Continued on inside back cover

Encyclopedia of Soil Science

edited by

Rattan Lal

School of Natural Resources
The Ohio State University
Columbus, Ohio, U.S.A.

MARCEL DEKKER

MARCEL DEKKER, INC. NEW YORK • BASEL

Set:
ISBN: 0-8247-0634-X print
ISBN: 0-8247-0518-1 electronic

This book is printed on acid-free paper.

Headquarters
Marcel Dekker, Inc.
270 Madison Avenue, New York, NY 10016
tel: 212-696-9000; fax: 212-685-4540

Eastern Hemisphere distribution
Marcel Dekker AG
Hutgasse 4, Postfach 812, CH-4001 Basel, Switzerland
tel: 41-61-261-8482; fax: 41-61-261-8896

World Wide Web
http://www.dekker.com

The publisher offers discounts on this book when ordered in bulk quantities. For more information, write to Special Sales/Professional Marketing at the headquarters address above.

Current printing (last digit)

ALD IBT 10 9 8 7 6 5 4 3

PRINTED IN THE UNITED STATES OF AMERICA

Encyclopedia of Soil Science

Rattan Lal, Editor
The Ohio State University

Topical Editors

Robert Ahrens / *United States Department of Agriculture, Natural Resources Conservation Service, National Soil Survey Center (USDA-NRCS-NSSC), Lincoln, Nebraska, U.S.A.*

Laj Ahuja / *United States Department of Agriculture-Agricultural Research Service (USDA-ARS), Great Plains Systems Research Unit (GPSR), Fort Collins, Colorado, U.S.A.*

Richard D. Bardgett / *Institute of Environmental and Natural Sciences, University of Lancaster, Lancaster, United Kingdom*

Richard W. Bell / *School of Environmental Science, Murdoch University, Murdoch, Western Australia, Australia*

Jagtar S. Bhatti / *Canadian Forest Service (CFS), Northern Forestry Centre, Edmonton, Alberta, Canada*

Virginie Bouchard / *School of Natural Resources, The Ohio State University, Columbus, Ohio, U.S.A.*

Mary Collins / *Soil and Water Science Department, University of Florida, Gainesville, Florida, U.S.A.*

Ram C. Dalal / *The Department of Natural Resources (DNR), Indooroopilly, Queensland, Australia*

David George Edwards / *School of Land and Food, The University of Queensland, St. Lucia, Queensland, Australia*

Delvin Fanning / *Department of Natural Resource Sciences and Landscape Architecture, University of Maryland, College Park, Maryland, U.S.A.*

Thomas E. Fenton / *Agronomy Department, Iowa State University, Ames, Iowa, U.S.A.*

Dennis Flanagan / *United States Department of Agriculture-Agricultural Research Service (USDA-ARS), National Soil Erosion Research Laboratory (NSERL), West Lafayette, Indiana, U.S.A.*

Serita Frey / *School of Natural Resources, The Ohio State University, Columbus, Ohio, U.S.A.*

Gavin Gillman / *Commonwealth Scientific and Industrial, Research Organisation (CSIRO), Land and Water, Davies Laboratory, Townsville, Queensland, Australia*

Sergey V. Goryachkin / *Institute of Geography, Moscow, Russia*

Robert C. Graham / *Department of Environmental Sciences, University of California, Riverside, California, U.S.A.*

Ed G. Gregorich / *Eastern Cereal and Oilseed Research Center, Agricultural and Agri-Food Canada, Ottawa, Ontario, Canada*

Chi-hua Huang / *United States Department of Agriculture-Agricultural Research Service (USDA-ARS), National Soil Erosion Research Laboratory (NSERL), West Lafayette, Indiana, U.S.A.*

Nilantha Hulugalle / *Co-operative Research Centre for Sustainable Cotton Production, Australian Cotton Research Institute, NSW Agriculture, Narrabri, New South Wales, Australia*

R. Cesar Izaurralde / *Pacific Northwest National Laboratory (PNL), Battelle Washington Operations, Washington, D.C., U.S.A.*

Selim Kapur / *Departments of Archaeometry and Soil Science, University of Cukurova, Balcali, Adana, Turkey*

Anastasios D. Karathanasis / *Department of Agronomy, University of Kentucky, Lexington, Kentucky, U.S.A.*

Eileen Kladivko / *Department of Agronomy, Purdue University, West Lafayette, Indiana, U.S.A.*

Rattan Lal / *School of Natural Resources, The Ohio State University, Columbus, Ohio, U.S.A.*

Dennis Linden / *United States Department of Agriculture-Agricultural Research Service (USDA-ARS), St. Paul, Minnesota, U.S.A.*

Mike Lindstrom / *United States Department of Agriculture-Agricultural Research Service (USDA-ARS), North Central Soil Conservation Research Laboratory, Morris, Minnesota, U.S.A.*

W. Dennis Nettleton / *United States Department of Agriculture-Natural Resource Conservation Service-National Soil Survey Center (USDA-NRCS-NSSC), Lincoln, Nebraska, U.S.A.*

Lee Nordt / *Department of Geology, Baylor University, Waco, Texas, U.S.A.*

Stephen Nortcliff / *Department of Soil Science, The University of Reading, Reading, United Kingdom*

Mark Peoples / *Commonwealth Scientific and Industrial Research Organisation (CSIRO), Plant Industry, Canberra, Australian Capital Territory, Australia*

Francis J. Pierce / *Center for Precision Agricultural Systems, Washington State University, Prosser, Washington, U.S.A.*

Gary W. Peterson / *Department of Agronomy, College of Agricultural Sciences, Pennsylvania State University, University Park, Pennsylvania, U.S.A.*

Joseph L. Pikul, Jr. / *United States Department of Agriculture-Agricultural Research Service (USDA-ARS), Northern Grain Insects Research Laboratory (NGIRL), Brookings, South Dakota, U.S.A.*

M.D. Ransom / *Department of Agronomy, Kansas State University, Manhattan, Kansas, U.S.A.*

Walter Rawls / *United States Department of Agriculture-Agricultural Research Service (USDA-ARS), Hydrology and Remote Sensing Laboratory, Beltsville, Maryland, U.S.A.*

Donald C. Reicosky / *United States Department of Agriculture-Agricultural Research Service (USDA-ARS), North Central Soil Conservation Research Laboratory, Morris, Minnesota, U.S.A.*

Pichu Rengasamy / *The University of Adelaide, Glen Osmond, South Australia, Australia*

Bal Ram Singh / *Department of Soil and Water Sciences, Agricultural University of Norway, Aas, Norway*

Brian Slater / *School of Natural Resources, The Ohio State University, Columbus, Ohio, U.S.A.*

Hwat Bing So / *School of Land and Food, The University of Queensland, St. Lucia, Queensland, Australia*

Moses Tenywa / *Department of Soil Science, University of Makerere, Kampala, Uganda*

Paul Tueller / *Department of Environmental and Resource Sciences, University of Nevada at Reno, Reno, Nevada, U.S.A.*

Larry T. West / *Department of Crop and Soil Sciences, University of Georgia, Athens, Georgia, U.S.A.*

Keith Wiebe / *United States Department of Agriculture-Economic Research Service (USDA-ERS), Resource Economics Division, Washington, D.C., U.S.A.*

Larry Wilding / *Department of Soil and Crop Sciences, Texas A&M University, College Station, Texas, U.S.A.*

William W. Wolfe / *Department of Civil and Environmental, Engineering and Geodetic Sciences, The Ohio State University, Columbus, Ohio, U.S.A.*

Ted M. Zobeck / *United States Department of Agriculture-Agricultural Research Service (USDA-ARS), Wind Erosion and Water Conservation Research Unit (WEWC), Lubbock, Texas, U.S.A.*

List of Contributors

Gayatri Acharya/*The World Bank, Washington, District of Columbia, U.S.A.*

L.R. Ahuja/*United States Department of Agriculture-Agricultural Research Service (USDA-ARS), Fort Collins, Colorado, U.S.A.*

Vasant A. Akala/*School of Natural Resources, The Ohio State University, Columbus, Ohio, U.S.A.*

Erhan Akça/*University of Çukurova, Department of Soil Science, Adana, Turkey*

Miguel Altieri/*University of California, Berkeley, California, U.S.A.*

Ronald Amundson/*University of California, Berkeley, California, U.S.A.*

D.W. Anderson/*Department of Soil Science, University of Saskatchewan, Saskatoon, Saskatchewan, Canada*

Samson D. Angima/*Department of Agronomy, Purdue University, West Lafayette, Indiana, U.S.A.*

Richard W. Arnold/*U.S. Department of Agriculture, Washington, D.C., U.S.A.*

Jeffrey G. Arnold/*United States Department of Agriculture-Agricultural Research Service (USDA-ARS), Temple, Texas, U.S.A.*

Paul A. Arp/*University of New Brunswick, Fredericton, New Brunswick, Canada*

K. Arulmozhiselvan/*Tamil Nadu Agricultural University, India*

L.A. Graham Aylmore/*The University of Western Australia, Western Australia, Australia*

Bryon W. Bache/*Cambridge University, Cambridge, United Kingdom*

John M. Baker/*United States Department of Agriculture (USDA), University of Minnesota, St. Paul, Minnesota, U.S.A.*

Kenneth A. Barbarick/*Colorado State University, Fort Collins, Colorado, U.S.A.*

Nikolaos I. Barbayiannis/*Aristotle University of Thessaloniki, Thessaloniki, Greece*

Richard Bardgett/*Lancaster University, Lancaster, United Kingdom*

N.J. Barrow/*Mt. Claremont, Western Australia, Australia*

C.D. Barton/*United States Department of Agriculture Forest Service, Aiken, South Carolina, U.S.A.*

Thomas Baumgartl/*Institute for Plant Nutrition and Soil Science, Kiel, Germany*

M.H. Beare/*New Zealand Institute for Crop & Food Research Limited, Christchurch, New Zealand*

Richard W. Bell/*School of Environmental Science, Murdoch University, Murdoch, Western Australia, Australia*

J.S. Bhatti/*Canadian Forest Service, Northern Forestry Centre, Edmonton, Alberta, Canada*

J.M. Bigham/*The Ohio State University, Columbus, Ohio, U.S.A.*

David E. Bignell/*Queen Mary, University of London, London, United Kingdom*

D.L. Bjorneberg/*Soil Scientist and Agricultural Engineer, Kimberly, Idaho, U.S.A.*

Elke Bloem/*Federal Agricultural Research Center, Braunschweig, Germany*

Wouter A. Blokhuis/*Agricultural University, Wageningen, The Netherlands*

John Boardman/*Environmental Change Institute, University of Oxford, Oxford, United Kingdom*

James G. Bockheim/*University of Wisconsin, Madison, Wisconsin, U.S.A.*

Pascal Boeckx/*Laboratory of Applied Physical Chemistry, Faculty of Agricultural and Applied Biological Sciences, Ghent University, Ghent, Belgium*

Janis L. Boettinger/*Utah State University, Logan, Utah, U.S.A.*

Patrick J. Bohlen/*Archbold Biological Station, Lake Placid, Florida, U.S.A.*

Julia Boike/*University of Alaska Fairbanks, Fairbanks, Alaska, U.S.A.*

J. Boixadera/*Departament d'Agricultura, Lleida, Spain*

N.S. Bolan/*Massey University, Palmerston North, New Zealand*

Getachew Boru/*United States Department of Agriculture, The Ohio State University, Columbus, Ohio, U.S.A.*

Virginie Bouchard/*The Ohio State University, Columbus, Ohio, U.S.A.*

E.M. Bridges/*Formerly of the International Soil Reference and Information Centre, Wageningen, The Netherlands*

María Jesús Iglesias Briones/*University of Vigo, Vigo, Spain*

Mary Ann Bruns/*The Pennsylvania State University, University Park, Pennsylvania, U.S.A.*

Graeme D. Buchan/*Lincoln University, Canterbury, New Zealand*

B.L. Bumb/*International Fertilizer Development Center, Muscle Shoals, Alabama, U.S.A.*

S.W. Buol/*North Carolina State University, Raleigh, North Carolina, U.S.A.*

Steven N. Burch/*University of Maryland, College Park, Maryland, U.S.A.*

W. Burghardt/*Soil Technology, University of Essen, Essen, Germany*

C. Lee Burras/*Iowa State University, Ames, Iowa, U.S.A.*

Robert H. Burris/*Professor Emeritus, Department of Biochemistry, University of Wisconsin-Madison, Madison, Wisconsin, U.S.A.*

David Burrow/*D.N.R.E. Agriculture Victoria, Tatura, Victoria, Australia*

R. Burt/*United States Department of Agriculture (USDA), Lincoln, Nebraska, U.S.A.*

Alan Busacca/*Washington State University, Pullman, Washington, U.S.A.*

Tarunjit Singh Butalia/*The Ohio State University, Columbus, Ohio, U.S.A.*

Cynthia A. Cambardella/*United States Department of Agriculture (USDA), Ames, Iowa, U.S.A.*

Keith C. Cameron/*Lincoln University, Canterbury, New Zealand*

Iain Campbell/*Land and Soil Consultancy Services, Nelson, New Zealand*

M.R. Carter/*Agriculture and Agri-Food Canada, Crops and Livestock Research Centre, Charlottetown, Prince Edward Island, Canada*

Elemar Antonino Cassol/*Federal University of Rio Grande do Sul, Porto Algre, Brazil*

Paul B. Cavers/*University of Western Ontario, London, Ontario, Canada*

Carlos C. Cerri/*Universidade de São Paulo, CENA, Piracicaba, São Pãulo, Brazil*

G.Y.S. Chan/*Department of Applied Biology and Chemical Technology, The Hong Kong Polytechnic University, Hung Hom, Hong Kong, China*

Kwong Yin Chan/*Wagga Wagga Agricultural Institute, NSW Agriculture, Wagga Wagga, New South Wales, Australia*

David Chandler/*Utah State University, Logan, Utah, U.S.A.*

S.S. Chernyanskii/*Moscow State University, Moscow, Russia*

Jon Chorover/*The University of Arizona, Tucson, Arizona, U.S.A.*

Torben Røjle Christensen/*Climate Impacts Group, Department of Ecology, Lund University, Lund, Sweden*

G.J. Churchman/*Commonwealth Scientific and Industrial Research Organisation (CSIRO) Land and Water, Glen Osmond, South Australia, Australia*

Jack Cline/*Department of Animal Sciences, The Ohio State University, Ohio, U.S.A.*

Matthew Cochran/*The Ohio State University, Columbus, Ohio, U.S.A.*

B. Cockroft/*Kialla, Victoria, Australia*

Neroli Pedro Cogo/*Federal University of Rio Grande do Sul (Department of Soils), Porto Alegre, Brazil*

Lisa Cole/*Lancaster University, Lancaster, United Kingdom*

David C. Coleman/*University of Georgia, Athens, Georgia, U.S.A.*

N.B. Comerford/*Soil and Water Science Department, University of Florida, Gainesville, Florida, U.S.A.*

R.D. Connolly/*URS Pty Ltd, East Perth, Western Australia, Australia*

M.K. Conyers/*New South Wales Agriculture, Wagga Wagga, New South Wales, Australia*

Ray Correll/*Commonwealth Scientific and Industrial Research Organisation (CSIRO), Mathematical and Information Sciences, Glen Osmond, Adelaide, South Australia, Australia*

Eric T. Craswell/*International Board for Soil Research and Management (IBSRAM), Jatujak, Bangkok, Thailand*

John W. Crawford/*University of Abertay, Dundee, Scotland, United Kingdom*

Pierre Crosson/*Resources for the Future, Washington, D.C., U.S.A.*

Seth M. Dabney/*United States Department of Agriculture, Oxford, Mississippi, U.S.A.*

Ram C. Dalal/*Department of Natural Resources, Indooroopilly, Queensland, Australia*

Jacob H. Dane/*Auburn University, Alabama, U.S.A.*

Eddy De Pauw/*International Center for Agricultural Research in the Dry Areas, Aleppo, Syria*

Jozef Deckers/*Institute for Land and Water Management, Leuven, Belgium*

M. Rifat Derici/*University of Çukurova, Department of Soil Science, Adana, Turkey*

Anthony R. Dexter/*Institute of Soil Science and Plant Cultivation, Pulawy, Poland*

M. Díaz-Zorita/*EEA INTA General Villegas, Buenos Aires, Argentina*

Warren Dick/*The Ohio State University, Wooster, Ohio, U.S.A.*

Peter Dillon/*Commonwealth Scientific and Industrial Research Organisation (CSIRO) Land and Water, Adelaide, South Australia, Australia*

Craig Ditzler/*United States Department of Agriculture-Natural Resources Conservation Service (USDA-NRCS), Soil Quality Institute, Ames, Iowa, U.S.A.*

Vsevolod V. Dobrovolsky/*Moscow State Pedagogical University, Moscow, Russia*

Douglas J. Dollhopf/*Department of Land Resources and Environmental Science, Montana State University, Bozeman, Montana, U.S.A.*

John W. Doran/*United States Department of Agriculture-Agricultural Research Service (USDA-ARS), University of Nebraska, Lincoln, Nebraska, U.S.A.*

L.R. Drees/*Texas A&M University, College Station, Texas, U.S.A.*

Paul Driessen/*Agricultural University, Wageningen, The Netherlands*

Sjoerd W. Duiker/*The Pennsylvania State University, University Park, Pennsylvania, U.S.A.*

Bahman Eghball/*United States Department of Agriculture-Agricultural Research Service (USDA-ARS), University of Nebraska–Lincoln, Lincoln, Nebraska, U.S.A.*

Reza Ehsani/*The Ohio State University, Columbus, Ohio, U.S.A.*

B.H. Ellert/*Agriculture and Agri-Food Canada, Lethbridge, Alberta, Canada*

William J. Elliot/*United States Department of Agriculture-Forest Service (USDA-FS), Rocky Mountain Research Station, Moscow, Idaho, U.S.A.*

M.M. Ellsbury/*United States Department of Agriculture (USDA), Brookings, South Dakota, U.S.A.*

Robert J. Engel/*United States Department of Agriculture (USDA), Lincoln, Nebraska, U.S.A.*

J.A. Entry/*Soil Scientist, Irrigation Engineer, and Soil Microbiologist, Kimberly, Idaho, U.S.A.*

G. Erpul/*Purdue University, West Lafayette, Indiana, U.S.A.*

Hari Eswaran/*United States Department of Agriculture-Natural Resources Conservation Service (USDA-NRCS), Washington, D.C., U.S.A.*

Christien H. Ettema/*Wageningen University, Wageningen, The Netherlands*

Robert O. Evans/*North Carolina State University, Raleigh, North Carolina, U.S.A.*

Delvin S. Fanning/*Department of Natural Resource Sciences and Landscape Architecture, University of Maryland, College Park, Maryland, U.S.A.*

Norman R. Fausey/*United States Department of Agriculture (USDA), Columbus, Ohio, U.S.A.*

David Favis-Mortlock/*Queen's University Belfast, Belfast, Northern Ireland, United Kingdom*

Thomas E. Fenton/*Iowa State University, Ames, Iowa, U.S.A.*

M.V. Fey/*University of Stellenbosch, Stellenbosch, South Africa*

Dennis C. Flanagan/*United States Department of Agriculture-Agricultural Research Service (USDA-ARS), West Lafayette, Indiana, U.S.A.*

G.N. Flerchinger/*United States Department of Agriculture-Northwest Watershed Research Center (USDA-NWRC), Boise, Idaho*

J.E. Foss/*University of Tennessee, Knoxville, Tennessee, U.S.A.*

Neil W. Foster/*Canadian Forest Service, Ontario, Sault Ste. Marie, Ontario, Canada*

Alan J. Franzluebbers/*United States Department of Agriculture-Agricultural Research Service (USDA-ARS), Watkinsville, Georgia, U.S.A.*

Donald P. Franzmeier/*Purdue University, West Lafayette, Indiana, U.S.A.*

Gary Frasier/*United States Department of Agriculture (USDA), Fort Collins, Colorado, U.S.A.*

Bruce E. Frazier/*Washington State University, Pullman, Washington, U.S.A.*

D.M. Freebairn/*Queensland Department of Natural Resources and Mines, Toowoomba, Queensland, Australia*

J.R. Freney/*Commonwealth Scientific and Industrial Research Organisation (CSIRO) Plant Industry, Canberra, Australian Capital Territory, Australia*

Serita Frey/*The Ohio State University, Columbus, Ohio, U.S.A.*

D. Gabriels/*Gent University, Gent, Belgium*

W.B. Gagliano/*The Ohio State University, Columbus, Ohio, U.S.A.*

John M. Galbraith/*Virginia Polytechnic Institute, State University, Blacksburg, Virginia, U.S.A.*

Patricia M. Gallagher/*Drexel University, Philadelphia, Pennsylvania, U.S.A.*

Gary J. Gascho/*University of Georgia, Tifton, Georgia, U.S.A.*

Harold R. Geering/*The University of Sydney, Sydney, New South Wales, Australia*

A.N. Gennadiyev/*Moscow State University, Moscow, Russia*

Argyrios Gerakis/*Arta, Greece*

Maria Gerasimova/*Moscow Lomonosov State University, Moscow, Russia*

Ken E. Giller/*Wageningen University, Wageningen, The Netherlands*

Dale Gillette/*Air Resources Laboratory, National Oceanic and Atmospheric Administration, Research Triangle Park, North Carolina, U.S.A.*

Daniel Giménez/*Rutgers, The State University of New Jersey, New Brunswick, New Jersey, U.S.A.*

Gunnar Gissel-Nielsen/*Risoe National Laboratory, Roskilde, Denmark*

Zitong Gong/*The Chinese Academy of Sciences, Nanjing, People's Republic of China*

Gerard Govers/*Katholicke Universiteit Leuven, Leuven, Belgium*

Robert C. Graham/*Soil and Water Sciences Program, Department of Environmental Sciences, University of California, Riverside, California, U.S.A.*

Timothy C. Granata/*The Ohio State University, Columbus, Ohio, U.S.A.*

Timothy R. Green/*United States Department of Agriculture (USDA), Fort Collins, Colorado, U.S.A.*

Richard S.B. Greene/*School of Resources, Environment and Society, Australian National University, Canberra, Australian Capital Territory, Australia*

D.J. Greenland/*University of Reading, Reading, United Kingdom*

E.G. Gregorich/*Agriculture and Agri-Food Canada, Central Experimental Farm, Ottawa, Ontario, Canada*

B.S. Griffiths/*Scottish Crop Research Institute, Invergowrie, Dundee, United Kingdom*

J.H. Grove/*University of Kentucky, Lexington, Kentucky, U.S.A.*

Noel J. Grundon/*Atherton, Queensland, Australia*

Georg Guggenberger/*Institute of Soil Science and Soil Geography, University of Bayreuth, Bayreuth, Germany*

S.C. Gupta/*University of Minnesota, St. Paul, Minnesota, U.S.A.*

Umesh C. Gupta/*Agriculture and Agri-Food Canada, Charlottetown, Prince Edward Island, Canada*

George Hall/*The Ohio State University, Columbus, Ohio, U.S.A.*

C.T. Hallmark/*Texas A&M University, College Station, Texas, U.S.A.*

Ardell D. Halvorson/*United States Department of Agriculture-Agricultural Research Service (USDA-ARS), Fort Collins, Colorado, U.S.A.*

L.L. Hammond/*International Fertilizer Development Center, Muscle Shoals, Alabama, U.S.A.*

Silvia Haneklaus/*Federal Agricultural Research Center, Braunschweig, Germany*

Jon Harbor/*Purdue University, West Lafayette, Indiana, U.S.A.*

R.F. Harris/*University of Wisconsin, Madison, Wisconsin, U.S.A.*

Jerry L. Hatfield/*United States Department of Agriculture-Agriculture Research Service (USDA-ARS), Ames, Iowa, U.S.A.*

S. Hauser/*International Institute of Tropical Agriculture, Mbalmayo, Cameroon*

Kris M. Havstad/*United States Department of Agriculture-Agricultural Research Service (USDA-ARS), Jornada Experimental Range, Las Cruces, New Mexico, U.S.A.*

J. Douglas Helms/*United States Department of Agriculture-Natural Resources Conservation Service (USDA-NRCS), Washington, D.C., U.S.A.*

Robert Hendershot/*United States Department of Agriculture-Natural Resources Conservation Service (USDA-NRCS), Lancaster, Ohio, U.S.A.*

J. Herrero/*Laboratorio asociado de Agricultura y Medioambiente (DGA-CSIC), Zaragoza, Spain*

David F. Herridge/*NSW Agriculture, Tamworth, New South Wales, Australia*

Karl Herweg/*University of Bern, Bern, Switzerland*

Dean Hesterberg/*North Carolina State University, Raleigh, North Carolina, U.S.A.*

A.E. Hewitt/*Landcare Reasearch, Lincoln, New Zealand*

Philippe Hinsinger/*INRA UMR Sol et Environnement, Montpellier, France*

Richard Hobbs/*Murdoch University, Murdoch, Western Australia, Australia*

John A. Holt/*James Cook University, Townsville, Australia*

Steve P. Hopkin/*University of Reading, Reading, United Kingdom*

Jeffrey W. Hopkins/*United States Department of Agriculture, Washington, D.C., U.S.A.*

Jan W. Hopmans/*University of California, Davis, California, U.S.A.*

D.J. Horne/*Massey University, Palmerston North, New Zealand*

L.R. Hossner/*Texas A&M University, College Station, Texas, U.S.A.*

Terry A. Howell/*United States Department of Agriculture-Agricultural Research Service (USDA-ARS), Bushland, Texas, U.S.A.*

Chi-hua Huang/*United States Department of Agriculture-Agricultural Research Service (USDA-ARS), West Lafayette, Indiana, U.S.A.*

J. Herbert Huddleston/*Oregon State University, Corvallis, Oregon*

A.R. Huete/*Department of Soil, Water, and Environmental Science, University of Arizona, Tucson, Arizona, U.S.A.*

Jeffrey Hughes/*University of Natal, Scottsville, South Africa*

Craig D. Idso/*Center for the Study of Carbon Dioxide and Global Change, Tempe, Arizona, U.S.A.*

Keith E. Idso/*Center for the Study of Carbon Dioxide and Global Change, Tempe, Arizona, U.S.A.*

Sherwood B. Idso/*U.S. Water Conservation Laboratory, Phoenix, Arizona, U.S.A.*

Samuel J. Indorante/*United States Department of Agriculture-Natural Resources Conservation Service (USDA-NRCS), Carbondale, Illinois, U.S.A.*

R.F. Isbell/*Landcare Reasearch, Lincoln, New Zealand*

R. César Izaurralde/*Joint Global Charge Research Institute, College Park, Maryland, U.S.A.*

T.J. Jackson/*United States Department of Agriculture-Agricultural Research Service (USDA-ARS), Beltsville, Maryland, U.S.A.*

Marc Jalbert/*Auburn University, Alabama, U.S.A.*

Bruce R. James/*University of Maryland, College Park, Maryland, U.S.A.*

H.H. Janzen/*Agriculture and Agri-Food Canada, Lethbridge, Alberta, Canada*

Philip M. Jardine/*Oak Ridge National Laboratory, Oak Ridge, Tennessee, U.S.A.*

David Jasper/*The University of Western Australia, Crawley, Western Australia*

J.D. Jastrow/*Argonne National Laboratory, Argonne, Illinois, U.S.A.*

Anna Joabsson/*Climate Impacts Group, Department of Ecology, Lund University, Lund, Sweden*

Jodi L. Johnson-Maynard/*University of Idaho, Moscow, Idaho, U.S.A.*

Uzi Kafkafi/*The Hebrew University of Jerusalem, Rehovot, Israel*

Yash P. Kalra/*Canadian Forest Service, Edmonton, Alberta, Canada*

Douglas L. Kane/*University of Alaska Fairbanks, Fairbanks, Alaska, U.S.A.*

Selim Kapur/*University of Çukurova, Department of Soil Science, Adana, Turkey*

A.D. Karathanasis/*University of Kentucky, Lexington, Kentucky, U.S.A.*

D.L. Karlen/*United States Department of Agriculture (USDA), Ames, Iowa, U.S.A.*

Lev O. Karpachevsky/*Moscow State University, Moscow, Russia*

Mikhail L. Karpachevsky/*Moscow State University, Moscow, Russia*

Thomas C. Kaspar/*United States Department of Agriculture (USDA), Ames, Iowa, U.S.A.*

Zulkuf Kaya/*University of Çukurova, Balcali, Adana, Turkey*

Jack Keller/*Utah State University, Logan, Utah, U.S.A.*

Katherine J. Kendrick/*U.S. Geological Survey, Pasadena, California, U.S.A.*

Vissarion Z. Keramidas/*Aristotle University of Thessaloniki, Thessaloniki, Greece*

J.M. Kimble/*United States Department of Agriculture (USDA), Lincoln, Nebraska, U.S.A.*

Kevin W. King/*United States Department of Agriculture-Agricultural Research Service (USDA-ARS), Temple, Texas, U.S.A.*

Peter I.A. Kinnell/*University of Canberra, Canberra, Australian Capital Territory, Australia*

Gunnar Kirchhof/*New South Wales Agriculture, Tamworth, New South Wales, Australia*

Eileen J. Kladivko/*Purdue University, West Lafayette, Indiana, U.S.A.*

Andreas Klik/*University of Agricultural Sciences Vienna—BOKU, Vienna, Austria*

Hans Klompen/*The Ohio State University, Columbus, Ohio, U.S.A.*

Rai Kookana/*Commonwealth Scientific and Industrial Research Organisation (CSIRO) Land and Water, Glen Osmond, Adelaide, South Australia, Australia*

Helmut H. Krause/*University of New Brunswick, Fredericton, New Brunswick, Canada*

Lars Krogh/*Institute of Geography, University of Copenhagen, Copenhagen, Denmark*

William F. Kuenstler/*United States Department of Agriculture-Natural Resources Conservation Service (USDA-NRCS), Fort Worth, Texas, U.S.A.*

William P. Kustas/*United States Department of Agriculture-Agricultural Research Service (USDA-ARS), Hydrology and Remote Sensing Lab, Beltsville, Maryland, U.S.A.*

John M. Laflen/*United States Department of Agriculture-Agricultural Research Service (USDA–ARS) (retired) West Lafayette, Indiana, U.S.A.*

Rattan Lal/*School of Natural Resources, The Ohio State University, Columbus, Ohio, U.S.A.*

Judith Lancaster/*Desert Research Institute, Reno, Nevada, U.S.A.*

W.E. Larson/*University of Minnesota, St. Paul, Minnesota, U.S.A.*

Robert J. Lascano/*Texas A&M University, Lubbock, Texas, U.S.A.*

Henry Noel Le Houérou/*Montpellier, France*

Jean-Claude Lefeuvre/*University of Rennes 1, Rennes, France*

Jean-Paul Legros/*INRA, Science du Sol, Montpellier, France*

Reynald Lemke/*Agriculture and Agri-Food Canada, Swift Current, Saskatchewan, Canada*

Renato Levien/*Federal University of Rio Grande do Sul (Department of Soils), Porto Alegre, Brazil*

John Leys/*Department of Land and Water Conservation, Gunnedah, New South Wales, Australia*

Michael J. Lindstrom/*United States Department of Agriculture-Agricultural Research Service (USDA-ARS), Morris, Minnesota, U.S.A.*

David A. Lobb/*University of Manitoba, Winnipeg, Manitoba, Canada*

Jau-Yau Lu/*National Chung-Hsing University, Taichung, Taiwan*

John Ludwig/*Commonwealth Scientific and Industrial Research Organisation, Australia*

A. Lufafa/*Makerere University, Kampala, Uganda*

Jian Feng Ma/*Kagawa University, Kagawa, Japan*

L. Ma/*United States Department of Agriculture-Agricultural Research Service (USDA-ARS), Fort Collins, Colorado, U.S.A.*

J.A. MacLeod/*Agriculture and Agri-Food Canada, Charlottetown, Prince Edward Island, Canada*

R.O. Maguire/*University of Delaware, Newark, Delaware, U.S.A.*

J.G.M. Majaliwa/*Makerere University, Kampala, Uganda*

D.D. Malo/*South Dakota State University, Brookings, South Dakota, U.S.A.*

Paul Mapfumo/*University of Zimbabwe, Harare, Zimbabwe*

Jay Martin/*The Ohio State University, Columbus, Ohio, U.S.A.*

José A. Martínez-Casasnovas/*University of Lleida, Lleida, Spain*

Christopher J. Matocha/*Agronomy Department, University of Kentucky, Lexington, Kentucky, U.S.A.*

Douglas G. Maynard/*Natural Resources Canada, Victoria, British Columbia, Canada*

G. Mazhitova/*Komi Science Center, Russian Academy of Sciences, Syktyvkar, Russia*

Ray McBride/*University of Guelph, Guelph, Ontario, Canada*

Donald K. McCool/*United States Department of Agriculture-Agricultural Research Service (USDA-ARS), Pullman, Washington, U.S.A.*

Paul A. McDaniel/*University of Idaho, Moscow, Idaho, U.S.A.*

Leslie McFadden/*University of New Mexico, Albuquerque, New Mexico, U.S.A.*

D. McGarry/*Natural Resource Sciences, Indooroopilly, Queensland, Australia*

Sean McGinn/*Agriculture and Agri-Food Canada, Lethbridge, Alberta, Canada*

Kevin McInnes/*Texas A&M University, College Station, Texas, U.S.A.*

David C. McKenzie/*Precision Land Management, Orange, New South Wales, Australia*

Ronald G. McLaren/*Lincoln University, Canterbury, New Zealand*

M.J. McLaughlin/*Commonwealth Scientific and Industrial Research Organisation-Land and Water, Glen Osmond, South Australia, Australia*

Mallavarapu Megharaj/*Commonwealth Scientific and Industrial Research Organisation (CSIRO) Land and Water, Glen Osmond, Adelaide, South Australia, Australia*

Michael D. Melville/*University of New South Wales, Sydney, New South Wales, Australia*

Rienk Miedema/*Department of Soil Science and Geology, Wageningen University, Wageningen, The Netherlands*

Fred P. Miller/*The Ohio State University, Columbus, Ohio, U.S.A.*

A.J. Mills/*University of Stellenbosch, Stellenbosch, South Africa*

Katsu Minami/*National Institute for Agro-Environmental Sciences, Tsukuba, Japan*

Douglas W. Ming/*NASA Johnson Space Center, Houston, Texas, U.S.A.*

Delbert Mokma/*Michigan State University, East Lansing, Michigan, U.S.A.*

H. Curtis Monger/*New Mexico State University, Las Cruces, New Mexico, U.S.A.*

L.E. Moody/*California Polytechnic State University, San Luis Obispo, California, U.S.A.*

David Mouat/*Desert Research Institute, Reno, Nevada, U.S.A.*

Brian Murphy/*Centre for Natural Resources, Cowra, Australia*

James A. Murphy/*Rutgers, The State University of New Jersey, New Brunswick, New Jersey, U.S.A.*

Stephanie L. Murphy/*Rutgers, The State University of New Jersey, New Brunswick, New Jersey, U.S.A.*

Freddy Nachtergaele/*FAO, Rome, Italy*

Ravendra Naidu/*Commonwealth Scientific and Industrial Research Organisation (CSIRO) Land and Water, Glen Osmond, Adelaide, South Australia, Australia*

M.A. Nearing/*Purdue University, West Lafayette, Indiana, U.S.A.*

Mark A. Nearing/*United States Department of Agriculture-Agricultural Research Service (USDA-ARS), West Lafayette, Indiana, U.S.A.*

John L. Nieber/*University of Minnesota, St. Paul, Minnesota, U.S.A.*

David C. Nielsen/*United States Department of Agriculture-Agricultural Research Service (USDA-ARS), Akron, Colorado, U.S.A.*

Niels Erik Nielsen/*The Royal Veterinary and Agricultural University, Copenhagen, Denmark*

T.M. Nissen/*University of Illinois, Urbana, Illinois, U.S.A.*

Egide Nizeyimana/*Environmental Resources Research Institute, The Pennsylvania State University, University Park, Pennsylvania, U.S.A.*

Lee Nordt/*Baylor University, Waco, Texas, U.S.A.*

L. Norgrove/*International Institute of Tropical Agriculture, Mbalmayo, Cameroon*

Stephen Nortcliff/*The University of Reading, Reading, United Kingdom*

L.D. Norton/*United States Department of Agriculture-Agricultural Research Service (USDA-ARS), National Soil Erosion Research Lab., West Lafayette, Indiana, U.S.A.*

T. Nyobe/*Institute de Recherche Agricole pour le Developpement, Nkolbisson, Yaounde, Cameroon*

R.D. O'Brien/*University of Stellenbosch, Stellenbosch, South Africa*

Robin O'Malley/*The H. John Heinz III Center for Science, Economics and the Environment, Washington, D.C., U.S.A.*

C.G. Olson/*United States Department of Agriculture-Natural Resources Conservation Service (USDA-NRCS), Lincoln, Nebraska, U.S.A.*

K.R. Olson/*University of Illinois, Urbana, Illinois, U.S.A.*

K.A. Olsson/*Tatura, Victoria, Australia*

Jacob Opadeyi/*The University of the West Indies, St. Augustine, Trinidad and Tobago*

Ibrahim Ortas/*University of Çukurova, Adana, Turkey*

Yakov A. Pachepsky/*United States Department of Agriculture-Agricultural Research Service (USDA-ARS), Hydrology and Remote Sensing Laboratory, Beltsville, Maryland, U.S.A.*

Eswaran Padmanabhan/*University of Malaysia Sarawak, Kota Samarahan, Malaysia*

Stefano Pagiola/*World Bank, Washington, D.C., U.S.A.*

P. Paramasivam/*Tamil Nadu Agricultural University, India*

S.J. Park/*University of Bonn, Bonn, Germany*

David R. Parker/*University of California, Riverside, California, U.S.A.*

Keith Paustian/*Colorado State University, Fort Collins, Colorado, U.S.A.*

Judith F. Pedler/*University of California, Riverside, California, U.S.A.*

Frits W.T. Penning de Vries/*International Board for Soil Research and Management (IBSRAM), Jatujak, Bangkok, Thailand*

Mark B. Peoples/*Commonwealth Scientific and Industrial Research Organization Plant Industry, Canberra, Australian Capital Territory, Australia*

E. Perfect/*University of Tennessee, Knoxville, Tennessee, U.S.A.*

G.W. Petersen/*The Pennsylvania State University, University Park, Pennsylvania, U.S.A.*

I.R. Phillips/*Griffith University, Queensland, Australia*

Joseph L. Pikul, Jr./*United States Department of Agriculture-Agricultural Research Service (USDA-ARS), Brookings, South Dakota, U.S.A.*

Rosa M. Poch/*University of Lleida, Lleida, Spain*

W.M. Post/*Oak Ridge National Laboratory, Oak Ridge, Tennessee, U.S.A.*

Sylvie A. Quideau/*University of Alberta, Edmonton, Alberta, Canada*

Timothy A. Quine/*University of Exeter, Exeter, Devon, United Kingdom*

Martin C. Rabenhorst/*Department of Natural Resource Sciences and Landscape Architecture, University of Maryland, College Park, Maryland, U.S.A.*

Mark Radosevich/*University of Delaware, Newark, Delaware, U.S.A.*

A. Subba Rao/*Indian Institute of Soil Science, Madhya Pradesh, India*

Walter J. Rawls/*United States Department of Agriculture-Agricultural Research Service (USDA-ARS), Hydrology and Remote Sensing Laboratory, Beltsville, Maryland, U.S.A.*

Randall C. Reeder/*The Ohio State University, Columbus, Ohio, U.S.A.*

Paul F. Reich/*United States Department of Agriculture-Natural Resources Conservation Service (USDA-NRCS), Washington, District of Columbia, U.S.A.*

José Miguel Reichert/*Federal University of Santa Maria, Santa Maria, Brazil*

D.C. Reicosky/*United States Department of Agriculture-Agriculture Research Service (USDA-ARS), North Central Soil Conservation Research Laboratory, Morris, Minnesota, U.S.A.*

Pichu Rengasamy/*Department of Soil and Water, Adelaide University, Glen Osmond, S.A., Australia*

Zdenko Rengel/*Soil Science and Plant Nutrition, The University of Western Australia, Crawley, Western Australia, Australia*

E. Danielle Rhine/*University of Delaware, Newark, Delaware, U.S.A.*

J.D. Rhoades/*Agricultural Salinity Consulting, Riverside, California, U.S.A.*

Charles W. Rice/*Kansas State University, Manhattan, Kansas, U.S.A.*

Jimmie L. Richardson/*Department of Soil Science, North Dakota State University, Fargo, North Dakota, U.S.A.*

Dominique Righi/*CNRS, Poitiers, France*

Joe T. Ritchie/*Belton, Texas, U.S.A.*

Philippe Rochette/*Agriculture and Agri-Food Canada, Sainte-Foy, Québec, Canada*

Dennis E. Rolston/*University of California, Davis, California, U.S.A.*

D.E. Romig/*Energy, Minerals and Natural Resources Department, Santa Fe, New Mexico, U.S.A.*

Calvin W. Rose/*Griffith University, Nathan, Queensland, Australia*

Stacey Rosen/*United States Department of Agriculture, Economic Research Service, Washington, D.C., U.S.A.*

Clayton Rubec/*Habitat Conservation Division, Canadian Wildlife Service, Environment Canada, Ottawa, Ontario, Canada*

Alain Ruellan/*Montpellier, France*

John Ryan/*International Center for Agricultural Research in the Dry Areas, Aleppo, Syria*

P.G. Saffigna/*The University of Queensland, Gatton, Queensland, Australia and Saffcorp Pty. Ltd, Tweed Heads, New South Wales, Australia*

K.L. Sahrawat/*West Africa Rice Development Association, Côte d'Ivoire (Ivory Coast), West Africa*

G.S. Saroa/*The Ohio State University, Columbus, Ohio, U.S.A.*

Ronald L. Sass/*Rice University, Houston, Texas, U.S.A.*

Thomas J. Sauer/*United States Department of Agriculture (USDA), Ames, Iowa, U.S.A.*

Brian M. Schafer/*The University of Queensland, Gatton, Queensland, Australia*

Sara J. Scherr/*University of Maryland, College Park, Maryland, U.S.A.*

David L. Schertz/*United States Department of Agriculture-Natural Resources Conservation Services (USDA-NRCS), Washington, D.C., U.S.A.*

Jan van Schilfgaarde/*United States Department of Agriculture (USDA), Fort Collins, Colorado, U.S.A.*

William H. Schlesinger/*Duke University, Durham, North Carolina, U.S.A.*

Ewald Schnug/*Federal Agricultural Research Center, Braunschweig, Germany*

Tom E. Schumacher/*South Dakota State University, Brookings, South Dakota, U.S.A.*

H.D. Scott/*University of Arkansas, Fayetteville, Arkansas, U.S.A.*

H. Shahandeh/*Texas A&M University, College Station, Texas, U.S.A.*

Shahla Shapouri/*United States Department of Agriculture, Economic Research Service, Washington, D.C., U.S.A.*

Brenton S. Sharratt/*United States Department of Agriculture (USDA), Morris, Minnesota, U.S.A.*

Joey N. Shaw/*Auburn University, Auburn, Alabama, U.S.A.*

J. Shen/*China Agricultural University, Beijing, Peoples Republic of China*

Hau-Jung Shieh/*National Chung-Hsing University, Taichung, Taiwan*

M.J. Shipitalo/*United States Department of Agriculture-Agricultural Research Service (USDA-ARS), Coshocton, Ohio, U.S.A.*

J.T. Sims/*University of Delaware, Newark, Delaware, U.S.A.*

Michael J. Singer/*Department of Land, Air and Water Resources, University of California, California, U.S.A.*

Bal Ram Singh/*Agricultural University of Norway, Aas, Norway*

J. Six/*Colorado State University, Fort Collins, Colorado, U.S.A*

Neil E. Smeck/*The Ohio State University, Columbus, Ohio, U.S.A.*

Keith R.J. Smettem/*The University of Western Australia, Nedlands, Western Australia, Australia*

G.D. Smith/*Queensland Department of Natural Resources and Mines, Toowoomba, Queensland, Australia*

Pete Smith/*University of Aberdeen, Aberdeen, United Kingdom*

Alvin J.M. Smucker/*Michigan State University, East Lansing, Michigan, U.S.A.*

Hwat Bing So/*The University of Queensland, St. Lucia, Queensland, Australia*

R.E. Sojka/*Soil Scientist and Agricultural Engineer, Kimberly, Idaho, U.S.A.*

Deborah A. Soukup/*Komex H_2O Science, Inc., Bakersfield, California, U.S.A.*

Otto Spaargaren/*International Soil Reference and Information Centre (ISRIC), Wageningen, The Netherlands*

Graham Sparling/*Landcare Research, Hamilton, New Zealand*

Victor R. Squires/*Adelaide University, Adelaide, South Australia, Australia*

Larry D. Stetler/*Department of Geology & Geological Engineering, South Dakota School of Mines and Technology, Rapid City, South Dakota, U.S.A.*

B.A. Stewart/*West Texas A&M University, Texas, U.S.A.*

Donald L. Suarez/*Salinity Laboratory, Agricultural Research Service-United States Department of Agriculture, Riverside, California, U.S.A.*

Matthew O. Sullivan/*The Ohio State University, Columbus, Ohio, U.S.A.*

David J. Susko/*University of Western Ontario, London, Ontario, Canada*

John P. Tandarich/*Hey and Associates, Inc., Chicago, Illinois, U.S.A.*

Charles Tarnocai/*Agriculture and Agri-Food Canada, Ottawa, Canada*

M.M. Tenywa/*Makerere University, Kampala, Uganda*

D.P. Thoma/*University of Minnesota, St. Paul, Minnesota, U.S.A.*

Grant W. Thomas/*Professor Emeritus, University of Kentucky, Lexington, Kentucky, U.S.A.*

Robin Thwaites/*The University of Queensland, St. Lucia, Queensland, Australia*

Gary L. Tibke/*Wind Erosion Research Unit, Throckmorton Hall, Kansas State University, Manhattan, Kansas, U.S.A.*

Ralph W. Tiner/*U.S. Fish and Wildlife Service, National Wetlands Inventory Program, Hadley, Massachusetts, U.S.A.*

Judy Tisdall/*La Trobe University, Bundoora, Australia*

David Tongway/*Commonwealth Scientific and Industrial Research Organisation, Australia*

Clinton C. Truman/*United States Department of Agriculture (USDA), Tifton, Georgia, U.S.A.*

Paul T. Tueller/*University of Nevada, Reno, Nevada, U.S.A.*

J.N. Tullberg/*University of Queensland, Gatton, Queensland, Australia*

Goro Uehara/*University of Hawaii, Honolulu, Hawaii, U.S.A.*

April L. Ulery/*New Mexico State University, Las Cruces, New Mexico, U.S.A.*

Paul W. Unger/*United States Department of Agriculture-Agricultural Research Service (USDA-ARS), Bushland, Texas, U.S.A.*

Oswald Van Cleemput/*Laboratory of Applied Physical Chemistry, Faculty of Agricultural and Applied Biological Sciences, Ghent University, Ghent, Belgium*

Ken Van Rees/*Department of Soil Science, University of Saskatchewan, Saskatoon, Saskatchewan, Canada*

George F. Vance/*University of Wyoming, Laramie, Wyoming, U.S.A.*

Tara T. VanToai/*United States Department of Agriculture, The Ohio State University, Columbus, Ohio, U.S.A.*

Joel T. Walker/*The Ohio State University, Columbus, Ohio, U.S.A.*

M.M. Wander/*University of Illinois, Urbana, Illinois, U.S.A.*

D. Wang/*University of Minnesota, St. Paul, Minnesota, U.S.A.*

Xingxiang Wang/*Institute of Soil Science, Chinese Academy of Sciences, Nanjing, People's Republic of China*

S.C. Ward/*Bicton, Australia*

E.D. Warner/*The Pennsylvania State University, University Park, Pennsylvania, U.S.A.*

Andrew Warren/*Department of Geography, University College London, London, United Kingdom*

Glenn A. Weesies/*United States Department of Agriculture-Natural Resources Conservation Service (USDA-NRCS), West Lafayette, Indiana, U.S.A.*

Walter Wenzel/*University of Agricultural Sciences Vienna—BOKU, Vienna, Austria*

Larry T. West/*University of Georgia, Athens, Georgia, U.S.A.*

Ian White/*Australian National University, Canberra, Australian Capital Territory, Australia*

Walter G. Whitford/*United States Department of Agriculture-Agricultural Research Service (USDA-ARS), Jornada Experimental Range, MSC 3JER, New Mexico State University, Las Cruces, New Mexico*

Keith Wiebe/*United States Department of Agriculture (USDA), Washington, D.C., U.S.A.*

Giles Wiggs/*University of Sheffield, Sheffield, United Kingdom*

Larry P. Wilding/*Texas A & M University, College Station, Texas, U.S.A.*

Jimmy R. Williams/*Texas A&M University, Temple, Texas, U.S.A.*

M.A. Wilson/*United States Department of Agriculture (USDA), Lincoln, Nebraska, U.S.A.*

William E. Wolfe/*Ohio State University, Columbus, Ohio, U.S.A.*

M.H. Wong/*Department of Biology and Institute of Natural Resources and Environmental Management, Hong Kong Baptist University, Kowloon Tong, Hong Kong, China*

J.H.M. Wösten/*Alterra Green World Research, Wageningen, The Netherlands*

Guohua Xu/*Nanjing Agricultural University, Nanjing, China*

Kazuyuki Yagi/*National Institute for Agro-Environmental Sciences, Tsukuba, Ibaraki, Japan*

D.F. Yule/*Natural Resource Sciences, Indooroopilly, Queensland, Australia*

F.S. Zhang/*China Agricultural University, Beijing, Peoples Republic of China*

Jianhuan Zhang/*United States Department of Agriculture, The Ohio State University, Columbus, Ohio, U.S.A.*

Taolin Zhang/*Institute of Soil Science, Chinese Academy of Sciences, Nanjing, People's Republic of China*

X.-C. (John) Zhang/*United States Department of Agriculture-Agricultural Research Service (USDA-ARS), El Reno, Oklahoma, U.S.A.*

Fenli Zheng/*Institute of Soil and Water Conservation, Yangling, Shaanxi, People's Republic of China*

Y-G. Zhu/*Research Centre for Eco-environmental Sciences, Chinese Academy of Sciences, Beijing, People's Republic of China*

Ted M. Zobeck/*United States Department of Agriculture, Lubbock, Texas, U.S.A.*

Michael A. Zoebisch/*Asian Institute of Technology (AIT), Bangkok, Thailand*

Contents

Preface

Six important challenges facing humanity at the start of the 21st century are: (1) a world population of 6.1 billion that is increasing at the rate of 1.3% per year, with most of the increase occurring in developing countries where natural resources are already under great stress; (2) a food-insecure population of 790 million, mostly in sub-Saharan Africa and South Asia; (3) severe and extreme forms of soil degradation affecting more than 300 million hectares (Mha) of agricultural land, continuing unabated in developing countries where the resource-poor farmers cannot afford to invest in soil restoration; (4) an anthropogenic increase in atmospheric concentration of greenhouse gases such as CO_2, with current levels at 370 ppm and increasing at the rate of 0.5% per year; CH_4 currently at 1.74 ppm and increasing at the rate of 0.75% per year; N_2O at 311 ppb and increasing 0.25% per year; and the atmospheric C pool now at 770 Pg and increasing at the global per capita rate of 1.1 Mg per year; (5) a projected 40% decrease in the global per capita arable land over the next 50 years; (6) the threat of a decreasing, renewable fresh-water supply, dropping to scarcity levels in many countries.

The solution to all of these challenges lies in developing management strategies based on an understanding of the nature, properties, and dynamics of the life-support processes of the most basic of all natural resources . . . the soil. This thin upper surface of Earth's crust supports life in numerous ways–moderating the purification of water, detoxification of pollutants, restoration and resilience of ecosystems, and cycling of basic elements such carbon, nitrogen, phosphorus, and sulfur. These life-supporting functions of soil are accurately summed up in an old Chinese proverb that states "Man . . . despite his artistic pretensions and many accomplishments . . . owes his existence to a six-inch layer of topsoil and the fact that it rains."

Traditionally, soil was regarded as a medium for plant growth, a repository of gene pool, an engineering foundation, an industrial raw material, and an archive of planetary history. However, meet the challenges of the post-modern era of the 21st century, world soils must also (1) maximize long-term productivity; (2) moderate moisture storage and distribution; (3) proxy for interpretation of past global changes and cultural heritage; and (4) minimize environmental pollution by functioning as a repository for waste disposal, an ameliorator of water quality through bioremediation, a sink for atmospheric carbon and other greenhouse gases, a component of biogeochemical cycles and other ecological processes, and as a medium for immersing and loading infrastructure. To enhance soil's capacity to perform these functions, it is important to understand factors and processes affecting soil quality under growing and conflicting modes of land uses (among them agricultural, forestry, urban, industrial, military, aesthetic, and cultural uses).

It is necessary that state-of-the-knowledge information about these complex attributes of soil and its multi-faceted functions of interest to humans is readily available to the scientific community, policy makers, and the public at large in an easily accessible compendium such as this encyclopedia.

The *Encyclopedia of Soil Science* has been complied by over 400 experts in their fields to meet this need. This first edition, to be updated periodically, addresses 40 thematic topics. Each of these topics was coordinated by a topic editor and includes discussions of related subjects prepared by soil scientists from around the world. The topics cover all branches of soil science including pedology, mineralogy, physics, soil mechanics, hydrology, chemistry, biology, ecology and management, and restoration of problem or degraded soils, and address numerous challenges including soil structure, tillage methods and mulch farming, irrigation, drainage and water table management, fertilizer and nutrient management, erosion control, and management of soil organic matter. Entries are alphabetically arranged, and an subject index is included for easy access.

Preparation of this state-of-the-knowledge compendium on soil science has been made possible by the hard work, vision, and commitment of hundreds of soil scientists from around the world who have contributed as members of the advisory board, as topic editors, or as authors. I thank all of them for their outstanding efforts to prepare this volume in record time. I am especially thankful to all the topic editors for identifying expert authors, for coordinating the writing with minimal duplication, and for carefully reviewing manuscripts to ensure high quality. I also thank all of the authors for their outstanding efforts in documenting and presenting, in a timely fashion, the scientific information supporting the current understanding of their field of specialization. Their efforts in the preparation of this volume provide up-to-date information on the nature, properties, functions, and dynamics of soil.

Thanks are also due to the staff of Marcel Dekker, Inc. for their efforts in making this information available to the world community. It has been a great pleasure working with Ms. Ellen Lichtenstein and Ms. Sapna Maloor. Their professionalism, commitment to excellence, and dedication to the mission is much appreciated. I also thank Ms. Brenda Swank of the School of Natural Resources of the Ohio State University for her efforts in handling the flow of manuscripts in the review process. These efforts have produced this volume on soil science that contains information useful to meeting the challenges of the 21st century and beyond.

Rattan Lal
Editor-in-Chief
School of Natural Resources
The Ohio State University
Columbus, Ohio, U.S.A.

ACID MINE DRAINAGE

W.B. Gagliano
J.M. Bigham
The Ohio State University, Columbus, Ohio, U.S.A.

A

INTRODUCTION

What Is Acid Mine Drainage?

Acid mine drainage refers to metal-rich sulfuric acid solutions released from mine tunnels, open pits, and waste rock piles (Table 1). Similar solutions are produced by the drainage of some coastal wetlands, resulting in the formation of acid sulfate soils. Acid mine drainage typically ranges in pH from 2 to 4; however, extreme sites like Iron Mountain, California have produced pH values as low as -3.6 (1). Neutral to alkaline mine drainage is also common in areas where the surrounding geologic units contain carbonate rocks to buffer acidity (Table 1).

Why Is Acid Mine Drainage a Problem?

Soils and spoils exposed to acid mine drainage do not support vegetation and are susceptible to erosion. When acid mine drainage enters natural waterways, changes in pH and the formation of voluminous precipitates of metal hydroxides can devastate fish populations and other aquatic life (Fig. 1). The corrosion of engineered structures like bridges is also greatly accelerated. There may be as many as 500,000 inactive or abandoned mines in the United States, with mine drainage severely impacting approximately 19,300 km of streams and more than 72,000 ha of lakes and reservoirs (2, 3). Once initiated, mine drainage may persist for decades, making it a challenging problem to solve.

What Causes Acid Mine Drainage?

Mine drainage results from the oxidation of sulfide minerals such as pyrite (cubic FeS_2), marcasite (orthorhombic FeS_2), pyrrhotite ($Fe_{1-x}S$), chalcopyrite ($CuFeS_2$) and arsenopyrite (FeAsS). These minerals are commonly found in coal and ore deposits and are stable until exposed to oxygen and water. Their oxidation causes the release of metals and the production of sulfuric acid. This process can occur as a form of natural mineral weathering but is exacerbated by mining because of the sudden, large-scale exposure of unweathered rock to atmospheric conditions.

MINE DRAINAGE CHEMISTRY

Mine drainage is a complex biogeochemical process involving oxidation-reduction, hydrolysis, precipitation, and dissolution reactions as well as microbial catalysis (1). The entire sequence is commonly represented by Reaction 1, which describes the overall oxidation of pyrite by oxygen in the presence of water to form iron hydroxide [$Fe(OH)_3$] and sulfuric acid.

$$FeS_{2(s)} + 3\,{}^3/_4O_{2(g)} + 3\,{}^1/_2H_2O_{(l)} \rightarrow Fe(OH)_{3(s)} + 2H_2SO_{4(aq)} \quad (1)$$

The actual oxidation process is considerably more complicated.

Pyrite and related sulfide minerals contain both Fe and S in reduced oxidation states. When exposed to oxygen and water the sulfur moiety is oxidized first, releasing Fe^{2+} and sulfuric acid to solution (Reaction 2). The rate of oxidation is dependent on environmental factors like temperature, pH, Eh, and relative humidity as well as mineral surface area and microbial catalysis.

$$FeS_{2(s)} + 3\,{}^1/_2O_{2(g)} + H_2O_{(l)} \rightarrow Fe^{2+}_{(aq)} + 2SO^{2-}_{4(aq)} + 2H^+_{(aq)} \quad (2)$$

Reaction 2 is most important in the initial stages of mine drainage generation and can be either strictly abiotic or mediated by contact with sulfur-oxidizing bacteria (4). The Fe^{2+} released by pyrite decomposition is rapidly oxidized by oxygen at $pH > 3$ as per Reaction 3.

$$Fe^{2+}_{(aq)} + {}^1/_4O_{2(g)} + H^+_{(aq)} \rightarrow Fe^{3+}_{(aq)} + {}^1/_2H_2O_{(l)} \quad (3)$$

If acidity generated by Reaction 2 exceeds the buffering capacity of the system, the pH eventually decreases. Below pH 3, Fe^{3+} solubility increases and a second mechanism of pyrite oxidation becomes important (5) (Reaction 4).

$$FeS_{2(s)} + 14Fe^{3+}_{(aq)} + 8H_2O_{(l)} \rightarrow 15Fe^{2+}_{(aq)} + 2SO^{2-}_{4(aq)} + 16H^+_{(aq)} \quad (4)$$

Table 1 Summary of mine drainage from 101 bituminous coal mine sites in Pennsylvania

	Range	Median	Mean
pH	2.7–7.3	5.2	3.6
Fe (mg/L)	0.16–512.0	43.0	58.9
Al (mg/L)	0.01–108.0	1.3	9.8
Mn (mg/L)	0.12–74.0	2.2	6.2
SO_4 (mg/L)	120–2000	580.0	711.2

(Unpublished data from C. Cravotta, III, 2001, USGS, Lemoyne, PA.)

In this case, pyrite is oxidized by Fe^{3+} resulting in the generation of even greater acidity than when oxygen is the primary oxidant. Pyrite decomposition is thus controlled by the rate at which Fe^{2+} is converted to Fe^{3+} at low pH (6). At pH < 3, Fe^{2+} oxidation is very slow unless it is catalyzed by populations of iron oxidizing bacteria like *Acidithiobacillus ferrooxidans* or *Leptospirillum ferrooxidans*. These acidophilic bacteria oxidize Fe^{2+} as a means of generating energy to fix carbon. In doing so they supply soluble Fe^{3+} at a rate equal to or slightly greater than the rate of pyrite oxidation by Fe^{3+} (5). Pyrite oxidation then regenerates Fe^{2+} (Reaction 4) creating a cyclic situation that leads to vigorous acidification of mine drainage water.

Fig. 1 Mixing of acid mine drainage (right) with a natural stream resulting in the formation of voluminous precipitates of iron minerals.

MINE DRAINAGE MINERALOGY

The hydrolysis of Fe^{3+} causes the precipitation of various iron minerals, generally represented as [$Fe(OH)_3$], that are often the most obvious indicators of mine drainage contamination (Reaction 5).

$$Fe^{3+}_{(aq)} + 3H_2O_{(l)} \rightarrow Fe(OH)_{3(s)} + 3H^+_{(aq)} \quad (5)$$

These precipitates are yellow-to-red-to-brown in color and have long been referred to by North American miners as "yellow boy." The actual mineralogy of the precipitates is determined by solution parameters like pH, sulfate, and metal concentration and can vary both spatially and temporally. Some of the most common mine drainage minerals are goethite (α-FeOOH), ferrihydrite ($Fe_5HO_8 \cdot 4H_2O$), schwertmannite [$Fe_8O_8(OH)_6SO_4$], and jarosite [$(H,K,Na)Fe_3(OH)_6(SO_4)_2$] (7).

Goethite is a crystalline oxyhydroxide that occurs over a wide pH range, is relatively stable, and may represent a final transformation product of other mine drainage minerals. Ferrihydrite is a poorly crystalline ferric oxide that forms in higher pH (> 6.5) environments. Schwertmannite is commonly found in drainage waters with pH ranging from 2.8 to 4.5 and with moderate to high sulfate contents. It may be the dominant phase controlling major and minor element activities in most acid mine drainage. Jarosite forms in more extreme environments with pH < 3, very high sulfate concentrations, and in the presence of appropriate cations like Na and K.

MINE DRAINAGE MICROBIOLOGY

The most studied bacterial species in mine drainage systems belong to the genus *Acidithiobacillus* (formerly *Thiobacillus*) (8). Species like *Acidithiobacillus thiooxidans* and *Acidithiobacillus ferrooxidans* are important in sulfur and iron oxidation in acid drainage; however, many other microorganisms may also be involved (9). Bacteria have been found in close association with pyrite grains and may play a direct role in mineral oxidation, but they most likely function indirectly through oxidation of dissolved Fe^{2+} as described previously. In low pH systems (< 3), *A. ferrooxidans* can increase the rate of iron oxidation as much as five orders of magnitude relative to strictly abiotic rates (6).

Iron oxidizing bacteria are chemolithotrophic, meaning they oxidize inorganic compounds, like Fe^{2+}, to generate energy and use CO_2 as a source of carbon. Iron oxidation, however, is a very low energy yielding process. It has been estimated that the oxidation of 90.1 moles of Fe^{2+} is required to assimilate one mole of C into biomass (10). Thus, large amounts of Fe^{2+} must be oxidized to achieve even modest growth.

In addition to mediating iron oxidation, bacteria may play an additional role in mineral formation. Bacteria in mine drainage systems have been shown to be partially encrusted with mineral precipitates (11). Bacterial cell walls provide reactive sites for the sorption of metal cations, which can accumulate and subsequently develop into precipitates using the bacterial surface (living or dead) as a template (12, 13).

ENVIRONMENTAL IMPACTS OF MINE DRAINAGE

Mine drainage is primarily released from open mine shafts or from mine spoil left exposed to the atmosphere. The drainage produced can have devastating effects on the surrounding ecosystem. Chemical precipitates can obstruct water flow, dramatically increase turbidity, and ruin stream aesthetics. Dissolved metals and acidity can also affect plant and aquatic animal populations.

Besides iron, Al is the most common dissolved metal in acid mine drainage. The primary source of Al is the acid dissolution of aluminosilicates found in soil, spoils, tailings deposits and gangue material (7). At high concentrations Al can be toxic to plants, and colloidal aluminum precipitates can irritate the gills of fish causing suffocation. Aluminum occurs as a dissolved species at low pH, but rapidly hydrolyzes at about pH 5 to form "amorphous" basaluminite [$Al_4(SO_4)(OH)_{10}{\cdot}4H_2O$] or gibbsite [$Al(OH)_3$] (7). Aluminum precipitates are white in color, but are readily masked by associated iron compounds.

Elevated levels of trace elements like As, Cu, Ni, Pb and Zn may be released during the oxidation of sulfide minerals. These elements can play a role in mineralization processes by forming co-precipitates but occur primarily as sorbed species (14). Mine drainage precipitates can retain both anions and cations, depending on pH. While co-precipitation and sorption function to immobilize trace elements by removing them from solution, this effect may not be permanent. Dissolution of precipitates and shifts in pH can result in the release of sorbed species, providing a latent source of pollution.

DEALING WITH MINE DRAINAGE

Successful control of mine drainage usually involves elements of both prevention and treatment.

Prevention

Prevention techniques include sealing mine shafts, burying or submerging spoil piles, and the addition of bactericides to limit the function of iron oxidizing bacteria. These techniques often have limited success. Sealing of mines is extremely difficult due to fractures and the permeability of surrounding rocks. Covering spoil with soil material can decrease the degree of sulfide oxidation by limiting exposure to oxygen, but establishment of a vegetative cover is necessary to prevent erosion from re-exposing the spoil. Inhibition of iron oxidizing bacteria with bactericides can decrease sulfide oxidation and reduce metal mobility; however, reapplication is necessary and adequate distribution to all affected areas is difficult. In addition, target bacteria may develop resistance and beneficial bacteria may be harmed (15).

Treatment

Solution pH usually underestimates the total acidity of mine drainage. Total acidity is the sum of "proton acidity" and "mineral acidity" generated upon oxidation and/or hydrolysis of metals like Fe^{2+}, Fe^{3+}, Mn, and Al^{3+}. The traditional approach to treatment of acid mine drainage involves neutralization of total acidity by the addition of alkaline agents like caustic soda (NaOH) or hydrated lime [$Ca(OH)_2$]. This method is effective in neutralizing acidity and precipitating dissolved metals; however, it requires continuous oversight and produces large amounts of waste sludge that require disposal. Newer remediation strategies focus on low cost, sustainable methods for treatment of drainage waters. For example, limestone drains coupled with compost wetlands have shown some promise as passive remediation technologies. In these systems, drainage is channeled through either oxic or anoxic limestone substrates to neutralize active acidity. Dissolved metals are then allowed to hydrolyze and precipitate in wetland cells. A major difficulty is the loss of reactive surface by armoring of limestone particles with precipitates of Fe and Al that eventually obstruct flow.

Compost wetlands are designed to stimulate the development of anaerobic microbial populations, particularly sulfate-reducing bacteria. The bacteria use the compost as an organic substrate and remove sulfate from solution either by converting it to H_2S, which is lost to the atmosphere, or by forming insoluble iron sulfides (Reactions 6 and 7).

$$2CH_3CHOHCOO^-_{(aq)} + SO_4^{2-}(aq) \rightarrow 2CH_3COO^-_{(aq)} + 2HCO_3^-(aq) + H_2S_{(g)} \quad (6)$$

$$H_2S_{(s)} + Fe^{2+}_{(aq)} \rightarrow FeS_{(s)} + 2H^+_{(aq)} \quad (7)$$

Bicarbonate is formed as a by-product of sulfate-reduction, and functions to buffer acidity. These systems have also shown limited success in the field. The sulfate removal rates are usually low (<10%), and pH often remains unchanged or decreases within the system (16).

REFERENCES

1. Nordstrom, D.K.; Alpers, C.N. Geochemistry of Acid Mine Waters. In *The Environmental Geochemistry of Mineral Deposits*; Plumlee, G.S., Logsdon, M.J., Eds.; Chpt. 6. Society of Economic Geologists Inc: Littleton, CO, 1999; 133–160.
2. Kleinmann, R.L.P. Acid Mine Drainage in the United States: Controlling the Impact on Streams and Rivers. *4th World Congress on the Conservation of Built and Natural Environments*; Toronto, Ontario, 1989; 1–10.
3. Lyon, J.S.; Hilliard, T.J.; Bethel, T.N. *Burden of Guilt*; Mineral Policy Center: Washington, DC, 1993; 68.
4. Rojas, J.; Giersig, M.; Tributsch, H. Sulfur Colloids as Temporary Energy Reservoirs for *Thiobacillus ferrooxidans* During Pyrite Oxidation. Archives of Microbiology **1995**, *163*, 352–356.
5. Nordstrom, D.K. Aqueous Pyrite Oxidation and the Consequent Formation of Secondary Iron Minerals. In *Acid Sulfate Weathering*; Kittrick, J.A., Fanning, D.S., Hossner, L.S., Eds.; Soil Science Society of America: Madison, WI, 1982; 37–57.
6. Singer, P.C.; Stumm, W. Acidic Mine Drainage: The Rate Determining Step. Science **1970**, *167*, 1121–1123.
7. Bigham, J.M.; Nordstrom, D.K. Iron and Aluminum Hydroxysulfates from Acid Mine Waters. In *Sulfate Minerals: Crystallography, Geochemistry and Environmental Significance*; Alpers, C.N., Jambor, J.L., Nordstrom, DK, Eds.; Reviews in Mineralogy and Geochemistry; The Mineralogical Society of America: Washington, DC, 2000; Vol. 40, 351–403.
8. Kelly, D.P.; Wood, A.P. Reclassification of Some Species of *Thiobacillus* to the Newly Designated Genera *Acidithiobacilus* gen. nov., *Halothiobacillus* gen. nov. and *Thermithiobacillus* gen. nov. International Journal of Systematic and Evolutionary Microbiology **2000**, *50*, 511–516.
9. Gould, W.D.; Berchard, G.; Lortie, L. The Nature and Role of Microorganisms in the Tailings Environment. In *The Environmental Geochemistry of Sulfide Mine-Wastes*; Jambor, J.L., Blowes, D.W., Eds.; Mineralogical Association of Canada Short Course; 1994; Vol. 22, 185–200.
10. Ehrlich, H.L. *Geomicrobiology*, 3rd Ed.; Marcel Dekker, Inc.: New York, 1996.
11. Clarke, W.; Konhouser, K.O.; Thomas, J.; Bottrell, S.H. Ferric Hydroxide and Ferric Hydroxysulfate Precipitation by Bacteria in an Acid Mine Drainage Lagoon. FEMS Microbiology **1997**, *20*, 351–361.
12. Schultze-Lam, S.; Fortin, D.; Davis, B.; Beveridge, T.J. Mineralization of Bacterial Surfaces. Chemical Geology **1996**, *132*, 171–181.
13. Konhauser, K.O. Diversity of Bacterial Iron Mineralization. Earth-Science Reviews **1998**, *43*, 91–121.
14. Winland, R.L.; Traina, S.J.; Bigham, J.M. Chemical Composition of Ocherous Precipitates from Ohio Coal Mine Drainage. Journal of Environmental Quality **1991**, *20*, 452–460.
15. Ledin, M.; Pedersen, K. The Environmental Impact of Mine Wastes—Roles of Microorganisms and Their Significance in Treatment of Mine Wastes. Earth-Science Reviews **1996**, *41*, 67–108.
16. Mitsch, W.J.; Wise, K.M. Water Quality, Fate of Metals, and Predictive Model Validation of a Constructed Wetland Treating Acid Mine Drainage. Water Resources **1998**, *32*, 1888–1900.

ACID RAIN AND N DEPOSITION

George F. Vance
University of Wyoming, Laramie, Wyoming, U.S.A.

A

INTRODUCTION

Air pollution has occurred naturally since the formation of the Earth's atmosphere; however, the industrial era has resulted in human activities greatly contributing to global atmospheric pollution (1, 2). One of the more highly publicized and controversial aspects of atmospheric pollution is that of acidic deposition. Acidic deposition includes rainfall, acidic fogs, mists, snowmelt, gases, and dry particulate matter (3). The primary origin of acidic deposition is the emission of sulfur dioxide (SO_2) and nitrogen oxides (NO_x) from fossil fuel combustion; electric power generating plants contribute approximately two-thirds of the SO_2 emissions and one-third of the NO_x emissions (4).

Acidic materials can be transported long distances, some as much as hundreds of kilometers. For example, 30–40% of the S deposition in the northeastern U.S. originates in industrial midwestern U.S. states (5). After years of debate, U.S. and Canada have agreed to develop strategies that reduce acidic compounds originating from their countries (5, 6). In Europe, the small size of many countries means that emissions in one industrialized area can readily affect forests, lakes, and cities in another country. For example, approximately 17% of the acidic deposition falling on Norway originated in Britain and 20% in Sweden came from eastern Europe (5).

The U.S. EPA National Acid Precipitation Assessment Program (NAPAP) conducted intensive research during the 1980s and 1990s that resulted in the "Acidic Deposition: State of the Science and Technology" that was mandated by the Acid Precipitation Act of 1980 (6). NAPAP Reports to Congress have been developed in accordance with the 1990 amendment to the 1970 Clean Air Act and present the expected benefits of the Acid Deposition Control Program (6, 7), http://www.nnic.noaa.gov/CENR/NAPAP/. Mandates include an annual 10 million ton or approximately 40% reduction in point-source SO_2 emissions below 1980 levels, with national emissions limit caps of 8.95 million tons from electric utility and 5.6 million tons from point-source industrial emissions. A reduction in NO_x of about 2 million tons from 1980 levels has also been set as a goal; however, while NO_x has been on the decline since 1980, projections estimate a rise in NO_x emissions after the year 2000. In 1980, the U.S. levels of SO_2 and NO_x emissions were 25.7 and 23.0 million tons, respectively.

Acidic deposition can impact buildings, sculptures, and monuments that are constructed using weatherable materials like limestone, marble, bronze, and galvanized steel (7, 8), http://www.nnic.noaa.gov/CENR/NAPAP/. While acid soil conditions are known to influence the growth of plants, agricultural impacts related to acidic deposition are of less concern due to the buffering capacity of these types of ecosystems (2, 5). When acidic substances are deposited in natural ecosystems, a number of adverse environmental effects are believed to occur, including damage to vegetation, particularly forests, and changes in soil and surface water chemistry (9, 10).

SOURCES AND DISTRIBUTION

Typical sources of acidic deposition include coal- and oil-burning electric power plants, automobiles, and large industrial operations (e.g., smelters). Once S and N gases enter the earth's atmosphere they react very rapidly with moisture in the air to form sulfuric (H_2SO_4) and nitric (HNO_3) acids (2, 3). The pH of natural rainfall in equilibrium with atmospheric CO_2 is about 5.6; however, the pH of rainfall is less than 4.5 in many industrialized areas. The nature of acidic deposition is controlled largely by the geographic distribution of the sources of SO_2 and NO_x (Fig. 1). In the midwestern and northeastern U.S., H_2SO_4 is the main source of acidity in precipitation because of the coal-burning electric utilities (2). In the western U.S., HNO_3 is of more concern because utilities and industry burn coal with low S contents and populated areas are high sources of NO_x (2).

Emissions of SO_2 and NO_x increased in the 20th century due to the accelerated industrialization in developed countries and antiquated processing practices in some undeveloped countries. However, there is some uncertainty as to the actual means by which acidic deposition affects our environment (11, 12), http://nadp.sws.uiuc.edu/isopleths/maps1999/. Chemical and biological

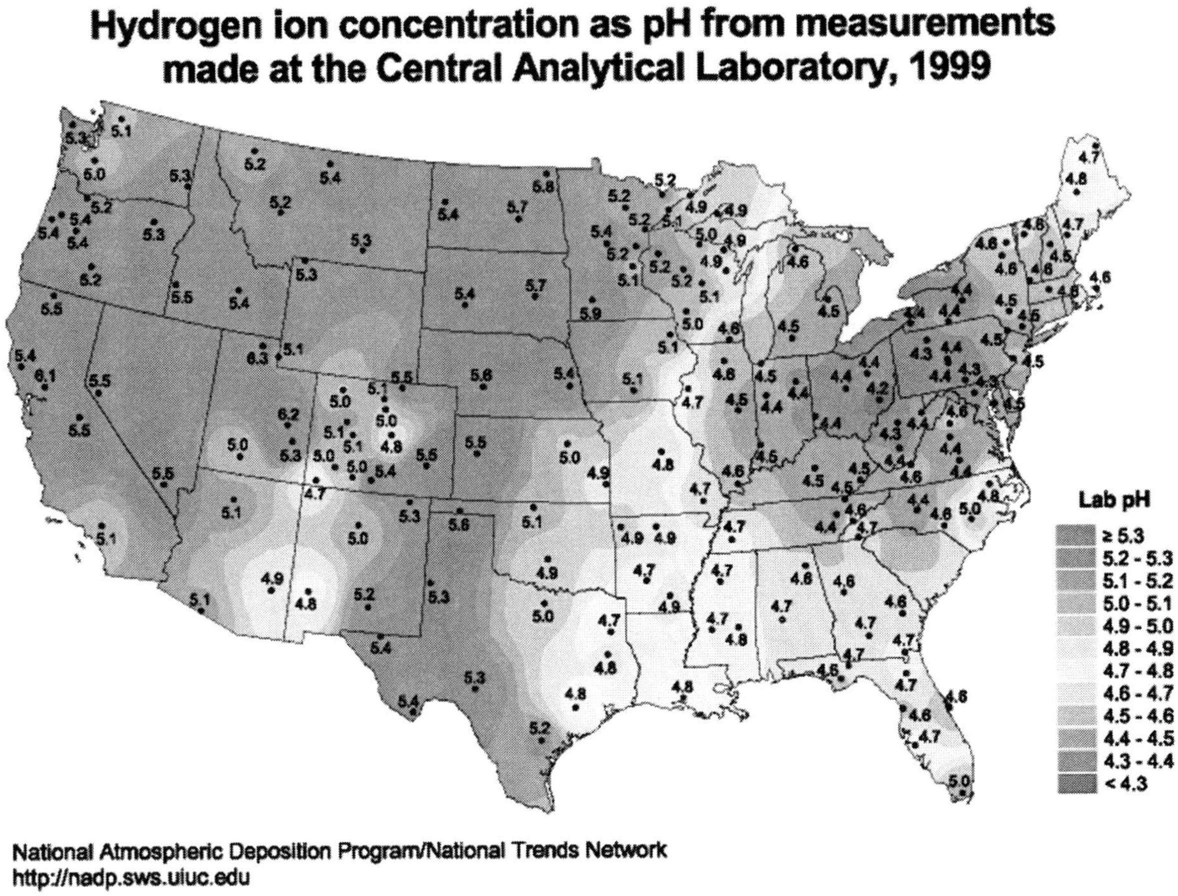

Fig. 1 Acidic deposition across the U.S. during 1999.

evidence, however, indicates that atmospheric deposition of H_2SO_4 caused some New England lakes to decrease in alkalinity (13, 14). Many scientists are reluctant to overgeneralize cause and effect relationships in an extremely complex environmental problem. Although, the National Acid Deposition Assessment Program has concluded there were definite consequences due to acidic deposition that warrant remediation (6, 7) http://www.nnic.noaa.gov/CENR/NAPAP/. Since 1995, when the 1990 Clean Air Act Amendment's Title IV reduction in acidic deposition was implemented, SO_2 and NO_x emissions have, respectively, decreased and remained constant during the late 1990s (4).

Both H_2SO_4 and HNO_3 are important components of acidic deposition, with volatile organic compounds and inorganic carbon also components of acidic deposition-related emissions. Pure water has a pH of 7.0, natural rainfall about 5.6, and severely acidic deposition less than 4.0. Uncontaminated rainwater should be pH 5.6 due to CO_2 chemistry and the formation of carbonic acid. The pH of most soils ranges from 3.0 to 8.0 (2). When acids are added to soils or waters, the decrease in pH that occurs depends greatly on the system's buffering capacity, the ability of a system to maintain its present pH by neutralizing added acidity. Clays, organic matter, oxides of Al and Fe, and Ca and Mg carbonates (limestones) are the components responsible for pH buffering in most soils. Acidic deposition, therefore, will have a greater impact on sandy, low organic matter soils than those higher in clay, organic matter, and carbonates. In fresh waters, the primary buffering mechanism is the reaction of dissolved bicarbonate ions with H^+ according to the following equation:

$$H^+ + HCO_3^- = H_2O + CO_2 \qquad (1)$$

HUMAN HEALTH EFFECTS

Few direct human health problems have been attributed to acidic deposition. Long-term exposure to acidic deposition precursor pollutants such as ozone (O_3) and NO_x, which are respiratory irritants, can cause pulmonary edema (5, 6). Sulfur dioxide (SO_2) is also a known respiratory irritant, but is generally absorbed high in the respiratory tract.

Indirect human health effects due to acidic deposition are more important. Concerns center around contaminated drinking water supplies and consumption of fish that contain potential toxic metal levels. With increasing acidity (e.g., lower pH levels), metals such as mercury, aluminum, cadmium, lead, zinc, and copper become more bioavailable (2). The greatest human health impact is due to the consumption of fish that bioaccumulate mercury; freshwater pike and trout have been shown to contain the highest average concentrations of mercury (5, 15). Therefore, the most susceptible individuals are those who live in an industrial area, have respiratory problems, drink water from a cistern, and consume a significant amount of freshwater fish.

A long-term urban concern is the possible impact of acidic deposition on surface-derived drinking water. Many municipalities make extensive use of lead and copper piping, which raises the question concerning human health effects related to the slow dissolution of some metals (lead, copper, zinc) from older plumbing materials when exposed to more acidic waters. Although metal toxicities due to acidic deposition impacts on drinking waters are rare, reductions in S and N fine particles expected by 2010 based on Clean Air Act Amendments will result in annual public health benefits valued at $50 billion with reduced mortality, hospital admissions and emergency room visits (16).

STRUCTURAL IMPACTS

Different types of materials and cultural resources can be impacted by air pollutants. Although the actual corrosion rates for most metals have decreased since the 1930s, data from three U.S. sites indicate that acidic deposition may account for 31–78% of the dissolution of galvanized steel and copper (7, 8), http://www.nnic.noaa.gov/CENR/NAPAP/. In urban or industrial settings, increases in atmospheric acidity can dissolve carbonates (e.g., limestone, marble) in buildings and other structures. Deterioration of stone products by acidic deposition is caused by: 1) erosion and dissolution of materials and surface details, 2) alterations (blackening of stone surfaces), and 3) spalling (cracking and spalling of stone surfaces due to accumulations of alternation crusts (8). Painted surfaces can be discolored or etched, and there may also be degradation of organic binders in paints (8).

ECOSYSTEM IMPACTS

It is important to examine the nature of acidity in soil, vegetation, and aquatic environments. Damage from acidification is often not directly due to the presence of excessive H^+, but is caused by changes in other elements. Examples include increased solubilization of metal ions such as Al^{3+} and some trace elements (e.g., Mn^{2+}, Pb^{2+}) that can be toxic to plants and animals, more rapid losses of basic cations (e.g., Ca^{2+}, Mg^{2+}), and the creation of unfavorable soil and aquatic environments for different fauna and flora.

Soils

Soil acidification is a natural process that occurs when precipitation exceeds evapotranspiration (2). "Natural" rainfall is acidic (pH of ~5.6) and continuously adds a weak acid (H_2CO_3) to soils. This acidification results in a gradual leaching of basic cations (Ca^{2+} and Mg^{2+}) from the uppermost soil horizons, leaving Al^{3+} as the dominant cation that can react with water to produce H^+. Most of the acidity in soils between pH 4.0 and 7.5 is due to the hydrolysis of Al^{3+} (17, 18), http://www.epa.gov/airmarkets/acidrain/effects/index.html. Other acidifying processes include plant and microbial respiration that produces CO_2, mineralization and nitrification of organic N, and the oxidation of FeS_2 in soils disturbed by mining or drainage (2). In extremely acidic soils ($pH < 4.0$), strong acids such as H_2SO_4 are a major component.

The degree of accelerated acidification depends both upon the buffering capacity of the soil and the use of the soil. Many of the areas subjected to the greatest amount of acidic deposition are also areas where considerable natural acidification occurs (19). Forested soils in the northeastern U.S. are developed on highly acidic, sandy parent materials that have undergone tremendous changes in land use in the past 200 years. However, clear-cutting and burning by the first European settlers have been almost completely reversed and many areas are now totally reforested (5). Soil organic matter that accumulated over time represents a natural source of acidity and buffering. Similarly, greater leaching or depletion of basic cations by plant uptake in increasingly reforested areas balances the significant inputs of these same cations in precipitation (20, 21). Acidic deposition affects forest soils more than agricultural or urban soils because the latter are routinely limed to neutralize acidity. Although it is possible to lime forest soils, which is done frequently in some European countries, the logistics and cost often preclude this except in areas severely impacted by acidic deposition (5).

Excessively acidic soils are undesirable for several reasons. Direct phytotoxicity from soluble Al^{3+} or Mn^{2+} can occur and seriously injure plant roots, reduce plant growth, and increase plant susceptibility to pathogens (21). The relationship between Al^{3+} toxicity and soil pH is complicated by the fact that in certain situations organic matter can form complexes with Al^{3+} that reduce its harmful effects on plants (18). Acid soils are usually less fertile because of a lack of important basic cations such as K^+, Ca^{2+}, and Mg^{2+}. Leguminous plants may fix less N_2 under very acidic conditions due to reduced rhizobial activity and greater soil adsorption of Mo by clays and Al and Fe oxides (2). Mineralization of N, P, and S can also be reduced because of the lower metabolic activity of bacteria. Many plants and microorganisms have adapted to very acidic conditions (e.g., pH<5.0). Examples include ornamentals such as azaleas and rhododendrons and food crops such as cassava, tea, blueberries, and potatoes (5, 22). In fact, considerable efforts in plant breeding and biotechnology are directed towards developing Al- and Mn- tolerant plants that can survive in highly acidic soils.

Agricultural Ecosystems

Acidic deposition contains N and S that are important plant nutrients. Therefore, foliar applications of acidic deposition at critical growth stages can be beneficial to plant development and reproduction. Generally, controlled experiments require the simulated acid rain to be pH 3.5 or less in order to produce injury to certain plants (22). The amount of acidity needed to damage some plants is 100 times greater than natural rainfall. Crops that respond negatively in simulated acid rain studies include garden beets, broccoli, carrots, mustard greens, radishes, and pinto beans, with different effects for some cultivars. Positive responses to acid rain have been identified with alfalfa, tomato, green pepper, strawberry, corn, lettuce, and some pasture grass crops.

Agricultural lands are maintained at pH levels that are optimal for crop production. In most cases the ideal pH is around pH 6.0–7.0; however, pH levels of organic soils are usually maintained at closer to pH 5.0. Because agricultural soils are generally well buffered, the amount of acidity derived from atmospheric inputs is not sufficient to significantly alter the overall soil pH (2). Nitrogen and S soil inputs from acidic deposition are beneficial, and with the reduction in S atmospheric levels mandated by 1990 amendments to the Clean Air Act, the S fertilizer market has grown. The amount of N added to agricultural ecosystems as acidic deposition is rather insignificant in relation to the 100–300 kg N/ha/yr required of most agricultural crops.

Forest Ecosystems

Perhaps the most publicized issue related to acidic deposition has been widespread forest decline. For example, in Europe estimates suggest that as much as 35% of all forests have been affected (23). Similarly, in the U.S. many important forest ranges such as the Adirondacks of New York, the Green Mountains of Vermont, and the Great Smoky Mountains in North Carolina have experienced sustained decreases in tree growth for several decades (6). Conclusive evidence that forest decline or dieback is caused solely be acidic deposition is lacking and complicated by interactions with other environmental or biotic factors. However, NAPAP research (6) has confirmed that acidic deposition has contributed to a decline in high-elevation red spruce in the northeastern U.S. In addition, nitrogen saturation of forest ecosystems from atmospheric N deposition is believed to result in increased plant growth, which in turn increases water and nutrient use followed by deficiencies that can cause chlorosis and premature needle-drop as well as increased leaching of base cations from the soil (24).

Acidic deposition on leaves may enter directly through plant stomates (1, 22). If the deposition is sufficiently acidic (pH ~ 3.0), damage can also occur to the waxy cuticle, increasing the potential for direct injury of exposed leaf mesophyll cells. Foliar lesions are one of the most common symptoms. Gaseous compounds such as SO_2 and SO_3 present in acidic mists or fogs can also enter leaves through the stomates, form H_2SO_4 upon reaction with H_2O in the cytoplasm, and disrupt many metabolic processes. Leaf and needle necrosis occurs when plants are exposed to high levels of SO_2 gas, possibly due to collapsed epidermal cells, eroded cuticles, loss of chloroplast integrity and decreased chlorophyll content, loosening of fibers in cell walls and reduced cell membrane integrity, and changes in osmotic potential that cause a decrease in cell turgor.

Root diseases may also increase in excessively acidic soils. In addition to the damages caused by exposure to H_2SO_4 and HNO_3, roots can be directly injured or their growth rates impaired by increased concentrations of soluble Al^{3+} and Mn^{2+} in the rhizosphere (2, 25), http://nadp.sws.uiuc.edu. Changes in the amount and composition of these exudates can then alter the activity and population diversity of soil-borne pathogens. The general tendency associated with increased root exudation is an enhancement in microbial populations due to an additional supply of carbon (energy). Chronic acidification can also alter nutrient availability and uptake patterns (8, 22).

Long-term studies in New England suggest acidic deposition has caused significant plant and soil leaching of base cations (1, 21), resulting in decreased growth of red spruce trees in the White Mountains (6). With reduction in about 80% of the airborne base cations, mainly Ca^{2+} but also Mg^{2+}, from 1950 levels, researchers suggest forest growth has slowed because soils are not capable of weathering at a rate that can replenish essential nutrients. In Germany, acidic deposition was implicated in the loss of soil Mg^{2+} as an accompanying cation associated with the downward leaching of SO_4^{2-}, which ultimately resulted in forest decline (2). Several European countries have used helicopters to fertilize and lime forests.

Aquatic Ecosystems

Ecological damage to aquatic systems has occurred from acidic deposition. As with forests, a number of interrelated factors associated with acidic deposition are responsible for undesirable changes. Acidification of aquatic ecosystems is not new. Studies of lake sediments suggest that increased acidification began in the mid-1800s, although the process has clearly accelerated since the 1940s (15). Current studies indicate there is significant S mineralization in forest soils impacted by acidic deposition and that the SO_4^{2-} levels in adjacent streams remain high, even though there has been a decrease in the amount of atmospheric-S deposition (24).

Geology, soil properties, and land use are the main determinants of the effect of acidic deposition on aquatic chemistry and biota. Lakes and streams located in areas with calcareous geology resist acidification more than those in granitic and gneiss materials (16). Soils developed from calcareous parent materials are generally deeper and more buffered than thin, acidic soils common to granitic areas (2). Land management decisions also affect freshwater acidity. Forested watersheds tend to contribute more acidity than those dominated by meadows, pastures, and agronomic ecosystems (8, 14, 20). Trees and other vegetation in forests are known to "scavenge" acidic compounds in fogs, mists, and atmospheric particulates. These acidic compounds are later deposited in forest soils when rainfall leaches forest vegetation surfaces. Rainfall below forest canopies (e.g., throughfall) is usually more acidic than ambient precipitation. Silvicultural operations that disturb soils in forests can increase acidity by stimulating the oxidization of organic N and S, and reduced S compounds such as FeS_2 (2).

A number of ecological problems arise when aquatic ecosystems are acidified below pH 5.0, and particularly below pH 4.0. Decreases in biodiversity and primary productivity of phytoplankton, zooplankton, and benthic invertebrates commonly occur (15, 16). Decreased rates of biological decomposition of organic matter have occasionally been reported, which can then lead to a reduced supply of nutrients (20). Microbial communities may also change, with fungi predominating over bacteria. Proposed mechanisms to explain these ecological changes center around physiological stresses caused by exposure of biota to higher concentrations of Al^{3+}, Mn^{2+}, and H^+ and lower amounts of available Ca^{2+} (15). One specific mechanism suggested involves the disruption of ion uptake and the ability of aquatic plants to regulate Na^+, K^+, and Ca^{2+} export and import from cells.

Acidic deposition is associated with declining aquatic vertebrate populations in acidified lakes and, under conditions of extreme acidity, of fish kills. In general, if the water pH remains above 5.0, few problems are observed; from pH 4.0 to 5.0 many fish are affected, and below pH 3.5 few fish can survive (23). The major cause of fish kill is due to the direct toxic effect of Al^{3+}, which interferes with the role Ca^{2+} plays in maintaining gill permeability and respiration. Calcium has been shown to mitigate the effects of Al^{3+}, but in many acidic lakes the Ca^{2+} levels are inadequate to overcome Al^{3+} toxicity. Low pH values also disrupt the Na^+ status of blood plasma in fish. Under very acidic conditions, H^+ influx into gill membrane cells both stimulates excessive efflux of Na^+ and reduces influx of Na^+ into the cells. Excessive loss of Na^+ can cause mortality. Other indirect effects include reduced rates of reproduction, high rates of mortality early in life or in reproductive phases of adults, and migration of adults away from acidic areas (16). Amphibians are affected in much the same manner as fish, although they are somewhat less sensitive to Al^{3+} toxicity. Birds and small mammals often have lower populations and lower reproductive rates in areas adjacent to acidified aquatic ecosystems. This may be due to a shortage of food due to smaller fish and insect populations or to physiological stresses caused by consuming organisms with high Al^{3+} concentrations.

REDUCING ACIDIC DEPOSITION EFFECTS

Damage caused by acidic deposition will be difficult and extremely expensive to correct, which will depend on our ability to reduce S and N emissions. For example, society may have to burn less fossil fuel, use cleaner energy sources and/or design more efficient "scrubbers" to reduce S and N gas entering our atmosphere. Despite the firm conviction of most nations to reduce acidic deposition, it appears that the staggering costs of such actions will delay implementation of this approach for many years. The 1990 amendments to the Clean Air Act are expected to reduce acid-producing air pollutants from electric power plants. The 1990 amendments established emission allowances based on a utilities' historical fuel use and SO_2 emissions, with each allowance representing 1 ton of SO_2 that can bought, sold or banked for future use (4, 6, 7), http://www.nnic.noaa.gov/CENR/NAPAP/. Short-term remedial actions for acidic deposition are available and have been successful in some ecosystems. Liming of lakes and some forests (also fertilization with trace elements and Mg^{2+}) has been practiced in European counties for over 50 years (16, 23). Hundreds of Swedish and Norwegian lakes have been successfully limed in the past 25 years. Lakes with short mean residence times for water retention may need annual or biannual liming; others may need to be limed every 5–10 years. Because vegetation in some forested ecosystems has adapted to acidic soils, liming (or over-liming) may result in an unpredictable and undesirable redistribution of plant species.

REFERENCES

1. Smith, W.H. Acid Rain. In *The Wiley Encyclopedia of Environmental Pollution and Cleanup*; Meyers, R.A., Dittrick, D.K., Eds.; Wiley: New York, 1999; 9–15.
2. Pierzynski, G.M.; Sims, J.T.; Vance, G.F. *Soils and Environmental Quality*; CRC Press: Boca Raton, FL, 2000; 459.
3. Wolff, G.T. Air Pollution. In *The Wiley Encyclopedia of Environmental Pollution and Cleanup*; Meyers, R.A., Dittrick, D.K., Eds.; Wiley: New York, 1999; 48–65.
4. U.S. Environmental Protection Agency, *Progress Report on the EPA Acid Rain Program*; EPA-430-R-99-011 U.S. Government Printing Office: Washington, DC, 1999; 20.
5. Forster, B.A. *The Acid Rain Debate: Science and Special Interests in Policy Formation*; Iowa State University Press: Ames, IA, 1993.
6. National Acid Precipitation Assessment Program Task Force Report, *National Acid Precipitation Assessment Program 1992 Report to Congress*; U.S. Government Printing Office: Pittsburgh, PA, 1992; 130.
7. National Science and Technology Council. *National Acid Precipitation Assessment Program Biennial Report to Congress: An Integrated Assessment.* 1998 (Accessed July 2001).
8. Charles, D.F., Ed. The Acidic Deposition Phenomenon and Its Effects: Critical Assessment Review Papers. *Effects Sciences*; EPA-600/8-83-016B U.S. Environmental Protection Agency: Washington, DC, 1984; Vol. 2.
9. McKinney, M.L.; Schoch, R.M. *Environmental Science: Systems and Solutions*; Jones and Bartlett Publishers: Sudbury, MA, 1998.
10. United Nations, World Bank and World Resources Institute, *World Resources: People and Ecosystems—The Fraying Web of Life*; Elsevier: New York, 2000.
11. *National Atmospheric Deposition Program (NRSP-3)/ National Trends Network. Isopleth Maps.* NADP Program Office, Illinois State Water Survey, 2204 Griffith Dr., Champaign, IL, 61820. 2000 (Accessed July 2001).
12. Council on Environmental Quality, *Environmental Quality, 18th and 19th Annual Reports*; U.S. Government Printing Office: Washington, DC, 1989.
13. Charles, D.F., Ed. *Acid Rain Research: Do We Have Enough Answers?* Proceedings of a Speciality Conference. Studies in Environmental Science #64, Elsevier: New York, 1995.
14. Kamari, J. *Impact Models to Assess Regional Acidification*; Kluwer Academic Publishers: London, 1990.
15. Charles, D.F., Ed. *Acidic Deposition and Aquatic Ecosystems*; Springer-Verlag: New York, 1991.
16. Mason, B.J. *Acid Rain: Its Causes and Effects on Inland Waters*; Oxford University Press: New York, 1992.
17. U.S. Environmental Protection Agency. Effects of Acid Rain: Human Health. *EPA Environmental Issues Website.* Update June 26, 2001 (accessed July 2001).
18. Marion, G.M.; Hendricks, D.M.; Dutt, G.R.; Fuller, W.H. Aluminum and Silica Solubility in Soils. Soil Science **1976**, *121*, 76–82.
19. Kennedy, I.R. *Acid Soil and Acid Rain*; Wiley: New York, 1992.
20. Reuss, J.O.; Johnson, D.W. *Acid Deposition and the Acidification of Soils and Waters*; Springer-Verlag: New York, 1986.
21. Likens, G.E.; Driscoll, C.T.; Buso, D.C. Long-Term Effects of Acid Rain: Response and Recovery of a Forest Ecosystem. Science **1996**, *272*, 244–246.
22. Linthurst, R.A. *Direct and Indirect Effects of Acidic Deposition on Vegetation*; Butterworth Publishers: Stoneham, MA, 1984.
23. Bush, M.B. *Ecology of a Changing Planet*; Prentice Hall: Upper Saddle River, NJ, 1997.
24. Alawell, C.; Mitchell, M.J.; Likens, G.E.; Krouse, H.R. Sources of Stream Sulfate at the Hubbard Brook Experimental Forest: Long-Term Analyses Using Stable Isotopes. Biogeochemistry **1999**, *44*, 281–299.
25. National Atmospheric Deposition Program. *Nitrogen in the Nation's Rain.* NADP Brochure 2000–01a (Accessed July 2001).

ACID SULFATE SOILS, DEFINITION AND CLASSIFICATION

Delvin S. Fanning
University of Maryland, College Park, Maryland, U.S.A.

INTRODUCTION

Pons (1) reported that acid sulfate soils have been known for ages and began to receive scientific attention in the 18th century when Linnaeus recognized them in the Netherlands with terms such as *argilla vitriolacea*, meaning "clay with sulfuric acid." Terms such as the Dutch *kattekleigronden*, or English "cat clay soils," or German *Maibolt* to imply hayfields affected by an evil spirit or *Gifterde* for poison earth, were applied to these soils to connote the mysterious and evil circumstances surrounding the difficulty in producing crops on them. The term acid sulfate soils is of more recent usage. It has gained popularity with the four international acid sulfate soils symposia (2–5) that have been held under the sponsorship of the International Land Reclamation Institute (ILRI), Wageningen, the Netherlands, and other organizations, and usage of the term is apparent in many other publications (e.g., 6–8) on these soils.

DEFINITION AND MAIN KINDS

In the broad sense, acid sulfate soils include all soils in which sulfuric acid may be produced, is being produced, or has been produced in amounts that have a lasting effect on main soil characteristics (1, 9). This definition includes potential, active, and postactive acid sulfate soils, which are broad genetic kinds as described in the succeeding three paragraphs.

Potential acid sulfate soils are anaerobic soils, commonly occurring in, or at one time formed in, coastal (tidal) sedimentary environments affected by sulfidization. Potential acid sulfate soils contain sulfide minerals at such levels in near surface horizons/layers that they are expected to generate, upon exposure to oxidizing conditions, sufficient sulfuric acid to drive the pH of these horizons/layers to ultralow levels. Under these conditions, most plants would be unable to grow on the soils and an active acid sulfate soil would then be recognized.

Active acid sulfate soils form where sulfide minerals (most typically iron sulfides and the mineral pyrite) have oxidized in near surface horizons and formed enough sulfuric acid, with insufficient neutralization, to have made the pH drop to ultralow levels. Most commonly the pH, as measured in water, is 3.5 or less, such that a *sulfuric horizon*, as defined later, is recognized. Some soil scientists would only require the pH to be 4.0 or below to recognize the active stage.

In post active acid sulfate soils weathering and pedogenesis have proceeded beyond the active stage to where sulfide minerals are no longer present in surface soil horizons and the pH in these horizons has risen to levels above that which would cause them to be recognized as *sulfuric horizons*.

A simpler definition of acid sulfate (8) is simply that they are soils that contain iron sulfides. This definition calls attention to the minerals that cause acid sulfate soils to become acid; however, it neglects to consider that how acidic the soils become depends on the balance between the acid forming substances (mainly iron sulfides) and the substances (minerals) that neutralize acidity, most commonly calcium carbonate minerals. This simpler definition also does not recognize the genetic distinction among the potential, active, and post-active stages of development of acid sulfate soils.

CLASSIFICATION IN SOIL TAXONOMY

Potential, active, and early postactive acid sulfate soils receive special recognition in *Soil Taxonomy* (10, 11). The taxonomic definition of *sulfidic materials* also permits the recognition of these acid-forming materials regardless of their depth in the soil-geologic column. This is useful for their recognition in construction activities such as mining, highway construction, and dredging operations, which can bring active acid sulfate soils into existence by exposing (to oxidation) *sulfidic materials* that previously occurred deep in the column or under deep water.

Diagnostic Characteristics

Sulfidic materials

Soil Taxonomy gives the following definition: Sulfidic materials contain oxidizable sulfur compounds. They are

Encyclopedia of Soil Science

mineral or organic soil materials that have a pH value of more than 3.5 and that, if incubated as a layer 1 cm thick under moist aerobic conditions (field capacity) at room temperature, show a drop in pH of 0.5 or more units to a pH value of 4.0 or less (1:1 by weight in water or in a minimum of water to permit measurement) within 8 weeks (10).

Sulfidic materials were defined to permit recognition of soil materials that lead to the formation of active acid sulfate soils if exposed to aerobic conditions at land surfaces.

Sulfidic materials when they are mineral soil materials typically have Munsell color chromas ≤1 and values ≤4 (12).

Sulfuric horizon

Following is the definition of sulfuric horizon as given in *Soil Taxonomy*: The sulfuric (L. *sulfur*) horizon is 15 cm or more thick and is composed of either mineral or organic soil material that has a pH value of 3.5 or less (1:1 in water or in a minimum of water to permit measurement) and shows evidence that the low pH value is caused by sulfuric acid. The evidence is *one or more* of the following: (a) Jarosite concentrations; (b) directly underlying sulfidic materials; *or* (c) 0.05 percent or more water-soluble sulfate (10).

The definition of the *sulfuric horizon* permits recognition of active acid sulfate soils and soil materials affected by active sulfuricization.

Classes

Potential and active acid sulfate soils are classified in *Soil Taxonomy* in the orders of *Entisols*, *Inceptisols*, and *Histosols*. Within these orders, acid sulfate soils are recognized at the great group and subgroup levels, which are intermediate categories—third and fourth in level from the top (order) level of the six categories of *Soil Taxonomy*. Mineral potential acid sulfate soils that have *sulfidic materials* within 50 cm of their surfaces without an overlying *sulfuric horizon* are classified in the great group of *Sulfaquents*. Organic potential acid sulfate soils that have *sulfidic materials* within 100 cm of their surfaces without a *sulfuric horizon* within 50 cm of their surface are classified as either *Sulfihemists* or *Sulfisaprists*.

Active acid sulfate soils are recognized in *Soil Taxonomy* as soils that have a *sulfuric horizon* with its upper boundary within 50 cm of the soil surface. Those that are mineral soils are classified as either *Sulfaquepts* or *Sulfudepts*, whereas those that are organic soils are classified as either *Sulfohemists* or *Sulfosaprists*. Most of these soils would be wet soils, belonging to an *Aqu*-suborder or are organic soils that formed under wet conditions, those with the *ist* stem; however, the *Sulfudepts* are better drained and would be the active acid sulfate soils of upland exposures, such as on mined land spoils that were not reclaimed properly.

Postactive acid sulfate soils are recognized in *Soil Taxonomy* only if they are early postactive. For example, *Sulfic Endoaquepts* are *Endoaquepts* that have either *sulfidic materials* or a *sulfuric horizon* or "a horizon 15 cm or more thick that has all of the characteristics of a sulfuric horizon, except that it has a pH between 3.5 and 4.0" within 150 cm of the mineral soil surface (10, 11). Many other soils beyond those recognized by *Soil Taxonomy* are considered to be postactive, such as certain *Alfisols* and *Ultisols* in Texas (13) and certain *Ultisols* in Maryland (14). It is important to recognize such soils because they do usually contain *sulfidic materials* at a depth that could give rise to new active acid sulfate soils if unearthed by land-moving operations and in order to better understand the morphology, genesis, and other aspects of these soils.

CLASSIFICATION IN OTHER SYSTEMS

Dent (6) has presented the ILRI system for the classification of acid sulfate soils, which considers acidity and potential acidity, salinity, soil composition and texture, degree of physical ripening, and profile form (the depth zone at which various properties occur). The (Food and Agriculture Organization) of the United Nations groups potential and active acid sulfate soils together in *Thionic* classes for purposes of showing these soils on the FAO/Unesco soil map of the world (15). Currently *Thionic Fluvisols*, *Thionic Gleysols* and *Thionic Histosols* are recognized (16).

REFERENCES

1. Pons, L.J. Outline of the Genesis, Characteristics, Classification and Improvement of Acid Sulphate Soils. In *Acid Sulphate Soils*; Proceedings of the International Symposium, Wageningen, The Netherlands, Aug 13–20, 1972; Dost, H., Ed.; International Land Reclamation Institute Pub. 18: Wageningen, The Netherlands, 1973; Vol. 1, 3–27.
2. Dost, H., Ed. *Acid Sulphate Soils*; Proceedings of the International Symposium, Wageningen, The Netherlands, Aug 13–29, 1972; International Land Reclamation Institute Pub. 18: Wageningen, The Netherlands, 1973.
3. Dost, H., van Breemen, N., Eds. *Proceedings of the Bangkok Symposium on Acid Sulphate Soils*; Second International Symposium on Acid Sulphate Soils, Bangkok, Thailand, Jan 18–24, 1981; International Land Reclamation Institute Pub. 31: Wageningen, The Netherlands, 1982.
4. Dost, H., Ed. *Selected Papers of the Dakar Symposium on Acid Sulphate Soils*; Dakar, Senegal, Jan 1986. International

Land Reclamation Institute Pub. 44: Wageningen, The Netherlands, 1982.

5. Dent, D.L., van Mensvoort, M.E.F., Eds. *Selected Papers of the Ho Chi Minh City Symposium on Acid Sulphate Soils*, Ho Chi Minh City, Vietnam, Mar, 1992; International Land Reclamation Institute Pub. 53: Wageningen, The Netherlands, 1993.
6. Dent, D. *Acid Sulphate Soils: A Baseline for Research and Development*; International Land Reclamation Institute Pub. 39: Wageningen, The Netherlands, 1993.
7. Kittrick, J.A., Fanning, D.S., Hossner, L.R., Eds. *Acid Sulfate Weathering*; Soil Science Society of America Special Pub. No. 10: Madison, WI. 1982; 234 pp.
8. Sammut, J. *An Introduction to Acid Sulfate Soils*; New South Wales Department of Agriculture: Wollongbar, New South Wales, Australia, 1997; 23 pp.
9. Fanning, D.S.; Burch, S.N. Coastal Acid Sulfate Soils. In *Reclamation of Drastically of Drastically Disturbed Lands*; Agronomy 41, Barnhisel, R.I., Darmody, R.G., Daniels, W.L., Eds.; American Society of Agronomy: Madison, WI, 2000; 921–937.
10. Soil Survey Staff *Soil Taxonomy*, 2nd Ed.; U.S. Dept. Agric. Handbook 436; U.S. Government. Printing Office: Washington, DC, 1999.
11. Soil Survey Staff, *Keys to Soil Taxonomy*, 8th Ed.; U.S. Government. Printing Office: Washington, DC, 1998.
12. Fanning, D.S.; Rabenhorst, M.C.; Bigham, J.M. Colors of Acid Sulfate Soils. In *Soil Color*; Soil Sci. Soc. Amer. Spec. Pub. No. 31, Bigham, J.M., Ciolkosz, E.J., Eds.; Soil Science Society of America: Madison, WI, 1993; 91–108.
13. Carson, C.D.; Fanning, D.S.; Dixon, J.B. Alfisols and Ultisols with Acid Sulfate Weathering Features in Texas. In *Acid Sulfate Weathering*; Soil Sci. Soc. Amer. Spec. Pub. No. 10, Kittrick, J.A., Fanning, D.S., Hossner, L.R., Eds.; Soil Science Society of America: Madison, WI, 1982; 127–146.
14. Wagner, D.P.; Fanning, D.S.; Foss, J.E.; Patterson, M.S.; Snow, P.A. Morphological and Mineralogical Features Related to Sulfide Oxidation Under Natural and Disturbed Land Surfaces in Maryland. In *Acid Sulfate Weathering*; Soil Sci. Soc. Amer. Spec. Pub. No. 10, Kittrick, J.A., Fanning, D.S., Hossner, L.R., Eds.; Soil Science Society of America: Madison, WI, 1982; 109–125.
15. Fanning, D.S.; Burch, S.N. Acid Sulphate Soils and Some Associated Environmental Problems. In *Soils and Environment*; Advances in Geoecology 30, Auerswald, K., Stanjek, H., Bigham, J.M., Eds.; Catena Verlag: Reiskirchen, Germany, 1997; 145–158.
16. FAO/Unesco, *Soil Map of the World, Revised Legend*; Food and Agriculture Organization of the United Nations: Rome, 1990.

ACID SULFATE SOILS, FORMATION

Martin C. Rabenhorst
Delvin S. Fanning
Steven N. Burch
University of Maryland, College Park, Maryland, U.S.A.

INTRODUCTION

Soils containing sulfide minerals that have not yet been oxidized through acid sulfate weathering are referred to as potential acid sulfate soils (potential acid SS). The processes involved in the formation and accumulation of sulfide minerals in soils leading to the formation of potential acid SS will be discussed first. Subsequently, processes related to the oxidation of sulfides in the formation of active acid SS will be examined. The extent of acid SS worldwide has been estimated to be approximately 12–15 MHa (1).

POTENTIAL ACID SULFATE SOILS AND SULFIDE MINERAL FORMATION—SULFIDIZATION

Biogeochemistry of Sulfide Mineral Formation

Several factors are required for sulfate reduction. These include a source of sulfate, a source of oxidizable carbon, reducing conditions and the presence of sulfate reducing bacteria (2). Any of these components could theoretically limit sulfate reduction. In saturated soil or sedimentary environments where the required factors are present, heterotrophic microbes utilize sulfate as an electron acceptor that becomes reduced to sulfide according to Eq. 1.

$$SO_4^{2-} + 10H^+ + 8e^- \rightarrow H_2S + 4H_2O \qquad (1)$$

Sulfate

Provided that the other required factors are met, the quantity of sulfate may limit the rate of sulfate reduction. Goldhaber and Kaplan (3) reported sulfate reduction to be independent of concentration when sulfate levels are above 10 mM (320 mg/l). Work by Haering (4), in Chesapeake Bay, indicated that sulfate levels may begin to limit sulfur accumulation in marsh soils when levels drop below 1 mM (32 mg/l). Some degree of sulfate reduction will continue as long as sulfate is present at minimal levels ($>$ 5–20 μM, 0.16–0.6 mg/l) (5). Because sulfate-reducing bacteria are better able to compete for electron-donating substrates than are methane-generating bacteria, methanogenesis is of minimal significance so long as sulfate levels are above 0.03–0.4 mM (6). Therefore, sulfate reduction dominates in brackish systems. In freshwater, sulfate reduction may become overshadowed by methanogenesis as sulfate is depleted.

Oxidizable organic carbon

The oxidation of organic matter provides the energy microorganisms need to facilitate sulfate reduction. Plant materials rich in labile components are more easily decomposed than humified soil organic matter or peat. In sediments low in organic matter, sulfate reduction may be limited by the paucity of oxidizable carbon. This can be demonstrated in thin sections from mineral horizons in tidal marsh soils where iron sulfide minerals have accumulated in pores occupied by decaying plant roots (Fig. 1). The intimate association of pyrite with the decomposing organic minerals, and its near absence from the surrounding soil matrix suggests that organic matter is limiting the formation of sulfide (7).

Reducing/saturated conditions

Because diffusion of gases through saturated soils and sediments is very slow, oxygen becomes depleted under saturated conditions and microbes which utilize other electron acceptors become active. Nitrate, Mn(IV) and Fe(III) are so utilized as the environment becomes progressively reduced (Fig. 2). If the conditions permit the entry of oxygen, then redox potentials may never become sufficiently low to foster sulfate reduction. More typically, diffusion of oxygen into a saturated soil or sediment is sufficiently slow, and if other necessary factors are present, sulfate reduction will occur. Fig. 2 illustrates that pH, as well as E_h, must be specified in assessing sulfur phase equilibria. For example, as the pH increases from 5 to 7, the minimum E_h at which sulfate reduction is expected decreases from approximately −50 to −200

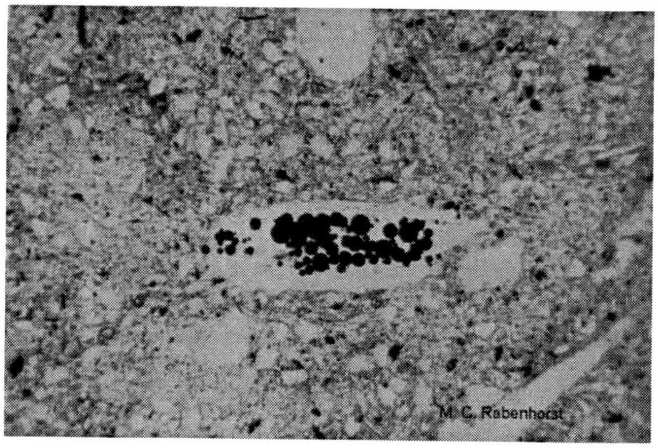

Fig. 1 Micrograph of a thin section from the mineral (Cg) horizon of a tidal marsh soil illustrating accumulation of pyrite framboids in the channel occupied by decaying plant roots. Plane polarized light; frame length 1.2mm.

(based on a SO_4^{2-} concentration of 10 mM and a pH_2S of 0.0001 atm).

Sulfate reducing bacteria

Some 15 genera of bacteria have been recognized as sulfate reducers including *Desulfovibrio*, *Desulfotomaculum*, and *Desulfobacter* (8, 9). These organisms thrive under strongly reducing conditions, but many are able to persist in aerobic conditions for significant periods of time. Thus, if the other factors necessary for sulfate reduction are present, sulfate reducing bacteria will also become active.

As with most heterotrophic bacteria, rates of sulfate reduction are temperature dependent. Optimum temperature for most sulfate reducers is 30–40°C (8), and the rate of sulfate reduction generally increase with temperature across this range. Some groups of sulfate reducers are thermophyllic and can function at temperature up to 85°C. Thus, in tropical coastal wetlands, sulfate reduction occurs all year round. In higher latitudes, where soil and sediment temperatures may approach biological zero, rates may become very slow during winter.

Reactive iron

Once formed, sulfide is available to form a variety of minerals provided there is adequate reactive iron present. Most of the iron enters coastal environments as iron oxides sorbed to the surface of clay and silt particles. When iron oxides in the sediments and marsh soils become reduced to Fe(II), they can form iron sulfide minerals. While monosulfide species may form first (Eq. 2), and minerals such as greigite (Fe_3S_4) may persist in recent sediments, disulfide forms such as pyrite are

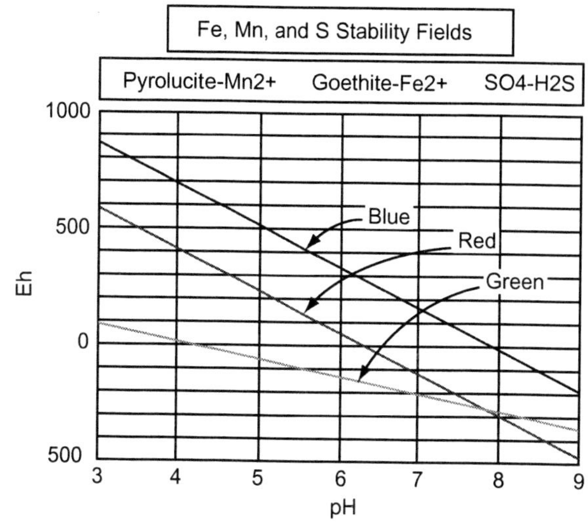

Fig. 2 pe-pH diagram illustrating location of stability fields for redox sensitive components. The sulfate-sulfide lines (Green) is based on a $(SO)_4^{2-}$ concentration of 10mM and a pH_2S of 0.001 atm. (Blue line separates the pyrolucite-Mn^{2+} stability fields; red line separates the goethite-Fe^{2+} stability fields.)

energetically more stable and will form at the expense of the monosulfides.

$$Fe^{2+} + S^{2-} \rightarrow FeS \qquad (2)$$

$$FeS \text{ and } (+S_X^{Y-}, \text{ loss of } e^-, \text{ or } +H_2S) \rightarrow FeS_2 + \text{various} \qquad (3)$$

Mechanisms for pyrite formation may follow several possible pathways including 1) reaction of monosulfide with polysulfide, 2) partial oxidation of monosulfide, and 3) reaction of monosulfides with H_2S (10) (Eq. 3). Sulfide itself has the ability to reduce Fe(III) to Fe(II) on the surface of iron oxides (11). Pyrite can occur either as small ($< 2\,\mu m$) individual crystals or as spherical clusters of crystals called framboids. In low organic mineral sediments, reactive iron is usually present in excess, resulting in a low degree of pyritization (12). However, in organic-rich soils iron may limit the accumulation of sulfide minerals, and the degree of pyritization is generally high. This has been demonstrated experimentally in salt marsh Histosols (13).

Environments of Sulfide Formation and Accumulation

It is clear that in environments which provide a source of oxidizable carbon and sulfate and which are sufficiently saturated to enhance reducing conditions, sulfate reducing bacteria will generate sulfide. If reactive iron is present, then solid phase ferrous minerals will accumulate. This process of *sulfidization* (14) is shown schematically in Fig. 3. The obvious settings for these processes are coastal marine and brackish environments, where sulfate is abundant. Under permanently submersed conditions, detrital carbon is added by flora and fauna to the sediment. In shallow water settings (< 3 m) where various pedogenic processes are at work, these accumulated sediments have been recognized as subaqueous soils (15) and are classified to reflect the sulfide components.

The soils of coastal marshes (in temperate environments) and mangroves (in tropical settings) also are ideal for sulfide formation and accumulation. The high primary productivity of these ecosystems (up to $3\,kg\,m^{-2}\,yr^{-1}$ in marshes and up to $5\,kg\,m^{-2}\,yr^{-1}$ in mangroves) (16) makes these an exceptionally good environment for sulfate reduction. Such soils may contain up to 20–30 g/kg of pyrite sulfur, and estimates of pyrite S accumulation rates in estuarine marshes are as high as $7\,g\,m^{-2}\,yr^{-1}$ (17).

Sulfate reduction can occur in other settings, so long as a source of sulfate is available. While generally small, atmospheric deposition of sulfate may be enough to induce sulfate reduction in the sediments of some interior freshwater lakes. Sulfate reduction has also been documented in prairie potholes where sulfate has apparently been contributed by the weathering of sulfur-bearing shales (18).

FORMATION OF ACTIVE ACID SULFATE SOILS—SULFURICIZATION

Chemistry of Sulfide Oxidation

Sulfides begin to oxidize once they are exposed to more oxidizing conditions. This occurs most often as a result of such human activities as drainage or dredging of sulfide-bearing soils or sediments, or the mining of sulfide bearing coal, but may also occur due to tectonic uplift or oceanic regression. Under humid or moist aerobic conditions, sedimentary sulfide minerals can oxidize chemically (19), but this is a slow process, probably due to particular rate-limiting reactions. Various microorganisms are adapted to oxidize sulfides either directly through sulfur transformations or by facilitating (catalyzing) such rate-limiting reactions as the oxidation of Fe(II) to Fe(III) (20). While there are many possible intermediate reactions in the oxidation of pyrite, the overall reaction is summarized in Eq. 4. One mole of pyrite eventually yields two moles of sulfuric acid and a mole of iron hydroxide.

$$FeS_2 + 3\tfrac{3}{4}O_2 + 3\tfrac{1}{2}H_2O \rightarrow 2H_2SO_4 + Fe(OH)_3 \qquad (4)$$

The oxidation of pyrite proceeds along two fronts (Fig. 4). First the S is oxidized (through intermediates) to sulfate yielding sulfuric acid and the remaining Fe(II).

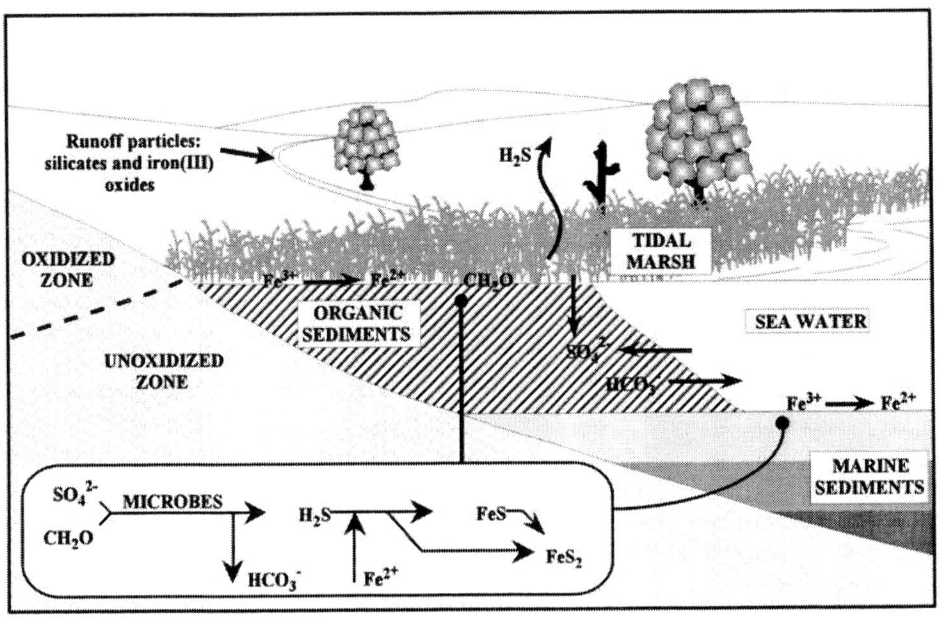

Fig. 3 Schematic diagram illustrating the generalized process of sulfidization which leads to the formation of iron sulfide minerals and potential acid SS. (Ref. 26.)

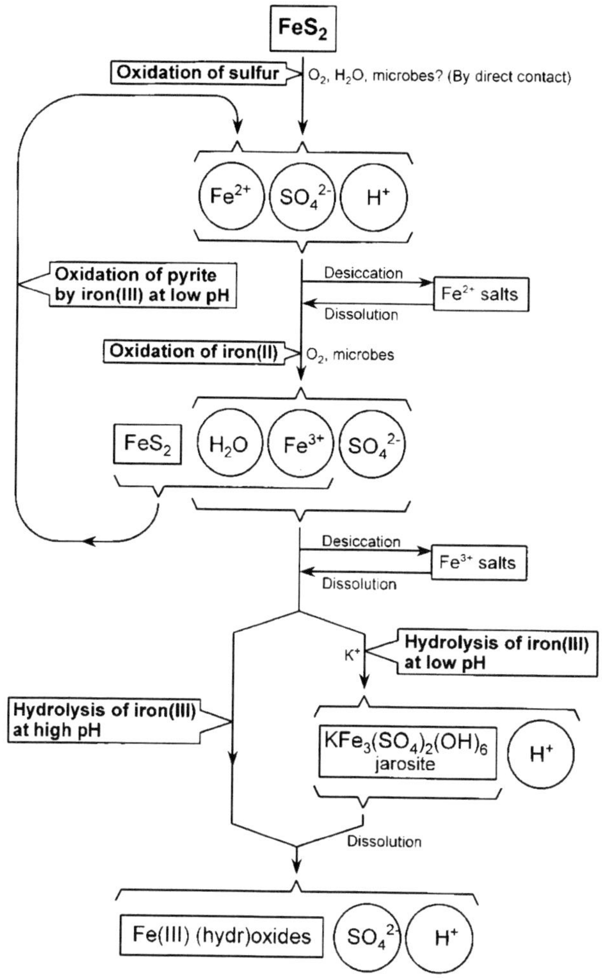

Fig. 4 Schematic diagram illustrating the generalized process of sulfuricization which involves the oxidation of iron sulfide minerals and the production of acidity and the formation of new sulfate and other minerals. (Ref. 26.)

The generated Fe(II) sulfate salts are very soluble and potentially mobile. Secondly Fe(II) is oxidized to Fe(III), which when hydrolyzed produces additional acid. At high pH, the oxidation of sulfide is accomplished with oxygen, but under low pH conditions sulfide is oxidized by Fe(III). Microorganisms such as *Thiobacillus ferrooxidans* facilitate this reaction by oxidizing Fe(II) to Fe(III). For more details refer to Gagliano and Bigham (21).

Other Aspects of Sulfuricization and Properties of Acid Sulfate Soils

Sulfuricization is the overall process by which sulfide-bearing minerals are oxidized, minerals are weathered by the sulfuric acid produced and new mineral phases are formed from the dissolution products (14, 21). When $CaCO_3$ minerals are present, the sulfuric acid reacts with them to form the mineral gypsum, according to Eq. 5.

$$CaCO_3 + H_2SO_4 + H_2O \rightarrow CaSO_4{\cdot}2H_2O + CO_2 \quad (5)$$

As long as sufficient $CaCO_3$ is present, the pH is prevented from becoming very low and the soil does not become acid. When insufficient acid neutralizing minerals are present, the oxidation of pyrite in soils will lower the pH. The pH of active SS commonly drops to below four and in extreme can go below two. As iron is oxidized and hydrolyzed, various iron minerals form in the soil including ferrihydrite, schwestmannite and goethite. If the soil pH falls below four while maintaining an oxidizing environment ($E_h > 400\,mV$) then the mineral jarosite ($KFe_3(SO_4)_2(OH)_6$) can form (23). Because jarosite forms under conditions of high E_h and very low pH, which can only develop from the generation of sulfuric acid, it is considered a diagnostic mineral for acid SS (24).

Jarosite has been reported in soils, which are not extremely acid and which may even contain carbonates (22). These are interpreted to be "postactive" acid SS, meaning that earlier in their pedogenic history, they had undergone acid sulfate weathering. Subsequently, the soil pH has risen due to weathering of silicate minerals or addition of eolian carbonates. Because the redox potential has remained strongly oxidized, the jarosite has persisted as a metastable species. A recent review of acid sulfate soils including a discussion of the modeling of associated processes was recently completed (25).

REFERENCES

1. Andriesse, W. Acid Sulfate Soils: Global Distribution. In *Encyclopedia of Soil Science*; Lal, R., Ed.; Marcel Dekker: New York, 2001.
2. Rabenhorst, M.C.; James, B.R.; Magness, M.C.; Shaw, J.N. Iron Removal from Acid Mine Drainage in Wetlands by Optimizing Sulfate Reduction. In *The Challenge of Integrating Diverse Perspectives in Reclamation*; Proc. Am. Soc. Surf. Mining Reclam. Spokane: Washington, 1993; 678–684.
3. Goldhaber, M.B.; Kaplan, I.R. Controls and Consequences of Sulfate Reduction Rates in Recent Marine Sediments. Soil Sci. **1975**, *119*, 42–55.
4. Haering, K.C. *Sulfur Distribution and Partitionment in Chesapeake Bay Tidal Marsh Soils*; M.S. Thesis University of Maryland: College Park, MD, 1986; 172.
5. Ingvorsen, K.; Zehnder, A.J.B.; Jorgensen, B.B. Kinetic of Sulfate and Acetate Uptake by Delulfobacter Postgatei. Appl. Environ. Microbiol. **1984**, *47*, 403–408.
6. Smith, D.W. Ecological Actions of Sulfate-Reducing Bacteria. In *The Sulfate-Reducing Bacteria: Contemporary*

Perspectives; Odom, J.M., Singleton, R., Jr. Eds.; Springer: New York, 1993; 161–188.

7. Rabenhorst, M.C.; Haering, K.C. Soil Micromorphology of a Chesapeake Bay Tidal Marsh: Implications for Sulfur Accumulation. Soil Sci. **1989**, *147*, 339–347.
8. Fauque, G.D. Ecology of Sulfate-Reducing Bacteria. In *Sulfate-Reducing Bacteria*; Barton, L.L., Ed.; Plenum Press: New York, 1995; 217–241.
9. Postgate, J.R. The Sulphate-Reducing Bacteria. *Ecology and Distribution*, 2nd Ed.; Ch. 7 Cambridge University Press: London, 1984; 107–122.
10. Rickard, D.; Schoonen, M.A.A.; Luther, G.W., III Chemistry of Iron Sulfides in Sedimentary Environments. In *Geochemical Transformations of Sedimentary Sulfur*; Vairavamurthy, M.A., Schoonen, M.A.A., Eds.; American Chemical Society: Washington, DC, 1995; 168–193.
11. Ghiorse, W.C. Microbial Reduction of Manganese and Iron. In *Biology and Anaerobic Microorganisms*; Zehnder, A.J.B., Ed.; John Wiley and Sons: New York, 1988; 305–331.
12. Griffin, T.M.; Rabenhorst, M.C. Processes and Rates of Pedogenesis in Some Maryland Tidal Marsh Soils. Soil Sci. Soc. Am. J. **1989**, *53*, 862–870.
13. Rabenhorst, M.C. Micromorphology of Induced Iron Sulfide Formation in a Chesapeake Bay (USA) Tidal Marsh. In *Micromorphology: A Basic and Applied Science*; Douglas, L.A., Ed.; Elsevier: Amsterdam, 1990; 303–310.
14. Fanning, D.S.; Fanning, M.C.B. *Soil Morphology, Genesis, and Classification*; John Wiley and Sons: New York, 1989; 395.
15. Demas, G.P.; Rabenhorst, M.C. Subaqueous Soils: Pedogenesis in a Submersed Environment. Soil Sci. Am. J. **1999**, *63*, 1250–1257.
16. Mitch, W.J.; Gosselink, J.G. *Wetlands*, 2nd Ed.; Van Nostrand Reinhold: New York, 1993; 722.
17. Hussein, A.H.; Rabenhorst, M.C. Modeling of Sulfur Sequestration in Coastal Marsh Soils. Soil. Sci. Soc. Am. J. **1999**, *63*, 1954–1963.
18. Arndt, J.L.; Richardson, J.L. Geochemistry of Hydric Soil Salinity in a Recharge-Throughflow-Discharge Prairie-Pothole Wetland System. Soil. Sci. Soc. Am. J. **1989**, *53*, 848–855.
19. Borek, S.L. Effect of Humidity of Pyrite Oxidation. In *Environmental Geochemistry of Sulfide Oxidation*; Alpers, C.N., Blowes, D.W., Eds.; American Chemical Society: Washington, DC, 1994; 31–44.
20. Nordstrom, D.K. Aqueous Pyrite Oxidation and the Consequent Formation of Secondary Iron Minerals. In *Acid Sulfate Weathering*; Kittrick, J.A., Fanning, D.S., Hossner, L.R., Eds.; Soil Sci. Soc. Am. Spec. Pub. No. 10: Madison, WI, 1982; 37–56.
21. Gagliano, W.B.; Bigham, J.M. Acid Mine Drainage. In *Encylopedia of Soil Science*; Lal, R. Ed.; Marcel Dekker: New York, 2001.
22. Carson, C.D.; Fanning, D.S.; Dixon, J.B. Alfisols and Ultisols with Acid Sulfate Weathering Features in Texas. In *Acid Sulfate Weathering*; Kittrick, J.A., Fanning, D.S., Hossner, L.R., Eds.; Soil Sci. Soc. Am. Spec. Pub. No. 10: Madison, WI, 1982; 127–146.
23. van Breeman, N. Genesis, Morphology, and Classification of Acid Sulfate Soils in Coastal Plains. In *Acid Sulfate Weathering*; Kittrick, J.A., Fanning, D.S., Hossner, L.R., Eds.; Soil Sci. Soc. Am. Spec. Pub. No. 10: Madison, WI, 1982, 95–108.
24. Fanning, D.S. Sulfate and Sulfide Minerals. In *Encyclopedia of Soil Science*; Lal, R., Ed.; Marcel Dekker: New York, 2001.
25. Ritsema, C.J.; van Mensvoort, M.E.F.; Dent, D.L.; Tan, Y.; van den Bosch, H.; van Wijk, A.L.M. Acid Sulfate Soils. In *Handbook of Soil Science*; Sumner, M.E., Ed.; CRC Press: Boca Raton, 1999; G121–G154.
26. Fanning, D.S.; Rabenhorst, M.C.; Burch, S.N.; Islam, K.R.; Tangren, S.A. Sulfides and Sulfates. In *Soil Mineralogy with Environmental Applications*; Dixon, J.B.; Schulze, D.G., Eds.; Soil Science Society of America Book Series #7: Madison, WI, 2002; 229–260.

ACID SULFATE SOILS, MANAGEMENT

Michael D. Melville
University of New South Wales, Sydney, New South Wales, Australia

Ian White
Australian National University, Canberra, Australian Capital Territory, Australia

A

INTRODUCTION

The definition, mineral characteristics, pedochemical processes, and distribution of acid sulfate soils are dealt with elsewhere in this publication. The biochemical processes in acid sulfate soils are in many ways analogous to those of acid mine drainage. Rain-induced acidic discharges from mine sites also have analogs in the hydrology of acid sulfate soils (1). Nevertheless, important differences exist between problems and management with acid mine drainage and with acid sulfate soil landscapes, mostly because of their respective scales, mineral and material contents, economics, and range of viable management options.

Much of the early research on acid sulfate soils was directed toward improving their agronomic usefulness (see successive International Symposia); however, recent research has concerned the environmental problems generated by their development and for their best management (1). There are three fundamental questions underlying the use of acid sulfate soil landscapes. Firstly, how best to use the existing environment productively while minimizing any increase in sulfide mineral oxidation; Secondly, how to neutralize effectively any existing or new acidity formed during a particular land use. Thirdly, how to eliminate or minimize downstream environmental impacts from acid sulfate drainage waters. The answers to these questions address the principles of ecological sustainable development.

EMERGING PROBLEMS

Many of the existing problems from acid sulfate soils has been discussed elsewhere (2). Only some of the more recent and emerging cases will be addressed here.

Ecological Impacts

Although the agronomic problems of acid sulfate soils have been recognized for centuries in Europe, in other parts of the world, such as Australia, widespread recognition of their existence, and their environmental impacts, have been more recent (3). Twenty-three kilometers of the Tweed River in eastern Australia were completely clarified and sterilized of fish, crustacea, and most benthic organisms after flooding in 1987 caused acidic drainage discharges from the drained acid sulfate soils of the river floodplain. It was the astute observations of the local government entomologist, published in an amateur fisherman's magazine, that identified the true cause of this devastating event that took up to 18 months to heal (4).

Acidic, aluminum-, and iron-rich drainage waters from acid sulfate soils cause gill tissue damage in fish that leads to their asphyxiation in waters containing adequate levels of dissolved oxygen (5). As well, these waters cause skin lesions in fish that allow the infection by the fungus *Aphanomyces invadans*, leading to Epizootic Ulcerative Syndrome (5). Such kills and diseases in fin fish, crustacea, and shellfish have important economic impacts for commercial and recreational fisheries and aquaculture in eastern Australia and southeast Asia (5).

Recently, elevated levels of dissolved iron from acid sulfate soil drainage have been implicated in blooms of the cyanobacterium *Lyngbya majascula* in coastal waters of eastern Australia (6). Toxins from these blooms have major impacts on most aquatic organisms but also pose a significant health risk to humans (6).

The use of acid sulfate soils as sites for brackish-water aquaculture in tropical coastal areas poses many problems with diseases and poor growth for shrimp and fin fish within the enterprises. More importantly, such enterprises can cause major downstream degradation from their effluent discharges, and a legacy of long-term impacts after aquaculture ventures become excessively acidified or disease prone and are abandoned.

Problems associated with saline scalding and acidification from contemporary sulfide mineral oxidation can occur in noncoastal areas (7). These contemporary inland acid sulfate soils are in part a product of human-induced, sulfate-rich groundwater hydrological changes, and the seasonality of the water balance associated with a Mediterranean climate.

An important outcome of increased acidification of drains and tidal tributaries associated with acid sulfate soils is the likely increase of acidophilic mosquito species, because of elimination of their natural predators such as fish. These insects pose a general nuisance problem for humans but more importantly, they may be vectors of existing or newly emerging, microbial and viral, human diseases such as malaria, dengue fever, or Japanese encephalitis (8). Since many humans already live near the coast, and are increasingly doing so, this is an important emerging issue.

Engineering Problems

Acidic pore water and its drainage from acid sulfate soils cause major problems with concrete and steel structures. Concrete in acidic, sulfate-rich conditions can require specialized concrete mixes, sacrificial surfaces, and especially pacified reinforcing steel. Nearly all steel products in acidic, chloride-rich conditions experience rapid and extensive corrosion. These problems can represent major costs from early replacement of existing infrastructure and in additional technologies for new installations.

Geomorphically in-filled estuarine mud basins often have a great thickness of sulfidic clay gels. These saturated materials are up to 80% volumetric water content and have bulk densities less than 0.7 t/m^3 (1). Such soft materials of small load-bearing capacity also occur in estuary dredging ponds (9). These saturated materials oxidize on excavation, and tend to flow under only moderate surface loads.

Recently in eastern Australia, such acid sulfate soil landscapes have been the site for extensive highway constructions. These engineering developments require the dewatering of the underlying soft material so as to support the road embankment above flood levels. This has been achieved with closely spaced (<2 m), geotextile, vertical wick drains and treatment of discharge water. This has required about 1000 km of wicks (in 10–20 m lengths) for approximately 30 km of the highway section. Some embankment subsidence and lateral shift have occurred with excessively rapid loading relative to drainage (Fig. 1). Not surprisingly such procedures and problems cause delays and have added approximately $100 M to construction costs.

Fig. 1 Failure of a major highway embankment constructed on unripe potential acid sulfate soils. The site is near Meleods Creek, on the Tweed River Floodplain NSW, Eastern Australia. (Photograph: A. Quirk.)

MANAGEMENT APPROACHES

The best management of acid sulfate soil involves a range of activities that addresses more than just the issues of minimizing the export of acidity into the downstream environment. This involves two issues, the creation of new acidity by sulfide mineral oxidation, and the management of the existing acidity in the landscape. The latter has been frequently ignored. Across the range of land uses of northern NSW, the existing acidity in the sulfuric layer averages approximately 50 t of sulfuric acid/hectare. Potential acidity represented in the sulfidic minerals of the deeper subsoil is many times this amount. Nevertheless, the annual discharge of acidity is <0.5 t/ha (10). The degree to which soil acidification has been caused by artificial drainage is uncertain, but natural processes are also involved. It is clear, however, that artificial drainage networks, which decrease by orders of magnitude the time of inundation of floodplain backswamps from their natural conditions, provide the conduit by which acidity is transferred rapidly to the estuary and potentially causes downstream impacts (1). Decreasing the density of drains has a major impact on the export of acidity. Laser leveling provides a technique for removing surface waters and decreasing drainage density and allows more land to be planted to crops (1).

The actual impact caused by rain-induced acidic discharges depends upon the magnitude of the discharge relative to the dilution by upland flows and neutralization capacity of the receiving waters. Land users on acid sulfate soil floodplains must appreciate that other parts of the estuarine ecosystem depend on the dissolved alkalinity of the receiving water, consumed during acidity neutralization. Floodplain land uses should be undertaken so as to avoid creating any new acidification and minimizing the export of any acidity.

Best management of acid sulfate soils requires knowledge of their distribution; the depth of the sulfidic layer from the soil-surface; the acidity stored in the sulfuric layer; the hydrological behavior of the soil profiles, landscape, and drainage system; the climatic regime; and the magnitude, tidal characteristics, and water quality of the receiving waters (1). These are infrequently considered.

Education and Assessment

Australian soil scientists were generally slow in appreciating both the presence and the environmental problems from acid sulfate soils, and in raising the awareness of the public, land managers, and policy makers. The first national symposium on acid sulfate soils was only as recent as 1993 (3). Nevertheless, in the past 10 years a major shift has occurred and now several states and the Commonwealth of Australia have included acid sulfate soil management in their environmental policies (11). Australia is one of the few countries with a National Strategy on Acid Sulfate Soils. This shift in NSW, has been greatly helped by publication of "Acid Sulphate Soil Risk Maps" (12) for all of the state's coastline (scale 1:25,000). These maps are being used as the basis for land-use planning instruments ("Local Environment Plans; LEPs") of local government authorities. These LEPs require submission of a Development Application for any activity that disturbs more than 1 tonne of acid sulfate soil.

The sugar industry, a major user of acid sulfate soils in NSW, has been granted statewide exemption from these requirements for normal farming and drain-cleaning activities because each of the 700 cane growers has signed a contract to comply with an approved code of best management practice for acid sulfate soils. Less than 10 years ago, the sugar industry was in a state of denial about the existence and problems of acid sulfate soils in its cane lands. The industry's new sense of land stewardship with acid sulfate soils is exemplary and indications are that economic rewards from up to 30% increase in yields are occurring. Education of the cane growers has been helped through an acid sulfate soil survey demonstration and assessment on each grower's land.

Oxidation Prevention

A primary preventative consideration with acid sulfate soils should be an avoidance strategy. Such a strategy includes the decision not to drain a potential acid sulfate soil wetland, and to divert or relocate a proposed land use to an alternative site. If use of the site is unavoidable then treatment to prevent sulfide mineral oxidation and export of any existing acidity is necessary. This might involve separation of the sulfidic material, and its capping with material that is impermeable to oxygen and rainfall infiltration. The use of other sulfidic clays as capping materials must be avoided (9). Bactericide treatments have limited application.

Neutralization

Neutralization of acid sulfate soil acidity by the alkalinity of seawater is the ultimate natural process in the geochemical sulfur cycle. The use of this approach with human-induced acidification and increased acidity discharge needs to be approached with caution. In eastern Australia where existing acidity concentrations can exceed 50 t/ha, application of sufficient crushed limestone or other neutralizing material is impracticable for agriculture. A far better option is for land-use practices that minimize acid export and for strategic application of crushed limestone to drains through which the acidity is exported. These practices can be incorporated into normal activities at the individual paddock scale of intensive agriculture such as with sugarcane. With low-value land uses, such as grazing, a more communal approach with some external funding may be necessary. The use of emerging technologies from acid mine drainage experience may be useful (13).

REFERENCES

1. White, I.; Melville, M.D.; Wilson, B.P.; Sammut, J. Reducing Acidic Discharges from Coastal Wetlands in Eastern Australia. Wetlands Ecol. Mgmt **1997**, *5*, 55–72.
2. Dent, D. *Acid Sulphate Soils: A Baseline for Research and Development*; ILRI Publ. No. 39, Institute for Land Reclamation and Improvement: Wageningen, The Netherlands, 1986.
3. Bush, R.T. *Proceedings of the National Conference on Acid Sulphate Soils*; Coolangatta, Australia, June 24–25, 1993; NSW Agriculture: Wollongbar, Australia, 1993.
4. Easton, C. The Trouble with the Tweed. Fishing World **1989**, 58–59, March.
5. Sammut, J.; Melville, M.D.; Callinan, R.B.; Frazer, G.C. Estuarine Acidification: Impacts on Aquatic Biota of Draining Acid Sulphate Soils. Aust. Geogr. Stud. **1995**, *33*, 89–100.
6. Dennison, W.C.; O'Neil, J.M.; Duffy, E.; Oliver, P.; Shaw, G. Blooms of the Alga *Lyngbya Majascula* In Coastal Waters of Queensland. In *Proceedings of the International Symposium on Marine Cyanobacterium*; Charpy, L., Larkum, A.W.D., Eds.; Bulletin de L'Institut Oceanographique: Monaco, 1997; 632.
7. Fitzpatrick, R.W.; Hudnall, W.H.; Self, P.G.; Naidu, R. Origin and Properties of Inland and Tidal Acid Sulphate

Soils in South Australia. In *Selected Papers from the Ho Chi Minh City Symposium on Acid Sulphate Soils*; ILRI Publ. No. 53, Dent, D.L., van Mensvoort, M.E.F., Eds.; Ho Chi Minh City, Vietnam, March, 1992; Institute for Land Reclamation and Improvement: Wageningen, The Netherlands, 1993; 71–80.

8. Garrett, L. *The Coming Plague: Newly Emerging Diseases in a World Out of Balance*; Penguin Books: New York, 1994.
9. Fanning, D.S.; Burch, S.N. Coastal Acid Sulfate Soils. In *Reclamation of Drastically Disturbed Lands*; Agronomy 41, Barnhisel, R.I., Darmody, R.G., Daniels, W.L., Eds.; American Society of Agronomy: Madison, WI, 2000; 921–937.
10. Wilson, B.P.; White, I.; Melville, M.D. Floodplain Hydrology, Acid Discharge and Change in Water Quality Associated with a Drained Acid Sulfate Soil. Mar. Freshwater Res. **1999**, *50*, 149–157.
11. Stone, Y., Ahern, C.R., Blunden, B., Eds. *Acid Sulfate Soils Manual*; Acid Sulfate Soil Management Advisory Committee, NSW Agriculture: Wollongbar, NSW, Australia, 1998.
12. Naylor, S.D., Chapman, G.A., Atkinson, G., Murphy, C.L, Tulau, M.J., Flewin, T.C.; Milford, H.B., Morand, D.T., Eds. *Guidelines for the Use of Acid Sulphate Soil Risk Maps*; NSW Soil Conservation Service, Department of Land and Water Conservation: Sydney, Australia, 1995.
13. Waite, T.D.; Desmier, R.; Melville, M.D.; Macdonald, B.C.T. Preliminary Investigations into the Suitability of Permeable Reactive Barriers for the Treatment of Acid Sulfate Soil Discharge. In *Handbook of Groundwater Remediation Using Permeable Reactive Barriers*; Naftz, D., Morrison, S.J., Davis, J.A., Fuller, C.C., Eds.; Ch 3, Academic Press: Amsterdam, 2002, 68–105.

ACID SULFATE SOILS, PROBLEMS

Martin C. Rabenhorst
Delvin S. Fanning
Department of Natural Resource Sciences and Landscape Architecture, University of Maryland, College Park, Maryland, U.S.A.

INTRODUCTION

What Are Acid Sulfate Soils?

Acid sulfate soils (acid SS) are, or have the potential to become, extremely acidic due to the oxidation of naturally occurring, reduced sulfur compounds. Oxidation forms sulfuric acid causing extreme soil acidity when there is inadequate buffering or neutralizing capability within the soil (1). In acid SS, pH values <4 are common, and in extreme instances, pH values below 2 may develop. Soils are sometimes referred to as potential acid SS prior to oxidation of the sulfide minerals, or active acid SS once the extreme acidity has developed (2).

In most acid SS, the reduced sulfur compound is the mineral pyrite (FeS_2), but less commonly elemental sulfur, monosulfide, or polysulfide minerals can cause the same problem. Therefore, acid SS may form wherever sulfide minerals accumulate. The environmental requirements for the formation of sulfide have been specified by numerous workers (1). Because its formation requires sulfate to be present, most acid SS occur in coastal areas although there are cases where acid SS have formed from older rocks or sediments, which at one time were near the coast but are now located in interior settings.

Sulfides within geological deposits remain stable indefinitely in anaerobic environments. Active acid SS form when environmental shifts or alterations expose sulfides to an oxidizing environment. This can be a natural process such as coastal regression or tectonic coastal uplift, or it may be the result of human activity such as drainage of coastal areas or earthmoving activities during construction or mining. Whatever the cause, once placed in an aerobic environment, particular bacteria oxidize sulfide minerals and produce sulfuric acid. When these soil materials lack minerals to adequately buffer or neutralize the acidity generated, extremely low pH conditions may develop leading to the formation of active acid SS (1).

ENVIRONMENTAL PROBLEMS ASSOCIATED WITH ACID SULFATE SOILS

Land Disturbance

Mining

Large-scale earthmoving activity is one important cause of acid sulfate soils. Mining activity was one of the first recognized (3). Many coal deposits formed in coastal environments contain pyrite either within the coal itself or in the surrounding rocks. Prior to recognition of the acid SS problem, pyrite-bearing soil was commonly left unattended resulting in the formation of active acid SS. Recent soil surveys in the Appalachian region of the U.S. include map units for active acid SS such as Sulfudepts. These soils have very low pH in upper horizons, large components of rock fragments, and a high density due to compaction by heavy equipment. For these reasons, extensive areas of abandoned mined lands remained nearly unvegetated for decades. In addition to the problem of acid soils, water [acid mine drainage (AMD)] (4) from such acid SS degrades aquatic environments.

Massive additions of lime are usually needed to neutralize soil acidity so that hardy plants may become established. In the last few decades, progress has been made in the U.S. in recognizing acid SS problems during mining, and both federal (Surface Mining and Control Act of 1977, SMCRA) and state regulations now require nonpyritic "topsoil" to be stored separately from the sulfide-bearing materials. During land reclamation, the topsoil is returned to the soil surface providing a better environment for plant establishment and growth.

New mineland reclamation strategies have largely eliminated the problem of barren acid SS on recently mined land, however, AMD still persists (4). In addition to sulfuric acid, drainage waters carry high levels of iron, aluminum and manganese, which detrimentally impact the microbial and invertebrate communities of streams (5, 6). Such waters often become uninhabitable by crustaceans, fish or other organisms further up the food chain.

Urban development

Geological weathering generally removes sulfides from the upper zone. However, pyrite may be found at depths ranging between 2 and 20 m in some portions of the mid-Atlantic region of the U.S. Excavation and leveling associated with building homes, commercial sites or the construction of highways may sometimes expose sulfide-bearing deposits. Dark gray colors associated with sulfide-bearing materials (7) and their near-neutral pH when unoxidized sometimes mislead people into viewing this potential acid SS material as good "topsoil." Unaware of the hazard, sulfide-bearing materials have sometimes been spread unintentionally over the land surface, leading to the formation of active acid SS.

Acid SS formed during urban development pose the same problems as mined lands. Extreme soil acidity presents difficulties in establishing turfgrass and ornamentals. In addition, corrosion of utility structures can be serious when concrete pipes, piers, drainage ways, foundations, streets and sidewalks are readily attacked by sulfuric acid. The combination of acidity and salinity is particularly corrosive to steel. These acid SS may accumulate ferrous sulfate salts during dry periods and release them into surface waters during rainfall events. There are anecdotal (if not documented) accounts of fish kills in creeks and streams in urbanizing watersheds where acidity from acid SS has been the likely cause.

Dredging

Opening and maintaining channels for marine traffic represent a nearly continuous operation in many port cities. In the eastern U.S., major cities including Richmond, Washington, Baltimore, Philadelphia; and New York were built near the upper limit of navigable waters. When the water bodies are saline or brackish (containing sulfate), materials dredged from the channels often contain sulfide minerals and represent potential acid SS. Where industry has contributed metals into the estuarine environment, acid SS problems in dredged materials (DM) may be further complicated by the presence of heavy metals in the acid soil environment.

Overboard disposal of DM can lead to release of sorbed and sulfide-bound metals (8). Application of DM to land, or creation of new land, may be viewed as good alternatives to overboard disposal of DM, but serious problems still remain. Acid SS formed in materials dredged from Baltimore Harbor, for example, had pH values of 2.5 and contained alarmingly high levels of Zn, Ni, and Cu. In the 1980s and 1990s, approximately 600 ha of land were created in the northern part of Chesapeake Bay through the deposition of DM, in an area known as Hart-Miller Island, with similar acid SS problems.

Agriculture

Effects on plant growth

Reclamation of coastal wetlands containing sulfides has been accomplished successfully in the Netherlands and northern Germany, especially where carbonates in the sediments neutralize the sulfuric acid generated through sulfide oxidation. However, there are also cases where the attempted reclamation of coastal wetlands has been disastrous, with soils developing both acidity and salinity problems (such as in Guinea-Bissau in West Africa) (9, 10). Portions of Southeast Asia also have extensive agriculture in areas plagued with acid SS problems, such as in the delta regions of Viet Nam, Thailand, Malaysia, Indonesia, China, and India (11), and similar problems also occur in Australia. The agricultural problems of acid SS are often associated with metal toxicity and nutrient deficiencies.

Acid drainage

Even in areas where agricultural use of coastal acid SS may not pose immediate problems for crops, there may still be environmental difficulties. For example, in areas of sugar cane production in Australia on acid SS, the cane survives by sending roots down into the unoxidized (and nonacidic) portion of the soil. Nevertheless, acidity and metals from these acid SS may be transported through drainage waters into the rivers and estuaries where dramatic fish kills have been reported (12). Strategies to minimize acid drainage from agricultural areas are under development (13, 14).

REMEDIATION STRATEGIES FOR ACID SULFATE SOIL PROBLEMS

Nondisturbance

The old adage "an ounce of prevention is worth a pound of cure" is especially applicable to acid SS, particularly where the problems are related to land disturbance. It is far better and far easier to avoid the exposure and oxidation of sulfide-bearing soil materials than it is to try to deal with the acidic aftermath. Using geomorphology and modeling techniques, recent efforts have been made to predict the depths at which sulfides are likely to be encountered within upland landscapes (15). This can help prevent the inadvertent exposure of sulfide-bearing materials during construction activity. In cases such as coal and lignite mining, where disturbance may be unavoidable, it has been effective (and is now mandated) to separate the sulfide-bearing material from that which is not sulfidic. During landscape reclamation, the nonsulfidic materials are replaced at the soil surface, preventing the development of acid SS in the upper rooting zone for plants.

Heavy Liming

Extremely high rates of liming have, in some cases, been proven effective in neutralizing the sulfuric acid in acid SS. In order to accomplish this, however, liming rates of 25–75 tons/acre may be needed and, in some cases, rates as high as 150 tons/acre have been used. It is important that sufficient lime be added to neutralize both the active acidity and also the potential acidity that remains in yet unoxidized sulfide minerals. A limitation of liming is that it usually can neutralize only acidity within the uppermost portion of the soil (15–30 cm), leaving the subsoil extremely acidic.

Natural Processes of Ripening or Flooding

Rather than trying to quickly neutralize acidity that is being generated through acid sulfate weathering, some have advocated permitting, or even enhancing, the natural oxidation of pyrite in acid SS. This process has been termed *ripening*. Allowing the sulfides to oxidize naturally for a period of time (estimates range from 10 to 50 years) makes the soil more easily neutralized by liming. On the other hand, some have advocated altering soil drainage conditions by flooding or saturation to cause the sulfate to revert to sulfide. While such strategies may be able to prevent or slow the oxidation of sulfides, generally there is insufficient energy (unoxidized organic matter) in the systems to induce significant sulfate reduction in soils that have already become acidic.

Strategies for Acid Drainage

Neutralization

One way to remediate acid drainage has been through direct neutralization of the surface water, either in ponds where precipitation products (usually iron oxides) may accumulate or within the streams themselves. The first efforts involved lining the streambed with limestone gravel. The high pH of the oxygenated, iron-rich water resulted in the precipitation of Fe and Al oxide coatings directly on the gravel. This *armoring* of the gravel quickly reduced their effectiveness in neutralizing acidity. Automatic lime dosers have been developed that periodically add pulverized limestone to a stream based on monitored water pH values. The fine-grained nature of the lime eliminates the problem of armoring, raises the pH of the water, and lowers the levels of soluble metals. The North Branch of the Potomac River, which was once terribly polluted by AMD, has been successfully restored as a recreational trout fishery through the use of lime dosers. Recent efforts have attempted to use some nontraditional or waste materials to neutralize acidified waters (16). Anoxic limestone drains, a variation on neutralization using limestone, were designed to intercept discharging groundwater before it emerged at the land surface. By isolating limestone gravel from oxygenated water below ground using impermeable plastic, it is thought that armoring of the gravel can be prevented (17).

Wetlands

Over the last 10–20 years, constructed wetlands have emerged as a treatment for AMD. Much of the published work emphasized treatment through such processes as neutralization and oxidation, which one might expect to be better accomplished in more aerobic environments than wetlands (18). Some have focused more on wetland processes such as sulfate reduction (19), and it has been shown that properly designed and engineered wetland systems can effectively remediate the negative effects of AMD through sulfidization and generation of bicarbonate (20).

REFERENCES

1. Rabenhorst, M.C.; Fanning, D.S.; Burch, S.N. Acid Sulfate Soils: Processes of Formation. In *Encyclopedia of Soil Science*; Lal, R., Ed.; Marcel Dekker: New York, 2001.
2. Fanning, D.S. Classification of Acid Sulfate Soils. In *Encyclopedia of Soil Science*; Lal, R., Ed.; Marcel Dekker: New York, 2001.
3. Barnhisel, R.I.; Powell, J.L.; Akin, G.W.; Ebelhar, M.W. Characteristics and Reclamation of Acid Sulfate Mine Spoils. In *Acid Sulfate Weathering. Kittrick*; Spec. Pub. Ser. No. 10, Kittrick, J.A., Fanning, D.S., Hossner, L.R., Eds.; Soil Sci. Soc. Am. Madison, WI, 1982; 225–234.
4. Gagliano, W.B.; Bigham, J.M. Acid Mine Drainage. In *Encyclopedia of Soil Science*; Lal, R., Ed.; Marcel Dekker: New York, 2001.
5. Niyogi, D.K.; Lewis, A.W.M.; McKnight, D.M. Litter Breakdown in Mountain Streams Affected by Mine Drainage: Biotic Mediation of Abiotic Controls. Ecol. Appl. **2000**, *11*, 506–516.

6. Cherry, D.S.; Currie, R.J.; Coucek, D.J.; Latimer, H.A.; Trent, G.C. An Integrative Assessment of a watershed Impacted by Abandoned Mined Land Discharges. Environ. Poll. **2001**, *111*, 377–388.
7. Fanning, D.S.; Rabenhorst, M.C.; Bigham, J.M. Colors of Acid Sulfate Soils. In *Soil Color*; Spec. Pub. Ser. No. 31, Bigham, J.M., Ciolkosz, E.J., Eds.; Soil Sci. Soc. Am: Madison, WI, 1992; 91–108.
8. Simpson, S.L.; Apte, S.C.; Batley, G.E. Effect of Short-Term Resuspension Events on the Oxidation of Cadmium, Lead, and Zinc Sulfide Phases in Anoxic Estuarine Sediments. Environ. Sci. Tech. **2000**, *34*, 4533–4537.
9. Fanning, D.S.; Burch, S.N. Acid Sulphate Soils and Some Associated Environmental Problems. Adv. GeoEcology. **1997**, *30*, 145–158.
10. Fanning, D.S.; Burch, S.N. Coastal Acid Sulfate Soils. In *Reclamation of Drastically Disturbed Lands*; Agron. Monogr. 41, Barnhisel, R.I., Darmody, R.G., Daniels, W.L., Eds.; ASA, CSSA, SSSA: Madison, WI, 2000; 921–937.
11. Mathew, E.K.; Panda, R.K.; Nair, M. Influence of Subsurface Drainage on Crop Production and Soil Quality in a Low-Lying Acid Sulphate Soil. Agric. Water Manage. **2001**, *47*, 191–209.
12. Cook, F.J.; Hick, W.; Gardner, E.A.; Carlin, G.D.; Froggatt, D.W. Export of Acidity in Drainage Water from Acid Sulphate Soils. Mar. Poll. Bull. **2000**, *41*, 319–326.
13. Yang, X.H.; Zhou, Q.M.; Melville, M. An Integrated Drainage Network Analysis System for Agricultural Drainage Management—Part 2: The Application. Agric. Water Manage. **2000**, *45*, 87–100.
14. Blunden, B.G.; Indraratna, B. Evaluation of Surface and Groundwater Management Strategies for Drained Sulfidic Soil Using Numerical Simulation Models. Austr. J. Soil Res. **2000**, *38*, 569–590.
15. Valladares, T.M. Estimating Depth to Sulfide-Bearing Sediments in the Maryland Coastal Plain: A Pedolgeomorphic Modeling Approach MS Thesis Univ. Maryland. College Park, MD, 1998; 232.
16. Chtaini, A.; Bellaloui, A.; Ballivy, G.; Narasiah, S. Field Investigation of Controlling Acid Mine Drainage Using Alkaline Paper Mill Waste. Water Air Soil Poll **2001**, *125*, 357–374.
17. Skousen, J.G.; Sextone, A.; Ziemkiewicz, P.F. Acid Mine Drainage Control and Treatment. In *Reclamation of Drastically Disturbed Lands*; Agron. Monogr. 41, Barnhisel, R.I., Darmody, R.G., Daniels, W.L., Eds.; ASA, CSSA, SSSA: Madison, WI, 2000; 131–168.
18. Henrot, J.; Wieder, R.K.; Heston, K.P.; Nardi, M.P. Wetland Treatment of Coal Mine Drainage: Controlled Studies of Iron Retention in Model Wetland Systems. In *Constructed Wetlands for Wastewater Treatment: Municipal, Industrial, and Agricultural*; Hammer, D.A., Ed.; Lewis Pub: Chelsea, MI, 1989; 793–800.
19. McIntire, P.E.; Edenborn, H.M. The Use of Bacterial Sulfate Reduction in the Treatment of Drainage from Coal Mines. In *Proceedings of the 1990 Mining and Reclamation Conference and Exhibition*; Skousen, J., Sencindiver, J., Samuel, D., Eds.; West Virginia University: Morgantown, WV, 1990; Vol. II., 409–415.
20. Rabenhorst, M.C.; James, B.R.; Magness, M.C.; Shaw, J.N. Iron Removal from Acid Mine Drainage in Wetlands by Optimizing Sulfate Reduction. *The Challenge of Integrating Diverse Perspectives in Reclamation*; Proc. 10th Ann. Meet. Am. Soc. Surf. Mining Reclam. ASSMR: Spokane, WA, 1993; 678–684.

AERATION MEASUREMENT

R.E. Sojka
Soil Scientist, Kimberly, Idaho, U.S.A.

H.D. Scott
University of Arkansas, Fayetteville, Arkansas, U.S.A.

A

INTRODUCTION

Soil oxygen enables aerobic respiration of plant roots and soil micro- and meso-flora and fauna. Its availability can be limited by soil wetness, compaction, discontinuous pores, or high respiration in moist soil due to elevated soil temperature or incorporation of fresh organic substrate. With oxygen depletion, soil redox potential shifts from oxidative to reducing conditions, hampering plant growth because of less efficient metabolic pathways and release into soil of toxic by-products of reduction chemistry or anaerobic respiration. Several texts are excellent sources for fundamental soil aeration concepts (1–3).

Measurements of soil aeration fall into three categories: "capacity," volume of gas-filled void space; "Intensity," partial pressure or concentration of oxygen (or other gases) in the voids; and "transport rate," the rapidity at which oxygen can be supplied to a point in the soil. Measurement difficulty increases in the order capacity $<$ intensity $<$ rate, as do the value and insight of the measurements.

CAPACITY MEASUREMENT

Capacity describes the ability of soil to contain air. Soil capacity parameters include total porosity, void ratio, relative saturation and air-filled porosity. These parameters are calculated from simple measurements of soil volume, particle density, bulk density and water content. Capacity has been used to understand plant growth and yield for over a century. A "rule of thumb" associates impaired plant growth with $<$10% soil air volume. Soil attributes affecting capacity include texture, structure, water content, clay mineralogy and sodium adsorption ratio (SAR). Sandy soils have less total porosity than clays, but in sands the pores are large and well-connected. Percent void space tends to increase in finer-textured (more clay), less-compact soils. Because clays have many small unconnected pores, and retain water more readily than coarser textured soils (sands), plants tend to suffer oxygen limitation more commonly in clays, despite their greater porosity. High soil sodium content or smectitic clay mineralogy (swelling clays) can exacerbate this tendency. Well-aggregated clays often avoid poor aeration because structure enhances macro-porosity. Inter-aggregate pores are larger, better connected and better drained than smaller intra-aggregate pores which usually contain more water.

INTENSITY MEASUREMENT

Intensity (partial pressure or concentration) measurements have been facilitated by instrumentation allowing rapid measurement of oxygen and other gas concentrations. The ambient atmosphere is 78% N and 21% O_2. The remaining gases total 1%. Ambient CO_2 is about 0.03%. In soil air, the O_2 concentration is $<$21%. The drop in O_2 below 21% generally corresponds to the CO_2 concentration increase, due to respiration of roots and soil organisms. In soil air, CO_2 is commonly 0.3–1%, but can be much higher. In warm wet soil with freshly incorporated organic matter, or where carbonates are abundant, CO_2 can rise above 10%, and, where drainage is restricted, can reach 20%. When water logging occurs and reducing conditions prevail, a few percent by volume of gaseous products of reducing chemistry or non-oxidative metabolism, e.g. methane or nitrous oxide, can be present.

Soil air oxygen concentration can be measured using various analytical techniques, depending mostly on whether air is withdrawn from the soil or analyzed in-situ (2, 4, 5). Withdrawn soil air samples have an interpretation problem stemming from over-representation of macro-pore composition and/or mixing during convective extraction from the heterogenous pore sites within the soil matrix. Buried diffusion cavities (artificial porous voids) are sometimes used for sampling points, but questions remain as to representativeness of cavity-equilibrated air. Once withdrawn, soil air samples can be analyzed by wet chemistry, paramagnetic or polarographic methods, or using gas chromatography (6, 7). Paramagnetic and polarographic instruments can be used in situ, but they still have limitations. Paramagnetic analyzers require gas flows of tens of cubic centimeters over the sensors. Polarographic soil oxygen sensors are usually membrane-covered. Double membrane probes are used in situ, to overcome calibration shifts caused by condensation on sensors. Sensors can be small enough to measure oxygen between small aggregates or within large aggregates (8). Samples withdrawn from soil allow for analysis of gases besides oxygen. Knowing the soil O_2 concentration, does not indicate if the O_2 consumption rate can be satisfied by the rate of O_2 convection and diffusion through soil.

Soil O_2 concentration $<10\%$ by volume indicates poor aeration. However, O_2 concentration per se is an imperfect predictor of plant response. The composition of other soil gases, such as carbon dioxide, ethylene, methane, etc., can affect response to a given oxygen concentration. Furthermore, O_2 concentration gives little information about the amount (mass) of O_2 (volume × concentration) in soil, or the rate at which it can move through tortuous soil pores or across water films and root membranes to reach metabolic sites where it is reduced (9). Specific composition of soil atmospheres vary with organic matter and mineral content, soil redox potential and pH, making it hard to make generalizations about soil air trace gas composition as soils become wetter and O_2 concentration decreases (1, 2).

TRANSPORT RATE MEASUREMENT

Measurements of the rate of gaseous transport in soil are of two types: diffusion and convection. To characterize diffusion, Lemon and Erickson (10) proposed placing a small platinum (Pt) wire micro-electrode in soil to electrically reduce oxygen. The Pt electrode simulates a respiring root to which oxygen diffuses through air- and water-filled pores and then through a water film surrounding the root. The micro-electrode measures the effect of restrictions along this pathway on the O_2 supply to the electrode surface. The current measured through an electrode tip of known geometry, supplied at steady potential, is related to the steady state reduction of oxygen supplied by diffusion to root surfaces of similar geometry. The technique became known as the soil ODR measurement (from "oxygen diffusion rate"). Lemon and Erickson's seminal concept, was perfected and field-adapted by others (11–13), providing a standardized approach to the technology and a unified interpretive framework. Eventually commercialization provided mass production of uniform reliable Pt electrodes and compact portable semi-automated multi-probe instrumentation.

ODR is probably the most common characterization of in situ soil oxygen status now conducted, and it is well correlated with plant physiological, nutritional and growth responses. ODR is the only soil oxygen status measurement suited to prolonged remote and nearly continuous measurement of dynamic soil oxygen status (14). Factors affecting ODR include water content, electrode contact with soil, presence of reducible compounds, salinity, temperature and oxygen concentration (5, 6). Numerous studies (1, 2) have identified an ODR value of $20\,\mu g\,m^{-2}s^{-1}$ as a threshold for a variety of plant physiological, nutritional, and growth responses to limited soil oxygen.

Subsequent ODR advances have been achieved. Electrode miniaturization allows oxygen flux measurements within roots and root microstructures, and within intact aggregates (15, 16). Also, due to potential poisoning of the Pt micro-electrodes during long term exposure to soil by oxides, alternatives to Pt have been developed. Wax-impregnated graphite electrodes (WIGEs) have greater current efficiencies and a wider current plateau than Pt electrodes (17). In moist soil, the current plateau region was a function of soil water content and the response was linear in atmospheres containing as much as 60% O_2. The WIGEs are less susceptible to oxide poisoning, can be fabricated easily and are less expensive.

Measurements of soil aeration based on convection of gas use simple flow permeameters to accurately measure mass flow through soil directly (18), or to measure the total air pressure or the difference in air pressure between the atmosphere and soil (19). Convection of soil air arises from spatial differences in total air pressures due to abrupt changes in air pressure of the atmosphere, the effects of temperature differences on gas properties, infiltration and redistribution of water in the soil profile, and microbial production of gases such as CO_2, NO, N_2O, and CH_4 (3). Flühler and Laser (8) developed a hydrophobic membrane probe (HMP) to measure total pressure and partial pressure in soil atmospheres. The HMP consisted of a membrane-covered chamber having a small volume and two teflon capillaries connected to a differential pressure transducer or to an oxygen electrode. A water-repellent non-rigid teflon membrane excluded wet soil from the continuous gas phase of the HMP. The total air pressure in the HMP is compared with the soil surface air pressure to determine pressure difference between the two locations. Renault et al. (20) developed an absolute pressure probe to measure in situ air pressure fluctuations at the soil surface and at varying depths within the profile. Their probe had negligible signal drift and an accuracy to 10 Pa. The sensitive component of the probe is a differential pressure sensor that functions from −14,000 to 14,000 Pa and senses a pressure differential of 2500 Pa. The signal results from resistance changes in an Arsenic-doped silicone membrane that functions as a Wheatstone bridge. The sensor requires a 1.5 mA direct current and its output is a −40 to 40 mV voltage. Circuitry converts the input voltage (24 V) into the stabilized current output (4–20 mA), providing a linear relationship between the differential air pressure and the output signal at a given temperature. The characteristics of the air pressure probe allow for in situ calibrations.

SUMMARY

Soil aeration is important to soil processes and plant growth. It can be characterized at various levels of complexity in terms of capacity, intensity or transport rate. Quantification of O_2 transport rates and measurement of concentrations of other important gases besides O_2 provide the best overall characterization of soil aeration for modern investigations, but simpler measurements can be valuable as rapid diagnostics for land managers. Modern instrumentation has made sophisticated characterization of soil aeration attainable at reasonable cost.

REFERENCES

1. Kozlowski, T.T., Ed. *Flooding and Plant Growth*; Academic Press, Inc: Orlando, FL, 1984; 356.
2. Glinski, J.; Stepniewski, W. *Soil Aeration and Its Role for Plants*; CRC Press, Inc: Boca Raton, FL, 1985; 229.
3. Scott, H.D. *Soil Physics: Agricultural and Environmental Applications*; Iowa State Press: Ames, IA, 2000.
4. Fatt, I. *Polarograhpic Oxygen Sensors*; CRC Press, Inc: Cleveland, OH, 1976.
5. Phene, C.J. Oxygen Electrode Measurement. In *Methods of Soil Analysis. Part 1. Physical and Mineralogical Methods. Monograph 9*, 2nd Ed.; Klute, A., Ed.; American Soc. Agron: Madison, WI, 1986; 1137–1159.
6. Tackett, J.L. Theory and Application of Gas Chromatography in Soil Aeration Research. Soil Sci. Soc. Amer. Proc. **1968**, *32* (3), 346–350.
7. Patrick, W.H. Oxygen Content of Soil Air by a Field Method. Soil Sci. Soc. Amer. J. **1977**, *41* (3), 651–652.
8. Flühler, H.; Laser, H.P. A Hydrophobic Membrane Probe for Total Pressure and Partial Pressure Measurements in the Soil Atmosphere. Soil Sci. **1975**, *120* (2), 85–91.
9. Hutchins, L.M. Studies on the Oxygen Supplying Power of the Soil Together with Quantitative Observations on the Oxygen-Supplying Power Requisite for Seed Germination. Plant Physiol. **1926**, *1* (2), 95–150.
10. Lemon, E.R.; Erickson, A.E. The Measurement of Oxygen Diffusion in the Soil with a Platinum Microelectrode. Soil Sci. Soc. Am. Proc. **1952**, *16* (2), 160–163.
11. Letey, J.; Stolzy, L.H. Measurement of Oxygen Diffusion Rates with the Platinum Microelectrode. I. Theory and Equipment. Hilgardia **1962**, *35*, 545–576.
12. Stolzy, L.H.; Letey, J. Characterizing Soil Oxygen Conditions with a Platinum Microelectrode. Advances in Agronomy **1964**, *16*, 249–279.
13. McIntyre, D.S. The Platinum Microelectrode Method for Soil Aeration Measurement. Advances in Agronomy **1970**, *22*, 235–283.
14. Phene, C.J.; Campbell, R.B.; Doty, C.W. Characterization of Soil Aeration In Situ with Automated Oxygen Diffusion Measurements. Soil Sci. **1976**, *122* (5), 271–281.
15. Hook, D.D.; McKevlin, M.A. Use of Oxygen Microelectrodes to Measure Aeration in the Roots of Intact Tree Seedlings. In *The Ecology and Management of Wetlands. Volume 1: Ecology of Wetlands*; Hook, D.D., Ed.; Croom Helm: London, 1988; 467–476.
16. Sextone, A.J.; Revsbech, N.P.; Parkin, T.B.; Tiedje, J.M. Direct Measurement of Oxygen Profiles and Denitrification Rates in Soil Aggregates. Soil Sci. Soc. Amer. J. **1985**, *49* (3), 645–651.
17. Shaikh, A.U.; Hawk, R.M.; Sims, R.A.; Scott, H.D. Graphite Electrode for the Measurement of Redox Potential and Oxygen Diffusion Rate in Soil. Nuclear and Chemical Waste Management **1985**, *5*, 237–243.
18. Evans, D.D. Gas Movement. In *Methods of Soil Analysis, Part 1. Physical and Mineralogical Properties, Including Statistics of Measurement and Sampling, Agronomy Monograph 9*; Black, C.A., Evans, D.D., White, J.L., Ensminger, L.E., Clark, F.E., Eds.; American Society of Agronomy: Madison, WI, 1965; 319–330.
19. Flühler, H.; Peck, A.J.; Stolzy, L.H. Air Pressure Measurement. In *Methods of Soil Analysis. Monograph 9. Part 1. Physical and Mineralogical Methods*; Second Edition; Klute, A., Ed.; American Soc. Agron: Madison, WI, 1986; 1161–1172.
20. Renault, P.; Mohrath, D.; Gaudu, J.-C.; Fumanal, J.-C. Air Pressure Fluctuations in a Prairie Soil. Soil Sci. Soc. Amer. J. **1998**, *62* (3), 553–563.

AERATION, TILLAGE EFFECTS ON

D.J. Horne
Massey University, Palmerston North, New Zealand

R.E. Sojka
Soil Scientist, Kimberly, Idaho, U.S.A.

INTRODUCTION

Few land management practices have the potential to impact upon soil aeration as directly or rapidly as tillage. Indeed, often, the reason for performing tillage is to modify or improve soil physical properties including aeration. The problems associated with inadequate aeration have been comprehensively reviewed elsewhere (1, 2). Important effects of limited soil aeration in crop production are: altered nutrient dynamics, a shift from oxidative to reductive chemical/biological reactions, impaired plant growth, and changes in gas equilibria affecting both soil and ambient atmospheres. For example, consider the soil nitrogen cycle which aeration effects via its influence on denitrification and gaseous nitrogen losses, decreased nitrogen mineralization rate and a reduction in nodulation and symbiotic fixation by leguminous plants (3). If the oxygen supply is sufficiently limited, and anerobosis sets in, then the products of reduction reactions may accumulate to toxic levels. In addition, a depleted oxygen supply may constrain root form and function, such as water and nutrient uptake, and therefore plant shoot performance even when many other soil physical factors are favorable (4). Unfortunately, relatively short periods of oxygen shortage can seriously compromise crop performance if they coincide with critical stages of crop growth (1). Finally, there are the effects of gas sources and sinks in the soil and transformations of soil gaseous components, and the exchange between soil and above ground air, on the atmosphere, e.g., diminished soil aeration may enhance the emission of greenhouse gases (5).

While the tillage-related literature is voluminous, little of it directly addresses soil aeration. Of necessity, this short article critiques only research which has measured aeration status directly—particularly indices of concentration and rate—and will make little or no attempt to draw inferences about the effect of tillage on soil aeration from studies reporting other related soil characteristics. Although bulk density, moisture content, and pore size distribution are related to soil aeration, and so may be indicative of aeration status, their direct relevance to a nuanced understanding of soil aeration is problematical. For instance, measurements of pore space convey little about pore continuity, tortuosity, or stability (6), whereas these effects are largely integrated de facto in measurements of oxygen diffusion rate (ODR).

EFFECTS OF SURFACE TILLAGE

Conventional tillage of virgin soil or soils growing permanent or long-term pasture for livestock grazing often improves soil aeration (7). In this situation, it is relatively straightforward to prepare a seedbed of good tilth and alleviate any compaction associated with animal or vehicular traffic. However, the timing of such plowing is important, as cultivation of wet soil leads to a dramatic reduction in soil aeration (8). If conventional tillage is poorly managed or executed in difficult circumstances (e.g., fine textured soils in humid climates) then relative to the permanent pasture datum, soil aeration may decline within a 3- or 4-year period (9). In the world's major cropping areas, tillage of pastureland is an infrequent occurrence; so a more interesting, if difficult, question is: what are the comparative effects of different tillage practices on soil aeration?

Accurately quantifying the effects of surface tillage on aeration has usually proved difficult. There are issues of both methodology and interpretation to contend with. For example, it has been argued that values for oxygen concentration in the seedbed say little about the quality of the soil air, or its rate of renewal where the roots or organisms are located, and that these factors are likely to be of greater importance than the average level for the whole soil (10). Also, some plants can compensate for low soil aeration by the supply of oxygen from one part of the root systems to another via internal diffusion. Another example relates to scale; what effect does a tillage practice have on intraaggregateo aeration in contrast to changes in aeration in the pore space between structural units where roots easily penetrate?

Interestingly, use of the ODR meter, arguably the best technique for field measurement of aeration status, is often unable to routinely and consistently discriminate between tillage treatments. This measure seems better able to reflect wet soil conditions associated with high rainfall and impeded drainage (11, 12). Yet a significant relationship between ODR and soybean yield has been observed, suggesting that it is only through critical stages of growth that low ODR will negatively impact on crop yield (12). Perhaps there are some implications here for the in-situ monitoring of the effects of tillage on aeration.

The difficulties mentioned above, and the sometimes conflicting research results that have been reported, mean that caution is required when attempting to draw broad conclusions about the effect of surface tillage on soil aeration. The impacts vary dramatically depending on: spatial heterogeneity, the depth in the soil profile under consideration (i.e., above or beneath the depth corresponding to plowing), soil type, the type or configuration of implements, climate, the cropping history of the field, the crop, the crop rotation, and the competency of the operator, especially regarding the timeliness of field work. The quantity and quality of crop residue may influence soil aeration; this aspect has received relatively little attention to date. Furthermore, interactions between aeration and other soil and crop properties make predictions about the exact effect of tillage on aeration risky. For example, aeration status may interact with earthworm populations and residue levels to effect seedling emergence (13).

The above discussion notwithstanding, most published literature has shown that surface soil is better aerated under conventional than other forms of tillage or that plowing enhances surface soil aeration to a greater extent than no tillage (14–16). The soil disturbance or loosening associated with tillage implements that invert or mix soil increases air movement in surface soil. In structurally degraded and/or fine textured soil, this increase in aeration may be confined to the inter-aggregate pore space. In an Indian study, tillage gave greater ODR than untilled (control) areas, and the greatest ODR was under moldboard plow and it was lowest with zero tillage (7). Likewise, in a compacted soil in New Zealand, rotary tillage of the soil surface improved ODR rates at 5-, 10-, and 15-cm depths compared with no-tillage (11). Enhanced rates of soil carbon oxidation are further evidence of improved aeration under cultivation.

The advantage that conventional tillage affords surface soil aeration may be somewhat transitory in nature both within seasons and across years. Enhanced soil aeration in a conventionally tilled seedbed may be short-lived as a result of soil reconsolidation (7), and may disappear by the time the crop becomes well established. The advantage of conventional tillage is likely to be more prominent early in the cropping cycle of a field: after many years of continuous cropping, no-tillage may be more beneficial to aeration status (17).

Although conventional tillage may generate more short-term macropores in the surface soil than no-tillage, they may not be as continuous down the profile and may be more tortuous (18–20). Therefore, no-tillage may improve aeration more uniformly throughout the soil profile particularly in the horizon just below the depth that corresponds to "plow level." Where sufficiently detailed measurements have been made, it would appear that not only may air permeability be more continuous or uniform down the profile under no-till, but it may also be more constant across "in rows" and between row positions (19).

Implement type will also have an important influence on the effect of tillage on soil aeration. In a summary of a number of New Zealand studies, the effects of some different drill openers on soil aeration are discussed (13). They demonstrate the benefits of some drill configurations (e.g., inverted T) relative to others (e.g., triple disk), and illustrate the importance of soil aeration to seedling establishment. In part, differences in tillage equipment, particularly direct-drilling machinery, may account for much of the disparity in the literature on the effects of tillage on aeration. For example, permeability measurements made under no-till treatments established using a triple disc are nearly always markedly inferior to those made under plowing (21).

EFFECTS OF SUBSURFACE TILLAGE

Subsurface tillage is practiced to improve aeration deeper in the profile where soils are naturally compact and/or have poor internal drainage. Where deep tillage is employed to improve drainage, it is important that there is an outlet for excess water or deep tillage may exacerbate aeration problems. In a study comparing deep tillage implements, the paraplow (angled shanks operating at 0.5 m depth) achieved more consistent improvement in air permeability at sowing, particularly in the important depth of 15–25 cm, than did subsoilers with straight shanks tilling at either shallow (0.25 m) or deeper (0.5 m) (11). In addition, all subsurface tillage treatments increased ODR values to a depth of 40 cm compared with the control. Due to reconsolidation in this humid climate, these advantages had disappeared by harvest.

The hypothesis that, as an alternative to conventional tillage, subsoiling in combination with no-tillage seeding would improve soil aeration, other physical properties, and crop yield in a soil that has become severely compacted following many years of cultivation has been tested (11). Some aspects of this study and a subsequent investigation (22) suggest that, for these New Zealand soils and conditions, despite the significant loosening achieved at depth by subsoiling implements, it was direct seeding that was the important component in the proposed system, and subsoiling contributed or added relatively little to the enhancement of crop yield. In part, this is due to the short-lived nature of the structural improvements generated by subsoiling (23).

TILLAGE, AERATION, AND GREENHOUSE GASES

In recent soil aeration research, there has been a shift in focus away from the study of the movement and use of oxygen and its effects on crop nutrition and growth to the role of soil aeration in environmental protection, e.g., greenhouse gas production and the susceptibility of nutrients to leaching. Indeed, concern about, and investigations of, the relationships between tillage and fluxes, composition and transformations of air into and out of soil is prominent on the environmental agenda. Of particular interest here is the effect of tillage practices on the release of the major greenhouse gases: carbon dioxide (CO_2), methane (CH_4), nitrous oxide (N_2O) (5, 24). In addition, there is the suggestion that conservation tillage practices may enhance carbon sequestration (25).

Relative to no-tillage, conventional cultivation may result in greater CO_2 fluxes from the soil (26). No-tillage may increase the frequency of N_2O emissions and the net fixation of carbon by decreasing CO_2 emissions (5, 27). Periods of low or zero CO_2 fluxes and very high N_2O fluxes under no-tillage may be associated with periodic reduced gas diffusivity and air-filled porosity, both of which are often caused by heavy rainfall. In general, peak N_2O emissions are mainly associated with heavy rainfall following fertilization particularly with no-tilled soils (5). Plowing may decrease the oxidation rate of atmospheric CH_4 in aerobic soils but this is unlikely to be sufficient to offset the lower N_2O emissions from the soil (5).

SUMMARY

Adequate soil aeration is important to crop production and environmental protection. Tillage may profoundly influence soil aeration. Although the effect of tillage on soil aeration is dependent on a range of factors and is, therefore, hard to predict, often surface soil aeration is more favorable under well-managed conventional cultivation than no-tillage. Often, these improvements are confined to shallow soil depths and relatively short time intervals. Where appropriate no-till equipment is used, direct-drilling may have greater potential to improve soil aeration, both in terms of uniformity throughout the soil profile and permanence in time. Tillage practices that result in anaerobic soil conditions are likely to promote N_2O production, while those that disturb larger volumes of soil and promote aeration may result in greater emissions of CO_2.

REFERENCES

1. Glinski, J.; Stepniewski, W. *Soil Aeration and Its Role for Plants*; CRC Press Inc.: Boca Raton, FL, 1985; 229.
2. Scott, H.D. *Soil Physics: Agricultural and Environmental Applications*; Iowa Sate Press: Ames, IA, 2000; 421.
3. Lipiec, J.; Stepniewski, W. Effects of Soil Compaction and Tillage Systems on Uptake and Losses of Nutrients. Soil Till. Res. **1995**, *35*, 37–52.
4. Letey, J.; Stolzy, L.H.; Valoras, N.; Szuszkiewicz, T.E. Influence of Soil Oxygen on Growth and Mineral Concentration of Barley. Agron. J. **1962**, *54*, 538–540.
5. Ball, B.C.; Scott, A.; Parker, J.P. Field N_2O, CO_2 and CH_4 Fluxes in Relation to Tillage, Compaction and Soil Quality in Scotland. Soil Till. Res. **1999**, *53*, 29–39.
6. Francis, G.S.; Cameron, K.C.; Kemp, R.A. A Comparison of Soil Porosity and Solute Leaching After Six Years of Direct Drilling or Conventional Cultivation. Aust. J. Soil Res. **1988**, *26*, 637–649.
7. Khan, A.R. Influence of Tillage on Soil Aeration. J. Agron. Crop Sci. **1996**, *177* (4), 253–259.
8. Hodgson, A.S.; McLeod, D.A. Oxygen Flux, Air-Filled Porosity and Bulk Density as Indices of Vertisol Structure. Soil Sci. Soc. Am. J. **1989**, *53*, 540–543.
9. Shepherd, T.G.; Dando, J.L. Physical Indicators of Soil Quality for Environmental Monitoring. In *Proceedings of Soil and Land Indicators*; Hawkes Bay Regional Council New Zealand Technical report EMT 96/3 pp 33–42.
10. Dexter, A.R. Physical Properties of Tilled Soils. Soil Till. Res. **1997**, *43*, 41–63.
11. Sojka, R.E.; Horne, D.J.; Ross, C.W.; Baker, C.J. Subsoiling and Surface Tillage Effects on Soil Physical Properties and Forage Oat Stand and Yield. Soil Till. Res. **1997**, *40*, 125–144.
12. Flowers, M.D.; Lal, R. Axle Load and Tillage Effects on Soil Physical Properties and Soybean Grain Yield on a Mollic Ochraqualf in Northwest Ohio. Soil Till. Res. **1998**, *48* (1–2), 21–35.
13. Baker, C.J.; Chaudhary, A.D.; Springett, J.A. Barley Seedling Establishment and Infiltration from Direct Drilling in a Wet Soil. Proceedings of Agronomy Society of New Zealand, **1987**, *17*, 59–66.

14. Ball, B.C.; O'Sullivan, M.F.; Hunter, R. Gas Diffusion, Fluid Flow and Derived Pore Continuity Indices in Relation to Vehicle Traffic and Tillage. J. Soil Sci. **1988**, *3*, 327–339.
15. Ball, B.C.; Campbell, D.J.; Douglas, J.T.; Henshall, J.K.; O'Sullivan, M.F. Soil Structural Quality, Compaction and Land Management. European J. Soil Sci. **1997**, *48*, 593–601.
16. Rasmussen, K.J. Impact of Ploughless Soil Tillage on Yield and Soil Quality: A Scandinavian Review. Soil Till. Res. **1999**, *53*, 3–14.
17. Voorhees, W.B.; Lindstrom, M.J. Long-Term Effects of Tillage Method on Soil Tilth Independent of Wheel Traffic Compaction. Soil Sci. Amer. J. **1984**, *48*, 152–156.
18. Boone, F.R.; Van der Werf, H.M.G.; Kroesbergen, B.; ten Hag, B.A.; Boers, A. The Effect of Compaction of the Arable Layer in Sandy Soils on the Growth of Maize for Silage I. Critical Matric Water Potential in Relation to Soil Aeration and Mechanical Impedance. Neth. J. Agric. Sci. **1986**, *34*, 155–171.
19. Herad, J.R.; Kladivko, E.J.; Mannering, J.V. Soil Macroporosity, Hydraulic Conductivity and Air Permeability of Silty Soils Under Long-Term Conservation Tillage in Indiana. Soil Till. Res. **1987**, *11*, 1–18.
20. Carter, M.R. Characterizing the Soil Physical Conditions in Reduced Tillage Systems for Winter Wheat on a Fine Sandy Loam Using Small Cores. Canadian J. Soil Sci. **1992**, *72*, 395–402.
21. Schjonning, P.; Rasmussen, K.J. Soil Strength and Soil Core Characteristics for Direct Drilled and Ploughed Soils. Soil Till. Res. **2000**, *57*, 69–82.
22. Hamilton-Manns, M. The Effects of No-Tillage and Subsoil Loosening on Soil Physical Properties and Crop Performance M.Sc. Thesis, **1998**, Massey University, New Zealand.
23. Busscher, W.J.; Sojka, R.E. Enhancement of Subsoiling Effect on Soil Strength by Conservation Tillage. Trans. ASAE **1987**, *30* (4), 888–892.
24. Soane, B.D.; van Ouwerkerk, C. Implications of Soil Compaction in Crop Production for the Quality of the Environment. Soil Till. Res. **1995**, *35*, 5–22.
25. Lal, R. Residue Management, Conservation Tillage and Soil Restoration for Mitigating Greenhouse Effect by CO_2 Enrichment. Soil Till. Res. **1997**, *43*, 81–107.
26. Reicosky, D.C.; Reeves, D.W.; Prior, S.A.; Runion, G.B.; Rogers, H.H.; Raper, R.L. Effects of Residue Management and Controlled Traffic on Carbon Dioxide and Water Loss. Soil Till. Res. **1999**, *52* (3–4), 153–165.
27. Aulakh, M.S.; Rennie, D.A.; Paul, E.A. Gaseous Nitrogen Losses from Soils Under Zero-Till as Compared to Conventional–Till Management Systems. J Environ. Qual. **1984**, *13* (1), 130–136.

AGGREGATION

Sjoerd W. Duiker
The Pennsylvania State University, University Park, Pennsylvania, U.S.A.

INTRODUCTION

Soil aggregation is the organization and bonding of soil particles into distinct units that can be separated by the application of energy to soil. When increasing levels of energy are applied, aggregates of most soils fall apart into smaller aggregates in a stepwise manner. Conceptual models have been formulated to explain this apparent hierarchy of soil aggregation. Different bonding agents have been found to play roles at specific hierarchical levels. In this article, the predominant model of aggregate hierarchy and known aggregating agents will be discussed.

AGGREGATE HIERARCHY

A common model of aggregation distinguishes four levels (1–3):

1. Domains, quasi crystals (also called clay tactoids), and assemblages (<2 μm);
2. clusters (2–20 μm);
3. microaggregates (20–250 μm); and
4. macroaggregates (250–2000 μm).

This classification is based on research in temperate and Mediterranean regions, and seems to apply also to moderately weathered tropical soils and volcanic ash soils (4). In highly weathered tropical soils with very stable microgranular structure, only one hierarchical level can be distinguished (2, 5).

MECHANISMS OF AGGREGATE FORMATION

Roots and VAM

Both plant roots and vesicular–arbuscular mycorrhizal hyphae (VAM) associated with roots are important binding agents at the scale of microaggregates and macroaggregates (6). Roots physically hold aggregates together, exert a drying effect, and may excrete organic substances that act as glues. The direct effect of roots on aggregation is greatest with perennial grass species due the enmeshment of their extensive fine root systems with soil. Annual crops have smaller root systems and therefore a less positive direct effect on aggregation (1).

Fungi and Bacteria

Fungi have a great direct effect on aggregate stability, more so than bacteria or actinomycetes. Fungal hyphae act as nets holding microaggregates together, and fungi and bacteria produce polysaccharides that favor aggregation (7). Dead fungi and bacteria act as centers of clusters (1). Clay particles adhere to the walls of living fungal hyphae, apparently due to the gluing action of fungal exudates. In sandy soils, roots and fungal hyphae may be the only components responsible for aggregation, because sand particles are large and inert and do not interact with charged organic ions (8).

Transient and Persistent Organic Matter

Transient organic matter is composed mostly of polysaccharides. Polysaccharides are produced by microorganisms or excreted by roots and bind clay-sized particles together into clusters and microaggregates (1). The effect of transient polysaccharides on aggregation is short-lived (generally less than two months). Persistent organic matter consists of highly humified organic molecules or polysaccharides protected against decomposition by aluminosilicate clays or metal oxides (9). Persistent organic matter is an important aggregating agent at the <2 μm level.

Earthworms

The casts of earthworms become highly stable aggregates that are resistant to slaking and dispersion. The casts contain more silt, clay, organic matter, nitrogen, and exchangeable nutrients, and have a higher porosity and stability

Encyclopedia of Soil Science

than the surrounding soil. Besides creating soil aggregates by their casts, earthworms create many pores in the soil as is explained elsewhere in the encyclopedia.

Termites

Termites bring large quantities of clay-sized particles from the subsurface to the surface of tropical soils to build their mounds. The clay particles are glued together by a sticky fluid excreted by the termites (10). The microaggregates thus formed are 0.03–1 mm in diameter and are stable to dispersion by raindrop impact (5). Termite activity may be a contributor to the "pseudosand" or "pseudosilt" texture of clayey tropical soils (2, 5).

Iron and Aluminum Oxides

Iron and aluminum (hydr)oxides have been thought to be responsible for the very stable structure of many highly weathered tropical soils. However, there is limited evidence to support this conclusion. There is some evidence that poorly crystalline iron oxides favor aggregation more than crystalline iron oxides (11–13). In some cases, aluminum oxides may be more important that iron oxides for aggregation (14). The role of these oxides needs more research before far-reaching conclusions can be drawn.

Inorganic Cations

Positively charged cations are attracted to negatively charged aluminosilicate clay surfaces, and form bridges between the clay particles, or between clay particles and organic matter (3). The role of cations in aggregation is primarily at the quasi-crystal and domain scale. Differences between cations in their effects on clay flocculation can be attributed to their charge, concentration, size, and hydration (15). The ability to flocculate clays decreases in the order: $Al^{3+} > Ca^{2+} > Mg^{2+} > K^{+} > Na^{+}$ (9).

Aluminosilicate Clays

Clay mineralogy determines to a large extent the basic soil structure of a soil. The most important clay minerals in soils are kaolinite, montmorillonite, and illite. Packs of montmorillonite clays have been termed "quasi crystals" (also called "clay tactoids"), those of illite "domains", and those of kaolinite "assemblages" (16). Clay particles in quasi crystals are more perfectly aligned than in domains, whereas they are randomly ordered in assemblages. Quasi crystals have a width: thickness ratio >100, domains <10, and assemblages <2. Montmorillonite plates in quasi crystals can come apart when water molecules enter the interlamellar space, explaining the shrink–swell behavior of these clays.

Wetting and Drying, Freezing and Thawing

Wetting and drying cycles are needed to form aggregates. The effect of wetting and drying may be due partly to its effect on microbial activity and to orientation of clay platelets (17). Freezing and thawing cycles can break up cloddy soil, if water contents are great enough to separate soil particles. A limited number of freeze–thaw cycles (up to 4) has been observed to increase aggregate stability because of localized drying and precipitation of binding agents (18).

CONCLUSION

Most soils are built up of a hierarchy of aggregates. Different components determine aggregation at different levels. At the macro- and microaggregate levels, roots and fungal hyphae appear to be the primary factors responsible for aggregation. Clusters are composed of fungal hyphae and bacterial remains encrusted with clay and silt particles. At the lowest level of aggregation (<2 μm), persistent organic matter and inorganic components are responsible for bonds between clay particles. Biological, chemical, and physical processes play a role in aggregation, including the crops grown, earthworms, termites, wetting and drying, and freeze–thaw cycles.

REFERENCES

1. Tisdall, J.M.; Oades, J.M. Organic Matter and Water-Stable Aggregates in Soils. J. Soil Sci. **1982**, *33*, 141–163.
2. Oades, J.M.; Waters, A.G. Aggregate Hierarchy in Soils. Aust. J. Soil Res. **1991**, *29*, 815–828.
3. Edwards, A.P.; Bremner, J.M. Microaggregates in Soils. J. Soil Sci. **1967**, *18*, 64–73.
4. Maeda, T.; Takenaka, H.; Warkentin, B.P. Physical Properties of Allophane Soils. Adv. Agron. **1977**, *29*, 229–264.
5. Trapnell, C.G.; Webster, R. Microaggregates in Red Earths and Related Soils in East and Central Africa, Their Classification and Occurrence. J. Soil Sci. **1986**, *37*, 109–123.
6. Thomas, R.S.; Franson, R.L. Bethelenfalvay. Separation of Vesicular–Arbuscular Mycorrhizal Fungus and Root Effects on Soil Aggregation. Soil Sci. Soc. Am. J. **1993**, *57*, 77–81.

7. Chaney and Swift. Studies on Aggregate Stability. II. The Effect of Humic Substances on the Stability of Re-Formed Soil Aggregates. J. Soil Sci. **1986**, *37*, 337–343.
8. Bond, R.D.; Harris, J.R. The Influence of the Microflora on Physical Properties of Soils. I. Effects Associated With Filamentous Algae and Fungi. Aust. J. Soil Res. **1964**, *2*, 111–122.
9. Emerson, W.W.; Greenland, D.J. Soil Aggregates—Formation and Stability. In *Soil Colloids and Their Association in Aggregates*; De Boodt, M.F., Hayes, M., Herbillon, A., Eds.; Plenum Press: New York, 1990; 485-511.
10. Jungerius, P.D.; Van Den Ancker, J.A.M.; Mücher, H.J. The Contribution of Termites to the Microgranular Structure of Soils on the Uasin Gishu Plateau, Kenya. Catena **1999**, *34*, 349–363.
11. Desphande, T.L.; Greenland, D.J.; Quirk, J.P. Changes in Soil Properties Associated with the Removal of Iron and Aluminum Oxides. J. Soil Sci. **1968**, *19*, 108–122.
12. Schahabi, S.; Schwertmann, U. Der Einfluß Von Synthetischen Eisenoxiden auf die Aggregation Zweier Löß Bodenhorizonte. Z. Pflanzenernaehr. Bodenkd. **1970**, *125*, 193–204.
13. Duiker, S.W. Soil Erodibility in Southern Spain and the Southern USA With Special Emphasis on the Role of Iron Oxides PhD Dissertation. The Ohio State University, Columbus, OH134.
14. El-Swaify, S.A.; Emerson, W.W. Changes in the Physical Properties of Soil Clays due to Precipitated Aluminum and Iron Hydroxides. I. Swelling and Aggregate Stability After Drying. Soil Sci. Soc. Am. Proc. **1975**, *39*, 1056–1063.
15. Sumner, M.E. The Electrical Double Layer and Clay Dispersion. In *Soil Crusting: Chemical and Physical Processes*; Sumner, M.E., Stewart, B.A., Eds.; Lewis Publishers: Boca Raton, FL, 1992; 1–31.
16. Quirk, J.P.; Aylmore, L.A.G. Domains and Quasi Crystalline Regions in Clay Systems. Soil Sci. Soc. Am. Proc. **1971**, *35*, 652–654.
17. Utomo, W.H.; Dexter, A.R. Changes in Soil Aggregate Water Stability Induced by Wetting and Drying Cycles in Non-Saturated Soil. J. Soil Sci. **1982**, *33*, 623–637.
18. Lehrsch, G.A. Freeze–Thaw Cycles Increase Near-Surface Aggregate Stability. Soil Sci. **1998**, *163*, 63–70.

AGGREGATION, FRAGMENTATION, AND STRUCTURAL STABILITY MEASUREMENT

A

M. Díaz-Zorita
EEA INTA General Villegas, Buenos Aires, Argentina

J.H. Grove
University of Kentucky, Lexington, Kentucky, U.S.A.

E. Perfect
University of Tennessee, Knoxville, Tennessee, U.S.A.

INTRODUCTION

Soil aggregation can be viewed as the arrangement of primary soil particles into hierarchical structural units, identified on the basis of varying failure zone strengths reflecting the characteristics of both the void and solid phases (1).

Direct characterization of soil aggregation can be performed by describing morphological features in the field (2), using image analysis techniques (3), or by measuring the size distribution and connectivity of pores (4). Other procedures are based on the partial breakdown of structural units by dispersion or fragmentation, and evaluation of the resulting fragment size distribution. In soils where dispersion–flocculation processes dominate the arrangement of secondary soil units, the use of fragmentation procedures to assess soil aggregation is less precise than the use of dispersion methods (5). This chapter focuses on fragmentation procedures and associated indices for assessing soil aggregation, based on the size distribution and stability of fragments after mechanical disruption.

Aggregates, Fragments, and Particles

Aggregate size and stability are interrelated concepts in the description of soil structure using fragmentation procedures. Assuming separation of aggregates occurs at planes of weakness surrounding coherent structural units, application of mechanical stress results in soil breakdown into fragments with stabilities greater than the applied stress (Fig. 1). With application of low energy stress, the resulting fragments will be similar in size to those naturally occurring aggregates. However, with increasing fragmentation energy, the rate of reduction in fragment size will be related to structural stability (bonding energy) between aggregation units (clay packages, microaggregates, etc.). With application of high-energy stress, the resulting fragment size distribution will be independent of further increments in stress and more closely related to the sizes of primary soil particles (e.g., sand, silt, clay) than aggregates. Soil fragmentation rarely implies complete disruption of aggregates and can be performed under dry, moist, or saturated conditions.

SAMPLING AND SAMPLE PREPARATION

The processes of breaking down involved in fragmentation methodologies are continuous, commencing at sampling and continuing until the final size distribution of fragments is determined. Thus, the purpose of any aggregation measurement must be well understood before sampling in order to determine the sampling and handling procedures to be employed. Recommended tools for taking undisturbed samples are shovels or soil core samplers (6, 7). Samplers with low wall area to volume ratios minimize compression. Air drying samples without initial fragmentation promotes the formation of clods which may not be present in the natural state, particularly when the sampled layer does not dry excessively under field conditions. Low energy fragmentation of moist samples before air drying is common. "Gentle" hand manipulation is widely used but provides a nonmeasurable energy input that is strongly operator dependent. We favor the application of a standardized low-energy fragmentation procedure like drop shatter (8). Fragmentation and/or compaction can continue during transportation and storage, so adequately sized containers are required. Cohesion of soil increases with drying/storage time (9). To be comparable, samples should be treated in the same manner and measurements to be

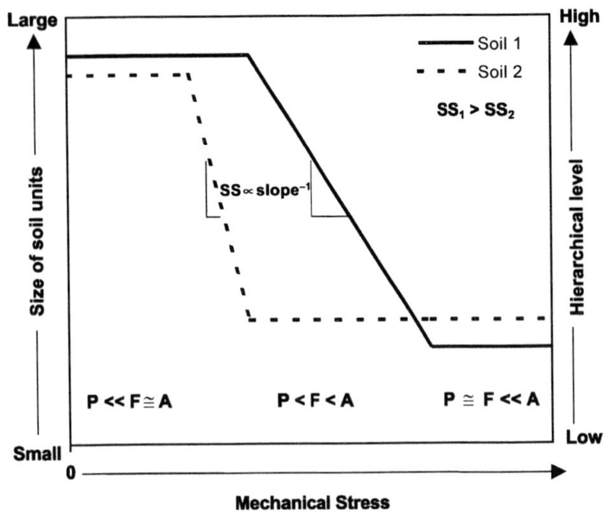

Fig. 1 Idealized changes in soil fragment size distribution as a function of applied mechanical stress. *P*, median particle size; *F*, median fragment size; *A*, median aggregate size; *SS*, soil structural stability, the subscript indicates the soil type 1 or 2.

done on samples with the same water potential whenever possible (10).

FRAGMENTATION PROCEDURES

Dry Sieving

Dry fragment separation is done with a nested set of rotary sieves, or with vibratory or oscillatory movements imposed on a nest of flat sieves. Rotary sieving is accepted as the standard for dry aggregate size distribution measurements for wind erosion assessment (11, 12). When flat sieves are used it is important to determine the sieving duration to avoid fragment abrasion. Several studies recommend 30 s of vibratory, horizontal or horizontal–vertical sieving for adequate fragment separation with minimal abrasion (13).

Wet Sieving

Wet sieving refers to the separation of soil fragments from an undisturbed sample or a single initial size fraction in the presence of free water or other liquids. Methods that involve sieving a single initial fragment size fraction are preferred because they are less time consuming. However, the selection of appropriately sensitive initial fragment sizes varies among soils. A better estimate of water stable fragments is obtained by sieving an undisturbed sample over a nest of sieves. The soil water content prior to immersion is a major factor in determining wet aggregate stability. Based on the wetting procedure, different mechanisms of aggregate breakdown can be identified: slaking due to fast wetting of dry aggregates, microcracking after slow wetting, and mechanical breakdown of prewetted aggregates (14). The use of "field moist" samples introduces another variable that may lead to erroneous conclusions (15).

Energy Input

Both dry and wet sieving subject soil to forces that are rarely found in the field and which cannot be easily

Table 1 Indices for soil fragment size distribution quantification. $P(x < X)$ is the fraction of soil structural units smaller than sieve size X

Distribution	Function	Parameters		
		Reference size[a]	Spread of distribution[b]	Reference
Normal	$P(x < X) = \frac{1}{\sigma\sqrt{2\pi}}\exp\left[-\frac{(x - \text{MD})^2}{2\text{SD}^2}\right]$	MD	SD	(20, 21)
Log-normal	$P(x < X) = \frac{1}{\log\ \text{GSD}\sqrt{2\pi}}\exp\left[-\frac{(\log(x) - \text{GMD})^2}{2(\log\ \text{GSD})^2}\right]$	GMD	log GSD	(20)
Rosin–Rammler	$P(x < X) = \exp\left[-\left(\frac{x}{\alpha}\right)^{\beta}\right]$	α	β	(22)
Gaudin–Schuhmann	$P(x < X) = \left(\frac{x}{x_0}\right)^m$	x_0	m	(23)

[a] MD, mean diameter; GMD, geometric mean diameter; α, fragment diameter corresponding to the 36.78 percentile of the cumulative probability function; x_0, diameter of the largest fragment.
[b] SD, standard deviation; log GSD, log of the geometric standard deviation; β, Rosin–Rammler exponent; m, Gaudin–Schuhmann exponent.

Table 2 Summary of soil structural/aggregate stability indices for fragmentation procedures

Index[a]	Equation[b]	Specific conditions	Reference
ΔMWD	$A-B$	Dry and wet mean weight diameters	(24)
Henin index, I_s	A/B	Water, ethanol and benzene stable aggregates	(25)
Stability index	B/A	Water stable aggregates ≥3.0 mm	(26)
WAS and DCF	B/A	Water stable aggregates and dispersible clay in 1–2 mm fraction	(27)
Stability index	B/A	Water stable aggregates ≥0.25 mm	(7)
Soil erodibility	B/A	Dry stable aggregates ≥0.84 mm	(11)
Dry structural stability	B/A	Mass fraction distribution after 1 and 5 min of sieving	(28)
Stability constant (k)	$(\ln B-\ln A)/E_s$	Ultrasonic dispersion	(16)

[a] ΔMWD, change in mean weight diameter; WAS, wet aggregate stability; DCF, dispersible clay fraction (dispersible clay/total clay).
[b] A, size of fragments under low energy stress; B, size of fragments under high-energy stress; E_s, sonic energy level.

quantified. Several attempts to quantify soil fragmentation energy have been described based on ultrasonic dispersion or drop shatter (8, 16).

SIZE DISTRIBUTION INDICES

The mean weight diameter is a commonly used index for expressing aggregation with a single value (17, 18). The literature also contains a variety of functions to parameterize the measured mass size distribution of soil fragments (Table 1). Indices based on a single fragment size class have been related to soil processes [e.g., wind erosion (11)], but are not useful in describing fragment size distributions. When performing the calculations, special care must be taken to avoid inclusion of primary particles (e.g., sand) with sizes similar to that of the measured fragments. Fragment sizes rarely follow a normal distribution so parameters from the log-normal or Rosin–Rammler distribution models are preferred. A new approach, using fractal theory, has been recently proposed (19).

SOIL STRUCTURAL STABILITY

Soil structural/aggregate stability is a measure of the ability of soil structural units to resist change (retain structure) in response to application of a mechanical stress (i.e., dry or moist fragmentation, disruption in water, etc.). Although there are a great variety of methods to assess aggregate stability, the ratio or difference in the fragment sizes before and after application of mechanical energy is generally used (Table 2). Defining the low- and high-energy input levels requires special attention, because fragment size depends on both energy input and structural stability (Fig. 1). Aggregate stability indices differ from indices used to describe the fragment size distribution because the former reflect the extent of breakdown relative to an initial condition, while the latter is independent of the soil's initial state.

Structural stability indices are empirical and comparisons among treatments, soil properties, and/or processes have meaning only when similar procedures are used. The choice of stability index should be carefully considered in relation to the type(s) of comparison desired.

CONCLUSIONS

Choosing between fragment size distribution and structural stability, procedures will depend on the use to be made of the indices. In general, if the procedure is to characterize the "field" soil condition without consideration of disruptive processes, then measurement of soil fragment size distribution after low-energy input is recommended. Structural stability measurements are most useful for characterizing soil susceptibility to wind and water erosion. In determining the stability of the soil to mechanical disruption, it is important to measure the distribution of soil fragments relative to an initial undisturbed (low-energy input) state.

REFERENCES

1. Kay, B.D.; Angers, D.A. Soil Structure. In *Handbook of Soil Science*; Sumner, M.E., Ed.; CRC Press: Boca Raton, FL, 1999; 229–276.
2. Manichon, H. Observation Morphologique De l'etat Structural Et Mise En Evidence D'effets De Compactage Des Horizons Travailles. In *Soil Compaction and Regeneration*; Monnier, G., Goss, M.J., Eds.; Balkema: Rotterdam, 1987; 39–52.

3. Dexter, A.R. Advances in Characterization of Soil Structure. Soil Till. Res. **1988**, *11*, 199–238.
4. Wilson, G.V.; Luxmoore, R.J. Infiltration, Macroporosity, and Mesoporosity Distributions on Two Forested Watersheds. Soil Sci. Soc. Am. J. **1988**, *52*, 329–335.
5. Emerson, W.W. Inter-particle Bonding. In *Soils: An Australian Viewpoint*; The Division of Soils CSIRO, Ed.; CSIRO/Academic Press: Melbourne, 1983; 477–498.
6. McIntyre, D.S. Soil Sampling Techniques for Physical Measurements. In *Methods for Analysis of Irrigated Soils*; Loveday, J., Ed.; Commonwealth Agricultural Bureaux: Victoria, Australia, 1974; 12–20.
7. Kemper, W.D.; Rosenau, R.C. Aggregate Stability and Size Distribution. In *Methods of Soil Analysis. Part 1. Physical and Mineralogical Methods*; Klute, A., Ed.; American Society of Agronomy and Soil Science Society of America: Madison, WI, 1986; 425–442.
8. Hadas, A.; Wolf, D. Refinement and Re-Evaluation of the Drop-Shatter Soil Fragmentation Method. Soil Till. Res. **1984**, *4*, 237–249.
9. Kemper, W.D.; Rosenau, R.C. Soil Cohesion as Affected by Time and Water Content. Soil Sci. Soc. Am. J. **1984**, *48*, 1001–1006.
10. Gollany, H.T.; Schumacher, T.E.; Evenson, P.D.; Lindstrom, M.J.; Lemme, G.D. Aggregate Stability of an Eroded and Desurfaced Typic Argiustoll. Soil Sci. Soc. Am. J. **1991**, *55*, 811–816.
11. Chepil, W.S. Improved Rotary Sieve for Measuring State and Stability of Dry Soil Structure. Soil Sci. Soc. Am. Proc. **1952**, *16*, 113–117.
12. Zobeck, T.M. Soil Properties Affecting Wind Erosion. J. Soil Water Conserv. **1991**, *46*, 112–118.
13. Braunack, M.V.; McPhee, J.E. The Effect of Initial Soil Water Content and Tillage Implement on Seedbed Formation. Soil Till. Res. **1991**, *20*, 5–17.
14. Le Bissonnais, Y. Soil Characteristics and Aggregate Stability. In *Soil Erosion, Conservation, and Rehabilitation*; Agassi, M., Ed.; Marcel Dekker: New York, NY, 1996; 41–60.
15. Yoder, R.E. A Direct Method of Aggregate Analysis of Soils and a Study of the Physical Nature of Erosion Losses. J. Am. Soc. Agron. **1936**, *28*, 337–351.
16. Fuller, L.G.; Goh, T.B. Stability–Energy Relationships and Their Application to Aggregation Studies. Can. J. Soil Sci. **1992**, *72*, 453–466.
17. Puri, A.N.; Puri, B.R. Physical Characteristics of Soils: II. Expressing Mechanical Analysis and State of Aggregation of Soils by Single Values. Soil Sci. **1939**, *47*, 77–81.
18. Youker, R.E.; McGuinness, J.L. A Short Method of Obtaining Mean Weight–Diameter Values of Aggregate Analyses of Soils. Soil Sci. **1957**, *83*, 291–294.
19. Perfect, E. Fractal Models for the Fragmentation of Rocks and Soils: A Review. Eng. Geol. **1997**, *48*, 185–198.
20. Gardner, W.R. Representation of Soil Aggregate Size Distribution by a Logarithmic-Normal Distribution. Soil Sci. Soc. Am. Proc. **1956**, *20*, 151–153.
21. Allen, T. *Particle Size Measurement. Volume 1. Powder Sampling and Particle Size Measurement*; Chapman & Hall: Cornwall, 1997; 525.
22. Irani, R.R.; Callis, C.F. *Particle Size: Measurement, Interpretation, and Application*; Wiley: New York, NY, 1963; 165.
23. Gaudin, A.M.; Meloy, T.P. Model and Comminution Distribution Equation for Single Fracture. Trans. AIME **1962**, *223*, 40–43.
24. De Leenheer, L.; De Boodt, M. Determination of Aggregate Stability by the Change in Mean Weight Diameter. *Proceedings of the International Symposium on Soil Structure, Ghent, Belgie*; Rijkslandouwhogeschool: Ghent, Belgie, 1959; 290–300.
25. Henin, S.; Monnier, G.; Combeau, A. Methode Pour l'etude De La Stabilite Structurale Des Sols. Ann. Agron. **1958**, *9*, 71–90.
26. Bryan, R.B. The Development, Use and Efficiency of Indices of Soil Erodibility. Geoderma **1968**, *2*, 5–26.
27. Pojasok, T.; Kay, B.D. Assessment of a combination of Wet Sieving and Turbidimetry to Characterize the Structural Stability of Moist Aggregates. Can. J. Soil Sci. **1990**, *70*, 33–42.
28. Buschiazzo, D.E.; Aimar, S.B.; Babinec, F.J.; Ferramola, L. Un método Para La Determinación De Estabilidad De Agregados En Seco En Suelos De La Región Semiárida Pampeana Central (Argentina). Ciencia del Suelo **1994**, *12*, 32–34.

AGGREGATION AND ORGANIC MATTER

Cynthia A. Cambardella
United States Department of Agriculture (USDA), Ames, Iowa, U.S.A.

A

INTRODUCTION

A soil aggregate is formed when closely packed sand, silt, clay, and organic particles adhere more strongly to each other than to surrounding particles (1). The arrangement of these aggregates and the pore space between them is referred to as soil structure. Soil structural stability is the ability of aggregates to remain intact when exposed to different stresses (2).

Soil organic matter (SOM) consists of plant, animal and microbial residues, and organic decomposition products that are associated with the inorganic soil matrix. Soil organic matter turnover is coupled to the agradation/degradation of soil aggregates through the activity of soil microorganisms. In order to understand how soil structure can mediate SOM turnover and formation, it is important to understand how soil primary particles and organic matter interact with each other.

AGGREGATE STRUCTURE

In the late 1950s, it was proposed that clay particles were organized into units called clay domains (3), which are defined as groups of clay crystals that behave in water as a single unit. Clay domains are joined to quartz particles through organic matter mediated bonds and these organo-mineral complexes can be further structured into larger discrete units held together by organic matter–polyvalent cation bridges. Based on experiments using ultrasonic vibration to disperse soils, Edwards and Bremner (4) suggested a model that classified aggregates into two groups based on their relative size and stability: macroaggregates (> 250-μm) and microaggregates (< 250-μm). They observed that the bonds which hold together macroaggregates were easily disrupted but that microaggregates were quite stable and their structure was disrupted only by applying mechanical energy (sonication) or by chemical treatment. They further postulated that the organic matter contained within microaggregates is inaccessible to microbial attack and, therefore, stabilized against decomposition.

In 1982 (5), two Australian scientists, Drs. Judy Tisdall and Malcolm Oades, presented a hierarchical conceptual model of aggregate formation and SOM turnover that further refined the ideas published in the late 1960s. The model was based on the observation that aggregates broke down in a stepwise fashion; macroaggregates > 250-μm broke down into microaggregates 20–250-μm before particles < 20-μm were released. These results have been confirmed by a number of researchers (6–8) and provide the basis for the concept of aggregate hierarchy. The idealized model of Tisdall and Oades (5) consists of three hierarchical levels of aggregates:

$$2\text{–}20\ \mu m \rightarrow 20\text{–}250\ \mu m \rightarrow\ > 250\ \mu m.$$

Microaggregates of 2–20-μm diameter may be formed by the flocculation of fine silt or clay particles. Flocculation of the negatively charged clay particles is increased in the presence of polyvalent cations such as Al^{+3} and Ca^{+2}. Organic matter may enhance aggregation at this level when it forms complexes with the clay and polyvalent cations. It is believed that the organic matter associated with aggregates < 20-μm is primarily of microbial origin and there is little evidence that any plant debris exists in aggregates of this size (9).

The next stage in the hierarchy is 20–250-μm microaggregates. The core of such aggregates may occasionally contain a fragment of plant debris, however Oades and Waters (9) reported that the cores of many of these aggregates, especially those between 20- and 90-μm, were void. They suggested that the empty core was due to the complete degradation of the plant debris they originally contained. The stability of this aggregate was maintained by the decomposition products of the organic core.

Macroaggregates (> 250-μm in diameter) consist of microaggregates, primary particles, and particulate organic matter (POM) held together by a network of fine roots and vesicular arbuscular mycorrhizae (VAM) hyphae. The surface of the roots and hyphae is covered with extracellular polysaccharides to which microaggregates may become bound. Macroaggregate stability is related to the amount of POM carbon (C) in the soil (10) and aggregate formation and stabilization processes in soil are directly related to the decomposition of roots and the dynamics of root-derived POM C in soil (11).

The model of Tisdall and Oades (5) is often interpreted to mean that microaggregates are gathered together by the ramifying effect of roots and stabilized by the enmeshing effect of roots and hyphae. It also suggests that microaggegates are very stable and the organic matter (OM) associated with them is biologically recalcitrant. A number of researchers have presented an alternative view in which microaggregates are formed within existing macroaggregates (12–15). According to this theory, macroaggregates form first through the ramifying effect of plant roots. Fragments of organic matter, originating as small pieces of plant root, fungal hyphae, or other organic debris may be included in the macroaggregate structure. During the decomposition process, the organic fragment becomes encrusted with clay particles and microbially produced mucilage. The decomposition of the organic core may be retarded. The result is a newly formed, stable microaggregate.

AGGREGATE STABILIZATION AND ORGANIC BINDING AGENTS

Tisdall and Oades (5) emphasized the importance of organic binding agents to soil structural stability and divided these agents into three broad classes based on their temporal persistence in the soil: persistent (humic material associated with polyvalent metal cations), transient (predominantly polysaccharides), and temporary (roots, root hairs, and fungal hyphae).

Humic substances are large, relatively immobile, aromatic molecules produced during the microbial decomposition of plant and animal residues. Persistent organic bonds occur when a water molecule forms a bridge between an anionic functional group (e.g., carboxyl or hydroxide) on the humic molecule and a polyvalent cation (Al^{+3}, Fe^{+3}, Ca^{+2}) attached to the clay surface. Because of their chemical nature and chelation with polyvalent cations, humic substances are resistant to decomposition, but not completely inert.

The origin of soil polysaccharides includes plant and animal tissue, microbial cells, and exudates secreted by roots, hyphae, and bacteria. Plant-originating polysaccharides may be easily decomposed by microorganisms, which in turn secrete polysaccharides themselves. Most polysaccharides are negatively charged and polyvalent cations serve as a bridge between polysaccharides and clay particles. Unlike humic substances, polysaccharides can be rapidly metabolized by microorganisms, hence they are considered to be transitory binding agents. If the binding effect of polysaccharides is to persist, they must be continually replenished or they must be protected from decomposition.

Plant roots, root hairs, mycorrhizae, and fungal hyphae may also serve as temporary binding agents within and between aggregates. Roots form an extensive network throughout the soil and as they grow, they may enmesh microaggregates and pull them together. In the rhizosphere, the hyphae of VAM may contribute further to the aggregating effect as they grow into small pores and bind

Table 1 Summary of the major organic binding agents in the stabilization of aggregates

Aggregating agents	Aggregation process	Major scale of aggregation
Humic substances	Form strong bonds with soil mineral components	Basis of microaggregate formation
Polysaccharides	Act as gelatinous gluing agents	Involved in the stabilization of both micro- and macroaggregates
	Form organo-mineral associations	
Plant roots	Enmesh soil aggregates Exude polysaccharides	Agents of macroaggregate formation and short-term binding
Fungal hyphae	Enmesh soil aggregates Exude polysaccharides	Agents of macroaggregate formation and short-term binding
Earthworms	Mix organic matter and clay colloids together Mix decaying detritus with the bulk soil	Agents of macroaggregate formation

(From Ref. 17.)

soil particles together (16). The hyphae of filamentous fungi associated with decomposing fragments of organic matter are likely to have a similar effect in the bulk soil.

A summary of the different binding agents and their role in soil aggregation is given in Table 1. It is generally agreed that the humic substances and polysaccharides are the primary organic binding agents for the formation of stable microaggregates. It is important to note that these binding agents are relatively immobile in the soil. The mobile component in the formation of microaggregates appears to be fine clay particles, which may be carried by water through soil pores toward plant roots by matrix or gravity suction. As the fine clays move, it appears that they align themselves parallel to and encrust roots or OM fragments. They may also surround fungal hyphae or encapsulate bacteria. Thus, it is envisioned that microaggregate formation occurs primarily in zones of high microbial activity because this is where the humic substances are being produced. Microaggregates formed in this way would have a core of plant- or microbially derived organic matter.

AGGREGATE STRUCTURE AND ORGANIC MATTER DECOMPOSITION

In general, it is thought that the humified organic matter which stabilizes microaggregates is fairly recalcitrant, while the temporary and transient binding agents are more easily mineralized. The actual rate at which organic materials are mineralized, however, depends not only on their chemical qualities, but also on their availability to microorganisms. The organic matter bound within soil aggregates may be physically or chemically protected from decomposition.

The physical protection of organic matter occurs when physical barriers exist between otherwise decomposable substrates and decomposers. The primary evidence that aggregate structure physically protects organic matter from microbial decomposition is provided by studies in which aggregates are crushed or ground. C and nitrogen (N) mineralization rates increase greatly when the aggregate structure is disrupted. This is attributed to the exposure of organic matter, which was previously inaccessible to microbial attack.

Chemical stabilization occurs by the chemical interaction of substrate with the mineral part of the soil. As polysaccharides dry, they become denatured and are more chemically resistant to decomposition. In addition, many aggregates are composed of organic cores encrusted by clay particles. During the decomposition process, oxygen is consumed and the cores may eventually become anaerobic. Under these conditions, the solubility of polyvalent cations (Cu^{+2}, Fe^{+3}, Zn^{+2} and Ca^{+2}) may increase with the result that polysaccharides form complexes with the polyvalent cations more easily. These structures may protect polysaccharides from decomposition.

Differences in the relative stability of macro- and microaggregates are a function of their relative size, as well as the location and lability of the organic binding agents within the aggregates. Because the OM binding microaggregates together is generally assumed to be recalcitrant and in smaller pores that cannot be accessed by microorganisms, it is expected that organic matter associated with microaggregates will decompose more slowly than that associated with macroaggregates. A number of research studies have supported the assertion that soil organic matter decompostion rates are higher in macroaggregates than in microaggregates for a variety of soils and land-use management protocols (6, 14, 18, 19).

A

REFERENCES

1. Martin, J.P.; Martin, W.P.; Page, J.B.; Raney, W.A.; De Ment, J.G. Soil Aggregation. Adv. Agron. **1955**, *7*, 1–37.
2. Kay, B.D.; Angers, D.A.; Baldock, J.A.; Groenevelt, P.H. Quantifying the Influence of Cropping History on Soil Structure. Can. J. Soil Sci. **1988**, *18*, 64–73.
3. Emerson, E.E. The Structure of Soil Crumbs. J. Soil Sci. **1959**, *10*, 233–244.
4. Edwards, A.P.; Bremner, J.M. Microaggregates in Soils. J. Soil Sci. **1967**, *18*, 64–73.
5. Tisdall, J.M.; Oades, J.M. Organic Matter and Water-Stable Aggregates in Soil. J. Soil Sci. **1982**, *33*, 141–163.
6. Elliott, E.T. Aggregate Structure and Carbon, Nitrogen and Phosphorus in Native and Cultivated Soils. Soil Sci. Soc. Am. J. **1986**, *50*, 627–633.
7. Cambardella, C.A.; Elliott, E.T. Carbon and Nitrogen Distribution in Aggregates from Cultivated and Native Grassland Soils. Soil Sci. Soc. Am. J. **1993**, *57*, 1071–1076.
8. Angers, D.A.; Giroux, M. Recently Deposited Organic Matter in Soil Water-Stable Aggregates. Soil Sci. Soc. Am. J. **1996**, *60*, 1547–1551.
9. Oades, J.M.; Waters, A.G. Aggregate Hierarchy in Soils. Aust. J. Soil Res. **1991**, *29*, 815–828.
10. Cambardella, C.A.; Elliott, E.T. Particulate Soil Organic Matter Changes Across a Grassland Cultivation Sequence. Soil Sci. Soc. Am. J. **1992**, *56*, 777–783.
11. Gale, W.J.; Cambardella, C.A.; Bailey, T.B. Root-Derived Carbon and the Formation and Stabilization of Aggregates. Soil Sci. Soc. Am. J. **2000**, *64*, 201–207.
12. Oades, J.M. Soil Organic Matter and Structural Stability: Mechanisms and Implications for Management. Plant Soil **1984**, *76*, 319–337.

13. Elliott, E.T.; Coleman, D.C. Let the Soil Work for Us. Ecol. Bull. **1988**, *39*, 23–32.
14. Beare, M.H.; Cabrera, M.L.; Hendrix, P.F.; Coleman, D.C. Aggregate-Protected and Unprotected Organic Matter Pools in Conventional- and No-Tillage Soils. Soil Sci. Soc. Am. J. **1994**, *58*, 787–795.
15. Gale, W.J.; Cambardella, C.A. Carbon Dynamics of Surface Residue- and Root-Derived Organic Matter Under Simulated No-Till. Soil Sci. Soc. Am. J. **2000**, *64*, 190–195.
16. Miller, R.M.; Jastrow, J.D. Hierarchy of Root and Mycorrhizal Fungal Interactions with Soil Aggregation. Soil Biol. Biochem. **1990**, *22*, 579–584.
17. Haynes, R.J.; Beare, M.H. Aggregation and Organic Matter Storage in Meso-Thermal, Humid Soils. In *Structure and Organic Matter Storage in Soils*; Carter, M.R., Stewart, B.A., Eds.; Lewis Publishers, CRC Press, Inc.: Boca Raton, FL, 1996.
18. Gupta, V.V.S.R.; Germida, J.J. Distribution of Microbial Biomass and Its Activity in Different Soil Aggregate Size Classes as Affected by Cultivation. Soil Biol. Biochem. **1988**, *20*, 777–786.
19. Jastrow, J.D.; Boutton, T.W.; Miller, R.M. Carbon Dynamics of Aggregate-Associated Organic Matter Estimated by Carbon-13 Natural Abundance. Soil Sci. Soc. Am. J. **1996**, *60*, 801–807.

ALFISOLS

Rienk Miedema

Department of Soil Science and Geology, Wageningen University, Wageningen, The Netherlands

INTRODUCTION

The definition (1) of the soil order of Alfisols is: Soils that do not have a plaggen epipedon and that have either 1) an argillic or kandic horizon (base saturation 35% or more) or a natric horizon, or 2) a fragipan with clay films of thickness 1 mm or more in some parts.

The argillic, kandic, and natric horizons are formed as illuvial subsurface horizons, characterized by a significantly higher percentage of clay than the overlying eluvial horizon. The fine fraction of clay is strongly involved in the clay illuviation process (the ratio fine clay/total clay is the highest in the argillic horizon).

The argillic and the kandic horizon differ because the kandic horizon has over more than half its thickness an apparent cation exchange capacity (CEC) of 16 cmol(+) or less per kg clay.

The natric horizon is an argillic horizon with: 1) columnar or prismatic structures in its upper part with or without uncoated silt and sandgrains in the top of the horizon, and 2) an exchangeable sodium percentage (ESP) of 15% or more, or a sodium adsorption ratio (SAR) of 13 or more in the upper part of the horizon, or more exchangeable magnesium plus sodium plus exchangeable acidity in the upper part of the horizon.

Alfisols, distinguished on the basis of a fragipan (firm rupture resistance, slaking of water-submerged air dry fragments), also show clear evidence of clay illuviation (2), but the relation between fragipans and argillic horizons remains controversial (3–5). Alfisols are very recently reviewed (6).

The objective of this short communication is to reflect, in a general way, on the (palaeo)-environmental conditions under which clay illuviation, the main soil forming process in Alfisols, takes place. The areal extent and geographical distribution of Alfisols are summarized.

THE CLAY ILLUVIATION PROCESS

Particulate dispersed (fine) clay is liberated in the surface soil, is transported down the profile and accumulates (precipitates) in the subsurface horizon. Which conditions promote the liberation of dispersed (fine) clay from (micro)aggregates in the surface soil?

These conditions can be considered from a (bio)physical and biochemical point of view. The clay mineralogy is also of importance: smectitic clays disperse to finer clay particles than kaolinitic clays (as in the kandic horizon).

From the (bio)physical point of view, the moisture content of the soil material is important. Suddenly wetted dry aggregated soil material may experience "air explosion." The wetting front proceeds to the center of the dry aggregate, the trapped air is compressed, the aggregate explodes and fine and coarse material is released. The porosity and chemical characteristics of the original aggregates determine whether air explosion happens. Very porous aggregates do not experience air explosion. Soil material with a lot of Ca^{2+} or Al^{3+} bonds forms stable aggregates that do not produce fine clay upon air explosion.

The ratio of water to solid is important (7). When the water content is extremely high this may promote the liberation of fine clay. From the (bio)chemical point of view, aggregate stability/clay flocculation is important. Soil material with a high percentage of exchangeable Ca^{2+} or Al^{3+} clay remains flocculated even with extremely high ratios of water to solid. Humic substances play a role in the liberation of fine clay, related to the clay mineralogy (8, 9).

It is possible to indicate pH ranges in the soil, where liberation of fine clay is possible (10–12). In soils with a pH below 5, the amount of Al^{3+} in the soil solution and the amount of exchangeable Al^{3+} is so high that the clay remains flocculated.

In soils with a pH of 7–8, the amount of Ca^{2+} in the soil solution and on the exchange complex is so high that the clay remains flocculated. In soils with a high pH (8.5 or more), Na^{+} is strongly dominant and the liberation of dispersed fine clay is promoted due to a very extended electric double layer (natric horizon).

Hence, the prerequisites for soils to become Alfisols are: 1) decalcification of originally calcareous parent materials, 2) not very acidic parent materials, and 3) strongly alkaline parent materials.

Downward transport of the fine clay occurs as laminar flow with water along the pore walls (13, 14). Turbulent flow moves coarser particles downward.

Accumulation (precipitation) of the liberated (fine) clay with adsorbed Fe may occur through a number of mechanisms.

Chemically, a calcareous subsurface layer (Fig. 2B) forces the fine clay to flocculate and precipitate on the pore wall.

Stratification may trigger the accumulation of the (fine) clay at boundaries of finer and coarser textured soil materials. In sandy soils, this leads to the characteristic "banded B horizons" with bridges of fine clay between the sandgrains (15, 16).

Water absorption in a dry subsoil from pores may also prompt precipitation of fine clay on the pore wall.

Recognition of Illuvial Clay

Macromorphologically, clay coatings in pores or on pedfaces, distinguished by their more shiny appearance, are noted in the profile description. To judge the composition of the coatings with the naked eye or with a hand lense is difficult.

The accumulation of fine clay in the argillic horizon leads to parallel arrangement of clay particles (continuous orientation) and distinctive optical properties such as strong birefringence and extinction phenomena (Fig. 1A and B) in thin sections (30 μm thick) of undisturbed soils (17, 18).

In macroscopically identified argillic horizons, and sometimes in micromorphologically very little clay coatings are observed (19, 20). Recognizing and counting fragments of clay coatings, and distinguishing between illuvial clay coatings and stress coatings is not always easy. Strong swell–shrink activity of smectitic clay in certain soils leads to argillic horizons "without clayskins" (21).

Quantification of Clay Illuviation

In micromorphological quantification of clay illuviation (22–24), different trained operators obtain different results on the same samples (coefficients of variance of 50%) (23, 24).

Volume percentages of micromorphologically illuviated clay in argillic horizons range from about 1 to 20%. Well-developed argillic horizons in loess typically contain 4–7% illuviated fine clay micromorphologically (Fig. 2A and B).

Pedological Translocations/Transformations of Clay Coatings

The accumulated fine clay is present as films coatings or pore walls. Subsequent biological activity may translocate fragments of the coatings into the groundmass of the soil.

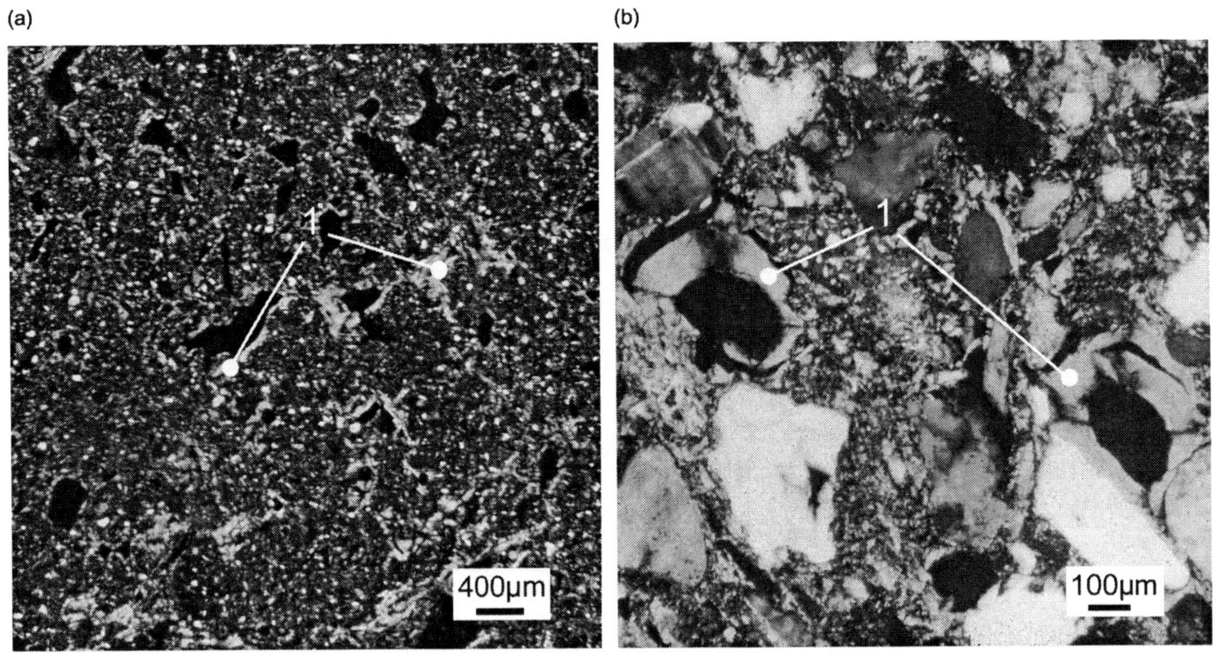

Fig. 1 Micromorphological images (crossed polarizers) of illuvial clay coatings (1) in (a) a Glossudalf from Russia (adapted from Ref. 38) and (b) an Ochraqualf from the Netherlands.

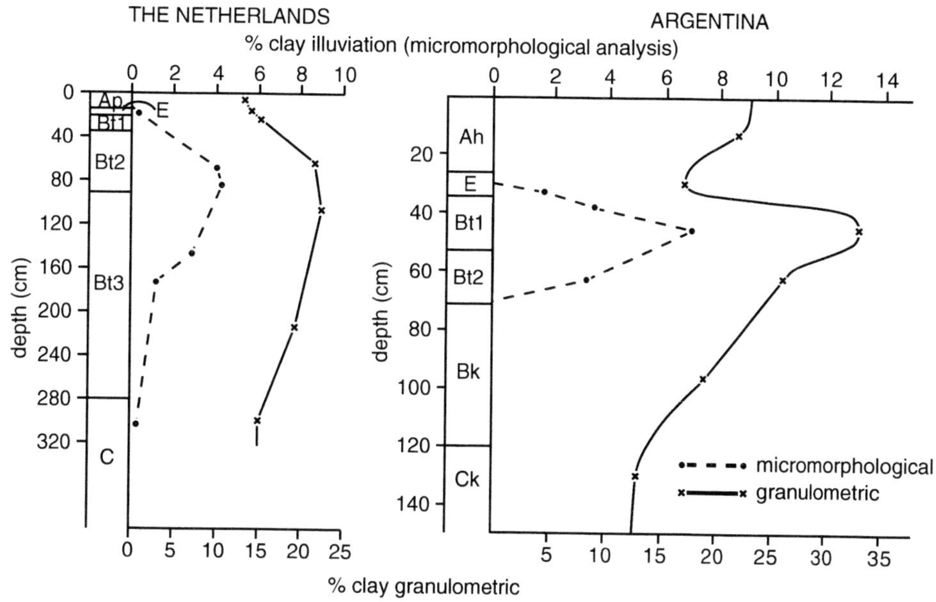

Fig. 2 Micromorphological quantification of clay coatings and granulometric clay content in an Udalf in loess from the Netherlands (adapted from Ref. 22.) and in an Ustalf in loess from Argentina.

The accumulation of clay in the subsurface horizon may lead to problems with the hydraulic conductivity in that horizon. Redox processes (pseudogley) in the upper part and along cracks and pores in the argillic horizon lead to degradation features (3, 25, 26) in the argillic horizon: clay coatings (partly) covered with Fe/Mn precipitates with less Fe content, or with grainy clay coatings (clay destruction) by ferrolysis (27). In Mediterranean (28) and semiarid areas, argillic horizons are found with clay illuviation and subsequent recarbonation from eolian dust (secondary calcite coatings covering fine clay coatings).

TIME OF FORMATION OF THE ARGILLIC HORIZON

The optimum conditions for the occurrence of clay illuviation (argillic horizon) are much disputed. One school of thought considers clay illuviation to occur in interglacial periods, such as the present-day Holocene period (4, 17, 29–31). Widespread semideciduous and mixed broadleaf-coniferous natural forest vegetation (through humic acids in their topsoil) is thought to create conditions optimally suitable for the liberation of fine clay in the topsoil.

Another school of thought (32–35) considers the optimum period for clay illuviation to occur at the end of a glacial period or the beginning of an interglacial or interstadial period. Following rapid Late Glacial decalcification of originally calcareous parent materials, the conditions for liberation of fine clay occur in the melting topsoil. This topsoil (active layer) is freeze-dried in frozen conditions and the melting of ice lenses produces sudden wetting, air explosion, and liberation of fine clay.

Relict argillic horizons occur in Late Pleistocene loess soils below present-day tidal marshes (submergence due to postglacial sealevel rise) in Maryland (36).

The Late Glacial conditions are also conducive for the formation of a fragipan or fragic characteristics (33, 37) in loess and glacial till in North America (2), Western Europe, (33) and Russia (38). Distinction is made between argillic horizons and paleo-argillic horizons (39). Paleo-argillic horizons are not associated with present-day soil-forming conditions (examples: 40–42).

GEOGRAPHICAL DISTRIBUTION AND AREAL EXTENT

Alfisols (total areal extent of which is 9.6% of the ice-free land surface of the earth) occur almost everywhere, provided that conditions suitable for clay illuviation are met, either presently or in the past (Fig. 3). The subdivisions of Alfisols are based on climate (temperature and precipitation). Suborders comprise the Cryalfs of Arctic areas, the Udalfs of temperate humid areas and the Ustalfs and Xeralfs of present-day semiarid and/or dry areas. A large proportion of the Alfisols occurs in the temperate and

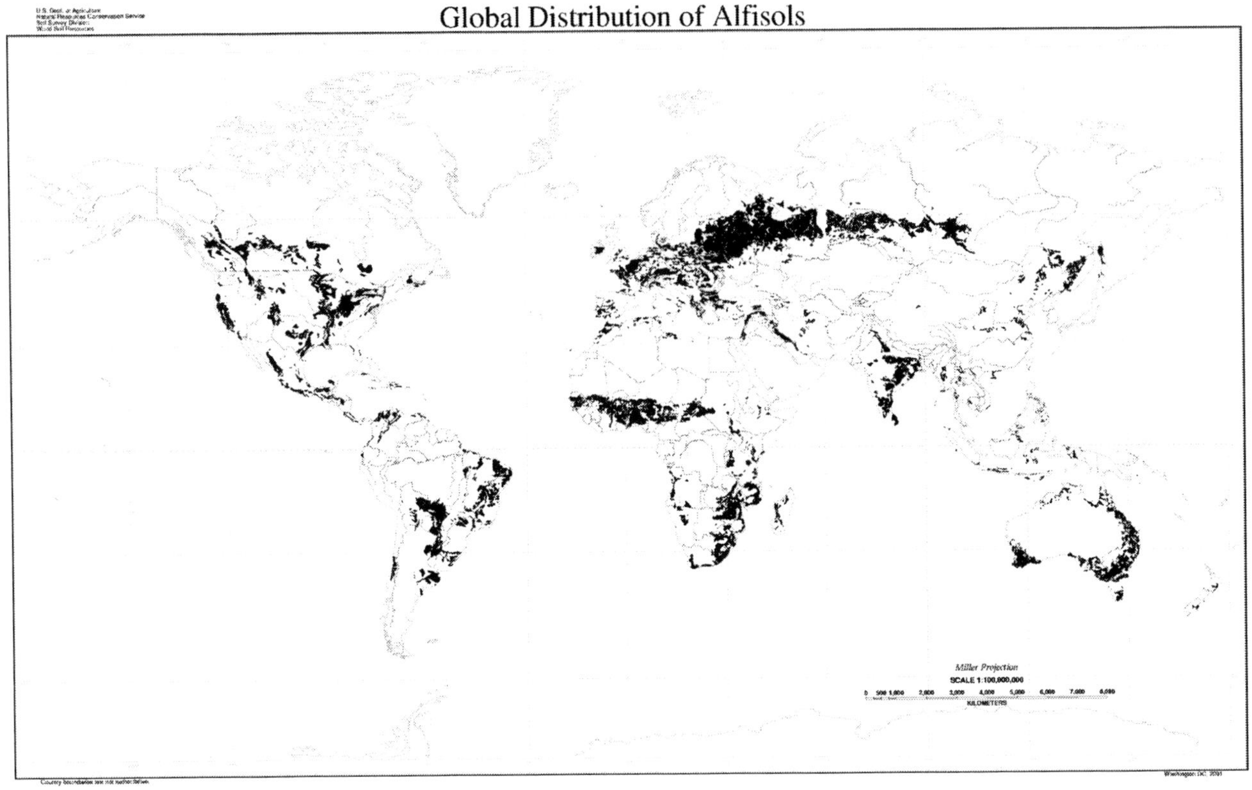

Fig. 3 Geographical distribution of Alfisols. (US Department of Agriculture, Nature Resources Conservation Service, Soil Survey Division, World Soil Resources, © US Department of Agriculture.)

Mediterranean zone. Medium textured, decalcified, weakly acidic parent materials like loess and glacial deposits provide suitable conditions for clay illuviation whilst the leaching of basic cations has not progressed too far and a high CEC is maintained.

Landscape position (degradation of Alfisols) is accommodated in the suborder of the Aqualfs (43, 44).

REFERENCES

1. Soil Survey Staff, *Keys to Soil Taxonomy*, 8th Ed.; USDA-NRCS US Government Printing Office: Washington, DC, 1998.
2. Smeck, N.E., Ciolkosz, E.J., Eds. *Fragipans: Their Occurrence, Classification and Genesis*; Soil Sci. Soc. Am. Spec. Publ. 24, Soil Sci. Soc. Am.: Madison, WI, 1989.
3. James, H.R.; Ransom, M.D.; Miles, R.J. Fragipan Genesis in Polygenetic Soils on the Springfield Plateau of Missouri. Soil Sci. Soc. Am. J. **1995**, *59*, 151–160.
4. Lindbo, D.L.; Rhoton, F.E.; Bigham, J.M.; Hudnall, W.H.; Jones, F.S.; Smeck, N.E.; Tyler, D.D. Loess Toposequences in the Lower Mississippi River Valley. I. Fragipan Morphology and Identification. Soil Sci. Soc. Am. J. **1995**, *59*, 487–500.
5. Lindbo, D.L.; Rhoton, F.E.; Hudnall, W.H.; Smeck, N.E.; Bigham, J.M. Loess Stratigraphy and Fragipan Occurrence in the Lower Mississippi River Valley. Soil Sci. Soc. Am. J. **1997**, *61*, 195–210.
6. Hallmark, C.T.; Franzmeier, D.P. Alfisols. In *Handbook of Soil Science*; Sumner, M.E., Ed.; CRC Press: Boca Raton, FL, 1999; E338–E359.
7. Gombeer, R.; d'Hoore, J. Induced Migration of Clay and Other Moderately Mobile Constituents. III. Critical Soil/Water Dispersion Ratio, Colloid Stability and Electrophoretic Mobility. Pédologie **1971**, *21*, 311–342.
8. Kaplan, D.I.; Bertsch, P.M.; Adriano, D.C.; Miller, W.P. Soil-borne Mobile Colloids as Influenced by Water Flow and Organic Carbon. Environ. Sci. Technol. **1993**, *27*, 1193–1200.
9. Kretzschmar, R.; Robarge, W.P.; Weed, S.B. Flocculation of Kaolinitic Soil Clays: Effect of Humic Substances and Iron Oxides. Soil Sci. Soc. Am. J. **1993**, *57*, 1277–1283.
10. Jenny, H.; Smith, G.D. Colloid Chemical Aspects of Clay Pan Formation in Soil Profiles. Soil Sci. **1935**, *39*, 377–389.
11. Beery, M.; Wilding, L.P. The Relationship Between Soil pH and Base Saturation Percentage for Surface and Subsoil Horizons of Selected Mollisols, Alfisols and Ultisols in Ohio. Ohio J. Sci. **1971**, *71*, 43–55.

12. Van Breemen, N.; Buurman, P. *Soil Formation*; Kluwer Academic Publishers: Dordrecht, 1998.
13. Hudson, B.D. Cohesive Water Films as a factor in Clay Translocation. Soil Surv. Horizons **1977**, *18*, 9–15.
14. Dalrymple, J.B. S.P. Theocharopoulos Intrapedal Cutans—An Experimental Production of Depositional (Illuviation) Channel Argillans. Geoderma **1984**, *33*, 237–243.
15. Dijkerman, J.C.; Cline, M.G.; Olson, G.W. Properties and Genesis of Textural Subsoil Lamellae. Soil Sci. **1967**, *104*, 7–16.
16. Torrent, J.; Nettleton, W.D.; Borst, G. Clay Illuviation and Lamella Formation in a Psammentic Haploxeralf in Southern California. Soil Sci. Soc. Am. J. **1980**, *44*, 363–369.
17. McKeague, J.A. Clay Skins and Argillic Horizons. In *Soil Micromorphology*; Bullock, P., Murphy, C.P., Eds.; AB Academic Publishers: Berkhamsted, UK, 1983; Vol. I, 367–388.
18. Bullock, P.; Thompson, M.L. Micromorphology of Alfisols. Soil Sci. Soc. Am. Spec. Publ. **1985**, *15*, 17–48.
19. McKeague, J.A.; Guertin, R.K.; Pagé, F.; Valentine, K.W. Micromorphological Evidence of Illuvial Clay in Horizons Designated Bt in the Field. Can. J. Soil Sci. **1978**, *58*, 179–186.
20. Bronger, A. Argillic Horizons in Modern Loess Soils in an Ustic Moisture Regime: Comparative Studies in Forest-Steppe and Steppe Areas from Eastern Europe and the United States. In *Advances in Soil Science*; Stewart, B.A., Ed.; Springer: New York, 1991; 15, 41–90.
21. Nettleton, W.D.; Flach, K.W.; Brasher, B.B. Argillic Horizons without Clay Skins. Soil Sci Soc Am. Proc. **1969**, *33*, 121–125.
22. Miedema, R.; Slager, S. Micromorphological Quantification of Clay Illuviation. J. Soil Sci. **1972**, *23*, 309–314.
23. McKeague, J.A.; Guertin, R.K.; Valentine, K.W.; Belisle, J.; Bourbeau, G.A.; Michalyna, W.; Hopkins, L.; Howell, L.; Pagé, F.; Bresson, L.M. Variability of Estimates of Illuvial Clay in Soils by Micromorphology. Soil Sci. **1980**, *129*, 386–388.
24. Murphy, C.P. Point Counting Pores and Illuvial Clay in Thin Sections. Geoderma **1983**, *31*, 133–150.
25. Bullock, P.; Milford, M.H.; Cline, M.G. Degradation of Argillic Horizons in Udalf Soils of New York State. Soil Sci. Soc. Am. Proc. **1974**, *38*, 621–628.
26. Ransom, M.D.; Smeck, N.E.; Bigham, J.M. Stratigraphy and Genesis of Polygenetic Soils on the Illinoian Till Plain of Southwestern Ohio. Soil Sci. Soc. Am. J. **1987**, *51*, 135–141.
27. Brinkman, R.; Jongmans, A.G.; Miedema, R.; Maaskant, P. Clay Decomposition in Seasonally Wet Acid Soils: Micromorphological, Chemical and Mineralogical Evidence from Indivual Argillans. Geoderma **1973**, *10*, 259–270.
28. Fedoroff, N. Clay Illuviation in Red Mediterranean Soils. Catena **1997**, *28*, 171–189.
29. Catt, J.A. Recent Work on Quaternay Paleosols in Britain. Quat. Int. **1996**, *34–36*, 183–190.
30. Mücher, H.J. Aspects of Loess and Loess-Derived Slope Deposits: An Experimental and Micromorphological Approach. PhD Thesis University of Amsterdam.
31. Hoffman, R.; Blume, H.P. Holocene Clay Migration As a Soil Forming Process in Loamy Soils of the Moraine Landscapes of North Germany. Catena **1977**, *4*, 359–368.
32. Hoeksema, K.J.; Edelman, C.H. The Role of Biological Homogenization in the Formation and Transformation of Gray–Brown Podzolic Soils. Transactions 7th Int. Congress Soil Sci.: Madison, WI, 1960; Vol IV, 402–405.
33. Van Vliet- Lanoë, B.; Langohr, R. Correlation Between Fragipans and Permafrost with Special Reference to Silty Deposits in Belgium and Northern France. Catena **1981**, *8*, 137–154.
34. Van Vliet-Lanoë, B. The Genesis and Age of the Argillic Horizon in Weichselian Loess of Northwestern Europe. Quat. Int. **1990**, *5*, 49–56.
35. Miedema, R. Processus De Formation Des Sols Tardiglaciaires Et Holocènes Sur Les Terrasses Alluviales Du Rhin Aux Pays Bas. Sci. du Sol **1992**, *30*, 149–168.
36. Stolt, M.H.; Rabenhorst, M.C. Micromorphology of Argillic Horizons in an Upland/tidal Marsh Catena. Soil Sci. Soc. Am. J. **1991**, *55*, 443–450.
37. FitzPatrick, E.A. An Indurated Horizon Formed in Permafrost. J. Soil Sci. **1956**, *7*, 248–254.
38. Miedema, R.; Koulechova, I.N.; Gerásimova, M.I. Soil Formation in Greyzems in Moscow District: Micromorphology, Chemistry, Clay Mineralogy and Particle Size Distribution. Catena **1999**, *34*, 315–347.
39. Bullock, P.; Murphy, C.P. Evolution of a Paleo-Argillic Brown Earth(Paleudalf) from Oxfordshire, England. Geoderma **1979**, *22*, 225–252.
40. Kemp, R.A. Micromorphology of Loess–Paleosol Sequences: A Record of Paleoenvironmental Change. Catena **1999**, *35*, 179–196.
41. Jongmans, A.G.; Feijtel, T.C.J.; Miedema, R.; van Breemen, N.; Veldkamp, A. Soil Formation in a Quaternay Terrace Sequence of the Allier, Limagne, France. Macro- and Micromorphology, Particle Size Distribution, Chemistry. Geoderma **1991**, *49*, 215–239.
42. Dorronsoro, C.; Alonso, P. Chronosequence in Almar River Fluvial Terrace Soil. Soil Sci. Soc. Am. J. **1994**, *58*, 910–925.
43. Smith, H.; Wilding, L.P. Genesis of Argillic Horizons on Ochraqualfs Derived from Fine-Textured Till Deposits in Ohio. Geoderma **1972**, *24*, 1–16.
44. Coventry, R.J.; Williams, J. Quantitative Relationships Between Morphology and Current Soil Hydrology in Some Alfisols in Semi-Arid Tropical Australia. Geoderma **1984**, *33*, 191–218.

ALLOPHANES

Jodi L. Johnson-Maynard
University of Idaho, Moscow, Idaho, U.S.A.

INTRODUCTION

Allophanes are poorly crystalline aluminosilicates that are especially prevalent in soils formed from volcanic ash. Allophane minerals impart unique chemical and physical properties to soils in which they are a predominant mineralogical component and their influence is largely responsible for the development of the Andisol soil order (1). The development of this soil order highlights the importance of allophane minerals in defining soil chemical and physical properties, and therefore the soil's response to management. The following discussion will focus on the structure and resulting properties of end-members, as defined by Al:Si ratio, of the allophane group. Despite having well-defined crystalline structure in one direction, imogolite will be included in this discussion because it is commonly found in association with allophane.

STRUCTURE

Imogolite

Imogolite exhibits tubular morphology with an outer diameter of about 2 nm (Fig. 1), an Al:Si ratio of about 2:1, and longer-range structure than that of allophane. The proposed structure of imogolite includes a gibbsite-like sheet that makes up the external surface of the tube and an interior that consists of SiO_4 groups coordinated with three Al atoms in the gibbsite-like sheet (2–4).

Allophanes

Due to their lack of long-range order, allophanes have been difficult to characterize by standard methods such as X-ray diffraction (5). Most of the information known about their chemical structures has been determined by methods such as IR and NMR (6). The precise details of methods suitable to study these short-ranged ordered minerals is beyond the scope of this discussion. For further information, the reader is directed to the excellent reviews of Wada and Dahlgren (2, 3). Although the exact structure of allophane has not been fully agreed upon (7), some similarities can be found among the various proposed structures.

The Al-rich allophanes (Al:Si = 2:1), also known as proto-imogolite or imogolite-like allophanes, are the most commonly encountered allophane minerals in soils and are similar in composition to imogolite (8). They are shaped like hollow spherules that range from 3.5 to 5 nm in diameter (Fig. 1). The structure is based on the "backbone" of a disordered single octahedral sheet. Each silicate tetrahedron is bound to the octahedral sheet through the sharing of three oxygens surrounding the octahedral sites (9). Similar IR and NMR spectra from Al-rich allophanes and imogolite indicate that the spherules contain imogolite-like fragments (4).

Si-rich allophanes (Al:Si close to 1:1) share the hollow-spherule morphology of the Al-rich allophanes. Structurally, they consist of Si tetrahedra polymerized into a disordered sheet and are sometimes likened to "defective kaolin structures" (4, 9, 10). Based on similarities between their NMR and IR spectra, Si-rich allophanes have areas similar to the proto-imogolite structure (4).

A computer model that accounted for structural data gathered by high resolution solid-state ^{29}Si and ^{27}Al NMR, thermal analysis, and X-ray powder diffraction, was used to develop a structure for allophanes with SiO_2/Al_2O_3 ratios between 0.93 and 1.73 (6). The modeled structure consists of a kaolinite-like structure with a curved tetrahedral sheet containing defects or holes through which imogolite-like orthosilicate units are anchored into the octahedral sheet (Fig. 2). This proposed structure was later supported by studies using ESCA and solid-state NMR (5). Together, the structural and chemical data show that in soils, allophanes form a continuum with respect to Al and Si concentrations and structure.

PROPERTIES

Probably the most well-noted property of allophane minerals is their high water-holding capacity. Pure imogolite, for example, was shown to have 1.5 times higher water retention than pure Na-montmorillonite at $-650\,J\,kg^{-1}$ (11). Due to their porous structures, allophanes have water contents

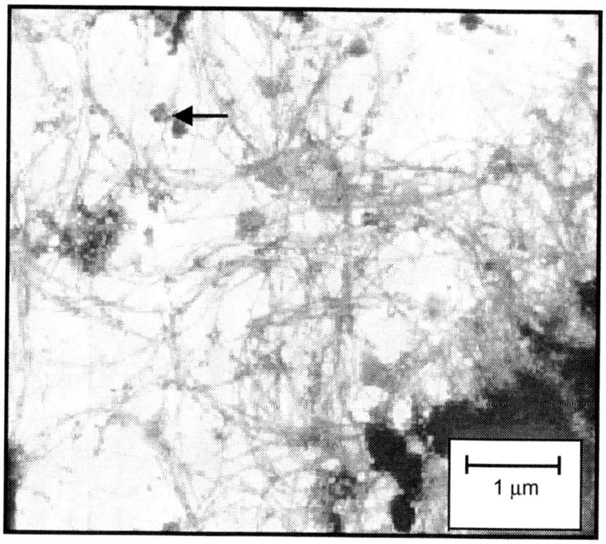

Fig. 1 Electron micrograph showing the fine tubular morphology of imogolite. The arrow indicates spherical allophane units. Photo courtesy of R. Dahlgren

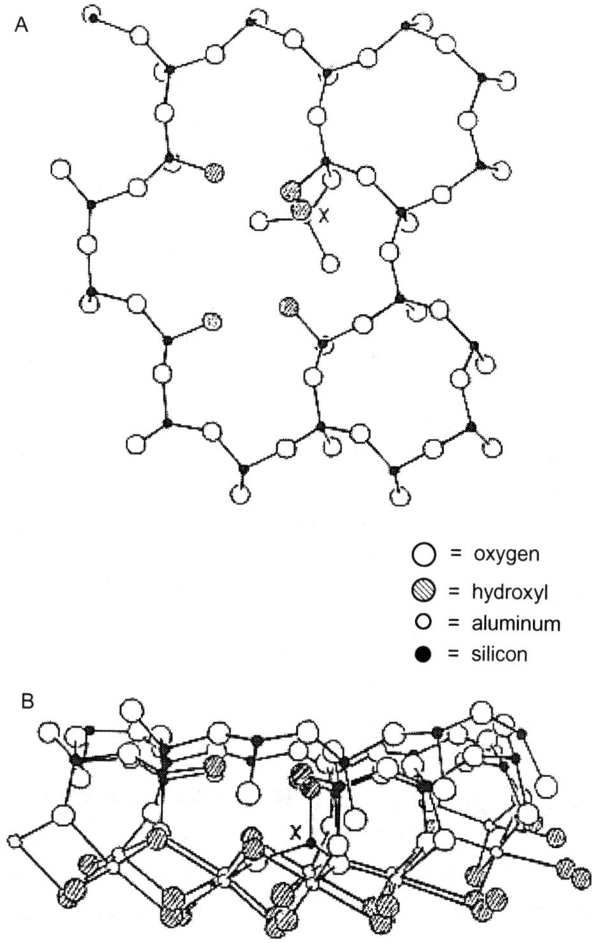

Fig. 2 Computer-generated diagram of A) Si–O sheet of proposed allophane structure showing a plan view from the center of the spherical particle, with a Si defect (hole) and an inserted imogolite unit, marked X; and B) curved cross-section of sphere wall with imogolite unit marked X. (From Ref. 3.)

(structural and hydration) ranging from 35.5 to 37.5% by weight (6). Allophanic soils, therefore, tend to have relatively high plant available water contents that are desirable for agricultural and agroforestry production systems. Andisols of the inland Pacific Northwest that support high productivity forests, have greater amounts of plant water available compared to nonandic soils with similar textures (Fig. 3).

Low bulk density is a definitive feature of the Andisol soil order. Since Andisols typically have particle densities ranging between 2.5 and 2.7 g cm^{-3}, the relatively low bulk density has been attributed to the development of porous structure (12). Bulk density values of allophanic Andisols with relatively low organic matter content (<3%) decrease with an increase in allophane content suggesting that allophane is the main noncrystalline material that results in porous structure (13).

It has been noted that Andisols exhibit irreversible changes in water retention, clay dispersibility, and liquid limit upon drying, with the degree of change strongly correlated to oxalate-extractable Al, Fe, and Si (13). Air-dried allophane, for example, has water contents that range from 23 to 29% of fresh allophane (11). The alteration of physical properties after drying has been attributed to irreversible aggregation (11). Allophane, even after treatment to enhance dispersion, tends to form chain-like floccules or domains that alter its physical properties (14).

Due to its charge characteristics and high surface area, allophane sorbs both P and organic compounds. Three possible P-sorption mechanisms are supported by experimental evidence (15). Phosphate is initially strongly adsorbed by ligand exchange on the most reactive AlOH defect sites. As these sites become saturated, P is held on less reactive defect sites. Under high P concentrations, phosphate reacts with Al, forming alumino-phosphate precipitates that cause a disruption of the allophane structure, thus creating new defect sites (15). These three types of bound P are bio-available to different degrees, indicating an important link between allophane and plant nutrition (16).

Allophanic soils accumulate relatively large amounts of organic matter relative to other soils. Allophane has been shown to stabilize organic matter through both direct and indirect mechanisms. Organic matter can be sorbed onto the surface of allophane minerals through ligand exchange

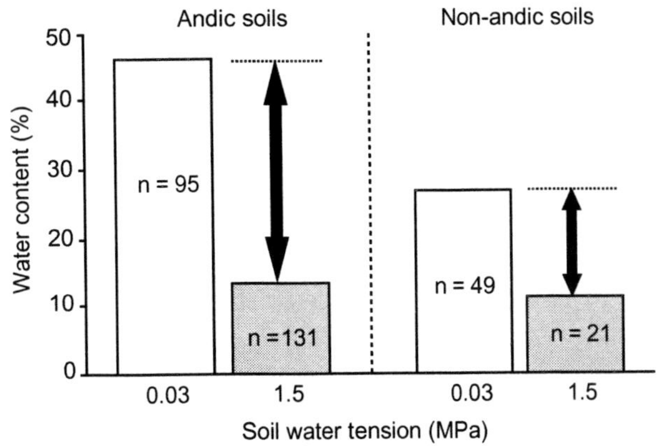

Fig. 3 Mean water content of inland Pacific Northwest Andisols and nonandic soils with similar textures at field capacity (0.03 MPa) and permanent wilting point (1.5 MPa). Vertical arrows indicate the range of plant available water (water $content_{0.03\,MPa} - water\ content_{1.5\,MPa}$) (17).

reactions. Indirectly, allophanes release Al into soil solution as a result of weathering. Subsequent formation of aluminum–humus complexes causes the retardation of organic matter decomposition (18). The accumulation of organic matter further contributes to the water retention capacity of Andisols.

The Anidsol order, because of the influence of allophane minerals, is one of the only two soil orders in the U.S. taxonomic system that are defined based on parent material characteristics. This demonstrates how unique the behavior of allophane is and emphasizes the need for careful consideration of land use decisions for allophanic soils.

REFERENCES

1. Soil Survey Staff, *Keys to Soil Taxonomy*; U.S. Government Printing Office: Washington, DC, 1998.
2. Dahlgren, R.A. Quantification of Allophane and Imogolite. In *Quantitative Methods in Soil Mineralogy*; Amonette, J.E., Zelazny, L.W., Eds. Soil Science Society of America: Madison, 1994; 360–408.
3. Wada, K. Allophane and Imogolite. In *Minerals in Soil Environments*, 2nd Ed.; Dixon, J.B., Weed, S.B., Eds. Soil Science Society of America: Madison, 1989; 1051–1081.
4. Parfitt, R.L. Allophane in New Zealand—A Review. Aust. J. Soil Res. **1990**, *28*, 343–360.
5. He, H.; Barr, T.L.; Klinowski, J. ESCA and Solid-State NMR Studies of Allophane. Clay Miner. **1995**, *30*, 201–209.
6. MacKenzie, K.J.D.; Bowden, M.E.; Meinhold, R.H. The Structure and Thermal Transformations of Allophanes Studied by ^{29}Si and ^{27}Al High Resolution Solid-State NMR. Clays Clay Miner. **1991**, *39*, 337–346.
7. Dahlgren, R.; Shoji, S.; Nanzyo, M. Mineralogical Characteristics of Volcanic Ash Soils. In *Volcanic Ash Soils: Genesis, Properties and Utilization*; Shoji, S., Nanzyo, M., Dahlgren, R., Eds. Elsevier: Amsterdam, 1993; 288.
8. Parfitt, R.L.; Kimble, J.M. Conditions for Formation of Allophane in Soils. Soil Sci. Soc. Am. J. **1989**, *53*, 971–977.
9. Childs, C.W.; Hayashi, S.; Newman, R.H. Five-Coordinate Aluminum in Allophane. Clays Clay Miner. **1999**, *47*, 64–69.
10. Childs, C.W.; Parfitt, R.L.; Newman, R.H. Structural Studies of Silica Springs Allophane. Clay Miner. Bull. **1990**, *25*, 329–341.
11. Karube, J.; Abe, Y. Water Retention by Colloidal Allophane and Imogolite with Different Charges. Clays Clay Miner. **1998**, *46*, 322–329.
12. Bielders, C.L.; DeBacker, L.W.; Delvaux, B. Particle Density of Volcanic Soils as Measured with a Gas Pyconometer. Soil Sci. Soc. Am. J. **1990**, *54*, 822–826.
13. Nanzyo, M.; Shoji, S.; Dahlgren, R. Physical Characteristics of Volcanic Ash Soils. In *Volcanic Ash Soils: Genesis, Properties and Utilization*; Shoji, S., Nanzyo, M., Dahlgren, R., Eds. Elsevier: Amsterdam, 1993; 189–207.
14. Karube, J.; Nakaishi, K.; Sugimoto, H.; Fujihira, M. Size and Shape of Allophane Particles in Dispersed Aqueous Systems. Clays Clay Miner. **1996**, *44*, 485–491.
15. Parfitt, R.L. Phosphate Reactions with Natural Allophane, Ferrihydrite, and Goethite. J. Soil Sci. **1989**, *40*, 359–369.
16. Parfitt, R.L.; Hart, P.B.S.; Meyrick, K.F.; Russell, M. Response of Ryegrass and White Clover to Phosphorus on an Allophanic Soil, Egmont Black Loam. NZ J. Agric. Res. **1982**, *25*, 549–555.
17. USDA–NR CS. 2001. Research Data. National Soil Survey Center-Soil Survey Laboratory.
18. Parfitt, R.L.; Fraser, A.R.; Farmer, V.C. Adsorption on Hydrous Oxides. III. Fulvic Acid and Humic Acid on Goethite, Gibbsite and Imogolite. J. Soil Sci. **1977**, *28*, 289–296.

AMENDMENTS/AMELIORANTS

David C. McKenzie
Precision Land Management, Orange, New South Wales, Australia

A

INTRODUCTION

Soil amendments/ameliorants can be used to overcome many physical problems in soil. For improving soil structural stability, the best-known ameliorants are gypsum (calcium sulfate) (1) and, to a lesser extent, finely ground limestone (calcium carbonate). Other useful inputs include natural and synthetic organic materials.

SOIL STRUCTURE LIMITATIONS

Key Processes

Good soil structure is vital for the growth of most plants (paddy rice is an exception), particularly where water application is either uncontrolled (rainfed), or not easily controlled (flood irrigated). Soil with excellent structural form accepts water readily, drains excess water quickly, and has a broad range of water contents over which waterlogging (poor aeration) and excessive hardness do not limit seedling emergence or root growth (2).

Structural form refers to the arrangement of the solid components of soil and the associated pore space (3). Structural stability is the ability of soil to retain its structural form when subjected to disruptive forces such as immersion in water.

There are two aspects of soil stability in water: slaking and dispersion (4). Slaking refers to the collapse of air-dry soil aggregates into subunits (diameter of about 0.2 mm) when placed in rainwater. It indicates that the bonds created by materials such as organic matter are not strong enough to withstand the forces associated with rapid wetting. If the sub-units created by slaking break down further, leading to a separation of clay, sand, and silt, dispersion is said to have occurred. An associated limitation is excessive subsoil swelling. Formation of a dispersed surface seal under wet conditions leads to poor aeration and inadequate water penetration; the runoff water may then cause erosion. Hard crusts develop when dispersed soil is dried, particularly where the shrink/swell potential is poor (3).

As the amount of sodium adsorbed on the surfaces of clay particles increases, so does dispersion. This problem is aggravated by a lack of electrolyte in soil solution. Clay type also is important. Illite disperses more readily than smectite clay, and kaolinite clay become more dispersive as pH increases. The presence of exchangeable magnesium on clay surfaces aggravates soil dispersion. Several direct measures of soil dispersibility are available (5, 6).

Management Options

Poor structure can be overcome indirectly by special management inputs. Options include:

- Raised beds and/or extra nitrogen fertilizer to deal with waterlogging limitations; and
- The use of drip irrigation systems to apply water in a way that maintains a very narrow range of soil water contents, where neither waterlogging nor soil strength will limit growth.

However, it is usually less expensive and less risky to improve structural stability by adding ameliorants to the soil. The benefits of gypsum application to sodic (dispersive) soil have been recognized since the early 1900s in the U.S. (7). The practice of applying lime (chalk) to improve the structure of clay soil has been carried out for thousands of years in England (8). In developing countries, ash and organic household waste often are used to fertilize nutrient-depleted soil (9), but may also improve soil structure.

Factors to consider before selecting an amelioration strategy include:

1. Soil condition should be measured so that amelioration is based on objective data about the severity of soil degradation.
2. The types and amounts of soil ameliorants that are applied will be strongly influenced by their cost in relation to the requirements and value of the plants being grown.

Secondary benefits, such as overcoming sulfur deficiency with gypsum and reducing acidity limitations with lime, should be considered when choosing an ameliorant. If the soil has been compacted by heavy machinery, deep ripping may also be required.

Encyclopedia of Soil Science

AMELIORANTS FOR IMPROVING SOIL STRUCTURE

Gypsum

Gypsum is the main compound used for the reclamation of soil that is dispersive and/or prone to excessive swelling. The benefits of using gypsum to improve soil structural stability are due to both electrolyte concentration and cation exchange effects (replacement of sodium ions attached to the clay surfaces with calcium ions) (10). Gypsum is available as a mined product, or can be obtained as a by-product of industrial processes (1). Mined gypsum, which has a coarser particle size than by-product gypsum, tends to be less soluble. This is beneficial in situations where a slow-release source of electrolyte and calcium is required. Gypsum can be either spread as a solid or dissolved in irrigation water. A typical application rate on sodic clay soil is 2.5 t/ha for dryland wheat and 7.5 t/ha for higher-value irrigated crops.

Soil that is most likely to show an economically viable response to gypsum application has a high clay content (greater than 30%) and is sodic with an exchangeable sodium percentage (ESP) greater than 5 (11). If the salt concentration in soil is very low, the critical ESP for dispersion may be as low as 2 (as found in the surface of some hardsetting soil types).

Overcoming a soil structure problem with gypsum may introduce new problems. For example, the correction of drainage limitations in a sodic clay soil may lead to a leaching of nitrates and other contaminants beyond the root zone. Therefore, soil water management needs to be adjusted after soil structural stability has been improved. Possible impurities in gypsum products also need to be considered. Mined gypsum often contains traces of clay, while some industrial gypsum may be contaminated by toxins such as cadmium and fluoride compounds. It is important to obtain data about the quality of ameliorants prior to application.

Lime

When the soil pH (measured in water) is less than approximately 7, the application of ground limestone is likely to release substantial amounts of electrolyte and calcium. Lime often is introduced to the soil naturally via irrigation water. Even in soil with an average pH >7, there may be pockets of soil adjacent to roots with a much lower pH that will encourage the dissolution of lime. Liming products (ground limestone, calcium oxide, and calcium hydroxide) are a more concentrated form of calcium than gypsum. The addition of acidifying materials (12) can release calcium by dissolving lime that occurs naturally in soil.

Another option is to apply a blend of gypsum and ground limestone (11). The gypsum provides an immediate source of calcium and electrolyte, while the less-soluble lime provides a longer term supply.

Organic Matter

For soil that is prone to slaking, the application of organic matter is likely to be beneficial (4). Organic ameliorants are particularly useful for nonswelling topsoil that is prone to hardsetting. Natural organic products include crop residues, animal and poultry manure and sewage sludge (biosolids). Synthetic soil conditioners such as polyacrylamide (PAM) are also effective, but cost and poor persistence of benefits limits their use. Organic matter application can reduce dispersion if accompanied by calcium ions (13). Organic mulches protect the soil-surface from the disruptive effects of raindrop impact. They also encourage soil fauna such as earthworms, which improve soil structure with their burrows and exudates.

CHALLENGES

Modelling the Effects of Ameliorants on Soil Condition and Plant Growth

Numerous field trials have been conducted throughout the world to determine the most effective rates and forms of gypsum (and related materials) for soil that is structurally unstable. However, the results generally are location-specific (5) and are strongly influenced by climatic conditions. A model is needed to predict the most profitable rates of gypsum and other ameliorants for various land use systems on soil that is prone to slaking and/or dispersion.

Integration of Soil Amelioration Strategies with "Precision Agriculture" Technology

The introduction of crop yield and quality mapping (14) will help land managers to identify soil structure problems within subsections of management units. This should allow ameliorants to be applied more efficiently via variable-rate application equipment, rather than as blanket applications using conventional machinery. Research is needed to optimize soil-sampling strategies for yield map interpretation, and to determine the extent to which within-field variability of soil properties can be minimized economically. More attention should also be given to the development of rapid and inexpensive field measurement techniques that allow the ameliorant requirements of soil to be mapped more accurately.

REFERENCES

1. Levy, G.J.; Sumner, M.E. Mined and By-Product Gypsum As Soil Amendments and Conditioners. In *Handbook of Soil Conditioners*; Wallace, A., Terry, R.E., Eds.; Marcel Dekker: New York, 1998; 187–215.
2. Letey, J. Relationship Between Soil Physical Properties and Crop Production. Advances in Soil Science **1985**, *1*, 277–294.
3. Kay, B.D. Rates of Change of Soil Structure Under Different Cropping Systems. Advances in Soil Science **1990**, *12*, 1–52.
4. Emerson, W.W. Structural Decline of Soils, Assessment and Prevention. Australian Journal of Soil Research **1991**, *29*, 905–921.
5. Rengasamy, P.; Churchman, G.J. Cation Exchange Capacity, Exchangeable Cations and Sodicity. In *Soil Analysis: An Interpretation Manual*; Peverill, K.I., Sparrow, L.A., Reuter, D.J., Eds.; CSIRO: Collingwood, Australia, 1999; 147–157.
6. Field, D.J.; McKenzie, D.C.; Koppi, A.J. Development of an Improved Vertisol Stability Test for SOILpak. Australian Journal of Soil Research **1997**, *35*, 843–852.
7. Richards, L.A., Ed. *Diagnosis and Improvement of Saline and Alkali Soils*; United States Department of Agriculture: Washington, DC, 1954.
8. Russell, E.W. *Soil Conditions and Plant Growth*, 9th Ed.; Longman: London, 1961.
9. Okigbo, B.N. *Development of Sustainable Agricultural Production Systems in Africa*; International Institute of Tropical Agriculture: Ibadan, Nigeria, 1991.
10. Loveday, J. Relative Significance of Electrolyte and Cation Exchange Effects when Gypsum is Applied to a Sodic Clay Soil. Australian Journal of Soil Research **1976**, *14*, 361–371.
11. Abbott, T.S.; McKenzie, D.C. *Improving Soil Structure with Gypsum and Lime*; NSW Agriculture: Orange, Australia, 1996.
12. Somani, L.L.; Totawat, K.L. Mined and Industrial Waste Products Capable of Generating Gypsum in Soil. In *Handbook of Soil Conditioners*; Wallace, A., Terry, R.E., Eds.; Marcel Dekker: New York, 1998; 257–291.
13. Muneer, M.; Oades, J.M. The Role of Ca-Organic Interactions in Soil Aggregate Stability II. Field Studies with ^{14}C-Labelled Straw, $CaCO_3$ And $CaSO_4 \cdot 2H_2O$. Australian Journal of Soil Research **1989**, *27*, 401–409.
14. Robert, P.C., Rust, R.H., Larson, W.E., Eds. *Proceedings of the Fourth International Conference on Precision Agriculture*; American Society of Agronomy: Madison, Wisconsin, 1999.

AMORPHOUS MINERALS

April L. Ulery
New Mexico State University, Las Cruces, New Mexico, U.S.A.

INTRODUCTION

The term amorphous literally means "without form" and refers to a class of materials that are noncrystalline or at best, poorly crystalline. Amorphous materials do not have a regular or repeating, ordered, internal atomic structure that is detectable by X-ray diffraction. Mineralogists use X-ray diffraction as the criterion for crystallinity and to identify most inorganic solids. Other terms used to define amorphous materials include short-range-order or X-ray zzamorphous minerals because the distances between repeating, internal patterns are too short to diffract X-rays.

Crystalline minerals, in contrast to amorphous or noncrystalline materials, have long-range-order, usually in three dimensions, and diffract X-rays from repeating atomic planes. Instead of a sharp boundary between amorphous and crystalline materials, a continuum exists between perfectly ordered, highly crystalline minerals, and disordered, poorly crystalline to amorphous, noncrystalline solids (Fig. 1).

VARIETIES OF AMORPHOUS MINERALS

Amorphous soil minerals include oxides and hydroxides of aluminum, silicon, iron, titanium, and manganese, and silicates of aluminum and iron. They may be hydrated weathering products of primary minerals or unweathered volcanic glass and biogenic silica phytoliths. Amorphous weathering products occur as coatings or gel-hulls on larger mineral grains and as aggregates or discrete particles in the soil clay fraction ($<2\,\mu m$ equivalent diameter). Glass particles and phytoliths are often larger than $2\,\mu m$ in diameter, and compared to the smaller amorphous soil components, are not as reactive.

Allophane, as a general term, has previously been used to describe any amorphous aluminosilicate in soils derived from volcanic material. Allophane ($1\text{-}2SiO_2Al_2O_3\ n\,H_2O^+$) is now recognized as a soil mineral group with a range in alumina and silica composition and specific diagnostic properties. Imogolite [$(OH)SiO_3Al_2\ (OH_3)$] is a mineral with threadlike or tubular morphology and long-range order in only one dimension. Although it exhibits unique diffraction characteristics, it is often included in the discussion of amorphous minerals because it has similar chemical and physical properties with allophane and is often associated with allophane in soils (1). For more detailed information on allophane and imogolite, please see the section entitled "Allophane" in this encyclopedia.

Poorly crystallized halloysite ($Al_2Si_2O_{10}OH_4\ n\,H_2O$) is another aluminosilicate that is included with the group of amorphous minerals because of its broad X-ray diffraction bands and tubular or curling morphology. Hisingerite ($Fe_2O_3\ 2SiO_2\ n\,H_2O$) is a "very finely crystalline" to amorphous variant of iron phyllosilicates or an iron silicate analog of allophane. Other iron minerals observed in soils but amorphous to X-rays includes ferrihydrite ($5Fe_2O_3\ 9H_2O$) and feroxyhite ($FeOOH$). Ferrihydrite may exist anywhere along the continuum between crystalline to poorly crystalline. It can range from having short-range order in three dimensions to less order in only two dimensions (2). Poorly crystalline varieties of aluminum and titanium are also identified as pseudomorphs of their corresponding well-ordered minerals. They include pseudoboehmite ($AlOOH$) and pseudorutile ($Fe_2O_3\ nTiO_2\ m\,H_2O$). Amorphous varieties of manganese oxides have not been specifically named.

Silica (SiO_2) may be present in soils as one or more mineral varieties along a continuum that ranges from well-ordered through disordered to no atomic order (Fig. 1). Quartz and cristobalite (opal-C) represent the well-ordered, crystalline phases of silica, whereas biogenic opal (opal-A) and volcanic glass are noncrystalline hydus silica compounds. Short-range ordered to disordered cristobalite (opal-CT) is also found along the silica continuum. Biogenic opaline phytoliths and diatoms are valuable indicators of paleoenvironments and are identified by the morphology of the biological structure in which they originated. However, both silica phytoliths and glass from pyroclastic deposits are amorphous (3). Silica is not as chemically reactive as other amorphous minerals, but is important as a cementing agent in arid zone soils.

A

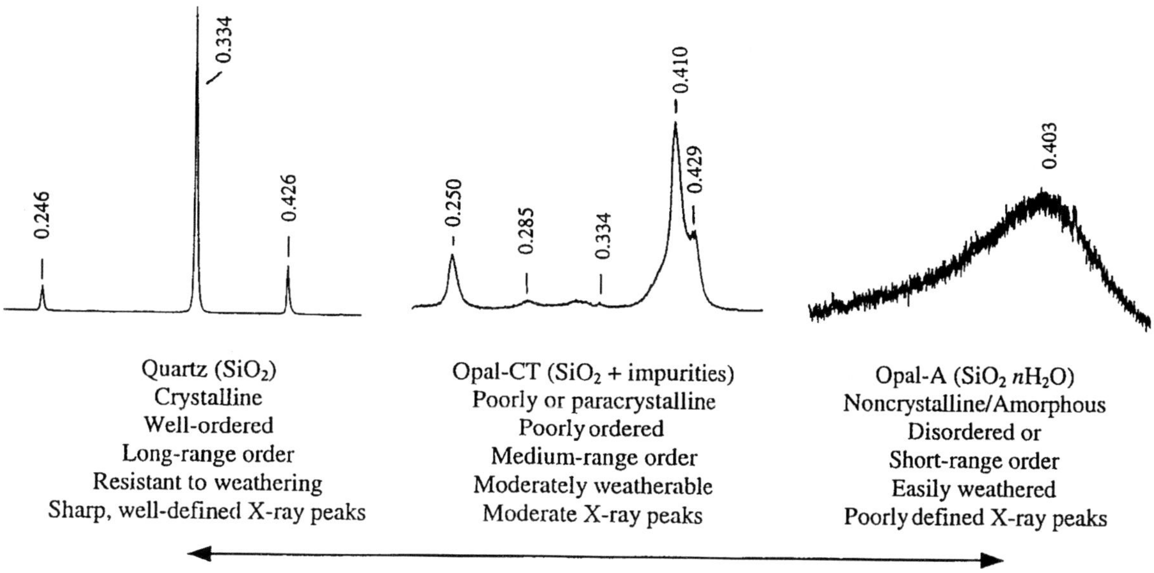

Fig. 1 A comparison of three silica varieties plotted along the crystallinity continuum. The X-ray peak positions are in nanometers and the spectra were made using CuKα radiation. Note that the well-ordered quartz is typically purer than either the opal-CT or opal-A, which both may contain numerous impurities in the lattice structure.

PROPERTIES OF SOILS CONTAINING AMORPHOUS MINERALS

Soils derived from volcanic ash are called Andisols and contain the highest amounts of amorphous components. Spodosols are soils with a diagnostic subsurface horizon containing up to 50% iron and aluminum hydrous oxides (a significant portion of which may be amorphous) and organic matter in the clay fraction. Hydrous iron and aluminum oxides, silica, and kaolin minerals, some of them amorphous or poorly crystalline, dominate in oxic horizons, diagnostic for Oxisols. Even though in most soils amorphous materials comprise a relatively small part of the total mass, they contribute significantly to the physical and chemical properties of the soil (4).

Physical Properties

Amorphous materials are extremely important both physically and chemically because of their small particle size, large surface area, and reactive surface chemistry (Table 1). The surface area of amorphous minerals represents a large pool of active sites that can interact with cations, anions, and water to aggregate soil particles by

Table 1 Ranges of selected physical and chemical properties of amorphous minerals

Mineral	Particle: Density ($Mg\,m^{-3}$)	Particle: Diameter (nm)	Morphology[b]	Surface area ($m^2\,g^{-1}$)	CEC[a] ($cmol_c\,kg^{-1}$)
Allophane	2.7–2.8	3–5	Spherical	100–800	5–350
Ferrihydrite	2.4–3.8	3–7	Coatings/Spherical	200–600	10–160
Halloysite	2.5–2.6	20–200	Tubular/Spherical	10–45	10–40
Immogolite	2.6–2.8	10–30	Threadlike	700–1100	19–37
Opaline Silica	1.5–2.3	10–50	Spherical[c]	40–120	<10

[a] CEC is the Cation Exchange Capacity, which for most amorphous minerals is strongly pH-dependent, increasing as the pH increases.
[b] Shape most commonly observed, others also may be reported.
[c] Amorphous silica also exists as phytoliths, spicules, or shards visible to the naked eye (mm in length).

forming bridges and coatings. Large surface areas result from the tiny diameters of individual tubes or spheres of many amorphous materials and are several times higher than most crystalline soil minerals.

Amorphous materials strongly influence bulk density, aggregate stability, water holding capacity, and plasticity. Soils containing large amounts of amorphous materials have bulk densities ranging from 0.3 to 1.0 Mg m^{-3} (5). In contrast, most mineral soils have bulk density values from 1.2 to 1.7 Mg m^{-3}. Low bulk density and high void ratios are largely due to the formation of stable, porous aggregates by amorphous materials and the inability of lightweight glass shards and porous minerals to be compacted. Soils high in amorphous materials have void ratios ranging from 2 to 5, whereas sandy alluvial soils have values of 0.8 to 1.0 and clayey soils range from 1.5 to 2.5. These soils also exhibit large preconsolidation loads and anomalous compaction behavior (5).

Stable aggregates are formed during the drying of soils rich in amorphous minerals. Amorphous or gel-like material coating soil particles cements those particles into larger aggregates (Fig. 2). The amorphous coatings behave as viscous bodies when moist and elastic bodies when dry (6). The high content of organic matter commonly associated with amorphous materials also contributes to enhanced aggregate formation and stability and darkens the soil (in Japanese *ando* = black).

The moisture content of amorphous-rich or volcanic soils is much higher than in other soils and ranges from 80 to 180% on an oven-dried basis (5). The high natural moisture content contributes to low bulk density values, which are based on dry soil weight per volume. The total porosity of many amorphous minerals and the aggregates they form provide numerous sites for water storage. Once dried however, these minerals and soils tend to rehydrate incompletely (5). The irreversible wetting of these soils and minerals affects several physical and engineering properties including liquid and plasticity limits, and water retention values.

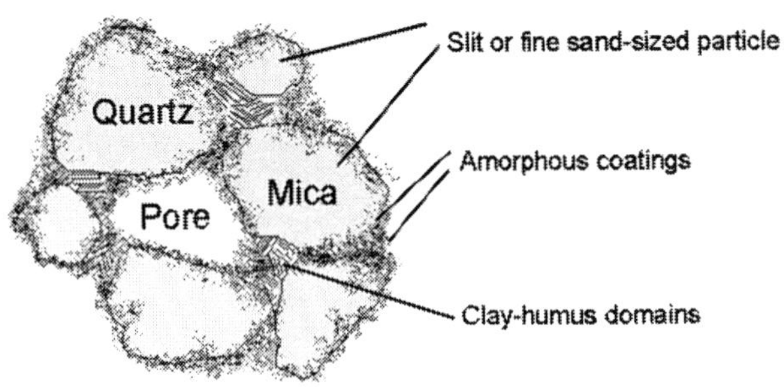

Fig. 2 Schematic representation of a soil aggregate stabilized by amorphous coatings surrounding larger primary minerals, biomass and connecting layers of clay-humus.

Chemical Properties

Amorphous minerals have numerous molecular units, or surface functional groups, that when charged can affect organic matter retention, soil fertility, and the mobility of metals and organic compounds. Anion and cation exchange capacities (AEC and CEC) are variable for most amorphous minerals depending on solution pH and ionic composition. The pH-dependent, or amphoteric charge tends to be positive at low pH and negative above the point of zero charge, which varies for each mineral but is usually between 3 and 7. Thus, anion exchange capacity is typically higher at low pH and CEC increases as the pH increases.

Because of the variable nature of surface charge on amorphous materials, the pH and method of analysis strongly influence the CEC and AEC values. Sample drying, solution concentration, temperature, pH, and mineral composition all affect CEC and AEC values (5). The CEC for individual amorphous minerals varies widely (Table 1). Amorphous components, even in small amounts, contribute to the variable CEC and AEC in soils. For example, smectite-rich clayey soils containing 4–32% amorphous material had CEC's ranging from 24 to 300 $cmol_c\,kg^{-1}$ (7).

Soil fertility and nutrient availability problems often arise in soils rich in amorphous materials. Strong covalent and ionic bonds form between oxyanions and amorphous mineral surfaces, especially under acidic conditions. The oxyanions, which include nutrients such as phosphate, sulfate, and molybdate, or contaminants like selenite and arsenate, may be fixed or permanently adsorbed on the solid phase, which removes them from solution and makes them unavailable to plants. Anion exchange, precipitation reactions, and physical adsorption are other mechanisms that also remove oxyanions from soil solution.

IDENTIFICATION AND CHARACTERIZATION OF AMORPHOUS MINERALS

There is not a specific method universally accepted for the chemical characterization of soil amorphous minerals (8). The techniques used to identify and quantify amorphous components include selective dissolution, electron microscopy, differential thermal analysis and loss on ignition, differential X-ray diffraction, X-ray fluorescence, and infrared spectroscopy.

The most common method used to measure the chemically active portion of the amorphous fraction is extraction with acid ammonium oxalate in the dark (9). Following the extraction of noncrystalline aluminosilicates and hydrous oxides, boiling in 0.5 M NaOH or KOH is used to remove amorphous silica. It is useful to pretreat soil samples using one or both of these methods to clear the crystalline minerals of amorphous materials that coat, dilute, or otherwise interfere with the crystalline components of soil (9). Among several methods estimating the amorphous content of soils, it appears that the Rietveld method is one of the best to determine the total amorphous content of the clay fraction (8). For a variety of quantitative methods in soil mineral analysis, including amorphous components, the reader is referred to the work of Amonette and Zelazny (10).

SUMMARY

Amorphous or noncrystalline minerals are atomic structures with short-range order to no order. Most amorphous minerals are extremely small and reactive both chemically and physically. They can exist as discrete particles and aggregates, but are most common as coatings on other soil particles. As coatings, they aid in the formation of aggregates, crusts, duripans, or other cemented features in soils. Soils high in amorphous minerals usually have high water holding capacity, low bulk densities, and high void ratios. They also have slippery but nonsticky consistence, high Atterberg limits, greater values for liquid and plastic limits on undried than on dried samples, large preconsolidation loads, and anomalous compaction behavior (5). Amorphous minerals have large organic matter retention capability and pH-dependent ion exchange capacities. Phosphate, sulfate, and other ions or metals are commonly fixed by amorphous soil minerals.

REFERENCES

1. Wada, K. Allophane and Imogolite. In *Minerals in Soil Environments*, Soil Sci. Soc. Am. Book Series No. 1, 2nd Ed.; Dixon, J.B., Weed, S.B., Eds.; SSSA: Madison, WI, 1989; 1051–1087.
2. Schwertmann, U.; Taylor, R.M. Iron Oxides. In *Minerals in Soil Environments*, Soil Sci. Soc. Am. Book Series No. 1, 2nd Ed.; Dixon, J.B., Weed, S.B., Eds.; SSSA: Madison, WI, 1989; 379–438.
3. Drees, L.R.; Wilding, L.P.; Smeck, N.E.; Senkayi, A.L. Silica in Soils: Quartz and Disordered Silica Polymorphs. In *Minerals in Soil Environments*, Soil Sci. Soc. Am. Book Series No. 1, 2nd Ed.; Dixon, J.B., Weed, S.B., Eds.; SSSA: Madison, WI, 1989; 913–974.
4. Fey, M.V.; LeRoux, J. Properties and Quantitative Estimation of Poorly Crystalline Components in Sesquioxidic Soil Clays. Clays Clay Miner. **1977**, *25*, 285–294.
5. Wada, K.; Harward, M.E. Amorphous Clay Constituents of Soils. Adv. Agron. **1974**, *26*, 211–260.
6. Jones, R.C.; Uehara, G. Amorphous Coatings on Mineral Surfaces. Soil Sci. Soc. Am. Proc. **1973**, *37*, 792–798.
7. Borchardt, G.A. Cation Exchange Capacity of Noncrystalline Clays in California Landslides. In Harward, M.E., and Wada, K., Eds. Amorphous and Poorly Crystalline Clays Nature, Properties and Magement: Extended Abstracts of; U.S.-Japan Seminar, Corvallis, OR, Oregon State University, 1976; 57–58.
8. Jones, R.C.; Babcock, C.J.; Knowlton, W.B. Estimation of the Total Amorphous Content of Hawaii Soils by the Rietveld Method. Soil Sci. Soc. Am. Proc. **2000**, *64* (3), 1100–1108.
9. Jackson, M.L.; Lim, C.H.; Zelazny, L.W. Oxides Hydroxides, and Aluminosilicates. In *Methods of Soil Analysis, Part 1. Physical and Mineralogical Methods*, Agronomy Monogr. No. 9, 2nd Ed.; Klute, A., Ed.; SSSA: Madison, WI, 1986; 101–150.
10. Amonette, J.E., Zelazny, L.W., Eds. *Quantitative Methods in Soil Mineralogy*; SSSA: Madison, WI, 1994; 1–462.

ANAEROBIC PROCESSES

Paul A. McDaniel
University of Idaho, Moscow, Idaho, U.S.A.

INTRODUCTION

Anaerobic processes are those that occur in soils when molecular oxygen (O_2) is absent or only present in small quantities. Because O_2 is so important to biological systems, its availability to soil organisms has a profound effect on the biogeochemical processes that take place in soils. As a result, anaerobic soils exhibit markedly different chemical and morphological properties than do aerobic soils in which O_2 is readily available.

ANAEROBIC CONDITIONS AND O_2 AVAILABILITY

A common means of describing O_2 status of soils is redox (Eh) potential. Eh can be thought of as a measure of the electron activity of a system (1). While the exact relationship between O_2 and Eh is dependent upon a number of specific soil properties, the general relationship is illustrated in Fig. 1: as O_2 levels decline, Eh also decreases. Redox status is dependent on soil pH, and ranges in Eh can be used to describe three general levels of O_2 availability (Fig. 2). Conditions where O_2 is readily available in soils are referred to as oxic or aerobic conditions, and exist at an Eh range of ~300–800 mV at pH 7. Oxygen is present in gaseous and dissolved forms at sufficient levels to be readily available to soil organisms. Hypoxic conditions or limited O_2 availability occur at Eh values of ~100–300 mV at pH 7, and anaerobic conditions may occur in localized zones of the soil. At Eh values less than ~100 mV at pH 7, strict anoxic or anaerobic conditions exist, and O_2 is absent or otherwise unavailable to organisms.

It is important to note that anaerobic conditions may occur on a microscale in soils. The O_2 content of soil aggregates may change dramatically across a distance of a few millimeters (7). Upon being allowed to air dry, the centers of soil aggregates as small as eight mm in diameter can remain anaerobic. Thus, anaerobic processes in soils are generally associated with the hypoxic and anoxic ranges in Eh values.

While the O_2 content of the Earth's atmosphere is approximately 21%, the O_2 content in soil air is generally lower. The main source of O_2 in soils is air, which can be found in pore spaces between mineral particles. Molecular O_2 is also dissolved in soil water. The gaseous and dissolved O_2 present in soils serves a critical biological function: it is consumed by many soil microorganisms and by the roots of most plants in the process of aerobic respiration. In aerobic respiration, organisms obtain energy through the oxidation of organic substrates, a process whereby electrons are passed through a series of carriers and ultimately accepted by O_2 (8).

DEVELOPMENT OF ANAEROBIC CONDITIONS

Selveral factors can contribute to the decrease in O_2 availability and the development of anaerobic conditions in soils. Aerobic respiration consumes O_2 and generates CO_2. Therefore, without rapid gas exchange between the soil and the atmosphere, the soil atmosphere will become depleted in O_2 and enriched in CO_2. The rate of gas exchange is dependent upon soil characteristics such as texture, structure, and the quantity, size, and connectivity of soil pores (9). In general, slower gas exchange occurs in fine-textured soils and those with poor soil structure. The amount of O_2 that can be dissolved in soil water also decreases as water temperature increases. As soils become very wet and approach saturation, air is physically excluded from soil pores, thereby reducing the amount of air present and inhibiting gas exchange.

ANAEROBIC CONDITIONS AND BIOLOGICAL ACTIVITY

Most plants are adversely affected by anaerobic conditions. Root respiration decreases in plants that lack special adaptations to anaerobic root environments and subsequent growth is limited. As a result of decreased root activity, uptake of water and nutrients also decreases (9).

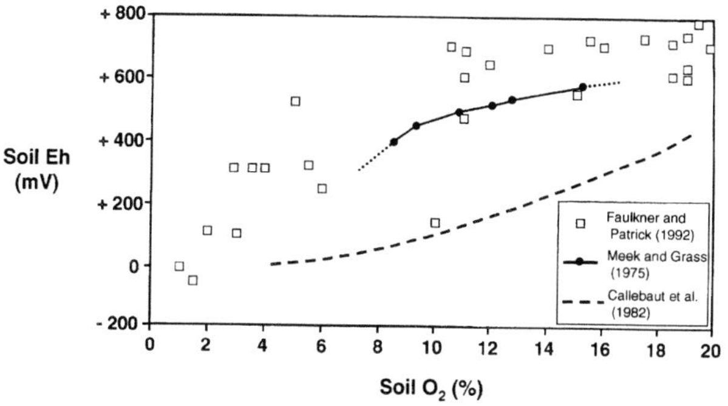

Fig. 1 General relationship between soil Eh and measured soil O_2 content. (Adapted from Refs. 2–4.)

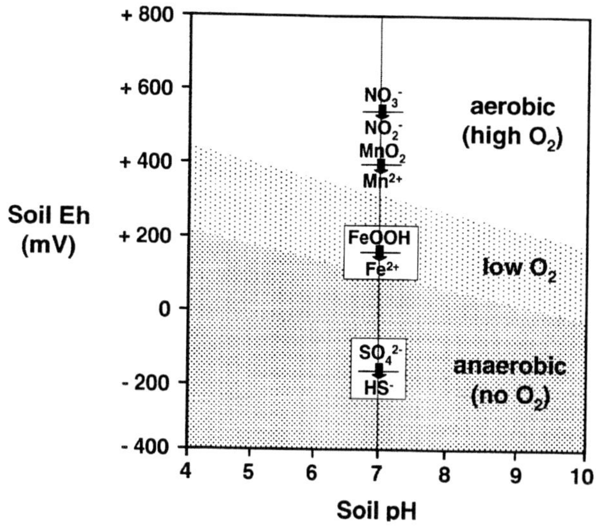

Fig. 2 Generalized relationship between soil O_2 availability, Eh, and pH. Shading patterns indicate the general aeration status of the soil (5) as adapted from (6). Eh values at which selected soil constituents are reduced at pH 7 are shown, based on data from (5). (Adapted from Refs. 5, 6.)

Most soil microorganisms are aerobes—this includes many common bacteria and fungi (8). Thus, when O_2 becomes limiting and anaerobic conditions develop, microbial activity is greatly affected. Development of anaerobic conditions results in a shift in the activity of microbial populations, with the activity of anaerobic and fermentative organisms increasing at the expense of aerobic organisms (8). This shift promotes the reduction of several important elements in soils that are used as alternative electron acceptors—these commonly include N, Mn, Fe, and S—in a process known as anaerobic respiration. The exact Eh at which these reduction reactions occur in soils is dependent on a number of variables, including pH and the form of the element that is present. The Eh values at which some of these reduction reactions occur at pH 7.0 are shown in Fig. 2. As Eh decreases, the general sequence of reduction observed in soils is: N→Mn→Fe→S.

Reduction of N in NO_3^-, may result in the formation of a variety of compounds, including NO_2^-, N_2, N_2O, or NH_4^+ (10). The reduction of insoluble mineral forms of Mn and Fe to soluble forms is characteristic of low-O_2 conditions (10). Reduction reactions involving Fe and Mn are of particular significance because they result in the formation of observable morphological features that can be used as indicators of anaerobic conditions. The reduction of SO_4^{2-} occurs at the lower Eh conditions found in soils and will typically give to rise to H_2S, HS^-, or $S_2O_3^-$ (10). Under extremely low Eh conditions, CO_2 and CH_4 are produced through the fermentation of organic compounds (5, 8).

A second consequence of the shift in microbial activity as soil O_2 becomes limiting is that the biodegradation of soil organic material is lessened. Under aerobic conditions, decomposition of organic materials can proceed rapidly (11). However, as the O_2 supply becomes limited, organic matter decomposition occurs via anaerobic respiration and fermentation. These are less efficient metabolic processes, resulting in slower decomposition of organic substrates. This leads to a greater net accumulation of organic materials in soils having limited O_2 availability.

ANAEROBIC CONDITIONS AND SOIL MORPHOLOGY

Anaerobic conditions associated with water tables and saturated conditions are often seasonal occurrences in soils. Field scientists therefore often rely on soil morphological features as indicators of seasonal anaerobic conditions. Two features commonly used for this purpose are the color patterns created by Fe and Mn compounds and the accumulation of soil organic matter (12).

The microbial reduction of Fe and Mn that occurs under conditions of limited O_2 is a critical step in the formation of redoximorphic features. Redoximorphic features result from alternating periods of reduction and oxidation of Fe and Mn compounds in soils (13). As Fe and Mn are reduced, they form soluble and generally colorless compounds. In the more mobile reduced state, Fe and Mn are subject to redistribution and, in some cases, removal from soils. Under oxidizing conditions, various oxide forms of Fe impart characteristic red, yellow, and brown colors to soils.

Under conditions of alternating oxidation and reduction, Fe and Mn become reduced and solubilized as

(a) (b)

Fig. 3 (a) Close-up view of redox concentrations and redox depletions in a seasonally anaerobic soil. The mottled color pattern consists of gray redox depletions and reddish-orange redox concentrations. Field of view is approximately 15 cm. (b) Morphological expression of anaerobic conditions in an Aquult from the southeastern U.S.A. The thick, dark surface layer reflects the accumulation of organic matter. Redox depletions and concentrations can be seen as gray zones and yellowish-orange zones in the subsoil. Scale shows depth in feet and centimeters.

soils become saturated. As soil water moves to more oxidizing zones or the soil dries out, Fe and Mn are re-oxidized and precipitate out of soil solution, giving rise to distinct zones of depletion and accumulation. These are referred to as redox concentrations and redox depletions (13). Redox depletions appear as low-chroma, grayish zones from which Fe and Mn have been at least partially removed. Redox concentrations appear as brightly colored soft masses or hard nodules composed of Fe and Mn oxide forms (Fig. 3a, b) (13).

Under strongly anaerobic conditions, the soil matrix is reduced and appears dull gray and may exhibit a bluish or greenish tint—these are referred to as gley colors. Upon exposure to air, these reduced matrices may develop brighter colors as Fe oxidizes (12).

Cycling and turnover of organic materials is directly related to O_2 availability in soils. Biodegradation of organic residues is inhibited under anaerobic conditions, and organic matter may accumulate rapidly under such conditions. This results in formation of a very dark surface horizon in mineral soils (Fig. 3b). Under more prolonged anaerobic conditions, organic materials may accumulate to the extent that soils are classified as organic soils (Histosols). Histosols contain a minimum of 20% organic matter on a weight basis to a depth of at least 40 cm (13). These soils typically occupy the wettest landscape positions, such as depressions and low-lying areas with high water tables.

REFERENCES

1. James, B.R.; Bartlett, R.J. Redox Phenomena. In *Handbook of Soil Science*; Sumner, M.E., Ed.; CRC Press: Boca Raton, FL, 2000; B169–B194.
2. Meek, B.D.; Grass, L.B. Redox Potential in Irrigated Desert Soils as an Indicator of Aeration Status. Soil Sci. Soc. Am. Proc. **1975**, *39*, 870–875.
3. Callebout, F.; Gabriels, D.; Minjauw, W.; De Boodt, M. Redox Potential, Oxygen Diffusion Rate, and Soil Gas Composition in Relation to Water Table Level in Two Soils. Soil Sci. **1982**, *134*, 149–156.

4. Faulkner, S.P.; Patrick, W.H., Jr. Redox Processes and Diagnostic Wetland Soil Indicators in Bottomland Hardwood Forests. Soil Sci. Soc. Am. J. **1992**, *56*, 856–865.
5. Evangelou, V.P. *Environmental Soil and Water Chemistry*; John Wiley & Sons: New York, NY, 1998.
6. Bass Becking, L.G.M.; Kaplan, L.R.; Moore, D. Limits of the Natural Environment in Terms of pH and Oxidation–Reduction Potentials. J. Geology **1960**, *68*, 224–284.
7. Sexstone, Alan J.; Revsbech, Niels P.; Parkin, Timothy B.; Tiedje James, M. Direct Measurement of Oxygen Profiles and Denitrification Rates in Soil Aggregates. Soil Sci. Soc. Am. J. **1985**, *49*, 645–651.
8. Coyne, M. *Soil Microbiology: An Exploratory Approach*; Delmar Publishers: Albany, NY, 1999.
9. Brady, N.C.; Weil, R.B. *The Nature and Properties of Soils*, 12th Ed.; Prentice Hall: Upper Saddle River, NJ, 1999.
10. Sposito, G. *The Chemistry of Soils*; Oxford University Press: New York, NY, 1989.
11. Duchaufour, P. *Handbook of Pedology*; A.A. Balkema: Rotterdam, The Netherlands, 1998.
12. USDA (U.S. Department of Agriculture), Field Indicators of Hydric Soils in the United States. Hurt, G.W., Whited, P.M., Pringle, R.F., Eds.; Version 4.0 USDA-NRCS: Fort Worth, TX, 1998.
13. Soil Survey Staff, *Soil Taxonomy. A Basic System of Soil Classification for Making and Interpreting Soil Surveys*, 2nd Ed.; Agric. Handbook No. 436, USDA-NRCS: U.S. Goverment Printing Office: Washington, DC, 1999.

ANDISOLS

Jon Chorover
The University of Arizona, Tucson, Arizona, U.S.A.

INTRODUCTION

Andisols are deep soils of low bulk density that are most often derived from volcanic parent materials. They are classified on the basis of unique mineralogical, chemical, and physical properties. This soil order, which is most prevalent along the tectonically active Pacific Ring of Fire, covers approximately 120 million hectares, or 1% of the earth's surface. Andisols are capable of supporting agricultural production and human populations that are large relative to the spatial extent of the soils. The prefix, *An*, refers to the dark color of the surface horizon(s) that results from high concentrations of humified organic matter stabilized by chemical interaction with poorly crystalline secondary minerals. The morphology of the soils typically includes multiple sequences of A and Bw horizons that are buried successively because of intermittent deposition events.

CLASSIFICATION OF ANDISOLS

The current classification of Andisols (1) originated with Smith (2) and was refined through the efforts of the International Committee on the Classification of Andisols. Parent material must be weathered to fall within a range of characteristics pertaining to bulk density, particle size, phosphorus adsorption capacity, and within a range of mass concentrations of organic carbon, volcanic glass, and oxalate extractable Fe and Al (1). Specifically, the soil material must contain no more than $250\,g\,kg^{-1}$ organic C and either: 1) oxalate extractable $Al+0.5\ Fe \geq 20\,g\,kg^{-1}$, bulk density (at 33 kPa) $\leq 0.9\,Mg\,m^{-3}$, and phosphate retention ≥85%, or 2) phosphate retention ≥25%, particles between 0.02 and 2 mm in size $\geq 300\,g\,kg^{-1}$, oxalate extractable $Al+0.5\ Fe \geq 4\,g\,kg^{-1}$, and volcanic glass in the 0.02–2 mm fraction $\geq 362-15.6\times$(oxalate extractable Al+0.5 Fe), where all units are $g\,kg^{-1}$. Soils classified as Andisols are, therefore, dominantly those developing in volcanic ash, pumice, cinders and other volcanic ejecta. Suborders are based on climate regimes, whereas great groups and subgroups are based on soil physical–chemical properties, such as clay and organic matter content.

PEDOGENIC FACTORS

Most Andisols develop from chemical weathering of volcanic ejecta in combination with organic matter humification (3). Common parent materials range in physical form from fine ash to viscous lava flows, and in mineralogical composition from andesitic and rhyollitic to basaltic rock types. Basalt weathers more rapidly than andesitic rock and, because of the availability of reactive surface, finely divided ash weathers more rapidly than lava. Sufficient weathering of parent material is required to produce the poorly crystalline Al, Fe, and Si compounds in the quantities needed to classify a soil as an Andisol. Thus, these soils commonly occur in humid regions (4). Andisols are formed from parent materials other than volcanic ejecta, provided that sufficiently high levels of reactive Al, Fe, and organic matter are present, but volcanic parent materials are most common. The high specific surface area and reactivity of poorly crystalline minerals result in greater organic matter retention than occurs in most other soil orders except Histosols. The sequence of soil horizons is typically A–Bw–C (Fig. 1). The surface A horizon often contains very high concentrations of organic matter (typically greater than $100\,g\,kg^{-1}$) that is stabilized by complexation with aluminum and iron. The B-horizon is normally dominated by poorly crystalline aluminosilicates and hydrous oxides (allophane, imogolite, ferrihydrite). Periodic eruptions in volcanically active areas can produce a series of buried A–Bw sequences, which is one cause of deep humus penetration in Andisols (5). This factor also contributes to the significant depth of Andisol profiles.

PHYSICAL–CHEMICAL PROPERTIES OF ANDISOLS

Mineralogy

The unique properties of Andisols are coupled tightly to the chemical properties of constituent solid phases (4, 6).

Encyclopedia of Soil Science

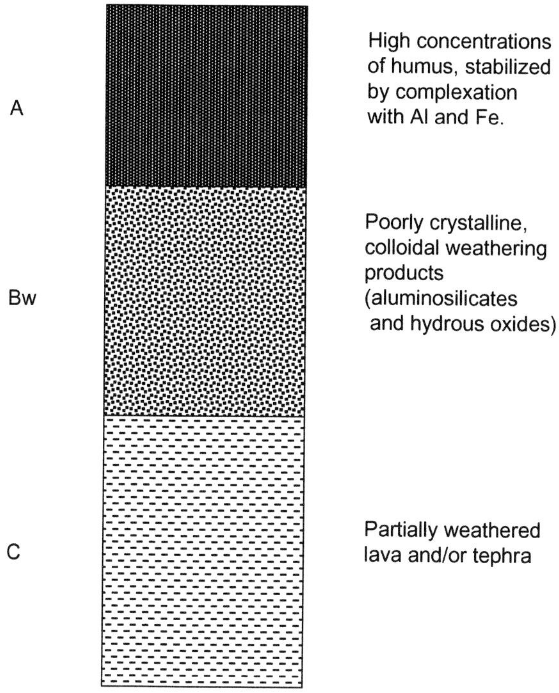

Fig. 1 Schematic of a typical Andisol profile.

Primary silicate minerals in andesitic and rhyollitic flows include quartz, hornblende, and biotite, whereas olivines, pyroxenes, and plagioclase are more common in basalts. Rapid weathering of volcanic ejecta (containing aluminosilicates, volcanic glass, and smaller amounts of ferromagnesian minerals) results in the accumulation of soluble silica, aluminum, and iron to high concentrations in the soil solution of young Andisols. In the near surface (e.g., A horizon), Fe and Al are often immobilized into humic complexes, whereas Si may polymerize to form opaline silica. At greater depth (e.g., Bw horizon), or if humus concentrations are low, hydrolysis and polymerization of Al, Fe, and Si results in precipitation of allophane ($x SiO_2 \cdot Al_2O_3 \cdot y H_2O$, where $x = 0.8$–2 and $y \geq 2.5$), ferrihydrite ($5Fe_2O_3 \cdot 9H_2O$), opaline silica ($SiO_2 \cdot n H_2O$), amorphous aluminum hydroxide ($Al(OH)_3$), and/or imogolite ($SiO_2 \cdot Al_2O_3 \cdot 2.5H_2O$), depending upon solution composition. Allophane and ferrihydrite, which do not exhibit the long-range crystal structure necessary to yield well-defined X-ray diffraction patterns, are commonly termed short-range-order (SRO) solids. These minerals and imogolite are "metastable" thermodynamically. That is, over the long-term, Ostwald ripening of soil solids transforms SRO phases into more crystalline forms, such as halloysite, goethite, and gibbsite. Since the accumulation of more crystalline phases causes soil characteristics to diverge from those that are diagnostic for the Andisol order, Andisols represent an intermediate stage of pedogenesis, the lifetime of which depends on interacting state factors that control mineral transformation rates.

The important properties of Andisols are generally ascribed to the preponderance of allophane and X-ray amorphous compounds of Al, Si and humus (2), but high concentrations of Al and Fe humus complexes can confer to nonallophanic soils the key diagnostic properties that are common in allophanic Andisols (7). It is, therefore, currently recognized that important SRO minerals of Andisols include not only allophane, but also imogolite, ferrihydrite, poorly crystalline aluminum hydroxides, and Al/Fe-humus complexes. The SRO minerals in Andisols exhibit very high specific surface areas (ca. $10^5-10^6\,m^2\,kg^{-1}$) and reactive site densities ($5-20\,\mu mol\,m^{-2}$). Unit particles of allophane are roughly spherical and 3.5–5.5 nm in diameter, whereas imogolite appears as smooth and curved threads of 10–30 nm diameter, and ferrihydrite is composed of spherical particles, 3–7 nm in diameter (8, 9).

Chemistry

Chemical properties of Andisols are dominated by the main pedogenic weathering products of volcanic debris in terrestrial ecosystems: X-ray amorphous SRO compounds of Al, Si and Fe, and their complexes with organic matter (10). These constituents are the target species measured during acid ammonium oxalate treatment of the soil and their prevalence is diagnostic for an Andisol. Humus concentrations correlate with SRO minerals and extend to depth accordingly. Allophane, imogolite, and ferrihydrite contain a prevalence of surface hydroxyl groups that: 1) retain and dissociate protons (H^+) in response to changes in soil solution pH, and 2) can form strong complexes with humic substances and oxyanions such as phosphate (4, 9). In particular, the strong binding of phosphate can lead to plant P deficiencies when these soils are used for crop production. These surface reactions confer significant variable charge on the soils. Cation adsorption increases and anion adsorption decreases with increasing pH. The pH value where moles of adsorbed cation and anion charge are equivalent is termed the point of zero net charge (PZNC). The total surface charge at any point is a function of solution chemistry, soil mineralogy, and organic matter chemistry. In general, increasing amounts of Fe and Al oxides will increase the PZNC, whereas increasing amounts of SRO silica and organic matter will decrease it. Andisols contain relatively low quantities of 2:1 layer-type clay minerals and, therefore, permanent (structural) charge represents a small fraction of the total surface charge (11).

Since SRO solids are more soluble than well-crystallized phases, they can support relatively high equilibrium concentrations of dissolved Al. The prevalence of Al in Andisols can maintain acidic soil pH (4.5–6.0) via Al hydrolysis reactions and diminish base saturation via displacement of nonhydrolyzing cations Ca^{2+}, Mg^{2+}, K^{+}, and Na^{+} from cation exchange sites. In addition, complexation with humic substances of monomeric, polymeric and colloidal forms of Al and Fe serves to coagulate and stabilize organic matter in Andisols, leading to long turnover times of organic C (12).

Physical Properties

Allophane, imogolite, ferrihydrite, and organic matter are responsible for the unique properties of soils derived from volcanic parent materials. The charge characteristics, size, and shape of these constituents are important determinants of soil physical properties. Andisols containing large quantities of SRO constituents have specific surface areas as high as $6 \times 10^5\,m^2\,kg^{-1}$. The density of allophane particles—ca. $2.7\,Mg\,m^{-3}$—is slightly higher than more crystalline aluminosilicates such as kaolinite and montmorillonite. Therefore, the much lower bulk densities encountered in Andisols ($<0.85\,Mg\,m^{-3}$) where SRO minerals are dominant, are attributed to the high porosity of microaggregates formed when unit particles interact with each other and with organic matter. Aggregation of SRO constituents and interaction with humic substances gives rise to large void volumes (25–45%) in the micropore and mesopore size ranges. In conjunction with the significant quantity of hydration water associated with primary particles, this gives Andisols the capacity to retain as much as 1.8 kg water per kilogram soil solids and to support lush vegetation even at low rainfall. Hydraulic conductivity values are high because of low bulk density values and granular structure. Hydraulic properties can be altered by changing chemical conditions to induce dispersion of variable-charge soil colloids. For example, saturated hydraulic conductivity values can be reduced dramatically by displacing soil pH several units from the point of zero charge. Because of the hydrated nature of SRO phases and microaggregates, soil drying to high tensions results in irreversible changes to many of these soil physical properties. Physical properties of Andisols have been reviewed by Maeda and co-workers (13, 14) and Wada (10).

LAND USE AND MANAGEMENT

Andisols can provide exceptional media for plant growth in comparison to other mineral soils (4). They tend to have a thick solum with unrestricted rooting zones, high organic matter contents, and abundant available water. The supply of lithogenic nutrients is provided through rapid weathering of residual volcanic ash and lava. In regions of high rainfall, however, loss of cationic nutrients may be facilitated by intense leaching. Under such conditions, and as a result of reactive SRO mineral surfaces, the sustainable management of Andisols for agriculture and forestry may also be limited by phosphorus deficiency and low cation exchange capacity, among other constraints. Nitrogen availability may also limit plant production to the extent that N-mineralization is limited by chemical stabilization of humus into metal–humus or mineral–humus complexes. The low bulk density, friable consistence, and granular structure, particularly in surface horizons, facilitate tillage operations such as seed-bed preparation and plowing. High porosity and plant available water favor seedling emergence, root development, and ramification. These soils resist compaction, even when cultivated continuously, and tend to regain physical properties after wetting and drying cycles once compacted by machinery. High hydraulic conductivities and water retention capacities promote infiltration and diminish erosive overland flow. Binding of SRO minerals and humic substances gives rise to strong soil aggregates that resist mechanical disruption by raindrops that could otherwise lead to particle detachment and erosion.

REFERENCES

1. Soil Survey Staff, *Keys to Soil Taxonomy*; 8th Ed. USDA–NRCS: Washington, DC, 1998.
2. Smith, G.D. *A Preliminary Proposal for the Reclassification of Andepts and Some Andic Subgroups (the Andisol Proposal 1978)*; New Zealand Soil Bureau Record, DSIR: Lower Hutt, 1978; 96.
3. Van Breeman, N.; Buurman, P. *Soil Formation*; Kluwer: Dordrecht, 1998.
4. Shoji, S.; Nanzyo, M.; Dahlgren, R.A. *Volcanic Ash Soils. Genesis, Properties and Utilization*; Elsevier: Amsterdam, The Netherlands, 1993.
5. Ping, C.L.; Shoji, S.; Ito, T. Properties and Classification of Three Volcanic Ash-Derived Pedons from Aleutian Islands and Alaska Peninsula, Alaska. Soil Sci. Soc. Am. J. **1988**, *52*, 455–462.
6. Wada, K. Mineralogical Characteristics of Andisols. In *Soils with Variable Charge*; Theng, B.K.G., Ed.; New Zealand Society of Soil Science: Lower Hutt, 1980; 87–108.

7. Shoji, S.; Ono, T. Physical and Chemical Properties and Clay Mineralogy of Andisols from Kitakami. Jpn. Soil Sci. **1978**, *126*, 297–312.
8. Schwertmann, U.; Taylor, R.M. Iron Oxides. In *Minerals in Soil Environments*; Dixon, J.B., Weed, S.B., Eds.; Soil Science Society of America: Madison, WI, 1989; 379–438.
9. Harsh, J.B.; Chorover, J.; Nizeyimana, E. Allophane and Imogolite in Soils and Their Effects on Environmental Processes. In *Environmental Soil Mineralogy*; Dixon, J.B., Schulze, D., Eds.; in press Soil Science Society of America: Madison, WI, 2002.
10. Wada, K. The Distinctive Properties of Andisols. Adv. Soil Sci. **1985**, *2*, 173–229.
11. Chorover, J.; DiChiaro, M.J.; Chadwick, O.A. Structural Charge and Cesium Retention in a Chronosequence of Tephritic Soils. Soil Sci. Soc. Am. J. **1999**, *63*, 169–177.
12. Torn, M.S.; Trumbore, S.E.; Chadwick, O.A.; Vitousek, P.M.; Hendricks, D.M. Mineral Control of Soil Organic Carbon Storage and Turnover. Nature **1997**, *389*, 170–173.
13. Maeda, T.; Takenaka, H.; Warkentin, B.P. Physical Properties of Allophane Soils. Adv. Agron. **1977**, *29*, 229–269.
14. Warkentin, B.P.; Maeda, T. Physical and Mechanical Characteristics of Andisols. In *Soil with Variable Charge*; Theng, B.K.G, Ed.; New Zealand Society of Soil Science: Lower Hutt, 1980; 281–302.

ANIMALS AND ECOSYSTEM FUNCTIONING

Alan J. Franzluebbers
United States Department of Agriculture-Agricultural Research Service (USDA-ARS), Watkinsville, Georgia, U.S.A.

INTRODUCTION

Soil animals (i.e., fauna) are represented by a diverse array of creatures living in or on soil for at least a part of their life cycle. Many animals have influences on soil properties, but should not be considered soil dwellers since only a minor portion of their life cycle is spent in the soil (Fig. 1).

Based on body size, soil animals can be divided into three categories:

1. microfauna (<200 μm length, <100 μm width) including protozoa, rotifers, and nematodes
2. mesofauna (0.2–10 mm length, 0.1–2 mm width) including tartigrades, collembola, and mites
3. macrofauna (>10 mm length, >2 mm width) including millipedes, spiders, ants, beetles, and earthworms

Soil animals can also be classified according to where they inhabit the soil. The aquatic fauna (e.g., protozoa, rotifers, tartigrades, and some nematodes) live primarily in the water-filled pore spaces and surface water films covering soil particles. Earthworms are divided into species that occupy the surface litter of soil (epigeic), that are found in the upper soil layers (endogeic), or that burrow deep into the soil profile (anecic).

A further classification of five groups of soil animals is based on feeding activity, which can be useful in distinguishing how different groups affect soil ecosystem functions:

1. Carnivores feed on other animals. This group can be subdivided into: i) predators (e.g., centipedes, spiders, ground beetles, scorpions, ants, and some nematodes), who normally engulf and digest their smaller prey and ii) parasites (e.g., some flies, wasps, and nematodes), who feed on or within their typically larger host organism.
2. Phytophages feed on living plant materials, including those that feed on above-ground vegetation (e.g., snails and butterfly larvae), roots (e.g., some nematodes, fly larvae, beetle larvae, rootworms, and cicadas), and woody materials (e.g., some termites and beetle larvae).
3. Saprophages feed on dead and decaying organic material and include many of the earthworms, enchytraeids, millipedes, dung beetles, and collembola (or springtails). Saprophages are often referred to as scavengers, debris-feeders, or detritivores.
4. Microphytic feeders consume bacteria, fungi, algae, and lichens. Typical microphytic feeders include mites, collembola, ants, termites, nematodes, and protozoa.
5. Miscellaneous feeders are not restrictive in their diet and consume a range of the previously mentioned sources of food. This group includes certain species of nematodes, mites, collembola, and fly larvae.

The arrangement of these feeding groups can be visualized as a soil food web with multiple trophic levels, beginning with the autotrophic flora (Fig. 2). Trophic levels describe the order in the food chain. The first trophic level is composed of photosynthetic organisms, including plants, algae, and cyanobacteria, which fix CO_2 from the atmosphere into organic compounds. Organisms that consume the photosynthesizers are in the second trophic level, which includes bacteria, actinomycetes, fungi, root-feeding nematodes and insects, and plant pathogens and parasites. The third trophic level feeds on the second trophic level, including many of the dominant soil animals, including bacterial- and fungal-feeding arthropods, nematodes, and protozoa. The soil food web can be continued to include various vertebrates, including amphibians, reptiles, and mammals.

SPATIAL DISTRIBUTION OF SOIL ANIMALS

Soil animals are not uniformly distributed in soil. Unlike the soil microflora, which could be considered ubiquitous, the proliferation of soil animal communities is more sensitive to environmental disturbances and ecological interactions. Gross climatic differences afford opportunities for unique assemblages of organisms. Even within a specific climatic region, large differences occur in the community of organisms present based upon type of

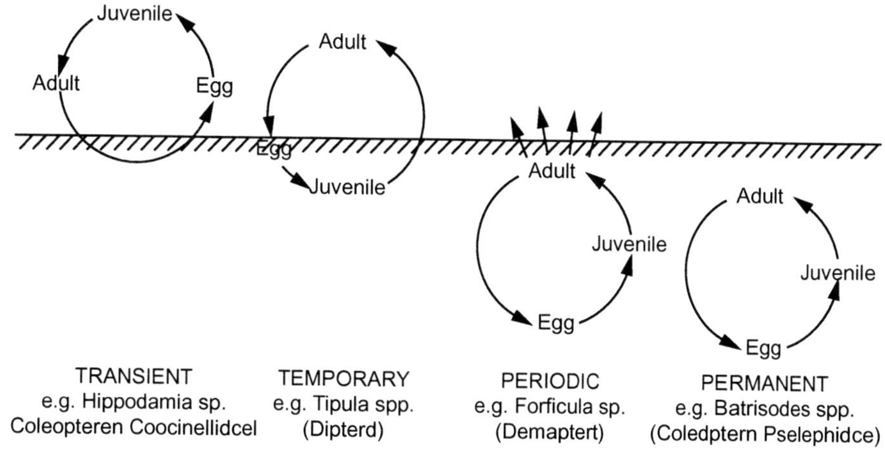

Fig. 1 Categories of soil animals defined according to degree of presence in soil, as illustrated by some insect groups. (From Ref. 1.)

vegetation, soil, availability of water, land use, and presence of xenobiotics. Within the confines of a seemingly uniform pedon, "hot spots" of soil organism activity can be isolated based on localized availability of resources and environmental conditions (Fig. 3).

INFLUENCE OF SOIL ANIMALS ON SOIL FUNCTIONS

Decomposition and Nutrient Cycling

Soil animals work directly and indirectly with the soil microflora (i.e., bacteria, actinomycetes, fungi, and algae) to decompose organic matter and mineralize nutrients (3). The primary consumers of organic materials are the soil microflora. Soil animals, like many of the microflora, are heterotrophs and therefore consume organic materials to

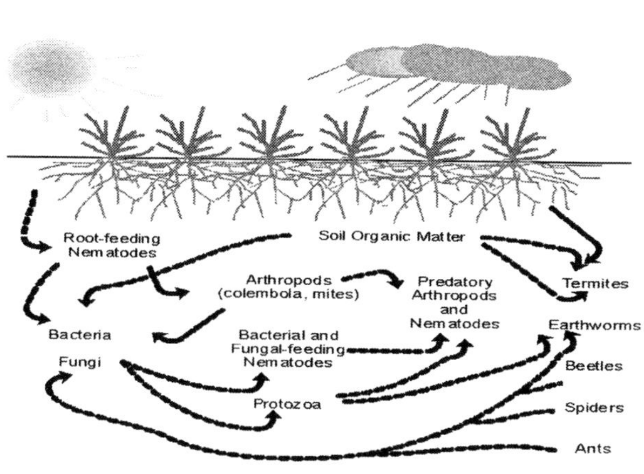

Fig. 2 Generalized diagram of a soil food web.

gain energy for growth and activity. Soil animals make important contributions to decomposition by

1. shredding organic materials, thereby exposing a greater surface area for enhancing the activities of other organisms, especially microorganisms;
2. consuming resistant plant materials that would decompose slowly otherwise, such as wood, roots, and dung, and transforming these materials into more decomposable constituents;
3. dispersing soil microorganisms (i.e., inoculation) within the soil profile by transporting them on their bodies and through their intestinal tracts;
4. creating burrows in soil to increase aeration, which stimulates microbial activity;
5. transporting organic materials from the soil surface to deeper in the soil profile, thereby improving environmental conditions for decomposition and increasing biological interactions deeper in the soil profile;
6. consuming bacteria and fungi, thereby releasing nutrients and stimulating the regeneration of microbial populations; and
7. providing unique food sources themselves for consumption by other soil fauna and microflora.

Water Cycling

Soil animals are active participants in the formation of soil structure, which is an important characteristic that influences water infiltration, soil water retention, and percolation (4). The biochemical activity of soil organisms transforms organic materials into soil-stabilizing cementing agents, which bind the primary soil particles (i.e., sand, silt, and clay) into aggregates. In addition, the burrowing

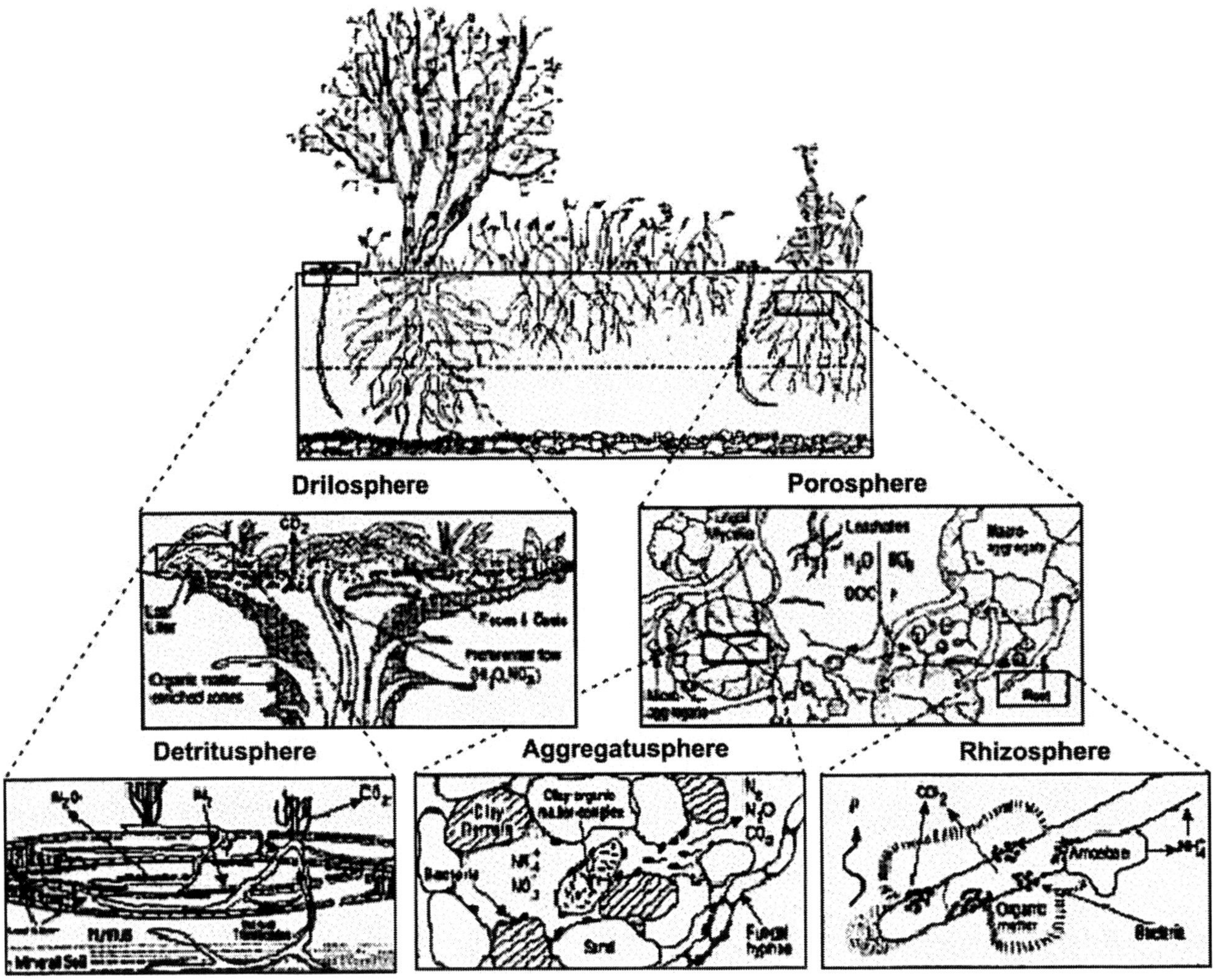

Fig. 3 Key locations of soil organism activity. (From Ref. 2. with kind permission from Kluwer Academic Publishers.)

activity of soil animals creates larger pores alongside water-stable aggregates to increase total porosity of soil, which aids water flow without decreasing overall water retention capacity and improves the plant rooting environment.

Both aggregates and porosity are important components of soil structure. Poor soil structure due to disruption of aggregates, which fills pores with disaggregated primary particles and causes crusting of the soil surface, results in more rainfall that runs off land (i.e., less infiltration), potentially carrying with it sediment, nutrients, and pesticides that can contaminate surface waters. Reduced infiltration with poor aggregation reduces available water for plant growth (i.e., reduces net primary productivity and the potential to fix atmospheric CO_2) and reduces percolation of water through the soil profile, essential for purification and recharge of groundwater.

Those animals that create burrows in soil also create conduits for water movement through the soil profile. These biopores can be important for improving water percolation and improving rooting below claypans and other restrictive soil layers.

Many different soil animals deposit fecal pellets, which become stable soil aggregates when the organic material is mixed with soil mineral particles. These aggregates are able to retain more water because of the high water-holding capacity of soil organic matter.

Pest Control

Intense competition among soil organisms keeps an ecosystem healthy by preventing one organism from becoming dominant. Potential plant pathogens, such as root-feeding nematodes, are often held below damaging levels because of consumption by predatory nematodes and arthropods. With a healthy food web rich in species diversity, the predatory activity of many arthropods can keep crop pests below economic thresholds.

Impact of Key Soil Animals

Earthworms

Earthworms are well-known soil animals inhabiting many environments, most prominently found in moist-temperate ecosystems. As earthworms ingest organic materials and mineral particles, they excrete waste as casts, which are a particular type of soil aggregate that is rich with organic matter and mineralizable nutrients. It is estimated that a healthy population of earthworms can consume and aerate a 15 cm surface of soil within one or two decades. Anecic or deep-burrowing species of earthworms can create relatively permanent vertical channels for improving root growth and water transport. Important attributes of earthworm activities are increased surface soil porosity, enhanced water infiltration and nutrient cycling, and distribution of organic matter within the soil profile to increase soil microbial activity.

Termites

Termites are important soil animals in grasslands and forests of tropical and subtropical regions. They often build mounds by excavating subsoil and depositing it above ground to build a city of activity with a complex social system. Termites are able to decompose cellulose in wood because they harbor various microorganisms (protozoa, bacteria, or fungi) to aid in decomposition. Better drainage and aeration of termite mounds may be beneficial to nearby plant growth in soils with a high water table. Stable macrochannels created by termites can improve water infiltration into soils that otherwise would form impermeable surface crusts.

Protozoa

Protozoa are single-celled animals that generally consume bacteria and soluble organic matter. Protozoa are more numerous in marine and freshwater environments, but do occur widely in water films of many soils (5). Their principal soil function is predation on soil bacteria, which releases nutrients for potential plant uptake; increases decomposition and soil aggregation by stimulating their bacterial prey; and prevents some bacterial pathogens from developing on plant roots.

SOIL BIODIVERSITY

There has been a great deal discovered about soils and the organisms that live in them, yet it is estimated that <10% of the species of soil organisms have been identified globally. A rich diversity of genetic information resides in soil. An awaiting challenge is to discover the ecological consequences of soil biodiversity. A critical understanding of how organisms interact has begun, yet there will be much more to learn about how species diversity interacts with functional diversity. There may be soil functions provided by organisms that we are unaware of even today.

How biodiversity relates to ecosystem functioning is an intensive area of current research in both above- and below-ground ecology (6, 7). Recent experimental evidence suggests that loss of species richness due to perturbations may not always lead to loss of ecosystem functioning, especially in initially species-diverse functional groups (8). Ecosystem functions can be performed by a number of different species within a trophic level, suggesting that functional redundancy is a mechanism to insure stability. However, loss of functional groups in trophic levels closest to the base of the detrital food web would be most detrimental to ecosystem stability. Future work on soil biodiversity should unravel the relative importance of species richness on the resistance and resilience of ecological processes under various short- and long-term stresses and environmental conditions.

REFERENCES

1. Wallwork, J.A. *Ecology of Soil Animals*; McGraw-Hill: London, 1970, 283.
2. Beare, M.H.; Coleman, D.C.; Crossley, D.A., Jr; Hendrix, P.F.; Odum, E.P. A Hierarchical Approach to Evaluating the Significance of Soil Biodiversity to Biogeochemical Cycling. Plant Soil **1995**, *170* (1), 5–22.
3. Coleman, D.C.; Crossley, D.A., Jr. *Fundamentals of Soil Ecology*; Academic Press: San Diego, CA, 1996; 205.
4. Brady, N.C.; Weil, R.R. *The Nature and Properties of Soils*, 12th Ed.; Prentice Hall: Upper Saddle River, NJ, 1999; 881.
5. Dabyshire, J.F., Ed. *Soil Protozoa*; CAB International: Wallingford, Oxon, UK, 1994; 209.
6. Tilmann, D.; Knops, J.; Wedin, D.; Reich, P.; Ritchie, M.; Siemann, E. The Influence of Functional Diversity and Composition on Ecosystem Processes. Science **1997**, *277*, 1300–1302.
7. Wardle, D.A. How Soil Food Webs Make Plants Grow. Trends Ecol. Evol. **1999**, *14*, 418–420.
8. Laakso, J.; Setälä, H. Sensitivity of Primary Production to Changes in the Architecture of Belowground Food Webs. Oikos **1999**, *87*, 57–64.

ANIMALS, MICROBIAL INTERACTIONS AND NUTRIENT CYCLING

Lisa Cole
Richard Bardgett
Lancaster University, Lancaster, United Kingdom

INTRODUCTION

The decomposer food web has a primary role in determining the mineralization of nutrients in soil, and hence plant nutrient acquisition and plant productivity (1). Although most mineralization of nutrients is governed directly by the activities of bacteria and fungi, their ability to do this is affected strongly by soil animals. Soil animals affect microbial communities either directly through selectively feeding on fungi and bacteria, or indirectly by comminution of organic matter, dissemination of microbial propagules, and the alteration of nutrient availability (2). Combined, these interactions between microorganisms and animals drive the ecosystem-level processes of energy flow and nutrient cycling, and, therefore, contribute to primary production (3–5). The effects of these biotic interactions on the structure and activity of microbial communities, and the consequences of this regards nutrient mineralization, will be discussed for the common groups of macro- and meso-biota found in soils, namely the earthworms, microarthropods, and enchytraeid worms.

SELECTIVE FEEDING BY SOIL FAUNA

Although soil animals are considered to be both generalist and opportunist in their feeding behaviour, two main groups of feeder that have been distinguished are bacterivorous and fungivorous. The structure of the microbial community is to some extent controlled by the selective feeding activities of these animal groups; selection of prey is based on their palatability, which varies with species, age, and physiological status. For example, the collembolan *Folsomia candida* has been shown to feed on metabolically active hyphae in preference to dead or inactive hyphae (6), and also selects regions of the fungal thallus with high N content (7). Preferential grazing is also thought to arise as a result of the avoidance of toxins that are produced by some fungal colonies (8) and some species of Collembola have been shown to locate and select their fungal food source by volatile compounds released from fungal mycelium (9).

Intermediate levels of grazing have been shown to enhance the activity and growth of selected fungi (10). Stimulation of fungal growth, which is often referred to as compensatory growth, is due to new fungal growth after senescent hyphae are grazed, and re-growth after periodic grazing of actively growing mycelia. For example, grazing of the fungus *Mortierella isabellina* by the collembolan *Onychiurus armatus* induced switching from a "normal" hyphal mode, with appressed growth and sporulating hyphae, to fan-shaped sectors of fast growing and non-sporulating mycelium with extensive areas of aerial hyphae (11). In addition, activities of specific amylase (starch degrading enzymes) were several times higher in grazed cultures than in those cultures that were ungrazed (11). Heavy grazing of fungal communities can reduce their activity. For example, heavy grazing of mycorrhizae has been shown to counteract the mutualistic relationship between mycorrhizae and their host plant (12).

Selective grazing by soil fauna can alter the composition of the soil microbial community by altering the competitive relationships between species of microorganisms. Selective grazing by the collembolan *O. latus* on the fungus *Marasmius androsaceus* in coniferous leaf litter, resulted in a reduction in the activity of this palatable fungi, and an increase in an unpalatable fungi *Mycena galopus* present in the litter (13, 14). Because *M. galopus* decomposes litter at a slower rate than *M. androsaceus*, the presence of *O. latus* reduced the decomposition of leaf litter (13, 14).

The ingestion and exposure of microbes to intestinal fluids can influence microbial activity. Microbial communities have been shown to be more abundant and active following passage through the gut of earthworms (15). The passage of microorganisms through earthworm (e.g., *Lumbricus terrestris*) guts has been shown to activate dormant bacteria due to the removal of endospore cell coats and subsequent germination of bacterial spores (16). Conversely, it has been suggested that the earthworm gut is a hostile environment for soil fungi (17); the germination

Encyclopedia of Soil Science

of fungal spores (*Fusarium lateritium*, *Agrocybe temulenta*, *Trichoderma sp.* and *Mucor hiemalis*) was reduced following ingestion by *L. terrestris* and *Aporrectodea longa*. It was demonstrated that the germination of fungal spores was enhanced by physical abrasion by soil particles in the gut, but that subsequent exposure to intestinal fluid significantly reduced germination (17).

EFFECTS ON NUTRIENT CYCLING AND PLANT GROWTH

The most studied interactions between invertebrates and the microbial community in soils involve those that result in the release of nutrients from the microbial biomass as a result of the feeding activities (grazing) of animals. In general, soil fauna have lower C:N ratios than the microbes upon which they graze, and therefore they excrete nutrients that are not required for production in forms which are biologically available (i.e. inorganic forms—"mineralized"). This release of nutrients into the soil system is effectively a re-mobilization of the nutrients that were bound up in the microbial biomass, and has been termed the "microbial loop" (18). In addition, this grazing of microorganisms often leads to changes in the structure, size, and activity of the microbial community in the soil; as discussed above, these changes in microbial communities indirectly affect processes of nutrient mineralization.

Many studies demonstrate the positive effects of soil fauna on carbon turnover and nitrogen mineralization in soil and litter materials. A study that used litter-bags buried in the field (19) demonstrated that the exclusion of enchytraeids and dipteran larva from the litter bags resulted in lower rates of decomposition of litter. Likewise, laboratory microcosm studies have shown that the presence of different groups of animals in soil enhances processes of nutrient mineralization. For example, the grazing activities of enchytraeids on microorganisms that colonise litter have been shown to result in increased rates of leaching of ammonium and dissolved organic carbon (20, 21) associated with increased microbial activity (21). The earthworm *A. caliginosa* has been shown to increase C and N mineralization in soils amended with birch litter (22), despite a reduction in the microbial biomass. The grazing of fungi on litter by the Collembola *F. candida* (23) and *Tomocerus minor* (24) also increased mineralization of N. These positive effects of soil fauna have been shown to translate into increased uptake of nutrients and growth of plants. It has been shown that Collembola and nematodes increased concentrations of ammonium in the soil solution of an upland organic soil, leading to increased plant nutrient uptake by an upland grass species, *Nardus stricta* (4). It has also been demonstrated that although grazing by the earthworm *A. caliginosa* reduced the root biomass of the grass *Hordelymus europaeus*, this was associated with an increased nitrogen content in its shoot and roots, due to increased availability of mineral N in the presence of the earthworm (5). The leaf, stem, and shoot biomass of birch seedlings (*Betula pendula*) was increased when grown in the presence of a diverse soil fauna (3). Further studies have demonstrated that reductions in microbial biomass through grazing activities of soil fauna can benefit plant growth, most likely as a result of increased activity of the microorganisms. For example, grazing of ectomycorrhizal fungi associated with *B. pendula* by a diverse soil fauna resulted in enhanced growth of and nutrient uptake by *B. pendula*, despite reductions in the biomass of ectomycorrhiza (25).

The effects of interactions between fauna and microorganisms on nutrient release are often regulated by resource quality. In studies of N-limited soil, Collembola did not increase the amount of nitrate and ammonium released into soil solution, because the extra N that was released as a result of animal grazing was rapidly re-utilized by the growing fungus that was nutrient limited (10). Similarly, a microcosm study showed that collembolan grazing on a leaf litter-inhabiting fungus had no effect on nutrient release due to efficient re-utilization of nutrients by the fungus (26). These studies, however, were conducted in the absence of plants, which are effective competitors with microbes for nutrients in N limited soil (27). Recent studies of ecosystem-scale N cycling point to the importance of these animal–microbial interactions in controlling seasonal patterns of plant–microbial competition for nutrients in nutrient limited ecosystems (27).

Recent work suggests that the effects of these biotic interactions on nutrient release are complex, and involve a diversity of species in more than one trophic group (28, 29). For example, the effects of feeding of fungal-feeding Collembola on nutrient cycling have been shown to become apparent only when they are interacting with another trophic group of soil fauna, namely microbial-feeding nematodes (4). Likewise, it has been demonstrated that combinations of soil animals, as opposed to a single group of soil fauna, had a synergistic effect on the microbial community in microcosms of coniferous forest humus, resulting in enhanced leaching of mineral nutrients (30). There is also evidence for top–down regulation of microbial biomass and nutrient mineralization rates in studies with microbes, microbivorous nematodes, and predatory organisms (31). However, recent microcosm studies (32) show that lower trophic levels in soil food webs are considerably more important than higher ones in regulating ecosystem productivity, as measured by plant growth. What is clear from these studies is that further testing of the effects of soil fauna on plant–microbe nutrient competition needs to examine the influence of more complex trophic relations.

Although not a direct interaction, the ability of soil fauna to comminute or fragment organic materials indirectly influences the activity of microorganisms that are involved in decomposition and nutrient cycling by increasing the surface area available for microbial colonisation (19, 33).

DISPERSAL OF MATERIALS

The movement of animals through the soil provides a passive means of transport for microbial propagules to new sites in the soil, both through the attachment and transport of microbes on the body surface, and through the ingestion and excretion of microbes at new sites in the soil. Mites, Collembola, and earthworms have been shown to carry propagules of several species of saprotrophic fungi and arbuscular mycorrhizal fungi (34, 35). Soil fauna also transport plant pathogens; the earthworm *L. terrestris* is known to enhance the dispersal of *Syntrichium endobioticum*, the causal agent of wart disease of potato (36). This has led to concerns that soil fauna may provide a means for the dispersal of genetically modified organisms. The beneficial role of soil fauna for the delivery of biocontrol agents to the rhizosphere is also a current research theme in the control of root pathogens.

REFERENCES

1. Wardle, D.A. How Soil Food Webs Make Plants Grow. Trends in Ecology and Evolution **1999**, *14*, 418–420.
2. Griffiths, B.S.; Bardgett, R.D. Interactions Between Microbe-Feeding Invertebrates and Soil Microorganisms. In *Modern Soil Microbiology*; van Elsas, J.D., Wellington, E., Trevors, J.T., Eds.; Marcel Dekker: New York, 1997; 165–182.
3. Setälä, H.; Huhta, V. Soil Fauna Increase *Betula pendula* Growth: Laboratory Experiments with Coniferous Forest Floor. Ecology **1991**, *72*, 665–671.
4. Bardgett, R.D.; Chan, K.F. Experimental Evidence that Soil Fauna Enhance Nutrient Mineralization and Plant Nutrient Uptake in Montane Grassland Ecosystems. Soil Biology & Biochemistry **1999**, *31*, 1007–1014.
5. Alphei, J.; Bonkowskim, M.; Scheu, S. Protozoa, Nematoda and Lumbricidae in the Rhizosphere of *Hordelymus europaeus* (Poaceae): Faunal Interactions, Response of Microorganisms and Effects on Plant Growth. Oecologia **1996**, *106*, 1111–1126.
6. Moore, J.C.; St. John, T.V.; Coleman, D.C. Ingestion of Vesicular-Arbuscular Mycorrhizal Hyphae and Spores by Soil Microarthropods. Ecology **1985**, *66*, 1979–1981.
7. Leonard, M.A. Observations on the Influence of Culture Conditions on the Fungal Preference of *Folsomia candida* (Collembola; Isotomidae). Pedobiologia **1984**, *26*, 361–367.
8. Parkinson, D.; Visser, S.; Whittaker, J.B. Effects of Collembolan Grazing on Fungal Colonization of Leaf Litter. Soil Biology & Biochemistry **1979**, *11*, 529–535.
9. Bengtsson, G.; Erlandsson, A.; Rundgren, S. Fungal Odour Attracts Soil Collembola. Soil Biology & Biochemistry **1988**, *20*, 25–30.
10. Bardgett, R.D.; Whittaker, J.B.; Frankland, J.C. The Effect of Collembolan Grazing on Fungal Activity in Differently Managed Upland Pastures—a Microcosm Study. Biology and Fertility of Soils **1993**, *16*, 255–262.
11. Hedlund, K.; Boddy, L.; Preston, C.M. Mycelial Responses of the Soil Fungus *Mortierella isabellina*, to Grazing by *Onychiurus armatus* (Collembola). Soil Biology & Biochemistry **1991**, *23*, 361–366.
12. Warnock, A.J.; Fitter, A.H.; Usher, M.B. The Influence of a Springtail *Folsomia candida* (Insecta, Collembola) on the Mycorrhizal Association of Leek *Allium porrum* and the Vesicular-Arbuscular Mycorrhizal Endophyte *Glomus fasciculatum*. New Phytologist **1982**, *90*, 285–292.
13. Newell, K. Interaction Between Two Decomposer Basidiomycetes and a Collembolan under Sitka Spruce: Distribution, Abundance and Selective Grazing. Soil Biology & Biochemistry **1984**, *16*, 227–233.
14. Newell, K. Interaction Between Two Decomposer Basidiomycetes and a Collembolan Under Sitka Spruce: Grazing and Its Potential Effects on Fungal Distribution and Litter Decomposition. Soil Biology & Biochemistry **1984**, *16*, 235–239.
15. Brown, G.G. How do Earthworms Affect Microfloral and Faunal Community Diversity? Plant and Soil **1995**, *170*, 209–231.
16. Fischer, K.; Hahn, D.; Hönerlage, W.; Zeyer, J. Effect of Passage Through the Gut of the Earthworm *Lumbricus terrestris* L. on *Bacillus megaterium* Studied by Whole Cell Hybridization. Soil Biology & Biochemistry **1997**, *29*, 1149–1152.
17. Moody, S.A.; Piearce, T.G.; Dighton, J. Fate of Some Fungal Spores Associated with Wheat Straw Decomposition on Passage Through the Guts of *Lumbricus terrestris* and *Aporrectodea longa*. Soil Biology & Biochemistry **1996**, *28*, 533–537.
18. Clarholm, M. Interactions of Bacteria, Protozoa and Plants Leading to Mineralization of Soil Nitrogen. Soil Biology & Biochemistry **1985**, *17*, 181–187.
19. Standen, V. The Influence of Soil Fauna on Decomposition by Micro-Organisms in Blanket Bog Litter. Journal of Animal Ecology **1978**, *47*, 25–38.
20. Förster, B.; Römbke, J.; Knacker, T.; Morgan, E. Microcosm Study of the Interactions Between Microorganisms and Enchytraeid Worms in Grassland Soils and Litter. European Journal of Soil Biology **1995**, *31*, 21–27.

21. Cole, L.; Bardgett, R.D.; Ineson, P. Enchytraeid Worms (Oligochaeta) Enhance Mineralization of Carbon in Organic Upland Soils. European Journal of Soil Science **2000**, *51*, 185–192.
22. Saetre, P. Decomposition, Microbial Community Structure and Earthworm Effects Along a Birch-Spruce Soil Gradient. Ecology **1998**, *79*, 834–846.
23. Ineson, P.; Leonard, M.A.; Anderson, J.M. Effect of Collembolan Grazing Upon Nitrogen and Cation Leaching from Decomposing Leaf Litter. Soil Biology & Biochemistry **1982**, *14*, 601–605.
24. Teuben, A. Nutrient Availability and Interactions Between Soil Arthropods and Microorganisms During Decomposition of Coniferous Litter: A Mesocosm Study. Biology and Fertility of Soils **1991**, *10*, 256–266.
25. Setälä, H. Growth of Birch and Pine Seedlings in Relation to Grazing by Soil Fauna on Ectomycorrhizal Fungi. Ecology **1995**, *76*, 1844–1851.
26. Visser, S.; Whittaker, J.B.; Parkinson, D. Effects of Collembolan Grazing on Nutrient Release and Respiration of a Leaf Litter Inhabiting Fungus. Soil Biology & Biochemistry **1981**, *13*, 215–218.
27. Jaeger, C.H.; Monson, R.K.; Fisk, M.C.; Schmidt, S.K. Seasonal Partitioning of Nitrogen by Plants and Soil Microorganisms in an Alpine Ecosystem. Ecology **1999**, *80*, 1883–1891.
28. Bardgett, R.D.; Cook, R. Functional Aspects of Soil Animal Diversity in Agricultural Grasslands. Applied Soil Ecology **1998**, *10*, 263–276.
29. Mikola, J.; Setälä, H. No Evidence of Trophic Cascades in an Experimental Microbial-Based Soil Food Web. Ecology **1998**, *79*, 153–164.
30. Setälä, H.; Tyynismaa, M.; Martikainen, E.; Huhta, V. Mineralization of C, N and P in Relation to Decomposer Community Structure in Coniferous Forest Soil. Pedobiologia **1991**, *35*, 285–296.
31. Brussaard, L.; Noordhuis, R.; Geurs, M.; Bouwman, L.A. Nitrogen Mineralization in Soil in Microcosms With and Without Bacterivorous Nematodes and Nematophageous Mites. Acta Zool. Fenn. **1995**, *196*, 15–21.
32. Laakso, J.; Setälä, H. Sensitivity of Primary Production to Changes in the Architecture of Belowground Food Webs. Oikos **1999**, *87*, 57–64.
33. Martin, A.; Marinissen, J.C.Y. Biological and Physico-chemical Processes in Excrements of Soil Animals. Geoderma **1993**, *56*, 331–347.
34. Behan-Pelletier, V.M.; Hill, S.B. Feeding Habits and Spore Dispersal of Oribatid Mites in the North American Arctic. Révue d'Écologie et de Biologie du Sol **1978**, *15*, 497–516.
35. Reddell, P.; Spain, A.V. Earthworms as Vectors of Viable Propagules of Mycorrhizal Fungi. Soil Biology and Biochemistry **1991**, *23*, 767–774.
36. Hampson, M.C.; Coombes, J.W. Pathogenesis of *Syntrichium Endobioticum* VII: Earthworms as Vectors of Wart Disease on Potato. Plant and Soil **1989**, *116*, 147–150.

ANTS

Walter G. Whitford
United States Department of Agriculture-Agricultural Research Service (USDA-ARS), Jornada Experimental Range, New Mexico State University, Las Cruces, New Mexico

INTRODUCTION

Ants are among the most ubiquitous insects on the planet. They occur in all biomes except for the extreme polar-regions. In the biomes where ants are abundant, they affect many soil processes that contribute to the creation of patch mosaics that characterize the soils and vegetation of many landscapes. The abundance and diversity of soil-nesting ants varies from as high as 7,000 colonies per hectare in tropical savanna to as few as 3–4 colonies per hectare on some periodically flooded, fine-textured soil, desert landscape units (1) Soil-nesting ants affect critical ecosystem processes such as nutrient cycling and water redistribution. Ant nest mounds vary from a few centimeters in height and diameter to >1 m in height and >2 m diameter (2). Ant nests consist of underground, branched networks of galleries and chambers. Surficial chambers are connected to lower chambers by vertical galleries with branching lateral galleries. Galleries and chambers vary in size and number depending upon the species of ant. For example, *Lasius neoniger*, an abundant ant species in temperate North America, constructs tubular galleries of 1.5–5.0 mm in diameter and chambers of 10–20 mm diameter and 30–50 mm in length. The volume of *L. neoniger* nests range from 20–250 cm^3 and are confined to the upper 70 cm of soil (3). Other species construct nests to depths ranging from 50 cm to greater than several meters depending upon species specific behavior, soil type, and landscape position. Soil profile mixing, texture, physical and chemical property modification of mound soils, soil macroporosity, and geomorphological attributes of ant nest mounds vary with species-specific colony longevity, body size, and numbers of workers of a colony, soil type, and landscape position. The pedturbation effects of ants are therefore dependent upon the species composition of the ant community, geomorphic history, soil properties, and topographic position of a landscape unit. Because most studies of the effects of ants on soils have focused on one or two species, a comprehensive analysis of the combined effects of all ant species on the soils of an ecosystem cannot be made.

MICROTOPOGRAPHY

In areas that are periodically flooded or where the water table is close to the surface, some species of soil-nesting ants build mounds that create favorable microhabitats for themselves and also a habitat for some species of plants that are confined to the aerated soils of the ant mounds. Soil-nesting ants create hummock microtopography in some wet meadow fens and tropical wet savannas (4). In the Chaco region of South America (parts of Paraguay, Bolivia, Argentina, and Brazil), nest mounds of *Camponotus punctulatus* occur at densities of between 200 and 1000 mounds ha^{-1}. These conical mounds average 0.62 cm high (up to 1.85 m high) with a mean basal diameter of 1.2 m. The mound soils are lighter textured than surrounding soils reflecting materials transported from surrounding sub-surface soil during mound construction (5). *Formica podzolica* mounds in a Montana fen are thought to contribute to the hummock-hollow microtopography of peat lands. Abandoned *F. podzolica* mounds provide drier, warmer microsites that are enriched with some soil nutrients (4). The mounds of *Lasius flavus* contribute to the microtopography of some European grasslands and salt marshes (6). Mima-type earth mounds up to 1.5 m in height and 20 m in diameter in the Buenos Aires Province, Argentina, are produced by horizontal translocation of soil to the colony sites of black fire ants, *Solenopsis richteri*. Continued occupation of the mounds by successive generations of ants gradually increases the mounds to mima-type size (7). Ants (*Formica* spp. and *Myrmica* spp.) are important agents in the process of development and maintenance of hummock microtopography of subarctic peatlands. Hummock retrogression is accelerated by the tunneling activity of ants (8).

HETEROGENEITY OF SOIL PHYSICAL AND CHEMICAL PROPERTIES

Many species of ants alter the texture and chemistry of the soil in the nest mounds. The nutrients most frequently

Encyclopedia of Soil Science

reported to be at higher concentrations in ant mound soils include nitrogen, phosphorus, potassium, calcium, magnesium, manganese, and iron (9). The effect of soil-nesting ants on soil nutrient patchiness and on vegetation varies as a function of landscape position, soil type, and the biology of the ant species. Nutrient enrichment of mound soils has been reported for several species of seed-harvesting ants and omnivorous species of ants that collect seeds, prey on insects, or collect insect carrion. Species of soil-nesting ants that enrich the nutrient content of mound soils are characterized by relatively long-lived colonies (>5 years) and the behavior of depositing chaff and unwanted insect parts on and around the nest mound or disk. Nutrient enrichment of mound soils by a species may not occur on all soils on a watershed or landscape. For example, *Pogonomyrmex rugosus* nest disks in desert shrubland and mixed shrub-grassland were nutrient enriched but the nest disks of this species in a piedmont grassland were not nutrient enriched (10). *Formica* spp. mounds in forest were nutrient enriched but *Formica* spp. mounds in meadows and grasslands were not (11).

The variability in soil nutrient enrichment of ant mounds has been documented in several species of leaf-cutting ants. In remnant Cerrado (woodland-savanna) in Brazil, leaf cutting ants (*Atta* spp.) had no detectable effect on nutrient enrichment (12). In northern Patagonia, soils associated with the leaf-cutting ant, *Acromyrmex lobicornis*, had higher concentrations of nitrogen, phosphorus, and organic matter than reference soils (13). The location of nutrient-rich organic refuse produced by leaf-cutting ant colonies varies among species. *Atta cephalotis* deposit organic refuse in subterranean chambers, whereas *A. colombica* place organic refuse on the soil surface near the nest. The location of organic refuse is a major factor affecting nutrient concentrations and the composition, abundance, and activity of soil microflora and microfauna (14). In the Orinoco Llanos savanna in Venezuela, *Atta laevigata* nests had higher concentrations of nitrogen, magnesium, calcium, and organic carbon, but other soil nutrients and properties were not affected by ant mounds (15).

In an Australia vertisol, ant nest soils had greater concentration of coarse and particulate organic matter, lower fine particulate soil organic matter (SOM)/coarse particulate SOM ratios, larger sand content, and lower clay content than surrounding soils (16). Nutrient enrichment of nest mound soils of funnel ants (*A. barbigula*) was attributed to entrapment of organic materials around the nest entrances. Re-excavation of nest chambers after rainfall buries trapped litter, resulting in higher concentrations of nitrogen, organic matter, and some cations compared to nest-free soils (17). In humid tropical savanna, ant mounds of *Camponotus* spp. had higher clay and coarse sand content than surrounding soils (9). Even exotic or alien species of ants change the chemical and physical properties of nest mound soils. Imported fire ants (*Solenopsis invicta*) mounds had higher concentrations of clay, phosphorus, and potassium, and lower concentration of soil organic matter, than reference soils. The effect of *S. invicta* on calcium concentrations relative to reference soils was dependent upon the characteristics of the unmodified soil (18).

Ants change the nutrient concentration of mound soils, but the physical and chemical properties of mound soils can also affect mineralization processes. Nitrogen mineralization rates were reduced in nest mound soils in moss-sedge, sedge, and alder peat habitats (19).

SOIL TURNOVER

The longevity and turnover rates of nests and nest mounds of species of ants in a community frequently follow a distribution gradient from high turnover (<3 months) to long-lived (>10 years). The importance of ants in the transport of subsurface horizon materials to the surface varies with the density and diversity of the ant community on a landscape unit. In Chihuahuan Desert grasslands, soil-nesting ants are an order of magnitude more abundant on sand and sandy loam soils than on fine-textured soils. Ants were estimated to move between 21.3 and 85.8 $kg\,ha^{-1}\,y^{-1}$ on sandy and sandy loam soils and between 0.1–3.4 $kg\,ha^{-1}\,y^{-1}$ on clay and clay-loam soils (20). The estimated annual soil turnover by ants in an *Atriplex vesicaria* shrubland in the semi-arid region of Australia was 350–420 $kg\,ha^{-1}y^{-1}$ (9). Soil that is excavated by ants in the construction of galleries and chambers and deposited on mounds around nest entrances are generally eroded by water and wind within a year unless the mound is protected from raindrop splash erosion by gravel, stones, or wood fragments. Nest mound soils may be replenished by the belowground expansion of galleries and chambers. Ant nest mounds in sparsely vegetated arid regions are prone to wind erosion. On an Australian aeolian soil, funnel ants (*Aphaenogaster barbigula*) nests were active for approximately 9 months and changed location approximately twice per year. Soil transport was estimated at 33.6 $kg\,ha^{-1}$ and it was estimated that 92% of the soil volume would be turned over by these ants in 100 years (21). In western Australia, ant communities on gray soils of semi-arid woodlands were estimated to turnover 46.5 $kg\,ha\,y^{-1}$ and on yellow soils, the soil-nesting ant community was estimated to turnover 22.3 $kg\,ha^{-1}\,y^{-1}$ (22). In a humid savanna environment, one abundant ant species, *Paltothyreus tarsatus*, was estimated to transport approximately $30\,g\,m^{-2}\,y^{-1}$ of sand particles and soil aggregates. This ant species increased the concentrations of clay, carbon, iron oxides, and coarse sand in the A horizon (9). The amount of soil transported to the surface by *Pognomyrmex occidentalis* in pinon-juniper woodland and ponderosa pine forest was estimated to be $650\,kg\,ha^{-1}$ (23). Soil turnover by the ant community in New England forest soil was estimated to be over $50\,kg\,ha^{-1}\,y^{-1}$. It was concluded that the translocation of B-horizon materials to the soil surface by soil-nesting ants was a important process in podzol formation in New England forest soils (24).

Some long-lived species of soil-nesting ants relocate their nests one or more times a year. Construction of new nests results in transport to the soil surface of a volume of soil equal to the volume of galleries and chambers. Most of that soil originates in lower soil horizons and contributes to soil profile homogenization. The relocation of nests by some species of ants results in lower estimates of soil turnover than occurs in some environments.

SOIL WATER RELATIONS

The structure of nests of soil-nesting ants provides extensive macroporosity to the soil in which the nests are constructed. The macropores constructed by ants affects rates of infiltration and rates of percolation. In some environments, extremely high densities of nest entrances can have a dramatic effect on infiltration. In semi-arid Western Australia, ant biopores were found to transmit water down the soil profile only when the soil was saturated and water was ponding on the surface (25). On aeolian sand soils in Australian semi-arid woodland, densities of nest entrances of funnel ants (*Aphaenogaster barbigula*) were estimated at $88{,}000\,ha^{-1}$. Steady-state water infiltration on soils with nest entrances averaged $23.3\,mm\,min^{-1}$, in comparison to an infiltration rate of $5.9\,mm\,min^{-1}$ on nest-entrance-free soil (26). In semi-arid woodland of eastern Australia on red earth soil, ponded steady-state infiltration averaged $1026\,mm\,h^{-1}$ on soil with nest entrances of *A. barbigula* but only $120\,mm\,h^{-1}$ on soils without nest entrances (27). Bulk flow along nest galleries provides an important route of recharge of deep soil moisture in arid and semi-arid environments.

Ant gallery macropores are not always avenues for bulk flow. In a study of a mesic Typic Quartzipsamment, there was no preferential flow down ant galleries. The lack of an effect on hydraulic conductivity was attributed to the sandy soil (28). In another study of a sandy soil, the estimated saturated soil matrix hydraulic conductivity of nest burrows was approximately eight times smaller than the soil matrix hydraulic conductivity of the bulk sandy soil. This reduction in hydraulic conductivity was attributed to the ants in-filling gallery walls with fine materials (28).

EFFECTS ON OTHER SOIL BIOTA

Soil around relatively long-lived ant colonies may be enriched with microflora, microfauna, and mesofauna. The soils of nest disks of western harvester ants, *Pogonomyrmex occidentalis*, are enriched with vesicular-arbuscular mycorrhizal fungi (29). In areas of North America dominated by the red imported fire ant, *Solenopsis invicta*, the species composition and abundance of soil yeast within mounds are altered by changes in soil properties produced by fire ants (30). Mound soils of *Formica aquilonia* are dominated by bacteria-feeding microfauna and have a higher microbial biomass than surrounding soils (31). Species specific differences in the effect of ants on soil microflora of mounds are related to the feeding strategies of the species and nest architecture. Three ant species, *Myrmica scabrinodis*, *Lasius niger*, and *L. flavus*, differ greatly in foraging strategies and methods of mound construction. Microbial functional diversity and evenness were higher in mound soils of *M. scabrinodis* and *L. niger* than in reference soils but were not different from reference soils in the mounds of *L. flavus*. Different functional groups of microorganisms were activated in the mounds of the different species. Carbon mineralization was higher in mound soils of all three species (32).

REFERENCES

1. Holldobler, B.; Wilson, E.O. *The Ants*; The Belknap Press of Harvard University Press: Cambridge, Mass, 1990.
2. Green, W.P.; Pettry, D.E.; Switzer, R.E. Formicarious Pedons, the Initial Effect of Mound-Building Ants on Soils. Soil Survey Horizons **1995**, *39* (2), 33–44.
3. Wang, D.; McSweeney, K.; Lowery, B.; Norman, J.M. Nest Structure of the Ant *Lasius neoniger* Emery and Its Implications to Soil Modification. Geoderma **1995**, *66* (3–4), 259–272.
4. Lesica, P.; Kannowski, P.B. Ants Create Hummocks and Alter Structure and Vegetation of a Montana Fen. Am. Midl. Nat. **1998**, *139* (1), 58–68.
5. Pire, E.F.; Torres, P.F.; Romagnoli, O.D.; Lewis, J.P. The Significance of Ant-Hills in Depressed Areas of the Great Chaco. Rev. Biol. Trop. **1991**, *39* (1), 71–76.

6. King, T.J. Ant-Hills and Grassland History. J. Biogeog. **1981**, *8*, 329–334.
7. Cox, G.W.; Mills, J.N.; Ellis, B.A. Fire Ants (Hymenoptera:Formicidae) as Major Agents of Landscape Development. Environ. Entomol. **1992**, *21* (2), 281–286.
8. Luken, J.O.; Billings, W.D. Hummock-Dwelling Ants and the Cycling of Microtopography in An Alaskan Peatland. Canadian Field Nat. **1986**, *100* (1), 69–73.
9. Lobry de Bruyn, L.A.; Conacher, A.J. The Role of Ants and Termites in Soil Modification a Review. Aust. J. Soil Res. **1990**, *28* (1), 55–95.
10. Whitford, W.G.; DiMarco, R. Variability in Soils and Vegetation Associated with Harvester Ant (*Pogonomyrmex rugosus*) Nests on a Chihuahuan Desert Watershed. Biol. Fertil. Soils **1995**, *20*, 169–173.
11. Culver, D.C.; Beattie, A.J. Effects of Ant Mounds on Soil Chemistry and Vegetation Patterns in a Colorado Montane Meadow. Ecology **1983**, *64* (3), 485–492.
12. Schoereder, J.H.; Howse, P.E. Do Trees Benefit from Nutrient Rich Patches Created by Leaf-Cutting Ants? Studies on Neotropical Fauna and Environment **1998**, *33* (2-3), 111–115.
13. FarjiBrener, A.G.; Ghermandi, L. Influence of Nests of Leaf-Cutting Ants on Plant Species Diversity in Road Verges of Northern Patagonia. J. Veg. Sci. **2000**, *11* (3), 453–460.
14. FarjiBrener, A.G.; Medina, C.A. The Importance of Where to Dump the Refuse: Seed Banks and Fine Roots in Nests of the Leaf-Cutting Ants *Atta cephalotes* and *A. colombica*. Biotropica **2000**, *32* (1), 120–126.
15. Brener, A.G.F.; Silva, J.F. Leaf-Cutting Ants and Forest Groves in a Tropical Parkland Savanna of Venezuela: Facilitated Succession. J. Trop. Ecol. **1995**, *11* (4), 651–669.
16. Hulugalle, N.R. Effects of Ant Hills on Soil Physical Properties of a Vertisol. Pedobiol. **1995**, *39* (1), 34–41.
17. Eldridge, D.J.; Myers, C.A. Enhancement of Soil Nutrients Around Nest Entrances of the Funnel Ant *Aphaenogaster barbigula* (Myrmicinae) in Semi-Arid Eastern Australia. Aust. J. Soil Res. **1998**, *36* (6), 1009–1017.
18. Green, W.P.; Pettry, D.E.; Switzer, R.E. Impact of Imported Fire Ants on the Texture and Fertility of Mississippi Soils. Comm. Soil Sci. and Plant Analysis **1998**, *29* (3-4), 447–457.
19. Petal, J. The Influence of Ants on Carbon and Nitrogen Mineralization in Drained Fen Soils. Appl. Soil Ecol. **1998**, *9* (1–3), 271–275.
20. Whitford, W.G.; Forbes, G.S.; Kerley, G.I. Diversity, Spatial Variability, and Functional Roles of Invertebrates in Desert Grassland Ecosystems. In *The Desert Grassland*; McClaran, M.P., Van Devender, T.R., Eds.; University of Arizona Press: Tucson, Az, 1995; 152–195.
21. Eldridge, D.J.; Pickard, J. Effects of Ants on Sandy Soils in Semiarid Eastern Australia. 2. Relocation of Nest Entrances and Consequences for Bioturbation. Aust. J. Soil Res. **1994**, *32* (2), 323–333.
22. Lobry de Bruyn, L.A.; Conacher, A.J. The Bioturbation Activity of Ants in Agricultural and Naturally Vegetated Habitats in Semiarid Environments. Aust. J. Soil Res. **1994**, *32* (3), 555–570.
23. Carlson, S.T.; Whitford, W.G. Ant Mound Influence on Vegetation and Soils in a Semiarid Mountain Ecosystem. Am. Midl. Nat. **1991**, *126* (1), 125–139.
24. Lyford, W.H. Importance of Ants to Brown Podzolic Soil Genesis in New England. Harvard Forest Paper No. 7, **1963**, 1–18.
25. Lobry de Bruyn, L.A.; Conacher, A.J. The Effect of Ant Biopores on Water Infiltration in Soils in Undisturbed Brushland and Farmland in a Semi-Arid Environment. Peodobiol **1994**, *38* (3), 193–207.
26. Eldridge, D.J. Effect of Ants on Sandy Soils in Semiarid Eastern Australia: Local Distribution of Nest Entrances and Their Effect on Infiltration of Water. Aust. J. Soil Res. **1993**, *31* (4), 509–518.
27. Eldridge, D.J. Nests of Ants and Termites Influence Infiltration in a Semiarid Woodland. Pedobiol. **1994**, *38* (6), 481–492.
28. Wang, D.; Lowery, B.; Norman, J.M.; McSweeney, K. Ant Burrow Effects on Water Flow and Soil Hydraulic Properties of Sparta Sand. Soil Till. Res. **1996**, *37* (2-3), 83–93.
29. Friese, C.F.; Allen, M.F. The Interaction of Harvester Ants and Vesicular Arbuscular Mycorrhizal Fungi in a Patchy Semiarid Environment: The Effects of Mound Structure on Fungal Dispersion and Establishment. Functional Ecology **1993**, *7* (1), 13–20.
30. Ba, A.S.; Phillips, S.A.; Anderson, J.T. Yeasts in Mound Soil of the Red Imported Fire Ant. Mycol. Res. **2000**, *104* (8), 966–973.
31. Laasko, J.; Setala, H. Composition and Trophic Structure of Detrital Food Web in Ant Nest Mounds of *Formica aquilonia* and in the Surrounding Forest Soil. Oikos **1998**, *81* (2), 266–278.
32. Dauber, J.; Wolters, V. Microbial Activity and Functional Diversity in Mounds of Three Ant Species. Soil Biol. Biochem. **2000**, *32* (1), 93–99.

ARCHAEOLOGY, SOIL AND

J.E. Foss
University of Tennessee, Knoxville, Tennessee, U.S.A.

INTRODUCTION

Geology has had a long period of interaction with archaeology, but the use of soil investigation in archaeology has a rather short history. In 1942, Nikiforoff (1) used the term *archeopedology* for those soil scientists working with fossil soils or paleosols. Early studies of soils at archaeologic sites were concerned mainly with soil chemical properties (e.g., 2–4). An early book by Cromwall (5) also played an important role in demonstrating the usefulness of soil–archaeologic interactions. The past 30 or 40 years have seen a substantial increase in the multidisciplinary effort between these two sciences and have involved more subdisciplines of soil science.

ROLE OF SOIL SCIENCE IN ARCHAEOLOGY

As archaeologists become more interested in a complete understanding of the chronology and environmental history of sites, a multidisciplinary effort is absolutely necessary. Team members commonly include scientists from soil science, geology, botany, zoology, palynology, and other specializations. Soil science, especially the pedology area (i.e., the study of soil formation and classification), has been particularly active within the past few decades in evaluation of archaeologic sites. Pedology, geology, and other earth sciences often work in the specialized field of geoarchaeology, which means using earth science principles to study archaeologic sites. Fig. 1 shows a landscape of Tikal, Guatemala (Mayan site); this site is one of the many important archaeologic sites that has required the expertise of pedologists to help interpret chronology and land use (6–9).

The study of soils and landscapes is an integral part of many archaeologic investigations. Some federal and state regulations that require geologic and soil input on archaeologic sites have also been responsible for including earth scientists in these studies. Publications in the past decade have indicated the interest of pedologists, geologists, and archaeologists in evaluating soils and landscapes as part of overall archaeologic investigations.

Publications such as "Soils in Archaeology" edited by V.T. Holliday (8), "Soil Science and Archaeology" by Scudder et al. (9), Soil Science Society of America Special Pub. No. 44 on "Pedological Perspectives in Archaeological Research," (10) and articles in the *Proceedings of Conferences on Pedo-archaeology* (11, 12) have raised pedologists' and archaeologists' awareness of the potential contributions of soil studies to site evaluation. The periodical *Geoarchaeology: An International Journal* has also been valuable in promoting earth science activity in archaeologic investigations.

Some of the major pedologic contributions to field archaeology have included the following (13):

- Determining site delimitation
- General pedologic stratigraphy
- Soil–landscape relationships
- Identification of geologic parent material
- Correlating soil morphology and archaeologic levels
- Identifying lithologic (parent material) and pedologic (soil weathering) discontinuities
- Approximating soil age
- Identifying paleosols (fossil soils)
- Contributing to the overall interpretation of site

In the past decade, many of the above pedologic contributions to archaeology were made during the final phase of archaeologic field work. More recently, pedologists have been more involved in phase 1 activity of archaeologic investigations. The early identification of major stratigraphic zones, preliminary analysis of landscape and soil age, and model of site development have resulted in more efficient archaeologic excavations and interpretations.

FIELD STUDIES

Archaeologic sites occur in many different geologic provinces and landscape positions. Determination of the site context is thus the most important initial stage in pedoarchaeology. Geologic maps can provide general knowledge of a region, but detailed soil surveys provide the most useful introduction to a study area. These maps

Fig. 1 Landscape view of Mayan city of Tikal, Guatemala.

produced by the National Resource Conservation Service (NRCS), in cooperation with the Land Grant Institutions, are usually on a county-wide basis using an air photo base with a scale of 1:15840 to 1:24000. At this scale, there is not sufficient detail to relate the morphology of individual soil mapping units to specific horizons encountered at an archaeologic site. The landscape and physiographic position of each soil mapping unit, however, are still useful in preliminary analysis of archaeologic sites.

The most important and informative archaeologic sites occur in landscapes where sediment is added to a preexisting surface, therby protecting the artifacts and soil horizons. Those buried sites may occur in the following areas or situations:

- Alluvial deposition
- Volcanic activity
- Eolian deposition
- Colluvial slopes
- Mass movement or slumping
- Seismic areas
- Artificial deposition or destruction

These situations provide the opportunity for soil burial (subsequently termed *paleosols*) and archaeologic levels. The buried surfaces (A horizons) of these paleosols are particularly good sources of artifacts and living surfaces when the events above took place in the Holocene. Holliday (8) provides an excellent background in the use of paleopedology in archaeology.

Soil Morphology

Soil morphology (e.g., a detailed description of soil profiles) provides the key to understanding and interpreting soils and landscapes at archaeologic sites. The unique soil morphology of a given region and site results from the weathering processes regulated by the interaction of soil-forming factors (14). These factors are climate, biotic, geology, topography, and length of time that the weathering processes have been operating. The morphologic properties of soils usually described in excavations and their interpretation for archaeologic sites are given in Table 1. A great deal of experience and technique is needed to provide an accurate and informative soil description. Evaluation of the age of soils, for example, requires an integration of all the morphologic features that are detailed in Table 1. Certain soil horizons provide general age estimates based on the length of time needed for weathering processes to develop specific features (e.g., argillic horizons). As noted in Table 1, a minimal argillic horizon can form in 4,000 years. Other age estimates of soil horizons have been published previously (9, 15).

One of the most useful applications of soil morphology in archaeologic site interpretation is that of distinguishing "natural" from "artificial" or "man-influenced" horizons. Some natural horizons—such as a spodic (Bh) with a dark-colored, organic-rich matrix—may appear as a buried surface or midden. Some albic E horizons could be interpreted as ash layers. Other characteristics that are related to soil genesis, such as redoximorphic features (i.e., mottling or gleying), result from water table fluctuations and often cause confusion in interpretation of color in archaeologic levels. Horizons with calcium carbonate filling (Bk or Ck) have sometimes been identified as plaster-filled.

LABORATORY

Laboratory soil characterization for archaeologic interpretations is used to verify and supplement field morphology. Laboratory analysis without complete soil morphology is generally of minimal value for archaeologic interpretation. Complete sampling of all soil horizons, columns, or archaeologic levels is also important to realize the full benefit of the additional cost and labor of soil analysis.

Those laboratory analyses that are frequently applied in pedoarchaeology are organic carbon (16), particle size distribution (17), and elemental composition (18). Other soil analysis may include pH, electrical conductivity, mineralogy, free iron, scanning electron microscopy (SEM), energy dispersive x-ray (EDAX), calcium carbonate, and micromorphology. The micromorphologic studies by Goldberg (19) and Macphail and Goldberg (20) have been especially useful in interpreting site stratigraphy and pedologic and geologic events.

Table 1 Morphologic properties of soils and their interpretive value at archaeologic sites

Soil property	Useful interpretative features
Texture	Lithologic and pedologic discontinuities; classification of geologic materials; determination of argillic horizons; determine relative energy of alluvial sedimentation
Structure	Relative abundance of macropores and potential artifact movement; degree of development is an indicator of soil age; development of clay or organic coatings on argillic horizons (e.g., continuous clay coatings on pedologic faces indicate 10,000 years of development while discontinuous coatings may indicate 4000–5000 years of weathering in southeastern U.S.)
Color	Indicator of organic matter and free iron content; classification of sediments; delineation of horizons; drainage characteristics (redoximorphic features)
Boundary	Abrupt boundary indicator of Ap (plow zone) or recent deposition; boundary becomes more diffuse with age
Consistence	Indicator of structural development, cementation, or consolidation (e.g., recent alluvium usually very friable or loose)
Clay coatings	Coatings on peds or in pores indicate state of development and age
Carbonate	Secondary $CaCO_3$ leaching, coatings, pore filling, and cementation can provide soil age estimates and climatic implications
Horizon identification	Indicates many weathering processes occurring in profile, e.g., A = organic matter accumulation; E = leached zone; Bt = argillic horizon with minimal 4,000 year age; distinguish natural vs. artificial horizons; horizon thickness (solum) is a measure of length of weathering time; diagnostic horizons useful indicator for archaeological interpretation (e.g., argillic, cambic, fragipan, spodic, etc.)

(Modified from Ref. 13.)

FUTURE

In the past few decades, pedologists have grown increasingly interested in work on archaeologic sites, and it is likely this trend will continue well into the future. Although we aid archaeologists in understanding the soils and pedologic features they carefully excavate, we have learned a great deal about weathering rates, horizon formation, and landscape development by teamwork with archaeologists and geologists. The use of additional techniques or applications by geoarchaeologists, such as x-ray diffraction (XRD), SEM, EDX, electrical resistivity, ground-penetrating radar, magnetic susceptibility, and micromorphology, will improve soil interpretation work in the future. Despite advances in analytical tools, the key to archaeologic site interpretation still remains the accurate, complete soil morphologic descriptions.

REFERENCES

1. Nikiforoff, C.C. Introduction to Paleopedology. Am. J. Sci. **1943**, *41*, 194–200.
2. Dietz, E.F. Phosphorus Accumulation in Soil of an Indian Habitation Site. Am. Antiquity **1957**, *22*, 405–409.
3. Cook, S.F.; Heizer, R.F. Studies on the Chemical Analysis of Archaeological Sites. Univ. Calif. Pub. Anthropol. **1965**, *2*.
4. Sokoloff, V.P.; Carter, G.F. Time and Trace Metals in Archaeological Sites. Science **1952**, *116*, 1–5.
5. Cromwall, I.W. *Soils for the Archeologist*; Phoenix House Ltd.: London, 1958.
6. Olson, G.W. *Soils and the Environment: A Guide to Soil Surveys and Their Application*; Chapman and Hall: New York, 1981.
7. Foss, J.E. Paleosols of Pompeii and Oplontis. In *Stvdia Pompeiana and Classics*; Curtis, R.L., Ed.; Orpheus Pub. Inc., 1988; 127–144.

A

8. Holliday, V.T., Ed. *Soils in Archaeology*; Smithsonian Institution Press: Washington, 1992.
9. Scudder, S.J.; Foss, J.E.; Collins, M.E. Soil Science and Archaeology. Advances in Agronomy **1996**, *57*, 1–76.
10. Collins, M.E.; Carter, B.J.; Gladfelter, B.G.; Southard, R.J., Eds. Pedological Perspectives in Archaeological Research. Soil Sci. Soc. Am. Special Pub. **1995**, *44*, 157.
11. Foss, J.E.; Timpson, M.E.; Morris, M.W., Eds. Proceedings of the First International Conference on Pedo-Archaeology. Univ. of Tennessee, Agr. Exp. Sta., Special Pub. **1993**, *93-03*, 210.
12. Goodyear, A.C.; Foss, J.E.; Sassaman, K.E., Eds. Proceedings of the Second International Conference on Pedo-Archaeology. South Carolina Institute of Archaeology and Anthropology, Univ. of South Carolina, Anthro. Studies **1994**, *10*, 157.
13. Foss, J.E.; Lewis, R.J.; Timpson, M.E.; Morris, M.W.; Ammons, J.T. Pedologic Approaches to Archaeological Sites of Contrasting Environments and Ages. In *Proceedings of the First International Conference on Pedo-Archaeology*; Foss, J.E., Timpson, M.E., Morris, M.W., Eds.; Univ. of Tenn, Agr. Exp. Sta. Spec. Pub. **1992**, *93–03*, 19–22.
14. Jenny, H. *Factors of Soil Formation*; McGraw-Hill: New York, 1941.
15. Foss, J.E.; Lewis, R.J.; Timpson, M.E. Soils in Alluvial Sequences: Some Archaeological Implications. In *Pedological Perspectives in Archaeological Research*; Collins, M.E., Carter, B.J., Gladfelter, B.G., Southard, R.J., Eds.; Soil Sci. Soc. Am. Special Pub. **1995**, *44*, 1–14.
16. Stein, Julie K. Organic Matter in Archaeological Contexts. In *Soils in Archaeology*; Holliday, V.T., Ed.; Smithsonian Institute Press: Washington DC, 1992, 193–216.
17. Timpson, M.E.; Foss, J.E. The Use of Particle-Size Analysis as a Tool in Pedological Investigations of Archaeological Sites. In *Proceedings of the First International Conference on Pedo-Archaeology*; Collins, M.E., Carter, B.J., Gladfelter, B.G., Southard, R.J., Eds.; Univ. of Tenn. Agr. Exp. Sta. Spec. Pub. **1992**, *93-03*, 69–80.
18. Schuldenrein, Joseph. Geochemistry, Phosphate Fractionation, and the Detection of Activity Areas at Prehistoric North American Sites. In *Pedologic Perspectives in Archaeological Research*; Collins, M.E., Ed.; Soil Sci. Soc. Am. Special Pub. **1995**, *44*, 107–132.
19. Goldberg, Paul. Micromorphology, Soils, and Archaeological Sites. In *Soils in Archaeology*; Holliday, V.T., Ed.; Smithsonian Institution Press: Washington DC, 1992, 45–167.
20. Macphail, Richard.; Goldberg, Paul. Recent Advances in Micromorphological Interpretation of Soils and Sediments from Archaeological Sites. In *Archaeological Sediments and Soils*; Barham, A.J., Macphail, R.I., Eds.; Institute of Archaeology, University College: London, 1995.

ARID SOILS

H. Curtis Monger
New Mexico State University, Las Cruces, New Mexico, U.S.A.

INTRODUCTION

Scarcity of rain is the dominant characteristic of arid soils. While age, parent material, carbonate, and salt content may vary from arid soil to arid soil, dryness is common to all. Of the total ice-free land area on Earth (130,797,000 km^2), about 22% or 28,703,000 km^2 is occupied by soils with aridic moisture regimes (1).

Although arid (L. *aridus*, dry) signifies lack of moisture, technical definitions of arid vary. In some cases, the arid–semiarid boundary is placed at 25 cm (10 in.) of annual rainfall (2). In other systems, such as the Köppen–Geiger–Pohl and Meigs systems, the arid (desert)–semiarid (steppe) boundary is based on a combination of rainfall and temperature (3, 4). Still other systems, such as those by Strahler and Soil Survey Staff, use soil moisture to define arid zones because the availability of moisture to plants is more important than annual precipitation itself (4, 5). In all cases, however, rainfall is insufficient to maintain perennial streams. Soils in these regions are unique because relatively little water percolates deep enough to reach groundwater. As a result, carbonates, gypsum, and more soluble salts acumulate in the profiles of many arid soils.

ARID SOILS OF RIVER FLOODPLAINS

Floodplain soils along rivers that flow through arid climates were sites of several ancient and eminent civilizations. Sumerian (ca. 3600 B.C.) and later Babylonian (ca. 2000 B.C.) civilizations grew into centers of trade and government as a result of irrigated agriculture on the Tigris and Euphrates River floodplains (6). Likewise, soils and irrigated agriculture along the Nile of ancient Egypt, the Indus of ancient India, and the Hoang-Ho (Yellow River) of ancient China made it possible for civilizations to create notable schools, calendars, armies, mathematics, medicine, literature, philosophy, science, and art. In the western hemisphere as well, Hohokam, Aztec, and Inca societies emerged in arid and semiarid environments (7).

These civilizations existed because floodplain soils are well suited for irrigated agriculture if groundwater tables are sufficiently deep and salts do not accumulate. In the case of the Nile prior to dam construction, the river would rise and spill over its banks, flood the adjacent plain, deposit sediment, and leach salts (7). In the case of the Tigris and Euphrates, however, drainage canals were needed to carry away leached salts, and with their demise soils became saline.

ARID SOILS ON UPLANDS

Most soils in arid regions are not on floodplains, but occur in vast upland areas composed of three major landforms: mountains, piedmont slopes, and basin floors (8). These major landforms, in turn, are composed of smaller, component landforms. Typically, soil boundaries correspond to component landforms. In mountains, for example, soil boundaries match the boundaries of colluvial wedges, valley fills, and pediments (9). On piedmont slopes, soil boundaries parallel the boundaries of alluvial fans, ballenas, and fan skirts. On basin floors, which characteristically have little topographic relief, soil boundaries generally follow landforms produced by wind, such as deflational blowouts, dunes, and eolian plains, or landforms produced by pluvial lakes, such as lake plains, playas, and beach plains.

Of the five soil-forming factors (climate, time, biota, topography, and parent material), climate is the defining factor of arid soils, although time is an important factor as well. The impact of time on arid soils is revealed by carbonate and clay accumulations in soils of progressively older geomorphic surfaces (Fig. 1). Carbonate in nongravelly soils, for example, progresses from carbonate filaments in middle Holocene soils to carbonate nodules in late Pleistocene soils to carbonate-indurated horizons in middle Pleistocene soils (10). Clay likewise accumulates with time to form argillic horizons. However, the correlation of clay accumulation with time is less robust than carbonate accumulation with time because many ancient soils that have calcretes do not have argillic horizons (11). This indicates that argillic horizons are more vulnerable to obliteration by erosion and bioturbation than calcic or petrocalcic horizons.

Fig. 1 Desert piedmont slope rising to a mountain chain in southern New Mexico. Progressively older geomorphic surfaces with progressively greater soil development are typical features of piedmont slopes. The younger soil on the right (about 3000 years old) has a small amount of carbonate (white zone in profile). The older soil to the left (25,000–150,000 years old) has substantially more carbonate. In addition, the older soil has an argillic horizon overlying the carbonate horizon.

Arid soils are not only unique because carbonate, gypsum, and soluble salt accumulate, but also because many have vesicular A-horizons covered by desert pavement (12) or microbiotic crust (13). In addition, inadequate water and nitrogen suppress biomass production on arid soils to about one-tenth the biomass of temperate forest soils (14). Nevertheless, soil animals such as rodents, ants, and termites are common. Ants, for example, can transfer 80 g/m^2 of desert soil to the land surface per year, which is as much as ants transfer in more mesic environments (15).

TYPES OF ARID SOILS

The main criterion for the classification of arid soils is soil dryness, or the aridic (torric) moisture regime, which is defined as soils too dry for agricultural crops unless irrigated (16). Further taxonomic subdivisions are based on diagnostic horizons. In contrast to the notion that arid soils are poorly developed, as written in some soil science books, many arid soils are strongly developed with a variety of diagnostic subsurface horizons (17). These horizons include the argillic, natric, salic, gypsic, petrogypsic, calcic, and petrocalcic horizons, and the duripan (5). Diagnostic surface horizons include the ochric epipedon with minor occurrences of the mollic and anthropic epidons.

Arid soils that have diagnostic subsurface horizons are generally classified as Aridisols. These include many of the older soils on piedmont slopes, basin floors, mountain uplands. Various types of Aridisols occur on the landscape because of lateral changes in particle size, truncation of diagnostic horizons, degradation of diagnostic horizons, moisture heterogeneity across the landscape, and age differences, which can range from Historical to Pliocene within small geographical areas (9, 18).

Arid soils that lack diagnostic subsurface horizons are generally classified as Entisols, which fall into the azonal concept of Sibirtsev (19). These include many of the younger soils on floodplains, dunes, and erosional surfaces. In Soil Taxonomy, floodplain soils are mainly classified as Fluvents or, more specifically, Torrifluvents (5). Arid soils associated with dunes are commonly Torripsamments and those associated with erosional surfaces are commonly Torriorthents.

Aridisols (14,942,000 km^2) and Entisols (12,682,000 km^2) are the dominant soil types in arid regions, although other soil types include Vertisols (889,000 km^2) and Oxisols (31,000 km^2) and very minor amounts of Mollisols, Andisols, Histosols, and Spodosols (1). Arid soils grade into semiarid soils across three climatic transects: laterally into wetter regions, upslope into wetter climates at higher elevations, or downslope into run-in areas with wetter microclimates. Taxonomically, changes in soil types from dry region aridic to wetter region ustic or xeric moisture regimes are expressed at the Suborder and Great Group level (Fig. 2). Linked to this climatic transition is a progressive change in vegetation—desert shrublands give way to grasslands that in turn give way to woodlands.

Also across this transition, soils have progressively deeper carbonate horizons. In the Chihuahuan Desert, for instance, carbonate zones are 50 cm deep at 230 mm of annual rainfall and 100 cm deep at 320 mm of annual rainfall (20). Likewise, gypsum zones progressively deepen from about 50 cm depth at 150 mm of annual rainfall to about 100 cm depth at 250 mm annual rainfall (21). Accompanying an increase in rainfall is an increase in soil organic matter. Although the amount of organic matter depends on the clay content, organic matter ordinarily increases from less than 0.5% in A-horizons of arid shrubland soils to 2–5% in semiarid and subhumid grassland soils (22).

ECOLOGICAL SIGNIFICANCE

Biodiversity in arid regions is linked to habitat diversity. Habitat diversity, in turn, is created by various microclimates caused by topographic factors and soil properties (23). Thus, soils help mold and are molded by ecosystems. In many arid regions, such as the southwestern United States, soils have been impacted by ecosystem changes of the Holocene and late Pleistocene when wetter climates alternated with drier climates (24–27). According to this model, landscape stability was greater during wetter climates because denser vegetative cover reduced erosion.

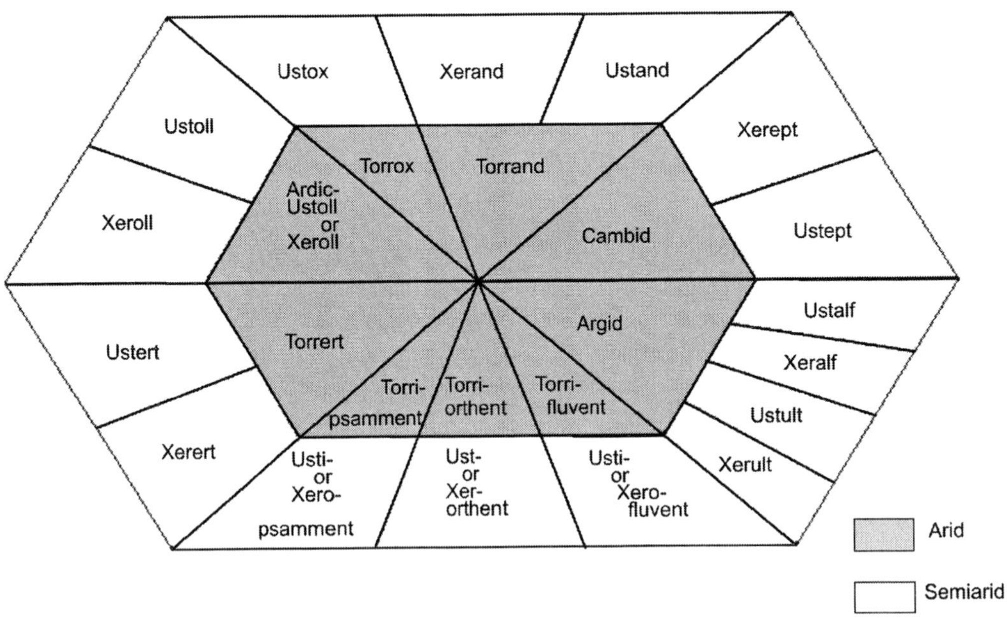

Fig. 2 Illustration of Soil Taxonomy Suborders and Great Groups that have aridic moisture regimes (shaded) and their moister counterparts that have ustic and xeric moisture regimes (From Ref. 5.)

With reduced erosion, soil formation occurred. In contrast, instability was greater during drier climates because sparse vegetative cover gave rise to more bare ground and increased erosion. With increased erosion, soil formation was inhibited. This oscillation between stability and instability is recorded as stacked sequences of buried paleosols in depositional environments and as stepped sequences of fan-terraces in areas that grade to fluctuating river base-levels.

Globally, arid soils affect atmospheric dust, rain chemistry, ocean fertilization, albedo, denitrification, and the carbon cycle as both sinks and sources of CO_2 (28, 29). Carbonate–carbon, for instance, is the second largest terrestrial carbon pool, totaling approximately 50–60 Pg C in the dryland zones of the U.S. (30) and approximately 750–950 Pg C in the dryland zones of the world (31).

Humans have lived on arid soils for millennia. In fact, the oldest known hominid tools are in arid East Africa and date back 2.5 million years (32). Today arid soils are important to humans for livestock grazing, irrigated agriculture, and urban development. In many arid regions of the world, human land use has resulted in desertification and diminishing groundwater supplies, both of which are increasingly important social and scientific issues as human population increases.

ACKNOWLEDGMENTS

Grateful acknowledgment is made to Haiyang Xing and Marco Inzunza for making the figures and Rebecca Kraimer for reviewing the manuscript.

REFERENCES

1. Wilding, L.P. Introduction: General Characteristics of Soil Orders and Global Distributions. In *Handbook of Soil Science*; Sumner, M.E., Ed.; CRC Press: Boca Raton, FL, 2000; E175–E182.
2. Bull, W.B. *Geomorphic Responses to Climatic Change*; Oxford University Press: New York, 1991; 326.
3. Dick-Peddie, W.A. Semiarid and Arid Lands: A Worldwide Scope. In *Semiarid Lands and Deserts*; Skujiφš, J., Ed.; Marcel Dekker: New York, 1991; 3–32.
4. Strahler, A.N.; Strahler, A.H. *Modern Physical Geography*, 3rd Ed.; Wiley: New York, 1987; 544.
5. Soil Survey Staff, *Soil Taxonomy—A Basic System of Soil Classification for Making and Interpreting Soil Surveys*, 2nd Ed.; USDA Agriculture Handbook Number 436; U.S. Govt. Printing Office: Washington, DC, 1999.
6. Durant, W. *Our Oriental Heritage*; Simon and Schuster: New York, 1935; 1049.
7. Dregne, H.E. *Soils of Arid Regions*; Elsevier: Amsterdam, 1976; 237.
8. Peterson, F.F. *Landforms of the Basin and Range Province Defined for Soil Survey*; Nevada Agricultural Experiment Station, Tech. Bull. 28, Univ. of Nevada: Reno, 1981; 52.
9. Gile, L.H.; Hawley, J.W.; Grossman, R.B. *Soils and Geomorphology in the Basin and Range Area of Southern New Mexico—Guidebook to the Desert Project*; New Mexico Bureau of Mines and Mineral Resources, Memoir 39, Socorro: New Mexico, 1981; 222.
10. Gile, L.H.; Peterson, F.F.; Grossman, R.B. Morphology and Genetic Sequences of Carbonate Accumulation in Desert Soils. Soil Sci. **1966**, *101*, 347–360.
11. Gile, L.H. Eolian and Associated Pedogenic Features of the Jornada Basin Floor, Southern New Mexico. Soil Sci. Soc. Am. J. **1999**, *63*, 151–163.
12. McFadden, L.D.; Wells, S.G.; Jercinovich, M.J. Influences of Eolian and Pedogenic Processes on the Origin and Evolution of Desert Pavements. Geology. **1987**, *15*, 504–508.
13. Kidron, G.J.; Yaalon, D.H.; Vonshak, A. Two Causes for Runoff Initiation on Microbiotic Crusts: Hydrophobia and Pore Clogging. Soil Sci. **1999**, *164*, 18–27.
14. Ludwig, J.A. Primary Productivity in Arid Lands: Myths and Realities. J. Arid Environ. **1987**, *13*, 1–7.
15. Whitford, W.G.; Schaefer, D.; Wisdom, W. Soil Movement by Desert Ants. The Southwestern Naturalist. **1986**, *31*, 273–274.
16. Smith, G.D. *The Guy Smith Interviews: Rationale for Concepts in Soil Taxonomy*; Soil Management Support Service Monograph No. 11, U.S. Government Printing Office: Washington, 1986.
17. Ahrens, R.J.; Eswaran, H. The International Committee on Aridisols: Deliberations and Rationale. Soil Surv. Horizons. **2000**, *41*, 110–117.
18. Gile, L.H. Causes of Soil Boundaries in an Arid Region; I. Age and Parent Materials. Soil Sci. Soc. Am. Proc. **1975**, *39*, 316–323.
19. Sibirtsev, N.M. *Selected Works, Vol. 1, Soil Science*; Issued in Translation by the Israel Program for Scientific Translation: Jerusalem, 1966; 354.
20. Gile, L.H. Holocene Soils and Soil-Geomorphic Relations in a Semi-Arid Region of Southern New Mexico. Quatern. Res. **1977**, *7*, 112–132.
21. Cooke, R.; Warren, A.; Goudie, A. *Desert Geomorphology*; UCL Press: London, 1993; 526.
22. Birkeland, P.W. *Soils and Geomorphology*, 3rd Ed.; Oxford University Press: New York, 1999; 430.
23. McAuliffe, J.R. Landscape Evolution, Soil Formation, and Ecological Patterns and Processes in Sonoran Desert Bajadas. Ecol. Monogr. **1994**, *64*, 111–148.
24. Ruhe, R.V. Age of the Rio Grande Valley in Southern New Mexico. J. Geol. **1962**, *70*, 151–167.
25. Gile, L.H.; Hawley, J.W. Periodic Sedimentation and Soil Formation on an Alluvial-piedmont in Southern New Mexico. Soil Sci. Soc. Am. Proc. **1966**, *30*, 261–268.
26. Monger, H.C.; Cole, D.R.; Gish, J.W.; Giordano, T.H. Stable Carbon and Oxygen Isotopes in Quaternary Soil Carbonates as Indicators of Ecogeomorphic Changes in

the Northern Chihuahuan Desert, USA. Geoderma **1998**, 137–172.
27. Buck, B.J.; Monger, H.C. Stable Isotopes and Soil-Geomorphology as Indicators of Holocene Climate Change, Northern Chihuahuan Desert. J. Arid Environ. **1999**, *43*, 357–373.
28. Schlesinger, W.H.; Reynolds, J.F.; Cunningham, G.L.; Huenneke, L.F.; Jarrell, W.M.; Virginia, R.A.; Whitford, W.G. Biological Feedbacks in Global Desertification. Science **1990**, *247*, 1043–1048.
29. Schlesinger, W.H. *Biogeochemistry: An Analysis of Global Change*, 2nd Ed.; Academic Press: New York, 1997.
30. Monger, H.C.; Martinez-Rios, J.J. Inorganic Carbon Sequestration in Grazing Lands. In *The Potential of U.S. Grazing Lands to Sequester Carbon and Mitigate the Greenhouse Effect*; Follette, R.F., Kimble, J.M., Lal, R., Eds.; CRC Press: Boca Raton, FL, 2001; 87–118.
31. Eswaran, H.; Reich, P.F.; Kimble, J.M.; Beiroth, F.H.; Radmanabhan, E.; Moncharoen, P. Global Carbon Sinks. In *Global Climate Change and Pedogenic Carbonates*; Lal, R., Kimble, J.M., Eswarar, H., Stewart, B.A., Ed.; CRC Press: Boca Raton, FL, 2000; 15–26.
32. Ambrose, S.H. Paleolithic Technology and Human Evolution. Science **2001**, *291*, 1748–1753.

ATTERBERG LIMITS

Thomas Baumgartl
Institute for Plant Nutrition and Soil Science, Kiel, Germany

A

INTRODUCTION

Soil is exposed to different states of stability depending on the amount of water that it contains. This characteristic is described as consistency (refer to entry on soil consistency and plasticity) and specifies the state of a remolded and cohesive soil in the range from the liquid (when wet) to plastic and finally solid (when dry) state.

Different soils contain a specific amount of water at these different states of stability. In 1911, the Swedish soil physicist Atterberg developed a classification system and method with which these states of consistency could be determined. The method is based on the determination of the water content [calculated as: (mass of water)/(dry mass of soil)] at distinct transitions between different states of consistency of soil. These transitions are defined as liquid limit, plastic limit and shrinkage limit, and are generally referred to as Atterberg limits. The values for these limits are dependent on various soil parameters, e.g., particle size, specific surface area of the particles which is able to take up water, and hence its particle size distribution. These limits are used to derive indices, e.g., index of plasticity and index of consistency, and are often used for the mechanical characterization of soils.

DEFINITION OF LIMITS AND THEIR DETERMINATION

Liquid Limit (Upper Plastic Limit) w_L

The liquid limit describes the transition from a viscous liquid to a plastic state. Soils with a water content at the liquid limit barely flow under an applied force. The associated capillary forces of the water menisci in the unsaturated pore system are equivalent to pF $\sim$ 0.5 ($\sim$0.3 kPa matric potential) (1).

The liquid limit is determined by a method and device developed by Casagrande. The principle is to find the water content (kg/kg) at which a soil sample starts to liquify under a small applied stress. In practice a groove is cut into soil samples with different water contents.

These soil samples are then exposed to a small standardized force by repeatedly dropping the Casagrande cup over a distance of 10 mm until the groove is close to ca. 10 mm. A semilogarithmic plot of the number of blows as a function of water content will result in a straight line. The liquid limit is defined as the water content at 25 blows (Fig. 1).

Plastic Limit (Lower Plastic Limit) w_P

The plastic limit determines the transition from a plastic (cohesive) to a semirigid or brittle state. Under an applied force cracking will occur. In sandy soils the plastic limit often cannot be determined. The matric potential at the plastic limit is the main cohesive stress and ranges between 63 and 200 kPa (pF 2.8–3.3) (1).

The plastic limit is determined by forming a moist ball from 2 to 3 g of soil, which is then rolled on a piece of frosted glass to a rod of thickness ca. 3 mm. The remolding and rolling is repeated until the 3 mm rod starts to break up into pieces of 10–20 mm. The gravimetric water content (kg/kg) at this point gives the plastic limit.

Shrinkage Limit w_S

Cohesive and remolded soils reduce their soil volume with the loss of water due to capillary forces. If in a drying process the reduction of the total soil volume equals the volume of water loss, then the soil shows normal shrinkage. Below a certain water content, the further shrinkage of the soil volume is restricted due to a high number of particle contact points and high effective stresses. This restricted shrinkage pattern is called residual shrinkage (Fig. 2). The transition from normal to residual shrinkage defines the shrinkage limit.

The shrinkage limit is often calculated by

$$w_S = 0.65 \times w_P$$

Index of Plasticity I_P

The index of plasticity is the amount of water between plastic and liquid limit and is calculated by

$$I_P = w_L - w_P$$

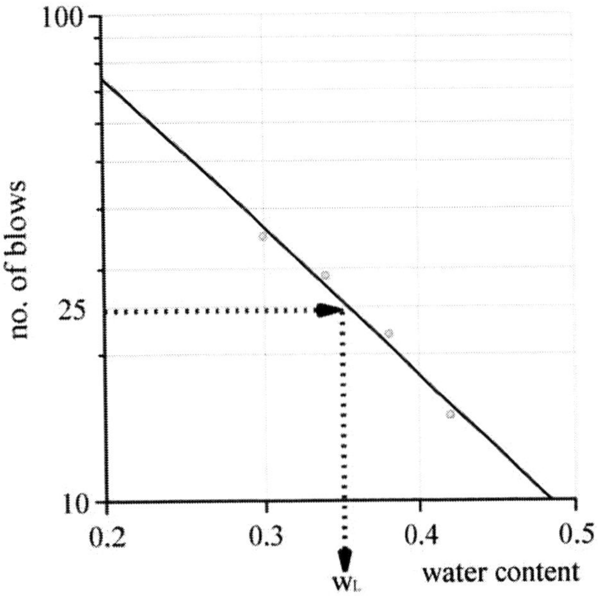

Fig. 1 Determination of the liquid limit.

It describes the sensitivity in the mechanical behavior of a soil towards changes in water content. However, it does not explain mechanical stability as hydraulic parameters are not included which are necessary as water flow becomes important when stresses are applied on a soil sample. Values for the index may range from 0 (no plastic behavior) for sandy material to 1 (100%) for clay.

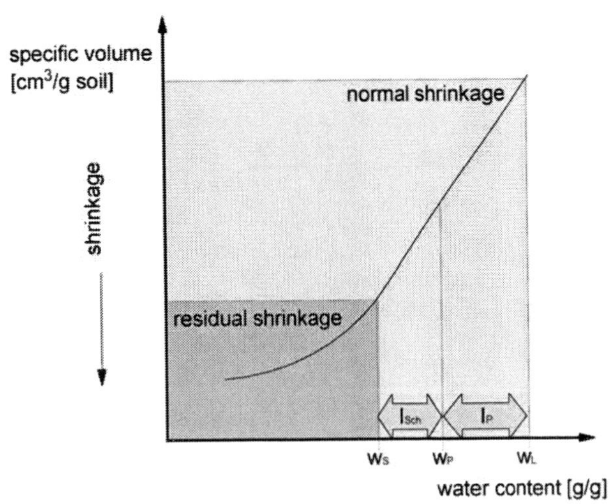

Fig. 2 Shrinkage behavior with change in water content; relations to Atterberg limits w_S, w_P, w_L and Atterberg indices. (From Ref. 9.)

Index of Consistency I_C

The index of consistency is determined as the ratio of the difference between the liquid limit and actual water content and the index of plasticity:

$$I_C = \frac{w_L - w}{w_L - w_P} = \frac{w_L - w}{I_P}$$

The water content w as such provides no information about the consistency of soils. The same water content in a sandy soil may reflect a liquid state, where as in a clay soil the behavior may be brittle. The index of consistency normalizes the water contents and characterizes whether the soil is close to the plastic limit (max $I_C = 1$) or the liquid limit (min $I_C = 0$).

Index of Shrinkage I_{Sch}

The difference between the plastic limit and the shrinkage limit results in the index of shrinkage:

$$I_{Sch} = w_P - w_S$$

Soils with water contents within this range are most suitable for cultivation (see Fig. 2). Table 1 summarizes some general values for the Atterberg limits and indices (8).

FACTORS INFLUENCING THE ATTERBERG LIMITS

Many mechanical processes are linked to hydrological properties of soils. Therefore, values of the limits and indices are influenced by factors which are generally important for the water retention curve, e.g., the capacity of swelling and shrinkage, clay content, type of clay minerals and organic matter.

Generally, the values of the limits and indices increase with their clay content. As the liquid limit increases in comparison to the plastic limit, the index of plasticity also

Table 1 Consistency limits (g water/g soil) for different soil textures

Texture	Sand	Silt	Clay
Consistency limits			
Liquid limit	0.15–0.20	0.30–0.40	0.40–1.50
Plastic limit	0	0.20–0.25	0.25–0.5
Index of plasticity	0	0.10–0.15	0.10–1.00
Shrinkage limit	0.12–0.18	0.14–0.25	0.08–0.25

(From Ref. 8.)

Table 2 Mechanical properties of soils and Atterberg limits

Index of consistency		0	0.25	0.5		0.75		1		1.3	
Symbol		w_L						w_P		w_S	
Index					Index of plasticity				Index of shrinkage		
Description	Slurry	Very soft		Soft	Deformable		Stiff		Medium hard		Hard
Unconfined compression strength (kPa)		<25	25–50		50–100		100–200		200–400		>400
Cultivation (pos., +; neg., −)	–	–			–		+			–	

(From Refs. 2, 9.)

increases. The swelling and shrinkage intensity is dependent not only on the amount but also on the type of clay mineral. Skempton introduced a factor described as the activity of clay (2, 9):

$$A = \frac{I_P}{\%\ \text{clay content}}.$$

The values of A can be classified as: 1) $A > 1.25$: active soil with high capacity of swelling and shrinking [Ca-montmorillonite ($A \sim 1.5$), Na-montmorillonite ($A \sim 7.5$), smectite, salt influenced clays]); 2) $0.75 < A < 1.25$: normal soils (illite); 3) $A < 0.75$: inactive clay with only little swelling–shrinking activity (kaolinite).

Organic matter increases both the plastic and liquid limits, but does not have a big effect on the index of plasticity. Organic substances in a soil matrix seem therefore to increase the surface hydration. Once this pool for water uptake is saturated, the soil shows the same mechanical behavior towards changes in water content as when organic matter is absent, only at a higher level of water content. Thus, the index of consistency is higher (1).

With the exception of organic matter and clays, the amount and type of exchangeable cations have a significant effect on the value of the limits (3, 5, 12). Na-saturated soils reduce the liquid limit, but increase the shrinkage limit. Soils therefore have the tendency to show crust formation at an earlier stage and will slake at lower water contents (1, 8).

APPLICATION

Soil charcteristics are inherent in the values of the Atterberg limits. Therefore, Atterberg limits are correlated with soil properties. For specific soils, investigations have been carried out which correlate the total particle surface with the plastic limit (4, 6). Soil strength can be defined by its compressibility and compressibility is correlated to the Atterberg limits (11). As soil strength is influenced to a great extent by the energy status of the capillary water, Atterberg limits reflect the soil water potential and show the dependency on texture and the water retention curve (10).

At a broader scale, the Atterberg limits can be used for the evaluation of the trafficability and cultivation of soils. Table 2 lists the limits and derived mechanical properties and qualities of soil substrates.

From this classification (Table 2), it is evident that an optimal range of water contents for agricultural use can be determined. This range is present when the soil is stiff and has a compression strength of >100 kPa and an index of consistency between 0.75 and 1. Drier soils increase the energy input needed for cultivation, which can be a serious problem for clay-rich soils as plowing can become difficult. In the case of lower than optimal plasticity indices, the soil structure can be destroyed easily when the

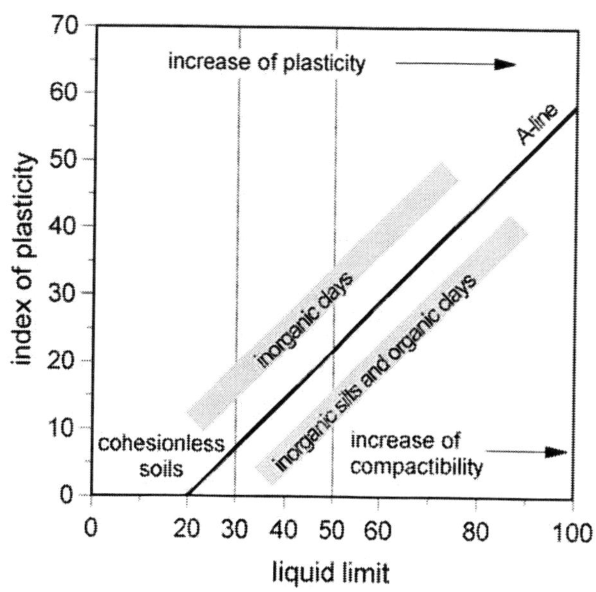

Fig. 3 Classification of soils according to Atterberg limits and Casagrande A-line.

Table 3 Mechanical parameters (angle of internal friction, cohesion) dependent on texture and index of plasticity

Texture	I_P	Angle of internal friction ϕ' (°)	Cohesion (kPa)
Clay of high plasticity: $w_L > 0.5$	0.50–0.75	17.5	0
	0.75–1.00	17.5	10
	1.00–1.30	17.5	25
Clay and silt of medium plasticity: $0.35 < w_L < 0.5$	0.50–0.75	22.5	0
	0.75–1.00	22.5	5
	1.00–1.30	22.5	10
Clay and silt of low plasticity: $w_L < 0.35$	0.50–0.75	27.5	0
	0.75–1.00	27.5	2
	1.00–1.30	27.5	5

(Adapted from Ref. 9.)

soil is kneaded by trafficking resulting in ecological problems. As a result, the hydraulic conductivity and gas flow as well as nutrient uptake of plants can be reduced. Hence, cultivation at index of consistency smaller than 0.75 can have a serious effect on plant growth and soil biological activity.

Although the values of the limits and indices are not independent values, they can be related to each other (e.g., I_P and w_L). Classifications with respect to particle size distribution, geological origin of material, and suitability under soil mechanical point of view can be derived thereafter. With the ratio of liquid limit and index of plasticity a linear relationship was found by Casagrande and described as A-line (13), following the equation:

$$I_P = 0.73(w_L - 0.2)$$

It distinguishes soil with content of organic matter of <5% above this line from those in the range of 5–30% below the line. Certain groups of soils and soil substrates can be classified as shown in Fig. 3.

Soil mechanical parameters including angle of internal friction and cohesion (the principle parameters for defining soil mechanical stability) can be as well related to the Atterberg limits (7, 14). Generally, the angle of internal friction increases with decreasing soil substrate plasticity, i.e., with decreasing liquid limits. Cohesion increases with increasing index of plasticity, the increase however is less with decreasing plasticities (see Table 3).

LIMITATIONS

Atterberg limits are empirical values developed for engineering purposes. They describe in approximation the change of the properties of soil material. The limits reflect not a sudden change in the mechanical state of soils but rather transitions. The determination of the limits do not describe processes, but only categories for the classification of the soil. As denoted, capillary forces, tensile stresses and volume change due to external and internal stresses are the parameters which determine the material behavior of the soil substrates. The determination of the Atterberg limits requires the destruction of structural units within the soil and remolding. Usually structure formation increases the strength of soils mainly by reorientation of particles and formation of bondings between particles. The interpretation of the Atterberg limits under agricultural–technical aspects should take these considerations into account.

REFERENCES

1. Baver, L.D.; Gardner, W.H.; Gardner, W.R. *Soil Physics*; Wiley: New York, 1972; 498.
2. Chancellor, W.J. Soil Physical Properties. ASAE Monograph. In *Advances in Soil Dynamics*, 12th Ed.; DeVore-Hansen, P., Ed.; American Society of Agricultural Engineers, 1994; Vol. 1.
3. Dexter, A.R.; Chan, K.Y. Soil Mechanical Properties as Influenced by Exchangeable Cations. J. Soil Sci. **1991**, *42* (2), 219–226.
4. Farrar, D.M.; Coleman, J.D. The Correlation of Surface Area with Other Properties of Nineteen British Clay Soils. J. Soil Sci. **1967**, *18*, 118–124.
5. Gill, W.R.; Reaves, C.A. Relationship of Atterberg Limits and Cation Exchange Capacity to Some Physical Properties of Soil. Soil Sci. Soc. Proc. **1957**, *21*, 491–497.
6. Hammel, J.E.; Sumner, M.E.; Burema, J. Atterberg Limits as Indices of External Surface Areas of Soils. Soil Sci. Soc. Am. **1983**, *47*, 1054–1056.
7. Kanji, M.A. The Relationship Between Drained Friction Angles and Atterberg Limits of Natural Soils. Geotechnique **1974**, *24* (4), 671–674.

8. Kezdi, A. Bodenphysik. *Handbuch der Bodenmechanik*; VEB Verlag fuer Bauwesen: Berlin, 1969; 259.
9. Kretschmer, H. Koernung Und Konsistenz. In *Handbuch der Bodenkunde*; Blume, H.P., Felix-Henningsen, P., Fischer, W.R., Frede, H.-G., Horn, R., Stahr, K., Eds.; Landsberg/Lech: Ecomed, 1997; 1–45.
10. McBride, R.A. A Re-Examination of Alternative Test Procedures for Soil Consistency Limit Determination: II. A Simulated Desorption Procedure. Soil Sci. Soc. Am. **1989**, *53*, 184–191.
11. McBride, R.A.; Bober, M.L. A Re-Examination of Alternative Test Procedures for Soil Consistency Limit Determination: I. A Compression-Based Procedure. Soil Sci. Soc. Am. **1989**, *53*, 178–183.
12. Smith, C.W.; Hadas, A.; Dan, J.; Koyumdjisky, H. Shrinkage and Atterberg Limits in Relation to Other Properties of Principal Soil Types in Israel. Geoderma **1999**, *35* (1), 47–65.
13. Terzaghi, K.; Peck, R.B. Soil Mechanics in Engineering Practice. Wiley: New York, 1967; 729.
14. Voight, B. Correlation Between Atterberg Plasticity Limits and Residual Shear Strength of Natural Soils. Geotechnique **1973**, *23* (2), 265–267.

A

BACTERIA

Mary Ann Bruns
The Pennsylvania State University, University Park, Pennsylvania, U.S.A.

INTRODUCTION

Soil bacteria are microscopic, single-celled organisms that inhabit terrestrial environments. They are prokaryotes (i.e., organisms having chromosomes not enclosed by nuclear membranes) belonging to the domain Bacteria, one of three fundamental domains of life (1). Bacteria coexist in soils with representatives of the other two domains, Archaea (also prokaryotic) and Eukarya (i.e., organisms having membrane-enclosed chromosomes), such as fungi and protozoans (2, 3). A handful of garden soil contains billions of bacterial cells, most ranging from 0.2–1 μm in size (4). When observed under the microscope, most soil bacteria appear round or rod-shaped, but such physical similarity masks the vast diversity in their genetic composition and biochemical capabilities (5). Less than 3% of bacteria observable by microscopy in soils can be grown in laboratory cultures (4, 6), so that most soil bacteria are thought to be dormant or dependent on other growth conditions. Analysis of deoxyribonucleic acids (DNA), extracted directly from soil after lysing organisms (7), indicates coexistence of thousands of different bacterial species (8). Under favorable conditions of moisture, temperature, and nutrients, bacterial numbers range from 10^8–10^{10} per gram of soil (2, 3). Bacteria are nonuniformly distributed in soil aggregates. They often reside in pores as microcolonies of typically 2–10 bacteria each (9), tightly associated with nutrient-rich clays and decaying organic matter (Fig. 1). The presence of active bacteria in soils can be demonstrated by oxygen consumption or carbon dioxide emission during respiration. Bacteria cultured from soils are frequently aerobic heterotrophs belonging to two major groups: Gram-positive bacteria and Proteobacteria (2, 3). Other, apparently abundant bacterial groups have been detected in soils with molecular methods (10, 11) but have yet to be cultured, including *Acidobacterium*, planctomycetes, and other less-studied taxa (12–15). Less abundant bacterial groups in soils include specialists and chemolithotrophs, which use inorganic compounds for energy (5). Soil bacteria therefore play important ecological roles by transforming, storing, and releasing elements required in primary and secondary production. The presence and abundance of different bacterial groups (Table 1) and their levels of activity in soils depend on specific soil properties and land-use history.

HISTORICAL BACKGROUND

Classical Microbiology

Soil bacteria are principal subjects of study in the field of *soil microbiology*, defined as "the branch of soil science concerned with soil-inhabiting microorganisms, their functions, and activities" (16). Two key developments made learning about bacteria in soils possible: the microscope, developed by Antonie van Leeuwenhoek (1632–1723); and laboratory culture methods, developed by Louis Pasteur (1822–1895) and Robert Koch (1843–1910). The latter methods were important, because the study of specific bacteria requires handling procedures (i.e., aseptic technique) that exclude other contaminating microorganisms (5). Such developments opened the way for extensive research on pure and mixed cultures by pioneers Sergei Winogradsky, known as the "Father of Soil Microbiology," and Martinus Beijerinck, of the Delft School of Microbiology in the Netherlands (2, 5). Winogradsky (1856–1953) elucidated many different biochemical processes performed by bacteria in soils, including fixation of atmospheric nitrogen to produce ammonia; nitrification (conversion of ammonia to nitrite and nitrate); utilization of carbon dioxide for cell synthesis (autotrophy); and oxidation-reduction of sulfur and iron. Beijerinck (1851–1931) isolated cultures of bacteria that could fix nitrogen either symbiotically in association with plant roots (e.g., *Rhizobium*) or as free-living organisms in the soil (e.g., *Azotobacter*).

Culture- and Process-Based Microbiology

Koch's techniques were originally used to isolate and grow bacteria that were pathogenic (disease-causing) to humans (5). Pure bacterial cultures were grown in petri dishes

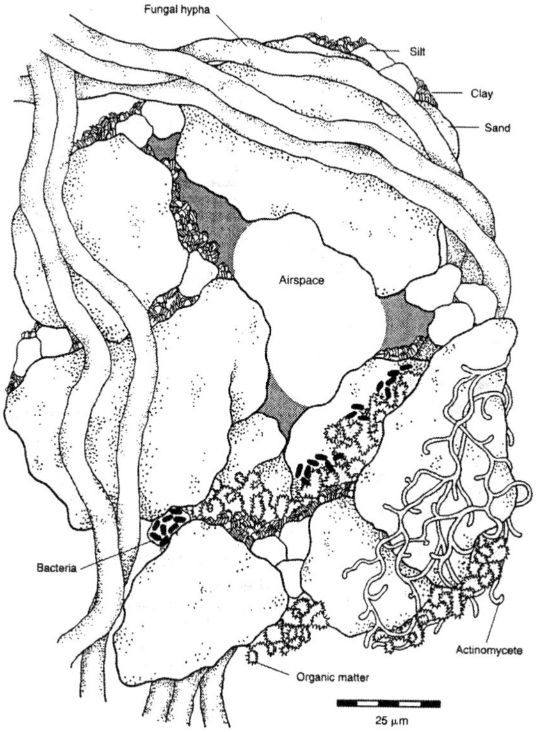

Fig. 1 Scale drawing of a typical soil aggregate. Sand (>50 μm), silt (2–50 μm), and clay (<2 μm) particles become cemented together with organic matter and microbial substances as a result of microbial activity. Note the pore in the center of the aggregate and the meniscus of water surrounding the airspace inside the pore. Bacteria are shown as dark rods and filaments (actinomycetes). Fungal filaments (hyphae) surround the aggregate. This illustrates how an aggregate can offer a diverse set of microsites for bacterial habitation over very short distances. (From Ref. 2.)

containing gels made with agar, a nondegradable seaweed material found to support bacterial growth better than starch or gelatin, both of which were consumed and liquefied by bacteria. Use of agar thus made it possible to separate and distinguish bacterial *colonies* (i.e., visible masses arising from one individual cell). Despite the early successes of Koch's techniques, Selman Waksman (1888–1973) at Rutgers University (2) was one of the first to recognize that not all soil bacteria could be grown in laboratory cultures. Discrepancies between plate counts and microscopic counts were at first attributed to dead and injured cells. Subsequent use of fluorescent stains to distinguish live from dead cells (17) indicated high proportions of viable, but nonculturable bacteria (18) in many environments, including soils (4, 6). Some of the species that do not grow in the laboratory may actually require unknown, alternative growth conditions (e.g., trace elements, pH, ionic strength, oxygen concentration). Other bacteria may require the activity of other species (i.e., syntrophs) in order to grow (18).

Recognition that culture-based methods were not comprehensive led to greater emphasis on measuring *in situ* processes in soils. Such measurements are conducted in the field or under controlled laboratory conditions in soil microcosms. In some cases, bacterial-process measurements require selective inhibition of fungi by antibiotics like cycloheximide (5). Process measurements relevant to carbon cycling include bacterial respiration (consumption of oxygen or emission of carbon dioxide); organic matter decomposition; biomass production (increases in carbon storage within cells); and methane production and consumption (5). Measurements of nitrogen-cycling processes include nitrogen fixation; mineralization (release of mineral forms of nitrogen during decomposition of organic matter); nitrification (oxidation of ammonium to nitrite and nitrate) (19); and denitrification (conversion of nitrate or nitrite to reduced forms of nitrogen, including the trace gases, NO and N_2O) (20). Chromatographic instruments, which provide fast and accurate measurements of trace gases, are also used to track products of pollutant biodegradation (21) and antibiotic synthesis in soils (22).

Molecular Microbiology

The discovery of DNA as the genetic information carrier in all organisms led to recognition that different species possess unique gene compositions. Vigdis Torsvik and coworkers at the University of Bergen, Norway, were the first to analyze DNA from soils to assess bacterial diversity. They estimated that 1 g of Norwegian forest soil harbored 4000 to 10,000 different bacterial species (8). Further advances in technologies to study bacteria without culturing resulted from the work of Carl Woese, who compared variation in ribosomal RNA (rRNA) genes to discern evolutionary relationships among all organisms (1, 23). Woese focused on these genes (which code for RNA components of ribosomes, the cell's protein-synthesizing organelles), because their sequences are *conserved* (i.e., slow-to-change over time) and *universal* (i.e., found in all organisms). The rRNA-based tree of life (5) consists of a single domain of eukaryotic organisms (Eukarya) and two prokaryotic domains (Archaea and Bacteria). Organisms formerly thought to be bacteria (e.g., methane-producing methanogens) are now classified as Archaea, a group that is as genetically distinct from bacteria as it is from animals and other eukaryotes (1, 23). Organisms representing all three domains coexist in soils. Approximately 1% of DNA or RNA extracted from soils

Table 1 Bacterial functional groups in soils and representative divisions/subdivisions and genera

Process in soil	Bacterial functional groups	Divisions/subdivisions and genera
Processes affecting primary production	Autotrophic blue-green "algae"	Cyanobacteria (*Anabaena*, *Nostoc*)
	Nitrogen-fixing plant symbionts	Alpha Proteobacteria (*Rhizobium*, *Bradyrhizobium*)
		High GC Gram-Positive (*Frankia*)
	Plant endophytes (reside in plant tissue)	Alpha Proteobacteria (*Agrobacterium*)
	Plant-associated beneficial bacteria	Alpha Proteobacteria (*Azospirillum*)
		Gamma Proteobacteria (*Pseudomonas*)
	Plant pathogens	High GC Gram-Positive (*Corynebacterium*)
		Gamma Proteobacteria (*Xanthomonas*, *Erwinia*)
Secondary production and carbon cycling	Cellulose degraders (possess cellulytic enzymes)	Cytophaga-Flexibacter-Bacteroides (*Cytophaga*)
		High GC Gram-Positive (*Cellulomonas*)
	Decomposers and mineralizers of organic matter	High GC Gram-Positive (*Arthrobacter*)
		Low GC Gram-Positive (*Bacillus*)
	Rhizosphere-associated	Beta Proteobacteria (*Comamonas*)
		Gamma Proteobacteria (*Pseudomonas*, *Klebsiella*)
	Predators of other bacteria	Alpha Proteobacteria (*Ensifer*)
		Beta Proteobacteria (*Cupriavidus*)
		Delta Proteobacteria (*Bdellovibrio*)
	Methanotrophs	Alpha Proteobacteria (*Methylosinus*, *Methylobacterium*)
		Beta Proteobacteria (*Methylobacillus*, *Methylophilus*)
		Gamma Proteobacteria (*Methylococcus*, *Methylomonas*)
Nitrogen cycling	Free-living nitrogen fixers	Gamma Proteobacteria (*Azotobacter*)
	Autotrophic nitrifiers (Chemolithotrophic ammonia and nitrite oxidizers)	Alpha Proteobacteria (*Nitrobacter*)
		Beta Proteobacteria (*Nitrosospira*, *Nitrosomonas*)
		Gamma Proteobacteria (*Nitrococcus*)
	Heterotrophic nitrifiers	High GC Gram-Positive (*Arthrobacter*)
	Denitrifiers	Beta Proteobacteria (*Alcaligenes*, *Azoarcus*)
		Gamma Proteobacteria (*Pseudomonas*)
Soil structure formation	Mucigel producers	Beta Proteobacteria (*Chromobacterium*)
		Gamma Proteobacteria (*Pseudomonas*)
	Mycelium (filament) producers	High GC Gram-Positive (*Agromyces*, *Actinomyces*)
	Commensals and symbionts in soil fauna	Cytophaga-Flexibacter-Bacteroides (*Bacteroides*)
Pollutant degradation	Pesticide degraders	Beta Proteobacteria (*Burkholderia*)
		Gamma Proteobacteria (*Pseudomonas*)
Specialized processes	Sulfur oxidizers	Gamma Proteobacteria (*Thiobacillus*)
	Hydrogen oxidizers	Beta Proteobacteria (*Hydrogenophaga*)
	Antibiotic producers	High GC Gram-Positive (*Streptomyces*)
Unknown processes	Groups detected by molecular analysis but with unknown functions	Acidobacterium
		Verrucomicrobiales

(From Refs. 2, 3, 13, 27, 28.)
GC stands for "Guanine + Cytosine" content in DNA.

appears to be derived from a subgroup of Archaea, known as the Crenarchaeota (24, 25), none of which have been isolated in the laboratory. Activities and functions of the Crenarchaeota in soils are currently unknown.

In addition to identifying three fundamental domains of life, Woese and coworkers provided a systematic classification (i.e., phylogeny) of Bacteria and Archaea on the basis of their rRNA sequences (23, 26). Previously, bacteria had been classified into taxonomic groups by their microscopic morphologies (i.e., cell sizes, shapes, and appendages); positive or negative Gram-stain reactions based on different cell wall compositions (5); and biochemical tests in laboratory cultures (27). Although the new rRNA phylogeny supported some previous taxonomic distinctions (e.g., between Gram-positive and Gram-negative bacteria), it placed some bacteria with extremely different biochemical characteristics into the same group (23, 26). Proteobacteria, a group that is common and abundant in soils, for example, encompasses more than 350 known genera (28, 29), ranging from chemolithotrophs (e.g., ammonia oxidizers, hydrogen oxidizers) to obligate symbionts, which can only reproduce inside specific host organisms (5, 27). Variations in rRNA genes also led to identification of *signature* sequences for narrow or broad groups (11, 23). These sequences can be synthesized and used as probes to detect complementary DNA of bacteria belonging to the various groups (10, 26). Signature sequences have also been used as specific *primers* to copy rRNA genes from soil DNA extracts with polymerase chain reaction, or PCR (11). Analysis of DNA extracted from soils has detected many novel soil bacteria that cannot be grown in the laboratory. Extensive phylogenetic studies in specific soils, based on hundreds of different rRNA sequences retrieved by PCR, indicate that uncultured bacterial groups are abundant and widespread in soils (e.g., *Acidobacterium*, planctomycetes, *Verrucomicrobium*) (12–15).

BACTERIAL LIFE IN SOILS

Soil Microenvironment

The finding that thousands of different bacterial species can coexist within 1 cm^3 of soil reflects a high degree of spatial heterogeneity in soil *microniches*. Such diversity may also reflect bacterial persistence and longevity. Bacteria are introduced into soil habitats by plant litter and particulate matter deposited by wind, water, and animals. By producing mucilaginous polysaccharides (i.e., mucigels), bacteria become enmeshed with clay particles ($<0.2\ \mu m$) and organic matter to initiate soil aggregation (2, 3). Bacteria adhering to soil aggregates inside pores (Fig. 1) are protected from desiccation and grazing by protozoan predators. Extensive formation of soil aggregates improves soil structure and productivity by increasing porosity, thereby enhancing air and water infiltration. Well-aggregated soils may provide more-varied niches for bacteria, because large aggregates can exhibit steep gradients in redox potential (2, 5). Despite their functional importance, bacteria occupy no more than 0.4% of the total pore space in arable soils (3). Table 1 lists key functional groups of soil bacteria and representative phyla and genera.

Growth Requirements and Survival Strategies

Like other organisms, bacteria require energy, carbon, nitrogen, adequate moisture, and favorable environmental conditions for growth and reproduction (5). Bacterial growth in most soils is limited primarily by soluble carbon (3). Although the bacterium's small size and high surface-to-volume ratio (5) make it highly efficient in taking up nutrients, its limited mobility renders it dependent on the immediate environment for nutrient supply. It has been estimated that total carbon inputs from plant litter and root exudates in an agricultural soil may support only 4 to 5 bacterial growth cycles per year (3). Bacterial distribution in soil tends to reflect organic matter distribution, with bacterial numbers being highest in upper-soil horizons and decreasing with depth (2, 3). Certain bacterial populations, such as *Pseudomonas* spp., are typically 10 to 100 times higher in zones immediately surrounding plant roots (i.e., rhizospheres rich in root exudates) than in soil zones not influenced by roots (2, 3). Bacteria may survive extended periods of soil nutrient deficit by invoking the following mechanisms: induction of low metabolic rates to reduce energy requirements (30); accumulation of intracellular polymers for carbon storage; production of extracellular polysaccharides to envelope and protect cells from desiccation; release of extracellular enzymes to solubilize nutrients; and formation of spores, cysts, and other resting-state structures (5). Some bacteria survive for decades in dried soils (3).

Ecological Interactions

Bacteria participate in many kinds of ecological interactions in soils: as predators of other bacteria (31); as food sources for protozoans; as pathogens, parasites, or antagonists of other organisms (e.g., antibiotic producers); and as plant commensals or symbionts (5, 27).

Cooperative interactions may even result from the transfer of genes from one bacterium to another via plasmids (e.g., for pollutant degradation or heavy metal resistance) (32). Diverse soil bacteria possess genes to transform and extract energy from a variety of substrates under different environmental conditions. Such genetic potential has been described as the "metagenome" of the soil (33).

REFERENCES

1. Woese, C.R.; Kandler, O.; Wheelis, M.L. Towards a Natural System of Organisms: Proposal for the Domains Archaea, Bacteria, and Eucarya. P. Natl. Acad. Sci. USA **1990**, *87*, 4576–4579.
2. Sylvia, D.M.; Fuhrmann, J.J.; Hartel, P.G.; Zuberer, D.A., Eds.; *Principles and Applications of Soil Microbiology*; Prentice Hall: Upper Saddle River, New Jersey, 1999.
3. Paul, E.A.; Clark, F.E. *Soil Microbiology and Biochemistry*, 2nd Ed.; Academic Press: New York, 1996.
4. Bakken, L.R.; Olsen, R.A. The Relationship Between Cell Size and Viability of Soil Bacteria. Microbial. Ecol. **1987**, *13*, 103–114.
5. Madigan, M.T.; Martinko, J.M.; Parker, J. *Brock Biology of Microorganisms*, 9th Ed.; Prentice Hall: Upper Saddle River, New Jersey, 2000.
6. Faegri, A.; Torsvik, V.L.; Goksoyr, J. Bacterial and Fungal Activities in Soil: Separation of Bacteria and Fungi by a Rapid Fractionated Centrifugation Technique. Soil Biol. Biochem. **1977**, *9*, 105–112.
7. Ogram, A. Isolation of Nucleic Acids from Environmental Samples. In *Techniques in Microbial Ecology*; Burlage, R.S., Atlas, R., Stahl, D., Geesey, G., Sayler, G., Eds.; Oxford University Press: New York, 1998; 273–288.
8. Torsvik, V.; Daae, F.L.; Goksoyr, J. High Diversity in DNA of Soil Bacteria. Appl. Environ. Microb. **1990**, *56*, 782–787.
9. Foster, R.C. Microenvironments of Soil Microorganisms. Biol. Fert. Soils **1988**, *6*, 189–203.
10. Stahl, D.A. Molecular Approaches for the Measurement of Density, Diversity, and Phylogeny. In *Manual of Environmental Microbiology*; Hurst, C.J., Knudsen, G.R., McInerney, M.J., Stetzenbach, L.D., Walter, M.V., Eds.; American Society for Microbiology Press: Washington, DC, 1997; 102–114.
11. Stackebrandt, E.; Goddfellow, M., Eds. *Nucleic Acid Techniques in Bacterial Systematics*; John Wiley and Sons, Ltd: New York, 1991.
12. Liesack, W.; Stackebrandt, E. Occurrence of Novel Groups of the Domain Bacteria as Revealed by Analysis of Genetic Material Isolated from an Australian Terrestrial Environment. J. Bacteriol. **1992**, *174*, 5072–5078.
13. Hugenholtz, P.; Goebel, B.M.; Pace, N.R. Impact of Culture-Independent Studies on the Emerging Phylogenetic View of Bacterial Diversity. J. Bacteriol. **1998**, *180*, 4765–4774.
14. Kuske, C.R.; Barns, S.M.; Busch, J.D. Diverse Uncultivated Bacterial Groups from Soils of the Arid Southwestern United States that are Present in Many Geographic Regions. Appl. Environ. Microbiol. **1997**, *63*, 3614–3621.
15. Tiedje, J.M.; Asuming-Brempong, S.; Nusslein, K.; Marsh, T.L.; Flynn, S.J. Opening the Black Box of Soil Microbial Diversity. Appl. Soil Ecol. **1999**, *13*, 109–122.
16. Soil Science Society of America. *Glossary of Soil Science Terms*; Soil Science Society of America: Madison, Wisconsin, 1998.
17. Kepner, R.L.; Pratt, J.R. Use of Fluorochromes for Direct Enumeration of Total Bacteria in Environmental Samples: Past and Present. Microbiol. Rev. **1994**, *58*, 603–615.
18. Roszak, D.B.; Colwell, R.R. Survival Strategies of Bacteria in the Natural Environment. Microbiol. Rev. **1987**, *51*, 365–379.
19. Hart, S.C.; Hart, J.M.; Davidson, E.A.; Firestone, M.K. Nitrogen Mineralization, Immobilization, and Nitrification. In *Methods of Soil Analysis: II. Microbiological and Biochemical Properties*; Weaver, R.W., Angle, J.S., Bottomley, P.S., Eds.; Soil Science Society of America: Madison, Wisconsin, 1994; 985–1019.
20. Tiedje, J.M. Denitrifiers. In *Methods of Soil Analysis: II. Microbiological and Biochemical Properties*; Weaver, R.W., Angle, J.S., Bottomley, P.S., Eds.; Soil Science Society of America: Madison, Wisconsin, 1994; 245–267.
21. Alexander, M. *Biodegradation and Bioremediation*, 2nd Ed.; Academic Press: London, 1999; 453.
22. Thomashow, L.S.; Bonsall, R.F.; Weller, D.M. Antibiotic Production by Soil Rhizosphere Microbes *In Situ*. In *Manual of Environmental Microbiology*; Hurst, C.J., Knudsen, G.R., McInerney, M.J., Stetzenbach, L.D., Walter, M.V., Eds.; American Society for Microbiology Press: Washington, DC, 1997; 493–499.
23. Woese, C.R. Bacterial Evolution. Microbiol. Rev. **1987**, *51*, 221–271.
24. Sandaa, R.-A.; Enger, O.; Torsvik, V. Abundance and Diversity Of *Archaea* In Heavy-Metal-Contaminated Soils. Appl. Environ. Microb. **1999**, *65*, 3293–3297.
25. Buckley, D.H.; Graber, J.R.; Schmidt, T.M. Phylogenetic Analysis of Nonthermophilic Members of the Kingdom *Crenarchaeota* and Their Diversity and Abundance in Soils. Appl. Environ. Microb. **1998**, *64*, 4333–4339.
26. Pace, N.R.; Stahl, D.A.; Lane, D.J.; Olsen, G.J. The Analysis of Natural Microbial Communities by Ribosomal RNA Sequences. Appl. Microb. Ecol. **1986**, *9*, 1–55.
27. Krieg, N.R.; Sneath, P.H.A.; Staley, J.T.; Williams, S.T.; Holt, J.G., Eds. *Bergey's Manual of Systematic Bacteriology*; Lippincott William and Wilkins: Philadelphia, Pennsylvania, 1984; Vol. *1–4*.

28. Maidak, B.L.; Cole, J.R.; Lilburn, T.G.; Parker, C.T., Jr.; Saxman, P.R.; Stredwick, J.M.; Garrity, G.M.; Li, B.; Olsen, G.J.; Pramanik, S.; Schmidt, T.M.; Tiedje, J.M. The RDP (Ribosomal Database Project). Nucleic Acids Res. **2000**, *28*, 173–174.

29. Center for Microbial Ecology, Michigan State University, Ribosomal Database Project, http://www.cme.msu.edu/RDP/html (accessed August 2000).

30. Morita, R.Y. Bioavailability of Energy and the Starvation State. In *Starvation in Bacteria*; Kjelleberg, S. Ed.; Plenum Press: New York, 1993; 20–33.

31. Casida, L.E. Minireview: Nonobligate Bacterial Predation of Bacteria in Soil. Microbial. Ecol. **1988**, *15*, 1–8.

32. Daane, L.L.; Molina, J.A.E.; Berry, E.C.; Sadowsky, M.J. Influence of Earthworm Activity on Gene Transfer from *Pseudomonas fluorescens* to Indigenous Soil Bacteria. Appl. Environ. Microb. **1996**, *62*, 515–521.

33. Handelsman, J.; Rondon, M.R.; Brady, S.F.; Clardy, J.; Goodman, R.M. Molecular Biological Access to the Chemistry of Unknown Soil Microbes: A New Frontier for Natural Products. Chem. Biol. **1998**, *5*, R245–R249.

B

BIOLOGICAL NITROGEN FIXATION

Robert H. Burris
Professor Emeritus, Department of Biochemistry, University of Wisconsin-Madison, Madison, Wisconsin, U.S.A.

INTRODUCTION

For centuries farmers practiced mixed cropping of leguminous and non-leguminous plants without knowing the basis of the empirically observed benefit from the practice. The first clear recognition that leguminous plants could utilize nitrogen (N_2) from the air is attributed to Boussingault's report in 1838 (1). Skepticism remained until Hellriegel and Wilfarth in Germany in 1886 reported definitive experiments demonstrating that pea plants (*Pisum sativum*) in association with bacteria borne in their root nodules were capable of utilizing N_2 (1).

SYMBIOTIC AND FREE-LIVING NITROGEN FIXERS

The symbiotic process of N_2 fixation by leguminous plants plus associated root nodule bacteria (*Rhizobium* species) is of greatest practical importance to agriculture, but there also are a number of free-living bacteria capable of N_2 fixation. The first of these recognized was the anaerobic organism *Clostridium pasteurianum* as reported by Winogradsky in 1893 (2). In 1901 Beijerinck recorded that the aerobic bacterium *Azotobacter chroococcum* also fixed N_2 (2). These early observations were followed by practical applications, and the practice of inoculation of leguminous seeds with cultures of the rhizobia became widespread. The bacteria were grown commercially and farmers applied these bacteria to leguminous seeds at the time of planting. There is a specificity between the plants and bacteria, and this gave rise to the recognition of cross-inoculation groups, i.e., groups of plants all of which are nodulated by the same species of rhizobia. For example, *Rhizobium leguminosarum* infects garden peas, sweet peas (*Lathyrus odoratus*), common vetch (*Vicia faba*), hairy vetch (*Vicia villosa*), and lentil (*Lens culinaris*), whereas *Rhizobium japonicum* is rather specific for soybeans (*Glycine max*). Not all strains of root nodule bacteria that infect a plant are equally effective in N_2 fixation, and this is referred to as strain variation. Because of strain variation, commercial inoculants often are a mixture of effective strains. Inoculants are grown as large batches of effective organisms in liquid culture, and then these cultures are mixed on a peat base. The finely-ground peat retains moisture, and the microorganisms remain viable for long periods. The peat culture is mixed with seeds before planting, and the viable organisms are in proximity when the seeds germinate and produce roots. The bacteria induce curling of root hairs and invade the plant through the root hairs. After invasion they induce the production of root nodules and proliferate in the nodules. The bacteria undergo morphological and metabolic changes in the nodule and are referred to as bacteroids. Bacteroids in the nodule are capable of fixing N_2 from the air.

Claims of fixation of N_2 by cultures of the rhizobia outside the host plant in general could not be verified. In 1975 three laboratory groups reported success in achieving such fixation, by furnishing succinic acid as the substrate for growth and by culturing the organisms under a very low pressure of oxygen (3).

The rhizobia function at a low oxygen concentration in the nodule. The nitrogenase enzyme system is labile to oxygen, but there must be sufficient oxygen to support the generation of ATP (adenosine triphosphate). Hemoglobin in nodules achieves this necessary balance in the oxygen level, keeping it adequate to support the action of cytochrome oxidase which generates ATP.

There are well over 10,000 legume plant species that fix N_2, and about 200 non-legumes. Only one of these non-legumes has been reported to be nodulated by a *Rhizobium* species such as nodulate the legumes. This is *Parasponia andersonii*, a tree of the elm group. The other non-leguminous plants, such as the alder (e.g., *Alnus glutinosa*), fix nitrogen in association with actinomycetes. It proved extremely difficult to culture these actinomycetes, but since methodology to isolate and culture these organisms was developed in 1977 (3), information on them has expanded substantially. Some of the actinorhizal associations have found extensive use, particularly in tropical and low rainfall areas.

BIOCHEMISTRY OF NITROGEN FIXATION

Early studies on biological nitrogen fixation emphasized practical applications of the process and the development of effective inoculants. Since 1925 there has been interest in the biochemistry of the N_2 fixation process. Wilson (1) utilized nodulated red clover plants (*Trifolium pratense*) to demonstrate that half-maximum fixation occurred at a pN_2 (partial pressure of N_2) of about 0.05 atmosphere (approximately 50 millibar). High levels of O_2 inhibited the fixation non-competitively. While growing clover plants under controlled atmospheres, Wilson (1) made the unexpected observation that fixation was inhibited by H_2 (hydrogen gas). H_2 is a specific and competitive inhibitor of N_2 fixation. The H_2 inhibition of N_2 fixation results from an ordered, sequential process in which the H_2 must bind to the nitrogenase enzyme before the N_2.

There was speculation about the first or key product of N_2 fixation. Winogradsky (2) suggested that ammonia (NH_3) was the first product. Virtanen (2), on the other hand, supported hydroxylamine as the key intermediate based on his claim that leguminous plants excreted aspartic acid, and that the hydroxylamine formed by fixation reacted with oxalacetic acid to form an oxime that then was converted to aspartic acid. Other investigators failed to confirm Virtanen's work. The use of the stable isotope ^{15}N as a tracer resolved the argument. ^{15}N-enriched N_2 was supplied to a culture of *Azotobacter vinelandii* fixing N_2 vigorously, and among the amino acids recovered from the hydrolysate of the cells, the highest enrichment in ^{15}N among the isolated amino acids was in glutamic acid (2) rather than in aspartic acid, which would have been supportive of Virtanen's hydroxylamine hypothesis. If ammonia were the key product of fixation, it could logically be incorporated into glutamic acid by the action of known enzymes. Similar experiments were performed with species of bacteria from the genera *Clostridium*, *Chromatium*, *Chlorobium*, and *Rhodospirillum*, as well as the blue-green alga *Nostoc muscorum* and excised soybean nodules. In all cases the highest concentration of ^{15}N among the isolated amino acids was in glutamic acid.

CELL-FREE NITROGEN FIXATION

As the metabolism of nitrogenous compounds in intact microorganisms can be rather complex, it was important to verify the path of fixation in cell-free preparations where reactions can be defined more specifically. ^{15}N as a tracer furnished a sensitive tool for seeking reliable cell-free fixation. Carnahan and coworkers in 1960 (3) developed a method for recovering active preparations consistently from *C. pasteurianum*. Such cell-free preparations supplied ^{15}N-enriched N_2 yielded ammonia with over 50 atom % ^{15}N excess, a very convincing verification that ammonia is the "key" intermediate in biological nitrogen fixation.

PURIFICATION OF NITROGENASE

Mortenson (3) demonstrated that the cell-free enzyme system consisted of two proteins. When the two proteins were purified, it turned out that one component was a molybdenum–iron (MoFe) protein and the other was an iron protein. In purification on a chromatographic column, the MoFe protein was eluted before the Fe protein, so the MoFe protein often was referred to as protein 1 or component 1, and the Fe protein as protein 2 or component 2. A number of other terminologies were used but abandoned. Descriptive terms are dinitrogenase for the MoFe protein, dinitrogenase reductase for the Fe protein, and nitrogenase for the complex of the two components. The primary function (it also has other functions) of the Fe protein is to transfer electrons to the MoFe protein, hence the designation dinitrogenase reductase.

REQUIREMENT FOR ADENOSINE TRIPHOSPHATE

In 1960 it was reported that ATP was inhibitory to N_2 fixation in cell-free preparations of nitrogenase, but in 1962 McNary (3) demonstrated that ATP is an absolute requirement for fixation by such preparations. The N–N bond is stable, so N_2 fixation is energy demanding whether it is accomplished chemically at high temperature and pressure, or whether it is accomplished by a biological system. A minimum of 16 ATPs are required for the reduction of one N_2 to $2NH_3$. Sixteen ATP are required under ideal conditions, and under most conditions in nature the requirement is 20 to 30 ATP.

SUBSTRATES OTHER THAN N_2

Nitrogenase is a versatile enzyme and is capable of reducing substrates other than N_2 (3). It was recognized early that H_2 is a by-product of the nitrogenase reaction. The next substrate to be recognized was nitrous oxide (N_2O) which is reduced to N_2. Cyanide and methyl isocyanide, and azide also can be reduced by the enzyme system. Nitrogenase can reduce acetylene to ethylene. Other substrates demonstrated include diazine, diazarine, cyanamide, and cyclopropene. Acetylene reduction to ethylene has become a widely used measure of nitrogenase activity. The substrate is easy to prepare, the ethylene formed as a product can be measured readily and accurately by gas chromatography, and the simple method has been applied to studies in the glasshouse and field.

CONTROL OF NITROGENASE

As nitrogen fixation is energy demanding, it is advantageous for the fixing organisms to turn off the system if they have access to a source of fixed nitrogen. Ludden (3) found that *Rhodospirillum rubrum* turned its nitrogenase off when supplied fixed nitrogen such as ammonia, or when it was placed in the dark and thus was deprived of light energy. When ammonia was exhausted or when light was restored, it turned nitrogenase back on. Inactivation is achieved by ADP ribosylation of arginine 100 of one of the two equivalent subunits of dinitrogenase reductase, and reactivation is achieved by removal of this ADP-ribosyl group.

TERTIARY STRUCTURE OF NITROGENASE

The tertiary structure of the nitrogenase components was established in 1992 (2). This aids one in visualizing how the dinitrogenase and dinitrogenase reductase fit together and how the active Fe and MoFe sites are aligned.

ASSOCIATIVE NITROGEN FIXATION

The leguminous plants and their associated rhizobia are of great practical importance in agriculture, but the non-leguminous actinorhizal systems often fix N_2 vigorously and occupy important niches in ecosystems. Certain non-legumes can benefit from association with bacteria that occupy their root surface or multiply within the plant without forming nodules (3, 4). In Brazil it has been demonstrated that sugarcane can derive nitrogen from fixation by *Acetobacter diazotrophicus* carried inside the cane. *A. diazotrophicus* is a remarkable bacterium, as it can grow in a 25% sucrose solution and fix N_2 at a pH as low as 3.0.

GENETICS OF THE NITROGENASE SYSTEMS

The genetics of biological N_2-fixing systems has occupied many investigators in recent years, as genetic manipulations may offer ways to improve the process or adapt it to new crops. Progress in our knowledge of the genetics of N_2-fixing organisms has led in recent years to a substantial reorganization of classification of these organisms.

REFERENCES

1. Wilson, Perry W. *The Biochemistry of Symbiotic Nitrogen Fixation*; The University of Wisconsin Press: Madison, WI, 1940; 302.
2. Triplett, Eric W., Eds. *Prokaryotic Nitrogen Fixation*; Horizon Scientific Press: Wymondham, England, 2000; 800.
3. Stacey, Gary, Burris, Robert H., Evans, Harold J., Eds. *Biological Nitrogen Fixation*; Chapman and Hall: New York, 1992; 943.
4. Postgate, John R. *The Fundamentals of Nitrogen Fixation*; Cambridge University Press: Cambridge, 1982; 252.

BIOLOGICAL NITROGEN FIXATION, CONTRIBUTIONS TO AGRICULTURE

B

Mark B. Peoples
Commonwealth Scientific and Industrial Research Organization Plant Industry, Canberra, Australian Capital Territory, Australia

INTRODUCTION

Although dinitrogen (N_2) gas represents almost 80% of the Earth's atmosphere, it is not a source of nitrogen (N) that is readily available to plants. However, a number of procaryotic microorganisms have evolved that utilize the enzyme nitrogenase to reduce atmospheric N_2 to ammonia, which can subsequently be used to support their growth. Biological N_2 fixation (BNF) by some diazotrophs can occur in a free-living state, or via associative relationships with plants, while others can fix N_2 only in symbiosis with specific plant hosts (1, 2). Although calculations of global contributions of BNF are subject to enormous approximations, annual inputs of fixed N in natural ecosystems, agriculture, and forestry have been conservatively estimated to be 140–170 million metric tonnes (t) of N, with a further 20 million tonnes of N per year occurring in marine systems. The terrestrial input of fixed N represents about twice the amount of N applied as fertilizer each year (2, 3).

SOURCES OF FIXED NITROGEN IN AGRICULTURAL SYSTEMS

The N_2 fixation process can contribute directly to agricultural production where the fixed N is harvested in grain or other food for human or animal consumption. However, BNF can also represent an important renewable source of N that can help maintain or enhance the fertility of many agricultural soils. Examples of experimental estimates of amounts of N fixed by various N_2-fixing organisms are presented in Table 1. Values for associative and symbiotic systems have almost always been determined from measures of plant shoot biomass. Below-ground contributions of fixed N have generally been ignored. However, research now suggests that N associated with roots might represent between 25% and 50% of the total N accumulated by crops and pastures (4, 5). Therefore, total inputs of fixed N could be 50–100% greater than has been traditionally determined from shoot-based measurements, such as those depicted in Table 1.

Contributions by Different N_2-Fixing Organisms

Free-living N_2-fixers probably contribute only small amounts of N to agricultural systems (Table 1). The data tend to be inconclusive concerning the role of diazotrophs associated with nonlegumes in temperate agriculture, but studies have demonstrated significant inputs of fixed N by tropical grasses and crops such as sugarcane (*Saccharum officinarum*) (1, 2, 6). Symbiotic associations between legumes and specific soil bacteria (*Rhizobium*, *Bradyrhizobium*, *Allorhizobium*, *Azorhizobium*, *Mesorhizobium*, or *Sinorhizobium* spp.) in specialized root structures (nodules) are generally responsible for the largest amounts of fixed N in farming systems (Table 1).

Inputs of Fixed Nitrogen by Legumes

That BNF by legume systems play a key role in world crop production is irrefutable. The ability of legumes to progressively improve the N status of soils has been utilized for thousands of years in crop rotations and traditional farming systems (1–3, 7). The 163 million ha of legume oilseeds (soybean—*Glycine max*; and groundnut—*Arachis hypogea*) and pulses sown globally each year, legume components of the 200 million ha under temporary pastures or fodder crops, and the 10–12 million ha of perennial legume cover crops in rubber (*Hevea brasiliensis*) and oil-palm (*Elaeis guineensis*) plantations all contribute fixed N to farming systems. Most modern methods used to quantify inputs of fixed N by legumes separate the plant N into fractions originating from soil N or N_2 fixation. (1, 5, 7). Once the legume N can be partitioned into that proportion derived from atmospheric N_2 (%Ndfa, sometimes also described as %Pfix) and that coming from the soil, the amounts of N_2 fixed can be calculated from measures of shoot dry matter (DM) and N content. The formation of the symbiosis between legume and rhizobia is dependent upon many factors and cannot

Table 1 Experimental estimates of the amounts of N_2 fixed in different agricultural systems

N_2-fixing organism	Range measured (kg N/ha per crop or per year)	Range commonly observed (kg N/ha per crop or per year)
Free-living		
• Crops	0–80	0–15
Associative		
• Tropical grasses	10–45	10–20
• Crops	0–240	25–65
Symbiotic		
• Azolla	10–150	10–50
• Green manure legumes	5–325	50–150
• Pasture/forage legumes	1–680	50–250
• Crop legumes	0–450	30–150
• Trees/shrubs	5–470	100–200

be assumed to occur as a matter of course. This is reflected in the range of values presented in Table 1. Such large variations in reported estimates of N_2 fixation make it difficult to generalize about how much N may be fixed by different legume species. Collectively the data suggest maximum rates of N_2 fixation of 3 to 4 kg shoot N/ha per day (5) and potential inputs of fixed N by many legumes of several hundred kg of shoot N/ha each year (Table 1). However, much of the information in Table 1 was derived from research trials in which specific treatments were imposed to generate differences in %Ndfa values and legume growth as an experimental means of studying factors which regulate BNF. Therefore, these data may be of little relevance to what might actually be occurring in farmers' crops and pastures. Fortunately, measurement procedures have now been developed that allow on-farm measures of legume N_2 fixation to be conducted with confidence (4, 5, 7).

Levels of Nitrogen Fixation Achieved in Farmers' Fields

Examples of the types of information that can be generated about BNF in farmers' fields are presented in Tables 2 and 3 for farming systems in different regions of the world. These on-farm data and observations can be used to develop a picture of N_2 fixation within an individual country or region and provide insights into contributions of BNF to agriculture on a global scale. Collectively, the results in Table 2 indicate that the potential for BNF inputs can differ between legumes and countries, but they also suggest many commonalities. Although wide ranges in %Ndfa values have been observed, it seems that, on average, most winter pulses (e.g., chickpea—*Cicer arietinum*; lentil—*Lens culinaris*; field pea—*Pisum sativum*; fababean—*Vicia faba*; lupin—*Lupinus albus*) satisfy relatively higher proportions of their growth requirements from N_2 fixation (>65%) than do the summer legumes (e.g., mungbean—*Vigna radiata*; mash-bean—*V. mungo*; soybean; groundnut) where %Ndfa values were commonly less than 60% (Table 2). The increased reliance on soil N by summer legumes may reflect greater N mineralization during summer (4).

In grazed pastures, the competition for mineral N between legumes and companion grasses or vigorous broadleaf weeds growing within the pasture sward results in low levels of plant-available soil N throughout much of the growing season. As a consequence, %Ndfa by the legume components of pastures tend to be high (Table 2). The slightly lower %Ndfa values detected in perennial legume species, such as alfalfa (*Medicago sativa*) and white clover (*Trifolium repens*), presumably result from a greater ability to scavenge soil mineral N from a larger rooting zone compared with annual pasture species (4).

Although the levels of %Ndfa are important, the amounts of N_2 fixed are usually regulated by legume growth rather than %Ndfa in most farming systems (Table 3), and many legumes appear to fix approximately 20 kg of shoot N for every metric ton of shoot DM accumulated (4, 8).

Impact of Management

Factors that either enhance or depress N_2 fixation (Table 3) can generally be summarized in terms of environmental or management constraints to crop growth (e.g., basic

B

Table 2 Summary of the proportion of plant N derived from N_2 fixation (%Ndfa) and the amounts of N_2 fixed by farmers' legume crops and pastures in different geographical regions

Country and legume	Number of fields	Mean Ndfa (%)	Total N fixed (kg N/ha)[a]
Winter pulse crops			
Pakistan	126	78	79
Nepal	27	79	78
Syria	46	67	na[b]
Australia	90	65	170
Summer legume crops			
Pakistan	63	47	42
Nepal	50	55	77
Thailand	13	75	78
Vietnam	45	48	125
South Africa	14	58	na[b]
Australia	33	53	267
Annual pastures			
Australia	300	75	na[b]
Perennial pastures			
Australia	110	64	na[b]

[a] Includes an estimate of fixed N from the roots and nodules which assumes that below-ground N represents 33% of total plant N.
[b] Data not available for all fields.
Ndfa = Nitrogen derived from atmospheric N_2.

Table 3 Key factors influencing inputs of fixed N by legumes in farmers' fields

Country	System	BNF regulated by		Primary factors
		DM	Soil nitrate	
Pakistan	Winter crop	+++		Rainfall, nutrition, weed control
	Summer crop	+++	+++	Fertilizer N, no inoculation, insects, disease
Nepal	Winter crop	+++		Rainfall, nutrition
	Summer crop	+++	+	Total soil N, mineralized N, available P, legume species
Syria	Winter crop	+++	++	Soil nutrients, insects, disease
Thailand	Summer crop	+++		Available P
Vietnam	Summer crop	+++		Plant density, soil pH, available P, legume species
South Africa	Summer crop	++	++	Effective inoculation, nutrition, rotation, water availability
Australia	Winter crop	+++	+++	Rainfall, fallowing, legume species
	Summer crop	+	+++	Crop rotation, tillage, rainfall
	Pasture	+++	+	Soil pH, available P, legume density, grazing management

BNF = biological N_2 fixation; DM = dry matter; P = phosphorus.

agronomy, nutrition, water supply, diseases, pests). A number of strategies can be employed that specifically enhance BNF through increased legume biomass. These include the use of legume genotypes adapted to the prevailing edaphic and environmental conditions, procedures to improve legume plant density, irrigation (if available), the amelioration of soil nutrient toxicities or deficiencies, and the control of weeds and pests (1, 7–9). However, since the formation of an active symbiosis is dependent upon the compatibility of both the diazotrophic microorganism and the legume host, local practices which limit the presence of effective rhizobia (no inoculation, poor inoculant quality) will also be crucial in determining the legume's capacity to fix N (Table 3), as will any management decisions that directly affect soil N fertility (excessive tillage, extended fallows, fertilizer N, rotations), since mineral N is a potent inhibitor of the N_2 fixation process (1, 7, 9).

REFERENCES

1. Giller, K.E.; Wilson, K.J. *Nitrogen Fixation in Tropical Cropping Systems*, 1st Ed.; CAB International: Wallingford, England, 1991.
2. Ledgard, S.T.; Giller, K.E. Atmospheric N_2 Fixation as an Alternative N Source. In *Nitrogen Fertilization in the Environment*; Bacon, P.E., Ed.; Marcel Dekker: New York, 1995; 443–486.
3. Peoples, M.B.; Herridge, D.F.; Ladha, J.K. Biological Nitrogen Fixation: An Efficient Source of Nitrogen for Sustainable Agricultural Production? Plant Soil **1995**, *174*, 3–28.
4. Peoples, M.B.; Herridge, D.F. Quantification of Biological Nitrogen Fixation in Agricultural Systems. In *Nitrogen Fixation: From Molecules to Crop Productivity*; Pedrosa, F.O., Hungria, M., Yates, M.G., Newton, W.E., Eds.; Kluwer Academic Publ: Dordrecht, The Netherlands, 2000; 519–524.
5. Unkovich, M.J.; Pate, J.S. An Appraisal of Recent Field Measurements of Symbiotic N_2 Fixation by Annual Legumes. Field Crops Res. **2000**, *65*, 211–228.
6. Boddey, R.M.; de Oliveira, O.C.; Urquiaga, S.; Reis, V.M.; de Olivares, F.L.; Baldani, V.L.D.; Döbereiner, J. Biological Nitrogen Fixation Associated with Sugar Cane and Rice: Contributions and Prospects for Improvement. Plant Soil **1995**, *174*, 195–209.
7. Peoples, M.B.; Herridge, D.F. Nitrogen Fixation by Legumes in Tropical and Sub-Tropical Agriculture. Adv. Agron. **1990**, *44*, 156–223.
8. Peoples, M.B.; Gault, R.R.; Scammell, G.J.; Dear, B.S.; Virgona, J.; Sandral, G.A.; Paul, J.; Wolfe, E.C.; Angus, J.F. Effect of Pasture Management on the Contributions of Fixed N to the N Economy of Ley-Farming Systems. Aust. J. Agri. Res. **1998**, *49*, 459–474.
9. Peoples, M.B.; Ladha, J.K.; Herridge, D.F. Enhancing Legume N_2 Fixation Through Plant and Soil Management. Plant Soil **1995**, *174*, 83–101.

BIOLOGICAL NITROGEN FIXATION, FORMS AND REGULATING FACTORS

B

Ken E. Giller
Wageningen University, Wageningen, The Netherlands

Paul Mapfumo
University of Zimbabwe, Harare, Zimbabwe

INTRODUCTION

Nitrogen fixation is the basis of the global N cycle. Therefore it is not surprising that the ability to fix atmospheric N_2 evolved in the "primeval soup" and is deeply rooted in the evolutionary tree of life. Despite this, nitrogenase remains an enzyme exclusive to prokaryotes; no eukaryote has been described that can fix N_2 except through a symbiotic relationship. Members of both the Archaea and Eubacteria can fix N_2 and even within the Eubacteria, nitrogenase is widely distributed among the various bacterial divisions (1).

THE MANY TYPES OF BIOLOGICAL NITROGEN FIXATION

The most widely distributed form of nitrogenase is known as the molybdenum nitrogenase, which, as its name suggests, contains molybdenum. Other nitrogenases in which molybdenum is substituted by iron or vanadium have been found in free-living bacteria such as *Azotobacter*. The structure and synthesis of nitrogenases is reviewed in detail by Smith (2). A novel mechanism for N_2 fixation was described recently in *Streptomyces thermoautotrophicus* (3). This nitrogenase is very unusual: it apparently uses superoxide as its reductant and contains a molybdo-pterin cofactor as the active site.

The diversity of types of N_2 fixation arises from the multiple modes by which bacteria and eukaryotes have co-evolved to meet the rather exacting demands of active N_2 fixation and these are considered below. Associations range from loose relationships of free-living heterotrophic bacteria with plant roots, to highly evolved symbioses that involve complex morphological differentiation of both partners (Table 1). Symbioses between N_2 fixing bacteria and eukaryotes range from the *Cyanobacteria* (often still referred to as blue–green algae) with fungi in lichens, Cycads and *Gunnera*; actinomycetes (generally placed in the genus *Frankia*) with a range of angiosperms; and the rhizobia with legumes (and the single non-legume genus *Parasponia*). N_2 fixing bacteria tend to occupy habitats abundant in C but, despite many claims, these cannot be considered to be true symbioses. Examples are N_2 fixing bacteria in plant rhizospheres (often referred to in the literature as "associative symbioses") and endophytic N_2 fixing bacteria present in the vascular tissues of graminaceous plants such as sugarcane (4).

INTRODUCTION OF NITROGEN FIXING ORGANISMS

Obviously if N_2 fixing organisms are not present within the ecosystem, they cannot contribute to the N cycle. In the case of agriculture, the identification of new niches for introduction of legumes into cropping systems is one of the most important means of increasing inputs from N_2 fixation (5). Another example of the introduction of N_2 fixing organisms that is practiced on a commercial scale is inoculation with rhizobia; this industry has existed for almost 100 years. Legumes that warrant rhizobial inoculation are fairly specific in their symbiosis, and often only require inoculation when they are transported to regions outside their centers of diversity, or to where they are not traditionally grown (6).

ENVIRONMENTAL FACTORS THAT REGULATE NITROGEN FIXATION

As with all biological processes, environmental conditions such as temperature and moisture availability regulate the rates of N_2 fixation achieved. There is a range of adaptation among N_2 fixing bacteria and symbioses. As indicated previously, many methanogenic archaebacteria can fix N_2 and some of these are found in hot springs. By contrast lichens are dominant primary producers in cold

Encyclopedia of Soil Science

Table 1 Examples of the range of N_2 fixing organisms

Status of organism	N_2 fixing organism	Symbiont
Heterotrophs		
Free-living		
Anaerobic	*Clostridium, Methanosarcina*	
Microaerophilic	*Frankia, Azospirillum, Bradyrhizobium*	
Aerobic	*Azotobacter, Derxia*	
Root-associated		
Microaerophilic	*Azospirillum, Paenibacillus*	
Endophytic	*Herbaspirillum, Acetobacter*	Sugar cane (*Saccharum*), tropical grasses
Symbiotic	*Frankia*	*Alnus, Myrica, Casuarina* etc.
	Bradyrhizobium, Mesorhizobium, Rhizobium, Sinorhzobium	Many legumes, *Parasponia*
	Azorhizobium	*Sesbania rostrata*
	Methylobacterium	*Crotalaria* spp.
Autotrophs		
Free-living		
Microaerophilic	*Rhodospirillum, Bradyrhizobium*	
Aerobic	*Cyanobacteria*	
Symbiotic	*Anabaena azollae*	*Azolla* spp.
	Cyanobacteria, Bradyrhizobium	Fungi (lichens), Cycads, *Gunnera*
		Aeschynomene spp.

(Adapted from Ref. 16.)

tundra. The optimum conditions for N_2 fixation therefore differ strongly, obviating generalizations in description of ideal environments. Extreme temperatures, whether hot or cold, and extremes of drought or waterlogging are important constraints to N_2 fixation in particular climates. The focus in this article is placed on environmental variables that commonly act as limiting factors for N_2 fixation at different scales.

Regulation at the Physiological Scale

The triple dinitrogen bond is highly stable and large amounts of energy are required for N_2 fixation in the form of ATP and electrons. Aerobic respiration is necessary to provide sufficient energy for high rates of N_2 fixation, yet the nitrogenase enzyme is denatured irreversibly on exposure to molecular oxygen. This is what has been termed the "oxygen paradox" of N_2 fixation and a diversity of mechanisms has evolved to circumvent this problem (7) (Table 1). In legume/rhizobium symbioses it is believed that the response of N_2 fixation to the majority of environmental stresses is regulated by altering the supply of oxygen to the N_2 fixing "bacteroids." This is mediated by the presence of an "oxygen-diffusion barrier" that has a variable permeability to oxygen (8).

A common feature is the repression of nitrogenase activity by the presence of available N in combined mineral forms (NH_4^+, NO_3^-). In cultures, active N_2 fixation by free-living bacteria is only strongly induced when the mineral N becomes limiting for bacterial growth. In N_2 fixing symbioses, availability of large amounts of mineral N can act to repress N_2 fixation in several ways. Nodule formation and development are inhibited by combined N (in what is sometimes referred to as "autoregulation"), or activity of N_2 fixing nodules can be suppressed (in what is termed N feedback regulation) (5). This can be viewed as an evolutionary demand of the symbiosis, as it demonstrates that the extent of development of colonization of plant tissues is under the firm control of the host—effectively preventing parasitism.

Nutrient Limitations

A range of nutrients is required for N_2 fixation (5, 9). Among the potential nutritional constraints to N_2 fixation, availability of phosphorus (P) is often overriding in both agricultural and natural systems. This is even true in aquatic systems where P limits the growth and N_2 fixation of Cyanobacteria and *Azolla* (10). Deficiency of P forms part of the "soil acidity complex" that limits N_2 fixation in acid soils, particularly in old, weathered soils of the tropics (11) or in volcanic soils (12). These are precisely the conditions that favor deficiency of molybdenum, which is an essential component of the nitrogenase enzyme. Molybdenum

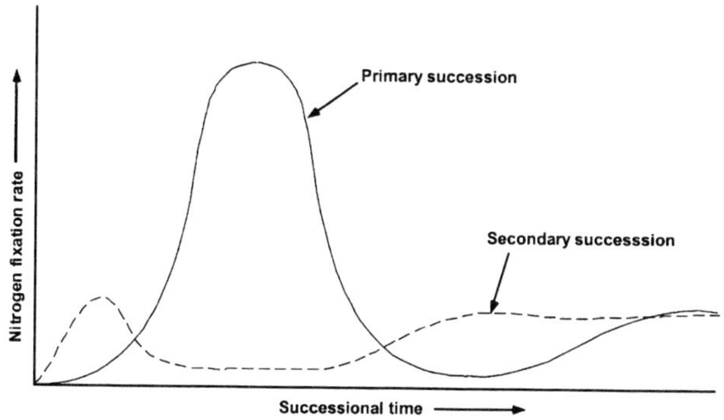

Fig. 1 Changes in the relative importance of N_2 fixation at different stages in vegetation succession. (After Ref. 15.)

deficiency is relatively uncommon except in older, weathered tropical soils but can be locally important. Many other major and micronutrients have special roles in N_2 fixation; for instance, calcium is important for infection and nodule formation in legumes and cobalt is required by N_2 fixing bacteria for electron transport (5, 9).

Toxicities

As with deficiencies, toxic effects can be manifested on both N_2 fixing bacteria or their symbionts. Toxicity of aluminium is of particular note: it appears to interfere specifically with the process of infection and nodule formation in legumes and is a major problem for survival of N_2 fixing bacteria in acid soils (5). Heavy metals appear to be particularly toxic to N_2 fixing bacteria in soil. This is true for free-living heteotrophic N_2 fixing bacteria, *Cyanobacteria*, and rhizobia that are all more sensitive to heavy metal toxicity than are host legumes (13).

Regulation of Nitrogen Fixation at the Ecosystem Scale

The feedback of abundant combined N on regulating rates of N_2 fixation by both free-living bacteria and N_2 fixing symbioses also acts to regulate rates of N_2 fixation at the scale of the (agro)ecosystem (14). In natural ecosystems, N_2 fixation is more important in early successional phases (unless other nutrients are strongly limiting) (Fig. 1), but becomes relatively unimportant as the amount of N in the soil gradually accumulates (15). A resurgence of the importance of N_2 fixation at later stages of primary or secondary succession is supported when the system becomes more abundant in C and limited in N. Agricultural systems are also influenced in the same way, rates of N_2 fixation are largest when N is limiting but other nutrients plentiful. In systems dominated by legumes, such as fodder banks, rates of N_2 fixation decline as the system accumulates N in the soil. Given that most agricultural production systems have high rates of N removal in harvested products, and hence slow rates of N accumulation, a large demand for high rates of N_2 fixation is the normal situation.

REFERENCES

1. Young, J.P.W. Phylogenetic Classification of Nitrogen Fixing Organisms. In *Biological Nitrogen Fixation*; Stacey, G., Burris, R.H., Evans, H.J., Eds.; Chapman and Hall: New York, NY, 1992; 43–85.
2. Smith, B.E. Structure, Function, and Biosynthesis of the Metallosulfur Clusters in Nitrogenases. Advances in Inorganic Chemistry **1999**, *47*, 159–218.
3. Ribbe, M.; Gadkari, D.; Meyer, O.J. N_2 Fixation by *Streptomyces Thermoautotrophicus* Involves a molybdenum Dinitrogenase and a Manganese Superoxide Oxidoreductase that Couple N_2 Reduction to the Oxidation of Superoxide Produced from O_2 by a Molybdenum-CO Dehydrogenase. J. Biol. Chem. **1997**, *272*, 26, 627–26, 633.
4. Boddey, R.M.; Döbereiner, J. Nitrogen Fixation Associated with Grasses and Cereals: Recent Progress and Perspectives for the Future. Fert. Res. **1995**, *42*, 241–250.
5. Giller, K.E. *Nitrogen Fixation in Tropical Cropping Systems*, 2nd Ed.; CAB International: Wallingford, UK, 2001; 352.
6. Eaglesham, A.R.J. Global Importance of *Rhizobium* as an Inoculant. In *Microbial Inoculation of Crop Plants*; Campbell, R., Macdonald, R.M., Eds.; IRL Press: Oxford, UK, 1989; 29–48.
7. Postgate, J. *Nitrogen Fixation*; Cambridge University Press: Cambridge, UK, 1998.
8. Minchin, F.R. Regulation of Oxygen Diffusion in Legume Nodules. Soil Biol. Biochem. **1997**, *29*, 881–888.
9. O'Hara, G.W. Nutritional Constraints on Root Nodule Bacteria Affecting Symbiotic Nitrogen Fixation. Aust. J. Exp. Agric. **2001**, *41*, 417–433.
10. Roger, P. *Biological Management of the Floodwater Ecosystem in Wetland Ricefields*; IRRI/ORSTOM: Los Baños, The Philippines/Paris, 1995; 250.
11. Sanchez, P.A. *Properties and Management of Soils in the Tropics*; John Wiley & Sons: New York, NY, 1976; 618.
12. Vitousek, P.M. Nutrient Limitation to Nitrogen Fixation in Young Volcanic Sites. Ecosystems **1999**, *2*, 505–510.
13. Giller, K.E.; Witter, E.; McGrath, S.P. Toxicity of Heavy Metals to Microorganisms and Microbial Processes in Agricultural Soils — A Review. Soil Biol. Biochem. **1998**, *30*, 1389–1414.
14. Hartwig, U.A. The Regulation of Symbiotic N_2 Fixation: A Conceptual Model of N Feedback from the Ecosystem to the Gene Expression Level. Perspect. Plant Ecol. Evol. Syst. **1998**, *1*, 92–120.
15. Gorham, E.; Vitousek, P.H.; Reiners, W.A. The Regulation of Chemical Budgets Over the Course of Terrestrial Ecosystem Succession. Ann. Rev. Ecol. Syst. **1979**, *10*, 53–84.
16. Ledgard, S.J.; Giller, K.E. Atmospheric N_2 Fixation as an Alternative Nitrogen Source. In *Nitrogen Fertilization and the Environment*; Bacon, P., Ed.; Marcel Dekker: New York, NY, 1995; 443–486.

BIOLOGICAL NITROGEN FIXATION, TECHNIQUES TO ENHANCE

David F. Herridge
NSW Agriculture, Tamworth, New South Wales, Australia

INTRODUCTION

Legume biological nitrogen fixation (BNF) is a key process in agriculture, with the crop legumes alone supplying about 17 million tons N annually for the world's grain production (1). Notwithstanding high-level achievements in legume and rhizobia technology, legume BNF, resulting from the symbiotic association between the two, remains essentially unmanaged by farmers. In many parts of the world, the legumes are considered secondary crops and are not provided the inputs and management given to major crops, usually cereals. In other situations, economic factors mitigate against the use of appropriate inputs. This lack of management, coupled with environmental influences beyond the control of farmers, results in wide variations in BNF, commonly between nil and approximately 400 kg N/ha (2).

FACTORS AFFECTING BNF

The factors accounting for variations in BNF are quite simple. Provided there are adequate numbers of highly effective rhizobia in the soil in which the legume is growing, BNF is essentially determined by the growth of the legume (i.e., the legume's N demand) and the proportion of legume N derived from BNF (% Ndfa), as follows:

$$\text{Legume BNF (kg/ha)} = \frac{[\text{Total N yield (kg/ha)} \times \%\text{Ndfa}]}{100}$$

Thus, legume BNF can be enhanced by increasing total N yield and/or increasing % Ndfa. With the majority of legumes, there is sufficient genetically based capacity for BNF to supply all of the N required for plant growth. For example, the data of Evans and colleagues (3) highlighted the very strong relationship between shoot dry matter and N_2 fixed for the temperate crop legumes lupin (*Lupinus angustifolius*) and pea (*Pisum sativum*), and indicated sufficient BNF capacity of lupin to support shoot dry matters of up to 14 t/ha.

There are cases, however, where insufficient BNF capacity limits legume yield, the most notable being common or dry bean (*Phaseolus vulgaris*). Redden and Herridge (4), after reviewing the genetic capacity for BNF of >1,000 genotypes of *P. vulgaris* concluded that BNF could supply only a fraction (<30%) of the N demand, with the remainder supplied from soil or fertilizer sources. Clearly, BNF of this species is weak, and management for yield will only have limited benefits for BNF.

LEGUME YIELD AND BNF

Just how much influence does legume yield have on BNF? The simple answer is a great deal. Researchers in Australia's northern grains belt (5) reported that 96% of the variation in BNF of alfalfa (*Medicago sativa*) was related to the variation in legume biomass with 28 kg N fixed for every ton of biomass produced. A study of BNF of pulse (pea) and pasture species [alfalfa and annual medic (*M. truncatula*)], showed that 99% of the variation in BNF was associated with differences in biomass production (23 kg N fixed/t biomass) (6).

Similar close relationships between biomass and BNF were reported by Evans and coworkers (3) in a network of 21 experiments at 10 locations of the cool-season pulses, lupin and pea, in southeastern Australia. They reported variations in BNF of 36–400 kg N/ha and 22–250 kg N/ha, for the two species. The original data were presented as fixed N in shoots only; whole-plant BNF was calculated by multiplying shoot N by 1.4 to account for N in below-ground parts (7). Values for %Ndfa varied 29–97% for lupin and 20–95% for pea. Almost 70% of the variation in BNF was associated with variation in legume biomass yield; an additional 18 kg N/ha was fixed for every extra ton of biomass produced.

Soil Water

In the rain-fed environments where the majority of agricultural legumes are grown, the overriding influence on plant growth (biomass yield) is the availability of water, either stored in the soil during the precrop fallow or falling as rain during growth. With the lupin and pea study referred to previously, 83% of variation in lupin biomass could be explained by in-crop rainfall (3). In an ancillary study, additional inputs of 75 and 125 mm water in the form of supplementary irrigation resulted in the enhancement of BNF at rates of 0.4–0.8 kg N/mm water. With the extra 125 mm water, BNF increased from 60–210 kg N/ha for lupin and from 47–150 kg N/ha for pea. The water-use efficiency values for lupin and pea BNF were similar to published values of 0.3–0.5 kg N/mm for chickpea (*Cicer arietinum*) (8).

These data highlight the close relationship between water availability, biomass production, and BNF. Irrigation is not often an option for farmers in dry-land agriculture, and they have no control over seasonal weather. What they have some control over is the efficiency with which water is infiltrated into and stored in the soil, coupled with the efficiency with which the water is used by the legume. The following sections discuss management practices that increase water infiltration and storage as well as the legume's utilization of water.

Rhizobial Inoculation

Legume inoculation with rhizobia is a long-established and successful practice, especially in countries in which legumes are widely grown, such as the U.S., Australia, Canada, Brazil, Thailand, and India. Available figures indicate that approximately 2,000 tons of inoculants are produced worldwide each year, sufficient to inoculate 20 million ha of legumes. By far, the largest producer of inoculant is the U.S., with an annual production of approximately 1,000 tons (9).

Many experiments have examined the need to inoculate in different parts of the world during the past 50 years. Results from 214 experiments of 10 legume species, conducted by the University of Hawaii's NifTAL project, indicated inoculation-enhanced yields in 53% of cases through enhanced BNF (10). Soybean (*Glycine max*) responded to inoculation in 65% of experiments, which is not surprising because of its specific requirements for rhizobia. In the same study, analysis of 305 soil samples from 17 countries indicated that only 14% of soils contained rhizobia effective on the highly bred American soybean.

These figures highlight the uncertainty regarding the benefits of and need for inoculation. In countries where inoculants are readily available, inoculation is generally recommended as a practice for all legume sowings. Unfortunately, few farmers in less-developed countries have access to high-quality inoculants or use inoculants as a normal part of their legume cultivating practices. In these situations, potential yield is lost unless soil N uptake is supplemented with fertilizer N (with additional cost to the farmer) or the soil already contains adequate numbers of effective rhizobia.

Legume Agronomy

Strategies to enhance legume biomass and legume BNF include optimizing crop use of growing season and space, and nutrient and water resources with careful attention to species and cultivar choice, planting time, row spacing and plant density, pest, disease and weed control, and crop nutrition. In the following sections, experiments are discussed in which legume BNF was enhanced through management for biomass yield. The examples are specific for particular sites and seasons, and are not meant to be prescriptive; rather, they serve to illustrate how management practices can be changed to enhance BNF.

Species/cultivar choice

Matching the species and cultivar of legume to the edaphic (soil) and environmental conditions can lead to enhanced BNF. In the same lupin and pea study referred to previously, BNF of lupin was approximately 80% greater than that of pea on the acidic soils because of substantially increased biomass yield and marginally improved %Ndfa (Table 1). The reverse was the case on the alkaline soils in the study; on those soils, pea outyielded lupin by approximately 60% and fixed approximately 80% more N.

Planting time

Optimizing planting time to take full advantage of growing season rainfall and temperatures and to minimize deleterious effects of pest and disease cycles can provide options for enhancing BNF. In the Australian pea experiment (Table 1), BNF was increased from 64 kg N/ha to 180 kg N/ha by planting 54 days earlier. Most of the increase was due to the larger biomass crops.

In a more radical departure from traditional agronomic practice, Beck and colleagues in Syria (11) increased BNF of chickpea, from 7 kg N/ha to 143 kg N/ha, by planting in winter rather than in spring. This was made possible by the introduction of cold-tolerant, ascochyta [*Ascochyta rabei* (Pass.) Labrousse] blight-resistant varieties. The longer growing season allowed the crop to utilize rainfall more efficiently and produce greater biomass, which resulted in the enhanced BNF.

Table 1 Examples of enhancement of legume BNF through practices to increase legume yield[a]

Agronomic practice	Crop N[b] (kg/ha)	Ndfa (%)	Total N fixed (kg/ha)	Reference
Choice of species				
Acid soil (pH 4.3)				
Lupin	253	91	230	3
Pea	166	78	130	
Alkaline soil (pH 7.8)				
Lupin	104	57	53	
Pea	166	57	95	
Time of planting				
Pea experiment (Australia)				
Early (0 days)	238	76	180	16
Mid (+26 days)	166	64	106	
Late (+54 days)	120	53	64	
Chickpea experiment (Syria)				
Winter sown	177	81	143	11
Spring sown	40	18	7	
Plant density (alfalfa plants/m^2)				
40	548	55	300	
20	442	53	234	13
10	346	51	176	
5	246	52	128	
Pest and weed control				
Control of sitona weevil in lentil				
+carbofuran	200	75	150	11
−carbofuran	174	71	124	
Control of grassy weeds in clover				
+herbicide	400	89	356	13
−herbicide	316	92	290	
Nutrient additions (subterranean clover)				
Nil	80	94	75	
+P	136	90	122	17
+lime	88	87	77	
+P+lime	182	90	164	

[a] In most cases, values shown are the means of several sites and/or seasons.
[b] Original values for shoot N recalculated as crop N using multiplication factors of 2.0 for chickpea, alfalfa, and subterranean clover, and 1.4 for pea, lupin, and lentil (7).

Plant density and row spacing

Within the limits of site and environment, narrow row spacing and/or high plant density can result in enhanced BNF. In on-farm surveys of 51 chickpea and faba bean (*Vicia faba*) crops in Australia's northern grains belt, Schwenke and colleagues (12) reported that crop biomass was greatest in narrow rows and that %Ndfa was positively correlated with plant density. In pasture systems in south-eastern Australia (13), increasing alfalfa numbers from 5 to 40 plants/m^2 more than doubled crop biomass (246–548 kg N/ha) and BNF (128–300 kg N/ha) (Table 1).

Pest control

Pests and diseases will reduce legume biomass. For example, a major constraint to lentil (*Lens culinaris*) production is the *Sitona* weevil, the larvae of which feed on the N_2-fixing root nodules. Destruction of nodules can be as high as 80%. Application of carbofuran (2,3-dihydro-2,2-dimethyl-7-benzofuranol methylcarbomate) at sowing controls the pest. In the Syrian study reported by Beck and colleagues (11), use of the insecticide resulted in a 20% increase in both biomass and BNF (Table 1).

Weeds compete with the legumes for light, water, and nutrients and their control benefits the production of legume biomass. Herbicide spraying of grassy weeds in subterranean clover (*Trifolium subterraneum*) pastures in southeastern Australia resulted in a 27% increase in biomass and a 23% enhancement of BNF (Table 1). Similar results were reported (6) for a mixed annual medic-alfalfa pasture in which winter cleaning of grassy weeds increased BNF by 110 kg N/ha.

Other soil constraints and plant nutrients

Other soil constraints include acidity, salinity, nutrient toxicities, and nutrient deficiencies. Such constraints must be addressed if potential legume biomass production is to be realized. Research has also established that the N_2-fixing legume may have additional edaphic and nutritional requirements, compared with the non–N_2-fixing plant (14). Examples are the higher requirement by N_2-fixing legumes for calcium, boron, iron, and molybdenum.

Soil acidity and phosphorus (P) deficiency are common constraints to legume BNF. In a three-year study at three sites in southeastern Australia, N yields and BNF of subterranean clover-dominated pastures were increased 65–70% with P fertilizer and 120–130% with a combination of lime and P (Table 1). Lime increased pH and reduced extractable aluminum and manganese, both of which are toxic to legumes and rhizobia at elevated concentrations. In an ancillary study, also involving subterranean clover at three sites, crop biomasses N and BNF were increased by 80 and 70 kg N/ha, respectively, with fertilizer P (13).

%NDFA AND BNF

At any level of yield, increasing %Ndfa results from practices to reduce soil nitrate suppression of BNF. This can be achieved in a number of ways, most dramatically by land use, and, to a lesser extent, by tillage practice and stubble and mulch management.

Land Use and Crop Sequence

Rain-fed chickpea grown immediately after sorghum (*Sorghum bicolor*) fixed 183 kg N/ha, compared with 54 kg N/ha by chickpea in previously fallowed soil (Table 2). The Ndfa values were 84% and 19%, respectively. The differences in BNF were due to the large differences in soil nitrate levels at sowing—17 and 226 kg N/ha—affecting the %Ndfa values.

Soybean BNF was similarly enhanced when grown on land that had previously grown un-nodulated soybean, rather than on land used for nodulated soybean cultivation or on fallowed land (Table 2). Effects on BNF were again mediated through soil nitrate levels affecting plant nodulation and %Ndfa. Un-nodulated soybean reduced soil nitrate levels more than nodulated soybean because the residues of the former were N-deficient and would have immobilized some of the mineral N as it was released from organic matter. Many other studies relating BNF to crop sequence and land use report similar findings,

Table 2 Examples of enhancement of legume BNF through practices to increase %Ndfa[a]

Agronomic practice	Soil nitrate (kg/ha; 1.2 m depth)	Nodulation (mg/plant)	Crop N (kg/ha)[c]	Ndfa (%)	Total N fixed (kg/ha)	Reference
Cropping sequence: Chickpea following						
Sorghum	17	n.d.	218	84	183	18
Fallow	226	n.d.	284	19	54	
Cropping sequence: Soybean following						
Un-nod soybean	70	306	310	59	183	19
Nod soybean	140	298	317	50	159	
Fallow	260	139	290	22	64	
Tillage practice: Soybean grown with						
No-tillage	n.d.[b]	139	363	88	320	20
Cultivated	n.d.	86	337	73	246	

[a] In most cases, values shown are the means of several sites and/or seasons.
[b] n.d.=Not determined
[c] Original values for shoot N recalculated as crop N using multiplication factors of 2.0 for chickpea and 1.5 for soybean (Ref. 7).

including on-farm surveys of 33 soybean crops (15) and 51 chickpea and fababean crops (12) in Australia's eastern grains belt.

Tillage

Cultivation accelerates the mineralization of organic N to plant-available mineral N in soils, with the result that cultivated soils often have an additional 20 to 25 kg nitrate-N/ha in the root zone, compared with untilled (or no-tilled) soils. For cereals under no-tillage, additional fertilizer N may be required to supplement the reduced soil nitrate N. For legumes, however, the lower nitrate N should result in enhanced BNF. In a series of experiments comparing effects of tillage practice on soybean production, the no-tilled soybean produced more nodules and more biomass, had higher %Ndfa, and fixed more N (320 vs. 246 kg/ha) than the cultivated crops (Table 2).

Although soil nitrate levels were not measured in this experiment, it is highly likely that reduced soil nitrate resulted in the higher %Ndfa value for the no-tilled soybean. It was interesting that the no-tilled crops produced 8% more biomass N than the cultivated crops. No-tilled soils are often wetter than cultivated soils, which may be crucial in arid to semiarid, rain-fed environments. An additional 40 mm plant-available water could mean an extra 25 kg biomass N/ha. Thus, although the extra 74 kg N fixed/ha should be mainly attributed to reduced nitrate suppression (and enhanced %Ndfa), the wetter no-tilled soils may have contributed to increased growth.

CONCLUSIONS

The previous examples show clearly that management options for enhancing BNF are available to farmers. The options roughly fall into two categories: those that increase biomass yield and those that reduce soil nitrate. As stated earlier, the examples are site and species specific and are not meant to be prescriptive. The means by which BNF is enhanced for a particular legume crop or pasture will depend on which of the crop or soil management practices can be used to increase legume biomass yield and/or reduce soil nitrate levels.

REFERENCES

1. Herridge, D.F.; Rose, I.A. Breeding for Enhanced Nitrogen Fixation in Crop Legumes. Field Crops Research **2000**, *65*, 225–248.
2. Peoples, M.B.; Ladha, J.K.; Herridge, D.F. Enhancing Legume N_2 Fixation Through Plant and Soil Management. Plant and Soil **1995**, *174*, 83–101.
3. Evans, J.; O'Connor, G.E.; Turner, G.L.; Coventry, D.R.; Fettell, N.; Mahoney, J.; Armstrong, E.L.; Walsgott, D.N. N_2 Fixation and Its Value to Soil N Increases in Lupin, Field Pea and Other Legumes in Southeastern Australia. Australian Journal of Agricultural Research **1989**, *40*, 791–805.
4. Redden, R.J.; Herridge, D.F. Evaluation of Genotypes of Navy and Culinary Bean (*Phaseolus Vulgaris* L.) Selected for Superior Growth and Nitrogen Fixation. Australian Journal of Experimental Agriculture **1999**, *39*, 975–980.
5. Hossain, S.A.; Waring, S.A.; Strong, W.M.; Dalal, R.C.; Weston, E.J. Estimates of Nitrogen Fixations by Legumes in Alternate Cropping Systems at Warra, Queensland, Using Enriched -^{15}N Dilution and Natural ^{15}N Abundance Techniques. Australian Journal of Agricultural Research **1995**, *46*, 493–505.
6. McCallum, M.H.; Peoples, M.B.; Connor, D.J. Contributions of Nitrogen by Field Pea (*Pisum Sativum* L.) in a Continuous Cropping Sequence Compared with Lucerne (*Medicago Sativa* L.)-Based Pasture Ley in the Victorian Wimmera. Australian Journal of Agricultural Research **2000**, *51*, 13–22.
7. Unkovich, M.J.; Pate, J.S. An Appraisal of Recent Field Measurements of Symbiotic N_2 Fixation by Annual Legumes. Field Crops Research **2000**, *65*, 211–228.
8. Herridge, D.F.; Marcellos, H.; Felton, W.L.; Turner, G.L.; Peoples, M.B. Chickpea Increases Soil-N Fertility in Cereal Systems Through Nitrate Sparing and N_2 Fixation. Soil Biology and Biochemistry **1995**, *27*, 545–551.
9. Singleton, P.W.; Boonkerd, N.; Carr, T.J.; Thompson, J.A. Technical and Market Constraints Limiting Legume Inoculant Use in Asia. In *Extending Nitrogen Fixation Research to Farmers' Fields*; Proceedings of an International Workshop on Managing Legume Nitrogen Fixation in the Cropping Systems of Asia, ICRISAT Asia Center, India, 1996; Rupela, O.P., Johansen, C., Herridge, D.F., Eds.; ICRISAT: Patancheru, AP, India, 1997; 17–38.
10. Singleton, P.W.; Bohlool, B.B.; Nakao, P.L. Legume Response to Rhizobial Inoculation in the Tropics: Myths and Mealities. In *Myths and Science of Soils of the Tropics*; Lal, R., Sanchez, P.A., Eds.; Soil Science Society of America and American Society of Agronomy Special Publication **1992**, *29*, 135–155.
11. Beck, D.P.; Wery, J.; Saxena, M.C.; Ayadi, A. Dinitrogen Fixation and Nitrogen Balance in Cool-Season Food Legumes. Agronomy Journal **1991**, *83*, 334–341.
12. Schwenke, G.D.; Peoples, M.B.; Turner, G.L.; Herridge, D.F. Does Nitrogen Fixation of Commercial, Dryland Chickpea and Faba Bean Crops in North-West New South Wales Maintain or Enhance Soil Nitrogen. Australian Journal of Experimental Agriculture **1988**, *38*, 61–70.
13. Peoples, M.B.; Gault, R.R.; Scammell, G.J.; Dear, B.S.; Virgona, J.; Sandral, G.A.; Paul, J.; Wolfe, E.C.; Angus, J.F. Effect of Pasture Management on the Contributions of Fixed N to the N Economy of Ley-Farming Systems. Australian Journal of Agricultural Research **1998**, *49*, 459–474.
14. O'Hara, G.W.; Boonkerd, N.; Dilworth, M.J. Mineral Constraints to Nitrogen Fixation. Plant and Soil **1988**, *108*, 93–110.

B

15. Peoples, M.B.; Gault, R.R.; Lean, B.; Sykes, J.D.; Brockwell, J. Nitrogen Fixation by Soybean in Commercial Irrigated Crops of Central and Southern New South Wales. Soil Biology and Biochemistry **1995**, *27*, 553–561.
16. O'Connor, G.E.; Evans, J.; Fettell, N.A.; Bamforth, I.; Stuchberry, J.; Heenan, D.P.; Chalk, P.M. Sowing Date and Varietal Effects on the N_2 Fixation of Field Pea and Implications for Improvement of Soil Nitrogen. Australian Journal of Agricultural Research **1993**, *44*, 151–163.
17. Peoples, M.B.; Lilley, D.M.; Burnett, V.F.; Ridley, A.M.; Garden, D.L. Effects of Surface Application of Lime and Superphosphate to Acid Soils on Growth and N_2 Fixation by Pasture Clover in Mixed Pasture Swards. Soil Biology and Biochemistry **1995**, *27*, 663–671.
18. Doughton, J.A.; Vallis, I.; Saffigna, P.G. Nitrogen Fixation in Chickpea 1. Influence of Prior Cropping or Fallow, Nitrogen Fertilizer and Tillage. Australian Journal of Agricultural Research **1993**, *44*, 1403–1413.
19. Herridge, D.F.; Bergersen, F.J.; Peoples, M.B. Measurement of Nitrogen Fixation by Soybean in the Field Using the Ureide and Natural ^{15}N Abundance Methods. Plant Physiology **1990**, *93*, 708–716.
20. Hughes, R.M.; Herridge, D.F. Effect of Tillage on Yield, Nodulation and N_2 Fixation of Soybean in Far North-Coastal New South Wales. Australian Journal of Experimental Agriculture **1989**, *29*, 671–677.

BIOTA

Lev O. Karpachevsky
Mikhail L. Karpachevsky
Moscow State University, Moscow, Russia

INTRODUCTION

Biota—the totality of living organisms—is an important factor of soil formation, which creates organic matter, loosens rock and modifies its composition by transforming into soil. In the absence of organisms, rocks are transformed into loose sedimentary materials similar to lunar regolith. Autotrophs, plants (their biomass on the Earth is 2.4×10^{12} ton) as well as sulfur bacteria, iron bacteria, etc., build-up organic matter. Heterotrophs, animals, and microorganisms (0.02×10^{12} ton), consume organic matter and decompose it to carbon dioxide and water. According to the type of feeding, they are divided into phytophages (feed on living plants), zoophages (predators), necrophages (eat animal corpses), and saprophages (consume decaying plants). The biota plays an essential role in creating the anisotropy of soils.

ROLE OF PLANTS

The biota is also responsible for the redistribution of nutrients within the pedosphere, soils and their accumulation in the upper layer of soil. This is especially true for such nutrients as C, N, P and S. Practically all biological cycles are primarily driven by plants. Plants uptake chemical elements from soil, water and atmosphere and build up phytomass—the source of soil humus. The phytomass deposits on or in soil. The annual fall of plant litter, mortmass, on the soil surface ranges from 1 to 30 ton/ha. Reserves of roots in soil may reach 20 ton/ha in tropical forest and 8–12 ton/ha in grasslands, temperate deciduous forests and southern taiga. Up to 30% roots die annually. Roots of herbaceous plants and partly fine tree roots promote humus formation in the soddy horizon, undershrub and root horizon and, partly, humic A horizon. These horizons are the soil factory of humus and it can form only in them.

The roots make up only 0.1–1% of soil by weight, while they directly affect soil in the vicinity up to 3 mm from the root surface, within the rhizosphere—the zone close to fine sucking roots. However, this generally has very little effect on the soil properties. Typically, the rhizosphere is densely populated with microorganisms, protozoans, and other invertebrates associated with these organisms.

If soil is continuously frozen, phytomass decomposition at the stage of the formation of litter (mortmass) slows down. The reserves of phytomass reach their maximum in forests—tropical, subtropical, temperate, deciduous and southern taiga. The reserves of mortmass—fresh and partly decomposed litter—generally decrease from tundra and northern taiga to steppes, accompanied by increasing humus content in soil (soil "mortmass").

The biogeochemical importance of plants is also due to the fact that they uptake nutrients. The comparison of coefficients of enrichment (biological uptake) of plants relative to lithosphere and soil shows that plants add in the trophic chain additional amounts of Na, Mg, K, Ca, Cu, B, P, S, Br, Cs, and Au. Plants, when compared with soils, contain noticeably less amounts of F, As, and Cd. At the same time, soil, when compared to lithosphere and sedimentary rocks, concentrates Be, S, Cr, Zn, As, Mo, Ag, Sb, Sn, I, Cs, Au, and Br. Soils on calcareous material contain much less Ca and Ba. Plants, when compared to soil, accumulate more Na, Mg, K, Ca, Mn, Co, Cu, Zn, Rb, Mo, Ag, and I. This accumulation is significant for B, P, S, Br, Cs and especially Au. The enrichment of a geological layer with these elements and impoverishing in F, Na, Mg, Cl, K, Ca, Cu, Rd, Hg, Pb, and U attests that this rock experienced a phase of soil formation.

Besides changing the chemical composition of rocks and accumulating humus, plants together with animals create the anisotropy of soils—a regular change in the system of soil horizons downward and spatial variation in properties of upper horizons (the soil pattern). Each particular plant and their group exert a specific effect on soil. Long-living immobile organisms, like trees, modify the soil during the whole life. However, their effect is not the same at various distances from the edifying plants, like large trees. Typically, soil pH in boreal ecosystems is lower near tree trunks, while litter reserves and humus content there are higher than under the canopy. The death of an individual tree or its group also has an important effect on soil formation. Thus, windthrow of trees mixes the soil,

Encyclopedia of Soil Science

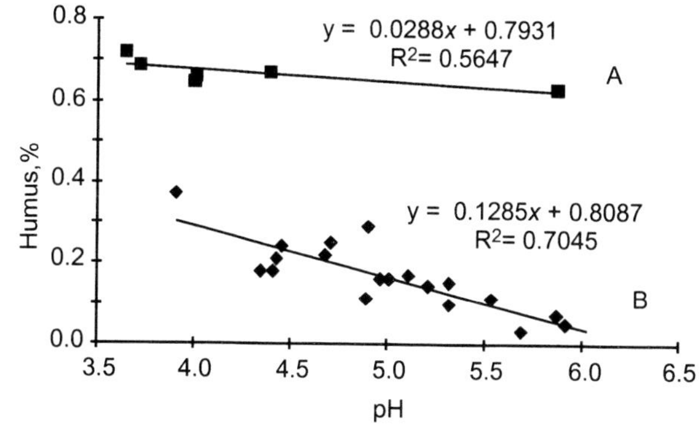

Fig. 1 The correlation between humus content and soil acidity in eluvial horizons of sandy Al–Fe podzols in pyrogenic pine forests of Surgut Woodland, Western Siberia. Samples from pink-colored eluvial horizons—the material from deeper horizons recently exposed by tree windthrow after wildfire and now being actively bleached. Samples from sugar-white eluvial horizons—long bleached material of upper horizons.

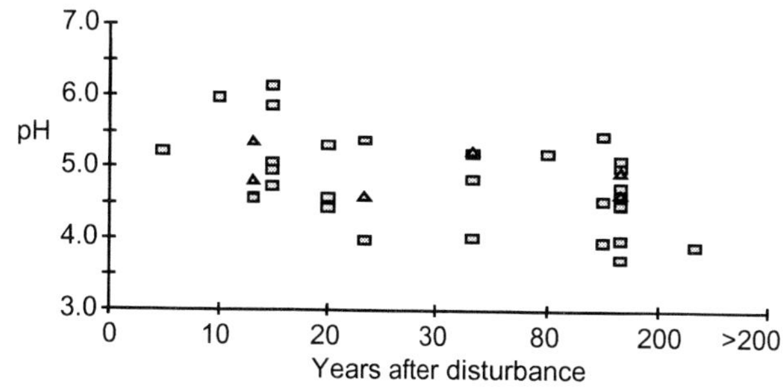

Fig. 2 Change in pH of drained and poorly drained Al–Fe podzols in Surgut Woodland, Western Siberia, in the course of successions after wildfire.

creating specific pit and mound topography. The traces of uprooting in soil can be recognized for decades. Fig. 1 shows the differences between two types of eluvial horizons of podzols found in pine forest, Western Siberia. Pink-colored eluvial horizons are the richer material from deeper horizons recently exposed by tree uprooting after wildfire and being actively bleached now. Sugar-white eluvial horizons are long bleached material of upper horizons.

Local disturbances, like the death of a tree, may maintain the stability of ecosystem, while dynamically changing its internal structure through gap mosaic. Extended disturbances, like wildfires, create an opportunity for new species to colonize the landscape. Such plant communities in the course of their development (succession) alter an environment, making it more suitable to newer species. Long-term changes in soil properties following wildfire can be traced for hundreds of years as illustrated by Fig. 2. Thus, plants change soil properties (Table 1), structure soil, and increase its porosity relative to a parent material. Different plants and plant communities transform soil material in different way.

ROLE OF ANIMALS

The distribution of animal weight is similar to plants. Animals living on the ground comprise less than 0.5% of the weight of all animals in ecosystem. In turn, animals dwelling in litter and organic horizons (mortmass) make up

Table 1 Properties of soils (0–30 cm) in forest, virgin grassland, and arable land (according to N.S. Oreshkina)

Land type	Soil solids density (g/cm^3)		Porosity (vol %)	Specific surface (m^2/g)	Infiltration (mm/min)	Humus (% of mass)
Soddy-podzolized soil, Malinki, Moscow Oblast						
Meadow	1.37	2.64	48	n.d.	2	3.4
Forest	1.19	2.62	54	46	130	1.1
Field	1.47	2.62	44	54	17	n.d.
Typical chernozems, Kursk Oblast						
Virgin grassland	1.08	2.62	56	114	68	4.5
Forest	1.05	2.59	58	94	148	5.4
Field	1.17	2.65	54	91	83	2.9
Dark chestnut soil, Dzhanibek, Western Kazakhstan						
Virgin grassland	1.06	2.58	60	122	2	3.7
Forest	1.08	2.62	58	n.d.	190	3.9
Field	1.30	2.66	51	137	12	3.1

n.d.: not determined.

about 90% of all animals by weight in tundra and northern taiga ecosystems. In steppe ecosystems, the weight of animals dwelling in mortmass decreases to 2–5% of the total zoomass as litter reserves decrease, while up to 90–98% of the zoomass occur in the soil. Most animals in grasslands and deserts live in soil, while in taiga they occur in forest litter. The distribution of microbial mass coincides with that of zoomass—in taiga its maximum occurs in litter, while further south it shifts to soil.

Major components of soil fauna are protozoans (amoeba, infusoria, etc.), worms, beetle larvas, arachnoids, slugs, etc. Saprophages play a prominent role in soil formation, especially burrowing animals—moles, ground squirrels, marmots, and mice. Soil-dwelling animals are subdivided into groups: geobionts (constantly live in soil), geophiles (spend a major part of their life cycle in soil), and geoxenes (occasionally dwell in soil or use it as a shelter).

Earthworms (*Oligochaeta* and a group of the families Megadrili) are represented by the families Lumbricidae, Criodrilidae, Ocnerodrilidae, Moniligastridae, Megascolecidae, etc. Most widespread are species of the family Lumbricidae. Earthworms are saprophages, part of them live in litter, others in soil. Litter-dwelling species, e.g., *Dendrobaena attemsi*, *D. hortensis*, *Dendrodrilus rubidus*, *Allobophora parva*, *Eisenia submontana*, also live in decaying trees, dispersing organic material and partially mixing it with mineral soil. Earthworms dwelling in soil (species *Nicodrilus caliginosus*, *Allobophora clorotica*, *A. biserialis*, *A. sturanyi*, *Kritidrilus calarensis*, etc.), pass soil through their gastrointestinal tract feeding on soil organic matter. Various ecological groups of the family Lumbricidae dwell on the ground; in soil and litter; in humus horizon; some large species make deep burrows and tunnels in soil. Earthworms produce coprolites—excrements rich in organic compounds, microorganisms, and nitrogen compounds; they often have high physical resistance. The upper layer of chernozems of old gullies and gray forest soils (Dnepropetrovsk Oblast, Ukraine) are 60–80% made up by earthworm coprolites. Groups of earthworms living in soil and litter and on the ground can penetrate far north; burrowing earthworms dwell even in dry forests of Mediterranean type. According to the type of feeding, earthworms are further subdivided into two ecological groups: feeding on leaf litter and detritus on the ground and proper soil forms feeding on raw humus. The number of earthworms in soils varies from one to hundreds earthworms per square meter. Earthworms feeding on litter on average consume up to 5 mg/g living mass per day. Earthworms may accelerate litter decomposition by 2–3 times. In the presence of earthworms (*N. caliginosus*, 20 worms and *L. terrestris*, three worms), the raw humus layer (H, or 0h) and humus layer, A, up to 5 cm thick develop. When earthworms are absent, only the F (0f) layer forms.

Other representatives of soil fauna are beetle larvae, ticks, annelids, collemboles, etc., basically dwelling in litter and the A horizon. Worms of the class Enchitraeidae, which ingest organic matter, are up to 10–25 mm long. In acid peat soils, there can be from 85–200 thousand worms per square meter of soil, with a total weight of 0.3–30 g. Nematodes are small (up to 1 mm) segmented worms, whose number may reach several million individuals per square meter, feeding on decaying remnants of animals and plants, soil microflora, alga, and fungal mycelium. Sloths, with a size of 0.2–0.3 mm, dwell in mosses, feeding on small animals and plant cells. Myriapods, e.g., small symphylids, eat decaying plants and few are predators. Diplopoda, including millepedes, *Rossiulus kessleri*, eating dead plants can be up to 10–17 cm long in the tropics. Their exoskeletons accumulate calcium, strontium, uranium, lead, etc. Few species are predators. Almost 95% of insects are dependent on soil during their life cycle. Some of them are phytophages—Elateridae, mole cricket, cutworm, phylloxera, May beetle larva, etc.—others are saprophages (larvae of Diptera—flies and mosquitoes: Bibionidae, Licoryidae, and Tipulidae). They decompose organic mass, transforming it into raw humus. Besides, beetle larvae (e.g., May beetle larvae), and beetles themselves live in soil.

Ants, which mix soil, are widespread in natural ecosystems (up to 50 kg/ha). They are partly predators and partly necrophages.

Wood louse, mainly saprophagous, is also common. Termites (5–22 g/m^2) live in southern soils under conditions of deep groundwater. With the help of microorganisms dwelling in their alimentary tracts, they digest cellulose and burrow channels down to the depth of 8 m from the soil surface.

Animals 0.1–3 mm long are referred to as microfauna. The most abundant are ticks (up to 50,000 species and especially those with exoskeleton, beetle mites) feeding on fungus hyphens and decaying plant debris. Their number in forest litter reaches 200,000–300,000 per square meter with the weight of 20 kg/ha. They annually discharge (in g/m^2): 1 in tundra, up to 6 in taiga, up to 8 in temperate deciduous forests, 2.2 in steppes, 1 in semi-deserts, 0.1 in deserts, and 13 in humid subtropics. Collemboles are lower wingless insects eating small plants (alga) dispersing plant debris. They give several generations per year with the total number of 1–50 million/m^2 and biomass of 0.2–6.4 g/m^2.

Protozoans are single-cell organisms inhabiting all soils, totaling up to 20 billion/m^2.

Rhizopoda, flagellates, and infusoria are 2–20 μm long and widely occur in soil. Their primary food is bacteria.

Microorganisms are also considered part of the biota. They dwell on plant leaves, in soil, litter, and alimentary tract of all animals, participating in transformation of organic matter of ecosystems. Microorganisms also promote nitrogen fixation in soil (*Azotobacter*, *Clostridium pasterianum*, and *Rhizobium*), methane emission from soil, etc. In soils, they are adsorbed on mineral particles. Microorganisms participate in all processes of organic matter transformation at all trophic levels of the food chain. The number of microorganisms in soils is estimated in tens of thousand to billions of cells per gram of soil.

More than 1500 alga species occur in soil—more than 600 species of green alga, Clorophyta, more than 400 species of cyanobacteria, Cyanophyta, nearly 300 species of diatomic alga, Bacillariophyta, and nearly 200 species of yellow green alga (Xantophyta). The total biomass of soil alga is estimated to be 0.5–1 ton/ha. Alga fix nitrogen from the atmosphere and accumulate carbon, thus making the primary trophic level.

Fungi and actinomycetes play a considerable role in organic matter decomposition. The length of the mycelium of actinomycetes can reach 100–300 m/g of soil, increasing to up to 6000 m/g in peatlands. The number of cells is estimated to be 2–300 million/g. Fungi decompose wood and litter, and accelerate carbon cycling in ecosystems.

Animals contain a noticeably less variety of elements in their bodies and the total amount of elements (except C, O, N, and H) is much less compared to plants. However, there is one part of animal activity, which is by scale comparable or even exceeds the biogeochemical activity of plants. This is the role of animals in reallocation and mixing of soil (burrows, disturbed grounds, passages in soil, etc.) and organic matter decomposition.

According to Abaturov (1), mole discards may occupy up to 50% of plot area, totaling to 50 ton/ha. Typically, animals discard the material from deeper horizons on the soil surface—E, EB, and B in forests, AB, B, and BC in steppes and forest–steppe. In taiga soils, discarded material is enriched with calcium and clay. In steppe soils, it is enriched with calcium carbonates, salts, and gypsum. The addition of elements to the upper layer of eluvial soils by animals often exceeds their input with plant litter. The discarded material becomes involved in soil formation. In the 0- to 7-cm layer, the humus content increases from 0.8% C during the first year after discard to 4% in 17 years. Animals like wild boar can actively reallocate soil matter—when abundant they may disturb 5–10% of the ground. Furthermore, they mix the upper and lower layers of soil down to the depth of 20 cm. Thus, in pine forests polluted during the Chernobyl accident, the radioactivity of soil decreased from 43 to 25 mR/h from 1987 to 1990, while in places disturbed by wild boar it decreased to 20 mR/h during the first year.

Dams made by beavers provoke soil waterlogging on midslopes and peat formation. Herbivores, including elephants trampling grasslands and savannas, make soil denser and decrease the content of humus in it. Excrements of mice in tundra and taiga lower the soil acidity permitting to survive pathogenic microorganisms infesting leptospirosis and pseudotuberculosis (*Hebdomadis*, *Grippotyphosa*, etc.).

Invertebrates and fungi are major destroyers of dead plant material. Fungi destroy wood and break down plant litter to trash. Observations show that trunks of beech, oak, and spruce with the diameter at breast height of 30–40 cm become fully decomposed on the soil surface for 100 years. Decomposing trunks, besides fungi and alga, are inhabited with mosses, higher plants, including woody species. In Northern European Russia, partly decomposed dead trunks are a place where seeds of spruce and fir may germinate, because large deadwood is not so quickly inhabited by mosses and undershrubs (e.g., bilberry). In Kamchatka, saplings of larch also prefer deadwood, because trunks of decomposing trees are characterized by high moisture content and nutrient reserve, thus promoting young trees to grow on these substrata. Decomposing trunks are often inhabited with ants. Saprophages can be divided into those which feed only on organic substratum and whose excrements consist of 80–90% organic matter and organisms, which together with organic matter consume mineral substratum (many species of earthworms). The excrements of earthworms (coprolites) contain much more nitrogen and organic matter than the A horizon. Thus, the coprolites of *Eisenia nordeskioldi* contain 29% organic matter and 0.95% nitrogen, while those of *Dendrobena octaedra* contain 53 and 1.21%, respectively. However, they contain more mineral matter than the excrements of wood louse (51–80% organic matter and 1.9% nitrogen), Bibionidae larvae (58–83 and 1.8%), and millipedes (76 and 1.5%). Fresh litter has about 90% organic matter and 1.5–1.9% nitrogen.

Although saprophages eating litter increase the ash content from 2–8 to 7–15%, they do not form soil humus. The humus matrix—humus tightly fixed on the surface of soil particles—is primarily formed by earthworms, which by three times accelerate decomposition of litter of various woody species. Soil worms mix organic matter with soil, thus promoting formation of the humus matrix in soil. Other animals and litter worms create raw humus, dispersed organic matter with higher ash content than fresh litter.

REFERENCES

1. Abaturov, B.D. *The Functional Role of Animals* (in Russian); Nauka: Moscow, 1980.
2. van der Drift, Doeksen J., Eds. *Soil Organisms*; North-Holland: Amsterdam, 1963.
3. Gilarov, M.S. Distribution Peculiarities of Soil-Dwelling Invertebrates in Different Zonal Genetic Soil Types. In *Proc. of 8th International Congress of Soil Science*; Bucharest: 1964; Vol. 3, 551–560.
4. Karpachevsky, L.O. *Ecological Soil Science (In Russ.)*; Mosk. Gos. Univ. Moscow, 1993.
5. Jenny, H. *Factors of Soil Formation*; McGraw-Hill: New York, 1941.
6. Sukachev, V.N. *The Principles of Forest Biogeocenology* (in Russian); Nauka: Moscow, 1964.
7. Vsevolodova-Perel, T.S. *Range and Regularities in the Distribution of Earthworms of the USSR Fauna* (in Russian); Nauka: Moscow, 1979.
8. Wilde, S.A. *Woodlands of Wisconsin*; Madison, 1976.
9. Zvyagintsev, D.G. Structure and Functions of the Soil Complex. In *Structural and Functional Role of Soil in the Biosphere* (in Russian); GEOS: Moscow, 1999.

BOREAL FOREST, SOILS OF THE

B

G. Mazhitova
Komi Science Center, Russian Academy of Sciences, Syktyvkar, Russia

INTRODUCTION

The boreal forest or taiga is a northern forest dominated by conifers (Fig. 1). The forest dominates the boreal biome, where it alternates with bogs and other ecosystems. The biome is circumpolar in the northern hemisphere (Fig. 2); It is subdivided into Low Boreal, Mid-Boreal and High Boreal. The transition between the High Boreal and the Arctic composed by forest-tundra/sparse taiga is called the Subarctic. Besides plains and lowlands, the boreal forest occupies a number of mountain systems: the Fennoscandian Range, the Urals, the mountains of southern and northeastern Siberia, and the northern extent of the North-American Cordillera. The boreal forest covers 58.6 Mha in Scandinavia, 709.0 Mha in the Former Soviet Union, 42.4 Mha in Alaska, and 221.8 Mha in Canada (1–6). In addition, the Subarctic comprises 340.0 Mha in Russia and 171.2 Mha in Canada (4, 6).

ENVIRONMENTAL CONDITIONS

Large boreal regions, with the exception of East Siberia, experienced multiple continental ice sheets in the Pleistocene, which predetermined the prevalence of geologically young parent materials and distribution of periglacial phenomena. Climate was highly dynamic during the Holocene, causing shifts of the northern treeline and southern border of the permafrost and the polygenetic nature of many Subarctic soils (7, 8). The northern treeline roughly corresponds to the July isotherm of 10°C. The climate of the boreal forest is characterized by long, cold, and snowy winters; however, it varies widely across the biome (Table 1) (9–13). East Canada is humid with relatively mild winters, whereas East Siberia is semiarid and reveals the highest thermal continentality on earth. Either continuous or discontinuous permafrost occurs everywhere in the Subarctic and in the closed taiga of East Siberia and Alaska. The rest of the biome is characterized by seasonal ground freezing. The most widely distributed boreal forests in the world are larch forests; in Russia they occupy 269 Mha (14). Other important tree species are spruce, pine, fir, and birch.

SOIL CLASSIFICATION

In terms of Soil Taxonomy, the boreal biome has the greatest area of Spodosols and Histosols, is the only biome but the Arctic and Alpine containing Gelisols, and is an important area of Alfisol formation. Inseptisols are also widespread. In terms of WRB, the boreal biome contains most of the Albeluvisols, Podzols, and Histosols and the second largest area of Cryosols after the Arctic. Cambisols, Gleysols, Regosols, Leptosols, and Fluvisols also occur.

MAJOR FEATURES OF SOIL FORMATION

Most soils in the boreal biome either are Gelisols or have cryic thermal regime. Aquic or udic soil moisture regimes are common. Ice-cemented permafrost, where present, adds to soil saturation. In East Siberia, however, salts accumulate in some soils as a result of the semiarid climate. The cold climate hinders chemical weathering. Combinations of moisture regime, low pH, and non-neutralized organic compounds foster leaching, clay illuviation, and podzolization in many soils. The latter, though not confined to the boreal biome, is best pronounced and widely expressed throughout it. Permafrost and, in many regions, seasonal frost cause various cryogenic features in soils. Nutrient turnover is slow. Only a small fraction of the ecosystem nutrients are contained in the vegetation; most are contained in soils (15, 16). The most biologically active nutrient fraction is cycled in litterfall to the forest floor. In the Subarctic, because the trees are shallow-rooted, the forest floor is the main source of nutrients for tree growth. However, boreal ecosystems are well adopted to the conditions of their growth. Conifers need a small amount of nutrients and are able to grow on highly leached acid soils.

Fig. 1 Mid-boreal spruce forest in European Russia. (Photo: E. Lopatin.)

About half of the world's wetland area is in the boreal biome; most wetlands are bogs; that is, their soils are Histosols and Histels. In the area of continuous permafrost peat plateaus and polygonal peatlands are typical. In the discontinuous permafrost area palsa peatlands develop, and in the zone of seasonal frost, string mires and domed bogs are characteristic (17, 18).

SOILS OF THE MAIN BOREAL REGIONS

Scandinavia

Spodosols, mostly Typic Haplocryods, and Histosols dominate the soil cover. Many Spodosols in Lapland are cryoturbated, possibly in the absence of permafrost. Frost heave takes place everywhere in the region of the seasonal frost in Finland. The only Gelisols recognized are Sphagnic or Terric Fibristels in the palsas of Lapland (17).

European Russia

Cryalfs and, in the northern taiga, Aqualfs (Albeluvisols in terms of WRB) on loamy materials (19) and Spodosols derived from coarse-textured parent materials (20) dominate the area with small inclusions of Histosols (Sphagnum domed bogs). Similar to Scandinavia, no mineral Gelisols form within the biome; however, Histels develop in the forest-tundra under peat plateaus or palsas. West-facing slope and foothills of the Urals support the least human-affected boreal forests in Europe, which grow on Cryalfs and Cryepts.

West Siberia

West Siberia contains one of the world's largest lowland areas (128 Mha). Continuous permafrost occurs in the forest-tundra and discontinuous in the northern taiga. Soils of the region have never been classified according to the

Fig. 2 Global boreal forest biome.

B

Table 1 Information from boreal weather stations

Station	Region	Mean annual temperature (°C)	Mean July temperature (°C)	Mean January temperature (°C)	Annual precipitation (mm)
Palojarvi[a]	Lapland	−2.1	12.1	−12.4	413
Syktyvkar[b]	European Russia	0.4	16.6	−15.1	514
Oimiakon[c]	East Siberia	−16.7	13.6	−47.5	233
Fairbanks[d]	Alaska	−2.7	16.0	−22.2	284
Moosonee[e]	Ontario, Canada	−1.3	15.1	−20.5	700
St. John's[f]	Newfoundland, Canada	4.7	15.4	−4.3	1482

[a]Period of 1991–1997, (9).
[b]1889–1989, (10).
[c]1943–1980, (10).
[d]1961–1983, (11).
[e]1951–1980, (12).
[f]1961–1990, (13).

Soil Taxonomy. Analysis of Russian literature (21, 22) indicates that not only Histels, but also mineral Gelisols develop in the Subarctic and in the east of the High Boreal. Cryalfs and Cryods dominate well-drained sites, whereas Cryaqualfs and Cryaquods with deep albic horizons and often ortstein develop in poorly drained sites affected by ground water. String-and-fen bogs with Fibrists and Hemists are typical for permafrost-free areas.

Central and East Siberia

The largest part of East Siberia has continuous low-temperature permaforst and is dominated by Gelisols (23–25). The recent Russian soil classification recognizes the soil type "Cryozem," which is widely distributed here (Oxyaqiuc Cryosols in WRB; Typic Haploturbels in the Soil Taxonomy). Oxyaquic features develop in fine-textured materials due to the presence of oxygen in the water of melting ground frost aided by the drainage provided by frost cracking (26). In Central-Yakutia unique Haplels showing evidence of solidification develop.

Alaska

Eighty percent of boreal Alaska has sparse or discontinuous permafrost. Turbels, with some Haplels and Histels, dominate permafrost-affected areas. Haplocryods and Humicryods are common for the Cordillera in southeastern Alaska, whereas in southern Alaska Andisols are the dominant soils (27).

Canada

About thirty percent of the land area in the Subarctic is covered with wetlands (18). In continental high Subarctic most wetlands are affected by permafrost and their soils include Histels; in the areas adjacent to Hudson Bay and in the east of the Canadian Subarctic, unfrozen fen and palsa complexes are common with Fibrists, Hemists, and Histels. Also, mineral Gelisols (both Haplels and Turbels) develop here as opposed to more southern boreal regions (28, 29). South from the Subarctic, Cryepts and, to a smaller extent, Cryalfs dominate the western part of the territory with more Spodosols and Cryalfs dominating the eastern part (29).

LAND USE

The main use of boreal lands is for forestry. In Scandinavia large areas of Histosols are artificially drained to cultivate forests. Agricultural use is limited. In European Russia in 1981, the area of arable lands commonly varied from 8 to 39 percent in major economic divisions (30). Most boreal soils are acid and nutrient poor and require liming and organic and inorganic fertilizers to grow crops.

GLOBAL ECOLOGICAL SIGNIFICANCE OF BOREAL SOILS

Global change ecological relationships within the taiga could assume global importance, because the biome's soils greatly affect the world's carbon balance. Boreal soils (excluding peat) contain 650 Pg of carbon, which corresponds to 25.6 percent of the world carbon. Of this, about one-third is contained in the forest floor and mineral soils (31). In the past, boreal regions, together with the Arctic ones, have been the net sink of carbon. However, they may become a major source of carbon due to anthropogenic activities and the increased rate of decomposition due to soil warming.

REFERENCES

1. *Statistics Finland/Yleisjulkaisut*; 1998.
2. *Statistics Norway/Natural Resources and Environment*; 1999.
3. *Statistics Sweden/Section: Agriculture, Forestry, and Fishery*; 2000.
4. Kolchugina, T.P.; Vinson, T.S.; Gaston, G.G.; Rozhkov, V.A.; Schlentner, S.F. Carbon Pools, Fluxes, and Sequestration Potential in Soils of the Former Soviet Union. In *Soil Management and Greenhouse Effect*; Lal, R., Kimble, J., Levine, E., Stewart, B., Eds.; Lewis Publishers: Boca Raton, London, Tokyo, 1995; 25–40.
5. Van Kleve, K.; Chapin, F.S., III; Dyrness, C.T.; Viereck, L.A. Element Cycling in Taiga Forest: State-Factor Control. BioScience **1991**, *41* (2), 78–88.
6. Tarnocai, C. Carbon Pools in Soils of the Arctic, Subarctic, and Boreal Regions of Canada. In *Global Climate Change and Cold Regions Ecosystems*; Lal, R., Kimble, J., Stewart, B., Eds.; Lewis Publishers: Boca Raton, London, New York, Washington, 2000; 91–103.
7. Aleksandrovsky, A.L. *Soil Evolution in the European Plain in Holocene*; (In Russian) Nauka: Moscow, 1983; 152.
8. Tarnocai, C. Paleosols of the Interglacial Climates in Canada. Geographie Physique et Quaternaire **1990**, *44* (3), 363–374.
9. *Meteorological yearbook of Finland*; Finnish Meteorological Institute: Helsinki, 1991 ... 1997 (annual issues).
10. *Scientific-Applied Climate Reference Book of the USSR. 1. Air and Soil Temperature*; Hydrometeoizdat: Leningrad, 1989; 1(2) and 24(3) (In Russian).
11. Ping, C.L. Soil Temperature Profiles of Two Alaskan Soils. Soil Science Society of America Journal **1987**, *51* (4), 1010–1018.
12. *Atmospheric Environment Service. Canadian Climate Normals 1951–1980. Temperature and Precipitation: Ontario*; Environment Canada, Atmospheric Environmental Service: Downsview, Ontario, 1982; 254.
13. *Atmospheric Environment Service. Canadian Climate Normals 1961–1990. Temperature and Precipitation: Atlantic Provinces*; Environment Canada, Atmospheric Environmental Service: Downsview, Ontario, 105.
14. Bukshtynov, A.D.; Groshev, B.I.; Krylov, G.V. *Forests*; Mysl: Moscow, 1981; 318 (In Russian).
15. Rodin, L.E.; Bazilevich, N.I. *Dynamics of the Organic Matter and Biological Turnover of Ash Elements and Nitrogen in the Main Types of the World Vegetation*; (In Russian) Nauka: Moscow, Leningrad, 1965; 254.
16. Powers, R.F.; Van Kleve, K. Long-term Ecological Research in Temperate and Boreal Forest Ecosystems. Agronomy Journal **1991**, *83* (1), 11–24.
17. Seppala, M. Palsas and Related Forms. In *Advances in Periglacial Geomorphology*; Clark, M.J., Ed.; John Wiley: Chichester, 1988; 247–278.
18. Zoltai, S.C.; Tarnocai, C.; Mills, G.F.; Veldhuis, H. Wetlands of Subarctic Canada. In *National Wetlands Working Group Canada Committee on Ecological Land Classification*, Ecological Land Classification Series, 1988; 24, 56–96.
19. Tonkonogov, V.D. *Texture-Differentiated Soils of European Russia*; V.V. Dokuchaev Soil Science Institute: Moscow, 1999; 156 (In Russian).
20. Aparin, B.F.; Zaboeva, I.V.; Lipkina, G.S.; Nogina, N.A.; Rudneva, E.N.; Rusakova, T.V.; Sloboda, A.V.; Urusevskaya, I.S. *Podzolic Soils of the Centre and East of the European Territory of the USSR*; Nauka: Leningrad, 1981; 200 (In Russian).
21. Karavaeva, N.A. *Paludification and the Evolution of Soils*; Nauka: Moscow, 1982; 296 (In Russian).
22. Avetov, N.A.; Trofimov, C.Y. Soil Cover of the Taiga and Floodplain Landscapes in the Pur River Basin, West Siberia. Pochvovedenie **1997**, (1), 31–35 (In Russian).
23. Elovskaya, L.G.; Petrova, E.I.; Teterina, L.V. *Soils of Northern Yakutia*; Nauka: Novosibirsk, 1979; 303 (In Russian).
24. Mazhitova, G.G. Soil Formation on the Yedomas of the Kolyma Lowland. Soviet Soil Science **1990**, *22* (3), 12–23.
25. Yershov, Yu.I. Features Proper to Forest Soil Formation in the North of Middle Siberia. Pochvovedenie **1995, (7), 805–810 (In Russian).**
26. Sokolov, I.A. Hydromorphic Non-Gley Soil Formation. Pochvovedenie **1980**, (1), 21–33 (In Russian).
27. Moore, J.P.; Swanson, D.K.; Fox, C.A.; Ping, C.L. *International Correlation Meeting on Permafrost-Affected Soils. Guidebook—Alaska Portion*; USDA Soil Conservation Service: Linkoln, NE, 1993.
28. Zoltai, S.C.; Tarnocai, C. Some Nonsorted Patterned Ground Types in Northern Canada. Arctic and Alpine Research **1981**, *13* (2), 139–151.
29. Lacelle, B.; Tarnocai, C.; Waltman, S.; Kimble, J.; Swanson, D.; Naumov, Ye.M.; Jacobson, B.; Broll, G. *Northern Circumpolar Soil Map*; Agriculture and Agrifood Canada, USDA, Dokuchaev Soil Institute, Institute of Geography—University of Copenhagen, Institute of Landscape Ecology—University of Muenster; Aug. 1998.
30. *Soil Map of the RSFSR, scale 1:2,500,000*. Fridland, V.M., Ed. GUGK: Moscow, 1988.
31. Bhatti, J.S.; Apps, M.J. Carbon and Nitrogen Storage in Upland Boreal Forests. In *Global Climate Change and Cold Region Ecosystems*; Lal, R., Kimble, J., Stewart, B., Eds.; Lewis Publishers: Boca Raton, London, New York, Washington, 2000; 79–89.

BORON AND MOLYBDENUM

Umesh C. Gupta
J.A. MacLeod
Agriculture and Agri-Food Canada, Charlottetown, Prince Edward Island, Canada

B

INTRODUCTION

Boron (B) and molybdenum (Mo) are among the chief micronutrients essential for plant growth that exist in soil solutions as anions. Boron deficiencies occur in many countries of the world with coarse-textured soils in humid regions where leaching and heavy cropping have diminished the soil B reserves. Boron toxicity occurs in soils which are inherently high in B or as a result of high B fertilization. Such may be the case, for example, if a B sensitive crop were to follow a crop that received a high rate of B, or through the use of irrigation water high in B. Crops sensitive to B, e.g., peanuts (*Arachis hypogaea* L.), could result in B toxicity (1) at B levels as low as 2 kg ha^{-1} (2).

Molybdenum deficiencies are common in many countries where soils are acidic and coarse in texture. Mo toxicities in crops tend to be rare in distribution and occur only when unusually high quantities of Mo are present. Soils high in Mo have been reported on some poorly drained soils, e.g., in Australia, Scotland, and the United States (3).

SOIL FACTORS AFFECTING PLANT UPTAKE

Boron

Soil reaction and pH: Generally, B becomes less available to plants with increasing soil pH. Visual symptoms of B deficiency were found to be accentuated by Ca deficiency and were less evident when Ca was added in excess (4).

Macronutrients: Nitrogen is of utmost importance in affecting B uptake by plants. Liberal N applications have been found to decrease plant B uptake. Increasing rates of P and K have also been found to decrease B uptake.

Soil properties: Higher quantities of available B are generally found in fine-textured than in coarse-textured soils. The lower amounts of B in sandy soils are associated with higher leaching of B. Organic matter is one of the chief sources of B in acid soils, as relatively little B adsorption occurs at low pH. Irrigation waters with high B can result in B toxicity in crops. Boron applied in bands is more effective than when applied broadcast in increasing B uptake by crops. Method of plowing (e.g., with moldboard plow) and ridge till resulted in higher B in plants than with no till and beds (5). Highly weathered acid soils are also responsible for causing B deficiency (6).

Other factors include environment, genotypes, and plant species, e.g., legumes can accumulate more B than grasses (7). Plant age and plant parts can affect the B composition of plants (8).

Molybdenum

Parent material and soil pH: Soils formed on sandstones that have experienced heavy leaching losses are most likely to be low in total Mo. Alfisols derived from mixed shale may be higher in Mo than spodosols formed from mixed schist. Lateritic soils derived from granite and basalt rocks are very high in total Mo. Soil solution MoO_4^{-2} is the most available form to plants. Its availability increases a hundred fold for each unit increase in pH. Alkaline soils have a relatively large proportion of Mo in the soil-solution phase. Some studies (9) have indicated that the severity of Mo deficiency in lentil (*Lens culinaris* Medic) can be reduced by liming on Mo deficient acid alluvial soils.

Soil properties: Coarse-textured loams and silty loams low in organic matter, as well as severely eroded and/or heavily weathered soils are particularly low in Mo. The amount of Mo adsorbed may be closely related to the organic matter content. Soil wetness is one of the chief factors affecting the availability of Mo. Peats and mucks are products of a wet environment and have been associated with high Mo in certain regions of the world. Well-drained soils, e.g., podzols are low in Mo. Availability of Mo to plants is also reduced by S fertilization.

PHYSIOLOGICAL ROLE, DEFICIENCY, AND TOXICITY SYMPTOMS

Boron

Boron plays a key role in seed production. Boron deficiency can severely impede seed production particularly in legumes (10). Boron is involved in the transport of sugars across cell membranes and in the synthesis of cell wall material. Some studies suggest that selection of transgenic cultivars with an increased sugar alcohol content can result in increased B uptake by translocating it as a complex sugar alcohol in the phloem (11). It promotes elongation of epicotyls and hypocotyls and increased height of seedlings grown under Al stress (12). Boron deficiency can inhibit the growth of seedlings (13). Boron plays an important role in maintaining the integrity of plasma membranes of leaf cells and in alleviating the damage of membrane caused by low temperatures (13).

Boron deficiency in crops is more widespread than deficiency of any other micronutrient. Because of its slow mobility in plants, B deficiency symptoms generally first appear on the young leaves at the top of the plant. Crops most sensitive to a B deficiency include rutabaga (*Brassica napobrassica* L.), other brassicas, forage legumes, and root crops.

The B toxicity symptoms are also similar among most plants. Generally they consist of marginal and tip chlorosis, which is quickly followed by a necrosis (8).

Molybdenum

Molybdenum is an essential constituent of several important molybdoenzymes. Nitrate reductase, which is involved in nitrate reduction in plants, and nitrogenase, which is involved in nitrogen fixation by legumes, are most important. Studies on beans in Brazil showed that foliar applications of Mo greatly enhanced nitrogenase and nitrate reductase activities resulting in increased N content in the shoots (14). Their results demonstrated that in certain soils, N fertilization may be replaced by small amounts of Mo as foliar application. Other roles of Mo include increased organic nitrogen, protein content, and drought resistance of plants. Lack of Mo can inhibit tasseling in corn and decrease sucrose and ascorbic acid concentrations in cauliflower (*Brassica oleracea* L. Botrytis Group) and legumes.

Deficiency symptoms for most micronutrients appear on the young leaves at the top of the plant, because most micronutrients are not readily translocated. Molybdenum is an exception in that it is readily translocated, and its deficiency symptoms generally appear as yellowing of the whole plant. Symptoms associated with Mo deficiency are closely related to N metabolism. In brassicas, Mo deficiency can result in cupped leaves, i.e., rolling or upward curling of leaves. Crops that commonly suffer due to Mo deficiency are forage legumes, brassicas, and soybeans (*Glycine max* [L.] Merr.). Molybdenum toxicity in crops is uncommon and is found only when unusually high concentrations of Mo are present. The most striking symptom of Mo toxicity as reported in alfalfa (*Medicago sativa* L.) is yellow or orange–yellow chlorosis, with some brownish tints that start in the youngest leaves. Further symptoms of Mo toxicity damage include moribund buds, thick stems, and development of axillary buds and some times succulent older leaves. Details of Mo deficiency and toxicity symptoms can be found elsewhere (3).

METHODS OF CORRECTING DEFICIENCY

Boron

Sodium borates of varying degrees of hydration and H_3BO_3 are principal sources of B fertilizers. Application rates of B generally range from 0.25 to 3.0 kg ha^{-1} depending upon crop requirements and the method of application. Higher rates are required for broadcast applications than for banded soil application or foliar sprays. On clay loam soils rates of up to 8.8 kg ha^{-1} are not toxic while on sandy loam soils this rate was toxic to cauliflower (15). However, on crops such as peanuts, sensitive to excess B, 2 kg B ha^{-1} was found to be toxic (1). Grasses are highly tolerant of B, e.g., 45 kg B ha^{-1} caused only 12% yield reduction (16). Most B sources are applied to soils with NPK fertilizers, and B may be incorporated or bulk blended with granular fertilizers or mixed with fluid fertilizers (17).

Molybdenum

Methods of controlling Mo deficiency in crops include band or broadcast application (generally of Mo contained

Table 1 Sufficiency levels of B (8) and Mo (3) in several crops

Crop	Plant part	Sufficiency level (mg kg^{-1}) Boron	Molybdenum
Alfalfa	Whole tops	20–40	0.12–1.29
Cauliflower	Leaves	11–97	0.19–0.25
Corn	Leaves	10	0.20
Red clover	Whole tops	21–45	0.3–1.59
Rutabagas	Leaves	40	—
Soybeans	Leaves	—	0.5–1.0
Wheat	Boot stage tissue	8	0.09–0.18

in P or NPK fertilizers) to soil, foliar sprays, and seed treatment. Amounts applied to soil range from 50 g to 1 kg ha^{-1} depending upon the crop and soil pH. However, seed treatment is the most common method because the recommended rates are low, ranging from 50 to 400 g ha^{-1}. Molybdenum sources can be applied to seeds as liquid or slurry, and some type of sticking or conditioning agents may be included. Because legume seeds, such as soybeans, peas, (*Pisum sativum* L.) alfalfa, and clovers (*Trifolium* spp.) are treated with a bacterial inoculant, the Mo source for seed treatment must be compatible with the inoculant. Foliar rates of application range from 60 to 218 g Mo ha^{-1} (18, 19). It should also be pointed out that on soils containing adequate Mo, liming soils to pH 6–6.5 can overcome a Mo deficiency in crops. Sufficiency levels of B and Mo for a few crops are shown in Table 1.

REFERENCES

1. Rashid, A.; Rafique, E.; Ali, N. Micronutrient Deficiencies in Rainfed Calcareous Soils of Pakistan II. Boron Nutrition of the Peanut Plant. Commun. Soil Sci. Plant Anal. **1997**, *28*, 149–159.
2. Gupta, U.C.; Jame, Y.W.; Campbell, C.A.; Leyshon, A.J.; Nicholaichuk, N. Boron Toxicity and Deficiency: A Review. Can. J. Soil Sci. **1985**, *65*, 381–409.
3. Gupta, U.C. *Molybdenum in Agriculture*; Cambridge University Press: U.K., 1997; 276.
4. Chatterjee, C.; Sinha, P.; Nautiyal, N.; Agarwala, S.C.; Sharma, C.P. Metabolic Changes Associated with Boron-Calcium Interaction in Maize. Soil Sci. Plant Nutr. **1987**, *33*, 607–617.
5. Lal, R.; Fausey, N.R.; Brown, L.C. Drainage and Tillage Effects on Leaf Tissue Nutrient Contents of Corn and Soybeans on Crosby-Kokomo Soils in Ohio. In *Drainage in the 21st Century: Food Production and the Environment;* 7th Intern. Drainage Symp., Orlando, FL, 1998; 465–471.
6. Fageria, N.K.; Baligar, V.C. Response of Common Bean, Upland Rice, Corn, Wheat, and Soybean to Soil Fertility of an Oxisol. J. Plant Nutr. **1997**, *20*, 1279–1289.
7. Adarve, M.J.; Hernandez, A.J.; Gil, A.; Pastor, J. Boron, Zinc, Iron, and Manganese Content in Four Grassland Species. J. Environ. Qual. **1998**, *27*, 1286–1293.
8. Gupta, U.C. *Boron and Its Role in Crop Production*; CRC Press: Boca Raton, FL, 1993; 237.
9. Mandal, B.; Pal, S.; Mandal, L.N. Effect of Molybdenum, Phosphorus, and Lime Application to Acid Soils on Dry Matter Yield and Molybdenum Nutrition on Lentil. J. Plant Nutr. **1998**, *21*, 139–147.
10. Mozafar, A. Role of Boron in Seed Production. In *Boron and Its Role In Crop Production*; Gupta, U.C., Ed.; CRC Press: Boca Raton, FL, 1993; 185–206.
11. Bellaloui, N.; Brown, P.H.; Dandekar, A.M. Manipulation of In Vivo Sorbitol Production Alters Boron Uptake and Transport in Tobacco. Plant Physiol. **1999**, *119*, 735–741.
12. Yang, Y.H.; Zhang, H.Y. Boron Amelioration of Aluminum Toxicity in Mungbean Seedlings. J. Plant Nutr. **1998**, *21*, 1045–1054.
13. Wang, Z.Y.; Tang, Y.L.; Zhang, F.S.; Wang, H. Effect of Boron and Low Temperature on Membrane Integrity of Cucumber Leaves. J. Plant Nutr. **1999**, *22*, 543–550.
14. Vieira, R.F.; Vieira, C.; Cardoso, E.J.B.N.; Mosquim, P.R. Foliar Application of Molybdenum in Common Bean. II. Nitrogenase and Nitrate Reductase Activities in a Soil of Low Fertility. J. Plant Nutr. **1998**, *21*, 2141–2151.
15. Batal, K.M.; Granberry, D.M.; Mullinix, Jr., B.G. Nitrogen, Magnesium, and Boron Applications Affect Cauliflower Yield, Curd Mass, and Hollow Stem Disorder. Hort. Sci. **1997**, *32*, 75–78.
16. Wilkinson, S.R. Response of Kentucky – 31 Tall Fescue to Broiler Litter and Composts Made from Broiler Litter. Commun. Soil Sci. Plant Anal. **1997**, *28*, 281–299.
17. Mortvedt, J.J.; Woodruff, J.R. Technology and Application of Boron Fertilizers for Crops. In *Boron and Its Role in Crop Production*; Gupta, U.C., Ed.; CRC Press: Boca Raton, FL, 1993; 157–176.
18. Mortvedt, J.J. Sources and Methods of Molybdenum Fertilization of Crops. In *Molybdenum in Agriculture*; Gupta, U.C., Ed.; CRC Press: Boca Raton, FL, 1997; 171–181.
19. Adams, J.F. Yield Responses to Molybdenum by Field and Horticultural Crops. In *Molybdenum in Agriculture*; Gupta, U.C., Ed.; CRC Press: Boca Raton, FL, 1997; 182–201.

BULK DENSITY

Kwong Yin Chan
Wagga Wagga Agricultural Institute, NSW Agriculture, Wagga Wagga, New South Wales, Australia

INTRODUCTION

Bulk density (ρ_b) is defined as the ratio of the mass of a given soil sample (M) to its bulk volume (V).

$$\rho_b = M/V \tag{1}$$

The mass is obtained by drying the sample to a constant weight at 105°C, and the bulk volume is that of soil particles plus pore space at the time of sampling (1). The commonly used SI unit for bulk density is $Mg\,m^{-3}$, which is numerically the same as $g\,cm^{-3}$. Bulk density is a key soil physical property and is required for: 1) determining the degree of compactness and hence as a measure of soil structure, 2) using as an indicator of soil aeration status (together with soil water content), and 3) converting soil water content and nutrient values from the gravimetric to the volumetric basis.

Bulk density of soil is dependent on the composition as well as the structural conditions. It therefore tends to vary with the soil's texture, which determines the packing of the soil particles. Sandy soils tend to have higher bulk density (ranged 1.4–1.9 $Mg\,m^{-3}$) than clays (0.9–1.4 $Mg\,m^{-3}$) (Table 1). Volcanic soils and peaty soils tend to have bulk density <1.0 $Mg\,m^{-3}$ because of the high organic matter levels and special clay minerals for the former and higher organic matter levels for the latter. Furthermore, as field soils are not just mixtures of primary particles in different degrees of packing, bulk density is also modified by soil aggregation (soil structure) and soil organic matter levels. Organic matter tends to reduce the degree of compaction. For 58 samples of surface soils in Scotland, a highly significant effect of organic matter is represented statistically by the relationship

$$\rho_{max} = 1.86\text{–}0.055 \times OM \tag{2}$$

where ρ_{max} is the maximum bulk density over a range from 1.3 to 1.8 $Mg\,m^{-3}$ and OM is the content of organic matter over a range of 2% to 10% was obtained (3).

For swelling soils, bulk density is a function of soil water content.

SOME RELATED TERMS

1. Wet bulk density (ρ_w)—is the wet mass of soil per unit volume and is related to ρ_b:

   $$\rho_w = \rho_b(1 + \theta g) \tag{3}$$

   where θg is gravimetric water content $kg\,kg^{-1}$ soil.
2. Porosity (ϵ)—is the total pore space per unit volume of soil, $m^3\,m^{-3}$ and is related to ρ_b:

   $$\epsilon = 1 - \rho_b/\rho_s \tag{4}$$

   where ρ_s is soil particle density $Mg\ m^{-3}$.
3. Air-filled porosity (ϵ_a)—is the air-filled pore space per unit volume of soil, $m^3\,m^{-3}$ and is related to ρ_b:

   $$\epsilon_a = \epsilon - \theta v = 1 - \rho_b/\rho_s\text{–}\rho_b\theta g \tag{5}$$

   where θv is volumetric water content, $m^3\,m^{-3}$.

METHODS FOR DETERMINATION

Broadly, methods commonly used for bulk density determination can be grouped into the following two categories.

Direct Methods

Direct methods involve measurement of the sample mass and volume, following Eq. (1). The most widely used procedure is the "core method" in which a cylindrical sampler is hammered into the soil (4, 5). Since the volume of the cylinder is known, trimming of the soil core flush with the ends of the cylinder and drying of the soil at 105°C for determining its weight allows the bulk density to be calculated. Accuracy of the core method is affected by compression and shattering, which are dependent on sample size and soil water content (4, 5). Samplers with too small a diameter are unsuitable because of the risk of compression. A thin-walled sampler with diameter of 75–100 mm has been proposed as a compromise (4). The core method works best in soft

Table 1 Range of bulk density and porosity values in different soils and particle mixtures[a]

Soil/particles mix	Bulk density ($Mg\,m^{-3}$)	Porosity (ϵ)
Surface soil of wet clay	1.12	0.58
Surface soil of loam	1.28	0.52
Sandy subsoil	1.61	0.39
Sandy loam compacted by heavy traffic	1.90	0.28
Spheres of uniform size in open packing	1.39	0.48
Spheres of uniform size in close packing	1.96	0.26
Sandstone	2.12	0.20
Peaty soil	0.3	–
Volcanic soil	<0.9	–

[a] Particle density (ρ_s) is taken as $2.65\,Mg\,m^{-3}$).
(Adapted from Ref. 2.)

cohesive soils sampled at water contents in the region of field capacity. The method is not suitable for non-cohesive soils, that is, sands, gravels, and gravelly soils. In the latter cases, excavation methods that involve digging a hole in the soil and determining the volume of the hole (using rubber balloon or sand displacement methods) and the oven-dried weight of the excavated soil are preferred (5, 6). For the sand displacement method, the volume of the hole is determined by filling it with sand from a pre-calibrated sand funnel (1).

For soil samples of irregular shape, for example, clods and aggregates, bulk density can be measured using the "clod method" which involves coating with paraffin wax or resin (e.g., Saran resin) and liquid displacement for determination of volume (4, 5). Saran resin coating has been used for the determination of bulk density changes of swelling soils over a range of soil water contents on a single clod (7). Since the resin has the unique property of being permeable to only water vapor molecules but not liquid water molecules, it allows the determination of shrinkage curve on one single clod, thus greatly reducing sampling variability.

With proper precautions taken and for uniform soils, field determination of bulk density using the core method has a relatively low coefficient of variation (<10%) (8). Therefore, about four samples should be sufficient to estimate the mean bulk density to within 10% of the true value, 95% of the time.

Radiation Methods

Radiation methods are based on the empirical relationships between radiation and soil wet bulk density. The most common methods involve the attenuation or scattering of gamma radiation by soil, both of which increase with bulk density (5). Because of the interaction of gamma radiation with water, simultaneous measurement of soil water content is required for the calculation of bulk density using Eq. (3). Radiation methods are potentially at least equal in accuracy to any of the direct methods (5, 9) and are simpler and quicker to use especially where measurements at depth are required. It has the added advantage of being nondestructive and therefore allows repeated measurements at the same location; however, the equipment is more expensive and involves safety issues associated with the use of radioactive radiation.

APPLICATION OF BULK DENSITY MEASUREMENTS

As Indicator of Soil Structure and Soil Quality

Bulk density has been included in a minimum data set for monitoring soil quality, as an indicator of both soil structure and soil strength (10). Changes in bulk density reflect changes in soil structure because of the relationship between bulk density and total porosity (Eq. 4). However, total porosity gives no indication of the pore size distribution and pore continuity, which are important attributes of soil structure and associated functions (11). Nevertheless, with the exception of swelling soils, soil compaction is often measured by the increases in bulk density. As bulk density of swelling soils changes with soil water content, comparison is only valid at the same soil water content.

Bulk density affects plant growth because of its effect on soil strength and soil porosity. With increasing bulk density, strength tends to increase and porosity tends to decrease; both tend to be limiting to root growth at some critical values. Critical bulk density for root growth was found to vary with different textures; for example, for sunflower, the critical value was 1.75 $Mg\,m^{-3}$ for sandy soil and 1.46–1.63 $Mg\,m^{-3}$ for clays (12, 13).

To account for the texture difference so that crop yield can be related to soil compaction (bulk density) over a whole range of soils, the concept of degree of compactness has been proposed (14). The latter refers to the ratio of the bulk density of the soil in the field to that of the same soil in a compacted reference state of 200 kPa. As an average of 100 field experiments on a range of soil types (% clay = 2–60%) the optimal degree of compactness for grain yield was 0.87, above which adverse effects start to occur.

As a Measure of Aeration Status

Together with gravimetric water content and using Eq. 5, air-filled porosity (ϵ_a) can be calculated. ϵ_a values ranging between 0.05 and 0.15 ($m^3\,m^{-3}$) are commonly used as the critical limit below which aeration is limiting to root growth (15).

Conversion of Gravimetric Data to Volumetric Data

Laboratory analyses for soil organic matter, nitrogen (N) contents, water contents, and so on are commonly expressed on a gravimetric or weight basis; however, for comparison in the field in terms of management practices, the amount per volume of soil or per area is preferred. For the conversion, soil bulk density, which frequently varies with management, depth of sampling, and time of the year, is needed. For example, to calculate the mass of organic carbon (OC) in t/ha to 0.10 m depth from gravimetric OC ($g\,kg^{-1}$),

$$\mathrm{OC}\ (t/ha) = \mathrm{OC}\ (g\,kg^{-1}) \times \rho_b\,(Mg\,m^{-3})$$

The inclusion of bulk density in a set of basic soil indicators for the proper interpretation of the change in magnitude in chemical and biochemical soil components has been proposed (10).

Estimate of Mass of Large Voluome of Soil

To estimate total mass of soil in 1 ha of land to 0.10 m depth, a knowledge of soil bulk density is needed:

$$\mathrm{Mass}\,(t/ha) = \rho_b\,(Mg\,m^{-3}) \times 1000$$

REFERENCES

1. Blake, G.R.; Hartge, K.H. Particle Density. In *Methods of Soil Analysis Part 1*; Klute, A., Ed.; Soil Science Society of America: Madison, WI, 1986; 363–375.
2. Marshall, T.J.; Holmes, J.W. Composition of Soil. In *Soil Physics* Chapter 1. Cambridge University Press: Cambridge, 1979; 10.
3. Soane, B.D. Studies on Some Physical Properties in Relation to Cultivation and Traffic. In *Soil Physical Conditions and Crop Production*; Min. Agr. Fish and Food: London, Tech. Bull. No. 29, 1975; 115–126.
4. McIntyre, D.S.; Loveday, J. Bulk Density. In *Methods for Analysis of Irrigated Soils*; Loveday, J., Ed.; Commonwealth Agricultural Bureaux: Farnham Royal, Bucks, England, 1974; 38–42.
5. Campbell, D.J.; Henshall, J.K. Bulk Density. In *Soil Analysis—Physical Methods*; Smith, K.A., Mulllins, C.E., Eds.; Marcel Dekker: New York, NY, 1991; 329–366.
6. Freitag, D.R. Methods of Measuring Soil Compaction. In *Compaction of Agricultural Soils*; Barnes, K.K., Carleton, W.W., Taylor, H.R., Throckmorton, R.I., Vanden Ber, G.E., Eds.; Am. Soc. Agric. Eng.: St Joseph, 1971; 47–103.
7. Chan, K.Y. Shrinkage Characteristics of Soil Clods from a Grey Clay Under Intensive Cultivation. Aust. J. Soil Res. **1982**, *20*, 65–74.
8. Warrick, A.W.; Neilson, D.R. Spatial Variability of Soil Physical Properties in the Field. In *Application of Soil Physics*; Hillel, D., Ed.; Academic Press: New York, NY, 1980; 319–344.
9. Soane, B.D.; Campbell, D.J.; Herkes, S.M. Hand Held Gamma-Ray Transmission Equipment for the Measurement of Bulk Density of Field Soils. J. Agric. Eng. Res. **1971**, *16*, 146–156.
10. Doran, J.W.; Parkin, T.B. Defining and Assessing Soil Quality. In *Defining Soil Quality for a Sustainable Environment*; Doran, J.W., Coleman, D.C., Bezdicek, D.F., Stewart, B.A., Eds.; Soil Sci Soc. Amer.: Madison, Wis., 1994; 3–21.
11. Kay, B.D. Rates of Change of Soil Structure Under Different Cropping Systems. In *Advances in Soil Science*; Stewart, B.A., Ed.; Springer-Verlag: Heidelberg, New York, 1990; 12, 1–52.
12. Veihmeyer, F.J.; Hendrickson, A.H. Soil Density and Root Penetration. Soil Sci. **1948**, *65*, 487–493.
13. Jones, C.A. Effect of Soil Texture on Critical Bulk Density for Root Growth. Soil Sci. Soc. Am. J. **1983**, *47*, 1208–1211.
14. Hakansson, I. A Method for Characterizing the State of Compactness of the Plough Layer. Soil & Tillage Research **1990**, *16*, 105–120.
15. Stepniewski, W.; Glinski, J.; Ball, B.C. Effects of Soil Compaction on Soil Aeration Properties. In *Soil Compaction in Crop Production*; Soane, B.D., van Ouwerkerk, C., Eds.; Elsevier Science B.V.: the Netherlands, 1994; 167–189.

CALCIFICATION

Janis L. Boettinger
Utah State University, Logan, Utah, U.S.A.

C

INTRODUCTION

Calcification is the general process by which calcium carbonate ($CaCO_3$) accumulates in soils. Most commonly, calcium carbonate accumulates in subsurface horizons of soils in subhumid, semiarid, or arid regions. The accumulation of calcium carbonate in subsoil horizons is often coupled with its removal from overlying horizons, linked by the downward translocation of the soluble ions that precipitate to form calcium carbonate. However, the properties and processes that govern water flow in soils also affect calcification. Calcification can operate at a rate and duration such that subsoil horizons become plugged with calcium carbonate, effectively limiting penetration by plant roots and reducing the conductivity of gases, solutions, and suspended solids. The objective of this entry is to describe the environmental factors, components, mechanisms, and manifestations of calcification in soils.

COMPONENTS OF CALCIFICATION

For calcification to occur in soil, there must first be sources of calcium carbonate ($CaCO_3$) or the ions that precipitate to form $CaCO_3$. Calcium (Ca^{2+}) is the cation required; bicarbonate (HCO_3^-) or, less commonly, carbonate (CO_3^{2-}) are the required anions.

The most direct source of $CaCO_3$ for calcification is calcareous soil parent materials, which already contain $CaCO_3$. Examples include limestone (1), calcareous shales and sandstones, and unconsolidated sediments derived from calcareous rocks (i.e., alluvium and lacustrine sediments). However, calcification can occur in soils that lack obvious calcareous parent materials. A likely external source of $CaCO_3$ in arid and semiarid regions is the eolian deposition of fine-grained calcareous sediments onto the soil surface ("calcareous dust") (2, 3). These calcareous eolian sediments originate from areas rich in calcareous sediments that can be easily entrained by the wind, such as playas or basins that contain dry lake beds (4).

Calcium-bearing, primary silicate minerals may also be a source for the Ca in the $CaCO_3$ that accumulates in soil (5). Silicates that can weather to provide a source of Ca for calcification include plagioclase feldspars (e.g., anorthite, $CaAl_2Si_2O_8$; oligoclase, $Na_{0.8}Ca_{0.2}Al_{1.2}Si_{2.8}O_8$; etc.) and Ca-bearing amphiboles (commonly hornblende) and pyroxenes.

Carbon dioxide gas (CO_2) is the ultimate source of bicarbonate and carbonate in soils. Carbon dioxide, which attains a higher partial pressure in soils than in the atmosphere because of respiration by plant roots and heterotrophic microorganisms, dissolves in water to form carbonic acid (H_2CO_3):

$$CO_2 + H_2O \leftrightarrow H_2CO_3$$

Carbonic acid (H_2CO_3) is a weak acid that dissociates to form bicarbonate (HCO_3^-) that dissociates very slightly to form carbonate (CO_3^{2-}) :

$$H_2CO_3 \leftrightarrow H^+ + HCO_3^- \leftrightarrow 2H^+ + CO_3^{2-}$$

The dominance of these species in soils depends on the pH. Bicarbonate is the dominant species in solutions of most soils, whereas carbonic acid dominates only in strongly acid soils and carbonate dominates only in strongly alkaline soils.

MECHANISMS OF CALCIFICATION

The first step in calcification is the chemical weathering of minerals in upper soil horizons. In soils that contain calcareous parent materials or receive calcareous eolian deposits, $CaCO_3$ must first dissolve. Some authors refer to this process as "decalcification" (6). Dissolution of $CaCO_3$ is accelerated by the presence of weak acids in soil solution because CO_3^{2-} associates readily with H^+ to form HCO_3^-. Carbonic acid is the most common solvent of $CaCO_3$ in natural systems (7):

$$CaCO_3 + H_2CO_3 \leftrightarrow Ca^{2+} + 2HCO_3^-$$

Carbonic acid also accelerates the dissolution of primary silicate minerals in soils and other natural systems. For example, diopside, a Ca-bearing pyroxene ($CaMgSi_2O_6$), can undergo carbonic acid weathering in soil solution to

Encyclopedia of Soil Science

produce Ca^{2+} and HCO_3^-, as well as Mg^{2+} and silicic acid (H_4SiO_4):

$$CaMgSi_2O_6 + 4H_2CO_3 + 2H_2O \leftrightarrow Ca^{2+} + Mg^{2+} + 4HCO_3^- + 2H_4SiO_4$$

Any process that increases CO_2 in the soil solution can increase the rate of carbonic acid weathering.

The second step in calcification involves the movement of Ca^{2+} and HCO_3^- ions in solution to the soil horizon(s) where the conditions allow the chemical precipitation of $CaCO_3$. Because atmospheric precipitation is limited in arid and semiarid climates, products of mineral dissolution are not effectively leached from soils. Instead, Ca^{2+} and HCO_3^- generally move downward with percolating solutions to deeper soil horizons. However, soil properties and processes that affect the magnitude and direction of saturated or unsaturated water flow in soils will influence where $CaCO_3$ ultimately precipitates and accumulates. For example, the downward movement of soil solution may be inhibited by an underlying slowly permeable layer (e.g., clay-rich horizon or bedrock) or by a textural discontinuity (e.g., silt loam or loam overlying loamy sand or gravelly sand). Alternatively, the soil solution may move upward in response to a water potential gradient set up by abundant water in the subsoil, such as a seasonally high water table, and evapotranspiration of water from near-surface soil in arid, semiarid, or seasonally dry climates.

The third step in calcification involves the chemical precipitation of $CaCO_3$; this step is defined exclusively by some authors as the process of "calcification" (6):

$$Ca^{2+} + 2HCO_3^- \leftrightarrow CaCO_3 + CO_2\ (\text{gas}) + H_2O\ (\text{liquid})$$

The removal of H_2O directly increases the concentration of Ca^{2+} and HCO_3^- in the soil solution and causes the precipitation of $CaCO_3$. Decreasing the partial pressure of CO_2 will also enhance the precipitation of $CaCO_3$. In the case of downward moving solution in arid and semiarid climates, deeper horizons of soils are generally drier and have lower partial pressures of CO_2 because of reduced biological activity. These conditions alone may favor the precipitation of $CaCO_3$. However, the removal of H_2O by plant roots via transpiration directly causes the precipitation of $CaCO_3$. Evapotranspiration of H_2O causes precipitation of $CaCO_3$ in soils in which the solution moves upward into drier horizons that have increasing concentrations of plant roots.

Soil microorganisms can play a direct role in the precipitation of pedogenic $CaCO_3$ (8). In soil horizons with Ca-rich solutions, some bacteria and fungi excrete excess Ca^{2+}. If the soil is sufficiently alkaline and moist, the CO_2 respired by heterotrophic microorganisms and plants dissolves and dissociates to form an abundance of HCO_3^-. Excreted Ca^{2+} combines with HCO_3^- in solution to precipitate $CaCO_3$ on the exterior surface of the microorganism.

The continued precipitation of $CaCO_3$ at about the same soil depth interval over hundreds or, perhaps, thousands of years results in the accumulation of $CaCO_3$. Calcium carbonate precipitates mainly in larger soil voids, which dry more quickly and have lower partial pressure of CO_2 than smaller micropores (9). Calcium carbonate also acts as a template for further precipitation of $CaCO_3$. The mineral form of pedogenic $CaCO_3$ is almost exclusively calcite, including that which is microbially precipitated (8).

MANIFESTATIONS OF CALCIFICATION

The morphology of $CaCO_3$ concentrations is varied (2, 10). Carbonates can be disseminated, or finely dispersed throughout the soil matrix such that specific carbonate features are not visible. Carbonates can also be segregated into visible morphological features such as filaments, soft masses, concentrically layered concretions, and irregularly shaped nodules. Carbonates can occur as coatings and pendants (coatings restricted to the undersides) on sand grains and rock fragments.

The manifestations of accumulating $CaCO_3$ with increasing time are different in gravelly and nongravelly soils (2) illustrated in this encyclopedia[a]. In gravelly soils, $CaCO_3$ first appears as thin, discontinuous coatings on rock fragments where water flows preferentially. With accumulation over time, coatings become continuous, followed by $CaCO_3$ filling in the interstices between rock fragments (Fig. 1). Eventually, $CaCO_3$ can plug the interstices and effectively cement the soil fabric, which reduces water, gas, and solute flow through the horizon. In nongravelly soils, which possess smaller pores and a larger total pore volume, $CaCO_3$ is first disseminated, then appears as filaments and coatings on sand grains. Nodules of $CaCO_3$ form, and increase in abundance and size with time. The soil fabric between the nodules eventually fills in with $CaCO_3$, effectively plugging and cementing the horizon. At this stage, in both gravelly and nongravelly soils, the downward moving solution is restricted and $CaCO_3$ accumulates at the upper boundary of the plugged horizon, forming a very hard, laminated layer.

Fig. 1 Profile of a soil with $CaCO_3$ accumulation in the subsoil. The soil formed in the seasonally dry, semiarid climate in late Pleistocene-aged deltaic sediments derived from pluvial Lake Bonneville and alluvium. Carbonates occur as coatings and pendants on rock fragments and are disseminated between rock fragments. The $CaCO_3$ is concentrated below the textural discontinuity between alluvial gravelly loam and the underlying deltaic very gravelly sandy loam and sand.

As mentioned before, the properties and processes that affect water flow in soils influence the patterns of $CaCO_3$ accumulation in soil. Similarly aged soils (ca. 13,000 years) formed in calcareous lacustrine sediments of Pleistocene pluvial Lake Bonneville in a seasonally dry climate exhibit different vertical distributions of $CaCO_3$ (Fig. 2) (11). The different patterns of $CaCO_3$ accumulation are strongly influenced by different depths to a seasonally high water table. The dominant direction of water flow is downward in soils that are not influenced by a water table, and $CaCO_3$ removed from upper horizons accumulates near the average maximum depth of wetting (e.g., Parleys, Fig. 2). Much of the water movement is upward in soils that have a seasonal high water table near the soil surface, and relatively insoluble $CaCO_3$ accumulates at about the depth

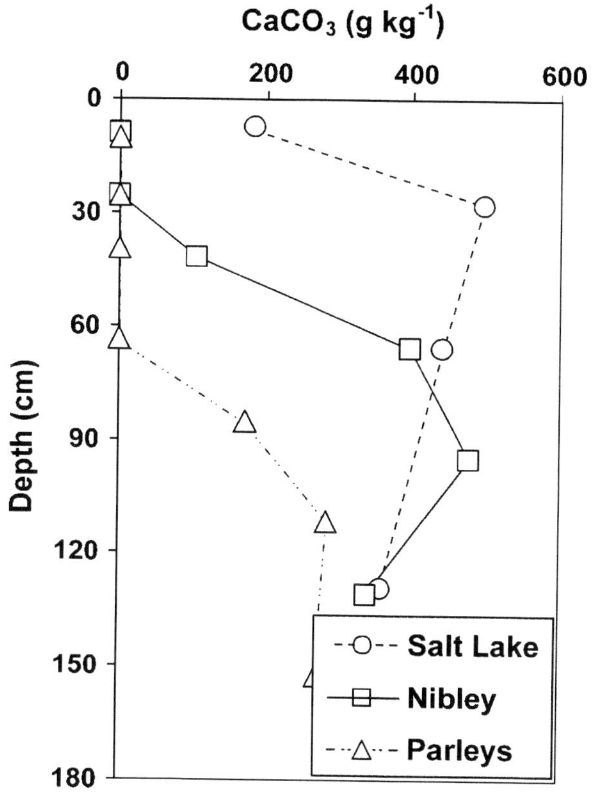

Fig. 2 Vertical distribution of $CaCO_3$ in three soils that have different depths to a seasonally high water table (11). The seasonal high water table is near the surface in the soil of the Salt Lake series and is between 75 and 100 cm in the soil of the Nibley series. The soil of the Parleys series is unaffected by a water table.

of the water table (e.g., Salt Lake, Fig. 2). Soils that experience downward percolation of water and have a seasonal water table present in the subsoil exhibit both removal of $CaCO_3$ from upper horizons and the precipitation and accumulation of $CaCO_3$ at about the depth that the downward moving solution and the water table meet (e.g., Nibley) (Fig. 2).

REFERENCES

1. Rabenhorst, M.C.; Wilding, L.P. Pedogenesis on the Edwards Plateau Texas. I. Nature and Continuity of Parent Material. Soil Sci. Soc. Am. J. **1986**, *50* (3), 678–687.
2. Gile, L.H.; Petersen, F.F.; Grossman, R.B. Morphological and Genetic Sequences of Carbonate Accumulation in Desert Soils. Soil Sci. **1966**, *101* (5), 347–360.
3. Reheis, M.C.; Goodmacher, J.C.; Harden, J.W.; McFadden, L.D.; Rockwell, T.K.; Shroba, R.R.; Sowers, J.M.; Taylor, E.M. Quaternary Soils and Dust Deposition in

Southern Nevada and California. Geol. Soc. Am. Bull. **1995**, *107* (9), 1003–1022.
4. McFadden, L.D.; Wells, S.G.; Dohrenwend, J.C. Influences of Quaternary Climatic Change on Processes of Soil Development on Desert Loess Deposits of the Cima Volcanic Field, California. Catena **1986**, *13* (4), 361–389.
5. Boettinger, J.L.; Southard, R.J. Silicate and Carbonate Sources for Aridisols on a Granitic Pediment, Western Mojave Desert. Soil Sci. Soc. Am. J. **1991**, *55* (4), 1057–1067.
6. Buol, S.W.; Hole, F.D.; McCracken, R.J.; Southard, R.J. *Soil Genesis and Classification*, 4th Ed.; Iowa State University Press: Ames, IA, 1997; 527.
7. Krauskopf, K.B. *Introduction to Geochemistry*, 2nd Ed.; McGraw-Hill: New York, 1979; 617.
8. Monger, H.C.; Daugherty, L.A.; Linderman, W.C.; Liddell, C.M. Microbial Precipitation of Pedogenic Calcite. Geology **1991**, *19*, 997–1000.
9. Chadwick, O.A.; Hendricks, D.M.; Nettleton, W.D. Silica in Duric Soils: A Depositional Model. Soil Sci. Soc. Am. J. **1987**, *51* (4), 975–982.
10. Schoeneberger, P.J.; Wysocki, D.A.; Benham, E.C.; Broderson, W.D. *Field Book for Describing and Sampling Soils, Version 1.1*; National Soil Survey Center, Natural Resources Conservation Service, US Department of Agriculture: Lincoln, NE, 1998; 185.
11. Erickson, A.J.; Mortensen, V.L. *Soil Survey of Cache Valley Area, Utah*; US Government Printing Office: Washington, DC, 1974; 192.

CALCIUM

C

Zdenko Rengel
Soil Science and Plant Nutrition, The University of Western Australia, Crawley, Western Australia, Australia

INTRODUCTION

Calcium has numerous roles in soils, plants, microflora, and fauna. In particular, Ca influences soil chemical and physical properties, maintains the structure and function of the cell plasma membrane, and is directly involved in cell signalling (secondary messenger).

CALCIUM IN SOIL

The average Ca content in the earth's crust is about 36 g/kg (1). In contrast, the average Ca content in soils is around 10 g/kg, which is about five times lower than the Ca content in igneous and sedimentary rocks, reflecting ease of Ca leaching in the weathering process as well as in well-drained soils. Calcium in igneous and metamorphic rocks occurs mainly in the plagioclase series of feldspars, with the high end of the plagioclase series weathering fairly rapidly. Noncalcareous soils contain Ca in the form of high sodium plagioclase, augite, hornblende, and epidote (2). These minerals weather slowly and therefore are not good sources of Ca. Calcareous soils have been formed from limestone or marl parent materials and frequently contain 50% or more calcium carbonate, with little Ca leaching over time due to arid climates.

Leaching of Ca from soils in wet climates can be substantial, amounting up to around 300 kg Ca/ha per year. Some fertilizers (e.g., ammonium) can increase Ca^{2+} leaching by displacing Ca^{2+} from the cation exchange complex. Losses of Ca^{2+} from soils by leaching are generally greater than the amounts taken out in farming products.

Exchangeable Calcium

The crystallographic (dehydrated) radius of Ca^{2+} cation is 0.099 nm, which is smaller than the radii of soil monovalent cations and many divalent cations (e.g., Sr^{2+}, Ba^{2+}), but not smaller than the Mg^{2+} radius (0.066 nm) (3). However, the hydrated radius of Ca^{2+} (0.43 nm) is greater than that of monovalents and also greater than radii of Mg^{2+} and Ba^{2+} (4).

Exchangeable cations are attracted to negative charges on the soil colloidal particles, forming an ion swarm in the diffuse double layer surrounding these particles. The specific adsorption increases with an increase in ion charge and with decreasing hydrated ion radius (because water molecules act as insulator). Exchangeable cations are present in the ratio $Ca^{2+} > Mg^{2+} > K^+ = Na^+$ in productive agricultural soils. Calcium occupies the major part of the cation exchange complex in the neutral soils (e.g., up to 84% in Mollisols in Russia at pH 7.0) but represents the most frequently occurring cation on the cation exchange complex even in the acidic soils (e.g., 48% of the exchange complex in Lanna soil in Sweden, see Ref. 3). Only highly acidic, sodic, or serpentine soils may not have exchangeable Ca^{2+} as the dominant cation on the exchange complex; instead, they may be dominated by H^+ (acidic), Na^+ (sodic), and Mg^{2+} (serpentine soils).

Cation exchange capacity (CEC) refers to the sum of negative sites on soil solid phase that can bind cations and therefore protect them from being leached but keep them in a state that will allow replenishment of cations depleted from the soil solution. With increased soil acidity, there is a decrease in CEC (5). As a consequence, greater amounts of cations (especially basic cations like Ca^{2+} and Mg^{2+}) are present in soil solution and are therefore prone to leaching into deeper layers of the soil profile [e.g., (5)] and eventually out of reach of roots. Liming of acidic soils results in an increase in effective CEC, with extra cation exchange sites mostly neutralized by Ca^{2+} (6).

Soil Solution Calcium

An average Ca^{2+} concentration in the soil solution of temperate region soils is between 0.8 and 19.4 mM (30 to 730 mg/L) (3, 7), with the median values around 1.5 mM (8). Optimal growth of monocots, however, is already achieved at $<10\,\mu M$ in flowing solution culture (9).

Concentrations between 10 and 1000 μM of Ca are generally needed for optimal growth of dicotyledonous species.

The soil solution Ca^{2+} is regulated by the exchangeable Ca^{2+}, but the relationship varies for different soils. The ratio between exchangeable Ca^{2+} on the soil solid particles and Ca^{2+} in soil solution represents Ca buffer power; it ranges from 19 to 107 for various soils (10), indicating a high capacity to replenish Ca^{2+} in soil solution.

Given relatively high concentration of Ca^{2+} in the soil solution, the predominant transport pathway toward roots is by mass flow rather than diffusion. However, diffusion of Ca^{2+} toward roots occurs when uptake of Ca^{2+} is faster than the resupply by mass flow. In contrast, when mass flow exceeds uptake by roots, Ca^{2+} accumulates in the rhizosphere, creating a diffusion gradient away from roots. Most plants would take up less Ca^{2+} than the amounts supplied via mass flow (e.g., ryegrass *Lolium rigidum*, subclover *Trifolium subterraneum*, and capeweed *Arctotheca calendula*), while some plant species can deplete Ca^{2+} in the rhizosphere by taking it up in excess of the supply by mass flow (e.g., lupin *Lupinus digitalus*) (11). Therefore, the relative contribution of diffusion to Ca^{2+} transport toward roots depends on the plant Ca requirement and on Ca^{2+} concentration in soil solution. If accumulation of Ca^{2+} takes place in the rhizosphere, precipitation of calcium carbonates or calcium sulfates may occur (10), especially in poorly leached, arid soils. Thus, soluble Ca^{2+} concentration in many arid soils is regulated not just by exchangeable Ca^{2+}, but also by solubility of calcium sulfate and/or calcium carbonate.

Calcium/Magnesium Ratio in Soil

Some authors have promoted the balance of soil exchangeable cations, particularly Ca^{2+} and Mg^{2+}, as an important soil test criterion. The recommended range of the exchangeable Ca:Mg ratio in soil is 3 to 7 (12). However, fertilization strategies based on cation saturation ratios (aimed at achieving a balance of soil cations instead of applying sufficient amounts of each nutrient) are not widely accepted (see Ref. 13). Although the concept has a good theoretical basis, it is undisputable that optimal plant growth may occur at a wide range of exchangeable Ca:Mg ratios (e.g., ratios ranging from 0.5 to 16 have produced constant yields for a variety of plant species and soil types (e.g., Refs. 14, 15)).

CALCIUM AS SOIL AMELIORANT

Amelioration of Acidic Soils

The widely accepted ameliorative measure for acid soils is liming, whereby Al held on the exchange complex is neutralized. Incorporating lime into the topsoil increases soil pH (3, 4). In contrast, incorporation of lime into deeper horizons to ameliorate acidic subsoil is technically difficult and frequently unprofitable. Alternatively, the surface-applied lime may in some cases allow sufficient leaching of Ca^{2+} down the profile to ameliorate subsoil acidity; this approach works well only in light-textured, sandy soils and requires lime rates sufficiently high not just to neutralize topsoil acidity, but to provide surplus alkalinity to be leached into deeper horizons.

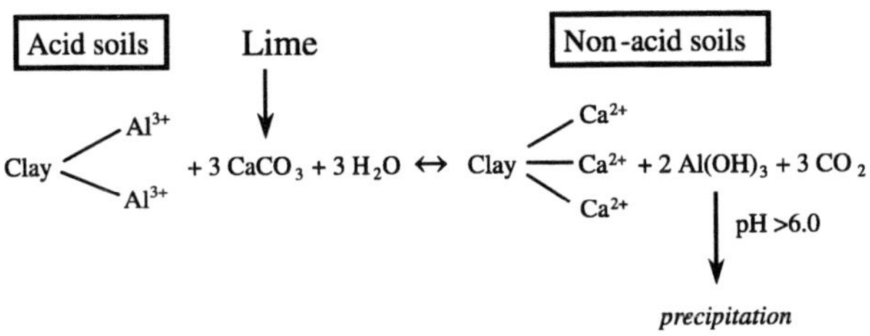

Lime reaction in soil is dependent on the nature and the fineness of the liming material. Dolomite [$CaMg(CO_3)_2$, with >15% Mg] has lower solubility but a higher surface area (smaller particle size) compared with calcite ($CaCO_3$); hence, the two types of material may have similar effectiveness in ameliorating acidic soil.

Amelioration of Sodic Soils

Sodic soils have high exchangeable Na^+ content and low hydraulic conductivity because Na promotes deflocculation and swelling of clay particles and thus dispersion of soils, especially those containing expanding montmorillonite clays (3, 4). Amelioration of sodic soils hinges on replacing exchangeable Na^+ by Ca^{2+} and the restoration of a granular physical structure. For noncalcareous sodic soils, gypsum is the Ca-containing material of choice.

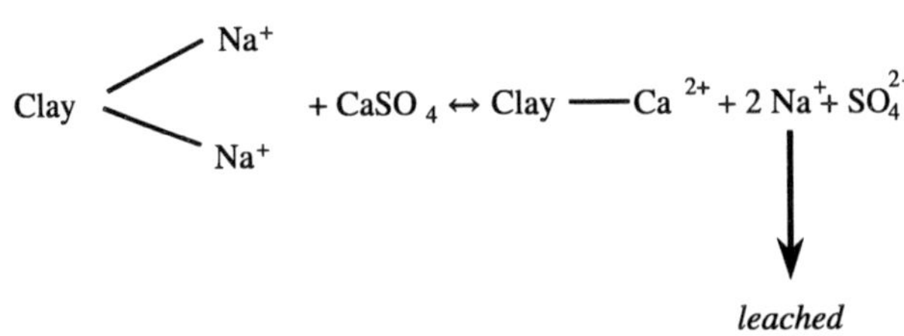

C

CALCIUM UPTAKE BY PLANTS

Calcium deficiency in soils is rare and occurs only on serpentine soils derived from Mg-rich materials or on highly leached acidic Al-saturated soils. Even under those conditions, Ca deficiency in plants may be due to Mg or Al repressing Ca uptake, rather than due to absolute Ca deficiency.

Symptoms of Ca deficiency are leaf distortion and petiole collapse (e.g., in *Glycine max*), rolled young leaves (e.g., in small grains), collapsed midrib, terminal dieback and defoliation (e.g., in *Prunus persica* seedlings), or reduced rate of wood formation and decreased amount of functional sapwood and live crown (see Ref. 16).

An average Ca concentration in the plant material is around 5 g/kg dry weight, which is about 125,000 times greater than for molybdenum as the least abundant nutrient and about 1/8 of nitrogen as the most abundant nutrient in plant tissue. Greater tissue concentrations of Ca are required in dicot compared to monocot plants [cf. (17)]. Most of Ca is bound in the cell wall (about 60%) or sequestered in different organelles (18). The activity of free cytosolic Ca^{2+} is therefore very low [10^{-7} to 10^{-6} M; (19)], preventing reaction of Ca^{2+} with inorganic phosphates and thus Ca^{2+} cytotoxicity.

Plant roots have cation exchange capacity similarly to soils (see Ref. 20). Therefore, one cation exchanger (root) is in contact with the other (soil), resulting in competition for cations based on physicochemical principles. Plant capacity to take up cations is at least partly influenced by the root CEC (e.g., Ref. 21) because ionic environment at the plasma membrane surface (where uptake of all ions takes place) depends on the strength of the electrical field developed in interaction of two charged surfaces (root and soil particles).

Calcium uptake is via an apoplastic pathway and is thus hindered by the Casparian strip or any suberization in the root cell walls in general. Therefore, the most effective transport of Ca^{2+} is at the root tip (22). Transport of Ca^{2+} across the root-cell plasma membrane into the cytosol is down the concentration gradient (high in the apoplast, low in the cytosol) via Ca^{2+}-selective channels (23).

Calcium is transported mainly in the xylem sap, while its mobility in the phloem sap is poor. As a consequence, concentration of Ca is greater in mature and old leaves than in the young ones (e.g., Ref. 24). Similarly, transport of Ca into fruits and grain is poor. The typical example of inadequate supply of Ca to developing fruits is blossom-end rot in capsicum and tomatoes (e.g., Ref. 25) and bitter pits in apples and pears (e.g., Ref. 26).

Calcium content in soils determines species distribution in the natural ecosystems. Calcicole species are adapted to soils rich in Ca (these species are also most sensitive to Al). In contrast, calcifuge species have a low Ca-demand and are therefore adapted to acid, low-Ca soils (these species are also less sensitive to Al) (8).

CONCLUSIONS

As the predominant soil exchangeable cation, Ca^{2+} determines chemical and physical properties of soil. Calcium can easily be leached, causing soils to acidify. Calcium lost by leaching and by export in farming products should be replenished. Liming is used to restore Ca content as well as to neutralize acidity.

REFERENCES

1. Glendinning, J.S. *Australian Soil Fertility Manual*; CSIRO Publishing: Collingwood, Vic, Australia, 2000; 154.
2. Bear, F.E. *Chemistry of the Soil*, 2nd Ed.; Van Nostrand Reinhold Company: New York, 1969; 515.
3. Bohn, H.L.; McNeal, B.L.; O'Connor, G.A. *Soil Chemistry*, 2nd Ed.; Wiley-Interscience: New York, 1985; 341.
4. Tan, K.H. *Principles of Soil Chemistry*, 3rd Ed.; Marcel Dekker Inc.: New York, 1998.
5. Blake, L.; Goulding, K.W.T.; Mott, C.J.B.; Johnston, A.E. Changes in Soil Chemistry Accompanying Acidification Over More Than 100 Years Under Woodland and Grass at Rothamsted Experimental Station, UK. Eur. J. Soil Sci. **1999**, *50*, 401–412.
6. Hochman, Z.; Edmeades, D.C.; White, E. Changes in Effective Cation Exchange Capacity and Exchangeable Aluminium with Soil pH in Lime-Amended Field Soils. Aust. J. Soil Res. **1992**, *30*, 177–187.
7. Adams, F. Soil Solution. In *The Plant Root and Its Environment*; Carson, E.W., Ed.; University Press of Virginia: Charlottesville, 1974; 441–481.
8. Runge, M.; Rode, M.W. Effects of Soil Acidity on Plant Associations. In *Soil Acidity*; Ulrich, B., Sumner, M.E., Eds.; Springer: Berlin/Heidelberg, 1991; 183–202.
9. Loneragan, J.F.; Snowball, K. Calcium Requirement of Plants. Aust. J. Agric. Res. **1969**, *20*, 465–478.
10. Barber, S.A. *Soil Nutrient Bioavailability. A Mechanistic Approach*, 2nd Ed.; John Wiley & Sons: New York, 1995.
11. Barber, S.A.; Ozanne, P.G. Autoradiographic Evidence for the Differential Effect of Four Plant Species in Altering the Ca Content of the Rhizosphere Soil. Soil Sci. Soc. Amer. Proc. **1970**, *34*, 635–637.
12. Kinsey, N.; Walters, C. *Neal Kinsey's Hands-on Agronomy*; Acres USA, 1995.
13. Dahnke, W.C.; Olson, R.A. Soil Test Correlation, Calibration and Recommendation. In *Soil Testing and*

Plant Analysis; Westerman, R.L., Ed.; Soil Science Society of America: Madison, WI, 1991; 65.
14. Simson, C.R.; Corey, R.B.; Sumner, M.E. Effect of Varying Ca:Mg Ratios on Yield and Composition of Corn (*Zea mays*) and Alfalfa (*Medicago sativa*). Commun. Soil Sci. Plant Anal. **1979**, *10*, 153–162.
15. McLean, E.O.; Hartwig, R.C.; Eckert, D.J.; Triplett, G.B. Basic Cation Saturation Ratios as a Basis for Fertilising and Liming Agronomic Crops. II. Field Studies. Agron. J. **1983**, *75*, 635–639.
16. Rengel, Z. Role of Calcium in Aluminium Toxicity. New Phytol. **1992**, *121*, 499–513.
17. Marschner, H. *Mineral Nutrition of Higher Plants*, 2nd Ed; Academic Press: London, 1995.
18. Clarkson, D.T. Calcium Transport Between Tissues and Its Distribution in the Plant. Plant Cell Environ. **1984**, *7*, 449–456.
19. Bush, D.S. Calcium Regulation in Plant Cells and Its Role in Signalling. Annu. Rev. Plant Physiol. Plant Mol. Biol. **1995**, *46*, 95–122.
20. Rengel, Z.; Robinson, D.L. Determination of Cation Exchange Capacity of Ryegrass Roots by Summing Exchangeable Cations. Plant Soil **1989**, *116*, 217–222.
21. Rengel, Z. Competitive Al^{3+} Inhibition of Net Mg^{2+} Uptake by Intact *Lolium multiflorum* Roots. II. Plant Age Effects. Plant Physiol. **1990**, *93*, 1261–1267.
22. Ferguson, I.B.; Clarkson, D.T. Ion Transport and Endodermal Suberization in the Roots of *Zea mays*. New Phytol. **1975**, *75*, 69–79.
23. Piñeros, M.; Tester, M. Calcium Channels in Higher Plant Cells: Selectivity, Regulation and Pharmacology. J. Exp. Bot. **1997**, *48*, 551–577.
24. Drossopoulos, J.B.; Bouranis, D.L.; Bairaktari, B.D. Patterns of Mineral Nutrient Fluctuations in Soybean Leaves in Relation to Their Position. J. Plant Nutr. **1994**, *17*, 1017–1035.
25. Ho, L.C.; Belda, R.; Brown, M.; Andrews, J.; Adams, P. Uptake and Transport of Calcium and the Possible Causes of Blossom-End Rot in Tomato. J. Exp. Bot. **1993**, *44*, 509–518.
26. Raese, J.T.; Drake, S.R. Effect of Calcium Spray Materials, Rate, Time of Spray Application, and Rootstocks on Fruit Quality of 'Red' and 'Golden Delicious' Apples. J. Plant Nutr. **2000**, *23*, 1435–1447.

CARBONATES

Douglas W. Ming
NASA Johnson Space Center, Houston, Texas, U.S.A.

INTRODUCTION

There are approximately 150 carbonate minerals that occur in nature; however, most of these carbonates are relatively rare. The most common rock-forming carbonates are calcite ($CaCO_3$) and dolomite ($CaMg(CO_3)_2$), which account for over 90% of natural carbonates (1). Calcite, dolomite, siderite ($FeCO_3$), and ankerite ($Ca(Mg,Fe)(CO_3)_2$) are the most common carbonates that occur in sedimentary environments (2). Carbonates are very common in soils and they either form during soil formation (i.e., pedogenic) or are inherited from parent materials (i.e., detrital or lithogenic). The most common carbonates that occur in soils are calcite, Mg-calcite ($Ca_xMg_{1-x}(CO_3)$), and dolomite. Aragonite ($CaCO_3$) occasionally occurs in soils and siderite has been reported to occur in some soils (3).

Carbonate dissolution and precipitation are among the most important chemical reactions in soils. Carbonates precipitate as coatings on soil particles and in soil pores that may result in cementation of particles. Other important reactions involving carbonate minerals are adsorption and/or precipitation of some anions and cations on their surfaces which has an impact on the mobility of essential plant growth elements (e.g., Fe^{2+}, Fe^{3+}, Zn^{2+}, HPO_4^{2-}, $H_2PO_4^-$) and elements of environmental concern (e.g., Cd^{2+}, Pb^{2+}). Carbonates (e.g., calcite, dolomite) are also added to raise the pH of acid soils (i.e., carbonates as liming agents) or to immobilize heavy metals (e.g., Pb^{2+}) in contaminated soils through precipitation of relatively insoluble carbonates (e.g., $PbCO_3$). These important processes have led to many studies of carbonate reactions and occurrences in soil environments.

STRUCTURE AND COMPOSITION

The carbonate ion $(CO_3)^{2-}$ is the basic structural unit for all carbonate minerals. The large planar CO_3 group occurs as a rigid structure with the C atom occupying the center, which is surrounded by three oxygen atoms arranged in an equilateral triangle. The linkage of the carbonate group to other cations has minimal effect on the ideal C–O bond angle of 120° in the CO_3 group (4). The nonhydrous carbonates can be grouped according to their structure into the calcite group, dolomite group, or the aragonite group (see Table 1). Hydrated carbonates have entirely different crystal structures and most crystallize in the monoclinic crystal system (Table 1).

Calcite group minerals crystallize in the hexagonal crystal system. The structure of calcite, which is the most common calcite-group mineral, is a distorted NaCl type structure where Ca alternates with CO_3. This distorted structure places Ca atoms in a face-centered rhombohedral unit cell (Fig. 1), which is the morphologic or cleavage cell. The smaller true unit cell is an elongated rhombohedral unit cell, which is easily compared to the hexagonal unit cell (Fig. 1). Magnesite, siderite, rhodochrosite ($MnCO_3$), and smithsonite ($ZnCO_3$) are isostructural with calcite (Table 1).

Aragonite-group minerals crystallize in the orthorhombic crystal system. In this group, large divalent cations (e.g., ionic radii >0.1 nm) do not permit a stable six-fold coordination and, therefore, result in a nine-fold coordination configuration. Aragonite, which is a higher tem-perature and pressure $CaCO_3$ polymorph (calcite is the stable polymorph at room temperature), has CO_3 groups in two structural planes rather than the single structural plane of calcite. Witherite ($BaCO_3$), strontianite ($SrCO_3$), and cerussite ($PbCO_3$) are isostructural with aragonite (Table 1).

The third major group of carbonate minerals, the dolomite group, crystallizes in the hexagonal system. Their structure is similar to calcite-group minerals, except that different cations (e.g., Ca–Mg, Ca–Fe, Ca–Mn) alternate along the *c*-axis. Because of this arrangement of alternating cations, the symmetry of dolomite-group minerals is reduced compared to the calcite-group minerals. Dolomite has Ca and Mg alternating along the *c*-axis. Ankerite and kutnahorite ($CaMn(CO_3)_2$) are isostructural with dolomite (Table 1).

Various cations may substitute into the carbonate structure. For example, Mg^{2+} may substitute for Ca^{2+} in the calcite structure. These Mg-substituted calcites are called magnesian calcites or Mg-calcites. Mg-calcites containing up to 20 mol% $MgCO_3$ are found in skeletal materials of marine organisms and even higher concentrations of Mg (e.g., 45 mol% $MgCO_3$) in the calcite structure have been synthesized at high pressure and temperature. Most soil Mg-calcites, however, contain less than 10 mol% $MgCO_3$

Table 1 Representative unit-cell formulae and selected properties of common carbonates

Carbonate group	Ideal unit-cell formula	Crystal system	Specific gravity
Calcite group			
Calcite	$CaCO_3$	Hexagonal (rhomobohedral)	2.71
Magnesite	$MgCO_3$	Hexagonal	3.00
Siderite	$FeCO_3$	Hexagonal	3.97
Rhodochrosite	$MnCO_3$	Hexagonal	3.70
Smithsonite	$ZnCO_3$	Hexagonal	4.43
Otavite	$CdCO_3$	Hexagonal	4.96
Gaspéite	$NiCO_3$	Hexagonal	4.39
Aragonite group			
Aragonite	$CaCO_3$	Orthorhombic	2.95
Witherite	$BaCO_3$	Orthorhombic	4.3
Strontianite	$SrCO_3$	Orthorhombic	3.7
Cerussite	$PbCO_3$	Orthorhombic	6.55
Dolomite group			
Dolomite	$CaMg(CO_3)_2$	Hexagonal	2.85
Ankerite	$CaFe(CO_3)_2$	Hexagonal	3.01
Kutnahorite	$CaMn(CO_3)_2$	Hexagonal	3.12
Minrecordite	$CaZn(CO_3)_2$	Hexagonal	3.45
Other carbonates			
Azurite	$Cu_3(CO_3)_2(OH)_2$	Monoclinic	3.77–3.89
Malachite	$Cu_2CO_3(OH)_2$	Monoclinic	3.7–4.1
Huntite	$Mg_3Ca(CO_3)_4$	Hexagonal	2.70
Landsfordite	$MgCO_3{\cdot}5H_2O$	Monoclinic	1.69
Artinite	$Mg_2(OH)_2CO_3{\cdot}3H_2O$	Monoclinic	2.04
Nesquehonite	$MgCO_3{\cdot}3H_2O$	Monoclinic	1.84
Hydromagnesite	$Mg_5(CO_3)_4(OH)_2{\cdot}4H_2O$	Monoclinic	2.25
Nyerereite	$Na_2Ca(CO_3)_2$	Orthorhombic	2.54
Gaylussite	$Na_2Ca(CO_3)_2{\cdot}5H_2O$	Monoclinic	1.99
Nahcolite	$NaHCO_3$	Monoclinic	2.19
Thermonatrite	$Na_2CO_3{\cdot}H_2O$	Orthorhombic	2.26
Soda (Natron)	$Na_2CO_3{\cdot}10H_2O$	Monoclinic	1.46
Trona	$NaCO_3{\cdot}NaHCO_3{\cdot}2H_2O$	Monoclinic	2.13

(Adapted from Ref. 5 and the JCPDS—International Centre for Diffraction Data.)

(3). Due to the smaller ionic radius of Mg^{2+} as compared to Ca^{2+}, the *a* and *c* unit cell axes decrease in length as the substitution of Mg^{2+} increases in the structure of calcite (7). Detailed discussion on the substitution of cations into carbonate structures can be found in Refs. 5, 8.

OCCURRENCE IN SOILS

Carbonates are common in many soils of the world, particularly in soils of subhumid to arid regions. Calcite is the most common carbonate in soils, but dolomite, aragonite, and siderite are also found in certain soils. Carbonates may be either inherited into soil or form directly via precipitation processes. The most common sources and mechanisms for carbonate occurrences in soils are: 1) inheritance from parent material, 2) precipitation following dissolution of soluble Ca-bearing minerals, 3) precipitation by reaction of CO_2-charged water with Ca^{2+} released by decomposition of plant residues, 4) deposition of windblown carbonates at the soil-surface and subsequent translocation within the soil by solution and precipitation, 5) precipitation by interaction of Ca^{2+} in rainwater entering the soil and combining with HCO_3^- from CO_2-charged water, 6) precipitation from CO_2-charged natural surface or irrigation water containing Ca^{2+}, and 7) precipitation in soils from groundwater that has moved through carbonate-containing soils or sediments (3).

Calcite and Mg-calcite are the two carbonates that commonly precipitate in soils. Although all of the carbonates commonly found in soils may be inherited,

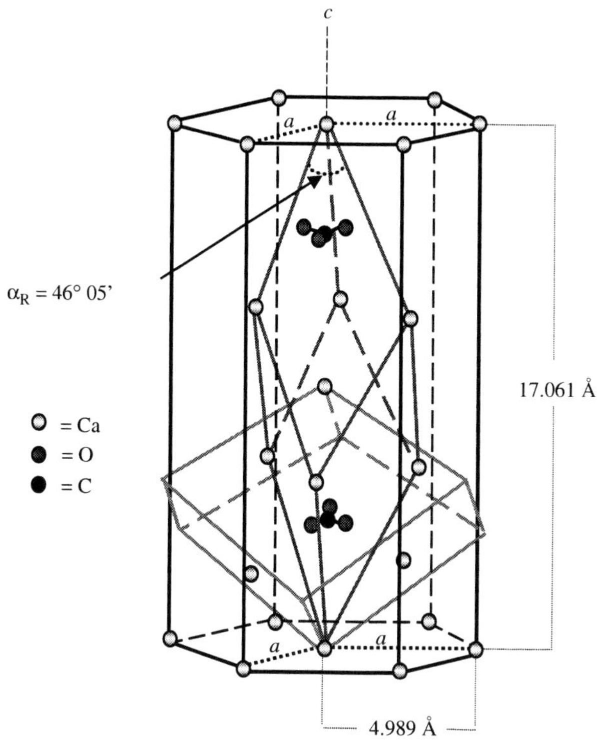

Fig. 1 Structure of calcite. Unit cell parameters for *a* and *c* are shown for the hexagonal unit cell. The elongated rhombohedral, true unit cell contains two $CaCO_3$; the two CO_3 groups are shown ($\alpha_R = \alpha_{\text{rhombohedral-cell}}$). The morphological or cleavage cell contains four $CaCO_3$ (face-centered rhombohedral cell). Note that ionic radii for Ca, C, and O are not drawn to scale. (Adapted from Ref. 6.)

aragonite and dolomite are almost always inherited. Pedogenic dolomite has been reported to occur in some saline, alkaline soils, but generally, its formation in soil is very rare (9). Mg-calcite with high $MgCO_3$ mole% (e.g., >10 mol%) is typically inherited in soils from marine sediments (i.e., from biogenic Mg-calcite). Siderite is relatively unstable in soil environments and is not known to form directly in soils. Magnesium carbonates are very rare in soils, but hydrated Mg carbonates (e.g., hydromagnesite, nesquehonite) may form at the surface during evaporation of solutions high in Mg^{2+} and HCO_3^- (10). Mechanisms by which carbonates precipitate in soils are provided in the following section.

Distribution of carbonates in a soil profile depends on the climate, soil chemical properties, and degree of leaching through the profile. In many arid soils, carbonates are common secondary products in the top horizons (e.g., A and B horizons). As a general rule, carbonates tend to be leached or removed from the upper horizons with increasing leaching or rainfall. Carbonates leached from upper horizons may precipitate as secondary carbonates in the lower horizons where they may form carbonate cemented layers (e.g., petrocalcic horizons).

Sodium carbonates may occur at the soil's surface in arid regions where soil salinity is high. Highly soluble sodium carbonates precipitate at the surface during periods of evaporation and such conditions produce soil pH values of higher than 8.5.

FORMATION

Carbonate dissolution or precipitation is dependent on several factors, such as the partial pressure of dissolved CO_2 in solution and temperature. A typical reaction for the dissolution of calcite may be expressed as a function of CO_2 as:

$$CaCO_3\ (\text{calcite}) + CO_2\ (\text{gas}) + H_2O \leftrightarrow Ca^{2+} + 2HCO_3^- \quad (1)$$

The equilibrium constant (K^0) for this reaction at 25°C is

$$K^0 = \frac{(Ca^{2+})(HCO_3^-)^2}{P_{CO_2}} = 10^{-5.90} \quad (2)$$

where P_{CO_2} is the partial pressure of CO_2 and brackets denote activities (11). The reaction may also be expressed as a function of H^+:

$$CaCO_3\ (\text{calcite}) + H^+ \leftrightarrow Ca^{2+} + HCO_3^- \quad (3)$$

with the respective equilibrium constant at 25°C being

$$K^0 = \frac{(Ca^{2+})(HCO_3^-)}{(H^+)} = 10^{+1.92} \quad (4)$$

Eqs. 2 and 4 indicate that the dissolution of calcite (and other carbonates) is favored with increasing CO_2 pressure and decreasing pH. Conversely, the degassing or removal of CO_2 from solution or an increase in pH will favor calcite participation. The equilibrium pH of a solution in contact with calcite at atmospheric CO_2 pressure ($10^{-3.5}$ bar P_{CO_2}) is 8.3. Detailed discussion on carbonate solution chemistry can be found in Refs. 11, 12.

PROPERTIES

Calcium carbonate has an influence on several soil properties, e.g., soil pH, adsorption–desorption, and soil cementation. Most soils that contain calcium carbonate will have a pH in the range of 7.1–8.5, in which the carbonate acts as a pH buffer. The surfaces of calcite are reactive and various ions may adsorb or interact at the crystal's surface.

For example, Mg^{2+}, Zn^{2+}, Cu^{2+}, Fe^{2+}, and Al^{3+} may replace Ca^{2+} on exposed surface lattice sites. The reactive surfaces of carbonates may adsorb essential plant ions such as HPO_4^{2-}, $H_2PO_4^-$, Fe^{2+}, Fe^{3+}, Zn^{2+}, and Mn^{2+} and adversely affect their availability for plant uptake. Iron deficiency chlorosis in plants has been attributed to the interaction of Fe and HCO_3^- in calcareous soils (13, 14). Other factors, such as quantity, mineralogy, and crystallinity of iron oxide phases in the soil, may also contribute to plant Fe chlorosis in calcareous soils (15).

Precipitation of carbonates on soil particles and in soil pores may form layers (i.e., calcic horizons) that impede the movement of water (16). Some layers may become indurated or cemented (i.e., petrocalcic horizons) and force water to move laterally in the soil.

Soil carbonates play an important role in the global carbon cycle, although their role in the greenhouse effect is not well understood (17). Soil inorganic carbon (i.e., lithogenic and pedogenic carbonates) accounts for about one-third of the total carbon in soils; the remaining two-thirds of soil carbon is comprised of organic carbon. Soil organic carbon is the primary C pool in humid regions, whereas carbonates are the predominant C pool in arid and semiarid regions. The role of pedogenic carbonates in atmospheric C sequestration is not clear; however, the flux of pedogenic carbonates with the atmosphere has been estimated as $0.023\,\mathrm{Pg\,C\,yr^{-1}}$, with a turnover time in the range of 30,000–90,000 years (18).

Carbonates such as calcite have been suggested for use in the remediation of Pb-contaminated soils where Pb^{2+} reacts with HCO_3^- to form insoluble Pb carbonates. Other soil contaminants (e.g., Ba^{2+}, Cd^{2+}) may also be immobilized through carbonate precipitation with the addition of more soluble carbonate amendments (19).

In summary, carbonates are of great importance in soils because they have considerable influence on the physical and chemical properties of soils, they have an important role in the global C cycle, and, occasionally, carbonates are the predo minant mineral phases that occur in soils of arid and semiarid regions.

REFERENCES

1. Reeder, R.J. Crystal Chemistry of the Rhombohedral Carbonates. In *Carbonates: Mineralogy and Chemistry*; Reeder, R.J., Ed.; Mineralogical Society of America: Washington, DC, 1983; Vol. 11, 1–47.
2. Lippmann, F. *Sedimentary Carbonate Minerals*; Springer: Secaucus, NJ, 1973; 228.
3. Doner, H.E.; Lynn, W.C. Carbonate, Halide, Sulfate, and Sulfide Minerals. In *Minerals in Soil Environments*; Dixon, J.B., Weed, S.B., Eds.; Soil Science Society of America: Madison, WI, 1989; 279–330.
4. Effenberger, H.; Mereiter, K.; Zemann, J. Crystal Structure Refinements of Magnesite, Calcite, Rhodochrosite, Siderite, Smithsonite, and Dolomite with Discussion of Some Aspects of the Stereochemistry of Calcite-Type Carbonates. Z. Krist. **1981**, *156*, 233–243.
5. Chang, L.L.Y.; Howie, R.A.; Zussman, J. Rock-Forming Minerals. *Non-Silicates: Sulphates, Carbonates, Phosphates, Halides*, 2nd Ed.; Longman: Essex, 1996; Vol. 5B, 383.
6. Hurlbut, C.S., Jr.; Klein, C., Jr. *Manual of Mineralogy (After James D. Dana)*, 19th Ed.; Wiley: New York, 1977; 532.
7. Paquette, J.; Reeder, R.J. Single-Crystal X-Ray Structure Refinements of Two Biogenic Magnesian Calcite Crystals. Am. Mineral. **1990**, *75*, 1151–1158.
8. Reeder, R.J. Carbonates: Mineralogy and Chemistry. *Reviews in Mineralogy*; Mineralogical Society of America: Washington, DC, 1983; Vol. 11, 394.
9. Kohut, C.; Muehlenbachs, K.; Dudas, M.J. Authigenic Dolomite in a Saline Soil in Alberta, Canada. Soil Sci. Soc. Am. J. **1994**, *59*, 1499–1504.
10. Ming, D.W.; Franklin, W.T. Synthesis and Characterization of Lansfordite and Nesquehonite. Soil Sci. Soc. Am. J. **1985**, *49*, 1303–1308.
11. Lindsay, W.L. *Chemical Equilibria in Soils*; Wiley: New York, 1979; 449.
12. Langmuir, D. *Aqueous Environmental Geochemistry*; Prentice-Hall: Upper Saddle Road, NJ, 1997; 600.
13. Lindsay, W.L.; Thorne, D.W. Bicarbonate Ion and Oxygen Level as Related to Chlorosis. Soil Sci. **1954**, *77*, 271–279.
14. Inskeep, W.P.; Bloom, P.R. Soil Chemical Factors Associated With Soybean Chlorosis in Calciaquolls of Western Minnesota. Agron. J. **1987**, *79*, 779–786.
15. Loeppert, R.H.; Hallmark, C.T. Indigenous Soil Properties Influencing the Availability of Iron in Calcareous Soils. Soil Sci. Soc. Am. J. **1985**, *49*, 597–603.
16. Stuart, D.M.; Dixon, R.M. Water Movement and Caliche Formation in Layered Arid and Semi-Arid Soils. Soil Sci. Soc. Am. Proc. **1973**, *37*, 323–324.
17. Lal, R., Kimble, J.M., Eswaran, H., Stewart, B.A., Eds. *Global Climate Change and Pedogenic Carbonates*; Lewis Publishers: Boca Raton, FL, 2000; 305.
18. Lal, R.; Kimble, J.M. Inorganic Carbon and the Global C Cycle: Research and Development Priorities. In *Global Climate and Pedogenic Carbonates*; Lal, R., Kimble, J.M., Eswaran, H., Stewart, B.A., Eds.; Lewis Publishers: Boca Raton, FL, 2000; 291–302.
19. Doner, H.E.; Grossl, P.R. Carbonates and Evaporates. In *Soil Mineralogy with Environmental Applications*; Dixon, J.B., Schulze, D.G., Eds.; Soil Science Society America: Madison, WI, 2002; in press.

CHEMICAL COMPOSITION

Goro Uehara
University of Hawaii, Honolulu, Hawaii, U.S.A.

C

INTRODUCTION

Every student of soil science learns to recognize and appreciate soil differences, and to understand how these differences affect soil behavior and performance. Soils can differ in particle size distribution or texture, in the way the individual particles are arranged and cemented to form compound particles or peds, in the amount and type of organic matter, and in the mineralogy of the inorganic particles. Variations in any of the above will determine whether a soil will behave like a sponge to absorb and hold water, as a sieve and filter to purify water as it flows through the soil, or as a plastic or elastic rheologic body to withstand the pressure of a tractor wheel. How a soil behaves will, in turn, determine whether it will perform well as a soybean field, building foundation, playground, or waste disposal site.

The soil differences just described generally go hand-in-hand with differences in chemical composition. A soil that is 95% quartz sand or silicon dioxide will clearly behave and perform differently in many ways from another that is 95% coral sand or calcium carbonate. These two soils would in turn behave and perform very differently from an organic soil, and all three would again differ greatly from a tropical soil rich in iron oxides. How do soils with such diverse chemical make-up originate? The purpose of this paper is to summarize the bio-geochemical processes and factors that lead to differences in the chemical composition of soils.

SOIL-FORMING PROCESSES

Changes in the chemical composition of the rock and the soil that forms from it are the result of four soil forming processes: transformations, translocations, additions, and losses.

Transformations

The process of weathering takes place because the minerals that make up the rock are thermodynamically unstable under the conditions prevailing at the site. Under these conditions, minerals exposed to the elements undergo hydration, hydrolysis, dissolution, and oxidation–reduction. Ions that go into solution, if not washed out to sea, seek new equilibria to form secondary or pedogenic minerals (1). In dry regions, one of the first secondary minerals to form is calcite. As rainfall and leaching intensity increase, Na, K, Ca, and Mg leave the system and secondary, layered alumino-silicates become more prominent. And with time and intense leaching, desilication transforms even the secondary silicates to metal oxides and hydrated oxides. It is no accident that bauxite deposits are more common in the tropics than elsewhere in the world. An example of concentration of aluminum through weathering is illustrated by the data in Table 1.

Translocations

Termites, earthworms, and burrowing animals move huge quantities of soil material to the surface. Surface soil materials can also fall between cracks in soils that shrink and fracture when dried. Repeated wetting and drying result in churning and inversion of the soil creating surface patterns called gilgai. These inverting soils are aptly named Vertisols. Particles suspended in water can also be transported to greater depth and deposited on walls of water-conducting pores. The wall coatings are called argillans or clay skins and can significantly increase the clay content of the subsoil horizons called argillic horizons.

Additions

Addition of water to soils is taken for granted but water enables most other soil processes to occur. Water is necessary for life and the production of soil organic matter. Weathering which may begin as a geochemical reaction soon turns into a bio-geochemical process with the first appearance of lichens on rock surfaces. Additions of dissolved gases and electrolytes play critical roles in soil formation. Rainfall also scrubs particulate matter from the atmosphere. Tropospheric dust from Asia containing quartz and mica is deposited over the northern Pacific but its concentration on land is highly dependent on rainfall (12). Close to their sources, aerosolic dusts can form thick loess deposits.

Table 1 Chemical compositions of (A) melilite-mepheline basalt and (B) soil formed from it

(A) *Melitite-mepheline basalt*

Chemical composition	%	Chemical composition	%	Chemical composition	%	Chemical composition	%
SiO_2	35.85	MgO	1398	CO_2	0.09	BaO	0.16
Al_2O_3	9.76	CaO	13.78	TiO_2	3.68	H_2O+	3.59
Fe_2O_3	5.09	Na_2O	2.22	P_2O_5	1.05	*Total*	99.76
FeO	9.59	K_2O	0.83	MnO	0.10		

(From Ref. 12.)

(B) *Soil*

Chemical composition	Depth (cm) 0–38	38–48	48–75	75–100	100–155
SiO_2	8.5	2.9	1.0	0.4	0.4
Al_2O_3	31.3	20.6	20.8	28.5	30.1
Fe_2O_3	45.5	48.2	46.4	42.3	41.3
MgO	0.0	0.0	0.0	0.0	0.0
CaO	0.0	0.0	0.0	0.0	0.0
Na_2O	0.6	0.7	0.7	0.7	0.0
K_2O	0.23	0.23	0.1	0.0	0.0
TiO_2	8.7	8.0	8.3	7.3	6.8
P_2O_5	0.70	0.63	0.60	0.61	0.64
MnO	0.07	0.07	0.04	0.09	0.08
H_2O+	5.0	19.7	20.5	20.7	20.2
Total	100.1	100.4	97.8	100.0	99.5

(From Ref. 13.)

Losses

One soil's loss can be another soil's gain. Wind erosion that generates aerosolic dust exposes underlying materials to the elements to renew the weathering processes. Landslides and water erosion do the same and create new parent material downstream.

Less obvious soil losses occur through solution in drainage waters as evidenced by the salty sea. Desilication or leaching of silica from soils into streams and oceans is evident in the data presented in Table 1. Less obvious are gaseous losses through microbial activity. Carbon dioxide, methane and nitrous oxide fluxes into and out of soil are measured with renewed interest because of their role as greenhouse gases.

SOIL-FORMING FACTORS

On a geologic time scale, soils are transient bodies on the landscape. What we see today may have little resemblance to what was there in the beginning, and what will remain in the end. The birth of a soil might begin with the solidification of molten lava to form solid rock or the exposure of sea floor from geologic uplift. The end could come with a landslide or during accelerated soil erosion on a cultivated hillside. But between beginning and end, major chemical transformation usually occurs so that the end product would be unrecognizable from the parent rock. In addition to time and the parent rock, three additional factors operate over time to produce differences in soil chemical composition (3). These three factors are climate, topography, and biota. All five of these factors have been discussed earlier, but will now be covered with emphasis on soil chemical composition.

Parent Rock

One would expect the chemical composition of the parent rock to have a significant effect on the chemical composition of the soil that forms from it. It does—but

C

the departure from initial rock composition is generally greater than one would expect (see Table 1). This departure from parent rock composition occurs because soils are open systems that gain new substances and gradually lose their original minerals primarily through erosion and dissolution and leaching. Rock-forming minerals, however, differ in their susceptibility to weathering (4). Acid igneous rocks containing potassium feldspars, muscovite and quartz tend to be more resistant to weathering than basic igneous rocks containing plagioclase feldspars, biotite, and amphiboles. Igneous rock erupted as lava or tefra having identical chemical compositions can and will, most likely weather into very different chemical end products. Tefra consisting mainly of glassy particles tends to weather more rapidly than lava rock and develop into soils that fall in the unique Andisol soil order.

Time

Mohr (5), a Dutch scientist investigating soils in Indonesia during colonial times marveled at the soils he found there, and to organize his findings, categorized soils as being either juvenile, mature, or senile. Mohr's juvenile soils still retained features of their parent rock and had not yet developed distinct features that set them apart from other soils. The mature soils were those favored by farmers. In mature soils, the parent rock had been weathered to clay but nutrient elements such as Ca, Mg, K, and P were still plentiful. But what intrigued Mohr most were the senile soils. Prolonged weathering under warm and humid conditions had stripped the soils of life sustaining nutrients and rendered them unsuitable for farming. What remained on the landscape were the insoluble residues of iron and

Table 2 Effect of climate on chemical composition of two soils developed from volcanic ash: (A) in a dry zone (mean annual rainfall: 800 mm; mean annual temperature 17°C) and (B) wet zone (mean annual rainfall: 3400 mm; mean annual temperature: 18°C)

(A) *Dry zone*

	Depth (cm)				
Chemical composition	**0–13**	**13–25**	**25–50**	**50–63**	**125–163**
SiO_2	32.3	32.9	36.1	38.6	55.7
Al_2O_3	21.1	23.9	24.6	24.0	21.1
Fe_2O_3	14.5	15.0	15.4	15.1	4.8
MgO	3.30	2.86	2.03	2.07	0.77
CaO	3.44	3.55	3.45	3.44	1.68
Na_2O	1.76	2.24	2.26	2.25	4.19
K_2O	0.80	0.87	0.88	0.82	2.66
H_2O+	18.3	14.7	11.7	10.0	7.6

(From Ref. 13.)

(B) *Wet zone*

	Depth (cm)				
Chemical composition	**0–18**	**18–35**	**35–50**	**50–63**	**148–163**
SiO_2	13.1	12.2	11.9	10.5	13.1
Al_2O_3	17.5	20.6	24.6	24.0	22.8
Fe_2O_3	23.1	24.3	19.5	21.4	29.7
MgO	0.50	0.24	0.29	0.27	trace
CaO	0.17	0.08	0.0	0.0	0.0
Na_2O	1.11	0.05	0.03	trace	0.10
K_2O	0.71	0.70	0.46	0.38	0.72
H_2O+	39.0	36.1	37.1	36.1	30.2

(From Ref. 13.)

aluminum oxides along with resistant rock farming minerals such as quartz. Senile soils are more likely to occur near the equator because soils age faster under warm tropical climates.

Climate

The influences of climate on soil chemical composition can best be observed along steep climate gradients located on a single parent rock of known geologic age. Such conditions occur on the Hawaiian Islands and researchers are now taking advantage of them to study the role of rock age (6), Island age (7–9) and climate (10). The islands offer temperatures associated with balmy beaches to snow-capped mountains and precipitation that range from rainforests to deserts, all compressed in a small area. A combination of steep rainfall gradient, warm tropical climate, and a basaltic parent rock containing easily weatherable minerals contributes to large soil chemical differences over short distances and in a relatively short period. The data in Tables 2 and 3 illustrate how differences in climate can produce soils of very different composition's and properties from identical parent rock.

Topography

Soil composition is affected by topography primarily through drainage and hydrology. Soil in depressions tend to be wetter, cooler, and richer in organic matter, clay and bases. These differences occur naturally but are amplified by cultivation. Soil on high points on the landscape tend to be net losers of fine particles and dissolved matter,

Table 3 Influence of climate on soil chemical properties as illustrated by a difference in (A) and (B) corresponding to soil (a) and (b) in Table 2

(A) *Dry zone*

	Depth (cm)				
Soil property	**0–13**	**13–25**	**25–50**	**50–63**	**125–163**
Organic carbon (%)	7.53	3.69	1.88	1.02	0.11
Nitrogen (%)	0.32	0.16	0.11	0.09	—
C:N ratio	24	23	17	11	—
Extractable Ca (cmol/kg)	27.9	28.0	32.6	34.6	33.3
Extractable Mg (cmol/kg)	7.8	6.4	7.1	9.8	14.4
Extractable Na (cmol/kg)	0.40	0.40	0.80	1.10	6.10
Extractable K (cmol/kg)	4.80	5.60	4.80	3.00	1.30
CEC (cmol/kg)	55.2	51.6	50.2	51.7	47.7
pH (H_2O)	6.6	7.1	7.3	7.5	8.2
pH (INKCI)	5.7	6.2	6.2	6.3	6.9

(B) *Wet zone*

	Depth (cm)				
Soil property	**0–18**	**18–35**	**35–50**	**50–63**	**148–163**
Organic carbon (%)	11.70	6.55	9.36	8.49	3.33
Nitrogen (%)	0.90	0.45	0.44	0.43	0.31
C:N ratio	13	15	21	20	11
Extractable Ca (cmol/kg)	9.4	0.8	0.5	trace	trace
Extractable Mg (cmol/kg)	2.6	0.3	0.3	0.1	0.6
Extractable Na (cmol/kg)	0.3	0.2	0.2	0.2	0.2
Extractable K (cmol/kg)	0.5	0.2	0.1	0.2	0.2
CEC (cmol/kg)	53.1	33.7	39.3	29.9	14.4
pH (H_2O)	5.4	5.2	5.4	5.4	5.2
pH (INKCI)	4.5	4.5	4.7	5.0	5.1

whereas bottom lands tend to be net accumulators of the same. Soils in between form a toposequence of composition which is repeated in a predictable pattern over the landscape. Topography and drainage control wetness, oxygen supply, oxidation–reduction potential, leaching, and pore water composition. All of the above affect biological diversity and activity and the bio-geochemistry of the system. Soil chemical compositional differences arising from topographic differences are the easiest to recognize in the landscape.

Biota

Plants clearly have an effect on mineral weathering (11). How different would a geochemically formed soil be from one formed bio-geochemically? The presence or absence of organic matter would make a difference but removing biota and organic matter from soils as we know them would not approximate one developed without living organisms. Not only do soil organisms such as termites and burrowing animals move huge quantities of soil material horizontally and vertically, but also, plants produce residue that serves as an energy source for other organisms and recycles inorganic nutrients. The organic residues also act as cementing agents to hold clay particles into stable compound particles or peds. It is unlikely that a soil can withstand the erosive impact of raindrops without the cementing action of humus or the high infiltration rate soils acquire through formation of large water-conducting pores between large water-stable peds. Organic matter also chelates metal ions such as iron and aluminum and facilitate their downward movement, but it is their contribution to the physical and hydraulic properties of soils that biota influence long-term stability of soil composition.

SUMMARY

Soil chemical composition is the product of the interactions of climate, topography, and biota operating over geologic time on a parent rock that is open to gains and losses of matter. Physical disintegration and chemical weathering transforms existing unstable minerals into secondary or pedogenic minerals that may differ in kind and composition with depth in the profile. Soil chemical composition is important only in so far as it can be related to the behavior and performance of soils.

REFERENCES

1. Garrels, R.M.; Christ, C.L. *Solutions, Minerals, and Equilibria*; Harper and Row: New York, 1965.
2. Jackson, M.L.; Level, T.W.M.; Syers, J.K.; Rex, R.W.; Clayton, R.M.; Sherman, G.D.; Uehara, G. Geomorphological Relationships of Tropospherically Derived Quartz in the Soils of the Hawaiian Islands. Soil Science Society of America Proceedings **1971**, *35*, 515–525.
3. Jenny, Hans. *Factors of Soil Formation: A System of Quantitative Pedology*; Dover: Mineola, New York, 1941.
4. Barshad, I. Chemistry of Soil Development. In *Chemistry of the Soil*; Bear, F.E., Ed.; Reinhold: New York, 1964.
5. Mohr, E.C.I. Tropical Soil Forming Processes and the Development of Tropical Soils. In *National Geological Survey of China*; National Geological Society of China: Peiping, China, 1930.
6. Chororor, J.; DiCharo, M.J.; Chadwick, O.A. Structural Charge and Cesium Retention in a Chronosequence of Tephritic Soils. Soil Science Society of America Proceedings **1999**, *63*, 169–177.
7. Vitousek, P.M.; Chadwick, O.A.; Crews, T.E.; Fownes, J.H.; Hendricks, D.M.; Hebert, D. Soil and Ecosystem Development Across the Hawaiian Islands. U.S.A. Today **1997**, *7*, 1–8.
8. Kennedy, M.J.; Chadwick, O.A.; Vitousek, P.M.; Derry, D.A.; Hendricks, D.M. Changing Sources of Base Cations During Ecosystem Development, Hawaiian Islands. Geology **1998**, *2G*, 1015–1018.
9. Chadwick, D.A.; Derry, L.A.; Witousek, P.M.; Huebert, B.J.; Hedin, L.O. Changing Sources of Nutrients During Four Million Years of Ecosystem Development. Nature **1999**, *397*, 491–497.
10. Chadwick, D.A.; Olson, C.G.; Hendricks, D.M.; Kelly, E.F.; Gavenda, R.T. *Quantifying Climatic Effects on Mineral Weathering and Neoformation in Hawaii*; Proceedings of the 15th International Soil Science Congress 8a, Acapulco, Mexico, 1994, 94–105.
11. Kelly, E.F.; Chadwick, O.A.; Hilinski, T. The Effect of Plants on Mineral Weathering. Bio-Geochemistry *42*, 53–72.
12. MacDonald, G.A.; Davis, D.A.; Cox, D.C. *Geology and Ground-Water Resources of the Island of Kauai, Hawaii*, Bulletin 13; Hawaii Division of Hydrography: Honolulu, HI, 1960.
13. Soil Conservation Service, U.S. Department of Agriculture. *Soil Survey Laboratory Data and Descriptions for Some Soils of Hawaii*, Soil Survey Investigations Report No. 29; HAES and HSPA: Honolulu, Hawaii, 1976.

CHEMISORPTION

N.J. Barrow
Mt. Claremont, Western Australia, Australia

INTRODUCTION

Many substances react with soil. Unless precipitation of ions from solution is involved, they initially react with the outside of soil particles in a process referred to as *adsorption*. When reaction continues and involves diffusion of the reactant into the interior of the soil particles, we use the more general word *sorption* to cover the whole process: adsorption plus diffusion. For some reactants, the attraction to the soil particles is largely electrical for example, the reaction of K^+ with a negatively charged surface. For other reactants, a strong inherent or chemical attraction is involved. This attraction may be so strong that ions overcome electrical repulsion and react with like-charged surfaces for example, reaction of phosphate ions with negatively charged surfaces. We use the word *chemisorption* to describe such reactions. Many nutrients such as phosphate, borate, copper (Cu), and zinc (Zn) are chemisorbed by soil, and this rations their supply to plants. Similarly, many pollutants, including selenite, arsenate, mercury (Hg), and lead (Pb), are chemisorbed, and this slows their entry into the food chain.

SPECIFICITY AND SORPTION

It follows that chemisorption involves specificity. That is, an ion may be sorbed despite the presence of a large excess of some other like-charged ion in solution. An example is sorption of phosphate from a solution containing chloride or nitrate. However, specificity does not always involve chemisorption. Preference may be a matter of valency or of differences in the tendency of ions to retain a sheath of water molecules. The terms also differ in that *specific absorption* describes an observation, whereas chemisorption requires some interpretation.

The strength of the chemical bond involved in chemisorption varies according to the ions involved. It therefore follows that specific adsorption is a graded property, with particular ions falling along a gradient of specificity.

CHEMICAL BONDING

For chemisorption to occur, there must be an attraction between the ion in solution and atoms on the surface of the soil particles. This attraction may show itself in other ways. For example, both iron (Fe) and aluminium (Al) form low-solubility compounds with phosphate. Similarly, metals, such as Zn, form hydroxides with low solubility. This has suggested to some that precipitates are important in the soil chemistry of such ions. However, discrete compounds are only formed when concentrated solutions leach out of fertilizer particles and have a limited life in soil. The ions eventually react with atoms, or groups of atoms, on the surface of soil particles: phosphate and several other anions with Fe and Al atoms; Zn and other metals with hydroxyls or with organic complexes.

SOIL COMPONENTS INVOLVED

Several kinds of soil particles may have appropriate groups of atoms on their surfaces. These include the edges of clay minerals, organic matter, and metal oxides. Fe oxides are the most widespread, with goethite being the most important and most frequently studied.

Even though chemisorption involves chemical bonds, these bonds must form between charged particles (ions) and charged surfaces. Two characteristics of the surfaces are important in this respect. One is that the charge on the surfaces varies with pH. In the case of metal oxides, this is because the metal atoms on the surface attempt to complete their electron shells by reacting with water molecules. These molecules lose or gain protons depending on the pH and the concentration of all ions in the solution in which reaction occurs. They are variable-charge surfaces. The charge is also variable in the spatial sense. Different particles, or different patches on the same particles, may carry different charge densities and therefore have different potentials. Direct evidence for this is that it is possible to measure both positive and negative charge in the one soil.

CHARACTERISTICS OF THE REACTING IONS

Two characteristics of the reacting ions are also important. First, many of the relevant acids have pK values in or near the range of soil pH values. Changes in pH therefore result in large changes in the ions present, with consequent effects on the reaction. Second, when the ions react with the surface, their charge must be conserved. Some of the charge is conveyed to the surface for reaction with phosphate, the surface becomes more negative. The remainder of the charge is returned to the solution for reaction with phosphate, hydroxide ions may be desorbed (or protons adsorbed). The balance between these two options changes in a complex way with changes in pH and in the amount of sorption. Because the surface charge is consequently more negative, the affinity of the surface for more phosphate ions decreases. Analogous effects occur for other ions.

This brief summary equips us to comprehend a great deal of the information on chemisorption. For example, the Langmuir equation, although still widely used to describe the relationship between sorption and concentration, is appropriate only in special circumstances. Soils do not provide one (or even two) uniform surfaces, and the feedback effect of the change in charge negates a basic assumption of this equation. A more realistic approach is to accept the range of surface characteristics and to apply a distribution to describe them. This distribution can be divided into slices, and a Langmuir-like equation that takes feedback into account can be applied to each slice. The net behavior is then given by the sum of the slices.

EXPLAINING THE EFFECTS OF CONCENTRATION

Consider how this model describes the most common observation in sorption studies: the relationship between sorption and concentration. In many studies, sorption curves are produced over a limited range of concentrations, about 100-fold. Over such a range, the Freundlich equation usually describes the data well. This means that a plot of log sorption against log concentration approximates a straight line. However, over a greater range, such as that illustrated in Fig. 1, curvature is obvious and is reproduced by the model outlined above. This curvature is inevitable because at low concentrations, sorption is confined to the most-favored end of the spectrum of sites and is therefore on a nearly uniform set of sites. Further, because the amount of sorption is small, the feedback effect is insignificant. This is one of the unusual situations in which the Langmuir equation is appropriate. However,

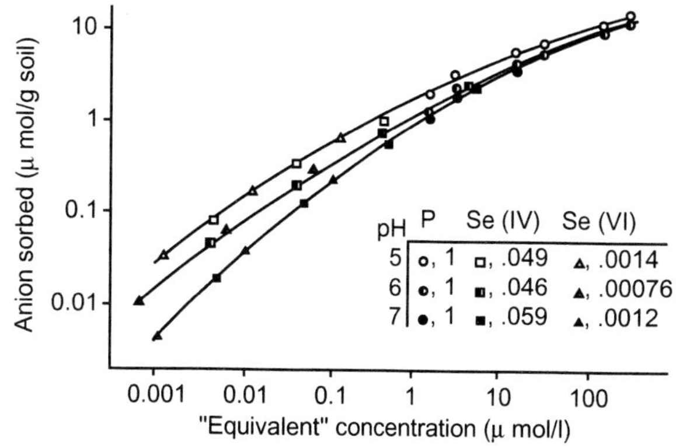

Fig. 1 Sorption of phosphate (P), selenite [Se(IV)], and selenate [Se(VI)] at pH 5, 6, and 7 plotted on common axes. The concentrations were multiplied by the values in the table to give equivalent concentrations. Thus, at pH 5, the value for phosphate is 1; for selenite, 0.049; and for selenate, 0.0014. (From Ref. 1.)

at very low concentrations, the form of the equation is such that sorption is virtually proportional to concentration. That means that on a log-log scale, the slope approaches unity. It is, therefore, much steeper than at higher concentrations.

EXPLAINING THE EFFECTS OF pH

Consider the interactions between pH and the concentration of the background solution on the sorption of some common inorganic anions (Fig. 2).

These interactions occur because there is an initial adsorption reaction of ions on a variable charge surface. An increase in salt concentration increases the number of counter ions close to the surface and therefore changes the way that electric potential changes as we move away from the surface. At any particular distance from the surface, an increase in salt concentration decreases the absolute value of the electric potential. For example, when a reaction is with a negatively charged surface, the potential becomes less negative and anion sorption increases. A convenient, though less precise, way of thinking about this is that an increase in the cation concentration near a negatively charged surface makes it easier for anions to react with the surface. The corollary is, when increasing salt concentration increases sorption of anions, the reacting surface must be negatively charged. In such cases, there must be a clear chemical component to sorption. By a similar argument, when sorption of anions is decreased by an increase in salt concentration, the reacting surface is positively charged. The results in Fig. 2, therefore, show

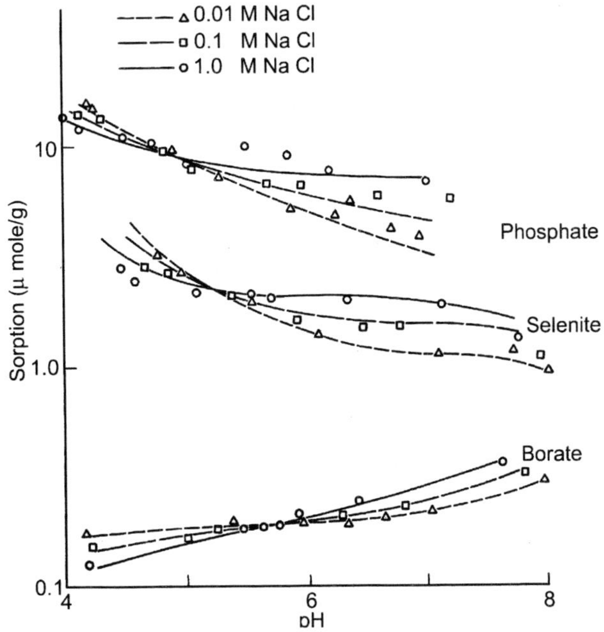

Fig. 2 The effect of pH and background electrolyte concentration on the sorption of phosphate, selenite, and borate by a soil. In each case, sorption was measured at a solution concentration of 100 μm. The lines were obtained by fitting the data into the model described in the text. (From Ref. 1.)

that reaction of borate, selenite, and phosphate at medium to high pH occurred with a negative surface and reaction at low pH occurred with a positive surface. Thus, all three species reacted with a variable-charge surface.

BEHAVIOR OF SOME IONS

Before considering the ions in detail, note that selenate, sulfate, phosphate, selenite, and arsenate all form bidentate, inner-sphere complexes with the surface. That is, two of the oxygen atoms provide direct chemical links to the surface atoms. When a reactant forms a bidentate link to the surface, it is appropriate to refer to the divalent ion in solution (1). Similarly, in the case of a monodentate link, as with borate, it is appropriate to refer to the monovalent ion.

Boric acid is fairly weak, with pH at about 9. Therefore, in the normal range of soil pH, the proportion of monovalent borate ion increases 10-fold for each unit increase in pH. The effects of pH on the charge and potential and the effects on acid dissociation therefore oppose each other: the increasingly negative electric potential favors decreased desorption; the increasing dissociation favors increased adsorption. Because the ion is monovalent, the effects of surface charge are not quite large enough to exceed the effects of the increasing value of the dissociation term. Thus, sorption increases with increasing pH.

Selenious acid is a diprotic acid with pK_1 at 2.7 and pK_2 at 8.5 in very dilute solution. The main species present in the range of soil pH values are $HSeO_3^-$ and SeO_3^{2-}, with the divalent ion increasing with increasing pH. However, because the relevant ion is divalent, the effects of the increasingly negative charge exceed the effects of increasing dissociation and sorption therefore decreases with increasing pH.

Phosphoric acid is a triprotic acid but the third dissociation (pH_3) is well beyond the range of soil pH. The species present in soil solutions are therefore $H_2PO_4^-$ and HPO_4^{2-}. The divalent ion increases 10-fold for each unit increase in pH up to just below pH_2, which is about 7. Its behavior is similar to that of selenite, and the explanation is analogous.

EXPLAINING THE EFFECTS OF TIME

We have thus far ignored the effects of the reaction period. However, many reactants continue to react with soil, and

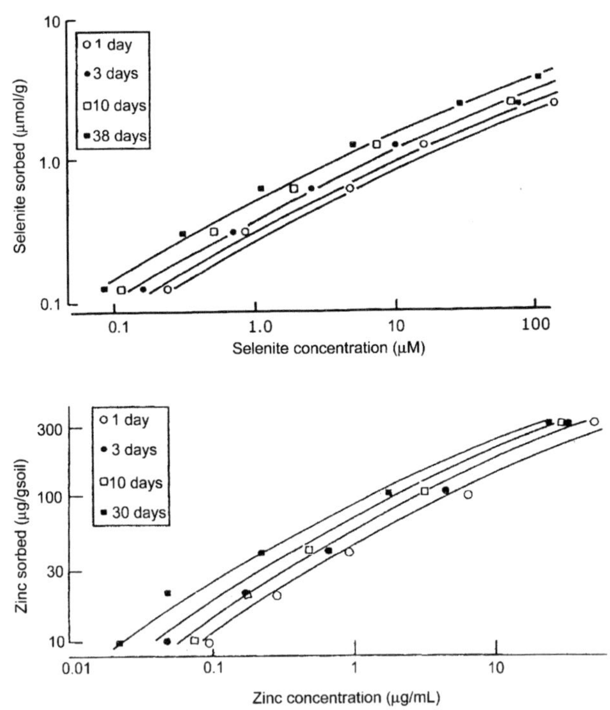

Fig. 3 The effect of period of reaction at 25°C for selenite and zinc on the relationship between solution concentration and amount of sorption. The lines were obtained by fitting the model described in the text to the data. (From Ref. 1.)

also with goethite, for a very long period. This continuing reaction is of great importance in agriculture where it has a large effect on the need to reapply fertilizer to land. Consequently, sorption is better described by a surface, with time as the extra dimension, rather than by a line (Fig. 3).

Ions differ in the extent to which they continue to react so that one can produce a sequence such as: phosphate> selenite>fluoride>selenate. For the anions, the greater the chemical attraction for the surface, the more marked the continuing reaction. The continuing reaction is caused by diffusion of adsorbed ions into the absorbing particle. This is a slow process but can be accelerated by raising the temperature. Consequently, increased temperature increases sorption and/or decreases solution concentration.

DESORPTION

The more marked the continuing reaction, the more deeply buried the ion becomes. Although desorption curves do not seem to follow the same track as sorption curves, they can be fully predicted if the continuing reaction is taken into consideration.

REFERENCE

1. Barrow, N.J. The Four Laws of Soil Chemistry: The Leeper Lecture 1998. Aust. J. Soil Res. **1999**, *37*, 787–829.

CHLORINE

Uzi Kafkafi
The Hebrew University of Jerusalem, Rehovot, Israel

Guohua Xu
Nanjing Agricultural University, Nanjing, China

INTRODUCTION

Chlorine is found in nature only in the form of negatively charged (Cl^-) highly water-soluble anion. Most of the world's Cl^- is found either in oceans or in salt deposits left by evaporation of old inland seas which are now found in deep quarries. Field and glasshouse studies in the mid-1800s to the early1900s have shown that the influence of Cl^- on plant growth varies with the plant variety (1). Lipman (2) demonstrated the beneficial effect of Cl^- on buckwheat (*Triticum asetivum* L.) growth. Arnon and Whatley (3) suggested that Cl^- is an essential co-factor in the O_2 evolution during photosynthesis. The first complete definition of Cl^- as a plant essential micronutrient was described by Broyer et al. (4).

The function of Cl^- in crop yield has been largely neglected (5), as it becomes a limiting factor for plant growth only in areas of high precipitation far from the sea. The negative effects of high Cl^- concentrations in soil and irrigation water on crop production are observed in coastal, arid and semi-arid areas, where freshwater sources are often scarce and the available groundwater is saline. The dependence of modern agriculture on irrigation and chemical fertilization causes more concern on excess of Cl^- rather than on its deficiency (6).

CHLORIDE IN SOIL

The concentrations of Cl^- in natural sources are listed in Table 1. The Cl^- content in the soil is not an intrinsic property of the soil but rather a result of soil management, rainfall and evaporation, and irrigation and fertilization.

The Cl^- concentration in rainwater ranges from about $20-50\,g\,m^{-3}$ close to seashore to $2-6\,g\,m^{-3}$ in inner continental areas (5, 7). The annual amount of Cl^- deposition on land ranges from $17-175\,kg\,ha^{-1}$. Mid-continental areas such as the Great Plains of North America receive less than $1.0\,kg\,Cl^-\,ha^{-1}$ annually through precipitation (5).

The chloride anion is not sorbed on soil particles at neutral and basic pH values, and therefore is easily leached. However, in acid soils containing variable-charge clays, a slight specific sorption of Cl^- is involved (8).

Plants can take up substantial amounts of Cl^- when soil Cl^- levels are high. At peak accumulation, the Cl^- content of spring wheat (*Triticum aestivum* L.) was 18 and $61\,kg\,ha^{-1}$ on sites testing low and high in Cl^-, respectively (5).

PLANT YIELD AND QUALITY RESPONSES TO CHLORIDE

The small amount of Cl^- in precipitation, combined with the limited use of KCl fertilizer in some K-rich inland soils results in low levels of Cl^- in soil testing (5). Substantial positive responses to Cl^- containing fertilizers have been reported for different crops in many parts of the world (5, 6, 9): barley (*Hordeum vulgare* L.), coconut (*Cocos nucifera* L.), kiwifruit (*Actinidia deliciosa* L.), oil palm (*Elaeis guineensis Jacq.*), potato (*Solanum tuberosum* L.), wheat (*Triticum aestivum* L.), tobacco (*Nicotiana tabacum* L.), and sugar beet (*Beta vulgaris* L.). Typical symptoms of Cl^- deficiency include wilting of leaves, curling of leaflets, bronzing and chlorosis, and severe inhibition of root growth. The so-called "Cl^- deficient leaf spot syndrome," the leaf spots and tissue necrosis in wheat, is caused by inadequate Cl^- nutrition (about $1\,g\,kg^{-1}$ DM) rather than pathogen infection (9). The concentration range of Cl^- deficiency in plants varies between 0.13 and $5.7\,g\,kg^{-1}$ (Table 2). Wheat yield benefits from Cl^- were attributed to either partial suppression of root or foliar diseases or to enhancement of the host tolerance to the "disease."

Sensitivity to high Cl^- concentrations varies widely between plant species and cultivars. Generally, many woody

Encyclopedia of Soil Science

Table 1 Chloride concentrations in some natural sources

Source	Chloride ($g\,kg^{-1}$)
Earth crust	1.50
Lithosphere	0.48
Basalt rocks	0.50
Syenite	0.98
Igneous rocks	0.23
Shale	0.16
Sandstone	0.02
Limestone	0.37
Dolomite	0.50
Soils	0.10
Plants	1.0–10.0
Ocean	19.0[a]
Dead Sea water	218.0[a]
Low to medium saline water	0.10–0.30[a]
High to very high saline water	0.30–1.20[a]
Table salt (NaCl)	607
Potassium chloride (KCl)	475
Potassium chloride fertilizers	450–570

[a] Unit: $kg\,m^{-3}$.
(Compiled from Refs. 7 and 16.)

plant species and legumes are susceptible to Cl^- toxicity, whereas most non-woody crops tolerate excessive Cl^- (10). The critical toxicity concentration of Cl^- (CTCC) in saturated soil extracts varies from 10–15 mM for sensitive crops like common bean (*Phaseolus vulgaris* L.), grape (*Vitis vinifera* L. ssp.*Vinifera*), lettuce (*Lactuca sativa* L.), potato, and strawberry (*Fragaria vesca*) and to 70–80 mM for crops, such as barley, cotton (*Gossypium hirsutum* L.), sorghum (*Sorghum bicolor* L.), and sugar beet (6, 10). The CTCC in plant is about 4–7 $g\,kg^{-1}$ and 15–50 $g\,kg^{-1}$ for Cl-sensitive and Cl-tolerant plant species, respectively (Table 2). It seems likely that factors associated with vigorous growth or Cl^- compartmentation within the cell could offset the inhibitory effects of Cl^- accumulation. The tolerance of a crop to Cl^- varies with the growth stage, as well as climatic conditions.

The specific influences of Cl^- on the quality of agricultural products are not clear. Chloride generally accumulates in the vegetative parts, mainly in the leaves, while its content in grains, fruits and seeds is very low and is hardly affected by the Cl^- concentration in the soil (5, 6). Therefore, the quality of leafy vegetables is relatively sensitive to Cl^- such as Chinese cabbage (*Brassica rapa* L.), lettuce, and tobacco, while fruit quality is generally tolerant to high Cl^- levels in the soil.

PLANT CHLORIDE UPTAKE AND DISTRIBUTION

Uptake of Cl^- by the plant roots is an active process that requires energy. The Cl^- transport through the cell membrane involves the nH^+:Cl^- symporter (11) or occurs via anti-port with hydroxyl ions energized by ATP (10). Specific protein channels energized by ATP for Cl^- transport are found both in the plasmalemma and in the tonoplast. The fluxes of Cl^- in intact plants are very diff erent from those measured in plants with excised roots. One of the schemes used to describe the control of Cl^- uptake and transport in plants is a homeostatic mechanism that senses vacuolar NO_3^- plus Cl^- or total anion concentration (12).

Chloride compartmentation appears to be highly regulated. In the chloroplast, Cl^- concentration remains relatively constant regardless of whether the plant growth medium is Cl^- deficient or excessive (10). Harmful effects of high leaf Cl^- concentrations could be avoided if salt excluding varieties are used as rootstocks.

BIOCHEMICAL AND PHYSIOLOGICAL FUNCTIONS OF CHLORIDE IN PLANT

Chloride has a stimulatory effect on asparagine synthethase, which uses glutamate as a substrate (13). The proton-pumping ATPase at the tonoplast is stimulated directly by Cl (14). The chlorinated indole-3-acetic acid (IAA) found in substantial amounts in some legume species seeds enhances hypocotyl elongation at 10 times the rate of IAA itself, probably because of the resistance of the former compound to degradation by peroxides (15). Chloride is required for the water-splitting system of Photosystem II (PSII) (3). Binding of Cl^- to membranes is needed for activation of the O_2-evolving enzyme. The preference for Cl^- over Br^-, NO_3^-, or I^- with regard to its ability to promote O_2-evolution is due to its specific ionic volume.

The ability of Cl^- to move rapidly across cell membranes against an electrochemical gradient and its relatively low biochemical activity are two important properties that make Cl^- particularly suited to serve as a key osmotic solute in plants. The accumulation of Cl^- by plants contributes greatly to an increase in cell hydration and turgor pressure, both of which are essential for cell elongation and stomata opening (17). The osmoregulatory function of Cl^- in plants seems to operate at different levels (16). The Cl^- concentration range usually found in plants (50–150 mM of tissue water) exceeds its critical deficiency level by 1 to 2 orders of magnitudes. At the low

Table 2 Chloride concentrations in plants

Plant family	Crop	Latin name	Plant part	Concentration ranges of tissue Cl (g kg^{-1} DM)			Reference
				Deficient	Normal	Toxicity[a]	
Actinidiaceae	Kiwifruit	*Actinidia deliciosa* L.	Leaves	2.1	6.0–13.0	>15.0	21, 23
Arecaceae	Coconut palm	*Cocos nucifera* L.	Leaves	2.5–4.5	>6.0–7.0		19
Asteraceae	Lettuce	*Lactuca sativa* L.	Leaves	>0.14	2.8–19.8	>23.0	24
Chenopodiaceae	Spinach	*Spinacia oleracea* L.	Shoot	>0.13			25
Chenopodiaceae	Sugarbeet	*Beta vulgaris* L.	Leaves	0.71–1.78			26
Chenopodiaceae	Sugarbeet	*Beta vulgaris* L.	Petioles	<5.7	>7.1–7.2	>50.8	6, 26
Fabaceae	Alfalfa	*Medicago sativa* L.	Shoot	0.65	0.9–2.7	6.1	27, 28
Fabaceae	Peanut	*Arachis hypogaea* L.	Shoot		<3.9	>4.6	29
Fabaceae	Red clover	*Trifolium pratense* L.	Shoot	0.15–0.21			30
Fabaceae	Soybean	*Glycine max* L. Merr.	Leaves		0.3–1.5	16.7–24.3	31, 32
Fabaceae	Subterranean clover	*Trifolium subterraneum* L.	Shoot	>1.0			27
Gramineae	Barley	*Hordeum vulgare* L.	Heading shoot	1.2–4.0	>4.0		5, 9
Gramineae	Corn	*Zea mays* L.	Ear leaves		1.1–10.0	>32.7	33
Gramineae	Corn	*Zea mays* L.	Shoots	0.05–0.11			24
Gramineae	Rice	*Oryza sativa* L.	Shoot	<3.0		>7.0–8.0	34
Gramineae	Rice	*Oryza sativa* L.	Mature straw		5.1–10.0	>13.6	6
Gramineae	Spring wheat	*Triticum aestivum* L.	Heading shoot	1.5	3.7–4.7	>7.0	5, 29
Gramineae	Wheat	*Triticum aestivum* L.	Heading shoot	1.2–4.0	>4.0		5, 9
Lauraceae	Avocado	*Persea americana* Mill.	Leaves		~1.5–4.0	~7.0	20
Malvaceae	Cotton	*Gossypium hirsutum* L.	Leaves		10.0–25.0	>25.0–33.1	6
Rosaceae	Apple	*Malus domestica* L.	Leaves	0.1		>2.1	28
Rosaceae	Peach	*Prunus persica* L.	Leaves		0.9–3.9	10.0–16.0	28, 35
Rosaceae	Pear	*Pyrus communis* L.	Leaves		<0.50	>10.0	35
Rosaceae	Strawberry	*Fragaria vesca* L.	Shoot		1.0–5.0	>5.3	29, 35
Rutaceae	Citrus	*Citrus sp.* L.	Leaves		~2.0	~4.0–7.0	20
Solanaceae	Potato	*Solanum tuberosum* L.	Mature shoot	<1.0	2.0–3.3	12.2	36
Solanaceae	Potato	*Solanum tuberosum* L.	Petioles	0.71–1.42	18.0	44.8	36, 37
Solanaceae	Sweet pepper	*Capsicum annuum* L.	Shoot		10–40		[b]
Solanaceae	Tobacco	*Nicotiana tabacum* L.	Leaves		1.2–10.0	>10.0	6, 28
Solanaceae	Tomato	*Lycopersicon esculentum* Mill	Shoot	0.25		~30.0	4, 38
Vitaceae	Grapevine	*Vitis vinifera* L. ssp. *vinifera*	Petioles		0.7–8.0	10.0–11.0	6, 28

[a] The plant yields decline or the plant shows visible scorching symptoms in leaves.
[b] Xu and Kafkafi, unpublished data.

concentration of Cl^- (1 mM or less in the whole plant tissue), the osmoregulatory functions of Cl^- are presumably confined to specialized compartments in tissues or cells. These are the elongation zones of roots and shoots, the pulvini, stigmata, and guard cells, where the concentrations of Cl^- might be much higher than the average Cl^- concentration in the bulk tissue (6).

Activation of a H^+ pump in the plasma membrane initiates K^+ and Cl^- influx, accompanied by malate synthesis, resulting in osmotic water flow into the guard cells, and consequent stomata opening. The relative contributions of Cl^- and malate may vary among species, and may depend on the availability of external Cl^- and on the plant growth environment (17). In plant species such as onion (*Allium cepa* L.), which lack the functional chloroplasts for malate synthesis in the guard cells, Cl^- is essential for stomata functioning (17, 18). Members of the *Palmaceae*, such as coconut and oil palm, which might possess starch-containing chloroplasts in their guard cells, also require Cl^- for stomata function (19).

INTERACTION OF CHLORIDE UPTAKE WITH THE UPTAKE OF OTHER NUTRIENTS

Ammonium is taken up by plant as a cation, and therefore relatively more anions have to be taken up to maintain the electrical neutrality of the uptake process. Plants fertilized with NH_4^+, usually contain higher Cl^- levels in the tissue than plants fertilized with NO_3^- or with both N sources, irrespective of the Cl^- concentration in the nutrient solution (6). Thus, when Cl^- is present in the root medium, NH_4^+ uptake may increases the salt sensitivity of plants.

The antagonism between NO_3^- and Cl^- uptake has been well demonstrated in many crops (6). When both Cl^- and NO_3^- anions are taken up by the root against their electrochemical gradient, Cl^- maintains its negative charge, while NO_3^- is metabolized and loose its negative charge. The accumulation of Cl^- reduces its further uptake since the Cl^- electrochemical potential gradient builds up during its accumulation in the cell (20). Nitrate can prevent Cl^- toxicity of avocado at a concentration of up to 16 mM in the root medium (20). On the other hand, Cl^- application may also be used as a strategy to decrease the NO_3^- content of leafy vegetables, as spinach (*Spinacia oleracea* L.), lettuce, and cabbage (*Brassica oleracea* L.), which are classified as NO_3^- accumulators (6).

Increasing concentrations of Cl^- in the root medium has generally no consistent effect on K concentrations in the plants. K concentrations in the leaves of kiwifruit are significantly higher for vines receiving KCl than for vines receiving K_2SO_4 as the kiwifruit uses Cl^- rather than organic anions for charge balance, and thus maintains a high K uptake (21).

CHLORIDE MANAGEMENT IN IRRIGATION AND FERTILIZATION

Large amounts of Cl^- enter the field through irrigation. The amount of Cl^- added to a field with 500 mm of irrigation water containing only $200\,g\,Cl\,m^{-3}$ is $1000\,kg\,ha^{-1}$. This is four times more than the amount of Cl^- applied by fertilization with KCl at $500\,kg\,ha^{-1}$. The irrigation system influences the distribution of Cl^- salts in the soil. Irrigation with saline water is managed by an excess of irrigation to meet the leaching requirement for avoiding salt accumulation in the root zone (22). The amount of water required to wash salts out of the root zone can be estimated from the electrical conductivity (EC) of the irrigation water and the mean EC of the saturated soil extract at which no crop yield reduction occurs under sprinkler or surface irrigation (22).

Fertilization with KNO_3 under saline conditions might reduce the toxic effect of salinity of some woody plants even if it is associated with increased soil osmotic potential (20). Under high salinity and Cl^- conditions, KNO_3 or K_2SO_4 as K fertilizers are preferred due to their lower salt index and absence of Cl^-. Addition of adequate P can also be helpful in alleviating salt stress (6).

REFERENCES

1. Tottingham, W.E. A Preliminary Study of the Influence of Chlorides on the Growth of Certain Agricultural Plants. J. Am. Soc. Agron. **1919**, *11*, 1–32.
2. Lipman, C.B. Importance of Silicon, Aluminum, and Chlorine for Higher Plants. Soil Sci. **1938**, *45*, 189–198.
3. Arnon, D.L.; Whatley, F.R. Is Chloride a Coenzyme of Photosynthesis? Science **1949**, *110*, 554–556.
4. Broyer, T.C.; Carlton, A.B.; Johnson, C.M.; Stout, P.R. Chloride—A Micronutrient Element for Higher Plants. Plant Physiol. **1954**, *29*, 526–532.
5. Fixen, P.E. Crop Responses to Chloride. Adv. Agron. **1993**, *50*, 107–150.
6. Xu, G.H.; Magen, H.; Tarchitzky, J.; Kafkafi, U. Advances in Chloride Nutrition of Plants. Adv. Agron. **2000**, *68*, 97–150.
7. Yaalon, D.H. The Origin and Accumulation of Salts in Groundwater and in Soils in Israel. Bull. Res. Counc. Israel **1963**, *11G*, 105–131.

8. Wang, J.H.; Yu, T.R. Release of Hydroxyl Ions During Specific Adsorption of Chloride by Variable-Charge Soils. Z Pflanzenernahr Bodenk **1998**, *161*, 109–113.
9. Engel, R.E.; Bruckner, P.L.; Mathre, D.E.; Brumfield, S.K.Z. A Chloride Deficient Leaf Spot Syndrome of Wheat. Soil Sci. Soc. Am. J. **1997**, *61*, 176–184.
10. Maas, E.V. Physiological Responses to Chloride. In *Special Bulletin on Chloride and Crop Production*, No. 2; Jackson, T.L., Ed.; Potash & Phosphate Institute: Georgia, 1996; 4–20.
11. Felle, H.H. The H^+/Cl^- Symporter in Root-Hair Cells of *Sinapsis alba*. An Electrophysiological Study Using Ion-Selective Microelectrodes. Plant Physiol. **1994**, *106*, 1131–1136.
12. Glass, A.D.M.; Siddiqi, M.Y. Nitrate Inhibition of Chloride Influx in Barley: Implications for a Proposed Chloride Homeostat. J. Exp. Bot. **1985**, *36*, 556–566.
13. Rognes, S.E. Anion Regulation of Lupin Asparagine Synthetase: Chloride Activation of the Glutamine-Utilizing Reaction. Phytochemistry **1980**, *19*, 2287–2293.
14. Churchill, K.A.; Sze, H. Anion-Sensitive, H^+-Pumping ATpase of Oat Roots. Plant Physiol. **1984**, *76*, 490–497.
15. Hofinger, M.; Bottger, M. Identification by GC-MS of 4-Chloroindolylacetic Acid and Its Methyl Ester in Immature *Vicia faba* Broad Bean Seeds. Phytochemistry **1979**, *18*, 653–654.
16. Flowers, T.J. Chloride as a Nutrient and as an Osmoticum. Adv. Plant Nutr. **1988**, *3*, 55–78.
17. Talbott, L.D.; Zeiger, E. Central Roles for Potassium and Sucrose in Guard-Cell Osmoregulation. Plant Physiol. **1996**, *111*, 1051–1057.
18. Schnabl, H.; Raschke, K. Potassium Chloride as Stomatal Osmoticum In *Allium cepa* L. (onion), a Species Devoid of Starch in Guard Cells. Plant Physiol. **1980**, *65*, 88–93.
19. von Uexkull, H.R. Chloride in the Nutrition of Coconut and Oil Palm. In *Trans. Int. Congr. Soil Sci.*, 14; 1990; Vol. IV, 134–139.
20. Bar, Y.; Apelbaum, A.; Kafkafi, U.; Goren, R. Relationship Between Chloride and Nitrate and Its Effect on Growth and Mineral Composition of Avocado and Citrus Plants. J. Plant Nutr. **1997**, *20*, 715–731.
21. Buwalda, J.G.; Smith, G.S. Influence of Anions on the Potassium Status and Productivity of Kiwifruit (*Actinidia deliciosa*) Vines. Plant Soil **1991**, *133*, 209–218.
22. Keller, J.; Bliesner, R.D. Trickle Irrigation Planning Factor. In *Sprinkler and Trickle Irrigation*; Keller, J., Bliesner, L.R.D., Eds.; Van Nostrand Reinhold: New York, 1990; 453–477.
23. Prasad, M.; Burge, G.K.; Spiers, T.M.; Fietje, G. Chloride Induced Leaf Breakdown in Kiwifruit. J. Plant Nutr. **1993**, *16*, 999–1012.
24. Johnson, C.M.; Stout, P.R.; Broyer, T.C.; Carlton, A.B. Comparative Chloride Requirements of Different Plant Species. Plant Soil **1957**, *8*, 337–353.
25. Robinson, S.P.; Downton, W.J.S. K, Na and Cl Content of Isolated Intact Chloroplasts in Relation to Ionic Compartmentation in Leaves. Arch. Biochem. Biophys. **1984**, *228*, 197–206.
26. Ulrich, A.; Ohki, K. Chloride, Bromine, and Sodium as Nutrients for Sugar Beet Plants. Plant Physiol. **1956**, *31*, 171–181.
27. Ozanne, P.G.; Woolley, J.T.; Broyer, T.C. Chloride and Bromine in the Nutrition of Higher Plants. Aust. J. Biol. Sci. **1957**, *10*, 66–79.
28. Eaton, F.M. Chapter: Chlorine. In *Diagnostic Criteria for Plants and Soils*; Chapman, H.D., Ed.; University of Califonia: Riverside, 1966; 98–135.
29. Wang, D.Q.; Guo, B.C.; Dong, X.Y. Toxicity Effects of Chloride on Crops. Chin. J. Soil. Sci. **1989**, *30*, 258–261.
30. Whitehead, D.C. Chlorine Deficiency in Red Clover Grown in Solution Culture. J. Plant Nutr. **1985**, *8*, 193–198.
31. Parker, M.B.; Gaines, T.P.; Gascho, G.J. The Chloride Toxicity Problem in Soybean in Georgia. In *Special Bulletin on Chloride and Crop Production*; 2; Jackson, T.L., Ed.; Potash & Phosphate Institute: Atlanta, Georgia, 1986; 100–108.
32. Yang, J.; Blanchar, R.W. Differentiating Chloride Susceptibility in Soybean Cultivars. Agron. J. **1993**, *85*, 880–885.
33. Parker, M.B.; Gaines, T.P.; Gascho, G.J. Chloride Effects on Corn. Commun. Soil Sci. Plant Anal. **1985**, *16*, 1319–1333.
34. Yin, M.J.; Sun, J.J.; Liu, C.S. Contents and Distribution of Chloride and Effects of Irrigation Water of Different Chloride Levels on Crops. Soil Fert. **1989**, *1*, 3–7, (Chinese).
35. Robinson, J.B. Fruits, Vines and Nuts. In *Plant Analysis—An Interpretation Manual*; Reuter, D.J., Robinson, J.B., Eds.; Inkata Press: Sydney, 1986; 120–147.
36. Corbett, E.G.; Gausman, H.W. The Interaction of Chloride and Sulfate in the Nutrition of Potato Plants. Agron. J. **1960**, *52*, 94–96.
37. James, D.W.; Weaver, W.H.; Reeder, R.L. Chloride Uptake by Potatoes and the Effects of Potassium Chloride, Nitrogen and Phosphorus Fertilization. Soil Sci. **1970**, *109*, 48–53.
38. Kafkafi, U.; Valoras, N.; Letay, J. Chloride Interaction with NO_3^- and Phosphate Nutrition in Tomato. J. Plant Nutr. **1982**, *5*, 1369–1385.

CLASSIFICATION, AUSTRALIA AND NEW ZEALAND

C

R.F. Isbell
A.E. Hewitt
Landcare Reasearch, Lincoln, New Zealand

INTRODUCTION

Although Australia and New Zealand are relatively close neighbors, both are independent island states and this probably accounts for some degree of isolation between the two countries. In terms of size, climate, geology, vegetation, and land form there are striking differences and it is not surprising, therefore, that the New Zealand soil pattern is mostly very different from that of much of Australia. It also follows for these factors, and other national differences, that separate soil classification schemes have always been in use.

It is widely accepted that classification is an essential part of any scientific discipline and is a necessary part of the language of science. No classification scheme can remain static; as new knowledge is gained, soil classifications need to be updated and improved. The history of national soil classifications in Australia (1, 2) is a good example of how benefits are obtained from modifications in the light of new knowledge. This, in effect, largely explains why Australia has rightly had a succession of national classification schemes. It is also of interest to note that two of the still commonly used schemes—the *Factual Key* of Northcote (3) and the so-called *Handbook of Australian Soils* (4) were a direct result of the decision to hold the 9th International Congress of Soil Science in Adelaide, Australia, in 1968. This meeting served as a catalyst to increase Australian soils knowledge relatively rapidly by means of a targeted approach to the mapping of the continent at a published scale of 1:2 million using the *Factual Key* (3). This was achieved on time in spite of the size and soil diversity of the Australian continent. The *Factual Key* is a bifurcating, hierarchical scheme with five categorical levels. All classes are mutually exclusive and the keying morphological attributes are determined in the field, including the presence of carbonate and pH as determined by a simple field test. Most class names are descriptive, e.g., Hard Pedal Mottled Red Duplex Soils.

The associated *Handbook of Australian Soils* (4) was a separate exercise to the soil mapping project and is a compendium of morphological and laboratory data of 94 soil profiles which were visited on the 1968 Congress field excursions, plus some extra representative soils (147) not visited in the field tours. The "Handbook" is not a formal classification but an assemblage of "great soil groups" largely derived from Stephens (5). This was more or less current at the time and was essentially based on the earlier United States Department of Agriculture schemes that were in vogue prior to the formal advent of the early *Soil Taxonomy Approximations* of the 1960s. Both the *Factual Key* and the Handbook are both still widely used in Australia, the former mainly because of the Australia—wide map coverage at 1:2 million; while the Handbook is still very useful because of the detailed accounts of the morphology, micromorphology, and laboratory data for a large number of widespread Australian soils. A disadvantage of the Handbook is the lack of a key to the soil groups.

The most comprehensive of the earlier New Zealand national soil classification systems is the New Zealand Genetic Soil Classification (6) reviewed by Hewitt (7) in 1992. In his classification (6), Taylor developed what would now be called soil landscape models in which soil groups were related to environmental factors in the zonal concepts of the Russians. The classification of Taylor was first published in 1948 as a legend to the first National Soil Map, but with time a number of weaknesses became apparent. Examples include the failure of the genetic scheme when applied at the scale of the farm paddock, and the difficulty of soil correlation between regions. It became obvious that a new model was required that made provision for important classes of New Zealand soils.

NATIONAL SOIL CLASSIFICATION SYSTEMS IN THE MODERN ERA

In the 1970s and 1980s there was an increase in soil surveys in some Australian States, particularly Queensland and the Northern Territory, where the then-current Northcote and Stace, et al. systems were often found to be inadequate to cater for many "new" Australian soils, particularly in the wet tropics and in the subtropics. A soil classification committee with an Australia-wide charter was set up in the early 1980s to remedy the lack of an adequate, up-to-date national soil classification. Extensive field travel was carried out in most parts of Australia in order to gather field and laboratory data to build up a database which would eventually contain some 14,000 published and unpublished profiles, many with laboratory data; mainly chemical. A "First Approximation" of a new national scheme was first compiled in 1989 and was widely circulated in Australia for comment. Two further approximations were produced and circulated before the published version appeared in 1996 titled *The Australian Soil Classification* (2).

It should be noted that due assessment was made of two so-called international systems viz, *Soil Taxonomy* (8) and the subsequent revisions of most of its Orders, and the other major "international" system, that of the FAO–UNESCO (1990) *Soil Map of the World* and its updated (1998) version titled *World Reference Base for Soil Resources* (9). While these two so-called international schemes are of considerable value and interest for comparative purposes, it is more

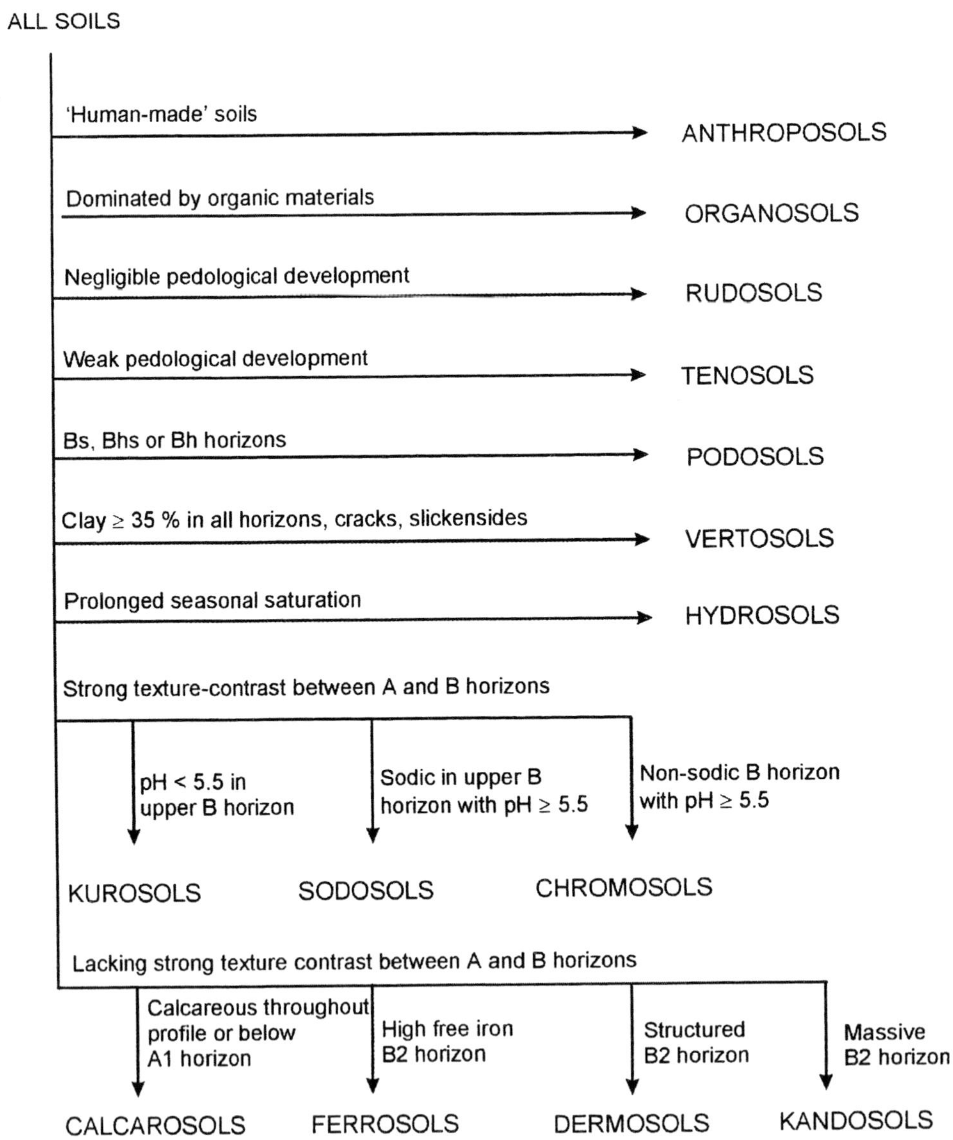

Fig. 1 Schematic summary of the 14 orders of the Australian Soil Classification. (From Ref. 10.) (Note: this figure is not to be used as a key.)

likely that most Australian pedologists would agree on the need for an appropriate national system that is regularly updated. The possibility of using methods of numerical classification was also examined but was thought to be inappropriate because of the great variability and inconsistencies in the soil datasets, e.g., lack of representation of important groups of soils.

The Australian Soil Classification (ASC)

This scheme (2, 10) is now widely used in most Australian soil surveys and pedological studies, primarily in an identification role. The scheme may be described as a multicategoric, hierarchical (order, suborder, great group, subgroup, family), general purpose scheme with classes defined on the basis of diagnostic horizons or materials and their arrangement in vertical sequence as seen in an exposed profile. The classes are mutually exclusive and the allocation of new or unknown individuals to the classes is by means of keys. The scheme is open-ended and new classes can be defined as knowledge increases, although these may not necessarily be added in sequence to the present list. Fourteen orders are currently defined (see Fig. 1). A recent innovation is the development of an interactive key to the ASC soon to be available on CD-ROM (11).

The structure of the ASC has strong similarities with that of *Soil Taxonomy*, and although the emphasis is on field morphology, laboratory data have been used as appropriate. All technical terms are defined in a glossary. In the companion publication (10) further descriptions of diagnostic horizons and materials are given, as well as the general occurrence of each order except for Anthroposols where data were unavailable. Some discussion is given of the major classes of each order and the use of major attributes in the subdivision of the orders. The main chemical properties used are pH and exchangeable cations, with pH commonly also determined in the field. A table is provided giving approximate correlations between ASC orders, three other Australian classifications, and *Soil Taxonomy* orders.

Perhaps the major benefit of the ASC is that it provides a means for efficient communication and a systematic framework for understanding and learning about the properties and interrelationships of soils. Experience with the *Factual Key* (3) suggested retention of the use of color and generalized texture profiles at high levels in the hierarchy of the new system, although formal justification of such decisions is often difficult to demonstrate. Another feature of the ASC that partly overcomes some of the problems associated with hierarchies is the use of family criteria at any level of the system.

New Zealand Soil Classification

With the advent of *Soil Taxonomy*, New Zealand soil scientists became heavily committed to its international development, involving a five year testing program in New Zealand. As in Australia, difficulties became apparent, particularly in regard to its complexity. Even so, the results of the investigations showed that *Soil Taxonomy* made inadequate provision for important classes of New Zealand soils, particularly in the case of Inceptisols. Much of the new order of Andisols was based on New Zealand, contributions in conjunction with Guy Smith who spent 12 months in the country studying these and other soils. The inadequacies of *Soil Taxonomy* and the older New Zealand genetic classification to serve as an up-to-date national classification led to the development and publication of new material, largely through the efforts of Hewitt (7, 12–15). The most recent version (13) has a helpful introductory section in which concepts, objectives, and principles are outlined. The scheme is a hierarchical one (order, group, subgroup, series), with class distinctions based on diagnostic horizons and materials. Keys are provided to enable easy class definitions and identification. Orders total 15 (Allophanic Soils, Anthropic Soils, Brown Soils, Gley Soils, Granular Soils, Melanic Soils, Organic Soils, Oxidic Soils, Pallic Soils, Podzols, Pumice Soils, Raw Soils, Recent Soils, Semiarid Soils, Ultic Soils). Accessory properties of the orders are also listed as well as concept, occurrence, and correlation with the earlier New Zealand Genetic Soil Classification and appropriate classes of *Soil Taxonomy*. The scheme is open-ended and hence new classes can usually be incorporated into the hierarchy.

CONCLUSION

This brief review of the development of two national soil classification systems indicates the need for some taxa which are required for one country but not the other. Of the 15 New Zealand soil orders that have been defined, at least two (Allophanic Soils and Pumice Soils) are virtually absent in Australian soil landscapes. Conversely some widespread Australian soils are inadequately defined in terms of the New Zealand system. Particular examples are the widespread Australian semiarid soils characterized by profiles which feature a clear or abrupt change to a textural B horizon which is frequently sodic. Finally, there are some soils common to both countries that probably could be satisfactorily classified at the order level by the other system. A few such examples include Arthropic soils and Arthroposols, Gley soils and Hydrosols, Organic soils and Organosols, Oxidic soils and Ferrosols, Podzols and Podosols.

REFERENCES

1. Isbell, R.F. A Brief History of National Soil Classification in Australia Since the 1920's. Australian Journal of Soil Research **1992**, *30*, 825–842.
2. Isbell, R.F. *The Australian Soil Classification*; CSIRO Publishing, 1996.
3. Northcote, K.H. *A Factual Key for the Recognition of Australian Soils*, 4th Ed.; Rellim Technical Publications: South Australia, 1979.
4. Stace, H.C.T.; Hubble, G.D.; Brewer, R.; Northcote, K.H.; Sleeman, J.R.; Mulcahy, M.J.; Hallsworth, E.G. *A Handbook of Australian Soils*; Rellim Technical Publications: South Australia, 1968.
5. Stephens, C.G. *A Manual of Australian Soils*, 3rd Ed.; CSIRO: Melbourne, Australia, 1962.
6. Taylor, N.H. *Soil Map of New Zealand, 1:2,027,520 Scale*; DSIR: Wellington, New Zealand, 1948.
7. Hewitt, A.E. Soil Classification in New Zealand: Legacy and Lessons. Australian Journal of Soil Research **1992**, *30*, 843–854.
8. Soil Survey Staff, *Soil Taxonomy, A Basic System of Soil Classification for Making and Interpreting Soil Surveys*; USDA Agriculture Handbook No. 436, U.S. Government Printing Office: Washington, DC, 1975.
9. FAO, *World Reference Base for Soil Resources*; FAO World Soil Resources Report 84, 1998.
10. Isbell, R.F.; McDonald, W.S.; Ashton, L.J. *Concepts and Rationale of the Australian Soil Classification, ACLEP*; CSIRO Land and Water: Canberra, Australia, 1997.
11. Jacquier, D.W.; McKenzie, N.J.; Brown, K.L.; Isbell, R.F.; Paine, T.A. *The Australian Soil Classification: An Interactive Key*; CSIRO Publishing, 2001.
12. Hewitt, A.E. *New Zealand Soil Classification (version 2.0)*; DSIR Division of Land and Soil Sciences Technical Record DN 2, Department of Scientific and Industrial Research: Wellington, 1989.
13. Hewitt, A.E. *New Zealand Soil Classification*; Landcare Research Science Series No. 1, Manaaki Whenua Press: Lincoln, New Zealand, 1998.
14. Hewitt, A.E. *Methods and Rationale of the New Zealand Soil Classification*; Landcare Research Science Series No. 2, Manaaki Whenua Press: New Zealand, 1993.
15. Clayden, B.; Webb, T.H. *Landcare Research Science Series No. 1*; Manaaki Whenua Press: Lincoln, New Zealand, 1994.

CLASSIFICATION, CHINESE

Zitong Gong
The Chinese Academy of Sciences, Nanjing, People's Republic of China

INTRODUCTION

Chinese Soil Taxonomy (CST) is a most recent taxonomic soil classification of China after a long history of related researches from ancient to modern times and peculiarized with several main points concerning *Anthrosols*, *Aridisols*, *Ferralisols*, *Alpine soils*, its nomenclature, and mapping. *CST* can be correlated to the worldwide popular systems like *Soil Taxonomy* and *WRB*.

HISTORY

Soil Classification in Ancient China

China is one of the most ancient civilized countries in the world with long history of agricultural activities. In about 2500 BP, the earliest soil classification in China appeared in two books, *Yugong* and *Guanzi Diyuan pian* (1).

Soil Classification in Modern China

It can be subdivided into three main stages, 1) In the 1930's the concept of soil great groups was adopted and more than 2000 soil series were established by combining C.F. Marbut's system with Chinese practice; 2) In 1954, a Chinese soil classification was formed by following the Russian system, that was further developed throughout the first National General Soil Survey (1958–1961), modified in 1978 and then accepted widely as the official system used in the Second Soil Survey (1979–1994); 3) With the increasing international exchanges, *Soil Taxonomy* and FAO/UNESCO *Soil Map Legend* were introduced into China since the 1980s. A Research Group of 35 universities and institutions led by ISSAS step by step, worked out a soil taxonomic classification of China—Chinese Soil Taxonomy (2).

PECULIARITY

Anthrosols (3–5)

Based on anthropogenic horizons as an anthrostagnic epipedon and hydragric horizon, a fimic epipedon and phos-agric horizon, and a irrgric epipedon *or* cumulic epipedon, Anthrosols are grouped into Stagnic Anthrosols, Fimic Anthrosols, Cumulic Anthrosols, and Irrgric Anthrosols.

Aridisols (6)

Aridisols are classified by the aridic epipedon *and* one or more of the following diagnostic horizons: salic, hypersalic, salipan, gypsic, hypergypsic, calcic, hypercalcic, calcicpan, argic, or cambic horizon whose upper boundaries are within 100 cm of the mineral soil surface.

Ferrosols

In tropical and subtropical China we have Ferrosols based on LAC-ferric subhorizon which has lower activity, clays, and rich in free iron oxides [CEC_7 16–24 cmol (+) kg^{-1}; free iron/total iron >40%] with a total area of 2 million km^2.

Alpine Soils

Tibet area is called the third polar, which is cold and dry. The soils there are classified as Cryic Aridsols that have a cryic soil temperature regime and Gelic Cambosols that have a cryic or cloder soil temperature regime, and that have frost-thawic features.

Nomenclature

Similar to *Soil Taxonomy*, we have adopted the principle of a segmental-continuous name. The first segment includes Order, Suborder, Group, and Subgroup. The Family name provides additional information on particle size class, mineral composition, and soil temperature regime. The Series is a second segment named independently.

Mapping

A 1:12M scale national soil map of China based upon *Chinese Soil Taxonomy* is already available.

Table 1 Correlation between Chinese soil taxonomy and ST, WRB systems

Chinese Soil Taxonomy, 1999	ST, 1999 (7)	WRB, 1998 (8)
Histosols	Histosols[a]	Histosols
Anthrosols	—	Anthrosols[a]
Spodosols	Spodosols	Podosols
Andosols	Andosols	Andosols[a], Cryosols[b]
Ferrosols	Oxisols	Ferralosols[a], Plinthosols[b], Acrisols[b], Lixisols[b]
Vertosols	Vertisols	Vertisols
Aridosols	Aridisols	Calcisols, Gypsisols
Halosols	Aridisols[b], Alfisols[b], Inceptisols[b]	Solonchaks, Solonetz
Gleyosols	Inceptisols, Gelisols, Entisols	Gleysols[a], Cryosols[b]
Isohumosols	Mollisols	Chernozems, Kastanozems, Phaeozems
Ferrosols	Ultisols[a], Alfisols[b], Inceptsols[b]	Acrisols[a], Lixisols[b], Plinthosols[b], Nitisols[b]
Argosols	Alfiisols[a], Ultisols[b], Mollisols[b]	Luvisols[a], Alisols[b]
Cambosols	Inceptsols[a], Mollisols[b], Gelisols[b]	Cambisols[a]
Primosols	Entisols[a], Gelisols[b]	Fluvisols, Leptisols, Arenosols, Regosols, Cryosols

[a] Mostly corresponding.
[b] Partly corresponding.

CORRELATION

Chinese Soil Taxonomy has become gradually known to the world after being suggested to its members by the Chinese Soil Society in 1996, translated into Japanese in Japan in 1997 (9), adopted by WRB especially for the Anthrosols in 1998, and introduced in *Handbook of Soil Science* (10) in 1999. The correlation between CST and ST, WRB systems is shown in (Table 1) .

REFERENCES

1. Gong, Z.T. *Chinese Soil Taxonomy (in Chinese)*; Science Press: Beijing, China, 1999.
2. Cooperative Research Group on Chinese Soil Taxonomy, *Chinese Soil Taxonomy*; Science Press: Beijing, China, 2001.
3. Gong, Z.T. Pedogenesis of Paddy Soil and Its Significance in Soil Classification. Soil Sci. **1983**, *35* (1), 5–10.
4. Gong, Z.T.; Zhang, G.; Luo, G. Diversity of Anthrosols in China. Pedosphere **1999**, *9* (3), 193–204.
5. Gong, Z.T.; Zhang, X.; Luo, G.; Shi, H.; Spaagaren, O. Extractable Phosphorus in Soils with Fimic Epipedon. Geoderma **1997**, *75*, 289–296.
6. Eswaran, H.; Gong, Z.T. Properties, Classification and Distribution of Soils with Gypsum. *Occurrence, Characteristics and Genesis of Carbonate, Gypsum and Silica Accumulation in Soil*; Special Publication No. 26, SSSA Madison, USA, 1991; 89–119.
7. USDA, *Soil Taxonomy*, 2nd Ed.; Agriculture Handbook No. 436, Government Printing Office, Washington, DC, 1999; 1–868.
8. FAO; ISRIC; ISSS, *World Reference Base for Soil Resources*; World Soil Resources Reports 84, FAO: Rome, 1998; 1–88.
9. Cooperative Research Group on Chinese Soil Taxonomy (Revised proposal) (in Japanese), JIRCAS Working Re No. 6, 1997; 1–128.
10. Sumner, Editior-in-Chief, *Handbook of Soil Science*; CRC Press: Boca Raton, FL, 1999; E161–E166.

CLASSIFICATION, FRENCH

Jean-Paul Legros
INRA, Science du Sol, Montpellier, France

INTRODUCTION

From the beginning of pedology, the French scientists used genetic classifications of soils in relation with the work of the Russian school of Dokuchaiev (1846–1903) and Glinka (1867–1927). For example, Valérien Agafonoff (1863–1955), a Russian soil scientist, came to Paris fleeing his country during the revolution of 1917. He built one of the first soil maps of France with the corresponding legend (created in 1928, but published mainly in 1936 after improvements) (1–3). Using the previous and successive works of Lagatu (1862–1942) (4, 5), of Demolon (1881–1954), of Kubiéna (1897–1970), of Erhart (1898–1982) and of several German authors, Georges Aubert (born in 1913) and Philippe Duchaufour (1912–2000), presented a first French classification in Paris, in 1956, during the Sixth International Congress of Soil Science. A second version of this system was given in 1962, known as "Aubert-Duchaufour classification." In 1960, Duchaufour published his famous "Précis de Pédologie" (6) that popularized his classification system to a much wider audience. Then, in 1967, the whole French scientific community collaborated and completed the system that became the "Classification of the CPCS" (*Commission de Pédologie et de Cartographie des Sols*). This document was used during 20 years and popularized a vocabulary nowadays used in the ordinary pedological language of French scientists, for example, *sol brun*, *ranker*, *rendzine*, etc.

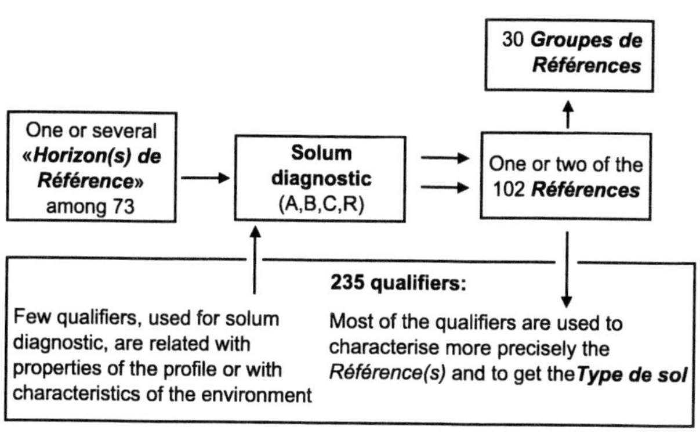

Fig. 1 Organization of the *Référentiel Pédologique*.

THE RP AND THE WRB

During this period, the American system and the FAO Soil Legend appeared (1960 and 1981). In parallel, the development of Computer Science and of Statistical Science demonstrated that the soils could be classified in a more objective way by searching for the mathematical similarities between taxa and objects to be classified rather than trying to enter a soil in a more or less adapted taxon of a rigid classification system. In this context, since 1986, the French community met, worked, and presented a first version of "Référentiel Pédologique" (RP) in 1992. A second version was produced in 1995 (7), with an English translation in 1998 (8). Because several French scientists were involved in the construction of both the new French trials and in the World Reference Base (WRB) (9), the two systems have great similarities in both philosophy and organization although not necessarily in the details of vocabulary. So, it is more simple to review the RP in comparison with the WRB:

- The 40 diagnostic horizons of the WRB are replaced by the 73 "Horizons de Référence" of the RP.
- The 30 reference soil groups of the WRB are replaced by the 30 "Grands Groupes de Références" (major groups of references or "GER"). But these 30 categories do not match exactly because, in the French system, the tropical soils are not considered. This last point was regretted (10), considering the large experience of French scientists in Africa. Theses 30 *Groupes* are divided into 102 "Sols de Référence."
- The WRB "Diagnostic properties," "Diagnostics materials," "Formative elements," and "Prefixes" correspond to the "Qualificatifs" (Qualifiers) of the RP. In the RP, all these elements are grouped in a single list of 235 terms.

The more original point of the RP is that a Diagnostic horizon alone is in general not sufficient to recognize a *Référence*. Several Diagnostic horizons are associated to identify a *Solum* (profile development as A, E, B, C including a part of the underlying rock) (11). For this reason, the "Horizons de Référence", in the RP, are defined considering strongly their relative development (A, B, C horizons, etc.). This indicates clearly that a part of

Table 1 Relationships between the WRB groups and the GER or the *Références* of the RP

WRB	Référentiel pédologique
Histosol	Histosol
Cryosol	Cryosol
Anthrosols	Anthroposols
Leptosols	Lithosols (superficial)
	Rankosols (on acid rocks)
	Organosols (rich in organic matter)
	Rendosols (on calcareous rocks)
Vertisols	Leptismectisols (A/C or A/R profile)
	Vertisols (with B horizon)
Fluvisols	Fluviosols
	Thalassosols (estuarine/marine soils)
Solonchaks	Salisols
Gley soils	Réductisols (dominant reduction)
	Rédoxisol (dominant oxidation)
Andosols	Andosols
	Vitrosols (with glass)
Podzols	Podzosols
Plinthosols	*Not already studied*
Ferralsols	*Not already studied*
Solonetz	Sodisols
Planosols	Planosols
Chernozems	Chernosols
Kastanozem	
Phaeozem	Phaeosols
(Greyzems)	Grisols
Gypsisols	Gypsosols
Durisols	*Not already studied*
Calcisols	Rendosol (rendzine)
	Rendisols (same morphol., saturated)
	Calcosols (calcareous with B)
	Dolomitosols (with $MgCO_3$)
	Calcisols (noncalcareous, saturated)
	Magnesisols (with Mg^{++} on clay)
	Calcarisols (calcaric within 25 cm)
Albeluvisols	Luvisols (for a part)
Alisols	*Not already studied*
Nitosols	Fersialsols
Acrisols	*Not already studied*
Luvisol	Luvisol
Lixisols	*Not already studied*
Umbrisols	Alocrisols humiques, rankosols
Cambisols	Brunisols
	Pelosols (rich in clay but not 2/1)
Arenosol	Arenosol
Regosol	Régosol

the previous genetic approach remains inside the RP. Its organization is presented in Fig. 1.

Even if it is not fully completed, the RP, richer in taxa and qualifiers for a smaller part of the world, may be

Table 2 Specific *Références* in the RP

Alocrisols	Acidic but without *argic* horizon (i.e different of the WRB "Acrisols")
Colluviosols	(From colluvium), i.e. on slopes, in parallel with fluviosols
Peyrosols	(From Peyre = stone in some local French languages), i.e. with important coarse fraction
Veracrisols	Sort of Acrisols rich in earth worms

considered to be more precise than the WRB. It is organized in three logical levels that are not truly hierarchical. It allows the user to identify a soil as an intermediate between two taxa, for example a "REGOSOL–CRYOSOL".

Table 1 shows the approximate links between WRB and RP at the level of GER, *Références* and Groups.

The most specific Références of the RP if we compare with WRB are presented in Table 2 with short explanations.

All this demonstrates that it is very easy to pass from one system to the other at the level of Groups and *Références* even if "Calcisol" is a false friend (Calcisol means calcareous in WRB but saturated and not calcareous in the RP).

The situation is more complicated at the level of soil types. Many of the qualifiers seem similar, but, this may not truly be the case. Table 3 shows some of the difficulties which may be encountered.

Table 4 characterizes the French soils using the RP system. The data were kindly provided by the *Service d'Etude des Sols et de la Carte Pédologique de France* (Orléans).

For the French scientific community, the construction of the RP was a good opportunity to work together on the concepts of the soil classifications and to develop the list of the soils identified in France. In the RP volume, the texts that present the Andosols (the soils with hydromorphic features, and the different kinds of humus) are valuable contributions to Pedology. Nevertheless, the interest of the RP, at an international level, seems limited by its great similarity with the WRB.

To pass from a purely genetic classification to an international reference system such as the WRB was rather a long journey for the French pedological community (12). The construction of the RP, initiated by Denis Baize and Michel-Claude Girard, was a good way to test the possibility of this large change in the French method of thinking. Now, the way is surely open for French scientists

C

Table 3 Differences between qualifiers in WRB and RP

Case	Problem	Examples	
		WRB terminology	RP terminology
1	Term specific of one of the two systems (scarce case)	For example: Carbic	or: Clinohumic (isohumic)
2	Same meaning but different terms (scarce case)	Alumic (Al sat>50%)	Aluminic (Al^{+++} dominant)
3	Same name but slightly different meanings (general case)	Magnesic $Ca^{++}/Mg^{++}<1$	Magnesic $0.2<Ca^{++}/Mg^{++}<2$
4	Same name but very different meaning (scarce case)	Ruptic	Ruptic
		Vertical discontinuity	Lateral discontinuity
5	Different orthography but same meaning (scarce case)	Bathi (prefixe)	Bathy
		Between 100 and 200 cm	Found at depth

to join the WRB community. Nevertheless, the international community certainly needs, in addition to the WRB and the U.S. Soil Taxonomy, a purely genetic system to standardize and to model the advances of knowledge in pedogenesis. The old classification of "Aubert-Duchaufour" could be a good starting point. Who will try to go further in this direction after Duchaufour himself (13)?

Table 4 Inventory of the French Soils using the RP (from INRA-SESCPF)

Référentiel Pédologique	% of France	Corresponding WRB group
Calcosols, Brunisols	48.5	Cambisols
Luvisols	14.5	Luvisols
Rendosols	8.3	Calcisols (p.p)
Fluviosols	7.8	Fluvisols
Podzolized luvisols	6.4	Albeluvisols
Podzosols	5.6	Podzols
Lithosols	2.3	Leptosols (p.p)
Rankosols	1.8	Leptosols (p.p)
Andosols, vitrosols	1.0	Andosols
Arénosols	0.8	Arenosols
Rédoxisols, Réductisols	0.4	Gleysols
Régosols	0.4	Regosols
Salisols	0.4	Solontchaks
Histosols	0.3	Histosols
Phaeosols	0.1	Phaeozems
Vertisols	0.1	Vertisols
Planosols	0.1	Planosols
Autres sols et surfaces	1.2	Others soils and surfaces

REFERENCES

1. Agafonoff, *Les sols de France au point de vue pédologique*; p. 154+carte Dunod: Paris, 1936.
2. Boulaine, J. *Histoire des pédologues et de la science des sols*; EditionsINRA: Paris, 1989; 285.
3. Legros, J.P. *Cartographies des sols*; De l'analyse spatiale à la gestion des territoires, Presses Polytechniques et Universitaires Romandes: Lausanne, 1996; 321.
4. Boulaine, J.; Legros, J.P.; Lagatu, Henri la pédologie Lui Doit Beaucoup. Revue Agriculture **1988**, *528*, 18–20.
5. Boulaine, J.; Legros, J.P. *D'olivier de Serres à René Dumont, portraits d'agronomes*; Coll. Tec/Doc, Lavoisier: Paris, 1998; 320.
6. Duchaufour, Ph. *Précis de Pédologie*; Masson et Cie: Paris, 1960; 438.
7. Baize, D., Girard, M.C., Coord. Référentiel pédologique, INRA-Editions: Paris, 1995; 332 pp.
8. Baize, D. *A Sound Reference Base for Soils*; The référentiel pédologique (text in English), INRA Editions: Paris, 1998; 322.
9. WRB. World Word Reference Base for Soil resources. ISSS, ISRIC, FAO. 1998, FAO Report No. 84, 88 pp.
10. Isbell, R.F. Book Review: A sound Reference Base for Soils—"référentiel pédologique" 1998. Catena **2001**, *43*, 157–159.
11. Baize, D. Place of Horizon in the New French "Référentiel Pédologique". Catena **1993**, *20*, 383–394.
12. Baize, D. Typologies Et Types En Pédologie. Science du Sol **1992**, *30* (2), 95–115.
13. Duchaufour, Ph. Réflexions Sur Les Classifications Des Sols. Etude et Gestion des Sols **1998**, *5* (3), 201–205.

CLASSIFICATION, NEED FOR

Stephen Nortcliff
The University of Reading, Reading, United Kingdom

INTRODUCTION

Soil varies across the Earth's surface. This variation occurs both laterally and vertically and is a result of the complex interactions of the processes of soil formation in the environmental context, particularly climate, parent material and topography. These processes have operated over varying time periods. As a consequence, some soils on the Earth's surface may have been developing for thousands and possibly millions of years and are well developed and in equilibrium with their environment; others will show only the initial stages of soil development and may be only a few years or tens of years old. Often the characteristics of this latter group of soils are very strongly influenced by the nature of the relatively unaltered parent material.

An important feature of the soil is that it is not static, rather it is a dynamic natural body interacting in a complex manner with its environment. As a consequence, the soil changes through time and in space as a response to environmental changes. The results of this are that the soil will show variation at different times within the development of a landscape but, of considerably greater importance, it will vary from place to place at any given time.

Taking this argument to its extreme conclusion, we might consider that no two places on the Earth's surface have identical soils! Rather than accept this extreme scenario it is generally agreed that it is possible to group soils together into classes, within which many properties of the soil may be expected to be similar. Soil classification seeks to group these soils into soil classes or taxonomic units, and the related practice of soil survey seeks to map the soils into spatially homogeneous groups so that they may be portrayed on maps. Before a map can be produced, however, the soil surveyor needs, at the minimum, a basic soil classification system. Frequently the soil surveyor producing soil maps will endeavor to produce soil-mapping units that correspond to the taxonomic units of the soil classification.

Soil classification is not new. In the 12th century, Yahya Ibn Mohammed, writing in his compendium entitled '*The Book of Agriculture*' (1) noted that "the first step in agriculture is recognizing the soil and knowing how to differentiate between the good and the poor one." It is interesting that there is very little documented consideration of soil classification from this time until the 19th century. In the second half of the 19th century, Dokuchaiev, working in Russia, highlighted the need to consider the nature and pattern of soils in the landscape and suggested the possibility of being able to group soils into distinct classes based on their properties and environmental context (2). From this time, throughout the 20th century and through to the 21st century, there have been numerous attempts to classify soils, some in the context of a local landscape of a few kilometers, others on a global scale.

WHICH PROPERTIES SHOULD BE CHOSEN FOR CLASSIFICATION?

It is unclear whether the approach to soil classification in the 12th century involved only the top few centimeters or greater thicknesses of soil material, but many of the early documented classification focused upon the upper soil layers, with little attention being given to underlying layers. As the knowledge concerning soils and their development has increased, so has the awareness of the need to consider the underlying layers. Today, we identify four broad groups of these layers, from the surface downwards the O, A, B and C, which we call soil horizons. The nature and significance of soil horizons are discussed by Bridges in this volume (3).

It has already been stated that part of the purpose of soil classification is to provide a degree of order to the considerable diversity of soils and soil properties found across the Earth's surface. The classes are normally defined with respect to one or a group of properties, with a particular soil having membership of the class if the values for the selected properties fall within defined ranges. The problem first arises in selecting the properties and secondly the ranges of properties to be used in determining class membership. In some cases, the properties selected for classification have not been properties of the soil, but properties of the environment. In using environmental factors in the classification of the soils, it is assumed that there are very strong environment–soil property relationships. This was a major feature of the classification produced by Dokuchaiev and co-workers in 19th century Russia.

Whilst the organization of our knowledge by means of a soil classification is a major step forward toward an understanding of the nature of soils and soil distributions, there still remains the often problematical decision of which properties of the soil are to be chosen to determine and identify the soil classes. As our knowledge of soils has increased, so has the range of properties upon which we might base a classification of soils. It is essential that there is a clear rationale for the selection of the often small number of soil properties chosen to differentiate the soil classes, and to determine the class limits.

Table 1 International and national general purpose soil classifications

Soil classification	Chapter in this volume
International classification	World Reference Base for Soil Resources (6)
International/national classification	USA—Soil Taxonomy (7)
National classification	Australia—The Australian Soil Classification (8)
	France—Référentiel Pédologique (9)
	Russia—Classification of Soils (10)
	South Africa—A taxonomic System for South Africa (11)

WHY CLASSIFY SOILS?

The principal reason for classifying soils is to organize the vast amount of information about soils, to produce classes that have either similar properties and/or similar responses to external inputs, both natural and man-made. The production of a classification should save the investigator time and simplify the description of the soil (4). If many of the soils that we encounter have three or four properties in common (for example, uniform texture down the profile, strong aggregated clay, and red color) it may be sensible to use one short name to describe them all. Having produced a framework for the classification of soils, this may provide the basis for the prediction of individual properties and behavior from knowledge of the defined class characteristics. In this respect, soil classification becomes an essential means of communication amongst soil scientists. It is also important when identifying the most appropriate use of the soil, estimating the production, extrapolating knowledge gained at one location to other often relatively little known locations, and providing a basis for future research needs.

SINGLE OR GENERAL PURPOSE CLASSIFICATION?

One of the matters that has concerned those involved with classification of soils is whether the classification should be with respect to a single objective, a narrow range of purposes, or whether the classifier should seek to produce a general purpose classification. The debate between those promoting either single or general purpose classifications has been widespread, yet the aims of these classifications are very different. Sneath (5) argues that classification based on as many features as possible is the most "natural" classification and that such a classification will be best for general purposes. Certainly a classification will have the greatest range of predictive powers, though for any specific purpose a special classification based on the features of importance to that purpose is likely to be more precise. Confusion frequently arises because the classifications have been undertaken according to the principles of general-purpose classification, but then used for very narrow ranges of specific purpose situations. Whilst the classifications may be very good at the general level, they will generally be less effective for specific needs. If the applications to specific purposes are inappropriate, they may result in erroneous predictions. Most widely used classifications, both international and national levels, are general purpose. Table 1 lists some of the general purpose soil classifications that are described elsewhere in this volume.

SOIL CLASSIFICATION AS A MEANS OF COMMUNICATION

Soil classification facilitates communication about soils, both between soil scientists and between soil scientists and nonsoil scientists. Because of this need for communication: it is essential that those persons using a classification are aware of the criteria used to define the classes, and the context within which these criteria were chosen. The outcomes of the classification must be understood in the context of the selection of the differentiating criteria chosen. If a soil classification is to be used as a means of communication, the principles upon which it is based and the criteria used to select the properties for class discrimination must be understood by those using it. Whilst this is important for communication amongst soil scientists, it is of particular importance when soil scientists communicate with nonsoil scientists. It is also important for the user to be aware whether there is any perceived co-variation between the properties chosen for discrimination and other soil properties, as frequently classifications based on one property or one set of properties are used by other people, although in the original specification of the classification no such co-variation had been assumed. Such misuses have the potential to result in substantial misunderstanding and major errors.

In summary, therefore, the key benefits that result from soil classification may be summarized as follows:

1. Soil classifications provide a framework for soil survey and for the observations made on soils.
2. Soil classifications provide an understanding of the relationships between (selected) soil properties and the significance of these properties in relation to use.
3. Soil classifications provide a means of organizing research information and the communication of research results (12).
4. The allocation of soils to classes should provide a framework within which the laws relating to the nature and development of the soil can be discovered (13).
5. Soil classifications provide a means of technology transfer, where the knowledge and behavior of soils in response to different management strategies, may be transferred to soils considered to be similar with respect to the classification (14, 15).

CONCLUSION

It is obvious that the vast amount of information that can be gathered for each soil requires some form of organization and management. To accomplish this, some form of allocation of soils into classes is essential. To simplify the wealth of information in this manner will greatly assist the communication between groups of soil scientists and between soil scientists and others. The key to a successful classification is to identify the purpose of classification and to select criteria selected as differentiating properties that are appropriate for this purpose. Frequently soil classifications are misused because no account is taken of why it was devised in the first place. Using an appropriate classification can provide greater insight into soil relationships whereas use of inappropriate classification may result in confusion.

REFERENCES

1. Johnson, W.M. Soil Classification and Design of Soil Surveys. In *Soil-Resource Data for Agricultural Development*; Swindale, L.D., Ed.; Hawaii Agricultural Experimental Station, University of Hawaii: Honolulu, 1978; 3–11.
2. Dokuchaiev, V.V. *The Russian Chernozem* (in Russian); PhD Thesis St. Petersburg, 1889.
3. Bridges, E.M. Diagnostic Horizons. In *Encyclopaedia of Soil Science*; Lal, R., Ed.; Marcel Decker: New York, 2001.
4. Leeper, G.W. The Classification of Soils. J. Soil Sci. **1956**, *7*, 59–64.
5. Sneath, P.H.A. The Application of Computers to Taxonomy. J. Gen. Microbiol. **1957**, *17*, 201–206.
6. Deckers, J.A., Nachtergaele, F.O., Spaargarer, O.C., Eds. *World Reference Base for Soil Resources*; 1st Ed. Introduction. ISSS/ISRIC/FAO. Acco: Leuven, Belgium, 1998; 165.
7. Soil Survey Staff, *Soil Taxonomy—A Basic System of Soil Classification for Making and Interpreting Maps*; Agriculture Handbook Number 436, USDA Natural Resources Conservation Service: Washington, DC, 1999; 869.
8. Isbell, R.F. *The Australian Soil Classification*; CSIRO Publishing: Melbourne, Australia, 1998; 143.
9. Baize, D. *A Sound Reference Base for Soils. The Référentiel Pédologique*; (Text in English) INRA Editions: Paris, 1998; 322.
10. Tonkonogov, V.D.; Lebedeva, I.I.; Shishov, L.L.; Gerasimora, M. Eds.; Russian Soil Classification System; English Version, Transl. by Gerasimora, M., ed. by Arnold, D., 2001; Dokuchaev Soil Science Institute: Moscow, 2001; 220 pp.
11. Soil Classification Working Group, *Soil Classification. A Taxonomic System for South Africa*; Memoirs on the Agricultural Natural Resources of South Africa No. 15, Department of Agricultural Development: Pretoria, South Africa, 1991; 1–257.
12. Runge, E.C.A.; McCracken, R.J. The Role of Soil Classification in Research Planning. *Soil Taxonomy—Achievements and Challenges*; SSSA Special Publication 14, SSSA: Madison, WI, 1984; 15–28.
13. Smith, G.D. Lectures on Soil Classification. Pedologie **1965**, 1–134, Special Edition 4.
14. Beinroth, F.H.; Uehara, G.; Silva, J.A.; Arnold, R.W.; Cady, F.B. Agrotechnology in the Tropics Based on Soil Taxonomy. Adv. Agron. **1980**, *33*, 303–309.
15. Buol, S.W.; Denton, H.P. The Role of Soil Classification in Technology Transfer. *Soil Taxonomy—Achievements and Challenges*; SSSA Special Publication 14, SSSA: Madison, WI, 1984; 29–43.

CLASSIFICATION, RUSSIAN

Maria Gerasimova
Moscow Lomonosov State University, Moscow, Russia

INTRODUCTION

The first soil classification systems appeared in Russia on the eve of the 20th century, and the imprint of their ideology and conceptual priorities have remained very strong in the later systems. Both early Russian schemes developed by Dokuchaev and Sibirtsev are traditionally regarded as factor-genetic, i.e., they comprise the notions on the origin of soils, soil-forming processes and soil position in the system of geographical zones. However, some importance was attributed to soil properties, moisture and temperature regimes, and the stage of soil development. These major features of pedogenetic manifestations and environment have proved to be key points in the further development of the classification schemes in Russia. Subsequent schemes differed in the priorities given either to the environment (zones), or to soil genesis, or to soil properties.

BACKGROUND INFORMATION

The diversity of approaches to soil classification was always inherent to Russian pedology. Unlike the early and recent American systems, all Russian schemes started from "above": the lowest taxa were derived from the upper ones, the names of which were retained in the low-category soil name; in other words, there was no similar concept to that of the American Series in the Russian approach. The majority of systems were hierarchical, with the soil type being the central unit. The definition of the soil type serves to illustrate the fundamental principles underlying the classification (Table 1). During the development of the various classification systems different sets of criteria for the upper taxonomic levels were proposed: soil genesis in terms of soil-forming processes, or soil age; soil chemical and/or morphological properties; soil horizons; and soil-forming agents. The categories higher in the system than soil type differed in definition depending upon the conceptual approach of the author(s), while the below-type categories have remained basically unchanged since they were first specified in the late 1950s. Not all classification systems have distinguished categories at levels higher than soil type.

The "discovery" of soil science by Dokuchaev (1) resulted in a simple and descriptive zonal soil classification system, which was the *first* in Russia. It was presented as a table with short comments in 1886. All soils—14 groups, were subdivided into normal—*zonal* soils, *transitional* (boggy-meadow soils, rendzinas, and solonetzes), and *abnormal* ones (e.g., alluvial and eolian). Groups of zonal soils were ascribed to zones, and characterized in terms of processes, sediments, climate, vegetation, fauna, and topography. The system of Sibirtsev (10) was slightly more advanced; the soils known in those times were grouped into *zonal, intrazonal, incomplete,* and *surface geologic formations*, and lower categories were outlined as related to soil texture and parent rocks.

The ecological–genetic system of Gerasimov et al. (4), under supervision of Prassolov, is widely recognized to be based on the strictly defined notion of "genetic soil type," with subdivision related to the potential drainage" and/or position in a catena. Soil sequences were identified as automorphic, hydromorphic, and semihydromorphic. The definition involves processes of substance migration–accumulation–transformation, geographic environments, soil evolution, and pedogenetic processes, which are specifically emphasized. A first categorization of subtypes and species was proposed; species—stages of pedogenetic process development—were later transformed into quantitative groupings. Units of the lowest category were identified in relation to land use and soil degradation (old-arable limed, old-irrigated, secondary saline, etc.).

In 1956 Ivanova (7) published a classification for world soils which was widely criticized by foreign and some Russian soil scientists because of its over reliance upon climatic subdivisions. Indeed, the highest taxa were differentiated with respect to climate and vegetation, the names of soil types were supplemented with landscape characteristics (for example, red soils of dry savanna). The degree of soil hydromorphism (automorphic, transitional and hydromorphic series) was also taken into account at the higher (above-type) level. Given this complexity it is perhaps not surprising that the number of genetic soil types reached 80. The system was termed evolutionary genetic–geographical and highly appreciated in the USSR for its geographical essence.

Table 1 Scheme of taxonomic units in some classification systems in USSR/Russia

	Author(s) (year)					
	Dokuchaev (1886)[a]	**Gerasimov et al. (1939)**	**Official systems (1967, 1977)**	**Kovda et al. (1967)**	**Glazovskaya (1972)**	**New system (1997)**
Criteria for the upper level	Zonality	No upper level		Soil age, budget of substances	Eh, pH, weathering, humus type	Pedogenesis vs. lithogenesis
(Genetic) soil type						
Criteria for lower levels	Subtype	Catenary and plants effects	Additional process	Not discussed		Diagnostic features
	Genus	–	Parent material			CEC, type of salinity
	Species	Stages	Quantitative characteristics			
	Variety	Texture				
	Phase	Human impacts	Degree of erosion	Not discussed		Parent material

[a] Only two levels were presented in this pioneer system, both related to zonality.

Two more classification systems illustrate the diversity of concepts which have prevailed in the former USSR. Both were concerned with the above-type categories, and they have common approaches to the choice of soil properties serving as differentiating criteria.

Perception of soils as results of broad geochemical mass–energy exchange processes interacting in time parallel to the evolution of landscapes served as background for the world system elaborated by Kovda et al. (8). The upper category is represented by broad soil geochemical formations differentiated by major trends of weathering and clay minerals, types of humus, acid–base properties, supplemented with characteristics of climate and vegetation.[a] The second entry—"stadial groups of soils"—corresponded to hypothetical evolution stages from submerged sites to excessively drained ones. The following stadial groups were proposed: hydroaccumulative, hydromorphic, mesohydromorphic, paleohydromorphic, proterohydromorphic, automorphic (including mountain soils), and paleoautomorphic. The number of stadial groups and of soil geochemical formations varied in different versions of this system; moreover, climatic zones or belts were sometimes introduced, and climatic facies appeared in the final version. The latter was published as the legend of the World Soil Map, scale 1:10M, 1975. The system was severely criticized for a high degree of clumsiness, insufficient paleogeographical substantiations, and internal inconsistency.

[a]Example: "Formation of acid, strongly freezing or cryogenic soils. Weak weathering. Clay of illite—hydromica type. Main processes: humus and peat accumulation, weak podzolization, gley, cryogenesis. Taiga vegetation. Cold boreal climate." (1975)

A similar geochemical approach was implemented by Glazovskaya (6) in the classification of world soils in late 1960s. Most of her genetic–taxonomic ideas were applied in the legend of the World Soil Map (produced for higher level educational use at universities, colleges, and advanced secondary schools) at a scale of 1:15 M, in 1981. This system is strictly hierarchical: there are three above-type levels: associations, generations (classes), and families. Soil types are arranged in families in accordance with distinct rules. Lower categories were preserved as traditional. Criteria for differentiation at the highest category are pH and soil features indicative of redox conditions; members of the second category are differentiated by a range of process-related properties, including broad processes of organic matter accumulation, formation of secondary minerals, translocation of substances, and effects of groundwater. The difference in the composition of pedogenetic accumulations (humus, illuvial horizons, concretions, etc.) allows the specification of families within classes that were the main soil units on the world map.

Official systems — an initial approximation produced in 1967 and the final version of 1977(a) derived from the system of Ivanova of 1956. Both systems lacked any above-type categories, whereas the lower ones were clearly and adequately defined. The version of 1977 is best known in Russia and abroad. It was translated into English (1986), and has been efficiently used in soil survey and small-scale mapping for more than 30 years. The ecological–genetic principle of the system was applied practically for all hierarchical categories and soil diagnostics. The upper category corresponds to genetic soil types, of which there are more than 100. The subtypes are very large, both conceptually and in number. Many subtypes were required to cover many and diverse soil features owing to their dual nature: there are subzonal and facies subtypes, both variants being more geographic than taxonomic units. The concept of subzonal subtypes as "central images" of types and intergrades corresponding to subzones did not contradict the traditional interpretation of subtypes as modifications of types by superimposed processes, while the facies subtypes are really virtual phenomena. They were recognized only by climatic parameters (total air and topsoil temperatures >10°C, depth and duration of soil freezing), and their names were like this: "very cold deeply (freezing to a depth in excess of 1 m) and long freezing… ."

The third—genus—category contained much information on parent rocks, groundwater, soil history, and its differentiating criteria were far from being universal for different types. Species (or kinds in some translations) and subspecies, unlike genera, were more consistent in specifying the quantitative parameters of pedogenetic processes (not horizon's properties!). For example, the thickness of humus-accumulative horizon and humus content in the 0–10 cm layer served to qualify the intensity of the "chernozemic process". Texture characteristics and erosion phases were also accommodated in this system.

The traditional logic and consistency of this system made it very popular in the Soviet Union. It really indicates the essence of soil formation in different zones (in accordance with the central images of types), and allows soil development in a changing environment to be forecast; it uses traditional nomenclature of Russian soils and its influence can be traced in many national classifications. However, this classification was considered by others to be out dated and is now considered by some to be inappropriate—given the new knowledge—on the soils of Russia. Moreover, the system embraced only soils of agriculturally suitable areas. Little attention was paid to human-modified soils. Using the basic factor approach, this classification system could not provide distinct morphological diagnostics of soil types and subtypes in these environments. Finally, it was considered to be based on a static understanding of the nature of soils and soil-forming environments and could not accommodate soils that did not fit the subtype criteria; therefore, recently described Siberian and northern soils could not be inserted into the system without considerable changes in its principles and structure. These drawbacks of the official system and the scarcity of appropriate information on the western classification systems provided the impetus for a new search in the field of soil classification in the late 1970s and 1980s.

The categorization of pedogenetic processes initiated by Neustruev in 1922 was further developed by his pupil Gerasimov (5), who, sticking to his "factors–processes–properties" triad, actively supported the idea of profile formulae [a similar profile formula approach using horizon sequences was suggested by Fitzpatrick (2)]. Such formulae presented sequences of horizons diagnostic for the majority of soil types. This was a change towards a substantive-genetic ideology in classifying soils.

Fridland (3) initiated the development of a new composite soil classification system based on three components: 1) regime, 2) petrographic–mineralogical and 3) profile-genetic component. The latter component presumed the assessment of soil properties via diagnostic horizons, in the definition of above-type categories. It evolved into the new soil classification of 1997.

Classification of Soils of Russia (11) is an open, substantive-genetic system, embracing soils of all regions. Soil properties related to genesis are used as criteria for the highest taxonomic categories (Trunks and Orders), and soil types, traditionally being central units of the system, are specified by sequences of diagnostic soil horizons. These horizons are defined by the integrity of substantive soil properties, whose choice is controlled by pedogenetic processes. Environmental agents, including climatic parameters, are virtually excluded from the diagnostics of most soil taxa. Special attention is paid to human-modified soils. Together with corresponding natural soils, they are perceived as a conceptual or spatial continuum: from natural soils to modified natural-anthropogenic soils and, finally, to nonsoil surface formations. The taxonomic position of human-modified soils does not take into account the goals and character of impacts on soil and the level of soil fertility; it is fully dictated by the morphology of soil profile. Traditional soil names were preserved, and supplemented by new constructions for human-modified soils and nonsoils.

The new Russian system has a number of features in common with the International and American systems in terms of methodology, in particular, in the attitude to soil horizons. However, in spite of many efforts, the correlation of soils between the systems remains broadly inadequate, and this may be a challenge in the future.

REFERENCES

1. Dokuchaev, V.V. Classification of Soils of Professor Dokuchaev, V.V. (Northern Hemisphere). Pochvovedeniye **1900**, *2*, Appendix.
2. Fitzpatrick, E.A. *Pedology, A Systematic Approach to Soil Science*; Oliver and Boyd: Edinburgh, 1971.
3. Fridland, V.M. *Major Principles and Elements of a Basic Soil Classification System and the Program of Its Development*; Dokuchaev Soil Institute: Moscow, 1982; 149.
4. Gerasimov, I.P.; Zavalishin, A.A.; Ivanova, E.N. A New Scheme of a General Soil Classification for the USSR. Pochvovedeniye **1939**, *7*, 10–43.
5. Gerasimov, I.P. Elementary Pedogenetic Processes as a Basis for Genetic Diagnostics of Soils. Pochvovedeniye **1973**, *5*, 102–113.
6. Glazovskaya, M.A. *Soils of The World*; Moscow Univ. Press: Moscow, 1972; vol. 1, 231.
7. Ivanova, E.N. Essai De Classification Générale Des Sols. *Rapports VI Congrés International de la Science du Sol, Paris, France*; Commission V, 1956; vol. E, 387–394.
8. Kovda, V.A.; Lobova, E.V.; Rozanov, B.G. Problem of World Soils Classification. II. Historical-Genetic Systematics of Soils. Pochvovedeniye **1967**, *7*, 3–16.
9. *Classification and Diagnostics of the Soils of the USSR*; Also Published by USDA and National Science Foundation: Washington, DC, 1986; Kolos Publ: Moscow, 1977; 222.
10. Sibirtsev, N.M. *Soil Classification as Applied to Russia* (Table), S. Peterburg, 1896.
11. Tonkonogov, V.D.; Lebedeva, I.I.; Shishov, L.L., Eds. *Classification of Soils of Russia*; Dokuchaev Soil Institute: Moscow, 1997; English Version: Arnold, R.W., Eds. Russian Soil Classification System, Moscow, 2001; 220.

CLASSIFICATION, SOUTH AFRICAN

Jeffrey Hughes
University of Natal, Scottsville, South Africa

INTRODUCTION

The present soil classification has its roots in a number of important field studies, the most influential of which covered 11,000 square miles of the Tugela (Thukela) River Basin in KwaZulu-Natal (1). Two of the co-authors of that survey, namely C.N. MacVicar (later, Director, Natal Region, Department of Agriculture) and J.M. de Villiers (later, Chair of Soil Science at the University of Natal) became two of the most influential figures in soil classification in South Africa. The classification presented in 1969 in the publication *Soils of the Tugela Basin* (1) laid the foundation for those that followed.

THE TUGELA BASIN CLASSIFICATION

The soils of the survey area were subdivided into 29 soil "forms" at the highest level in the classification and into 110 soil series at the most detailed level. Thus, just two levels in the hierarchy were recognized—a feature that has remained in all subsequent changes. The concept of the soil form was "developed to include all soils that have the same kind and number of diagnostic horizons, arranged in the same vertical order." (1). The soil series were distinguished from the forms by using soil properties that were either not used to define the forms or, where properties were used, a narrower range of variation was employed at the series level. The major criteria used for defining the soil series were texture, base status, and calcareousness. In individual soil forms, characteristics such as pH and surface crusting were used. The classification also recognized that not all the possible soil series were present within the Basin and that there would be others found elsewhere in the country. Gaps were therefore left so that those series could be slotted in as they were identified.

Eighteen diagnostic horizons were defined and their general connotation given, i.e., the meaning behind the definition and where they were most likely to be found. All the soil forms and series were named after the geographic location where the "type" soil was identified. One of the major features of this classification was its simplicity, since almost all the properties needed to classify a soil were identifiable in the field. The main exception to this was base status (either dystrophic, mesotrophic, or eutrophic), although this often could be inferred from other characteristics such as the vegetation, climate, and soil structure. Base status, together with other field-estimated or measured properties such as pH and texture, needed confirmation by laboratory measurement to finalize the classification of some soils. The degree of success that this system achieved can be measured by so many of its features having been retained in future editions of the classification.

THE BINOMIAL SYSTEM

Soil Classification. A Binomial System for South Africa was the first detailed system that allowed for the classification of the soils of South Africa (2), as opposed to individual classifications developed for specific areas of the country. Its publication resulted only after considerable consultations with the major role players in South Africa and extensive in-field testing. The authors also recognized in the Preface that "This is the first edition; it is certain that the system as presented here is not the final word." (2). The classification in many ways followed the same format as the Tugela Basin survey. The name "binomial" came from the fact that soils were classified into two classes, i.e., soil form and soil series. In total, 41 soil forms and 504 soil series were recognized and described. As in the Tugela Basin survey, the soil forms were defined on the basis of a unique vertical sequence of diagnostic horizons. A total of 5 topsoil and 15 subsoil horizons were defined, and, as before, each had a connotation in terms of presumed genesis and provenance. The soil series (and soil forms) had unique geographic names and the series were also given a number within their soil form, which in turn had a unique two-letter code. Thus, e.g., the Griffin form, Farmhill series has a notation Gf 13; the Cartref form, Cranbrook series a notation of Cf 22. This greatly facilitated soil mapping, soil survey, and communication. The main soil series criteria were again texture, base status, and calcareousness, with horizon color (especially the distinction between defined red and non red colors) and pH important in some soil forms. The format followed that begun in the Tugela survey in that each soil form was represented by a full-color plate and the series criteria were given on the facing page. This classification also included some useful features, notably that each soil form was given possible correlations to the FAO (3) and U.S. Department of Agriculture (4) classifications. Although, of necessity, these were only broadly equivalent, they did introduce the local users of the classification to these two more widely known classifications and, internationally, allowed readers of the binomial system to relate to terminology with which they would be familiar. The system also included a full glossary. Perhaps most importantly, the system retained its user-friendly nature in that it remained possible to classify almost all soils to series level *in the field* (with the same provisos as earlier), without resorting to expensive laboratory analyses. A criticism leveled at the binomial system was that, because it was based on the evidence of previous soil surveys, it contained no differentiae that those surveys had not recognized (5). Thus, it lacked the previous flexibility. However, as stated earlier, the authors recognized that it was just "the first edition" and that as knowledge expanded so the classification would have to be modified. Perhaps a more important limitation to the binomial system was that it was based strongly on the knowledge built-up within the Province of KwaZulu-Natal, and drew heavily on the Tugela Basin survey information. This resulted in the (very different) soils of the Cape Province, especially, being largely restricted to two podzol forms and to the calcareous series of a few of the other soil forms. These strictions made it inevitable that a second edition would be needed.

SOIL CLASSIFICATION. A TAXONOMIC SYSTEM FOR SOUTH AFRICA

The current soil classification (6) is the second (revised) edition of the binomial system and retains many of the features of that classification although there are some fundamental differences. As stated in the Preface, the main pressure for a second edition came from soil scientists in the Cape Province (see above). The major change to the classification itself was the decision to "include only those classes which accommodate similar, naturally occurring, more or less uniform soil bodies (polypedons), and to exclude arbitrarily chosen classes (mainly texture) which cause uniform soil bodies to be split artificially by class boundaries" (6). This reflects a change in philosophy from the Tugela survey where, in explaining the use of the term "soil individual," the authors stated "The U.S.D.A. Soil Survey Staff has defined discrete individuals of the soil population in terms of three-dimensional bodies of soil (polypedons). We fail to see the necessity, or logic of this" (1). The decision to exclude texture as a differentiating criterion, given that it was the major feature that distinguished the soil series in the binomial system, meant that soil series were omitted from the classification, and the lowest level became the soil family. It was also felt that the information needed to define the series satisfactorily was not generally available. By excluding the most detailed level of the classification, the soil series, the authors also tacitly recognized that the criticism leveled at the binomial system by Butler (5) had some basis. Despite texture being excluded from the current classification there is a recommendation that the texture of the *topsoil* should be appended to the formal communication of any particular soil. This emphasis on the topsoil texture only, makes this classification unusual (unique?), since others generally use the texture of the subsoil as a differentiating feature. The classification thus consists of two levels, soil form and soil family. As in the first edition, all form and family names are from geographic locations and none of the earlier series names is used in the second edition. Similarly, the forms are designated by unique two-letter codes and the families by four-digit numbers so that, e.g., the Estcourt form, Zastron family is designated Es 1100; the Clovelly form, Brereton family by the code Cv 1200. The number of forms recognized has increased to 73, with a total of 404 soil families. There are now 5 diagnostic topsoil horizons and 25 diagnostic subsoil horizons or materials (Fig. 1).

Concepts and characteristics brought in from other systems include the luvic character of B horizons, the placic pan, and the use of the plasticity index to distinguish between vertic and non-vertic topsoils. However, the system has retained its local character in that the terminology of many of the diagnostic horizons is different from that used elsewhere. Examples include podzol B (not spodic), hardpan carbonate horizon (not calcrete), and a particular usage of the term "saprolite." The classification book also includes as appendices the methods of analysis used for the distinguishing criteria, an outline of the diagnostic horizons from the FAO (7) and the U.S. Department of Agriculture (8) soil classifications, and a full glossary. However, lacking from the second edition is any attempt to correlate the South African system with either of these more generally used international systems. The increase in the number of forms is mainly accounted for by the increase in the podzolic soils

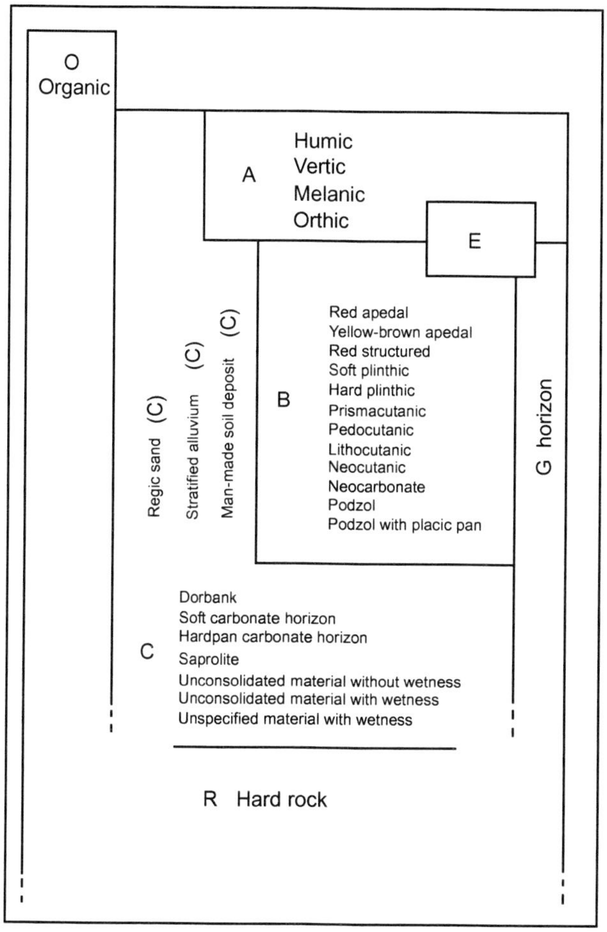

Fig. 1 Diagnostic horizons and materials. (From Ref. 6. Courtesy of The Director, Institute for Soil, Climate and Water, Pretoria.)

from 2 in the first edition to 8, the recognition of 19 arid-zone soils [either with calcareous B horizons or underlain by dorbank (a hard subsurface horizon cemented by silica)]—these two categories largely to cater for the soils of the Cape Province—and three more soils underlain by wet materials. The families are distinguished by a total of 19 characteristics of which color, base status, calcareousness, luvic character in the B horizon, and signs of wetness in the subsoil are the most important. The classification has retained many of the positive features of the earlier ones remaining one of the most user-friendly classifications, with it still being very largely a field classification. However, the addition of measurable characteristics such as the plasticity index to distinguish vertic soils and the addition of a cation exchange capacity criterion to distinguish borderline structured from apedal subsoils means that laboratory data are more necessary now than in the past.

THE FUTURE?

Discussions are currently underway among the South African soil science community with a view to decide on the next edition of the soil classification and what changes are required. The main item of contention is whether we are in a position yet to define soil series satisfactorily and, if so, what differentiae should be used. The decision not to reuse any of the earlier soil series names in the current classification means that they could be brought back into any future edition, providing that their characteristics can be agreed upon. What is certain is that the present classification will not be the final word as South African soil science develops and more knowledge is gained about the country's soil mantle.

REFERENCES

1. Van der Eyk, J.J.; MacVicar, C.N.; de Villiers, J.M. *Soils of the Tugela Basin. A Study in Subtropical Africa*; Natal Town and Regional Planning Reports, Town and Regional Planning Commission, Natal: Pietermartizburg, South Africa, 1969; Vol. 15, 1–263.
2. MacVicar, C.N.; de Villiers, J.M.; Loxton, R.F.; Verster, E.; Lambrechts, J.J.N.; Merryweather, F.R.; Le Roux, J.; Van Rooyen, T.H.; Von M. Harmse, H.J. *Soil Classification. A Binomial System for South Africa*; Department of Agricultural Technical Services: Pretoria, South Africa, 1977; 1–150.
3. Dudal, R. Definitions of Soil Units for the Soil Map of the World. *World Soil Resources Reports No. 33*; World Soil Resources Office, Land and Water Development Division, FAO: Rome, 1968; 1–72.
4. Soil Survey Staff. *Soil Classification. A Comprehensive System (Seventh Approximation)*; SCS, U.S. Department of Agriculture; U.S. Government Printing Office: Washington, DC, 1960; 1–265.
5. Butler, B.E. *Soil Classification for Soil Survey*; Monographs on Soil Survey, Clarendon Press: Oxford, 1980; 120–122.
6. Soil Classification Working Group. *Soil Classification. A Taxonomic System for South Africa*; Memoirs on the Agricultural Natural Resources of South Africa No. 15, Department of Agricultural Development: Pretoria, South Africa, 1991; 1–257.
7. FAO-UNESCO. Soil Map of the World, 1:5,000,000. Vol. 1. Legend. UNESCO: Paris, 1974.
8. Soil Survey Staff. *Soil Taxonomy. A Basic System of Soil Classification for Making and Interpreting Soil Surveys*; U.S. Department of Agriculture Handbook No. 436, U.S. Government Printing Office: Washington, DC, 1975; 1–754.

CLASSIFICATION SYSTEMS, MAJOR

Lars Krogh
Institute of Geography, University of Copenhagen, Copenhagen, Denmark

INTRODUCTION

Soil classification is a tool for stratifying, generalizing, remembering, predicting, mapping, and communicating information on soil resources for a specific purpose and, in addition, it serves as a basis for further enquiry into soil processes.

Soil cover varies continuously vertically and horizontally at many scales in response to the interactive effects of climate, vegetation, relief, parent material, time, and the influence of man. Soils have many users and the use of soils also varies according to climate, technology, agricultural systems, trade possibilities, and traditions, and therefore the concepts of soils are very diverse. In addition, the very nature of soils; slowly evolving three-dimensional bodies with no distinct borders in combination with a strong symbolic meaning and prestige attached to a nation's soil or land, probably explains why many national systems of soil classification have evolved.

The American classification system *Soil Taxonomy* (1) and the joint FAO, ISRIC, and ISSS system *World Reference Base for Soil Resources* (2), which is a further development of the FAO–UNESCO *Soil Map of the World* (3) and its update from 1988 (4), are currently the two most widely accepted and used soil classification systems. In addition to these, the Australian (6, 7), Brazilian (8), Canadian (9, 10), Chinese (11), French (12), and Russian (13–15) systems might also be considered major systems in that they are used on a large proportion of the land surface of the Earth. However, these systems have all been developed for national purposes only.

SOIL TAXONOMY

Building upon existing soil concepts and classification systems in the U.S.A., *Soil Taxonomy* was started in 1951 and through a series of modifying steps the Seventh Approximation was published in 1960 (16) followed by the publication of the definitive system for the first time in 1975, under the name *Soil Taxonomy* (17, 18). The system has been amended eight times since then (published as keys) and is continuously undergoing modifications to encompass results of new research and changing soil concepts, and the second comprehensive edition was published in 1999 (1, 19, 20). The original purpose of the system was to serve as a tool for making and interpreting soil surveys in the U.S.A., predominantly in relation to agricultural production, but in recent years it has evolved into a means of communication in soil science. *Soil Taxonomy* differs from most other systems by its rather special names for describing soils, which were created in order to avoid older traditional names that now have ambiguous meanings. At the time of publication, it was among the very few systems that separated soils on the basis of precise quantitative definitions of observable and measurable properties.

Soil Taxonomy is a hierarchical, six level, morphogenetic classification system, in which soils can be arranged—with decreasing rank and with increasingly narrower variation in properties—in 12 orders, 64 suborders, ~300 great groups, ~2400 subgroups, ~8000 families, and (in the U.S.) ~19,000 series.

Soil Taxonomy has defined a number of diagnostic horizons (Table 1) and diagnostic soil characteristics, which are used to identify and assign soils into its categories by the means of a key. Diagnostic horizons are recognized as reflecting common and widespread results of major pedogenetic processes and are defined by observable and measurable morphological and chemical properties. Diagnostic horizons present at the surface are termed epipedons and those present below the surface are termed subsurface horizons. Diagnostic soil characteristics, some unique to mineral soils and some to both mineral and organic soil, are also defined by specific morphological and chemical properties. They do, however, not necessarily make up a coherent horizon, and they also include soil water and temperature characteristics as defined by several soil moisture and soil temperature regimes. For a soil to be assigned into a category at any level of the system, a soil must meet the criteria specified in the definitions of the category.

Soil orders (Table 2) are separated on the basis of stable and long-term development processes and properties

Table 1 Brief description of the 8 diagnostic epipedons and the 20 diagnostic subsurface horizons in *Soil Taxonomy*

	Description
Epipedons	
Anthropic (Gr. *anthropos*, human being)	Manmade, thick, dark, formed by long continued residence or application of organic residues. Often with high phosphorus content.
Folistic (L. *folia*, leaf)	More than 20 cm thick, high in organic matter of cool humid regions under forest vegetation.
Histic (Gr. *histos*, tissue)	Between 20 and 40 cm thick, high in organic matter that is saturated with water for long periods.
Melanic (Gr. *melas*, black)	Thick, black, high in organic matter of soils formed from volcanic ash. The organic matter is associated with short-range order minerals or Al–humus complexes.
Mollic (L. *mollis*, soft)	Thick, dark, high in organic matter, well structured with over 50% base saturation, predominantly from divalent cations. Typical for steppe soils.
Ochric (Gr. *ochros*, pale)	Thin, light or low in organic matter, thereby failing to be mollic or umbric.
Plaggen (Ger. *plaggen*, sod)	Manmade, more than 50 cm thick, high in organic matter, and raised above the original soil surface. Formed by spreading of a mixture of manure and sod, the latter used for bedding livestock in medieval times.
Umbric (L. *umbra*, shade = dark)	Thick, dark, high in organic matter, well structured with under 50% base saturation.
Subsurface horizons	
Agric (L. *ager*, field)	Illuvial horizon below the plough layer containing silt, clay, and organic matter. Formed under (old-world) cultivation.
Albic (L. *albus*, white)	Eluvial, bleached and light-colored horizon from which clay and Fe-oxides have been removed.
Argillic (L. *argilla*, white clay)	Horizon with significantly higher clay content than horizon above. Formed through clay illuviation and/or in situ weathering.
Calcic (L. *calcis*, lime)	Illuvial horizon with secondary $CaCO_3$ as powder or concretions. Mostly found in arid environments.
Cambic (L. *cambiare*, to exchange)	Horizon in which weak structure formation, weathering of primary minerals, and pecipitation of Fe-oxides have taken place.
Duripan (L. *durus*, hard)	Horizon cemented by illuvial silica thereby impeding root and water movement.
Fragipan (L. *fragilis*, brittle)	Compact, high bulk density, noncemented horizon impeding root and water movement.
Glossic (Gr. *glossa*, tongue)	Transitional horizon with vertical bleached tongues formed by movement and destruction of clay and Fe-oxides in existing argillic horizon.
Gypsic (L. *gypsum*, gypsum)	Illuvial horizon with secondary gypsum. Mostly found in arid environments.
Kandic (from kandite)	Horizon with significantly higher clay content than horizon above and dominated by LAC clays. Formed through clay illuviation and/or in situ weathering.
Natric (L. *natrium*, sodium)	Horizon with significantly higher clay content than horizon above and 15% or more exchangeable Na. Formed through clay illuviation and/or in situ weathering.
Ortstein (Ger.)	Cemented horizon consisting of complexes of Al and organic matter with or without Fe (spodic materials).
Oxic (F. *oxide*)	Fine textured, thoroughly weathered horizon with clay consisting of LAC and sesquioxides and having a low CEC. Mostly found in old tropical environments.
Petrocalcic (Gr. *petra*, rock and calcic)	Indurated illuvial horizon with secondary $CaCO_3$. Mostly found in arid environments. Barrier to roots and water.

(*Continued*)

Table 1 Brief description of the 8 diagnostic epipedons and the 20 diagnostic subsurface horizons in *Soil Taxonomy (Continued)*

	Description
Petrogypsic (Gr. *petra*, rock and gypsic)	Indurated illuvial horizon with secondary gypsum. Mostly found in arid environments. Barrier to roots and water.
Placic (Gr. *plax*, flat stone)	Thin (<25 mm), black to reddish horizon cemented by Fe (and Mn) and organic matter.
Salic (L. *sal*, salt)	Horizon with accumulation of salts more soluble than gypsum and in quantities harmful to roots. Mostly found in arid environments.
Sombric (F. *sombre*, dark)	Horizon with illuvial concentrations of organic matter that is not in combination with Al. Mostly found in cool, mountainous environments.
Spodic (Gr. *spodos*, wood ash)	Illuvial horizon with amorphous material consisting of organic matter and Al with or without Fe.
Sulfuric (L. *sulphur*,sulfur)	More than 15 cm thick mineral or organic horizon with low pH(≤3.5) formed by artificial drainage of sulfide rich materials.

Table 2 Brief description of the 12 orders in *Soil Taxonomy* and their surface area in percent

Order	Formative name element and meaning	Breif description of order	% of ice-free land surface[a]
Alfisols	Alf (M[b]. *pedalfer*)	Weakly weathered soils with argillic or kandic horizon and high base saturation in subsoil.	9.6
Andisols	And (Jap. *an*, dark)	Soils formed from volcanic ash and with short-range order minerals.	0.7
Aridisols	Id (L. *aridus*, dry)	Weakly developed soils of dry regions.	12.0
Entisols	Ent (M. *recent*)	Soils with very weak pedogenic development.	16.2
Gelisols	El (L. *gelare*, to freeze)	Soils with perennial permafrost within 100 cm of the surface.	8.6
Histosols	Ist (Gr. *histos*, tissue)	Periodically wet soils with more than 30% organic matter to a depth of 30 cm.	1.2
Inceptisols	Ept (L. *inceptum*, beginning)	Soils with incipient development in terms of weathering, color, and structure.	9.8
Mollisols	Oll (L. *mollis*, soft)	Soils with mollic horizon and high base saturation.	6.9
Oxisols	Ox (F. *oxide*, oxide)	Highly weathered red soils of the tropics.	7.5
Spodosols	Od (Gr. *spodos*, wood ash)	Acids soils with spodic horizon.	2.5
Ultisols	Ult (L. *ultimus*, last)	Strongly weathered soils with argillic or kandic horizon and low base saturation in subsoil.	8.5
Vertisols	Ert (L. *vertere*, to turn)	Soils with more than 30% expandable clays in all horizons and development of cracks during dry periods.	2.4

[a] Ice-free land surface = 129,788,231 km^2.
[b] M = meaningless syllable.

C

Table 3 Brief description of the 30 reference soil groups in *World Reference Base for Soil Resources* and their surface area in percent

Reference soil group	Formative element	Brief description of reference soil groups	% of ice-free land surface[a]
Histosols	Gr. *histos*, tissue	Periodically wet soils with more than 30% organic matter to a depth of 30 cm.	2.2
Cryosols	Gr. *kraios*, cold ice	Soils with perenial permafrost.	12.3
Anthrosols	Gr. *anthropos*, man	Soils transformed by human activity through deep working, application of plaggen manure, or sediment-rich irrigation water.	?
Leptosols	Gr. *leptos*, thin	Weakly developed shallow soils with hard rock within 25 cm of surface.	11.5
Vertisols	L. *vertere*, to turn	Soils with more than 30% expandable clays in all horizons and development of cracks during dry periods.	2.3
Fluvisols	L. *fluvius*, river	Soils developed on stratified floodplain deposits.	2.4
Solonchaks	Russ. *sol*, salt; *chak*, salty area	Soils with accumulation of soluble salts.	2.1
Gleysols	Russ. *gley*, wet soil	Poorly drained waterlogged soils with mottles and anaerobic conditions.	5.0
Andosols	Jap. *an*, dark; *do*, soil	Soils formed from volcanic ash and with short-range order minerals.	0.8
Podzols	Russ. *pod*, under; *zola*, ash	Acid soils with accumulation of amorphous material consisting of organic matter and Al with or without Fe in the subsoil (spodic horizon).	3.4
Plinthosols	Gr. *plinthos*, brick	Soils with mottled appearance due to segregation of ironoxides, which harden upon exposure to the air.	0.4
Ferralsols	L. *ferrum*, iron; *alumen*, aluminium	Highly weathered red soils of the tropics with high content of Fe/Al oxides and LAC clays.	5.2
Solonetz	Russ. *sol*, salt; *etz*, strongly	Soils with high content of exchangeable sodium.	0.9
Planosols	L. *planos*, flat	Soils on flat terrain with abrupt textural change from sandy surface to clayey subsoil, which causes seasonal water stagnation.	0.9
Chernozems	Russ. *chern*, black; *zemlja*, earth	Dark-colored loess soils of the grass steppe regions with high content of organic matter and accumulation of calcium carbonate in subsoil.	1.6
Kastanozems	L. *kastanea*, chestnut; Russ. *zemlja*, earth	Brown-colored soils of the grass steppe regions with high content of organic matter and accumulation of calcium carbonate or gypsum in subsoil.	3.2

(Continued)

Table 3 Brief description of the 30 reference soil groups in *World Reference Base for Soil Resources* and their surface area in percent *(Continued)*

Reference soil group	Formative element	Brief description of reference soil groups	% of ice-free land surface[a]
Phaeozems	Gr. *phaios*, dusky; Russ. *zemlja*, earth	Dark-colored soils of the steppe regions with high content of organic matter and leaching of calcium-carbonate.	1.3
Gypsisols	L. *gypsum*, gypsum	Soils with accumulation of secondary gypsum.	0.6
Durisols	L. *durus*, hard	Well-drained, arid region soils with compact and silica-cemented layer.	?
Calcisols	L. *calx*, lime	Soils with accumulation of secondary calcium carbonate as soft powder or as cemented layers.	5.6
Albeluviols	L. *albus*, white; *eluere*, to wash out	Soils with clay accumulation in subsoil penetrated by tongues of coarse-textured and lighter material of eluvial horizon.	2.2
Alisols	L. *alumen*, alum	Acid soils with clay accumulation in subsoil, high CEC, and low base saturation.	0.7
Nitisols	L. *nitidus*, shiny	Low lattitude, deep, well-drained soils with clay accumulation in subsoil and shiny ped-surfaces.	1.4
Acrisols	L. *acer*, strong acid	Acid, weathered soils with clay accumulation in subsoil, low CEC, and low base saturation.	7.0
Luvisols	L. *luere*, to wash	Soils with clay accumulation in subsoil, high CEC, and high base saturation.	4.5
Lixisols	L. *lixia*, washing	Strongly weathered soils with clay accumulation in subsoil, low CEC, and high base saturation.	3.0
Umbrisols	(L. *umbra*, shade = dark	Well-drained soils with dark-colored, organic-rich, acid surface horizon (umbric horizon).	0.7
Cambisols	L. *cambiare*, to change	Soils with incipient weathering and slight development of color and structure.	10.5
Arenosols	Gr. *arena*, sand	Soils with sandy texture throughout.	6.3
Regosols	Gr. *rhegos*, blanket	Weakly developed fine-textured soils.	1.8

[a] Ice-free land surface = 129,788,231 km^2.

reflected by diagnostic horizons; suborders are separated on the basis of mostly dynamic properties controlling the current soil-forming processes, such as soil moisture regime and presence or absence of water or vegetation; great groups are separated on the basis of generally static properties with minor or additional control on current soil forming processes; subgroups are separated on the basis of a central concept and divergence from this; families are separated on the basis of properties which relate to plant growth or the potential for further change; series, which are the mapping unit carrying local names, are separated on the basis of properties that reflect relatively narrow ranges of soil forming factors. To illustrate the special nomenclature of *Soil Taxonomy*, the name Typic Argiaquolls (subgroup level) signifies that the soil is a typical base rich soil with a deep, organic matter-rich A-horizon (Mollisol), that is relatively wet (Aquic conditions) and with clay accumulation in the subsoil (argillic subsurface horizon).

Table 4 Nearest WRB equivalents to the 12 orders in *Soil Taxonomy*

***Soil Taxonomy* orders**	**WRB reference soil groups**
Alfisols	Luvisols, Lixisols, Planosols, and (Solonetz)
Andisols	Andosols
Aridisols	Leptosols, Gypsisols, Durisols, Calcisols, Solonchaks
Entisols	Arenosols, Fluvisols, Regosols, and (Gleysols)
Gelisols	Cryosols
Histosols	Histosols
Inceptisols	Cambisols and (Fluvisols, Gleysols, Solonschaks)
Mollisols	Chernozems, Kastanozems, Phaeozems
Oxisols	Ferralsols, Plintosols
Spodosols	Podzols
Ultisols	Alisols, Acrisols, and (Albeluvisols, Nitisols)
Vertisols	Vertisols

THE *WORLD REFERENCE BASE FOR SOIL RESOURCES*

The *World Reference Base for Soil Resources* (WRB) was developed jointly by the ISSS, FAO, and ISRIC through a series of steps and published in 1998 (2, 21–23). It is not an entirely new system, but rather a revision of the FAO–UNESCO *Soil Map of the World* from 1974 (3) and the so-called FAO Revised Legend from 1988 (4, 5), to some extent also inspired by *Soil Taxonomy* and the French *Référentiel Pédologique* (12). WRB was launched in order to provide scientific depth and background to the FAO system, as it became clear that important new insight and knowledge about the world's soils had been generated. Additional objectives are to serve as a link to and between national classification systems, to serve as a basic language in soil science, to strengthen the application of soil science, and to facilitate the use of soil data for other professionals in soil and land management. Together with the now superseded FAO system, WRB is the only system specifically designed to accommodate global soil resources.

The WRB system is a two level morpho-genetic system. At the highest level it has 30 reference soil groups that represent the major soil regions of the world (Table 3). Table 4 shows the nearest WRB equivalents to the 12 soil orders in *Soil Taxonomy*.

At the second level the WRB is a classification system where combinations of 121 qualifiers may be added to the reference groups, thus allowing very precise characterization of individual soil profiles. Traditional soil names have been retained to the greatest extent possible, e.g., Podzols and Solonetz, but in case of ambiguity of traditional names, new names derived from various linguistic roots were selected, e.g., Albeluvisols instead of Podzoluvisols.

For describing the reference soil groups, WRB has defined diagnostic horizons, diagnostic properties (assemblages of soil characteristics, e.g. texture, color, soil reaction, or CEC), and diagnostic soil materials (original parent materials that are only slightly altered, e.g., anthropogenic material, fluvic material, or sulfidic material). Diagnostic horizons and properties are the expression of pedogenic processes and have been defined as representing generally agreed pathways of soil formation and importance for soil management and use. Many of the differentiating criteria have been adopted from *Soil Taxonomy* either directly and simplified, e.g., the mollic, umbric and ochric horizon, or modified and simplified, e.g., the argic B horizon (from the argillic subsurface horizon) and the ferralic B horizon (from the oxic subsurface horizon). In contrast to *Soil Taxonomy*, the use of soil climate is not part of the system, except for the effects it may have on soil properties. Assignment of soils into the WRB categories is by the means of a key. To illustrate how the the second level classification systems of WRB works, a reddish-colored Vertisol with a surface-near calcic horizon would be termed a Chromi-Epicalcic Vertisol.

REFERENCES

1. Soil Survey Staff, *Soil Taxonomy*. In *A Basic System of Soil Classification for Making and Interpreting Soil Surveys*, 2nd Ed.; United States Department of Agriculture, Natural Resources Conservation Service, Agriculture Handbook No. 436; U.S. Government Printing Office: Washington, DC, 1999.
2. FAO–ISRIC–ISSS, *World Reference Base for Soil Resources*; World Soil Resources Report 84, FAO: Rome, Italy, 1998.
3. FAO–UNESCO. *FAO–UNESCO Soil Map of the World. 1:5,000,000*; Vol. I, Legend; UNESCO: Paris, France, 1974.
4. FAO. *FAO–UNESCO Soil Map of the World, Revised Legend, with Corrections and Updates*; World Soil Resources Report 60; FAO: Rome, 1988; Reprinted with Updates as Technical Paper 20; ISRIC: Wageningen, The Netherlands, 1997.
5. Driessen, P.M., Dudal, R., Eds. *The Major Soils of the World*; Agricultural University Wageningen, The Netherlands and Katholieke Universiteit Leuven, Belgium; K. Wöhrmann B.V, Zutphen: The Netherlands, 1991.
6. Isbell, R.F. *The Australian Soil Classification*; Australian Soil and Land Survey Handbook, CSIRO Publishing: Collingwood, Australia, 1996.
7. Isbell, R.F.; McDonald, W.S.; Ashton, L.J. *Concepts and Rationale of the Australian Soil Classification*; ACLEP, CSIRO Land and Water: Canberra, Australia, 1997.
8. DoPrado, H. *Manual de Classificacão de Solos do Brazil*, 3rd Ed.; FUNEP: Brasilia, Brazil, 1996.
9. Canada Soil Survey Committee, Subcommittee on Soil Classification. In *The Canadian System of Soil Classification*; Canadian Department of Agriculture, Publication No. 1646, Supply and Services Canada: Ottawa, Canada, 1978.
10. Soil Classification Working Group. In *The Canadian System of Soil Classification*, 3rd Ed.; Research Branch, Agriculture Canada, Publication 1646; NRC Research Press: Ottawa, Canada, 1998.
11. Gong, Z. *Chinese Soil Taxonomic Classification*; Institute of Soil Science, Academica Sinica: Nanjing, China, 1994.
12. Baize, D.; Girard, M.C., Eds. *Référentiel Pédologique*; Institut National de la Recherche Agronomique: Paris, France, 1995.
13. VASKhNIL (Dokuchaev Institute of Soil Science). In *Classification and Diagnostics of Soils of the USSR*, Translated by S.Viswanathan; Amerind Publishing Co.: New Delhi, India, 1986.
14. Shishov, L.L.; Sokolov, I.A. Genetic Classification of Soils in the USSR. In *Soil Classification*; Reports of the International Conference on Soil Classification, Rozanov, B.G., Ed.; USSR State Committee for Environmental Protection: Moscow, Russia, 1990; 77–93.
15. Stolbovoi, V. *Soils of Russia: Correlated with the Revised Legend of the FAO Soil Map of the World and World Reference Base for Soil Resources*; IIASA: Laxenburg, Austria, 2000.
16. Soil Survey Staff. In *Soil Classification—A Comprehensive System—7th Approximation*; Soil Conservation Service, United States Department of Agriculture, Agriculture Handbook, U.S. Government Printing Office: Washington, DC, 1960.
17. Soil Survey Staff, *Soil Taxonomy. A Basic System of Soil Classification for Making and Interpreting Soil Surveys*; Soil Conservation Service, United States Department of Agriculture, Agriculture Handbook No. 36, U.S. Government Printing Office: Washington, DC, 1975.
18. Smith, G.D. *The Guy Smith Interviews: Rationale for Concepts in Soil Taxonomy*; Soil Management Support Services, Department of Agronomy, Cornell University: Ithaca, NY, 1986.
19. http://www.statlab.iastate.edu/soils/nssc (Accessed February 2001)
20. http://www.nhq.nrcs.usda.gov/wsr (Accessed February 2001)
21. ISSS Working Group RB, In *World Reference Base for Soil Resources: Atlas*; Bridges, E.M., Batjes, N.H., Nachtergaele, F.O., Eds.; ISRIC–FAO–ISSS–Acco: Leuven, Belgium, 1998.
22. ISSS Working Group RB, In *World Reference Base for Soil Resources: Introduction*, 1st Ed.; Deckers, J.A., Nachtergaele, F.O., Spargaren, O.C., Eds.; ISRIC–FAO–ISSS–Acco: Leuven, Belgium, 1998.
23. http://www.fao.org/waicent/Faoinfo/agricult/agl/agll/wrb/wrbhome.htm (Accessed February 2001)

CLAY, CHARGE PROPERTIES OF

C

Christopher J. Matocha
Agronomy Department, University of Kentucky, Lexington, Kentucky, U.S.A.

INTRODUCTION

Surfaces of soil particles are classified as having either permanent charge or pH-dependent charge. A permanent charge is derived from isomorphic substitutions in the mineral structure and is indifferent to the conditions surrounding the mineral. Isomorphic substitution involves replacement of constituent metal ions of the lattice by cations of lower charge, generating a net negative charge on the mineral, which is compensated by the positive charge of adsorbed (exchangeable) cations. The substitution can occur during or after crystal formation. Surface functional groups that may be exposed on the periphery of minerals protonate and deprotonate depending on the solution pH, generating pH-dependent surface charge. Most soils carry a net negative charge because of negative charges on phyllosilicate minerals, but some weathered soils can develop a net positive charge at sufficiently low pH due to the presence of hydrous iron and aluminum oxides. The net negative charge on soil minerals produces a positive electrostatic attraction for cations in soil solution that are retained on the surface. The total quantity of cations retained in this way is called the cation exchange capacity (CEC) because they are easily exchanged with other cations. The focus of the present article will be the surface charge properties of 1:1 and 2:1 phyllosilicate minerals because of their abundance and significant influence on agricultural and environmental processes.

ORIGINS OF CHARGE

Phyllosilicate minerals consist of aluminosilicate tetrahedral and octahedral sheets stacked one above the other and differentiated by the number and combination of sheets and the cation substitution in each sheet (1, 2). In 1:1 phyllosilicate minerals, such as kaolinite, the full-cell chemical formula is $Si_4Al_4O_{10}(OH)_8$ where one Si tetrahedral sheet is combined with one Al octahedral sheet and individual layers stack by hydrogen bonding (O–H–O) in the *c*-dimension (perpendicular to the plane of the sheets) with a combined thickness of 0.7 nm. Generally, no isomorphic substitution occurs in either sheet, thus kaolinite has no significant permanent negative structural charge on basal sites and little CEC (1). Where crystal growth stops and no more unit cells are added, broken bonds exist on the edges of clay minerals and a surface is obtained (Fig. 1). These "edge" sites exhibit pH-dependent charge that results from protonation/deprotonation of aluminol (>AlOH) and silanol (>SiOH), functional groups that are exposed on the periphery or edges of the mineral (3). The acidity or dissociation constants (pK_as) for the aluminol edge sites are $pK_{a1} = 6.3$ and $pK_{a2} = 8.7$, meaning that at pH<6, these edge sites are positively charged (4). This explains the early findings by Schofield and Samson (5), who documented Cl^- sorption at low pH. Kaolinite has a significant concentration of edge sites because it tends to stack in the *c*-dimension (3).

The 2:1 phyllosilicate minerals consist of two tetrahedral sheets that sandwich one octahedral sheet (Fig. 1). The ideal full cell chemical formula for a 2:1 clay would be $Si_8^{IV}Al_4^{VI}O_{20}(OH)_4$, where the roman numerals refer to the tetrahedral (IV) and octahedral (VI) sheets, respectively (6). This formula is represented by the mineral pyrophyllite, which is neutral because no cation substitution occurs in either sheet. More common 2:1 phyllosilicate minerals include the smectite and vermiculite groups. The number of cation positions occupied within the octahedral sheet determines whether a clay mineral is di- or trioctahedral. If Al^{3+} is present in the octahedral sheet, then the mineral is dioctahedral and trioctahedral if Mg^{2+} is present. Within the smectite group of minerals, dioctahedral montmorillonites have the majority of structural negative charge. These are derived primarily from octahedral site substitutions of Fe^{2+} or Mg^{2+} for Al^{3+} with a structural formula of $Si_8^{IV}M_{0.66}H_2OAl_{3.34}^{VI}\ (Fe^{2+}, Mg^{2+})_{0.66}O_{20}(OH)_4$, where M refers to the exchangeable cation balancing the permanent charge produced by substitution of Fe^{2+} or Mg^{2+} for Al^{3+} (7). The octahedral layer charge for montmorillonite, or the number of moles of excess electron charge per formula unit (1), would be −0.66, balanced by $M_{0.66}$ exchangeable cations adsorbed near the basal plane of the tetrahedral sheet. In dioctahedral beidellites, tetrahedral substitutions

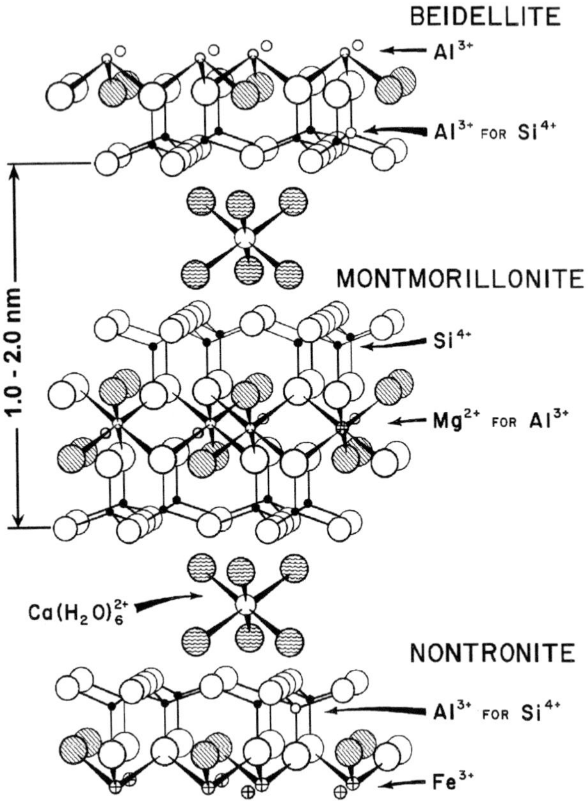

Fig. 1 Representative 2:1 phyllosilicate minerals indicating charge sites and different locations of substitution. (From Ref. 7.)

of Al^{3+}-for-Si^{4+} exceed substitutions in the octahedral sheet (Fig. 1). Nontronite is similar to beidellite except it has redox-active Fe^{3+} in octahedral coordination (Fig. 1). The vermiculite group differs from the smectite group because the primary source of negative charge arises from a large degree of Al^{3+}-for-Si^{4+} substitution in the tetrahedral layer, giving rise to a layer charge greater than that in smectites, ranging from 1.2–1.8.

The basal plane of a tetrahedral sheet composed of oxygen atoms is called the siloxane surface and its reactivity depends on the location and type of isomorphic substitution (8). In a tetrahedrally-substituted clay such as beidellite, the excess negative charge is localized and functions as a strong electron donor because of the lone pair, nonbonding electrons on the siloxane oxygens (8, 9). Siloxane oxygens not adjacent to the site of substitution, as is the case for octahedrally substituted montmorillonite, are weaker electron donors because the negative charge is distributed over at least ten surface oxygens (9).

One characteristic that the smectite and vermiculite groups have in common with kaolinite is the presence of edge sites, although in lower concentration because stacking occurs predominantly in the horizontal dimensions (a, b) rather than the vertical c-dimension. A distinguishing feature of the smectite and vermiculite groups from kaolinite is the presence of an interlayer, occupied by exchangeable cations such as Ca^{2+} and water (Fig. 1). Thus, the 2:1 clay mineral particles have external planar sites, internal sites, and edge sites, while kaolinite contains mainly external edge sites. The active interlayer and the higher level of isomorphic substitution give rise to greater CEC values for smectite and vermiculite minerals (CEC$\sim$0.80–1.50 mEq g^{-1}) when compared to kaolinite (CEC$\sim$0.02–0.05 mEq g^{-1}).

SURFACE CHARGE DENSITY OF CLAY MINERALS

It is important to note at this point that the clay-size fraction is, by definition, any mineral with a particle size $<2\,\mu$m. Thus, the clay-size fraction has a large amount of surface area per unit mass when compared to the silt and sand fractions. This is clearly shown when one derives the specific surface area, or the area per unit mass. For uniform, spherical particles this ratio equals

$$S = 3/\rho R$$

where S is the specific surface area (cm^2 g^{-1}) having a radius R (cm) and particle density ρ (g cm^{-3}). For a given particle size, an equal mass of smectite or vermiculite would have a greater specific surface area when compared to kaolinite because of the presence of the interlayer. A typical value of S for montmorillonite would be 700 m^2 g^{-1}, while for kaolinite S is considerably less ($\sim$10 m^2 g^{-1}). The surface density of charge, obtained by normalizing the CEC to S, is similar irrespective of clay mineral type because S tends to scale with CEC (10). Higher S values usually indicate greater potential for reactivity.

AGRICULTURAL AND ENVIRONMENTAL IMPLICATIONS OF SURFACE CHARGE

High Activity and Low Activity Clays

Small amounts of smectite minerals can greatly influence overall soil charge properties due to their greater specific surface area. In the context of charge, clay activity classes have been defined as the ratio of CEC/clay content (11). High activity clay minerals, such as smectite or vermiculite, have ratios >0.7 while low activity clays like kaolinite have ratios <0.3. These indices are useful in estimating mineralogy when no specific mineralogy data are available (11).

Adsorption of Plant Nutrients and Contaminants

As mentioned earlier, the cations balancing the net negative charge generated from isomorphic substitutions in order to maintain electroneutrality are exchangeable. They maintain their inner hydration shell, thus forming the so called outer-sphere complexes (8). An example of an outer-sphere complex would be exchangeable Ca^{2+} in the interlayer space of 2:1 minerals (Fig. 1). Exchangeable cations are important agronomically because they are available to plants. In contrast, cations that lose H_2O from their inner hydration shell to form a direct bond with a mineral surface functional group are inner-sphere surface complexes (8). The type of surface complex formed influences the fate of plant nutrients and contaminants in the soil environment.

Tetrahedrally-substituted clay minerals can form stronger adsorption complexes with cations than octahedrally substituted clay minerals (1). Cations with the largest ionic radius and lowest hydration energy are preferentially adsorbed on permanent charge sites (10). Vermiculites and beidellites, with more tetrahedral charge sites, can fix K^+ more readily by forming inner-sphere bonds with siloxane oxygens. The fixation of K^+ makes it unavailable to plants because formation of these bonds causes the collapse of adjacent silicate layers (6). The proximity of tetrahedral charge to the interlayer space of 2:1 phyllosilicates explains the greater K^+-fixing ability of tetrahedrally-charged clays (12).

Varying pH and background electrolyte concentration is an approach used to assess the contribution of pH-dependent charge edge sites and permanent charge sites in controlling metal contaminant solubility on montmorillonite. Spectroscopic investigations confirmed that Pb^{2+} and Cu^{2+} adsorbed as inner-sphere surface complexes on hydroxyl edge sites in montmorillonite, particularly as pH and ionic strength increased, and as outer-sphere surface complexes on permanent charge sites at low pH and ionic strength (13, 14). The inner-sphere Cu^{2+} complexes coordinated to the hydroxyl edge sites were resistant to desorption, whereas outer-sphere Cu^{2+} surface complexes on permanent charge sites were exchangeable and potentially more mobile. Therefore, the practical benefit achieved in considering the difference between outer-sphere and inner-sphere surface complexation would be in predicting the fate of plant nutrients and contaminants in the soil environment.

Flocculation Processes

Clay particle–particle interactions regulate dispersion and flocculation processes and are important to soil quality. Flocculation, by definition, is the formation of particle aggregates from association of individual particles in water that settle by gravity and promote aggregation and optimal soil structure (10). Positively-charged clay mineral edges contribute to flocculation, because of the electrostatic attraction between positive edges and negative planar surface faces of adjacent minerals. Studies of the Schofield group (5) showed that kaolinite can be deflocculated by addition of montmorillonite particles. Other factors influencing the degree of flocculation include charge of exchangeable cation, ionic strength of the soil solution, and cation size. High charge exchangeable cations and high ionic strength promote flocculation of clay mineral suspensions dominated by permanent charge minerals (10). High Ca^{2+} status and near neutral pH are associated with stable aggregation in less weathered soils carrying a net negative charge. There is a direct relationship between cation size and ability to form inner-sphere surface complexes with clay mineral sites. For instance, high concentrations of exchangeable Na^+ promote dispersion and poor structure because it is weakly adsorbed, likely bound as an outer-sphere surface complex with octahedral charge sites on smectites (1). The greater hydration of Na^+, a function of its small ion size, explains the weaker retention of Na^+ when compared to larger cations like Cs^+ (10).

Clay Swelling

Water penetrates the interlayer of 2:1 phyllosilicates and forces them apart causing the clay to swell, which is an important natural soil phenomenon. High activity clay minerals, such as montmorillonite, influence the engineering behavior of soils because of their ability to take up and adsorb large amounts of water, when compared to low activity clay minerals like kaolinite (11). Water adsorption is an exothermic process, thus "heat of wetting" values are typically greater for montmorillonite than kaolinite. In addition, the type of exchangeable cation and location and distribution of charge strongly influence hydration characteristics of the clay mineral. The amount of water adsorbed and internal surface hydrated on montmorillonite is positively correlated with hydration energy of the exchangeable cation (15). Sodium-montmorillonite exhibited greater adsorption–desorption hysteresis of water when compared to Ca- and Mg-montmorillonite, which was ascribed to differences in hydration energy for the cations (16). Well-hydrated interlayer Ca and Mg serve to prop open the interlayer when compared to Na-montmorillonite. The ability of montmorillonite to swell in water more than vermiculite, despite the greater layer charge in the latter mineral, is explained by the location of the charge. The localized charge in tetrahedrally-substituted vermiculite apparently causes the exchangeable cations to pull the layers together and limit the expansion (10).

REFERENCES

1. Sposito, G.; Skipper, N.T.; Sutton, R.; Park, S.H.; Soper, A.K.; Greathouse, J.A. Surface Geochemistry of the Clay Minerals. Proc. Natl Acad. Sci. **1999**, *96*, 3358–3364.
2. Swartzen-Allen, S.L.; Matijevic, E. Surface and Colloid Chemistry of Clays. Chem. Rev. **1974**, *74*, 385–400.
3. White, G.W.; Zelazny, L.W. Analysis and Implications of the Edge Structure of Dioctahedral Phyllosilicates. Clays Clay Miner. **1988**, *36*, 141–146.
4. Wieland, E.; Stumm, W. Dissolution Kinetics of Kaolinite in Acidic Aqueous Solutions at 25 Degrees C. Geochim. Cosmochim. Acta **1992**, *56*, 3339–3355.
5. Schofield, R.D.; Samson, H.R. The Deflocculation of Kaolinite Suspensions and the Accompanying Change Over from Positive to Negative Chloride Adsorption. Clay Miner. Bull. **1953**, *2*, 45–51.
6. Sparks, D.L. *Environmental Soil Chemistry*; Academic Press: San Diego, CA, 1995; 267.
7. Borchardt, G. Smectites. In *Minerals in Soil Environments*; Dixon, J.B., Weed, S.B., Eds.; Soil Science Society of America: Madison, WI, 1989; 439–465.
8. Sposito, G. *The Surface Chemistry of Soils*; Oxford University Press: New York, 1984.
9. Farmer, V.C.; Russell, J.D. Interlayer Complexes in Layer Silicates. The Structure of Water in Lamellar Ionic Solutions. Water on Particle Surfaces. Trans. Faraday Soc. **1971**, *67*, 2737–2749.
10. McBride, M.B. *Environmental Chemistry of Soils*; Oxford University Press: New York, 1994; 406.
11. Olson, C.G.; Thompson, M.L.; Wilson, M.A. Phyllosilicates. In *Handbook of Soil Science*; Sumner, M.E., Ed.; CRC Press: Boca Raton, FL, 2000; F77–F123.
12. Bouabid, R.; Badraoui, M.; Bloom, P.R. Potassium Fixation and Charge Characteristics of Soil Clays. Soil Sci. Soc. Am. J. **1991**, *55*, 1493–1498.
13. Strawn, D.G.; Sparks, D.L. The Use of XAFS to Distinguish Between Inner- and Outer-Sphere Lead Adsorption Complexes on Montmorillonite. J. Colloid Interface Sci. **1999**, *216*, 257–269.
14. Morton, J.D.; Semrau, J.D.; Hayes, K.F. An X-Ray Absorption Spectrometry Study of the Structure and Reversibility of Copper Adsorbed to Montmorillonite Clay. Geochim. Cosmochim. Acta **2001**, *65*, 2709–2722.
15. Sposito, G.; Prost, R. Structure of Water Adsorbed on Smectites. Chem. Rev. **1982**, *82*, 553–573.
16. Xu, W.; Johnson, C.T.; Parker, P.; Agnew, S.F. Infrared Study of Water Sorption on Na-, Li-, Ca-, and Mg-Exchanged (SWy-1 and SAz-1) Montmorillonite. Clays Clay Miner. **2000**, *48*, 120–131.

CLAY MINERALS

C

C.D. Barton

United States Department of Agriculture Forest Service, Aiken, South Carolina, U.S.A.

A.D. Karathanasis

University of Kentucky, Lexington, Kentucky, U.S.A.

INTRODUCTION

Clay minerals refers to a group of hydrous aluminosilicates that predominate the clay-sized (<2 μm) fraction of soils. These minerals are similar in chemical and structural composition to the primary minerals that originate from the Earth's crust; however, transformations in the geometric arrangement of atoms and ions within their structures occur due to weathering. Primary minerals form at elevated temperatures and pressures, and are usually derived from igneous or metamorphic rocks. Inside the Earth these minerals are relatively stable, but transformations may occur once exposed to the ambient conditions of the Earth's surface. Although some of the most resistant primary minerals (quartz, micas, and some feldspars) may persist in soils, other less resistant minerals (pyroxenes, amphiboles, and a host of accessory minerals) are prone to breakdown and weathering, thus forming secondary minerals. The resultant secondary minerals are the culmination of either alteration of the primary mineral structure (incongruent reaction) or neoformation through precipitation or recrystallization of dissolved constituents into a more stable structure (congruent reaction). These secondary minerals are often referred to as phyllosilicates because, as the name implies (Greek: *phyllon*, leaf), they exhibit a platy or flaky habit, while one of their fundamental structural units is an extended sheet of SiO_4 tetrahedra.

STRUCTURE OF CLAY MINERALS

The properties that determine the composition of a mineral are derived from its chemical foundation, geometric arrangement of atoms and ions, and the electrical forces that bind them together (1). Given that there are eight elements that constitute over 99% of the Earth's crust (Table 1), the inclusion of these in the elemental makeup of soil minerals is understandable. Notwithstanding, the prevalence of silicon and oxygen in the phyllosilicate structure is logical. The SiO_4 tetrahedron is the foundation of all silicate structures. It consists of four O^{2-} ions at the apices of a regular tetrahedron coordinated to one Si^{4+} at the center (Fig. 1). An interlocking array of these tetrahedral connected at three corners in the same plane by shared oxygen anions forms a hexagonal network called the tetrahedral sheet (2). When external ions bond to the tetrahedral sheet they are coordinated to one hydroxyl and two oxygen anion groups. An aluminum, magnesium, or iron ion typically serves as the coordinating cation and is surrounded by six oxygen atoms or hydroxyl groups resulting in an eight-sided building block termed an octohedron (Fig. 1). The horizontal linkage of multiple octahedra comprises the octahedral sheet. The minerals brucite $Mg(OH)_2$ and gibbsite $Al(OH)_3$ are similar to the octahedral sheets found in many clay minerals; however, phyllosilicates may contain coordinating anions other than hydroxyls. Cations in the octahedral layer may exist in a divalent or trivalent state. When the cations are divalent (Mg, Fe^{2+}), the layer exhibits a geometry similar to brucite, such that electrical neutrality is maintained. In this arrangement the ratio of divalent cations to oxygens is 1:2 and all three possible cation sites in the octahedron are occupied. This configuration and the respective sheet formed from an array of such as octahedral are referred to as trioctahedral. When the cations are trivalent (Al, Fe^{3+}), the charge balance is maintained by leaving one of every three octahedral cation sites empty. Under this configuration, the ratio of trivalent cations to oxygens is 1:3 and the layer exhibits a gibbsite-like dioctahedral arrangement. A combination of tetrahedral and di- or trioctahedral sheets bound by shared oxygen atoms forms aluminosilicate layers that comprise the basic structural units of phyllosilicates (Fig. 2). Sheet arrangement within the aluminosilicate layers varies between clay mineral types resulting in variable physical and chemical properties that differentiate the clay mineral classes.

Table 1 Common elements in Earth's crust and ionic radius

Element	Crustal average ($g\,kg^{-1}$)	Ionic radius (nm)	Volume (%)
O^{2-}	466.0	0.140	89.84
Si^{4+}	277.2	0.039	2.37
Al^{3+}	81.3	0.051	1.24
Fe^{2+}	50.0	0.074	0.79
Mg^{2+}	20.9	0.066	0.60
Ca^{2+}	36.3	0.099	1.39
Na^{2+}	28.3	0.097	1.84
K^{+}	25.9	0.133	1.84

(Adapted from Ref. 1.)

ISOMORPHOUS SUBSTITUTION

The structural arrangement of the elements described above forms the template for the silicate clay minerals. However, the composition varies frequently due to substitution of ions within the mineral structure. Weathering allows for the substitution of Si^{4+}, Al^{3+}, and Mg^{2+} with cations with comparable ionic radii in their respective tetrahedral and octahedral sheets (Table 1). Consequently, Si^{4+} may be replaced by Al^{3+} in the center of the tetrahedron without changing the basic structure of the crystal. Moreover, cations such as $Fe^{3+/2+}$ and Zn^{2+} (ionic radius = 0.074 nm) may replace Al^{3+} and Mg^{2+} in the octahedra. The process of replacing one structural cation for another of similar size is referred to as isomorphous substitution. This replacement represents the primary source of both negative and positive charges in clay minerals. For example, the substitution of one Al^{3+} for a Si^{4+} in the tetrahedron results in a gain of one negative charge.

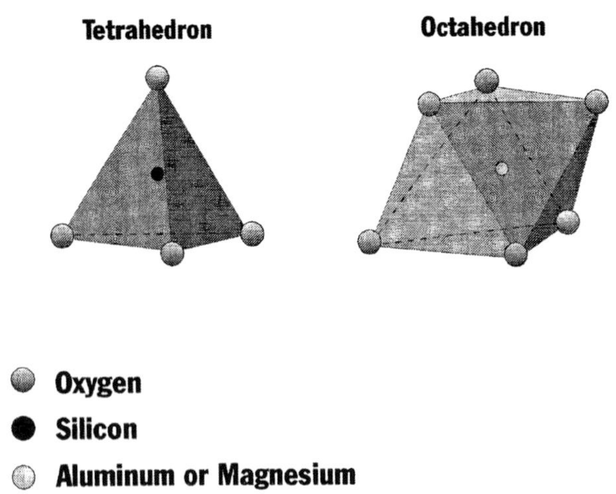

Fig. 1 The basic structural components of clay minerals; a single four-sided tetrahedron, and a single eight-sided octahedron. (From Ref. 6.)

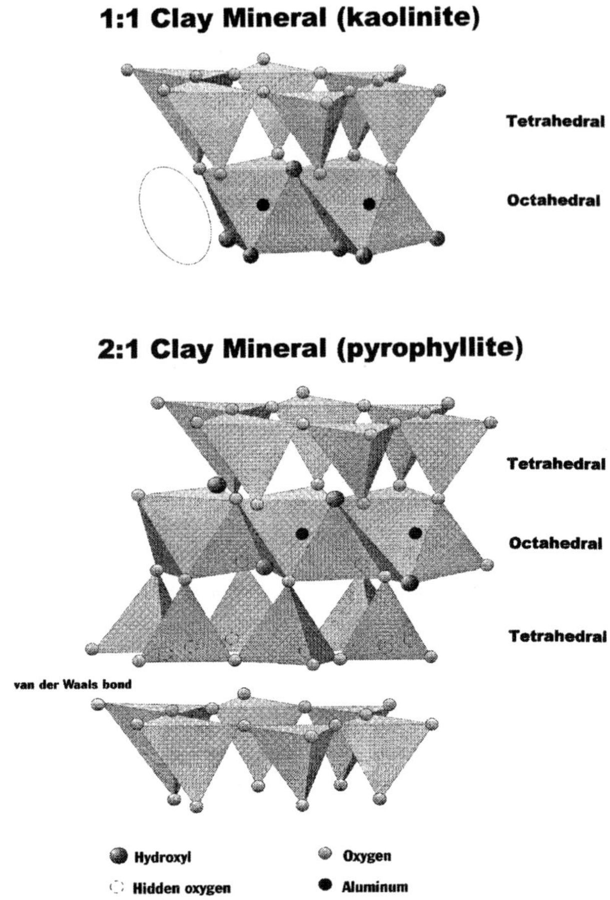

Fig. 2 Diagrammatic sketch of a 1:1 clay mineral consisting of one tetrahedral sheet bonded to an octahedral sheet (kaolinite); and a 2:1 clay mineral consisting of an octahedral sheet bound between two tetrahedral sheets (pyrophyllite). (From Ref. 6.)

Alternatively, replacement of a lower valence cation by one with a higher valence (Fe^{2+} by Fe^{3+}) results in a gain of one positive charge. Some clay minerals exhibit substitutions that result in both positive and negative charges. A balance of electron loss and gain within the structure determines the net charge of the mineral. In most soils, however, substitutions that result in net negative charge exceed those producing a positive charge.

Table 2 Properties of clay mineral groups

Group	Layer type	Net negative charge (cmol kg^{-1})	Surface area ($m^2 g^{-1}$)	Basal spacing (nm)
Kaolinite	1:1	2–5	10–30	0.7
Fine-grained mica	2:1	15–40	70–100	1.0
Smectite	2:1	80–120	600–800	1.0–2.0
Vermiculite	2:1	100–180	550–700	1.0–1.5
Chlorite	2:1:1	15–40	70–100	1.4

(Adapted from Ref. 5.)

CLAY MINERAL CLASSIFICATION

Clay minerals are generally classified into three layer types based upon the number and arrangement of tetrahedral and octahedral sheets in their basic structure. These are further separated into five groups that differ with respect to their net charge (Table 2).

1:1 Clay Minerals

The 1:1 layer minerals contain one tetrahedral and one octahedral sheet in their basic structural unit (Fig. 2.) This two-sheet mineral type is represented by the kaolin group, with the general formula $Al_2Si_2O_5(OH)_4$. Kaolinite, the most common mineral in this group, is dioctahedral, exhibiting Al^{3+} octahedral and Si^{4+} tetrahedral coordination. The sheets are held together by van der Waals bonds between the basal oxygens of the tetrahedral sheet and the hydroxyls of the octahedral sheet. Layers are held together tightly by hydrogen bonding, which restricts expansion and limits the reactive area to external surfaces. Isomorphic substitution for Si^{4+} and Al^{3+} in this mineral is negligible. As such, soils dominated by 1:1 minerals exhibit a low capacity for adsorbing cations and have low fertility. The serpentine group, with the general formula $Mg_3Si_2O_5(OH)_4$, represents the trioctahedral version of the 1:1 layer minerals.

2:1 Clay Minerals

The joining of two tetrahedral sheets (one from each side) to one octahedral sheet produces a three-sheet mineral type, which is called 2:1 and is represented by the mica, smectite, and vermiculite groups. Talc $[Mg_3Si_4O_{10}(OH)_2]$ and pyrophyllite $[Al_2Si_4O_{10}(OH)_2]$ are typical representatives of electrically neutral 2:1 type minerals in which adjacent layers are joined to each other by van der Waals bonds (Fig. 2). Although these two minerals are found infrequently in soils (2), their structure serves as a model for discussing transitions leading to the formation of other more common 2:1 clay minerals.

The true micas have a similar structure to that of talc and pyrophyllite, except that substitution of Al^{3+} for Si^{4+} in every fourth tetrahedral site results in an excess of one negative charge per formula unit. The negative charge is satisfied by monovalent cations, primarily K^+, that reside on interlayer sites between the 2:1 layers. The interlayer cation forms a strong bond between adjoining tetrahedral sheets, which limits expansion of the mineral. The mica group is subdivided into tri- and dioctahedral minerals according to cation substitutions in the octahedral sheet and within the interlayer. The trioctahedral group of micas contains interlayer K^+ cations and is represented by phlogopite $[KMg_3(AlSi_3O_{10})(OH)_2]$, with Mg^{2+} occupying the octahedral sites, and biotite, which contains both Fe^{2+} and Mg^{2+} in the octahedron. Muscovite $[KAl_2(AlSi_3O_{10})(OH)_2]$ is a dioctahedral mica containing Al^{3+} in the octahedral sheet and K^+ in the interlayer, while paragonite exhibits a similar dioctahedral coordination with interlayer K^+ and Na^+ cations. In most soils, micas are generally inherited from the parent material and occur in a relatively unweathered state in the sand and silt fractions. Mica in the clay fraction usually exhibits poorer crystallinity, lower K^+ content, higher water content, and possible substitutions of Fe^{2+} and Fe^{3+} in the octahedral sheets and Ca^{2+} in the interlayer. Manganese, vanadium, lithium, chromium, titanium, and several other cations are also known to occur in varying amounts in these fine-grained or clay-sized micas (1). Illite and glauconite (dioctahedral iron illites) are commonly associated with the clay-sized micas; however, the structures of these minerals are poorly defined and likely to be representative of a mixture of weathered micas. Expandable 2:1 clay minerals exhibit a similar layer structure to that described for mica, but vary widely in layer charge and interlayer spacing due to the presence of weakly bound cations, water, or polar organic molecules in the interlayer region.

Smectites generally refer to a group of expandable dioctahedral 2:1 minerals with a charge of 0.2–0.6 per formula unit. Montmorillonite, the most common member of this group, derives its charge from the octahedral substitution of Mg^{2+} for Al^{3+}. Beidellite and nontronite, which are less abundant in soils, derive much of their charge from tetrahedral substitutions. Nontronite is distinguished from beidellite by the presence of iron in the octahedral sheet. The 2:1 layers in smectites are held together by van der Waals bonds and weak cation-to-oxygen linkages. The presence of exchangeable cations located between water molecules in the interlayer allows for expansion of the crystal lattice as the mineral hydrates. When the mineral is saturated with water, the basal spacing between layers can approach 2 nm, while under dry conditions, the basal spacing may be reduced to less than 1 nm (Fig. 3). This expansion and contraction trait found in smectites, often referred to as shrink–swell potential, is problematic to engineers and farmers alike due to the propensity for crack formation and general instability of the soil surface.

The weathering of mica, via replacement of interlayer K^+ with hydrated exchangeable cations, results in the formation of vermiculite. In soils, vermiculite exists as an Al^{3+} dominated dioctahedral and, to a lesser extent, Mg^{2+} dominated trioctahedral mineral. A charge of 0.6–0.9 per formula unit is derived in these minerals by tetrahedral substitution of Al^{3+} for Si^{4+}. Consequently, vermiculites exhibit a high cation exchange capacity, particularly for weakly hydrated cations such as K^+, NH_4^+, and Cs^+ (3). Because of the tetrahedral charge origin, water molecules and exchangeable cations—primarily Mg^{2+} and Ca^{2+}—are strongly adsorbed within the interlayer space of vermiculites. Unlike in smectites, the strong bonding of the interlayer cations holds the 2:1 layers together in vermiculites, thus limiting expansion of the basal spacing to 1.5 nm (Fig. 3).

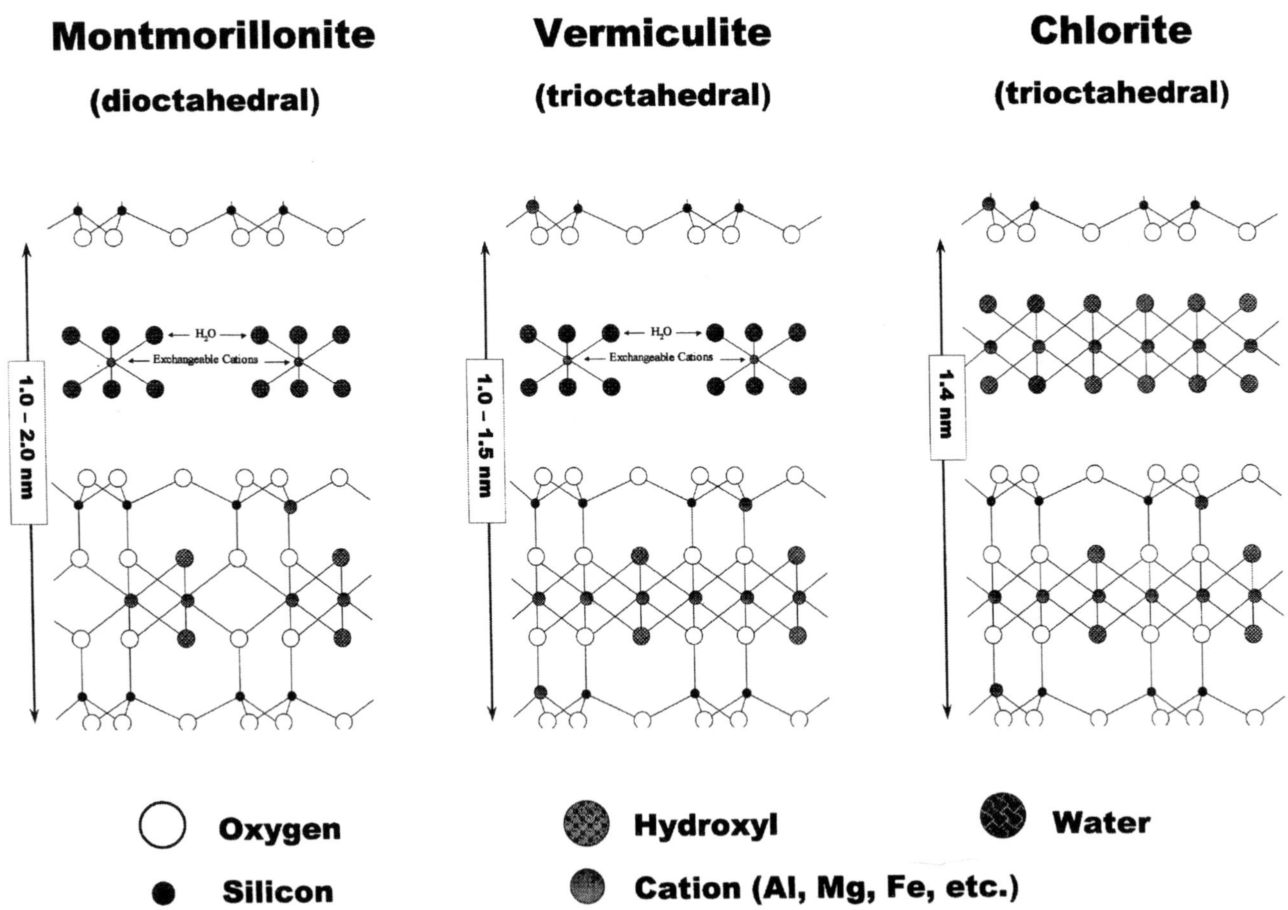

Fig. 3 Schematic representation of expandable dioctahedral montmorillonite and trioctahedral vermiculite, and nonexpandable trioctahedral chlorite.

C

2:1:1 Clay Minerals

Chlorites are a group of minerals that exhibits a basic 2:1 layer structure similar to that described for talc or pyrophyllite, but with an interlayer brucite- or gibbsite-like sheet, which forms a 2:1:1 structural arrangement. Isomorphic substitutions within the interlayer hydroxide sheet create a net positive charge that balances the negative charge arising from the 2:1 layers. A typical formula for the interlayer sheet is $(MgFeAl)(OH)_6^+$; however, a variety of cation species may exist in these brucite or gibbsite-like islands that contribute to a large number of mineral species within this group. There is no water adsorption within the interlayer space; thus, chlorites are considered nonexpansive minerals (Fig. 3).

Hydroxy-Al Interlayered Vermiculites and Smectites

Chlorites exist in soils as primary minerals that weather to form vermiculite and smectite. In contrast, hydroxy-Al interlayered vermiculites or smectites are considered to be exclusively secondary minerals forming as intermediate mineral weathering products, or from deposition of hydroxy-Al polymeric components within the interlayer space of expanding minerals (4). These hydroxy-Al polycations—with the general formula $Al_6(OH)_{15}^{3+}$—balance a portion of the charge but they are not exchangeable. Because the level of hydroxy-Al occupancy within the interlayer space is variable, the CEC of the expandable 2:1 clays is reduced as a function of the quantity and valence of the hydroxy-Al polymer residing within the interlayer.

Mixed Minerals

In the soil environment, clay minerals rarely occur as homogeneous mixtures of single groups, types, or mineral phases. Instead, they comprise complex assemblages of primary minerals, and weathering intermediates of multiple structural and compositional combinations. The potential also exists for a discrete mineral grain to be composed of more than one clay type or contain components that are intermediate between two known minerals. These minerals are referred to as mixed-layer or interstratified minerals. Examples of such mixed-layer mineral sequences include mica–vermiculite, mica–smectite, chlorite–vermiculite, and kaolinite–smectite, among others. The sequence of differing layers making up the mixed mineral occurs at both regular and random intervals. Consequently, efforts to fully characterize clay mineralogy within certain soils may be hindered considerably.

Other Clay Minerals

In addition to the phyllosilicates described, certain oxides, hydroxides, and hydroxy-oxides (sesquioxides) of Si, Al, Fe, as well as some poorly crystalline aluminosilicates, may also be found in small quantities in the soil clay fraction (7, 8). Typical representatives of Si-oxide minerals that are often found in the clay fraction are quartz—due to its high resistance to weathering—and opal—a poorly crystalline polymorph of quartz, which precipitates from Si-supersaturated solutions or is of volcanic or biogenic origin. Gibbsite ($Al(OH)_3$) is the main Al-hydroxide representative in clay fractions of highly weathered soils, while goethite (FeOOH) is the most common clay fraction Fe-mineral. The poorly crystalline aluminosilicate clay minerals allophane and imogolite are typically found in clay fractions of soils derived from volcanic materials. Extensive discussions about these minerals and their properties are provided in other soil mineral sections of this encyclopedia.

Clay Mineral Identification and Quantification

The basic methods used for clay mineral identification and quantification involve X-ray diffraction (XRD) and thermal analyses (TA) (9). Electron microscopy (EM) is also employed for complementary mineral characterizations. In XRD analysis the basal spacing (or *d*-spacing) between equivalent crystal planes is measured from the specific angle at which they diffract X-rays of known wavelength according to Bragg's Law: ($d/n = \lambda/2\sin\theta$) where d = basal spacing, n = order of diffraction, usually set to 1, λ = X–ray wavelength, and θ = diffraction angle. Therefore, diffraction peaks at specific angles corresponding to fixed *d*-spacing characteristic of specific minerals are used for mineral identification, while diffraction peak intensities are utilized for quantitative estimations. The identification of minerals with overlapping *d*-spacing regions (smectite, vermiculite, chlorite) is based on specific diffraction peak shifts, following solvation with specific ions and/or heating treatments.

Thermal methods are based on characteristic temperature regions in which specific minerals lose their structural OH water (thermogravimetric analysis or TGA) and the enthalpy associated with the dehydroxylation reation (differential scanning calorimetry or DSC). Mineral quantities are estimated from comparisons of the OH- or enthalpy fractions that have been measured and standard quantities of representative pure minerals.

REFERENCES

1. Klein, C.; Hurlbut, C. Systematic Mineralogy Part IV: Silicates. *Manual of Mineralogy*, 20th Ed.; John Wiley & Sons: New York, 1985; 366–467.
2. Schultz, D. An Introduction to Soil Mineralogy. In *Minerals in Soil Environments*; Dixon, J., Weed, S., Eds.; Soil Science Society of America: Madison, WI, 1989; 1–34.
3. Douglas, L. Vermiculites. In *Minerals in Soil Environments*; Dixon, J., Weed, S., Eds.; Soil Science Society of America: Madison, WI, 1989; 635–676.
4. Barnhisel, R.; Bertsch, P. Chlorites and Hydroxy-Interlayered Vermiculite and Smectite. In *Minerals in Soil Environments*; Dixon, J., Weed, S., Eds.; Soil Science Society of America: Madison, WI, 1989; 729–788.
5. Brady, N. Soil Colloids: Their Nature and Practical Significance. *The Nature and Properties of Soils*, 10th Ed.; Macmillan Publishing Co.: New York, 1990; 177–212.
6. Grim, R. *Clay Mineralogy*, 2nd Ed. McGraw-Hill Book Co.: New York, 1968.
7. Amonette, J.E., Zelazny, L.W., Eds. *Handbook of Soil Science*; CRC Press: Boca Raton, Fl, 2000.
8. Amonette, J.E., Zelazny, L.W., Eds. *Origin and Mineralogy of Clays: Clays and the Environment*; Springer-Verlag: New York, NY, 1995.
9. Amonette, J.E., Zelazny, L.W., Eds. *Quantitative Methods in Soil Mineralogy*; Soil Science Society of America Miscellaneous Publication: Madison, WI, 1994.

CLAY MINERALS, WEATHERING AND ALTERATION OF

C

A.D. Karathanasis
University of Kentucky, Lexington, Kentucky, U.S.A.

INTRODUCTION

Mineral weathering encompasses the physical and chemical breakdown of minerals, that are unstable under the conditions prevailing at the Earth's surface. It is driven by physical, chemical and biological gradients, which cause gradual structural and compositional mineral alterations that allow them to attain more stable states. Most weathering reactions lead into complete dissolution of the original mineral structure to soluble components, which re-synthesize to form a new mineral of different composition. A significant number of mineral alterations, however, result in modified residual mineral structures without going through a complete dissolution phase. Water is the main mineral weathering agent. Under its dominant influence physical forces (wetting and drying, freezing and thawing, wind and water erosion, thermal expansion, salt crystallization) combine with chemical forces (hydration–dehydration, acidification, oxidation–reduction, chelation) to produce new mineral phases more conforming to the current environment.

MINERAL RESISTANCE TO WEATHERING

Weathering affects both primary and secondary minerals. Primary minerals having formed under high temperature and pressure conditions are less stable under ambient conditions and tend to transform into secondary minerals. Their resistance to weathering is inversely related to the temperature of their crystallization from cooling magmas. The higher the temperature of formation, the lower their stability at the surface of the earth. Based on this principle, Goldich (1) proposed a stability series of common primary minerals (Table 1), which was experimentally confirmed recently with long-term mineral dissolution comparisons (2). Mineral stability is also a function of structural bond strength with which atoms and ions are held together in the crystal lattice of each mineral. Most primary minerals are silicates or aluminosilicates, with strong Si–O bonds, moderately strong Al–O bonds, and weaker metal–O bonds in their crystal lattice. Their susceptibility to weathering depends on the number of weak bonds and the strength of the Si–O bond, which is a function of the Si:O ratio in the structure. The lower the ratio (increased sharing of oxygens between adjacent silica tetrahedrons) the stronger the overall mineral structure. Quartz, for example, having the maximum sharing of tetrahedral oxygens, is one of the most resistant minerals. Olivine is one of the most weatherable minerals because there is no sharing of oxygens in its structure. The stability of other minerals is controlled by the number of shared oxygens, and the type of cation-bonds satisfying isomorphic substitutions within the structure. In addition, the presence of Fe and Mn in the lattice usually induces higher weatherability because oxidation-reduction transitions create charge imbalances within the structure. Keller (3) and Birkeland (4) used the following mineral stability sequence to rank the silicate mineral groups according to bond strength considerations: nesosilicates ($Si:O = 1/4$) $<$ single-chain inosilicates ($Si:O = 1/3$) $<$ double-chain inosilicates ($Si:O = 4/11$) $<$ phyllosilicates ($Si:O = 2/5$) $<$ tectosilicates ($Si:O = 1/2$). Even though some discrepancies exist between the crystallization temperature and the Si–O bond strength sequences, these two schemes are used extensively for making mineral weathering stability interpretations.

Weathering continues after the transformation of primary to secondary minerals because some of them may be stable only within a certain range of solution ionic gradients. As the concentration of soluble Si, Al, and other cations decreases through leaching processes, initially stable secondary minerals may weather further to more stable states. Therefore, secondary minerals may be products of alteration of primary or secondary minerals subjected to multiple weathering cycles. Jackson and Sherman (5) proposed a mineral weathering sequence based on the relative degree of soil development with continuous leaching and the types of minerals commonly found in the soil clay fraction (Table 2). This weathering scheme or stages of it may be applicable to many soils, but it does not specify the physicochemical gradients that could trigger a certain set of transformations, or whether a given soil under specific environmental conditions will go through every one of the predicted stages. Nevertheless, it has been an effective tool in describing soil development

Encyclopedia of Soil Science

Table 1 Mineral stability table[a]

Mafic minerals	Felspathic minerals
Olivine	Calcic plagioclase
Augite	Calcic-alkalic plagioclase
Hornblende	Alkali-calcic plagioclase
Biotite	Alkalic plagioclase
Potassium felspar	
Muscovite	
Quartz	

[a] Stability increasing from top to bottom.
(From Ref. 1.)

stages and processes, mainly because the clay fraction minerals play a greater role and are more actively involved than their sand and silt size counterparts in the changes occurring in the soil over time. Furthermore, the clay minerals due to their smaller size, high surface area and chemical activity are in closest agreement with the surrounding soil solution composition, which is the driving force for mineralogical changes. The mineral groups of the clay size fraction listed in Table 2 indicate progressively increasing stages of soil development. A certain mineral suite may not include all the minerals of a specific weathering stage or index or may not necessarily dominate the clay fraction in order to be a fairly reliable indicator of the overall weathering or maturity status of a soil. A given soil clay fraction often contains a mixture of minerals covering several adjacent weathering stages. A quantitative weathering mean representing the weighted average of mineral composition by soil horizon depth is used to estimate the overall weathering stage of soil development (6). In most soils this mean generally shifts from lower to higher numerical weathering stages as the particle size decreases, as we move from the parent material to the soil surface, and as the soil advances in maturity. With continuous weathering and sufficient time, the effect of distinct climatic and geographic settings should be very evident in the distribution of soil clay minerals. Indeed, these relationships have been the foundation for developing various soil classification schemes, grouping soils with similar behavior, including Soil Taxonomy (7).

Table 2 Weathering sequence for clay size minerals[a]

(1)	Gypsum (*also* halite, sodium nitrate, ammonium chloride, etc.)
(2)	Calcite (*also* dolomite, aragonite, apatite, etc.)
(3)	Olivine-hornblende (*also* pyroxenes, diopside, etc.)
(4)	Biotite (*also* glauconite, magnesium chlorite, antigorite, nontronite, etc.)
(5)	Albite (*also* anorthite, stilbite, microcline, orthoclase, etc.)
(6)	Quartz (*also* cristobalite, etc.)
(7)	Muscovite (*also* illite, etc.)
(8)	Vermiculite (*also* interstratified 2:1 layer silicates)
(9)	Montmorillonite (*also* beidellite, saponite, etc.)
(10)	Kaolinite (*also* halloysite, hydroxyinterlayered vermiculite and smectite, etc.)
(11)	Gibbsite (*also* boehmite, allophane, etc.)
(12)	Hematite (*also* goethite, limonite, etc.)
(13)	Anatase (*also* zircon, rutile, ilmenite, leucoxene, corundum, etc.)

[a] Stability increasing from 1 to 13.
(From Ref. 5.)

MINERAL WEATHERING PATHWAYS

Regardless of the mineralogical and structural composition of the minerals undergoing weathering, the degree of alteration and mineral weathering products depend upon the intensity and extent of leaching by water. Differences in the chemical composition between the leaching water and the mineral structures provide the chemical energy to initiate and sustain mineral changes. The rate of leaching controls the composition of the aqueous solution by dictating the rate of removal of the dissolved solutes and hence, the course of the mineral alteration. In regions of limited rainfall, where evapotranspiration exceeds precipitation, the downward flow of water through a soil containing soluble salts and carbonates is sufficient only in removing the most soluble weathering products, such as Na salts. Less soluble minerals tend to form and accumulate in lower soil depths because of limited leaching. Soluble K and Mg may form secondary aluminosilicates, while soluble Ca may re-precipitate as calcite or gypsum. Secondary calcite precipitation is induced by lower CO_2 pressure and higher pH below the soil surface. Accumulation of secondary carbonates in the soil results in considerable inhibition of silicate mineral weathering because of their control of pH and soluble Ca concentrations.

The main weathering pathway for silicate minerals is desilication. Part of the Si released from the mineral structure reacts with Al and to a lesser extent Fe or Mg to form other clay minerals, while the remainder is subject to leaching. Consequently, most soils experience a loss of Si and basic cations during weathering, which results in a relative increase and residual accumulation of Al- and Fe-enriched mineral structures (ferrallitization) (Fig. 1).

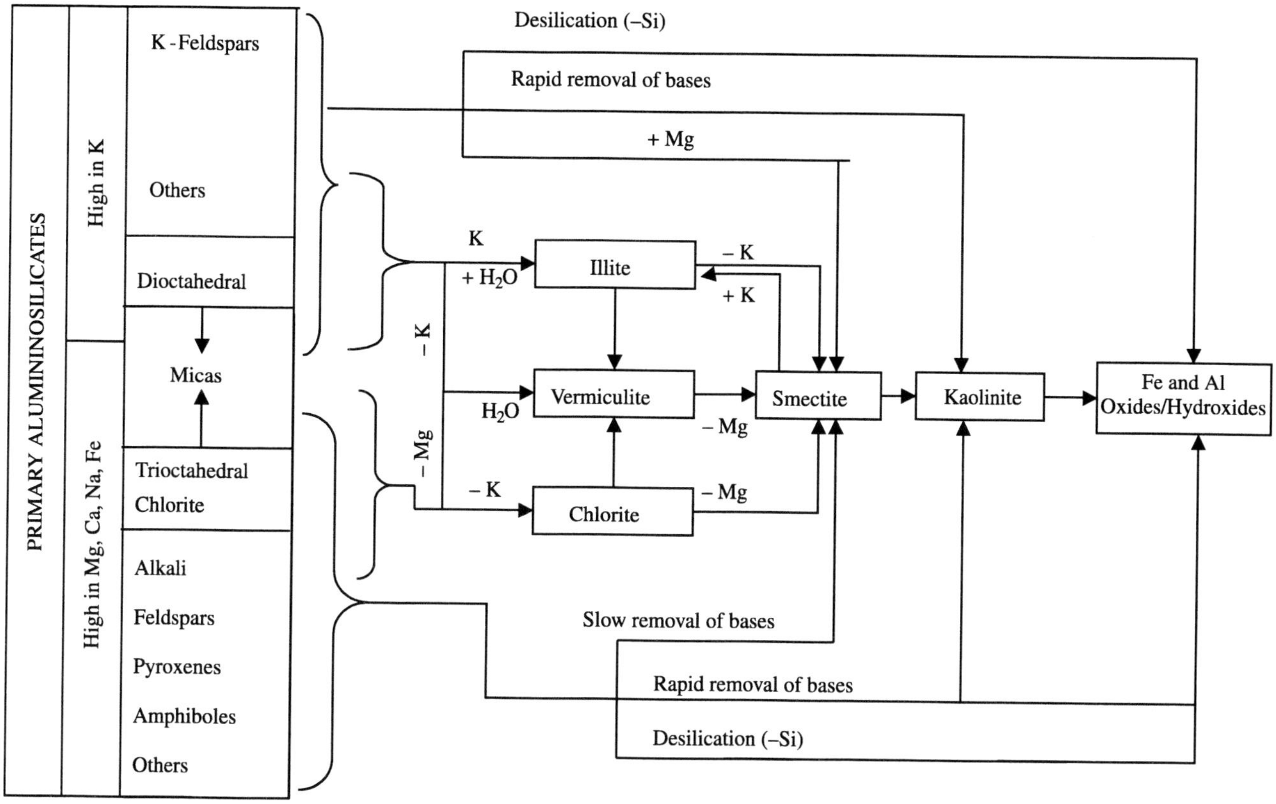

Fig. 1 A generalized mineral weathering/formation pathway scheme.

Desilication is more pronounced in humid tropical environments, but it occurs with lesser intensity in temperate regions as well. Upon desilication and release of Mg and Fe, olivines weather mainly to serpentine and subsequently smectite and goethite (8). Similarly, pyroxenes and amphiboles weather through loss of Mg, Fe and Ca to chlorite and smectite. Under stronger leaching conditions all three of these mineral groups may produce more advanced weathering mineral products, such as kaolinite, halloysite, and various oxyhydroxides (7). Trioctahedral micas (biotite, phlogopite) weather more readily than dioctahedral micas (muscovite, illite) through K release and charge reduction to form interstratified mica-vermiculite, vermiculite, and subsequently smectite (8, 9). Upon stronger weathering under tropical and subtropical conditions, trioctahedral micas may ultimately form kaolinite and halloysite, as well as goethite and hematite. The weathering stages of trioctahedral chlorites are analogous to those of trioctahedral micas, producing interstratified chlorite-vermiculite, vermiculite, smectite, halloysite, kaolinite, and Fe-oxyhydroxides (7–9). Dioctahedral micas (muscovite, illite) undergo similar transformations upon weathering to those discussed for trioctahedral micas. However, their intermediate and advanced weathering products are clearly dioctahedral. Furthermore, under conditions of active leaching, moderately acidic conditions and low organic carbon content, their vermiculite and smectite weathering products may transform into Al-hydroxyinterlayered versions as a result of Al-hydroxide polymerization within their interlayer spaces (7). These minerals sometimes known as pedogenic chlorites, have a lower charge but display considerably greater stability in the weathering environment (10).

Feldspar weathering could lead into a variety of mineral products, including sericite (a micaceous mineral), interstratified mica-smectite, smectite, and a variety of 2:1 aluminosilicates (7, 11, 12). Most often, however, feldspars weather to kaolinite, halloysite, gibbsite and even quartz (9, 12). Slow rates of leaching through the feldspar grains usually form 2:1 phyllosilicates, while high rates of leaching induce gibbsite formation.

While vermiculite is considered mostly an intermediate product of transformation of primary or secondary mica minerals, smectites can be inherited from parent materials, or form either by transformation of other 2:1 phyllosilicates (micas, vermiculite) or precipitation from saturated solutions (7, 9). Transformations from other 2:1 minerals usually occur under high leaching acidic environments, whereas neoformation is induced by poor drainage conditions and accumulation of the necessary cations and silica. Under prolonged acidic conditions, vermiculites may transform to smectites through interlayering, while smectites may alter to kaolinite or halloysite following various stages of interstratification (9, 13). Smectite to kaolinite transformations may involve several stages of interlayering of smectite with various kaolinite precursor phases (7, 13). Both vermiculite and smectite may also form Al-hydroxyinterlayer phases, which enhance their stability in intense weathering environments. Kaolinite may also form via precipitation from solution under warm or temperate humid conditions, while halloysite neoformation has been often associated with weathering of volcanic materials (7). Under extreme desilication conditions and low residence time of Si in the soil solution gibbsite instead of kaolinite may form. Reverse transformations of gibbsite to kaolinite and kaolinite to smectite in environments experiencing resilication have also been reported. The nature of Fe oxides and oxyhydroxides formed through mineral weathering processes depends mostly on the environmental conditions and less on the structure of the dissolved mineral. Accumulation occurs in well oxidized environments in the Fe(III) form, while remobilization occurs under anaerobic conditions in the Fe(II) form (14). Goethite formation is usually favored under cool temperate conditions with moderate organic carbon content, while hematite prefers low organic carbon, warmer, and drier environments. Finally, the most stable titanium and zirconium minerals are the ultimate products of weathering, which accumulate residually in the soil as a result of dissolution of silicate minerals containing Ti and Zr mainly as impurities in their structure (7). Because of their high stability, these minerals are used as indices of soil development stages and soil age.

THERMODYNAMIC FORMULATIONS OF MINERAL WEATHERING

Based on the principle that the stability of a mineral in the weathering environment depends on the balance or imbalance of its structural ions with the activities of the common ions in the soil solution, Kittrick (15) developed a thermodynamic approach in predicting mineral stability. The approach involves the formulation of mineral stability lines in terms of major ionic constituents from dissolution-precipitation reactions and plotting them in a stability diagram. For relatively simple aluminosilicate minerals such diagrams are plotted as functions

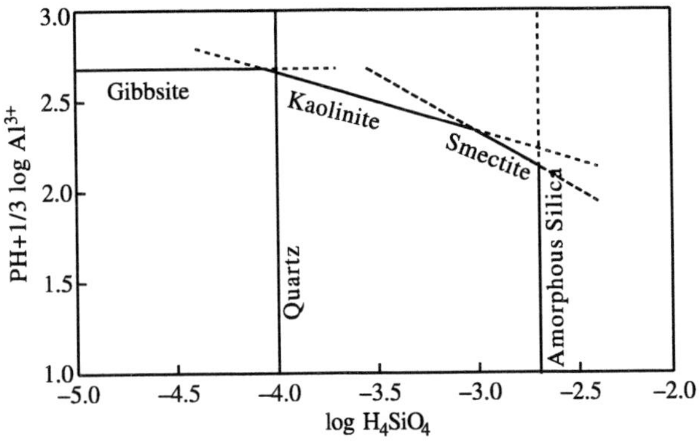

Fig. 2 A typical thermodynamic mineral stability diagram for common soil clay minerals.

of $pH + 1/3 \log Al^{3+}$ and $\log H_4SiO_4$ supported by different minerals from which their stability is deduced under specific ranges of solution ionic conditions (Fig. 2). For minerals with additional structural elements a fixed nominal range in their activity is assumed so that their stability lines are expressed in terms of major structural ions only (Al, Si) and pH, and be comparable to those of other minerals. The stability of a given mineral under a specific soil environment, is determined from the solution composition coordinates in terms of $pH + 1/3 \log Al^{3+}$ and $\log H_4SiO_4$ plotted in the diagram. Solution points plotted along a mineral stability line are considered in equilibrium with the mineral, while points plotted above (also to the right) or below (also to the left) the line represent environments in which the specific mineral is stable or unstable, respectively. Since desilication generally produces more stable minerals, as a convention, the most stable mineral in a stability diagram is the one supporting the lowest level of $pH + 1/3 \log Al^{3+}$ for a given level of $\log H_4SiO_4$. In the stability diagram of Fig. 2, at $\log H_4SiO_4 = -3.5$ the mineral supporting the lowest $pH + 1/3 \log Al^{3+}$, and, therefore, most stable is kaolinite. Similar thermodynamic stability concepts have been used for non-silicate minerals in a variety of geochemical and environmental applications (16).

MECHANISMS AND RATES OF MINERAL WEATHERING

Important mechanisms of mineral weathering include dissolution–precipitation, hydrolysis, oxidation–reduction, hydration–dehydration, chelation and ion exchange (4). Dissolution reactions could be congruent or incongruent. During congruent dissolution the mineral goes into solution completely with no concurrent precipitation of other minerals. In contrast, with incongruent dissolution, all or some of the ions released by weathering precipitate to form new minerals. Relatively soluble minerals (salts, gypsum, carbonates) and most silicate minerals dissolve congruently, while most aluminosilicate minerals weather through incongruent dissolution reactions. Hydrolysis involves the interaction and replacement of structural cations with H^+ ions dissociated from H_2O. This exchange has a disrupting effect on the crystal surface because of the high charge to ion size ratio of H^+ and weakens the rigidity of the mineral structure. Hydrolysis is the main mechanism in the weathering of feldspars and other aluminosilicates. Minerals containing Fe, Mn, or S are especially susceptible to oxidation–reduction reactions. Oxidation involves the loss of electrons, and therefore, changes in charge. The disruption of the crystal electrostatic neutrality along with the size change of the same element at different oxidation states cause considerable instability in the crystal lattice and make the mineral more susceptible to weathering. Hydration and dehydration involve the addition or removal of water molecules in the crystal structure of a mineral. They are not as important weathering mechanisms because they affect very few minerals. Examples are the formation of gypsum and anhydrite or halloysite and kaolinite following hydration–dehydration. Complexation is the process by which chelating agents produced through soil biological processes interact with structural cations to form stable ligands and remove them into solution. The process, which affects many clay minerals, destabilizes the mineral structure and facilitates further decomposition. Ion exchange between the solution and mineral surfaces does not produce drastic alterations, since it leaves the mineral structure unchanged, but hydration differences between exchanged ions may cause considerable changes in the physicochemical properties of the interlayer space.

Temperature, moisture flux, and solution gradients are the major environmental variables affecting mineral weathering rates. Additional factors include mineral particle surface area and biological activity. Generally, weathering rates increase with temperature, intense leaching and removal of soluble products, surface area, and biological activity. The rate of weathering of silicate minerals is very slow, with average lifetimes ranging from 2.3×10^3 for olivine to 2.4×10^7 for quartz considering 1-mm diameter crystals under pH = 5.0 and 25°C (17). This suggests that the bulk mineralogy of a soil is not expected to change considerably during the lifespan of any given individual. Relative weathering rates are generally consistent with the mineral stability sequences reported earlier, but natural rates are usually several orders of magnitude lower than laboratory estimated rates (17). One of the main reasons is that natural systems experience variable moisture fluxes and periodic replenishment compared to the steady flows used during laboratory experiments. Also, the presence and distribution of structural defects and crystal imperfections enhances significantly mineral dissolution rates. Therefore quite variable dissolution rates may be obtained for the same mineral depending on its degree of crystallinity.

PRACTICAL IMPLICATIONS OF MINERAL WEATHERING

Mineral weathering sequences have been important models in explaining soil development stages over time and understanding respective pedochemical processes. Thermodynamic approaches of mineral weathering have allowed assessments of current mineral stability trends and predictions of future changes in soil ecosystems as they are impacted by shifts in use and management (16). Mineral weathering is also the process contributing several macronutrients, such as Ca, Mg, K, P, and S and most micronutrients to the soil as they are released from dissolving minerals. Ultimately these elements are used and recycled by plant communities or leached out of the rooting zone, thus providing an important link between the soil fertility status and degree of mineral weathering. Weathering of silicate minerals also plays an important role in buffering watershed acidification from atmospheric inputs. This is the only process by which long-term neutralization of the produced acidity is achieved through continuous release and replenishment of basic cations from mineral weathering. Global climatic changes are also affected by mineral weathering processes (18). On geological time scales, atmospheric CO_2 levels have been primarily controlled by the balance of volcanic inputs form the earth's interior and silicate weathering outputs at the earth's surface. This balance has been disturbed in recent years by burning large quantities of fossil fuel. It will be interesting to see whether the greenhouse effect will be able to trigger high enough silicate weathering rates able to moderate or buffer larger global climatic changes.

REFERENCES

1. Goldich, S.S. A Study in Rock Weathering. J. Geol. **1938**, *46*, 17–58.
2. Franke, W.A.; Teschner-Steinhardt, R. An Experimental Approach to the Sequence of the Stability of Rock-Forming

Minerals Towards Chemical Weathering. Catena **1994**, *21*, 279–290.
3. Keller, W.D. Environmental Aspects of Clay Minerals. J. Sed. Petrol. **1970**, *40*, 788–854.
4. Birkeland, P.W. *Pedology, Weathering and Geomorphological Research*; Oxford University Press: New York, 1974.
5. Jackson, M.L.; Sherman, G.D. Chemical Weathering of Minerals in Soils. In *Advances in Agronomy*; Norman, A.G., Ed.; Academic Press: New York, 1953; 221–317.
6. Marshall, C.E. *The Physical Chemistry and Mineralogy of Soils: Soils in Place*; John Wiley and Sons: New York, NY, 1977.
7. Dixon, J.B., Weed, S.B., Eds. *Minerals in Soil Environments*; Soil. Sci. Soc. Am. Madison, WI, 1989.
8. Loughnan, F.C. *Chemical Weathering of the Silicate Minerals*; Elsevier: New York, NY, 1969.
9. Churchman, G.J. The Alteration and Formation of Soil Minerals by Weathering. In *Handbook of Soil Science*; Sumner, M.E., Ed.; CRC Press: Boca Raton, FL, 2000; F-3–76.
10. Karathanasis, A.D. Compositional and Solubility Relationships Between Al-Hydroxyinterlayered Soil Smectites and Vermiculites. Soil Sci. Soc. Am. J. **1988**, *52*, 1500–1508.
11. Millot, G. *Geology of Clays*; Springer-Verlag: New York, NY, 1970.
12. Eswaran, H. The Alteration of Plagioclases and Augites Under Different Environmental Conditions. J. Soil Sci. **1979**, *30*, 547–555.
13. Karathanasis, A.D.; Hajek, B.F. Transformation of Smectite to Kaolinite in Naturally Acid Systems: Structural and Thermodynamic Considerations. Soil Sci. Soc. Am. J. **1983**, *47*, 158–163.
14. Schwertmann, U.; Fitzpatrick, R.W. Iron Minerals in Surface Environments. Catena Suppl. **1992**, *21*, 7–30.
15. Kittrick, J.A. *Soil Mineral Weathering*; Van Nostrand Reinhold: New York, 1986.
16. Karathanasis, A.D. Mineral Equilibria in Environmental Soil Systems. In *Environmental Soil Mineralogy*; Dixon, J.B., Schulze, D. Eds.; Soil Sci. Soc. Am. Madison, WI, 2001 (in press).
17. White, A.F.; Brantley, S.L. Chemical Weathering Rates of Silicate Minerals. *Reviews in Mineralogy*; Miner. Soc. Am. Washington, D.C., 1995; Vol. 31.
18. Lasaga, A.C.; Soler, J.M.; Canor, J.; Burch, T.E.; Nagy, K.L. Chemical Weathering Rate Laws and Global Geochemical Cycles. Geochim. Cosmochim. Acta. **1994**, *58*, 2361–2386.

CLIMATE

A. N. Gennadiyev
S. S. Chernyanskii
Moscow State University, Moscow, Russia

INTRODUCTION

Climate is a versatile factor that affects virtually all pedogenetic phenomena. Climatic conditions specify the amounts of solar radiation and precipitation getting to the soil surface. Solar radiation is the main source of energy for most of the processes in the pedosphere. The degree of transformation of soil mineral mass, the rate and character of decomposition of organic remains in the soil, and the nature and functioning of soil biota depend on the supply of solar radiation to the soil, provided that the soil gets a sufficient amount of moisture. Annual precipitation and its distribution by separate seasons together with the evapotranspiration value dictate the degree of soil moistening, the depth of water percolation, the reserves of plant-available water, and the water supply of soil microbiota. The dependence of soils on climate is manifested by the climatic zonality of the soil cover, i.e., by the distribution of particular soils over the Earth's surface in agreement with thermal belts and regions of different moisture levels. Climatic zonality is one of the main regularities in the development in soil cover patterns.

HISTORICAL BACKGROUND

Since the end of the nineteenth century, special in-depth studies of the role of climate as a factor that controls the supply of energy and moisture to soil have been conducted by soil scientists. In 1883, Dokuchaev advanced the concept of soil forming factors in his classical monograph *Russian Chernozem* (1). Along with climate, he considered parent rocks, living organisms, relief, and time to be equally important obligatory factors for soil formation. At the same time, it was climate that received Dokuchaev's primary attention. The student and follower of Dokuchaev, N.M. Sibirtsev considered climate as the most important soil forming factor. The system of soil classification developed in 1898 by Sibirtsev (2) stressed the priority of climatic factors. At approximately the same time the works of E.W. Hilgard on the role of climate in soil formation appeared in the United States. Hilgard paid special attention to the study of soil temperature and moisture conditions and their effect on soil processes. The results of these studies were summarized in his works *A Report on the Relations of Soil to Climate* (3) and *Soils: Their Formation, Properties, Composition, and Relation to Climate and Plant Growth in the Humid and Arid Regions* (4). Later, numerous empirical relationships between various soil properties (the contents of humus, carbonates, nitrogen, clay, saturation capacity, acidity, silica-to-alumina ratio, etc.) and the main climatic parameters (annual precipitation and mean annual temperature) were analyzed by H. Jenny. Jenny suggested the equation of soil forming factors in which the contributions of the inner (soil) and outer (atmosphere) climate to soil formation were considered, respectively, as the components of the internal and external potentials of pedogenesis (5). V.R. Volobuev suggested the concept of hydrothermic sequences of soils. He revealed global relationships between the distribution of soils and ecosystems and several climatic parameters (annual precipitation, mean annual temperature, radiation balance, and evaporation capacity) (6). V.A. Kovda stressed that, in contrast to all the other factors of soil formation having terrestrial origin, the climate is governed by cosmic factors (7). He emphasized the need to study soil evolution as related to climate changes in the Earth's history. More recently, S.W. Buol, F.D. Hole, and R.J. McCracken considered climate (together with living organisms) in the group of *flux factors* responsible for the transfer across a surface of interface of energy, fluids, particles, and other entities (8). M.A. Glazovskaya and A.N. Gennadiyev analyzed the contribution of climate to the energy state of soils, their material basis, and the dynamics of soil forming processes (9).

At present, it is firmly established that the great diversity of soils on our planet is connected to the diversity of climatic conditions.

DISCUSSION

The annual sum of active (above 10°C) air temperatures (accumulated active temperature), mean annual temperatures, and the indices of radiation balance increases by tens and even hundreds of times from polar regions to the equator. This results in considerable differences between polar and tropical regions with respect to the thermal state of soils and the climatic potential of pedogenesis. According to Van't Hoff's Law, a 10°C rise in temperature accelerates the rate of chemical reactions by two to three times. It means that the intensity of chemical reactions, including the weathering of minerals, decomposition of organic matter, and the synthesis of new organomineral compounds, in the soils of tropical regions is tens of times higher than in cold polar regions. Therefore, the thickness of soil profiles and weathering crusts in the tropics is much greater than that in temperate and cold regions of the Earth.

The input of solar radiation to the Earth's surface proceeds with varying intensity, displaying diurnal, seasonal, annual, and long-term fluctuations (cycles). Therefore, periods of soil heating alternate with periods of soil cooling; in many regions, the alternation between soil thawing and freezing is observed. Various combinations of these processes together with particular temperature conditions in the soil specify the *heat (or temperature) regimes of soils*.

In soils that remain in a frozen state for a long period, the possibilities for the development of soil forming processes are strongly restricted in time: at low temperatures, chemical reactions proceed very slowly or cease, the activity of living organisms is decelerated, and the migration of soil solutions stops. At the same time, soil material may be subjected to cryogenic churning (*cryoturbation*). In contrast, long periods with relatively high temperature favor the development of diverse pedogenetic processes.

The soil temperature regime is conditioned by the direct supply of solar radiation, as well as by the peculiarities of atmospheric circulation, including the advection of warm or cool air masses.

The pedogenetic and biological effects of the heat transmitted from the Sun to the Earth's surface can only manifest themselves in full measure if the moisture supply of the soil is sufficient. Atmospheric precipitation is the main factor controlling the solution, leaching, and migration of mobile compounds in the soils. The hydrolysis of primary minerals and the synthesis of secondary clay minerals in soils occurs due to the supply of water. Atmospheric water that reaches the soil is retained in soil horizons and consumed by plants in the course of photosynthesis. The biomass produced by plants serves as a source of energy and nutrients for animals and microorganisms. Thus, precipitation is directly and indirectly related to the processes of humus formation in the soils. Eventually the degree of soil water supply dictates the development of the main genetic horizons of soils, i.e., humus-accumulative, eluvial (horizons with predominant leaching), metamorphic, and illuvial (whose accumulation of substances leached from the above-lying horizons takes place) horizons. Different types of *soil water regime* can be distinguished on the basis of data on the depth and intensity of water migration and the predominance of ascending or descending water flows in the soil. The many-sided role of atmospheric precipitation in the pedogenetic process is illustrated in Figs. 1 and 2. The effect of precipitation on soil properties is so significant that it can be indirectly reflected in soil classification. Thus, C. Marbut suggested that all soils could be subdivided at the highest level into two groups, Pedalfers and Pedocals (10). Pedalfers included the soils of tundra, taiga, and tropical rainforests, i.e., territories with sufficient or excessive moistening. Intensive leaching of soluble compounds in Pedalfers leads to residual accumulation of less soluble sesquioxides (Al and Fe). Pedocals included the soils of steppes, semideserts, and deserts with insufficient moistening and accumulation of calcium carbonates in the profile (Fig. 3). The leading role of the moisture factor in soil development is especially stressed by I.A. Sokolov (11). According to Sokolov, two fundamentally different worlds of soils can be distinguished: the world

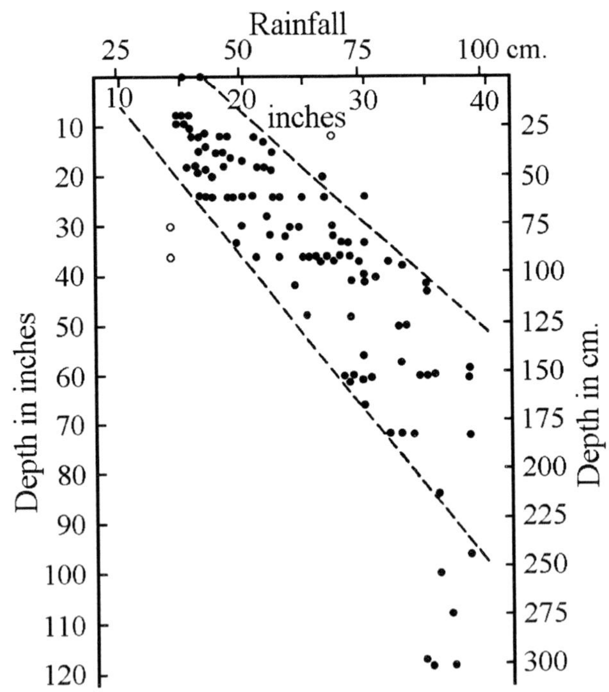

Fig. 1 Relation between depth of carbonate accumulation and rainfall in loessial soils. Every point represents the depth of the beginning of the concretion zone in a soil cut investigated. (From Ref. 5.)

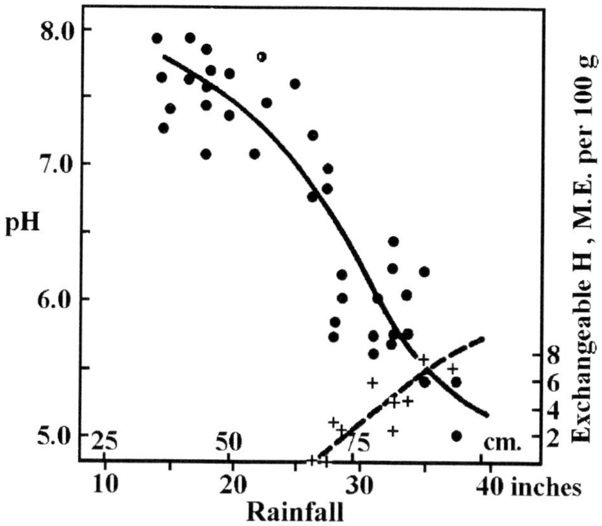

Fig. 2 Relation between soil acidity and rainfall for surface soils derived from similar parent material (loess). Circles stand for pH, crosses for exchangeable hydrogen ions. (From Ref. 5.)

(pedocosmos) of arid pedogenesis and the world of humid pedogenesis. Arid pedocosmos consists of accumulative, saturated, neutral or alkaline, calcareous and/or saline xeromorphic soils. Humid pedocosmos is the world of eluvial, leached, acid, and unsaturated soils.

Variations in annual precipitation across the globe are very considerable. A general trend toward an increase in precipitation from polar regions to the equator (i.e., in the same direction as mean annual air and soil temperatures) is observed. However, the input of solar radiation, and, hence, the heat supply of soils, has a rather distinct latitudinal (zonal) pattern, whereas the zonal distribution of precipitation is strongly altered by the particular geomorphic conditions (the distance from the sea, the location of mountain chains), general circulation of the atmosphere, ocean currents, etc. The overall effect of precipitation and temperature on soil formation has an extremely complex character.

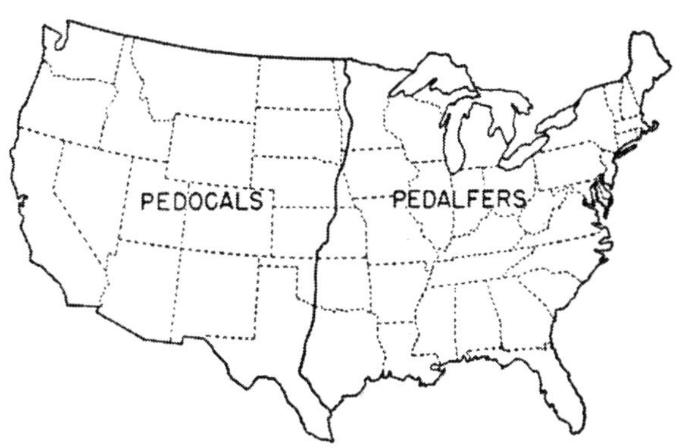

Fig. 3 Marbut's classification of soils into pedocals and pedalfers. (From Ref. 5.)

Initially, the most general regularities of soil distribution dependent on spatial changes in moisture and temperature conditions were formulated by Dokuchaev in the form of the Law of Horizontal (for plain territories) and Vertical (for mountains) Soil Zonality. Later, with acquisition of new factual data, this initial perception of soil zonality was refined and shaped into a more complex form. In 1933, I.P. Gerasimov formulated the concept of bioclimatic soil facies. He stressed that local provincial (facial) peculiarities of the climate conditioned by local thermodynamic processes in the atmosphere complicate the latitudinal pattern of soil zonality.

In works devoted to geographic division (regionalization) of the soil cover of the planet and particular regions, much attention has been paid to climatic criteria of such division. Thus, M.A. Glazovskaya (12) separated the pedosphere into *pedobioclimatic belts* encompassing horizontal soil zones and vertical soil sequences (mountainous soil provinces) with similar radiation and thermal conditions (e.g., polar, subboreal, and tropical belts). In turn, these belts were subdivided into *pedobioclimatic regions* composed of horizontal soil zones and mountainous soil provinces that not only had similar radiation and thermal conditions but also similar moistening and degree of climatic continentality (e.g., the regions of boreal forests, subboreal steppes, subtropical semideserts, etc.).

The economic activity of humans is an important factor affecting microclimatic conditions. Some human impacts are directly aimed at the transformation of soil climate, i.e., soil water and temperature regimes. Thus, drainage and irrigation, shading the soil surface with a dense canopy of crops, destruction of the root mat on the soil surface by plowing, and changes in the direction and velocity of local winds caused by human constructions or artificial planting of trees (windbreaking strips or shelter belts) lead to considerable changes in the provision of soils with heat and moisture. All these measures are aimed at maintaining the optimum parameters of soil water and heat supply during the growing season.

REFERENCES

1. Dokuchaev, V.V. Russian Chernozem. In *Selected Works*; USSR Academy of Sciences Publishers: Moscow, Russia, 1948.
2. Sibirtsev, N.M. Pochvovedeniye (Soil Science). In *Selected Works*; Sel'khozgiz: Moscow, Russia, 1951; Vol. 1.

3. Hilgard, E.W. A Report on the Relations of Soil to Climate. U.S. Department Agric. Weather Bulletin **1892**, *3*, 1–59.
4. Hilgard, E.W. *Soils: Their Formation, Properties, Composition, and Relation to Climate and Plant Growth in the Humid and Arid Regions*; MacMillan: New York, NY, 1906.
5. Jenny, H. *Factors of Soil Formation*; McGraw-Hill: New York, NY, 1941. Jenny, H. Derivation of State Equations of Soil and Ecosystems. Soil Sci. Soc. Am. Proc. **1961**, 25, 385–388.
6. Volobuev, V.R. *Pochvy i Klimat (Soils and Climate)*; Izd. Akad. Nauk Azerb SSR: Baku, 1953; 486.
7. Kovda, V.A. *Osnovy Ucheniya o Pochvakh (Basics of Soil Science)*; Book 1 Nauka: Moscow, Russia, 1973; 448.
8. Buol, S.W.; Hole, F.D.; McCracken, R.J. *Soil Genesis and Classification*; Iowa State University Press, IA, 1989; 446.
9. Glazovskaya, M.A.; Gennadiev, A.N. *Geografiya Pochv s Osnovami Pochvovedeniya (The Geography of Soils with the Basics of Soil Science)*; Izd. Mosk. Gos. Univ. Moscow, Russia, 1995; 246.
10. Marbut, C.F. *Soils of the United States*; Atlas of American Agriculture, Part III, Advance Sheets, No. 8, US Department of Agriculture, 1935.
11. Sokolov, I.A. *Pochvoobrazovanie i Ekzogenez (Soil Formation and Exogenesis)*; Dokuchaev Soil Science Institute: Moscow, Russia, 1997; 244.
12. Glazovskaya, M.A. *Obshcheye Pochvovedeniye i Geografiya Pochv (General Soil Science and Soil Geography)*; Vysshaya Shkola: Moscow, Russia, 1981; 400.

COBALT AND IODINE

Ronald G. McLaren
Lincoln University, Canterbury, New Zealand

C

INTRODUCTION

Cobalt (Co) and iodine (I) are two trace elements that are generally considered to be nonessential for the growth of higher plants (1). However, both elements are essential nutrients for animals, particularly in the case of Co for ruminants (sheep and cattle), and deficiencies of Co and I in grazing animals are not uncommon. Cobalt is also essential for nitrogen (N) fixation by micro-organisms such as rhizobium and blue-green algae, and at high concentrations in the soil can actually be toxic to plants. Most investigations of Co and I in the soil have concentrated on factors affecting the plant availability of these elements, with the aim of diagnosing potential deficiencies.

SOIL COBALT

The Co concentration in soils depends primarily on the parent materials (rocks) from which they were formed and on the degree of weathering undergone during soil development (2). Cobalt tends to be most abundant as a substituent ion in ferromagnesian minerals, and therefore has relatively high concentrations in mafic and ultra-mafic rocks (rocks containing high or extremely high proportions of ferromagnesian minerals). Conversely, Co concentrations are relatively low in felsic rocks (rocks containing large amounts of silica-rich minerals) such as granite, and in coarse-textured quartz-rich sedimentary rocks (sandstones). Higher concentrations of Co may be associated with finer textured sediments (shales) in which Co has become surface adsorbed by, or incorporated into, secondary layer silicates by isomorphous substitution (3). Typical concentrations of Co reported in different rock types are shown in Table 1. As a result of the large range in Co concentrations of soil parent materials, and variation in the degree of weathering, total soil Co concentrations also vary widely. However, the mean values reported for agricultural soils from many countries appear to have a somewhat restricted range of between approximately 2 and 20 mg/kg (Table 1).

Forms of Cobalt in Soils

Cobalt in soils, whether released from parent materials during soil development, or derived from anthropogenic contaminant sources, occurs in several different forms or associations. Cobalt may be present: 1) as the simple Co^{2+} ion, or as complexes with various organic or inorganic ligands in the soil solution, 2) as exchangeable Co^{2+} ions, 3) as specifically adsorbed Co, bound to the surfaces of inorganic soil colloids (clays, oxides/hydrous oxides of Al, Fe, and Mn), 4) as Co complexed by soil organic colloids, 5) as Co occluded by soil oxide materials, and 6) as Co present within the crystal structures of primary and secondary silicate minerals (6). In some soils there appears to be a particularly strong association between Co and manganese (Mn) oxides, particularly in soils where Mn oxides occur as distinct nodules or coatings (7, 8).

Plant Availability of Soil Cobalt

The immediate source of Co for plant uptake is the soil solution. However Co concentrations in the soil solution are extremely low, generally much less than 0.1 μg/mL (1). Soil solution Co appears to be in equilibrium with Co adsorbed at the surfaces of soil colloids and the distribution between solution and surface phases is strongly influenced by soil pH. As pH increases, soluble Co decreases (6). Similarly, soils with high capacities to adsorb Co, particularly those soils with high Mn oxide contents, also have low solution Co concentrations (9). Thus, in addition to soils with low total Co concentrations, the plant availability of Co may be low in soils with high pH or high Co adsorption capacities. Cobalt availability is also influenced by soil moisture status, availability increasing under waterlogged conditions (10).

Determination of soil Co dissolved by various extractants is the most common way of assessing the plant availability of soil Co. Ideally, the forms of Co extracted should include any soluble and exchangeable Co, together with any solid-phase forms of Co able to move readily into the soil solution in response to changes in solution Co concentrations (6). The two extractants used most commonly for this purpose are 2.5% acetic acid and solutions of EDTA. However, the ability of these

Table 1 Cobalt and iodine concentrations in rocks and soils

	Co concentration (mg/kg)	I concentration (mg/kg)
Rock type		
Ultramafic (e.g., serpentinite)	100–300	0.01–0.5
Mafic (e.g., basalt, gabbro)	30–100	0.08–50
Intermediate (e.g., diorites)	1–30	0.3–0.5
Felsic (e.g., granites, gneiss)	<1–10	0.2–0.5
Sandstones	0.3–10	0.5–1.5
Shales/argillites	11–40	2–6
Limestones	0.1–3.0	0.5–3.0
Soils		
Complete range	0.1–300	<0.1–25.4
Range of mean values	2–21.5	1.1–13.1

(From Refs. 1, 4, and 5.)

extractants to accurately assess Co availability appears to be somewhat limited (11, 12).

COBALT DEFICIENCY

Cobalt deficiency in grazing sheep and cattle was first diagnosed in the 1930s, initially in New Zealand, Australia, and Scotland (13–15). The condition causes animals to loose their appetite, become weak and emaciated, suffer severe anaemia, and eventually die. Subsequently, it was shown that Co is a constituent of both vitamin B_{12} and a closely related coenzyme, and that Co deficiency is in effect a deficiency of vitamin B_{12} (16). It has been concluded from field studies that pasture containing below 0.08 mg Co/kg for sheep or below 0.04 mg/kg for cattle is unlikely to meet nutritional requirements in terms of maintaining adequate serum and liver vitamin B_{12} concentrations, and healthy growth rates (17, 18).

Cobalt deficiency is commonly treated by applying Co-containing fertilizers to pastures, usually at very low rates—e.g., 350 g ($CoSO_4.7H_2O$)/ha/yr. However, on some soils such treatments appear to be ineffective (11). Alternative treatments for Co deficiency involve either injecting the animal with vitamin B_{12}, the use of Co drenches, or the insertion of slowly dissolving Co "bullets" in the animal's rumen.

SOIL IODINE

The range of I concentrations found in rocks and soils is shown in Table 1. Iodine occurs as a minor constituent in various minerals, where it can replace anions such as OH^-, SiO_4^{4-}, and CO_3^{2-}, and has relatively low concentrations in most types of rock (19). Highest I concentrations are generally found in fine-grained sedimentary rocks (Table 1). However, soil I concentrations are generally higher than in the rocks from which they have been derived, a fact attributed predominantly to atmospheric accessions of I (20). Iodine is known to be present in the atmosphere in vapor form and associated with particles of dust. In coastal areas, accession of I may also be related to sea spray (1).

Forms of Iodine in Soils

Information of the forms of I in soils is limited, most published analyses of soils have determined total I concentrations only. Of the three most common forms of I, elemental (I_2), iodide (I^-) and iodate (IO_3^-), it seems likely that most I in soils is present as iodide or possibly as elemental I. The presence of iodate in soils has also been postulated but would require high oxidation conditions in neutral or alkaline soils (20). Indeed, there is some evidence that when iodate is added to soils it is rapidly reduced to elemental I or iodide by soil organic matter (21). There is also evidence that, under some conditions, elemental I can be volatilized from soils (22).

Most I in soils appears to be associated with soil organic matter and iron (Fe) and aluminium (Al) oxides, materials by which both iodide and elemental I are known to be strongly adsorbed (21, 23). Indeed the distribution of I in soil profiles appears to follow the distribution of these soil components (24). The importance of atmospheric accessions of I and the affinity between I and organic matter often results in maximum concentrations of I in the surface horizons of soils (20, 24).

Plant Availability of Soil Iodine

Interest in soil analysis as a means of assessing plant availability of I has been minimal (20), and relationships between the I status of soils and the concentrations of I in plants appear to be poor (1). Even soil extractants designed to determine the most soluble forms of I in soils do not provide a good indication of I availability to plants (25). Plants are capable of absorbing I directly from the atmosphere (1), and plant species or varietal differences appear to have a greater influence on plant I concentration than soil I status (20, 26). Dicotyledonous pasture species (e.g., clovers) generally have higher I contents than grasses (26). Concentrations of I in plants may be reduced by liming (25), by the application of N fertilizer (26), and by the application of farmyard manure (25). Seasonal effects on pasture I concentrations have also been observed, with decreases in the summer, and slight increases in the autumn (26).

IODINE DEFICIENCY

Low concentrations of I in food and water have been associated with the occurrence of endemic goitre (enlargement of the thyroid gland) in humans and farm livestock (27). Early work suggested a close relationship between goitre incidence and low soil levels of I, however, it is now recognized that other factors are also involved. In particular, the presence of a group of substances known as goitrogens, occurring in various plant species, has been shown to reduce thyroid hormone synthesis and metabolism (20). In the absence of goitrogens, it is considered that diets containing 0.5 mg I/kg DM will more than adequately meet the I requirements of all classes of animals, while levels as low as 0.15 mg/kg might be sufficient to meet the requirements for growing animals (28). In the presence of goitrogens, I requirements may be as high as 2 mg/I kg DM (28). Attempts to increase pasture I concentrations with I-containing fertilizers have been generally ineffective, with very low recoveries of the applied I (25, 26). Direct supplementation of livestock is normally the preferred way to increase I intakes.

REFERENCES

1. Kabata-Pendias, A.; Pendias, H. *Trace Elements in Soils and Plants*; CRC Press: Boca Raton, Florida, 1984; 238–246.
2. Mitchell, R.L. Cobalt in Soil and Its Uptake by Plants. In *Atti del IX Simposio Internazionale di Agrochimica su La Fitonutrizione Oligominerale*; Punta Ala: Italy, 1972; 521–532.
3. Hodgson, J.F. Chemistry of Micronutrient Elements in Soils. Adv. Agron. **1963**, *15*, 119–159.
4. Aubert, H.; Pinta, M. *Trace Elements in Soils*, Developments in Soil Science 7; Elsevier Scientific Publishing Co.: Amsterdam, 1977; 395 pp.
5. Swaine, D.J. *The Trace Element Content of Soils*; CAB: Harpenden, England, 1955; 157 pp.
6. McLaren, R.G.; Lawson, D.M.; Swift, R.S. The Forms of Cobalt in Some Scottish Soils as Determined by Extraction and Isotopic Exchange. J. Soil Sci. **1986**, *37*, 223–234.
7. Taylor, R.M.; McKenzie, R.M. The Association of Trace Elements with Manganese Minerals in Australian Soils. Aust. J. Soil Res. **1966**, *4*, 29–39.
8. Jarvis, S.C. The Association of Cobalt with Easily Reducible Manganese in Some Acidic Permanent Grassland Soils. J. Soil Sci. **1984**, *35*, 431–438.
9. Tiller, K.G.; Honeysett, J.L.; Hallsworth, E.G. The Isotopically Exchangeable Form of Native and Applied Cobalt in Soils. Aust. J. Soil Res. **1969**, *7*, 43–56.
10. Adams, S.N.; Honeysett, J.L. Some Effects of Soil Waterlogging on Co and Cu Status of Pasture Plants Grown in Pots. Aust. J. Agric. Res. **1964**, *15*, 357–367.
11. McLaren, R.G.; Lawson, D.M.; Swift, R.S.; Purves, D. The Effects of Cobalt Additions on Soil and Herbage Cobalt Concentrations in Some S.E. Scotland Pastures. J. Agric. Sci., Camb. **1985**, *105*, 347–363.
12. McLaren, R.G.; Lawson, D.M.; Swift, R.S. The Availability to Pasture Plants of Native and Applied Soil Cobalt in Relation to Extractable Soil Cobalt and Other Soil Properties. J. Sci. Food Agric. **1987**, *39*, 101–112.
13. Grange, L.I.; Taylor, N.H. Bush Sickness, Part IIA. The Distribution and Field Characteristics of Bush-Sickness Soils. Bull. N. Z. Dept. Sci. Ind. Res. **1932**, *32*, 21.
14. Underwood, E.J.; Filmer, J.F. The Determination of the Biologically Potent Element (Cobalt) in Limonite. Aust. Vet. J. **1935**, *11*, 84–92.
15. Corner, H.H.; Smith, A.M. The Influence of Cobalt on Pine Disease in Sheep. Biochem. J. **1938**, *32*, 1800–1805.
16. Smith, R.M.; Gawthorne, J.M. The Biochemical Basis of Deficiencies of Zinc, Manganese, Copper and Cobalt in Animals. In *Trace Elements in the Soil-Plant-Animal System*; Nicholas, D.J.D., Egan, A.R., Eds.; Academic Press: New York, 1975; 243–258.
17. Andrews, E.D. Observations on the Thrift of Young Sheep on a Marginally Cobalt Deficient Area. N. Z. J. Agric. Res. **1965**, *8*, 788–817.
18. Gardner, M.R. *Cobalt in Ruminant Nutrition. A Review*, Technical Bulletin 36; Department of Agriculture: Western Australia, 1977.

19. Goldschmidt, V.M. *Geochemistry*; Oxford University Press: Oxford, 1954.
20. Fleming, G.A. Essential Micronutrients II: Iodine and Selenium. In *Applied Soil Trace Elements*; Davies, B.E., Ed.; John Wiley and Sons, Ltd: Chichester, 1980; 199–234.
21. Whitehead, D.C. The Influence of Organic Matter, Chalk, and Sesquioxides on the Solubility of Iodide, Elemental Iodine, and Iodate Incubated with Soil. J. Soil Sci. **1974**, *25*, 461–470.
22. Whitehead, D.C. The Volatilisation, from Soils and Mixtures of Soil Components, of Iodine Added as Potassium Iodide. J. Soil Sci. **1981**, *32*, 97–102.
23. Whitehead, D.C. The Sorption of Iodide by Soils as Influenced by Equilibrium Conditions and Soil Properties. J. Sci. Food. Agric. **1973**, *24*, 547–556.
24. Whitehead, D.C. Iodine in Soil Profiles in Relation to Iron and Aluminium Oxides and Organic Matter. J. Soil Sci. **1978**, *29*, 88–94.
25. Whitehead, D.C. Uptake by Perennial Ryegrass of Iodide, Elemental Iodine and Iodate added to Soil as Influenced by Various Amendments. J. Sci. Food. Agric. **1975**, *26*, 361–367.
26. Hartmans, J. Factors Affecting the Herbage Iodine Content. Neth. J. Agric. Sci. **1974**, *22*, 195–206.
27. Underwood, E.J. *Trace Elements in Human and Animal Nutrition*, 3rd Ed.; Academic Press: New York, 1971; 543 pp.
28. Agricultural Research Council, *The Nutrient Requirements of Ruminant Livestock*; Commonwealth Agricultural Bureau: Slough, UK, 1980; 351 pp.

COLLEMBOLA

Steve P. Hopkin
University of Reading, Reading, United Kingdom

INTRODUCTION

Collembola (commonly known as "springtails") are currently considered to be a monophyletic Class of the Phylum Arthropoda (1) although their exact taxonomic position is still the subject of some debate. Many authors treat Collembola as insects. However, many of the old ideas on arthropod evolution are being reassessed in the light of modern evolutionary theory and molecular phylogeny so their placement may change.

There are three main Orders of Collembola. Members of the **Arthropleona** (about 5500 species) have a more or less elongated body shape and range from highly active surface-dwelling species to those that live out all their lives in the depths of the soil. An example of this Order is *Folsomia candida* (Fig. 1), which belongs to the Family Isotomidae. It is familiar to ecotoxicologists as one of a suite of "standard test soil organisms," which are used to assess the toxicity of new chemicals before they are released into the environment. The **Symphypleona** (about 1000 species) have a much more rounded body shape and are mostly attractively patterned, surface-living species. A typical example is *Dicyrtomina ornata* (Fig. 2) (Subfamily Dicyrtominae). The **Neelipleona** are very small soil inhabiting springtails (typically 0.5 mm in length) with only about 25 species known in the world. They have a rounded body shape and bear a superficial resemblance to Symphypleona.

MORPHOLOGY

The oldest fossil Collembola (indeed the oldest fossil insects) are specimens preserved in Rhynie Chert of more than 400 million years in age. Collembola are present in amber in which remarkable detail is preserved (Fig. 3). The most obvious feature of Collembola is the jumping organ or furca, which is clearly visible in the specimen shown in Fig. 1 projecting behind the animal in its "sprung" state. The furca evolved through the basal fusion of a pair of appendages on the fourth abdominal segment and is capable of propelling some springtails many times their own body length in a fraction of a second. The spring developed as an escape mechanism to avoid predators. Species of Collembola confined to the soil have a reduced furca to ease their movement between soil particles and tightly packed leaf litter. Some have lost the jumping organ altogether. The maximum number of ocelli in each eye is eight, but these are often reduced and many soil-dwelling and cave species are blind.

All Collembola have a *ventral tube* that consists of eversible sacs derived from a pair of appendages on the first abdominal segment. This organ is extremely important in fluid balance but can also function as a sticky appendage to enable springtails to adhere to slippery surfaces. In some species, the vesicles of the ventral tube may extend more than twice the length of the body and be used for self-righting after a jump. Lubbock introduced the scientific name for springtails in 1873 (2). He rightly considered the ventral tube to be the most characteristic feature of the group and gave them the name Collembola based on the Greek *Colle* (= glue) and *embolon* (= piston).

Collembola are small animals. Most are only a few millimeters long although the Central European species *Tetrodontophora bielanensis* can reach 9 mm in length, and some members of the Subfamily Uchidanurinae grow to 10 mm and bear brightly colored "spines." The tiniest and least-pigmented species tend to be those that live permanently in the spaces between particles of soil or sand. The blood of some Collembola possesses chemicals that act as powerful feeding deterrents to predators (3).

ECONOMIC IMPORTANCE

The majority of springtails feed on fungal hyphae or decaying plant material. In the soil, they may influence the growth of mycorrhizae (4) and control fungal diseases of some plants (5). In general, these effects are beneficial; however, there are a few species, including *Sminthurus viridis* the "Lucerne flea," which feed directly on plant material. In Australia, a country to which *S. viridis* has been introduced, the species can cause significant

Fig. 1 Adults and juveniles of *Folsomia candida*. The largest specimen is 2 mm in length. (Photograph by Steve Hopkin.)

Fig. 2 *Dicyrtomina ornata* (1.5 mm in length). (Photograph by Steve Hopkin.)

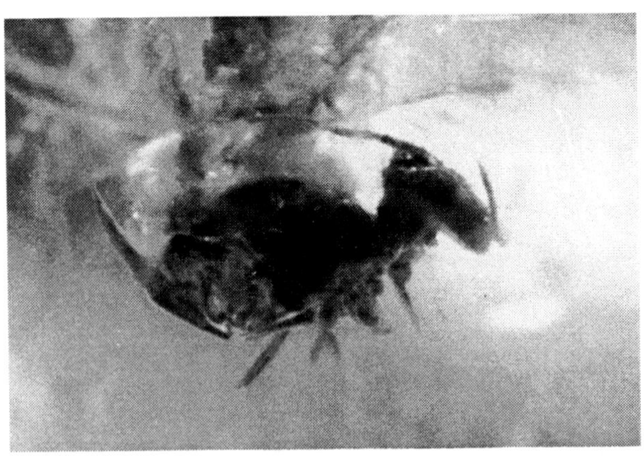

Fig. 3 Symphypleonid springtail (1 mm in length) preserved in Baltic amber of 40 million years in age. (Photograph by Steve Hopkin. Specimen supplied by Andrew Ross, Natural History Museum, London.)

economic damage (6). Some springtails are carnivorous and eat nematodes, rotifers, and even other Collembola.

LONGEVITY

Springtails are generally short-lived. Few survive as adults for more than a year or two. The documented longevity record is held by *Pseudosinella decipiens*, which survived for up to 67 months in the laboratory, although some cave species, or those in very cold climates, may live longer. Most Collembola continue to moult after reaching reproductive maturity and may alternate reproductive and "sexually dormant" instars through several cycles. Several species are parthenogenetic (males absent) including *Folsomia candida* (Fig. 1).

HOW MANY SPECIES?

Approximately 6500 species of Collembola have been described although it is difficult to give an exact figure as there are many species yet to be discovered, especially in countries such as Australia and New Zealand where there is a high level of endemism. The most recent estimate of the total number of species of all insects on the earth quotes a figure of between 5 and 10 million, less than 20% of which have been described (7). In view of their cryptic behavior and relative lack of scientific study, it is likely that at least 50,000 species of springtail exist on our planet.

DISTRIBUTION

Collembola have a very wide global distribution. They are abundant on every continent, including Antarctica, where *Biscoia sudpolaris* and *Antarctophorus sudpolaris* have been found crawling among lichen at a latitude of 84° 47′S, the most southerly location for any invertebrate (8). *Aackia karakoramensis* occurs on newly fallen snow in the Himalayas at an altitude of 7742 meters (9). *Folsomides arnoldi* is abundant in Australian deserts (10).

Collembola are common on the seashore. *Anurida maritima* is a marine species and is one of the most familiar invertebrates of the littoral zone in Europe (11). Several species live almost permanently on the surface of freshwater, including the common and widespread *Podura aquatica*, a frequent sight on puddles after rain, sometimes in huge numbers. Species of *Hypogastrura* are abundant

C

and are important in clearing growths from the percolation filters of sewage beds.

Many species live all their lives in the soil, where they penetrate more than 150 cm below the surface. Others live on trees and are abundant in rain forest canopies. In one study, about one million Collembola of 16 species were collected from 100 m^2 of dry forest in Mexico by insecticide fogging (12). Some Collembola are specialized for living in sand (13).

DIVERSITY

Habitats with extreme climates such as deserts and polar regions support few species of Collembola, but sites with many niches have a diverse springtail fauna. Collembola seem to follow the general rule that diversity is inversely related to latitude; that is, there are more species in tropical than in temperate zones. In tropical rain forests, more than 130 species m^{-2} have been found in soil, leaf litter, and aboveground vegetation (14). In more temperate forests, diversity is lower, but it is not unusual to find more than 40 species in deciduous woodland (15). Collembola exhibit dominance patterns typical of most groups of terrestrial arthropods. The majority of individuals are usually represented by a small number of common species. In most populations, a large fraction of the species (usually >50%) is rare with dominance values of <1%.

"SWARMING"

Collembola are most obvious when they swarm. Most reports are of species in the family Hypogastruridae. They occur following synchronized reproduction in conditions of ideal humidity and temperature and abundant food supply. There are numerous references in the literature to swarming, particularly on snow and glaciers (e.g., 16). Swarms certainly can be huge, often comprising several millions of individuals. The reasons for this behavior are not completely understood although in most cases the Collembola are probably searching for a more favorable habitat after becoming overcrowded.

ABUNDANCE

Collembola are extremely abundant in soil and leaf litter. In most terrestrial ecosystems they occur in high numbers, typically between 10^4 and $10^5\,m^{-2}$. Densities of springtails of more than $10^5\,m^{-2}$ have been found in pine forests in India and Japan, moorland in England, and dry meadows in Norway. Collembola are particularly abundant in agricultural soils that are farmed "organically" (17). In the rain forests of Seram, Indonesia, Collembola comprise about 20% of the total number of arthropods on tree trunks and 50% and 60% of the total from soil and leaf litter, respectively (18). However, because of their small size the contribution of Collembola to total soil animal biomass and respiration is low, typically between 1% and 5% in temperate ecosystems, but up to about 10% in some arctic sites and as much as 33% of total soil fauna respiration in ecosystems in early stages of succession. Typical values for the dry weight of springtails in temperate ecosystems are $0.15\,g\,m^{-2}$ in deciduous woodland and $0.3\,g\,m^{-2}$ in limestone grassland.

EFFECTS ON SOIL STRUCTURE

Despite their relatively low biomass, springtails are extremely important in influencing the structure of some soils. For example, "alpine pitch rendzinas" on limestone are composed mainly of a deep black humus layer of 15 to 20 cm in depth that is formed almost entirely of Collembola feces (19). Most soils contain millions of collembolan fecal pellets m^{-2}, and these must be beneficial in slowly releasing essential nutrients to plant roots as the pellets are broken down by microbes.

ROLE IN DECOMPOSITION

The main effect of Collembola on decomposition and "soil respiration" is through feeding on fungal hyphae. At certain densities of Collembola, grazing of mycorrhizae on roots can stimulate growth of the symbiont and improve plant growth (4). In other situations, Collembola may reduce disease by consuming pest fungi (5).

Selective grazing by springtails may be an important factor limiting the distribution of certain species of basidiomycete fungi in the field. However, many of these effects are density-dependent, and too little information is available for quantifying accurately the specific contribution of Collembola to "indirect" or "catalytic" decomposition. Nevertheless, the influence of springtails on decomposition and nutrient availability must be significant in many ecosystems.

FURTHER READING

Further information can be found in the review by Hopkin (20), which contains a list of the most important references on Collembola published before 1996. An up-to-date list of Collembola publications from 1995 onward is available at http://www.ams.rdg.ac.uk/zoology/collembola/. A very detailed Collembola site is maintained by Frans Janssens at http://www.geocities.com/CapeCanaveral/Lab/1300/. This includes a world checklist of species together with identification keys to genus level for most families. Since the publication of Hopkin (20) a number of important monographs have been published that should be consulted for recent developments in taxonomy and biology. These are

- a major revision of the Iberian Collembola by Jordana et al. (21)
- a second edition of the key to North American Collembola by Christiansen and Bellinger (22)
- the first volume of a key to the Nordic Collembola by Fjellberg (23)
- an identification guide to South American Collembola by Heckman (24)
- a review of Collembola found associated with streams and lakes in Europe by Palissa (25).

REFERENCES

1. Bitsch, C.; Bitsch, J. The Phylogenetic Interrelationships of the Higher Taxa of Apterygote Hexapods. Zoologica Scripta **2000**, *29*, 131–156.
2. Lubbock, J. *Monograph of the Collembola and Thysanura*; Ray Society: London, 1873.
3. Messer, C.; Walther, J.; Dettner, K.; Schulz, S. Chemical Deterrents in Podurid Collembola. Pedobiologia **2000**, *44*, 210–220.
4. Gange, A. Arbuscular Mycorrhizal Fungi, Collembola and Plant Growth. Trends in Ecology and Evolution **2000**, *15*, 369–372.
5. Sabatini, M.A.; Innocenti, G. Effects of Collembola on Plant-Pathogenic Fungi Interactions in Simple Experimental Systems. Biology and Fertility of Soils **2001**, *33*, 62–66.
6. Bishop, A.L.; Harris, A.M.; McKenzie, H.J. Distribution and Ecology of the Lucerne Flea, *Sminthurus viridis* (L.) (Collembola: Sminthuridae), in Irrigated Lucerne in the Hunter Dairying Region of New South Wales. Australian Journal of Entomology **2001**, *40*, 49–55.
7. Odegaard, F. How Many Species of Arthropods? Erwin's Estimate Revised. Biological Journal of the Linnean Society **2000**, *71*, 583–597.
8. Block, W. Terrestrial Microbiology, Invertebrates and Ecosystems. In *Antarctic Ecology*; Laws, R.M., Ed.; Academic Press: London, 1984; Vol. 1, 163–236.
9. Yosii, R. Snow Collembola of the Siachen Glacier in Karakoram. Results of the Kyoto University Scientific Expedition to the Karakoram and Hindukush **1966**, *8*, 407–410.
10. Suhardjono, Y.R.; Greenslade, P. *Folsomides arnoldi* n.sp. (Isotomidae): A New Collembolan Abundant in Arid Australia, with a Redescription of *Folsomides denisi* (Womersley). Proceedings of the Linnean Society of New South Wales **1994**, *114*, 21–27.
11. McMeechan, F.K.; Manica, A.; Foster, W.A. Rhythms of Activity and Foraging in the Intertidal Insect *Anurida maritima*: Coping with the Tide. Journal of the Marine Biological Association of the United Kingdom **2000**, *80*, 189–190.
12. Palacios-Vargas, J.G.; Gonzalez, V. Two New Species of *Deuterosminthurus* (Bourletiellidae), Epiphytic Collembola from the Neotropical Region with a Key for the American Species. Florida Entomologist **1995**, *78*, 286–294.
13. D'Haese, C. Is Psammophily an Evolutionary Dead End? A Phylogenetic Test in the Genus *Willemia* (Collembola: Hypogastruridae). Cladistics **2000**, *16*, 255–273.
14. Deharveng, L.; Bedos, A.; Leksawasdi, P. Diversity in Tropical Forest Soils: The Collembola of Doi Inthanon (Thailand). In *Third International Seminar on Apterygota*; Dallai, R., Ed.; University of Siena: Siena, 1989; 317–328.
15. Lauga-Reyrel, F.; De Conchat, M. Diversity Within the Collembola Community in Fragmented Coppice Forests in South-western France. European Journal of Soil Biology **1999**, *35*, 177–187.
16. Hagvar, S. Navigation and Behaviour of Four Collembola Species Migrating on the Snow Surface. Pedobiologia **2000**, *44*, 221–233.
17. Axelsen, J.A.; Kristensen, K.T. Collembola and Mites in Plots Fertilised with Different Types of Green Manure. Pedobiologia **2000**, *44*, 556–566.
18. Stork, N.E.; Blackburn, T.M. Abundance, Body Size and Biomass of Arthropods in Tropical Forest. Oikos **1993**, *67*, 483–489.
19. Kubiena, W.L. *The Soils of Europe*; Thomas Murby: London, 1953.
20. Hopkin, S.P. *Biology of the Springtails (Insecta: Collembola)*; Oxford University Press: Oxford, 1997.
21. Jordana, R.; Arbea, J.I.; Simón, C.; Luciáñez, M.J. Collembola, Poduromorpha. In *Fauna Ibérica*; Ramos, M.A., Ed.; Museo Nacional de Ciencas Naturales, CSIC: Madrid, 1997; Vol. 8.
22. Christiansen, K.; Bellinger, P. *The Collembola of North America, North of the Rio Grande*; 2nd Ed.; Grinnell College: Grinnell, Iowa, 1998.
23. Fjellberg, A. *The Collembola of Fennoscandinavia and Denmark Part I. Poduromorpha*; Fauna Entomologica Scandinavica: Brill, Leiden, 1998; 35.
24. Heckman, C.W. *Encyclopedia of South American Aquatic Insects*; Kluwer: Collembola, 2000.
25. Palissa, A. Süsswasserfauna von Mitteleuropa. Band 10. *Collembola*; Spektrum Akademischer Verlag: Heidelberg, Berlin, 2000.

COLOR

Robin Thwaites
The University of Queensland, St. Lucia, Queensland, Australia

C

INTRODUCTION

Soil color is one of the most evident characteristics observed on viewing a soil exposure. It is often used in lay terms as a descriptor for any soil type, e.g., "red soils," "black soil," "mottled soils," and it is always recorded as part of a soil description for inventory purposes. The pioneering Russian school of pedology noted variations in soil color before the turn of the 19th century but it was not until the end of the 1920s before a satisfactory attempt had been made in the U.S. to standardize colors of dry soil samples (1). Nowadays soil color, defined through a standard system, plays an important role in all major soil classification systems and for all means of soil interpretation. Although soil color per se has no known direct effect on soil behavior (except for heat absorption at the surface), it provides a valuable indication of soil conditions, composition and genesis. It can indicate conditions of aeration (therefore drainage), organic matter content, fertility, and degree of weathering. Various minerals at different stages of weathering, particularly the iron oxides, lend their color to the soil matrix and to individual soil grains. Color is also a major criterion to differentiate horizons within a soil as a means to its classification and interpretation. It can occur as whole-colored matrix of the soil, as mottles (blotches, specks), in marbled or streaked patterns, as precipitates around mineral grains (lining root holes), or as nodules and other segregations.

MEASURING SOIL COLOR

Soil colors vary from very pale blues and greens and grays, even white, through browns, yellows, oranges and reds, to dark forms of these as well as completely black. There is a general perception that soil color cannot be measured with any precision or accuracy, but color has been recorded consistently and assiduously in both the field and the laboratory for many years. Routine determination of soil color in the field is the less precise and less accurate form of measurement, although consistent results can be achieved with experience. It is determined through visual comparison of a small, fresh soil sample with color chips in standard color charts. The most familiar and widely used are the Munsell Soil Color Charts (2). The Munsell color scheme is an objective three-coordinate system that defines color through its *hue* (shade), *value* (lightness), and *chroma* (intensity/saturation) (Fig. 1). For soil colors the Munsell chart ranges in value from 0 to 10, in chroma from 0 to 8, and in hue (in units of principal colors—which are primarily red and yellow for soils) as 10R (red), 2.5YR (yellow–red), 5YR, 7.5YR, 10YR, 2.5Y (yellow), and 5Y. Red soil colors are usually 5YR or redder (with chromas >1) and yellow soils are hues 7.5YR or yellower. Dark soil colors have values <3 and chromas <2 for any hue. Some soil colors have been recorded outside this standard range (e.g. 5R in central Australia) but these occurrences are rare. Some "gley" hues (5Y, 5GY, 5G, 5BG, 5B) are also included in the standard charts to account for pale yellow, green and blue colors from hydromorphic, or waterlogged, soils with reduced ferrous iron conditions. Color matching is usually undertaken on moist, or moistened, soil samples, although dry soil colors are needed for some classification purposes. Moistening a soil can darken the value by up to three units and up to two units in chroma (3).

Soil colors measured this way rarely match with the standard color system perfectly, and training is recommended to perform consistent and accurate color matching (4). The factors that most affect visual perception of soil color are: 1) the light source characteristics illuminating the sample, 2) the surface characteristics of the sample, and 3) the variable spectral response characteristics of the human eye (5). Even among professional soil scientists agreement on the same color "chip" in the Munsell chart for all three color components is around 50% for a variety of soil color samples, although this rises to around 70% for individual color components (4).

In the laboratory soil color can be ascertained best by spectral reflectance from prepared, disturbed soil samples via a diffuse reflectance spectrophotometer. The reflectance data provided by this process can be converted to a three-figure "tristimulus" value that defines the visually perceived color. This, in turn, can then be converted to a standard color system such as the Munsell notation (6). Photoelectric colorimeters can also be used but results are deemed to be less accurate (6).

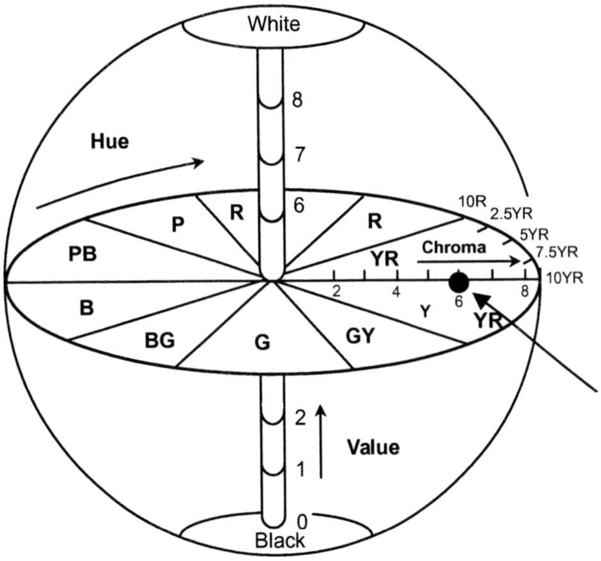

Fig. 1 The Munsell three-dimensional color concept. The hue and chroma ranges for a value of 5 are shown. Thus, the point arrowed represents a color of 10YR 5/6.

WHAT CAUSES SOIL COLORS?

Variations in soil color may be due to soil forming processes or inherited from the parent material. It was recognized in the earliest known study devoted to soil color in 1911 (7) that the color of soil depends upon the content of organic matter and the type of iron oxides. Several iron oxide mineral types can be present in soil and they each possess distinctive colors (Table 1) (8). The type of iron mineral present is influenced by the parent material and the immediate soil environment. The type and amount of organic matter will influence the darkness of the topsoil. It must also be realized that some soil colors may be relics from prior climates and landscape processes that currently do not affect the soil.

There is a broad relationship between the darker color of surface soils and organic matter content. However, the relative darkness does not reflect absolute levels of organic matter content. Higher levels of humification will produce darker colors as will increasing wetness: the near black color of some peats and mucks is due to intense humification. The dark color of many black soils (e.g., dark clays with high calcium carbonate contents) may not be attributable to high levels of organic matter but rather due to the clay mineralogy and stage of humification (9). The nature of the organic compounds will usually be clay–iron–(manganese)–organic matter complexes. Intense black coloration, often as mottles, is due to the presence of manganese oxides.

The bright brown, red, and yellow colors of many subsoils are caused by oxidized (often hydrated) iron products, Fe(III) oxides (or "Fe oxides"), in the fine clay-sized fraction. Contrary to common belief, it is not the *amount* of Fe oxides in the soil that determines the nature of the color but the *type* and *nature* of Fe oxides. The iron mineral responsible for most nonorganic soil coloration is goethite (α-FeO–OH), which imparts colors from reddish-brown through brown to yellow with increasing hydration. Geothite, the most common soil Fe oxide, gives Munsell hues of 10YR–7.5YR if it is finely dispersed, variations in this color come from variations in crystal size and

Table 1 The soil colors and soil environments associated with the common iron oxides

Iron oxide: formula	Munsell color range	Soil environment
Goethite: α-FeOOH	7.5YR–2.5YR (brown–reddish yellow–reddish brown–dark red)	Weathering of all soils in all regions with Fe release
Lepidocrocite: γ-FeOOH	5YR–7.5YR, Value≥6 (light brown–reddish yellow–light reddish brown–pink)	Noncalcareous, aerobic/anaerobic conditions in temperate regions
Hematite: α-Fe_2O_3	7.5YR–5YR (brown–reddish yellow–yellowish red–reddish brown)	Aerobic conditions of the tropics and subtropics. High Fe release rate, high soil temperature, rapid biomass turnover, low leaching activity
Maghemite: γ-Fe_2O_3	2.5YR–5YR (red–reddish brown–reddish yellow–yellowish red)	Tropical and subtropical regions. May be a product of fires.
Ferrihydrite: $Fe_5HO_8 \cdot 4H_2O$	5YR–7.5YR Value≤6 (reddish brown–yellowish red–brown–dark brown)	Rapid oxidation in humic environments in gleys and podzols (spodosols).

(Adapted from Ref. 8.)

cementation (8). Cementation of goethite into nodules and ferricretes darkens the appearance. The strong red colors of tropical and subtropical soils under aerobic conditions are due to the dominance of hematite (α-Fe_2O_3). This mineral gives Munsell hues of 5YR–5R but, again, variations are due to crystal size: larger hematites, in the micrometer range, tend to purple colors and may even reach Munsell RP hue (8). Other Fe oxides that may impart distinctive colors to the soil are: 1) lepidocrocite (γ-FeO–OH), which gives dull yellow to orange mottle colors in anaerobic/aerobic, gleyed conditions, 2) ferrihydrite ($Fe_5HO_8 \cdot 4H_2O$) giving rusty red colors to specks, mottles and root holes in otherwise wet or waterlogged (anaerobic) conditions, including the reddish tinge to the podzol (spodic) B horizon (Bs, Bhs), 3) maghemite (γ-Fe_2O_3), often the result of fire affecting the soil transforming goethite, produces colors between those of goethite and hematite (8).

Other iron compounds impart distinctive colors to subsoils. A special case is that of jarosite [K Fe_3 $(OH)_6$ $(SO_4)_2$] which occurs commonly as yellow (5Y) mottles in acid sulfate soils. Reduced Fe compounds (through anaerobic, often waterlogged, conditions) lack the red and yellow pigmentation. The soil matrix under these conditions attains a range of shades between gray, green, and blue as the Fe oxides are completely removed after mobilization by microbial reduction. These "gley" colors form in ferriferous clay minerals through partial reduction of structural Fe (8). It is possible that some of the greenish tinge is also due to the presence of the so-called "green rusts", which are a group of Fe(II, III) hydroxy

Table 2 Simple indications given by some soil color characteristics and examples of possible soil types, for illustrative purposes only

Color characteristic	Indication	Soil type example
Brown, whole-colored profile	Well drained, moderately weathered, high organic matter, high fertility	ST: Haplustoll; FAO: Chernozem
Red, whole-colored profile	Well drained, highly leached, intensely weathered (high proportion of sesquioxides), low fertility	ST: Durargid, Humox; FAO: Ferralsol
Yellow, whole-colored profile	Moderately drained, moderately leached, moderately weathered, (geothite dominant), moderate fertility	ST: Haplustalf, Paleustult; FAO: Xanthic Ferralsol, Ferric Luvisol
Black, whole-colored profile	Moderately to imperfectly drained, little weathered, little leached, high organic carbon, high fertility	ST: Ustert, Haploxeroll; FAO: Pellic Vertisol, Rendzina
Black topsoil, whole-colored	Imperfectly to poorly drained, high organic matter, unweathered, unleached, high fertility	ST: Umbrept; FAO: Humic Cambisol, Humic Podzol
Gray topsoil, whole-colored	Well drained, leached, low organic carbon, low fertility	ST: Palexeralf; FAO: Dystric Cambisol
Pale-white subsoil	Intensely leached, no organic carbon and low cations, very low fertility	ST: Haplorthod; FAO: Podzol, Albic Luvisol
Yellow subsoil, whole-colored	Moderately drained, moderately weathered, high iron content (probably goethite), moderate fertility	ST: Paleustalfs; FAO: Luvisol, Acrisol
Green–gray mottled subsoil, Green–blue mottled subsoil	Poorly drained, often saturated by groundwater, anaerobic/reducing conditions (Fe(II) oxides), poor fertility	ST: Aquult; FAO: Gleyic Acrisol, Gleysol
Red–yellow mottled subsoil	Imperfectly drained, fluctuating water status (may be relict), regular oxidizing of wet soils (Fe (III) oxides), mod-poor fertility	ST: Paleustult, Palexeralf; FAO: Plinthic Luvisol, Chromic Luvisol
Orange–rusty mottled topsoil	Regular surface waterlogging and draining of surface, oxidizing mineral matter, high organic matter, mod-high fertility	ST: Pellustert; FAO: Chromic Vertisol

ST: Soil Taxonomy, FAO: FAO/Unesco Soil Map of the World

salts. If this is the case, then the minerals lose their color rapidly upon exposure to air (8).

In some cases, subsoil colors may be inherited from distinctively colored parent materials, e.g. red–brown soils from hematite-rich red sandstone, green tinges from glauconite-rich substrates, recent red soils from pre-weathered volcanic ash deposits, sooty black soils from manganese-rich, ore-bearing substrates.

SOIL CONDITIONS INDICATED BY COLOR

Dark brown colors near the surface are associated with high levels of organic matter, thus higher than average nutrient fertility.

Bright yellows and reddish colors in the subsoil signal the presence of Fe(III) oxides, therefore indicating good drainage and aeration conditions. Red soils are usually better aerated/drained than yellow soils. The strong surface charge of these Fe oxides also contributes to the good aggregation properties of the soil (9).

Very pale gray to white colors, particularly in earthly fabric topsoils, indicate significant leaching and low organic matter. Fertility status is, therefore, low. A distinctively paler sub-topsoil horizon, characterized by leaching, is known as an E (eluviated), or A_2, horizon and is distinctive in podzols (spodosols).

Orange or dull yellow mottles (lepidocrocite) in gray, bluish or greenish (reduced) ferriferrous clay matrix (gley) signifies prolonged anaerobic conditions, permanent or seasonal waterlogging.

Black matrix material or mottles in wet conditions with a hydrogen sulfide (rotten eggs) odor (the anaerobic decay of organic matter) indicates severe permanent waterlogging to the surface (either perched or from groundwater) (9).

Red, orange and yellow mottles in pale (sometimes colored) matrix, often in a marbled pattern, may well be relict features of previous weathering environments associated with fluctuating water tables. Such features are common to the so-called "lateritized" soil profiles on older geomorphic surfaces where current weathering environments are not responsible for the observed, dominant color patterns.

Other indications are given in Table 2.

REFERENCES

1. Simonson, R.W. Soil Color Standards and Terms for Field Use—History of Their Development. In *Soil Color*; Special Publication 31. Bigham, J.M., Ciolkosz, E.J., Eds.; SSSA: Madison, WI, 1993; 1–20.
2. Munsell Color Company, *Munsell Soil Color Charts*; Munsell Color Company: Baltimore, MD, 1975.
3. Hunt, C.B. *Geology of Soils. Their Evolution, Classification, and Uses*; W.H. Freeman: San Francisco, CA, 1972.
4. Post, D.F.; Bryant, R.B.; Batchily, A.K.; Huete, A.R.; Levine, S.J.; Mays, M.D.; Escadafal, R. Correlations Between Field and Laboratory Measurements of Soil Color. In *Soil Color*; Special Publication 31, Bigham, J.M., Ciolkosz, E.J., Eds.; SSSA: Madison, WI, 1993; 35–49.
5. Melville, M.D.; Atkinson, G. Soil Color: Its Measurement and Its Designation in Models of Uniform Color Space. J. Soil Sci. **1985**, *36*, 495–512.
6. Torrent, J.; Barrón, V. Laboratory Measurement of Soil Color: Theory and Practice. In *Soil Color*; Special Publication 31, Bigham, J.M., Ciolkosz, E.J., Eds.; SSSA: Madison, WI, 1993; 21–33.
7. Robinson, W.O.; McCaughey, W.J. *The Color of Soils*; USDA Bureau of Soils Bulletin 79, U.S. Govt. Printing Office: Washington, DC, 1911.
8. Schwertmann, U. Relations Between Iron Oxides, Soil Color, and Soil Formation. In *Soil Color*; Special Publication 31, Bigham, J.M., Ciolkosz, E.J., Eds.; SSSA: Madison, WI, 1993; 51–69.
9. Fitzpatrick, R.W.; McKenzie, N.; Maschmedt, D.J. Soil Morphological Indicators and Their Importance to Soil Fertility. In *Soil Analysis. An Interpretation Manual*; Peverill, K.I., Sparrow, L.A., Reuter, D.J., Eds.; CSIRO: Collingwood, Vic.

COMPACTION

William E. Wolfe
Ohio State University, Columbus, Ohio, U.S.A.

C

INTRODUCTION

A determination of the behavior of a soil mass subjected to structural loads is performed by a geotechnical engineer who evaluates soil properties and makes design recommendations. Satisfactory performance is usually achieved when ground movements are small enough that the structure placed on the prepared soil behaves the way the owners want it to. This is accomplished by specifying minimum performance requirements. Often, onsite soils are not ideally suited for the proposed project and the specified performance cannot be achieved unless the soil's properties are modified.

Modification can be accomplished in a number of ways. The two most commonly used classes of modification are chemical stabilization and mechanical stabilization. Chemical methods involve the addition of a stabilizing agent, usually lime or cement, to the soil. The chemically modified soil has a higher strength and lower compressibility than the untreated soil. Chemical additives have also been effective in reducing the swelling potential in clays. A mechanically modified soil has had its properties improved either by the inclusion of foreign material (usually reinforcing elements) or by densification. Densification is most commonly achieved through preloading or compaction. The two methods differ in the means employed to achieve the necessary increase in density. Preloading is used primarily to force water out of a saturated soil. Because preloading causes a reduction in volume by forcing water to flow away from areas of load, preloading a site requires the imposition of large loads over extended periods of time. The principles governing the rate and quantity of volume decrease accomplished by preloading are discussed in the section on consolidation.

In the compaction process, the volume of void space is reduced by applying high loads over small areas to force the air out of an unsaturated soil mass. Because the area covered by the compaction load is much smaller than the one used to preload a site, the volume of soil stabilized during each application of the compaction load is much less. So, to be successful, the treatment must be moved from place to place and applied to thin layers. The most common method involves driving heavy vehicles, a specified number of times (passes) over the soil surface as the soil is being placed. Typically, each 8–12 in. (20–30 cm) thick layer of the soil is compacted and the amount of compaction measured. Since the procedure specifically targets the air fraction of the soil mass, the water content is not affected by compaction (1).

COMPACTION PRINCIPLES

Although compaction may not change the water content of a soil, the maximum density obtainable does depend on the amount of water present in the soil mass. Although the engineering properties of all soils are affected by the soil density, the response of a coarse grained soil to compaction is different and needs to be considered separately from the response of a fine grained soil.

Coarse grained soil—because the individual particles in a coarse grained soil are chemically inert, there is little interaction between the solid and liquid fractions of the soil mass. The effect of water in the deposit is largely through the formation of capillary tensions, which result in the particles being tightly bound in a matrix that resists rearrangement. As the degree of saturation increases, the capillary forces are destroyed and compaction energy becomes more effective in densifying the soil. An example of a compaction curve obtained for a coarse grained soil is presented as Fig. 1. Vibratory loading has been found to be the most effective method for densifying coarse grained soils (2, 3).

Fine grained soil—compaction tests performed on fine grained soil samples have clearly shown that for every soil and compactive effort, an optimum water content yielding the greatest density of the solid fraction (dry density) can be identified. At water contents both above and below the optimum, the same compactive effort results in lower soil densities. Typical curves showing the dependence of density on water content and compactive effort are presented as Fig. 2.

Although the density decreases above and below the optimum water content, the behavior of the soil is much different on the two sides of the optimum condition. On

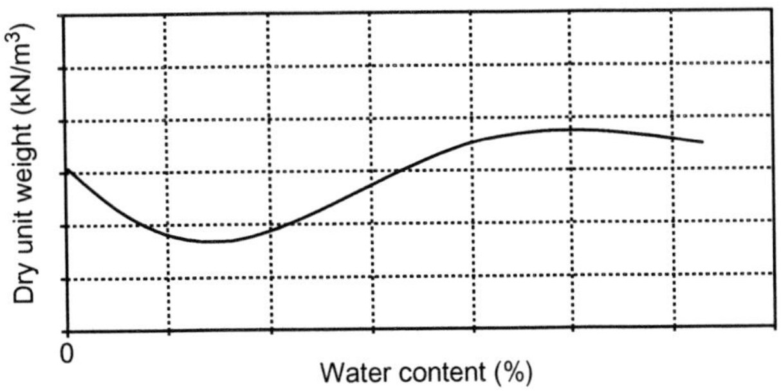

Fig. 1 Typical compaction curve for a coarse grained soil.

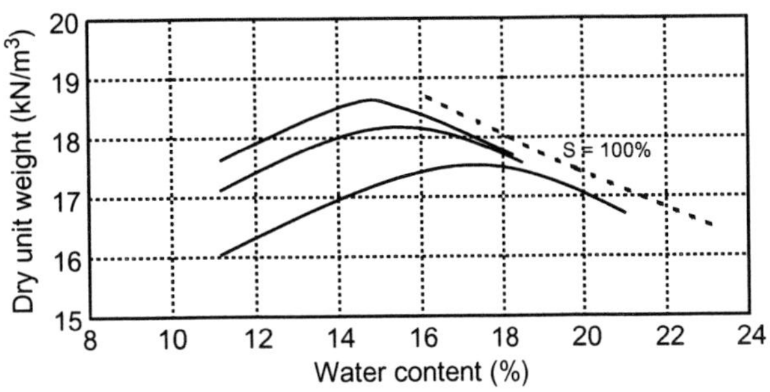

Fig. 2 Typical compaction curves for a fine grained soil.

the dry side of optimum, the electrical charges on the sheet like solid particles encourage an edge to face orientation of the individual soil grains with the resultant structure looking something like a honeycomb. This particle arrangement results in a fairly open structure, but one in which the high negative pressures in the water resulting from capillary tension cause the individual particles to be tightly bound together. As water content increases, the capillary tensions decrease. This decrease in capillarity reduces the force holding grains together and increases the effectiveness of the compactive energy. The particles are forced into more orderly patterns. As the water content continues to increase, the air volume fraction becomes small enough that air voids are discontinuous and so air cannot be easily expelled. The maximum density is quickly reached and further increases in water content merely increase the water adsorbed onto the solid grains reducing contact strength and density (3).

TESTING

It is the job of the geotechnical engineer to determine if the soil, whether onsite or imported, is strong or stiff enough to support the design loads. The most direct measure of the ability of a soil to perform as designed would be to measure the strength, stiffness, and/or hydraulic conductivity at many locations in the field. In most cases, field testing of these important engineering properties is not practical and an indirect measure of how the soil will behave when loaded is appropriate. Density and moisture content are the most common substitutes for the actual soil properties upon which design was based.

Laboratory Tests

The actual combination of maximum dry density and optimum moisture content is different for each soil. Therefore, a laboratory obtained moisture content versus density relationship such as one of those shown in Fig. 2 should be developed for each different soil or mixture of soils to be used on the site. A moisture-density curve is generated by plotting the results of five separate compaction tests performed on different samples of the same soil at five different moisture contents. Several test methods have been standardized by the American Society for Testing and Materials (ASTM) (4). The most commonly used are referred to as the Standard Proctor Test (ASTM D-698) and the Modified Proctor Test (ASTM D-1557), after R.R. Proctor, who developed the procedures for conducting repeatable compaction tests in the 1920s and 30s.

Ideally, the engineer then performs the appropriate engineering tests on laboratory samples compacted to the optimum moisture content and density. Once the data showing a correlation between soil density and its strength and stiffness have been generated for each soil type, the density and water content become surrogates for the true design parameters. Because these two parameters are easy to determine and can be related to the significant engineering properties, the measurement of density and moisture content has become the preferred method for verifying adequate preparation of the soil.

Field Tests

With the lab data obtained as described above, field verification of the appropriate density and moisture content is now the extent of the field testing necessary to be confident that the required strength and stiffness have been achieved. The two methods most commonly used to check density are the sand cone method (ASTM D-1556) and the nuclear gauge (ASTM D-2922).

Sand Cone

In this method, a test hole is excavated in the soil whose density is to be determined and all the excavated material

Fig. 3 Troxler Model 3450 moisture/density gauge in the direct transmission mode.

weighed. The hole is filled with a clean, uniform sand of a known density and the amount of sand used to fill the hole is determined. A fraction of the sample of the soil from the hole is dried in an oven at 105°C to determine the water content. The field density and water content can then be compared with the optimum conditions obtained in the laboratory.

Nuclear Gauge

The principle behind the use of the nuclear gauge is that measuring the attenuation of gamma radiation can accurately determine the total density and water content of a near surface soil. In the direct transmission method, a radiation source is placed in a pre-drilled hole, usually about 300 mm below the surface. A gamma radiation detector is placed on the ground surface. In the backscatter method, both the radiation source and the detector are placed on the ground surface (5). Figure 3 depicts a nuclear gauge (Troxler Model 3450) measuring the density of clay material in the direct transmission mode. In recent years, the nuclear gauge has become a very popular way to make density and water content determinations because it is much faster than other methods, less prone to operator error and, in the backscatter mode, is nondestructive, since no excavation needs to be made. However, since the method is an indirect measure of density, the instrument must be calibrated at the start of each day's work and the calibration must be kept in the permanent record for the gauge. In addition, the precision of even properly calibrated gauges may be adversely affected by a number of elements in the soil including organic matter. The U.S. Army Corps of Engineers (6) specifies readings from a nuclear gauge must be frequently compared with direct measurements for density (sand cone) and water content (conventional oven).

SUMMARY

Compaction is perhaps the oldest technique for enhancing the engineering properties of a soil mass. Experience in compacting a large number of soil types has shown that, in many soils, adequate engineering properties can be expected, if a minimum density that can be related to the laboratory Proctor density is achieved. This knowledge allows the geotechnical engineer to reduce the testing that must be done to establish the performance of compacted soil profiles.

REFERENCES

1. Hilf, J.W. Compacted Fill. In *Foundation Engineering Handbook*, 2nd Ed.; Fang, H.Y, Ed. Van Norstrand Reinhold Company: New York, 1991; 249–316.
2. Schroeder, W.L.; Dickenson, S.E. *Soils in Construction*, 4th Ed. Prentice Hall: Englewood Cliffs, New Jersey, 1996.
3. Lambe, T.W.; Whitman, R.V. *Soil Mechanics*; John Wiley and Sons: New York, 1969.
4. American Society for Testing and Materials ASTM Standards, Book of Standards Volume 04.08, Soil and Rock (I): D420–D5779, March 2000.
5. Department of the Army, US Army Corps of Engineers, Engineer Manual 1101-2-1906, Appendix VI, Compaction Tests, 20 August 1986.
6. Department of the Army, US Army Corps of Engineers, Engineer Manual 1110-2-1911, Construction Control for Earth and Rock-Fill Dams, 30 September 1995.

COMPUTER MODELING OF WATER MANAGEMENT

L.R. Ahuja
L. Ma
United States Department of Agriculture-Agricultural Research Service (USDA-ARS), Fort Collins, Colorado, U.S.A.

INTRODUCTION

Water is a vital resource for food and fiber production. Irrigated farmlands in the semi-arid to arid regions of the world contribute about 36 percent of the total production. As the human population increases, more irrigation water will be needed to meet the greater food demand. However, the supplies of useable and renewable fresh water are scarce. Agriculture is already facing a stiff competition for fresh water supplies from drinking, industrial, and recreational uses, even in the U.S. that has vast resources. The only answer is to conserve water through judicious management for all uses. In agriculture, the management of soil water based on current scientific principles can help to increase water use efficiency, reduce losses, and prevent secondary salinization of groundwater resources, a common problem in irrigated areas of the world.

WATER FLOW

The theories of water flow in the integrated soil–plant–atmosphere system or continuum have vital field applications for soil water management under both rainfed and irrigated agriculture. During the last 40 years, the increase in our basic understanding of soil water entry (infiltration), redistribution, retention, and uptake by plants has changed the old concepts of field capacity (water holding capacity) of soils and the water available for plant growth ingrained in the traditional soil water management for plant production. The field capacity is no longer considered a constant intrinsic property of soil, but rather a continually decreasing soil water content, albeit at a slow rate of internal drainage (1, 2). Factors such as the impeding or coarse-textured layers in the soil profile, pre-infiltration wetness, depth of the groundwater table, and surface evaporation rate, influence the field capacity value (3–5). Similarly, the concept that soil water is equally available to plants in the range of soil water contents from field capacity to permanent wilting (6) has been shown to be invalid (7) and replaced by a dynamic concept of water availability to plants. It is postulated that the rate of water uptake that would sustain normal plant growth at any time depends not only upon soil water status but also upon the atmospheric conditions and properties of the plants (8–10). Atmospheric conditions dictate the evapotranspiration (ET) demand, or the rate at which the plant is required to transpire and absorb water from the soil to maintain its turgidity. Rooting depth, density, proliferation, and extension, physiological adjustment of plant to water stress, and soil's unsaturated conductivities determine the soil water content at which the actual ET falls below the demand ET. These conditions also determine the water content or soil water suction at which the plants will wilt. This new dynamic concept of soil water availability has tremendous implications for soil water management. This concept requires that irrigation applications be more flexible, based on meteorological conditions, plant growth stage, root system, and soil properties.

The knowledge of water flow in the soil–plant–atmospheric continuum, needs to be combined with the knowledge of heat flow, nutrient processes, plant growth, and management effects in order to attain the best scientific management of soil water. However, the interactions among these components of the agricultural system are highly complex, dynamic, and difficult to comprehend. Fortunately, the modern technology has allowed us to develop computer models of the entire agricultural systems that incorporate these interactions (11–14). Therefore, computer models of this nature are useful as research and management guides for efficient use of water and nutrients under varying weather and climatic conditions. Modeling may be used for all areas of soil water management in a variety of ways—from designing an irrigation and drainage system for uniformity and effectiveness, comparison of soil and water conservation practices, to simulation of the entire soil–plant–atmosphere continuum on a day-to-day basis for making decisions on water, fertilizer, or pesticide applications. Although there are many gaps in our knowledge at present, the models are developed well enough to use them as guides for optimum use of resources. This article illustrates the applications of models in soil water management for plant production through examples. (Readers interested in the theories and assumptions

Encyclopedia of Soil Science

Table 1 Management practices, timing, and application options in RZWQM

Management practice	Timing options	Application options
Tillage operations	Plant growth cycle Specified date	29 Different implements User specified intensity
Manure applications	Plant growth cycle Specified date	Surface broadcast Broadcast incorporated Injected fumigation Chemigation 15 Different types of manure available
Fertilizer applications	Plant growth cycle Specified date Preplant with topdress Preplant and topdress with plant demand trigger	Surface broadcast Broadcast incorporated Injected NH_3-N Fertigation Nitrification inhibitor additives
Pesticide applications	Plant growth cycle Specified date	Surface broadcast Broadcast incorporated Injected fumigation Chemigation Slow release formulations Foliar or soil surface only
Crop planting and harvesting	Multiple year same crop Multi-species rotations	5 Harvesting operations External crop models can be used
Irrigation applications	Fixed interval Specified dates Root zone depletion	Fixed amount Varying amounts Refill root zone

Table 2 Average yearly summary for manure and water management based on RZWQM

Management option	Manure	Water applied (cm/event)	Silage yield (Mg ha^{-1} yr^{-1})	Nitrogen uptake (kg N ha^{-1} yr^{-1})	N-seepage (kg N ha^{-1} yr^{-1})	Denitrification (kg N ha^{-1} yr^{-1})	Volatilization (kg N ha^{-1} yr^{-1})
1	Full rate every year	20	25.2	254	196	130	11.1
2	Half rate every year	20	21.9	182	106	48	3.5
3	Full rate every two years	20	22.8	192	91	57	4.7
4	Full rate every three years	20	19.0	153	83	34	3.2
5	Full rate, split application	20	25.1	250	188	140	7.1
6	Full rate every year	10	25.8	307	116	141	11.1
7	Half rate every year	10	22.7	228	49	50	3.5
8	Full rate every two years	10	22.9	232	39	59	4.7
9	Full rate every three years	10	20.1	186	40	35	3.2
10	Full rate, split application	10	25.6	302	109	150	7.1

Full manure application rate was 44.8 Mg ha^{-1} and half rate was 22.4 Mg ha^{-1}.
Manure was applied on October 15 for one application and October 15/April 1 for split application (22.4 Mg ha^{-1} each).
(From Ref. 16.)

underlying these models, and the extent of their validation should see the cited references.)

EXAMPLES OF APPLICATION

The USDA-ARS Root Zone Water Quality Model (RZWQM) is an agricultural system model with a Windows-based graphical user interface. It has been developed and used for a wide range of agricultural management problems, including both water quantity and quality (14, 15). Table 1 shows the various management practices simulated in RZWQM that affect soil water management for plants. The model can be used to address questions such as the amounts and timings of irrigation water, fertilizer, and pesticides applications under various tillage and cropping schemes in different soil types (Table 1). Table 2 shows the results of an example application of RZWQM for water and manure management in eastern Colorado from Ma et al. (16). The field has been used for corn silage production for more than a decade with manure application every fall after harvesting. Irrigation water was applied four to six times during the months of July and August at a rate of 20 cm per event. Management Option 1 was the simulation result under current management practices. As seen in Table 2, reducing water irrigation to 10 cm per event not only increased yield and N uptake, but also decreased N losses.

Another example application of RZWQM is to answer questions on when and how much to irrigate. A common practice is to apply irrigation at a fixed time interval, and a fixed amount of water for each irrigation. This practice disregards the water actually used by the crop in response to the atmospheric evapotranspiration demand. A more scientific way is to link irrigation timing and amount to soil water contents in the root zone. For example, the irrigation is triggered when the soil water storage falls below a certain lower limit of the so-called available water, arbitrarily defined as soil water content in equilibrium with 33 kPa (1/3 bar) water suction minus the soil water content at 1500 kPa suction. The irrigation is stopped when the soil water storage is filled to a certain upper limit of available water. What the lower and upper limits of available water for triggering and stopping irrigation should be depends upon the soil type, crop, and climatic conditions in the area. A system model will be a good guide to determine optimal limits for an area. For the eastern Colorado site noted previously, on a Vona sandy loam soil, RZWQM simulated average corn silage yields, N (nitrogen) uptake and N leaching at different irrigation amounts are shown in Fig. 1. The lower limit of available water to trigger irrigation was varied from 10% to 90%, and the upper limit to stop irrigation from 50% to 90%. The irrigation period lasted up to 100 days after planting. These results indicate that the irrigations triggered when the soil water content falls to about 20% of the available water and stopped when the soil water storage is filled to about 50% of available water gave the best results—maximum N uptake and silage yield, and minimum nitrate-N leached.

A third example is the use of the Gossym model for cotton crop management from Lemmon (12). Gossym was

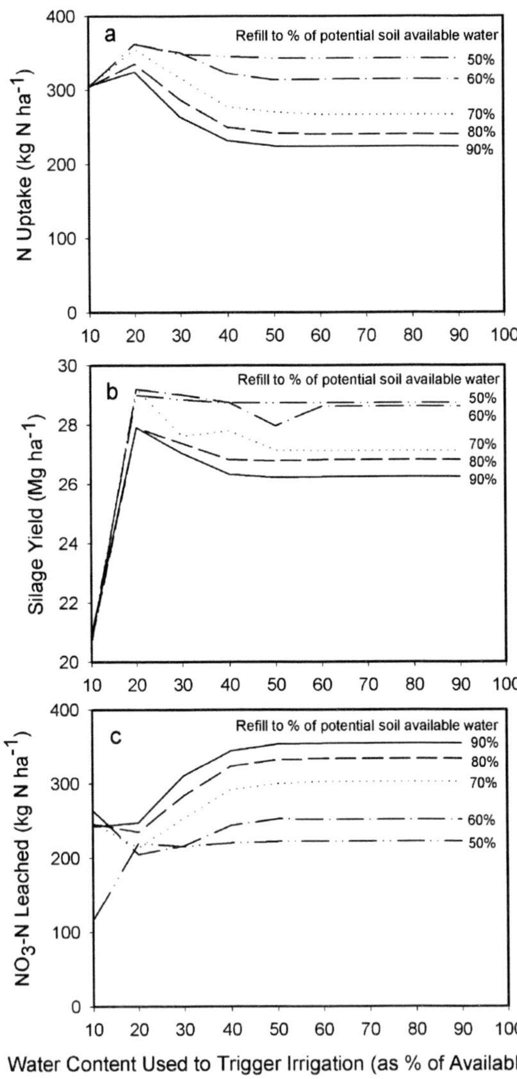

Fig. 1 The responses of yearly average N uptake, silage yield, and nitrate-N leached to irrigation according to degree of root zone water depletion, as simulated by RZWQM. (RZWQM was calibrated by Ma et al. From Ref. 16.)

developed by a team of scientists in Mississippi and South Carolina, USA (17). It simulates the growth and development of cotton on an organ-by-organ basis (roots, stems, leaves, blooms, squares, and bolls), as well as the transport and uptake of water and nutrients through the soil profile. The model is driven by daily weather variables, such as solar radiation, rainfall, and maximum-minimum air temperatures, and requires input data on soil properties and soil fertility. Gossym is used to determine irrigation schedules, N requirements, and crop maturity dates. An example of how Gossym simulation results are used is shown in Fig. 2 (12). The N and water stress values are ratios of the amount of N or water used to the amount needed for full growth at different times. The first row of graphs was produced after the third N application. These results show N and water stresses and their projected effects on yield organs and yield of cotton. The second row of graphs is the simulation results of increasing irrigation

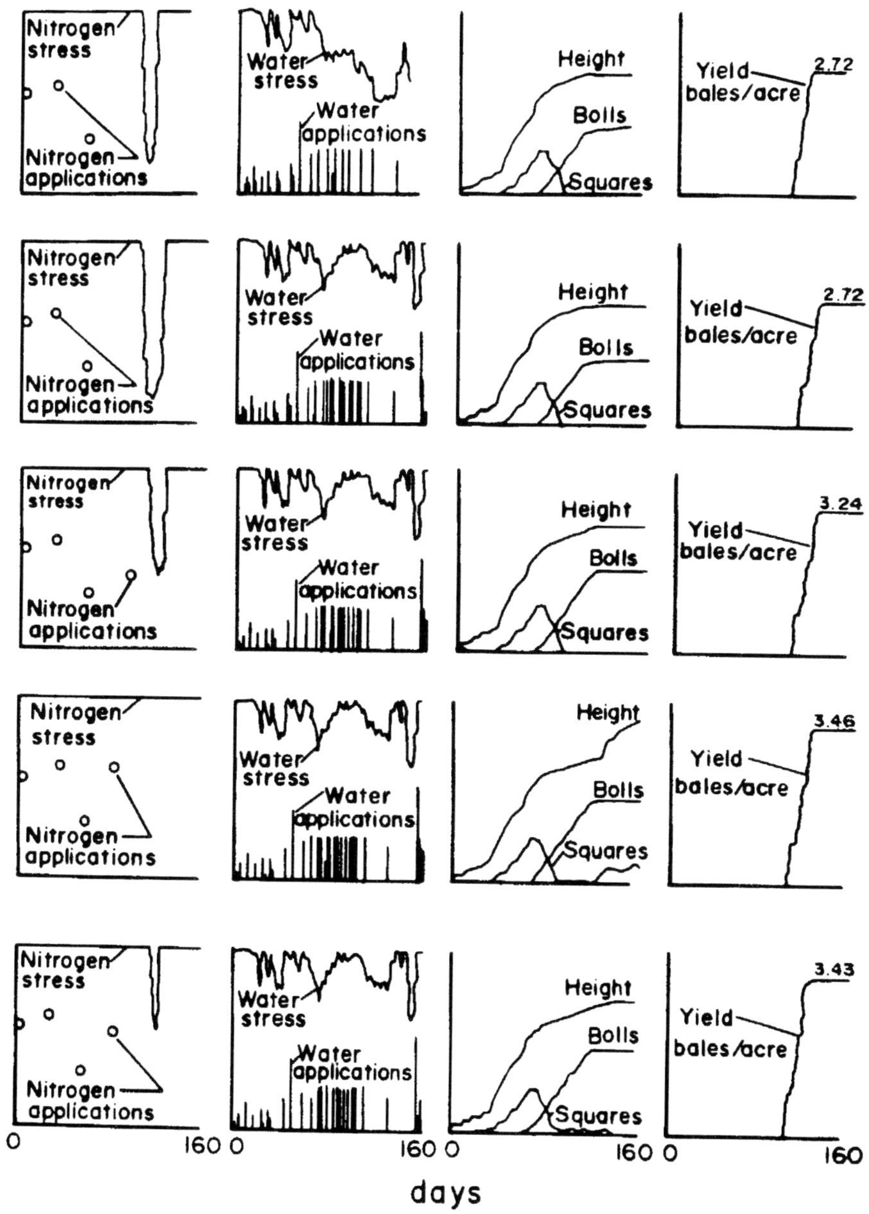

Fig. 2 Graphs produced by Commax from the results of Gossym simulations, showing the process whereby Commax reduces the water stress and then the N stress. (From Ref. 12.)

schedule to reduce water stress. However, increased irrigation results in increased N stress. The third row shows the simulation result of adding a small N application (fourth N application), which reduces N stress and increases yield. In the fourth row, N application was increased for the fourth N application, which eliminated N stress. However, this increased application also resulted in some new vegetative growth and squares near the harvest time. Thus, the fifth row presents optimal results with a moderate N application rate. These simulation results have convinced Mississippi cotton growers to apply additional N that was not planned originally, and resulted in increased yield.

THE FUTURE

The results presented show the value of the models to guide soil water management for plant production in complex agricultural systems. At the dawn of the 21st century, computers are playing a central role in managing industries, businesses, and many other aspects of life. Their use in managing water and other natural resources and environmental quality is on the increase. It is time that computer models of agricultural systems be integrated with all field research, so that the models continually capture new knowledge as it becomes available. The models should then be used to guide agricultural management under varying climatic and weather conditions.

REFERENCES

1. Richards, L.A.; Gardner, W.R.; Ogata, G. Physical Processes Determining Water Loss from Soils. Soil Sci. Soc. Am. Proc. **1956**, *20*, 310–314.
2. Ogata, G.; Richards, L.A. Water Content Change Following Irrigation of Bare Field Soil that is Protected from Evaporation. Soil Sci. Soc. Am. Proc. **1957**, *21*, 355–356.
3. Miller, D.E. *Flow and Retention of Water in Layered Soils*, Conserv. Res. Rep. 13; USDA-ARS: Washington, DC, 1969.
4. Biswas, T.D.; Nielsen, D.R.; Biggar, J.W. Redistribution of Soil Water After Infiltration. Water Resour. Res. **1966**, *2*, 513–514.
5. Ratliff, L.F.; Ritchie, J.T.; Cassel, D.K. Field-Measured Limits of Soil Water Availability as Related to Laboratory Measured Properties. Soil Sci. Soc. Am. J. **1983**, *47*, 770–775.
6. Veihmeyer, F.J.; Hendrickson, A.H. Soil Moisture Conditions in Relation to Plant Growth. Plant Physiol. **1927**, *2*, 71–78.
7. Richards, L.A.; Wadleigh, C.H. Soil Water in Plant Growth. In: Soil Physical Conditions and Plant Growth. Agronomy 1952, *2*, 13.
8. Penman, H.L. The Dependence of Transpiration on Weather and Soil Conditions. J. Soil Sci. **1949**, *1*, 74–89.
9. Philip, J.R. The Physical Principles of Soil Water Movement During the Irrigation Cycle. 3rd Congr. Int. Comm. Irrig. Drain., 1957, Question 8:8, 125–8.154.
10. Gardner, W.R. Dynamic Aspects of Water Availability to Plants. Soil Sci. **1960**, *89*, 63–73.
11. Watts, D.C.; Hanks, R.J. Soil-Water-Nitrogen Model for Irrigated Corn on Sandy Soils. Soil Sci. Soc. Am. J. **1978**, *42*, 492–499.
12. Lemmon, H. Commax: An Expert System for Cotton Crop Management. Science **1986**, *233*, 29–33.
13. Shaffer, M.J., Larson, W.E., Eds. *NTRM: A Soil-Crop Simulation Model for Nitrogen, Tillage, and Crop Residue Management*, Conserv. Res. Rep. 34–1; USDA-ARS: Washington, DC, 1987.
14. Shaffer, M.J., Larson, W.E., Eds. *Root Zone Water Quality Model: Modeling Management Effects on Water Quality and Crop Production*; Water Resources Publications, LLC, Littleton, CO, 2000, 372
15. Ma, L.; Ahuja, L.R.; Ascough, J.C., II; Shaffer, M.J.; Rojas, K.W.; Malone, R.W.; Cameira, M.R. Integrating System Modeling with Field Research in Agriculture: Applications of the Root Zone Water Quality Model (RZWQM). Adv. Agron. **2000**, *71*, 233–292.
16. Ma, L.; Shaffer, M.J.; Boyd, J.K.; Waskom, R.; Ahuja, L.R.; Rojas, K.W.; Xu, C. Manure Management in an Irrigated Silage Corn Field: Experiment and Modeling. Soil Sci. Soc. Am. J. **1998**, *62*, 1006–1017.
17. Baker, D.N.; Lambert, J.R.; McKinion, J.M. A Simulator of Cotton Crop Growth and Yield. South Carolina Agric. Exp. Stn. Tech. Bull. 1983, 1089.

CONSERVATION TILLAGE

C

Jack Cline

Department of Animal Sciences, The Ohio State University, Columbus, Ohio, U.S.A.

Robert Hendershot

United States Department of Agriculture-Natural Resources Conservation Service (USDA-NRCS), Lancaster, Ohio, U.S.A.

INTRODUCTION

History of Conservation Tillage

Conservation tillage has been around since humans first planted a seed by making a hole in the ground or spreading seeds on the surface of the soil. Conservation tillage can greatly reduce water- and wind-driven soil erosion. In spite of its long history, the science of conservation tillage is relatively recent. The relationship between plant residues and the forces of soil erosion and soil quality parameters has been studied in depth only during this generation. As used in this article, conservation tillage is an agronomic production system that leaves at least 30% of the soil surface covered with plant residue after the planting process. Laflen (2) showed that 30% crop residue cover reduces soil erosion by 60%. This conservation practice can be utilized on both cropland and grazing lands. Protecting the soil from erosion is important for the maintenance of long-term productivity of the soil resource base. Conservation tillage also improves the quality of the soil by potentially increasing the soil organic matter and generally improving the soil quality.

Concept of Conservation Tillage

Conservation tillage or crop residue management systems range from no-till or direct-seeding methods to reduced tillage systems that disturb the soil but leave the required amount of crop residue on the soil surface after planting. Equipment used for no-till planting till only a narrow strip of soil or push aside the residue so the planting can be completed with good soil-to-seed contact. Forage and cereal crops can be broadcast on the soil surface where they rely on the forces of nature such as frost or rain to incorporate the seeds into the soil. Reduced tillage equipment utilizes chisel plows, field cultivators or disks that incorporate a portion of previous crop residue. The number of trips, speed, depth and type of point used on the cultivation tool all affects the amount of crop residue incorporated. Sweeps and straight chisel points leave more residue on the surface than twisted chisel points.

SOIL EROSION

Soil erosion is a natural process that has been a major problem for all civilizations since recorded time began. Fifty years ago, Lowdenmilk (3) pointed out the tremendous impact soil erosion has had on great civilizations around the world. The loss of soil productivity and the effect the resulting sediment from the eroding fields of both cropland and grazing lands has, in many cases, hastened or even caused the demise of some of these past societies. Modern governments have begun to recognize that many practices lead to the accelerated soil erosion that affects the productivity of the land unit and the displaced sediment causes many other problems. Many now recognize the importance of sustaining the soil resources not only for food and fiber production, but also for the economic savings to communities. Sediment removal is a costly endeavor for many communities and governments in order to maintain drainage ditches, irrigation canals, shipping and boating channels, water storage reservoirs, and lakes for flood control. Sediment also lowers water quality and increases water treatment costs for drinking and other uses. Deposition of eroded soil interferes with the utilization of the land resources and often requires removal or incorporation into the existing soil.

THE ECONOMICS OF CONSERVATION TILLAGE

Conservation tillage has been a major part of the conservation program in the U.S. since the 1970s. It is

Encyclopedia of Soil Science

the primary conservation practice used to control soil erosion on cropland. Conservation tillage is also used to control soil erosion on grazing lands when establishing a forage crop or introducing a new forage species to an existing pasture or rangeland. Conservation tillage systems generally require more management and have less opportunity for adjustment in decisions, but they are considered to be the most cost-effective tillage systems due to the often significant reduction in fuel and labor costs as well as investment in machinery.

CONSERVATION TILLAGE EFFECTS ON SOIL QUALITY

Conservation tillage improves soil quality by improving soil structure and soil tilth by increasing the soil organic matter. Tilling the soil oxidizes organic matter and exposes it to erosion. The less tillage performed the more organic matter will be in the soil profile. Humus is the stable portion of the soil organic matter. Humus aids in binding soil aggregates together and increases the cation exchange capacity. These strong soil aggregates create a porous soil that allows both water and air to move freely. This porous condition is ideal for plant root growth. The increase in soil organic matter increases the water holding capacity of the soil, increases the water infiltration rate, and decreases runoff (7). Soil organisms flourish under conservation tillage systems (5). The more plant residue left in the field the more soil organisms there are in the soil. Bacteria, fungi, algae, protozoa, nematodes, earthworms, and insects are all working in the soil profile breaking down organic matter and providing a biodiverse community. The plant residues provided by conservation tillage feed the soil biology.

SOIL CONSERVATION BY GRASSLANDS

In addition to the many techniques described above for conserving soil and water resources, the use of forages, either as a permanent soil cover or in some rotational system with crops for direct human use, may be the most effective of all techniques for soil and water conservation. Historically and prehistorically, natural grasslands and shrub lands have supported large populations of plant eaters (herbivores), which have grazed and/or browsed these resources. These plant eaters have then been the source of food for a variety of carnivores (meat eaters) and omnivores (mixed diet of plant and meat). Grazing lands include the tundra of the arctic regions, the great temperate region grasslands, and the savannas of the tropical regions. These growing plants convert solar energy by the process of photosynthesis into chemical energy (primarily carbohydrates) and then some is transformed by complex biochemical processes into lipids, amino acids and proteins, and vitamins, which are the nutrients needed by all multicellular life.

GRASSLAND UTILIZATION

The most abundant carbohydrates in nature are the structural compounds of cellulose and hemicellulose and in mature plants increasing levels of the derived lignin. Higher animals do not digest these compounds, but the anaerobic intestinal microorganisms can degrade cellulose and hemicellulose to usable end products; while lignin though somewhat degraded, cannot be utilized at all. Herbivores have an enormous population of these anaerobic microorganisms either in the forestomachs (before the true gastric stomach therefore called pregastric digesters) or in the caecum and colon and therefore called postgastric digesters. These microorganisms produce short-chain fatty acids, gases, and microbial cells. The fatty acids provide 60–80% of the host animal's energy, the gases are waste products and the microbial cells are digested to amino acids, carbohydrate, fats, minerals, vitamins or excreted in feces, depending on pre or postgastric anatomy. Thus, these structural carbohydrates not digestible by humans and higher animals are converted into usable products like meat, milk, cheese, hides, and wool for human utilization and benefit.

GRASSLAND HISTORIC BENEFITS

In the natural environment, wild herbivores were migratory and many species traveled as much as 1200 to 1500 miles on their migratory routes. This meant that grazing pressure was heavy for a relatively short period as the animals passed through, but that extensive periods without grazing allowed plants to recover and maintain high levels of root-energy reserves. These grasslands therefore accumulated very significant organic matter in the soils and most of the high organic matter soils of the world developed under grasslands (which is an important mechanism in sequestering carbon). The notable exception to this statement is a forest growing in swamps or bogs where the leaves and small branches, etc., fall into water, is an anaerobic or reduced-oxygen environment, as opposed to the normal dry aerobic environment where oxidative processes re-release the sequestered carbon with significantly reduced yearly gain. Arctic areas have low oxidation rates for much of the year because of low temperatures and therefore reduced rates of organic matter degradation.

OVERGRAZING PROBLEMS

Domestication and fencing of land have destroyed migratory patterns in much of these grasslands and range lands. This is especially true in those areas of sufficient rainfall for relatively high productivity. The tendency therefore has been to graze continually, inevitably resulting in the depletion of root energy reserves, reduced soil organic matter during much of the year, and, therefore, reduced rates of organic matter production. The density of plant cover is reduced and the time needed for plant growth to recover after grazing is increased. These processes increase the probability and quantity of soil erosion and reduce the efficiency of water infiltration into the soil. Continuous overgrazing is one of the major causes of desertification and either reduction of animal concentration or reduction in the length of the grazing period is necessary to prevent overgrazing.

RESEARCH

Long-term research has been conducted at the USDA North Appalachian Experimental Watershed near Coshocton, Ohio, where precipitation, surface water runoff, and sediment loss in that runoff have been measured. In Table 1, which is composed of data from two articles by Owens et al. (1, 4), precipitation, surface runoff and sediment load were measured. Data from an unimproved pasture ungrazed (2 years), grazed only during the grazing season (3 years), and grazed during the growing season and winter hay feeding (6 years) were taken and summarized here. A similar sloped wooded watershed nearby was also sampled for the same measures. Note that by comparing the 3-year data, it is apparent that less surface runoff occurred from unfertilized pasture grazed during the growing season only (203 mm/yr) than from a similar sloped, wooded watershed nearby (261 mm/yr). Also note for the 6-year data presented by Owens et al. (1, 4) there was less annual surface runoff (137 mm/yr) from pasture (grazed and imported winter fed hay) than from the wooded watershed (173 mm/yr). The lower values for runoff recorded in the pasture watershed as compared to the wooded watershed occurred even though precipitation levels in the pasture watershed were higher than the wooded watershed. Note also that the data from the storm of September 13, 1979, record a significantly higher sediment load was carried in surface water runoff from the wooded watershed (1032 kg/ka) as compared to the load from the grazed pasture watershed (739 kg/ha), even though surface runoff from the pasture was 6 mm greater.

MANAGEMENT INTENSIVE GRAZING (MIG)

Subsequent research studied moderately fertilized and heavy fertilized pasture with rotational grazing as well as continual grazing. Many interesting conclusions reached from these studies and others in the eastern and midwestern areas of the U.S. have led to the development of what are commonly called Management Intensive Grazing (MIG) systems. MIG systems are described below. The principles followed in MIG are designed to easily meet the animals' nutrient needs and thus their productivity levels while at the same time to keep pasture plants in the immature or boot-pre bud stage of growth; when root reserves have been restored, rapid regrowth can be expected. To accomplish this goal, sufficient cells or grazing paddocks are needed so that animals remain in individual paddocks no more than 3 days to prevent animals from grazing new regrowth, and so that at least 3–4 in. of leaf remain, although regrowth is rapid. The plant species used are a function of moisture supply,

Table 1 Sediment transport and supporting hydrologic data

	Unimproved pasture watershed				Wooded watershed		
	2 year	3 year	6 year	13/9/79[a]	3 year	6 year	13/9/79[a]
Precipitation, mm/yr	991	1204	1108	116	1120	1034	121
Surface runoff, mm/yr	123	203	137	55	261	173	49
Sediment, average annual, kg/ha	228	934	2088	739	347	301	1032

[a] Data from the storm of September 13, 1979—the worst storm in more than a century.

latitude, fertility, etc., but the MIG system increases plant diversity and maximizes ground cover over time.

MIG BENEFITS—CONSERVATION AND ECONOMICS

The advantages of MIG include greatly reduced soil erosion with the improved air and water quality this entails. MIG greatly improves plant vigor, production, and diversity and improved fish, game bird, and wildlife habitat. The increased organic matter accumulating in soils allows a higher percentage of rainfall to infiltrate where it can be used to stimulate plant growth rather than runoff and increase sedimentation problems in streams. The animal wastes (manure and urine) are more evenly distributed by MIG than other grazing systems, allowing those nutrients to be recycled through the plants to the animals without getting into runoff water. In research performed at the North Appalachian Experimental Watershed, water quality in pastured watershed was found to be as good or better than both surface and groundwater from an adjacent woodland watershed. In total, grazing lands properly used are one of the best tools to conserve water and soil resources, to sequester significant quantities of atmospheric CO_2 (6) and to provide an excellent economic return to the producer by converting forages into higher valued animal products for human use.

REFERENCES

1. Owens, L.B.; Edwards, W.M.; Van Keuren, R.W. Surface Run-Off Water Quality Comparisons Between Unimproved Pasture and Woodland. Journal of Environmental Quality **1983**, *12*, 518–522.
2. Laflen, J.M. Tillage and Residue Effect on Erosion from Cropland. In *Proceedings of the Natural Resources Modeling Symposium, 1983*; DeCoursey, D.G., Ed.; October 16–20 USDA–ARS: Pinegree Park, CO, 1985; 438–441.
3. Lowdenmilk, W.C. *Conquest of the Land Through 7000 Years*; Agriculture Information Bulletin Number 99, Reprint of the 1953 Report; Natural Resources Conservation Service, USDA, 1999.
4. Owens, L.B.; Edwards, W.M.; Van Keuren, R.W. Sediment and Nutrient Losses from an Unimproved, All Year Grazed Watershed. Journal of Environmental Quality **1989**, *18*, 232–238.
5. Soil and Water Conservation Society, Farming for a Better Environment a White Paper Soil and Water Conservation Society, 1995.
6. Owen, L.B.; Hothem, D.L. Carbon Stored in Soils Under Eastern Grasslands. *Proceedings of the Second Eastern Native Grass Symposium, Baltimore MD*; 1999; 231–234.
7. Swan, J.B.; Lowery, B.; Cruse, R.; Kasper, T.; Lindstrom, M.J.; Moncrief, J.F.; Staricka, J. Interactions of Crop Residue with Soil and Climate. In *Crop Residue Management to Reduce Erosion and Improve Soil Quality*; USDA–Agricultural Research Service Conservation Research Report Number 42, North Central Region, Moldenhauer, W.C., Mielke, L.N., Eds.; 1995; 73–77.

CONSISTENCE

Ray McBride
University of Guelph, Guelph, Ontario, Canada

C

INTRODUCTION

Soil consistence and consistency are closely related terms used by pedologists and soil engineers, respectively, to describe the degree of resistance to deformation or rupture exhibited by a structured soil when subjected to externally-applied mechanical stresses. This aspect of a soil's rheological behavior, however, has only ever been characterized with semiquantitative criteria. One of these criteria is that the test soil should be under one or more prescribed (and stated) conditions of initial soil wetness (e.g., dry, moist, wet), as these properties are highly dependent on soil water content (Fig. 1). These soil properties are a manifestation of the combined forces of adhesion/cohesion that contribute to: 1) the attraction between soil particles and pore water, and 2) the attraction of soil particles to one another. As a result, particle-size distribution, organic matter content, clay mineralogy, soil solution chemistry (adsorbed cations), soil fabric, sample pretreatment, etc. also have a large influence on these properties. Classes or ratings of soil consistence are commonly determined by pedologists with simple manual field tests that involve application of compressive forces to a structurally-intact soil specimen (e.g., clod) until *rupture* occurs. Conversely, soil consistency classes or ratings are routinely used by soil engineers in the classification of soils used for engineering purposes,[a] but in this case the field tests usually involve defining the soil resistance to *penetration* by a blunt object (e.g., thumb/thumbnail; blunt end of a pencil).

SOIL CONSISTENCE

Soil profiles are routinely characterized for their consistence by horizon as part of a soil morphological description, which can aid in making pertinent soil survey interpretations for common land uses (e.g., agriculture) where knowledge of soil strength is an important consideration (e.g., tillage, traffic). Simple field tests for soil consistence at prescribed water contents have been developed by the USDA-SCS (1), and have been adopted by pedologists in many other countries including Australia (2) and Canada (3).

The principal field test involves compressing an unconfined, block-like specimen between the extended thumb and forefinger, between both hands, or between the foot and a nonresilient flat surface (Table 1) (1). Rupture resistance to compressive stress is indirectly related to the tensile strength (i.e., the load per unit area at which an unconfined cylindrical specimen will fail in a simple tension test) (4). Table 1 includes approximate values of applied force/energy for the various classes. If induration with cementing agents is suspected (e.g., organic substances, calcite, silica, sesquioxides), the specimen is air-dried and then submerged in water for at least one hour before manual compression testing takes place (Table 1). Apart from brittle fracture (rupture), however, soil specimens can fail in different manners (e.g., plastic deformation, flow, smearing) and meaningful categories can also be determined manually in the field (1) (Fig. 1).

Other related tests that can be performed on structured, confined soil include resistance to penetration with a commercially-available, inexpensive "pocket penetrometer" (i.e., the pressure required to push a 6.4 mm diameter, cylindrical rod a distance of 6.4 mm into the soil within 1 second) (1). Related field tests on puddled, unconfined soils include plasticity (permanent deformation at constant volume without rupturing as force is applied), toughness (manual force needed to perform the plastic limit roll test), and stickiness (capacity of a soil to adhere to other objects). The plasticity and toughness tests now used by pedologists were originally derived from soil engineering practice (5).

SOIL CONSISTENCY

Soil consistency (or Atterberg) "limits" are gravimetric soil water contents that correspond to arbitrary boundaries defining the liquid, plastic, and semisolid states as measured by simple laboratory tests (6) (Fig. 1). The term

[a]See Plastic Properties on pages 1005–1007 and Atterberg Limits on pages 89–93.

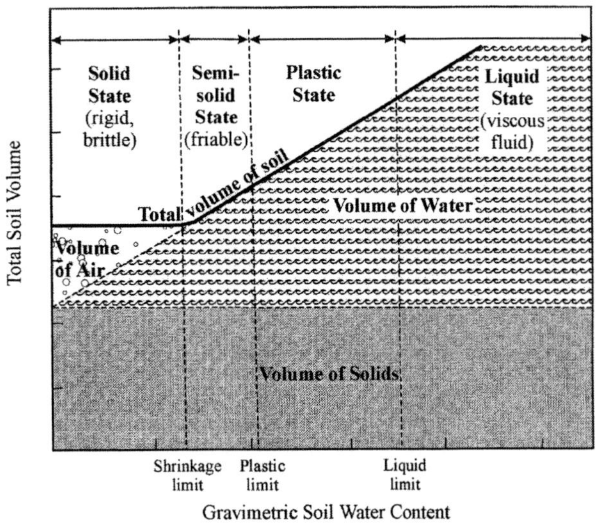

Fig. 1 Generalized schematic of consistency states and shrinkage behaviour of clay-rich soils.

consistency, however, was originally used by soil engineers with reference to a simple field test of soil resistance to penetration by a thumb or thumbnail when the specimen was at field water content (5). To date, there has been no attempt by the ASTM to standardize their test method using a "pocket penetrometer" or similar device (5), but approximate values of undrained shear strength have been assigned to consistency classes of cohesive soils (Table 2). Soil engineering tests for cementation (5) are performed manually by applying pressure on structured, unconfined specimens as in Ref. 1, but do not involve submersion in water prior to testing.

Only soils with significant clay contents (i.e., "plastic" or "cohesive" soils) have measurable Atterberg limits, and they are amongst the most meaningful and widely interpreted of soil engineering test indices. Applications include being used to estimate the shear strength and bearing capacity, compressibility, permeability, swelling potential, and specific surface of soils.

Table 1 Terms for describing consistence (rupture resistance) of block-like specimens

Soil consistence classes at prescribed soil wetness conditions			Field test description	
Moderately dry and very dry	**Slightly dry and wetter**	**Air dried, then submerged**	**Operation**	**Approx. force/energy applied[a]**
Loose	Loose	Not applicable	Specimen not obtainable	—
Soft	Very friable	Noncemented	Fails under very slight force applied slowly between thumb and forefinger	<8 N
Slightly hard	Friable	Extremely weakly cemented	Fails under slight force applied slowly between thumb and forefinger	8–20 N
Moderately hard	Firm	Very weakly cemented	Fails under moderate force applied slowly between thumb and forefinger	20–40 N
Hard	Very firm	Weakly cemented	Fails under strong force applied slowly between thumb and forefinger (80 N is about the maximum force that can be applied)	40–80 N
Very hard	Extremely firm	Moderately cemented	Cannot be failed between thumb and forefinger but can be between both hands or by placing on a nonresilient surface and applying gentle force underfoot	80–160 N
Extremely hard	Slightly rigid	Strongly cemented	Cannot be failed in hands but can be underfoot by full body weight (approx. 800 N) applied slowly	160–800 N
Rigid	Rigid	Very strongly cemented	Cannot be failed underfoot by full body weight but can be by <3 J blow	800 N–3 J
Very rigid	Very rigid	Indurated	Cannot be failed by blow of <3 J	≥3 J

[a] Both units of force (Newtons, N) and energy (joules, J) are used.
(From Ref. 1, pp. 174–175.)

Table 2 Terms for describing consistency (penetration resistance) of cohesive material at field water content

Soil consistency term	Field test description	Approx. undrained shear strength ($kN\,m^{-2}$)
Very soft	Easily penetrated several cm by fist	<20
Soft	Easily penetrated several cm by thumb	20–40
Soft to firm	–	40–50
Firm	Moderate effort needed to penetrate by thumb	50–75
Firm to stiff	–	75–100
Stiff	Great effort needed to penetrate by thumb	100–150
Very stiff	Readily indented by thumbnail	150–200
Hard	Very difficult to indent with thumbnail	>200

(Adapted from Ref. 5.)

REFERENCES

1. Soil Survey Division Staff, *Soil Survey Manual*; USDA-SCS Agricultural Handbook 18, U.S. Government Printing Office: Washington, DC, 1993.
2. McDonald, R.C.; Isbell, R.F.; Speight, J.G.; Walker, J.; Hopkins, M.S. *Australian Soil and Land Survey Field Handbook*, 2nd Ed.; CSIRO Land and Water Division, Inkata Press: Melbourne, Australia, 1998.
3. Expert Committee on Soil Survey, *The Canada Soil Information System (CanSIS): Manual for Describing Soils in the Field*; Land Resource Research Institute (LRRI) Contribution No. 82–52, Day, J.H., Ed.; Research Branch, Agriculture Canada, Supply and Services Canada: Ottawa, Ontario, Canada, 1982.
4. SSSA, *Glossary of Soil Science Terms*, 1996 Ed.; Soil Science Society of America: Madison, WI, 1997.
5. ASTM, Standard Practice for Description and Identification of Soils (Visual–Manual Procedure) (D2488–93). *2000 Annual Book of ASTM Standards*; American Society for Testing and Materials: West Conshohocken, PA, 2000; Vol. 04.08, 249–259.
6. ASTM, Standard Test Methods for Liquid Limit, Plastic Limit, and Plasticity Index of Soils (D 4318–98). *2000 Annual Book of ASTM Standards*; American Society for Testing and Materials: West Conshohocken, PA, 2000; Vol. 04.08, 546–558.

CONSOLIDATION

Tarunjit Singh Butalia
The Ohio State University, Columbus, Ohio, U.S.A.

INTRODUCTION

Consolidation is the reduction in volume of soil caused by movement of water out of a soil mass. The term is commonly referred to as primary compression. The rate of consolidation is an important property of a soil mass because water has to move in the direction of decreasing potential for consolidation to occur. It is the movement of water rather than the compression of air filled voids that distinguishes consolidation from compaction. The magnitude of consolidation for coarse-grained soils (sand and gravel) is much less compared to fine-grained soils (clay and silt). However, because fine-grained soils exhibit low coefficient of permeability, the rate of consolidation in fine-grained soils is much slower than coarse-grained soils.

THEORIES AND ANALYSES

Consolidation analysis generally focuses on saturated soils. When an external load is applied on soil or if the vertical stresses increase for saturated soils, the pore water pressure increases immediately by an amount equal to the increase in applied stress. Gradually, some of the pore water moves out of the soil mass in the direction of decreasing potential, thus transferring the stresses from the water particles to the soil skeleton. This results in reduced volume of the soil. During the consolidation process, the volume of the soil particles remains unchanged, while the volume of the water filled voids is reduced.

One of the first soil consolidation laboratory experiments was presented by Frontard (1). He placed clay samples in metal containers and loaded the samples with a piston. Meantime, Forchheimer (2) was the first to propose a mathematical model for consolidation analysis. While the mathematical formulation was simple, it was not very accurate due to incomplete understanding of the concept of effective stress. Karl Terzaghi (3), one of Forchheimer's former students, was the first person to recognize clearly the concept of effective stress and its application to the soil consolidation problem.

Terzaghi's theory of consolidation, relating soil compression and effective stress as well as the rate of consolidation, is considered to be the advent of modernsoil mechanics. Terzaghi's simple theory assumed a saturated clayey soil, the void ratio–logarithm of effective stress relationship to be linear, and that the soil properties do not change during the consolidation process. The analysis was carried out for one-dimensional (axial) settlement with no radial deformations. For a homogeneous soil layer of thickness H with initial void ratio e_0, the amount of one-dimensional consolidation settlement due to Δe decrease in void ratio is

$$\Delta H = \frac{\Delta e H}{1 + e_0} \tag{1}$$

In most cases, the decrease in void ratio is generally expressed in terms of compression index or coefficient of compressibility and change in effective stress. The rate of consolidation is affected by several factors including soil, stratum, and loading properties. Soil properties of interest include soil permeability, void ratio, and compressibility. Stratum factors include the thickness of the layer under consideration as well as the drainage conditions surrounding the layer. Loading factors include the ratio of the new loading to original loading. The soil properties are commonly modeled through the coefficient of consolidation (c_v) as

$$c_v = \frac{k(1 + e)}{a_v \gamma_w} \tag{2}$$

where k is the coefficient of permeability, e is the in-place void ratio, a_v is the coefficient of compressibility, and γ_w is the density of the pore fluid (typically water). The governing equation of one-dimensional diffusion as applied to transient water flow

$$\frac{\partial u}{\partial t} = c_v \frac{\partial^2 u}{\partial z^2}$$

is solved to obtain the rate at which consolidation occurs.

Odometer or one-dimensional consolidation laboratory test (ASTM D2435) (4) is used to evaluate the consolidation characteristics of a soil sample. The test involves applying varying levels of stress on a soil sample that is confined in the radial direction and allowed to drain and compress axially (refer to Fig. 1). Data collection for

Fig. 1 Consolidation test apparatus.

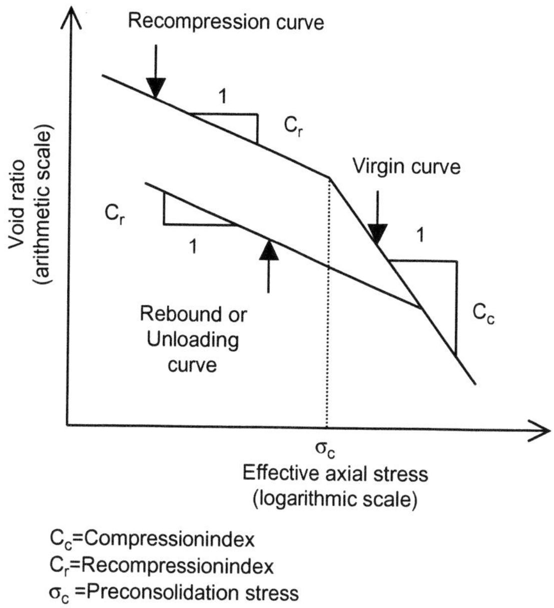

Fig. 3 Idealized void ratio vs. axial effective stress plot.

the test mainly consists of axial compression vs. time readings for each stress loading and unloading process (refer to Fig. 2). The ultimate void ratio for each loading and unloading step can then be plotted against the logarithm of the vertical (axial) effective stress (refer to Fig. 3). The slope of the virgin curve is referred to as the compression index (C_c) while the slope of the recompression or unloading curve is referred to as the recompression

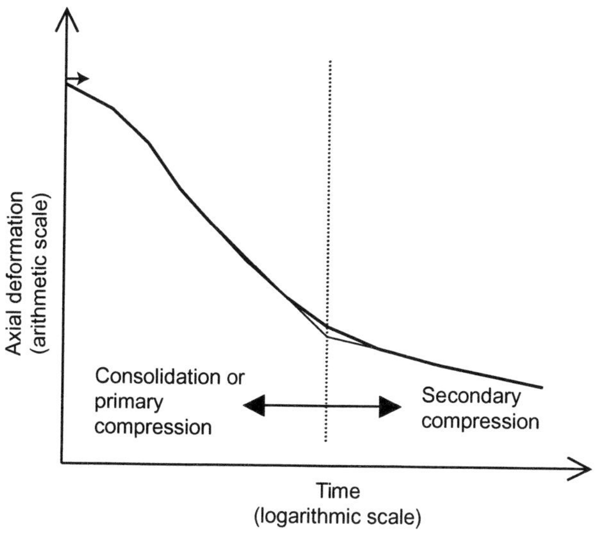

Fig. 2 Idealized axial compression vs. time plot.

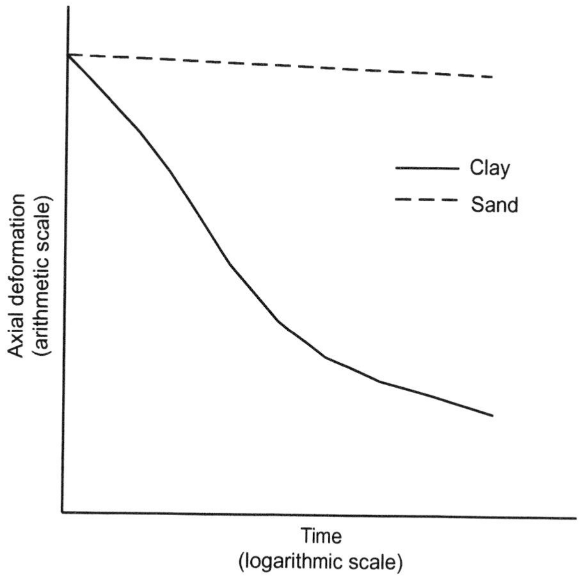

Fig. 4 Typical axial compression vs. time plot for clay and sand at specified effective axial stress.

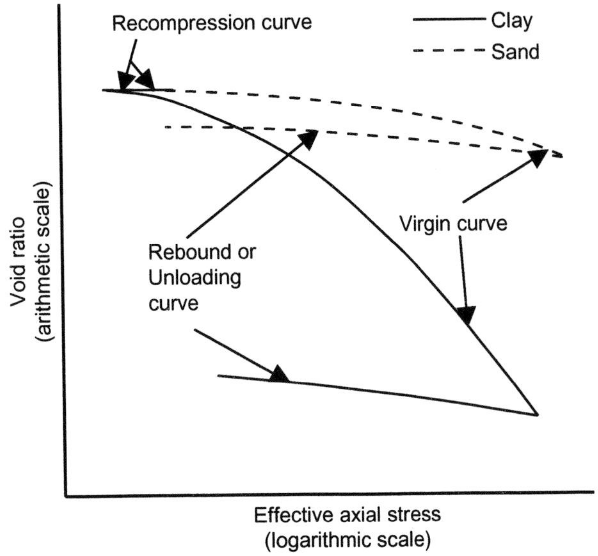

Fig. 5 Typical void ratio vs. axial effective stress plot for clay and sand.

index (C_r). Typical consolidation test curves for clay and sand samples are shown in Figs. 4 and 5. Compression and recompression index values less than 0.2 represent soils with low compressibility. Values between 0.2 and 0.4 indicate moderately compressible soils, while values larger than 0.4 represent highly compressible soils. Accurate laboratory evaluation of coefficient of consolidation values can be challenging due to its dependence on permeability coefficient (k), coefficient of compressibility (a_v), and potential sample disturbance. The average coefficient of consolidation varies from 10^{-8} to 10^{-6} m^2/s. For controlled-strain loading testing of soil samples, test method ASTM D4186 (5) should be used.

Terzaghi's theory and the one-dimensional laboratory experiment based on it are used extensively by engineers for consolidation analysis. However, the stringent soil behavior and modeling considerations in Terzaghi's theory limit the use of the theory for predicting pore water pressure responses and settlements (amount and rate) in complex situations. Alternatively, nonlinear consolidation models proposed by Schiffman (6), Gibson (7), Tse (8), Scott (9), and Leroueil (10) can be used. Many of these models allow for a nonlinear variation of void ratio with logarithmic effective stress and the change in soil permeability as consolidation occurs. Additional technical information on soil consolidation can be obtained from Wu (11), Mitchell (12), Terzaghi (13), Coduto (14), McCarthy (15), and DeBoer (16).

REFERENCES

1. Frontard, J. Notice Sur l'accident de la Digue de Charmes. Ann. Ponts et Chausees **1914**, *23* (9), 173–280, in French.
2. Forchheimer, P. *Hydraulik*; Leipzig, 1914; 494–495 (in German).
3. Terzaghi, K. *Erdaumechanik auf Bodenphysikalischer Grundlage*; Deuticke: Vienna, 1925.
4. ASTM D2435, *Standard Test Method for One-Dimensional Consolidation Properties of Soils*; Book of Standards, American Society for Testing and Materials, 2001; Vol. 04.08.
5. ASTM D4186, *Standard Test Method for One-Dimensional Consolidation Properties of Soils Using Controlled-Strain Loading*; Book of Standards, American Society for Testing and Materials, 2001; Vol. 04.08.
6. Schiffman, R.L. The Use of Visco-Elastic Stress–Strain Laws in Soil Testing. ASTM Spec. Tech. Publ. **1959**, *254*, 131–155.
7. Gibson, R.E.; Schiffman, R.L.; Cargill, F.W. The Theory of One-Dimensional Consolidation of Saturated Clays. II. Finite Non-Linear Consolidation of Thick Homogenous Layers. Can. Geotech. J. **1981**, *18*, 280–293.
8. Tse, E.C. Influence of Structure Change on Pore Pressure and Deformation Behavior of Soft Clays Under Surface Loadings PhD Dissertation Department of Civil Engineering, University of California, Berkley.
9. Scott, R.F. Consolidation of Sensitive Clay as a Phase Change Process. J. Geotech. Engng, ASCE **1989**, *115* (10), 1439–1458.
10. Leroueil, S.; Magnan, J-P.; Tavenas, P. *Embankments on Soft Clays*; Ellis Horwood Series in Civil Engineering, D.M. Wood (Trans.), Ellis Horwood: New York, 1990.
11. Wu, Tien H. *Soil Mechanics*; Allyn and Bacon, Inc.: Worthington, OH, 1982.
12. Mitchell, James F. *Fundamentals of Soil Behavior*; Wiley: New York, 1993.
13. Terzaghi, Karl; Peck, Ralph B.; Mesri, Gholamreza *Soil Mechanics in Engineering Practice*; Wiley: New York, 1996.
14. Coduto, Donald P. *Geotechnical Engineering: Principles and Practices*; Prentice Hall: Upper Saddle River, NJ, 1999.
15. McCarthy, David F. *Essentials of Soil Mechanics and Foundations: Basic Geotechnics*; Prentice Hall: Upper Saddle River, NJ, 1998.
16. DeBoer, R.; Schiffman, R.L.; Gibson, R.E. The Origins of the Terzaghi–Fillunger Disputet. Geotechnique **1996**, *46* (2), 175–186.

CONTROLLED TRAFFIC

Randall C. Reeder
The Ohio State University, Columbus, Ohio, U.S.A.

C

INTRODUCTION

Controlled traffic is a method to manage soil compaction. Compaction is "managed" by assuring that all heavy traffic is confined to specific lanes through the crop, year after year. The lanes become compacted, and the soil between lanes is never driven on.

Controlled traffic is an ideal match for most conservation tillage systems. However, if the entire soil surface is tilled, a GPS-based guidance system would be required because the operator cannot see where the lanes were. With controlled traffic, field efficiency and crop yields increase and the cost of inputs (pesticides, seed, labor, fuel) decrease. Many farmers convert all of their machinery to controlled traffic at one time while others choose to reduce or eliminate the extra cost by moving to controlled traffic gradually as equipment is replaced. Maximum advantages of controlled traffic occur when the operating width of all equipment is matched and tire spacings are set to minimize the trafficked area.

The first step to controlled traffic is to make all equipment cover the same width, or multiples of that width. For big grain farms, the combine harvester will likely be the determining factor. For example, if the grain head is 36 ft (11 m), the drill or air seeder might be 35 ft (10.7 m), and the sprayer either 70 ft (21.3 m) or 105 ft (32 m) wide. For a farm with row crops (for example, corn in 30-in. (750 mm) rows), the combine could be 12-row, the planter either 12- or 24-row, and the sprayer either 60 ft (18.3 m) or 90 ft (27.4 m) wide. If the same farm raises small grains or soybeans, the drill, air seeder, and grain head must match these widths. For smaller operations, the basic width might be 15 ft (4.5 m) or 20 ft (6 m).

The second step is to look at the tires (or tracks) of all vehicles in the field. The goal is to minimize the number of traffic lanes and the width of those lanes. Any equipment (sprayer, cultivator) driven through a growing crop must use the traffic lanes and the tires should fit between the rows. A combine or four-wheel drive tractor with wide tires is satisfactory, but the goal would be to go to narrower tires (maybe split duals) using the same traffic lanes (Table 1).

Controlled traffic improves traction, flotation, and timeliness of planting, spraying, and harvesting while minimizing potential yield losses from compaction. Compaction is managed, not eliminated. Controlled traffic takes advantage of compaction where it is beneficial. Compacted soil under wheel tracks provides better flotation and improved traction, especially on wet soil. Controlled traffic eliminates compaction where it is a detriment (under and close to rows) and leads to better water infiltration, root development, and fertilizer uptake.

Controlled traffic eliminates overlaps and skips during pesticide and fertilizer application and while seeding crops. Overlaps often waste 10–15% of chemicals, drilled seed, fuel, and all other machine operation expenses. Precisely placed traffic paths can eliminate the waste and more than offset the expense of establishing and maintaining the controlled traffic system.

NO TILL

One concern farmers have about no till is the potential for compaction and ruts caused by driving on wet soil. Controlled traffic removes most of that concern. Controlled traffic provides for potential yield increases by restricting compaction to trafficked lanes. The stalks or skip rows remaining from the previous year's crop provide visual guidance at planting (Fig. 1).

RIDGE TILL

Controlled traffic is an integral part of a ridge till system and is responsible for part of the yield advantages. All tires must be narrow enough and spaced properly to fit between the rows. Driving on the row with a combine or other heavy load significantly reduces yield in those rows the following year. If the ground is frozen solid, driving on the row does not harm the soil.

MULCH TILL

Controlled traffic is less convenient with mulch till systems, unless the controlled traffic pattern is maintained

Table 1 Example traffic patterns for row crops

No. of rows	Tractor (in.)	Combine (in.)	No. of paths	% Trafficked
30-in. (750 mm) row spacing				
6	60	120	4	44
6	120	120	2	22
8	120	120	2	17
8	60 and 120	120 and 180	6	50
12	60	120 (6-row)	4	22
16	60 and 120	120 and 180 (8-row)	8	33
24	60 and 120	120 and 180 (12-row)	12	33
36-in. (900 mm) row spacing				
6	72	144	4	37
8	72	144	4	28
12	72	144 (6-row)	4	18

Example wheel spacings for tractor and combine, single and dual tire options. Percent of the field trafficked assumes each path is 20 in. (500 mm) wide.

with a GPS-based guidance system. Otherwise, a type of strip tillage is required to leave evidence of the previous year's traffic paths. Tillage implements could be adapted to controlled traffic by removing shanks that follow the tire tracks. The tillage implement would have to be the same width as the planter or drill.

BENEFITS

Timeliness of planting and harvesting are major advantages of controlled traffic. Compacted paths provide a firm base for tractor and combine tires. Flotation and traction are both improved. Although not recommended, farmers with controlled traffic sometimes drive through standing

Fig. 1 Permanent traffic lanes are easy to follow at planting.

water to plant without sinking or destroying soil structure. With controlled traffic, the absorption of water by the soil is greater except in the traffic lanes. Because the soil can store more water in the root zone, yields may improve in dry seasons, and runoff and erosion may be reduced.

Benefits related to chemical application may outweigh any other. Precisely spaced traffic lanes eliminate overlaps and uneven guess rows. Traffic lanes make it easy to spray at night, which usually improves the efficiency of the plant protection product. Eliminating waste from overlaps, and reduced yields from any gaps in application, may quickly pay for the extra investment in controlled traffic.

Controlled traffic helps retain any long-term benefits of subsoiling to alleviate compaction. With controlled traffic, repeated subsoiling is not needed because root zones under row crops are not compacted. Otherwise, as few as two passes of a tractor in the spring for shallow tillage and planting after fall subsoiling may recompact soil in the tire tracks to the same porosity as in nonsubsoiled, nontrafficked areas. Wheel tracks become compacted even when relatively light machinery is used. With a typical plow-based cropping system, random trafficking would cover 75–90% of a field by the time planting is completed the second year, nearly eliminating the benefits of subsoiling.

Cotton research with no till has shown yield reductions caused by shallow compaction, especially in a layer 2–4 in. (50–100 mm) below the surface (1). This surface compaction resulting from random wheel traffic could be minimized by controlled traffic.

With an automatic guidance system, the operator can devote attention to a machine's performance instead of steering. The operator could literally look backwards 90% of the time. Ease of driving minimizes fatigue allowing for longer work days.

IMPLEMENTING CONTROLLED TRAFFIC

Controlled traffic can work with, but is not limited to the following systems:

- continuous row crops;
- row crops in rotation with drilled beans or grain (requires skip rows);
- continuous small grains (using skip rows).

Because the goal of controlled traffic is to drive on the least amount of soil, it is important that tire width be minimized and that wheels run in the fewest possible row middles (Table 1) (2).

Corn and Other Row Crops

Getting all wheels spaced properly is the major challenge with existing equipment. For corn and other crops planted in rows, machinery could include a grain cart, manure spreader, fertilizer spreader and sprayer. For continuous cotton, or continuous small grains, there are usually fewer machines to get aligned.

Cotton

Typical cotton rows are spaced at 38–40 in. (1 m). The wide rows allow adequate space for large tires. Where cotton is grown on beds or ridges, the maximum tire width would be about 10 in. less than the row spacing. A tire spacing of 80 in. (2 m) is a fairly easy adjustment for all equipment. Bigger tractors may have split duals spaced at 80 and 160 in. (2 and 4 m). Some cotton is grown in 30-in. (750 mm) rows making the wheel alignments similar to most corn production systems.

Cotton is often planted in the same row position year after year, so controlled traffic is ideal. With controlled traffic, planters and cultivators would likely be twice as wide as the cotton harvester. Harvesters are usually 4-row, with some 6-row units.

Small Grains

For small grain production, a major benefit from controlled traffic is in sprayer operation efficiency. Controlled traffic lanes allow the sprayer operator to drive accurately, even at night.

Fig. 2 An 8-row combine with single tires is a key part of one controlled traffic system.

Examples

An example of an optimum system is the tractor and combine on single tires with the same spacing, making only two tracks with 8-row equipment (Figs. 2 and 3). The section width of the tires should be several inches less than the row spacing (for example, 22 in. (550 mm) for 30-in. (750 mm) rows). Less than 20% of the soil would be trafficked, leaving more than 80% of the soil in good condition for root growth. Even a combine with split duals and a tractor with narrow wheel spacing (six tracks with 8-row equipment) would result in only about half of the soil being driven on. The rows between adjacent tracks would have a strip of uncompacted soil about 8 in. (20 mm) wide.

With narrow row spacings of 20 or 15 in. (50 or 38 mm), installing tires that can fit between the rows and still carry the load is a challenge. For example, a tractor with split duals for 30 in. (750 mm) rows would likely need split triples for the narrower row spacing. This results in four rows with tire traffic on both sides. An option with 15 in.

Fig. 3 The tractor and grain cart (and a sprayer) all have single tires spaced 10 ft on centers, matching the combine.

rows is to leave 30 in. spaces for tractor tires. Another is to keep the wide tires, realizing that for any operations after the crop emerges some damage will occur to the plants driven over. This last option may be better than the wide skip rows because of poor weed control in the unshaded space. Even a very low yield from the plants driven over is better than no yield (and weed problems) from a skip row.

All field equipment needs to cover multiples of the same width. Lighter implements, such as a sprayer or rotary hoe may cover double or triple the width of the planter or drill. The combine or cotton harvester might be one-half, or in some cases one-third, the planter width.

Break long fields in the middle for crop hauling unless a grain cart is used for on-the-go unloading. Any trip through the field should continue to the end, or to a cross road, instead of turning around a fraction of the way through.

PLANNING AHEAD

Adopting controlled traffic is not a simple change, but rather a transition that can take several years to complete. Consider controlled traffic in all major machinery buying decisions. Selecting wheels and tires to avoid running on rows adds little to the cost and can provide immediate benefits. Better yet, design a controlled traffic plan appropriate for your operation. Perhaps tires spaced 120 in. (3 m) would fit a farm with large machinery, and 80- or 90-in. (2 or 2.3 m) spacing would be ideal for a smaller farm. The combine header may define the operating width, with the planter being the same or, perhaps, double, the width of the combine. (With a precise controlled traffic pattern, the combine can conveniently harvest half the planter width, centered on guess rows.)

Here are tips to consider when buying new equipment:

- Choose a combine with tires that match the row spacing. If standard single tires are too wide, use split duals or tall single tires. Costs vary widely by manufacturer. Some combines require several mechanical modifications to accommodate tire changes, others do not. For ridge till, the maximum recommended tire width is 8 in. (20 mm) less than the row width. For no till or mulch till, tires could be wider, perhaps 4 in. (10 mm) less than row width, but rows with traffic on both sides may lose yield.
- Select a small-grain platform the same width as the corn head on combines.
- Buy a drill the same width as the planter.
- Choose pesticide sprayers and fertilizer applicators that cover the planter width, or double or triple that width.
- Match grain cart tire spacings with the combine. Plan to run the cart in the same tracks as the previous pass of the combine. This may require an extension on the combine unloading auger. In some cases, the grain cart could be towed by the combine.

REFERENCES

1. Burmester, C.H.; Patterson, M.G.; Reeves, D.W. *Challenges of No-Till Cotton Production on Silty Clay Soils in Alabama*; Conservation Tillage Systems for Cotton Special Report 169, University of Arkansas: Fayetteville, AR, 1995; 5–7.
2. Grisso, R.D.; Jasa, P.J.; Jones, A.J.; Peterson, T.A. *Equipment Wheel Spacing for Ridge-Till and No-Till Row Crops*; University of Nebraska Cooperative Extension EC96-780, University of Nebraska: Lincoln, NE, 1996; 1–6.

COPPER

Judith F. Pedler
David R. Parker
University of California, Riverside, California, U.S.A.

INTRODUCTION

Copper (Cu) is a metallic trace element with two naturally occurring isotopes (^{63}Cu and ^{65}Cu), and is essential for the growth of all life. Naturally present in all soils, Cu is found in excessive and potentially toxic concentrations in mine spoils, in waste products from industrial and agricultural activities, and in some agricultural soils due to historic use of Cu sprays (e.g., Bordeaux mix) for disease control. Copper is deficient in many soils around the world, and addition of Cu fertilizer is required for productive crop growth.

COPPER IN SOILS

The average total concentration of Cu in the Earth's crust is estimated to be 70 mg/kg, although levels of 20–30 mg/kg are prevalent in average soils (1). Common primary minerals include Cu sulfides, with Cu largely in the +I oxidation state, which dissolve by weathering processes. Secondary minerals of Cu(II) include oxides, carbonates (malachite), silicates, sulfates, and chlorides, most of which are relatively soluble. Copper(II) may substitute for Fe, Mg, and Mn in an assortment of minerals, especially silicates and carbonates (1).

The Cu^{2+} ion can form strong inner-sphere complexes, and is thus immobilized by carboxylic, carbonyl, or phenolic functional groups, even at low pH. Exchangeable and weak acid-extractable Cu represent a small percentage of total Cu in most soils. The bulk of the Cu is complexed by organic matter, occluded in oxides, and substituted in primary and secondary minerals. Organic matter and Mn oxides are the most likely materials to retain Cu in a nonexchangeable form in soils. Alkali extraction techniques that remove organic matter from soils usually release large fractions of the total soil Cu (2). Addition of organic matter to soils, and biological exudation of organic acids may both increase dissolved organic carbon, thus solubilizing Cu from mineral forms, increasing the total dissolved Cu in soil solution (3), but predictive models of humic acid binding of Cu in soil solution are generally inadequate (4). Overall, Cu is one of the least mobile of the trace elements, maintained in a form sufficiently available to plants but relatively resistant to movement by leaching (1).

Free Cu^{2+} in soil solution decreases with increasing pH, reaching a minimum above pH 10. In the absence of organic ligands, Cu speciation is dominated by free Cu^{2+}, and increasingly by carbonate and hydroxy complexes as pH rises above 6.5 (1). The dissolved organic carbon found in most surface soils has a strong affinity for Cu, but estimates of the percentage of soluble Cu that is organically complexed can vary widely.

COPPER AND PLANTS

Plant uptake of Cu appears to be directly related to the concentration of the free ion, Cu^{2+}, but may also be influenced by the total concentration in soil solution, including organic complexes (5). As with most trace metals, it is not known whether Cu is passively absorbed or actively taken up across the root-cell membrane (6). Rates of absorption are generally low, on the order of 1 nmole h^{-1} (g root dry weight)$^{-1}$ (7f). The activity of free Cu^{2+} required in nutrient solution for optimal plant growth is just 10^{-14}–10^{-16} M (5). Copper absorption is generally halted by metabolic inhibitors and uncoupling agents which disrupt the normal transmembrane potential (7).

Uptake of Cu is strongly affected by pH: increasing concentration of hydrogen ions decreases the absorption of Cu ions by plant roots (8). Uptake is also affected by the presence of Ca, and to a lesser extent by Mg, both of which compete with Cu for binding sites at the root plasmalemma (9). The effects of other trace metals on Cu uptake have frequently been seen as inhibitory (Zn), or stimulatory (Mn), but not necessarily under well-defined conditions (2).

COPPER AS AN ESSENTIAL ELEMENT

Copper is an essential element for plant growth and is a component of many enzymes, including plastocyanin, and thus is an indispensable prosthetic group in Photosystem 2. Cu-containing proteins are also important in respiration (cytochrome c oxidase is the terminal oxidase of the mitochondrial electron chain), in detoxification of superoxide radicals (superoxide dismutase), and in lignification (polyphenol oxidase). Ascorbate oxidase, which contains eight Cu^{2+} ions, has been proposed as an indicator of plant Cu status, although its relevant biological function has yet to be determined (10).

The critical concentration of Cu in shoot tissue for optimal growth does not vary greatly between plant species, ranging from 1 to 6 μg/g dry weight of young leaf tissue (11). Most crops are recorded as having a requirement of 3–5 μg/g (11). The average concentration of Cu in plant parts varies with age and with the level of Cu and N supply. Translocation of Cu to plant shoots increases with an increasing supply of N. In the xylem and phloem saps, Cu is probably complexed by amino acids (2). Copper is usually described as having "variable" phloem mobility in plants (2), as the retranslocation of Cu from older tissues is regulated by both Cu-supply and by N-status. Lack of a sufficiently long-lived radioisotope makes study of Cu transport and translocation problematic.

PLANT GROWTH ON COPPER-DEFICIENT SOILS

Copper deficiency most often occurs on organic soils where excessive leaching has occurred, or on calcareous sands. In general, crops grown on mineral soils with Cu contents less than 4 to 6 μg/g, or on organic soils with less than 20 to 30 μg/g are the most likely to suffer Cu deficiency (13), although this varies with specific soil type and the crop grown.

Symptoms of Cu-deficiency include chlorosis, necrosis, leaf distortion and terminal dieback, and are most evident in new leaf growth. Wilting can also occur, indicating structural weaknesses due to reduced lignification of the xylem elements. These symptoms are not entirely specific to Cu deficiency, and can be observed in plants under a variety of stresses. The most profound symptoms of Cu deficiency are those seen in the reproductive cycles of many sensitive species: delay of flowering, and/or reductions in seed and fruit yield as a consequence of sterile pollen or reduced floret numbers. Because these latter symptoms are not observed until maturity or harvest, Cu deficiency is often termed a "hidden hunger." Rice, citrus and cassava are sometimes referred to as indicator species that are more sensitive to Cu deficiency, but are still not reported to require more than 5–6 μg/g Cu to avoid Cu deficiency. Cereal rye and canola are crops more tolerant of Cu deficiency, requiring only 1–2 μg/g Cu (11).

Copper deficiency in legumes depresses nodulation and N_2 fixation, leading to N deficiencies. Unlike Mo and Co, there seems to be no specific Cu requirement for N_2 fixation in nodules beyond that required for plant growth and the production of carbohydrates (12).

Copper deficiency decreases polyphenol oxidase activity and thus lignification (14). Susceptibility of Cu-deficient plants to pathogenic attack may be increased due to reduced lignification of xylem elements, or due to impaired lignification in response to pathogenic invasion (wounding response). Application of Cu to soil, at rates too low to directly affect the pathogen, has controlled powdery mildew in wheat (15). It has also been suggested that, where Cu in soils is more than sufficient, the accumulation of additional Cu in roots provides a fungistatic defense against pathogens (16). Conversely, where Cu is deficient, roots are more vulnerable to pathogenic invasion.

There are genetic differences in the absorption of Cu by plant roots. Rye is able to take up significantly more Cu from soil than wheat, and is thus viewed as being more Cu-efficient. Tritcale, the wheat–rye hybrid, inherits the efficiency factor (7). Copper efficiency could be a useful trait in breeding crops for regions where soils are commonly Cu-deficient (17). However, the mechanism of the efficiency factor is not clear.

PLANT GROWTH ON HIGH-COPPER SOILS

As some plant species have adapted to soils of low Cu status, others have evolved tolerance to Cu-toxic conditions. The 16th century author, Agricola, wrote of indicator plants that grow on naturally Cu-rich soils. Natural revegetation of mine spoils has been shown to reflect rapid genetic evolution of Cu tolerance by grasses and other plants (14). There seem to be several possible mechanisms of Cu tolerance, although exclusion from shoots is a common feature. The exceptions are a few Cu-accumulator species which may contain in excess of 1000 μg/g Cu in shoot tissue (12). In other Cu-tolerant species, root compartmentation or immobilization of Cu may be achieved by immobilization in cell walls, by complexation with intracellular proteins, or by removal of Cu to the vacuole (14).

With nontolerant taxa, plant growth is likely to be depressed when Cu concentrations in the whole shoots exceed ~20 μg Cu g^{-1}. Symptoms of Cu toxicity include poorly developed and discolored root systems, reduced shoot vigor, and leaf chlorosis. Toxicity thresholds (e.g., for a 10% yield reduction) seem to vary widely, probably because of the low translocation of Cu from roots to shoots. Only when roots are overwhelmed by Cu rhizotoxicity does sufficient Cu reach shoots to affect growth and function. Other syndromes, such as Fe deficiency, can readily occur as secondary consequences of excess Cu (14).

Exclusion of Cu from the shoot protects photosynthetic activity, which is highly sensitive to excess Cu. Photosynthetic electron transport is blocked by high levels of free Cu, at the oxidizing side of Photosystem 2, and at the reducing side of Photosystem 1. Excess Cu supply results in reduced lipid content and noticeable changes in the fatty-acid composition of tomato roots and primary leaves, indicating enhanced activity of enzymes which catalyze lipid peroxidation (18).

Concentrated Cu sprays have historically been used to control foliar pathogens, especially in vineyard and orchard crops. These fungicidal sprays often included limestone to reduce their phytotoxicity, and to make them more rain-fast. The accumulation of Cu in the soils under these crops has not regularly caused Cu toxicity, which indicates the remarkable ability of most plant roots to accumulate Cu, while regulating its flow to the shoots. Both Cu deficiency and toxicity may result in non-specific symptoms of plant stress. Assessment of the Cu status of a soil, or of crop plants, is most accurate when soil type, soil history, and soil and plant analyses for Cu are all considered.

REFERENCES

1. McBride, M.B. Forms and Distribution of Copper in Solid and Solution Phases of Soil. In *Copper in Soils and Plants*; Loneragan, J.F., Robson, A.D., Graham, R.D., Eds.; Academic Press: Sydney, Australia, 1981; 25–45.
2. Loneragan, J.F. Distribution and Movement of Copper in Plants. In *Copper in Soils and Plants*; Loneragan, J.F., Robson, A.D., Graham, R.D., Eds.; Academic Press: Sydney, Australia, 1981; 165–188.
3. Sanders, J.R.; McGrath, S.P. Experimental Measurements and Computer Predictions of Copper Complex Formation by Soluble Soil Organic Matter. Environ. Pollut. **1988**, *49*, 63–79.
4. Robertson, A.P.; Leckie, J.O. Acid/base Copper Binding, and Cu^{2+}/H^{+} Exchange Properties of a Soil Humic Acid, an Experimental and Modelling Study. Environ. Sci. Technol. **1999**, *33*, 786–795.
5. Bell, P.F.; Chaney, R.L.; Angle, J.S. Free Metal Activity and Total Metal Concentrations as Indices of Micronutrient Availability to Barley (*Hordeum vulgare* L. Cv Klages). Plant Soil **1991**, *130*, 51–62.
6. Strange, J.; Macnair, M.R. Evidence for a Role for the Cell Membrane in Copper Tolerance of *Mimulus guttatus* Fischer Ex DC. New Phytol. **1991**, *119*, 383–388.
7. Graham, R.D. Absorption of Copper by Plant Roots. In *Copper in Soils and Plants*; Loneragan, J.F., Robson, A.D., Graham, R.D., Eds.; Academic Press: Sydney, Australia, 1981; 141–163.
8. Lexmond, Th.M.; van der Vorm, P.D.J. The Effect of pH on Copper Toxicity to Hydroponically Grown Maize. Neth. J. Agric. Sci. **1981**, *29*, 217–238.
9. Parker, D.R.; Pedler, J.F.; Thomason, D.T.; Li, H. Alleviation of Copper Rhizotoxicity by Calcium and Magnesium at Defined Free Metal-Ion Activities. Soil Sci. Soc. Am. J. **1998**, *62*, 965–972.
10. Maksymiec, W. Effect of Copper on Cellular Processes in Higher Plants. Photosynthetica **1997**, *34*, 321–342.
11. Reuter, D.J.; Robinson, J.B. *Plant Analysis: an Interpretation Manual*; CSIRO Press: Melbourne, Australia, 1997; 83–566.
12. Römheld, V.; Marschner, H. Function of Micronutrients in Plants. In *Micronutrients in Agriculture*, 2nd Ed.; Mortvedt, J.J., Cox, F.R., Shuman, L.M., Welch, R.M., Eds.; SSSA: Madison, WI, 1991; 297–328.
13. Jarvis, S.C. Copper Concentrations in Plants and Their Relationship to Soil Properties. In *Copper in Soils and Plants*; Loneragan, J.F., Robson, A.D., Graham, R.D., Eds.; Academic Press: Sydney, Australia, 1981; 265–285.
14. Woolhouse, H.W.; Walker, S. The Physiological Basis of Copper Toxicity and Copper Tolerance in Higher Plants. In *Copper in Soils and Plants*; Loneragan, J.F., Robson, A.D., Graham, R.D., Eds.; Academic Press: Sydney, Australia, 1981; 235–262.
15. Graham, R.D. Susceptibility to Powdery Mildew (*Erisiphe graminis*) of Wheat Plants Deficient in Copper. Plant Soil **1980**, *56*, 181–185.
16. Graham, R.D.; Webb, M.J. Micronutrients and Disease Resistance and Tolerance in Plants. In *Micronutrients in Agriculture*; Mortvedt, J.J., Cox, F.R., Shuman, L.M., Welch, R.M., Eds.; Soil Science Society of America: Madison, WI, 1991; 329–370.
17. Owuoche, J.O.; Briggs, K.G.; Taylor, G.J. The Efficiency of Copper Use by 5A/5RL Wheat-Rye Translocation Lines and Wheat (*Triticum aestivum* L.) Cultivars. Plant Soil **1996**, *180*, 113–120.
18. Ouariti, O.; Boussama, N.; Zarrouk, M.; Cherif, A.; Ghorbal, M.H. Cadmium- and Copper-Induced Changes in Tomato Membrane Lipids. Phytochemistry **1997**, *45*, 1343–1350.

CROP ROOT RESPONSE TO SOIL TEMPERATURE IN TEMPERATE REGIONS

Thomas C. Kaspar
United States Department of Agriculture (USDA), Ames, Iowa, U.S.A.

INTRODUCTION

In temperate regions, the average temperature of the soil profile warms from winter minimums to summer maximums as the growing season progresses. At the time of planting for most summer annual crops, a temperature gradient has developed in the soil profile with temperature decreasing with depth. As the soil surface absorbs thermal energy throughout the growing season, progressively deeper soil layers become warmer (1). Because temperature influences a wide variety of plant functions, it is not surprising that soil temperature affects root growth in many ways (1–5) and that expansion of crop root systems in temperate regions is limited by cool soil temperatures (1, 6). Root system expansion is a function of three temperature-dependent processes: growth, development, and orientation. Temperature affects root growth through its influence on root weight, root length, and root diameter. Root development is influenced by the effect of temperature on root initiation and root turnover. Lastly, temperature controls root orientation through its impact on the direction of root growth.

ROOT GROWTH

Root dry weight is the most commonly measured growth parameter. For most plant species the temperature response of root dry weight follows a typical temperature response curve (2; Fig. 1). The optimum, minimum, and maximum growth temperatures of the response curves differ among plant species and among genotypes within species. Generally, crop species that are winter annuals or cool-season species, such as oat (*Avena sativa* L.), rye (*Secale cereale* L.), wheat (*Triticum aestivum* L.), and rape (*Brassica napus* L.), have lower optimum, minimum, and maximum growth temperatures than crops such as cotton (*Gossypium hirsutum* L.) or maize (*Zea mays* L.), which are summer annuals.

Crop root length is an important factor in determining water and nutrient uptake. Root elongation responds to temperature in much the same way as root dry weight (7; Fig. 2). Similarly, plant species and genotypes within species differ in their response to soil temperature. Root length, however, often responds more strongly to soil temperature than root dry weight. For example, Loffroy et al. (8) found that cotton taproot length at a root temperature of 17°C was less than 40% of its length at 27°C, whereas root dry weight was 68% of that at 27°C. Presumably, this occurred because the mean taproot diameter was greater at 17°C than at 27°C.

Root diameter along with root length determines root surface area. Additionally, the amount of carbohydrate required for root system expansion is determined by root diameter. Unfortunately, the effect of temperature on root diameter is not clear. Part of the confusion arises because of the differences between mean root diameter of an entire root system and mean root diameter of the roots in a branching order of roots within a root system. Root branching orders are determined by where a root originates. For example, a first-order root originates from the seed, whereas second-order roots originate from first-order roots. Generally, higher-order roots have smaller diameters than the lower-order roots from which they originated. Thus, an increase in root initiation or branching results in more roots with smaller diameters. If temperature increases root initiation, then mean root diameter of the root system is likely to decrease. Another complicating factor is that root diameters of individual roots tend to decrease as they elongate. Therefore, the generally shorter roots at cooler temperatures would most likely have greater mean diameters than the longer roots grown at warmer temperatures. In summary, the mean root diameter of an entire root system will decrease with increasing temperature, if root elongation and root branching increase with increasing temperature. In contrast, the root diameter of individual roots from the same branching order and of equal length may have similar root diameters at both the higher and lower temperatures (1).

ROOT DEVELOPMENT

Root initiation is a developmental process that responds to soil temperature. The number of lateral roots of cotton and

Encyclopedia of Soil Science
Published 2002 by Marcel Dekker, Inc.

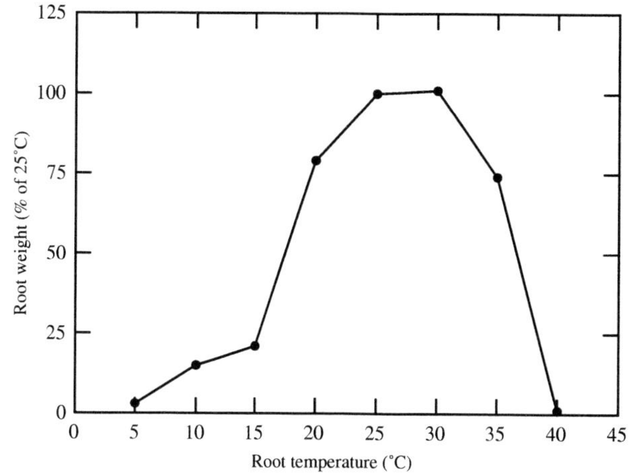

Fig. 1 Root temperature response of maize (*Zea mays* L.) root dry weight at 24 days after germination. (Redrawn from Ref. 2.)

sunflower (*Helianthus annus* L.) seedlings changed in response to soil temperature in a manner similar to that of root length or root dry weight (5). Number of lateral roots increased from zero at low temperatures to maxima at 27°C for sunflower and 35°C for cotton. Above the optimum temperature, the numbers of laterals decreased rapidly. Root initiation can also be described in terms of accumulated degree days or thermal time for temperatures below the optimum (9; Fig. 3), which is normally the case in temperate regions. Accumulated degree days are calculated by summing over a period of days the difference for each day between the daily mean temperature and the minimum temperature required for root initiation. Studies

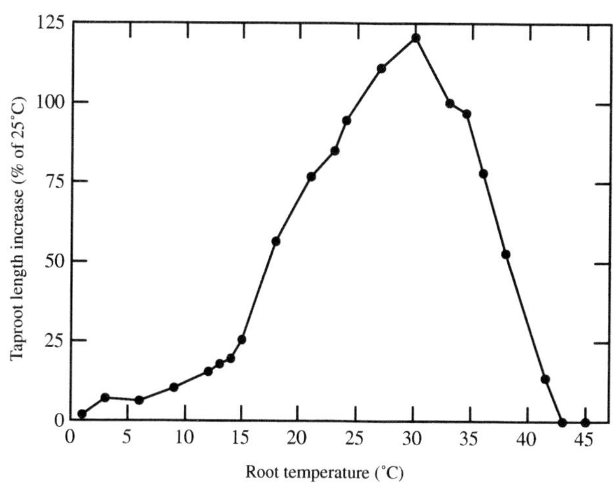

Fig. 2 Increase in taproot length of pecan (*Carya illinoensis*) seedlings after 4 days of growth at the treatment soil temperature. (Drawn from data from Ref. 7.)

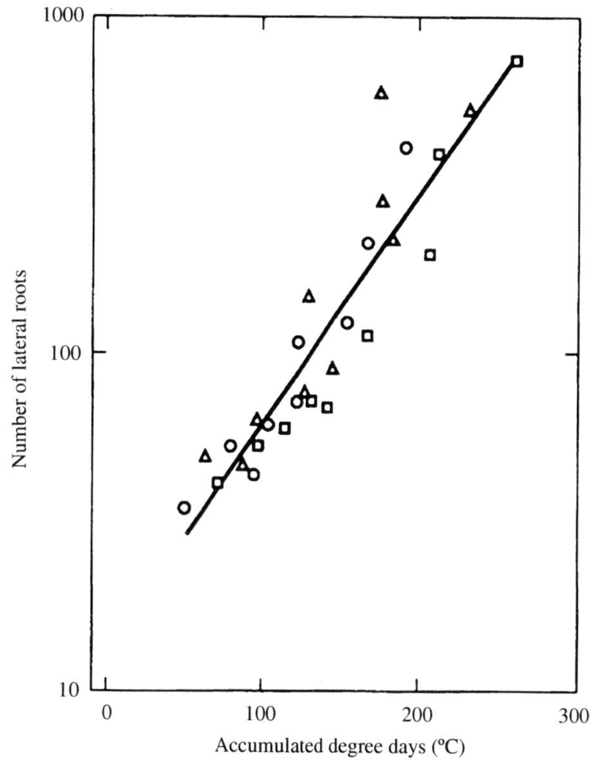

Fig. 3 The relationship between the number of lateral roots of pearl millet (*Pennisetum typhoides* S. & H.) and accumulated degree days (base 12°C) at the shoot meristem during the period 4–22 days after planting. (From Ref. 9.)

of pearl millet (*Pennisetum typhoides* S. & H.), for which the minimum temperature of initiation is 12°C, have found that the numbers of lateral roots and adventitious roots were strongly correlated to accumulated degree days (9). In other words, both the length of time and the temperature during the time period controlled the number of lateral and adventitious roots. Interestingly, the number of seminal roots of oat, wheat, and maize are not determined by soil temperature because the seminal root initials are already present in the seed embryos at planting (1).

There have been few studies that have examined the effect of soil temperature on root maturation and turnover. In general, roots at low temperatures remain white and succulent longer than roots grown at high temperatures, which become brown and shriveled more quickly. In one study, root longevity of several grass species was observed to be greater at a site with cooler soil temperatures than at a similar site with warmer soil temperatures (10). Root production and standing numbers of roots, however, were greater at the warmer site. Another indication of root maturation is the suberization of endodermal cells. For maize roots of similar length suberization occurred closer to the apex at higher soil temperatures, indicating that low temperatures delayed root maturation (1).

ROOT ORIENTATION

Roots of many plant species grow out from the plant main axis at a specific angle from the vertical rather than straight down. The angle at which these roots grow is temperature sensitive (1). In general, cooler temperatures tend to cause more horizontal growth and warmer temperatures result in more downward growth. Additionally, maize and soybean (*Glycine max* [L.] Merr.) genotypes differ in the temperature response and sensitivity of the growth direction of their roots (1).

ROOT RESPONSE TO SOIL TEMPERATURE GRADIENTS

In temperate regions, soil temperatures below the 30 cm depth are usually less than optimum for root growth of summer annual crops like cotton, maize, or soybean at the start of the growing season. As the growing season continues, the temperatures of the deeper soil layers continue to increase until canopy closure and decreasing air temperatures halt the process. Because warming of soil layers progresses downward from the surface, some deep soil layers may never reach the optimum temperature for root growth and at greater depths the soil may only warm to a few degrees above the minimum root growth temperature. As a result, the downward extension of crop roots may be limited by the cool temperatures of the subsoil layers (1, 6). A controlled environment experiment in which temperature gradients were established in soil columns has shown that the rate of downward growth of cotton and soybean roots is limited by temperature gradients and the rate at which soil warming progresses from the surface (6). These results corroborate field observations of soybean rooting depth increases that seemed to follow the 17°C isotherms downward from the surface (1).

REFERENCES

1. Kaspar, T.C.; Bland, W.L. Soil Temperature and Root Growth. Soil Science **1992**, *154*, 290–299.
2. Brouwer, R. *Influence of Temperature of the Root Medium on the Growth of Seedlings of Various Crop Plants*; Jaarb. Inst. Biol: Scheik, Wageningen, 1962; 11–18.
3. Cooper, A.J. *Root Temperature and Plant Growth A Review*. Research Review No. 4; Commonwealth Bureau of Horticulture and Plantation Crops; Agricultural Bureau: Farnham Royal, England, 1973; 1–73.
4. Bowen, G.D. Soil Temperature, Root Growth, and Plant Function. In *Plant Roots: The Hidden Half*, 1st Ed.; Waisel, Y., Eshel, A., Kafkafi, U., Eds.; Marcel Dekker, Inc.: New York, 1991; 309–330.
5. McMichael, B.L.; Burke, J.J. Temperature Effects on Root Growth. In *Plant Roots: The Hidden Half*, 2nd Ed.; Waisel, Y., Eshel, A., Kafkafi, U., Eds.; Marcel Dekker, Inc.: New York, 1996; 383–396.
6. Bland, W.L. Cotton and Soybean Root System Growth in Three Soil Temperature Regimes. Agron. J. **1993**, *85*, 906–911.
7. Woodroof, J.G.; Woodroof, N.C. Pecan Root Growth and Development. J. Agric. Res. **1934**, *49*, 511–530.
8. Loffroy, O.; Hubac, C.; da Silva, J.B.V. Effect of Temperature on Drought Resistance and Growth of Cotton Plants. Physiol. Plant. **1983**, *59*, 297–301.
9. Gregory, P.J. Response to Temperature in a Stand of Pearl Millet (*Pennisetum typhoides* S. & H.). III. Root Development. J. Exp. Bot. **1983**, *34*, 744–756.
10. Eissenstat, D.M.; Yanai, R.D. The Ecology of Root Lifespan. In *Advances in Ecological Research*, Begon, M., Fitter, A.H., Eds.; Academic Press, Ltd.: New York, 1997; *27*, 1–60.

CROP ROTATION AND FARMING SYSTEMS: SMALLHOLDER FARMING SYSTEMS AND AGROFORESTRY SYSTEMS

C

S. Hauser
L. Norgrove
International Institute of Tropical Agriculture, Mbalmayo, Cameroon

T. Nyobe
Institute de Recherche Agricole pour le Developpement, Nkolbisson, Yaounde, Cameroon

INTRODUCTION

Soil structure is the manner in which soil mineral particles and organic matter are spatially arranged into groups or aggregates plus the arrangement of voids between the solid material. It is a function of soil texture, clay type, the quality and quantity of soil organic matter inputs (1), and the way they are modified by the activity of soil biota (microflora, plant roots, and soil fauna). Soil structure is characterized by porosity, pore size distribution, and pore continuity. These can be assessed by measurements of penetrometer resistance, bulk density, infiltration rate, transmissivity, and aggregate stability.

SMALLHOLDER AGRICULTURE AND THE IMPORTANCE OF SOIL STRUCTURE

Most smallholders in the tropics practice shifting cultivation or other forms of "slash and burn" agriculture so these shall be the focus of this article. Slash and burn agriculture is a generic term for agricultural systems in which the fallow vegetation is manually slashed, left to dry and cleared from the field by burning before crop cultivation. After a cropping phase, the land is abandoned to a fallow phase. Later, the cycle is repeated (2).

Relative to undisturbed forest or fallow land, soil structure is generally degraded under slash and burn agriculture. In Brazil, compared with undisturbed forest, continuous sugar cane production for 20 years on both Oxisols and Alfisols resulted in reductions in wet aggregate stability at low water contents, 3-fold reductions in pore tortuosity, and thus higher unsaturated hydraulic conductivity. In the Oxisol, total porosity was reduced under cultivation but in the Alfisol, land use had no significant effect (3).

However, such structural changes may not be the limiting factors in crop production. Many soils in the humid tropics are constrained by low nutrient reserves rather than soil physical properties. For example, there were few differences in soil physical properties between "productive" and "non-productive" land as classified by smallholder farmers in central Kenya, although there were substantial differences in soil chemical properties (4). In most tropical soils, the majority of nutrients is concentrated in the shallow topsoil, subtended by nutrient-poor and acid subsoil. Thus the retention of the topsoil is of paramount importance for sustained crop production. Land use practices that cause structural collapse reduce the topsoil's ability to absorb and conduct water. On sloping land this leads to surface run-off and soil erosion, which can completely remove the topsoil layer, exposing infertile subsoil unsuitable for agriculture and unlikely to be reclaimed by vegetation.

The Importance of Fallow

Slash and burn smallholder systems can be subdivided by the fallow length employed and the climatic zone in which they are practised. The basis of the processes leading to soil aggregation during fallow phases is biomass production as an energy supply. Litterfall and root turn-over apply and incorporate biomass into the soil. Therefore, generally soil organic matter levels increase with time under fallow (5). The fallow vegetation covers the land, preventing particle detachment by wind and rain. Shade plus the increased soil organic matter levels encourage macrofaunal activity (6). Mites, collembola, earthworms, termites, and ants burrow in the soil generating macropores connected to the soil surface. These pores will provide for rapid water infiltration during intensive rainstorms (7).

The Importance of Fallow Length

In some areas of the humid tropics, farmers still clear primary or secondary forest, and return to the same land after a forest fallow phase of 10 to 30 years. These systems are perceived to be sustainable as fallow recovery is complete (2). However, many smallholder farmers face a shortage of land so fields are cleared after shorter fallow periods, with insufficient regeneration.

The Impact of Climatic Zone

In the moist savanna, the semi-arid and arid tropics, forest fallow does not exist. With decreasing rainfall, the climax vegetation shifts from forest to savanna, shrubland, and grassland. However, in either ecozone, a minimum amount of time is required to re-establish the climax vegetation after clearing and cropping. If the fallow phase is repeatedly shorter than this minimum, "arrested succession" will occur and other types of vegetation will be found by the start of another cropping cycle.

In areas with low rainfall, biomass production is limited. Soil protection relies on other, mainly physical measures preventing or reducing soil exposure to forces degrading soil structure. Windbreaks, stone bunds, ridges, and mulching are measures to reduce wind erosion, some of them indirectly through water harvesting and retention.

HOW SMALLHOLDER FARMERS CAN INFLUENCE SOIL STRUCTURE

Smallholder agriculture in the tropics is thus typified by the clearing of fallow vegetation, the use of burning, manual labour rather than mechanization, and the use of fallow. Human activity is not able to improve directly soil aggregation. Smallholders can modify soil structure only through the promotion of soil aggregating factors and processes increasing soil porosity. These factors are largely biological processes, for example, plant root growth and above and below ground plant biomass decomposition, the breakdown of organic matter by soil microbes, forming humic substances, and soil meso and macrofauna, such as collembola, mites, termites (8), and earthworms (9) which ingest and mix mineral soil particles and organic matter. The selection of the land, clearing method employed, whether the land is burned or mulched, tilled or not, the use of inputs and the management of fallow are the smallholder's tools to balance soil structural degradation versus structural regeneration.

Site Selection: Avoiding Slopes and Soil With Low Structural Resilience

Slash and burn farmers have topographic preferences. In southern Cameroon, on structurally unstable Ultisols, farmers prefer the plateaux and avoid cropping even on gentle slopes (2). However, farmers' choice is increasingly limited by the type of land available so unsuitable land is taken into crop production, increasing the risk of degradation.

Clearing Method

Smallholder farmers generally clear bush or forest manually, using machetes and in some cases chainsaws. Compaction, which would occur with mechanized clearing by bulldozer, is avoided. For example, in the Amazon basin, the effects of mechanical and manual clearance were compared (10). Prior to clearance, infiltration rates were $420\,\mathrm{mm\,h^{-1}}$. Manual clearance reduced infiltration rates to $304\,\mathrm{mm\,h^{-1}}$, whereas mechanical clearing reduced infiltration rates to $32\,\mathrm{mm\,h^{-1}}$. All clearing methods increased topsoil bulk density to 15 cm depth, however, only mechanical clearing caused significant increases to 25 cm depth. There was no effect upon soil texture. Similar results were obtained in analagous experiments on an Ultisol in southern Nigeria (11).

If forest is cleared without a chainsaw, farmers tend to retain trees that are either too big or too hard to fell. These trees may be retained or are killed by burning around their base rather than being felled. Retention of living trees, as compared to clear felling them, maintained lower soil bulk density and higher infiltration rates on a southern Cameroonian Ultisol (12).

Burning Versus Mulching

In burning experiments conducted in southern Cameroon, burning had no effect on bulk density and penetrometer resistance (2). It has been reported in the literature that the burn loosens the soil and facilitates tillage and planting. Direct measurements in southern Cameroon could not confirm these reports and the reports might refer to the lack of litter and shallow root biomass, rather than any changes in the soil. Where hot burns are attained, such as under a log, compaction and soil structural collapse can occur due to a relative excess of monovalent cations leading to a breakdown of cation bridges between clay minerals, the combustion of soil organic carbon and thus substances binding soil particles, and the elimination of soil macro and micro-fauna mixing soil particles and organic materials to form stable aggregates.

After land clearing and burning, the soil surface is exposed to the impact of raindrops. The retention of biomass would reduce the impact of raindrops and through decomposition would contribute to soil structural maintenance. Mulch layers prevent detachment of soil particles and are conducive for soil biota such as earthworms which produce stable soil aggregates by casting.

In semi-arid zones, where mulch is applied to crusted soil, termite activity can be triggered and this can have positive effects upon soil structure. In Burkina Faso, in plots mulched with a wood–straw mixture, saturated soil hydraulic conductivity was $3.6 \times 10^{-5}\,\mathrm{m\,s^{-1}}$ when termites were present yet $0.34 \times 10^{-5}\,\mathrm{m\,s^{-1}}$ when termites were excluded (13).

However, mulching is not a generally applicable solution, because not all crops respond well to a mulch layer, some cannot be grown at all unless there is a "clean" soil surface. In systems without fallow phases mulch has to be produced during cropping phases by a cover crop or intercropped tree or shrub hedgerows (see the section on *Improved Planted Fallows*, following).

Tillage Versus No-tillage

Farmers have various reasons for tillage, depending on site conditions and the intended crop. Tillage can loosen crusted and compacted soil, increase the depth of the loose soil layer, mix ash and residue into the soil to avoid nutrient losses by run-off, and concentrate topsoil through mounding. Breaking up of soil aggregates increases the decomposition rate of soil organic matter, previously physically protected, and thus contributes to increased nutrient availability. Accordingly, the purposes of tillage range from water harvesting to nutrient supply and accumulation in small areas. It can generally be assumed, that tillage does not contribute to soil aggregation. Its loosening effect is temporary and rain may disintegrate soil aggregates further, leading to more severe crusting and compaction. Some crops, however, do require tillage and this exposes the soil and can lead to structural degradation.

Under humid conditions tillage is mainly serving the requirements of certain crops. Groundnut (*Arachis hypogaea*) for instance, requires its pegs (fertilized ovules) to penetrate the soil. To form grains the crop has to take up Ca^{2+} through the pod walls. Therefore the soil is tilled with handhoes, mixing Ca^{2+}-rich ash with the soil at seeding. Other crops may require a deeper layer of loose soil than naturally formed. This is achieved by mounding for crops such as yam (*Dioscorea spp.*), which forms a tuber that grows with a blunt end vertically into the soil. Any obstacle will stop tuber growth and lead to deformed, unmarketable tubers. Other crops planted in somewhat smaller mounds are sweet potatoes (*Ipomoea batatas*) and tannia (*Xanthosoma sagittifolium*), which may not require a deeper layer of loose soil but may positively respond to an increased amount of nutrients. Furthermore, tubers are more easily harvested from mounds than flat soil, especially during dry seasons, when soils are hard.

Manuring, Compost, and Fertilizer Application

Although primarily used to apply nutrients, the high organic matter content in manures and composts has positive effects on soil aggregation. This effect is mainly caused by processes of decomposition and ingestion and egestion by soil meso- and macrofauna. In drier areas, manure is usually incorporated by tillage (see the section on *Tillage*, earlier) to facilitate biological activity which would be hampered at the surface due to drought. The major constraints are the low quantities of manure and compost available and the high labor requirement for transport and incorporation. Organic matter inputs will increase if crop biomass production increases. Of most crops only some part or proportion is harvested and various amounts of biomass will remain in the field (unless required for livestock feeding). Judicious fertilizer use can increase biomass production, crop yields and thus the amount of organic matter returned to the soil. Furthermore, a fertilized crop may grow more vigorously and cover the soil earlier and more intensively, reducing the negative impact of rain.

Improved Planted Fallows

If shortened natural regrowth fallow is not capable to restore soil properties to a level at which adverse effects on long term suitability of the soil for crop production can be avoided, planted fallows may be an alternative. There are two main types of planted fallow: herbaceous legume fallow and tree or shrub based fallow. In south western Nigeria, under *Pueraria phaseoloides* live mulch and *Leucaena leucocephala* alley cropping, aggregate stability, measured as mean weight diameter (MWD) was higher than in natural regrowth system when continuously cropped (14). The differences between the fallow types diminished with increasing fallow length. MWD variation over time was less under *P. phaseoloides* live mulch. On southern Cameroonian Ultisol, increased infiltration rates, reduced soil temperature, and reduced penetrometer resistance were determined in a tilled alley cropping system using *Senna spectabilis* compared with a tilled no-tree control (15). However, planted fallows do not always have more positive effects on soil physical properties than the natural regrowth (14).

EXAMPLES OF INDIGENOUS SYSTEMS OF SOIL STRUCTURAL MANAGEMENT

An example of how tillage, manuring, and soil biological processes interact is the Sahelian "Zaï" system. *Zaï* ("water pockets") is an intensive technique for the management of manure and the preservation of water and is used to regenerate the poorest parts of fields (16). Farmers dig small holes of 15–30 cm diameter, about 10–20 cm deep. During the dry season, fine sand and litter is trapped in these holes. Farmers apply animal manure, compost, and sometimes rockphosphate into the hole, attracting termites. The termites build stable channels in the soil under the manure to access it and distribute some of the organic material. Once the rains commence, the holes collect water, rapidly conducted in the termite channels, thus avoiding run-off. The fine sand can retain more water than coarse sand while the organic layer reduces evaporation thus reducing the risk of seeds failing to germinate. Sorghum, millet and cowpea are commonly planted in the zaï (16). In Tanzania a similar system was developed for steep slopes. The "Ngoro" system is a regular series of pits, 2.4 m long by 2.1 m wide, aligned down the slope and surrounded by ridges, under which buried plant residues are decomposing. Pits are 0.15–0.3 m deep. Although run-off and erosion occur, the losses are small because most of the soil is deposited in the pits.

REFERENCES

1. Feller, C.; Albrecht, A.; Tessier, D. Aggregation and Organic Matter Storage in Kaolinitic and Smectitic Tropical Soils. In *Structure and Organic Matter Storage in Agricultural Soils*; Carter, M.R., Stuart, B.A., Eds.; CRC Press: Boca Raton, FL, 1996; 309–359.
2. Hauser, S.; Norgrove, L. Slash and Burn Agriculture, Effects of. In *Encyclopedia of Biodiversity*; Levin, S., Ed.; Academic Press: San Diego, USA, 2001; *5*, 269–284.
3. Caron, J.; Espindola, C.R.; Angers, D.A. Soil Structural Stability During Rapid Wetting: Influence of Land Use on Some Aggregate Properties. Soil Science Society of America Journal **1996**, *60* (3), 901–908.
4. Murage, E.W.; Karanja, N.K.; Smithson, P.C.; Woomer, P.L. Diagnostic Indicators of Soil Quality in Productive and Non-Productive Smallholders' Fields of Kenya's Central Highlands. Agriculture, Ecosystems and Environment **1999**, *79* (1), 1–8.
5. Kass, D.C.L.; Somarriba, E. Traditional Fallows in Latin America. Agroforestry Systems **1999**, *47*, 13–36.
6. Hauser, S. Distribution and Activity of Earthworms and Contribution to Nutrient Recycling in Alley Cropping. Biology and Fertility of Soils **1993**, *15*, 16–20.
7. Hauser, S.; Asawalam, D. Effects of Fallow System and Cropping Frequency Upon Quality and Composition of Earthworm Casts. Zeitschrift für Pflanzenernährung und Bodenkunde **1998**, *161*, 23–30.
8. Mando, A. The Impact of Termites and Mulch on the Water Balance of Crusted Sahelian Soil. Soil Technology **1997**, *11* (2), 121–138.
9. Hauser, S.; Vanlauwe, B.; Asawalam, D.O.; Norgrove, L. Role of Earthworms in Traditional and Improved Low-Input Agricultural Systems in West Africa. In *Soil Ecology in Sustainable Agricultural Systems*; Brussaard, L., Ferrera-Cerrato, R., Eds.; CRC Press: Boca Raton, 1997; 113–136.
10. Alegre, J.C.; Cassel, D.K.; Bandy, D.E. Effects of Land Clearing and Subsequent Management on Soil Physical Properties. Soil Science Society of America Journal **1986**, *50* (6), 1379–1384.
11. Ghuman, B.S.; Lal, R.; Shearer, W. Land Clearing and Use in the Humid Nigerian Tropics: I. Soil Physical Properties. Soil Science Society of America Journal **1991**, *55* (1), 178–183.
12. Hulugalle, N.R.; Ndi, J.N. Contributory Factors to Soil Spatial Variability in an Ultisol. II. Retention of Living Trees *In Situ* Following Land Clearing. Commun. Soil Sci. Plant Anal. **1993**, *24*, 1409–1419.
13. Mando, A.; Van Rheenen, T. Termites and Agricultural Production in the Sahel: From Enemy to Friend? Netherlands Journal of Agricultural Science **1998**, *46* (1), 77–85.
14. Salako, F.K.; Babalola, O.; Hauser, S.; Kang, B.T. Soil Macroaggregate Stability Under Different Fallow Management Systems and Cropping Intensities in Southwestern Nigeria. Geoderma **1999**, *91*, 103–123.
15. Hulugalle, N.R.; Ndi, N.J. Effects of No-Tillage and Alley Cropping on Soil Properties and Crop Yields in a Typic Kandiudult of Southern Cameroon. Agroforestry Systems **1993**, *22*, 207–220.
16. Maatman, A.; Sawadogo, H.; Schweigman, C.; Ouedraogol, A. Application of Zaï and Rock Bunds in the Northwest Region of Burkina Faso; Study of Its Impact on Household Level by Using a Stochastic Linear Programming Model. Netherlands Journal of Agricultural Science **1998**, *46* (1), 123–136.

CROP ROTATION AND FARMING SYSTEMS: TEMPERATE ZONES

C

M.R. Carter
Agriculture and Agri-Food Canada, Crops and Livestock Research Centre, Charlottetown, Prince Edward Island, Canada

INTRODUCTION

Crop rotation and farming systems play major roles in soil structure management, especially in temperate zones where moderate climates provide a potential for accumulation of soil organic matter and related improvements in soil structure. *Soil structure* can be defined as the spatial arrangement of both soil particles and voids, and improvements in its quality are positively related to crop and soil management practices, such as organic matter inputs (e.g., continuous cropping, organic amendments), soil nutrient management (e.g., fertilizers, calcium amendments), and conservation tillage practices. Increasing organic matter inputs through crop rotation, which enhances the supply of structure-forming agents (e.g., humic substances, plant roots, fungal hyphae, soil fauna), is the major contributor to soil structure management in temperate soils.

SOIL STRUCTURE MANAGEMENT IN TEMPERATE AGRICULTURE

Soil structure can be described in the broadest sense as the spatial arrangement of both soil particles (aggregates) and voids (pores), which has significant effects on air–water relationships in soil (1–3). The shape, size, and stability of aggregates in cultivated soils can be greatly modified by crop rotation and farming practices.

Soil Structure in Temperate Zones

A useful approach to understanding the management of soil structure in temperate agriculture, which accommodates both aggregates and pores, characterizes structure on the following basis (4):

- Structural form—arrangement and size of the pore space
- Structural stability—ability to retain the distribution and size of aggregates after exposure to various stresses (e.g., external forces of impact, shear, abrasion, slaking)
- Structural resiliency—ability of a soil to recover its pore-space arrangement after the removal of a specific stress (e.g., compaction)

On a macroscopic scale, soil structural form can be characterized by soil bulk density. Pore-size distribution, however, provides a more detailed measure of structural form. Pore space can be categorized on the basis of the dominant water process (i.e., water transmission or storage pores).

Although both the pore and aggregate viewpoints are useful, the latter usually dominates through ease of measurement. It assesses soil structural stability on the basis of the stability of a certain aggregate size. The level of dispersed clay can also provide a useful index of soil structural stability, especially for processes involving reduced permeability, sodicity, and soil crusting (2).

Soil structural resiliency is characterized by assessing a soil's ability to self-generate structure through natural processes. Biological processes, such as the activity of soil fauna; and physical processes, such as freezing and thawing, and wetting and drying, are the main agents involved in structure recreation at the soil surface (4). These "self-mulching" soils are mainly clay to clay loams that undergo sufficient swelling to allow structure formation.

Soil Structure-forming Agents in Crop Rotation and Farming Systems

The formation of soil aggregates is dependent upon both abiotic and biotic factors, the former being related mainly to soil clay content and the capacity for natural structure-forming processes (e.g., alternating shrink–swell, freeze–thaw, wet–dry). In the temperate zone, organic matter (i.e., biotic factor) plays the major role in soil structure formation and management. The structure-forming process occurs with some degree of order (2, 4). This ordering or arrangement of particles and aggregates can be viewed conceptually as a hierarchy consisting of three main orders (2): clay microstructures ($<2\,\mu m$ diameter), microaggregates (2–$250\,\mu m$ diameter), and macroaggregates ($>250\,\mu m$ diameter).

Table 1 Benefit of crop rotation on soil structure in temperate soils: Agents, processes, and scale of structure formation

Structure-forming agent	Structure-forming process	Scale of structure
Humic substances; hydroxides of Fe and Al; polyvalent metal cations (e.g., Ca); clays	Allow bonding between soil mineral and organic components	Microaggregation
Polysaccharides	Gelatinous glue; organomineral bonding	Micro- and macroaggregation
Plant roots; fungal hyphae	Enmesh soil aggregates; exude polysaccharides	Macroaggregation
Soil fauna (e.g., earthworms)	Mix organic matter with soil colloids; form large pore or gallery networks	Macroaggregation

(From Ref. 2.)

In clay microstructures, clay–organic matter complexes are stabilized by humic acids and inorganic ions (e.g., Ca). Microaggregates are stabilized directly by microbial materials, such as polysaccharides, hyphal fragments, and bacterial cells or colonies (2, 5). In comparison, the formation of macroaggregates and their temporary stabilization are the result of a combination of mechanisms related to plant roots and the activity of soil fungi and fauna (5) (Table 1). Soil fauna, strongly associated with structure formation, is a major determinant of soil processes influencing nutrient cycling, aggregate formation, and permeability of soil (6). Macrofauna (e.g., earthworms) influence the formation of large soil pores and play an important role in preferential flow.

Importance of Soil Structural Quality in Temperate Farming Systems

Management of soil structure is needed to ensure the maintenance of soil structural quality, which is important for many soil functions, including serving as a medium for root growth and development, and regulation of water and air (1). Inherent soil properties, such as clay content, influence the magnitude of the soil water holding capacity. This is also augmented by the organic matter content. Dynamic soil structural properties, such as soil density and porosity, control the soil storage capacity for water and air. Pore size distribution, considered a good indicator of the soil structural condition, has proven useful in predicting water infiltration rates, water availability to plants, soil water storage capacity, and soil aeration status. Macroporosity, air permeability, and/or oxygen diffusion rate provide a measure of soil macropore continuity and organization. Soil strength and aggregate stability reflect a soil's ability to resist compaction and other stresses that can lead to loss of structure.

BENEFITS OF CROP ROTATION ON SOIL STRUCTURE

In the temperate zone, soil structure and organic matter accumulation are interrelated: organic matter (or fractions thereof) is basic to the structure-forming process, while organic matter sequestered within microaggregates is protected against decomposition. However, crop rotation, land use and farming system, and soil and crop management practices influence the degree of soil structure.

Perennial Forages and Pastures

Water-stable macroaggregation increases rapidly when perennial forages and grasses are utilized in crop rotations. Maximum stability is often achieved after 3 to 5 years (7, 8). Grass roots and fungal hyphae are the main aggregating agents involved in forming and stabilizing macroaggregates (9). Early improvement in water-stable aggregates is related to soluble carbohydrates and microbial biomass (2). Choice of perennial forage crop species and cultivar can also influence the extent of aggregate formation and stabilization.

Short-term Rotations and Cover Crops

Crop rotation, an ordered sequence of crops on the same land, and the use of cover crops mainly in row crop systems [i.e., maize (*Zea mays* Linnaeus), potato (*Solanum tuberosum* L.)], provide important benefits to soil structure in temperate agriculture. The degree of structure modification, especially improvement of water-stable aggregation, is dependent on crop species and, in particular, the amount of residue that each crop of the rotation returns to the soil. Carbon inputs vary widely among crop species. Generally, an annual input of 2 to 3 Mg C ha^{-1} is required in temperate soils to prevent a decline in soil organic matter concentration and soil structure (2).

Organic Additions, Fertilization, and Calcium

Many organic materials are applied to agricultural soils, such as manure, compost, and sewage sludge. The most widespread is farmyard manure. Organic additions and inputs can lead to the formation of water-stable macroaggregates (2, 10). However, organic materials vary widely in composition and, consequently, differ in their rate of decomposition and resultant benefit on soil structure and organic matter content.

Mineral fertilizer, nitrogen (N) in particular, can have both positive and negative effects on soil structure and organic matter. The potential improvement in soil aggregation associated with increased production of roots due to fertilization may not be fully realized if fertilization also causes increased rates of mineralization of the binding agents (4).

Common calcium (Ca) amendments (e.g., lime, gypsum) can lead to soil structure formation and stabilization. The Ca functions as an agent that promotes bridging of microbial cells and clay minerals to form organomineral microstructures (2).

Soil Cultivation and Conservation Tillage

Long-term studies have shown that cultivation of native soils has resulted in a loss of soil organic carbon and a subsequent decline in soil structure (11). Some of this loss would be associated with a change in vegetation or cropping (e.g., from native grassland to annual grains) and the resulting decrease in organic matter inputs to the soil. Reducing the amount and degree of tillage intensity, and retaining a soil cover of crop residues (i.e., conservation tillage; >30% soil residue cover) can lead to an accumulation of organic matter in the surface layer (5–10 cm). This increase in organic matter yields major benefits for improved soil structural stability and soil physical quality (2, 5). However, in comparison, the mixing effect of tillage increases the association of mineral and organic particles, resulting in the beneficial formation of organomineral microstructures (12). Although in many cropping practices (especially conservation tillage), the mixing activity of soil macrofauna may complement, to some degree, the role of tillage (6).

C

REFERENCES

1. Gregorich, E.G., Carter, M.R., Eds. *Soil Quality for Crop Production and Ecosystem Health*; Elsevier: Amsterdam, The Netherlands, 1997.
2. Carter, M.R., Stewart, B.A., Eds. *Structure and Organic Matter Storage in Agricultural Soils*; Lewis/CRC Press: Boca Raton, Florida, 1996.
3. Burke, W., Gabriels, D., Bouma, J., Eds. *Soil Structure Assessment*; A.A. Balkema: Rotterdam, The Netherlands, 1986.
4. Kay, B.D. Rates of Change of Soil Structure Under Different Cropping Systems. Advances Soil Sci. **1990**, *12*, 1–52.
5. Carter, M.R., Gregorich, E.G., Angers, D.A., Beare, M.H.; Sparling, G.P., Wardle, D.A., Voroney, R.P. Interpretation of Microbial Biomass Measurements for Soil Quality Assessment in Humid Temperate Regions. Canadian J. Soil Science **1999**, *79*, 507–520.
6. Lavelle, P., Bignell, D., LePage, M., Wolters, V., Roger, P., Ineson, P., Heal, O.W., Dhillion, S. Soil Function in a Changing World: The Role of Invertebrate Ecosystem Engineers. European J. Soil Biology **1997**, *33*, 159–193.
7. Low, A.J. Improvements in the Structural State of Soil Under Leys. J. Soil Science **1955**, *6*, 177–199.
8. Angers, D.A. Changes in Soil Aggregation and Organic Carbon Under Corn and Alfalfa. Soil Sci. Society America J. **1992**, *56*, 1244–1249.
9. Tisdall, J.M., Oades, J.M. Organic Matter and Water Stable Aggregates in Soils. J. Soil Science **1982**, *33*, 141–163.
10. Aoyama, M., Angers, D.A., N'dayegamiye, A. Particulate and Mineral-Associated Organic Matter in Water-Stable Aggregates As Affected by Mineral Fertilizer and Manure Applications. Canadian J. Soil Science **1999**, *79*, 295–302.
11. Mann, L.K. Changes in Soil Carbon Storage After Cultivation. Soil Science **1986**, *142*, 279–288.
12. Angers, D.A., Bolinder, M.A., Carter, M.R., Gregorich, E.G., Drury, C.F., Liang, B.C., Voroney, R.P., Simard, R.R., Donald, R.G., Beyaert, R.P., Martel, J. Impact of Tillage Practices on Organic Carbon and Nitrogen Storage in Cool, Humid Soils of Eastern Canada. Soil Tillage Research **1997**, *41*, 191–201.

CROP ROTATION AND FARMING SYSTEMS: TROPICS AND SUBTROPICS AND SWELLING SOILS

D.M. Freebairn
G.D. Smith
Queensland Department of Natural Resources and Mines, Toowoomba, Queensland, Australia

R.D. Connolly
URS Pty Ltd, East Perth, Western Australia, Australia

J.N. Tullberg
University of Queensland, Gatton, Queensland, Australia

INTRODUCTION

Large areas of swelling clay soils (vertisols) occur in arid and semiarid tropical and subtropical regions of Africa, India and Australia, and are increasingly being relied on to produce food for the growing population. Farmers of vertisols are often resource-poor and face many risks and constraints associated with soil and weather interactions. These regions usually have wet and dry seasons but rainfall may be highly variable between years and within seasons. Intense rains may follow long dry periods. In some areas, a high proportion of annual rainfall may be in light falls that do not wet below the surface and are quickly lost by evaporation. In other areas, there may be long wet periods in which waterlogging and weed growth present major problems for farmers.

Vertisols have long been recognized as difficult to farm, because they are soft and sticky when wet and hard, and intractable when dry. The swelling characteristics arise from the high proportion of montmorillonitic clays. Forces associated with the interaction between water and the highly active clay particles are extremely powerful and usually predetermine soil structural relations in the soils. Management inputs will have the best effect if they work in harmony with these forces. For example, the aggregation characteristics strongly influence the effects of tillage and the nature of pore spaces available for roots and water entry.

Cropping vertisols alter water and carbon balances. For example, changes in soil cover and hydrology as a result of cultivation can result in doubling of annual runoff and increasing soil erosion by a factor of 100 (1). Also, tillage reduces organic matter in the soil-surface (2), while the use of large tractors compacts subsurface soil (3).

Fallowing is a key management tactic in rainfed farming. A distinguishing feature of vertisols is their very high infiltration capacity when dry and cracked, in contrast to low infiltration rates when wet (3). Low infiltration rates are generally a result of: a crusted or saturated surface, a wet profile (4), or poor internal drainage due to compaction or loss of structural voids. High erosion rates associated with bare soil fallows are clearly unsustainable, far exceeding likely soil formation rates (1). Loss of productivity associated with shallow soils and nutrient depletion result in lower biomass production and declines in organic stores in soil.

While all soils are vulnerable to compaction, the ability of clay soils to store water for long periods makes them particularly susceptible, with subsoils being above the plastic limit for much of the time (5). The hidden nature of subsoil compaction, and variable crop responses make the diagnosis and quantification of impacts of compaction difficult.

OPTIONS FOR MANAGING SOIL STRUCTURE

Retaining Soil Cover

Soil cover (from growing crops and their residues) and reduced tillage maintain infiltration by reducing aggregate disruption by raindrops (6, 7). The amount of crop residue available to provide surface cover is determined by crop type, yield and tillage method (8). Reduced tillage and no till practices result in less stubble breakdown and minimize the deleterious features of tillage while maximizing water storage.

Maintenance of soil cover can reduce erosion rates by an order of magnitude (1), thus protecting the soil resource base. Tillage systems with high nutrition levels and crop residue retention can also maintain higher organic carbon levels (9).

Rotations to Manage Soil Moisture Deficit

Soil moisture content is the most important factor in determining infiltration into cracking clay soils. Although seasonal conditions have a strong and uncontrollable influence on soil moisture, the sequence and number of crops grown have a major effect on the timing of soil water deficits during the year and hence on runoff (10). Rotations that grow more crops, or crops grown during the wetter season have lower runoff and deep drainage. One element of a system to improve water use efficiency includes planting of crops before the onset of the wet season, to make better use of incident rainfall (11). Increasing cropping intensity can also result in greater total productivity although the yields per crop may be reduced (12). The concept of "opportunity" or "response" cropping involves matching cropping intensity with rainfall expectancy and current moisture conditions. Crop rotations are based more on current conditions rather than a fixed pattern, thus exploiting above-average rainfall conditions when they occur. Rotations that maintain a higher soil water deficit are also less prone to compaction damage due to being below the plastic limit for a greater proportion of the time.

Exploiting Soil Cracking

The cracking nature of vertisols offers potential to capture intense falls of rain that might otherwise be lost as runoff. Rapid water movement to depth via cracks can occur during high intensity rainfall, and provides a means for improving the efficiency of water storage. Tillage should be avoided as it may close cracks that would otherwise channel water below the highly evaporative surface soil.

Tillage and Roughness

Tillage, although superimposed on structural arrangements produced by wetting and drying, influences aggregate and pore size distribution, roughness, residue cover and strength of the surface soil (13). Tillage can increase infiltration by breaking surface crusts, increasing surface porosity and surface storage capacity (14). In some situations, smooth soil-surfaces associated with no till systems can result in higher runoff compared to tilled soil (6). Depression storage can be created using specialized tillage equipment. Its function is to store excess rainfall, allowing greater time for infiltration. Such storage is variously referred to as furrow dikes, tied ridges or pits. A particular example of surface configuration used to improve productivity and reduce risk is the broad bed and furrow system developed to allow double-cropping on vertisols at ICRISAT (14). This system reduces the risk of waterlogging, and by controlling runoff, reduces soil erosion (15).

Deep tillage (20–40 cm depth) has been practiced to reduce runoff, an improve water storage and root growth, but results have been variable. Deep ripping (35 cm) prior to wheat and sorghum on a poorly structured clay in southern Queensland has only short-term benefits, not sufficient to justify the cost (16). In contrast, deep tillage reduced runoff and erosion on a steep (14%) cracking soil in southern Italy (17). The variability in results between studies suggests caution in applying deep tillage.

Zonal Wheel Traffic to Reduce Compaction

The possibilities of conserving energy and time associated with confining wheel traffic to permanent laneways is being explored to determine whether there were improvements in soil physical conditions through removal of tractor wheel loads and compaction (18). The system involves creating permanent tracks for most machinery operations to follow. Global positioning system technology is being employed to guide machinery operations. Advantages of this "controlled traffic" system are: less overlap of implements leading to greater efficiency, lower draft requirements (3), more timely planting and spray operations and less compaction of the majority of the crop growing soil mass. A trafficked micro-catchment yielded considerably more runoff (30–60 mm) compared to an untrafficked area (19). The energy implications of controlled traffic also make the system attractive by improving the efficiency in the application of seed and chemicals. Soil benefits are not always expressed as improvements in yield (19), although consistent yield advantages have been observed on a clay loam (20).

Pasture Phases to Rebuild Organic Carbon and Structure

Soil organic matter declines with cultivation, with associated loss of nutrient supply capacity and soil structure (2). Incorporating a pasture ley in a cropping system has the capacity to increase soil organic matter, and improve infiltration characteristics (21, 22). The use of pasture leys in Mediterranean cropping systems is well established, but is not common in clay soils in the subtropics.

With many interactions between soil, weather and crops in cropping systems, the design of improved cropping systems is difficult to explore when based wholly on experimentation. Results from a sequence of seasons may not be representative, and options are restricted by logistics. Cropping system models can be valuable tools in the exploration of new options for improving both the economic and natural resource sustainability of alternative rotations and management systems. As an example, the APSIM–SWIM model (23, 24) was used to explore interactions between changes in soil properties resulting from rotations of crops and pastures, providing predictions of variables such as soil organic carbon and wheat crop yield (25). Such models allow extrapolation of experimental results beyond the experience and locations where detailed experiments have been conducted. In general, cropping systems that have high fertilizer inputs and less tillage result in improvements in soil organic matter, infiltration characteristics, and yield. Analysis of time series of costs and returns then allow for a robust analysis of economic performance.

CONCLUSION

Soil-structure in swelling soils is closely related to soil–weather–plant interactions. These interactions need to be taken into consideration when designing improved soil–crop strategies. Manipulation of soil cover and moisture content can be used to improve infiltration and moisture conservation, leading to better production and reduced soil erosion. Tillage systems that reduce soil compaction both improve production efficiency and soil function. With the incorporation of pasture leys in cropping systems, an improvement in the organic carbon status can be achieved. Relatively slow soil responses to pasture phases make such management options economically difficult for many land managers. System models provide a means to explore some of the biophysical interactions in cropping–tillage systems. When used in conjunction with field experimentation, new management systems can be efficiently developed.

REFERENCES

1. Freebairn, D.M.; Wockner, G.H. A Study of Soil Erosion on Vertisols of the Eastern Darling Downs, Queensland. I. The Effect of Surface Conditions on Soil Movement Within Contour Bay Catchments, Aust. J. Soil Res. **1986**, *24*, 135–158.
2. Dalal, R.C.; Mayer, R.J. Long-Term Trends in Fertility of Soils Under Continuous Cultivation and Cereal Cropping in Southern Queensland. II. Total Organic Carbon and Its Rate of Loss from the Soil Profile. Aust. J. Soil Res. **1986**, *24*, 281–292.
3. Freebairn, D.M.; Loch, R.J.; Cogle, A.L. Tillage Methods and Soil and Water Conservation in Australia. Soil Till. Res. **1993**, *27*, 303–325.
4. Freebairn, D.M.; Loch, R.J.; Glanville, S.F.; Boughton, W.C. Use of Simulated Rain and Rainfall-Runoff Data to Determine Final Infiltration Rates for a Heavy Clay. In *Properties and Utilisation of Cracking Clay Soils*; Reviews in Rural Science No. 5, McGarity, J., Hoult, E.H., So, H.B., Eds.; Univ. of New England: New South Wales, 1984; 348–351.
5. McGarry, D.; Sharp, G.; Bray, S.G. *The Current Status of Soil Structural Degradation in Queensland Cropping Soils*; Project Report DNRQ990092, Department of Natural Resources: Queensland, 1999.
6. Hewitt, J.S.; Dexter, A.R. Effects of Tillage and Stubble Management on the Structure of a Swelling Soil. J. Soil Sci. **1980**, *31*, 203–215.
7. Loch, R.J. Effects of Fallow Management and Cropping History on Aggregate Breakdown Under Rainfall for a Range of Queensland Soils. Aust. J. Soil Res. **1994**, *32*, 1125–1139.
8. Sallaway, M.M.; Lawson, D.; Yule, D.F. Ground Cover During Fallow of Wheat, Sorghum and Sunflower Stubble Under Three Tillage Practices in Central Queensland. Soil Till. Res. **1988**, *12*, 347–364.
9. Marley, J.M.; Littler, J.W. Winter Cereal Production on the Darling Downs — An 11 Year Study of Fallowing Practices Aust. J. Exp. Agric. **1989**, *29*, 807–827.
10. Adams, J.E.; Henderson, R.C.; Smith, R.C. Interpretations of Runoff and Erosion from Field-Scale Plots on Texas Blackland Soil. Soil Sci. **1959**, *87*, 232–238.
11. Krantz, B.A. Technological Packages and Approaches to Management for Rainfed Agriculture in Semi-Arid Tropics Proceedings of the Agricultural Sector Symposia, January 7–11, 1980, pp. 1–29.
12. Berndt, R.D.; White, B.J. A Simulation-Based Evaluation of Three Cropping Systems on Cracking Clay Soils in a Summer–Rainfall Environment. Agric. Meteorol. **1976**, *16*, 211–229.
13. Smith, G.D.; Yule, D.F.; Coughlan, K.J. Soil Physical Factors in Crop Production on Vertisols in Queensland, Australia. Pages 87–104 in Proceedings of the International Workshop on Soils, 12–16 September 1983, Townsville, Queensland, Australia. Australian Centre for International Agricultural Research, Canberra, Australia.
14. El-Swaify, S.A.; Pathak, P.; Rego, T.J.; Singh, S. Soil Management for Optimized Productivity Under Rainfed Conditions in the Semi-Arid Tropics. Adv. Soil Sci. **1985**, *1*, 1–64.
15. Kampen, J.; Hari Krishna, J.; Pathak, P. Rainy Season Cropping on Deep Vertisols in the Semi-Arid Tropics—Effects on Hydrology and Soil Erosion. In *Tropical Agricultural Hydrology*; Lal, R., Russell, E.W., Eds.; Wiley: New York, 1981; 257–271.

16. Radford, B.J.; Gibson, G.; Nielsen, R.G.H.; Butler, D.G.; Smith, G.D.; Orange, D.N. Fallowing Practices, Soil Water Storage, Soil Nitrate Accumulation and Wheat Performance in South West Queensland. Soil Till. Res. **1992**, *22*, 73–93.
17. Postiglione, L.F.; Basso, M.; Amato, M.; Carone, F. Effects of Soil Tillage Methods on Soil Losses, Soil Characteristics and Crop Productions in Hilly Areas of Southern Italy in Proc. NATO Conf. In *Advanced Soil Mechanics and Related Soil Properties*; Series E: Applied Science Series, Larson, W.E., Blake, G.R., Allmaras, R.R., Voorhes, W.B., Gupta, S.C., Eds.; Kluwer Academic: London, 1988; Vol. 172.
18. Proceedings of a National Controlled Traffic Conference, 13–14 Sept, 1995, Yule, D.F., Tullberg, J.N., Eds.; Department of Primary Industries: Rockhampton, Queensland,1985; 209.
19. Tullberg, J.N.; Ziebarth, P.J.; Li, Y. Tillage and Traffic Effects on Runoff. Aust. J. Soil Res. **2001**, *39*, 249–257.
20. Ziebarth, P.D.; Tullberg, J.N. Wheeltrack Compaction Effects on Runoff, Infiltration and Crop Yield. Proceedings of National Controlled Traffic Conference, September 1995, Rockhampton. pp. 103–108.
21. Connolly, R.D.; Freebairn, D.M.; Bridge, B.J. Change in Infiltration Characteristics Associated with Cultivation History of Soils in Southeastern Queensland. Aust. J. Soil Res. **1997**, *35*, 1341–1358.
22. Connolly, R.D.; Freebairn, D.M.; Bell, M.J. Change in Soil Infiltration Associated with Leys in South-Eastern Queensland. Aust. J. Soil Res. **1998**, *36*, 1057–1072.
23. McCown, R.L.; Hammer, G.L.; Hargreaves, J.N.G.; Holzworth, D.P.; Freebairn, D.M. APSIM: A Novel Software System for Model Development, Model Testing and Simulation in Agricultural Systems Research. Agric. Syst. **1996**, *50*, 255–271.
24. Verberg, K., Ed. *Methodology in Soil Water and Solute Balance Modelling: An Evaluation of the APSIM-SoilWat and SWIMv2 Models*; Divisional Report No. 131, CSIRO Div. of Soils: Canberra, 1996.
25. Connolly, R.D.; Freebairn, D.M.; Bell, M.J.; Thomas, G. Effects of Rundown in Soil Hydraulic Condition on Crop Productivity in South Eastern Queensland — A Simulation Study. Aust. J. Soil Res. **2001**, *39*, 1111–1129.

CROPLANDS

B.A. Stewart
West Texas A&M University, Texas, U.S.A.

Robin O'Malley
The H. John Heinz III Center for Science, Economics and the Environment, Washington, D.C., U.S.A.

INTRODUCTION

Croplands are used to produce food, fiber, and other products. Depending upon one's definition, the term "croplands" can include harvested croplands, land with perennial crops such as orchards and vineyards, pasture, land with failed crops[a] and areas idled in a particular year either for production reasons[b] or through government programs. Pastureland is intensively managed land that supports forage such as improved grasses or a mixture of grasses and legumes. Pastureland is very different from rangeland, which that refers to unplowed areas where native grasses, forbs, shrubs, and trees are used for forage.

CROPLAND IN THE U.S.

Cropland occupies about 158 million ha, or 17%, of the nation's land, with pasture adding between 20 and 40 million acres more.[c] Total cropland (not including pasture) in the U.S. has remained fairly stable since the 1950s (Fig. 1).

Government programs have often had the most significant effect on how much cropland is actively used and, in some cases, how it is used. Acreage of cropland idled by U.S. government programs has ranged from 0 to nearly 20% of the nation's total cropland acreage over the past 50 years, with a high of 31 million ha (over 19%) in 1988. In 1997, almost 9% of the nation's croplands—more than 13.4 million ha—were out of production under the Conservation Reserve Program (CRP). The U.S. Congress established the CRP to pay producers to take highly erodible land out of production; these acres are under 10 year leases and are seeded to grass or planted to trees. In previous years, such acreage set-asides had, as their explicit purpose, the control or stabilization of the amounts of crops produced.

The amount of crops and livestock produced in the U.S. has more than doubled since 1948 (1). However, the amount of cropland used for this production changed very little and was actually even less in 1997 than in 1948 (Fig. 1). The average agricultural productivity increase of 1.9% per year during this period was the result of farm inputs such as fertilizers, improved breeding, pesticides, and other capital investment. Irrigation has also played a major role. Only about 15% of the harvested cropland is irrigated, but accounts for approximately 40% of the total value of crops produced (2). Irrigated areas dominate the production of several major crops, including rice with 100% irrigation, orchards (76%), Irish potatoes (71%), and vegetables (65%). Irrigated cropland area has expanded by a third since 1969, but nearly all of this expansion occurred prior to 1982. This expansion in the area irrigated was also accompanied by increases in water use efficiency—while the total area irrigated increased, total irrigation water applied was roughly the same in 1995 as it was in 1969.

CROPLAND IN THE WORLD

According to the Food and Agriculture Organization of the United Nations, there are about 1.5 billion ha of cropland in the world (Table 1). Turner et al. (3) estimated that the global extent of cropland increased from about 265 million ha in 1700 to around 1.2 billion ha in 1950, with the addition of about 315 million ha since that time. Almost 60% of the world's cropland is dedicated to the production

[a]Not all acreage listed as "failed" actually had a crop failure. Some cropland not harvested due to lack of labor or low market prices is also considered as crop failure.

[b]For example, in certain semiarid and subhumid region areas of the U.S., farmers cultivate or treat lands with herbicides for one or more seasons to control weeds and accumulate soil water before a crop is grown. This practice is known as summer fallow.

[c]USDA estimates for the amount of pasture vary significantly. Before 1997, different agency estimates range from 64.5 million acres (Census of Agriculture) to 120 million acres (Natural Resources Inventory).

Encyclopedia of Soil Science

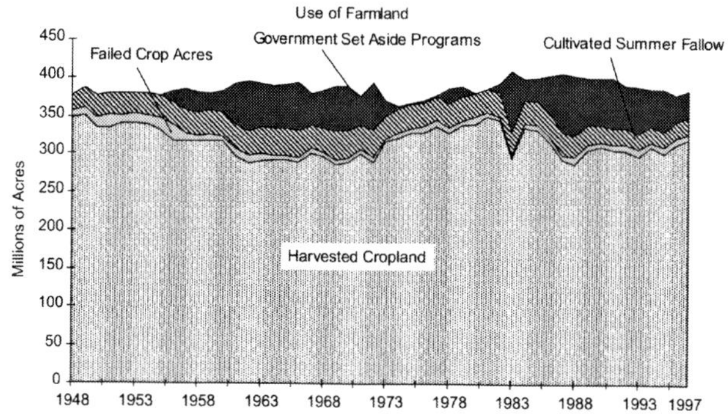

Fig. 1 Source: Adapted from Ref.1; data are from USDA Economic Research Service (2.54 acres = 1 ha).

of cereals. Wheat, rice, and maize are the dominant cereals, occupying 32%, 21%, and 20% of the cereal crop area, respectively. Only in Latin America, with some 43% of cropland devoted to cereals, does the cereal share fall below half of all harvested land (4).

Irrigated cropland makes up about 17% of the total cropland but accounts for 30–40% of the crop output. Irrigated areas vary greatly among regions. Less than 4% of the cropland is irrigated in sub-Saharan Africa while more than 32% of the cropland is irrigated in Asia. This high proportion of irrigated cropland has helped Asia to cope with a high and growing population. Asia has only 9.5% of the world's cropland but is home to 55.6% of the world's population (4). The people in Asia also consume a high percentage of their calories directly from cereal grains while most developed countries use much of their cereal grains for animal feed and then consume meat products. Maize is the dominant cereal grain fed to animals and on average, about 66% of the global maize production is used for animal feed. This figure is higher for developed countries, which use about 76% of the maize for animal feed, and lower (about 56%) for developing countries. Rice is almost exclusively consumed by people and Asia contains a staggering 89% of the world's harvested rice area (4).

Cropland is also used for producing fiber crops (such as cotton, flax, sisal), medicines, dyes, chemicals, and other nonfood industrial raw materials. About 7% of the world's harvested cropland is used for these crops (4). Some food crops are grown primarily for fuel such as ethanol and more wood biofuels are also becoming widespread.

CROPPING INTENSITY

A simple measure of how intensive cropland is used is the cropping intensity index. This is defined as the annual harvested area expressed as a proportion of the total cropland area (land in use plus fallow). For example, some cropping systems require a portion of the production area to be placed in fallow every year resulting in a cropping intensity index of less than 1. In comparison, some irrigated areas can produce up to three crops per year from the same physical area and have a cropping intensity of 3. On a global basis, the average cropping intensity for annual crops is about 0.8, but regional variations are again very large (4). Asia has a cropping intensity index of about 1.1 that is in sharp contrast to temperate, developed-country agriculture, such as Western Europe and North America, where temperature-limited growing seasons and proportionately less irrigation give rise to cropping intensities of 0.6–0.7.

CONDITION OF CROPLANDS

The potential for expanding suitable cropland is limited by the availability of suitable topography, soils, and climate,

Table 1 Cropland in various regions of the world

Region	Total land (million ha)	Cropland (million ha)	Irrigated cropland (million ha)
North America	1872	231	22
Latin America and the Caribbean	2018	153	17
Europe	472	136	17
Former Soviet Union	1788	230	20
West Asia/North Africa	1189	92	24
Sub-Saharan Africa	2268	166	6
Asia	2470	444	143
Oceania	849	52	2
Total	12,926	1504	251

(Adapted from Ref. 5.)

and by competition with other land uses. Cropland area has expanded at the expense of grasslands and forests, with attendant losses of habitat, and engineering projects have altered the hydrological regime of most of the world's major rivers to provide irrigation water.

While there have been only modest increases in the land devoted to agriculture in recent decades, existing croplands are being used more intensively. Irrigated area has increased, fallow time has decreased, and the use of purchased inputs such as fertilizer and equipment has increased. This intensification has enabled food production to more than keep pace with the demands of a growing global population. On average, food supplies are 24% higher per person than in 1961 and real prices are 40% lower.

There are many signs that the capacity of ecosystems to continue to produce many of the goods and services on which we depend may be declining (4). These include instances—some widespread, some not—of soil degradation from erosion, soil organic matter depletion, increased salinity from irrigation, acidity, and soil compaction. For example, one recent survey indicates that about 40% of the world's agricultural areas have moderately or extremely degraded soils, with attendant losses in crop productivity (4). Depletion of soil organic matter and increases in salinization resulting from irrigation are thought to be widespread. One recent estimate indicates that 20% of irrigated land worldwide suffers from salinization (4).

In addition to increasing food production, intensification of agriculture through increased use of inputs such as nitrogen fertilizer can have off-farm impacts. Nitrogen concentrations in U.S. streams and shallow groundwater in agricultural areas are higher—2–6 times as high—as areas with forest cover or residential land use. In agricultural areas, more than three-fourths of the streams and 40% of groundwater had at least one pesticide, while many streams and wells had multiple pesticide compounds.

Public policies can reduce or even halt such degradation. For example, the area of U.S. cropland that is most likely to erode decreased by 28 million acres from 1987 to 1992. This improvement is widely attributed to improvements in farm practice, encouraged by various public incentives and education programs, and retirement of highly erodible lands under programs such as the Conservation Reserve, previosly noted.

There is, however, a startling lack of information on the condition of croplands. Even for such a basic resource as soil, there are few sources of consistent data that are comparable over large areas to confirm these trends. Particularly at the global level, facts about resource conditions are sketchy and often not comparable. As an example, the estimate of 20% of irrigated lands having elevated salinity is accompanied by the caveats that "salinization data are poor" and the estimate itself is described as "rough." In the U.S., there are no reliable, survey-based data on soil organic matter, compaction, salinization, or biological condition, and very thin data on the amount of nutrients present in agricultural soils nationwide. And there is very little consistent broad-scale information on whether agricultural areas are losing wildlife habitat or gaining it, on whether the species that inhabit these areas are increasing or declining, or whether the quality of habitat, such as stream conditions, is improving or degrading.

REFERENCES

1. The Heinz Center, *Designing a Report on the State of the Nation's Ecosystems: Selected Measurements for Croplands, Forests and Coasts and Oceans*; The H. John Heinz III Center: Washington, DC, 1999.
2. USDA, *Agricultural Resources and Environmental Indicators, 1996–97*; Agricultural Handbook Number 712, Economic Research Service, U.S. Department of Agriculture: Washington, DC, 1997.
3. Turner, B.I., II, Clark, W.C., Kates, R.W., Richards, J.F., Mathews, J.T., Meyer, W.B., Eds. *The Earth as Transformed by Human Action: Global and Regional Change in the Biosphere Over the Past 300 Years*; Cambridge University Press: Cambridge, 1990.
4. Wood, S.; Sebastian, K.; Scherr, S.J. *Pilot Analysis of Global Ecosystems*; World Resources Institute: Washington, DC, 2000.
5. FAOSTAT. 1999. Food and Agriculture Organization of the United Nations, Statistical Databases. Online at: http://apps.fao.org.

DEBRIS FLOW

Jau-Yau Lu
National Chung-Hsing University, Taichung, Taiwan

INTRODUCTION

Debris flow is a rapidly moving two-phase gravity flow with a high content of wide gradation solid particles. It usually occurs in gullies or on sloping land near the upstream areas. Debris flow has a strong erosive force that erodes the gully bed or banks as it moves downstream. Debris flow-hazard mitigation is very important because the flow often causes heavy losses of lives and properties.

The debris flow problem has been recognized as a very critical problem worldwide (1–3). Figure 1 is an example showing the disaster caused by the debris flow of April 29, 2000, at Chia-Yi County in central Taiwan. This debris flow event was induced by a devastating earthquake (100-year return period), 7.3 on the Richter scale, in central Taiwan on September 21, 1999. The earthquake caused the exposure of a large amount of loose soil on the sloping land. With a dry season from October through February, debris flow did not occur until the wet season the following spring. The total amount of rainfall for the storm (April 28–29, 2000) that triggered this particular debris flow event was 152 mm.

CLASSIFICATION OF DEBRIS FLOW

Based on materials contained in the flow, debris flow can be classified into the following three types (4, 5): 1) viscous debris flow, 2) nonviscous (stony, water-borne) debris flow, and 3) mud flow (muddy debris flow).

Viscous debris flow, which carries material ranging from fine clay to large boulders, is the most common type of debris flow. The debris flow shown in Fig. 1 is an example of the viscous debris flow. Both nonviscous (stony) debris flow and the mud flow can be considered as two extreme cases of the viscous debris flow. Nonviscous (stony) debris flow has very few fine particles. It usually runs smoothly as a single and continuous event. On the contrary, viscous debris flow is characterized by the intermittent pulsing. A single event may include several dozen pulses.

Mud flow mainly carries sediment finer than coarse sand (2 mm). However, the flow sometimes also carries a small amount of rock fragment. After an intense thunderstorm on August 25, 1967, a mud flow of more than 50,000 cubic yards occurred in the 1.5 square mile Second Creek Basin in Nevada. It damaged properties and roadways and polluted Lake Tahoe (6).

INITIATION OF DEBRIS FLOW

In general, physical processes of debris flow consist of initiation, transportation, and deposition of solid particles. Conditions that trigger the initiation of debris flow include: 1) availability of loose solid material, 2) steep slope gradient, and 3) heavy rainfall (4). The first two are the potential factors, and the third one is the triggering factor for the formation of a debris flow.

The loose solid material may be due to the prior landfalls, landslides, typhoons, earthquakes (e.g., Fig. 1), glaciers, or rock weathering. The production of loose material is also related to the geological conditions. The gravitational force associated with steep slopes is an important driving force for the initiation of debris flow. An analysis of 150 debris flow gullies in Tibet showed that the debris flow occurred in gullies with slopes of 10–30%. In Japan, most of the debris flow occurred in gullies with slopes greater than 15 or 27%. These observations have been incorporated into a model that predicts the critical slope for initiation of debris flow (7).

An intense rainstorm is the triggering factor for the initiation of debris flow as illustrated in Fig. 1. Figure 2 shows the debris flow of July 31, 1996, near Hsin-Yi bridge (180 m long), Nantou County, Taiwan. The debris flow was caused by a 24-h rainfall of 745 mm from Typhoon Herb, which broke the local 200-year maximum daily rainfall record. The amount of deposited material was so large that the surface of the deposited material almost reached the lower surface of the bridge deck (5–6 m high from the original channel bottom).

Essentially, three types of debris flow initiation predominate (5). The first two types are due to erosion of channel bed and landslide. The third type is the destruction

Fig. 1 Disaster caused by the debris flow of April 29, 2000, at Chia-Yi County in central Taiwan. (Photo by I.Y. Wu.)

Fig. 2 View looking upstream of the 1996 debris flow near Hsin-Yi bridge. (Photo by S.H. Tseng.)

of a natural dam by the overtopping of river water or by the collapse of the dam body itself under the effects of the seepage water and the hydraulic pressure.

Several landslides occurred near the Tsao-Ling area in central Taiwan, which is approximately 40 km south of the epicenter for the earthquake (7.3 on the Richter scale) of September 21, 1999. An upstream impoundment of water, with an estimated total storage volume of about 43 million m^3, was produced. In fact, this location has a long history of landslides induced by either earthquake or heavy rainfall, resulting in the impoundment of water and subsequent breach outflow causing downstream damage and loss of life. For example, landslides in December 1941 and August 1942 produced a total impoundment about 170 m deep with a volume of about 120 million m^3 by October 1942 (after heavy rainfall), which eventually failed during overtopping (4 m depth) in May 1951, resulting in 437 deaths and associated damage.

MOVEMENT AND DEPOSITION OF DEBRIS FLOW

Debris flow usually occurs near the upland areas (upper reach of a watershed) with steep-slope gradients, moves along gullies or channels (middle reach of a watershed), and decelerates rapidly once it reaches a gully or channel outlet (mouth) where the slope abruptly levels off. The depositional area below the channel outlet is usually called "debris cone" or "debris fan."

One of the interesting characteristics during the transportation of debris flow is the accumulation of large boulders at the front of the flow. The phenomenon can be explained by Bagnold's (8) dispersive pressure concept, i.e., the dispersive force is proportional to the square of particle diameter. With this vertical sorting mechanism, the coarse particles gradually move toward the water surface. Since the maximum flow velocity occurs near the water surface in an open channel flow, the large boulders proceed faster than the smaller particles and gradually accumulate at the front of debris flow.

The mechanics of debris flow are very complex. Models that account for the rheological properties of the debris flow material and processes have been developed by researchers (9–11). The debris flow has been modeled as a Bingham substance, as a viscoplastic fluid, or as a dilatant fluid. The flow velocities of a steady two-dimensional debris flow can be obtained using various models with the assumption of a uniform mixture (continuum).

Mathematical models have also been developed to predict the snout profiles of constantly advancing debris flows on hillslopes (12, 13). The deposition process of debris flow can be modeled based on the theory of mass and momentum conservation (14, 15). The proper prediction of the deposition areas (e.g., evolution of debris cones) and deposition depths is very useful for disaster and damage prevention.

HAZARDS MITIGATION

Absolute control over debris flows is rarely feasible either physically or economically. There are many different methods for debris flow-hazard mitigation. The most commonly accepted measures for reducing debris flow damage can be divided into two categories, i.e., structural and nonstructural measures.

The structural measures include different types of dams (e.g., consolidation dams, dams with holes such as grid-type dam and slit dam), retarding basins, and diverting and deflecting walls (16). Selection of a proper structure depends on the type of debris flow (e.g., stony debris flow vs. mud flow), and the area to be protected (i.e., upper reach, middle reach, or debris cone of a watershed).

The watershed management practices, such as revegetation of barren areas and reforestation, are nonstructural debris flow countermeasures. Other nonstructural methods include hazard zoning, relocation of endangered properties, and temporary evacuation of people in the debris-flow threatened areas on the basis of warning systems.

Debris flow mitigation projects often utilize a combination of several measures based on both physical and economical considerations. For example, many landslides occurred on the hillslopes and mountainous areas in central Taiwan due to the devastating earthquake of September 21, 1999 (7.3 on the Richter scale). As it was a dry season from October 1999 through next February, the revegetation and reforestation were not immediately effective after the earthquake. The cost and the construction time also prohibited the selection of structural measures. Therefore, hazard zoning, relocation, and emergency evacuation were the most suitable means of debris flow mitigation until other long-term measures became effective or could be implemented.

REFERENCES

1. Scott, K.M. Precipitation-Triggered Debris-Flow at Casita Volcano, Nicaragua: Implications for Mitigation Strategies in Volcanic and Tectonically Active Steeplands. In *Proceedings of the Second International Conference on Debris-Flow Hazards Mitigation, Taipei, Taiwan, August 16–18, 2000*; Wieczorek, G.F., Naeser, N.D., Eds.; Balkema: Netherlands, 2000; 3–13.
2. Chen, C.L. *Proceedings of the First International Conference on Debris-Flow Hazards Mitigation: Mechanics, Prediction, and Assessment*; ASCE: New York, 1997; 817.
3. Wieczorek, G.F.; Naeser, N.D. *Proceedings of the Second International Conference on Debris-Flow Hazards Mitigation: Mechanics, Prediction, and Assessment*; Balkema: Netherlands, 2000; 608.
4. Chien, N.; Wan, Z. *Mechanics of Sediment Transport*; (in Chinese) Science Publication: Beijing, 1991; 656.
5. Takahashi, T. Initiation and Flow of Various Types of Debris-Flow. In *Proceedings of the Second International Conference on Debris-Flow Hazards Mitigation, Taipei, Taiwan, August 16–18, 2000*; Wieczorek, G.F., Naeser, N.D., Eds.; Balkema: Netherlands, 2000; 15–25.
6. Glancy, P.A. *A Mudflow in the Second Creek Drainage, Lake Tahoe Basin, Nevada, and Its Relation to Sedimentation and Urbanization*, U.S. Department of Interior Geological Survey Professional Paper 650-C C195-C200, 1969.
7. Takahashi, T. Mechanical Characteristics of Debris-Flow. J. Hydr. Div. ASCE **1978**, *104* (8), 1153–1169.
8. Bagnold, R.A. Experiments on a Gravity Free Dispersion of Large Solid Sphere in a Newtonian Fluid Under Shear. Proc. R. Soc. Lond. Ser. A **1954**, *225*, 49–63.
9. Jan, C.D.; Shen, H.W. Review Dynamic Modeling of Debris Flows. Lect. Notes Earth Sci. **1997**, *64*, 93–116.
10. Chen, C.L. Comprehensive Review of Debris Flow Modeling Concepts in Japan. *Debris Flows/Avalanches: Process Recognition, and Mitigations*; (Rev. Eng. Geol., *7*, 13–29) Geological Society of America: Boulder, CO, 1987.
11. Chen, C.L. Generalized Viscoplastic Modeling of Debris Flow. J. Hydr. Eng. ASCE **1988**, *114* (3), 237–258.
12. Takahashi, T. *Debris Flow*; Monograph Series of IAHR, Balkema: Rotterdam, 1991; 165.
13. Chen, C.L.; Ling, C.H. Fully Developed Snout Profiles of Noncohesive Debris Flows with Internal Friction. In *Proceedings of the Second International Conference on Debris-Flow Hazards Mitigation, Taipei, Taiwan, August 16–18, 2000*; Wieczorek, G.F., Naeser, N.D., Eds.; Balkema: Netherlands, 2000; 335–344.
14. Takahashi, T.; Nakagawa, H.; Harada, T.; Yamashiki, Y. Routing Debris Flows with Particle Segregation. J. Hydr. Eng. ASCE **1992**, *118* (11), 1490–1507.
15. Hsieh, C.L. *On the Analysis and Prediction of Hazardous Zone of Debris Flows*; Research Report NSC80-0410-E006-29, National Science Council: Taiwan, 1991; 1–71.
16. Heumader, J. Technical Debris-Flow Countermeasures in Austria—A Review. In *Proceedings of the Second International Conference on Debris-Flow Hazards Mitigation, Taipei, Taiwan, August 16–18, 2000*; Wieczorek, G.F., Naeser, N.D., Eds.; Balkema: Netherlands, 2000; 553–564.

DEGRADATION

Rosa M. Poch
José A. Martínez-Casasnovas
University of Lleida, Lleida, Spain

INTRODUCTION

Soil degradation is defined as *the loss of the soil's capacity to develop its functions; e.g., support for plant growth, hydrological regulation of watersheds, environmental filtering, and support for buildings, among others.* Former approaches to soil degradation, from a productivity or agronomic point of view, have evolved to an environmental one in the last decades. A more comprehensive approach considers the soil to be a part of the land and part of a socioeconomic, historical, and cultural reality. The environmental point of view enables other degradation processes to be accepted—not only those processes affecting soil's intrinsic characteristic (changes of the physical, chemical, biological soil properties, or agricultural use), but also those processes due to externalities. The best known of these degradation processes are soil erosion, compaction, alkalinization, salinization, pollution, acidification, nutrient depletion, and organic matter loss. There are other, less common factors that limit soil use, such as the lack of accessibility, natural disasters, and war and even cultural, socio-economical, and historical limitations to proper use of the land. The increasing interest in sustainable land management is a result of the scarcity of land resources, together with the probable effects of global climatic changes. The soil in a sustainable system must keep its qualities by means of adequate use.

SOIL DEGRADATION: CONCEPTS AND EVOLUTION

Historical Perspective

The earlier definitions of soil degradation were based on the zonality concept (1–3); today, they are replaced by definitions related to land use. Degradation now means the loss of the soil's agricultural productivity, which according to Charman (4) includes soil erosion, salinization, loss of chemical fertility, acidification, and structural degradation as the main soil degradation processes. Aveyard and Charman (5) also stress the importance of the off-site effects and human impact on the degradation processes, which are the result of the mismanagement of the soil. Within the framework of land evaluation, degradation is then defined as *the decline in soil quality caused through its improper use by humans* (6). The Food Agricultural Organization (FAO) (7) also points out the importance of the rate at which the process occurs: Degradation may not necessarily be continuous in time but can occur during a short period between two stages of ecological equilibrium. Six types of soil degradation that follow this concept are presented, together with the units of measurement that assess the rate of the process: 1) water erosion ($Mg\,ha^{-1}\,year^{-1}$); 2) wind erosion ($Mg\,ha^{-1}\,year^{-1}$); 3) excess of ions: salinization (ECs) and alkalinization (ESP); 4) chemical degradation: acidification ($\Delta V\,year^{-1}$) and toxicity ($\Delta ppm\,year^{-1}$); 5) physical degradation ($\Delta^b\rho\,year^{-1}$; $\Delta Ks\,year^{-1}$); and 6) biological degradation ($\Delta OM\,year^{-1}$).

At the end of the 1970s, society's general awareness of the importance of the environment permitted an expansion of the concept of soil degradation, which is finally defined as *the loss of the soil's capacity to develop its functions in the environment.* These functions are based on soil being the natural substrate for plant growth, the regulator of the water regime of hydrological units, and the natural environmental filter (8). This definition allows an acceptance of processes affecting the land and its socioeconomic, historical, and cultural framework other than those related to agricultural aspects. Degradation processes include those limiting soil use in the short, medium, and long term, not merely those caused by inappropriate agricultural use. Polluting soils, urbanization, mining activities, watershed mismanagement, flooding, global climatic change, loss of accessibility, or even minefield installation in war zones are also recognized as human activities or natural phenomena leading to land degradation. All of these processes have been better assessed with the increase of soil knowledge, the use of models to foresee its behavior, and land information management, particularly geographic information systems.

Nevertheless, sometimes the functions developed by the soil are mutually exclusive. It is not expected that soil along a river can be used for agriculture and act as a green filter simultaneously, or serve as the support for human settlement and become an area of water table recharge as well. Is a saline soil with a specific halophytic community a degraded soil? Or, must a value be given to the genetic background of the plants growing there? In fact, soil degradation processes are often assessed from an anthropocentric point of view, by the evaluation of the possible use humankind can make of it, instead of considering its impact on the environment in the long run.

Soil Conservation and Degradation

The concept of soil conservation has also evolved along with soil degradation. Although defined as the group of measures used to prevent degradation, conservation was originally only used as a synonym for anti-erosion measures. It was formerly defined as *measures to maintain the soil layer* (1) or *the techniques to maintain the productivity of the land, reducing erosion* (9). Now, other management techniques, such as liming or salt-control practices in irrigated land, are also included when defining soil conservation in the broadest sense.

Other environmental definitions of soil conservation implicitly include the concept of sustainability and refer more to the conservation of a land devoted to a specific use than to the intrinsic conservation of the soil itself. Soil conservation consists of the *optimal rational use of the natural resources and the environment, taking into account the different use demands and the need to save and maintain them for the future* (10); it also consists of *the promotion of the optimum use of the land in accordance with its capability to assure its maintenance and improvement. In the same way, conservation means also halting degradation and restoring productivity where it has been reduced* (11). This approach assumes that soil conservation is not only a group of management practices, but also that land should be used in a manner adequate to the land's qualities and capacities. Therefore, soil conservation by definition requires a previous exercise of land evaluation. The FAO framework for land evaluation (12, 13) allows the matching of soil conservation within land evaluation, in the sense that management systems preserve its qualities (14).

PROCESSES AND CAUSES IN SOIL DEGRADATION

When listing the different types of soil degradation processes, there is confusion regarding causes, processes involved, and effects on soil and environment. This is a result of the inherent complexity and diversity of the phenomena, which makes the accordance between them difficult. Moreover, the result of one degradation process may be the cause of another one.

Two processes lead to the loss of the soil's capacity to perform its functions: those that change their physical, chemical, and biological properties (intrinsic processes), and those that prevent their use by other causes (extrinsic processes) (Table 1). Some processes correspond partly to soil formation processes (acidification, salinization,

Table 1 Soil degradation processes

	Intrinsic Processes	
Degradation of the physical fertility:		
Compaction		
Sealing		
Crusting		
Structural degradation		
Soil loss:	Water and wind erosion	
	Land movements by civil engineering or urbanization (levelings, landfills, landcuts)	
	Mining	
	Resource consumption; urbanization on highly productive soils	
Water excess		
Hydromorphism		
Impermeabilization by urban use		
Sodification; alkalinization		
Degradation of chemical fertility:		
Loss of nutrients:	Leaching	
	Extraction by plants (*nutrient mining*)	
	Runoff	
	Immobilization	
Acidification		
Salinization		
Sodification; alkalinization		
Pollution		
Degradation of biological fertility		
Loss of organic matter		
	Extrinsic Processes	
Loss of accessibility:	Loss or damage of roads	
	Change of ownership regime	
	Loss of material or human resources	
Conversion to risk area:	Natural disasters	
	Volcanic and seismic activity	
	Floods	
	Fires	
	Human diseases	
	Wars	
Development of inadequate agricultural policies		
Climatic fluctuations		
Illiteracy		
Cultural traumas		

D

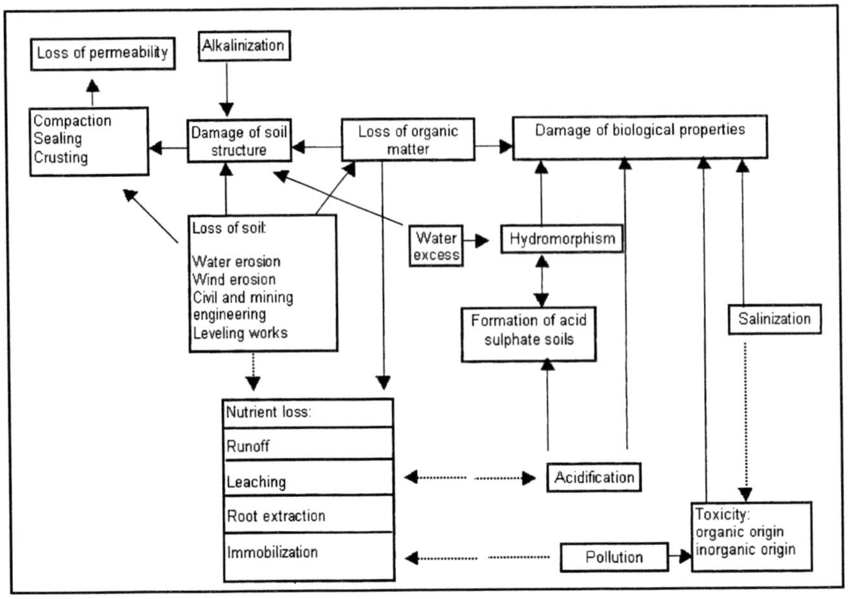

Fig. 1 Some cause–effect relations in soil degradation processes.

sodification) that, because of their rate of progress, can be observed during the course of one generation and can be modified by humans. Other soil formation processes have similar effects, such as podsolization, cementation, or ferralitization, but because they are not observable on a human scale and have few possibilities of modification, they are normally not considered to be soil degradation processes. Fig. 1 shows some of the cause–effect relationships between the intrinsic processes. Table 2 displays some of the causes of soil degradation. The list is not exhaustive because of the complexity of the relationship between the agents involved.

The term desertification has often been used as a synonym for soil degradation. It was introduced in 1949 by A. Aubréville and has been applied since then to the degradation of arid and semi-arid environments. During the 1950s, its meaning expanded to include ecological degradation under different climates, even in tropical or temperate areas.

The use of the term desertification was generalized after the Nairobi Conference in 1977, when desertification was defined as *the diminution or destruction of the biological potential of the land, and can lead ultimately to desert-like conditions. It is an aspect of the widespread deterioration of ecosystems, and has diminished or destroyed the biological potential, i.e., plant and animal production, for multiple use purposes at a time when increased productivity is needed to support growing populations* (15).

The previous definition has a clear, economical approach applicable to developing countries, and assumes anthropic causes in the process, derived from a population

Table 2 Causes of soil degradation

Agricultural practices
Improper agricultural practices
New cultures in former rangeland
Monocultures
Traffic of machinery
Excessive irrigation schemes and schemes without drainage systems
Inadequate or overfertilization
Inappropriate or excess of tillage
Overgrazing
Deforestation
Generation and application of residues to the soil
Natural causes
Climatic changes
Earthquakes and active volcanoes
Floods
Fires
Industrial activities
Mining
Residue applications
Buildings and civil engineering structures
Socio-economical causes
Human pressure
Urbanization of cropland and marginal agroecosystems
Lack of education
Inadequate ownership system
Tourism
Political causes
Environmental policies
Social and cultural policies
Economical and market policies
Wars

increase causing a high demographic pressure in the system. It does not specifically refer to arid environments, although the lack of water or the lack of the environment's storage capacity are conditions similar to those of a desert. Thus, because soil is the component of the ecosystem with the largest water storage capacity, it is evident that soil degradation is one of the main causes of desertification. In this sense, Porta and colleagues (16) indicate that desertification may be better defined as the *soil degradation that goes beyond the tolerance limits, generally due to a human intervention*. The authors advise to restrict the term's use to degradation processes taking place in marginal areas near deserts. The use of the general term *land degradation* would be more appropriate for environments other than arid or semi-arid.

REFERENCES

1. Plaisance, G.; Cailleux, A. *Dictionaire des Sols*; La Maison Rustique: Paris, 1958.
2. SSSA. *Glossary of Soil Science Terms*; Soil Science Society of America: Washington, DC, 1979.
3. SSSA. *Glossary of Soil Science Terms*; Soil Science Society of America: Washington, DC, 1982.
4. Charman, P.E.V. Other Forms of Soil Degradation. In *Soils: Their Properties and Management. A Soil Conservation Handbook for New South Wales*; Charman, P.E.V., Murphy, B.W., Eds.; Sydney University Press: Sydney, 1991; 48–53.
5. Aveyard, J.M.; Charman, P.E.V. Soil Degradation & Productivity: A Concluding Perspective. In *Soils: Their Properties and Management. A Soil Conservation Handbook for New South Wales*; Charman, P.E.V., Murphy, B.W., Eds.; Sydney University Press: Sydney, 1991; 317–321.
6. Houghton, P.D.; Charman, P.E.V. *Glossary of Terms Used in Soil Conservation*; Soil Conservation Service of New South Wales: Sydney, 1986.
7. FAO. *Método Provisional para la Evaluación de la Degradación de Suelos*; FAO: Rome, 1980.
8. Pla, I.; Ovalles, F., Eds. *Efecto de los Sistemas de Labranza en la Degradación y Productividad de los Suelos*. II. Reunión Bienal de la Red Latinoamericana de Labranza Conservacionista, Guanare, Acarigua, Venezuela, November, 14–19, 1993.
9. Agence de Coopération Culturelle et Technique. *Dictionaire D'Agriculture et des Sciences Annexes*; La Maison Rustique: Paris, 1997.
10. Dalal-Clayton, D.B. *Black's Agricultural Dictionary*, 2nd Ed.; A&C Black: London, 1985.
11. Dudal, R. An Evaluation of Conservation Needs. In *Soil Conservation: Problems and Prospects*; Morgan, R.P.C., Ed.; Wiley: New York, 1981; 3–12.
12. FAO. *Framework for Land Evaluation*, FAO Soils Bulletin 32, FAO: Rome, 1976.
13. FAO. *Planning for Sustainable Use of Land Resources: Towards a New Approach*; FAO Land and Water Bulletin 2; FAO: Rome, 1995.
14. FAO. *Land Quality Indicators and Their Use in Sustainable Agriculture and Rural Development*, FAO Land and Water Bulletin 5; FAO: Rome, 1997.
15. United Nations. *UN Conference on Desertification: Roundup, Plan of Action and Resolutions*; United Nations: New York, 1978.
16. Porta, J.; López-Acevedo, M.; Roquero, C. *Edafología Para la Agricultura y el Medio Ambiente*, 2nd Ed.; Mundiprensa: Madrid, 1999.

DEGRADATION, BIOLOGICAL

Ibrahim Ortas
University of Çukurova, Adana, Turkey

INTRODUCTION

The soil component of the agrosystem consists of plant roots, microflora, fauna, organic matter, and the abiotic geochemical matrix. In this system so far, soil has taken second place with respect to plants. Bethlenfalvay and Schüepp (1) stated that the importance of soil is recognized not only as an agricultural resource base, but also as a complex, living, and fragile system that must be protected and managed for its long-term stability and productivity. Soil degradation has been defined as the loss of a soil's capacity to produce crop and also as the antithesis of soil resilience and quality (2). Rao and Dadarwal (3) describe biological degradation as the decline in soil organic matter and biomass C, a decrease in diversity and activity of soil flora and fauna, or the indiscriminate use of chemicals and pollutants.

The human relationship with soil started more than 8000 years ago with the ancient civilizations in Anatolia and Mesopotamia, located in the Fertile Crescent of the Tigris and Euphrates river basins. Throughout the early history of agriculture, human use of land for farming and grazing had little effect on the depletion of soil properties. As the human population increased and an agriculture became more settled over the last two centuries, however, cultivation practices such as tillage, intensified conversion of native soil to agriculture, and a large amount of native micro-organisms have disappeared, resulting in depletion of organic matter and decreasing soil fertility. Harvesting of residues and deforestation gradually accelerated the problem. Increasing loss of species in the past few decades has been a consequence of habitat destruction and direct exploitation of plant populations. As a result, most endemic plants have disappeared simultaneously with most rhizosphere organisms, such as mycorrhizal fungi and *Rizobium* bacteria, causing the decline in soil biological diversity.

Mycorrhizal fungi are the largest existing symbiotic plant communities in plant roots, which are the essential part of healthy plant growth, survival, and soil biological quality, particularly in desertified landscapes, because of their essential role in sustaining the vegetation cover. In natural soils, mycorrhizae are abundant and readily available to plants for their health and soil quality. However, loss of mycorrhizal fungi is indispensable in mismanaged soils with excess use of chemicals and practices of tillage and irrigation. Desertified and mismanaged ecosystems are very fragile and subject to progressive disturbance of the vegetation cover and rapid degradation of soils. High disturbance and degradation generally results in the loss or reduction of mycorrhizal propogules present in the soil, consequentially the subsequent reduction in the inoculum potential of mycorrhizal development (4). If the mycorrhizal inoculum potential is low or ineffective, it may limit the successful reestablishment of native plants as a result of the decline of soil productivity and increase in biological degradation.

FACTORS THAT AFFECT SOIL BIOLOGICAL DEGRADATION

Soil Management

Agricultural machinery often causes soil compaction, which increases bulk density and alters soil pore-size distribution, and movement of air, water, and nutrients. Recently, North American farmers have been widely using a technique known as *conservation tillage* or *zero tillage* to reduce the degradation problem. This technique employs special machinery and herbicides to plant crops with minimal disturbance to the soil surface. Studies since the 1980s have indicated that soil disturbance by tillage reduced the effectiveness of mycorrhizae symbiosis (5). Mycorrhizal symbionts produce extraradical hyphae that may extend several centimeters out into the soil and exude organic material, which are the substrates for micro-organisms. The extraradical or external soil hyphae of arbuscular mycorrhizal (AM) association is an important contributor to the process of creating a stable soil aggregate structure. These hyphae associates frequently produce sticky materials that cause soil particles to adhere, creating small aggregates that impart structural properties to soil, allowing for improved aeration and water percolation, as well as stability (6).

The infectivity of the external hyphae network becomes inactive following soil disturbance (7). Soil disturbance can reduce the infective potential of AM fungi in several ways: 1) propoaguls may be physically damaged, spores may be crushed, the external hyphae network and colonized root fragments may be disrupted; 2) disturbance may alter the physical, chemical, or biological environment of the soil, which in turn prevents the colonization by or germination of AM propaguls; 3) soil disturbance may eliminate host plants, leading to changes in the carbon supply available to the fungi.

Tisdall's (6) studies clearly document the role of mycorrhizal hyphae on facilitating the flow of C compounds to the bulk soil, which is an essential process for soil stability. The author has outlined this contribution in three steps, namely the hyphae growth into the soil matrix creating the skeletal structure that holds the primary soil particles together through physical entanglement, the plant roots and hyphae together creating particular physical and chemical conditions and producing organic and amorphous materials for the binding of particles, and the plant roots and hyphae enmeshing micro-aggregates into macro-aggregate structures. After the development of the soil structure, the aggregates enhance carbon and store nutrients, thus providing microhabitats for soil organisms.

Recent discoveries of copious production of glycoprotein (glomalin) by arbuscular mycorrhizal fungi and the apparent recalcitrance of this material in soils led to the comparison between concentration of glomalin and aggregate stability. Aggregate stability was linearly correlated with all measures of glomalin (mg/g of aggregates) in the soil (8).

Recent research conducted on soil organisms has defined its influence on the increase of soil fertility through the development of soil structure (9). The best example of this are the earthworms and termites that transport large amounts of soil through the soil profile, mixing organic and mineral compounds, building up resistant soil structural units, and changing micro- and macro-pore volumes. Conservation tillage has also been proven to stimulate rhizosphere micro-organisms in different soil layers (9).

The Effect of Chemicals on SBD

Long-term applications of P and N and micronutrient fertilizers may reduce the abundance of indigenous AM fungi (or AMF). The highest concentrations of P, Zn, and Cu in maize were observed in nonreduced and reduced tillage treatments concurrently with the highest hyphal densities (10), preserved against destruction by tillage. The results have clearly shown that the greater rates of P fertilizers were not required in reduced tillage systems, compared to conventional causing greater soil disturbance (5, 11).

High concentrations of heavy metals were determined to adversely affect the size, diversity, and activity of microbial populations in soil and total AMF spore numbers decreased with increasing amounts of heavy metals in the soil (12). Pesticide, herbicide, and nematicides are used to control a wide variety of organisms on many crops; unfortunately, often the control they provide is erratic. This erratic behavior is not always predictable and has been associated with the chemical, physical, and biological degradation of soils.

Crop Management

Fallowing

Crop rotation and fallow systems have been observed to significantly affect the diversity and the function of mycorrhizal fungi (13) and the loss of root colonizing AM fungi that benefit the plant by increasing uptake of P in the semi-arid cropping systems of Australia and the United States (14). Decline in AM colonization and soil propogule densities were the results of fallowing and correlated with plant P and Zn deficiency (14). Content of P, Zn, and Cu in test plants was reduced by about 40%, 63%, and 70%, respectively, by 90 days of fallow (10).

Crop rotation

Humankind's knowledge of the benefit of crop rotation dates back more than 3000 years. Recent literature links its effect on yield to increasing AM colonization. The so-called green revolution inducing the trend of monoculture has been responsible for the increased use of chemical and the control of weeds, insects, and plant diseases.

It has been demonstrated that rotation not only affects mycorrhiza formation in general, but also can affect species of AM fungi associated with different crops (15). Mader and colleagues (16) studied the effect of the conventional rotation and two low-input sustainable systems on mycorrhizal spores. They revealed that the low-input systems tended to have higher populations of spores and a greater diversity of mycorrhizal fungi than did the conventional farming systems.

Stubble burning and grazing

Harvest residues are usually burned on the surface or used as forage, household fuel wood, or construction material before soil cultivation for second crop sowing. This, together with overgrazing, removes large quantities of organic matter from the soil, causing biological degradation by altering the turnover rate of the organic matter. Serious negative environmental impacts of harvesting and overgrazing result, inducing land degradation reflected by increased soil compaction and erosion and decreased soil fertility. This leads to the biological degradation of the soil, with decreased indigenous mycorrhizae—the "must" for success for sustainable agriculture, conventionally restored by pasture development (17). Forest clearing followed by maize/cassava cropping in Ghana is a good example with a decrease in the carbon content of the soil from 2.19% to 1.50% in eight years (18).

Heavy grazing strongly affected soil properties indirectly through a change in shrub composition and cultivation, resulting in a decrease in soil nitrogen and organic matter. This loss was still detectable after 20 years of fallow in Australia and South Africa (19).

SOIL RESILIENCE

Soil biological degradation can be restored by several methods: 1) recycling of organic wastes; 2) suitable soil and crop management reducing tillage; 3) increasing crop rotation; 4) better water management; 5) use of cover crops for enhancing soil organic matter; and 6) use of beneficial soil organisms, such as mycorrhizal fungi.

Agroforestry is being used to control soil biological degradation, for reforestation and to establish pasture. The latter, especially with deep-rooted perennial grassland, allows the sustainability of the soil to be regained by increasing the number of indigenous micro-organism populations and thereby increasing carbon levels in the soil through organic matter.

CONCLUSION

The use of native and genetically engineered micro-organisms holds considerable promise for increasing soil biological productivity. An understanding of the impact of agronomic practices on communities of mycorrhizal fungi would help to ensure an opportunity for the utilization of the symbiosis and would contribute to the success of sustainable agriculture. It is important to appreciate that AMF are present and are an integral part of most conventional agriculture production systems. It is also important to identify how those benefits can be maximized to enhance agricultural sustainability.

REFERENCES

1. Bethlenfalvay, G.J.; Schüepp, H. Arbuscular Mycorrhizaes and Agrosystem Stability. In *Impact of Arbuscular Mycorrhizas on Sustainable Agriculture and Natural Ecosystems*; Gianinazzi, S., Schüepp, H., Eds.; Birkhauser Verlag: Basel, 1994; 117–132.
2. Elliott, L.F.; Lynch, J.M. Biodiversity and Soil Resilience. In *Soil Resilience and Sustainable Land Use*; Greenland, D.J., Szabolcs, I., Eds.; CAB International: Oxon, United Kingdom, 1994; 353–364.
3. Rao, D.L.N.; Dadarwal, K.R. *Biological Amelioration of Degraded and Contaminated Soils. Biotechnological Approaches in Soil Micro-Organisms for Sustainable Crop Production*; Central Soil Salinity Research Institute: Karnal, India, 1997; 261–275.
4. Requena, N.; Jeffries, P.; Barea, J.M. Assessment of Natural Mycorrhizal Potential in a Desertified Semiarid Ecosystem. Applied Environmental Microbiol. **1996**, *62*, 842–847.
5. Miller, M.H.; McGonigle, T.P.; Addy, H.D. Functional Ecology of Vesicular Arbuscular Mycorrhizas as Influenced by Phosphate Fertilization and Tillage in an Agricultural Ecosystem. Critical Revive Biotechnology **1995**, *15*, 241–255.
6. Tisdall, J.M. Fungal Hyphae and Structural Stability of Soil. Australia Journal Soil Research **1991**, *29*, 723–743.
7. McGonigle, T.P.; Miller, M.H.; Young, D. Mycorrhizae, Crop Growth, and Crop Phosphorus Nutrition in Maize-Soybean Rotations Given Various Tillage Treatments. Plant and Soil **1999**, *210* (1), 33–42.
8. Wright, S.F.; Upadhyaya, A.A. Survey of Soils for Aggregate Stability and Glomalin, a Glycoprotein Produced by Hyphae of Arbuscular Mycorrhizal Fungi. Plant and Soil **1998**, *198* (1), 97–107.
9. Aslam, T.; Choudhary, M.A.; Saggar, S. Tillage Impacts on Soil Microbial Biomes C, N and P, Earthworms and Agronomy After Two Years of Cropping Following Permanent Pasture in New Zealand. Soil and Tillage Research **1999**, *51* (1-2), 103–111.
10. Kabir, Z.; O'Halloran, I.P.; Hamel, C. Combined Effects of Soil Disturbance and Fallowing on Plant and Fungal Components of Mycorrhizal Corn (*Zea mays* L.). Soil Biology and Biochemistry **1999**, *31* (2), 307–314.
11. Miller, M.H. Arbuscular Mycorrhizae and the Phosphorus Nutrition of Maize: A Review of Guelph Studies. Canadian Journal of Plant Science **2000**, *80* (1), 47–52.
12. Del Val, C.; Barea, J.M.; Azon-Aguilar, C. Diversity of Arbuscular Mycorrhizal Fungus Populations in Heavy-Metal-Contaminated Soils. Applied Environmental Microbial **1999**, *65* (2), 718–723.
13. Thompson, J.P. Correction of Dual Phosphorus and Zinc Deficiencies of Linseed (*Linum usitatissimum* L.) with Cultures of Vesicula-Arbuscular Mycorrhizal Fungi. Soil Biol Biochem. **1996**, *28* (7), 941–951.
14. Ellis, J.R. Post Flood Syndrome and Vesicular-Arbuscular Mycorrhizal Fungi. Prod. Agriculture **1998**, *11* (2), 200–204.

D

15. Johnson, N.C.; Pflenger, R.K.; Crookston, R.K.; Simmons, S.R.; Copeland, P.J. Vesicular-Arbuscular Mycorrhizas Respond to Corn and Soybean Cropping History. New Phytology **1991**, *117*, 657–663.
16. Mader, P.; Edenhofer, S.; Boller, T.; Wiemken, A.; Niggli, U. Arbuscular Mycorrhizae in a Long-Term Field Trial Comparing Low-Input (Organic, Biological) and High-Input (Conventional) Farming Systems in a Crop Rotation. Biology and Fertility of Soils **2000**, *31* (2), 150–156.
17. Crush, J. Changes in Effectiveness of Soil Endomycorrhizal Fungal Populations During Pasture Development. New Zealand Journal Agriculture Research **1978**, *21*, 683–685.
18. Ney, P.H.; Greenland, D.J. *The Soil Under Shifting Cultivation*, Technical Communication No. 51; Common-Wealth Bureau of Soils: Herpenden, 1960.
19. Malinde, D.K.; Fawcett, R.K.; Bligh, K.; Cock, G.; Darling, W. The Effect of Grazing on Erosion and Nutrient Depletion. In *Towards Sustainable Land Use*, 9th Conference of the International Soil Conservation Organization (ISCO), Bon, Germany, 26–30 August 1996.

DEGRADATION, CHEMICAL

M. Rifat Derici
University of Çukurova, Department of Soil Science, Adana, Turkey

INTRODUCTION

Soil chemical degradation can be described as an undesirable change in soil chemical properties such as pH, size and composition of cation exchange complex, contents of organic matter, mineral nutrients, and soluble salts. Changes in one or more of these properties often have direct or indirect adverse effects on the chemical fertility of soils, which can lead to a decrease in soil productivity. Soils are continually undergoing natural chemical changes as a result of weathering. The combination of the weathering process with other factors, such as parent material, climate, biota, and topography, are responsible for the evolution of soil variety. The forces and factors affecting soil formation are always operational in a manner such that a static equilibrium state is never attained. However, the balance at any given time is highly sensitive to these factors and a new equilibrium towards soil degradation may be the outcome when there are natural or man-made changes.

The most widespread types of chemical degradation in soils are an excessive decrease or increase in pH (acidity or alkalinity), an increase in soluble salt content (salinity), a decrease in organic matter content, and a loss of mineral nutrients through leaching or crop offtake (Table 1). Salinity and alkalinity are closely related and thus will be discussed together.

SOIL PH AND DEVELOPMENT OF SOIL ACIDITY

Chemically fertile soils typically have a pH range of 5.5–7.5. Soil pH is determined by the mineralogical make-up (clay minerals, various metal oxides and hydroxides, lime, etc.), organic matter content of the soils, and dissolved CO_2 in the aqueous phase. In spite of small diurnal or seasonal fluctuations, soil pH is strongly buffered by these factors. Any measure of pH below 7 is defined as the active acidity, whereas the ability of the soil to maintain a low pH level is referred to as the potential acidity (1). The active acidity represents the H^+ ions in the soil solution. Potential acidity includes exchange and titratable acidities where the former constitutes most of the latter in acid soils. In most of the acid soils, the exchangeable acidity is approximately equal to the strong acid component, and the nonexchangeable but titratable acidity to the weak acid component (2). The exchange acidity includes the protons associated with the cation exchange sites on the clay mineral and organic fractions. The exchange acidity as a portion of total acidity varies with the nature of the soil and the percentage base saturation. There is an equilibrium between active and exchange acidities and as the H^+ ions in soil solution is neutralized the cation exchange phase brings new H^+ ions into solution. The sources of soil acidity are humus, aluminosilicates, hydrous oxides and soluble salts (1).

Humus or humic matter causes acidity through dissociation of H^+ in its carboxylic, phenolic, and similar H^+ yielding functional groups. The humic fraction is considered the weak acid component of the total acidity. Furthermore, the complexes of humus with iron and aluminum can produce H^+ ions upon hydrolysis (1). The charged sites associated with aluminosilicate clay minerals are occupied by various cations present in the solution phase. As the portion of basic cations such as Ca, K, Mg, Na are reduced through leaching or by plant uptake, the portion of the total charge occupied by H^+ ions increases. This process is accompanied by a reduction in pH as the dominating exchangeable H^+ controls the solution phase. When the soil pH falls below 6, Al in octahedral sheets dissociates and is adsorbed in exchangeable form by clays (3), thereby increasing the Al saturation. Exchangeable Al is the major cause of exchange acidity. When dissociated from the exchange complex as the Al^{3+} ion, it produces H^+ ions through the following hydrolysis reactions (1).

$$Al^{3+} + H_2O \rightarrow Al(OH)^{2+} + H^+$$

$$Al(OH)^{2+} + H_2O \rightarrow Al(OH)_2^+ + H^+$$

$$Al(OH)_2^+ + H_2O \rightarrow Al(OH)_3 + H^+$$

Hydrous oxides of Al and Fe either in free crystalline and amorphous forms or as coatings or interlayers in clay

Table 1 Extent of chemical degradation of soils on continental basis (million ha)

Degradation type	Africa	Australasia	Europe	North America	North and Central Asia	South and Central America	South and Southeast Asia	Total
Acidity[a]	423,306	40,941	217,794	335,583	170,066	770,036	268,987	2,226,713
Salinity[a]	48,574	16,565	2308	127	46,895	24,344	48,512	187,325
Sodicity[a]	13,800	38,099	7906	10,748	30,062	34,652	–	135,267
Nutrient depletion[b]	45,000	–	3000	–	–	72,000	14,000	134,000

[a]FAO/AGL, Land and plant nutrition management service, prosoil problem soils database (www/fao.org/ag/agll/prosoil).
[b]GLASOD; Global assessment of the status of human-induced soil degradation (www/fao.org/docrep/w4745E/w4745e0a.htm).

minerals are also subjected to stepwise hydrolysis and thus produce H^+ ions.

Soluble salts stemming from dissolution of soil minerals or from fertilizers may directly or indirectly cause acidity. When only the cation of a salt is taken up by plants, H^+ ions released from roots can contribute acidity. Fertilizers, especially ones of ammonium, indirectly cause acidity by production of H^+ ions as a by-product of nitrification. It is reported that continuous application of ammonium sulfate, when not accompanied by liming, decreased the soil pH from 5.8 to 4 in profiles with depth to at least 1 m in less than 70 years (4).

Soil acidity limits the plant growth by H^+ toxicity and decreases the macronutrient base cation content. Furthermore, the solubility of Fe^-, Mn^-, and Al^-containing minerals are enhanced at low pH levels and the toxicity of these elements becomes a major problem (5). The activities of soil organisms, including nitrifying bacteria, are severely restricted at pH levels lower than 5.5.

Human activities may accelerate or initiate acidification of soils. A monoculture with an excessive use of fertilizers on sensitive soils often results in the depletion of bases and an increase in soil acidity. Soil acidification can also arise as a result of changes in land use pattern, mostly due to differential rooting patterns and nutrient requirements of plants. For example, establishment of coniferous forest on a former grassland in New Zealand increased exchangeable acidity in the top 20–30 cm of soil (6). Oxides of N and S released in the atmosphere by industrial activity and washed onto soils with rainfall (acid rains) is an ever growing problem. Acid rains are not confined to industrial areas, but are transported long distances by mass air movements and affect both agricultural areas and forestland. Forests are especially vulnerable to atmospheric deposition of acids (7, 8).

SALINITY AND ALKALINITY

Salinity

Salinity is a common problem in arid and semiarid regions where evapotranspiration exceeds rainfall. Under these conditions (salinization regime) there is not enough water to wash the soluble salts down the profile below the rooting zone. Thus, soluble salts, originating from various sources, accumulate in the soil profile at certain depths—known as the salic horizon—or at the soil surface (Fig. 1), depending on the water regime. The potential for soil salinization may be greater in irrigated agricultural areas since even good quality irrigation water contains appreciable amounts of salts. If not washed from the soil profile to a drainage system by the leaching fraction of water, chloride, sulfate, carbonate, and bicarbonate salts of Na may concentrate in the soil profile and cause salinity in a very short period of time (9). Another cause of salinity is the absence of a drainage system or poor drainage especially in lowlands. A raised (high) water table as a result of an ineffective discharge system is still another cause of salinity. For example, clearing of perennial species and conversion of this land to production of annual crop species with much lower transpiration rates has led to rising water tables across vast areas of Australia. The upward capillary movement of water carried the dissolved salts, previously present at depth, to the rooting zone. The sources of soluble salts, besides irrigation water, are mineral weathering, fertilizers, salts used on frozen roads, atmospheric transfer of sea spray, and lateral movement of groundwater from salt-containing areas. The contribution from each of these sources depends on the geographical position and topography of the area. In this respect, coastal areas, especially delta plains and regions lacking natural

Fig. 1 Salt crusts on soil surface caused by irrigation with low quality water in Faisalabad, Pakistan. (Photo courtesy of Dr. C. Kirda, University of Çukurova, Turkey.)

water outlets, are the most vulnerable areas. It is strongly emphasized that unless the salts are properly discharged to the sea by an effective drainage system, the salinity problem cannot be eliminated, but will only transferred to another location (10).

Salinity affects plant growth by affecting water and nutrient uptake and through specific toxicity of Na, Cl, and B. The dissolved salts in water increase the osmotic potential, thereby creating the so-called physiological drought. Toxicity develops when the ions taken up from soil solution accumulate in leaves. As water is lost in transpiration, the concentration of the toxic ions increases and causes damage to various degrees, depending on the sensitivity of plants (11). Salinity affects the mineral nutrition of plants by reducing the availability and uptake of nutrients through the interaction of Na and Cl with nutrient cations and anions and by interfering with the transport of elements within the plant (12).

The degree of salinity is measured as electrical conductivity of a soil saturation paste or extract and is reported as deci Siemens per meter (dS^{m-1}). Plant sensitivity to salinity is expressed as a percent decrease in yield at a given level of salinity (11).

Alkalinity

Addition of salts to soils increases the concentration of Na in the soil solution more than those of Ca and Mg, since the Na ion is more weakly bound to the soil surface than the divalent cations. This increase in exchangeable Na (Na_x) is called sodication, and soils degraded in this manner are referred to as sodic soils. The measure of sodicity is exchangeable sodium percentage, which is the ratio of Na_x to cation exchange capacity. This parameter is sometimes expressed as the exchangeable sodium ratio, which is the ratio of Na_x to other exchangeable cations. If the soil solution contains CO_3^{2-} and HCO_3^- in excess of Ca^{2+} and Mg^{2+}, highly soluble Na salts of these anions hydrolyze and the soil pH rises above 8.5. This process is termed as alkalinization. Sodic soils do not necessarily have high pH, but in well-aerated soils alkalinization often follows sodicity. Sodicity, although a chemical property, has adverse effects on soil structure. As Na_x increases, the binding effects of divalent Ca and Mg ions on clay particles are overcome by the dispersing action of Na ions. Dispersed particles move with water and quickly clog the soil pores, causing drastic reductions in water and air permeability in sodic nonsaline soils.

DEPLETION OF SOIL ORGANIC MATTER

The accumulation of soil organic matter, namely humus, in soils starts with the production of biomass and approaches an equilibrium dependent on the effects of factors such as climate, type of vegetation, topography, soil texture, and drainage conditions. At equilibrium, when additions by biomass and removal by mineralization are in balance, organic matter contents range from less than 1% in arid regions to over 20% in organic deposits of cool humid climates. Any change in these factors may disturb the system and a new equilibrium towards depletion of soil organic matter results. A highly disturbing factor in this respect is cultivation. Less organic material is returned to soil at harvest in most cropping systems, and tillage accelerates decomposition of soil organic matter (3, 13–15). Man-made changes in plant cover, such as the conversion of forests and grasslands to croplands, promotes rapid decomposition of the organic matter present in soils (13, 15). Soil erosion is another factor that causes significant reductions in soil organic matter content, first, by decreasing the overall productivity and thus the production of biomass some of which is returned to soil and second, by carrying away the organic matter present in the lighter fraction of the surface soils (13).

LOSS OF PLANT NUTRIENTS

Plant nutrient elements are continuously lost from soils by crop removal (nutrient mining), erosion, leaching, and volatilization at rates determined by the type of vegetative cover, cropping system, and climatic conditions. In natural systems such as grasslands and forests, nutrient losses from soils are somewhat compensated by returns from falling plant debris and mineral weathering. On the other hand, in intensive agriculture much larger amounts of nutrients are taken away from soil with little return in crop residues, in many cases exhausting the nutrient reserves in soils (16). Erosion permanently removes the readily available nutrients from the top soil, which is the most productive layer.

Leaching losses in agricultural lands can be significant for anionic nutrients, especially for nitrate, under various farming systems (17). Basic nutrient cations such as Ca, Mg, and K may be leached from soils acidified by atmospheric deposition (18) or from those under shifting cultivation (19).

REFERENCES

1. Seats, L.F.; Peterson, H.B. Acid, Alkaline and Sodic Soils. In *Chemistry of the Soil*; Bear, F.E., Ed.; Reinhold: New York, 1967; 124–145.
2. Coleman, N.T.; Thomas, G.W. The Basic Chemistry of Soil Acidity. In *Soil Acidity and Liming*; Agronomy no. 12, Pearson, R.W., Adams, F., Eds.; American Society of Agronomy: Madison, WI, 1967.
3. Foth, H.D.; Turk, D.M. *Fundamentals of Soil Science*, 5th Ed.; Wiley: New York, 1972.
4. Goulding, K.W.T.; Blake, L. Land Use Liming and the Mobilization of Potentially Toxic Metals. Agri. Ecosys. Environ. **1998**, *67*, 135–144.
5. Marschner, H. *Mineral Nutrition of Higher Plants*; Academic Press: New York, 1995.
6. Alfredsson, H.; Condron, L.M.; Clarholm, M.; Davis, M.R. Changes in Soil Acidity and Organic Matter Following the Establishment of Conifers on Former Grassland in New Zealand Forest. Ecol. Mgmt. **1998**, *112*, 245–252.
7. Ulrich, B. Waldsterben: Forest Decline in West Germany. Environ. Sci. Technol. **1990**, *24*, 436–441.
8. Van Breemen, N.; Driscoll, C.T.; Mulder, J. Acidic Deposition and Internal Proton Sources in Acidification of Soils and Waters. Nature **1984**, *307*, 599–604.
9. Haw, M.; Cocklin, C.; Mercer, D. A Pinch of Salt: Landowner Perception and Adjustment to Salinity Hazard in Victoria, Australia. J. Rural Stud. **2000**, *16*, 155–169.
10. Kamphorst, A.; Bolt, G.H. Saline and Sodic Soils. In *Soil Chemistry A. Basic Elements*; Bolt, G.H., Bruggenwert, M.G.M., Eds.; Elsevier: New York, 1978; 171–191.
11. Ayers, R.S.; Westcot, D.W. *Water Quality for Agriculture*; FAO Irrigation and Drainage Paper 29, FAO: Rome, 1985.
12. Grattan, S.R.; Grieve, C.M. Salinity–Mineral Nutrient Relations in Horticultural Crops. Sci. Hort. **1999**, *78*, 127–157.
13. Gregoricha, E.G.; Greerb, K.J.; Andersonb;, D.W.; Liange, B.C. Carbon Distribution and Losses: Erosion and Deposition Effects. Soil Till. Res. **1998**, *47*, 291–302.
14. Chan, K.Y.; Hulugalle, N.R. Changes in Some Soil Properties Due to Tillage Practices in Rainfed Hardsetting Alfisols and Irrigated Vertisols of Eastern Australia. Soil Till. Res. **1999**, *53*, 49–57.
15. Riezebos, H.Th; Loerts, A.C. Influence of Land Use Change and Tillage Practice on Soil Organic Matter in Southern Brazil and Eastern Paraguay. Soil Till. Res. **1998**, *49*, 271–275.
16. Kirchmann, H.; Throvaldson, G. Challenging Targets for Future Agriculture. Eur. J. Agron. **2000**, *12*, 145–161.
17. Hansen, B.; Kristensen, E.S.; Grant, R.; Hogh-Jensen, H.; Simmelgaard, S.E.; Olesen, J.E. Nitrogen Leaching from Conventional Versus Organic Farming Systems—A System Modelling Approach. Eur. J. Agron. **2000**, *13*, 65–82.
18. Thimoier, A.; Dupouey, J.L.; Tacon, F. Recent Losses of Base Cations from Soils Of *Fagus Sylvatica* L. Stands in Northeastern France. Ambio **2000**, *29*, 314–321.
19. Brand, J.; Pfund, J.L. Site and Watershed-Level Assessment of Nutrient Dynamics Under Shifting Cultivation in Eastern Madagascar. Agric. Ecosys. Environ. **1998**, *71*, 169–183.

DEGRADATION, CRITICAL LIMITS OF SOIL PROPERTIES AND IRREVERSIBLE

Anthony R. Dexter
Institute of Soil Science and Plant Cultivation, Pulawy, Poland

Michael A. Zoebisch
Asian Institute of Technology, Pathumthani, Thailand

INTRODUCTION

Major causes of soil degradation are compaction, loss of organic matter, salinization, and pollution. Soil properties that are important for crop growth or for soil management must often lie within certain critical limits which are summarized in Table 1. Degradation of topsoils is usually reversible, whereas degradation of subsoils is much more difficult to reverse or may even be permanent.

SOIL STRUCTURE

Virgin soils have structures that took hundreds or thousands of years to form as a result of pedogenic processes. When such soils are brought under cultivation, some stuctural features are destroyed permanently. However, if the soil is managed carefully and appropriately, then it can be productive and can form part of a sustainable ecosystem. Soil with a good structure has a wide distribution of pore sizes and the structure is stable (1). A wide distribution of pore sizes is necessary for soil function and arises when the soil primary particles are arranged into compound particles of different sizes. The structure will be stable if the clay and other colloidal particles, which glue the soil together, remain flocculated and do not disperse when in contact with water. Therefore, a low value of clay dispersibility is required for a stable soil structure. Factors associated with low clay dispersibility include organic matter and calcium. Factors that cause clay dispersion include the presence of sodium and mechanical energy inputs, e.g., from excessive tillage, especially when the soil is wetter than the lower plastic (or lower Atterberg) limit (2).

Soil Bulk Density

The bulk density of the soil (or mass per unit volume) is not important as such. Plant roots, for example, do not sense soil density. What they do sense is mechanical impedance to elongation, which is positively correlated with bulk density, although the correlation is different for different soils. Effects of bulk density on root growth for soils of different clay contents are shown in Fig. 1. The values in this figure are indicative only because mechanical impedance depends on soil water potential, clay mineralogy, soil physical chemistry, and other factors. Roots may also bypass the dense parts of a soil by growing in cracks and biopores. Therefore, an exact relationship between bulk density and root growth is not to be expected. Bulk density is often used as an indicator of the soil physical condition, although it is not a measure of soil function and its use is subject to limitations.

HYDRAULIC PROPERTIES

Water Retention

Soil must be able to store useful amounts of water for crop growth. The maximum amount that can be held is known as the *field capacity*, which is the water content of a soil after at least 2 days of drainage following heavy rain or a large irrigation. In the laboratory, it may be measured as the water content of soil samples when drained from saturation to a matric water potential of -100 h Pa. The dry limit for crop use is the *permanent wilting point,* which is a plant characteristic. This is the water content of the soil below which plants lose turgor pressure and cease growing, even under conditions of zero transpiration. For many crops (e.g., cereals) it corresponds to a matric water potential of -1.5 MPa. The *available water capacity* or *plant available water* are defined as the difference between field capacity and permanent wilting point, and quantify how much water a soil can store for use by crops. The higher this quantity, the better. However, the soil must be aerobic at field capacity. Soil compaction reduces the value of the available water

Table 1 Critical limits and optimum values of soil properties for most soils under typical agricultural use

	Soil water/solution			Aeration			Mechanical	
	Matric potential (kPa)	EC (mS cm^{-1})	Acidity (pH)	Air-filled porosity (%)	Redox potential Eh (mV)	ODR (μg m^{-2}s^{-1})	Penetrometer resistance (kPa)	Friability
Maximum	−5 to −10[a]	1.5–12[b]	7.5[b]	–	700	–	2000	–
Optimum	−20	<0.5	7.0	15	450	>50	1000	0.4[c], 0.8[d]
Minimum	−1000 to −1500[b]	–	6.5[b]	10	330	20–30[b]	500	0.2[c], 0.4[d]

[a] Soil dependent.
[b] Plant dependent.
[c] By the method of size dependence.
[d] By the method of coefficient of variation.

capacity and also the air-filled porosity through preferential destruction of the larger pores.

Infiltration and Hydraulic Conductivity

Rain or irrigation water must infiltrate into soil where it can be stored for subsequent use by crops, otherwise, runoff will occur with adverse consequences (e.g., erosion, flooding, transport of nutrients, etc.). Infiltration occurs in two main stages. At short times, infiltration is rapid because water is being "sucked" into the soil by capillary forces. At longer times, infiltration reaches a steady rate equal to the saturated hydraulic conductivity of the soil. Whereas irrigation rates can be reduced to match the ability of the soil to absorb water, no such control over rainfall is possible. To avoid runoff, the rate at which the soil can absorb water should be greater than the highest rainfall rates that may be expected. Because of the two-stage nature of water infiltration, the onset of runoff depends not only on rainfall intensity but also on storm duration and "storm profile" (how intensity varies with time through a storm). Subsoil compaction may reduce hydraulic conductivity by factors of 10 or even 100. Therefore, once the pore space in the topsoil has been filled-up with water, runoff may occur even at low rainfall rates.

SOIL AERATION

Air-Filled Porosity

Adequate soil aeration requires a network of continuous air-filled pores throughout the soil volume. For soil to be aerobic at field capacity (see earlier), these pores need to be larger than 30 μm. Air-filled porosity is the simplest measure of soil aeration (4, 5). The soil air should contain at least 5% oxygen.

Electrochemical Measurements

Redox potential, E_h, may be measured by the electrical potential between a saturated calomel reference electrode and a platinum electrode (4). E_h ranges from +700 mV in fully aerobic conditions to −200 mV in completely

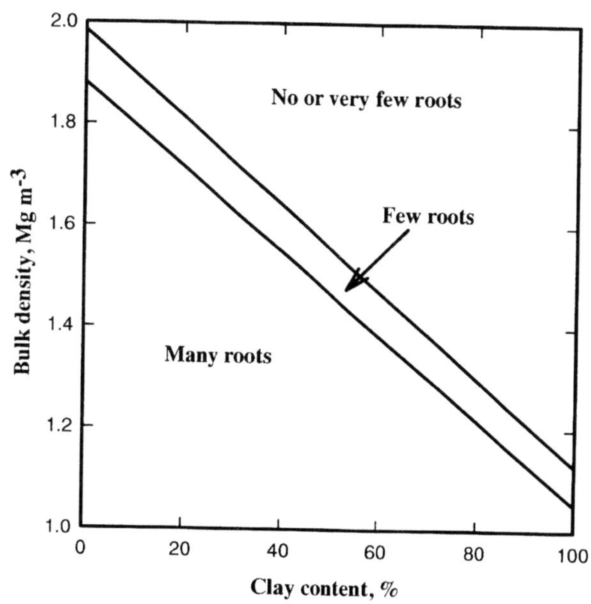

Fig. 1 Critical values of soil bulk density for root growth at a water potential of −33 kPa as a function of soil clay content. (Adapted from Ref. 3.)

anaerobic and reduced conditions. A value of $E_h = +300$ mV corresponds to the initiation of iron reduction, which causes the blue mottling that is characteristic of inadequately aerated soils.

The oxygen diffusion rate (ODR) is a measure of the rate at which oxygen molecules impact with a simulated root that is inserted into the soil. Typically, a root is simulated by a platinum wire of 0.5 mm diameter at an electrical potential of −650 mV relative to a saturated calomel reference electrode (4). When an oxygen molecule impacts with the electrode, it is reduced and four electrons flow in an electrical circuit. The electrical current in the circuit is proportional to the number of molecules per second which are impacting with the Pt wire electrode.

SOIL STRENGTH

Penetrometer Resistance

Penetrometers are used to measure soil strength in the field (6). Penetrometer resistance is measured by the force required to push a conical metal probe into the soil. The probe is typically of about 20 mm in diameter (or smaller in strong soils) and 30° total included angle. Penetrometer resistance is expressed as the force per unit projected area of the cone, in kPa or MPa. When penetrometer resistance >5 MPa, crop development is seriously affected. Penetrometer resistance increases sharply with decreasing soil water content.

Tillage becomes difficult when penetrometer resistance >2 MPa. There is no real upper limit of soil strength for tillage; it is a question of how much energy the farmer is prepared to use to produce the desired soil structure.

The lower limit for soil strength arises from the requirement for the soil to support traffic and from the need for the soil to be strong enough to prevent plants from falling over. It has been found that passage of a tractor across the soil was not possible when penetrometer resistance was <250 kPa, was possible when it was in the range 250–500 kPa, and was easy when it was >500 kPa (7). Trafficability of soil depends very much on the design of the vehicle. Similarly, the ability of the soil to hold plants upright depends not only on the soil but also on the geometry and mechanical properties of the plants.

Precompaction Stress

When soil is subjected to compressive stresses, it may compact (i.e., its volume will reduce) until the stress required for further compaction is equal to the applied stress (8). Compaction is an irreversible deformation, and the soil will not return to its original volume when the applied stress is removed. Further compaction will not occur until the previous maximum value of applied stress or *precompaction stress* of the soil is exceeded. Precompaction stress can be measured in the laboratory, and can then be used to determine the vehicle parameters (e.g., total weight, axle loads, tire inflation pressures), which may be used without causing further compaction damage. It is not possible to give a maximum value for precompaction stress, as each case needs to be considered separately. For example, some soils even after light traffic may have extremely low hydraulic conductivities or infiltration rates which would then give rise to water runoff. For every soil, there is a critical value of precompaction stress which should not be exceeded and which must be determined in relation to all aspects of soil function (8).

Friability

Friability describes the ease with which soil can be tilled to produce a seedbed. This can be evaluated qualitatively in the field by feeling how easily the soil crumbles when rubbed between the fingers. Friability depends on the existence of air-filled pores which provide surfaces of weakness along which the soil can break readily. Such pores are typically microcracks that are wider than 30 μm, and therefore air-filled at field capacity. Friability can be quantified either from the decrease of tensile strength with increasing sample size or from the coefficient of variation (standard deviation divided by the mean) of tensile strength values (9). A well-structured soil has a high friability, whereas a structureless soil has a small (in the extreme case, zero) friability. Friability is greater when the content of readily dispersible clay in the soil is smaller; in noncompacted soil relative to compacted soil, and in direct-drilled (zero-tilled) soil relative to conventional (plow) tilled soil.

SALINITY AND SODICITY

The term salinity refers to the amount of salt (usually sodium chloride) in the soil, whereas sodicity refers to the amount of sodium adsorbed on the surfaces of the soil particles (most of the surface area exists on the clay particles). Increase of salinity or of other sodium salts (e.g., Na_2CO_3 and $NaHCO_3$) produces an increase of sodicity. Salinity of soil water is usually measured by electrical conductivity, EC, and sodicity is usually expressed as the exchangeable sodium percentage, ESP.

Salinization of soil occurs when the input of salt is greater than the output. This can occur in irrigation systems where there is some salt in the irrigation water and where there is inadequate drainage to remove these salts. It can also occur as dryland salinization in which clearance of natural deep-rooted vegetation and its replacement by crop or pasture species reduces water extraction from deep layers. The water table can then rise and can bring salt of geological origin toward or to the soil surface.

Sodicity causes soil particles to repel and to disperse in water. Therefore, sodic soils are highly unstable. They dry into extremely dense, strong blocks and their friability is essentially zero. Tillage of such soils, if possible at all, can be done only over a very narrow range of water contents. Sodic soils may exhibit a complete loss of soil function. The critical value of ESP beyond which soil physical degradation becomes apparent is 15%.

ACIDITY (pH)

The normal range of soil pH values is from 4 to 10. However, most plants prefer conditions between pH 6.5 and 7.5. Some specialized plants tolerate values outside this range. In highly acidic conditions (pH <5), some heavy metals and other elements such as iron, copper, zinc, cobalt, and aluminum become soluble and can become toxic to plants and may be mobilized in the environment with the potential to cause pollution. In agriculture, there is a natural tendency for pH values to drop. A typical rate of decrease of pH might be 0.4 unit for every 10 years although, of course, this depends on the soil type and on conditions. Therefore, soil pH must be adjusted periodically by additions of lime (as $Ca(OH)_2$ or $CaCO_3$).

DEGRADATION: REVERSIBLE OR IRREVERSIBLE?

All types of soil degradation can be caused very easily. However, a cure may be either very slow and expensive or completely impracticable.

Symptoms of degradation of the topsoil (i.e., the arable layer in agricultural soils) include slumping, crusting, low friability, etc. These problems can usually be alleviated by increasing the soil organic matter content and by adding calcium compounds such as gypsum ($CaSO_4$) or lime if the soil is acidic. Organic matter can be increased by changing crop rotations to include a pasture phase, by keeping plant cover all the year round, by adding manures, and also by minimizing losses of soil organic matter, which can occur as a result of excessive or high-intensity tillage. Organic matter increases soil aggregation and reduces clay dispersion in water. Calcium compounds flocculate the clay, thereby reducing dispersion and increasing soil stability. Clay floccules form the building blocks for larger compound particles such as aggregates. The use of such practices over a number of years can produce a soil structure that is better for agricultural production and is more stable. Compaction damage of topsoils is usually not permanent because of the ameliorative effects of tillage, biological activity, and wetting and drying cycles.

Degradation of the subsoil often occurs through compaction (8). Subsoil tillage may make a temporary improvement, but often the soil will recompact to be as bad as or worse than it was before subsoiling. However, recompaction can be minimized if vehicle weights or axle loads after subsoiling are kept significantly below those that caused the original compaction. Any attempt at subsoil tillage should be done only when the whole soil layer to be tilled is drier than the plastic (lower Atterberg) limit; otherwise; the soil will be destabilized. In many soils, the effects can be considered to be permanent.

Soil salinity can be cured in principle by leaching. However, dryland salinity usually cannot be reversed but may be slowed or halted by the planting of deep-rooted plants (e.g., salt-tolerant trees) that can transpire enough water to stop further water table rise. Sodicity can be cured by displacing the sodium with calcium. However, this is difficult because sodic soils have an extremely low permeability to water. Displacement and leaching may be accelerated by not using pure water but by using water with a high enough electrolyte concentration to keep the clay flocculated and hence the permeability high (10, 11). It is necessary to consider where the displaced sodium will go in order to avoid degradation of other areas or pollution of water courses.

Heavy metals can be removed slowly from topsoils by "hyperaccumulator plants," which can be harvested and taken to a safe place for disposal. However, heavy metals are often best controlled by keeping them "fixed" in the soil (adsorbed on the exchange complex). This may require liming of the soil periodically for all time, otherwise the naturally occurring pH drop of the soil may eventually mobilize the heavy metals in the environment at some time in the future. This is the so-called "chemical time bomb" effect.

REFERENCES

1. Dexter, A.R. Advances in the Characterization of Soil Structure. Soil Till. Res. **1988**, *11*, 199–238.
2. Watts, C.W.; Dexter, A.R. The Influence of Organic Matter in Reducing the Destabilisation of Soil by Simulated Tillage. Soil Till. Res. **1997**, *42*, 253–275.
3. Jones, C.A. Effect of Soil Texture on Critical Bulk Densities for Root Growth. Soil Sci. Soc. Am. J. **1983**, *47*, 1208–1211.
4. Glinski, J.; Stepniewski, W. *Soil Aeration and Its Role for Plants*; CRC Press: Boca Raton, FL, 1985.
5. Ball, B.C.; Smith, K.A. Gas Movement and Air-Filled Porosity. In *Soil and Environmental Analysis: Physical Methods*; Smith, K.A., Mullins, C.E., Eds.; Marcel Dekker: New York, 2000; 499–538.
6. Bengough, A.G.; Campbell, D.J.; O'Sullivan, M.F. Penetrometer Techniques in Relation to Soil Compaction and Root Growth. In *Soil and Environmental Analysis: Physical Methods*; Smith, K.A., Mullins, C.E., Eds.; Marcel Dekker: New York, 2000; 377–403.
7. Japanese Agriculture, Forestry and Fisheries Research Council Secretariat, Paddy Field Engineering for Mechanization of Agriculture, Bulletin of Major Research Task No. 40, Tokyo, Japan, 1969 (in Japanese).
8. Horn, R., van den Akker, J.J.H., Arvidsson, J., Eds. *Subsoil Compaction: Distribution, Processes and Consequences*; Catena: Reiskirchen, 2000.
9. Dexter, A.R.; Watts, C.W. Tensile Strength and Friability. In *Soil and Environmental Analysis: Physical Methods*; Smith, K.A., Mullins, C.E., Eds.; Marcel Dekker: New York, 2000; 405–433.
10. Quirk, J.P. Soil Permeability in Relation to Sodicity and Salinity. Phil. Trans. Roy. Soc. Lond. **1986**, A*316*, 297–317.
11. Reeve, R.C.; Bower, C.A. Use of High-Salt Waters As a Flocculant and Source of Divalent Cations for Reclaiming Sodic Soils. Soil Sci. **1960**, *90*, 139–144.

DEGRADATION, FOOD AID NEEDS IN LOW-INCOME COUNTRIES AND

D

Shahla Shapouri
Stacey Rosen
United States Department of Agriculture, Economic Research Service, Washington, D.C., U.S.A.

INTRODUCTION

Global food production has grown faster than the world's population in the past forty years. As a result, enough food is available worldwide so that countries with production shortfalls can import food as long as they have adequate financial resources. Yet many poor countries and millions of poor people do not share in this abundance. They suffer from food insecurity and hunger that are conditions associated with low-quality resources and inadequate purchasing power. For these countries, demand for food aid is high and is likely to increase unless steps are taken to implement improved policies and raise productivity. Gains in domestic production will likely stem from improved yields that are, in part, dependent upon soil quality.

The Economic Research Service (ERS) estimates that about 33 percent—roughly 900 million—of the developing world's population suffers from insufficient food intake (7). For 67 lower income countries, the amount of food needed to provide nutritionally adequate diets was estimated at 13.7 million tons in 2000 and is projected to increase 70 percent to more than 22 million tons in the next decade. Sub-Saharan Africa is the most vulnerable region and is projected to account for almost 80 percent of this "nutritional" gap (of the 67 study countries) in 2010 (Table 1).

The donor countries have demonstrated an interest in providing food assistance to low-income countries; however, there are budget pressures that limit these transfers. The question at hand is what the factors are leading to food insecurity in these countries and what kind of changes can be initiated to compensate for the decline in the quality of resources without reducing the benefits of increased productivity.

IMPROVING AGRICULTURAL PRODUCTIVITY

A common characteristic among the food-deficit countries is the large contribution of domestic food production to food consumption, with only a very small share of domestic consumption provided by imports. Agricultural products also are very important sources of foreign exchange earnings to support the growing import needs. Agricultural production growth depends on the availability and quality of resources. In low-income, food-deficit countries, this means land and labor because of limited use of new technologies.

Constraints in Expanding Agricultural Areas

In many low-income countries, most increases in agricultural output have stemmed from area expansion (1, 8) (Fig. 1). In Sub-Saharan Africa, area expansion accounted for more than 80 percent of the region's grain output growth between 1980 and 1999. This means that yield growth contributed to less than 20 percent of the growth. In Latin America, area expansion accounted for 68 percent of the growth in grain production. In Asia, however, the reverse was true; area expansion accounted for less than 5 percent of the growth in grain output. Stated another way, gains in yields contributed to nearly all of the growth experienced in the region's grain output.

The long-term prospects for acreage expansion are not bright, because, in most countries, a large part of land that could be used for farming is unfit to cultivate without major investment. In Latin America and Sub-Saharan Africa, continued expansion of cropland means converting range and forestland to crop production, a process with high economic and environmental costs. According to FAO estimates, about half of the land that could be used to produce food in Sub-Saharan Africa has poor quality soil (8). Sub-Saharan Africa has a vast and diverse land area, but the region faces a number of resource constraints (such as lack of water) to sustainable agricultural growth (5). Some countries, such as Sudan and the Democratic Republic of Congo, have vast areas of rainfed land with crop potential, while other countries such as Kenya and Madagascar have already exhausted their high-potential land. Demographic changes are

Table 1 Food gaps by region

	2000	2010
	(million tons)	
Sub-Saharan Africa	10.70	17.50
Latin America	0.63	0.99
Asia	1.70	3.63
Other	0.17	0.18
Total: 67 countries	13.20	22.30

placing increased pressure on land in the region. More than 20 percent of all vegetative land is degraded due to human causes; however, water and wind erosion still account for a majority of the affected hectares. Much of this degraded area is in the Sahel, Sudan, Ethiopia, Somalia, Kenya, and southern Africa. Historically, farmers adjusted to resource constraints by following several years of planting with several fallow years. However, population pressures have reduced the practice of these sustainable agricultural techniques, leading to declines in land productivity.

Given the constraints to area expansion, changes were made in the assumptions for area growth relative to the base scenario for Sub-Saharan Africa, Latin America, and Asia using the structural framework of ERS's food security model. For Sub-Saharan Africa and Latin America, we cut the projected area growth rates used for the baseline scenario in half for each country in the two regions. In Asia, we assumed the area to remain constant during the entire projection period under the new scenario as population pressures will intensify competition for land. In Sub-Saharan Africa, production in the baseline scenario was projected to grow at a rate of 2.1 percent per year;

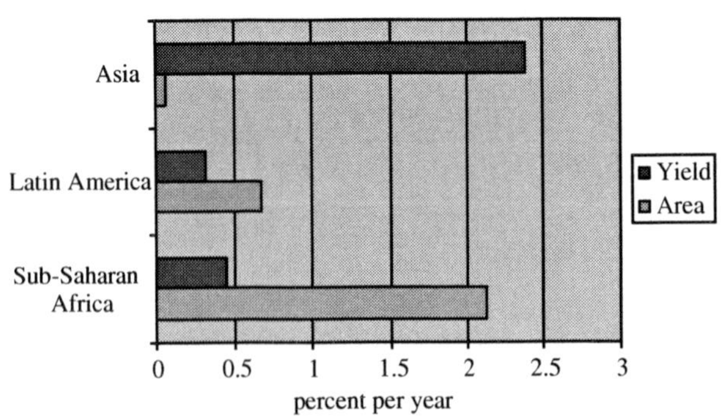

Fig. 1 Sources of grain production growth, 1980–99.

Table 2 Food gaps under alternative scenarios

	Baseline	Reduced area growth
	(million tons)	
Sub-Saharan Africa	17.5	25.6
Latin America	1.0	1.6
Asia	3.6	4.7

under the reduced area growth scenario, this rate is projected to fall to 1.7 percent. As a result of the slower production growth, the region's nutritional food gap in 2010 jumps more than 40 percent to more than 25 million tons (Table 2). In Latin America, the decline in production growth is much less significant—from 2.3 percent to 2.2 percent. The impact is less in this region because gains in yields are expected to be a more significant contributor to production growth than area expansion. In Sub-Saharan Africa, both area and yield were estimated to contribute almost evenly to production gains; therefore, a cut in area growth is going to have a greater impact. This small change in production growth will still have strong implications for the region's nutrition gap that is projected to rise 40 percent under this new scenario. In Asia, production growth is projected to slow from 1.7 percent per year to 1.6 percent when area is held constant. This slowdown translates into a 30 percent increase in the nutrition gap for 2010.

Limited Use of New Technologies

The only option to sustain production growth is to increase yields. Within the developing world, average regional grain yields are the highest in North Africa and the lowest in Sub-Saharan Africa. Yields are highly dependent on the use of improved inputs, particularly fertilizer use. However, yields do not increase commensurately with fertilizer use. A cross-country estimate for developing countries showed that a 1 percent increase in fertilizer use results only in a 0.1–0.3 percent increase in yields (2, 5). The principal factor limiting yield response to fertilizer use is the inadequate supply of water during the growing season. Although water availability varies considerably across regions, it has been a serious problem in many countries. According to a World Bank study, depletion and degradation of water resources have become major problems facing many low-income countries (2, 3). Within 10 years, if population grows at projected rates, per capita water availability will decrease by an average of 20 percent in developing countries, on

D

Table 3 Land use and agricultural productivity

	Irrigated land (% of crop land) 1995–97	Arable land (ha per capita) 1995–97	Fertilizer use (Kg per hectare arable land) 1979–81	Fertilizer use (Kg per hectare arable land) 1996–98	Agricultural productivity: Agr. value added per agr. worker (1995 dollars) 1979–81	Agricultural productivity: Agr. value added per agr. worker (1995 dollars) 1996–98
North Africa	13.2	0.24	91	117	1255	1982
Sub-Sahara Africa	4.2	0.25	12	10	244	128
Asia	36[a]	0.14[a]	54[b]	135[b]	521[a]	618[a]
Latin America	13.2	0.17	56	99	1528[c]	1779[c]

[a] Regional average without Afghanistan, North Korea, and Vietnam.
[b] Regional average without North Korea.
[c] Regional average without Bolivia.

average, and by 34 percent in African countries. The agricultural sector consumes over half the annual freshwater withdrawals in most of the countries and could face greater competing demands from household and industrial uses in the future.

The sparse rainfall that characterizes much of Sub-Saharan Africa affects response to and demand for fertilizer (4, 6). Farmers are reluctant to risk fertilizer use until rain begins to fall because inadequate moisture will fail to dissolve the nutrients in fertilizer (especially nitrogen) and crops can "burn." Irrigation can make the use of fertilizer profitable and can increase agricultural output. However, in Sub-Saharan Africa, only 4.2 percent of arable land is irrigated (Table 3). This is low, even when compared with other developing regions. In Latin America, 13 percent of arable land is irrigated, and 36 percent is irrigated in Asia (9). The world average is 19 percent. There is potential for expanding irrigated area in Sub-Saharan Africa, but it is costly and requires investment. Increasing the use of fertilizer raises production costs. In many low-income countries, particularly those in Sub-Saharan Africa and Latin America, almost all fertilizer is imported, and the lack of adequate foreign exchange constrains availability.

CONCLUSION

Uneven distribution of the world's resources means that the poor, low-resource countries are vulnerable to food insecurity. Sub-Saharan Africa, the most food-insecure region, faces the challenge of achieving multiple goals: meeting basic human needs, stabilizing population, encouraging economic development, and conserving natural resources for future growth. In most low-income, food-deficit countries, gains in production have been due to area expansion. However, population pressures and poor farming practices that have contributed to soil erosion and nutrient-deficient soils have already pushed farmers to marginal lands. Therefore, the only option for sustainable production growth is to increase yields. So far, most countries in the region remain unsuccessful in adopting new technologies to raise crop yields and increase productivity.

REFERENCES

1. Brown, M.; Goldin, I. *The Future of Agriculture: Developing Country Implications*; Development Center of the Organization for Economic Cooperation and Development: Paris, France, 1992.
2. Cleaver, K.; Schreiber, G. *Reversing the Spiral: The Population, Agricultural, and Environment Nexus in Sub-Saharan Africa*; The World Bank: Washington, D.C., 1994.
3. Crosson, P.; Anderson, J. *Resources and Global Food Prospects: Supply and Demand for Cereals*, World Bank Technical Paper No. 184; The World Bank: Washington, D.C., 1992.
4. Harold, Courtney; Larson, Bruce; Scott, Linda. Fertilizer Consumption Remains Low. *International Agricultural and Trade Reports, Africa and Middle East Situation and Outlook Series*, WRS-94-3; United States Department of Agriculture, Economic Research Service: Washington, D.C., 1994.
5. Ingram, Kevin; Frisvold, George. Productivity in African Agriculture: Sources of Growth and Stagnation. *International Agricultural and Trade Reports, Africa and Middle East Situation and Outlook Series*, WRS-94-3; United States Department of Agriculture, Economic Research Service: Washington, D.C., 1994.
6. Seckler, D.; Gollin, D.; Antoine, P. Agricultural Potential of "Mid-Africa": A Technological Assessment. *Agricultural*

Technology in Sub-Saharan Africa, World Bank Discussion Papers 126; The World Bank: Washington, D.C., 1991.
7. Shapouri, Shahla; Rosen, Stacey. Global Food Security: Overview. *Food Security Assessment, International Agricultural and Trade Reports, Situation and Outlook Series*; United States Department of Agriculture, Economic Research Service: December 2000.
8. United Nations, Food and Agriculture Organization. *Agriculture: Towards 2010*; FAO: Rome, 1993.
9. United Nations, Food and Agriculture Organization. http://apps.fao.org/cgi-bin/nph-db.pl (accessed June 2000).
10. Wiebe, Keith D.; Soule, Meredith J.; Schimmelpfennig, David E. Agricultural Productivity and Food Security in SubSaharan Africa. *Food Security Assessment, International Agricultural and Trade Reports, Situation and Outlook Series*, GFA-10; United States Department of Agriculture, Economic Research Service: December, 1998.

DEGRADATION, FOOD SECURITY ON A GLOBAL SCALE AND

D

Michael A. Zoebisch
Asian Institute of Technology, Klong Luang, Pathumthani, Thailand

Eddy De Pauw
International Center for Agricultural Research in the Dry Areas, Aleppo, Syria

INTRODUCTION

Growing populations lead to an increasing demand for food. With decreasing per-capita areas of arable land, this leads to additional pressure and stresses on the limited land resources, especially the soils. As a consequence, large parts of the world are affected by soil degradation, and this has a direct effect on food security.

Food security is commonly defined as the access by all people at all times to enough food for an active, healthy life. Extended concepts include food-quality aspects, cultural acceptability of the food, equitable distribution among the different social groups, and gender balance (1–4). These concepts imply that, in addition to producing enough food, people must also have access to it. This opens the arena for the socioeconomic, cultural, policy, and political dimensions of food production and supply. Thus, food security—and consequently food insecurity—are multi-dimensional. It is not possible to separate clearly the biophysical and socioeconomic dimensions of food security, because they are closely interrelated and interdependent (2). However, the most limiting of the natural resources needed for food production are soil and water. If these resources are depleting, land productivity—the key factor for crop production—declines, and the land is no longer capable of producing the biomass needed for direct and indirect human consumption. If soil degrades, its plant-life supporting functions are lost, together with its role in the hydrological cycle. Soil is a limited resource and, because of the long duration of soil-forming processes, can be considered as nonrenewable. The production of food and feed is directly linked to the productive capacity of the soil. The degree and extent of soil degradation and, consequently, its effect on food production are dependent on how the land is used.

UNDERNOURISHMENT—A MAJOR INDICATOR OF FOOD INSECURITY

The proportion of chronically undernourished people is a manifestation of the degree of food insecurity. Worldwide, more than 830 million people are undernourished; 800 million of these people live in the developing countries, i.e., about 18% of the population of these countries (1). Table 1 shows that the most seriously affected region is Sub-Saharan Africa, with 33% of the population chronically undernourished. In Asia and the Pacific chronic undernourishment has decreased significantly over the past two decades.

Yields of the major food crops in many countries of these regions have risen remarkably. But in most cases the yield increases have not been able to keep pace with the needs of growing populations. Subregional trends, especially from Africa, China, South Asia, and Central America, indicate large yield declines due to soil degradation (2, 5, 6). Countries of the less degraded temperate regions can substitute for the losses in other regions. However, in the affected areas, food prices will increase and hence the incidence of malnutrition. Densely populated countries and regions that depend solely on agriculture will be most affected. As input requirements—and their costs—grow with increasing degradation, these countries will probably not be able to maintain adequate levels of soil productivity without special efforts to stabilize the system.

VULNERABLE AREAS

Only about one tenth of the world's arable areas are not endangered by degradation. Most soils are vulnerable to degradation and have a low resilience to fully recover from stresses. The soils in the tropics and subtropics are more vulnerable to degradation than those in temperate areas, mainly due to the climatic conditions (7). The total land area available for agricultural production is limited. Different types of land have different capabilities and limitations. Not all land is suitable for cultivation. The major natural limiting factors for soil productivity are steep slopes, shallow soils, low levels of natural fertility, poor soil drainage, sandy or stony soil, salinity, and sodicity.

Table 1 Undernourishment in developing countries

Region/subregion	Undernourished population: Proportion of total population 1979/81 (%)	Proportion of total population 1995/97 (%)	Number of people affected 1995/97 (millions)
Total Developing World	*29*	*18*	*791.5*
Asia and Pacific	*32*	*17*	*525.5*
East Asia	29	14	176.8
Oceania	31	24	1.1
Southeast Asia	27	13	63.7
South Asia	38	23	283.9
Latin America and Caribbean	*13*	*11*	*53.4*
North America	5	6	5.1
Caribbean	19	31	9.3
Central America	20	17	5.6
South America	14	10	33.3
Near East and North Africa	*9*	*9*	*32.9*
Near East	10	12	27.5
North Africa	8	4	5.4
Sub-Saharan Africa	*37*	*33*	*179.6*
Central Africa	36	48	35.6
East Africa	35	42	77.9
Southern Africa	32	44	35
West Africa	40	16	31.1

(Modified from Ref. 10.)

Different agro-ecosystems are typically susceptible to different types of soil degradation (6–8). In mountainous areas and steeplands, water erosion is the dominant form of soil degradation. For arid areas, both water and wind erosion and the loss of soil organic matter are typical. In humid areas, soil acidification and fertility decline are of importance. In irrigated areas, the hazards of soil salinization and waterlogging are of special significance. Knowledge of these hazards can help understand the limitations of the land and introduce appropriate land-use practices that reduce soil degradation and loss of soil productivity.

UNDERLYING CAUSES

Misuse of Land

The main causative factors of human-induced soil degradation, in a broad sense, are overuse and inappropriate management of agricultural land, deforestation and the removal of natural vegetation, and overgrazing as well as industrial activities that pollute the soil (6). Worldwide, soil erosion by wind and water are the most important forms of soil degradation, accounting for between 70% and 90% of the total area affected by soil degradation (5, 6). High population pressure and the resultant increased need for higher land productivity is the main factor leading to inappropriate and exploitative land use.

In high-input systems, loss of soil quality due to degradation can be compensated to a certain extent by inputs, such as fertilizers and irrigation water. In low-input systems—usually practiced by resource-poor farmers (in poor countries)—soil quality contributes relatively more to agricultural productivity (9). Therefore, soil degradation in these systems has more drastic and immediate effects on soil productivity and, hence, food production.

Pressure on the Land

When the land is not enough, i.e., when land scarcity becomes a limiting factor for food production, people usually resort to intensification of their land use and opening up new, often unsuitable, land for cultivation (4, 9). Both pathways will most certainly lead to soil degradation, if unsuitable land-use technologies are used. The majority of land users in the developing world do not have adequate access to appropriate land-use technologies and the financial means to invest in their land (1). The area available for crop production therefore puts a definite limit to sustainable food production in many parts of the developing world. Alternative, i.e., nonagricultural means of income generation would enable the people to supplement their food requirements on the market and could relieve immediate pressure on the land. This would save soil and land resources.

Losses and Gains of Agricultural Land

Estimates show that more than 10 million hectares of agricultural land worldwide is lost per year, most of it due to soil degradation but also due to urbanization (i.e., between 2 and 4 million hectares) (4). Annual deforestation and clearing of savanna land accounts for about 20 million hectares, of which 16 million hectares are converted to cropland. Thus, there is an annual net gain of cropland of about 5–6 million hectares. Most of these converted lands can be considered marginal, i.e., less suitable for cropping. Loss of traditional sources of grazing and firewood, together with the ecological functions of forests, i.e., their role in regulating watershed hydrology, are factors inducing the degradation of soil.

MINIMUM PER-CAPITA CROPLAND REQUIREMENT

On a global scale, it is estimated that the minimum per-capita area of cropland required to produce an adequate quantity of food is around 0.1 ha (2, 5). This rough estimate is dependent on many different location-specific factors and circumstances, such as soil, terrain, climate, farming system, and land-use technology. However, it permits a suitable comparison at the overall global scale. Estimates predict that a large number of countries will have reached the 0.1 ha limit of per-capita cropland by the year 2025 (1, 2, 7). In the absence of more land for cultivation, future food needs of the people in these countries will have to depend on increased soil productivity through intensified land use. Resource-poor farmers with no option to expand their cultivated area usually cannot invest in necessary soil-conservation and soil-fertility maintenance measures. Over time, this leads to nutrient depletion and other forms of soil degradation that affect soil productivity (4, 9, 10).

Although some developing countries have indeed increased their yields on a per-unit area basis significantly, these increases cannot compensate for the increased demands by growing populations. Table 2 illustrates past and future trends in per-capita available cropland in selected countries with high population growth and gives examples of yield-level trends. The table shows that, to a large extent, the gains in yields are offset by the decrease in available cropland to grow the required total quantities of crops to feed the population. In these countries, food insecurity can be directly related to the pressure on the land, or land scarcity.

The growing pressure on the land is not only a reflection of population growth and the direct needs for food, but also of changed diets and increasing cash needs (by the land users) that have to be met from land cultivation, putting additional stress on the soil resources.

ESTIMATING SOIL DEGRADATION ON A GLOBAL SCALE

Limitations of Methodology

The scientific relationships between different soil-degradation processes and soil productivity have been established mainly on field and plot level (9, 11). Although reliable data are available for a wide range of different agro-ecologies and farming systems, they cannot give a precise picture of degradation-dependent soil productivity for larger land units. The increasing complexity of the determining factors and processes on larger land units, the endless array of location-specific conditions, and the lack of adequate data make upscaling extremely difficult for country and regional levels. Estimates of the extent and severity of soil degradation—and their effects on soil productivity—on a country, regional, continental, and global level therefore bear a substantial degree of uncertainty. Most estimates use the methodology developed by the Global Assessment of Human-Induced Soil Degradation (GLASOD) Project (5, 6). The GLASOD estimates are based on expert assessment and are therefore largely subjective. However, the methodology has been widely adopted, and GLASOD estimates of soil degradation are accepted as the best estimates available. Global estimates are, therefore, only rough approximations and should not be taken literally.

Table 2 Trends of available per-capita cropland and average yield levels of main staple crops for selected countries

	Per-capita cropland (ha)						Average yield (kg/ha)			
Country	1961	1980	1990	2000	2015[a]	Crop	1961	1980	1990	2000[b]
Syria	1.40	0.62	0.46	0.34	0.17	Wheat	575	1,536	1,544	1,882
						Barley	461	1,312	310	381
Kenya	0.21	0.15	0.10	0.08	0.04	Wheat	1,090	2,156	1,864	1.400
						Maize	1,253	1,200	1,580	1,321
Pakistan	0.34	0.25	0.17	0.13	0.07	Wheat	822	1,568	1,825	2,492
						Rice	1.391	2.423	2,315	2.866
Bangladesh	0.17	0.10	0.09	0.07	0.05	Wheat	573	1,899	1,504	2.258
						Rice	1,700	2.019	2,566	2.851
Nepal	0.19	0.17	0.14	0.11	0.07	Wheat	1.227	1.199	1,415	1.820
						Rice	1.937	1.932	2,407	2.600
Zambia	1.52	0.92	0.65	0.57	0.28	Wheat	1.600	4.068	4,399	5.454
						Maize	882	1,688	1,432	1,431
Côte d'Ivoire	0.69	0.44	0.31	0.24	0.10	Rice	757	1.166	1,155	1.548
						Sorghum	667	583	575	352
Sudan	0.97	0.62	0.52	0.43	0.22	Rice	1.409	634	1,250	563
						Sorghum	970	712	428	634
Peru	0.20	0.20	0.17	0.14	0.10	Wheat	1.001	939	1,085	1.289
						Potato	5.287	7.196	7,881	10.818

[a] UN Projection, medium level
[b] FAO Estimates
(Modified from Refs. 2 and 12.)

Effects on Soil Productivity and Food Production

Table 3 shows estimates of degradation and cumulative losses in soil productivity on a continental level for different land-use types. The figures show that, overall, Africa and Central America are the most severely affected continents, both in terms of soil degradation and reduction of productivity. Sixty-five percent of African and 75% of Central American cropland are affected by

Table 3 Estimates of degradation and losses in soil productivity for different land-use types in the major regions

	Degraded areas									
	Agricultural land		Pasture land		Forests		Total degraded area		Loss in productivity[a]	
Region	(Mha)	(%)	(Mha)	(%)	(Mha)	(%)	(Mha)	(%)	Cropland (%)	Pasture (%)
Africa	121	65	243	31	130	19	494	30	25	6.6
Asia	206	38	197	20	344	27	747	27	12.8	3.6
South America	64	45	68	14	112	13	244	16	13.9	2.2
Central America	28	74	10	11	25	38	63	32	36.8	3.3
North America	63	26	29	11	4	1	96	9	8.8	1.8
Europe	72	25	54	35	92	26	218	27	7.9	5.6
Oceania	8	16	84	19	12	8	104	17	3.2	1.1
World	562	38	685	21	719	18	1,966	23	12.7	3.8

[a] Cumulative loss in productivity since 1945. Adjusted figures according to type and degree of degradation
(From Refs. 5, 6, 13.)

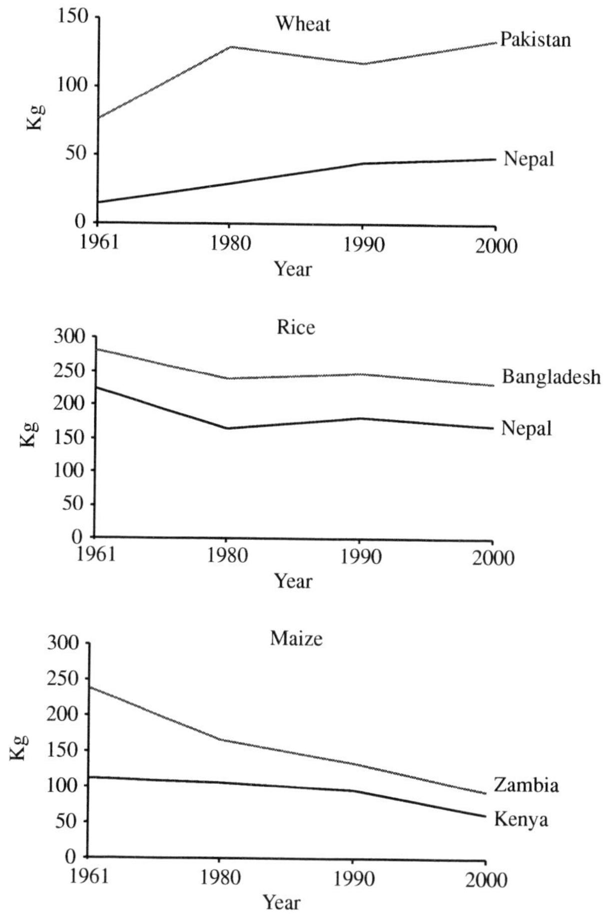

Fig. 1 Per-capita production of selected crops in countries with high population pressure. (From Refs. 2 and 12.)

soil degradation. The overall loss in productivity in these two regions over the last 50 years is estimated between 25% and 37%. On a continental basis, Africa still has land reserves, but large areas of the continent are marginal for crop production, and their food production is therefore most seriously affected by degradation. Europe also shows relatively high trends in degradation, but mainly of pastures and forests. Productivity losses generally are low, as for North America and Oceania (i.e., mainly Australia), and these can most probably be compensated for by improvements in technology and input supply.

A more detailed picture of trends in food productivity for the main food cereals in selected countries with high population growth rates and agriculture-based economies is given in Fig. 1. The overall past trends in the different countries show a consolidation of per-capita food production. However, the projections until 2025 indicate clear decreases. This suggests that long-term food security in these countries is at stake.

SOIL DEGRADATION AND DECLINE IN PRODUCTIVITY

Over the past decades, the cumulative loss of productivity of cropland has been estimated at about 13%. However, aggregate global food security does not appear to be under a significant threat (2, 5). There is no conclusive evidence that soil degradation has in the past affected global food security. We cannot conclude to such relationship because 1) the database on the extent and magnitude of various types of soil degradation is inadequate and 2) the impact of soil degradation on land productivity is very site-specific, often anecdotal and difficult to quantify at regional and global levels. In fact, globally, per-capita food production has increased by about 25% since 1961 when the first production surveys were conducted by FAO (Food and Agriculture Organization of the United Nations) (12). The yields per unit area of the major cereals (i.e., wheat, rice, maize) have steadily increased and are still rising (1, 12). This, however, does not imply that sufficient food is—and will be—available in all countries and regions. Food is not necessarily produced where it is most needed. A top priority should therefore be to improve inventories of land degradation at regional, national, and subnational levels.

However, the evidence is conclusive that soil degradation affects food security at the subnational and national levels in the developing countries by the gradual decline of the land's productive capacity and that this trend will continue in the future. We therefore have to assume that, eventually, land degradation will constitute a serious threat to global food security by its particular impact on the developing countries. According to most scenarios, these countries are most vulnerable to degradation induced by increasing pressure on their land resources, the effects of climate change and their inability to finance programs to rehabilitate affected areas and prevent further degradation and decline of productivity.

REFERENCES

1. FAO. *The State of Food Insecurity in the World*; Food and Agriculture Organization of the United Nations (FAO): Rome, Italy, 1999.
2. Engelman, R.; LeRoy, P. *Conserving Land: Population and Sustainable Food Production*; Population and Environment

Program, Population Action International: Washington, D.C., USA, 1995.
3. Saad, M.B. *Food Security for the Food-Insecure: New Challenges and Renewed Commitments*; Position Paper: Sustainable Agriculture. Centre for Development Studies, University College Dublin: Dublin, Ireland, 1999.
4. Kindall, H.W.; Pimentel, D. Constraints on the Expansion of the Global Food Supply. Ambio **1994**, *23* (3), 198–205.
5. Scherr, S.J. *Soil Degradation—A Threat to Developing-Country Food Security by 2020?* Discussion Paper 27; International Food Policy Research Institute: Washington, D.C., 1999.
6. van Lynden, G.W.J.; Oldeman, L.R. *The Assessment of the Status of Human-Induced Soil Degradation in South and Southeast Asia*; International Soil Reference and Information Centre: Wageningen, The Netherlands, 1997.
7. Greenland, D.; Bowen, G.; Eswaran, H.; Rhoades, R.; Valentin, C. *Soil, Water, and Nutrient Management Research—A New Agenda*, IBSRAM Position Paper; International Board for Soil Research and Management: Bangkok, Thailand, 1994.
8. Lal, R. Soil Management in the Developing Countries. Soil Science **2000**, *165* (1), 57–72.
9. Stocking, M.; Murnaghan, N. *Land Degradation—Guidelines for Field Assessment*; UNU/UNEP/PLEC Working Paper, Overseas Development Group, University of East Anglia: Norwich, UK, 2000.
10. Lal, R.; Singh, B.R. Effects of Soil Degradation on Crop Productivity in East Africa. J. Sustain. Agricul. **1998**, *13* (1), 15–36.
11. Stocking, M.; Benites, J. *Erosion-Induced Loss in Soil Productivity: Preparatory Papers and Country Report Analyses*; Food and Agriculture Organization of the United Nations: Rome, Italy, 1996.
12. FAO. FAOSTAT. Statistical Database of the Food and Agriculture Organization of the United Nations. Internet Web-Source http://www.fao.org (accessed January 2001): Rome, Italy, 2000.
13. Oldeman, L.R. *Soil Degradation: A Threat to Food Security?* Report 98–01; International Soil Reference and Information Centre (ISRIC): Wageningen, The Netherlands, 1998.

DEGRADATION, FOOD SECURITY ON A LOCAL/ECO-REGIONAL SCALE AND

Sara J. Scherr
University of Maryland, College Park, Maryland, U.S.A.

D

INTRODUCTION

An estimated 16% of all agricultural land in developing countries, and a much higher proportion of cropland, has experienced significant soil degradation since the middle of the twentieth century. Deterioration in the physical or chemical attributes of these soils has reduced agricultural yields and increased production costs. Large land areas—an estimated 5 to 7 million hectares (ha) per year—have gone out of production entirely (1, 2). Degradation appears to have had little impact on food availability or prices in international markets because of the global capacity for substitution and the dominance of temperate producers (2). However, international trade plays a minor role in food supplies in most poor countries, and especially in most poor rural areas, largely because of a lack of income to purchase imported food and, in some cases, poor transport and market infrastructure. Thus, many areas undergoing serious soil degradation are experiencing reduced food supply and farm income, in some cases on a scale large enough to slow economic development. Degradation especially threatens the livelihoods of poor producers, who have limited access to capital inputs and for whom the underlying quality of agricultural soil is the principal determinant of farm productivity.

The period since World War II has seen remarkable growth in rural population, agricultural land area, and agricultural productivity in the developing world. Although the annual rural population growth rate declined from 2.2% between 1960 and 1965 to 1% between 1990 and 1995, the absolute number of rural dwellers grew by almost 40%, from 2.0 to 2.8 billion. Total growth rates of agricultural production in developing countries have rivaled or surpassed historical growth rates in industrialized countries, though not on a per capita basis. This growth came in part from extensive land clearing, but primarily through yield increases. It is not surprising that rural population increase, area expansion, and widescale intensification were associated with widespread soil resource degradation (2).

NATURE OF SOIL DEGRADATION

The nature of soil degradation has varied according to the particular pathways of agricultural intensification, reflecting different pressures and resource endowments. *Irrigated lands* have expanded and become highly productive as beneficiaries of Green Revolution technologies. However, poor water management has led to widespread salinization and waterlogging, while subtle nutrient management problems associated with multiple cropping have slowed yield increases in recent years. In *high-quality rainfed lands*, with naturally fertile, less weathered soils and lower climatic risks, cropping intensity also increased greatly. But inappropriate machinery use has sometimes led to soil compaction, and poor vegetation management has exposed soils to erosion. Substitution of organic inputs with chemical fertilizers has led to declining organic matter, acidification of vulnerable soils, and changes in soil fauna. In *densely populated, marginal areas*, low-quality and degradation-prone soils, which were traditionally managed through moderate- to long-fallow systems, are now under continuous cultivation. Erosion from poor soil cover, and nutrient depletion from inadequate management of organic matter and fertilizers, are common results. In *extensively managed, marginal lands*, degradation has been caused principally by the land-clearing process itself, by nutrient depletion due to declining fallow periods, by widespread burning to control weeds and pests and provide ash for plant nutrition, and in rangelands by overgrazing. The importance of *urban agriculture* accelerated dramatically in the 1980s, contributing to food security, income generation, and recycling of urban wastes, but also to some environmental problems. Soil contamination with urban pollutants may pose a health hazard to consumers and reduce production, while insecure land access and tenure discourage sustainable grazing and soil-management practices.

The concerns of policymakers regarding agricultural soil degradation relate not to its extent or severity, but to its demonstrated impact on the following policy objectives[a]:

- *aggregate supply, stability, or price of agricultural output* (i.e., if lands with degrading soils are a significant source of market supply for national consumers or export markets and alternative sources of supply not available or economic);
- *agricultural income or economic growth* (i.e., if soil degradation reduces agricultural income and its economic multiplier effects, through lower production or higher costs, and alternative sources of economic growth are limited or expensive to develop);
- *consumption by poor farm households* (i.e., if lands with degrading soils are a critical source of food security for subsistence or semisubsistence producers with few alternative livelihood options); or
- *national wealth* (i.e., if degradation reduces the long-term productive capacity of soil resources deemed to be of future economic or environmental significance, threatening the resource base and food security of future generations) (2).

Considerable evidence of the economic and food security impacts of soil degradation has been produced since the late 1980s, although results must be interpreted cautiously (2). Weaknesses include uneven geographic coverage and numerous methodological and data problems (3). For many soil types, there is little information about the productivity impacts of degradation, the costs of rehabilitation, and the thresholds for soil quality below which future investment in restoration will be uneconomic (4). Most studies still use simplistic models of degradation–yield–income relationships, ignoring the dynamics of markets and farmer response that have been widely observed in field studies (5, 6). Only a few studies explicitly document the impacts of soil degradation on consumption by the rural poor. Still, the weight of the evidence (summarized in the following section) suggests that the magnitude of agricultural supply and economic growth impacts of soil degradation justify more active policy engagement.

AGRICULTURAL SUPPLY

Asia

In Asia during the 1980s, it was estimated that over one-third of irrigated areas and over one-half of dry rainfed areas had experienced at least a 10% loss in productive potential, and 8% of irrigated and 10% of rainfed dry areas had suffered more than a 25% loss. Soil productivity losses of at least 20% from human-induced water erosion was well confirmed in eight countries, and presumptive evidence of such losses was documented in five other countries. Wind erosion was found to have little effect (7). The 1995 expert survey found that soil degradation had had moderate or worse impacts on agricultural productivity on a tenth of all lands in South and Southeast Asia, with serious fertility decline or salinization affecting 15% of arable land (8).

Field-based evidence confirms supply effects. A synthesis of survey and experimental data in India concluded that 48% of soils had suffered a productivity loss of over 33% (9). Salinization and waterlogging in four villages of Uttar Pradesh, India reduced paddy yield by 61% and wheat yield by 68% over 10 years (10). In Bangladesh, deteriorating nutrient balance and organic matter decline led to an annual 0.36-ton decrease in total factor productivity in an intensive cropping system (11). In irrigated districts of the Pakistani Punjab, soil degradation lowered aggregate economic growth from 1966 to 1997 in four farm production systems by 0.58% per year; in the wheat–rice (*Triticum sativum-Oryza sativa*) system, resource degradation more than cancelled the effects of technological change (12). In China, grain yields would have been 30% higher between 1983 and 1989 in the absence of erosion, salinity, and degradation due to increased cropping intensity (13).

Africa

Soil experts estimated that in the 1980s over one-half of dry rainfed areas in Africa had experienced at least a 10% loss in productive potential, with irreversible productivity losses of at least 20% due to erosion in large parts of 11 countries (14). Cropland productivity loss from all types of soil degradation between the mid-twentieth century and 1990 was estimated to be 25% (15). In 1989, crop yield losses resulting from past erosion were calculated to range from 2% to 40%, with a mean of 6.2% for sub-Saharan Africa (16).

Evidence from field studies and cropping system models is variable. A Malawi study found crop-specific soil erosion impacts ranging from 4% to 11% per year yield losses (16). Models of maize production in four highland farming systems predicted significant yield declines, even with fertilizer use, although cotton (*Gossypium* spp.) and coffee (*Coffea* spp.) yields were more stable. Fertilizer-induced soil acidification reduced highland maize yield to zero in

[a] Environmental effects of soil degradation, — including off-site damages from sedimentation and diminished watershed function, as well as habitat alteration and reduced carbon sequestration, — are also of interest to policymakers.

20 years; even with lime, yields were projected to drop by half in 30 years (17). National- and district-level regressions for maize (*Zea mays*) and sorghum (*Sorghum vulgare*) yields in Lesotho and for maize yields in Zimbabwe found no statistically significant relationship with erosion, possibly due to the dominant effect of climatic variation (18, 19).

Central America

The cumulative loss of agricultural productivity due to soil degradation was estimated by soil experts to be especially high—nearly 37%—in Central American croplands (15). Most cost–benefit models in the region predict large yield declines without soil conservation measures (20). Bioeconomic-model simulations predicted that between 1991 and 2000, soil erosion in Nicaragua would result in 16–21% reductions in output for sesame (*Sesamum indicum*), beans (*Phaseolus* spp.), maize, cattle, and cotton; large declines in coffee, cotton, and sesame exports; and price increases of over 25% for maize, beans, and cattle (21). Because they find ways to manage their soils, farmers generally report lower estimates of production impact when surveyed. For example, a national sample of smallholders in hilly El Salvador reported current erosion-related productivity problems in less than one-fifth of sloping farm fields (22).

ECONOMIC GROWTH

Asia

The annual cost of soil degradation in South Asia in the early 1990s was estimated to be $9.8–$11.0 billion, the equivalent of 7% of aggregate agricultural gross domestic product (AGDP). Water and wind erosion accounted for over two-thirds of the loss, salinization and waterlogging for one-fifth, and soil fertility decline the rest (23). Other studies of China, India, Java, and Pakistan calculated impacts ranging from less than 1% to 5% of AGDP (2). In Java, one-year costs of erosion in 1984 equaled 4% of the annual value of rainfed farm output—the same order of magnitude as annual recorded growth in uplands production (24). In the irrigated wheat–rice system of the Pakistani Punjab, degradation also entirely offset productivity gains from other sources (12). The economic loss from degradation in China in the late 1980s was equal to China's entire budget for rural infrastructure investment, though less than 1% of AGDP (25). An econometric analysis of soil erosion and salinization using district-level data in China showed more serious impacts in poorer and more densely populated areas (26). Yet despite nutrient depletion, the economic value of topsoil rose by nearly 8% between the 1950s and 1980s as a result of shifts to soil-preserving products, yield increases, and actions to reduce soil alkalinity (27).

Africa

Bojö (28) evaluated evidence on the economic losses due to soil erosion from 12 studies completed in eight countries in sub-Saharan Africa. He found low gross annual immediate losses for hill agriculture in Madagascar, the Ethiopian highlands, Mali, and South Africa. Losses were significantly greater in Ghana, in another Ethiopian study, and in Zimbabwe for nutrient replacement alone. The gross discounted future loss from degradation varied from under 1% to 18%, and the gross discounted cumulative loss (which assumed a continued process of degradation over time) for five countries ranged from under 1% to 44%, of AGDP. More recent estimates (29), based on replacement cost of depleted nutrients (calculated from nutrient balances predicted for 2000), are much higher than others based on productivity loss. Farm monitoring in Kenya found that the average cost of replacing depleted soil nutrients equaled 32% of average net farm income (30). A model was developed for Ghana to calculate the impacts of soil erosion, acidification, and nutrient depletion on different crops and agroecozones, and trace their effects through the national economy. Comparing the baseline scenario over 8 years, with and without soil degradation, showed agricultural productivity losses of 2.9% per year. Even with greater fertilizer use, such losses would lower economic growth by one percentage point. In some scenarios, real GDP is reduced as much as 4.8% in the eighth year (31).

Central America

The few estimates of agricultural income effects in Central America are high. The annual cost of replacing nutrients lost through soil erosion in Costa Rica was estimated at 5.3–13.3% of annual value-added in agriculture that year (32). The average cost of soil erosion in areas planted to maize in Mexico represented 2.7% of AGDP, reaching 12.3% in the areas with highest erosion effects. Economic losses were nine times higher in the highlands and semiarid regions than in the lowland tropics and four times higher without than with soil conservation (33). A model for Nicaragua projected that annual productivity loss due to erosion in 1991 would reduce GDP, imports, exports, and consumption in 2000 by 4–7% from the baseline scenario without erosion, and investment by 9%. Erosion would raise the producer and domestic price indexes by 1.7% and 2.1%, respectively, compared with the baseline, while consumer price indexes would rise 4.0–5.8% for different social groups (2).

CONCLUSION

Soil degradation is widespread, and its pace has certainly accelerated in developing countries in the past 50 years. Major productivity declines have resulted in some parts of Africa, Latin America, and Asia. The principal impacts appear to be on food consumption by the rural poor and agricultural income in areas where alternative livelihood options, sources of food supply, or nonagricultural development potential have been limited. Current and future soil degradation "hot spots" include areas with degradation-prone soils (particularly in sub-Saharan Africa), inadequately managed irrigation (particularly in South Asia), and rapidly intensifying production without the technology (many densely populated marginal lands) or the economic incentives (extensively managed marginal lands) for good resource husbandry (34). Thus, while posing particular problems for the poor, soil degradation is likely to have far-reaching impacts on economic development in many agriculture-dependent countries.

REFERENCES

1. Oldeman, L.R. The Global Extent of Soil Degradation. In *Soil Resilience and Sustainable Land Use*; Greenland, D.J., Szaboles, T., Eds.; Commonwealth Agricultural Bureau International: Wallingford, 1994; 99–118.
2. Scherr, S.J. *Soil Degradation: A Threat to Developing Country Food Security in 2020?*, Food, Agriculture and the Environment Discussion Paper No. 27; International Food Policy Research Institute: Washington, DC, 1999.
3. Scoones, I.; Toulmin, C. *Policies for Soil Fertility Management in Africa*, A Report Prepared for the Department for International Development; International Institute for Environment and Development, Drylands Programme and the Institute of Development Studies; The Environment Group: London and Brighton, 1999.
4. Tengberg, A.; Stocking, M. Land Degradation, Food Security and Agro-biodiversity—Examining an Old Problem in a New Way. In *Response to Soil Degradation*; Bridges, E.M., Hannam, I., Oldeman, L.R., Penning de Vries, F., Scherr, S.J., Sombatpanit, S., Eds.; Oxford & IBH Publishing Co. Pvt. Ltd, New Delhi, 2001, 171–185.
5. Templeton, S.; Scherr, S.J. The Effects of Demographic and Related Microeconomic Change on Land Quality in Hills and Mountains of Developing Countries. World Dev. **1999**, *27* (6), 903–918.
6. Tiffen, M.; Mortimore, M.; Gichuki, F. *More People, Less Erosion: Environmental Recovery in Kenya*; John Wiley and Sons: Chichester, England, 1994.
7. Dregne, H.E. Erosion and Soil Productivity in Asia. J. Soil Water Conserv. **1992**, *47* (1), 8–13.
8. Van Lynden, G.; Oldeman, R. *Soil Degradation in South and Southeast Asia*; International Soil Reference and Information Centre for the United Nations Environment Programme: Wageningen, The Netherlands, 1997.
9. Seghal, J.; Abrol, I.P. *Soil Degradation in India: Status and Impact*; Oxford and IBH: New Delhi, 1994.
10. Joshi, P.K.; Jha, D. *Farm-Level Effects of Soil Degradation in Sharda Sahayak Irrigation Project*, Working Papers on Future Growth in Indian Agriculture No. 1.; Central Soil Salinity Research Institute, ICAR, and International Food Policy Research Institute, 1991.
11. Pagiola, S. *Environmental and Natural Resource Degradation in Intensive Agriculture in Bangladesh*, Environmental Economics Series Paper 15, Environment Department; The World Bank: Washington, DC, 1995.
12. Ali, M.; Byerlee, D. Productivity Growth and Resource Degradation in Pakistan's Punjab. In *Response to Land Degradation*; Bridges, E.M., Hannam, I.D., Oldeman, L.R., Penning de Vries, F., Scherr, S.J., Sombatpanit, S., Eds.; Oxford & IBH Publishing Co. Pvt. Ltd, New Delhi, 2001, 171–185.
13. Huang, J.; Rozelle, S. Environmental Stress and Grain Yields in China. Am. J. Agr. Econ. **1994**, *77*, 246–256.
14. Dregne, H.E. Erosion and Soil Productivity in Africa. J. Soil Water Conserv. **1990**, *45* (4), 432–436.
15. Oldeman, L.R. *Soil Degradation: A Threat to Food Security?*, Report 98/01; International Soil Reference and Information Centre: Wageningen, The Netherlands, 1998.
16. Lal, R. Erosion-Crop Productivity Relationships for Soil of Africa. Soil Sci. Soc. Am. J. **1994**, *59* (3), 661–667.
17. Aune, J.B.; Kullaya, I.K.; Kilasara, M.; Kaihura, F.S.B.; Singh, B.R.; Lal, R., Consequences of Soil Erosion on Productivity and Productivity Restoration by Soil Management in Tanzania. In *Soil Quality and Sustainable Agriculture*; Lal, R., Ed.; Ann Arbor Press: Ann Arbor, 1997.
18. Böjo, J. *The Economics of Land Degradation: Theory and Applications to Lesotho*; The Stockholm School of Economics: Stockholm, 1991.
19. Grohs, F. *Economics of Soil Degradation, Erosion and Conservation: A Case Study of Zimbabwe*; Wissenschaftsverlag Vauk Kiel KG: Kiel, Germany, 1994.
20. Lutz, E.; Pagiola, S.; Reiche, C., Eds.; *Economic and Institutional Analyses of Soil Conservation Projects in Central America and the Caribbean*, A CATIE-World Bank Project, World Bank Environment Paper No. 8; The World Bank: Washington, DC, 1994.
21. Alfsen, K.M.; Defranco, M.A.; Glomsrod, S.; Johnsen, T. The Cost of Soil Erosion in Nicaragua. Ecol. Econ. **1996**, *16*, 129–145.
22. Pagiola, S.; Dixon, J. *Land Degradation Problems in El Salvador*, Annex 7, El Salvador Rural Development Study Report 16253-ES; World Bank: Washington, DC, 1997.

23. Young, A. *Land Degradation in South Asia: Its Severity, Causes and Effects upon the People*, Final Report for the Economic and Social Council of the United Nations; Food and Agriculture Organization of the United Nations, United Nations Development Programme, and United Nations Environment Programme: Rome, 1993.
24. Repetto, R.; Magrath, W.; Welk, M.; Beer, C.; Rossini, F. *Wasting Assets*; World Resources Institute: Washington, DC, 1989.
25. Rozelle, S.; Veeck, G.; Huang, J. The Impact of Environmental Degradation on Grain Production in China's Provinces. Econ. Geogr. **1997**, *73* (June), 44–66.
26. Rozelle, S.; Huang, J.; Zhang, L. Poverty, Population, and Environmental Degradation in China. Food Policy **1997**, *22* (3), 229–251.
27. Lindert, P. *The Bad Earth? China's Agricultural Soils Since the 1930s*, Working Paper Series No. 83; Agricultural History Center, University of California: Davis, CA, 1996.
28. Bojö, J. The Costs of Land Degradation in Sub-Saharan Africa. Ecol. Econ. **1996**, *16*, 161–173.
29. Dreschel, P.; Gyiele, L. *The Economic Assessment of Soil Nutrient Depletion: Analytical Issues for Framework Development*; International Board for Soil Resource Management: Bangkok, 2000.
30. Shepherd, K.D.; Soule, M.J. Soil Fertility Management in West Kenya. Agr., Ecosyst. Environ. **1998**, *71* (1–3), 133–147.
31. Alfsen, K.M.; Bye, T.; Glomsrod, S.; Wiig, H. Soil Degradation and Economic Development in Ghana. Environment and Development Economics **1997**, *2*, 119–143.
32. Solorzano, R.; De Camino, R.; Woodward, R. *Accounts Overdue: Natural Resource Depreciation in Costa Rica*; World Resources Institute: Washington, DC, 1991.
33. Mcintire, J. A Review of the Soil Conservation Sector in Mexico. In *Economic and Institutional Analyses of Soil Conservation Projects in Central America and the Caribbean*; Lutz, E., Pagiola, S., Reiche, C., Eds.; A CATIE-World Bank Project, World Bank Environment Paper No. 8, The World Bank: Washington, DC, 1994; 107–130.
34. Scherr, S.J.; Yadav, S. *Land Degradation in the Developing World: Implications for Food, Agriculture, and the Environment to 2020*, Food, Agriculture and Environment Discussion Paper 14; International Food Policy Research Institute: Washington, DC, 1995.

DEGRADATION, FOOD SECURITY, POVERTY AND

Pierre Crosson
Resources for the Future, Washington, D.C., U.S.A.

INTRODUCTION

Food-insecure people are defined here as those who are considered undernourished by the Food and Agriculture Organization of the United Nations (FAO). These are people who do not consume enough calories over the course of a year to maintain normal body weight and support light physical activity. According to the FAO, there were approximately 780 million people who, by this definition, were food insecure in 1990 (1). Almost all were from the less developed countries (LDCs) of Asia, Africa, and Latin America.

POVERTY AND FOOD SECURITY

The consensus among students of world agricultural development is that, where people are food-insecure, it is mainly because of poverty. People are too poor to purchase the food needed for adequate nourishment. This means that the food-insecurity problem lies primarily on the demand side of world food markets. Soil degradation reduces the capacity of the land to produce food, which means that soil degradation is on the supply side of world food markets. In principle, it could also be on the demand side because degradation-induced production losses reduce farmers' incomes, and thus their ability to buy food.

This implies that reducing soil degradation would help solve the problem of global food insecurity to the extent that degradation significantly reduces farm income. Investigation of present information on soil degradation and its productivity consequences suggests, with some caveats, that the effect of soil degradation on farm income is not significant on a global scale.

FAO data on the number of undernourished people in developing countries show that the 780 million in this category (in 1990) represented 20% of the total population of these countries. This was a marked decline from the 36% of the LDC population that was food-insecure in the late 1960s. It must be noted, however, that this improvement occurred in only Asia and Latin America. In Africa, the number of food-insecure people increased over the 20-year period, both as a percentage of the population and in absolute number.

In the 1990s, per-capita food production and per-capita income increased in Asia and Latin America, so it is likely that the population percentage and the absolute amount of food-insecure people in those areas declined. In Africa, however, the trends in food production and per-capita income continued to be unfavorable in that time frame, suggesting that, in percentage and absolute number, food-insecure people increased further.

These numbers indicate that despite the relatively good performance in Asia and Latin America over the past 30 years, the global community continues to confront a serious problem of food insecurity in the LDCs, especially in Africa.

The consensus mentioned previously that poverty is the main reason for food insecurity includes the view that if food-insecure people had sufficient income and the global food trading system were no less robust than it is now, farmers worldwide could increase food production enough to meet the higher demand at prices only moderately higher than those now prevailing. It also was mentioned that because soil degradation is a supply factor in agricultural production, reducing soil degradation would not significantly contribute toward world food security. However, this assertion cannot be accepted on its face; it must be investigated. Specifically, we must review estimates of the extent and productivity consequences of soil degradation worldwide.

SOIL DEGRADATION

It should be noted that over the past 60 or 70 years, most soil scientists and the general public have viewed soil degradation as a serious threat to agricultural production capacity. In 1928, Hugh Hammond Bennett—regarded as the father of the soil conservation movement in the United States—and a colleague published *Soil Erosion: A National Menace* (2). In the United States, this view of erosion was widely shared and still persists among many soil scientists and others. The same view was more recently expressed, with respect both to the United States, and the world as a whole, by Brown and Wolf (3), and Pimentel et al. (4).

The view has held despite the lack of strong scientifically reliable information about the amounts and extent of soil erosion or its productivity impacts, not only in the

Encyclopedia of Soil Science

United States but around the world. In the United States, it was not until the U.S. Soil Conservation Service (an agency of the Department of Agriculture) conducted a comprehensive nation-wide survey in 1977 that such information became available. On a global scale, no such data existed, even as late as 1988. Nelson (5) expressed the consensus after conducting a comprehensive literature review; he described worldwide evidence regarding the rate, extent, and severity of soil degradation as "extraordinarily skimpy." Other students of the problem came to the same conclusion as well (6, 7).

MEASUREMENTS OF SOIL DEGRADATION

In 1982, 1987, 1992, and 1997, the U.S. Soil Conservation Service (now the Natural Resources Conservation Service) conducted nationwide surveys of erosion amounts comparable with those done in 1977. The surveys showed that over the period covered the amounts declined, both in total and per hectare.

In addition, studies done of the productivity consequences of erosion in the United States indicate that in the past they were, and in the future would continue to be, small. (In these studies, productivity is measured as crop output per hectare.) Crosson (8) found that from 1950 to 1980, erosion in more than 600 counties in the Corn Belt and Great Plains reduced increases in corn and soybean yields by a cumulative 1.5–2%. That is, in 1980, yields of those crops were smaller by those percentages than they would have been without erosion. Crosson also found that erosion in those counties had no statistically significant effect on wheat yields.

Soil scientists at the University of Minnesota developed a Productivity Index (PI) model to estimate the future effects of soil erosion on crop yields. They applied the model to 39 million hectares of cropland in the Corn Belt and found that if 1977 rates of erosion were to continue for 100 years, corn yields at the end of the period would be approximately 4% less than without erosion (9).

In yet another study, a team of soil scientists, agronomists, and economists at the Department of Agriculture facility in Temple, Texas, developed the Erosion Productivity Impact Calculator (EPIC), a model designed to estimate the effects of erosion on crop yields (10). The team used the model to estimate the erosion-induced loss of yield in the nation as a whole that would occur if the 1982 crop production and management practices were to continue for 100 years. They found that at the end of that time period yields would be 2–3% lower than without erosion.

In a comprehensive review of 90 micro-scale studies of erosion-induced productivity losses in the United States and Canada, den Biggelaar et al. (11) came to a conclusion consistent with that of the three previous studies.

Two studies done in the early 1990s provide the only reasonably reliable global scale estimates of the extent and severity of soil degradation on agricultural land. One of the studies, by a team headed by Roel Oldeman (12) at the Agricultural University at Wageningen, The Netherlands, was based on a survey of more than 200 worldwide soil specialists. The specialists were asked to classify as degraded or not degraded the soils in their areas that were in permanent or annual crops, permanent pasture, or in forest and woodland. They were then asked to categorize each type of land into four degrees of degradation: light, moderate, strong, and extreme. To translate these estimates to the global scale, Oldeman et al. used FAO data showing that some 8,735 million hectares were in the four land uses. Responses from the specialists were used to allocate these hectares into degraded or not degraded; if degraded, they were placed into each of the four degrees of degradation. The results showed that 1,965 million of the 8,735 million hectares (22%) were degraded. Of the 1,965 degraded hectares, 33% were lightly so, 46% moderately, 20% strongly, and 1% severely.

Another global study of productivity effects of soil degradation was done by Dregne and Chou (13), specialists in dryland agriculture at Texas Tech University. They estimated the degradation-induced percentage losses of yield on irrigated cropland, rain-fed cropland, and range in dry regions of the world. (Dry regions are those in arid, semi-arid, and dry sub-humid agro-climatic zones.) Using FAO data, Dregne and Chou found approximately 5,100 million global hectares in the three categories of land use. Drawing on the published literature and other available evidence, as well as their experience in dryland agriculture, Dregne and Chou classified land in irrigated and rain-fed crop production into four categories by severity of degradation-induced losses of productivity. Slightly degraded land was estimated to have lost 0–10% of its undegraded level of productivity. The estimated loss for moderately degraded land was 10–25%, for severely degraded land the loss was 25–50%, and very severely degraded land was estimated to have lost more than 50%. For rangeland, the estimated losses of productivity in each of the four categories were assumed to be somewhat higher (e.g., slightly degraded rangeland was estimated to have lost 0–25% of its undegraded productivity). For each of the three land uses, Dregne and Chou divided the amount in each among the four degree-of-degradation categories.

Crosson (14) used the data in the Dregne and Chou study to estimate that the weighted average cumulative global loss of soil productivity on the three classes of land use in dry areas was 12%. This is the loss over some period, which Dregne and Chou did not specify, but it must have been averaged over approximately three to four decades. Over three decades, the average annual rate of productivity loss would be 0.4%.

Crosson (14) combined parts of the Dregne and Chou study with results of the Oldeman et al. study (12) and calculated the global scale cumulative loss of productivity on land in all agro-climatic zones to be 5%. Oldeman et al. specifically note that their estimate is for the period from 1945 to 1990. Over that 45-year period, the average annual rate of loss would be 0.1%.

In the latter calculation, Crosson assumed that the degradation-induced loss of productivity on the undegraded land in the Oldeman et al. study (78% of the total land) was zero. With so much of the land in an undegraded state, the extent of degradation-induced losses of productivity on *all* the land necessarily would be small.

These results suggest that over the past several decades, the impact of soil degradation on United States and global agricultural productivity has been small. Results that estimate future erosion-induced losses of productivity in the United States suggest that with current rates of erosion, the losses would continue to be small.

CONCLUSION

The conclusion for both the United States, and global scales is subject to five important caveats. First, other students of the U.S. and global soil erosion issue do not agree with the conclusion (3, 4); however, it must be noted that neither of these authors has directly confronted the argument made here.

Another caveat is that the data used in the Dregne and Chou and Oldeman et al. studies are subject to wide margins of error. Both sets of authors recognize this and warn against careless use of their data. If and when more reliable data become available, they may contradict these initial findings.

The third caveat, regarding global scale estimates, is that the estimates obscure the fact that in many areas, particularly in the LDCs, there are "hot spots" where soil degradation has been severe—with significant impact on the food security of people in those areas (15). In parts of sub-Saharan Africa, for example, the productivity effects of soil degradation may be particularly severe on a local scale because of shallow depths of topsoil and a high concentration of nutrients in the surface soil layer. In these areas, the erosion-induced nutrient loss is particularly serious because supplies of inorganic fertilizers are either not available or are very expensive. Thus, to say that soil degradation on a global scale does not seem to be a significant threat to food security does not mean the problem is not serious for the millions of people in particular locations affected by it.

Fourth, with respect to the global estimates, the estimates reflect experience over the past several decades. While soil degradation has not greatly reduced agricultural productivity over this period, this does not guarantee that, in the future, degradation could not increase enough to significantly threaten world food security.

Much will depend on whether land tenure systems in the LDCs move in the direction of providing farmers more secure property rights in the land. The evidence around the world is clear that where these rights are protected, farmers have incentive to invest in soil conservation because they can be confident that they will reap the gains of the investments. Several studies in Asia and Africa suggest that land tenure systems in those areas are slowly shifting toward more secure property rights in the land (16–18). If this trend continues, it will have much impact on the size of the future threat of soil degradation to world food security.

Finally, while soil degradation does not presently appear to be a serious threat to world food security, this does not mean that measures to protect the soil are unimportant. On the contrary, they are very important. For most farmers, land is by far their most important asset. The apparently low cumulative and present rates for erosion-induced losses of soil productivity are strong indirect evidence that farmers worldwide recognize this and have taken measures to protect their land against erosion damages. If this practice continues, soil degradation will not likely threaten future global food security.

Continued protection of the soil against the inroads of degradation is essential to maintaining and increasing the global capacity to produce food. Using that capacity to achieve global food security requires an increase in the incomes of the poor worldwide. If that is done on an adequate scale, the problem of global food insecurity will be solved.

REFERENCES

1. *World Agriculture: Toward 2000*; Food and Agriculture Organization: Rome, 1995.
2. Bennett, H.; Chapline, W. *Soil Erosion: A National Menace*; Circular Number 13; Soil Conservation Service, U.S. Department of Agriculture, Government Printing Office: Washington, DC, 1928.

3. Brown, L.; Wolf, E. *Soil Erosion: Quiet Crisis in the World Economy*; Worldwatch Paper 60, Worldwatch Institute: Washington, DC, 1984.
4. Pimentel, D.; Harvey, C.; Resosudarmo, P.; Sinclair, K.; Kurz, D.; McNair, M.; Crist, S.; Shpritz, L.; Fitton, L.; Saffouri, R.; Blair, R. Environmental and Economic Costs of Soil Erosion and Conservation Benefits. Science **1995**, *267*, 1117–1123.
5. Nelson, R. *Dryland Management: The Soil Erosion Problem*; Environment Department Working Paper No. 8, The World Bank: Washington, DC, 1988.
6. Dregne, H. Desertification of Drylands. In *Challenges in Dryland Agriculture: A Global Perspective- Proceedings of the International Conference on Dryland Agriculture*; Unger, P., Sneed, T., Jordan, W., Jensen, R. Eds.; Texas Agricultural Experiment Station: Bushland/Amarillo, Texas, 1988.
7. El-Swaify, S.; Dangler, E.; Armstrong, C. *Soil Erosion by Water in the Tropics*; University of Hawaii Press: Honolulu, 1982.
8. Crosson, P. *Productivity Consequences of Cropland Erosion in the United States*; Resources for the Future: Washington, DC, 1983.
9. Pierce, F.; Dowdy, R.; Larson, W.; Graham, W. Soil Erosion in the Corn Belt: An Assessment of Erosion's Long-Term Effect. Journal of Soil and Water Conservation **1984**, *39*, 131–136.
10. *U.S. Department of Agriculture. The Second RCA Appraisal: Soil, Water, and Related Resources on Nonfederal Land in the United States*; U.S. Department of Agriculture: Washington, DC, 1989.
11. den Biggelaar, C.; Lal, R.; Wiebe, K.; Breneman V. Impact of Soil Erosion on Crop Yields in North America. *Advances in Agronomy*, **2001**, *72*, 1–52.
12. Oldeman, R.; Hakkeling, R.; Sombroeck, W. *World Map of the Status of Human-Induced Soil Degradation: An Explanatory Note*; International Soil Information and Reference Center, Wageningen, The Netherlands, and United Nations Environment Programme: Nairobi, Kenya, 1990.
13. Dregne, H.; Chou, N.T. Global Desertification Dimensions and Costs. In *Degradation and Restoration of Arid Lands*; Dregne, H. Ed.; Texas Tech University: Lubbock, Texas, 1992, 249–382.
14. Crosson, P. *Soil Erosion and Its On-Farm Productivity Consequences: What Do We Know?* Resources for the Future: Washington, DC, 1995.
15. Sheer, S.; Yadev, S. *Land Degradation in the Developing World: Implications for Food and the Environment*; International Food Policy Research Institute: Washington, DC, 1996.
16. English, J.; Tiffen, M.; Mortimore, M. *Land Resource Management in Machakos District, Kenya*; Environment Department Working Paper No. 5, World Bank: Washington, DC, 1994.
17. Pingali, P. Institutional and Environmental Constraints to Agricultural Intensification. Population and Development Review **1989**, *15*, 243–260.
18. Migot-Adholla, S.; Hazell, P.; Blarel, B.; Place, F. Indigenous Land Rights Systems in Sub-Saharan Africa: A Constraint on Productivity? World Bank Economic Review **1991**, *5* (1), 155–175.

DEGRADATION, GLOBAL ASSESSMENT OF

Selim Kapur
Erhan Akça
University of Çukurova, Department of Soil Science, Adana, Turkey

INTRODUCTION

One of the major challenges facing humanity in the coming decades is the need to increase food production to cope with the ever-increasing population, which is indeed the greatest threat for food/soil security (1). The worldwide loss of 5–7 M ha of agricultural land per year was reflected in the average annual loss of 120 ha in Germany in 1997 (2) in spite of the strict laws for environmental protection enacted there. This is the unpalatable but important indicator of a steady decline in agricultural production and increase in soil damage, reflecting the "impaired productivity" in a quarter of the world's agricultural land (3, 4).

BACKGROUND

Soil can degrade without actually eroding. It can lose its nutrients and soil biota and can become damaged by waterlogging and compaction. Erosion is only the most visible part of degradation, where the forces of gravity, water flow, or wind actively remove soil particles.

Rather than taking the classical view that soil degradation was, is, and will remain an ongoing process mainly found in countries of the developing world, this phenomenon should be seen as a worldwide process that occurs at different scales and different time-frames in different regions. The causes of biophysical and chemical soil degradations are enhanced by socioeconomic interventions, which are the main anthropogenic components of this problem, together with agricultural mismanagement, overgrazing, deforestation, overexploitation, and pollution as reiterated by Lal and Cummings (5), UNEP/ISRIC (6), Lal (7), Eswaran et al. (8, 9), Kapur et al. (10), and Cangir et al. (11), as the main reasons for erosion and chemical soil degradation.

Soil degradation, the threat to "Soil Security," is ubiquitous across the globe in its various forms and at varying magnitudes, depending on the specific demands of people and the inexorably increasing pressures on land. Europe provides many telling examples of the fragile nature of soil security and the destructive consequences of a wide range of soil degradation processes. Asia, Africa, and South and North America all are partly affected by the nonresilient impacts of soil degradation but are also experiencing more subtle destruction of soils via political developments, which seek to provide temporary relief and welfare in response to the demands of local populations.

Europe

The major problems concerning the soils of Europe are the loss of such resources due to erosion, sealing, flooding, and large mass movements as well as local and diffuse soil contamination especially in industrial and urban areas and soil acidification (12, 13). Salinity is a minor problem in some parts of Western and Eastern Europe, but with severe effects at the northern and western parts of the Caspian Sea (with low salinity; 14) mainly due to the shift to irrigated agriculture and destruction of the natural vegetation (Fig. 1).

Urbanization and construction of infrastructure at the expense of fertile land is widespread in Europe, particularly in the Benelux countries, France, Germany, and Switzerland, and such effects are most conspicuously destructive along the misused coasts of Spain, France, Italy, Greece, Turkey, Croatia, and Albania. The drastic increase in the rate of urbanization since the 1980s is now expected to follow the Blue Plan (2), which seeks to create beneficial relationships between populations, natural resources, the main elements of the environment, and the major sectors of development in the Mediterranean Basin and to work for sustainable development in the Mediterranean region. The very appropriate term "industrial desertification" remains valid for the once degraded soils of East Europe, under the pressure of mining and heavy industry, as in the Ukraine where such land occupies 3% of the total land area of the country (15).

Encyclopedia of Soil Science

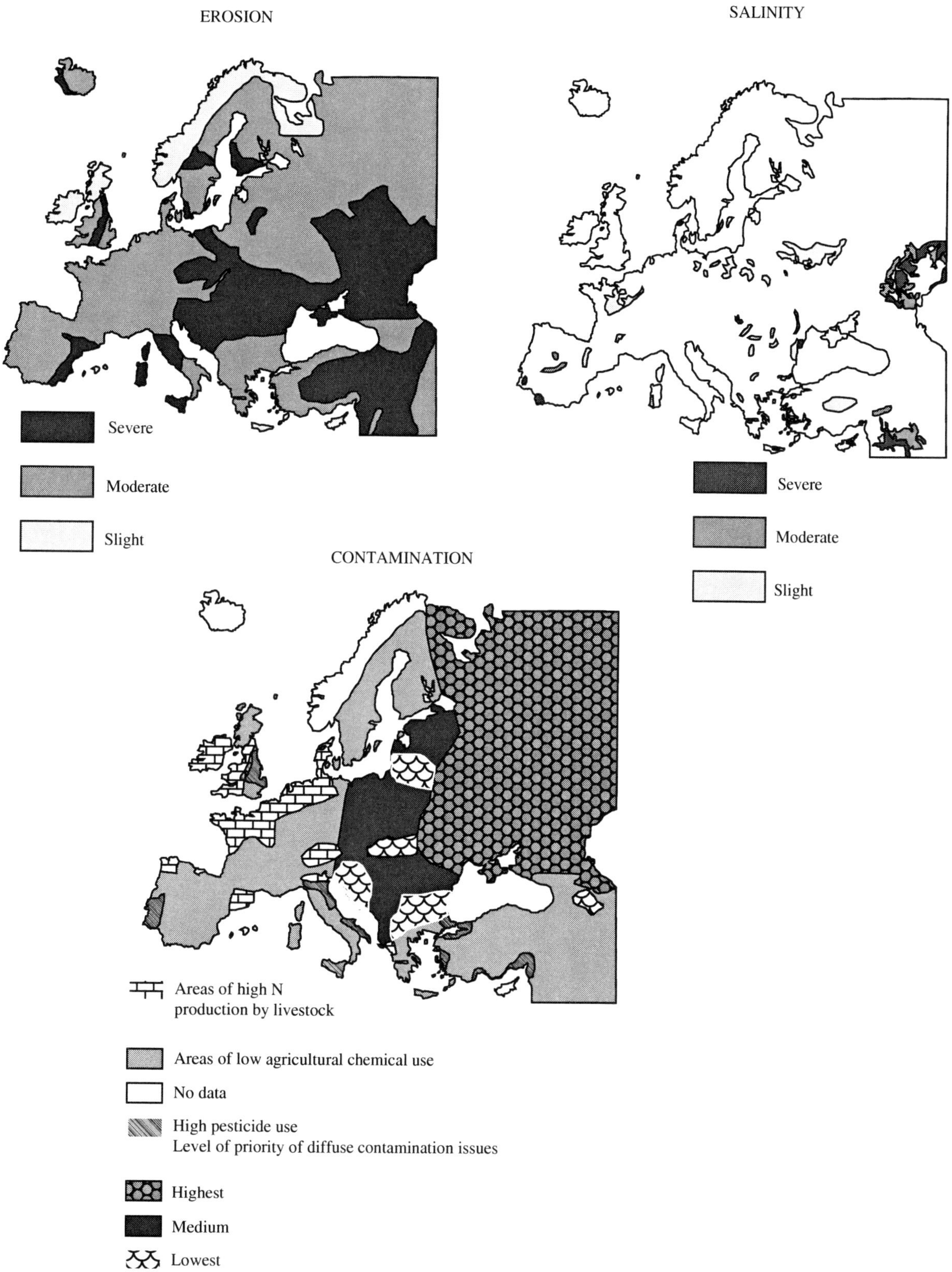

Fig. 1 Soil degradation in Europe. (From Refs. 6, 17–22.)

There are three broad zones of "natural" erosion across Europe, including Iceland: 1) the southern zone (the Mediterranean countries); 2) a northern loess zone comprising the Baltic States and part of Russia; and 3) the eastern zone of Slovenia, Croatia, Bosnia-Herzegovina, Rumania, Bulgaria, Poland, Hungary, Slovakia, the Czech Republic, and Ukraine (Fig. 1). Seasonal rainfalls are responsible for severe erosion due to overgrazing and the shift from traditional crops. Erosion in southern Europe is an ancient problem and still continues in many places, with marked on-site impacts and with significant decreases in soil productivity as a result of soil thinning. The northern zone of high-quality loess soils displays moderate effects of erosion with less intense precipitation on saturated soils. Local wind erosion on light textured soils is also responsible for the transportation of agricultural chemicals used in the intensive farming systems of the northern zone to adjacent water bodies, along with eroded sediments. The high erodibility of the soils of the eastern zone is exacerbated by the presence of large state-controlled farms that have introduced intensive agriculture at the expense of a decrease in the natural vegetation. Contaminated sediments are also present in this zone, particularly in the vicinity of former industrial operations/deserts, with high rates of erosion in Ukraine (41% of the total land area) and Russia (57% of the total land area) whose agricultural land has been subjected to strong water and wind erosion ever since the beginning of industrialization.

Localized zones of likely soil contamination through the activities of heavy industry are common in northwestern and central Europe as well as Northern Italy, together with more scattered areas of known and likely soil contamination due to the intensive use of agricultural chemicals. Sources of contamination are especially abundant in the "hot spots" associated with urban areas and industrial enclaves in the northwestern, southern, and central parts of the continent (Fig. 1). Acidification through deposition of windborne industrial effluents and aerosols has been a long-standing problem for the whole of Europe; however, this is not expected to increase much further, especially in Western Europe, as a result of the successful implementation of emission-control policies over the recent years (2).

The desertification of parts of Europe has been evident for some decades, and the parameters of the problem are now becoming clear, with current emphasis on monitoring of the environmentally sensitive areas (16, 21) on selected sites, seeking quality indicators for: 1) soil, 2) vegetation, 3) climate, and 4) human management throughout the Mediterranean basin. Apart from the human factor, these indicators are inherent.

Asia

The most severe aspect of soil degradation on Asian lands has been desertification due to the historical, climatic, and topographic character of this region as well as the political and population pressures created by the conflicts of the past 500 years or so. Salinization caused by the rapid drop in the level of the Aral Sea and the water-logging of rangelands in Central Asia due to the destruction of the vegetation cover by overgrazing and cultivation provide the most striking examples of an extreme version of degradation—desertification caused by misuse of land. Soil salinity, the second colossal threat to the Asian environment has occurred through the accumulation of soluble salts, mainly deposited from saline irrigation water or through mismanagement of available water resources, as in the drying Aral Sea and the Turan lowlands as well as the deterioration of the Oases in Turkmenistan, with excessive abstraction of water in Central Asia (Fig. 2) (22).

The drylands of the Middle East have been degrading since the Sumerian epoch, with excessive irrigation causing severe salinity and erosion/siltation problems (23), especially in Iraq, Syria, and Saudi Arabia. Iraq has been unique in the magnitude of the historically recorded build-up of salinity levels, with 8.5 million ha saline land, which is 64% of the total arable land surface (i.e., 90% of the land in the southern part of that country). The historical lands of Iran, Pakistan, Afghanistan, India, and China are also subject to ancient and ongoing soil degradation processes, which are subtle in some areas but evident and drastic in others (Fig. 2).

Africa

Africa's primary past and present concern has been the loss of soil security by nutrient depletion, i.e., the decreasing NPK levels (kg ha^{-1}) in cultivated soils following the exponential growth in population and the resulting starvation and migrations (Fig. 3). Intensification of land use to meet the increased food demands combined with the mismanagement of the land leads to the degradation of the continental soils. This poses the ultimate question of how the appropriate sustainable technologies can be identified that will permit increased productivity of soils. This problem is illustrated by the example of the Sudan, where nutrient depletion has steadily increased through more mechanized land preparation, planting and threshing without the use of inorganic fertilizers, and legume rotations. Thus, aggregate yields have been falling as it became more difficult to expand the cultivated area without substantial public

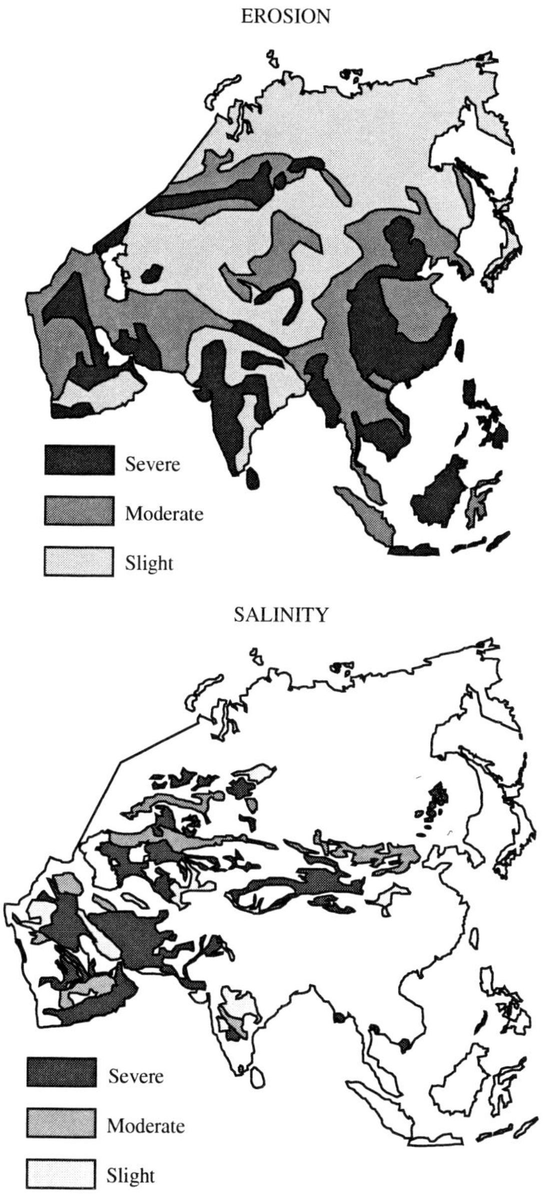

Fig. 2 Soil degradation in Asia. (From Refs. 6, 20, 23.)

investments in infrastructure. This decline in yield has occurred at enormous rates in the Vertisols of the Sudan, with the mean annual sorghum yields decreasing from 1000 to 500 $kg\,ha^{-1}$ (24). In Burkina Faso the decreased infiltration and increased runoff causing erosion is a further consequence of repeated cultivation. Thus, the technological measures to be identified for these two African examples must include development of water retention technologies in Burkina Faso, while polyculture/rotations with proper manuring and fertilization for cost-efficient provisions of N and P and preferably green manuring are all needed in the Sudan to permit the balanced management of soil moisture, nutrients, and organic matter (and to enhance C-sequestration, a main goal for sustainability based on the earth sciences—to ensure the security of both the soil and the global climate) (24, 25).

Central and South America

Water and wind erosion are the dominant soil degradation processes in these regions and have caused the loss of the topsoil at alarming rates due to the prevailing climatic and topographic conditions. Almost as important is the loss of nutrients from the Amazon basin (26) (Fig. 4). These effects are mainly attributable to deforestation and overgrazing, the former being responsible for the degradation of 576 million ha of potential agricultural land. Another important factor has been the ever-increasing introduction of inappropriate agricultural practices derived from the so-called imported technology, which has not been properly adapted to indigenous land-use procedures. The traditional methods of permitting the land to recover naturally has been almost totally abandoned and has been replaced by unsuitable technological measures designed to maintain production levels (temporarily) and to overcome the loss of soil resilience, thus increasing chemical inputs.

The rapid industrialization/urbanization of the limited land resources in the Caribbean region has been expelling agricultural communities to remote and marginal regions that are at present rich in biodiversity and biomass—a major global C sink. Moreover, large-scale livestock herding of Central and South America is also a major threat to soil security and has been responsible for degrading 35% of the Argentinian pasture lands.

North America

The most prominent outcome of soil degradation (or more correctly desertification) in the United States is exemplified by the accelerated dust storm episodes of the 1930s—the Dust Bowl years, marked by the "Black Blizzards" which were caused by persistent strong winds, droughts, and overuse of the soils. These resulted in the destruction of large tracts of farmland in the south and central United States. Recently, salinization has become an equally severe problem in the western part of the country (Fig. 5) through artificial elevation of water tables by extensive irrigation, with associated acute drainage problems. An area of about 10 million ha in the west of the United States has been suffering from

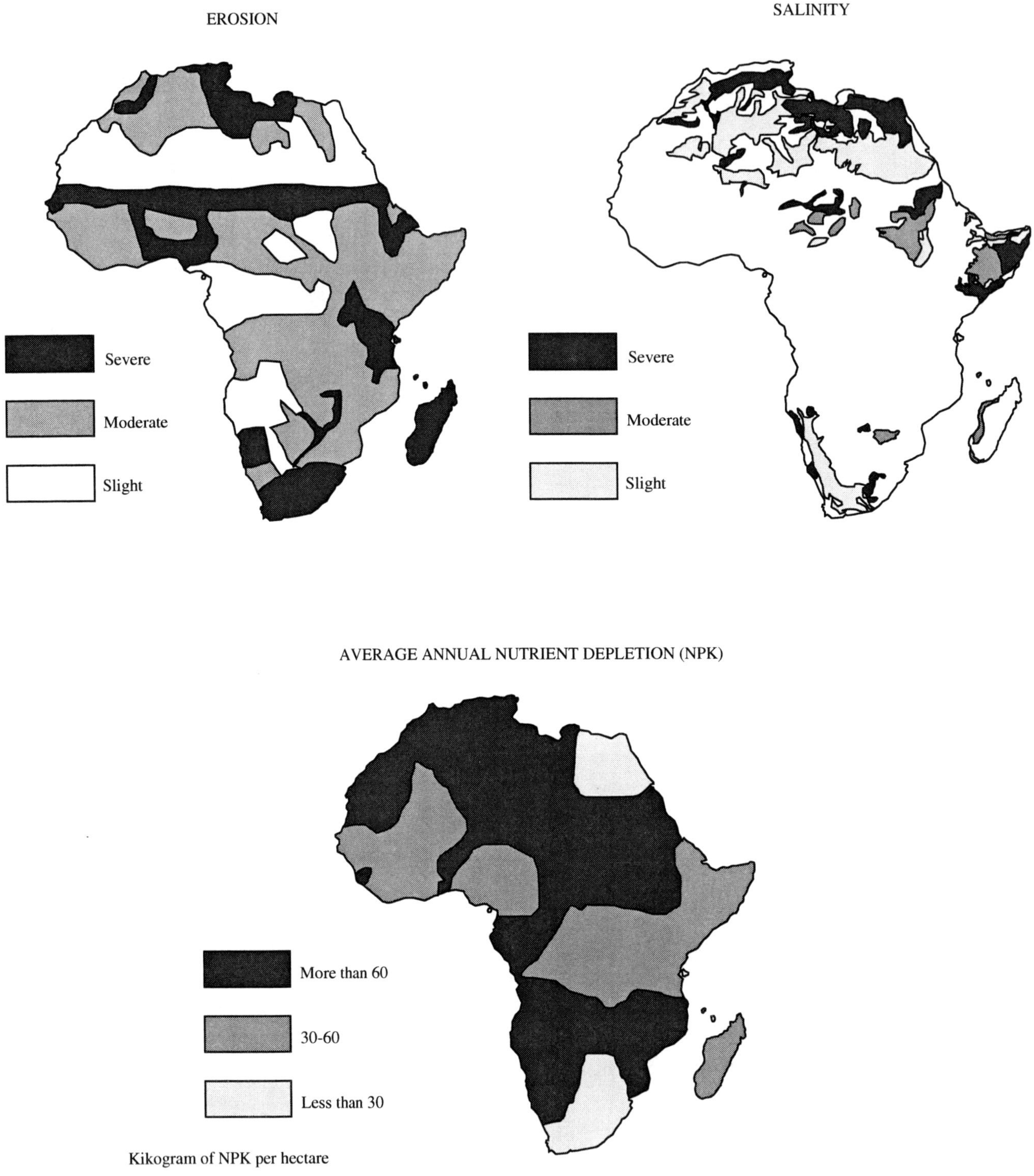

Fig. 3 Soil degradation in Africa. (From Refs. 6, 23.)

salinity-related reductions in yields, coupled with very high costs in both the Colorado River basin and the San Joaquin Valley (27). Unfortunately, new irrigation technologies such as the Center-Pivot irrigation system (developed as an alternative to the conventional irrigation systems causing the salinity problems) has

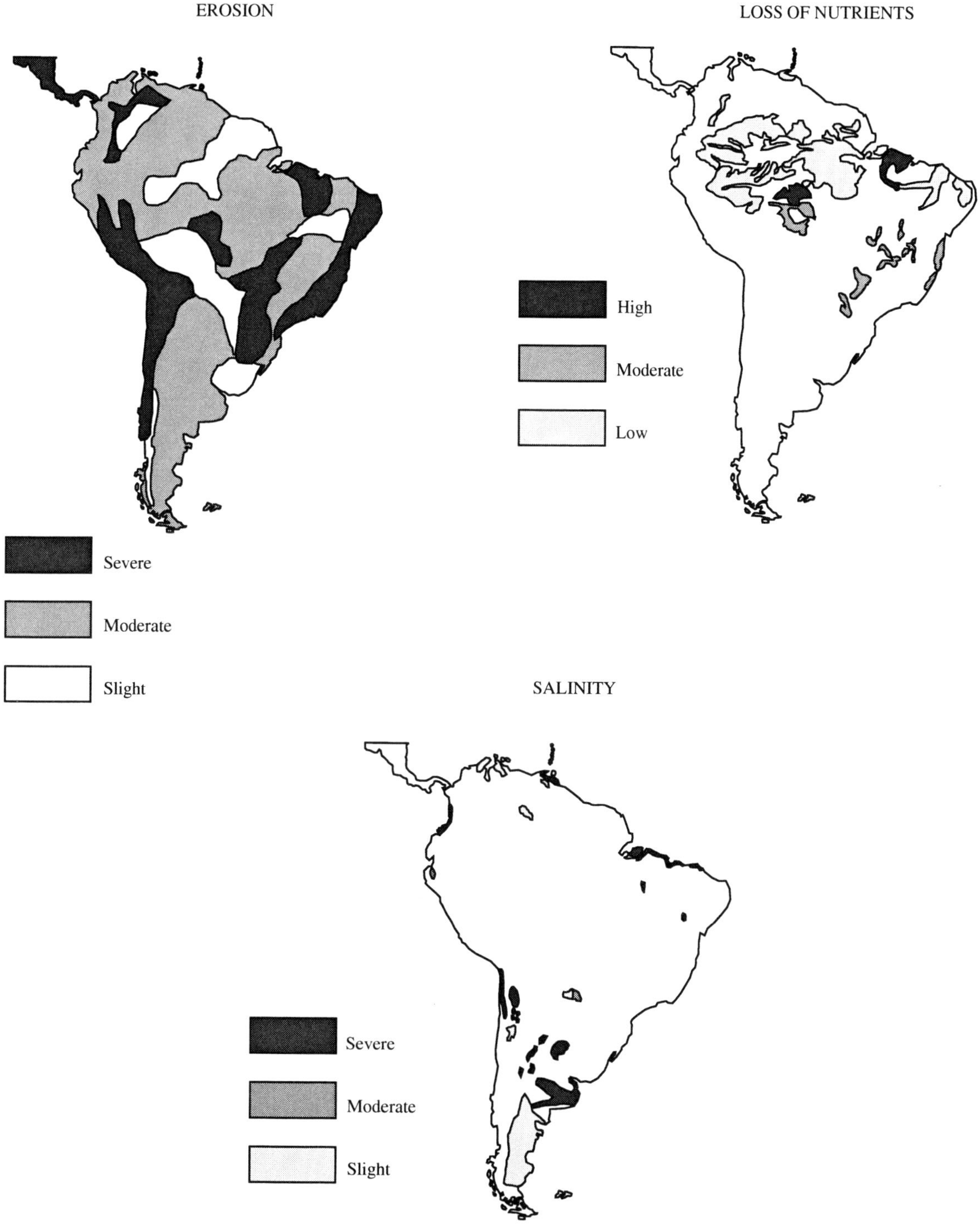

Fig. 4 Soil degradation in Central and South America. (From Refs. 6, 22.)

caused a decline of the watertable levels in areas north of Lubbock, Texas, by around 30–50 m, leading to a dramatic decrease in the thickness of the well-known Ogallalla aquifer by 50%. In some areas this has been followed by ground subsidence, which is an extreme form of soil structure degradation, that is, loss of the physical integrity of the soil.

Loss of topsoil, as the result of more than 200 years of intensive farming in the United States, is estimated to vary between 25% and 75% and exceeds the upper

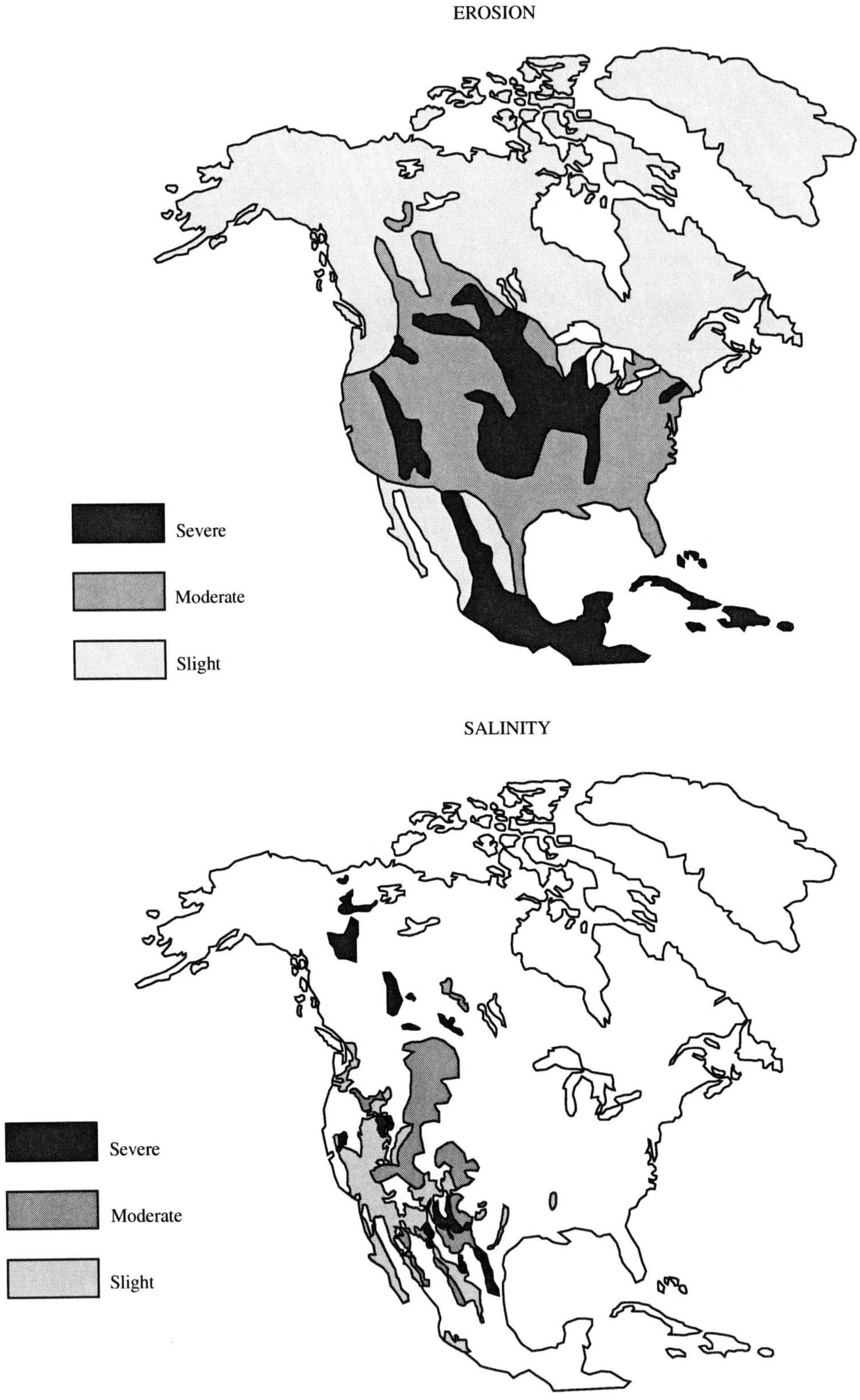

Fig. 5 Soil degradation in North America. (From Refs. 6, 22.)

limit in some parts of the country (19, 20). The United States provides good examples of the difficulties involved in erosion control, with its large-scale intensive agriculture—deteriorating soil structure and increasing erosion of its susceptible soils. This problem could be overcome primarily by the strict introduction of the no-till system. No-till areas have increased from 4 million ha in 1989 to 18 million ha in 1997, and they are forecast to follow a linear extrapolation until 2012, when they will attain 48 million ha out of the total 81 million ha of cultivated land (28). Conservation farming is practiced in only about half of all U.S. agricultural land and on less than half of the country's most erodible cropland. Conservation farmers are encouraged to use only the basic types of organic fertilizers, such as animal and green manure together with compost, mulch farming, improved pasture management, and crop rotation to conserve soil nutrients.

Canada is a large country where half of the 68 million ha of available land is cultivated, with an average farm size of 450 ha. It is reported that Canada has experienced annual soil losses on the prairies, through wind and water erosion, that are similar to the Asian steppes, amounting, respectively, to 60 and 117 million tonnes. These annual rates are much higher than the rate of soil formation, resulting in an annual potential grain production loss of 4.6 million tonnes of wheat. With regard to primary soil salinity, during historic times the prairies have experienced steady increases, related partially to increasing groundwater levels. Major problems of secondary salinity are estimated to affect 2.2 million ha of land in Alberta, Saskatchewan, and parts of Manitoba, with an immense economic impact each year.

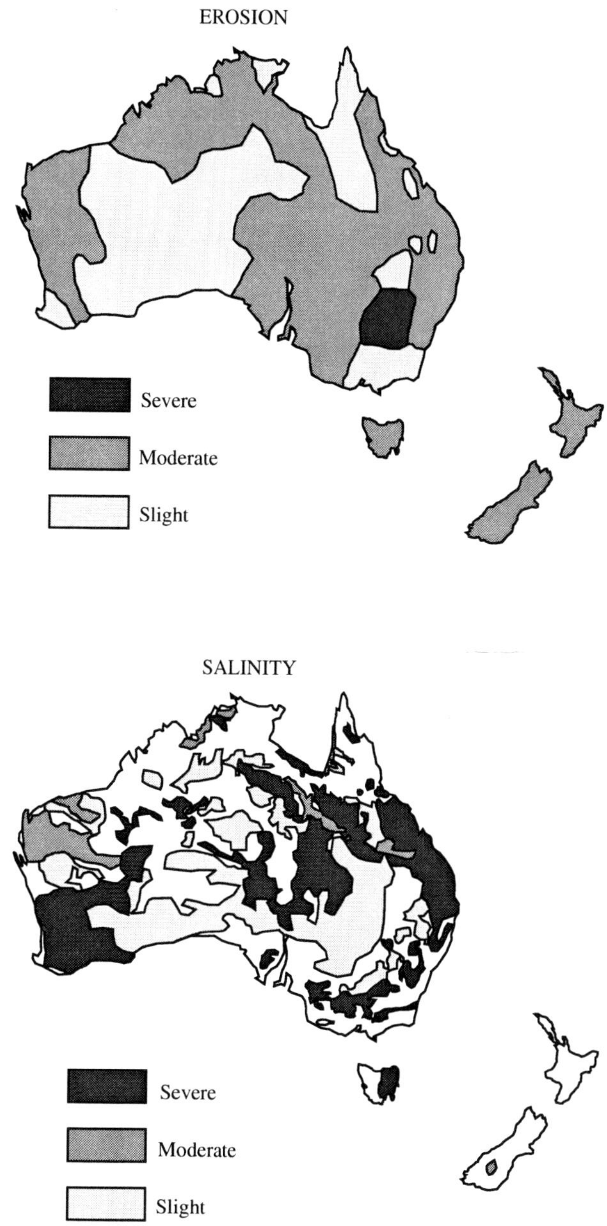

Fig. 6 Soil degradation in Australia. (From Refs. 6, 22.)

Australia

The Australian agricultural/soil resource base has been endangered since the "business as usual" concept was adopted on the continent to achieve temporary economic betterment. Identification of the different types of soil degradation in Australia reveals that erosion has been the main component, primarily via dust storms, which still are a serious problem, especially where cropping practices do not include retention of cover and minimum tillage methods. Water erosion effects are also particularly severe in areas of summer rainfall and topographic extremities (Fig. 6). Remedial actions for this include the well-known measures of maintaining adequate cover and changing prevailing attitudes towards stock management, storage feed, redesign of watering sites, and management of riparian areas.

Part of the excess salinity in Australia is of primary origin and was retained in the subsoil by trees, which have now been cleared to create soil surfaces for cropping and pastures, allowing penetration of water to the saline subsoil, then followed by abstraction from the water table, thus leading to the ultimate disaster. About 30% of Australia's agricultural land is sodic, creating poor physical conditions and impeded productivity. This problem can only be alleviated by massive revegetation programs and by taking extra care of the water table and plant cover. Despite the introduction of costly conventional measures for reclamation, salinity levels continue to increase across Australia in the dry and irrigated soils. The dryland salinity in the

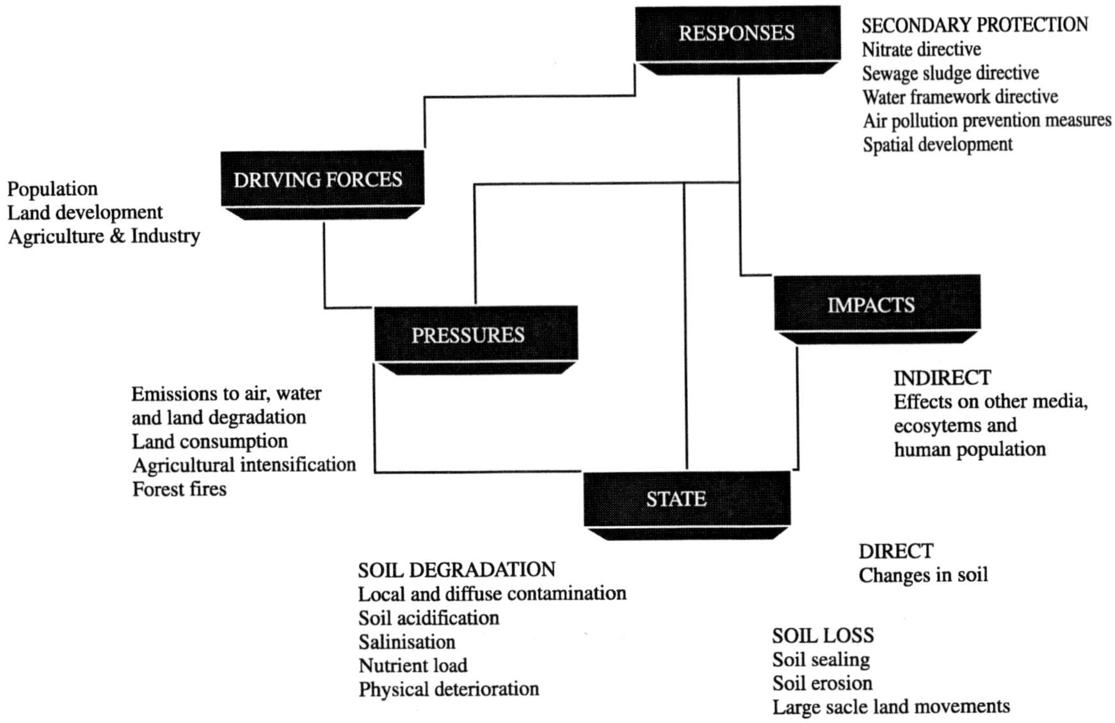

Fig. 7 DPSIR framework applied to soil. (From Ref. 2.)

continent affects about 2.5 million ha of farmland and is expanding at a rate of 3–5% $year^{-1}$ (29, 30).

The retardation of organic matter levels also requires remediation measures, with economically justified fertilizer use strategies to be utilized throughout the continent. Moreover, overgrazing has resulted in the impoverishment of plant communities and loss of habitats as well as the decline in the chemical fertility of the soil by progressive depletion of organic matter in the topsoil, followed by deterioration in the soil structure.

Acidification caused by legume-based mixed farming plus use of ammonia-based fertilizers threatens 55 million ha of Australian land. Liming seems to be the most effective present remedy, but this is costly, does not lead to rapid recovery, and is impractical for subsoil acidity. Thus, the precise remedies have yet to be developed for the conditions on this continent, utilizing careful, long-term monitoring and the experience of farmers to devise specific treatment and conservation procedures.

CONCLUSIONS

The state of soil degradation and its remedies as a multifunction–multi impact approach, has been identified through a Driving Force–Pressure–State–Impact–Response (DPSIR) matrix by the European Environmental Agency (2) (Fig. 7) leading to Sustainable Land Management (SLM; 31–33) measures to be taken for the future. SLM is concerned with more soil-friendly farming practices that minimize the erosion potential of soils, together with the adoption by landholders of property management planning procedures that involve community actions, such as the Landcare Programme of Australia (34). Moreover, as Smyth and Dumanski (35) have stated, it combines socioeconomic principles with environmental concerns so that the production is enhanced, together with the reduction of its level of risk with the protection of natural resources, which would prevent degradation of soil and water quality to be successfully accepted by the farmer. The methods to be adapted for SLM via community actions include contour farming, terracing, vegetative barriers, and other land-use practices amalgamated with Indigenous (traditional) Technical Knowledge (ITK) (36) as applied to farming and landscape preservation. The impetus to the use of ITK by scientists and local communities in creating new strategies for sustainable resource management was initiated in the United Nations Conference on Environment and Development (UNCED) held in Rio de Janeiro (Brazil) in 1992.

REFERENCES

1. Alexandratos, N. World Food and Agriculture: The Outlook for the Medium and Longer Term. *NAS Colloquium. Plants and Population: Is There Time?* 5–6 December 1998. Beckman Center of the National Academy of Sciences: UC Irvine, 1998.
2. UNEP/EEA. Soil Degradation and Sustainable Development in Europe: A Challenge for the 21st Century. *4th Conference to the Parties*. Bonn. 19 December 2000. UNEP/EUROPEAN ENVIRONMENTAL AGENCY. 2000a;.
3. Oldeman, L.R.. *The Global Extent of Soil Degradation*; Soil Resilience and Sustainable Land Use, CAB International: Oxon, U.K., 1994; 115.
4. Steiner, K.G. *Causes of Soil Degradation and Development Approaches to Sustainable Soil Management*; Steiner, K.G., Ed.; GTZ. Weikersheim. Margraf Verlag: Filderstadt, Germany, 1996; 50.
5. Lal, R.; Cummings, D.J. Clearing a Tropical Forest. I. Effect on Soil and Micro-Climate. Field Crop Res. **1979**, *2*, 91–107.
6. UNEP/ISRIC. *GLASOD World Map of the Status of Human-Induced Soil Degradation, 1:10 M. 3 Sheets*, 2nd Revised Ed.; (130×270 cm) folded in A-4 format, with explanatory note, ISRIC: Wageningen, 1991.
7. Lal, R. Methods and Guidelines for Assessing Sustainable Use of Soil and Water Resources in the Tropics. N.21. *SMSS Technical Monograph*; US Agency for International Development Washington, D.C., 1994; 79.
8. Eswaran, H.; Reich, P. Desertification: A Global Assessment and Risks to Sustainability. *Proc. of the 16th Int. Cong. Soil Science*; Montpellier: France, 1998, CD-ROM.
9. Eswaran, H.; Beinroth, F.H.; Reich, P. Global Land Resources and Population Supporting Capacity. Am. J. Alternative Agric. **1999**, *14*, 129–136.
10. Kapur, S.; Atalay, İ.; Ernst, F.; Akça, E.; Yetiss, C.; İssler, F.; Öcal, A.D.; Üzel, İ; Ssafak, Ü. A Review of the Late Quaternary History of Anatolia. In *1st International Symposium on Cilician Archaeology*; Durugönül, S., Ed.; Mersin University: Turkey, 1999; 253–272.
11. Cangir, C.; Kapur, S.; Boyraz, D.; Akça, E.; Eswaran, H. Land Resource Consumption in Turkey. Journal of Soil and Water Conservation **2000**, *3*, 253–259.
12. UNEP. *Magazine for Sustainable Development. Our Planet*. Nairobi, Kenya, 2000b; *Vol. 6*. No. 5.
13. Oldeman, L.R.; Hakkeling, R.T.A.; Sombroek, W.G. *World Map of the Status of Human-Induced Soil Degradation: An Explanatory Note*; International Soil Reference Center: Wageningen, Netherlands, 1992; 34.
14. Karpinsky, M.G. Aspects of the Caspian Sea Benthic Ecosystem. Marine Pollution Bulletin **1992**, *24*, 3849.
15. Mnatsakanian, R. *Environmental Legacy of the Former Soviet Republics*; Center for Human Ecology; University of Edinburgh: Edinburgh, 1992.
16. Denisov, N.B.; Mnatsakanian, R.A.; Semichaevsky, A.V. Environmental Reporting in Central and Eastern Europe: A Review of Selected Publications and Frameworks. Central European University/United Nations Environment Program. 1997. CEU/50-97.1. UNEP/DEIA/TR.97-6.
17. Kobza, J. Monitoring and Contamination of Soils in Slovakia in Relation to Their Degradation. In *Proceedings of the 1st International Conference on Land Degradation*, 10–14 June 1996, Adana, Turkey; Kapur, S., Akça, E., Eswaran, H., Kelling, G., Vita-Finzi, C., Mermut, A.R., Öcal, A.D., Eds.; 1999; 68–76.
18. USDA-NRCS. Risk of Human Induced Desertification. NRSC. Soil Survey Division, World Soil Resources, 1999.
19. USDA-NRCS. Global Desertification Vulnerability. NRSC. Soil Survey Division, World Soil Resources, 1999.
20. Erol, O. Misuse of the Mediterranean Coastal Area in Turkey. In *Proceedings of the 1st International Conference on Land Degradation*, 10–14 June 1996, Adana, Turkey, Kapur, S., Akça, E., Eswaran, H., Kelling, G., Vita-Finzi, C., Mermut, A.R., Öcal, A.D., Eds.; 1999; 195–203.
21. Kosmas, C.; Ferrara, A.; Briassouli, H.; Imeson, A. *Methodology for Mapping Environmentally Sensitive Areas (ESAs) to Desertification*. The MEDALUS Project. Mediterranean Desertification and Land Use Report (Eds. C. Kosmas, M. Kirkby, N. Geeson). European Commission, Belgium. 1999; 31–47.
22. Kharin, N.G.; Tateishi, R.; Harahsheh, H. *Degradation of the Drylands of Asia*; Center for Environmental Remote Sensing Chiba Univ. Japan, 1999; 81.
23. Lowdermilk, W.C. Conquest of the Land Through Seven Thousands Years. No. 99. USDA. Soil Conservation Service. U.S. Government Printing Office, Washington, D.C., 1997; 1–30.
24. Sanders, J.H.; Southgate, D.D.; Lee, J.G. The Economics of Soil Degradation: Technological Change and Policy Alternatives. No.22. *SMSS Technical Monograph*. Dept. of Agricultural Ecomomics. Purdue University. 1995; 8–11.
25. Lal, R. Soil Degradation and Future of Agriculture in Sub-saharan Africa. Journal of Soil and Water Conservation **1988a**, *43*, 445–451.
26. Lal, R. Monitoring Soil Erosion's Impact on Crop Productivity. In *Soil Erosion Research Methods*; Lal, R., Ed.; Soil and Water Conservation Society Ankeny: Iowa, 1988b; 187–200.
27. Barrow, C.J. *Land Degradation*; Cambridge University Press: Cambridge, 1994.
28. El-Ashry, M.T.; Gibbons, D.C. *Water and Arid Lands of the Western United States*; Cambridge University Press: Cambridge, 1988.
29. Hamblin, A.; Kyrieur, G. *Trends in Wheat Yields and Soil Fertility in Australia*; Department of Primary Industries and Energy, Bureau of Resource Sciences, AGPS: Canberra, 1993.
30. Hamblin, A.; Williams, J. Alarming Erosion of Australia's Soil and Land Base. 1996. ATSE Focus, No. 90. www.atse.org.au/publications/focus/focus-hamblin.htm. (accessed Apr 2001).
31. Dumanski, J.; Craswell, E. Resource Management Domains for Evaluation and Management of Agro-Ecological Systems. In *Proceedings of the IBSRAM International Workdshop on Resource Management Domains*, Kuala Lumpur, 26–29 August 1996; Leslie, R.N., Ed.; The International Board for Soil Research and Management (IBSRAM): Bangkok, Thailand, 1996; 1–16.
32. Eswaran, H.; Beinroth, F.; Reich, P. Biophysical Considerations in Developing Resource Management

Domains. In *Proceedings of the IBSRAM International Workdshop on Resource Management Domains*, Kuala Lumpur, 26–29 August 1996; Leslie, R.N., Ed.; The International Board for Soil Research and Management (IBSRAM): Bangkok, Thailand, 1996; 61–78.
33. Kapur, S.; Eswaran, H.; Akça, E.; Dingil, M. Developing a Sustainable Land Management (SLM) Research Strategy for the GAP (Southeastern Anatolian Irrigation Project). Workshop Abstracts on the Challenge of Production System Sustainability: Long-term Studies in Agronomic Research in Dry Areas, 1997. 8–11 DEC 1997, Aleppo, Syria. 1997; 24–25.
34. Landcare Australia. Annual Report and Yearbook. Landcare Australia. Australia, 1998; 70.
35. Smyth, A.J.; Dumanski, J. *FESLM: An International Framework for Evaluating Sustainable Land Management;* World Soil Resources Report. No. 73; FAO: Rome, 1993.
36. Prasad, R.N.; Arya, S.L. Indigenous Technical Knowledge of Soil and Water Conservation in India. In *Indigenous Technical Knowledge for Land Management in Asia*; Leslie, R.N., Ed.; 28 January–2 February 1997. The International Board for Soil Research and Management (IBSRAM): Bangkok, Thailand, 1998; 1–24.

DEGRADATION, GREENHOUSE EFFECT AND

Eddy De Pauw
International Centre for Agricultural Research in the Dry Areas (ICARDA), Aleppo, Syria

Michael A. Zoebisch
Asian Institute of Technology (AIT), Bangkok, Thailand

INTRODUCTION

Soil degradation refers to a decline of soil quality which reduces either the soil's productive capacity, its sustainability, its resistance to degradation, or the biodiversity of the supported natural ecosystems. Soil degradation is a global phenomenon and its most important expressions are water and wind erosion, fertility decline, salinization, waterlogging, or lowering of the groundwater table. Being both a cause and effect of rural poverty, soil degradation is one of the most serious threats to the prosperity of rural populations worldwide, and is accelerated by nonsustainable land management induced by rural poverty.

The greenhouse effect refers to the trapping of heat in the lower levels of the atmosphere. The atmosphere allows a large percentage of the rays of visible light from the Sun to reach the Earth's surface and heat it. Part of this energy is re-radiated by the Earth's surface in the form of long-wave infrared radiation, much of which is absorbed in the atmosphere by water vapor and greenhouse gases, in particular carbon dioxide (CO_2), but also other trace gases, notably methane (CH_4), nitrous oxide (N_2O), and chlorofluorocarbons (CFCs). This process of heat-trapping causes the Earth's surface and lower atmospheric layers to warm to higher temperatures than would otherwise be the case. In fact, this natural greenhouse effect makes life on Earth possible.

Atmospheric concentrations of greenhouse gases have increased because of human activities, primarily the combustion of fossil fuels (coal, oil, and gas), deforestation, and agricultural practices. Since the beginning of the pre-industrial era around 1750, CO_2 has risen by nearly 30%, methane by more than a factor of two, and N_2O by about 15%. Their concentrations are higher now than at any time during the last 420,000 years. The human-induced greenhouse gas emissions are responsible for the rise of average global temperature of about 0.7°C over the last 100 years, with the most rapid rise occurring since the mid 1970s. If current greenhouse gas emission trends continue, an increase in global mean surface temperature in the range 1.5–6°C is expected by 2100 (1) (http://www.ipcc.ch/press/sp-cop6.htm). In addition, there is evidence that precipitation patterns are changing, sea level is increasing, glaciers are retreating world-wide, Arctic Sea ice is thinning, and the incidence of extreme weather events is increasing in some parts of the world.

The overwhelming majority of scientific experts, while recognizing that scientific uncertainties exist, now believe that global warming is at least partially linked to the increased concentration of greenhouse gases by human activities and that climate change is inevitable. This will have potentially major repercussions on water resources, agriculture, forestry, fisheries and human settlements, ecological systems and human health with developing countries being the most vulnerable (1) (http://www.ipcc.ch/press/sp-cop6.htm).

LINKAGES BETWEEN CARBON, GREENHOUSE EFFECT, AND SOIL DEGRADATION

Soil degradation and greenhouse effect are both linked to human activities and are likely to interact in the context of global climate change. The expected shift in climate patterns, which will involve both changes in temperature and rainfall patterns, has the potential to either accelerate degradational trends in regions where they already prevail, or to initiate certain types of degradation in new areas where they do not exist at present. On the other hand, current or future soil degradation is likely to enhance the greenhouse effect through a reduced ability of the soil to retain carbon.

The global carbon cycle, showing the major reservoirs and flows of carbon in gigatonnes of carbon or GtC (10^{12} kg) per year, is outlined in Fig. 1. The main stores or "sinks" of carbon are the oceans, the lithosphere with geological deposits containing fossil fuels, the terrestrial

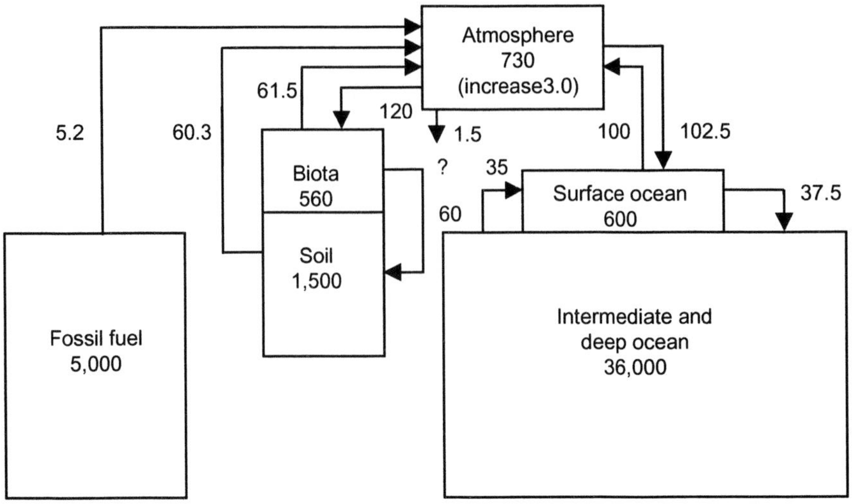

Fig. 1 The global carbon cycle in about 1980. (From Ref. 2.)

ecosystems, composed of living biomass and soil, and the atmosphere. Of these, the oceanic carbon reservoir is the largest but least reactive, and the atmospheric reservoir, although the smallest, is the most critical one since it drives global warming. With the exception of the fossil fuel reservoir, each reservoir at times behaves as either a source or a sink of carbon, and it is the rate of exchange of carbon between different reservoirs that determines whether they act as net sinks or sources of carbon. Fig. 1 represents an estimate and refers to the situation of fossil fuel use in about 1980. An estimate of carbon flows and reservoir sizes around 1995, based on data provided by the Intergovernmental Panel on Climate Change, is given in Table 1 and compared with the estimates of 1980.

Table 1 shows that the terrestrial ecosystems have both an important carbon source and sink role. They act as a source of atmospheric carbon through land use change and tropical deforestation, at the same time they act as an important sink through the regrowth of forests in temperate regions. 1.5–1.7 GtC yr^{-1} is not accounted for. The "missing" carbon is probably going into the terrestrial biosphere in the northern hemisphere where it contributes to an increasing capture of carbon in terrestrial ecosystems through the CO_2 fertilization effect (3).

Table 1 Global carbon flux budget

Carbon flows	1980	1995
Annual atmospheric increase of CO_2	3.0	3.4
Sources		
Terrestrial ecosystems	1.8	2.7
Fossil fuels	5.2	6.4
Sinks		
Terrestrial ecosystems		2.0
Oceans	2.5	2.0
"Missing"	1.5	1.7

Note: Units are *GtC* (gigaton carbon, 10^{12} kg).
(From Refs. 3, 4.)

Within the terrestrial ecosystems, soils constitute the largest reservoir of carbon, which is retained in the form of soil organic matter. All soil types of the world have an inherent capacity to store organic carbon, which is determined by soil temperature and moisture regime, internal drainage, textural composition, and clay mineralogy. World soils show great variation in soil organic matter content, with values of 4–6% (weight basis) in the most fertile to less than 0.5% in others (5).

Figure 1 shows that currently soils are a net source of carbon to the atmosphere. This is attributed to the massive land use change resulting from the conversion of forests and other natural ecosystems into farmland. From a terrestrial carbon balance model, Esser (6) estimated for the period 1860–1980 the net decline of the carbon pool in plant biomass at 22 Gt, and in the soil at 5 Gt. If adjustment was made for increased carbon mineralization resulting from the global temperature increase, the loss of carbon in the soil reservoir would have increased to 12 Gt.

Apart from land use change, soil degradation has been considered a major contributing factor to the loss of soil carbon to the atmosphere and therefore to an increase in the greenhouse effect. The main reason is that soil degradation, through the processes of erosion, salinization, and fertility decline affects the productive capacity of land,

and as a result less carbon is returned to the soil. In fact, due to degradation many soils throughout the world do not achieve their potential carbon-carrying capacity. This is particularly the case in most arid and semiarid regions of the developing world, where overgrazing, fuel–wood collection, and opportunity cropping have seriously depleted the limited stock of organic matter. The resulting erosion, soil structural degradation and fertility loss have often all but prevented the reestablishment of the natural vegetation, and in order to regain the original productive capacity, expensive reclamation or conversion into protected lands is now required in huge parts of the dry areas.

Also in the (sub)humid tropics, particularly in Africa, soil carbon stocks have been much depleted due to a generally negative nutrient balance resulting from inadequate fertilizer use and reductions of fallow periods in shifting cultivation systems.

The scale of soil degradation worldwide is certainly sufficient to have impact on the global terrestrial carbon stocks. Oldeman et al. (7) estimated that 38% of the agricultural land, 21% of the permanent pastures, and 18% of the forests and woodlands worldwide are affected by one form or another of soil degradation. However, an accurate assessment of the contribution of soil degradation to the greenhouse effect is, at the moment, difficult because current inventories of the actual extent of different soil types, their carbon carrying capacities in different eco-regions, the extent of various types of land degradation, and the carbon contents of degraded soils are inadequate.

In addition, some of the degradation observed does not necessarily lead to carbon loss. In the case of erosion, carbon is transported from one place to another without involving an increase in mineralization. There is also evidence from the geological past that an increased atmospheric CO_2, coupled with warmer than current temperatures, higher rainfall, and increased biological activity—the scenario predicted by global change—could lead to higher concentrations of reactive bicarbonate (HCO_3^-) in soil solutions, producing more deeply weathered soil profiles rather than atmospheric carbon (8).

Global warming is also likely to increase the risk of soil degradation. In its assessment of regional vulnerability to climate change (9), the Intergovernmental Panel on Climate Change (IPCC) anticipates an increase of global mean precipitation and of the frequency of intensive rainfall, entailing an increased risk of soil erosion by water and of soil acidification in humid areas. However, currently arid and semiarid regions in southern and northern Africa, southern Europe, the Middle East, and parts of Latin America and Australia are expected to become even drier. Such areas would then become affected by more frequent drought and associated wind erosion, salinization, and carbon mineralization. Most regions of the world are expected to experience an increase in floods because of the projected increase in heavy precipitation events. Whereas the IPCC study outlines likely global trends and can be used for policy guidance at regional level, a quantified assessment of the impact of global warming at national or subnational level is highly speculative owing to major uncertainties about the geographical location and magnitude of the expected changes.

Whether the soil ultimately serves as a source or sink of carbon very much depends on the way it is managed. Even agricultural land is capable of storing more carbon if appropriate land management practices are adopted, and thus to reclaim some or all of the carbon lost in the transformation from natural ecosystems. In this respect the potential to increase the carbon sink is particularly high in the dry areas, because their current carbon content is far below their potential and the size of desertified and degraded lands worldwide, estimated at some 2 billion hectares (3). Soil management principles advocated to increase carbon sequestration include the increase of the total and subsoil organic matter content, soil microaggregation, and soil biodiversity (10). Among the technological options proposed are conservation tillage practices, crop residue management, shelterbelts and windbreaks, intercropping, optimal crop rotations, winter cover crops, fertility improvement through green manuring, fallowing, precision farming, and improvement of the soil moisture regime through irrigation and salinity management (2, 10).

Generally speaking, land use practices designed to prevent or remedy soil degradation will contribute to carbon sequestration. Given the high spatial variability of soils, which will persist under global warming, it is essential that the proposed principles are fine-tuned for specific soils and eco-regions, and the practices adapted to local conditions.

REFERENCES

1. Watson, R.T. *Presentation of the Chair Intergovernmental Panel on Climate Change at the 6th Conference of Parties to the United Nations Framework Convention on Climate Change,* November 13, 2000. Available on URL: http//www.ipcc.ch/press/spcop6.htm.
2. Rosenzweig, C.; Hillel, D. Soils and Global Climate Change: Challenges and Opportunities. Soil Science **2000**, *165* (1), 47–56.

3. CAST, *Storing Carbon in Agricultural Soils to Help Mitigate Global Warming*; Issue Paper 14, April 2000, Council for Agricultural Science and Technology: Ames, Iowa, 2000.
4. IPCC, *Climate Change 1995: The Science of Climate Change*; Working Group 1, Cambridge University Press: New York, 1996.
5. Lal, R.; Logan, T.J. Agricultural Activities and Greenhouse Gas Emissions from Soils of the Tropics. In *Soil Management and Greenhouse Effect. Advances in Soil Science*; Lal, R., Kimble, J., Levine, E., Stewart, B.A., Eds.; CRC Press: Boca Rato, USA, 1995.
6. Esser, Modelling Global Terrestrial Sources and Sinks of CO_2 with Special Reference to Soil Organic Matter. In *Soils and the Greenhouse Effect*; Bouwman, A.F., Ed.; Wiley: Chichester, UK, 1990; 247–262.
7. Oldeman, L.R.; Hakkeling, R.T.A.; Sombroek, W.G. *World Map of The Status of Human-Induced Soil Degradation: An Explanatory Note*; 3 maps ISRIC/UNEP: Wageningen, The Netherlands/Kenya, 1990; 27.
8. Bird, M.; Fyfe, B.; Chivas, A.; Longstaffe, F. Deep Weathering at Extra-Tropical Latitudes: A Response to Increased Atmospheric CO_2. In *Soils and the Greenhouse Effect*; Bouwman, A.F., Ed.; Wiley: Chichester, UK, 1990; 383–389.
9. Watson, R.T., Zinyawera, M.C., Moss, R.H., Dokken, D., Ed. *The Regional Impacts of Climate Change. An Assessment of Vulnerability*; Special report of IPCC Working Group II. Intergovernmental Panel on Climate Change, ISBN 92-9169-110-0, 1997.
10. Lal, R.; Kimble, J.; Stewart, B.A. Towards Soil Management for Mitigating the Greenhouse Effect. In *Soil Management and Greenhouse Effect*; Advances in Soil Science, Lal, R., Kimble, J., Levine, E., Stewart, B.A., Eds.; CRC Press: Boca Rato, USA, 1995; 373–381.

DEGRADATION, PHYSICAL

Michael A. Zoebisch
Asian Institute of Technology, Klong Luang, Pathumthani, Thailand

Anthony R. Dexter
Institute of Soil Science and Plant Cultivation (IUNG), Pulawy, Poland

D

INTRODUCTION

There is no accepted standard terminology concerning the definitions of terms used to describe and classify soil degradation. It is difficult to draw clear boundaries between the different types of soil degradation, because, as with any complex natural system, there are a lot of interrelationship and interdependencies between chemical, biological and physical degradation processes in the soil. Different forms of soil physical degradation go along with each other—they are interlinked and not always clearly distinguishable from each other. The International Union of Soil Science (1) has proposed a comprehensive terminology for soil physical processes and characteristics but the use of terms is still not coordinated and sometimes confused. The classification of the different types of soil physical degradation is, therefore, not clearly differentiated. For example, it may be argued that waterlogging is not a degradation type on its own. It is rather a consequence of other physical degradation processes, and caused by a reduction in permeability due to soil-structure breakdown or compaction. Because of impeded water-entry and movement, waterlogging can also lead to other types of degradation, such as salinization. These examples illustrate the fuzziness of boundaries between soil physical degradation and other forms of degradation. Different types of soil physical degradation cannot be distinguished uniquely in terms of simple cause-and-effect relationships. The causes and effects are often the same for several types of degradation. Fig. 1 depicts the cause-and-effect system typical for soil physical degradation.

The main types of soil physical degradation described here are characterized by the different sets of processes involved and responsible for the degradation. Deterioration of soil structure is the overriding and more general form of soil physical degradation. Compaction, sealing and crusting, and hardsetting are important special types of soil-structure breakdown with important implications for soil management and land use.

LOSS OF SOIL STRUCTURE

There are a number of definitions of soil structure in the literature. Many of these describe the arrangement of primary particles into aggregates of different sizes and shapes and the associated pore spaces between them. Therefore, a structured soil is heterogeneous; whereas a degraded, structureless soil is homogeneous. A more general definition is "the spatial heterogeneity of the different components or properties of soil" (2). Soil structure significantly influences all processes that take place in the soil. It influences water infiltration (and hence runoff), the movement of water within the soil, and the amount of water that can be stored in the soil. Therefore, it has a direct influence on soil erosion. Soil structure also determines aeration levels in the soil, which are essential for the oxygen supply to roots, soil fauna, and for aerobic microbial activity. High soil strength impedes seedling emergence and root growth, and also makes tillage more difficult and more energy demanding. These functions demonstrate the role and importance of the size and continuity of pores in the soil for the movement of soil and water into and within the soil (3). Not only the soil structure but also the stability of the structure is of major importance. Structural stability determines the ability of a soil to withstand imposed stresses without changes in its geometric structure and functions. These stresses may be due to rapid wetting, raindrop impact, wheel traffic and excessive tillage.

Soil physical degradation reflects soil-structure breakdown or homogenization, when soil aggregates are destroyed by internal or external forces (4). Internal forces are applied when entrapped air breaks out of soil aggregates upon flooding. External forces appear in the form of rain impact or pressure and shearing forces exerted by animal trampling, wheel traffic and tillage implements. Depending on the water content, this results either in pulverization or compaction of the soil. Soil physical degradation, however, depends not only on the degrading forces and stresses but also on the stability of a soil to

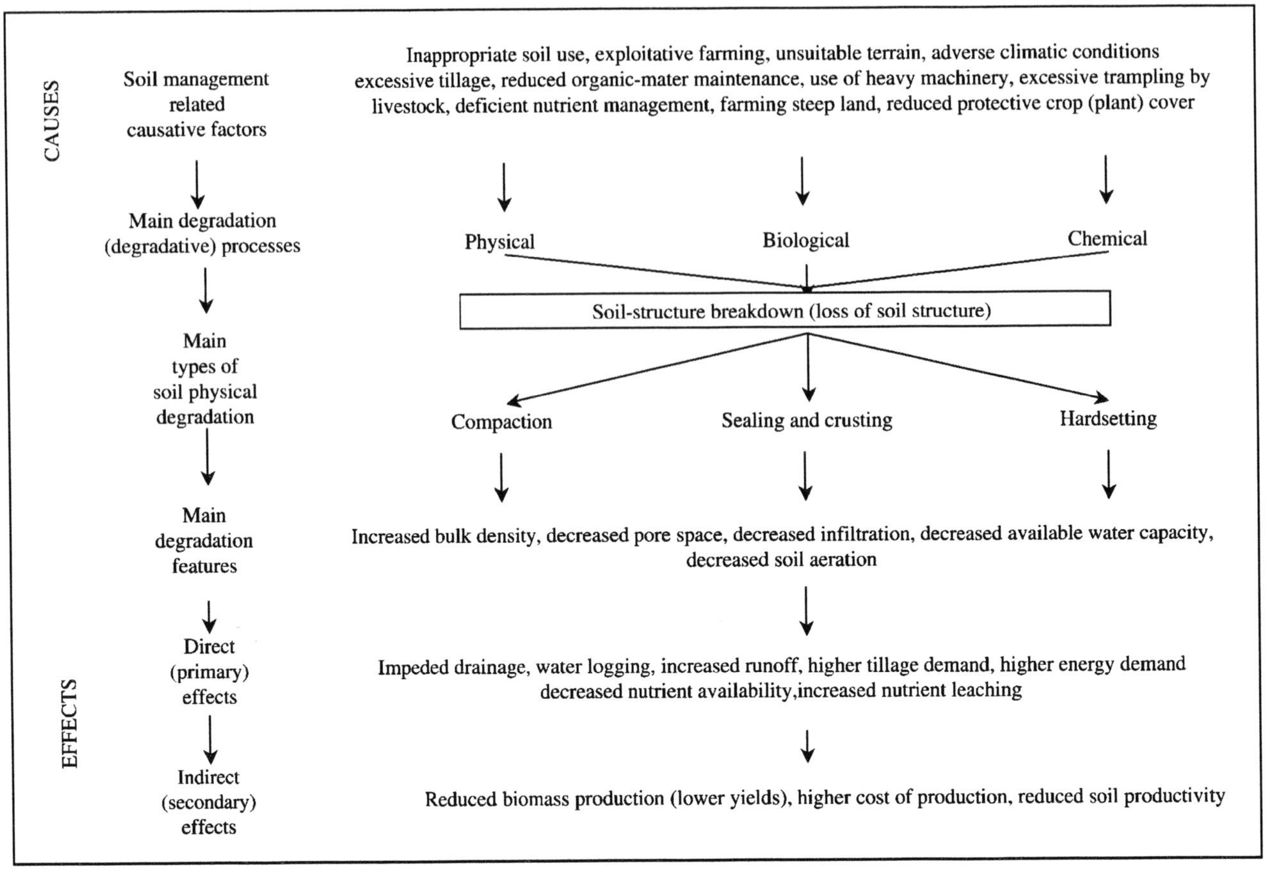

Fig. 1 Simplified cause-effect system of soil physical degradation. (Adapted from Refs. 2–4, 8, 9.)

withstand these stresses, and its resilience to recover from different levels of short-term and long-term degradation.

Not all activities that break down soil are necessarily degrading. Appropriate use and management of the soil also includes measures to maintain—and improve—the resilience of soils. Appropriate soil tillage aims at producing soil structural conditions that improve plant establishment and growth. However, excessive tillage with high-energy inputs can result in losses of soil organic matter and can increase the amount of readily dispersible clay. Both of these consequences can result in decreased structural stability and hence structure breakdown. Such a degraded soil will be weaker when wet but stronger when dry, and will be more difficult to manage under all conditions.

Compaction

Compaction describes the state, or the increase, of "compactness," i.e., bulk density, of a soil. Compared with its undisturbed condition, a compacted soil exhibits reduced total pore space, especially because of a drastic reduction of the macropores, and a pronounced discontinuity of the pore system within the profile (3). This affects the conductive properties of the soil and reduces its ability to retain air and water. Hence, plants growing under high evaporative demand suffer more from compaction than plants growing under low evaporative demand (5). Compaction also inhibits root penetration and development, thus affecting nutrient uptake and, consequently, plant growth. Table 1 shows examples of the effects of soil compaction on the yields of different crops, due to tillage and heavy machinery.

The most important cause of compaction is off-road wheel traffic and the use of heavy machinery in mechanized agriculture (5, 10). In the upper part of the soil, mainly the contact pressure of moving vehicles and the shearing effects of the wheels determine compaction. At greater soil depth, the axle load is more important. The bulk density of the soil increases with the weight, i.e., pressure, exerted by vehicles and machinery, and also

Table 1 Crop yield decline due to soil compaction

Crop	Soil type	Selected control parameter	Yield trend	Country	Source
Wheat	Sandy loam	Penetration resistance increased from 0.83 to 1.29 MPa; bulk density increased from 1.68 to 1.78 Mg m^{-3} after two crops	Decrease between 12–38%	Pakistan	(6)
Sorghum (fodder)	Sandy loam	As above, but after one crop	Decrease between 14–22%		
Spring barley Oilseed rape Winter wheat	Loam and sandy loam	Saturated hydraulic conductivity reduced by about 90% after one crop	Average overall decrease of only 1%, nonsignificant	Sweden	(7)
Corn (grain)	Silt loam	Bulk density was correlated with tillage type, i.e., lowest under chisel plowing (1.25 Mg m^{-3}) and highest under moldboard plowing (1.31 Mg m^{-3}) after 11 cropping seasons	Under regular machine traffic, yield decline was 5% for 7.5 Mg axle load and 10% for 7.5 Mg axle load over 11 cropping seasons	Ohio, U.S.A.	(8)
Barley	Clay loam	Bulk density increased from 1.15 to 1.50 g cm^{-3} after one crop depending on depth, type of tillage, and equipment used	Yield declined from around 1300–1000 kg ha^{-1}	Jordan	(9)

increases with increasing soil water content and the number of passes of the vehicles and machinery over the soil surface. Soils with high clay contents and well-developed pore systems are generally more compressible than sandy soils. Tillage practices also influence the degree of topsoil compaction. Within the annual cropping cycle, the topsoil undergoes changes in its bulk density. To a large extent, plowing alleviates the effects of topsoil compaction produced by the crop-husbandry measures and harvesting activities carried out during and after the cropping season. However, seedbed quality may be severely reduced if the topsoil was severely compacted in the previous year. A cumulative effect of compaction caused by heavy machinery is often observed after several years. Soil compaction increases the demand for tillage. It has been shown that compaction reduces the range of water content over which tillage can satisfactorily be done and therefore reduces the opportunities for tillage (11).

For practical purposes, two main types of soil compaction can be distinguished, namely, the surface-layer compaction and subsoil compaction.

Surface-layer compaction describes the compaction in the upper part of the soil profile, i.e., in arable soils usually the plow layer. Compaction in the surface layer is dynamic and changes significantly over the cropping season, increasing with increasing machinery passes over the field and decreasing again with primary tillage for seedbed preparation for the following season. In reduced tillage systems, the effects of soil compaction build up and tend to be more lasting than in plow systems (5). These effects seem to be more pronounced in sandy soils than in clay soils, because of their lower capacity to maintain a continuous macropore system (2). Adequate tillage effectively reduces soil compaction in the surface layer and its effects.

Sub-soil compaction affects soils beyond the surface layer at depths >30 cm. It is caused by heavy machinery, especially those with high axle loads (5, 10). Swelling–shrinking, freezing–thawing and biological activities can alleviate compaction to a certain extent. However, these processes that are not usually effective beyond a depth of 35 cm. Sub-soiling using specially designed equipment can in some cases alleviate sub-soil compaction, but it is very energy demanding. Sub-soiling should be attempted only when the whole soil profile is drier than the plastic limit (or lower Atterberg limit) of the soil, otherwise more harm than good may be done. Soil at these depths can be dried sufficiently only by plant roots. However, the compaction may prevent the necessary root growth. In many situations after sub-soiling, the soil recompacts readily to be more dense than it was before sub-soiling because of the destabilization of the soil caused by the mechanical energy input from the sub-soiling operation. Compaction should be considered to be an irreversible, permanent form of degradation.

Sealing and Crusting

Seals and crusts are consequences of rain and flooding on unprotected soil-surfaces. Under the impact of raindrops and the "soaking" effect of water, the bonds that hold the particles together become weak and the soil aggregates tend to fall apart. Individual particles become separated. These particles become rearranged and the finer particles tend to be washed into the cavities of the surface. There they form a very thin (1–5 mm) and dense layer that clogs the soil pores and seals the surface. This process is also referred to as capping. These seals are usually very elastic. Typical characteristics of soil seals are that they do not crack and cannot be removed from the surface.

Soil crusts are formed by the same processes that form seals. They are much thicker than seals (usually 5–20 mm) and can be separated easily from the soil-surface, and they crack upon drying. Crusts are typically formed on soils with high contents of nonswelling clay susceptible to dispersion. Seals that become hard upon drying are frequently also termed crusts (12). Soils with a high contents of fine or very fine sand or silt are especially prone to sealing and crusting. The presence of exchangeable sodium in the soil can enhance clay dispersion and thus contribute to seal and crust formation.

Depending on the dominant processes, three types of seals or crusts are distinguished (4, 13). *Structural crusts* are a result of physical forces, i.e., raindrop impact, animal trampling, wheel traffic and flooding. *Slaking crusts* are formed as a result of the disintegration of the soil structure when in contact with water. This can take the form of chemical dispersion of the clay due to, e.g., the presence of exchangeable sodium, mainly in sodic soils, or of the slaking of the soil into microaggregates. *Depositional crusts* are formed by the translocation and deposition of fine particles by, e.g., erosion processes and on irrigated fields.

The main direct effects of seals and crusts are a decrease in infiltration and hence increase in surface runoff and soil erosion, and an obstruction to seedling emergence.

Hardsetting

Hardsetting is a result of structural breakdown and reconsolidation of soil under repeated wetting and drying cycles (12). The result is a hard, homogeneous mass difficult—if not impossible—to till and therefore very restrictive for soil use and management. Hardsetting takes place in two stages. First, the soil structure breaks down under wetting. This process is, in principle, the same for all forms of soil physical degradation and is influenced, as described before, by a complex set of interacting physical, chemical, and biological conditions and processes. The second stage is characterized by the drying of the soil—without a significant regain of structure—and hardening of the soil mass. During this second stage, effective stresses due to the menisci of water films and the water matric potential contribute to the consolidation process. The main effect of hardsetting—an increase of soil tensile strength—increases with each wetting and drying cycle. Soils that tend to hardset usually are low in nonswelling clays and have moderate levels of silt and fine sand (12). In soils that release soluble salts when moist, the effects of hardsetting upon drying are usually increased. Hardsetting can affect soils to depths of 20–30 cm. Because of their extremely high tensile strength, hardsetting soils are very difficult to till when dry. To offset the effects of drying, tillage has to be timed after rainfall or irrigation before the hardening of the soil occurs. This restricts the time available for effective soil tillage.

Hardsetting processes affect many soils, but this does not necessarily make them hardsetting soils. More sandy soils with a weaker structure do not usually become too hard to till upon drying, although they also undergo hardsetting processes. These soils develop features and problems related to hardsetting, but they can still be tilled without major limitations and constraints. Typical effects of hardsetting are the obstruction to seedling emergence and increased resistance to root growth.

CAUSES AND EFFECTS OF SOIL PHYSICAL DEGRADATION

The main causes of soil physical degradation, i.e., the breakdown of the soil structure, are inappropriate land use and soil-management practices. All "exploitative" practices will ultimately lead to degradation and hence reduced soil productivity. The basic detrimental effects of certain practices—or of their neglect—on different soil properties are known. But their specific contribution to the breakdown of the soil structure in a specific location also depends on the other prevalent practices and factors.

The same set of circumstances applies to the effects that soil physical degradation has on soil productivity and the environment.

In highly mechanized systems, heavy machinery and excessive tillage are important factors of degradation. In the developing world, land not suitable for cultivation, such as drylands or steep terrain, is increasingly being cropped. The cultivation and husbandry practices associated with these land-use systems are largely responsible for degradation. The systems themselves are, in the long term, also largely affected by them. Fig. 1 depicts the causative factors and effects that lead to soil physical degradation. It shows that inappropriate soil and land management leads to an array of different types of soil physical degradation. Ultimately, soil physical degradation leads to reduced plant growth, crop yields, and soil productivity.

The soil environment is very dynamic and changes in one property may affect the response of another property to different degrading stresses. Also, climatic conditions, which play an important role in degradation, vary from season to season (5).

MEASURES TO MITIGATE AND REDUCE SOIL PHYSICAL DEGRADATION

The best way to prevent and control soil degradation is by using the land and soil according to their capability, and observe natural constraints, such as terrain and climatic conditions. This is best done by what is commonly referred to as "appropriate" soil and land use. However, the conditions determining the suitability of soil use are always location specific. Therefore, there is no single universally valid system of appropriate farming. All

Table 2 Examples of simple (indicative) tests to assess soil characteristics important for the evaluation of soil physical degradation

Soil characteristic (indicator)	Field assessment (test)	Laboratory assessment (test)
Compaction	Bulk density (e.g., sand-replacement method) Penetration resistance (cone penetrometer) Shear vane test	Bulk density (using soil cores) Precompaction stress
Total porosity	From bulk density (above) assuming a particle density of 2.65 g cm^{-3}	Pyknometer test Calculation from bulk density
Pore space distribution	N.P.	Water-retention at various potentials
Clay dispersion	"Emerson (1967) test" (11)	Gravimetric method (Pipette test) Turbidimetry
Aggregate stability	"Emerson (1967) test" (11)	Wet sieving
Aggregate strength	Feel (crushing between thumb and first finger)	Tensile strength test (crushing force)
Infiltration	Double-ring infiltrometer Tension (disk) infiltrometer	
Permeability	Constant-head permeameter	Constant-head permeameter
Hydraulic conductivity	Tension (disk) infiltrometer	Falling-head permeameter
Runoff generation	Rainfall simulation	Rainfall simulation
Soil organic matter content	N.P.	Loss on combustion Walkley Black method
Texture (particle-size distribution)	By feel	Sedimentation test
Crust strength	Cone penetrometer test Crop-emergence count	Cone penetrometer test Modulus of rupture test
Water-holding capacity	Neutron probe or TDR 2 days after heavy rain	Soil cores
Aeration status	Air permeability Redox potential Oxygen diffusion rate	Air-filled porosity Air permeability Redox potential Oxygen diffusion rate
Soil-particle detachment	Rainfall simulation and sediment sampling	Rainfall simulation and sediment sampling

N.P. = Not possible.
(Adapted from Refs. 1, 4, 5, 7–9.)

measures that lead to an optimization of plant cover and plant population, and maintain a stable structure, are suited to prevent soil physical degradation.

ASSESSMENT AND MEASUREMENT OF SOIL PHYSICAL DEGRADATION

Because of the high degree of interaction between the different physical, biological and chemical factors and processes involved, the quantification of soil physical degradation in its totality and its effects on plant growth are extremely difficult. There are no standard, objective methods to assess soil physical degradation in its entirety (2, 4, 5, 13). In practice in most cases, it is usually a subjective assessment by specialists, based on their individual experience. A well-known example of this is the methodology used by the Global Assessment of Human-Induced Soil Degradation (GLASOD) project (14).

However, many basic soil properties that are altered by physical degradation can be measured objectively and quite easily. Changes over time in these characteristics can be used to estimate the influence of a single or a set of properties on plant growth and soil productivity. Table 2 lists a number of simple field and laboratory methods that can be used to measure and characterize these properties.

One of the two common basic concepts is the assessment of the severity of past and ongoing degradation. This is usually a comprehensive assessment of actual field conditions. The other concept is the assessment of the susceptibility of a certain soil to degradation under a variety of soil-use and management practice scenarios. Often, treatments are simulated in the laboratory. Another aspect of current concern is the impact of climate change on the sensitivity of soils to physical degradation.

REFERENCES

1. ISSS, *Terminology for Soil Erosion and Conservation*; International Society of Soil Science Special Publication, International Soil Reference and Information Centre (ISRIC): Wageningen, The Netherlands, 1998.
2. Dexter, A.R. Advances in the Characterization of Soil Structure. Soil Tillage Res. **1988**, *11*, 199–238.
3. Shaxson, F. *New Concepts and Approaches to Land Management in the Tropics with Emphasis on Steeplands*; Fao Soils Bulletin 75, Food and Agriculture Organization of the United Nations: Rome, 1999.
4. Gabriels, D.; Horn, R.; Villagra, M.M.; Hartmann, R. Assessment, Prevention, and Rehabilitation of Soil Structure Caused by Soil-Surface Sealing, Crusting, and Compaction. In *Methods for the Assessment of Soil Degradation*; Lal, R., Blum, W.H., Valentin, C., Stewart, B.A., Eds.; CRC Press: New York, 1998; 129–165.
5. Håkansson, I.; Voorhees, W.B. Soil Compaction. In *Methods for the Assessment of Soil Degradation*; Lal, R., Blum, W.H., Valentin, C., Stewart, B.A., Eds.; CRC Press: New York, 1998; 167–179.
6. Ishaq, M.; Hassan, A.; Saeed, M.; Ibrahim, M.; Lal, R. Subsoil Compaction Effects on Crops in Punjab, Pakistan. I. Soil Physical Properties and Crop Yield. Soil Tillage Res. **2001**, *59*, 57–65.
7. Arvidsson, J. Subsoil Compaction Caused by Heavy Sugarbeet Harvesters in Southern Sweden. I. Soil Physical Properties and Crop Yield in Six Field Experiments. Soil Tillage Res. **2001**, *60*, 67–78.
8. Lal, R.; Ahmadi, M. Axle Load and Tillage Effects on Crop Yield for Two Soils in Central Ohio. Soil Tillage Res. **2000**, *54*, 111–119.
9. Abu-Hamdeh, N.H.; Al-Widyan, M.I. Effect of Axle Load, Tire Pressure, and Tillage System on Soil Physical Properties and Crop Yield of a Jordanian Soil. Trans. ASAE **2000**, *43* (1), 13–21.
10. Horn, R., van den Akker, J.J.H., Arvidsson, J., Eds.; *Subsoil Compaction: Distribution, Processes and Consequences*; Catena Verlag: Reiskirchen, 2000.
11. Dexter, A.R.; Bird, N.R.A. Methods for Predicting the Optimum and the Range of Soil Water Contents for Tillage Based on the Water Retention Curve. Soil Tillage Res. **2001**, *57*, 203–212.
12. Mullins, C.E. Hardsetting. In *Methods for the Assessment of Soil Degradation*; Lal, R., Blum, W.H., Valentin, C., Stewart, B.A. Eds.; CRC Press: New York, 1998; 109–127.
13. Valentin, C.; Bresson, L.M. Soil Crusting. In *Methods for the Assessment of Soil Degradation*; Lal, R., Blum, W.H., Valentin, C., Stewart, B.A., Eds.; CRC Press: New York, 1998; 89–107.
14. ISRIC-UNEP, *World Map of the Status of Human-Induced Soil Degradation*; Explanatory Note International Soil Reference and Information Centre (ISRIC): Wageningen, The Netherlands, 1990.

DEGRADATION, QUALITY AND

Rattan Lal
The Ohio State University, Columbus, Ohio, U.S.A.

INTRODUCTION

Soil degradation is defined as diminution of soil quality in terms of its current and potential productivity and/or reduction in its ability to be a multipurpose resource by both natural and anthropogenic causes (1, 2). A decrease in soil quality implies adverse changes in soil properties and processes with a negative effect on the soil's life-support systems (2–4). Soil degradation has both agronomic and ecologic consequences. Agronomically, soil degradation leads to loss of production and renders an ecosystem unsustainable under the specific land use (2). Ecologically, soil degradation leads to pollution of natural waters, emission of radiatively active gases into the atmosphere, decrease in soil biodiversity, etc. Soil degradation is the result of interactive effects of processes, factor, and causes or management practices.

PROCESSES OF SOIL DEGRADATION

The processes of soil degradation involve mechanisms, actions, and interactions that affect the soil's: 1) resilience or capacity of self-regulation, 2) life-support processes or productivity, and 3) ability to moderate environments. The principal processes of soil degradation (Fig. 1, Table 1) are briefly described as follows.

Physical Processes

These processes lead to adverse changes in soil physical, mechanical, or hydrological properties. Principal among these processes is the decline in soil structure, which manifests in crusting, compaction, impeded surface and subsoil drainage, and the poor or limited aeration. A reduction in aggregation and the dispersion of clay lead to a loss of clay through erosion and eluviation. Accelerated soil erosion is a major soil degradative process, has potentially severe ecological and economic consequences, and has local, regional, or global impacts.

Chemical Processes

These processes lead to undesirable changes in soil chemical (reaction) and nutritional (fertility) properties. Soil chemical degradation involves adverse changes in: 1) properties and processes that regulate nutrient capacity and intensity, 2) soil's ability to inactivate or denature toxic substances, 3) the soil's chemical balance, leading to elemental toxification (Al, Mn, salts) and nutrient deficiencies, and 4) the quality of natural waters, atmosphere, and the general environments.

Biological Processes

These processes affect soil biodiversity. The principal effect of biological degradation is a decline in soil organic matter content, a reduction in biomass carbon, and a decrease in the activity and species diversity of soil fauna and flora.

NATURAL VS. ANTHROPOGENIC PROCESSES

Soil degradative processes may be natural or anthropogenic. Natural processes are generally slow, soil forming or pedological mechanisms, and their effect is felt over the geological time-span. In contrast, anthropogenic processes are rapid, highly destructive, and their effect is felt over human time-span. Some important natural and anthropogenic processes of soil degradation are listed in Table 2 and are briefly described as follows.

Natural Processes

One of the natural degradative processes is "laterization", which is a process of iron accumulation in soils. At its extreme, the process involves intense weathering and intense leaching. Intense weathering leads to the breakdown of all minerals except quartz. Intense leaching of soil removes all the soluble salts and some of the Fe and Al. These processes lead to a predominance of kaolinite and a depletion of plant-available nutrients. As a result, high

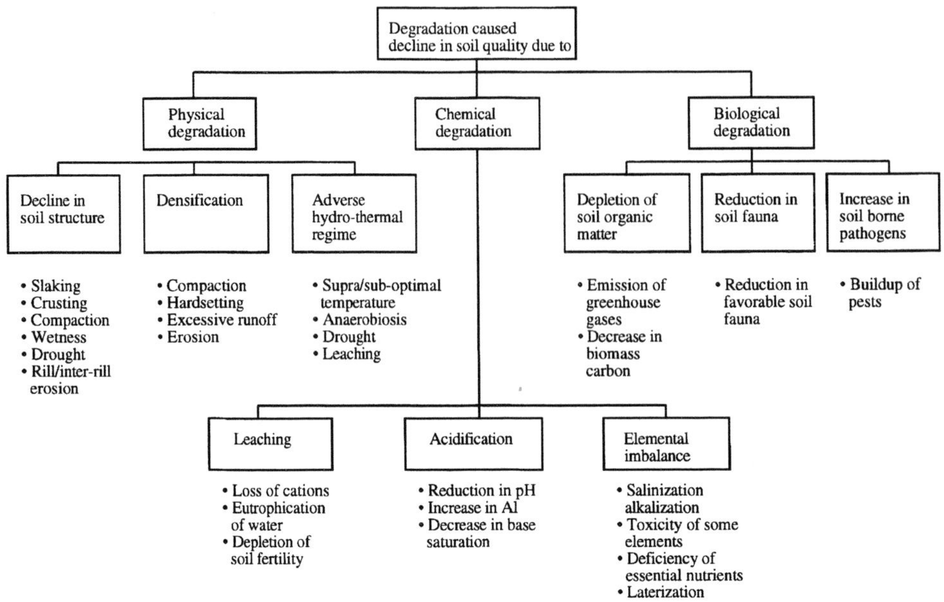

Fig. 1 Principal processes of soil degradation. (Adapted from Ref. 5.)

concentrations of Fe and Al cause complex fertility management problems.

Hard-setting is another natural process. Hard-setting soils are characterized by weakly cemented surface horizons in deeply weathered soils of low inherent fertility and low soil organic matter content (6). Such soils have massive, single-grain or weak pedality of the surface. These soils set hard on drying, and have a narrow range of moisture content at which they can be managed. Hard-setting characteristics are accentuated by physical and biological processes of degradation.

Pan formation within the soil profile is another natural degradative process. There are two principal types of pans. Fragipans are pedogenic subsurface horizons characterized by their medium texture; a high content of silt, very fine sand, and low-to-moderate clay content; high bulk density; low organic matter content; and brittleness when moist and hardness when dry (1). These pans affect rooting depth and plant-available water reserves. In contrast to Fragipans, clay pans are not brittle and may not have high density. In situ formation of a clay pan in the B-horizon happens due to translocation of fine clay to the subsoil. The physical barrier formed by the clay pan is root-restrictive and can also lead to impeded drainage.

Anthropogenic Processes

Human activities have created formation of "anthropic soils." Anthropic soils may be due to urbanization, waste disposal, or other drastic perturbation.

Table 1 Processes of soil degradation

Physical processes	Chemical processes	Biological processes
Decline in soil structure, e.g., compaction, crusting, etc.	Leaching and acidification	Decline in soil biodiversity
Accelerated erosion by water, wind, gravity, glaciation	Nutrient depletion	Decrease in soil organic matter content
Decrease in effective rooting depth	Nutrient imbalance: i) toxicity, e.g., Al, Mn, Fe, ii) deficiency, e.g., P	Reduction in soil biomass carbon
Anaerobiosis or excessive wetness	Sodication	Increase in pests and pathogens
Excessive drainage or droughtiness	Laterization	

Table 2 Natural and anthropogenic processes of soil degradation

Natural processes	Anthropogenic processes
Set-in-motion by natural causes	Set-in-motion by human intervention
Laterization	Accelerated erosion
Fragipan formation	Impeded surface drainage
Clay pan formation	Soil compaction
Hard setting	Fertility depletion

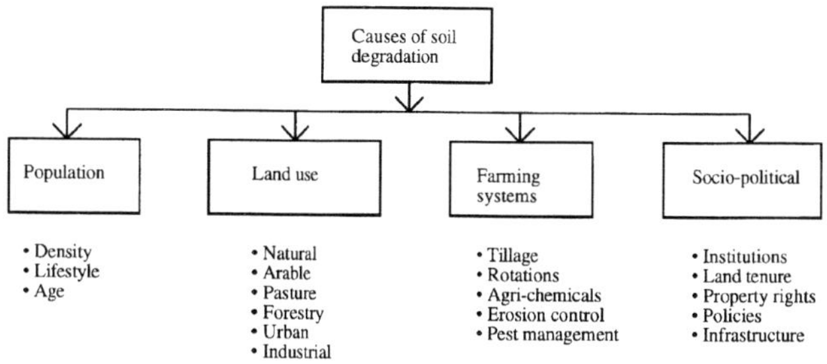

Fig. 3 Principal causes of soil degradation. (Adapted from Ref. 5.)

FACTORS AND CAUSES OF SOIL DEGRADATION

Factors of soil degradation are ecosystem characteristics. Notable among these are climate (e.g., rainfall, temperature), terrain characteristics and vegetation (Fig. 2). Factors are agents and catalysts, and accentuate the rate of different processes of soil degradation. Causes of soil degradation are anthropogenic activities, which grossly alter the influence of processes and factors. Important causes of soil degradation include agricultural activities, industrial activities, and urbanization (Fig. 3, Table 3). Causes of degradation are driven by demographic pressure, human needs, and other issues related to the human dimension. Consequently, increase in demographic pressure aggravates the effects of processes and factors of soil degradation.

EFFECTS OF SOIL DEGRADATION

Soil degradative processes are influenced strongly by biophysical factors and socioeconomic causes (Fig. 4). The effects of physical processes of soil degradation are exacerbated by anthropogenic activities in ecologically sensitive ecoregions. The soil erosion problem is extremely severe when sloping lands and shallow soils are cultivated for row crop production in ecologically sensitive ecoregions such as the Himalayan–Tibetan ecosystem, Andean region, the Caribbean, and the West African Sahel. Deforestation in ecologically sensitive regions can cause severe degradation of soil and water resources, e.g., tropical rainforest ecosystem.

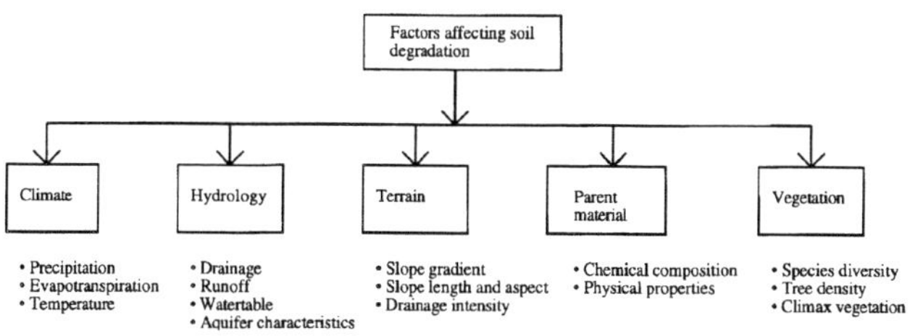

Fig. 2 Major natural and anthropogenic factors affecting soil degradation. (Adapted from Ref. 5.)

Degradation of soil has both economic and ecologic consequences. The most direct and economic effect of degradation of soil resources is the loss of agricultural productivity. The adverse effects of soil degradation on productivity are accentuated by a decline progressive in the per capita arable land area at national and global scales. The per capita arable land is rapidly decreasing due to an increase in population and the degradation of soil resources. Global per capita arable land area of 0.3 ha in 1985 is expected to progressively decline to 0.23 ha by the year 2000, 0.15 ha by 2050, and to 0.14 ha by 2100. Countries with per capita arable land area of less than 0.15 ha were 5 in 1960, 23 in 1990, and will be 50 in 2025 (9). Mismanagement, neglect, and exploitation can deplete or even ruin the fragile soil resources.

While the per capita arable land area is decreasing, its productivity is also being reduced by several processes of soil degradation, e.g., erosion, desertification, leaching, etc. The effects of soil degradation on productivity are not easily understood because they are often cumulative and may be observed long after the degradation has occurred. Furthermore, productivity is the net result of all factors of production, including climate, and the effects are masked by management and the prevailing weather conditions. In addition to the direct effects of degradation on productivity, there are many indirect or hidden effects due to alterations in soil properties. In fact, there are no reliable estimates of worldwide loss in productivity due to different processes of soil degradation (7, 8).

Ecologic consequences of soil degradation are major environmental concerns. The annual discharge of chemicals to world rivers is phenomenal and rapidly increasing.

Table 3 Causes of soil degradation

Anthropogenic causes		Natural causes
Agricultural activities	**Industrial activities and urban development**	**Climate and land**
Deforestation: felling trees, biomass burning	Mining: surface or strip mining, subsurface mining	Rainfall: intensity, amount energy load
Cultivation: plowing, leveling, land forming	Waste disposal: industrial and urban wastes, by-products	Wind: velocity direction
Farming/cropping systems: cropping intensity, grain crops vs. tree crops, pasture, stocking rate	Acid rain: intensity, chemical agents	Temperature: maximum, minimum seasonal and diennal fluctuations length aspect
Agricultural chemicals: fertilizers, pesticides	Nonagricultural land uses: urban development civil structures, mining	Terrain: slope gradient, length aspect
		Parent material: geological formation
		Vegetation: age, species composition

Chemical discharge into natural waters from agricultural land may be from fertilizers and pesticides, feedlots, and from sediment. Risks of agrochemical contamination of natural waters are extremely high in both developing and developed economies.

Emission of radiatively active gases from soil-related processes is another principal ecologic consequence of soil degradation. The emission of CO_2 and other greenhouse gases (e.g., CH_4, CO_2, N_2O, NO_x) by soil erosion can be enormous (10). Yet, world soils can be a major source or sink for radiatively active gases depending on the land use and management system adopted (11).

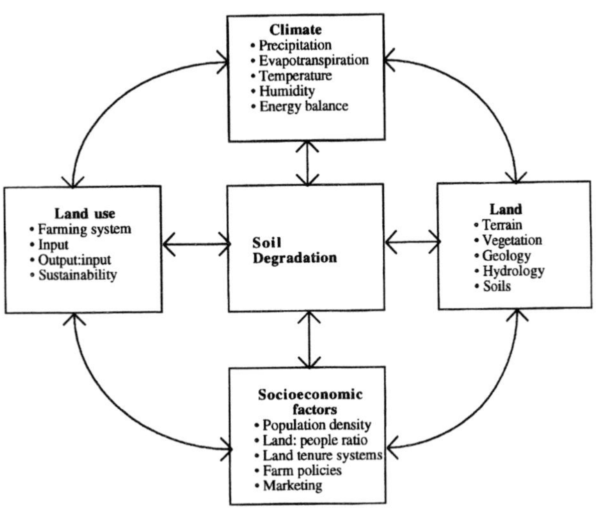

Fig. 4 Interdependence of soil degradation on biophysical and socioeconomic factors. (Adapted from Ref. 5.)

GLOBAL EXTENT OF SOIL DEGRADATION

The problem of soil degradation has been accentuated drastically by changes in land use witnessed since the 18th century. Despite the widespread awareness about the problem, exact estimates of the extent and severity of soil degradation are not known. According to some estimates, about 2.0 billion ha (Bha) of land is degraded to some degree out of the world's total land areas of 13.4 Bha (12, 13). Furthermore, Asia and Africa combined account for a total of 1.24 Bha of degraded land. The most predominant degradative processes are the accelerated erosion by wind and water, chemical degradation and soil fertility depletion, and physical degradation resulting from a decline in soil structure (Table 4). Estimates of soil

Table 4 Global estimates of soil degradation by different processes

Process of soil degradation	World's total degraded area (10^6 ha)	Regional degraded area (10^6 ha) Asia	Africa
Water erosion	1100.0	433.2	227.3
Wind erosion	550.0	224.1	187.3
Chemical degradation	235.8	74.7	59.3
Physical degradation	78.6	19.8	15.0
Total	1964.4	747.0	494.2

(Adapted from Refs. 12, 13.)

D

Table 5 Global estimates of desertification in arid, semiarid and subhumid regions

		Land affected by desertification	
Land use	Total area (10^6)	Area (10^6 ha)	% of total
Rangeland	3700	3100	80
Rainfed cropland	570	335	60
Irrigated land	131	40	30
Total	4409	3475	70

(Adapted from Ref. 14.)

degradation in arid climates, due to wind erosion and desertification, are equally alarming (Table 5).

DATA ACCURACY AND RELIABILITY

Statistics, such as those shown in Tables 4 and 5, play an important role in creating awareness about the hazard, and in formulating a global strategy in addressing it. Indeed, if these statistics are correct, the challenge they present to the human race is one of the greatest because soil resources are finite and nonrenewable over the human time-span (15, 16). Even though the statistics can be approximately correct, it is a matter of greatest urgency for decision-makers to do something about it. However, careful analyses of the data may reveal several problems:

1. There may be a problem with the definition. The term soil degradation is vague and highly subjective. To avoid ambiguity, soil degradation needs to be defined quantitatively (2). To do so is to delineate threshold values or critical limits or soil properties beyond which soil's life support processes are severely jeopardized. Important soil properties whose critical limits need to be defined are rooting depth, plant-available water capacity, soil organic matter content, soil structural attributes, and capacity and intensity factors and limiting levels of principal nutrients. These limits are not known and vary among soils, climates, farming systems, land uses, and management.
2. There is a problem with the methodology. There have been only few attempts to standardize the methodology and validate the estimates by ground truthing representative soils and ecoregions. Consequently, estimates of soil degradation by the same process vary widely because of the differences in methodology and criteria used.
3. Soil degradation is estimated on the basis of visual observations, reconnaissance surveys, and opinions of regional/local organizations. It is rarely related to actual and potential productivity.
4. There is a problem with double accounting. The same soil prone to more than one degradative process may be counted as many times.
5. Productivity is affected by land use, inputs, improved technology, and management. Estimating productivity without considering of important technology and management processes leads to erroneous data.

CONCLUSIONS

Understanding principal processes, factors, and causes of soil degradation is an important prerequisite to our ability to reverse the degradative trends and to preserve a healthy state of global environment. Being a complex problem, it requires a systematic and coordinated effort to effectively address the major issues involved. These issues include: 1) obtaining reliable estimates of the land area affected by different soil degradative processes, 2) standardizing definitions and methods of assessment of soil degradation, 3) evaluating on-site and off-site impacts of soil degradation, 4) developing appropriate measures of soil restoration, and 5) assessing economic and environmental impacts of restoring degraded soils and ecosystems.

REFERENCES

1. Lal, R.; Hall, G.F.; Miller, F.P. Soil Degradation. I. Basic Processes. Land Degrad. Rehab. **1989**, *1*, 51–69.
2. Lal, R. Degradation and Resilience of Soils. Phil. Trans. R. Soc. Lond. B **1997**, *352*, 997–1010.
3. Lal, R. Soil Quality and Agricultural Sustainability. In *Soil Quality and Agricultural Sustainability*; Lal, R., Ed.; Ann Arbor Press: Chelsea, MI, 1998; 3–12.
4. Lal, R. Soil Quality and Food Security. In *Soil Quality and Food Security: The Global Perspective*; Lal, R., Ed.; CRC Press: Boca Raton, FL, 1999; 3–16.
5. Lal, R., Stewart, B.A., Eds. Need for Action: Research and Development Priorities. Adv. Soil Sci. **1990**, *11*, 331–336.
6. Mullins, C.E.; MacLeod, D.A.; Northcote, K.H.; Tisdall, J.M.; Young, T.M. Hard-Setting Soil's Behavior, Occurrence and Management. Adv. Soil Sci. **1990**, *11*, 37–108.
7. Lal, R. Soil Erosion Impact on Agronomic Productivity and Environment Quality. Crit. Rev. Plant Sci. **1998**, *17*, 319–464.
8. Oldcman, L.R. *Soil Degradation: A Threat to Food Security*; ISRIC: Wageningen, The Netherlands, 1998.

9. Engehnan, R.; LcRoy, P. *Conserving Land: Population and Sustainable Food Production*; Population and Environment Program, Population Action International: Washington, DC, 1995; 48.
10. Lal, R. Global Soil Erosion by Water and Carbon Dynamics. In *Soils and Global Change*; Lal, R., Kimble, J., Levine, E., Stewart, B.A., Eds.; CRC/Lewis Publishers: Boca Raton, FL, 1995; 131–142.
11. Lal, R. Soil Management and Restoration for C Sequestration to Mitigate the Accelerated Greenhouse Effect. Prog. Environ. Sci. **1999**, *1*, 307–326.
12. World Resources Institute, Towards Sustainable Development. *A Guide to the Global Environment*; WRI: Washington, DC, 1992; 385.
13. Oldcman, L.R. The Global Extent of Soil Degradation. In *Soil Resilience and Sustainable Land Use*; Greenland, D.J., Szabolcs, I., Eds.; CAB International: Wallingford, UK, 1994; 99–118.
14. UNEP, *Status of Desertification, and Implementation of United Nations Plan of Action to Combat Desertification*; UNEP: Nairobi, Kenya, 1991; 77.
15. Lal, R. Land Degradation and Its Impact on Food and Other Resources. In *Food and Natural Resources*; Pimentel Ed.; Academic Press: New York, 1989; 85–140.
16. Pierce, F.J.; Lal, R. Soil Management in the 21st Century. In *Soil Management for Sustainability*; Lal, R., Pierce, F.J., Eds.; Soil and Water Conservation Society: Ankeny, IA, 1991; 175–179.

DESERTIFICATION

Paul T. Tueller
University of Nevada, Reno, Nevada, U.S.A.

INTRODUCTION

The concept of "desertification" has been heavily discussed in the literature since the 1977 UNEP Desertification Conference held in Nairobi, Kenya. It is believed that the term itself was first used by Lavauden (1, 2) to describe forest decline in the Sahara (3), though most researchers credit Aubreville (4). Both were French ecologists who eyewitnessed the land degradation occurring in north and west Africa (5). Aubreville noted that land degradation seemed to be extending north into semiarid and subhumid regions of North Africa from the more arid zones of the Sahara (4). Land degradation had been observed in the Mediterranean area nearly two millennia before by the Roman senator Cicero who spoke of the destruction of the north African forests and their replacement by barren, desert-like areas. The devastating Sahel droughts of 1968–1973 and the apparent accelerated southward advance of the Sahara Desert led to extensive international discussion of the problem and the formation of the United Nations Conference on Desertification (UNCOD). An early and short definition of desertification came from the Desertification Control Program within the United Nations Environment Program: "The diminution or destruction of the biological potential of the land, and can lead ultimately to desert-like conditions" (Fig. 1).

DEFINITIONS

The definitions of desertification have changed considerably over the years (6) and at least 100 have appeared in the literature. The United Nations conference on Environment and Development (UNCED) has defined desertification as "land degradation in arid, semiarid and dry subhumid areas resulting from various factors including climatic variations and human activities" (7). This definition has the advantage of being concise but it makes no explicit mention of two features commonly associated with desertification. One is the irreversibility of the process. The second is the loss of biodiversity that often accompanies land degradation, which is important "because species and genes well adapted to the drier areas are so few, our loss is great" (8). An earlier definition by UNEP ignored the role of climate entirely (9).

In 1977 in Nairobi, Kenya, UNCOD defined desertification as follows: "Desertification is the diminution or destruction of the biological potential of land, and can lead ultimately to desert-like conditions. It is an aspect of the widespread deterioration of ecosystems, and has diminished or destroyed the biological potential, i.e., plant and animal production, for multiple use purposes at a time when increased productivity is needed to support growing populations in quest of development" (10).

During the following years, various agencies, scientific institutions, and individual scientists found this definition to be inadequate (11). Cyclic oscillations of vegetation productivity related to climate fluctuations had been observed in satellite data, and there was a need to differentiate between desertification and these cyclic climatic oscillations (11).

"Desertification" has also been defined as the reduction or spatial reorganization of net primary productivity of arid and semiarid lands. This definition tends to ignore the many examples where good arid land management reverses the process. Likewise, desertification was often defined and experienced as the irreversible loss of the production and ecological functions. Generally agreed indicators of desertification include soil erosion, loss of biodiversity, and lower productivity. This definition is now used less because it is generally recognized that in many cases the process of desertification is not irreversible.

Desertification is not, as sometimes described, an invasion of nondesert areas from a desert core. Desertification has been compared to guerrilla warfare with no real "front line" (12). Desertification has also been described as a patch of land degraded through human abuse that then spreads outward if the abuse continues (13). The spread of desertification has been described as follows: "These degraded patches, like a skin disease, link up to carry the process over extended areas" (10). Mainguet stated, "The theory of the encroaching desert, which has now been scientifically rejected, is still a fixed idea in the minds of governments, donors, and journalists; this must change" (14).

Fig. 1 Children collecting firewood from one of the few surviving trees on the edge of the desert in the Sahel. (Taken from "Save The Earth," J. Porritt).

HUMAN ACTIVITY OR CLIMATE?

The early debate centered on whether degraded land actually irreversibly become desert or was simply in exceptionally poor condition. Because the lands were often marginal, great emphasis was placed on expansion into marginal areas of the Sahel, which occurred in wetter years. When drought struck, researchers emphasized changing agriculture practices, overgrazing, and the many hydrologic improvement projects that had gone awry. By the 1980s, researchers from Lund University in Sweden were challenging the emphasis on human impact. Better understanding of the Southern Oscillation and its cycles explained in part the droughts (9). The often cited work of Lamprey (not actually published until 1988) concerning desert encroachment in Sudan was called into question. The Lund researchers produced several Sahelian studies that utilized precipitation and agricultural statistics in combination with remote sensing information to conclude that the climate was the main driving force behind the degradation (15–18).

Perhaps the most significant aspect of many definitions is the focus on "human impact." This differentiates the issue of desertification from simple climatic fluctuations such as drought, but it should be noted that drought can cause an exacerbation of damage derived from human activities. One definition of desertification states that desertification is "revealed by drought," but "caused by human activities" (14). Nearly one-quarter of the vegetated land area of the Earth has been highly disturbed by human activity and an additional 28% has been moderately disturbed (19). This massive level of human disturbance suggests that there are large areas of the Earth's surface that have either been degraded through human activity or have a significant potential of being degraded through human activity.

Desertification reduces the land's resilience to natural climate variability. Soil, vegetation, freshwater supplies, and other dryland resources tend to be resilient. They can eventually recover from climatic disturbances, such as drought, and even from human-induced impacts, such as overgrazing. When land is degraded, however, this resilience is greatly weakened. This has both physical and socio-economic consequences (20).

The occupation of humans in lands described as desert or near-desert conditions is often associated with long habitation of a region. Two major factors are believed to account for the growth of human-induced deserts. In the first place, semiarid to semihumid regions proved the most favorable sites for the early development of human culture. Such areas, however, stand in a condition of delicate ecological balance between humid and true desert climates. Processes of soil erosion are accelerated by the exposure of soil surfaces previously protected by complete mantles of vegetation, whether grass or forest, or by heavy grazing and cultivation. It is only within the past decade that experimental studies of these processes have been made. So enormous have been the differences in soil wastage and superficial runoff of rainwater from bared sloping lands, as compared with similar surfaces protected by a complete coverage of vegetation, that new light is thrown on the problem of the decadence of former civilizations" (21).

REVERSIBILITY OF DESERTIFICATION

Vegetation changes were documented in the Sahara from 1980 to 1990 (22). Their results showed increased greenness with increased rainfall that followed the dry period of 1980–1984. This was later combined with ground photography to demonstrate the "reversibility" of the degradation (23, 24). These conclusions led the authors to believe that desertification, as earlier defined, was simply a myth.

The view has been expressed that "the extent of desertification as an irreversible state has probably been exaggerated, although it is correct to classify it as a serious problem" (25). Topsoil lost to erosion can usually be rebuilt in the given time, and the well-known processes of biological succession shows that a forest or rangeland ecosystem can replace an abandoned field in a few centuries. However, on the scale of people's need to use the land for sustenance, a few centuries is forever. Human effort can accelerate the regeneration of topsoil or ecosystems, shortening the time need for regeneration, but this may require considerable expense. In extreme cases of degradation where the clay fraction of the soil is nearly completely lost or invasive weeds become dominant, natural recovery to the original ecosystem will not occur. There is a lack of knowledge required to assess the resilience or recoverability of soils and plant communities.

Desert land degradation has been characterized as having the following symptoms: declining groundwater tables, salinization of topsoil and water, reduction of surface waters, unnaturally high soil erosion, and the desolation of native vegetation (12). The major processes of desertification in arid regions are: water erosion, wind erosion, salinization, waterlogging, and soil compaction (13). Additionally, the mechanical destruction of desert pavements and calcrete (caliche) makes the soil more vulnerable to erosion. Through these different processes, the land is made more barren, and it may become effectively like a desert, even without climate changes. When the desert pokes a hot finger into the border regions, the people speak of a drought; when it pulls the finger back, they say "the country is getting more seasonable." "At the heart of the desert there is no drought, there is only an occasional mitigation of dryness" (26).

In summary it is reasonable to conclude that the concept of desertification is still evolving and that a clear definition favored by all is not really available. Desertification occurs only in arid and semiarid areas and is mostly associated with land degradation caused by human activities. In some cases it may be reversible and in some cases it is not reversible especially if considerable topsoil is lost. The concern for desertification will hopefully lead to better management, protection and use of our arid land resources throughout the world.

REFERENCES

1. Lavauden, L. Les Forets Du Sahara. Rev. des Eaux et Forets **1927**, LXV (*6*), 265–277.
2. Lavauden, L. Les Forets Du Sahara. Rev. des Eaux et Forets **1927**, LXV (*7*), 329–341.
3. Le Houerou, H.N. Land Degradation in Mediterranean Europe: Can Agroforestry be a Part of the Solution? A Prospective Review. Agroforestry Systems **1993**, *21*, 43–61.
4. Aubreville, A. *Climats, Forets et Desertification de l'Afrique Tropicale*; Geogr. Marit. & Col. Paris, 1949; 351.
5. Dregne, H.E. Forum for Applied Research and Public Policy **1996**, *11* (*3*), 5–11.
6. Verstraete, M.M. Another Look at the Concept of Desertification. In *Origin and Evolution of Deserts*; Wells, S., Haragan, D., Eds.; University of New Mexico Press: 1983.
7. United Nations. Managing Fragile Ecosystems: Combating Desertification and Drought, Chapter 12 of Agenda 21. New York: United Nations, 1992.
8. Bie & Imevbore. http://www.fao.org/waicent/faoinfo/sustdev/EPdirect/EPan0005.htm.
9. Hulme, M.; Kelly, M. Exploring Links Between Desertification and Climate Change. Environment **1993**, *35* (*4–11*), 3945.
10. UNEP. Desertification Control Bulletin United Nations Environment Programme. DC/PAC 1978–1985.
11. UNEP, *Convention on Biological Diversity, Opened for Signature,* June 5, 1992; 31 I.L.M. 822 United Nations Environment Programme: Nairobi, Kenya, 1992; 27.
12. Sheridan, D. *Desertification of the United States*; Council on Environmental Quality: Washington, DC, 1981; 142.
13. Dregne, H.E. *Desertification of Arid Lands*; Harwood Academic Publishers: New York, 1983; 242.
14. Mainguet, Monique, Université de Reims, France, *Desertification Natural Background and Human Mismanagement*; Springer: Berlin, 1991.
15. Olsson, L. Desertification or Climate? Lund Studies in Geography Ser. A Physical Geography No. 60, Lund, Sweden, 1983; 36 pp.
16. Hellden, U. *Drought Impact Monitoring*; Rapporter Och Notiser 61, Lunds Universitets Naturgeografiska Institution: Lund, Sweden, 1984; 61.
17. Olsson, L. An Integrated Study of Desertification. Lund Studies in Geography Ser. C General and Mathematical Geography No. 13, Lund, Sweden, 1985; 170 pp.
18. Hellden, U. Desertification Monitoring: Is the Desert Encroaching? Desertification Control Bulletin **1988**, *17*, 8–12.
19. World Resources Institute, *World Resources* 1992–1993; Oxford University Press: Oxford, 1993; 385.
20. Westoby, M.; Walker, B.; Noy-Meir, I. Opportunistic Rangeland Management for Rangelands Not at Equilibrium. Journal of Range Management **1989**, *42*, 266–274.
21. Lowdermilk, W.C. Man Made Deserts. Public Affairs **1935**, *8* (*4*), 409–419.
22. Tucker, C.J.; Dregne, H.E.; Newcombe, W.W. Expansion and Contraction of the Sahara Desert from 1980 to 1990. Science **1991**, *253*, 299–301.
23. Hellden, U. Desertification—Time for an Assessment. Ambio **1992**, *20*, 372–383.
24. Olsson, L. On the Causes of Famine: Drought, Desertification and Market Failure in the Sudan. Ambio **1993**, *22*, 395–403.
25. Nelson, R. *Working Paper 8, for ENVST*; World Bank: Washington, DC, 1988.
26. Webb, Walter Prescott The American Desert: Perpetual Mirage. Harper's Magazine **1957**, *214*, 25–31.

DESERTIFICATION, EXTENT OF

Victor R. Squires
Adelaide University, Adelaide, South Australia, Australia

INTRODUCTION

Desertification, as a *concept*, is one of the most complex to consider objectively. The word desertification unfortunately conjures up images of advancing sand dunes, which is only part of the problem. There is now a general acceptance of the term *desertification* to encompass a process of land degradation in drylands (those regions with a growing season of about 75–120 days per year) induced by climatic factors and human use. Fig. 1 shows the broad distribution of the world's drylands.

Whatever the definition, it is clear that large areas of the world's drylands are affected by land degradation and that this adversely affects the lives of about 1 billion people on almost every continent. To assess the global extent of desertification implies that there is universal acceptance of the term and that the monitoring and evaluation is somewhat uniform. Neither supposition is true. Until the decision by the United Nations Conference on Desertification (1) there was only broad agreement as to what was meant by it. It is only as signatories to the Convention to Combat Desertification (CCD) have been preparing and implementing their respective Action Plans that more accurate assessments have been made of the current status of desertification in each country. Prior to this, the maps and other documents produced by the UN through its various agencies were the product of expert panels who developed maps of the areas of potential desertification (2).

Every inhabited continent has a problem with desertification but in some it is quite acute. Land degradation induced by human activities (anthropogenic) is believed to be currently lowering the productivity of at least one fifth of the world's agricultural lands. Soil erosion from water and wind is a principal but by no means the only soil degrading process. Sub-humid and semi-arid areas in crops or pastures are particularly vulnerable to erosion. Land degradation in such areas leads to desertification by definition. Anthropogenic land degradation, however, also affects large areas of arable land and pasture in humid regions as well.

Depending on how it is measured exactly (3) up to 75 percent of some continents are affected. The impacts are greatest in those regions with high population density and where the level of economic development is low. For example, Australia is reported as having more than 75 per cent of its land surface affected to a greater or lesser extent by desertification. Yet its low population density and the low level of dependence on these drier regions for its national prosperity means that desertification is perceived as a lesser threat in Australia than in say regions of comparable climate in the Indian subcontinent or in Africa.

ASSESSING THE EXTENT OF THE DESERTIFICATION PROBLEM

Desertification comprises two main types of degradation: vegetation degradation and soil degradation.

These can occur anywhere in dry areas and not just on desert fringes. Vegetation degradation involves a temporary or permanent reduction in the density, structure, species composition or productivity of vegetation cover. It has been argued that degradation of vegetation should be given less weight in assessment of desertification because there is clear evidence that most change in vegetation is reversible over reasonable time scales whereas soil degradation is unlikely to occur within a human life span.

Desertification might be thought of as a situation where a landscape is stressed beyond its resilience. Most scientists would agree that, despite their apparent fragility, dryland ecological systems are quite resilient. The spatial and temporal variation in conditions means that coping mechanisms have developed to allow the system to continue in the face of adversity. Under the combined force of human-induced pressure and climatic forces, such as recurrent drought, major changes in soils and vegetation can occur. This is what has come to be recognized as desertification. This is what the Convention on Combating Desertification is designed to combat.

Desertification is a product of complex interactions between the social and economic systems (disease, poverty, hunger and unreliable economy), and natural factors (drought, water erosion, soil salinization, degradation and loss of vegetation cover). It is not surprising

Encyclopedia of Soil Science

D

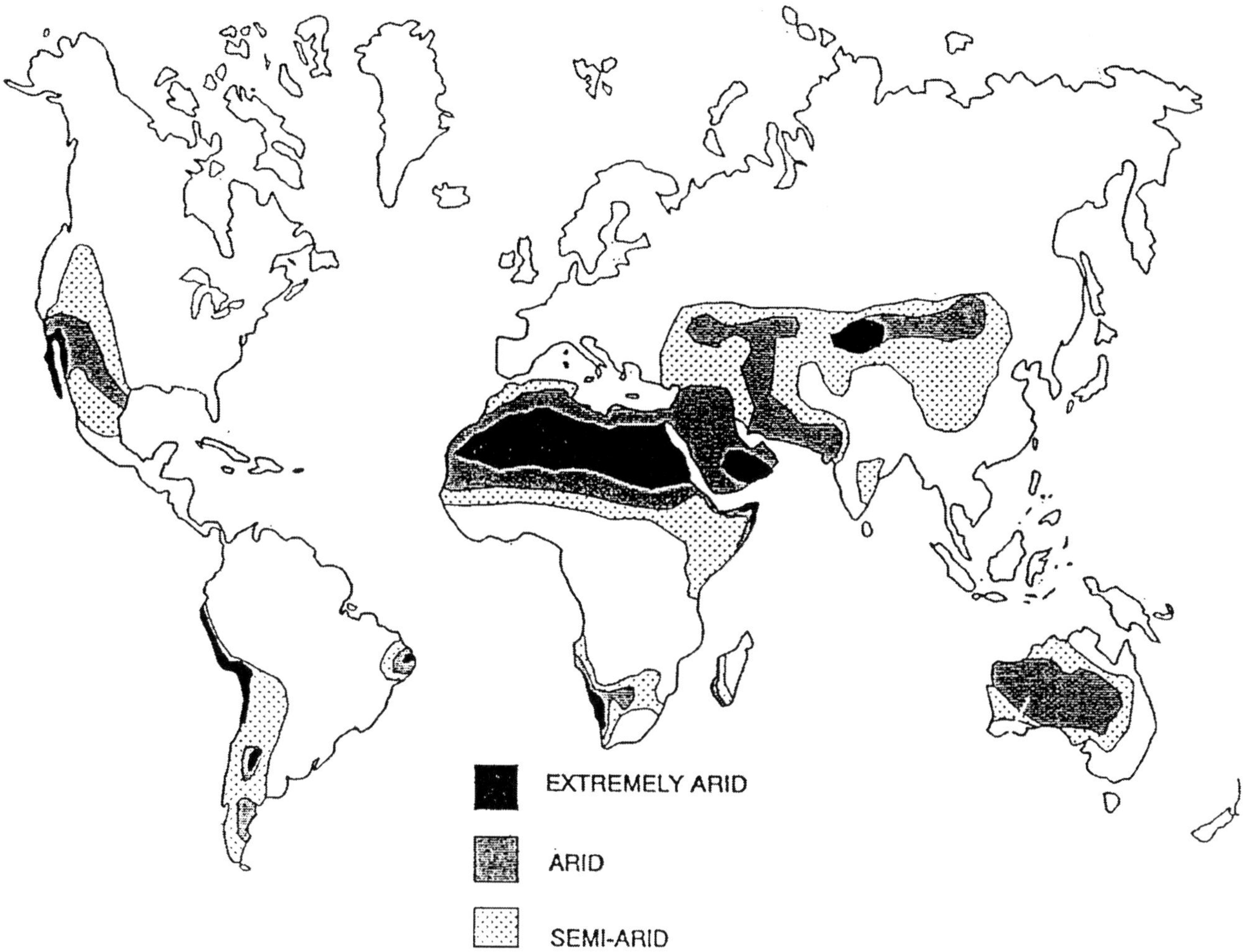

Fig. 1 Distribution of world's drylands.

then that the criteria used to objectively assess the areal extent varies between regions. It is clear that a definition based purely on climatic indices does not give a true picture of the full extent of the problem.

No one has been able so far to monitor and document desertification and the resulting land cover change through reliable, verifiable and repeated observation on a global, continental, regional or even national scale. There is not full agreement on a single indicator of dryland degradation or an approach to assess and study desertification. Dryland degradation has many faces (driving forces as well as symptoms). It can only be assessed and understood through an interdisciplinary study of the changing characteristics and integrated trends of a variety of biological, agricultural, physical and socioeconomic indicators over a long time period and at a variety of spatial scales.

Because of the scale of mapping and the fact that most desertification occurs as a mosaic of small, often-isolated, areas mapping defies the use of synoptic techniques such as remote sensing from Landsat and other satellites. Land cover/land use maps are being prepared by a number of countries and on a global scale land cover is being assessed by analysis of satellite data. These may help to clarify and quantify the extent of desertification, especially where local input helps in verification and ground truthing.

The use of the low-resolution NOAA/AVHRR satellite to generate a greenness index and to develop the Normalized Difference Vegetation Index (NVDI) might, on a broad scale, assist in monitoring change in plant cover and hence be an early warning of incipient change. But such change may only be a reflection of variability in space and time that characterizes the world's dryland regions. It is likely to be more useful in studies of drought and vegetation mapping. Its potential for use in desertification monitoring may be limited (3).

Table 1 Some common manifestations of desertification

Economic manifestations	Ecological manifestations	Social manifestations
Economic loss in cash	Loss of diversity in terms of wildlife, plants and ecosystems	Migration of population off affected areas
Decreased crop yield	Loss of inland lakes	Rural poverty
Loss of farmland due to desertification	Loss of topsoil in terms of organic matter, N, P, and K plant nutrients	Influx of ecological refugees into urban areas
Loss of rangeland due to desertification	Decreased ground water level, increasing salinity of water	
Decreased grazing capacity in terms of the number of livestock	Increased frequency of sand storms and associated losses of human lives and livestock	
Abandoned farmland		
Abandoned rangeland		
Drifting sand affects railway lines and highways		
Increase in suspended load raises river height and increases flood problems		

The most immediately detectable changes can be quite misleading.

AREAL EXTENT OF DESERTIFICATION

The area affected by desertification in many developing countries is on the increase although the lack of sound baseline data make any attempt at quantification difficult. Because a lot of desertification is characterized by a more dispersed, patch-like process of degradation it is not amenable to remote sensing. It has long been recognized that operational monitoring of patchy soil and vegetation degradation in dry areas is inherently difficult, even with medium resolution satellite technology like Landsat MSS and that improved techniques are needed to make reliable monitoring and trend detection viable.

What is clear is that most of the more than 170 signatories to the CCD report growing concern about loss of productivity of their land in the face of rapid and continuing land degradation. The impacts are reflected in:

- a loss of land resource through erosion and transportation by wind and water
- a loss of productivity through land degradation
- eco-environmental change impacting on living conditions and health
- increased poverty and social instability in the rural hinterland
- flow-on impacts on cities and on industry
- effects on infrastructure and economic development

Table 2 Distribution of the area of land susceptible to desertification (by bioclimatic zones)

	Bioclimatic zone (000's km^2 %)						
Susceptibility to desertification	**Arid**		**Semi-arid**		**Sub-humid**		
Moderately	1144.5	6.6	12713.8	68.4	3345.6	23.4	
Highly	14585.8	82.5	2686.5	14.2		589.8	4.0
Very highly	1040.3	6.4	2157.5	12.4		173.5	1.3
Total	16770.6		17557.8		4108.9		

(Data from UNEP and other sources. Used with permission.)

Table 3 Area of desertified lands by continent/region

Degree of desertification	Africa	Asia	Australia	North America	South America	Europe	World total
Slight	12430	7980	2330	440	1340	–	24520
Moderate	1870	4480	3510	2720	1050	140	13770
Severe	3030	3210	520	1200	680	60	8700
Very severe	–	–	–	67	8	–	73
Total	17330	15670	6360	4427	3076	200	47063

(Data from UNEP, CCD and other sources. Used with permission.)

The principal manifestations of severe desertification are shown in Table 1.

UNEP's 1991 (4) estimate of the area of desertified land is 3.6 billion ha, including 1 billion ha of dryland suffering from soil degradation and another 2.6 billion ha of rangeland with degraded vegetation (5). But an external review claimed that its accuracy was limited by considerable subjectivity (so observations were not repeatable); lack of resolution (so comparisons through time were not possible); and the use of point assessments were unrepresentative of larger areas (6).

As was indicated above, there are few objective measures of the areas of land affected by desertification. The most reliable maps are those published by UNEP in the revised edition of the *World Atlas of Desertification* (2).

Dr. Harold Dregne of Texas Tech University calculated the area of potential desertification (Table 2). Dregne determined the desertification status according to three factors taken as indicators, namely,

- changes in the composition of the vegetation
- extent of erosion
- presence of soil salinity

He singled out four degrees of desertification: slight, moderate, strong, and very strong. Strong desertification is an irreversible process when it is impossible to restore the land. This classification comprises mobile sand dunes, heavily salt-affected land, and badlands. In Dregne's opinion such areas are not extensive and their area is about 50,000 km^2 (7).

The areas subject to severe desertification (Table 3) are in Africa, Asia and South America and are mainly occupied by developing countries. There are a number of drivers of change. These include rapid population growth, the availability of modern technologies for land conversion, as well as loss of traditional land use controls. Political and economic policy decisions, market and trade arrangements, lack of environmental awareness and lack of capacity to combat the problem also contribute.

CONCLUSIONS

The evidence from many sources is that desertification is a global problem. UNEP estimate that about 35 percent of the earth's land surface and about 20 percent of its population are affected by it. Clearly, according to these figures and the data presented here, desertification is a major environmental and social problem. The problem is worse for heavily populated countries on the desert margins. When the extent of desertification is broken down according to major types of land use we see that grazing land and rainfed cropping are most severely affected, each being desertified on over three-quarters of its area. Irrigated land suffers less so, with about one-fifth of the irrigated area in drylands affected by desertification (salinization and waterlogging).

The global overview of desertification provides a summary but recognition must be given to regional differences and impacts. A global summary of the kind presented here is a useful first step to take when looking at a large-scale issue. However, when working out a plan of action to combat desertification at the regional or local level, more detailed information needs to be gathered on the way desertification is happening and on its root causes. Such information can only be gathered by long term monitoring of a particular area.

REFERENCES

1. UNCOD. International Convention to Combat Desertification in Those Countries Experiencing Serious Drought and/or Desertification, Particularly in Africa, I.L.M. Content Summary. International Legal Materials, **1995**, *33*, 1328–1382.

2. Middleton, N., Thomas, D., Eds. *World Atlas of Desertification*, 2nd Ed; UNEP/Edward Arnold: London, 1997.
3. Hellden, U. Desertification Monitoring—is the Desert Encroaching? Desertification Control Bulletin No. 17, **1998**, 8–12.
4. UNEP. *Status of Desertification and Implementation of the United Nations Plan of Action to Combat Desertification*; UNEP: Nairobi, 1991.
5. Dregne, H.; M. Kassas, M.; Rosanov, B. A New Assessment of the World Status of Desertification. Desertification Control Bulletin **1991**, *20*, 6–18.
6. UNEP. *Draft Report of the Expert Panel Meeting on Development of Guidelines for Assessment and Mapping of Desertification/Land Degradation in Asia/Pacific*. United Nations Environment Programme, Nairobi, 1994.
7. Kassas, M. Desertification: A General Review. Journal of Arid Environments **1995**, *30*, 15.

DESERTIFICATION, GREENHOUSE EFFECT AND

D

Sherwood B. Idso
U.S. Water Conservation Laboratory, Phoenix, Arizona, U.S.A.

Craig D. Idso
Center for the Study of Carbon Dioxide and Global Change, Tempe, Arizona, U.S.A.

INTRODUCTION

What is the impact of the ongoing rise in the air's carbon dioxide (CO_2) concentration on the ecological stability of the world's deserts and the shrubs and grasslands that surround them? This question weighs heavily on the minds of many, as the nations of the Earth debate the pros and cons of the prodigious CO_2 emissions produced by the burning of fossil fuels.

On the downside, there is concern that more CO_2 in the air will exacerbate the atmosphere's natural greenhouse effect, producing changes in climate that lead to desertification. On the upside, the aerial fertilization effect of additional atmospheric CO_2 may enhance plant prowess, increasing plant water use efficiency and enabling vegetation to reclaim great tracts of desert. The challenge, therefore, is to determine the relative merits of these competing phenomena.

THE WORLD IN TRANSITION

Climatical Changes

Questions surrounding the climatical effects of the ongoing rise in the air's CO_2 content are contentious. There is evidence the Earth has warmed significantly over the course of what is deemed the "Age of Fossil Fuels." Yet, most of this warming occurred well before the largest increases in the air's CO_2 content were recorded, peaking in the 1930s. Thereafter, the atmospheric CO_2 concentration rose at a much greater rate than it had previously, while temperatures stagnated before staging a hotly contested comeback in the 1980s and 1990s—a comeback more virtual than real. It is also possible that the atmospheric warming in the late 19th and early 20th centuries was nothing more that a natural recovery from the global chill of the Little Ice Age. Hence, although many scientists believe there has been a discernable human influence on the global climate of the past century, other scientists take serious issue with that contention.

Biological Changes

Climatical effects in the biological arena are also complicated, but not so much that certain facts cannot be used to draw some broad conclusions. Carbon dioxide, for example, is one of two main raw materials (the other being water) used by plants to produce the organic matter from which they construct their tissues. Thus, increasing the air's CO_2 content typically enables plants to grow better, as has been demonstrated in literally thousands of laboratory and field experiments (1). In addition, higher concentrations of atmospheric CO_2 cause many plants to reduce the apertures of the small pores in their leaves, through which water vapor escapes to the air. Consequently, with more biomass production per unit of water lost, plant water use efficiency is significantly enhanced. In fact, it approximately doubles with a doubling of the air's CO_2 content (2), allowing many species of plants to grow and reproduce where it had been too dry for them previously. Furthermore, the enhanced degree of groundcover resulting from this phenomenon reduces the magnitude of soil erosion caused by the ravages of wind and rain. And with greater plant growth, both above and below ground, there are significant increases in the amounts of organic matter that enter the soil. That matter enhances the soil's ability to sustain the more productive shrub and grassland ecosystems that come into being via this process of reverse desertification.

Although supported by a plethora of scientific studies, this scenario of vegetative transformation has been challenged on the assumption that resource limitations and environmental stresses encountered in nature might overpower the ability of atmospheric CO_2 enrichment to significantly enhance the vitality of plants. However, in a massive literature review designed to investigate this question, it was found that the percentage increase in plant growth produced by an increase in atmospheric CO_2 is generally not reduced by less-than-optimal levels of light, water, or nutrients, or by high temperatures, salinity, or gaseous air pollution (3). In fact, the data demonstrate that the relative growth-enhancing effects of atmospheric CO_2 enrichment are typically greatest when resource limitations and environmental stresses are most severe.

Greening of the Earth Hypothesis

In light of these observations, there is reason to believe that the historical trend in the air's CO_2 concentration, which rose from a value of 280 parts per million (ppm) at the start of the 19th century to 370 ppm at the turn of the millennium, may already be producing an ubiquitous "greening" of the Earth. This is especially true for bushes, shrubs, and trees, since woody plants are typically more positively affected by increases in the air's CO_2 content than herbaceous plants (4). Consequently, the most readily documented aspect of this biological transformation should be seen in the spreading of woody vegetation onto grasslands.

WOODY PLANT RANGE EXPANSIONS

From Prehistory to Industrial Revolution

The savannas, grasslands, and deserts of the American southwest and the southern Great Plains got their start approximately 25 million years ago. About that time, the climate began to dry, and it continued to become more arid, particularly over the past 15 millennia. But when the engines of the Industrial Revolution began to pump CO_2 into the air at rates that exceeded natural geological processes, things began to change. As early as 1844, in fact, a trader from Santa Fe, New Mexico wrote in his memoirs (5) "there are parts of the southwest now thickly set with trees of good size, that, within the remembrance of the oldest inhabitants, were as naked as the prairie plains." He summarized the situation by saying "we are now witnessing the encroachment of timber upon the prairies." And in surveying the land a century later, Malin (6) would verify that trader's prescience by referring to the ecosystem it supported as a "tangled jungle."

Modern Studies

An especially good history of the vegetative transformation of the American Southwest was developed by Blackburn and Tueller (7) from a study of the growth rings of juniper and pinyon communities in east-central Nevada, where they determined that juniper began expanding its coverage of the land well before 1800, with more rapid increases in the densities and sizes of both species occurring subsequent to that time. But just as the rise in the air's CO_2 content developed most dramatically in the 20th century, the most remarkable increases in the presence and abundance of trees at these sites manifest themselves after 1920.

A parallel example involving other woody species has been documented on the Jornada Experimental Range in New Mexico (8). When surveyed by the U.S. Land Office in 1858, no shrubs were evident on 60% of the land and only 5% of the range contained what could be described as dense stands of brush. However, 105 years later, none of the area was free of woody plants and 73% of it was dominated by thick stands of mesquite and creosote. Texas, too, had been thus transformed. By 1963, 88 million acres of former grasslands had been replaced by

Fig. 1 Photographs of the U.S.–Mexico border just east of Sasabe, Arizona. *Top*: Taken in 1893 by D.R. Payne, the area was devoid of shrubs. *Bottom*: Same areas, taken in 1984 by R.R. Humphrey, was dominated by velvet mesquite, ocotillo, velvetpod mimosa, snakeweed and burroweed. (From Humphrey, R.R. *90 Years and 535 Miles: Vegetation Changes Along the Mexican Border*; University of New Mexico Press: Albuquerque, 1987.)

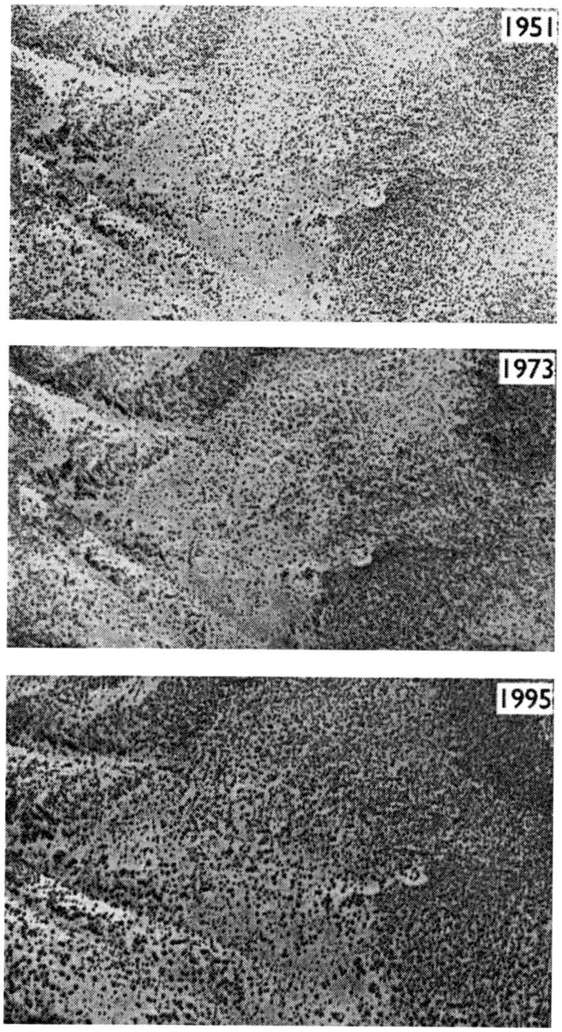

Fig. 2 Aerial photographs of the Horse Ridge Research Natural Area in central Oregon depicting an increase in western juniper cover and density between 1951 and 1995. (From Knapp P.A.; Soule, P.T. Recent *Juniperus occidentalis* (Western Juniper) expansion on a protected site in central Oregon. Global Change Biology **1998**, *4*, 347–357.)

brush and shrubs. By 1982, the figure had risen to 105 million acres (Figs. 1 and 2).

In a comprehensive review of this subject, Idso (8) assembled many examples of woody-plant range expansions onto grasslands, citing studies conducted in California, Idaho, Kansas, Missouri, Montana, Nebraska, North Dakota, Oklahoma, Oregon, South Dakota, and a number of New England states. He also cited examples of the same phenomenon in South America, Europe, Asia, Africa, Australia, and New Zealand, augmenting this evidence with equally numerous and widespread reports of ever-accelerating woody-plant growth-rate increases.

Although a number of different hypotheses have been proposed to account for this ongoing vegetative transformation of the planet, many of which play significant roles in specific locations and circumstances, the ubiquitous nature of the phenomenon argues strongly for a single worldwide forcing factor that dominates the effects of most other influences. The ongoing rise in the air's CO_2 concentration is the only phenomenon that would appear to meet this global-scale criterion.

CONCLUSION

What does the future hold? It will probably be more of the same. There are no signs that humanity's appetite for fossil-fuel energy will abate any time soon. Neither are there any indications the nations of the Earth will be able to reverse this trend by regulatory or legislative fiat. Therefore, if we do not completely cover the globe with concrete and asphalt before the Age of Fossil Fuels ends, we will probably see woody plants continue their invasion of grasslands, while grasslands intensify their assault upon the world's deserts.

With respect to global warming, it's anybody's guess. Even if it does occur, CO_2-induced increases in plant water use efficiency (a doubling for a doubling of atmospheric CO_2) and plant thermal tolerance (a 5°C increase in optimum growing temperature for a doubling of CO_2) should enable Earth's ecosystems to keep on responding as they have over the past two centuries. Thus, the greening of the Earth should continue, increasing vegetative productivity, the biomass of higher food-chain trophic levels, and ecosystem biodiversity concomitantly (9).

REFERENCES

1. Idso, K.E. *Plant Responses to Rising Levels of Atmospheric Carbon Dioxide: A Compilation and Analysis of the Results of a Decade of International Research into the Direct Biological Effects of Atmospheric CO_2 Enrichment*, Climatological Publications Scientific Paper No. 23; Office of Climatology, Arizona State University: Tempe Arizona, 1992; 1–186.
2. Idso, S.B. *Carbon Dioxide and Global Change: Earth in Transition*; IBR Press: Tempe, Arizona, 1989; 1–292.
3. Idso, K.E.; Idso, S.B. Plant Responses to Atmospheric CO_2 Enrichment in the Face of Environmental Constraints: A Review of the Past 10 Years' Research. Agric. For. Meteorol. **1994**, *69*, 153–203.
4. Idso, S.B.; Kimball, B.A. Tree Growth in Carbon Dioxide Enriched Air and Its Implications for Global Carbon Cycling

and Maximum Levels of Atmospheric CO_2. Global Biogeochem. Cycles **1993**, *7*, 537–555.
5. Gregg, J. *Commerce of the Prairies: Or, the Journal of a Santa Fe Trader, During Eight Expeditions Across the Great Western Prairies, and a Residence of Nearly Nine Years in Northern Mexico*; Henry C. Langley: New York, 1844; 2, 202.
6. Malin, J.C. Soil, Animal, and Plant Relations of the Grassland, Historically Reconsidered. Sci. Mon. **1953**, *76*, 207–220.
7. Blackburn, W.H.; Tueller, P.T. Pinyon and Juniper Invasion in Black Sagebrush Communities in East-central Nevada. Ecology **1970**, *51*, 841–848.
8. Idso, S.B. *CO_2 and the Biosphere: The Incredible Legacy of the Industrial Revolution*; Department of Soil Water & Climate, University of Minnesota: St. Paul, Minnesota, 1995; 1–60.
9. Idso, K.E.; Idso, S.B. Atmospheric CO_2 Enrichment: Implications for Ecosystem Biodiversity. Technology **2000**, *7S*, 57–69.

DESERTIFICATION IMPACTS ON LANDSCAPES, BIODIVERSITY, AND HUMAN POPULATIONS

David Mouat
Judith Lancaster
Desert Research Institute, Reno, Nevada, U.S.A.

INTRODUCTION

The United Nations defines desertification as: "Land degradation in arid, semi-arid and dry sub-humid areas resulting from various factors, including climatic variations and human activities" (1). These dryland regions comprise 41% of the global land area, or 6150 million ha (2) and are present in every continent. Hyper-arid regions are almost rainless, natural deserts and not susceptible to desertification processes. Areas of concern are those that are desertified, that are at risk for increases or acceleration of desertification processes, and that are on the threshold of change but where sustainable human life is still possible.

Economic and political factors have changed the patterns and strategies of human occupation of dryland regions, where previous lifestyles of transhumance or nomadism and minimal involvement in a market economy permitted human occupation of arid and semiarid areas even in times of drought, without causing ecosystem stress.

MANIFESTATIONS OF THE DESERTIFICATION PROCESS

Dryland ecosystems are fragile, highly vulnerable to natural climatic fluctuations, and susceptible to desertification. Desertification impacts not only human populations and their livelihood, but also the landscape—vegetation, soils, hydrology—as well as insect, animal, and bird populations and biodiversity as a whole. Dryland soils are particularly at risk—their low levels of biological activity, organic matter, and aggregate stability can easily result in a breakdown or decline of soil structure, accelerated soil erosion, reduction in moisture retention, and increase in surface runoff (3). Reduction in plant cover will exacerbate these processes, leading to a change of scale in the spatial distribution of soil resources (4), an increasingly "patchy" landscape, and a decline in sustainability that is difficult or impossible to reverse (5).

Dryland vegetation communities include grassland, shrubland, woodland, savanna, and steppe (Fig. 1) and vary according to climatic, edaphic, hydrologic, and anthropogenic factors (3). Change in vegetation community structure and reduction in cover may be a catalyst for desertification processes (6). A shift from grassland-dominated to shrub-dominated rangeland seems to be ubiquitous (7, 8) and is generally associated with increased soil runoff. The main causes of this vegetation change seem to be grazing, fencing, and alterations to the natural fire regime, in association with climatic variability (9).

The World Atlas of Desertification (10) suggests that overgrazing, agricultural practices, overexploitation of vegetation, and deforestation are the four most significant causes of soil degradation although bioindustry plays a locally important role in some countries. Soil degradation may result from displacement (by wind and water erosion) or internal deterioration by chemical or physical variables (11). Erosion of soil by water results in loss of topsoil as well as changes to the landscape such as gullying and rilling. Loss of topsoil has an impact on both natural vegetation and cultivated crops (12), and is the dominant erosion process in Africa (10). Wind erosion is most severe in areas that have been disturbed by human activity—for example, by grazing pressure that reduces plant cover. Chemical deterioration can be divided into three classes: a loss of nutrients, salinization, and acidification (10). A change in soil nutrients and a patchy distribution of nutrients at the landscape scale is a consequence of desertification (5) and in some cases may be partially due to erosion of fine particles by wind and water (13). Physical changes in soil structure result from compaction, crusting and waterlogging, and—like salinization and acidification—are exacerbated by pressure of human land use and widespread irrigation (10).

Desertification is linked to poverty (12, 14) and tends to result in out-migration, initially of young people and men. This changes family structure and places burdens and constraints on the women, children, and old people remaining in the settlements—which in turn changes patterns of land use and may exacerbate desertification (12). Loss of human dignity is perhaps the most insidious consequence of the desertification process—it is difficult to measure and difficult to reverse.

Measurement of the extent and severity of desertification has been an ongoing process since the 1930s and is

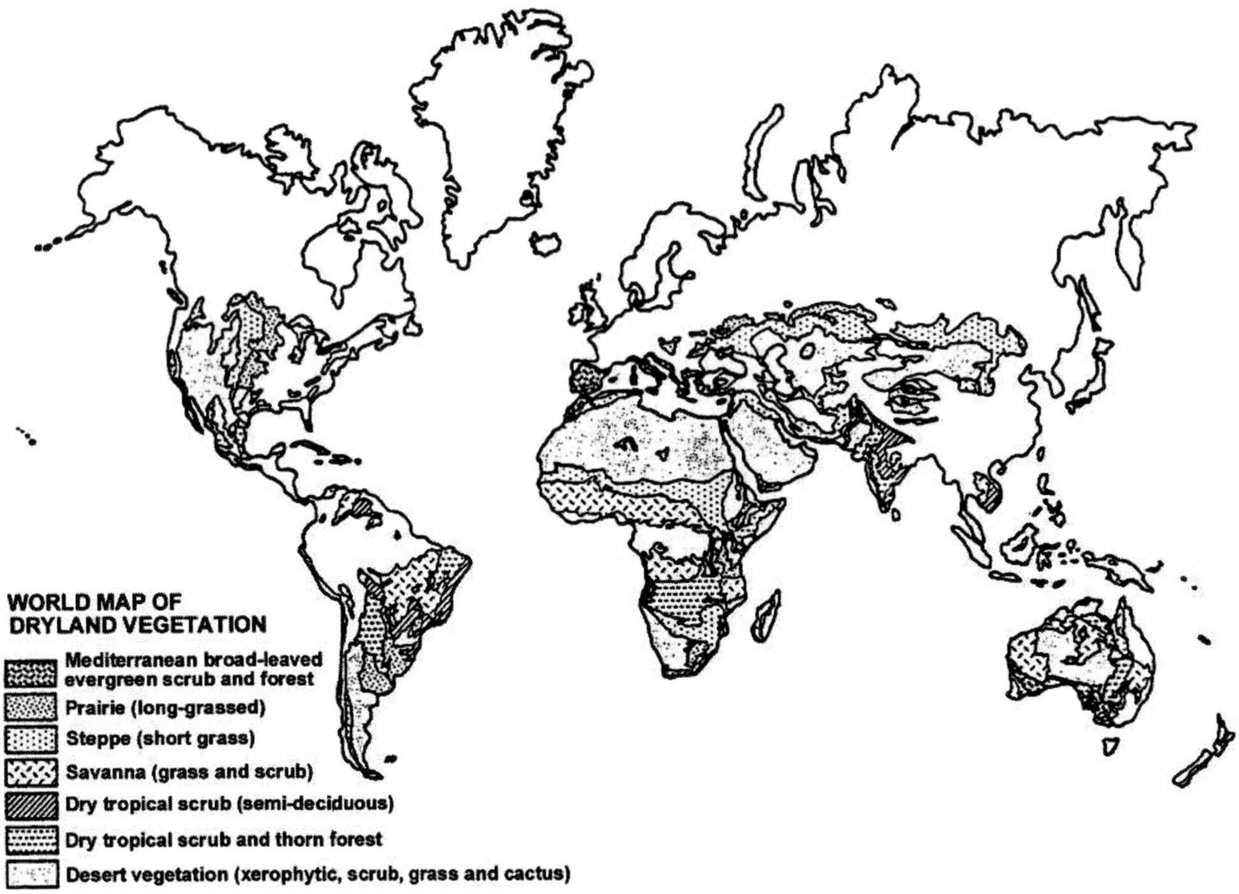

Fig. 1 Global distribution of vegetation communities. (From Ref. 3.)

now well documented in many countries. However, in order to comply with UN treaties, management plans must be developed and implemented, education programs initiated, and local communities involved in decision making. To do this effectively, it is necessary to identify those areas that will most benefit from intervention, and where desertification processes may be stabilized or reversed. International conferences addressing these issues were held in Tucson, Arizona, U.S.A. in 1994 and 1997 (15, 16) at which indicators of desertification were much discussed.

INDICATORS OF DESERTIFICATION

Although some indicators may be relevant universally (17), a standardized list of indicators is not possible as the climatic, geographic, and anthropomorphic conditions of desertification are variable, as are technology, human, and financial resources. Most indicators tend to be of a "snapshot" nature that provide measures of condition and trend, and a measure of what is "normal" only when compared in time and space. Process-based studies in semiarid and arid ecosystems that appear to be structurally and functionally intact (18) may provide baseline data on the expected and inherent variability of systems, to permit the establishment of thresholds indicative of irreversible change (19).

Reporting on environmental indicators for the land, to meet Australia's UN obligations, Hamblin (20) selected indicators that would distinguish between anthropogenic interventions and natural causes, and recommended "key indicators" of condition, pressure, and response that would describe trend and impact upon the environment. The indicators are grouped by process in relation to change in erosion, habitat, hydrology, biota, nutrients, and pollution.

The 1994 International Desertification Conference in Arizona identified nutrient availability, water budget, energy balance, and biological diversity as key ecosystem processes affected by desertification which apply across international boundaries, and again recommended indicators that were grouped by process—in this case, nutrient availability, water budget, energy balance, and biological diversity. Out of a total of 28 recommended indicators, the following may be most significant.

- Infiltration, both under and between plants
- Cover type and distribution
- Erosion as indicated by rills, gullies, pedestalling, litter movement, flow patterns, and fetch length
- Depth to water table
- Normalized Difference Vegetation Index (NDVI) as a measure of greenness
- Surface temperature
- Ratio of native to exotic species for flora and fauna

There is a paucity of demographic or socioeconomic indicators in this list. Of the numerous indicators of desertification in general use, those concerning socioeconomic factors are the least developed, and there is a disconnect in many areas between the communities affected by desertification processes, the scientists conducting research, and policy makers and managers (21). The disconnect between communities, researchers, and policy makers has been a subject of concern in South Africa and Namibia for some time, and has been addressed by several recent projects (22–24). The Karoo veld (rangeland) assessment handbook produced by Milton and Dean (25) is a successful example of orienting scientific data and their interpretation to a specific nonscience audience.

Soil is variable and complex, and the methods for assessing its condition are slow, tedious, and expensive; and to compound the issue—differences in soil type may be mistaken for differences in soil condition (26). Also targeted to ranchers and managers, Tongway's rangeland soil condition assessment handbook (27) proposes an approach to assessing soil condition based on recognizing and classifying processes related to erosion, infiltration, and nutrient cycling.

The world's drylands provide critical habitats for migratory birds, and other wildlife (12), and the maintenance of biodiversity in arid and semiarid areas is a major concern. Measures of floral, faunal, and avian diversity and reproductive success must be included in the list of indicators used to assess and monitor the status and extent of desertification. However, like demographic indicators, those involving biodiversity can be difficult to define, measure, and analyze and are frequently site specific. They may, however, provide the elusive "early warning" of a susceptibility for desertification that researchers and managers are seeking.

TECHNOLOGY, MEASUREMENT, AND ANALYSIS

The field data needed for desertification assessment and monitoring, such as soils and vegetation measurements, are typically collected at the plot scale, and many researchers, managers, and policy makers have questioned the use of remote sensing, because it provides a generalized measurement at scales of tens of meters to kilometers. However, there are numerous remote-sensing systems, operating at different spectral, spatial, temporal, and radiometric scales. The latest in the Earth Observing System (EOS) series of missions launched in December 1999 carried new sensors, two of which are the Advanced Spaceborne Thermal Emission and Reflection Radiometer (ASTER) and the Moderate Resolution Imaging Spectrometer (MODIS). The ASTER mission objectives focus on understanding the Earth as a system and the construction of models of Earth's global dynamics—including desertification (28).

The long-term perspective critical to desertification assessment is also an essential component of the evaluation of future management options and development of remediation, land use, and management plans. A recently developed technique, alternative futures assessment, models potential changes to the landscape in a Geographic Information System (GIS) over several time scales, including projections for population growth. Models for basic environmental processes are then operated on the results of the change scenarios to show cumulative effects on the system as a whole (29). Workshops and questionnaires are used to ascertain current and future attributes of the area stakeholders consider most desirable, to identify plausible change scenarios, and evaluate results.

Long-term ecological studies at plot, regional and local scales, remote sensing data, GIS, and alternative futures assessments are components of an integrated assessment, monitoring, and management strategy that is evolving in response to the need for greater interaction and communication between scientists, managers, and communities. This is a global concern, and the following case study for southeastern Arizona, discusses previous and ongoing research in the context of a strategy that involves local communities, land owners, and managers in the decision-making process.

CASE STUDY—THE SAN PEDRO RIVER BASIN

The San Pedro River rises in Mexico and flows north through the semiarid shrub steppe of southeastern Arizona.

The Bureau of Land Management (U.S. Department of Interior) manages the San Pedro Riparian National Conservation Area immediately bordering the river, where there is one of the highest animal biodiversity totals in North America. Vegetation change, especially the deterioration of grasslands due in part to grazing and the suppression of fire has been a concern in the area for some years (30). This change was quantified using aerial photography, remote sensing, and GIS to compare the extent and change over time between the eight vegetation classes in the region (9). The most striking change between 1974 and 1987 was the considerable fragmentation of vegetation classes. This patchiness of the landscape is indicative of loss of biodiversity, and could be an early warning of desertification in the area (9).

A multi-agency, multi-national, global change initiative to investigate the consequences of natural and human-induced environmental change in semiarid regions was started in 1995 in the San Pedro River Basin. Several multi-disciplinary projects have been initiated, and ASTER imagery is being acquired for the region. Among the studies currently being conducted in the San Pedro River Basin is an alternative futures assessment. This latter study has two objectives: to develop an array of plausible alternative patterns of land use and to assess their impacts on biodiversity, vegetation dynamics, fire regimes, hydrology, and aesthetics; and as a pilot study to develop a methodology applicable to other areas (31).

SUMMARY

Desertification impacts on landscapes, biodiversity, and human populations are moderately well documented, and there are numerous suites of indicators in use for assessment and monitoring (20). Studies of the actual processes involved are valuable, particularly in areas that are marginal for desertification susceptibility, and are contributing to our understanding and therefore to our ability to meaningfully intervene to reduce or mitigate the effects of desertification on both the environment and human populations (18).

In the long term, we still do not know what "desertification" really means, nor what the impacts to humans, flora, fauna, and the landscape will be. Desertification is a phenomenon that will not disappear as a result of human intervention, but may be reduced in severity or extent by a strategy including increasing understanding of the processes involved, monitoring, and investigating alternative management options. In particular, the societal impacts of desertification may be mitigated by involving local communities in the policy and decision making process.

REFERENCES

1. *Report of the United Nations Conference on Environment and Development*; United Nations Conference on Environment and Development (UNCED). United Nations: New York, 1992; Vol. 1, 486.
2. Kassas, M. Desertification: A general Review. J. Arid Environ. **1995**, *30*, 115–128.
3. Williams, M.A.J.; Balling, R.C., Jr. *Interactions of Desertification and Climate*; Arnold: London, 1996; 270.
4. de Soyza, A.G.; Whitford, W.G.; Herrick, J.E.; Van Zee, J.W.; Havstad, K.M. Early Warning Indicators of Desertification: Examples of Tests in the Chihuahuan Desert. J. Arid Environ. **1998**, *39* (2), 101–112.
5. Schlesinger, W.H.; Reynolds, J.F.; Cunningham, G.L.; Huenneke, L.F.; Jarrell, W.M.; Virginia, R.A.; Whitford, W.G. Biological Feedbacks in Global Desertification. Science **1990**, *247*, 1043–1048.
6. Le Houérou, H.N. Climate Change, Drought and Desertification. J. Arid Environ. **1996**, *34*, 133–185.
7. Ludwig, J.A.; Tongway, D.J. Desertification in Australia: An Eye to Grass Roots and Landscapes. In *Desertification in Developed Countries*; Mouat, D.A., Hutchinson, C.F., Eds.; Kluwer Academic Publishers: Dordrecht, 1995; 231–237.
8. Milton, S.J.; Dean, W.R.J. South Africa's Arid and Semiarid Rangelands: Why are They Changing and Can They be Restored? In *Desertification in Developed Countries*; Mouat, D.A., Hutchinson, C.F., Eds.; Kluwer Academic Publishers: Dordrecht, 1995; 245–264.
9. Mouat, D.A.; Lancaster, J. Use of Remote Sensing and GIS to Identify Vegetation Change in the Upper San Pedro River Watershed, Arizona. Geocarto Int. **1996**, *11* (2), 55–67.
10. Middleton, N.; Thomas, D. *World Atlas of Desertification*, 2nd Ed; Middleton, N., Thomas, D., Eds.; Arnold: London, 1997; 182.
11. Oldeman, L.R. *Guidelines for General Assessment of the Status of Human-Induced Soil Degradation*; International Soil Reference and Information Center: Wageningen, 1988.
12. http://www.unccd.int/main.php (Accessed Dec. 2000).
13. Schlesinger, W.H.; Ward, T.J.; Anderson, J. Nutrient Losses in Runoff from Grassland and Shrubland Habitats in Southern New Mexico: II. Field Plots. Biogeochemistry **2000**, *49*, 69–86.
14. Glantz, M.H. *Drought Follows the Plow: Cultivating Marginal Areas*; Glantz, M.H., Ed.; Cambridge University Press: Cambridge, 1994; 197.
15. Mouat, D.A.; Hutchinson, C.F. *Desertification in Developed Countries*; Mouat, D.A., Hutchinson, C.F., Eds.; Kluwer Academic Publishers: Dordrecht, 1995; 363.
16. Mouat, D.A.; McGinty, H.K. *Combating Desertification: Connecting Science with Community Action*; J. Arid Environ., Mouat, D.A., McGinty, H.K., Eds.; (Special Issue) Academic Press: London, 1998; 340.
17. Mouat, D.A.; Lancaster, J.; Wade, T.; Wickham, J.; Fox, C.; Kepner, W.; Ball, T. Desertification Evaluated Using an

Integrated Environmental Assessment Model. Environ. Monit. Assess. **1997**, *48*, 139–156.
18. Dean, W.R.J.; Milton, S.J. *The Karoo: Ecological Patterns and Processes*; Dean, W.R.J., Milton, S.J., Eds.; Cambridge University Press: Cambridge, 1999; 374.
19. Tausch, R.J.; Wigand, P.E.; Burkhardt, J.W. Plant Community Thresholds, Multiple Steady States, and Multiple Successional Pathways: Legacy of the Quaternary? J. Range Manag. **1993**, *46* (5), 439–447.
20. Hamblin, A. *Environmental Indicators for National State of the Environment Reporting: The Land*; Department of the Environment: Canberra, 1998; 124.
21. Mouat, D.A.; McGinty, H.K.; McClure, B.C. Introduction. In *Combating Desertification: Connecting Science with Community Action*; Mouat, D.A., McGinty, H.K., Eds.; Special Issue: J. Arid Environ. Academic Press: London, 1998; 97–99.
22. Seely, M.K. Can Science and Community Action Connect to Combat Desertification? J. Arid Environ. **1998**, *39* (2), 267–278.
23. van Rooyen, A.F. Combating Desertification in the Southern Kalahari: Connecting Science with Community Action in South Africa. J. Arid Environ. **1998**, *39* (2), 285–298.
24. Milton, S.J.; Dean, W.R.J.; Ellis, P.R. Rangeland Health Assessment: A practical Guide for Ranchers in Arid Karoo Shrublands. J. Arid Environ. **1998**, *39* (2), 253–266.
25. Milton, S.J.; Dean, W.R.J. *Karoo Veld: Ecology and Management*; Agricultural Research Council Range and Forage Institute: Lynn East, 1996; 94.
26. Tongway, D. Monitoring Soil Productive Potential. J. Arid Environ. **1998**, *39* (2), 303–318.
27. Tongway, D.J. *Rangeland Soil Condition Assessment Manual*; Commonwealth Scientific and Industrial Research Organization: Melbourne, 1993; 69.
28. http://asterweb.jpl.nasa.gov (Accessed Dec. 2000).
29. Steinitz, C.; Binford, M.; Cote, P.; Edwards, T.; Ervin, S.; Forman, R.T.T.; Johnson, C.; Kiester, R.; Mouat, D.; Olson, D.; Shearer, A.; Toth, R.; Wills, R. *Biodiversity and Landscape Planning: Alternative Futures for the Region of Camp Pendleton, California*; Harvard University, Graduate School of Design: Cambridge, MA, 1996; 142.
30. Hastings, J.R.; Turner, R.M. *The Changing Mile: an Ecological Study of Vegetation Change with Time in the Lower Mile of an Arid and Semi-Arid Region*; University of Arizona Press: Tucson, AZ, 1965.
31. http://www.gsd.harvard.edu/faculty/steinitz/sanpedroriver.html (Accessed Dec. 2000).

DESERTIFICATION, PRODUCTIVITY AND

Kris M. Havstad

United States Department of Agriculture-Agricultural Research Service (USDA-ARS), Jornada Experimental Range, Las Cruces, New Mexico, U.S.A.

INTRODUCTION

The United Nations Convention to Combat Desertification (UN-CCD) defined desertification as the degradation of land in arid, semi-arid, and dry sub-humid areas resulting from various factors, including climatic variations and human activities (1). Desertification is not a new problem, but one that traces back to 2200 BC and the degradation of Mesopotamia (2). However, the estimated extent of degradation of the world's dry lands at the beginning of the 3rd millennium is staggering (3) (Table 1). Though estimates of the magnitude of degradation vary widely, it is gen-erally accepted that at least 60% of the world's 2556 m ha of rangelands are at least moderately desertified (4). It should also be noted that these estimates often do not consider lands outside the world's subtropics (5). Globally, about 1 of every 6 ha of land has degraded soils as a result of disturbances such as overgrazing, deforestation, and overexploitation for domestic uses (6). The UN-CCD (1) estimates that over 250 million people are affected by desertification with about one-sixth of the world's population at risk. Generally, the effects of this land degradation are reductions in desirable plant production, alterations in biomass, lowered carrying capacities for livestock, increased soil erosion, and an overall increase in environmental deterioration (7). It is usually assumed that desertification has direct and measurable effects on either primary or secondary production from these lands. In very severely desertified situations, such as southern Kazakhstan, productivity has been dramatically reduced (8). For other regions of the world where desertification is certainly serious but not severe the relationships between degradation and productivity are less obvious and direct. These relationships need to be understood so that measures other than simply productivity reductions are employed to accurately assess rates, extent, and threats of degradation to these land resources.

PRIMARY PRODUCTIVITY

Primary productivity is defined as the fixation of carbon dioxide into organic molecules by photosynthetic organisms. Typically, agriculturally and ecologically based estimates of primary productivity focus on some aspect of annual above ground net primary productivity (ANPP). For crop lands this may be annual grain yields, and reductions in yield of 25% or more have been reported for dry land and irrigated crops (3). Often, these losses in productivity are coupled to increased soil erosion due to degradation. For rangelands, primary production is often expressed as biomass production on a dry matter basis per some unit area. Plant processes affected by degradation include reproduction, germination, establishment, survival, and competition, and adverse effects on these processes can result in reductions in ANPP. As a result, there are a number of documented reports from different continents where these shifts have resulted in large declines in primary productivity (9). Yet, the initial effects of chronic disturbances on these processes may result in structural changes in vegetation such as a shift from herbaceous to woody dominated plant communities, and these structural changes do not always result in reductions in primary productivity (10, 11). For example, in the northern Chihuahuan Desert of North America, ANPP does not differ among major vegetation types, including shrub land types that are regarded as degraded from the historic desert grasslands (12). These types do differ in other vegetation attributes including species richness and in their seasonality of production. Irrespective, seasonal measures of primary production in heterogeneous environments are extremely variable both temporally and spatially (13). Thus, it is difficult to consistently and directly equate reductions in primary productivity in all rangeland environments with desertification.

SECONDARY PRODUCTIVITY

Secondary productivity is the transfer of energy from primary producers to consumers, and is often expressed as animal production. For the world's rangelands secondary production is often measured in terms of livestock productivity. There are specific instances where deserti-

Encyclopedia of Soil Science

Table 1 Global desertification in drylands (arid, semiarid, and dry subhumid regions)

	% Degraded of total area		
Continent	**Irrigated land**	**Rainfed cropland**	**Range land**
Africa	18	61	74
Asia	35	56	76
Australia and New Zealand	13	34	55
Europe	16	54	72
North America	28	16	85
South America	17	31	76

(From Ref. 3.)

fication has resulted in a documented decline in livestock stocking rates, such as in the Western Cape of South Africa (14). However, in many instances losses of stocking rates have not been directly linked to desertification, and loss of secondary productivity can be an ineffective indicator of degradation (3). For example, looking more broadly at South Africa during the last half of the 20th century livestock numbers have been fairly stable (15). A more effective indicator of desertification may not be the total number of livestock, but the shift in livestock species used as vegetation types typically change from herbaceous to more woody species with degradation (16).

CONCLUSIONS

Despite the increase in desertified lands during the 20th century, the world's populations of grazing livestock, primarily cattle, sheep, and goats, have collectively increased 1% annually since World War II (17, 18). This is an increase of 1.1 billion head of livestock over the past five decades, and the increase has been spatially heterogeneous. For example, cattle numbers have more than doubled in Africa, South America, and Eastern Europe, a rate more than 50% higher than the global average. Seventy-seven percent of the global increase in sheep numbers has occurred in Asia. Yet, these continents have reported extensive desertification during this century. Declines in numbers of livestock may be misleading as an indicator of desertification at regional, national, and continental scales.

Hoffman (19) has described five phases of vegetation change during the degradation of the Karoo in South Africa. These were: 1) primary degradation with a loss of cover, primarily palatable perennial grasses and shrubs, 2) denudation with further loss of palatable plants, an increase in unpalatable species, and accelerated rates of erosion, 3) revegetation with an increase in unpalatable species, particularly woody plants, 4) secondary degradation with a relatively stable community of undesirable plants, and 5) desertification with loss of most plant species other than a few hardy shrubs, a soil surface widely exposed to accelerated erosion, and invasion by exotics. Similar vegetation dynamics have been described for other dry land environments in response to overgrazing and prolonged drought. Throughout several of these phases primary and secondary productivity may or may not be reduced. In general, effects of desertification on primary or secondary productivity of rangelands will be related to resulting species alterations and/or changes on water and nutrient economics (20).

REFERENCES

1. Secretariat for the Convention to Combat Desertification. United Nations Convention to Combat Desertification: An Explanatory Leaflet. http://www.unccd.int/convention/text/leaflet.php(accessed February 2001).
2. Burns, W.C. The International Convention to Combat Desertification: Drawing a Line in the Sand? Mich. J. Int. Law **1995**, *16*, 831–882.
3. Dregne, H.E.; Chou, N. Global Desertification Dimensions and Costs. In *Degradation and Restoration of Arid Lands*; Dregne, H.E., Ed.; International Center for Arid and Semiarid Land Studies, Texas Tech University: Lubbock, Texas, 1992; 249–282.
4. Mabbutt, J.A. Desertification of the World's Rangelands. Desert. Control Bull. **1985**, *12*, 1–11.
5. Arnalds, O. Desertification: An Appeal for a Broader Perspective. In *Rangeland Desertification*; Advances in Vegetation Science 19, Arnalds, O., Archer, S., Eds.; Kluwer Academic Publishers: Dordrecht, 2000; 5–15.
6. United Nations Environment Program, *World Atlas of Desertification*; Edward Arnold Publishing Agencies: London, 1992.
7. Dregne, H.E. Desertification: Man's Abuse of the Land. J. Soil Water Conserv. **1978**, 11–15.
8. Shaukhamanov, S.S. The Influence of Anthropogenic Desertification on the Productivity of Pasture Lands. Probl. Desert Dev. **1996**, *2*, 81–83.
9. Milchunas, D.G.; Lauenroth, W.K. Quantitative Effects of Grazing on Vegetation and Soils Over a Global Range of Environments. Ecol. Monogr. **1993**, *63*, 327–366.
10. Whitford, W.G.; Reynolds, J.F.; Cunningham, G.L. How Desertification Affects Nitrogen Limitations of Primary Production on Chihuahuan Desert Watersheds. In *Strategies for Classification and Management of Native Vegetation for Food Production in Arid Zones*; U.S. For. Serv. Gen Tech. Rep. RM-150, Aldon, E.F., Gonzales-Vincente, C.E., Moir,

W.H., Eds.; United States Department of Agriculture: Fort Collins, Colorado, USA, 1987; 143–153.
11. Grover, H.D.; Musick, H.B. Shrub Land Encroachment in Southern New Mexico, U.S.A.: An Analysis of Desertification Processes in the American Southwest. Clim. Change **1990**, *17*, 305–330.
12. Huenneke, L.F. Shrublands and Grasslands on the Jornada Long-Term Ecological Research Site: Desertification and Plant Community Structure in the Northern Chihuahuan Desert. In *Proceedings: Shrub Land Ecosystem Dynamics in a Changing Environment*; U.S. Dept. Agric. Forest Service Gen. Tech. Rep. INT-GTR-338, Barrow, J.R., McCarthur, E.D., Sosebee, R.R., Tausch, R.J., Eds.; United States Department of Agriculture: Ogden, Utah, USA, 1996; 48–50.
13. Huenneke, L.F.; Clason, D.; Muldavin, E. Spatial Heterogeneity in Chihuahuan Desert Vegetation: Implications for Sampling Methods in Semi-Arid Ecosystems. J. Arid Environ. **2001**, *47*, 257–270.
14. Dean, W.R.J.; MacDonald, I.A.W. Historical Changes in Stocking Rates of Domestic Livestock as a Measure of Semi-Arid and Arid Rangeland Degradation in the Cape Province, South Africa. J. Arid Environ. **1994**, *26*, 281–298.
15. Milton, S.J.; Dean, R.J. South Africa's Arid and Semiarid Rangelands: Why are they Changing and Can they be Restored? Environ. Monit. Assess. **1995**, *37*, 245–264.
16. Babiker, M.M. Nomad Settlements and Deterioration of Rangelands in Sudan. Environ. Conserv. **1985**, *12*, 356–358.
17. Food and Agriculture Organization, *Production Yearbook*; Food and Agricultural Organization of the United Nations: Rome, 1956; 2.
18. Food and Agriculture Organization, *Production Yearbook*; Food and Agricultural Organization of the United Nations: Rome, 1996; 50.
19. Hoffman, M.T. Agricultural and Ecological Perspectives of Vegetation Dynamics and Desertification. In *Rangeland Desertification*; Advances in Vegetation Science 19, Arnalds, O., Archer, S., Eds.; Kluwer Academic Publishers: Dordrecht, 2000; 115–130.
20. Illius, A.W.; O'Connor, T.G. When is Grazing a major Determinant of Rangeland Condition and Productivity? In *People and Rangelands: Proceedings of the VIth International Rangeland Congress*; Eldridge, D., Freudenberger, D., Eds.; VIth International Rangeland Congress, Inc.: Queensland, 1999; 1, 419–424.

DESERTIFICATION, REVERSING

David Tongway
John Ludwig
Commonwealth Scientific and Industrial Research Organisation, Australia

INTRODUCTION

The term *desertification* was coined to graphically represent the state of the Sahelian lands in the 1970s when major drought accompanied by big increases in the human population served to cause the desert margins to apparently move into formerly more productive land (1, 2). The image of an encroaching desert is powerful and evocative and resulted in major international efforts to understand and deal with the problem. Since that time, the notion of desertification has been reworked in the light of additional information to the extent that the desert is no longer seen as inexorably increasing in size nor restricted to the Sahel (3–5). Most rangeland areas in the world have suffered some sort of degradation, and recent reviews (6) have shown the process to be not at all restricted to hot deserts or areas of high population density. This is not to deny, however, the major effects on the human populations using these lands, and no doubt, much human hardship has been endured.

DESERTIFICATION REDEFINED

This article focuses on the biophysical aspects of desertification. Traditionally, easy to measure plant structural attributes, such as species composition and productivity, were the means by which desertification was assessed. These methods served to show the effect of desertification but did not provide a predictive understanding of how to combat it. However, recent advances in landscape and restoration ecology (7–9) have proposed generic approaches to study the basic nature and reversibility of desertification. Principally, this involves treating the affected landscapes as biogeochemical systems, with logical sequences of processes and feedback loops.

Vital resources such as water, nutrients, and topsoil are distributed, utilized, and cycled in space and time by processes such as runoff/runon, erosion/deposition, and decomposition (8, 10, 11). Landscapes are said to be "functional" or nondesertified if resources are retained within the system and utilization and cycling processes are efficient. "Dysfunctional" or desertified landscapes are characterized by the depletion of the stock of some vital resources and the continued flow of these resources out of the system. This mindset emphasizes the system attributes of processes acting in space, over time, in relation to applied stress and disturbance, rather than just the change in lists of species or yields of commodities. The role of vegetation as a significant regulator of energy and resources is integral to this approach (12). Desertification should be viewed as a continuum ranging from slight to severe, rather than as a simple step function (Fig. 1).

ASSESSING THE DEGREE OF DESERTIFICATION

If field sites were characterized according to "resource regulation" capacity, not only could the degree of desertification be assessed, but also the critical pathway of resource loss might be identified. Tongway and Hindley (13, 14) have designed and implemented monitoring programs to quickly provide information about biophysical processes and edaphic properties related to plant habitat favorability at both landscape and plot scales. The data gathered needs an interpretational framework to facilitate generic application. Graetz and Ludwig (15) proposed that system response to desertification be represented by a 4-parameter sigmoidal curve. Their field data analysis suggested that this shape was appropriate to relate landscape function to the degree of stress and disturbance causing the dysfunction. The curve form acknowledges upper and lower asymptotic plateaus at the non- and highly desertified ends of the spectrum, respectively, and a transition between those plateaus. This approach permits questions about whether a system was "fragile"—easily made dysfunctional, with low restoration potential—or "robust"—able to withstand stress and disturbance with only small biogeochemical attenuation (Fig. 2) to be asked. Importantly, this curve type enables critical thresholds to be detected and quantified using field indicators (16).

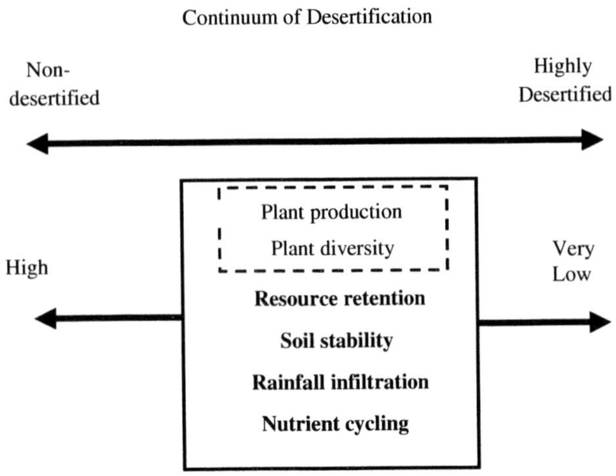

Fig. 1 Desertification as a continuum. The four new biophysical parameters (bold) are added to the two existing desertification descriptors (dotted box) to locate any given landscape on the continuum.

PROCEDURES TO REVERSE DESERTIFICATION

Rehabilitation of desertified landscapes, under the biogeochemical system mindset, requires that processes that accumulate resources be reintroduced, thus providing a rational procedure in restoration of desertified landscapes. The approach explicitly seeks to retain resources by ecological processes (9, 17, 18). Once the analysis of resource regulation system has been completed and the most affected process identified, remedial efforts can begin. For example, Rhodes and Ringrose-Voase (19) deduced that ponding water on swelling clay soils would eventually result in open, friable soils with high infiltration and water store through swelling and shrinking processes. Tongway and Ludwig (17) used piles of branches arranged on the contour line to trap water, sediment, and organic matter to effect substantial improvements to a range of soil properties, permitting perennial grasses to self-establish. Both these procedures were effective because each had identified the critical biophysical process to be ameliorated in their respective landscapes. Attempting to revegetate desertified areas by simply reseeding without understanding both the current and required edaphic properties needed for the desired vegetation mix frequently results in unexplained failure. In some instances, where the system function is close to the lower asymptote (Fig. 2), simple treatments such as exclusion from grazing will be too slow or ineffective, and active intervention may be needed to improve one or more functional processes.

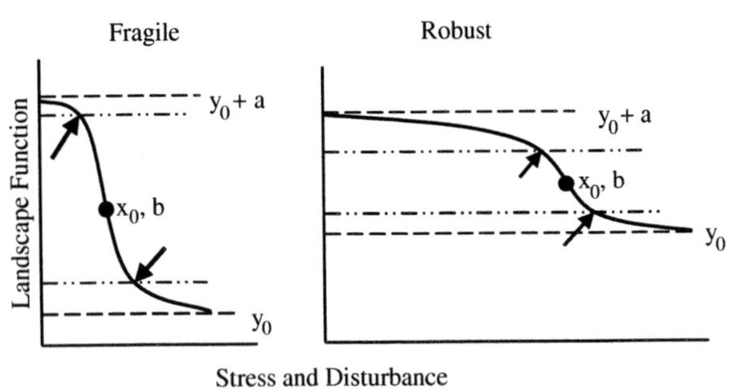

Fig. 2 Examples of response curves for fragile and robust landscapes. The initial response of landscape function to stress and/or disturbance is markedly different. The fragile landscape deteriorates with low applied stress and has a much lower base value (y_0) than the more robust landscape. Four-parameter sigmoid curves of the form $y = (y_0 + a)/1 + e^{-(x-x_0)/b}$ provide four practical values reflecting the nature of the landscape. Critical thresholds (arrows) for each index of desertification can be determined by curve analysis.

MONITORING REHABILITATION

It will be important to monitor the progress of processes set in train by the rehabilitation activities. Essentially, the degradation curves such as those in Fig. 2 need to be driven "in reverse." The procedure proposed by Tongway and Hindley (20) can be used to follow the trajectory of the system. The procedures use simple, easily observed indicators of processes of resource regulation. Biota establishment and development will also need to be monitored and interpreted in terms of the response of the whole system to restoration. It is important that the monitoring process should provide accurate information quickly and at low cost. Remotely sensed data, related to landscape function is a cost-effective procedure (21). Also, the effect of rare stochastic events such as fire may need to be assessed to see if the resultant stress and disturbance has set the system back beyond a critical threshold or not.

SUMMARY

We have described an ecologically based set of data-gathering processes by which desertification can be assessed and combatted, using a framework that characterizes the biogeochemical status of the system. This can be simply expressed as "assessing the regulation of vital resources in space and time." The procedure is able to deliver biogeochemical solutions appropriate to the problem, rather than a set of recipes with varying degrees of potential success. The information-gathering procedure can be adapted to a wide range of bioclimatic situations, because it deals with the basic processes controlling the availability of vital resources to biota.

REFERENCES

1. UNEP. *United Nations Conference on Desertification: An Overview*; United Nations Environment Program: Nairobi, Kenya, 1977.
2. UNEP. *General Assessment of Progress in the Implementation of the Plan of Action to Combat Desertification 1978–84*; United Nations Environment Program: Nairobi, Kenya, 1984.
3. Arnalds, O. Desertification: An Appeal for a Broader Perspective. In *Rangeland Desertification*; Arnalds, Olafur, Archer, Steve, Eds.; Kluwer Academic: Dordrecht, 2000; 5–15.
4. Chen, Z.; Li, X. In *People and Rangelands: Building the Future*; Proceedings of the VI International Rangeland Congress, Townsville, Australia, July 19–23, 1999; Eldridge, D., Freudenberger, D., Eds.; VI International Rangeland Congress: Townsville, 1999; 105–107.
5. Tongway, D.; Whitford, W. *In People and Rangelands: Building the Future*; Proceedings of the VI International Rangeland Congress, Townsville, Australia, July 19–23, 1999; Eldridge, D., Freudenberger, D., Eds.; VI International Rangeland Congress: Townsville, 1999; 89–142.
6. Archer, S.; Stokes, C. Stress, Disturbance and Change in Rangeland Ecosystems. In *Rangeland Desertification*; Arnalds, O., Archer, S., Eds.; Kluwer Academic: Dordrecht, 2000; 17–38.
7. Breedlow, O.A.; Voris, P.V.; Rogers, L.E. Theoretical Perspective on Ecosystem Disturbance and Recovery. In *Shrub-Steppe: Balance and Change in a Semi-Arid Terrestrial Ecosystem*; Rickard, W.H., Rogers, L.E., Vaughn, B.E., Liebetrau, S.F., Eds.; Elsevier: New York, 1988; 257–269.
8. Ludwig, J.A.; Tongway, D.J. A Landscape Approach to Rangeland Ecology. In *Landscape Ecology Function and Management: Principles from Australia's Rangelands*; Ludwig, J., Tongway, D., Freudenberger, D., Noble, J., Hodgkinson, K., Eds.; CSIRO: Melbourne, Australia, 1997; 1–12.
9. Whisenant, S.G. *Repairing Damaged Wildlands: A Process-Oriented, Landscape-Scale Approach*; Cambridge Univ. Press: Cambridge, England, 1999; 312.
10. Jorgenson, S.E.; Mitsch, W.J. Ecological Engineering Principles. In *Ecological Engineering: An Introduction to Ecotechnology*; Mitsch, W.J., Jorgensen, S.E., Eds.; John Wiley and Sons: New York, 1989; 21–37.
11. Schlesinger, W.H.; Reynolds, J.F.; Cunningham, G.L.; Huenecke, L.F.; Jarrel, W.M.; Virginia, R.A.; Whitford, W.G. Biological Feedbacks in Global Desertification. Science **1990**, *247*, 1043–1048.
12. Farrell, J. The Influence of Trees in Selected Agroecosystems in Mexico. In *Agroecology: Researching the Ecological Basis for Sustainable Agriculture*; Gliessman, S.R., Ed.; Springer-Verlag: New York, 1990; 167–183.
13. Tongway, D.J. *Rangeland Soil Condition Assessment Manual*; CSIRO: Melbourne, Australia, 1994; 69.
14. Tongway, D.; Hindley, N. *Assessment of Soil Condition of Tropical Grasslands*; CSIRO Division of Wildlife and Ecology: Canberra, Australia, 1995; 60.
15. Graetz, R.D.; Ludwig, J.A. A Method for the Analysis of Piosphere Data Applicable to Range Assessment. Australian Rangeland Journal **1978**, *1*, 126–136.
16. Tongway, David; Hindley, Norman. *Ecosystem Function Analysis of Rangeland Monitoring Data*, 2000; http://www.nlwra.gov.au/atlas
17. Tongway, D.J.; Ludwig, J.A. Rehabilitation of Semi-Arid Landscapes in Australia. I. Restoring Productive Soil Patches. Restoration Ecology **1996**, *4*, 388–397.
18. Ludwig, J.A.; Tongway, D.J. Rehabilitation of Semi-Arid Landscapes in Australia. II. Restoring Vegetation Patches. Restoration Ecology **1996**, *4*, 398–406.
19. Rhodes, D.W.; Ringrose-Voase, A.J. Changes in Soil Properties During Scald Reclamation. J. Soil Conservation Service NSW **1987**, *43*, 84–90.
20. Tongway, D.; Hindley, N. Assessing and Monitoring Desertification with Soil Indicators. In *Rangeland Desertification*; Arnalds, O., Archer, S., Eds.; Kluwer Academic: Dordrecht, 2000; 89–98.
21. Kinloch, J.E.; Bastin, G.N.; Tongway, D.J. Measuring Landscape Function in Chenopod Shrublands Using Aerial Videography. In *Proceedings of the 10th Australasian Remote Sensing and Photogrammetry Conference*, Adelaide, Australia, 21–25 August 2000, Causal Productions: Adelaide, South Australia, 2000; 480–491.

DESERTIZATION[a]

Henry Noel Le Houérou
Montpellier, France

INTRODUCTION

Desertization is the irreversible extension of desert-like landscapes and landforms to areas where they did not occur in the recent past (1, 2). Desertization results, to a large extent, from land abuse by humans and livestock; it is not a consequence of climate fluctuation, albeit drought may worsen and accelerate a phenomenon that could have been triggered during high rainfall periods (there are many examples of such seemingly paradoxical situations). It is estimated (3–6) that over 20% of the 42 million km^2 of dry lands are submitted to a greater or lesser extent to desertization processes.

The causes of desertization are both direct and indirect. Direct causes (3, 7) are those that have a strong immediate impact on site soil characteristics and properties. The major one is the reduction of perennial plant cover and biomass that contributes to soil organic matter depletion. This reduction of soil organic matter content triggers a chain of actions and reactions affecting both soil and microclimate, leading to lowered biological activity, to compacted soils with high runoff, and fast erosion processes, both by water and wind; soil may thus ultimately be turned sterile or almost so in a few years of abuse.

The indirect causes are those that bring about the depletion of natural resources, such as long-standing overcultivation, overstocking, poor farming practices, deforestation, destructive collection of fuel wood, and other mismanagement practices (1, 2); all these re sult from excessive and permanent pressure on the land by exceedingly numerous human and livestock populations: continuous high densities and stocking rates.

Combatting desertization first requires the discontinuation, or at least the serious mitigation of previous destructive activities that brought the situation about. Biological recovery may not be feasible any more whenever degradation is extreme; artificial human-induced intervention may then be deemed necessary. On the other hand, in less extreme cases, artificial rehabilitation and restoration may be desirable to hasten the desired goal. Biological recovery may be achieved via centuries-old techniques of soil and water conservation, water harvesting, runoff farming, agroforestry, appropriate farming, stocking, range, and forest management. But such activities often necessitate a deliberate policy and adequate strategies of conservative land use; this, in turn, may require political and legal actions encouraging responsible landuse management to achieve sustainability, i.e., a profound land reform. This has actually been achieved by various centuries-old peasant civilizations in various parts of the world arid lands.

DEFINITION

Deserts are defined as regions where rainfed agriculture is not feasible because of excessive aridity (8). Unlike the word "desertification," which has at least four distinct meanings, the term "desertization" is clearly defined. It implies irreversible extension of desert land-forms and conditions to regions where they did not occur in the recent past (1, 2). The desertization is primarily due to destructive human and livestock activities in arid environments. In spite of the UNEP Plan to Combat Desertification (as a follow-up to the UNCOD Conference of Nairobi in 1977, the Earth Summit of Rio de Janeiro in 1992, and the Convention of Paris in 1994), little has been done on a large scale, so desertization is likely to continue over the next 50 years as it has in the past 50 years (9).

FACTS AND FALLACIES

Origin of Deserts

Deserts are regions where, because of excessive aridity, no agriculture is feasible without supplemental irrigation (8, 10). On the basis of their origin, there are five types of deserts (10, 11). True climatic deserts result from general atmospheric circulation whereby the adiabatic subsidence of dried-up air masses occurs on the limits between the Ferrel and Hadley cells along the 30° N and 30° S parallels (11). The origin of true deserts varies with local conditions (e.g., tropical and subtropical deserts). Rain-shadow deserts

[a] The text of the present entry is a 75% reduced version of an article to be published in *Arid Lands Research and Management* 16:1–36, 2002.

Encyclopedia of Soil Science

(a particular case of climatic deserts) are formed due to orographic obstacles cutting the trajectory of rain-bearing frontal depressions (e.g., the Death Valley and Colorado River Delta east of the Sierra Nevada of California, Patagonia). Coastal deserts are formed due to the atmospheric stability resulting from the upwellings of coastal cold currents, such as along the Western Continental shores of America and Africa (e.g., the Humboldt current in Chile and Peru, the Benguela current of Southwest Africa). Edaphic deserts result from geological and soil conditions that are inappropriate for agricultural activities and the development of natural vegetation, such as unweathered hard or toxic rocks, salinity, and alkalinity, etc. Man-made deserts result from anthropgenic activities and destructive land abuse. The present article is concerned with the formation of the human-induced desert.

Fluctuations and Trends of Climatic Variables in Historical Times

Rainfall

There is a consensus among scientists on the absence of trends in long-term rainfall since the beginning of history (3, 12–14). There is, however, evidence of considerable changes in the course of the Pleistocene and Holocene periods beginning 3000 and 10,000 years B.P., respectively, with some 10 alternating periods of dry and humid climates in the Sahara and Arabia (9, 12, 15, 16). This evidence comes essentially from tree-ring studies, pollen analysis, and lake-level fluctuations. Minor fluctuations took place, however, in the course of history such as the recent 25-year period of drought in the African Sahel (1968–1983) and East Africa (1975–1984) (15, 17), in Chile from ca. 1900 to date, and in Argentina between 1900 and 1950.

Temperature

Variations in temperature are more complex to analyze. There is a consensus on a global increase of 0.5°C over the past 100 years (13, 14). But this includes the effect of urbanization that may globally account for up to 0.2°C (18). Fluctuations during the historical period are fairly well documented. The causes of these "random" fluctuations are not known with any certainty. The magnitude of increase over the past 100 years is thus of the same dimension or smaller than previous random fluctuations.

Air Pollution

There is general agreement on the fact that the atmospheric content CO_2 and other pollutants is increasing. The CO_2 content of the atmosphere was 280–300 ppmv (parts per million in volume) at the beginning of the industrial era some 150 years ago (19). It is now 370 ppm, essentially due to the burning of fossil fuel (14). Similarly, the content of methane has doubled from 800 to 1750 ppbv (parts per billion in volume) over the same period (14). The main sources of CH_4 are swamps, rice fields, ruminants, forest fires, garbage disposal, oil industries, and termites.

But the present situation is different in as much as it is the increasing content of CO_2 and CH_4 in the atmosphere that is causing a temperature rise, while in the past 160,000 years, and presumably all the Quaternary, the situation was the opposite. It is the increase in temperature from orbital forcing that resulted in a rise of CO_2 and CH_4, by oxidation of soil organic matter content, melting of tundra permafrost soils, and release of CH_4 from the thaw of frozen swamps. The present undisputed 0.5°C/100 years of global temperature increase thus cannot, strictly speaking, be attributed to CO^2 or other atmospheric pollutants.

Land Surface Albedo

As vegetation is increasingly depleted and perennial plant cover recedes, the proportion of bare ground increases, enhancing its reflectivity, or the surface albedo (20). Higher albedo reduces temperature, decreases evaporation, and creates a thermic depression, impeding cloud formation and possibly decreasing rainfall. This so-called surface albedo feedback theory is, however, highly debated—actual facts have not, so far, proved it; rather the opposite (21, 22).

DESERTIZATION

Definition

Desertization is the irreversible degradation of arid land resulting in desert-like land-forms in areas where they did not occur in a recent past (1, 2). This definition differs from the concept of desertification as accepted by UNEP in the 1977 UNCOD Conference of Nairobi (4, 23). It was then defined as "land degradation under arid, semiarid and dry subhumid climates that may ultimately lead to desert-like conditions." The term "desertification" is presently used for two distinct meanings. It implies the *irreversible* degradation of arid lands leading to formation of man-made deserts. In western Europe and EEC, desertification is synonymous with land abandonment, resulting in natural reafforestation and expansion of woodlands (24, 25). Land abandonment may be referred to as "land desertion" because the rural population migrates to towns (25).

Causes of Desertization: Degradation Processes

The causes of desertization are both direct and indirect.

Direct causes

The direct causes pertain essentially to the destruction of the perennial plant cover and biomass and its consequences on soil characteristics through erosion, sedimentation, water logging, and salinization (1, 2, 16, 26–28). The destruction of perennial biomass and cover results in the following degradative chain reaction:

1. Reduced litter production and return to the soil particles decrease soil organic matter content.
2. Reduced soil organic matter weakens or destroys soil structure and aggregate stability.
3. Unstable soil aggregates increase surface soil sealing from raindrop splash on silty or loamy soils, and biological crusting on coarse sandy soils (29).
4. Sealed soil surfaces increase runoff, decrease soil water intake and storage, and create a drier environment.
5. Increased runoff causes flooding, water logging, and mass die-off of shrubs, trees, and crops downstream and in closed depressions.
6. High wind speed on the soil surface, a consequence of reduced perennial plant cover, leads to higher albedo, increased potential evapotranspiration, increased water and wind erosion, and therefore drier microclimatic conditions.
7. Higher soil surface daily maximum temperatures and therefore high evapotranspiration, due to the lack of shading, enhances oxidation of soil organic matter, a rapid depletion of water reserves, and a shorter growing season. Lower soil-surface night temperature causes a larger diurnal and seasonal temperature range (thermal amplitude), leading to less favorable conditions for germination, emergence, and establishment of seedlings, and to the restoration of the natural vegetation.
8. Decreases in the number, biomass, and activity of soil microflora (e.g., bacteria, actynomycetes, fungi, algae, and rhizospheric symbionts) as a result of decreasing organic matter content in the soil slow down nutrient turnover and reduce availability of major and trace geobiogene elements, leading to lower fertility and productivity.
9. Decreased productivity reduces the number, biomass, and activity of soil microfauna, mesofauna, macrofauna, and megafauna.
10. Wind erosion (creeping, saltation, suspension, corrosion) results in the development of desert pavements (regs, serirs, hamadas, gobis) rock mushrooms, and yardangs.
11. Wind deposition leads to the formation of sand clay dunes (lunettes), loess, nebkas, ramlas, shamo, ergs, edeyen, nefuds, sand sheets, and sand veils. Gravity erosion leads to dry colluvium deposits.
12. Water erosion (sheet, rill, gully, pipe, landslides, etc.) leads to flooding, sedimentation, and alluvium deposition.
13. Rising water tables and salinity result from higher runoff and lesser pumping by shrubs and trees.
14. Rising salinity and soil sterilization are caused by poor drainage and other mismanagements.

Indirect causes

The indirect causes of desertization are those that exacerbate the development of the direct causes.

1. Rapid human population growth and quick increase of population density lead to high human and livestock pressure on the land. The human population in the arid lands of the developing countries has grown at an annual rate of 3.2% since 1950 (2, 3, 16, 30).
2. Livestock populations in Africa and Southwest Asia have increased at an average annual rate of 2.2% since 1950 (3, 12, 16).
3. Soil degradation is accentuated by overcultivation, inappropriate farming methods, poor land management (e.g., cultivation of land which is too dry, too steep, too shallow), inappropriate tillage (up and downhill plowing), quasi-exclusive utilization of disc-plows, inappropriate crop rotation patterns, inadequate field size, careless harvesting, and postharvesting land management, etc.
4. Deforestation is exacerbated by excessive collection of firewood (1.5 kg per person per day). As arid land steppic vegetation of dwarf shrubs bears some 300–600 kg dry woody matter per hectare, each person depending on this type of fuel destroys about 1 ha of steppic vegetation per annum (8).
5. Overcollection of medicinal and other useful plants contribute to the decline of plant cover and biomass.
6. Wildfires are particularly harmful in the tropical savannas. As they occur in the dry season, they leave a soil surface that remains bare and therefore vulnerable to erosion during the next rainy season (17, 31).
7. Vegetation arcs are particular patterns of distribution whereby plants are concentrated on contour strips alternating with bare and sealed interstrips. Seen from above, as on air photos or satellite images, the land looks like a tiger skin, hence the expression of "tiger

bush" (17, 24, 27). This is a case of edaphic aridity when, owing to the thinning of vegetation from woodcutting and/or overbrowsing, and the consecutive runoff and erosion, the soil moisture regime is too low for the vegetation. Thus, vegetation is confined in strips that permit higher soil moisture due to runon.

8. Inadequate legislation or lack of enforcement of legislation on land use and erosion control measures.
9. Inappropriate land-ownership systems under which land, grazing, and water are free communal resources while livestock are privately owned. Thus, it is in each stockman's interest to draw the maximum from the common resource without investing anything in it, the "tragedy of the commons."
10. Inappropriate marketing facilities and inadequate incentive to de-stock and cull animals.
11. Poverty and lack of education and inefficient extension services.

Diagnosis and evaluation of desertization

Desertization may be evaluated via various criteria, namely, measuring soil erosion and sedimentation, assessing perennial plant canopy cover and biomass or the basal cover of perennial species, and evaluating the rain-use efficiency (RUE = Kg DM/ha/yr/mm) and computing the production to rain variability ratio (PRVR = CV Ann. Prod./CV Ann. rainf.) (32). The evaluation of RUE is a particularly easy, quick, and clean method.

Low perennial plant cover is the principal factor of soil surface erosion by water and wind. In the Mediterranean steppelands there is virtually no soil-surface erosion when perennial plant canopy cover reaches 25% of soil-surface. In such cases, erosion is compensated by deposition and sedimentation (26). In the North African steppes, for instance, each 1% of perennial canopy cover corresponds with a biomass of 30 to 60 kg DM per ha (33). It is then easy to determine the stage of degradation and desertization from evaluation of perennial plant canopy cover (Table 1). The world regional distribution of arid lands is shown in Table 2.

Table 1 Desertization risk assessment from perennial plant canopy cover evaluation

	Perennial plant canopy projection cover in percent of ground surface					
MAR nmm	0–1	1–5	5–15	15–25	25–50	>50
50–100	5	5	4	3	2	1
100–200	5	4	3	2	1	0
200–300	4	3	2	1	0	0
300–400	3	2	1	0	0	0
400–500	2	1	0	0	0	0
500–600	1	0	0	0	0	0

0 — Present hazard nonexistent or very low.
1 — Immediate risk low.
2 — Immediate hazard moderate
3 — Immediate risk serious.
4 — Immediate hazard severe.
(From Ref. 33.)

Severity and Extent of Desertization

Can desertization occur without the impact of drought?

True climatic deserts cannot exist without permanent and extreme climatic aridity. But such is not the case with the occurrence and spread of man-made deserts. Contracted vegetation occurs in all world deserts with a mean annual rainfall isohyets of between 50 and 150 mm depending on the substrate, the anthropozonic impact, and other local factors (e.g., high atmospheric humidity and occult precipitation, or, conversely, continentality, low RH, and high PET). Desertization is often caused by droughts, but the primary causes have often been operating long before drought occurred. The effects may have been hidden by a period of better than average climatic conditions. This was the case in the West African Sahel.

The analyses of available information lead to the following conclusions:

1. Desertization may be triggered during nondrought periods.
2. Desertization cannot result from drought alone.
3. The main causative factor appears to be land mismanagement from excessive anthropogenic pressure on the land (land abuse), beyond the ecosystems' resilience.

Severity and extent

The global extent of land degradation has been assessed by ISRIC/UNEP (4–6, 23). The conclusions are in good agreement with other independent studies (3). Tables 3, 4 and 5 show that overstocking and overgrazing are the major causes of arid land degradation worldwide, followed by overcultivation and poor farming methods. The regional and local causes, however, may considerably alter this pattern. Overcultivation and shrubland clearing, for instance, are by far the prime causes in northern Africa, southwestern Asia, and northwestern China, while overgrazing is the foremost cause in the Sahel, East Africa, North and South America, and Australia (Table 5). Salinization affects 1.5 million ha annually out of a world

Table 2 Regional distribution of world arid lands

	Continental distribution (10^3 km^2)						
Geographic area	**Eremitic**	**Hyperarid**	**Arid**	**Semiarid**	**Total**	**%[a]**	
MAR mm[b]	<50	50–100	100–400	400–600			
100 P/PET mm[c]	<3	3–5	5–28	28–45			
Budyko's Aridity Index[d]	>50	10–50	3–10	2–3			
World	130,737	7,500	7,059	14,330	12,651	41,440	100
Africa	30,312	6,232	3,017	3,570	2,951	15,770	52
N. America	21,322	10	90	1,025	1,935	3,060	14
S. America	17,818	275	105	972	1,274	2,626	14
Asia	43,770	1,595	3,225	5,415	4,817	15,042	34
Europe (Spain)	500	NA[e]	NA	100	300	400	80

[a]Percent of geographic area; [b]MAR = mean annual rainfall; [c]100 P/PET = precipitation to potential evapotranspiration ratio, × 100 for easier handling. PET being evaluated via Penman Equation; [d]Budyko's Index is the ratio between net energy budget and actual annual rainfall, in other words it evaluates how many times the energy budget could evaporate the MAR (which does not take into consideration of the aerodynamic parameter of Penman's equation); [e]NA = not applicable.
(From Ref. 3.)

irrigated land area of 2 million km^2. The total area lost to secondary salinity over the last century is about 25 million ha (3, 32).

Natural Amelioration

Natural amelioration of land management to restore or rehabilitate ecosystems and environment implies the elimination or substantial reduction of the causes that leads to degradation. This goal can be achieved by reducing stocking rates to carrying capacity; restricting farming to appropriate areas, such as depressions with deep soils benefiting from runoff or having a water table; improving grazing systems through control of livestock, fencing, rotational/controlled grazing, and installing temporary exclosures. These activities are usually cheap and efficient tools for restoring and rehabilitating arid lands (34–36).

Managed Restoration

Human-assisted rehabilitation includes a large array of techniques from the simplest and cheapest to the most sophisticated and costly but not necessarily efficient.

Table 3 Extent and severity of arid and semiarid land degradation

	Degradation intensity (10^3 km^2 and %)									
Bioclimatic zones	**Light**		**Moderate**		**Strong**		**Very strong**		**Total**	
	S	**%**	**S**	**%**	**S**	**%**	**S**	**%**	**S**	**%[a]**
Arid and semiarid zones	3,601	42	3,964	45	1,096	12	62	1	8,723	20
Water erosion	1,475	37	1,757	45	666	17	40	1	3,938	9
Wind erosion	1,662	46	1,815	50	152	4	15	0	3,644	8
Chemical Degradation	373	44	266	31	203	24	7	1	849	2
Physical Degradation	91	31	126	43	75	26	0	0	292	1

[a]Percent of dry lands (SA+A+HA.Er).
(From Refs. 5, 6.)

Table 4 Causes of degradation in dry lands ($10^3\,km^2$)

	Deforestation	Overgrazing	Overcultivation	Other	Total degradation	Nondegradation	Total dry lands
S	1,700	4490	2,450	83	8,723	34,832	43,554
%	20	51	28	1	100/20	80	100

(From Refs. 3, 12, 23.)

Surface roughness

Techniques to create surface roughness break the surficial crust (e.g., a silt seal or a biological crust), increase permeability to air and rainwater, and facilitate seed germination and seedling emergence. A simple technique for breaking crust involves spreading branches from spiny trees and shrubs on the soil surface (*Acacia* spp., *Balanites* spp., *Ziziphus* spp., *Prosopis* spp., *Rhus* spp., *Parkinsonia* spp., *Commiphora* spp., etc.). Soil-surface roughening may also be achieved with the plow harrow, hoe, cultivator, subsoiler, ripper, basin-lister, etc. (17).

Contour lining

This may be achieved by using stones or pieces of hardpan or duricrusts (e.g., gypsic pans, calcrete, iron pans, or lateritic pans). Light contour lining aims at reducing sheet erosion and improving water intake. It may also be achieved via mechanical equipment used for terracing and building contour banks, contour benches, or ditches.

Ripping and subsoiling

These techniques use heavy equipment to break deep, hard lime, gypsum, silicium (silcrete), or iron pans or to loosen hard compacted soils to improve water intake and storage and deepen the root zone of protective plants, particularly shrubs and trees.

Termite-assisted hand-pitting across a surficial pan

The southern Sahelian technique of the "Zai" is a cheap and efficient practice. It involves digging pits (40 L × 40 W × 15 D) about 80 cm apart at a density of 2,500 pits/ha. These pits are dug a few months prior to the onset of the rainy season, and 1,000–1,500 kg/ha of organic matter (e.g., litter, straw, stovers, etc.) is placed at the bottom of these pits. The organic matter attracts termites that dig deep galleries below the pan. The microenvironment thus created is then manured at a rate of 1,000 kg/ha of dry corral dung. Pearl millet or sorghum planted in these pits may reach maturity even in a dry year, owing to the better soil water regime. Each pit is a green island amid a devastated environment landscape, and the beneficial effect may last up to 30 years. The cost of making Zai is 150–300 human-days per hectare for a pit density of 2,500/ha, 1,000 kg of organic matter, and 1,000 kg of corral manure (37).

Range reseeding

Range reseeding and establishing improved pasture are also feasible in the semiarid zone, but not so in the arid zone (7, 28).

Agroforestry

Agroforestry is practiced on 3.2 million hectares in world arid lands. Agroforestry, using either native or exotic species, is a very potent and efficient tool for biological recovery, allowing productivity 3 to 10 times higher than pristine vegetation under the same ecological conditions. These lands are distributed as shown in Table 6 (35, 36).

Table 5 M-Causes of desertization, % of desertized land

Regions	Overcultivation	Overstocking	Fuelwood collection	Salinization	Urbanization	Other	Total
N.W. China	45	18	18	2	3	14	100
N. Africa and S.W. Asia	50	26	21	2	1	–	100
Sahel and E. Africa	25	65	10	–	–	–	100
Middle Asia	10	62	–	9	10	9	100
Australia	20	75	–	5	?	–	100
U.S.A.	22	73	–	5	1	–	100

(From Refs. 7, 30.)

Table 6 World areas under agroforestry management

Fodder cacti	1,200,000 ha	N. Africa, S. Africa, N. and S. America
Saltbushes	1,000,000 ha	N. Africa, S. Africa, S.W. Asia, N. and S. America, Australia
Acacias spp	600,000 ha	N. Africa, Sahel, S. Africa, Australia
Prosopis spp	50,000 ha	N. and S. America, India
Agave americana	100,000 ha	S. Africa
Miscellaneous (*Saxaouls, Tamarix, Bohemia olive, poplars*)	250,000 ha	Africa, Australia, N. and S. America, Europe, Central Asia

Wildlife management, wildlife husbandry, tourism

Wildlife management and husbandry are actively pursued in an increasing number of arid land countries: U.S.A., Australia, East Africa, Southern Africa, New Zealand, Argentina and Western Europe. Privately owned game ranches or game farms are quickly developing in Africa.

Adaptive arid land strategy and sustainable development policies

Mitigating desertization is a problem of sustainable arid land development. Sustainable development policies include many facets (e.g., technical, social, economic, political, legislative, organizational, and educational). An important step toward sustainable development is reforming the land tenure system. This would require a fundamental land reform in order to shift from communal or tribal land ownership to secure private systems whereby long-term investment becomes possible and long-term planning can be implemented.

REFERENCES

1. Le Houérou, H.N. La Désertisation Du Sahara Septentrional Et Des Steppes Limitrophes. Ann. Algér. de Géogr. **1968**, *6*, 2–27.
2. Le Houérou, H.N. The Nature and Causes of Desertization. Arid Lands Newsletter **1976**, *3*, 1–7.
3. Le Houérou, H.N. An Overview of Vegetation and Land Degradation in World Arid Lands. In *Degradation and Restoration of Arid Lands*; Dregne, H.E., Ed. Texas Technical University: Lubbock, 1992; 127–163.
4. UNCOD, United Nations Conference on Desertification; UNEP, Nairobi & Pergamon Press, New York, 1977.
5. Oldeman, L.R., Hakkeling, R.T.A. and Sombroek, W.G., World Map of Status of Human-Induced Soil Degradation. An Explanatory Note, ISRIC, Wageningen and UNEP, Nairobi, 1990.
6. Middleton, N.; Thomas, D. *World Atlas of Desertification*, 2nd Ed; Edward Arnold Ed. London and UNEP: Nairobi, 1997.
7. Le Houérou, H.N. Drought-tolerant and Water-Efficient Trees and Shrubs for the Rehabilitation of Tropical and Subtropical Arid Lands. Journal of Land Husbandry **1996**, *1* (1–2), 43–64.
8. Meigs, P. *World Distribution of Arid and Semiarid Homoclimates*; 5 Maps. Reviews of Research on Arid Zone Hydrology, UNESCO: Paris, 1952.
9. Le Houérou, H.N. *Climate Change Drought and Desertification*; Report to IPCC, Washington, 1995.
10. McGinnies, W.G.; Goldman, B.G.; Paylore, P. *Deserts of the World: An Appraisal of Research into Physical and Biological Environments*; University of Arizona Press: Tuscon, 1968.
11. Monod, Th. *Les Déserts*; Horizons de France, Publ. Paris, 1973.
12. Le Houérou, H.N. Vegetation and Land-Use in the Mediterranean Basin by the Year 2050: A Prospective Study. In *Climatic Change and the Mediterranean*; Jeftic, L., Milliman, J.D., Sestini, G., Eds. Edward Arnold: London, 1992; 175–232.
13. Houghton, J.T.; Callander, B.A.; Varney, S.K. *Climate Change, 1992: The Supplementary Report to IPCC Scientific Assessment*; Cambridge University Press: London, 1992.
14. Houghton, J.T.; Jenkins, G.J.; Ephraums, J.J. *Climate Change: The IPCC Scientific Assessment*; Cambridge University Press: London, 1990.
15. Le Houérou, H.N. Changements Climatiques Et désertisation. Sècheresse **1993**, *4*, 95–111.
16. Le Houérou, H.N., La Méditerranée En 2050: Impacts Respectifs D'une Éventuelle Évolution Climatique et de la demographie Sur La Végétation, Les Écosystèmes Et L'utilisation Des Terres. Etude Prospective. *La Météorologie*, VII Série, No. 36: pp. 4-37, 1991.
17. Le Houérou, H.N. *The Grazing Land Ecosystems of the African Sahel*; Ecological Studies n. 75, Springer Verlag: Heidelberg, 1989.
18. Duplessy, J-C. Quand l'Ocean Se fâche. In *Histoire Naturelle du Climat*; Odile, Jacob Ed.; Paris, 1996.
19. Keeling, C.D.; Bacastow, R.B.; Bainbridge, A.E.; Ekdahl, C.A.; Guenther, P.R.; Waterman, L.S. Carbon Dioxide Variation At Mauna Loa Observatory Hawaii. Tellus **1976**, *28*, 28.
20. Charney, J. Dynamics of Deserts and Drought in the Sahel. Quat. J. Roy. Meteor. Soc. **1975**, *101*, 193–202.
21. Williams, M.A.J.; Balling, R.C., Jr *Interactions of Desertification and Climate WMO*; Geneva & UNEP: Nairobi, 1994.

D

22. Courel, M.F. *Etude de L'évolution Recente Des Milieux Sahéliens à Partir Des Measures Fournies Par Les Satellites*; Thése Doct., Univ. Paris I, 1985.
23. UNEP, World Atlas of Desertification. UNEP, Nairobi and Edward Arnold, London, 1992.
24. Le Houérou, H.N. Impact of Man and his Animals on Mediterranean Vegetaion. In *Mediterranean-Type Shrublands*; Ch. 25. Di Castri, F., Goodall, D.W., Specht, R.L., Eds. Ecosystems of the World, Elsevier: Amsterdam, 1981; Vol. 11, 479–521.
25. Le Houerous, H.N. Land Degradation in Mediterranean Europe: Can Agroforestry be a part of the Solution? A Prospective Review. Agroforestry Systems **1993**, *21*, 43–61.
26. Le Houérou, H.N., Recherches Écologiques et Floristiques Sur La Végétation De La Tunisie Méridionale. Mém. H.S no 6., Instit. Rech. Sahar., Univ. d'Alger, **1959**.
27. Le Houérou, H.N. The Sahara from the Bioclimatic View Point. Definition and Limits. Ann. of Arid Zones **1995**, *34* (1), 1–16.
28. Le Houérou, H.N. Bioclimatologie Et Biogéographie Des Steppes Arides Du Nord De l'Afrique. *Options Mediterranéennes*; Sér. B (10): 1–396, CIHEAM: Paris, 1995.
29. Verrecchia, E.; Yaïr, A.; Kidron, G.J.; Verrecchia, K. Physical Properties of the Psammophile Cryptogarnic Crust and Their Consequences to the Water Regime of Sandy Soils, North-Western Negev Desert, Israel. J. Arid Environments **1995**, *29* (4), 427–438.
30. Le Houérou, H.N. Dégradation, Régénération Et Mise En Valeur Des Terres Séches D'Afrique. In *L'homme Peut-il Refaire ce Qu'il a Défait?* Pontanier, R., M'Hiri, A., Aronson, J., Akrimi, N., Le Floc'h, E., Eds. John Libbey Eurotext: Montrouge, 1995; 65–104.
31. Goldammer, J.G., Ed. Fire in Tropical Biota. Ecological Studies no 84, Spring Verlag: Heidelberg, 1990.
32. Le Houérou, H.N. A Probabilistic Approach to Assessing Arid Rangelands Productivity, Carrying Capacity and Stocking Rates. In *Drylands: Sustainable Use of Rangelands into the Twenty-First Century*; Squires, V.R., Sidahmed, A., Eds.; Ch. 12, IFAD: Rome, 1998; 159–172.
33. Le Houérou, H.N., Aspects méteorologiques De la Croissance et du Dévelopment Végétal Dans Les Déserts et Les Zones Menacées De Désertisation. UNEP, Nairobi and WMO, Report no WMO/TD-No 194., Geneva, 1987.
34. Le Houérou, H.N. The Role of Cacti (*Opuntia* Spp.) in Land Rehabilitation in the Mediterranean Basin. J. of Arid Environments **1996**, *32*, 1–25.
35. Le Houérou, H.N. Utilization of Fodder Trees and Shrubs for the Arid and Semiarid Zones of West Asia and North Africa. Arid Soil Research and Rehabilitation **2000**, *14*, 101–135.
36. Le Houérou, H.N. Restoration and Rehabilitation of Arid and Semiarid Mediterranean Ecosystems in North Africa and West Asia. A Review Arid Soil Research and Rehabilitation **2000**, *14*, 3–14.
37. Roose, E., Introduction à la Gestion Conservatoire de L'eau, de la Biomasse et de la fertilité Des Sols (GCES). Bulletin Pédologique no 70, FAO, Rome, 1994.
38. Le Houérou, H.N. Man-Made Deserts: Desertization Process and Threats. Arid Land Research and Management **200** *16*, 1–36.

DIAGNOSTIC HORIZONS

E.M. Bridges
Formerly of the International Soil Reference and Information Centre, Wageningen, The Netherlands

INTRODUCTION

In layman's terms, the concept of topsoil and subsoil is widely understood, but pedologists have built upon this idea and given it a scientific basis, using the term soil horizons for these layers of soil. A soil horizon may be defined as a layer of soil lying approximately parallel to the Earth's surface, having pedological characteristics. A soil profile includes the collection of soil horizons revealed in a section of the soil downward from the surface. As soil horizons develop by means of the processes of soil formation acting upon weathered geological materials, they are often referred to as genetic horizons. Definition of these genetic horizons was effectively qualitative in approach, reflecting a perceived means of genesis. Toward the end of the 19th century and throughout the first part of the 20th century, various systems of nomenclature arose in different countries allocating letters O, A, B, C, etc.; to genetic horizons, but their use was not uniformly applied and so they became discredited in the eyes of some authorities. Such shorthand designations are "an interpretive symbol," based upon horizon morphology and implied genesis that are used to identify and label a soil horizon (1). In the U.S., as demonstrated in Refs. (10, 12), certain organic O horizons and mineral A, B, and C horizons were designated master horizons and were in vogue for the description of soil profiles. A similar system was developed in the then Soviet Union as described in Pochvennia Syemka (15), and many other countries followed their example. Although the ABC system of labeling worked reasonably well in the areas where it developed, e.g., Russia, U.S., soil scientists working in tropical areas, where soils had a longer and more complex history of development, found it difficult to use (6, 17), who strived to develop systems of their own. Prior to the 1960s, the presence of certain horizons was essential for the classification of soils; podzols, for example, had to possess a bleached horizon and below it a horizon of accumulation of organic matter, iron and aluminum, and gley soils had to have the features of hydromorphism. As the definitions of soil horizons in the early 20th century were cast in general, rather than in specific, quantitative terms, there were overlaps and gaps that caused confusion in the recognition of taxonomic units employed in soil classification systems.

The idea of diagnostic horizons first appeared in the 6th Approximation of 1957 (9), when the earlier concept of soil horizons was taken and developed into a basis for the quantitative identification of horizons and their unambiguous incorporation into systems of soil classification.

DIAGNOSTIC HORIZONS IN THE USDA SOIL TAXONOMY

From 1951, soil scientists in the U.S. began to develop a new approach to soil classification and after a number of preliminary drafts, the 7th Approximation was published as a discussion document. In this approach to soil classification, named diagnostic horizons were selected as a key to the classification system. Originally, six surface diagnostic horizons, called epipedons, and 12 subsurface diagnostic horizons were proposed. Although the conceptual basis of what was a diagnostic horizon or how it was chosen were not defined or described, it amounted to a soil horizon, or group of closely related horizons that possess a set of quantitatively defined properties. By fulfilling their criteria, and with the use of a key, soils may be placed in their correct position in one of the 10 orders of the system.

The quantitative definition of these diagnostic horizons is such that they cannot be related in all cases to the master horizons of other systems of genetic horizon designation. In discussion about the development of the Soil Taxonomy system, Smith (9) explained that the concept of master horizons and the ABC system was not used in Soil Taxonomy because of a lack of agreement about its use among pedologists. Therefore, he adopted the diagnostic horizon concept, which did not carry with it any inherited controversy. The diagnostic horizons proposed in the 7th Approximation are presented in Table 1.

Since its publication in 1975, Soil Taxonomy has been continually updated and improved through correlation meetings and advice received from soil scientists around the world. As a result, two new epipedons have been introduced: a Melanic epipedon to provide for the surface

Table 1 Diagnostic horizons of the USDA 7th Approximation

Epipedons	Subsurface horizons	Other horizons
Mollic	Agric, Albic, Anthropic, Argillic	Calcic
Umbric	Cambic	Gypsic
Histic	Natric	Salic
Ochric	Oxic,	Pans: Duripan, Fragipan
Plaggen	Spodic	

(From Ref. 11.)

Table 2 Reference horizons proposed by FitzPatrick

Alkalon	Fermenton	Kastanon	Proxon
Alton	Ferron	Krasnon	Pseudofibron
Anmorphon	Fibron	Kuron	Rosson
Andon	Flambon	Limon	Rubon
Anmooron	Flavon	Lithon	Rufon
Arenon	Fragon	Litter	Sapron
Argillon	Gelon	Luton	Seron
Buron	Gleyson	Luvon	Sesquon
Calcon	Glosson	Marblon	Sideron
Candon	Gluton	Minon	Solon
Celon	Gypson	Modon	Sulphon
Cerulon	Gyttjon	Mullon	Tannon
Chernon	Halon	Oron	Thion
Chloron	Hamadon	Pallon	Verton
Clamon	Hudepon	Pelon	Veson
Crumon	Humifon	Pesson	Zhelton
Cryon	Husesquon	Placon	Zolon
Cumulon	Hydromoron	Planon	
Dermon	Ison	Primon	

(From Ref. 6.)

horizons of soils on tephra and volcanic glass, and a Folistic epipedon to provide for organic soil materials, saturated for more than 30 days and consisting of more than 75% sphagnum. Six of the subsurface diagnostic horizons that first appeared in the 7th Approximation remain with improved definition in Soil Taxonomy. These are the agric, argillic, cambic, natric, oxic, and spodic horizons, together with horizons referred to as pans: duripan and fragipan. "Other horizons" that have persisted throughout the development of Soil Taxonomy are the calcic, gypsic, and salic horizons. New horizons that have been introduced are the glossic (to provide for degraded argillic) kandic, and natric horizons; the kandic to provide for a more clayey subsurface horizon with an apparent CEC of 16 cmol or less per kg of clay; orstein has been separated from the spodic horizon, and placic has been introduced for thin iron pans. Petrocalcic and Petrogypsic have been introduced for carbonate and gypsic-cemented horizons, respectively. Sombric caters for dark-colored, freely drained subsurface horizons with illuvial humus. Additionally, there is a sulfuric horizon with low pH and containing jarosite. Other diagnostic soil characteristics, e.g., slickensides and lamellae, and soil conditions, e.g., andic soil properties, are strictly defined. Full definitions of all these horizons and characteristics are given in Soil Taxonomy (14) and summaries are provided in Ref. (13) and various textbooks. Since the development of Soil Taxonomy, the concept of diagnostic horizons has been widely used as the basis of several national systems of soil classification (e.g., Brazil, South Africa), and in the legend for Ref. (2).

REFERENCE HORIZONS OF FITZPATRICK

In parallel with developments in the U.S., FitzPatrick (6, 7, 18) independently proposed a system of classification with a greater number of defined horizons, called reference horizons. The soil is viewed as a three-dimensional continuum, whereas a soil profile is in reality only a two-dimensional section through the soil. As the soil reflects the conditions in which it was formed, most importantly the effects of climate, parent material, and relief, a pattern of soils occurs over the landscape that is three-dimensional. The boundaries between soils are rarely abrupt, but are zones of gradual change. Thus, it is necessary to consider soils as three-dimensional entities that vary in space and time. Within this three-dimensional system, both conceptual space and observed field characteristics are accommodated in reference segments where there is a single unique dominating property or combination of properties formed principally by a single set of processes. Intergrade segments contain properties that gradually change between two reference segments. Within each segment, it is possible to recognize reference horizons, intergrade horizons, compound horizons, and composite horizons. The reference horizons proposed are presented in Table 2.

THE FAO–UNESCO SOIL MAP OF THE WORLD LEGEND

The FAO–UNESCO Soil Map of the World was completed during the decade 1971–1981 (2, 3) with the aim of making a first scientific appraisal of the soil resources of the world, providing a basis for transfer of knowledge, and promoting a generally accepted system of

Table 3 Diagnostic horizons used in the legends of the Soil Map of the World

Histic H horizon	Albic E horizon	Calcic horizon
Mollic A horizon	Argillic B horizon	Gypsic horizon
Umbric A horizon	Cambic B horizon	Sulfuric horizon
Ochric A horizon	Natric B horizon	
Spodic B horizon		
Oxic B horizon		

(From Ref. 2.)

soil classification, especially for educational, research, and development activities. Following the use of diagnostic horizons in Soil Taxonomy, a similar limited number of horizons with measurable and observable properties was employed by the editors of the 1974 legend of the Soil Map of the World (Table 3).

Reflecting the increasing impact of human activity in modifying, in particular, the A horizon of soils, a fimic A horizon was added in the 1988 revision to provide for horizons with a man-made surface layer 50 cm or more thick, produced by long continued manuring with earthy mixtures. It commonly contains artifacts such as brick and pottery and includes the plaggen epipedon and the anthropic epipedon of the Soil Taxonomy system. The difficulties, which had been experienced in the field, especially in tropical regions, where oxic and argillic horizons were difficult to separate, resulted in a redefinition into the argic and the ferralic B horizons, respectively. The petrocalcic and petrogypsic phases became diagnostic horizons and other definitions were amended, but Calcic, Gypsic, Histic A, Mollic A, Ochric A, Spodic B, Sulfuric, and Umbric A horizon definitions were unchanged. As with the system of Soil Taxonomy, a number of diagnostic properties, which are not considered as horizons, are also used in the definitions of the diagnostic horizons. These refer to andic soil material and ferralitic, ferric, fluvic, hydromorphic, nitic, and vertic properties. With their sets of quantitatively defined properties, produced by the soil-forming processes, the use of diagnostic horizons has made it possible to base a classification on general principles of soil genesis but objectivity was retained because the processes themselves are not used as criteria, only their effects expressed in terms of morphological properties, which have measurable values.

THE WORLD REFERENCE BASE OF THE IUSS

Within the international community, it has long been appreciated that there has not been a generally accepted system of soil classification. In 1982, the International Society of Soil Science initiated the program to develop an international reference base for soils, but it languished until 1992. In that year, the decision was taken to make a world reference base (WRB) by developing the revised (1988) FAO–UNESCO legend for the Soil Map of the World (4). In this way, advantage could be taken of the international correlation work that had taken place since the Soil Map of the World had been published.

The objectives of WRB were to develop the FAO–UNESCO map legend into a comprehensive, internationally acceptable system for delineating soil resources and to provide a sound scientific basis to support it. The WRB is based upon morphological characterization of soils and is supported by laboratory analyses where necessary. Its authors claim that it provides an easy means of communication about soils, including the transfer of data and technology relevant to soils both for soil scientists and for related environmental fields. As with other soil classification systems it purports to show the relationships within and between the different soils of the world.

WRB adopted the use of diagnostic horizons as reflecting the results of widely recognized processes of soil development, and improved the FAO definitions in terms of morphological characteristics and/or analytical criteria. Of the 16 diagnostic horizons of the 1988 FAO–UNESCO revised legend, the fimic A horizon has been discarded and replaced by hortic, plaggic, and terric horizons, respectively. The histic horizon has been broadened as its minimum thickness has been reduced to 10 cm and has now been used to define Histosols. Experience with Chinese soils has shown that the former P_2O_5 requirement for the mollic and umbric A horizons of the revised FAO–UNESCO system is untenable as a criterion. The unique properties of Chernozems were not effectively reflected in the definition of the mollic A horizon and so a chernic horizon has been introduced for the deep, blackish, porous surface horizons typical of Chernozems. Increasing the percentage of clay skins from 1 to 5% on both horizontal and vertical ped faces and pores is expected to give an improved correlation between field and micromorphological evidence.

The cambic horizon has always suffered from being a negative concept, without many pedological criteria. Its positive points such as depth, texture, structure, color, and presence of weatherable minerals, etc., have now given it a more positive image. In the case of the Spodic horizon, it has been supplemented with recent (1996) developments in Soil Taxonomy. In addition to the diagnostic horizons mentioned, an additional 19 new horizons have been proposed. A full list is given in Table 4 and in the World Reference Base for Soil Resources (5, 8).

D

Table 4 Diagnostic horizons of the World Reference Base for Soil Resources

Albic	Ferric	Mollic	Salic
Andic	Folic	Natric	Spodic
Anthraquic	Fragic	Nitic	Sulfuric
Argic	Fulvic	Ochric	Takyric
Calcic	Gypsic	Petrocalcic	Terric
Cambic	Histic	Petroduric	Umbric
Chernic	Hydragric	Petrogypsic	Vertic
Cryic	Hortic	Petroplinthic	Vitric
Duric	Irragric	Plaggic	Yermic
Ferralic	Melanic	Plinthic	

(From Refs. 5, 8.)

The adoption of the concept of diagnostic horizons by American Soil Survey staff has led to their acceptance in systems of soil classification in many countries of the world. Subsequent refinements and work for the World Reference Base indicate the necessity for a greater number of diagnostic horizons to cover the nature and range of soil morphological features.

The use of diagnostic horizons, focused upon observable and measurable soil morphology, rather than ill-defined genetic processes, helped greatly in reaching agreement for the World Reference Base as a single international system of soil classification. In the revised legend for the Soil Map of the World (4), it is stated that the use of diagnostic horizons has proved to be most appropriate. In both theory and practice it has been one of the most useful developments in soil classification.

REFERENCES

1. Bridges, E.M. *Soil Horizon Designations*; Technical Paper 19, International Soil Reference and Information Centre: Wageningen, 1990.
2. FAO *FAO–UNESCO Soil Map of the World, 1:5,000,000, Legend*; UNESCO: Paris, 1974; Vol. I.
3. FAO, *FAO–UNESCO Soil Map of the World, 1:5,000,000*; UNESCO: Paris, 1971–1981; Vols. 2–10.
4. FAO, *FAO–UNESCO Soil Map of the World, 1:5,000,000, Revised Legend*; World Soil Resources Report No. 60, FAO: Rome, 1988.
5. FAO, *World Reference Base for Soil Resources*; World Soil Resources Report No. 84, (see also Ref. 8) FAO: Rome, 1998.
6. FitzPatrick, E.A. Soil Nomenclature and Classification. Geoderma **1967**, *1*, 91–105.
7. Fitzpatrick, E.A. *Soil Horizon Designation and Classification*; Technical Paper 17, International Soil Reference and Information Centre: Wageningen, 1988.
8. ISSS Working Group RB, *World Reference Base for Soil Resources*; Vol. 1. Introduction (Deckers, J.A., Nachtergaele, F.O., Spaargaren, O.C., Eds.), Vol. 2. Atlas (Bridges, E.M., Batjes, N.H., Nachtergaele, F.O., Eds.), (see also Ref. 5 for Full Technical Details) ISRIC/FAO: Acco, Leuven, 1998.
9. Smith, G.D. *The Guy Smith Interviews: Rationale for Concepts in Soil Taxonomy*; Technical Monograph No. 11, Soil Management Support Services, Soil Conservation Service, USDA/Agronomy Department, Cornell University: Washington, DC, 1986.
10. Soil Survey Staff, *Soil Survey Manual*; Agriculture Handbook 18, USDA: Washington, DC, 1951.
11. Soil Survey Staff, *Soil Classification: A Comprehensive System (7th Approximation)*; Soil Conservation Service, USDA: Washington, DC, 1960.
12. Soil Survey Staff, *Supplement to Soil Survey Manual*; Agriculture Handbook 18, USDA: Washington, DC, 1962; 173–188.
13. Soil Survey Staff, *Keys to Soil Taxonomy*; USDA–Natural Resources Conservation Service: Washington, DC, 1998.
14. Soil Survey Staff, *Soil Taxonomy. A Basic System of Soil Classification for Making and Interpreting Soil Surveys*, 2nd Ed.; Agriculture Handbook 436; USDA–Natural Resources Conservation Service: Washington, DC, 1999.
15. Tiurin, L.V., Gerasimov, I.P., Ivanova, E.N., Nosin, V.A., Eds. *Pochvennia Syemka*; Akademia Nauk: Moscow, 1959.
16. Nye, P.H. Some Soil-Forming Processes in the Humid Tropics. Journal of Soil Science **1954**, *5*, 7–21.
17. Watson, J.P. A Soil Catena on Granite in Southern Rhodesia. 1. Field Observations. Journal of Soil Science **1964**, *15*, 238–250.
18. FitzPatrick, E.A. *Soils: Their Formation, Classification and Distribution*; Longman: London and New York, 1980.

DIURNAL FREEZE/THAW: TEMPERATURE AND WATER DISTRIBUTION

Joseph L. Pikul, Jr.
United States Department of Agriculture-Agricultural Research Service (USDA-ARS), Brookings, South Dakota, U.S.A.

INTRODUCTION

In northern climates throughout the world, periodic freezing and thawing of soil is common. Sharratt et al. (1) estimated that approximately 30% of the earth's landmass is periodically frozen, and that an additional 20% of the landmass is underlain by permafrost. Some of the world's most productive soil lies within regions where soil may be seasonally or diurnally frozen. Freezing and thawing of soil exerts a profound effect on soil water distribution, solute movement, soil physical and chemical condition, biological processes, and hydrology (1–4). There are few land-management options available to lessen the effect of freezing weather on the soil thermal environment. However, maintaining surface cover can reduce the incidence and penetration depth of frozen soil (5) on croplands. This article discusses the diurnal soil freeze–thaw and soil–water movement associated with nighttime radiative freezing of the soil surface.

SOIL FREEZING AND WATER MOVEMENT

Nocturnal soil freezing and daytime thawing cycles are numerous in late autumn and early spring (6). About 100 diurnal freezing and thawing cycles, recorded at a 1-cm depth, have been observed in a year in mountain regions (7). Nighttime freezing, termed *radiation frost*, (8) is a consequence of rapid loss of long-wave radiation from the earth to the atmosphere on cloudless nights.

When moist soil begins to freeze, as during a radiation frost, water migrates upward to the freezing front (surface) and is held there as ice. During subsequent daytime thaw, near-saturated soil conditions may develop at the surface (9). Differences in soil water potential within the soil profile are responsible for water movement within soil. Water moves from regions of high water potential (wet soil) to low water potential (dry soil). How fast water moves within soil depends upon the water potential gradient (driving force) and the ability of soil to transmit water (hydraulic conductivity). In the case of soil freezing, very large water potential gradients develop as a consequence of rather small temperature gradients.

As the surface cools and soil temperature dips below 0°C, a fraction of the water freezes within the soil pores (10). Ice particles remain separated from solid particles by a film of water; the thickness of which depends upon temperature, pore size distribution, solutes in pore water, and freezing and thawing history (11). Ice formation in soil pores effectually dries the soil and results in a reduction of water potential. A water potential gradient is then established across the freezing front, creating the conditions whereby water flows from unfrozen soil toward the freezing front. Water flows towards the region of low potential.

Repeated saturation and subsequent drying of the soil surface, as occurs during diurnal freeze and thaw cycles, may accelerate surface crusting. However, identification of direct consequences of freeze–thaw on soil structural stability is confounded due to soil type, soil water content at freezing, and characteristics of cyclic freezing and thawing (12).

FIELD STUDIES

Microenvironment created by the presence or absence of surface residue strongly affects surface energy exchange and, consequently, soil thermal environment. Pikul and Allmaras (9) revealed effects of crop residue cover on soil water movement during consecutive freezing and thawing cycles. Their field studies included both residue-covered and bare plots. Water redistribution during one freeze–thaw cycle described by Pikul and Allmaras (9) is illustrated in Fig. 1. The soil did not freeze on plots with residue cover, but froze to about 1.5 cm on plots having a bare surface. Soil water content under residue cover remained essentially unchanged throughout the night (Fig. 1A). On bare plots, soil water content of the surface 0.5-cm layer (Fig. 1A) increased from 0.28 kg kg^{-1} at 2400 h to 0.45 kg kg^{-1} at 0600 h. Soil water content (Fig. 1A) includes both liquid

Encyclopedia of Soil Science

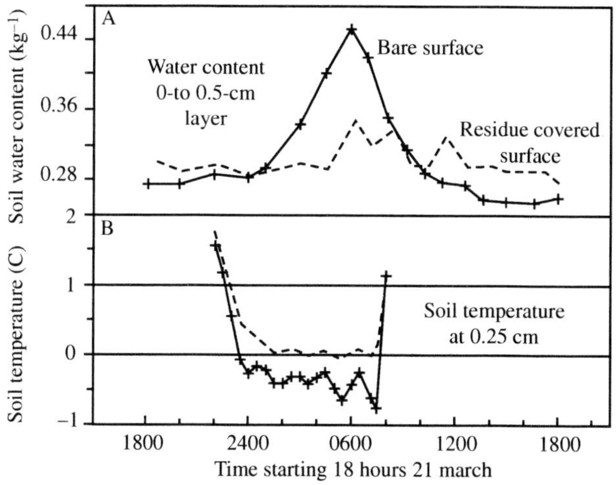

Fig. 1 Water redistribution during one freeze–thaw cycle: (A) soil water content (ice and liquid) of the surface- to 0.5-cm depth; (B) soil temperature at 0.25 cm. Soil froze to a depth of 1.5 cm on the bare surface treatment, but did not freeze under residue cover.

and ice. The quantity of water redistributed, as a consequence of soil freezing, depends on soil water content at the onset of soil freezing (13). In this example, both the bare surface and residue-covered plots entered the freezing cycle at about the same water content (Fig. 1A). Therefore, the sharp increase in soil water on the bare surface plot was a result of steep water potential gradients that developed as a consequence of soil water freezing.

Concurrent with the decrease in soil temperature to below freezing (Fig. 1B), was an increase in soil water content in the bare surface plot (Fig. 1A). The onset of soil freezing in the bare surface plot was at 23:30 h, and significant water redistribution began shortly thereafter. Minimum soil temperature at 0.25 cm was −0.8°C. Soil temperature in residue-covered plots hovered near 0.0°C (Fig. 1B) during the night.

SIMULATION STUDIES

Heat and water flow in freezing soil have been modeled using several approaches with varying degrees of complexity. All models require detailed knowledge of physical, chemical, hydrologic, and thermal properties of the soil. Generally, the most complex models include provisions for frost heave and overburden pressures (14). However, heat and water flow during daily freeze and thaw cycles have been successfully simulated by excluding frost heave (15). Cary (16) provided an example of a simulation model that included solute transport and frost heave. The Simultaneous Heat and Water (SHAW) model (17), one of the more detailed models of snowmelt and soil freezing, has been used to simulate soil freezing, thawing, snowmelt, and water runoff.

CONCLUSION

Diurnal freezing and thawing of soils is a worldwide phenomenon impacting many biological and physical aspects of agricultural soils. Available management options to reduce soil-freezing events may be limited to manipulation of crop residues. Surface cover creates a unique microclimate, affording thermal protection to soil and thereby reducing the incidence and severity of frozen soil. Reduction of soil-freezing events could be expected to reduce soil water evaporation, soil surface crusting, and undesirable movement of agricultural chemicals.

REFERENCES

1. Sharratt, B.S.; Radke, J.K.; Hinzman, L.D.; Iskandar, I.K.; Groenevelt, P.H. Physics, Chemistry, and Ecology of Frozen Soils in Managed Ecosystems: An Introduction. In *Proceedings of the International Symposium on Physics, Chemistry, and Ecology of Seasonally Frozen Soils*, Fairbanks, Alaska, June 10–12, 1997; Iskandar, I.K., Wright, E.A., Radke, J.K., Sharratt, B.S., Groenevelt, P.H., Hinzman, L.D., Eds.; CRREL Special Report 97–10; U.S. Army Cold Regions Research and Engineering Lab: Hanover, New Hampshire, 1997; 1–7.
2. Williams, P.J. The Seasonally Frozen Layer—Geotechnical Significance and Needed Research. In *Proceedings of the International Symposium on Physics, Chemistry, and Ecology of Seasonally Frozen Soils*, Fairbanks, Alaska, June 10–12, 1997; Iskandar, I.K., Wright, E.A., Radke, J.K., Sharratt, B.S., Groenevelt, P.H., Hinzman, L.D., Eds.; CREEL Special Report 97–10; U.S. Army Cold Regions Research and Engineering Lab: Hanover, New Hampshire, 1997; 9–15.
3. Ellsbury, M.M.; Pikul, J.L., Jr.; Woodson, W.D. A Review of Insect Survival in Frozen Soils with Particular Reference to Soil Dwelling Stages of Corn Rootworms. Cold Reg. Sci. Technol. **1998**, *27*, 49–56.
4. Kane, D.L. Snowmelt Infiltration into Seasonally Frozen Soils. Cold Reg. Sci. Technol. **1980**, *3*, 153–161.
5. Pikul, J.L., Jr.; Zuzel, J.F.; Greenwalt, R.N. Formation of Soil Frost as Influenced by Tillage and Residue Management. J. Soil Water Conserv. **1986**, *41*, 196–199.
6. Hershfield, D.M. The Frequency of Freeze-Thaw Cycles. J. Appl. Meteorol. **1974**, *13*, 348–354.
7. Fahey, B.D. An Analysis of Diurnal Freeze-Thaw and Frost Heave Cycles in the Indian Peaks Region of the Colorado Front Range. Arctic Alpine Res. **1973**, *5*, 269–281.
8. Oke, T.R. *Boundary Layer Climates*; Methuen and Co.: London, 1978.
9. Pikul, J.L., Jr.; Allmaras, R.R. Hydraulic Potential in Unfrozen Soil in Response to Diurnal Freezing and Thawing of the Soil Surface. T. ASAE. **1985**, *28*, 164–168.

10. Farouki, O.T. *Thermal Properties of Soils*, CRREL Monograph 81-1; U.S. Army Cold Regions Research and Engineering Lab: Hanover, New Hampshire, 1981.
11. Anderson, D.M.; Tice, A.R. Predicting Unfrozen Water Contents in Frozen Soils from Surface Area Measurements. Highw. Res. Rec. **1972**, *393*, 12–18.
12. Dagesse, D.F.; Groenevelt, P.H.; Kay, B.D. The Effect of Freezing Cycles on Water Stability of Soil Aggregates. In *Proceedings of the International Symposium on Physics, Chemistry, and Ecology of Seasonally Frozen Soils*, Fairbanks, Alaska, June 10–12, 1997 Iskandar, I.K., Wright, E.A., Radke, J.K., Sharratt, B.S., Groenevelt, P.H., Hinzman, L.D. Eds.; CREEL Special Report 97-10; U.S. Army Cold Regions Research and Engineering Lab: Hanover, New Hampshire, 1997; 177–181.
13. Cary, J.W.; Papendick, R.I.; Campbell, G.S. Water and Salt Movement in Unsaturated Frozen Soil: Principles and Field Observations. Soil Sci. Soc. Am. J. **1979**, *43*, 3–8.
14. O'Neill, K. The Physics of Mathematical Frost Heave Models: A Review. Cold Reg. Sci. Technol. **1983**, *6*, 275–291.
15. Pikul, J.L., Jr.; Boersma, L.; Rickman, R.W. Temperature and Water Profiles During Diurnal Soil Freezing and Thawing: Field Measurements and Simulation. Soil Sci. Soc. Am. J. **1989**, *53*, 3–10.
16. Cary, J.W. A New Method for Calculating Frost Heave Including Solute Effects. Water Resour. Res. **1987**, *23*, 1620–1624.
17. Flerchinger, G.N.; Seyfried, M.S. Modeling Soil Freezing and Thawing, and Frozen Soil Runoff with the SHAW Model. In *Proceedings of the International Symposium on Physics, Chemistry, and Ecology of Seasonally Frozen Soils*, Fairbanks, Alaska, June 10–12, 1997 Iskandar, I.K., Wright, E.A., Radke, J.K., Sharratt, B.S., Groenevelt, P.H., Hinzman, L.D., Eds.; CRREL Special Report 97-10; U.S. Army Cold Regions Research and Engineering Lab: Hanover, New Hampshire, 1997; 537–543.

DRAINAGE, AERATION, AND TRAFFICABILITY

Norman R. Fausey
United States Department of Agriculture (USDA), Columbus, Ohio, U.S.A.

INTRODUCTION

Soil drainage is a natural process by which water moves through and out of the soil as a result of the force of gravity. This natural process provides the water that supports seeps, springs, stream base flow, and aquifer recharge. As water leaves the soil, air moves into the space previously occupied by the water. This process is called aeration. Soil aeration is vital for plant roots and for many beneficial organisms that live in the soil and require oxygen for respiration. As the proportion of water and air in the soil changes as a result of drainage, the ability of the soil to provide support and traction for animals and vehicles (trafficability) is altered because the strength of the soil changes with water content. The natural drainage of the soil can be accelerated by the use of surface and subsurface drainage practices. Surface drainage diverts excess water from the soil surface directly to streams, thereby reducing the amount of water that will move into and possibly through the soil. Subsurface drainage, provided by ditches and drain pipes, collects and diverts water from within the soil directly to streams.

DRAINAGE (ACCELERATED DRAINAGE)

A complete modern reference for drainage is provided by the American Society of Agronomy (1). This monograph provides a comprehensive treatment of the need for, and consequences of, drainage as well as the methods and materials used for drainage, and the design of drainage systems.

Soil is comprised of between 50 and 70% solid material. The solid phase of the soil is predominately mineral particles in varying proportions of sand, silt, and clay sizes. A small amount of the solid phase is organic material, and this amount decreases with depth. [While the amount of the soil organic matter is quite small on a weight basis, its impact on drainage, aeration, and trafficability is substantial (2).] This leaves 30–50% of the soil volume that is comprised of void space that can be filled with either air or water. The relative amounts of air and water that are present in the soil determine the ability of the soil to bear the weight of machines and animals, provide traction, sustain the biological life in the soil, and maintain productive and healthy plants. Drainage plays an important role in managing the relative amounts of air and water that are present in the soil.

FACTORS THAT AFFECT SOIL–AIR–WATER RELATIONSHIPS

Bulk Density

The greater the proportion of mineral particles to void spaces, the higher the bulk density. Rocks, boulders, stones, sand grains, silt particles, and even clay particles all typically have a specific weight (density), compared to water, in the range of 2.6 to 2.7 $Mg\,m^{-3}$ because they are all made up of mineral particles. If you were to pour dry soil into a bucket, with the individual soil particles having a specific density of 2.65 $Mg\,m^{-3}$, the overall or bulk density of the soil in the bucket would typically be about 1.4 $Mg\,m^{-3}$ because there are void spaces between the soil particles. The bulk density of soil varies depending on the parent material from which it formed, the organic matter content, the depth below the soil surface, and the soil management. Certain processes and activities can decrease the bulk density and thereby create more void space in the soil, while other processes and activities can increase the bulk density and cause loss of void space. The growth of ice crystals in the soil during freezing can result in soil heaving and a decrease in bulk density. Cultivation loosens soil, resulting in more voids and a lower bulk density. Heavy machine traffic compacts the soil, resulting in fewer voids and a higher bulk density.

Soil Texture

Soil texture refers to the relative amounts of sand, silt, and clay size mineral particles in the soil. Sand particles are the largest, ranging from 0.05 to 2.0 mm equivalent diameter; silt particles are smaller, ranging from 0.002 to 0.05 mm equivalent diameter; and clay particles are

Encyclopedia of Soil Science

smallest, less than 0.002 mm equivalent diameter. Equivalent diameter is used to characterize the size because soil particles are not spherical but rough and random in shape. The clay size mineral particles have unique and important characteristics including an electrical charge and tremendous surface area per unit volume. Most of the electrical charges on clay size particles are negative, causing the particles to repel each other like similar magnetic poles and establishing void spaces between the particles. When considering the sand and silt particles, a useful illustration is to consider how many basketballs (representing sand particles) can fit into a bathtub. Between the basketballs there are large void spaces. Then consider how many marbles (silt particles) would fit into the same bathtub. The void spaces between the marbles are smaller, but the total volume of the void space is similar in both cases. By putting small particles into the large voids between the large particles, it is possible to end up with more solids and less voids. This happens in soils where the small silt and smaller clay size particles fill the spaces between the large sand size particles. Even when smaller particles occupy the spaces between larger particles, some void space will remain—usually 30 to 40% of the total space will remain unfilled. However, all the void spaces will be small.

Sandy soils tend to have uniformly distributed large void spaces, resulting in rapid movement of fluids through the soil. Silty soils tend to have uniformly distributed small void spaces and slow movement of fluids. Clay soils tend to have only very small void spaces and very slow movement of fluids.

Soil Structure

Mixtures of the sand, silt, and clay size particles and the organic matter in the soil become bonded together—a result of the electrical charges on the clays and sticky organic materials in the soil—forming larger units that have a visible structure and are referred to as aggregates. Because they are large units, these aggregates can be thought of again as larger size spheres having larger voids between them. Within the aggregates there are small voids; between the aggregates, the voids are larger. Aggregates are dynamic rather than permanent arrangements of particles. Accelerated drainage tends to reduce the general size of aggregates, but increase their stability. Any soil tillage activities tend to disrupt aggregates leading to smaller voids. Water moves more rapidly through soils with durable aggregates or structural units, and these soils will also typically retain more water that can be utilized by crop plants.

Pore Size Distribution

The interconnected void spaces in soil are typically referred to as pores. Most of these pores are small and can be visualized as capillary tubes, while some pores are as large as drinking straws. Pores are not usually straight and uniform in diameter, however, being more tortuous as they twist and turn around particles and through or between aggregates. The flow and retention of fluids (air and water) in pores is ultimately what controls the drainage, aeration, and trafficability of the soil. Fluids move more easily through larger pores than through small and more tortuous pores. Burrowing insects or earthworms usually create the largest pores, called macropores. Well-aggregated soils and sandy soils also have large, well-connected pores that transmit air and water easily. Silt and clay soils with many small pores experience slow movement of fluids and entrapment of air within pores during rewetting events. Soils with a range of pores sizes, some macropores, some large pores between aggregates, and some smaller pores within aggregates are usually better suited for agricultural use and respond economically to accelerated drainage installation.

AERATION

Aerobic respiration of plant root cells and microorganisms in the soil is dependent upon an adequate supply of oxygen. This oxygen is supplied to a great extent by air moving into the soil through the soil pores. When it rains, some of the rainwater enters the soil through the soil pores and displaces the air from the pores. If enough rainwater enters the soil, all the pores become filled with water and the soil becomes saturated. In the saturated pores, air movement is blocked and the oxygen needed for respiration by plant roots and microorganisms cannot be supplied through the soil pores.

Under natural drainage conditions, some of the water that saturates the soil may slowly percolate downward through the soil to great depths or may leave the soil by evaporation. In some soils the natural drainage of water occurs rapidly enough to prevent the death of plant roots and microorganisms from lack of oxygen. In other soils, accelerated drainage is needed to provide a means to get the water out of the pores (and get oxygen in) quickly enough so that the plant roots and microorganisms can survive.

TRAFFICABILITY

The ability of the soil to support the weight of machines and animals and provide traction for moving about on the soil is called trafficability. Trafficability is dependent on both bearing strength and slippage. The surfaces of soil particles are typically not smooth and friction develops between soil particles when force or weight is applied, such as when a machine drives across the surface. Organic matter also increases the resistance of the soil to deformation and increases elasticity (2). It is this friction that gives the soil (bearing) strength to support the weight of the machine and (shear) strength to move the machine forward as the wheels turn. Water acts as a lubricant between soil particles, reducing the friction and allowing the soil particles to move past one another when force or weight is applied. This results in a loss of strength and poor trafficability. With the loss of strength, sinkage and slippage occur in response to the weight of the machine and the rotational force on the tires. Similar sinkage and slippage occurs under the feet of animals and humans. Drainage accelerates the removal of water from the soil pores between the soil particles, and allows friction between soil particles to occur so that the bearing and shear strength can support trafficability.

Trafficability can also be impeded by stickiness whereby clay particles stick to the tires of machines or the feet of animals and humans. This phenomenon occurs in very wet but not saturated soils with a high proportion of clay size particles.

CONCLUSION

Accelerated drainage, using ditches or drain pipes to collect the water that is free to be moved by gravity and to transport this water out of the soil, promotes the desired states of aeration and trafficability that are important for efficient and economical management of soil for agriculture, highways, recreation, and many other uses.

REFERENCES

1. Skaggs, R.W., van Schilfgaarde, J., Eds. *Agricultural Drainage*; Monograph No. 38, American Society of Agronomy: Madison, WI, 1999.
2. Soane, B.D. The Role of Organic Matter in Soil Compactibility: A Review of Some Practical Aspects. Soil and Tillage Research **1990**, *16*, 179–201.

DRAINAGE AND ANAEROBIOSIS

Robert O. Evans
North Carolina State University, Raleigh, North Carolina, U.S.A.

INTRODUCTION

Agricultural crops are obligate aerobes. As such, most plant metabolic processes require oxygen; in particular, aerobic respiration provides energy from the degradation of glucose (derived from sugars and starches in the tissue) to pyruvic acid which is further oxidized to CO_2 and water. Anaerobiosis implies life in the absence of oxygen. Anaerobiosis is most commonly caused by waterlogging of the root zone. Under anaerobic conditions (lack of oxygen), pyruvic acid produced from glucose in the root cells of higher plants is converted to CO_2 and ethyl alcohol. Ethyl alcohol tends to accumulate within root cells, and if anaerobic conditions persist, may accumulate at concentrations that can become toxic. Anaerobiosis thus creates an environment that impairs the growth rate of most plants that require oxygen for respiration. Drainage is the practice of removing excess water from land to minimize the occurrence of waterlogging. It has been an important management practice for centuries (1), and the consequences of poor drainage have been studied extensively over the past century (2–8).

SOIL AERATION

Either air or water occupies the pore volume of the soil. Excess water displaces air and affects the root zone environment and subsequent plant survival, growth, and yield. Processes affected by excess soil water include: soil oxygen and carbon dioxide exchange with the atmosphere, oxidation reduction reactions in the soil and subsequent diffusion of gaseous metabolites, nutrient availability and uptake, and accumulation of carbon dioxide, ethylene, and other toxic substances (Fig. 1). These conditions can impact crop production through such processes as suppressing seed germination, restricting root growth, death of root cells, shoot wilting, chlorosis (yellowing of leaves due to failure of leaves to produce normal amounts of chlorophyll), and slowed growth and dry matter accumulation.

The most basic consequence of excessive soil water is restriction of the supply of oxygen to roots by displacing air in the pore space and by reducing the exchange of soil air with the atmosphere. These processes are often lumped together and referred to as soil aeration (10). Soil aeration status has often been reported on a percentage basis of water-free pore space. Air porosities in the range of 5–20% were reported in earlier literature to be adequate for crop growth. A minimum value of 10% was reported in an early study (11) although the optimum air filled porosity may range from 4 to 50% depending on soil texture, type and age of crop, and other growing conditions (7).

Soil aeration can also be expressed as the oxygen content of air samples extracted from the soil (12). In well-aerated soil, the composition of soil air closely approximates that of the atmosphere, namely, 20% O_2 and 0.03% CO_2. Oxygen contents in the range of 8–14% have been reported to be minimum for unrestricted root respiration; however, critical values are affected by many factors including soil physical properties such as texture and structure, depth in the soil, type and growth stage of crop, soil water content, organic matter present and, of course, temperature. Numerous studies have documented declines in O_2 content to less than 2% within 24–48 h under waterlogged conditions (Fig. 2). The rate of decline of O_2 is closely correlated to metabolic activity of microbes and roots and to soil temperature (13, 14).

The oxygen diffusion rate (ODR) is the rate of exchange of soil air with the atmosphere. At the soil-surface, the composition of soil air is in equilibrium or nearly so with the atmosphere. The O_2 content of soil air generally declines and CO_2 content increases with soil depth resulting from the production of CO_2 during aerobic respiration. At any depth, the rate of exchange of soil air with the atmosphere decreases as soil water content increases (Fig. 3). Under waterlogged conditions, the O_2 content declines to less than 1% and ODR typically approaches zero (12). The exchange of soil air with the atmosphere occurs by two mechanisms: convection (often referred to as mass flow) and diffusion. Early work (16) reported that oxygen transport into the soil by mass flow (convective flow) was insignificant compared to diffusion. The diffusion rate of oxygen is roughly 10,000 times less through water than through contiguous air filled soil pores.

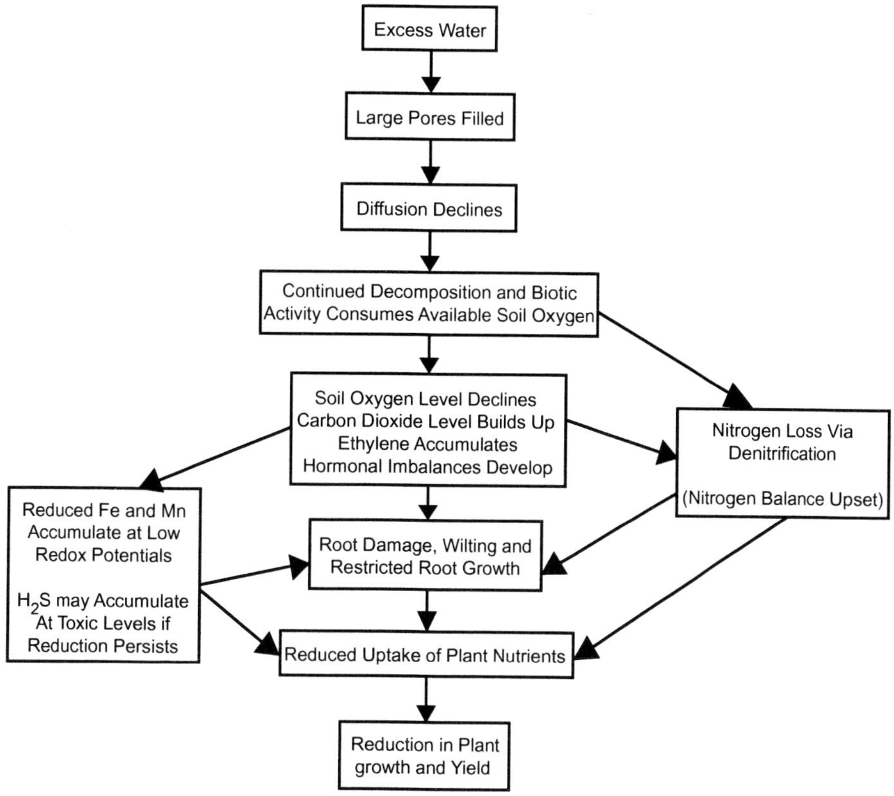

Fig. 1 Processes causing reduction in plant growth and yields due to excessive soil water and anaerobiosis. (Adapted from Refs. 7, 9.)

Recently, convective transport was considered to be important under some conditions such as shallow depths or in soils with large pores (7, 10). Theoretical relationships for describing airflow processes in soil have been reported (7, 12).

Oxygen diffusion rates required for normal plant growth have been reported for several crops. In early work, plants like sunflowers, cotton, and snapdragons stopped growing when the soil oxygen diffusion rate dropped below 3.3×10^{-8} kg/m^2/s (17). Plant functions such as photosynthesis may be suppressed when the ODR drops below 2×10^{-8} kg/m^2/s (18). A value of $6.8-5\times10^{-8}$ kg/m^2/s was found for soybean (19) with a critical value of 3.3×10^{-8} kg/m^2/s. The required rate of diffusion is constantly changing due to constant changes in growing conditions (12).

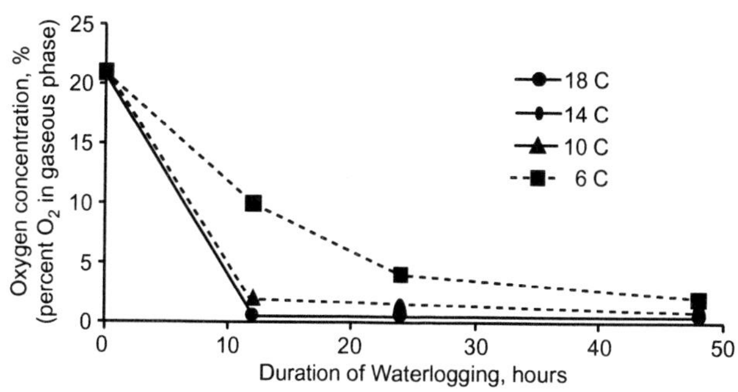

Fig. 2 Oxygen content of soil air as influenced by duration of waterlogging at four soil temperatures. (Adapted from Ref. 13.)

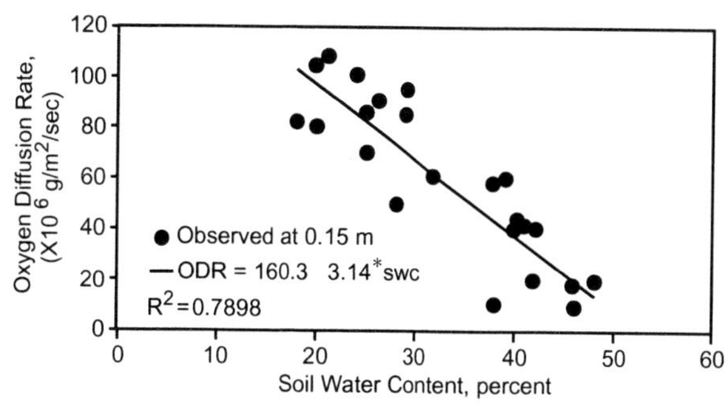

Fig. 3 Relationship between oxygen diffusion rate (ODR) and soil water content (SWC). (Adapted from Ref. 15.)

PLANT RESPONSES TO POOR AERATION

As noted earlier, there are many crop responses to poor aeration resulting from waterlogged or excessive soil water conditions. A prolonged lack of oxygen suppresses and in many cases prevents germination of most seeds. While seeds of many species can tolerate short periods of limited aeration, cell division and growth rate are reduced when the supply of oxygen is cutoff, even for a few hours (13, 20, 21). Fausey and McDonald (21) observed that emergence suppression was much greater at warmer temperatures.

Low oxygen diffusion limits root respiration and the exchange of gaseous metabolites such as carbon dioxide and ethylene. There is extensive evidence that the rapid decline in oxygen content is the primary mechanism of injury to roots (2, 22–24). Although ethylene accumulation was once thought to be a direct cause of root injury (25), Cannell and Jackson (2), citing work of others, concluded that ethanol accumulation in roots likely did little more than modestly depress root growth. Root injury tends to be in the form of slow or terminated growth during waterlogged conditions due to a lack of energy normally obtained through aerobic respiration. Oxygen content within the root zone often drops below critical levels within hours of waterlogging (26). Again, growth retardation responses can be rapid, especially at higher temperatures (27). In studies involving static water tables, investigators have reported little or no root penetration below the water table (28). McDaniel (29) observed that corn roots responded quickly to saturation with all growth ceasing within 24 h. He observed that 3 days of waterlogging significantly affected root mass, maximum root depth, and consumptive water use for the remainder of the growing season. Like roots, plant shoots can respond quickly to flooded conditions. Frequently cited symptoms of above-ground plant parts to waterlogging include shoot wilting (30), increased resistance to water flow (31–35), elevated plant temperature (36), and chlorosis that leads to slowed growth and dry matter accumulation (12, 23, 24, 31, 36–38).

In soils with a high water table (water table less than 2 m from the soil surface during the growing season), yield of most crops can be related to the water table depth. The yield response is generally not a direct response to the position of the water table, but rather to the indirect effects the water table has on aeration status, nutrient availability, and plant available soil moisture. At very shallow water table positions, most soil pores are filled with water and the crop experiences aeration or oxygen stress (anaerobiosis), which in turn suppresses yield. A common method

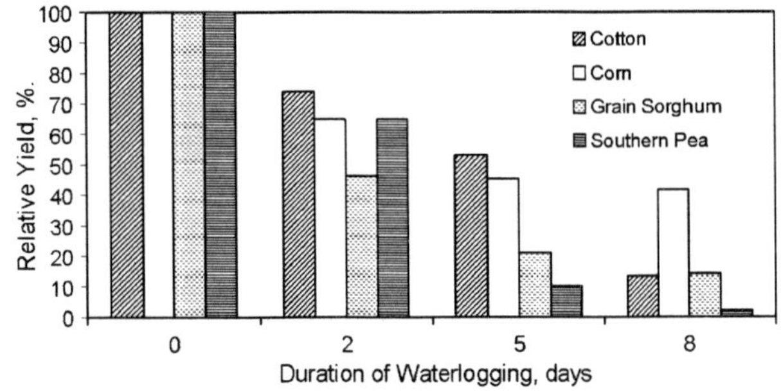

Fig. 4 Relationship between crop yield and flooding duration. (Adapted from Refs. 39, 40.)

for quantifying the stress imposed on a crop by excess soil water or lack of oxygen relies on experimental determination of crop yield in response to an elevated water table. The most frequently applied approach has been to grow crops in lysimeters and examine the effect of inundation or waterlogging on crop yield. This approach does not directly account for stress caused by lack of oxygen, excess carbon dioxide, or accumulation of toxins within the soil profile; but it does provide one method of relating crop yield to a condition of obvious stress (7).

Several general conclusions can be developed from inundation studies. Crop yield reductions increase as the flooding duration increases, as demonstrated in Fig. 4. This conclusion is consistently supported by most studies for most crops. Many crops can tolerate short-term periods of waterlogging; but the tolerance duration depends on many factors. One important factor affecting crop yield response to excess water and poor aeration is the crop stage of development. Most crops studied tend to be more susceptible during earlier stages of development (pre-flowering) (26, 41–43). Another important factor is accumulation of toxic substances found in the soil associated with anaerobic respiration. Nitrite is the first potentially toxic product of anaerobic respiration. At low pH, incomplete conversion of nitrate to N_2 gas can favor the formation of nitrous acid (44). Under extended reducing conditions where E_h values drop below 100 mV, iron and manganese can be reduced and accumulate at toxic concentrations. If anaerobic conditions persist, hydrogen sulfide can be produced that is reported to kill roots at concentrations greater than 2.5 ppm (45).

DRAINAGE TO REDUCE ANAEROBIOSIS

A primary goal of agricultural drainage in humid regions is to lower the water content of the root zone to provide adequate aeration following excessive rainfall or irrigation so as to provide a soil condition that is favorable for crop growth (46). Several techniques are available that improve drainage and reduce excess water-related stress. Drainage methods and materials are described in detail in another section of this encyclopedia. Surface drainage (47) can be achieved by a system of open ditches typically installed on 100–200 m intervals parallel to each other in the general direction of the prevailing slope. To encourage surface runoff and reduce surface ponding, fields are often graded and sometimes crowned into a turtleback shape. Field ditches are typically 0.5–1.5 m deep and discharge into collector canals typically laid out on 1 or 2 km grids. Irregularly spaced ditches are often used to drain depressional areas. Bedded rows parallel to the field ditches are sometimes used to elevate the seedbed and help reduce water stress while the plants are small. Surface drainage systems encourage aeration of the root zone by quickly removing surface water. Once surface ponding is eliminated, the water table declines rapidly during periods with moderate to high potential evapotranspiration (PET), which is usually the case during the growing season. Thus, a surface drainage system encourages reaeration of the root zone by reducing the volume of water entering the soil profile, while the water table is actually lowered by some means other than direct drainage, i.e., evapotranspiration. As the water table declines, air replaces the water that is removed.

Excess water can also be removed by a system of subsurface drains (48–50) comprised clay tile, perforated plastic tubing, or unlined mole drains. In general, 100–150 mm diameter tile or tubing is buried 1–1.5 m deep at intervals of 10–50 m. The subsurface drainage pipes generally outlet to an open ditch or stream. In the U.S., subsurface drainage systems have typically been designed to remove 125 mm (0.5 in.) of water per day for grain crops. Higher design rates up to 250 mm/day have been used for vegetable crops grown in organic soils. Subsurface drainage systems directly improve root zone aeration by lowering the water table following heavy rainfall. In many cases, crop protection is also provided as a result of the water table having been lowered in advance of the rainfall. In this case, subsurface drainage provides adequate soil storage to prevent the water table from rising into the root zone and as a result, aerobic conditions may persist throughout the rainfall/drainage event. As a general rule, aeration and oxygen diffusion into the root zone are associated with increased water table depth and greater drained porosity (51, 52). No degree of site drainage will provide the "optimal" soil water environment in all situations. Drainage requirements depend on many interrelated factors including rainfall duration and intensity, soil physical properties, soil and air temperature, type of crop being grown, and the growth stage of the crop.

CONCLUDING REMARKS

Excess water in the root zone displaces oxygen and leads to a condition of anaerobiosis, also known as oxygen stress. Oxygen stress leads to many adverse effects on plants including discoloration of leaves (chlorosis), accumulation of toxic substances in the root zone, yield loss, and in extreme cases, plant death. Some plants can tolerate longer periods of excess soil water or waterlogging than others. The mechanisms of tolerance are varied and may include the ability to respire anaerobically, slowed growth or self-imposed dormancy, development of adventitious roots, and development of aerenchyma cells. Recent work using molecular genetics techniques has shown that specific genes are responsible for flooding tolerance (53) and provides future promise as a means of improving the flooding tolerance of major agronomic crops. Drainage can, as it has for centuries, be used to remove excess water from land and reduce the occurrence of waterlogging and subsequent anaerobiosis. Improved knowledge of the physiological mechanisms of plant tolerance may lead to alternatives for overcoming the effects of poor soil aeration (2). At present, however, drainage continues to be the most viable alternative for reducing the frequency and duration of anaerobiosis resulting from excess soil water.

REFERENCES

1. Beaucamp, K.H. A History of Drainage and Drainage Methods. In *Farm Drainage in the United States: History, Status, and Prospects*; Misc. Publ. No. 1455, Pavelis, G.A., Ed.; Economics Research Service, U.S. Department of Agriculture: Washington, D.C, 1987; 13–29.
2. Cannell, R.Q.; Jackson, M.B. Alleviating Aeration Stresses. In *Modifying the Root Environment to Reduce Crop Stress*; Arkin, G.F., Taylors, H.M., Eds.; ASAE: St Joseph, MI, 1981; 141–192.
3. Kozlowski, T.T., Ed. *Flooding and Plant Growth*; Academic Press: Orlando, FL, 1984; 356.
4. Glinski, J.; Stepniewski, W. *Soil Aeration and Its Role for Plants*; CRC Press: Boca Raton, FL, 1985; 229.
5. Heritage, A.D. Temporary Waterlogging, Poor Soil Aeration and Root Susceptibility to Fungal Infection—A Review. CSIRO **1985**, 117–125.

6. Kowalik, P.J. *Influence of Land Improvement on Soil Oxidation*; Report 42, Swedish University of Agricultural Science: Uppsala, 1985; 145.
7. Patwardhan, A.S.; Nieber, J.L.; Moore, I.D. Oxygen, Carbon Dioxide, and Water Transfer in Soils: Mechanisms and Crop Response. Trans. ASAE **1988**, *31* (5), 1383–1395.
8. Evans, R.O.; Fausey, N.R. Effects of Inadequate Drainage on Crop Growth and Yield. In *Agricultural Drainage*; Skaggs, R.W., vanSchilfgaarde, J., Eds. Agronomy Monograph 38. American Society of Agronomy, Inc.: Madison, WI, 1999; 13–54.
9. Ravelo A Rational Approach for Incorporating Crop Needs into Drainage System Design. PhD Dissertation: Texas A&M University, College Station, TX, 1978; 159.
10. Hillel, D. *Fundamentals of Soil Physics*; Academic Press: New York, 1980.
11. Wesseling, J.; van Wijk, W.R. Soil Physical Conditions in Relation to Drain Depth. In *Drainage of Agricultural Lands*; Monograph No. 7, Luthin, J.N., Ed.; Chapter VI ASA: Madison, WI, 1957; 461–504.
12. Wesseling, J. Crop Growth and Wet Soils. In *Drainage for Agriculture*; Monograph. No. 17, vanSchilfgaarde, J., Ed.; Chapter 3 ASA: Madison, WI, 1974; 39–90.
13. Trought, M.C.T.; Drew, M.C. Effects of Waterlogging on Young Wheat Plants (*Triticum Aestivum* L.) and on Soil Solutes at Different Soil Temperatures. Plant Soil **1982**, *69*, 311–326.
14. Belford, R.K.; Cannell, R.Q.; Thomson, R.J. Effects of Single and Multiple Waterlogging on the Growth and Yield of Winter Wheat on a Clay Soil. J. Sci. Food Agric. **1985**, *36*, 142–156.
15. Bornstein, J.; Benoit, G.R.; Scott, F.R.; Hepler, P.R.; Hedstrom, W.E. Alfalfa Growth and Soil Oxygen Diffusion as Influenced by Depth to Water Table. Soil Sci. Soc. Am. J. **1984**, *48*, 1165–1169.
16. Buckingham, E. *Contributions to Our Knowledge of the Aeration of Soils*; Bulletin 38, U.S. Bureau Soils Washington, D.C., 1904.
17. Stolzy, L.H.; Letey, J. Characterizing Soil Oxygen Conditions with a Platinum Microelectrode. Adv. Agron. **1964**, *16*, 249–279.
18. Sojka, R.E.; Stolzy, L.H. Soil Oxygen Effects on Stomatal Response. Soil Sci. **1980**, *130*, 350–358.
19. VanToai, T.T.; Desmond, E.D.; Fausey, N.R. Lysimeter Study of Soybean Responses to Excess Water. In *Drainage Design and Management*; Proceedings of the Fifth National Drainage Symposium, Johnston, W.R., Ed.; ASAE: St Joseph, MI, 1987; 213–216.
20. Cannell, R.Q.; Belford, R.K. Crop Growth After Transient Waterlogging. In *Advances in Drainage*; Proceedings of the Fourth National Drainage Symposium, Kriz, G.J., Ed.; ASAE: St Joseph, MI, 1982; 163–170.
21. Fausey, N.R.; McDonald, M.B. Emergence of Inbred and Hybrid Corn Following Flooding. Agron. J. **1985**, *77*, 51–56.
22. Box, J.E., Jr. The Effects of Waterlogging on Rooting of Soft Red Winter Wheat. Dev. Agric. Mgmt For. Ecol. **1991**, *24*, 418–430.
23. Huang, B.; Johnson, J.W. Root Respiration and Carbohydrate Status of Two Wheat Genotypes in Response to Hypoxia. Ann. Bot. **1995**, *75* (4), 427–432.
24. Huang, B.; NeSmith, D.S.; Bridges, D.C.; Johnson, J.W. Responses of Squash to Salinity, Water Logging, and Subsequent Drainage. II. Root and Shoot Growth. J. Plant Nutr. **1995**, *18* (1), 141–152.
25. Crawford, R.M.M. Tolerance of Anoxia and Ethanol Metabolism in Germinating Seeds. New Phytol. **1977**, *79*, 511–517.
26. Mukhtar, S.; Baker, J.L.; Kanwar, R.S. Corn Growth As Affected by Excess Soil Water. Trans. ASAE **1990**, *33* (2), 437–442.
27. del Castillo Torres, D. Soybean Root Growth and Recovery Under Temporary Anaerobic Soil Conditions. PhD Dissertation: Mississippi State University, MS, 1983.
28. Baser, D.K.; Jaggi, I.K.; Sinha, S.B. Root and Shoot Growth and Transpiration from Maize Plant as Influenced by Various Depths of Water Table. Jawaharlal Nehru Agric. Univ. Jabalpur, India **1981**, 390–399.
29. McDaniel, V. Effects of Shallow Water Tables on Corn Roots, Crop Yield and Hydrology. PhD Dissertation: North Carolina State University, Raleigh, 1995; 166.
30. Parsons, L.R.; Kramer, P.J. Diurnal Cycling in Root Resistance to Water Movement. Physiol. Plant **1974**, *30*, 19–23.
31. Galloway, R.; Davidson, N.J. The Response of Atriplex Amnicola to the Interactive Effects of Salinity and Hypoxia. J. Exp. Bot. **1993**, *44* (260), 653–663.
32. Timsina, J.; Garrity, D.P.; Pandey, R.K. Water Table Gradient Effects on the Performance of Diverse Cowpea Cultivars. Agron. J. **1993**, *85* (2), 359–367.
33. Huang, B.; Johnson, J.W.; Nesmith, S.; Bridges, D.C. Growth, Physiological and Anatomical Responses of Two Wheat Genotypes to Waterlogging and Nutrient Supply. J. Exp. Bot. **1994**, *45* (271), 193–202.
34. Huang, B.; Johnson, J.W.; NeSmith, D.S.; Bridges, D.C. Root and Shoot Growth of Wheat Genotypes in Response to Hypoxia and Subsequent Resumption of Aeration. Crop Sci. **1994**, *34* (6), 1538–1544.
35. Pezeshki, S.R.; Pardue, J.H.; DeLaune, R.D. Leaf Gas Exchange and Growth of Flood-Tolerant and Flood-Sensitive Tree Species Under Low Soil Redox Conditions. Tree Physiol. **1996**, *16* (4), 453–458.
36. Timsina, J.; Garrity, D.P.; Pandey, R.K. Cowpea Water Relations Growth Response on a Toposequence Water Table Gradient. Agron. J. **1993**, *85* (2), 368–378.
37. Cannell, R.Q.; Belford, R.; K, .; Blackwell, P.S.; Govi, G.; Thompson, R.J. Effects of Waterlogging on Soil Aeration and on Root and Shoot Growth and Yield of Winter Oats (*Avena Sativa* L.). Plant Soil **1985**, *85*, 361–373.
38. Ahmad, N.; Kanwar, R.S.; Kaspar, T.C.; Bailey, T.B. Effect of Soil Surface Submergence and a Water Table on Vegetative Growth and Nutrient Uptake of Corn. Trans. ASAE **1992**, *35* (4), 1173–1177.
39. Howell, T.A.; Hiler, E.A. Effects of Inundation Period on Seedling Growth. Trans. ASAE **1974**, *17* (2), 286–288, See Also Page 294.
40. Hiler, E.A. Drainage Requirements of Crops. In *Proceedings of the Third National Drainage Symposium*; Hoffman, G.J., Ed.; ASAE: St Joseph, MI, 1977; 127–129.
41. Griffin, J.L.; Saxton, A.M. Response of Solid-Seeded Soybean to Flood Irrigation. II. Flood Duration. Agron. J. **1988**, *80* (6), 885–888.

D

42. Scott, H.D.; DeAngulo, J.; Daniels, M.B.; Wood, L.S. Flood Duration Effects on Soybean Growth and Yield. Agron. J. **1989**, *81* (4), 631–636.
43. Evans, R.O.; Skaggs, R.W.; Sneed, R.E. Normalized Crop Susceptibility Factors for Corn and Soybean to Excess Water Stress. Trans. ASAE **1990**, *33* (4), 1153–1161.
44. Lee, R.B. The Effect of Nitrite on Root Growth of Barley and Maize. New Phytol. **1979**, *83*, 615–622.
45. Culbert, D.L.; Ford, H.W. The Use of a Multi-Celled Apparatus for Anaerobic Studies of Flooded Root Systems. Hort. Sci. **1972**, *71*, 29–31.
46. Fausey, N.R.; Doering, E.J.; Palmer, M.L. Purposes and Benefits of Drainage. In *Farm Drainage in the United States: History, Status, and Prospects*; Misc. Publ. No. 1455, Pavelis, G.A., Ed.; Economics Research Service, U.S. Department of Agriculture, 1987; 48–51.
47. Carter, C.E. Surface Drainage. In *Agricultural Drainage*; Skaggs, R.W., vanSchilfgaarde, J., Eds. Agronomy Monograph 38. American Society of Agronomy, Inc.: Madison, WI, 1999; 1023–1050.
48. Schwab, G.O.; Fouss, J.L. Drainage Materials. In *Agricultural Drainage*; Skaggs, R.W., vanSchilfgaarde, J., Eds.; Agronomy Monograph 38. American Society of Agronomy, Inc.: Madison, WI, 1999; 911–926.
49. Broughton, R.S.; Fouss, J.L. Subsurface Drainage Installation Machinery and Methods. In *Agricultural Drainage*; Skaggs, R.W., vanSchilfgaarde, J., Eds. Agronomy Monograph 38. American Society of Agronomy, Inc.: Madison, WI, 1999; 967–1004.
50. Spoor, G.; Leeds-Harrison, P. Nature of Heavy Soils and Potential Drainage Problems. In *Agricultural Drainage*; Skaggs, R.W., vanSchilfgaarde, J., Eds. Agronomy Monograph 38. American Society of Agronomy, Inc.: Madison, WI, 1999; 1051–1182.
51. Skaggs, R.W.; Wells, L.G.; Ghate, S.R. Predicted and Measured Drainable Porosities for Field Soils. Trans. ASAE **1978**, *21* (3), 522–528.
52. Silins, U.; Rothwell, R.L. Spatial Patterns of Aerobic Limit Depth and Oxygen Diffusion Rate at Two Peatlands Drained for Forestry in Alberta. Natl Res. Council Can. **1999**, *29* (1), 53–61.
53. VanToai, T.T.; Beuerlein, J.E.; Schmitthenner, A.F.; St. Martin, S.K. Genetic Variability for Flooding Tolerance in Soybeans. Crop Sci. **1994**, *34* (4), 1112–1115.

EARTHWORMS

Patrick J. Bohlen
Archbold Biological Station, Lake Placid, Florida, U.S.A.

INTRODUCTION

Earthworms are perhaps the most important soil organisms in terms of their influence on organic matter breakdown, soil structural development, and nutrient cycling, especially in productive ecosystems. Aristotle called them the "intestines of the earth" and the eminent nineteenth century biologist, Charles Darwin, spent many years observing their major influence on the formation of humus and transport of soil (1). Despite the vast increase in scientific literature on earthworms in recent years, much remains to be known of their basic biology and ecology (2, 3).

EARTHWORM BIOLOGY AND CLASSIFICATION

Earthworms are terrestrial annelids with bilateral symmetry and corresponding external and internal segmentation. They have a thinly pigmented cuticle bearing setae on all segments except the first two. All earthworms are hermaphroditic. Their gonads are situated in specific segments, which vary among different taxa. When sexually mature, they develop a swollen area of the epidermis called a clitellum. This region produces a cocoon in which one or more eggs are deposited and then the cocoon is passed over the anterior segments and deposited in or on the soil. The young develop within the cocoon and newly hatched worms resemble adults. The time to hatching and reproductive maturity varies widely among different earthworm species and is influenced by environmental factors.

Earthworms are classified within the phylum Annelida and the class Oligochaeta, which consists of as many as 36 families worldwide. About two-thirds of oligochaete families comprise aquatic or semi-aquatic worms, and the remaining families comprise mostly or exclusively terrestrial worms or earthworms. There are over 3500 known earthworm species and it is estimated that the global total may be twice that number. Distinct taxonomic groups have arisen on every continent except Antarctica, and some groups are now distributed throughout the world.

EARTHWORM ECOLOGY

Ecological Groups

Earthworm species can be grouped according to behavioral, morphological or physiological adaptations that enable them to partition available resources in the soil. The three main life history strategies are termed epi, anecic, and endogeic (2, 3). Epigeic worms feed on plant litter, dwell on the soil surface or within the litter layer, tend to be heavily pigmented, and are small- to medium-sized. Anecic worms feed on plant litter and soil, live in nearly vertical permanent burrows, are dorsally pigmented, and large. Endogeic species are soil-feeders, are not heavily pigmented, form extensive horizontal burrow systems, and range in size from small to large. Endogeic worms have been further divided into polyhumic, mesohumic, and oligohumic groups, which are separated, respectively, by the descending importance of organic rich mineral soil in their diet and increasing size. Earthworm species do not always fall clearly into these three main categories and may even exhibit traits of different groups at different life stages or under different environmental conditions.

Earthworm Communities

Earthworm communities generally consist of from one to six species. The relative abundance and species composition of earthworm communities depends upon soil type, topography, and vegetation, and is also influenced by land use history and earthworm biogeography. Earthworms account for the majority of animal biomass in soil in a wide range of productive ecosystems, from temperate grasslands, pastures and forests to tropical pastures, savannas and rainforests, and many temperate and tropical agroecosystems. They generally do not occur in deserts and arid grasslands or in extreme alpine or boreal habitats. Earthworms are often absent from strongly acidic forest soils with poor litter quality, such as some northern coniferous forests.

Many earthworm communities consist of invasive exotic species. In North America, where approximately 100 native earthworm species have been described, at least 45 exotic species have been introduced. Lumbricid earthworms of European origin dominate many North American agroecosystems. A worldwide survey of earthworms in tropical regions reported that a total of 51 exotic and 151 native species commonly occur in tropical agroecosystems (4).

EFFECTS OF EARTHWORMS ON SOIL PROPERTIES

The influence of a given earthworm species on soil properties depends upon that species' life history strategy. For example, the large vertical burrows of anecic earthworms, such as the common nightcrawler *Lumbricus terrestris*, can facilitate preferential flow of water through the soil profile, increasing the transport of water, nutrients, and agricultural chemicals into deeper soil layers (5, 6). Epigeic species facilitate the breakdown and mineralization of surface litter, whereas anecic species incorporate surface litter deeper into the soil profile. Anecic species also bring soil from deeper soil horizons to the surface, which over a long period of time can change the mineralogy of surface soil. Endogeic species feed on fragmented organic matter and mix it thoroughly through the surface mineral soil.

Effects on Organic Matter Breakdown and Nutrient Cycling

Earthworm activity accelerates the decomposition of plant litter, increases rates of nutrient transformation and plant nutrient uptake, improves soil aggregation and porosity, and enhances water infiltration and solute transport. Earthworms can consume and incorporate large amounts of organic matter into soil (Table 1). Such mixing is largely responsible for the formation of mull soils in which surface organic horizons are thoroughly mixed with underlying mineral soil.

Earthworms are a major influence on nutrient cycling in many ecosystems. Although they generally increase the mineralization of soil carbon, earthworms can also decrease mineralization of carbon by contributing to the formation of stable soil aggregates in which carbon is protected from further breakdown. Direct fluxes of nutrients through earthworm biomass can be considerable. For example, the turnover of nitrogen (N) through earthworm tissues can be up to $150\,kg\,N\,ha^{-1}\,yr^{-1}$ (3). Earthworm casts contain elevated amounts of inorganic N relative to surrounding soil. As a consequence, earthworms can greatly enhance the mineralization of N and can stimulate other N transformations such as denitrification (7). By increasing bypass flow of infiltrating water, earthworm burrows can increase the amount N and other nutrients leaching from the soil profile (5). Alternatively earthworms can reduce the amount of nutrients lost in surface runoff by increasing rates of water infiltration into the soil (8).

Effects on Physical Properties of Soil

The effect of earthworms on soil structure results from the net outcome of their feeding and burrowing activities. Earthworms ingest soil particles and organic matter, mixing these two fractions together and egesting them as surface or subsurface casts. Estimates of annual rates of production of earthworm casts range from less than 5 to over 250 Mg ha^{-1} in various ecosystems (Table 2). Once egested, soil in casts can be eroded due to impact of rainfall or can form stable soil aggregates through a variety of stabilizing mechanisms (16).

Earthworms generally improve the aeration and porosity of soil through formation of burrows and by increasing the proportion of large aggregates in the soil, and their effects are especially important in poorly structured or reclaimed soil. By increasing rates of water infiltration, earthworms can reduce the amount of surface runoff. Alternatively, earthworms can increase erosion by

Table 1 Amount of organic matter ingested or incorporated into soil by earthworm populations in different environments

Ecosystem	Location	Type of organic matter	Amount consumed or incorporated (kg ha^{-1})	Reference
Maize field	United States	Maize residues	840	(9)
Orchard	England	Apple leaves	2,000	(10)
Mixed forest	England	Canopy tree leaves	3,000	(11)
Oak forest	Japan	Oak leaves	1,071	(12)
Alfalfa field	United States	Alfalfa residues	1,220	(13)
Tallgrass prairie	United States	Total organic matter	740–8,980	(14)
Savanna	Ivory Coast	Total organic matter	1,300	(15)

Table 2 The amount of earthworm casts produced and in different environments in various locations around the world

Environment	Location	Earthworm casts ($Mg\ ha^{-1}$)
Arable	Germany	92
Arable	Switzerland	18–81
Arable	Nigeria	50
Arable/floodplain	Egypt	268
Pasture	England	19–40
Pasture	England	28
Pasture	Germany	91
Pasture	Switzerland	22–42
Grassland/tallgrass prairie	United States	24–94
Grassland	India	4–78
Savanna	Columbia	10–50
Tropical forest	Ivory Coast	32–50
Temperate forest	Germany	7–60

(Data from various sources)

removing the protective cover of surface litter, increasing surface sealing and depositing surface casts, which can be carried downslope during heavy rains. Some tropical species actually increase soil bulk density and decrease infiltration by producing compact soil casts. Overall, the effects of earthworms on soil structure are considered to improve soil fertility.

Effects on Plant Growth

The majority of studies examining the influence of earthworms on plant growth have reported that earthworms stimulate plant growth, although some studies have reported no effect or even a negative effect of earthworms on plant growth (2, 17). Earthworms have been shown to increase production of shoots and grain in a variety of field trials and greenhouse experiments. Introduction of earthworms into reclaimed polders in the Netherlands and in pastures in New Zealand resulted in large increases in forage quantity and quality (18, 19). Beneficial effects of earthworms on plant growth may be due to increased nutrient and water availability, improved soil structure, stimulation of microorganisms or formation of microbial products that enhance plant growth, or possibly through direct production of plant growth promoting substances.

Undesirable Effects of Earthworms

Despite the documented and putative beneficial effects of earthworms on nutrient dynamics, soil structure, and fertility, some aspects of earthworm activities are considered undesirable (2). These include removing and burying surface residues that would otherwise protect soil surfaces from erosion; producing fresh casts that increase erosion and surface sealing; depositing castings on the surface of lawns and golf greens or irrigation ditches where they are a nuisance, or in pastures where they interfere with haying operations; dispersing weed seeds in gardens and agricultural fields; transmitting plant or animal pathogens; increasing losses of soil nitrogen through leaching and denitrification; and increasing soil carbon loss through enhanced microbial respiration. It is the net outcome of their positive and negative effects that determines whether earthworms are to be considered pests or benefactors in any given situation, and they are generally considered to be of great benefit to soil properties and fertility.

EARTHWORMS AND SOIL MANAGEMENT

Earthworms and Soil Reclamation

Introduction of appropriate earthworm species or encouraging natural population through the addition of suitable amendments can increase the rate of soil improvement and genesis of soil structure of reclaimed land. Introduction of European earthworms into pasture in New Zealand and Australia, as well as in Dutch polders, greatly facilitated improvements in soil structure and plant productivity (18, 19). There has also been some success in introducing earthworms into reclaimed mining sites and in reclaimed peat with beneficial effects on soil structural development, nutrient cycling, and productivity (2).

Influence of Agricultural Practices on Earthworm Populations

Earthworms can be affected favorably or adversely by different agricultural practices. Whereas heavy cultivation is detrimental to earthworm populations, reduced and no-tillage agricultural practices promote the growth of earthworm populations. Organic amendments, such as animal or green manures, also stimulate the growth of earthworm populations. Inorganic fertilizers may benefit earthworm populations by increasing the production of crop residues, but their effects are not as great as the effects of organic fertilizers. Long-term application of inorganic fertilizers may adversely affect earthworm populations due to soil acidification or other changes in the soil environment. Liming can benefit earthworm populations in some instances.

E

Effects of Pesticides and Soil Pollutants on Earthworms

Pesticides have widely varying effects on earthworm populations, with many chemicals having little or no toxicity toward earthworms and others exhibiting acute toxicity (20). In general, the carbamate insecticides and soil fumigants are very or highly toxic to earthworms. Herbicides in general show low toxicity toward worms, although there are some exceptions. Organochlorine and organophosphorous insecticides have varying levels of toxicity.

Soil contamination with organic pollutants, heavy metals, and acid precipitation can be detrimental to earthworm populations. Most toxic organic soil pollutants are highly toxic to earthworms. Heavy metals vary in their toxicity toward earthworms. In addition to acute negative affects of some heavy metals on earthworms, sublethal heavy metal concentrations can cause declines in earthworm growth and reproduction. Heavy metals can accumulate in earthworm tissues, which is potentially problematic for movement of heavy metals into higher trophic levels, because of the wide array of animals that prey on earthworms. Earthworm species vary in their tolerance of acid soil conditions but some reports have noted a decline in earthworm populations in response to large amounts of acid deposition.

REFERENCES

1. Darwin, C. *The Formation of Vegetable Mould, Through the Action of Worms, with Observations on Their Habits*; John Murray: London, 1881.
2. Edwards, C.A.; Bohlen, P.J. *Biology and Ecology of Earthworms*, 3rd Ed. Chapman and Hall: London, 1996.
3. Lee, K. *Earthworms: Their Ecology and Relationships with Soils and Land Use*; Academic Press: New York, 1985.
4. Fragoso, C.; Kanyonyo, J.; Moreno, A.; Senapati, B.K.; Blanchart, E.; Rodriguez, C. A Survey of Tropical Earthworms: Taxonomy, Biogeography and Environmental Plasticity. In *Earthworm Management in Tropical Agroecosystems*; Lavelle, P., Brussaard, L., Hendrix, P., Eds. CABI Publishing: New York, 1999; 1–26.
5. Subler, S.; Baranski, C.M.; Edwards, C.A. Earthworm Additions Increased Short-Term Nitrogen Availability and Leaching in Two Grain-Crop Agroecosystems. Soil Biology and Biochemistry **1997**, *29*, 413–421.
6. Edwards, W.M.; Shipitalo, M.J.; Traina, S.J.; Edwards, C.A.; Owens, L.B. Role of *Lumbricus Terrestris* (L.) Burrows on Quality of Infiltrating Water. Soil Biology and Biochemistry **1992**, *24*, 1555–1561.
7. Elliot, P.W.; Knight, E.; Anderson, J.M. Denitrification in Earthworm Casts and Soil from Pastures Under Different Fertilizer and Drainage Regimes. Soil Biology and Biochemistry **1990**, *22*, 601–605.
8. Sharpley, A.N.; Syers, J.K.; Springett, J.A. Effect of Surface-Casting Earthworms on the Transport of Phosphorus and Nitrogen in Surface Runoff from Pasture. Soil Biology and Biochemistry **1979**, *11*, 459–462.
9. Bohlen, P.J.; Parmelee, R.W.; McCartney, D.A.; Edwards, C.A. Effects of Earthworms (*Lumbricus Terrestris*) on the Carbon and Nitrogen Dynamics of Decomposing Surface Litter in Corn Agroecosystems. Ecological Applications **1997**, *7*, 1341–1349.
10. Raw, R. Studies on Earthworm Populations in Orchards. I. Leaf Burial in Apple Orchards. Annals of Applied Biology **1962**, *50*, 389–404.
11. Satchell, J.E. Lumbricidae. In *Soil Biology*; Burgess, A., Raw, F., Eds.; Academic Press: London, 1967, 259–322.
12. Sugi, Y.; Tanaka, M. Population Study of an Earthworm, *Pheretima Sieboldi*. In *Biological Production in a Warm-Temperate Evergreen Oak Forest of Japan*; J.I.B.P Synthesis, no. 18, Kira, T., Ono, Y.H., Hosokawa, T., Eds.; University of Tokyo Press: Tokyo, 1978, 163–171.
13. Gallagher, A.V.; Wollenhaupt, N.C. Surface Alfalfa Residue Removal by Earthworms *Lumbricus Terrestris* L. In a No-Till Agroecosystem. Soil Biology and Biochemistry **1997**, *29*, 477–480.
14. James, S.W. Soil, Nitrogen, Phosphorus, and Organic Matter Processing by Earthworms in Tallgrass Prairie. Ecology **1991**, *72*, 2101–2109.
15. Lavelle, P. Consommation Annuelle d'une Population Naturelle De Vers De Terre (*Millsonia Anomala* Omodes, Acanthodrilidae: Oligochaetes) Dans La Savanne De Lamto (Côte d'Ivoire). In *Progress in Soil Zoology*; Vanek, J., Ed.; Academia Publishing House: Prague, 1975, 299–304.
16. Tomlin, A.D.; Shipitalo, M.J.; Edwards, W.M.; Protz, R. Earthworms and Their Influence on Soil Structure and Infiltration. In *Earthworm Ecology and Biogeography in North America*; Hendrix, P.F., Ed.; Lewis Publishers: Chelsea, MI, 1995; 159–184.
17. Brown, C.; Pashanasi, B.; Villenave, C.; Patron, J.C.; Senapati, B.K.; Giri, S.; Barois, I.; Lavelle, P.; Blanchart, E.; Blakemore, R.J.; Spain, A.V.; Boyer, J. Effects of Earthworms on Plant Production in the Tropics. In *Earthworm Management in Tropical Agroecosystems*; Lavelle, P., Brussaard, L., Hendrix, P., Eds.; CABI Publishing: New York, 1999; 87–148.
18. Hoogerkamp, M.; Rogaar, H.; Eijsackers, H.J.P. Effect of Earthworms on Grassland on Recently Reclaimed Polder Soils in the Netherlands. In *Earthworm Ecology: From Darwin to Vermiculture*; Satchell, J.E., Ed.; Chapman and Hall: London, 1983; 85–106.
19. Stockdill, S.M.J. Effect of Introduced Earthworms on the Productivity of New Zealand Pastures. Pedobiologia **1982**, *24*, 29–35.
20. Edwards, C.A.; Bohlen, P.J. The Effects of Toxic Chemicals on Earthworms. Reviews of Environmental Contamination and Toxicology **1992**, *125*, 23–99.

ECOLOGY AND THE CYCLING OF CARBON AND NITROGEN

Alan J. Franzluebbers
United States Department of Agriculture-Agricultural Research Service (USDA-ARS), Watkinsville, Georgia, U.S.A.

INTRODUCTION

Carbon (C) and nitrogen (N) are two of the most important elements that affect the soil's productivity and environmental quality (1). Carbon is found throughout nature in a wide variety of forms and particularly in soil as 1) complex organic compounds (e.g., $C_xH_{2x}O_x$) derived from living organisms, 2) carbonate minerals such as calcite ($CaCO_3$) and dolomite ($CaMg(CO_3)_2$), and 3) carbon dioxide (CO_2) and methane (CH_4) as decomposition endproducts. Nitrogen is an essential element of plants, animals, and microorganisms that is a part of chlorophyll, enzymes, amino acids, and proteins, which are necessary for growth and development of organisms. In soil, the quantity of N in organic matter and as clay-fixed ammonium (NH_4^+) far exceeds quantities in plant-available forms of nitrate (NO_3^-) and NH_4^+.

CARBON AND NITROGEN CYCLES

Carbon and N occur in various forms and undergo transformations from one form to another, primarily through biochemical manipulations involving enzymes (2, 3). Enzymes are catalysts of very specific reactions that function either 1) intracellularly within plants, microorganism, or soil animals or 2) extracellularly in soil solution or attached to soil colloids.

The forms and fluxes of an element are commonly illustrated in a cycle following the principles of conservation of mass (i.e., elements are transferred from one molecule to another). The C and N cycles have global dimensions with terrestrial, aquatic, and atmospheric components of major significance (4, 5). The sun initiates a chain of energy reactions, which drive elemental cycles. The elemental cycles of C and N interact closely with the water cycle, as water is a fundamental internal component of life and a major transport mechanism of nutrients.

Autotrophic fixation of atmospheric CO_2 by plants captures the energy of the sun within organic compounds via the process of photosynthesis (Fig. 1). Inorganic N is taken up by plant roots and synthesized into amino acids and proteins during plant development. Plants are eventually consumed by animals or microorganisms, transferring portions of this stored energy through biochemical processes into various cellular components. Once in soil, the C cycle is dominated by the heterotrophic process of decomposition, i.e., the breakdown of complex organic compounds into simple organic constituents. Mineralization is the complete decomposition of organic compounds into mineral constituents:

$$C_6H_{12}O_6 + 6O_2 \rightarrow 6CO_2 + 6H_2O + \text{energy}$$

$$R\text{–}NH_2 + H_2O \rightarrow R\text{–}OH + NH_3$$

Immobilization of N occurs simultaneously with N mineralization when soil organisms consume inorganic N to meet the demands for new body tissue. Net N mineralization occurs when gross N mineralization exceeds that of N immobilization.

ENVIRONMENTAL INFLUENCES ON SOIL MICROBIAL ACTIVITY

The dominant organisms responsible for decomposition of organic matter and associated mineralization of C and N are soil microorganisms, composed of bacteria, actinomycetes, fungi, and protozoa (6, 7). Soil fauna also indirectly affect C and N cycling by 1) comminuting plant residues and exposing a greater surface area to soil microorganisms, 2) transporting plant and animal residues to new locations in the soil to facilitate decomposition, interaction with soil nutrients, or isolation from environmental conditions, 3) inoculating partially digested organic substrates with specific bacteria and enzymes, and 4) altering physical characteristics of soil by creating burrows, fecal pellets, and distribution of soil particles that influence water, air, nutrient, and energy retention and transport. With suitable environmental conditions, soil microorganisms grow rapidly in response to the availability of organic substrates rich in C and N.

Soil Temperature

Temperature controls both plant and soil microbial activity, although not at the same level (Fig. 2). Plant and

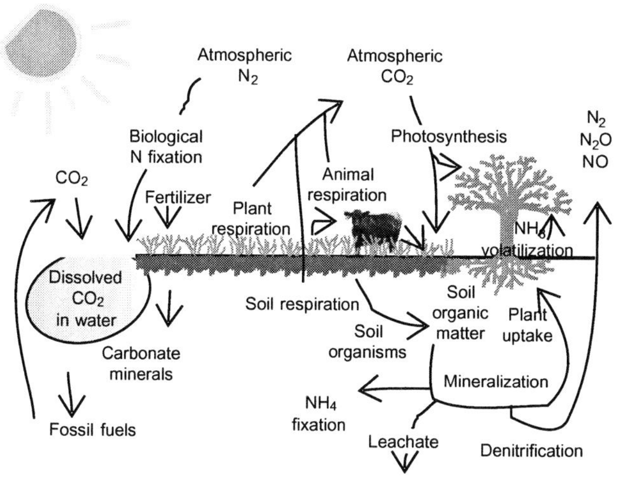

Fig. 1 Generalized diagram of the C and N cycles in soil.

soil microbial activity are limited by low temperature resulting in low photosynthetic potential, as well as low decomposition potential. For many plants, net photosynthetic activity is optimized between 20 and 30°C, because at higher temperatures plant respiration consumes energy for maintenance. In many temperate soils, microbial activity is maximum near 30°C and decreases at higher temperatures. An intermediate temperature is often ideal for maximizing C retention in soil, where optimum plant activity outdoes soil microbial activity.

Soil Water Content

The greatest diversity of soil microorganisms is found under aerobic conditions, where maximum energy is obtained. However, there are a number of soil bacteria that thrive under anaerobic conditions, in which alcohols, acetic acid, lactic acid, and CH_4 become C endproducts via fermentation, and nitrate is converted to N gases (e.g., N_2, N_2O, NO) via the process of denitrification. Soil C and net N mineralization are maximized at an optimum balance between soil moisture and oxygen availability (Fig. 3). Significant denitrification occurs at water-filled pore space >70%, resulting in what appears as reduced net N mineralization.

Soil Texture

Soil texture can influence both the quantity of C and N accumulation in soil and their potential mineralization. Potential C mineralization is often greater in coarse- than in fine-textured soils, which may be due to both increased microbial predation by soil fauna and greater accessibility of organic substrates in coarse-textured soils. Organic C and N can also be protected from decomposition when bound within soil aggregates, which are a coherent assemblage of primary soil particles (i.e., sand, silt, clay) cemented through natural forces and substances derived from root exudates and soil microbial activity.

Spatial Distribution of Organic Substrates

Distribution of organic substrates in soil has a major impact on potential C and N mineralization. Potential C mineralization is often several-fold greater in the rhizosphere (i.e., 0–5 mm zone surrounding roots) than in bulk soil. However, because of the high demand for N by plant roots and the stimulated soil microflora, net N mineralization is often lower in the rhizosphere because of immobilization of N.

The soil-surface often contains greater quantities of organic matter than at lower depths due to surface deposition of plant residues, as well as highest plant root activity. The soil-surface usually undergoes the most extreme drying/wetting cycles and has the greatest exchange of gases, both of which contribute to enhanced soil microbial biomass and activity. Tillage of soil with traditional agriculture redistributes the organic substrates uniformly within the tillage layer, often resulting in stimulated soil microbial activity from disruption of organic substrates protected within stable soil aggregates. Minimum soil disturbance with conservation tillage practices can reduce oxidation of soil organic matter and preserve more C within soil, which can have implications for potentially mitigating the greenhouse effect (8).

ORGANIC SUBSTRATE QUALITY

The quality of organic substrates has a major influence on the rate of decomposition and the transformations that occur in soil. Plant residues do not vary greatly in total C concentration on a dry-weight basis (e.g., 37–47 mg g^{-1}), but do vary in the suite of C compounds, which determine its quality or conversely its resistance to degradation. The diversity of organic compounds attacked by soil microorganisms is extensive (e.g., organic acids, polysaccharides, lignins, aromatic and aliphatic hydrocarbons, sugars, alcohols, amino acids, purines, pyrimidines, proteins, lipids, and nucleic acids). Almost all naturally occurring organic compounds, and even most synthetic organic compounds, are susceptible to decomposition, given the appropriate environment, microbial community, and time (9, 10). Generally, the primary components of

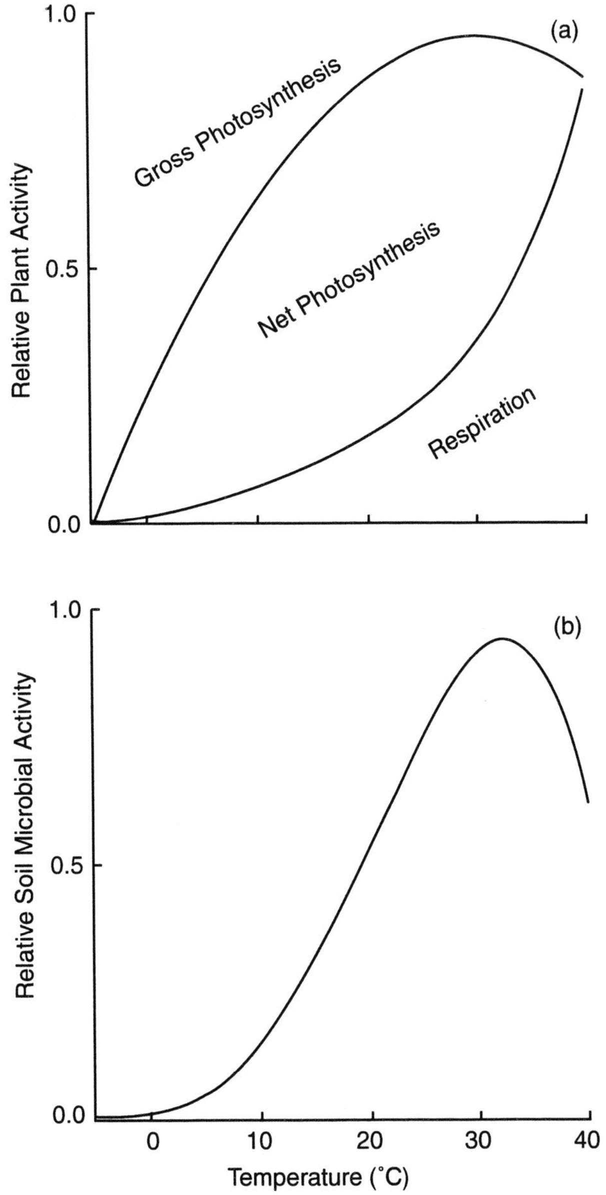

Fig. 2 Typical responses of plant and soil microbial activities to temperature.

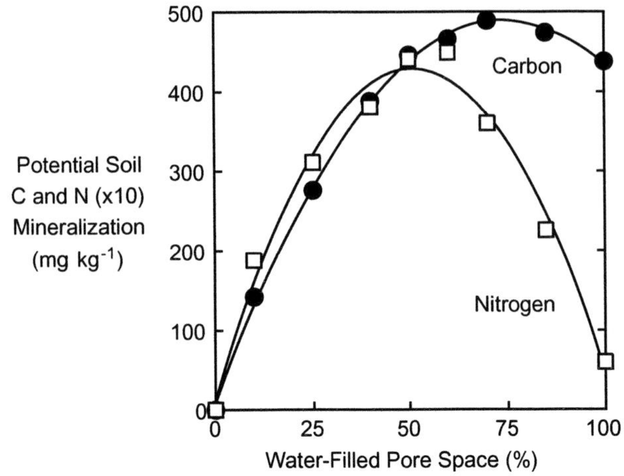

Fig. 3 Responses of potential soil C and N mineralization to water-filled pore space in Typic Kanhapludults in Georgia, U.S. Air-filled pore space is 100−(water-filled pore space).

plants can be categorized according to the relative rate of decomposition: rapid (sugars, starches, fats, and proteins), intermediate (cellulose and hemicellulose), and slow (lignin and lignocellulose). Young plants are of high quality and low resistance to decomposition, whereas with aging, lignin and polyphenolic concentrations increase, resulting in greater resistance to decomposition. Low N concentration of organic amendments usually results in temporary net N immobilization into microbial biomass, which grows rapidly in response to the availability of organic C. Soil microbial biomass maintains a C:N ratio of 10 ± 5. Following a proliferation of microbial biomass that depletes the source of readily decomposable organic C:N in excess of microbial demands becomes mineralized and available for plant uptake (Fig. 4). In general, plant residues with C:N ratio >40 will result in longer periods of net N immobilization.

Soil Organic Matter

Soil organic matter is composed of a large variety of organic compounds that can be characterized in many ways. A useful separation of soil organic matter for modeling is based on the turnover times, whereby at least three pools can be defined: 1) active (composed of microbial biomass and light fraction material with a turnover time of <1 yr), 2) passive (composed of macroorganic matter and protected organic matter with a turnover time of 3–10 yr), and 3) slow (stable humus fraction with a turnover time of >100 yr).

LOSSES OF NITROGEN FROM SOIL

Nitrogen cycling in soil is different from that of C because of the more numerous transformations that can occur upon mineralization to an inorganic form (11). Mineralization of N from organic matter results in NH_4^+ released into soil solution. In the presence of nitrifying bacteria, NH_4^+ is converted to NO_3^-; a process called nitrification. The fate of NO_3^- in soil depends upon environmental conditions. Active plant growth in natural and agricultural systems would provide enough demand to reimmobilize N into

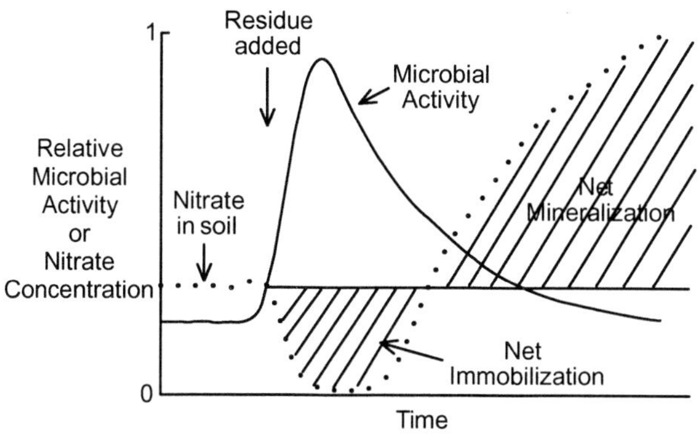

Fig. 4 Typical responses in soil microbial activity and soil nitrate concentration with the addition of plant residue relatively low in N concentration.

organic forms. However, NO_3^- can be used as an electron acceptor in place of O_2 under anaerobic conditions with a supply of soluble organic C, resulting in gaseous loss of N to the atmosphere via denitrification. In temperate soils with a net negative charge on colloidal surfaces, NO_3^- can readily leach into the vadose zone and contaminate groundwater. In tropical soils with a net positive charge, NO_3^- can be retained on anion exchange sites. Contradicting behavior of the cations, NH_4^+ and NO_3^-, occurs with respect to clay mineralogy.

Humans, as well as roving animals, impose great demands on the C and N cycles. Management of agricultural and forest land for food and fiber removes nutrients from soil for consumption and utilization elsewhere. Return of these nutrients to soil is possible when municipal and agricultural solid wastes and wastewater are applied to land. Losses of C and N from managed lands also occur through soil erosion, which transports nutrients via 1) water from overland flow into streams, lakes and oceans, and 2) air from exposed land.

Volatilization of ammonia (NH_3) to the atmosphere is possible when NH_4^+ is exposed to alkaline soil conditions. Significant ammonia volatilization can occur with surface application of urea fertilizer to nonacidic soils, from animal manures, and from green plant materials not incorporated into the soil.

FIXATION OF CARBON AND NITROGEN

Both C and N are biologically fixed from inorganic atmospheric forms to organically bound plant and microbial forms. As mentioned earlier, photosynthesis converts CO_2 from the atmosphere into organically bound forms in plants, algae, and cyanobacteria. Biological N_2 fixation is a unique transformation carried out by a number of bacteria, which convert N_2 gas into NH_3 for biological uptake. These bacteria are most prevalent in symbiotic relationships with plants, such as *Rhizobium* that forms nodules on the roots of clovers where the nitrogenase enzyme catalyzes the reaction. Fertilizer manufacturing converts N_2 gas into NH_3 in a similar manner without an enzyme, but rather large quantities of energy necessary to create the pressure required for the transformation.

Under certain conditions, both inorganic C and N can be chemically fixed in the subsoil. Carbon dioxide forms carbonic acid in water, which can precipitate with the basic cations, Ca^{2+}, Mg^{2+}, and Na^+, to form pedogenic carbonates. Inorganic C is most abundant in soils of the semiarid and arid regions. Ammonium can be fixed as nonexchangeable components of the lattice structure of 2:1-type clay minerals, which are especially prevalent in the subsoil of many younger soils.

REFERENCES

1. Follett, R.F., Stewart, J.W.B., Cole, C.V., Ed. *Soil Fertility and Organic Matter as Critical Components of Production Systems*; Spec. Publ. No. 19, Soil Science Society of America: Madison, WI, 1987; 166.
2. Burns, R.G., Ed. *Soil Enzymes*; Academic Press: London, 1978; 380.
3. Tabatabai, M.A. Soil Enzymes. In *Methods of Soil Analysis, Part 2. Microbiological and Biochemical Properties*; Book Series No. 5, Weaver, R.W., Angle, J.S., Bottomley, P.S., Eds.; Soil Science Society of America: Madison, WI, 1994; 775–833.
4. Stevenson, F.J.; Cole, M.A. *Cycles of Soil: Carbon, Nitrogen, Phosphorus, Sulfur, Micronutrients*, 2nd Ed.; Wiley: New York, 1999; 427.
5. Schlesinger, W.H. *Biogeochemistry: An Analysis of Global Change,* 2nd Ed.; Academic Press: San Diego, CA, 1997; 588.
6. Alexander, M.A. *Introduction to Soil Microbiology,* 2nd Ed.; Krieger: Malabar, FL, 1991; 467.
7. Sylvia, D.M.; Fuhrmann, J.J.; Hartel, P.G.; Zuberer, D.A. *Principles and Applications of Soil Microbiology*; Prentice-Hall: Upper Saddle River, NJ, 1998; 550.
8. Lal, R., Kumble, J.M., Follett, R.F., Cole, C.V., Ed. *The Potential of U.S. Cropland to Sequester Carbon and Mitigate the Greenhouse Effect*; Ann Arbor Press: Chelsea, MI, 1998; 128.
9. Tate, R.L. III *Soil Organic Matter: Biological and Ecological Effects*; Wiley: New York, 1987; 291.
10. Alexander, M. *Biodegradation and Bioremediation*; Academic Press: San Diego, CA, 1994; 302.
11. Stevenson, F.J., Ed. *Nitrogen in Agricultural Soils*; Agronomy Monograph 22, American Society of Agronomy: Madison, WI, 1982; 940.

E

ECONOMICS OF SOIL MANAGEMENT IN DEVELOPING COUNTRIES

Stefano Pagiola
World Bank, Washington, D.C., U.S.A.

INTRODUCTION

Agriculture plays a vital role in the economies of developing countries and in the welfare of their populations, including many of their poorest members. Inappropriate soil management is thought to threaten the sustainability of agricultural production by undermining the soil resource base on which agriculture depends (1–4). Developing-country farmers often have limited capacity to substitute fertilizers and other inputs for soil fertility, making the problem particularly acute. Addressing such problems is not easy, however, because decisions about soil management are made not by governments but by individual land users according to their own priorities and constraints. Economic analysis of soil-management problems in developing countries has focused increasingly on understanding why farmers use soils in the way they do, and in particular on understanding their reasons for not using them sustainably.

ASSESSING THE PROBLEM

Soil has value because of its role in crop production. Cultivation, however, can damage the soil. For example, clearing vegetation cover exposes soil to water and wind erosion; repeated tillage weakens soil structure; crop production removes nutrients; and use of machinery leads to soil compaction. The economic impacts of land degradation can be divided into on-site and off-site effects. On-site, degradation can result in lower yields (and, hence, lower revenues), or in the need for higher input levels (and, hence, higher costs) to maintain yields (5). Off-site, degradation can cause problems such as sedimentation and changes in hydrological flows (6,7). Adverse off-site effects can also occur in the absence of on-site effects (e.g., if agrochemicals contaminate water supplies). More recently, land degradation has also been seen as contributing to global problems, such as climate change, by reducing soil's ability to store carbon (8,9). On-site problems have usually been thought to be most important in developing countries (10), although more recently attention has focused on off-site effects, particularly in Asia and Latin America (11).

Assessing the absolute and relative magnitude of soil degradation and its consequences is difficult. Valuing on-site effects is straightforward in principle, but it has proved empirically difficult in practice because of the lack of appropriate data, particularly on the yield impact of degradation. The site-specificity of soil-management relationships limits the applicability of data collected in one location to analysis of problems at another. Efforts to value off-site effects are also constrained by insufficient data, exacerbated by unclear cause–and–effect relationships between events that are widely separated in both space and time (12). In sub-Saharan Africa, estimated annual on-site losses from degradation range from under 1% of agricultural gross domestic product (GDP) in Madagascar, Mali, and South Africa, to as much as 8% in Zimbabwe (13). Three different studies of Ethiopia, however, showed estimated annual losses of less than 1% (14), 4% (15), and 5% (16) of agricultural GDP, demonstrating the weakness of the data and the dependence on the assumptions made.

Although data weaknesses make it difficult to arrive at strong conclusions, fears of catastrophic damage from soil degradation do not appear to have been realized. A review of 26 global and regional studies and 54 national studies concludes that degradation is not likely to threaten aggregate global food supply by 2020, although degradation problems can be acute at the local level (4). An analysis of El Salvador, for example, concludes that while the oft-cited estimate that "75% of the country is degraded" is almost certainly exaggerated, degradation does affect about one-third of farm households—a significant social problem by any standard (17).

UNDERSTANDING FARMER BEHAVIOR

Because farmers adopt soil-management practices that degrade the soil, many researchers have concluded that farmers are ignorant or tradition-bound. Beginning with the work of T.W. Schultz (18), however, it has been increasingly recognized that developing-country farmers are, in fact, rational decision makers (19, 20). Recent research demonstrates that they are quite aware of the properties and behavior of their soils, and of the problems that inappropriate soil management can cause (21–25).

Farmers experience the effects of any on-site problems directly, so they generally have a direct incentive to respond to them. But even though sustainable practices may bring long-term benefits, they can also be costly, both directly in terms of investment requirements and maintenance costs, and indirectly in terms of forgone production. The critical question farmers face is whether the long-term benefits of sustainable practices make these costs worth bearing. In many cases, the answer is clearly yes. Almost all farmers in the Kitui/Machakos area of Kenya, for example, have adopted some form of conservation practice (26, 27), while small farmers in El Salvador have done so on over one-half the fields on moderate and steep slopes (17). In some instances, however, the costs of adopting particular conservation techniques exceed the benefits. Analysis of conservation measures in six Central American and Caribbean countries, for example, found that many measures promoted by conservation projects had negative returns from the farmers' perspective (28).

Many factors can affect the profitability of conservation measures from the farmers' perspective. These include the rate and severity of damage to the soil; the consequent effects on productivity; the value of lost production (for sale or subsistence consumption); the extent to which fertilizers and other inputs can cost-effectively substitute for lost fertility; the effectiveness of available conservation measures and their costs, including initial investment and maintenance; the relative riskiness of production, with and without conservation; and the farmers' own preferences, including—crucially—their preference for present as opposed to future consumption (i.e., their *discount rate*) (27–30). Since all of these factors can vary substantially, even in small areas, it should not be surprising that the extent of adoption of conservation measures also varies substantially. Adoption can also vary over time: When initial productivity losses are small, it makes sense to delay adopting conservation practices until later (31, 32).

Poverty is often thought to play a significant role in conservation-adoption decisions of developing-country farmers, by causing them to emphasize short-term over long-term benefits and limiting their ability to undertake investments (33, 34). Conversely, it has been argued that poorer households may have greater incentives to conserve soil than better-off households, because they face greater penalties for failing to do so (35). Empirically, one can also observe that poor farmers do often undertake a variety of long-term investments—in livestock, tree crops, the education of their children, and other assets with long-term returns. Thus, in the case of Kitui/Machakos, even expensive conservation measures such as terraces have been widely adopted by poor farmers lacking access to formal credit markets (26, 27). Like many other aspects of the soil-conservation debate, the exact role played by poverty is likely to vary from case to case.

Government policies can often affect the perceived incentives to adopt conservation practices. Until recently, many developing countries' policies discriminated heavily against agriculture (20, 36), making investments in agriculture, including conservation investments, less attractive. The policy environment has been identified as the main cause of degradation in Colombia, for example (37). The relationship between government policies and conservation is complex, however, and could go in either direction depending on the details of the policy and the specific conditions encountered at any given site (38, 39). By keeping prices low, for example, Kenyan maize price policy tended to reduce returns to conservation in Kitui/Machakos, and thus discourage its adoption, but even in that area the opposite effect was also observed (39).

Institutions often play a critical role in conservation decisions. A key institutional factor is the security of rights to land. Given the long-term nature of conservation investments, incentives to undertake them will be lower if farmers fear losing their land (40, 41). Lack of titles is not, however, always synonymous to tenure insecurity. A substantial literature demonstrates that many traditional African land tenure systems do not result in tenure insecurity, even in the absence of titles (42). Institutions can also play other roles. In Kitui/Machakos, women's labor exchange groups (*mwethya*) substituted for missing credit markets and allowed farmers to undertake costly conservation measures (27).

When the effects of soil degradation are felt primarily off-site, the situation is very different. Because these problems affect others rather than themselves, farmers have no incentive to address them. Many soil conservation projects have been motivated by the desire to protect downstream infrastructure, such as dams, from the effects of upstream degradation. Because the practices promoted by these projects are not necessarily profitable from the farmers' perspective, their success rate has been very low (43). Subsidies and/or compulsion can sometimes lead to temporary adoption of the measures being promoted, but they tend to be rapidly abandoned once the project is completed (28, 44).

CONCLUSIONS

Land degradation problems seem unlikely to cause the catastrophic agricultural declines some fear, but they can have acute consequences at the local level, particularly for poor farmers. Farmers have strong incentives to respond to the on-site effects of soil degradation, and the evidence shows they often do so. The changing understanding of farmers' incentives to undertake appropriate soil-management practices has shifted the emphasis of interventions away from top-down projects focusing on technical solutions. Overall policy reform is now seen as a necessary precondition for any intervention. Where on-site effects dominate, participative projects in which responses appropriate to local conditions are developed in collaboration with the affected land users, have come to play an important role. Where off-site effects dominate, recent efforts have turned to the development of systems of payments for environmental services.

REFERENCES

1. Eckholm, E. *Losing Ground: Environmental Stress and World Food Prospects*; W.W. Norton: New York, 1976.
2. Brown, L.R.; Wolf, E.C. *Soil Erosion: Quiet Crisis in the World Economy*, Worldwatch Paper No. 60; Worldwatch Institute: Washington, DC, 1984.
3. Pimentel, D.; Harvey, C.; Resosudarmo, P.; Sinclair, K.; Kurz, D.; McNair, M.; Crist, S.; Shriptz, L.; Fitton, L.; Saffouri, R.; Blair, R. Environmental and Economic Costs of Soil Erosion and Conservation Benefits. Science **1995**, *267*, 1117–1123.
4. Scherr, S.J. *Soil Degradation: A Threat to Developing Country Food Security by 2020*, Food, Agriculture and Environment Discussion Paper No. 27; IFPRI: Washington, DC, 1999.
5. Lal, R.C. Effects of Erosion on Soil Productivity. CRC Critical Reviews in Plant Sciences **1987**, *5* (4), 303–367.
6. Hamilton, L.S.; King, P.N. *Tropical Forest Watersheds: Hydrologic and Soils Response to Major Uses and Conversions*; Westview Press: Boulder, Colorado, 1983.
7. Chomitz, K.; Kumari, K. The Domestic Benefits of Tropical Forests: A Critical Review. World Bank Research Observer, **1998**, *13* (1), 13–35.
8. Pagiola, S. *The Global Environmental Benefits of Land Degradation Control on Agricultural Land*, Environment Paper No.16; World Bank: Washington, DC, 1999.
9. Lal, R.; Logan, T.J. Agricultural Activities and Greenhouse Gas Emissions from Soils in the Tropics. In *Soil Management and Greenhouse Effect: Advances in Soil Science*; Lal, R., Kimble, J., Levine, E., Stewart, B.A., Eds.; CRC Press: Boca Raton, Florida, 1995.
10. Magrath, W.B.; Arens, P. *The Cost of Soil Erosion on Java: A Natural Resource Accounting Approach*, Environment Department Working Paper No. 18; World Bank: Washington, DC, 1989.
11. Asian Productivity Organization (APO). *Soil Conservation and Watershed Protection in Asia and the Pacific*; Asian Productivity Organization: Tokyo, 1995.
12. Walling, D.E. Measuring Sediment Yield from River Basins. In *Soil Erosion Research Methods*; Lal, R., Ed.; Soil and Water Conservation Society: Ankeny, Iowa, 1988.
13. Bojö, J. The Costs of Land Degradation in Sub-Saharan Africa. Ecological Economics **1996**, *16* (2), 161–173.
14. Food and Agriculture Organisation (FAO). *Highlands Reclamation Study: Ethiopia*; FAO: Rome, 1986.
15. Bojö, J.; Cassells, D. *Land Degradation and Rehabilitation in Ethiopia: A Reassessment*, AFTES Working Paper No.17; World Bank: Washington, DC, 1995.
16. Sutcliffe, J.P. *Economic Assessment of Land Degradation in the Ethiopian Highlands: A Case Study*. National Conservation Strategy Secretariat, Ministry of Planning and Economic Development: Addis Ababa, 1993.
17. Pagiola, S.; Dixon, J.A. Land Degradation Problems in El Salvador. In *El Salvador: Rural Development Study*, Report No.16253-ES; World Bank: Washington, DC, 1997.
18. Schultz, T.W. *Transforming Traditional Agriculture*; University of Chicago Press: Chicago, 1964.
19. Popkin, S.L. *The Rational Peasant*; University of California Press: Berkeley, 1979.
20. Timmer, C.P.; Falcon, W.P.; Pearson, S.R. *Food Policy Analysis*; Johns Hopkins University Press for the World Bank: Baltimore, Maryland, 1983.
21. Forsyth, T. The Use of Cesium-137 Measurements of Soil Erosion and Farmers' Perceptions to Indicate Land Degradation Amongst Shifting Cultivators in Northern Thailand. Mountain Research and Development **1994**, *14* (3), 229–244.
22. Kiome, R.M.; Stocking, M. Rationality of Farmer Perception of Soil Erosion. Global Environmental Change **1995**, *5* (4), 281–295.
23. Pender, J.; Kerr, J. *Determinants of Farmers' Indigenous Soil and Water Conservation Investments in the India's Semi-Arid Tropics*, EPTD Discussion Paper No.17; IFPRI: Washington, DC, 1996.
24. Reij, C. Building on Tradition: The Improvement of Indigenous Soil and Water Conservation Techniques in the West African Sahel. In *Adopting Conservation on the Farm*; Napier, T.L., Camboni, S.M., El-Swaify, S.A., Eds.; Soil and Water Conservation Society: Ankeny, Iowa, 1994.
25. Ryder, R. Farmer Perception of Soils in the Mountains of the Dominican Republic. Mountain Research and Development **1996**, *14* (3), 261–266.
26. Tiffen, M.; Mortimore, M.; Gichuki, F. *More People, Less Erosion: Environmental Recovery in Kenya*; John Wiley: Chichester, England, 1994.
27. Pagiola, S. Soil Conservation in a Semi-Arid Region of Kenya: Rates of Return and Adoption by Farmers. In *Adopting Conservation on the Farm*; Napier, T.L., Camboni, S.M., El-Swaify, S.A., Eds.; Soil and Water Conservation Society: Ankeny, Iowa, 1994.
28. Lutz, E.; Pagiola, S.; Reiche, C. Cost-Benefit Analysis of Soil Conservation: The Farmers' Viewpoint. The World Bank Research Observer **1994**, *9* (2), 273–295.

29. Bojö, J. *The Economics of Land Degradation: Theory and Applications to Lesotho*; Stockholm School of Economics: Stockholm, 1991.
30. Shively, G.E. Consumption Risk, Farm Characteristics, and Soil Conservation Adoption Among Low-Income Farmers in the Philippines. Agricultural Economics **1997**, *17*, 165–177.
31. Walker, D.J. A Damage Function to Evaluate Erosion Control Economics. American Journal of Agricultural Economics **1982**, *64* (4), 690–698.
32. Pagiola, S.; Bendaoud, M. *Long-Run Economic Effects of Erosion on Wheat Production in a Semi-Arid Region of Morocco: A Simulation Analysis*, Agricultural Economics Staff Paper No. AE 95-12; Washington State University: Pullman, 1994.
33. Holden, S.T.; Shiferaw, B.; Wik, M. Poverty, Market Imperfections and Time Preferences: Of Relevance for Environmental Policy? Environment and Development Economics **1998**, *3* (1), 105–130.
34. Reardon, T.; Vosti, S.A. Poverty-Environment Links in Rural Areas of Developing Countries. In *Sustainability, Growth, and Poverty Alleviation: A Policy and Agroecological Perspective*; Vosti, S.A., Reardon, T., Eds.; Johns Hopkins University Press for IFPRI: Baltimore, Maryland, 1997.
35. Pagiola, S. The Effect of Subsistence Requirements on Sustainable Land Use Practices. Annual Meetings of the American Agricultural Economics Association, Indianapolis, Indiana, Aug 6–9, 1995.
36. Schiff, M.; Valdès, A. *The Political Economy of Agricultural Pricing Policy*; Johns Hopkins University Press: Baltimore, Maryland, 1992.
37. Heath, J.; Binswanger, H. Natural Resource Degradation Effects of Poverty and Population Growth are Largely Policy-Induced: The Case of Colombia. Environment and Development Economics **1996**, *1* (1), 65–83.
38. LaFrance, J.T. Do Increased Commodity Prices Lead to More or Less Soil Degradation? Australian Journal of Agricultural Economics **1992**, *36*, 57–82.
39. Pagiola, S. Price Policy and Returns to Soil Conservation in Semi-Arid Kenya. Environmental and Resource Economics **1996**, *8*, 255–271.
40. Ervin, D.E. Constraints to Practicing Soil Conservation: Land Tenure Relationships. In *Conserving Soil: Insights from Socioeconomic Research*; Lovejoy, S.B., Napier, T.L., Eds.; Soil and Water Conservation Society: Ankeny, Iowa, 1986.
41. Wachter, D. Land Titling: Possible Contributions to Farmland Conservation in Central America. In *Economic and Institutional Analyses of Soil Conservation Projects in Central America and the Caribbean*; Lutz, E., Pagiola, S., Reiche, C., Eds.; Environment Paper No. 8; World Bank: Washington, DC, 1994.
42. Place, F.; Hazell, P. Productivity Effects of Indigenous Land Tenure Systems in Sub-Saharan Africa. American Journal of Agricultural Economics **1993**, *75* (1), 10–19.
43. Pagiola, S. Economic Analysis of Incentives for Soil Conservation. In *Using Incentives for Soil Conservation: From Theory to Practice*; Sanders, D.W., Huszar, P.C., Sombatpanit, S., Enters, T., Eds.; Science Publishers: Enfield, 1999.
44. Enters, T. The Token Line: Adoption and Non-Adoption of Soil Conservation Practices in the Highlands of Northern Thailand. In *Soil Conservation Extension: From Concepts to Adoption*; Sombatpanit, S., Zöbisch, M.A., Sanders, D.W., Cook, M.G., Eds.; Science Publishers: Enfield, 1997.

E

ECONOMICS OF SOIL MANAGEMENT IN THE UNITED STATES

Jeffrey W. Hopkins
United States Department of Agriculture, Washington, D.C., U.S.A.

INTRODUCTION

The economics of soil management differs from the physical science of soil management—it looks at maximizing long-run economic return, rather than maximizing soil performance and quality. With regard to maximizing soil performance, the prescription from soil scientists is clear and can be summarized as follows: Add organic matter, avoid most tillage and other activities that might erode or compact the soil, apply chemicals on an as-needed basis, and keep the ground covered between growing seasons; increase crop diversity across a landscape; and systematically monitor soil chemical and physical properties (1).

Economists have a long and continuing tradition of evaluating soil management practices as they affect soil fertility and soil degradation. But economists realize that, in practice, soil management techniques are applied only to the point where the farmer (or society) is better off economically. Because economics considers all inputs and outputs of soil management, including off-site impacts, the environmental impacts of soil management demand constant study as well. Therefore, an economic assessment of soil management must extend beyond the farm field to consider off-site effects on water, air quality, and wildlife habitat.

ECONOMIC EVALUATION OF SOIL MANAGEMENT PRACTICES

Soil Fertility Management

Starting in the 1950s, economists used static production functions to formulate soil fertility recommendations, relying heavily on data generated by soil scientistzs and agronomists on crop yield response to soil management practices (2–4). Economists have since sought improvement over the single-period crop soil management recommendation by searching for solutions that increase both current and future profits. In cases in which a dynamic biological process is considered, such as fertilizer carry-over (5), or crop response to management is highly variable (6), multiperiod analysis has proven to be particularly important. In the 1990s, with advances in computer, positioning systems, and application technology, many producers, as well as scientists, have become interested in managing spatial variability within a field (7). Economic management of site-specific soil conditions requires significant knowledge about the variability of soil properties and how these impact crop response. While the relative profitability of site-specific soil management is still unproven, some studies have shown that modest improvements in economic return can be achieved (8, 9).

Soil Degradation Management

Soil fertility research finds ways to increase yields; soil degradation research looks for soil management techniques to keep crop yield growth from leveling off. In the United States, water and wind erosion and soil compaction remain the primary soil degradation concerns (10) despite a 38% decrease in cropland wind and water erosion from 1982 to 1997 (11). Soil erosion decreases potential crop yield by depleting soil nutrients and topsoil depth. Some have argued that it is the latter effect that poses the greatest threat to future food security (12). Because the rate of soil genesis is so low, topsoil losses are often considered irreversible and, therefore, of particular concern. This argument has apparently been particularly persuasive in the United States, where conservation policy abandoned an earlier recommendation to limit degradation to "tolerable" nutrient losses, and, instead, focused exclusively on limiting degradation to "tolerable" losses of soil depth (13). However, U.S. irreversible topsoil losses have been shown to be slight in aggregate (14), especially when viewed in relation to past yield increases from improvements in technology and management.

Off-Site Impacts of Soil Management

Except for doubts regarding the ability to continue to achieve productivity gains that doubled world grain harvests since 1960 (15), conventional thinking among economists is that the downstream costs of poor soil management (borne by users of downstream surface water) are larger than the combined costs of on-site productivity losses and future food security (16). To decrease the likelihood of both social and private costs from nonoptimal soil management, U.S. conservation policy has promoted farmer adoption of soil-enhancing practices. These *best management practices* hold the promise of a win–win solution for farmers and society at large. However, farmer adoption is far from universal for any single practice.

ADOPTION OF SOIL MANAGEMENT PRACTICES IN THE UNITED STATES

Adoption of most agricultural innovations has been shown to depend on both *farm* and *farmer* characteristics. The experience of U.S. hybrid corn adoption is instructive in understanding why best management practice adoption may not be universal. While hybrid corn was especially quick to catch on in Iowa, adoption of open-pollinated corn elsewhere was in part limited by the availability (different hybrids appeared in different parts of the country) and acceptability (some farmers were reluctant initially) of the technology (17). Adoption of soil-based best management practices is likewise limited to factors related to the farm (climate and soils) and the farmer (firm management and household characteristics). The following material examines soil performance characteristics and on-farm adoption patterns of three best management practices—conservation tillage, monitoring soil properties, and crop rotations.

Conservation Tillage

Conservation tillage is a system of crop production that leaves additional crop residue remaining on the field after harvest to decompose naturally. Conservation tillage can increase soil organic matter and, therefore, increase soil productivity by increasing moisture retention and improving soil tilth. It can also drastically decrease soil erosion levels compared to conventional tillage methods, and similarly reduce some rates of nutrient loss, particularly phosphorus. While adoption of the practice (defined strictly as 30% or more of the soil surface covered with flat and standing residue) grew steadily through the 1970s until the mid 1990s, and is still growing for some crops, overall adoption has hovered over the past 5 years at 35–37% of U.S. planted cropland, according to field surveys (18, 20).

The adoption rate of conservation tillage follows a regional distribution, with the highest rate (about 50%) among corn and soybeans producers in the Corn Belt, and lower adoption rates (about 10%) by cotton and small grain producers in the West (19). While adoption of conservation tillage can result in stable or increased crop yields compared to conventional systems, as well as decreased input costs (primarily machinery, chemical, fuel and labor costs) these savings are not always seen in practice (21). Any absolute economic advantage is likely derived only after several years of use of the system, and may still be slight unless multiplied over many acres. In 1995, conservation tillage adoption rates on farms with sales greater than $100,000 were more than 50% higher than on smaller farms (22).

Monitoring Soil Properties

Monitoring soil chemical and physical properties through soil testing can provide valuable information to farmers, allowing them to more closely match soil conditions with soil requirements for the growing crop. Optimal use of monitoring technology allows farmers to match soil conditions with crop-growing requirements both temporally (such as splitting nutrient applications into two smaller doses rather than one large dose), and spatially (dividing a field into different management units depending on soil conditions). Adoption of soil monitoring varies by the soil property monitored (20% of all cropland is tested for nutrient levels) and is greatest in high-value crops, particularly fruit, vegetables, and cotton production (20). Site-specific soil management systems include spatially monitoring crop yield and soil quality as well as spatially applying inputs. On-farm adoption of site-specific management has been less widespread than the soil performance gains suggest, with less than 1% adoption across all farms in the United States. Adoption of some components, such as intensive soil sampling, has appealed to certain types of farms, however, with corn and soybean producers (11% adoption) and high sales farms (18% adoption for farms with sales greater than $500,000) among the early adopters (23).

Crop Rotations

Crop rotations allow producers to increase soil organic matter and reduce soil erosion, although these soil performance improvements are most likely to occur when row crops are rotated with hay, meadow, or pasture. In 1997, while 80% of cropland was grown in some sort of rotation, only 3% of acreage followed the hay, meadow, or pasture rotation reported to maximize soil performance (20). The economic role of crop rotations may be different from the physical role. For corn producers, crop rotations are particularly popular with both very large farms (with more than $250,000 in sales) and with part-time farmers (22). For large farms, rotations could play a role in managing busy planting and harvesting periods, while for part-time farmers the role may be in managing on-farm and off-farm employment. Low use of hay rotations for row crop producers may be an indication that few producers are willing to forego any row crop sales in favor of greater soil performance. Cover crops, which do not occupy a full growing season, can effectively increase soil organic matter and decrease erosion, while still allowing farmers to produce the crop giving highest economic return. Cover crops, however, are used rarely outside of the South, where the practice is common with soybean production. In this system, some cash gain can be achieved in double-cropping winter wheat or rye (20).

IMPLEMENTING SOIL MANAGEMENT PROGRAMS IN THE UNITED STATES

In cases in which the off-site costs of sub-optimal soil management are great, such as where soil and nutrient runoff cause water quality problems, additional economic incentives have been used to increase adoption of a best management practice. For example, the Environmental Quality Incentive Program, administered by the U.S. Department of Agriculture, provides technical assistance and extension education, as well as cash payments to producers adopting certain best management practices. The program is unlike earlier soil management initiatives because it targets participation to maximize the environmental benefit per dollar spent on the program. In most cases, encouraging adoption first in those areas where off-site damages can be decreased cheaply is an improvement over programs that solely attempt to increase adoption in general. No matter how programs are targeted, however, increasing the existing size of the pool of adopters has come at increasing cost.

Because improved practices have costs as well as benefits, economists advise a farmer who invests in soil performance to do so selectively, with an eye toward balancing the costs of improving soil performance with the long-term economic opportunities afforded by the practice. In general, however, the cost to producers of an improvement in soil performance is usually more clear than the benefits, which are realized in the future. To address the uncertainty of benefits relative to costs, producers will continue to need information on how to profitably improve soil management.

REFERENCES

1. University of Minnesota Extension Service. Soil Management; BU-7390-GO. 2000.
2. Swanson, Earl D. The Static Theory of the Firm and Three Laws of Plant Growth. Soil Science **1963**, *95*, 338–343.
3. Heady, Earl O.; Dillon, John L. *Agricultural Production Functions*; The Iowa State University Press: Ames, 1961.
4. Hall, Hary H. Economic Evaluation of Crop Response to Lime. American Journal of Agricultural Economics **1983**, *65*, 811–817.
5. Kennedy, John O.S.; Whan, I.F.; Jackson, R.; Dillon, J.L. Optimal Fertilizer Carryover and Crop Recycling Policies for a Tropical Grain Crop. Australian Journal of Agricultural Economics **1973**, *17*, 104–113.
6. Antle, John M. Sequential Decision Making in Production Models. American Journal of Agricultural Economics **1983**, *75*, 282–290.
7. Sawyer, John E. Concepts of Variable Rate Technology with Consideration for Fertilizer Application. Journal of Production Agriculture **1994**, *7*, 195–201.
8. Schnitkey, Gary D.; Hopkins, Jeffrey W.; Tweeten, Luther G. An Economic Evaluation of Precision Fertilizer Applications on Corn-Soybean Fields. In *Proceedings of the Third International Conference on Precision Agriculture, Minneapolis, Minnesota,* June 23–26, 1996; Robert, P.C., Rust, H.R., Larson, W.E., Eds.; American Society of Agronomy, Crop Science Society of American, and Soil Science Society of America: Madison, Wisconsin, 1996; 977–987.
9. Babcock, Bruce A.; Pautsch, Gregory R. Moving from Uniform to Variable Fertilizer Rate on Iowa Corn: Effects on Rates and Returns. Journal of Agricultural and Resource Economics **1998**, *23* (2), 385–400.
10. Syers, J.K. Managing Soils for Long-Term Productivity. In *Land Resource on the Edge of the Malthusian Precipice*; Greenland, D.J., Gregory, P.J., Nye, P.H., Eds.; CAB International: New York, 1997; 151–160.
11. Summary Report: 1997 National Resources Inventory; Natural Resources Conservation Services, U.S. Department of Agriculture: Washington, D.C., December, 1999.
12. Carter, V.G.; Dale, T. *Topsoil and Civilization*; University of Oklahoma Press: Norman, Oklahoma, 1974.
13. Hall, G.F.; Logan, T.J.; Young, K.K. Criteria for Determining Tolerable Erosion Rates. In *Soil Erosion and Crop Productivity*; Follett, R.F., Stewart, B.A., Eds.; ASA-CSSA-SSSA: Madison, Wisconsin, 1985; 173–188.
14. Crosson, Pierre. Future Supplies of Land and Water for World Agriculture. In *Population and Food in the Early Twenty-First Century: Meeting Future Food Demand of an Increasing Population*; Islam, N., Ed.; International Food Policy Research Institute: Washington, 1995; 143–159.
15. Mann, Charles. Crop Scientists Seek a New Revolution. Science **1999**, January 15, *283*, 310–314.

16. Trimble, Stanley W.; Crosson, Pierre. U.S. Soil Erosion Rates—Myth and Reality. Science **2000**, July 14, *289*, 248–250.
17. Griliches, Zvi. Hybrid Corn: An Exploration in the Economics of Technological Change. Econometrica **1957**, *25* (4), 501–522.
18. Lal, Rattan; Kimble, J.M.; Follett, R.F.; Cole, C.V. *The Potential of U.S. Cropland to Sequester Carbon and Mitigate the Greenhouse Effect*; Sleeping Bear Press: Chelsea, Michigan, 1998.
19. CTIC, *Crop Residue Management Executive Summary*; Conservation Tillage Information Center: West Lafayette, Indiana, November, 1998.
20. Padgitt, Merritt; Newton, Doris; Penn, Renata; Sandretto, Carmen. *Production Practices for Major Crops in U.S. Agriculture, 1990–1997*; Agricultural Statistical Bulletin No. 969, June Economic Research Service, USDA: Washington, DC, June 2000.
21. Day, John C.; Sandretto, Carmen L.; McBride, William D.; Breneman, Vincent E. *Conservation Tillage in U.S. Corn Production: An Economic Appraisal*, American Agricultural Economics Association Annual Meeting, Salt Lake City, Utah, August 2–5, 1998.
22. Soule, Meredith *Soil Management and the Farm Typology: Do Small Family Farms Manage Soil and Nutrient Resources Differently than Large Family Farms?* Agricultural and Resource Economics Review, **2001**, *30* (2), 179–188
23. Daberkow, Stan; McBride, William. *Adoption of Precision Agriculture Technologies by U.S. Farmers*, 5th International Conference on Precision Agriculture, Minneapolis, Minnesota, July 16–19, 2000.

ENCHYTRAEIDAE

María Jesús Iglesias Briones
University of Vigo, Vigo, Spain

INTRODUCTION

Enchytraeidae (potworms) constitute a little known family of aquatic (both marine and freshwater) and terrestrial worms within the class Oligochaeta, phylum Annelida. They are less conspicuous in size than other soil Annelids like earthworms, being between 10 to 50 mm long (Fig. 1). They are regarded as "mesofauna," together with Acari and Collembola, as they occupy pore spaces with a diameter of less than 2 mm. Like the rest of Oligochaetes, enchytraeids are hermaphroditic with amphimictic reproduction, although there are some species which can reproduce parthenogenetically and asexually by fragmentation and also by self-fertilization.

IDENTIFICATION

The lack of taxonomic information about enchytraeids is due to their small size and the confusing identification criteria. Only after the revisions published by Nielsen and Christensen (1–3) was it possible to resolve the taxonomy of this group by providing a standard set of criteria for identification of genera and species. More than 600 species have now been recognised, although the list is still increasing.

Taxonomical criteria include external characteristics such as colour (although the majority of the species are grey or whitish and usually gut contents, lymphocytes, chloragogen cells and blood vessels are responsible for other colorations), size, number of segments, cutaneous glands (size, shape and arrangement), setae (shape and arrangement) or setae follicles (except in the genus *Achaeta* where they are absent), dorsal pores (presence and position) and clitellum (position), and internal ones like the number and location of oesophageal appendages, salivary glands and septal glands, the origin of the dorsal vessel and the color of the blood, the shape of the nephridia and several reproductive organs (e.g., size and shape of seminal vesicles, length and shape of the sperm funnel, size and shape of the spermatheca or the number of eggs). The analysis of these features requires the careful study of living, mature specimens because fixative liquids turn the tegument opaque making it impossible to see the internal organs. Dózsa-Farkas (4) provided a list of 20 necessary characteristics for the identification and description of enchytraeids; this emphasises the difficulty of assigning specimens to the species level with certainty.

Futhermore, modern tools including spermatozoal ultrastructure, chemical composition of lipids, protein analysis and isoenzymes, immunological methods, and restriction fragmentation patterns have been successfully employed in identifying new species of enchytraeids.

SAMPLING

Sampling of field populations of enchytraeids is performed by taking soil cores of 5–6 cm diameter which gives an area of 20 cm^2 for accurate counting of the worms and easy calculation of the numbers per square meter. The depth of the core depends on the soil type; as enchytraeids are mainly concentrated in the surface organic horizons, sampling depth has to be determined from preliminary trials in the studied area. In relation to the number of sampled cores, O'Connor (5) established that a minimum number of 10 units would ensure a representative estimate of the population numbers in a given ecosystem. The extraction of enchytraeids from the soil cores is based on the movement of the worms in response to a gradient of light and temperature. Although several methods can be employed, the most widely used is "wet extraction" (Fig. 2), which has been shown to extract more than 95% of the total population of organic soils in three hours (6). A modified wet extractor has been proposed (7) in which heating is omitted and the length of the extraction time is increased to several days. This method proved to be the best in terms of efficiency, but the extremely long extraction time could be inconvenient when processing a large set of samples.

SPATIAL DISTRIBUTION AND POPULATION DYNAMICS

Terrestrial enchytraeids are distributed globally although the majority of the studies on their distribution have been

A

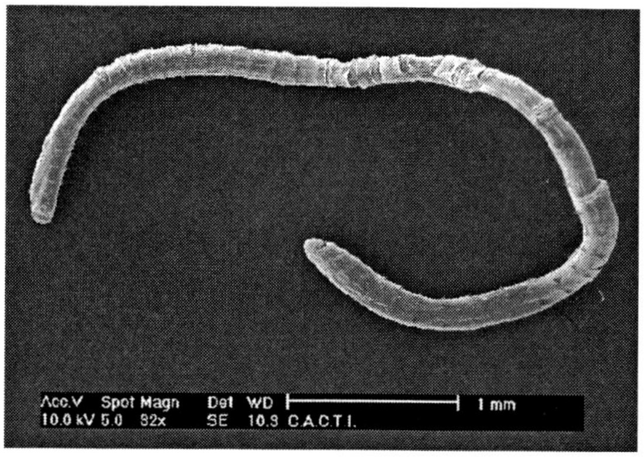

B

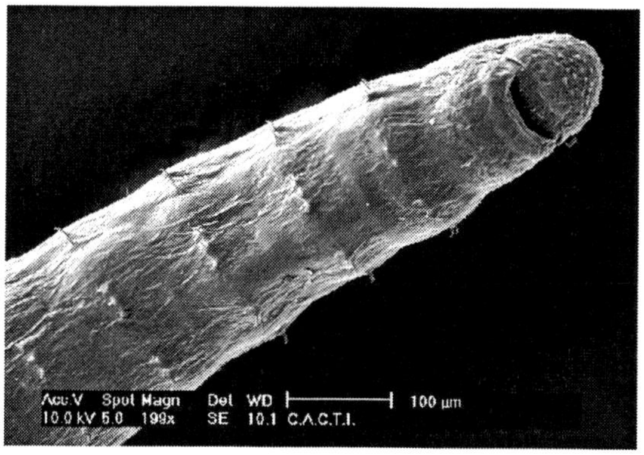

Fig. 1 Scanning electron micrograph (SEM) of *Cognettia sphagnetorum*: A. whole worm, B. ventral view of the head region showing the prostomium, the mouth and the setae follicles.

carried out in Europe. Climatic events, competition with other species, linkage with certain vegetation or soil types, and historical events are thought to be the main controlling factors of enchytraeid distribution (8).

Present knowledge of the distribution of enchytraeids is highly variable because most studies are based on a low number of samples from few localities, and that these samples are taken over a short period of time with different core sizes, sampling depths, and extraction methods. For these reasons it is very difficult to give accurate estimates of enchytraeid population sizes in different habitats. This results in the observed variability of abundance, with numbers ranging from zero through to an average of 350,000 individuals per m^2. Didden (9) reviewed the available literature and concluded that highest population densities are found in cold to temperate moist habitats, such as moorland soils, coniferous forests and grasslands; however, low numbers have been recorded in these habitats when other soil characteristics are unsuitable (e.g., soil type).

Besides the methods employed to obtain the mean density values of enchytraeids in various habitats, seasonal climatic fluctuations play an important role in controlling the population dynamics. Temperature and/or moisture content, as well as other biotic activities, seem to be the main factors responsible for the observed variation in enchytraeid population sizes. Summer droughts can drastically affect total numbers as these organisms cannot survive if the soil moisture content is less than 10% of field capacity, and their reproduction rate can be reduced at water contents of 15% (10). Severe cold winters can also profoundly affect reproduction with frequent frost periods leading to higher enchytraeid mortality.

Briones et al. (11) carried out a transplant experiment in order to determine whether the responses to climate were species dependent or whether the community as a whole was responding to changes in environmental conditions. The data suggested that temperature was the main controlling factor determining population sizes and that response was species specific: some species were positively influenced by the warmer temperatures and responded by increasing their reproduction rate (*Cognettia sphagnetorum*), whilst other species were either resistant to (*Achaeta eiseni*), and/or were dramatically reduced in numbers as result of the higher temperature regimes (*Cernosvitoviella atrata*). Differences in reproduction strategies, including asexual reproduction by fragmentation by the former species (Fig. 3), and physiological adaptations such as a thicker cuticle in several species of the genus *Achaeta*, could have conferred more resistance to the adverse conditions.

Enchytraeids mainly concentrate in the upper soil layers where organic matter is accumulated, although different species show varying vertical distributions. It is also known that some enchytraeid species can also migrate to deeper soil layers in response to moisture changes (11, 12), but numbers in deeper layers depend greatly on other factors such as stage of decomposition and differences in food supply (13). Vertical migration seems to be the short-term strategy to overcome adverse temporal climatic conditions (11, 12), at least for some species, whereas others are unable to burrow deeper (11). However, we lack information about similar responses in other species and whether this vertical movement would be a long-term survival strategy or whether food availability at deeper layers would be limiting their existence.

FEEDING BEHAVIOR

Culture methods suggested that enchytraeids are fungivorous rather than bacteriovorous (14) and ultrastructural studies of the intestinal content of several species of

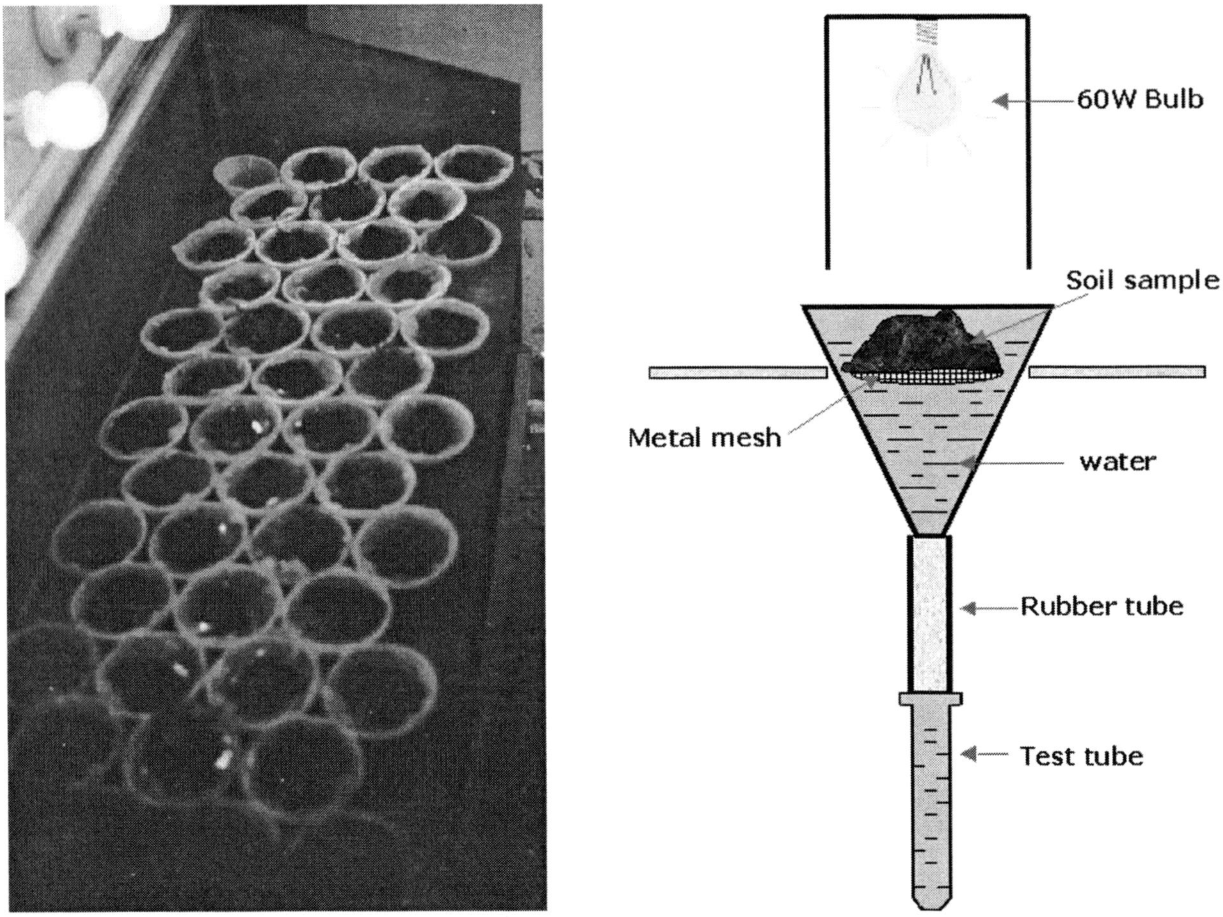

Fig. 2 Wet extractor: The soil sample placed on 1 mm stainless metal mesh is submerged in a water-filled funnel. The temperature is gradually increased to a maximum of 40°C; after three hours the worms are collected alive in the test tube.

enchytraeids show that they are saprovores consuming plant debris (15). Moreover, some studies suggest that enchytraeids ingest mineral particles, fungi, bacteria, oats and yeasts, algae, and even dead bodies of lumbricids and arthropods. However, recent evidence suggests that microbivory has been underestimated and that enchytraeids are 80% microbivorous and 20% saprovorous (9).

Enzymatic studies have been performed to gain a better understanding of the digestive processes, and suggest that enchytraeids are able to process polysaccharides and that they preferentially digest starch, xylan, and laminarin (16).

ECOLOGICAL IMPORTANCE

The contribution of the enchytraeids to energy and nutrient fluxes in soil ecosystems is the result of their consumption and assimilation efficiency, their influence on soil properties, and feedbacks to other trophic levels (9). This contribution is difficult to assess because it depends on their activity in relation to other decomposer invertebrates and microorganisms, and on the amount and quality of the organic matter available. The assimilation efficiency of enchytraeid populations is very low; as a result they have to ingest large quantities of organic material and considerable amounts of undigested material are produced, creating a perfect substrate for microbial growth. Together with tunneling and burrowing activities, the high production of faecal pellets, which later fuse to form a soil matrix, are key influences of enchytraeids on soil structure (17).

The feeding activities of enchytraeids also have strong effects on soil microflora. Hedlund and Augustsson (18) demonstrated that the grazing activities of *C. sphagnetorum* increased the hyphal length and respiration of the

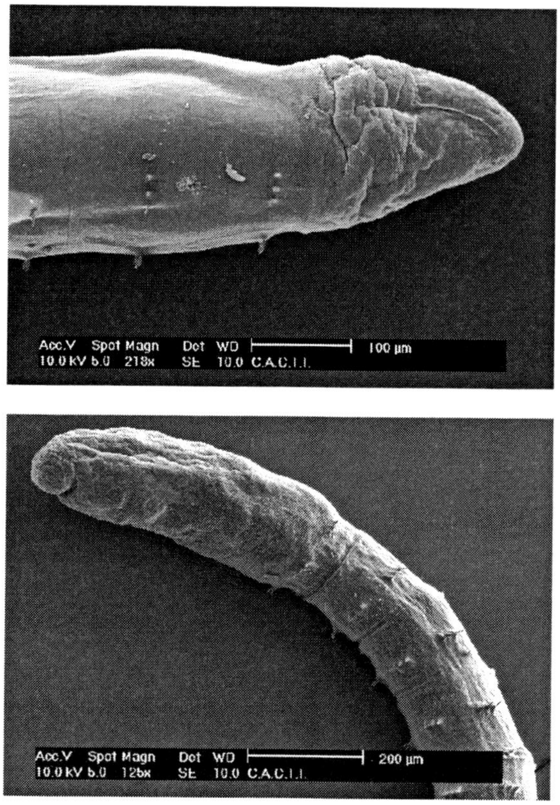

Fig. 3 SEM of a tail fragment of *Cognettia sphagnetorum* regenerating its head: A. early stage, B. later stage.

fungus *Mortierella isabellina* and this could have important implications for organic matter turnover. Similarly, Cole et al. (19) showed that enchytraeids significantly enhanced microbial activity by 35% in the surface horizon of blanket peat.

The contribution of enchytraeids to total soil respiration is generally no more than 2% (9). In a broad-leaved forest in England (20) the respiration of the total soil fauna was calculated to be 361 Kcal/m^2 per year of which enchytraeids were responsible for 167 Kcal/m^2.

Enchytraeidae can also influence strongly the immobilization or excretion of Ca^{++} (21) and the leaching of nutrients, particularly of N and P (22). Recent studies have demonstrated the effect of these organisms on the mineralization of carbon and on the release of dissolved organic carbon (DOC) (19, 23) into soil solution. The positive effect of enchytraeids on the release of DOC is thought to be controlled by the effect of climate on their total abundance and vertical distribution (23).

Despite all the information that is available the functional role of enchytraeids in terrestrial ecosystems has not yet been fully described. It is clear that in organic soils they are key organisms in terms of their effect on carbon and nitrogen dynamics. However, unlike with other soil fauna groups like earthworms (24) and nematodes (25), there still is not a functional classification of enchytraeids (functional group = a grouping of organisms which affect a process in a similar way) with respect to the different functional roles of enchytraeid species in ecosystems. An attempt at an ecological classification of these organisms has been performed by Graefe and Schmelz (26) in which species are classified in relation to pH, soil moisture, salinity, reproductive strategy, stress tolerance, and their occurrence in the gradient of humus forms.

REFERENCES

1. Nielsen, C.O.; Christensen, B. The Enchytraeidae, Critical Revision and Taxonomy of European Species. Nat. Jutl. **1959**, *8–9*, 7–160.
2. Nielsen, C.O.; Christensen, B. The Enchytraeidae, Critical Revision and Taxonomy of European Species (suppl. 1). Nat. Jutl. **1961**, *10*, 1–23.
3. Nielsen, C.O.; Christensen, B. The Enchytraeidae, Critical Revision and Taxonomy of European Species (suppl. 2). Nat. Jutl. **1963**, *10*, 1–19.
4. Dózsa-Farkas, K. List of Parameters Useful for Identification of Enchytraeids. Newslett. Enchytraeidae **1992**, *3*, 49–50.
5. O'Connor, F.B. The Enchytraeids. In *Methods of Study in Quantitative Soil Ecology*; Philipson, J., Ed.; Blackwell: Oxford, Edinburgh, 1971; 83–106.
6. O'Connor, F.B. Extraction of Enchytraeid Worms from a Coniferous Forest Soil. Nature **1955**, *175*, 815–816.
7. Graefe, U. Eine Einfache Methode der Extraktion von Enchytraeiden aus Bodenproben. In *Protokoll des Workshops zu Methoden der Mesofaunaerfassung (Moderation H. Koehler) und zu PCP-Wirkungen auf Collembolen und Andere Mesofauna-Gruppen (Moderation L. Beck)*; 1984; 17.
8. Römbke, J. Contribution to the Biogeography of Some Species of Terrestrial Enchytraeidae (Oligochaeta, Annelida). Soil Biol. Biochem. **1992**, *24* (12), 1283–1290.
9. Didden, V.A.M. Ecology of Terrestrial Enchytraeidae. Pedobiologia **1993**, *37*, 2–29.
10. Beylich, A.; Achazi, R.K. Influence of Low Soil Moisture on Enchytraeids. Newslett. Enchytraeidae **1999**, *6*, 49–58.
11. Briones, M.J.I.; Ineson, P.; Piearce, T.G. Effects of Climate Change on Soil Fauna; Responses of Enchytraeids, Diptera Larvae and Tardigrades in a Transplant Experiment. App. Soil Ecol. **1997**, *6*, 117–134.
12. Springett, J.A.; Brittain, J.E.; Springett, B.P. Vertical Movement of Enchytraeidae (Oligochaeta) in Moorland Soils. Oikos **1970**, *21*, 16–21.
13. Didden, V.A.M.; Fluiter, R. Dynamics and Stratification of Enchytraeidae in the Organic Layer of a Scots Pine Forest. Biol. Fertil. Soils **1988**, *26*, 305–312.
14. Standen, V.; Latter, P.M. Distribution of a Population of *Cognettia sphagnetorum* (Enchytraeidae) in Relation to Microhabitats in a Blanket Bog. J. Anim. Ecol. **1977**, *46*, 213–229.

15. Toutain, F.; Villemin, G.; Albrecht, A.; Reisinger, O. Etude Ultrastructurale des Processus de Biodégradation II. Modèle Enchytraeides-Litière de Feuillus. Pedobiologia **1982**, *23*, 145–156.
16. Urbášek, F.; Chalupský, J. Activity of Digestive Enzymes in 4 Species of Enchytraeidae (Oligochaeta). Rev. Écol. Biol. Sol **1991**, *28* (2), 145–154.
17. Dawood, V.; FitzPatrick, E.A. Some Population Sizes and Effects of the Enchytraeidae (Oligochaeta) on Soil Structure in a Selection of Scottish Soils. Geoderma **1993**, *56*, 173–178.
18. Hedlund, K.; Augustsson, A. Effects of Enchytraeid Grazing on Fungal Growth and Respiration. Soil Biol. Biochem. **1995**, *27* (7), 905–909.
19. Cole, L.; Bardgett, R.D.; Ineson, P. Enchytraeid Worms (Oligochaeta) Enhance Mineralization of Carbon in Organic Upland Soils. Eur. J. Soil Sci. **2000**, *51*, 185–192.
20. Satchell, J.E. Feasibility Study of an Energy Budget for Meathop Wood. In *Productivity of Forest Ecosystems*; Duvigneaus, P., Ed.; Unesco: Paris, 1971; 619–630.
21. Anderson, J.M.; Ineson, P.; Huish, S.A. The Effects of Animal Feeding Activities on Element Release from Deciduous Forest Litter and Soil Organic Matter. In *New Trends in Soil Biology*; Lebrun, P., André, H.M., de Medts, A., Grégoire-Wibo, C., Wauthy, G., Eds.; Louvain-La-Neuve; 1983; 87–100.
22. Briones, M.J.I.; Carreira, J.; Ineson, P. *Cognettia sphagnetorum* (Enchytraeidae) and Nutrient Cycling in Organic Soils: A Microcosm Experiment. Appl. Soil Ecol. **1988a**, *9*, 289–294.
23. Briones, M.J.I.; Ineson, P.; Poskitt, J. Climate Change and *Cognettia sphagnetorum*: Effects on Carbon Dynamics in Organic Soils. Funct. Ecol. **1998b**, *12*, 528–535.
24. Bouché, M.B. *Lombriciens de France. Écologie et Systématique*; Institut de la Récherche Agronomique: Paris, France, 1972; 671.
25. Yeates, G.W.; Bongers, T.; de Goede, R.G.M.; Freckman, D.W.; Georgieva, S.S. Feeding Habits in Soil Nematode Families and Genera—An Outline for Soil Ecologists. J. Nematol. **1993**, *25*, 315–331.
26. Graefe, U.; Schmelz, R.M. Indicator Values, Strategy Types and Life Forms of Terrestrial Enchytraeidae and Other Microannelids. Newslett. Enchytraeidae **1999**, *6*, 59–67.

ENTISOLS

L.T. West
University of Georgia, Athens, Georgia, U.S.A.

E

INTRODUCTION

Entisols are soils that have minimal pedogenic development and no diagnostic horizons other than an ochric epipedon (1). Because of limited pedogenesis, subsoil characteristics are those of the parent material. Entisols are the most extensive soil order comprising about 16% of the Earth's ice-free land area. About 60% of the Entisols occur in desert regions (2), and the largest area of Entisols occurs in Africa (Table 1).

Most Entisols occur on unstable geomorphic surfaces that are subject to frequent deposition, high rates of geologic erosion and truncation, or drastic disturbance by humans. Common Entisol landscapes include flood plains of rivers and streams, sand dunes, steep mountains, and mined or otherwise disturbed lands. Entisols may also be found on older more stable landscapes if parent material or hydrologic characteristics retard pedogenesis.

Because of flooding, steep slopes, arid conditions, and other environmental factors, Entisols often have use limitations and can be easily degraded. Even so, Entisols have been and continue to be important for food production and other human activities. Entisols along river floodplains are often intensively farmed and are some of the most agriculturally productive soils in the world. Many towns and cities are located where rivers or oceans provide transportation, and these landscapes have abundant Entisols. In addition, human land disturbance in and near urban centers have created Entisols, and these comprise the soil resource that must be managed for the future.

CLASSIFICATION

Entisols are divided into five suborders; Aquents, Arents, Psamments, Fluvents, and Orthents. Aquents, the wet Entisols, are not extensive worldwide covering less than 0.1% of the ice-free land surface (Table 1). However, they are widely distributed in landscape segments that accumulate water and receive sediment additions, including coastal tidal zones, fresh-water marshes and swamps, and flood plain backswamps. Except for seasonal saturation with water at or near the surface and associated redoximorphic features, properties of Aquents are widely variable. Aquents in coastal marshes may contain sulfide minerals (Sulfaquents) and/or have low bearing capacity (Hydraquents). Aquents saturated at or above the soil surface for long periods may accumulate sufficient organic matter to develop O horizons. Unless sulfidic materials are present within 50 cm of the surface (Sulfaquents), however, a histic epipedon results in classification as Inceptisols.

Arents include those Entisols that have fragments of diagnostic horizons mixed with other soil material because they have been physically disturbed by plowing, spading, or other human activities (1). Properties of Arents vary widely and depend on properties of the soil present before disturbance.

Often soil materials deposited or exposed by human activities do not contain recognizable fragments of diagnostic horizons and thus classify as Orthents rather than Arents. Many consider this placement unacceptable, and many proposals have been made to modify Entisol classification to recognize various types of soil materials and deposits that result from human activity (3–8). At the time of writing, classification of human modified or formed soils is under discussion by an international scientific committee (ICOMANTH). Recommendations from this committee will likely alter the classification of Entisols and other parts of *Soil Taxonomy* to better accommodate and recognize properties of human modified soils.

Typically, Fluvents are Entisols formed in recent water-deposited sediments and are commonly found on flood plains, fans, and deltas of rivers and streams where the sediment deposition rate exceeds the rate of pedogenesis. About 2% of soils worldwide are Fluvents (Table 1). Fluvents are defined as having an irregular decrease in organic C with depth or as having more than 0.2% organic C at a depth of 1.25 m (1). Because clayey and loamy strata commonly have more organic C than sandy strata, organic C decreases irregularly in stratified sediments. If the sediment has uniform texture, organic C content in lower subsoil horizons of Fluvents is more than 0.2% because of limited time for organic C decomposition. Physical and chemical properties of Fluvents are variable and depend

Encyclopedia of Soil Science

Table 1 Land area of Entisols suborders by continent

	Land area ($km^2 \times 10^3$)							
Suborder	**North America**	**Central America**	**South America**	**Europe**	**Africa**	**Asia**	**Australia/ Oceania**	**Global**
Aquents	–[a]	–	1	–	17	57	–	75
Arents	–	–	–	–	–	–	–	–
Psamments	–	–	787	22	2257	117	1277	4460
Fluvents	77	19	620	197	647	1202	3	2765
Orthents	1013	10	631	149	6992	4417	487	13,699
Total	**1090**	**29**	**2039**	**368**	**9913**	**5793**	**1767**	**20,999**
Total ice-free land area	20,753	696	17,546	5741	29,808	47,826	7897	130,268

[a] The suborder occurs but individual areas are too small to show and be tallied from global scale inventories.
(From Ref. 18.)

largely on the source of the alluvial sediment serving as parent material for the soils. Fluvent landscapes are commonly subject to frequent flooding, which limits their suitability for many uses. Many Fluvents are fertile and productive, however, and are used extensively for crop production if flood cycles do not coincide with cropping periods.

Orthents are common in deserts and mountainous areas and are the most extensive of the Entisols suborders (Table 1). Pedogenesis is limited in deserts because of the lack of water to drive pedogenic processes. In mountainous landscapes, erosion rates are high, and the rate at which weathered material is removed exceeds the rate of pedogenesis. In colluvial landscapes at the base of steep slopes, the rate of sediment deposition may exceed the soil development rate, which will also promote the occurrence of Orthents. Soils in large gullies formed by cultural erosion are commonly Orthents as are soils associated with mine pits and spoils that do not have fragments of diagnostic horizons (Arents). Extremely slow weathering of resistant parent materials such as quartzite-rich bedrock retards horizon formation and results in landscapes dominated by Orthents (9). Physical and chemical properties of Orthents are inherited from the parent material and vary as widely as the properties of the unconsolidated sediments in which the soils occur.

Psamments are the Entisols with sandy parent materials and cover about 3% of the Earth's surface (Table 1). Specifically, Psamments must have texture of loamy fine sand or coarser throughout the control section (25–100 cm for soils deep to rock), and many Psamments have sandy textures to depths of 2 m or more. Common Psamment landscapes include dunes, natural levees, sand sheets, and sandy outwash plains. These landscapes can be young or ancient. On ancient landscapes, the materials in which Psamments occur are commonly quartzitic sands that do not readily form horizons because of resistance to weathering. Psamments are by definition sandy and thus, are commonly low in plant nutrients, have low water-holding capacity, and rapid permeability. These characteristics limit crop production without irrigation and may impose limitations for other uses. Psamments are subject to wind erosion if bare, but with irrigation and proper management are often used for citrus, melon, and vegetable production.

GENESIS

Although the state factor equation for soil formation, $S=f(Cl, PM, R, B, T)$, where S = soil, Cl = climate, PM = parent material, R = relief, B = the biotic factor, and T = time (10), has many shortcomings, it is a convenient way to discuss the influence of various environmental factors on genesis of soils, including Entisols. Limited time for soil development because of unstable landscapes is the most common and most easily understood reason for occurrence of Entisols. Entisols may occur, however, on relatively stable landscapes on parent materials that have been in place for a relatively long time. Conditions that lead to the occurrence of Entisols on more stable landscapes include arid climates that limit water movement through the soil, parent materials resistant to weathering, parent materials toxic to plant growth, and permanent, or almost permanent, saturation with water.

Climate

Entisols occur in all climates and there are no climatic restrictions on Entisol classification. Because low rainfall limits rates of leaching and other pedogenic processes, formation of diagnostic horizons and evolution of Entisols to other orders is slower in arid climates than in semiarid and humid areas (11). In humid climates, mollic epipedons and Mollisols have been reported to form in as little as 100 years (12), and cambic horizons and Inceptisols may develop within 500–600 years or less (13, 14). To form cambic or calcic horizons and Aridisols in arid environments, however, may take more than 1000 and possibly as much as 4000 years (15).

Retardation of soil development by limited leaching partially explains the high proportion of Entisols that occur in arid regions. In addition, sparse vegetation and intense rainfall events enhance erosion and fluvial deposition, which limits soil development and contributes to occurrence of Entisols in desert areas (15, 16).

Parent Material

Entisols are found on all types and compositions of parent material. Entisols are often associated with alluvial parent materials since many of these soils are found in floodplains with active fluvial deposition. Extremely resistant bedrock weathers slowly, which tends to maintain Entisol landscapes. Sediments dominated by resistant minerals, such as quartzitic sand, do not have or produce clay, Fe, or other mobile components needed to form diagnostic horizons that would result in the evolution of Entisols to other orders (17).

Relief

Entisols are often associated with two differing types of relief: steep eroding landscapes and nearly level to gently sloping depositional landscapes. On steep convex slopes, erosion may result in soils that are rapidly truncated, which destroys diagnostic horizons that may be forming. On low relief depositional areas, such as stream floodplains, deltas, fans, and colluvial footslopes, the rate of deposition may be such that the soil accretes faster than diagnostic horizons are formed. Entisols also occur in areas where the local relief results in soils that are permanently saturated, such as coastal marshes and freshwater swamps. Permanent saturation inhibits leaching, translocation, and reduction–oxidation cycles that form diagnostic horizons.

Biota

Entisols can be found associated with any type of vegetation. Thus, little can be said about vegetation effects on genesis of Entisols. If humans are included, however, biota may have a profound impact on development and properties of Entisols. Earth moving activities and/or deep tillage often destroys existing soils and creates Entisols.

Time

Time or age of the soil, specifically limited time over which soil development processes have proceeded, is the environmental factor most often associated with Entisol landscapes. If the landscape is "young" because of continual erosion or deposition, time for soil development will be limited, and Entisols will dominate. If parent materials are inert, however, Entisols may occur in old landscapes if weathering does not produce mobile components that are necessary for formation of diagnostic horizons.

SUMMARY

Entisols are soils that are minimally developed and lack diagnostic horizons other than an ochric epipedon. Thus, properties of Entisols are those of the parent material and vary as widely as the properties of the materials in which they occur. Most Entisols occur on unstable geomorphic surfaces that are subject to frequent deposition, high rates of geologic erosion and truncation, or drastic disturbance by humans, although Entisols may occur in old stable landscapes if environmental or parent material characteristics retard pedogenesis. Entisols along river floodplains are commonly fertile and productive and are important for food production. Soil disturbance in contemporary urban landscapes is continually creating Entisols, and Entisols are the soil resource in these areas that must be managed for the future.

REFERENCES

1. Soil Survey Staff, *Soil Taxonomy: A Basic System of Soil Classification for Making and Interpreting Soil Surveys*, 2nd Ed.; USDA Handbook 436, U.S. Department of Agriculture Natural Resources Conservation Service: Washington, DC, 1999.
2. Wilding, L.P. Introduction: General Characteristics of Soil Orders and Global Distributions. In *Handbook of Soil Science*; Sumner, M.E., Ed.; CRC Press: Boca Raton, FL, 1999; E-175–E-183.

3. Sencindiver, J.A. *Classification and Genesis of Minesoils* Ph.D. Dissertation West Virginia University: Morgantown, WV, 1977.
4. Smith, R.M.; Sobek, A.A. Physical and Chemical Properties of Overburden, Spoils, Wastes, and New Soils. In *Reclamation of Drastically Disturbed Land*; Schaller, F.W., Sutton, P., Eds.; American Society of Agronomy: Madison, WI, 1978; 149–172.
5. Short, J.R.; Fanning, D.S.; McIntosh, M.S.; Foss, J.E.; Patterson, J.C. Soils of the Mall in Washington, D.C. II. Genesis, Classification, and Mapping. Soil Sci. Soc. Am. J. **1986**, *50*, 705–710.
6. Fanning, D.S.; Fanning, M.C.B. *Soil Morphology, Genesis, and Classification*; John Wiley & Sons: New York, NY, 1989.
7. Fanning, D.S. Human-Influenced and Disturbed Soils: Overview with Emphasis on Classification. In *Human-Influenced and Disturbed Soils*; Evans, C.V., Ed.; University of New Hampshire Department Natural Resources: Durham, NH, 1991; 3–14.
8. Strain, M.R.; Evans, C.V. Map Unit Development for Sand- and Gravel-Pit Soils in New Hampshire. Soil Sci. Soc. Am. J. **1994**, *58*, 147–155.
9. Buol, S.W.; Hole, F.D.; McCracken, R.J.; Southard, R.J. *Soil Genesis and Classification*, 4th Ed.; Iowa State University Press: Ames, IA, 1997.
10. Jenny, H. *Factors of Soil Formation*; McGraw-Hill: New York, NY, 1941.
11. Nordt, L.C.; Collins, M.E.; Fanning, D.S.; Monger, H.C. Entisols. In *Handbook of Soil Science*; Sumner, M.E., Ed.; 1999; E-224–E-242.
12. Ruhe, R.V.; Fenton, T.E.; Ledesma, L.L. *Missouri River History, Flood Plain Construction, and Soil Formation in Southwestern Iowa*; Research Bulletin 580, Iowa Agricultural and Home Economics Experiment Station: Ames, IA, 1975; 38–791.
13. Bilzi, A.F.; Ciolkosz, E.J. Time as a Factor in the Genesis of Four Soils Developed in Recent Alluvium in Pennsylvania. Soil Sci. Soc. Am. J. **1977**, *41*, 122–127.
14. Scully, R.W.; Arnold, R.W. Holocene Alluvial Stratigraphy in the Upper Susquehanna River Basin, New York. Quat. Res. **1981**, *15*, 327–344.
15. Gile, L.H. Holocene Soils and Soil-Geomorphic Relations in an Arid Region of Southern New Mexico. Quat. Res. **1975**, *5*, 321–360.
16. Gile, L.H.; Grossman, R.B. *The Desert Project Soil Monograph*; U.S. Department of Agriculture Natural Resources Conservation Service: Washington, DC, 1979.
17. Cabrera-Martinez, F.; Harris, W.G.; Carlisle, V.W.; Collins, M.E. Evidence for Clay Translocation in Coastal Plain Soils with Sandy/loamy Boundaries. Soil Sci. Soc. Am. J. **1989**, *53*, 1108–1114.
18. Reich, P.; Eswaran, H. *Personal Communication*; U.S. Department of Agriculture Natural Resources Conservation Service World Soil Resources: Washington, DC, 2001.

EROSION

Dennis C. Flanagan
United States Department of Agriculture-Agricultural Research Service (USDA-ARS), West Lafayette, Indiana, U.S.A.

E

INTRODUCTION

Soil erosion is the detachment or breaking away of soil particles from a land surface by some erosive agent, most commonly water or wind, and subsequent transportation of the detached particles to another location. Usually erosion occurs when a fluid (air or water) moves into and/or across a soil surface. Fluid and sediment particle impact forces, shear forces, and turbulence act to detach and lift soil into the fluid flow that then transports the particles away (Fig. 1). The force of gravity moves detached soil particles downwards, while cohesive forces between soil particles resist detachment and transport. Physical and chemical dispersion can disrupt cohesion and break soil aggregates into smaller and more easily transported particles. At some time and location away from the initial point of detachment, the sediment particles will eventually move back down to a state of rest on a soil surface, in a process known as deposition or sedimentation.

Erosion is a natural process and is a critical factor in soil formation from rock parent material. However, once productive agricultural soils have been formed over periods of thousands or millions of years, erosion of the soil material is then usually very low or negligible because of the impacts of protective natural plant and residue cover. Human activities are responsible for greatly accelerating erosion rates, usually by reducing or eliminating plant and residue cover. This exposes the soil to wind and water erosive forces, weakening the soil cohesive forces by tillage disturbance, and increasing the erosive agents, particularly by activities that increase surface runoff.

EROSION BY WIND

Erosion by wind occurs when wind speed exceeds a certain critical or threshold value. Soil particles can be detached and moved through suspension, saltation, or creep (Fig. 1). Suspension usually lifts the smallest soil particles (clays, silts, organic matter) so high into the air mass that they are easily kept in motion and can travel for long distances. Soil particles that move by creep are larger sand grains and aggregates that stay in contact with the soil surface at almost all times—their motion is often through rolling and bouncing. Saltating soil particles are usually moderate in size and once detached move in trajectories up into the air and then back down to the soil surface. Saltating particles often cause further detachment through abrasion, by striking the soil surface with sufficient momentum to dislodge additional soil particles from the in situ soil mass.

EROSION BY WATER

The most common types of soil erosion by water are sheet and rill erosion on upland areas, channel and gully erosion in small watersheds, and stream channel and bank erosion in larger catchments. Sheet erosion is caused by the action of raindrops (Fig. 2) and shallow overland flows that remove a relatively uniform depth (or sheet) of soil. Because of the uniform nature of the soil loss, it is often difficult to detect and gauge the extent of damage due to sheet erosion. On the other hand, rill erosion occurs in well-defined and visible flow concentrations, or rills (Fig. 3). Soil detachment in rills is largely due to flow shear stress forces acting on the wetted perimeter of the rill channel (Fig. 4). Once detached, larger sediment particles move as bedload, rolling and bouncing down slope with the flow, and are almost always in contact with the soil (or bed) surface. Smaller sediment particles (silts, clays) are much easier to transport and travel in the rill channels as suspended load. Rills are also the major pathway for transporting away sediment that is detached by sheet erosion (also known as interrill detachment).

By definition, rill channels are small enough to be obliterated by tillage and will not reform in exactly the same location. As one moves from smaller hillslopes to larger fields and watersheds, additional erosion processes come into play, because of the increasing amounts of runoff water. Gullies are incised erosion channels that are larger than rills and form in regions of large runoff flow concentration. Ephemeral gullies are a common type of

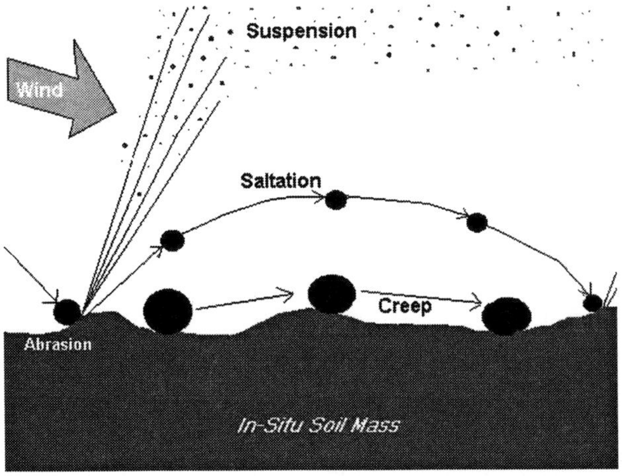

Fig. 1 Soil erosion by wind, showing the three modes of movement (creep, saltation, suspension).

Fig. 3 Rill erosion, caused by concentration of flowing water, forms easily recognizable regions of detachment on a soil surface (photograph courtesy M. Huhnke, Oklahoma State University).

erosion feature in many fields (Fig. 5). They are small enough to be tilled over, but will re-form in the same location due to convergent topography in small catchments. Runoff flows from large events can erode down through tilled soil layers until a nonerodible layer is reached, then the ephemeral gully channel will widen and soil detachment will decrease. Classical gullies are larger erosion features that cannot normally be tilled across (Fig. 6). The physical processes in classical gullies include other factors such as headcutting, seepage, sidewall sluffing and clean-out of fallen sidewall materials.

As the size of watersheds increases further, and streams increase in size and become perennial (due to subsurface water flows from springs and aquifers), the erosion processes in play change as well. Stream and channel erosion at these larger scales can include scouring of the channel beds, as well as contributions from the channel banks. Areas in streams may be in states of degradation, in which active detachment is lowering the level of the channel bed, or they may be in states of aggradation, in which sediment deposition is raising the bed level.

GRAVITY-INDUCED EROSION

There are also less frequent but more extreme forms of gravity-induced erosion on steep slopes from saturated soils that can be exacerbated by events such as earthquakes.

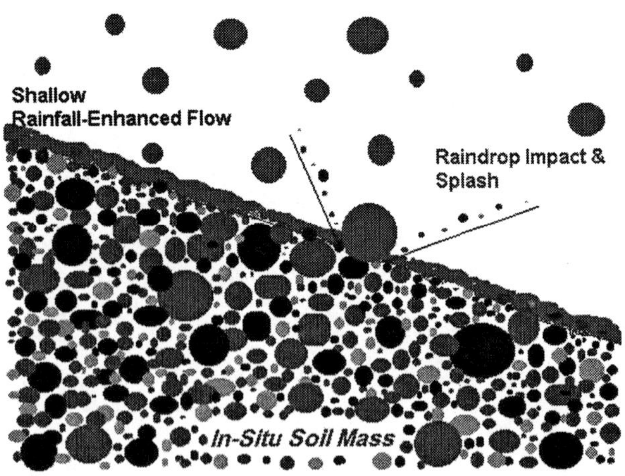

Fig. 2 Soil detachment by raindrop impact and shallow flow transport.

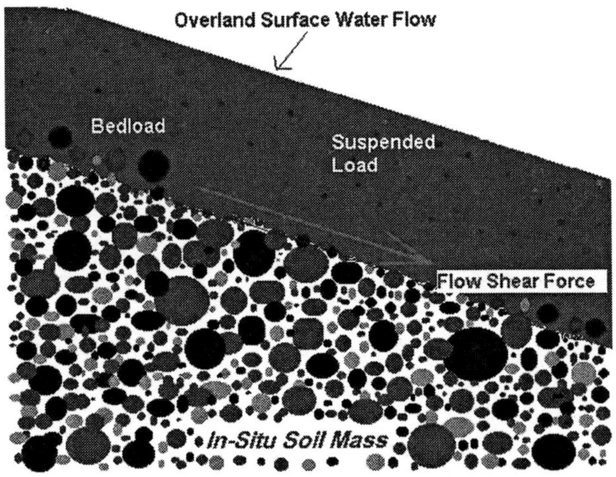

Fig. 4 Soil detachment and transport in rills are largely due to flow shear forces.

Fig. 5 Typical ephemeral gully, located in a soybean field in Indiana.

Large masses of land can slowly or rapidly slide down hills when the cohesive forces holding them in place fail (landcreep, landslide, debris flow, etc.). These types of erosion events typically occur when large rainfall or snowmelt water depths saturate soil profiles and weaken their resistance to slip.

EROSION ASSESSMENT

Erosion is a serious problem within the U.S. and throughout the world. In 1997, the USDA Natural Resources Conservation Service estimated that about 1.9 billion tons of soil are lost each year from nonfederal rural lands due to sheet and rill erosion by water and erosion by wind (1). Also, 29% of cropland in the U.S. is eroding at excessive rates. These estimates are on the low side, since erosion of other types (gully, etc.) and at other locations (urban lands, federal lands) were not included in this inventory. Throughout the world, FAO estimates that 16% of the total land area (21,960,000 km^2 of 134,907,000 km^2) is subject to significant risk of soil erosion (2). In Asia, South America, and Africa soil erosion rates are highest at an estimated average of 30–40 tons/ha/yr, while in Europe and North America average rates are somewhat lower at about 17 tons/ha/yr (3, 4). A sustainable rate of soil loss (rate of soil loss is equal to rate of soil formation) is thought to be about 1 ton/ha/yr (3).

Fig. 6 Classical gullies in the Loess Plateau of China. Terrace farming is being used to stabilize some of the hillslopes.

Erosion assessment can be a difficult task to perform in the field, and monitoring of soil lost and transported by wind or water can be expensive and prone to measurement errors. Gullies are easy to recognize, while soil lost to sheet and rill erosion is hard to gauge. Sheet and rill erosion may be occurring on hillslopes adjacent to a gully and may actually contribute more sediment to runoff water than the gully itself. Visual assessment of rates of wind erosion losses can be even more difficult to perform. Mathematical equations or sets of equations have been developed and used since the mid-1900s to estimate the rates of soil loss due to wind [(Wind Erosion Equation) (5)] or water [(Universal Soil Loss Equation) (6, 7)]. More recently, computer models are now being applied to simulate soil erosion processes and to estimate detachment, transport, and deposition of sediment (8, 9).

EROSION IMPACTS

Erosion has a range of impacts, both on-site as well as off-site. Soil loss removes fertile topsoil, organic matter, and nutrients, thus decreasing the tilth, water-holding capacity, and general productivity of a soil for on-site agricultural production. Regions of detachment can expand to dislodge and remove small crop seedlings, while regions of deposition can bury and kill small plants. In the case of wind erosion, the erosion process can damage fragile young seedlings through abrasion of plant tissue. When excessive detachment occurs, such as is the case with gully erosion, whole sections of fields may be destroyed, or may become inaccessible to farmers and their equipment. Eroded sediment can cause a number of off-site problems, including deposition along windbreaks, ditches, and waterways. The deposited sediment may require costly dredging and removal operations. Nutrients and pesticides associated with sediments can also contaminate air and water bodies. Erosion by wind can cause massive dust storms that blind drivers and cause traffic accidents, and sand particles can abrade and damage painted surfaces on buildings and vehicles. Some recent estimates are that the cost of combined on-site and off-site effects from soil erosion in the U.S. is as high as $44 billion per year (3).

EROSION CONTROL AND SOIL CONSERVATION

Many nations have created government agencies or organizations to specifically deal with soil erosion problems and to interact with landowners to get conservation practices implemented on the landscape. In the U.S., the Natural Resources Conservation Service assists in implementation of soil conservation practices on agricultural lands, the Forest Service manages sediment delivery from forests and timber harvest roads, and the Bureau of Land Management manages soil loss on range and grazing lands. The Department of Defense is responsible for managing erosion and off-site sediment delivery from lands that it uses for military training activities. However, in some countries efforts to address and minimize erosion problems are nonexistent or severely limited due to poor economic conditions, failure to recognize the erosion threat, and/or the extreme magnitude of the soil erosion.

A variety of soil conservation methods are available that can be applied on a landscape to minimize erosion problems due to wind or water. Wind erosion can be controlled through the use of windbreaks, crop residues, and tillage to induce significant surface roughness. Control procedures for erosion by water need to be determined based upon the types of active erosion processes. For example, if sheet and rill erosion are a major problem, then some type of conservation tillage practice that leaves large amounts of crop residues intact on the soil surface may be appropriate. However, if the water erosion problem is due to large amounts of surface runoff concentrating in a field and forming an ephemeral gully, then crop residues may not be adequate; instead, permanent vegetative cover in a grass waterway may be necessary, along with appropriate engineering structures (drop-box, etc.). Erosion prediction models can be used to assist in selecting and designing appropriate conservation practices.

Through the use of proper conservation planning and application of appropriate soil conservation methods, most erosion problems can be minimized or eliminated. This is critically important if the soil resource is to be preserved for continuous use in food and fiber production for current and future generations.

REFERENCES

1. United States Department of Agriculture National Resources Conservation Service. 1997. National Resource Inventory (Accessed 5/2001 http://www.nhq.nrcs.usda.gov/NRI/1997).
2. Food and Agriculture Organization of the United Nations, *World Soil Resources Report*; Accessed 5/2001 http://www.fao.org/ag/agl/agll/terrastat FAO Land and Water Development Division: Rome, Italy, 2000.
3. Pimental, D.; Harvey, C.; Resosudarmo, P.; Sinclair, K.; Kurz, D.; McNair, M.; Crist, S.; Shpritz, L.; Fitton, L.; Saffouri, R.; Blair, R. Environmental and Economic Costs of Soil Erosion and Conservation Benefits. Science **1995**, *267*, 1117–1122.
4. Barrow, C.J. *Land Degradation*; Cambridge University Press: Cambridge, 1991.
5. Woodruff, N.P.; Siddoway, F.H. A Wind Erosion Equation. Proc. Soil Sci. Soc. Am. **1965**, *29*, 602–608.
6. Wischmeier, W.H.; Smith, D.D. *Predicting Rainfall Erosion Losses—A Guide to Conservation Planning*; Agricultural Handbook No. 537, USDA: Washington, DC, 1978.
7. Renard, K.G.; Foster, G.R.; Weesies, G.A.; McCool, D.K.; Yoder, D.C. *Compilers Predicting Soil Erosion by Water: A Guide to Conservation Planning with the Revised Universal Soil Loss Equation (RUSLE)*. Agricultural Handbook No. 703; USDA: Washington, DC 1997; 384
8. Flanagan, D.C., Nearing, M.A. Eds. USDA-Water Erosion Prediction Project (WEPP) Hillslope Profile and Watershed Model Documentation, NSERL Report No. 10; National Soil Erosion Research Laboratory, USDA-Agricultural Research Service, West Lafayette, IN. 1995. 298 pp. Available Online: http://topsoil.nerl.pwdue.edu/nserlweb/weppmain/docs/readme. htm (Accessed October 2001).
9. Hagen, L.J. A Wind Erosion Prediction System to Meet User Needs. J. Soil Water Conserv. **1991**, *46* (2), 106–111.

EROSION ASSESSMENT

John Boardman
Environmental Change Institute, University of Oxford, Oxford, United Kingdom

INTRODUCTION

Soil erosion is the loss of soil from the surface of the Earth. It is a two-stage process with detachment of soil particles preceding transport by the agency of water and wind. It is a natural process occurring at relatively low rates but which may be accelerated by human actions. Most erosion is associated with the formation of rills and gullies. Erosion has serious impacts on the fertility of soils, on water quality and on reservoir storage capacity. Assessment of actual and potential erosion risk is necessary so that measures can be taken to prevent the loss of soil.

IMPACT OF EROSION

Natural rates of erosion are usually below 1 t ha^{-1} yr^{-1} whereas soil losses from farmland in temperate areas may exceed 100 t ha^{-1} in storms. Rates of soil formation are very low—similar to the natural rates of erosion—so that erosion means the loss of a finite resource. Erosion removes the organic-rich topsoil with its water and nutrient-holding capacity. On the Loess Plateau of China, 7.2 Mha is reported to have erosion rates in excess of 100 t ha^{-1} yr^{-1} (1).

Agents of erosion are wind, water and gravity (causing mass movement). Feet of people and animals may contribute, and soils will be more susceptible if weakened by weathering (wetting and drying, frost, etc.).

Erosion is an important contributor to global soil degradation. It is also a major component of desertification. Under semiarid conditions, vegetation damage typically precedes erosion. If climate variability is a major influence then degradation may be cyclic, interspersed with periods of improved land quality. Frequently, because of population pressure, degraded land does not recover.

The major impact of soil erosion is on agricultural land, both arable and grazing. Other areas of concern are construction sites and protected natural areas (e.g., footpath erosion in National Parks).

Impacts of erosion are 1) on-site, which includes loss of crop, seeds, fertilizer and thinning of the soil layer thus directly affecting present and future productivity and 2) off-site, including dam sedimentation, pollution of water courses, and property damage by muddy floods.

EROSION ASSESSMENT

Erosion assessment may describe the *actual* risk of erosion: GLASOD is an example of this type of assessment applied globally (2). Similarly, Canada has published maps of the risk of wind and water erosion at the provincial level. New Zealand has combined, on single maps at 1:250,000 scale, the actual and *potential* risk of erosion. Potential in this case refers to a change of land use to pastoral or arable. Maps from the CORINE project show actual and potential soil erosion in southern Europe; the latter is defined as the worst possible situation that might be reached.

Erosion assessment may focus on changes through time in the same area. Trimble has modeled changes in erosion and deposition rates in the Coon Creek catchment in Wisconsin since the beginning of European settler farming in the 1850s. This study relates erosion to land use change, management and conservation techniques, as well as to the capacity of rivers to transport eroded materials. It therefore links on-site and downstream effects through time and space (3). Erosion rates since Neolithic clearance have been modeled on the loessic soils of the South Downs, U.K. This assessment is based on changing land use, climate and farming practices (4). Assessment of erosion and its impacts may be combined in complex scenarios of change involving population pressures. Assessments of erosion through time have also been carried out for eastern U.S., England and Wales and Slovakia. Assessments over short time scales may be carried out using event-based models (see below).

The spatial scale of modeling may vary from a few m^2 to thousands of km^2. The global change and terrestrial ecosystems (GCTE) model evaluation exercise divides models into field, catchment and landscape, based on spatial scale (5). Resolution of erosion models is determined by grid size, e.g., km^2 grid size used in CORINE for southern Europe and the 250 m grid used in the maps of seasonal erosion risk in France.

Assessments are frequently expressed in map form (6). Landscape features indicative of erosion may be mapped as a record of an actual event or series of events. In arable landscapes features may be temporary; thus rills, ephemeral gullies and fans are usually plowed out annually. In semiarid, grazed landscapes erosional features may become permanent, e.g., gullies and badlands (7). Assessments should distinguish features active at the present time from those that are inactive and represent erosion under former conditions.

Several methods of erosion assessment are commonly used (Table 1)

1. Small experimental plots in the field; these may operate with natural or simulated rainfall (9). They are assumed to represent larger parts of the landscape. Results are frequently extrapolated to fields or catchments and used as input to models (8). Extrapolation from small plots to continental size areas is not acceptable (10).
2. Sediment yield of rivers represents the net loss of soil from a catchment (11). Erosion rates on fields may be very different due to storage of soils between the field and the river. Highest rates of sediment yields from a large river are from the Yellow River, China, with an average silt content of 38 kg m^3 (12). Rivers of the east and south Asia deliver about 67% of the sediment reaching the world's oceans; rates are highest in this region due to tectonic uplift, high rainfall, deforestation and farming activities (13).
3. Field monitoring is defined as "field-based measurement of erosional and/or depositional forms over a significant area (e.g., $> 10\,km^2$) and for a period of >2 years." Regular measurements of volumes of soil loss from rills and gullies or deposited in fans are made. Notable examples are the monitoring of 17 localities in England and Wales 1982–1986 (14), a 10-year program on the South Downs, England 1982–1991, and monitoring of ephemeral gullies in Belgium 1982–1993 (15).
4. Remote sensing is used increasingly in erosion assessment. For rill and gully erosion, it may be used as a locational technique to allow detailed field measurements to be made at selected sites. Gullies and badlands may be identified on black and white air photographs of 1:20,000 scale; in the Karoo of South Africa comparisons have been made between the extent of gullying in 1945 and 1980 (16). Also in South Africa, in a pioneering study, Talbot mapped gullying in the Swartberg resulting from land conversion to wheat farming. More recently, the extent of Icelandic land degradation has been assessed using Landsat imagery.
5. Cesium-137, derived from weapons testing since 1954, is used as a tracer to assess amounts of erosion and deposition. Annual average rates for the last ca. 45 years are estimated by measurement of the amount and distribution of cesium and comparison with an uneroded reference site (17). Studies have been carried out in Australia, Canada, China, The Netherlands, Poland, Thailand, U.K. and U.S.
6. Total amounts of soil loss since a historical baseline such as woodland clearance or the beginning of European settlement have been assessed by comparison of existing soil profiles with uneroded ones. Pedological and stratigraphic approaches have also been used to assess the effect of past extreme events, for example, the major storms and floods in Germany in the early 14th century (18).

Table 1 Assessment methods: their merits and limitations

Assessment method	Merits	Limitations
Experimental plots	Control over erosion factors; ease of measurement of runoff and soil	Small size: difficult to extrapolate to larger areas
Sediment yield of rivers	Ease of measurement	Measures total soil lost from watershed, not from specific areas
Field monitoring	Records soil lost from rills and gullies; covers large areas; inexpensive	Gives very approximate amounts of soil lost
Remote sensing	Covers large areas and may cover long time-span	Limited availability for many areas; may not be suitable for small scale features
Cesium-137	Gives 45-year average	Many unknowns: unreliable without careful calibration
Stratigraphy and pedology	Deposited soil may contain artifacts; suitable for long-time-period studies	Scarcity of suitable sites; low precision of some dating methods
Expert opinion	May reveal unrecorded information	Subjectivity: difficult to compare different areas
Models	Objectivity; ease of use even by "nonexperts"; numerical output	Availability of data; complexity of process descriptions; lack of validation

7. Expert opinion has been used to assess erosion on a regional or global scale (2). GLASOD assesses erosion at a global scale (1:10 M). A total of 12 degradation types are recognized under the broad headings of water erosion, wind erosion, chemical deterioration and physical deterioration. The *degree* of degradation is assessed for each mapping unit, as is its *relative extent*. A combination of degree and relative extent gives an assessment of *severity* which is expressed in *severity classes* and mapped in four colors.
8. Assessment of erosion has frequently been based on the application of models. The universal soil loss equation (USLE) is most widely used for reason of its simplicity and low-level data requirements (19). Recent developments in modeling have concentrated on simulation of erosional processes and computerization. Models may be used to estimate average, long term, erosion rates under specified conditions, or to assess the effect of particular rainfall events. Erosion models have also been used to assess chemical losses in solution or attached to soil particles.

REFERENCES

1. Dazhong, Wen Soil Erosion and Conservation in China. In *World Soil Erosion and Conservation*; Pimental, D., Ed.; Cambridge University Press: Cambridge, 1993; 63–85.
2. Oldeman, L.R.; Hakkeling, R.T.A. and Sombroek, W.G. World Map of the Status of Human-Induced Soil Degradation: An Explanatory Note, ISCRIC and UNEP. 1990.
3. Trimble, S.W. A Sediment Budget for Coon Creek Basin in the Driftless Area, Wisconsin, 1853–1977. Am. J. Sci. **1983**, *283*, 454–474.
4. Favis-Mortlock, D.; Boardman, J.; Bell, M. Modelling Long-Term Anthropogenic Erosion of a Loess Cover: South Downs, UK, The Holocene **1997**, *7* (1), 79–89.
5. Boardman, J.; Favis-Mortlock, D.T. *Modelling Soil Erosion by Water*; Modelling Soil Erosion by Water, NATO ASI Series, Boardman, J., Favis-Mortlock, D.T., Eds.; Springer: Berlin, 1998; 3–6.
6. Morgan, R.P.C. Chapter 4: Erosion Hazard Assessment. In *Soil Erosion and Conservation*; Longman: London, 1995.
7. Torri, D.; Calzolari, C.; Rodolfi, G. Badlands in Changing Environments: An Introduction. Catena **2000**, *40*, 119–125.
8. Hudson, N.W. *Field Measurement of Soil Erosion and Runoff*; FAO Soils Bulletin 68, FAO: Rome, 1993.
9. Lal, R., Ed. *Soil Erosion Research Methods*; Soil and Water Conservation Society: Ankeny, IA, 1988.
10. Boardman, J. An Average Soil Erosion Rate for Europe: Myth or Reality? J. Soil Water Conserv. **1998**, *53* (1), 46–50.
11. Walling, D.E. Measuring Sediment Yield from River Basins. In *Soil Erosion Research Methods*; Lal, R., Ed.; Soil and Water Conservation Society: Ankeny, IA, 1988; 39–73.
12. Dazhong, Wen. Soil Erosion and Conservation in China. In *World Soil Erosion and Conservation*; Pimental, D., Ed.; Cambridge University Press: Cambridge, 1993; 63–85.
13. Milliman, J.D.; Meade, R.H. World-Wide Delivery of River Sediment to the Oceans. J. Geol. **1983**, *91*, 751–762.
14. Evans, R. Extent, Frequency and Rates of Rilling of Arable Land in Localities in England and Wales. In *Farm Land Erosion: In Temperate Plains Environment and Hills*; Wicherek, S., Ed.; Elsevier: Amsterdam, 1993; 177–190.
15. Poesen, J. Gully Typology and Gully Control Measures in the European Loess Belt. In *Farm Land Erosion: In Temperate Plains Environment and Hills*; Wicherek, S., Ed.; Elsevier: Amsterdam, 1993; 221–239.
16. Boardman, J.; Parsons, A.J.; Holmes, P.J.; Holland, R.; Washington, R. Development of Badlands and Gullies in the Sneeuberg, Great Karoo, South Africa. Catena, in press.
17. Quine, T.A.; Walling, D.E. Assessing Recent Rates of Soil Loss from Areas of Arable Cultivation in the UK. In *Farm Land Erosion: In Temperate Plains Environment and Hills*; Wicherek, S., Ed.; Elsevier: Amsterdam, 1993; 357–371.
18. Bork, H.-R. The Importance of Catastrophic Events in Past Erosion. In *Climate Change and Soil Erosion*; Boardman, J., Favis-Mortlock, D., Eds.; Imperial College Press: London, in press.
19. Wischmeier, W. H.; Smith, D.D. *Predicting Rainfall Erosion Losses*; Agricultural Research Service Handbook 537, U.S. Department of Agriculture: Washington, DC, 1978.

E

EROSION CONTROL BY AGRONOMIC PRACTICES

Glenn A. Weesies
United States Department of Agriculture-Natural Resources Conservation Service (USDA-NRCS), West Lafayette, Indiana, U.S.A.

David L. Schertz
United States Department of Agriculture-Natural Resources Conservation Service (USDA-NRCS), Washington, D.C., U.S.A.

William F. Kuenstler
United States Department of Agriculture-Natural Resources Conservation Service (USDA-NRCS), Fort Worth, Texas, U.S.A.

INTRODUCTION

Soil erosion by wind and water degrades our soils by removing material from the fertile topsoil. Windstorms and rainstorms detach and transport soil particles high in nutrients and organic matter. The result is a soil surface with reduced plant-available nutrients, high in bulk density, and low in porosity and capacity for water intake.

SOIL EROSION AND ITS CONTROL

In regions with adequate rainfall, nature's way of protecting the soil is for vegetation to grow during periods of adequate rainfall and temperature, die during periods of low rainfall or temperature, and emerge again to start the next cycle. Either growing plants or plant residue would cover the soil to protect it from erosion. During the process of soil formation, the cycle of growing plants and decaying residue would provide protection for the soil while new soil developed (1).

Mechanization has, over the last half-century, allowed farmers to till the soil and plant row crops continuously under intensive cultivation. This has seriously altered nature's system of producing topsoil and protecting the soil from erosion. Today, under intensive, conventional tillage, soils are left unprotected and exposed to wind, rainfall, and runoff. Erosion has reduced the depth of topsoil, altered its natural tilth and structure, and reduced its productivity. Recent advances in technology have made it possible to employ techniques similar to nature's methods of protecting the soil during crop production by leaving the previous crop's residue on the soil surface, adding nutrients to the soil, and maintaining its tilth and structure (1). Of all the tools that technology has made available to farmers to control erosion, the most efficient and economical are agronomic practices—those that involve management practices that protect the soil during production of crops.

AGRONOMIC PRACTICES THAT CONTROL EROSION

Types of Agronomic Practices

The more common agronomic practices used to control erosion on cropland are described as follows:

Residue management

Residue management is managing the amount, orientation, and distribution of crop and other plant residue on the soil surface year-round while growing crops (2). Crops are grown in narrow slots, in residue-free strips in previously untilled fields, or in fields where the entire surface is tilled prior to planting. It includes limiting or reducing tillage to leave more residues on the surface. An increase in ground cover of just 10% can reduce soil loss by as much as 20%.

Conservation tillage

Conservation tillage is any tillage and planting system that covers 30% or more of the soil surface with crop residue, after planting, to reduce soil erosion by water. Where soil erosion by wind is the primary concern, any system that maintains the equivalent of at least 1120 kg/ha (1000 lb/acre) of flat, small grain residue on the soil

Encyclopedia of Soil Science

Fig. 1 A field with a contour stripcropping system. Strips of corn are alternated with strips of a hay crop. Planting is on the contour. (Photo courtesy of USDA, Natural Resources Conservation Service.)

surface throughout the critical wind erosion period qualifies as conservation tillage (3). Types of conservation tillage include no-till, strip-till, ridge-till, and mulch-till. With this agronomic practice, tillage tools are selected and operated to maximize the amount of residues left on the surface. Weed control is accomplished with a combination of herbicides and cultivation.

Contour farming

Contour farming is tillage, planting, and other farming operations performed on or near the contour of the field slope (2). Crops are planted around the hill nearly on the level, rather than up and down the hill. This makes many short slopes out of one long slope. Contouring can reduce erosion by 50% in many instances.

Contour stripcropping

This is growing row crops, forages, small grains, or fallow in a systematic arrangement of equal width strips on or near the contour of the field slope (2). This practice involves a combination of crop rotation and contouring. Contour stripcropping can reduce soil loss by 75% in many instances (Fig. 1).

Contour buffer strips

These are narrow strips of permanent, herbaceous vegetative cover established across the slope. These strips are alternated down the slope with parallel, wider cropped strips (2). The permanent vegetation in the buffers is typically grass, legumes, or a grass–legume mix. Contour buffer strips can reduce erosion by 60% in many instances.

Fig. 2 Corn growing in a no till system. The crop was planted in residue from a cover crop that has been chemically killed prior to planting corn. The cover crop and its residue protect the soil until the corn becomes established. (Photo courtesy of USDA, Natural Resources Conservation Service.)

Filter strips

A strip or area of herbaceous vegetation situated between cropland, grazing land, or disturbed land (including forest land), and environmentally sensitive areas (2). Typically, a filter strip is placed at the base of a slope to filter sediment and other contaminants from runoff water before they reach offsite areas or bodies of water.

Crop rotation

This involves growing crops in a recurring sequence on the same field. Erosion control is achieved by changing the crops year by year to alternate high residue-producing crops with lower residue-producers, close-growing crops with row crops, or perennial crops with annual crops. Some rotations can reduce erosion by 60%.

Cover crops

Grasses, legumes, forbs, or other herbaceous plants established for seasonal cover and conservation purposes are known as cover crops (2). Cover crops temporarily protect the soil during seasons when major crops are not growing (Fig. 2).

Grassed waterways

These are natural or constructed channels that are shaped or graded to required dimensions and established with suitable vegetation (2). This practice is used where water usually concentrates from terraces, diversions, concentrated flow channels, or natural waterways as it runs off a field. Grassed waterways will slow water and guide it off the field, significantly reducing gully erosion.

Erosion Control with Surface Cover

Each of the agronomic practices listed previously utilizes surface cover to control erosion. Surface cover is the single most important agronomic factor in determining soil loss. Surface cover affects erosion by water by reducing the transport capacity of runoff water, by causing deposition in ponded areas, and by decreasing the surface area susceptible to raindrop impact. It also protects the soil surface from erosive winds. Surface cover includes the live plant canopy, as well as plant residue.

CROP RESIDUE MANAGEMENT WITH CONSERVATION TILLAGE

The remaining section discusses in greater detail the two most important agronomic practices for controlling erosion—crop residue management and conservation tillage. Crop residue management is an "umbrella" term encompassing several conservation tillage systems. In the 1960s, the terms *minimum tillage* and *reduced tillage* were used to refer to less extensive tillage operations during the production of agronomic crops. In the 1980s, the term *conservation tillage* was used to refer to any tillage and planting system that left at least 30% of the soil surface covered by plant residue. In recent years, the term *crop residue management* has evolved to address the various methods and benefits of surface residue in reducing erosion.

Crop residue management is any tillage and planting system that uses conservation tillage (preferably no-till) or another system designed to retain all or a portion of the previous crop's residue on the soil surface.

Conservation tillage, defined earlier, is any tillage and planting system that leaves more crop residue on the soil surface after planting to reduce soil erosion. Types of conservation tillage include:

- No-till/strip-till—the soil is left undisturbed from harvest to planting except for strips up to one-third of the row width (strips may involve only residue disturbance or may include soil disturbance). Planting or drilling is accomplished using disc openers, coulter(s), row cleaners, in-row chisels, or rotary tillers. Weed control is accomplished primarily with herbicides. Cultivation may be used for emergency weed control (4). Other common terms used to describe no-till include direct seeding, slot planting, zero-till, row-till, and slot-till.
- Ridge-till—the soil is left undisturbed from harvest to planting except for strips up to one-third of the row width. Planting is completed on the ridge and usually involves the removal of the top of the ridge. Planting is completed with sweeps, disk openers, coulters, or row cleaners. Residue is left on the surface between ridges. Weed control is accomplished with herbicides (frequently banded) and/or cultivation (4). Ridges are rebuilt during row cultivation (Fig. 3).
- Mulch-till—full-width tillage involving one or more tilage trips which disturbs all of the soil surface and is done prior to and/or during planting. Tillage tools such as chisels, field cultivators, disks, sweeps, or blades are used. Weed control is accomplished with herbicides and/or cultivation (4).

Fig. 3 Corn growing in a ridge till system. During planting, the ridgetop was cleared and the seed planted. Residue is left on the surface between ridges. The ridges will be rebuilt during cultivation as the crop grows. (Photo courtesy of USDA, Natural Resources Conservation Service.)

Table 1 shows average amounts of residue cover left after various tillage operations in corn and soybeans in several Iowa studies (5). The residue left after harvesting soybeans was 83%; that left after corn was 93%. Both levels gave very good erosion control compared with bare soil. The loss of erosion control because of the reduction in residue cover is evident, especially for fragile residues like soybean residue.

Crop residue management utilizes agronomic practices for the entire cropping year. A crop rotation sequence is chosen to include high residue producing crops. A crop is planted that will provide residue to meet a specified goal, such as reducing erosion to a specified level. Planting in narrower rows and at higher plant populations will create more residue. Cover crops can be used when residue from the major crop is insufficient. When no-till is not utilized, tillage operations are delayed until near planting time instead of tilling soon after harvest. With mulch-till, tillage is performed at shallower depths, with fewer passes across the field. Tillage tools are used that bury less residue

Table 1 Percentage of soybean and corn residue cover remaining after harvest in different tillage operations in Iowa

Crop and type of operation	Average % residue cover remaining	Erosion reduction% (compared to fall moldboard plowing)
Harvest		
Soybeans	83	
Corn	93	
Fall moldboard plowing		
Soybeans	2	
Corn	9	
Fall chisel plowing		
Soybeans	15	37
Corn	49	75
Spring disking		
Soybeans	18	43
Corn	56	81
No till		
Soybeans	58	86
Corn	86	93

compared to conventional tillage. At harvest, residue is distributed uniformly over the soil surface, and baling, grazing, burning, and other operations that remove residue are eliminated.

The effectiveness of residue cover in reducing erosion varies, as shown in Fig. 4. However, as this figure illustrates, even at relatively low residue cover levels, residue cover is highly effective. In Fig. 4, residue effectiveness is expressed as a *b* value. Fields easily eroded by flow, such as thawing soils where rill erosion is the primary mechanism of soil loss, have a *b* value of 0.05. Fields with highly cohesive soils, such as with long-term no-till where interrill erosion is the primary mechanism of soil loss, have a *b* value of 0.025. Fields that are regularly tilled have an intermediate *b* value of 0.035 (6).

Residues vary in their effectiveness, depending on their size and weight, as shown in Fig. 5 (6). For a given weight, different kinds of residue provide varying amounts of surface cover. For example, 2690 kg/ha (2400 lb/acre) of

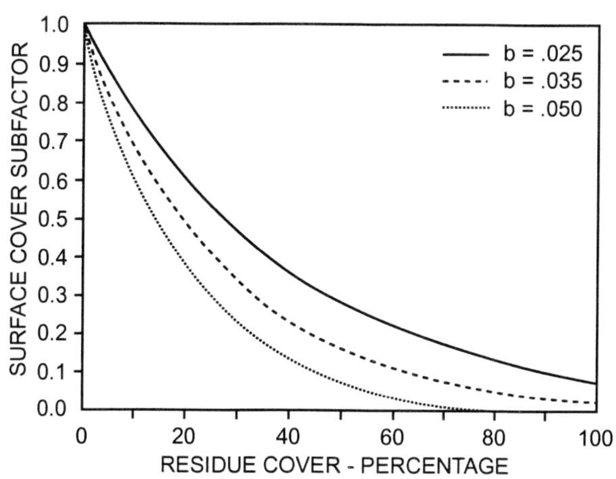

Fig. 4 Effect of residue cover and *b* values on the surface cover subfactor for smooth field surfaces in the Revised Universal Soil Loss Equation. (From Ref. 6.)

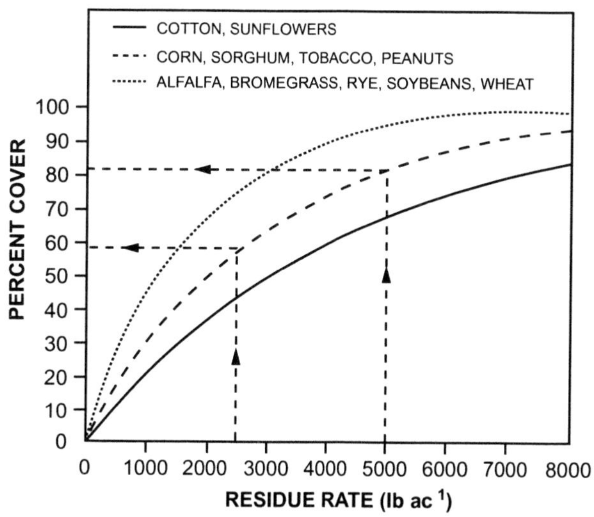

Fig. 5 Relationship of residue amount to percent residue cover for various crops. (From Ref. 6.)

cotton residue provides 45% surface cover. The same weight of corn provides 60% surface cover, and that weight of soybean residue provides nearly 85% cover.

Leaving crop residue on the soil surface has a number of advantages over tillage that leaves the soil surface nearly bare. The most significant advantage is the greatly reduced erosion. Other advantages include the potential for increased yield due to water conserved by surface residue, lower soil temperatures in warmer climates, improved soil quality over time due to increased organic matter levels, and reduced input of time, labor, and fuel (7).

Success of agronomic practices, such as crop residue management and conservation tillage to control erosion has been proven by the increasing rate of acceptance and use by farm operators. Success is due in part to the improved effectiveness and reduced cost of herbicides. Success is also due to improvements in tillage tools that leave more residue on the surface and planting equipment that works well in high-residue conditions. Finally, farm operators measure success through economics. The reduction in the number of tillage operations, the size and number of tractors, fuel used, and maintenance costs are important considerations in making the decision to use crop residue management practices.

REFERENCES

1. *No-Till User's Manual*; Agronomic Development Center, BASF Corporation: Research Triangle Park, NC, January 1990.
2. USDA–Natural Resources Conservation Service. *Field Office Technical Guide*; Section IV.
3. Conservation Technology Information Center (CTIC). 2000. Core4/Conservation Tillage [Online]. Available at http://www.ctic.purdue.edu/Core4/CT/Definitions.html (verified October 5, 2000).
4. USDA–Soil Conservation Service, *Conservation Catalog for the 1990s*; USDA–Soil Conservation Service: Des Moines, IA, October 1991.
5. Colvin, T.S.; Laflen, J.M.; Erbach, D.C. A Review of Residue Reduction by Individual Tillage Implements. *Crop Production with Conservation in the 80s*; Siemens, J.C., Ed.; American Society of Agricultural Engineers Publication: St. Joseph, MI, 1980; 7–81.
6. Renard, K.G.; Foster, G.R.; Weesies, G.A.; McCool, D.K.; Yoder, D.C. Coord.; *Predicting Soil Erosion by Water: A Guide to Conservation Planning with the Revised Universal Soil Loss Equation (RUSLE)*; US Department of Agriculture: Agriculture Handbook No. 703, 1997.
7. USDA–Soil Conservation Service. *Crop Residue Systems for Conservation and Profit*; USDA–Soil Conservation Service: Washington, DC:USDA–SCS, December 1992.

EROSION CONTROL, SOIL CONSERVATION AND

Eric T. Craswell
International Board for Soil Research and Management, Jatujak, Bangkok, Thailand[a]

INTRODUCTION

The conservation of soil and water resources is a growing challenge to humankind because these resources are under stress to produce food and fiber to meet the needs of a burgeoning population. Consequently, poor land management associated with land clearing and agricultural intensification has caused widespread land degradation. Most natural ecosystems with climax vegetation display a dynamic equilibrium between soil formation and erosion. This equilibrium is displaced when vegetation is cleared or overgrazed, and when fire and mechanical cultivation expose soil to water and/or wind erosion. Apart from agricultural intensification, construction of roads and buildings causes serious, but commonly localized, problems of soil erosion. Loss of the soil-surface layer, which contains the highest level of nutrients and organic matter reduces soil fertility and hence vegetative growth and yields. The broad scale of human-induced land degradation due to soil loss on agricultural lands is indicated by the estimates of Oldeman et al. (1). Globally, the total area affected by moderate to serious soil erosion is 1028 million hectares, of which 748 million hectares is due to water erosion and the rest by wind erosion. Asia and Africa together make up 673 million hectares of the total affected area. Oldeman also estimates that 186 million hectares are affected by chemical and physical degradation, which reduce vegetative cover and exacerbate soil losses by erosion.

SOIL CONSERVATION—THE EVOLVING PARADIGM

Soil conservation may be defined as the protection of soil against physical loss by erosion or against chemical deterioration. This article emphasizes soil erosion and its control. Human efforts to combat soil erosion on steepland areas date back to the dawn of recorded history. Indigenous technologies, documented recently (2)[b], include a wide range of methods that comprise terraces, minimum tillage, handmade ridges, natural vegetation strips, stone bunds, logs, etc. (Table 1). Some of these technologies, conceived over the centuries in tune with the diverse belief systems of the people, are similar to those based on modern science (see later for more details). Soil erosion affects not only the productivity of the individual farm, but also the long-term capacity of a society to produce food for its population. The importance of the social dimensions of soil erosion are reinforced by the off-site impacts of transported soil or sediment in the landscape in the case of water erosion; and the environmental impacts on air quality and dust deposition of airborne soil particles, in the case of wind erosion. These off-site impacts may affect urban populations at long distances from the site of the erosion, and in some spectacular cases—such as the dustbowl in the U.S. in the 1930s—have drawn the attention of policy makers to the need for more investment in soil conservation in rural areas. Models for the most effective community and the government's involvement in soil conservation continue to evolve.

In the 20th century, governments in countries with mechanized modern agriculture, such as the U.S., recognized the serious problems created by erosion and established institutions to promote and subsidize soil conservation. A land classification scheme was developed to identify high-risk erosion-prone land that was in some cases set aside as natural rangeland. On other classes of vulnerable land, agronomic practices, and/or conservation structures such as contour banks, were introduced on an individual farm basis with the help of government agents. Some developing countries introduced similar schemes. However, these efforts have not been successful in many countries. In the U.S., despite massive investments in soil conservation since the 1930s, some experts believe that the rates of soil erosion have not declined (3). Many practitioners now consider that traditional top-down soil conservation programs should be replaced by systems

[a]IBSRAM's programs merged with those of the International Water Management Institute on 1 April 2001.

[b](http://www.wocat.net/ftp/qtqalist.pdf (accessed June 2001)).

Table 1 Some examples of indigenous technologies for soil conservation and erosion control

Country	Technology	Description	Source (Reference)
China	Ditches along hillsides	Grassed contour ditches between rows of fruit trees	(2)
Ethiopia	Stone bunds	Stones collected and placed in lines on contours	a
India	Kana bandi	Windbreaks of dead wood	(2)
Kenya	Fanya-juu terraces	Soil thrown uphill to form contour bunds	a
Laos	Lao Theung systems	Sustainable shifting cultivation on mid-altitude steeplands	(2)
Nepal	Terraces	Manually built on steep slopes; terrace risers sliced regularly	(2)

[a](From http: //www.wocat.net/ftp/qtqalist.pdf (accessed june 2001).)

utilizing community-based approaches. Movements such as the land-care program in Australia [www.landcareaustralia.com.au/ (accessed February 2001] and conservation districts in the U.S. (www.nacdnet.org/ (accessed February 2001)) are based on a bottom-up approach in which members of rural communities work together to conserve not only the soil and water resources but also, on the basis of a more holistic approach, the biodiversity and landscape integrity.

The limited successes of soil conservation schemes contributed to concerns about global environmental security that in the mid-1980s led to widespread acceptance of the concept of sustainability. The World Commission on Environment and Development defined sustainable development as meeting the needs of the present without compromising the ability of future generations to meet their own needs (4). This new thinking also recognized that technological innovation is a necessary, but not a sufficient requirement for successful soil and water conservation. A new paradigm that shifted from an emphasis on soil and water conservation to the more holistic goal of sustainable land management was developed. The shift entailed a change from a technology-based approach to an approach designed to preserve and enhance the quality and productivity of global natural and environmental resources, with a view to the dynamic and constantly changing needs of all of the Earth's inhabitants (5). The scales of intervention ranged from the household through the community to the international level, involving treaties and conventions.

At the farm and field level, a mechanism is needed for assessing sustainable land management. The Framework for Evaluating Sustainable Land Management (FESLM) was designed for this purpose (6). The FESLM sets the goal of a system that combines technologies, policies, and activities aimed at integrating socioeconomic principles with environmental concerns so as to simultaneously: maintain or enhance production/service; reduce the level of production risk; protect the potential of the resource base; be economically viable and socially acceptable. These objectives—productivity, security, protection, viability, and acceptability—constitute the five pillars of the FESLM.

This broad view of sustainable land management, involving social and economic as well as biophysical dimensions, requires a new community-based approach to institutional support. Farmer participation is the key to success in both developed and developing countries although the most appropriate frameworks for community, institution, and government involvement vary. In developing countries, there is increased adoption of participatory technology development that involves farmers in the planning, collection, selection, testing, and dissemination of technologies. The Landcare approach in Australia is an example of the new approach in the context of a developed country. Landcare was initiated to address the failure of the conventional government soil conservation approach to halt the disturbing rate of natural resource degradation in Australia; in 1996 there were 3000 Landcare groups (7). The model is based on government support of these land care groups consisting of a collection of farmers and interested citizens joined together to reverse natural resource degradation. The roles of communities and governments at different levels in underpinning the Landcare movement are still evolving and are critically important to the ultimate success of these soil conservation efforts. The policy interventions needed in developing countries must take account of credit, land tenure, and pricing policies that strongly influence the actions of farmers in marginal areas (8).

EROSION CONTROL

The control of soil erosion can be achieved using a wide range of technologies, both vegetative and mechanical (Table 2). The principles underlying these control

Table 2 Technologies for soil erosion control on agricultural land

Vegetative	Mechanical
Cultivated crop land • Mulching • Crop management Dense planting Multiple cropping/agroforestry (contour strips or rotations) Cover cropping Forest land • Tree planting • Agri silviculture Grazing land • Plant grasses • Plant shrubs Wasteland • Plant trees, shrubs, grasses • Biological engineering Wattling and staking Brush matting • Hydro seeding	Cultivated crop, forest and grazing land • Conservation tillage Contour tillage Ridging and ridge tying Minimum till and no-till • Terracing • Waterways • Structures

(Adapted from Ref. 10.)

measures have been summarized in models such as the USLE, RUSLE, WEPP [http://topsoil.nserl.purdue.edu/nserlweb/ (accessed February 2001)] and GUEST (9). Rainfall impact is a key factor in dispersing and detaching the surface soil, hence soil cover that protects the surface is an important factor in selecting technologies to control erosion. However, since surface runoff plays an overriding and critical role in most erosion, control measures should generally be selected to increase infiltration and thereby reduce runoff, or to slow the surface runoff rate by reducing the degree of slope or slope length. In tropical steeplands with high rainfall rates, high surface runoff rates create high stream power leading to unacceptable erosion rates higher than 100 t/ha/yr. Some soils with initially high infiltration rates have no runoff until the profile is saturated by tropical storms, then upwelling and runoff occur leading to high soil losses (9). For erosion control, the choice of technologies from lists such as those in Table 2, depends on the biophysical, social and economic characteristics of the site. Use of schemes such as the FESLM discussed before helps ensure that the technology(ies) chosen lead to sustainable land management.

Lal (11) recently presented the rationale to consider the watershed as the basic unit for sustainable land and water management. Since erosion processes of detachment, transport, and deposition operate in the landscape, effective erosion control requires an understanding of those processes, and the hydrological processes and their impact in the watershed scale. The trend toward integrated watershed management as the basis for natural resource management is consistent and adds weight to the shift towards community-based land management discussed in the previous section on soil conservation.

Control of wind erosion presents a difficult challenge, particularly in the developing world. The underlying causes are overgrazing and removal of crop residues in cultivated fields in dry areas. The first line of defense is to reduce stocking rates or retain residues in the field. However, these techniques, and long-term solutions such as the planting of windbreaks, are not within the realm of economic possibilities for poor farmers in regions such as West Africa. Nevertheless, Stirk (12) showed that farmer-managed regeneration of woody species was a promising and effective approach to reduce wind erosion without reducing income or requiring too much investment of labor or capital. This illustrates once more the principle that all the dimensions of sustainable land management must be considered to achieve successful erosion control.

REFERENCES

1. Oldeman, L.R.; Hakkeling, R.T.A.; Sombroek, W.G. *World Map of the Status of Human-Induced Soil Degradation: An Explanatory Note*; International Soil Reference and Information Centre: Wageningen, 1991; 1–34.
2. Pongsapich, A. *Indigenous Technical Knowledge for Land Management in Asia*; Issues in Sustainable Land Management No. 3 International Board for Soil Research and Management: Bangkok, 1998.
3. Trimble, S.W.; Crosson, P. U.S. Soil Erosion Rates—Myth and Reality. Science **2000**, *289* (5477), 248–250.
4. World Commission on Environment and Development, *Our Common Future*; Oxford University Press: Oxford, 1987; 1–400.
5. Hurni, H. *Precious Earth: From Soil and Water Conservation to Sustainable Land Management*; International Soil Conservation Organization and Centre for Development and Environment: Berne, 1996; 1–89.
6. Smyth, A.J.; Dumanski, J. *FESLM: An International Framework for Evaluating Sustainable Land Management*; Soil Bulletin No. 32, FAO: Rome, 1993; 1–74.
7. Prior, J.C. From Technology Transfer to Community Development: The Policy Implications of the Australian Landcare Movement. In *Soil Conservation Extension: From Concepts to Adoption*; Sombatpanit, S., Zoebisch, M.A., Sanders, D.W., Cook, M.G., Eds.; Oxford and IBH Publishing: New Delhi, 1997; 77–86.

8. Anderson, J.R.; Thampapillai, J. *Soil Conservation in Developing Countries: Project and Policy Intervention*; Policy and Research Series No. 8, World Bank: Washington, DC, 1990; 1–45.
9. Coughlan, K.J.; Rose, C.W. *A New Soil Conservation Methodology and Application to Cropping Systems in Tropical Steeplands*; Technical Report No. 40, Australian Centre for International Agricultural Research: Canberra, 1997; 1–147.
10. El-Swaify, S.A.; Dangler, E.W.; Armstrong, C.L. *Soil Erosion by Water in the Tropics*; Research Extension Series No. 24, College of Tropical Agriculture and Human Resources, University of Hawaii: Honolulu, 1982; 1–173.
11. Lal, R. Rationale for Watershed as a Basis for Sustainable Management of Soil and Water Resources. In *Integrated Watershed Management in the Global Ecosystem*; Lal, R., Ed.; CRC Press: Boca Raton, FL, 2000; 3–16.
12. Stirk, G. *Wind Erosion in the Sahelian Zone of Niger: Processes. Models, and Control Techniques*; Tropical Resource Management Papers, No. 15, Wageningen Agricultural University: Wageningen, 1997; 1–151.

EROSION, CONTROLLING IRRIGATION-INDUCED

E

R.E. Sojka
D.L. Bjorneberg
Soil Scientist and Agricultural Engineer, Kimberly, Idaho, U.S.A.

INTRODUCTION

Erosion is the greatest threat to agricultural sustainability. Most irrigation is on fragile arid soils that have enormous crop yield potential when irrigated. However, that yield potential is easily lost if the thin veneer of "topsoil" is eroded (1). Erosion prevention on irrigated land is, arguably, more important than on rainfed land. Yields from irrigated land are more than double those from non-irrigated land, with nearly triple the crop value per hectare (2). In addition, runoff and irrigation return flows (necessary in many surface irrigation schemes) deliver sediment; human, animal and plant pathogens; nutrients and pesticides to downstream fields and riparian waters. These pollutants accumulate in runoff primarily as a result of erosion.

IRRIGATION'S UNIQUE EROSION CHARACTERISTICS

Irrigation-induced erosion and rainfall-induced erosion result from the same physical and chemical processes. However, the processes come together and interact very differently in each case (2–4). The magnitude of the differences depends upon the type of irrigation system and on soil and water properties[a]. Briefly, the most important differences stem from soil and water chemistry, wetting rate, water application and infiltration patterns, and, for surface irrigation, absence of water drop impact. These factors are the basis for many erosion control practices unique to irrigation (4, 5–7). Since 1990, advances in irrigation erosion control have resulted from improved understanding of water quality and antecedent soil condition effects on erosion and from development of polyacrylamide (PAM) use.

The key to controlling erosion is controlling runoff. Runoff is controlled in two ways. It is minimized by scheduling irrigation to meet, but not exceed, crop water and salt leaching requirements (i.e. avoid over-irrigation), and it is managed by using application rates during each scheduled irrigation that minimize runoff and erosion for that event. Systems should be designed and operated to minimize over-irrigating some areas in order to adequately irrigate others. In furrow irrigation this is accomplished by reducing the length of furrows; managing inflow rates and advance times, and where possible, cutting back inflow rates once runoff begins; or through use of surge irrigation (surge irrigation sometimes erodes near the inlet because of higher flows during initial pulsing of water). Sprinkler systems can reduce runoff with variable rate emitters that match application rates to soil infiltration rates at specific field locations.

Erosion reduction from improved scheduling and application management is usually proportional to runoff reduction. Reducing over application also reduces pumping costs and losses of applied nutrients and agri-chemicals. In surface irrigation systems, where 20 to 40% runoff is often required to achieve field application uniformity, erosion remediation can be integrated into water supply enhancement by pumping sediment-laden drain water back onto fields. This does not prevent erosion, but does replace most of the eroded soil along with the saved water, for only the pump-back cost.

METHODS OF CONTROL

Conversion to Sprinklers

One effective way to prevent irrigation-induced erosion is conversion from surface to sprinkler irrigation. Again, the soil conservation benefit from conversion to sprinklers derives from and is proportional to the reduction of runoff. Sprinkler irrigation has higher technical, capital, energy and infrastructure requirements than surface irrigation. Therefore, sprinklers are used on only a small fraction of global irrigated area, whereas, nearly 60% of US irrigated land uses sprinklers. Properly designed and managed sprinkler systems eliminate 100% of off site sediment losses. However, with sprinklers, there is a tendency to extend irrigation to steeper slopes or otherwise more erosive lands. On steep land, when sprinkler systems are poorly designed or managed, erosion can occur.

[a] See *Irrigation Erosion* on page 742.

Center pivots can cause erosion problems due to water running in wheel ruts, down steep slopes, or because of high application rates at outer reaches of the pivot (8), especially when using extendable booms and high volume end-guns to reach corners. Erosion from high application areas, or where runoff concentrates, can be reduced using tillage, pitting and mulching between rows to increase surface roughness storage and reduce runoff (9–11).

Soil Protection and Tillage

Many approaches developed to control rainfall-induced erosion can prevent irrigation-induced erosion, particularly under sprinklers, e.g. no-till and conservation tillage, which rely on crop residue to protect the soil surface. Yet, despite typical erosion reductions > 90%, often with increased yields (12), no-till and conservation tillage are rarely practiced by surface irrigators. Floating residue often migrates along and clogs irrigation furrows, washing out adjacent beds and furrows, while under-irrigating the blocked furrow. In basin flood irrigation, floating debris can interfere with water spreading, sometimes concentrating initial flows, eroding some areas and elsewhere burying emerging plants with sediment or debris. No-till farming with furrow irrigation is further complicated by crop rotations that require different row (and furrow) spacings each season.

Sojka et al. (13) demonstrated 60% reduction in field sediment loss from furrow-irrigated potatoes that were paratilled (subsoiled) following planting. Slight yield increases and significant tuber grade improvements raised profitability under both furrow and sprinkler irrigation with paratilling (14). Because irrigation assures crop water availability, yield benefits from improved root development are not consistently seen with subsoiling in irrigated crops (15). Subsoiling is practiced commonly with sprinkler irrigation to enhance infiltration and decrease runoff, thereby reducing erosion. However, farmers are cautious about subsoiling furrow-irrigated crops because of the potential for irregular water flows in subsurface cracks to interfere with irrigation uniformity. Field preparation or land forming practices that reduce water application uniformity or increase runoff, are avoided by irrigators.

Placing mulch or growing sod in irrigated furrows reduces erosion. Sod nearly eliminated runoff sediment (16). Straw mulching reduced sediment loss 52 to 71% (17–20). Drawbacks of these techniques relate to management of sodded furrows, the added operations and equipment needed to place straw, and debris migrating along and clogging mulched furrows.

Site Modification

Various "engineering" approaches have been used to reduce field sediment losses from surface irrigation. The most common is use of settling basins. Large quiescent pools to facilitate particulate settling from runoff collected from fields up to 20 hectares are fairly typical. Settling pond size depends upon the area served, rate and volume of runoff, sediment concentrations expected and particle size distribution. Small settling basins along the bottoms of surface irrigated fields, serving a few rows per basin, are sometimes easier to manage at season's end, when trapped sediment can be spread back onto the field using farm equipment. Big ponds require large scale equipment for construction, cleaning and soil redistribution. For medium-textured soils about 60% of suspended mass entering settling ponds is retained. The non-retained soil is in the clay size range (21). Since clay carries most of sediment's adsorbed nutrient and chemical load, failure of ponds to retain clay impedes retention of agricultural chemical pollutants, despite the high percentage of sediment mass captured. Furthermore, effectiveness declines as ponds fill with sediment, reducing water residence time. Another variation on ponds is installation of buried drains and stand pipes to regulate water level in tail ditches (22). The stand pipes force ponding and prevent gradual concaving of field tail ends. They do not, however, prevent loss into the drain of sediment entrained in runoff from upper field reaches.

Altering canopy configuration can reduce erosion. Sojka et al. (23) halved field sediment loss using narrow or twin row plantings. Water ran between closely placed furrows, reducing irrigation duration (and runoff) and allowing root systems and canopy debris to reduce soil detachment in the furrow. Filter strip crops drilled at right angles into the final three to six meters of furrow-irrigated row crops also remove entrained sediments from runoff (6), but do not prevent sediment migration from field inlet to tail end. Because filter strip management is a compromise between two crops, yield from the strips is typically half that expected for either crop alone.

Water Properties

Both the physical and chemical properties of irrigation water affect erosion. Erosion is greatly reduced by reducing sprinkler droplet size or energy (24, 25) or by reducing stream flow in furrows (26). These physical changes require adjustments in application timing, furrow lengths and irrigation durations to properly match water application constraints with crop water needs.

Water electrolyte chemistry greatly affects the erosiveness of irrigation water (27–30). High sodium adsorption ratio (SAR) and low electrical conductivity (EC) contribute to soil aggregate detachment, disruption and dispersion of fine primary soil particles in runoff. The effect of low EC and high SAR are synergistic. Increasing electrolyte concentration with a calcium source lowers SAR, shrinks the ionic diffuse double layer around charged soil particles, and prevents dispersion, thereby maintaining aggregate stability and resisting erosion. The conjunctive use of waters from different sources or the addition of calcium can raise EC and/or lower SAR to reduce erosion potential and improve infiltration by stabilizing surface-soil structure.

Adding large polymeric compounds to irrigation water is an effective erosion prevention technology (31–33). These compounds, when delivered in dilute concentrations (typically 1 to 10 ppm) by the irrigation stream, increase aggregate stability and inter-aggregate cohesion as water infiltrates. Erosion reduction of 95% is typical for application of 1 to 2 $kg\,ha^{-1}$ per treated irrigation. Adoption has been greatest for furrow irrigation erosion reduction, but interest in extending the technology to sprinklers is growing, as much to improve infiltration uniformity as to reduce erosion (34–36). The most successful class of polymers has been anionic polyacrylamide (PAM), allowing safe, easy and effective erosion prevention for seasonal application rates of 3 to 5 $kg\,ha^{-1}$, or under \$35 ha^{-1} per season (37).

REFERENCES

1. Carter, D.L.; Berg, R.D.; Sanders, B.J. The Effect of Furrow Irrigated Erosion on Crop Productivity. Soil Sci. Soc. Am. J. **1985**, *49* (1), 207–211.
2. Bjorneberg, D.L.; Kincaid, D.C.; Lentz, R.D.; Sojka, R.E.; Trout, T.J. Unique Aspects of Modeling Irrigation-Induced Soil Erosion. Intern. J. Sed. Res. **2000**, *15* (2), 245–252.
3. Trout, T.J.; Neibling, W.H. Erosion and Sedimentation Processes on Irrigated Fields. J. Irr. Drain. Eng. ASCE. **1993**, *119* (6), 947–963.
4. Sojka, R.E. Understanding and Managing Irrigation-Induced Erosion. In *Advances in Soil and Water Conservation*; Pierce, F.J., Frye, W.W. Eds.; Ann Arbor Press: Chelsea, MI, 1998, 21–37.
5. Carter, D.L. Soil Erosion on Irrigated Lands. In *Irrigation of Agricultural Crops*; Agronomy Monograph no. 30, Stewart, B.A., Nielson, D.R. Eds.; Am. Soc. Agron.: Madison, WI, 1990, 1143–1171.
6. Carter, D.L.; Brockway, C.E.; Tanji, K.K. Controlling Erosion and Sediment Loss from Furrow-Irrigated Cropland. J. Irrig. Drain. Engr. ASCE. **1993**, *119* (6), 975–988.
7. Sojka, R.E.; Carter, D.L. Constraints on Conservation Tillage Under Dryland and Irrigated Agriculture in the United States Pacific Northwest. In *Conservation Tillage in Temperate Agroecosystems*; Carter, M.R. Ed.; Lewis Publishers: Boca Raton, FL, 1994, 285–304.
8. Gilley, J.R.; Mielke, L.N. Conserving Energy With Low-Pressure Center Pivots. J. Irrig. Drain. Div. ASCE. **1980**, *106* (1), 49–59.
9. Aarstad, J.S.; Miller, D.E. Soil Management to Reduce Runoff Under Center-Pivot Sprinkler Systems. J. Soil Water Cons. **1973**, *28* (1), 171–173.
10. Kranz, W.L.; Eisenhauer, D.E. Sprinkler Irrigation Runoff and Erosion Control Using Interrow Tillage Techniques. Applied Eng. in Ag. **1990**, *6* (6), 739–744.
11. Oliveira, C.A.S.; Hanks, R.J.; Shani, U. Infiltration and Runoff as Affected by Pitting, Mulching and Sprinkler Irrigation. Irr. Sci. **1987**, *8* (1), 49–64.
12. Carter, D.L.; Berg, R.D. Crop Sequences and Conservation Tillage to Control Irrigation Furrow Erosion and Increase Farmer Income. J. Soil Water Conserv. **1991**, *46* (12), 139–142.
13. Sojka, R.E.; Westermann, D.T.; Brown, M.J.; Meek, B.D. Zone-Subsoiling Effects on Infiltration, Runoff, Erosion, and Yields of Furrow-Irrigated Potatoes. Soil & Till. Res. **1993**, *25* (4), 351–368.
14. Sojka, R.E.; Westermann, D.T.; Kincaid, D.C.; McCann, I.R.; Halderson, J.L.; Thornton, M. Zone-Subsoiling Effects on Potato Yield and Grade. Am. Pot. J. **1993**, *70* (6), 475–484.
15. Aase, J.K.; Bjorneberg, D.L.; Sojka, R.E. Zone-Subsoiling Relationships to Bulk Density, Cone Index, Infiltration, Runoff and Erosion on a furrow Irrigated Soil. Trans. ASAE. **2001**, *44* (3), 577–583.
16. Cary, J.W. Irrigating Row Crops from Sod Furrows to Reduce Erosion. Soil Sci. Soc. Am. J. **1986**, *50* (5), 1299–1302.
17. Aarstad, J.S.; Miller, D.E. Effects of Small Amounts of Residue on Furrow Erosion. Soil Sci. Soc. Am. J. **1981**, *45* (1), 116–118.
18. Brown, M.J. Effect of Grain Straw and Furrow Irrigation Stream Size on Soil Erosion and Infiltration. J. Soil Water Conserv. **1985**, *40* (4), 389–391.
19. Miller, D.E.; Aarstad, J.S.; Evans, R.G. Control of Furrow Erosion with Crop Residues and Surge Flow Irrigation. Soil Sci. Soc. Am. J. **1987**, *51* (2), 421–425.
20. Brown, M.J.; Kemper, W.D. Using Straw in Steep Furrows to Reduce Soil Erosion and Increase Dry Bean Yields. J. Soil Water Conserv **1987**, *42* (4), 187–191.
21. Brown, M.J.; Bondurant, J.A.; Brockway, C.E. Ponding Surface Drainage Water for Sediment and Phosphorus Removal. Trans. ASAE. **1981**, *24* (6), 1478–1481.
22. Carter, D.L.; Berg, R.D. A Buried Pipe System for Controlling Erosion and Sediment Loss on Irrigated Land. Soil Sci. Soc. Am. J. **1983**, *47* (4), 749–752.
23. Sojka, R.E.; Brown, M.J.; Kennedy-Ketcheson, E.C. Reducing Erosion from Surface Irrigation Using Furrow Spacing and Plant Position. Agron. J. **1992**, *84* (4), 668–675.
24. Bubenzer, G.D.; Jones, B.A. Drop Size and Impact Velocity Effects on the Detachment of Soils Under Simulated Rainfall. Trans. of the ASAE **1971**, *14* (4), 625–628.
25. Kinnell, P.I.A. Laboratory Studies on the Effect of Drop Size on Splash Erosion. J. Agric. Eng. Res. **1982**, *27* (3), 431–439.
26. Trout, T.J. Furrow Irrigation Erosion and Sedimentation: On-Field Distribution. Trans. of the ASAE **1996**, *39* (5), 1717–1723.

27. Le Bissonais, Y.; Singer, M.J. Seal Formation, Runoff, and Interill Erosion from Seventeen California Soils. Soil Sci. Soc. Am. J. **1993**, *57* (1), 224–229.
28. Lentz, R.D.; Sojka, R.E.; Carter, D.L. Furrow Irrigation Water Quality Effects on Soil Loss and Infiltration. Soil Sci. Soc. Am. J. **1996**, *60* (1), 238–245.
29. Levy, G.J.; Levin, J.; Shainberg, I. Seal Formation and Interill Soil Erosion. Soil Sci. Soc. Am. J. **1994**, *58* (1), 203–209.
30. Shainberg, I.; Laflen, J.M.; Bradford, J.M.; Norton, L.D. Hydraulic Flow and Water Quality Characteristics in Rill Erosion. Soil Sci. Soc. Am. J. **1994**, *58* (4), 1007–1012.
31. Brown, M.J.; Robbins, C.W.; Freeborn, L.L. Combining Cottage Cheese Whey and Straw Reduces Erosion While Increasing Infiltration in Furrow Irrigation. J. Soil Water Conserv. **1998**, *53* (2), 152–156.
32. Lentz, R.D.; Sojka, R.E. Field Results Using Polyacrylamide to Manage Furrow Erosion and Infiltration. Soil Sci. **1994**, *158* (4), 274–282.
33. Orts, W.J.; Sojka, R.E.; Glenn, G.M. Biopolymer Additives to Reduce Erosion-Induced Soil Losses During Irrigation. Industrial Crops and Products. **2000**, *11* (1), 19–29.
34. Aase, J.K.; Bjorneberg, D.L.; Sojka, R.E. Sprinkler Irrigation Runoff and Erosion Control with Polyacrylamide – Laboratory Tests. Soil Sci. Soc. Am. J. **1998**, *62* (6), 1681–1687.
35. Bjorneberg, D.L.; Aase, J.K. Multiple Polyacrylamide Applications for Controlling Sprinkler Irrigation Runoff and Erosion. Applied Eng. in Ag. **2000**, *16* (5), 501–504.
36. Sojka, R.E.; Lentz, R.D.; Ross, C.W.; Trout, T.J.; Bjorneberg, D.L.; Aase, J.K. Polyacrylamide Effects on Infiltration in Irrigated Agriculture. J. Soil Water Conserv. **1998**, *53* (4), 325–331.
37. Sojka, R.E.; Lentz, R.D. Reducing Furrow Irrigation Erosion With Polyacrylamide (PAM). J. Prod. Ag. **1997**, *10* (1), 47–52.

EROSION ON DISTURBED LANDS

William J. Elliot
United States Department of Agriculture-Forest Service (USDA-FS), Rocky Mountain Research Station, Moscow, Idaho, U.S.A.

INTRODUCTION

Disturbed lands include construction sites, mine sites, roads, or forests that have been disturbed by fire or other operations, and military maneuver sites. Erosion rates on disturbed lands are highly dependent upon factors such as climate conditions immediately following the disturbance, the type and degree of disturbance, local topography, and the natural vegetation regeneration or mitigation measures. Soil properties also influence erosion rates, but tend to be overshadowed by the other factors. A high degree of variability is associated with each of these factors; thus, the erosion rates resulting from disturbances are highly variable.

CLIMATE AND HYDROLOGY

Vegetation regeneration on disturbed sites depends upon precipitation. Too little precipitation means that the site takes longer to regenerate, whereas too much precipitation will likely lead to excessive soil erosion that may also hinder re-vegetation. However, high levels of low-intensity precipitation will generally result in rapid regeneration and minimal erosion risk.

DISTURBANCES

In the absence of disturbances, land surfaces are often covered in forests or rangelands. Disturbances can accelerate erosion rates as much as one thousand times more quickly than on undisturbed sites. Disturbances may be natural, such as major storm or flood events and fire (Fig. 1), or human caused, such as roads (Fig. 1), construction (Fig. 2), forestry, or mining (Fig. 3). Erosion rates of undisturbed rangelands and forests are low (Table 1). However, occasional natural disturbances are part of the ecosystem, and any increased erosion rates from human disturbances should be compared with both the undisturbed and naturally disturbed erosion rates.

Fire and Water Repellency

Fire may be natural or intentional. Generally, natural or "wild" fires tend to be more severe because the forest or range must be in a drier condition for a wildfire to initiate. Intentional or "prescribed" fires are applied in early summer or after autumn rains have begun, to minimize the amount of litter that is burned, while reducing the amount of above-ground woody debris. Managers use fire to provide improved grazing for livestock and wildlife and to try to improve forest or range health. Fires are commonly used in forests following a harvesting operation to reduce the amount of remaining woody material, or to improve the health of the forest and the success of regeneration. Such practices—when used with a suitable undisturbed buffer between the fire and a stream channel—can minimize upland erosion and sediment delivery risks.

Fire may cause soils to become water repellent. Water repellency is caused by the condensation of hydrocarbons on soil particles and aggregates when organic material is combusted. It can reduce the hydraulic conductivity of soil from 30–80 $mm\,h^{-1}$ to 15–40 $mm\,h^{-1}$ in forests (1). Rangelands tend to be less susceptible to repellent conditions following fire, although some range shrub communities as well as some forest soils can be naturally repellent. The reduced hydraulic conductivity values associated with repellency may still be greater than rainfall intensities or snowmelt rates in the year following the fire. The chemicals causing repellency are water soluble, and repellent soils tend to recover to pre-fire conditions within three or four years, if there is sufficient precipitation. If there is a major storm or runoff event that occurs when soils are highly repellent, erosion rates can be large (20–100 $Mg\,ha^{-1}$), but if storms are of low to average intensity, erosion rates are likely to be moderate, typically under 10 $Mg\,ha^{-1}$. Water repellency can be reduced through tillage, but burying remaining surface residue tends to offset any benefits of reduced repellency.

Anthropogenic Soil Disturbances

Anthropogenic or human-caused disturbances can cause minor changes in the landscape, such as a road (Fig. 1),

Fig. 1 Two disturbances that may cause increased erosion are wildfire, and roads. This photograph was taken several weeks after forest fires ravaged western Montana in 2000.

Fig. 2 A construction site exposing large areas to soil erosion until vegetation can be re-established. Silt fences and straw bales have been installed to reduce sediment delivery.

Fig. 3 A surface mine spoil heap many years after mining in northern Idaho. The lack of vegetation increases susceptibility to erosion, but the large rock content reduces actual erosion risks.

a footpath, a logging skid trail, or off-road vehicle tracks from recreational or military uses. Human activities can also cause major landscape disturbances, such as a construction site (Fig. 2) or a surface mine spoil heap (Fig. 3). The same general erosion principles, however, can be applied to all disturbances. The main impacts of these disturbances are reduced live vegetation, loss of surface cover by decaying vegetation, and soil compaction. In addition, roads and mining may expose soils low in organic matter, and high in rock content or undesirable trace elements. These conditions are not conducive to vegetation regeneration, further aggravating long-term erosion risks.

Runoff rates are much higher on most disturbed areas than undisturbed areas. On roads, for example, the hydraulic conductivity can be reduced to near zero (under $0.1\,\mathrm{mm\,hr^{-1}}$) (2). Bulk densities are generally increased from under $1.0\,\mathrm{Mg\,m^{-3}}$ in forests and range to over $1.4\,\mathrm{Mg\,m^{-3}}$ on roads. Compaction reduces the water-holding capacity of the soil, which increases runoff and erosion. The increased runoff reduces the amount of water held in storage for the summer, subsequently reducing vegetation regeneration rates. Large rocks (Fig. 3) and acidic soils are common to many mining sites. The rocky surface tends to reduce runoff and erosion, but increases the difficulty in establishing vegetation.

In many cases, it is not possible to prevent erosion on a road, a highly disturbed construction site, or mining site. In these cases, managers frequently incorporate practices to reduce off-site erosion using various sediment control techniques such as silt fences, straw bales (Fig. 2) or sediment basins. Vegetated buffer strips are also commonly used to reduce the amount of runoff and sediment leaving such sites.

Reclamation of highly disturbed sites generally involves techniques such as subsoiling, mulching, and fertilizing to improve soil productivity. Vegetation is then established that will likely provide rapid surface cover until natural cover returns. Once the site is revegetated, erosion risk is minimal.

VARIABILITY

Soil erosion varies greatly with time and in space on disturbed sites. Soil properties and subsurface geologic properties vary along all hillslopes. Roads, skid trails, and other trails can be quite different than the surrounding areas in hydrology and erodibility. Spatial variation is very high following both prescribed and wildfire, leading to major variability in the erosion risk that may follow (3). Seasonal variability in climate can have major impacts on erosion rates. Large soil erosion events tend to occur only from major runoff events, or

E

Table 1 Typical erosion rates in forests, rangelands, and disturbed sites

Condition	Estimated area in the U.S. (millions of ha)[a]	Typical erosion rate ($Mg\,ha^{-1}$)
Undisturbed forests	200	0.1
Undisturbed rangelands	200	1
Harvested forests	4	1
Conservation agriculture	80	5
Forests after low-severity wildfire	1	10
Overgrazed rangelands	50	10
Rangelands after fire	10	10
Forest roads	3	20
Intensive agriculture	80	20
Surface mines in need of reclamation	0.02	50
Forests after high-severity wildfire	1	100
Construction sites	–	100

[a] From Refs. 7–11.

when large runoff events occur on areas that have been disturbed. "Average annual erosion" has little meaning under conditions dominated by temporal variability.

Following disturbance, hydrologic recovery also varies greatly. North-facing slopes may recover faster than south-facing slopes. Topsoils recover faster than subsoils or mine spoil. A dry year following a disturbance may result in little water repellent recovery, a moderate year significant recovery, but a wet year may cause catastrophic flooding and erosion. Large events could disturb a watershed for decades, while stream channels and upland surfaces return to hydrologic stability. Under these conditions, it may be more appropriate to consider the probability of a given amount of erosion occurring, rather than an average value.

Because of the spatial variability following disturbances, a given landscape will have sites of net runoff and sediment generation, and sites of net infiltration and deposition. The distribution and occurrence of these sites will determine the net landscape response to an erosion event. On landscapes with a few scattered areas of severe disturbance, there may be no net runoff and erosion. This is particularly true with roads, where sediment delivery is minimal if the roads are located a reasonable distance from streams. Only severe storm events that exceed the infiltration capacity of the entire hillslope will cause any net off-site sedimentation.

SURFACE COVER

The amount and distribution of surface cover will determine the erosion rate following most disturbances. Surface cover is generally decomposing vegetation, but can also include animal waste, gravel, rocks, and fabricated materials like erosion control blankets. As surface cover increases from zero to 100%, erosion rates will drop by 99% or more. With fires and grazing, natural spatial variability is extremely high, whereas with military activities, forest harvesting, and roads, the distribution of disturbance on the landscape may be more predictable and more easily managed. Increasing surface cover with vegetation regeneration, mulch, or erosion-control fabrics is usually the most effective mitigation method for disturbed sites, although costs can be high.

EROSION PREDICTION

Erosion on disturbed sites can be predicted with both empirical models [such as the Universal Soil Loss Equation (USLE)] and physically based models [such as the Water Erosion Prediction Project (WEPP) model] (12). Factors have been developed for construction sites for the USLE (4), and that technology has been incorporated into many predictive software programs. Interfaces for predicting erosion following fire and forest roads have been developed for the WEPP model (5), as have input soil files for construction sites (6). When using any of these technologies, model users must carefully consider the period of time that the site is disturbed, the chosen mitigation practices, and how fast the site is re-vegetated.

REFERENCES

1. Robichaud, P.R. Fire Effects on Infiltration Rates After Prescribed Fire in Northern Rocky Mountain Forests, USA. J. Hydrology **2000**, *231–232*, 220–229.
2. Elliot, W.J.; Hall, D.E. *Water Erosion Prediction Project (WEPP) Forest Applications*, General Technical Report INT-GTR-365; USDA Forest Service, Rocky Mountain Research Station: Ogden, Utah, 1997. Available online at http://forest.moscowfsl.wsu.edu/4702/forestap/forestap.pdf (Accessed October 2001).
3. Robichaud, P.R.; Miller, S.M. Spatial Interpolation and Simulation of Post-Burn Duff Thickness After Prescribed Fire. International Journal of Wildland Fire **2001**, *9* (2), 137–143.
4. Wischmeier, W.H.; Smith, D.D. *Predicting Rainfall Erosion Losses–A Guide to Conservation Planning*, USDA Agriculture Handbook No. 537; U.S. Department of Agriculture: Washington, DC, 1978.
5. Elliot, W.J.; Hall, D.E.; Scheele, D.L. *FS WEPP Forest Service Interfaces for the Water Erosion Prediction Project Computer Model*; USDA Rocky Mountain Research Station: Fort Collins, Colorado, 1999. Available online at http://forest.moscowfsl.wsu.edu/fswepp/. (Accessed January 2001).
6. Laflen, J.M. *Application of WEPP to Construction Sites*, Proceedings of the International Symposium for Soil Erosion Research for the 21st Century, Honolulu, Hawaii, Jan 3–5, 2001; American Society of Agric. Engrs., St. Joseph, Michigan, 2001; 135–138.
7. USDA Forest Service. National and Regional Areas Summary. Available online at http://www.fs.fed.us. 2001. (Accessed January 2001).
8. USDA Natural Resource Conservation Service. Land Cover/Use of Nonfederal Rural Land, by State and Year, 2001. Available online at http://www.nrcs.usda.gov. (Accessed January 2001).
9. USDI Bureau of Land Management. Bureau of Land Management Facts. 2001. Available online at http://www.blm.gov/nhp/index.htm. (Accessed January 2001).
10. USDI National Park Service. The National Park System Acreage, 2001. Available online at http://www.nps.gov. (Accessed January 2001).
11. USDI Office of Surface Mining. Unreclaimed Public Health and Safety Coal Related problems, 2001. Available online at http://www.osmre.gov. (Accessed January 2001).
12. Flanagan, D.C.; Nearing, M.A. Eds. *USDA-Water Erosion Prediction Project (WEPP) Hillslope Profile and Watershed Model Documentation*, NSERL Report No. 10; National Soil Erosion Research Laboratory, USDA–Agricultural Research Service, West Lafayette, IN. 1995. 298 pp. Available online: http://topsoil.nserl.purdue.edu/nserlweb/weppmain/docs/readme.htm (Accessed October 2001).

EROSION AND GLOBAL CHANGE

Taolin Zhang
Xingxiang Wang
Institute of Soil Science, Chinese Academy of Sciences, Nanjing, People's Republic of China

INTRODUCTION

Soil erosion is the movement and transport of soil by various agents, which leads to a loss of soil, particularly by water, wind, and mass movement. It is the most serious and widespread form of soil degradation (1). Global change is becoming a great concern since it was put forward by ICSU in 1986. Global change, as a result of physical, chemical, and biological processes in the Earth's system—particularly interactions between humans and the environment—refers to changes in global environmental components that are vital to human survival, such as climate, land, water and air in the form of greenhouse effect, diminution of forest, depletion of biodiversity, degradation of land (desertification) and deficiency in water resources. Soil erosion is an important form of global change, and it severely threatens the quality of the soil, land, air and water resources, and hence the whole ecosystem and agricultural development. Soils and global changes are related by bi-directional interaction (2), and global change may exacerbate soil erosion problems (3).

THE CONTRIBUTION OF SOIL EROSION TO GLOBAL CHANGE

Extent of Soil Erosion in China and in the World

China is one of the nations that is most severely affected by soil erosion in the world. The land area affected by accelerated soil erosion is estimated as 367 million ha, including 179 million ha by water erosion and 188 million ha by wind erosion, and agriculture land affected by soil erosion is estimated as 45.4 million ha in China (4) (Table 1). The annual soil loss due to accelerated soil erosion is estimated at 5000 million tons, including 3300 million tons from agricultural land. The region most severely affected by soil erosion in China is the watershed of Yellow River, with 45 million ha area and 6821 $t\,km^{-2}\,y^{-1}$ erosion modulus, especially in the Loess plateau with more than 10,000 $t\,km^{-2}\,y^{-1}$ erosion modulus (5) (Fig. 1). On a global scale, the land area affected by accelerated soil erosion is estimated at about 1640 million ha, including 1100 million ha by water erosion and 550 million ha by wind erosion (1, 6) (Table 2). The annual soil loss due to soil accelerated erosion is estimated at 19,000 million tonnes (7).

Soil Erosion Degrades Soil Quality and Leads to Loss of Agricultural Lands

Unlike soil formation, the destruction of fertile soil can proceed very quickly. Soil erosion causes loss of nutrients and organic matter, structural deterioration, and soil desertification, and has negative effects on soil quality and biological productivity. Soil erosion also leads to loss of agricultural lands. It is estimated that the annual global loss of agricultural land due to accelerated soil erosion is 3 million ha (8). This has been exerting a profound effect on the balance of world food supply, although the on-site impacts of soil erosion on productivity are easily minimize through additional external agricultural input and adoption of improved agricultural technology. A decline in productivity depends on the quality of soil and the crop planted. Soil erosion and its effects on crop production have been of concern to western agricultural scientists since the turn of the 20th century. Yield reductions due to soil erosion have been measured on plot and field scales. Simulation models, such as the productivity index (PI), and the erosion productivity impact calculator (EPIC), have been developed to predict the effects of erosion on soil properties and crop responses (9). Unfortunately, little is known about the impact of soil erosion on productivity at the regional, national, and global scales.

Soil Erosion Deteriorates Global Environment and Influences Global Change

Soil erosion deteriorates the global environment, through flood disasters, environmental pollution, desertification, etc. It aggravates flood disasters through raising riverbeds, silting lakes, and shrinking water bodies. In the U.S., it is

Table 1 Agricultural land area affected by soil erosion in China

	Area affected (million ha)	Percentage in agricultural land (%)
Northeast	8.00	37.36
North China	7.00	26.85
Loess plateau	11.28	71.30
Arid region in northwest	2.03	15.34
Qingzang plateau	0.19	19.75
Middle and low reach of Yangtse Basin	4.62	17.75
South China	2.12	21.97
Southwest	10.17	52.53
China	45.41	34.20

(From Ref. 4.)

Table 2 Global area affected by accelerated soil erosion

	Area affected (million ha)	
Region	Water erosion	Wind erosion
Africa	227.3	187.8
Asia	433.2	224.1
Europe	113.9	41.6
North and Central America	106.6	38.8
Oceania/Pacific Islands	83.4	16.4
South America	124.1	41.4
World	1088.5	550.1

(From Refs. 1, 6.)

estimated that sediment reduces the water-storage capacity of lakes and reservoir by 0.22% per year. In addition, the annual economic cost of reservoir capacity losses by sediment deposition is $819 million (10). In China, flood disasters increased from once per 20 years 1300 years ago to once per 1.6 years in the 1990s in the Yangtse River Basin due to soil erosion. Soil erosion is also an important nonpoint source pollution. With an enrichment of nutrients, heavy metals, and pesticides in the sediments and a large amount of rapidly available nutrients in the surface runoff, soil erosion can cause serious damage to the water quality and aquatic ecosystems. Meanwhile, soil erosion can also have an adverse effect on transportation, communication, and human health through increasing dust in the air. In addition, soil erosion can also be a driving force to soil desertification.

Fig. 1 Landscape of severe soil erosion in the Loess Plateau of China (the soil has been blown or washed away and only naked rocks remain).

Soil erosion influences global climatic change by changing the carbon, nitrogen and water cycles. World soils are an important pool of active carbon and nitrogen, and play a major role in the global carbon and nitrogen cycles. Small changes in the pools of carbon and nitrogen in the world soils could have large effects on atmospheric concentration of CO_2, CH_4, and NO_2. Soil erosion also has a profound impact on both quality and quantity of organic matter in soils, and can be a significant factor in local carbon losses and redistribution. It is estimated that 5.7×10^{15} g C is displaced, and 1.14×10^{15} g C flux into the atmosphere annually due to global soil erosion (11).

Soil erosion may cause a shortage of freshwater through deteriorating soil physical structure, increasing surface runoff, and changing global water cycling. Moreover, it may also directly lead to changes in microclimate, regional and/or zonal landscapes and ecosystems, and it may even alter the global climatic pattern. In addition, soil erosion may also lead to a change in land uses and land covers through degrading soil quality and land suitability.

GLOBAL CHANGE MAY ACCELERATE SOIL EROSION

The main factors influencing soil erosion are rainfall (amount, frequency, duration, and intensity), wind speed (direction, strength, and frequency of high-intensity events), land use and management, topography, and soils and their properties. Since the 1950s, significant advances have been made in predicting erosion risks, particularly with the development of modeling technology (3, 12, 13). However, the response of soil erosion to global change is very complex temporally and spatially due to uncertainties in general circulation models (GCMs) and the complex relationships between soil erosion processes and climate, land use, and land cover.

Future changes in atmospheric CO_2 concentration and climate will change the hydrologic cycle, and hence affect the soil–plant–water interactions, which in turn affect soil erosion. Climatic erosivity is a major determinant of soil erosion and sediment transport, and is directly affected by change in climate. In the future, climatic change is expected to increase the atmospheric temperature, and the weather patterns will have an apparent change and become less predictable (14). Climate warming will enhance evaporation, and lead to a more vigorous hydrological cycle. In some regions, any increase in rainfall, whether amount, intensity, or frequency, which occurs as a result of climate change, may directly exacerbate erosion. On the contrary, some regions will not experience the increase in precipitation, and instead may experience a decrease in rainfall and hence a decline in soil moisture because of the enhanced evaporation. In this case, drought soil conditions and less vegetation can make soils more vulnerable to wind erosion. Moreover, soil erosion tends to be dominated by extreme events, which might occur only rarely. An increase in the possibility of extreme rainfall events in the future, which is suggested by several climatic change models, may accelerate soil erosion. While a large emphasis has been focused on climate change and how it may impact soil erosion, other relevant global change factors may have a more profound influence on soil loss. Global change may influence soil erosion through changing temporal and spatial evolution of global soil cover, such as change in soil-forming processes and soil properties. The response of soil to the erosion process is complex, and is also influenced by soil properties such as texture, structural stability, organic matter content, clay mineralogy, and chemical constituents. All these factors are likely to change with change in climate and/or land use. Soil organic matter (SOM), which is a key element of the soil properties, is susceptible to change with changing climate. Higher temperature will usually increase rates of decomposition of organic matter, and decrease its content in soils. A decline in SOM level would cause a decrease in soil aggregate stability, lower infiltration rates, increased runoff, and hence increase the likelihood of erosion. Whether enhanced atmospheric CO_2 concentrations will give rise to sufficient increases in plant growth to return the organic matter to soils and compensate for this loss is still not established.

The processes of land use and culture changes are also strongly impacted by the climatic changes. With climatic change and soil degradation, land use and land cover may be changed. A decline in forest cover due to deforestation and wildfire, especially tropical rainforests, and the conversion from forest or grassland to arable land in some areas under global change will accelerate soil erosion. Moreover, a rising sea level could also increase coastal erosion.

REFERENCES

1. Oldeman, L.R. The Global Extent of Soil Degradation. In *Soil Resilence and Sustainable Land Use*; Greenland, D.J., Szabolcs, I., Eds.; CAB International: Wallingford, UK, 1994; 99–118.
2. Mosier, A.R. Soil Processes and Global Change. Biol. Fert. Soils **1998**, *27*, 221–229.
3. Ingram, J.; Lee, J.; Valentin, C. The GCTE Soil Erosion Network: A Multi-Participatory Research Program. J. Soil Water Conserv. **1996**, *51* (5), 377–380.
4. Yang, R.Z. Cultivated Land Soil and Water Loss for Our Country and Control Measures. Bull. Soil Water Conserv. **1994**, *14* (2), 32–36, (in Chinese).
5. Chen, H.S.; Shao, M.A. Soil Erosion and Its Control Measures in China. Science **2000**, *2*, 55–57, (in Chinese).
6. World Resources Institute, *Towards Sustainable Development—A Guide to the Global Environment*; World Resources Institute: Washington, DC, 1992–1993.
7. Walling, D.E. Rainfall, Runoff and Erosion of the Land: A Global View. In *Energetics of the Physical Environment*; Gregory, K.J., Ed.; Wiley: UK, 1987; 89–117.
8. Buringh, P. *An Assessment of Losses and Degradation of Productive Agricultural Land in the World*; FAO Working Group on Soil Policy: Rome, Italy, 1981.
9. Lal, R. Agronomic Impact of Soil Degradation. In *Methods for Assessment of Soil Degradation*; Lal, R., Blum, W.H., Valentine, C., Stewart, B.A., Eds.; CRC Press: Boca Raton, FL, 1998; 459–473.
10. Crowder, B.M. Economics Costs of Reservoir Sedimentation: A Regional Approach to Estimating Cropland Erosion Damages. J. Soil Water Conserv. **1987**, *42* (3), 194–197.
11. Lal, R. Global Soil Erosion by Water and Carbon Dynamics. In *Soils and Global Change*; Lal, R., Kimble, J., Levine, E., Stewart, B.A., Eds.; CRC press: Boca Raton, FL, 1995; 131–142.
12. Rose, C.W. Modeling Erosion by Water and Wind. In *Methods for Assessment of Soil Degradation*; Lal, R., Blum, W.H., Valentine, C., Stewart, B.A., Eds.; CRC Press: Boca Raton, FL, 1998; 57–88.
13. Williams, J.; Nearing, M.; Nicks, A.; Skidmore, E.; Valentin, C.; King, K.; Savabi, R. Using Soil Erosion Models for Global Change Studies. J. Soil Water Conserv. **1996**, *51* (5), 381–385.
14. Watson, R.T., Zinyowera, M.C., Moss, R.H., Dokken, D.J., Eds. *Climate Change 1995—Impacts, Adaptations and Mitigation of Climatic Change: Scientific–Technical-Analyses*; Cambridge University Press: Cambridge, 1996.

EROSION IMPACTS, ON/OFF SITE IMPACTS

Jerry L. Hatfield
United States Department of Agriculture-Agriculture Research Service (USDA-ARS), Ames, Iowa, U.S.A.

INTRODUCTION

Erosion, either by wind or water, moves large amounts of soil each year. It has been estimated that over 80% of the world's agricultural lands are affected to some degree by erosion (1, 2). Continued loss of topsoil removes the most productive portion from the soil profile and although nutrients and water can be resupplied to the soil, the productivity levels remain affected. The major on-site impact of erosion is the loss of crop productivity and this is mirrored in reductions in crop yield and water and nutrient use efficiency. In regions of the world with variable rainfall the loss of topsoil increases the vulnerability of human food supplies. Off-site impacts of erosion relate to the economic and ecological costs of sediment, nutrients, or agricultural chemicals being deposited in streams, rivers, and lakes. Adoption of soil management practices to reduce erosion can have profound effects on future world food supplies.

Erosion is the process of moving soil by the action of water or wind from one location to another in the landscape and often over large distances as off-site impacts. Redistribution of soil within the field is often overlooked as a major impact of erosion processes compared to the more visible off-site movement that is often seen as alluvial fans in streams and rivers or the dunes of sand in highly erodible arid regions. Erosion processes are responsible for the shaping of the canyons, creation of the Palouse areas in Washington, and the loess deposits in both China and the United States. Erosion is a powerful force at work in the agricultural lands of the world. It has been estimated that almost 56% of the world's agricultural lands have been degraded by water erosion while 28% have been degraded by wind erosion. Continued degradation of world agricultural lands by erosion threatens the stability of food production (1–3).

ON-SITE IMPACTS

Topsoil in agricultural fields has been removed since humans began farming. Evidence of this effect is seen by the level of agricultural fields compared to adjacent fence rows or native pastures. Removal of this topsoil has a large impact on the ability of the soil to produce food, feed, or fiber. On-site impacts of erosion are evident in the productivity levels of the soil. Erosion removes the topsoil that is highest in organic matter content, has the most stable soil structure, and offers the most optimal seedbed for germinating and emerging plants. The upper layers of soil are those that reflect the influence of anthropogenic activities. Removal of the topsoil reduces the water-holding capacity of the soil and further reduces the available rooting depth of the crop. All of these changes have a negative impact on soil productivity. Soil productivity levels represent the potential capability of the soil to produce a crop under optimal meteorological conditions. Therefore, any removal of soil would reduce this potential (Fig. 1).

There is a large variation among soils in their response to erosion. Temperate soils are less affected by erosion than tropical soils; however, in both soils productivity is negatively affected. In tropical soils the on-site impacts vary among crops. Cereal crops are more affected than root crops (3). Similar responses are seen for other areas of the world. Continued erosion will cause a shift in the types of plants that can be grown on a given soil. Erosion effects can be modified by the use of inputs (nutrients, tillage, supplemental supplies of water). However, adding soil nutrients is not often a replacement for the natural productivity of the soil. Rooting depth of the crop is restricted more in eroded than noneroded soils. Problems with rooting depth and extensive root development reduce the efficiency of water and nutrient use by the crop (4–6).

On-site impacts of soil erosion are difficult to quantify. Crop production levels (yields) are often the most common measure of soil performance. However, measurement of water or nutrient use efficiency provides a more quantitative method of being able to compare among locations or crop species. Water use and nutrient use efficiency (expressed as the amount of grain or total biomass produced relative to the amount of water used or nutrient applied) provides a measure of the capability of the soil to produce a crop. In wheat, removal of the upper 0.18 m of the soil profile reduced the water use efficiency by 20% but didn't affect the amount of water used by the crop (7, 8). Nutrient use efficiency in the wheat was

Fig. 1 Erosion dissects productive cropland with water flow and sediment movement through fields. (Courtesy of USDA–NRCS.)

affected and it was found that additions of either nitrogen or phosphorus fertilizers to eroded soils did not restore productivity to the noneroded levels (4). In tropical soils, erosion reduced the water use efficiency by 60% with 0.35 m of erosion and reduced nutrient use efficiency of the noneroded soil by over 500% (2). These examples are typical of the responses that have often been observed when crop production efficiency has been studied. Efficiency of crop production diminishes with decreasing topsoil depth.

In Africa—where much of the crop production is by subsistence agriculture—the loss of 0.4 m of topsoil could reduce crop yields by over 75% (3). A conceptual diagram of the potential changes in productivity and required production inputs is shown in Fig. 2. This level of reduction would increase the potential vulnerability of food supplies to weather variation within the growing season or climate changes across years. In arid and semiarid environments—where water is a limiting component in crop or forage production—erosion would have a larger impact on productivity than in the temperate or tropical climates.

In addition to removal of the topsoil from the field, an on-site impact is often the disruption of cultural operations because of gullies or rills in the field. The generation of eroded areas within the field can cause problems with all cultural operations that result in damage to machinery or increased labor costs to repair these problems (2). Deposition of soil within a field can create problems by burying emerging crops, and often the moving water transports crop residues from previous crops to low-lying areas of the field. Another on-site impact that occurs is the deposition of soil that contains agricultural chemicals onto adjacent fields. These chemicals can cause problems with crop growth, particularly if the crop in the adjacent field is sensitive to the particular chemical. This effect is seen in localized sites but the impact is often large if there are high-value crops being grown (Fig. 3).

OFF-SITE IMPACTS

Off-site impacts are those linked with the transport and deposition of soil from production fields or paddocks onto adjacent or downstream areas. These impacts can be classified into two major components: economic and environmental. Economic impacts of either water or wind erosion are difficult to quantify. Costs associated with the removal of eroded soil from roads after floods or windstorms are billions of dollars each year (1). This amount does not include the amount of money spent on

Fig. 2 Conceptual diagram of the effect of topsoil removal on the input and output changes in agricultural systems.

Fig. 3 Surface runoff occurs in fields and removes large amounts of soil and crop material. (Courtesy of USDA–NRCS.)

repairing damage caused by sediment deposited in homes after floods or the costs associated with temporary housing. These off-site impacts are often the ones that are most noticeable due to water and wind erosion. Other economic impacts of erosion are the removal of sediment from lakes and rivers to maintain either navigation, recreation, or flow.

Environmental impacts from erosion may take a longer time to be noticeable. Sediment deposition into streams, rivers, and lakes impacts the biological health of the ecosystem. The presence of excessive sediment in the water column disrupts the ecological balance and it is difficult to place a monetary value on this impact. Sediment that is moved off-site may contain nutrients, in particular phosphorus, agricultural chemicals, or pathogens. Each of these components has an impact on the ecosystem. Excess phosphorus can cause eutrophication in water bodies that leads to diminished productive capability and water-quality problems (9). Phosphorus movement is mainly through attachment to sediment; and movement of soil from fields often transports large amounts of sediment-bound materials (10). Likewise, many agricultural chemicals are tightly bound to sediments and only move off-site when soil is transported. Transport of agricultural chemicals (herbicides or insecticides) poses health hazards to both fish and humans if the concentration in the water supply exceeds toxic levels. Treatment of drinking water supplies to remove agricultural chemicals varies across regions; however, annual costs that exceed $500,000 in water treatment facilities have been reported (L.D. McMullen, Des Moines Water Works, Personnel Communication, 2000). Environmental quality aspects of erosion are more evident in the effect of materials bound to the moving sediment than sediment itself.

Transport of pathogens from agricultural lands to adjacent waterbodies has become one of the most important topics in agricultural watersheds. Pathogens may impact fish and human populations and the recent tragedy in Milwaukee, Wisconsin, with the deaths of over 20 people from waterborne pathogens, shows the potential of off-site movement to affect human life. Movement of sediment and nutrients has been considered to be the largest off-site impact and the most costly (9, 10). However, transport of pathogenic material from fields or stream banks into water supplies is emerging as one of the critical problems in watershed management and scientific papers to evaluate the cause and effect relationships will emerge from these observations.

Off-site impacts from wind erosion are often characterized as nuisances because of the problems with deposition of wind-blown materials. The cost of removing these deposits is small compared to the costs of removing sediment. There is also a nuisance factor caused by reduced visibility; however, there are direct human health concerns associated with increased particulate loads and soil-borne pathogens in the air.

CONCLUSIONS

Erosion has a major impact on both on-site productivity and efficiency of production. It is difficult to develop an exact cost figure for erosion. However, continual degradation of the soil through erosion threatens the stability of future food supplies. Off-site impacts of erosion affect both water and air quality. Increasing concern about the potential impacts of off-site movement of nutrients, chemicals, and pathogens on water quality will continue to be a major concern in how we view the soil resource and implement practices to protect soil. Decreasing erosion will have positive on-site and off-site impacts.

REFERENCES

1. El-Swaify, S.A. State of the Art for Assessing Soil and Water Conservation Needs and Technologies. In *Adopting Conservation on the Farm*; Napier, T.L., Camboni, S.M., El-Swaify, S.A., Eds.; Soil and Water Conservation Society: Ankeny, IA, 1994; 13–27.
2. Larson, W.E.; Pierce, F.J.; Dowdy, R.H. The Threat of Soil Erosion to Long-Term Crop Production. Science **1983**, *219*, 458–465.
3. Lal, R. Erosion–Crop Productivity Relationships for Soils of Africa. Soil Science Society of America Journal **1995**, *59*, 661–667.
4. Massee, T.W.; Waggoner, H.O. Productivity Losses from Soil Erosion on Dry Cropland in the Intermountain Area. Journal of Soil and Water Conservation **1985**, *40*, 447–450.
5. Martin, C.K.; Cassel, D.K. Soil Loss and Silage Yield for Three Tillage Management Systems. Journal of Production Agriculture **1992**, *5*, 581–586.
6. McGregor, K.C.; Cullum, R.F.; Mutchler, C.K. Long-Term Management Effects on Runoff, Erosion, and Crop Production. Transactions of the ASAE **1999**, *42*, 99–105.
7. Tanaka, D.L. Topsoil Removal Influences on Spring Wheat Water-Use Efficiency and Nutrient Concentration and Content. Transactions of the ASAE **1990**, *33*, 1518–1524.
8. Tanaka, D.L.; Aase, J.K. Influence of Topsoil Removal and Fertilizer Application on Spring Wheat Yields. Soil Science Society of America Journal **1989**, *53*, 228–232.
9. Chambers, B.J.; Garwood, T.W.D.; Unwin, R.J. Controlling Soil Water Erosion and Phosphorus Losses from Arable Land in England and Wales. Journal of Environmental Quality **2000**, *29*, 145–150.
10. Sauer, T.J.; Daniel, T.C.; Nichols, D.J.; West, C.P.; Moore, P.J., Jr.; Wheeler, G.C. Runoff Water Quality from Poultry Litter Treated Pasture and Forest Sites. Journal of Environmental Quality **2000**, *29*, 515–521.

EROSION AND LANDSCAPE DEVELOPMENT, SCALE (SPACE AND TIME FACTORS)

E

Jon Harbor
Purdue University, West Lafayette, Indiana, U.S.A.

INTRODUCTION

Landscapes develop as a result of the interaction between rates and patterns of uplift and denudation. Therefore styles of landscape development depend in part on rates and patterns of soil formation and soil erosion (1). Although the similarity of some aspects of landscape form across spatial scales (Fig. 1) (2) research suggests that fundamental balances of erosional and depositional processes are scale independent, scale issues are extremely important in understanding how soil erosion and empirical data on soil erosion relate to landscape development.

TIMESCALES

Landscape development occurs over timescales that are typically extremely large ($\geq 10^5$ years) from the perspective of soil scientists, and the landscape itself is a spatial scale far greater than those tackled in most soil studies. Thus scientists interested in present-day soil erosion processes and controls usually consider the landscape and landscape features as fixed boundary conditions (slope angles, lengths, and shapes do not typically change over months or years). Jenny's (3) often-quoted list of factors of soil formation (including climate, organic matter, relief, and parent material) is in fact a list of independent variables controlling spatial patterns of soil erosion over short timescales (Table 1).

However, over thousands to millions of years, rates and patterns of soil formation and erosion are important in controlling the long-term development of landscape form. Over these long timescales if soil accumulation at a site (soil production plus deposition) exceeds soil loss by erosion, then the denudational balance (5) results in increasing soil thickness, decreasing sub-soil weathering, and a situation where the erosion rate is limited by the effectiveness of the processes transporting soil on the slope. In such transport-limited situations, long-term slope evolution is towards convexo-concave forms that become progressively gentler over time (6). In contrast, if the soil accumulation rate is less than the potential soil erosion rate, then weathered material is removed as soon as it is produced and the rate and pattern of weathering limits the amount and pattern of soil erosion. Under such weathering-limited situations slopes have prominent straight sections and develop over time by parallel retreat in which slope angles are maintained in dynamic equilibrium (6). Thus, over long timescales, soil erosion rates and patterns are not only affected by landscape form but are also a determinant of landscape form.

More generally, as discussed in detail by Schumm and Lichty (4), as the timescale of interest changes, what we consider to be relevant, dependent, and independent variables in soil erosion also change (Table 1). Over short timescales it is irrelevant what the landscape was like when it was first exposed by uplift, or how much time has elapsed since then, and morphometric variables such as relief and slope shape are independent variables in a study of soil erosion. However, on the timescale of landscape development, variables such as relief and valley morphology change over time as a function of the pattern of geomorphic processes, including soil erosion, and so must be treated as dependant variables. Thus feedback relationships develop between changes in slope form and soil erosion patterns: soil erosion patterns depend on slope form, but on long timescales soil erosion patterns change slope form, and thus slope form and soil erosion patterns evolve interdependently.

Timescales are also important in soil erosion measurement programs and in the ways in which we use observations of soil erosion processes to make inferences about landscape development. Over short timescales, our understanding of spatial and temporal patterns of soil erosion is based on monitoring erosion plots and on determining sediment yields from watersheds across a range of spatial scales (7). Over geologic timescales, our understanding of spatial and temporal patterns of soil erosion is based on depositional records from lake and ocean sediments, and more recently on cosmogenic nuclide techniques (8). The cosmogenic nuclide approach is based on the observation that reactions between cosmic rays and minerals within about 2–3 m of the soil or rock surface produce a range of cosmogenic nuclides within the mineral

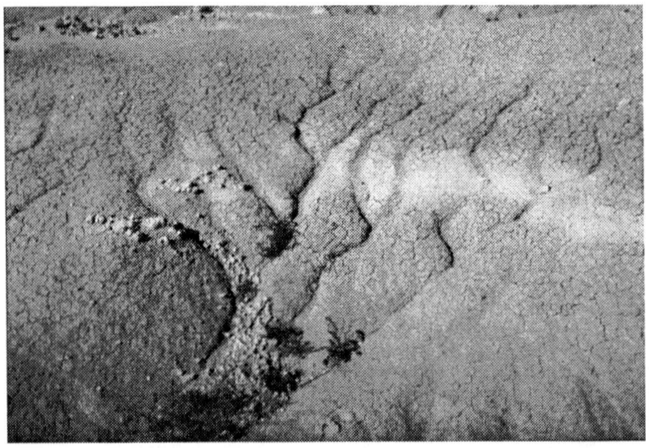

Fig. 1 Scale and soil erosion features. Topography is approximately fractal (2) so, without scale indicators, this landscape photo could have been taken from an airplane of a landscape element on the order of $10^2\,km^2$, or it could be a close up of a microscale feature on the order of a $10^0\,m^2$. (Photo courtesy of Christian Renschler, USDA National Soil Erosion Laboratory, Purdue University. Soil-surface ~ $1\,m^2$, Badlands National Park, U.S.)

structure, such as the radionuclides ^{26}Al, ^{10}Be, and ^{36}Cl. These nuclides accumulate at known rates, and the production rate decreases exponentially with depth below the soil or rock surface. Concentrations of radionuclides in eroded sediments can thus be used to back calculate the amount of time it took for a sample to go from ≥~3 m deep to the surface, and thus is a measure of the surface lowering (soil erosion) rate. Applications of this technique and comparison with other sources of data allow evaluation of the scale of variations in soil erosion rates at different timescales. For example, current sediment yields for a hyperarid drainage basin in southern Israel, based on a 33-year sediment budget, exceed long-term sediment generation rates based on ^{10}Be and ^{26}Al by 53–86% (9). In comparison, over a timescale of several decades, different averaging periods in data from 189 gauging stations in the eastern U.S. and Canada produced sediment yield variations of 10–50% (10). At the shortest timescales variations in erosion rates increase dramatically; because soil erosion is an episodic process, measurements over periods of days, hours, or less fluctuate wildly depending on whether or not an erosional event occurs during the measurement period. Thus soil erosion rate variability decreases as a function of increasing timescale.

SPATIAL SCALES

Sediment yield data also demonstrate the importance of linkages between temporal scales and spatial patterns in soil erosion. In the data from gauging stations in the eastern U.S. and Canada (10), in addition to temporal variation in sediment yield as a function of the timescale used for averaging, the spatial pattern of sediment yield varied dramatically between averaging periods. If only one averaging period had been used to derive inferences about spatial patterns of soil loss, this could lead to misleading conclusions concerning longer-term patterns of landscape development. Thus understanding spatial patterns of soil erosion for application to landscape development requires use of information on soil erosion patterns at the appropriate temporal scale.

Spatial scale also has a significant impact on soil erosion process and rate data. Soil erosion rates, as indicated by sediment yields back calculated to derive denudation rates (Table 2), typically are greatest at small spatial scales.

Table 1 Impact of timescale on the status of some variables in soil erosion

	Status during the designated timescale, I: independent; D: dependant	
Variable	**Short timescale ≲ 10^2 years**	**Long timescale ≳ 10^4 years**
Time of landscape formation	n.a.	I
Initial relief	n.a.	I
Geology	I	I
Paleoclimate	I	I
Paleohydrology	I	I
Relief	I	D
Valley dimensions	I	D
Climate	I	
Vegetation	I	

(Adapted from Ref. 4.)

Table 2 Impact of spatial scale on denudation rates

Basin area (km^2)	Mean denudation rate (mm/1000 yr)
3×10^{-1}	12,600
3×10^{0}	2550
8×10^{1}	220–60
3.9×10^{3}	100–30
3.7×10^{4}–3.3×10^{6}	60–30

(Adapted from Ref. 11.)

This important observation is due in large part to differences in the relative magnitudes of processes operating at different spatial scales, and on sampling techniques. At larger spatial scales, soil erosion, as measured by sediment yield, includes both erosion and deposition. Loss of sediment to deposition on hillslope sections with decreasing gradient and on floodplains significantly reduces net soil loss rates from a landscape (7), and this pattern of erosion and deposition is important for landscape form development. In contrast, soil erosion studies at plot scales rarely include areas of deposition and are dominated by studies of small-scale soil detachment and transport processes. Thus, from a landscape development perspective, extrapolation of rates from small spatial scales to predict landscape change rates is likely to result in significant overestimation of rates and styles of landscape change.

REFERENCES

1. Kneupfer, P.; McFadden, L. *Soils and Landscape Evolution*; Elsevier: Amsterdam, 1990.
2. Chase, C. Fluvial Landsculpting and the Fractal Dimension of Topography. Geomorphology **1992**, *5*, 39–57.
3. Jenny, H. *Factors of Soil Formation*; McGraw-Hill: New York, 1941.
4. Schumm, S.A.; Lichty, R.W. Time, Space and Causality in Geomorphology. Am. J. Sci. **1965**, *263*, 110–119.
5. Jahn, A. Denudational Balance of Slopes. Geogr. Polonica **1968**, *13*, 9–29.
6. Carson, M.; Kirkby, M. *Hillslope Form and Process*; Cambridge University Press: Cambridge, 1972.
7. Lane, L.; Hernandez, M.; Nichols, M. Processes Controlling Sediment Yield from Watersheds as a Function of Spatial Scale. Environ. Model. Software **1998**, *12*, 355–369.
8. Granger, D.; Kirchner, J.; Finkel, R. Spatially Averaged Long Term Erosion Rates Measured from In Situ Produced Cosmogenic Nuclides in Alluvial Sediment. J. Geol. **1996**, *104*, 249–257.
9. Clapp, E.; Bierman, P.; Schick, A.; Lekach, J.; Enzel, Y.; Caffee, M. Sediment Yield Exceeds Sediment Production in Arid Region Drainage Basins. Geology **2000**, *28*, 995–998.
10. Conrad, C.; Saunderson, H. Temporal and Spatial Variability in Suspended Sediment Yields from Eastern North America. In *Uplift, Erosion and Stability: Perspectives on Longterm Landscape Development*; Smith, B.J., Whalley, W.B., Warke, P.A. Eds. Special Publication, 162. The Geological Society of London: London, 1999; 219–228.
11. Chorley, R.; Schumm, S.; Sudgen, D. *Geomorphology*; Methuen: London, 1984.

EROSION AND PRODUCTIVITY, HUMAN LIFE

Neroli Pedro Cogo
Renato Levien
Federal University of Rio Grande do Sul (Department of Soils), Porto Alegre, Brazil

INTRODUCTION

Soil is, and long will be, a vital resource and an essential medium for life on Earth. Plants growing in the soil provide food for human and animal needs. However, because human-induced water and wind erosion can greatly exceed natural rates of soil formation and erosion (1, 2), the soil may become depleted and severely affect crop production. Therefore, the soil must be considered a non-renewable natural resource to be protected and preserved for future generations. Maintaining soil productivity for sustainable food production is an ongoing challenge for farmers, agricultural and environmental professionals, and governments worldwide.

HISTORICAL EVIDENCE OF EROSIONAL EFFECTS

Decline of civilizations as a result of soil degradation (particularly soil erosion) is well-documented (3–7). Soil degradation and civilization interact in both directions, meaning erosion-induced land degradation affects human lives and human activities often enhance soil degradation (6).

Historically, three major human-induced erosion periods can be identified (7). The first period, between 1,000 and 3,000 years ago, was caused by excessive timber cutting and expanded cultivation. The second period, during the 19th and early 20th centuries, occurred in countries colonized by European immigrants. Through the establishment of an exportation-type of agriculture, the immigrants forced the native population to move to and explore the more erodible, marginal lands while holding the better lands for a marketing economy. The third period, beginning in the early 1920s and extending to the present day, occurs mainly in developing countries with a growing population pressure for more land and food production. This population growth forces farmers to develop new lands, some good and some only marginal for agricultural production, and to adopt intensive, non-conserving cropping and management systems which cause spectacular levels of soil erosion.

The present situation varies throughout the world. In many regions, the seriousness of the erosion problem has been recognized and effective erosion control measures adopted. In other regions, erosion still is a major crop production impairment and continuously causes extensive sedimentation and water pollution problems.

HOW SOIL EROSION AFFECTS CROP PRODUCTIVITY AND HUMAN LIFE

The primary effect of soil erosion on crop productivity is a reduced yield (less $kg\,ha^{-1}$ of grain, dry matter biomass, tubers, and fruits), that comes from the removal of fertile topsoil and the contained nutrients. In a review paper, Langdale and Schrader (8) prepared a table showing soil erosion–crop productivity relationships for soils in the southeastern and midwestern United States; some of their findings are detailed in Table 1. Because it reduces animal feeds, erosion also results in a reduced livestock yield (less $kg\,ha^{-1}$ of meat, milk, and wool). This means erosion affects both vegetal- and animal-originated food productions. In addition, products of erosion (e.g., soil particles, water-runoff, dissolved chemicals in the runoff-water, and chemicals adsorbed to the soil particles) may cause serious damage to the environment. Decreased food production and increased sedimentation and water pollution caused by soil erosion mean a diminished quality of life for humans (8–13).

Erosion affects crop yield through degradations in the soil's physical, chemical, and biological properties. Physical degradations caused by soil erosion are: 1) loss of plant-available water reserve and water-storage capacity; 2) reduced rooting depth; 3) increased soil crusting; 4) reduced aggregate stability and soil tilth; 5) formation of rills and gullies that fragmentize the land; and 6) exposure of a subsurface soil layer that is inadequate for plant growth. Chemical degradations from erosion include: 1) reduced plant nutrients; 2) reduced cation-exchange

Table 1 Examples of how soil erosion affected crop productivity

	Crop yield ($kg\,ha^{-1}$)				
Degree of erosion	**Corn**	**Soybeans**	**Cotton**	**Small grains**	**Forage**
			Southeastern USA		
Grenada silt loam (Glossic Fragiudalfs), 0–5% slope:					
None	6000	2700	840	3600	7200
Eroded	5300	2000	784	3100	6700
Severe	4400	1600	672	2700	6000
Cecil sandy clay (Typic Hapludults), 2–10% slope:					
Deposition (local alluvium)	6200	–	–	–	–
Eroded	5800	2100–3100	1389	2400	17400
Severe	1900	1500–2400	866	1600	13700
			Midwestern USA		
Marshall silty clay loam (Typic Hapludolls), 2.5–6.0% slope:					
None	–	–	–	–	–
Slight	6700	2800	–	2200	9000
Moderate	6200	2600	–	2000	8500
Ida silt loam (Typic Udorthents), 6.0–9.0% slope:					
None	–	–	–	–	–
Moderate	5200	2200	–	2100	6900
Severe	4300	1700	–	1700	5800

(From Ref. 8.)

capacity from the removal of fine soil particles and associated organic matter; and 3) increased toxicity of aluminum and manganese, and soil acidity due to the exposure of a less-favorable subsurface soil layer. Biological factors affected by erosion include loss of carbon in the biomass and reduced micro- and macro-fauna activity.

Production costs on degraded lands are also elevated because of increased fertilizer needs to offset soil nutrient loss; machine power to till and plant more compacted subsoil; and additional seed cost in replanting stand loss from soil crusting and seeds washing off. These effects also cause an increase in machine wear, fuel consumption, time and labor, and consequently, increased production cost.

The effects of soil erosion on crop productivity and the environment may be strongly felt by society as a whole, but it essentially affects the human's quality of life (Fig. 1). In

Fig. 1 A family from Rio Grande do Sul State, Brazil, suffer from severe soil erosion on their farm.

addition to rural society stricken by reduced crop yield and increased production costs, urban society also suffers from higher cost for food and reduced environmental quality. There is evidence of this in developing countries with extreme cases of erosion. In these instances, farmers cannot maintain a basic subsistence level, so they abandon their land, move to the cities and cause what is known as rural exodus. The society also bears the costs of restoring reservoir and river capacities and cleaning chemically polluted surface waters caused by eroded sediments.

AWARENESS OF SOIL EROSION: CROP PRODUCTIVITY RELATIONSHIP

Despite historical evidence that human-induced erosion has occurred worldwide wherever land has been cultivated, awareness of the erosion-induced problem developed slowly. The study of soil erosion as a science started only approximately 100 years ago, between 1877 and 1895, by German soil scientist Ewald Wollny (14).

It is difficult to assess the cost of erosion-induced damages to society in terms of loss of food production, water storage and quality, and quality of life. This may partly explain why soil erosion, even though universally recognized as the greatest threat to food security, environmental quality, and human well-being, is continuously uncontrolled in many parts of the world. If soil erosion and crop productivity relationships were better understood, people might be more willing to accept erosion-control measures. The society could determine whether erosion control regulations should be imposed for land use purposes (15). Perhaps a change must be made to the customary form of reporting soil erosion loss, from mass of soil loss per hectare per year to an economic loss expressed in monetary values and associated with production and natural resource losses (11, 15).

The effects of erosion on crop productivity are difficult to evaluate independently because of various interactions (8–12, 15, 16). Interactions occur among soil properties, crop characteristics, and climate, which should be accounted for when relating soil erosion and crop yields. Much of the soil's water and nutrient supply potential may be lost before the productivity loss is realized, because erosion effects are gradual, cumulative, not felt from year to year, and significant only over a long time period.

Furthermore, the combined effect on crop yield varies with the intrinsic characteristics of the soil and the specific soil management systems used. This is especially true because improved technology can offset erosion removal of plant nutrients and soil-water storage capacity and greatly mask erosion effects on crop yields, especially in the short-term. Krauss and Allmaras (16) stated that: "sustained productivity of an eroding soil cannot be determined unless yield increases from technology advances are separated from soil productivity changes due to erosion." Even with great obstacles, efforts to quantify erosion's effects on crop productivity must continue. This information is fundamental in developing soil conservation plans and effective land-use policies.

TECHNIQUES FOR EVALUATING EROSION IMPACT ON CROP PRODUCTIVITY

Several techniques have been proposed to evaluate the cause–and–effect relationship between crop yield and soil erosion (8, 11). The direct method involves field experiments of short and long duration in which the soil of the test plot is naturally eroded by rain or artificially eroded by mechanical removal of the topsoil layer. With the direct method, productivity losses can be measured under field conditions where crop performance, over a period of time, is related to soil loss or erosion-induced alterations in soil properties.

Overall, the direct method is more accurate than the indirect method, but requires more capital, time, and labor to accomplish. The indirect method relates measured crop yields with estimated erosion losses from changes in physical and chemical soil properties occurring in the field. The crop productivity–soil erosion relationship is evaluated through mathematical, geomorphological, and crop productivity models. These models are less accurate than the direct method, but require less time and labor and are cheaper to execute. Although satisfactory techniques are available to monitor soil erosion effects on crop productivity, (i.e., the on-site damages), there is a lack of effective procedures to quantitatively evaluate sedimentation and water pollution impacts on the environment and human life (i.e., the off-site damages).

CONCLUSION

Soil erosion is a problem associated with human lives on Earth. It threatens soil productivity, food production, the environment, and quality of life. Fortunately, erosion-induced problems have been well recognized and sound control measures have been developed. To maintain the natural resources and environment for a sustainable civilization, soil erosion from agricultural lands must be controlled.

REFERENCES

1. Shumm, S.A.; Harvey, M.D. Natural Erosion in the USA. In *Determinants of Soil Loss Tolerance*; Kral, M.K., Ed.; American Society of Agronomy, Soil Science Society of America: Madison, Wisconsin, 1982; 15–22.
2. Hall, G.F.; Daniels, R.B.; Foss, J.E. Rate of Soil Formation and Renewal in the USA. In *Determinants of Soil Loss Tolerance*; Kral, M.K., Ed.; American Society of Agronomy, Soil Science Society of America: Madison, Wisconsin, 1982; 23–29.
3. Bennett, H.H. *Soil Erosion*; McGraw-Hill: New York, 1939.
4. Lowdermilk, W.C. *Conquest of the Land Through Seven Thousand Years*, Agricultural Information Bulletin 99; United States Department of Agriculture: Washington, DC, 1953; 1–12.
5. Stallings, J.H. *Soil Conservation*; Prentice Hall: New Jersey, 1957.
6. Hudson, N.W. *Soil Degradation and Civilization*, Occasional Paper 9; Cranfield Institute of Technology: Cranfield, United Kingdom, 1980; 1–13.
7. Dregne, H.E. Historical Perspective of Accelerated Erosion and Effect on World Civilization. In *Determinants of Soil Loss Tolerance*; Kral, M.K., Ed.; American Society of Agronomy, Soil Science Society of America: Madison, Wisconsin, 1982; 1–14.
8. Langdale, G.W.; Schrader, W.D. Soil Erosion Effects on Soil Productivity of Cultivated Cropland. In *Determinants of Soil Loss Tolerance*; Kral, M.K., Ed.; American Society of Agronomy, Soil Science Society of America: Madison, Wisconsin, 1982; 41–51.
9. Stocking, M. *Erosion and Soil Productivity — A Review*; Food and Agriculture Organization: Rome, 1984.
10. Crosson, P. Impact of Erosion on Land Productivity and Water Quality in the United States. In *Soil Erosion and Conservation*; El-Swaify, S.A., Moldenhauer, W.C., Lo, A., Eds.; Soil Conservation Society of America: Ankeny, Iowa, 1985; 217–236.
11. Lal, R. Soil Erosion and Its Relation to Productivity in Tropical Soils. In *Soil Erosion and Conservation*; El-Swaify, S.A., Moldenhauer, W.C., Lo, A., Eds.; Soil Conservation Society of America: Ankeny, Iowa, 1985; 237–247.
12. Lal, R. Monitoring of Soil Erosion's Impacts on Crop Productivity. In *Soil Erosion Research Methods*; Lal, R., Ed.; Soil and Water Conservation Society: Ankeny, Iowa, 1988; 187–200.
13. Young, K.K. The Impact of Erosion on the Productivity of Soils in the United States. In *Assessment of Erosion*; DeBoodt, M., Gabriels, D., Eds.; John Wiley & Sons, Ltd.: London, 1980; 295–303.
14. Hudson, N. *Soil Conservation*; 3rd Ed; Cornell University Press: New York, 1995.
15. Moldenhauer, W.C. Soil Erosion — A Global Problem. In *Assessment of Erosion*; DeBoodt, M., Gabriels, D., Eds.; John Wiley & Sons, Ltd.: London, 1980; 3–8.
16. Krauss, H.A.; Allmaras, R.R. Technology Masks the Effects of Soil Erosion on Wheat Yields — A Case Study in Whiteman County, Washington. In *Determinants of Soil Loss Tolerance*; Kral, M.K., Ed.; American Society of Agronomy, Soil Science Society of America: Madison, Wisconsin, 1982; 75–86.

EROSION AND SEDIMENTATION CONTROL, AMENDMENT TECHNIQUES FOR

X.-C. (John) Zhang
United States Department of Agriculture-Agricultural Research Service (USDA-ARS), El Reno, Oklahoma, U.S.A.

INTRODUCTION

Any material that is added to soil to benefit the soil or plant is called a soil amendment (1). For example, lime, gypsum, animal manure, green manure, sewage sludge, compost, crop residue, woodchips, leaves, sawdust, synthetic polymers, and industrial waste such as paper sludge are all soil amendments. Soil amendments, though arbitrary, may be grouped by their origins into five different categories (Table 1). Technically, chemical fertilizers are soil amendments, but conventionally they are not referred to as such. A closely related term to soil amendment is soil conditioner. A soil conditioner is a substance that has the ability to improve the physical condition of a soil for better aeration, water-holding capacity and infiltration, porosity, structural stability, erodibility (susceptibility to erosion), and tilth. Actually, soil conditioners are also soil amendments that are used to create favorable physical environments for plant growth. Soil amendments are either incorporated into the soil or applied on the surfaces. Incorporated materials ameliorate soil properties, while surface-applied materials mostly affect the atmosphere–soil interface. Soil amendments, such as animal waste, farmyard manure, green manure, crop residues, and composts, have been used for many centuries to recycle plant nutrients and to improve soil structure and tilth. Crop residues, preserved by no-till and conservation tillage, have been widely used to reduce soil erosion during the past half century (2). Due to increasing landfill costs, applications of municipal and industrial wastes and by-products such as sewage sludge, paper mill sludge, wood products, and gypsum to agricultural lands have increased dramatically in past decades (3). More recently, interests in using synthetic polymers to reduce water runoff and soil erosion have been revived. Overall, soil amendments have beneficial effects on the physical, chemical, and biological properties of soil and, therefore, on plant growth (1–3). In relation to erosion control, most soil amendments are effective in increasing soil aggregation and soil structural stability (4), decreasing erosive forces of raindrop impact and flowing water, reducing surface crusting (a thin, dense surface layer with very low water permeability), and improving water infiltration. These beneficial effects not only reduce soil erodibility but also rainfall-runoff erosivity. Thus, most amendments are effective in reducing soil erosion by water, though only a few (e.g., standing residue cover) are effective in controlling erosion by wind.

TYPES OF SOIL AMENDMENTS

Lime

Liming materials such as quicklime (CaO) and limestones are the most widely used mined or mineral soil amendments. Liming soil has a long history in the United States. Early settlers started to lime soil from the beginning of the nineteenth century. By the early twentieth century, liming had become a routine agricultural practice to raise soil pH and increase crop productivity. Liming soil generally improves the physical and chemical properties of soil (1, 5) by neutralizing soil acidity, reducing phytotoxicity of Al, Mn, and H ions, increasing plant availability of many nutrients, promoting microbial activity, and increasing soil structural stability. However, over-liming can result in soil dispersion and aggregation disintegration in highly weathered, high Al^{3+} soils. Beneficial effects of liming on soil erosion result from reduction of soil erodibility due to improved soil aggregation and structural stability, and increase of surface cover due to enhanced plant growth. Lime has also been successfully applied as a stabilizer and cementing agent to control side sloughing of canal embankments and pipe erosion on earthen dams, which are composed of highly dispersive and expansible clays (1, 6).

Gypsum

Gypsum (mined or industrial by-products) has been widely used for many decades to ameliorate unstable, high Na soils (e.g., sodic soils). In recent years, research has demonstrated that soils with low Na and electrolyte

Table 1 Type and example of commonly used soil amendments

Type	Example
Mined or mineral	Lime, gypsum
Synthetic polymer	Polyacrylamide (PAM)
Manure and compost	Animal manure, green manure
Mulches	Crop residue, woodchips
Municipal and industrial wastes	Sewage sludge, paper mill sludge

(From Refs. 12 and 13)

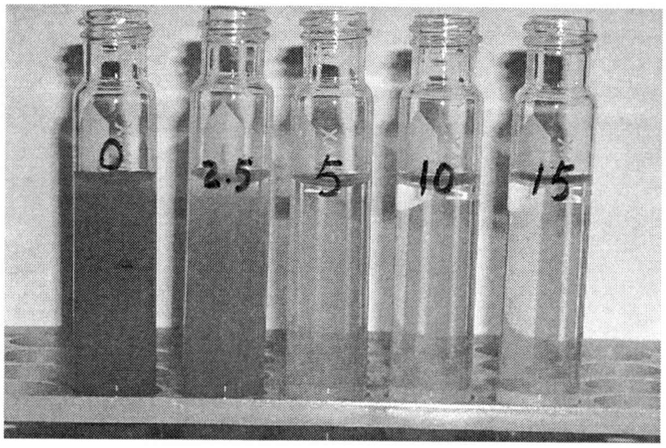

Fig. 1 Flocculation test series after 6-h settling using Cecil soil clay. (From left: 0, 2.5, 5, 10, and 15 mmol$_c$ gypsum L^{-1}.)

concentration (EC) can also be dispersive, and gypsum addition is effective in reducing surface sealing and crusting, increasing water infiltration, and reducing soil erosion by water (7, 8). The beneficial effects of gypsum addition on soil erosion reduction result from increased soil aggregate and structural stability due to elevated EC in soil solution and the predomination of calcium ions on the exchange sites. Cation valence and EC are two major factors influencing clay behavior, i.e., clay flocculation or dispersion (9). When two clay particles approach each other, their electric double layers (a negatively charged clay surface surrounded by a compensating cation cloud) begin to overlap or repel each other. The cation cloud is pushed toward the particle surface when EC is increased or when a divalent cation replaces a monovalent cation. The reduction in the cloud thickness results in a decrease in this repulsive force. When a critical EC is reached, when the van der Waals attractive force overcomes the repulsive force, rapid coagulation takes place. This concentration is termed critical flocculation concentration (CFC). It should be pointed out that cation valence has a dominant effect on clay flocculation. This is why Al salts such as Alum have been widely used to flocculate suspended particles in many water-treatment plants. Among the common soil cations (e.g., Al^{3+}, Ca^{2+}, Mg^{2+}, K^+, Na^+, and H^+), Ca^{2+} has the greatest flocculation power excluding Al^{3+} (which is only significant in acid soils), while Na^+ has the least. This is why sodic soils have poor soil structure and gypsum addition helps improve it. Based on this theory, clay particles flocculate when EC is greater than CFC, otherwise dispersion takes place. Clay flocculation is clearly demonstrated in Fig. 1. After 6-h settling, more dispersed clay particles have coagulated into micro-aggregates and settled out to the bottom as gypsum concentrations increase from left to right. Clay flocculation promotes soil aggregation and structural stability, while clay dispersion often leads to deterioration of soil structure. This is because clay is one of the primary cementing agents for soil aggregation. Increased soil aggregate stability decreases soil erodibility, improves water infiltration, and reduces soil erosion by water (1, 7, 8).

Synthetic Polymer

Synthetic polymers with various charge and chemical properties have been used to improve soil physical properties for more than a half century. In early studies, polymers were typically incorporated into the entire plow layer to promote structural stability and water permeability. The enthusiasm reached its peak in the 1950s upon reports that polymers were effective in reducing erosion, improving water infiltration, and increasing crop yields. However, the enthusiasm was subdued by high cost and large quantities needed to amend deep soil layers. In recent years, interest has been revived by new polymer products and new application strategies. Only a small amount of polymer, when sprayed on the soil surface or dissolved in irrigation water, can reduce surface seal and crust formation, and, therefore, reduce runoff and soil erosion (1). Surface sealing, formed by disintegration of soil aggregates and dispersion of clay particles, causes extremely low water permeability. The threadlike large polymer molecules or interwoven polymer nets are strongly (often irreversibly) adsorbed by soil aggregates and clay particles. The adsorbed polymer nets that surround soil aggregates physically protect them from disintegration by raindrop impact or shear force imparted by flowing water. This is demonstrated in Fig. 2, where soil aggregates, which

Fig. 2 Soil aggregates of a Miami silt loam soil both without (left) and with (right) an anionic PAM treatment after exposure to a simulated rainfall event.

have been treated with an anionic polyacrylamide, are more stable than untreated aggregates after exposure to a 65-mm/h rain for 10 min. Polymers can also chain clay particles together to prevent clay dispersion or promote flocculation. Polymers have been used to flocculate clay particles in sedimentation ponds. For anionic polymers, the effectiveness is usually improved by adding an electrolyte such as gypsum, because cations are basic ingredients for building cation bridges between negatively charged clay platelets and polymer molecules. Due to the ability to stabilize soil aggregates and improve soil structure, polymers are effective in controlling surface seal and crust formation, increasing water infiltration, and reducing soil erosion. Polymers have also been successfully used to control soil erosion on steep slopes on construction sites and road banks.

Manure, Compost, and Organic Sludge

Animal, farmyard, and green manures have been used as organic soil amendments to improve soil productivity for many centuries. In recent years, application of municipal biosolids (sewage sludge) and industrial waste compost such as paper sludge to agricultural lands has increased dramatically. These materials have the ability to increase aggregate stability, porosity, and water-holding capacity, reduce runoff and erosion, improve soil tilth, increase plant nutrient availability and soil cation exchange capacity, and elevate soil microbial activities (7). Decomposed organic substances such as humus along with soil clays are primary binding agents for soil aggregation. An increase in soil organic matter content often leads to increased soil aggregation and structural stability. Thus, addition of organic materials such as manure and compost reduces soil erosion by water by improving soil aggregation and reducing runoff volumes. In addition, improved soil conditions enhance rooting and plant growth, which increases plant cover and root biomass that further reduce soil erosion. It should be pointed out that substantial rates of application over time are needed to considerably increase soil humus levels.

Mulches

Most mulch materials used for runoff and erosion control are naturally occurring soil amendments. Both inorganic and organic mulches have been used. Inorganic mulches include stone, gravel, sand, and many others, while organic mulches are normally undecomposed materials derived from living matter such as crop residue, hay, leaves, woodchips, and sawdust. These materials dissipate raindrop impact energy, protect surface soil aggregates from disintegration by raindrop impact, prevent surface seal formation, reduce raindrop-impact-induced turbulence, obstruct overland flow, and enhance water infiltration. These processes not only reduce soil detachment by raindrop impact but also suppress sediment transport by thin overland flow. Therefore, surface mulches are very effective in reducing surface runoff and soil erosion by water. Many studies have shown that both runoff volumes and soil loss rates decrease rapidly (often exponentially) with increase of surface mulches (2, 10, 11). Buried or incorporated residues are also effective in reducing soil erosion, especially rill erosion (erosion by concentrated overland flow in small channels). However, the reduction is generally less as compared to surface residues. Moreover, organic mulches eventually decompose into humus and impart to the soil all the beneficial

Table 2 Effects of gypsum, anionic PAM, and canopy cover on water infiltration and soil loss in a Cecil sandy loam soil during an initial 31-min simulated rain at an intensity of 88 mm h^{-1}

Treatment	Total infiltration (mm)	Percent runoff (%)	Erosion rate (kg/m^2)
Untreated	20.5	54.7	1.63
Canopy cover	27.2	39.9	0.67
Gypsum	26.5	41.4	1.13
PAM	43.2	4.5	0.01

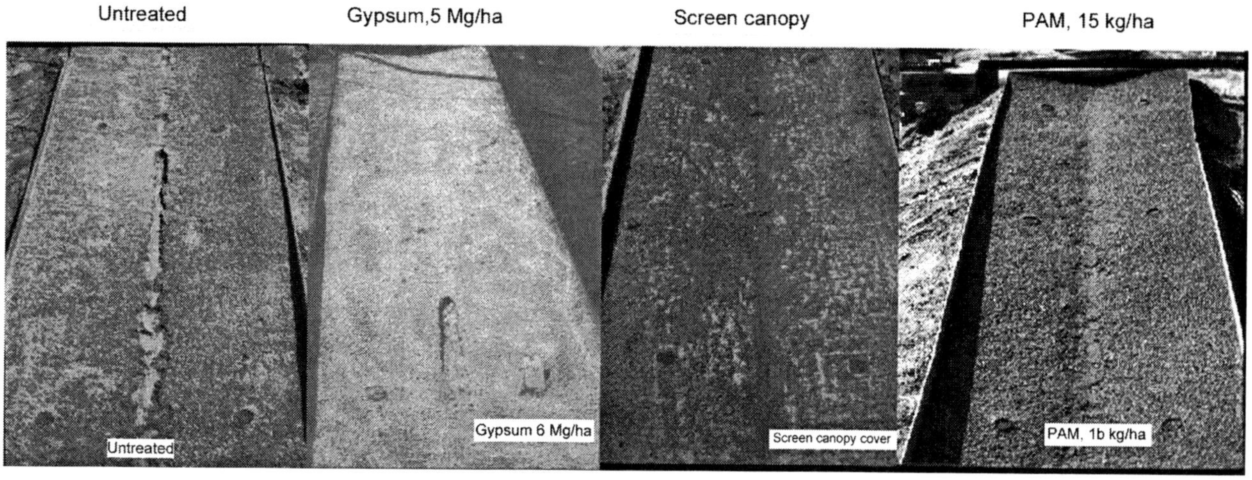

Fig. 3 Rill development and surface conditions of four surface treatments (from left: untreated, 5 Mg ha^{-1} gypsum, screen canopy cover, and 15 kg ha^{-1} PAM) after exposed to an 88-mm h^{-1} rain for 31 min (soil: Cecil, sandy loam; plot size: 3.5 by 0.9 m; slope: 11%).

qualities derived from traditional organic matter additions as discussed earlier.

CONCLUSION

An example from field studies (12, 13) is summarized in Table 2 to illustrate the general effects of soil amendments on erosion control. Compared with an untreated plot, the canopy cover, gypsum, and polyacrylamide (PAM) treatments were all effective in reducing runoff and erosion. In this example, PAM was the most effective treatment, followed by the cover treatment. This was also illustrated by the surface conditions in Fig. 3. A full rill was developed in the untreated plot during a 31-min rain; however, only a small rill was initiated in the gypsum plot, and no rill incision occurred in the cover and PAM plots. Also, the resultant surface conditions differed among different treatments. A smoother surface, on the untreated and the cover plots, indicates somewhat of a disintegration of soil aggregates and seal formation. In contrast, a well-aggregated, rough surface was exhibited on the PAM plot. This explains why almost all rainfall water has infiltrated into the soil on this plot. In general, the effectiveness of these treatments declined with time and the treatment longevity varied from weeks to months. To maintain the effect, PAM and gypsum must be reapplied at some intervals, which depend largely upon soil and environmental conditions.

REFERENCES

1. Wallace, A.; Terry, R.E., Eds. *Handbook of Soil Conditioners: Substances that Enhance the Physical Properties of Soil*; Marcel Dekker: New York, NY, 1998.
2. Oschwald, W.R., Ed. *Crop Residue Management System*, ASA Spec. Publ. No. 31; ASA, CSSA, and SSSA: Madison, WI, 1978.
3. Rechcigl, J.E.; MacKinnon, H.C., Eds. *Agricultural Uses of By-Products and Wastes*, ACS Symp. Series 668; ACS: Washington, DC, 1997.
4. Emerson, W.W.; Bond, R.D.; Dexter, A.R., Eds. *Modification of Soil Structure*; John Wiley & Sons: New York, 1978.
5. Adams, F., Ed. *Soil Acidity and Liming*, Agron. Monogr. 12; ASA, CSSA, and SSSA: Madison, WI, 1984.
6. Gutschick, K.A., Ed. *Lime for Environmental Uses*; ASTM: Philadelphia, PA, 1985.
7. Sumner, M.E.; Stewart, B.A., Eds. *Soil Crusting: Chemical and Physical Processes*; Lewis: Boca Raton, FL, 1992.
8. Shainberg, I.; Sumner, M.E.; Miller, W.P.; Farina, M.P.W.; Pavan, M.A.; Fey, M.V. Use of Gypsum on Soils: A Review. Adv. Soil Sci. **1989**, *9*, 1–111.
9. Van Olphen, H. *An Introduction to Clay Colloid Chemistry*; John Wiley & Sons: New York, 1977.
10. Unger, P.W., Ed. *Managing Agricultural Residues*; Lewis: Boca Raton, FL, 1994.
11. Wischmeier, W.H.; Smith, D.D. *Predicting Rainfall Erosion Losses: A Guide to Conservation Planning*, USDA Agric. Handb. 537; U.S. Gov. Print. Office: Washington, DC, 1978.
12. Zhang, X.C.; Miller, W.P. Physical and Chemical Crusting Processes Affecting Runoff and Erosion in Furrows. Soil Sci. Soc. Am. J. **1996**, *60* (3), 860–865.
13. Zhang, X.C.; Miller, W.P. Polyacrylamide Effect on Infiltration and Erosion in Furrows. Soil Sci. Soc. Am. J. **1996**, *60* (3), 866–872.

EROSION AND SEDIMENTATION CONTROL, ENGINEERING TECHNIQUES FOR

José Miguel Reichert
Federal University of Santa Maria, Santa Maria, Brazil

Elemar Antonino Cassol
Federal University of Rio Grande do Sul, Porto Algre, Brazil

INTRODUCTION

Water erosion is caused basically by raindrop impact and runoff of excess water (Fig. 1), thus erosion and sedimentation control strategies must be based on covering the soil against raindrop impact, increasing water infiltration to reduce runoff generation, and increasing surface roughness to reduce overland flow velocity. To pursue these objectives, several strategies (Fig. 2) can be adopted, including agronomic measures, soil management, and mechanical or engineering methods (1, 2) as discussed and illustrated in this topic.

Erosion is a natural process which cannot be completely controlled, but engineering methods can help in reducing erosion rates to maintain a maximum sustained level of production while keeping soil loss below a threshold (soil loss tolerance). Erosion control is also needed to reduce nutrient and pesticide losses, to prevent pollution of surface water, and avoid sedimentation or siltation of water bodies.

It is largely accepted that soil management and agronomic measures are more suitable for erosion control due to lower cost and greater effect on reducing soil detachment due to surface protection, increased water infiltration, and runoff reductions, and are easily incorporated into existing and improved farming systems (3), with characteristics of sustainability.

Engineering techniques for soil erosion/sedimentation control involve manipulation of surface topography, and usually are complex and costly to install and maintain, thus requiring proper planning and implementation. However, they are essential for controlling and orienting the flow of excess water generated in agricultural, forest, pasture, or urban land, and should be used in conjunction with agronomic and management practices.

ENGINEERING TECHNIQUES

Practices of engineering erosion control include contouring, terraces, contour barriers and stone terraces, waterways, conservation structures, gully stabilization structures, and geotextiles (1–3).

Contouring

Contouring consists of performing all field practices such as plowing, disking, planting, and cultivation on the contour (Fig. 3). It is effective for low slopes and short slope length, and for rainstorms of low intensity. The effectiveness of this practice is about 50% (2) depending upon slope grade and length, and this effectiveness can be increased by establishing a series of depressions or pits (Fig. 4) which fill with water and sediment during rain.

Terraces

Terraces are mechanical structures constructed across a slope and consist of a channel and an earth or stone bank to intercept surface runoff, allow it to infiltrate and evaporate or divert the excess to a stable outlet at a nonerosive velocity. Depending upon their function or type of construction, terraces are classified as retention or absorption, diversion or graded, and bench terraces. Retention or absorption terraces are used when water is to be stored on the hillside of slopes less than 4.5°, having level channels able to accumulate runoff volume generated by a 10-year return period rainfall. Diversion or graded terraces aim at intercepting runoff and conduct it to a proper outlet through a channel of a slight grade, usually 1:250 (Fig. 5) on slopes less than 7°. Bench terraces are used on steep slopes, up to 30°, alternating a series of shelves (cuts) and raises (fills) (Fig. 6). Research data from several authors (1) indicate an effectiveness in soil loss control of bench terraces up to 93%. Terraces must be properly designed to avoid even greater soil erosion damage (2), and should be planned considering the whole farm, a group of neighboring farms, or even a small hydrographic catchment (3), to ensure optimization of land use.

Encyclopedia of Soil Science

Fig. 1 Rill and interrill erosion and sedimentation due to improper soil conservation in southern Brazil in the early 70s.

Fig. 2 Agricultural field with agronomic and engineering soil conservation methods in southern Brazil.

Fig. 4 Sedimentation pits in a horticultural field in Costa Rica.

Contour Ridges and Stone Terraces

Contour ridges are earth banks, similar to narrow-based terraces, constructed across the slope using hand tools, animal traction implements, and sometimes motorized farm machinery. These structures are frequently used on small-scale farms and hilly regions. Their effectiveness in reducing soil loss varies, but research data from several authors indicate values up to 80 and even 100% control (1). Contour stone bunds (Fig. 7) are an alternative technique

Fig. 3 Contouring in an agricultural field in southern Brazil.

Fig. 5 Diversion or graded terraces in southern Brazil.

Fig. 6 Bench terraces on high-grade hillslope in Guatemela.

Fig. 8 Grass waterway draining to a concrete drop inlet structure in southern Brazil.

on stony lands, where stone walls are set up in shallow trenches. These allow slow retention of sediment and are permeable to water. With time, up-slope sedimen tation will generated a gradual development into bench terraces. The construction of stone barriers requires considerable labor and time, but is more permanent and easier to maintain than Earth barriers. Spacing and construction details are presented in Ref. (3).

Waterways

A system of graded terraces requires waterways to receive the excess drained runoff and safely conduct at a nonerosive velocity to lower parts of the landscape. Waterways must be carefully established (considering location, construction, and stabilization) and maintained (managing grass species and controlling incipient erosion) to avoid serious erosion problems, even causing gully formation. Grass waterways (Fig. 8), tile drains, or on steep slopes, concrete structures (Fig. 8) are located in natural depressions or, if needed, reshaped channels. The grass waterways reduce water-flow velocity because of the retardance effects of vegetation and increased soil resistance to erosion due to soil coverage and aggressive rooting system holding soil particles together. To work properly, waterway design follows defined principles from channel hydraulics (1, 4) and the flow ways must be maintained and cared for (3).

Fig. 7 Contour stone barriers in an agricultural field in Cape Verde (photo by L.D. Norton).

Conservation Structures

In certain situations, the only feasible way to convey safely large amounts of runoff water from a higher elevation to a lower one is through the use of conservation structures that are made of some type of nonerodible material, such as concrete or large rocks. Examples of these structures include chute drop inlets (Fig. 8), flumes and drop spillways (Fig. 9), and pipe spillways, which must follow standard functional requirements to maintain channel stabilization (4). Selection and design of these structures is based upon existing or anticipated erosion problem, slope steepness, estimated maximum water flow discharge, and cost of alternative systems. Conservation structures are usually used in conjunction with other erosion-control techniques such as grass waterways (Fig. 8).

Gully Stabilization Structures

Gullies represent an advanced stage of rill erosion (Fig. 10), and are most important as a source of sediment in streams. If the gully is small, recovery might be suitable by filling it with soil, but in many cases recovery is

E

Fig. 9 Conservation structures showing drop spillway and flumes to convey water downslope, along with stone barriers, in Spain.

technically and economically not viable. In such cases, gully stabilization is recommended by constructing small dams using diverse materials depending on availability. These materials may include stones, wood planks or branches, earth, etc., to trap sediment, thus reducing gully depth and slope. Along with these stabilization structures, plants must be established to reclaim the affected areas

Fig. 10 Gully development on sandy soil in southern Brazil.

Fig. 11 Gully stabilization with used tires, stones, and vegetation in Costa Rica.

(Fig. 11). Sometimes, to control the overfall of water on the headwall on large gullies, more permanent structures are required (1, 2).

Geotextiles

Erosion on road banks or urban areas is best controlled by vegetation, but establishment on these highly erodible and usually infertile soil surfaces is difficult. Thus, other materials including straw and hay, to help establishing vegetation are used, but these do not hold tight to the soil and are washed away easily (2). Alternative materials include geotextiles that are permeable textiles for erosion control interacting with soil and vegetation. The geotechnical material is in the form of a mat, sheet, grid, or web of natural or synthetic fiber (1). The first type is biodegradable while the second type gives a more permanent protection to a slope, interacting with roots to improve soil cohesion and slope stability.

SUMMARY

Engineering techniques vary in their purpose and suitability. For a given field condition a variety of methods should be considered and adopted in an integrated manner, depending on whether water or wind erosion is the most prominent problem. Technology, money, and labor availability are also constraints in adopting a given method; thus, for different locations around the world, myriad ways of adaptation are developed and used by professionals and farmers.

REFERENCES

1. Morgan, R.P.C. *Soil Erosion and Conservation*, 2nd Ed.; Wiley: New York, 1995.
2. Hudson, N. *Soil Conservation*, 2nd Ed.; Cornell University Press: Ithaca, NY, 1995.
3. FAO, *Manual on Integrated Soil Management and Conservation Practices*; FAO Land and Water Bulletin 8, IITA and FAO: Rome, Italy, 2000.
4. Schwab, G.O.; Fangmeier, D.D.; Elliot, W.J. *Soil and Water Conservation Engineering*, 4th Ed.; New York: Wiley, 1992.

EROSION AND SEDIMENTATION CONTROL, VEGETATIVE TECHNIQUES FOR

E

Samson D. Angima
Department of Agronomy, Purdue University, West Lafayette, Indiana, U.S.A.

INTRODUCTION

Soil detachment and erosion occur when soil is disturbed by either human activity or natural conditions, such as extreme weather. This results in soil being moved from one point to another by either wind or water. Practices such as road construction, suburban and industrial developments, stream channel and other types of construction on sloping lands, inadequate drainage facilities, poor grading practices, deforestation, cultivation on sloping lands, and general lack of adequate planning by land users cause soil erosion. Consequences of soil erosion include water and air pollution, reduced land productivity from loss of topsoil and nutrients, and degradation of the environment. Soil erosion by water is controlled mainly by earthworks and engineered constructions such as terracing aimed at reducing the slope and at collecting and storing moisture while reducing runoff to acceptable limits. Vegetative barriers are grass, shrubs, or small trees grown in close rows that can be used to control both water and wind erosion by providing protection from soil, dislodging sources such as rainfall, and offering a semi-permeable barrier to erosion agents resulting in soil deposition. The resultant vegetation also shields the soil surface from overland flow and decreases the erosive capacity of water flow by reducing its velocity.

VEGETATIVE MATERIALS FOR SOIL EROSION AND SEDIMENTATION CONTROL

Vegetative cover is essential for the design and stabilization of many structural erosion-control devices (Fig. 1). Proper vegetative cover provides excellent erosion protection and sedimentation control. Vegetative barriers are planted in close rows along contours on slopes to intercept water runoff, or are planted perpendicular to the direction of wind to retard wind movement, resulting in soil deposition. Plant roots and lateral stems provide a structure to hold soil particles in place. These features also improve the soil's physical properties and increase infiltration rate, thereby decreasing runoff. Plant transpiration reduces soil moisture levels, which increases soil absorption capabilities. Not every plant material can serve as a soil erosion control agent. Those plants that do must possess some bioengineering characteristics in both the root and shoot system that encompasses both living plants and organic materials as construction elements for erosion control. Such properties include non-competition with adjacent crops or fruit trees for moisture or nutrients, a rooting system that reaches deep down to anchor the plant and also extract leached nutrients from the subsoil, and shade tolerance. The plants must be perennials with high seed vigor, have the ability to increase soil organic matter and reduce surrounding soil bulk density to allow faster water infiltration, and have strong woody stalks to withstand pressure from erosive agents.

Vegetative materials used for soil erosion control include grass species, legumes, trees, shrubs, vetches, and sods. Common grass species that bunch during growth include Tall fescue, Perennial ryegrass, Orchardgrass, Timothy, Switchgrass, Weeping lovegrass, Deertongue, and Big bluestem. Bunching legumes include Birdsfoot trefoil and Serecia lespedeza. Cereals such as Winter wheat, Winter rye, Spring oats, Sudangrass, and Japanese millet are also used. Sod-forming species include grasses such as Redtop, Fine fescue, Kentucky bluegrass, and Smooth bromegrass, and legumes such as Crownvetch and Flatpea (1). Shrubs and small trees used as barriers in erosion control include *Calliandra calothyrsus, Sesbania sesban, Leucaena leucocephala, Gliricidia sepium, Cassia siamea, Eucalyptus* spp., *Casuarina* spp., *Acacia* spp., *Azadirchta indiaca*, and *Grevillea robusta* (2). Tree species selection depends on local landuse and climatic conditions that favor establishment.

MASS PLANTING OF VEGETATIVE COVER

Mass planting of vegetative cover is used where land that is susceptible to erosion is converted from cropland to

Fig. 1 Vegetative barriers/strips hold the terraces intact for agricultural production in Kabale, Uganda. (Picture by Samson Angima.)

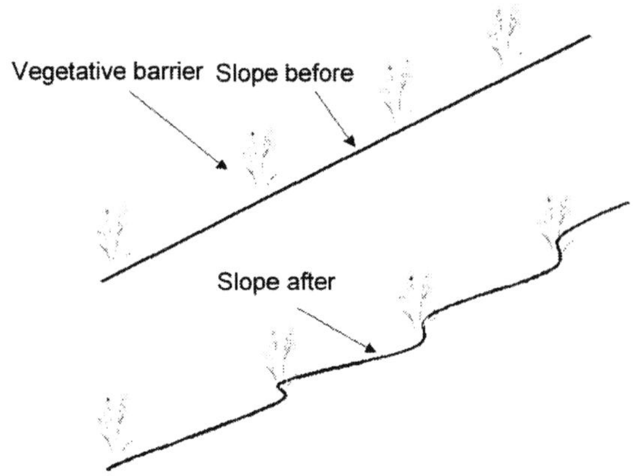

Fig. 2 Schematic sketch of vegetative barriers illustrating expected change in land slope over time resulting from tillage and erosion deposition process.

vegetative cover such as in the Conservation Reserve Program (CRP). In this program, trees as well as sod are planted in strategic places where they will provide maximum protection to the soil resource, and the landowner agrees to leave the areas under vegetation for a given period of time for land stabilization to occur. The vegetative cover also reduces water pollutants such as fertilizer nutrients, pesticides, and herbicides in the runoff water; increases oxygen levels; reduces greenhouse gases; reduces evaporation rates; provides shade; and buffers against high winds (2).

VEGETATIVE STRIPS

Vegetative strips constitute different types of filter strips that reduce runoff velocity and provide differing degrees of filtering action depending on the species used. Larger soil particles tend to settle out readily, leaving only clay particles suspended and thereby reducing pollutant transfer to ponds, rivers, and lakes. Filter strips can be classified as grass strips, vegetated waterways, filter strip terraces, buffer and riparian strips, and settling basins (3).

Grass Strips

Grasses are by far the most important vegetative materials used to control erosion. Bands of grass about 1–2 m in width are planted along the contours and spaced every 10 to 100 m depending on the slope (Fig. 2). Prior to planting, the land is roughened by disking, harrowing, or raking, then limed, fertilized, and seeded with mixtures of adapted grasses and legumes to enhance a good stand (4). Vetiver grass (*Vetiveria zizanioides*), Napier grass (*Pennisetum purpureum*), and Stiff grass (*Miscanthus sinesis*) are the most commonly used species (2). These are upright, tufted, deep-rooted, and very dense grasses that are by far the most important in erosion control, particularly in tropical countries. Vetiver is a bunch grass with very rapid growth in warm, moist conditions. It grows to more than 2 m in height and has a remarkably dense and vertical rooting system, growing deep (3 to 5 m). Napier grass is a tall perennial reaching over 3 m high. It is resistant to drought, and grows at altitudes up to 2400 m. Stiff grass has dense roots and coarse stems that withstand erosive agents. These grass hedges present a virtually impenetrable barrier through which soil can hardly pass and through which only water, with much reduced velocity, passes (5). As silt builds up behind the grass, other grass shoots arise from the nodes above the deposited silt to form a natural terrace. Weeds and undesirable foreign grasses are unable to penetrate through a well-established grass hedge. Studies carried out in Mississippi, using stiff grass (*Miscanthus sinesis*) on no-till and conventional tillage (6), showed that grass strips help reduce sediment losses by up to 88% on conventional tillage and 57% on no-till cotton plots (Table 1). Other grass species include Buffalo grass (*Buchloe dactyloides*), grama or mesquite grasses (genus *Bouteloua*), Switch grass (*Panicum virgatum*), and *Phalaris aquatica*.

Vegetated Waterways

Vegetated waterways are channels or waterways that transfer runoff from a higher to a lower elevation over a short distance without allowing erosion to occur (Fig. 3). In

Table 1 Sediment loss with and without grass strip (*Miscanthus sinesis*) under no-till and conventional tillage cotton at 5% slope

Period	No-till with grass strip	No-till without grass strip	Conventional tillage with grass strip	Conventional tillage without grass strip	No-till with winter wheat cover
			Sediment loss (t/ha)		
1992	2.6	4.5	12.3	60.3	2.9
1993	0.9	1.6	5.4	21.9	1.2
1994	3.2	9.6	18.6	63.2	1.9
Average	2.2	5.2	12.1	48.5	2.0

(From McGregor, K.C.; Dabney, S.M.; Johnson, J.R. USDA-ARS, National Sedimentation Laboratory, Oxford, MS.)

addition to dissipating flow energy, some structures also act to retain soil. They are especially effective in arresting gully development, a situation that might need both structure(s) and some vegetated channel. They have many uses in comprehensive conservation plans, but they primarily collect and concentrate flows and then safely transport the water to major drainage systems. Dense vegetation is used to minimize the area required, i.e. the protective action of the vegetation permits higher flow velocities and thus smaller waterway cross-sections. Most vegetated waterways run directly down a slope; however, they can also be constructed somewhat across the slope as diversions or sometimes just to reduce channel slope. Vegetated channels are not used where continuous flow occurs because the vegetation will die out. Tillage near a waterway is accomplished in a direction across the waterway. Sod-forming, cool-season grasses such as smooth brome or western wheatgrass are used in grass waterways. Grass waterways are usually designed to carry runoff of a 24 h storm of the intensity that happens once every 10 years. In areas with prolonged water flows, high water tables, or seepage problems, a rock-lined center is added (1).

Filter Strip Terraces

Filter strip terraces are strips of grass sod, legumes, and other vegetation on the contour that surface water runoff crosses as it runs downhill. They serve as an alternative to earthen terraces but do not have a channel to conduct water along the contour, as earthen terraces do. Filter strip terraces are excellent removers of sediment, pesticides, organic matter, and other pollutants. They are better than grass waterways because water enters the strip uniformly and over a wide area (3). The width and type of vegetation established in the filter area are determined by site conditions including soil type, land slope, and type of runoff entering the filter.

Small trees and shrubs are also used as conservation hedges in erosion control. These are planted close together and are periodically pruned to maintain height while the cut branches are inserted upslope of the hedge to trap more sediment. Trees that have been used successfully in erosion control, especially in agroforestry systems in the tropics, include *Calliandra calothyrsus*, *Sesbania sesban*, *Leucaena leucocephala*, *Gliricidia sepium*, and *Cassia siamea*. Trees used for windbreaks include *Eucalyptus* spp., *Casuarina* spp., *Acacia* spp., Neem tree (*Azadirchta indiaca*), and *Grevillea robusta* (7). In the Temperate Zone, thorny hedge plants include barberry, Osage orange, buckthorn, and hawthorn. Evergreen hedge plants are box, privet, azalea, yew, arborvitae, rhododendron, mountain laurel, and holly (8). Decorative deciduous shrubs often used are lilac, forsythia, mock orange, spiraea, euonymus, and viburnum. *Rosa rugosa* can be planted along highway embankments and rows of poplars, hemlocks, and other trees can be used as shelterbelts. Vegetative or biological measures may include log bundles anchored to the stream bank or the planting of herbaceous or woody plants, which can withstand high velocity flow while the roots form a protective net for the soil.

Fig. 3 Grass waterways are the most common type of vegetative filter strip. (Courtesy of Regehr, D.L., Devlin, D.L., and Barnes, P.L., Kansas State University.)

Fig. 4 Buffer strips at lower elevations of fields intercept surface runoff water from crop fields. (Courtesy of Regehr, D.L., Devlin, D.L., and Barnes, P.L., Kansas State University.)

Fig. 5 Riparian strips along stream banks intercept surface runoff water from crop fields. (Courtesy of Regehr, D.L., Devlin, D.L., and Barnes, P.L., Kansas State University.)

Buffer Strips and Riparian Strips

Buffer strips at lower elevations of fields (Fig. 4), and Riparian strips along stream banks, ponds, and lakes (Fig. 5), intercept surface runoff water from crop fields. Buffer strips may constitute ordinary grassed fencerows or strips of grasses, shrubs, and trees lining hillsides or banks of rivers. Runoff water must flow in a shallow, even layer across the buffer strip to remove sediments. Most common grasses used in buffer strips are Bluestem and Indiangrass. Riparian strips are planted so that surface and subsurface runoff must filter through them before it reaches a pond, lake, or stream (1). The body of water can be permanent or temporary. Riparian strips can also be placed next to wetlands, such as marshy or swampy areas, and additional vegetation can be placed uphill if excessive amounts of sediments enter the waters.

Settling Basins

Settling basins are constructed around inlets to tile-outlet terraces. These are important in reducing soil sediment loads and also act as setback zones for herbicides such as Atrazine. The basins are designed to retain water for up to 24 h, giving most of the larger sediments time to settle out (3). These are common where tile lines have been installed.

REFERENCES

1. Landschoot, P. *Erosion Control and Conservation Plantings on Non Cropland*; Penn State College of Agricultural Sciences Extension Publication, 1997.
2. Young, A. Agroforestry for Soil Conservation. *Agroforestry for Soil Conservation*; CAB International: Wallingford, UK, 1989; 197–229.
3. Regehr, D.L.; Devlin, D.L.; Barnes, P.L. *Using Vegetative Strips in Crop Fields*; Kansas State University, 1996, (http://www.oznet.ksu.edu).
4. Dabney, S.M.; Liu, Z.; Lane, M.; Douglas, J.; Zhu, J.; Flanagan, D.C. Landscape Benching from Tillage Erosion Between Grass Hedges. Soil Tillage & Res. **1999**, *51*, 219–231.
5. Dabney, S.M.; McGregor, K.C.; Meyer, L.D.; Grissinger, E.H.; Foster, G.R. Vegetative Barriers for Runoff and Sediment Control. In *Integrated Resources Management and Landscape Modification for Environmental Protection*; Mitchell, J.K., Ed.; Amer. Soc. of Agri. Eng.: St. Josephs, MI, 1993; 60–70.
6. McGregor, K.C.; Dabney, S.M.; Johnson, J.R. Runoff and Soil Loss from Plots with and without Stiff-Grass Hedges. Trans. Amer. Soc. of Agri. Eng. **1999**, *42* (2), 361–368.
7. Angima, S.D.; O'Neill, M.K.; Omwega, A.; Stott, D.E. Use of Tree/Grass Hedges for Soil Erosion Control in the Central Kenyan Highlands. J. Soil & Water Conserv. **2000**, *55* (4), 478–482.
8. Nair, P.K.N. *An Introduction to Agroforestry*; Kluwer Academic Publishers: Boston, USA, 1993; 325–343.

EROSION, SNOWMELT (WINTER EROSION PROCESSES)

E

Donald K. McCool
United States Department of Agriculture-Agricultural Research Service (USDA-ARS), Pullman, Washington, U.S.A.

INTRODUCTION

In many areas of the world, winter hydrology is an important part of the annual erosion process; in some regions it is the primary cause of erosion. Sharratt et al. (1) indicate that about half of the Earth's land surface is frozen at some time during the year. In this topic, we will consider winter processes to include a range of conditions from those where the soil freezes and thaws diurnally and is subjected mainly to rainfall to those where the soil remains frozen for several months and precipitation occurs as snow or in other solid form during all or part of the winter season. In areas where daily minimum temperatures are rarely below 0°C, winter erosion processes are of relatively minor importance as compared to spring, summer and autumn erosion. At the other end of the spectrum are areas where the soil freezes in the autumn and stays frozen until spring and the ground is blanketed with snow for the entire winter. Erosion is then confined to a period of spring snowmelt. Some areas such as the Palouse region of eastern Washington, northern Idaho, and northeastern Oregon are subjected to several runoff and erosion events each winter because the winter temperatures and precipitation patterns lead to multiple soil freezing and thawing occurrences each winter with accompanying rain or snowmelt events. The result of a particularly severe rain and snowmelt event when the soil had thawed at the surface on a fall-seeded field is shown in Fig. 1.

PRECIPITATION

Precipitation in areas where winter processes are important can occur as rain, or in solid form as sleet or snow. Rain can cause erosion due to splash and runoff detachment whereas sleet and snow cause detachment by runoff as they melt. In general, winter rainfall intensities are lower than summer rainfall intensities because of the lack of thunderstorm activity in the winter. However, the frequency and duration of precipitation is commonly greater. Since water loss from the soil is considerably lower during the winter due to lower mean temperatures and the absence of actively growing crops, even low intensity rainfall can increase the moisture content of soil rather quickly with a resultant dramatic increase in its susceptibility to erosion. The kinetic energy and intensity associated with rainstorms is not a direct factor in snowmelt erosion. Snowmelt erosion is more closely related to volume and peak rate of runoff, which can be strongly influenced by rainfall when it occurs concurrently with the snowmelt.

SOIL

The phase of the water in the surface layer of the soil is important in the winter erosion process. When soil water is frozen, erodibility, the susceptibility of the soil to erosion, is very low and erosion rates under sheet flow conditions are generally quite low. However, when runoff concentrates in small channels, the flow may cut into and through the frozen layer, leaving gullies and deeply incised channels. Soils that have been frozen and are thawing from the surface are weakened due to water expansion during the freezing process. Lee (2) cited a number of studies that indicate an increase in water content as water moves to the freezing front from deeper in the soil because the area where the soil is freezing is at very high water tension. The soil will regain strength after the frost has thawed and the soil reconsolidates as water drains from the soil (3).

Water content of the soil at the time of freezing is important to the permeability of the soil and the opportunity for infiltration and deep percolation. Lee (2) found a linear relationship with a negative slope between the ratio of frozen soil infiltration hydraulic conductivity to the unfrozen soil infiltration hydraulic conductivity. Soil frozen at water content near saturation is frequently impervious and runoff can be nearly 100% of the rainfall or snowmelt (2), thus leading to severe concentrated flow and gully erosion (Fig. 2).

Fig. 1 Erosion on fall seeded winter wheat in eastern Washington caused by rain and snowmelt on thawing soil. Measured rill erosion was 200 metric tons per hectare.

EFFECT OF SNOW AND FREEZING CONDITIONS

During the nonwinter period or in climatic regions where winter erosion processes are not a factor, the only type of water erosion that occurs is rainfall on unfrozen soil. Winter erosion is more complex because there are seven basic types of erosion events:

1. Rain on frozen soil,
2. Rain on thawing soil,
3. Rain on unfrozen soil,
4. Snowmelt on frozen soil,
5. Snowmelt on unfrozen soil,
6. Rain on snowmelt on frozen soil, and
7. Rain on snowmelt on unfrozen soil.

Snowmelt on thawing soil and rain on snowmelt on thawing soil are not listed because, as discussed later, these conditions generally do not occur due to the insulating properties of snow. Further complicating these various types of events is that they typically do not stand alone but occur in some combination and how they combine will dictate the severity of any resultant erosion.

A blanket of snow on the soil surface can have positive as well as negative effects with regard to soil erodibility. Snow is an excellent insulator; the deeper the snow the better the insulation. Thus, a snow layer prior to or concurrent with freezing conditions can retard or prevent frozen soil. However, this insulating property also works extremely well at preventing frozen soil from thawing from the surface down (though thawing will occur in an upward direction due to deeper, warmer unfrozen layers if above ground temperatures are favorable). Snow cover on frozen soil typically results in the soil surface remaining frozen until bare soil begins to appear as the snow melts with warming temperatures, at which time the exposed soil begins to thaw from the surface down (Fig. 3). The potential for severe erosion can be quite high with the presence of snow and depends on how all the contributing factors come together. If rainfall accompanies the warming temperatures required for snowmelt, runoff can be initiated over the frozen soil beneath the snowpack. If these conditions persist as patches of soil begin to be exposed and surface thawing follows, soil erosion will begin. The nature and severity of the erosion will be determined largely by the intensity and duration of the rainfall and to a lesser degree by temperature which affects the rapidity of the thawing process.

Fig. 2 Gully erosion from snowmelt runoff in northern Idaho.

Fig. 3 Melting snow exposes bare soil to thaw and increases erodibility.

E

RUNOFF EVENTS

It is difficult to ascribe a single set of characteristics to describe winter hydrology events on hillslopes or small watersheds. Events from rainfall on frost-impacted soil can be of very short duration whereas events resulting from extended periods of snowmelt can be of much longer duration. Events involving rainfall without snowmelt are usually brief although erosion rates can be high if the soil is thawing from the surface. Events involving snowmelt alone are extended in nature; runoff may fluctuate diurnally as air temperature dips below freezing at night. Erosion resulting from snowmelt alone is generally not severe. The highest runoff rates and flooding result from a combination of rainfall and snowmelt with high air temperatures. As long as the soil is frozen, sheet and rill erosion rates may be quite small, although concentrated flow erosion may be significant. When substantial quantities of bare soil appear, and the soil frost starts to thaw from the surface, erosion rates may increase dramatically. Sediment concentration graphs may lead or lag hydrographs under winter conditions. When there is no snow and the soil has thawed at the surface, a small amount of rainfall may lead to very high sediment concentration that occurs early in the event. When snow melts and the soil is not frozen, the sediment concentration and hydrographs may coincide. When snow melts over frozen soil, there may be little sediment until a portion of the soil is bare and melts. Then the runoff from the melting snow will detach and transport sediment, leading to a significant lag of peak sediment concentration.

MODELING

Erosion models are used to estimate location specific soil losses under various management systems within different land uses. They are also used to estimate expected amounts of sediment and chemicals that may be associated with those sediments. Models may be empirical in nature, such as the universal soil loss equation (USLE) (4) or the revised universal soil loss equation (RUSLE) (5), or they may be process based such as WEPP (6). The inclusion of winter erosion processes even in empirical models such as USLE or RUSLE complicates the models. Erosion relationships based on runoff characteristics rather than rainfall characteristics may be more appropriate, creating the need to estimate runoff. Process based models require inclusion of soil freezing and thawing, snowfall and snowmelt, and movement of water into frost impacted soil. Simple soil frost models are driven by air temperature, but more complex models consider radiation and other methods of energy transfer. It is difficult to strike a balance between excessive data requirement and adequate performance for applied use.

AMELIORATION

Preventing erosion damages from winter processes is similar to preventing erosion damages in other seasons. Cropland and rangeland are best protected by crop and surface cover. Snow collected in standing residue can prevent soil freezing and the creation of an impermeable condition. Surface cover protects from splash detachment, insulates the soil, and slows runoff. No till practices produce root channels and other pores that enable infiltration when the soil is frozen. Likewise, for areas where frost depth seldom exceeds 30–40 cm, deep ripping, chiseling, or slot mulching to expected frost depth can prevent formation of a continuous frost layer and lead to increased infiltration.

REFERENCES

1. Sharratt, B.S.; Radke, J.K.; Hinzman, L.D.; Iskandar, I.K.; Groenevelt, P.H. Physics, Chemistry, and Ecology of Frozen Soils in Managed Ecosystems: An Introduction. In *Intern. Symposium Physics, Chemistry, and Ecology of Seasonally Frozen Soils*; U.S. Army Cold Regions Res. Engin. Lab Spec. Rep. 97-10, Iskandar, I.K., Ed.; 1997; 1–7.
2. Lee, H.W. *Determination of Infiltration Characteristics of a Frozen Palouse Silt Loam Soil Under Simulated Rainfall.* Dissertation, Doctor of Philosophy in Agricultural Engineering, University of Idaho, Moscow, 1983.
3. Formanek, G.E.; McCool, D.K.; Papendick, R.I. Freeze–Thaw and Consolidation Effects on Strength of a Wet Silt Loam. Trans. ASAE **1984**, *26* (6), 1749–1752.
4. Wischmeier, W.H.; Smith, D.D. Predicting Soil Erosion by Water—A Guide to Conservation Planning. USDA Agriculture Handbook No. 537, 1978, 58.
5. Renard, K.G.; Foster, G.R.; Weesies, G.A.; McCool, D.K.; Yoder, D.C.; (Coordinators). *Predicting Soil Erosion by Water: A Guide to Conservation Planning with the Revised Universal Soil Loss Equation (RUSLE)*, USDA Agriculture Handbook No. 703, 1997; 404 pp.
6. Flanagan, D.C., Nearing, M.A., Eds. *USDA–Water Prediction Project: Hillslope Profile and Watershed Model Documentation*; NSERL Report No. 10, USDA–ARS National Soil Erosion Research Laboratory: West Lafayette, IN, 1995.

EROSION TOLERANCE/SOIL LOSS TOLERANCES

David L. Schertz

United States Department of Agriculture-Natural Resources Conservation Services (USDA-NRCS), Washington, D.C., U.S.A.

Mark A. Nearing

United States Department of Agriculture-Agricultural Research Service (USDA-ARS), West Lafayette, Indiana, U.S.A.

INTRODUCTION

Scientists first defined the *Soil Loss Tolerance* (T) value in the 1940s as "the amount of soil that could be lost without a decline in fertility, thereby maintaining crop productivity indefinitely" (1–3). The Natural Resources Conservation Service (NRCS, formerly Soil Conservation Service [SCS]) has used T values in conservation planning since the mid-1960s. Today, a T value is most commonly defined as "the maximum rate of annual soil erosion that may occur and still permit a high level of crop productivity to be obtained economically and indefinitely" (4).

Although not a component of current erosion prediction technologies, the T value is used as a basis for comparing estimated soil losses by prediction models to determine whether or not a particular cropping and management system is sustainable, and therefore acceptable. For example, if a particular soil under a specific cropping and management system is estimated to be eroding at 22.4 tonnes ha^{-1} $year^{-1}$, and the T value for that particular soil is 9.0 tonnes ha^{-1} $year^{-1}$, soil erosion would be considered excessive, and additional conservation treatment would be recommended.

Because tremendous technological advances have occurred in the area of soil fertility and amendments, on which the original definition was based, the T value concept may now be outdated. In other words, if any loss in soil fertility due to soil erosion can be replenished with commercial fertilizer, is there cause for concern about soil erosion? The answer to this question is obviously, yes, especially if maintenance of our soil resource, as well as other resource concerns, are considered.

BASIS FOR SOIL LOSS TOLERANCE VALUES

A primary reason for setting T values was to maintain soil fertility and productivity levels. Organic matter was used to indicate the level of soil fertility present. Other soil characteristics considered included soil loss limits to prevent formation of gullies, variation in soil type, and whether soils contained restrictive layers (e.g., claypan or heavy infertile subsoil) (5). Early work (3) assigned a T value of 6.7 tonnes ha^{-1} $year^{-1}$ for Putnam soil, and 9.0 tonnes ha^{-1} $year^{-1}$ for Marshall and Shelby soils, taking into account that if soil loss exceeded these rates, soil fertility levels would decline. Using available research on a limited set of soils, T values were then assigned to other soils based on interpolation.

The rate of soil formation for each soil type must also be considered in establishing a soil erosion limit. There is even less scientific data available to address this area. Estimates suggest that under natural conditions soil forms at a rate of about 2.5 cm in 300 to 1000 years (6). Under cultivation, however, soil may form at a rate of about 2.5 cm in 100 years. More recent research suggests that formation of the A horizon may occur at a rate of 2.5 cm in 30 years for medium to moderately coarse-textured soils but would be slower in finer-textured soils (8). However, increasing the depth of a favorable rooting zone through weathering takes much longer than the development of the A horizon. One estimate indicates we might expect to develop 1.1 tonnes ha^{-1} $year^{-1}$ of soil for unconsolidated parent material, and much longer for consolidated parent material (9).

Assuming a soil density of 1320 kg m^{-3}, if the A horizon renews itself at the rate of 2.5 cm every 30 years, an average erosion rate of 11.2 tonnes ha^{-1} $year^{-1}$ would maintain the A horizon. However, assuming that rooting depth is renewed at the rate of 1.1 tonnes ha^{-1} $year^{-1}$ from unconsolidated parent material, total rooting depth will continue to decrease even with soil erosion occurring at 11.2 tonnes ha^{-1} $year^{-1}$ (5). Therefore, if soils erode for several centuries at present soil loss tolerance levels, considerable loss in soil productivity may result (10). However, reducing the allowable soil loss to 1.1 tonnes ha^{-1} $year^{-1}$ (the level estimated to be the rate of rooting

Encyclopedia of Soil Science

depth development for unconsolidated parent material), would likely cause significant production constraints. Hugh Hammond Bennett, the first Chief of the Soil Erosion Service, stated in 1939, "As nearly as can be ascertained, it takes nature, under most favorable conditions, including a good cover of trees, grass, or other protective vegetation anywhere from 300 to 1000 years or more to build a single inch of topsoil. When 7 inches of topsoil is allowed to wash away, therefore at least 2,000 to 7,000 years of nature's work goes to waste" (7).

A 1956 joint conference involving the U.S. Department of Agriculture, Agricultural Research Service, and the Soil Conservation Service addressed the issues surrounding the establishment of T values (11). As a result of this conference, a committee established the following specific factors as being important in establishing T values:

1. Maintenance of adequate soil depth
2. Value of nutrients lost
3. Maintenance of water-control structures and control of floodplain sedimentation
4. Prevention of gullies
5. Yield reduction per inch of topsoil lost
6. Water losses
7. Seeding losses

This committee also set the caveat that soil loss tolerances should not exceed 11.2 tonnes ha^{-1} $year^{-1}$. The SCS held six regional workshops in 1961 and 1962 that resulted in T values from 2.2 to 11.2 tonnes ha^{-1} $year^{-1}$ being established in each region (9). Table 1 shows the percent distribution of T values in twelve Soil Orders occurring in the United States. Although some adjustments have been made to the criteria for establishing T values, the basic structure is still in use today.

EXISTING CONCERNS

Since the late 1950s and early 1960s, soil erosion prediction technology has significantly improved. The Universal Soil Loss Equation was replaced by the Revised Universal Soil Loss Equation in 1995. In addition, considerable work has been completed on process-based models for both wind and water erosion. As erosion prediction models become more sophisticated and better represent site-specific conditions, questions arise as to whether traditionally defined T values are outdated.

Because T values were developed primarily for use on cropland soils, there is concern that they may not be applicable to permanent pasture or rangeland. For the very

Table 1 Soil loss tolerance values expressed in percent of soils mapped for each soil order in the United States

		T value (t/ha)					
Soil order	**Order code**	**2.2**	**4.5**	**6.7**	**9.0**	**11.2**	**Total area (ha)[a]**
Alfisol	A	1	6	26	16	51	103,829,046
Andisol	C	2	18	33	15	32	9,307,531
Aridisol	D	21	19	12	3	44	53,041,017
Entisol	E	10	15	5	2	68	78,667,763
Gelisol	G	23	61	9	1	7	141,601
Histisol	H	16	32	30	4	19	7,711,917
Inceptisol	I	8	19	37	5	32	54,910,918
Mollisol	M	10	9	11	6	64	166,850,655
Oxisol	O	0	2	2	8	89	143,521
Spodosol	S	12	15	17	9	48	20,028,476
Ultisol	U	1	5	24	24	46	71,305,466
Vertisol	V	0	3	19	8	70	12,546,385

[a]Total hectares shown represent approximately 68% of the total land area in the United States stored in the Soil Survey Database.

ha = hectares; t = tonnes; T value = soil loss tolerance value.

(Data from Soil Survey Staff. *National Map Unit Interpretation Record (MUIR)*, U.S. Department of Agriculture, Natural Resources Conservation Service; National Soil Survey Center: Lincoln, NE and Iowa State University Statistical Laboratory: Ames, IA, 1997.)

fragile landscapes on which rangeland often occurs, tolerable soil losses might need to be significantly lower than those established for similar, more fertile soils that are cropped. For example, where a maximum of 11.2 tonnes ha^{-1} $year^{-1}$ is recommended for soils that are cropped, a limit of 5 to 7 tonnes ha^{-1} $year^{-1}$ might be more appropriate for rangeland. In addition, sheet and rill erosion occurs less often and concentrated flow erosion is often of greater concern on rangeland than on cropland. T values, which address only sheet and rill erosion, are not applicable to concentrated flow or gully erosion. Therefore, consideration should be given to establishing T values based on the specific use of the land. For example, it is not appropriate to use a T value of 11.2 tonnes ha^{-1} $year^{-1}$ for a range site that would likely be devastated if more than 5 tonnes ha^{-1} of soil erosion occurred on an annual basis. Productivity of the site, potential loss of desired vegetation, and soil rooting depth must therefore be taken into consideration.

In contrast to on-site effects caused by sheet and rill erosion, off-site impacts of soil erosion—such as sediment damage, water and air quality degradation, and emission of greenhouse gases—are of great concern. Society's increasing urban nature and environmental consciousness have made off-site concerns more important now than they were at the time that soil tolerance values were established. Formally incorporating the societal costs of off-site damages into the concept of soil loss tolerance is a major challenge for the future. Some of those costs are direct and relatively easy to quantify, such as the cost of dredging navigable waterways, ditches, and reservoirs. Other costs, such as those associated with the degradation of water quality in lakes and streams, impact their use by humans, fish, and wildlife, and are therefore less easily quantified.

The sediment discharge and amount of deposition in sensitive areas resulting from eroding fields depends greatly on the distance and the transport conditions between both areas. For example, excessive erosion on a field several kilometers from a sensitive area is less of a potential concern than erosion on a field adjacent to that same sensitive area. Therefore, proximity to sensitive areas, as well as transport conditions, may need to be considered when establishing or revising T values. In addition, fish and other wildlife vary in their ability to tolerate sediment and associated pollutants. Some species are very sensitive while others are not, indicating that a single tolerance value is inadequate when a broader set of concerns is evaluated.

If site-specific T values become an option, they must consider these concerns in order to be effective in protecting and enhancing all of our natural resources. In addition, a more modern approach to traditional productivity losses from erosion is needed. The short- and long-term cost and benefit trade-offs of soil conservation can and should be addressed using state-of-the-art methods of establishing soil loss tolerances.

CONCLUSION

T values have proven tremendously beneficial in developing and evaluating conservation systems that maintain or improve the soil resource. Without them, soil conservationists, farmers, and ranchers would have little basis for evaluating the adequacy of a particular conservation system. Several national programs, including Conservation Compliance, Conservation Reserve Program, and the Environmental Quality Incentives program, have been implemented based, in part, on the T value concept.

Because of its widespread use in program activities, it is time to reevaluate the T value concept. Perhaps a T value assigned to an individual soil should be viewed as a generic T value or basic starting point. Tables, equations, or process-based models could be developed to adjust the generic T value based on various conditions, such as land use, proximity to sensitive areas, susceptibility of wildlife species to sediment and associated contaminants, and other off-site concerns. For example, an assigned T value for a cropped soil may need to be reduced to maintain the ecological integrity of a rangeland site. This same T value may need to be reduced further, depending on off-site considerations. Or a T value for a cropped site may need to be adjusted downward if excessive sedimentation would be detrimental to the proper functioning of an adjacent wetland.

With the additional scientific knowledge existing today, we have the basis to refine T values to make them more responsive to site-specific conditions.

REFERENCES

1. Browing, G.M.; Parish, G.L.; Glass, J. A Method for Determining the Use and Limitation of Rotation and Conservation Practices in the Control of Soil Erosion in Iowa. J. Am. Soc. Agron. **1947**, *39*, 65–73.
2. Smith, D.D. Interpretation of Soil Conservation Data for Field Use. Agr. Eng. **1941**, *22*, 173–175.
3. Smith, D.D.; Whitt, D.M. Evaluating Soil Losses from Field Areas. Agr. Eng. **1948**, *29*, 394–396, 398.
4. Wischmeier, W.H.; Smith, D.D. *Predicting Rainfall Erosion Losses—A Guide to Conservation Planning*, Science and Education Administration Agriculture Handbook

No. 537; U.S. Department of Agriculture: Washington, DC, 1978; 1–58.

5. Schertz, D.L. The Basis for Soil Loss Tolerances. J. Soil Water Conserv. **1983**, *38*, 10–14.

6. Pimental, D.; Terhune, E.C.; Dyson-Hudson, R.; Rochereau, S.; Samis, R.; Smith, E.A.; Denman, D.; Reifschneider, D.; Shepard, M. Land Degradation: Effects on Food and Energy Resources. Science **1976**, *94*, 149–155.

7. Bennett, H.H. *Soil Conservation*; McGraw-Hill: New York, 1939.

8. Hall, G.F.; Daniels, R.B.; Foss, J.E. Rate of Soil Formation and Renewal in the USA. In *Determinants of Soil Loss Tolerance*, ASA Publication No. 45; Am. Soc. Agron.: Madison, Wisconsin, 1979.

9. McCormack, D.E.; Young, K.K.; Kimberlin, L.W. Current Criteria for Determining Soil Loss Tolerance. In *Determinants of Soil Loss Tolerance*, ASA Publication Number 45; Am. Soc. Agron.: Madison, Wisconsin, 1979.

10. McCormack, D.E.; Young, K.K. Technical and Societal Implications of Soil Loss Tolerance. In *Soil Conservation, Problems and Prospects*; Morgan, R.P.C., Ed.; John Wiley & Sons: New York, 1981; 365–376.

11. Paschall, A.H.; Klingebiel, A.A.; Allaway, W.H.; Bender, W.H.; Carpenter, W.W.; Glymph, L.M. *Permissible Soil Loss and Relative Erodibility of Different Soils*, Committee Report; U.S. Department of Agriculture, Agricultural Research; Service and Soil Conservation Service: Washington, DC, 1956.

EROSION BY WATER

David Favis-Mortlock
Queen's University Belfast, Belfast, Northern Ireland, United Kingdom

INTRODUCTION

Erosion by water is the redistribution and removal of the upper layers of the soil, both by the action of falling rain and by water flowing over the soil during and after rain or following snow melt. It occurs on both agricultural and undeveloped landscapes. While erosion by water is a natural phenomenon, removal of the soil's protective cover of vegetation at times of heavy rains greatly increases erosion rates, so that they exceed rates of soil formation. This is known as accelerated erosion, and is always the result of human actions such as unwise agricultural practices, overgrazing, or construction activity. Accelerated water erosion is a present-day problem over much of the Earth's surface and has both on-site and off-site impacts. The main on-site impact is the reduction in soil quality which results from the loss of the nutrient-rich upper layers of the soil, and the reduced water-holding capacity of many eroded soils. In affluent areas of the world, accelerated water erosion's on-site effects upon agricultural soils can be mitigated by increased the use of artificial fertilizers; however, this is not an option for much of the Earth's population. Water erosion's main off-site effect is the movement of sediment and agricultural pollutants into watercourses, leading to the silting-up of dams, disruption of the ecosystems of lakes, and contamination of drinking water. In some cases, increased downstream flooding may also occur due to the reduced capacity of eroded soil to absorb water.

PROCESSES OF EROSION BY WATER

Water erosion is driven by the gravitational energy of rainfall and flowing water. When rain falls upon unprotected soil, its kinetic energy may detach soil particles: this is one of the subprocesses (1) of water erosion (RD-ST in Fig. 1) and is often referred to as "rainsplash erosion" or "splash erosion." A more accurate term is "rainsplash redistribution," since although considerable quantities of soil can be moved by splash, almost all of it is merely redistributed on the soil's surface, i.e., the net downslope movement of splashed soil is generally small. Rainsplash redistribution is most effective where rainfall intensities are high (2), e.g., as a result of convective rainstorms in the world's equatorial regions. Low-intensity rainfall is often of frontal origin. Where such rainfall is common (e.g., northwest Europe), rainsplash is ineffective.

Some of the rain may infiltrate into (i.e., be absorbed by) the body of the soil. In dry conditions, all rain will infiltrate: the result is then rain with no runoff (Fig. 1). Runoff, that part of the rainwater which has not infiltrated, tends to flow downhill under the action of gravity. This thin diffuse film of water has lost virtually all the kinetic energy which it possessed as falling rain, thus it moves only slowly, has low flow power, and is generally incapable of detaching or transporting soil particles ("no erosion" in Fig. 1).

The microtopography (i.e., small-scale pattern of irregularities) of the soil's surface tends to cause this overland flow to concentrate in closed depressions which slowly fill; This is known as "detention storage" or "ponding". Both the flowing water and the water in detention storage protect the soil from raindrop impact, so that rainsplash redistribution usually decreases over time within a storm as the depth of surface water increases. There are, however, complex interactions between rainsplash and overland flow (Fig. 1).

If the rain continues, the increasing depth of water will eventually overtop the depressions. Overland flow that is released in this way is likely to flow downhill more quickly and in greater quantities (i.e., possess more flow power as a result of its kinetic energy) and so may be able to begin transporting (RD-RIFT, RD-FT and FD-FT in Fig. 1) and even detaching (FD-FT in Fig. 1) soil particles. Where it does so, the soil's surface will be lowered slightly. Lowered areas form preferential flow paths for subsequent flow, and these flow paths are in turn eroded further. Eventually, this positive feedback (3) results in small, well-defined linear concentrations of overland flow ("microrills" or "traces").

In many cases, individual microrills become ineffective over time due to sedimentation. A subset, however, grow further to become rills (4) and a smaller subset may go on to develop into gullies. This process of "competition" between microrills and rills leads to the self-organized

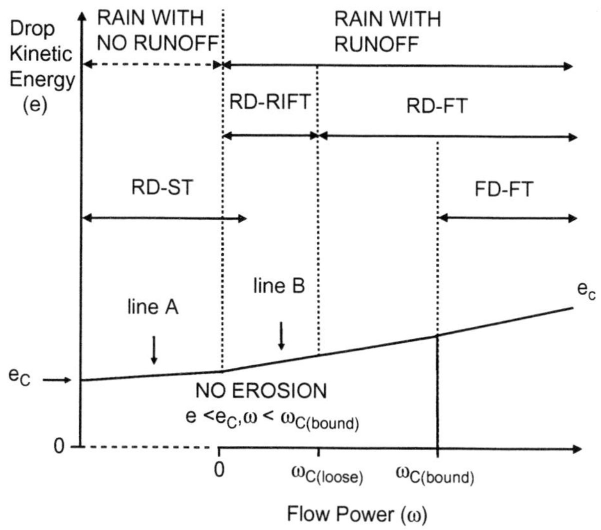

Fig. 1 Detachment and transport processes associated with variations in raindrop and flow energies. Line e_c: critical raindrop energy to cause erosion. Line A: raindrop energy prior to flow (this often increases slightly with time, as a soil crust develops). Line B: raindrop energy when flow occurs (increasing with time, as more drop energy is used to penetrate the deepening flow). $\omega_{c(loose)}$: critical stream power for transporting loose material. $\omega_{c(bound)}$: critical stream power for detaching soil from surface of soil matrix. RD-ST: raindrop detachment, splash transport. RD-RIFT: raindrop detachment, raindrop induced flow transport. RD-FT: raindrop detachment, flow transport. FD-FT: flow detachment, flow transport. (From Ref. 1.)

formation (3) of networks of erosional channels (dendritic on natural soil surfaces (5); constrained by the direction of tillage on agricultural soils (6)), which form efficient pathways for the removal of water from hillslopes. It is in such erosional channels that water erosion also operates most effectively to detach and remove soil by its kinetic energy (7). In most situations erosion by concentrated flow is the main agent of erosion by water. The flow-dominated erosional channels are separated by interrill areas (8) where the dominant processes are rainsplash and diffuse overland flow; however, boundaries between rill and interrill areas are both ill-defined and constantly shifting.

In some circumstances subsurface flow may be important in determining where channel erosion will begin and develop (e.g., at the base of slopes (9), and in areas of very deep soils such as tropical saprolites (10)). Meltwater from thawing snow operates in a broadly similar way to rain-derived overland flow (11), detaching and transporting unfrozen soil in areas of concentrated flow; however, snowmelt erosion is less well studied and less well understood.

As erosional channels increase in size (i.e., become large rills and gullies) (12), processes such as the gravitational collapse of channel walls and heads increase in importance (13). Runoff and sediment from rills and gullies may be moved into ditches, stream and rivers, and so transported well away from the point of origin. However, sediment may also be deposited within the rill or gully or beyond the rill or gully's confines in a depositional fan, at locations where the gradient slackens. Here it may be stored for a variable period of time (14), possibly being reworked by tillage activity until a subsequent erosion event is of sufficient size to re-erode the stored sediment. It may then be redeposited further downstream or make its way into a permanent watercourse and thence to lake or ocean.

EROSION BY WATER: SPATIAL AND TEMPORAL SCALE

Water erosion's complex hierarchy of subprocesses (Fig. 1) mean that erosion by water operates (and is studied) over a wide range of spatial scales (15). Rainsplash redistribution and the initiation of microrills and rills occur at a scale of millimeters (Fig. 2). Rill erosion on agricultural hillslopes operates at a scale of meters to tens of meters (Fig. 3), while gully erosion can occur on a scale of hundreds of meters or even kilometers. The off-site impacts of erosion can affect very large areas, sometimes hundreds or even thousands of square kilometers (Fig. 4).

Fig. 2 Impacts of erosion by water at the microscale: overland flow between millimeter-scale soil aggregates. The finger is pointing at an area where concentrated flow is just beginning to incise a microrill. (Photo from a rainfall simulation experiment by A.J.T. Guerra and D.T. Favis-Mortlock, 1997.)

Fig. 3 Impacts of erosion by water at the field scale: a large rill on agricultural land in Germany. (Photo from Katharina Helming, 2001.)

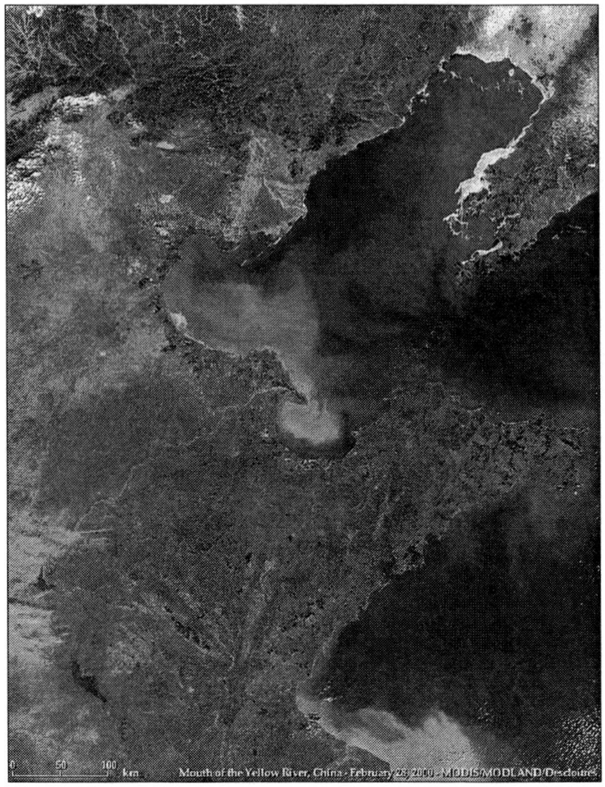

Fig. 4 Impacts of erosion by water at the global scale: a sediment plume covering tens of thousands of square kilometers at the mouth of the Yellow River, China. This results from erosion by water on the Chinese Loess Plateau. (Photo from NASA, 2000 (http://modis.gsfc.nasa.gov/MODIS/IMAGE_GALLERY/MODIS1000152_md.html.))

At every spatial scale, however, erosion is highly patchy. Even in areas of severe erosion, rates of soil loss can vary greatly from point to point on the landscape as the vagaries of topography and land use concentrate erosive flows on a wide range of spatial scales (16). Obvious erosion in one field can be found side-by-side with virtually untouched areas, and within an eroded field, the severity of erosion can vary markedly.

Erosion by water is often also variable across a range of temporal scales. Soil loss from water erosion occurs both incrementally, as a result of many small rainfall events, and more dramatically, as a result of large but relatively rare storms. Dramatic erosion events can produce large gullies and create flooding and property damage, which hits the news headlines. However, a significant proportion of total soil loss may be due to small but frequent erosion events, which nonetheless have a notable cumulative impact (17).

THE GLOBAL PROBLEM OF EROSION BY WATER

While soil erosion by water has been occurring naturally for some 450 million years (since the first land plants formed the first soil), accelerated erosion is of much more recent origin. Yet on a human timescale, accelerated erosion is old. There is considerable archaeological evidence from many parts of the world that accelerated erosion by water is often associated with early agriculture (18, 19). Water erosion's association with unwise agricultural practices was first noted within a scientific context during the second and third decades of the 20th century by pioneers of soil conservation such as Hugh Hammond Bennett in the U.S., and subsequently by workers in other parts of the globe. During the period of colonialism, the imposed adoption of European agricultural methods frequently led to accelerated erosion in developing countries, a problem which continues to the present day (20, 21). During the last few decades of the 20th century, a move towards intensive agricultural technologies which leave the soil bare during times of heavy rainfall meant that previously problem-free areas of the world, such as northwest Europe, began to experience notable increases in water erosion (22).

Yet despite the global nature of the problem of erosion by water, even today we do not have good information regarding the global extent of erosion by water (23, 24). Data on the severity of erosion are also often limited (25).

IMPACTS OF EROSION BY WATER

Impacts of erosion by water can be categorized into on-site (i.e., affecting only the place where the erosion occurs) and off-site (i.e., affecting locations other than that at which the erosion occurs) problems.

Erosion's removal of the upper horizons of the soil results in a reduction in soil quality (26), i.e., a diminution of the soil's suitability for agriculture or other vegetation. This is because the eroded upper horizons tend to be the most nutrient-rich. Also, because the finest constituents of eroded soil tends to be transported furthest, eroded soils become preferentially depleted of their finer fraction over time; this often reduces their water-holding capacity (26). In other words, "Erosion removes the cream of the soil". Increased use of artificial fertilizers may to an extent, and for a time, compensate for erosion-induced loss of soil quality where economic circumstances are favorable. This is not usually feasible in developing countries, however (20, 21). Loss of soil quality is a long-term problem; globally, it is water erosion's most serious impact (27, 28).

Movement of sediment and associated agricultural pollutants into watercourses is the major off-site impact resulting from erosion. This leads to silting-up of dams, disruption of the ecosystems of lakes, and contamination of drinking water (22). Rates of erosion do not have to be high for significant quantities of agricultural pollutants to be transported off-site (29). This is a shorter-term impact than loss of soil quality; in the more affluent areas of the world it can be the main driver for present-day soil conservation policy initiatives (24). A more minor off-site effect can occur in situations where eroded soil has a decreased capacity to absorb water; increased runoff may lead to downstream flooding and local damage to property (30).

REFERENCES

1. Kinnell, P.I.A. *A Discourse on Rainfall Erosion Processes and Modelling on Hillslopes*; Centre for Australian Regolith Studies: Canberra, ACT, 2000.
2. Nearing, M.A.; Bradford, J.M.; Holtz, R.D. Measurement of Waterdrop Impact Pressures on Soil Surfaces. Soil Science Society of America Journal **1987**, *51* (5), 1302–1306.
3. Favis-Mortlock, D.T.; Boardman, J.; Parsons, A.J.; Lascelles, B. Emergence and Erosion: A Model for Rill Initiation and Development. Hydrological Processes **2000**, *14* (11–12), 2173–2205.
4. Slattery, M.C.; Bryan, R.B. Hydraulic Conditions for Rill Incision Under Simulated Rainfall: A Laboratory Experiment. Earth Surface Processes and Landforms **1992**, *17*, 127–146.
5. Helming, K.; Roth, C.H.; Wolf, R.; Diestel, H. Characterization of Rainfall–Microrelief Interactions with Runoff Using Parameters Derived from Digital Elevation Models (DEMs). Soil Technology **1993**, *6*, 273–286.
6. Planchon, O.; Esteves, M.; Silvera, N.; Lapetite, J.-M. Raindrop Erosion of Tillage-Induced Microrelief: Possible Use of the Diffusion Equation. Soil and Tillage Research **2000**, *15–16*, 1–14.
7. Nearing, M.A.; Norton, L.D.; Bulgakov, D.A.; Larionov, G.A.; West, L.T.; Dontsova, K. Hydraulics and Erosion in Eroding Rills. Water Resources Research **1997**, *33* (4), 865–876.
8. Govers, G.; Poesen, J. Assessment of the Interrill and Rill Contributions to Total Soil Loss from an Upland Field Plot. Geomorphology **1988**, *1*, 343–354.
9. Huang, C.; Laflen, J.M. Seepage and Soil Erosion for a Clay Loam Soil. Soil Science Society of America Journal **1996**, *60* (2), 408–416.
10. Fernandes, N.F.; Coelho Netto, A.L. Subsurface Hydrology of Layered Colluvium Mantles in Unchannelled Valleys—South-Eastern Brazil. Earth Surface Processes and Landforms **1994**, *19*, 609–626.
11. Sharratt, B.S.; Lindstrom, M.J.; Benoit, G.R.; Young, R.A.; Wilts, A. Runoff and Soil Erosion During Spring Thaw in the Northern U.S. Corn Belt. Journal of Soil and Water Conservation **2000**, *55* (4), 487–494.
12. Poesen, J.; Vandaele, K.; van Wesemael, B. Gully Erosion: Importance and Model Implications. In *Modelling Soil Erosion by Water*; NATO–ASI Series I-55, Boardman, J., Favis-Mortlock, D.T., Eds.; 1998; 285–312.
13. De Ploey, J. A Model for Headcut Retreat in Rills and Gullies. Catena Supplement **1989**, *14*, 81–86.
14. Trimble, S.W. A Sediment Budget for Coon Creek Basin in the Driftless Area Wisconsin, 1853–1977. American Journal of Science **1983**, *283*, 454–474.
15. Kirkby, M.J.; Imeson, A.C.; Bergkamp, G.; Cammerat, L.H. Scaling up Processes and Models from the Field Plot to the Watershed and Regional Areas. Journal of Soil and Water Conservation **1996**, *51* (5), 391–396.
16. Favis-Mortlock, D.T.; Boardman, J.; MacMillan, V.J. The Limits of Erosion Modeling: Why We Should Proceed with Care. In *Landscape Erosion and Evolution Modeling*; Harmon, R., Doe, W.W., III Eds.; Kluwer Academic/Plenum Publishing: New York, 2001; in press.
17. Boardman, J.; Favis-Mortlock, D.T. Frequency-Magnitude Distributions for Soil Erosion, Runoff and Rainfall—A Comparative Analysis. Zeitschrift Für Geomorphologie N.F. Supplementband **1999**, *115*, 51–70.
18. Bell, M.G., Boardman, J., Eds. *Past and Present Soil Erosion*; Oxbow Monograph 22, Oxbow Books: Oxford, UK, 1992.

E

19. O'Hara, S.L.; Street-Perrott, F.A.; Burt, T.P. Accelerated Soil Erosion Around a Mexican Highland Lake Caused by Prehispanic Agriculture. Nature **1993**, *362*, 48–51.
20. Hallsworth, E.G. *Anatomy, Physiology and Psychology of Soil Erosion*; Wiley-Interscience: New York, 1987.
21. Stocking, M. Soil Erosion in Developing Countries: Where Geomorphology Fears to Tread! Catena **1995**, *25* (1–4), 253–267.
22. Evans, R. *Soil Erosion and Its Impacts in England and Wales*; Friends of the Earth: London, 1996.
23. United Nations Environment Program; Global Land Degradation map; http://www.unep.no/db/maps/prod/level3/id-1238.htm; (accessed July 2001).
24. European Environment Agency, *Down to Earth: Soil Degradation and Sustainable Development in Europe*; Office for Official Publications of the European Communities: Luxembourg, 2000; http://reports.eea.eu.int/Environmental_issue_series_16/en; (accessed July 2001).
25. Boardman, J. An Average Soil Erosion Rate for Europe: Myth or Reality? Journal of Soil and Water Conservation **1998**, *53* (1), 46–50.
26. Lal, R. Monitoring Soil Erosion's Impact on Crop Productivity. In *Soil Erosion Research Methods*; Lal, R., Ed.; Soil and Water Conservation Society: Ankeny, IA, 1989.
27. Larson, W.E.; Pierce, F.J.; Dowdy, R.H. The Threat of Soil Erosion to Long-Term Crop Production. Science **1983**, *219*, 458–465.
28. Pimentel, D.; Allen, J.; Beers, A.; Guinand, L.; Hawkins, A.; Linder, R.; McLaughlin, P.; Meer, B.; Musonda, D.; Perdue, D.; Poisson, S.; Salazar, R.; Siebert, S.; Stoner, K. Soil Erosion and Agricultural Productivity. *World Soil Erosion and Conservation*; Cambridge University Press: Cambridge, UK, 1993; 277–292.
29. Harrod, T.R. Runoff, Soil Erosion and Pesticide Pollution in Cornwall. In *Conserving Soil Resources: European Perspectives*; Rickson, R.J., Ed.; CAB International: Wallingford, UK, 1994; 105–115.
30. Boardman, J.; Ligneau, L.; de Roo, A.; Vandaele, K. Flooding of Property by Runoff from Agricultural Land in Northwestern Europe. Geomorphology **1994**, *10*, 183–196.

EROSION BY WATER, EMPIRICAL MODELS

John M. Laflen
United States Department of Agriculture–Agricultural Research Service (USDA–ARS) (retired) West Lafayette, Indiana, U.S.A.

INTRODUCTION

Empirical models of soil erosion by water are used to determine the extent of soil erosion, to identify areas needing soil erosion control measures, to select appropriate practices to minimize damages resulting from soil erosion, and to estimate downstream delivery of sediment and other materials. Examples of damages might include reduced productivity, exposure of landfill materials, obstacles to farming, living plant damages, increased maintenance of locks and dams, reduced reservoir capacity, reduced aquatic habitat, deposited nuclear materials, and increased eutrophication of surface waters.

An empirical model is a representation of data (1). Although the models described here were developed based on relationships from data, and while these relationships are explainable to a certain extent based on our understanding of the erosion process, they came from data. The models described do not include simulation models, even those that might include empirical components, nor the application of empirical models in Geographical Information Systems.

EMPIRICAL EROSION MODEL DEVELOPMENT

Early erosion research literature describes the effect of various factors—climatic, topographic, soil, management and conservation practices, on soil erosion, and empirical models include most of these factors. Early models were relatively simple relationships with factor values generally determined based on a limited set of experimental observations and involving only a few factors. As the science progressed, the complexity increased, with relationships of factor values to factor characteristics being developed based on experimental data.

One of the early developments was establishment of erosion plots at numerous locations in the United States. These plots provided an immense data set that is still used today for evaluating soil erosion and hydrologic models. Over 10000 plot years of data were collected on these plots (2). Additionally, many plots were located on experimental farms that contained small watersheds on which conservation practices were installed, monitored, and evaluated. These experimental areas provided the necessary data for developing the relationships in water erosion empirical models.

Erosion plots have also been established in other regions of the world. Hudson (3) described field experiments involving erosion plots in Rhodesia. Erosion plots can be found on every continent (except Antarctica) and are frequently used to evaluate the applicability of empirical and non empirical erosion models to specific regions, and to develop appropriate factor values for practices common to those regions.

The development of empirical models has been a classic example of the operation of science and its application to solving real world problems. The steady progression of science is illustrated in Table 1 where the steps in development of the United States empirical erosion modeling are given chronologically. The work initiated with Zingg continues even today, each step built on top of earlier work. Meyer (4) gives an excellent review of the development of the Universal Soil Loss Equation in the United States.

EMPIRICAL MODELS

Universal Soil Loss Equation (USLE)

The USLE is written as

$$A = RKLSCP \tag{1}$$

where, A is soil loss per unit area (units depend on units used for R and K); R is the rainfall and runoff factor, the number of rainfall erosion index units, plus factors for runoff from snowmelt or applied water. K is the soil erodibility factor, the soil loss rate per erosion index unit on a plot 72.6 feet long of uniform 9% slope maintained in a clean tilled fallow. L is the slope-length factor, the ratio of soil loss for a given slope length to a 72.6 foot long slope. S is the slope-steepness factor, the ratio of soil loss

Table 1 Evolution of United States empirical water erosion models

Reference	Soil loss	Coefficient or rainfall factor	Soil erodibility factor	Length–slope factor	Cropping management factor	Conservation practice factor
Zingg 1940 (25)	A	C'		$L^{0.6} S^{1.4}$		
Smith 1941 (26)	A	C''		$L^{0.6} S^{1.4}$		P
Browning 1947 (27)	A	C'''	K'	$L^{0.6} S^{1.4}$		P
Musgrave 1947 (28)	A'	$P_{30}^{1.75}$	K''	$L^{0.35} S^{1.35}$	C^*	P
USLE 1961 (6)	A	EI_{30}	K	$(L/72.6)^{0.5}(.065+.045 S+.0065 S^2)$	C	P
USLE 1978 (2)	A	EI_{30}	K	$(L/72.6)^{0.5}(65.4 \text{ Sin}^2 \Theta + 4.56 \text{ Sin } \Theta +.065)$	C	P
MUSLE 1975 (11)	A^*	$95(Qq_p)^{0.56}$	K	$(L/72.6)^{0.5}(65.4 \text{ Sin}^2 \Theta + 4.56 \text{ Sin } \Theta +.065)$	C	P
OF 1975 (14)	A	$0.5\, EI_{30} + 15 Q_o q_o^{0.33}$	K	$(L/72.6)^{0.5}(65.4 \text{ Sin}^2 \Theta + 4.56 \text{ Sin } \Theta +.065)$	C	P
RUSLE 1997 (24)	A	EI_{30}	K	$(L/72.6)^{m} (a \text{ Sin } \Theta + b)$	C	P

A = Soil loss (tons/acre); A' = Soil loss (inches/year); A^* = Sediment delivery (tons/year).
P_{30} = Maximum precipitation amount (inches) falling in 30 minutes in a storm.
Q = Storm runoff volume (acre feet); q_p = peak runoff rate (cubic feet per second);
Q_o = Storm runoff volume (inches); q_o = peak runoff rate (inches per hour).
E = Storm rainfall energy (hundreds of foot-tons per acre).
I_{30} = Maximum rainfall intensity in a 30 minute period within a storm (inches per hour).
K', K'', K = Soil erodibility factors.
L = Slope length (feet); S = Slope (%); Θ = Slope angle (degrees).
m = Exponent on length term; values depend on slope or slope and rill/interrill ratio.
a,b = coefficients in function making up slope term; values depend on slope.
C', C'', C''' = coefficients. C^* = vegetal cover factor. C = Cropping and management factor (before 1978) and cover and management factor (after 1978).
P = Conservation practice factor.

for a given slope to soil loss from a 9% slope. C is the cover and management factor, the ratio of soil loss for a given land use and management to soil loss from a continuous clean tilled fallow. P is the support practice factor, the ratio of soil loss with a support practice, such as contouring, terracing or stripcropping, to soil loss for up and down the slope.

Much more detail on factor values, and assistance in computing these values can be found in Wischmeier and Smith (2).

The USLE was the culmination of over 2 decades of work in developing a national model of soil erosion by water. As shown in Table 1, the form and factors were long established before the USLE was developed. The defining scientific finding in the development of the USLE was that of Wischmeier and Smith (5) when they isolated a single rainfall variable that could be used over the United States to model soil erosion from rainstorms. This allowed the application of factor values in one region to be used in another region. Climatic databases were developed for the entire United States, and tables and charts were published to allow use of the USLE for most US conditions (2, 6, 7). The USLE technology was the water erosion model of choice over much of the world through the last century. It is being replaced by RUSLE in the United States (8, 24).

While the USLE is a good predictor of average annual erosion, it is a poor predictor of individual storm soil erosion because it does not contain variables associated with the hydrologic condition of the soil at the time the storm occurs. Additionally, the USLE does not consider deposition on a field, and does not predict sediment yield, a needed estimate for many applications. The USLE's area of application is from the top of a hillslope (or divide) to where deposition begins, or runoff enters a channel.

Modified Universal Soil Loss Equation (MUSLE)

MUSLE is written as

$$A^* = 95(Qq_p)^{0.56}KLSCP \tag{2}$$

where A^* is sediment delivery, Q is total volume of runoff and q_p is peak rate of surface runoff, and K, L, S, C, and P are the same as for the USLE.

MUSLE was developed to predict individual storm sediment yield. A major shortcoming of the USLE for many applications was that while it estimated soil erosion, it did not predict sediment yield, and it was a poor predictor of individual storm soil erosion. Williams et al. (9) determined that for individual storms, sediment concentration for 5 small watersheds near Riesel, Texas was highly correlated with the ratio of peak runoff rate per square mile to total volume of runoff. Williams and Berndt (10) applied the USLE to average annual sediment yield prediction for the same 5 small watersheds using a delivery ratio based on watershed characteristics. They weighted the various factors in the USLE (except for the rainfall factor) based on drainage area of the various factor values, channel lengths, channel slopes, and cultivated area.

Williams (11) examined 3 forms of a runoff factor, and determined that the runoff factor $95(Qq_p)^{0.56}$ (with Q, the volume of runoff given in acre-feet, and q_p, the peak flow rate given in cubic feet per second) was the best form for estimating individual storm sediment yield from small watersheds. In the analysis, he used the procedures of weighting the USLE factors pioneered in 1972. The 18 watersheds ranged in size from 2.7 to 4380 acres. Two watersheds were located at Hastings, Nebraska; the remainder were located at Riesel, Texas and included the 5 small watersheds used in earlier analyses.

MUSLE requires inputs—storm volume of runoff and peak rate of runoff, not used in the USLE. This information can be generated by simulation models or it can be estimated for design storms.

MUSLE continues to be heavily used in various modeling activities. It is an option in the EPIC model (12) for computing sediment yield. While MUSLE is widely used, there are few publications that detail the accuracy of MUSLE, outside of the original work performed by Williams in its development.

ONSTAD–FOSTER MODIFICATION OF THE USLE (OF)

The Onstad–Foster modification of the USLE is written as:

$$A = (0.5R + 15Q_o q_o^{0.33})KLSCP \tag{3}$$

where A, R, K, L, S, C, and P are the same as for the USLE, and Q_o is runoff volume in inches and q_o is peak flow rate in inches per hour.

Foster et al. (13), using basic erosion principles, derived a soil loss equation that included a runoff parameter and a rainfall parameter to replace the rainfall factor in the USLE. This factor was $0.5R+15Q_o q_o^{0.33}$. Onstad and Foster (14) found good results when the factor was tested using watershed data from Ohio and Iowa.

Again, as for MUSLE, hydrologic information must be available to estimate peak rate and total volume of storm runoff. A major difference between MUSLE and OF are the approaches used in developing the equations—MUSLE variables were selected and tested, while OF variables were determined based on analysis of fundamental erosion processes. Hence, the difference in forms. It has never been established that one is a superior predictor to the other.

OF has received limited use, although it is also an option for use in some modeling activities, including the EPIC model.

SLEMSA (Soil-Loss Estimation Model for Southern Africa)

Elwell (15) introduced SLEMSA as a reliable management tool developed from limited resources. The development team evaluated the USLE for adoption in southern Africa, but were concerned about adopting factor values developed in America (16). Additionally, there was considerable concern about differences in farming practices. They felt that they did not have the resources to implement the USLE, and a simpler approach was needed.

SLEMSA is written as

$$Z = KCX \tag{4}$$

where Z is mean annual soil loss, K is mean annual soil loss from a standard field plot 30 m × 10 m at 4.5% slope for a soil of known erodibility under a weed-free bare fallow, C is the ratio of soil lost from a cropped plot to that lost from bare fallow land and X is the ratio of soil lost from a plot length L and slope percent S to that lost from the standard plot.

K values are computed as

$$Ln\ K = b\ln\ E + a \tag{5}$$

where E is seasonal rainfall energy, and a and b are given by

$$a = 2.884 - 8.1209\ F \tag{6}$$

$$b = 0.4681 + 0.7663\ F \tag{7}$$

where F is soil erodibility. Based on soil texture, F is assigned a value between 4 (light textured, or sandy soil) and 6 (heavy textured, or clay soil). Additional adjustments are made based on hydrologic characteristics of the soil, tillage, soil structure and crusting, giving a range of F values from 1 to 9.

C is computed as $C = e^{(-.06i)}$ when $i < 50\%$, and as $C = (2.3 - 0.01i)/30$ when $i \geq 50\%$, where i is the percentage rainfall energy intercepted by the crop (used crop cover %).

X is computed as:

$$X = (l)^{1/2}(0.76 + 0.53s + 0.076s^2)/25.65 \tag{8}$$

This is derived from the USLE LS factor in use in the 1970's. The equation above converted the USLE LS factor to the metric system, and adjusted it to a 4.5%, 30 m long slope. The variables l and s are length (m) and slope (%).

The USLE and SLEMSA are very similar in many respects. The major differences are in the climate variables and the cropping-management approaches. The USLE climate factor contains both rainfall energy and rainfall intensity where SLEMSA contains only rainfall energy. The USLE C value incorporates both tillage and cropping while the C value for SLEMSA is a function of the interception of rainfall energy by crops. However, SLEMSA does capture some of the tillage component in the F value.

ABAG (German USLE)

Auerswald (17) described the application of the USLE in Germany. Experimentally, most of the USLE factors were validated, and few if any modifications were needed to apply it under German conditions. One modification was that soils highly enriched in potassium were found to be more erodible than expected, based on the soil erodibility nomogram of Wischmeier et al. (18).

While no modification of the USLE was found necessary, ABAG factors were necessary for German conditions, specifically the R and C factors. Schwertmann et al. (19) published a book for extension workers in Germany detailing the use of the USLE in Germany.

SOIL LOSS

Rosewell and Edwards (20) developed a computer program that used the USLE approach to estimate average annual soil losses, and to give recommendations on ways to reduce soil loss through changes in land and crop management practices. SOILOSS was developed for use in New South Wales, Australia. It used a slightly different equation for computing rainfall energy than that used in the USLE. It also used both the USLE nomograph for soil erodibility, plus experimentally derived values for NSW. In a later revision (21), Rosewell adopted LS values from RUSLE. From the earliest version of SOILOSS, the subfactor approach of Laflen et al. (22) that was implemented in RUSLE was used to estimate the C value. It also contained P value estimates similar to those in the USLE, but some were from experimental work in NSW. It also included a benefit from sediment deposited in bank (terrace) channels.

Keats and Rosewell developed a SOILOSS user guide (23) for high school students. It included the program, an introduction to the Universal Soil Loss Equation, and complete parameterization for most Australian conditions. It included worksheets and examples.

In a number of ways, SOILOSS could be considered an early RUSLE, using much of the science and approaches that were eventually used in RUSLE. It was also developed in the time frame where RUSLE was developed.

Revised Universal Soil Loss Equation (RUSLE)

RUSLE (24) maintains the form of the USLE in terms of the factors. But, since it is operated on a computer, many improvements in computations, particularly those including interactions, have been incorporated. Major improvements in the revised USLE include improved R value map for the United States, time variation in soil erodibility, improved LS factor that incorporates recent science, a subfactor approach for computing the cover and management factor, and P values based on modeling analyses that include storm severity, ridge height, drainage, and off-grade contouring.

RUSLE has been implemented for use in erosion prediction by the Natural Resource Conservation Service in the United States. It is also used in several simulation models. RUSLE development continues, with improvements in science, ease of use, supporting databases, and applications related to sediment deposition and delivery.

SUMMARY

Empirical erosion modeling has been applied on much of the Earth's land surface. The approaches have almost exclusively been to quantify factors that influence soil erosion, and then to combine these factors, usually as products of the factors. The development of empirical models has clearly been an evolutionary process, and almost all models owe all or much of their technology to the Universal Soil Loss Equation.

REFERENCES

1. Haan, C.T. *Statistical Methods in Hydrology*; Iowa State Press: Ames, Iowa, 1977; 394.
2. Wischmeier, W.H.; Smith, D.D. *Predicting Rainfall-Erosion Losse—A Guide to Conservation Farming*, Agricultural Handbook No. 537; U.S. Department of Agriculture: Washington DC, 1978; 58.
3. Hudson, N.W. Erosion Control Research. Rhodesian Agricultural J. **1957**, *54* (4), 297–323.
4. Meyer, L.D. Evolution of the Universal Soil Loss Equation. J. Soil and Water Conserv. **1984**, *39* (2), 99–104.
5. Wischmeier, W.H.; Smith, D.D. Predicting Rainfall Erosion Losses—A Guide to Conservation Planning. Trans. Am. Geophys. Union. **1958**, *39* (2), 285–291.
6. Wischmeier, W.H.; Smith, D.D. *A Universal Equation for Predicting Rainfall–Erosion Losses—An Aid to Conservation Farming in Humid Regions*, ARS Special Report 22–66; Agricultural Research Service, U.S. Department of Agriculture: Washington, DC, 1961; 11.
7. Wischmeier, W.H.; Smith, D.D. *Predicting Rainfall–Erosion Losses from Cropland East of the Rocky Mountains—Guide for Selection of Practices for Soil and Water Conservation*, Agricultural Handbook No. 282; U.S. Department of Agriculture: Washington DC, 1965; 47.
8. Renard, K.G.; Foster, G.R.; Weesies, G.A.; Porter, J.P. RUSLE: Revised Universal Soil Loss Equation. J. Soil and Water Conserv. **1991**, *46* (1), 30–33.
9. Williams, J.R.; Hiler, E.A.; Baird, R.W. Prediction of Sediment Yields from Small Watersheds. Trans. Am. Soc. Agric. Eng. **1971**, *14* (6), 1157–1162.
10. Williams, J.R.; Berndt, H.D. Sediment Yield Computed With Universal Equation. J. Hydr. Div., Proc. Am. Soc. Civil Eng. **1972**, *98* (HY12), 2087–2098.
11. Williams, J.R. Sediment Yield Prediction with Universal Equation Using Runoff Energy Factor. In *Present and Prospective Technology for Predicting Sediment Yields and Sources*, Sediment-Yield Workshop, USDA Sedimentation Laboratory, Oxford, Mississippi, November 28–30, 1975; ARS-S-40, Agricultural Research Service, U.S. Department of Agriculture: Washington DC, 1975; 244–252.
12. Sharpley, A.N.; Williams, J.R. *EPIC–Erosion/Productivity Impact Calculator: 1. Model Documentation*, Technical Bulletin No. 1768; U.S. Department of Agriculture: Washington DC, 1990; 235.
13. Foster, G.R.; Meyer, L.D.; Onstad, C.A. An Erosion Equation Derived from Basic Erosion Principles. Trans. Am. Soc. Agric. Eng. **1977**, *20* (4), 678–682.
14. Onstad, C.A.; Foster, G.R. Erosion Modeling on a Watershed. Trans. Am. Soc. Agric. Eng. **1975**, *18* (2), 288–292.
15. Elwell, H. A. Modeling Soil Losses in Zimbabwe Rhodesia. Am. Soc. of Agr. Eng. Paper Number 79–2051, 1979; Am. Soc. Agr. Eng., St. Joseph MI.
16. Elwell, H.A. Modeling Soil Losses in Southern Africa. J. Agr. Eng. Res. **1978**, *23*, 117–127.
17. Auerswald, K. Empirical Soil Erosion Models. Personal Letter, 4-12-2000, 2000.
18. Wischmeier, W.H.; Johnson, C.B.; Cross, B.V. A Soil Erodibility Nomograph for Farmland and Construction Sites. J. Soil and Water Conserv. **1971**, *26* (5), 189–193.
19. Schwertmann, U.; Vogl, W.; Kainz, M. *Bodenerosion durch Wasser-Vorhersage des Abtrags und Bewertung von GegenmaBnahmen*; Ulmer Verlag: Stuttgart, Germany, 1987; 64.
20. Rosewell, C.J.; Edwards, K. *SOILOSS—A Program to Assist in the Selection of Management Practices to Reduce*

Erosion, Technical Handbook No. 11; Soil Conservation Service of New South Wales: Sydney, NSW, 1988; 71.

21. Rosewell, C.J. *SOILOSS-A Program to Assist in the Selection of Management Practices to Reduce Erosion*, 2nd Ed.; Technical Handbook No. 11; Soil Conservation Service of New South Wales: Sydney, NSW, 1993; 108.
22. Laflen, J.M.; Foster, G.R.; Onstad, C.A. Simulation of Individual Storm Soil Loss for Modeling Impact of Soil Erosion on Crop Productivity. In *Soil Erosion and Conservation*; El-Swaify, S.A., Moldenhauer, W.C., Lo, A., Eds.; Soil Conserv. Soc. Am.: Ankeny, IA, 1985; 285–295.
23. Keats, J.; Rosewell, C. *SOILOSS-High School User Guide*; Department of Conservation and Land Management: NSW, Australia; 87.
24. Renard, K.G.; Foster, G.R.; Weesies, G.A.; McCool, D.K.; Yoder, D.C. *Predicting Soil Erosion by Water: A Guide to Conservation Planning with the Revised Universal Soil Loss Equation (RUSLE)*, Agricultural Handbook No. 703; U.S. Department of Agriculture: Washington DC, 1997; 384.
25. Zingg, A.W. Degree and Length of Land Slope as it Affects Soil Loss in Runoff. Agr. Eng. **1940**, *21* (2), 59–64.
26. Smith, D.D. Interpretation of Soil Conservation Data for Field Use. Agr. Eng. **1941**, *22* (5), 173–175.
27. Browning, G.M.; Parish, C.L.; Glass, J. A Method for Determining the Use and Limitations of Rotation and Conservation Practices in the Control of Soil Erosion in Iowa. J. Am. Soc. Agron. **1947**, *39* (1), 65–73.
28. Musgrave, G.W. The Quantitative Evaluation of Factors in Water Erosion—A First Approximation. J. Soil and Water Conserv. **1947**, *2* (3), 133–138, 170.

EROSION BY WATER, HYBRID MODELS

Jeffrey G. Arnold
Kevin W. King
United States Department of Agriculture-Agricultural Research Service (USDA-ARS), Temple, Texas, U.S.A.

Jimmy R. Williams
Texas A&M University, Temple, Texas, U.S.A.

INTRODUCTION

Each year, soil erosion causes billions of dollars of damage in the world (1). Simulation models are among the best tools available for analyzing soil erosion and conservation management. Models can project the consequences of alternative management, conservation planning, or policy-level activities and substantially reduce the cost of managing soil and water resources. Numerous models have been developed to estimate soil erosion and the impact of land management on erosion. Early models such as the universal soil loss equation (USLE) (2) were empirical and predicted average annual erosion rates using simple nomograph procedures. The USLE and its subsequent improvements have been modified and enhanced (3, 4). These equations have been integrated into comprehensive hybrid models that directly simulate hydrology, plant-growth, nutrient-cycling, and land-management impacts on soil erosion. Then the models use predicted soil erosion to estimate nutrient, pesticide and bacteria transport; change in soil productivity; and carbon dynamics.

EROSION EQUATIONS

The USLE has been used throughout the world to estimate average annual sheet and rill erosion. Several improved forms of the USLE have been developed that include the Onstad–Foster equation (4) and modified universal soil loss equation (MUSLE) (3). The equations are identical except for their energy components. The USLE strictly depends on rainfall as an indicator of erosive energy. The MUSLE and its variations use only runoff variables to simulate erosion and sediment yield. Runoff variables increase the prediction accuracy, eliminate the need for a delivery ratio, and enable the equation to give single-storm estimates of sediment yields. The USLE gives only average annual estimates. The Onstad–Foster equation contains a combination of the USLE and MUSLE energy factors. The empirical erosion model equations are of the form

$$Y = \chi(K)(C)(P)(LS) \tag{1}$$

$$\chi = \mathrm{EI} \quad \text{for USLE}$$

$$\chi = 0.646\mathrm{EI} + 0.45(Qq_{\mathrm{p}})^{0.33} \quad \text{for Onstad–Foster}$$

$$\chi = 1.586(Qq_{\mathrm{p}})^{0.56}A^{0.12} \quad \text{for MUSLE}$$

where Y is the sediment yield (t/ha), K is the soil erodibility factor, C is the crop management factor, P is the erosion control practice fact, LS is the slope length and steepness factor, Q is the runoff volume (mm), q_{p} is the peak runoff rate (mm/h), and A is the watershed area (ha).

COMPREHENSIVE HYBRID MODELS

Numerous hybrid models have been developed including RUSLE (5), GLEAMS (6), and AGNPS (7). Three comprehensive hybrid models with varying spatial scales will be discussed in more detail: field scale—environmental policy impact calculator (EPIC), small watershed scale—agricultural policy/environmental extender (APEX), and river basin scale—soil and water assessment tool (SWAT). All three models use modified versions of USLE for erosion and link with components for plant growth, hydrology, and management. APEX and SWAT also include components for channel sediment routing and pond and reservoir deposition.

The EPIC model (8) was developed in the early 1980s to assess the effect of erosion on productivity. The model has since been expanded and refined to allow simulation of many processes important in agricultural management (9). EPIC is a continuous simulation model (daily time step) that can be used to determine the effect of management strategies on water quality. The drainage area considered by EPIC is generally a field-sized area, up to 100 ha, where weather, soils, and management systems are assumed to be homogeneous. The major components in EPIC are weather simulation, hydrology, erosion, nutrient cycling, pesticide fate, plant growth, soil temperature, tillage, economics, and plant-environment control. EPIC can be used to compare management systems and their effects on sediment nitrogen, phosphorus, and pesticides. The management operations that can be simulated are crop rotations, tillage operations, irrigation scheduling, drainage, furrow diking, liming, grazing, manure handling, and nutrient and pesticide application rates and timing.

The APEX model (10) was developed to extend the EPIC model capabilities to whole farms and small watersheds. In addition to the EPIC functions, APEX has components for routing water, sediment, nutrients, and pesticides across complex landscapes and channel systems to the watershed outlet. A watershed can be subdivided to assure that each subarea is relatively homogeneous in terms of soil, land use, management, etc. The routing mechanisms provide for evaluation of interactions between subareas involving surface runon, return flow, sediment deposition and degradation, and nutrient transport. Water quality in terms of nitrogen (ammonium, nitrate, organic), phosphorus (soluble and adsorbed/mineral and organic), and pesticide concentrations may be estimated for each subarea and at the watershed outlet. APEX is limited in watershed size because of the detailed crop-management system and because daily rainfall is uniformly distributed over the entire watershed.

SWAT (11) is a complex, conceptual model with spatially explicit parameterization. It is a continuous time model that operates on a daily time step. The objective in model development was to predict the impact of management on water, sediment, and agricultural chemical yields in ungauged basins. To satisfy this objective, the model is physically based (as calibration is not possible on ungauged basins) on readily available inputs, computationally efficient to operate on large watersheds, in a reasonable time, and capable of continuous simulation over long time periods necessary for computing the effects of management changes. Major model components include weather, hydrology, erosion/sedimentation, soil temperature, plant growth, nutrients, pesticides, groundwater flow, land management, stream routing, and pond/reservoir routing. SWAT allows a large basin to be subdivided into grid cells or subwatersheds and allows point source inputs and reservoirs to be included in the drainage network. Basins ranging in size from 100 to 100,000 km^2 have been simulated (12). GIS interfaces have been developed using GRASS and ArcView to automate model-input development and spatially display model outputs (13).

MODEL COMPONENTS THAT INTERACT WITH EROSION EQUATIONS

Fig. 1 shows major model components and their interaction with other components and management. The major process components that influence erosion are hydrology, plant growth, residue decay, and management (rotations, tillage, irrigation, fertilizer, and manure).

Hydrology

The primary driver of soil erosion by water is surface water runoff that is caused by rainfall, snowmelt, and irrigation. To simulate surface runoff accurately, an estimate of soil moisture at the start of the runoff event is critical. For continuous time models, the components of the soil water balance must be simulated between events. The main components of the soil water balance are canopy storage, surface runoff, infiltration, snow fall and melt, evapotranspiration (ET), percolation, and lateral soil flow (Fig. 2). The hybrid models discussed here use a daily time step for most water balance processes with the option of sub-daily for the rainfall/runoff process. Hybrid models using empirical erosion models such as MUSLE and Onstad–Foster, require peak rate and total runoff volume from the hydrology routines.

Plant Growth

Simulation of plant growth is crucial for accurate prediction of soil erosion. Plant growth impacts: water extraction from the soil (ET), canopy and residue cover, and nutrient uptake. In most climates, the largest component of the water balance is ET. ET indirectly affects erosion by modifying soil moisture between rainfall events and thus impacting soil erosion. Canopy and residue cover have direct impact on the erosion equations. For the hybrid models, a relationship between cover (canopy and residue) and the USLE C-factor was developed (8).

Residue Decay

Decay of plant residue (including standing, flat, buried, and roots) reduces the amount of residue protecting the soil-surface from erosion. Decay is normally simulated as a function of temperature, soil moisture, and soil carbon, nitrogen and phosphorus contents. The hybrid models simulate nutrient cycling and track nutrient pools in the soil and nutrient concentrations in the plants and residue. The decay rates can then be calculated as a function of the soil C:N and C:P ratios and the composition of plant residue.

E

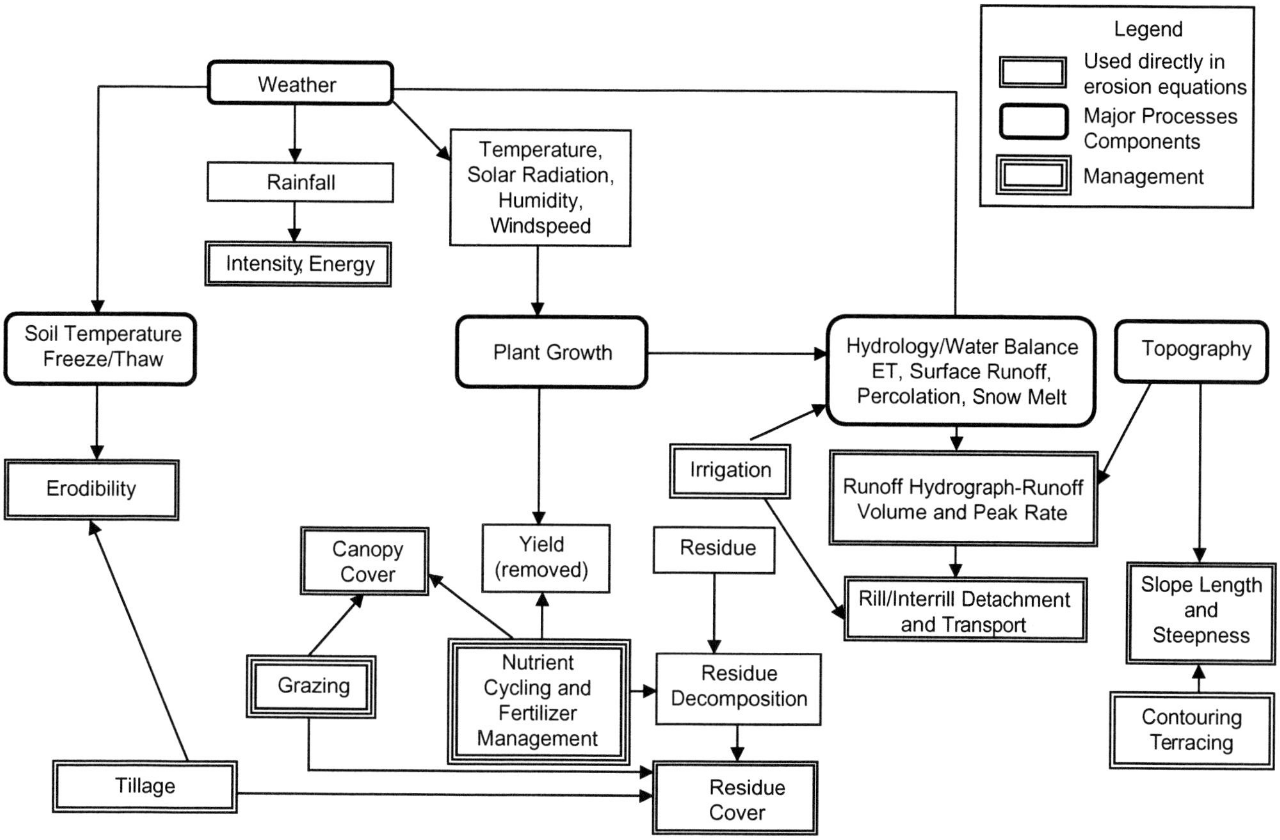

Fig. 1 Major hybrid model components and their interactions with management practices.

Management

Plant-growth management is simulated by most comprehensive models. Crop rotations, planting and harvest dates, length of growing season, cuttings, and grazing all impact the water balance and the amount of canopy and ground cover.

Tillage

Tillage operations remove plant residue from the soil-surface (standing and flat) and incorporate residue into the soil, reducing cover. Tillage also mixes nutrients and changes soil properties such as erodibility, bulk density, and hydraulic conductivity.

Irrigation

Surface irrigation often produces surface runoff and erosion. Irrigation also affects the soil water balance by increasing soil moisture and thus increasing surface runoff and erosion during rainfall events.

Fertilizer and manure management

Application of manure and commercial fertilizers influences soil carbon, nitrogen, and phosphorus and thus influences residue decomposition. Increased fertilizer application can increase plant growth and potentially increase canopy and residue cover. Soil properties such as erodibility and bulk density are altered by fertilizer and manure applications.

MODEL APPLICATIONS

Applications of hybrid soil erosion models include farm planning, erosion–productivity (14), sustainable farming systems, carbon sequestrations (15), climate change (16), and total maximum daily loads.

EPIC Applications

1985 RCA appraisal

The USDA used EPIC to simulate the soil–climate–plant-management processes in agricultural production and to estimate the impact of soil erosion on resource productivity and fertilizer requirements for the 1985

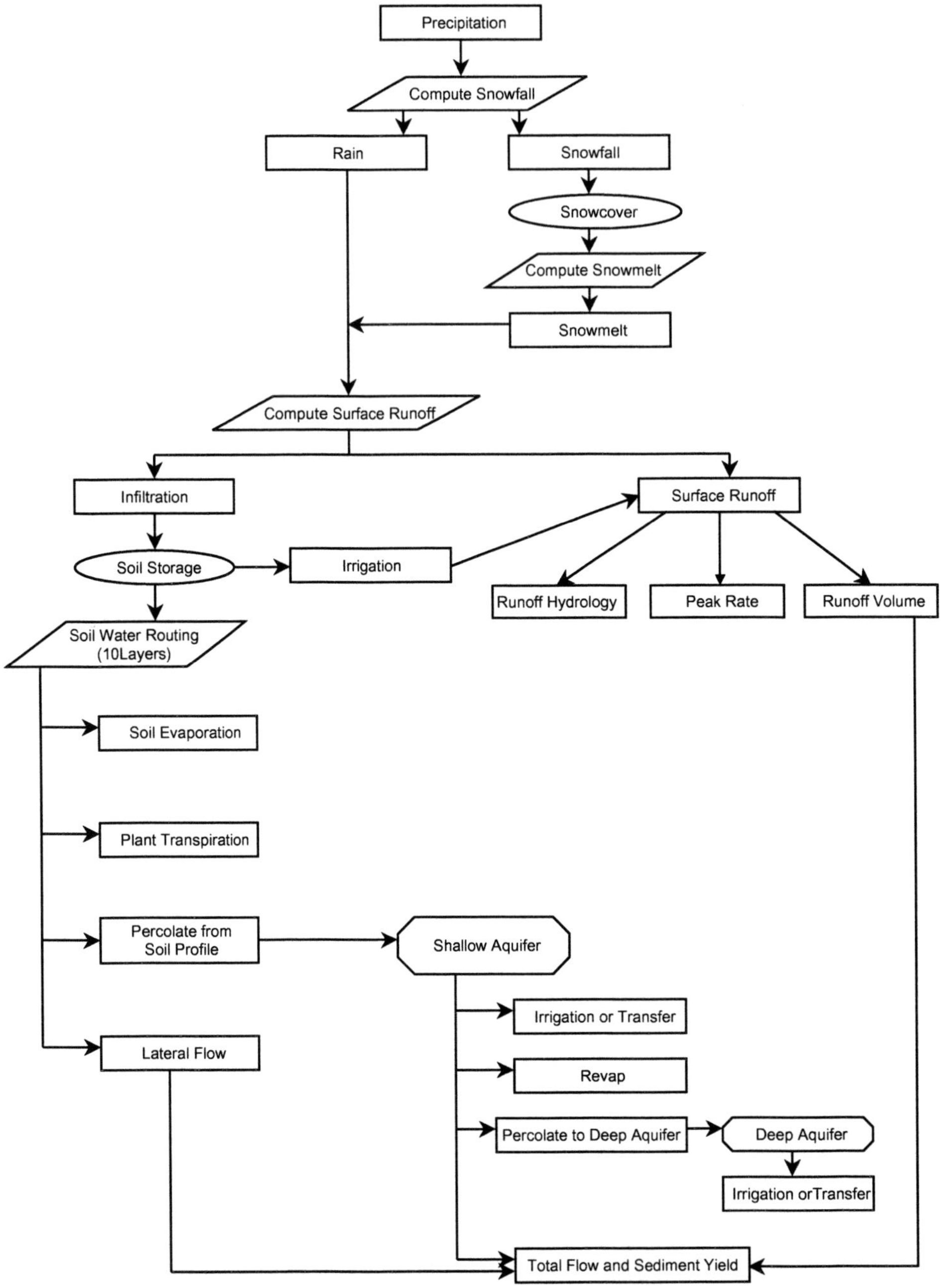

Fig. 2 Major components of the hydrologic balance in hybrid models.

Resources Conservation Act appraisal (14). EPIC results showed that if cropping patterns and the mix of management, tillage, and conservation practices continued for 100 years, sheet and rill erosion will exceed soil loss tolerance on 127 million acres. Annual fertility requirements are estimated to increase by 798 and 672 million pounds of nitrogen and phosphate, respectively.

EPA nutrient movement study

EPA used EPIC to examine the impacts of tillage on nitrogen and phosphorus movement. Four tillage strategies were examined using 100-year EPIC simulations at 100 sites in Illinois. Simulated crop yields closely corresponded to expected yields. No till corn/soybean rotations were found to reduce sediment attached nitrogen and phosporus in runoff relative to conventional till continuous corn (17).

APEX Applications

Animal waste recycling

The APEX model has been used to study animal water disposal systems in the Upper North Bosque watershed in Texas. Various forage production systems and soils were simulated to illustrate the interactions among crops, manure recycling, and nutrient losses.

Buffer analysis

APEX was used to evaluate buffer grass strip efficiency under a variety of soils, climate, slope, and management across the U.S. Relationships of buffer trapping efficiency of sediment and nutrients were developed for use in regional buffer analysis studies and for use in watershed models (18).

SWAT Applications

Hydrologic unit model of the U.S. (HUMUS)

The NRCS used the SWAT model for the 1997 Resource Conservation Assessment. The model was linked to national economic models and used for national policy planning of water supply and quantity of the 18 major river basins of the U.S. Geographic information systems were utilized to integrate SWAT with national soils, land use, and elevation databases (19).

Total maximum daily load (TMDL) analysis

There are approximately 15,000 water bodies in the U.S. identified by EPA as impaired for various uses. For each of these, states must estimate the severity of the problem (develop a TMDL) and determine potential solutions. SWAT has been selected by EPA for use in TMDL analysis and is being used to determine the impact of proposed management practices on attaining water-quality standards (20).

REFERENCES

1. Committee on Conservation Needs and Opportunities, *Soil Conservation: Assessing the National Resource Inventory*; National Academy Press: Washington, DC, 1986; Vol. 1, 114.
2. Wischmeier, W.H.; Smith, D.D. *Predicting Rainfall Erosion Losses, A Guide to Conservation Planning*; Agric. Handbook No. 537, U.S. Dept. Agric: Washington, DC, 1978.
3. Williams, J.R.; Berndt, H.D. Sediment Yield Prediction Based on Watershed Hydrology. Trans. ASAE **1975**, *20* (6), 1100–1104.
4. Onstad, C.A.; Foster, G.R. Erosion Modeling on a Watershed. Trans. ASAE **1975**, *18* (2), 288–292.
5. Renard, K.G.; Farreira, V.A. RUSLE Model Description and Data Base Sensitivity. J. Environ. Qual. **1993**, *22* (3), 458–466.
6. Leonard, R.A.; Knisel, W.G.; Still, D.A. GLEAMS: Groundwater Loading Effects of Agricultural Management Systems. Trans. ASAE **1987**, *30* (5), 1403–1418.
7. Young, R.A.; Onstad, C.A.; Bosch, D.D.; Anderson, W.P. *AGNPS, Agricultural Nonpoint Source Pollution Model—A Watershed Analysis Tool*; Conservation Research Report 35, U.S. Dept. of Agriculture: Washington, DC, 1987.
8. Williams, J.R.; Jones, C.A.; Dyke, P.T. A Modeling Approach to Determining the Relationship Between Erosion and Soil Productivity. Trans. ASAE **1984**, *27*, 129–144.
9. Sharpley, A.N.; Williams J.R. 1990. *EPIC—Erosion/Productivity Impact Calculator: 1. Model Documentation*. U.S. Dept. of Agriculture Tech. Bulletin No. 1768.
10. Williams, J.R.; Arnold, J.G.; Srinivasan, R.; Ramanarayanan, T.S. 1998. APEX: A New Tool for Predicting the Effects of Climate Change and CO_2 Changes on Erosion and Water Quality. In: Boardman, J., Favis Mortlock, D., Eds.; Modelling Soil Erosion by Water, NATO ASI Ser. I(55), 1998, 441–449.
11. Arnold, J.G.; Srinivasan, R.; Muttiah, R.S.; Williams, J.R. Large Area Hydrologic Modeling and Assessment. Part I Model Development. J. Am. Water Resour. Assoc. **1998**, *34* (1), 73–89.
12. Arnold, J.G.; Muttiah, R.S.; Srinivasan, R.; Allen, P.M. Regional Estimation of Base Flow and Groundwater Recharge in the Upper Mississippi Basin. J. Hydrol. **2000**, *227* (2000), 21–40.
13. DiLuzio, M.; Srinivasan, R.; Arnold, J.G. Watershed Oriented Non-Point Pollution Assessment Tool. *7th Annual Conference on Computers in Agriculture*; ASAE: Orlando, FL, 1998; 7.
14. Putnam, J.; Williams, J.R.; Sawyer, D. Using the Erosion-Productivity Impact Calculator (EPIC) Model to Estimate the Impact of Soil Erosion for the 1985 RCA Appraisal. J. Soil Water Conserv. **1985**, *43* (4), 321–326.
15. Izaurralde, R.C.; Haugen-Kozyra, K.H.; Jans, D.C.; McGill, W.B.; Grant, R.F.; Hiley, J.C. Soil Dynamics: Measurement, Simulation and Site-To-Region Scale-Up. Adv. Soil Sci. **2001**, (in press).
16. Rosenberg, N.J.; Epstein, D.L.; Wang, D.; Vail, L.; Srinivasan, R.; Arnold, J.G. Possible Impacts of Global Warming on the Hydrology of the Ogallala Aquifer Region. J. Climate **1999**, *42*, 677–692.
17. Phillips, D.L.; Hardin, P.D.; Benson, V.W.; Baglio, J.V. Using the National Resources Inventory and the EPIC Model to Evaluate the Impact of Alternative Agricultural Management Practices in Illinois. J. Soil Water Conserv. **1993**, *48* (5), 449–457.
18. Rodriguez, O.; Williams, J.R.; Arnold, J.G. 2001. Buffer Grass Strip Efficiency Evaluated with APEX as Input for Regional Applications (Draft).
19. Srinivasan, R.S.; Arnold, J.G.; Jones, C.A. Hydrologic Modeling of the United States with the Soil and Water Assessment Tool. Water Resour. Dev. **1998**, *14* (3), 315–325.
20. Saleh, A.; Arnold, J.G.; Gassman, P.W.; Hauck, L.W.; Rosenthal, W.D.; Williams, J.R.; McFarland, A.M.S. Application of SWAT for the Upper North Bosque Watershed. Trans. ASAE **2000**, *43* (5), 1077–1087.

EROSION BY WATER, MODELING

Calvin W. Rose
Griffith University, Nathan, Queensland, Australia

INTRODUCTION

Modeling soil erosion by water was initially developed in the context of agricultural productivity and land degradation (1–4). Modeling was subsequently stimulated by off-site concerns for the water quality of streams, lakes, and oceans (5, 6). Soil erosion by water is also recognized as one process in the (mostly) descriptive long-term, land-forming models of geomorphologists and geographers (7, 8), which include mass movement and other forms of erosion. More recently, concern that soil erosion might ultimately expose buried contaminants has encouraged modeling possible landscape change over thousands of years (9, 10).

Early erosion models used statistical techniques to summarize large bodies of data, giving the effect of environmental and agricultural management factors on net soil loss from agricultural plots (1, 2). The Revised Universal Soil Loss Equation (RUSLE) (11), a familiar example, developed from the USLE, which summarizes erosion experiments in U.S. field plots (1). RUSLE allows prediction of annual expected net soil loss from a plot of uniform slope throughout an entire crop rotation, using databases for climate, vegetation, and field-management operations which are available for the United States. The annual soil loss per unit area, A, is given by

$$A = RKLSCP$$

where R is a rainfall-runoff factor, K is soil erodibility factor, L is slope length factor, S is slope steepness factor, C is cover management factor, and P is support practices factor.

Later soil-erosion modeling developments primarily use deterministic models, which seek to represent the physical, hydrological, and sediment transport–deposition processes involved.

SOIL EROSION PROCESSES

Water-induced soil erosion involves downslope flow of water with a sediment concentration that depends on the net balance of two rapid but opposing types of processes (12–14). Sediment removed from the soil surface adds to the sediment concentration, a process countered by deposition, which continuously returns sediment to the soil surface at a rate depending predominantly on the size and density of the sediment (13). The rate of soil removal depends on an interaction of rainfall and overland-flow characteristics with soil-strength-related properties, which can be affected by subsurface hydrology (15–21).

Raindrop impact on unprotected soil is commonly the initial soil-removal mechanism (13, 22). However, when an excess of rainfall over infiltration rate occurs, the resulting overland flow both transports sediment and exerts a shear stress on the soil, τ. This shear stress arises chiefly from turbulence (15, 17, 18) and its rate of action is called the stream power, Ω, where

$$\Omega = \tau V$$

with V the velocity of flow.

For cultivated soil, overland flow-driven erosion processes usually become dominant over rainfall-driven processes when Ω exceeds approximately 0.2 $W\,m^{-2}$ (23).

DYNAMIC STOCHASTIC AND DETERMINISTIC MODELS OF SOIL EROSION

The rates of sediment removal from, and return to, the soil surface in deposition can be described stochastically (24). The probability density of resting on the soil surface is a function of time alone, but the probability to describe motion involves both space and time. This leads to two simultaneous partial differential equations, one related to the deposited or stationary mass, the second to sediment in motion (24). These two equations are formally identical with the two partial differential equations derived using deterministic mass-conservation considerations (16, 24).

Fig. 1 gives a deterministic representation of rainfall-driven erosion, recognizing that soil consists of a range of particle sizes. For a typical size class (i), (e_i) is the rate of detachment from the original soil matrix. Deposition at rate (d_i) from sediment in the overland flow forms a deposited layer of mass per unit area (m_{di}), from which

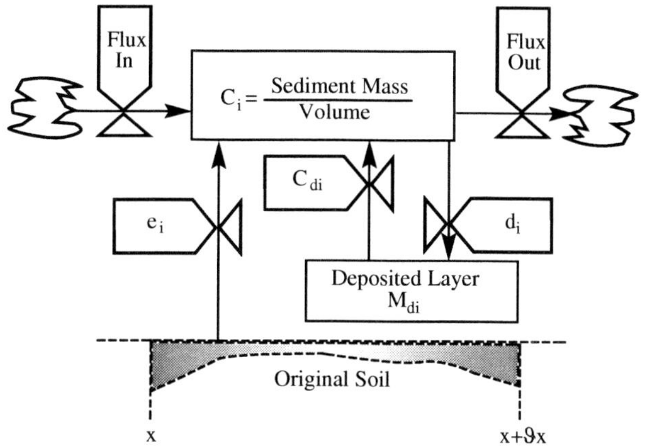

Fig. 1 Forrester-style flow diagram describing the interaction of rainfall detachment and re-detachment processes among the original soil, the layer formed by deposition, and the sediment flux. The water layer with sediment concentration c_i (for size class i), and the deposited layer of mass per unit area m_{di} are shown artificially elevated for clarity. Rates of processes exchanging sediment are shown by valve symbols. (Adapted from Ref. 16.)

sediment is redetached at rate e_{di}. Let c_i denote the concentration of sediment in i at any time (t) in depth of water (D), with volumetric flow per unit width (q) in the x-direction. Then mass conservation of sediment of i flowing over the soil surface requires that (16, 25)

$$\frac{\partial}{\partial t}(Dc_i) + \frac{\partial}{\partial x}(qc_i) = e_i + e_{di} - d_i \tag{1}$$

Mass conservation of sediment in the deposited layer (Fig. 1) requires that

$$\frac{\partial}{\partial t}(m_{di}) = d_i - e_{di} \tag{2}$$

Solution of Eqs. 1 and 2 (or corresponding equations when flow-driven processes are also important) involves numerical or approximate analytical methods (25–28). A major purpose of such models is to test the physical understanding of basic dynamic soil erosion and deposition processes. Dynamic models also help in choosing approximations for simpler steady-rate models, which are more suitable for application in field-predictive applications. Use of steady-rate models avoids the need to solve partial differential equations in space and time.

Especially at long time scales and catchment space scales, there is interest in the change in land surface elevation that results from net erosion or net deposition, an issue ignored in equations such as Eqs. 1 and 2, but crucial in seeking to describe landscape evolution (7, 8). A quantitative general model of channel network growth and hillslope evolution has been developed whose output resembles many observed landscape features (29–32). In this model, surface denudation is represented as the sum of fluvial transport processes, dependent on discharge and slope, and a slower analog of diffusive processes dependent on slope alone. Digital terrain modeling is used to determine drainage areas, and the landform is adjusted through time in response to erosion processes represented.

Three physically based steady-rate models which have received significant application in land-use contexts will now be outlined.

STEADY-RATE MODELS OF WATER EROSION

Water Erosion Prediction Project (WEPP)

Developed by the U.S. Department of Agriculture, the hillslope version of WEPP is a comprehensive, continuous simulation model that draws on U.S. databases describing climate, soils, tillage, and crop parameters (33, 34). Runoff is generated from rainfall input using an infiltration equation (34). The peak predicted runoff rate is the steady-rate value assumed for the erosion event.

Sediment sourced from the interrill area is delivered to rills, and net erosion within rills can add to transported sediment. If predicted total sediment transported exceeds the maximum transport limit, a separate equation acknowledges that net deposition will occur, representing a mass conservation of transported sediment. A flow chart for the WEPP profile model is given in Fig. 2.

U.S. databases provide the 64 parameters needed to run the model. However, experience outside the United States indicates that infiltration and erodibility parameters, in particular, must be obtained from in-country experiments if predictions of soil and water loss are to be of useful accuracy (35). More-recent versions of the WEPP model allow for simulation of small watersheds (36).

Griffith University Erosion System Template (GUEST)

Program GUEST (37) is based on physical theory (13, 16–18, 38). The simpler of two options in GUEST defines an empirical erodibility parameter β and uses the mean velocity of sediment settling as a soil characteristic (13). GUEST also addresses rilling in a physical manner (13, 38). If total runoff amount alone is measured, an effective runoff rate for an erosion event can be calculated from rainfall rate (39, 40). If β has been previously determined

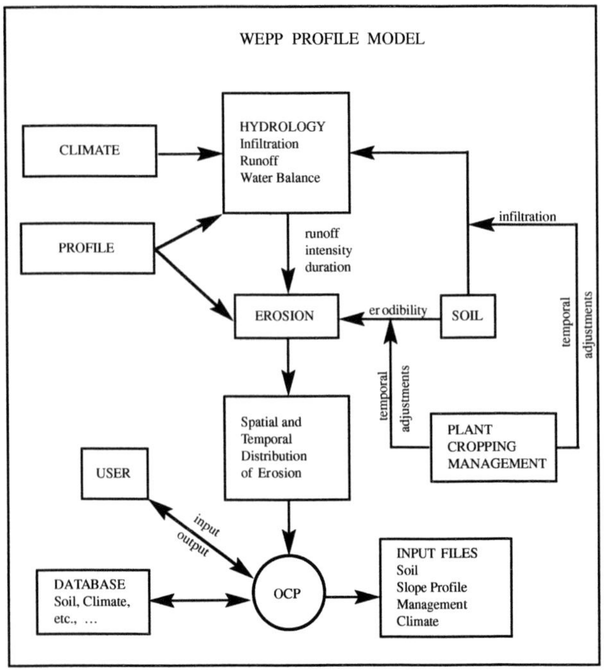

Fig. 2 Flow chart illustrating the components of the hillslope profile version of WEPP. (Adapted from Ref. 34.)

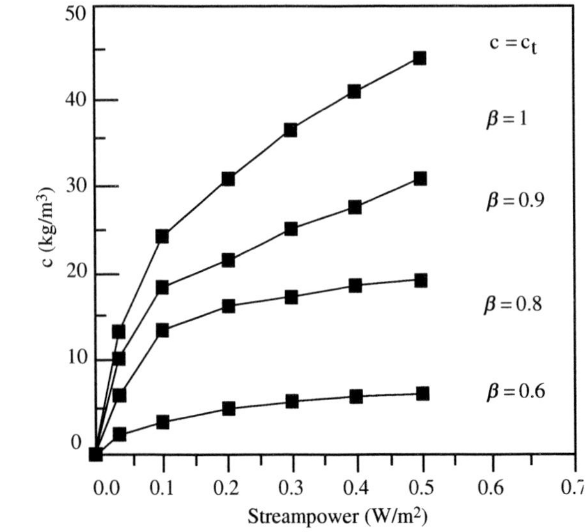

Fig. 3 The relationship between sediment concentration, c, and streampower for various values of the erodibility parameter β given in Eq. 3. For β = 1 (the upper curve), $c = c_t$, the sediment concentration at the transport limit. (After Ref. 14, where assumed values of other parameters are given.)

or can be estimated, then the program can be used to predict soil loss.

GUEST assumes that a constant fraction of the excess stream power ($\Omega - \Omega_0$, where Ω_0 is a threshold value) is used in flow-driven erosion (13, 18). The maximum sediment concentration or transport limit (c_t) is derived theoretically (13, 18). The erodibility parameter β is given by

$$\beta = (ln\,\bar{c})/(ln\,\bar{c}_t) \tag{3}$$

where $\bar{c}$ is the average sediment concentration, and $\bar{c}_t$ the average value of c_t during an erosion event. How the relation between $\bar{c}$ and Ω is controlled by the value of β is illustrated in Fig. 3.

European Soil Erosion Model (EUROSEM)

EUROSEM (41, 42) describes erosion for single events on single-slope segments (as does GUEST), uses an experimentally based relationship for the transport limit (43), and is linked to the KINEROS hydrologic model (44). The numerical-solution procedure in KINEROS provides the basis for sediment discharge calculation. Net deposition occurs if the calculated sediment concentration exceeds the transport limit. Sediment removed by rainfall impact is delivered to rills, which transport the sediment (as in WEPP).

EUROSEM represents canopy protection against rainfall impact, adding drop impact due to throughfall. The effect of different cultivation practices on surface water storage is also represented.

REFERENCES

1. Wischmeier, W.H.; Smith, D.D. *Predicting Rainfall Erosion Losses—A Guide to Conservation Planning*, Agriculture Handbook No. 537; U.S. Department of Agriculture: Washington, DC, 1978.
2. Romkens, M.J.M. The Soil Erodibility Factor: A Perspective. In *Soil Erosion and Conservation*; El-Swaify, S., Moldenhauer, W., Lo, A. Eds.; Soil Conservation Society of America: Ankeny, Iowa, 1985; 237–347.
3. El-Swaify, S.; Moldenhauer, W.; Lo, A., Eds. *Soil Erosion and Conservation*; Soil Conservation Society of America: Ankeny, Iowa, 1985.
4. Hurni, H.; Tato, K., Eds. *Erosion, Conservation, and Small-Scale Farming*; Geographica Bernesia Wadsworth Publishing Company: Missouri, 1992.
5. Knisel, W., Jr. (Ed.) *CREAMS: A Field-Scale Model for Chemicals, Runoff and Erosion from Agricultural Management Systems*, Conser. Res. Rep. No. 26; U.S. Department of Agriculture: Washington, DC, 1980.
6. Ghadiri, H.; Rose, C.W. *Modeling Chemical Transport in Soils: Natural and Applied Contaminants*; Lewis Publishers: London, 1992.

7. Kirkby, M.J. Hillslope Process-Response Models Based on the Continuity Equation. In *Slopes: Form and Process, Spec. Publ. 3*; Institute of British Geographers: London, 1971; 15–30.
8. Huggett, R.J. Dissipative Systems: Implications for Geomorphology. Earth Surface Processes and Landforms **1998**, *13*, 45–49.
9. Evans, K. Methods for Assessing Mine Site Rehabilitation Design for Erosion Impact. Australian Journal of Soil Research **2000**, *38* (2), 231–247.
10. Hancock, G.; Evans, K.; Willgoose, G.; Moliere, D.; Saynor, M.; Loch, R. Medium-Term Erosion Simulation of an Abandoned Mine Site Using the SIBERIA Landscape Evolution Model. Australian Journal of Soil Research **2000**, *38* (2), 249–263.
11. Renard, K.; Foster, G.; Weesies, G.; Porter, J. RUSLE: Revised Universal Soil Loss Equation. J. Soil and Water Conservation **1991**, *46*, 30–33.
12. Rose, C.W. Developments in Soil Erosion and Deposition Models. In *Advances in Soil Science*; Stewart, B.A., Ed.; Springer-Verlag: New York, 1985; Vol. *2*, 1–63.
13. Rose, C.W. Erosion and Sedimentation. In *Hydrology and Water Management in the Humid Tropics—Hydrological Research Issues and Strategies for Water Management*; Bonell, M., Hufschmidt, M.M., Gladwell, J.S., Eds.; Cambridge University Press: Cambridge, 1993; 301–343.
14. Rose, C.W. Research Progress on Soil Erosion Processes and a Basis for Soil Conservation Practices. In *Soil Erosion Research Methods*, 2nd Ed.; Lal, R., Ed.; Soil and Water Conservation Society: Ankeny, Iowa, 1994; 159–178.
15. Foster, G.R. Modeling the Erosion Process. In *Hydrologic Modeling of Small Watersheds*; Hann, C.T., Ed.; Am. Soc. Agr. Eng. Monograph No. 5: St. Joseph, Michigan, 1982; 297–379.
16. Hairsine, P.B.; Rose, C.W. Rainfall Detachment and Deposition: Sediment Transport in the Absence of Flow-Driven Processes. Soil Sci. Soc. Am. J. **1991**, *55*, 320–324.
17. Hairsine, P.B.; Rose, C.W. Modeling Water Erosion Due to Overland Flow Using Physical Principles: I. Uniform Flow. Water Resources Research **1992**, *28*, 237–243.
18. Hairsine, P.B.; Rose, C.W. Modeling Water Erosion Due to Overland Flow Using Physical Principles: II. Rill Flow. Water Resources Research **1992**, *28*, 245–250.
19. Proffitt, A.P.B.; Hairsine, P.B.; Rose, C.W. Modeling Soil Erosion by Overland Flow: Application Over a Range of Hydraulic Conditions. Trans. ASAE **1993**, *36*, 1743–1753.
20. Rose, C.W.; Hairsine, P.B.; Proffitt, A.P.B.; Misra, R.K. Interpreting the Role of Soil Strength in Erosion Processes. Catena Supp. **1990**, *17*, 153–165.
21. Huang, C.; Gabbard, D.S.; Norton, L.D.; Laflen, J.M. Effects of Hillslope Hydrology and Surface Condition on Soil Erosion. In *Advances in Geology 31*; Rohdenburg, M., Ed.; Catena Verlag GMBH: Reiskirchen, Germany, 1998; 257–262.
22. Proffitt, A.P.B.; Rose, C.W.; Hairsine, P.B. Rainfall Detachment and Deposition: Experiments with Low Slopes and Significant Water Depths. Soil Sci. Soc. Am. J. **1991**, *55*, 325–332.
23. Proffitt, A.P.B.; Rose, C.W. Soil Erosion Processes I. The Relative Importance of Rainfall Detachment and Runoff Entrainment. Aust. J. Soil Res. **1991**, *29*, 671–683.
24. Lisle, I.G.; Rose, C.W.; Hogarth, W.L.; Hairsine, P.B.; Sander, G.C.; Parlange, J.-Y. Stochastic Sediment Transport in Soil Erosion. Journal of Hydrology **1997**, *204*, 217–230.
25. Sander, G.C.; Hairsine, P.B.; Rose, C.W.; Cassidy, D.; Parlange, J.-Y.; Hogarth, W.L.; Lisle, I.G. Unsteady Soil Erosion Model, Analytical Solutions and Comparison with Experimental Results. Journal of Hydrology **1996**, *178*, 351–367.
26. Rose, C.W.; Hogarth, W.L.; Sander, G.; Lisle, I.; Hairsine, P.B.; Parlange, J.-Y. Modeling Processes of Soil Erosion by Water. Trends in Hydrology **1994**, *1*, 443–452.
27. Rose, C.W.; Hogarth, W.L. Process-Based Approaches to Modeling Soil Erosion. In *Modeling Soil Erosion by Water*; Boardman, J., Favis-Mortloch, D., Eds.; Springer: Berlin, 1998; 259–270.
28. Parlange, J.-Y.; Hogarth, W.L.; Rose, C.W.; Sander, G.C.; Hairsine, P.; Lisle, I. Addendum to Unsteady Soil Erosion Model. Journal of Hydrology **1999**, *217*, 149–156.
29. Willgoose, G.R.; Bras, R.L.; Rodriguez-Iturbe, I. A Physically Based Coupled Network Growth and Hillslope Evolution Model. 1. Theory. Water Resources Research **1991**, *27*, 1671–1684.
30. Willgoose, G.R.; Bras, R.L.; Rodriguez-Iturbe, I. A Physically Based Coupled Network Growth and Hillslope Evolution Model. 2. Applications. Water Resources Research **1991**, *27*, 1685–1696.
31. Willgoose, G.R.; Bras, R.L.; Rodriguez-Iturbe, I. A Physical Explanation of an Observed Link Area-Slope Relationship. Water Resources Research **1991**, *27*, 1697–1702.
32. Willgoose, G.R.; Bras, R.L.; Rodriguez-Iturbe, I. Results from a New Model of River Basin Evolution. Earth Surface Processes and Landforms **1991**, *16*, 237–254.
33. Nearing, M.A.; Foster, G.R.; Lane, L.J.; Finkner, S.C. A Process Based Erosion Model for USDA Water Erosion Prediction Project Technology. Trans. ASAE **1989**, *32*, 1587–1593.
34. Lane, L.J.; Nearing, M.A. *USDA Water Erosion Prediction Project: Hillslope Profile Model Documentation*; NSERL Report No. 2, National Soil Erosion Laboratory, USDA-ARS: West Lafayette, Indiana, 1989.
35. Yu, B.; Ciesiolka, C.A.A.; Rose, C.W.; Coughlan, K.J. A Validation Test of WEPP to Predict Runoff and Soil Loss from a Pineapple Farm on a Sandy Soil in Subtropical Queensland, Australia. Aust. J. Soil Res. **2000**, *38*, 537–554.
36. Flanagan, D.C.; Nearing, M.A., Eds. *USDA Water Erosion Prediction Project: Hillslope Profile and Watershed Model Documentation*, NSERL Report No. 10; National Soil Erosion Research Laboratory, USDA-ARS: West Lafayette, Indiana, 1995.
37. Misra, R.K.; Rose, C.W. Application and Sensitivity Analysis of Process-Based Erosion Model GUEST. European Journal of Soil Science **1996**, *47*, 593–604.
38. Ciesiolka, C.A.A.; Coughlan, K.J.; Rose, C.W.; Escalante, M.C.; Hashim, G.M.; Paningbatan, E.P.; Sombatpanit, S. Methodology for a Multi-Country Study of Soil Erosion Management. Soil Technology **1995**, *8*, 179–192.

39. Yu, B.; Rose, C.W.; Ciesiolka, C.A.A.; Coughlan, K.J.; Fentie, B. Toward a Framework for Runoff and Soil Loss Prediction Using GUEST Technology. Aust J. Soil Res. **1997**, *35*, 1191–1212.
40. Yu, B.; Rose, C.W. Application of a Physically Based Soil Erosion Model, GUEST, in the Absence of Data on Runoff Rate. I. Theory and Methodology. Aust. J. Soil Res. **1999**, *37*, 1–11.
41. Morgan, R.P.C.; Quinton, J.N.; Rickson, R.J. *EUROSEM: Documentation Manual*; Silsoe College: Silsoe, United Kingdom, 1992.
42. Morgan, R.P.C. The European Soil Erosion Model: An Update on Its Structure and Research Base. In *Conserving Soil Resources, European Perspectives*; Rickson, R.J., Ed.; CAB International: Wallingford, 1994.
43. Govers, G. Empirical Relationships on the Transporting Capacity of Overland Flow. International Association of Hydrological Sciences Publication **1990**, *189*, 45–63.
44. Woolhiser, D.A.; Smith, R.E.; Goodrich, D.C. *KINEROS: A Kinematic Runoff and Erosion Model: Documentation and User Manual*, USDA Agricultural Research Service ARS-677, 1990.

EROSION BY WATER, PROCESS-BASED MODELING

E

M.A. Nearing
Purdue University, West Lafayette, Indiana, U.S.A.

INTRODUCTION

Soil erosion models can play a critical role in addressing problems associated with soil protection and conservation. They are used for assessment and inventory purposes when financial and time-costs of obtaining useful soil erosion measurements are prohibitive. They provide assessments of both on-site loss of topsoil and off-site delivery of sediments from fields and catchments. Thus they have value for addressing both declining soil productivity on agricultural lands and problems associated with nonpoint source pollution. Models are used for purposes of conservation planning, primarily for selecting the most appropriate conservation measures from a range of options for a particular field or geographical region. Models also are used with increasing frequency by governmental agencies at all levels to set regulations for erosion control practices for agriculture, construction, and forestry operations. Finally, models are used to increase and synthesize our knowledge of soil erosion and conservation science.

Historically models of soil erosion by water have been primarily either empirically based or process-based (sometime referred to as physically based). The first models of soil erosion were empirically based. The prime example of the empirically based model is the Universal Soil Loss Equation (USLE) (1, 2). More recent models have been based on equations which describe the physical, biological, and/or chemical processes which cause or affect soil erosion. It is important to understand that both the process-based and the empirically based models possess a major empirical component, in the sense that the constitutive equations use parameters based on experimental data.

PROCESS-BASED MODELS

Process-based erosion models address soil erosion on a relatively fundamental level using mass balance differential equations for describing sediment continuity on a land surface. The fundamental equation for mass balance of sediment in a single direction on a hillslope profile is given as:

$$\partial(cq)/\partial x + \partial(ch)/\partial t + S = 0 \quad (1)$$

where c (kg/m^3) is sediment concentration, q (m^2/s) is unit discharge of runoff, h (m) is depth of flow, x (m) is distance in the direction of flow, t (s) is time, and S [kg/(m^2 s)] is the source/sink term for sediment generation. Eq. 1 is an exact one-dimensional equation. It is the starting point for development of all existing process-based models. The differences in various erosion models are primarily: a) whether the partial differential with respect to time is included, and b) differing representations of the source/sink term, S. If the partial differential term with respect to time is dropped then the equation is solved for the steady state, whereas the representation of the full partial equation represents a fully dynamic model. The source/sink term for sediment, S, is generally the greatest source of differences in soil erosion models. It is this term that may contain elements for soil detachment, transport capacity terms, and sediment deposition functions. It is through the source/sink term of the equation that empirical relationships and parameters are introduced.

One of the earliest process-based erosion models was presented by Meyer and Wischmeier (3). This model considered soil detachment by rain and by runoff, and then compared the sediment load generated from this detachment to the sediment transport capacity of the flow. If sediment load exceeded sediment transport capacity, then deposition was calculated. The ideas from this early work were expanded and revised by Foster and Meyer (4), and ultimately were worked into the Chemicals, Runoff, and Erosion from Agricultural Systems (CREAMS) model (5), which is a hybrid model in the sense that it uses process-based equations with parameters from the USLE. Experience from developing CREAMS then led to the development of the fully process-based Water Erosion Prediction Project (WEPP) model (6).

CONTINUOUS SIMULATION MODELS

Process-based models for soil erosion by water may be either continuous simulation or event models. Continuous simulation models predict soil erosion for series of individual storms, and have auxiliary components for updating system parameters between storms. Event models simply predict erosion for individual storm events, but require the user to provide system information for each event. As such, the continuous simulation model is useful for analyzing the effects of crop management systems on erosion, whereas the event model is relatively limited in that regard.

The Water Erosion Prediction Project (WEPP) model is an example of a continuous simulation model. WEPP includes nine major components: climate, infiltration, water balance, winter process and snowmelt erosion, plant growth and residue decomposition, irrigation, surface runoff, rainfall erosion, and channel routing for watersheds (6). System parameters which are updated on a daily basis within the continuous simulation include soil moisture, soil density, soil resistance to erosion, plant canopy, surface residue, soil surface roughness, soil surface sealing and crusting, buried residue amounts, frozen soil layers, snow cover, and roots. All of these system parameters define the antecedent conditions for each erosion event.

The WEPP climate model has been tested for erosion and well parameterized for the United States (7). The infiltration component of the hillslope model is based on the Green and Ampt equation as modified by Mein and Larson (8), with the ponding time calculation for an unsteady rainfall (9). The water balance and percolation component of the profile model is based on the water balance component of SWRRB (Simulator for Water Resources in Rural Basins) (10, 11), with some modifications for improving estimation of percolation and soil evaporation parameters. The plant growth component of the model simulates plant growth and residue decomposition for cropland and rangeland conditions. The residue and root decomposition model simulates decomposition of surface residue (both standing and flat), buried residue, and roots for the annual crops specified in the WEPP User Requirements (12) plus perennial crops of alfalfa and grasses. Surface runoff is calculated using a kinematic wave equation. Flow is partitioned into broad sheet flow for interrill erosion calculations and concentrated flow for rill erosion calculations. The erosion component of the model uses a steady-state sediment continuity equation that calculates net values of detachment or deposition rates along the hillslope profile (13). The erosion process is divided into rill and interrill components where the interrill areas act as sediment feeds to the rills, or small channel flows. The model is applicable to hillslopes and small watersheds.

Because the model is based on all of the processes described above, and more, it is possible with a continuous simulation, process-based model to have an enormous array of possible system interactions represented in the simulations. Just to name a very few examples, slope length and steepness effects are functions of soil consolidation, surface sealing, ground residue cover, canopy cover, soil water content, crop type and many other factors. Ground residue cover is a function of biomass production rates, tillage implement types, residue type, soil moisture, temperature and solar radiation, previous rainfall, and many other factors. Rill erosion rates are a function of soil surface roughness, ground cover, consolidation of the soil, soil physical and chemical properties, organic matter, roots, interrill erosion rates, slope, and runoff rates, among other factors. The lists continue ad infinitum. These are interactions which are simply not possible to represent with an empirical model. The continuous simulation, process-based model is quite complex in this sense.

The disadvantage of the process-based model is also the complexity of the model. Data requirements are huge, and with every new datum element comes the opportunity to introduce uncertainty, as a first order error analysis would clearly indicate. Model structure interactions are also enormous in number, and with every structural interaction comes the opportunity for error, as well. In a sense, the goal in using the process-based model is to capture the advantages of the complexity of model interactions, while gaining the accuracy and dependability associated with the simpler empirically based model. This can be done, and was done with the WEPP model, using a combination of detailed sensitivity analyses and calibration of the model to the large database of natural runoff plot information used to develop the USLE. Without the tie between model and database, and without knowledge of the sensitive input variables so as to know where to focus efforts, turning a complex model such as WEPP into a useful conservation tool would not be possible. Thus in a sense, even though WEPP routines are process-based descriptors of various components of the erosional system, ultimately the model must be empirically based on the same type of data as was used to develop the USLE, along with additional experimental data collected specifically for WEPP.

Examples of process-based event models of soil erosion by water include EUROSEM (14), the Hairsine and Rose model (15, 16), and GUEST (17). EUROSEM is a fully dynamic model which simulates the erosion, transport and deposition of sediment over the land surface by interrill and rill processes. The model requires break-point rainfall data, ideally of one-minute resolution, as input along with a detailed description of the soils, slopes and land cover of the watershed. The model computes the interception of rainfall by the plant cover, the generation of runoff as infiltration-excess, soil detachment by raindrop impact and runoff, the transport capacity of runoff and the deposition of sediment. EUROSEM uses the runoff generator and the water and sediment routing routines of KINEROS (18). The model of Hairsine and Rose (15, 16) develops the ideas presented earlier by Rose and colleagues (19, 20), where erosion is calculated as a net balance of instantaneous processes of sediment entrainment, deposition, and re-entrainment. The GUEST model (17) is a practical conservation tool developed from these same concepts.

REFERENCES

1. Wischmeier, W.H.; Smith, D.D. A Universal Soil-Loss Equation to Guide Conservation Farm Planning. Trans. Int. Congr. Soil Sci. **1960**, 418–425.
2. Wischmeier, W.H.; Smith, D.D. *Predicting Rainfall Erosion Losses. A Guide to Conservation Planning*, Agriculture Handbook No. 537; USDA-SEA, U.S. Govt. Printing Office: Washington, DC, 1978.
3. Meyer, L.D.; Wischmeier, W.H. Mathematical Simulation of the Process of Soil Erosion by Water. Trans. Am Soc. Agric. Eng. **1969**, *12* (6), 754–758, 762.
4. Foster, G.R.; Meyer, L.D. A Closed-Form Soil Erosion Equation for Upland Areas. In *Sedimentation (Einstein)*; Shen, H.W., Ed.; Colorado State Univ. Ft. Collins, CO, 1972.
5. Knisel, W.G., Ed. *CREAMS: A Field Scale Model for Chemicals, Runoff, and Erosion from Agricultural Management Systems*, USDA, Conservation Research Report No. 26.; USDA-ARS: Washington, DC, 1980; 643.
6. Flanagan, D.C.; Nearing, M.A. *USDA-Water Erosion Prediction Project: Hillslope Profile and Watershed Model Documentation*, NSERL Report No. 10; USDA-ARS National Soil Erosion Research Laboratory: West Lafayette, IN, 1995; 47097–1196.
7. Baffaut, C.; Nearing, M.A.; Nicks, A.D. Impact of Climate Parameters on Soil Erosion Using CLIGEN and WEPP. Trans. Am. Soc. of Agricultural Eng. **1996**, *39*, 447–457.
8. Mein, R.G.; Larson, C.L. Modeling Infiltration During a Steady Rain. Water Resources Research **1973**, *9* (2), 384–394.
9. Chu, S.T. Infiltration During an Unsteady Rain. Water Resources Research **1978**, *14* (3), 461–466.
10. Williams, J.R.; Nicks, A.D. SWRRB, a Simulator for Water Resources in Rural Basins: An Overview. Proceedings of the Natural Resources Modeling Symposium, Pingree Park, CO, October 16–21, 1983; DeCoursey, D.G., Ed.; USDA-ARS, Washington, DC, 1985; 17–22.
11. Arnold, J.G.; Williams, J.R.; Nicks, A.D.; Sammons, N.B. *SWRRB: A Basin Scale Simulation Model for Soil and Water Resource Management*; Texas A&M University Press: College Station, TX, 1990; 142.
12. Flanagan, D.C.; Livingston, S.J. *USDA-Water Erosion Prediction Project: WEPP User Summary*, NSERL Report No. 11; USDA-ARS National Soil Erosion Research Laboratory: West Lafayette, IN, 1995; 47097–1196.
13. Nearing, M.A.; Foster, G.R.; Lane, L.J.; Finkner, S.C. A Process-Based Soil Erosion Model for USDA-Water Erosion Prediction Technology. Trans. Am Soc. Agric. Eng. **1989**, *32* (5), 1587–1593.
14. Morgan, R.P.C.; Quinton, J.N.; Smith, R.E.; Govers, G.; Poesen, J.W.A.; Auerswald, K.; Chisci, G.; Torri, D.; Styczen, M.E. The European Soil Erosion Model (EUROSEM): A Dynamic Approach for Predicting Sediment Transport From Fields and Small Catchments. Earth Surface Processes and Landforms **1998**, *23*, 527–544.
15. Hairsine, P.B.; Rose, C.W. Modeling Water Erosion Due to Overland Flow Using Physical Principles: 1. Sheet Flow. Water Resource Res. **1992**, *28* (1), 237–243.
16. Hairsine, P.B.; Rose, C.W. Modeling Water Erosion Due to Overland Flow Using Physical Principles: 2. Rill Flow. Water Resource Res. **1992**, *28* (1), 245–250.
17. Misra, R.K.; Rose, C.W. Application and Sensitivity Analysis of Process-Based Erosion Model GUEST. Eur. J. Soil Sci. **1996**, *47*, 593–604.
18. Woolhiser, D.A.; Smith, R.E.; Goodrich, D.C. KINEROS: A Kinematic Runoff and Erosion Model: Documentation and User Manual. USDA Agricultural Research Service ARS-77, 1990.
19. Rose, C.W.; Williams, J.R.; Sander, G.C.; Barry, D.A. A Mathematical Model of Soil Erosion and Deposition Processes: I. Theory for a Plane Land Element. Soil Sci. Soc. Am. J. **1983**, *47*, 991–995.
20. Rose, C.W. Developments in Soil Erosion and Deposition Models. *Advances in Soil Science*; Springer-Verlag: New York, 1985; Vol. 2.

EROSION AND WATER QUALITY

Andreas Klik
University of Agricultural Sciences Vienna—BOKU, Vienna, Austria

INTRODUCTION

The adjustments and modernization of agriculture in the last three decades led to considerable changes of the agricultural landscape. To increase economic efficiency, changes were made mainly in land use, crop production, crop rotation, fertilizer management, and pest control. These human impacts affected the stability of agricultural ecosystems. Soil erosion by wind and water increased in many parts of the world, resulting in negative changes of soil and water resources.

Soil and water quality problems caused by agricultural production practices are receiving increased attention and are now partly perceived by society as important environmental problems. Agricultural land use has a great impact on soil erosion and sediment deposition as well as on water quality. Sediment input is still the most polluting source of aquatic ecosystems worldwide. Severe soil degradation from soil erosion by wind and water can destroy the productive capacity of the soil. Sediments as well as nutrients, pesticides, salts, other trace elements, pathogens, and toxic substances (which are either attached to sediments or dissolved in surface runoff) lead to contamination of surface and groundwater bodies and diminish water quality.

EFFECTS OF SEDIMENTS

Within a watershed, various soil erosion processes including detachment, transport, and deposition occur. Their magnitude mainly depends on slope, soil, topographic, and land use conditions and can greatly vary within an area. Although high erosion rates can be reached in a watershed, not all sediments may reach surface water bodies due to deposition nearby.

According to estimates, on a global scale sediments carried into the oceans increased from 9 billion tons per year before introduction of intensive agriculture, grazing, and other activities to between 23 and 45 billion tons thereafter (1). Of the total 0.9 billion tons of sediments carried by rivers from the continental U.S., about 60% is estimated to come from agricultural lands (2). In the Chinese Loess Plateau, total soil loss in recent years is estimated to be about 2200 million tons annually. Three-quarters of the total soil loss is transported to the lower reaches of the Yellow river (3). Therefore, average sediment concentration of Yellow river is 38 kg m^{-3}, 20 times higher than of the Nile River in Egypt and 38 times higher than of the Mississippi River. During periods of floods, silt content in the Yellow River can rise to more than 650 kg m^{-3}.

Different erosion processes produce different sediment qualities. Sheet or interrill erosion usually produces fine-textured sediment from top soil layers. These layers contain the bulk of agriculturally applied chemicals that attach to and move with the eroded soil. Channel erosion produces sediment from all soil layers incised by this erosion process. Streambanks erode into previously deposited alluvial sediments that normally do not contain significant amounts of agrochemicals.

High concentration of suspended sediments in streams diminishes their recreational uses because pathogens and toxic substances commonly associated with suspended sediments are threats to public health. High sediment concentrations reduce water clarity and the esthetic appeal of streams. Suspended sediment is also harmful to stream biota; it inhibits respiration, diminishes the transmission of light needed for plant photosynthesis, and promotes infections. Sediment depositions can lead to suffocation of benthic organisms.

EFFECTS OF NUTRIENTS

For growth of plants and animals, macronutrients like nitrogen (N), phosphorus (P), potassium (K), sodium (Na), sulfur (S), magnesium (Mg), and calcium (Ca), and micronutrients like boron (B), molybdenum (Mo), chlorine (Cl), iron (Fe), manganese (Mn) and others are essential. When input in surface water bodies, nitrogen and phosphorus especially can increase their biological productivity. Although nitrogen and carbon (C) are essential to the growth of aquatic biota, most attention has focused on P inputs, because of the difficulty in controlling the exchange of N and C between the atmosphere and water, and fixation of atmospheric N by some blue-green

Table 1 Dissolved load from different continents to world's oceans

Continent	Dissolved load (10^6 Mg yr^{-1})
Africa	201
Asia	1592
Europe	425
North and Central Europe	758
Oceania/Pacific Islands	293
South America	603
Total	3872

(From Ref. 4.)

algae. Thus, P is often the limiting element and its control is of prime importance in reducing the accelerated eutrophication of fresh waters. Walling (4, 5) calculated that globally the transport of a dissolved P load from different continents to the oceans is about 4 billion tons per year (Table 1).

In undisturbed rivers, total phosphorus concentrations are generally lower than 25 μg P l^{-1}. Concentrations higher than 50 μg P l^{-1} result from human activities. Investigations show that considerably more than half of all European river stations exceed that level. In Europe the highest levels are found in a band stretching from the northwest across the middle part of the continent, reflecting the intensity of agriculture and livestock production in these regions (6). From 1980 to 1989, average P concentrations in U.S. freshwater ecosystems were 0.1 mg l^{-1} or greater. Concentrations greater than 0.5 mg l^{-1} were especially common in the central and southcentral regions where extensive agricultural use of phosphorus and highly erodible soils combine to create large nonpoint source loadings (7). Nationwide measurements since 1991 showed that annual amounts of total P and total N measured in agricultural streams were equivalent to less than 20% of phosphorus and less than 50% of the nitrogen that was applied to the land (8). In more than one-half of sampled streams and in three-fourths of agricultural and urban streams, average annual concentrations of total phosphorus exceeded the USEPA desired goal for prevention of nuisance plant growth. The highest total N and total P concentrations were found in small streams draining watersheds with large proportions of agricultural or urban land.

Nitrogen

Most of the nitrogen in the soil is stored in soil organic matter. This nitrogen is transformed through mineralization into ammonium ions (NH_4) and released into the soil. Ammonium adsorbs to clay minerals and organic matter and can be transported to surface water attached to sediment or suspended matter. Nitrification transforms ammonium ions to nitrite (NO_2) and nitrate (NO_3). Nitrate is totally soluble and moves freely in solution.

The dominant factors in the loss of nitrogen in runoff are the amount and timing of rainfall, and soil properties. Much of the nitrogen which enters rivers is associated with eroded sediments and eroding soil organic matter or it is dissolved in surface runoff. Runoff during rainfall or snowmelt may have a high concentration of organic N attached to sediments, but is typically low in nitrate concentration. Most rains have a significant wetting period before runoff starts. Soluble fertilizer is carried down into the soil during these first few minutes, so little is lost in the runoff. Exceptional losses can occur if intense rain strikes very suddenly or if rain cannot infiltrate because of frozen or impermeable soil, or if water exfiltrates at the foot of the hillslope due to near-surface hydraulic gradient (9).

High nitrate levels in water are a critical concern for human beings because ingestion of such water has the potential to reduce the oxygen-carrying capacity of the blood. When nitrate is reduced to NO_2^- it oxidizes the Fe of the hemoglobin in the blood and methemoglobin is formed, which cannot carry oxygen (10). Infants are more endangered to suffer these ill effects than adults.

Phosphorus

Soil P exists in inorganic and organic forms. In most agricultural soils, 50–75% of P is inorganic, although this fraction can vary from 10 to 90%. Inorganic P forms are dominated by hydrous sesquioxides and amorphous and crystalline Al and Fe compounds in acidic, noncalcareous soils and by Ca compounds in alkaline, calcareous soils. Organic forms include relatively labile phospholipids, nucleic acids, inositols and fulvic acids, while more resistant forms are comprised of humic acids. The lability of these forms of P is based on the extent to which extractants of increasing acidity or alkalinity, applied sequentially, can dissolve soil P (11).

During erosion processes, P can be transported by surface runoff in dissolved (DP) and particulate (PP) forms. Particulate P includes P attached to eroded soil particles and absorbed by organic matter. Particulate P constitutes the major fraction of P transported from cultivated land with values between 60 and 90% (12, 13). Runoff from grass or forest land carries little sediment and is therefore generally dominated by DP. While DP is immediately available for biological uptake, PP can provide a long-term source of P for aquatic biota (14). Not all P transported by sediments is bioavailable (15). Sediment transported P not presently bioavailable may become bioavailable (16), with the release of P depending on the nature of the sediment and surrounding environment. Under regular pH and aerobic conditions in aquatic systems P is present as secondary or tertiary phosphate and combines with Fe as a heavily soluble substance which is deposited in the sediment. Under anaerobic conditions (= oxygen shortage) iron has a higher affinity to sulfur and combines with it under production of ironsulfide; P is then released and again available as nutrient for aquatic biota (17).

As P is usually the limiting factor in surface water bodies, each increase of P concentration leads to increased growth of undesirable algae and aquatic weeds, oxygen shortages, and subsequently to problems with fisheries, and water for recreation, industry and drinking. Massive surface blooms of cyanobacteria (blue-green algae) lead to fish kills, make drinking water unpalatable, and contribute to the formation of trihalomethane during water chlorination (18). Consumption of algal blooms or of the water-soluble neurotoxins and hepatotoxins released when the algae die can kill livestock and may pose a serious health hazard to humans (19).

Most of the annual nutrient losses measured in surface runoff are associated with sediment losses. Data in Table 2 show that losses of soluble N and P in runoff water are usually much less than the losses of these nutrients transported attached to sediments. Soil erosion protection measures like contouring or no till management are able to diminish nutrient losses by reducing soil loss.

EFFECTS OF PESTICIDES

Pesticides are used to protect food and fiber from damage by weeds, insects, diseases, nematodes, and rodents. When used properly, modern pesticides can perform their functions without causing significant hazards to humans or to the environment. The most serious long-term effects of pesticides on humans are cancer, genetic defects, and birth defects. Effects on nontarget organisms may lead to a decreasing biodiversity.

The application of pesticides in the European Community fell between 1985 and 1995, but this does not necessarily indicate a decrease in environmental impact since the range of pesticides in use has changed (6). In the U.S., the potential for pesticide runoff loss is greatest for watersheds in the Midwest and the Mississippi Embayment (27).

The persistence of a pesticide in the environment and three physical properties of pesticides—soil sorption propensity, solubility, and vapor pressure—determine the tendency of pesticides to move from the application site (28, 29).

Persistence is a pesticide's resistance to decomposition through chemical, photochemical, and microbial action. It is expressed in terms of half-life (in days). It varies from a few hours to months depending on the pesticide and the local environment. Pesticides that persist are more likely to move off-site than less persistent ones. Most pesticides decompose into less toxic chemical forms, but sometimes the metabolites retain pestical properties.

Sorption is the binding of the chemical to the soil. Some pesticides are preferentially transported, adsorbed to entrained sediments. Wauchope (30) found that those pesticides with solubilities below 2 $mg\,l^{-1}$ are transported primarily by sediment. Since soil organic matter is mainly responsible for adsorption of nonionic organic compounds, sorption is commonly evaluated by use of a sorption partitioning coefficient (Koc, expressed in $ml\,g^{-1}$) based on the organic carbon content of the soil.

Water solubility (expressed in $mg\,l^{-1}$) determines how easily it goes into solution with water. Simply being water soluble does not mean that a pesticide will leach into groundwater or runoff into surface water. However, solubility means that if a soluble pesticide somehow gets into water, it will probably stay there and go where the water flows. Although the relation between soil sorptivity and water solubility is complex, it is generally true that given a particular soil adsorption level, the greater the solubility, the more potential there is for losses from the field when it rains.

Vapor pressure (expressed in mmHg) is the measure of a pesticide's tendency to evaporate. Wind speed, air temperature, soil temperature, humidity, and equipment determine the amount of losses to the atmosphere.

Soil, pesticide, and rainfall characteristics influence the timing and amount of pesticide loss (31, 32) and the dominant transporting agent of that pesticide. Pesticide persistence and sorption properties influence the time they stay near their application site and whether they will be adsorbed to sediment or remain in solution phase. Pesticide persistence and sorption determine, in part, the probability of loss by runoff, and in what form the loss will occur (in solution/runoff water or adsorbed to sediment). Percentage of pesticide in runoff and on sediment will depend not only on sorption properties but also on the time elapsed since application (33) and on the processes

Table 2 Nitrogen and phosphorus losses associated with surface runoff (R) and soil loss (S) as well as total losses (T)

Study description	Land use/crop	Fertilization ($kg\,ha^{-1}$)	Runoff ($mm\,yr^{-1}$)	Soil loss ($Mg\,ha^{-1}\,yr^{-1}$)	N losses associated with ($kg\,ha^{-1}$) R	S	T	P losses associated with ($kg\,ha^{-1}$) R	S	T	Reference
Minnesota field plot study	Fallow	Recommended fertilizer application	12.72	37.00	3.43	146.86	150.29	0.19	33.15	33.34	(20)
	Continuous corn		8.61	16.47	2.42	75.56	77.98	0.41	18.19	18.60	
	Corn rotation		4.74	7.54	1.18	34.77	35.95	0.23	8.43	8.66	
	Hay rotation		13.21	3.96	4.01	0.09	4.10	0.66	0.02	0.68	
England field plot study	Contour	No information	7.4	2.43	0.14	5.85	5.99	0.09	3.32	3.41	(21)
potatoes–winter wheat–Winter barley	Up/down slope		9.0	3.51	0.21	5.09	5.30	0.11	4.30	4.41	
Nigeria field plot study	Bare fallow		504.1	232.6	11.5	310	321.5	3.7	20	23.7	(22)
	Maize–maize (mulch)		29.3	0.2	0.6	t	0.6	1.0	t	1.0	
	Maize–maize (plow till)		88.4	7.2	1.6	14	15.6	0.6	1	1.6	
Alabama watershed study	Cotton conventional tillage	67 N, 36 P	182	2.98	4.95	n.a.		0.75	0.24	0.96	(23)
	Cotton conservation tillage	78 N, 25 P	273	1.31	6.39	n.a.		2.62	0.30	2.92	
Mississippi watershed study	Soybeans conventional tillage	45 N at beginning of experiment	532	n.a.	3.47	11.60	15.04	0.48	12.21	12.69	(24)
	Soybeans no till		392	n.a.	3.33	1.54	4.87	2.06	0.92	2.98	
North Carolina field plot study	Conventional till	120 N and 60 P	313.5	6.84	8.49	22.76	31.25	2.51	0.17	2.68	(25)
	Strip till		348.5	3.59	12.81	12.27	25.08	4.63	0.06	4.69	
	No till		284.5	2.04	11.47	8.77	20.24	4.2	0.03	4.23	
Southern China field plot study	Peanut	309 N and 197 P at beginning of experiment	160	9.0	17.2	84.0	101.2	3.1	63.6	66.7	(26)
	Corn–soybean–buckwheat–rape conventional		160.5	6.0	24.6	75.4	100.0	3.4	63.5	66.0	
	Corn–soybean–buckwheat–rape min.till+mulch		72.2	1.16	9.6	13.2	22.8	1.4	10.1	11.5	

t, traces; n.a., not available.

E

Table 3 Pesticide losses associated with runoff

Study description	Compound	Application rate ($kg\,ha^{-1}\,yr^{-1}$)	Location	Crop/treatment	Total seasonal losses by runoff in % of application	Reference
Comparison between field and watershed scale 1997–1998			France	Vineyards		(36)
	Diuron	2.0		No till field	3.28	
				Tilled field	0.92	
				Watershed	0.52	
	Simazine	1.0		No till field	2.96	
				Tilled field	0.54	
				Watershed	0.24	
Field study 1993–1996	Endosulfan	2.25–3.0	Australia	Irrigated cotton	1.5–1.9	(37)
Field study natural rainfall	Atrazine	1.45–4.03	Georgia	Corn	0.2–1.9	(38)
	Paraquat	1.52–15.3		Corn	3.4–10.9	
	Trifluralin	1.12		Soybean	0.1–0.3	
	Cyanazine	1.35–1.61		Corn	0.07–1.0	
	Diphenamid	2.31–3.52		Soybean	0.1–7.2	
Field study, natural rainfall	Alachlor	2.24	Iowa	Corn	0.96 (average)	(39)
	Atrazine	2.24		Corn	2.1 (average)	
	Cyanazine	2.24		Corn	2.1 (average)	
	Fonofos	1.12		Corn	0.36 (avgerage)	
Field study, natural rainfall 1994–1999	Rimsulfuron	0.014	Austria	Corn–winter wheat–sugar beet–summer barley–sunflower rotation		(40)
	Bromoxynil	0.225		Conventional tillage	5.63 (average)	
	Metamitron	0.49		Conservation tillage	1.74 (average)	
	Pendimethalin	1.40		No till	2.58 (average)	
Field study, rainfall simulation	Alachlor	3.36	Illinois	Up/down slope	6.2	(41)
				Contour	2.4	
	Tebufos	1.12		Up/down slope	0.5	
				Contour	0.4	
Field study, rainfall simulation	Atrazine	0.56	Oklahoma	Continuous winter wheat		(42)
				No till	7.9 (average)	
				Chisel tillage	10.5 (average)	
				Disk tillage	7.5 (average)	

Table 4 Pesticides in stream and water bodies resulting from agricultural applications

Watershed stream system	Location	Pesticide residues	Concentration	Comments	Reference
Rivers and agricultural drainage	Japan	CNP	0.01–16.67	Highest levels found above 4 weeks after rice planting and when flood waters released from paddies	(43)
Parena River, 600 km upstream from month	Argentina	Lindane Parathion ABHC	0.009 (average) 0.022 (average) 0.009 (avgerage)	Sediment transported pesticides were positively correlated with discharge as was sediment concentrations	(32, 44)
Rhein River	Germany	Six main herbicides (bentzon, chloridazon, chlortoluron, isoproturon, MCPP, terbuthylazin	$10.9\,t\,yr^{-1}$ total load	$2.7\,g\,ha^{-1}\,yr^{-1}$ from nonpoint source = 0.15% of applied amount; main part from point sources	(45)

controlling runoff and sediment production during rainfall (34, 35). Those pesticides that are strongly sorbed to soil clays and organic matter may be subject to removal by surface runoff or wind. Other pesticides that are weakly sorbed and have high water solubilities may be lost in the dissolved state.

Table 3 summarizes results from field experiments with natural and simulated rainfall about losses of different pesticides associated with runoff. In many studies, the bulk of measured pesticide losses occurred in a single storm soon after pesticide application. Therefore, pesticide runoff losses from simulated rainfall may be higher than from natural rainfall studies. Examples of reports of pesticides in surface waters are shown in Table 4.

SUMMARY AND CONCLUSIONS

In future, it will be difficult to prevent contamination of water bodies by sediments, nutrients, pesticides, salts and other pollutants if soil degradation is not controlled. Worldwide, there already exists various effective soil erosion control practices and measures. They are able to reduce soil erosion by wind and/or water and therefore to prevent negative on-site and off-site impacts. Additionally, predictive tools are available to identify problem areas and to perform and evaluate environmental sound watershed management to achieve sustainable land and soil use. We have to keep in mind that soil and water quality are inherently linked; conserving or enhancing soil quality is a fundamental step to improve the quality of our surface and groundwater bodies. Improvement of the quality of both resources (soil and water) will be essential for future generations.

REFERENCES

1. Judson, S. Whats's Happening to Our Continents? In *Use and Misuse of Earth's Surface*; Skinner, B.J., Ed.; William Kaufmann: Los Altos, CA, 1981; 12–139.
2. National Research Council, *Productive Agriculture and a Quality Environment*; National Academic Press: Washington, DC, 1974.
3. Wen, Dazongh Soil Erosion and Conservation in China. In *World Soil Erosion and Conservation*; Pimentel, D., Ed.; Cambridge University Press: Cambridge, UK, 1993; 63–85.
4. Walling, D.E. Rainfall, Runoff, and Erosion of the Land. A Global Review. In *Energetics of Physical Environment*; Gregory, K.J., Ed.; Wiley: Chichester, 1987; 89–117.
5. Lal, R. Global Overview of Soil Erosion. In *Soil and Water Science: Key to Understanding Our Global Environment*; SSSA Special Publication Number 41; Baker, R.S., Gee, G.W., Rosenzweig, C., Eds.; Soil Science Society of America: Madison, WI, 1994; 39–51.
6. European Environment Agency (EEA). Report of the EEA: Europe's Environment—The Second Assessment. http://www.mem.dk/aarhus-conference/issues/dobris+3/second.htm (Accessed October 2000).
7. Smith, R.A.; Alexander, R.B.; Lanfear, K.J. Stream Water Quality in the Conterminous United States: Status and Trends of Selected Indicators During the 1980s. *National*

Water Summary 1990–91: Hydrologic Events and Stream Water Quality; Water Supply Pap. 2400, U.S. Geological Survey: Washington, DC, 1993.
8. U.S. Geological Survey, *The Quality of Our Nation's Waters—Nutrients and Pesticides. Circular 1225*; U.S. Geological Survey: Washington, DC, 1999; 82.
9. Zheng, F.; Huang, C.; Norton, L.D. How Near-Surface Moisture Gradients Affect Phosphorus and Nitrate Losses. *Soil Erosion Research for the 21st Century*; Proceedings of the International Symposium, Honolulu, Hawaii, Jan 3–5, 2001; American Society of Agricultural Engineers: St. Joseph, MI, 2001; 649–652.
10. Keeney, D.R. Nitrogen Management for Maximum Efficiency and Minimum Pollution. In *Nitrogen in Agricultural Soils*; Stevenson, F.J., Ed.; American Society of Agronomy: Madison, WI, 1982; 22, 605–649.
11. Sharpley, A.N.; Halverson, A.D. The Management of Soil Phosphorus Availability and Its Impact on Surface Water Quality. In *Soil Processes and Water Quality*; Lal, R., Stewart, B.A., Eds.; Advances in Soil Science: Lewis: Boca Raton, FL, 1994; 7–90.
12. Sharpley, A.N.; Smith, S.J.; Jones, O.R.; Berg, W.A.; Coleman, G.A. The Transport of Bioavailable Phosphorus in Agricultural Runoff. J. of Environ. Qual. **1992**, *21*, 30–35.
13. Sharpley, A.N.; Rekolainen, S. Phosphorus in Agriculture and Its Environmental Implications. In *Phosphorus Losses from Soil to Water*; Tunney, H., Carton, O.T., Brookes, P.C., Johnston, A.E., Eds.; CAB International: Wallingford, UK, 1997; 1–54.
14. Sharpley, A.N.; Menzel, R.G. The Impact of Soil and Fertilizer Phosphorus on the Environment. Adv. Agron. **1987**, *4*, 297–324.
15. Sonzogni, W.C.; Chapra, S.C.; Armstrong, D.E.; Logan, T.J. Bioavailability of Phosphorus Inputs to Lakes. J. of Environ. Qual. **1982**, *11*, 555–563.
16. Wildung, R.E.; Schmidt, R.L.; Gahler, A.R. The Phosphorus Status of Eutrophic Lake Sediments as Related to Changes in Limnological Conditions—Total, Inorganic and Organic Phosphorus. J. of Environ. Qual. **1974**, *3*, 133–138.
17. Hartmann, L. *Oekologie und Technik. Analyse, Bewertung und Nutzung von Oekosystemen*; Springer: Berlin, 1992; 269.
18. Kotak, B.G.; Kenefick, S.L.; Fritz, D.L.; Rousseau, C.C.; Prepas, E.E.; Hrudey, S.E. Blue-Green Algal Toxins in Drinking Water Supplies. Research in Alberta. Lake Line **1994**, *14*, 37–40.
19. Martin, A.; Cooke, G.D. Health Risks in Eutrophic Water Supplies. Lake Line **1994**, *14*, 24–26.
20. Burwell, R.E.; Timmons, D.R.; Holt, R.F. Nutrient Transport in Surface Runoff as Influenced by Soil Cover and Seasonal Periods. Soil Sci. Soc. Am. Proc. **1975**, *39*, 523–528.
21. Catt, J.A.; Quinton, J.N.; Rickson, R.J.; Styles, P. Nutrient Losses and Crop Yields in the Wooburn Erosion Reference Experiment. In *Conserving Soil Resources. European Perspectives*; Rickson, R.J., Ed.; CAB International: Wallingford, 1994; 94–104.
22. Lal, R. Soil Erosion and Conservation in West Africa. In *World Soil Erosion and Conservation*; Pimentel, D., Ed.; Cambridge University Press: Cambridge, UK, 1993; 7–25.
23. Soileau, J.M.; Touchton, J.T.; Hajek, B.F.; Yoo, K.H. Sediment, Nitrogen, and Phosphorus Runoff with Conventional- and Conservation-Tillage Cotton in a Small Watershed. J. Soil Water Conserv. **1994**, *49* (1), 82–89.
24. Schreiber, J.D.; Cullen, R.F. Tillage Effects on Surface and Groundwater Quality in Loessial Upland Soybean Watersheds. Trans. ASAE **1998**, *41* (3), 607–614.
25. Reddy, G.B.; Raczkowski, C.W.; Reyes, M.R.; Gayle, G.A. Surface Losses of N, P, and Herbicides from a Long-Term Tillage Study at North Carolina A & T State University. *Soil Erosion Research for the 21st Century*; Proceedings of the International Symposium, Honolulu, Hawaii, Jan 3–5, 2001; American Society of Agricultural Engineers: St. Joseph, MI, 2001; 665–668.
26. Zhang, T.; Wang, X.; Zhang, B. Soil Erosion and Nutrient Cycling on Sloping Upland Ecosystems of Red Soil in Southern China. In *Soil Erosion and Dryland Farming*; Laflen, J.M., Tian, J., Huang, C., Eds.; CRC Press: Boca Raton, FL, 2000; 203–216.
27. United States Department of Agriculture-National Resources Conservation Service, *Water Quality and Agriculture. Status, Conditions, and Trends*; Working Paper #16, USDA-NRCS: Washington, DC, 1997; 20013, 125.
28. Hornsby, A.G.; Wauchope, R.D.; Herner, A.E. *Pesticide Properties in the Environment*; Springer: New York, 1996; 227.
29. Linders, J.B.H.J.; Jansma, J.W.; Mensink, B.J.W.G.; Otermann, K. *Pesticides: Benefaction or Pandora's Box?* Report No. 679101014, National Institut of Public Health and Environmental Protection: Bilthoven, Netherlands, 1994; 204.
30. Wauchope, R.D. The Pesticide Content of Surface Water Draining from Agricultural Fields: A Review. J. of Environ. Qual. **1978**, *7*, 459–472.
31. Leonard, R.A. Herbicides in Surface Waters. In *Environmental Chemistry of Herbicides*; Grover, R., Ed.; CRC Press: Boca Raton, FL, 1988; 45–89.
32. Leonard, R.A. Movement of Pesticides into Surface Waters. In *Pesticides in the Soil Environment: Processes, Impact and Modeling*; Soil Sci. Soc. Am. Book Ser. 2, Cheng, H.H., Ed.; Soil Science Society of America: Madison, WI, 1990; 303–349.
33. Auerswald, K. Seasonal Variation of Erosive Rains in Southern Germany. Zeitschrift Fuer Kulturtechnik und Landentwicklung **1996**, *37* (2), 81–84.
34. Zhang, X.C.; Norton, L.D.; Hickman, M. Rain Pattern and Soil Moisture Content Effects on Atrazine and Matolachlor Losses in Runoff. J. of Environ. Qual. **1997**, *26*, 1539–1547.
35. Truman, C.C.; Steinberger, P.; Leonard, R.A.; Klik, A. Laboratory Determination of Water and Pesticide Partitioning. Soil Sci. **1998**, *163* (7), 556–568.
36. Louchart, X.; Voltz, M.; Andrieux, P.; Moussa, R. Herbicide Transport to Surface Waters at Field and Watershed Scales in a Mediterranean Vineyard Area. J. of Environ. Qual. **2001**, *30*, 982–991.
37. Kennedy, I.R.; Sánchez-Bayo, F.; Kimber, S.W.; Hugo, L.; Ahmad, N. Off-Site Movement of Endosulfan from Irrigated Cotton in New South Wales. J. of Environ. Qual. **2001**, *30*, 683–696.
38. Leonard, R.A.; Langdale, G.W.; Fleming, W.G. Herbicide Runoff from Upland Piedmont Watersheds—Data and

Implications for Modeling Pesticide Transport. J. of Environ. Qual. **1979**, *8*, 223–229.
39. Baker, J.L.; Johnson, H.P. The Effect of Tillage Systems on Pesticides in Runoff from Small Watersheds. Trans. ASAE **1979**, *22*, 554–559.
40. Klik, A.; Zartl, A.S.; Rosner, J. Tillage Effects on Soil Erosion, Nutrient, and Pesticide Transport. *Soil Erosion Research for the 21st Century*; Proceedings of the International Symposium, Honolulu, Hawaii, Jan 3–5, 2001; American Society of Agricultural Engineers: St. Joseph, MI, 2001; 71–74.
41. Kenimer, A.L.; Mitchell, J.K.; Felsot, A.S.; Hirschi, M.C. Pesticide Formulation and Application Technique Effects on Surface Pesticide Losses. Trans. ASAE **1997**, *40* (6), 1617–1622.
42. Basta, N.T.; Huhnke, R.L.; Stiegler, J.H. Atrazine Runoff from Conservation Tillage Systems: A Simulated Rainfall Study. J. Soil Water Conserv. **1997**, *52* (1), 44–48.
43. Suzuki, M.; Yamato, Y.; Akiyama, T. Fate of Herbicide 2,4,6-Trichlorphenyl-4-Nitrophenyl Ether in Rivers and Agricultural Drainages. Water Resour. **1978**, *12*, 777–782.
44. Lenardon, A.M.; Hevia, M.I.M.; Fuse, J.A.; DeNochetto, C.B.; Depetris, P.J. Organochlorine and Organophosphorus Pesticides in Paran River, Argentina. Sci. Total Environ. **1984**, *34*, 289–298.
45. Bach, M.; Huber, A.; Frede, H.-G.; Mohaupt, V.; Zullei-Seibert, N. *Schaetzung der Eintraege von Pflanzenschutzmitteln aus der Landwirtschaft in die Oberflaechengewaesser Deutschlands*; Berichte/Umweltbundesamt. Erich Schmidt: Berlin, 2000; 273.

EROSION BY WIND, CLIMATE CHANGE AND

Alan Busacca
Washington State University, Pullman, Washington, U.S.A.

David Chandler
Utah State University, Logan, Utah, U.S.A.

INTRODUCTION

Wind erosion has been driven by cycles of climate change over geologic time, resulting in the transport and accumulation of eolian sediments. As an important process in the Pleistocene Epoch, or Ice Ages, it left conspicuous eolian deposits worldwide. Wind erosion, driven by shorter time-scale changes in climate, has attended the present interglacial period of the last 12,000 years. This time period has also seen the rise of human societies, the development of agriculture, and, in the last 200 years, the technological age in which human population increased from less than 1 billion to more than 6 billion. The methods we use today to study climate change and wind erosion shift at the major glacial–interglacial climatic transition 12,000 years ago. That is, to reconstruct wind erosion events during the Pleistocene Epoch and their causes, such as climate change, we have relied until recently on geologic studies of the deposits resulting from wind erosion. To reconstruct events since 12,000 years ago, however, we have, in addition, variably detailed archeological and written records of wind erosion and climate change. In the last several decades we have added increasingly detailed observations of meteorological and wind erosion events from global networks of satellites and remote sensors. Since the 1980s, we have added regional- to global-scale computer models to investigate both ancient and modern wind erosion events and their causes (1).

Wind erosion is a common phenomenon today in areas of arid and semiarid climate and sparse vegetation. As much as 30% of the Earth's land area is susceptible to wind erosion today. The total annual production of mineral aerosols from wind erosion is estimated at 1 to 2 billion metric tons (2). Human impacts, such as farming, grazing, deforestation, urbanization, and withdrawal of groundwater, may contribute as much as 20–30% of the total. Strikingly, during the cold phases or glacials of the Pleistocene Epoch, wind erosion, as judged by reconstructions from dust in polar ice caps and in deep-sea sediments, may have been one order of magnitude or more greater than that of today (1).

WIND EROSION AND CLIMATE CHANGE IN THE GEOLOGIC RECORD

Eolian deposits associated with ancient wind erosion events have been identified back to the Proterozoic (600–2500 Ma) (3). It is especially difficult to reconstruct ancient environments and the paleoclimatic significance of eolian deposits before the Devonian Period (360–410 Ma), when land surfaces were first colonized by plants (3).

Today, about 10% of the land area of the Earth is covered by loess (i.e., deposits of windblown silt) (Fig. 1) in several different climatic settings (3–5), and perhaps a similar area is covered by sand dunes and sand seas in semiarid and arid climate areas. To a great extent, these are the tangible products of very active wind-erosion processes triggered by global climatic changes beginning in the Pleistocene Epoch and continuing through the present interglacial period. Eolian dust, an important component of soils downwind of areas of wind erosion, may play a role in renewing plant nutrients in areas of rapid leaching in the tropics (e.g., the input of Saharan dust to soils of the Amazon basin) (1). Thus, global pedological and biogeochemical cycles may be affected by climatically induced dust aerosols. Deep-ocean sediments dating from the last glacial maximum (LGM) in many places contain up to 20% quartz from eolian dust (Fig. 2) (3, 4).

Wind erosion, driven by climate change, has played an important role in human history and development over the last 12,000 years. By some measures of dust aerosols, however, eolian sediment fluxes during the LGM about 18,000 years ago were *10 to 40 times* higher than during the present day (5, 6). The principal means by which geologists and paleoclimatologists have reconstructed past episodes of wind erosion and linked them to changing climates has been the stratigraphy, geomorphology, and chronology of eolian deposits associated with arid,

E

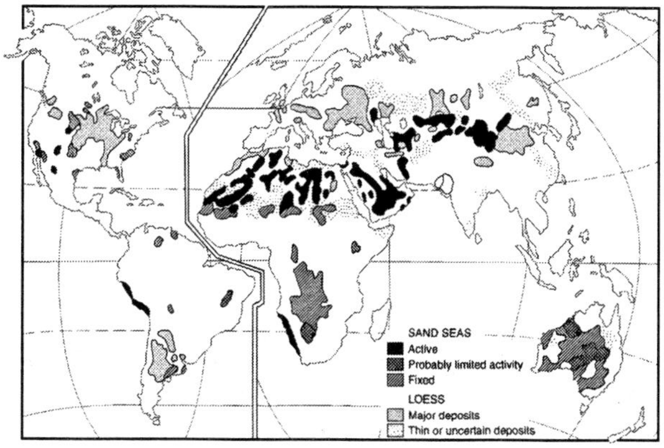

Fig. 1 The global distribution of major eolian deposits. (From Ref. 3.)

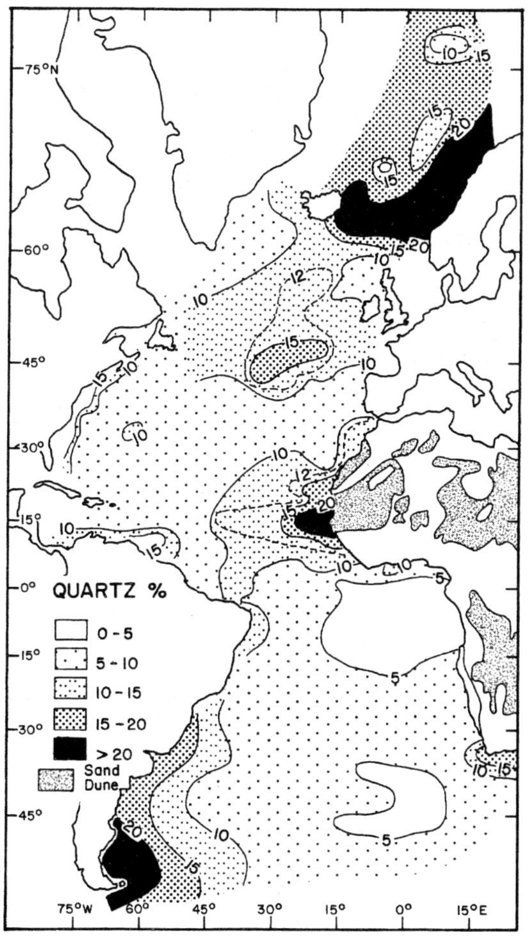

Fig. 2 Distribution of weight percent quartz (carbonate-free) in Atlantic sediments of the LGM (18,000 years B.P.). (From Ref. 4.)

semiarid, and periglacial areas. These studies began in the mid-nineteenth century and accelerated with new age-dating techniques after the World War II. Starting in the 1960s, more refined reconstructions of past wind erosion and links to climate change were made possible by examination of eolian materials, mainly dust, in cores from Arctic and Antarctic glacial ice and in deep-sea sediments (1, 6).

Stratigraphic studies of ice cores, deep-sea sediments, and loess demonstrate that atmospheric dust deposition changed markedly over glacial to interglacial, as well as over shorter, time scales. Dust deposition shows millennial-scale variation that tracks the patterns of other paleoenvironmental indicators, such as atmospheric CO_2 concentrations (1), suggesting a direct, but not necessarily simple, link to climate change over several time scales. Stratigraphic studies confirm that the LGM had much greater wind erosion than today, with periglacial and peridesert zones of the temperate latitudes and peridesert zones of Africa showing dramatic increases in dust fluxes (1, 5), while other regions showed decreases in dust flux.

Three causes may explain why wind erosion and dust loading into the atmosphere were higher during glacials than interglacials: increased wind speeds, reduced intensity of the hydrologic cycle, and expansion of eroding areas (1). Increased wind speeds would increase the probability of exceeding the threshold velocity for sediment entrainment. Reduced intensity of the hydrologic cycle (less precipitation) could allow dust to remain longer in the atmosphere and be carried farther from the source before rainout would occur. Source areas probably expanded by: 1) exposure of sediments on continental shelves with lowered sea level; 2) deposition of large volumes of outwash from continental and alpine glaciers, especially in the Northern Hemisphere; and 3) reduction in soil moisture and/or reduction in plant cover. Simulations of the atmosphere during glacial climates show an increase in wind intensities, but direct evidence of higher wind speeds is equivocal (1). Some models predict that there were strong cyclonic winds in areas of periglacial outwash in front of major Northern Hemisphere ice sheets (1). Temperature depression at the LGM is estimated to have been 10°C or more in temperate latitudes. These estimates are based on pollen evidence for changes in vegetation. Such decreases in temperature would have resulted in decreased intensity of the hydrologic cycle. Evidence regarding lake levels suggests that runoff was less than it is today over much of the globe, including the Sahel in northern Africa (5). Drier soils and colder temperatures also drove the expansion of desert, grassland, and shrubland at the expense of forests, resulting in a worldwide increase in the area susceptible to wind erosion. Widespread loess deposits in North America, Europe, Asia, and South America (4, 5) are of glacial age, which also suggests that the LGM was more arid than today or that sediment production was more robust (e.g., glacial erosion).

Detailed paleoenvironmental studies of Saharan sand seas have produced process–response models of environmental change driven by arid–humid cycles that result from orbital forcing of climate (Fig. 3) (7). For example, in Fig. 3, the model predicts that during humid phases, sediment production by chemical weathering and alluviation dominates and eolian systems are stabilized by soil moisture and vegetation. In contrast, during arid phases, transport capacity and sediment availability increase, resulting in time-lagged deflation of sediments accumulated during the humid phase with construction of eolian deposits (7).

The deserts and the loess plateau in northern China (Fig. 1) form perhaps the most famous example of an eolian system coupled to glacial–interglacial changes in the global climate system. As many as 40 loess units are interstratified with paleosols there, representing 2.5 million years of record. The cyclical alternation of loess and paleosols is thought to represent fluctuations in the dominance of the Asian winter (dust-transporting) and summer (precipitation-bringing) monsoons associated with variations in Earth's orbital parameters that drove the Ice Ages. At the desert margins, sand beds within the loess suggest that the desert margin advanced many kilometers to the south by active dune migration during cold glacial periods of intensified winter monsoon. Conversely, loess or paleosols in desert sites suggest northern movement of the boundary (8).

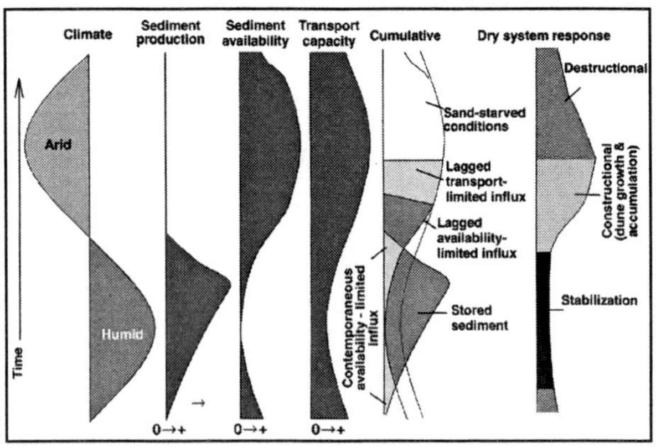

Fig. 3 A climate-change driven process–response model based on sediment production, sediment availability (supply), and transport capacity of wind for Saharan sand seas. (From Ref. 7.)

The results of simulation modelling of radiative forcing today and during the LGM are intriguing. They suggest that dust aerosols in the atmosphere from wind erosion, anthropogenic sources, and volcanic eruptions can *effect* climate change as well as *result from* climate change (e.g., the cooling of global temperatures after the eruption of Mount Pinatubo in 1991). The dustiness of the LGM atmosphere could have produced an additional cooling of 3°C beyond predictions, based on orbital forcing, ice-sheet albedo, and CO_2 forcing alone (1), although different models produce different latitudinal distributions of predicted temperature change.

WIND EROSION AND CLIMATE CHANGE IN HUMAN HISTORY AND TODAY

Climate change refers to statistical changes in climatic variables such as temperature, evaporation, or precipitation over a long period, relative to human records, such as decades to centuries. *Climate variablility* refers to statistical changes in climate variables over shorter time scales, such as months to years. In recorded human history, wind erosion has related both to dry periods within the natural cycles of climate variability and to progressive, long-term climate change. Throughout history and around the world, humankind has responded to the agricultural challenge posed by climate variability and climate change by developing hydraulic infrastructure, especially in areas of arid and semiarid climates. The fall of ancient hydraulic systems, such as those in Mesopotamia and along the Indus River, and the subsequent wind erosion, soil degradation, and societal collapse have often been attributed to climate change (9). Chinese records document quasi-periodic variations in rainfall, temperature, and dustfall for the past 2000 years, with periods ranging from 2–100 years (10). Periods of high dustfall were strongly correlated with periods of drought.

When human impacts to the natural and agricultural environment are added to natural climate variability and climate change, greater wind erosion is likely to occur than would be the case under climate change alone. The world's arid and semiarid areas have the greatest interannual variability of rainfall (Fig. 4) (9). Although ancient sand seas continue to shift by dune migration, generation of dust aerosols is generally low because deflation has already occurred (5). Areas at greatest risk of wind erosion with soil degradation are those at desert margins. The fine- to medium-textured soils deposited previously by deflation at desert margins are often ideal for agriculture, but particularly susceptible to entrainment by wind. Human

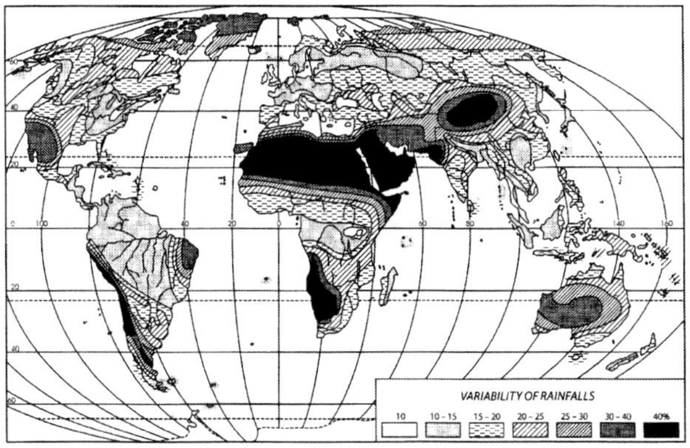

Fig. 4 Interannual variability of rainfall on the surface of the Earth. [From Ref. 9; this figure originally appeared in the French edition of this book (Ref. 15).]

disturbance of eolian soils, such as fallow rotations in dryland agriculture, can lead to accelerated wind erosion, as in the U.S. Great Plains during the Dust Bowl years of the 1930s.

Drought can persist for decades in some arid and semiarid areas, with dramatic effects on plant cover, soil moisture, and consequent wind erosion: In the Sahel of West Africa, where decadal-scale fluctuations in climate are particularly strong (11), mean annual rainfall decreased by 25–40% from 1968 to 1997. West Africa is the source of nearly half of the mineral dust in the atmosphere today. Analysis of dust storms over West Africa shows a steady increase in the frequency of dust occurrence since the early 1970s, which parallels negative rainfall anomalies during this period. The source is principally the semiarid Sahel, not the Sahara desert as once was assumed (11). Paradoxically, some measurements and models suggest that atmospheric dust over the Sahel has an important climatic feedback effect that may prolong drought cycles by scattering or absorbing solar radiation or even by modifying the African Easterly Jet (11). Dust from the Sahel is carried into the upper atmosphere and transported across the Atlantic Ocean to Bermuda, the Caribbean islands, and even the Amazon Basin (1).

Redistribution of surface water also can drive local and regional changes in surface hydrology that result in wind erosion. The diversion of the Owens River to Los Angeles, California, in the early 1900s turned Owens Lake into a 300-km^2 dry playa, now known for wind erosion and dust generation. A recent agreement to flood parts of the lake bed avoids decades of litigation and sets a goal to meet federal air quality standards. In central Asia, large-scale river diversions for irrigation led to the desiccation of Lake Aral. The dramatic reduction in the lake's surface area diminished its effect as a regulator of regional climate and exposed over 2 million ha of salinized lake bottom sediments. The combined effect of a shorter growing season, degraded soils, higher winds, and exposed sediments results in erosion of millions of tons of dust and salts annually (9). For the lands above the rapidly depleting Ogallala aquifer in southwestern Kansas and the Oklahoma–Texas Panhandle, deep well irrigation has been key to limiting the recurrence of the Dust Bowl (12, 14).

Individual severe wind erosion events can also be triggered by extreme synoptic weather patterns, with or without antecedent drought. For example, in 1998, on April 15 and 19, two intense dust storms were generated over the Gobi Desert by springtime cold-weather systems. The April 15 dust cloud was removed by precipitation over East Asia. The April 19 dust cloud was transported across the Pacific in 5 days. Part of the dust subsided to the surface between British Columbia and California, producing average Asian dust concentrations over the West Coast of about 20–50 $\mu g/m^3$. The 1998 Asian dust event was larger than any other similar event in western North America in the previous decade.

WIND EROSION AND CLIMATE CHANGE IN THE FUTURE

There is considerable interest in the prediction of future climate change, especially in light of the increasing rate of global warming over the past several decades. Several General Circulation Models (GCMs) have been applied to estimate climate change driven by rising atmospheric CO_2 levels. These models have more recently been coupled with the effects of atmospheric aerosols, such as sulfate and anthropogenic dust (1). The best estimate for the year 2100, under a doubled concentration of CO_2 and incorporating aerosols, is a temperature increase of 1.0–3.5°C (13, 14), with an increase in global average precipitation ranging from 5–15%. Paradoxically, the GCMs also predict that midcontinental land areas in the middle latitudes of the Northern Hemisphere may undergo a drying trend in summer (14). This could have the effect of increasing the potential for wind erosion in critical areas, such as the Great Plains of North America and the desert basins of Asia, especially if human land use results in lesser plant or crop cover. Different predictions come from models that take into account the physiological effects of increased CO_2 on plant growth and climatic warming. These models predict substantial reductions in the area of arid and semiarid vegetation and, thus, reductions of about 20% in source areas for potential wind erosion (1).

Under greenhouse warming, changes will occur not only in the mean values of climate variables, but also in the frequency and severity of extreme weather events (14) due to increased energy in the atmosphere. The predicted increase in variability of temperature, rainfall, and wind speed, already high in arid and semiarid areas (Fig. 4), may exacerbate wind erosion of midcontinental agricultural, range, and desert lands. Although smaller changes in mean values of climatic variables are predicted for tropical latitudes, wind erosion could increase in these areas as well.

Under greenhouse-warming scenarios, the severity of drought in the Great Plains of North America may well exceed that experienced during the Dust Bowl (14). In the midlatitudes and in other arid and semiarid areas at low latitudes, increased incidence of high-wind events could begin to "pirate" or erode the large deposits of eolian sands and loess that were emplaced during the LGM, if plant cover is reduced. According to recent estimates, present wind speeds exceed the threshold wind velocity 30–60% of the time (5), thus the eolian soils of the Great Plains are poised for renewed wind erosion if existing plant cover or land use patterns are significantly disturbed. Dropping water tables in the Ogallala aquifer may eventually force the abandonment of up to 6 million ha of irrigated cropland on soils sensitive to wind erosion (12). A similar situation may occur in the irrigated lands around Lake Aral. The legacies of ancient hydraulic societies, which rose to high levels of achievement but then collapsed upon climate change, raise questions about the future of today's hydraulic societies (14). If the competition for freshwater resources continues to increase and the availability of groundwater resources declines, especially under climate-change scenarios, a reduction or relocation of irrigated croplands is possible.

Improved stochastic climate modeling, coupled with proper land-use management of the agricultural lands at the desert margins, may provide the means to limit wind erosion of those lands under future climates. As human land use encompasses ever greater areas of Earth's arid and semiarid lands in the future, the control of wind erosion will increasingly require decision making that accounts fully for climate change and climate variability.

REFERENCES

1. Harrison, S.P.; Kohfeld, K.E.; Roelandt, C.; Claquin, T. The Role of Dust in Climate Changes Today, at the Last Glacial Maximum and in the Future. Earth Sci. Rev. **2001**, *54* (1–3), 43–80.
2. D'Almeida, G.A. Desert Aerosol: Characteristics and Effects on Climate. In *Palaeoclimatology and Palaeometeorology: Modern and Past Patterns of Global Atmospheric Transport*; Leinen, M., Sarnthein, M., Eds.; Kluwer Academic Press: Dordrecht, 1989; 311–338.
3. Thomas, D.J.T., Ed.; *Arid Zone Geomorphology*, 2nd Ed.; John Wiley and Sons: New York, 1997.
4. Pye, Kenneth *Aeolian Dust and Dust Deposits*; Academic Press: London, 1987.
5. Goudie, A.S., Livingstone, I., Stokes, S., Eds.; *Aeolian Environments, Sediments, and Landforms*; John Wiley and Sons, Ltd. West Sussex, England, 1999.
6. Maggi, V. Mineralogy of Atmospheric Microparticles Deposited Along the Greenland Ice Core Project Core. J. Geophys. Res. **1997**, 102 (C12), 26, 725–26, 734.
7. Kocurek, G. Aeolian System Response to External Forcing Factors–A Sequence Stratigraphic View of the Saharan Region. In *Quaternary Deserts and Climatic Change*; Alsharhan, A.S., Glennie, K.W., Whittle, G.L., Kendall, C.G. St. C., Eds.; Balkema: Rotterdam, 1998; 125–153.
8. Ding, Z.; Sun, J.; Rutter, N.; Rokosh, D. Changes in Sand Content of Loess Deposits Along a North-South Transect of the Chinese Loess Plateau and the Implications for Desert Variations. Quat. Res. **1999**, *52*, 56–62.
9. Mainguet, M. *Aridity, Droughts, and Human Development*; Springer-Verlag: Berlin, Germany, 1999.
10. De'er, Z. Historical Records of Climate Change in China. Quat. Sci. Rev. **1991**, *10*, 551–554.
11. Nicholson, S. Land Surface Processes and Sahel Climate. Reviews of Geophysics **2000**, *38* (1), 117–139.
12. Opie, J. The Drought of 1988, the Global Warming Experiment, and Its Challenge to Irrigation in the Old Dust Bowl Region. Agricultural History **1992**, *66* (2), 279–336.
13. Houghton, J.T., Meira Filho, L.G., Callander, B.A., Harris, N., Kattenberg, A., Maskell, K., Eds.; Intergovernmental Panel on Climate Change. *Climate Change 1995: The Science of Climate Change*; Cambridge University Press: Cambridge, England, 1996.
14. Rozensweig, C.; Hillel, D. *Climate Change and the Global Harvest*; Oxford University Press: New York, 1998.
15. Mainguet, M. *L'homme et la Sécheresse*; Masson: Paris, France, 1995.

EROSION BY WIND, CONTROL MEASURES

Gary L. Tibke
Wind Erosion Research Unit, Throckmorton Hall, Kansas State University, Manhattan, Kansas, U.S.A.

INTRODUCTION

Wind erosion can be controlled by reducing the wind forces at the soil surface or by creating surface conditions more resistant to wind forces (1). This can be accomplished by maintaining vegetative residues on the soil surface, reducing field width, utilizing stable soil aggregates, roughing of the land surface and reshaping the land (i.e., the five basic principles of wind erosion control). While individual conservation practices can be successful in controlling erosion, a combination of practices should generally be considered when a wind erosion control system is being planned. Droughts will cause a shortage of residue in the stubble mulch program, erosive winds will not always blow in the prevailing direction when barriers are used, and soil clods and ridging for erosion control are temporary control measures and best serve as supplemental practices. Conditions conducive to wind erosion exist when the soil surface is loose, dry, and finely granulated, when the soil surface is smooth and vegetative cover is sparse or absent, and when the susceptible area is sufficiently large. Wind erosion control is usually needed in areas with low and variable precipitation and frequent droughts, and where high winds, high temperatures, and consequent high evaporation prevail (2). Control is also needed in areas where crops that are easily damaged by moving soil particles are grown on highly erodible soils. Research has shown that it is possible to alter some of the components of the erosion process and that this can be accomplished with one or more of the five basic principles of wind erosion control (3).

ESTABLISHING AND MAINTAINING VEGETATION OR VEGETATIVE RESIDUES TO PROTECT THE SOIL

On unprotected fallow fields, applying the principle of vegetative cover can best be done by practicing no-till or stubble-mulch farming (4). The importance of vegetative protection on the land cannot be overstressed.

Stubble Mulching

Stubble mulching is a form of subsurface conservation tillage where the soil is tilled without inversion. Tillage tools, such as chisels, field cultivators, sweeps or blades, are used (Fig. 1).

Weed control is accomplished with cultivation and/or herbicides. The level of protection afforded by stubble mulching is proportional to the kind, quantity and orientation of crop residue maintained on the soil surface. Excessive use of tillage implements that bury crop residues are the major cause of inadequate vegetative cover for erosion control on cropland. Also, tillage operations that flatten crop residues must be minimized. Standing residues are 5 to 10 times more effective in controlling wind erosion than flat residue. Conservation tillage systems must be designed to avoid excessive loss of surface residues if effective wind erosion control is to be accomplished.

No Tillage

Since 1990, no-till farming has grown from 6% to more than 16% of the planted acres in the United States (Fig. 2). This is an increase of over 32 million acres (5). No-till is a procedure whereby a crop is planted directly into a seedbed that has not been tilled since harvest of the previous crop. Only the immediate seed zone is disturbed and no additional tillage occurs. Many studies have shown that a no-till program is a way to virtually eliminate erosion problems (Table 1) (6).

The bodies of plants and animals that have finished their life cycle are involved in the processes of soil formation and help to generate fertile and productive soils. No-till largely approximates this ingenious work of nature since it in fact makes use of crop residue for this purpose.

In many farming systems, the plow can be replaced efficiently by biological activities. Plants, roots, worms, arthropods, and other living beings are natural cultivators of the soil. In one study, after 7 years of no-till, the earth worm population was 36 times greater than the population in a conventionally tilled field (7). The switch to

Fig. 1 Stubble mulch tillage. (Courtesy of USDA.)

no-till causes many soil changes. The type and extent of changes depend on soil type, climate, and farming history. Farmers who have no-tilled for some years usually notice more soil moisture, better seedbed tilth, more organic matter and earthworms, less soil erosion, and improved trafficability (8).

Cover Crops

The purpose of a cover crop is to produce vegetative protection for the land against wind and water erosion. Cover crops are usually planted when protective residues are inadequate and winter and spring winds are high. Cover crops are also planted between rows to provide protection for vegetable or other crops that are highly susceptible to abrasive injury in the seedling stage. Cover crops are best suited to humid and higher rainfall areas or irrigated land. In many of the semiarid and arid areas of the Great Plains and intermountain valleys, cover crops are not practical because they compete for nutrients and limited soil moisture. Generally, cover crops are used in these dry non-irrigated areas only to control erosion on erosion-susceptible knolls or where low residue crops have been grown.

Table 1 Estimated soil losses with four tillage systems in Nebraska (6)

Tillage system	Estimated soil loss (t ha^{-1} year^{-1})[a]
Conventional	20.2
Disk and plant	10.5
Till plant	4.5
No-tillage	1.8

[a] $t\,ha^{-1} \times 0.446 = t\,acre^{-1}$. (From Ref. 6.)

REDUCING FIELD WIDTHS BY ESTABLISHING BARRIERS OR STRIP CROPPING

Windbreaks and windbarriers contribute to wind erosion control by reducing windspeed, on their leeward side, below the threshold required for initiation of soil movement and by decreasing the field length along the erosive wind direction (Fig. 3).

The main effect of reducing downwind field length is a reduction in breakdown of clods and crust by abrasion from saltating (hopping) particles. Abrasion of the soil surface components and breakage of saltating aggregates

Fig. 2 No-till corn. (Courtesy of USDA.)

Fig. 3 Windbreaks and strip cropping. (Courtesy of USDA.)

increase the soil flow with distance downwind. This in turn reduces surface roughness and permits even higher rates of soil transport for a given wind speed. The downwind rate of soil movement varies directly with erodibility of the surface and the width of the eroding field. The effect of any barrier in reducing the rate of soil movement depends on the wind velocity and direction, the threshold velocities needed to initiate soil movement, and the barrier shape, width, height and porosity. For planning purposes, the distance of 10 barrier heights (h) has been widely adopted as the minimum protected area on the leeward side of barriers.

The rate of soil movement by wind varies directly as the cube of the windspeed (wind erosion force), above the threshold, and inversely with surface soil water. Considering this, it is apparent that a small decrease in windspeed, some distance from the barrier, could still result in a significant decrease in wind erosion forces (9). In one test, the windspeed at 12 h leeward of a barrier was 62% of open field velocity, whereas windspeed cubed (wind erosion force) was only 25% of that in the open field (10).

Tree Windbreaks

One advantage of windbreaks over most other types of wind erosion control is their relative permanency. Siddoway and Fenster (11) made the point that during a series of drought years, windbreaks may be the only effective and persistent control measure on cropland. Single-row plantings are most common in field windbreaks and 2- to 5-row plantings are used for farmstead and wildlife protection (Fig. 4). Chepil (12) found single-row field windbreaks more suitable than double or multiple rows because they occupied the least amount of land area for the amount of protection derived from them.

Fig. 4 Single row field windbreak. (Courtesy of USDA.)

Fig. 5 Herbaceous wind barriers. (Courtesy of USDA.)

It usually takes many years for tree windbreaks to reach sufficient height and density for good protection, especially in semiarid areas. Therefore, if used, windbreaks require a combination of other conservation practices if they are to be successful.

Herbaceous Wind Barriers

Herbaceous barriers (annual and perennial) have worked well for controlling wind erosion, trapping snow, and reducing evaporation in dryland cropping areas (Fig. 5).

Additional advantages of perennial grass barriers, as Black and Bauer (13) explained, are ease of establishment, low cost, and lack of potential injury from herbicides used for broadleaf weed control. Research in Montana by Aase et al. (14) on tall wheatgrass (thinopyrum ponticum) barriers showed that the barrier-protected land would only experience 6.6% of open-field erosion for the entire year.

Artificial Barriers

Artificial barriers such as snow fencing, board walls, bamboo (bambusa bambos) and willow (salix spp.) fences, earthen banks, hand-inserted straw rows, and rock walls have been used for wind erosion control, but only on a very limited basis. There is usually a very high cost in material and labor to construct these barriers and their use generally restricted to high-value crops.

Strip Cropping

Strip cropping is dividing a large field into strips that are narrow enough to help control wind erosion. Usually strips of erosion-resistant crops or standing stubble are alternated with other crops or unprotected fallow fields. Strip cropping reduces the ever increasing downwind abrasion caused by wind eroding fields and is a very effective erosion control method used extensively in the northern Great Plains of the United States, Canada and the spring wheat (triticum aestivum) region of Russia.

MAINTAINING STABLE AGGREGATES OR CLODS ON THE SOIL SURFACE

Texture is the dominant soil characteristic relative to inherent erodibility. In general, susceptibility to wind erosion increases as soil texture becomes coarser. Sandy or coarse-textured soil classes lack sufficient amounts of silt and clay to bind sand particles together and form soil aggregates or clods resistant to wind erosion. Consequently, such soils form a "single grain" structure or weakly cemented clods, a condition that is quite susceptible to erosion by wind. A soil surface of stable aggregates of about 70% clods >0.84 mm in diameter, resists erosion from all but the highest winds (9). Clods of this size are not easily moved by wind, and they protect smaller clods and particles in their lee. Loams, silt loams, and clay loams tend to form the most stable aggregates and are, therefore, the least affected by erosive winds. The type of tillage equipment used has a definite influence on soil cloddiness and surface roughness. Smika and Greb (15) found that tillage by machines other than the chisel tend to reduce the non-erodible soil aggregation. One-way, offset or tandem disks leave a smooth surface. Subsurface sweeps, because they do not disturb the soil surface, do not create a rough, ridged soil surface, but they do maintain a greater vegetative roughness by allowing some of the vegetation to remain erect (16).

ROUGHENING THE LAND SURFACE

Soil surface roughness is composed of anchored vegetative material, soil ridges, soil clods, or combinations of all three. All help to control wind erosion by lowering the wind velocity near the soil surface and by sheltering erodible soil fractions (17). Tillage implements form ridges and depressions which alter wind velocity. The depressions behind the ridges trap saltating soil particles and stop the normal build-up of eroding material downwind.

Fig. 6 Field following emergency tillage. (Courtesy of USDA.)

Emergency Tillage

Emergency tillage is a last-resort wind erosion control practice that can provide a rough, cloddy surface. It is usually carried out when vegetative cover is depleted by excessive grazing, drought, improper or excessive tillage, or by growing crops that produce little or no residue (Fig. 6).

Emergency tillage is an inadequate wind erosion control measure and its only purpose is to create a temporary erosion-resistant soil surface. Implements such as listers, chisels, shovels and "sandfighters" should traverse fields at right angles to erosive winds to roughen the soil surface and bring clods to the surface (9). Listers and narrow chisels were found by Chepil and Woodruff (16) to have the most effective tillage points for emergency tillage. Listers provide a high degree of roughness, and in extremely sandy soils, where clods can be produced only by deep tillage, they are the most effective tools available. Chisel cultivators are more widely used because they require less power and destroy less growing crop than listers.

RESHAPING THE LAND TO REDUCE EROSION ON KNOLLS

Reshaping the land by leveling knolls and benching slopes to shorten the unsheltered distance is an option in wind erosion control, but is usually not economical or practical. Because land reshaping is very costly, other effective control measures, such as no-till and seeding to permanent grass, are usually options that are more viable. Hills and knolls affect tillage system requirements indirectly by influencing wind shear stress. When the wind blows over a hill, streamlines of airflow are squeezed together, which increases the wind velocity and shear stress, thereby increasing the erosion potential on the windward slope and hilltop. Consequently, this increases the amount of residue, cloddiness, or roughness needed to control wind erosion on the knoll.

OTHER WIND EROSION SITUATIONS

Wind Erosion on Irrigated Land

Wind erosion on irrigated land can be a serious problem in areas characterized by variable high-wind velocities, where the soils are organic or quite sandy and low in organic matter or where crop residues are inadequate. Under certain conditions it is impractical and wasteful of water to irrigate frequently enough to prevent a finely pulverized surface soil from blowing. The depth of drying may only be a fraction of an inch and the soil below this may be wet, but if the immediate surface is dry and the wind is strong enough, the top layer can erode unless the soil particles are consolidated into clods or protected by vegetation. The basic element in erosion control by tillage on irrigated land, as on dryland, is the creation of a rough, cloddy surface which will resist the force of the wind, decrease its velocity at the ground level and trap moving soil. Sandy soils, usually found in irrigated areas, are far more difficult to protect by emergency measures than fine-textured soils.

Wind Erosion Control on Sand Dunes and Other Problem Areas

Wind erosion control on sand dunes and other problem areas require measures that are more intensive to get sand dunes in check. Dunes lack a soil profile because they are unstable and underdeveloped. The sand is fine, loose and easily moved by wind. It has no organic matter, and consequently retains little moisture for plants and has inherently low fertility. Sand dunes and drift areas often require artificial barriers or cover for stabilization before vegetation can be established. These include oil, clay gravel, picket fence, brush, straw, and hay. Clay is effective, but is expensive. Hay or straw can be used as temporary mulches on blowout or small areas of dune sand at road cuts, around dwellings and other disturbed areas. They provide some organic matter, which is critical for successful dune plantings. The establishment of permanent vegetation is the final objective in the stabilization of dunes.

Other Non-Vegetative Erosion Protection

Some of the non-vegetative and processed vegetative materials used are gravel and crushed rock, various surface films such as resin-in-water emulsion (petroleum origin), rapid-curing cutback asphalt, asphalt-in-water emulsion, starch compounds, latex-in-water emulsion (elastomeric polymer emulsion), by-products of the paper pulp industry, and wood cellulose fiber (2). Several of these spray-on adhesives are available for temporary wind erosion control of vegetable seedlings on mineral soils. Some of the adhesives are relatively expensive, but a few are economically feasible on high-value crops threatened by serious blowing that cannot be controlled by other methods.

REFERENCES

1. Lyles, L. Basic Wind Erosion Processes. Agric. Ecosystems Environ. **1988**, *22/23*, 91–101.
2. Woodruff, N.P.; Lyles, L.; Siddoway, F.H.; Fryrear, D.W. *How to Control Wind Erosion*, Agric. Inf. Bull. No. 354; U.S.D.A., A.R.S.: Washington, D.C., 1972; 22.
3. Greb, B.W. *Reducing Drought Effects on Cropland in the Western-Central Great Plains*, U.S.D.A., SEA Agric. Inf. Bull. No. 420: Washington, D.C., 1979; 31.
4. Haas, H.J.; Willis, W.O.; Bond, J.J. *Summer Fallow in the Western United States*, Agriculture Research Service Conservation Research Report No. 17; U.S.D.A.: Washington, D.C., 1974; 149–160.
5. Watson, S., Ed. *Kansas No-Till Handbook*; Kansas State University: Manhattan, KS, 1999; 72.
6. Rice, R.W. *Fundamentals of No-Till Farming*; American Association for Vocational Instructional Materials: Winterville, 1983; 148.
7. Crovetto, C. Stubble Over the Soil. *The Vital Role of Plant Residue in Soil Management to Improve Soil Quality*, Amer. Soc. Agronomy: Madison, WI, 1996; 245.
8. Domitruk, D., Crabtree, B., Eds. *Zero Tillage. Advancing, the Art*, Manitoba–North Dakota Zero Tillage Farmers Association: Brandon, Manitoba 1997; 40.
9. Tibke, G. Basic Principles of Wind Erosion Control. Agric. Ecosystems Environ. **1988**, *22/33*, 103–122.
10. Skidmore, E.L.; Hagen, J.L. Reducing Wind Erosion with Barriers. Trans. ASAE **1977**, *20*, 911–915.
11. Siddoway, F.H.; Fenster, C.R. Soil Conservation: Western Great Plains. In *Dryland Agriculture*; Dregne, H.E., Willis, W.O., Eds.; Monograph No 23; Amer. Soc. Agronomy: Madison, WI, 1983; 231–246.

12. Chepil, W.S. Wind Erosion Control with Shelterbelts in North China. Agron. J. **1949**, *41*, 127–129.
13. Black, A.L.; Bauer, A. Soil Conservation: Northern Great Plains. In *Dryland Agriculture*; Dregne, H.E., Willis, W.O., Eds.; Monograph No 23; Amer. Soc. Agronomy: Madison, WI, 1983; 247–257.
14. Aase, J.K.; Siddoway, F.H.; Black, A.L. Effectiveness of Grass Barriers for Reducing Wind Erosiveness. J. Soil Water Conserv. **1985**, *40*, 354–360.
15. Smika, D.E.; Greb, B.W. Nonerodible Aggregates and Concentration of Fats, Waxes and Oils in Soils as Related to Wheat Straw Mulch. Soil Sci. Soc. Am. Proc. **1975**, *39*, 104–107.
16. Chepil, W.S.; Woodruff, N.P. The Physics of Wind Erosion and Its Control. Adv. Agron., Water Conserv. **1963**, *15*, 211–302.
17. Armbrust, D.V.; Chepil, W.S.; Siddoway, F.H. Effects of Ridges on Erosion of Soil by Wind. Soil Sci. Soc. Am. Proc. **1964**, *28*, 557–560.

EROSION BY WIND-DRIVEN RAIN

E

G. Erpul
Purdue University, West Lafayette, Indiana, U.S.A.

L.D. Norton
United States Department of Agriculture-Agricultural Research Service (USDA-ARS), National Soil Erosion Research Lab., West Lafayette, Indiana, U.S.A.

D. Gabriels
Gent University, Gent, Belgium

INTRODUCTION

Wind-driven rain can be described as raindrops falling through a wind field at a vertical angle under the effects of both gravitational and drag forces. Schematic representation of wind-driven rain incidental to a sloping soil surface is given in Fig. 1.

A study of wind-driven rain erosion is an attempt to investigate the combined effect of wind and rain on soil erosion processes in situations where wind and rain occur at the same time.

EFFECTS OF WIND ON PHYSICAL CHARACTERISTICS OF RAINS

Raindrop Size Distribution

In the assessment of the distribution of small simulated raindrops in a wind tunnel, Erpul et al. (1) obtained a narrower raindrop distribution and observed a distinct increase in mean drop diameter under wind-driven rain compared to windless rain. Collisions between small drops occurred more frequently as a result of their greater number per unit volume in air leading to an increase in mean drop size. For large drops, however, this would not occur as large drops are less stable and the wind caused some of them to break up into smaller drops (2, 3). Basically, the effect of wind on raindrop size distributions is a potentially important effect that needs to be considered when estimating the rainfall erosivity.

Raindrop Impact Energy

Wind-driven raindrops gain some degree of horizontal velocity, which increases their resultant impact velocity. The kinetic energy load of the rainfall may be expected to change as a result of increased velocity and altered size of the raindrops. The exponential relationship between the horizontal wind velocity and the kinetic energy of raindrops was found in natural rains (4). The effect of the wind on the horizontal component of small raindrops would be greater, so a greater percentage increase in kinetic energy would be expected for small raindrops than large raindrops (5).

Raindrop Impact Angle

Wind-driven raindrops strike the soil surface with an angle deviated from the vertical because of their horizontal and vertical velocities. This inclination depends on the magnitude of wind velocity. In mid latitudes rain mostly falls at considerable inclination from the vertical, and resultant angles of 40–70° have been found in rains driven by wind velocity of $10\,ms^{-1}$ (6). Gabriels et al. (7) reported that the mean angles of rain inclination were 52, 66, and 67° for the simulated rains in a wind tunnel driven by 6, 10, and $12\,ms^{-1}$ of wind velocitites, respectively. For these rains, a median drop size was approximately 1.50 mm. Little is known about the physical impact of raindrops on a soil in situations where this impact is not vertical. It is also not known if inclined raindrops have stronger erosive effects than vertical ones. The extent and magnitude of the rainsplash detachment increased as the angle of deviation increased within the range of 5–30° (8). Whereas, with greater deviations, the impact angle could be so small that raindrops would hardly hit the soil surface. Especially when impact angles were less than 30°, rainsplash detachment rate highly decreased (Table 1).

Raindrop Impact Frequency

The distribution and intensity of rain on sloping surfaces differs depending on wind direction and velocity. In

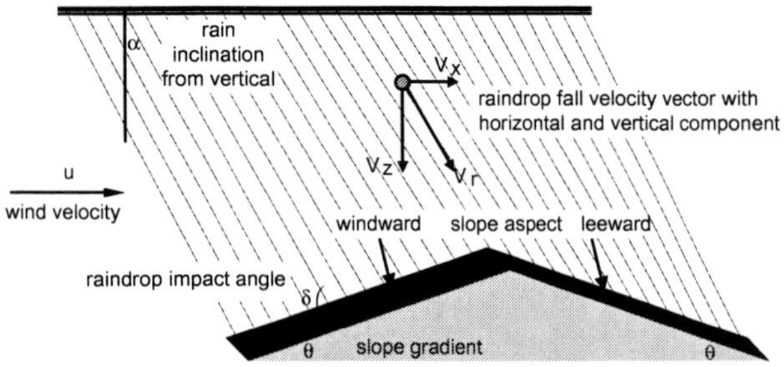

Fig. 1 Schematic representation of wind-driven rain with an angle from vertical and incident on sloping soil surface.

fact, the angle of rain incidence (ϕ), which is a function of rain inclination (α), slope gradient (θ), and slope aspect, determines the rain intensity in wind-driven rains. As an example, a windward facing slope can receive two times more rain intensity than a leeward facing slope, or even exceed it in extreme cases for rain inclinations of 40–70° (6). When the rain inclination and the slope gradient increase, the discrepancies in the rain intensity between wind- and leeward slopes become greater (Table 1).

EFFECTS OF WIND ON RAINSPLASH DETACHMENT AND TRANSPORT

Rainsplash Detachment

Similar to effects on rain characteristics, wind movement and velocitites can have a profound effect on some aspects of the soil erosion process. When wind accompanies rain, rainsplash detachment tends to increase due to the increased energy of raindrops. Pedersen and Hasholt (4) obtained a better correlation between the rainsplash detachment and erosivity indexes when wind velocity was taken into account in the kinetic energy calculation. However, the wind not only increased the raindrop impact energy but also altered the angle of rain incidence, resulting in the variations in the raindrop impact frequency and the raindrop impact angle. In other words, raindrop impact energy, impact frequency, and impact angle determine the magnitude of rainsplash detachment under wind-driven rainfall. The influence of each factor on the process is not exclusive and not clearly distinguished because all are closely related and are a function of the wind speed. The highest rainsplash rate occurred with the highest impact frequency and the highest impact angle at a given raindrop impact energy under wind-driven rains (Table 1).

Rainsplash Transport

Wind, as well as slope and overland flow, is another possible factor capable of transporting detached particles by raindrop impact. It is possible to find similarities between the movement and trajectory of sand saltation by wind and the movement and trajectory of soil particles by rainsplash under wind-driven rainfall. In saltation, the hitting sand grains, once ejected, initiate the motion of uplift (9), whereas hitting raindrops on the soil surface with an angle initiate a jumping movement of soil particles in wind-driven rainfall (10). A threshold is given by impacting raindrops, and wind does not account for the upward movement except that it changes energy, frequency, and angle of raindrop impact. Once soil particles are entrained in the splash droplets that have risen into the air by raindrop impact, wind velocity gradient will transport these particles. It is found that, together with wind velocity gradient, the impact angle is also playing a significant role in determining the extent of the process. At the same wind speed, longer mean rainsplash distances (X_c) were observed as the impact angle became greater (Table 1). There was a very discernible decrease in X_c in leeward slopes, which was mostly associated with the impact angles of less than 30°. These results showed that the raindrop impact angle might determine the soil particle ejection angle.

As mentioned above, the amount of soil particles to be transported by the wind will depend upon the raindrop impact energy, impact frequency and impact angle. Subsequently, the soil particle ejection angle and the wind velocity gradient will determine the extent of the process (Fig. 2).

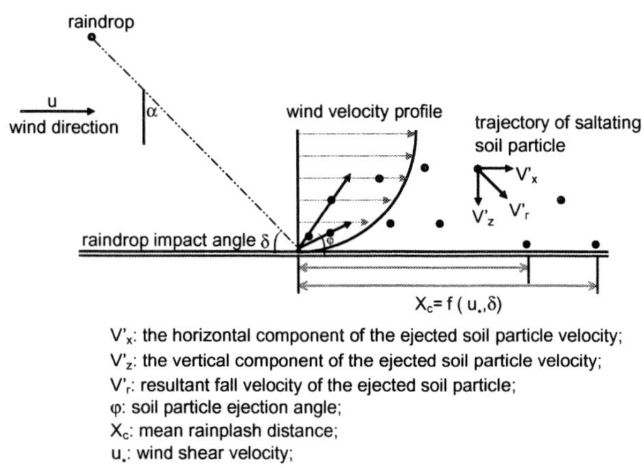

Fig. 2 Rainsplash transport: raindrop-induced and wind-driven splash trajectories of soil particles falling through a wind profile.

Table 1 The effect of wind on rain energy, rain intensity, and rainsplash detachment and transport of a silt loam soil

			Windward						Leeward				
u (ms^{-1})	KE ($Jm^{-2}mm^{-1}$)	α (°)	θ (°)	ϕ (°)	I (mmh^{-1})	D ($gm^{-2}s^{-1}$)	X_c (m)	q_s ($gm^{-1}s^{-1}$)	ϕ (°)	I (mmh^{-1})	D ($gm^{-2}s^{-1}$)	X_c (m)	q_s ($gm^{-1}s^{-1}$)
0	2.23	0	4	4	143	0.23	0.35	0.08	4	161	0.47	0.34	0.16
			9	9	141	0.32	0.38	0.12	9	172	0.75	0.35	0.26
			11	11	132	0.41	0.45	0.18	11	180	0.98	0.36	0.35
6	4.37	52	4	48	92	0.41	0.61	0.25	56	126	0.82	0.73	0.60
			9	44	106	0.51	0.61	0.31	61	107	0.28	0.74	0.21
			11	41	112	0.77	0.62	0.48	63	97	0.30	0.70	0.21
10	13.20	66	4	62	125	2.01	1.26	2.53	70	90	0.78	1.12	0.87
			9	53	137	2.01	1.37	2.75	75	61	0.27	1.05	0.28
			11	55	141	2.91	1.40	4.07	78	50	0.29	0.95	0.28
12	17.94	67	4	63	93	2.26	1.92	4.34	71	66	0.94	1.66	1.56
			9	59	114	4.39	1.86	8.17	76	42	0.28	1.50	0.42
			11	56	120	5.47	1.88	10.28	78	32	0.24	1.17	0.28

u: horizontal wind velocity
KE: kinetic energy. The exponential relationship found between KE and u is $E(u) = 1.9723e^{0.1796u}$ (the relationship between columns 1 and 2).
α: rain inclination from vertical
θ: slope gradient
ϕ: angle of rain incidence calculated using rain inclination, slope gradient and slope aspect by cosine law: $\cos(\phi) = \cos(\alpha - \theta) = \cos\alpha\cos\theta + \sin\alpha\sin\theta$ for windward slopes. $\cos(\phi) = \cos(\alpha + \theta) = \cos\alpha\cos\theta - \sin\alpha\sin\theta$ for leeward slopes
I: rain intensity
D: rainsplash detachment rate
X_c: mean rainsplash distance calculated by center of gravity of mass distribution curves
q_s: rainsplash transport estimated by: $q_s = DX_c$

E

CONCLUSIONS

Wind could make significant changes in physical characteristics of rains and hence in soil erosion processes. The rainsplash detachment and transport under wind-driven rainfall would differ from that under windless rain. The rainsplash transport could be a significant process to the extent that it may not be negligible in accurately predicting water erosion, and this process could result in a net transportation in the prevailing wind direction.

REFERENCES

1. Erpul, G.; Gabriels, D.; Janssens, D. Assessing the Drop Size Distribution of Simulated Rainfall in a Wind Tunnel. Soil and Tillage Research **1998**, *45*, 455–463.
2. Lyles, L.; Disrud, L.A.; Woodruff, N.P. Effects of Soil Physical Properties, Rainfall Characteristics, and Wind Velocity on Clod Disintegration by Simulated Rainfall. Soil Sci. Soc. Am. Proc. **1969**, *33* (2), 302–306.
3. Disrud, L.A.; Lyles, L.; Skidmore, E.L. How Wind Effects the Size and Shape of Raindrops. Agricultural Engineering **1969**, *50* (10), 617.
4. Pedersen, H.S.; Hasholt, B. Influence of Wind Speed on Rainsplash Erosion. Catena **1995**, *24*, 39–54.
5. Erpul, G.; Gabriels, D.; Janssens, D. The Effect of Wind on Size and Energy of Small Simulated Raindrops: A Wind Tunnel Study. International Agrophysics **2000**, *14*, 1–7.
6. Sharon, D. The Distribution of Hydrologically Effective Rainfall Incident on Sloping Ground. Journal of Hydrology **1980**, *46*, 165–188.
7. Gabriels, D.; Tack, K.; Erpul, G.; Cornelis, W.M.; Norton, L.D.; Biesemans, J. *Effect of Wind-Driven Rain on Splash Detachment and Transport of a Silt Loam Soil: A Short Slope Wind Tunnel Experiment*; Proceedings of the International Workshop on Technical Aspects and Use of Wind Tunnels for Wind-Erosion Control, and Combined Effect of Wind and Water on Erosion Processes. Gent–Belgium, November 17–18, 1997; Gabriels, D., Cornelis, W.M., Eds.; I.C.E. Special Report No. 1998/1 International Center for Eromology, University of Ghent: Belgium, 1997; 87–93.
8. Van Heerden, W.M. Splash Erosion as Affected by the Angle of Incidence of Raindrop Impact Unpublished Ph.D. Thesis Purdue University, Lafayette, Ind., 1961.
9. Bagnold, R.A. *The Physics of Blown Sand and Desert Dunes*; 1973 Reprint, Chapman & Hall: London, 1941; 265.
10. De Lima, J.L.M.P.; Van Dick, P.M.; Spaan, W.P. Splash-Saltation Transport Under Wind-Driven Rain. Soil Technology **1992**, *5*, 151–166.

EROSION BY WIND, EFFECTS ON SOIL QUALITY AND PRODUCTIVITY

John Leys
Department of Land and Water Conservation, Gunnedah, New South Wales, Australia

INTRODUCTION

Accelerated erosion on agricultural lands has adverse effects on soil quality and productivity through the removal of soil particles and nutrients. The majority of reviews of this topic have concentrated on the impact of water erosion on soil and crop productivity (1, 2). However, there is a growing body of research on the impact of wind erosion on soil quality and productivity, and that is the focus of this article.

Wind erosion reduces soil quality and production using a mechanism different from that of water erosion. Water erosion removes the soil *en masse*, while wind erosion winnows the finer/lighter particles from the surface, leaving the larger (generally inert) particles behind. As a result, wind erosion removes topsoil (3), reduces soil clay and silt content (4), and reduces organic matter (5). Wind erosion also impacts on crop productivity. It sandblasts emerging crops (6), reshapes the landsurface, thereby making it difficult to traverse with wide agricultural implements, buries or undermines infrastructure such as fences and roads, and buries adjacent land with sand drift. This results in limiting the drifted land's production in the short term (7). Although wind erosion also has considerable off-site impact—reduces visibility (8), deposits unwanted dust and associated contaminates off-farm, and raises airborne particulate levels (9), with particle sizes less than 10 micrometers (PM_{10}), which can have adverse health effects (10)—off-site impacts will not be discussed here.

SOIL QUALITY

Soil quality is generally adversely affected by wind erosion via the removal of soil fines (clay, silt, and organic fractions). Soil texture changes are largely irreversible, unless topsoil is imported to the site. Quantification of the changes in soil texture, as well as identifying the eroded fractions in the eroded sediments, indicates the magnitude of the decline in soil quality brought about by wind erosion. Descriptions of soil texture changes and the eroded sediments will be presented in the next two sections.

Soil Texture Changes

The impact of erosion on soils has been measured over many years, using many different methods. Long-term analysis of soils exposed to wind erosion (11) showed a decline in the fertility and particle size distribution (PSD) of the surface soil. Over a 36-year period, increases in the sand fraction of the top 0–10 cm of 6.5% have been reported for the midwestern United States (12). Comparison of the PSD of a soil that had been farmed for 30 years and had suffered periodic erosion with an adjacent soil under native vegetation (13), revealed a loss of fines in the 10–100 μm fraction, as well as an increase in the coarse 350–1000 μm fraction (Fig. 1). Increases in the saltation fraction (>250 μm), and reduction in the 75–210 μm fraction, have been reported for the top 1 cm soil layer over a 15-week period for southeastern Australia (3). Further research at the same site using a portable field wind tunnel indicated an increase in the dominant sediment population of PSD of the surface soil layer (approx. 0–500 μm depth) after a 30-minute simulated erosion event. The analysis indicates that for the soil cultivation ridges, the proportion of the 300 μm sediment population increased from 60% in the parent soil to 86% after the simulated erosion. In the cultivation furrows, the 420 μm population increased from 0 to 85%. Analysis of the eroded sediments indicated that >78% of material being eroded fell within the 180 μm population, compared to 29% in the parent soil. The increase in the coarser fractions of the surface soil and high proportion of finer fractions in the eroded sediment, implies that wind erosion is winnowing the fines.

Eroded Sediments

A large number of studies in North America (14), Belgium (15), Nigeria (16), Australia (3), and China (17) show

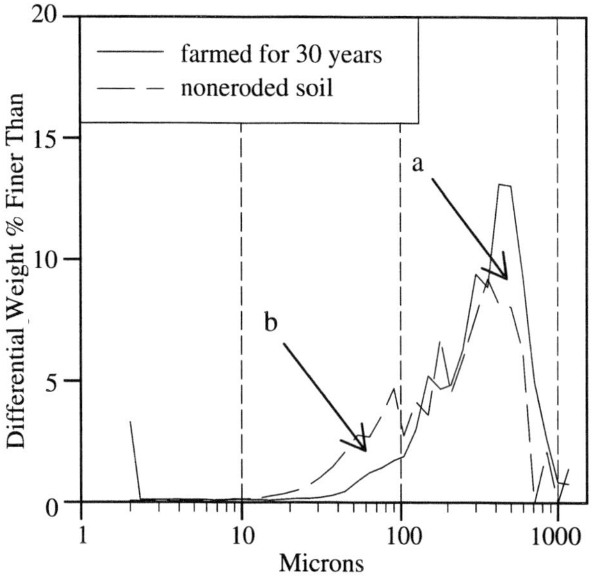

Fig. 1 Particle size distribution of two adjacent soils. One has been farmed for 30 years and has undergone repeated erosion; the other is under native vegetation. The PSDs show an increase in the coarse fraction (*arrow a*) and a decrease in the finer fraction (*arrow b*) in the eroded soil.

a decreasing particle size with height, indicating a selective sorting of eroded particles. When the vertical component of the wind exceeds the fall velocity of the eroded particles, they and their associated nutrients are removed from the site. Australian results indicate that for a one-week period of monitoring, 27% of the total eroded sediment is in the suspension fraction and thus, are removed from the eroded field (3). These studies further highlight the winnowing action of the wind erosion processes and the potential loss of soil and nutrients.

SOIL PRODUCTIVITY

When soil is eroded from a site, there is often a subsequent decrease in rooting depth and available water holding capacity (13). There is also a loss of nutrients and a subsequent decline in soil productivity. So if there is wind erosion, where does the nutrient go and how much impact does it have on crop production?

Soil Nutrient Loss

There is considerable evidence that wind-eroded sediments are enriched with nutrients. The nutrient loss has been measured at the site of dust emission, at various heights above the eroded surface, and downwind of the eroded site, both immediately adjacent in drift banks of soil and from deposited dust. It is important to examine the magnitude of this nutrient loss.

The nutrient content and particle size of eroded sediments collected at 0–0.5 m height during wind tunnel tests were similar to source sediments (18). However, if the sediments are sieved and the <90 μm fraction of the sediments analyzed, then there are significant enrichment ratios, where the enrichment ratio is the ratio of the nutrient concentration in the eroded sediment compared to the nutrient concentration of the source soil. The sieve size of 90 μm was chosen because the modal particlesize of sediments collected, using wind vane samplers (19) from field trials in Australia, where the wind tunnel studies were undertaken, were all less than 90 μm for sediments collected at or greater than 0.5 m (3).

The mean enrichment ratio (standard deviation in parentheses) of eroded sediments sampled in a wind tunnel with a 4 m working section, for 25 sites, each replicated 10 times, for a range of farming systems and soil types in the drier agricultural areas of South Australia, were 3.43 (1.89) for total nitrogen (N), 2.53 (1.09) for total phosphorous (P), and 5.00 (2.46) for organic carbon after one minute duration of 75 km h^{-1} wind (20). Therefore, there is significant winnowing of nutrients and enrichment of eroded sediments, even over short distances, i.e., 4 m.

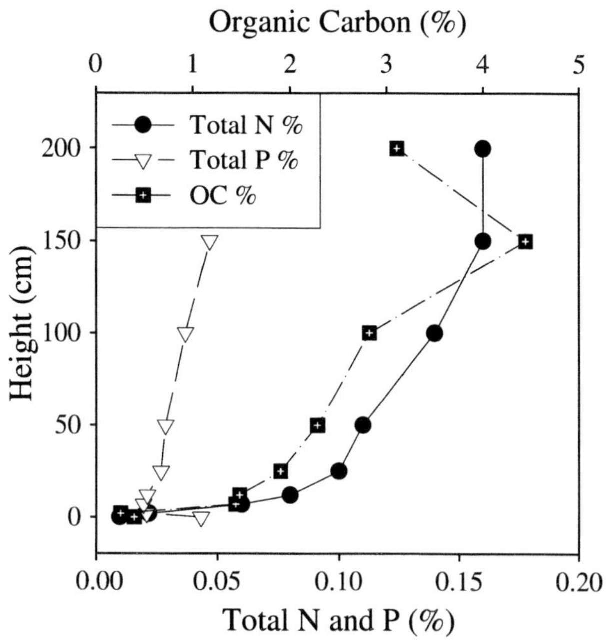

Fig. 2 Relationship between nutrient concentration and height for eroded sediments. (Adapted from Ref. 13.)

The mean (standard deviation in parentheses) emission rates of nutrients from the previously mentioned 25 sites (in $g\,m^{-2}\,s^{-1}$), were 0.0007 (0.0008) for total N, 0.0002 (0.0002) for total P, and 0.0086 (0.0108) for organic carbon after one minute duration of a $75\,km\,h^{-1}$ wind.

Analysis of dust collected in passive dust samplers at the source area, shows that with increasing height above the ground, nutrient concentration increases, along with a decrease in particle size (13) (Fig. 2). This indicates, the greater the sorting of the sediments, the greater the nutrient concentration in the eroded sediments.

Enrichment ratios of deposited sediment increase with distance from the erosion source because the sediments become more sorted. Measurement of wind-eroded material deposited near eroding fields have enrichment ratios of the order 1.21 for total N and 1.25 organic carbon (21); therefore, some nutrient is redistributed locally. For one study in southeastern Australia, enrichment ratios of 7.84 for total N and 2.96 for total P have been reported with corresponding deposition rates of $0.0034\,g\,m^{-2}\,day^{-1}$ for total N and $0.0008\,g\,m^{-2}\,day^{-1}$ for total P (22).

The previous discussion reveals that significant enrichment of eroded sediments occurs and that this enrichment increases with distance from source, both vertically and horizontally. The magnitude of this loss is large enough to influence subsequent crop production.

Crop Production Decline

There have been many experiments to test the assumption that soil loss causes crop production decline. The basic method has been to scalp off the surface soil and measure the subsequent changes in crop production. However, this approach is not entirely applicable to wind erosion because of the selective nature of the wind erosion processes as previously described. The problem is compounded because plant response to erosion is complex.

Wind erosion winnows the finer fractions from the soil surface. When these sediments are analyzed, they do show considerable enrichment in nutrients compared to the soil from which they were derived. But is this nutrient loss significant?

Erosion can impact crop production because there is a reduction in the soil depth (23); there is a direct reduction in the amount of nutrients and water availability (13); or a combination of both.

Trial results are inconclusive, with some crops in some years showing no yield effects (24). Canadian research shows a linear decrease with increasing wind erosion severity (25). For spring wheat the results demonstrate that for every 1 m across a 200 m field, there was a decline in grain yield of $3.6\,kg\,ha^{-1}$. When the experiment was repeated for canola the following year, no yield depression was observed. Similarly, results in the United States revealed yield depression for two consecutive years for grain sorghum and kenaf yields, but not in each year for cotton and forage sorghum, because of climatic variations between years (24). It has also been suggested that above-average rainfall may also compensate for nutrient loss (26). This would further mask the effects of nutrient loss on soil production.

Nutrient Restoration

The obvious solution to nutrient loss is to replace it with the next crop. Previous studies demonstrate that additional fertilizers only partially compensate for nutrient loss from erosion (27). The effects of fertilizer are highly dependent on soil type, soil depth, the level of the erosion, the initial fertilizer level, and the chemistry of the soil (i.e., P can be bound up in soils with high calcium carbonate).

CONCLUSION

Wind erosion of agricultural land has negative erosion effects on soil quality and soil productivity. The winnowing of the clay, silt, and organic fractions from the surface soil, and its transport off-site during wind erosion, result in nutrient decline and a reduction in water-holding capacity of the soil. Sandy soils are more prone to wind erosion and suffer the greatest impact of soil and nutrient loss. The loss of soil productivity through wind erosion impacts on subsequent crop production. The level of crop production decline depends on crop, soil, and climatic factors. The use of inorganic fertilizers to replace lost nutrients assists in maintaining crop production, but does not fully compensate for eroded particles and nutrients. Minimizing erosion is the best way to maintain soil quality and productivity.

REFERENCES

1. Lal, R. *Soil Quality and Agricultural Sustainability*; Ann Arbor Press: Chelsea, Michigan, 1998; 1–378.
2. Williams, J.R.; Allmaras, R.R.; Renard, K.G.; Lyles, L.; Moldenhauer, W.D.; Langdale, G.W.; Meyer, L.D.; Rawls, W.J. Soil Erosion Effects on Soil Productivity: A Research Perspective. Journal of Soil and Water Conservation **1981**, *36*, 82–90.
3. Leys, J.F.; McTainsh, G.H. Sediment Fluxes and Particle Grain-Size Characteristics of Wind-Eroded Sediments in

Southeastern Australia. Earth Surface Processes and Landforms **1996**, *21*, 661–671.
4. Zobeck, T.M.; Fryrear, D.W. Chemical and Physical Characteristics of Windblown Sediment Quantities and Physical Characteristics. Transactions of the ASAE **1986**, *29* (4), 1032–1036.
5. Daniel, H.A.; Langham, W.H. The Effect of Wind Erosion and Cultivation on the Total Nitrogen and Organic Matter Content of Soil in the Southern High Plains. Journal of the American Society of Agronomy **1936**, *28* (8), 587–596.
6. Armbrust, D.V. Wind and Sandblast Injury to Field Crops: Effect of Plant Age. Agronomy Journal **1984**, *76*, 991–993.
7. Gaynor, J.D.; MacTavish, D.C. Movement of Granular Simazine by Wind Erosion. Horticultural Science **1981**, *16*, 756–757.
8. Hagen, L.J.; Skidmore, E.L. *Wind Erosion and Visibility Problems*; No. 76-2019 American Society of Agricultural Engineers: St. Joseph, Michigan, 1976; 1–15.
9. Saxton, K.E. Wind Erosion and Its Impact on Off-Site Air Quality in the Columbia Plateau-An Integrated Research Plan. Transactions of the ASAE **1996**, *38*, 1031–1038.
10. Choudhury, A.H.; Gordian, M.E.; Morris, S.S. Associations Between Respiratory Illness and PM10 Air Pollution. Archives of Environmental Health **1997**, *52* (2), 113–117.
11. Lyles, L. *Wind Erosion: Processes and Effect on Soil Productivity*; 1976 Annual Meeting, American Society of Agricultural Engineers; American Society of Agricultural Engineers, Michigan, 1976; No. 76–2016; 23.
12. Lyles, L.; Tatarko, J. Wind Erosion Effects on Soil Texture and Organic Matter. Journal of Soil and Water Conservation **1986**, *41*, 191–193.
13. Leys, J.F.; McTainsh, G.H. Soil Loss and Nutrient Decline by Wind Erosion-Cause for Concern. Australian Journal of Soil and Water Conservation **1994**, *7* (3), 30–40.
14. Vories, E.D.; Fryrear, D.W. Vertical Distribution of Wind-eroded Soil Over a Smooth, Bare Field. Transactions of the ASAE **1991**, *34* (4), 1763–1768.
15. Goossens, L.D. The Granulometrical Characteristics of a Slowly Moving Dust Cloud. Earth Surface Processes and Landforms **1985**, *10*, 353–362.
16. Sterk, G.; Raats, P.A.C. Comparison of Models Describing the Vertical Distribution of Wind-Eroded Sediments. Soil Science Society of America Journal **1996**, *60*, 1914–1919.
17. Chen, W.; Fryrear, D.W. Sedimentary Characteristics of Drifting Sediments Above an Eroding Loessal Sandy Loam Soil as Affected by Mechanical Disturbance. Journal of Arid Environments **1998**, *39*, 241–440.
18. Leys, J.F.; Heinjus, D.R. *Simulated Wind Erosion in the South Australian Murray Mallee*; Research Report, Soil Conservation Service of NSW: Sydney, 1991.
19. Shao, Y.; McTainsh, G.H.; Leys, J.F.; Raupach, M.R. Efficiencies of Sediment Samplers for Wind Erosion Measurement. Australian Journal of Soil Research **1993**, *31*, 519–532.
20. Leys, J.F.; Packer, S.; Butler, P. *Wind Erosion Research at Keith in the Upper South East of South Australia*; Department of Land and Water Conservation: Sydney, 1997; 28.
21. Cihacek, L.J.; Sweeny, M.D.; Deibert, E.J. Characterization of Wind Erosion Sediments in the Red River Valley of North Dakota. Journal of Environmental Quality **1993**, *22* (2), 305–310.
22. Leys, J.F.; McTainsh, G.H. Dust and Nutrient Deposition To riverine Environments of Southeastern Australia. Zeitschrift für Geomorphologie Supplementband **1999**, *116*, 59–76.
23. Larney, F.J.; Izaurralde, R.C.; Janzen, H.H.; Olson, B.M.; Solberg, E.D.; Lindwall, C.W.; Nyborg, M. Soil Erosion — Crop Productivity Relationships for Six Alberta Soils. Journal of Soil and Water Conservation **1995**, *50* (1), 87–91.
24. Zobeck, T.M.; Bilbro, J.D. Crop Productivity and Surface Soil Properties of a Severely Wind-Eroded Soil. In *Sustaining the Global Farm*; Nearing, M., Ed.; Purdue University, West Lafayette, Indiana Soil Conservation Organization: Ankeny, 2000.
25. Larney, F.J.; Bullock, M.S.; Janzen, H.H.; Ellert, B.H.; Olson, E.C.S. Wind Erosion Effects on Nutrient Redistribution and Soil Productivity. Journal of Soil and Water Conservation **1998**, *53* (2), 133–140.
26. Arce-Diaz, E.; Featherstone, A.M.; Williams, J.R.; Tanaka, D.L. Substitutability of Fertilizer and Rain Fall for Erosion in Spring Wheat Production. Journal of Production Agriculture **1993**, *6*, 72–76.
27. Larney, F.J.; Janzen, H.H.; Olson, B.M. Efficiency of Inorganic Fertilizers in Restoring Wheat Yields on Artificially Eroded Soils. Canada Journal of Soil Science **1995**, *75*, 369–377.

EROSION BY WIND, FIELD MEASUREMENT OF

Ted M. Zobeck
United States Department of Agriculture, Lubbock, Texas, U.S.A.

INTRODUCTION

Accurate, reliable and direct measurements of windblown sediment are needed to confirm and validate erosion models, assess the intensity of aeolian (wind related) processes in the environment (1), determine source and origin of pollutants, and for many other applications. Field measurements of wind erosion consist of sensing or collecting samples of windblown sediment that may be rolling (creep), or bouncing along the ground (saltation), or suspended in the atmosphere (suspension). The sediment may be of local or distant origin and be moving horizontally, particularly during intense dust storms, or vertically, as dust fall. This section will discuss factors important in sampler design and describe samplers commonly used in wind erosion research.

SAMPLING CONSIDERATIONS

The type of sampler used to collect windblown sediment during a wind erosion event may vary depending upon the mode of particle transport. Particles moving in the creep or saltation mode are larger particles that travel close to the soil surface and move primarily in a horizontal direction. Suspended particles are small particles that may travel horizontally great distances at great heights and return vertically as dust fall.

Because the sediment is moved by the wind, any changes in the wind field caused by a sampler may also cause inaccuracies or errors in sample collection. A good sampler should be isokinetic and highly efficient over a wide range of incident wind angles (2). An isokinetic sampler maintains the same wind speed through the orifice of the sampler as the ambient wind at the same height. The sampling efficiency describes how well samplers collect sediment compared with the actual amount of sediment in the wind stream. Samplers that collect the same amount of sediment that is actually in the path of the sampler are 100% efficient. Samplers that only collect half of the amount actually present are 50% efficient.

Many different kinds of wind erosion sediment samplers and sensors have been used in the past (3–5). The sampling process may be passive or active. Passive samplers rely on natural ambient wind conditions when collecting samples. Active samplers are most often used for collecting fine suspended particles and use vacuum pumps to ensure the sampler is isokinetic or to draw particles into the samplers.

CREEP, SALTATION, AND OTHER TRANSPORT MODE COMBINATION SAMPLERS

Particles moving in creep and saltation have been measured by using both horizontally and vertically configured samplers. Horizontally configured creep and saltation samplers, also called sand traps, have horizontal openings and are generally constructed flush to the soil surface. Horizontal sand traps present very little interference with the wind stream and are passive and highly efficient for rolling or bouncing sediment if the sample area is longer than the bounce distance (3). Sand traps usually consist of a series of pits dug parallel to the expected wind direction. The main problems associated with pit sand traps include maintaining an adequate depth to prevent scour of the samples deposited in the traps, large volumes of sample if the pits are large, and inability to change sampling direction as wind direction changes. Sand traps have most often been used in studies of sand movement in coastal environments where wind direction tends to be less variable.

Vertically configured samplers have been used for many years to study sand and dust movement. Perhaps the earliest application of these samplers was in a wind erosion study of dunes in Western Australia in 1908 (6). In this study, the carrying capacity of the wind was estimated by measuring sediment collected at 20 mm intervals over a total height of 80 mm using five sheets of corrugated iron held together in a frame. Perhaps the most well-known device was developed by Bagnold to measure creeping and saltating particles (7). The Bagnold creep and saltation sampler has a vertical slot 760 mm high and 13 mm wide connected to a buried sample reservoir (Fig. 1). The slot is only open on the end facing the wind and has downward

Encyclopedia of Soil Science

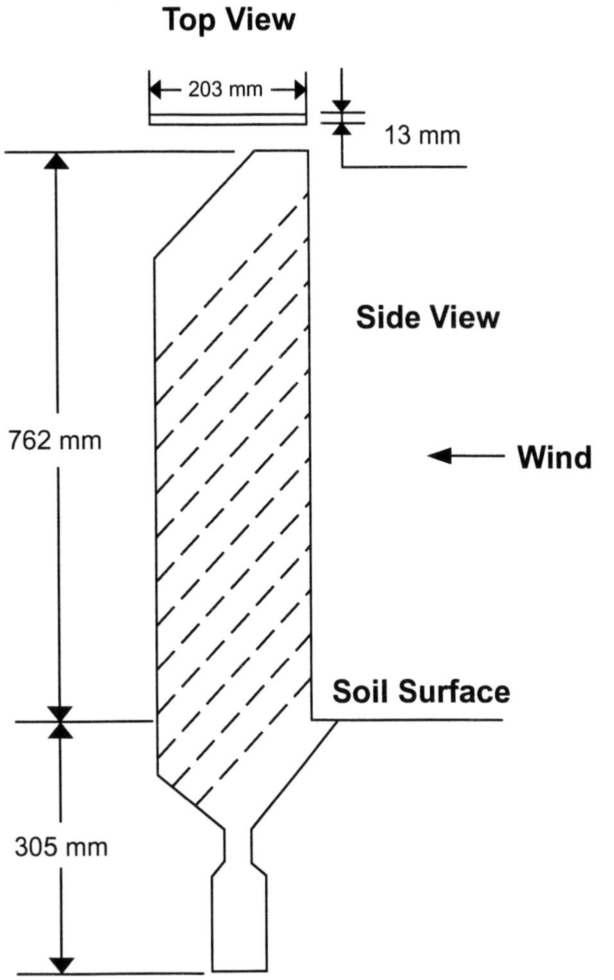

Fig. 1 Schematic diagram of Bagnold sampler. (Adapted from Ref. 3.)

slanting partitions to direct the sample down to the buried reservoir. This type of sampler measures all of the sediment collected throughout its height.

Many modifications have been made to improve the efficiency and utility of the basic Bagnold design. Chepil (8) modified the Bagnold design by adding a wind vane and mounting the sampling slot on bearings to allow it to change direction with the wind. Because the Bagnold sampler is closed at the back and top, a back-pressure develops, yielding a sampling efficiency of about 20% (3). Tests have shown that simply loosening the connection of the sampling reservoir allows air to pass through the sampler and increases the efficiency to about 50% (3). Other innovations included adding baffled openings (3) or screens with various opening sizes to the backs or sides of the sampler to allow relief of the back-pressure. Attempts to further improve sampling efficiency were made

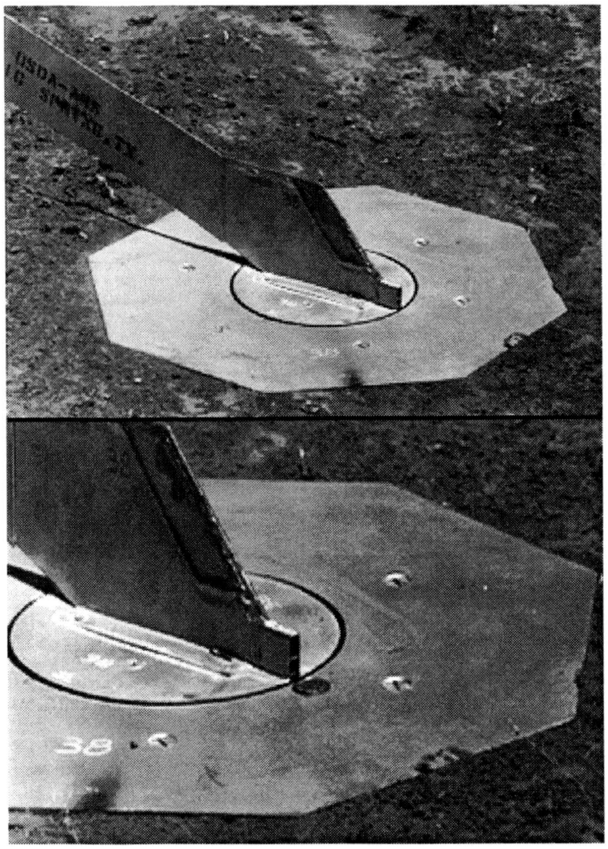

Fig. 2 Creep-saltation sampler (*top*) and close-up of sampler openings (*bottom*); a dime is placed in front of the sampler openings for scale.

by using a wedge-shaped design. The location and size of screen used with a wedge-shaped sampler directly affect the inlet sampling velocity and subsequent sampling efficiency (9). A recent passive wedge-shaped design, with a 500 mm high, 20 mm wide vertical slot and a stainless steel 200 mesh screen (62.5 micron openings) covering the back, had a sampling efficiency exceeding 90% over a wide range of wind velocities (2).

Vertical samplers have also been developed to sample over specific height intervals rather than integrating over the entire sampler height. A creep and saltation sampler developed by USDA (10) collects creep-saltation material from three inlets at 0–3 mm, 3–9 mm, and 9–20 mm above the surface (Fig. 2). The sampler rotates into the wind and vents air out of a 60 mesh screen. A rotating wedge-shaped saltation and suspension combination sampler, called BSNE (Big Spring Number Eight), has been developed to efficiently sample saltation sediment at greater heights above the soil surface (10, 11). In typical applications (12) samplers are mounted at heights of 0.05,

Fig. 3 BSNE sampling cluster (*top*) and close-up of three lowest samplers (*bottom*); a dime is placed above the lowest sampler opening for scale.

Fig. 4 Wilson and Cook-type sampling cluster (*left*) and close-up (*right*) of the four lowest samplers. (Photo by Geert Sterk.)

0.10, 0.25, 0.50 and 1.0 m (Fig. 3). The efficiency of the BSNE sampler for saltation-sized sediment exceeds 90% (11). Another very inexpensive and simple saltation and suspension combination sampler (Wilson and Cooke samplers) used to collect sediment at various heights is constructed from plastic bottles with an inlet and outlet made of glass tube bent at a 90 degree angle (13). These samplers can be mounted on wind vanes similar to the BSNE (Fig. 4).

The devices described above are usually used to collect sediment over a considerable period of time, often including several storms. Recent advances in sampler design have focused on measuring erosion at very short time intervals. A horizontal sand trap has been developed with a measurement frequency of once per second (1 Hz) (14). The trap uses a circular, horizontal opening with a load cell and tipping bucket to give continuous measurement of sediment flux. A similar device (15) uses a wedge-shaped vertical slot sampler (2) connected to a load cell and tipping bucket apparatus (Fig. 5).

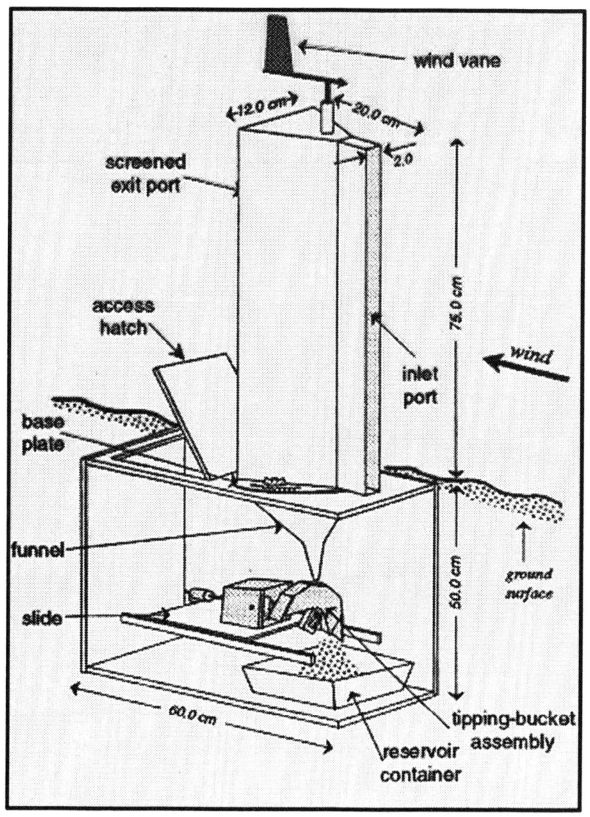

Fig. 5 Wedge-shaped, vertical slot sampler with tipping bucket assembly. (From Ref. 15.)

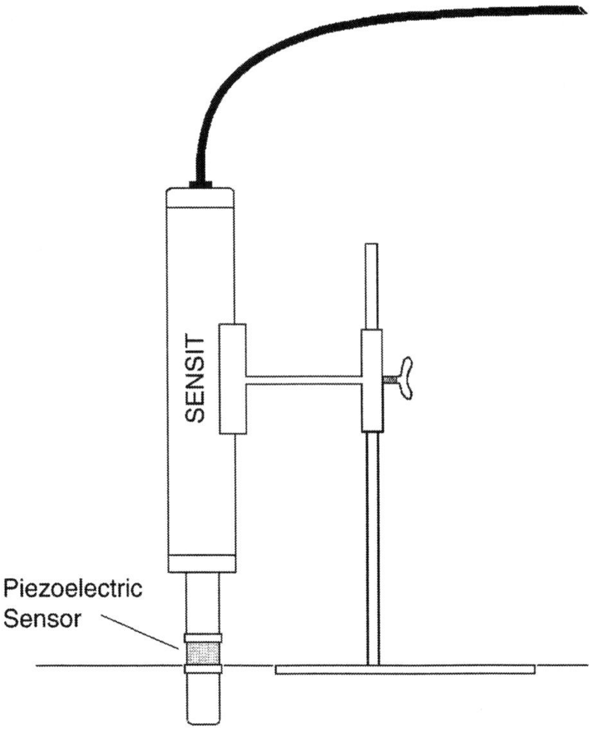

Fig. 6 SENSIT particle impact sensor.

Fig. 7 Pan-type dust collector. The pan diameter is 250 mm. The metal bands crossing over the top discourage bird perching.

Two electronic devices have been used to detect the onset of erosion by detecting particles moving in saltation. The SENSIT (16) uses a piezoelectric crystal sensor to detect the impact of saltating grains (Fig. 6). The SENSIT is not sensitive to very small suspended particles but can be easily used to estimate the duration of erosion events by estimating saltating particle impacts at a frequency of 1 Hz (17). The Saltiphone (18) is an acoustic sensor that records the impacts of saltating particles striking a 200 mm^2 diameter microphone mounted in a tube with a wind vane. The Saltiphone can measure at a frequency of 1000 Hz but movement of the vane is somewhat limited by the connecting cable.

SUSPENSION SAMPLERS

Particles moving in suspension are very small (<100 microns) and are easily transported horizontally and vertically by the wind. A wide variety of devices have been designed to measure suspended particulate matter for wind erosion and air quality studies (4, 5). This discussion will be limited to devices typically used in field wind erosion studies.

Although many types of devices and materials, including growing moss and moss bags (4), have been used to collect vertically falling suspended dust, pan-type samplers have been most commonly employed. When mounted at a height of about 2 m or at lower heights over stabilized vegetation-covered surfaces, the amount of saltating grains captured is very small. Regional studies of suspended dust (19) use a simple aluminum cake pan 250 mm in diameter and 100 mm deep mounted on a pole 2 m above the soil surface (Fig. 7). Several layers of glass marbles are suspended above the bottom of the pan on hardware cloth to reduce sample scouring. Metal bands covered with a sticky substance cross over the sampler in an attempt to repel birds.

Horizontally moving, suspended sediment also may be captured by BSNE or Wilson and Cooke samplers. Wind tunnel studies for suspended sediment using loess soil as a dust source have indicated the BSNE is only about 40% efficient and the Wilson and Cooke sampler is over 80% efficient (1). This study indicated that while the BSNE sampler sampling efficiency was constant, the Wilson and Cooke sampling efficiency varied somewhat with wind speed.

Perhaps the best method of collecting horizontally moving suspended sediment is with an active impinger-type device. In these devices, a vacuum source is connected to filtered sampling tubes pointed into the wind. The vacuum is usually controlled to ensure the sampler is isokinetic. The dust drawn into the sampler is then caught on some type of filter material. These devices have been successfully used for many years. Chepil and Woodruff (20) used 48 mm diameter sampling tubes mounted at four heights up to 6 m to measure dust in Kansas and Colorado. More recent studies in Africa employed 13 mm diameter tubes that automatically oriented into the wind mounted at four height to 10 m (21).

REFERENCES

1. Goosens, D.; Offer, Z.Y. Wind Tunnel and Field Calibration of Six Aeolian Dust Samplers. Atmospheric Environment **2000**, *34*, 1043–1057.
2. Nickling, W.G.; McKenna Neuman, C. Wind Tunnel Evaluation of a Wedge-Shaped Aeolian Sediment Trap. Geomorphology **1997**, *18*, 333–345.
3. Horikawa, K.; Shen, H.W. *Sand Movement by Wind Action: On the Characteristics of Sand Traps*, Tech. Memorandum No. 119; U.S Army Corps of Engineers, Beach Erosion Board: Vicksburg, MS, 1960; 1–51.
4. Knott, P.; Warren, A. Aeolian Processes. In *Geomorphological Techniques*; Goudie, A., Lewin, J., Richards, K., Anderson, M., Burt, T., Whalley, B., Worsley, P., Eds.; George Allen and Unwin: London, 1981.
5. Mark, D.; Hall, D.J. Recent Developments in Airborne Dust Monitoring. Clean Air **1994**, *23*, 193–217.
6. Olsson-Seffer, P. Relation of Wind to Topography of Coastal Drift Sands. J. Geol. **1908**, *16*, 549–564.
7. Bagnold, R.A. *The Physics of Blown Sand and Desert Dunes*; William Morrow & Co.: New York, 1941; 1–265.
8. Chepil, W.S. *Width of Field Strips to Control Wind Erosion*, Tech. Bull. 92; Kansas State College Agricultural Experiment Station: Manhattan, KS, 1947; 1–16.
9. Stout, J.E.; Fryrear, D.W. Performance of a Windblown-Particle Sampler. Trans. ASAE **1989**, *32* (6), 2041–2045.
10. http://www.csrl.ars.usda.gov/wewc/tooltech.htm.
11. Fryrear, D.W. A Field Dust Sampler. J. Soil and Water Conserv. **1986**, *41* (2), 117–120.
12. Stout, J.E.; Zobeck, T.M. The Wolfforth Field Experiment: A Wind Erosion Study. Soil Sci. **1996**, *161* (9), 616–632.
13. Wilson, S.J.; Cooke, R.U. Wind Erosion. In *Soil Erosion*; Kirkby, M.J., Morgan, R.P.C., Eds.; John Wiley and Sons: New York, 1980; 217–251.
14. Jackson, D.W.T. A New, Instantaneous Aeolian Sand Trap Design for Field Use. Sedimentology **1991**, *43*, 791–796.
15. Bauer, B.O.; Namikas, S.L. Design and Field Test of a Continuously Weighing, Tipping-Bucket Assembly for Aeolian Sand Traps. Earth Surface Processes and Landforms **1998**, *23*, 1171–1183.
16. Stockton, P.; Gillette, D.A. Field Measurements of the Sheltering Effect of Vegetation on Erodible Land Surfaces. Land Degradation and Rehabilitation **1990**, *2*, 77–85.
17. Stout, J.E.; Zobeck, T.M. Intermittent Saltation. Sedimentology **1997**, *44*, 959–970.
18. Spaan, W.P.; van der Abeele, G.D. Wind Borne Particle Measurements with Acoustic Sensors. Soil Technology **1991**, *4*, 51–63.
19. Rehies, M.C. Dust Deposition in Southern Nevada and California, 1984–1989: Relations to Climate, Source Area, and Source Lithology. J. Geophysical Res. **1995**, *100* (D5), 8893–8918.
20. Chepil, W.S.; Woodruff, N.P. Sedimentary Characteristics of Dust Storms: II. Visibility and Dust Concentration. Am. J. Sci. **1957**, *255*, 104–114.
21. Nickling, W.G.; Gillies, J.A. Dust Emission and Transport in Mali, West Africa. Sedimentology **1993**, *40*, 859–868.

EROSION BY WIND, GLOBAL HOT SPOTS OF

Andrew Warren
Department of Geography, University College London, London, United Kingdom

INTRODUCTION

Some distinctions, assumptions: first, "natural" can be differentiated from "induced" erosion (this entry will concentrate on the latter); second, hot spots vary both in time and space (they come and go and their intensity varies from place to place); third, a hot spot is assumed to be where the rate of wind erosion is well above the mean, even though it may not have serious economic impact. Hence, the Dust Bowl of the Great Plains in the 1930s was probably a hot spot, yet, today, agriculture is prosperous there. There and elsewhere, as in the West African Sahel, the Mallee of southeastern Australia, a great deal of soil may be lost without much impact on yield, until the last few centimeters of soil are left.

NATURAL EROSION

The Sahara, the Central Asian Deserts, and the Chinese deserts are the sites of very high rates of "natural" erosion. Much more sediment leaves the Sahara in wind than in water. Contemporary outputs from the Sahara are of the order of 10^9 tons per year (1). Dry Australia, the southwestern United States, and southern South America are less important, but still significant natural hot spots. The dust reaches far into the oceans and beyond, as from northern African to the United States, and from Australia to New Zealand (Fig. 1). Marine and land sediments (loess and stabilized sand dunes) show that the hot spots were hotter in the Pleistocene, and that they got hotter as it progressed. In the central Sahara, Egypt, western Argentina, eastern Iran, and parts of western China, great grooves were cut at those times by wind erosion, some in hard rock. There were also more hot spots, as in parts of northern Europe and central North America. Ice-core data from Greenland, the Tibetan Plateau, and the Antarctic corroborate the picture. The contemporary global source areas are all very dry. Explanation of the temporal change is more disputed (2).

Seen at a finer scale, most "natural" erosion in these deserts occurs in soft sediments, particularly those in ancient, dry lake basins. The northern Chad Depression in the central Sahara and the Lake Eyre basin in central Australia are major natural hot spots (3). In the late Pleistocene, the bare outwash plains round glaciers were the sources of loess, and of sand that formed dunes (now mostly stabilized) in northern Europe, Asia and northern Canada. People have little influence on rates of wind erosion in deserts. There are exceptions. Off-road vehicles create a minor hot spot in the Mojave of southern California; and wars and training for wars has created others in North Africa during the 1940s, and during the Gulf War (2).

INDUCED EROSION

Identifying global hot spots of induced wind erosion is more problematic. Measurements of emissions are very few indeed, measurements of rates of erosion are even fewer, and measurements of economic impact are the rarest of all. This leaves some results from modeling, historical accounts built on disparate evidence, and informed guesses. Modeling and measurement suggest short-term rates of the order of $30-60\,t\,ha^{-1}\,yr^{-1}$ on dry High Plains sites, depending on winds, rainfall, and land use. Caesium-based measurements in the Sahel of West Africa, on sandy soils under fallow-cultivation cycling, and with mean annual rainfalls of about 550 mm give rates of up to $40\,t\,ha^{-1}\,yr^{-1}$ over a 30-year period (4).

Some historical accounts are well documented and compelling. The Dust Bowl of the American Great Plains in the 1930s was an undisputed hot spot (5). Dust storms were more frequent and violent and many farmers went out of business, though more from drought than erosion. There were renewed hot spots here in the 1950s, 1960s, and 1970s, on the evidence of agricultural surveys and dust emissions. Probably more serious, though less well documented, at least in English, were the hot spots in the Soviet Union, specially northern Kazakhstan and neighboring parts of Russia, at about the same time (6) (although the earliest recorded paper in Russian on wind erosion control was written in 1768). More surprising, even to some experts, are histories of wind erosion

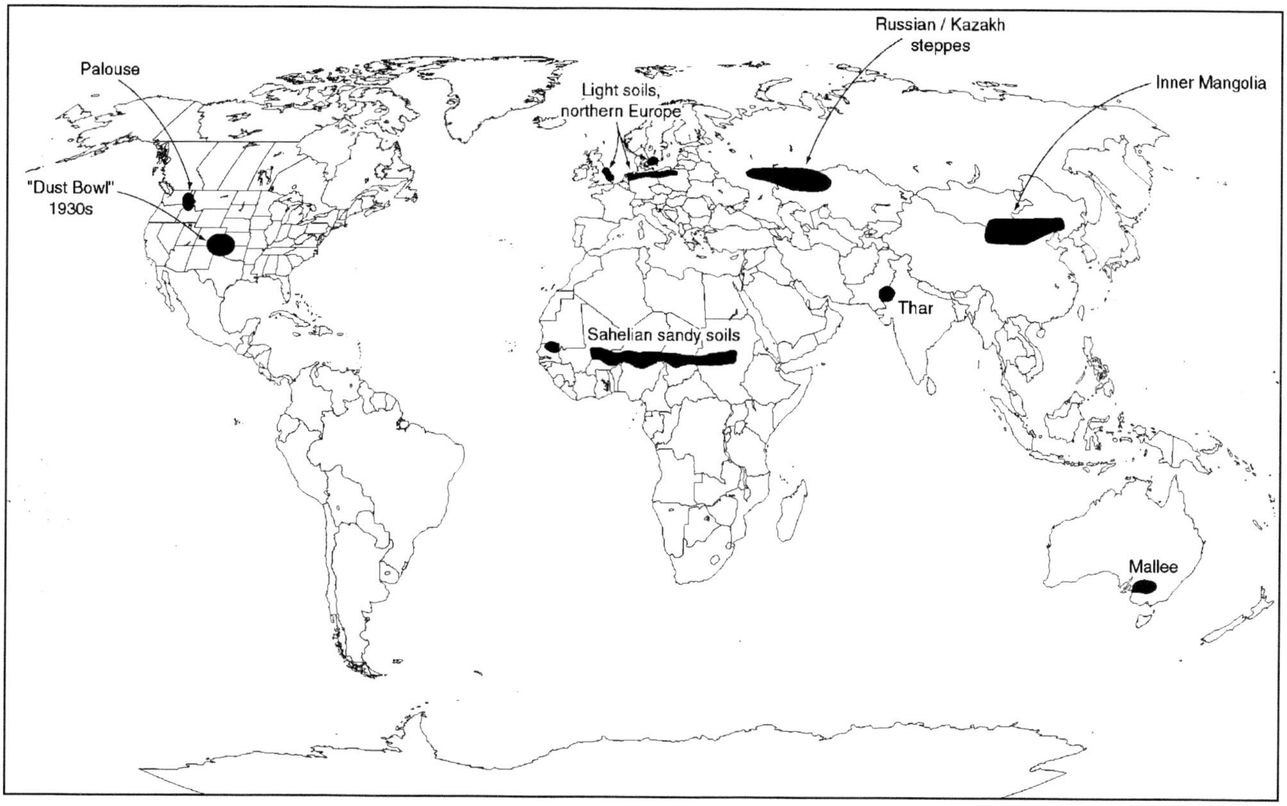

Fig. 1 Global hot spots for wind erosion.

in western Europe. An even earlier paper on wind erosion, titled "A curious and exact relation of a sand cloud which hath lately overwhelmed a great tract of land in the county of Suffolk" appeared in the *Philosophical Transactions of the Royal Society of London* in 1669. In Sweden, no less a person than the great naturalist Linnaeus recorded destitution because of wind erosion in Skania in the 18th century. There is similar evidence in northern Germany, The Netherlands, Denmark, Poland and Hungary. Archaeology now shows that there were hot spots as far back as the Neolithic when ancient farms were buried in sand in southeastern England and The Netherlands. In all these areas, it was and is sandy soils, inherited from glacial outwash, that are the most susceptible. Much more recently, the period of desiccation in the Sahel of West Africa is said to have been a hot spot for wind erosion. Some evidence lies in dust-storm data (1), and, as in the Great Plains of the 1930s, there was undoubtedly great hardship. However, here too, the exact role of wind erosion is debatable. Much of the dust may have come from quite small perimeters round dust-measuring stations, and the hardship was to do more with drought than erosion.

Informed guesses are a more dubious source. The literature is no great help, even with the wind erosion bibliography (listing works till 1995) and its 2700 odd entries (7). It is undoubtedly biased toward work in English, toward areas where funds have been more available and to places where there has been a strong scientific tradition. For example, research in the USA, and the Great Plains in particular, where massive funds for research were released and maintained after the 1930s, overwhelms the bibliography, yet the aforementioned historical evidence suggests that the problem may have been more serious in the USSR. Indeed, Petrov's bibliography of relevant work in Russian (1768–1950) has 852 entries. Germany, with its long scientific tradition, apparently dominates research in Europe, yet it is known that the problem is as serious in neighboring counties. Southeastern Australia is also, by this evidence, a hot-spot (8), yet in terms of damage to cropland, the Sahel of West Africa is probably more of one (the Sahel is only belatedly receiving scientific attention). Using the literature, Crosson (9) estimated that the Sahel was a major wind erosion hot spot, yet there are few data on which to base this conclusion (4).

Only one systematic attempt at informed guessing on a worldwide scale has been made. This is GLASOD (The *GL*obal *A*ssessment of *SO*il *D*egradation). It is most accessible in the World Atlas of Desertification, which is now in its second edition (10). The assessment was for "polygons" of the order of size of southeast England, the province of Skania in Sweden, or a quarter of an average US State. Experts who knew each polygon were asked a set of carefully constructed questions about the extent and severity of soil degradation, including wind erosion. The result is the best guess we have of the severity of wind erosion globally. But it has serious flaws: in Africa and many other parts of the world, there were no measurements at the time of the questionnaire against which to compare the guesses; the distinction between induced at natural erosion is unclear; some experts have interpreted the questions differently from others; the identifications are generally of potential rather than actual erosion; the polygons are huge.

The expert opinion used in GLASOD is based on our present scientific knowledge about wind erosion. If, as a simple framework, we take the factors in the famous Wind Erosion Equation of Woodruff and Siddoway, we can follow it in predicting the hot spots (11). Three factors are almost equally important. First is the Climatic (C) factor. Most induced wind erosion occurs in areas with semi-arid climates, as corroborated by maps of the C factor in the Sahel of West Africa, and in the United States. This is also true in the USSR, Australia, and northwestern Europe. However, wind erosion can and does take place in much wetter climates. In Wales and in the Pennine Hills of England peat is blown away after fires and in long dry spells. Erosion occurs in parts of England where the mean annual rainfall is of the order of 1500 mm. At roughly the same mean annual rainfall, there is also significant wind erosion on the coastal plain of South Carolina. Wind erosion is reported in the Austrian Tyrol, where rainfalls approach 2000 mm. Even in Iceland, where evapotranspiration is very low, there is some serious wind erosion. The C factor is also important in the temporal dimension: when there are dry years, wind erosion hot spots appear, as in the Dust Bowl and the Sahel.

The K, or soil, factor in the Wind Erosion Equation is the second important control, for wind erosion rarely affects fine-texture soils, although a few may be "pelletized" into aggregates of sand size, which may then be moved by the wind. Areas with agriculture on sandy soils are, worldwide, the common locations of hot spots, because they lack binding agents for aggregation. Peats, which also may be loose are also eroded in western Europe and Florida, when they are drained, cleared, and cultivated. Light loess soils, as in the Palouse of Washington State, parts of the Great Plains and the Russian steppes, are other hot spots.

The last crucial factor is vegetation cover, and this is the most susceptible to management. The plowing, or harrowing of a field can raise considerable quantities of dust (12), but it is clearance of vegetation in dry and windy seasons that is the more generally damaging. The reasons why farmers clear susceptible fields are exceedingly complex. Ignorance may sometimes produce a minute hot spot, but it is an inexperienced farmer indeed who does not realize that erosion follows clearance. In the Sahelian hot spot, for example, the laying of millet (*Pennisetum glaucum*) stalks to counter erosion is standard practice. Economic forces produce the main hot spots for induced erosion. In Europe, at least, the short-term economic impact of wind erosion does not provide a strong enough incentive to persuade farmers to manage it. Worster believed that the huge hot spot in the US Dust Bowl was mainly due to farmers being driven to cultivate by other economic forces, in this case the need to pay off loans from banks and from machinery salesmen. The massive Soviet hot spots are said to have been the result of yet another kind of economic force: centralized and ill informed attempts to reach US standards of economic output. In our study of a village in Niger, economics takes yet another guise: wind erosion is the outcome of a gamble by those farmers who can afford to lose a crop, because they have other economic options (13).

REFERENCES

1. Middleton, Nicholas J. Climatic Controls on the Frequency, Magnitude and Distribution of Dust Storms: Examples from India/Pakistan, Mauritania and Mongolia. In *Palaeoclimatology and Palaeometeorology: Modern and Past Processes of Global Atmospheric Transport*; Leinen, Margaret, Sarnthein, M., Eds.; Kluwer: Dordrecht, **1989**; 97–132.
2. Livingstone, Ian; Warren, Andrew. *Aeolian Geomorphology: An Introduction*; Addison-Wesley Longman: Harlow, **1996**.
3. McTainsh, Grant H. Dust Processes in Australia and West Africa: A Comparison. Search **1985**, *16* (3–4), 104–106.
4. Chappell, Adrian; Warren, Andrew; Oliver, Margaret A.; Charlton, M. The Utility of ^{137}Cs for Measuring Soil Redistribution Rates in South-West Niger. Geoderma **1998**, *81* (3–4), 313–338.
5. Worster, Donald. *Dust Bowl: The Southern High Plains in the 1930s*; Oxford University Press: Oxford, **1979**.
6. Alayev, E.B.; Badenkov, Yu. P.; Karavaeva, N.A. The Russian Plain. In *The Earth as Transformed by Human*

Action; Turner, Billy Lee II, Clark, Bill, Kates, Bob W., Richards, J.F., Matthews, J.T., Meyer, W.B., Eds.; Cambridge University Press: Cambridge, **1990**; 453–560.

7. Wind Erosion Bibliography:http://www.csrl.ars.usda.gov/wewc/apm.htm.
8. McTainsh, Grant H., Boughton, W.C., Eds. *Land Degradation Processes in Australia*; Longman Cheshire: Melbourne, **1993**.
9. Crosson, P.R. Will Erosion Threaten Agricultural Productivity? Environment **1997**, *39* (8), 4–31.
10. United Nations Environment Programme (UNEP). *World Atlas of Desertification*, 2nd Ed.; Arnold: London, **1997**.
11. Woodruff, N.P.; Siddoway, F.H. A Wind Erosion Equation. Proceedings of the Soil Science Society of America **1965**, *29* (5), 602–608.
12. Lee, Jeff A.; Tchakerian, Van T. Magnitude and Frequency of Bowing Dust on the Southern High Plains of the United States, 1947–1989. Annals of the Association of American Geographers **1995**, *85* (4), 684–693.
13. Warren, Andrew, Batterbury, Simon.P.J.; Osbahr, Henny, Soil Erosion in the Sahel of West Africa: A Review of Research Issues and an Assessment of New Approaches. Global Environmental Change-Human and Policy Dimensions **2000**, *11* (1), 79–95.

EROSION BY WIND, MICROMETEOROLOGY OF

Giles Wiggs
University of Sheffield, Sheffield, United Kingdom

INTRODUCTION

Wind erosion of soils and sediments can be considered at a number of temporal scales ranging from sub-second to annual or greater. Agricultural scientists often refer to amounts of erosion in terms of tonnes per hectare per year or equivalent but, whilst such terms are useful for comparison between fields, they mask many of the dynamic processes operating on the soil at much smaller spatial and temporal scales. In order to fully appreciate wind erosion it is important to have an understanding of the smaller-scale processes that are operating at time-scales ranging from seconds to minutes or hours. It is processes operating at this micro-scale that control the local erosivity (erosive power) of the atmosphere and can have an impact on the erodibility (susceptibility to erosion) of soils. Such meteorological factors include wind shear stress, wind turbulence, thermal stability, and humidity.

VELOCITY PROFILE, WIND SHEAR STRESS, AND SURFACE ROUGHNESS

As wind moves over a soil surface it is retarded at its base by friction from the roughness of that surface. This creates a velocity profile with a region very close to the surface of zero velocity and wind high in the profile (or boundary layer) unaffected by the surface friction (called free stream velocity). The depth of this zero-velocity layer (termed z_0) is called the aerodynamic roughness length. It is an important parameter because it partly controls the gradient of the velocity profile, which is proportional to the shear stress (or force) of the wind on the surface. When wind shear stress at the surface reaches a critical level, wind erosion will occur (1).

Under normal atmospheric conditions a wind velocity profile plots as a straight line on a logarithmic chart, the gradient of which is proportional to the surface shear stress. It is simpler to measure velocity profiles than it is to measure surface shear stresses, and so the gradient of a velocity profile (equivalent to shear velocity, u_*) is often used as a surrogate for surface shear stress.

The relationship between surface shear stress (τ_0) and shear velocity (u_*) is given by the following expression:

$$u_* = \sqrt{\tau_0/\rho}$$

where: ρ = air density.

From measurements of a simple mean velocity profile, an assessment can therefore be made of the shear stress imparted by the wind to the surface. High shear stresses are associated with high mean wind speeds, but many other factors also have an influence.

CHANGES IN SURFACE ROUGHNESS

One of the most important controls on shear stress is the roughness of the surface and its effect on the value of aerodynamic roughness, z_0. Any change in the value of z_0 has a subsequent effect on the velocity profile gradient because of the alteration in the depth of the zero-velocity region. With all other factors remaining the same, an increase in z_0 will lead to an increase in u_* (and hence surface shear stress) and vice versa (see Fig. 1).

The wind will respond to surface roughness at a variety of spatial scales. At a small scale, the individual soil clods create a drag on the airflow, as do ploughed furrows or crops. At a larger scale, shelterbelts, fences, and hummocky terrain also affect the velocity profile (see Table 1). As the wind blows across a single field, it will respond to these changing surface roughnesses and so too will the shear velocity, surface shear stress, and the ability of the wind to entrain and transport soil particles. Rough surfaces should therefore be more susceptible to wind erosion than smooth surfaces. However, the relationship is complicated by the fact that increases in surface shear stress caused by a larger value of z_0 may not necessarily lead to increases in sediment entrainment and transport. This is because any increase in surface roughness may result in more grains lying within the depth of z_0 and hence protected from erosion in a layer of zero-velocity flow. This is the reason why fields ploughed perpendicular to the mean wind direction can often be stable and resistant to wind erosion as the aerodynamic roughness height caused

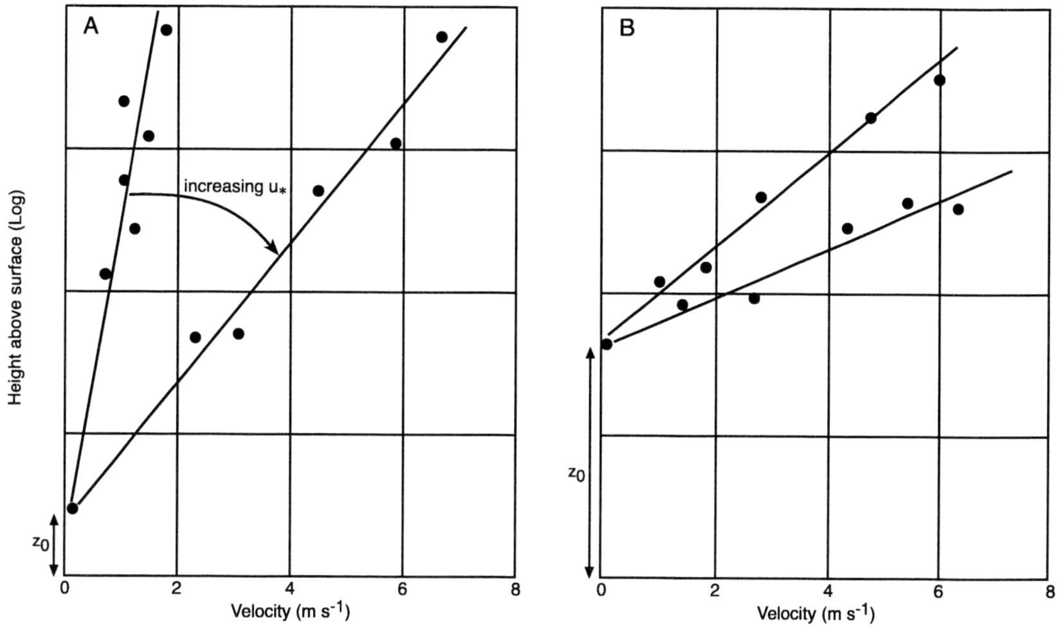

Fig. 1 Velocity profiles showing a focus at height z_0, the aerodynamic roughness length. The aerodynamic roughness in (**A**) is less than in (**B**), hence shear velocities (u_*) in (**B**) are greater. (From Ref. 1.)

by the furrows allows protection of fine surface soil. Typical values of z_0 for different surfaces are shown in Table 1.

ROLE OF TURBULENCE

The semi-logarithmic wind velocity profile (Fig. 1) only occurs under mean wind conditions, averaged over time periods ranging from perhaps 30 seconds or greater. However, under most natural conditions, wind in the environment is turbulent and consists of high velocity gusts and low velocity troughs. For this reason, velocity profiles measured over timescales of the order of 1 second do not display a semi-logarithmic relationship with height.

Table 1 Typical aerodynamic roughness (z_0) values for different surfaces

Surface	z_0 (m)
Sand	0.0003
Rippled sand	0.003
Soil	0.001–0.01
Grass	0.003–0.1
Agricultural crops	0.04–0.2
Forests	1.0–6.0

(Adapted from Ref. 2.)

In a turbulent velocity profile momentum transfer through the profile is achieved by turbulent eddies of varying sizes. These operate in a cascade from eddies as large as the boundary layer to viscous energy dissipation at the surface. These eddies move between different layers of fluid and impart or absorb energy depending on whether the eddy is faster or slower moving than the surrounding air (3). This type of momentum transfer is very efficient and results in high mean shear stresses being imparted to a soil surface.

Research (3 and 4) suggests that sediment entrainment and transport can be greatly influenced by peak values in instantaneous shear stress provided by these turbulent flow structures (see Fig. 2).

Three types of important turbulent structure have been identified: 1) eddies of low momentum fluid at the surface elevated into faster air above ("burst" events); 2) high momentum fluid from high in the velocity profile impacting the surface ("sweeps"); and 3) flow-parallel longitudinal vortices ("streaks").

Each of these flow structures may have periodicities of around 20 Hz and are considered to be critical in wind erosion because they may provide instantaneous peaks in shear stress which might exceed the threshold of grain entrainment, even where the mean wind velocity is just below that threshold. In these circumstances, turbulent flow structures may provide high instantaneous shear stresses which could entrain a few grains into saltation. These

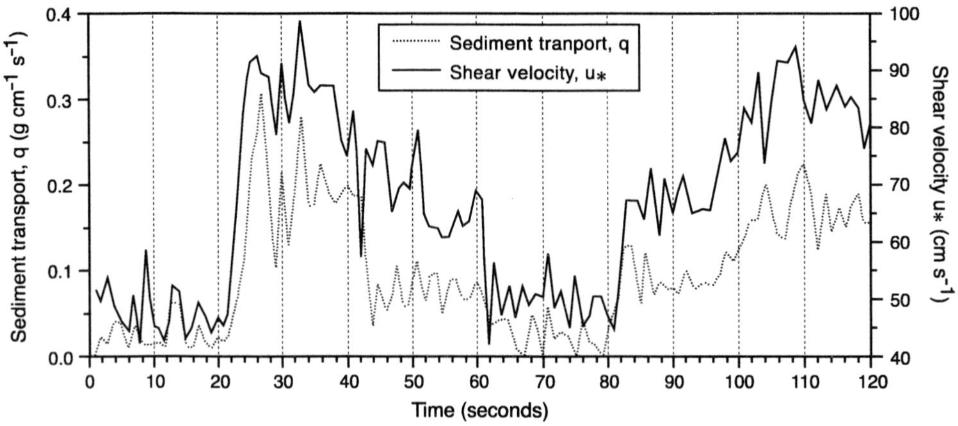

Fig. 2 The response of sediment transport (q) to fluctuations in instantaneous shear velocity (u_*). (From Ref. 4.)

grains may then be kept in transport by the mean wind velocity (as the energy required for sediment transport is less than that required for entrainment) and, on impacting the surface, splash other grains into transport. In this way, a few saltating grains entrained by turbulent motion may provide a downwind cascade of entrainment and result in significant wind erosion (1).

THERMAL STABILITY

The logarithmic relationships shown in Fig. 1 only arise in atmospheric conditions with neutral buoyancy. Such conditions are found with cloudy skies (which reduce radiative heating) and strong winds (which promote atmospheric mixing and prevent temperature stratification) (2). On clear and sunny days (especially in arid or semi-arid areas), strong radiative heating may result in thermal instability (with a steep temperature gradient), which increases buoyancy effects and vertically stretches turbulent eddies. This results in a reduction of the wind gradient (becoming more vertical with height, in Fig. 1) with atmospheric turbulence driven by buoyancy effects rather than mechanical forces. Conversely, atmospheric stability (often occurring at night with radiative cooling of the surface) tends to squeeze turbulent eddies vertically resulting in a strong wind gradient (becoming more horizontal with height in Fig. 1) with little vertical mixing.

Both atmospheric stability and instability therefore have an impact on the structure of the velocity profile and hence on the shear stresses evident at the surface. The impact of thermal stability or instability on wind erosion is a function of the role of turbulent eddies. Atmospheric instability is associated with greater vertical mixing of air and so fast-moving upper-air is commonly brought closer to the surface resulting in an increase in shear stress. In contrast, a stable atmosphere is associated with relatively low surface shear stresses because turbulent mixing is damped down (5).

Sediment transport and wind erosion are therefore more likely to be evident where the atmosphere displays thermal instability (rather than stability) because buoyancy forces result in strong vertical mixing and increased turbulence. However, atmospheric instability is also associated with low mean wind speeds and such conditions are unlikely to result in significant erosion. Wind erosion is more closely associated with neutral atmospheric buoyancy where a reasonable amount of turbulent mixing is combined with high mean wind speeds. As noted, such conditions are generally associated with overcast skies and high winds during the daytime.

CHANGES IN HUMIDITY

Humidity refers to the water vapour content of the atmosphere and it is an important variable in wind erosion because of the two-way transfer of water vapour between the soil surface and the near-surface atmosphere. This transfer has large implications for the erodibility of a soil.

Grain entrainment is a function of the ratio of resisting forces to dynamic forces (1). Whilst grain size is the most important of the resisting forces, considerable resistance to movement can be provided by moisture in the upper layers of the soil. The exact physical nature of the relationship between grain entrainment threshold and moisture content has yet to be quantitatively defined but there is a theoretical basis for suggesting that the threshold rises as a function of surface tension associated with pore moisture (6). Humidity has an important role to play here because an unsaturated air mass (low humidity) will have a significant drying effect on a soil and thus increase its erodibility (lowering the threshold of grain entrainment), whilst a saturated air mass will have the opposing effect. Research has suggested that the threshold of grain entrainment might be very sensitive to changes in moisture content below about 4–8% (6). Fairly small changes in atmospheric humidity may therefore have a critical impact on grain entrainment and wind erosion.

FUTURE DIRECTIONS FOR MICROMETEOROLOGICAL RESEARCH

A major challenge in wind erosion research is to successfully link the differing scales of investigation and applied activity. It is clear that an understanding of erosion processes at the micro-scale is necessary to successfully comprehend the wind erosion system, and many advances have been made toward this end. However, the application of this understanding is often at the field-scale or larger and it is in the transference of scales where failure has occurred in the past. Process measurements operating over timescales of seconds or hours cannot easily be linked to changes in field erosion which may occur over timescales of months to years. As scales of interest increase, the reliability and appropriateness of small-scale process work on wind erosion decreases. It has been argued that, where appropriate, small-scale process measurements should become check-points for model calibration (7). In this respect, micrometeorology is likely to become increasingly important in soil erosion models for the prediction and assessment of wind erosion.

REFERENCES

1. Wiggs, Giles F.S. Sediment Mobilisation by the Wind. In *Arid Zone Geomorphology*; Thomas, D.S.G., Ed.; Wiley: London, 1997; 351–372.
2. Oke, T.R. *Boundary Layer Climates*, 2nd Ed.; Routledge: London, 1987.
3. Clifford, Nicholas J.; French, Jon R.; Hardisty, Jack *Turbulence: Perspectives on Sediment Transport*; Wiley: Chichester, 1993.
4. Butterfield, Graeme R. Sand Transport Response to Fluctuating Wind Velocity. In *Turbulence: Perspectives on Sediment Transport*; Clifford, N.J., French, J.R., Hardisty, J., Eds.; Wiley: Chichester, 1993; 305–335.
5. Frank, Andy; Kocurek, Gary. Effects of Atmospheric Conditions on Wind Profiles and Aeolian Sand Transport with an Example from White Sands National Monument. Earth Surface Processes and Landforms **1994**, *19*, 735–745.
6. Sherman, Douglas J. Discussion: Evaluation of Aeolian Sand Transport Equations Using Inter-Tidal Zone Measurements. Sedimentology **1990**, *37*, 385–392.
7. Sherman, Douglas J. Problems of Scale in the Modelling and Interpretation of Coastal Dunes. Marine Geology **1995**, *124* (1–4), 339–349.

EROSION BY WIND, PRINCIPLES OF

Larry D. Stetler
Department of Geology & Geological Engineering, South Dakota School of Mines and Technology, Rapid City, South Dakota, U.S.A.

INTRODUCTION

Erosion of sediment and soil by wind arises when the kinetic energy in the passing airstream exceeds the energy employed in adhering the grain to the surface. Initial grain motion is achieved at this instant and the grain will be transported in saltation as a series of short hops along the surface. Each time the grain impacts the surface, additional grains are released into the flow until at some distance downwind the carrying capacity of the wind is reached. From this point on, the mass of sediment in transport remains relatively constant; e.g., it is self-balanced. However, mass is lost from the process primarily through deposition of heavy particles behind surface roughness elements. In addition, suspendable particles of fine dust are released vertically through the saltation layer. Many additional factors, such as climate and surface conditions, complicate the simple mechanics involved.

EROSION PROCESSES

Erosion of fine-grained sediment and soil particles by wind is a natural geologic process occurring worldwide on a nearly continuous basis. Throughout the geological record evidence of wind processes eroding, transporting, and depositing sediment exists primarily as vast accumulations of eolian (wind-blown) sand—both modern sand dune fields and ancient eolian sandstone formations—and loess (wind-blown silt). Currently, about 30% of the total land surface of the planet consists of semi-arid and arid lands which give rise to most of the transported sediment and dust. Modern agriculture and industrial activities are also a significant contributor to transported sediment and dust due to induced or mechanical destabilization of the land surface.

Conceptually, mobilization of loose sediment and soil particles occurs during wind storms or wind "events," where the mean wind speed exceeds a critical threshold value based on some erodible fraction of the sediment. Individual wind events may be of short (less than one to a few hours) or long (multiple hours to days) duration. In terms of distance and total mass moved, a typical erosion event will transport the greatest mass a short distance and a significantly lesser mass a greater distance. Perception of moving sediment, however, may make this statement appear to be contradictory since there is, generally, an abundance of the smaller and lighter particles which are visually acute when airborne. Exceptional, or non-typical, wind events are able to transport fine particles across entire continents and oceans to distant lands where visual impacts to air quality may be readily observed.

Knowledge of particle entrainment and transport studies were greatly advanced through the work of Bagnold who used field observations in the deserts of Egypt and later wind tunnel studies to detail the mechanics of the erosional and transportation process. This work resulted in his well-known and widely used treatise *The Physics of Blown Sand and Desert Dunes* (1). He observed that once a particle begins to move it will be transported by one of three modes: creep, saltation, or suspension. The primary difference in these three transport modes is a relatively simple relation between the particle size and the wind force which, unfortunately, becomes a complex mechanical problem that is today still not fully understood.

WIND DYNAMICS

Wind is by nature a random process and as such, the internal structure of a natural wind contains random variations known as turbulence (except in possibly very stable conditions). Turbulence develops in natural flows as inertial forces, which provide forward momentum (motion), react to viscous forces, which impede motion due to fluid properties. This results in a flow that is strongly sheared.

The unit of measure for wind speed is the mean wind velocity, $\overline{U}$, defined as an average value for a continuous time series of instantaneous velocities (Fig. 1):

$$\overline{U} = \frac{\sum u_1 + u_2 + \ldots u_i}{T} \qquad (1)$$

Encyclopedia of Soil Science

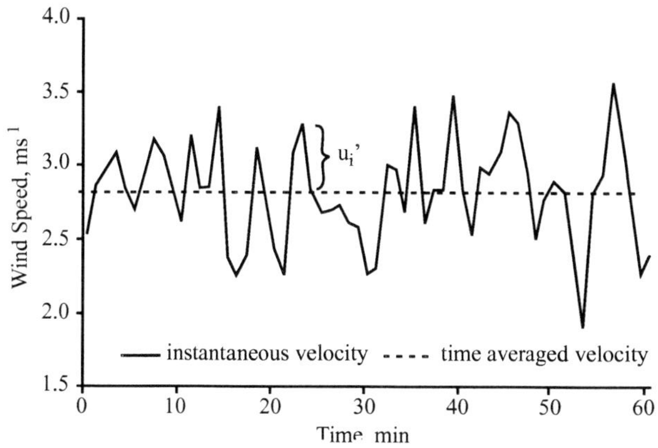

Fig. 1 Time series of one-min wind speeds illustrating principles of determining mean wind speed, instantaneous and fluctuating values. Unpublished data from author.

where

$u_1 + u_2 + \ldots u_i$ = components of instantaneous velocity
T=time interval over which velocity values are summed

As shown in Fig. 1, at every instant a fluctuating component of velocity, u_i' exists defined as the difference between the instantaneous and the mean velocity values:

$$u_i' = u_i - \overline{U} \tag{2}$$

It is these velocity fluctuations, whether positive or negative in regard to the mean velocity, that give rise to flow instability and thus, turbulence. As such, turbulence is not a property of the fluid but of the flow of the fluid. Turbulent flows are, therefore, unique in their ability to mix momentum and kinetic energy and to erode and transport particles.

PARTICLE ENTRAINMENT

Bagnold described the instant of initial particle motion by a threshold fluid velocity which represents the wind speed at which the most susceptible particle becomes entrained. As a flow passes over a stationary particle, turbulent kinetic energy is imparted to the grain which builds a reservoir of stored potential energy (2). Further increases in this energy reservoir cause the grain to oscillate and at the critical point where potential energy exceeds the energy of adherence, the grain lifts, or more accurately is ejected in a near vertical trajectory away from the surface (2, 3) initiating saltation. For a single grain of diameter d, Bagnold showed that the threshold fluid velocity, v_t,

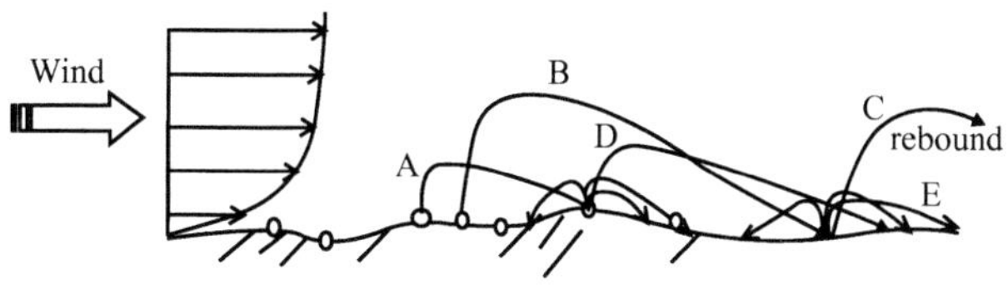

Fig. 2 Particle transport paths in response to the wind profile (at left). A, B, C, and D are high-energy impacting grains moving in saltation. E are low-energy reptating particles including those moving by creep.

required to initiate saltation motion is related to particle diameter by:

$$v_t = 5.75A\sqrt{\frac{\rho_p - \rho_f}{\rho_f} - gd}\,\log\frac{z}{k} \tag{3}$$

where

A = empirical constant = 0.1 for fully turbulent flows
ρ_p = particle density
ρ_f = fluid density
g = gravity force
z = height for which threshold velocity is being calculated
k = surface roughness height equal to about 1/30 the surface grain diameter.

For the case of a naturally irregular surface, such as a field or a sand dune, fluid threshold is not defined by a single wind speed or particle size but by a continuum of wind speeds over a length of time where grains of various diameter become entrained almost simultaneously (4).

Once these grains are ejected into the overriding flow they attain a high-energy ballistic trajectory that depends upon particle size and wind velocity (1, 5, 6) (Fig. 2). Heavier and larger particles will travel relatively close to the surface (A) contrasted to lighter and smaller particles that initially are ejected higher into the wind profile (B). At some distant point downstream these grains (impactors) will impact on the sediment bed and either rebound upward again (C), eject new impactors (D), or lodge in the loose sediment. In either case, it is probable that upon impact sufficient kinetic energy will be transferred to one or several additional grains that are "splashed," or dislodged a short distance (E) known as reptation (6, 5), a low-energy process that includes a small number of grains traveling upstream. Reptation includes what Bagnold referred to as creep—grains that are too large to be entrained aerodynamically and move by rolling or sliding along the surface. Impact force is also a primary method leading to disaggregation of larger particle agglomerates by breaking particle bonds or the particles themselves into smaller pieces, enabling aerodynamic transport.

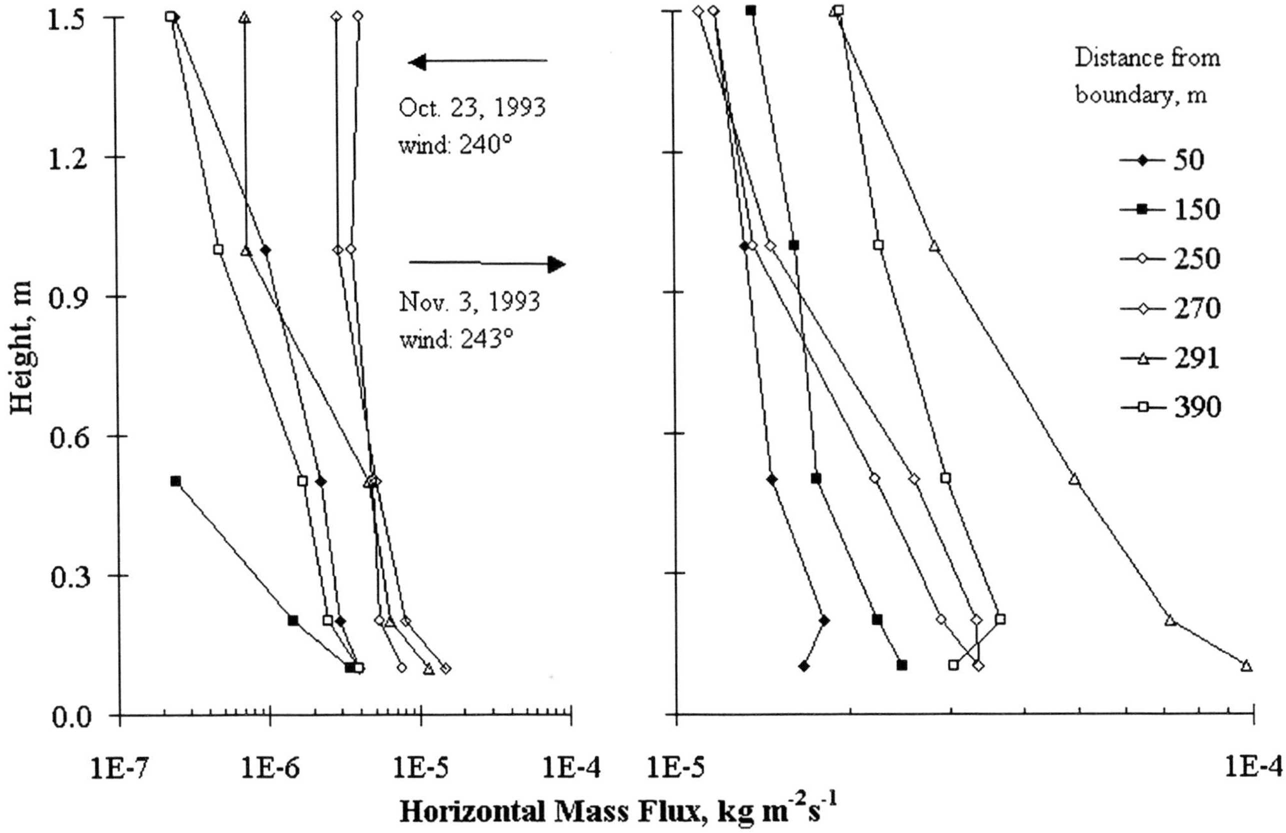

Fig. 3 Vertical and horizontal mass flux profiles for two wind events illustrating the decrease and increase of mass with height and downstream distance, respectively. (Data from Ref. 10.)

Saltation, which is defined as movement in a series of short hops, includes particles in the high-energy population (A, B, C, and D). These particles are initially moved by wind speeds exceeding the fluid threshold and subsequently through surface impacts that eject new particles at a point known as the impact threshold. Bagnold suggested that the impact threshold involves the same mechanics included in Eq. 3 except $A = 0.08$. In other words, after saltation is initiated, additional grains are placed in motion at the threshold impact velocity that is about 80% of the threshold fluid velocity. This infers that once motion has begun, further motion can be sustained at wind speeds less than those required for initial grain motion.

The third transport mode, suspension, occurs in grains that are small enough to respond directly to turbulent fluctuations and thus, are lofted above the surface saltation layer.

SELF-BALANCING CONCEPT

Energy required to transport particles comes from extraction of turbulent momentum contained in the flow and a reduction in shearing forces near the surface (7, 8). Thus, as the flow adjusts to the presence of particles, a self-balancing condition in mass flux develops downstream of the point of initial grain motion (9). In other words, the carrying capacity of the wind has been realized. This does not imply additional particles are not eroded but that as larger grains become deposited behind surface roughness elements, the loss of mass is balanced by erosion of new grains from the surface, primarily by impact and abrasion forces.

The self-balancing concept for maximum horizontal mass flux, f_{max}, has been analyzed by Stout (8), reducing to:

$$f_x = f_{max}\left(1 - e^{-\frac{x}{b}}\right) \tag{4}$$

where

f_x = horizontal mass flux at downstream distance, x, from a non-eroding boundary
b = length scale for stability in horizontal mass flux

and is applied to height-specific mass flux measurements. In utility, Eq. 4 states that beginning from a non-eroding boundary at $x = 0, f_x$ increases with x until f_{max} is attained. Downstream of this point, horizontal mass flux is relatively unchanged. Typical vertical and horizontal mass flux profiles from two wind events are shown in Fig. 3 (10). Two components require discussion. First, vertical mass flux profiles, from 0.1 to 1.5 m height, show mass decreases with height. This is consistent with the discussion above, e.g., saltating particles having less mass loft higher into the flow. For any single profile, mass is greatest near the surface and decreases vertically. Second, horizontal mass flux increases downstream until the maximum transport capacity is reached. In the figure, mass continues to increase downstream not yet having attained f_{max}.

SUMMARY

Complications to the above processes arrive due to numerous natural and human-induced conditions including: soil type, particle size distribution and fraction of erodible particles, climate, and land use. In cultivated fields, surface roughness is a primary consideration complicating the entrainment process as well as the establishment of the saltation layer. Rough surface elements have the effect of reducing wind shear forces thereby limiting erosion contrasted to a smoother surface that is favorable for a higher erosion rate (11). Surface roughness can be either from soil clods, vegetation, or a combination of both (12). Additionally, given all the factors discussed above, increases in soil moisture will cause an increase in the wind speed required to initiate grain motion (13).

Prediction of wind erosion is, therefore, an inexact science, even with the advent of sophisticated computer models. The numerous and complex relations between all the critical parameters imply a degree of uncertainty, even in the best of calculations. The remaining articles will address some of these parameters more specifically.

REFERENCES

1. Bagnold, R.A. *The Physics of Blown Sand and Desert Dunes*; Methuen and Co., Ltd.: London, 1941; 265.
2. Reeks, M.W.; Hall, D. Deposition and Resuspension of Gas-Borne Particles in Recirculating Turbulent Flows. J. Fluids Engr. **1988**, *110*, 165–171.
3. Greely, R.; Iverson, J.D. *Wind as a Geological Process on Earth, Mars, Venus and Titan*; Cambridge University Press: New York, 1985; 333.
4. Nickling, W.G. The Initiation of Particle Movement by Wind. Sedimentology **1988**, *35*, 499–510.
5. Anderson, R.A.; Sorenson, M.; Willets, B.B. A Review of Recent Progress in Our Understanding of Aeolian Sediment Transport. Acta Mechanica Suppl. **1991**, *1*, 1–19.
6. Unger, J.E.; Haff, P.K. Steady State Saltation in Air. Sedimentology **1987**, *34*, 289–299.
7. Anderson, R.S.; Haff, P.K. Simulation of Eolian Saltation. Science **1988**, *241*, 820–823.
8. Stout, J.E. Wind Erosion within a Simple Field. Trans ASAE **1990**, *33* (5), 1597–1600.
9. Owen, P.R. Saltation of Uniform Grains in Air. Jour. Fluid Mech **1964**, *20*, 225–242.
10. Stetler, L.D.; Saxton, K.E. Wind Erosion, PM_{10} Emissions, and Dry-Land Farming on the Columbia Plateau. Earth Surface Processes and Landforms **1996**, *21*, 673–685.
11. Stout, J.E.; Zobeck, T.M. The Wolfforth Field Experiment: A Wind Erosion Study. Soil Sci. **1996**, *161* (9), 616–632.
12. Horning, L.B.; Stetler, L.D.; Saxton, K.E. Surface Residue and Soil Roughness for Wind Erosion Protection. Trans. ASAE **1998**, *41* (4), 1061–1065.
13. Pye, K. *Aeolian Dust and Dust Deposits*; Academic Press: London, 1987; 334.

EROSIVITY AND ERODIBILITY

Peter I.A. Kinnell
University of Canberra, Canberra, Australian Capitol Territory, Australia

INTRODUCTION

In water erosion, erosivity and erodibility are terms used to identify rain-associated factors driving erosion and the susceptibility of the soil to erosion. Their application in modeling erosion in empirical and process-based models will be described and discussed.

THE UNIVERSAL SOIL LOSS EQUATION

Historically, the terms erosivity and erodibility were originally associated with the *R* and *K* factors in the Universal Soil Loss Equation (USLE):

$$A = RKLSCP \tag{1}$$

where *A* is the annual average soil loss per unit area, *R* the rainfall (*erosivity*) factor, *K* the soil (*erodibility*) factor, *L* the slope length factor, *S* the slope gradient factor, *C* the crop and crop management factor, and *P* the support practice factor (1, 2). Fundamental to the USLE is the concept of the unit plot—a bare fallow plot having a slope length of 72.6 ft (22.1 m), a slope gradient of 9%, and cultivation up and down the slope. For the unit plots

$$C = L = S = P = 1.0$$

and

$$A = RK \tag{2}$$

In the USLE, *R* is defined as the average annual value of the product of the total storm kinetic energy and the maximum 30 min intensity for rainfall events producing at least 0.5 in. (12.5 mm) of rain or at least 0.25 in. (6.35 mm) of rain in 15 min. In the Revised USLE (RUSLE) (3), all rainfall events were considered in calculating *R* in the western part of the USA. Customary U.S. units for *R* are 10^2 ft T_f in A^{-1} hr^{-1} y^{-1} (hundreds of foot–Tonf–inch per Acre–hour–year) and *K* are T A hr 10^{-2} ft^{-1} T_f^{-1} in^{-1} (Ton–Acre–hour per hundreds of foot–Tonf–inch). Customary metric units are MJ mm ha^{-1} hr^{-1} y^{-1} (maga-joule–millimeter per hectare–hour–year) and t hrMJ^{-1} mm^{-1} (metric ton–hour per megajoule–millimeter) (3).

K is essentially the regression coefficient for the direct relationship between *R* and *A* under unit plot conditions. Experimental plot data were used originally to determine values of *K*. Subsequently, equations have been developed to predict *K* from soil properties in various geographic locations (2, 3, 5).

The USLE model works reasonably well in some locations (Fig. 1A) but does not in others (Fig. 1B). The USLE is less effective when the soil has a high capacity to accept rainfall (4). In the latter situation, there is a tendency for the USLE to overestimate the annual soil loss. Including runoff in the storm erosivity factor by multiplying the EI_{30} index by the runoff coefficient (Q_R, runoff amount divided by rainfall amount) provides a model known as the USLE M (4) which, as can be seen by comparing Fig. 2 with Fig. 1B, is better able to deal with the effect of differences in soil infiltration characteristics.

MORE PROCESS-BASED APPROACHES

The USLE is designed to predict erosion from sheet and rill eroded areas. Modern understanding of erosion processes recognizes that energy required for soil detachment associated with rill erosion comes from flow energy while detachment in sheet and interrill erosion is dominated by energy derived from raindrop impact. The USLE does not take these different energy sources into account, more process-based models have been developed which do. Rill and interrill erosion equations have been developed within the USDA Water Erosion Prediction Project (6). In the rill equation

$$D_r = K_r(\tau - \tau_0)(1 - q_s T_{cF}^{-1}) \tag{3}$$

D_r is rill detachment, K_r the rill erodibility, τ the flow shear stress, τ_0 the critical shear stress, q_s the sediment load, and T_{cF} is the sediment load when the transport capacity is reached. τ_0 is a soil dependent parameter. However, it is not considered to be an erodibility factor. In the basic interrill equation

$$D_i = K_i I_e I_x S_f \tag{4}$$

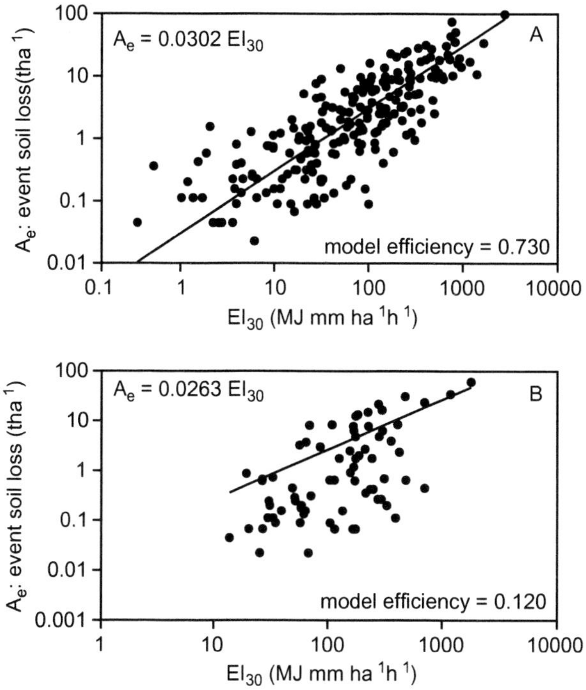

Fig. 1 Relationships between event soil losses from bare fallow and the EI_{30} index on (A) plot 5 at Holly Springs, MS during 1961–1968 and (B) plot 13 at Morris, MN during 1961–1971. The logarithmic form of the Nash–Sutcliffe efficiency factor (4) provides the measure of model efficiency.

D_i is interrill detachment, K_i the interrill erodibility, I_e the effective rainfall rate during the period of rainfall excess, I_x is the excess rainfall rate, and S_f is a function of interrill slope. Base values for K_r and K_i have been determined from experiments undertaken on a number of U.S. soils (6). In practice, the base values for K_r and K_i are adjusted to account for various effects including roots, sealing, and crusting to predict soil loss. Surface conditions of freshly tilled soils impacted by raindrops change over time and this causes the susceptibility of the soil to erosion to also change with time. The base values for K_i were determined from data usually collected after rain at 62 mm hr^{-1} had been applied to short (250 mm long slopes) inclined slopes for 30 min or more to ensure reasonably stable surface conditions.

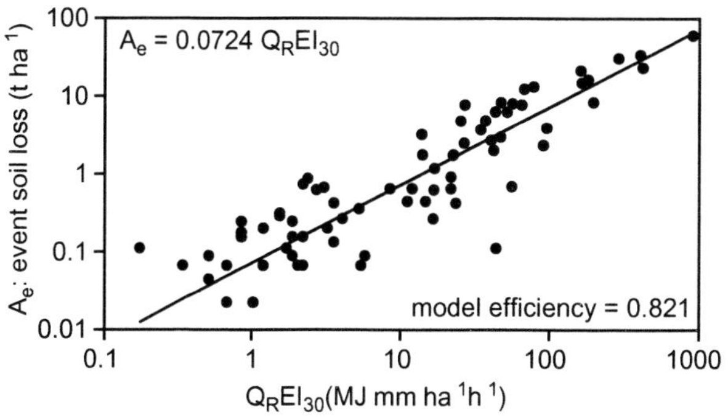

Fig. 2 Relationships between event soil losses from bare fallow and $Q_R EI_{30}$ index on plot 13 at Morris, MN during 1961–1971.

Although interrill erosion is dominated by detachment caused by raindrop impact there are a number of detachment and transport processes that can occur during a rainfall event. At the onset of rain on a dry surface, material detached by raindrop impact is splashed down slope. This is shown as RD–ST (raindrop detachment–splash transport) in Fig. 3. The RD–ST system controls what is commonly known as splash erosion. Before detachment can occur, a raindrop must posses sufficient energy to detach soil material from the cohesive forces that hold that material within the surface layer. This critical energy is indicated by e_c. e_c tends to increase during rain as, for example, a surface crust develops and this contributes to a decrease in soil erodibility associated with splash erosion. As

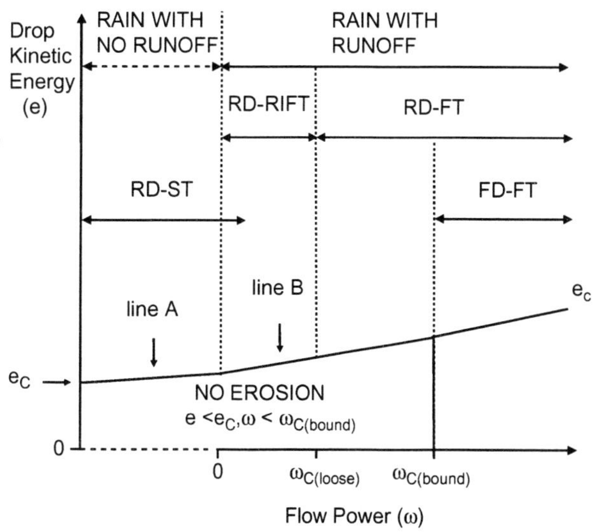

Fig. 3 Detachment and transport processes associated with variations in raindrop and flow energies. e_c = critical raindrop energy to cause erosion. Line A = e_c prior to flow (increasing through crust development). Line B = e_c when flow occurs (increasing drop energy used to penetrate flow). $\tau_{c(loose)}$ = critical shear stress for transporting loose material. $\tau_{c(bound)}$ = critical shear stress for detaching soil from surface of soil matrix. RD–ST = raindrop detachment, splash transport. RD–RIFT = raindrop detachment, raindrop induced flow transport. RD–FT = raindrop detachment, flow transport. FD–FT = flow detachment, flow transport. (Adapted from Ref. 7.)

runoff develops, raindrops penetrating the flow lift soil material into the flow and this material moves down stream in the flow. This detachment–transport system is called the raindrop detachment–raindrop induced flow transport system (RD–RIFT) and requires both raindrop impact and flow to operate. On short slopes, if the slope gradient is sufficiently high, flow can entrain loose material sitting on the surface without the aid of raindrop impact. This produces a raindrop detachment–flow transport system (RD–FT). The flow detachment–flow transport system (FD–FT) is normally associated with rills, but can also occur on short high-gradient slopes. In principle, each detachment–transport system can be represented by an equation that has an erosivity term and associated erodibility term. Since all the detachment–transport systems may operate to various degrees in any given area depending on the prevailing conditions, interrill erodibilities are less predictable than rill erodibilities.

In the USLE, soil erodibility (K) was kept constant during the calculation of the annual soil loss while interaction between short-term variations in erosivity and crop factors were considered in determining C values. In RUSLE, K is considered to vary during the year. At the process model level, changes in the soil surface conditions associated with, for example, the development of soil crusts have been noted already as having an effect on erodibility. However, erosion is limited by either detachment or transport processes. Crusts influence detachment but transport limiting conditions need to be considered in modeling erosion. In the WEPP rill model (Eq. 3), this is achieved by the term $1 - q_s \mathrm{T}_{\mathrm{cF}}^{-1}$. The RD–ST and RD–RIFT are inherently transport-limited systems because material detached by drop impact tends to fall back to the soil surface and is remobilized by numerous drop impacts before being discharged from the eroding area. The material on the surface near the end of the eroding area contributes directly to the material being discharged and also protects the underlying surface to some degree. This influences erodibility (8). For the RD–RIFT systems, erodibility is given by

$$K_{\mathrm{i}} = H_{\mathrm{PD}}K_{\mathrm{iPD}} + (1 - H_{\mathrm{PD}})K_{\mathrm{iM}} \qquad (5)$$

where H_{PD} is the degree of protection provided by the predetached material, K_{iPD} the erodibility when the predetached material fully protects the underlying surface, and K_{iM} is the erodibility of the soil surface when there is no predetached material sitting upon it. However, because H_{PD} and the composition of the predetached material vary with flow conditions and particle characteristics, H_{PD} and K_{iPD} vary in time and space. Relatively complex mathematical models that consider particle characteristics (9) have been developed with the aim of dealing with this variation.

SIMPLE CONCEPT FOR COMPLEX SYSTEMS

Originally, the concept of erosivity and erodibility as considered in the USLE was quite simple because the model was designed to predict long term average annual soil loss based on experimental data. Equations developed to predict K and C enabled the model to be applied to areas and situations not covered in the original experiments. However, the USLE and the RUSLE do not capture the effects of some of the fundamental erosion processes well in some cases. In order to overcome this deficiency, more process-based models have been developed. However, because the susceptibility of the soil to erosion can vary in space and time as a result of an interaction between the soil surface and the detachment and transport mechanisms, erodibility can vary in time and space in a complex manner. Complex process-based mathematical models may be needed to account for this variation, but less complex models may produce sufficiently accurate results to make reasonable decisions on land management options. Runoff is an important factor in rainfall erosion. Models that include it as part of the erosivity factor have the potential to account for erosion better than those that do not.

REFERENCES

1. Wischmeier, W.H.; Smith, D.D. *Predicting Rainfall–Erosion Losses from Cropland East of the Rocky Mountains: Guide for Selection of Practices for Soil and Water Conservation*; Agriculture Handbook 282, U.S. Department of Agriculture: Washington, DC, 1965.
2. Wischmeier, W.H.; Smith, D.D. *Predicting Rainfall Erosion Losses: A Guide to Conservation Planning*; Agriculture Handbook 537, U.S. Department of Agriculture: Washington, DC, 1978.
3. Renard, K.G.; Foster, G.R.; Weesies, D.K.; McCool, D.K.; Yoder, D.C. *Predicting Soil Erosion by Water: A Guide to Conservation Planning with the Revised Universal Soil Loss Equation (RUSLE)*; Agriculture Handbook 703, U.S. Department of Agriculture: Washington, DC, 1997.
4. Kinnell, P.I.A.; Risse, L.M. USLE-M: Empirical Modelling Rainfall Erosion Through Runoff and Sediment Concentration. Soil Sci. Soci. Am. J. **1998**, *62*, 1667–1672.

E

5. Loch, R.J.; Slater, B.K.; Devoil, C. Soil Erodibility Values (K_m) for Some Australian Soils. Aust. J. Soil Res. **1998**, *36*, 1045–1055.
6. Flanagan, D.C.; Nearing, M.A. *USDA—Water Erosion Prediction Project: Technical Documentation*; NSERL Report No. 10, National Soil Erosion Research Laboratory: West Lafayette, IN, 1995.
7. Kinnell, P.I.A. *A Discourse on Rainfall Erosion Processes and Modelling on Hillslopes*; Occasional Paper No.6, Centre for Australian Regolith Studies. University of Canberra: Australia, 2000.
8. Kinnell, P.I.A. The Effect of Slope Length on Sediment Concentrations Associated with Side-Slope Erosion. Soil Sci. Soc. Am. J. **2000**, *64*, 1004–1008.
9. Hairsine, P.B.; Sander, G.C.; Rose, C.W.; Parlange, J.-Y.; Hogarth, W.L.; Lisle, I.; Rouhipour, H. Unsteady Soil Erosion Due to Raindrop Impact: A Model of Sediment Sorting on a Hillslope. J. Hydrol. **1999**, *220*, 115–128.

EVAPORATION

William P. Kustas

United States Department of Agriculture-Agricultural Research Service (USDA-ARS), Hydrology and Remote Sensing Lab, Beltsville, Maryland, U.S.A.

INTRODUCTION

Soil evaporation is a critical component for many vegetated, as well as bare, soil surfaces because the resulting surface soil moisture and temperature can have a significant impact on the microclimate in partially vegetated canopies, indirectly influencing plant transpiration (1). Over a growing season, soil evaporation can be a significant fraction of total water loss for agricultural crops (2). On a seasonal basis in semiarid and arid regions, soil evaporation can have a significant influence on the relative fraction of runoff to rainfall, which in turn has a major impact on the available water for plants (3).

The measurement of soil evaporation over bare soil fields can be performed using standard micro-meteorological techniques, namely Bowen ratio and eddy covariance methods. However, these techniques cannot be applied for partial canopy cover conditions due to the fetch requirements. The only exception to this is when vegetation is clumped and spacing is on the order of 10's of meters, which allows for a micro-Bowen ratio approach where gradients of temperature and humidity are measured very close to the soil surface (3, 4). Otherwise soil evaporation is measured using micro-lysimeters (5), chambers (e.g., Ref. 6) or a combination of micro-lysimetry and Time Domain Reflectometers (TDR) (7). Clearly, when dealing with plant canopies, these point-based measurements are difficult to extrapolate to the field scale, due to the significant micro-scale variability in surface soil moisture. In addition, these techniques have other associated problems (e.g., the soil in micro-lysimeters being hydraulically isolated).

Therefore, models have been developed to estimate the contribution of soil evaporation to the total evapotranspiration process. Models of varying degrees of complexity are reviewed, but the primary focus is on relatively simple analytical models, some of which provide daily estimates and can be implemented operationally. The potential application of models using remote sensing data for large-scale estimation is also briefly discussed.

METHODOLOGIES

Numerical Models

Numerous mechanistic/numerical models of heat and mass flows exist and are primarily based on the theory of Philip and DeVries (8) (see also Ref. 9). However, they continue to be refined through improved parameterization of the moisture and heat transport through the soil profile (e.g., Refs. 9–11). Some of these mechanistic models have been used recently to explore the utility of bulk transfer approaches used in weather forecasting models (e.g., Ref. 12) and in soil-vegetation-atmosphere models (e.g., Ref. 13) computing field to regional scale fluxes. These bulk transport approaches are commonly called the "alpha" and "beta" methods defined by the following expressions:

$$LE_S = \frac{\rho C_P}{\gamma}\left(\frac{\alpha e_*(T_S) - e_A}{R_A}\right) \tag{1}$$

and

$$LE_S = \frac{\rho C_P}{\gamma}\beta\left(\frac{e_*(T_S) - e_A}{R_A}\right) \tag{2}$$

ρ is the air density ($\sim 1\,\mathrm{kg\,m^{-3}}$) and C_P is the heat capacity of air ($\sim 1000\,\mathrm{J\,kg^{-1}\,K^{-1}}$), γ is the pyschrometric constant ($\sim 65\,\mathrm{Pa}$), $e_*(T_S)$ is the saturated vapor pressure (Pa) at soil temperature T_S(K), e_A is the vapor pressure (Pa) at some reference level in the atmosphere, and R_A is the resistance ($\mathrm{s\,m^{-1}}$) to vapor transport from the surface usually defined from surface layer similarity theory (14). For α, several different formulations exist (see Refs. 15, 16) with one of the first relating α to the thermodynamic relationship for relative humidity in the soil pore space, h_R (17):

$$h_R = \exp\left(\frac{\psi g}{R_V T_S}\right) \tag{3}$$

where P is the soil matric potential (m), g is the acceleration of gravity ($9.8\,\mathrm{m\,s^{-2}}$), R_V is the gas constant for water vapor ($461.5\,\mathrm{J\,kg^{-1}\,K^{-1}}$). From Eq. 2, β can be defined as a ratio of aeodynamic and a soil resistance to vapor transport from the soil layer to the surface, R_S, namely

$$\beta = \frac{R_A}{R_A + R_S} \quad (4)$$

Both modeling (16) and observational results (18, 19) indicate that more reliable results are obtained using the "beta method." In fact Ye and Pielke (16) formulate an expression similar to Camillo and Gurney (20), which combines Eqs. 1 and 2 and provides more reliable evaporation rates for a wider range of conditions,

$$LE_S = \frac{\rho C_P}{\gamma} \beta \left(\frac{\alpha e*(T_S) - e_A}{R_A} \right) \quad (5)$$

Unfortunately, there is no general consensus concerning the depth in the soil profile to consider in defining the α and β terms. In particular, studies evaluating the soil resistance term R_S use a range of soil moisture depths from 0–1/2 cm (20), 0–1 cm (6), 0–2 cm (18), and 0–2 and 0–5 cm (11). Chanzy and Buckler (10) appear to be one of the few studies that attempt to determine the most useful soil moisture depth for modeling soil evaporation by considering the penetration depth of passive microwave sensors of varying wavelengths. Using field data and numerical simulations with a mechanistic model, they find that the 0–5 cm depth, which can be provided by the L-band microwave frequency, appears to be the most adequate frequency for evaluating soil evaporation.

Besides the depth of the soil layer to consider, the equations relating soil moisture to R_S have ranged from linear to exponential (Table 1). Furthermore, observations and numerical models have shown that R_S varies significantly throughout the day and that its magnitude is also affected by climatic conditions (10, 11).

These are not the only complicating factors that make the use of such a bulk resistance approach somewhat tenuous. From detailed observations of soil moisture changes and water movement, Jackson et al. (21), found that the soil water flux in the 0 to 9 cm depth to be very dynamic with fluxes at all depths continually changing in magnitude and sometimes direction over the course of a day. These phenomena observed by Jackson et al. (21) are due in part to a process that occurs during soil drying where dry surface soil layer forms, significantly affecting the vapor transport through the profile (22). Yamanaka

Table 1 Bulk soil resistance formulations, R_S, from previous studies

R_S formula ($s\,m^{-1}$)	Value of coefficients	Soil type	Depth (cm)	Reference
$R_S = a\left(\frac{\theta_S}{\theta}\right)^n + b$	$a = 3.5$ $b = 33.5$ $n = 2.3$	Loam[a]	0–1/2	40
$R_S = a(\theta_S - \theta) + b$	$a = 4140$ $b = -805$	Loam[a]	0–1/2	20
$R_S = R_{SMIN}\ \exp\ (a[\theta_{MIN} - \theta])$	$R_{SMIN} = 10\,s\,m^{-1}$ $\theta_{MIN} = 15\%$ $a = 0.3563$	Fine sandy loam	0–1	6
$R_S = \frac{a(\theta_S - \theta)^n}{2.3 \times 10^{-4}(T_S/273.16)^{1.75}}$	$a = 2.16 \times 10^2$ $n = 10$ $\theta_S = 0.49$	Loam	0–2	18
	$a = 8.32 \times 105$ $n = 16.16$ $\theta_S = 0.392$	Sand	0–2	18
$R_S = a\theta + b$	$a = -73420$ to -51650 $b = 1940$ to 3900	Sand	0–2	11
$R_S = \exp\left(b - a\left(\frac{\theta}{\theta_S}\right)\right)$	$a = 4.3$ $b = 8.2$	Silty clay loam[b]	0–5	13
	$a = 5.9$ $b = 8.5$	Gravelly sandy loam	0–5	38

[a] Soil type for the data from Jackson et al. (21) was determined from texture-dependent soil hydraulic conductivity and matric potential equations of Clapp and Hornberger (41) evaluated by Camillo and Gurney (20).

[b] Soil type was determined from texture-dependent soil hydraulic conductivity and matric potential equations of Clapp and Hornberger (41) evaluated by Sellers et al. (13).

et al. (23) recently developed and verified, using wind tunnel data, a simple energy balance model in which the soil moisture available for evaporation is defined using the depth of the evaporating/drying front in the soil. This approach removes the ambiguity of defining the thickness of the soil layer and resulting moisture available for evaporation. However, the depth of the evaporating surface is not generally known *a priori*, nor can it be measured in field conditions; hence this approach at present is limited to exploring the effects of evaporating front on R_S type formulations.

Analytical Models

To reduce the effect of temporal varying, R_S, Chanzy and Buckler (10) developed a simple analytically based daily LE_S in $\mathrm{mm\,d^{-1}}$ (E_D) model using simulations from their mechanistic model for different soil texture, moisture, and climatic conditions as quantified by potential evaporation (E_{PD}) using Penman (24). The analytical daily model requires midday 0–5 cm soil moisture, θ, daily potential evaporation and daily average wind speed, U_D. The simple model has the following form:

$$\frac{E_D}{E_{PD}} = \left[\frac{\exp(A\theta + B)}{1 + \exp(A\theta + B)}\right] C + (1 - C) \tag{6a}$$

$$A = a + 5\ \max(3 - E_{PD}, 0) \tag{6b}$$

$$B = b - 5(-0.025b - 0.05)\ \max(3 - E_{PD}, 0) + \alpha(U_D - 3) \tag{6c}$$

$$C = 0.90 - 0.05c(U_D - 3) \tag{6d}$$

where the coefficients *a*, *b*, and *c* depend on soil texture (Table 2) and were derived from their detailed mechanistic model simulations for loam, silty clay loam, and clay soils (10). In Fig. 1 a plot of Eq. 6a is given for two soil types, loam (Fig. 1a), and silty clay loam (Fig. 1b) under two climatic conditions, namely a relatively low evaporative demand condition with $U_D = 1\,\mathrm{m\,s^{-1}}$ and $E_{PD} = 2\,\mathrm{mm\,d^{-1}}$ and high demand $U_D = 5\,\mathrm{m\,s^{-1}}$ and $E_{PD} = 10\,\mathrm{mm\,d^{-1}}$. Notice the transition for $E_D/E_{PD} \sim 1$ to $E_D/E_{PD} < 1$ as a function of θ varies not only with the soil texture, but also the evaporative demand. The simplicity of such a scheme outlined in Eq. 6 needs further testing for different soil textures and under a wider range of climatic conditions.

The ratio E_D/E_{PD} as a function of θ illustrated in Fig. 1 also depicts the effect of the two "drying stages" typically used to describe soil evaporation (14). The "first stage" (S_1) of drying is under the condition where water is

Table 2 Values of desorptivity, D_E, evaluated from various experimental sites

Desorptivity D_E ($\mathrm{mm\,d^{-1/2}}$)	Soil type	Reference
4.96 to 4.30	Sand	26
5.08	Clay loam	27
4.04	Loam	27
3.5	Clay	27
~4 to ~8[a]	Loam[b]	28
5.8	Clay loam	29
4.95[c]	Silty clay loam[d]	31
2.11	Gravelly sandy loam	3

[a] The magnitude of D_E was found to have a seasonal dependency.
[b] Soil type was determined from texture-dependent soil hydraulic conductivity and matric potential equations of Clapp and Hornberger (41) evaluated by Camillo and Gurney (20).
[c] This value was evaluated for a vegetated surface.
[d] Soil type was determined from texture-dependent soil hydraulic conductivity and matric potential equations of Clapp and Hornberger (41) evaluated by Sellers et al. (13).

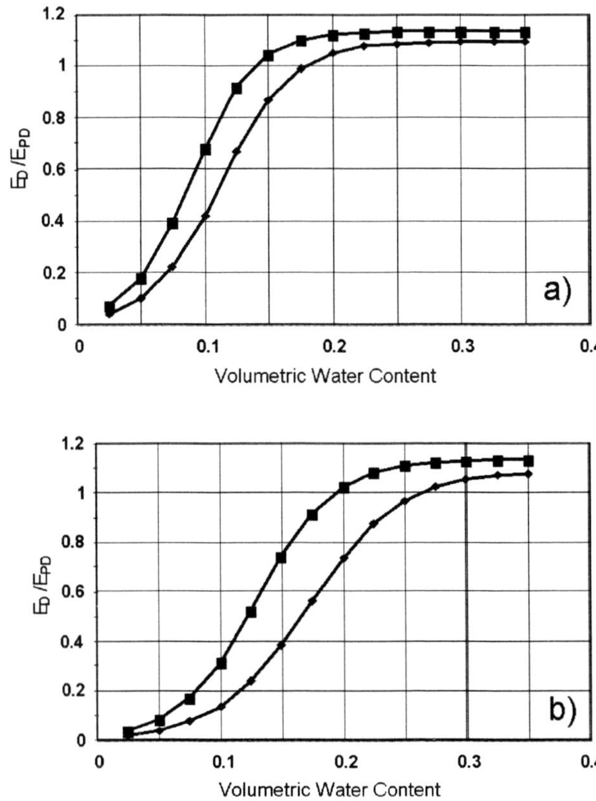

Fig. 1 A plot of E_D/E_{PD} estimated using Eq. 6 from Chanzy and Buckler (10) versus volumetric water content for a) loam and b) silty clay loam soil under two evaporative demand conditions: $U_D = 1\,\mathrm{m\,s^{-1}}$ and $E_{PD} = 2\,\mathrm{mm\,d^{-1}}$ (squares) and $U_D = 5\,\mathrm{m\,s^{-1}}$ and $E_{PD} = 10\,\mathrm{mm\,d^{-1}}$ (diamonds).

available in the near-surface soil to meet atmospheric demand, i.e., $E_D/E_{PD} \sim 1$. In the "second stage" (S_2) of drying, the water availability or θ falls below a certain threshold where the soil evaporation is no longer controlled by the evaporative demand, namely $E_D/E_{PD} < 1$. Under S_2, several studies find that a simple formulation can be derived by assuming that the time change in θ is governed by desorption, namely as isothermal diffusion with negligible gravity effects from a semi-infinite uniform medium. This leads to the rate of evaporation for S_2 being approximated by (14, 25)

$$E_D = 0.5D_E t^{-1/2} \tag{7}$$

where the desorptivity D_E (mm d$^{-1/2}$) is assumed to be a constant for a particular soil type and t is the time (in days) from the start of S_2. Although, both numerical models and observations indicate the soil evaporation is certainly a more complicated process than the simple analytical expression given by Eq. 7, a number of field studies (e.g., Refs. 26–30) have shown that for S_2 conditions, reliable daily values can be obtained using Eq. 7. In many of these studies for determining D_E, the integral of Eq. 7 is used, which yields the cumulative evaporation as a function of $t^{1/2}$

$$\sum E_D = D_E(t - t_O)^{1/2} \tag{8}$$

where t_O is the number of days where $E_D/E_{PD} \sim 1$ or is in S_1. In practice, observations of $\sum E_D$ are plotted versus $(t - t_O)^{1/2}$ and in many cases the choice of the starting point of S_2 is $t_O \approx 0$ or immediately after the soil is saturated. As shown by Campbell and Norman (30) the course of evaporation rate for three drying experiments (see their Fig. 9.6) indicates that for a loam soil t_O depends on the evaporative demand or E_{PD} with $t_O \sim 2$ days when E_{PD} is high versus $t_O \sim 5$ days when E_{PD} is low. On the other hand, for a sandy soil, there is almost an immediate change from S_1 to S_2 conditions with $t_O \approx 1$ day. As suggested by the analysis of Jackson et al. (28) and stated more explicitly by Brutsaert and Chen (31), the value of t_O can significantly influence the value computed for D_E. Jackson et al. (28) also show that for the same soil type the value of D_E has a seasonal dependency (ranging from 4 to 8 mm d$^{-1/2}$) most likely related to the evaporative demand, which they correlate to daily average soil temperature (see their Fig. 2). Values of D_E from the various studies are listed in Table 2. Brutsaert and Chen (31) modified Eq. 7 for deriving D_E by rewriting in terms of a "time-shifted" variable $T = t - t_O$ and expressing it in the form

$$E_D^{-2} = \left(\frac{2}{D_E}\right)^2 T \tag{9}$$

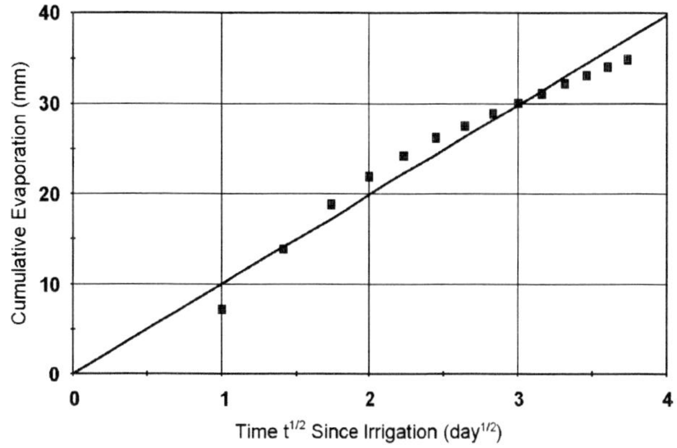

Fig. 2 The desorptivity D_E (mm d$^{-1/2}$) estimated from a least squares linear fit to the data from Jackson et al. (28) assuming $t_O = 0$ (i.e., stage two drying occurs immediately after irrigation/precipitation).

where D_E and t_O will come from the slope and intercept (see their Fig. 1). It follows that $\sum E_D$ under S_2 will start at $T = T_O$ and not at $T = 0$ so that Eq. (8) is rewritten as

$$\sum E_D = D_E\left(T^{1/2} - T_O^{1/2}\right) \tag{10}$$

They evaluated the effect on the derived D_E using this technique with the data from Black et al. (26). The value of D_E using Eqs. 9 and 10 was estimated to be approximately 3.3 mm d$^{-1/2}$, which is smaller than D_E values reported by Black et al. (26), namely, 4.3 to 5 mm d$^{-1/2}$. However, Brutsaert and Chen (31) show that this technique yields a better linear fit to the data points that were actually under S_2 conditions.

Eqs. 9 and 10 were used with the September 1973 data set from Jackson et al. (28) and compared to using Eq. 8 with $t_O = 0$. The plot of Eq. 8 with the regression line (Fig. 2) yields D_E. 10 mm d$^{-1/2}$, which is significantly larger than any previous estimates (Table 2). Moreover, it is obvious from the figure that Eq. 8 should not be applied with $t_O = 0$, since this relationship is not linear over the whole drying processes. With Eq. 9 applied to the data, t_O is estimated to be approximately 4.3 days and thus a linear relationship should start at the shifted time scale $T = t - 4.3$; this means $\sum E_D$ should start on day 5 or $T_O \approx 5 - 4.3$ (Fig. 3a). With Eq. 10, a more realistic $D_E \approx 4.6$ mm d$^{-1/2}$ is estimated for the linear portion of daily evaporation following the S_2 condition (Fig. 3b).

While this approach is relatively easy to implement operationally, D_E will likely depend on climatic factors as well as soil textural properties. However, it might be feasible to describe the main climate/seasonal affect on D_E from soil temperature observations (28). These might

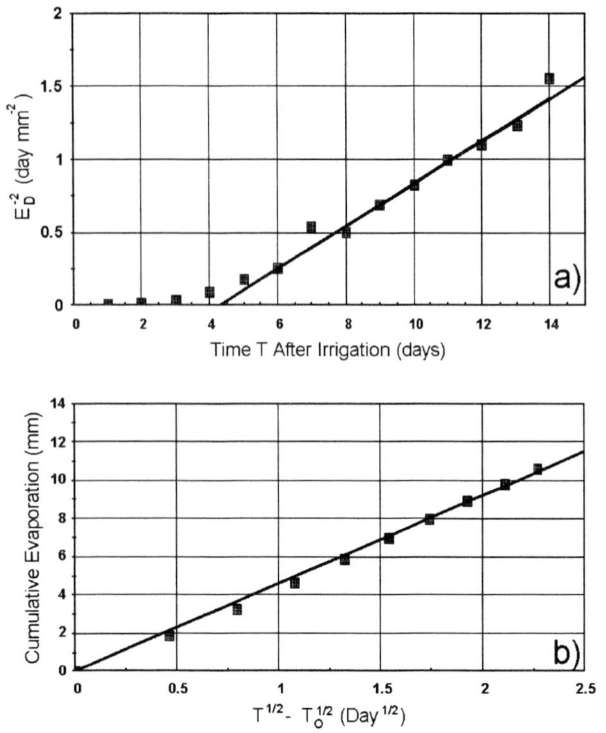

Fig. 3 Estimation of a) D_E and t_O with the data from Fig. 2 using Eq. 9 from Brutsaert and Chen (31); and b) the resulting cumulative evaporation $\sum E_D$ curve under second stage drying using Eq. 10.

come from weather station observations or possibly from multitemporal remote sensing observations of surface temperature.

The difficulty in developing a formulation for R_S, which correctly describes the water vapor transfer process in the soil, was recognized much earlier by Fuchs and Tanner (32) and Tanner and Fuchs (33). They proposed instead a combination method that involves atmospheric surface layer observations and remotely sensed surface temperature, T_{RS}. Starting with the energy balance equation

$$R_N = H + G + LE \tag{11}$$

where R_N is the net radiation, H is the sensible heat flux, and G the soil heat flux all in W m^{-2}, and assuming the resistance to heat and water vapor transfer are similar yielding,

$$H = \rho C_P\left(\frac{T_{RS} - T_A}{R_A}\right) \tag{12a}$$

$$LE = \frac{\rho C_P}{\gamma}\left(\frac{e_{RS} - e_A}{R_A}\right) \tag{12b}$$

an equation of the following form can be derived

$$LE = \left(\frac{\gamma + \Delta}{\Delta}\right)LE_P - \frac{\rho C_P}{\Delta}\left(\frac{e_*(T_{RS}) - e_A}{R_A}\right) \tag{13}$$

where

$$LE_P = \rho C_P\left(\frac{e_*(T_A) - e_A}{R_A(\Delta + \gamma)}\right) + \Delta\left(\frac{R_N - G}{\Delta + \gamma}\right) \tag{14}$$

The difference $(e_*(T_A) - e_A)$ is commonly called the saturation vapor pressure deficit and the value of soil surface vapor pressure e_{RS} is equal to $h_{RS}\ e*(T_{RS})$ where h_{RS} is the soil surface relative humidity. Substituting Eq. 14 into Eq. 13 yields

$$LE = R_N - G - \frac{\rho C_P}{\Delta}\left(\frac{e_*(T_{RS}) - e_*(T_A)}{R_A}\right) \tag{15}$$

This equation has the advantage over the above bulk resistance formulations using R_S in that there are no assumptions made concerning the saturation deficit at or near the soil surface or how to define h_{RS}. Instead, this effect is accounted for by T_{RS} because as the soil dries T_{RS} increases and hence $e_*(T_{RS})$, which generally results in the last term on the right-hand side of Eq. 15 to increase, thus causing LE to decrease. In a related approach, the magnitude of LE is simply computed as a residual in the energy balance equation, Eq. (11), (e.g., Ref. 34) namely

$$LE = R_N - G - \rho C_P\left(\frac{T_{RS} - T_A}{R_A}\right) \tag{16}$$

Particularly crucial in the application of either Eq. 15 or 16 is a reliable estimate of R_A and T_{RS}. Issues involved in correcting radiometric temperature observations for surface emissivity, viewing angle effects, and so on are summarized in Becker and Norman (35). The aerodynamic resistance R_A is typically expressed in terms of Monin-Obukhov similarity theory (14),

$$R_A = \frac{\left\{\ln\left[\frac{z - d_O}{z_{OM}}\right] - \psi_M\right\}\left\{\ln\left[\frac{z - d_O}{z_{OS}}\right] - \psi_S\right\}}{k^2 U} \tag{17}$$

where z is the observation height in the surface layer (typically 2–10 m), d_O is the displacement height and z_{OM} the momentum roughness length, z_{OS} is the roughness length for scalars (i.e., heat and water vapor), k (~0.4) is von Karman's constant, P_M is the stability correction function for momentum, and P_S is the stability correction function for scalars. Both d_O and z_{OM} are dependent on the height and density of the roughness obstacles and can be considered a constant for a given surface while the magnitude of z_{OS} can vary for a given bare soil surface since it is also a function of the surface friction velocity (14). Experimental evidence suggests existing theory with possible modification to some of the "constants" can still be used to determine z_{OS} providing acceptable estimates of H for bare soil surfaces. However, application of Eq. 16 in partial canopy cover conditions has not been successful in general because z_{OS} is not well defined in Eq. 17, exhibiting large scatter with existing theory (e.g., Ref. 36).

For this reason, estimating soil evaporation from partially vegetated surfaces using T_{RS} invariably has to involve "two-source" approaches whereby the energy exchanges from the soil and vegetated components are explicitly treated (e.g., Ref. 37). Similarly, when using remotely sensed soil moisture for vegetated surfaces, a two-source modeling framework needs to be applied (38). In these two-source approaches, there is the added complication of determining aerodynamic resistances between soil and vegetated surfaces and the canopy air space. Schematically, the resistance network and corresponding flux components for two-source models is shown in Fig. 4. An advantage with the two-source formulation of Norman et al. (37) is that R_S is not actually needed for computing LE_S since it is solved as a residual. Nevertheless, the formulations in such parameterizations that are used (such as the aerodynamic resistance formulations) are likely to strongly influence LE_S values (1).

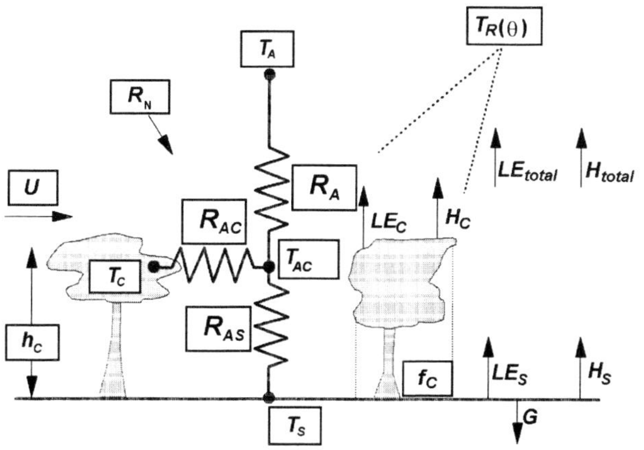

Fig. 4 A schematic diagram illustrating the resistance network for the two-source approach where the subscript *c* refers to the vegetated canopy and *s* refers to the soil surface. The symbol $T_R(\theta)$ refers to a radiometric temperature observation at a viewing angle θ, T_{AC} is the model-derived within canopy air space temperature, h_C is the canopy height, f_C is the fractional vegetation/canopy cover, and R_{AC} and R_{AS} are the aerodynamic resistances to sensible heat flux from the canopy and soil surface, respectively. The main meteorological inputs required for the model are also illustrated, namely wind speed, U, air temperature, T_A, and the net radiation, R_N. (Adapted From Ref. 37. Permission from Excerpta Medica Inc.)

CONCLUDING REMARKS

For many landscapes having partial vegetative cover, the contribution of soil evaporation to the total evapotranspiration flux cannot be ignored, particularly with regard to the influence of surface soil moisture and temperature on the microclimate in the canopy air space (see Fig. 4). Numerical models for simulating soil evaporation have been developed and are probably the most reliable. However, the required inputs for defining model parameters often limit their application to field sites having detailed soil profile information and are well instrumented with ancillary weather data.

For many operational applications where detailed soils and ancillary weather data are unavailable or where daily evaporation values are only needed, some of the analytical models described may provide the necessary level of accuracy. Moreover, in the application of weather forecast and hydrologic models, the use of simplified approaches is necessitated by the computational requirements and/or the lack of adequate data for defining more complex numerical model parameters and variables.

For large area estimation, the use of remotely sensed soil moisture and surface temperature offer the greatest potential for operational applications. The development of modeling schemes that can incorporate this remote sensing information and readily apply it on a regional scale basis have been proposed and show promise (e.g., Refs. 38, 39).

REFERENCES

1. Kustas, W.P.; Norman, J.M. Evaluation of Soil and Vegetation Heat Flux Predictions Using a Simple Two-Source Model with Radiometric Temperatures for Partial Canopy Cover. Agricultural and Forest Meteorology **1999**, *94*, 13–29.
2. Tanner, C.B.; Jury, W.A. Estimating Evaporation and Transpiration from a Row Crop During Incomplete Cover. Agron. J. **1976**, *68*, 239–243.
3. Wallace, J.S.; Holwill, C.J. Soil Evaporation from Tiger-Bush in South-West Niger. J. of Hydrol. **1997**, *188–189*, 426–442.
4. Ashktorab, H.; Pruitt, W.O.; Paw U, K.T.; George, W.V. Energy Balance Determinations Close to the Soil Surface Using a Micro-Bowen System. Agric. and For. Meteorol. **1989**, *46*, 259–274.

5. Allen, S.J. Measurements and Estimation of Evaporation from Soil Under Sparse Barley Crops in Northern Syria. Agric. and For. Meteorol. **1990**, *49*, 291–309.
6. van de Griend, A.A.; Owe, M. Bare Soil Surface Resistance to Evaporation and Vapor Diffusion Under Semiarid Conditions. Water Resour. Res. **1994**, *30*, 181–188.
7. Baker, J.M.; Spaans, E.J.A. Measuring Water Exchange Between Soil and Atmosphere with TDR-Microlysimetry. Soil Sci. **1994**, *158*, 22–30.
8. Philip, J.R.; de Vries, D.A. Moisture Movement in Porous Materials Under Temperature Gradients. Eos Trans. AGU **1957**, *38*, 222–232.
9. Camillo, P.J.; Gurney, R.J.; Schmugge, T.J. A Soil and Atmospheric Boundary Layer Model for Evapotranspiration and Soil Moisture Studies. Water Resour. Res. **1983**, *19*, 371–380.
10. Chanzy, A.; Bluckler, L. Significance of Soil Surface Moisture with Respect to Daily Bare Soil Evaporation. Water Resour. Res. **1993**, *29*, 1113–1125.
11. Daamen, C.C.; Simmonds, L.P. Measurement of Evaporation from Bare Soil and Its Estimation Using Surface Resistance. Water Resour. Res. **1996**, *32* (5), 1393–1402.
12. Noilhan, J.; Planton, S. A Simple Parameterization of Land Surface Processes for Meteorological Models. Mon. Weather Rev. **1989**, *117* (3), 536–549.
13. Sellers, P.J.; Heiser, M.D.; Hall, F.G. Relations Between Surface Conductance and Spectral Vegetation Indices at Intermediate (100 m^2 to 15 km^2) Length Scales. J. Geophys. Res. **1992**, *97* (D17), 19033–19059.
14. Brutsaert, W. *Evaporation into the Atmosphere. Theory, History, and Applications*; Kluwer Academic: Norwell, MA, 1982.
15. Mahfouf, J.F.; Noilhan, J. Comparative Study of Various Formulations of Evaporation from Bare Soil Using In Situ Data. J. Appl. Meteorol. **1991**, *30*, 1354–1365.
16. Ye, Z.; Pielke, R.A. Atmospheric Parameterization of Evaporation from Non-Plant-Covered Surfaces. J. Appl. Meteor. **1993**, *32*, 1248–1258.
17. Philip, J.R. Evaporation, and Moisture and Heat Fields in the Soil. J. Meteor. **1957**, *14*, 354–366.
18. Kondo, J.; Saigusa, N.; Sato, T. A Parameterization of Evaporation from Bare Soil Surfaces. J. Appl. Meteorol. **1990**, *29*, 385–389.
19. Cahill, A.T.; Parlange, M.B.; Jackson, T.J.; O'Neill, P.; Schmugge, T.J. Evaporation from Nonvegetated Surfaces: Surface Aridity Methods and Passive Microwave Remote Sensing. J. Appl. Meteor. **1999**, *38*, 1346–1351.
20. Camillo, P.J.; Gurney, R.J. A Resistance Parameter for Bare Soil Evaporation Models. Soil Sci. **1986**, *141*, 95–105.
21. Jackson, R.D.; Kimball, B.A.; Reginato, R.J.; Nakayama, F.S. Diurnal Soil Water Evaporation: Time-Depth-Flux Patterns. Soil Sci. Soc. Am. Proc. **1973**, *37*, 505–509.
22. Hillel, D. *Applications of Soil Physics*; Academic: San Diego, CA, 1980; 383.
23. Yamanaka, T.; Takeda, A.; Sugita, F. A Modified Surface-Resistance Approach for Representing Bare-Soil Evaporation: Wind Tunnel Experiments Under Various Atmospheric Conditions. Water Resour. Res. **1997**, *33* (9), 2117–2128.
24. Penman, H.L. Natural Evaporation from Open Water, Bare Soil, and Grass. Proc. R. Soc. London A. **1948**, *193*, 120–146.
25. Gardner, W.R. Solution of the Flow Equation for the Drying of Soils and Other Porous Media. Soil Sci. Soc. Am. Proc. **1959**, *23*, 183–187.
26. Black, R.A.; Gardner, W.R.; Thurtell, G.W. The Prediction of Evaporation, Drainage, and Soil Water Storage for a Bare Soil. Soil Sci. Soc. Am. Proc. **1969**, *33*, 655–660.
27. Ritchie, J.T. Model for Predicting Evaporation from a Row Crop with Incomplete Cover. Water Resour. Res. **1972**, *8*, 1204–1213.
28. Jackson, R.D.; Idso, S.B.; Reginato, R.J. Calculation of Evaporation Rates During the Transition from Energy-Limiting to Soil-limiting Phases Using Albedo Data. Water Resour. Res. **1976**, *12*, 23–26.
29. Parlange, M.B.; Katul, G.G.; Cuenca, R.H.; Kavvas, M.L.; Nielsen, D.R.; Mata, M. Physical Basis for a Time Series Model of Soil Water Content. Water Resour. Res. **1992**, *28*, 2437–2446.
30. Campbell, G.S.; Norman, J.M. *An Introduction to Environmental Biophysics*; Springer-Verlag: New York, 1998.
31. Brutsaert, W.; Chen, D. Desorption and the Two Stages of Drying of Natural Tallgrass Prairie. Water Resour. Res. **1995**, *31* (5), 1305–1313.
32. Fuchs, M.; Tanner, C.B. Evaporation from a Drying Soil. J. Appl. Meteor. **1967**, *6*, 852–857.
33. Tanner, C.B.; Fuchs, M. Evaporation from Unsaturated Surfaces: A Generalized Combination Method. J. Geophys. Res. **1968**, *73* (4), 1299–1304.
34. Jackson, R.D.; Reginato, R.J.; Idso, S.B. Wheat Canopy Temperature: A Practical Tool for Evaluating Water Requirements. Water Resour. Res. **1977**, *13*, 651–656.
35. Norman, J.M.; Becker, F. Terminology in Thermal Infrared Remote Sensing of Natural Surfaces. Remote Sensing Rev. **1995**, *12*, 159–173.
36. Verhoef, A.H.; de Bruin, H.A.R.; van den Hurk, B.J.J.M. Some Practical Notes on the Parameter kB^{-1} For Sparse Vegetation. J. Appl. Meteor. **1997**, *36*, 560–572.
37. Norman, J.M.; Kustas, W.P.; Humes, K.S. A Two-Source Approach for Estimating Soil and Vegetation Energy Fluxes from Observations of Directional Radiometric Surface Temperature. Agric. Forest Meteorol. **1995**, *77*, 263–293.
38. Kustas, W.P.; Zhan, X.; Schmugge, T.J. Combining Optical and Microwave Remote Sensing for Mapping Energy Fluxes in a Semiarid Watershed. Remote Sens. Environ. **1998**, *64*, 116–131.
39. Mecikalski, J.R.; Diak, G.R.; Anderson, M.C.; Norman, J.M. Estimating Fluxes on Continental Scales Using Remotely-Sensed Data in an Atmospheric-Land Exchange Model. J. Appl. Meteorol. **1999**, *38*, 1352–1369.
40. Sun, S.F. Moisture and Heat Transport in a Soil Layer Forced by Atmospheric Conditions M.S. thesis.
41. Clapp, R.B.; Hornberger, G.M. Empirical Equations for Some Soil Hydraulic Properties. Water Resour. Res. **1978**, *14*, 601–604.

EVOLUTION: THE ROLE OF THE SOIL SEED BANK

Paul B. Cavers
David J. Susko
University of Western Ontario, London, Ontario, Canada

INTRODUCTION

The development of the concept of the soil seed bank is generally attributed to Harper (1). He separated the typical "bank" into a "deposit account," consisting of seeds in a dormant state and a "current account," consisting of nongerminated seeds "in which the only hindrance to immediate germination is a shortage of water and a favorable temperature." The first use of the term "seed bank" is attributed to Sarukhán (2). Over the past century, there have been hundreds of studies of seed banks made in a variety of terrestrial and aquatic habitats. Seed banks can be incredibly complex; often with hundreds of different species distributed heterogeneously, with few to hundreds of thousands of seeds per species in each square meter of soil and with great variability in dormancy and longevity, even among the seeds of a single species. Our understanding of this complexity is far from complete (3).

SIZES OF SEED BANKS IN DIFFERENT HABITATS AND REGIONS

The numbers of seeds in seed banks can be determined by a) extraction of seeds followed by identification and counting, or b) identification and counting of seedlings arising from soil samples put under controlled conditions in greenhouses or growth rooms. There are problems with both approaches (4). Seed extraction is laborious and may miss small or cryptically colored seeds; additionally, some viable seeds can die before subsequent germination tests. Counts of seedlings, the preferred method in most studies (5), will miss seeds that are alive but unable to germinate under the conditions provided. For the most abundant species, there is usually a good correlation between the two procedures (4).

Fenner (6) summarized the reports on the numbers of seeds found in a wide range of communities; 10^2–$10^3\,m^{-2}$ for forest soils, 10^3–$10^6\,m^{-2}$ for grasslands and 10^3–$10^5\,m^{-2}$ for arable soils. Arable and disturbed soils often have huge seed banks (5), but seed numbers in arable soils have declined as use of chemical herbicides has increased (7). Wet sites and marshlands can have enormous seed banks as well (8). Baskin and Baskin (9) presented data on seed banks for 78 disparate communities. In general, their values fell into the ranges described by Fenner, but they cited several forests and grasslands with seed banks having < 10 seeds m^{-2} vs. up to 646,000 seeds m^{-2} in arable land.

SEED LONGEVITY IN THE SOIL

Two well-known experiments, one started by Beal in 1879 (10) and the second by Duvel in 1905 (11), have shown that the seeds of some plant species can remain dormant for decades, and that crop seeds have the shortest longevity, on average, and weed seeds the greatest. Since then, there have been numerous reviews of seed longevity in seed banks. Recently, Baskin and Baskin (9) separated seed bank studies into a) those with freshly collected seeds that are buried for specified periods before testing and b) those that infer the age of buried seeds by dating material in the soil around them.

Studies in which fresh seeds have been buried for various lengths of time are numerous but very few of them have extended beyond ten years (9). Beal's experiment is still continuing (10), but numbers of seeds per sample are small and only one set of storage conditions was used. None of the other studies exceeds 40 years and only two continuing studies are designed to reach 50 years (9). There is a great need for large, long-term, comprehensive studies of this type.

In studies of known-age seeds, the decrease in seed numbers over time generally follows the pattern of an exponential decay curve; a constant percentage being lost each year through death and germination (5). Seed dormancy is overcome and germination initiated by many factors. Amongst the most important are light (including quality, quantity and daylength), diurnally alternating temperatures, wet stratification at chilling temperatures (0–6°C), and soil nutrients, especially nitrogen (nitrate) levels (9).

VARIABILITY IN SEED BANKS IN TIME AND SPACE

Seed banks are more complex than communities of growing plants, primarily because they contain species that are not present as growing plants and because species found as growing plants have additional genotypes present in the seed bank. In addition, seed banks are variable over time and space.

Variability in Time

When viable seeds are retrieved from the seed bank they can be classed as either a) able to germinate when exposed to favorable conditions or b) remaining dormant when exposed to a favorable germination regime (12). Many fresh seeds fall into the latter category but this dormancy is often lost after a few weeks or months (12). Furthermore, unfavorable environmental conditions, such as high summer temperatures or drought, can return seeds from the former to the latter category.

Thompson and Grime (12) made an important distinction by classifying the seed banks of individual species as either: 1) "transient," lasting for less than one year and consequently with periods each year when no viable seeds are available to replace or reinforce the plant population, or 2) "persistent" lasting for more than one year, with remaining seeds overlapping fresh inputs each year. Thompson and Grime (12) produced four models of seed banks, two transient and two persistent, which have been used as the basis for comparison in many subsequent articles.

Variability in Space

One of the greatest problems facing an investigator of seed banks is how to estimate accurately their sizes, given the unknown and notoriously heterogeneous distribution of seeds, both laterally and at different depths in the soil (13, 14). For example, the number of seeds per unit area declines rapidly with increasing depth in the soil (9, 15). Further, seeds buried at greater depths tend to remain dormant longer, but fewer of them can germinate and emerge successfully to become established seedlings (15).

SEED BANK GENETICS

The offspring of plant genotypes that existed in the recent and far past may be preserved together in the soil, since seed banks can retain nongerminating but viable seeds for many years, or even centuries. Hence, the majority of genetic variation present in a plant population (i.e., the gene pool) is often contained within the seed bank, particularly in short-lived species with long-lived persistent seed banks, while only a fraction of this total variation is present in surface plants at any given date.

Levin (16) summarized the genetic consequences of a persistent seed bank on a population's gene pool. First, a seed bank maintains the genetic composition of a population and buffers it from extinction by increasing the effective population size, even when plant population size fluctuates dramatically over time (17). For instance, the genotypes of surface plants that fail to reproduce before they die may still be present in the seed bank. Second, the seed bank slows the rate of response of seedling or adult plant characters to selection, since gene loci not directly related to seed dormancy are constrained as much as those that influence dormancy. The degree of evolutionary retardation is dependent on seed longevity in the seed bank; retardation is greater in species with long-lived seeds than in those with short-term seed viability, because of the increased genetic load in seed banks containing long-lived seeds (18). Third, selection is biased toward traits associated with years of high seed production. As a consequence, the seed bank can serve as an evolutionary filter, where seeds from years of large seed crops may be favored by selection over seeds from years of small seed crops (19). Lastly, seed banks may be sources of novel genetic variation. Levin (16) noted that as seeds age, seed viability declines, whereas chromosome damage and the fre-quency of mutations within seeds increase. If such mutations are not lethal and can be transmitted to future generations, then the potential for plant populations to evolve may be increased due to the presence of a persistent seed bank.

Recent empirical studies have compared the amount and distribution of genetic variation in seed banks and surface plants. Typically, such comparisons among seeds of different ages within the seed bank, or among seeds in the soil vs. surface plants are made either: a) by examining morphological or physiological traits of genotypes when grown in common and multiple en-vironment experiments (20), or b) by analyzing poly-morphic allozyme and isozyme loci using starch gel electrophoresis (21–23). From the former technique, consistent phenotypic differences were found among seedlings arising from young (recently buried) and old seeds in the seed bank when the plants were grown at different densities, temperatures, and nutrient levels (20). These results indicated genetic differences between seeds of different ages, but our understanding of the causal mechanisms of such differences remains incomplete.

Studies using electrophoretic techniques have shown, in general, that allele frequencies in seed banks differ significantly from those in surface plants (21–23), but one study found little differentiation (17). Furthermore, many of these studies have found greater homozygosity in the seed bank than in later stages of the life cycle (21–23). These results suggest that the genotypes of surface plants constitute only a genetic subpopulation of the gene pool present in the soil seed bank. Hence, seed banks may be particularly important for the persistence and/or recovery of rare and endangered plant species. Presumably, susceptibility to genetic drift and inbreeding depression in small, scarce populations could be mediated by the presence of a homogeneously distributed, genetically diverse reserve of seeds. Examples of such mediation are few, but a study on the rare annual *Clarkia springvillensis* by McCue and Holtsford (23) supports this argument.

CONCLUSION

The number and composition of plant seeds found within the soil in a given unit of area comprises the soil seed bank. Seed banks can be incredibly large, often consisting of hundreds to hundreds of thousands of seeds per square meter in a variety of different habitats and communities around the world. Also, these seeds can be short- or long-lived. Seed banks provide a "memory" of recent prevailing site conditions, as well as climatic and edaphic conditions present many years or decades earlier. Hence, much of the genetic diversity of plant populations is preserved in seed banks. Furthermore, as seeds age in the soil, mutations accumulate and novel variation is introduced into a population. It is clear from the accumulated information in this article that soil seed banks play a pivotal role in the formation, maintenance and evolution of plant communities.

REFERENCES

1. Harper, John L. *Population Biology of Plants*; Academic Press: New York, 1977.
2. Sarukhán, José. Studies on Plant Demography: *Ranunculus repens* L., *R. bulbosus* L. and *R. acris* L. II. Reproductive Strategies and Seed Population Dynamics. Journal of Ecology **1974**, *62* (1), 151–177.
3. Vázquez-Yanes, C.; Orozco-Segovia, A.; Sánchez-Coronado, M.E.; Rojas-Arjchiga, M.; Batis, A.I. Seed Ecology at the Northern Limit of the Tropical Rain Forest in America. In *Seed Biology: Advances and Applications*; Black Michael, Bradford, Kent J., Vázquez-Ramos, Jorge, Eds.; CAB International: Wallingford, UK, 2000; 375–388.
4. Gross, Katherine L. A Comparison of Methods for Estimating Seed Numbers in the Soil. Journal of Ecology **1990**, *78* (4), 1079–1093.
5. Roberts, Harold A. Seed Banks in Soil. Advances in Applied Biology **1981**, *6*, 1–55.
6. Fenner, Michael. *Seed Ecology*; Chapman and Hall: London, UK, 1985.
7. Cavers, Paul B.; Benoit, Diane L. Seed Banks in Arable Land. In *Ecology of Soil Seed Banks*; Leck, Mary A., Parker, V. Thomas, Simpson, Robert L., Eds.; Academic Press: San Diego, 1989; 309–328.
8. Staniforth, Richard J.; Griller, Natalie; Lajzerowicz, Cleo. Soil Seed Banks from Coastal Subarctic Ecosystems of Bird Cove, Hudson Bay. Ecoscience **1998**, *5* (2), 241–249.
9. Baskin, Carol C.; Baskin, Jerry M. *Seeds: Ecology, Biogeography and Evolution of Dormancy and Germination*; Academic Press: San Diego, CA, 1998.
10. Kivilaan, A.; Bandurski, Robert S. The One Hundred-Year Period for Dr. Beal's Seed Viability Experiment. American Journal of Botany **1981**, *68* (9), 1290–1291.
11. Toole, E.H.; Brown, E. Final Results of the Duvel Buried Seed Experiment. Journal of Agricultural Research **1946**, *72* (6), 201–210.
12. Thompson, Ken; Grime, J. Phillip. Seasonal Variation in the Seed Banks of Herbaceous Species in Ten Contrasting Habitats. Journal of Ecology **1979**, *67* (3), 893–921.
13. Benoit, Diane L.; Derksen, Douglas A.; Panneton, B. Innovative Approaches to Seed Bank Studies. Weed Science **1992**, *40* (4), 660–669.
14. Dessaint, Fabrice; Barralis, G.; Caixinhas, M.L.; Mayor, J-P.; Recasens, J.; Zanin, G. Precision of Soil Seedbank Sampling: How Many Soil Cores? Weed Research **1996**, *36* (2), 143–151.
15. Cavers, Paul B. Seed Demography. Canadian Journal of Botany **1983**, *61* (12), 3578–3590.
16. Levin, Donald A. The Seed Bank as a Source of Genetic Novelty in Plants. American Naturalist **1990**, *135* (4), 563–572.
17. Mahy, Grégory; Vekemans, Xavier; Jacquemart, Anne-Laure. Patterns of Allozymic Variation within *Calluna vulgaris* Populations at Seed Bank and Adult Stages. Heredity **1999**, *82* (4), 432–440.
18. Templeton, Alan R.; Levin, Donald A. Evolutionary Consequences of Seed Pools. American Naturalist **1979**, *114* (2), 232–249.

19. Venable, D. Lawrence. Modeling the Evolutionary Ecology of Seed Banks. In *Ecology of Soil Seed Banks*; Leck, Mary A., Parker, V. Thomas, Simpson, Robert L., Eds.; Academic Press: San Diego, 1989; 67–87.
20. Vavrek, Milan C.; McGraw, James B.; Bennington, Cynthia C. Ecological Genetic Variation in Seed Banks. III. Phenotypic and Genetic Differences Between Young and Old Seed Populations of *Carex bigelowii*. Journal of Ecology **1991**, *79* (3), 645–662.
21. Tonsor, Stephen J.; Kalisz, Susan; Fisher, Jill; Holtsford, Timothy P. A Life-History Based Study of Population Genetic Structure: Seed Bank to Adults in *Plantago lanceolata*. Evolution **1993**, *47* (3), 833–843.
22. Cabin, Robert J.; Mitchell, Randall J.; Marshall, Diane L. Do Surface Plant and Soil Seed Bank Populations Differ Genetically? A Multipopulation Study of the Desert Mustard *Lesquerella fendleri* (Brassicaceae). American Journal of Botany **1998**, *85* (8), 1098–1109.
23. McCue, Kimberly A.; Holtsford, Timothy P. Seed Bank Influences on Genetic Diversity in the Rare Annual *Clarkia springvillensis* (Onagraceae). American Journal of Botany **1998**, *85* (1), 30–36.

FAUNA AND MICROFLORA: MACROFAUNA

Judy Tisdall
La Trobe University, Bundoora, Australia

F

INTRODUCTION

Soil macrofauna (> 2 mm) live all or part of their life below ground (Table 1). Through turnover (or bioturbation) of soil, the macrofauna contribute significantly to the structure, texture, and nutrient cycling of soil. Tillage destroys macrofauna habitat and food, hence the number of macrofauna is usually higher in undisturbed soils than in tilled soils.

This article discusses earthworms, termites, and ants and their effect on soil structure. One or more of these groups generally dominate the soil fauna.

EARTHWORMS

About 3000 species in 20 families of earthworms have been identified and their distribution and number of species vary with climate and soil management. Earthworms affect soil structure by producing casts and burrows.

Casts

Earthworms daily ingest and mix in the gut about half their own mass of soil plus large volumes of organic matter and deposit the material as casts. Each year, earthworms deposit 0.3–27.0 kg m^{-2} of casts on the surface of the soil, with more below the surface (13). Earthworms remould the soil in the gut (21) so that fresh casts can be up to 15% more dispersive, i.e., less stable (8), than the surrounding soil and can easily be eroded during rain. As they age and dry, the casts often become more stable and stronger than the surrounding soil, due to age-hardening, cohesive forces of water and cements, clay-bound carbohydrates and microorganisms (5, 7, 17). By preventing the formation of crusts, stable casts on the surface of soil often increase infiltration and seedling emergence. Physical properties of casts may depend on the properties (content of clay, water or organic matter, or compaction) of the original soil (1, 8, 21). Some casts produced in very wet soil are so dense and hard that the movement of water and air, mineralization of carbon, and root growth are decreased (2, 5).

Burrows

The main types of burrows are: a) semipermanent vertical burrows from the surface to at least 3 m depth, b) almost horizontal and some vertical burrows to the surface, and c) deep temporary burrows (13). Extensive networks of horizontal and vertical burrows develop in soils with several species of earthworms, but may vary seasonally, being filled with casts or destroyed by tillage or mammals (12).

Earthworms can produce burrows in soil at least as hard as 3 MPa, whereas roots are often limited in soil at 2 MPa (4). Many burrows are stable and may persist long after the earthworms have left, possibly because the burrows are lined with orientated clay or humic materials, carbonates and iron oxides (13). The burrows can be up to at least 10 mm wide depending on the size of the earthworm. The burrows affect the continuity and tortuosity of pores, and often increase the macroporosity and infiltration in the soil (12). The matrix of the soil may be wetted laterally from open burrows rather than from the surface of the soil (22). Water that is absorbed into the matrix may later move back to burrows. Although water runs down the burrows during rainfall in temperate climates, the burrows always contain air and do not need to be linked for better aeration of the matrix (12).

Roots can grow along the earthworm burrows (9), possibly through hard, dry, waterlogged or acidic soil to reach nutrients and water in soft, moist, drained or neutral soil. However, water, nutrients, and pesticides may follow burrows below the root zone and pollute the groundwater (5).

TERMITES

Termites consist of several thousand species, with several castes in a colony (6). Termites turn over 0.02–470 t soil $ha^{-1} yr^{-1}$, affecting over 0.5–20% of the soil surface (15). Termites can increase erosion by removing protective litter from the surface, and bringing dispersive particles to the surface.

Termites build a wide range of nests (mounds), in which they live and store food, and sometimes runways and subterranean galleries (i.e., the nests contain large

Table 1 Climatic zone, feeding habit and food of invertebrate macrofauna in soil

Fauna	Climatic zone	Feeding habit (food)
Oligochaeta		
Earthworms	All but driest and coldest	Saprovores, geovores, (organic litter, soil)
Macroarthropods		
Isopods	Various including desert	Saprovores, predators, (organic litter, roots and leaves of seedlings)
Millipedes (Diplopoda)	Temperate, Tropical, arid, semiarid	Saprovores, (leaf litter, dung)
Centipedes (Chilopoda)	Forest to desert	Saprovores, predators
Scorpions	Mainly desert; also warm dry tropics, temperate	Predators, cannibals
Spiders (Araneae)	All but Arctic/Antarctic	Carnivores, predators, (arthropods, small vertebrates)
Termites (Isoptera)	Mainly tropics, subtropics; also temperate and deserts	Saprovores, (wood, plants, lichens, humus, fungi)
Ants (Hymenoptera)	Arctic to tropics	Saprovores, carnivores, predators (organic litter, seeds, secretions of plants and aphids, plants)
Wasps (Hymenoptera)	Temperate, tropics	Saprovores, (nectar, sap)
Beetles (Coleoptera)	Temperate	Saprovores, predators, carnivores, (dung, animal carcasses, seeds, roots)

pores). The nests and runways consist of soil, excreta, plant residues and sometimes saliva, often with higher concentrations of clay, exchangeable cations and microbial activity, than the surrounding soil (10, 15).

Mounds are usually up to at least 30 cm wide and 9 m high (6, 15), with 200–1000 small mounds ha^{-1}, or 2–10 large mounds ha^{-1}; there are few mounds on sands, self-mulching clays and shallow soils. Mounds last from 8–80 years, surviving while the mound is active. Once inactive, mounds are quickly eroded, especially when dug by mammals (6). The surface becomes covered with fine particles of subsoil, adding 25–1000 mm depth of soil every 1000 years. The runways are quickly eroded by rain and rebuilt just as quickly.

The few reports available suggest that termites have less overall effect on soil structure than do earthworms. Some termites slightly increase the bulk density of soil, probably because they repack the soil particles and select fine particles. Reports on the infiltration of water in termite galleries differ from that of the surrounding soil, although water-holding capacity is usually higher in the mounds (15).

ANTS

There are at least 15,000 species of ants, all of which are social insects living in colonies with up to 500,000 individuals, with a biomass of 0.1–0.8 $g\,m^{-2}$ (15). Many ants burrow extensively, producing nests ranging from simple excavations on the surface or under stones, to craters and domes (18). The mounds are either Type I, which are small and ephemeral, or Type II, which are large, dense and can last for more than 100 years (20). Ants move between 3–11,360 kg of soil $ha^{-1}\,yr^{-1}$ (agricultural land) and 21–8,410 kg $ha^{-1}\,yr^{-1}$ (natural vegetation) depending on the species (14). Ants mix soil and litter, or decompose organic litter or the surface of soil, deposit subsoil on the surface (11), and appear eventually to homogenize the texture to a depth of at least 260 mm (16).

The effect of ants on soil structure has not been greatly studied. However, compared with the surrounding soil, the mounds may have a lower plastic limit (11), be more stable and denser (15) (at least at the top of the mound), or less dense (3, 11), with higher macroporosity due to the galleries (19). The infiltration rate into an ant nest was over three times that of the surrounding soil (15), probably through the galleries. Compared with the surrounding soil, infiltration into ant mounds was increased as the liquid moved to a depth of 2 m through vertical and lateral galleries (19). However, ants expose bare soil around their burrows, which could impede infiltration and encourage soil erosion.

REFERENCES

1. Buck, C.; Langmaack, M.; Schrader, S. Influence of Mulch and Soil Compaction on Earthworm Cast Properties. Appl. Soil Ecol. **2000**, *14*, 223–229.
2. Cockroft, B.; Olsson, K.A. Degradation of Soil Structure Due to Coalescence of Aggregates in No-Till, No-Traffic Beds in Irrigated Crops. Aust. J. Soil Res. **2000**, *38*, 61–70.

F

3. Cowan, J.A.; Humphreys, G.S.; Mitchell, P.B.; Murphy, C.L. An Assessment of Pedoturbation by Two Species of Mound-Building Ants. *Camponotus Intrepidus* (Kirby) and *Iridomyrmex Purpureus* (F. Smith). Aust. J. Soil Res. **1985**, *23*, 95–107.
4. Dexter, A.R. Tunnelling in Soil by Earthworms. Soil Biol. Biochem. **1978**, *10*, 447–449.
5. Edwards, W.M.; Shipitalo, M.J. Consequences of Earthworms in Agricultural Soils: Aggregation and Porosity. In *Earthworm Ecology*; Edwards, C.A., Ed.; CRC Press: Boca Raton, 1998; 147–161.
6. Goudie, A.S. The Morphological Role of Termites and Earthworms in the Tropics. In *Biogeomorphology*; Viles, H., Ed.; Basil Blackwell: Oxford, 1988; 166–192.
7. Hindell, R.P.; McKenzie, B.M.; Tisdall, J.M.; Silvapulle, M.J. Relationships Between Casts of Geophagous Earthworms (*Lumbricidae*, *Oligochaeta*) and Matric Potential. Biol. Fertil. Soils **1994**, *18*, 119–126.
8. Hindell, R.P.; McKenzie, B.M.; Tisdall, J.M. Influence of Drying and Ageing on the Stabilization of Earthworm (*Lumbricidae*) Casts. Biol. Fertil. Soils **1997**, *25*, 119–126.
9. Hirth, J.R. Interactions Between Plant Roots (*Lolium Perenne* L.) *and Earthworm Burrows* (*Aporrectodea* Species) PhD Thesis La Trobe University, 1996.
10. Holt, J. Microbial Activity in the Mounds of Some Australian Termites. Appl. Soil Ecol. **1998**, 187–191.
11. Hulugalle, N.R. Effects of Ant Hills on Soil Physical Properties of a Vertisol. Pedobiologia **1995**, 34–41.
12. Kretzschmar, A. Earthworm Interactions with Soil Organization. In *Earthworm Ecology*; Edwards, C.A., Ed.; CRC Press: Boca Raton, 1998; 163–176.
13. Lee, K.E. *Earthworms—Their Ecology and Relationships with Soils and Land Use*; Academic Press: Sydney, 1985.
14. Lobry de Bruyn, L.A. Ants as Bioindicators of Soil Function in Rural Environments. In *Invertebrate Biodiversity as Bioindicators of Sustainable Landscapes*; Paoletti, M.G., Ed.; Elsevier: Amsterdam, 1999; 425–441.
15. Lobry de Bruyn, L.A.; Conacher, A.J. The Role of Termites and Ants in Soil Modification: A Review. Aust. J. Soil Res. **1990**, *28*, 55–93.
16. Lobry de Bruyn, L.A.; Conacher, A.J. The Bioturbation Activity of Ants in Agricultural and Naturally Vegetated Habitats in Semi-Arid Environments. Aust. J. Soil Res. **1994**, *2*, 555–570.
17. McKenzie, B.M.; Dexter, A.R. Physical Properties of Casts of the Earthworm *Aporrectodea Rosea*. Biol. Fertil. Soils **1987**, *9*, 163–167.
18. Mitchell, P. The Influence of Vegetation. In *Animals and Micro-Organisms on Soil Processes. Biogeomorphology*; Viles, H., Ed.; Basil Blackwell: Oxford, 1988; 43–82.
19. Nkem, J.N.; de Bruyn, L.A.L.; Grant, C.D.; Hulugalle, N.R. The Impact of Ant Bioturbation and Foraging Activities on Surrounding Soil Properties. Pedobiologia **2000**, *44*, 609–621.
20. Paton, T.R.; Humphries, G.S.; Mitchell, P.B. *Soils: A New Global View*; Yale University Press: New Haven, 1995.
21. Schrader, H.; Zhang, H. Earthworm Casting: Stabilization or Destabilization of Soil Structure. Soil Biol. Biochem. **1997**, *29*, 469–475.
22. Smettem, K.R.J. Analysis of Water Flow from Cylindrical Macropores. Soil Sci. Soc. Am. J. **1986**, *50*, 1139–1142.

FERTILITY, ENVIRONMENTALLY COMPATIBLE MANAGEMENT OF

Bal Ram Singh
Agricultural University of Norway, Aas, Norway

INTRODUCTION

Soil Fertility in the Past

In writings dating back to 2500 B.C. the fertility of land is mentioned. Herodotus, the Greek historian, traveling through Mesopotamia some 2000 years later reported the phenomenal yields obtained by the inhabitants of this land. Later Theophrastus (372–287 B.C.) recommended the abundant manuring of thin soils but suggested that rich soils be manured sparingly. During the seventeenth and eighteenth centuries, agricultural writings reflected that plants consisted of one substance, and most of the workers were searching for this principle of vegetation during this period. It was not until the later half of nineteenth and the beginning of the twentieth century that some progress was made to understand the subject of plant nutrition and crop fertilization. It was Justus von Liebig (1803–1873), a German chemist, who effectively deposed the humus myth and eventually developed the law of minimum. The law says that the growth of plants is limited by the plant–nutrient element present in the smallest quantity, all others being present in adequate amounts. These developments led to a rapid increase in chemical fertilizer's use.

Soil Fertility in Modern Times

With advances in our knowledge with regards to various processes affecting the nutrient dynamics in soils, the definition of soil fertility is also refined. Soil fertility integrates the basic principles of soil biology, chemistry and physics to develop the practices needed to manage nutrients in a profitable and environmentally sound manner. The main focus of soil fertility is to manage nutrient status in soils to create optimum conditions for plant growth. Two fundamental principles underlay the study of soil fertility. First is the recognition that optimum nutrient status alone will not ensure soil productivity. Other factors, such as soil moisture and temperature, soil physical conditions, soil acidity and salinity, and biotic stress can reduce the productivity of even more fertile soils. Second is the realization that modern soil fertility practice must stress soil productivity and environmental protection (1). Taking the second realization in perspective, it is imperative that soil fertility in relation to agricultural sustainability and environmental protection will be the main focus of this paper.

SOIL FERTILITY AND AGRICULTURAL SUSTAINABILITY

In recent years, a new dimension to soil fertility in relation to agricultural production has been added. It is the concept of "sustainability," which has been defined and interpreted differently by different workers. Okigbo (2) after analyzing the various definitions of sustainable agriculture by different workers defined "a sustainable agricultural production system as one that maintains an acceptable and increasing level of productivity that satisfies prevailing needs and is continuously adapted to meet the future needs for increasing the carrying capacity of the resource base and other worthwhile human needs" (2).

The sustainability of a production system is location specific and is determined to a greater degree by an interaction among several production factors, viz. physiochemical (soil, climate, radiation etc.) biological (crop species, weeds and pests etc.), management and socioeconomic elements. Maintenance and management of soil fertility is central to the development of sustainable food production systems. Sustainability is dependent to a large degree on recycling the inputs into a production system, thereby increasing efficiency of output per unit of resource input. Soil fertility management is concerned with the essential plant nutrients, their amounts, availability to crop plants, chemical reactions in soil, loss mechanisms, processes making them unavailable or less available to crop plants, and ways and means of replenishing them in these soils (3).

ESSENTIAL NUTRIENTS

Because soil fertility involves management of nutrients required for plant production, it is important to describe

Encyclopedia of Soil Science

briefly the elements, which are considered essential for plant growth. The essential nutrients required by higher plant are exclusively of inorganic nature, and Arnon and Stout (4) proposed the term essential nutrient (element). The essential element must meet three criteria to be considered essential: 1) a given plant must be unable to complete its life cycle without the presence of mineral element in question, 2) the function of the element cannot be replaced by another element, and 3) the element is directly involved in the nutrition of the plant—for example as a component of an essential plant constituent such as enzyme—or must be required for a distinct metabolic step such as an enzyme reaction

Some elements, which either compensate for the toxic effect of other elements or they simply replace mineral elements in some of their less specific functions, such as maintenance of osmotic pressure, can be described as beneficial elements. Out of a large number of elements found in plants, 14 mineral elements are recognized as essential, whereas the requirement of chlorine and nickel is yet restricted to a limited number of plant species. The plant nutrients may be divided into macronutrients (N, P, S, K, Mg, Ca), micronutrients (Fe, Mn, Zn, Cu, B, Mo, Cl, Ni) and beneficial elements (Na, Si, Co). The last three elements have been found essential only for some plant species, for example, Na for plants with C4 photosynthetic pathway, Si for rice, and Co for fixation of atmospheric N by rhizobia and blue green algae, but they have not yet been included in the list of essential nutrients (3).

MANAGEMENT OF SOIL FERTILITY AND THE ENVIRONMENT

As pointed out by Sims (1) that modern soil fertility in addition to involving soil productivity, must also include protection of the environment and thus the environmental aspects related to soil fertility are described under this section. Long-term use of organic and inorganic fertilizers to agricultural soils has led to slow build up of nutrient reserves and especially under temperate conditions. The same nutrients, which are considered essential for plant growth and crop production, if lost from the system, can create a concern for the environment. Increased use of fertilizers for meeting the world food demand and recycling of on farms (manures and slurries) and off farms organic wastes (municipal and industrial sludges) on agricultural lands in the last three decades have resulted in some undesirable effects on the environment in intensively cultivated areas. This has created greater awareness for environmental issues not only among scientists and policy makers but also among general public. Although there are a number of issues related to soil fertility and the environment, emphasis is placed on four main concerns of environmental protection, viz. soil quality, water quality, air quality, and heavy metals and food concerns.

Soil Quality

The export of agricultural produce takes with it large amounts of nutrients (especially N, P, K, and S) and if these nutrients are not replenished, soil organic matter (SOM) and the fertility of the soil will decline and the soil will degrade. A number of such examples from the subsistence farming of the tropics are available, where "nutrient mining" of soils has been responsible for decline in soil fertility and soil quality leading to loss of productivity (5). The rapid loss of SOM of the topsoil after clearing of the natural vegetation or under continuous cultivation is a common phenomenon in Africa. In Zambia it was found that the SOM content dropped from 59 to 32 $Mg\,ha^{-1}$ after 17 years of cultivation of a newly cleared land, which represented an average loss of 1.6 $Mg\,ha^{-1}\,y^{-1}$ (6).

On the other hand, long-term studies carried out in temperate regions have shown that organic matter of soil can be maintained or raised modestly by proper fertilization and especially by organic manures and crop residue management (7). Any management practice that results in an increase in the crop residues returned to soil will have a positive influence on soil organic matter (7), which in turn will affect soil aggregation and related properties such as soil structure, erodibility, workability, and water infiltration. Growing of legumes is an effective way for promoting good soil aggregation and for reversing degradative trends.

Water Quality

The major concerns with regards to water quality are accelerated eutrophication of surface waters and nitrate (NO_3) content of drinking water. Eutrophication, the rapid growth and decay of aquatic vegetation, is most often limited by P and sometimes by N concentration in water. The NO_3 concentration of 10 $mg\,L^{-1}$ in drinking water is considered safe but higher concentration can cause methaemoglobinaemaia (reduced carrying capacity of blood for O_2) in children.

Both over fertilization and under fertilization can lead to N losses. Losses occur either as runoff to surface water or as leaching to underground waters. Under normal conditions, runoff losses in watersheds are low (e.g., 10 mg L^{-1}). Much of the N transported in runoff is particulate N associated with the sediments and it can range from 0 to 7 kg N ha^{-1} (8). These losses can be reduced by incorporation of fertilizer and providing a good vegetation cover. Much of N losses occur through leaching but there is a large variation in the quantities of N lost in leaching depending on the amount of N fertilizer applied, soil type, crop grown and climatic conditions. Letey et al. (9) reviewed NO_3 concentration in tile effluents from 55 sites of California; the concentration ranged from 1 to 196 mg L^{-1}, with only one-quarter of the sites averaging <10 mg L^{-1}. High nitrate levels in drinking water have also been observed where animal or poultry manures are applied. In Sussex county, Delaware, it was found that 32% of the water wells had high (>10 mg L^{-1}) nitrate levels (10).

From surface water viewpoint, P is the element of primary concern, since it is considered limiting for eutrophication. Excessive application of organic and inorganic fertilizers can result in building of P near the soil surface, which is subjected to soil erosion leading to P losses to surface waters. In aquatic ecosystems of southwestern Australia an excess of nutrients has caused serious eutrophication, which was manifested by excessive growth and accumulation of green and blue green algae (11). Phosphorus is generally the limiting nutrient for algae growth and phosphatic fertilizers applied to nutrient-deficient, leaching, sandy soils were the main source of P in these ecosystems of Australia.

Air Quality

The primary goal of good nutrient management and especially N should be to maximize N uptake at critical growth stages and to minimize transformation processes, which lead to formation of "greenhouse gases". The main N-containing greenhouse gas is N_2O. The N_2O produced during nitrification and denitrification contributes to global warming and stratospheric ozone depletion. Annual losses of N as N_20 from fertilized field soil can be as high as 40 kg h^{-1} compared to that from unfertilized soils being <2 kg h^{-1} (12). Based on a summary of 104 field experiments conducted in temperate climate, it was reported that almost all experiments showed evidence of fertilizer derived N_2O emissions (13). Legumes, like fertilizers have been shown to contribute to increased N_2O emissions (7) caused by enhanced denitrification due to the addition of easily decomposable organic matter. On the other hand, it was found that gaseous losses via denitrification and N_2O emissions from irrigated corn in Shelton, NE, USA represented only 1 to 5% of the N applied as fertilizer and thus it was concluded that under good irrigation and good N management practices gaseous N losses from denitrification and N_2O emissions do not pose additional threat to air quality (14).

Under tropical climate, N loss to air takes place both through denitrification and volatilization of NH_3. High losses of NH_3 in rice fields have been attributed to high concentration of $NH_4 - N$ following urea application, high pH, high temperature and high wind speed. Ammonia loss of applied urea-N ranged from 7 to 54% in Philippines and the loss depended on time and the method of urea-N application. For example, incorporation of urea before transplanting with no standing water reduced urea-N loss as NH_3 from 42 to 11% at IRRI, Philippines (15). Direct field measurement of denitrification has been made difficult by the lack of suitable methodology. Buresh and Austin (16) used a floating chamber technique to directly collect $(N_2 + N_2O)^{-15}$ N from labeled urea applied to flooded soil and found that the total N loss as unaccounted-for ^{15}N by the ^{15}N balance study ranged from 26 to 46% of the applied N.

Heavy Metals and Food Concerns

Contamination sources

The major sources of anthropogenic inputs of metals to soils are either primary sources as fertilizers and amendments (e.g., lime), sewage sludge, and pesticides or secondary source, where metals are added to the soil as a result of nearby activities such as mining or industry or aerosol deposition from more distant sources. However, there are also rural areas, where the content of metal in soils is many fold higher due to natural processes and geological formation of soils. A good example is alum shale soils (developed from sulphide-bearing rocks formed in anoxic environment) in south-eastern Norway (17).

It is well documented that the use of commercial P fertilizers is one of the main sources of metals and especially of Cd addition to agricultural soils (18–20). Besides fertilizers, certain animal waste products may also contain high amount of metals. Pig manure contained more than twice the amount of Zn (158 mg/kg) and Cu (15 mg/kg) than cow manure because of supplementation of these metals in their diets (21). Disposal of biosolids (e.g., sewage sludge) is a potential source to contaminate soils. In many developed countries, soil metal loading through sewage sludge is regulated through "maximum permitted concentration (MPC)."

F

Table 1 Dutch reference values and interventional values (mg kg^{-1}) in soils based on ecotoxicity and human health

Metal	Reference value for standard soil[a]	Ecotoxicity intervention value	Human health intervention value
Cd	0.8	12	680
Cu	36	190	16000
Pb	85	290	300
Ni	35	210	6600
Zn	140	720	56000

[a] Standard soil with 25% clay and 10% organic matter.
(From Ref. 20.)

Soil contamination with metals and guidelines for soil quality

That long-term use of fertilizers can lead to increased content of metals in soils is well documented (20, 22, 26) and especially Cd enrichment of soils after long-term use of P fertilizers is reported in many countries (19, 22–24). Many studies based on long-term fertility experiments in USA and western Europe have shown a net accumulation of Cd in soils (19, 23, 24). The balance calculations based on 70 years data indicated that Cd accumulation in the soil was <1 g Cd ha^{-1} y^{-1} but increasing the doses of commercial fertilizers or farm yard manure would likely result in increased accumulation of this element (19). In some countries guidelines to protect soil quality has been developed. A best example of such values is found in The Netherlands. They have developed "reference or target values," which represent soils with negligible risks and "intervention values," which represent soils with the maximum possible risk levels (Table 1). In Australia and New Zealand environmental soil quality guidelines to protect plant production, animal health, and food quality have been prepared. These soil quality guidelines for Cd, Cu, Pb, Ni and Zn are 3, 60, 300, 60, and 200 mg/kg, respectively (20).

Phytouptake and phytotoxicity

Soil conditions related directly or indirectly to soil fertility such as soil pH, organic matter and clay content, metal content, moisture content and soil aeration are determining factors for metal. For these reasons some workers have found increased plant uptake of Cd added through commercial fertilizers (18, 24–26), whereas others did not find any increase (23). The application of P fertilizer containing increasing amount of Cd resulted in increased concentration of Cd in different crops but the magnitude of Cd uptake differed considerably among crops. For example, spinach contained almost 25 times more Cd than oat grain (Table 2) (25). Furthermore, increasing disposal of sludges and other organic wastes to agricultural lands has also resulted in increased uptake of metal in plants (27, 28). The uptake of Zn, Cu and Cd in field pea increased significantly in sludge or pig manure treated plots but liming of these plots reduced the concentration of all metals significantly (Table 3). (28).

Metals in foods and their regulations

Among the metals, Cd in food stuff is considered a greater risk and hence regulatory limits for food, so called

Table 2 Cadmium concentration (mg kg^{-1}) in different crop species as affected by varying levels of Cd in P fertilizer applied

Cd applied in fertilizer (μg kg^{-1})	Oat grain	Ryegrass	Carrot root	Spinach
0.0	0.06	0.21	0.14	—
2.7	0.07	0.21	0.15	1.52
12.5	0.07	0.24	0.18	1.80

(From Ref. 22.)

Table 3 Zinc, Cu and Cd concentration in field pea (*Pisum arvense*) in sludge and pig manure treated soils with and without liming

Treatment	Zn (mg/kg)	Cu (mg/kg)	Cd (μg/kg)
Unlimed:			
Control	73.3 cd	9.2 c	63.2 d
Sludge	77.9 bc	7.8 d	81.8 dc
Pig manure	75.3 cd	7.4 bd	60.9 d
Limed:			
Control	49.5 e	7.0 be	37.3 e
Sludge	67.5	7.1 b	36.7 e
Pig manure	61.1 f	7.0 be	33.5 e

Mean values within the same column followed by the same letter are not significantly different at ($P = 0.05$).
(Adapted and modified from Krebs et al. 1998.)

"maximum permissible concentrations (MPC)" have been worked out in several countries (e.g. The Netherlands, Germany, Denmark, Sweden. Australia and New Zealand). These values do not pose health risks but they are taken as target concentrations for initiating corrective measures to minimize Cd entry into the food chain. Cadmium concentration in leafy vegetable is much higher than seed grains and fruits (Table 3). For metals like Pb and Hg, their entry through roots is rather restricted, resulting in their low concentration in food crops.

Management practices for minimizing metal transfer to food crops

Elevated levels of metals in soils and food crops have created a need to search for management practices, which can reduce metal transfer from soils to food crops. A number of practices have been tested with varying rates of success. However, the mechanisms involved are far from fully understood. One practice, which has been more often used, is liming, but the results obtained have shown both decrease and increase in metal uptake and especially that of Cd (25, 29). Liming raises soil pH and thus reduces metal concentration in soil but it also raises the Ca concentration in soil solution. Increased concentration of Ca further complicates the effect of liming. Addition of Zn is shown to reduce Cd uptake by crops and especially under Zn deficiency conditions (30), but the mechanism is not known. In metal-rich alum shale soils, increasing rates of cow and pig manures decreased the exchangeable fractions of Cd and Ni. (21). Other practices, such as crop rotation showed marked effect, where wheat grown after lupins had higher concentration of Cd than wheat grown after a cereal crop (31).

REFERENCES

1. Sims, T. Soil Fertility Evaluation. In *Handbook of Soil Science*; Sumner, M.E., Ed.; CRC Press: Boca Raton, 1999; D113–D153.
2. Okigbo, B.N. *Development of Sustainable Agricultural Production Systems in Africa*; International Institute of Tropical Agriculture, Ibadan, Nigeria 1991; 66.
3. Prasad, R.; Power, J.F. *Soil Fertility Management for Sustainable Agriculture*; Lewis Publishers: Boca Raton, 1995; 1–4.
4. Arnon, D.I.; Stout, P.R. The Essentiality of Certain Elements in Minute Quantity for Plant with Special Reference to Copper. Plant Physiol. **1939**, *14*, 371–375.
5. Stoorvogel, J.J.; Smaling, E.M.A.; Janssen, B.H. Calculating Soil Nutrient Balances in Africa at Different Scales: I. Supernational Scale. Fert. Res. **1993**, *35*, 227–235.
6. Singh, B.R.; Goma, H.C. Long-Term Soil Fertility Management in Eastern Africa. In *Soil Management-Experimental Basis for Sustainability and Environmental Quality*; Lal, R., Stewart, B.A., Eds.; Lewis Publishers: Boca Raton, 1995; 347–382.
7. Campbell, C.A.; Myers, R.J.K.; Curti, D. Managing Nitrogen for Sustainable Crop Production. Fert. Res. **1995**, *42*, 277–296.
8. Smith, S.J.; Schepers, J.S.; Porter, L.K. Assesseing and Managing Agricultural Nitrogen Losses to the Environment. Adv. Agron. **1990**, *14*, 1–43.
9. Letey, J.; Blair, J.W.; Devitt, D.; Lund, L.J.; Nash, P. Nitrate-Nitrogen in Effluent from Agricultural Tile Drains in California. Hilgardia **1977**, *45*, 289–319.
10. Ritter, W.E.; Chirnside, A.E.M. *Ground Water Quality in Selected Areas of Kent and Sussex Counties, Delaware*; Delaware Agriculture Experiment Station, Project Completion Report, Moore, P.A.; Daniel, T.C.; Sharpley, A.N.; Wood, C.W. Poultry Manure Management: Environmentally Sound Options. J. Soil Water Cons. **1995**, *50* (3), 321–327.
11. Hodgkin, E.P.; Hamilton, B.H. Fertilizers and Eutrophication in Southwestern Australia: Setting the Scene. Fert. Res. **1993**, *36*, 95–103.
12. Sahrawat, K.L.; Keeny, D.R. Nitrous Oxide Emission from Soils. Adv. Soil Sci. **1986**, *4*, 143–148.
13. Eichner, M.J. Nitrous Oxide Emissions from Fertilized Soils. Summary of Available Data. J. Environ. Qual. **1990**, *19*, 272–280.
14. Qian, J.H.; Doran, J.W.; Weier, K.L.; Mosier, A.R.; Peterson, T.A.; Power, J.E. Soil Denitrification and Nitrous Oxide Losses Under Corn Irrigated with High-Nitrate Ground Water. J. Environ. Qual. **1997**, *26*, 348–360.
15. De Datta, S.K.; Buresh, R.J. Integrated Nitrogen Management in Irrigated Rice. Adv. Agron. **1989**, *10*, 143–169.
16. Buresh, R.J.; Austin, E.R. Direct Measurement of Dinitrogen and Nitrous Oxide Flux in Flooded Rice Fields. Soil Sci. Soc. Am. J. **1988**, *52*, 681–688.
17. Mehlum, H.K.; Arnesen, A.K.M.; Singh, B.R. Extractability and Plant Uptake of Heavy Metals in Alum Shale Soils. Commun. Soil Sci. Plant Anal. **1998**, *29*, 183–198.
18. Singh, B.R. Trace Element Availability to Plants in Agricultural Soils, with Special Emphasis on Fertilizer Inputs. Environ. Reviews. **1994**, *2*, 133–146.

19. Jeng, A.J.; Singh, B.R. Cadmium Status of Soils and Plants from a Long-Term Fertility Experiment in Southeast Norway. Plant Soil **1995**, *175*, 67–74.
20. McLaughlin, M.J.; Hamon, R.E.; McLaren, R.G.; Speir, T.W.; Roger, S.L. Review: A. Bioavailability-Based Rationale for Controlling Metal and Metalloid Contamination of Agricultural Land in Australia and New Zealand. Aust. J. Soil Res. **2000**, *38*, 1037–1086.
21. Narwal, R.P.; Singh, B.R. Effect of Organic Materials on Partitioning, Extractability and Plant Uptake of Metals in an Alum Shale Soil. Water, Air, and Soil Pollut. **1998**, *103*, 405–421.
22. He, Q.B.; Singh, B.R. Plant Availability of Cd in Soils: I. Extractability of Cd in Newly- and Long-Term Cultivated Soils. Acta. Agric. Scand. **1993**, *43*, 134–141.
23. Mordvedt, J.J. Cadmium Levels in Soils and Plants from Some Long-Term Soil Fertility Experiments in the USA. J. Environ. Qual. **1987**, *16*, 137–142.
24. Jones, K.C.; Johnston, A.E. Cadmium in Cereal Grain and Herbage from Long-Term Experimental Plots at Rothamstead., U.K. Environ. Pollut. **1989**, *67*, 75–89.
25. He, Q.B.; Singh, B.R. Crop Uptake of Cd from Phosphate Fertilizers: I Yield and Cd Content. Water Air Soil Pollut. **1994**, *74*, 251–265.
26. McLaughlin, M.J.; Tiller, K.G.; Naidu, R.; Stevens, D.P. Review: The Behaviour and Environmental Impact of Contaminants in Fertilizers. Aus. J. Soil Res. **1996**, *34*, 1–56.
27. Brown, S.L.; Chaney, R.L.; Angle, J.S.; Ryan, J.A. The Phytoavailability of Cd to Lettuce in Long-Term Biosolids-amended Soils. J. Environ. Qual. **1998**, *27*, 1071–1078.
28. Krebs, R.; Gupta, S.K.; Furrer, G.; Schulin, R. Solubility and Plant Uptake of Metals With and Without Liming of Sludge-Amended Soils. J. Envir. Qual. **1998**, *27*, 18–23.
29. Maier, M.A.; McLaughlin, M.J.; Heap, M.; Butt, M.; Smart, M.K.; William, C.M.J. Effect of Current-Season Application of Calcitic Lime on Soil pH, Yield and Cd Concentration in Potato (*Solanum tuberosum* L.) Tubers. Nutr. Cycling in Agroecos **1997**, *47*, 29–40.
30. Oliver, D.P.; Hamon, R.; Tiller, K.G.; Wilhelm, N.S.; Merry, R.H.; Cozens, G.D. The Effect of Zinc Fertilization on Cd Concentration in Wheat Grain. J. Environ. Qual. **1994**, *23*, 705–711.
31. Oliver, D.P.; Schultz, J.E.; Tiller, K.G.; Merry, R.H. The Effect of Crop Rotations and Tillage Practices on Cd Concentration in Wheat Grain. Aust. J. Agric. Res. **1993**, *44*, 1221–1234.

F

FERTILITY EVALUATION SYSTEMS

Ram C. Dalal
Department of Natural Resources, Indooroopilly, Australia

A. Subba Rao
Indian Institute of Soil Science, Madhya Pradesh, India

INTRODUCTION

The soil fertility evaluation is aimed at providing an adequate supply of essential plant nutrients to ensure optimum plant productivity while maximizing economic benefits and minimizing environmental degradation. The essential plant nutrients include: N, P, K, S, Ca, Mg, Na, Cu, Zn, Mn, Fe, Mo, B, and Cl. The soil fertility evaluation systems include biological assessment, plant nutrient analysis, soil testing (chemical, biological, and physical), and spectral sensing for site-specific soil fertility.

BIOLOGICAL ASSESSMENT

The significance of the supply of an essential plant nutrient for crop production was contained in the Liebig's Law of the Minimum (1), which states that by the deficiency or absence of one necessary constituent, all others being present, the soil is rendered barren for all those crops to the life of which that one constituent is indispensable. When crop response to the quantity of an essential nutrient present in soil or added to it increases in direct proportion, it can be expressed by the Linear equation:

$$y = a + mx \tag{1}$$

where y is the crop yield, x is the quantity of nutrient added, a is the intercept, and m is the slope. Eq. 1 applies to low soil fertility conditions and for nutrient uptake. In such cases, the value of a, the intercept, is equivalent to the plant available quantity of nutrient in soil.

However, as the quantity of a nutrient increases, the unit increment in crop yield becomes smaller, and eventually, the crop yield reaches a maximum value. The crop response curve is then represented by the Mitscherlich equation (2):

$$y = A(1 - e^{-Cx_t})$$

or

$$y = A(1 - e^{-C(x+b)}) \tag{2}$$

where x_t or $x + b$ is the total nutrient and b is the plant available quantity of nutrient in soil (and seed, considered if present in significant quantity), A is the maximum yield attainable when $x + b$ is present in sufficient quantity, and C is the efficiency factor.

When quantity of a nutrient present in, as well as added to, soil is in excess of crop requirements, the nutrient becomes toxic to plants or other factors become yield limiting, then the crop yield response curve obeys polynomial equations such as the quadratic equation:

$$y = a_1 + b_1x + c_1x^2 \tag{3}$$

where a_1, b_1 and c_1 are constants. The maximum yield is given by the value of $-b_1/2c_1$. When y is expressed as the nutrient uptake, a_1 estimates the quantity of available nutrient in soil.

Because maximum yield, A value in Eq. 2 and $-b_1/2c_1$ in Eq. 3, varies across crops, soils, seasons, and years, it is usually preferable to express plant productivity as relative yield in a field experiment. The relative yield at a given nutrient supply is: $100 \times y_n/A$ from Eq. 2, and $100 \times y_n/(-b_1/2c_1)$ from Eq. 3, where y_n is the crop yield at a given nutrient supply in soil.

This biological assessment of soil fertility evaluation is shown in Fig. 1.

Although this biological assessment in the field is the most appropriate system for soil fertility evaluation, it is expensive and time consuming and subject to the limitations and constraints imposed on crop growth due to other factors such as pests, diseases, water stress, spatial variability. It is primarily used for benchmarking other soil fertility evaluation systems such as plant analysis and tissue testing, soil testing, and remote sensing.

PLANT ANALYSIS

When nutrient deficiency is severe enough, plants exhibit visual symptoms (3), and this provides a rapid but qualitative assessment of soil fertility. Conversely,

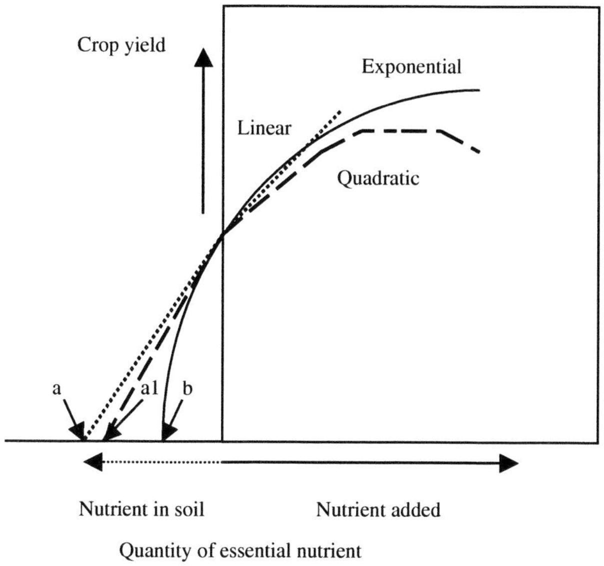

Fig. 1 Crop yield response curve with quantities of a nutrient added and in soil (and seed, if contributing significant quantity); the quantity in soil is estimated from *a* in Eq. 1, *b* in Eq. 2, and a_1 in Eq. 3.

nutrient toxicity in the presence of excess nutrients and heavy metals in soil can be readily assessed from visual symptoms on plants in the field (3). However, it is often too late to take corrective measures to avert the decline in crop yields. Other limitations include disease, undefined symptoms, environmental factors, and multi-nutritional disorders.

Because critical nutrient concentration range (≈90% maximum yield) in plants at which crop yield is affected is relatively narrow, plant analysis for critical nutrient concentration has been used to evaluate soil fertility and recommend nutrient applications to correct nutrient deficiency (3).

The further application of plant analysis for critical nutrient concentration is in tissue testing of a crop growing in the field. This involves a rapid analysis of a nutrient, a form of the nutrient such as nitrate for nitrogen, enzyme activity such as nitrate reductase activity for nitrate, and ribonuclease activity for zinc, and greenness index for chlorophyll associated with critical nitrogen concentration in the leaf. It usually provides only a qualitative or semi-quantitative assessment of soil fertility because the critical concentration differs among crop cultivars, crop types, incidence of disease, time of sampling, nature of nutrient interactions, water supply, temperature, dry matter yield level, physiological maturity of the leaf or plant part sampled, and other undefined factors.

The balance of nutrients within a plant is often more important than the concentration of any individual nutrient. The Diagnosis and Recommendation Integrated System (DRIS) focuses on nutrient balance as an alternative to the critical nutrient concentration range approach to plant analysis interpretation. One advantage of the DRIS system is that nutrient ratios in plants tend to be less variable throughout the growing season than individual nutrient concentrations, especially for mobile nutrients such as N, P, and S, but there is no one universal benchmark nutrient ratio for plants (4).

SOIL TESTING

Soil testing provides a rapid chemical and/or biological analysis to assess the quantity of plant-available nutrients in soil and, hence, for evaluation of soil fertility (4–6). This is usually done by analyzing an extract of the soil obtained by adding water, chemical extractants, ion-exchange resin and membranes, and electroultrafiltration. Short-term incubation, especially for assessing nitrogen mineralization, fungal growth such as *Aspergillus niger*, or even short-term plant growth in the greenhouse is employed as biological methods for soil testing. However, biological methods are time-consuming and are rarely used in soil testing except for short-term anaerobic nitrogen mineralization (4).

Unfortunately, most values of available plant nutrients are expressed on a soil's mass basis rather than the volume basis; the latter requires soil's bulk density and depth of sampling. Furthermore, for most plant-available nutrients, soil from only the top 10 cm or 30 cm depth is sampled although crop roots access nutrients and water from much deeper layers, often deeper than 100 cm in many soils.

SOIL TEST–CROP RESPONSE CORRELATION

Soil test values are used for evaluating soil fertility of a group of soils based on correlation with crop response such as dry matter yield, crop yields, nutrient yield, or nutrient concentration. It is essential that the soil tests be calibrated against crop response to applied nutrients in field experiments conducted over a wide range of soils, crops, and climate. A properly calibrated soil test correctly identifies the degree of deficiency or sufficiency of a nutrient. This is shown in Fig. 2.

Most soil tests are based on theoretical consideration of some aspects of plant nutrient availability (intensity, quantity, buffer power, and diffusion of nutrients in soil).

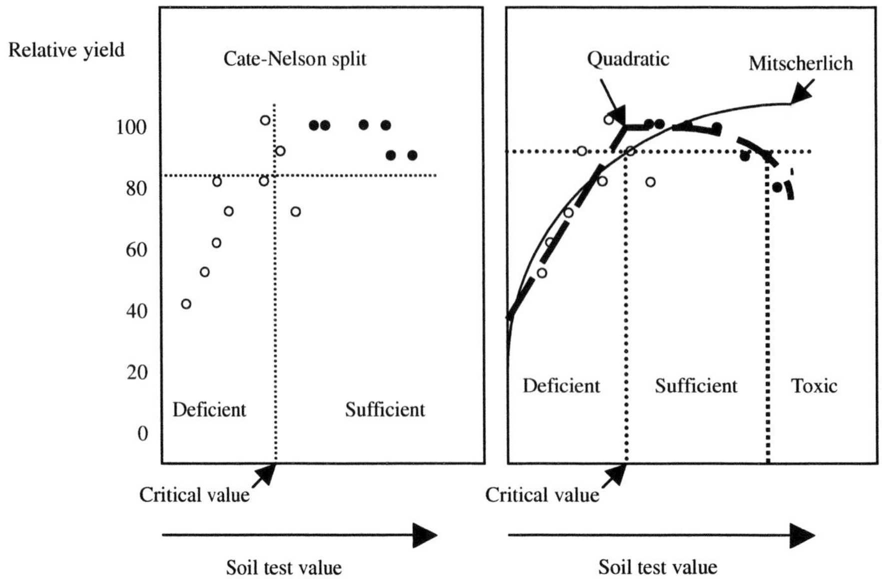

Fig. 2 A diagrammatic representation of soil test–crop response correlation to evaluate soil fertility of a group of soils. In the deficient range, soil fertility status is further classified as very low (<50% relative yield), low (50–75% relative yield), medium (75–90% relative yield), and high (>90% relative yield). Toxic range is of special interest for environmental considerations of soil fertility.

Isotopic dilution techniques are used to label the plant-available nutrient in the soil, such as phosphorus in greenhouse or field, and then trace its availability through crops, thus avoiding the artifact effects of soil extractants (7). In some soil test–crop response correlation procedures, soils are grouped according to clay content, CEC, or soil pH, and separate calibrations are established for each group of soils. In fact, the Fertility Capability Soil Classification System groups soils according to surface soil texture, subsoil texture, and 15 other soil characteristics (8).

SPECTRAL SENSING AND SITE-SPECIFIC SOIL FERTILITY EVALUATION

Soil fertility variability in conventional soil testing, which aims at obtaining an average value for a field or experimental site, is not evaluated to make site-specific management decisions. The two outstanding benefits of site-specific soil fertility evaluations are: 1) increased economic efficiency of nutrient application, that is, maximum profits; and 2) consequent reduced environmental nutrient pollution. Spectral sensing techniques (physical soil testing) such as static and portable near- and mid-infrared, gamma ray, and satellites using hyperspectral sensors (9) are increasingly being used to map soil fertility for multiple nutrients and crop yields at $<1\,m^2$ scale to $>1\,km^2$ scales, while using precision technologies for site-specific nutrient applications.

ENVIRONMENTAL IMPACT

Usually maximum economic yield is obtained with soil nutrient supply just enough for achieving about 90–95% of relative yield when nutrients are fully taken up by plants. In crops when quality of the produce attracts a premium such as protein concentration of wheat >11.5% protein, nitrogen supply at the higher level of sufficiency and luxury range is economically preferred. Yet high soil nitrate-N fertility may lead to excess of nitrate-N in soil and, consequently, to denitrification (greenhouse effect) and leaching. Excess nitrate-N in surface water bodies causes algal blooms and in drinking waters causes methemoglobinemia in infants. Also, adverse environmental and health impacts occur when soil nutrient fertility is in toxic range (Fig. 2), especially for heavy metals, either due to human consumption of produce or on ecosystems.

PERSPECTIVES

Soil fertility evaluation has evolved in the last two centuries, from a major concern for enhancing crop production to that of maintaining environmental integrity. With increasing population, crop production will remain the main objective of managing the soil fertility. However, with the recent advances in multi-element extractants and analysis and remote sensing and precision technologies for site-specific soil fertility evaluation, it is possible to meet the twin goals of economic optimum yields and minimal environmental pollution. By managing soil fertility this way, sustainable use of land, water, and nutrients is also ensured, resulting in enhanced soil and water quality for the current and future generations. In spite of the advances in sensor and analytical technologies, however, the challenge lies ahead in developing soil nutrient tests that closely mimic nutrient uptake by a crop.

REFERENCES

1. Russell, E.W. *Soil Conditions and Plant Growth*, 10th Ed.; Longman: London, 1973; 49–50.
2. Mitscherlich, E.A. Das Gesetz des Minimums und das Gesetz des Abnehmenden Bodenertrages. Landwirtschaftliche Jahrbücher **1909**, *38*, 537–552.
3. Reuter, D.J., Robinson, J.B., Eds.; *Plant Analysis: An Interpretation Manual*, 2nd Ed.; CSIRO Publishing: Melbourne, 1997; 1–572.
4. Black, C.A. *Soil Fertility Evaluation and Control*; Lewis Publishers: Boca Raton, FL, 1993; 1–746.
5. Havlin, J.L.; Beaton, J.D.; Tisdale, S.L.; Nelson, W.L. *Soil Fertility and Fertilizers*, 6th Ed.; Prentice Hall: Upper Saddle River, New Jersey, 1999; 1–499.
6. Peverill, K.I., Sparrow, L.A., Reuter, D.J., Eds.; *Soil Analysis: An Interpretation Manual*; CSIRO Publishing: Melbourne, 1999; 1–369.
7. Dalal, R.C.; Hallsworth, E.G. Evaluation of the Parameters of Soil Phosphorus Availability Factors in Predicting Yield Response and Phosphorus Uptake. Soil Science Society of America Journal **1976**, *40*, 541–546.
8. Sanchez, C.A.; Couto, W.; Buol, S.W. The Fertility Capability Soil Classification System: Interpretation, Application and Modification. Geoderma **1982**, *27*, 282–309.
9. Sudduth, K.A.; Hummel, J.W.; Birrel, S.J. Sensors for Site-Specific Management. In *The State of Site-Specific Management for Agriculture*; Pierce, F.J., Sadler, E.J., Eds.; ASA/CSSA/SSSA: Madison, 1997; 183–210.

FERTILITY, LOW INPUT AND TRADITIONAL METHODS OF MAINTAINING

Miguel Altieri
University of California, Berkeley, California, U.S.A.

INTRODUCTION

There are thousands of farmers throughout the world (mostly organic farmers and resource-poor farmers) who, due to economic constraints or environmental reasons, cannot or do not want to use chemical fertilizers to enhance or maintain soil fertility (1). These farmers who can neither afford nor rely on a regular supply of inorganic fertilizers must find alternative sources of nutrients. These sources are often cheaper, more efficient than inorganic compounds and their use focuses on recycling of nutrients rather than on supplying nutrients on a regular basis (2). The techniques that these farmers use to maintain soil fertility and to improve biological, chemical, and physical soil properties tend to enhance organic matter content, increase the efficiency of nutrient use by closing the nutrient cycles (by returning exported nutrients to cropland) and by minimizing nutrient loss from the agroecosystem (3).

PRINCIPLES OF LOW INPUT SOIL FERTILITY MANAGEMENT

Although many farmers have developed alternative fertilizing techniques through trial and error, research shows that there are several agroecological principles underlying such low input methods of maintaining soil fertility (4, 5). Such principles can be summarized as follows.

Securing Arable Soil Conditions for Plant Growth

In order to grow healthy and productive plants, farmers must create and/or maintain the following soil conditions (6): 1) timely availability of water, air, and nutrients in balanced and buffered quantities; 2) soil structure which enhances root growth, exchange of gaseous elements, water availability, and storage capacity; 3) soil temperature which enhances soil biology and plant growth; 4) absence of toxic elements such as pesticide residues; 5) increased addition of organic residues by using varied sources of organic materials.

Optimizing Nutrient Availability and Cycling

Nutrient availability depends greatly on the general soil condition, soil life, and organic matter content. However, deliberate attention must also be given to providing the nutrients required for plant growth. There is a constant flow of nutrients through the farm. Some of the nutrients are lost or exported, e.g., by export of products, erosion, leaching, and volatilization. These nutrients have to be replaced. Nutrient losses can be limited by (7): 1) recycling organic wastes, e.g., manure, night soil, and crop residues by returning them to the field, either directly or treated (composted, fermented, etc.); 2) handling organic fertilizers in such a way that nutrients are not leached by excessive rain or volatilized by high temperature or solar radiation; 3) reducing runoff and soil erosion, which removes nutrients and organic matter; 4) reducing burning of vegetation when farming is intensified, as this leads to losses of organic matter; 5) reducing leaching by maintaining a high humus content in the soil, soil cover and intercropping plant species with different rooting depths; 6) pumping up partly leached nutrients from deeper soil layers and bringing them back to the topsoil by using litter from trees or other deep-rooting plants.

Clearly, nutrient export from the farm to the market cannot be completely avoided, nor can all nutrient losses resulting from erosion and leaching. However, soil cover and efficient organic management can minimize losses to acceptable levels. In fact, nutrients can be captured on the farm by (8): 1) fixing nitrogen by microorganisms living in symbiosis with leguminous trees, shrubs, or cover crops; 2) harvesting nutrients by capturing wind or water sediments from outside the farm; this can be done with special soil conservation practices; 3) integrating livestock into the farming system.

TECHNIQUES FOR SUSTAINABLE SOIL FERTILITY MANAGEMENT

The above principles are universal in nature but when applied by different farmers under a diversity of environmental and socioeconomic conditions, principles take various technical forms which can be grouped as follows (9).

Soil Conservation

The most effective way to improve the productivity of small, resource-poor farmers is to conserve and utilize the resources that they already possess. Soil conservation thus aims to maintain soil fertility by preserving soil structure, organic matter, and nutrients. These practices include techniques that decrease rates of leaching, erosion, and organic matter destruction. Also important are methods that capture and conserve water, especially in low rainfall areas, for water and soil fertility act synergistically on crop production (3).

Conservation techniques are often advantageous because they require little outlay of cash and they can usually be undertaken during the fallow season, when farmers have less work. As a result, they do not put the farmer in debt risk and they do not interrupt or interfere with traditional work schedules. These techniques include the use of windbreaks, terraces, and deep-rooted plants. Windbreaks are usually composed of tree hedgerows planted at regular intervals in a crop field. They reduce erosion by reducing wind velocity. Terraces can be used to prevent soil erosion on cultivated steeplands. Although large-scale terracing programs may be expensive and require much labor, there are several techniques for low-resources terrace building, in which the terraces actually build themselves (10). Deep-rooted plants, such as perennial shrubs and trees, can be used to recycle nutrients that have leached to deeper layers in the subsoil, beyond the reach of the roots of most crop species. This strategy is particularly important in certain forms of agroforestry such as alley farming (11).

Nutrient Recycling

The nutrients that are removed from agricultural fields as crops eventually become wastes and thus can be reused in the field. Some wastes, such as manure, can be used directly on cropland, but most organic materials must first be "prepared" before they are used as a fertilizer. The variety of materials that have been used to maintain soil fertility include animal wastes, bones, blood, seaweed, seashells, peat, ashes, sawdust, leaves, sewage, sludge, etc. (2).

In temperate regions, one of the most common recycling practices is the use of "mixed" farming, where livestock or chickens are the chief product of the farm. On these farms, grains and hay are grown and fed to the livestock. Their manure is used to fertilize the soil. Legumes are grown in rotation with grain crops in order to resupply the soil with nitrogen. Opportunities are endless, but the main approaches include the following.

Cover crops and green manures

Some crops are grown solely for the purpose of enhancing soil fertility. These crops are not harvested, but rather they are plowed into the soil, while they are still growing, which is why they are referred to as green manures. Typically, green manures or cover crops are grown in rotation or as "improved fallow" and are mixed into the soil before seeding the subsequent crop, allowing enough time for the decomposition of residues and the mineralization of nutrients. Green manures serve many different purposes in soil fertility. They can improve the soil physical structure by adding organic matter. They can be used to prevent erosion, as when they are used as a fallow season cover crop. They can improve soil nutrient status when they are used as "nutrient pumps" to recycle nutrients that have leached to deeper horizons (12). When a legume is used as a green manure crop, large amounts of soil nitrogen may be added.

One of the most remarkable cover crops is the velvetbean (*Mucuna pruriens*). This has been widely promoted as part of the work of World Neighbors in Central America, though its effectiveness is attested by its spontaneous spread from village to village without outside intervention. Establishing *Mucuna* requires very little investment in labor. It grows rapidly, is palatable to animals and people, fixes large amounts of nitrogen, and can produce as much as 60 ton/ha of organic matter. It can grow on most soils and its spreading habit suppresses weed growth. Incorporating such green manures into cropping systems can increase yields substantially. Honduran farmers are able to harvest some 2.5–3.2 ton/ha of maize when grown after velvetbean. This compares with an average for the country of just 0.6 ton/ha (13).

Azolla and *Anabaena*

Blue-green algae are another important source of nitrogen, the most widely exploited being the alga *Anabaena azollae* which fixes atmospheric nitrogen while living in cavities in the leaves of a small fern, *Azolla*, which grows in the water of rice fields in both tropical and temperate regions. *Azolla* quickly covers the water surface in the ricefield, but does not interfere with the normal cultivation of the rice crop.

Very high nitrogen production is possible following *Azolla* inoculation in rice fields. In the Philippines, 57 tons of fresh weight *Azolla* can be harvested after 100 days yielding more than 120 kg/ha of nitrogen. Over the whole year, *Azolla* can fix more than 400 kg N/ha, a rate in excess of most tropical and subtropical legumes. This nitrogen is only available to the rice crop after *Azolla* has decomposed and so exploitation consists of incorporating the ferns into the soil while wet as a green manure or removing them for drying and then reapplying them to the ricefields.

The results of at least 1500 studies in China, Philippines, Vietnam, India, Thailand, and the U.S.A. have shown that when *Azolla* is grown in paddy fields, rice yields increase by on average 700 kg/ha, with a range of 400–1500 kg/ha (11).

Mulches

Soil fertility can often be enhanced simply by covering the soil's surface with a layer of organic material (mainly straw), a process that is known as mulching. In the humid tropics, where cultivation of any sort is often deleterious to soil fertility and crop productivity, mulching is a highly effective method to maintain soil health. Mulches are particularly useful in preventing soil damage because they reduce the destructive effect of raindrop impact on soil aggregation and prevent soil erosion. Mulches also improve soil water status by increasing infiltration rates. Mulches also stimulate the growth of soil biota (9).

Agroforestry and intercropping

There is a huge diversity of agroforestry systems throughout the world, in which the incorporation of bushes and trees result in many benefits. Trees with nitrogen fixing capacity have beneficial effects on plants growing beneath them. In many arid parts of Africa, millet is planted under *Acacia albida* trees. The acacias, which conveniently drop their leaves at the start of the growing season, provide nutrients for the growing crop; millet yields under the tree's canopy are 56–64% higher than yields obtained away from the tree (14). Most positive effects of trees are the result of the fixed nitrogen, but significant quantities of N can also be supplied in the leaf litter or from deliberate pruning. In Africa, leguminous shrubs (*Leucaena*) are grown as hedgerows 2–4 m apart, and their prunings added to crops grown in the alleys. This proved a remarkably efficient agroforestry technique to stabilize maize yields at about 2 t/ha without the use of fertilizers (15).

Trees also improve the microclimate by acting as windbreaks, by improving the water-holding capacity of the soil and by acting as shade trees for livestock—so focusing the deposition of manure (16).

Intercropping, the process of growing more than one crop simultaneously in the same plot, is used throughout the world and has many advantages over monoculture, including a more efficient use of space and nutrients, higher total yields, and less vulnerability to crop failure due to pests (17). As in the case of maize–bean polycultures, legumes can be used in the intercrop to add nitrogen and thereby reduce competition between the intercropped species. In fact, bean-fixed N is made directly available to the maize through mycorrhizal fungi connections between root systems. Intercropping also helps prevent erosion and soil-surface damage (i.e., degradation of surface pores) by keeping the soil covered. Due to these and other effects, polycultures generally overyield corresponding monocultures (18).

Integration of livestock

Livestock are a critical component of sustainable agricultural systems. The nutrient value of manures largely depends on how it is handled, stored, and applied. Losses of nitrogen tend to be highest when liquid systems of storage are used and when the manure is broadcast without incorporation. Livestock manures from cattle, pigs, and chickens are important, as they positively affect soil structure and water retention, and benefit soil organisms. Manure itself is an excellent fertilizer and soils usually require 5–20 t/ha of fresh manure to remain in continuous productivity (19).

It is becoming more common for small farmers to keep their animals permanently penned in zero-grazing or stall-feeding units rather than permit livestock to graze freely. In many rural areas, zero-grazing units are a central part of efforts to improve soil and water conservation. Fodder grown on the farm in the form of improved grasses, tree fodder and the residues of cultivated crops are cut and carried to the animals. Because of the proximity to the crops, manures can be returned directly to the land, so improving nutrient supply and soil structure (20).

Reduced tillage

The use of reduced tillage or no till systems can maintain or enhance soil organic matter more than the conventional moldboard plow and disk system. The decreased soil disturbance under reduced tillage slows the rate of organic matter decomposition and helps maintain a soil structure that allows rainfall to infiltrate rapidly. Leaving residue on the surface encourages the development of earthworm populations, which also improves soil structure. Compared with conventional tillage, soil erosion is greatly reduced under minimum tillage systems, which helps keep the organic matter and rich topsoil in place. The only drawback of these systems is their high dependence on herbicides.

Composting

Organic wastes can also be concentrated through the process of composting, in which microorganisms attack the waste before it is used as fertilizer. There are many methods of composting, each combining the use of animal manures, green material and household wastes in different ways and proportions. The materials are headed or placed in a pit in such a fashion that anaerobic decomposition occurs. Harmful substances and toxic products of metabolism are broken down, while pathogens, and the seeds and roots of weeds are destroyed by the heat generated within the compost heap. Compost created by a variety of methods has proven to be an effective fertilizer in many agricultural settings.

Inoculating the soil with microorganisms

Soil microbes play many beneficial roles in maintaining soil fertility. In some situations, however, these beneficial microbes might be present in very low numbers. In such situations, the soil can be "inoculated" with beneficial microorganisms thus enhancing crop productivity. These introductions may be done with any species of microbe that is known to benefit plant productivity. Those species that fix nitrogen, enhance phosphorus availability, increase mineralization of soil organic matter or prevent diseases have attracted the attention of researchers. There are many studies that have examined the use of improved strains of the nitrogen fixing bacterium *Rhizobium*, the nitrogen fixing soil bacteria *Azotobacter* and *Azospirillum* and Mycorrhizal fungi that enhance plant phosphorus nutrition (21).

INTEGRATED APPROACHES TO SOIL FERTILITY MAINTENANCE

As discussed earlier, there are many options for making soils more healthy and fertile. Combining the various options is challenging, but usually implies the following (7).

1. Using rotations that utilize grass, legume, or a combination of grass and legume sod crops: or crops with large amounts of residue as important parts of the system.
2. Using of cover crops when soils would otherwise be bare to add organic matter, capture residual plant nutrients, and reduce erosion. Cover crops also help maintain soil organic matter in resource-scarce regions that lack possible substitutes to using crop residues for fuel or building materials.
3. Leaving residues from annual crops in the field or, if they were removed for use as bedding for animal, returning them to the soil as manure or compost.
4. Raising animals or having access to animal wastes from nearby farms gives farmers wider choices of economically sound rotations. Rotations that include perennial forages make hay or pasture available for use by dairy and beef cows, sheep, and goats. In addition, on mixed crop–livestock farms, animal manures can be applied on cropland. It is easier to maintain organic material on a diversified crop-and-livestock farm, where sod crops are fed to animal and manures returned to the soil. However, growing crops with high quantities of residues plus frequent use of green manures and composts from vegetable residues helps maintain soil organic matter even without animals.

Throughout the world, more and more farmers are learning that the combination of reduced tillage, cover crops, and better rotations can have a dramatic effect on their soil and the health of crops. They are finding that, by combining practices, they are reducing pest damage, improving soil tilth, vastly reducing runoff and erosion, increasing soil organic matter and producing better crop growth (22). Among resource-poor farmers in developing countries, adoption of sustainable soil fertility management practices can have important impacts on crop yields. An analysis of 208 agroecological projects showed clear increases in food production over 29 million hectares, with nearly 9 million households benefiting (11). In North America and Europe, organic farmers who adopt low-input soil fertility practices typically obtain yields 6–11% lower than conventional farmers who rely on chemical fertilizers. Such lower yields are, however, offset by the economic and environmental performance of organic agriculture (2).

REFERENCES

1. Conway, G.R. *The Doubly Green Revolution*; Penguin Books: London, 1997.
2. Lampkin, N. *Organic Farming*; Farming Press: Ipswich, 1990.
3. Lal, R.; Pierce, F.J. *Soil Management for Sustainability*; Soil and Water Conservation Society: Amkeng, 1991.

4. Altieri, M.A. *Agroecology: The Science of Sustainable Agriculture*; Westview Press: Boulder, 1995.
5. Gliessman, S.R. *Agroecology: Ecological Processes in Agriculture*; Ann Arbor press: Chelsea, 1998.
6. Reinjntjes, C.; Haverkort, B.; Waters-Bayer, A. *Farming in the Future*; McMillan Press: London, 1992.
7. Magdoff, F.; van Es, H. *Building Soils for Better Crops*; Sustainable Agriculture Network: Beltsville, 2000.
8. Kotschi, J. *Ecofarming Practices for Tropical Smallholdings*; Masgraf: Weikersheim, 1990.
9. Gershuny, G.; Smillie, J. *The Soul of Soil: A Guide to Ecological Soil Management*; GAIA Services: St. Johnsburg, 1986.
10. Dover, M.; Talbot, L.M. *To Feed the Earth: Agroecology for Sustainable Development*; World Resources Institute: Washington, DC, 1987.
11. Pretty, J.N. *Regenerating Agriculture*; Earthscan: London, 1995.
12. Sarrontino, M. *Northeast Cover Crop Handbook*; Soil Health Series, Rodale Institute: Kutztown, 1997.
13. Buckles, D.; Triomphe, B.; Sain, G. *Cover Crops in Hillside Agriculture*; International Development Research Center: Ottawa, 1998.
14. Poschen, P. An Evaluation of the *Acacia albida*-Based Agro-Forestry Practices in the Hararghe Highlands of Eastern Ethiopia. Agrofor. Syst. **1986**, *4*, 129–143.
15. Kang, B.T.; Wilson, G.F.; Lawson, T.L. *Alley Cropping: A Stable Alternative to Shifting Cultivation*; International Institute of Tropical Agriculture: Ibadan, Nigeria, 1986; 22.
16. Nair, P.K. *Agroforestry Systems in the Tropics*; Kluwer Academic Pub.: Dordrecht, 1989.
17. Francis, C.A. *Multiple Cropping Systems*; Wiley and Sons: New York, 1986.
18. VanderMeer, J. The Ecological Basis of Alternative Agriculture. Annu. Rev. Ecol. Syst. **1995**, *5*, 201–224.
19. Follet, R.F. *Soil Fertility and Organic Matter as Initial Components of Production Systems*; SSSA Special Pub. No. 19, Soil Sci. Soc. Am. Madison, WI, 1987.
20. Pearson, C.J.; Ison, R.L. *Agronomy of Grassland Systems*; Cambridge University Press: Cambridge, 1995.
21. McGuinnes, H. *Living Soils: Sustainable Alternatives to Chemical Fertilizers for Developing Countries*; Consumers Policy Institute: New York, 1993.
22. Edwards, C.A., Ed. *Sustainable Agricultural Systems*; SWCS, Amkeng, 1990.

FERTILIZER APPLICATION, METHODS OF

John Ryan
International Center for Agricultural Research in the Dry Areas, Aleppo, Syria

INTRODUCTION

Since the beginning of civilization, soil fertility has been a dominant factor influencing human destiny. It took millennia before humans could unravel the mysteries of soil nutrients in relation to crops. In the 19th century, the invention of superphosphate laid the basis for the chemical fertilizer industry, an essential component of modern agriculture. Developments in fertilizer technology have provided an array of commercial fertilizer types and formulations, including solids, liquids, gases, suspensions; single-elements to complex multi-nutrients; soluble to slow-release materials; pelleted and powder forms; and bagged and bulk materials (1). Soil chemistry–plant nutrition research has provided the technical basis for fertilizer application methods. Both research and legislation have evolved to address changing fertilizer technology and cropping systems. The chemical fertilizer industry, now a complex global one, is the cornerstone of agricultural production and world food supplies through the provision of crop nutrients (2). Today, the broad goals of nutrient management are: 1) cost-effective production of quality crops, 2) efficient use of fertilizers, 3) maintenance of soil quality, and 4) protection of the environment. This brief review highlights fertilizer application methods within such a context.

FERTILIZER APPLICATION: BASIC CONSIDERATIONS

Decisions about sustainable and economical fertilizer use must consider many crop, soil, and environmental factors. Soil testing is now a well-established practice for identifying the fertility status of the soil, nutrient constraints, and the need for fertilization, especially in the West (3); however, progress has also been made in developing countries in promoting soil testing, correlation and calibration (6). It involves representative field sampling; analyzing the sample in the laboratory using appropriate tests; interpreting the test data using established norms; and making fertilizer recommendations. Criteria vary with the nutrient in question, soil type, or management. Apart from nutrients, crop growth is indirectly influenced by soil factors such as acidity, salinity, sodicity, toxic levels of elements, e.g. boron, and by high nutrient "fixation" capacity. Soil physical constraints such as compaction and increased bulk density must also be considered.

Crop characteristics such as yield goal or predicted yield and specific nutrient uptake are considered in fertilizer recommendations. Thus, fertilizer nutrient needs vary widely between short- and long-season crops and within crops; the higher the uptake or demand at maximum yield, the more fertilizer is needed. The architecture and extent of the crop's root system, including mycorrhizal colonization, dictate the relative use of soil and fertilizer nutrients and fertilizer application methods.

Environmental factors govern fertilizer-use efficiency. For example, frozen soil leads to poor nutrient penetration and runoff loss; rain after application promotes leaching losses; high ambient temperature promotes both crop growth and nutrient uptake; warm, windy conditions can increase ammonia volatilization; and increasing water use, from rainfall or irrigation, enhances the uptake of fertilizer nutrients. Strategies to adapt to such conditions include splitting the fertilizer rates to coincide with crop growth demand, with due consideration of rainfall, and soil incorporation of nitrogen fertilizers to minimize loss.

FERTILIZER APPLICATION METHODS

Given the variability of chemical fertilizers (physical state, type of nutrient, solubility, release rates) and range of crops (field, greenhouse, annual, perennial), fertilizer management conditions (rainfed, irrigated), and economic considerations, fertilizer application methods have correspondingly changed (1). The application methods used should be accurate, efficient in terms of nutrient use, and economic in terms of comparative investment. Fertilizers can be applied during soil preparation and before planting the crop, i.e., the more traditional approach, and/or after crop emergence (Table 1). Nowadays, with increased fertilizer use and greater emphasis on efficiency, fertilizers

Table 1 Fertilizer application methods

Application time/method	Attribute (+/−)	Remarks
1. Pre-plant		
Broadcast	+	Traditional, suitable for solids/liquids
	−	Less efficient for P, could cause NH_3 loss
Injection	+	Reduces NH_3 loss, suitable for immobile nutrients
	−	More expensive, elaborate equipment needed
2. Planting time		
With seed	+	Efficient, economic
	−	Needs machines, possible damage to germination, especially with N
Band placement	+	More efficient for P in high-fixing soils
	+	Can use liquids and solids
3. Postemergence		
Side dressing	+	Suitable for P and K
	−	Needs care with N
Topdressing	+	Suitable for cereals, perennial crops
	+	Solids and liquids can be used
4. Fertigation	+	High water and N-use efficiency
	−	High initial costs and maintenance
	−	Need soluble nutrient sources
	−	In-line precipitation, poor soil mobility with P
5. Foliar sprays		
	+	Suitable, effective where soil application is not possible
	+	Good for micronutrients, especially in perennial crops
	−	May not supply full N, P, K requirement
	−	Several sprayings needed
	−	Limited N concentration (1–2%), also biuret

can be applied by broadcasting, banding, or with water, i.e., foliar sprays and in irrigation water.

Broadcasting

The essence of broadcasting fertilizer involves spreading the material evenly on the bare or cropped soil. Prior to the age of mechanization, solid-form fertilizer was spread by hand and is still done so in many developing countries. Mechanized application represented a major improvement in speed, accuracy, and uniformity of spreading. The range of such equipment varies from small to large, and from pull-type drop-spreaders to self-propelled machines with power-driven, disc spreaders. Generally, the application rate of fertilizer in such machines is determined by their ground speed. Large, bulk fertilizer spreaders have flotation types to mitigate soil compaction.

Topdressing is essentially broadcasting over the established crop; it is the only feasible means for pasture and perennial forage crops and flooded rice. Fertilizer-use efficiency and reduced adverse environmental effects are promoted with splitting the fertilizer application between planting and a later crop growth stage. Thus, depending on the standing crop, its nutrient needs, and growing conditions, application of fertilizer to the crop stand is a common practice. Topdressing may be done several times with long-season crops. Plant analysis can help in determining the need for such applications.

The use of liquid fertilizers is a relatively recent development, mainly in the U.S. for N fertilizers; liquids can be sprayed or dribbled on the soil surface in bands between rows with gravity feed. Where spraying is involved, there is need for a tank, pump, pressure gauge, nozzles, fittings, and a spreading boom; it can have variable traction mechanisms, many tractor mounted. Now in the age of precision agriculture, sophisticated and expensive machinery is used to variably apply the fertilizer, liquid or solid, across a given field having different fertility levels, as determined by soil analyses, whose spatial variability is determined by global positioning systems.

Placement/Banding

The main approach to fertilizer application at planting time is with the seed or by banding. When applied with the seed, application rates must be limited for avoiding seedling damage and poor emergence, especially with N fertilizers and those with a high "salt index." Banding implies application of fertilizer a few centimeters to the side and below the seed row. Fertilizer and seed are usually in the same drilling machine, but in parallel hoppers; both are delivered by metered gravity feed through tubes behind the shoe or disc. Banding reduces P-fixation, and hence increases fertilizer-use efficiency. The efficiency of N fertilizer, alone or with P, is also increased.

The placing of the fertilizer, usually liquids, at some depth in the soil, via tool bar-mounted knives in shank openers, is referred to as injection or "knifing in." It helps reduce or eliminate volatile loss of N fertilizers such as urea and ammonium compounds, and, to a lesser extent, it places immobile nutrients like P closer to the root zone, and also reduces nutrient losses due to erosion and runoff. This can be done before planting or after crop emergence.

As with preplant application, liquids such as anhydrous ammonia can be injected under pressure into the soil between or close to the seed rows; use of solid materials is less common by this method. Placement can be combined with row cultivation. The type of fertilizer must be considered; both P and K should be close to the roots as they are immobile. Plant damage due to high localized ammonia concentrations, as well as disruption of plant roots, must be avoided.

Fertigation

The practice of applying water and fertilizers together, fertigation, involves technical and economic advantages, e.g., savings in labor, time, and equipment and fuel costs (7). Open fertigation systems involve the metering of soluble solid or liquid fertilizer, mainly N, into irrigation ditches. However, volatile NH_3 gas loss can occur when the N source is urea or anhydrous ammonia. Simultaneous application of H_2SO_4 can counteract the alkalinity and rise in pH following ammonia application (4); it also prevents calcite precipitation (5), thus reducing the water's sodium-absorption ratio (SAR) and increasing infiltration. Phosphorus is less suitable for open-channel application because it can be readily precipitated and thus be unevenly spread within the field. Where soils are sodic and have poor hydraulic characteristics, amendments such as polysulfides, gypsum, and H_2SO_4 can be readily metered into the irrigation system.

Closed fertigation systems involve fertilizer applied to irrigation water in pipes under pressure with pumping, e.g., sprinklers and trickle or drip irrigation, with many variations of the latter system; fertilizer solubility and subsequent nutrient movement in the soil are considered. As most N fertilizers are water-soluble, N is easily used in both systems; K sources are soluble but relatively immobile and remain in the top few centimeters of soil or at the point of dripper. The primary concern with P fertilizers is poor solubility, precipitation, and limited mobility at the dripper site. With Ca-rich water, soluble P forms precipitates in the tubing that can block the tiny orifices, thus rendering the system ineffective. Flushing with acids such as urea phosphate and H_3PO_4 can dissolve such precipitates. Acidic solutions promote greater P mobility in the wetted soil volume and by dissolving insoluble soil P sources, viz. $Ca_3(PO_4)_2$ (8). Acidic P solutions may also enhance micronutrient availability in calcareous soils.

While close-system irrigation is a sophisticated technology, there are many obstacles to using fertilizers in such systems. However, these obstacles are now well understood by commercial farmers and irrigation company personnel, and workable solutions are readily available. As a consequence, fertigation is now part of high-intensity production systems in field and under greenhouse conditions.

Foliar Feeding

The concept of applying nutrients as foliar sprays is relatively new and is a practice adapted in situations where soil application of fertilizers is ineffective and where a quick growth response is required (9). Aerial application of nutrients in sprays is used for broadacre crops where the equipment has problems entering or crossing the field and where large areas need to be sprayed, or where nutrient uptake is limited by cold. Such nutrients can be normally combined with pesticides using ordinary pesticide sprayers; however, compatibility and stability of the ingredients must be considered.

However, as with any technology, there are pitfalls and obstacles. Fertilizer solution is a salt solution and as such can injure growing plants. Thus, nutrient concentration in the sprayed solution should be no greater than 1–2%. While urea is suitable for foliar sprays, the concentration of biuret, a toxic component, should be limited. Other practical considerations associated with foliar application involve varying the nutrient application relative to the irrigation spray and applying nutrient sprays frequently where the permissible concentrations of nutrients are low with respect to the growing crop needs, especially for major nutrients. However, one or two sprayings with micronutrients are usually sufficient to alleviate the deficiency.

CONCLUSIONS

As crop yields increase, even more fertilizer than used at present will be needed. The growth in demand will come particularly from developing countries where fertilizer usage is currently low and where populations are expanding rapidly. The twin concerns—efficient economic use and mitigating environmental damage, as well as maintaining nutritional quality of crops—underscore the continued importance of precise, accurate, and efficient application of chemical fertilizers. A major part of world food production is attributable to chemical fertilizers. This will continue—there is no alternative. But the industry must be sensitive to societal concerns. While the emphasis will continue to be on the environment in the developed world and on production in the developing countries, efficient fertilizer application technology can accommodate both concerns. Fertilizer application will, in future, be part of strategies of "positive soil fertility management", where the broader perspective will apply, i.e., consideration of the whole cropping system—not just one year's crop—and the residual or carryover value of the added fertilizer. Indeed, its implications for the entire food chain cannot be ignored. While research will continue to achieve even greater efficiencies of application, the greatest challenge will be to transfer what is known to those who need it most, especially in the developing world.

REFERENCES

1. UNIDO/IFDC, *Fertilizer Manual*; United Nations Industrial Development Organization, and International Fertilizer Development Center Kluwer: Dordrecht, The Netherlands, 1998.
2. IFA/UNEP, *The Fertilizer Industry, World Food Supplies and the Environment*; International Fertilizer Industry Association and United Nations Environment Programme: Paris, France, 1998.
3. Matar, A.; Torrent, J.; Ryan, J. Soil and Fertilizer Phosphorus and Crop Responses in the Dryland Mediterranean Zone. Adv. Soil Sci. **1992**, *18*, 82–146.
4. Miyamoto, S.; Ryan, J.; Stroehlein, J.L. Sulfuric Acid for The Treatment of Ammoniated Irrigation Water. I. Reducing Ammonia Volatilization. Soil Sci. Soc. Am. Proc. **1975**, *30*, 544–548.
5. Miyamoto, S.; Ryan, J. Sulfuric Acid for the Treatment of Ammoniated Irrigation Water. II. Reducing Calcite Precipitation and Sodium Hazard. Soil Sci. Soc. Am. J. **1976**, *40*, 305–309.
6. Ryan, J., Ed. *Accomplishments and Future Challenges in Dryland Soil Fertility Research in the Mediterranean Area*; Proceedings of the International Soil Fertility Workshop, November 19–23, ICARDA: Aleppo, Syria, 1997.
7. Ryan, J., Ed. *Plant Nutrient Management Under Pressurized Irrigation Systems in the Mediterranean Region*; World Phosphorus Institute, Casablanca, Morocco, and International Center for Agricultural Research in the Dry Areas: Aleppo, Syria, 2000.
8. Ryan, J.; Eisa, M.A.; Tabbara, H.; Baasiri, M. Phosphorus Movement Following Water-Applied Urea Phosphate and Phosphoric Acid in a Calcareous Clay Soil. J. Fert. Issues **1998**, *5* (3), 89–96.
9. Soil Improvement Committee, California Fertilizer Association, *Western Fertilizer Handbook*, 7th Ed.; The Interstate Printers and Publishers: Danville, IL, 1985.

FERTILIZERS, LEACHING LOSSES

P.G. Saffigna
The University of Queensland, Gatton, Queensland, Australia and Saffcorp Pty. Ltd, Tweed Heads, New South Wales, Australia

I.R. Phillips
Griffith University, Queensland, Australia

F

INTRODUCTION

Leaching is the downward movement of fertilizer or waste in soil with the drainage water and is a critical process in global nutrient cycling (1). To be available for plant uptake, fertilizer/waste (solute) must dissolve in the soil water. The combination of soil water and dissolved solute is the soil solution. The soil solution and soil air are present in the pores between soil particles. If water fills all the pores and no air is present, the soil is saturated and leaching is rapid. If water and air are both present, the soil is unsaturated and leaching is slower.

Efficient nutrient use occurs when the plant demand coincides with the presence of solute in the root zone. When solute leaches below the root zone, it is unavailable for plant uptake and is lost from the soil–plant system. Depending on the amount of water draining below the root zone, the leached solute may simply accumulate at depth in the soil or it may pollute the underlying groundwater (2). Pollution of groundwater by fertilizers and waste is a major environmental problem worldwide (1). Although substantial progress has been made in understanding the movement of fertilizers and wastes in soil, the words of Leonardo da Vinci are equally applicable today as they were 500 years ago: "We know more about the movement of celestial bodies than about the soil underfoot."

THE MECHANISMS AND PATTERN OF LEACHING

Water movement due to gravity causes leaching (3). When there is no mixing between the solute and the draining water, the solute is pushed ahead of the advancing water like a piston and remains in a narrow zone or band. This somewhat idealized leaching pattern is referred to as *piston flow* (Fig. 1b). Commonly, mechanisms such as dispersion, ion exchange, and ion exclusion modify this leaching pattern. *Dispersion* is a result of solute mixing with the drainage water by molecular diffusion (3) and/or hydrodynamic dispersion (3, 4), and causes solute to be more "spread out" in the soil than if only piston flow occurred (Fig. 1c). The process of *cation exchange* with negatively charged soil colloids (5) slows the rate of cation leaching in soil, reduces the number of cations in the soil solution, and skews the fertilizer leaching pattern (Fig. 1d). *Exclusion* of negatively charged solute ions such as nitrate and chloride by the negatively charged soil colloids (3) confines them to the middle of the soil pores. Here water drains faster due to reduced friction, causing solute to leach more quickly than if piston flow alone was occurring (compare Fig. 1e with Fig. 1b).

MEASUREMENT AND MODELING OF LEACHING

Although the pattern of leaching is well represented by the examples shown in Fig. 1, this does not provide sufficient information to measure leaching losses. By multiplying the concentration of solute by the volume of drainage water, the amount of solute leached from the soil can be quantified. Representative samples of soil solution are critical for quantifying solute leaching (6). Column displacement and centrifugation are commonly employed to extract soil solution in the laboratory, while tension lysimeters are commonly used in the field. The volume of water draining through soil can be measured using repacked soil cores, intact (undisturbed) soil cores, and field lysimeters (undisturbed or refilled). These methods can be used to quantify the key mechanisms controlling the rate and magnitude of solute leaching in the laboratory and field (7).

Modeling solute leaching in and below the plant root zone is important for predicting soil water quality, and for developing fertilizer and waste management practices

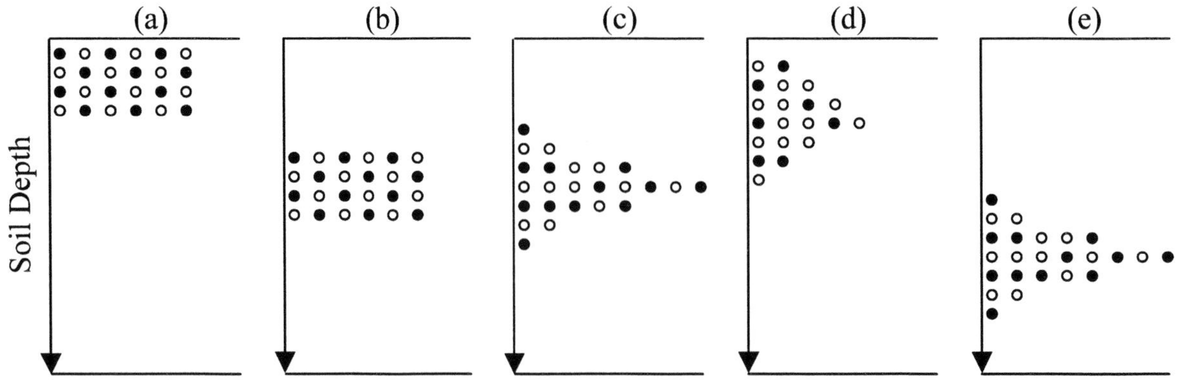

Fig. 1 Schematic representation of solute leaching in soil. a) Solute distribution pattern prior to leaching—note that the solute band contains 12 cations (•) and 12 anions (○) (24 ions); b) Solute distribution pattern after leaching involving piston flow only—note that the solute band still contains 24 ions and has only been moved deeper in the soil; c) Solute distribution pattern after leaching involving piston flow and dispersion—note that the solute band has become more spread out but still contains 24 ions; d) Solute distribution pattern after leaching involving piston flow, dispersion, and exchange—note that the solute band only contains seven cations, is skewed in shape and is at a shallower depth than b); e) Solute distribution pattern after leaching involving piston flow, dispersion, and exclusion—note that the solute band still contains 24 ions but has moved deeper in the soil than b).

which minimize adverse environmental impacts. Numerous analytical and numerical models have been developed which include key physical and chemical processes that affect solute leaching (8). Water flow models, solute transport models, and aqueous chemical equilibrium models were often developed independently, and solute leaching models mostly considered only a single ion and simplified chemical processes. Recently, there has been a conscious effort to couple leaching models for water flow and solute transport with chemical equilibrium models (9).

MAGNITUDE OF LEACHING

Not all nutrient elements are susceptible to leaching, and the amount of a nutrient lost by leaching varies widely between climates, land use, soils, and nutrients (Table 1) (10). Generally, fertilizer-leaching losses are greatest in sandy soils, moderate in loamy soils, and least in clay soils for two reasons. Firstly, water drains faster in sandy soils, which have a low capacity to retain cations. Secondly, anions tend to leach faster than cations because soils carry

Table 1 Leaching losses from agriculture under different climates and land uses

Country	Land use	Nutrient	Form	Rate applied (kg/ha/y)	Amount leached (kg/ha/y)	Reference
UK	Grazing	N	Fertilizer	250	4–184	(10)
UK	Grazing	N	Fertilizer	500	88–590	(10)
Europe	Cropped	N	Fertilizer	80	47–74	(10)
USA	Potatoes	N	Fertilizer	265	200	(12)
USA	Potatoes	Cl	Fertilizer	360	300	(12)
Australia	Bare	P	Wastewater	50	2–6	(13)
Australia	Bare	N	Wastewater	180–1500	113–315	(13)
Australia	Bare	K	Wastewater	300–2040	1–280	(13)
Australia	Cropped	Ca	Wastewater	185–530	127–330	(13)
New Zealand	Grazing	N	Wastewater	200–400	6–10	(14)
New Zealand	Grazing	N	Fertilizer	200–400	8–17	(14)
New Zealand	Bare	K	Fertilizer	50	1–8	(11)
New Zealand	Bare	K	Fertilizer	200	6–22	(11)
New Zealand	Bare	K	Wastewater	1000	20–240	(11)
New Zealand	Bare	Mg	Fertilizer	62	0.6–17	(11)

mainly negative electrical surface charge. For example, after adding 50, 200, or 1000 kg/ha of KCl to a clay soil, about 7, 8, and 6 kg/ha of K was leached below the root zone, whereas in a sandy soil receiving the same fertilizer rates, 7, 20, and 240 kg/ha of K was leached (11). In both of these soils, nearly all of the added Cl leached below the root zone. Nitrate (200 kg/ha of nitrate–nitrogen) and chloride leaching to depths >5 m and into the underlying groundwater have been found in sandy soils within 1 yr of fertilizer application (1, 12).

Although there are many benefits from using organic wastes as nutrient sources, their long-term environmental impact is a major concern. Sewage sludge can contain heavy metals (copper, lead, cadmium, and zinc) which can contaminate soils, while the use of wastewater can introduce very high concentrations of unwanted nutrients (nitrogen, chloride, and potassium), which can increase the likelihood of groundwater pollution. Generally, leaching of inorganic (fertilizer) phosphorus (P) is negligible; however, P leaching (ca. 10% of applied) has been observed in a wide range of soils treated with wastewater. The major source of wastewater P is dissolved organic P, and organic P compounds are relatively stable and mobile in soils. Consequently, the sustainability of long-term wastewater application to soil is a major environmental concern for many countries.

MITIGATION AND MANAGEMENT OF FERTILIZER NUTRIENT LOSSES

The amount of solute leached from soil is highly dependent on dissolved nutrient type (cation or anion), concentration and form (ionic or complexed), chemical, physical, and biological properties of the soil, timing and placement of fertilizer application, rainfall, and irrigation practice. For example, large leaching losses of fertilizer nitrogen occur if it is present as nitrate, there is limited plant uptake, and the soil is permeable (well drained) and located in a high rainfall environment. Mitigation of fertilizer leaching is site-specific, and highly dependent on site management practices. Recommended management strategies include 1) synchronization of fertilizer supply with crop demand, and 2) employing irrigation strategies to minimize the drainage of soil water and loss of nutrients from the plant root zone. Other strategies include the use of nutrient release inhibitors, slow release fertilizers (maintains low-nutrient concentrations in the soil solution) and incorporation of amendments (zeolite, iron oxides, gypsum, fly ash) into soil that remove fertilizer nutrients from the soil solution. There is an urgent need to develop best-management practices for land disposal of organic wastes as these nutrient sources are likely to have a greater effect on soil and water pollution than fertilizers.

REFERENCES

1. Smil, V. Nitrogen in Crop Production: An Account of Global Flows. Global Biogeochem. Cycles **1999**, *13*, 647–662.
2. Saffigna, P.G.; Keeney, D.R. Nitrate and Chloride in Groundwater Under Irrigated Agriculture in Central Wisconsin. Groundwater **1977**, *15*, 170–177.
3. Biggar, J.W.; Nielsen, D.R. Miscible Displacement and Leaching Phenomenon. In *Irrigation of Agricultural Lands*; Am. Soc. Agron. Monog. 11, Hagen, R.M., Ed.; Am. Soc. Agron. Madison, WI, 1967; 254–274.
4. Fried, J.J.; Combarnous, M.A. Dispersion in Porous Media. Adv. Hydrosci. **1971**, *7*, 169–282.
5. Thomas, G.W. Historical Developments in Soil Chemistry: Ion Exchange. Soil Sci. Soc. Am. J. **1977**, *41*, 230–238.
6. Wolt, J. *Soil Solution Chemistry. Applications to Environmental Science and Agriculture*; Wiley: New York, 1994; 345.
7. Phillips, I.R. Collection and Automation of Large Undisturbed Soil Cores for Laboratory Leaching Studies. Commun. Soil Sci. Plant Anal. **2001**, *32* (5, 6), 843–862.
8. Ghadiri, H.; Rose, C.W. *Modelling Chemical Transport in Soils. Natural and Applied Contaminants*; Lewis Publishers: London, 1992; 217.
9. Suarez, D.L.; Simunek, J. UNSATCHEM: Unsaturated Water and Solute Transport Model with Equilibrium and Kinetic Chemistry. Soil Sci. Soc. Am. J. **1997**, *61*, 1633–1646.
10. Wild, A.; Cameron, K.C. Soil Nitrogen and Nitrate Leaching. In *Soils In Agriculture*; Tinker, P.B., Ed.; Blackwell: Oxford, 1980; 36–70.
11. Phillips, I.R.; Black, A.S.; Cameron, K.C. Effect of Cation Exchange on the Distribution and Movement of Cations in Soils with Variable Charge. II—Effect of Lime or Phosphate on Potassium and Magnesium Leaching. Fert. Res. **1988**, *17*, 31–46.
12. Saffigna, P.G.; Keeney, D.R.; Tanner, C.B. Nitrogen, Chloride, and Water Balance with Irrigated Russet Burbank Potatoes in a Sandy Soil. Agron. J. **1977**, *69*, 251–257.
13. Phillips, I.R. Nutrient Leaching Losses from Undisturbed Soil Cores Following Application of Piggery Wastewater. Aust. J. Soil Res. **2001**, *6*, (in press).
14. Silva, R.G.; Cameron, K.C.; Di, H.J.; Hendry, T. A Lysimeter Study of the Impact of Cow Urine, Dairy Shed Effluent, and Nitrogen Fertilizer on Nitrate Leaching. Austr. J. Soil Res. **1999**, *37*, 357–369.

FERTILIZERS, MINERAL

B.L. Bumb
L.L. Hammond
International Fertilizer Development Center, Muscle Shoals, Alabama, U.S.A.

INTRODUCTION

Mineral fertilizers, also known as chemical fertilizers (1), refer to the products manufactured by the chemical fertilizer industry through chemical reactions of different elements or products. Nitrogen, phosphate, and potash fertilizers are primary mineral fertilizers produced by the fertilizer industry. Nitrogen fertilizer products are mostly manufactured from ammonia (which is derived from chemical reaction of nitrogen and hydrogen). Phosphate fertilizers are derived by reacting phosphate rock with sulfuric or phosphoric acid, and potash fertilizers are refined from mined potash ores or sea brines.

Harvested crops remove nutrients. Unless the removed nutrients are replenished, soil fertility declines and the capacity of soils to produce additional crops degrades. To maintain soil health and productivity, nutrients must be continuously replaced. Because natural processes can replenish only limited quantities of the nutrients removed, these nutrients must be supplied from external sources including organic materials, biological fixation, and mineral fertilizers. Although organic and biological sources are important in supplying nutrients, these are not sufficient to meet the nutrient requirements of the food and fiber production needed for an ever-growing world population. World population increased from less than 2 billion at the turn of the century to over 6 billion at the beginning of the 21st century and may reach over 9 billion by 2050 (2, 3, 4). Since the mid-1960s, mineral fertilizers have played an important role in promoting cereal production, especially in developing countries (Fig. 1).

MAIN FERTILIZERS—NUTRIENTS AND PRODUCTS

There are more than 100 chemical elements in nature but only 16 are considered essential for plant growth. If any of these 16 elements are lacking, plants cannot complete their vegetative or reproductive cycles. Thirteen of these are "mineral" nutrients (Table 1). Some of these nutrients combine to form compounds that compose cells, enzymes, etc. Others must be present in order for certain plant chemical and physiological processes to occur. A productive soil should contain all the essential plant nutrients in sufficient quantity and in balanced proportions. Soils differ widely in their ability to provide the essential nutrients for plant growth, depending on the total amount of nutrient native to the soil and the mobilization or fixation reactions in the soil that affect the chemical form and resulting availability of the nutrients. When the nutrient supply from the soil is insufficient for crop needs, additional nutrient can be supplied by fertilizer to compensate for the difference. The type and amount of fertilizer required depends on the crop to be grown and the soil characteristics.

The nutrients taken up by plants from mineral fertilizer products are in the same form as the nutrients taken up from organic sources, but many different products have been developed to meet specific needs for crop production.

Simple Fertilizers Versus Complex Fertilizers

Simple fertilizers contain only one primary plant nutrient. In contrast, complex fertilizers contain two or more plant nutrients. Complex fertilizers are produced by processes involving chemical reaction between the constituents containing the primary plant nutrients (ammonium phosphates, nitrophosphate NPK fertilizer, etc.). Fertilizers produced by granulating together single-nutrient materials or mixing granules of single-nutrient fertilizers are referred to as compound or mixed fertilizers. The nutrient content of fertilizer products, or "grade" expressed as a percentage, may refer either to the total or available nutrient content, and may be expressed traditionally for some nutrients in the oxide form (P_2O_5, K_2O) (6). Thus, 100 kg of 12-6-6 fertilizer contains 12 kg of N, 6 kg of P_2O_5, and 6 kg of K_2O.

Nitrogen Fertilizers

Nitrogen fertilizers can broadly be classified into four groups depending on the chemical form in which

F

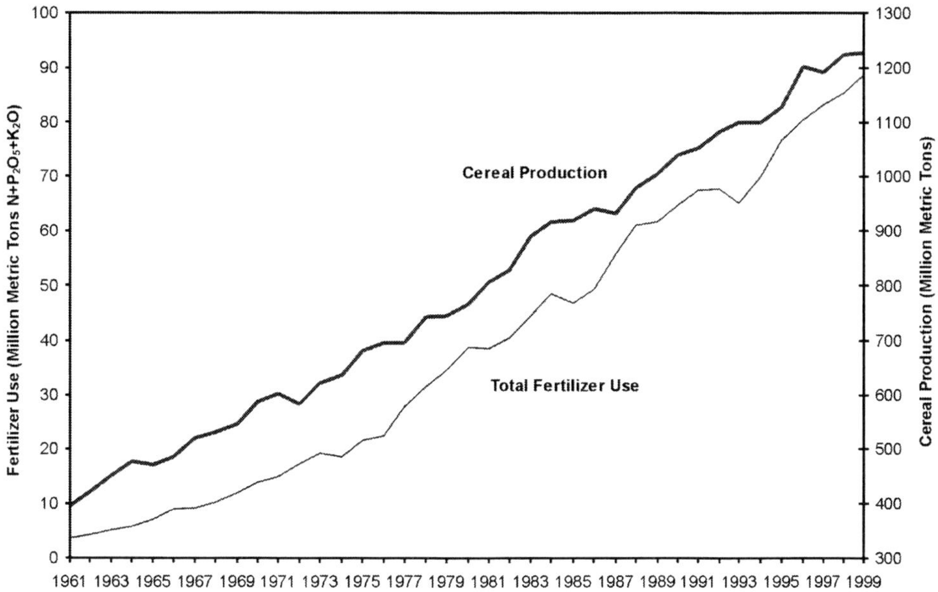

Fig. 1 Fertilizer use and cereal production in developing countries 1961–1999. (Derived from data in Ref. 5.)

Table 1 Essential mineral nutrients and their chemical symbols

	Principal chemical form taken up by plants	Nutrient uptake by wheat (5 mt/ha) (kg nutrient/ha)
Primary nutrients		
Nitrogen (N)	NH_4^+, NO_3^-	105
Phosphorus (P)	$H_2PO_4^-$	18
Potassium (K)	K^+	15
Secondary nutrients		
Sulfur (S)	SO_4^{2-}	8
Magnesium (Mg)	Mg^{2+}	6
Calcium (Ca)	Ca^{2+}	2
Micronutrients		
Chlorine (Cl)	Cl^-	3
Iron (Fe)	Fe^{2+}	0.2
Manganese (Mn)	Mn^{2+}	0.2
Zinc (Zn)	Zn^{2+}	0.2
Copper (Cu)	Cu^{2+}	0.03
Boron (B)	H_3BO_3	0.02
Molybdenum (Mo)	MoO_4^{2-}	
Essential non-mineral nutrients		
Carbon (C)		
Hydrogen (H)		
Oxygen (O)		

the nitrogen is present in them: ammonium fertilizers, nitrate fertilizers, combined ammonium and nitrate fertilizers, and amide fertilizers (Table 2). Also, there are NPK complex fertilizers, mixed compounds such as granular compound fertilizers, and bulk blends.

Nitrogen fertilizers are completely water-soluble and most sources are essentially equal in effectiveness for crops if incorporated into well-drained soils. If the nitrogen source applied is in the nitrate (NO_3^-) form, it is immediately available for plant uptake. If it is applied in the ammonium form, (NH_4^+) is initially absorbed by the soil organic matter and clay particles. Under soil conditions suitable for good plant growth, however, the ammonium form is rapidly converted by nitrifying bacteria to the nitrate form (nitrification). Because the nitrate is not absorbed strongly by soil particles, it moves freely in soil water. For maximum uptake of N fertilizer, plants must be in a vigorous growth stage soon after the N is applied. Otherwise, the nitrate may move below the root zone.

Phosphate Fertilizers

Phosphate fertilizers vary widely in solubility. Phosphate rock is essentially insoluble in water and is only slowly available to plants if applied directly to the soil. Phosphate rock, however, is the basic raw material from which the most typically used phosphate fertilizers are produced.

Table 2 Fertilizer materials

Fertilizer products	Grade (%) N-	Grade (%) P_2O_5-	Grade (%) K_2O
Nitrogen fertilizers			
Ammonium fertilizers			
Anhydrous ammonia	82		
Ammonium sulfate (AS)	21		
Ammonium chloride	26		
Nitrate fertilizers			
Calcium nitrate	16		
Sodium nitrate	16		
Ammonium nitrate fertilizers			
Ammonium nitrate (AN)	34		
Calcium ammonium nitrate (CAN)	21		
Ammonium nitrate-sulfate	26		
Amide fertilizers			
Urea	46		
S-coated urea	35		
Calcium cyanamide	22		
Multinutrient fertilizers			
Nitrophosphate (NP)	21	21	
Mono-ammonium phosphate (MAP)	11	52	
Diammonium phosphate (DAP)	18	46	
Liquid ammonium polyphosphate	12	40	
Urea ammonium phosphate	28–32	30	
Phosphate fertilizers			
Water-soluble types			
Single superphosphate (SSP)		18	
Triple superphosphate (TSP)		46	
Partly water-soluble types			
Partially acidulated phosphate		(23–26)	
Slow-acting types			
Dicalcium phosphate (citrate soluble)		40	
Basic slag		9	
Calcium metaphosphate		62	
Very slow-acting types			
Rock phosphates		<32	
Potassium fertilizers			
Muriate of potash (Potassium chloride) (MOP)			60
Sulfate of potash (Potassium sulfate) (SOP)			48
K-Mag / Sulpomag (K, Mg sulfate)			23
Nitrate of potash (potassium nitrate)	13		44
Potassium polyphosphate		59	39

The main deposits and production of phosphate rock are located in the United States, Morocco, Togo, Senegal, Tunisia, Russia, Jordan, China, and Oceania.

The rock can be solubilized by direct acidulation with sulfuric acid to form single superphosphate or to produce phosphoric acid. The phosphoric acid can then be used to directly acidulate more phosphate rock to form triple superphosphate or the acid can be reacted with ammonia to form ammonium phosphates. The availability of the phosphorus to the plant is typically expressed as the sum of the water- and citrate-soluble P in the fertilizer source. Water-soluble P is easily available to plants while citrate-soluble P is more slowly available to plants. The soluble P, however, can react with other minerals in the soil, thus reducing the immediate availability to plants. This transformation can be retarded somewhat through granulation and placement of the fertilizer. The P products formed in the soil, however, limits the movement of P in the soil and results in a longer-term "residual value" not observed with nitrogen. Generally, the agronomic performance of phosphate fertilizers containing 60% or more water-soluble P is essentially equal to that of fertilizers containing 100% water-soluble P (7).

Potash Fertilizers

Canada has the world's largest known deposit of K. Potash is also mined in France, Germany, Italy, Spain, England, Israel, Jordan, Russia, Belarus, Ukraine, Brazil, and China. Potassium chloride (KCl) accounts for more than 90% of the K sold in North America. As with most of the common potash fertilizers, KCl is water-soluble. Whereas KCl is very concentrated in K (60%), the concentration of chloride in KCl is about 45% and, therefore, potassium sulfate is often used for chloride-sensitive crops such as tobacco and to supply S. Sulfate of potash-magnesia contains not only potash but also about 11% Mg and 22% S. It is a good source of water-soluble K and Mg and is important where Mg and/or S is deficient, or with Cl-sensitive crops. Potassium nitrate contains little or no Cl or S and is completely water-soluble. It is widely used in foliar spray applications for fruit, vegetables, and cotton.

After application, most of the K from mineral fertilizer is absorbed near the application point in the soil by clay minerals and organic matter like ammonium N. It thus becomes "exchangeable" K. However, unlike ammonium N, which converts to a mobile form, K continues to be held in the soil and is relatively immobile.

Secondary and Micronutrients

While the nomenclature of secondary nutrients (Ca, Mg, and S) and micronutrients (Cl, Fe, Mn, Zn, Cu, B, and Mo) may indicate a lower degree of importance relative to N, P, and K, these nutrients are just as important to plant nutrition as the primary nutrients. They are referred to in this manner because plants do not usually require as great a quantity of them. Calcium is generally supplied in the limestone (calcitic or dolomitic) used to reduce soil acidity or in basic slag from the steel industry or gypsum. The most common source of magnesium is dolomitic limestone. Other sources include potassium-magnesium sulfate, magnesium sulfate, magnesium oxide, and basic slag. Most fertilizer S sources are sulfates (ammonium, potassium, or magnesium) that are moderately to highly water-soluble and are quickly available to plants. Elemental S (>85% S) can also be used as a fertilizer, but it results in a slower crop response because it is water insoluble. If well incorporated into the soil in advance of planting, however, elemental S is an agronomically effective and economically efficient S source.

While there are many fertilizer sources available to correct micronutrient deficiencies, some of the more common are borax (11% B), copper sulfate (23% Cu), iron sulfate (19%–23% Fe), iron oxide (69%–73% Fe), manganese sulfate (26%–30% Mn), manganese oxide (41%–68% Mn), ammonium molybdate (54% Mo), sodium molybdate (40% Mo), zinc sulfate (23%–36% Zn), and zinc oxide (78% Zn). Cu, Fe, Mn, and Zn-chelates are also common fertilizer sources.

FERTILIZER USE: HISTORY AND RECENT TRENDS

Although the first use of mineral fertilizers was introduced during the early 20th century, large-scale use of mineral fertilizers started only after World War II—initially as a result of the use of surplus ammonia capacity remaining from the production of explosives during the war and then as a result of the introduction of fertilizer-responsive improved crop varieties during the Green Revolution era.

At the turn of the century, fertilizer use was insignificant. By 1950 it increased to 14 million nutrient tons, and by 1960 total nutrient use was about 28 million tons (8,9). After the mid-1960s, due to the adoption of the Green Revolution technologies, fertilizer use increased rapidly—reaching 85 million tons in 1973/74 and 146 million nutrient tons in 1988/89 (Fig. 2). Thereafter, global fertilizer use decreased due to a steep decline in fertilizer use in the reforming markets (Eastern Europe and Eurasia). Policy reforms in Africa and Latin America and agricultural policy changes in the developed markets also contributed to this process (Fig. 3). After declining to a low of 120 million tons in 1993/94, global fertilizer use recovered slowly to reach 141 million tons in 1999/2000. Yet, it was lower than the peak of 146 million nutrient tons that was reached in 1988/89.

During the 1990s, while fertilizer use was declining in the reforming markets and was nearly stagnant in the developed markets, it continued to grow in the developing markets—mostly in Asian economies. Consequently, the developing markets' share of global fertilizer use increased from 34% in 1980 to 64% in 2000. China, United States, India, Brazil, and France are five main users of mineral fertilizers in the world. These countries account for 60% of the global fertilizer use.

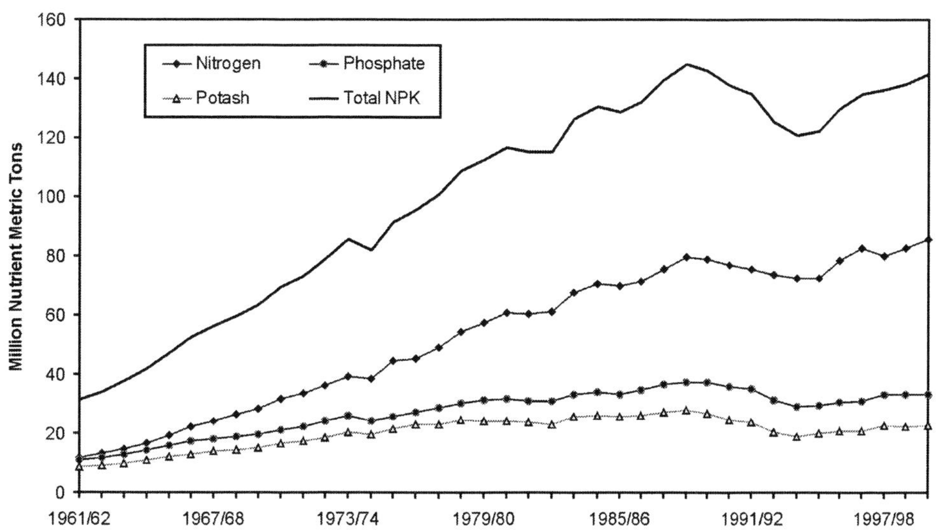

Fig. 2 World: nitrogen, phosphate, potash, and total NPK consumption, 1961/62–1999/00. (Derived from data in Ref. 5.)

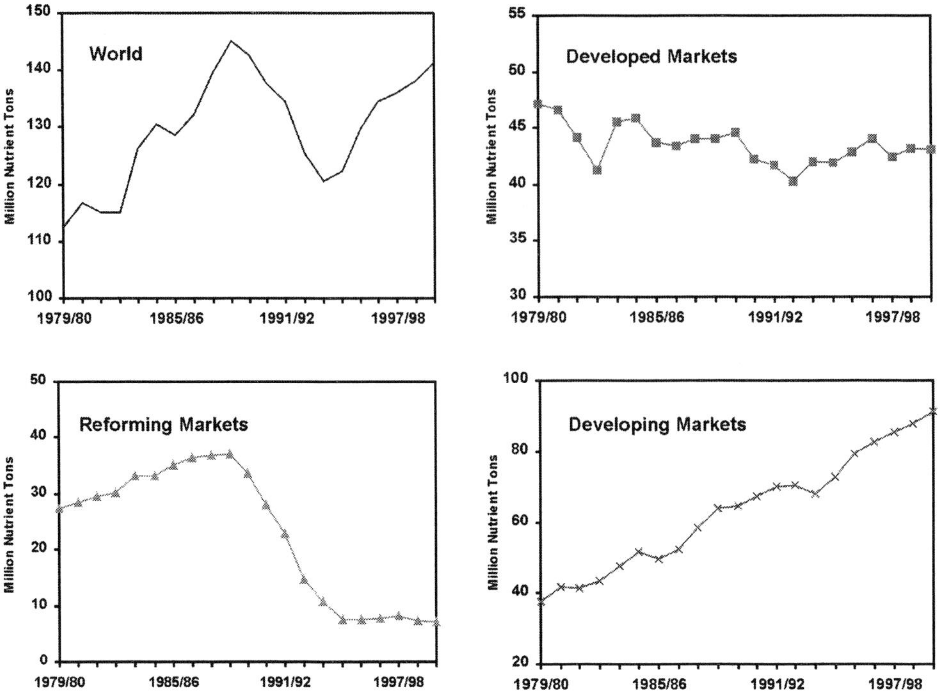

Fig. 3 Total fertilizer use by markets, 1979/80–1999/00. (Derived from data in Ref. 5.)

Fertilizer Use by Nutrients and Products

In 1999/2000, global use of N, P_2O_5, and K_2O was estimated to be 85.5, 33.1, and 22.7 million nutrient tons. Nutrient use by different regions is shown in Table 3. East Asia, South-West Asia, and North America are dominant users of N and P_2O_5. In K_2O use, Latin America dominates South-West Asia. Reforming markets account for less than 5% of global fertilizer use.

Urea, diammonium phosphate (DAP)/monoammonium phosphate (MAP), and muriate of potash (MOP) are the main fertilizer products manufactured by the fertilizer industry. These products accounted for over 50% of global fertilizer use in 1998/99 (Table 4). Other commonly used products are ammonium sulfate (AS), ammonium nitrate (AN)/calcium ammonium nitrate (CAN), single superphosphate (SSP), triple superphosphate (TSP), and NPKs.

Rapid growth in fertilizer use during the 1970s and 1980s also induced growth in fertilizer production and trade. Total fertilizer production increased from 28 million nutrient tons in 1959/60 to 90 million nutrient tons in 1973/74 and to 158.5 million tons in 1988/89 and then decreased until 1994/95 and recovered slowly thereafter (11,12,13). In 1999/2000, 148.9 million tons of nutrient production consisted of 90.8 million tons of N, 32.6 million tons of P_2O_5, and 25.4 million tons of K_2O (5).

Fertilizer Use Efficiency

Because of rapid growth in fertilizer use in the past, little attention was paid to fertilizer use efficiency. Overall, fertilizer efficiency may not exceed 40%. Nitrogen use

Table 3 Fertilizer use by regions and markets, 1999/2000

Region/market	N	P_2O_5	K_2O	Total
	(million nutrient tons)			
I. Developed markets	24.4	9.5	9.3	43.2
North America	12.9	4.6	5.0	22.5
Western Europe	10.2	3.5	3.9	17.7
Oceania	1.3	1.4	0.4	3.0
II. Reforming markets	4.6	1.0	1.5	7.1
Eastern Europe	2.0	0.5	0.6	3.1
Eurasia	2.6	0.5	0.9	4.0
III. Developing markets	56.5	22.6	11.9	91.1
Africa	2.4	1.0	0.5	4.0
East Asia	30.6	11.3	6.0	47.9
South & West Asia	18.5	7.0	2.3	27.8
Latin America	5.0	3.3	3.1	11.4
IV. World	85.5	33.1	22.7	141.4

(From Ref. 5.)

Table 4 Global fertilizer use by products, 1998/99

Product	N	P_2O_5	K_2O	Total
	(million nutrient tons)			
Urea	39.6			39.6
AS	2.6			2.6
AN/CAN	10.7			10.7
TSP		2.2		2.2
SSP		7.0		7.0
DAP/MAP	4.9	14.1		19.0
MOP			13.4	13.4
SOP			0.8	0.8
NP/NPKs	8.0	8.5	7.7	24.2
Others	17.2	1.5	0.2	18.9
Total	83.0	33.3	22.1	138.4

(From Ref. 10.)

efficiency may vary from 30% to 50% for rice, maize, and sorghum in developing countries (14). Phosphorus and potassium use efficiencies are also low. Improving fertilizer use efficiency by proper placement and timing of fertilizer products and soil- and crop-specific recommendations could lead to a win-win situation by increasing crop productivity and reducing nutrient losses to the atmosphere thereby protecting the environment. New technologies of precision agriculture may prove useful in improving nutrient use efficiency by targeting fertilizer applications more accurately.

CONCLUSION

At the turn of the 21st century, the world faces many challenges. First, food production must be increased to eliminate hunger and malnutrition and ensure food security for all. Second, natural resources, such as rain forests, wildlife habitats, and biodiversity, must be protected. Third, soil degradation and nutrient depletion must be halted. Fourth, global warming and pollution must be reduced. In confronting all these challenges, increased and environmentally sound use of mineral fertilizers will be essential for enhancing the productivity of cultivable lands, so that the increased food could be produced while protecting the environment by sparing forests and marginal lands, replenishing the removed nutrients, and sequestering carbon in the biomass production and the soils.

Because of uncertainties associated with future events, it is difficult to make precise projections. However, one study estimated that annual fertilizer use should increase to 263 million tons to satisfy the food production requirements in 2020 (15). This means that fertilizer use should be nearly doubled in the next 20 years. Given the successes of the past (over fivefold increase in fertilizer use during the 1960–90 period), this may not be a difficult target to achieve but would require a strong commitment from all stakeholders including donors and policymakers.

REFERENCES

1. Hereinafter Mineral Fertilizers Are Also Referred to as Fertilizers.
2. United Nations (UN). *The State of World Population, 1988*; New York, 1988.
3. UN. *Long Range World Population Projections: Two Centuries of Population Growth, 1950–2150*; New York, 1992.
4. UN. World Population Prospects: The 2000 Revision—Highlights; Draft. New York, 2001.
5. FAO. FAOSTAT-Statistical Database; http://apps.fao.org/ (Accessed July 2001).
6. Conversion Factors from Oxides to Elements Are as Follows: $P = 0.436\ P_2O_5$, $K = 0.830\ K_2O$.
7. Engelstad, O.P.; Hellums, D.T. *Water Solubility of Phosphate Fertilizers: Agronomic Aspects—A Literature Review*; IFDC: Muscle Shoals, Alabama, U.S.A., 1993; 17.
8. Food and Agriculture Organization of the United Nations (FAO). *Fertilizers: An Annual Review of World Production, Consumption, and Trade*; Rome, Italy, 1961.
9. FAO. *Fertilizers: A World Report on Production and Consumption*; Rome, Italy, 1952.
10. International Fertilizer Industry Association. IFADATA Statistics, 1973/74 to 1998/99, Compact Disk (CD); Paris, France, 2000.
11. Bumb, B.L. *Global Fertilizer Perspective, 1960–95: The Dynamics of Growth and Structural Change*; IFDC: Muscle Shoals, Alabama, U.S.A., 1989; T-34, Executive Summary T-35.
12. Bumb, B.L. *Global Fertilizer Perspective, 1980–2000: The Challenges in Structural Transformation*; IFDC: Muscle Shoals, Alabama, U.S.A., 1995; T-42.
13. International Fertilizer Development Center (IFDC). Global and Regional Data on Fertilizer Production and Consumption, 1961/62–1998/99; Unpublished Report. Muscle Shoals, Alabama, U.S.A., 1999.
14. Peoples, M.B.; Freney, J.R.; Mosier, A.R. Minimizing Gaseous Losses of Nitrogen. In *Nitrogen Fertilization in the Environment*; Bacon, P.E., Ed.; Marcel Dekker, Inc.: New York, 1995.
15. Bumb, B.L.; Baanante, C.A. *The Role of Fertilizer in Sustaining Food Security and Protecting the Environment to 2020*; Food, Agriculture, and the Environment Discussion Paper 17, International Food Policy Research Institute (IFPRI): Washington, DC, U.S.A., 1995.

FERTILIZERS, ORGANIC

Gary J. Gascho
University of Georgia, Tifton, Georgia, U.S.A.

INTRODUCTION

By one definition, organic fertilizers may be nutrient sources that can be used in organic farming. With such a definition, inorganic materials that are deemed to be "natural," such as mined gypsum, untreated salts, unprocessed lime, etc., can be included (1). For this article, we define organic fertilizers as those that are organic, from a chemical standpoint. Only an overview of the subject can be provided in this brief contribution. More comprehensive recent publications are available (1, 2).

ANCIENT APPLICATIONS OF ORGANIC FERTILIZER MATERIALS

The first fertilizers were organic. They have been used for thousands of years and the literature is voluminous. Ancient people principally used animal wastes to enhance the growth of plants, usually with great success. More recently, Guano, a commercial organic material from the excreta of sea birds and bats, was an important fertilizer in the southern United States in the early 1900s. Application of organic fertilizers has now become much more complicated as modern municipal and industrial products, byproducts, and wastes must be considered.

GENERAL ADVANTAGES AND DISADVANTAGES OF ORGANIC FERTILIZERS

Agriculture occupies a great proportion of the land mass in comparison to industrial operations. It is quite natural that a disposal philosophy exists. Unwanted wastes are often promoted as beneficial for farm crops, sometimes only to remove the wastes from manufacturing sites or to save costly land-filling. Agriculture should make careful assessments of all fertilizer materials applied, but especially of industrial and municipal wastes. That assessment should include, not only the nutrient value in relation to need, but also the concentration of elements and compounds that could lead to crop and/or soil toxicity. Concentrations of heavy metals are a primary concern in many cases. When applying wastes containing acceptable concentrations of heavy metals, careful maintenance of soil pH at >6.0 can help limit their release in the soil and availability to plants.

A general advantage of organic fertilizers can be cost of the nutrient, but that is not always the case. It largely depends on the proximity of the organic fertilizer to the site of application and the value of the organic. In cases where a material must be placed in a landfill, the material may have a negative value due to the cost and maintenance of that landfill. If so, a material may be obtained at little or no cost, or even at a negative cost. As organic fertilizers generally have lesser nutrient values than modern manufactured inorganic fertilizers, the cost of transportation per unit of nutrient is often the controlling factor in value of the fertilizer. A second potential advantage of organic fertilizers is that they may provide nutrient release over an extended period of time; in particular, nitrogen (N) can be mineralized over a season to eliminate the need for repeated applications. Soil organic matter can be increased when organic fertilizers are applied at sufficient rates. The value of soil organic matter is well recognized, but the ability of added organic matter to substantially add to the pool of soil organic matter is debatable. It greatly depends on many factors, especially climatic conditions.

General disadvantages of organics are their fertilizer ratios, that are not likely to be matched to plant needs, their nonuniform composition that can prohibit precise uniform applications, and their unpredictable release of nutrients, that may not coincide with the time a particular nutrient is most required by a particular plant. Some organic fertilizers may also contain unwanted, harmful, and/or toxic elements or compounds making them unsatisfactory for utilization. Some may be objectionable due to strong smells or irritants to humans. This latter disadvantage is now important to the quality of life in residential developments near sites of application.

ANIMAL RESIDUES

Manures

Animal manures were the first fertilizers used in agriculture. Feces and urine from large animals and feces from

Encyclopedia of Soil Science

poultry are the main sources. Most are applied, together with the bedding material, directly to the soil and usually in near proximity to the confined animals. Some is composted and/or pelleted to provide a material that is more agreeable for handling and transport. Nutrient composition can vary widely depending on the animal, its feed, the bedding, and the proportion of manure to bedding, and handling or storage prior to application (Table 1).

Because of the importance of animal agriculture, manures will continue to be important sources of fertilizers and pose problems for agriculture and society. As farmers move toward precision applications (that provide improved nutrition without over-application that can result in pollution of soil or water), it becomes increasingly important to analyze manures and apply them in accordance with plant needs. Analyses provided indicate that phosphorus (P) composition is high relative to N for most plant needs. Therefore, application according to N needs provides too much P, that may be bioavailable and add to *eutrophication* of water via runoff. In some areas with high soil P, the recommendation is to apply manures on the basis of P needs. Such a philosophy will greatly limit applications in fields that are close to confined animals. It will also require N application and possibly other fertilizer nutrients from commercial fertilizers for balanced fertilization of most crops.

One attempt to provide more uniform applications and products that are more environmentally friendly to transport is to compost or pelletize manures. Such operations concentrate the manure and reduce transportation costs. Many composting processes have been developed. To date the economics are favorable for some specialty plants but not for broad-scale crop production.

Other Animal Residues

Bone meal, dried blood, and other animal processing wastes are important fertilizers, but now account for little of the total fertilizer used in the USA. Bone meal is a good source of P, often used by the homeowner but seldom used in commercial agriculture. Dried blood is a rapid releasing organic N source. Other animal wastes are variable in composition and are often considered as disposal problems rather than desirable fertilizers. Applications may be difficult due to poor and nonuniform physical characteristics. Chemical analysis should be known before application.

PLANT RESIDUES

Leaving plant residues on the soil surface provides both nutrients for a following crop and soil protection from wind and rain erosion. Plant residue fertilization may also be obtained from the application of wastes or byproducts from many agricultural processing plants. Trash from cotton gins, filter mud from sugarcane, and processing wastes from fruits and vegetables are but a few of the plant residues utilized. Soil conservationists have long recommended keeping the soil protected by maintaining cover crops during the seasons when the main crops are not planted and returning those plant residues to the soil. The modern emphasis on conservation tillage includes planting directly into cover crops. If the cover crop is a legume, N may be supplied to the planted crop due to the N_2-fixation from the legume. A good stand of alfalfa, clover, birdsfoot trefoil or vetch plowed into the soil may replace a fertilizer N application of 80–125 kg N/ha for a following crop of corn (3, 4). Presently, it is believed that such cover crops used in conservation tillage and not incorporated would supply nearly as much N to the following crop.

MUNICIPAL WASTES

Sewage treatment plants produce fertilizers in the forms of effluent (the liquid portion) and as biosolids that are settled from the effluent. In some cases, biosolids are placed in a landfill, but costs of maintaining and monitoring seepage from landfills is now making the economics favor

Table 1 Concentrations of N, P, and K on a dry-weight basis in commonly applied wastes

Waste	N	P	K	References
Livestock manures	1–3%	0.4–2.0%	1–2.5%	1, 2
Poultry manures	3–5%	1–3%	1–2%	1, 2
Plant residues	1–7%	0.1–1.7%	0.1–9%	1
Municipal biosolids	2–9%	1.5–5%	0.2–0.8%	9
Municipal effluents	1.6–2.7 mg/L	0.2–1.2 mg/L	1.1–1.7 mg/L	7, 8

transportation and land application. Proper handling in the treatment plant removes most human and animal health concerns, but not the stigma of such applications.

Effluents

Ammonium and nitrate-N should be monitored to avoid over-application of effluents, as N leaching can be a problem in sandy soils. When application is made year around, it is necessary to keep the soil covered with an actively growing crop to minimize leaching (5). Phosphorus fixation capacity of the soil is an important determinant of the amount of effluent that should be applied without promoting a soluble-P problem. High fixing soils will handle greater amounts of P than soils with little fixing capacity (6). Heavy metals are not a problem as they are concentrated in the solids. Nitrogen, P, and K concentrations of effluents vary quite widely (Table 1).

Biosolids

Because of advances in technology, treatment plants are now able to remove nearly all heavy metals and bacteria, allowing the nutrient-rich organic material byproducts or biosolids to be recycled and applied as fertilizer. Elemental concentrations in biosolids from different sources have extremely wide variances (Table 1). They are generally good sources of both N and P for crops, but as for most manures, P application will be too great if enough is applied to satisfy N requirements. Application according to P requirements will allow little or no use in many cases.

OTHER BYPRODUCTS AND WASTES

Many other organic byproducts and wastes may be suitable and even valuable as fertilizers. Included would be paper and pulp, vegetable and fruit process byproducts, food wastes, etc.—the list is too extensive to be included here. All wastes should be analyzed for nutrient elements and for potential pollutants prior to application.

REFERENCES

1. Huntley, E.E.; Barker, A.V.; Stratton, M.L. Composition and Uses of Organic Fertilizers. In *Agricultural Uses of Byproducts and Wastes*; Rechcigl, J.E., MacKinnon, H.C., Eds.; American Chemical Society: Washington, DC, 1997; 120–139.
2. Miller, D.M.; Miller, W.P. Land Application of Wastes. In *Handbook of Soil Science*; Sumner, M.E., Ed.; CRC Press: Boca Raton, FL, 2000; G-217–G-242.
3. Bruulsema, T.W.; Christie, B.R. Nitrogen Contribution to Succeeding Corn from Alfalfa and Red Clover. Agron. J. **1987**, *79*, 96–100.
4. Fox, R.H.; Piekielek, W.P. Fertilizer Nitrogen Equivalence of Alfalfa, Birdsfoot Trefoil, and Red Clover for Succeeding Corn Crops. J. Prod. Agric. **1988**, *1*, 313–317.
5. Hook, J.E. Comparison of the Crop Management Strategies Developed from Studies at Pennsylvania State University. University of Minnesota, and the Muskegon County Land Treatment System. In *Land Treatment of Municipal Wastewater*; D'Itri, F., Ed.; Ann Arbor Science: Ann Arbor, MI, 1982; 65–78.
6. Sumner, M.E. Beneficial Use of Effluents, Wastes, and Biosolids. Commun. Soil Sci. Plant Anal. **2000**, *31* (11–14), 1701–1715.
7. Ellis, B.G.; Erickson, A.E.; Jacobs, L.W.; Knezek, B.D. Crop Management Studies at the Muskegon County Michigan Land Treatment System. In *Land Treatment of Municipal Wastewater*; D'Itri, F., Ed.; Ann Arbor Science: Ann Arbor. MI, 1982; 49–63.
8. Dowdy, R.H.; Clapp, C.E.; Marten, G.C.; Linden, D.R.; Larsen, W.E. Wastewater Crop Management Studies in Minnesota. In *Land Treatment of Municipal Wastewater*; D'Itri, F., Ed.; Ann Arbor Science: Ann Arbor, MI, 1982; 35–47.
9. Forste, J.B. Biosolids Processing, Products and Uses. In *Agricultural Uses of Byproducts and Wastes*; Rechcigl, J.E., MacKinnon, H.C., Eds.; American Chemical Society: Washington, DC, 1997; 50–62.

FERTILIZERS, URBAN WASTE AND SLUDGES AS

R.O. Maguire
J.T. Sims
University of Delaware, Newark, Delaware, U.S.A.

INTRODUCTION

Environmental quality is a major issue in many parts of the world and includes important topics such as sustainability and environmental protection. Sustainability and environmental protection often overlap and cover issues such as soil conservation, careful use of finite resources such as inorganic fertilizers, and the responsible recycling of urban wastes and animal manures by agriculture. The beneficial re-use of urban wastes and by-products, e.g., sewage sludge (biosolids), has become an increasingly important issue in order to attain sustainability as the world's population increases. This article describes the uses of urban wastes and biosolids as fertilizers primarily for agricultural settings and the major trends, issues, options, and regulations involved. The main focus is on biosolids, since they constitute the largest portion of urban wastes that are currently land applied.

TYPES OF URBAN WASTES THAT ARE USED AS FERTILIZER

Legislation in most countries requires treatment of wastewater from combined residential, commercial, and industrial sources. Treatment of wastewater produces a semisolid by-product that is commonly known as "sewage sludge" or "biosolids." There are several disposal pathways for biosolids including incineration, landfilling, and application to land as a fertilizer. Use of biosolids in agriculture is a well-established and regulated process in many parts of the world, including the United States and Europe. To a lesser extent, other urban and industrial by-products, e.g., paper waste, and municipal solid waste (MSW), such as leaves, grass clippings, or other organic materials, are also applied to agricultural land, often following composting. Sometimes biosolids can be co-composted with industrial by-products or MSW.

PRODUCTION OF URBAN WASTES USED AS FERTILIZER

Biosolids Production and Quality

There are several treatment processes that are routinely carried out in wastewater treatment plants. These determine the quality and properties of the biosolids produced, which in turn affects the options for land application. Following initial screening, wastewater can undergo primary, secondary, and tertiary treatments (Table 1). The treatment used at any particular wastewater treatment plant depends on the effluent discharge limits (e.g., biological oxygen demand (BOD) or nutrient content of the treated wastewater to be discharged into surface waters), the proposed use for the biosolids and resources available. In addition to these wastewater treatment processes, de-watering of biosolids is frequently used to reduce biosolids volume, which can in turn reduce transportation and disposal costs.

Composting of Biosolids and Municipal Solid Waste

Composting (often called co-composting when two or more materials are composted together) is the decomposition of organic matter by microorganisms in a controlled environment that has optimum moisture and oxygen contents. The increase in temperature during composting can destroy most pathogens. Composting of biosolids or the organic fraction of MSW may cause offensive odors and odor control systems, such as scrubbers and biofilters, are usually required in populated areas. Composting of biosolids involves mixing de-watered biosolids with a bulking agent, such as MSW, wood chips, or straw, followed by aerobic decomposition. Only a small proportion of biosolids and MSW are composted worldwide, but in certain areas composting accounts for a large proportion of the urban wastes generated. For example, Edmonton, Canada, is able to divert 70% of its residential waste from landfill using recycling and co-composting (1).

Table 1 Wastewater treatment processes

Treatment name	Treatment process
Screening and grit removal	Generally a combination of sedimentation and passing through a large screen to remove stones, grit, and any large debris, such as branches, that may have entered the sewage system. The product of this process is nearly always regarded as a solid waste rather than biosolids and is usually landfilled
Primary	Follows screening and grit removal and usually involves sedimentation to remove suspended solids prior to secondary treatment
Secondary	Normally a biological treatment in which microorganisms are used to reduce suspended solids and BOD thereby eliminating fish kills when the wastewater is discharged. This is the minimum wastewater treatment process required in the United States
Tertiary	Used where higher standards are required for effluent quality or to improve biosolids quality for land application. Examples include addition of lime for pathogen and odor control, iron or aluminum salts for precipitation of P, and polymers for removal of suspended sediments

Where the organic fraction of MSW is composted, it can either be source separated by the resident or screened and separated from regular residential garbage, and composted alone or mixed with biosolids before being composted.

REGULATIONS COVERING LAND APPLICATION OF URBAN WASTES

Land application of biosolids normally depends on the wastewater treatment process used to produce the biosolids. As specific rules vary between countries, it is impossible to describe them all here (2, 3). However, it is possible to make some generalizations about land application of biosolids. For example, secondary and/or tertiary treatment, mainly to control pathogens and odors, is normally required before land application is permitted.

In the United States, biosolids applications are governed by Title 40 of the Code of Federal Regulations, Part 503, commonly referred to as the "503 rule," which sets limits for toxic metals in biosolids, and for both annual and cumulative loadings to land (Table 2). Limits for chromium and molybdenum are currently under consideration. The 503 rule also classifies biosolids as either "Class A" or "Class B" for pathogen control. The content of polychlorinated biphenyls in biosolids to be land applied is limited to a maximum of 50 mg/kg, under Title 40 of the Code of Federal Regulations, Part 761. Similar rules are in effect in the European Community, under Council Directive 86/278/EEC, implemented in 1986 and currently under revision.

AGRICULTURAL MANAGEMENT OF URBAN WASTES AS FERTILIZERS

The benefits of biosolids and compost applications to soil quality are many and well documented. Biosolids and composts are good sources of nitrogen (N), phosphorus (P), and potassium (K). Typical biosolids contain 4.0, 2.0,

Table 2 Pollutant limits set by the 503 rule for toxic metals in biosolids applications to land

Pollutant	Ceiling concentration limits for all biosolids applied to land[a] (mg/kg)	Cumulative pollutant loading rate limits (kg/ha)	Annual pollutant loading rate limits (kg/ha/yr)
Arsenic	75	41	2
Cadmium	85	39	1.9
Copper	4300	1500	75
Lead	840	300	15
Mercury	57	17	0.85
Nickel	420	420	21
Selenium	100	100	5
Zinc	7500	2800	140

[a] Maximum concentration of pollutant permitted in any biosolids to be land applied.

and 0.4% of N, P, and K, respectively, on a dry weight basis, while the equivalent values for MSWs are 0.7, 0.2, and 0.3%, respectively. Composted biosolids contain less total N than uncomposted biosolids, due to the addition of other materials to aid composting and loss of ammonia during the composting process. The N in composted biosolids is released more slowly, which decreases nitrate leaching. Thus, the N is available to plants over a longer time period, which is more consistent with crop N uptake patterns (2). Biosolids and composts can also be good sources of Ca, Mg, and S, and of the micronutrients Fe, Cu, B, Zn, Mn, and Mo. Biosolids and composts can promote beneficial microbial activity and diversity, that suppress plant diseases and the need for costly pesticides. Biosolids and composts can also improve water infiltration, water retention and soil structure, which in turn increases resistance to wind and water erosion. In the United States in 1998, land application of biosolids accounted for 41% of the total produced, while advanced treatments such as composting accounted for 12% (2). The annual production of biosolids (7 million $Mg\,yr^{-1}$) is small in comparison to animal manure production (174 million $Mg\,yr^{-1}$) in the United States. However, the land application of biosolids constitutes a significant economic saving for many biosolids producers and saves landfill space, which is an increasingly expensive, finite resource in many areas.

POTENTIAL PROBLEMS ASSOCIATED WITH LAND APPLICATION OF BIOSOLIDS

Legitimate concerns about the need to prevent toxic metals and pathogens from affecting human and ecosystem health have been addressed through regulation and record keeping of applications of biosolids to land. However, there is currently a debate as to whether the limits set for toxic metals and pathogens are strict enough. Some scientists think that metal bioavailability will increase as the organic matter added with the biosolids is mineralized, while others argue that the evidence to support this hypothesis is inconclusive.

Biosolids and composts have a low N:P ratio compared to crop requirements. Biosolids applications are generally carried out according to N-based nutrient management plans that over-apply P, and can lead to a buildup of P in agricultural soils (4). Buildup of soil P in many areas has been linked to increasing losses of P from agriculture to surface waters, with a corresponding decrease in water quality. In the future, biosolids may have to be applied according to P based nutrient management plans. This will decrease the amount of biosolids that can be applied per unit area of land, introduce the necessity for inorganic N fertilizers, and increase other associated costs. Nuisance issues such as odors and attraction of pests are of public concern when application sites are close to residential areas (5).

THE FUTURE

Efforts to recycle municipal wastes will likely continue, driven by human population growth, the continuing rise in cost of landfill, and the goal of sustainability in agriculture and the environment as a whole. However, the future of land-application programs for biosolids is uncertain. For example, the U.S. EPA has forecast an increase in the beneficial use of biosolids in land-application programs because of increasing costs associated with landfill, and an increase in biosolids quality due to stricter regulations covering biosolids production. Despite increasing biosolids quality, the argument over the long-term environmental impact of pollutants in biosolids will likely continue. Further regulation of the application of P to agricultural land may also increase costs associated with land-application programs for biosolids and municipal wastes. However, if the potential negative side effects associated with the land application of biosolids and composted waste are properly managed, then the beneficial use of these by-products will continue.

REFERENCES

1. www.gov.edmonton.ab.ca (accessed May 2001).
2. U.S. Environmental Protection Agency. Biosolids Generation, Use, and Disposal in the United States. EPA530-R-99-009, September 1999.
3. Commission of the European Communities. Council Directive on the Protection of the Environment, and in Particular Soil, When Sewage Sludge is Used in Agriculture (86/278/EEC). In Official Journal of the European Communities, No. 2, 181/6-12, 1986.
4. Maguire, R.O.; Sims, J.T.; Coale, F.J. Phosphorus Solubility in Biosolids-Amended Farm Soils in the Mid-Atlantic Region of the U.S. J. Environ. Qual. **2000**, *29* (4), 1225–1233.
5. Sims, J.T.; Pierzynski, G.M. Assessing the Impacts of Agricultural, Municipal, and Industrial By-Products on Soil Quality. In *Land Application of Agricultural, Industrial, and Municipal By-Products*; Power, J.F., Warren, A.D., Eds.; Soil Sci. Soc. Am. Madison, WI, 2000; 237–261.

F

FLOODING TOLERANCE OF CROPS

Tara T. VanToai
Getachew Boru
Jianhuan Zhang
United States Department of Agriculture, The Ohio State University, Columbus, Ohio, U.S.A.

INTRODUCTION

Periodic flooding during the growing season adversely affects crop growth and productivity, with the exception of flooded rice, in many areas of the U.S. and the rest of the world. Soil can become flooded when it is poorly drained or when rainfall or irrigation is excessive. Other terms, such as soil saturation, waterlogging, anoxia, and hypoxia, are also commonly used to describe flooding conditions. Flooding causes premature senescence, which results in leaf chlorosis, necrosis, defoliation, reduced nitrogen fixation, cessation of growth, and reduced yield. The severity of the flooding stress is affected by many factors, including flooding duration, crop variety, growth stage, soil type, fertility levels, pathogens, and flooding conditions (1). In general, stream flooding, characterized by the overflow of rivers or creeks into a flood plain, is more damaging than lowland flooding, characterized by inadequate surface drainage and slow soil permeability of depressional areas. Sediments carried by stream flooding, when deposited on the leaves of flooded plants, can cause severe wilting and plant death within 24 hours of the stress. Flooding can be further divided into either waterlogging, where only the roots are flooded, or complete submergence, where the entire plants are under water. While plants develop adaptive mechanisms to allow them to survive long-term waterlogging, most plants die within one or two days of submergence (1).

The lack of oxygen has been proposed as the main problem associated with flooding (2). Indeed, tolerance of anoxia and hypoxia has been used synonymously with tolerance of flooding stress. During the last two decades, a great deal more information has accumulated from research on the molecular, biochemical, and physiological responses of plants to the lack of oxygen rather than to flooding per se (3–5). However, tolerance of field flooding appears to be much more complex than tolerance of artificially induced hypoxia and anoxia. Contrary to the injury seen in flooded fields, soybeans can thrive in stagnant water in the greenhouse and soybeans grown in hydroponic medium continuously bubbled with nitrogen gas, where the dissolved oxygen level was not detectable, showed no symptoms of stress (6). Soybean, therefore, is much more tolerant to excessive water and a lack of oxygen than previously expected. The reasons underlying the dramatic differences between responses to flooding in the greenhouse and flooding in the field are not known. However, growth reduction and yield loss in flooded fields could arise from root rot diseases, nitrogen deficiency, nutrient imbalance, and/or the accumulation of toxic levels of CO_2 in the root zone. Indeed, the levels of CO_2 commonly found in flooded soil (30%) severely caused leaf chlorosis and reduced plant biomass of soybean, a flood-susceptible crop, but not of rice, a flooding-tolerant crop (VanToai, unreported data).

MORPHOLOGICAL AND ANATOMICAL ADAPTATION TO FLOODING STRESS

One important morphological change associated with flooded roots is the formation of aerenchyma tissue, which contains continuous gas-filled channels connecting the root with the shoot. Other morphological changes, including hypertrophy and the formation of lenticels, adventitious roots, and pneumatophores, have also been observed in many plant species (7). Flooding can also change the direction of root growth. Roots of tomato and sunflower plants become disgeotropic or negatively geotropic under flooding conditions instead of positively geotropic (7). The changes in orientation of roots under flooding conditions enables them to escape stress from the reduced oxygen availability by growing closer to the better aerated soil surface.

PHYSIOLOGICAL, BIOCHEMICAL, AND MOLECULAR ADAPTATION TO FLOODING STRESS

Rice cultivars that showed rapid leaf and sheath growth during submergence did not survive as well as cultivars that did not elongate. Tolerant cultivars appeared to conserve carbohydrates in the shoots and roots during submergence (8). Upon removal of the stress, tolerant rice cultivars are able to recover more rapidly and suffer less plant mortality (9). An adequate supply of sugar is needed for corn root tips to survive anoxic and hypoxic stresses (10). The lack of oxygen induces a set of anaerobic proteins in roots to allow plants to cope with the stress (11). These stress proteins are enzymes of either glycolysis, glucose metabolism, or fermentation (10, 12).

GENETIC VARIABILITY IN FLOODING TOLERANCE

Flooding tolerance is usually defined as minimal or no yield loss. According to VanToai et al. (13), waterlogging for four weeks during the early flowering stage reduced the average grain yield of 84 U.S. soybean cultivars by 25%. Yield reduction, however, varied from 9% in the most flooding tolerant cultivar to 75% in the most flooding susceptible cultivar. Flooding tolerance can also be defined as high yield under flooding stress. According to this definition, the most flooding tolerant variety in this study produced 3.7 Mg ha^{-1}, while the least produced 1.27 Mg ha^{-1}. When the cultivars were ranked for flooding tolerance based on both definitions, seven of the ten most flooding tolerant cultivars were the same; and seven of the ten least flooding tolerant cultivars were also the same. Thus, the two definitions of flooding tolerance, either high yield under flooding or minimal yield difference between nonflooded and flooded conditions, appear to be compatible. Flooding tolerance is independent from nonflooded yield indicating that genetic variability for flooding tolerance exists and could be improved through plant breeding and selection.

Studies of submergence tolerance of rice showed that the unimproved land races FR13A, Janki, and FR43B had survival values ranging from 41 to 51% after 10 days of submergence, while only 2–4% of elite cultivars (IR74, IR48 and IR68) survived (9).

IMPROVING FLOODING TOLERANCE BY TRADITIONAL PLANT BREEDING

Tolerance of flooding in wheat (*Triticum aestivum* L.) and rice (*Oryza sativa* L.) is a quantitative trait controlled by a small number of genes (9, 14). Using the submergence tolerant land races FR13A, Janki, and FR43B as donor parents, rice breeders at the International Rice Research Institute (IRRI) at Los Banos in the Philippines have developed an experimental rice line (IR49830-7-1-2-2) from crosses with the short stature, high-yield IR lines, which produced as much as 4880 kg ha^{-1} (9). The result showed that submergence tolerance can be incorporated into improved, high-yielding cultivars to raise the productivity in submergence-prone areas.

IDENTIFYING FLOODING TOLERANCE LOCI AND IMPROVING FLOODING TOLERANCE BY MOLECULAR PLANT BREEDING

Xu and Mackill (15) identified a single submergence tolerance locus, *Sub*1, that controls about 50% of the variation in submergence tolerance of rice. During the last few years, molecular marker aided selection has been used successfully for the breeding of crops with improved quantitative traits.

VanToai et al. (16) identified a single DNA marker that was associated with improved plant growth (from 11 to 18%) and grain yields (from 47 to 180%) of soybean in waterlogged environments. The identified marker was uniquely associated with waterlogging tolerance and was not associated with maturity, normal plant height or grain yield. Near isogenic lines with and without the flooding tolerant marker have been developed and are being field tested under waterlogging conditions to confirm the association of the marker with the tolerance of soybean to waterlogging stress.

IMPROVING FLOODING TOLERANCE BY GENETIC TRANSFORMATION

Flooding induces or accelerates plant senescence in tobacco, tomato, sunflower, carrot, barley, peas, wheat, maize, and soybean. The most obvious visual symptom of flooded plants under stress is the yellowing of leaves followed by necrosis due to premature senescence. Within one day of flooding, the concentration of the antiaging hormone, cytokinin, in sunflower xylem sap declined sharply to a very low level (17). In order to test if enhanced endogenous cytokinin production could improve flooding tolerance, Zhang et al. (18) generated transgenic plants containing a gene coding for cytokinin biosynthesis. Four transgenic *Arabidopsis* lines were chosen for cytokinin and flooding tolerance determinations. The levels of cytokinin were similar between wild-type and transgenic plants in the unflooded treatment. After 5 days of waterlogging, the cytokinin increased 3–10 times in transgenic plants as compared to wild-type plants. In three independent experiments, all four transgenic lines were consistently more tolerant to soil waterlogging and complete submergence than wild-type plants. The results indicated that endogenously produced cytokinin can regulate senescence caused by flooding stress, thereby increasing plant tolerance of flooding. This study provides a novel mechanism to improve flooding tolerance in plants (18).

In summary, while the lack of oxygen has been used interchangeably with flooding stress, tolerance of field flooding is more complex than tolerance of anoxia and hypoxia. The use of molecular plant breeding and genomic transformation to improve flooding tolerance in crops is promising to be successful.

REFERENCES

1. Sullivan, M.; VanToai, T.; Fausey, N.; Beuerlein, J.; Parkinson, R.; Soboyejo, A. Evaluating On-Farm Flooding Impacts on Soybean. Crop Science **2001**, *41*, 1–8.
2. Kozlowski, T.T. Extent, Causes, and Impacts of Flooding. In *Flooding and Plant Growth*; Kozlowski, T.T., Ed.; Academic Press Inc.: Orlando, FL, 1984; 1–7.
3. Kennedy, R.A.; Rumpho, M.E.; Fox, T.C. Anaerobic Metabolism in Plants. Plant Physiology **1992**, *84*, 1204–1209.
4. Perata, V.M.; Alpi, A. Plant Responses to Anaerobiosis. Plant Science **1993**, *93*, 1–17.
5. Ricard, B.; Couee, I.; Raymond, P.; Saglio, P.H.; Saint-Ges, V.; Pradet, A. Plant Metabolism Under Hypoxia and Anoxia. Plant Biochemistry **1994**, *32*, 1–10.
6. Boru, G.; VanToai, T.; Alves, J.D. Flooding Injuries in Soybean are Caused by Elevated Carbon Dioxide Levels in the Root Zone. Fifth National Symposium on Stand Establishment **1997**, 205–209.
7. Hook, D.D. Adaptations to Flooding with Fresh Water. In *Flooding and Plant Growth*; Kolowski, T.T., Ed.; Academic Press: Orlando, FL, 1984; 265–294.
8. Jackson, M.B.; Waters, I.; Setter, T.; Greenway, H. Injury to Rice Plants Caused by Complete Submergence; A Contribution by Ethylene (Ethene). Journal of Experimental Botany **1987**, *38*, 1826–1838.
9. Mackill, D.J.; Amante, M.M.; Vergara, B.S.; Sarkarung, S. Improved Semidwarf Rice Lines With Tolerance to Submergence of Seedlings. Crop Science **1993**, *33*, 749–753.
10. Ricard, B.; Saglio, P.; VanToai, T.; Chourey, P. Evidence for the Critical Role of Sucrose Synthase for Anoxic Tolerance of Maize Roots Using a Double Mutant. Plant Physiology **1998**, *116*, 1323–1331.
11. Sachs, M.M.; Freeling, M.; Okimoto, R. The Anaerobic Proteins of Maize. Cell **1980**, *20*, 761–767.
12. Sachs, M.M.; Subbaiah, C.C.; Sabb, I.N. Anaerobic Gene Expression and Flooding Tolerance in Maize. Journal of Experimental Botany **1996**, *47*, 1–15.
13. VanToai, T.; Beuerlein, J.E.; Schmitthenner, A.F.; St. Martin, S.K. Genetic Variability for Flooding Tolerance in Soybeans. Crop Science **1994**, *34*, 1112–1115.
14. Boru, G.; Ginkel, V.M.; Kronstad, W.E.; Boersma, L. Expression and Inheritance of Tolerance to Waterlogging Stress in Wheat. Euphytica **2001**, *117*, 91–98.
15. Xu, K.; Mackill, D.J. A Major Locus for Submergence Tolerance Mapped on Rice Chromosome 9. Molecular Breeding **1996**, *2*, 219–224.
16. VanToai, T.T.; St. Martin, S.K.; Chase, K.; Boru, G.; Schnipke, V.; Schmitthenner, A.F.; Lark, K.G. Identification of a QTL Associated with Tolerance of Soybean to Soil Waterlogging. Crop Science **2001**, *41*, 1247–1252.
17. Burrows, W.J.; Carr, D.J. Effects of Flooding the Root System of Sunflower Plants on the Cytokinin Content of the Xylem Sap. Physiol Plant. **1969**, *22*, 1105–1112.
18. Zhang, J.; VanToai, T.; Huynh, L.; Preiszner, J. Development of Flooding-Tolerant *Arabidopsis Thaliana* by Autoregulated Cytokinin Production. Molecular Breeding **2000**, *6*, 135–144.

FLUID FLOW, CHALLENGES MODELING

John L. Nieber
University of Minnesota, St. Paul, Minnesota, U.S.A.

INTRODUCTION

The unsaturated zone (vadose zone) of the Earth's crust is an important interface for both the underlying groundwater and the overlying surface water resources and atmosphere. Quantifying fluid flow, mass transport, and energy transport processes in the unsaturated zone have become a focus for researchers, government agencies, and consultants during the past three decades because it is found that the outcome of these processes have an impact on the sustainability of modern social structures. While sophisticated sensors and instrumentation seems to have been developed to provide data from the field on a real-time basis, these cannot be used to directly predict the possible outcomes. Instead, this prediction requires the use of mathematical models representing the flow and transport processes. The following describes some of the applications of models for flow in the unsaturated zone, past challenges and achievements in improving modeling methods, and future challenges to modeling flow.

APPLICATION OF MODELS

Models for simulating flow processes in soil and groundwater find applications in many environmentally oriented disciplines including geography, soil science, agricultural engineering, civil engineering, geoengineering, hydrogeology, and meteorology. These models are generally based on numerical solutions of governing equations and require the use of digital computers to complete the calculation task. There are many current and potential applications for such models, but a short list of the common applications includes: 1) estimation of groundwater recharge volume and contaminant loading; 2) design of measures to remediate contaminated soils and groundwater; 3) design of efficient drainage and irrigation systems for efficient crop production; 4) estimation of runoff production from land in response to rainfall and/or snowmelt; and 5) assessment of the impact of global climate change on surface and subsurface water resources.

Models developed for these applications need to satisfy two criteria to be put to use by a practitioner. They have to be fairly easy to use and dependable. The first criterion, ease-of-use, only requires a good team of programmers, and does not pose a challenge for flow modeling. The second criterion stipulates that the model provides accurate results in a timely manner. This is a direct challenge to flow modeling. The following sections present information on the past and future challenges associated with the development of dependable models.

PAST CHALLENGES AND ACHIEVEMENTS

During the past three decades there has been substantial progress in the development of numerical models for simulating relevant unsaturated zone processes. During this time predominant attention in modeling flow processes was given to the Richards equation (1). In the past, many of the difficulties in model development involved the determination of the best ways to solve this equation.

Due to the highly nonlinear character of the Richards equation, analytical solutions (2) to the equation were possible only for simplified conditions, and therefore numerical solution methods were necessary to treat realistic field conditions. Numerical methods such as the finite difference method and the finite element method were adopted from other engineering and science applications and applied to discretize the Richards equation into systems of nonlinear algebraic equations. Two major numerical solution problems were faced by researchers in solving this equation. One was the highly nonlinear nature of the equation and associated boundary conditions, leading to problems of slow convergence or even non-convergence of the solution methods. This problem was handled by applying nonlinear equation solvers classified within the broad class of Newton methods and Picard methods (3).

A second problem was the need to be able to solve large systems of algebraic equations, which in earlier years involved hundreds or perhaps thousands of equations. Due to the relatively small memory available on early computers it was necessary to use iterative methods to solve even moderate-sized problems. Conventional iterative methods like Jacobi, Gauss–Seidel, and successive over-relaxation methods (4) were used to solve these problems. While these methods were satisfactory for relatively homogeneous systems, for strongly heterogeneous systems it was found that these methods led to poor convergence or even nonconvergence. Computer memory did increase substantially during the 1980s and 1990s, and this allowed the use of direct equation solvers for the types of problems solved in the earlier years. However, with the increase of computer memory storage, the problems tackled have also increased memory requirements (tens of thousands to millions of equations need to be solved) and iterative methods are once again back in vogue. Fortunately, the efficiency of iterative methods has also increased substantially with the development of conjugate gradient methods (5) and multigrid methods (6).

During the last 30 years there has been an ongoing effort to find the recipe for a numerical method or set of methods that will be robust for solving the Richards equation (3, 4, 7–12). The desire has been to develop computationally efficient methods applicable to a broad range of practical problems, especially for large-scale, three-dimensional, heterogeneous flow systems. Associated problems involved assuring that the solutions were mass conservative and that the iterative methods used to solve the nonlinear algebraic equations would converge even for conditions where the soil is very dry and highly heterogeneous. Detailed analysis of how to assure a mass conservative solution has been given in Ref. (8). Assurance of nonlinear iteration convergence has been found to be more problematic and recent improvements have been made using techniques involving primary variable switching (9, 10), higher order time integration (11), and variable transformation (12).

Aside from the obvious problems associated with solving the Richards equation there has been the need to assign (spatially) the equation parameters for field scale applications of the numerical solutions. For instance, the catchment scale model of Ref. (13), like the model of Ref. (4), was developed to facilitate the simulation of three-dimensional variably saturated flow over an entire catchment. While the solution of the large system of algebraic equations for such a problem offers a significant challenge to modeling, an equally if not larger challenge is the problem of assigning parameters to the cells in the numerical grid.

The problem of parameter assignment is two-fold. For one, there is the problem of determining the spatial distribution of the parameters at the scale of the numerical grid cell. Involved with this problem is the uncertainty in predicting these values, given limited field data. Geostatistical methods (14) were developed to facilitate such prediction and to estimate limits of uncertainty. A second problem arises because small scale features, and even small scale processes, occurring below the sampling scale of the field scale grid cell can significantly influence the actual outcome at the field scale, but are not directly predictable by the governing equation(s) discretized at the field scale grid. There has been some success (15) in developing techniques that provide parameters that account for these subgrid scale features and processes. These parameters, called *effective* parameters, are those that yield *effectively* the same outcome as would occur if the more detailed parameter distribution were used.

FUTURE CHALLENGES TO MODELING

Two of the greatest future challenges to modeling include the need to more completely describe the flow and transport processes, and the need to incorporate multiscale phenomena into the modeling analysis. The first challenge involves the expansion of the governing equations to include coupled physical and chemical processes. The second challenge involves the assignment of equation parameters and the incorporation of subgrid features and processes. To assure the success of future modeling, this second challenge is the most critical to address.

Governing Equations

As awareness of environmental problems has increased, and environmental regulations have become more stringent, the scope for modeling has expanded from modeling the flow of water alone to modeling coupled multiphase fluid flows, mass (solute) transport and energy (thermal) transport. The coupled equations for isothermal multiphase fluid flow have been reviewed in detail (16). Numerical solutions of coupled two-phase flow equations for applications to environmental problems have advanced considerably in the last 20 years. One of the earlier multiphase flow solutions is given in Ref. (17), while methods representing the latest advances in computational efficiency are given in Ref. (18).

Equations for nonisothermal conditions, which until recently have received much less attention, have been presented in Ref. (19). Solutions of these equations offer new challenges to those developing numerical solutions (20, 21). Recent techniques such as those mentioned before for the solution of the Richards equation and the coupled multiphase flow equations should be tested to improve the efficiency of such solutions.

Process Scale

Numerical solutions of flow and transport processes for field-scale applications are generally performed on relatively large numerical grids. The reason for this is two-fold. First, the computational effort to simulate field-scale problems can easily exceed the capabilities of today's computers (using current numerical methods) if too fine a grid is used. Second, one generally does not know what values of parameters to assign to very fine grids due to inadequate spatial resolution in field data. As a result, it is necessary to assign effective parameters to the field-scale grid cells. One approach is to use stochastic methods (15) to derive effective parameters. Renormalization methods (22), which rely on numerical methods, are another means to derive effective parameters.

In some instances the governing equations may behave differently at the small scale than at the large scale. For instance, in the case of finger flow (23, 24) or funnel flow (25), the flow occurs on a scale of about 10 cm. To simulate these flow features directly it is necessary to use relatively small grid cells. Using large grid cells without considering these small-scale processes leads to a diffuse solution and the small-scale features are missed. Effectively capturing the dynamics of these small-scale processes into a field-scale grid requires an appropriate procedure for upscaling of the governing equations from the small-scale to the field-grid scale. Such a procedure has been demonstrated (26) for viscous fingering in multiphase flow and is an area of active research.

The discussion about process scale poses the question about whether the governing equations maintain a constant form in the progression from the small scale to the large scale. Addressing this question involves the principles currently developed in the field of multiscale science (27). While these developments originated in the fields of solid mechanics and fluid mechanics, they are currently receiving much attention in hydrology, soil physics, and hydrogeology.

REFERENCES

1. Richards, L.A. Capillary Conduction of Liquids Through Porous Mediums. Physics **1931**, *1*, 318–333.
2. Philip, J.R. Steady Infiltration from Spheroidal Cavities in Isotropic and Anisotropic Soils. Water Resour. Res. **1986**, *22*, 1874–1880.
3. Paniconi, C.; Putti, M.A. A Comparison of Picard and Newton Iteration in the Numerical Solution of Multi-dimensional Variably Saturated Flow Problems. Water Resour. Res. **1994**, *30*, 3357–3374.
4. Freeze, R.A. Three-Dimensional, Transient, Saturated-Unsaturated Flow in a groundwater Basin. Water Resour Res. **1971**, *7*, 347–366.
5. Saad, Y. *Iterative Methods for Sparse Linear Systems*; PWS Publ. Co.: Boston, MA, 1996.
6. Brandt, A. Multi-Level Adaptive Solutions to Boundary Value Problems. Math. Comput. **1977**, *31*, 333–390.
7. Huyakorn, P.S.; Thomas, S.D.; Thompson, B.M. Techniques for Making Finite Elements Competitive in Modeling Flow in Variably Saturated Media. Water Resour. Res. **1986**, *20*, 1099–1115.
8. Rathfelder, K.; Abriola, L.M. Mass Conservative Numerical Solutions of the Head-Based Richards Equation. Water Resour. Res. **1994**, *30*, 2579–2586.
9. Forsyth, P.A.; Wu, Y.S.; Pruess, K. Robust Numerical Methods for Saturated-Unsaturated Flow with Dry Initial Conditions in Heterogeneous Media. Adv. Water Resour. **1995**, *18*, 844–856.
10. Diersch, H.-J.G.; Perrochet, P. On the Primary Variable Switching Technique for Simulating Unsaturated-Saturated Flows. Adv. Water Resour. **1999**, *23*, 271–301.
11. Tocci, M.D.; Kelley, C.T.; Miller, C.T. Accurate and Economical Solution of the Pressure-Head Form of Richards' Equation by the Method of Lines. Adv. Water Resour. **1997**, *20*, 1–14.
12. Williams, G.A.; Miller, C.T. An Evaluation of Temporally Adaptive Transformation Approaches for Solving Richards' Equation. Adv. Water Resour. **1999**, *22*, 831–840.
13. Paniconi, C.; Wood, E.F. A Detailed Model for Simulation of Catchment Scale Subsurface Hydrologic Processes. Water Resour. Res. **1993**, *29*, 1601–1620.
14. Webster, R. Quantitative Spatial Analysis of Soil in the Field. In *Advances in Soil Science*; Stewart, B.A., Ed.; 1985; Springer Verlag: New York, Vol. 3, 1–70.
15. Mantoglou, A.; Gelhar, L.W. Stochastic Modeling of Large-Scale Transient Unsaturated Flow Systems. Water Resour. Res. **1987**, *23*, 37–46.
16. Miller, C.T.; Christakos, G.; Imhoff, P.T.; McBride, J.F.; Pedit, J.; Trangenstein, J.A. Multiphase Flow and Transport Modeling in Heterogeneous Porous Media: Challenges and Approaches. Adv. Water Resour. **1999**, *21*, 77–120.
17. Kaluarachchi, J.J.; Parker, J.C. An Efficient Finite Element Method for Modeling Multiphase Flow. Water Resour. Res. **1989**, *25*, 43–54.
18. Bastian, P.; Helmig, R. Efficient Fully-Coupled Solution Techniques for Two-Phase Flow in Porous Media; Parallel Multigrid Solution and Large Scale Computations. Adv. Water Resour. **1999**, *23*, 199–216.
19. Nassar, I.N.; Horton, R. Heat, Water and Solute Transfer in Unsaturated Porous Media: I. Theory Development and Transport Coefficient Evaluation. Trans. Por. Med. **1997**, *27*, 17–39.
20. Thomas, H.R.; Missoum, H. Three-Dimensional Coupled Heat, Moisture, and Air Transfer in a Deformable Unsaturated Soil, Int. J. Num. Meth. Engng. **1999**, *44*, 919–943.

21. Chounet, L.M.; Hilhorst, D.; Jouron, C.; Kelanemer, Y.; Nicolas, P. Simulation of Water Flow and Heat Transfer in Soils by Means of a Mixed Finite Element Method. Adv. Water Resour. **1999**, *22*, 445–460.
22. King, P.R. The Use of Renormalization for Calculating Effective Permeability. Transp. Por. Med. **1989**, *4*, 37–58.
23. Glass, R.J.; Steenhuis, T.S.; Parlange, J.-Y. Wetting Front Instability As a Rapid and Far-Reaching Hydrologic Process in the Vadose Zone. J. Contam. Hydrol. **1988**, *3*, 207–226.
24. Nieber, J.L. Modeling Finger Development and Persistence in Initially Dry Porous Media. Geoderma **1996**, *70*, 209–229.
25. Ju, S.-H.; Kung, K.-J.S. Steady-State Funnel Flow: Its Characteristics and Impact on Modeling. Soil Sci. Soc. Am. J. **1997**, *61*, 416–427.
26. Blunt, M.; Christie, M. How to Predict Viscous Fingering In Three-Component Flow. Transp. Por. Med. **1993**, *12*, 207–236.
27. Glimm, J.; Sharp, D.H. Multiscale Science: A Challenge for the Twenty-First Century, SIAM News **1997**; *30*, 4,17,19.

FOREST ECOSYSTEMS, NUTRIENT CYCLING IN

F

Neil W. Foster
Canadian Forest Service, Ontario, Sault Ste. Marie, Ontario, Canada

J.S. Bhatti
Canadian Forest Service, Northern Forestry Centre, Edmonton, Alberta, Canada

INTRODUCTION

Nutrients are elements or compounds that are essential for the growth and survival of plants. Plants require large amounts of nutrients such as nitrogen (N), phosphorus (P), carbon (C), hydrogen (H), oxygen (O), potassium (K), calcium (Ca), and magnesium (Mg), but only small amounts of others such as boron (B), manganese (Mn), iron (Fe), copper (Cu), zinc (Zn) and chlorine (Cl) (micronutrients). Forest nutrient cycling is defined as the exchange of elements between the living and nonliving components of an ecosystem (1). The processes of the forest nutrient cycle include: nutrient uptake and storage in vegetation perennial tissues, litter production, litter decomposition, nutrient transformations by soil fauna and flora, nutrient inputs from the atmosphere and the weathering of primary minerals, and nutrient export from the soil by leaching and gaseous transfers.

Each nutrient element is characterized by a unique biogeochemical cycle. Some of the key features of the major nutrients are shown in Table 1. Forest trees make less demand on the soil for nutrients than annual crops because a large proportion of absorbed nutrients are returned annually to the soil in leaf and fine root litter and are reabsorbed after biological breakdown of litter materials. Also, a large portion of nutrient requirement of trees are met through internal cycling as compared with agricultural crops.

Nutrient cycling in forest ecosystems is controlled primarily by three key factors: climate, site, abiotic properties (topography, parent material), and biotic communities. The role of each factor in ecosystem nutrient dynamics is discussed and illustrated with selected examples from boreal, temperate, and tropical zones. The importance of ecosystem disturbance to nutrient cycling is examined briefly, since some nutrients are added or lost from forest ecosystems through natural (e.g., fire, erosion, leaching) or human activity (harvesting, fertilization).

INFLUENCE OF CLIMATE

Large-scale patterns in terrestrial primary productivity have been explained by climatic variables. In aboveground vegetation, nutrient storage generally increases in the order: boreal$<$temperate$<$tropical forests (Table 2). In contrast, forest floor nutrient content and residence time increases from tropical to boreal forests, as a result of slower decomposition in the cold conditions of higher latitudes.

In subarctic woodland soils and Alaskan taiga forests, nutrient cycling rates are low because of extreme environmental conditions (2). Arctic and subarctic forest ecosystems have lower rates of nutrient turnover and primary production because of low soil temperature, a short growing season, low net AET and the occurrence of permafrost. Low temperature reduces microbial activity, litter decomposition rates, and nutrient availability and increases C accumulation in soil.

In contrast with high latitudes, conditions in a tropical forest favor microbial activity throughout the year, which generally results in faster decomposition except in situations with periodic flooding, soil dessication, and low litter quality (3). Rates of plant material decay are an order of magnitude higher in tropical soils than in subarctic woodland soils. The low storage of C and high amount of litter production in highly productive tropical forests contrasts with the high C storage and low litter production in boreal forests (Table 2).

INFLUENCE OF BIOTIC FACTORS

Nutrient cycles are modified substantially by tree species-specific controls over resource use efficiency (nutrient use per unit net primary production). Species vary widely in their inherent nutrient requirements and use (4). These

Table 1 Features of the major nutrient cycles

Element	Uptake by the trees	Major sources for tree uptake	Limiting situations
Carbon	Atmosphere	Atmosphere	Atmospheric concentration may limit growth
Oxygen	Atmosphere	Atmosphere	Waterlogged soils
Hydrogen	Atmosphere	Atmosphere	Extremely acidic and alkaline conditions
Nitrogen	Soluble NO_3 and NH_4; N_2 for nitrogen fixing species	Soil organic matter; atmospheric N_2 for nitrogen fixing species	Most temperate forests, many boreal forests and some tropical forests
Phosphorus	Soluble phosphorus	Soil organic matter; adsorbed phosphate and mineral phosphorus	Old soils high in iron and aluminum, common in subtropical and tropical environment
Potassium calcium magnesium	Soluble K^+/soluble Ca^{2+}/ soluble Mg^{2+}	Soil organic matter; exchange complex and minerals	Miscellaneous situations and some old soils

effects can be split into two categories: accumulation into living phytomass and production of various types of nutrient-containing dead phytomass. Rapid accumulation of phytomass is associated with a net movement of nutrients from soil into vegetation. More than half of the annual nutrient uptake by a forest is typically returned to forest floor (litterfall) and soil (fine-root turnover). The subsequent recycling of these nutrients is a major source of available nutrients for forest vegetation. The mean annual litterfall from above-ground vegetation increases from boreal regions to the tropics following the gradient of productivity (Table 2).

Nutrient availability is strongly influenced by the quantity and quality of litter produced in a forest. A high proportion of the variation in foliar N concentrations at the continental scale has been explained by differences between forest types, which in turn has large impact on litter quality and the nutrient content of forest floors. In many temperate and boreal forest ecosystems, microbial requirement for N increases or decreases with labile

Table 2 Nutrient distribution in different forest ecosystems

	Vegetation ($Mg\,ha^{-1}$)	Forest floor ($Mg\,ha^{-1}$)	Soil ($Mg\,ha^{-1}$)	Residence time (year)
Carbon				
Boreal coniferous	78–93	37–113	41–207	800
Temperate deciduous	103–367	42–105	185–223	200
Tropical rain forest	332–359	7–72	2–188	120
Nitrogen				
Boreal coniferous	0.3–0.5	0.6–1.1	0.7–2.87	200
Temperate deciduous	0.1–1.2	0.2–1.0	2.0–9.45	6
Tropical rain forest	1.0–4.0	0.03–0.05	5.0–19.2	0.6
Phosphorus				
Boreal coniferous	0.033–0.060	0.075–0.15	0.04–1.06	300
Temperate deciduous	0.06–0.08	0.20–0.10	0.91–1.68	6
Tropical rain forest	0.2–0.3	0.001–0.005	0.06–7.2	0.6
Potassium				
Boreal coniferous	0.15–0.35	0.3–0.75	0.07–0.8	100
Temperate deciduous	0.3–0.6	0.050–0.15	0.01–38	1
Tropical rain forest	2.0–3.5	0.020–0.040	0.05–7.1	0.2

supplies of soil C. Increased microbial demand for N may temporarily decrease the N availability to trees during the initial decomposition of forest residues with a wide C/N ratio. Microbes immobilize N from the surrounding soil, relatively rapid for readily decomposable organic matter (needle litter), and more slowly for recalcitrant material (branches, boles).

Rates of net N mineralization are higher and retention of foliar N is lower in temperate and tropical than in boreal forest soils. Nitrogen limitation of productivity, therefore, is weak in tropical forests and increases from temperate to boreal and tundra forest systems. Trees may obtain organic N and P from the soil via mycorrhizae or by relocation from older foliage prior to abscission, and thereby, partly reduce their dependence on soil as a source of inorganic nutrients. Increased understanding of the fundamental relationships between soil properties and plant nutrients requirements will most likely come from examination of plant–fauna–microbe interactions at root surfaces (rhizosphere), rather than in the bulk soil.

INFLUENCE OF ABIOTIC FACTORS

Forests have distinctive physiographic, floristic, and edaphic characteristics that vary predictably across the landscape within a climatically homogeneous region. Differences in the elemental content of parent material influence the tree species composition between and within a landscape unit. For example, wind deposited soils, which support hardwood or mixed wood forest, are likely to be fine textured with high nutrient supplying capacity. In contrast, outwash sands that often support pine forests are coarse textured and infertile.

Heterogeneity within the landscape results in sites differing in microclimatic conditions, and physical and chemical properties, which produces different geochemical reaction rates and pools of available nutrients in soil. Soil type and topographic–microclimate interactions are important feedbacks that influence biological processes, such as the rate of N mineralization in soil. Low P availability is a characteristic of geomorphically old, highly weathered tropical, subtropical, and warm temperate soils (3). The type and age of parent material from which the soil is derived can influence the base status and nutrient levels in soil. Soils in glaciated regions are relatively young and rich in weatherable minerals. Mineral weathering is an important source of most nutrients for plant uptake, with the exception of N. Nutrient availability is regulated by the balance between weathering of soil minerals and precipitation, adsorbtion, and fixation reactions in soil.

Edaphic conditions can exert a strong influence on forest productivity and produce considerable variation in nutrient cycling processes. Soils with low N, P, or pH support trees with low litter quality (high in lignin and tannins that bind N) that decomposes slowly. Edaphic limitations on growth may be compensated for by an increase in rooting density and depth. Some late-succession or tolerant species have a shallower root distribution relative to intolerant pioneer species and are adapted to sites where nutrients and moisture are concentrated at the soil surface. In contrast, nutrient uptake from sub-soil horizons is more important in highly weathered warm temperate soils where nutrient depletion takes place deeper in the soil.

ROLE OF DISTURBANCE

Disturbances such as fire, harvesting, hurricanes, or pests affect nutrient cycling long after the event. In fire-dominated ecosystems, intensive wildfire results in a horizontal and vertical redistribution of ecosystem nutrients. Redistribution results from the combined effects of the following processes: 1) oxidation and volatilization of live and decomposing plant material; 2) convection of ash particles in fire generated winds; 3) water erosion of surface soils; and 4) percolation of solutes through and out of the soil. The relative importance of these processes varies with each nutrient and is modified by differences in fire intensity, soil characteristics, topography, and climatic patterns. Expressed as a percentage of the amount present in vegetation and litter before fire, the changes often follow the order:

$$N > K > Mg > Ca > P$$

Harvesting removes nutrients from the site and interrupts nutrient cycling temporarily. The recovery of the nutrient cycle from harvest disturbance is dependent partly on the rate of re-establishment of trees and competing vegetation. Re-vegetation may occur within months in the tropics, 2–5 years in temperate regions, and longer in boreal and tundra regions (5). Recovery assumes that the soil's ability to supply nutrients to plant roots has not been altered by disturbance. If nutrients cannot be supplied by the soil at rates sufficient to at least maintain the rate of growth of the previous forest then fertilization may be necessary to maintain site productivity.

Nutrient cycling and the impacts of disturbance on nutrient cycling, have been examined thoroughly in many representative world forests. The impact of natural disturbances and management practices on nutrient cycling processes are generally characterized of the stand or occasionally on a watershed basis. There is a growing demand from policy makers and forest managers for spatial estimates on nutrient cycling at local, regional, and national scales.

The availability of N, P, and K in soil largely determines the leaf area, photosynthetic rate, and net primary production of forest ecosystems. Forest management practices that produce physical and chemical changes in the soil that accentuate the cycle of nutrients between soil and trees, may increase forest productivity. Clear-cut harvesting and site preparation practices (mechanical disturbance, slash burning) remove nutrients from soil in tree components and by increased surface runoff, soil erosion, and off-site movement of nutrients in dissolved form or in sediment transport. In the tropics, potential negative impacts associated with complete forest removal and slash burning are greatest because a larger proportion of site nutrients are contained in the living biomass. Environmental impacts associated with clear-cutting and forest management in general, are confounded by climatic, topographic, soil, and vegetation diversity associated with the world's forests. Best forest management practices can be utilized to control negative impacts on nutrient cycling.

REFERENCES

1. *Glossary of Soil Science Terms*; Soil Science Society of America: Madison, WI, 1996; 134 pp.
2. Van Cleve, K.; Chapin, F.S., III; Dyrness, C.T., III; Viereck, L.A., Element Cycling in Taiga Forests: State Factor Control. Bioscience **1991**, *41*, 78–88.
3. Vitousek, P.M.; Stanford, R.L., Jr. Nutrient Cycling in Moist Tropical Forest. Ann. Rev. Ecol. Syst. **1986**, *17*, 137–167.
4. Cole, D.W.; Rapp, M. Elemental Cycling in Forest Ecosystems. In *Dynamic Properties of Forest Ecosystems*; Reichle, E.D., Ed.; Cambridge University Press: London, 1981; 341–409.
5. Keenan, R.J.; Kimmins, J.P. The Ecological Effects of Clear-Cutting. Environ. Rev. **1993**, *1*, 121–144.

FOREST ECOSYSTEMS, SOILS ASSOCIATED WITH MAJOR

F

Ken Van Rees

Department of Soil Science, University of Saskatchewan, Saskatoon, Saskatchewan, Canada

INTRODUCTION

Soil is defined as the unconsolidated mineral or organic material that occurs at the Earth's surface and is capable of supporting plant growth. Forest soils are those soils that have developed underneath forest vegetation (1). Forests cover approximately 29% of the world's land surface and play a key role in most ecosystems except for the tundra, deserts, some grasslands, and wetlands. Forests and forest soils contain about 60% of the carbon contained in the Earth's land surface and thus are an integral part in the global carbon cycle (2). Soil development or genesis beneath forest ecosystems is influenced by a number of different factors. These factors include the type and nature of the forest vegetation, parent material and topography, climate, human or organism influence, and the amount of time that these factors have been influencing soil development (3). All these factors combined together can result in soils with different physical, chemical, and biological properties that are unique to those conditions and provides the framework for how soils are classified. Generally there are two main classification systems used in the world: the U.S. Soil Taxonomic (4) and the FAO systems (5).

Soils types will vary around the world, but some generalizations can be made based on the major forest biomes that exist today. The major forest biomes include boreal and coniferous forests, temperate deciduous/mixed forests, scrub and woodlands, temperate rain forest, tropical rain forests, and tropical monsoon/deciduous forests. A generalized vegetation map of the world showing most of these forest biomes is provided in Fig. 1. Typical soil types associated with these forests are summarized in Table 1 and discussed below for each major forest biome.

BOREAL AND OTHER CONIFEROUS FORESTS

Coniferous forests, otherwise known as boreal forests are comprised mainly of spruce and pine trees with birch and poplar appearing at the southern range. These forests, located in the northern hemisphere, extend southward from the tundra treeline and occur across Canada, Europe, and Asia. The soils for these forests are generally acidic and form on sandy deposits from glaciation and range in thickness from >1 m to very shallow (<1 m) underlain by bedrock (precambium shield). These forests receive about 90 cm/year of precipitation or less and have low rates of evapotranspiration, which results in water moving through the soil profile (1). Because these forests have adapted to cooler climates, the decomposition of litter material is slow, resulting in thick forest floors compared to other forests as well as the accumulation of mosses and lichens. Soils of the boreal forest are relatively young compared to tropical soils because of glaciation; consequently, there are soils that have little to no development or no distinct soil horizons and are known as Entisols or Regosols. Sandy soils that have undergone more soil development, such as the development of a reddish colored B horizon, are called Inceptisols and are associated with pine forests growing on outwash plains. More soil development can result when organic acids that are released from slowly decomposing litter material bind with Fe and Al in the soil causing them to be leached deeper into the soil profile. This process is called podzolization and results in the development of a dark colored B or spodic horizon (the accumulation of Fe, Al, and organic matter), which belongs to the Spodosol or Podzol soil groups. Often these soil profiles have a bleached-looking (light grey) surface horizon (E or Ae horizon) underlain by a darker colored (reddish-brown) spodic horizon. Boreal forests also have extensive low-lying areas in the landscape that have high water tables resulting in organic soils known as Histisols or Organic soils. These organic layers can be thicker than 40 cm and in some cases are meters thick. These soils are generally unproductive for tree growth but are still important for wildlife and water quality.

TEMPERATE DECIDUOUS FORESTS

Temperate deciduous forests are found in North America, Europe, and China and are comprised of oaks, beech, ash, maple, basswood, and yellow-poplar trees. These forests

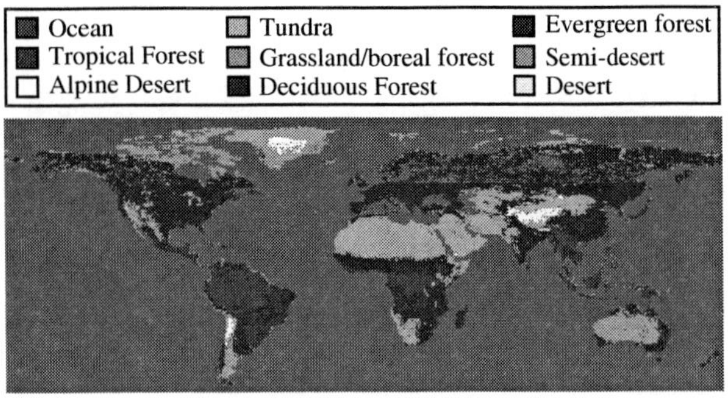

Fig. 1 World map of major vegetation types. (Modified from Ref. 7.)

occur on a wide diversity of parent materials, but the soils are generally mildly acidic. Spodosols are common under oak/birch forests in western Europe. In central Europe and China, forests have developed on more neutral, clayey Mollisols or Chernozemic soils, which are characterized by a surface mineral horizon containing high amounts of well-decomposed organic matter. This surface horizon enriched with organic matter makes these soils very productive for tree growth, and the accumulation of organic matter is believed to have originated from grassland conditions. In North America, deciduous forests have developed on Alfisols in the north and Ultisols in the south. Alfisols develop in cool to warm humid climates and are characterized by an accumulation of clay in the subsurface. This clay was eluviated or transported from upper horizons and deposited or illuviated in the B horizon to form an argillic horizon. These argillic horizons have more than 35% of their cation exchange sites saturated with bases such as calcium (Ca) and magnesium (Mg) (3). The degree of soil development for Alfisols is greater than that for Inceptisols but less than that for Spodosols. Ultisols are similar to Alfisols in that they have developed an argillic horizon; however, Ulitisols develop in more humid tropical areas with greater precipitation resulting in more weathering and thus $<35\%$ base cations with more acidic conditions compared to Alfisols. Subsurface horizons in Ultisols tend to be redder in color, indicating an accumulation of Fe oxides, and are less productive than Alfisols.

Mixed forests are transition forests between boreal and deciduous forests. Mixed forests of aspen/spruce or birch, maple, and oaks growing with conifers occur in the northern hemisphere. In the southern hemisphere in Chile, southern Brazil, South Africa, and New Zealand one can also find mixed forests of broad-leaved evergreens and conifers. These forests will occur on Mollisols, Inceptisols, and Alfisols.

TEMPERATE RAIN FORESTS

Temperate rain forests consisting of conifers are located on the northwest coast of North America and broadleaf evergreens in southern New Zealand, Chile, and southeastern Australia. These areas have milder, marine climates with relatively higher rainfalls. Spruce, cedar, hemlock, Douglas-fir, and redwoods are the dominant species in North America whereas beech and eucalyptus are found in the southern hemisphere. These forests have thick forest floors, and soil types range from Spodosols and Inceptisols to Alfisols and Ultisols. These forests are highly productive and produce rather tall and large trees.

SHRUB AND WOODLAND

These forests are limited by the amount of rainfall and are less productive than most forests in the world. They range from sclerophyllous communities with Mediterranean-type climates to xerophytic shrublands where droughts can persist throughout the year (6). The sclerophyllous communities consisting of evergreen oaks or Eucalyptus species are located in southwestern North America, Australia, Chile, South Africa, and Mediterranean regions.

Table 1 Soil types associated with the major forest biomes of the world

Major forest biomes	Area (ha $\times 10^8$)	Soil orders
Boreal and other coniferous forests	18.1	Spodosols, Histisols, Inceptisols, Entisols, Gelisols
Temperate deciduous forests	11	Spodosols, Alfisols, Mollisols, Ultisols
Temperate rain forests	3.2	Spodosols, Alfisols, Inceptisols, Ultisols
Shrub and woodland	8.5	Aridisols, Mollisols, Alfisols
Tropical rain forests	17.4	Ultisols, Inceptisols, Oxisols
Tropical monsoon/deciduous forests	4.6	Ulitisols, Oxisols

(From Refs. 2, 6, and 8.)

The xerophytic shrublands (oaks, pinyon, and ponderosa pines) are located in the interior of the southwestern United States, central Australia, Eurasia, southern Sahara, and the western regions of the Andes in central and southern South America. Soils that have developed under these conditions range from Mollisols and Alfisols to Aridisols. Aridisols, also known as dry soils, are common when evapotranspiration exceeds precipitation inputs, resulting in accumulations of calcium carbonate, gypsum, soluble salts, or exchangeable sodium in the soil profile. The accumulation of soluble salts makes these soils more akaline and are termed saline soils; however, if there are high concentrations of sodium on the exchange sites, these soils are classified as sodic soils, which can disperse organic matter and clay resulting in a black surface crust (3).

TROPICAL RAIN FORESTS

There are three distinct tropical rain forests in the world—the American, African, and Indo-Malaysian (1). These evergreen forests have high air temperatures and humidity, resulting in rapid decomposition of the forest floor, which is the opposite of that seen for boreal forests. Higher precipitation in these areas results in highly weathered soils, with soil development occurring deeper in the soil profile. The range of soil types is quite variable under tropical forests, and they have been studied less than temperate forests. Soils range from Inceptisols and Ultisols to Oxisols and Andisols. Oxisols are the oldest and hence most weathered soils, with intense leaching that removes a large part of the silica from silicate materials, resulting in more 1:1 type silicate clays or nonexpanding clays. The subsurface horizon characteristic of Oxisols is the oxic horizon, which is very high in clay particles dominated with oxides of Fe and Al (3).

TROPICAL MONSOON/DECIDUOUS FORESTS

Tropical monsoon forests experience significant seasonal droughts (5 months with <9 cm of rain) and are dominated by seasonal deciduous trees such as teak. These forests are found in Central and South America, Indo-Malaysia, and Africa, with the largest area of tropical deciduous forests located in southeastern Asia. The soils are not as highly weathered as tropical rain forests but consist of soils from the Ultisols and Oxisols orders.

SUMMARY

Soils vary around the world and reflect the types of vegetation and other soil-forming factors/processes that have been acting upon them. Soils are much younger and less developed in boreal forests compared to those soils associated with tropical forests. Regardless of the degree of soil development, forest soils will continue to play an important role in ecosystem function.

REFERENCES

1. Fisher, R.F.; Binkley, D. *Ecology and Management of Forest Soils*; John Wiley & Sons, Inc.: New York, 2000; 489.
2. Landsberg, J.J.; Gower, S.T. *Applications of Physiological Ecology to Forest Management*; Academic Press: San Diego, 1997; 354.
3. Brady, N.C.; Weil, R.R. *The Nature and Properties of Soils*; Prentice Hall: Upper Saddle River, New Jersey, 1996; 740.
4. United States Department of Agriculture. *Keys to Soil Taxonomy*, Eighth Edition; Natural Resources Conservation Service; U.S. Government Printing Office: Washington, D.C., 1998; 326.
5. Food and Agriculture Organization of the United Nations. *FAO-UNESCO Soil Map of the World*; Vol. 1, Legend. UNESCO, Paris, 1974; 59.
6. Pritchett, W.L.; Fisher, R.F. *Properties and Management of Forest Soils*; John Wiley & Sons, Inc.: New York, 1987; 494.
7. Murai, S.; Honda, Y.; Asakura, K.; Goto, S. *An Analysis of Global Environment by Satellite Remote Sensing, Institute of Industrial Science*; (http://www-cger.nies.go.jp/griddata/grid22.gif accessed June 2001) University of Tokyo: Tokyo, Japan, 1990.
8. Waring, R.H.; Schlesinger, W.H. *Forest Ecosystems. Concepts and Management*; Academic Press: San Diego, CA, 1985; 340.

F

FOREST SOILS

N.B. Comerford
Soil and Water Science Department, University of Florida, Gainesville, Florida, U.S.A.

INTRODUCTION

Forest soil science is a well-established field of study, having been a concern in Europe as early as the late 1800s. This is best illustrated by both the number of textbooks and the dates of the texts published in this area (Table 1). Other soil scientists have asked: "If there is a subdiscipline called forest soils, why not also study corn or tomato soils?" The justification for forest soil science is twofold. First, a long-lived forest cover imposes a unique set of characteristics on the soil in which it grows. Second, and probably of equal importance, is that the soil characteristics and processes important to the forest soil scientist are not of great concern to cropland soil scientists for whom the biological and economic time frames of food production are relatively short (Table 2). In order to understand "Why study forest soils?" it is useful to first define the science and then discuss the soil/ecosystem characteristics that define the uniqueness of a forest soil.

The Soil Science Society of America defines soil science as "the science dealing with soils as a natural resource on the surface of the earth including soil formation, classification and mapping, physical, chemical, biological, and fertility properties of soils per se; and these properties in relation to the use and management of soils" (1).

While the Soil Science Society of America does not define forest soil science, the following definition is now offered. Forest soil science is "the science dealing with soils as a natural resource on the earth's surface which supports a forest cover. It includes how the forest cover influences, and is influenced by, soil formation and soil physical, chemical, and biological properties. It addresses how soil properties affect the sustainable use of the soil for the production of multipurpose forest outputs including, but not limited to, wood and pulp, wildlife, water, and the maintenance of environmental quality."

WHY FOREST SOILS?

By the above definition, a forest soil is one that developed under, and is currently supporting, a forest cover. Realistically, forest soils today include soils that have been continually developing under a native forest cover, soils that are supporting a forest cover different from the native cover, and soils that were once in cropland and have been converted back to a forest. Therefore, the chemical and physical properties of a forest soil are determined by the conditions under which it was formed as well as its more recent historical use.

Steep slopes, stoniness, inherent low fertility, remoteness from markets, excessive or limited moisture, land ownership patterns, and environmental policy are all reasons why agriculture has not claimed these soils. Forest soils can be highly variable; fertile or infertile; occur on level or steep slopes; composed of marine sands or too stony for a plow. What they all have in common is that they support a forest ecosystem, managed or unmanaged.

Historically, forest soil management encompassed a range of intensities depending on the product to be produced and the economic value of that product (2) (Fig. 1). The intensity of management represents a significant difference in perspective between forest soil scientists and cropland soil scientists. At its most intensive, the time between investment and harvest is about 7 years. At its least, the forest is preserved with no intent to harvest wood products. The relatively long management time frame means that soil processes and properties, which have minimal importance in cropland systems, can be controlling factors in forest soils (Table 2).

A closed, multiyear nutrient cycle is the most significant process that separates a forest soil from an agricultural soil. The time between management interventions, if any, allows organic soil horizons (forest floor) to form above the mineral soil surface. The degree of organic matter incorporation into the surface mineral soil will produce a range of forest floor types. These organic horizons not only influence root distribution, but are also zones for rapid cycling of some nutrients, such as K, and immobilization or metered release of others, such as N and P.

Because of the perennial nature of forest cover, and the low intervention during stand development, roots are capable of growing, surviving and functioning at depth, allowing the forest stand to define a greater soil depth and to exploit water resources that short-lived agricultural

Table 1 Textbooks in forest soils

1893	Ramann, E. *Forstliche Bodenkunde und Standortslehre*; Julius Springer: Berlin.
1908	Henry, E. *Les Sols Forestiers*; Berger-Levrault & Co.: Paris
1946	Lutz, H.J.; Chandler, R.F., Jr. *Forest Soils*; John Wiley & Sons: New York.
1946	Wilde, S.A. *Forest Soils and Forest Growth*; Chronica Botanica Co.
1950	Tamm, Olof. *Northern Coniferous Forest Soils*; The Scrivener Press: Oxford
1951	Hartmann, F. *Der Waldboden: Humus-, Boden- und Wurzeltypen als Standortsanzeiger*. \sterreichisches Produktivit-ts-Zentrum. Wein, Germany
1958	Wilde, S.A. *Forest Soils: Their Properties and Relation to Silviculture*; Ronald Press: New York.
1969	Remezov, N.P.; Pogrebnyak, P.S. *Forest Soil Science*; Israel Program for Scientific Translations, Jerusalem
1971	Stefanovits, P. *Brown Forest Soils of Hungary*; Akadϑmiai Kiad; Budapest, Hungary
1977	Armson, K.A. *Forest Soils: Properties and Processes*; University of Toronto: Toronto.
1979	Pritchett, W.L. *Properties and Management of Forest Soils*; Wiley: New York
1986	Binkley, D. *Forest Nutrition Management*; Wiley: New York.
1987	Pritchett, W.L.; Fisher, R.F. *Properties and Management of Forest Soils*, 2nd Ed.; Wiley: New York.
1987	Attiwill, P.M. *Forest Soils and Nutrient Cycles*; Melbourne University Press: Carlton, Victoria.
1987	Salas, Gonzalo de las. *Suelos y ecosistemas forestales: con enfasis em America tropical*; Instituto Interamericano de Cooperacion para la Agricultura: San Jose, Costa Rica.
2000	Fisher, R.F.; Binkley, D.; Pritchett, W.L. *Ecology and Management of Forest Soils*; Wiley and Sons: New York.

crops might not find. Where cropland is most commonly defined by monocultures and fertilization to control soil fertility, forest soils generally support a diverse plant community where groups of species may have their own strategy for using the soil's resources.

Wood products generally have low economic value compared to agricultural crops. Therefore, low intensity management is justified in order to gain an economic return on the investment. In some circumstances, wood products are not the primary goal of management, which might be water yield and quality, wildlife, or maintenance of biodiversity. The latter objectives require understanding of how the entire forest ecosystem functions as well as of the multitude of consequences (on-site and off-site) that might come from human intervention.

The forest soil scientist is interested in deep rooting, nutrient cycling, mineralization, the plant's ability to thrive under native soil fertility, soil properties that influence both the occurrence and productive potential of a forest stand, soil erosion following forest harvesting, and the influence of natural inputs and pollutants on forest ecosystems. Many of these are of little concern to cropland soil scientists.

With the advent of high intensity management strategies, such as eucalypts in South America, modeling of soil/plant processes, and a more ecosystematic approach to agriculture, the uniqueness of forest soils is not as strong as it once was. As forest management becomes more intensive, forest soil management moves toward the time scale of other cropland ventures and the uniqueness separating forest soil science from other areas of soil

Table 2 Contrast of soil properties, processes, and management strategies between forest soils and cropland soils. (Management contrasts are for intensive forest management versus intensive cropland management)

Property/process/management	Forest soil	Cropland soil
Time between harvests	7 yr to indefinite	Annual to biannual
Nutrient cycling	Closed	Open
Erosion	Very low	High
Topography	Level to very steep	Level to moderate
Fertilization practices	Low amounts applied one to several times per rotation	Annual high amounts
Irrigation	Not employed	Employed
Soil temperature	Buffered by a forest canopy	More variable
Utilized soil depth	Deeper	Shallower
Stoniness	None to high	Low
Remoteness from markets	Close to remote	Close
Surface soil organic horizons	Present	Absent

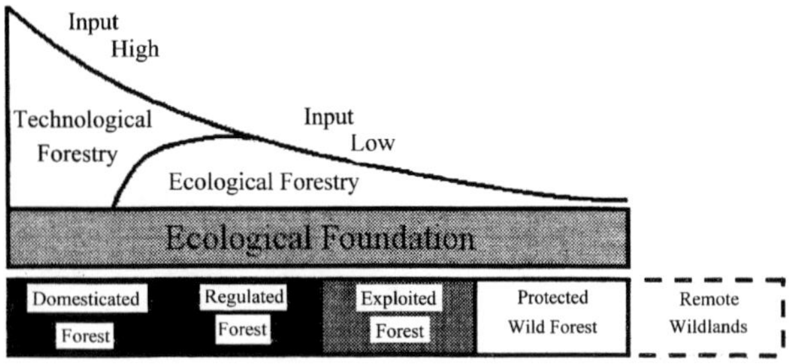

Fig. 1 The range of management intensity for a forest soil and forest stand. (Adapted from Ref. 2.)

science becomes less clear. However, that should not undermine the importance of the past discoveries made in forest soil science or diminish the importance of future work where forest management remains different from agriculture in time scale and management intensity.

CONTRIBUTIONS OF FOREST SOILS TO SOIL SCIENCE

Like many scientific fields, the discipline is changing with changing societal needs. Yet the accomplishments in forest soil science during the 1900s testifies to its importance in ecology, forest management, and forest conservation. The following is not meant to be a compilation of research in the area, but is offered to show this field's breadth and depth.

Forest Ecosystem Nutrient Cycling and Long-Term Productivity Decline

Evaluating and developing forest management strategies based on nutrient cycling has been a collaborative effort of ecologists, silviculturalists, tree physiologists, and forest soil scientists. Nutrient cycling is often the basis for both soil management and forest harvesting schemes. A problem that constantly haunts forest managers is whether their harvesting regimes allow for sustainable forest productivity (3). Defining the soil's role in nutrient cycling as related to mineralization, exchange reactions, water regime, and root depth has been crucial in defining a site's ability to maintain sustainable forest growth.

Soil Development and Soil-Site Relationships

One of the five soil forming factors is the influence of vegetation. The influence of different species on soil formation is well known, as is the influence of the soil on forest ecosystems. This is best seen in the surface 20–40 cm of mineral soil. The study of forest soils has defined soil changes that occur under different types of forest vegetation. The early interest in the type of forest floor [mor vs. mull (4)] is indicative of this type of influence. The ability of low-nutrient-demanding conifers to ameliorate a degraded soil and make it more productive for nutrient-demanding hardwood species has been well established (5). Podzolization has been linked with certain types of conifers that provide the acidic litter and soluble organic compounds that are responsible for sesquioxide movement in soils as well as soil acidification.

Since forest ecosystems are related to the soils on which they grow, species have been matched to sites that meet their autecological requirements. Likewise, the productivity of the same species will change with changes in soil characteristics. Soil-site studies have done little to resolve these relationships, most likely due to the highly empirical nature of the studies (6). However, such work provides an appreciation of the range of site characteristics of economically important forest species, and provides clues as to why their growth patterns vary on different soils.

Fertility of Forest Soils

Soil fertility is probably the most studied topic in forest soil science in recent years. Evaluation of nutrient deficiencies and testing of fertilization practices to correct deficiencies have provided productive forest management strategies throughout the world. Given the fact that many soils are not in cropland because of their inherent low fertility, commercially productive forests require nutrient management strategies that will allow large growth responses to minimal fertilizer inputs. Models such as NUTRICALC (7) are combining our knowledge of nutrient cycling with soil chemical properties to evaluate nutrient deficiencies and recommend fertilization regimes. More process-oriented models such at TREGRO (8) and ECOSYS (9) have the potential to do the same thing, but at a more mechanistic level.

Off-Site Effects of Forest Management

Probably the most controversial environmental debates of the late 1900s in which forest soils played a role concerned the off-site effects of forest harvesting on water quality. Studies at the Hubbard Brook Experimental Forest (10) forced scientists to evaluate whether harvesting could be responsible for serious nutrient losses and detrimental off-site effects. These articles set off a flurry of soil research on mineralization rates, water flow in forest soils, leaching losses through forest soils, and soil recovery from degradation. The stimulation was healthy in that it produced a compendium of information, from a wide range of landscapes, showing how soil treatments influence water quality.

The influence of acid deposition, both in the U.S. and Europe, stimulated another round of research on forest soils. Soil's natural buffering against acidification, leaching of aluminum from soils, nitrate loading of soils by pollution, and amelioration practices to combat the effect of acid deposition were all topics of interest that stimulated research and provided a better understanding of the soil resource.

CONCLUSION

In conclusion, forest soil science is a crucial discipline in the management of our natural forest resources. Tall trees use the soil for physical support, as well as a source of water and nutrients, so we will always need to understand all processes involved. As our information becomes more mechanistic, we should find it harder to separate the study of forest soil science from cropland systems because the only real difference is the dominance of different sets of processes during a different management time scale. Since our forests are considered a source of wood and pulp products as well as water, wildlife, and recreation, forest soil science will continue to contribute to both the knowledge base and the management of these natural, renewable resources.

REFERENCES

1. Soil Science Society of America, *Glossary of Soil Science Terms*; Soil Science Society of America: Madison, WI, 1996.
2. Stone, E.L. Soils and Man's Use of Forest Land. In *Forest Soils and Forest Land Management*; Bernier, B., Winget, C.H., Eds.; Laval University Press: Quebec, 1975; 1–9.
3. Powers, R.F. On the Sustainability of Planted Forests. New Forests **1999**, *17*, 263–306.
4. Tamm, O. *Northern Coniferous Forest Soils*; The Scrivener Press: Oxford, UK, 1950.
5. Stone, E.L. Effects of Species on Nutrient Cycles and Soil Change. Philosophical Transactions of the Royal Society of London. B. **1975**, *271*, 149–162.
6. Carmean, W.H. Forest Site Quality Evaluation in the United States. Advance. Agron. **1975**, *27*, 209–269.
7. Barros, N.F.; Novais, R.F.; Teixeira, J.L.; Fernandes Filho, E.I. NUTRICALC: Sistema Para Calculo del Balance Nutricional Y Recomendación de Fertilizantes Para El Cultivo de Eucalipto. Bosque, (Chile) **1995**, *16*, 129–131.
8. Weinstein, D.A.; Yanai, R.D. Integrating the Effects of Simultaneous Multiple Stresses on Plants Using the Simulation-Model TREGRO. Journal of Environ. Quality **1994**, *23*, 418–428.
9. Grant, R.F. ECOSYS. *Global Change and Terrestrial Ecosystems Focus 3 Wheat Network. Model and Experimental Meta Data*; NERC Centre for Ecology and Hydrology: Oxon, UK, 1993; 13–23.
10. Bormann, F.H.; Likens, G.E.; Fisher, D.W.; Pierce, R.S. Nutrient Loss Accelerated by Clear-Cutting of a Forest Ecosystem. In *Symposium on Primary Productivity and Mineral Cycling in Natural Ecosystems*; Young, H.D., Ed.; University of Maine Press: Orono, ME, 1968; 187–193.

FOREST SOIL PROPERTIES AND SITE PRODUCTIVITY

Paul A. Arp
Helmut H. Krause
University of New Brunswick, Fredericton, New Brunswick, Canada

INTRODUCTION

Site productivity is a key variable in forest management and yield: it depends on site quality—the inherent capability of the site to produce—and on management input (1). Site quality—in turn—is largely determined by soil, topography, and climate. Maximum production occurs when: macronutrients (N, P, K, Ca, Mg, S) and micronutrients (e.g., Zn, B, Cu) are consumed in certain amounts and proportions to each other for every unit of biomass produced, soil moisture and nutrients are available in sufficient quantities, and an extensive root system exists for efficient water and nutrient uptake. Related soil and site research throughout the world (e.g., (2)) has, inter alia, produced simple guidelines for interpretations of soil survey data, and complex site classification systems based on climate, geology and soil properties (3).

In practice, the site index (SI = average height of dominant trees in a stand at a certain reference age, e.g., 50 years) is a local measure of site quality. This index is presumed to integrate the effects of climate, topography and soil. Unfortunately, SI also varies with stand history, stocking level, and species composition. Soil properties that affect site quality, and, therefore, forest growth and ecosystem functioning, are highlighted in Table 1. Examples of soils with contrasting productivity are shown in Fig. 1.

INFLUENCE OF SOIL PHYSICAL PROPERTIES

Highest site productivity generally occurs on deep, friable, and organically enriched soils with well-functioning root systems. In other soils, root growth is often restricted by, e.g., mechanical impedance, lack of aeration, low temperature, and/or the presence of toxic substances. Aside from shallow bedrock, roots encounter mechanical impedance in soils with high soil strength. High soil strength arises naturally in hardpans and fragipans, beneath the plough layer on old fields, and on soils that may have been compacted by logging equipment. In such cases, tree growth generally decreases with decreasing depth of root-permeable soil. In mineral soils, roots usually encounter mechanical impedance at bulk density values of 1.3–1.75 Mg m^{-3}, depending mainly on texture.

SOIL MOISTURE

A crucial requirement to maximize site utilization is the soil's ability to store water to compensate for precipitation deficits during periods of high evapotranspiration (4). For example, annual basal area increments of southern pines were shown to increase directly with number of days of adequate soil water quality. Available water is customarily estimated as the difference between the field capacity (FC) and the permanent wilting point (PWP) of the root-accessible portion of each soil. These parameters—in turn—are primarily determined by soil texture and organic matter.

SOIL AERATION

Good air supply is essential for good root growth (5). Insufficient aeration occurs in soils with high water tables, and in fine-textured, structureless soils that are water-saturated throughout most of the growing season. Insufficient aeration is usually indicated by gleyed and mottled soil layers. Most tree roots do not penetrate such soil layers. However, some species are tolerant of low soil aeration. For example, Douglas fir does not tolerate low soil aeration, but red alder, western cedar, sitka spruce, and western hemlock are adapted to shallow water tables (1). Controlled drainage to remove excess water often improves productivity and postdisturbance tree regeneration, especially when the roots continue to be constantly supplied with water by way of capillary rise from the water table below, as is often the case for sandy soils (6).

Encyclopedia of Soil Science

F

Table 1 Soil properties related to site quality

Soil property	Relation to site quality	Optimal conditions
Mineralogical composition of parent material	Affects K, Mg, and Ca supply	Abundance of dark (mafic) minerals
pH	Affects nutrient availability	5–7
SOM content	Source of N and P, promotes structure	Presence of forest floor, accumulations in A and B horizons
Texture	Affects drainage, water, and nutrient storage	Sandy loam, loam, silt loam[a]
Structure	Promotes aeration, counteracts compaction, and high soil strength	Granular[a]
Consistency	Affects soil strength	Friable[a]
Drainage	Determines soil moisture regime and aeration status	Well and moderately well drained[a]
Depth of A horizon	Location of high biological activity and nutrient availability	>10 cm
Stone content	Improves thermal regime, limits soil volume for rooting	<15%
Depth to least permeable horizon	Delimits rooting space, improves moisture regime in coarse soils	>80 cm
Depth to mottling	Reveals presence of excess water and lack of aeration	>75 cm
Thickness of B horizon	Clay, SOM, and nutrient enriched zone	>30 cm

[a]Classes as defined in Agriculture Canada Expert Committee on Soil Survey, 1987. The Canadian System of Soil Classification.

On fine-textured soils, however, this benefit might not occur, because the same capillary rise may reduce soil aeration.

SOIL TEMPERATURE

Soil temperature is an important parameter because it influences root growth, and microbial activities, which—in turn—influence SOM decomposition and nutrient availability. Literature values for minimal (0–7°) and optimal (10–25°C) soil temperatures vary over wide ranges, thereby reflecting the climatic adaptation of species. Soil temperatures generally decrease with increasing elevation and latitude, and are generally lower under coniferous forests than deciduous forests: dense conifer canopies intercept much of the incoming solar radiation, thereby reducing the solar heating of the soils underneath, especially of wet soils. Delayed soil heating often implies delayed root growth, thereby shortening the effective growing season and hence forest growth. Fine roots are also sensitive to soil acidification (7): slowly soluble soil Al may convert to toxicologically active Al as the forest soil acidifies either naturally during the course of their development, or on account of acid deposition.

INFLUENCE OF NUTRIENT SUPPLIES AND AVAILABILITIES

Matters dealing with nutrient supplies, retention and availability are of particular importance to sustainable forest production, as follows:

- Long-term nutrient supplies of forest ecosystems need to be sustainable in principle. Additions of nutrients to each site come from external sources (atmospheric accretions, liming, and fertilizer), and from internal sources (weathering of primary soil minerals, biochemical cycling). Under natural conditions, origin and type of soil parent material are strong determinants of rate of weathering and related nutrient supplies.
- Nutrient retention and release involves numerous reactions, including ion sorption and desorption on mineral and organic colloids; precipitation by metal oxides; complexation by humic substances; immobilization of nutrient ions by microorganisms and SOM mineralization. All these processes are influenced by soil texture, SOM content, cation exchange capacity (CEC), pH, base saturation, aeration, and prevailing soil temperature and moisture regimes.

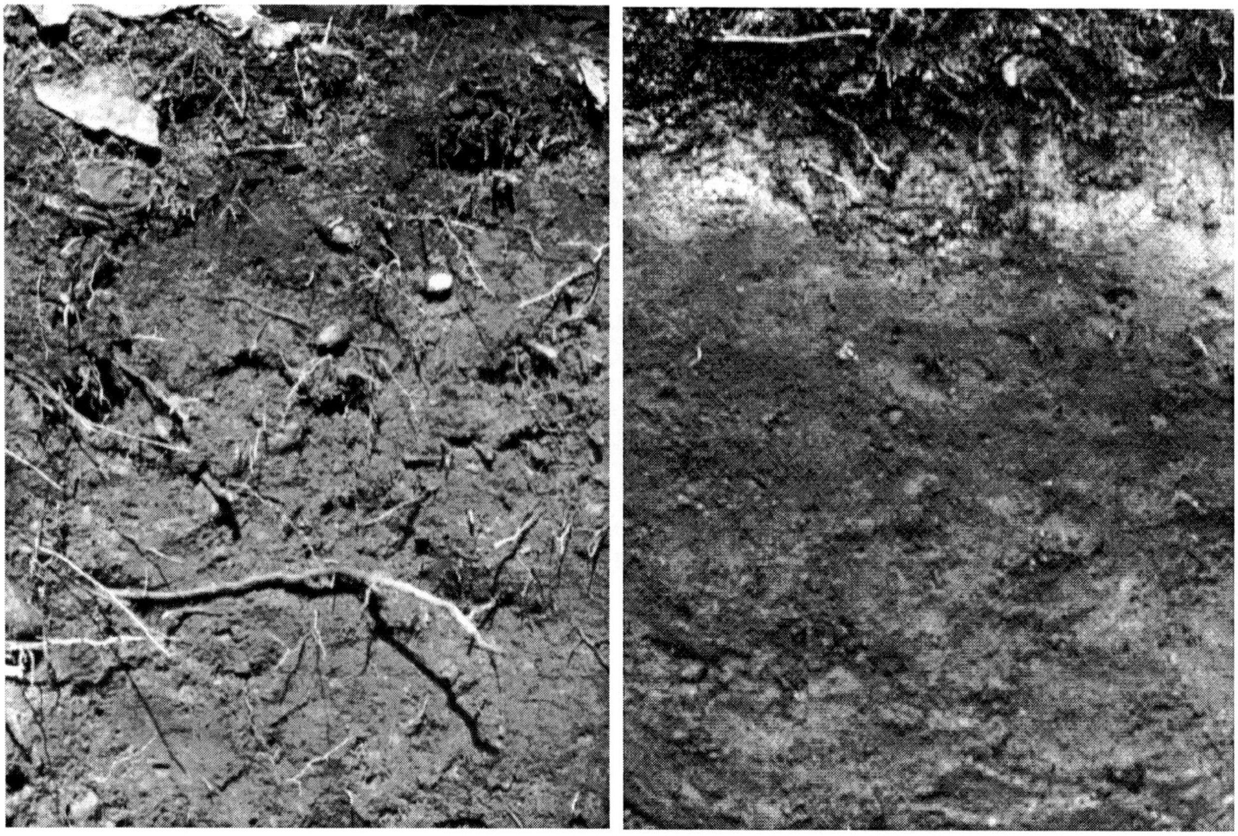

Fig. 1 Soils with contrasting productivity in the Atlantic Region of Northeastern North America. Left: brunisolic podsol supporting vigorous forest of northern tolerant hardwoods. Right: gleyed podsol supporting a pure black spruce stand. Note uniformly brown color, structure (granular to subangular blocky), and abundance of roots throughtout.

- Nutrient availability results from the net cumulative difference between supplies, immobilization, leaching, mineralization, and past uptake, and is also controlled by chemical equilibria. Potential fertility of a soil can provisionally be appraised from simple field observations. For example, favorable contents and distribution of SOM and clay imply improved N, P and base cation supply, and the presence of black minerals in the parent material may indicate K and Mg sufficiency. To be reliable, such observations must be supplemented by appropriate soil analysis.

The effect of soil parent material, via the continuing weathering of primary minerals, is often reflected in the species composition of natural forests. For example, in central regions of eastern North America, soils derived from calcareous shales often support, e.g., basswood, white ash, yellow-poplar and hickory, while soils derived from acid sedimentary rocks support high percentages of beech, yellow birch, red maple and certain oaks. As well, white cedar and eastern red cedar grow better on calcareous soils than on acid soils. In Sweden, site productivity is grouped by calcium content of the soil parent material: soils derived from calcium-poor substrates commonly support Scots pine at low productivity; soils with intermediate calcium availability support pine and mixed conifer forests with high productivity; soils derived from basic igneous sedimentary substrates support productive stands of Norway spruce and hardwoods.

CA, MG, K, P AND TRACE ELEMENTS

In areas where soils are very old and deeply weathered (e.g., tropical and subtropical regions in Africa, Asia, Australia, and Central and South America), or where soils have been heavily cropped, forest productivity is often limited because of limited supplies of Ca, Mg, K, P and/or micronutrients such as Zn, Cu, and B (8). Such soils often consist of abundant accumulation of Al and Fe sesquioxides, which typically produce soils with high anion exchange capacities

(AEC), but low CEC. As a result, nutrient elements such as K, Mg and Ca are generally in short supply, and severe losses of forest productivity are likely to occur when such soils are stripped of their natural forest vegetation. Such soils, furthermore, may loose their originally friable consistency with gradual SOM loss and become cementation when exposed to air and allowed to dry.

Areas that have been covered by glaciers in the recent past and currently have a temperate to boreal climate can support productive forests set within the limits set by climate and soil depth. This can be attributed to the continuing release of Ca, Mg, K and P through natural weathering of the ground-up bedrock (till). Soils in these areas generally have low accumulations of Fe and Al sesquioxides, except in areas of high soil acidity where forest productivity may decrease as a result of advanced podsolization, i.e., the formation of Fe- and organic matter cemented hardpans within the B horizon. Except for N, mineral deficiencies are infrequently encountered in these areas.

N AVAILABILITY

In boreal and sub-boreal climates, forest productivity is commonly restricted by low N availability, which is the result of slow SOM turn-over rates. Under these conditions, the forest floor is the principal source of available N: the A horizon is—more often than not—strongly leached, poorly developed or absent, and N in the B layers is essentially unavailable. In these soils, N availability is further adversely affected by high soil acidity, impeded drainage, and restricted soil aeration. In contrast, soils with mesic or warmer temperature regimes have SOM-enriched A layers with high levels of available N. As shown in many studies (2), site quality on these soils rises with increasing depth of the A layer.

SUMMARY

Forest growth restrictions occur on soils with nutrient-poor parent materials, inefficient water and nutrient storage and retrieval capacities, unfavorable pH, low SOM contents, and extreme texture (sand, clay). Some of these soil restrictions can—in part—be corrected by silvicultural means such as change of forest cover type, control of stand density, fertilization, irrigation, and artificial drainage. All of these growth restrictions have been considered in recent efforts to develop soil quality indices for monitoring sustainability of current forest management practices (9). Additional efforts are being made to quantify the relationships between soil and forest production by way of modeling (10, 11).

REFERENCES

1. Pritchett, L.W.; Fisher, F.R. *Properties and Management of Forest Soils*, 2nd Ed.; Wiley: New York, 1987; 494.
2. Carmean, W.H. Forest Site Quality Evaluation in the United States. Adv. Agron. **1975**, *27*, 209–269.
3. Klinka, K.; van der Horst, W.D.; Nuszdorfer, F.C.; Harding, R.G. An Ecosystematic Approach to Forest Planning. For. Chron. **1980**, *56*, 97–103.
4. Zahner, R. Water Deficit and Growth of Trees. In *Water Deficits and Plant Growth*; Kozlowski, T.T., Ed.; Academic Press: New York, 1968; 191–254.
5. Drew, M.C. Soil Aeration and Plant Root Metabolism. Soil Sci. **1992**, *154*, 259–268.
6. Wilde, S.A.; Iyer, J.G.; Tanzer, C.; Trautman, W.L.; Watterston, K.G. *Growth of Wisconsin Coniferous Plantations*; Research Bulletin 262, University of Wisconsin: Madison, WI, 1965; 80.
7. Reuss, J.O.; Walthall, P.M.; Roswall, E.C.; Hopper, R.W.E. Aluminum Solubility, Calcium–Aluminum Exchange, and pH in Acid Forest Soils. Soil Sci. Soc. Am. J. **1990**, *54*, 374–380.
8. Stone, E.L. Microelement Nutrition of Forest Trees: A Review. *Forest fertilization—Theory and Practice*; Tennessee Valley Authority: Knoxville, TN, 1968; 132–175.
9. Burger, J.A.; Kelting, D.L. Using Soil Quality Indicators to Assess Forest Stand Management. For. Ecol. Mgmt **1999**, *122*, 155–166.
10. Kimmins, J.P.; Scoullar, K.A. *FORCYTE 10*; Faculty of Forestry, University of British Columbia: Vancouver, BC, 1983; 112.
11. Bhatti, J.S.; Foster, N.W.; Oja, T.; Moayeri, M.H.; Arp, P.A. Modelling Potentially Sustainable Biomass Productivity in Jack Pine Forest Stands. Can. J. Soil. Sci. **1998**, *78*, 105–113.

F

FUNGI

Serita Frey
The Ohio State University, Columbus, Ohio, U.S.A.

INTRODUCTION

Soil fungi are a highly diverse group of microscopic, eukaryotic organisms belonging to the Kingdom *Fungi* (*Mycota*). While some are single-celled (e.g., yeast), most are multicellular with a mycelial morphology comprising a network of tube-like strands (hyphae). They are indeterminate organisms, growing and extending indefinitely as long as sufficient resources are available to sustain their growth (1). The mycelial growth form is well suited to the heterogenous soil environment where nutrient resources are spatially separated over long distances at the microbial scale. Fungi are heterotrophs, obtaining carbon, nutrients, and energy through the extracellular degradation and absorption of organic matter from their external environment. Oxygen is generally required for growth. Fungi are an integral part of the soil biotic community, significantly contributing to the decomposition of organic matter, the release and turnover of nutrients, the formation and maintenance of soil structure, the extension of plant root systems via the formation of mycorrhizal networks, and the promotion and suppression of plant disease (2).

FUNGAL CLASSIFICATION

Fungi obtained kingdom status in 1969 when Whittaker (3) proposed a five-kingdom classification system that recognized five distinct organism groups: *Plantae*, *Animalia*, *Fungi*, *Monera* (bacteria and cyanobacteria), and *Protoctista* (algae, slime molds, protozoa). This classification scheme has been debated for the past two decades, largely due to advances in molecular biology which allow the direct analysis of genetic material (i.e., RNA and DNA). These molecular analyses have greatly enhanced biologists' ability to discern evolutionary relationships and have provided extraordinary insight into the wealth of biological diversity present on Earth. Several new classification schemes containing a greater number of kingdoms have been proposed to adequately represent this diversity (4, 5). Fungi are currently assigned to four subdivisions within Kingdom *Fungi*: *Basidiomycotina*, *Ascomycotina*, *Zygomycotina*, and *Mastigomycotina* (6).

Classification within these groups is based on the characteristics of both vegetative and reproductive structures. The oomycetes, an important group of organisms that include several highly destructive plant pathogens (*e.g.*, *Pythium* and *Phytophthora* species), were traditionally classified as fungi within the *Mastigomycotina*, but are now generally considered to represent a separate evolutionary line from the "true fungi" and are placed with the brown algae in the Kingdom *Chromista* (7).

GROWTH AND REPRODUCTION

Fungi—often incorrectly labeled as plants—superficially resemble plants in that they possess cell walls, are generally non-motile, and reproduce by microscopic, seed-like structures (i.e., spores). However, fungi differ from plants in several key features, including being heterotrophic and having cell walls composed mainly of chitin rather than cellulose. Heterotrophy results in fungi obtaining nutrients and energy by producing and releasing extracellular enzymes which break complex organic compounds down into simpler constituents (e.g., sugars, amino acids) which are then absorbed back through the fungal cell wall and plasmalemma. Recent molecular evidence suggests that fungi are, in fact, more closely related to animals than plants (7, 8).

The vegetative structure of most fungi consists of a filamentous network of hyphae that extends outward by growing at the hyphal tips. Interconnected hyphae are collectively referred to as the *fungal mycelium* or *thallus*. Hyphae may be branched or unbranched and may or may not contain septa, the rigid but porous crosswalls that provide structural support and regulate the flow of cytoplasm. Fungi reproduce either sexually or asexually by producing microscopic propagules called *spores* (2). Spores may be produced singly or in visible fruiting bodies (e.g., mushrooms). In addition to serving as dispersal agents, spores also provide an important survival mechanism, often resistant to low temperatures, low nutrient availability, desiccation, UV irradiation, and other adverse environmental conditions (2, 6). Dormant spores may remain viable for long periods of time, germinating when conditions again become favorable for vegetative growth (2).

ABUNDANCE AND DIVERSITY

Fungi, with bacteria, comprise the majority of the total living biomass of soil biota (9, 10). In some soils, fungal biomass equals or exceeds that of plant root biomass (11). The hyphae of filamentous fungi are usually 2 to 10 μm in diameter, but can reach extensive lengths and cover large areas. A gram of soil typically contains several hundred meters of fungal hyphae and may easily contain several hundred different fungal species. Despite their microscopic size, a single fungal mycelium can cover an area of several hectares (12, 13). Approximately 70,000 fungal species have been described worldwide, but an estimated 1.5 million species exist, indicating that 95% of all extant species have yet to be identified (14). Current limitations to the adequate characterization of fungal diversity includes the methodological constraints historically associated with the identification of microscopic organisms, the lack of trained fungal taxonomists, and the economic cost associated with the large research efforts required for this type of work (15).

FUNGI AND ECOSYSTEM FUNCTION

Soil fungi can be grouped into three general functional groups (Table 1). While plant pathogenic fungi, such as *Fusarium* and *Rhizoctonia*, cause significant losses of agricultural crops each year, most soil fungi are beneficial and perform a number of critically important ecological functions. A primary role of fungi in any soil is the decomposition of plant residues. Saprophytic fungi produce a suite of extracellular enzymes capable of depolymerizing plant cell constituents such as cellulose, hemicellulose, and lignin. During the decomposition process, fungi immobilize (i.e., retain in their biomass in organic form) and mineralize (i.e., release to the environment in inorganic form) nutrients simultaneously, with the balance between these two processes determining the plant availability of nutrients such as nitrogen, phosphorus, potassium, and sulfur. In addition to nutrient immobilization, fungi are known to accumulate potentially toxic substances within the mycelium, including radionuclides and heavy metals (9). The branching of fungal hyphae around soil particles, combined with the production of extracellular polysaccharides which serve as binding agents, promotes the formation of stable soil aggregates (15). This fungal-mediated process modifies air and water relations by altering soil permeability and may be an important mechanism for the physical protection of soil organic matter (16). Many saprophytic fungi promote the suppression of plant disease, either by producing antibiotics inhibitory to disease-causing organisms or by outcompeting pathogens for available resources (11). For example, the fungus *Chaetomium globosum* produces an antibiotic suppressive to *Fusarium*, the agent of seedling blight on maize (11).

Table 1 Functional groups of soil fungi

Group	Function
Saprotrophs	Decomposition of organic matter Nutrient immobilization and release Accumulation of toxic materials Aggregate formation and stabilization Suppression of pathogens
Mutualists	Transport of nutrients and water to plant roots Protection from root pathogens and heavy metals
Pathogens	Cause plant and animal disease

Mutualistic associations involving fungi include lichens, endophytes, and mycorrhizas. In all cases, a fungus establishes a mutually beneficial relationship with an autotrophic organism. Lichens are fungus–alga or fungus–cyanobacterium associations in which the alga or cyanobacterium partner captures energy through photosynthesis and the fungal partner provides structural support, supplies mineral nutrients, and helps regulate water relations (11). Endophytic fungi grow within living plants, obtaining nutrients from the plant without causing noticeable damage and, in at least some cases, providing protection from insects and plant pathogens (2). For example, some grasses (e.g., ryegrass) can be infected with a fungus that produces toxins thought to inhibit insect attack (2).

Mycorrhizal fungi, which form an intimate association with plant roots, enhance the uptake and transfer of mineral nutrients to the plant host in exchange for carbon. "Indeed, mycorrhizas, not roots, are the chief organs of nutrient uptake by land plants" (17). Phosphorus is often the nutrient of primary importance, but nitrogen, zinc, and sulfur have also been shown to be absorbed and transported by mycorrhizal fungi (11). Some mycorrhizal species participate in water as well as nutrient absorption, and may also provide protection against root pathogens

(11). Mycorrhizas are currently grouped into seven types, including arbuscular mycorrhizas, ectomycorrhizas, ectendomycorrhizas, arbutoid mycorrhizas, monotropoid mycorrhizas, ericoid mycorrhizas, and orchid mycorrhizas (17). Arbuscular mycorrhizas are the most common, and are formed by fungi in the order *Glomales* (*Zygomycetes*) in association with roots of a wide range of plants (17). Recent fossil evidence suggests that mycorrhizal-like structures, resembling modern arbuscular mycorrhizas formed by fungi in the genus *Glomus*, were present 460 million years ago and may have played an important role in the successful establishment of early terrestrial plants (18, 19). Ecological studies indicate that mycorrhizal species diversity in soil can influence the diversity and productivity of plant communities (20).

Fungi are sensitive to disturbance, and human-induced modifications of the soil environment often have negative impacts on the soil fungal community. Tillage of agricultural land disrupts the establishment and development of hyphal networks, thereby significantly reducing fungal hyphal lengths and biomass within the plow layer (21, 22). Increased nitrogen inputs, through fertilization (23, 24) and atmospheric deposition (25) have been implicated in the decline of both fungal biomass and fungal species diversity. Exposure to UV-B radiation, which has been increasing due to thinning of the protective stratospheric ozone layer, has been shown to influence fungal growth and the relative competitive abilities of different fungal species (26). Reductions in fungal biomass, activity, and diversity could have significant implications for the functioning of terrestrial ecosystems.

REFERENCES

1. Andrews, J.H. Fungal Life-History Strategies. In *The Fungal Community*; Carroll, G.C., Wicklow, D.T., Eds.; Marcel Dekker: New York, 1992; 119–144.
2. Deacon, J.W. *Modern Mycology*, 3rd Ed.; Blackwell Science: Oxford, 1997.
3. Whittaker, R.H. New Concepts of Kingdoms of Organisms. Science **1969**, *163*, 150–160.
4. Cavalier-Smith, T. Eukaryote Kingdoms: Seven or Nine? BioSystems **1981**, *14*, 461–481.
5. Woese, C.R.; Kandler, O.; Wheelis, M.L. Towards a Natural System of Organisms: Proposal for the Domains Archaea, Bacteria, and Eucarya. Proc. Natl. Acad. Sci. USA **1990**, *87*, 4576–4579.
6. Cooke, R.C.; Whipps, J.M. *Ecophysiology of Fungi*; Blackwell Scientific Publications: London, 1993.
7. Baldauf, S.L.; Palmer, J.D. Animals and Fungi are Each Other's Closest Relatives: Congruent Evidence from Multiple Proteins. Proc. Natl. Acad. Sci. USA **1993**, *90*, 11558–11562.
8. Baldauf, S.L. A Search for the Origins of Animals and Fungi: Comparing and Combining Molecular Data. American Naturalist **1999**, *154*, S178–S188.
9. Killham, K. *Soilecology*; Cambridge University Press: Cambridge, 1994.
10. Beare, M.H. Fungal and Bacterial Pathways of Organic Matter Decomposition and Nitrogen Mineralization in Arable Soils. In *Soil Ecology Insustainable Agricultural Systems*; Brussaard, L., Ferrera-Cerrato, R., Eds.; Lewis Publishers: Boca Raton, Florida, 1997; 37–70.
11. Paul, E.A.; Clark, F.E. *Soil Microbiology and Biochemistry*, 2nd Ed.; Academic Press: San Diego, California, 1996.
12. Smith, M.L.; Bruhn, J.N.; Anderson, J.B. The Fungus *Armillaria Bulbosa* is Among the Largest and Oldest Living Organisms. Nature **1992**, *356* (6368), 428–432.
13. Bonello, P.; Bruns, T.D.; Gardes, M. Genetic Structure of a Natural Population of the Ectomycorrhizal Fungus *Suillus Pungens*. New Phytol. **1998**, *138*, 533–542.
14. Hawksworth, D.L. The Fungal Dimension of Biodiversity: Magnitude, Significance, and Conversation. Mycol. Res. **1991**, *95* (6), 641–655.
15. Cannon, P.F. Strategies for Rapid Assessment of Fungal Diversity. Biodiversity and Conservation **1997**, *6*, 669–680.
16. Beare, M.H.; Hu, S.; Coleman, D.C.; Hendrix, P.F. Influences of Mycelial Fungi on Soil Aggregation and Organic Matter Storage in Conventional and No-Tillage Soils. Applied Soil Ecology **1997**, *5*, 211–219.
17. Smith, S.E.; Read, D.J. *Mycorrhizal Symbiosis*, 2nd Ed.; Academic Press: San Diego, California, 1997.
18. Blackwell, M. Terrestrial Life–Fungal from the Start? Science **2000**, *289*, 1884–1885.
19. Redecker, D.; Kodner, R.; Graham, L.E. Glomalean Fungi from the Ordovician. Science **2000**, *289*, 1920.
20. van der Heijden, M.G.A.; Klironomos, J.N.; Ursic, M.; Moutoglis, P.; Streitwolf-Engel, R.; Boller, T.; Wiemken, A.; Sanders, I.R. Mycorrhizal Fungal Diversity Determines Plant Biodiversity, Ecosystem Variability and Productivity. Nature **1998**, *396*, 69–72.
21. Beare, M.H.; Parmelee, R.W.; Hendrix, P.F.; Cheng, W.; Coleman, D.C.; Crossley, D.A. Microbial and Faunal Interactions and Effects on Litter Nitrogen and Decomposition in Agroecosystems. Ecological Monographs **1992**, *62* (4), 569–591.
22. Frey, S.D.; Elliott, E.T.; Paustian, K. Bacterial and Fungal Abundance and Biomass in Conventional and No-Tillage Agroecosystems Along Two Climatic Gradients. Soil Biology and Biochemistry **1999**, *31*, 573–585.
23. Bardgett, R.D.; Hobbs, P.J.; Frostegard, A. Changes in Soil Fungal: bacterial Biomass Ratios Following Reductions in the Intensity of Management of an Upland Grassland. Biology and Fertility of Soils **1996**, *22*, 261–264.
24. Bardgett, R.D.; Lovell, R.D.; Hobbs, P.J.; Jarvis, S.C. Seasonal Changes in Soil Microbial Communities Along a Fertility Gradient of Temperate Grasslands. Soil Biology and Biochemistry **1999**, *31*, 1021–1030.
25. Arnolds, E. The Changing Macromycete Flora in the Netherlands. Trans. Br. Mycol. Soc. **1988**, *90*, 391–406.
26. Duguay, K.J.; Klironomos, J.N. Direct and Indirect Effects of Enhanced UV-B Radiation on the Decomposing and Competitive Abilities of Saprobic Fungi. Applied Soil Ecology **2000**, *14*, 157–164.

GAS AND VAPOR PHASE TRANSPORT

G

Dennis E. Rolston
University of California, Davis, California, U.S.A.

INTRODUCTION

Diffusion is the principal mechanism in the interchange of gases between the soil and the atmosphere. The interchange results from concentration gradients established within soil by respiration of micro-organisms and plant roots; by production of gases associated with biological reactions such as fermentation and nitrogen transformations; and by soil incorporation of materials such as fumigants, anhydrous ammonia, pesticides, and various volatile organic chemicals in toxic waste sites. The diffusion of water vapor within the soil also occurs due to differences in vapor pressure gradients induced by temperature differences or by evaporative conditions at the soil surface.

TRANSPORT PHYSICS AND EQUATIONS

The diffusion velocities of gas mixtures in porous media are related to each other in a complex manner dependent upon the mole fraction of each gas, the molar fluxes of each gas, and the binary diffusion coefficient of each gas pair. If gravity effects are ignored or diffusion occurs only horizontally, the well-known Stefan–Maxwell equations provide the theoretical framework for diffusion of gases in soils. Fick's law for diffusion is a restrictive case of the Stefan–Maxwell equations and is generally applicable for only a few special cases (1). One of these cases is for the diffusion of a trace gas in a binary mixture, meaning that the mole fraction of the tracer gas is small. The second special case is for diffusion of two gases in a closed system (total pressure remains constant). In this case, neither gas needs to be in trace amounts. A third case where Fick's law is applicable is for a three-component system where one of the gases exists in trace amounts and the binary diffusion coefficients of the other two pairs do not differ much from one another (basically the first case). Examples of this case would be for the gas pairs of N_2-O_2 and N_2-Ar.

Assuming that the special case conditions are met, Fick's law is given by

$$\frac{M_g}{At} = f_g = -D_p \frac{dC_g}{dx} \tag{1}$$

where M_g is the amount of gas diffusing (g gas), A is the cross-sectional area of the soil (m^2 soil), t is time (s), f_g is the gas flux density (g gas m^{-2} soil s^{-1}), C_g is concentration in the gaseous phase (g gas m^{-3} soil air), x is distance (m soil), and D_p is the soil–gas diffusion coefficient (m^3 soil air m^{-1} soil s^{-1}).

MEASUREMENT OF THE SOIL–GAS DIFFUSION COEFFICIENT

The soil–gas diffusion coefficient, D_p, is the fundamental property that must be known to use Eq. 1 to calculate gas transport in soils. Values of D_p change with soil–air content and tortuosity (crookedness of the diffusion path).

The standard laboratory method for measuring the soil–gas diffusion coefficient is based upon establishing gas concentration, C_o, within a chamber (Fig. 1). One end of a soil core of concentration C_s is placed in contact with the gas within the chamber (2). The other end of the soil core is maintained at concentration C_s. The gas of interest diffuses either into or out of the chamber depending upon the concentration C_o compared to that outside the core. Obviously, the other gases making up the atmosphere will diffuse in an opposite direction to that of the gas of interest. The time rate of change of concentration in the chamber is related to the soil–gas diffusion coefficient and can be described by equations for unsteady diffusion of gas. Several investigators (see Ref. 2) have used similar procedures.

The unsteady diffusion of a gas, which is nonreactive (physically, biologically, and chemically) is described by the combination of Fick's first law (Eq. 1) and the continuity equation (conservation of mass)

$$\epsilon \frac{\partial C_g}{\partial t} = D_p \frac{\partial^2 C_g}{\partial x^2} \tag{2}$$

where ϵ is the soil–air content (m^3 air m^{-3} soil).

In developing Eq. 2, it is assumed that the soil is uniform with respect to the diffusion coefficient and that ϵ is constant in space and time. A simple solution of Eq. 2 that allows for the determination of D_p from laboratory measurements is available (2, 3).

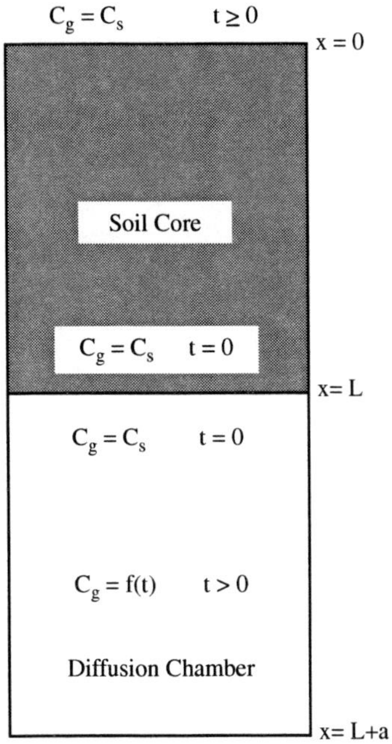

Fig. 1 Schematic diagram of the apparatus design used for measuring the soil–gas diffusion coefficient of a soil core.

PREDICTIVE EQUATIONS FOR THE SOIL–GAS DIFFUSION COEFFICIENT

Gas transport and fate simulation studies often depend on accurately estimating gas diffusivity (D_p/D_0) as a function of soil–air content (ϵ) from easily obtainable physical parameters of the soil, because measured data of D_p (ϵ) are often not available. Many empirical models for predicting the gas diffusivity have been proposed, such as the well-known Millington–Quirk equation (4), with varying degrees of prediction accuracy (5). Recently, a series of papers (5–9) have offered new equations that appear to greatly increase the predictive accuracy. Separate equations are needed for undisturbed soils compared to sieved, repacked soils. To accurately predict diffusivity in undisturbed soils, the effects of texture and structure on the diffusivity must be considered.

Undisturbed Soil

It has been shown that the Campbell (10) pore-size distribution (water retention) parameter, b, is an effective parameter to describe effects of soil type (soil texture and structure) on $D_p(\epsilon)$ in undisturbed soils (5–8) to give

$$\frac{D_p}{D_0} = \Phi^2 \left(\frac{\epsilon}{\Phi}\right)^{2+3/b} \tag{3}$$

where D_0 is the gas diffusion coefficient in air (without soil) (m^2 air s^{-1}), Φ is soil total porosity (m^3 voids m^{-3} soil), and b is the Campbell pore size distribution parameter, corresponding to the slope of a plot of the log of the soil–water pressure potential versus the log of the volumetric soil–water content.

It has also been shown that there is a significant effect of macro-porosity on D_p (8). In this respect, macro-porosity is defined as the air-filled porosity at a soil–water pressure head of -100 cm H_2O (-10 kPa), corresponding to the volumetric content of soil pores with an equivalent pore diameter larger than 30 μm. The macro-porosity (ϵ_{100}) is found by subtracting the volumetric soil–water content measured at a soil–water pressure head of -100 cm from the soil total porosity. Using this concept results in an additional equation

$$\frac{D_p}{D_0} = (2\epsilon_{100}^3 + 0.04\epsilon_{100})\left(\frac{\epsilon}{\epsilon_{100}}\right)^{2+3/b} \tag{4}$$

To summarize, Eq. 4 appears slightly more accurate and is recommended if both ϵ_{100} and b are known, whereas Eq. 3 is recommended if only b is known.

If the Campbell pore size distribution parameter, b, is not measured, clay fraction (CF) is a good indicator of b (11, 12), and is given by

$$b = 13.6\ \mathrm{CF} + 3.5 \tag{5}$$

Thus, Eq. 5 can be used to determine b to be subsequently used in either Eqs. 3 or 4.

Sieved, Repacked Soil

Soil-type effects, such as texture and structure (manifested through pore size distribution), on gas diffusivity in sieved and repacked soils appear to be minor and can likely be neglected (9,13,14). A simple, physically based model (9) for $D_p(\epsilon)/D_0$ in sieved and repacked soils is given by

$$\frac{D_p}{D_0} = \epsilon^{1.5}\left(\frac{\epsilon}{\Phi}\right) \tag{6}$$

The reduction term (ϵ/Φ) describes the increased tortuosity in a wet soil, compared to a dry soil at the same soil–air content, due to interconnected water films. This equation has been shown to give accurate predictions of the soil–gas diffusion coefficient in sieved, repacked soil samples (9).

SUMMARY

Diffusion is the principal mechanism in the interchange of gases between the soil and the atmosphere. Fick's Law of diffusion can describe the diffusion of gases in soils as long as the restrictive conditions for the applicability of the equation apply. The soil–gas diffusion coefficient, D_p, is the fundamental property that must be known to use Fick's Law to calculate gas transport in soils. Values of D_p change with soil–air content and tortuosity. Techniques for accurately measuring D_p in soil cores are well established. However, because measured data for D_p are often not available, equations to predict D_p from easily obtainable soil–physical parameters have been developed. It is now well established that different equations are needed to estimate D_p for undisturbed and for disturbed soil samples. Equations for undisturbed soil must include parameters that take into account the effects of soil type, such as texture, structure, and pore size distribution, on D_p. Soil-type effects on D_p in sieved and repacked soils appear to be minor.

REFERENCES

1. Amali, S.; Rolston, D.E. Theoretical Investigation of Multicomponent Volatile Organic Vapor Diffusion: Steady-state Fluxes. J. Environ. Qual. **1993**, *22*, 825–831.
2. Rolston, D.E. Gas Diffusivity. In *Methods of Soil Analysis, Part1. Physical and Mineralogical Methods*, 2nd Ed; Klute, A. Agronomy Monograph No. 9; American Society of Agronomy, Soil Science Society of American: Madison, WI, 1986; 1089–1102.
3. Currie, J.A. Gaseous Diffusion in Porous Media. Part 1. A Non-steady State Method. Br. J. Appl. Phys. **1960**, *11*, 314–317.
4. Millington, R.J.; Quirk, J.M. Permeability of Porous Solids. Trans. Faraday Soc. **1961**, *57*, 1200–1207.
5. Moldrup, P.; Kruse, C.W.; Rolston, D.E.; Yamaguchi, T. Modeling Diffusion and Reaction in Soils: III. Predicting Gas Diffusivity from the Campbell Soil–Water Retention Model. Soil Sci. **1996**, *161*, 366–375.
6. Moldrup, P.; Olesen, T.; Rolston, D.E.; Yamaguchi, T. Modeling Diffusion and Reaction in Soils: VII. Predicting Gas and Ion Diffusivity in Undisturbed and Sieved Soils. Soil Sci. **1997**, *162*, 632–640.
7. Moldrup, P.; Olesen, T.; Yamaguchi, T.; Schjønning, P.; Rolston, D.E. Modeling Diffusion and Reaction in Soils: IX. The Buckingham-Burdine-Campbell Equation for Gas Diffusivity in Undisturbed Soil. Soil Sci. **1999**, *164*, 542–551.
8. Moldrup, P.; Olesen, T.; Schjønning, P.; Yamaguchi, T.; Rolston, D.E. Predicting the Gas Diffusion Coefficient in Undisturbed Soil from Soil Water Characteristics. Soil Sci. Soc. Am. J. **2000**, *64*, 94–100.
9. Moldrup, P.; Olesen, T.; Gamst, J.; Schjønning, P.; Yamaguchi, T.; Rolston, D.E. Predicting the Gas Diffusion Coefficient in Repacked Soil: Water-induced Linear Reduction Model. Soil Sci. Soc. Am. J. **2000**, *64*, 1588–1594.
10. Campbell, G.S. A Simple Method for Determining Unsaturated Conductivity from Moisture Retention Data. Soil Sci. **1974**, *117*, 311–314.
11. Clapp, R.B.; Hornberger, G.M. Empirical Equations for Some Soil Hydraulic Properties. Water Resour. Res. **1978**, *14*, 601–604.
12. Olesen, T.; Moldrup, P.; Henriksen, K.; Petersen, L.W. Modeling Diffusion and Reaction in Soils: IV. New Models for Predicting Ion Diffusivity. Soil Sci. **1996**, *161*, 633–645.
13. Xu, X.; Nieber, J.L.; Gupta, S.C. Compaction Effects on the Gas Diffusion Coefficient in Soils. Soil Sci. Soc. Am. J. **1992**, *56*, 1743–1750.
14. Jin, Y.; Jury, W.A. Characterizing the Dependency of Gas Diffusion Coefficient on Soil Properties. Soil Sci. Soc. Am. J. **1996**, *60*, 66–71.

GELISOLS

James G. Bockheim
University of Wisconsin, Madison, Wisconsin, U.S.A.

INTRODUCTION

Gelisols, the permafrost-affected soils, comprise 18 million km^2 or about 13% of the Earth's's land surface. They occur in polar regions, the Arctic and Antarctic, and in some alpine regions (Fig. 1). Gelisols are of global concern because they contain many protected areas, support indigenous people who depend on the land and surrounding oceans for sustenance, may be subject to considerable impacts from human development (oil, coal, gas, and gas hydrite exploitation), and are already experiencing global warming (1).

Gelisols are defined as having permafrost within 100 cm of the soil surface, or gelic materials within 100 cm of the surface and permafrost within 200 cm of the surface (2). Permafrost is commonly defined as a thermal condition in which a material remains below 0°C for two or more years in succession; it may be ice-cemented or, in the case of insufficient interstitial water, dry. The zone above permafrost that is subject to seasonal thawing is known as the active layer. Gelic materials are defined as seasonally or perennially frozen mineral or organic soil materials that have evidence of cryoturbation (frost churning), ice segregation, or cracking from cryodesiccation.

THE PEDON AS A BASIC SOIL UNIT FOR GELISOLS

The pedon is the basic soil unit for sampling in *Soil Taxonomy* (2) and is especially important for describing, classifying, and sampling Gelisols, which often occur in areas with patterned ground. Patterned ground is a general term for any ground surface that is ordered into polygons, nets, circles, or stripes as a result of freezing and thawing processes. Fig. 1 shows Earth hummocks in an alpine region resulting from such processes. Each hummock is 1 – 1.5 m across and is about 1 m high. The pedon is defined so as to encompass the full cycle of patterned ground with a 2 m linear interval or a half cycle with a 2–7 m cycle. This interval is suitable for most patterned ground features such as Earth hummocks. In the case of large-scale (>7 m) patterned ground, such as ice-rich, low-center polygons on the North Slope of Alaska, two pedons are established: one within the center of the polygon and the other within the ice wedge. Therefore, each hummock in the figure represents a single pedon for descriptive and sampling purposes.

CLASSIFICATION OF GELISOLS

There are three suborders within the Gelisol order, Histels, Turbels, and Orthels; they are distinguished on the basis of organic matter content and for mineral soils whether or not there is evidence for cryoturbation.

Histels have 80% or more organic materials by volume within the upper 50 cm or to a restricting layer. They are subdivided into five great groups based on the nature of the underlying material and the degree of decomposition: Glacistels, Folistels, Fibristels, Hemistels, and Sapristels (3). The Glacistel illustrated in Fig. 2 contains 60 cm of organic material directly overlying ground ice. Disturbance to the surface-insulating layer may cause the ice layer to melt and the soil to collapse, a condition known as thermokarst (Fig. 3). The key properties of Histels are abundant organic materials ranging from woody to highly amorphous materials, a high moisture holding capacity, and a pH that for an array of Histels may range from as low as 2.5 to above 7.0. Although Histels comprise less than 3% of the Gelisols globally, they cover large areas in North America and Russia (>15% of the area of Gelisols).

Turbels represent a second suborder of the Gelisols; they are mineral soils subject to cryoturbation. Turbels comprise about 87% of the Gelisols on a global basis. Cryoturbation is evidenced by irregular or broken horizons, involutions, organic matter accumulated on the surface of the permafrost, oriented rock fragments, and silt coatings and silt-enriched subsoil horizons that result from freezing and thawing, frost heaving, and cryogenic sorting (4). Cryoturbation is caused primarily by differential frost heave, but its action can be enhanced by cryostatic pressure, differential swelling, and load casting on poorly drained sites (5).

Figure 4 shows an Aquiturbel on Earth hummocks (see also Fig. 1) in northern Canada. This pedon contains an

Fig. 1 Earth hummocks, a type of patterned ground, in the North Cascades of Washington State, U.S.A. Each hummock is 1–1.5 m wide and about 1 m high. This landform features a type of soil known as Turbels.

organic layer along the rim of the Earth hummocks and intensely cryoturbated mineral soil horizons in the center.

Although the underlying concept of Turbels is that they are subject to cryoturbation, they may contain diagnostic horizons that are common to soils not having permafrost. However, these horizons are recognized at the great group level. There are seven great groups within the Turbels (Histoturbels, Aquiturbels, Anhyturbels, Molliturbels, Umbriturbels, Psammoturbels, and Haploturbels) that link them with other soils not containing permafrost (2).

Fig. 3 This landform in the Russian Far East contains abundant ground ice that has melted following disturbance of the surface organic materials, resulting in a condition known as thermokarst.

It should be emphasized that cryopedogenic processes such as cryoturbation, thermal cracking, and ice segregation are soil-forming processes characteristic of soils with permafrost. They should not be viewed as operating against the other soil-forming processes in low-latitude soils; rather, they are distinctive processes producing horizons and properties that are uncommon to other soil orders. Processes common to the other soil orders operate in Gelisols but at a lesser magnitude because of the dominance of cryopedogenic processes.

The third suborder within the Gelisols is the Orthels, which are other mineral soils containing permafrost within the upper 100 cm. These soils comprise less than 10% of the Gelisols globally and are the smallest suborder within Gelisols in the Northern Hemisphere. The Orthels are subdivided into eight great groups that are more or less parallel

Fig. 2 A Hemic Glacistel in northern Canada. The organic soil materials, which are intermediate in decomposition from fibric and sapric (muck), are underlain by ground ice at 60 cm.

Fig. 4 An Aquiturbel developed on an Earth hummock in northern Canada (see also Fig. 1).

Fig. 5 A Spodic Psammorthel in northern Russia. The soil is developed in excessively drained, sandy floodplain deposits and contains weakly developed spodic materials in the upper part (photo provided by L. Huber).

to those in the Turbels and that link them to soils not containing permafrost (Historthels, Aquorthels, Anhyorthels, Mollorthels, Umbrorthels, Argiorthels, Psammorthels, and Haplorthels). Fig. 5 shows a Spodic Psamm-orthel in northern Russia. The soil is derived from sandy floodplain deposits and contains weakly developed spodic materials.

GELISOLS AND LAND USE

Gelisols offer special challenges to land management for interpretation and practices, including construction (structures and pipelines), mining, forestry, and agriculture. For example, structures must be built on refrigerated pilings or above ground (Fig. 6) so that heat from the structure does not melt the permafrost and cause subsidence. A main concern with Gelisols is that they are large C sinks (Table 1). The concern is that warming in the Arctic could increase the thickness of the seasonal thaw layer and enhance heterotrophic respiration, releasing additional CO_2 to the atmosphere. In this case, Gelisols would become a C source.

Fig. 6 An apartment complex in the upper Kolyma River region of the Russian Far East that is built on stilts to avoid heat transfer, melting of permafrost, and subsidence.

CONCLUSIONS

Gelisols occur in the polar and in some alpine regions and are soils containing permafrost within 100 cm of the soil surface, or gelic materials within 100 cm and permafrost within 200 cm of the surface. Gelisols originate from cryopedogenic processes that include cryoturbation, freez-

Table 1 Carbon storage in two Gelisols

Horizon	Depth (cm)	Organic C (%)	Bulk density ($g\,cm^{-3}$)	C storage ($kg\,m^{-3}$)
Pedon A96-29: Typic Aquiturbel; Barrow, Alaska (6):				
Oi	1	25.3	0.42	1.1
Bg	32	9.9	1.05	33.3
Oejj	2	18.8	0.38	1.4
Bgf	15	7.3	0.83	9.1
BCgf	12	3.9	1.05	4.9
Cgf	38	4.4	1.02	17.1
Total				66.9
Pedon 94P0668: Typic Molliorthel; lower Kolyma River, Russian Far East (2):				
A	7	7.5	1.04	5.5
AB	5	0.65	1.65	0.5
Bw1	26	0.56		2.4
Bw2	21	0.48		1.7
Bw3	19	0.44	1.66	1.4
Bw4	16	0.44	1.79	1.3
Cf	6	0.39		0.4
Total				13.2

ing and thawing, cryogenic sorting, ice segregation, and cryodesiccation. The low temperatures in Gelisols give rise to cryostatic pressure development and migration of water with resultant ice buildup. Therefore, Gelisols are differentiated from other soils primarily on the basis of thermal characteristics and physical properties that are readily observed in the field. The three main types of Gelisols are Histels (organic soils with permafrost), Turbels (mineral soils with cryoturbation), and Orthels (other mineral soils with permafrost in the upper 100 cm). The pedon concept is especially important for sampling Gelisols and is determined by the size of the repeating units of patterned ground features, e.g., <2 m, 2–7 m, or >7 m.

REFERENCES

1. Bockheim, J.G.; Ping, C.L.; Moore, J.P.; Kimble, J.M. Gelisols: A New Proposed Order for Permafrost-Affected Soils. In *Proceedings of the Meeting on the Classification, Correlation, and Management of Permafrost-Affected Soils, Fairbanks, Alaska*; Kimble, J.M., Ahrens, R.J., Eds.; USDA, Soil Conservation Service, National Soil Survey Center: Lincoln, NE, 1994; 25–44.
2. Soil Survey Staff, *Soil Taxonomy: A Basis System of Soil Classification for Making and Interpreting Soil Surveys*, 2nd Ed.; USDA Natural Resources Conservation Service. Agric. Handbook No. 436; U.S. Government Printing Office: Washington, DC, 1999.
3. Soil Survey Staff, *Keys to Soil Taxonomy*, 8th Ed.; USDA Natural Resources Conservation Service, 1998.
4. Bockheim, J.G.; Tarnocai, C. Recognition of Cryoturbation for Classifying Permafrost-Affected Soils. Soil Sci. **1998**, *162*, 927–939.
5. Van Vliet-Lanoe, B. Differential Frost Heave, Load Casting and Convection: Converging Mechanisms; A Discussion of the Origin of Cryoturbations. Perm. Periglac. Proc. **1991**, *2*, 123–139.
6. Bockheim, J.G.; Everett, L.R.; Hinkel, K.M.; Nelson, F.E.; Brown, J. Soil Organic Carbon Storage and Distribution in Arctic Tundra, Barrow, Alaska. Soil Sci. Soc. Am. J. **1999**, *63*, 934–940.

GEOLOGY, SOIL AND

K.R. Olson
University of Illinois, Urbana, Illinois, U.S.A.

INTRODUCTION

Geology is the study of Earth, its internal structure, its materials, its chemical and physical processes, and its physical and biological history (1). One of the most important geological discoveries was that rocks could form by crystallization of molten material. Rocks do change from one kind to another. As rocks exposed at the surface weather, particles move downstream, eventually to be deposited as sediments that lithify into sedimentary rocks. By tracing distributions of rock materials, early geologists were able to link sediments to highly consolidated, mineralogically distinct metamorphic rocks.

ATMOSPHERE, HYDROSPHERE, AND LITHOSPHERE

The crust, mantle, and core account for 99.97% of the mass of the Earth (1). The lithosphere includes the Earth's crust and upper mantle. The term has previously been used to describe the entire portion of the Earth that is composed of rocks. The remaining 0.03% comprises the atmosphere and the hydrosphere. The hydrosphere includes the portion of the Earth's surface that is covered by water. The hydrosphere behaves as an intermediate reservoir for carbonates, silicates, and other mineral groups leached from the rock of the continents and carried to the oceans by rivers (1). Both silicates and calcium carbonate follow involved paths from the time they are weathered from continental rock, until they are deposited on the sea floor.

The composition of the atmosphere is strongly influenced by the cycling of water from the oceans to the continents, and by its return to the oceans in rivers and subsurface flow. The amount of free oxygen in the atmosphere seems to remain essentially constant. Volcanic gases contain only traces of molecular oxygen but eject large quantities of molecular hydrogen, carbon monoxide, and sulfur dioxide. These gases react with atmospheric oxygen to produce carbon dioxide, water vapor, and sulfur dioxide. The lithosphere, which consumes free oxygen through weathering of rock, and the biosphere, which produces oxygen through photosynthesis, maintain this equilibrium. The control over this equilibrium may be a feedback mechanism involving the organisms whose by-products and remains become constituents of sedimentary rock.

CYCLING OF ELEMENTS

Oxygen, silicon, aluminum, and iron contribute 96% by volume of the elemental composition of the Earth's crust (2). The other elements only occupy the remaining 4%. Individual atoms or molecules of such elements as carbon, nitrogen, oxygen, and sodium change form countless times as they cycle through the atmosphere, biosphere, and lithosphere (1). Oxygen, for instance, may be converted from a neutral atom to dissolved gas, to combined molecule to ionized particle, to protoplasmic components, and, perhaps, back to the ionic state as a constituent of rock (1). Other atoms, such as sodium or silicon, may be more constrained in the variety of chemical forms they may assume. These and other representative routes are indicated in a revised and modified schematic shown in Fig. 1.

CHEMICAL WEATHERING PROCESSES

Weathering is a collective term for the combined effects of all the physical and chemical processes that break down and transform pre-existing rock materials near the Earth's surface, to products that are more stable under the physical and chemical conditions at the surface. Weathering constitutes the response of rock materials to several forms of energy as a function of time (1). The products of weathering include solids (i.e., sediments and soils) and liquids (including the solutions of salts present in rivers and the ocean). Each physical and chemical factor of weathering affects the outcome of the weathering process. These factors and related variables of the rock cycle, including erosion, transportation, and sedimentation, combine to operate as a complex chemical sorting system

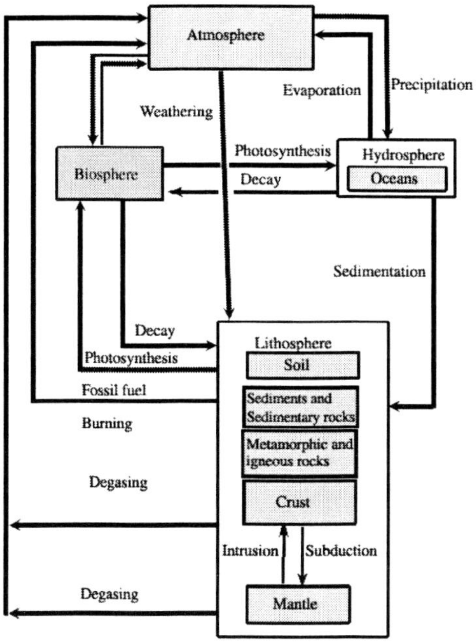

Fig. 1 Cycling of elements. (From Ref. 1.)

which distributes products of different composition to various sites of deposition.

Soils are the result of leaching, oxidation, and dissolution of surface materials by the percolation of groundwater and by humic acids derived from oxidized organic materials. New minerals, such as clays, are formed by the chemical alteration of the original mineral of the bedrock (1). The chemical weathering process develops a soil profile ranging from heavily weathered surface materials down to unaltered rock. The soil zone is the zone of transition between solid rock and the atmosphere. The solid rock, or bedrock, usually has numerous tiny cracks or joints. Chemical weathering caused by water, which fills the cracks, attacks the joints' surfaces and enlarges the cracks.

PARENT MATERIAL

Parent material is the initial or starting material from which a soil develops. This initial material can include both consolidated rock or unconsolidated material deposited by gravity, wind, or water, and consists of specific minerals of different sizes, or even plant materials of various plant types (2). Mineral matter inherited from rocks is referred to as soil parent material because it is the principal ingredient from which soils are formed (3). However, the principal parent materials of organic soils are decomposing plants.

In many cases, relief prior to and during soil formation is related to the nature of the initial soil material. In broad river deltas, crests of natural levees near the stream channels have more coarse material than the areas beyond the levees that are nearly level, and have the finer textured initial material (4). In steeper topography, where the valleys below the mountain ranges are characterized by broad alluvial–colluvial fans, the initial material near the mountain range contains more coarse and angular material than areas farther away from the mountain range (5).

After unconsolidated parent material is deposited on a stable landscape, or after bedrock has been exposed at the Earth's surface, soil formation begins and continues over time. The rate of soil formation depends on the climate, including temperature and rainfall. It also depends on vegetation and the activity of other organisms, which live on or in the parent material. These organisms help convert parent materials to soil.

Russian pedologists (6) identified parent material as one of the five significant soil-forming factors. Early approaches to soil survey and classification were based on the geology and composition of the soil-forming material (7, 8). The geologic origin and composition of the initial material was identified by the terms *Agranite soils* or *Aglacial soils*. Soils that originate for a parent material inherit the mineral types found in them. Over time, these original minerals are weathered (dissolved) and new minerals form and accumulate in the soils (2).

Russian soil scientists (9) showed the controlling effects of parent material on soil properties. Jenny (10) conducted a systematic analysis of the relationship between soil properties and parent materials from which the soils developed. Jenny proposed parent materials as an independent soil-forming factor, defining parent material as *the state of the soil system at time zero of soil formation.* The physical body of soil and its associated mineralogical and chemical properties are the starting point for the interaction of other soil-forming factors. Previous weathered rock—even a previous soil—could be considered parent material.

The properties of a modern soil are the result of the composition of the surficial layer, which existed when the other factors started to impact, and the alterations resulting from the effect of these factors over time. Properties of younger soils are greatly influenced by parent material. Weathering and pedogenic processes result in characteristics of the original parent material being eliminated. Extremely resistant initial material, such as quartz or sand, may still exist in old, weathered soils. It can be difficult to separate the nature (or characteristics) of the initial material and its influence on soil, the kind of "pre-weathering" of the initial material before becoming parent material for the soil, and the effects of the other soil-forming factors on the parent material of this soil. The complex environmental history (climatic and vegetation changes in the recent geologic past) makes it difficult to separate parent material influences on soil properties from other factor effects.

Rock types influence the soil properties of the modern soil. The impact of rock type on soil properties was organized by Buol and colleagues (11) into the following subdivisions: sedimentary, igneous, and combinations of mineralogically similar metamorphic and igneous rocks. Sedimentary rocks include unconsolidated glacial deposits, loessial deposits, and unconsolidated coastal plain sediments, as well as consolidated rock, such as limestone and dolomites, sandstones, and shales.

Siliceous crystalline rocks include more "acidic" quartzose, igneous, and metamorphic rocks including granites, granite gneiss, and schists. Dark-colored ferromagnesian rocks include andosites, diorites, basalt, and hornblende gneiss. Volcanic ash parent materials are composed of noncrystalline materials, any glass fragments, bits of the easily weatherable feldspars, ferromagnesian minerals, and varying amounts of quartz.

Mineral components of many soils are inherited almost exclusively from parent materials, while others are developed mostly in situ during the course of weathering and pedogenesis. Primary minerals are formed at high temperature igneous and metamorphic rocks. Secondary minerals are formed at lower temperature in sedimentary rocks and soils (12).

CONCLUSIONS

The basic model of soil implies that soils are dynamic and geographical. In soil systems, the processes or driving forces are best described as dynamic, rather than static. Morphological properties of soil develop as a result of processes acting on parent materials. In addressing the influence of parent materials in soil genesis. Chesworth (13, 14) stated that "time has the result of modifying the parent material effects so that only in young or relatively immature soils will the parent material exert its strongest influence on the soil-forming process. That influence will be an inverse function of time."

REFERENCES

1. Jackson, C. *Geology Today*; CRM Books, Communications Research Machines: Del Mar, California, 1973.
2. Singer, M.J.; Munns, D.N. *Soils, An Introduction*, 3rd Ed.; Prentice Hall: Upper Saddle River, New Jersey, 1996.
3. Troeh, F.R.; Thompson, L.M. *Soils and Soil Fertility*, 5th Ed.; Oxford University Press: New York, 1993.
4. Russell, R.J. *River Plains and Sea Coasts*; University Calif. Press: Berkeley, 1967.
5. Birot, P. *The Cycle of Erosion in Different Climates*; Translated by C.I. Jackson and K.M. Clayton; Univ. Calif. Press: Berkeley, 1960.
6. Dokuchaev, V.V. Russian Chernozem (Russkii Chernozen). In *Collected Writings (Sochineniya)*; Acad. of Sci.: Moscow, USSR, 1883; Vol. *3*.
7. Richthofen, F.F. von. Fuhrer fur Forschungsreisende; Berlin, 1886.
8. Thaer, A.D. Grundsatze der Rationaellen Landwirtschaft. (Cited in J.S. Joffe, 1949: Pedology Publ., New Brunswick, N.J.), 1809, 1810, 1812; Vol. 1–4.
9. Polynov, B.B. Das Muttergestein als Faktor de Bodenbildung und als Kriterium fur die Bodenklassification. Soil Res. **1930**, *2*, 165–180.
10. Jenny, H. *Factors of Soil Formation*; McGraw-Hill: New York, 1941.
11. Buol, S.W.; Hole, F.D.; Mc Cracken, R.J. *Soil Genesis and Classification*, 2nd Ed.; The Iowa State University Press: Ames, 1980.
12. Jackson, M.L. Chemical Composition of Soils. In *Chemistry of Soil*, 2nd Ed.; Bear, F.E., Ed.; Reinhold: New York, 1964.
13. Chesworth, W. The Parent Rock Effect in the Genesis of Soil. Geoderma **1973**, *10*, 215–225.
14. Chesworth, W. Conceptual Models in Pedogenesis: A Rejoinder. Geoderma **1976**, *16*, 257–260.

GLOBAL RESOURCES

G

Paul F. Reich
Hari Eswaran
United States Department of Agriculture-Natural Resources Conservation Service (USDA-NRCS), Washington, D.C., U.S.A.

INTRODUCTION

A biome is defined as "a community of organisms interacting with one another and with the chemical and physical factors making up their environments" (1). For most purposes, the term biome is used to identify the natural habitat conditions around the world. Depending on the purpose, the global ecosystem is divided into units each characterized by a specific combination of climatic factors. Two major determinants of biome type are precipitation (total and its distribution) and air temperature. These two elements of climate have been commonly used to define the major biomes of the world. A third variable that affects the habitat type is the soil. This section presents a general overview of the soils characterizing the major biomes of the world.

Detailed maps showing the biomes of the world are not available due to conceptual differences of definitions and reliable global databases. There are many excellent and detailed studies of specific habitats around the world, and using these and the global soils and climate database of the world (2), a map showing the distribution of the major biomes was drawn. The terms used to describe the biomes are common in use, but their subdivisions are based on important differentiating factors.

MAJOR BIOMES

The five major biomes of the world and their subdivisions are listed in Table 1 and their distribution is shown in Fig. 1. Some geographers have used elevation as a differentiating factor and recognized a "montane biome." This biome's small extent precludes its description in this section. Few geographers recognize the Mediterranean biome as presented here as a subdivision of the Temperate biome. Within each biome, some major distinction is made. In the Polar biome, an attempt is made to differentiate areas with permafrost and warmer areas with intermittent permafrost. The latter is termed "interfrost." The Boreal Forest biome has a humid and a semiarid counterpart similar to the temperate and tropical zones. The semiarid vegetation is generally grass or shrubs. The desert areas are divided into hot, cool, and cold equivalents.

Deserts occupy about 30% of the global ice-free landmass. Potential evapotranspiration exceeds precipitation during most days of the year, and the biome shows the greatest contrast in temperature, both diurnal and seasonal. Desert vegetation is very sparse with a few isolated shrubs dominated by succulents and annuals. Grasses and forbs become more dominant at the margins. The tropics occupy about 27% of the landmass and are characterized by only a small variation in temperature during the year. The soil moisture conditions range from semiarid to humid. The availability of soil moisture determines the biota. The temperate grassland and forest biome occupies about 15%. The cold biomes, the boreal and polar, are mainly in the northern latitudes with small areas in South America. The low temperatures control the flora and fauna, and vegetation is scarce to nonexistent in the extreme cold polar regions. Each of these areas has specific kinds of soils that have developed as a response to the specific soil moisture and temperature conditions. The diversity in vegetation has its equivalent in fauna with animal species adapting to the specific bio-climatic conditions. Within each of the biomes, there are specific conditions that promote unique flora and fauna. Examples of such localized systems are volcanic hot springs, wetlands, and oases.

Soils of the Major Biomes

Soil temperature and moisture, with their seasonal and annual variations, are an integral part of the soil classification system called Soil Taxonomy (3) which is used here. As Fig. 1 and Table 1 are based on soil moisture and temperature conditions, there is an implied link between the biomes and their soil resource endowments.

Table 1 Major global biomes with subdivisions and their areas

			Global land area		
Biome	**Subdivision**	**Remarks**	**In thousand km^2**	**In%**	**In%**
Polar	Permafrost	Mosses and lichens	10,550	8.06	15.46
	Interfrost	Shrubs and stunted trees	9686	7.40	
Boreal	Semi arid	Shrub	3555	2.72	9.94
	Humid	Forest	9446	7.22	
Temperate	Semi-arid	Grassland	7352	5.62	15.14
	Humid	Forest	12,458	9.52	
	Mediterranean warm	Shrubs and forest	3625	2.77	3.37
	Mediterranean cold	Forbs and shrubs	790	0.60	
Desert	Hot	Barren	4419	3.38	29.46
	Cool	Barren	28,525	21.80	
	Cold	Barren	5599	4.28	
Tropical	Semi-arid	Grassland, savanna	20,348	15.55	26.65
	Humid	Forests	14,521	11.10	
Total			130,879	100.00	

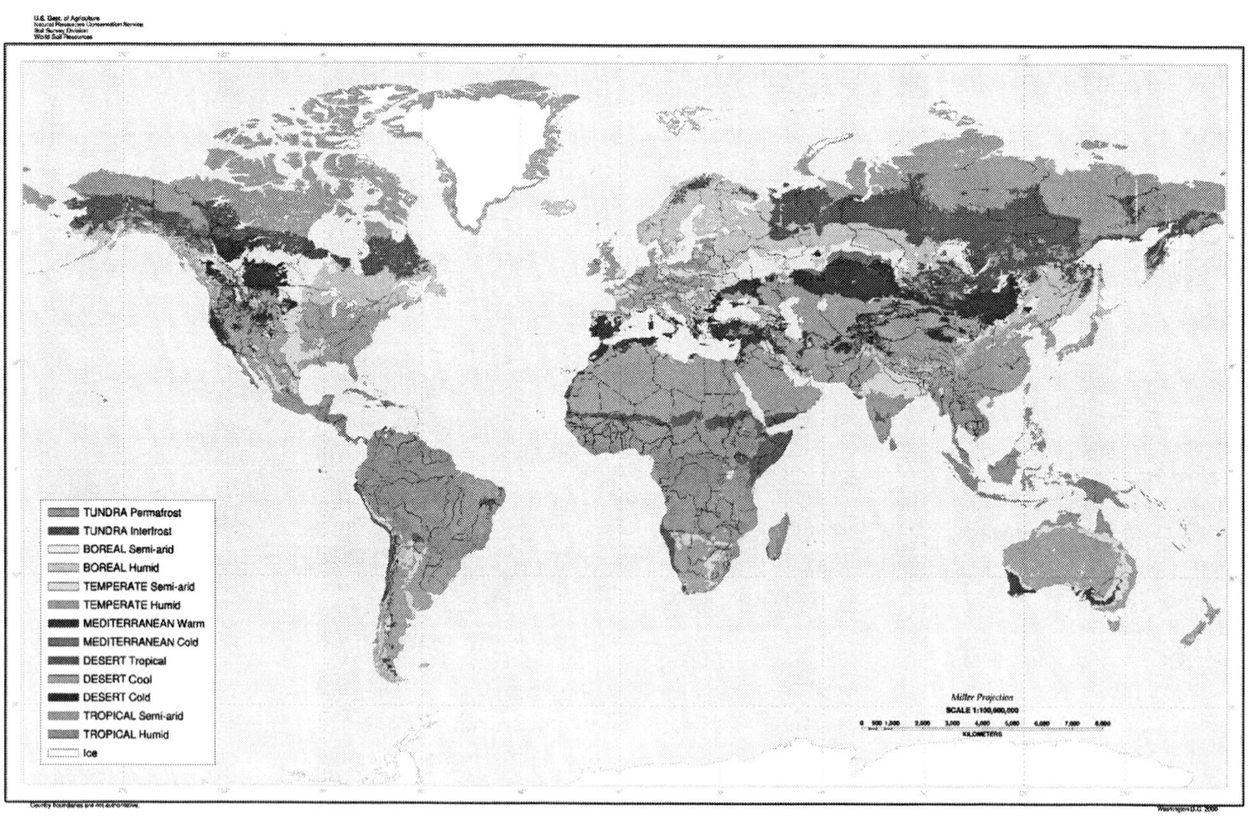

Fig. 1 Major biomes.

The purpose of this introductory section is to present the general geographic distribution of the biomes and the major soils characterizing each. The major soil orders are listed in Table 2. The subsequent sections elaborate on the soil resources of each of the six biomes listed in Table 1.

Polar Biome

Characterized largely by a mean annual soil temperature of less than 0°C, the Polar biome occupies a large land mass adjacent to the Arctic circle to about 60°N latitude. This is generally the ice-free land of the northern latitudes. The sub-zero soil temperatures that prevail during most of the year are conducive to the formation of permafrost. Apart from bare rock, which occupies about 44% of the area, the dominant soils of the region are Gelisols. Such soils freeze and thaw once or more during a year which triggers cryoturbation or physical mixing of the soil material. Soils with such intense mixing are called Turbels, and are the most extensive kind of soils, occupying about 31% of the Tundra zone. The Orthels are generally shallower or drier soils and have little or no cryoturbation. The low temperatures and periodic moisture saturation promote the accumulation of organic matter. The Histels are characterized by the high organic matter and occupy about 1.01 million km^2 (5%). These organic soils form the largest contiguous extent of such soils in the world and serve as an important sink of CO_2.

Boreal Forest Biome

Bordering the southern flank of the Polar biome is the Boreal Forest biome in the northern hemisphere. The high volcanic area between Chile and Argentina has similar climatic regimes but the nature of the soil and physiography may result in a different set of habitat conditions. There are about 243,000 km^2 of Andisols or volcanic ash soils characterizing the Boreal Forest biome in the southern hemisphere. Such soils only occur as a small area in the Kamchatca Peninsula in the Northern Hemisphere. In Northern Europe and Siberia, the Boreal Forest biome has a humid part and is bordered in its southern periphery by a semiarid to arid part. In Canada, the semiarid part is in the middle of the continent. A range of soils are present and the most extensive are the Spodosols, which occupy 2.7 million km^2 (20.6%). The Spodosols are mostly in the humid part of the Boreal Forest biome and form under acid vegetation. Histosols are present in the depressions in this cold region.

Temperate Grassland and Forest Biome

The Temperate Grassland and Forest biome extends from about 25 to 55°N with counterparts in the Southern Hemisphere. Large areas in this belt are also deserts. Due to the favorable climate and soil endowments of this biome, much of the land is used for agriculture and native habitats are local and sporadic. About 50% of the zone (9.9 million km^2) is occupied by Mollisols and Alfisols (grasslands), and Ultisols in the forests. Their general good fertility and tilth have made them in great demand for grain production. Due to the long history of civilization in the Temperate Grassland and Forest biome of Europe and China, pristine ecosystems are rare. In large areas, there have been successive replacements and changes in the floral and faunal composition. Within the Temperate biome, are areas characterized by moist winters and dry summers. These Mediterranean conditions are conducive to unique ecosystems. These areas occur around the Mediterranean Sea and small areas in the western U.S. and southern Australia.

Desert Biome

Deserts occupy about 38.6 million km^2 and may be distinguished as warm (or tropical), cool (or temperate) and cold (or boreal) kinds of habitats. Though the biome is characterized by lack of moisture for normal vegetative growth of most plants, the ambient temperature conditions further distinguish the habitat conditions. About 38.6% of the Desert biome is occupied by Aridisols, which by definition have some kind of subsurface horizon. Shifting sands or moving dunes occupy about 13.7% and Entisols, which are very recent deposits, occupying about 33% of the Desert biome. The harshness of the environment has resulted in plants and animals with special adaptive features. Many of the Aridisols are increasingly used for agriculture when irrigation facilities are made available. In this fragile ecosystem, both soil and habitat conditions are drastically altered when irrigation is introduced.

Tropics Biome

An absence of winter and summer temperature extremes characterize the Tropics biome, which occupy about 34.8 million km^2. Availability of moisture separates the semiarid from the humid tropics. As a corollary to the desert, in

Table 2 Dominant soils in the major biomes

Soil order	Ice-free land	Polar		Boreal		Temperate				Desert			Tropics	
		Permafrost	Interfrost	Semiarid	Humid	Semiarid	Humid	Mediterranean warm	Mediterranean cold	Warm	Cool	Cold	Semiarid	Humid
Gelisols	11,260	5413	5847	0	0	0	0	0	0	0	0	0	0	0
Histosols	1526	0	0	299	686	17	82	1	8	0	23	93	37	281
Spodosols	3354	0	0	180	2500	105	450	39	6	0	0	12	0	61
Andisols	912	0	0	47	196	26	135	32	0	17	10	14	198	235
Oxisols	9811	0	0	0	0	9	184	0	0	27	4	0	3387	6199
Vertisols	3160	0	0	2	15	595	303	99	0	239	651	0	1170	87
Aridisols	15,728	1	0	143	76	353	18	203	23	1824	11,014	2029	44	1
Ultisols	11,053	0	0	1	14	266	3122	20	18	0	0	0	4371	3240
Mollisols	9005	0	0	1059	511	1280	1232	874	323	49	1110	2255	136	176
Alfisols	12,621	0	0	1002	1816	1867	2149	860	125	0	0	0	4165	638
Inceptisols	12,830	0	0	181	1095	1636	3078	683	89	0	0	0	3566	2502
Entisols	21,137	0	0	122	232	1173	1697	807	45	1708	10,693	342	3225	1093
Miscellaneous		397,693	0	0	0	0	0	0	0	0	0	0	0	0
Total	130,797	10,547	9685	3552	9446	7348	12,453	3624	791	4418	28,521	5599	20,299	14,514

the humid tropics there are the perhumid areas where the potential evapotranspiration never exceeds the precipitation during any month of the year. This is a biomic condition that deserves much greater detailed studies. The Tropics biome is the home to the Oxisols (9.6 million km^2), which are unique to this biome. These are highly weathered soils where the original vegetation is closed or open forests. Most of the plant nutrients are concentrated in the top 5 cm of the soil and recycled through plant uptake and leaf fall. Ultisols also occur in this biome. Apart from these soils, there are small areas of most of the other soil orders. The wetlands of the tropics present a separate and unique habitat condition and those along the coast have some special soils.

REFERENCES

1. Tootil, E., Ed. *Dictionary of Biology*; Intercontinental Kook Productions, Ltd.: Maidenhead, Berkshire, England, 1980.
2. Eswaran, H.; Beinroth, F.H.; Kimble, J.; Cook, T. Soil Diversity in the Tropics: Implications for Agricultural Development. In *Myths and Science of Soils of the Tropics*; Lal, R., Sanchez, P.A., Eds.; Soil Sci. Soc. Am. Spec. Publ., 1992; *29*, 1–16.
3. Soil Survey Staff, *Soil Taxonomy: A Basic System of Soil Classification for Making and Interpreting Soil Surveys*, 2nd Ed.; Natural Resources Conservation Service. US Department of Agriculture, Handbook 436; US Government Printing Office: Washington, DC, 1999.

GLOBAL WARMING, CARBON SEQUESTRATION TO MITIGATE

Sherwood B. Idso
U.S. Water Conservation Laboratory, Phoenix, Arizona, U.S.A.

Keith E. Idso
Center for the Study of Carbon Dioxide and Global Change, Tempe, Arizona, U.S.A.

INTRODUCTION

Concomitant with mankind's growing numbers and the progression of the Industrial Revolution, there has been a significant increase in the burning of fossil fuels (coal, gas, and oil) over the past 200 years, the carbon dioxide emissions from which have led to ever-increasing concentrations of atmospheric CO_2. This "large-scale geophysical experiment," to borrow the words of two of the phenomenon's early investigators (1), is still ongoing and expected to continue throughout the current century. Furthermore, this enriching of the air with CO_2 is looked upon with great concern, because CO_2 is an important greenhouse gas, the augmentation of which is believed by many to have the potential to produce significant global warming. Therefore, and because of perceived serious consequences, such as the melting of polar ice, rising sea levels, coastal flooding, and more frequent and intense droughts, floods, and storms (2), a concerted effort is underway to slow the rate at which CO_2 accumulates in the atmosphere, with the goal of stabilizing its concentration at a level that would prevent dangerous anthropogenic interference with the planet's climate system.

One of the more promising ways of reducing the rate of rise of the air's CO_2 content is to encourage land management policies that promote plant growth, which removes CO_2 from the atmosphere and sequesters its carbon, first in vegetative tissues and ultimately in soils. Some of these policies deal with managed forests and agro-ecosystems, while others apply to natural ecosystems, such as unmanaged forests and grasslands. In all instances, however, questions abound. Can carbon inputs to soils really be enhanced or carbon losses reduced? Can carbon storage in recalcitrant fractions of soil organic matter be increased, making it possible to successfully maintain new stores of sequestered carbon for long periods? And what if global warming runs wild? Will the ensuing rise in temperature stimulate plant and microbial respiration rates, returning even more CO_2 to the air than is removed by photosynthesis and leading to a negative net ecosystem exchange of carbon? These important questions rank high on the priority lists of many research organizations concerned about the planet's future climate and the sustainability of the biosphere.

REMOVING CARBON FROM THE AIR

The Role of Man

There are only two ways to significantly increase the natural flux of carbon from the atmosphere to the biosphere within the time frame required for effective ameliorative action if the ongoing rise in the air's CO_2 content is indeed a bona fide global warming threat: 1) increase the rate of vegetative CO_2 assimilation (photosynthesis) per unit leaf area and/or 2) increase the total plant population of the globe, i.e., leaf area per unit land area. Additionally, these things must be done without increasing the rate at which carbon is lost from the soil.

Man can do certain things to promote both of these phenomena while meeting the latter requirement as well. He can, for example, increase the rate of CO_2 assimilation per unit leaf area in agro-ecosystems by supplying additional nutrients and water to his crops. As has recently been noted, however, there are significant carbon costs associated with the production and application of fertilizers, as well as the transport of irrigation water; and factoring the CO_2 emissions of these activities into the equation often results in little net CO_2 removal from the atmosphere via these intensified agricultural interventions (3).

Man can also draw more CO_2 out of the air by increasing the acreage of land devoted to growing crops, but this approach simultaneously releases great stores of soil carbon built up over prior centuries. When the plow exposes buried organic matter and it is oxidized, for example, prodigious amounts of CO_2 are produced and released to the atmosphere. But if a transition to less intensive tillage is made on fields that have a long history of conventional management and have thus been largely depleted of carbon, there is a good opportunity for nature to rebuild previously lost stores of soil organic matter (4).

This approach to carbon sequestration is doubly beneficial for it results in a net removal of CO_2 from the atmosphere at the same time that it enhances a whole host of beneficial soil properties (4, 5). Also, abandoned farmlands will gradually replenish their carbon stores, both above- and below-ground, as native vegetation gradually reestablishes itself upon them. And, of course, the process can be hastened and made even more effective if trees are planted on such lands. Even without trees, it has been estimated that agricultural "best management practices" that employ conservation tillage techniques have the potential to boost the current U.S. farm and rangeland soil carbon sequestration rate of 20 million metric tons of carbon per year to fully 200 million metric tons per year (6), which is approximately 13% of the country's yearly carbon emissions (7).

Commercial forests also offer excellent opportunities for CO_2 removal from the air for considerable periods of time, especially when harvested wood is used to produce products that have long lifetimes. In addition, since some species of trees, such as many of those found in tropical rainforests (8), can live in excess of a thousand years, CO_2 can be removed from the atmosphere and sequestered within their tissues—if man protects the trees from logging—until either long after the Age of Fossil Fuels has run its course or until significant changes in energy systems have reduced our dependence on fossil fuels and the CO_2 content of the air has returned to a level no longer considered problematic. Furthermore, carbon transferred to the soil beneath the trees via root exudation and turnover has the potential to remain sequestered even longer.

The Role of Nature

The fact that the biosphere has maintained itself over the eons in the face of a vast array of environmental perturbations (albeit with significant modifications) suggests that earth's plant life has great resiliency and may even be able to exert a restraining influence on climate change.

A particularly important negative feedback of this type is the biosphere's ability to intensify its rate of carbon sequestration in the face of rising atmospheric CO_2 concentrations, as this phenomenon slows the rate of rise of the air's CO_2 content and thereby reduces the degree of intensification of the atmosphere's greenhouse effect. This particular climate-moderating influence of atmospheric CO_2 enrichment was first described in quantitative terms by Idso (9, 10). It begins when the aerial fertilization effect produced by the rising CO_2 content of the atmosphere elicits an increase in plant CO_2 assimilation rate per unit leaf area and when the concomitant plant water use efficiency-enhancing effect of the elevated CO_2 leads to an increase in the total plant population of the globe, due to the ability of more water-use-efficient plants to live and successfully reproduce in areas where it was formerly too dry for them to survive. In fact, these two effects are so powerful, they may actually be able to stabilize the CO_2 content of the atmosphere sometime during the current century, but only if anthropogenic CO_2 emission rates do not rise by an inordinate amount in the interim (9, 10). At the very least, together with the things man can do, they have the potential to "buy time" until other less-CO_2-emitting technologies become available (11).

KEEPING CARBON IN THE SOIL

As more carbon is added to soils via CO_2-enhanced root growth, turnover and exudation, as well as from CO_2-induced increases in leaf litter and other decaying plant parts, the trick of significantly augmenting soil carbon sequestration is to keep at least the same percentage of this carbon in the soil as has historically been the case and to do so in the face of potential global warming.

A number of studies have addressed various aspects of this subject in recent years, with most of them finding that atmospheric CO_2 enrichment has little to no significant effect on plant litter decomposition rates. Furthermore, in nearly all of the cases where elevated CO_2 was observed to impact this phenomenon, the extra CO_2 was found to actually slow the rate of plant decomposition (12). Much the same results have been obtained when analogous studies have used temperature as the independent variable. Warming has had either no effect on CO_2 evolution from the soil, or it has led to an actual decrease in CO_2 loss to the atmosphere (13). Hence, the balance of evidence obtained from these studies suggests that the same—or a greater—percentage of plant material produced in a world of elevated atmospheric CO_2 concentration (and possibly higher mean air temperature) would indeed be retained in the soils of the terrestrial biosphere.

Even more compelling are the results of experiments where scientists have made direct measurements of changes in soil carbon storage under conditions of elevated atmospheric CO_2. Nearly every such study has observed increases in soil organic matter. In a Free-Air CO_2 Enrichment (FACE) experiment where portions of a cotton field were exposed to a 50% increase in atmospheric CO_2, for example, Leavitt et al. (14) found that 10% of the organic carbon present in the soil below the CO_2-enriched plants at the conclusion of the three-year experiment came from the extra CO_2 supplied to the FACE plants. In addition, some of the stored carbon had made its way into a very recalcitrant portion of the soil organic matter that had an average soil residence time of 2200 years.

Here, too, most experiments indicate that concomitant increases in temperature do not negate the increased carbon storage produced by atmospheric CO_2 enrichment. In a two-year study of perennial ryegrass grown at ambient and twice-ambient atmospheric CO_2 concentrations, as well as ambient and ambient +3°C temperature levels, for example, Casella and Soussana (15) determined that the elevated CO_2 increased soil carbon storage by 32% and 96% at low and high levels of soil nitrogen supply, respectively, "with no significant increased temperature effect." Hence, as in the case of studies of plant decomposition rates, the balance of evidence obtained from these studies also suggests that the same—or a greater—percentage of plant material produced in a world of elevated atmospheric CO_2 concentration (and possibly higher mean air temperature) would indeed be retained in the soils of the terrestrial biosphere.

Table 1 Potential rates of carbon sequestration (kilograms carbon per hectare per year) due to land management practices that could be employed for this purpose

Improved rangeland management	50 to 150
Improved pastureland management	
Commercial fertilizer applications	100 to 200
Manure applications	200 to 500
Use of improved plant species	100 to 300
Improved grazing management	300 to 1300
Nitrogen fertilization of mountain meadows	100 to 200
Restoration of eroded soils	50 to 200
Restoration of mined lands	1000 to 3000
Conversion of cropland to pasture	400 to 1200
Conversion of cropland to natural vegetation	600 to 900
Conversion from conventional to conservation tillage	
No till	500
Mulch till	500
Ridge till	500

Adapted from data reported by Follett, R.F.; Kimble, J.M.; Lal, R. *The Potential of U.S. Grazing Lands to Sequester Carbon and Mitigate the Greenhouse Effect*; Lewis Publishers, Boca Raton, FL, 2001; 1–442, and by Lal, R.; Kimble, J.M.; Follett, R.F.; Cole, C.V. *The Potential for U.S. Cropland to Sequester Carbon and Mitigate the Greenhouse Effect*; Sleeping Bear Press, Chelsea, MI, 1998; 1–128.

CONCLUSIONS

As the air's CO_2 content continues to rise, there will almost certainly be a significant upward trend in the yearly production of terrestrial vegetative biomass, due to the growth-enhancing aerial fertilization effect of atmospheric CO_2 enrichment and the concomitant CO_2-induced increase in plant water use efficiency that enables plants to grow where it is currently too dry for them. Experimental evidence further suggests that at least the same percentage—but in all likelihood more—of this yearly-increasing mass of plant tissue will be sequestered in earth's soils. Consequently, it is almost impossible to conclude that the carbon sequestering prowess of the planet will not be greatly enhanced in the years ahead, even without any overt actions on the part of man. Hence, if the nations of the earth were to implement even a modicum of carbon-conserving measures—such as 1) using minimum tillage techniques wherever possible in agricultural settings, 2) allowing abandoned agricultural land to revert to its natural vegetative state, 3) allowing stands of trees that can grow to very old age to actually do so, and 4) employing wise forestry practices to produce wood for making products that have long lifetimes—it is possible that the antiwarming feedback produced by the subsequent removal of CO_2 from the atmosphere would be sufficient to keep the risk of potential greenhouse gas-induced global warming at an acceptable level. Estimates of the carbon-sequestering power of some of these "best management practices" are given in Table 1.

REFERENCES

1. Revelle, R.; Suess, H.E. Carbon Dioxide Exchange Between Atmosphere and Ocean and the Question of an Increase of Atmospheric CO_2 During the Past Decades. Tellus **1957**, *9*, 18–27.
2. Intergovernmental Panel on Climate Change. *Climate Change 2001: The Scientific Basis, Summary for Policy Makers and Technical Summary of the Working Group I Report*; Cambridge University Press: Cambridge, UK, 2001; 1–98.
3. Schlesinger, W.H. Carbon Sequestration in Soils: Some Cautions Amidst Optimism. Agric. Ecosys. Environ. **2000**, *82*, 121–127.
4. Lal, R.; Kimble, J.M.; Follett, R.F.; Cole, C.V. *The Potential for U.S. Cropland to Sequester Carbon and Mitigate the Greenhouse Effect*; Sleeping Bear Press: Chelsea, MI, 1998; 1–128.

5. Idso, S.B. *Carbon Dioxide and Global Change: Earth in Transition*; IBR Press: Tempe, AZ, 1989; 1–292.
6. Jawson, M.D.; Shafer, S.R. Carbon Credits on the Chicago Board of Trade? Agric. Res. **2001**, *49*(2), 2.
7. Comis, D.; Becker, H.; Stelljes, K.B. Depositing Carbon in the Bank. Agric. Res. **2001**, *49*(2), 4–7.
8. Chambers, J.Q.; Higuchi, N.; Schimel, J.P. Ancient Trees in Amazonia. Nature **1998**, *391*, 135–136.
9. Idso, S.B. The Aerial Fertilization Effect of CO_2 and its Implications for Global Carbon Cycling and Maximum Greenhouse Warming. Bull. Amer. Meteorol. Soc. **1991**, *72*, 962–965.
10. Idso, S.B. Reply to Comments of L.D. Danny Harvey, Bert Bolin, and P. Lehmann. Bull. Amer. Meteorol. Soc. **1991**, *72*, 1910–1914.
11. Izaurralde, R.C.; Rosenberg, N.J.; Lal, R. Mitigation of Climatic Change by Soil Carbon Sequestration: Issues of Science, Monitoring, and Degraded Lands. Adv. Agron. **2001**, *70*, 1–75.
12. Nitschelm, J.J.; Luscher, A.; Hartwig, U.A.; van Kessel, C. Using Stable Isotopes to Determine Soil Carbon Input Differences Under Ambient and Elevated Atmospheric CO_2 Conditions. Global Change Biol. **1997**, *3*, 411–416.
13. van Ginkel, J.H.; Whitmore, A.P.; Gorissen, A. *Lolium perenne* Grasslands May Function as a Sink for Atmospheric Carbon Dioxide. J. Environ. Quality **1999**, *28*, 1580–1584.
14. Leavitt, S.W.; Paul, E.A.; Kimball, B.A.; Hendrey, G.R.; Mauney, J.R.; Rauschkolb, R.; Rogers, H.; Lewin, K.F.; Nagy, J.; Pinter, P.J., Jr.; Johnson, H.B., Jr. Carbon Isotope Dynamics of Free-Air CO_2-Enriched Cotton and Soils. Agric. For. Meteorol. **1994**, *70*, 87–101.
15. Casella, E.; Soussana, J.-F. Long-Term Effects of CO_2 Enrichment and Temperature Increase on the Carbon Balance of a Temperate Grass Sward. J. Exp. Bot. **1997**, *48*, 1309–1321.

GRASSLANDS, SOILS OF THE

D.D. Malo
South Dakota State University, Brookings, South Dakota, U.S.A.

INTRODUCTION

Soil science developed in recent times in response to problems. In Europe during the 1800s, food shortages, social upheavals, and declining soil productivity brought about the need to study soil to improve and increase its productivity. At the same time, in Russia, a need arose to administer and manage geographically diverse soil resources. Russia had large areas of fertile, productive soils unlike those in Europe. As a result, Russian scientists developed an inventory of agricultural resources and determined the factors causing soils to vary across Russia. Soils were found to have relationships with climatic and vegetative zones (1). It is from these concepts that the living soil individual was developed. Soil is a natural body and not a geologic formation. With time, soil develops from the parent material under the influence of the climate, vegetation, and topography (relief). These five factors of soil formation are interdependent and not independent. Changing one soil-forming factor often changes other soil-forming factors. Changing any one, some, or all of the soil-forming factors causes differences in soils and soil profiles (2).

Soils and plants interact and evolve together forming different ecosystems. These soil–plant relationships are often most strongly expressed when native vegetation is present. The soils in grasslands, a major ecosystem, are very different from other soils due to this interaction.

IMPORTANCE, LOCATION, AND EXTENT OF GRASSLAND SOILS

Grassland soils have thick, dark-colored, humus-rich, surface horizons or layers and are some of the most productive soils in the world. These soils occupy about 7% of the world's land area (Table 1). The majority of grassland soils occur in temperate (3.4%) and boreal (3.2%) regions (3). These soils are the dominant food- and fiber-producing soils in subhumid and semiarid regions because of favorable soil properties and climatic conditions. The world's breadbaskets (e.g., U.S. Midwest and Great Plains prairies; Canadian prairie provinces of Manitoba and Saskatchewan; the pampas of Argentina, Paraguay, and Uruguay; and the steppe regions of Europe, Russia, Mongolia, and Northern China) are dominated by grassland soils (Fig. 1). The largest grassland area in the world is found in Kazakhstan, Russia, and the Ukraine. Generally, small grains and grain sorghum are raised in the drier grassland regions. The warmer, humid grasslands are better suited for row crops like maize (corn) and soybeans. Where slopes are too great for cultivation or the climate is not favorable, grassland soils are used for pasture and rangeland. Native grassland types include: desert grasslands, <15 cm high; short-grass prairie, 15–30 cm high; mid-grass prairie, 30–100 cm high; and tall-grass prairie, 1–3 m high (4).

Grassland soils (e.g., Mollisols) are very extensive in the U.S. (Fig. 2), especially the Midwest and Great Plains. They occupy about 21.5% of the U.S. land area (3), more than any other soil ecosystem.

Grasslands tend to occur in subhumid to semiarid regions, where average annual precipitation ranges from 20 to 90 cm and average annual temperatures range from 0 to 32°C (5). Grasslands tend to be located in areas between deserts (e.g., Aridisols) and forested (e.g., Alfisols) areas (3). Different climates with similar effective annual precipitation levels (e.g., prairies of California and western Kansas) give rise to grassland soils with similar properties (6). Soil texture and topography does modify the boundaries between prairie, forest, and desert areas (e.g., sandy, south-facing slopes are drier than loamy, north-facing slopes) (7).

PROPERTIES OF GRASSLAND SOILS

Soils that form under grass vegetation have unique soil properties and characteristics (Table 2). Grassland soils (e.g., Houdek soil—fine-loamy, mixed, superactive, mesic Typic Argiustoll) have thick, soft, dark-colored, moderately high organic carbon-containing A horizons (e.g., 0.5–10% organic carbon) with developed structure (e.g., mollic surface). Granular structure is most common in surface horizons with blocky (humid areas)

Encyclopedia of Soil Science

G

Table 1 Global grassland soil distribution (based on ice-free land)

Continent/region	Area (km^2)	Area (% of region)	Area (% of world's grasslands)	Area (% of world's total area)
Africa	76,911	0.26	0.84	0.06
Asia	4,041,017	8.45	43.90	3.10
Australia/Oceania	103,569	1.31	1.13	0.08
Central America	32,302	4.64	0.35	0.02
Europe	724,280	12.62	7.87	0.56
North America	3,230,829	15.57	35.10	2.48
South America	994,431	5.67	10.81	0.76
Total	9,203,339		100.00	7.06

(From Refs. 3, 20, and unpublished data from P. Reich, USDA–NRCS, World Soil Resources Division.)

and prismatic (semiarid areas) structure present in B horizons. Cultivation often causes granular surface soil structure to change to subangular blocky (9). Grassland soils have sufficient strength to support grazing livestock and normal cultivational activity. The dark-colored A horizon slowly changes into a B horizon in most prairie-derived soils (Fig. 3).

The Houdek Bt horizon has developed structure and evidence of clay illuviation (Table 2). Clay illuviation is much less than that found in comparable forested soils. Base saturation and soil pH are high in all layers of most grassland soils. Divalent cations, especially Ca^{2+}, dominate the soil-cation exchange sites. Most grassland soils have smectite (2:1 lattice silicate clay minerals) present in the clay fraction, and textures are usually finer than loamy fine sand (10). The average cation-exchange capacity of the soil solum is at least $15\,cmol_c\,kg^{-1}$ but averages $25–35\,cmol_c\,kg^{-1}$. Grassland soils can have many types of B horizons. Soils of the steppes tend to have horizons of lime accumulation (Bk in Fig. 3), whereas the grassland soils in humid areas tend to lack horizons of lime accumulation but have argillic, Bt, horizons. Some grassland soils have pans (e.g., duripans, petrocalcic layers), and some have strongly developed eluvial, E, horizons (e.g., albic). Still other grassland soils have only weakly developed B horizons (Bw and Bg) or no B horizon at all.

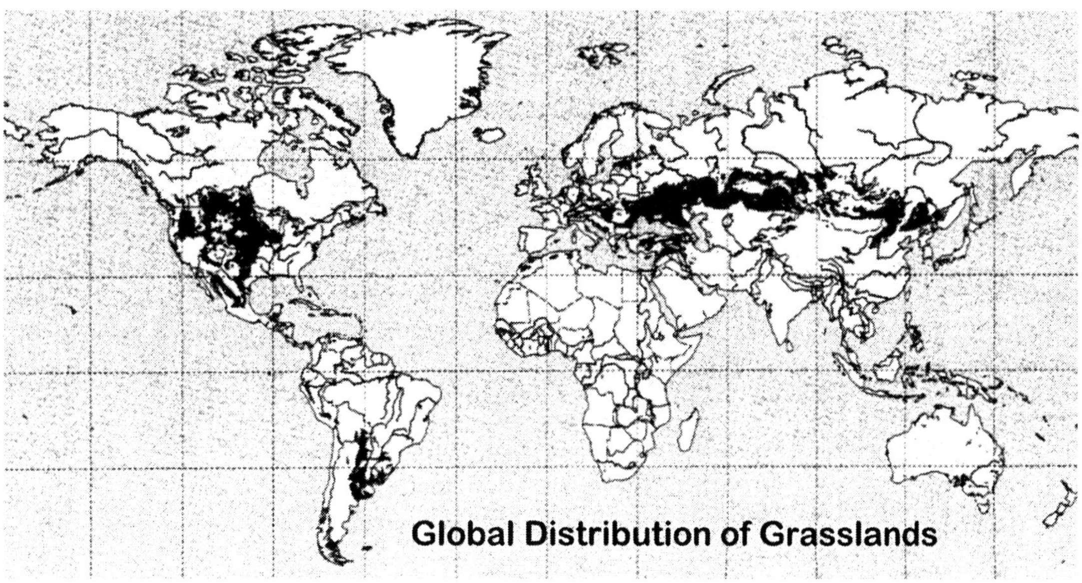

Fig. 1 Distribution of grassland soils (Mollisols) in the world. (From Refs. 3, 20.)

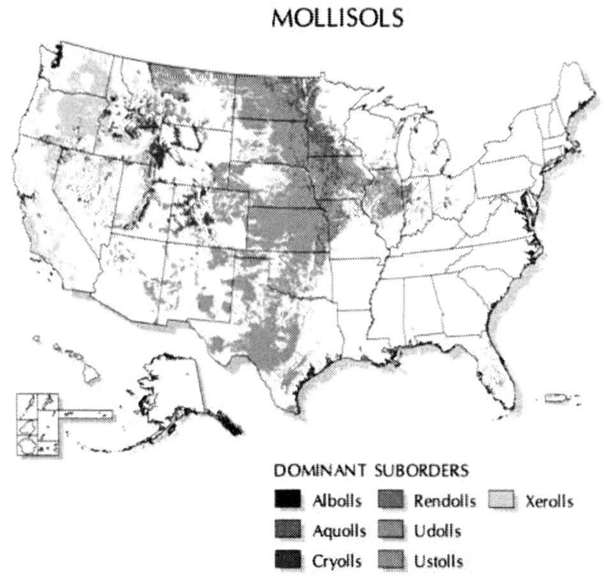

Fig. 2 Distribution of grassland soils (Mollisols) in the U.S. (From Ref. 21.)

GENESIS OF GRASSLAND SOILS

The genesis of grassland soils is strongly dependent on the vegetation. Grasses, a source of organic matter and nutrient cycling, provide these two key components to the soil body. The principal process in grassland genesis is the accumulation of high base-content humus in the soil from the dense grass root systems near the soil surface (2). Through the microbial decay of plant roots and plant tissue, organic matter is added to the soil surface and the soil profile in grasslands. Soil macrofauna (e.g., earthworms) incorporate the above-ground biomass tissue into the soil surface and help increase surface organic-carbon levels. Most grass roots are found in the top 30 cm of grassland soils (11). Each year more than 50% of the biomass produced by unharvested grasses (nearly all of the above-ground biomass and about 30% of the below-surface biomass) is added to the soil (12). The

Fig. 3 Typical Great Plains grassland soil (Houdek—fine-loamy, mixed, superactive, mesic Typic Argiustoll). Scale is in feet.

Table 2 Selected soil properties of typical grassland (Houdek) soil

Horizon	Depth (in.)	Clay (%)	pH	Org. C (%)	Base saturation (%)	Structure: Grade	Structure: Shape
Houdek (grassland derived)							
Ap	0–6	22.0	6.2	4.6	84	Strong	Granular
Bt1	6–10	29.2	6.4	1.8	86	Strong	Subangular blocky
Bt2	10–18	26.5	7.0	1.1	95	Moderate	Prismatic
Bk1	18–28	25.9	8.2	0.5	100	Moderate	Prismatic
Bk2	28–40	24.8	8.4	0.3	100	Weak	Prismatic
C	40–80	23.2	8.4	0.2	100	Structureless	Massive

(From Ref. 8, http://www.statlab.iastate.edu/cgi-bin/osd/osdname.cgi?-P (accessed January 2001), and unpublished data from D.D. Malo, South Dakota State University.)

amount of above-ground grassland-biomass production can range from 1500 to 3500 kg ha^{-1} yr^{-1}, dry weight (2). The amount of air-dry organic matter added per year can vary depending on climate conditions and vegetation type (e.g., 1250 kg ha^{-1} for humid prairie areas to 600 kg ha^{-1} for short- to mid-grass prairies) (13).

The distribution of organic carbon in grassland soils shows a gradual decline with increasing soil depth due to the gradual decline of fibrous grass roots and microbial activities with increasing soil depth (Table 2). This pattern of organic-carbon distribution is strikingly different from that in forest-derived soils where the organic carbon is high in a thin surface horizon and low in the rest of the soil. As a result, the structural stability of grassland soil peds tends to be stronger than nongrassland soil peds. Humus is not very water soluble, so it tends to stay in one location unless there is suspension or some mechanism for mechanical movement. The organic-carbon decline continues until the native rooting depth of the grasses is reached, and there is little or no organic carbon.

Grasses are large users of bases, especially calcium (Ca^{2+}). As a result, when grass residue is added to the soil surface, large amounts of cations (Ca^{2+}, Mg^{2+}, and K^{1+}) are brought to the surface, replacing cations lost by leaching and weathering activities. This results in higher soil pH values and base saturation throughout the profile (Table 1). The high concentration of bases help, along with the high content of shrinking–swelling silicate clay minerals and high humus levels, to form the granular structure that is common in grassland soils. In general, leaching under grass vegetation is minimal when compared to other types of vegetation (e.g., forest). Deep percolation of precipitation to the water table is not common, and deep leaching is not usually evident in most grassland soils. Grasses tend to utilize the water in the soil, preventing the deep leaching of nutrients. The grass plants keep recycling bases to the soil surface and this limits soil acidity, clay illuviation, and base-saturation reductions.

The development of textural horizons and silicate clay illuviation in grassland soils tends to happen in three steps (14).

1. Removal of free carbonates—as long as free lime is present, the soil remains flocculated and little or no silicate clay will translocate.
2. Silicate clay formation and alteration—silicate clays are formed and altered as a result of weathering and soil genesis.
3. Silicate clay eluviation and illuviation—the fine silicate clays ($<0.2\ \mu m$) move and precipitate out lower in the soil profile. The place where silicate clay illuviation occurs depends on pH, lime content, microbial activity, and moisture levels. Often, joint illuviation of both silicate clays and humus in grassland soils forms organo-argillans.

Another key indicator of soil genesis in grassland soils is phosphorus. Significant differences in phosphorus species have been noted in different grassland soils. Semiarid grasslands (Ustolls or Chestnut soils) had a larger percentage of calcium phosphates when compared to humid grasslands (Udolls or Chernozems), whereas humid grasslands had a higher percentage of iron phosphate (15). Phosphorus distributions (total, organic, and inorganic) within the soil profile depend on horizon, lime illuviation, vegetation, and drainage (14).

With cultivation in grassland soils, significant changes occur in the soil. Significant organic-carbon losses (>20% reduction in organic C levels after 20 years of cultivation) and soil quality reductions occur (5). The losses are most dramatic in the soil surface and diminish with increasing soil depth (14).

Table 3 Soil classification of grassland soils

Soil or environmental conditions	Soil taxonomy	1938/1949 equivalents	FAO equivalents
Albic horizon	Alboll	Planosol	Planosol
Wet, hydric conditions	Aquoll	Gley	Gleysol
Very cold	Cryoll	Chernozem	Greyzem
Highly calcareous parent materials	Rendoll[a]	Rendzina	Rendzina
Humid	Udoll	Brunizem	Phaeozem
Moist spring/dry summer	Ustoll	Chernozem, Chestnut	Chernozem, Kastanozem
Dry summer/moist winter	Xeroll	Chestnut, Brown	Chernozem, Kastanozem

[a] Mostly formed under forest vegetation.

(From Refs. 3, 16–18. http://www/fao.org/waicent/faoinfo/agricult/agl/agll/key2soil.htm accessed January 2001.)

CLASSIFICATION OF GRASSLAND SOILS

The classification of grassland soils with various soil-classification systems is given in Table 3.

SUMMARY

Throughout the world where grasslands are located, developed cultures exist. These soils are extremely productive and have allowed societies to flourish. About 40% of the grasslands are tall-grass prairies and 60% are short- and mid-grass steppes (4). Grasslands have higher organism activity than most other soils (5). Carbon sequestration in grasslands is critical to the global warming process. Grasslands are not only important for food and fiber production, but they have the potential to help solve one of the major problems facing human survival, global warming (19). Grasslands are one of the most important ecosystems in the world.

REFERENCES

1. Kellogg, C.E. Soil in Society. *Soils and Men, Yearbook of Agriculture*; U.S. Department of Agriculture, U.S. Government Printing Office: Washington, DC, 1938; 863–886.
2. Jenny, H. *Factors of Soil Formation*; McGraw-Hill: New York, NY, 1941.
3. Soil Survey Staff, *Soil Taxonomy*, 2nd Ed.; Agriculture Handbook 436, U.S. Department of Agriculture, Natural Resources Conservation Service, Superintendent of Documents, U.S. Government Printing Office: Washington, DC, 1999.
4. Buol, S.W.; Hole, F.D.; McCracken, R.J.; Southard, R.J. *Soil Genesis and Classification*, 4th Ed.; Iowa State University Press: Ames, IA, 1997; 317–329.
5. Brady, N.C.; Weil, R.R. *The Nature and Properties of Soils*, 12th Ed. Prentice Hall: Upper Saddle River, NJ, 1999; 29–116.
6. Barshad, I. Chemistry of Soil Development. In *Chemistry of the Soil*, 2nd Ed.; American Chemical Society Monograph 160, Bear, F.E., Ed.; Reinhold: New York, NY, 1964; 1–70.
7. Jenny, H. *The Soil Resource*; Ecological Studies 37, Springer: New York, NY, 1980; 337–360.
8. Natural Resources Conservation Service, Houdek Series, Official Series Description Sheet, U.S. Department of Agriculture, Natural Resources Conservation Service: Ames, IA, 1997.
9. Soil Survey Division Staff, *Soil Survey Manual*; U.S. Department of Agriculture Handbook 18, U.S. Department of Agriculture, Superintendent of Documents, U.S. Government Printing Office: Washington, DC, 1993; 59–196.
10. Kononova, M.M. Humus of Virgin and Cultivated Soils. In *Soil Components*; Gieseking, J.E., Ed.; Springer: New York, NY, 1975; Vol. 1, 475–576.
11. Douglas, C.L.; Fehrenbacher, J.B.; Ray, B.W. The Lower Boundary of Selected Mollisols. Soil Sci. Soc. Am. Proc. **1967**, *31*, 795–800.
12. Kononova, M.M. *Soil Organic Matter*; Pergamon Press: Oxford, 1966.
13. Thorp, J. How Soils Develop Under Grass. *Yearbook of Agriculture*; U.S. Department of Agriculture, U.S. Government Printing Office: Washington, DC, 1948; 55–66.
14. Fenton, T.E. Mollisols. In *Pedogenesis and Soil Taxonomy, Part II, The Soil Orders*; Developments in Soil Science 11B, Wilding, L.P., Smeck, N.E., Hall, G.F., Eds.; Elsevier: Amsterdam, 1983; 125–163.
15. Westin, F.C.; Buntley, G.J. Soil Phosphorus in SD: III. Phosphorus Fractions of Some Borolls and Ustolls. Soil Sci. Soc. Am. Proc. **1967**, *17*, 287–293.
16. Baldwin, M.; Kellogg, C.E.; Thorp, J. Soil Classification. *Soils and Men*; Yearbook of Agriculture, U.S. Department of Agriculture, U.S. Government Printing Office: Washington, DC, 1938; 979–1001.
17. Thorp, J.; Smith, G.D. Higher Categories of Soil Classification. Soil Sci. **1949**, *67*, 117–126.
18. Food and Agriculture Organization (FAO) of the United Nations Land and Plant Nutrition Management Service, *Key to the FAO Soil Units in the FAO/UNESCO Soil Map of the World*; FAO: Rome, 2000.
19. Lal, R.; Kimble, J.M.; Folet, R.F.; Cole, C.V. *The Potential of U.S. Cropland to Sequester Carbon and Mitigate the Greenhouse Effect*; Sleeping Bear Press: Chelsea, MI, 1998; 35–51.
20. U.S. Department of Agriculture–Natural Resources Conservation Service, *Global Distribution of Mollisols*; Soil Survey Division, World Soil Resources: Washington, DC, 2000.
21. U.S. Department of Agriculture–Natural Resources Conservation Service, *Percent of Land Area in Mollisols, Map m4034*; U.S. Department of Agriculture, Natural Resources Conservation Service, National Soil Survey Center: Lincoln, NE, 1999.

GREENHOUSE EFFECT, WORLD SOILS AND

Rattan Lal
The Ohio State University, Columbus, Ohio, U.S.A.

INTRODUCTION

Earth's vast reservoir of carbon (C) is contained in four distinct pools (Table 1). The soil C pool comprises both organic and inorganic forms. Soil organic carbon (SOC) is estimated at 1580 Pg (1 Pg = petagram = 10^{15} g = 1 billion metric tons) and soil inorganic carbon (SIC) at 748 Pg. The atmospheric pool contains 750 Pg of C, and the vegetation or the biotic pool amounts to about 610 Pg of C. The pedologic or soil C pool, with a total reserve of about 2300 Pg in both organic and inorganic form, is about 3.1 times the atmospheric pool and 3.8 times the biotic pool (1–9).

In addition to being a pool, world soils also play a major role in the global C cycle (Fig. 1). The SOC pool is highly dynamic, is concentrated near the soil-surface, and comprises material of plant and animal origin at various stages of decay and decomposition. A large part of the SOC pool is held in the organic soils of tundra and boreal forest ecosystems. It is this dynamic SOC pool that is intimately connected with the projected global warming or the accelerated greenhouse effect. Change in soil C pool by 1 Pg is equivalent to 0.47 ppm change in atmospheric concentration of CO_2.

These global pools are interconnected through fluxes of C between them (Fig. 1). The atmospheric pool has been increasing at the expense of other pools. Some known sources and sinks are listed in Table 2. Two principal sources are fossil-fuel combustion (6.8 Pg/yr), deforestation, and soil disturbance (1.6 Pg/yr). Two proven sinks are atmosphere (3.5 Pg/yr) and ocean (2.0 Pg/yr). The unknown sink for the remaining 2.7 Pg/yr is most likely the world soils and other terrestrial ecosystems (11). The largest exchanges occur between: 1) world biota and the atmosphere with an annual flux of 120 Pg photosynthesized by plants and returned to the atmosphere in equal amounts through plant and soil respiration, and 2) oceans and the atmosphere amounting to an annual flux of 107 Pg from atmosphere to the ocean and return of 105 Pg making the ocean a net sink of about 2 Pg/yr.

There has been a steady increase in the atmospheric concentration of greenhouse gases (GHGs, CO_2, CH_4, and N_2O) since the beginning of the industrial revolution in about the 1850s. This increase is primarily due to human activities, e.g., industry-based fossil-fuel consumption, and agriculture-based deforestation and conversion to cultivable land involving intensive use of fertilizers and manures. The CO_2 concentration in the atmosphere has increased from 280 ppmv in about 1850 to 370 ppmv in 2000 (4), and is increasing at the rate of 0.5% per yr (Table 2).

AGRICULTURAL ACTIVITIES AND GREENHOUSE GAS EMISSIONS

In addition to industrial sources, GHGs are also emitted by numerous agricultural activities (Fig. 2). Agricultural sources of CO_2 are deforestation, land use conversion, biomass burning, soil plowing and other mechanical disturbances of the topsoil, and fossil-fuel consumption. About 50% of the total increase in atmospheric CO_2 is attributed to agricultural activities. Among principal sources of CH_4 are enteric fermentation in ruminant animals, cultivation of rice paddies, anaerobiosis in poorly drained soils, and biomass burning. Use of nitrogenous fertilizers and manures is the principal source of N_2O emission. Soil management plays a major role in the type and amount of emission of GHGs.

The rate of oxidation or mineralization of the SOC pool depends on its composition and numerous other factors. Factors that increase oxidation are soil disturbance, high soil temperature, lack of protective vegetal cover, and risks of soil erosion and degradation. In contrast, factors that decrease oxidation are minimal soil disturbance, low soil temperature, minimal soil disturbance, presence of protective vegetal cover, and high soil resilience. The annual flux of CO_2 from soil to the atmosphere, about 60 Pg, is due to soil respiration, whose rate increases with increase in soil temperature. The mean resident time of SOC in soil is about 32 years, but some resistance humic substances can stay in soil for thousands of years. Losses of SOC pool by oxidation are accentuated by soil erosion by water and wind (12) and other degradative processes.

Introduction of mechanized agriculture (e.g., tractor based plowing and other farm operations causing soil disturbance) have accentuated losses of SOC pool from

Table 1 Global pools of C

Pool	Amount (Pg)
1. Ocean	
i) Intermediate and deep ocean	38,100
ii) Surface ocean	1,020
iii) Dissolved organic carbon	700
iv) Surface sediment	150
v) Marine biota	3
2. Soil (1-m depth)	
i) SOC	1,580
ii) SIC	748
3. Vegetation	610
4. Atmosphere	750

(From Ref. 4.)

Table 2 Sources and sinks of atmospheric CO_2

Source/sink	Flux (Pg)
A. Source	
1) Fossil fuel combustion	6.8
2) Deforestation, land use and world soils	1.6
Total	8.4
B. Sinks	
1) Increase in atmospheric concentration	3.5
2) Ocean uptake	2.0
Total	5.7
Balance of unknown sink most likely to be world soils	2.7

(From Refs. 1, 4 and 5.)

cultivated soils. Some soils lose as much as 25% of their SOC pool within 1 year after deforestation and cultivation (13). The rate of loss of SOC content is often more in the tropics than in the temperate climate. Some tropical soils, cultivated for millennia in Asia, have extremely low levels of remaining SOC content (as low as 0.2%). These soils have probably lost 75–80% of their initial SOC content. It is likely that as much as 66–90 Pg of C in the atmospheric pool has been contributed by world soils (14).

In addition to being a source of atmospheric C (as CO_2 CH_4 or CO), world soils are also a major source of N_2O. The relative contribution of N_2O has also increased because of the intensive use of nitrogenous fertilizers since World War II. The magnitude of N_2O emission

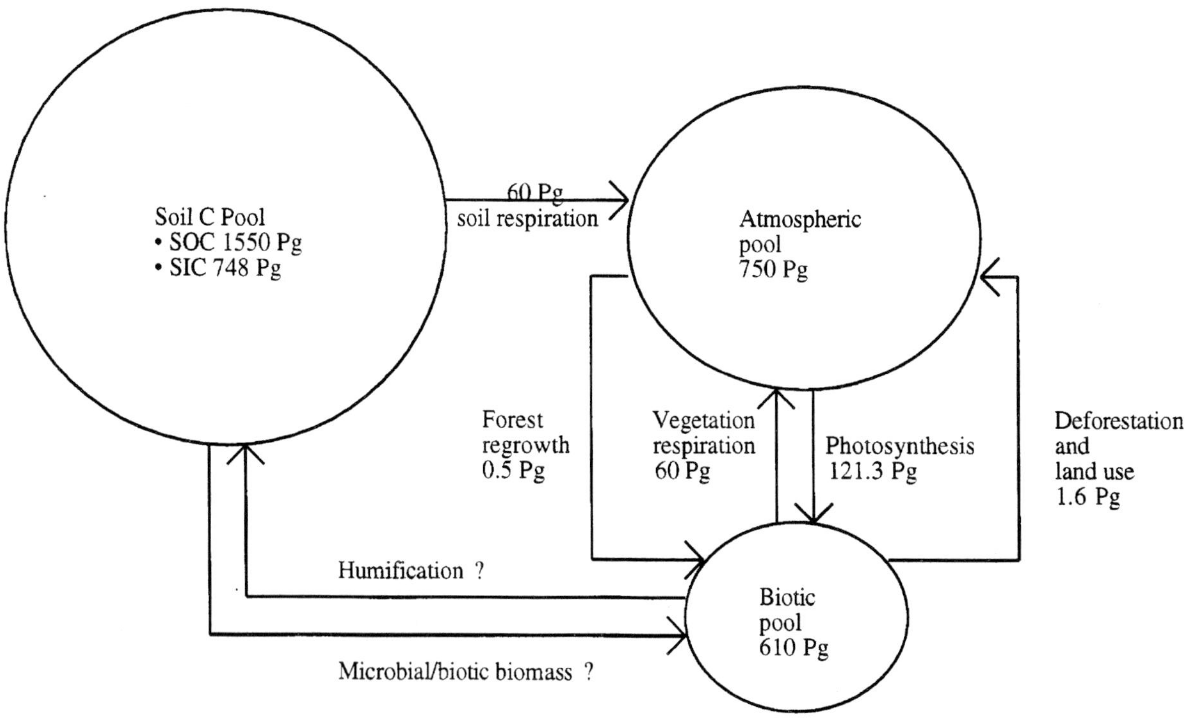

Fig. 1 Interactive effects among three C pools (soils, biota, and atmosphere) have lead to a steady increase in the atmospheric pool at the expense of the soil and biotic pools. Values in the arrow indicate direction and magnitude of the known fluxes, while the question mark indicates the lack of information. (From Refs. 1, 4, and 10.)

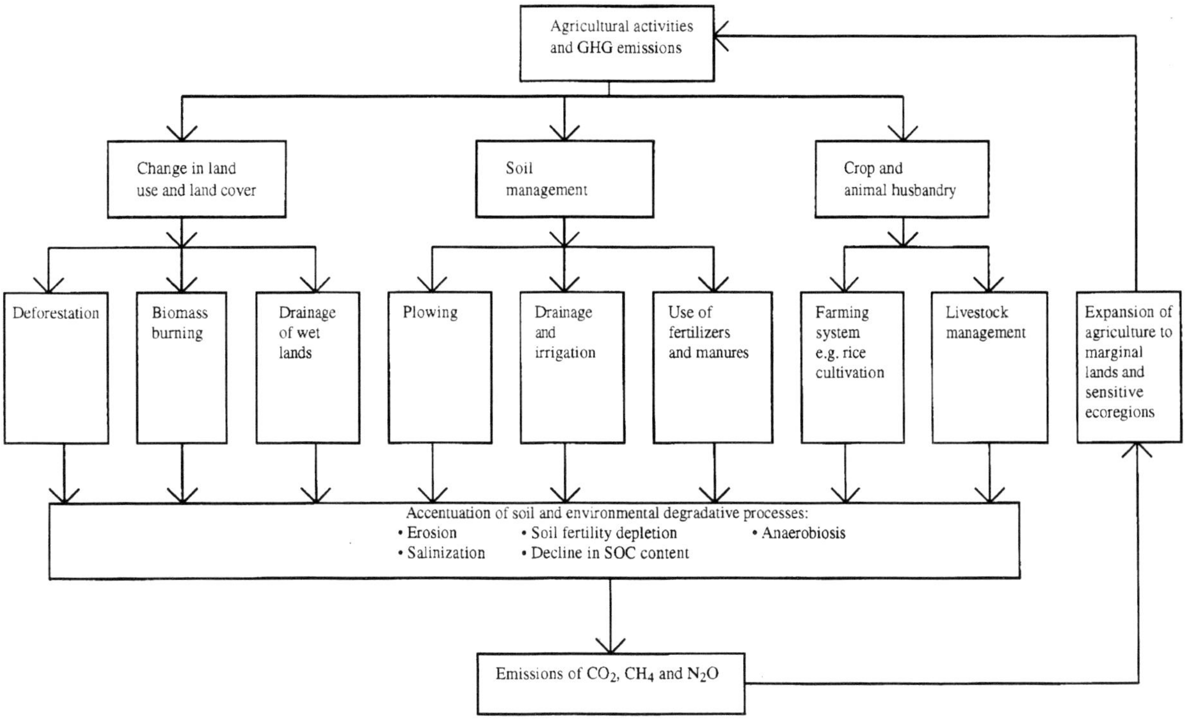

Fig. 2 Agricultural activities and emission of GHG from soil to the atmosphere.

varies among soils, type of nitrogenous fertilizer used (formulation), climate, land use, and soil and crop management (3).

POTENTIAL OF WORLD SOILS AS C SINK

The SOC in most agricultural soils is below their potential capacity. Since a considerable part of atmospheric CO_2 has come from world soils, it is possible to reverse the trend and store C back into the soil. Increase in SOC content, through change in land use and farming practices, could slow the rate of increase of atmospheric CO_2 and provide a negative feedback to the greenhouse effect. Because of their vast pool, it is very likely that world soils are the unknown sink for 2.7 Pg of C/yr (Table 2).

Principal mechanisms by which soils can be used as an effective sink for some of the atmospheric CO_2 through increase in SOC content include: 1) humification or conversion of biomass into resistant humic substances with long turnover time, 2) aggregation or formation of stable organo-mineral complexes that render SOC inaccessible to microbes, and 3) deep incorporation of SOC below the plow layer through deep root system development and translocation by leaching. These mechanisms can be set-in-motion through adoption of proper land use, choice of appropriate farming/cropping systems, and use of judicious soil and crop management practices.

The potential of C sequestration in world soils depends on the current land use, present level of SOC content, the initial SOC content prior to soil disturbance by deforestation and conversion to other land uses, climatic or ecoreginal characteristics, and farming systems. Among all land uses, the largest potential of increase in SOC content is in arable lands. If the SOC content of the world's arable land can be increased at the low rate of 0.01% per yr to 1 m depth, which may be difficult to achieve for soils of arid and semiarid regions, the C sequestration potential of such a possibility is about 2 Pg/yr (Eq. 1). These calculations are based on the assumption that the mean bulk density of soils is 1.33 Mg/m^3.

$$\begin{aligned} &\text{C sequestration potential of } 1500\times10^6\text{ ha of worldarable land} \\ &= (1500\times10^6\text{ ha})\left(10^4\,\frac{\text{m}^2}{\text{ha}}\right)\times(1\text{ m}) \\ &\times\left(1.33\,\frac{\text{Mg}}{\text{m}^3}\right)\left(\frac{1\times10^{-4}}{\text{yr}}\right) \\ &= 2\text{ Pg/yr} \end{aligned} \tag{1}$$

MANAGING WORLD SOILS FOR MITIGATION OF THE GREENHOUSE EFFECT

There are several soil management practices that can be used to increase SOC content and reduce the risks of accelerated greenhouse effect (Fig. 3). These practices can be grouped under two broad categories: 1) those which decrease losses due to soil degradative processes, and 2) those which increase SOC content through soil restorative processes (Table 3). Principal strategy is of agricultural intensification through adoption of appropriate land use and recommended agricultural practices.

A tremendous potential of mitigation of the greenhouse effect exists through restoration of degraded and drastically disturbed soils. It is estimated that the area of the degraded soils of the world is about 2.0 billion ha (15). These soils include the presently cultivated land area, pastoral land areas, and other managed and natural ecosystems. Restoration to improve soil quality can lead to an increase in below ground and above ground biomass production, increase in SOC content, and strengthening of a negative feedback to the greenhouse effect. The C sequestration potential of restoration of degraded soils is about 3 Pg/yr (Eq. 2)

$$\begin{aligned}&\text{C sequestration potential of restoration}\\&\quad\text{of degraded soils}\\&\qquad= (2\times 10^9\ \text{ha})\times(10^4\ \text{m}^2/\text{ha})\\&\qquad\quad\times(1\ \text{m})(1.5\ \text{Mg/m}^3)\times(10^{-4}/\text{yr})\\&\qquad= 3\ \text{Pg/yr}\end{aligned} \tag{2}$$

These calculations are based on the assumption that mean soil bulk density to 1 m depth is 1.5 Mg/m^3. The potential of 3 Pg/yr is almost equivalent to the current rate of annual increase in atmospheric C at 3.5 Pg/yr (Table 2), and can be realized over 25–50 years. Realization of this potential requires development of strict landcare ethics and global soil policy that must be strictly adhered to. While attempts at decreasing the emissions due to fossil-fuel combustion must be continued, it is important that the vast potential of world soils in C sequestration is realizable through adoption of known and proven technologies.

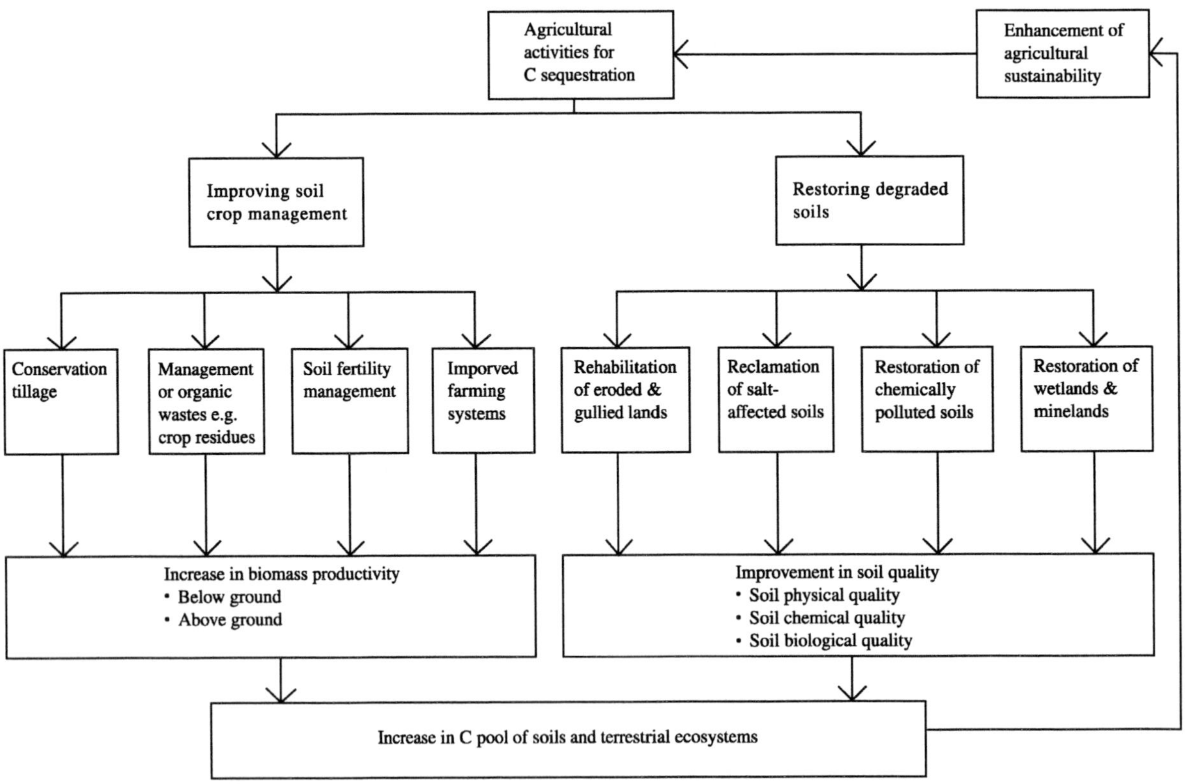

Fig. 3 Agricultural activities with a potential of sequester C in world soils and decrease the risks of the greenhouse effect.

G

Table 3 Strategies for increasing SOC content for mitigation of the greenhouse effect

Practices that decrease loss of SOC content	Practices that enhance SOC content
1) Soil erosion control techniques	1) Improved soil fertility and integrated nutrient management
i) Conservation tillage	i) Precision farming
ii) Residue management	ii) Nutrient cycling
iii) Terraces and contour barriers	iii) Application of biosolids and biological nitrogen fixation
iv) Improved cropping systems	2) Restoration of degraded soils
v) Afforestation	i) Eroded soils
2) Decreasing losses of dissolved organic carbon through leaching	ii) Mineland and drastically disturbed soils
	iii) Salt-affected soils
	3) Water table management (drainage and subirrigation)

CONCLUSIONS

World soils constitute a major terrestrial carbon pool, and even a small change in the C pool of world soils can lead to a substantial change in the atmospheric C pool. A considerable part of the increase in atmospheric C pool since the industrial revolution has come from the soil and biotic pools. Agricultural activities that lead to emissions of CO_2 and CH_4 from soil to the atmosphere include deforestation, biomass burning, and plowing. Gaseous emissions are accentuated by soil degradative processes including accelerated erosion, structural decline, etc.

Conversion to an appropriate land use and adoption of judicious management can reverse the trend by making world soils a major sink for C. The SOC content of cropland soils can be maintained or enhanced by use of conservation tillage, application of crop residue mulch, and judicious use of chemical fertilizers and organic amendments which enhance soil quality. Restoration of the world's degraded soils and desertification control have a potential to sequester C at the rate of 3.0 Pg/yr.

REFERENCES

1. Schlesinger, W.H.; Lal, R.; Kimble, J.; Levine, E.; Stewart, B.A. An Overview of the Carbon Cycle. In *Soils and Global Change*; CRC/Lewis Publishers: Boca Raton, FL, 1995, 9–25.
2. Schlesinger, W.H. *Biogeochemistry: analysis of global change*; Academic Press: San Diego, CA, 1991.
3. Lal, R.; Kimble, J.; Levine, E.; Whitman, C. World Soils and Greenhouse Effect: An Overview. In *Soils and Global Change*; Lal, R., Kimble, J., Levine, E., Stewart, B.A. Eds.; CRC/Lewis Publishers: Boca Raton, FL, 1995, 1–7.
4. Kennel, C. *UC Revalle Program on Climate, Science and Policy*; Scripps Institution of Oceanography: La Jolla, CA, 2000.
5. Bouwman, A.F., (Ed.) *Soils and the Greenhouse Effect*; Wiley: London, 1990, 575.
6. Post, W.M.; Emmanuel, W.R.; Zinke, P.J.; Stangenberger, A.G. Soil Carbon Pool in World Life Zones. Nature **1982**, *298*, 156–159.
7. Post, W.M.; Peng, T.H.; Emmanuel, W.R.; King, A.W.; Dale, V.H.; De Angelis, D.L. The Global C Cycle. Am. Sci. **1990**, *78*, 310–326.
8. Sandquist, E.T. The Global CO_2 Budget. Science **1993**, *259*, 934–941.
9. Eswaran, H.; Vandenberg, E.; Reich, P.; Kimble, J. Global Soil Carbon Resources. In *Soils and Global Change*; Lal, R., Kimble, J., Levine, E., Stewart, B.A. Eds.; CRC/Lewis Publishers: Boca Raton, FL, 1993, 27–43.
10. Batjes, N.H. Total Carbon and Nitrogen Pools in Soils of the World. Eur. J. Soil Sci. **1996**, *47*, 151–163.
11. Tans, P.P.; Fuang, I.Y.; Takahashi, T. Observational Constraints on the Global Atmospheric CO_2 Budget. Science **1990**, *247*, 1431–1438.
12. Lal, R. Global Soil Erosion by Water and Carbon Dynamics. In *Soils and Global Change*; Lal, R., Kimble, J., Levine, E., Stewart, B.A. Eds.; CRC/Lewis Publishers: Boca Raton, FL, 1995, 131–141.
13. Lal, R. Deforestation and Land Use Effects on Soil Degradation and Rehabilitation in Western Nigeria. III. Runoff, Soil Erosion and Nutrient Loss. Land Degrad. Dev. **1996**, *7*, 99–119.
14. Lal, R. Soil Management and Restoration for C Sequestration to Mitigate the Accelerated Greenhouse Effect. Prog. Env. Sci. **1999**, *1*, 307–326.
15. Oldeman, L.R. The Global Extent of Soil Degradation. In *Soil Resilience and Sustainable Land Use*; Greenland, D.J., Szabolcs, I. Eds.; CAB International: Wallingford, UK, 1994, 99–118.

GREENHOUSE GAS FLUXES, MEASUREMENT OF

Oswald Van Cleemput
Pascal Boeckx
Laboratory of Applied Physical Chemistry, Faculty of Agricultural and Applied Biological Sciences, Ghent University, Ghent, Belgium

INTRODUCTION

Increasing concentrations of atmospheric gases, such as carbon dioxide (CO_2), methane (CH_4), and nitrous oxide (N_2O), contribute to radiative forcing. Activities related to land use are importantly responsible for the buildup of CO_2, CH_4, and N_2O in the atmosphere. A survey of techniques will be described for measuring fluxes of greenhouse gases (GHG) from soils. Subsequently, sampling techniques, sample handling, and means of detection will be presented.

SAMPLING TECHNIQUES, SAMPLE HANDLING, AND ANALYSIS

Chamber Techniques

Flux chambers are simple inverted containers, which form an enclosure for gases emitted from the soil surface (1). Both closed (static) and open (dynamic) chambers can be used. Advantages and disadvantages of chamber techniques are listed in Table 1.

Closed and Open Chambers

Emissions of CO_2, CH_4, and N_2O are very variable, both spatially and temporally. It has not been determined whether the size of the flux chambers has an influence on this variability. Nevertheless, at least six chambers should be used per campaign. Flux chambers (cylindrical, or square or rectangular and box-like) can either be installed as a complete assembly, eventually for a short period and then removed until the next sampling occasion (Fig. 1A), or they can exist out of a basal part, which is installed for the entire duration of the experiment with a gas-tight chamber attached to it for short periods (Fig. 1B). This last variant is often used in flooded systems (e.g., paddy soils). The normal procedure for installation is to make a slit in the soil with a metal cutting edge, the same size and shape as the collar of the chamber, and to insert the chamber collar for about 3 cm into the slit.

When a closed chamber is fixed in place, gas samples can be taken from the headspace at different time intervals. The change in concentration in the chamber over time is used to calculate the gas flux. The calculation of the flux goes through a linear regression analysis of the gas concentration increase with time (corrected for eventual temperature changes) and a calculation of the chamber volume and area. Typical expressions of fluxes are g GHG $ha^{-1} d^{-1}$. A minimum of three measurements should be made to check the linearity of concentration increase in the chamber. A non-linear increase could indicate an inadequate sealing of the chamber or an important increase of pressure in the chamber due to temperature increases (2). However, venting of the closed chambers can create large errors. The chamber cover should be removed once the final sample has been taken to minimize disturbance of the environmental conditions of the area covered by the chamber.

Open chambers can be used as well (Fig. 2). In open chambers outside air flows into the chamber via an inlet and is forced to flow over the enclosed soil surface before leaving the chamber via an outlet. The concentration of the respective gas is measured at the in- and outlet sides. The gas flux from the soil surface can be calculated from the concentration difference between the in- and outlet, gas flow rate, and volume and area covered by the chamber.

Manual or Automated Sampling and Analysis

Using closed chambers, the headspace can be sampled by syringe. The gas samples are transferred to the laboratory into sealed containers (evacuated vials fitted with rubber septa) for analysis. To obtain a representative sample of the headspace, a diffusive mixing over 30 min is adequate, unless a mixing fan is used. Gas samples should be taken from the headspace immediately after sealing and at equal time intervals, thereafter, during which, gas concentrations increase linearly.

Table 1 Advantages and disadvantages of the closed chamber, open chamber, and micrometeorological methods to measure gas fluxes from soils

Method	Advantage	Disadvantage
Closed chamber	Simple and low cost Multiple gases can be sampled Small fluxes can be measured Manual and automated gas sampling can be used	Labor intensive Small area is covered Only a short-term emission event is monitored (1–2 h) Disturbance of the emitting surface upon installation Altered conditions of temperature and soil atmosphere exchange Different functioning of plants in the chamber
Open chamber	Relatively simple Environmental condition close to uncovered field Continuous long-term monitoring possible	Small area is covered Disturbance of the emitting surface upon installation Pressure deficits can cause artificially high fluxes Automated sampling is required
Micrometeorological	Useful for diurnal and seasonal variations Large areas can be monitored (aggregate flux) Minimal disturbance of the emitting surface	Expensive and sophisticated instrumentation needed Dependence on a uniform, large surface and constant atmospheric conditions

With automatic sampling, manual gas sampling from the chamber is replaced by a gas flow system providing a periodic sample transfer to a detector. The basic elements of an automatic system consist of closed chambers (equipped with lids that open and close automatically) or open chambers, gas flow systems (tubing and pump); a sampling unit; an analytical unit (detector); a time controller; and a data acquisition system. Automatic sampling is more expensive and it needs to be done in the vicinity of the analysis device. However, this technique is helpful when extensive data sets need to be collected over longer periods of time. It increases the reliability of the emission data obtained, because the number of manipulations is reduced. In most cases open chambers are sampled automatically.

ANALYTICAL ASPECTS

Gas Chromatography (2)

Gas samples are injected into the gas chromatograph (GC) either manually or through the use of a sample loop (automated). Depending on the type of gas to be analyzed, specific GC settings and columns are used. Carbon dioxide is detected using a thermal conductivity detector (TCD). Methane is detected using a flame ionization detector (FID). Nitrous oxide is detected using a ^{63}Ni electron capture detector (ECD). For analysis of N_2O, care must be taken to remove both CO_2 and water vapor. In experiments with acetylene, to block the last step in the denitrification process, one may need to bypass the ECD with the acetylene in the sample after it exits from the chromatographic column. Acetylene alters both the sensitivity and stability of some ECDs.

Photo-Acoustic-Infrared Detector (PAID) (4)

The measuring principle is based on photo-acoustic detection of the absorption of infrared light. This means that any gas that absorbs infrared light of a specific wavelength can be measured. The most common PAID is the *Brüel and Kjær multi-gas monitor.* The PAID is equipped with a pump, which draws air from the flux chamber into an analysis cell inside the gas monitor. This cell can be sealed hermetically. Light from an infrared source is pulsated by a mechanical chopper and then passes through one of the optical filters of the filter carousel.

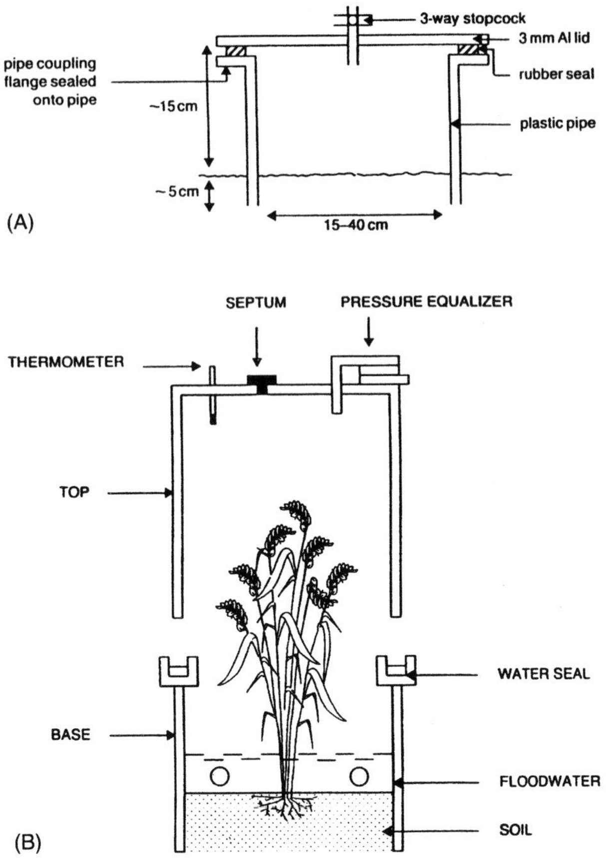

Fig. 1 (A) Typical closed chamber used for greenhouse gas flux measurements from aerobic soils. (Adapted from Ref. 2.) (B) Schematic drawing of closed chamber used for greenhouse gas flux measurements in flooded soil. (Adapted from Ref. 3.)

Fig. 2 Field set-up of an open chamber with removable lid to measure NO fluxes from forest soils: In the front, the outlet with the air sampling tube; in the back, the air inlet with the air sampling tube and the NO analyzer with data acquisition system.

The filter produces an infrared wavelength, which is selectively absorbed by the gas being monitored. Absorption of infrared light by the gas in the closed analysis cell causes the temperature to increase. Because the infrared light is pulsating, the gas temperature increases and decreases. In the closed cell this results in an increase and decrease of the pressure, which can be measured via 2 microphones. This acoustic signal is proportional to the concentration of the gas monitored. The *Brüel and Kjær multi-gas monitor* allows the analysis of 5 different gases and water vapor from 1 sample in approximately 120 seconds. The device has the ability to compensate for temperature fluctuations and water vapor interference. Using an automated sampling system, careful calibration of the multi-gas monitor and the sampling unit is required (4).

Micrometeorological Methods (5) (6)

The basic concept of micrometeorological methods for measuring trace gas fluxes to or from the soil surface is that gas transport is accomplished by the eddying motion of the atmosphere which displaces parcels of air from one level to another. Transport of a gas through the free atmosphere is provided by turbulent diffusion and convection in which the displacement of individual eddies is the basic transport process. Micrometeorological methods are based on the assumption that the flux to or from the soil surface is identical to the vertical flux measured at the reference level some distance above the surface. Therefore, a flat and homogeneous terrain is needed. The flux measured at the reference level provides than the average flux over the upwind area (fetch), provided that sampling point at the reference level is in the height range in which the vertical flux is constant with height (fetch $\approx$ 100H, where H = height of the sampling point). In the simplest of the micrometeorological methods,

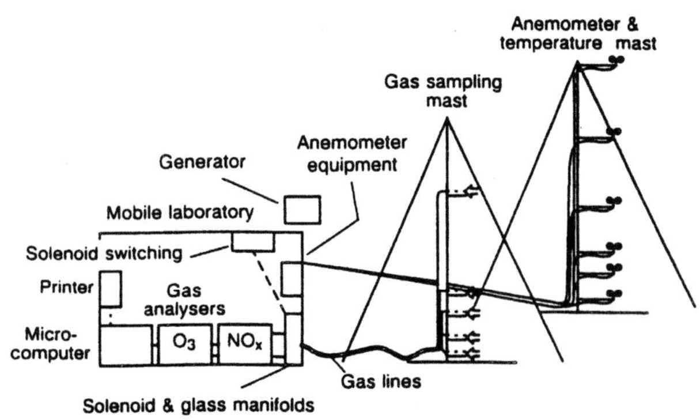

Fig. 3 Typical instrument for flux gradient measurements of greenhouse gas fluxes in the field. (Adapted from Ref. 5.)

the flux may be measured by sensing the concentrations and velocities of components of the turbulence.

Two general micrometeorological techniques are used to measure trace gas fluxes: the eddy correlation and flux-gradient technique (Fig. 3). Application of both approaches is limited to situations in which the air analyzed has passed over a homogeneous exchange surface for a long distance so that profiles of gas concentration in the air are in equilibrium with the local rates of exchange. These methods also require that horizontal concentration gradients are negligible. Eddy correlation methods require a fast response detector. The tuneable diode laser (TDL) technique is based on infrared absorption spectrometry, whereby the absorption depends upon path length, line strength, and absorber concentration. Liquid nitrogen temperature diodes are commercially available to cover the infrared spectrum from about 2–10 μm, the region where most trace gases have absorption spectra. Detailed information about these methods can be found in Refs. 5 and 6. Advantages and disadvantages of micrometeorological methods are listed in Table 1.

Non-Isotopic Tracer Methods (2)

Tracer methods involve the release of an inert tracer gas, most commonly sulphur hexafluoride (SF_6), from an emitting surface. The tracer gas is released at a known rate in a pattern similar to the release pattern of the GHG, perpendicular to the direction of the prevailing wind. This method can be applied when a definite plume of the GHG can be readily detected in the ambient environment. Under these conditions the plume of the dispersed emission is located based on analyses of upwind and downwind air samples. The flux rate is computed, using the ratio of the plume concentration of the tracer and the GHG and the known release rate of the tracer. The advantage of this technique is that aggregate gas emissions can be collected from heterogeneous areas, such as landfills, circumventing the problem of spatial heterogeneity. However, the high costs, dependence on meteorological conditions, and the potential for interfering sources limit its application.

Ultra-Large Chambers with Long-Path Infrared Spectrometers (2)

Infrared absorption spectrometers are available that can give an average value for the gas concentration over distances of tens or hundreds of meters. They are useful for measurements of average emissions from a whole experimental plot, by covering the plot temporarily with a large canopy to act as a chamber and retain the gas emitted from the soil. Two systems are available: 1) the Fourier Transform Infrared (FTIR) spectrometer with a mirror system, which allows multiple reflections and thus a total path of up to 1 km that is capable of measuring GHG concentration changes down to a fraction of 1 ppb and 2) a simpler, less-sensitive IR spectrometer with the capacity to detect a concentration change of about 25 ppbv of N_2O and 10 ppbv of CH_4.

REFERENCES

1. Hutchinson, G.L.; Livingstone, G.P. Use of Chamber Techniques to Measure Trace Gas Fluxes. In *Agricultural Ecosystems on Trace Gases and Global Change*, ASA Special Publication 55; ASA: Madison, 1993; 63–78.
2. *Manual on Measurement of Methane and Nitrous Oxide Emissions from Agriculture*. IAEA-TECDOC-674; IAEA: Vienna, 1992; 91.
3. Lindau, C.W.; Bollich, P.K.; De Laune, R.D.; Patrick, W.H. Jr.; Law, V.J. Effect of Urea Fertilizer and Environmental Factors on CH_4 Emissions from a Louisiana USA Rice Field. Plant and Soil **1991**, *136*, 195–203.
4. De Visscher, A.; Goossens, A.; Van Cleemput, A. Calibration of a Multipoint Sampling System Connected with a Photoacoustic Detector. Intern. J. Environ Anal. Chem. **2000**, *76*, 115–133.
5. Denmead, O.T. Micrometeorological Methods for Measuring Gaseous Losses of Nitrogen in the Field. In *Gaseous Loss of Nitrogen from Plant-Soil Systems*; Freney, J.R., Simpson, J.R., Eds.; Nyhoff Junk Publ.: The Hague, 1983; 133–157.
6. Fowler, D.; Duyzer, J. Micrometeorological Techniques for the Measurement of Trace Gas Exchange. In *Exchange of Trace Gases Between Terrestrial Ecosystems and the Atmosphere*; Andreae, M.O., Schimel, D.S., Eds.; John Wiley and Sons: Chichester, 1989; 189–207.

GULLY EROSION

Fenli Zheng
Institute of Soil and Water Conservation, Yangling, Shaanxi, China

Chi-hua Huang
United States Department of Agriculture-Agricultural Research Service (USDA-ARS), West Lafayette, Indiana, U.S.A.

INTRODUCTION

Gully erosion is defined as erosion in channels where runoff water accumulates and removes soils from this channel area (1). Gully erosion is prevalent throughout the world. In 1939, Bennett (2) estimated that 200 million active gullies were in existence in the U.S. These gullies ranged in depths from 0.3 to 0.6 m to as much as 50 to 100 m. The severe erosion in the 1930s prompted the U.S. government to take actions to control erosion and successful results are now well known all over the world. Gullies remove portions of land completely from production, and they fragmentize the landscape making it difficult for vehicles or farm machinery to move across. Gully erosion significantly reduces land quality and value, and also contributes to downstream sedimentation problems (3). Despite the fact that a gully can be viewed as a part of the natural drainage system, it is different from a stream channel due to its intermittent flow only triggered by rainfall events.

Gullies often develop from intense erosion caused by flow over a steep overfall at the top of the gully. This overfall, called a headcut, moves upstream in a natural drainage way, and it can be initiated off-site and move into a field. Gullies can also be enlarged by lateral erosion, sloughing of their sidewalls and cleaning out of debris by flow in the gullies. Subsurface flow through the gully walls can significantly reduce soil strength and accelerate gully erosion (3).

Depending on the scale, gullies are divided into two types: ephemeral and classical gullies (Figs. 1 and 2). Ephemeral gullies are wider and deeper than rills, but they can be tilled across and filled in partially or completely. Although the ephemeral gully can be filled in or obliterated by tillage operations, it tends to reappear later at the same location because the depression it has formed on the landscape will concentrate the runoff. Classical gullies are eroded channels too large to cross and obliterate with tillage equipment. Table 1 shows the comparative characteristics of rill, ephemeral gully and classical gully erosion (4).

EPHEMERAL GULLY EROSION

Ephemeral gully erosion is referred to as concentrated-flow erosion, mega-rill erosion (4) or shallow gully erosion (5). The ephemeral gully erosion is the process of overland flow converging into a few major natural waterways, detaching soil particles from the waterway boundary, and transporting sediments downstream. Ephemeral gullies are main drainage systems for a field, and most water and sediment are discharged from fields through these channels. Contribution of ephemeral gully erosion varies geographically. A recent USDA-Natural Resources Conservation Service estimate from 19 U.S. states (Table 2) showed ephemeral gully erosion contributed from 17% of total soil loss at New York State to 73% at Washington State (6, 7). Ephemeral gully erosion also affects the loess belt and Mediterranean region of Western Europe and the Loess Plateau of P. R. China. Ephemeral gully erosion contributes at least 10% of the total loss in Europe (8) and 35% of the total soil loss in China. Most erosion in ephemeral gully channels is from rainstorms soon after seedbed preparation. As time progresses through the growing season, soil consolidates and these channels become less erodible. Many ephemeral gullies begin at a third to half way down the slope, and then extend and grow in both directions.

The following factors have been identified for the development of ephemeral gullies: 1) critical slope length and slope gradient that are dependent on slope characteristics and crop row direction; 2) occurrence and depth of a fragipan; 3) agricultural practices, principally row direction and timing of cultivation; and 4) timing and total amount of precipitation (7, 9).

CLASSICAL GULLY EROSION

Classical gullies are those that are too large to be filled in or obliterated through tillage operations. The quantities of

Fig. 1 Ephemeral gully.

sediment moved from gully systems in loessial regions are highly variable. Gully erosion contributed 10–30% of average annual sediment yield in the deep loess of western Nebraska from 1937 to 1952 (10). At the Loess Plateau of P. R. China, gully erosion contributes about 60% of total sediment yield in the hilly-gully region and 90% of total sediment yield in the high plateau with deep-cutting and large gully regions (Fig. 2). The contribution of gully erosion to total sediment yield also changes with year, dependent upon annual precipitation and the timing of storms. Gully channel formation is very rapid during the period of gully initiation, when morphological characteristics of a gully (length, width, area, and volume) are far from stable. This period is relative short, about 5% of a gully's lifetime. For the most part of a gully's lifetime, its size is nearly stable, except near the headcut area.

Main factors affecting gully development in loessial regions are watershed size, runoff, the distance from the top of the gully to the watershed boundary, gully channel gradient, surface material, sub-stratum moisture, and vegetation.

Fig. 2 Classical gullies.

TECHNIQUES OF MEASURING AND PREDICTING GULLY EROSION

Aerial photography is widely used for making accurate and repetitive measurements of gully development (11). Sequential stereo-photographs taken at critical intervals provide excellent information for assessing gully development. Field survey is also used to quantify gully geometry. Nevertheless, measuring gully geometry, especially at different times to monitor the gully development, is laborious.

The ephemeral gully erosion model (EGEM), a computer program to provide estimates of the average annual soil loss removed from a single ephemeral gully, was developed by the USDA Agricultural Research Service (ARS) and USDA Soil Conservation Service (12).

Two types of models have been developed to predict gully development:

1. Dynamic models to predict rapid changes of gully morphology at the initial period of gully development with active headcut and channel expansion, which is based on the solution of the equations of mass conservation and gully bed deformation.
2. Static models to estimate final geometric parameters of stable gullies, which is based on the assumption of final morphological equilibrium of a gully (13).

The dynamic and static gully models were verified on the data on gully morphology and dynamics from Yamal Peninsula, Russia and New South Wales, Australia.

CONTROL OF GULLY EROSION

There are different practices to prevent or mitigate ephemeral gully formation and erosion. Conservation tillage is an excellent soil conservation practice for preventing ephemeral gully erosion formation in agricultural fields. In areas where ephemeral gullies have formed, vegetative control, such as grassed waterways and vegetative barriers are often used to reduce their further development. Sometimes, a retention structure is built within gully channels to trap sediment (7).

Because many classical gullies have become a permanent landscape feature, stabilizing the channel banks and headcut areas is the primary goal. Establishment

Table 1 Comparative characteristics of rill erosion, ephemeral gully erosion and classical gully erosion (4)

Rill erosion	Ephemeral gully erosion	Classical gully erosion
Rills are normally erased by tillage; usually do not reoccur in the same place.	Ephemeral gullies are temporary features, usually obscured by tillage; recur in the same location.	Gullies are not obscured by normal tillage operations.
May be of any size but are usually smaller than ephemeral gullies.	May be of any size but are usually lager than rills and smaller than permanent gullies.	Usually larger than ephemeral gullies.
Cross-sections tend to be narrow relative to depth.	Cross-sections tend to be wide relative to depth. Sidewalls frequently are not well defined. Headcuts are usually not readily visible and are not prominent because of tillage.	Cross-sections of many gullies tend to be narrow relative to depth. Sidewalls are steep. Headcuts are usually prominent.
Flow pattern develops as many small-disconnected parallel channels ending at ephemeral gullies, terrace channels, or where deposition occurs. They are generally uniformly spaced and sized.	Usually forms a dendritical pattern along depression water courses, beginning where overland flow including rills, converge. Flow patterns may be influenced by tillage, crop rows, terraces, or other man-related features.	Tend to form a dendritical pattern along natural watercourses. Non-dendritical patterns may occur in road ditches, terraces, or diversion channels.
Occurs on smooth side slopes above drainage ways.	Occurs along shallow drainage ways upstream from incised channels or gullies.	Generally occurs in well-defined drainage ways.
Soil is removed in shallow channels but annual tillage causes the soil profile to become thinner over the entire slope.	Soil in removed along a narrow flow path, typically to the depth of the tillage layer where the untilled layer is resistant. Soil is moved into the voided area from adjacent land by mechanical action (tillage) and sheet and rill erosion, damaging area wider than the eroded channels.	Soil may be eroded to depth of the profile, and can erode into soft bedrock.

(From Ref. 4.)

G

Table 2 Assessment of ephemeral gully erosion in selected areas of the U.S.

Location	Estimated sheet and rill erosion (tons $ha^{-1} yr^{-1}$)	Measured ephemeral gully erosion (EGE[a]) (tons $ha^{-1} yr^{-1}$)	EGE as a percentage of total (sheet, rill and gully erosion) (%)
Alabama	5.73	3.42	37
Delaware	0.38	0.93	71
Illinois	2.61	1.91	42
Iowa	3.53	1.1	24
Kansas	8.07	2.94	27
Louisiana	6.54	2.22	25
Maine	4.12	1.89	31
Maryland	1.95	1.47	43
Michigan	1.72	0.45	21
Mississippi	6.46	2.75	30
New Jersey	2.46	1.91	44
New York	8.73	1.85	17
North Dakota	2.77	1.3	32
Pennsylvania	0.93	0.65	41
Rhode Island	3.31	1.36	29
Vermont	1.65	2.24	58
Virginia	4.77	4.7	50
Washington	0.25	0.69	73
Wisconsin	2.89	1.54	35

[a] EGE is ephemeral gully erosion.
(From Refs. 6 and 7.)

of vegetation along the channel banks and headcut areas are most effective in preventing classical gully growth. For large gullies such as those at the Loess Plateau of China, terracing on the gully slopes and building retention structures in the gully channels are widely used for controlling classical gully erosion (Fig. 3).

Fig. 3 Classical gully control.

REFERENCES

1. Hutchinson, D.E.; Pritchard, H.W. Resource Conservation Glossary. J. of Soil and Water Conservation. **1976**, *31* (4), 1–63.
2. Bennett, H.H.; *Soil Conservation*; Mcgraw-Hall: New York, 1939; 993 pp.
3. Piest, R.F.; Bradford, J.M.; Wyatt, G.M. Soil Erosion and Sediment Transport from Gullies. J. of the Hydraulics Division **1975**, *101* (HY1), 65–80.
4. Foster, G.R.; Lane, L.J.; Mildner, W.F. Seasonally Ephemeral Cropland Gully Erosion. *Proceedings of Natural Resources Modeling Symposium, Pingree Park, CO., USA, Oct. 16–21*; 1983; 463–465.
5. Zhu, X. Soil Erosion Classification at the Loessial Region. ACTA Pedological Sinica. **1956**, *4* (2), 99–116.
6. USDA-NRCS. *America's Private Land, a Geography of Hope*; USDA-NRCS: Washington, DC, 1977.
7. Robinson, K.M.; Bennett, S.J.; Casali, J. Headcut Dynamics and Ephemeral Gully Erosion. *ASAE Annual International Meeting, Orlando, Florida, USA, July 12–16, 1998*; American Society of Agricultural Engineering: St Joseph, USA, 1998.
8. Wijdenes, D.J.; Poesen, J.; Vandakerckhove, L.; Ghesquiere, M. Spatial Distribution of Gully Head Activity and Sediment Supply Along an Ephemeral Channel in a

Mediterranean Environment. Catena. **2000**, *39* (3), 147–167.

9. Smith, L.M. *Investigation of Ephemeral Gullies in Loessial Soils in Mississippi*; U.S. Technical Report GL-93-11; Army Corps of Engineers, 1990.
10. Piest, R.F.; Spomer, R.C. Sheet and Gully Erosion in the Missouri Valley Loessial Region. Trans. ASAE. **1968**, *11* (6), 850–853.
11. Thomas, A.W.; Welch, R. Measurement of Ephemeral Gully Erosion. Trans. ASAE. **1988**, *31* (6), 1723–1728.
12. Woodward, D.E. Method to Predict Cropland Ephemeral Gully Erosion. Special Issue: Soil Erosion Modeling at the Catchment Scale. Catena. **1999**, *37* (3–4), 393–399.
13. Sidorchuk, A. Dynamic and Static Models of Gully Erosion. Special Issue: Soil Erosion Modeling at the Catchment Scale. Catena. **1999**, *37* (3–4), 401–414.

GYPSIC SOILS

J. Herrero
Laboratorio asociado de Agricultura y Medioambiente (DGA-CSIC), Zaragoza, Spain

J. Boixadera
Departament d'Agricultura, Lleida, Spain

G

INTRODUCTION

Gypsic is here applied in a broad sense to those soils whose behavior and appearance are dependent on their gypsum ($CaSO_4 \cdot 2H_2O$) content either in the entire profile or in a particular horizon. High amounts of gypsum in soils are more likely to occur in dry climates.

GYPSIC SOILS IN THE WORLD

The world distribution of gyprock outcrops and the solubility of gypsum restrict the extensive occurrence of gypsic soils to the dry regions (Figs. 1 and 2). These soils are distinct and recognized by local populations, as well as by early pedologists, in areas such as northern Africa, the Middle East, and Spain. Now most soil classification systems have the formative element *Gyps*. The extent of soils with gypsic or petrogypsic diagnostic horizons (1) in Africa, Asia, the Near East, Europe, and the United States has been estimated at 207 million ha (2), and broad gypseous areas also occur in Australia, Mexico, and South America. These soils and other gypsiferous surface formations interact in global cycles through calcium and the bicarbonate equilibria with the atmosphere and water (3, 4), processes also involving respiration. Gypsic soils deserve an environmental valuation due to the pedodiversity protection and to the plant endemisms produced by selective forces not yet established (5).

OCCURRENCE OF GYPSUM IN SOILS AND THEIR RELATED PROPERTIES

The semi-solubility of gypsum in water ($\approx 2.6\,g\,L^{-1}$) controls the occurrence of gypsum in soils. The soil solution or the water extracts of gypsic soils have an electrical conductivity of 2.2 dS/m or more if soluble salts with no common ions are present. The leaching of gypsum out of the profile is common in wet climates. The low solubility of gypsum enables its transport from gyprock or other primary sources and its redistribution in the soil by dissolution/precipitation, and its permanent presence as a significant soil component in dry environments or in specific geomorphic positions. Moreover, the size of some pedogenic or sedimentary gypsum crystals allows their transport by wind from bare ground surfaces. Gypsum in soils can also result from other natural or artificial materials containing sulfur, for example rain, industrial products or mine spoils, but in these cases gypsum is a minor soil component.

Gypsum, even in small quantities, can prevent clay dispersion and soil degradation. The effect is due to the displacement of Na^+ from the cation exchange complex of the soil by the Ca^{2+} released by the gypsum. This is one of the reasons for the use of gypsum as a soil amendment not only for sodic soils but also to avoid soil sealing and crusting under irrigation with sodic or with low electrical conductivity waters (6). Gypsum is also a soil acidity ameliorant (7, 8). Plants are not stressed by the osmotic potential generated by gypsum, and this fact together with the beneficial effect on crops with high calcium or sulfate requirements must be taken into account when land evaluation deals with gypsiferous soils (9). In gypsic soils a constant osmotic potential generated by the gypsum exists over all the range of soil moisture contents; it may reach up to -80 kPa and it has a mild effect on plants quite different from the osmotic effect in saline soils because the last one is several times higher even for the very slightily saline soils (electrical conductivity in the saturation extract = 2–4 dS/m at 25°C).

HYPERGYPSIC SOILS

The name of hypergypsic was coined for a subsurface diagnostic soil horizon with 60% or more gypsum (2),

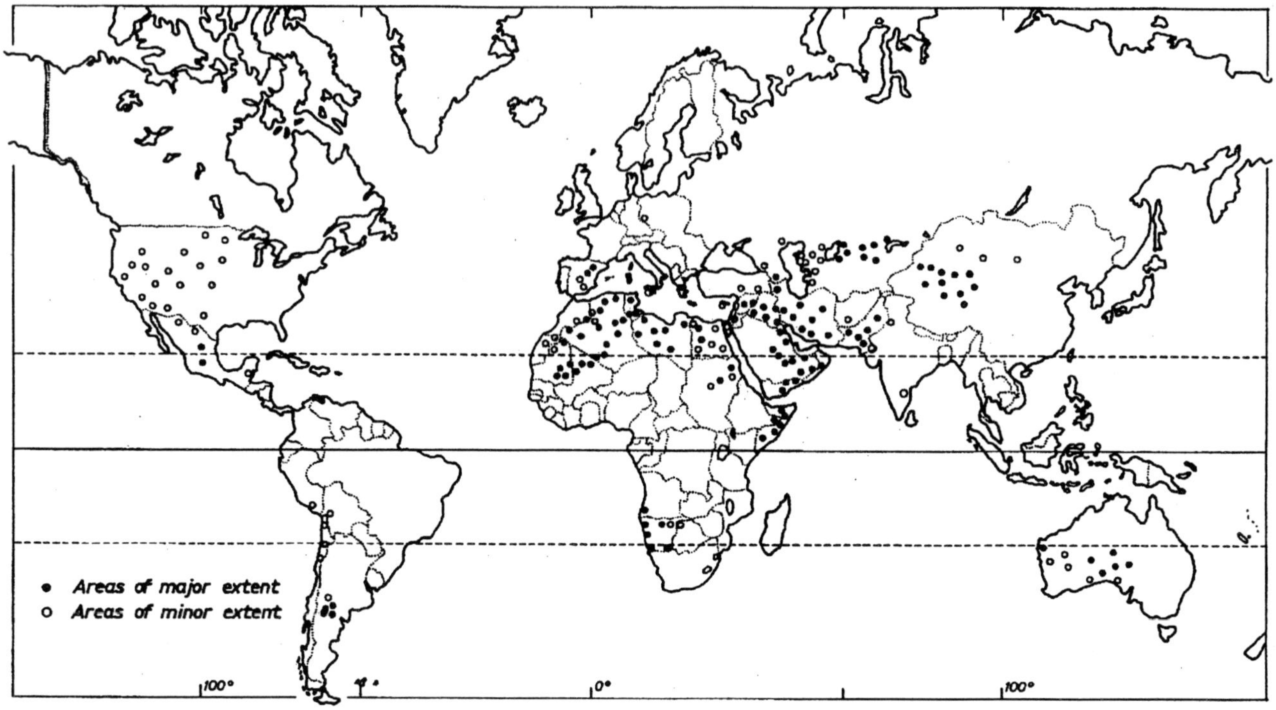

Fig. 1 World distribution of gypsic soils. (Prepared from data contained in Refs. 2, 19, 24–35; many of the documents quoted in Ref. 4; and from authors' studies.)

and later reused with genetic meaning (10). Those soils having gypsum as a major component can also be named hypergypsic (11). These types of soils appear on gyprock outcrops or close to some other gypsum source, commonly in arid climates (4). In these conditions gypsum is not transitory, and can constitute the soil groundmass, supporting pedological processes such as the redistribution of silty or clay-size carbonatic materials in the profile (12). The growth of gypsum crystals is displacive, resulting in a mixing of the soil components: This can be illustrated by the disturbance and comminution of lutitic geological materials leading to C horizons (13) or by the mechanical weathering of the parent gyprock (14). Many hypergypsic horizons show an isles fabric under the microscope, i.e., masses of nongypsic fine material embedded in a gypsic pedofeature, or the gypsic pedofeature can even be continuous producing a horizon with gypsic groundmass (11). In the field, these horizons feel either "gritty," "flour-like," or like "hard-bread crumbs," composed of sandy or coarser gypsum lentils, microcrystalline gypsum, or travertinic gypsum, respectively (15, 16). Some of these horizon can qualify for petrogypsic (1, 10) if a certain degree of cementation is attained. Commonly gypsic soils occur in Holocene surfaces, but the ones with cemented horizons have been found in Pleistocene surfaces.

For sand-sized or smaller grains, the distinction between pedogenic or geologic gypsum is often unfeasible even under the microscope (17). Moreover, this distinction lacks agricultural or environmental interest, at least in the gypsic soils in areas with ubiquitous gypsum. This makes it possible to argue against the incorporation of the distinction between primary and secondary gypsum into the concept of hypergypsic (10), and even in the definitions of the gypsic and petrogypsic diagnostic horizons (1, 10).

CONSTRAINTS TO THE USE OF GYPSIC SOILS

The presence of gypsum must not be confused with the osmotic effect of soil salinity, even though gypsic soils can also be saline, as is the case in playa- or sabkha-like (18) environments or in some alluvial plains with incomplete drainage. In fact, gypsum is often used as an amendment in sodic soils to allow salt leaching.

Physical and Engineering Constraints

A large amount of gypsum in soils implies minor contents of clay minerals and other components having high water

Fig. 2 Aerial view of the valleys carved with a fingered pattern in the gyprock surrounding Zaragoza (Spain). These valleys are flat-bottomed because of their filling during the Quaternary with materials from the soils on the slopes. Transversal dry-stone walls were built to control the erosion and to capture water in the bottoms. Deep gypsic soils there only permit poor barley yields even in the favorable years, because of the aridic soil moisture regime. The hills have shallow discontinuous gypsic soils, with shrub cover in the shaded slopes. (Photo by O. Artieda, 1997.)

and nutrient retention capacity. Water is the main limitant to plants' life, thus in arid climates the soil water holding capacity is decisive for life. In gypsic soils this capacity is controled by the gypsum grain size. Moreover, the soil exploration by roots can be hampered by massive gypsic layers (3). Drought is severe for plants in gypsic soils grazed or sown with barley or other annuals, except where irrigation or rain is frequent during the growing season.

Sparse shrubs or trees, also ants or other animals, often take advantage of cracks, or shallow karstic cavities in gyprock where lower temperatures and some moisture are available in arid environments.

Dissolved gypsum attacks concrete and iron, hindering, together with karstic phenomena, the construction of irrigation canals and tunnels. These problems plus the sinkholes by dissolution make flood irrigation of gypsic soils difficult or impossible (19); although sprinkler or drip irrigation techniques and the use of plastics in agriculture have now overcome some of the obstacles to the irrigation of these soils. The water return flows from irrigated gypsic soils are saturated with calcium and sulfate, limiting the urban or industrial reusability of these waters.

Fertility Constraints

Gypsic soils support forests or shrubs under appropriate climatic and management conditions, and have been grazed and cultivated from times immemorial in a sustainable manner, but fertility is a matter of concern for intensive agriculture in gypsic soils. One favorable point is that, in contrast with other anions, no luxury consumption of sulfate by plants occurs, as was earlier reported (20) from pot experiments. Sulfur is an essential element, and sulfate its main source for plants, but many different sulfur-based anions can occur in the soil depending on local or temporal conditions of pH, redox, microbial action, etc. (21) making the assessment of the role of gypsum in soil fertility difficult. The relationship between sulfate and phosphate adsorption (22) must be remembered, but probably the main fertility problems in gypsic soils are related to their low nutrient retention capacity and to the high amounts of Ca^{2+} in the soil solution (gypsum is over 100 times more soluble than calcite), reducing the availability of P for plants or inducing chlorosis in sensitive fruit trees. These and other constraints related to N and other nutrients in gypsiferous soils have been reviewed by F.A.O. (3), this work also gives an overview on the effects of gypsum on several crops. Many knowledge gaps in this subject are due to the limited research conducted about fertility in gypsic soils, and to the failure of some experiments to meet the analytical procedures (23) required for gypsiferous soils.

REFERENCES

1. Soil Survey Staff. *Soil Taxonomy: A Basic System of Soil Classification for Making and Interpreting Soil Surveys*, 2nd Ed.; USDA Agriculture Handb. 436; USDA: Washington, DC, 1999; 869.
2. Eswaran, H.; Zi-Tong, G. Properties, Genesis, Classification and Distribution of Soils with Gypsum. In *Properties, Characteristics and Genesis of Carbonate, Gypsum and Silica Accumulations in Soils*; Nettleton, W.D., Ed.; SSSA Spec. Publ. 26; SSSA: Madison, WI, 1991; 89–119.
3. FAO. *Management of Gypsiferous Soils*, FAO Soils Bulletin 62; FAO: Rome, 1990; 81.
4. Herrero, J.; Porta, J. The Terminology and the Concepts of Gypsum-Rich Soils. Geoderma **2000**, *96*, 47–61.
5. Escudero, A.; Carnes, L.F.; Pérez-García, F. Seed Germination of Gypsophytes and Gypsovags in Semi-Arid Central Spain. Journal of Arid Environments **1997**, *36*, 487–497.
6. Amézketa, E. Soil Aggregate Stability: A Review. Journal of Sustainable Agriculture **1999**, *14*, 83–151.
7. Sumner, M.E. Gypsum and Acid Soils: The World Scene. Advan. Agron. **1993**, *51*, 1–32.
8. Toma, M.; Sumner, M.E.; Weeks, G.; Saigusa, M. Long-Term Effects of Gypsum on Crop Yield and Subsoil Chemical Properties. Soil Sci. Soc. Am. J. **1999**, *39*, 891–895.
9. Laya, D.; Van Ranst, E.; Herrero, J. A Modified Parametric Index to Estimate Yield Potentials for Irrigated Alfalfa on Soils with Gypsum in Quinto (Aragón, Spain). Geoderma **1998**, *87* (1–2), 111–122.
10. FAO. *World Reference Base for Soil Resources*, Word Soil Resources Report 84; FAO: Rome, 1998; 88.
11. Herrero, J.; Porta, J.; Fédoroff, N. Hypergypsic Soil Micromorphology and Landscape Relationships in N.E. Spain. Soil Sci. Soc. Am. J. **1992**, *56*, 1188–1194.
12. Herrero, J. *Morfología y Génesis de Suelos Sobre Yesos*, Monografía 77; I.N.I.A.: Madrid, 1991; 447.
13. Artieda, O.; Herrero, J. Micromorphological Features of Lutitic Cr Horizons in Gypsiferous Soils. In *Extended Abstracts of 6th Internat. Meeting on Soils with Mediterranean Type of Climate*, Barcelona, Spain, July 4–9, 1999; Bech, J., Ed.; University of Barcelona: Spain, 1999; 527–529.
14. Artieda, O. Decimetric Weathering Features in the Gyprock Surface (Ebro Basin, Spain). In *Extended Abstracts of 6th Internat. Meeting on Soils with Mediterranean Type of Climate*, Barcelona, Spain, July 4–9, 1999; Bech, J., Ed.; University of Barcelona: Spain, 1999; 524–526.
15. Artieda, O. *Génesis y Distribución de Suelos en un Medio Semiárido. Quinto (Zaragoza)*; Ministerio de Agricultura, Pesca y Alimentación: Madrid, 1996; 223 + map.
16. Herrero, J.; Artieda, O. Gypsum-Rich Horizons as Seen in the Field and Under the Microscope. Annual Meeting Abstracts, A.S.A.-C.S.S.A.-S.S.S.A., Salt Lake City, UT, Oct. 31-Nov 4, 1999; 269.
17. Stoops, G.; Poch, R.M. Micromorphological Classification of Gypsiferous Soil Materials. In *Soil Micromorphology: Studies in Management and Genesis*; Ringrose-Voase, A.J., Humphreys, G.S., Eds.; Developments in Soil Science, 22; Elsevier: Amsterdam, 1994; 327–332.
18. Briere, P.R. Playa, Playa Lake, Sabkha: Proposed Definitions for Old Terms. Journal of Arid Environments **2000**, *45*, 1–7.
19. Alphen, J.G.; Ríos, F. *Gypsiferous Soils Notes on Their Characteristics and Management*; ILRI: Wageningen, The Netherlands, 1971; 44.
20. Barbier, G. Action Des Chlorures Et Sulfates Sur La Nutrition Minérale Des Plantes. Comptes Rendues de l'Académie d'Agriculture **1937**, *123*, 699–706.
21. Pédro, G.; Robert, M. Les Minéraux Contenant Du Soufre, Nature, Cristalochimie, Stabilité. *In International Symposium on Sulphur in Agriculture*, Versailles, France, December 1970; Institut National de la Recherche Agronomique: Paris, 1972; 137–180.
22. Kamprath, E.J.; Nelson, W.L.; Fitts, J.W. Influence of PH, Sulfate and Phosphate Concentrations on the Adsorption of Sulfate by Soils. Soil Sci. Soc. Am. Proceedings **1956**, *20* (4), 463–466.
23. Porta, J. Methodologies for the Analysis and Characterization of Gypsum in Soils: A Review. Geoderma **1998**, *87* (1–2), 31–46.
24. Boyadgiev, T.G.; Verheye, W.H. Contribution to a Utilitarian Classification of Gypsiferous Soils. Geoderma **1996**, *74*, 321–338.
25. Bridges, E.M.; Batjes, N.H. *Nachtergaele World Reference Base for Soil Resource*: Atlas. Acco: Leuven, Belgium, 1998, 79.
26. FAO. World Soil Resources. An Explanatory Note on the FAO World Soil Resources Map at 1/25.000.000 Scale. FAO: Rome. 1993; 64.
27. Fierotti, G.; Dazzi, C.; Raimondi, S. Isuoli Della Serie Gessoso Solfifera: Bosco Di Mustigarufi. Guia Alla Escursione Pedologica. Bollettino della Società della Scienza del Suolo, Nuova Serie **1993**, *4*, 63–131. Palermo, Italia.
28. Heinze, M.; Fiedler, H.J. Chemische Eigenschoaften Von Gips-Rendzinen und Begleitbodenformen des Kyffhäuïser-Gebirges (DDR). Chemie und Erde **1984**, *43*, 65–75.
29. Jacob, J.S. Archaeological Pedology in the Maya Low Lands. In *Pedological Perspectives in Archaeological Research*; Collins, M.E., Carter, B.J., Gladfelter, B.G., Southard, R.J., Eds.; SSSA Sp. Publ. 44; SSSA: Madison, WI, 1995; 51–80.
30. Laya, H. Génesis Y Propiedades Tecnológicas Y Posibilidades De Utilización De Los Suelos Salinos De La Cuenca Del Duero Y del Tajo Con Énfasis En Los

Suelos Yesíferos PhD. Thesis Universidad Politécnica de Madrid. 1989; 350.
31. Macau, F.; Riba, O. Situación, Características Y Extensión De Los Terrenos Yesíferos En España. *Primer Coloquio Internacional Obras Públicas en Terrenos Yesíferos*; Servicio Geológico de Obras Públicas: Madrid, Spain, 1965; 28.
32. Mashali, A.M. Soil Management Practices for Gypsiferous Soils. In *Proceedings of the International Symposium on Soils with Gypsum*; Poch, R., Ed.; Lleida: Spain, 1996; 34–52.
33. Mees, F. Distribution Paterns of Gypsum and Kalistronite in a Dry Lake Basin of the Southwestern Kalahari (Omongwa Pan Namibia). Earth Surface Processes and Landforms **1999**, *24*, 731–744.
34. Nettleton, W.D.; Nelson, R.E.; Brasher, B.R.; Derr, P.S. *Gypsiferous Soils in the Western United States*; Kittrick, J.A., Fanning, D.S., Hossner, L.R., Eds.; SSSA Spec. Publ. 10; SSSA: Madison, WI, 1982; 147–167.
35. Watson, A. Structure, Chemistry and Origin of Gypsum Crusts in Southern Tunisia and the Central Namib Desert. Sedimentology **1985**, *32*, 855–875.

HARD SETTING SOILS

Richard S.B. Greene
School of Resources, Environment and Society, Australian National University, Canberra, Australian Capital Territory, Australia

INTRODUCTION

Hardsetting is a characteristic of soil horizons, usually cultivated seedbeds, that contain unstable soil aggregates. Under rapid wetting by either rainfall or irrigation, the aggregates collapse, the seedbed slumps, and upon drying a hard, structureless mass of soil results. The change of structure is due to the breakdown of the soil aggregates under rapid wetting into microaggregates $< 250\,\mu m$ diam. by a process called slaking (1). Hardsetting soils may also have poor physical properties due to breakdown of their soil aggregates by the process of clay dispersion (2). Hardsetting is a widely occurring problem, as most soils (apart from well-structured clays, organic soils, and sands) can undergo hardsetting. Even though hardsetting is usually associated with the A horizon, it can occur in any horizon, thus causing a range of management problems. Mullins et al. (3) and Blackwell (4) have reviewed the topic of hardsetting. This article discusses the occurrence, formation, assessment, and management of hardsetting soils.

OCCURRENCE AND PROCESSES OF FORMATION

The term hardsetting has generally only been used in Australia, and was first used by Northcote (5) in his soil-classification system. It is a widespread phenomena in Australian soils: e.g., in the new Australian Classification developed by Isbell (6), of the 14 orders recognized, only the Organosols (soils with organic horizons), Podosols (deep sands), and probably the Calcarosols (calcareous soils), can be assumed to be generally nonhardsetting. Internationally, the conditions also exist for hardsetting soils, but due to the lack of a common definition, they are harder to recognize.

Even though the soil properties that predispose a soil to surface crusting and hardsetting may be similar and involve the same processes (such as slaking and dispersion), the phenomena apply to different thicknesses of the seedbed. Crusting occurs when surface aggregates breakdown on wetting (usually under droplet impact) to form a hard, dense, stuctureless mass at the soil surface, often only a few millimeters thick. Hardsetting, however, involves a much greater thickness of material, which commonly not only includes the A1 or Ap horizon, but also may include A2 (E) horizons which may be bleached. Therefore in some soils the depth affected by hardsetting may be up to 0.3 m or more (7). Thus, hardsetting can cause different problems at different depths. For example, when it occurs at the surface horizon, infiltration and aeration are reduced, and seedling emergence is inhibited (Fig. 1). When hardsetting occurs lower in the profile, it causes problems of root elongation and water logging.

The processes required for hardsetting are: 1) slaking of aggregates under rapid wetting, and/or 2) dispersion of the clay fraction. Although the worst possible situation results when both slaking and dispersion occur, slaking or dispersion alone will result in a soil developing hardsetting properties. The soil mass consisting of these breakdown products then increases in strength as the soil dries. In addition, in some subsoil layers, soluble materials such as silica are released on wetting and act as cementing agents as the soil dries (8). These cementing agents further exacerbate the problem of hardsetting. The processes of slaking and dispersion are now discussed in detail. The interparticle bonding in hardsetting soils is also outlined.

Slaking Under Rapid Wetting by Irrigation or Rainfall

Under the forces involved in rapid wetting, soil aggregates may disintegrate into smaller particles called microaggregates (1). The extent of breakdown depends on how the strength of the aggregate can withstand the stresses due to rapid wetting. The stresses include swelling of the clays, compression of entrapped air, release of the heat of wetting, and the mechanical action of the wetting (9, 10). Slaking is therefore affected by the amount and type of clay minerals, the organic matter content, the initial moisture content, and the rate of wetting (11). For example, silty surface horizons that are frequently cultivated are commonly low in organic matter. Under rapid wetting, they undergo severe slaking and result in

Fig. 1 A silty loam surface soil that has hardset following irrigation. Note the problems of seedling emergence.

hardsetting horizons. The hardsetting soil in Fig. 1 is typical of such a soil. It has a silty clay loam texture, organic carbon content of approx. 1.5%, and is used for row cropping.

Dispersion

Dispersion is the process whereby the microaggregates formed by slaking undergo further breakdown into individual clay particles or packets of clay particles known as quasi-crystals (12). The amount of dispersion depends on the exchangeable sodium percentage (ESP) of the soil, the total electrolyte concentration of the soil solution, and the amount of mechanical agitation the soil receives during wetting (2).

Interparticle Bonding in Hardsetting Soils

The material produced by slaking of hardsetting soils consists of a mixture of sand grains and fragments of aggregated material, as well as clay particles or packets of clay particles (11). In soils that have undergone both slaking and dispersion, the released clays bridge between the sand grains and fragments of disrupted material to form a dense, structureless mass. Greene et al. (13) demonstrated that when soils were sodium (Na^+) saturated, the extent of hardsetting increased as the smectite content increased. However, when calcium (Ca^{2+}) saturated, hardsetting decreased as the smectite content increased. The difference was due to the formation of quasi-crystals in the Ca^{2+} system, which formed an open, less dense structure with less hardsetting properties than a Na^+ system.

ASSESSMENT OF HARDSETTING SOILS

It is important to be able to assess the likelihood of a soil to set hard in the field and to be able to characterize the severity of that hardsetting. Some of the measurements used are: 1) modulus of rupture (MOR), 2) tensile strength, 3) extent of slaking into microaggregates, and 4) amount of clay dispersion. The first two measurements involve strength measurements on reconstituted soil discs, while the later two involve subjecting natural soil aggregates to wetting and characterising the amount of structural disintegration into microaggregates and clay particles respectively. Details of the four measurements are as follows.

Modulus of Rupture

Aylmore and Sills (14) used a combination of MOR and hydraulic conductivity measurements to classify hardsetting sandy loam soils of the Western Australian wheatbelt. Soils were classified as hardsetting if their MOR had a value of $\geq$60 kPa. Hardsetting soils also exhibited a more rapid increase in MOR with increasing ESP (sodium sensitivity) compared with nonhardsetting soils.

Tensile Strength

Tensile strength measurements on remoulded cores of soil are usually carried out after saturating the exchange complex of the soil with Na^+. Greene et al. (13) used measurements of tensile strength to characterize the effect of clay content, cation exchange capacity, and clay type on the hardsetting properties of a range of hardsetting soils used for horticulture in the Carnarvon district of western Australia.

Extent of Slaking (Wet Sieving Measurements)

Wet sieving techniques are used to measure the portion of soil aggregates that disintegrate on wetting due to slaking into microaggregates <250 μm in size. These microaggregates are important in clogging interaggregate porosity and initiating problems of hardsetting and crusting. As the extent of slaking depends upon conditions such as the initial aggregate size, moisture content, and rate of wetting, it is important to standardize these parameters in the test. It is also important to relate the size range of the products of aggregate disintegration formed by laboratory wet sieving procedures to actual processes of structural decline and soil strength development in the field.

Measurement of Soil Dispersion

Rengasamy et al. (2) used measurements of clay dispersion to predict if a soil was likely to undergo surface crusting processes in the field. The test can also be used to predict the likelihood of a soil undergoing hardsetting. Because clay dispersion was measured under different conditions, i.e., spontaneous and mechanical situations, it was also possible to use this test to predict the effects of different management procedures, e.g., tillage techniques and types of rotation, on soil behavior. The test was particularly useful in predicting the likely response of the soil to ameliorants such as gypsum and/or lime and to irrigation with different qualities of effluent.

MANAGEMENT

As rapid and prolonged wetting of air dried cultivated soils leads to severe slaking and formation of hardsetting soils, any management techniques that avoid these practices will minimize structural breakdown and hence the extent of hardsetting. Therefore, the use of surface mulches, low-application-rate micro-irrigation, and raised beds, all of which slow the rate of wetting (11), are important.

Soil organic matter is critical in controlling the structural stability of a soil, and hence its tendency to hardset. Tillage practices such as intensive, continuous cropping leads to low soil organic matter concentrations and concomitant low structural stability. Therefore, pasture phases, which increase soil organic matter, are critical for structural improvement of hardsetting soils (11).

Sodic soils, i.e., those containing a level of exchangeable Na ≥6%, are prone to dispersion and formation of crusting surfaces and hardsetting properties in the profile. The application of gypsum to dispersive, sodic soils, is a very efficient way of preventing dispersion, and thus the problems of hardsetting (2).

CONCLUSIONS

Hardsetting is shown to be one of the most serious forms of soil structural decline that can occur. Its effects on a range of soil processes such as infiltration, emergence, air–gas exchange and waterlogging are critical in determining plant performance and productivity. Only by correct assessment, followed by appropriate amelioration and management practices, can these potential limitations to productivity be avoided. Further work is urgently needed to develop an international system for classifying hardsetting soils.

REFERENCES

1. Oades, J.M. Soil Organic Matter and Structural Stability: Mechanisms and Implications for Management. Plant and Soil **1984**, *76*, 319–337.
2. Rengasamy, P.; Greene, R.S.B.; Ford, G.W.; Mehanni, A.H. Identification of Dispersive Behaviour and the Management of Red-Brown Earths. Aust. J. Soil Res. **1984**, *22*, 413–431.
3. Mullins, C.E.; MacLeod, D.A.; Northcote, K.H.; Tisdall, J.M.; Young, I.M. Hard-Setting Soils: Behaviour, Occurrence, and Management. In *Soil Degradation*; Lal, R., Stewart, B.A., Eds.; Adv. Soil Sci., 11, Springer-Verlag: New York, 1990; 37–108.
4. Blackwell, P.S. Slaking and Hardsetting Soils: Some Research and Management Aspects. Soil and Tillage Res. **1992**, *25*, 111–261.
5. Northcote, K.H. *A Factual Key for the Recognition of Australian Soils*; Rellim Technical Publications Pty. Ltd.: Coffs Harbour, NSW, 1960; 124.
6. Isbell, R.F. *The Australian Soil Classification*; CSIRO Publishing: Melbourne, 1996; 144.
7. Isbell, R.F. Sealing, Crusting and Hardsetting Conditions in Australian Soils. In *Proceedings of the Second International Symposium on Sealing, Crusting and Hardsetting Soils: Productivity and Conservation*; So, H.B., Smith, G.D., Raine, S.R., Schafer, B.M., Loch, R.J., Eds.; ASSSI: Queensland Branch, 1995; 15–30.
8. Chartres, C.J.; Kirby, J.M.; Raupach, M. Poorly Ordered Silica and Aluminosilicates as Temporary Cementing Agents in Setting Soils. Soil Sci Am. J. **1990**, *54*, 1060–1067.
9. Emerson, W.W. Physical Properties and Structure. In *Soil Factors in Crop Production in a Semi-Arid Environment*; Russell, J.S., Greacen, E.L., Eds.; Queensland University Press: Brisbane, 1977; 78–104.
10. Collis-George, N.; Greene, R.S.B. The Effect of Aggregate Size on the Infiltration Behaviour of a Slaking Soil and Its Relevance to Ponded Irrigation. Aust. J. Soil Res. **1979**, *17*, 65–73.
11. Chan, K.Y.; Mullins, C.E. Slaking Characteristics of Some Australian and British Soils. Eur. J. Soil Sci. **1994**, *45*, 273–283.
12. Greene, R.S.B.; Posner, A.M.; Quirk, J.P. A Study of the Coagulation of Montmorillonite and Illite Suspensions by Calcium Chloride Using the Electron Microscope. In *Modification of Soil Structure*; Emerson, W.W., Bond, R.D., Dexter, A.R., Eds.; Wiley: New York, 1978; 35–40.
13. Greene, R.S.B.; Eggleton, R.A.; Rengasamy, P. Relationships Between Clay Mineralogy and Hardsetting Properties of Soils in the Carnarvon Horticultural District of Western Australia. Appl. Clay Sci. **2001**, (in press).
14. Aylmore, L.A.G.; Sills, I.D. Characterization of Soil Structure and Stability Using Modulus of Rupture-Exchangeable Sodium Percentage Relationships. Aust. J. Soil Res. **1982**, *20*, 213–224.

HEALTH

H

R.F. Harris
University of Wisconsin, Madison, Wisconsin, U.S.A.

D.E. Romig
Energy, Minerals and Natural Resources Department, Santa Fe, New Mexico, U.S.A.

INTRODUCTION

Historically, the "healthy soil–healthy people" concept made an important grass roots contribution to our reawakened interest in the key role that soil plays in the Earth's life support system (1). Soil health may be viewed as the component of soil quality and ecosystem health that reflects the properties of soil as a living system. Whereas a healthy soil is an integral part of a healthy ecosystem, soil health plays a variable and sometimes negative role in soil quality, depending on the specific homocentric use of soil under consideration. The terms soil quality and soil health converge but are not conceptually identical when the specific use of soil is to support ecosystem health.

SOIL HEALTH AND SOIL QUALITY

In recent years, the terms soil health and soil quality have been used synonymously, at least from the standpoint that objective differences have not been clearly established (2–6). Soil quality is now used almost exclusively by the scientific community, in part for brevity and in part because soil quality, concisely defined as "the degree of fitness of soil for a specific use" or "capacity of soil to function" (6–8)[a] is felt to be less subjective, more quantifiable, and thus more defendable and less controversial than soil health (e.g., 8). In addition, integrating the concise and broader definitions of soil quality (1, 8)[b] and the six landscape functions of soil (9), identifies that soil quality is broader than soil health. Accordingly, for the three ecosystem functions, soil quality becomes the degree of fitness of soil to sustainably 1) support biomass production and diversity, 2) protect environmental water and air quality by filtering, buffering and transforming, and 3) maintain the Earth's genetic heritage; and, for the three "nonecosystem" functions, soil quality is the degree of fitness of soil to act as sustainably as possible as a 1) physical medium for infrastructures, 2) source of mining materials, and 3) keeper of the Earth's paleontological and our archeological heritage. The concept of soil health has little or no relevance to the three "nonecosystem" functions, unless to rationalize a negative relationship between a healthy biologically active soil and the quality of soil to support these functions.

Despite the shift away from soil health to soil quality, there is general agreement that soil health appropriately conveys a more "soil as a living system" image as compared to the animate and/or inanimate "soil as a resource for human development" focus of soil quality (3–9), (3–8)[a,b]. Rather than force synonimity between soil health and soil quality, we suggest that soil health should be considered a component of soil quality that reflects the properties of soil as a living system, and, within the context of ecosystem functions, can be defined separately as the degree of fitness of soil as a living system, within its natural means, to sustainably 1) support biological production and promote plant and animal health, 2) maintain the Earth's genomic heritage, and 3) act as a living filter protector of water and air quality. The "within its natural means" qualification recognizes that the fitness of a soil to perform its ecosystem functions may be limited because of its inherent make up, yet the soil may still be a resilient living system as long as it is managed within its capabilities. In practice, this definition of soil health converges with, but is not conceptually identical to, an exclusively ecosystem-based definition of soil quality[b].

SOIL HEALTH AND SOIL QUALITY WITHIN THE CONTEXT OF ECOSYSTEM HEALTH

Conflicts over the pros and cons of health as compared to quality concepts exist for managing rivers and watersheds

[a] See *Quality* on page 1058.
[b] See *Quality and Sustainable Agriculture* on page 1088.

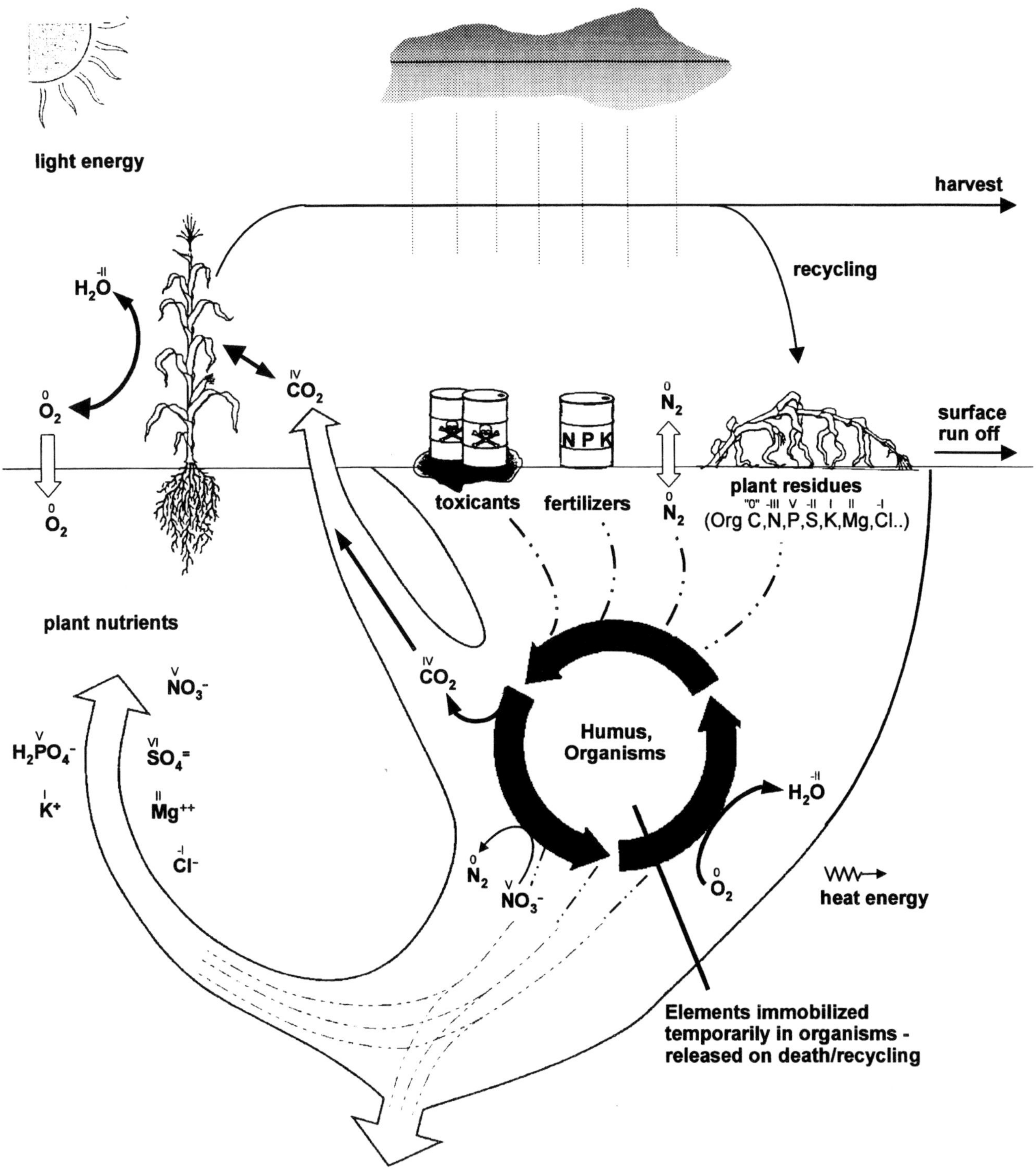

Fig. 1 Role of soil health in biomass production, biogeochemical cycling, energy flux and environmental protection in an agricultural ecosystem. (Adapted from Ref. 14.) A Roman numeral above an element is the oxidation number of the element.

as well as soil. In this regard, Karr and Chu (10) persuasively rationalize that the concept of "ecological health," as difficult as it is to define and measure, is an important policy goal with a greater opportunity to engage public support and interest than "ecological quality" because people more readily extend the idea of their own health to that of a watershed or ecosystem. As a component of ecosystem health (11), soil health may be defined as the fitness of soil as a living system to sustainably help: 1) prevent ecosystem distress syndrome, 2) support ecosystem sustainability, and 3) protect the environmental quality of interfacing ecosystems. Ecosystem distress syndrome comprises a group of signs by which ecosystem breakdown is generally recognized, and for terrestrial ecosystems, includes leaching of soil nutrients, reduced species diversity, shifts in species composition to opportunistic species, reduced productivity, and increased pest and disease loads (11). Fig. 1, with its focus with its is on the key role of soil as a biological reactor (12), illustrates the role of soil health in supporting biological production, biogeochemical cycling, energy flux, and environmental protection in an agroecosystem. For example, for a healthy soil: 1) agroecosystem distress syndrome is prevented by optimized biogeochemical cycling and a stable soil tilth restricting leaching and surface runoff loss; 2) agroecosystem sustainability is promoted by minimal external input of chemicals to the soil; and 3) contamination of interfacing ecosystems is attenuated by the presence of an effective living filter of soil and non-leakage of nutrients (such as nitrate) into groundwater by tightly coupled mineralization and plant root uptake. Similarly, contaminants entering the soil, either as a function of agronomic use or by accident, would ideally be integrated into the natural nutrient cycles, and converted to harmless by products for metabolism by soil organisms and higher plants (Fig. 1). Finally, it should be recognized that as a component of ecosystem health, soil health implicitly has built-in checks and balances. For example, because of its contribution to ecosystem distress syndrome and environmental contamination, rapid organic turnover and nitrogen mineralization resulting in soil organic matter depletion and excessive nitrate loss to the groundwater is not a characteristic of a healthy soil from an ecosystem health standpoint.

HOW DO YOU RECOGNIZE A HEALTHY SOIL?

As for other subsets of soil quality, the institutional approach to recognizing or assessing soil health involves describing each function on which soil health is to be based, selecting characteristics or properties that influence the capacity of the soil to carry out each function, choosing indicators of these characteristics, and using methods that can accurately measure these indicators (6). Many of the results of the application of this approach to soil quality (1–8)[a] also apply to soil health and are not reviewed here.

Alternatively, building on tacit knowledge of how farmers recognize a healthy soil (2), a soil health scorecard was developed from structured interviews with Wisconsin farmers (13, 14). The Wisconsin Soil Health Scorecard uses an ordinal, qualitative scoring mechanism for 43 soil health indicator properties to evaluate and monitor a soil's ability to support crop production. In addition to soil properties *per se*, farmers used plant, animal, human health, and to a lesser extent water properties to judge the health of their soils, and relied heavily on sensory, descriptive criteria expressing how a healthy soil looked, felt, and smelled. Farmers commonly associated a sense of well-being and harmony with nature, with a healthy soil characterized by dark, rich earthy smelling humus, and friable, crumb structured soil tilth. However, it should be noted that the top 10 choices of farmers were more scientifically acceptable indicators of soil health, specifically, soil organic matter content, crop appearance, erosion, earthworms, drainage, tillage ease, soil structure, pH, soil nutrient test, and crop yield (13, 14). The usefulness of integrating farmer-based, descriptive soil health indicators with analytical indicators in assessment of agroecosystem health remains to be seen.

REFERENCES

1. Harris, R.F.; Karlen, D.L.; Mulla, D.J. A Conceptual Framework for Assessment and Management of Soil Quality and Health. In *Handbook of Methods for Assessment of Soil Quality*; Special Pub. No. 49, Doran, J.W., Jones, A.J., Eds.; American Society Agronomy: Madison, WI, 1996; 61–82.
2. Harris, R.F.; Bezdiceck, D.F. Descriptive Aspects of Soil Quality/Health. In *Defining Soil Quality for a Sustainable Environment*; Special Pub. No. 35, Doran, J.W., Coleman, D.C., Bezedick, D.F., Stewart, B.A., Eds.; American Society Agronomy: Madison, WI, 1994; 23–35.
3. Acton, D.F.; Gregorich, L.J. Understanding Soil Health. In *The Health of Our Soils Towards Sustainable Agriculture in Canada*; Acton, D.F., Gregorich, L.J., Eds.; Center for Land and Biological Resources Research, Research Branch, Agriculture and Agr-Food Canada: Ottawa, Ontario, Canada, 1995; 5–10.
4. Doran, J.W.; Sarrantonio, M.; Leibig, M.A. Soil Health and Sustainability. Adv. Agron. **1996**, *56*, 1–54.
5. Pankhurst, C.E.; Doube, B.M.; Gupta, V.V.S.R. Biological Indicators of Soil Health: Synthesis. In *Biological*

Indicators of Soil Health; Pankhurst, C.E., Doube, B.M., Gupta, V.V.S.R., Eds.; CAB International: Wallingford, UK, 1997; 419–435.

6. Carter, M.R.; Gregorich, E.G.; Anderson, D.W.; Doran, J.W.; Janzen, H.H.; Pierce, F.J. Concepts of Soil Quality and Their Significance. In *Soil Quality for Crop Production and Ecosystem Health*; Gregorich, E.G., Carter, M.R., Eds.; Elsevier: Amsterdam, The Netherlands, 1997; 1–9.
7. Singer, M.J.; Ewing, S. Soil Quality. In *Handbook of Soil Science*; Sumner, M.E., Ed.; CRC Press: Boca Raton, FL, 2000; G271–298.
8. Karlen, D.L.; Mausbach, M.J.; Doran, J.W.; Cline, R.G.; Harris, R.F.; Schuman, G.E. Soil Quality: A Concept, Definition, and Framework for Evaluation. Soil Sci. Soc. Am. J. **1997**, *61*, 4–10.
9. Blum, W.E.H.; Santelises, A.A. A Concept of Sustainability and Resilience Based on Soil Functions. In *Soil Resilience and Sustainable Land Use*; Greenland, D.J., Szabolcs, I., Eds.; CAB International: Wallingford, UK, 1994; 535–542.
10. Karr, J.R.; Chu, E.W. *Restoring Life in Running Waters*; Island Press: Washington, DC, 1998.
11. Rapport, D.J.; McCullum, J.; Miller, M.H. Soil Health: Its Relationship to Ecosystem Health. In *Biological Indicators of Soil Health*; Pankhurst, C.E., Doube, B.M., Gupta, V.V.S.R., Eds.; CAB International: Wallingford, UK, 1997; 29–47.
12. Harris, R.F.; Arnold, S.M. Redox and Energy Aspects of Soil Bioremediation. In *Bioremediation: Science and Applications*; Special Pub. No. 43, Skipper, H.D., Turco, R.F., Eds.; American Society Agronomy: Madison, WI, 1995; 55–87.
13. Romig, D.E.; Garlynd, M.J.; Harris, R.F.; McSweeney, K. How Farmers Assess Soil Health and Quality. J. Soil Water Conserv. **1995**, *50*, 229–236.
14. Romig, D.E.; Garlynd, M.J.; Harris, R.F. Farmer Based Assessment of Soil Quality: A Soil Health Scorecard. In *Handbook of Methods for Assessment of Soil Quality*; Special Pub. No. 49, Doran, J.W., Jones, A.J., Eds.; American Society Agronomy: Madison, WI, 1996; 39–60.

HEAT FLUX

Thomas J. Sauer
United States Department of Agriculture (USDA), Ames, Iowa, U.S.A.

INTRODUCTION

The range of climatic conditions existing on the Earth's land surface are related to physical position (elevation and latitude) and large-scale meteorological forces (such as ocean and air currents). Soil plays an integral role in influencing the climate near the ground. Properties of the soil surface affect the amount of sunlight energy reflected back to the atmosphere and the amount used to evaporate water, grow plants, warm the air above the ground, or warm the soil. The amount of thermal energy that moves through an area of soil in a length of time is referred to as the soil heat flux or heat flux density. *Soil heat flux* is a measure of the amount of energy moving into or out of the soil, which determines soil temperature and the rate of daily and seasonal temperature change. Soil temperature is one of the key factors in determining the rate of chemical and biological processes in the soil.

SOIL HEAT FLUX AND SURFACE ENERGY BALANCE

Early research on soil thermal regimes focused on soil temperature and its effects on crop growth (1, 2). Soils subjected to different tillage practices or with contrasting physical properties were observed to warm at different rates in the spring. However, the processes controlling heat flow in soils were not adequately understood. Patten (3) provided the first quantitative, comprehensive treatment of heat-transfer processes in soil by measuring the thermal properties of several soils under controlled laboratory conditions. By the mid-twentieth century, meteorologists became interested in measuring soil heat flux as a component of the surface energy balance:

$$R_n - G = LE + H \tag{1}$$

where R_n is the net radiation, G is the soil heat flux, LE is the latent heat flux, and H is the sensible heat flux (all in watts per square meter, W m^{-2}). *Net radiation* represents the net difference between incoming shortwave and longwave (thermal) radiation and reflected shortwave and terrestrial longwave radiation. *Latent heat flux* is the amount of energy consumed by evaporating water or released during dew formation. *Sensible heat flux* refers to the energy involved in heating or cooling the air layer near the soil surface.

The magnitude of G can vary, from near zero on a daily basis in temperate regions with dense crop cover, to 50% of hourly R_n in semiarid regions with no vegetation. Energy balance investigations (4, 5) have found complex interactions between atmospheric transport processes and soil thermal properties because both affect the dynamics of energy flow in surface soil layers. Fig. 1 presents an example of diurnal patterns of energy balance terms at a pasture site in northwestern Arkansas. These data are for 2 days in autumn (October) when the soil was moist and the grass canopy was 0.35 m tall. The sum of G for both days was negative (i.e., the soil was cooling) and equivalent to less than 1% of the magnitude of the daily R_n.

FACTORS AFFECTING SOIL HEAT FLUX

The partitioning of R_n at the soil surface and the soil thermal properties influence diurnal and annual patterns of soil heat flux and determine the ability of the soil to transfer energy to and from the surface. The amount of reflected shortwave radiation, which depends on soil color, soil wetness, surface cover, and surface roughness, is one component of the partitioning of R_n at the surface. In general, plants or plant residues reflect more shortwave radiation than do bare soils. Smooth surfaces with lighter-colored or dry soils also tend to reflect more shortwave radiation than do uneven surfaces with dark-colored or wet soils. Surface roughness and soil wetness also affect LE and H in that roughness enhances turbulent mixing of the air near the surface and the availability of soil water affects evaporation.

Soil heat capacity and thermal conductivity are the key thermal properties affecting soil heat flux (6). The heat capacity of a soil is the amount of energy (megajoules, MJ) required to change the temperature of a volume of soil (m^3) 1 degree Kelvin (K). A soil's thermal conductivity

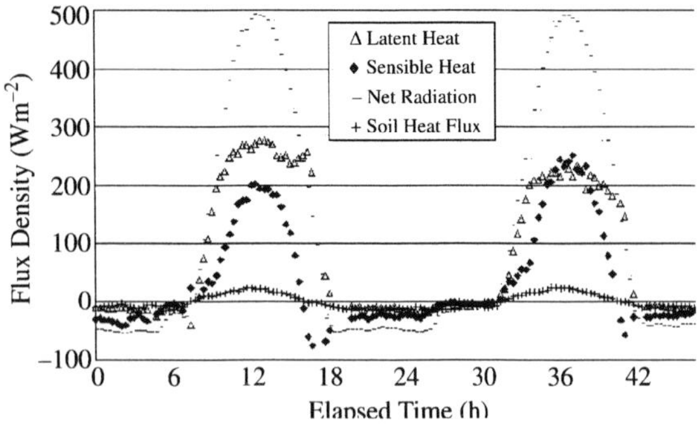

Fig. 1 Energy balance components for a pasture site in northwestern Arkansas for 2 days in October 1997.

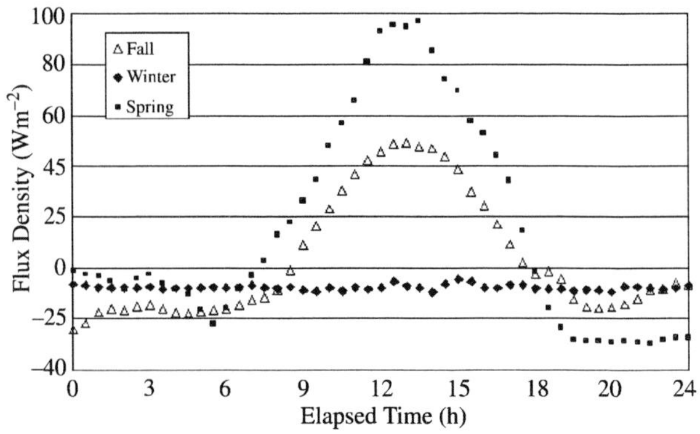

Fig. 2 Seasonal variation in soil heat flux at 0.05 m in a no-till corn field in central Iowa.

is the constant of proportionality relating the temperature gradient across the soil to the soil heat flux, as defined by Fourier's equation:

$$G = -k\ dT/dz \tag{2}$$

where k is the thermal conductivity ($W\,m^{-1}\,K^{-1}$), T is the soil temperature (K), and z is the distance between the soil temperature measurements (m). Table 1 lists typical heat capacity and thermal conductivity values for different soils and soil constituents. Because of the much greater (approximately 3500 times) heat capacity of water as compared with air (4.18 vs. 0.0012 $MJ\,m^{-3}\,K^{-1}$), a moist soil has a much higher heat capacity than a dry one. However, the thermal conductivity of water is approximately 23 times greater than air (0.57 vs. 0.025 $W\,m^{-1}\,K^{-1}$). Thus, a moist soil conducts heat more efficiently than a dry one, but it takes much more energy to develop the same temperature gradient in a moist soil as compared with a dry one.

Table 1 Observed thermal properties of various moist soils and soil components at ambient temperatures

Material	Heat capacity ($MJ\,m^{-3}K^{-1}$)	Thermal conductivity ($W\,m^{-1}K^{-1}$)
Sandy soil	2.09	1.8
Silt loam soil	1.02	1.2
Peat soil	3.14	0.29
Quartz	2.13	8.8
Other soil minerals	2.39	2.9
Organic matter	2.50	0.25
Water	4.18	0.57
Air	0.0012	0.025

K = Kelvin; m = meters; MJ = megajoules; W = watts.
(From Refs. 6–8.)

In field settings, G can be limited by the ability of the soil to conduct heat or by the amount of available energy at the surface. Soil heat flux measured over a 24-hour period at a depth of 0.05 m in a no-till corn field in central Iowa in fall, winter, and spring is presented in Fig. 2. In fall, the fresh corn residue effectively insulated the dry soil from solar radiation inputs. In winter, a layer of snow 0.23 m thick covered the wet residue and soil. Because of the insulating effect of the snow layer, the soil heat flux remained negative (i.e., the soil was slowly losing heat to the atmosphere) and showed little diurnal trend. By spring, the increase in decayed residue exposed more of the wet soil to incoming radiation, causing the peak soil heat flux to nearly double that of the peak measured during the fall.

MEASUREMENT TECHNIQUES

Because soil heat flux is such an important parameter in agrometeorology and soil science research, several measurement methods have been developed (9, 10). The choice of a specific method depends on the intended use of the data, the resources available, and the degree of accuracy required. One category of techniques uses measured and/or estimated soil thermal properties combined with soil heat flow theory to estimate G. These techniques, which include calorimetric, gradient, and combination, generally require numerous and accurate measurements of soil temperature, heat capacity, and thermal conductivity at various depths in the soil. Depending on the accuracy desired, some or all of these parameters must be measured at several locations, at frequent intervals, and for each depth throughout the measurement period. Such requirements make their use relatively tedious and time consuming. The second category uses a sensor, called a heat flux plate or heat flow transducer, designed specifically to measure soil heat flux. Soil heat flux plates are small (often <50 mm per side), thin (<5 mm thick), rigid sensors placed horizontally in the soil. Most employ a thermopile located inside a waterproof housing to measure the temperature difference between the top and bottom of the plate. Temperature difference across the plate is used to estimate heat flux through the soil. Although this technique is much simpler to use, the plate itself can disturb soil heat and water flow, thereby reducing accuracy in some situations. Corrections must be made for heat storage in the soil above the plate in order to calculate G at the soil surface accurately.

CONCLUSION

The rates of chemical and biological processes in soil are strongly affected by temperature. Soil heat flux, the amount of heat moving into or out of the soil, can be altered to optimize the soil thermal environment. Irrigation, drainage, tillage, and crop residue management can all be used to affect energy transfer at the soil surface by modifying the surface characteristics and soil thermal properties. Knowledge of the factors affecting soil heat flux is essential for optimizing plant growth through modification of the soil thermal regime and microclimate near the ground.

REFERENCES

1. King, F.H. *The Soil*; Macmillan: New York, 1906.
2. Keen, B.A. *Physical Properties of the Soil*; Longmans, Green and Co.: London, 1931.
3. Patten, H.E. *Heat Transference in Soils*, Bureau of Soils Bulletin No. 59; U.S. Department of Agriculture: Washington, DC, 1909.
4. Staley, R.C.; Gerhardt, J.R. Soil Heat Flux Measurements. In *Exploring the Atmosphere's First Mile*; Lettau, H., Davidson, B., Eds.; Pergamon Press: New York, 1957; *1*, 58–80.
5. Sellers, W.D. *Physical Climatology*; The Univ. of Chicago Press: Chicago, 1965.
6. de Vries, D.A. Thermal Properties of Soils. In *Physics of Plant Environment*, 2nd Ed.; van Wijk, W.R., Ed.; North-Holland Publishing Co.: Amsterdam, 1966; 210–235.
7. van Wijk, W.R.; de Vries, D.A. Periodic Temperature Variations in a Homogeneous Soil. In *Physics of Plant Environment*, 2nd Ed.; van Wijk, W.R., Ed.; North-Holland Publishing Co.: Amsterdam, 1966; 102–140.
8. Al Nakshabandi, G.; Kohnke, H. Thermal Conductivity and Diffusivity of Soils as Related to Moisture Tension and Other Physical Properties. Agr. Meteorol. **1965**, *2*, 271–279.
9. Fuchs, M. Heat Flux. In *Methods of Soil Analysis: I. Physical and Mineralogical Methods*, 2nd Ed.; Klute, A., Ed.; Soil Science Society of America: Madison, Wisconsin, 1986; 957–968.
10. Kimball, B.A.; Jackson, R.D. Soil Heat Flux. In *Modification of the Aerial Environment of Plants*; Barfield, B.J., Gerber, J.F., Eds.; American Society of Agricultural Engineers: St. Joseph, Michigan, 1979; 211–229.

HEAVY METALS

M.J. McLaughlin
Commonwealth Scientific and Industrial Research Organisation Land and Water, Glen Osmond, South Australia, Australia

INTRODUCTION

It is generally accepted that metals having a specific gravity (weight per unit volume) greater than $5\,Mg\,m^{-3}$ are termed heavy metals. In soils, these elements include cadmium (Cd), cobalt (Co), chromium (Cr), copper (Cu), iron (Fe), mercury (Hg), manganese (Mn), molybdenum (Mo), nickel (Ni), lead (Pb) and zinc (Zn) (1) (http://www.soils.org/sssagloss/ (accessed February 2001)). The term "heavy metals" is often used synonymously with the term "trace elements," but this is incorrect as trace elements are generally defined as those elements normally occurring in soil at concentrations less than $100\,mg\,kg^{-1}$ (2), which precludes several heavy metals, e.g., Cr, Fe, and Mn. Arsenic (As) is often included in the group "heavy metals," but is more correctly classified as a metalloid.

GENERAL CHEMISTRY

Some heavy metals are essential for either plant or animal survival on land (Co, Cr, Cu, Fe, Mn, Mo, Ni, and Zn) while others are nonessential and toxic at low concentrations (Cd, Hg, and Pb). All the heavy metals, except Pb, are transition elements, belonging to the d-block in the periodic table. Many of these elements differ from the alkaline earth metals (e.g., Na, Ca, and Mg) in that they can exist in several valence states in soil (Table 1). In particular, Cr, Fe, and Mn are markedly affected by soil redox potential and undergo both oxidation and reduction depending on soil conditions. This has important implications for the availability and toxicity of many heavy metals. As both Fe and Mn are major structural metals in soil minerals, reduction of the Fe^{3+} and Mn^{4+} ions may lead to a change in soil mineral surfaces important for retention of many elements, including other metals. For Cr, oxidation converts the nontoxic Cr^{3+} ion to the toxic and carcinogenic Cr^{6+} ion. This reaction has even more significance in soils as a strongly sorbed or precipitated cation (Cr^{3+}) is converted into a poorly sorbed or soluble anion (CrO_4^{2-}). However, even in aerobic soils, Cr^{3+} is the thermodynamically stable state, so added Cr^{6+} ion is rapidly converted to Cr^{3+} in most soils.

Solubility and availability/toxicity to organisms of heavy metal cations (Cd^{2+}, Cr^{3+}, Fe^{n+}, Pb^{2+}, Mn^{n+}, Hg^{2+}, Ni^{2+}, and Zn^{2+}) decrease as soil pH increases. This is due to the increase in negative charge on variable charge surfaces in soil (4) and the propensity for these metals to precipitate as sparingly soluble compounds (phosphates, carbonates, and hydroxides) as soil pH increases (5). On the other hand, solubility and availability/toxicity to organisms of anionic heavy metals (CrO_4^{2-}, MoO_4^{2-}) may increase as soil pH increases, again due to increases in surface negative charge on soil particles affecting sorption. Under reducing conditions in soil, many of the metals form insoluble metal sulfides, therefore reducing availability to plants and animals.

Heavy metals are also subject to complexation reactions in soil with both inorganic and organic ligands, which may markedly increase mobility in soil. This may be used to good effect by plants in scavenging essential metals from soil, e.g., phytochelatins (see below). However, for nonessential metals, complexation and increased mobility also may lead to increased environmental risks through leaching and plant uptake, e.g., chloride-induced uptake of Cd by food crops in saline soils (6).

ABUNDANCE IN ROCKS AND SOILS

Heavy metals become incorporated into primary minerals in igneous rocks through isomorphous substitution. In sedimentary rocks, heavy metals are incorporated as constituents of minerals, or are removed from the water column and trapped in sediments by adsorption or precipitation processes (7). Apart from Cr, Fe, and Mn, heavy metals are generally present at trace concentrations ($<100\,mg\,kg^{-1}$) in most soils, with the exception of soils developed over mineralized parent materials or soils developed through biological enrichment (8). Background concentrations in soil usually reflect the composition of the parent rock material. Background concentrations of heavy metals are usually determined where no known anthropogenic inputs have occurred, but this assessment is often problematic due to the global spread of anthropogenic emissions through atmospheric transport processes (9).

Encyclopedia of Soil Science

Table 1 Physical and chemical properties of the heavy metals

Element	Symbol	Atomic no.	Molecular weight	Valence	Natural isotopes	Density ($Mg\,m^{-3}$)	Melting point (°C)	First ionization potential (eV)	Dominant species in soil[a]	Dominant species in soil solution[b] pH 3.5–6.0	Dominant species in soil solution[b] pH 6.0–8.5
Cadmium	Cd	48	112.41	2	8	8.65	321	8.96	Cd^{2+}	Cd^{2+}, $CdCl^{+}$, $CdSO_4^0$	Cd^{2+}, $CdCl^{+}$, $CdSO_4^0$
Chromium	Cr	24	52.01	2, 3, 6	4	7.19	1875	6.76	Cr^{3+}, CrO_4^{2-}	Cr^{3+}, $CrOH^{2+}$	$Cr(OH)_4^-$
Cobalt	Co	27	58.94	2, 3	1	8.90	1493	7.86	Co^{2+}	–	–
Copper	Cu	29	63.54	1, 2	2	8.94	1083	7.73	Cu^{2+}	Cu^{2+}, Cu-org.	Cu-hydroxy species, $CuCO_3^0$, Cu-org.
Iron	Fe	26	55.85	2, 3	4	7.87	1536	7.87	Fe^{2+}, Fe^{3+}	Fe-hydroxy species, Fe-org.	Fe-hydroxy species, Fe-org.
Lead	Pb	82	207.19	2, 4	4	11.35	327	7.42	Pb^{2+}	Pb^{2+}, $PbSO_4^0$, Pb-org.	Pb-hydroxy and carbonate species, Pb-org.
Manganese	Mn	25	54.94	2, 3, 4, 7	1	7.44	1244	7.44	Mn^{2+}, Mn^{4+}	Mn^{2+}, $MnSO_4^0$, Mn-org.	Mn^{2+}, $MnSO_4^0$, $MnCO_3^0$,
Mercury	Hg	80	200.61	1, 2	7	13.54	−39	10.44	Hg^{2+}, $(CH_3)_2Hg$	–	–
Molybdenum	Mo	42	95.94	6	7	10.22	2610	7.10	MoO_4^{2-}	–	–
Nickel	Ni	28	58.71	2, 3	5	8.91	1453	7.64	Ni^{2+}	Ni^{2+}, $NiSO_4^0$, Ni-org.	Ni^{2+}, $NiHCO_3^+$, $NiCO_3$
Zinc	Zn	30	65.37	2	5	7.14	420	9.39	Zn^{2+}	Zn^{2+}, $ZnSO_4^0$, Zn-org.	Zn^{2+}, Zn-hydroxy and carbonate species, Zn-org.

[a] From Ref. 2.
[b] From Ref. 3.

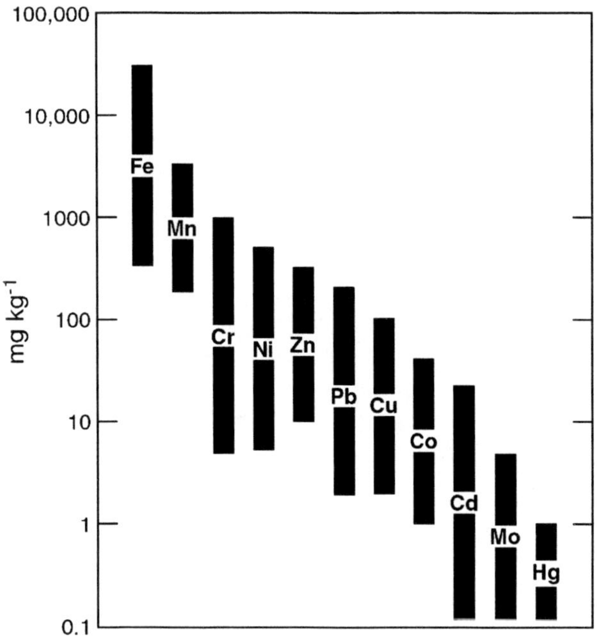

Fig. 1 Range of heavy metal concentrations of soil. Bars represent commonly found values. (Adapted from Refs. 2, 7–12.)

The typical range in abundances of heavy metals in unpolluted soils is shown in Fig. 1.

Concentrations of heavy metals in soil can be significantly increased through human activity. A number of primary and secondary sources have been identified as contributing to enhanced concentrations of heavy metals in soil (12) (Table 2). Many countries have introduced legislation to minimize the amount of metals accumulating in soil, and set ceiling concentrations above which further metal additions must stop.

Table 2 Sources of heavy metal contamination in soils

Source	Main heavy metals
Primary sources	
Fertilizers	Cd, Cu, Mo, Pb, Zn
Irrigation water	Cd, Fe
Manures and composts	Cd, Cr, Cu, Fe, Hg, Mn, Mo, Ni, Pb, Zn
Pesticides	Cu, Hg, Pb, Zn
Sewage biosolids (sludges)	Cd, Cr, Cu, Fe, Hg, Mn, Mo, Ni, Pb, Zn
Soil amendments (lime, gypsum, etc.)	Cu, Mn, Pb, Zn
Secondary sources	
Automobile aerosols	Pb
Coal combustion	Pb
Mine waste and effluents	Cd, Cu, Fe, Hg, Mn, Ni, Pb, Zn
Nonferrous smelter waste	Cd, Cu, Hg, Mn, Ni, Pb, Zn
Paint dispersal	Cd, Pb
Tire wear	Cd, Zn
Waste combustion	Cd, Pb

(Adapted from Ref. 8.)

BIOLOGICAL EFFECTS

Co, Cr, Cu, Fe, Mn, Mo, Ni, and Zn are all essential for either healthy plant or animal functioning in soil and all these metals, except Ni, are used in fertilizers or animal stock supplements to ensure efficient agricultural production on soils deficient in essential metals. Due to the low solubility of essential heavy metals in neutral and alkaline soils, many plants have developed strategies to mobilize solid phase forms to facilitate uptake by roots (e.g., Fe, Mn, and Zn). Nonproteinogenic amino acids, or phytochelatins, secreted by actively growing roots are important in acquisition of Cu, Fe, Zn, and possibly Mn from deficient soils (13). At high concentrations in soil, all the essential elements may pose risks to microorganisms, plants, animals, or humans (Fig. 2). Nonessential heavy metals in soil (Cd, Hg, and Pb) have no beneficial effects at low concentrations (Fig. 2) and may be toxic at even trace concentrations (e.g., Cd, Hg).

Critical exposure pathways for expression of heavy metal toxicity in soil have been examined as part of regulation governing re-use of biosolids (sewage sludge) on soil (14), but act as a general risk analysis template for all types of pollution of soil by heavy metals (15). For some elements, e.g., Cd and Co, food chain transfer is the main risk pathway as these elements are easily accumulated by plants in edible tissues. For other elements, sorption in soil is strong, bioaccumulation by

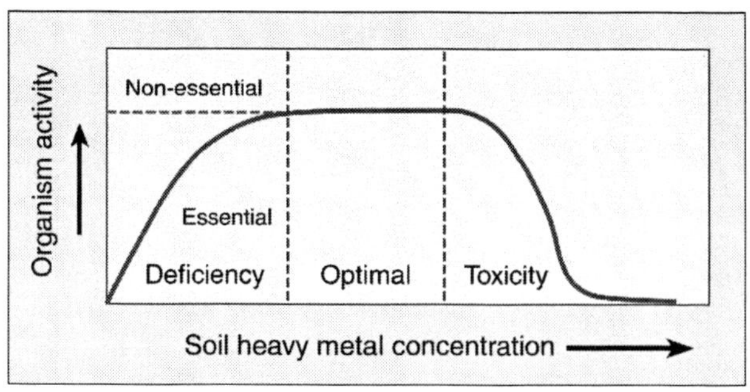

Fig. 2 Typical concentration–response relationship for (a) essential and (b) nonessential heavy metals in soil.

H

Table 3 Critical risk pathway assessments for heavy metal pollution of soils

Metal	Dominant risk pathway	Secondary risk pathway	Most important predictor required
Cd	Food chain transfer	Phyto- and ecotoxicity	Soil–plant uptake
Co	Food chain transfer	Phyto- and ecotoxicity	Soil–plant uptake
Cr	Phyto- and ecotoxicity	Metal leaching	Toxic threshold definition, speciation
Cu	Phyto- and ecotoxicity	Soil ingestion by animals/humans	Toxic threshold definition
Fe	Phyto- and ecotoxicity[a]	None	Toxic threshold definition
Hg	Soil ingestion by animals/humans	Metal leaching	Toxic threshold definition, speciation
Mn	Phyto- and ecotoxicity	Soil ingestion by animals/humans	Toxic threshold definition
Ni	Phyto- and ecotoxicity	Soil ingestion by animals/humans	Toxic threshold definition
Pb	Soil ingestion by animals/humans	Phyto- and ecotoxicity	Oral bioavailability assessment
Zn	Phyto- and ecotoxicity	Food chain transfer	Soil–plant uptake

[a] Only in acid soils under reducing condition.
(Adapted from Ref. 15.)

microorganisms low, and plant uptake and translocation so low that the dominant risk pathway is (for higher animals and humans) through direct ingestion of soil, e.g., Hg and Pb. For the other heavy metals, behavior in soil and bioaccumulation characteristics result in toxicity to plants and microorganisms (phyto- and ecotoxicity) being the dominant risk pathway at high concentrations, e.g., Cr, Cu, Mn, Ni, and Zn. Thus, prediction of risks from heavy metal pollution of soils through soil testing requires a different emphasis depending on the metal considered (Table 3).

REFERENCES

1. *Internet Glossary of Soil Science Terms*; Soil Science Society of America: Madison, WI.
2. Logan, T.J. Soils and Environmental Quality. In *Handbook of Soil Science*; Sumner, M.E., Ed.; CRC Press: Boca Raton, FL, 2000, G155–G163.
3. Ritchie, G.S.P.; Sposito, G. Speciation in Soils. In *Chemical Speciation in the Environment*; Ure, A.M.E., Davidson, C.M., Eds.; Blackie Academic and Professional: London, 1995; 201–233.
4. Barrow, N.J. *Reactions with Variable Charge Soils*; Martinus Nijhoff: Dordrecht, The Netherlands, 1997; 191.
5. Lindsay, W.L. *Chemical Equilibria in Soils*; Wiley: New York, 1979; 449.
6. McLaughlin, M.J.; Tiller, K.G.; Smart, M.K. Speciation of Cadmium in Soil Solution of Saline/Sodic Soils and Relationship with Cadmium Concentrations in Potato Tubers. Aust. J. Soil Res. **1997**, *35*, 1–17.
7. Alloway, B.J. The Origins of Heavy Metals in Soils. In *Heavy Metals in Soils*, 2nd Ed.; Alloway, B.J., Ed.; Blackie Academic and Professional: London, 1995; 38–57.
8. Barry, G.; Rayment, G.E. Heavy Metals and Nutrients in Soils and Sediments of Raine Island, Great Barrier Reef. Land Contam. Reclam. **1997**, *5* (4), 281–285.
9. Lantzy, R.J.; Mackenzie, F.T. Atmospheric Trace Metals: Global Cycles and Assessment of Man's Impact. Geochim. Cosmochim. Acta **1979**, *43*, 511–525.
10. Kabata-Pendias, A.; Pendias, H. *Trace Elements in Soils and Plants*; 2nd Ed.; CRC Press: Boca Raton, FL, 1992; 365.
11. Swaine, D.J. *The Trace Element Content of Soils*; Commonwealth Bureau of Soil Science Technical Communication No. 48, Commonwealth Agricultural Bureaux: Farnham Royal, UK, 1955; 157.
12. Ferguson, J.E. *The Heavy Elements: Chemistry, Environmental Impact and Health Effects*; Pergamon Press: Oxford, UK, 1990; 614.
13. Römheld, V. Different Strategies for Iron Acquisition in Plants. Physiol. Plant. **1987**, *70*, 231–234.
14. USEPA—United States Environment Protection Agency. *Part 503—Standards for the Use and Disposal of Sewage Sludge*, Federal Register 58, 1993; 9387–9404.
15. McLaughlin, M.J.; Zarcinas, B.A.; Stevens, D.P.; Cook, N. Soil Testing for Heavy Metals. Commun. Soil Sci. Plant Anal. **2000**, *31* (11–14), 1661–1700.

HISTORY

Fred P. Miller
The Ohio State University, Columbus, Ohio, U.S.A.

INTRODUCTION

The Development of Soil Science as a Discipline

The year 1927 serves as a logical divide between the early history of soil science and its twentieth century counterpart. That year the First International Congress of Soil Science convened in Washington, D.C. By this point in time, soil science's fragmented, multidisciplinary origins had begun to coalesce through national and international conferences, the formation of scientific societies, and institutional organization of its subject matter. For example, in 1924, the International Society of Soil Science (ISSS) had been founded in Rome, linking the major sub-disciplines of soil science (soil physics; soil chemistry; soil biology; soil genesis, classification, cartography, and pedology) in a single, structured international organization (1). Also by this time, pedology was seen as an independent sub-discipline separated from its earlier ties to the discipline of geology (1–8).

While chemistry (mineralogy), physics, and biology (microbiology) constitute major pillars upon which the discipline of soil science was founded and still rests, there is a hard core of science unique to soil science and that has no place in other more basic sciences, namely, pedology (7). Pedology constitutes the fourth pillar of the discipline of soil science—the study of the relationship of soils to landscapes and ecosystems as well as their genesis and morphology.

DEVELOPMENT OF SOIL SCIENCE IN THE UNITED STATES

The First Half of the Twentieth Century

In another article, John Tandarich has set forth the early history of soil science, including the development of soil science in the United States up to 1927. The early development of soil science in the United States was influenced by the soil survey program, eventually called the National Cooperative Soil Survey of the United States (9, 10). This soil survey program had a substantial impact on soil science and its literature in the United States (10, 11).

In 1920 those working in the soil survey program organized themselves into the American Association of Soil Survey Workers, later known as the American Soil Survey Association (1923). It was the American Soil survey program that sponsored the First International Congress of Soil Science in 1927 in Washington, D.C. In 1936, this Association merged with the Soil Section of the American Society of Agronomy, which had been founded in 1907, to form the Soil Science Society of America. Although this merger was not without detractors in the soil survey program, it did unite various sub disciplines of soil science [soil physics, soil chemistry (fertility), soil biology–microbiology, and pedology] under a single professional–scientific society. As recorded in an increasing literature about soil science, much progress had been made in soil physics, chemistry, mineralogy, and microbiology by this time (5, 6, 10).

After 1929, the Great Depression coupled with the Dust Bowl of the 1930s stimulated demand for soil information in general and soil survey information in particular. Soil erosion had long been the Achilles heel of American agriculture. More and better information about the problem was needed by farmers and land managers to manage land better and minimize the ecological scourges it faced. Such economic and natural forces prompted the Roosevelt administration to form the Soil Erosion Service in 1933 and later the Soil Conservation Service in 1935. After a rocky fiscal start, the soil survey program was a major component underpinning these agencies' missions (9).

In the fifteen years preceding his death in 1935, Curtis F. Marbut, the American pioneer of pedology, devoted himself to developing a comprehensive natural system for classifying all known soils (9). Marbut's final soil classification system (1935) was revised in 1938 by his successor, Dr Charles E. Kellogg, and published in the 1938 Yearbook of Agriculture (12). This classification system was revised again in 1949 and published in a special issue of the journal *Soil Science* (13).

At the same time all subdivisions of soil science were making advances and refinements in soil classification were continuing, field soil mapping was greatly enhanced by aerial photography. Although the feasibility of mapping on air photos was well established by 1929, aerial photo base maps did not come into general use in soil surveys until several years later (9). In 1937, Kellogg issued the first Soil Survey Manual (14) setting forth guidelines, standards, and techniques for soil surveys.

Midway through the twentieth century, the major paradigms of soil science had been established. These included:

1. The pedogenesis concepts of Dochuchaev–Hilgard (vs. the earlier geology-based concepts of soil formation), refined by Marbut, Jenny, Cline, Simonson, and others (1, 2, 5, 6, 9, 11, 15).
2. The chemical basis of plant nutrition, begun by Sprengel and von Liebig in Germany in the mid-1800s (vs. earlier focus on humus as the essence of soil) (1, 6) and essential concepts such as the chemistry of ion exchange, solute transport, and colloidal chemical phenomena (5, 6).
3. The theoretical basis of soil physics, such as soil hydraulic conductivity (Darcy in 1856) and soil mechanics (Terzaghi in the 1920s and 1930s) (1, 5).
4. The understanding and characterization of soil mineralogy, particularly clay mineralogy (with the advent of x-ray techniques and thermic analysis) and its roles in soil chemical phenomena and the physical behavior of soils (6).
5. The basis of soil microbiology and the role microorganisms play in nitrification and the cycling of elements and compounds, the practical relationship of soil microbiology to plant growth, the decomposition of plant residues, the character and dynamics of humus, and an inventory of microbial species in the soil and the study of their roles (1, 5, 6).

As Boulaine (6) noted, the end of World War II and the close of the first half of the twentieth century brought much attention to the need for soil science, including "the expansion of irrigation, the conquest of newly cultivated land, attempts at rehabilitating salty soils, the drainage of marshes, and the fight against erosion requiring technical inventories, periodic surveillance, and predictions concerning the eventual evolution of the land. These needs had repercussions on research programs and on ways of defining pedological variables." Boulaine concluded that "a remarkable flowering of soil science" marked the next half century to follow (6).

The Second Half of the Twentieth Century

McDonald (16) noted that the year 1950 is a logical starting point for an analysis of modern trends in soil science and its literature. In 1949, the first *Advances in Agronomy* had been issued under the auspices of the American Society of Agronomy and included a number of articles concerning soils. In 1951, the revised and expanded *Soil Survey Manual* was published (16, 17). This book refined the standardization of soil survey procedures and guidelines as well as established the primary functions of the United States Soil Survey program (9). Research on soils was taking place in many areas. For example, in 1952, Selman A. Waksman, a microbiologist–biochemist at Rutgers University, received a Nobel Prize for his discovery and isolation of the soil microbe streptomycin (1, 5, 6).

The year 1950 also marked the beginning of an effort to develop an entirely new natural system for classifying the known soils of the world. Since the 1938 major revision of the U.S. classification system (12) and its less intensive revision in 1949 (13), expanded knowledge of soils, demands on the use of soils, and new needs of soil surveys could no longer be reconciled with the established soil classification schemes of 1938 and 1949 (9, 18). Guy D. Smith, co-author of the 1949 revision (13) and a soil scientist in the United States Division of Soil Survey, assumed leadership of this international effort to develop a new comprehensive system of soil classification (9).

This project resulted in the publication of *Soil Taxonomy* in 1975 (18, 19). Subsequently, international committees were established by the United States National Cooperative Soil Survey to work on problems and concerns arising from the use of *Soil Taxonomy* (8). Revisions of this work resulted in further refinement of this new classification system, now totaling twelve major global soil orders. The first level in the classification system, with the most narrowly prescribed ranges of soil properties, is known as the soil series. In the United States, there are more than 14,000 established soil series, a 24-fold increase in the eighty-year period between 1912 and 1992. By the end of 1992, virtually all arable land in the United States had been analyzed by soil surveys under the mandates of the Food Security Act of 1985 (18).

Advances in technology including isotopic tracers, computers, satellites, fractal mathematics, geo-statistics, and biotechnology have been applied rapidly to many areas of soil science. Computers and sophisticated software have allowed complex soil-system behavior to be modeled. The remote sensing of landscapes via satellites and rapidly processed information gained through geographic information systems provide soil scientists with powerful tools to investigate spatial soil relationships as well as predict soil behavior under various ecosystem manipulations. Biotechnology provides tools for assessing, understanding, and manipulating the soil microbes that drive many soil and ecological processes. These and many other opportunities for soil science research have been recently reviewed (20).

In the 1960s and 1970s, agricultural interests emphasized economic and resource efficiency in production. During the same period, the environmental movement became more influential. Soil scientists and their colleagues in agronomy, engineering, forestry, range management, and other scientific areas began to research not only the cropping system responses, but environmental leakages and impacts. Both interest in and demand for such research, coupled with available funding, drove the research agenda, as reflected in the mainline soil science journals (*Journal of Environmental Quality, Soil Science Society of America Journal, Agronomy Journal, Soil Science*) for this period reflect these research interests and demands.

The discipline of soil science, particularly pedology, has also had significant influence on American assessments of land and water resources inventories during the twentieth century. After Bennett's first comprehensive soil erosion inventory in 1934 (21), a soil-and-water conservation needs inventory was conducted and published in 1945 (22). Later inventories of the nation's soil and water resources were conducted using scientific and statistically reliable protocols. A series of conservation needs inventories were carried out between 1958 and 1967 (23, 24). Subsequent inventories of the condition of the nation's land and water resources, known as the National Resource Inventories (NRI), were completed in response to the requirements of the Rural Development Act of 1972 and the Soil and Water Conservation Act of 1977 (RCA). The first RCA appraisal (25) was conducted in 1977. The second and more comprehensive RCA–NRI appraisal (26) reported conditions for the nation's land for 1982. The 1987, 1992, and 1997 NRI updates have been completed (27–29) and provide information to guide policy formulation in managing and sustaining the nation's natural resources.

During the 1990s, the discipline of soil science focused on a variety of agricultural and non-agricultural problems and issues. An assessment of the literature of soil science as it pertains to these issues can be found in McDonald (30). Soil scientists in the 1990s assessed soils as ecosystem components and concentrated on processes that drive ecosystem functions. Several references for this article provide examples of the research initiatives of the 1990s, including isotopic signatures in soil development, computer-assisted tomography, thermo-mechanical processes in soils, fractal mathematics and chaotic flow analytical methodologies, aqueous and surface chemical speciation, microbial habitat assessment and genetic characterization of soil microbes, and extraterrestrial pedology (20, 31–35). The number of issues and problems addressed by American soil scientists is also illustrated by the topics presented at the 2000 Annual Meeting of the Soil Science Society of America. These topics were not of interest to soil scientists fifty years ago. These include:

- ecosystem manipulation and response to mitigation strategies;
- pushing the limits of soils and the environment for food production;
- carbon sequestration potential of managed soil-cropping systems;
- remediation of contaminated soils and landscapes;
- indicators of soil and environmental quality;
- sustaining ecological integrity under high-output agriculture;
- computer modeling and computer-assisted decision making; and
- riparian buffer management as a water quality protection practice.

THE STATUS OF SOIL SCIENCE AS A DISCIPLINE IN 2000

Born out of the basic and natural science disciplines, but nurtured throughout most of the twentieth century by agricultural institutions (U.S. land grant colleges and the United States Department of Agriculture agencies, such as the Soil Conservation Service/Natural Resources Conservation Service and the Agricultural Research Service), soil science now seeks professional–scientific parity with its earth science and natural science counterparts (36, 37). Soil science is still perceived by many to be an agriculturally-oriented discipline. But since World War II, and particularly since the 1970s and 1980s, soil science research and application have been broadened well beyond the boundaries of agriculture. Evidence of the discipline's growth as a bona fide earth science and natural science discipline include:

- greater identity of the Soil Science Society of America (SSSA) from its long professional partnership with the American Society of Agronomy and the Crop Science Society of America;
- SSSA's membership in the American Geological Institute;
- the publication of soil science research in international science journals such as *Science, Nature,* and others;
- research addressing global, national, regional, and local issues in both agricultural and nonagricultural domains;

- soil scientists becoming members of the United States National Academy of Science and other prestigious science-oriented affiliations and awards;
- collaboration in interdisciplinary research;
- successfully competing for research grants;
- membership on political boards, advisory panels, and other political organizations guiding public policy;
- a long history of international exposure and participation; and
- contributions to state-of-the-art research on a natural resource critical to the sustenance of humankind and its well being.

At the threshold of the twenty-first century, soil science is positioned to attain both recognition and legitimacy within the scientific community as a natural science discipline. Soil plays a major role in accommodating and driving ecological processes. It is the planetary medium in which our human sustenance is literally rooted, upon which much of civilization's activities take place, and through which most of society's waste loads are disposed of and recycled. To paraphrase Boulaine's observation a half century ago (6), the "flowering of soil science" should continue into the twenty-first century.

REFERENCES

1. Yaalon, D.H. History of Soil Science in Context: International Perspective. Adv. GeoEcol. **1997**, *29*, 1–13.
2. Tandarich, J.P.; Sprecher, S.W. The Intellectual Background for the Factors of Soil Formation. In *Factors of Soil Formation: A Fiftieth Anniversary Retrospective*; Special Publication No. 33, Soil Science Society of America: Madison, WI, 1994; 1–13.
3. Tandarich, J.P. Pedology: Disciplinary History. In *Sciences of the Earth, An Encyclopedia of Events, People, and Phenomena*; Good, G.A., Ed.; Garland Publishing Inc.: New York, NY, 1998; Vol. 2, 666–670.
4. Tandarich, J.P. Agricultural Geology: Disciplinary History. In *Sciences of the Earth, an Encyclopedia of Events, People, and Phenomena*; Good, G.A., Ed.; Garland Publishing Inc. New York, NY, 1998; Vol. 1, 23–29.
5. Warkentin, B.P. Trends and Developments in Soil Science. In *Literature of Soil Science*; McDonald, P., Ed.; Cornell University Press: Ithaca, NY, 1994; 1–19.
6. Boulaine, J. Early Soil Science and Trends in the Literature. In *The Literature of Soil Science*; McDonald, P. Ed.; Cornell University Press: Ithaca, NY, 1994; 20–42.
7. Simonson, R.W. Soil Classification in the United States. Science **1962**, *137*, 1027–1034.
8. McCraken, R.J.; Helms, D. Soil Surveys and Maps. In *The Literature of Soil Science*; McDonald, P., Ed.; Cornell University Press: Ithaca, NY, 1994; 275–311.
9. Gardner, D.R. *The National Cooperative Soil Survey of the United States*; Historical Notes No. 7, U.S. Department of Agriculture, Natural Resources Conservation Service, Soil Survey Division: Washington, DC, 1998; 1–270.
10. Simonson, R.W.; McDonald, P. Historical Soil Science Literature of the United States. In *The Literature of Soil Science*; McDonald, P., Ed.; Cornell University Press: Ithaca, NY, 1994; 379–434.
11. Simonson, R.W. *Historical Highlights of Soil Survey and Soil Classification with Emphasis on the United States, 1899–1970*; Technical Paper No. 18, International Soil Reference and Information Center: Wageningen, The Netherlands, 1989.
12. Baldwin, M.; Charles, E.K.; Thorp, J. Soil Classification. *Soils and Men*; Yearbook of Agriculture, U.S. Department of Agriculture, U.S. Government Printing Office: Washington, DC, 1938; 979–1001.
13. Thorp, J.; Smith, G.D. Higher Categories of Soil Classification: Order, Suborder, and Great Soil Groups. Soil Sci. **1949**, *67*, 117–126.
14. Kellogg, C.E. *Soil Survey Manual*; Misc. Publication No. 274, U.S. Department of Agriculture: Washington, DC, 1937; 1–116.
15. Jenny, H. *Factors of Soil Formation: A System of Quantitative Pedology*; McGraw-Hill: NY, 1941; 1–281.
16. McDonald, P. Characteristics of Soil Science Literature. In *The Literature of Soil Science*; McDonald, P., Ed.; Cornell University Press: Ithaca, NY, 1994; 43–72.
17. Soil Survey Staff, *Soil Survey Manual*; Agricultural Handbook No. 18, U.S. Department of Agriculture, U.S. Government Printing Office: Washington, DC, 1951; 1–503.
18. Simonson, R. Evolution of Soil Series and Type Concepts in the United States. Adv. GeoEcol. **1997**, *29*, 79–108.
19. Soil Survey Staff, *Soil Taxonomy: A Basic System of Soil Classification for Making and Interpreting Soil Surveys*; USDA Handbook No. 436, U.S. Department of Agriculture, U.S. Government Printing Office: Washington, DC, 1975.
20. Sposito, G.R., Reginato, J., Frankenberger, W.T., Jr., Sims, R.C., Jr., Eds. *Opportunities in Basic Soil Science Research*; Soil Science Society of America, Inc.: Madison, WI, 1992; 1–109.
21. Bennett, H.H. *Report of the Chief of the Soil Conservation Service*; Mimeograph, (to the Secretary of Agriculture) U.S. Department of Agriculture, Soil Conservation Service: Washington, DC, 1935.
22. U.S. Department of Agriculture, *Soil and Water Conservation Needs Estimates for the U.S., by States*; Soil Conservation Service, U.S. Government Printing Office: Washington, DC, 1945.
23. U.S. Department of Agriculture, *Basic Statistics of the National Inventory of Soil and Water Conservation Needs*; Agriculture Statistical Bulletin No. 317, U.S. Government Printing Office: Washington, DC, 1962.
24. U.S. Department of Agriculture, *National Inventory of Soil and Water Conservation Needs, 1967*; Statistical Bulletin No. 461, Soil Conservation Service, U.S. Government Printing Office: Washington, DC, 1971; 1–211.
25. U.S. Department of Agriculture, *Basic Statistics: 1977 National Resources Inventory*; Statistical Bulletin No. 686, Soil Conservation Service, U.S. Government Printing Office: Washington, DC, 1982.
26. U.S. Department of Agriculture, *Basic Statistics: 1982 National Resources Inventory*; Statistical Bulletin No. 756,

Soil Conservation Service, U.S. Government Printing Office: Washington, DC, 1987.
27. U.S. Department of Agriculture, *Summary Report, 1987 National Resources Inventory*; Statistical Bulletin No. 790, Soil Conservation Service, U.S. Government Printing Office: Washington, DC, 1989.
28. U.S. Department of Agriculture, *Summary Report: 1992 National Resources Inventory*; Soil Conservation Service, U.S. Government Printing Office: Washington, DC, 1994; 1–54.
29. U.S. Department of Agriculture. *1997 National Resources Inventory*, www.nhq.nrcs.usda.gov/NRI/, 1999.
30. McDonald, P. *The Literature of Soil Science*; Cornell University Press: Ithaca, NY, 1994; 1–448.
31. Boersma, L.L. *Future Developments in Soil Science Research*; Soil Science Society of America, Inc.: Madison, WI, 1987; 1–537.
32. Doran, J.W., Coleman, D.C., Bezdicek, D.F., Stewart, B.A., Eds.; *Defining Soil Quality for a Sustainable Environment*; SSSA Special Publication No. 35, Soil Science Society of America, Inc., and the American Society of Agronomy, Inc.: Madison, WI, 1994; 1–244.
33. Wagenet, R.J., Bouma, J., Eds.; *The Role of Soil Science in Interdisciplinary Research*; SSSA Special Publication No. 45, American Society of Agronomy, Inc. and the Soil Science Society of America, Inc. Madison, WI, 1996; 1–143.
34. Huang, P.M., Eds.; *Soil Chemistry and Ecosystem Health*; SSSA Special Publication No. 52, Soil Science Society of America, Inc.: Madison, WI, 1998; 1–386.
35. Adriano, D.C., Bollag, J.M., Frankenberger, W.T., Jr., Sims, R.C., Jr., Eds.; *Bioremediation of Contaminated Soils*; Agronomy Monograph No. 37, American Society of Agronomy, Inc., Crop Science Society of America, Inc., and Soil Science Society of America, Inc.: Madison, WI, 1999; 1–772.
36. Simonson, R.W. Soil Science—Goals for the Next 75 Years. Soil Sci. **1991**, *151*, 7–18.
37. Miller, F.P. Soil Science: A Scope Broader Than Its Identity. Soil Sci. Soc. Am. J. **1993**, *57*, 299–564.

HISTORY TO EARLY-MID 20TH CENTURY

John P. Tandarich
Hey and Associates, Inc., Chicago, Illinois, U.S.A.

H

INTRODUCTION

Soil science is the scientific study of soil as a unique independent phenomenon of nature, in all its complex interrelationships. The interrelationships of soil are those interactions of physical, chemical, biological, geological, geomorphological, climatological and temporal factors that are combined to create the soil and are responsible for the geographical arrangement of soil on the shallow waterscapes and landscapes of the Earth.

This article briefly traces the development of soil science in the Western World from its origins in Greek thought, evolving as a separate agricultural interest of classical chemistry and classical mineralogy. The agricultural interests crystallized as agricultural chemistry and agricultural geology. The science evolved within the last 150 years from the interactions with other disciplines, principally chemistry and geology.

The historiography of soil science has been increasing within the last 25 years. Some of the more recent works are by Boulaine (1, 2), Jenny (3), Krupenikov (4), Simonson (5), Strzemski (6), Tandarich (7–11), Warkentin (12), and Yaalon (13). This article ends at 1927 when the First International Congress of Soil Science was held. Miller treats soil science history subsequent to this point.

AGRICULTURAL CHEMISTRY

In the Greek world, it was Empedocles of Agrigentum who proposed four elements as the basis for all existence: earth, air, fire, and water. The four elements of Empedocles, developed and refined by Aristotle, became the accepted "Classical Model" for almost 2000 years (14). The Classical Model of four fundamental elements was the basis for understanding the physical world from the eighth through the 14th centuries, when knowledge centers were within monastery walls and courts of royalty. The rediscovery of the ancient Greek and Roman philosophers by Renaissance scholars in monasteries and newly established universities revitalized scientific inquisitiveness and allowed progress in knowledge to move forward in the 15th and 16th centuries (10).

By the late 17th century, the Royal Society of London, and particularly Robert Boyle, recognized the agricultural value of the knowledge of soil types, their properties, and condition (15). Simultaneously, northern European chemists were also analyzing and classifying soils; prominent among these were Johann Becher and his student Georg Stahl, and Johann Wallerius and his student Count Gustavus Adolphus Gyllenborg (14). By the late 18th century, Scottish chemist Joseph Black had analyzed and classified calcareous marls.

With the rise in France of the new experimental chemistry of Antoine Lavoisier in the late 18th and early 19th centuries, scientists in many countries became involved in soil analyses, particularly focused on the organic matter component that for them was the soil. Some of these individuals were Jean Chaptal and Jean-Baptiste Boussingault in France; Humphry Davy in England; and Albrecht Thaer, Heinrich Einhof and their student Carl Sprengel, Jons Berzelius and his students Heinrich Rose, Eilhard Mitscherlich and Gerardus Mulder, and Gustav Schubler in northern Europe (10, 14).

Scientists in northern Europe were particularly active. In 1837 Sprengel (16) first used the term Bodenkunde (soil knowledge) in his text. At the same time Justus von Liebig, his colleagues and students initiated an academic tradition that continued to the early 20th century and produced many agricultural chemists (14). Many of these scientists were also outstanding teachers, such as Pieter Muller, Emil Ramann, Hermann Stremme, and Georg Wiegner and his student Hans Jenny in Europe; Joseph H. Gilbert in England; Alexandr Khodnev, Paval Ilienkov, Alexandr Voskresenskii and his student Dmitri Mendeleev, and Nikolai Zinin and his student Alexandr Butlerov in Russia; and Eben Horsford, Eugene Hilgard, John Norton, William H. Brewer, Samuel W. Johnson, and George Caldwell in the U.S. (10, 14, 17, 18).

AGRICULTURAL GEOLOGY

Simultaneously with the development of agricultural chemistry, knowledge of soils was accumulating, expressed within what today would be called geology and geography.

In the late 17th and early 18th centuries, Royal Society of London members Martin Lister, John Aubrey, and William Stukeley proposed that soils be analyzed and mapped. The 1684 work by Lister (19) included the first soil classification scheme. In 1743 Christopher Packe (20) published a large scale physiographic and soils map of east Kent, England. Pioneer soil mapping of the late 18th and early 19th centuries in Great Britain and Ireland culminated in a series of county-based agricultural surveys, called "General Views of Agriculture" made for the British Board of Agriculture under the leadership of Arthur Young (21).

By the end of the 18th century, a renewed interest in soils developed from a redefinition of classical mineralogy by German classical mineralogist Abraham Werner into Geognosie (geognosy—eventually replaced by the name geology), Oryctognosie (modern mineralogy), and Mineral Geography (22, 23). Geognosie was defined as "the abstract systematic knowledge of the solid earth" (22). That part of geognosy regarding the Earth's surface, and as applied to agriculture, was spread through Werner's students. Werner's student Alexander von Humboldt and his student and colleague Carl Ritter spread new geographic and cartographic methods and techniques through Europe and Russia. They influenced their academic colleagues and successors, and government officials to adopt these new methods and ideas. Some of the inspired in northern Europe were Friedrich Fallou, who coined the name Pedologie in 1862 to represent the true science of the soil elevated from Sprengel's soil knowledge (24), and Albert Orth (25, 26), who developed the soil profile concept; Alexandre Brongniart, Georges Cuvier and his Swiss student Louis Agassiz in France; and members of the Imperial Free Economic Society of St. Petersburg in Russia such as Vasilli Dokuchaev.

In Scotland, Robert Jameson's interpretation of that part of geognosy concerned with the Earth's surface influenced William Maclure. In the early 19th century, Maclure brought this knowledge, much of which would become known as agricultural geology, to Benjamin Silliman and his students and academic successors in the U.S.; some of whom were Amos Eaton, Edward Hitchcock, Ebenezer Emmons, David Dale Owen, George H. Cook, Nathaniel Shaler, William Morris Davis, Alexander Winchell, Eugene Hilgard, Thomas Chamberlin, W. J. McGee, Frank Leverett, and Curtis F. Marbut (7–9, 11). The agricultural geologists' concept of soil focused on soil as separate formation in its own right deserving of study (27). This discipline became concerned with processes leading to soil formation, such as the weathering of rocks and minerals, and how these processes influenced soil fertility.

Practicing agricultural geologists (or agrogeologists as they called themselves by the turn of the 20th century) began meeting in 1909 (28). Van Baren (29) summarized the state of the science of agrogeology in an introduction to a comprehensive bibliographic work on the subject by Wulff (30). The agricultural geology, or agrogeology, period lasted until the 1920s (11, 13).

THE EMERGENCE OF SOIL SCIENCE

When did modern soil science emerge from agricultural chemistry and agricultural geology? It appears that emergence took place during the period from 1837 to 1927, and particularly after 1860, as shown by the indicators that Greene (31) proposed: 1) the establishment of major organizations representing soil science, 2) review articles and bibliographies published, and 3) textbooks published on the subject.

Rossiter (17) forcefully argued that agricultural chemistry was already accepted, taught, and practiced in the U.S. by the mid-19th century. Simultaneously, agricultural geology was already establishing itself in Europe and the U.S. with its own literature (11, 30). Modern soil science developed from agricultural chemistry and agricultural geology through a complex series of interactions among individual scholars, academic and governmental institutions, and scientific societies of England and Scotland, France, Denmark, Germany, Hungary, Romania, Sweden, Russia, and the U.S.

Many of the major organizations representing soil science were created during the early 20th century: the American Society of Agronomy (ASA), Soils Section in 1908; the International Conferences on Agrogeology in 1909 that became the International Society of Soil Science (ISSS) in 1924 (28); and the American Association of Soil Survey Workers in 1920 that became the American Soil Survey Association in 1922. Between 1924 and 1927, the ISSS created an organizational structure that recognized commissions (soil science subdisciplines) of soil physics, soil chemistry, soil bacteriology (biology), nomenclature, classification and cartography (what today is known as pedology), soil fertility, and soil technology (13).

Greene (31) noted that the appearance of historical review articles of disciplines indicate that the particular discipline is in a transitional or significant developmental state. In Russia, Dokuchaev (32) reviewed the literature on soils in 1879. There were review articles and bibliographies on soils published immediately after the formation of the ASA (33, 34), prior to the establishment of the ISSS (29,30), and in conjunction with the First International Congress of Soil Science (35).

Textbooks on soil science began to appear in Europe and Russia beginning in the early to mid-19th century (2, 12, 14, 17, 36). Briefly, some of the principal works were by Sprengel (16), Fallou (24), Orth (25, 26), and Dokuchaev (32, 37, 38). Dokuchaev's student Konstantin Glinka (39) brought the Russian ideas to the West with the assistance of Hermann Stremme and Marbut (40).

In England, significant publications included those by John Bennett Lawes and Joseph Henry Gilbert, who pioneered in soil fertility research at the Rothamsted Experiment Station, Lawes established in 1843 (41, 42), and by Arthur Hall (43) and Edward J. Russell (44). In Germany, the books of Emil Ramann (45, 46), who originated forest soil science, and Hermann Stremme (47) built upon the work of Fallou and Orth.

Significant books on soils in the U.S. began with Nathaniel Shaler's (48, 49) in the early 1890s. Others that followed included those by Franklin King (50) in soil physics, Hilgard (51), Milton Whitney (52), Thomas L. Lyon, Elmer Fippin, and Harry Buckman (53), George Coffey (54), Marbut (55), Lyon (56), and Lyon and Buckman (57) that has gone through 11 editions.

The First International Congress of Soil Science was held in the U.S. in Washington, D.C., in 1927 and brought together scientists from the Eastern and Western Worlds and from all subdisciplines of soil science for meetings and a transcontinental field trip (58). This meeting effectively launched modern soil science. Miller treats the history of soil science from this turning point to the present (this volume).

REFERENCES

1. Boulaine, J. *Histoire des Pedologues et de la Science du Sols*; INRA: Paris, 1989.
2. Boulaine, J. Early Soil Science and Trends in the Literature. In *The Literature of Soil Science*; McDonald, P., Ed. Cornell University Press: Ithaca, New York, 1994; 20–42.
3. Jenny, H. *E.W. Hilgard and the Birth of Modern Soil Science*; Collana Della Rivista Agrochimica No. 3, CDRA: Pisa, Italy, 1961.
4. Krupenikov, I.A. *History of Soil Science from Its Inception to the Present*; Academy of Sciences of the USSR, translated from the Russian Amerind Publishing: New Delhi, 1992.
5. Simonson, R.W. Soil Classifications in the Past—Roots and Philosophies. *Annual Report 1984*; International Soil Reference and Information Centre: Wageningen, The Netherlands, 1985; 6–18.
6. Strzemski, M. *Ideas Underlying Soil Systematics*; Pub. TT 73-54013 (Translated from Polish) Foreign Scientific Publications Department of the National Center for Scientific, Technical and Economic Information: Warsaw, Poland, 1975.
7. Tandarich, J.P.; Darmody, R.G.; Follmer, L.R. The Development of Pedological Thought: Some People Involved. Phys. Geog. **1988**, *9*, 162–174.
8. Tandarich, J.P.; Darmody, R.G.; Follmer, L.R. Some International and Interdisciplinary Connections in the Development of Pedology. *Transactions of the 14th International Congress of Soil Science, Kyoto, Japan, August 8–15, 1990*; International Society of Soil Science: Kyoto, Japan, 1990; Vol. 5, 191–196.
9. Tandarich, J.P.; Sprecher, S.W. The Intellectual Background for the Factors of Soil Formation. In *Factors of Soil Formation: A Fiftieth Anniversary Retrospective*; Soil Sci. Soc. Am. Spec. Pub. No. 33, Amundson, R., Harden, J., Singer, M., Eds.; Soil Science Society of America: Madison, WI, 1994; 1–13.
10. Tandarich, J.P. Agricultural Chemistry—Disciplinary History. In *Sciences of the Earth: An Encyclopedia of Events, People, and Phenomena*; Good, G., Ed. Garland Publishing: New York, 1998; Vol. 1, 19–23.
11. Tandarich, J.P. Agricultural Geology—Disciplinary History. In *Sciences of the Earth: An Encyclopedia of Events, People, and Phenomena*; Good, G., Ed. Garland Publishing: New York, 1998; Vol. 1, 23–29.
12. Warkentin, B.P. Trends and Developments in Soil Science. In *The Literature of Soil Science*; McDonald, P., Ed. Cornell University Press: Ithaca, New York, 1994; 1–19.
13. Yaalon, D.H. History of Soil Science in Context: International Perspective. In *History of Soil Science: International Perspectives*; Adv. Geoecology 29, Yaalon, D.H., Berkowicz, S., Eds.; Catena Verlag GMBH: Reiskirchen, Germany, 1997; 15–46.
14. Browne, C.A. A Source Book of Agricultural Chemistry. Chron. Bot. **1944**, *8* (1), 1–144.
15. Georgical Committee; Enquiries Concerning Agriculture.... Phil. Trans. Royal Soc. Lond. **1665**, *1*, 91–94.
16. Sprengel, C.S., *Die Bodenkunde oder die Lehre vom Boden Nebst Einer Vollstandigen Anleitung zur Chemischen Analyse der Ackererden*; Leipzig, 1837.
17. Rossiter, M.W. *The Emergence of Agricultural Science: Justus Liebig and the Americans, 1840–1880*; Yale University Press: New Haven, 1975.
18. Vucinich, A. *Science in Russian Culture 1861–1917*; Stanford University Press: Stanford, CA, 1970.
19. Lister, M. An Ingenious Proposal for a New Sort of Maps of Countrys, Together with Tables of Sands and Clays, Such Chiefly as are Found in the North Parts of England, Drawn up About 10 Years Since, and Delivered to the Royal Society Mar. 12, 1683. Phil. Trans. Royal Soc. Lond. **1684**, *14*, 739–746.
20. Packe, C. *Ankographia; Sive, Convallium Descripto. In Which are Briefly but Fully Expounded the Origine, Course and Insertion; Extent Elevation and Congruity of all the Valleys and Hills, Brooks and Rivers (As an Explanation of a New Philosophico-Choreographical Chart) of East Kent*; J. Abre: Canterbury, England, 1743.
21. Young, A. *General View of the Agriculture of the County of Sussex*; McMillan: London, 1793.
22. Ospovat, A.M. *Short Classification and Description of the Various Rocks by Abraham Gottlob Werner*; Translator Hafner Publishing Company: New York, 1971.

23. Laudan, R. *From Mineralogy to Geology: The Foundations of a Science, 1650–1830*; The University of Chicago Press: Chicago, 1987.
24. Fallou, F.A. *Pedologie oder Allgemeine und Besondere Bodenkunde*; Verlag Schonfeld: Dresden, 1862.
25. Orth, A. *Das Geologische Bodenprofile Nach Seiner Bedautung fur den Bodenwert und die Landeskultur*; Berlin, 1873.
26. Orth, A. *Die Geognostisch-Agronomische Kartierung*; Verlag von Ernst & Korn: Berlin, 1875.
27. Emmons, E.; Prime, A.J. Agricultural Geology. Am. Q. J. Agr. Sci. **1845**, *2* (1), 1–13.
28. Szabolcs, I. The 1st International Conference on Agrogeology, April 14–24, 1909, Budapest, Hungary. In *History of Soil Science: International Perspectives*; Adv. Geoecology 29, Yaalon, D.H., Berkowicz, S., Eds.; Catena Verlag GMBH: Reiskirchen, Germany, 1997; 67–78.
29. Van Baren, J. Agrogeology as a Science. Mededeelingen van de Landbouwhoogeschool **1921**, *20*, 5–9.
30. Wulff, A. Bibliographia Agrogeologia. Mededeelingen van de Landbouwhoogeschool **1921**; *20*, 1–285.
31. Greene, M.T. History of Geology. Osiris (2nd series) **1985**, *1*, 97–116.
32. Dokuchacv, V.V. *Cartography of Russian Soils*; In Russian Imperial University of St. Petersburg: St. Petersburg, Russia, 1879.
33. Coffey, G.N. The Development of Soil Survey Work in the United States with a Brief Reference to Foreign Countries. Proc. Am. Soc. Agron. **1912**, *3*, 115–129.
34. Fippin, E.O. The Practical Classification of Soils. Proc. Am. Soc. Agron. **1912**, *3*, 76–89.
35. McDonald, P., Ed. *A Classified List of Soil Publications of the United States and Canada*; Bibliographical Contributions No. 13, U.S. Department of Agriculture Library: Washington, DC, 1927.
36. McDonald, P., Ed. *The Literature of Soil Science*; Cornell University Press: Ithaca, NY, 1994.
37. Dokuchaev, V.V. *Tchernozeme (Terre Noire) de la Russie D'Europe*; Societe Imperiale Libre Economique Imprimeric Trenke & Fusnot: St. Petersburg, Russia, 1879.
38. Dokuchaev, V.V. *Russian Chernozem*; In Russian Imperial University of St. Petersburg: St. Petersburg, Russia, 1883.
39. Glinka, K.D. *Die Typen der Bodenbildung, Ihre Klassification und Geographische Verbreitung*; Gebrueder Borntraeger: Berlin, 1914.
40. Marbut, C.F. *The Great Soil Groups of the World and Their Development By K.D. Glinka*; Translator Edwards Brothers: Ann Arbor, MI, 1927.
41. Russell, E.J. *A History of Agricultural Science in Great Britain: 1620–1954*; George Allen & Unwin: London, 1966.
42. Catt, J.A.; Henderson, I.F. Rothamsted Experiment Station—150 Years of Agricultural Research/the Longest Continuous Scientific Experiment? Interdisc. Sci. Revs. **1993**, *18* (4), 365–378.
43. Hall, A.D. *The Soil*; E. P. Dutton: New York, 1907.
44. Russell, E.J. *Soil Conditions and Plant Growth*; Longmans, Green and Co. New York, 1912.
45. Ramann, E. *Forstliche Bodenkunde und Standortslehre*; Verlag von Julius Springer: Berlin, 1893.
46. Ramann, E. *Bodenkunde*, 3rd Ed.; Verlag von Julius Springer: Berlin, 1911.
47. Stremme, Hermann *Grundzuge der Praktischen Bodenkunde*; Verlag von Gebruder Borntraeger: Berlin, 1926.
48. Shaler, N.S. *Aspects of the Earth*; Charles Scribner's Sons: New York, 1890.
49. Shaler, N.S. The Origin and Nature of Soils. *Twelfth Annual Report of the United States Geological Survey 1890–91*; U.S. Government Printing Office: Washington, DC, 1891; 219–345.
50. King, F.H. *The Physics of Agriculture*; F.H. King: Madison, WI, 1901.
51. Hilgard, E.W. *Soils*; Macmillan: New York, 1906.
52. Whitney, M. *Soils of the United States*; U.S. Dept. Agr. Bur. Soils Bull. 55, U.S. Government Printing Office: Washington, DC, 1909.
53. Lyon, T.L.; Fippin, E.O.; Buckman, H.O. *Soils: Their Properties and Management*; Macmillan: New York, 1909.
54. Coffey, G.N. *A Study of Soils of the United States*; U.S. Dept. Agr. Bur. Soils Bull. 85, U.S. Government Printing Office: Washington, DC, 1913.
55. Marbut, C.F.; Bennett, H.H.; Lapham, J.E.; Lapham, M.H. *Soils of the United States*; U.S. Dept. Agr. Bur. Soils Bull. 96, U.S. Government Printing Office: Washington, DC, 1913.
56. Lyon, T.L. *Soils and Fertilizers*; Macmillan: New York, 1917.
57. Lyon, T.L.; Buckman, H.O. *The Nature and Properties of Soils*; Macmillan: New York, 1922.
58. Tedrow, J.C.F.; Simonson, R.W. The Transcontinental Excursion of the First International Congress of Soil Science. Bull. Intl. Soc. Soil Sci. **1997**, *91*, 75–78.

HUMAN SOCIETY, SOIL AND

Rattan Lal
The Ohio State University, Columbus, Ohio, U.S.A.

INTRODUCTION

Humans are the most highly evolved animals who, in addition to flesh and skeleton, have intellectual capacity to perceive, rationalize, and develop technological innovations to overcome nature's constraints to meet their needs. There is a strong interdependence between soil and humans. Soil is the three-dimensional layer of Earth's crust that, through numerous biophysical and chemical interactive processes, is capable of supporting plant and animal life and moderating environment (air and water) quality. Rather than a mere combination of inorganic and organic substances to varying proportions represented among different soil types, soil is a living entity. It is full of life, highly dynamic, and ceases to produce biomass and purify water and air when life in it ceases to exist due to natural or human-induced causes. Thus, there has existed a strong inter-relationship between soil and humans.

Ever since the evolution of humans on Earth, the quality of life and number of humans in any ecoregion or biome have depended on the quality and resilience characteristics of the soils of that region. While human societies rose and vanished, along with the quality of soil that supported them, they also changed soil quality in a manner that no other species of the animal kingdom has ever done throughout Earth's history. The fate of humans and the soil that supports its society is so closely intertwined that it is difficult to understand the nature of one without knowing properties and dynamics of the other. Humans have manipulated soils to meet the nutritional needs of their flesh and the spiritual needs of their minds. In that process, they have left an indelible mark on soil quality. As Lowdermilk (1) wrote, "Individual nations and civilizations write their records on the land—a record that is easy to read by those who understand the simple language of the land." The extent and severity of change in soil quality due to human-induced perturbation depends on the demands based on human needs and expectations.

Soil is the most basic of all natural resources, because most processes governing terrestrial life depend on it. Important among these processes are: 1) biomass productivity, 2) purification of water, 3) detoxification of pollutants, 4) recycling of elements, and 5) resilience and restoration of ecosystems. Hugh Hammond Bennett, father of the soil conservation movement in the U.S., wrote "Out of the long list of nature's gifts to man, none is perhaps so utterly essential to human life as soil." The human history is intimately linked to soil and its quality. That is why "soil" or "Mother Earth" has been held sacred and worshiped in many ancient cultures.

Ancient Hindus and Buddhists worshiped Earth as Dharra, Vasundharra, Buhumi, Prithvi as sacred. The word Dharra has the same roots as the Latin word Terra. The ancient Greeks worshiped the Earth as Gaea or the Mother Earth. Water, another important constituent of the land and essential for life, is also held sacred in many cultures. The Koran, sacred to the followers of Islam, states that "with water we have made all living things." Similar beliefs exist in Hebrew, Christian, and other cultures (2).

Soil for modern societies is even more important than ever before. The increase in importance lies in the fact that soil resources of the world are 1) finite, 2) nonrenewable over the human time frame, 3) liable to misuse and mismanagement, 4) unevenly distributed among biomes and ecoregions, and 5) highly variable over time and space. Area under prime soils, with few or no limitations to intensive use for agricultural production, is rapidly shrinking because of soil degradation, conversion to nonagricultural uses, and urbanization. This article describes the dependence of human society on soil in both ancient and modern eras, and the need for maintaining and enhancing soil quality for achieving food security and improving environment quality.

SOIL AND ANCIENT SOCIETIES

The history of human civilization is based on the relation between humans and soil, humans' influence on soil quality, and humans' perception of how to manage soil resources. Hyams (3) outlined three types of interactions between soil and ancient societies, depending on the renewability of the soil and the managerial skills of the evolving human society.

Soil Parasitism

Parasitism implies the life of an organism through exploitation (and in extreme cases, eventual debilitation) of its host organism. Humans have existed as parasites on soils since the dawn of civilization and evolution of structured society. Some alluvial soils, due to their natural rejuvenation and renewability, have supported human societies for millennia. Examples of such constantly renewed alluvial soils include those in the valleys of the Nile, the Euphrates–Tigris, the Indus, the Yangtze, and the Hwang-Ho. Similar to alluvium, some loess (wind-borne) soils also have the renewability and capacity to withstand human exploitation for a long time. In arid and semiarid regions where loess is formed, soils must have adequate supply of good quality water to support and sustain human society for an extended period of time. Because of their natural renewability, caused by annual floods, societies living off the alluvial soils did not have to discover ameliorative tools (e.g., plow, manure, and other amendments).

Soil Occupation or Militaries

In contrast to the deltaic soils of the Nile, those of the Euphrates–Tigris river system were easily eroded and degraded because of undulating terrain. For example, the soils of ancient Babylon were desolated by misuse and abuse over an extended period of time. Sustaining production on these soils implied input of energy, nutrients, and water. Therefore, it was in this region where innovations such as the plow, irrigation, and manuring were introduced to alleviate soil-related constraints to crop production. Yet irrigation and soil erosion cannot be taken for granted. After all, it was the neglect of ancient canals, which created waterlogging and salinity on the one hand, and erosion in the catchment, which led to siltation of waterways and reservoirs on the other. Agriculture that was thriving about 10,000 B.C. in Mesopotamia, present-day Iraq, was converted to desert and shifting sand dunes by deforestation and erosion of the surrounding hills (1, 4). There is a lot of similarity in ecosystems of Euphrates–Tigris and Indus valleys. Salinity, erosion (by water and wind), and siltation also played havoc with human-occupied soils of the Indus. Widespread deforestation for wood for brick-making in the Indus Valley (3) may have caused severe problems of erosion and degradation. The collapse of the 1700-year old Mayan civilization in Guatemala around 900 A.D. is also attributed to accelerated soil erosion (5). Increase in population caused soil depletion, and the Mayan culture paid the ultimate price (6). Military occupation of any form, soil or society, cannot survive/sustain for an extended period.

Humans as Soil Pathogens

There are many examples where humans have been the "afflicting" and the soil, the "afflicted" species in functions rather than in form. A disease is caused by imbalance between the activities of afflicting (parasite) over that of the afflicted (host) species, leading to eventual collapse of both host and parasite. The extinction of many ancient societies (e.g., Mesopotamia in Iraq, ancient kingdoms of Sardis and Lydia in Turkey, the Mayan kingdom in Guatemala, etc.) are but a few examples of the severe cases of humans afflicting soil, leading to an eventual collapse of both. In contrast to the vanquished, ancient societies that realized that soil is to be respected and worshiped as a living entity have thrived and flourished.

THE VANISHING SOIL AND MODERN SOCIETY

The per capita arable land area is rapidly decreasing due to an ever-increasing human population. In 1995, the per capita land area was 0.23 ha worldwide, 0.23 ha in Africa, 0.20 ha in Latin America, 0.89 ha in North America, 0.12 ha in Asia, 0.17 ha in Europe, and 1.8 ha in Oceania. By 2050, even if there is no soil degradation and conversion to nonagricultural uses, the per capita land area will change to 0.14 ha worldwide, 0.08 ha in Africa, 0.11 ha in Latin America, 0.69 ha in North America, 0.07 ha in Asia, 0.19 ha in Europe and 1.1 ha in Oceania. Similar to the arable land area, world grain harvested area per person declined from 0.23 ha in 1950 to 0.11 ha in 1998 (7). These vanishing soil resources imply that the basic necessities of life (e.g., food, feed, fuel, and fiber) will have to be met from low per capita land area, and likely with adverse moisture/rainfall and temperature regimes due to the projected greenhouse effect.

The importance of soil resources on global food security cannot be overemphasized. Despite impressive gains in food production brought about by the Green Revolution technology, the problem of food insecurity persists in several less-endowed and densely populated regions of the world. The food insecure population of the world was 960, 938, 831, and 791 millions in 1970, 1980, 1990, and 1996 respectively, and is projected to be 680 million in 2010 (8, 9). There is also a widespread problem of malnutrition (10), which is accentuated by food grown on nutrient-deficient and degraded soils.

Table 1 Challenges posed by human society at the threshold of the 21st century that world soils must meet

Challenge	Current status	Rate of change
World population	6 billion	+1.3%/yr
Food insecure population	790 million	−1.0%/yr
Soil degradation	1966 Mha	+5−10 Mha/year
Desertification	1016 Mha	+5.8 Mha/year
Global irrigated area per person	0.045 ha	−1.3%/yr
World grain harvested area per person	0.11 ha	−0.55%/yr
Forested area per capita	0.59 ha	−0.78%/yr
Atmospheric concentration of GHGs		
i) CO_2	370 ppmv	+0.5%/yr
ii) CH_4	1.74 ppmv	+0.75%/yr
iii) N_2O	311 ppbv	+0.25%/yr

(Adapted from Refs. 7, 12.)

The threat of pending famine was voiced by Thomas Malthus in 1798 when the world population was merely 940 million. Since then, the world population has multiplied geometrically, as Malthus predicted, to 1.26, 1.65, 2.52, and 6.06 billions in 1850, 1900, 1950, and 2000 respectively (11, 12). In 1000 years, from year 1000 to 2000, the human race has multiplied 20-fold, and is still increasing at the rate of 73 million or 1.3% per yr. The population is projected to be 7.5 billion by 2020 and 9.4 billion by 2050. The welfare of the present and future population depends on soil quality and resilience under an intensive agricultural use. Several global challenges facing human society at the onset of the 21st century (Table 1) will have to be met through sustainable use of the world's soil resources.

SOIL–HUMAN SYMBIOSIS

Soil resources must be managed, improved and restored if the challenges of the 21st century (Table 1) are to be successfully met. The large human population (6 billion now and 9.4 billion by 2050) cannot survive by parasitism, militaries, or pathogenic strategies. Humans have to live in harmony or symbiosis with soil. Humans, through their inquisitive nature and innovative instincts, have the capacity to meet the challenges of the growing population. As Adlai E. Stevenson wrote, "Nature is neutral. Man has wrested from nature the power to make the world a desert or to make the desert bloom." The power to make the desert bloom lies in the realization of maintaining and enhancing soil quality. The concept of soil quality in one form or another has been used by agricultural philosophers and writers for some 2500 years. An agriculturist of Moorish Spain, Ibn-Al-Awam wrote several volumes during the 12th century on agricultural issues. The book *Kitab al-Felhah* or *Book of Agriculture* was translated into Spanish in 1802, and was brought to public attention in the *Encyclopedia of Islam* (1760–1777). In this book, the author writes, "The first step in the science of the agriculture is the recognition of soils and of how to distinguish that which is of good quality and that which is of inferior quality" (vol. 1, p. 23) (1). The concept of soil quality, now well developed and understood for its management and enhancement, has to be used in a "symbiotic" manner for meeting the challenge of the modern era. These challenges have to be met through: i) restoration of degraded soils, ii) enhancement of soil organic matter content and soil quality, iii) intensification of agriculture on prime soils, iv) minimizing risks of erosion and soil degradation by other processes, and v) enhancing efficiency of all inputs (water, fertilizers, energy). Humans have to learn to live in symbiosis with soil for their survival.

CONCLUSIONS

In the old Roman Empire, all roads led to Rome. In the long-term survival of human society, all roads lead back to the soil (13). Human society must respect soil, the media that feeds the world and purifies the environment. World soils have the capacity to feed the present and future population, provided that soils are used, improved, and restored. Judicious management of soil resources is a solution to the environmental issues (water quality and the greenhouse effect) and also to achieving global food security.

REFERENCES

1. Lowdermilk, W.C. *Conquest of the Land Through Seven Thousand Years*; Agric. Inform. Bull. 99, SCS: Washington, DC, 1953.
2. Hillel, D.J. *Out of the Earth: Civilization and the Life of the Soil*; The Free Press: New York, 1991; 321.
3. Hyams, E. *Soil and Civilization*; Thames and Hudson: London, 1952; 312.
4. Beasley, R.P. *Erosion and Sediment Pollution Control*, 1st Ed.; Iowa State University Press: Ames, IA, 1972.
5. Olson, G.W. Archaeology: Lessons on Future Soil Use. J. Soil Water Conserv. **1981**, *36*, 261–264.

6. Hardin, G. The Tragedy of the Commons. Science **1968**, *162*, 1243–1248.
7. Lal, R. *Controlling Greenhouse Gases and Feeding the Globe Through Soil Management*; University Distinguished Lecture, 17 February, The Ohio State University: Columbus, OH, 2000.
8. FAO, The State of Food and Agriculture. Thirtieth Session of the FAO Conference, 12–23 November, Rome, Italy, 1999.
9. FAO, Assessment of the World Food Security Situation. Report #CFS: 99/2 of the 25th Session of the Committee on World Food Security, 31 May–2 June, Rome, Italy, 1999.
10. Mesham, A.R.; Chatterjee, M. *Wasting Away: The Crisis of Malnutrition in India*; The World Bank: Washington, DC, 1999; 78.
11. Anonymous Like Herrings in a Barrel. The Economist **1999**, 13–14.
12. Brown, L.R.; Gardner, G.; Halweil, B. Beyond Malthus. Sixteen Dimensions of the Population Problem, Worldwatch Paper 143, Washington, DC, 1998; 87 pp.
13. Hambridge, G. Soils and Men: A Summary. *Soils and Men*; Yearbook of Agriculture, USDA: Washington, DC, 1938; 1–44.

HYDRAULIC CONDUCTIVITY

H

Jacob H. Dane
Marc Jalbert
Auburn University, Alabama, U.S.A.

Jan W. Hopmans
University of California, Davis, California, U.S.A.

INTRODUCTION

Transport of chemicals and flow of water in soil depend on, among other things, the soil's hydraulic properties. One of these hydraulic properties is the hydraulic conductivity function. In general terms, one could say that the hydraulic conductivity ($m\,s^{-1}$) is a measure of the ability of a soil to conduct water.

BACKGROUND INFORMATION

The flow of liquid water through soil is generally viscous and laminar, i.e., nonturbulent. This is due to the small sizes of the pores in which the water movement takes place. Under these conditions, the water flux density q ($m\,s^{-1}$; vector quantity) is proportional to the driving force, which is equal to the negative of the hydraulic head gradient ∇H (dimensionless; vector quantity). The magnitude of q is defined as the volume of water V (m^3) passing through a cross-sectional area of soil A (m^2) during time t (s). The hydraulic head H (m; scalar) is the potential energy (J) of water on a weight (N) basis. It incorporates the influences of forces such as gravity, pressure, and capillarity. The proportionality factor relating the flux density to the driving force is called the hydraulic conductivity K ($m\,s^{-1}$; tensor). In mathematical form the flux density can be expressed as

$$q = -K \cdot \nabla H \tag{1}$$

K varies from soil to soil and even for a given soil, as it depends on soil properties such as fineness, clay content, solid particle orientation, organic matter content, and water content. Eq. 1 was first observed by the French engineer Darcy (1) and is often referred to as Darcy's law. As the hydraulic head H (m) has units of length, the hydraulic gradient, ∇H, is dimensionless. Thus, the units for the water flux density and the hydraulic conductivity are the same, i.e., $m\,s^{-1}$. Other, less frequently used, units for the hydraulic potential, result in different units of the hydraulic conductivity. The term hydraulic conductivity should not be confused with permeability (m^2) as the former includes liquid and porous medium properties, while the latter should only reflect porous medium properties.

In general, the proportionality relation, Eq. 1, between the two vectors q and ∇H is tensorial, which means that K needs to be expressed as a matrix. This is necessary when the soil's structure forces water to flow in a direction different from the direction of the hydraulic head gradient. These soils, called anisotropic, occur frequently in nature. Most soils resulting from sedimentary processes exhibit preferential flow pathways, often perpendicular to the deposition direction that occurred during soil formation. Also, charged clay particles often arrange themselves in sheet-like configurations that create anisotropy. Although important, consideration of the soil's anisotropy involves complications that are beyond the scope of this article, and the reader is referred to the literature for further information (2).

If the soil possesses the same conductive properties in all directions, it is called isotropic. In this case, the water flow takes place parallel to and in the opposite direction of the hydraulic head gradient, and K can be considered a scalar. For a simple application of Darcy's law (Eq. 1) to an isotropic soil, such as a well-packed sand, let us consider a horizontal soil column of constant crosssectional area subjected to a hydraulic head gradient established by two constant-level water reservoirs (Fig. 1). The column is located beneath the water levels, and the soil's entire pore volume is filled with water. The hydraulic head at each end of the column is then simply equal to the water levels H_1 and H_2 in the water reservoirs as measured with respect to an arbitrary elevation reference. The hydraulic head gradient, uniform throughout the soil column, is equal to the difference between the hydraulic head values H_1 and H_2 divided by the length of the column L, i.e.,

$$\nabla H = \frac{H_2 - H_1}{L} \tag{2}$$

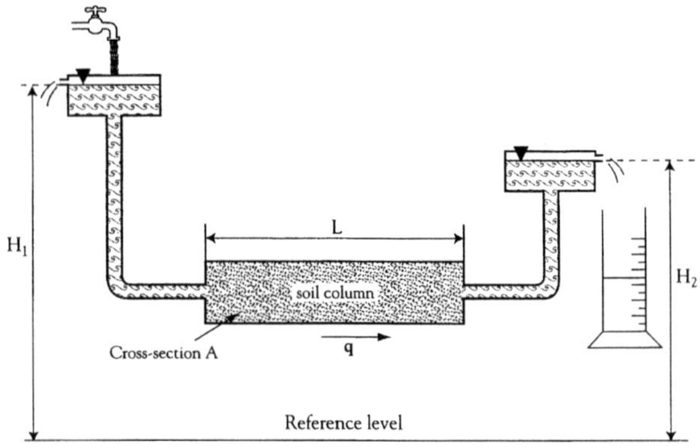

Fig. 1 Schematic diagram for application of Darcy's law.

which is negative. This implies that water flows from reservoir 1 to reservoir 2, i.e., in the direction of decreasing hydraulic head values. For a soil with hydraulic conductivity K, the magnitude of the water flux density (q; scalar) can then be calculated from

$$q = \frac{V}{At} = K\left|\frac{H_2 - H_1}{L}\right| \tag{3}$$

As an example calculation, consider a 1-m long column ($L = 1\,\mathrm{m}$) with a cross-sectional area $A = 10^{-2}\,\mathrm{m}^2$, containing a water-saturated sand with a hydraulic conductivity value $K = 10^{-4}\,\mathrm{m\,s^{-1}}$. The water level in the left reservoir is 0.06 m higher than in the right reservoir. This results in a water flux density of $6 \times 10^{-6}\,\mathrm{m\,s^{-1}}$, oriented from left to right, and hence a discharge rate, V/t, of $6 \times 10^{-8}\,\mathrm{m^3\,s^{-1}}$, i.e., 60 μl of water crosses any section of the column during 1 s. The value for the water flux density, which is expressed as a velocity and is sometimes referred to as the Darcy flux or Darcy velocity, should not be mistaken for the average velocity of the water in the pores, known as the seepage or pore-water velocity v ($\mathrm{m\,s^{-1}}$; vector quantity). The difference between the water flux density and the pore-water velocity is due to the fact that water only occupies a limited fraction of the soil's total volume. In the case of a water-saturated soil, the soil's volumetric water content, θ, is equal to the porosity n of the soil, i.e., the ratio of the pore volume and the soil's bulk volume. Consequently, water flows only through the fraction nA of a cross-section with area A. The true water velocity, or the pore-water velocity, is therefore (3)

$$v = \frac{q}{n} \tag{4}$$

Assuming a porosity $n = 0.3$, the average pore-water velocity or seepage velocity of the water in the column is $2 \times 10^{-5}\,\mathrm{m\,s^{-1}}$ or $7.2\,\mathrm{cm\,h^{-1}}$.

Darcy's law can also be applied to unsaturated soils, i.e., to soils containing water and air (4). In this case, the soil's hydraulic conductivity is a function of its volumetric water content θ (defined as the volume of water per unit volume of bulk soil), and the seepage velocity is expressed as

$$v = \frac{q}{\theta} \tag{5}$$

Hydraulic conductivity is difficult to measure in situ, especially when the influence of θ needs to be quantified. Consequently, a great deal of research has been conducted to estimate K from more easily available soil data. Most of these approaches rely on assuming a certain geometry for the soil pore space, the simplest of them being a bundle of circular capillary tubes, in which the well-defined Hagen–Poiseuille flow occurs. Pioneering work in this area was conducted by Kozeny (5) and later by Carman (6), whose work led to a formula for estimating K of saturated soils. Applying the Kozeny–Carman concepts, it can be shown that saturated hydraulic conductivity is approximately proportional to the square of a mean soil particle diameter. Hence, if the mean particle diameter for a given soil is twice as small as for another soil, its saturated K value will be four times smaller. Notably, clayey soils are often very restrictive to water flow and have saturated K values ranging from 10^{-6} to $10^{-9}\,\mathrm{m\,s^{-1}}$. Sandy soils, on the other hand, offer much less resistance to water flow and have saturated K values in the range of 10^{-4} to $10^{-5}\,\mathrm{m\,s^{-1}}$. Saturated K values (K_{sat}) for various soils are listed by soil texture in Table 1. Although the Kozeny–Carman theory provides good results for sandy, fully saturated porous

Table 1 Saturated hydraulic conductivity values (K_{sat}) for various soils by soil texture

Soil texture	Silt fraction	Clay fraction	$K_{sat} \times 10^5$ ($\mathrm{m\,s^{-1}}$)
Sand	0.05	0.05	1.4 (0.70–7)[a]
Loamy sand	0.10	0.07	1.0[b] or 1.9[c]
Sandy loam	0.25	0.10	0.7 (0.4–2)[a]
Silt loam	0.65	0.15	0.07[b] or 0.14[c]
Loam	0.40	0.18	0.4 (0.2–0.6)[a]
Sandy clay loam	0.13	0.27	0.36[b] or 0.42[c]
Silty clay loam	0.55	0.34	0.05[b] or 0.06[c]
Clay loam	0.35	0.34	0.2 (0.1–0.4)[a]
Silty clay loam	0.47	0.47	0.07 (0.01–0.1)[a]
Clay	0.20	0.60	0.1 (0.04–0.3)[a]

[a] From Israelsen and Hansen (7); numbers in parentheses indicate range.
[b] Calculated according to Eq. 6.11 in Ref. 8.
[c] Calculated according to Eq. 6.12 in Ref. 8.

media, its application to soil science is somewhat limited, because the predictions for fine-textured or structured soils are usually not very good and most soils contain entrapped air when "saturated." For this reason, soil scientists sometimes use the term satiated θ, with its corresponding satiated K. Under unsaturated conditions in the presence of a continuous air phase, capillary forces cause the water to occupy the smaller soil pores, while air invades the larger pores. Because the ability to conduct water decreases with decreasing pore size, the unsaturated hydraulic conductivity decreases rapidly with decreasing θ. Corey (9) observed that in sedimentary rocks, K is often approximately proportional to the fourth power of θ. A twofold reduction in θ will therefore result in a 16-fold decrease in K. In fact, K values easily decrease five orders of magnitude when θ decreases from its satiated to its residual value. Besides the Brooks–Corey type of empirical relationship between K and θ, more elaborate techniques are often used to estimate unsaturated K values from water retention data. Most of these techniques assume a certain form for the soil water retention curve, and use parameters associated with the water retention relationship to express the hydraulic conductivity as a function of water content or matric head. The most commonly used models are those of Brooks and Corey (10) and van Genuchten (11), using the capillary models of Burdine (12) and Mualem (13), which are extensions of the Kozeny–Carman theory. Based on the pore size distribution factor of their water retention function, λ, Brooks and Corey (10) derived

$$K_{\text{rel}} = \frac{K}{K_{\text{sat}}} = \left(\frac{h_{\text{d}}}{h_{\text{m}}}\right)^{\eta} \tag{6}$$

where K_{rel} and K_{sat} are the relative and saturated hydraulic conductivity values, respectively, h_{m} (m) is the matric head, h_{d} (m) is the displacement head, i.e., the value for h_{m} at which water in a saturated soil is first displaced by air, and

$$\eta = 2 + 3\lambda$$

Based on the dimensionless constant m appearing in his functional relationship of the water retention curve, van Genuchten (11) derived

$$K_{\text{rel}} = S_{\text{e}}^{1/2}[1 - (1 - S_{\text{e}}^{1/m})]^2 \tag{7}$$

where the effective saturation

$$S_{\text{e}} = (\theta - \theta_{\text{r}})/(\theta_{\text{s}} - \theta_{\text{r}})$$

The subscripts s and r refer to saturation and residual, respectively. It was found that the applicability of these methods, based on retention information, is greatly increased when at least one value for K at a θ different from saturation (mostly satiated) is measured. This is due to the fact that it is difficult to extrapolate the soil's conductive behavior from the situation near water saturation, where most water is conducted through macropores or fissures, which barely influence the soil's water retention relationship. In addition, more detailed information regarding the above and other relationships can be found in Kosugi et al. (14). Techniques for direct or indirect determinations of saturated and/or unsaturated hydraulic conductivity values are described in Reynolds and Elrick (constant head on soil cores, falling head on soil cores, measurement techniques above the water table, pressure (single-ring) infiltrometer, constant head well-permeameter); Booltink and Bouma (steady flow/piezometers, suction crust infiltrometer); Youngs (double/multiple ring infiltrometer); Amoozegar (auger hole, piezometer); Scotter and Clothier (one-dimensional infiltration equations, horizontal absorption, three-dimensional infiltration); Corey (long column); Nimmo et al. (steady-state centrifuge); Arya (wind and hot air methods, plane of zero flux); Vachaud and Dane (instantaneous profile); Parkin and Kachanoski (constant flux vertical TDR); Hopmans et al. (inverse methods) (14).

Complementary estimation of hydraulic conductivity can be obtained through statistical, empirical analysis (see section on Pedotransfer Functions; and Nimmo; Ref. 14). When numerous samples of known hydraulic conductivity have been taken over an area presenting some general characteristics, the hydraulic conductivity of a new sample can be roughly estimated from more easily available data. Particle-size data are often chosen (e.g., Refs. 15, 16) because they are routinely obtained by sedimentation analysis, and because hydraulic conductivity is very much influenced by the amounts of clay contained in the soil.

Because hydraulic conductivity is very much dependent on soil structure, i.e., on the occurrence of aggregates and/or cracks, either due to natural causes (soil erosion, shrink/swell, biological activity) or human factors (tillage, compaction), its value can vary greatly within short distances in the field. Microbial and fungal activity can also have quantitative effects on the hydraulic conductivity by pore clogging, as does temperature by its influence on the water viscosity (17). Finally, soil water chemistry influences the hydraulic conductivity of soils containing shrink/swell clay minerals (18). If the soil water does not contain sufficient quantities of mainly polyvalent cations, it can cause the clay to swell or even destroy the soil structure. This can greatly reduce the soil's conductive properties to water. As a result of these chemical, biological, and human factors, the hydraulic conductivity of a soil is not only likely to vary greatly over distance, but also over time.

REFERENCES

1. Darcy, H. Détermination des Lois d'Écoulement de l'Eau à Travers le Sable. *Les Fontaines Publiques de la Ville de Dijon*; Victor Dalmont: Paris, 1856.
2. Bear, J. *Dynamics of Fluids in Porous Media*; Elsevier: New York, 1972; 764.
3. Dupuit, J. *Études Théoriques et Pratiques sur le Mouvement des Eaux dans les Canaux Découverts et à Travers les Terrains Perméables*; Dunod: Paris, 1863.
4. Buckingham, E. *Studies on the Movement of Soil Moisture*; U.S. Dept. Agric. Bur. Soils Bull. 38; U.S. Gov. Print. Office: Washington, DC, 1907; 29–61.
5. Kozeny, J. Über Kapillare Leitung des Wassers im Boden, Sitzungsber. Akad. Wiss. Wien **1927**, *136*, 271–306.
6. Carman, P.C. Fluid Flow Through a Granular Bed. Trans. Inst. Chem. Eng., London **1937**, *15*, 150–166.
7. Israelsen, O.W.; Hansen, V.E. *Irrigation Principles and Practices*; Wiley: New York, 1962.
8. Campbell, G.S. *Soil Physics with Basic*; Elsevier: New York, 1985; 150.
9. Corey, A.T. The Interrelation Between Gas and Oil Relative Permeabilities. Producer's Monthly **1954**, *19* (1), 38–44.
10. Brooks, R.H.; Corey, A.T. Properties of Porous Media Affecting Fluid Flow. J. Irrig. Drain. Div., Proc. ASCE **1966**, *92*, 61–87.
11. van Genuchten, M.Th. A Closed-Form Equation for Predicting the Hydraulic Conductivity of Unsaturated Soils. Soil Sci. Soc. Am. J. **1980**, *44*, 892–898.
12. Burdine, N.T. Relative Permeability Calculations from Pore-Size Distribution Data. Trans. AIME **1953**, *198*, 71–77.
13. Mualem, Y. A New Model for Predicting the Hydraulic Conductivity of Unsaturated Porous Media. Water Resour. Res. **1976**, *12*, 513–522.
14. Dane, J.H.; Topp, G.C., Eds. *Methods of Soil Analysis, Part 4, Physical Methods*; Agronomy 9; Soil Sci. Soc. Amer.: Madison, WI, 2002.
15. Puckett, W.E.; Dane, J.H.; Hajek, B.F. Physical and Mineralogical Data to Determine Soil Hydraulic Properties. Soil Sci. Soc. Am. J. **1985**, *49*, 831–836.
16. Dane, J.H.; Puckett, W.E. Field Soil Hydraulic Properties Based on Physical and Mineralogical Information. In *Indirect Methods for Estimating the Hydraulic Properties of Unsaturated Soils*; van Genuchten, M.Th., Leij, F.J., Lund, L.J., Eds. University of California: Riverside, 1992; 389–404.
17. Hopmans, J.W.; Dane, J.H. Temperature Dependence of Soil Hydraulic Properties. Soil Sci. Soc. Am. J. **1986**, *50*, 4–9.
18. Dane, J.H.; Klute, A. Salt Effects on the Hydraulic Properties of a Swelling Soil. Soil Sci. Soc. Am. J. **1977**, *41*, 1043–1049.

HYDROLOGIC CYCLE, SOILS ROLE IN THE

H

Keith R.J. Smettem
The University of Western Australia, Nedlands, Western Australia, Australia

INTRODUCTION

Climate, soil, vegetation, and topography all play important roles in maintaining the water balance of catchments (1). Changes in even one of these factors will affect the hydrological response to rainfall, in terms of both water quality and quantity. At a broad regional scale, Milly (2) showed that the annual water balance is controlled primarily by climate (as measured by a simple climatic index of dryness, given by the ratio of annual potential evaporation to annual precipitation) and to a lesser extent by the soil storage capacity of the landscape. Reggiani et al. (3) also show that at catchment scale, total annual evaporation and runoff are controlled by climate (through a dryness index) and soil type. Topographic or landscape factors only become significant when consideration is given to partitioning total runoff into surface and subsurface contributions (3).

A REGIONAL SCALE PERSPECTIVE FOR LARGE SPACE AND TIME SCALES

At large space and annual time scales the role of the soil in the hydrological cycle is therefore, primarily as a transient store for water. At continental scales, Manabe (4) simulated the water balance using a simple bucket representing soil water storage and an evaporative threshold. Milly and Dunne (5) presented similar results showing that at continental and regional scales, simple lumped storage representation of surface-hydrology proved to be sufficient to capture the salient features of catchment behavior.

A LOCAL PERSPECTIVE AT SMALLER SPACE AND TIME SCALES

Soils and Runoff Generation Processes

When spatial and/or temporal scales of interest are reduced, consideration must be given to a more detailed understanding of processes leading to runoff generation and groundwater response. In particular, water flux rates over and through the soil assume increasing importance. Surface runoff may be generated as a "top down" or "bottom up" process. The "top down" process is referred to as infiltration excess runoff and occurs when the intensity of rainfall exceeds the rate of infiltration at the soil-surface. The "bottom up" process is referred to as "saturation excess" and occurs when initially unsaturated soil water stores are filled to overflowing so that additional rainfall simply runs off the saturated area.

Dynamics of Runoff Source Areas

The area of soil contributing to saturation excess runoff from a catchment is temporally dynamic, increasing as a storm progresses and diminishing thereafter. This "variable source area" (6) for runoff generation is dependent on storm characteristics (duration and volume), soil type and position in the landscape. Topographic indices of wetness have long been used to identify likely runoff producing areas (7) because the general tendency of water to flow downhill is amenable to macroscopic conceptualization. Soils tend to remain wetter as the upslope contributing area increases, particularly in areas of topographic convergence (8), so it is perhaps not surprising to find that soils close to watercourses tend to saturate first during rainfall. The concept of an upslope contributing area for saturation-excess flow is most readily applied to soils that are shallow, relative to the hillslope scale, under conditions that are wet enough to generate lateral downslope saturated flow through the soil. Lateral downslope saturated flow can of course, also be generated over impeding layers in the soil profile.

Predicting Infiltration Excess Runoff

Prediction of infiltration excess runoff requires knowledge of particular soil hydraulic properties that characterize infiltration behavior and the rainfall rate must be specified at short timesteps (typically minutes). One typical quasi-analytic equation for predicting the time to incipient ponding (the time at which a free water film first appears on the soil-surface) is (9)

$$t_p = M(S^2/K_S R_p)\ln\{R(t_p)/[R(t_p) - K_S]\} \quad (1)$$

where S is the sorptivity ($LT^{-0.5}$), K_S is the saturated hydraulic conductivity (LT^{-1}), $R(t_p)$ is the rainfall rate at incipient ponding (LT^{-1}), R_p is the mean rainfall rate up to ponding and M is a constant, generally set to 0.55.

The sorptivity (10) is a characteristic of the soil's capillarity properties and varies with soil texture, structure and initial water content. The saturated hydraulic conductivity is dependent upon soil texture and can be significantly affected by the degree of soil structure development. Both hydraulic properties can exhibit a high degree of spatial variability and frequency distributions of saturated hydraulic conductivity are often log-normal (11). Rawls et al. (12) provide summary data on saturated hydraulic conductivity for each of the major United States Department Agency textural classes and point out that soil structure effects can override any differences between texture classes.

Factors Affecting the Time of Runoff Generation

The time at which surface runoff commences after incipient ponding depends on a number of complex interacting factors. Surface microtopography is particularly important because it controls the amount of surface (detention) storage that can occur. Flow concentration into rills can have a major effect on reducing effective wetted areas subject to infiltration after rainfall ceases. If the saturated hydraulic conductivity increases downslope then this in turn, can have a major impact on runoff peaks, volumes and the time to peak runoff (13).

Over time scales that are typically of interest in process hydrology (one year or less), soil texture can be regarded as a fixed property in space and therefore patterns of textural distribution within catchments are amenable to mapping. Maps of soil hydraulic properties can then be derived from this basic information using indirect functional relations (14, 15). In contrast, soil structure arising from biotic or pedogenic activity can be highly variable temporally, particularly if the land surface is subjected to agronomic manipulation. Agronomic practices that expose soil to the beating action of raindrops often result in the destruction of surface soil structure and the consequent formation of surface skins or seals with low hydraulic conductivity. Physically, within the seal layer both porosity and pore size distribution are altered. Large conducting pores are destroyed and the total porosity is reduced. These changes lead to greatly enhanced surface runoff and possible removal of soil by water erosion.

The effects of soil structure on hydrologic response are often incorporated within descriptions of land use and treatment. Although land use and treatment are not primary physical variables, they are easy to map and have a long history of use as classificatory discriminators of runoff behavior. For example, the SCS curve number procedure (16) represents the soil-cover complex and a five-day antecedent rainfall that is referenced to dry, normal, or wet conditions. This approach is also to be found in well known hydrological models such as CREAMS (17).

Partial Wetting and Other Preferential Flow Processes

Soils that exhibit hydrophobic (nonwetting) behavior can yield considerably more surface runoff under dry antecedent conditions than would be predicted from a consideration of soil texture alone. Nonwetting behavior can also generate "finger flow", which is a form of physical nonequilibrium leading to preferential wetting of only a small fraction of the total soil volume. Preferential flow can also occur through soil macropores that are connected to a source of free water or by funnel-type flow due to layering structures in sandy soils (18).

Identification of soils and field conditions that result in preferential flow is the subject of considerable ongoing research effort because of the potential contamination risk to groundwater systems posed by the rapid movement of dissolved chemicals through the unsaturated zone.

CONCLUSIONS

The role of soils in the hydrological cycle becomes progressively more complex as time and space scales are reduced. The requirement for detailed soils information is essential for understanding runoff generation and preferential flow processes at local scale. However, for understanding annual or seasonal hydrological trends at large catchment or regional scales information on soil texture and soil depth can provide sufficient information for estimating the soil water storage capacity.

REFERENCES

1. Eagleson, P.S. Climate Soil, and Vegetation 1. Introduction to Water Balance Dynamics. Wat. Resour. Res. **1978**, *14*, 705–712.
2. Milly, P.C. Climate, Inter-Seasonal Storage of Soil Water and the Annual Water Balance. Wat. Resour. Res. **1994**, *17*, 19–24.
3. Reggiani, P.; Sivapalan, M.; Hassanizadeh, S.M. Conservation Equations Governing Hillslope Responses: Exploring the Physical Basis of Water Balance. Wat. Resour. Res. **2000**, *36*, 1845–1863.

4. Manabe, S. Climate and Ocean Circulation. 1. The Atmospheric Circulation and the Hydrology of the Earth's Surface. Mon. Weather Rev. **1969**, *97*, 739–774.
5. Milly, P.C.; Dunne, K.A. Sensitivity of the Global Water Cycle to the Water Holding Capacity of Land. J. Climate **1994**, *7*, 506–526.
6. Dunne, T.; Black, R.D. Partial Area Contributions to Storm Runoff in a Small New England Watershed. Wat. Resour. Res. **1970**, *6*, 1296–1311.
7. Beven, K.J.; Kirkby, M.J. A Physically-Based Variable Contributing Area Model of Basin Hydrology. Hydrol. Sci. Bull. **1979**, *24*, 43–69.
8. O'Loughlin, E.M. Saturation Regions in Catchments and Their Relation to Soil and Topographic Properties. J. Hydrol. **1981**, *53*, 229–246.
9. White, I.; Sully, M.J.; Melville, M.D. Use and Hydrological Robustness of Time-To-Incipient Ponding. Soil Sci. Soc. Am. J. **1989**, *29*, 1343–1346.
10. Philip, J.R. The Theory of Infiltration. 1. The Infiltration Equation and Its Solution. Soil Sci. **1957**, *77*, 153–157.
11. Nielsen, D.R.; Biggar, J.W.; Erh, R.T. Spatial Variability of Field-Measured Soil-Water Properties. Hilgardia **1973**, *42*, 215–259.
12. Rawls, W.J.; Brakenseik, D.L.; Logsdon, S.D. Predicting Saturated Hydraulic Conductivity Utilizing Fractal Principles. Soil Sci. Soc. Am. J. **1993**, *57*, 1193–1197.
13. Woolhiser, D.A.; Smith, R.E.; Giraldez, J.-V. Effects of Spatial Variability of Saturated Hydraulic Conductivity on Hortonian Overland Flow. Wat. Resour. Res. **1989**, *32*, 671–678.
14. Rawls, W.J.; Brakensiek, D.L.; Saxton, K.E. Estimation of Soil Water Properties. Trans. ASAE **1982**, *25*, 1316–1320.
15. Smettem, K.R.J.; Oliver, Y.M.; Heng, L.K.; Bristow, K.L.; Ford, E.J. Obtaining Soil Hydraulic Properties for Water Balance and Leaching Models from Survey Data. 1. Water Retention. Aust. J. Agric. Res. **1999**, *50*, 283–289.
16. USDA, Soil Conservation Service, *National Engineering Handbook, Hydrology*; U.S. Government Printing Office: Washington, DC, 1972; 548.
17. Kinsel, W.G., Ed. *CREAMS: A Field-Scale Model for Chemicals, Runoff, and Erosion from Agricultural Management Systems*; Conservation Research Report No. 26, USDA, Science and Education Administration, 1980; 643.
18. Ju, S.-H.; Kung, K.-J.S. Finite Element Simulation of Funnel Flow and Overall Flow Property Induced by Multiple Soil Layers. J. Environ. Qual. **1993**, *22*, 432–442.

INCEPTISOLS

John M. Galbraith
Virginia Polytechnic Institute & State University, Blacksburg, Virginia, U.S.A.

Robert J. Engel
United States Department of Agriculture (USDA), Lincoln, Nebraska, U.S.A.

INTRODUCTION

Inceptisols[a] are a unique soil order. They are essentially intermediate between soils of any other order and soils that have no diagnostic subsurface horizon (*Entisols*). Inceptisols are mineral soils that have significant subsoil or surface alteration but do not have the diagnostic horizon, material, *permafrost*, or other features definitive of the more developed soil orders that precede them in the *Keys to Soil Taxonomy* (1). Their development has been or is limited by one or more of the five soil forming factors (parent material, climate, relief, organisms, and time). If they have more than 8% clay, they must have dewatered enough following aqueous deposition to be able to support the weight of humans and most livestock (have a "*n*" value of more than 0.7) to avoid being Entisols. Note that there are also almost 200 soil series in the U.S. that have the morphology of *Inceptisols* but have sandy subsoils and are considered *Entisols* because of limited potential for further development.

Inceptisols can form in any type of parent material except thick organic deposits. They occur on any slope or landform from floodplains to mountaintops and are found under all vegetation types. *Inceptisols* may occur in any *soil moisture regime* other than *aridic* or *torric* and any *soil temperature regime* that is too warm to contain *permafrost*. *Inceptisols* range in age from recent to very old.

Inceptisols are concentrated in four major settings: cold regions, dry regions (*ustic* and *xeric*), wetlands, and geologically young surfaces (such as floodplains, volcanic and glacial deposits, and steep or highly erosive slopes). About one-fourth of the land area of Europe and Central America and less than one-tenth of the other continents consist of *Inceptisols* (Table 2). Fig. 1 shows *Inceptisol* distribution in the U.S.

The major categories (*suborders*) of Inceptisols are: *Aquepts* (aquic conditions), *Anthrepts* (human-altered), *Cryepts* (cold), *Ustepts* (alternating moist and dry with moist summers or no defined winter season), *Xerepts* (alternating moist and dry with dry summers and moist winters), and *Udepts* (moist in all seasons). Table 1 correlates Soil Taxonomy *suborders* (1) with World Reference Base reference groups (2). These categories have been established because of major differences in parent material or climatic factor that affect soil genesis. Approximate areas of each suborders (3) by continent are listed in Table 2. Inceptisols make up about one-quarter of Europe and Central America.

[a]Terms from *Soil Taxonomy* (1) are italicized text.

AQUEPTS

The *Aquepts* are the suborder of *Inceptisols* that formed with near-surface aquic conditions during biologically active seasons. Some have accumulated sodium or soluble salts in the surface. Aquepts occur in low-lying coastal areas, glacial lake plains, floodplains, playas, and in other nearly level areas where shallow groundwater or perched water exists. Major areas are in the in sub-Arctic areas, in humid areas that were covered with the most recent glaciation, in wet floodplains and lake plains, and on wet, low terrace deposits of major rivers (1, 2).

ANTHREPTS

The *Anthrepts* are mainly long-term agricultural soils with significant surface deposits of human-transported material such as manure, kitchen refuse, and fish bones or oyster shells. They occur on any slope and landform because humans controlled the parent material and deposition. These rare soils formed in areas where long-term (a few thousand years or more) human habitation and deposition of manure has occurred (such as in northwestern Europe), in regions of long-term irrigated agriculture in Asia, northern Africa, and North and South America, and near coastal villages where oysters and fish were the major food source (1).

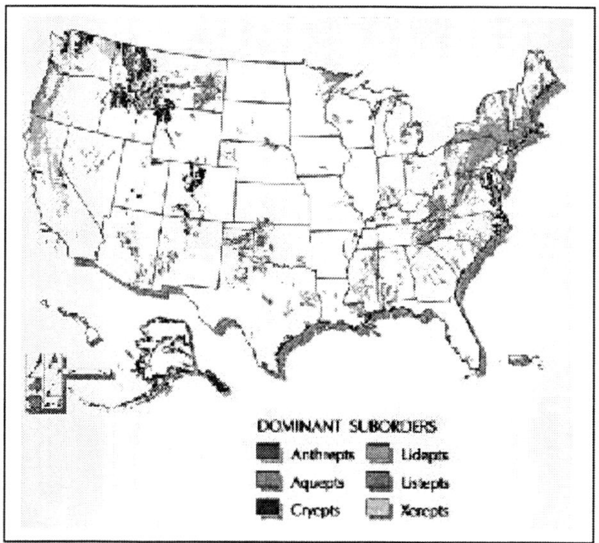

Fig. 1 Distribution of *Inceptisols* in the U.S. (From Ref. 5.)

CRYEPTS

Cryepts are the *Inceptisols* with persistently cold climates, but they do not have permanently frozen ground (*permafrost*) within 1 m. They occur on any landform in northern latitudes just south of the areas with *permafrost* and in high mountains around the globe (1, 2).

USTEPTS

The *Ustepts* are the *suborder* of Inceptisols that receive most of their precipitation in the summer when evapotranspiration is highest or they do not have a defined winter. Native vegetation is typically grassland, shrubland, or savanna. They occur on any freely drained landform and that are not persistently cold. Major areas of *Ustepts* occur on major river floodplains, terraces, glacial till plains, hills, and foothills (1, 2). *Ustepts* form where precipitation is sufficient to cause weathering and leaching but is insufficient to completely leach the byproducts. The effectiveness of the precipitation is severely limited either by slope, high evapotranspiration, or both. Cemented layer or bedrock prevents deep leaching in some of these soils.

Table 1 Correlation of *Inceptisol suborders* with WRB reference groups

WRB reference group	Inceptisol suborders
Gleysols	Aquepts
Anthrosols	Anthrepts
Cryosols	Cryepts
Cambisols, Durisols, Calcisols, Phaeozems	Ustepts
Cambisols, Durisols, Calcisols, Phaeozems	Xerepts
Cambisols, Umbrisols, Phaeozems	Udepts

(From Refs. 1, 2.)

XEREPTS

The *Xerepts* are the suborder of Inceptisols that are moist in the winter and dry in the summer. They receive most of their precipitation in the winter season when evapotranspiration is lowest, but experience a summer moisture deficit in most years when evapotranspiration is highest (Mediterranean climate). Native vegetation ranges from grassland or chaparral to forest. They occur on any freely drained landform if the temperature is not persistently cold or hot. Major areas of *Xerepts* occur in mid-latitude coastal areas and nearby zones with Mediterranean climate (1, 2). *Xerepts* occur in areas where precipitation is sufficient to cause weathering and leaching but it is insufficient to completely leach the byproducts in some areas. The effectiveness of the precipitation is enhanced by low evapotranspiration rates in winter. These soils are dry in summer when evapotranspiration rates are highest.

UDEPTS

The *Udepts* are the suborder of *Inceptisols* that do not experience more than short periods of moisture deficit in the growing season for many years. The moisture is distributed evenly enough to keep up with evapotranspiration losses most of the time. Native vegetation is typically forest. *Udepts* occur on freely drained landforms in climates where the temperature is not persistently cold. Figure 2 shows an acidic *Udept* from the humid mountains of the eastern U.S. (4). Major areas of *Udepts* occur in humid areas on floodplains and low terraces, in areas that had recent glacial or periglacial conditions, and in humid hills and mountains (1, 2). *Udepts* occur in areas where precipitation is sufficient to cause abundant weathering and leaching, unless steep slopes decrease the effectiveness of the moisture or some cemented or dense subsurface horizon or feature restricts deep leaching of the weathering byproducts.

Table 2 Area (km^2) of Inceptisols on different continents

Suborder	North America	Central America	South America	Europe	Africa	Asia	Australia/ Oceania	Global total
Aquepts	361,062	38,231	623,873	164,499	554,277	1,386,338	1882	3,130,161
Anthrepts	–	–	–	–	–	–	–	–
Cryepts	271,137	–	36,442	139,495	–	206,342	7075	660,491
Ustepts	331,262	83,255	611,443	377,338	1,675,118	832,200	257,463	4,168,080
Xerepts	109,246	–	8342	310,697	163,793	178,312	264	770,654
Udepts	526,222	79,498	892,932	772,928	214,883	1,336,508	154,169	3,977,140
Total Inceptisols	1,237,867	162,753	1,549,159	1,600,458	2,053,794	2,553,361	418,971	9,576,365
Total ice free land area	20,752,934	696,418	17,546,182	5,741,378	29,808,032	47,825,795	7,897,311	130,268,050
Percentage Inceptisols	6	23	9	28	7	5	5	7

(From Ref. 3.)

SOIL GENESIS

Organic matter accumulates in the surface if the water tables stay very near the surface round the year and mineralization of organic matter is restricted, forming organic or humus-rich mineral surfaces (*histic, mollic* or *umbric epipedons*). Some *Inceptisols* have thick surface deposits of human-transported soil mixed with barnyard manure and livestock bedding material (*Plaggen epipedons*). This practice was used on fields with low fertility before commercial fertilizer became available. *Anthropic* surfaces formed where crop residues, manure (human and livestock), bones, oyster shells and kitchen waste were repeatedly spread on fields as fertilizers. All of these materials decomposed (but did not leach) to form surface soils enriched in carbon, calcium and phosphorus compared to surrounding soils. Some *Inceptisols* in marine estuaries or dry climates have accumulated sodium and other soluble salts in the surface (*salic horizon*) through upward movement of an element-rich or saline water table. Evapotranspiration removes water but leaves precipitated salts behind as the water table recedes. In dry climates, incomplete leaching leaves subsoil water enriched in soluble compounds that precipitate and cement the soil fabric with silica (*duripans*), gypsum (*gypsic* or *petrogypsic horizons*), or lime (*calcic* or *petrocalcic horizons*). Sandy parent materials may accumulate very thin bands (*lamellae*) of layered clay in the subsoil where the soil water resides and dries. In cool and humid areas, compounds with organic carbon and either aluminum or iron may accumulate in the acidic subsoil (*spodic

Fig. 2 Hazleton series, state soil of Pennsylvania. (From Ref. 4.) Scale in feet.

materials). *Cambic horizons* form from incomplete accumulation of these materials.

Inceptisols with *aquic conditions* are saturated in the upper soil for more than a few consecutive days during seasons of biological activity. Water fills all the pores in a horizon when rainfall exceeds evapotranspiration and either some subsoil horizon or layer restricts downward movement, groundwater rises in the soil, or porewater moves very slowly through a horizon. In tidal marshes or areas where reduced parent materials high in sulfides (*sulfidic materials*) are exposed to oxygen by mining or artificial drainage, a *sulfuric horizon* quickly forms and pH drops to very low levels. Water- and root-restricting subsoil horizons form in moist climates when restricted leaching leaves subsoil water enriched in soluble iron compounds that precipitate and cement the soil fabric to form continuous *plinthite, or petroferric* (*ironstone*) *horizons*. In moist climates, noncemented root-restricting subsoils (*fragipans*) may form through internal desiccation or compaction or partial cementation. The persistently cold (*cryic*) soils in northern latitudes or high elevations have a slower rate of soil forming processes than do warmer soils. On gentle slopes, volume changes from alternating freezing and thawing may cause churning (*cryoturbation*). *Cryic* soils on steep slopes have shallow profiles because of reduced effectiveness of precipitation, increased runoff, and accelerated losses from erosion and mass movement. Volcanic materials ejected into the air and some other physically and chemically similar materials may weather to form amorphous clays instead of layered clays. Soils rich in amorphous clays have *andic soil properties* such as high surface area, high retention of charged elements and compounds, high water-holding capacity, and very low bulk density. Deep, wide cracks and large volume changes occur in soils that wet and dry often and contain a high amount of layered clay that has high shrink–swell potential. *Inceptisols* form on major floodplains and low terraces that rarely flood or are very young deposits. Some of these soils still contain evidence of sediment deposition from flooding in the subsoil that has not been completely destroyed by soil forming processes. *Cambic horizons* form from incomplete action of the alteration processes mentioned before.

In extremely old, highly weathered and continuously leached *Inceptisols* have *cambic horizons* that are dominated by low chemical- and physical-activity clays and oxides. *Cambic horizons* may also form from removal of gypsum and lime by percolating soil water over time. Often the leached horizon develops soil structure and becomes lighter or brighter in color or accumulates red colors from the release of oxidized iron compounds. *Inceptisols* have formed on some shoulder positions from more developed soils in intensively farmed areas that have had long-term accelerated erosion. These soils may have had diagnostic horizons removed by accelerated erosion because of farming practices and their convex landscape position.

USE AND MANAGEMENT

Gently sloping, deep Inceptisols are well suited to use as cropland. Moderately sloping areas that are not terraced are better suited for biofuels, forages, grazing, forest products, or agroforestry because of the erosion hazards. Areas which are very steep or shallow can be used for forest products, rangeland, wildlife habitat, or recreational uses.

Udepts used for cropping or pasture seldom need irrigation to supplement growing season soil moisture. Native areas are used for forest products, recreation, and wildlife habitat. However, irrigation is often needed to grow annual crops on *Ustepts*. Without irrigation, moisture conservation farming practices and terracing are needed for growing forages and drought-tolerant crops. Perennial crops such as biofuels, root crops, and trees are better able to take advantage of stored moisture. *Xerepts* are important soils for growing small grains and other winter annual crops. Summer crops such as grapes, peas, lentils, and mustards are able to use soil moisture stored during the previous winter season. Native areas are used for timber production, grazing, and wildlife habitat. *Ustepts* and *Xerepts* with a shallow cemented root-limiting layer can be mechanically ripped to allow deeper moisture percolation and increase the rooting zone. Note that if the cemented layer is mechanically ripped for agriculture, these soils become *Entisols*. However, ripping is not financially practical unless irrigated specialty crops can be grown.

Acidic *Inceptisols* with low quantity of nutrients (low base saturation) are in separated from high base saturation soils in all but the *Aquepts* and *Anthrepts* suborders. These soils occur on uplands and on minor floodplains that drain mostly acidic rocks and soils. They may be used to grow native crops but must be limed to raise the pH to grow most introduced crops, especially where free aluminum is present in the soil solution and aluminum toxicity may occur. These acidic soils are inherently infertile and must be fertilized to sustain crops or forages, provided the soil mineralogy has sufficient capacity to adsorb the added nutrients. Agroforestry and perennial biofuel cropping may be used in combination with organic fertilizers and lime on some of these soils. Extremely weathered and leached rainforest soils have most of their nutrients tied up in the above- and belowground organic matter and cannot sustain continuous cropping without the addition of fertilizer and other soil amendments.

Inceptisols on major floodplains typically have high quantity of nutrients (high base saturation). These soils are some of the most important cropland soils in the world. They may be used to grow almost any type of vegetation, depending on their texture, climate, and available moisture.

Aquepts and *Udepts* with *sulfuric horizons* have extremely low pH and are not suitable for agriculture. They can be remediated only if the remaining sulfide-bearing materials are prevented from oxidizing, either by plugging drainage ditches or deeply burying the layer with better soil. *Aquepts* in warm climates have value for raising rice or water-tolerant specialty crops. Those that have been drained can be used for many types of forages, agricultural crops, biofuels, or forest products (timber or pulpwood). Saline and sodic *Inceptisols* must be drained to prevent the saline water table from rising near the surface if they are used for growing crops. Rainfall or nonsaline irrigation water can then leach the excess sodium and salts from the surface. These soils are then suitable for growing a wide variety of crops, depending on their climate and texture.

Cryepts are not suited to many crops but may be used for small grains, berries, potatoes, forest products, wildlife habitat, and recreation. The *Cryepts* that are in high latitudes are more suited to cropland because of the long day lengths than those at high elevations. The *Cryepts* that occur at high elevations are important sources of streams and fresh water aquifers because they typically have high runoff from rain and melting snowfall and decrease evapotranspiration.

Plaggen surfaces have many uses for agriculture because they are fertile and have high organic carbon and water-holding capacity in the surface. *Anthropic* surfaces are chemically suitable for agricultural fields or gardens but may need irrigation. Soils with high accumulation of organic carbon in the surface have better structure, and higher water nutrient-holding capacity than other surface soils lower in carbon. Natural fertility is variable.

REFERENCES

1. Soil Survey Staff, *Soil Taxonomy: A Basic System of Soil Classification for Making and Interpreting Soil Surveys*, 2nd Ed.; USDA–Natural Resources Conservation Service Agricultural Handbook #436; U.S. Govt Printing Office: Washington, DC, 1999.
2. Food and Agriculture Organization of the United Nations, International Society of Soil Science, and International Soil Reference and Information Centre. World Reference Base for Soil Resources. World Soil Resources Report No. 84. Rome, 1998.
3. USDA–Natural Resources Conservation Service, Soil Survey Division, World Soil Resources. Washington, DC, 2001.
4. Soil Survey Staff. State Soil Photograph Gallery. USDA–Natural Resources Conservation Service, National Soil Survey Center. Photos and Data Available at National Soil Data Access Facility, Iowa State University, Ames, IA, 1998.
5. Soil Survey Staff, *Keys to Soil Taxonomy*, 6th Ed.; USDA–Soil Conservation Service, Pocahontas Press: Blacksburg, VA, 1994.

INDIGENOUS SOIL MANAGEMENT

Karl Herweg
University of Bern, Bern, Switzerland

INTRODUCTION

Indigenous or *local* knowledge is unique to a given culture or society, in contrast with external knowledge that is generated within the international system of universities, research institutes, and private firms (1). Nowadays, however, virtually all indigenous knowledge systems have been exposed to external influence, be it during the eras of colonization or development cooperation, due to migration and trading, international training of local experts, or through the mass media. What is today termed indigenous knowledge has already incorporated suitable components from external knowledge systems, or in other words, introduced technologies have been indigenized (2). Thus, in this article, the term *indigenous soil management* refers to agro-pastoral or mixed smallholder farming systems with low external agricultural inputs (3), rather than to the rare cases that may be absolutely isolated from the outside world.

INTEGRATED SOIL MANAGEMENT

The term indigenous soil management reflects a typical scientific approach. In the reality confronted by land users, an equivalent term for soil management as such may not exist, because soil is always integrated in an overall farming system, e.g., to produce food and fodder. Thus, indigenous soil management should be seen in the framework of a land management system (4).

Indigenous knowledge systems are highly site-specific (4, 5). They undergo permanent adaptation in response to local socioeconomic and environmental changes, and thus always imply an aspect of innovation. Apart from its site-specificity, indigenous soil management may best be characterized as multi-functional, flexible, and dynamic (2, 6, 7). Table 1 provides examples of various indigenous soil management options.

Multi-Functionality

Each household pursues its own livelihood strategy and land users apply a range of land management practices to achieve their goals. Each practice may involve the management of soils besides other components such as water, crops, animals, forests, and human resources. Each practice usually serves more than one purpose, e.g., crop production and resource protection (6, 7).

Flexibility and Dynamism

Indigenous soil and land management usually is flexible and dynamic in several aspects. A practice responds to one or more specific problems (soil erosion, low production, pests, etc.) at a specific location (steep slope, valley bottom, communal land, etc.), as perceived by a land user with specific experience, capabilities, and options to react. In addition, the positive impact and the side effects of a practice are permanently observed, and the practice is modified, if necessary, over time. This means that, depending on the variations in these factors, a practice will always look different, from one field to another, and from one farmer to another (6, 7).

INDIGENOUS SOIL CLASSIFICATIONS AND FERTILITY INDICATORS

Indigenous soil classifications reflect land users' perceptions of land and soils, their potential for a specific purpose, and their spatial variability. Such classifications are usually well categorized and based on well-identified criteria, which typically incorporate diverse properties of soil or soil-related factors, e.g., texture, color, water holding capacity, irrigation requirements, soil softness, drainage, fertility, and crop suitability (8). Soils are often classified according to their suitability for growing certain crops, based on vegetative, topographic, and microclimatic indicators.

Soil fertility is considered a dynamic characteristic, not an inherent quality (7, 8). In local terms, soil fertility may, for instance, be connected with manure requirements, but interestingly, it is not globally attributed to nutrient availability. For example, water-holding capacity (in semiarid areas), and permeability and drainage of the soils (in subhumid areas) may be considered more important than the overall nutrient status. Also, socioeconomic terms such as workability are used to describe a fertile soil, e.g., if the soil can be prepared with a few tillage operations only, if tillage is relatively easy, or if labor requirements are low. Although indigenous soil classifications and indicators of soil fertility are highly site-specific, soil texture and color are the most common factors in indigenous knowledge systems to distinguish soils of different quality (8).

Local knowledge of soil–plant–fauna relationships is used to select crop varieties or crop rotations suitable to a specific soil. Soil heterogeneity is acknowledged and advantageously used, and not eliminated by fertilization. The choice of crops for each soil is not necessarily unanimous; but depends largely on the individual experience and interest of the land user (8). Local plants that indicate soil fertility are as common as the knowledge of plants that are beneficial for improving soil fertility. The range of indigenous biological indicators is wide, e.g., the appearance of mauls in the accumulation of indigenous terraces, in combination with grass production on the terraces, is known to be an indicator of improved soil fertility (7).

INDIGENOUS SOIL MANAGEMENT PRACTICES

Continuous farming or animal grazing affects soil in many ways. For example, mono-cropping and soil erosion diminish nutrients and organic matter; soil erosion changes soil depth, texture, structure, and the capacity to store water and nutrients; irrigation in semiarid areas increases the salt content; and so on. Indigenous soil management practices try to prevent or minimize degradation processes and focus on managing labor, soil fertility (soil nutrients, organic matter), and both the excess and shortage of water (9).

Managing Limited Labor

Relatively wealthy farmers have a greater labor force, more draught power, and a greater range of soil management options compared to relatively poor land users. Limited labor and transport facilities usually result in intensively used land around the homesteads and decreasing intensity toward the most distant plots. Shortage of labor often forces farmers to rely on each other and make sharecropping, animal fattening, and other arrangements (6, 7). Decisions are often made together with neighbors or in the community, because individual strategies can lead to severe problems. For example, a single farmer who decides to plant early will attract all the pests to his fields, or a single field that is left fallow will be a refuge for pests that affect all neighboring plots (9). Local societies have a number of working arrangements to overcome labor shortages, resolve conflicts over scarce resources, and use optimum timing for land preparation, weeding, and other farming activities (6, 7).

Managing Soil Fertility and Soil Water

Probably the most common way of recycling nutrients and reclaiming land after a period of nutrient and water exploitation is to leave it *fallow* until its fertility has recovered. However, increasing population and animal pressure, fragmentation of the land, insecure land tenure, and other factors force land users to shorten their fallow periods and to look for alternatives (2, 4, 8–10). For example, crop rotation (e.g., cereals and legumes), selection of crops corresponding with the potential of the soil, and the use of nitrogen-fixing trees are common methods for preventing exploitation of nutrients. A number of indigenous practices contribute to improved nutrient levels and physical soil properties that provide better aeration, infiltration, water-holding capacity, and soil moisture. For example, the organic matter content can be improved by manure, if the herd size is sufficient, if manure can be purchased from pastoralists, or if the manure is not used for other purposes (fuel, construction, to protect seeds from diseases, etc.). Application of manure requires improved weed control. The highest concentration of manure is found around the homesteads (fixed pens). Distant plots can be improved through mobile pens and open grazing. Other sources of organic matter can include incorporation of compost, household waste, leaf litter, weeds (uprooted and burned) in the soil, and crop residues left after harvesting or disseminated as mulch. Again, availability is the limiting factor, since crop residues are frequently used as fodder. It is also common to carry fertile silt from community ponds back to improve the cultivated fields. Soil burning—together with dung and other organic matter—may improve water infiltration and availability of nutrients on a rather short-term basis.

Particularly in arid and semiarid areas, the shortage of water is the greatest constraint on production. Effective indigenous water harvesting practices include ridging and infiltration pits (zaï, tassa), that break hard layers in the topsoil, crescent or rectangular embankments that allow

Table 1 Examples of indigenous soil management options

Option	Main purposes	Area (reference source)
Working arrangements (sharecropping, animal fattening)	Overcome labor shortage, optimize timing of farming operations	Ethiopia (6, 7)
Gradual build-up of terraces	Effective use of labor	Ethiopia (2, 6, 7)
Focusing only on locations with severe erosion	Effective use of labor	Ethiopia (6, 7)
Joint decision-making among neighbors	Avoid pests as result of individual cropping patterns	Nepal (9)
Fallow	Recycling of nutrients, recovering of soil fertility	Africa (2, 10), Ethiopia (4), South America (8), Nepal (9)
Crop rotation, selection of crops to correspond with soil potential, use of nitrogen-fixing trees	Prevent exploitation of nutrients	Africa (2, 10), Ethiopia (4), South America (8) Nepal (9)
Manure (from fixed pens, open grazing, compost, crop residues, household waste, leaf litter, weeds)	Improve nutrient level and physical soil properties to provide better aeration, infiltration, water holding capacity and soil moisture	Africa (2, 10), Ethiopia (4), Kenya (5), South America (8), Nepal (9)
Mulching	Improve soil fertility, reduce evaporation	South America (8)
Soil burning	Improve infiltration and availability of nutrients	Ethiopia (6)
Trapping eroded soil	Improve soil fertility	Mexico (8)
Application of fertile silt from ponds and rivers	Improve soil fertility	Africa (2)
Temporary structures (trash-lines)	Collect and redistribute fertile accumulation	Ethiopia (2, 6, 7)
Indigenous terraces, grass and vegetation strips	Collect and redistribute fertile accumulation	Ethiopia (2, 6, 7)
Water harvesting (drainage ditches, gully reclamation)	Conserve water/moisture	Africa, semiarid areas (2)
Crescent-shaped or rectangular embankments	Slow down runoff, increase infiltration	Africa, semiarid areas (2)
Ridging and infiltration pits (zaï, tassa)	Break hard topsoil layers, increase infiltration	Africa, semiarid areas (2)

slow movement of runoff and better infiltration, and mulch that reduces evaporation. Drainage ditches and gully reclamation measures help control erosion in subhumid areas and harvest water in semiarid areas. Indigenous terraces, grass and vegetation strips are used to enforce accumulation of fertile topsoil and to prevent further decline of soil fertility (2). But unlike introduced soil and water conservation (SWC) structures such as soil bunds, indigenous SWC measures are built from material that can best be spared, such as stones or brushes, instead of fertile soil. In addition, indigenous SWC measures rarely cover the entire slope but concentrate on "hot spots," where severe soil erosion is expected or where soils are particularly shallow [sedimentation management (8)]. Indigenous SWC is often a mixture of temporary structures (e.g., trash-lines)—i.e., scheduled for removal to redistribute the fertile accumulation—and permanent structures (e.g., stone bunds serving at the same time as field borders or fences). Probably the greatest difference between introduced and indigenous measures is that the latter are rarely constructed in one go, but are gradually built up in the course of other farming activities such as ploughing, weeding, etc. (2, 6, 7).

CONCLUSIONS

Former approaches to technology transfer tried to identify a problem, such as declining soil fertility, soil degradation, pests, etc., and to respond to it, for example, through on-station research and dissemination of technologies. This has not only improved soil management but has also been accompanied by major setbacks in the sense that adoption of alien technologies often ignored socioeconomic aspects and turned out to be costly or difficult to accept at the social level. More recent approaches, therefore, focus on incremental adaptation of technologies rather than adoption, i.e., understanding indigenous farming practices to guide sustainable land use development (8). This does not mean that indigenous knowledge always provides the solution. Rapid changes in the socioeconomic, political, and biophysical environments put indigenous knowledge systems under tremendous pressure and do not give them time to adapt themselves to the changes. But learning about local knowledge, recognizing indigenous practices and their functions, and then trying to improve soil management together with local farmers, researchers, and external advisors means making the best use of all available knowledge systems.

REFERENCES

1. Van Marrewijk, A. Indigenous Knowledge and Development Monitor **2000**, *8* (3), 48.
2. Scoones, I; Reij, C; Toulmin, C. Sustaining the Soil. *Indigenous Soil and Water Conservation in Africa;* 11ED Drylands Programme, issue paper No. 67, London; 1996; 260.
3. Eyasu, E. Soil Enrichment and Depletion in Southern Ethiopia. In *Nutrients on the Move*; Hilhorst, T., Muchena, F., Eds.; 11ED Drylands Programme, London; 2000; 65–82.
4. Corbeels, M.; Shiferaw, A.; Haile, M. Farmers' Knowledge of Soil Fertility and Local Management Strategies in Tigray, Ethiopia. Manag. Africa's Soils **2000**, *10*, 24.
5. Nandwa, S.M.; Ondurru, D.D.; Gachimbi, L.N. Soil Fertility Regeneration in Kenya. In *Nutrients on the Move*; Hilhorst, T., Muchena, F., Eds.; 11ED Drylands Programme, London; 2000; 119–132.
6. Gebre Michael, Y. *The Use, Maintenance, and Development of Soil and Water Conservation Measures by Small-Scale Farming Households in Different Agro-Climatic Zones of Northern Shewa and Southern Wello, Ethiopia.* Soil Conservation Research Project 2000; Research Report 44, 188 pp.
7. Gebre Michael, Y.; Herweg, K. *Soil and Water Conservation—From Indigenous Knowledge to Participatory Technology Development*; CDE, Bern; 2000; 52.
8. Talawar, S.; Rhoades, R.E. Scientific and Local Classification and Management of Soils. Agric. Human Values **1998**, *15*, 3–14.
9. Tamang, D. Living in a Fragile Ecosystem: Indigenous Soil Management in the Hills of Nepal. IIED Gatekeeper Ser. **1993**, *41*, 23.
10. Hilhorst, T., Muchena, F., Eds. Nutrients on the Move. *Soil Fertility Dynamics in African Farming Systems*; 2000; 146.

INDUSTRIAL WASTE

Warren Dick
The Ohio State University, Wooster, Ohio, U.S.A.

I

INTRODUCTION

There is no universally accepted definition of industrial waste. Common sense suggests industrial wastes are any materials created or left over as a result of industrial activities to produce durable and nondurable goods. In a utopian world, we would be 100% efficient in use of raw materials and products, once manufactured, would serve their purpose and then simply vanish with no need for disposal.

Unfortunately, although we are not operating in the same grossly inefficient way we did during the peak of the Industrial Revolution, we are still generating millions of metric tons of industrial wastes every year. We will never achieve the utopian dream, but we can make it our goal to reduce the amount of industrial wastes and to properly dispose those materials that cannot be beneficially recycled. The first step towards achieving this goal is to identify the types of industrial activities and the wastes they create. Then processes and technologies can be designed that will either reduce the amounts of wastes created or recycle the wastes in ways that are environmentally safe and economically beneficial.

Currently, large amounts of industrial wastes are disposed in either a planned manner or simply allowed to disperse into the environment (e.g., carbon dioxide due to combustion of fossil fuels). However, there are only a few places where disposal can be done on (or in) the land, dumped in water bodies such as the oceans, or incinerated. Ocean dumping has been outlawed or severely curtailed in many of the more developed countries. Incineration is also often opposed because during combustion of the waste there is potential to create even more toxic materials such as dioxins. Therefore, the only remaining disposal area is the land. This can be in dedicated landfills. Alternatively, many industrial by-products are being viewed as having potential beneficial properties which can be captured either as a recycled material or land applied as a soil amendment. Each of these options must be carefully researched to avoid the creation of a new problem by solving an existing problem.

TYPES OF INDUSTRIAL PRODUCTS

Our modern lifestyle has generated an insatiable appetite for all types of manufactured products. Table 1 provides a listing of industrial products. These products all generate, at some point in their product life cycle, industrial waste. The type of wastes may be unique or be common to many different types of products. The major variables that affect the amounts and types of wastes for each product are the raw materials used, the amount of processing required of the raw materials before a finished product is created, the relative toxicity of the processing materials, and the total amount of energy required during product manufacture.

TYPES OF INDUSTRIAL BY-PRODUCTS

The types of industrial by-products are as varied as there are industries to produce materials and goods for human consumption. However, there are several broad categories that can be identified and the information presented below is taken from Ref. (2).

Metal Refining and Processing

Slags are produced by iron and steel making and in 1995, 21 million metric tons were produced in the U.S. Slags are produced by addition of lime to furnaces to remove silicon impurities and are composed primarily of $CaSiO_3$. About 60% of the slags produced are iron slags and 40% are steel slags. *Fine dusts* are fine airborne particulates collected from metal processing, refining and fabricating industries. The most commonly encountered dusts are fines from the metal processing industry (especially from electric furnaces) and are primarily ferrous in nature with admixture of heavy metals. Approximately 5 million metric tons per year are produced in the U.S. and most are either recycled within the facility or used as land fill.

Table 1 Census Bureau's industrial products overview covering manufactured durable and nondurable goods

Industrial product category	Industrial product subcategory	Industrial product category	Industrial product subcategory
Building materials	Clay construction		Internal combustion engines
	Lumber and mill stocks		Metalworking machinery
	Plumbing fixtures		Mining machinery
	Refractories	Industrial equipment	Air pollution control
Chemicals	Fertilizer materials		Antifriction bearings
	Industrial gases		Fluid power
	Inorganic chemicals		Motors and generators
	Paint and allied products		Pumps and compressors
	Paint, varnish, and lacquer		Refrigeration and heating
	Pharmaceutical preparations		Steel drums and pails
Consumer goods	Consumer electronics		Vending machines
	Electric housewares	Primary metals	Aluminum mill products
	Household appliances		Insulated wire and cable
Electronics	Communication equipment		Iron and steel castings
	Computers and office machines		Nonferrous castings
	Electromedical equipment		Steel mill products
	Fluorescent lamp ballasts		Steel mill inventories
	Lighting fixtures	Textile and apparel	Apparel
	Measurement instruments		Broadwoven fabrics
	Semiconductors and components		Carpets and rugs
	Switchgear and industrial		Cotton consumption
	controls		Footwear (annual)
	Wiring devices		Footwear (quarterly)
Food	Confectionery		Gloves and mittens
	Fats and oils		Knit fabric production
	Flour milling		Bed and bath furnishings
	Oilseed crushings		(sheets, pillowcases, and towels)
Glass	Flat glass		Woolen consumption
	Glass containers		Yarn production
	Glassware	Transportation	Aerospace industry
Heavy machinery	Construction machinery		Civil aircraft and engines
	Farm equipment		Truck trailers

(Adapted from Ref. 1. http://www.census.gov/econ/www/industry.html site confirmed April 20, 2001.)

Mineral Processing

Gypsum ($CaSO_4{\cdot}2H_2O$) is produced as a by-product by a number of industries, most notably the phosphate fertilizer industry, the pigment processing and electroplating industries, and the electricity generating industry. By far the largest production, however, is the gypsum produced during the processing of rock phosphate to produce phosphate fertilizer. Nearly 1 billion metric tons of "phosphogypsum" are stockpiled in the phosphate district of central Florida, where most of the commercial rock phosphate deposits are found in the U.S. A major concern related to phosphogypsum is the radioactivity that is due to radioactive elements originating in the sedimentary phosphate deposits. *Cement kiln dust* is a fine material produced during manufacture of Portland cement, which involves heating lime and siliceous materials (clay, shale) to a molten state in rotary ovens or kilns. Dust in the exhaust is removed by filter units and is composed of fine feed material, cement, and sloughed refractory materials. In 1995 there were 118 facilities in 37 states producing 80 million metric tons of cement and approximately 3–5 million metric tons of kiln dust. *Tailings* are fines remaining after crushing and extraction of ores during mineral separation and refining. Many different separation and refining processes result in a wide range of materials such as the clayey phosphate slimes and the red muds of Al metal processing.

Coal Combustion By-products

Fly ash is derived when coal is burned for energy and is produced in larger amounts than any other single industrial waste. In 1998 approximately 60 million metric tons of fly ash as well as 20 million tons of *bottom ash* and *boiler slag* materials and 25 millions metric tons of *flue gas desulfurization* by-products were produced in the U.S. All of the above named materials are derived when coal is burned and are, in their conglomerate, referred to as coal combustion by-products. Fly ash is a fine (typically less that 0.05 mm in diameter) largely siliceous material derived from coal. It is removed from the flue gases by special collectors and about 25% is recycled as a cement additive, structural fill, or for other purposes. Fly ashes are classified as Class C or Class F and the properties of each type of fly ash are due primarily to the coal source and furnace operating conditions. Class C fly ashes are higher in alkali metals and Class F ashes have high Si and Al contents. Bottom ashes are sand-like granular substances, which are typically used as aggregate in cement blocks, for construction backfill, and for treatment of snow or icy roads. Boiler slag is coarse textured and shiny-black in appearance and resembles crushed glass. Its major applications include use as blasting grit, aggregate, granules for roofing shingles asphalt, and snow and ice abrasive. Flue gas desulfurization by-products are created as a result of the Clean Air Acts, which mandate removal (or scrubbing) of sulfur from the flue gases when coal is burned. They are often a mixture of coal ash, gypsum, and unreactive lime, although the composition can vary tremendously based on the type of technology used to scrub the sulfur.

Pulp and Paper Manufacture

The production of pulp for papermaking and as an industrial feedstock by the Kraft and related processes is the major industry in several regions of the U.S. Total production of pulp and paper manufacture materials in 1989 was approximately 13 million metric tons. Material properties are highly variable due to the wide range of processes used in wood digestion and in handling of the resultant wastes.

Construction Wastes

A large amount of construction waste is generated each year in the U.S. waste wallboard, lumber, waste fill material, and other construction materials. For example, in new home construction, approximately 0.5 kg of wallboard waste is generated per square foot of floor space resulting in the annual production of about 1 million metric tons of waste. Typical construction waste of a 2000 square foot home is estimated to be 3600 kg (3) (http://www.greenbuilder.com/sourcebook/ConstructionWaste.html#CSI, site confirmed April 26, 2001.)

LAND APPLICATION OF INDUSTRIAL BY-PRODUCTS

The physical form of industrial wastes can vary greatly ranging from liquid to fine powders to large-sized pieces of construction wastes. Not all industrial wastes are destined to be entombed in landfills. Many have properties that make them beneficial as raw materials for subsequent manufactured products (4). For example, coal combustion by-products are commonly included in the preparation of high quality, high strength concrete. When industrial wastes are used in such a manner, they are really no longer considered wastes, but are valued by-products used as raw materials to create another type of industrial activity.

Many industrial wastes also have properties that make them valuable as soil amendments (4). Organic by-products from some food processing plants, for example, when applied to soil can affect a large number of properties and lead to an increase in soil quality. A summary of the more important physical, chemical and biological properties that can be affected by land application of by-products is provided in Table 2.

A large number of studies have been conducted to determine the best management practices for use of by-products as a soil amendment. This includes information such as presented in Table 2, but also involves methods of application and techniques related to the application process itself.

ROLE OF REGULATORY AGENCIES REGARDING DISPOSAL AND REUSE OF INDUSTRIAL WASTES

Many states require a plan be submitted to state regulatory agencies which provides information on the type of material to be land applied, the properties of the material, the rates to be applied, and how the material will be transported and applied. The by-product must be applied without creating odors, must often be mixed into the soil with farm machinery to reduce risk of runoff if a heavy rain should occur soon after application, and often cannot be applied within certain distances from wells, streams,

Table 2 Properties of soil that can be either improved or degraded by the application of by-products to soil. All of the properties listed in the table below are useful in models that seek to integrate soil, landscape, and geographic variability for the purpose of assessing and predicting impacts of applying by-products to soil

Soil properties affected	Reasons these properties are important
Soil physical properties	Indicators of retention and transport of water and chemicals, soil erosion, leaching, surface and subsurface water runoff, and soil productivity
Texture	
Topsoil depth	
Rooting zone depth	
Water infiltration rate	
Soil bulk density	
Water holding capacity	
Erodibility	
Compaction	
Soil chemical properties	Defines the fertility of the soil and the potential to either positively or negatively affect the soil's quality and the quality of the air and water that interact with the soil
Soil organic matter concentration	
pH	
Electrical conductivity	
Available plant nutrients	
Available heavy metals	
Sorption capacity	
Soil biological properties	Indicates the soil's microbial ability to degrade organic compounds in soil and the soil's potential to cycle nutrients and to purify the soil and water. The more microbial diversity maintained in an ecosystem, the better the ecosystem's ability to adjust to changes
Microbial biomass size	
Microbial biomass activity (i.e., respiration)	
Microbial diversity	
Redox potential	
Ability to biodegrade organic materials	

(Adapted from Ref. 5.)

houses, or other more sensitive areas. This is not to imply that all industrial by-products are toxic. Like medicines, properly handled by-products can also be beneficial. Guidelines are established, however, to ensure good stewardship of our soil–water–air resources. Many states have guidelines for land application of by-products posted on the Internet.

The Resources Conservation and Recovery Act (RCRA), enacted on October 21, 1976, regulates the most hazardous of industrial wastes "from cradle to the grave." The statute requires EPA to establish minimum acceptable requirements for all aspects of hazardous wastes for generators and transporters as well as for treatment, storage, and disposal facilities. The regulatory framework established under Subtitle C of RCRA was designed to protect human health and the environment from the effects of improper management of hazardous waste. Determining what is hazardous waste is therefore a key activity because only those wastes that meet the definition of hazardous waste are subject to Subtitle C. This is also a complex issue. In general, a material is classified as hazardous if it exhibits one of the EPA-defined characteristics of a hazardous waste (i.e., corrosive, ignitable, reactive, toxic). A hazardous waste, however, must also meet the definition of a solid waste, and a solid waste is "any discarded material that is not specifically excluded" from a listing of solid wastes. Materials excluded by the various definitions of wastes are indicated in Table 3.

Solid waste does not refer to the materials physical state per se as it is a regulatory term only. Nonhazardous wastes, as defined by RCRA, are not necessarily guaranteed to be safe for dispersal into the environment. Potential hazards could include soil degradation, surface and groundwater contamination, direct exposure during handling, or indirect exposure such as plant uptake into the food chain can all result in harmful environmental and human health effects.

MINIMIZING PRODUCTION OF INDUSTRIAL WASTES

In the late 1980s, many groups argued for a defined approach to prevention of wastes (including industrial wastes). These efforts translated into the Pollution

Table 3 Exclusions from Subtitle C of RCRA

Excluded from solid waste definition	Excluded from hazardous waste definition	Excluded materials requiring special management
Domestic sewage	Household wastes	Product storage wastes
Mixtures of domestic sewage and wastes going to publicly owned treatment works	Agricultural wastes used as fertilizers	Waste identification samples
Industrial point source discharges under Section 402CWA	Mining overburden returned to site	Treatability samples
Irrigation return flows	Discarded wood treated with arsenic	Residues remaining in empty containers
Source, special nuclear, or by-product material under AED	Specific ore and mineral beneficiating wastes	Conditionally exempt small-quantity generator wastes
		Farm wastes (pesticides)
In situ mining waste	Fossil fuel combustion wastes	PCB-mixed wastes[a]
Reclaimed pulling liquors from Kraft paper process	Oil and gas exploration, development and production wastes	Low-level radioactive mixed waste[a]
Spent sulfuric acid used to produce new acid	Cement kiln dust	
Secondary materials returned to the original process under certain conditions	Petroleum-contaminated media and debris from underground tank cleanup	
Spent reused wood preservatives	Spent chlorofluorocarbon refrigerant	
Certain coke by-products	Used oil filters	
Splash condenser dross residue	Still bottoms from the re-refining of used oil	
Recovered refinery oil wastes		

[a] These wastes are not explicitly excluded from RCRA, but require special management.
(Adapted from Ref. 6.)

Prevention Act of 1990 and development of a qualitative hierarchy as a means to evaluate various approaches to pollution reduction in terms of their adherence to a pollution provision theme. A simple version of this hierarchy (from most to least desirable) is as follows: 1) source reduction—use of cleaner feed materials and processes, 2) reuse and recycle—especially in-process recycling, 3) treatment, and 4) disposal (7).

The Chemical Manufacturers Association (8) have developed a 5-step approach which is applicable to the design and analysis of chemical manufacturing plants and can be used in various stages of design. This approach can be extended to many different industries which generate waste:

Step 1: Characterize process streams
Step 2: Evaluate environmental impacts and issues
Step 3: Identify pollution prevention opportunities
Step 4: Analyze alternatives
Step 5: Document results

CONCLUSIONS

Increasing human population and expanded industrial activities make it imperative that we pay attention to the creation, recycling, and disposal of industrial wastes. Our quality of life will largely depend on how we address this issue. While progress has been made, there is still a long way to go before we will achieve a truly sustainable way of life.

REFERENCES

1. U.S. Census Bureau. *Industrial Products Overview.*
2. Miller, D.M.; Miller, W.P.; Dudka, S.; Sumner, M.E. Characterization of Industrial By-Products. In *Land Application of Agricultural, Industrial, and Municipal By-Products*; SSSA Book Series 6, Power, J.F., Dick, W.A., Eds.; Soil Science Society of America: Madison, WI, 2001; 107–126.

3. Sustainable Sources. *Sustainable Building Workbook—Construction Waste.*
4. Power, J.F.; Dick, W.A. *Land Application of Agricultural, Industrial, and Municipal By-Products*; SSSA Book Series 6, Soil Science Society of America: Madison, WI, 2000.
5. Sims, J.T.; Pierzynski, G.M. Assessing the Impacts of Agricultural, Municipal, and Industrial By-Products on Soil Quality. In *Land Application of Agricultural, Industrial, and Municipal By-Products*; SSSA Book Series 6, Power, J.F., Dick, W.A., Eds.; Soil Science Society of America: Madison, WI, 2001; 237–261.
6. Wagner, T.P. *The Complete Guide to the Hazardous Waste Regulations—RCRA, TSCA, OSHA, and Superfund*; Wiley: New York, 1999.
7. Radecki, P.P.. In *Proceedings of the 51st Purdue Industrial Waste Conference, West Lafayette, IN*, May 6–8, 1996; Wukasch, R.F., Ed.; Ann Arbor Press: Chelsea, MI, 1997; 5–22.
8. Chemical Manufacturers Association, *Designing Pollution Prevention into the Process: Research, Development and Engineering*; Chemical Manufacturers Association: Washington, DC, 1993.

INFILTRATION PROPERTIES

Walter J. Rawls
United States Department of Agriculture-Agricultural Research Service (USDA-ARS), Beltsville, Maryland, U.S.A.

INTRODUCTION

Infiltration is commonly defined as the process of water entry at the land surface into a soil from a source such as rainfall, irrigation, and snowmelt. The rate of infiltration is generally controlled by the rate of soil-water movement below the surface. Factors which affect infiltration have been divided into categories of soil, surface, management, and natural properties (1). These factors should be taken into account in the infiltration model parameters. Soil factors encompass soil particle size, morphological, chemical and soil water properties (2). Surface factors are those that affect the movement of water through the air–soil interface such as soil crust. Management systems involve different types of tillage, vegetation, and surface cover (2, 3). Natural factors include such things as precipitation, soil temperature, residue, and soil moisture which vary with time and space, and interact with other factors in their effect on infiltration.

INFILTRATION MODELS

Models to characterize infiltration for field applications usually employ simplified concepts, which predict the infiltration rate or cumulative infiltration volume and assume that surface ponding begins when the surface application rate exceeds the soil-surface infiltration rate. The evolution of infiltration modeling has taken three directions: the empirical, the approximation to the physically based models, and the physically based approach. Most of the empirical and approximate models treat the soil as a semi-infinite medium with the soil saturating from the surface down. Physically based models specify appropriate boundary conditions and normally require detailed data input. The Richards equation is the physically based infiltration equation primarily used for describing water flow in soils; however, it has not been used in operational infiltration models.

The empirical models generally relate infiltration rate or volume to elapsed time modified by certain soil properties (4). Parameters used in these models are commonly estimated from measured infiltration rate–time relationships for a given soil condition. The three most common empirical models were proposed by: Kostiakov (5), Horton (6), and Holtan (7). The most common physically based models were proposed by: Green and Ampt (8), Phillip (9), Morel-Seytoux and Khanji (10), and Smith and Parlange (11).

MODEL PARAMETERS

The most common ways to determine infiltration parameters are to fit models to experimental infiltration data (infiltration rate as it varies with time) or to develop relationships between readily available data such as soils and land-use data and the infiltration model parameters.

Experimental infiltration measurements are normally made by applying water at a specific site to a finite area and measuring the intake rate of the soil. There are four types of infiltrometers: the *ponded-water ring* or *cylinder* type, the *sprinkler* type, the *tension* type, and the *furrow* type. When choosing an infiltrometer one must choose the one that replicates the system being investigated. For example, ring infiltrometers should be used to determine infiltration rates for inundated soils such as flood irrigation or pond seepage. Sprinkler infiltrometers should be used where the effect of rainfall on surface conditions influences the infiltration rate. Tension infiltrometers are used to determine the infiltration rates of the soil matrix in the presence of macropores. Furrow infiltrometers are used when the effect of flowing water is important such as furrow irrigation (12).

The Green–Ampt (8) model is typical of physically based models and has probably had the most effort directed toward estimating the parameters from readily available data. Also, as shown in Table 1, the Green–Ampt parameters can be used to estimate the parameters for most of the physically based models (13).

The Green–Ampt (8) model is an approximate model utilizing Darcy's law. The original model was developed for ponded infiltration into a deep homogenous soil with

Table 1 Physically based infiltration-capacity models, assuming immediate ponding

Infiltration rate equation	Parameters	Parameter estimation	Ref.
$i = k_s\left(1 + \frac{(\phi - \theta_i)^{H_f}}{I}\right)$	K_s, saturated hydraulic conductivity; H_f, wetting front suction; φ, porosity		
$i = St^{-1/2} + A$	S, sorptivity; A, parameter with dimension of conductivity	$H_f = \frac{2 + 3\lambda}{1 + 3\lambda}\left(\frac{h_b}{2}\right)$	(8)
$i = \frac{\tilde{K}}{B}\left(\frac{(\varphi - \theta_i)(H_o + H_c) + 1}{I}\right)$	$\tilde{K}$, hydraulic conductivity at natural saturation; B, viscous resistance correction factor; H_c, effective capillary drive; φ, porosity	$S = (2(H_0 + H_f)(\phi - \theta_i)(K_s))^{1/2}$ $A = \alpha(K_s)$ α ranges 0.33–1, (recommend 1)	(9)
$i = K_s\left(\frac{C}{K_s I} + 1\right)$	C, sorptivity; K_s, saturated hydraulic conductivity; φ, porosity; θ_i, initial water content	$\tilde{K} = K_s$ $1 \leq B = \leq 1.7$ $H_c = H_f$	(10)
$i = \frac{K_s e^{(I(K_s)/C)}}{e^{(I(K_s)/C-1)}}$	i = infiltration rate (cm/hr); I = cumulative infiltration depth (cm); θ_i = initial water content (vol. Fraction); H_o = depth of ponded water (cm); λ = Brooks-Corey pore size index; h_b = Brooks-Corey bubbling capillary pressure	$C = [2(\phi - \theta_i)H_f K_s]^{1/2}$	(11)

(From Ref. 13.)

a uniform initial water content. Water is assumed to infiltrate into the soil as piston flow resulting in a sharply defined wetting front that separates the wetted and unwetted zones. Neglecting the depth of ponding at the surface, the Green–Ampt rate equation is

$$f = K(1 + (\varphi - \theta_i)H_f/F)$$

where K is the effective hydraulic conductivity (cm/h), H_f is the effective suction at the wetting front (cm), φ is the soil porosity (cm^3/cm^3), θ_i is the initial water content (cm^3/cm^3), F is accumulated infiltration (cm), and f is the infiltration rate (cm/h).

The Green–Ampt parameters can be estimated from readily available data such as soils and land-use data (14). For example, average values for the Green–Ampt wetting front suction, S_f, saturated hydraulic conductivity, K_s, and porosity, φ, are given in Table 2 for the 11 USDA soil textures. These values can be used as a first estimate; however, if more detailed soil properties are available, more refined estimates can be made.

Porosity

Table 2 gives estimates of porosity for soil textures; however, the porosity, φ, can be estimated from measured bulk density or from soil particle size data (%sand, %clay), % organic matter and cation-exchange capacity of the clay (an indicator of the shrink–swell capacity of the clay) (16). Coarse fragments (>2 mm in size) in the soil affect the porosity and adjustments should be made (17).

Initial Water Content

The initial water content (θ_i) should be measured or it can be estimated from moisture-retention relationships (14). Good estimates of wet, average and dry initial water contents are the water content held at −10, −33, and −1500 kPa, respectively.

Wetting Front Suction

The Green–Ampt wetting front suction parameter (H_f) can be estimated from the parameters of the Brooks–Corey water retention equation (14) as

$$H_f = ((2 + 3\lambda)/(1 + 3\lambda))(h_b/2)$$

where H_f is the Green–Ampt wetting front suction (cm), λ is the Brooks–Corey pore size distribution index, and h_b is the Brooks–Corey bubbling pressure.

Rawls and Brakensiek further simplified the above equation by relating the Green–Ampt wetting front suction parameter to soil properties in the following equation

Table 2 Green–Ampt parameters

Soil texture class	Porosity, φ	Wetting front soil suction head, S_f	Saturated hydraulic conductivity, K_s
Sand	0.437 (0.374–0.500)	4.95 (0.97–25.36)	23.56
Loamy sand	0.437 (0.363–0.506)	6.13 (1.35–27.94)	5.98
Sandy loam	0.453 (0.351–0.555)	11.01 (2.67–45.47)	2.18
Loam	0.463 (0.375–0.551)	8.89 (1.33–59.38)	1.32
Silt loam	0.501 (0.420–0.582)	16.68 (2.92–95.39)	0.68
Sandy clay loam	0.398 (0.332–0.464)	21.85 (4.42–108.0)	0.30
Clay loam	0.464 (0.409–0.519)	20.88 (4.79–91.10)	0.20
Silty clay loam	0.471 (0.418–0.524)	27.30 (5.67–131.50)	0.20
Sandy clay	0.430 (0.370–0.490)	23.90 (4.08–140.2)	0.12
Silty clay	0.479 (0.425–0.533)	29.22 (6.13–139.4)	0.10
Clay	0.475 (0.427–0.523)	31.63 (6.39–156.5)	0.06

Numbers in parentheses are the 25% and 75% ranges.
(Adapted from Ref. 15.)

$$
\begin{aligned}
H_f = \exp[&6.53 - 7.326(\varphi) + 0.00158(C^2) \\
&+ 3.809(\varphi^2) + 0.000344(S)(C) - 0.04989(S)(\varphi) \\
&+ 0.0016(S^2)(\varphi^2) + 0.0016(C^2)(\varphi^2) \\
&- 0.0000136(S^2)(C) - 0.00348(C^2)(\varphi) \\
&- 0.000799(S^2)(\varphi)
\end{aligned}
$$

where S is the % sand, C is the % clay, and φ is the porosity (vol.%).

Effective Hydraulic Conductivity

The effective hydraulic conductivity can be obtained from measurements or from saturated hydraulic conductivity prediction techniques developed from physical and hydraulic soil properties and modified for management and natural conditions. If only soil texture classes are available, the saturated hydraulic conductivity can be obtained from Table 2 and, as a rule of thumb, the effective hydraulic conductivity can be taken as half of the saturated hydraulic conductivity. There are various techniques for estimating saturated hydraulic conductivity from soil properties (18, 19) or characteristics of the water retention curve (15, 20, 21). Other techniques are given in the hydraulic conductivity section.

The effective hydraulic conductivity is determined by modifying the saturated hydraulic conductivity to incorporate the effects of such factors as coarse fragments (>2 mm in size) (17), soil freezing (14), soil crusts (22), and macropores (23).

A practical approach to handling the various conditions is to set the effective hydraulic conductivity equal to the saturated hydraulic conductivity (K_s) times a macroporosity factor (12). Brakensiek and Rawls (1) developed two macroporosity factors for areas that do not undergo mechanical disturbance on a regular basis (for example, rangeland), and one for areas that do (for example, agricultural areas). For the bare area outside a plant canopy, the soil is assumed to be crusted and the effective hydraulic conductivity is equal to the saturated hydraulic conductivity (K_s) times a crust factor.

CONCLUSIONS

Defining infiltration parameters that represent the conditions at a point are well established; however, defining how these parameters vary both temporally and spatially have only been defined in a very elementary way and an elaborate research is needed in these areas.

REFERENCES

1. Brakensiek, D.L.; Rawls, W.J. Effects of Agricultural and Rangeland Management Systems on Infiltration. *Modeling Agricultural, Forest, and Rangeland Hydrology*; Am. Soc. Agric. Engrs: St. Joseph, MI, 1988; 247.
2. Rawls, W.J.; Brakensiek, D.L.; Soni, B. Agricultural Management Effects on Soil Water Processes: Part I. Soil Water Retention and Green-Ampt Parameters. Trans. Am. Soc. Agric. Engrs **1983**, *26* (6), 1747–1752.
3. Rawls, W.J.; Brakensiek, D.L.; Savabi, R. Infiltration Parameters for Rangeland Soils. J. Range Mgnt. **1989**, *42* (2), 139–142.

4. Skaggs, R.W.; Khaleel, R. Infiltration. In *Hydrologic Modeling of Small Watersheds*; Monogr. No. 5, Haan, C.T., Ed.; Am. Soc. Agric. Engrs: Joseph, MI, 1982; 4–166.
5. Kostiakov, A.N. On the Dynamics of the Coefficient of Water Percolation in Soils and on the Necessity for Studying it from a Dynamic Point of View for Purposes of Amelioration. Trans. Sixth Commun. Int. Soil Sci. Soc. **1932**, 17–21, Russian Part A.
6. Horton, R.E. An Approach Toward a Physical Interpretation of Infiltration-Capacity. Soil Sci. Soc. Am. J. **1940**, *5*, 399–417.
7. Holtan, H.N. A Concept for Infiltration Estimates in Watershed Engineering. U.S. Dept. of Agric. Bull. **1961**, *41–51*, 25.
8. Green, W.H.; Ampt, G.A. Studies on Soil Physics: 1. Flow of Air and Water Through Soils. J. Agric. Sci. **1911**, *4*, 1–24.
9. Philip, J.R. The Theory of Infiltration: 1. The Infiltration Equation and Its Solution. Soil Sci. **1957**, *83*, 345–357.
10. Morel-Seytoux, H.J.; Khanji, J. Derivation of an Equation of Infiltration. Water Resour. Res. **1974**, *10* (4), 795–800.
11. Smith, R.E.; Parlange, J.Y. A Parameter-Efficient Hydrologic Infiltration Model. Water Resour. Res. **1978**, *14* (3), 533–538.
12. Rawls, W.J.; Ahuja, L.R.; Brakensiek, D.L.; Shirmohammadi, A. Chapter 5: Infiltration and Soil Water Movement. In *Handbook of Hydrology*; Maidment, D.R., Ed.; McGraw-Hill: New York, 1993.
13. Rawls, W.J.; Brakensiek, D.L. Estimation of Soil Retention and Hydraulic Properties. In *Unsaturated Flow in Hydrologic Modeling*; Morel-Seytoux, H.J., Ed.; NATO ASI Series C: Mathematical and Physical Sciences; Vol. 275: Kluwer Academic Publishers: Dordrecht, The Netherlands, **1989**; 275–300.
14. Rawls, W.J.; Brakensiek, D.L. A Procedure to Predict Green Ampt Infiltration Parameters. *Advances in Infiltration*; Am. Soc. Agric. Engrs: St. Joseph, MI, 1983; 102–112.
15. Rawls, W.J.; Brakensiek, D.L.; Logsdon, S.D. Predicting Saturated Hydraulic Conductivity Utlizing Fractal Principles. Soil Sci. Soc. Am. J. **1993**, *57*, 1193–1197.
16. Rawls, W.J. Estimating Soil Bulk Density from Particle Size Analysis and Organic Matter Content. Soil Sci. **1983**, *135* (2), 123–125.
17. Brakensiek, D.L.; Rawls, W.J.; Stephenson, G.R. Determining the Saturated Hydraulic Conductivity of Soil Containing Rock Fragments. Soil Sci. Soc. Am. J. **1986**, *50* (3), 834–835.
18. Saxton, K.E.; Rawls, W.J.; Romberger, J.S.; Papendick, R.I. Estimating Generalized Soil Water Characteristics for Texture. Soil Sci. Soc. Am. J. **1986**, *50*, 1031–1036.
19. Rawls, W.J.; Brakensiek, D.L. Prediction of Soil Water Properties for Hydrologic Modeling. *Watershed Management in the Eighties*; ASCE: New York, 1985; 293–299.
20. Ahuja, L.R.; Bruce, R.R.; Cassel, D.K.; Rawls, W.J. Simpler Field Measurement and Estimation of Soil Hydraulic Properties and Their Spatial Variability for Modeling. *Proc. Modeling Agricultural Forest and Rangeland Hydrology*; ASAE Pub. 07-88, Am. Soc. Agr. Engrs: St. Joseph, MI, 1988; 19–33.
21. Rawls, W.J.; Gimenez, D.; Grossman, R. Use of Soil Texture, Bulk Density, and the Slope of the Water Retention Curve to Predict Saturated Hydraulic Conductivity. Trans. Am. Soc. Agric. Engrs **1998**, *41* (4), 983–988.
22. Rawls, W.J.; Brakensiek, D.L.; Simanton, J.R.; Kohl, K.D. Development of a Crust Factor for the Green–Ampt Model. Trans. Am. Soc. Agric. Engrs **1990**, *33*, 1224–1228.
23. Rawls, W.J.; Brakensiek, D.L.; Logsdon, S.D. Estimation of Macropore Properties for No Till Soils. Trans. Am. Soc. Agric. Engrs **1996**, *39* (1), 91–95.

INORGANIC CARBON ANALYSIS

I

L.R. Drees,
C.T. Hallmark
Texas A&M University, College Station, Texas, U.S.A.

INTRODUCTION

Soil inorganic carbon (SIC) exists primarily as carbonates of calcium (calcite, $CaCO_3$) and calcium plus magnesium (dolomite, $CaMg(CO_3)_2$). Carbonates may be pedogenic (i.e., formed in the soil) or lithogenic (i.e., inherited from calcareous parent materials). Calcite is dominant in pedogenic environments, although dolomite has been reported to occur under certain conditions (1). Many animal shells are also calcareous—often aragonite, which is another form of $CaCO_3$. Other unique carbonate minerals may exist in soil, but regional geologic settings or unique environmental conditions govern their existence.

Soil is a major reservoir of inorganic carbon, containing more carbon than the atmosphere and biosphere combined (2).[a] Estimates of SIC are important in evaluating carbon reserves, rates of carbon sequestration, or loss from the soil system. Also, the presence of carbonates has a major influence on soil chemical and physical reactions.

There are numerous procedures for SIC determination, and many excellent reviews exist (3–6). Each technique uses unique physical, chemical, or mineralogical properties of carbonates. It is not our intent to repeat the details of these procedures, but rather to present the fundamental basis of analysis and to focus on commonly used procedures in soil investigations.

CHEMICAL REACTIONS

Carbonates dissolve in strong acids. Ideal reaction equations are:

$$2H^+ + CaCO_3 \rightarrow Ca^{2+} + H_2O + CO_2(\uparrow) \qquad (1)$$

or

$$4H^+ + CaMg(CO_3)_2 \rightarrow Ca^{2+} + Mg^{2+} + 2H_2O + 2CO_2(\uparrow) \qquad (2)$$

[a]See also "Soil Inorganic Carbon, Global Stocks."

These reactions can be used to determine carbonates by measuring the weight loss due to carbonate dissolution, amount of CO_2 produced, or the amount of acid consumed in the reaction. The reaction rate of calcite (Eq. 1) is much greater than dolomite (Eq. 2) when particle size and surface area are standardized (6), which permits quantitative determination of the two minerals. For techniques that cannot differentiate calcite from dolomite, results are reported as $CaCO_3$ equivalent.

Gravimetric Techniques

The reaction of carbonates with a strong acid releases CO_2, resulting in a weight loss (Eqs. 1 and 2). The weight loss is a direct measure of carbonates (3, 5).

$$\%CaCO_3 \text{ equivalent} = \frac{\text{g } CO_2 \text{ lost}}{\text{g oven–dried soil}} \times 227.3 \qquad (3)$$

Weight gained by trapping evolved CO_2 gas with an absorbent is another common technique for carbonate determination (6). There are two methods of generating CO_2 for gravimetric techniques: 1) treating carbonates with acid (Eqs. 1 and 2); and 2) heating the sample (dry combustion) to 1000°C after organic carbon removal to release carbonate CO_2 (7) in the reaction

$$CaCO_3 \xrightarrow{\Delta} CaO + CO_2(\uparrow) \qquad (4)$$

In both instances, evolved CO_2 is passed through a series of traps to absorb water and other reaction constituents before the gas passes through a Nesbitt absorption bulb containing NaOH to absorb CO_2 and Mg perchlorate to retain the water generated in the reaction (6)

$$2NaOH + CO_2 \rightarrow Na_2CO_3 + H_2O \qquad (5)$$

The calculations are the same as in Eq. 3, except that weight gain replaces weight loss. For the acid reaction, if the weight gain/loss is monitored over time, dolomite content can be determined due to its slower reaction rate. However, it is not possible to differentiate calcite and dolomite by the dry-combustion method (7).

Volume/Pressure Techniques

Measurements of gas evolution are relatively rapid and are often used to determine soil carbonates. The common techniques used are a volumetric measure of evolved CO_2 (constant pressure) and an increase in pressure (constant volume). For the constant pressure procedure (commonly called the Chittick procedure), soil is treated with a strong acid, and the volume of released CO_2 is directly correlated with carbonate content (6, 8). Fig. 1 illustrates a constant pressure apparatus. A finely ground sample is weighed to the nearest 0.1 mg and placed into a decomposition flask. By a stopcock (*S* in Fig. 1), the reaction vessel is closed to the atmosphere just before introducing the acid.

The leveling bulb and burette contain a saturated salt solution with a red indicator dye for easy reading (3). As the acid is added, the sample is stirred to provide a uniform reaction, and the leveling bulb is constantly adjusted to maintain a slight negative pressure. After carbonates have reacted, the volume of solution displaced is measured.

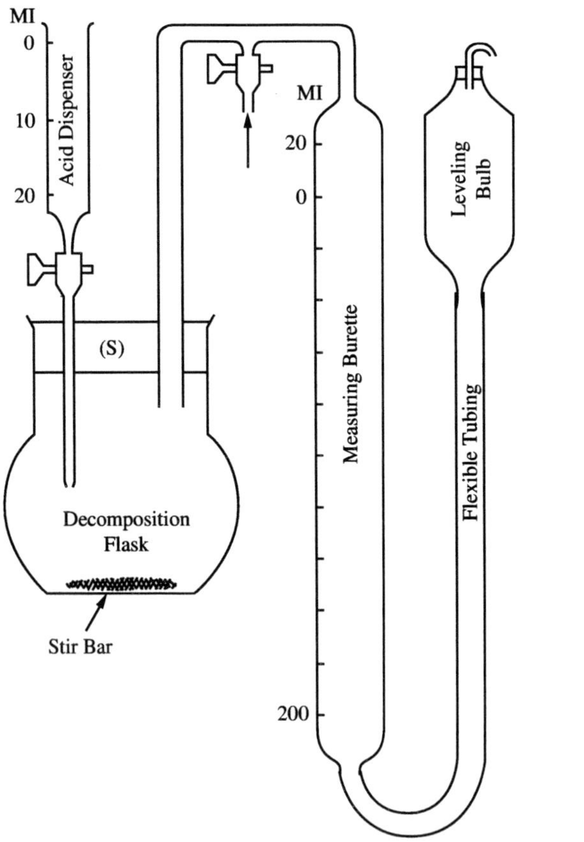

Fig. 1 Constant pressure apparatus for carbonate determination. (From Ref. 8.)

Timing of the reaction is important because calcite and dolomite have different reaction rates. The volume change after 30 seconds is often used as a measure of calcite and any volume increase between 30 seconds and 2 hours is attributed to dolomite. Adjustments are made for changes in temperature and atmospheric pressure during the reaction. Calculations to convert gas volume to weight percentage carbonates are given by several authors (5, 6).

When carbonates react with acid at constant temperature and volume, the gas released causes an increase in gas pressure that is directly related to carbonate content (9–13). A pressure gauge, pressure transducer, or manometer is used to monitor change in pressure over time. Pressure transducers can be connected to a personal computer for instantaneous data acquisition. Pressure change over time can be used to quantify calcite and dolomite and has been shown to be highly sensitive and rapid (13).

The change of volume or pressure with CO_2 evolution has certain limitations. The solubility of CO_2 in acid varies with temperature, type of acid used, acid concentration, and barometric pressure (3, 6). Calibration with pure calcite and dolomite standards is required in both the constant pressure and constant volume system for accurate results.

Titrimetric Techniques

Carbonates are the only major soil constituents that readily react with acids. The amount of acid consumed (Eqs. 1, 2) during reaction with carbonates can be determined by titrating the unreacted acid to a common endpoint with a base. Normally, 50 ml of standardized 0.5N HCl is carefully introduced to 5–25 g of soil, and the soil–acid mixture is gently boiled for 5 minutes, then allowed to cool. The suspension is filtered and washed to remove all acid. The filtrate is then titrated with standardized 0.25N NaOH to the phenolphthalein endpoint (3, 14) to determine the amount of acid consumed:

$$\%CaCO_3 \text{ equivalent} = 100 \times \frac{(\text{Meq HCl added} - \text{Meq NaOH used})}{\text{g oven-dried soil}} \times 0.05\,\text{g } CaCO_3/\text{Meq} \qquad (6)$$

X-RAY DIFFRACTION

Carbonates are crystalline minerals that yield characteristic x-ray diffraction peaks. The intensity of the diffraction peak should be proportional to the quantity of carbonate present. Accuracy is dependent on carbonate crystallinity, particle size, chemical composition, and sample orientation (15). Excellent results can be attained using mineral standards (calcite or dolomite) and an internal reference standard (16). A calibration curve is constructed that plots the internal standard:mineral standard intensity ratio against the weight percentage of the mineral standard (16). In most instances, carbonate-containing samples and reference standards must be ground to produce a uniform, small particle size. Increased grinding time generally yields a decrease in error (17).

X-ray diffraction has proved to be a reliable semi-quantitative method for the determination of dolomite in carbonate rocks and sediments even without an internal standard (18). Peak intensities corrected for absorption also achieved accurate results in carbonate rocks (19), but it is unclear how this method would work for samples of low carbonate concentration. Due to slight differences in diffraction intensity between carbonates, it is recommend that reference carbonates be as similar as possible to those contained in the soil being analyzed (15).

Magnesium may substitute for Ca within the calcite structure up to approximately 10 mole %. X-ray diffraction can differentiate pure calcite from Mg-substituted calcite whereas other techniques cannot. Small amounts of Mg substituted within the calcite structure result in slight shifts in diffraction peaks that has been calibrated with the amount of Mg substitution (20).

THIN SECTIONS

Carbonate minerals can usually be readily identified in thin sections by the high birefringence of carbonates and the microstructure of shells. Thin section techniques are time consuming because sample preparation is lengthy and quantification relies on point count determinations (21), which are laborious. Accuracy is dependent on carbonate quantity, distribution, particle size, and how well the section is representative of the sample as a whole. As particle size decreases, positive identification becomes more difficult (17) with a lower practical limit of detection of approximately 20 μm. For low concentrations (<10%), quantification becomes problematic. Despite limitations, microscopic examinations of soil material may differentiate pedogenic from lithogenic carbonates (22), which can have implications in soil genesis studies.

Calcite and dolomite often look similar in thin sections. Dolomite typically has rhombohedral habits, but this cannot be relied upon as a differentiating criterion. Chemical staining techniques have been used successfully to differentiate carbonate species (15, 17, 23–25).

X-RAY FLUORESCENCE AND MICROPROBE

X-ray fluorescence determines the chemical composition of a bulk sample, whereas microprobe is more site specific. Interpretation is based on assigning chemical composition to specific minerals. There are limitations in this assignment, since other minerals contain Ca and Mg, and carbon approaches the detection limit of many instruments. In microprobe analyses, common stoichiometric correction factors may result in errors up to 20% (26). These are probably not the best techniques for carbonate determination, especially in low concentrations, for they are indirect and do not employ characteristic properties of carbonate minerals. However, if carbonate concentrations are above 10–20% and other analyses indicate that there are none to few other Ca- or Mg-bearing minerals, x-ray fluorescence and microprobe analysis of total Ca and/or Mg may be good indicators of carbonates.

Microprobe analysis is particularly useful in determining Ca/Mg ratios in dolomite or Mg-substituted calcite of individual crystals, as it is site specific. However, it is not as efficient in bulk sample analysis as other techniques relying on carbonate reaction.

CONCLUSION

Numerous methods for determining SIC exist. The nature of the soil investigation and the intended use of carbonate data govern the technique of choice. Although highly sophisticated instruments can be used, the simple titrimetric technique (Eq. 6) and the change in gas volume or pressure upon reacting carbonates with acid (Fig. 1) remain popular. Such techniques are simple and accurate, requiring commonly available laboratory apparatus and reagents. The change in gas pressure or volume has the advantage of differentiating between calcite and dolomite. The change in gas pressure technique may gain some favor because a pressure transducer can be connected to a personal computer for instantaneous results.

REFERENCES

1. Kohut, C.; Muehlenbachs, K.; Dudas, M.J. Authigenic Dolomite in a Saline Soil in Alberta, Canada. Soil Sci. Soc. Am. J. **1995**, *59*, 1499–1504.
2. Drees, L.R.; Wilding, L.P.; Nordt, L.C. Reconstruction of Inorganic Carbon Sequestration Across Broad Geoclimatic Regions. In *Soil Carbon Sequestration and the Greenhouse Effect*; Lal, R., and McSweeney, K., Eds.; SSSA Special

Pub. No. 57. Soil Sci. Soc. Am.: Madison, Wisconsin, 2001; 155–172.

3. Allison, L.E.; Moodie, C.D. Carbonate. In *Methods of Soil Analysis. Part 2. Chemical and Microbiological Properties*; Black, C.A., Ed.; Soil Sci. Soc. Am.: Madison, Wisconsin, 1965; 1379–1396.
4. Nelson, R.E. Carbonate and Gypsum. In *Methods of Soil Analysis. Part 2. Chemical and Microbiological Properties*; Page, A.L., Ed.; Soil Sci. Soc. Am.: Madison, Wisconsin, 1982; 181–197.
5. Goh, T.B.; St. Arnaud, R.J.; Mermut, A.R. Carbonates. In *Soil Sampling and Methods of Analysis*; Carter, M.R., Ed.; Can. J. Soil Sci. Lewis Pub.: Boca Raton, Florida, 1993; 177–185.
6. Loeppert, R.H.; Suarez, D.L. Carbonate and Gypsum. In *Methods of Soil Analysis. Part 3. Chemical Methods*; Sparks, D.L., Page, A.L., Eds.; Soil Sci. Soc. Am.: Madison, Wisconsin, 1996; 437–474.
7. Rabenhorst, M.C. Determination of Organic and Carbonate Carbon in Calcareous Soils Using Dry Combustion. Soil Sci. Soc. Am. J. **1988**, *52*, 965–969.
8. Dreimanis, A. Quantitative Gasometric Determination of Calcite and Dolomite by Using Chittick Apparatus. J. Sediment. Petrol. **1962**, *32*, 520–529.
9. Presley, B.J. A Simple Method for Determining Calcium Carbonate in Sediment Samples. J. Sediment. Petrol. **1975**, *45*, 745–746.
10. Schink, J.C.; Stockwell, J.H.; Ellis, R.A. An Improved Device for Gasometric Determination of Carbonate in Sediment. J. Sediment. Petrol. **1979**, *49*, 651–653.
11. Dunn, D.A. Revised Techniques for Quantitative Calcium Carbonate Analysis Using the Karbonat-Bombe and Comparisons to Other Quantitative Carbonate Analysis Methods. J. Sediment. Petrol. **1980**, *50*, 631–636.
12. Jones, G.A.; Kaiteris, P.A. Vacuum-Gasometric Technique for Rapid and Precise Analysis of Calcium Carbonate in Sediments and Soils. J. Sediment. Petrol. **1983**, *53*, 655–660.
13. Evangelou, V.P.; Whittig, L.D.; Tanji, K.K. An Automated Monometric Method for Quantitative Determination of Calcite and Dolomite. Soil Sci. Soc. Am. J. **1984**, *48*, 1236–1239.
14. Richards, L.A., Ed. *Diagnosis and Improvement of Saline and Alkali Soils*, U.S. Dept. Agr. Hbk. 60; U.S. Salinity Laboratory: Washington, DC, 1954; 159.
15. St. Arnaud, R.J.; Mermut, A.R.; Goh, T.B. Identification and Measurement of Carbonate Minerals. In *Soil Sampling and Methods of Analysis*; Carter, M.R., Ed.; Can. J. Soil Sci. Lewis Pub.: Boca Raton, Florida, 1993; 737–743.
16. Ulas, M.; Sayin, M. Quantitative Determination of Calcite and Dolomite in Soils by X-Ray Diffraction. J. Soil Sci. **1984**, *35*, 685–691.
17. Gensmer, R.P.; Weiss, M.P. Accuracy of Calcite/Dolomite Ratios by X-Ray Diffraction and Comparisons with Results from Staining Techniques. J. Sediment. Petrol. **1980**, *50*, 626–629.
18. Royse, C.F.; Wadell, J.S.; Petersen, L.E. X-Ray Determination of Calcite-Dolomite: An Evaluation. J. Sediment. Petrol. **1971**, *41*, 483–488.
19. Fang, J.H.; Zevin, L. Quantitative X-Ray Diffractometry of Carbonate Rocks. J. Sediment. Petrol. **1985**, *55*, 611–613.
20. Hutchison, C.S. *Laboratory Handbook of Petrographic Techniques*; John Wiley & Sons: New York, 1974; 527.
21. Drees, L.R.; Ransom, M.D. Light Microscopic Techniques in Quantitative Soil Microscopy. In *Quantitative Methods in Soil Mineralogy*; Amonette, J.E., Zelazny, L.W., Eds.; SSSA Misc. Pub. Soil Sci. Soc. Am.: Madison, Wisconsin, 1994; 137–176.
22. West, L.T.; Drees, L.R.; Wilding, L.P.; Rabenhorst, M.C. Differentiation of Pedogenic and Lithogenic Carbonate Forms in Texas. Geoderma **1988**, *43*, 271–287.
23. Warne, S.St.J. A Quick Field or Laboratory Staining Scheme for the Differentiation of the Major Carbonate Minerals. J. Sediment. Petrol. **1962**, *32*, 29–38.
24. Dickson, J.A.D. A Modified Staining Technique for Carbonates in Thin Section. Nature **1965**, *205*, 587.
25. Lindholm, R.C.; Finkelman, R.B. Calcite Staining: Semiquantitative Determination of Ferrous Iron. J. Sediment. Petrol. **1972**, *42*, 239–242.
26. Lane, S.J.; Dalton, J.A. Electron Microprobe Analysis of Geologic Carbonates. Am. Mineral. **1994**, *79*, 745–749.

INORGANIC CARBON: CLIMATE AND TIME

Lee Nordt
Baylor University, Waco, Texas, U.S.A.

I

INTRODUCTION

Soil inorganic carbon (SIC) is contained in solid, solution, and gaseous phases with the regional and global distribution dependent, in large part, on the influence of climate and time (1). Common sources for the solid SIC phase are limestone (calcite and dolomite), carbonate-rich airborne dust, and unconsolidated calcareous parent materials such as alluvium, loess and till. The gaseous phase consists of soil-respired CO_2 that is generated by plant and animal respiration, and by the influx of atmospheric CO_2 into the upper part of the soil profile. The CO_2 phase combines with water to produce a solution enriched in H_2CO_3 and HCO_3^-. Carbonic acid also enhances weathering of calcareous parent materials and calcium-bearing silicates that generates Ca^{2+} and CO_3^{2-} [or $Ca(HCO_3^-)_2$] by-products in the solution phase. Dissolution is enhanced in the upper soil profile where there is a greater flux of water, higher CO_2 production from respiration, and a lower pH.

Recombination of ions from solution into a solid SIC phase leads to the formation of pedogenic carbonate. Formation typically occurs below the zone of maximum biological activity where CO_2 production is lower and pH is higher. Further, because of evapotranspiration and drying at the terminus of the wetting front, the ionic concentration of Ca_2^+ and CO_3^{2-} may exceed the solubility product of $CaCO_3$. In association with a shallow groundwater table, capillary rise in soils can also contribute to the concentration of Ca^{2+} and CO_3^{2-} ions, and, ultimately, to the formation of pedogenic carbonate. Also, the role of micro-organisms and roots in the formation of calcium carbonate is now recognized, but its importance and extent are largely unknown (2). Under well-drained conditions, and in soils with pHs greater than 6.5, $CaCO_3$ is the dominant SIC solid species.

When pedogenic forms of carbonate are visible in the field, the associated soil horizons are commonly described as Bk (filaments, soft masses, nodules) or Bkm (indurated). Although not formally recognized, there is evidence that disseminated pedogenic carbonate accumulates in near-surface horizons not visible in the field (3).

SIC AND CLIMATE

The global distribution of soil CO_2, HCO_3^-, and $CaCO_3$ is determined, in part, by climate, particularly in response to rainfall and temperature. The concentration of soil CO_2 is commonly 10 to 100 times greater than at the soil-atmosphere interface (365 ppmV) because of biotic respiration and organic matter decomposition (4). Soil CO_2 concentrations are greater in warm-humid climates where rainfall and respiration rates are higher, and decrease systemically to hot- or cold-arid climates where rainfall and respiration rates are lower. Soil respiration rates and CO_2 concentrations typically increase with elevation in arid to semiarid regions in response to more rainfall, cooler temperatures, and less evapotranspiration (5). In areas where soil CO_2 concentrations are relatively low (arid), atmospheric CO_2 can penetrate into the soil to depths of as much as 50 cm (6). Increased greenhouse gases and global warming could in some areas increase soil respiration rates (7).

As CO_2 dissolves in soil water, the inorganic carbon species H_2CO_3 is formed, which rapidly dissociates to HCO_3^- in soils with pH above 6.5 (8). The production of HCO_3^- will be greater in warm-humid climates, especially where there are high respiration rates or rapid weathering of carbonate-rich parent materials. The fate of HCO_3^- in leaching soils is typically to deep groundwater and to rivers (9). In arid to semiarid regions where evapotranspiration is high and rainfall is low, HCO_3^- may become part of the SIC pool as pedogenic carbonate. Elevated soil CO_2 levels generated in some areas by global warming could enhance HCO_3^- production from increased interaction of water and CO_2, leading to lower soil pH and more intense weathering of primary and other secondary minerals (7).

Retallack (10) recently provided a robust data set quantifying the relationship between rainfall and depth to the top of Bk horizons in warm-temperate climates (Fig. 1). This relationship indicates that at less than 100 mm of mean annual precipitation, insufficient soil moisture is available for weathering and for the production of visible forms of pedogenic carbonate. In contrast, most soils are leached of carbonate within the upper 1.5 to 2 m when

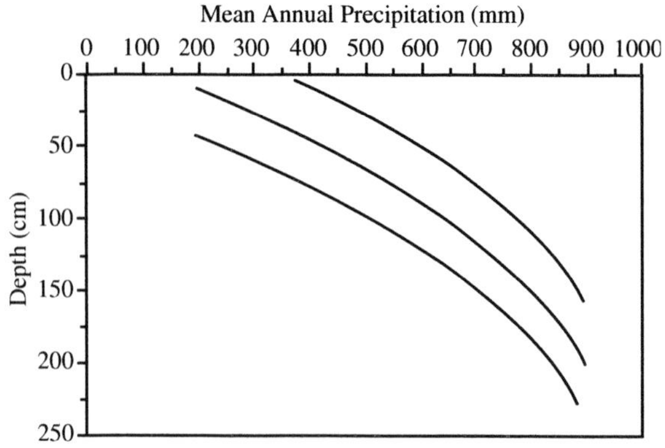

Fig. 1 Relationship between mean annual precipitation (P) in mm and depth to top of Bk horizon (D) in cm. The relationship is described by the equation $P = 139.6 - 6.388D - 0.01303D^2$, with a correlation coefficient of 0.79 and a standard error of ±33 cm (10). Results derived from well-drained and loamy, calcareous soils in warm-temperate climates principally for Stages II and III carbonate morphologies.

mean annual precipitation exceeds 1000 mm. Using a national soil taxonomic data base, the rainfall to depth of pedogenic carbonate relationship has been brought into question (11). Careful control is needed to ensure that climate is the primary soil-forming factor in using this relationship for paleoenvironmental reconstruction.

SIC AND TIME

In the initial stages of soil formation, the production and concentration of CO_2 likely increases over decades to centuries as pioneer plant communities evolve into climax plant communities. Soil CO_2 production may decrease with time in response to vegetation disturbances created by increased aridity, soil erosion, over-grazing, or plowing (7).

In general, soil HCO_3^- in solution will increase or decrease in proportion to the production of soil CO_2. However, if continued weathering leads to greater concentration of H^+, a more acidic soil environment may produce more H_2CO_3 at the expense of HCO_3^-. Another long-term mechanism for decreasing HCO_3^- in soil solution is related to the weathering and depletion of parent material carbonate.

Gile and colleagues (12) proposed a model of carbonate formation recognized as morphological stages for desert climates. For the model, it is assumed that a sufficient supply of Ca^{2+} and HCO_3^- ions are provided to form all stages of carbonate development. In addition, minimal climatic change during the interval of pedogenesis is ideal. The rate at which carbonate development passes through the individual stages is also dependent on parent material texture.

With these caveats in mind, the carbonate stages are designated Stage I, II, III, and IV from least to most developed. In non-gravelly parent materials, it can take 25,000 to 75,000 years to progress from Stage I (filaments) to Stage III (weakly indurated mass) in the formation of Bk horizons (Fig. 2). Stage IV represents nearly complete plugging of macropores forming a laminar cap and Bkm horizon. In desert environments this process can take as long as 400,000 years (Fig. 2).

Machette (13) proposed two additional morphological stages reflecting the further passage of time and enhanced geomorphic activity. Stage V is also recognized as a Bkm horizon, but with a laminar cap greater than 1 cm thick. Stage VI is characterized by multiple episodes of brecciation and recementation and is thought to have formed over a period of 2 million years.

Similar stages of carbonate development form in soils with gravelly parent materials (12). Stage I results from the accumulation of carbonate pendants on the bottom of gravels. Stage II forms with nearly complete encasement of the gravels with a coat of carbonate. The more advanced stages have similar morphologies to the carbonate that accumulates in non-gravelly parent materials. However,

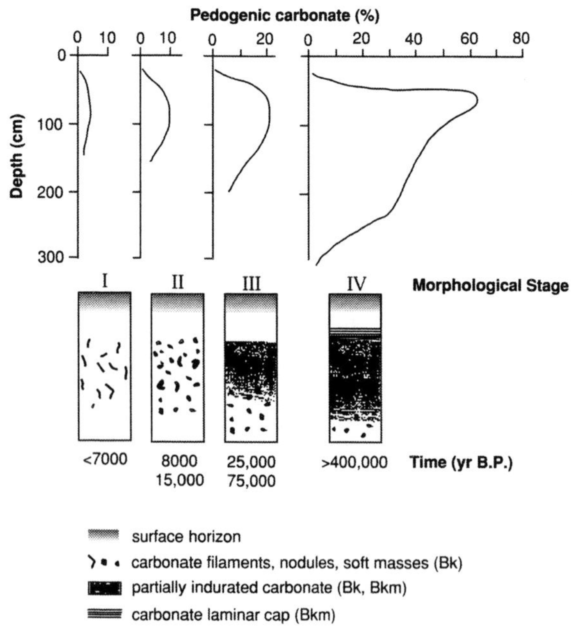

Fig. 2 Morphological stages of pedogenic carbonate with time in non-gravelly parent materials of arid to semiarid regions of the southwestern United States. (From Ref. 12.) Not shown are advanced carbonate morphologies typifying Stages V and VI.

because of reduced inherent porosity, the first four stages can be attained within 25,000 to 50,000 years of soil formation.

In soils formed from indurated and soft limestone, the upper surface tends to undergo numerous in situ dissolution-precipitation cycles that generate limestone porosity, unique pedogenic carbonate morphological fabrics, and eventually a laminar cap. Rabenhorst and colleagues (14) present a model showing 5 stages of carbonate development in this system for semiarid to subhumid climates. Initial stages involve the accumulation of pedogenic carbonate around macrovoids either weathered into limestone or along fracture planes. Advanced stages culminate in complete plugging of both macro and micropores along with the formation of a laminar cap.

SIC: INTEGRATION OF CLIMATE AND TIME

As shown in Fig. 3, the morphological characteristics of soil carbonate generally reflect the combined effects of climate and time. In this idealized example, it is assumed that the soils are loamy, well drained, calcareous, and formed in warm-temperate climates. Horizontally, the effects of time on soil development for a constant climate can be observed. In contrast to arid to semiarid climates discussed previously, carbonate development may never pass through Stage I in humid climates before net leaching of carbonate by-products begins to intensify. Here, A-E-Bt horizon sequences quickly form at the expense of incipient Bk and Bw horizons. In transitional subhumid climates Bk horizons may eventually evolve into Bt horizons in the upper profile with Bk horizons persisting at depth.

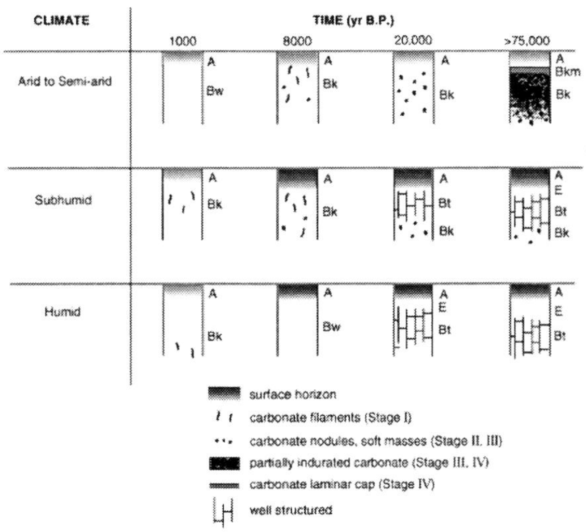

Fig. 3 Integrative effects of climate and time on horizonation and pedogenic carbonate development in soils formed from calcareous parent materials in well-drained, loamy soils. (From Ref. 9.)

Observing Fig. 3 vertically across climatic regions while holding time conzstant provides another view of these complex relationships. At 1000 years, incipient Bk horizons form quickly in subhumid and humid climates. At the endpoint of weathering at >75,000 years, A-Bkm horizons may form and persist in arid to semiarid climates with well-developed A-E-Bt horizons forming in subhumid and humid climates.

CONCLUSION

Knowing the formation and distribution of SIC components is important for the mapping and classification of soils, understanding carbonate equilibria reactions, better interpreting of the role of soils in the global carbon cycle, and understanding feedback mechanisms that soil may provide in a future greenhouse world.

REFERENCES

1. Birkeland, P.W. *Soils and Geomorphology*; Oxford University Press: New York, 1999.
2. Monger, H.C.; Daugherty, L.A.; Lindemann, W.C.; Liddell, C.M. Microbial Precipitation of Pedogenic Calcite. Geology **1991**, *19*, 997–1000.
3. Nordt, L.C.; Hallmark, C.T.; Wilding, L.P.; Boutton, T.W. Quantifying Pedogenic Carbonate Accumulations Using Stable Carbon Isotopes. Geoderma **1998**, *82*, 115–136.
4. Schlesinger, W.H. *Biogeochemistry: Analysis of Global Change*, 1st Ed.; Academic Press: San Diego, California, 1991; 1–443.
5. Amundson, R.G.; Chadwick, O.A.; Sowers, J.M. A Comparison of Soil Climate and Biological Activity Along An Elevation Gradient in the Eastern Mojave Desert. Oecologia **1989**, *80*, 395–400.
6. Cerling, T.E. The Stable Isotopic Composition of Modern Soil Carbonate and Its Relationship to Climate. Earth and Planetary Science Letters **1984**, *71*, 229–240.
7. Suarez, D.L. Impact of Agriculture on CO_2 as Affected by Changes in Inorganic Carbon. In *Global Climate Change and Pedogenic Carbonates*; Lal, R., Kimble, J.M., Eswaran, H., Stewart, B.A., Eds.; Lewis Publishers: Boca Raton, Florida, 2000; 257–272.
8. Schwab, A.P. The Soil Solution. In *Handbook of Soil Science*; Sumner, M.E., Ed.; CRC Press: Boca Raton, Florida, 2000; B85–B120.
9. Nordt, L.C.; Wilding, L.P.; Drees, L.R. Pedogenic Carbonate Transformations in Leaching Soil Systems: Implications for

the Global C Cycle. In *Global Climate Change and Pedogenic Carbonates*; Lal, R., Kimble, J.M., Eswaran, H., Stewart, B.A., Eds.; Lewis Publishers: Boca Raton, Florida, 2000; 43–64.

10. Retallack, G.J. The Environmental Factor Approach to the Interpretation of Paleosols. In *Factors of Soil Formation: A Fiftieth Anniversary Retrospective*; Amundson, R., Harden, J., Singer, M., Eds.; Special Publication Number 33; Soil Science Society of America: Madison, Wisconsin, 1994; 31–64.
11. Royer, D.L. Depth to Pedogenic Carbonate as a Paleoeprecipitation Indicator? Geology **1999**, *27*, 1123–1126.
12. Gile, L.H.; Hawley, J.W.; Grossman, R.B. *Soils and Geomorphology in the Basin and Range Area of Southern New Mexico-Guidebook to the Desert Project*, Memoir 39; New Mexico Bureau of Mines and Mineral Resources: Socorro, New Mexico, 1981; 1–222.
13. Machette, M.N. Calcic Soils of the Southwestern United States. In *Soils and Quaternary Geomorphology of the Southwestern United States*; Weide, D.L. Ed.; Special Paper 203; Geological Society of America: Boulder, Colorado, 1985; 1–21.
14. Rabenhorst, M.C.; West, L.T.; Wilding, L.P. Genesis of Calcic and Petrocalcic Horizons in Soils Over Carbonate Rocks. In *Occurrence, Characteristics, and Genesis of Carbonate, Gypsum, and Silica Accumulations in Soils*; Nettleton, W.D., Ed.; Publication Number 26; Soil Science Society of America: Madison, Wisconsin, 1991; 61–74.

INORGANIC CARBON: COMPOSITION AND FORMATION

H. Curtis Monger
New Mexico State University, Las Cruces, New Mexico, U.S.A.

Larry P. Wilding
Texas A&M University, College Station, Texas, U.S.A.

INTRODUCTION

It has become increasingly common for "soil inorganic carbon" to mean soil carbonate mineral carbon, mainly $CaCO_3$. In the strict sense, however, inorganic carbon not only encompasses carbon in carbonate minerals, but also carbon in the carbonic acid system (1). The carbonic acid system includes gaseous carbon dioxide ($CO_{2(g)}$), aqueous carbon dioxide ($CO_{2(aq)}$), carbonic acid ($H_2CO_{3(aq)}$), bicarbonate ion ($HCO^-_{3(aq)}$), and carbonate ion ($CO^{2-}_{3(aq)}$).

COMPOSITION

In the soil solution, as with other solutions, the interaction of these species can be represented by the following reaction (1):

$$CO_{2(g)} + H_2O_{(l)} = H_2CO_{3(aq)} = HCO^-_{3(aq)} + H^+_{(aq)} = CO^{2-}_{3(aq)} + 2H^+_{(aq)} \qquad (1)$$

Cations, such as Ca^{2+}, Mg^{2+}, Fe^{2+}, Mn^{2+}, and Na^+, precipitate with HCO_3^- (which is the dominant anion between pH 6.5 and 10.5) and CO_3^{2-} (which is the dominant anion above pH 10.5) to form a variety of carbonate minerals. The reaction of $Ca^{2+}_{(aq)}$ with $HCO^-_{3(aq)}$ to form calcite is illustrated below:

$$Ca^{2+}_{(aq)} + 2HCO^-_{3(aq)} = CaCO_{3(s)} + CO_{2(g)} + H_2O_{(l)} \qquad (2)$$

There are about 60 carbonate minerals, which in addition to calcite, include aragonite ($CaCO_3$), dolomite [$CaMg(CO_3)_2$], siderite ($FeCO_3$), magnesite ($MgCO_3$), rhodocrosite ($MnCO_3$), cerussite ($PbCO_3$), and malachite [$CuCO_3Cu(OH)_2$]. In soil, the overwhelmingly abundant carbonate mineral is calcite (2). In unique soil environments, however, other carbonate minerals have been found, such as pedogenic siderite and dolomite (3).

Individual carbonate crystals of pedogenic origin are generally too small to be seen with the unaided eye. Yet when concentrated together, their combined presence takes on a white color with a variety of macroscopic forms. These forms include carbonate filaments (also called mycelia, pseudomycelia, and threads), films, coatings, soft spheroidal segregations (white eyes), nodules, cylindroids, concretions, glaebules, and veins. Soil fabric which is impregnated with carbonate to the point that it occurs as an essentially continuous medium has been termed "k-fabric" (4). Stages of morphogenetic carbonate accumulation, in which progressively greater amounts of carbonate occur in progressively older soils, are important chronologic indicators. The calcic and petrocalcic horizon are diagnostic horizons in Soil Taxonomy (5). Calcic horizons generally contain greater than 15% carbonate by weight. Petrocalcic horizons are indurated forms of calcic horizons. Examples of these horizons are shown in Figs. 1a and b. Examples of carbonate crystals as viewed with optical microscopy and scanning electron microscopy are shown in Figs. 1c and d.

Dissolution of carbonates in soil systems can be represented by the following reaction (Eq. 3). In humid regions, soluble products of this weathering reaction flux through the vadose zone into groundwater, or precipitate as pedogenic carbonates deep in the soil or geologic system. In arid regions, soluble products precipitate at relatively shallow depths as a result of sparse rainfall and insufficient leaching.

$$HCO^-_{3(aq)} + H^+_{(aq)} + CaCO_{3(s)} = Ca(HCO_3)_{2(aq)} \qquad (3)$$

FORMATION

Being located in an arid, semiarid, or subhumid climate is the primary factor that controls carbonate formation. In many areas, the boundary between carbonate-accumulating soil and noncarbonate-accumulating soil is about 500 mm (20 in.) mean annual rainfall (6). This relationship

Fig. 1 Examples of inorganic carbon as it exists in the field and under the microscope. The white horizon is a calcic horizon in (a), and petrocalcic horizon in (b). The small golden crystals in (c) are calcite crystals coating sand grains; the black region is a pore space as it appears in cross-polarized light. A calcified fungal filament (cf) viewed with scanning electron microscopy is shown in (d).

is confounded, however, by the effects of soil temperature, soil drainage, nature and properties of the parent material (e.g., soil texture, carbonate content, carbonate mineralogy, and porosity), soil drainage, landform position, geomorphic stability, and effectiveness of precipitation (rainfall intensity and duration). Hence, there are many examples in humid and subhumid environments where soil carbonate persists in the soil system at depths inconsistent with regional models. In humid regions, for example, inorganic carbon persists as calcite or dolomite detritus in soils derived from certain parent materials (e.g., calcareous loess, till, outwash, alluvial deposits, sedimentary and metamorphic rocks). In seasonally wet soils, carbonate can accumulate in upper subsoils from capillary rise of bicarbonates via evaporative pumping from shallow groundwater (7). In addition, carbonate minerals can occur in wetland soils which commonly contain soluble carbonates, bicarbonates or carbonic acid depending on the pH of the local environment.

Pedogenic vs. Geogenic Carbonate

Many soils develop in calcareous parent materials. For these soils, it has been a challenge quantifying carbonate that formed in the soil profile vs. carbonate mechanically inherited from parent material. Carbonate formed in the soil profile has been termed “secondary,” “authigenic,” or “pedogenic” (4, 8). On the other hand, carbonate mechanically inherited from parent material has been termed “primary”, “geogenic”, or “lithogenic” (2, 9). Criteria for distinguishing pedogenic from geogenic carbonates involve the scrutiny of both field and laboratory evidence. Field evidence, for example, includes differences in the presence of marine fossils, carbonate morphology (such as nodules, pendants, and laminar caps which indicate pedogenic), and distribution patterns with depth, where, for example, a carbonate horizon of pedogenic origin is overlain and underlain by noncalcareous soil. Laboratory evidence includes comparing mineralogy, particle size, microfabric, and $^{13}C/^{12}C$ ratios of carbonate with unknown origin to those of carbonate with known geogenic origin (10–12).

Models of Carbonate Formation

There are several processes that cause carbonates to form in soil. Excluding geologic processes, such as lacustrine and deep groundwater cementation that preserves the original sedimentary structure, the formation pedogenic carbonate can broadly be placed into four models—per descensum, per ascensum, in situ, and biogenic models.

The per descensum model

The per descensum model accounts for carbonate formation resulting from downward moving meteoric water and can be subdivided into three types. First is the dissolution of pre-existing carbonates in the upper profile, their vertical translocation, and their precipitation in the subsoil. This model was invoked to explain why progressively shallower carbonates occur in progressively drier climates (13). Later, this per descensum model was used as the basis for calculating the number of wetting-fronts required to leach carbonates to a particular depth (14). In both cases, it was assumed that carbonate was uniformly distributed in parent material at the beginning of pedogenesis.

Second is the case in which pedogenic carbonate forms in soils with noncalcareous parent materials. Unlike the model described before, noncalcareous parent material does not have carbonate uniformly distributed throughout the profile at the beginning of pedogenesis. In southern New Mexico, for example, prominent calcic and petrocalcic horizons occur in soils with rhyolite alluvium as parent material. This alluvium would yield low amounts of calcium if the rhyolite particles were thoroughly decomposed, which they were not (15). Therefore, atmospheric additions, another per descensum model, was judged to be the source of carbonates (15). Initially calcareous dust was measured and considered to be the source of carbonate. Later it was realized that Ca^{2+} in rain was an additional, and even larger source of Ca^{2+} for reaction with soil HCO_3^- to form carbonates (15). Building on these per descensum concepts, compartmental models have been constructed that compute the depth, amount, and distribution of pedogenic carbonate as a function of climate and time (16, 17).

Third, in addition to vertical illuviation within a soil profile, lateral, downslope migration of the soil solution containing soluble products of carbonate is another per descensum model. In this case, carbonate is thought to precipitate after carbonate-charged waters migrate from upslope positions to lower landscape positions (18).

The per ascensum model

The per ascensum model accounts for carbonate formation resulting from bottom–up movement. A primary example is the capillary rise of Ca^{2+} and bicarbonate from shallow water tables by evaporative pumping, which leads to the precipitation of carbonates in the upper subsoil (7). Moreover, chemical studies have shown that in some environments plants promote carbonate formation by transporting Ca^{2+} upward to the land surface from subsoil, rock, and groundwater (19).

The in situ model

Third, in the in situ model, pedogenic carbonate is the result of in-place dissolution and reprecipitation of bedrock composed of marine carbonate (20). Limestone, for instance, is progressively transformed into pedogenic carbonate as a result of short-range carbonate dissolution and reprecipitation proximal to the depth of the upper contact with limestone. This is a rather unique method to form pedogenic carbonates where the total carbonate content of the zone of enriched pedogenic products is less than the carbonate content of the limestone originally. These pedogenic zones have a much higher macro- and micro-porosity than the limestone.

In addition to marine carbonates, the in situ model also includes carbonate formation resulting from in-place chemical weathering of Ca-bearing igneous rock. Upon release into the soil solution by weathering, Ca^{2+} precipitates with bicarbonate formed from the reaction of water with CO_2 generated by root and microbial respiration. In many cases, however, igneous parent material has been considered as an insufficient source of Ca^{2+} and hence external sources, such as atmospheric additions of Ca^{2+}, have been sought (6).

The biogenic model

Fourth, some plants, microorganisms, and termites produce calcium carbonate. Evidence that various plants play a direct role in carbonate formation comes from the presence of euhedral calcite crystals on plant roots (21). Moreover, several references in the Russian literature note carbonate formation by plant tissue and the downward translocation of these carbonates with wetting fronts (22). Evidence that some microorganisms precipitate carbonates is based on observations of calcified bacteria and fungal hyphae with electron microscopy and in vitro laboratory experiments (23, 24). Evidence that termites precipitate carbonate in certain environments is based on the studies of termite mounds in Africa and southeast Asia (25). Such mounds can be calcareous even though surrounding soils are noncalcareous, making the mounds attractive to native farmers who spread them over their agricultural fields (25).

CONCLUSION

The formation of pedogenic carbonate may be dominated by one of the models listed before or may involve several of the models working together in different magnitudes. Understanding pedogenic carbonate formation has been extremely useful for understanding relative ages of geomorphic surfaces and landscape evolution (15). A knowledge of pedogenic carbonate formation has also been useful for soil classification. Marbut (26), for instance, used the presence of carbonate as a criterion for the highest category of his soil classification—Pedocal (soils with carbonate accumulation) and Pedalfers (soil with Al and Fe accumulation). Today, studies of pedogenic carbonate have expanded to include questions about paleoclimate, paleoecology, paleoatmospheric composition, global carbon cycles and the greenhouse effect.

REFERENCES

1. Morse, J.W.; Mackenzie, F.T. *Geochemistry of Sedimentary Carbonates. Developments in Sedimentology 48*; Elsevier: Amsterdam, 1990.
2. Doner, H.E.; Lynn, W.C. Carbonate, Halide, Sulfate, and Sulfide Minerals. In *Minerals in Soil Environments*, 2nd Ed.; Dixon, J.B., Weed, S.B., Eds.; Soil Science Society of America: Madison, WI, 1989; 279–330.
3. Capo, R.C.; Whipkey, C.E.; Blachère, J.R.; Chadwick, O.A. Pedogenic Origin of Dolomite in a Basaltic Weathering Profile, Kohala Peninsula, Hawaii. Geology **2000**, *28*, 271–274.
4. Gile, L.H.; Peterson, F.F.; Grossman, R.B. The K Horizon—A Master Soil Horizon of Carbonate Accumulation. Soil Sci. **1965**, *99*, 74–82.
5. Soil Survey Staff, *Soil Taxonomy—A Basic System of Soil Classification for Making and Interpreting Soil Surveys*, 2nd Ed.; USDA Agriculture Handbook Number 436, U.S. Govt. Printing Office: Washington, DC, 1999.
6. Birkeland, P.W. *Soils and Geomorphology*, 3rd Ed.; Oxford University Press: New York, 1999, 430.
7. Sobecki, T.M.; Wilding, L.P. Formation of Calcic and Argillic Horizons in Selected Soils of the Texas Coast Prairie. Soil Sci. Soc. Am. J. **1983**, *47*, 707–715.
8. Pal, D.K.; Dasog, G.S.; Vadivelu, S.; Ahuja, R.L.; Bhattacharyya, T. Secondary Calcium Carbonate in Soils of Arid and Semiarid Regions of India. In *Global Climate Change and Pedogenic Carbonates*; Lal, R., Kimble, J.M., Eswaran, H., Stewart, B.A., Eds.; Lewis Publishers: London, 2000; 149–185.
9. West, L.T.; Wilding, L.R.; Rabenhorst, M.C. Differentiation of Pedogenic and Lithogenic Carbonate Forms in Texas. Geoderma **1987**, *43*, 271–287.
10. Rabenhorst, M.C.; Wilding, L.P.; West, L.T. Identification of Pedogenic Carbonates Using Stable Carbon Isotopes and Microfabric Analysis. Soil Sci. Soc. Am. J. **1984**, *48*, 125–132.
11. Drees, L.R.; Wilding, L.P. Micromorphic Record and Interpretations of Carbonate Forms in the Rolling Plains of Texas. Geoderma **1987**, *40*, 157–175.
12. Nordt, L.C.; Hallmark, C.T.; Wilding, L.P.; Boutton, T.W. Quantifying Pedogenic Carbonate Accumulations Using Stable Carbon Isotopes. Geoderma **1998**, *82*, 115–136.
13. Jenny, H.; Leonard, C.D. Functional Relationships Between Soil Properties and Rainfall. Soil Sci. **1934**, *38*, 363–381.
14. Arkley, R.J. Calculation of Carbonate and Water Movement in Soil from Climatic Data. Soil Sci. **1963**, *96*, 239–248.
15. Gile, L.H.; Hawley, J.W.; Grossman, R.B. *Soils and Geomorphology in the Basin and Range Area of Southern New Mexico—Guidebook to the Desert Project*; New Mexico Bureau of Mines and Mineral Resources, Memoir 39: Socorro, New Mexico, 1981; 222.
16. McFadden, L.D.; Tinsley, J.C. Rate and Depth of Pedogenic-Carbonate Accumulation in Soils: Formation and Testing of a Compartment Model. In *Soils and Quaternary Geology of the Southwestern United States*; Special Paper 203, Weide, D.L., Ed.; Geological Society of America: Boulder, CO, 1985; 23–42.
17. Marion, G.M.; Schlesinger, W.H. Quantitative Modeling of Soil Forming Processes in Deserts: The CALDEP and CALGYP Models. In *Quantitative Modeling of Soil-Forming Processes*; Bryant, R.B., Arnold, R.W., Eds.; Soil Sci. Soc. Am. Spec. Publ. 39: Madison, WI, 1994; 129–145.
18. Scharpenseel, H.W.; Mtimet, A.; Freytag, J. Soil Inorganic Carbon and Global Change. In *Global Climate Change and Pedogenic Carbonates*; Lal, R., Kimble, J.M., Eswaran, H., Stewart, B.A., Eds.; Lewis Publishers: London, 2000; 27–42.
19. Goudie, A. *Duricrusts in Tropical and Subtropical Landscapes*; Oxford Univ. Press: London, 1973; 174.
20. Rabenhorst, M.C.; Wilding, L.P. Pedogenesis on the Edwards Plateau, Texas: III. A New Model for the Formation of Petrocalcic Horizons. Soil Sci. Soc. Am. J. **1986**, *50*, 693–699.

21. Monger, H.C.; Gallegos, R.A. Biotic and Abiotic Processes and Rates of Pedogenic Carbonate Accumulation in The Southwestern United States—Relationship to Atmospheric CO_2 Sequestration. In *Global Climate Change and Pedogenic Carbonates*; Lal, R., Kimble, J.M., Eswaran, H., Stewart, B.A., Eds.; Lewis Publishers: London, 2000; 273–290.
22. Labova, E. *Soils of the Desert Zone of the USSR*; Israel Program for Scientific Translation: Jerusalem, 1967.
23. Phillips, S.E.; Milnes, A.R.; Foster, R.C. Calcified Filaments: An Example of Geological Influences in the Formation of Calcretes in South Australia. Aust. J. Soil Res. **1987**, *25*, 405–428.
24. Monger, H.C.; Daugherty, L.A.; Lindemann, W.C.; Liddell, C.M. Microbial Precipitation of Pedogenic Calcite. Geology **1991**, *19*, 997–1000.
25. Thorp, J. Effects of Certain Animals that Live in Soils. Sci. Monthly **1949**, *68*, 180–191.
26. Marbut, C.F. Subcommission II. Classification, Nomenclature, and Mapping of Soils in the United States. Soil Sci. **1928**, *25*, 61–71.

I

INORGANIC CARBON AND THE GLOBAL C CYCLE

William H. Schlesinger
Duke University, Durham, North Carolina, U.S.A.

INTRODUCTON

The *global C cycle* is defined as the exchange of carbon (C) among the atmosphere, seawater, land vegetation, and soil reservoirs (Fig. 1). Each year dead plant materials entering the soil are decomposed by soil microbes that return carbon dioxide (CO_2) to the atmosphere. If the amount of land vegetation remains the same, the amount of CO_2 removed from the atmosphere by plant growth each year is balanced by the amount of plant death and decomposition. Such a perfect balance, however, is seldom seen. Changes in the quantity of C in vegetation and soils play a major role in determining short- and long-term fluctuations in the concentration of CO_2 in Earth's atmosphere. A portion of the current atmospheric increase in CO_2, for example, is due to the destruction of vegetation and the disturbance of soils by humans.

MEAN RESIDENCE TIME

For comparative purposes, biogeochemists calculate the *mean residence time* (MRT), or the amount of time C resides in each pool of the global C cycle before circulating to the others. For instance, the average molecule of CO_2 spends about 5 years in the atmosphere before it enters the terrestrial biosphere or the oceans. The average atom of C spends about 10 years in vegetation and 35 years in soil organic matter (SOM) before it returns to the atmosphere as CO_2. In comparison, the circulation of C in the oceans is rather sluggish; the average C atom spends hundreds of years in the sea, where it is found predominantly as dissolved bicarbonate (HCO_3^-). Human activities have the greatest impact on pools with short mean residence times.

WEATHERING

Most studies of soil C focus on SOM, which globally contains nearly 2300×10^{15} g of C (2). The release of CO_2 by the microbial decomposition of SOM is one of the largest fluxes in the global C cycle (Fig. 1). However, soils also contain C in various inorganic forms—CO_2 held in the soil pore spaces, bicarbonate dissolved in soil waters, and calcium carbonate ($CaCO_3$) as a soil mineral. CO_2 in the soil pores is largely derived from the respiration of plant roots and soil microbes, which varies as a function of soil moisture and temperature. However, even when the respiration rate is very high, the amount of C in soil gases and the soil solution is not large enough to contribute materially to the total C content of soils globally. In contrast, the amount of C held in soil carbonates is quite large, totaling about 750 to 1000×10^{15} gC (Fig. 1) (3, 4). The vast majority of this carbonate is found in the world's arid and semiarid lands, where inorganic C may exceed soil organic C by a factor of 10 (5).

In some soils, $CaCO_3$ and other carbonates (e.g., dolomite) are derived from the parent rocks from which the soils have formed. However, additional carbonate may form as a result of the release of Ca from the chemical weathering of rocks and the precipitation of $CaCO_3$ when water is lost from the soil by evaporation and plant uptake. In the case of silicate parent minerals, such as plagioclase (calcium feldspar), the relevant reaction of weathering is

$$CaAl_2Si_2O_8 + 3H_2O + 2CO_2 \rightarrow Ca^{2+} + 2HCO_3^- + Al_2Si_2O_5(OH)_4 \quad (1)$$

Here Ca is released from the silicate mineral by the weak solution of carbonic acid that forms when CO_2 dissolves in water.

Declining soil concentration of either CO_2 or water precipitates carbonate via the reaction

$$Ca^{2+} + 2HCO_3^- \rightarrow CaCO_3 \downarrow + H_2O + CO_2 \quad (2)$$

While the weathering process occurs most rapidly when plant activity increases the concentration of CO_2 in the soil pore space, the precipitation reaction occurs during seasonal periods of drought. Carbonate formed in the soil, known as secondary or *pedogenic* carbonate, can be distinguished from carbonate inherited from parent materials by examining thin sections and the isotopic ratio (^{13}C vs. ^{12}C) of the C in carbonate minerals.

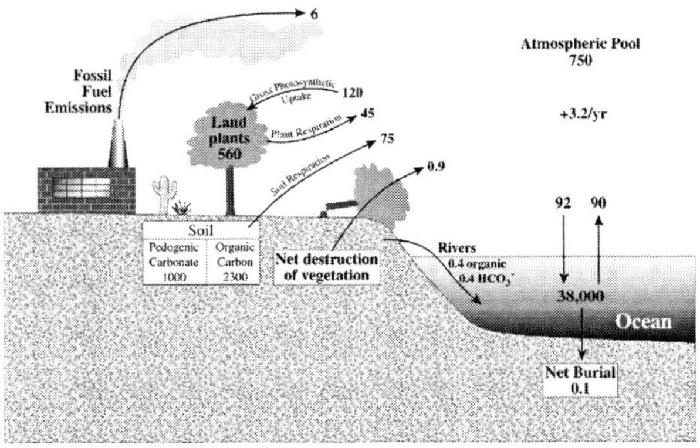

Fig. 1 The global C cycle, showing the size of reservoirs (10^{15} gC) and the annual flux between them (10^{15} gC/yr). (Modified from Ref. 1.)

ATMOSPHERIC DERIVATION

In most areas, the Ca content of pedogenic carbonate is derived from the atmosphere and is deposited as a constituent of rain and dust (6, 7). However, because the Ca carried in the atmosphere is ultimately derived from rock weathering in some upwind area, the two afore–mentioned reactions are general (8). Across deserts of the southwestern United States, the formation of pedogenic carbonate is closely related to mean annual rainfall (Fig. 2).

If we know the rate of formation of pedogenic carbonate, usually expressed in grams per square meter per year ($g/m^2/yr$), and the total amount of carbonate in the soil profile, we can calculate the length of time it has taken the current quantity of soil pedogenic carbonate to accumulate. In the Mojave Desert of California, radiocarbon and

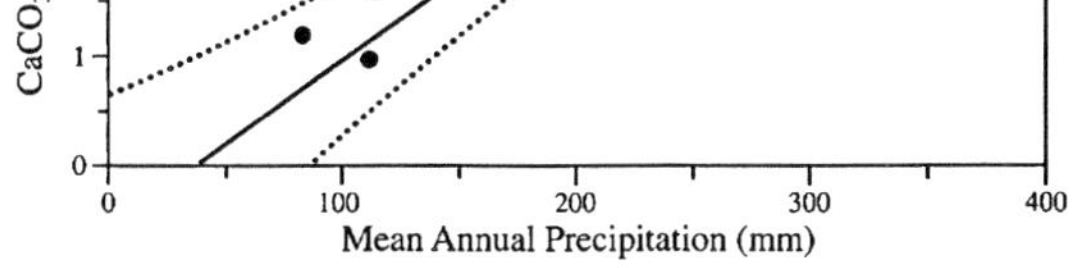

Fig. 2 The rate of formation of pedogenic carbonate in arid soils of the southwestern United States as a function of the modern precipitation at each site. (Modified from Ref. 9.)

uranium-thorium dating show that pedogenic carbonate found in the upper 1.5 meters of soils, derived from silicate materials, accumulated over a period of up to 20,000 years (10). During that period, pedogenic carbonate formed at rates ranging from 1.0 to 3.5 $g/m^2/yr$, with some indications of greater rates during the Pleistocene, when rainfall was greater in this region. The global MRT of soil pedogenic carbonate is about 85,000 years, making this C pool much less dynamic than SOM, in which the global MRT for C is about 35 years (Fig. 1).

While CO_2 is sequestered from the atmosphere during the formation of pedogenic carbonate from silicate parent materials, no such net storage occurs when pedogenic carbonate forms from carbonate parent materials (11). Overall, the formation of soil pedogenic carbonates is less effective than the formation of SOM in the storage of atmospheric CO_2 (12). This is disappointing, of course, to those who view increasing the formation of pedogenic carbonate in desert soils as a means of slowing the rise of CO_2 in Earth's atmosphere and reducing global warming.

EFFECT OF HUMAN ACTIVITIES

Human activities, such as the irrigation of agricultural soils in arid regions, can alter the accumulation of pedogenic carbonate in soils. The groundwater used for irrigation is often extracted from subsurface environments, where CO_2 concentration is much higher than in Earth's atmosphere, and often contains high concentrations of dissolved Ca. Applying such water to arid lands, precipitates dissolved Ca in the soil, forming $CaCO_3$, and releases CO_2 to the atmosphere, via the reaction in Eq. 2. Precipitation of calcite is also favored when large amounts of gypsum ($CaSO_4 \cdot 2H_2O$), a ready source of Ca, are used to remediate dryland soils. The formation of pedogenic carbonate in arid, agricultural soils, as a result of these human activities makes a small contribution to the current increase in atmospheric CO_2 (13).

Human activities leading to the formation of acid rain also affect soil carbonates. If the acidity in rainfall is derived from sulfuric acid (H_2SO_4), then CO_2 is released when the rain falls on carbonate-rich soils. The reaction is

$$CaCO_3 + H_2SO_4 \rightarrow Ca^{2+} + SO_4^{2-} + CO_2\uparrow + H_2O \quad (3)$$

and gypsum ($CaSO_4 \cdot 2H_2O$) precipitates as the soil dries. Although most fossil fuels contribute small amounts of sulphur to the atmosphere, the global amount of CO_2 derived from this reaction is relatively small compared to the direct release of CO_2 from fossil fuel combustion (Fig. 1).

CONCLUSION

Although the amount of pedogenic carbonate in world soils is quite large, perhaps as much as 1000×10^{15} gC, this pool of C is relatively sluggish. The long MRT of the pedogenic carbonate pool ensures that it will not become a major sink or source of atmospheric CO_2 over the next several centuries.

REFERENCES

1. Schlesinger, W.H. *Biogeochemistry: An Analysis of Global Change*; Academic Press: San Diego, California, 1997.
2. Jobbagy, E.G.; Jackson, R.B. The Vertical Distribution of Soil Organic Matter and Its Relation to Climate and Vegetation. Ecol. Appl. **2000**, *10*, 423–436.
3. Eswaran, H.; Reich, P.F.; Kimble, J.M.; Beinroth, F.H.; Padmanabhan, E.; Moncharoen, P. Global Carbon Stocks. In *Global Climate Change and Pedogenic Carbonates*; Lal, R., Kimble, J.M., Eswaran, H., Stewart, B.A., Eds.; Lewis Publishers: Boca Raton, Florida, 2000; 15–25.
4. Batjes, N.H. Total Carbon and Nitrogen in Soils of the World. Eur. J. Soil Sci. **1997**, *47*, 151–163.
5. Schlesinger, W.H. Carbon Storage in the Caliche of Arid Soils: A Case Study from Arizona. Soil Sci. **1982**, *133*, 247–255.
6. Capo, R.C.; Chadwick, O.A. Sources of Strontium and Calcium in Desert Soil and Calcrete. Earth Planet. Sci. Lett. **1999**, *170*, 61–72.
7. Chiquet, A.; Michard, M.; Nahon, D.; Hamelin, B. Atmospheric Input vs. In Situ Weathering in the Genesis of Calcretes: An Sr Isotope Study at Gálvez (Central Spain). Geochimica et Cosmochimica Acta. **1999**, *63*, 311–323.
8. Naiman, Z.; Quade, J.; Patchett, P.J. Isotopic Evidence for Eolian Recycling of Pedogenic Carbonate and Variations in Carbonate Dust Sources Throughout the Southwest United States. Geochimica et Cosmochimica Acta. **2000**, *64*, 3099–3109.
9. Marion, G.M. Correlation Between Long-Term Pedogenic $CaCO_3$ Formation Rate and Modern Precipitation in Deserts of the American Southwest. Quaternary Res. **1989**, *32*, 291–295.
10. Schlesinger, W.H. The Formation of Caliche in Soils of the Mojave Desert, California. Geochimica et Cosmochimica Acta. **1985**, *49*, 57–66.
11. Monger, H.C.; Martinez-Rios, J. Inorganic Carbon Sequestration in Grazing Lands. In *The Potential of U.S. Grazing Lands to Sequester Carbon and Mitigate the Greenhouse Effect*; Follett, R.F., Kimble, J.M., Lal, R., Eds.; CRC Press: Boca Raton, Florida, 2000; 87–118.
12. Chadwick, O.A.; Kelley, E.F.; Merritts, D.M.; Amundson, R.G. Carbon Dioxide Consumption During Soil Development. Biogeochemistry **1994**, *24*, 115–127.
13. Suarez, D.L. Impact of Agriculture on CO_2 as Affected by Changes in Inorganic Carbon. In *Global Climate Change and Pedogenic Carbonates*; Lal, R., Kimble, J.M., Eswaran, H., Stewart, B.A., Eds.; Lewis Publishers: Boca Raton, Florida, 2000; 257–272.

INORGANIC CARBON: GLOBAL STOCKS

I

L.P. Wilding
Texas A&M University, College Station, Texas, U.S.A.

L.C. Nordt
Baylor University, Waco, Texas, U.S.A.

J.M. Kimble
United States Department of Agriculture (USDA), Lincoln, Nebraska, U.S.A.

INTRODUCTION

In recent years more work has focused on the role of soil organic carbon (SOC) (1–5), than soil inorganic carbon (SIC) (6, 7) in the global carbon cycle. One of the constraints to quantifying SIC stocks is the incomplete global dataset. Thus, we can estimate global SIC stocks, but with considerable uncertainty. Further, there is difficulty differentiating lithogenic from pedogenic carbonates because most routine chemical methods determine the total quantity of carbonates in a soil sample without regard to origin. The amount of pedogenic carbonate formed in soils is typically estimated from visible segregations in the field, from chemical analysis in the laboratory under the assumption that the carbonate is pedogenic (8), or from stable carbon isotope analysis to differentiate lithogenic from disseminated pedogenic carbonate components (9). Gile and Grossman (10) consider that much of the calcium in carbonates results from carbonate dusts or from calcium in rainwater. Chadwick and Capo (11) believe that at least 95% of the Ca in carbonates of arid and semiarid regions is derived from dusts. Many soils also have lithogenic carbonate inherited from alluvial, till, and bedrock parent materials. It is possible that as little as 10% of the total SIC reported in soils is formed from the authigenic weathering of calcium-bearing minerals.

Perhaps even less well understood are the flux rates of SIC stocks (e.g., soluble carbonates, bicarbonates, and carbonic acid) from soil systems into the vadose zone, geologic substrata, ground water aquifers, rivers, lakes, and oceans. The flux rates of inorganic carbon stocks require knowledge of soil chronology coupled with mass balance reconstruction analyses to determine the magnitude of gains or losses (12, 13). This long-term transitory SIC stock is believed to have important impacts on soil carbon sequestration over geologic time periods up to 10 times longer than the SOC stocks (12–14). Also, the biogenesis of SIC stocks by various organisms (e.g., plant roots, blue green algae, termites, plankton, etc.), while well documented, is not as well quantified.

MORPHOLOGY OF SIC STOCKS

The major SIC stocks are in petrocalcic and calcic horizons. It is estimated that calcic horizons contain 209 Pg C (10^{15} g C), petrocalcic horizons 104 Pg C, and calcareous soils not meeting the taxonomic criteria for a named carbonate horizon about 636 Pg C (6). Disseminated carbonates (micritic or fine-grained carbonate particles) and segregated forms of soft masses, threads, films, or filaments are often present in soils at quantities insufficient to be recognized as calcic horizons (8). Such disseminated and segregated carbonates may be the most chemically reactive form of SIC.

MAGNITUDE OF SIC STOCKS

The amount of soil global inorganic carbon is large, with estimates ranging from 1576 to 695 Pg C (3, 15, 16). Eswaran et al. (17) recently revised the estimates down to between 949 to 940 Pg C, which appear to be more reasonable (see Tables 1 and 2). Uncertainties in these numbers result because the depth of the soil represented has been arbitrarily set at 1 m, values represent combined lithogenic and pedogenic sources, and data for SIC stocks that are partitioned by soil taxa and global areal extent are approximate. Discrepancies in areal extents of soil orders/suborders, and the SIC stocks in Tables 1–4 are

Table 1 Soil inorganic carbon (SIC) and total soil carbon (TC) in soil orders of soil taxonomy

	TC		SIC			
Order	**Pg**	**% Global**	**Pg**	**% Global**	**SIC/TC % of Order**	**SIC/TC % Global**
Alfisols	201	8.1	43	4.5	21.4	1.7
Andisols	20	0.8	0	0.0	0.0	0.0
Aridisols	515	20.9	456	48.5	88.5	18.5
Entisols	353	14.3	263	28.0	74.5	10.6
Gelisols	323	13.1	7	0.8	2.2	0.3
Histosols	180	7.2	0	0.0	0.0	0.0
Inceptisols	224	9.1	34	3.6	15.2	1.4
Mollisols	237	9.5	116	12.3	48.9	4.7
Oxisols	126	5.1	0	0.0	0.0	0.0
Spodosols	64	2.6	0	0.0	0.0	0.0
Ultisols	137	5.6	0	0.0	0.0	0.0
Vertisols	64	2.6	21	2.3	32.8	0.9
Miscellaneous	24	1.0	0	0.0	0.0	0.0
Total	2468	100.0	940	100.0		38.1

(Adapted from Ref. 17.)

due to rounding errors and different sources of published information.

SIC STOCKS RELATED TO SOIL TAXONOMY ORDERS AND SUBORDERS

Areal stocks of SIC in soil orders are presented in Table 1 and partitioned by the suborders of soil taxonomy in Table 2 (8). About 38% of the total carbon (TC) stocks are accounted for by SIC, with Aridisols, Entisols, and Mollisols contributing the most (Table 1). Nearly half of the SIC occurs in Aridisols (Table 1) with over 85% distributed among Argids, Calcids, and Cambids (Table 2). Entisols contain 28% of SIC (Table 1) with 77% associated with Orthents (Table 2). Mollisols contain the next largest amount of SIC (≈12%) with over 80% stored in Ustolls. The Alfisols contribute nearly 5% of the SIC stocks with over 95% stored in Ustalfs and Xeralfs. These distributions of SIC stocks reflect heavily on the semiarid and drier climates, coupled with carbonate-rich parent materials.

Five of the soil orders contain no SIC stocks. These orders are the Andisols, Histosols, Oxisols, Spodosols, and Ultisols. They represent soils in subhumid to humid environments where carbonates have been leached into lower vadose zones, geologic substrata, and/or ground water aquifers. In these orders one would expect the parent materials to have been either noncalcareous or low in carbonate content.

The ratios of SIC/TC as a percentage of the order and as a global percentage are given in Table 1. It is noteworthy that nearly 90% of the TC is SIC in Aridisols, 75% of the TC is SIC in Entisols, and about 50% of the TC is SIC in Mollisols. This reflects environments that do not favor sequestration of SOC because the soils are youthful, well oxidized, and have primary productivity constrained by long periods of soil moisture stress.

SIC STOCKS RELATED TO SOIL MOISTURE CONDITIONS

Table 3 presents SIC and TC stocks according to global soil moisture and temperature conditions. Two soil moisture conditions account for 92% of the SIC stocks, namely arid (≈78%) and semiarid (≈14%). This is explained by the fact that these environments exhibit little soluble transport of pedogenic and lithogenic carbon stocks. In these environments, the proportion of TC that is SIC is ≈84% and ≈29%, respectively, indicating that accumulation of SOC is not as important in these dry climates as in humid and wetter regimes where SOC comprises ≈97% of the TC stock (Table 3). It is interesting to note that in Mediterranean climates with winter rain, SIC contributes ≈56% of the TC stock. Again, because of

Table 2 Listing of soil taxonomy orders, suborders, ice-free land areas, and soil inorganic carbon (SIC) stocks

		Area ice-free land[a]		SIC[b]	
Soil orders	**Soil suborders**	**10^6 km^2**	**%**	**Pg**	**%**
Alfisols		12.620	9.65	43	4.6
	Aqualfs	0.836	0.64	0	0.0
	Cryalfs	2.518	1.92	1	0.1
	Ustalfs	5.664	4.33	29	3.1
	Xeralfs	0.896	0.69	12	1.3
	Udalfs	2.706	2.07	1	0.1
Andisols		0.91	0.70	0	0.0
Aridisols		15.728	12.02	456	48.5
	Cryids	0.943	0.72	17	1.8
	Salids	0.890	0.68	21	2.2
	Gypsids	0.682	0.52	23	2.4
	Argids	5.407	4.13	135	14.4
	Calcids	4.872	3.73	165	17.6
	Cambids	2.931	2.24	95	10.1
Entisols		21.137	16.16	263	28.0
	Aquents	0.116	0.09	0	0.0
	Psamments	4.428	3.39	33	3.5
	Fluvents	2.860	2.18	27	2.9
	Orthents	13.733	10.50	203	21.6
Gelisols		11.26	8.60	7	0.8
	Histels	1.01	0.77	4	0.4
	Turbels	6.33	4.84	3	0.3
	Orthels	3.91	2.99	1	0.1
Histosols		1.53	1.17	0	0.0
Inceptisols		12.829	9.81	34	3.6
	Aquepts	3.199	2.45	2	0.2
	Cryepts	0.456	0.35		
	Ustepts	4.241	3.24	18	2.0
	Xerepts	0.685	0.52		
	Undepts	4.247	3.25	14	1.4
Mollisols		9.005	6.89	116	12.3
	Albolls	0.028	0.02	0	0.0
	Aquolls	0.118	0.09	1	0.1
	Rendolls	0.266	0.20	4	0.4
	Xerolls	0.924	0.71	15	1.6
	Cryolls	1.164	0.89	3	0.3
	Ustolls	5.244	4.01	94	9.9
	Udolls	1.261	0.96	0	0.0
Oxisols		9.810	0.75	0	0.0
Spodosols		3.350	2.56	0	0.0
Ultisols		11.052	8.45	0	0.0
Vertisols		3.160	2.42	21	2.3
	Aquerts	0.054	0.00	0	0.0
	Cryerts	0.014	0.01	0	0.0
	Xererts	0.098	0.08	1	0.1
	Torrerts	0.889	0.68	12	1.3
	Usterts	1.767	1.35	8	0.9
	Uderts	0.384	0.29	0	0.0
Miscellaneous land		18.405	10.0	0	0.0
Total		130.796	100.0	940	100.0

[a] (From Ref. 8.)

Table 3 Soil inorganic carbon (SIC) and total carbon (TC) stocks according to moisture conditions

	TC		SIC		SIC/TC	SIC/TC
Moisture condition	**Pg**	**% Global**	**Pg**	**% Global**	**% Region**	**% Global**
Permafrost	405	16.4	18	1.9	4.4	0.7
Arid	877	35.5	732	77.8	83.5	29.7
Mediterranean	90	3.6	50	5.4	55.6	2.0
Semiarid	471	19.1	134	14.2	28.5	5.5
Humid	539	21.9	4	0.5	0.7	0.2
Perhumid	85	3.4	2	0.2	2.4	0.1
Total	2472	100.0	946	100.0		38.2

(From Ref. 17.)

the semiarid condition the SOC pool is less important than SIC. From the above comments, it is not surprising that a little over 38% of the TC stock (2472 Pg C) is SIC (946 Pg C), with arid (30%) and semiarid (6%) regions serving as major contributors.

SIC STOCKS RELATED TO ECOLOGICAL REGIONS

Table 4 illustrates SIC and TC stocks partitioned by ecological regions. Boreal and temperate regions contribute ≈82% of the global SIC, with tropical regions ≈16% and tundra regions ≈2%. The SIC is much higher in the temperate region (55%) due to the larger extent of deserts. The small contributions of SIC to TC in tundra regions reflect cold temperatures favorable to SOC accumulation and dissolution of lithogenic SIC. Many of the tropical regions are more humid, and thus any SIC in the soil parent materials has been leached.

FLUX OF SIC STOCKS

An estimate of soluble carbonate fluxes can be made from mass balance reconstruction analysis of the soil carbonate phase. Two problems with this method are the lack of reliable estimates of soil chronology and assumptions of flux rates. The rates of soil genesis, and especially the more labile carbonate components, are known to be exponential functions, but the precise geometric rate function applicable is not known and would vary with different amounts of carbonate, carbonate mineralogy, leaching potential, soil temperature, soil permeability, and vegetative cover.

The magnitude of leaching of soluble carbonates and bicarbonates has been estimated globally from calcareous soils (12, 13). These projections are based on a flux rate of between 7 and $8\,g\,m^{-2}\,yr^{-1}$ from soil great groups likely to have leaching potentials sufficient to transport soluble products of carbonate weathering from the soil. Estimated results indicate that between 0.25 and nearly 0.40 Pg of SIC is lost per year globally in leaching calcareous environments. The major contributors to these

Table 4 Soil inorganic carbon (SIC) and total carbon (TC) stocks according to ecological regions

	TC		SIC		SIC/TC	SIC/TC
Ecological region	**Pg**	**% Global**	**Pg**	**% Global**	**% Region**	**% Global**
Tundra	405	16.4	18	1.9	4.4	0.7
Boreal	632	25.6	256	27.2	40.5	10.4
Temperate	873	35.4	518	55.1	59.3	21.0
Tropical	557	22.6	149	15.9	26.7	6.0
Total	2466	100.0	940	100.0		38.1

(From Ref. 17.)

leachates are the Entisol, Mollisol, and Alfisol soil orders. This value probably underestimates the actual flux of SIC stocks because it does not consider the amount of carbonic acid used to weather soil minerals or the amount that is flushed from soils of high leaching potentials, such as, Oxisols, Ultisols, and Spodosols, which occupy about 20% of the ice-free Earth's surface.

CONCLUSION

There is a very large global pool of SIC, but its formation, quantity, and dynamics are still uncertain. The major problems in determining SIC stocks are the incomplete global database and the difficulty in separating pedogenic from lithogenic carbonate sources. To provide better estimates of SIC stocks, more sampling and analysis are needed that involve good soil chronology, stable carbon isotopes to differentiate pedogenic from lithogenic carbonates, and techniques to quantify soluble SIC stocks.

REFERENCES

1. Lal, R.; Kimble, J.M.; Eswaran, H.; Stewart, B.A., Eds. *Soil and Global Change*; Advances in Soil Science, CRC Press, Inc.: Boca Raton, FL, 1995.
2. Lal, R., Kimble, J.M., Eswaran, H., Stewart, B.A., Eds. *Soil Processes and C Cycles*; Advances in Soil Science, CRC Press, Inc.: Boca Raton, FL, 1998.
3. Bouwman, A.F., Ed.; *Soils and the Greenhouse Effect*; John Wiley and Sons: New York, 1990.
4. Wisniewski, J.; Sampson, R.N.; Eds. Terrestrial Biospheric Carbon Fluxes: Quantification of Sinks and Sources of CO_2, 1993.
5. Lal, R., Kimble, J.M., Eswaran, H., Stewart, B.A., Eds. *Soil Organic Matter in Temperate Agroecosystems*; CRC Press: Boca Raton, FL, 1997.
6. Lal, R., Kimble, J.M., Eswaran, H., Stewart, B.A., Eds. *Global Climate Change and Pedogenic Carbonates*; Lewis Publishers: Boca Raton, FL, 2000.
7. Batjes, N.H. Total Carbon and Nitrogen in the Soils of the World. European Journal of Soil Science **1996**, *47*, 151–163.
8. Soil Survey Staff, *Soil Taxonomy: A Basic System of Soil Classification for Making and Interpreting Soil Surveys*, 2nd Ed.; USDA-NRCS Agriculture Handbook 436, U.S. Government Printing Press: Washington, DC, 1999.
9. Nordt, L.C.; Hallmark, T.C.; Wilding, L.P.; Boutton, T.C. Quantifying Pedogenic Carbonate Accumulations Using Stable Carbon Isotopes. Geoderma **1998**, *82*, 115–136.
10. Gile, L.H.; Grossman, R.B. *The Desert Project Soil Monograph*; USDA-SCS, National Technical Information Service: Springfield, VA, 1979.
11. Chadwick, O.A.; Capo, R.C. Partitioning Allogenic and Authigenic Sources of Carbonate in New Mexico Calcrete. Agronomy Abstracts, American Society of Agrononmy, Cincinnati, Ohio, November 7–12, 1993; Madison, WI, 1993; 295.
12. Nordt, L.C.; Wilding, L.P.; Drees, L.R. Pedogenic Transformations in Leaching Soil Systems: Implications for the Global Carbon Cycle. In *Global Climate Change and Pedogenic Carbonates*; Lal, R., Kimble, J.M., Eswaran, H., Stewart, B.A., Eds.; Lewis Publishers: Boca Raton, FL, 2000; 43–64.
13. Drees, L.R.; Wilding, L.P.; Nordt, L.C. Reconstruction of Inorganic and Organic Carbon Sequestration Across Broad Geoclimatic Regions. In *Soil Carbon Sequestration and the Greenhouse Effect*; Lal, R., Ed.; Soil Science Society of America Special Publication No 57: Madison, WI, 2001; 155–172.
14. Chadwick, O.A.; Kelly, E.F.; Merritts, D.M.; Amundson, R.G. Carbon Dioxide Consumption During Soil Development. Biogeochemistry **1994**, *24*, 115–127.
15. Eswaran, H.; Van den Berg, E.; Reich, P.; Kimble, J. Global Soil Carbon Resources. In *Soils and Global Change*; Advances in Soil Science, Lal, R., Kimble, J., Levine, E., Steward, B.A., Eds.; CRC Press, Inc.: Boca Raton, FL, 1995; 27–44.
16. Sombroek, W.G.; Nachtergaele, F.O.; Hebel, A. Amounts, Dynamics and Sequestering of Carbon in Tropical and Subtropical Soils. Ambio. **1993**, *22*, 417–426.
17. Eswaran, H.; Reich, P.F.; Kimble, J.M.; Beinroth, F.H.; Padmanabhan, E.; Moncharoen, P. Global Carbon Stocks. In *Global Climate Change and Pedogenic Carbonates*; Lal, R., Kimble, J.M., Eswaran, H., Stewart, B.A., Eds.; Lewis Publishers: Boca Raton, FL, 2000; 15–26.

I

INORGANIC CARBON: LAND USE IMPACTS

Donald L. Suarez
Salinity Laboratory, Agricultural Research Service-United States Department of Agriculture, Riverside, California, U.S.A.

INTRODUCTION

Land use affects soil inorganic carbon, but the changes are significant only over the long term. Both increases and decreases in carbon storage occur as a result of various management practices. In irrigated lands, the primary factors causing a decrease in inorganic carbon are high leaching and maintenance of elevated water content at or near the soil surface. These factors result in elevated carbon dioxide concentrations and thus increased dissolution of soil carbonates. Use of acidifying fertilizers such as ammonia and sulfur also act to reduce soil inorganic carbon. Practices that favor accumulation of carbonates in the soil include efficient irrigation with surface waters in arid and semiarid regions (leaching less than 30% of the applied water), irrigation with ground waters at elevated CO_2 concentrations, application of gypsum to alkaline soils, and use of nitrate fertilizer. Other factors that affect soil carbonate content include land clearing, cropping practices, and erosion.

Soil inorganic carbon constitutes a major carbon pool in the near surface environment. In arid regions, the inorganic carbon can comprise more than 90% of the total C in the soil. The major inorganic C mineral phases are calcite and dolomite. Both minerals are relatively insoluble, however, dolomite dissolution is much slower than calcite dissolution at the intermediate pH values relevant to soils. Also, dolomite does not readily precipitate under Earth surface conditions. As a result, dolomite content in soils will remain constant or decrease due to dissolution, while calcium carbonate content may either increase or decrease.

LAND CLEARING AND CROPPING

Land clearing generally results in increased water runoff and soil erosion. This process or any other process such as tillage that increases erosion serves to remove the surface soil horizons. Since these horizons are generally depleted in inorganic carbon relative to less weathered, deeper horizons, there is an apparent increase in the inorganic carbon content of the surface soil as a result of erosion. In terms of carbonate dissolution the impact of land clearing is not certain. After clearing there is increased runoff, thus decreased infiltration, favoring less dissolution of carbonates. This effect may be compensated by the decreased water consumption (lower evapotranspiration) after clearing, resulting in increased deep recharge (greater carbonate dissolution). Depending on how much biomass remains after clearing, there is likely a short-term increase in soil CO_2 followed by a longer-term reduction, favoring less carbonate dissolution in the soil. Land clearing and overgrazing in arid lands serve to increase wind erosion and to redistribute soil in the landscape. In this manner noncalcareous soils receive inputs of carbonates. This process increases net dissolution of carbonates in the landscape.

In humid environments inorganic carbon is leached from the soil. The elevated CO_2 concentrations in the soil enhance calcite solubility relative to Earth surface conditions. In humid environments carbonates are successively leached from the upper portions of the soil profile. Agricultural practices may serve to enhance or reduce the net removal of carbonates. Removal of vegetation from a site with practices such as tree harvesting or harvesting of forages serves to remove base cations and causes net acidification of the upper portions of the soil profile. If carbonates are present deeper in the soil, this acidification increases dissolution. The impact of removal of vegetation in humid environments with carbonates in the subsoil can be calculated by assuming that the net harvested alkalinity is compensated by an equal increase in carbonate dissolution in the subsurface.

FERTILIZATION

Since optimum plant growth is generally at a pH lower than that observed in untreated calcareous soils, acid fertilizers are commonly applied. Use of sulfur with

Encyclopedia of Soil Science

subsequent oxidation to sulfate results in acid release to the soil (2 mol of protons per mole of sulfur). Application of ammonia salts, with subsequent fixation into organic matter or oxidation to nitrous oxide or elemental nitrogen, also releases protons (2 and 1 mol of protons per mole of ammonia ions, respectively). This acidification will increase carbonate dissolution proportionately, and has a significant effect since ammonia salts are widely used.

Application of urea or ammonia gas should have no net effect on carbonate dissolution (upon oxidation to nitrous oxide or elemental nitrogen). In contrast, use of nitrate fertilizers serves to increase pH and thus reduce carbonate dissolution. Generally nitrate is not utilized on calcareous soils so the impact on inorganic carbon storage in soil is slight. The quantitative impact of fertilization on changes in inorganic carbon is not easily calculated, as it depends on the extent of N incorporation into organic matter, mineralization, the extent to which the harvested biomass is removed from the site, and the occurrence of carbonates in the subsurface. The addition of liming products, primarily calcite, are reported as 3.7 Tg C yr^{-1} in the U.S. for 1978 (1). This is a significant but temporary addition to the soil carbon pool as it is assumed that the majority of the material is applied to acid soils and thus it is readily dissolved.

HUMID REGION IRRIGATION

Irrigation in humid environments serves to increase the net recharge through the soils and thus should increase the removal of carbonates. These changes may be relatively difficult to detect in view of the limited amount of irrigation water added and the fact that irrigation in humid environments, although increasing rapidly, was very limited in the past. Field studies are needed to determine the impact of irrigation on changes in inorganic carbon storage in humid environments.

ARID REGION SOILS

Arid zone soils usually contain at least minor amounts of carbonates, even if classified as noncalcareous. In the absence of irrigation, there may be redistribution of carbonates within the soil but little net precipitation. The majority of the pedogenic calcite is reprecipitated calcite with relatively small amounts added as a result of mineral weathering. Significant amounts of carbonates are also added to the surface of arid land soils as dust. Calcite is leached from the upper part of the soil profile by dissolution into the infiltrating rain, and is mostly reprecipitated at depth after plant extraction of the available water.

Irrigation in arid and semiarid environments may result in a net increase or decrease in soil carbonate, depending on the water source and fraction of water applied that is leached (leaching fraction). There are two opposing effects. First, elevated CO_2 concentrations in the root zone relative to the atmospheric condition results in enhanced calcite solubility and dissolution. Second, plant water extraction and evaporation concentrate the salts into a smaller volume of water and enhance calcite precipitation. At low leaching fractions, the effect of concentration of salts due to plant water extraction and evaporation is greater than the enhanced CO_2 effect and there is net precipitation of calcite.

For a calcite saturated surface water such as the Colorado River, it is estimated (2) that at a leaching fraction of 0.1 there is net precipitation of 125 kg ha^{-1} yr^{-1} of C, based on water consumption of 1.2 m yr^{-1} and a CO_2 partial pressure of 3 kPa. Model simulations indicated that net precipitation of calcite occurred as the sum of the loss of carbonates in the upper portion of the root zone and precipitation of calcite in the lower portion. At high leaching fractions there is net dissolution of carbonates. Using a calcite equilibrium model it is predicted that at a leaching fraction of 0.4 there will be a net dissolution of carbonates of 70 kg ha^{-1} yr^{-1} of C, again, based on water consumption of 1.2 m yr^{-1} and CO_2 partial pressure of 3 kPa (3). In all instances there is a prediction of net dissolution in the upper portion of the soil profile and net precipitation in the lower portions of the profile. Using average leaching fractions for the western U.S., it is estimated that irrigation with surface waters on 12 million ha results in an increase in inorganic carbon of 1 Tg yr^{-1} (4), or 80 kg ha^{-1} yr^{-1}.

Irrigation with groundwater saturated with respect to calcite will result in precipitation of carbonates at almost all leaching fractions, since the irrigation water is equilibrated at the groundwater CO_2 partial pressure and is supersaturated upon degassing and application to the soil. Calcite saturated groundwater is used for irrigation on an estimated 3.12 million ha in the U.S. It is estimated that irrigation on these soils results in a net inorganic C precipitation of 1.3 Tg yr^{-1} or 420 kg ha^{-1} yr^{-1} (3). The above calculations are dependent on several assumptions regarding calcite equilibrium and soil CO_2 concentrations. Using a kinetic model for calcite precipitation and the measured CO_2 partial pressure in the groundwater in Palo Verde Valley there is prediction of no net change in inorganic carbon at a leaching fraction of 0.5. Consistent with these predictions the groundwater composition in Palo Verde Valley shows no evidence for net precipitation or dissolution of carbonates in the soil.

There is limited direct field evidence for the influence of irrigation on inorganic soil carbon and it is difficult to be certain that differences among sites are only related to changes in management. Researchers (4) observed a net decrease in the calcium carbonate content of three pairs of soil profiles taken from sites in Israel irrigated for approximately 40 yr as compared to nonirrigated sites. The estimated input of 4.40 m of water per year at those sites is contrasted with the yearly potential evapotranspiration of 1.93 m. The observed trend is qualitatively consistent with model predictions if we account for the input of rain and the estimated leaching fraction of 0.56. Isotopic evidence indicated that there was precipitation of pedogenic carbonate at depth despite a net decrease in carbonate content at depth.

In a study in the San Joaquin Valley in California researchers compared samples of a soil taken from irrigated and native vegetation sites (5). They also measured a net loss of carbonates attributed to 8 yr of irrigation. Net carbonate loss was estimated as 7×10^3 kg ha^{-1} y^{-1} (800 kg C ha^{-1} yr^{-1}). Leaching fractions at the site were not reported but this value corresponds to approximately 10 times greater dissolution than expected based on model simulations. However, another study by the same author (6) found no change in total carbonate when comparing pedons with native vegetation and those irrigated for 5–25 yr. In this instance both gypsum and sulfur were applied as amendments for reclamation. Gypsum would tend to increase precipitation of carbonates while sulfur would acidify the soil and cause net dissolution of carbonates. In a recent, more extensive study (7) on paired soil cores (irrigated and adjacent nonirrigated sites) from the lower Colorado River basin there were no observed changes in net inorganic carbon storage after 90 yr of irrigation, and no isotopic C shifts indicative of recrystallization.

SODIC SOIL RECLAMATION

Reclamation of sodic soils can result in either an increase or decrease in inorganic carbon in the soil. Gypsum application to a sodic and alkaline soil will increase the soil carbonate content, as the increased Ca will precipitate most of the soluble bicarbonate and carbonate. Application of sulfuric acid, sulfur, or green manuring all serve to dissolve soil carbonates. Green manuring as a reclamation practice consists of incorporating plant residues into the soil and leaching with water. The high CO_2 production is combined with restricted gas transport creating very high CO_2 concentrations in the soil, dissolving large amounts of calcium carbonate. It is estimated that this process can dissolve in the order of 400–800 kg ha^{-1} during a year of reclamation. Use of acid is currently a widespread and generally recommended practice to prevent emitter clogging in drip irrigation systems. This practice may result in total removal of carbonates within 10–20 yr, for soils with less than 3% carbonate content.

IMPACT ON ATMOSPHERIC CARBON DIOXIDE

Dissolution of carbonates in neutral to alkaline environments results in consumption of CO_2 gas and formation of aqueous HCO_3^-, while precipitation of carbonates results in release of CO_2. The net effect of dissolution or precipitation of soil carbonate on atmospheric CO_2 depends on the solution flow path. In regions irrigated with surface water the dissolution of carbonates results in a net C sink. However, the high alkalinity drainage water usually flows back to the river. The resultant degassing of carbonic acid and reprecipitation of carbonate in the river or reservoir releases CO_2 back into the atmosphere. If the water is recharged into deep aquifers the net soil flux is preserved. In acid environments, liming of soils results in CO_2 release to the atmosphere as there is little or no net alkalinity produced.

CONCLUSION

Land use practices have a long-term impact on soil inorganic carbon. Due to the large C pools in the soil these impacts are not generally observed in short-term studies. In humid environments the major anthropogenic impacts on inorganic C are liming of surface soils, use of NH_4 vs. nitrate fertilizer, removal of vegetation, and erosion. In semiarid and arid environments increased inorganic C is favored by the use of groundwater for irrigation and application of gypsum. Decreased inorganic C is favored by inefficient irrigation with surface water and application of NH_4 fertilizer. The net effect of irrigation on a global scale, neglecting the effects of fertilizer addition, is to increase soil inorganic C by 30 Tg C yr^{-1} as well as to release an equal amount of C to the atmosphere. Liming practices in humid regions throughout the world are estimated to have no net effect on inorganic soil C but release up to 85 Tg C yr^{-1} to the atmosphere.

REFERENCES

1. Voss, R.D. What Constitutes an Effective Liming Material. In *National Conference on Agricultural Limestone*; National Fertilizer Development Center: Muscle Shoals, AL, 1980; 52–61.
2. Suarez, D.; Rhoades, J. Effect of Leaching Fraction on River Salinity. J. Irr. Drainage Div. ASCE **1977**, *103* (2), 245–257.
3. Suarez, D. Impact of Agriculture on CO_2 as Affected by Changes in Inorganic Carbon. In *Global Climate Change and Pedogenic Carbonates*; Lal, R., Kimble, J., Eswaran, H., Stewart, B., Eds.; Lewis: Boca Raton, FL, 1999; 257–272.
4. Magaritz, M.; Amiel, A. Influence of Intensive Cultivation and Irrigation on Soil Properties in the Jordan Valley, Israel: Recrystallization of Carbonate Minerals. Soil Sci. Soc. Am. J. **1981**, *45*, 1201–1205.
5. Amundson, R.G.; Smith, V.S. Effects of Irrigation on the Chemical Properties of a Soil in the Western San Joaquin Valley, California. Arid Soil Res. Rehab. **1988**, *2*, 1–17.
6. Amundson, R.D.; Lund, L. The Stable Isotope Chemistry of a Native and Irrigated Typic Natrargid in the San Joaquin Valley of California. Soil Sci. Soc. Am. J. **1987**, *51*, 761–767.
7. Suarez, D.L. Impact of Agriculture on Soil Inorganic Carbon. Agron. Abstr., Madison, WI, 1998; 258–259.

I

INORGANIC CARBON MODELING

Leslie McFadden
University of New Mexico, Albuquerque, New Mexico, U.S.A.

Ronald Amundson
University of California, Berkeley, California, U.S.A.

INTRODUCTION

Virtually all carbon in soils of arid and semiarid regions of the world accumulates as pedogenic calcium carbonate (referred to subsequently as carbonate). The carbonate usually accumulates in layers that eventually attain the status of calcic horizons and, in much older soils, petrocalcic horizons. Numerous mechanisms for the accumulation of pedogenic carbonate in soils are recognized, but the most fundamental reason for accumulation is limited depth of soil-water movement and seasonally high evapotranspiration that favors precipitation of carbonate within the soil (1). Many studies of calcic soils in the past few decades demonstrate a close correspondence between the depth of pedogenic carbonate accumulation and modern annual precipitation (2–4), although recent studies show the relationship may be more complicated (5). Other studies also show progressive, time-dependent accumulation in many environments (1); these studies have led to the now generally accepted conceptual models of calcic soil development (6).

Development of a numerical model of carbonate accumulation, however, is a more challenging proposition, given the remarkably complex character of the soil system. Fortunately, certain aspects of calcic soils facilitate formulation of such numerical models. For example, the observed soil depth–climate relationship implies that, utilizing a sound strategy for simulation of water movement, determination of carbonate movement via solution transport is a reasonable proposition. In addition, a significant body of research shows that the majority of carbonate is derived from accumulated entrapped dust and Ca in rainwater (1). Finally, data pertaining to calcite geochemistry and dissolution rates in different environments are available and show that, in soils associated with typical ranges in soil CO_2, pH, and salinity, calcite is far more soluble than virtually all silicate minerals and has more rapid dissolution rates (7). Consequently, a relatively simple model for carbonate movement in a soil based on relations in the $CaCO_3$–H_2O–CO_2 system can be formulated that essentially ignores the more complex chemical reactions involving aluminosilicates. Research on the nature and composition of stable and unstable isotopes in pedogenic carbonate has also helped elucidate the nature of calcic soil development and improve the design for testing the results of numerical modeling.

THE COMPARTMENT MODEL AND SIMULATIONS OF PEDOGENIC CARBONATE ACCUMULATION

The compartment-model, or "box-model," approach to modeling of calcic soils accommodates continuously changing values among the interdependent variables that influence soil development. It enables integration of several factors that influence pedogenic carbonate accumulation and that can be explicitly considered in this model. These include soil–water movement and soil–water balance, changing soil CO_2 concentrations and temperature with depth and season, initial parent material composition, carbonate and soluble salt additions from external sources, and calcite reactant surface area. The soil profile is represented by a vertical sequence of compartments of arbitrary dimensions, with the initial characteristics of each compartment specified (i.e., texture, available water-holding capacity, pCO_2). A series of equations that enable forward modeling and simulation of evolving carbonate depth functions using the box-model approach can be derived on the basis of consideration of the factors indicated above. For example, the solubility of calcite is derived from the following equation, after Drever (8):

$$m^3Ca^{2+} = (pCO_2K_1K_{cal}K_{CO_2})/(4K_2\gamma Ca^{2+} + \gamma^2HCO_3^-) \quad (1)$$

where K_{cal} is the calcite solubility product and K_1, K_2 and K_{CO_2} are dissociation constants in the carbonate system, and γCa^{2+} and γHCO_3^- are the activity coefficients of Ca^{2+} and HCO_3^-. A gravelly, permeable calcic soil probably best approximates an open system-weathering environment,

Encyclopedia of Soil Science

in which case calcite dissolution rates are probably surface-area controlled rather than diffusion controlled. Also, the dissolution rate is defined ultimately by the rate-limiting conversion of dissolved carbon dioxide (CO_2*) to H_2CO_3. At a very low solution volume to surface area ratios, and with fast, surface-controlled calcite dissolution rates, H_2CO_3 is rapidly depleted. In such circumstances, a commonly used rate equation that enables determination of dissolution rates in the CO_2–$CaCO_3$–H_2O system (9) is:

$$dC/dt = (A'k/V)(1 - C/C^*)^n \text{ mg l}^{-1} \text{ s}^{-1} \quad (2)$$

where A' is the surface area of rock in contact with water (cm^2), V the water volume (cm^3), k the reaction coefficient (mg cm l^{-1} s^{-1}), n the reaction order, C the moles of calcite in solution, and C^* is the solubility of calcite. Values of n and k vary with saturation ratio, temperature, and pCO_2. In the model, A'/V can be specified depending on observed soil features. Eqs. 1 and 2 show that soil CO_2 content is a very important variable, but soil CO_2 contents may be highly variable (10). Fortunately, studies show that a depth function for pCO_2 that reflects prolonged seasonal respiration levels can be estimated, assuming the concentration of soil CO_2 is described by mass transport of CO_2 by gas diffusion (11, 12). The following diffusion-reaction equation, essentially Fick's Second Law for a one-dimensional case, is used in the model:

$$\partial C_s/\partial t = D_s(\partial^2 C_s/\partial z^2) + \phi_s(z) \quad (3)$$

where C_s is the concentration of CO_2 in the soil (mol cm^{-3}), t the time (s), D_s the diffusion coefficient for CO_2 in the soil ($cm^2 s^{-1}$), z the depth in the soil (cm), and $\phi_s(z)$ is the production rate of CO_2 as a function of depth (mol $cm^{-3} s^{-1}$). At steady state, when $\partial C_s/\partial t = 0 = D_s\partial^2 C_s/\partial z^2 + \phi_s$, the general solution to this equation to produce a simple production function is:

$$C_s(z) = \phi/D_s(Lz - z^2/2) + C_0 \quad (4)$$

where C_0 is the concentration of CO_2 in the atmosphere (ppm) and L is the depth to the lower, no-flux boundary. Soil CO_2 contents with depth calculated using this method are used to calculate carbonate solubility and dissolution rates with depth.

Available water-holding capacity, infiltration, and percolation rates can be specified on the basis of laboratory soil measurements or estimated from field measurements or theoretical considerations.

Earlier versions of the simulation model included certain assumptions that simplified numerical calculations, such as simple vertical saturated flow and constant soil temperature with depth. The lack of certain types of data (e.g., variation of pCO_2 with depth and time of year) also constituted a limitation on utility of the model. The model did enable simulation of 1) realistic depths and magnitudes of carbonate accumulation over thousands of years, and 2) the range of effects of large climatic changes on calcic soils (13–16). Model results emphasized the critical roles of external Ca^{2+} influx and influence of soil CO_2 concentrations on carbonate accumulation. Model-simulated bimodal concentrations of carbonate based on theoretical, late-Pleistocene climatic conditions resembled those observed in late-Pleistocene, polygenetic soils; however, incompletely understood changes in the magnitude of climate changes, dust flux, and vegetation change in the Quaternary complicate attempts to simulate polygenetic soils (7, 14, 16).

Later versions of the model utilized important new inputs and employed routines that reflected improved understanding of key processes that strongly influence calcic soils. Studies of dust accumulation rates in the American Southwest (17), C, O, and Sr isotopes in carbonate, and development of more sophisticated models for unsaturated flow in calcic soils (18) have allowed development of improved compartment models that can address new and more challenging research problems. For example, such numerical simulations demonstrate how climate changes in the Holocene might have dramatically influenced the rates and temporal patterns of soluble salt leaching and accumulation relative to pedogenic carbonate (19). A more recent modeling study addressed the problem of how carbonate can occasionally accumulate at much shallower depths than those expected from the depth—annual leaching depth relationship (20). This study showed how carbonate can be preferentially removed from depths of a few cm to a few dm below the soil surface, while carbonate simultaneously accumulates either as collars on surface pavement clasts or in the vesicular A horizon. Model results also explain how a significant change in climate or soil erosion rates could cause the dissolution of carbonate rinds on the tops and sides of boulders and/or the tops of limestone boulders at depths of up to several dm, unusual features observed in some calcic soils (21).

ISOTOPES IN CALCIC SOILS

During weathering, parent material carbonate undergoes dissolution and reprecipitation in the soil. The carbon ($^{13}C/^{12}C$, $^{14}C/^{12}C$) and oxygen ($^{18}O/^{16}O$) isotope ratios of pedogenic carbonate that forms from dust or parent material carbonate, or from Ca^{2+} derived from silicate weathering, are determined by isotopic composition of soil CO_2 and H_2O. These are the primary carbon and oxygen reservoirs, respectively, for the carbonate. Therefore, pedogenic carbonate reflects only isotopic conditions of the soil and bears no memory of the isotopic composition of the rock or mineral from which it was derived.

Soil CO_2 is derived primarily from decomposition of soil organic matter and root respiration. The C isotope composition of soil CO_2 reflects: 1) the isotopic composition of these CO_2 sources; 2) the effects of the diffusion of this CO_2 toward the atmosphere; and 3) the isotopic composition of atmospheric CO_2. In the 1980s, researchers recognized that fairly simple, steady state, diffusion models could be used to reasonably explain the observed depth patterns of C isotopes in both soil CO_2 and pedogenic carbonate. The solution to the mathematical model that encompasses the forementioned processes describes the abundance of $^{12}CO_2$ in soils. A related equation can be derived for $^{13}CO_2$, and the ratio of the two models then describes the ratio of C isotopes at any given soil depth. A similar approach can also be used to model the ^{14}C composition of CO_2 with soil depth, with the additional complication that the two main sources of soil CO_2 (humus decomposition and root respiration) have different ^{14}C contents, making the solution to the model slightly more complex (22). The C isotope diffusion model of soil CO_2 has provided the opportunity to quantitatively use pedo-genic carbonates in a number of applications: 1) paleovegetation studies (23), 2) paleo-atmospheric CO_2 studies (24), and 3) radiocarbon dating of pedogenic carbonate and geomorphic surfaces (25, 26).

The O isotopic composition of soil water is determined by the O isotope composition of precipitation and the evaporation of soil water. It has been observed that the ^{18}O content of modern precipitation is generally correlated with mean annual temperature on a global scale (27), but regional differences due to storm sources can obscure these patterns (28). If precipitation water (once stored in the soil) is subject to evaporation, an enrichment of the remaining soil water in ^{18}O occurs because water vapor depleted in the "heavy" isotope is preferentially removed during evaporation. Models have been made that successfully explain the key components of this process (29) and the fact that soils subject to evaporation commonly have generally decreasing ^{18}O contents of soil water with depth (30). These models have two components. The first is a vapor transport layer (describing the flow of evaporating soil water to the atmosphere through a dry soil layer), and the second is an evaporating front layer. The evaporating front layer exists below the vapor transport zone. At the evaporating front ^{18}O enrichment of soil water occurs as water is transferred to a vapor phase, and the remaining ^{18}O-enriched soil water at the evaporating front then undergoes diffusional mixing with the less ^{18}O-enriched water at greater depths. In general, these models have been more difficult to use than C isotope models due to the dynamic nature of soil water (steady state assumptions are difficult to apply) and the array of model parameters, many of them not known with certainty for most soils.

Oxygen isotopes in pedogenic carbonate have been used less extensively in paleoclimate work than C isotopes because of concern over possible evaporation of soil water that formed the carbonate. Amundson et al. (28) demonstrated that, except for hyperarid regions, the O isotope composition of pedogenic carbonate appears to reasonably reflect that of the local precipitation. There is a growing list of studies using carbonate O isotopes in Quaternary (28) and Tertiary (31) paleoclimate applications.

CONCLUSION

Numerical models and isotope studies have proven to be valuable tools in the study of calcic soil development. They have helped elucidate the relation of climate, vegetation, and geomorphic processes to carbonate accumulation. The models are not able to explain the observed character of certain aspects of calcic soils, such as patterns of pedogenic carbonate development in some soil chronosequences (32), or the somewhat enigmatic formation of calcic soils in humid, monsoonal climates. Also, these models are not designed to simulate the evolution of very old, morphologically complex soils with petrocalcic horizons. Future models must be designed to address locally abundant calcic soils. Additional fieldwork and application of recently developed field and laboratory techniques will provide the basis for development of the next generation of numerical models.

REFERENCES

1. Birkeland, P.W. *Soils and Geomorphology*; Oxford University Press: New York, 1999; 720.
2. Jenny, H. *Factors in Soil Formation*; McGraw-Hill: New York, 1941; 281.
3. Arkley, R.J. Calculation of Carbonate and Water Movement in Soil from Climatic Data. Soil Sci. **1963**, *96*, 239–248.
4. Retallack, G.J. The Environmental Factor Approach to the Interpretation of Paleosols. In *Factors of Soil Formation: A Fiftieth Anniversary Retrospective*; Special Publication 33, Amundson, R. Ed.; SSSA: Madison, WI, 1994; 31–64.
5. Royer, D.L. Depth to Pedogenic Carbonate Horizon as a Paleoprecipitation Indicator? Geology **1999**, *27*, 1123–1126.
6. Machette, M.A. Calcic Soils of the Southwestern United States. In *Soils and Quaternary Geology of the Southwestern United States*; Special Paper 203, Weide, D.L., Faber, M.L.

Eds.; Geological Society of America: Boulder, CO, 1985, 1–21.
7. McFadden, L.D.; Tinsley, J.C. The Rate and Depth of Accumulation of Pedogenic Carbonate Accumulation in Soils: Formation and Testing of a Compartment Model. In *Soils and Quaternary Geology of the Southwestern United States*; Special Paper 203, Weide, D.L., Faber, M.L., Eds.; Geological Society of America: Boulder, CO, 1985, 23–42.
8. Drever, J.I. *The Geochemistry of Natural Waters*, 2nd Ed.; Prentice-Hall: Englewood Cliffs, NJ, 1991; 437.
9. Palmer, A.N. Origin and Morphology of Limestone Caves. Geol. Soc. Am. Bull. **1991**, *103*, 1–21.
10. Kiefer, R.H.; Amey, R.G. Concentrations and Controls of Soil Carbon Dioxide in Sandy Soil in the North Carolina Coastal Plain. Catena **1992**, *19*, 539–559.
11. Solomon, D.K.; Cerling, T.E. The Annual Carbon Dioxide Cycle in a Montane Soil: Observations, Modeling, and Implications for Weathering. Water Resour. Res. **1987**, *23*, 2257–2265.
12. Cerling, T.E.; Quade, J.; Yang, W.; Bowman, J.R. Carbon Isotopes in Soils and Palaeosols, and Ecology and Palaeoecology Indicators. Nature **1989**, *341*, 138–139.
13. McFadden, L.D.; Amundson, R.G.; Chadwick, O.A. Numerical Modeling, Chemical, and Isotopic Studies of Carbonate Accumulation in Soils of Arid Regions. Soil Sci. Soc. Am. Spec. Publ. **1991**, *26*, 17–35.
14. McFadden, L.D. The Impacts of Temporal and Spatial Climatic Changes on Alluvial Soils Genesis in Southern California PhD Dissertation University of Arizona Tucson430.
15. Marion, G.M.; Schlesinger, W.H.; Fonteyn, P.J. Caldep: A Regional Model for Soil $CaCO_3$ (Caliche) Deposition in Southwestern Deserts. Soil Sci. **1985**, *139*, 468–481.
16. Mayer, L.; McFadden, L.D.; Harden, J.W. Distribution of Calcium Carbonate in Desert Soils: A Model. Geology **1988**, *16*, 303–306.
17. Reheis, M.C.; Kihl, R. Dust Deposition in Southern Nevada and California, 1984–1989: Relations to Climate, Source Area, and Source Lithology. J. Geophys. Res. **1995**, *100*, 8893–8918.
18. McDonald, E.V.; Pierson, F.B.; Flerchinger, G.N.; McFadden, L.D. Application of a Soil–Water Balance Model to Evaluate the Influence of Holocene Climatic Change on Calcic Soils, Mojave Desert, California, USA. Geoderma **1996**, *74*, 167–192.
19. McFadden, L.D.; Crossey, L.J.; McDonald, E.V. Predicted Response to Calcic Soil Development to Periods of Significantly Wetter Climate During the Late Holocene (Abstr.). Geol. Soc. Am. Abstr. Progr. **1990**, *24* (6), A252.
20. McFadden, L.; McDonald, E.; Wells, S.; Anderson, K.; Quade, J.; Forman, S. The Vesicular Layer and Carbonate Collars of Desert Soils and Pavements: Formation, Age and Relation to Climate Change. Geomorphology **1998**, *24*, 101–145.
21. Treadwell-Steitz, C.; McFadden, L.D. Influence of Parent Material and Grain Size on Carbonate Coatings in Gravelly Soils, Palo Duro Wash, New Mexico. Geoderma **2000**, *94*, 1–22.
22. Amundson, R.; Stern, L.; Baisden, T.; Wang, Y. The Isotopic Composition of Soil CO_2. Geoderma **1998**, *82*, 83–114.
23. Cerling, T.E. Development of Grasslands and Savannas in East Africa During the Neogene. Palaeogeogr. Palaeoclimatol. Palaeoecol. **1992**, *97*, 241–247.
24. Cerling, T.E. Use of Carbon Isotopes in Paleosols as an Indicator of the $P(CO_2)$ of the Paleoatmosphere. Global Biogeochem. Cycles **1992**, *6*, 307–314.
25. Amundson, R.; Wang, Y.; Chadwick, O.; Trumbore, S.; McFadden, L.D.; McDonald, E.; Wells, S.; DeNiro, M. Factors and Processes Governing the ^{14}C Content of Carbonate in Desert Soils. Earth Planet. Sci. Lett. **1994**, *125*, 385–405.
26. Wang, Y.; McDonald, E.; Amundson, R.; McFadden, L.D.; Chadwick, O. An Isotopic Study of Soil in Chronological Sequences of Alluvial Deposits, Providence Mountains, California. Geol. Soc. Am. Bull. **1996**, *108*, 379–391.
27. Rozanski, K.L.; Araguás-Araguás, L.; Gonfiantini, R. Isotopic Patterns in Modern Global Precipitation. In *Climate Change in Continental Isotopic Records*; American Geophysical Union Monograph 78, Swart, P.K., Ed.; Washington, D.C., 1998; 1–36.
28. Amundson, R.; Chadwick, O.; Kendall, C.; Wang, Y.; DeNiro, M. Isotopic Evidence for Shifts in Atmospheric Circulation Patterns During the Late Quaternary in Mid-North America. Geology **1996**, *24*, 23–26.
29. Barnes, C.J.; Allison, G.B. The Distribution of Deuterium and ^{18}O in Dry Soils. I. Theory. J. Hydrol. **1983**, *60*, 141–156.
30. Allison, G.B.; Hughes, M.W. The Use of Natural Tracers as Indicators of Soil Water Movement in Temperate Semiarid Regions. J. Hydrol. **1983**, *60*, 157–173.
31. Quade, J.; Cerling, T.E.; Bowman, J.R. Development of the Asian Monsoon Revealed by Marked Ecological Shift During the Latest Miocene in Northern Pakistan. Nature **1989**, *343*, 163–166.
32. Holliday, V.T.; McFadden, L.D.; Bettis, E.A.; Birkeland, P.W. Soil Survey and Soil Geomorphology. In *Profiles in the History of the U.S. Soil Survey*; Helms, D., Efflard, A., Dwara, P., Eds.; Iowa State University Press: Ames, IO, 2001.

INSECT SURVIVAL, EFFECT OF SUBZERO TEMPERATURE ON

M.M. Ellsbury
United States Department of Agriculture (USDA), Brookings, South Dakota, U.S.A.

INTRODUCTION

Insects that inhabit soil in regions where winter temperatures fall below freezing have necessarily evolved behavioral and physiological mechanisms for surviving and developing in seasonally cold environments. Reviews of literature relating to the physiological, ecological, and behavioral adaptations that allow insects to survive cold temperature appear in Baust et al. (1), Cannon and Block (2), Block (3), and Danks (4). A recent review by Danks (5) considers relationships between dehydration and cold hardiness in dormant insects. Danks et al. (6) and Block (7) also have compiled concise glossaries of scientific terms associated with cold-hardiness in insects.

BEHAVIORAL ADAPTATION

Behavioral mechanisms for avoidance of low temperatures may entail movement of freezing-susceptible overwintering stages deeper into the soil or to sites protected from cold temperatures (8, 9). Snow cover has an insulating effect that mediates winter soil temperatures, producing nearly isothermal soil temperature regimes when early snow cover occurs (9, 10). Where snow cover is sparse or lacking soil temperature more closely follows extremes of air temperature (11) and cold-hardiness adaptation becomes more important for overwinter survival of soil-dwelling insects. This is particularly true of soil-dwelling invertebrates of arctic regions that cannot avoid exposure to cold temperature and thus have necessarily evolved cold hardiness traits and diapause capability that enable them to survive during exposure to subfreezing temperature. Where short warm seasons occur, modified, sometimes very prolonged, life cycles extending over more than one overwinter season are found in insects of cold environments (3, 6, 9).

PHYSIOLOGICAL ADAPTATION

Cold hardiness involves physiological and metabolic adaptations that may not always be concomitant with a depressed diapause metabolism. Diapause has been defined as hormonally controlled metabolic dormancy that occurs during a specific stage in the life cycle of an insect (12–14). Thus, cold hardiness may occur independently of diapause, but often is an integral component of the diapause trait and diapause expression may improve the cold hardiness capability of the insect (15).

FREEZE TOLERANCE VS. FREEZE SUSCEPTIBILITY

Cold-hardy insects have been characterized into two groups: those that are tolerant to freezing and those that are susceptible to freezing (8, 16, 17). The first of these, insects that are freeze tolerant, survive ice formation in body tissues but avoid damage to intracellular components because ice formation usually is confined to extracellular fluids. Protective mechanisms for freezing resistance include elevated solute levels, presence of nucleating agents, and accumulation of cryoprotectants in body fluids. The second broad category includes insects capable of undergoing supercooling without freezing of body fluids at temperatures that are below the true freezing point or melting point (6). Insects that have the ability to supercool are considered freezing-susceptible since they avoid freezing damage to body tissues only at temperatures above the supercooling point. Below the supercooling point, ice formation in body tissues results in death. These relationships are summarized in Fig. 1. It should be noted however, that a strict dichotomous categorization of the cold-hardy stages of insects as either freeze-tolerant or freeze-susceptible may be an oversimplification (7). Kostal and Havelka (18) observe that in soil environments where freezing events occur unpredictably, overwintering stages of some insects display both freeze susceptibility (supercooling) and freeze tolerance. In many insects, particularly those of more temperate environments, cold hardiness is associated with ability to survive low temperature during chilling events but is not necessarily associated with extremes of subfreezing temperature.

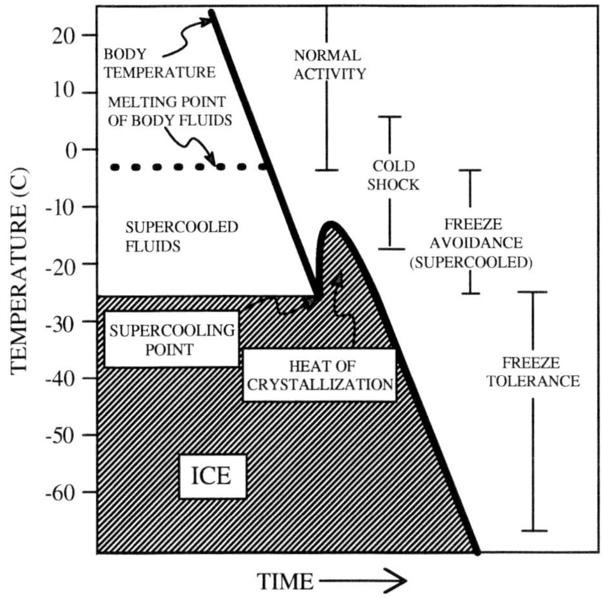

Fig. 1 Generalized diagram showing insect responses to cold temperature. Heavy line indicates insect body temperature and shaded area is a zone where ice nucleation occurs. (From Ref. 6.)

The capacity of insects to supercool may be enhanced by acclimation at low temperatures, still above the supercooling point, prior to exposure to subfreezing temperatures. However, where acclimation does not occur, exposure to rapidly falling temperature, even for short durations, may induce cold shock or direct chilling injury. Cold shock is defined as stress associated with short duration or rapid exposure to low, but non-freezing temperature (19). The rate of cooling is more significant in producing cold shock injury than are the duration or the level of low temperature attained during the chilling event. Insects subjected to cold shock sustain injury at temperatures well above the supercooling point. The mechanism of injury caused during cold shock probably involves changes in structure of lipids associated with membranes or thermoelastic stress induced by rapidly falling temperature (19–21).

Lee et al. (22) described a process for rapid cold-hardening of insects during exposure to low temperature that otherwise would result in cold shock. The phenomenon has been associated with glycerol accumulation (23) and production of stress proteins similar to those produced in response to heat shock (24). The rapid cold-hardening phenomenon was observed in nonoverwintering stages and probably provides insects with means to avoid direct chilling injury during rapid exposure to low temperatures above the supercooling point. This adaptation could be particularly important for soil-dwelling organisms of temperate regions that may be exposed to rapid freeze–thaw events during the autumn or spring. The association of heat shock protein production with cold-hardiness suggests similarities between rapid cold hardening and insect response to heat stress by production of heat-shock proteins.

Mortality during cold exposure may result from dehydration during the freezing process, even in insects, which are freeze-tolerant (8). Indeed, Ring and Danks (25) suggested that physiological and behavioral traits considered adaptive to cold-hardiness may also be adaptive to survival in dry winter conditions. Block (26) concluded that resistance to desiccation may be considered preadaptive to cold-hardiness in terrestrial arthropods.

CORN ROOTWORMS

The ability of certain *Diabrotica* beetle species (Coleoptera: Chrysomelidae) to survive winter desiccation in tropical climates may have preadapted them for survival in more northern climates. Pest *Diabrotica*, of the corn rootworm complex, are thought to have expanded their range northward from tropical regions into areas of more temperate climate concomitantly with the agricultural development of corn, *Zea mays* L., as a primary host plant. Western corn rootworms, *Diabrotica virgifera virgifera* LeConte, considered by Krysan (27) to be of tropical origin, survive cold winter conditions in temperate regions as a diapausing egg stage in the soil. The diapause mechanism that enabled survival of eggs during seasonal desiccation in the tropics probably also preadapted the soil-dwelling egg stage of this insect to survive in frozen soils (28). A closely related subspecies, the Mexican corn rootworm, *D. virgifera zeae*, also overwinters as a diapausing egg stage in the soil but is adapted to survive dry winters in more southern climates (29). These observations are consistent with the suggestion by Block (7) that insects tolerant of desiccation also are probably preadapted for survival of cold temperatures.

In the northern Great Plains the *Diabrotica* pest complex consists of two species, the western corn rootworm, *D. virgifera virgifera*, and the northern corn rootworm, *D. barberi*. Because both species overwinter as eggs in the soil, the occurrence of an economic infestation during the following growing season depends in part on the overwintering survival of the egg. Chiang (30) Chiang et al. (31), Calkins and Kirk (32), and Gustin (33, 34) determined that winter soil temperatures reach levels low enough to cause mortality in overwintering western and northern corn rootworm eggs in the northern Great Plains.

Nonetheless a significant proportion of eggs laid each fall frequently survive winter conditions in corn-growing areas of North America to produce economic infestations of these insects each year.

Ellsbury et al. (35, 36) have shown that adult emergence patterns of both rootworm species have highly variable spatial distributions in the field. These distributions are probably the result of behavioral and mortality factors operating on eggs in the soil at two points in the seasonal life cycle of these insects. Firstly, ovipositional patterns may vary spatially in three dimensions with adult response to corn maturity, and soil texture and moisture conditions. Secondly, distributions of eggs in the soil may change from those that exist just after the oviposition period because of differential overwinter mortality. The influence of fall tillage, soil characteristics, snow cover, and soil temperature are known to mediate egg mortality, but the interactions between these factors that produce spatial and temporal variation in surviving rootworm populations are poorly understood.

The effect of low winter soil temperature on corn rootworm eggs has been the subject of intense study. Some early research on this subject was driven by the need for laboratory rearing of large quantities of postdiapause eggs for artificial infestation and thus was focused on the biology of egg diapause, temperature limits for development, and optimal chilling conditions for storage of eggs. Much research effort has been directed toward determining developmental thresholds and diapause requirements of rootworms in cold temperature environments; yet little information is available about cold hardiness, and the actual cold hardening mechanism of the overwintering egg stage, nor is it known whether corn rootworm eggs can be considered freezing tolerant or freeze-susceptible.

CONCLUSIONS

Much of the current knowledge about how soil-dwelling insects survive cold soil temperature has come from scientific studies done in arctic or alpine environments under extreme conditions. The basic concepts arising from these studies have been summarized in this contribution. It is evident that insects have adapted to cold soil environments through a variety of mechanisms and that no one strategy suffices for survival in cold environments for all insects. The discussion of survival of corn rootworm eggs was intended to illustrate how some of this knowledge can be applied to an insect pest in a temperate agricultural setting. There are of course other soil-dwelling insects that overwinter in various life stages in both agricultural and natural settings that, for want of space, could not be considered here. However, it can said that these basic concepts and references provided herein should serve as a point of departure for the reader wishing to undertake more in-depth study or research on the survival of insects in frozen soil environments.

REFERENCES

1. Baust, J.G.; Lee, R.E., Jr.; Ring, R.A., The Physiology and Biochemistry of Low Temperature Tolerance in Insects and Other Terrestrial Arthropods: A Bibliography. Cryo-Letters **1982**, *3*, 191–212.
2. Cannon, R.J.C.; Block, W. Cold Tolerance of Microarthropods. Biol. Rev. **1988**, *63*, 23–77.
3. Block, W. Cold Tolerance of Insects and Other Arthropods. Phil. Trans. R. Soc. Lond. B **1990**, *326*, 613–633.
4. Danks, H.V. Winter Habitats and Ecological Adaptations for Winter Survival. In *Insects at Low Temperature*; Lee, R.E., Jr., Denlinger, D.L., Eds.; Chapman and Hall: New York, 1991; 31–259.
5. Danks, H.V. Dehydration in Dormant Insects. J. Insect Physiol. **2000**, *46*, 837–852.
6. Danks, H.V.; Kukal, O.; Ring, R.A. Insect Cold-Hardiness: Insights from the Arctic. Arctic **1994**, *47* (4), 391–404.
7. Block, W. Insects and Freezing. Science Progress **1995**, *78* (4), 349–372.
8. Lee, R.E., Jr. Insect Cold-Hardiness: To Freeze or Not to Freeze. Bioscience **1989**, *39* (5), 308–312.
9. Coulson, S.J.; Hodkinson, I.D.; Block, W.; Webb, N.R.; Worland, M.R. Low Summer Temperatures: A Potential Mortality Factor for High Arctic Soil Microarthropods. J. Insect Physiol. **1995**, *41*, 783–792.
10. Sharratt, B.S.; Baker, D.G.; Wall, D.B.; Skaggs, R.H.; Ruschy, D.L. Snow Depth Required for Near Steady State Soil Temperatures. Agric. Forest Meteorol. **1992**, *57*, 243–251.
11. Coulson, S.J.; Hodkinson, I.D.; Strathdee, A.T.; Block, W.; Webb, N.R.; Bale, J.S.; Worland, M.R. Thermal Environments of Arctic Soil Organisms During Winter. Arct. Alp. Res. **1995**, *27* (4), 364–370.
12. Beck, S.D. *Insect Photoperiodism*, 2nd Ed.; Academic Press: New York, 1980; 387.
13. Saunders, D.S. *Insect Clocks*, 2nd Ed.; Pergamon Press: Oxford, 1982; 288.
14. Tauber, M.J.; Tauber, C.A.; Masaki, S. *Seasonal Adaptations of Insects*; Oxford University Press: New York, 1986; 411.
15. Denlinger, D.L. Relationship Between Cold Hardiness and Diapause. In *Insects at Low Temperature*; Lee, R.E., Jr., Denlinger, D.L., Eds.; Chapman and Hall: New York, 1991; 174–198.
16. Salt, R.W. Principles of Insect Cold-Hardiness. Ann. Rev. Entomol. **1961**, *6*, 55–74.
17. Block, W. Cold Hardiness in Invertebrate Poikilotherms. Comp. Biochem. Physiol. A **1982**, *73*, 581–593.

18. Kostal, V.; Havelka, J. Diapausing Larvae of the Midge *Aphidoletes aphidimyza* (Diptera: Cecidomyiidae) Survive at Subzero Temperatures in a Supercooled State but Tolerate Freezing if Inoculated by External Ice. Eur. J. Entomol. **2000**, *97* (3), 433–436.
19. Denlinger, D.L.; Joplin, K.H.; Chen, C.-P.; Lee, R.E., Jr. Cold Shock and Heat Shock. In *Insects at Low Temperature*; Lee, R.E., Jr., Denlinger, D.L., Eds.; Chapman and Hall: New York, 1991; 131–148.
20. Quinn, P.J. A Lipid-Phase Separation Model of Low-Temperature Damage to Biological Membranes. Cryobiology **1985**, *22*, 128–146.
21. McGrath, J.J. Cold Shock: Thermoelastic Stress in Chilled Biological Membranes. In *Network Thermodynamics, Heat and Mass Transfer in Biotechnology*; Diller, K.R., Ed.; United Engineering Center: New York, 1987; 57–66.
22. Lee, R.E., Jr.; Chen, C.-P.; Denlinger, D.L. A Rapid Cold-Hardening Process in Insects. Science **1987**, *238*, 1415–1417.
23. Chen, C.-P.; Denlinger, D.L.; Lee, R.E., Jr. Cold Shock Injury and Rapid Cold Hardening in the Flesh Fly, *Sarcophaga crassipalpis*. Physiol. Zool. **1987**, *60* (3), 297–304.
24. Burdon, R.H. Heat Shock and the Heat Shock Proteins. Biochem. J. **1986**, *240*, 313–324.
25. Ring, R.A.; Danks, H.V. Desiccation and Cryoprotection: Overlapping Adaptations. Cryo-Letters **1994**, *15*, 181–190.
26. Block, W. Cold or Drought—the Lesser of Two Evils for Terrestrial Arthropods? Eur. J. Entomol. **1996**, *93*, 325–339.
27. Krysan, J.L. Diapause in the Nearctic Species of the *Virgifera* Group of *Diabrotica*: Evidence for Tropical Origin and Temperate Adaptation. Ann. Entomol. Soc. Amer. **1982**, *75* (2), 136–142.
28. Krysan, J.L. Introduction: Biology, Distribution, and Identification of Pest *Diabrotica*. In *Methods for the Study of Pest Diabrotica*; Krysan, J.L., Miller, T.A., Eds.; Springer-Verlag: New York, 1986; 1–24.
29. Krysan, J.L.; Branson, T.F.; Castro, G.D. Diapause in *Diabrotica virgifera* (Coleoptera: Chrysomelidae): A Comparison of Eggs from Temperate and Subtropical Climates. Ent. Exp. Appl. **1977**, *22*, 81–89.
30. Chiang, H.C. Survival of Northern Corn Rootworm Eggs Through One and Two Winters. J. Econ. Entomol. **1965**, *58* (3), 470–472.
31. Chiang, H.C.; Mihm, J.A.; Windels, M.B. Temperature Effects on Hatching of Western and Northern Corn Rootworms. Proc. North Cent. Br. Entomol. Soc. Amer. **1972**, *27*, 127–131.
32. Calkins, C.O.; Kirk, V.M. Effect of Winter Precipitation and Temperature on Over-Wintering Eggs of Northern and Western Corn Rootworms. J. Econ. Entomol. **1969**, *62* (3), 541–543.
33. Gustin, R.D. Soil Temperature Environment of Over-Wintering Western Corn Rootworm Eggs. Environ. Entomol. **1981**, *10* (4), 483–487.
34. Gustin, R.D. *Diabrotica longicornis barberi* (Coleoptera: Chrysomelidae): Cold Hardiness of Eggs. Environ. Entomol. **1983**, *12* (3), 633–634.
35. Ellsbury, M.M.; Woodson, W.D.; Clay, S.A.; Carlson, C.G. Spatial Characterization of Corn Rootworm Populations in Continuous and Rotated Corn. In *Precision Agriculture*; Robert, P.C., Rust; R.H., Larson, W.E., Eds.; Proceedings of the 3rd International Symposium on Precision Agriculture, Bloomington, Minnesota, June 23–26, 1996; American Society of Agronomy: Madison, Wisconsin, 1996; 487–494.
36. Ellsbury, M.M.; Woodson, W.D.; Clay, S.A.; Malo, D.; Schumacher, J.; Clay, D.E.; Carlson, C.G. Geostatistical Characterization of the Spatial Distribution of Adult Corn Rootworm (Coleoptera: Chrysomelidae) Emergence. Environ. Entomol. **1998**, *27* (4), 910–917.

ION EXCHANGE

Bryon W. Bache
Cambridge University, Cambridge, United Kingdom

INTRODUCTION

Nature and Origin of Cation Exchange

Many common substances exist as ions. The simplest example is common salt, sodium chloride (NaCl). This innocuous and life-sustaining compound is made of two dangerous and highly reactive elements, sodium and chlorine. When these two elements react together, the sodium atom loses an electron, giving it a positive charge (Na^+), while the chlorine atom gains an electron, giving it a negative charge (Cl^-). The positively charged ions are *cations*; the negative ones are *anions*. In a salt solution, the sodium and chloride ions are not physically bonded together but are free to diffuse independently in solution, attracted to each other only by electrostatic charges. Hence, the original derivation of the term *ion* (from the Greek for "wanderer") is now applied generally to any charged particle.

Many substances in the soil solution exist as ions, but in a typical agricultural or garden soil most of the anions (the negatively charged particles) exist as large, insoluble macroions. Thus, the solid matrix of soil has a negative charge, and to maintain electrical neutrality, positively-charged cations are adsorbed onto the surface of the insoluble anions in a diffuse-ion swarm. This scenario facilitates *cation exchange*, whereby some of the adsorbed cations are exchanged for others.

CATION EXCHANGE

Cation exchange is best illustrated by conducting a simple experiment. Place a layer of dry soil about 5 mm deep into a funnel, supported by a filter paper. Leach the soil slowly with demineralized water to remove the bulk of any soluble salts present, and discard this leachate. Then leach with 50 ml of water, but this time collect the leachate. (Although this will take some time, mixing the soil with coarse sand will speed up the process.) Next, leach the soil with 50 ml of a solution containing 2 to 3 g of ammonium chloride, and again collect the leachate. Test both the water leachate and the ammonium chloride leachate for calcium by adding ammonium oxalate solution. The water leachate will be free of calcium but the ammonium chloride leachate will give a white precipitate of calcium oxalate, showing that ammonium NH_4^+ has displaced calcium Ca^{2+} from the soil surfaces into solution. Ammonium has *exchanged* for calcium. A diagrammatic representation of this exchange reaction is shown in Fig. 1A.

Humus and clay are the two components of the soil matrix that contain the macroions responsible for cation exchange. Humus contains proton-donating functional groups on the surface of its molecules, which are undissociated at low pH but dissociate as the pH rises, to give a negatively charged surface. The most important of these is the carboxy group:

$$R - COOH \rightarrow R - COO^- + H^+ \qquad (1)$$

where R is the organic core. Clay consists of fine platy crystals of aluminosilicate minerals. The lattice of these minerals has a positive-charge deficiency caused by the isomorphous substitution of Al^{3+} for Si^{4+}, or Mg^{2+} for Al^{3+} in the structure of the mineral, which gives the crystal a net negative charge.

EXCHANGEABLE CATIONS, CATION EXCHANGE CAPACITY, AND BASE CATION SATURATION

In most agricultural soils, the dominant exchangeable cation is calcium (Ca^{2+}), with lesser amounts of magnesium (Mg^{2+}), sodium (Na^+), and potassium (K^+). These four are often termed *base cations*—i.e., the cations of the bases $Ca(OH)_2$, $Mg(OH)_2$, NaOH, and KOH. These should not to be confused with *basic cations*, of which $MgOH^+$ is an example. In contrast, the acid cation aluminium Al^{3+} is dominant in acid mineral soils, and the hydrogen ion H_3O^+ is dominant in acid organic soils.

The total amount of cations that may be exchanged is the *cation exchange capacity* (CEC) of the soil. The sum of the positive charge on these cations equals the negative charge on the soil surface. The convention for CEC units most consistent with the International System of Units (SI) is "millimoles of charge per kilogram of dry soil"

Encyclopedia of Soil Science

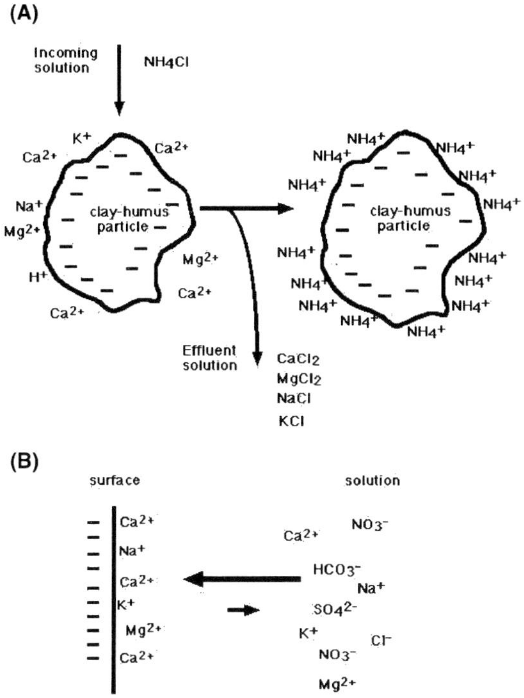

Fig. 1 Diagrammatic representations of cation exchange: (**A**) the experiment described in the text; (**B**) the equilibrium between the surface and the solution.

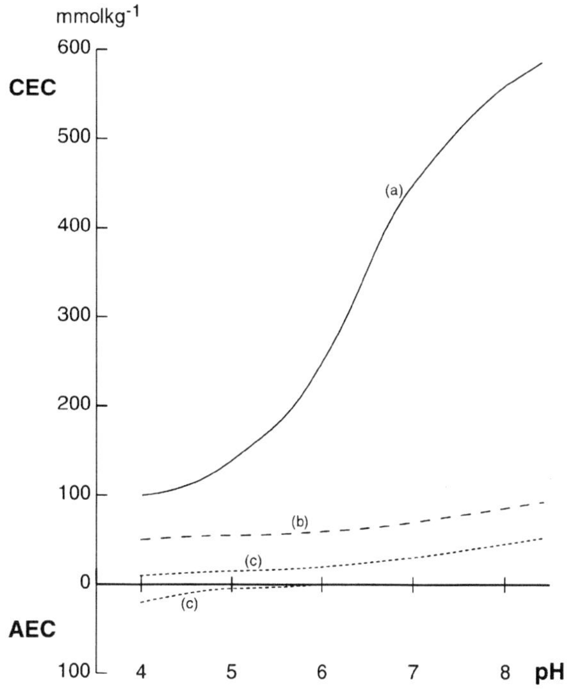

Fig. 2 The effect of pH on CEC: (**A**) a humic topsoil; (**B**) a mineral subsoil; (**C**) a strongly weathered Oxisol, showing both CEC and AEC.

i.e., mmol(+)kg^{-1} for the cations and mmol(−)kg^{-1} for the surface charge. A variety of experimental procedures has been developed to measure exchangeable cations and CEC (1). CEC values in mmolkg^{-1} may vary for clay minerals from 30 for kaolinite to 1200 for smectite, and for soils from 50 for Oxisols to 2000 for Histosols.

The negative charge on humus is strongly pH-dependent, so that apart from the nature and amounts of clay and humus, the most important variable affecting CEC is the pH of the soil (Fig. 2). Thus, when comparing different soils, CEC is conventionally measured at a standard pH, normally at the neutral point of pH 7.0 with an ammonium acetate buffer and ammonium NH_4^+ as the displacing cation, or at pH 8.2 with a triethanolamine buffer and barium Ba^{2+} as the displacing cation.

Conventions for measuring CEC at a standard pH are appropriate for near-neutral soils dominated by permanent-charge clays. However, it is clear from Fig. 2 that they result in a highly inflated CEC value for acid soils when compared with the natural field soil, and they are even less appropriate for variable-charge soils, which acquire pH-dependent positive and negative charges (see Fig. 2C and Anion Exchange below). In these circumstances, CEC measured with unbuffered salt solutions at field pH provides a better interpretative value. This *effective* CEC (ECEC) equals the sum of the exchangeable cations extracted from the soil, including both base and acid cations. Thus, the pH of measurement should always be stated for CEC data.

An important concept for acid soils is the *base cation saturation* (V). This is usually expressed as a percentage, and is then given by

$$V = 100\ \Sigma M^+/CEC \qquad (2)$$

where ΣM^+ is the sum of the exchangeable base cations, ΣM^+ and CEC are measured in the same units, and CEC is at pH 7.0.

CATION EXCHANGE EQUILIBRIA AND CATION SELECTIVITY

Cation exchange was first discovered by H.S.Thompson in 1850; a review of early related work was compiled by Kelley (2). Good recent accounts are given by Mott (3), Sposito (4), and McBride (5). Apart from the source of cation exchange (discussed earlier), most interest has focused on the relationship between the relative concentrations of different cations in solution and the amounts of surface-adsorbed cations, and how this varies for different exchange materials (e.g., silicate clay minerals, humus, and synthetic resins).

Fig. 1B illustrates the equilibrium between the adsorbed cations and those in the surrounding solution, all of which are free to diffuse within the constraints of electrical neutrality. This equilibrium is established rapidly, allowing the exchange reactions to occur when the solution concentrations are altered, as in the experiment described above, or when fertilizers, irrigation water, or environmental pollutants are added to soils. Theoretical approaches to quantify this equilibrium are important to predict cation behavior in soil processes. While rigorous thermodynamic treatments are complicated (6), simplified empirical treatments are often more satisfactory. This author has found the most useful to be the *corrected rational selectivity coefficient* (7), which incorporates the activities of the solution ions and the equivalent fraction of the adsorbed ions. *Activity* has been described as effective concentration and is usually a little lower than the actual concentration measured in mole.dm^{-3}. *Equivalent fraction* is the fraction of the total negative charge occupied by the adsorbed cation. In a simple homovalent exchange, such as potassium displacing sodium,

$$\text{soil Na} + K^+ \leftrightarrow \text{soil K} + Na^+ \quad (3)$$

the selectivity coefficient $\boldsymbol{K}$ is given by

$$\boldsymbol{K} = \frac{X_K}{X_{Na}} \frac{(Na^+)}{(K^+)} \quad (4)$$

where X is the equivalent fraction of ion on the exchanger and the parentheses indicate the activity of that ion in solution. This selectivity coefficient shows the preference of the exchanger for one ion over another. In this example, if $\boldsymbol{K}$ is less than 1, sodium would be preferred by the exchanger, but in practice $\boldsymbol{K}$ is greater than 1 and potassium is preferred.

For a heterovalent exchange reaction, such as calcium displacing sodium,

$$\text{soil } Na_2 + Ca^{2+} \leftrightarrow \text{soil Ca} + 2Na^+ \quad (5)$$

the selectivity coefficient is given by

$$\boldsymbol{K} = \frac{X_{ca}}{[X_{Na}]^2} \frac{(Na+)2}{(Ca^{2+})}, \quad (6)$$

using the same conventions as previously.

In general, cations with higher charge are more strongly adsorbed by exchangers (e.g., $Al^{3+} > Ca^{2+} > Na^+$) and for cations with similar charge, those with higher radius and lower hydration energy are preferred (e.g., $Ba^{2+} > Ca^{2+} > Mg^{2+}$). Structural considerations may also be involved. For example, the size of the potassium ion enables it to fit neatly into the holes on the surface of silicate clays, thereby increasing the preference of the surface for potassium.

PRACTICAL IMPORTANCE OF CATION EXCHANGE

This section outlines three of the many practical applications of cation exchange.

Plant Nutrition (8)

Many plant nutrients exist as cations: the four base cations mentioned above, ammonium NH_4^+, and the heavy metal cations iron (Fe^{2+}), copper (Cu^{2+}), zinc (Zn^{2+}) and manganese (Mn^{2+}). Plants take up their nutrients as simple ions from solution, but when the solution is depleted, *exchange desorption* allows the adsorbed cations to provide a reserve from which the solution may be replenished (Fig. 1B). To maintain electrical neutrality, desorption of one ion must be accompanied by adsorption of another, usually either Ca^{2+} (the most abundant cation) or the hydrogen ion H_3O^+, which plants excrete when they take up cation nutrients.

Soil Development and Acidification (5, 9)

In a humid climate the course of soil development in a freely drained environment is always towards acidification. The hydrogen ion H_3O^+ is strongly adsorbed and it exchanges for the base cations on the soil surfaces, which are then leached out, so that the soil gradually becomes more acid. The main sources of hydrogen ions are the dissolution of carbon dioxide, the release of organic acids when plant residues decompose, the deposition of atmospheric pollution, and the addition of fertilizer.

Amelioration of Saline and Alkali Soils (5)

An excess of sodium salts and/or exchangeable sodium in these soils causes problems. Amelioration is effected by adding gypsum ($CaSO_4$) prior to leaching the soil with excess water. Calcium dissolves from the $CaSO_4$ and exchanges for sodium (as in the example of heterovalent exchange 5, above), which is then leached in the drainage water.

I

ANION EXCHANGE

Anion exchange (3–5) can occur on positively charged surfaces in a manner similar to cation exchange on negative surfaces (e.g., as with synthetic resins based on quaternary ammonium groups (7). Although there are no soil minerals that possess permanent positive charges and humus does not acquire a positive charge, strongly weathered acid soils containing amorphous hydrated iron and aluminum oxides can acquire a pH-dependent positive charge:

$$FeOOH + H^+ \rightarrow FeO^+ + H_2O \qquad (7)$$

This allows a small *anion exchange capacity* (AEC) to develop at a pH below 5, as illustrated in Fig. 2C, in addition to cation exchange on the negative surfaces of these variable-charge soils (3, 10).

The main practical importance of anion exchange is that it allows the anions chloride Cl^- and nitrate NO_3^- to be resistant to leaching from the soil. This is particularly important for nitrate, which provides a reserve for plant nutrition. Phosphate has often been associated with anion exchange in the popular soil literature, but because its bonding to both neutral and positively charged sites is by hydroxyl ligand exchange rather than by diffuse-ion swarming, it is not genuinely "exchangeable" within the meaning of this article.

REFERENCES

1. Page, A.L.; Miller, R.H.; Keeney, D.R. *Methods of Soil Analysis, Part 2, Chemical and Microbiological Properties*; American Society of Agronomy: Madison, Wisconsin, 1982.
2. Kelley, W.P. *Cation Exchange in Soils*, American Chemical Society, Monograph No.109; Reinhold: New York, 1948.
3. Mott, C.J.B. Surface Chemistry of Soil Particles. In *Russell's Soil Conditions and Plant Growth*, 11th Ed.; Wild, A., Ed.; Longman: Harlow, 1988; 239–281.
4. Sposito, G. *The Chemistry of Soils*; Oxford Univ. Press: New York, 1989.
5. McBride, M.B. *Environmental Chemistry of Soils*; Oxford Univ. Press: New York, 1994.
6. Sposito, G. *The Thermodynamics of Soil Solutions*; Clarendon Press: Oxford, 1981.
7. Helfferich, F. *Ion Exchange*; McGraw Hill: New York, 1962.
8. Mengel, K.; Kirkby, E.A. *Principles of Plant Nutrition*, 3rd Ed. International Potash Institute: Bern, 1982.
9. Bache, B.W. Soil Acidification and Aluminium Mobility. Soil Use and Management **1985**, *1*, 10–14.
10. Gillman, G.P.; Sumpter, E.A. Surface Charge Characteristics and Lime Requirement of Soils Derived from Basaltic, Granitic and Metamorphic Rocks in High-Rainfall Tropical Queensland. Aust. J. Soil. Res. **1986**, *24*, 173–192.

IRRIGATION

Jack Keller
Utah State University, Logan, Utah, U.S.A.

INTRODUCTION

The basic idea behind irrigation is to supply water to and store water in soils for later use by plants where or when insufficient water for the desired level of plant growth is supplied by natural precipitation. There are three basic methods used for supplying irrigation water to the surface of soils and two for applying water beneath the soil surface.

IRRIGATION METHODS

The three application methods for applying water to the surface of soils are: 1) surface irrigation, where the soil surface is used to convey the irrigation water to where it will be stored in the soil; 2) sprinkle irrigation, where the water is sprinkled on the soil surface where it is to be stored; and 3) drip irrigation, where the water is directly applied by dripping or being sprayed from small holes in the pipe network to where it is to be stored in the soil. The two methods for applying irrigation water beneath the soil surface are: 1) subsurface drip irrigation, where the network of pipe and outlets are buried beneath the soil surface; and 2) subirrigation, where the water table is controlled and the water reaches the plant root system through capillary action.

These methods of irrigation and irrigation efficiency issues are discussed in detail in other sections of this encyclopedia.

WATER STORAGE CAPACITY OF SOILS

Understanding basic soil–water–plant relations is central to the ability to design and manage irrigation systems. The tables related to soil water presented in this section provide useful guidelines for estimating irrigation requirements and for preliminary irrigation design purposes. However, more site-specific information should be used for designing and managing irrigation systems and evaluating irrigation system performance.

Soil Water Holding Capacity

Soils of various textures have varying abilities to retain water. Except for required periodic leaching, any irrigation beyond the field capacity of the soil is an economic loss. Table 1 gives typical ranges of available water holding capacities (field capacity minus permanent wilting point) of soils of different textures. Table 1 should only be used as a guide for preliminary system designs and management purposes. Final designs and management decisions should be based on actual field data.

Root Depth

The total amount of soil water available for plant use in any soil is the sum of the available water-holding capacities of all soil horizons occupied by plant roots. Typical plant feeder root and total root depth are given in many references; however, the actual depths of rooting of the various crops are affected by soil conditions. Therefore, the root depth for any set of crop and site should be checked. Where local data are not available and there are no expected root restrictions, Table 2 can be used as a guide to estimating the effective root depths. The values given are taken from Ref. 2 and are based on the author's estimate of averages selected from a large number of references. They represent the depth at which crops will obtain the major portion of the water they need when grown in a deep, well-drained soil that is adequately irrigated.

CONSUMPTIVE USE AND DESIGN

Deciding how much water an irrigation system should be able to deliver to a crop over a given period is ultimately a question of selecting a capacity that will maximize profits to the farmer. To begin to address this question of system capacity, it is necessary to know how much water a crop will use over the entire growing season and during the part of the season when water use is at its peak. The rate of water use during this peak period provides the basis for determining the rate at which irrigation water should be

Table 1 Range in available water-holding capacity of soils of different texture

Soil texture	Water-holding capacity Range (mm/rn)	Average (mm/rn)
Very coarse texture—very coarse sands	33–62	42
Coarse texture—coarse sands, fine sands, and loamy soils	62–104	83
Moderately coarse texture—sandy loams	104–145	25
Medium texture—very fine sandy loams, loams, and silt loams	125–192	167
Moderately fine texture—clay loams, silty clay loams, and sandy clay loams	145–208	183
Fine texture—sandy clays, silty clays, and clays	133–208	192
Peats and mucks	167–250	208

(Adapted from Ref. 1.)

delivered to the field. Examples of typical seasonal and peak daily crop water requirements are given in Table 3.

SOIL MOISTURE MANAGEMENT

A general rule of thumb for many field crops in arid and semiarid regions is to maintain the soil moisture deficit (SMD) within the root zone above 50% of the total available water-holding capacity. This is a management-allowed deficit (MAD = 50%) because it is also desirable to bring the moisture level back to field capacity with each irrigation. (In humid regions, it is necessary to allow for rains during the irrigation period. However, the 50% limitation on soil moisture depletion should be followed as a general guide for field crops.)

Soil management, water management, and economic considerations determine the amount of water used in irrigating and the rate of water application. The standard approach has been to determine the amount of water needed to fill the entire root zone to field capacity and then to apply a sufficiently larger amount to account for evaporation, leaching, and efficiency of application. The traditional approach to the frequency of application has been to take the depth of water in the root zone reservoir that can be extracted assuming MAD = 50% and, using the daily consumptive use rate of the crop, to determine how long this supply will last. Such an approach is useful only as a guide to irrigation requirements, as many factors affect the volume and timing of applications for optimal design and operation of a system.

Table 4 is presented as a guide for selecting the appropriate MAD or for near optimum production of various crops. As indicated in Table 4 for crops with a high market value, it is often profitable to irrigate well before half of the Soil moisture in the root zone has been depleted, i.e., SMD = 50%.

Irrigation Depth

The maximum net depth of water to be applied per irrigation, d_x, is the same as the maximum allowable depletion of soil water between irrigations. Assuming no additional water is necessary for leaching purposes, it is computed by:

$$d_x = \frac{\text{MAD}}{100} W_a Z \tag{1}$$

where

d_x Maximum net depth of water to be applied and stored in the root zone per irrigation event [mm (in.)].

MAD Management-allowed deficit, which can be estimated from Table 4 (%).

W_a Available water-holding capacity of the soil, which can be estimated from Table 1 [mm/m (in./ft)].

Z Effective root depth, which can be taken from Table 2 [mm (ft)].

Irrigation Interval

The appropriate irrigation interval, which is the time that should elapse between the beginning of two successive irrigation events, is determined by:

$$f' = \frac{d_n}{U_d} \tag{2}$$

where

f' = Irrigation interval or frequency (days).

d_n = Net depth of water applied and stored in the soil

Table 2 Effective crop root depths that would contain approximately 80% of the feeder roots in a deep, uniform, well-drained soil profile[a]

Crop	Root depth (M)	Crop	Root depth (M)
Alfalfa	1.2–1.8	Lettuce	0.2–0.5
Almonds	0.6–1.2	Lucerne	1.2–1.8
Apple	0.8–1.2	Oats	0.6–1.1
Apricot	0.6–1.4	Olives	0.9–1.5
Artichoke	0.6–0.9	Onion	0.3–0.6
Asparagus	1.2–1.8	Parsnip	0.6–0.9
Avocado	0.6–0.9	Passion fruit	0.3–0.5
Banana	0.3–0.6	Pastures	0.3–0.8
Barley	0.9–1.1	Pea	0.4–0.8
Bean (dry)	0.6–1.2	Peach	0.6–1.2
Bean (green)	0.5–0.9	Peanuts	0.4–0.8
Bean (lima)	0.6–1.2	Pear	0.6–1.2
Beet (sugar)	0.6–1.2	Pepper	0.6–0.9
Beet (table)	0.4–0.6	Plum	0.8–1.2
Berries	0.6–1.2	Potato (Irish)	0.6–0.9
Broccoli	0.6	Potato (sweet)	0.6–0.9
Brussel sprouts	0.6	Pumpkin	0.9–1.2
Cabbage	0.6	Radish	0.3
Cantaloupe	0.6–1.2	Safflower	0.9–1.5
Carrot	0.4–0.6	Sorghum (grain & sweet)	0.6–0.9
Cauliflower	0.6	Sorghum (silage)	0.9–1.2
Celery	0.6	Soybean	0.6–0.9
Chard	0.6–0.9	Spinach	0.4–0.6
Cherry	0.8–1.2	Squash	0.6–0.9
Citrus	0.9–1.5	Strawberry	0.3–0.5
Coffee	0.9–1.5	Sugarcane	0.5–1.1
Corn (grain & silage)	0.6–1.2	Sudan grass	0.9–1.2
Corn (sweet)	0.4–0.6	Tobacco	0.6–1.2
Cotton	0.6–1.8	Tomato	0.6–1.2
Cucumber	0.4–0.6	Turnip (white)	0.5–0.8
Eggplant	0.8	Walnuts	1.7–2.4
Fig	0.9	Watermelon	0.6–0.9
Flax	0.6–0.9	Wheat	0.8–1.1
Grapes	0.5–1.2		

[a] Approximately 80% of the feeder roots are in the top 60% of the soil profile. Soil and plant environmental factors often offset normal root development; therefore, soil density, pore shapes and sizes, soil–water status, aeration, nutrition, texture and structure modification, soluble salts, and plant root damage by organisms should all be taken into account.
(Adapted from Ref. 2.)

per irrigation event, to meet consumptive use requirements [mm (in.)].

U_d = Conventionally computed average daily crop water requirement, or use rate, during the peak-use month, which can be estimated from Table 3 [mm/day (in./day)].

The value selected for d_n will depend upon system design and environmental factors and should be equal to or less than d_x. When d_n is replaced by d_x in Eq. 2, f' becomes the maximum irrigation interval, f_x.

IRRIGATION SYSTEM DESIGN, MANAGEMENT AND SCHEDULING

General Design Concepts

Design of all irrigation systems is process of synthesis, where properties, such as a soil's intake rate and crop water requirements; items, such as canals, pipes, and pumps; and processes, such as trenching, coaxing water down furrows, or moving pipe, must be integrated to

Table 3 Typical peak daily and seasonal crop water requirements in different climates

	Type of climate and water requirements (mm)									
	Cool		Moderate		Hot		High desert		Low desert	
Crop	Daily	Seas	Daily	Seas	Daily	Seas	Daily	Seas	Daily	Seas
Alfalfa	5.1	635	6.4	762	7.6	914	8.9	1016	10.2	1219
Pasture	4.6	508	5.6	610	6.6	711	7.6	762	8.9	914
Grain	3.8	381	5.1	457	5.8	508	6.6	533	5.8[a]	508[a]
Beets	4.6	584	5.8	635	6.9	711	8.1	732	9.1	914
Beans	4.6	330	5.1	381	6.1	457	7.1	508	7.6	559
Corn	5.1	508	6.4	559	7.6	610	8.9	660	10.2	762
Cotton	–	–	6.4	559	7.6	660	–	–	10.2	813
Peas	4.6	305	4.8	330	5.1	356	5.6	356	5.1[b]	356[b]
Tomatoes	4.6	457	5.1	508	5.6	559	6.4	610	7.1	660
Potatoes	4.6	406	5.8	457	6.9	553	8.1	584	6.9[b]	533[b]
Truck vegetables	4.1	305	4.6	356	5.1	406	5.6	457	6.3[b]	508[b]
Melons	4.1	381	4.6	406	5.1	457	5.6	508	6.4[b]	559[b]
Strawberries	4.6	457	5.1	508	5.6	559	6.1	610	6.6	660
Citrus	4.1	508	4.6	559	5.1	660	–	–	5.6	711
Citrus (w/cover)	5.1	635	5.6	711	6.4	813	–	–	6.9	889
Dec orchard	3.8	483	4.8	533	5.8	584	6.6	635	7.6	762
Dec orchard (w/cover)	5.1	635	6.4	711	7.6	813	8.9	914	10.2	1016
Vineyards	3.6	356	4.1	406	4.8	457	5.6	508	6.4	610

[a] Winter planting.
[b] Fall or winter planting.
(From Ref. 2.)

form a good irrigation system. The irrigation designer's art is to know the kinds of hardware and management techniques appropriate for a given cropping system and site and to have a clear mental image of what the system can accomplish and how the completed system will appear. Careful site analysis is essential as it provides data that lead to an understanding of the physical, economic, and social resources that determine what can and ought to be accomplished by a proposed irrigation system.

Table 4 Guide for selecting management-allowed deficit (MAD), values for various crops

MAD (%)	Crop and root depth
25–40	Shallow-rooted, high-value fruit, and vegetable crops
40–50	Orchards,[a] vineyards, berries, and medium-rooted row crops
50	Forage crops, grain crops, and deep-rooted row crops

[a] Some fresh fruit orchards require lower multiwavelength anomalous diffraction or values during fruit finishing for sizing.
(From Ref. 2.)

Generally irrigation systems are designed to meet average peak-water-use requirements. Sometimes, to reduce costs or to stretch limited water supplies, systems are designed to optimize production per unit of water applied. In such cases systems can be designed to apply only about 80% of peak water requirements and still obtain up to 95% of optimum yields. For deep-rooted crops in fine-textured soils, an appreciable amount of water can be stored prior to the critical peak-use periods. By drawing on this stored water, peak system delivery requirements can be reduced without reducing yield potential providing the quality of the irrigation water is high.

Salinity Control

All irrigation water contains some dissolved salts that are pushed downward by rainfall and application of irrigation water. By applying more water than the plants consume, most of the salts can be pushed or leached below the root zone. The first step in computing the additional water

Table 5 Values of EC_e that will give 10% yield reduction for various crops[a]

Crop	EC_e − dS/m	Crop	EC_e − dS/m
Field crops		*Vegetable crops*	
Barley	10	Beets	5.1
Cotton	9.6	Broccoli	3.9
Sugar beets	8.7	Tomato	3.5
Wheat	7.4	Cucumber	3.3
Soybean	5.5	Cantaloupe	3.6
Sorghum	5.1	Spinach	3.3
Groundnut	3.5	Cabbage	2.8
Rice	3.8	Potato	2.5
Corn	2.5	Sweet corn	2.5
Flax	2.5	Sweet potato	2.4
Broadbeans	2.6	Pepper	2.2
Cowpeas	2.2	Lettuce	2.1
Beans	1.5	Radish	2
Fruit and nut crops		Onion	1.8
Date palm	6.8	Carrot	1.7
Fig, olive	3.8	Beans	1.5
Pomegranate	3.8	*Forage crops*	
Grapefruit	2.4	Tall wheat grass	9.9
Orange	2.3	Bermuda grass	8.5
Lemon	2.3	Barley (hay)	7.4
Apple, pear	2.3	Rye grass	6.9
Walnut	2.3	Crested wheat grass	6
Peach	2.2	Tall fescue	5.8
Apricot	2	Sudan grass	5.1
Grape	2.5	Wild rye grass	4.4
Almond	2	Vetch	3.9
Plum	2.1	Alfalfa	3.4
Blackberry	2	Corn (forage)	3.2
Boysenberry	2	Berseem clover	3.2
Avocado	1.8	Orchard grass	3.1
Raspberry	1.4	Clover	2.3
Strawberry	1.3		

[a](Adapted from Ref. 3.)

required for leaching is to determine the leaching requirement by:

$$LR = \frac{EC_w}{5\ EC_e - EC_w} \tag{3}$$

where

LR = Leaching requirement ratio for sprinkle or surface irrigation.

EC_w = Electrical conductivity of the irrigation water [dS/m (mmhos/cm)].

EC_e = Estimated electrical conductivity of the average saturation extract of the soil root zone profile for an appropriate yield reduction [dS/m (mmhos/cm)].

Unless more specific information on specific crop cultivars is available the EC_e values presented in Table 5, which are taken from (3), can be used in Eq. 3. These are values that will give an approximate 10% yield reduction, as presented by (3). (For conversion purposes: 1.0 ppm = 640 × EC in dS/cm.)

Under full irrigation, where $LR < 0.1$, the annual deep percolation losses—even in most of the least watered areas—will normally be sufficient to provide the necessary leaching. However, under deficit irrigation or when $LR < 0.1$, water in addition to the consumptive use should be applied or available at some time during the year to satisfy leaching requirements. The ratio of the total depth of irrigation water required with and without leaching is equal to $1/(1 - LR)$.

I

Drainage

Drainage of the excess water from the soil profile and conveying it to a sump for reuse or disposal is as important as irrigation. In fact natural or man-made subsurface drainage is essential for sustaining irrigated agricultural production. Without it leaching will not be possible, causing salts to accumulate until they become toxic to plant growth, and the water table will rise and literally drown the plants. Furthermore, the productive capability of the land itself may be severely damaged and require major reclamation to become productive again. Details regarding drainage needs are presented elsewhere in this encyclopedia.

Application Depth and Frequency

For periodic-move, and low-frequency, continuous-move systems, such as traveling sprinklers, it is desirable to irrigate as infrequently as practical to reduce labor costs and Eq. 1 will apply. For drip solid-set and center-pivot sprinkle systems labor costs are not a major consideration and the irrigation frequency can be selected to provide the optimal environment for plant growth, water conservation, and economic production within the physical limitations of the system.

Systems are usually designed so that their discharge, depths of application, and irrigation frequency meet crop water requirements during the peak consumptive use period. Therefore, systems must be managed to avoid wasting water, labor, and energy, and leaching nutrients from the soil during periods of the crops' growth cycle when water requirements are less, when the crops' roots may not have penetrated to their full depth, and during rainy periods. For optimum efficiency, irrigation applications should be scientifically scheduled from water budgets based on crop evapotranspiration estimates or soil moisture observations.

REFERENCES

1. Soil Conservation Service (SCS), Irrigation Water Requirements. *Technical Release 21. Chapter 1, Section 15, Soil Conservation Service's (SCS) National Engineering Handbook*; U.S. Department of Agriculture: Washington, DC, 1970.
2. Keller, J.; Bliesner, R.D. Soil–Water–Plant Relations. In *Sprinkle and Trickle Irrigation*, 1st Ed.; Van Norstrand Reinhold: New York, 1990.
3. Ayers, R.S.; Westcott, D.W. Salinity Problems. *Water Quality for Agricultures*; Irrigation and Drainage Paper 29, Rev. 1, Food and Agricultural Organization of the United Nations: Rome, 1985; 31–33.

IRRIGATION EFFICIENCY[a]

Terry A. Howell
United States Department of Agriculture-Agricultural Research Service (USDA-ARS), Bushland, Texas, U.S.A.

INTRODUCTION

Irrigation efficiency is a critical measure of irrigation performance in terms of the water required to irrigate a field, farm, basin, irrigation district, or an entire watershed. The value of irrigation efficiency and its definition are important to the societal views of irrigated agriculture and its benefit in supplying the high quality, abundant food supply required to meet our growing world's population. "Irrigation efficiency" is a basic engineering term used in irrigation science to characterize irrigation performance, evaluate irrigation water use, and to promote better or improved use of water resources, particularly those used in agriculture and turf/landscape management (1–4). Irrigation efficiency is defined in terms of: 1) the irrigation system performance, 2) the uniformity of the water application, and 3) the response of the crop to irrigation. Each of these irrigation efficiency measures is interrelated and will vary with scale and time. Fig. 1 illustrates several of the water-transport components involved in defining various irrigation performance measures. The spatial scale can vary from a single irrigation application device (a siphon tube, a gated pipe gate, a sprinkler, a microirrigation emitter) to an irrigation set (basin plot, a furrow set, a single sprinkler lateral, or a microirrigation lateral) to broader land scales (field, farm, an irrigation canal lateral, a whole irrigation district, a basin or watershed, a river system, or an aquifer). The timescale can vary from a single application (or irrigation set), a part of the crop season (preplanting, emergence to bloom or pollination, or reproduction to maturity), the irrigation season, to a crop season, or a year, partial year (pre-monsoon season, summer, etc.), or a water year (typically from the beginning of spring snow melt through the end of irrigation diversion, or a rainy or monsoon season), or a period of years (a drought or a "wet" cycle). Irrigation efficiency affects the economics of irrigation, the amount of water needed to irrigate a specific land area, the spatial uniformity of the crop and its yield, the amount of water that might percolate beneath the crop root zone, the amount of water that can return to surface sources for downstream uses or to groundwater aquifers that might supply other water uses, and the amount of water lost to unrecoverable sources (salt sink, saline aquifer, ocean, or unsaturated vadose zone).

The volumes of the water for the various irrigation components are typically given in units of depth (volume per unit area) or simply the volume for the area being evaluated. Irrigation water application volume is difficult to measure, so it is usually computed as the product of water flow rate and time. This places emphasis on accurately measuring the flow rate. It remains difficult to accurately measure water percolation volumes, groundwater flow volumes, and water uptake from shallow ground water.

IRRIGATION SYSTEM PERFORMANCE EFFICIENCY

Irrigation water can be diverted from a storage reservoir and transported to the field or farm through a system of canals or pipelines; it can be pumped from a reservoir on the farm and transported through a system of farm canals or pipelines; or it might be pumped from a single well or a series of wells through farm canals or pipelines. Irrigation districts often include small to moderate size reservoirs to regulate flow and to provide short-term storage to manage the diverted water with the on-farm demand. Some on-farm systems include reservoirs for storage or regulation of flows from multiple wells.

Water Conveyance Efficiency

The conveyance efficiency is typically defined as the ratio between the water that reaches a farm or field and that

[a] The use of trade, firm, or corporation names in this publication is for the information and convenience of the reader. Such use does not constitute an official endorsement or approval by the United States Department of Agriculture or the Agricultural Research Service of any product or service to the exclusion of others that may be suitable. This work was done by a U.S. Government employee and cannot be copyrighted. It may be reproduced freely for any noncommercial use.

Encyclopedia of Soil Science

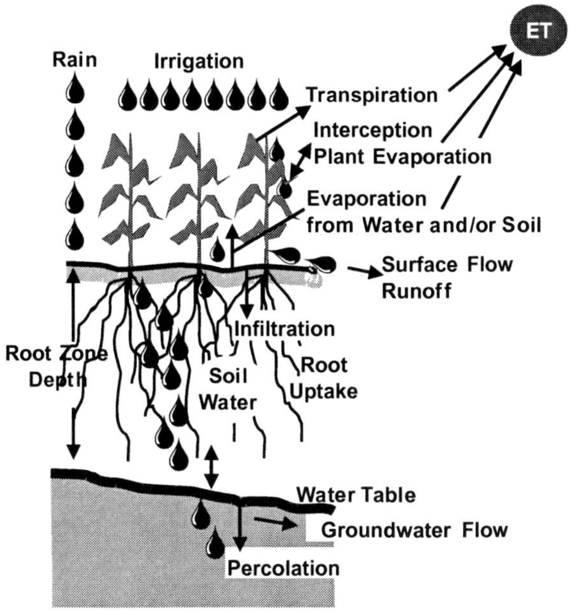

Fig. 1 Illustration of the various water transport components needed to characterize irrigation efficiency.

diverted from the irrigation water source (1, 3, 4). It is defined as

$$E_c = 100\frac{V_f}{V_t} \tag{1}$$

where E_c is the conveyance efficiency (%), V_f is the volume of water that reaches the farm or field (m^3), and V_t is the volume of water diverted (m^3) from the source. E_c also applies to segments of canals or pipelines, where the water losses include canal seepage or leaks in pipelines. The global E_c can be computed as the product of the individual component efficiencies, E_{ci}, where i represents the segment number. Conveyance losses include any canal spills (operational or accidental) and reservoir seepage and evaporation that might result from management as well as losses resulting from the physical configuration or condition of the irrigation system. Typically, conveyance losses are much lower for closed conduits or pipelines (4) compared with unlined or lined canals. Even the conveyance efficiency of lined canals may decline over time due to material deterioration or poor maintenance.

Application Efficiency

Application efficiency relates to the actual storage of water in the root zone to meet the crop water needs in relation to the water applied to the field. It might be defined for individual irrigations or parts of irrigations (irrigation sets). Application efficiency includes any application losses to evaporation or seepage from surface water channels or furrows, any leaks from sprinkler or drip pipelines, percolation beneath the root zone, drift from sprinklers, evaporation of droplets in the air, or runoff from the field. Application efficiency is defined as

$$E_a = 100\frac{V_s}{V_f} \tag{2}$$

where E_a is the application efficiency, (%), V_s is the irrigation needed by the crop (m^3), and V_f is the water delivered to the field or farm (m^3). The root zone may not need to be fully refilled, particularly if some root zone water-holding capacity is needed to store possible or likely rainfall. Often, V_s is characterized as the volume of water stored in the root zone from the irrigation application. Some irrigations may be applied for reasons other than meeting the crop water requirement (germination, frost control, crop cooling, chemigation, fertigation, or weed germination). The crop need is often based on the "beneficial water needs" (5). In some surface irrigation systems, the runoff water that is necessary to achieve good uniformity across the field can be recovered in a "tailwater pit" and recirculated with the current irrigation or used for later irrigations, and V_f should be adjusted to account for the "net" recovered tailwater. Efficiency values are typically site specific. Table 1 provides a range of typical farm and field irrigation application efficiencies (6–8) and potential or attainable efficiencies for different irrigation methods that assumes irrigations are applied to meet the crop need.

Storage Efficiency

Since the crop root zone may not need to be refilled with each irrigation, the storage efficiency has been defined (4). The storage efficiency is given as

$$E_s = 100\frac{V_s}{V_{rz}} \tag{3}$$

where E_s is the storage efficiency (%) and V_{rz} is the root zone storage capacity (m^3). The root zone depth and the water-holding capacity of the root zone determine V_{rz}. The storage efficiency has little utility for sprinkler or microirrigation because these irrigation methods seldom refill the root zone, while it is more often applied to surface irrigation methods (4).

Seasonal Irrigation Efficiency

The seasonal irrigation efficiency is defined as

$$E_i = 100\frac{V_b}{V_f} \tag{4}$$

Table 1 Example of farm and field irrigation application efficiency and attainable efficiencies

	Field efficiency (%)			Farm efficiency (%)		
Irrigation method	**Attainable**	**Range**	**Average**	**Attainable**	**Range**	**Average**
Surface						
Graded furrow	75	50–80	65	70	40–70	65
w/tailwater reuse	85	60–90	75	85	–	–
Level furrow	85	65–95	80	85	–	–
Graded border	80	50–80	65	75	–	–
Level basins	90	80–95	85	80	–	–
Sprinkler						
Periodic move	80	60–85	75	80	60–90	80
Side roll	80	60–85	75	80	60–85	80
Moving big gun	75	55–75	65	80	60–80	70
Center pivot						
Impact heads w/end gun	85	75–90	80	85	75–90	80
Spray heads wo/end gun	95	75–95	90	85	75–95	90
LEPA[a] wo/end gun	98	80–98	95	95	80–98	92
Lateral move						
Spray heads w/hose feed	95	75–95	90	85	80–98	90
Spray heads w/canal feed	90	70–95	85	90	75–95	85
Microirrigation						
Trickle	95	70–95	85	95	75–95	85
Subsurface drip	95	75–95	90	95	75–95	90
Microspray	95	70–95	85	95	70–95	85
Water Table Control						
Surface ditch	80	50–80	65	80	50–80	60
Subsurface drain lines	85	60–80	75	85	65–85	70

[a] LEPA is low energy precision application.
(From Refs. 6, 7, 11.)

where E_i is the seasonal irrigation efficiency (%) and V_b is the water volume beneficially used by the crop (m^3). V_b is somewhat subjective (4, 5), but it basically includes the required crop evapotranspiration (ET_c) plus any required leaching water (V_l) for salinity management of the crop root zone.

Leaching requirement (or the leaching fraction)

The leaching requirement (9), also called the leaching fraction, is defined as

$$L_r = \frac{V_d}{V_f} = \frac{EC_i}{EC_d} \tag{5}$$

where L_r is the leaching requirement, V_d is the volume of drainage water (m^3), V_f is the volume of irrigation (m^3) applied to the farm or field, EC_i is the electrical conductivity of the irrigation water ($dS\,m^{-1}$), and EC_d is the electrical conductivity of the drainage water ($dS\,m^{-1}$). The L_r is related to the irrigation application efficiency, particularly when drainage is the primary irrigation loss component. The L_r would be required "beneficial" irrigation use

$$V_l \equiv L_r V_i,$$

so only V_d greater than the minimum required leaching should reduce irrigation efficiency. Then, the irrigation efficiency can be determined by combining Eq. (4) and (5)

$$E_i = 100\left(\frac{V_b}{V_f} + L_r\right) \tag{6}$$

Burt et al. (5) defined the "beneficial" water use to include possible off-site needs to benefit society (riparian needs or wildlife or fishery needs). They also indicated that V_f should not include the change in the field or farm storage of water, principally soil water but it could include field (tailwater pits) or farm water storage (a reservoir) that was not used within the time frame that was used to define E_i.

IRRIGATION UNIFORMITY

The fraction of water used efficiently and beneficially is important for improved irrigation practice. The uniformity of the applied water significantly affects irrigation efficiency. This uniformity is a statistical property of the applied water's distribution. This distribution depends on many factors that are related to the method of irrigation, soil topography, soil hydraulic or infiltration characteristics, and hydraulic characteristics (pressure, flow rate, etc.) of the irrigation system. Irrigation application distributions are usually based on depths of water (volume per unit area); however, for microirrigation systems they are usually based on emitter flow volumes because the entire land area is not typically wetted.

Christiansen's Uniformity Coefficient

Christiansen (10) proposed a coefficient intended mainly for sprinkler systems based on the catch volumes given as

$$C_U = 100\left[\frac{1 - (\sum | X - \bar{x} |)}{\sum X}\right] \tag{7}$$

where C_U is the Christiansen's uniformity coefficient (%), X is the depth (or volume) of water in each of the equally spaced catch containers (mm or ml), and $\bar{x}$ is the mean depth (volume) of the catch (mm or ml). For C_U values >70%, Hart (11) and Keller and Bliesner (8) presented

$$C_U = 100\left[1 - \left(\frac{\sigma}{x}\right)\left(\frac{2}{\pi}\right)^{0.5}\right] \tag{8}$$

where σ is the standard deviation of the catch depth (mm) or volume (ml). Eq. 8 approximates the normal distribution for the catch amounts.

The C_U should be weighted by the area represented by the container (12) when the sprinkler catch containers intentionally represent unequal land areas, as is the case for catch containers beneath a center pivot. Heermann and Hein (12) revised the C_U formula (Eq. 8) to reflect the weighted area, particularly intended for a center pivot sprinkler, as follows:

$$C_{U(H\&H)} = 100\left\{1 - \left[\frac{\left(\sum S_i \left| V_i - \left(\frac{\sum V_i S_i}{\sum S_i}\right)\right|\right)}{\sum (V_i S_i)}\right]\right\} \tag{9}$$

where S_i is the distance (m) from the pivot to the ith equally spaced catch container and V_i is the volume of the catch in the ith container (mm or ml).

Low-Quarter Distribution Uniformity

The distribution uniformity represents the spatial evenness of the applied water across a field or a farm as well as within a field or farm. The general form of the distribution uniformity can be given as

$$D_{Up} = 100\left(\frac{\bar{V}_p}{\bar{V}_f}\right) \tag{10}$$

where D_{Up} is the distribution uniformity (%) for the lowest p fraction of the field or farm (lowest one-half $p = 1/2$, lowest one-quarter $p = 1/4$), $\bar{V}_p$ is the mean application volume (m^3), and $\bar{V}_f$ is the mean application volume (m^3) for the whole field or farm. When $p = 1/2$ and $C_U > 70\%$, then the D_U and C_U are essentially equal (13). The USDA-NRCS (formerly, the Soil Conservation Service) has widely used D_{Ulq} ($p = 1/4$) for surface irrigation to access the uniformity applied to a field, i.e., by the irrigation volume (amount) received by the lowest one-quarter of the field from applications for the whole field. Typically, D_{Up} is based on the post-irrigation measurement (5) of water volume that infiltrates the soil because it can more easily be measured and better represents the water available to the crop. However, the post-irrigation infiltrated water ignores any water intercepted by the crop and evaporated and any soil water evaporation that occurs before the measurement. Any water that percolates beneath the root zone or the sampling depth will also be ignored.

The D_U and C_U coefficients are mathematically interrelated through the statistical variation (coefficient of variation, $\sigma/\bar{x}$, C_v) and the type of distribution. Warrick (13) presented relationships between D_U and C_U for normal, log-normal, uniform, specialized power, beta-, and gamma-distributions of applied irrigations.

Emission Uniformity

For microirrigation systems, both the C_U and D_U concepts are impractical because the entire soil surface is not wetted. Keller and Karmeli (14) developed an equation for microirrigation design as follows

$$E_U = 100\left[1 - 1.27(C_{vm})n^{-1/2}\right]\left(\frac{q_m}{\bar{q}}\right) \tag{11}$$

where E_U is the design emission uniformity (%), C_{vm} is the manufacturer's coefficient of variability in emission device flow rate (l hr^{-1}), n is the number of emitters per plant, q_m is the minimum emission device flow rate (l hr^{-1}) at the minimum system pressure, and $\bar{q}$ is the mean emission device flow rate (l hr^{-1}). This equation is based on the D_{Ulq} concept (4), and includes the influence of multiple emitters per plant that each may have a flow rate from a population of random flow rates based on the emission device manufacturing variation. Nakayama et al. (15) developed a design coefficient based more closely on the C_U concept for emission device flow rates from a normal distribution given as

$$C_{Ud} = 100\,(1 - 0.798(C_{vm})n^{-1/2}) \qquad (12)$$

where C_{Ud} is the coefficient of design uniformity (%) and the numerical value, 0.798, is

$$\left(\frac{2}{\pi}\right)^{0.5}$$

from Eq. 8.

Many additional factors affect microirrigation uniformity including hydraulic factors, topographic factors, and emitter plugging or clogging.

WATER USE EFFICIENCY

The previous sections discussed the engineering aspects of irrigation efficiency. Irrigation efficiency is clearly influenced by the amount of water used in relation to the irrigation water applied to the crop and the uniformity of the applied water. These efficiency factors impact irrigation costs, irrigation design, and more important, in some cases, the crop productivity. The water use efficiency has been the most widely used parameter to describe irrigation effectiveness in terms of crop yield. Viets (16) defined water use efficiency as

$$\mathrm{WUE} = \frac{Y_g}{\mathrm{ET}} \qquad (13)$$

where WUE is water use efficiency (kg m^{-3}), Y_g is the economic yield (g m^{-2}), and ET is the crop water use (mm). WUE is usually expressed by the economic yield, but it has been historically expressed as well in terms of the crop dry matter yield (either total biomass or aboveground dry matter). These two WUE bases (economic yield or dry matter yield) have led to some inconsistencies in the use of the WUE concept. The transpiration ratio (transpiration per unit dry matter) is a more consistent value that depends primarily on crop species and the environmental evaporative demand (17), and it is simply the inverse of WUE expressed on a dry matter basis.

Irrigation Water Use Efficiency

The previous discussion of WUE does not explicitly explain the crop yield response to irrigation. WUE is influenced by the crop water use (ET). Bos (3) defined a term for water use efficiency to characterize the influence of irrigation on WUE as

$$\mathrm{WUE} = \frac{(Y_{gi} - Y_{gd})}{(\mathrm{ET}_i - \mathrm{ET_d})} \qquad (14)$$

where WUE is irrigation water use efficiency (kg m^{-3}), Y_{gi} is the economic yield (g m^{-2}) for irrigation level i, Y_{gd} is the dryland yield (g m^{-2}; actually, the crop yield without irrigation), ET_i is the evapotranspiration (mm) for irrigation level i, and $\mathrm{ET_d}$ is the evapotranspiration of the dryland crops (or of the ET without irrigation). Although Eq. 14 seems easy to use, both Y_{gd} and $\mathrm{ET_d}$ are difficult to evaluate. If the purpose is to compare irrigation and dryland production systems, then dryland rather than nonirrigated conditions should be used. If the purpose is to compare irrigated regimes with an unirrigated regime, then appropriate values for Y_{gd} and $\mathrm{ET_d}$ should be used. Often, in most semiarid to arid locations, Y_{gd} may be zero. Bos (3) defined irrigation water use efficiency as

$$\mathrm{IWUE} = \frac{(Y_{gi} - Y_{gd})}{\mathrm{IRR}_i} \qquad (15)$$

where IWUE is the irrigation efficiency (kg m^{-3}) and IRR_i is the irrigation water applied (mm) for irrigation level i. In Eq. 15, Y_{gd} may be often zero in many arid situations.

SUMMARY

Irrigation efficiency is an important engineering term that involves understanding soil and agronomic sciences to achieve the greatest benefit from irrigation. The enhanced understanding of irrigation efficiency can improve the beneficial use of limited and declining water resources needed to enhance crop and food production from irrigated lands.

REFERENCES

1. Israelsen, O.R.; Hansen, V.E. *Irrigation Principles and Practices*; 3rd Ed.; Wiley: New York, 1962; 447.
2. ASCE; Describing Irrigation Efficiency and Uniformity. J. Irrig. Drain. Div., ASCE **1978**; *104* (IR1), 35–41.
3. Bos, M.G. Standards for Irrigation Efficiencies of ICID. J. Irrig. Drain. Div., ASCE **1979**; *105* (IR1), 37–43.
4. Heermann, D.F.; Wallender, W.W.; Bos, M.G. Irrigation Efficiency and Uniformity. In *Management of Farm Irrigation Systems*; Hoffman, G.J., Howell, T.A., Solomon, K.H., Eds.; Am. Soc. Agric. Engrs: St. Joseph, MI, 1990; 125–149.

5. Burt, C.M.; Clemmens, A.J.; Strelkoff, T.S.; Solomon, K.H.; Bliesner, R.D.; Hardy, L.A.; Howell, T.A.; Eisenhauer, D.E. Irrigation Performance Measures: Efficiency and Uniformity. J. Irrig. Drain. Eng. **1997**; *123* (3), 423–442.
6. Howell, T.A. Irrigation Efficiencies. In *Handbook of Engineering in Agriculture*; Brown, R.H., Ed.; CRC Press: Boca Raton, FL, 1988; Vol. I, 173–184.
7. Merriam, J.L.; Keller, J. *Farm Irrigation System Evaluation: A Guide for Management*; Utah State Univ. Logan, UT, 1978; 271.
8. Keller, J.; Bliesner, R.D. *Sprinkle and Trickle Irrigation*; The Blackburn Press: Caldwell, NJ, 2000; 652.
9. U.S. Salinity Laboratory Staff, *Diagnosis and Improvement of Saline and Alkali Soils*; Handbook 60, U.S. Govt. Printing Office: Washington, DC, 1954; 160.
10. Christiansen, J.E. *Irrigation by Sprinkling*; California Agric. Exp. Bull. No. 570, Univ. of Calif. Berkeley, CA, 1942; 94.
11. Hart, W.E. Overhead Irrigation by Sprinkling. Agric. Eng. **1961**; *42* (7), 354–355.
12. Heermann, D.F.; Hein, P.R. Performance Characteristics of Self-Propelled Center-Pivot Sprinkler Machines. Trans. ASAE **1968**; *11* (1), 11–15.
13. Warrick, A.W. Interrelationships of Irrigation Uniformity Terms. J. Irrig. Drain. Eng., ASCE **1983**; *109* (3), 317–332.
14. Keller, J.; Karmeli, D. *Trickle Irrigation Design*; Rainbird Sprinkler Manufacturing. Glendora, CA, 1975; 133.
15. Nakayama, F.S.; Bucks, D.A.; Clemmens, A.J. Assessing Trickle Emitter Application Uniformity. Trans. ASAE **1979**; *22* (4), 816–821.
16. Viets, F.G. Fertilizers and the Efficient Use of Water. Adv. Agron. **1962**; *14*, 223–264.
17. Tanner, C.B.; Sinclair, T.R. Efficient Use of Water in Crop Production: Research or Re-Search? In *Limitations to Efficient Water Use in Crop Production*; Taylor, H.M., Jordan, W.R., Sinclair, T.R., Eds.; Am. Soc. Agron., Crop Sci. Soc. Am., Soil Sci. Soc. Am. Madison, WI, 1983; 1–27.

IRRIGATION EROSION

D.L. Bjorneberg
R.E. Sojka
United States Department of Agriculture-Agricultural Research Service (USDA-ARS), Kimberly, Idaho, U.S.A.

INTRODUCTION

Irrigation is important to global food production. About 15% of cropland (1) and 5% of food production land, which includes rangeland and permanent cropland (2), are irrigated. However, irrigated land produces more than 30% of the world's food (3), which is 2.5 times as much per unit area compared with nonirrigated production (1). In the United States, approximately 15% of the harvested cropland is irrigated, however, almost 40% of the total crop value is produced on irrigated land (4).

Although sprinkler- and drip-irrigated areas are increasing, most of the world's irrigated land uses surface or flood irrigation. The countries with the largest irrigated areas are India—59,000,000 hectares (ha), China—52,580,000 ha, United States—21,400,000 ha, and Pakistan—18,000,000 ha (2). These countries account for 55% of the world's irrigated land; all other countries have less than 10 million ha each of irrigated land (2). About 50% of the irrigated land in the United States is surface irrigated (5) although 95 to 99% of the irrigated land in India, China, and Pakistan is surface irrigated (6).

Soil erosion from irrigated fields has been discussed previously (7, 8) and we focus on unique aspects of irrigation-induced soil erosion that are important when managing and simulating soil erosion on irrigated lands.

Soil erosion mechanics can be divided into three components: detachment, transport, and deposition. Water droplets and flowing water detach soil particles; flowing water then transports these detached particles downstream; deposition occurs when flowing water can no longer transport the soil particles because flow rate decreases as water infiltrates or as rill slope or roughness change. Some particles are deposited within a few meters, although others are transported off the field with runoff water. These mechanisms are the same for surface irrigation, sprinkler irrigation, and rainfall; however, there are some systematic differences between irrigation and rainfall erosion, especially between surface irrigation and rainfall.

SURFACE IRRIGATION

Soil erosion is often a serious problem on surface-irrigated land (Figs. 1 and 2). Erosion rates as high as 145 Mg/ha in 1 h (9) and 40 Mg/ha in 30 min (10) were reported in some early surface irrigation erosion studies. These extreme losses do not represent a sustained seasonal rate. Annual soil losses of 1 to 141 Mg/ha from surface irrigated fields were reported in a 1980 southern Idaho study (11). Within-field erosion rates on the upper quarter of a furrow-irrigated field can be 10 to 30 times more than the field average erosion rate (12). Some soil eroded from the upper end of a field is deposited on the lower end, whereas some soil leaves the field with runoff.. Losing topsoil from the upper end of the field can decrease crop yields 25% compared with the lower end of the field (13).

Sediment cannot be transported without runoff. Runoff is planned with many surface irrigation schemes in order to irrigate all areas of the field adequately. Under ideal conditions, properly designed and managed sprinkler irrigation systems will not have any runoff from the irrigated area. However, economic and water supply constraints, along with variable slope and soil conditions, often force compromises in sprinkler irrigation design.

SPRINKLER IRRIGATION

Runoff is rarely a problem with solid-set sprinkler irrigation systems because stationary sprinklers uniformly apply water at low rates (e.g., 2 mm/h). At the other end of the spectrum are systems with continuously moving laterals (center-pivot and lateral-move systems), which apply water to smaller areas (5–20 m wide) at higher rates than solid-set systems (e.g., 80 mm/h). Traveling lateral systems must irrigate large fields to reduce cost per unit area; this necessitates high instantaneous application rates to meet crop water requirements over the entire field. Application rates for center-pivot and lateral-move irrigation systems often exceed the soil infiltration rate, therefore runoff is

Encyclopedia of Soil Science

Fig. 1 White area in the field caused by erosion from more than 80 years of surface irrigation.

almost always a potential problem. Sprinkler type, nozzle pressure, and nozzle size influence runoff and soil erosion by affecting application rate, wetted area, and droplet size. Low pressure sprinklers, which reduce energy costs, have smaller pattern widths and therefore greater application rates. Lower pressure also produces larger drops with greater impact energy on the soil.

Sprinkler systems, particularly center pivots, operate on variable slopes and topography. Slope direction relative to the lateral affects how runoff accumulates. If the lateral is perpendicular to the slope direction, runoff will tend to move away from the lateral where water is being applied, allowing water to infiltrate before traveling very far. However, if the slope is parallel to the lateral, runoff can accumulate down slope and begin flowing in erosive

Fig. 2 Eroded irrigation furrows near the inflow end of a field.

streams. Furthermore, if the lateral is traveling up slope, runoff will flow onto a previously wetted area; whereas with down slope travel, runoff can flow onto dry soil. These factors are further complicated by wheel tracks from moving sprinkler systems that create compacted channels for water flow.

SURFACE IRRIGATION AND RAINFALL EROSION DIFFERENCES

The most obvious difference between soil erosion from rain or sprinkler irrigation and from surface irrigation is the lack of water droplets impacting the soil during surface irrigation. This fundamental difference is important because droplet kinetic energy affects both erosion and infiltration (14). When rain begins, droplets wet the soil surface and detach soil particles; as runoff begins, rills form in wet soil. Water flowing in rills is also exposed to falling raindrops, which affects detachment, transport, and deposition in the rills.

For furrow irrigation, rills are mechanically formed in dry soil before irrigation begins. Water is applied to only a small portion of the soil surface. As water advances down the field, it flows over dry, loose soil on the first irrigation and dry, consolidated soil on subsequent irrigations. Irrigation water instantaneously wets the soil, rapidly displacing air adsorbed on internal soil particle surfaces (15). The rapid replacement of air with water breaks apart soil aggregates (7), increasing the erodibility of the soil. Preliminary results from a southern Idaho field study showed that soil erosion from initially dry furrows was greater than erosion from furrows that were pre-wet by drip irrigation.

The hydraulics of rill flow from rain differ from furrow irrigation. Rill flow rate tends to increase downstream as additional rain water plus sheet and rill flow combine. During furrow irrigation, flow rate decreases with distance down the furrow as water infiltrates and increases with time as infiltration rate decreases which changes sediment detachment and transport capacities with distance and time. The duration of furrow irrigation runoff (typically 12 h or more) is generally longer than most rain runoff events. Temporal changes in infiltration, soil and water temperature, rill size and shape, and soil erodibility become more important for longer runoff events. Sediment concentration tends to decrease with time during furrow irrigation. Flow rate, however, increases with time, which should increase sediment detachment and transport. This indicates that soil erodibility decreases during furrow irrigation by phenomena such as armoring, surface sealing, or other unrecognized processes.

Predicting small erosion events is important for irrigation. Seasonal irrigation-induced erosion occurs during numerous controlled and often small events rather than during one or two large erosion events. In southern Idaho for example, a cornfield may be sprinkler irrigated 15 to 20 times or furrow irrigated six to eight times during the growing season. The magnitude of a single irrigation erosion event is usually much smaller and less dramatic than erosion from a single 50-mm thunderstorm occurring on freshly tilled soil without an established crop. However, the cumulative soil loss from irrigation during the growing season may be substantial.

Chemical quality of rainfall varies less from location to location than surface water and groundwater quality. Irrigation water quality can also vary during the season as return flow is added to surface water sources or as groundwater and surface water sources are mixed. Water quality can significantly impact erosion from furrow and sprinkler-irrigated fields. Increasing electrical conductivity (EC) tends to decrease erosion whereas increasing sodium adsorption ratio (SAR) tends to increase erosion (16, 17). Interactions among EC, SAR, clay flocculation, soil chemistry, rainfall application rate, etc. influence the effects of water quality on infiltration and erosion.

REFERENCES

1. Kendall, H.W.; Pimentel, D. Constraints on the Expansion of the Global Food Supply. Ambio **1994**, *23* (3), 198–205.
2. Food and Agriculture Organization. FAOSTAT–Agriculture Data. Food and Agriculture Organization On-line Database, apps.fao.org (accessed Sept. 2000).
3. Tribe, D. *Feeding and Greening the World, the Role of Agricultural Research*; CAB International: Wallingford, Oxon, United Kingdom, 1994.
4. National Research Council, *A New Era for Irrigation*; National Academy Press: Washington, DC, 1996, 203.
5. United States Department of Agriculture. 1998 Farm and Ranch Irrigation Survey. National Agricultural Statistics Service. www.nass.usda.gov/census/ (accessed Sept. 2000).
6. Food and Agriculture Organization. AQUASTAT—Country Profiles. Food and Agriculture Organization On-line Database; www.fao.org/WAICENT/FAOINFO/AGRICULT/AGL/AGLW/aquastat (accessed Sept. 2000).
7. Carter, D.L. Soil Erosion on Irrigated Lands. In *Irrigation of Agricultural Crops*; Agronomy Monograph no. 30, Stewart, B.A., Nielson, D.R., Eds.; Am. Soc. Agronomy: Madison, WI, 1990, 1143–1171.
8. Koluvek, P.K.; Tanji, K.K.; Trout, T.J. Overview of Soil Erosion from Irrigation. J. Irr. Drain. Eng. **1993**, *119* (6), 929–946.
9. Israelson, O.W.; Clyde, G.D.; Lauritzen, C.W. *Soil Erosion in Small Irrigation Furrows*; Bull. 320 Utah Agr. Exp. Sta. Logan, UT, 1946.
10. Mech, S.J. Effect of Slope and Length of Run on Erosion Under Irrigation. Agr. Eng. **1949**, *30*, 379–383.
11. Berg, R.D.; Carter, D.L. Furrow Erosion and Sediment Losses on Irrigated Cropland. J. Soil Water Cons. **1980**, *35* (6), 267–270.
12. Trout, T.J. Furrow Irrigation Erosion and Sedimentation: On-field Distribution. Trans. of the ASAE **1996**, *39* (5), 1717–1723.
13. Carter, D.L.; Berg, R.D.; Sanders, B.J. The Effect of Furrow Irrigation Erosion on Crop Productivity. Soil Sci. Soc. Am. J. **1985**, *49* (1), 207–211.
14. Thompson, A.L.; James, L.G. Water Droplet Impact and Its Effect on Infiltration. Trans. of the ASAE **1985**, *28* (5), 1506–1510.
15. Kemper, W.D.; Rosenau, R.; Nelson, S. Gas Displacement and Aggregate Stability of Soils. Soil Sci. Soc. Am. J. **1985**, *49* (1), 25–28.
16. Lentz, R.D.; Sojka, R.E.; Carter, D.L. Furrow Irrigation Water-Quality Effects on Soil Loss and Infiltration. Soil Sci. Soc. Am. J. **1996**, *60* (1), 238–245.
17. Kim, K.-H.; Miller, W.P. Effect of Rainfall Electrolyte Concentration and Slope on Infiltration and Erosion. Soil Technology **1996**, *9*, 173–185.

IRRIGATION: AN HISTORICAL PERSPECTIVE

R.E. Sojka
D.L. Bjorneberg
J.A. Entry
Soil Scientist, Irrigation Engineer, and Soil Microbiologist, Kimberly, Idaho, U.S.A.

INTRODUCTION

Irrigation can be broadly defined as the practice of applying additional water (beyond what is available from rainfall) to soil to enable or enhance plant growth and yield, and, in some cases, the quality of foliage or harvested plant parts. The water source could be groundwater pumped to the surface, or surface water diverted from one position on the landscape to another. Development of irrigation water often entails development of large-scale, geographically significant dams and water impoundments and/or diversions that can provide additional functions apart from crop growth enhancement, e.g., flood control, recreation, or generation of electricity. In many cases sustainable irrigation development requires concomitant development of surface and/or subsurface drainage.

ANCIENT ORIGINS AND IMPORTANCE

Irrigation may be the single most strategically important intentional environmental modification humans have learned to perform. While irrigation's impact has not always been as critical to the global agricultural economy and food supply as it is today, it has always had major local impacts and profound historical and social consequences. In the Bible's book of Genesis, we are told that God's creation of humans was accompanied shortly thereafter by His assignation to Adam of the stewardship of the irrigated orchard that was Paradise. The four life-giving water heads of Judeo-Christian Paradise are also mentioned in the 47th Sura of the Koran (1). Some anthropologists and historians point to the development of irrigation as the catalyst for the interaction of engineering, organizational, political and related creative or entrepreneurial skills and activities which produced the outcome referred to as "civilization" (2–5). In the ancient Persian language, the word *abadan*, civilized, is derived from the root word *ab*, water (1). Fundamental differences in social, cultural, religious, political, esthetic, economic, technological, and environmental outlook have been attributed to modern groupings of humankind related to their use of irrigation (5).

The earliest archeological evidence of irrigation in farming dates to about 6000 B.C. in the Middle East's Jordan Valley (1). It is widely believed that irrigation was being practiced in Egypt at about the same time (6), and the earliest pictorial representation of irrigation is from Egypt around 3100 B.C. (1). In the following millennia, irrigation spread throughout Persia, the Middle East and westward along the Mediterranean. In the same broad time frame, irrigation technology sprang up more or less independently across the Asian continent in India, Pakistan, China, and elsewhere. In the New World the Inca, Maya, and Aztec made wide use of irrigation. The technology migrated as far North as the current southwestern U.S., where the Hohokam built some 700 miles of irrigation canals in what is today central Arizona to feed their emerging civilization, only to mysteriously abandon it in the 14th century A.D. (3).

In the ancient world, the level of irrigation sophistication varied from one setting to the next. The differences, however, stemmed mostly from variations in understanding of both large- and small-scale hydraulic principles, as well as the capabilities to construct feats of hydraulic engineering. The Assyrians, for example, built an inverted siphon into the Nineveh Aqueduct 700 years before the birth of Christ, an engineering feat unrivaled until the 1860 construction of the pressurized siphons of the New York Aqueduct (3).

Some ancient irrigation schemes have survived to the present day where geologic, soil, and climatic conditions were favorable and where then-known management principles were adequate for the prevailing conditions. However, some ancient schemes failed. In the Mesopotamian Valley, Syria, Egypt, and other areas throughout the Middle East, there were many cases where the principles of salt management and drainage were insufficiently understood, resulting in eventual permanent impairment of the land (1).

Siltation of ancient dams and reservoirs is a testament to inadequate soil conservation measures that eventually reduced the productivity of the land as well as destroyed the capacity of reservoirs to provide an adequate supply of water (3). Erosion of irrigation channels, in geologically unstable areas like the Chilean deserts, and catastrophic failure of irrigation channels after earthquakes often defeated the best efforts of ancient engineers to maintain water supplies (3).

Modern irrigation technology probably began with the Mormon settlement of the Utah Great Salt Lake Basin in 1847, and their eventual cultivation of nearly 2.5 million ha irrigated across the inter-mountain western U.S. by the turn of the century. Whereas relationships of mass, energy, and turbulence of flow were mastered at remarkably high levels of proficiency in ancient cultures, understanding of chemistry and physico-chemical interactions of soil and salt-bearing water was relatively meager even into the 19th century.

MODERNIZATION OF IRRIGATION

The mid-19th century marked a conjunction of several ascending areas of scientific learning, including chemistry, physical chemistry, physics, mineralogy, and biology. These were adapted, blended, and applied in important emerging new sub-disciplines of soil chemistry, soil physics, plant physiology, and agronomy, whose fundamental principles were to prove essential for sustainable irrigation system design and operation.

In ancient irrigation developments, soils, climate, and water quality came together in more forgiving combinations at some locations than at others. Where seasonal rains provided leaching, where soils were permeable and well drained, and/or where irrigation water had favorable combinations of electrolyte concentrations and specific cations, irrigation has continued to the present day, even without sophisticated management. In other areas, salinization, increased soil sodicity, and elevated water tables have limited the life spans of irrigation schemes or impaired their productivity. As irrigation moved into more marginal settings, with less productive soils, poorer drainage, and greater salinity and sodicity problems, the success or failure and ultimate longevity of the schemes became more dependent on knowledgeable application and adaptation of scientific principles. America's Mormon pioneers, choosing to settle in a remote salt-impaired desert habitat, were forced of necessity to use trial and error and the enlightened application of all available new knowledge to reclaim their lands from the desert and to practice a sustainable irrigated crop husbandry. They were so successful in their efforts that their approaches to irrigation and salt-threatened arid land reclamation and management provided the guiding principles for development of irrigation throughout the western U.S. from 1902 (with passage of the Reclamation Act) to the close of the 20th century (3). The science of irrigated agriculture and arid zone soil science in general relied heavily on the foundation and contributions stemming from these mid-19th century origins (7). Development of irrigation in the western U.S. was further spurred by passage of the Desert Land Act of 1877 and the Carey Act of 1894, which provided land for settlement and governmental infrastructure for development. The first university level irrigation course is believed to have been taught by Elwood Mead (Lake Mead's namesake) at the Agricultural College of Colorado in Fort Collins, Colorado (8). Mead later took positions with the United State Department of Agriculture and eventually was a commissioner for the Bureau of Reclamation. Worldwide, many of the practical modern principles of irrigation system design and irrigated soil management can be traced to the lessons learned in the settling of the American West from 1847 to the close of World War II, when the total U.S. irrigated area had grown to 7.5 million ha (6).

Following World War II, irrigation development worldwide entered a heady period of rapid expansion. World populations were growing, in part because of increased life expectancies resulting from new medicines and use of DDT to control malaria and other disease carrying insects. The advances in technology spurred by the first and second world wars were being applied to all avenues of life including agriculture. Electrical, steam and internal combustion power sources became available to pump and pressurize water. New pump designs, the patenting of the center pivot and other sprinkler delivery systems came together in a few short decades between and immediately following the wars to revolutionize the ability to deliver water (7).

CURRENT STATUS

In the U.S., Soviet Union, Australia, and Africa huge government-sponsored programs were initiated in the 1930s, 40s and 50s to build dams for hydropower, flood control, irrigation, and to encourage settlement and stabilization of sparsely populated frontiers. The worldwide total irrigated area was about 94 million ha in 1950 and grew to 198 million ha by 1970 (9). In contrast, the world total irrigated area grew to only about 220 million ha by 1990

(9) and to 263 million ha by 1996 (10). Not surprisingly, the easiest, least technically challenging, least expensive irrigation developments occurred first, and more difficult, more technically challenging, more expensive projects dominate the remaining potential for water development. In some instances, dams and large-scale water development projects have been hampered by poor economies and the instability of the countries in the potential development areas, rather than by the cost or technical challenges per se.

Today 60% of the earth's grain production and half the value of all crops harvested result from irrigation (10). Perhaps most remarkable is the agricultural production efficiency that irrigation provides worldwide. Some 50 million ha of the earth's most productive irrigated cropland (4% of the earth's total cropland) produces a third of the entire planet's food crop (11). Hectare for hectare, irrigated land produces two to two and a half times the yield and three times the crop value per hectare compared with non-irrigated land (10, 12, 13). Yet, the irrigated portion amounts to only about one sixth of the world's total cropped area (14) and about 5% of the world's total production area, which includes cropland, range and pasture (15). In America, most fresh fruits and vegetables in grocery stores come from irrigated agriculture. Beyond survival and economic impact, even our entertainment and esthetics rely heavily on irrigation. Nearly all garden nursery stock in America is propagated and maintained under irrigation and today's parks, play fields, golf courses, and commercial landscaping are seldom established and maintained without irrigation.

To put the global production impact of irrigated agriculture in perspective, it would require over a quarter billion hectares of new rainfed agricultural land (an area the size of Argentina) to supply the average additional production that irrigation's high yield and efficiency provides. Actually, this estimate is conservative. If the land currently irrigated was no longer irrigated but left in production, its output would be well below the mean of existing rainfed land; this is because the lion's share of irrigation occurs in arid or semiarid environments. Furthermore, additional rainfed land brought into production to replace irrigated agriculture, would be well below the current rainfed average productivity; this is because the rainfed land with greatest yield potential has already been brought into production. A more realistic estimate might be double or triple the quarter billion hectare nominal replacement estimate.

In a world of six billion people, irrigation has become essential by providing yet another benefit that cannot be immediately quantified, but which is as important as or more important than production efficiency or economic gain, or even the often uncredited benefits in many irrigation development schemes of hydropower, flood control, transportation, and rural development. The overriding benefit is security—security derived from food production stability. Substantial portions of the world food supply are subject to precipitous and often unpredictable yield reductions due to drought. Irrigation was a key component of the "Green Revolution" of the 1960s and 70s, which stabilized food production in the developing world, providing a new tier of nations the opportunity to turn some of their monetary and human resources to nonagricultural avenues of economic and social development. Much of the drop in the rate of increase of worldwide food production in the last two decades relates to the decrease in the rate of irrigation development since 1980.

ISSUES AFFECTING THE FUTURE

Although there are large projects currently underway or planned for the near future, notably in China, Pakistan, Brazil, Canada, Spain, and Portugal, the equal of the great dam building era from 1930 to 1970 will likely never be seen again. Much of the development of irrigation in the last decade has been achieved through exploitation of groundwater or by smaller scale entrepreneurial surface water developments. In Australia, for example, with the disastrous deflation of the world wool market in the 1990s, substantial numbers of individual sheep stations ceased raising animals and developed their surface water supplies to grow vast hectares of irrigated cotton and rice.

Worldwide, further expansions in irrigated area are unlikely to be large because of the limited remaining surface water sources to exploit and because of the growing environmental concerns, especially related to soil waterlogging, salinization, and sodication problems. Future increases in irrigated area will likely come mainly from the development of the so-called "supplemental" irrigation in humid rainfed areas, from improvements in water use efficiencies associated with utilization of existing irrigation resources, and from improvements in the reuse of municipal, industrial and agricultural wastewaters. Howell (10) noted that improved efficiencies have resulted in a reduction in the mean applied depth of water in the U.S. from about 650 mm annually in 1965 to 500 mm currently. These increased efficiencies have come in great part from the improved understanding of the energy physics of water which led to modern evapotranspiration (ET) theory and ET-based crop irrigation scheduling (9, 10). Many other water conservation practices were developed in the last half of the 20th century, including drip and microirrigation, which has spread from the hyper-xeric conditions of Israel in the early 1950s (1) to nearly every climate and rainfall environment where there is a need, for one reason or another, to conserve water.

Loss of productive capacity caused by soil salinization, sodication, and waterlogging, as well as runoff contamination, riparian habitat impairment, and species losses, are often cited by critics of irrigation as evidence of fundamental drawbacks to irrigated agriculture. Surveys have indicated that of the existing irrigated lands, some 40–50 million ha show measurable degradation from waterlogging, salinization, and sodication (16, 17). Erosion and sedimentation of reservoirs and channels caused the failures of ancient irrigation schemes and have limited the life expectancy of some modern dams to only a few decades as well (3, 18). These problems should not be trivialized. They demonstrate the need for intensified research and conservation, as well as improved dissemination and use of known prophylactic and remedial technologies. However, neither should they be overstated nor presented without due consideration of mitigating factors.

If rates of production loss from these problems are weighted by relative yield or economic value of irrigation compared to rainfed agriculture, and if other positive effects of irrigation are considered, the relative magnitude of negative impacts of irrigation is greatly diminished. For example, runoff contamination from irrigated land would have to be three times the mean for rainfed land, on a crop value basis, or two to two and a half times the mean, on a yield basis, to be "comparable" to problems from nonirrigated agriculture because of the respective relative efficiencies of irrigated agriculture. Both the absolute and relative area of impaired production, plus the degree of impairment, need to be compared on a global basis to rainfed losses, as well as the potential for remediation and production expansion under either circumstance. Positive impacts of irrigation water development include many social and economic benefits such as hydropower, flood control, transportation, recreation, and rural development. Positive environmental effects result from crops, field borders, canals, ditches, and reservoirs that provide significant expansions of habitat for a variety of wildlife compared to undeveloped arid land.

As with all agriculture in recent years, irrigated agriculture has greatly improved its ability to provide humanity's essential needs in closer harmony with environmental needs. This remains a key priority in modern irrigated agricultural research along with continued improvement of production potential to meet the needs of a growing population. Population growth is occurring mostly in underdeveloped nations, where there is an added expectation of improved diet and standard of living. This expectation raises the need for improved production per capita above a simple linear extrapolation based on population. Only high yield intensive production from irrigated agriculture has shown the potential to meet these projected needs.

The knowledge and technology exist to design and operate irrigated agricultural systems sustainably, and without environmental damage or irreversible soil impairment (9, 16, 19). The problem lies in implementing known scientific principles and technologies in a timely fashion as part and parcel of irrigation project and system design and management. This is true both on a regional or project basis and at the farm or field level. Politics and economics play pivotal roles in how well known science and technology are applied. In this respect irrigated agriculture is no different than the myriad manifestations of rainfed agriculture, or any other environmentally impacting activity. Because of modern political and economic considerations, there is usually great pressure, when designing and developing a large-scale irrigation project, to allocate resources for development of as many irrigated hectares as possible at the outset. This often occurs without provision of an adequate technical or social support network to the farming community making the transition to irrigated agriculture. Many schemes fail to provide sufficient financial or technical resources to install drainage systems, to educate farmers or to include them in policy formulation. The resouces are needed to help guarantee prudent water application and salinity or drainage management compatible with the social, technical and financial capabilities of the water users. These are not failures of irrigation. They are failures of human institutions. In this respect, human political, economic, and institutional considerations rather than technical advances or water availability may represent the real challenges for irrigation in the 21st century. These obstacles must be overcome if irrigated agriculture is to provide the production advantage required to satisfy future human needs and meet improved dietary and living standard expectations.

REFERENCES

1. Hillel, D. *Rivers of Eden: The Struggle for Water and the Quest for Peace in the Middle East*; Oxford University Press: New York, 1994; 355.
2. Mitchell, W. The Hydraulic Hypothesis. Curr. Anthropol. **1973**, *14*, 532–534.
3. Reisner, M. *Cadillac Desert: The American West and Its Disappearing Water*; Penguin Books: New York, 1986; 582.

4. Wittfogel, K.A. *Oriental Despotism: A Comparative Study of Total Power*; Yale University Press: New Haven, CT, 1956.
5. Worster, D. *Rivers of Empire: Water, Aridity & The Growth of The American West*; Pantheon Books: New York, 1985; 402.
6. Hoffman, G.J.; Howell, T.A.; Solomon, K.H. Introduction. In *Management of Farm Irrigation Systems*; Hoffman, G.J., Howell, T.A., Solomon, K.H., Eds.; American Society of Agricultural Engineers: St. Joseph, MI, 1990; 5–10.
7. Morgan, R.M. *Water and the Land, a History of American Irrigation*; The Irrigation Association: Fairfax, VA, 1993; 208.
8. Heerman, D.F. Where We Have Been, What We Have Learned and Where We Are Going. In *National Irrigation Symposium: Proceedings of the 4th Decennial Symposium*; Evans, R.G., Benham, B.L., Trooien, T.P., Eds.; American Society of Agricultural Engineers: St. Joseph, MI, 2000; 40–51.
9. Jensen, M.E.; Rangeley, W.R.; Dieleman, P.J. Irrigation Trends in World Agriculture. In *Irrigation of Agricultural Crops, Agronomy Monograph 30*; Stewart, B.A., Nielsen, D.R., Eds.; American Society of Agronomy: Madison, WI, 1990; 31–67.
10. Howell, T.A. Irrigation's Role in Enhancing Water Use Efficiency. In *National Irrigation Symposium: Proceedings of the 4th Decennial Symposium*; Evans, R.G., Benham, B.L., Trooien, T.P., Eds.; American Society of Agricultural Engineers: St. Joseph, MI, 2000; 66–80.
11. Tribe, D. *Feeding and Greening the World, the Role of Agricultural Research*; CAB International: Wallingford, UK, 1994; 274.
12. Bucks, D.A.; Sammis, T.W.; Dickey, G.L. Irrigation for Arid Areas. In *Management of Farm Irrigation Systems*; ASAE Monograph, Hoffman, G.J., Howell, T.A., Solomon, K.H., Eds.; American Society of Agricultural Engineers: St. Joseph, MI, 1990; 499–548.
13. Kendall, H.W.; Pimentel, D. Constraints on the Expansion of the Global Food Supply. Ambio **1994**, *23* (3), 198–205.
14. Gleick, P.H. *Water in Crisis: A Guide to the World? Fresh Water Resources*; Oxford University Press: New York, 1993; 473.
15. Food and Agriculture Organization. FAOSTAT—Agriculture Data. Food and Agriculture Organization On-Line Database, http://apps.fao.org (accessed Sept. 2000).
16. Rhoades, J.D. Sustainability of Irrigation: An Overview of Salinity Problems and Control Strategies. Pp1–42. IN CWRA 1997 Annual Conf. Footprints of Humanity: Reflections on Fifty Years of Water Resource Developments, Lethbridge, Alta. June 3–6 1997.
17. Ghassemi, F.; Jakeman, A.J.; Nix, H.A. *Salinization of Land and Water Resources, Human Causes, Extent, Management, and Case Studies*; CAB International: Wallingford, UK, 1995.
18. Fukuda, H. *Irrigation in the World*; University of Tokyo Press: Tokyo, 1976.
19. Sojka, R.E. Understanding and Managing Irrigation-Induced Erosion. In *Advances in Soil and Water Conservation. Pages*; Pierce, F.J., Frye, W.W., Eds.; Sleeping Bear Press: Ann Arbor, MI, 1998; 21–37.

I

IRRIGATION AND SOIL SALINITY

J.D. Rhoades
Agricultural Salinity Consulting, Riverside, California, U.S.A.

INTRODUCTION

Irrigation is an ancient practice that predates recorded history. While irrigated farmland comprises only about 15% of the worlds' total farmland, it contributes about 36% of the total supply of food and fiber, and it stabilizes production against the vagaries of weather (1). In 30 years, time, irrigated agriculture is expected to have to supply 50% of the worlds' food production requirements (1). However, over the last 20 years, irrigation growth has actually slowed to a rate that is now inadequate to keep up with the projected expanding food requirements (1). Furthermore, irrigation has resulted in considerable salination of associated land and water. It has been estimated variably that the salinized area is as low as 20 and as high as 50% of the worlds' irrigated land (2–4). Worldwide, about 76.6 Mha of land have become degraded by human-induced salination over the last 45–50 years (3). It has been estimated that the world is losing at least three hectares of arable land every minute to soil salination (about 1.6 Mha per year), second only to erosion as the leading worldwide cause of soil degradation (5–7). These data imply that the rate of salinization in developed irrigation projects now exceeds the rate of irrigation expansion (8).

Surviving the salinity threat requires that the seriousness of the problem be recognized more widely, the processes contributing to salination of irrigated lands be understood, effective control measures be developed and implemented that will sustain the viability of irrigated agriculture, and that practical reclamation measures be implemented to rejuvenate the presently degraded lands (9, 10).

DELETERIOUS EFFECTS OF SALTS ON PLANTS, SOILS, AND WATERS

Salt-affected soils have reduced value for agriculture because of their content and proportions of salts, consisting mainly of sodium, magnesium, calcium, chloride, and sulfate and secondarily of potassium, bicarbonate, carbonate, nitrate, and boron. Saline soils contain excessive amounts of soluble salts for the practical and normal production of most agricultural crops. Sodic soils are those that contain excessive amounts of adsorbed sodium in proportion to calcium and magnesium, given the salinity level of the soil water. An example of a salt-affected irrigated soil is shown in Fig. 1

Soluble salts exert both general and specific effects on plants, both of which reduce crop yield (11). Excess salinity in the seedbed hinders seedling establishment and in the crop root zone causes a general reduction in growth rate. In addition, certain salt constituents are specifically toxic to some plants. For example, boron is highly toxic to susceptible crops when present in the soil water at concentrations of only a few parts per million. In some woody crops sodium and chloride may accumulate in the tissue over time to toxic levels. These toxicity problems are, however, much less prevalent than is the general salinity problem.

Salts may also change soil properties that affect the suitability of the soil as a medium for plant growth (12). The suitability of soils for cropping depends appreciably on the readiness with which they conduct water and air (permeability) and on their aggregate properties (structure), which control the friability (ease with which crumbled) of the seedbed (tilth). In contrast to saline soils, which are well aggregated and whose tillage properties and permeability to water and air are equal to or higher than those of similar nonsaline soils, sodic soils have reduced permeabilities and poor tilth. These problems are caused by the swelling and dispersion of clay minerals and by the breakdown of soil structure (slaking and crusting), which results in loss of permeability and tilth. Sodic soils are generally less extensive but more difficult to reclaim than saline soils.

Beneficial use of water in irrigation consists of transpiration and leaching for salinity control (the leaching requirement). Plant growth is directly proportional to water consumption through transpiration (13). From the point of view of irrigated agriculture, the ultimate objective of irrigation is to increase the amount of water available to support transpiration. Salts reduce the fraction of water in a supply (or in the soil profile) that can be consumed beneficially in plant transpiration (14). In considering the use of a saline water for irrigation and in

Fig. 1 Photograph of salt-affected irrigated field.

selecting appropriate policies and practices of irrigation and drainage management, it is important to recognize that the total volume of a saline water supply cannot be consumed beneficially in crop production (i.e., transpired by the plant). A plant will not grow properly when the salt concentration in the soil water exceeds some limit specific to it under the given conditions of climate and management (11). This is even true for halophytes (15). Thus, the practice of blending or diluting excessively saline waters with good quality water supplies should be undertaken only after consideration is given to how it affects the volumes of consumable (usable) water in the combined and separated supplies (14).

CAUSES OF SALINATION INDUCED BY IRRIGATION AND DRAINAGE

While salt-affected soils occur extensively under natural conditions, the salt problems of greatest importance to agriculture arise when previously productive soils become salinized as a result of agricultural activities (the so-called secondary salination). The extent and salt balance of salt-affected areas has been modified considerably by the redistribution of water (hence salt) through irrigation and drainage. The development of large-scale irrigation and drainage projects, which involves diversion of rivers, construction of large reservoirs, and irrigation of large landscapes, causes large changes in the natural water and salt balances of entire geohydrologic systems. The impact of such developments can extend well beyond that of the immediate irrigated area. Excessive water diversions and applications are major causes of soil and water salination in irrigated lands. It is not unusual to find that less than 60% of the water diverted for irrigation is used in crop transpiration (9). This implies that about 40% of the irrigation water eventually ends up as deep percolation. This drainage water contains more salt than that added with the irrigation water because of salt dissolution and mineral weathering (14) within the root zone. It often gains additional salt-load as it dissolves salts of geologic origin from the underlying substrata through which it flows in its down-gradient path. This drainage water often flows laterally to lower lying areas, eventually resulting in shallow saline groundwaters of large areas of land (waterlogging). Salination occurs in soils underlain by saline shallow groundwater through the process of "capillary rise" as groundwater (hence, salt) is driven upwards by the force of evaporation of water from the soil surface. Correspondingly, saline soils and waterlogging are closely associated problems.

Seepage from unlined or inadequately lined delivery canals occurs in many irrigation projects and is often substantial. Law *et al.* (16) estimated that 20% of the total water diverted for irrigation in the United States is lost by seepage from conveyance and irrigation canals. Biswas (17) estimated that 57% of the total water diverted for irrigation in the world is lost from conveyance and distribution canals. Analogous to on-farm deep percolation resulting from irrigation, these seepage waters typically percolate through the underlying strata (often dissolving additional salts in the process), flow to lower elevation lands or waters, and add to the problems of waterlogging and salt-loading associated with on-farm irrigation there. A classic example of the rise in the water table following the development of irrigation has been documented in Pakistan and is described by Jensen et al. (9) and Ghassemi et al. (2). The depth to the water table in the irrigated landscape located between three major river-tributaries rose from 20 to 30 m over a period of 80–100 years, i.e., from preirrigated time (about 1860) to the early 1960s, until it was nearly at the soil-surface. In one region, the water table rose nearly linearly from 1929 to 1950, demonstrating that deep percolation and seepage resulting from irrigation were the primary causes. Ahmad (18) concluded that about 50% of the water diverted into irrigation canals in Pakistan eventually goes to the groundwater by seepage and deep percolation.

The role of irrigated agriculture in salinizing soil systems has been well recognized for hundreds of years. It is of relatively more recent recognition that salination of water resources from agricultural activities is a major and widespread phenomenon of likely equal concern to that of soil salination. The causes of water salination are essentially the same as those of soils, only the final reservoir of the discharged salt-load is a water supply in the former case (14). The volume of the water supply is reduced through irrigation diversions and irrigation; thus, its capacity to assimilate such received salts before reaching use-limiting levels is reduced proportionately. Only in the past 15 years has it become apparent that trace toxic constituents, such as selenium, in agricultural drainage waters can also cause serious pollution problems (19).

IRRIGATION AND DRAINAGE MANAGEMENT TO CONTROL SOIL SALINITY

The key to overall salinity control is strict control that maintains a net downward movement of soil water in the root zone of irrigated fields over time while minimizing excess irrigation diversions, applications, and deep percolation (20). The direct effect of salinity on plant growth is minimized by maintaining the soil-water content in the root zone within a narrow range at a relatively high level, while at the same time avoiding surface-ponding and oxygen deletion and minimizing deep percolation. Combined methods of pressurized, high-frequency irrigation and irrigation scheduling have been developed that permit substantially the desired control to be achieved (21, 22). These systems transfer control of water distribution and infiltration from the soil to the irrigation equipment. This results in less excess water (and hence, less salt) being applied overall to the field to meet the needs of a part of the field area having lowest intake rate, as done in the more traditional gravity irrigated systems. However, gravity irrigation systems can be designed to achieve good irrigation efficiency and salinity control, even though surface ponding still does occur. The so-called level-basin, multi-set, cablegation, surge, and tailwater-return systems are among them (21, 22). The need for irrigation and the amount required to meet evapotranspiration and leaching requirement is determined from plant stress measurements, calculations of evapotranspiration amounts, measurements of soil-water depletion, measurements of soil (or soil-water) salinity, or a combination of them (21, 22).

In addition to effective methods of irrigation scheduling and application, appropriate irrigation and salinity management also require an effective delivery system. Delivery systems have generally been designed to provide water on a regular schedule. Efficient irrigation systems require more flexible deliveries that can provide water on demand as each crop and particular field have need of it. Delivery systems can be improved by lining the canals, by containing the water within closed conduits, and by implementing techniques that increase the flexibility of delivery.

As briefly discussed earlier, irrigated agriculture is a major contributor to the salinity of many rivers and groundwaters, as well as soils. Reducing deep percolation generally lessens the salt load that is returned to rivers or groundwater and their pollution (14). Additionally, saline drainage waters should be intercepted before being allowed to mix with water of better quality. The intercepted saline drainage water should be desalted and reused, disposed of by pond evaporation or by injection into some suitably isolated deep aquifer, or better yet it should be used for irrigation in a situation where brackish water is appropriate. Various irrigation and drainage strategies have been developed for minimizing the pollution of waters from irrigation and for using brackish waters for irrigation (14, 23). Desalination of agricultural drainage waters is not now economically feasible, but improved techniques for doing this exist and some are being implemented. However, more needs to be done in this regard.

Traditionally, the concepts of leaching requirement and salt-balance index have been used to plan and judge the appropriateness of irrigation and drainage systems, operations and practices with respect to salinity control, water use efficiency, and irrigation sustainability. However, these approaches are inadequate. The recommended method is to monitor directly the root-zone salinity levels and distributions across fields as a means to evaluate the effectiveness of salinity, irrigation, and drainage management practices, to detect problems (current and developing), to help determine the underlying causes of problems, and to determine source areas of major water and salt-load contributions to the underlying groundwater. Theory, equipment, and practical technology have been developed for these purposes (24). More information about irrigation and drainage management to control soil and water salinity is found elsewhere (25–27).

REFERENCES

1. FAO, In *World Agriculture Toward 2000: An FAO Study*; Alexandratos, N., Ed. Bellhaven Press: London, 1988; 338.
2. Ghassemi, F.; Jakeman, A.J.; Nix, H.A. *Salination of Land and Water Resources. Human Causes, Extent, Management and Case Studies*; CAB International: Wallingford, UK, 1995; 526.
3. Oldeman, L.R.; van Engelen, V.N.P.; Pulles, J.H.M. The Extent of Human-Induced Soil Degradation. In *World Map of the Status of Human-Induced Soil Degradation: An Explanatory Note*; Oldeman, L.R., Hakkeling, R.T.A., Sombroek, W.G., Eds. International Soil Reference and Information Center (ISRIC): Wageningen, The Netherlands, 1991; 27–33.
4. Adams, W.M.; Hughes, F.M.R. Irrigation Development in Desert Environments. In *Techniques for Desert Reclamation*; Goudie, A.S., Ed. Wiley: New York, 1990; 135–160.

5. Buringh, P. Food Production Potential of the World. In *The World Food Problem: Consensus and Conflict*; Sinha, R., Ed. Pergamon Press: Oxford, 1977; 477–485.
6. Dregne, H.; Kassas, M.; Razanov, B. A New Assessment of the World Status of Desertification. Desertification Control Bull. **1991**, *20*, 6–18.
7. Umali, D.L. Irrigation-Induced Salinity. *A Growing Problem for Development and Environment*; Technical Paper World Bank: Washington, DC, 1993.
8. Seckler, D. *The New Era of Water Resources Management: From "Dry" to "Wet" Water Savings*; Consultative Group on International Agricultural Research: Washington, D.C., 1996.
9. Jensen, M.E.; Rangeley, W.R.; Dieleman, P.J. Irrigation Trends in World Agriculture. *Irrigation of Agricultural Crops*; American Society of Agronomy Monograph No. 30, ASA: Madison, WI, 1990; 31–67.
10. UNEP, *Saving Our Planet: Challenges and Hopes*; Nairobi, United Nations Environment Program: Nairobi, Kenya, 1992; 20.
11. Maas, E.V. Crop Salt Tolerance. ASCE Manuals and Reports on Engineering No. 71. In *Agricultural Salinity Assessment and Management Manual*; Tanji, K.K., Ed. ASCE: New York, 1990; 262–304.
12. Rhoades, J.D. Principal Effects of Salts on Soils and Plants. In *Water, Soil & Crop Management Relating to the Use of Saline Water*; FAO (AGL) Misc. Series Publication 16/90, Kandiah, A., Ed. Food and Agriculture Organization of the United Nations: Rome, 1990; 1933.
13. Sinclair, T.R. Limits to Crop Yield? In *Physiology and Determination of Crop Yield*; Boone, K.J., Ed. American Society of Agronomy: Madison, WI, 1994; 509–532.
14. Rhoades, J.D.; Kandiah, A.; Mashali, A.M. *The Use of Saline Waters for Crop Production*; FAO Irrigation and Drainage Paper 48, FAO: Rome, Italy, 1992; 133.
15. Miyamoto, S.; Glenn, E.P.; Oslen, M.W. Growth, Water Use and Salt Uptake of Four Halophytes Irrigated with Highly Saline Water. J. Arid Environ. **1996**, *32*, 141–159.
16. Law, J.P.; Skogerboe, G.V.; Denit, J.D. The Need for Implementing Irrigation Return Flow Control. p. 1–17. In *Managing Irrigated Agriculture to Improve Water Quality*; Proc. Math. Conf. Manag. Irrig. Agric. Improve Water Avail., Denver, CO, May, 1972. Graphics Manage. Corp., Wash., DC; 16–18.
17. Biswas, A.K. Conservation and Management of Water Resources. In *Techniques for Desert Reclamation*; Goudie, A.S., Ed.; Wiley: New York, 1990; 251–265.
18. Ahmad, N. Planning for Future Water Resources of Pakistan. *Proceedings of Darves Bornoz Spec. Conference, National Committee of Pakistan*; ICID: New Delhi, India, 1986; 279–294.
19. Letey, J.; Roberts, C.; Penberth, M.; Vasek, C. *An Agriculturl Dilemma: Drainage Water and Toxics Disposal in the San Joaquin Valley*; Special Publication 3319, University of California: Oakland, 1986.
20. Rhoades, J.D. Soil Salinity—Causes and Controls. In *Techniques for Desert Reclamation*; Goude, A.S., Ed. Wiley: New York, 1990; 109–134.
21. Hoffman, G.J.; Rhoades, J.D.; Letey, J.; Sheng, F. Salinity Management. In *Management of Farm Irrigation Systems*; Hoffman, G.J., Howell, T.A., Solomon, K.H., Eds. ASCE: St. Joseph, MI, 1990; 667–715.
22. Kruse, E.G.; Willardson, L.; Ayars, J. On-Farm Irrigation and Drainage Practices. In *Agricultural Salinity Assessment and Management Manual*; ASCE Manuals and Reports on Engineering No. 71, Tanji, K.K., Ed. ASCE: New York, 1990; 349–371.
23. Rhoades, J.D. Use of Saline Drainage Water for Irrigation. In *Agricultural Drainage*; ASA Drainage Monograph 38, Skaggs, R.W., van Schilfgaarde, J., Eds. American Society of Agronomy: Madison, WI, 1999; 619–657.
24. Rhoades, J.D.; Chanduvi, F.; Lesch, S. *Soil Salinity Assessment: Methods and Interpretation of Electrical Conductivity*; FAO Irrigation and Drainage Paper 57, FAO, United Nations: Rome, Italy, 1999; 152.
25. Rhoades, J.D. Use of Saline and Brackish Waters for Irrigation: Implications and Role in Increasing Food Production, Conserving Water, Sustaining Irrigation and Controlling Soil and Water Degradation. In *Proceedings of the International Workshop on "The Use of Saline and Brackish Waters for Irrigation: Implications for the Management of Irrigation, Drainage and Crops" at the 10th Afro-Asian Conference of the International Committee on Irrigation and Drainage, Bali, Indonesia, July 23–24, 1998*; Ragab, R., Pearce, G., Eds. International Committee on Irrigation and Drainage: Bali, Indonesia, 1998; 261–304.
26. Rhoades, J.D.; Loveday, J. Salinity in Irrigated Agriculture. In *Irrigation of Agricultural Crops*; Agron. Monograph. No. 30, Stewart, B.A., Nielsen, D.R., Eds. American Society of Agronomy: Madison, Wisconsin, 1990; 1089–1142.
27. Tanji, K.K. Nature and Extent of Agricultural Salinity. In *Agricultural Salinity Assessment and Management*; ASCE Manuals and Reports on Engineering No. 71, Tanji, K.K., Ed. ASCE: New York, 1990; 1–17.

LAND CAPABILITY ANALYSIS

Michael J. Singer
Department of Land, Air and Water Resources,
University of California, California, U.S.A.

INTRODUCTION

All soils are not the same and they are not of the same capability for every use. Land capability implies that the choice of land for a particular use contributes to the success or failure of that use. It further implies that the choice of land for a particular use will determine the potential impact of that use on surrounding resources such as air and water. To make the best use of land and to minimize the potential for negative impacts on surrounding lands, land capability analysis is needed. The assessment of land performance for specific purposes is land evaluation (1). A system that organizes soil and landscape properties into a form that helps to differentiate among useful and less useful soils for a purpose is land capability classification. Land capability is a broader concept than soil quality, which has been defined as the degree of fitness of a soil for a specific use (2). Bouma (3) points out that land capability or land potential needs to be evaluated via various scales and gives the example of precision agriculture, which requires land capability analysis in more detail than that required to determine if an investment should be made to initiate agriculture.

Land capability or suitability classification systems have been designed to rate land and soil characteristics for specific uses (Table 1). Huddleston (4) has reviewed many of these systems. They may also rate land qualities. Land qualities have been defined by the United Nations Food and Agricultural Organization (1) as "attributes of land that act in a distinct manner in their influence on the function of land for a specific kind of use." An example of a land quality is the plant available water stored in soil, and an example of a soil characteristic is the clay content that contributes to the plant available water holding capacity.

It is generally recognized that a single soil characteristic is of limited use in evaluating differences among soils (5), and that use of more than one quantitative variable requires a system for combining the measurements into a useful index (6). Gersmehl and Brown (7) advocate regionally targeted systems.

KINDS OF SYSTEMS

Land rating systems for agriculture include those that are used to evaluate the potential for agricultural development of new areas, and others that evaluate the potential for agriculture in already developed areas. Many other land capability assessments exist to help planners rate suitability of agricultural lands for nonagricultural uses. Some examples of agricultural land capability systems are described in this chapter. All systems have, in common, a set of assumptions on which the analysis is based and each system answers the question "capability for what use."

Development Potential

Examples of systems designed to determine the potential for agricultural development include the FAO framework for land evaluation and the U.S. Bureau of Reclamation (USBR) irrigation suitability classification. The FAO framework combines soil and land properties with a climatic resources inventory to develop an agro-ecological land suitability assessment.

The USBR capability classification was frequently used to evaluate land's potential for irrigation in the Western U.S. during the period of rapid expansion of water delivery systems (8–11). It combines social and economic evaluations with soil and other ecological variables to determine whether the land has the productive capacity, once irrigated, to repay the investment necessary to bring water to an area. It recognizes the unique importance of irrigation to agriculture and the special qualities of soils that make them irrigable.

Land Capability

The USDA Land Capability Classification (see entry by Fenton) is narrower in scope than either the FAO or USBR capability rating systems. The purpose of the Land Capability Classification (LCC) is to place arable soils into groups based on their ability to sustain common cultivated crops that do not require specialized site conditioning or

Table 1 Examples of land capability systems

System	Purpose	Property	References
FAO framework	Development potential	Used for large-scale development of agriculture	(1, 10)
USBR irrigation suitability	Potential for irrigation development	Used for determining potential to repay costs of developing irrigation	(8, 9)
USDA land capability classification	Land capability for agriculture	Uses 13 soil, climate, and landscape properties to determine agricultural capability	(4, 12)
Storie index	Land capability for agriculture	Uses nine soil and management factors to determine agricultural capability	(17–19)
Soil potential	Soil suitability for specific uses	Uses a cost index to rate land for any potential use	(16)
Soil quality	Determining status of soil profile	Used for determining the status of selected soil properties	(6, 16, 20)

treatment (12). Nonarable soils, unsuitable for long term, sustained cultivation, are grouped according to their ability to support permanent vegetation, and according to the risk of soil damage if mismanaged.

Several studies have shown that lands of higher LCC have higher productivity with lower production costs than lands of lower LCC (13–15). In a study of 744 alfalfa-, corn-, cotton-, sugar beet- and wheat-, growing fields in the San Joaquin Valley of California, those with LCC ratings between 1 and 3 had significantly lower input/output ratios than fields with ratings between 3.01 and 6 (15). The input/output ratio is a measure of the cost of producing a unit of output and is a better measure of land capability than output (yield) alone. This suggests that the LCC system provides an economically meaningful assessment of agricultural soil capability.

Quantitative systems result in a numerical index, typically with the highest number being assigned to the land or soil with the highest capability for the selected use. The final index may be additive, multiplicative, or more complex functions of many land or soil attributes. Quantitative systems have two important advantages over nonquantitative systems: 1) they are easier to use with GIS and other automated data retrieval and display systems, and 2) they typically provide a continuous scale of assessment (16). No single national system is presently in use but several state or regional systems exist.

One example of a quantitative system is the Storie index rating (SIR). Storie (17) determined that land productivity is dependent on 32 soil, climate, and vegetative properties. He combined only nine of these properties into the SIR, to keep the system from becoming unwieldy. The nine factors are soil morphology (A), surface texture (B), slope (C), and management factors drainage class (X_1), sodicity (X_2), acidity (X_3), erosion (X_4), micro-relief (X_5), and fertility (X_6). Each factor is rated from 1 to 100%. These are converted to their decimal value and multiplied together to yield a single rating for a soil map unit (17–19) (Eq. 1):

$$\mathrm{SIR} = \left(A \times B \times C \times \prod_{i=1}^{6} X_i \right) \times 100 \qquad (1)$$

An area-weighted SIR can be calculated by multiplying the SIR for each soil map unit within a parcel by the area of the soil unit within the parcel, followed by summing the weighted values and dividing by the total area (Eq. 2).

$$\text{Area-weighted SIR} = \frac{1}{\text{total area}} \times \sum_{i=1}^{n} \mathrm{SIR\ soil_i} \times \mathrm{area\ soil_i}. \qquad (2)$$

Values for each factor were derived from Storie's experience mapping and evaluating soils in California, and in soil productivity studies in cooperation with California Agricultural Experiment Station cost-efficiency projects relating to orchard crops, grapes and cotton. Soils that were deep, which had no restricting subsoil horizons, and held water well had the greatest potential for the widest range of crops.

Reganold and Singer (15) found that area-weighted average SIR values between 60 and 100 for 744 fields in the San Joaquin Valley had lower but statistically insignificant input/output ratios than fields with indices <60. The lack of statistical significance is scientifically meaningful, but may not be interpreted the same by planners or farmers for whom small differences in profit can be significant.

Designers of these systems have selected thresholds or critical limits that separate one land capability class from another. Frequently, little justification has been given for the critical limit selected. The issue of critical limits is a difficult one in soils because of the range of potential uses and the interactions among variables (20).

Soil Potential

Soil potential is an interpretive system that is used for both agricultural and nonagricultural capability assessments. It is an example of a land capability system that is both quantitative and highly specific. Soil potential is defined as the usefulness of a site for a specific purpose using available technology at an indexed cost. Experts in the location where the system will be used determine the soil properties that most influence the successful use of a soil for a particular purpose. The well-suited soils for the use are given a rating of 100. Soil limitations that can be removed or reduced are identified and continuing limitations that occur even after limitations have been addressed are also identified. The cost of available practices and technologies used to remove limitations and the cost of continuing limitations are indexed and these index values are subtracted from 100 to yield a soil potential rating for each soil (Eq. 3).

$$\mathrm{SPI} = \mathrm{PI} - (\mathrm{CI} + \mathrm{CL}). \qquad (3)$$

The best soils in an area are those that have the lowest indexed cost for operating or maintaining use, while ensuring the lowest possible environment.

CONCLUSIONS

All soils are not the same and are not equally capable of sustained use. Various systems have been created that rate soil properties for these uses. Examples include FAO land suitability, USBR irrigation suitability, USDA land capability classification, Storie index, and soil potential. Each has strengths and weaknesses that the user must recognize before using the system.

REFERENCES

1. Food and Agricultural Organization, *A Framework for Land Evaluation*; Soils Bulletin 32, FAO: Rome, Italy, 1976.
2. Gregorich, E.G.; Carter, M.R.; Angers, D.A.; Monreal, C.M.; Ellert, B.H. Towards a Minimum Data Set to Assess Soil Organic Matter Quality in Agricultural Soils. Can. J. Soil Sci. **1994**, *74*, 367–386.
3. Bouma, J. Land Evaluation for Landscape Units. In *Handbook of Soil Science*; Sumner, M.E., Ed.; CRC Press: Boca Raton, FL, 2000; E-393–E-412.
4. Huddleston, J.H. Development and Use of Soil Productivity Ratings in the United States. Geoderma **1984**, *32*, 297–317.
5. Reganold, J.P.; Palmer, A.S. Significance of Gravimetric Versus Volumetric Measurements of Soil Quality Under Biodynamic, Conventional, and Continuous Grass Management. J. Soil Water Conserv. **1995**, *50*, 298–305.
6. Halvorson, J.J.; Smith, J.L.; Papendick, R.I. Integration of Multiple Soil Parameters to Evaluate Soil Quality: A Field Example. Biol. Fert. Soils **1996**, *21*, 207–214.
7. Gersmehl, P.J.; Brown, D.A. Geographic Differences in the Validity of a Linear Scale of Innate Soil Productivity. J. Soil Water Conserv. **1990**, *45*, 379–382.
8. USBR, *Land Classification Handbook*; Bur. Recl. Pub. V, Part 2, USDI: Washington, DC, 1953.
9. Maletic, J.T.; Hutchings, T.B. Selection and Classification of Irrigable Lands. In *Irrigation of Agricultural Lands*; Hagen, R.M., Ed.; Soil Science Society of America: Madison, Wisconsin, 1967; 125–173.
10. Food and Agricultural Organization, *Land Evaluation Criteria for Irrigation. Report of an Expert Consultation*; World Soil Resourc. Rep. 50, FAO: Rome, Italy, 1979.
11. McRae, S.G.; Burnham, C.P. *Land Evaluation*; Clarendon Press: Oxford, UK, 1981.
12. Klingebiel, A.A.; Montgomery, P.H. *Land-Capability Classification*; Agriculture Handbook No. 210, Soil Conservation Service USDA: Washington, DC, 1973.
13. Patterson, G.T.; MacIntosh, E.E. Relationship Between Soil Capability Class and Economic Returns from Grain Corn Production in Southwestern Ontario. Can. J. Soil Sci. **1976**, *56*, 167–174.
14. Van Vliet, L.J.; Mackintosh, E.E.; Hoffman, D.W. Effects of Land Capability on Apple Production in Southern Ontario. Can. J. Soil Sci. **1979**, *59*, 163–175.
15. Reganold, J.P.; Singer, M.J. Comparison of Farm Production Input/Output Ratios of Two Land Classification-Systems. J. Soil Water Conserv. **1984**, *39*, 47–53.
16. Singer, M.J.; Ewing, S. Soil Quality. In *Handbook of Soil Science*; Sumner, M.E., Ed.; CRC Press: Boca Raton, Florida, 2000; 9271–9298.
17. Storie, R.E. *An Index for Rating the Agricultural Value of Soils*; California Agr. Exp. Sta. Bull. 556. Berkeley, CA, 1932.
18. Wier, W.W.; Storie, R.E. *A Rating of California Soils*; Bull. 599, California Agr. Exp. Sta. Berkeley, CA, 1936.
19. Storie, R.E. *Handbook of Soil Evaluation*; Associated Students Store: U. Cal. Berkeley, CA, 1964.
20. Arshad, M.A.; Coen, G.M. Characterization of Soil Quality: Physical and Chemical Criteria. Am. J. Altern. Agric. **1992**, *7*, 25–31.

LAND CAPABILITY CLASSIFICATION

L

Thomas E. Fenton
Iowa State University, Ames, Iowa, U.S.A.

INTRODUCTION

A common definiton of land is the surface of the Earth and all its natural resources. This is interpreted to include the atmosphere, the soil and underlying geology, hydrology, and plants on the Earth's surface (1). In the 1938 Yearbook of Agriculture, land is defined as the total natural and cultural environment within which production must take place. Its attributes include climate, surface configuration, soil, water supply, subsurface conditions, etc., together with its location with respect to centers of commerce and population. It should not be used as synonymous with soil or in the sense of the Earth's surface only (2). Other definitions included the results of the past and present human activities as well as the animals within this area when they exert a significant influence on the present and future uses of land by man. Thus, the concept of land is much broader than soil. However, since soils are a major component of terrestrial ecosystems (3), they integrate and reflect internal and external environmental factors. They have been considered an important factor—if not the major factor—in many land classification systems.

CLASSIFICATIONS

Land classification may be defined as a classification of specific bodies of land according to their characteristics or to their capabilities for use. A natural land classification system is one in which natural land types are placed in categories according to their inherent characteristics. A land classification according to their capabilities for use may be defined as one in which the bodies of land are classified on the basis of physical characteristics with or without economic considerations according to their capabilities for man's use (2).

Olson (4) defined land classification as the assignment of classes, categories, or values to the areas of the Earth's surface (generally excluding water surfaces) for immediate or future practical use. The project, product, or proposal resulting from this activity may be also generally referred to as land classification. Any land classification involves two parts or phases: resource inventory and analysis and categorization. The inventory consists of gathering data and delineating land characteristics on maps. The analysis and categorization put the basic data into a form that can be used generally for a specific use. Thus, there are potentially many land classification systems that could be developed depending on the objectives of the system.

HISTORICAL PERSPECTIVE

Some sort of land classification system was used in the transfer of public lands to private owners and users as the U.S. settled. However, a scientific approach was not possible until soil surveys, topographic maps, economic analyses of production and other activities became available in the early 1900s. These sources provided essential data for a scientific approach to land classification. This approach gained added momentum in the mid-1930s due to the need for land classification for many new government programs.

A report submitted to the President of the U.S. by the National Resources Board dated December 31, 1934 (5) included a section that dealt with the physical classification of productivity of the land. The total land area of the U.S. was rated and divided into five grades based on the factors thought to affect productivity—soil type, topography, rainfall, and temperature. Norton (6, 7) defined five classes of land in the regions of arable soils according to their use capability. Two other classes were defined for land that should not be cultivated. A National Conference on Land Classification (NCLC) was held in the campus of the University of Missouri in October 1940. The proceedings of this conference were published in the bulletin 421 of the Missouri Agricultural Experiment Station (8). At this conference it was recognized that land classification in the U.S. had begun many years prior to 1940. Simonson (9) summarized the purpose of the NCLC meeting as follows: to encourage and provide opportunity for the discussion of land classification in all of its different aspects.

In the years following the conference, there were many papers published that examined questions related to land classification. General principles associated with technical groupings (10) or solution of soil management problems (11) using natural soil classification systems were discussed in detail. Hockensmith and Steele (12) presented a land capability classification developed for conservation and development of land and for land use adjustments. The system used eight land capability classes with the first four suited for cultivation and the remaining four not suited for cultivation. Within a capability class, the subclasses were determined by the kind of limitation. Within each subclass, the land that required the same kind of management and the same kind of conservation treatment was called a land capability unit. The highest level of abstraction, the capability class, provided a quick, easy, understanding of the general suitability of the land for cultivation. Hockensmith (13) discussed the use of soil-survey information in farm planning. A land capability map was an essential part of the farm plan. This map was an interpretative map of the soil map together with other pertinent facts. It was used to help the farmers understand the limitations of their soils and to aid in selection of those practices that would preserve their soils and keep them productive. Klingebiel (14) defined capability classification as the grouping of individual kinds of soils, called soil mapping unit, into groups of similar soils, called capability units, within a framework of eight general capability classes that were divided into subclasses representing four kinds of conservation problems. This land classification system has been used as a basis for all the conservation farm plans developed by the Soil Conservation Service (now Natural Resources Conservation Service) in the U.S. since the 1950s and continues to be used today. Because of its widespread use, this system is discussed in more detail in the following section.

LAND CAPABILITY CLASSIFICATION

One of the most commonly used land classification systems for land management is the land capability classification. Hockensmith (15) defined this classification as a systematic arrangement of different kinds of land according to those properties that determine the ability of the land to produce permanently. Classification was based on the detailed soil survey (scale of 1:15,840 or 1:20,000). The system was described in detail in the Agriculture Handbook No. 210 (16) and continues to be in use, with no significant changes except that presently Arabic numerals are used to identify the capability classes rather than Roman numerals. There are a number of land capability classifications used throughout the world but all these methods are patterned after the U.S. system (17).

The capability grouping of soils is designed to:

1. help landowners and others use and interpret the soil maps;
2. introduce users to the detail of the soil map itself;
3. to make possible broad generalizations based on soil potentialities, limitations in use, and management problems.

There are three major groupings: 1) capability classes, 2) capability subclasses, and 3) capability units.

Capability Classes

The broadest unit is the capability class and there are eight classes. The probability of soil damage or limitations in use become progressively greater from Class I to VIII. There are no additional subdivisions of soils in Class I because by definition the soils in this category have no limitations in use. The proper grouping of the soils assumes that the recommended use will not deteriorate the soil over time.

- Class I soils have few limitations that restrict their use.
- Class II soils have some limitations that reduce the choice of plants or require moderate conservation practices.
- Class III soils have severe limitations that reduce the choice of plants or require special conservation practices or both.
- Class IV has very severe limitations that restrict the choice of plants, require very careful management, or both.
- Class V has little or no erosion hazard but has other limitations that are impractical to remove and limit their use largely to pasture, range, woodland, or wildlife food and cover.
- Class VI soils have severe limitations that make them generally unsuited to cultivation and limit their use largely to pasture or range, woodland, or wildlife food and cover.
- Class VII soils have very severe limitation that make them unsuited to cultivation and that restrict their use largely to grazing, woodland, or wildlife.
- Class VIII soils and landforms limitations that preclude their use for commercial plant production and restrict their use to recreation, wildlife, or water supply or to esthetic purposes.

Capability Subclasses

Subclasses are groups of capability units within classes with the same kind of major limitations for agricultural use. The capability subclass is a grouping of capability units having similar kinds of limitations and hazards. Four general kinds of limitations or hazards are recognized:

1. erosion hazard
2. wetness
3. rooting-zone limitations
4. climate

The limitations recognized and the symbols used to identify them are: risk of erosion designated by the symbol e; wetness, drainage, or overflow, w; rooting-zone limitations, s; and climatic limitations, c. There is a priority of use among the subclasses if the limitations are approximately of the same degree. The order of use is e, w, s. For example, if a group of soils has both erosion and excess water hazard, e takes precedence over w. However, for some uses, it may be desirable to show two kinds of limitations. This combination is rarely used but when used, the higher priority one is shown first, i.e., IIew.

Capability Units

Capability units provide more specific and detailed information than class or subclass. They consist of soils that are nearly alike in suitability for plant growth and responses to the same kinds of soil management. However, they may have characteristics that place them in different soil series or soil map units. Soil map units in any capability unit are adapted to the same kinds of common cultivated and pasture plants, and require similar alternative systems of management for these crops. Another assumption is that the longtime estimated yields of adapted crops for individual soil map units within the unit under comparable management do not vary more than about 25%.

CONCLUSION

The grouping of soil map units as used in the land-capability classification is a technical classification. The basis for this classification is the natural soil classification system in which the soils and soil map units are defined based on their characteristics and interpretations made based on those properties that affect use and management. For conservation planning, environmental quality, and generation of interpretive maps, it is important to know the kind of soil, its location on the landscape, its extent, and its suitability for various uses.

REFERENCES

1. Brinkman, R.; Smyth, A.J. *Land Evaluation for Rural Purposes*; Publication 17, International Institute for Land Reclamation and Improvement: Wageningen, The Netherlands, 1973; 116.
2. United States Department of Agriculture, In *Soils and Men. Yearbook of Agriculture*; U.S. Govt. Printing Office: Washington, DC, 1938; 1232.
3. Jenny, H. *The Soil Resource*; Ecological Studies 37, Springer: New York, 1980; 377.
4. Olson, G.W. *Land Classification*. Cornell Univ. Agri. Exp. Station Agronomy 4. Ithaca, New York, 1970.
5. National Resources Board Report. Part II. Report of the Land Planning Committee. Section II. U.S. Govt. Printing Office: Washington, DC, 1934; 108–152.
6. Norton, E.A. Classes of Land According to Use Capability. Soil Sci. Soc. Am. Proc. **1939**, *4*, 378–381.
7. Norton, E.A. *Soil Conservation Survey Handbook*; Misc. Publ. 352, U.S. Dept. Agric. Soil Conservation Service: Washington, DC, 1939; 40.
8. Miller, M.F. *The Classification of Land*. Proc. First International Conference on Land Classification, Bulletin 421, Missouri Agric. Exp. Station, 1940; 334.
9. Simonson, R.W. The National Conference on Land Classification. Soil Sci. Soc. Am. Proc. **1940**, *5*, 324–326.
10. Orvedal, A.C.; Edwards, M.J. General Principles of Technical Grouping of Soils. Soil Sci. Soc. Am. Proc. **1941**, *6*, 386–391.
11. Simonson, R.W.; Englehorn, A.J. Interpretation and Use of Soil Classification in the Solution of Soil Management Problems. Soil Sci. Soc. Am. Proc. **1942**, *7*, 419–426.
12. Hockensmith, R.D.; Steele, J.G. Recent Trends in the Use of the Land-Capability Classification. Soil Sci. Soc. Am. Proc. **1949**, *14*, 383–387.
13. Hockensmith, R.D. Using Soil Survey Information for Farm Planning. Soil Sci. Soc. Am. Proc. **1953**, *18*, 285–287.
14. Klingebiel, A.A. Soil Survey Interpretation-Capability Groupings. Soil Sci. Soc. Am. Proc. **1958**, *23*, 160–163.
15. Hockensmith, R.D. Classification of Land According to Its Capability as a Basis for a Soil Conservation Program. Reprinted from Proceedings of the Inter-American Conference on Conservation of Renewable Natural Resources, Denver, CO, September 7–20, 1948.
16. Klingebiel, A.A.; Montgomery, P.H. *Land Capability Classification*; Agric. Handbook 210, U.S. Dept Agri. Soil Conservation Service: Washington, DC, 1961; 21.
17. Hudson, N. Land Use and Soil Conservation. In *Soil Conservation*, 3rd Ed.; Iowa State University Press: Ames, IA, 1995; 391.

LAND EVALUATION

J. Herbert Huddleston
Oregon State University, Corvallis, Oregon, U.S.A.

INTRODUCTION

Land refers to a specific area of the earth's surface that is characterized by its soil, climatic, geologic, topographic, hydrologic, floral, and faunal resources, as well as by the kinds of human activities that take place on it (1–4). *Land use* refers to those biological and technological activities that manage and improve land resources for economic and social purposes (2). *Land evaluation* refers to an assessment of the properties of the several resources that characterize land as they bear on the requirements of one or more specific land uses (3). Land evaluation is undertaken both to assess the *performance* of land under specified uses (3), and to *communicate* information about alternative uses of the same tract of land (5).

Soil, strictly speaking, is only one of the natural resources that characterize the land resource. But the same set of natural resources that define land also define the environment in which soil forms (6), and soil expresses the integrated effects of both its internal properties and the combined effects of its surroundings. Soil resource evaluation, therefore, often serves as a surrogate for land evaluation. In fact, the only real difference between comprehensive land evaluation and soil resource evaluation is that land evaluation may include assessments of social and economic factors (3, 4, 7), whereas soil resource evaluation includes such factors only in a very general way, if at all. Readers who are interested in the comprehensive land evaluation approach will find excellent resources in FAO (3) and McRae and Burnham (4).

HISTORICAL OVERVIEW

Land evaluation in some form has been going on for as long as human societies have been able to observe their environment and adapt their activities to the resource qualities available to them. Over four millennia ago the Chinese were making soil maps as a basis for taxation and agricultural management, and the early Romans clearly understood that different regions had different soils, and these differences were manifest in different crop yields.

Land evaluation in the United States began in earnest as the Division of Soils initiated soil surveys to direct agricultural development onto the more productive soils (8) and the Bureau of Reclamation began classifying land suitability for irrigation development. The Bureau of Reclamation later wrote a comprehensive land evaluation manual (7) in which all related natural resources, as well as the economic factors affecting the development and use of the land, were assessed. The U.S. Department of Agriculture developed the Land Capability Classification (9, 10), which was based on an interpretation of soil survey information and defined capability explicitly in terms of limitations to use of soil and risk of damage to soil when cultivated.

After World War II soil survey information was used increasingly to develop interpretations for land use planning (11) and other nonagricultural uses of land (12). Specific procedures for making these interpretations are spelled out in the National Soil Survey Handbook (13). In the 1980s, two new interpretations, the soil potential rating (SPR) and the land evaluation and site assessment (LESA) program were developed, both of which incorporate a degree of economic analysis in their evaluation criteria. SPRs (13, 14) are based on net returns from specific agricultural or nonagricultural uses on specific soils. They are determined by expressing the yield, output, or performance of a soil in economic terms and subtracting from the output value the costs of inputs or management practices needed to achieve the desired output. LESA combines an evaluation of resource quality with an evaluation of various social, cultural, and economic impacts on land use (15–17). The land evaluation part is really soil resource evaluation because it is based on soil potential ratings, soil productivity ratings, land capability classes, or some combination of these. Site assessment considers impacts associated with parcel size and shape, access to supporting services and markets, historical or cultural values, and compatibility among adjacent and nearby land uses.

STEPS IN THE DEVELOPMENT OF LAND EVALUATION MODELS

The first and most important task is to *specify the objective* as thoroughly and completely as possible. Land evaluation

Encyclopedia of Soil Science

is always undertaken for some specific land use, and because different land uses require different sets of soil and other resource attributes, the objective must be stated clearly to properly develop evaluation criteria. An important corollary of objective specification is that a land evaluation developed for one specific purpose should not be used as a surrogate to classify land for a different use. Land capability classes, for example, are neither measures of soil productivity nor classes of agricultural suitability and should not be used in land use planning programs that require evaluations of productivity or suitability.

The second step is to *identify those resource properties that most affect the desired use.* This is best done by assembling a committee of local people—farmers, farm service advisers, planners, developers, realtors—who know both the resources of the area and the issues associated with the intended use. Involving local people ensures that the widest possible range of ideas and issues is represented and gives credibility to the land evaluation system that emerges from the process. The committee should first make a long list of all factors that might influence the land use of interest, then, through discussion and debate, narrow the list to a few important factors that are the least correlated with each other as possible.

The third step is to *develop specific criteria for evaluating the impact of each factor selected.* Land evaluation models may be either qualitative or quantitative, deductively or inductively derived (18). Qualitative criteria assign ratings such as good, fair, or poor suitability to soil property characteristics. Quantitative criteria assign numerical ratings on a scale of 10, or 100, to the values associated with each property. Deductive models use actual data, such as crop yields, to create classes of land suitability. Inductive models use values of various soil properties that are highly correlated with performance to develop empirical ratings of land suitability. In all cases criteria should be specified as completely as possible to maximize consistency among all users. Soil and landscape complexity, however, may defy absolute prescription of criteria that are applicable to all possible situations, in which case the model should leave some flexibility for users to subjectively determine exactly how a criterion should be evaluated.

The fourth step is to *combine all of the individual factor ratings into an overall evaluation* of land quality for the specified objective. Additive models, multiplicative models, or combination models (18) may all be appropriate, depending on the objective and the consensus of the committee undertaking the land evaluation. Additive models simply sum up all of the individual factor ratings, or subtract them from an ideal score of, say, 100. They are easy to use, but these models risk minimizing the full impact of one or more particularly limiting factors, especially if the model includes a large number of factors. Multiplicative models, which create a product of all individual factor ratings (19), do allow single factors to control the overall evaluation, but the final rating generated can be significantly lower than each of the individual factor ratings, even when all factor ratings are high. Combined models capture the advantages of both additive and multiplicative models by using additive techniques to combine property ratings within major factors, then multiplying major factor ratings to generate the overall rating.

The final steps in the process are to *set thresholds for creating land suitability classes* and to *validate the results of the model.* Setting class limits is another committee effort, and it may be done using either overall ratings only, ratings of component factors, or both (20). Model validation is necessary to ensure the usefulness and credibility of the final product. Several test cases are selected, all model factors are evaluated, and the committee determines subjectively whether the results are realistic and "make sense." If not, the model is revised—by adding or deleting factors, changing factor ratings, or adjusting criteria for setting thresholds—then tested again, until consensus is reached.

ROLE OF SOIL SURVEYS IN LAND EVALUATION AND LAND CLASSIFICATION

Soil surveys are the backbone of any land evaluation system for three reasons: 1) soils are essential land resources; 2) soil survey interpretations are often sufficient to meet the land evaluation objective; and 3) soil surveys provide a cartographic base from which maps of resultant land evaluation classes can be made. Two issues must be fully understood to use soil surveys properly for land evaluation: composition of soil map units and limitations of scale.

Most soil map units are defined by a single kind of soil. Each delineation of these map units represents a single element of the landscape within which soil properties and behavior are reasonably homogeneous. Because soils are naturally variable, however, each soil-landscape element contains small areas whose soils are different from the dominant soil for which the map unit is named. These areas are called *inclusions*, and their soils are identified in map unit descriptions.The problem is that, although inclusions may behave quite differently from the dominant soil for a given kind of land use, it is common practice to treat each delineation as though it were composed entirely of the dominant, named soil, classify its land use suitability accordingly, and map it as such. Users must understand that some error is inherent in such interpretive maps and avoid the temptation to assume that, once mapped, all areas shown in a particular land use suitability class are 100% pure.

Most soil surveys used for land evaluation purposes are order 2 or order 3 surveys having scales ranging from 1:15,840 to 1:63,360. Because landscape scale soil variability occurs within meters, order 2 and 3 surveys necessarily entail generalizations that result in inclusions, particularly in the vicinity of soil boundaries. Some detailed land use planning projects may invoke detailed soil surveys at scales of 1:1200 or 1:4800, which do permit higher degrees of accuracy, but it is more common to enlarge order 2 or 3 maps to overlay the results of a land evaluation on larger scale planning or plat maps. This is particularly easy to do when a digitized soil map is brought into a computerized GIS system and scales can be changed easily at the click of a mouse. Users must understand that simply changing the scale at which a soil or land evaluation map is displayed does not change its accuracy, and all the inclusions present in the original delineations are still present in their representation at a larger scale.

Soil surveys, when used appropriately with full understanding of their limitations, are excellent tools for conducting land evaluations for many kinds of uses. Many land evaluations can be based entirely on existing soil survey interpretations. Others may be developed by identifying relevant factors, evaluating each one according to precisely specified criteria, combining factor ratings into a land evaluation model, and setting thresholds for land suitability classes. Existing soil surveys may then serve as templates for mapping the results of land evaluation efforts.

REFERENCES

1. Brinkman, R.; Smyth, A.J. *Land Evaluation for Rural Purposes*; Publication 17, International Institute for Land Reclamation and Improvement: Wageningen, The Netherlands, 1973.
2. Vink, A.P.A. *Land Use in Advancing Agriculture*; Advanced Series in Agricultural Sciences 1, Springer-Verlag: New York, 1975.
3. FAO. *A Framework for Land Evaluation*; Soils Bulletin 32, Food and Agriculture Organization of the United Nations: Rome, 1976.
4. McRae, S.G.; Burnham, C.P. *Land Evaluation*; Oxford Science Publications Monographs on Soil Survey, Oxford University Press: New York, 1981.
5. Beek, K.J. From Soil Survey Interpretation to Land Evaluation Part 1. From the Past to the Present. Soil Survey and Land Eval. **1981**, *1*, 6–12.
6. Jenny, H. *Factors of Soil Formation. A System of Quantitative Pedology*; Dover Publications, Inc.: New York, 1994.
7. USBR. Land Classification. *Bureau of Reclamation Manual*, Vol. V, Part 2, U.S. Dept. of Interior: Washington, DC, 1953.
8. Whitney, M. *Announcement*; Circ. 1, U.S. Dept. Agr. Bureau of Soils: Washington, DC, 1894.
9. Norton, E.A. Classes of Land According to Use Capability. Soil Sci. Soc. Am. Proc. **1939**, *4*, 378–381.
10. Klingebiel, A.A.; Montgomery, P.H. *Land Capability Classification*; Agr. Handbook 210, U.S. Dept. Agr. Soil Conservation Service: Washington, DC, 1961.
11. Bartelli, L.J., Klingebiel, A.A., Baird, J.V., Heddleson, M.R., Eds. *Soil Surveys and Land Use Planning*; Soil Science Society of America: Madison, WI, 1966.
12. Simonson, R. W., Ed. Non-Agricultural Applications of Soil Surveys; Developments in Soil Sci. 4, Elsevier: Amsterdam, 1974.
13. Soil Survey Staff, U.S. Natural Resources Conservation Service. *National Soil Survey Handbook*; Title 430-IV, U.S. Govt. Printing Office: Washington, DC, 1999.
14. McCormack, D.E.; Stocking, M.A. Soil Potential Ratings: I. An Alternative Form of Land Evaluation. Soil Survey and Land Eval. **1986**, *6*, 37–42.
15. Wright, L.E. Agricultural Land Evaluation and Site Assessment (LESA): A New Agricultural Land Protection Tool in the U.S.A. Soil Survey and Land Eval. **1984**, *4*, 25–38.
16. Pease, J.R.; Coughlin, R.E. *Land Evaluation and Site Assessment. A Guidebook for Rating Agricultural Lands*; Soil and Water Conservation Society Ankeny: IA, 1996.
17. Huddleston, J.H.; Pease, J.R.; Forrest, W.G.; Hickerson, H.J.; Langridge, R.W. Use of Agricultural Land Evaluation and Site Assessment in Linn County, Oregon, USA. Env. Management **1987**, *11*, 389–405.
18. Huddleston, J.H. Development and Use of Soil Productivity Ratings in the United States. Geoderma **1984**, *32*, 297–317.
19. Storie, R.E. *Storie Index Rating*; Spec. Pub. 3203, University of California Division Agriculture Science: Davis, CA, 1976.
20. Huddleston, J.H. Importance of the LESA Objective in Selecting LE Methods and Setting Thresholds for Decision Making. In *a Decade with LESA. The Evolution of Land Evaluation and Site Assessment*; Steiner, F.R., Pease, J.R., Coughlin, R.E., Eds.; Soil and Water Conservation Society: Ankeny, IA, 1994; 78–93.

LANDFORMS

L

C.G. Olson
United States Department of Agriculture-Natural Resources Conservation Service (USDA-NRCS), Lincoln, Nebraska, U.S.A

INTRODUCTION

Landforms are features of the earth that together comprise the land surface. They may be large features such as plains, plateaus, or mountains, or smaller features such as hills, eskers or hillslopes. The glossary of geology defines a landform as "any physical, recognizable form or feature of the Earth's surface having a characteristic shape and produced by natural causes" (1). A similar definition for landform can be found in almost any geology or geomorphology textbook. The word "landform" is derived from the Old English or Old German word *land* and the Latin *forma* (2).

DESCRIPTION AND MORPHOMETRY

There are several ways to describe a landform. Surface form is the most common method of description and includes an analysis of shape factors. Shape is characterized by breaks and changes of slope, directions and angles of maximum slope, and other similar information. A minimum of three parameters are used to describe slope shape: 1) slope length or vertical curvature, 2) slope width or horizontal curvature, and 3) gradient. Slope length is measured by a radial that crosses topographic contour lines at right angles. This dimension is sometimes referred to as a longitudinal profile. Slope width, measured perpendicular to the slope length, parallels topographic contour lines. The curvature or shape of a simple landform such as a hillslope may be defined by a series of forms (Fig. 1). Ruhe represented changes in hillslope curvature by a matrix of nine basic forms (3). Huggett added surface flow lines to basic slope shapes (4) and Pennock et al. others combined curvature with landform elements to identify seven different hillslope positions (5).

Since there are hundreds of landforms, it might be useful to choose a commonly recognized landform such as a dune to discuss some of the aspects of shape. Characteristically, many dunes have a steeper, shorter, leeward side and a longer, lower gradient or windward side and in cross section have a distinctive asymmetry of slope length. Slope width is dependent on the amount of material composing the dune. The curvature of the windward side is often linear to linear convex and that of the leeward side almost any possible combination of forms. In the case of dunes, each distinctive slope also provides information on the wind direction during formation. The map view on topographic maps and aerial photographs, of a barchan dune with its commonly recognized crescentic patterns illustrates its direction of movement. Lancaster provides an excellent series of drawings and photographs illustrating several types of dune landforms (6).

The third slope-shape factor is gradient, the angle of inclination of the slope with respect to the horizontal. On a topographic map, it is measured at right angles to the contour lines. Soil scientists measure gradient in percentage. Geologists and engineers generally measure gradient in degrees.

In addition to shape, certain landforms are best described by waveforms which include the dimensions of length, height, and periodicity. Examples of landforms for which waveform description is useful are meander curves and dune fields.

Internal form or structure is often included in a landform description. This is an important distinguishing criterion for identification of eroded or degraded landforms in which characteristic surface morphology has been obscured. For example, many landforms such as dunes, eskers, or barrier islands are composed primarily of sand. However, the distribution of sands and the cross-bedding in a dune are internal structural elements unique to a landform developed by eolian processes.

The shape and size of a landform are not random. These parameters are determined by the processes that build landforms and by the properties of their materials. In the case of a depositional feature such as a dune, the size of the feature is also closely related to the amount of depositional material available during formation. A dune can form that is a few centimeters to several hundred meters in height. The former can occur in the dry bed of a small ditch or in intermittent streams as the wind blows sandy bed material into dune forms. Dunes in the central deserts of China or the Sahara of Northern Africa are examples of the latter.

Encyclopedia of Soil Science

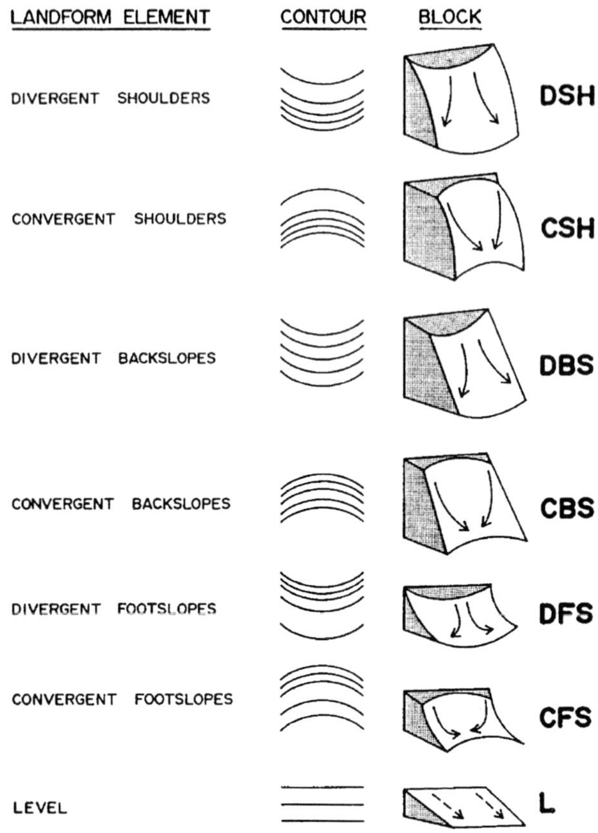

Fig. 1 A matrix of representative hillslope shapes. (From Ref. 5.)

Landforms have a natural relationship to each other and to the surrounding landscape. Fractal geometry has been used to produce realistic computer-generated reproductions of landforms (7). The mathematical concept of fractal geometry also eliminates the problems associated with scale of observation. A landform is observed and defined differently by an individual on the ground than is a feature observed from Earth orbit.

CLASSIFICATION

There are numerous approaches to the classification of a landform (8). A summary of early classification schemes (9) first presented in Ref. (10) generally fall into five categories: genetic, process, size, topographic expression (morphology), and degree of erosion.

Features have been classified by size (11). Landforms were grouped into numbered orders. First order relief features included continents and ocean basins. Second order relief features were tectonic mountain belts, plateaus, and plains, and third order relief features were erosional and constructional features.

Groups of landforms have been organized into physiographic provinces (12, 13). The provinces are subdivided into sections and grouped into divisions. Each grouping was based on landforms resulting from a set of similar processes or a succession of processes. Similar regional groupings have been produced for the coterminous U.S. (14–16), and for Alaska (17).

A genetic classification, or the mode of origin, has some overlap with descriptive systems because different manner of origin may produce different forms. Examples of genetic systems include the classification of volcanic rocks (18), the classification of glacial landforms (19, 20), and landslide types and processes (21).

LANDFORM DEVELOPMENT

Landforms may be erosional or constructional in form, or a composite of both. Most landforms are the products of erosion, but many are formed by deposition of sediments, by volcanic activity, or by movements from within the Earth's crust. Examples of depositional features include dunes, moraines, and spits. Examples of erosional features are water gaps, deflation basins, or arêtes. Lahars, lava plains, and volcanoes result from volcanic activity. The formation of many alluvial fans and subsequent episodes of erosion and sedimentation in the basin and range of the western U.S. are triggered by tectonic uplift.

LANDFORMS AND MAPPING

Landforms are part of a continuum of near surface features, soils, and surficial sediments. Ref. (22) describes a landform in a continuum as a unit of systematic analysis. When mapping, distinctive landform populations need to be delineated. The patterns are identifiable in the field with aids such as topographic maps, aerial photography, geophysical surveys, land use maps, and other available information.

Delineating landforms can sometimes simplify soil mapping. For example, in traversing a hillslope from its summit to its toe, one might recognize several differing soil map units. The sequence may repeat itself on adjacent hillslopes resulting in a predictable pattern of soils that can be related back to a distinctive landform population.

In mapping large tracts of land or in remote areas, it is theoretically possible to systematically subdivide surficial deposits into textural, morphological, or genetic components. In reality, it is often difficult. Instead, the composition and nature of map unit materials over wide areas are generally inferred from landforms (23, 24). For example, in flying over a region, drumlins are observed, then drumlins and eskers, then disintegration features, followed by moraines at right angles blocking valleys. If the only advance information one has is that the area had been glaciated, the sequence of landforms becomes the key to interpreting that one is crossing the periphery of a former ice sheet, possibly its maximum extent. One can then map the active zone, the transition to the proglacial, ice-contact zone, and the outer edge of the glacial advance from the landform distribution.

REFERENCES

1. Goldthwaite, R.P.; Matsch, C.L., Eds. *Glossary of Geology*, 4th Ed.; American Geological Institute: Alexandria, VA, 1997; 769.
2. Goldthwaite, R.P.; Matsch, C.L., Eds. *The Random House Dictionary of the English Language, Unabridged*, 2nd Ed.; Random House, Inc.: New York, 1987.
3. Ruhe, R.V. *Geomorphology, Geomorphic Processes and Surficial Geology*; Houghton Mifflin: Boston, 1975; 246.
4. Huggett, R.J. Soil Landscape Systems: A Model of Soil Genesis. Geoderma **1975**, *13*, 1–22.
5. Pennock, D.J.; Zebarth, B.J.; deJong, E. Landform Classification and Soil Distribution in Hummocky Terrain Saskatchewan, Canada. Geoderma **1987**, *40*, 297–315.
6. Lancaster, N. *Geomorphology of Desert Dunes*; Routledge: New York, 1995; 290.
7. Mandelbrot, B. *The Fractal Geometry of Nature*; W.H. Freeman: New York, 1983; 468.
8. Small, R.J. *The Study of Landforms, a Textbook of Geomorphology*, 2nd Ed.; Cambridge University Press: New York, 1978; 502.
9. Goldthwaite, R.P.; Matsch, C.L., Eds. *The Encyclopedia of Geomorphology, Encyclopedia of Earth Sciences*; Dowden, Hutchinson and Ross: Stroudsburg, PA, 1968; Vol. 3, 1295.
10. Howard, A.D.; Spock, L.E. A Classification of Landforms. J. Geomorphol. **1940**, *3*, 332–345.
11. Salisbury, R.D. *Physiography*, 3rd Ed.; Henry Holt: New York, 1919; 676.
12. Fenneman, N.M. *Physiography of Western United States*; McGraw-Hill: New York, 1931; 534.
13. Fenneman, N.M. *Physiography of Eastern United States*; McGraw-Hill: New York, 1938; 714.
14. Hammond, E.H. Classes of Land-Surface Form in the Forty-Eight States, USA, 1963: Map Supp. No. 4. Assoc. Am. Geograph. Ann. **1964**, 54.
15. Thornbury, W.D. *Regional Geomorphology of the United States*; Wiley: New York, 1965; 609.
16. Hunt, C.B. *Natural Regions of the United States and Canada*; Freeman W.H. San Francisco, CA, 1974; 725.
17. Wahrhaftig, C. *Physiographic Divisions of Alaska*; US Geol. Survey Prof. Paper 482, US Department of Interior Geological Survey: Washington, DC, 1965; 52.
18. Rittman, A. *Volcanoes and Their Activity*; Wiley–Interscience: New York, 1962; 305.
19. Sugden, D.E.; John, B.S. *Glaciers and Landscape, A Geomorphological Approach*; Routledge/Chapman, & Hall: New York, 1991; 376.
20. Goldthwaite, R.P.; Matsch, C.L., Eds. *Genetic Classification of Glaciogenic Deposits; Final Report of the Commission on Genesis and Lithology of Glacial Quaternary Deposits of the International Union for Quaternary Research (INQUA)*; Balkema: Rotterdam, 1988; 294.
21. Cruden, D.M.; Varnes, D.J. Landslide Types and Processes. National Research Council Transportation Research Board Special Report 247; Turner, A.K., Schuster, R.L., Eds.; National Academy Press: Washington, DC, 1996; 673.
22. Bloom, A.L. *Geomorphology, A Systematic Analysis of Late Cenozoic Landforms*, 2nd Ed.; Prentice-Hall: Engle-Wood Cliffs, NJ, 1991; 532.
23. Fulton, R.J. Surficial Geology Mapping at the Geological Survey of Canada: Its Evolution to Meet Canada's Changing Needs. Can. J. Earth Sci. **1993**, *30*, 232–242.
24. McMillan, A.A.; Powell, J.H. *British Geological Survey Rock Classification Scheme, 4: classification of Artificial (Man-Made) Ground and Natural Superficial Deposits—applications to Geological Maps and Datasets in the UK*; British Geological Survey Research Report Resonance Raman 99-04, NERC: Nottingham, UK, 1999; 65.

LAND RESTORATION, PRINCIPLES OF

Richard W. Bell
School of Environmental Science, Murdoch University, Murdoch, Western Australia, Australia

INTRODUCTION

Soils are essentially a finite and nonrenewable resource (1) but are subject to various forms of degradation. Recent monographs by Lal and co-workers (1–3, and papers therein) have dealt in-depth with soil and land degradation processes, their measurement, impact, and management. Conacher and Conacher (4) define land degradation broadly as "alteration to all aspects of the natural (or biophysical) environment by human actions, to the detriment of vegetation, soils, landforms and water (surface and subsurface, terrestrial and marine) and ecosystems." The challenge for sustainable use of land resources embraces both the minimization of the degrading processes, and the restoration of previously degraded land. In this paper, the author focuses on restoration of degraded land from a biophysical perspective.

The term restoration is used in the present review as a generic term after the usage of Hobbs and Norton (5) who suggest "that restoration occurs along a continuum and that different activities are simply different forms of restoration." Given that part of the paper also concerns restoration of land degraded by mining, the term rehabilitation as defined by Aronson et al. (6) is also appropriate. Restoration will usually focus on restoring ecosystem functions such as nutrient cycling, hydrological balance, and ecosystem resilience (7), although restoring the original flora may on occasions be a realistic and appropriate goal. Reclamation is the term that describes the replacement of ecological functions by planting a different vegetation to that which previously grew; it is commonly used in the U.S.

LAND RESTORATION PRINCIPLES

Land restoration comprises: determination of end land use, determining the main limiting factors for restoration and means of alleviating them, and finally planning and implementation of the restoration program. This article will deal mostly with the first two of these components.

End Land Use

A clear definition of the end land use is a prerequisite for effective restoration of highly disturbed land. This will determine the stakeholders, the scope for restoration, the key constraints that have to be alleviated, and help to define the goals for determining success. End land use has a major bearing on the degree of difficulty of the task of restoration and hence its cost. Restoration of a complete ecosystem, and its function, is a markedly more complex task than restoration of a stable surface or creating a pasture for low intensity grazing (5). The former may require decades of systematic research before restoration goals can be reliably accomplished and demonstrably sustainable.

Prior land use and current use of the surrounding areas will generally determine the end land use except where the substrate or landscape is so radically altered by degrading processes as to make this impossible. Apart from restoring natural ecosystems, agriculture, forestry, housing, industry, amenity and recreation, wetlands, and waste disposal are all possible end land uses (8). Economic and socio-political factors have a major bearing on end land use, particularly on production lands, protected landscapes, and in densely settled areas.

Whilst setting end land use is necessary in order to develop tangible goals for restoration, there are also merits in an adaptive rather than a prescriptive approach. As practice improves, new benchmarks for completion can be set and new possibilities for end land use emerge, as shown in the changes with time in bauxite mine rehabilitation in southwest Australia (9).

Problem Diagnosis and Definition: Biophysical Factors Limiting Land Restoration

Land degradation has many different forms, and hence different consequences for restoration (3). Diagnosing the main form(s) of degradation at a site or in a landscape is a prerequisite for appropriate land restoration. The main biophysical factors limiting land restoration are: climate, landform, hydrology, substrate properties, and plant establishment. These are described briefly later and in Table 1. Many examples of their importance and treatment can be found elsewhere in this volume.

Table 1 Biophysical factors limiting land restoration, their consequences for site stability and plant growth, and common treatment methods

Factor	Constraint	Consequence	Treatment
Climate	Drought	Failure to germinate, poor emergence or establishment, plant death	Irrigation, drought tolerant species, adjust time of sowing
	High temperature	Poor germination and emergence, plant death	Mulching
	Low temperature/frost	Delayed emergence, plant death, poor seed set	
	Extreme rainfall/wind events	Water or wind erosion episodes	Contour banks, soil cover by foliage, mulch or stubble, wind breaks
Landform	Slope	Land slippage, soil creep	Deep rooted plants, reshape to lower slope, contour banks, engineering design
	Runoff	Water erosion	Contour banks, reshape to lower slope
	Exposure	Drought, high winds, extreme temperatures	Tolerant plant species
	Aspect	High winds, extreme temperatures	Tolerant plant species
Hydrology	Runoff	Erosion, waterlogging, flooding	Contour banks, increased drainage intensity
	Limited profile water storage	Drought, runoff, increased groundwater recharge	Tolerant plant species, deep ripping
	Groundwater discharge	Water-filled voids, acid mine drainage	Containment of water, wetland treatment ponds
Substrate properties	Acidity	Poor plant growth especially roots, plant death	Lime, acid tolerant species, P fertilizer
	Alkalinity	Poor plant growth, plant death	Gypsum, leaching, alkaline-tolerant species, acidifying materials
	Salinity	Poor plant growth, plant death	Leaching, salt tolerant species, drainage
	Sodicity	Soil dispersion, crusting, poor seedling emergence, erosion	Gypsum, leaching, organic matter addition
	Nutrient deficiency	Poor plant growth	Fertilizer
	Metal toxicity	Poor plant growth, plant death	Lime, tolerant species, phytoremediation
	Low water availability	Poor plant growth, plant death	Irrigation, mulching, organic matter, deep ripping
	Waterlogging	Poor plant growth, plant death	Drainage, tolerant plant species
	Poorly structured soils	Crusting, poor water holding capacity	Mulching, organic matter, gypsum
	Mycorrhiza	Poor plant growth, nutrient deficiency	Topsoil management, inoculation of nursery plants
	Rhizobium	Nitrogen deficiency	Topsoil management, inoculation
	Soil microbes	Poor mineralization of soil organic matter	Topsoil management

L

Climate sets the underlying growing conditions for plants in land restoration and in a broad sense determines the range of species able to survive and grow in a particular location. Landform is a prime consideration in restoration because of the need to achieve slope stability and effective erosion control. The important components of landform that limit restoration are slope angle, elevation, aspect, and drainage. Any clearing or significant disturbance of vegetation alters hydrology. Increased runoff is usually the first on-site response, followed by erosion, and then downstream consequences such as flooding. Other changes in hydrology include altered profile water storage and increased groundwater recharge.

Substrate properties

Physical properties: Land degradation often alters the soil physical conditions in ways that decrease its suitability for plant establishment and growth, particularly by changing water storage and availability. Water erosion may strip away topsoil; decrease soil depth; and expose subsoil material with different texture, lower organic matter levels, and poorer soil structure. Wind erosion selectively removes clay and humus from the soil surface, increasing the prevalence of coarse materials. Sand deposits from wind erosion may bury topsoil and vegetation.

Chemical properties: Nutrient levels, acidity, alkalinity, salinity, sodicity, and toxic metals and organic compounds are the chemical factors in soils and substrates that can limit plant growth and therefore hamper effective land restoration. Deficiencies of nutrient elements can be diagnosed or predicted by several means (10). Once identified, deficiencies can generally be corrected by fertilizer application. Selection of plants with high tolerances of the toxicities from acidity or heavy metals is often the most cost-effective strategy of revegetation even if it means planting a different species mix to that which existed before disturbance (11). However, the preferred modern practice in mining is to identify adverse materials and selectively place them in waste dumps where there is no contact with the root zone.

Biological properties: Mining activities and topsoil storage have quite profound negative effects on microbial biomass and activity (12). Restoration of microbial biomass may take 7–10 years on revegetated mine sites. Careful handling and reuse of topsoil can often minimize the problems associated with decreased biological activity of soil after land restoration. Maximum activity is retained if topsoil is reused immediately without a period of storage.

Plant establishment

The prerequisite for effective plant establishment is to ameliorate substrate conditions most likely to limit plant growth. In mining, the aim is to plan mine operations so that the most benign materials are placed on top of surfaces to be revegetated. This minimizes the need for expensive amelioration of adverse soil conditions.

Plant establishment relies on three strategies, either individually or in combination. Topsoil spreading will often be the most reliable strategy if it is available, because it contains seed native to the area, adds soil organisms, and creates a seedbed for seed germination and establishment. However, in some plant communities, seed is stored in the plant canopy rather than in the soil. For restoration of these communities, cutting branches and spreading them on the soil surface has proved effective. Nursery-raised seedlings will generally be reliable for plant establishment but are relati-vely expensive. In addition, there are often many species that are difficult to propagate from seed in the nursery, or they generally regenerate by vegetative means. Direct seeding is usually cheaper than other options, although there is still significant cost associated with collection of native seed. Establishment of seedlings especially from direct seeding is hampered by weed competition, and grazing by herbivores.

Sustainability

After plant establishment, some measure of the success of restoration is needed. According to Hobbs and Norton (5), ecosystem characteristics should be considered when setting goals for land restoration: composition, structure, pattern, heterogeneity, function, species interactions, and dynamics and resilience. This raises the need for reliable low-cost indicators of success (12). To date, most indicators have relied on composition and structure of vegetation. Microbial biomass has been advocated as an indicator of nutrient cycling functions. Indicators for pattern, heterogeneity, dynamics, and resilience are less advanced in their development, in part because few land restoration activities have run for long enough to measure these characteristics let alone develop indicators for them.

CONCLUSIONS

Land is a finite resource. Every attempt must be made to prevent or minimize land degradation. In addition, restoration of already degraded land needs to be undertaken. Land restoration is still a relatively new science that is increasingly informed by the emerging discipline of restoration ecology. Land restoration occurs at a range of different scales. Mine rehabilitation has developed many successful practices of land restoration at a site-specific scale. The challenge ahead is to develop effective land restoration practices at a regional or landscape scale. Most success at both scales has, to date, been concerned with restoring key ecosystem functions such as nutrient cycling and water balance.

REFERENCES

1. Lal, R. Blum, W.H., Valentine, C., Stewart, B.A., Eds.; *Soil Restoration*; Advances in Soil Science, CRC Press: Boca Raton, FL, 1992; Vol. 17, 456.
2. Lal, R., Blum, W.H., Valentine, C., Stewart, B.A., Eds.; *Soil Degradation*; Advances in Soil Science, CRC Press: Boca Raton, FL, 1990; Vol. 11, 345.
3. Lal, R., Blum, W.H., Valentine, C., Stewart, B.A., Eds.; *Methods for Assessment of Soil Degradation*; CRC Press: Boca Raton, FL, 1998.
4. Conacher, A.J.; Conacher, J. *Rural Land Degradation in Australia*; Oxford Univ. Press: Melbourne, Australia, 1995; 170.
5. Hobbs, R.J.; Norton, D.A. Towards a Conceptual Framework for Restoration Ecology. Restor. Ecol. **1996**, *4*, 93–110.
6. Aronson, J.; LeFoc'h, E.; Floret, C.; Ovalle, C.; Pontanier, R. Restoration and Rehabilitation of Degraded Ecosystems in Arid and Semiarid Regions. II. Case Studies in Chile, Tunisia and Cameroon. Restor. Ecol. **1993**, *1*, 168–187.
7. Hobbs, R.J. Restoration Ecology. In *The Encyclopedia of Soil Science Section XXXVI, Rehabilitation of Highly Disturbed Soils*; Bell, R.W., Ed.; (This Volume) Marcel Dekker: New York, 2001.
8. McRae, S.G. Land Reclamation After Open-Pit Mineral Extraction in Britain. In *Remediation and Management of Degraded Lands*; Wong, M.H., Wong, J.W.C., Baker, A.J.M., Eds.; Lewis Publishers: Boca Raton, FL, 1999; 47–62.
9. Koch, J.M.; Ward, S.C. The Technology of Bauxite Mine Rehabilitation in the Jarrah Forest of Western Australia. In *Proceedings of Remade Lands 2000, The International Conference on Remediation and Management of Degraded Lands, Fremantle, 30 Nov–2 Dec 2000*; Brion, A., Bell, R.W., Eds.; Promaco Conventions: Canning Bridge, WA, 2000; 38–39.
10. Bell, R.W. Chapter 16 Diagnosis and Prognosis of Soil Fertility Constraints for Land Restoration. In *Remediation and Management of Degraded Lands*; Wong, M.H., Wong, J.W.C., Baker, A.J.M., Eds.; Lewis Publishers: Boca Raton, FL, 1999; 163–173.
11. Ho, G.E.; Samaraweera, M.K.S.A.; Bell, R.W. Revegetation Directly on Salt-Affected Gold Ore Refining Residues—a Case Study. In *Remediation and Management of Degraded Lands*; Wong, M.H., Wong, J.W.C., Baker, A.J.M., Eds.; Lewis Publishers: Boca Raton, FL, 1999; 123–135.
12. Jasper, D.A.; Sawada, Y.; Gaunt, E.; Ward, S.C. Indicators of Reclamation Success- Recovery Patterns of Soil Biological Activity Compared to Remote Sending of Vegetation. In *Land Reclamation: Achieving Sustainable Benefits*; Fox, H.R., Moore, H.M., McIntosh, A.D., Eds.; A. A. Balkema: Rotterdam, 1998; 21–24.

LAND USE, HISTORIC

J. Douglas Helms
Hari Eswaran
Paul Reich
United States Department of Agriculture-Natural Resources Conservation Service (USDA-NRCS), Washington, D.C., U.S.A.

INTRODUCTION

Land, as a fundamental factor in production, has always been integrally related to the economic growth of a country. Until or about the time of the industrial revolution, land resources and their productivity constituted much of a society's wealth. Societies use land differently based on their culture, technology, and political and economics aspirations. European colonization of the much of the world since Columbus' explorations radically altered global land use. In addition to mineral exploration and mining, the colonial powers introduced new agricultural production systems to serve their political and economic goals. The current land use patterns in the developing countries were, to a large extent, determined by the resources and raw material needs of the colonial powers. The soil and climatic endowments and the cultural habits of the people, of course, control the specific land use within a region or country.

There were many events and reasons that resulted in land use changes. It is not the intent here to examine these in detail. The purpose of this entry is to evaluate, on a global perspective, some major changes that took place in the last few centuries and evaluate how these have resulted in the current global land use patterns. This article will focus on agricultural land uses, although it is recognized that increasing urbanization is a serious issue for preserving valuable agricultural land.

ORIGINS OF AGRICULTURE

The term "land use" implies human choices and actions that change and alter the landscape. Initially, agriculture was the main use of the land, and land was valued for its agricultural productivity. Agriculture altered land use in multiple, reinforcing ways. First, humans began altering the land surface as more of it was devoted to growing domesticated plants and raising domesticated animals. Next, agriculture supported an increasing population, which in turn brought more land into production. While it is true that some hunter-gatherers lived in communities, agriculture freed people to pursue other livelihoods and to congregate in towns and cities, and to be supplied with food from the countryside.

Scholars have debated the origins and spread of agriculture. Jared Diamond's *Guns, Germs, and Steel* argues for five centers where food production arose independently: Southwest Asia (also referred to as the Near East or Fertile Crescent); China; Mesoamerica (central and southern Mexico and areas of Central America); the Andes of South America, and the eastern U.S. There are four other possibilities for the independent agricultural origins: the Sahel of Africa, tropical West Africa, Ethiopia, and New Guinea (1).

Wheat had its origins in Mesopotamia in the Tigris and Euphrates river valleys about 10,000 years ago. By about 1000–3000 BC, Egyptians were already consuming yeast-leavened bread. Corn is considered to have its origin in the Mexican plateau. Native Americans knew corn, squash and beans as the three sisters. Recent archeological studies in Hunan province, China suggest that cultivation of rice dates back 6000–7000 years and quickly spread across the East Asian crescent from Assam in Northeast India to Yunnan, China. Soon it became a dominant cultivar, occupying the major river valleys and making it the most important cereal grain in Asia. Today, it is a staple food for more than three billion people, with most Asians depending on it for their caloric intake. These—wheat, corn, and rice—are the three basic grain crops that together dominate the agricultural scene of the world.

FLUCTUATIONS IN AGRICULTURAL USES

But expansion of cropland and population did not proceed unabated at all times and places. The Black Death of 1348–1349 was followed by repeated, smaller epidemics. During the 15th century, there were fewer people in

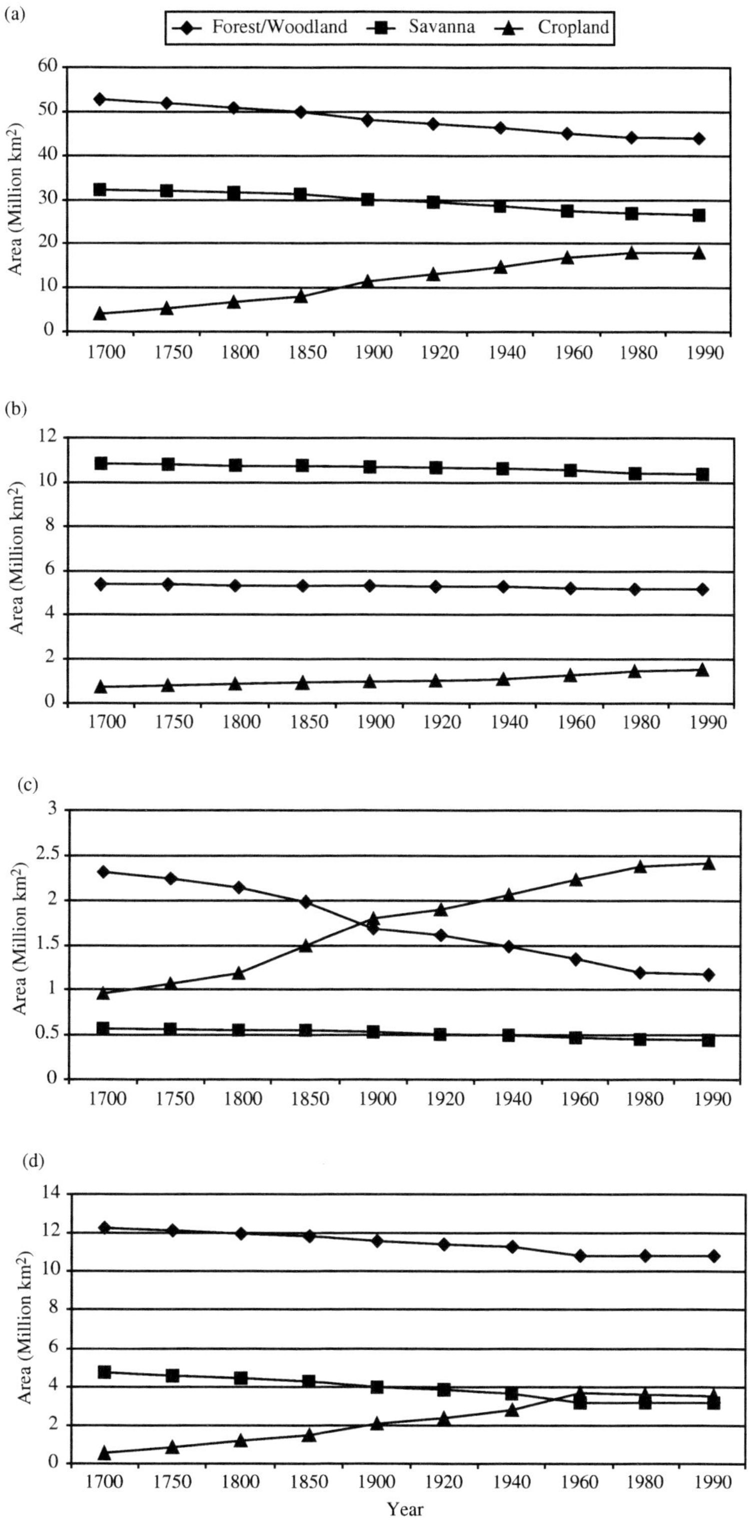
(a)
Forest/Woodland
Savanna
Cropland
Area (Million km²)
60
50
40
30
20
10
0
1700
1750
1800
1850
1900
1920
1940
1960
1980
1990
(b)
Area (Million km²)
12
10
8
6
4
2
0
1700
1750
1800
1850
1900
1920
1940
1960
1980
1990
(c)
Area (Million km²)
3
2.5
2
1.5
1
0.5
0
1700
1750
1800
1850
1900
1920
1940
1960
1980
1990
(d)
Area (Million km²)
14
12
10
8
6
4
2
0
1700
1750
1800
1850
1900
1920
1940
1960
1980
1990
Year

Fig. 1

the European countryside than at any time since about 1150. By the mid-15th century, probably 60% of the plowed cropland in the Parisian basin was no longer planted. Forests reclaimed some land, and pasture replaced some cropland. As a result of less intensive land use, peasants enjoyed more meat and higher quality breads in their diet due to the depopulation of the countryside (2). Native Americans succumbed to disease such as smallpox after European contact (3). The "Indian old fields" of the Eastern woodlands, the Indian mounds of urban Mississippian settlements, and the abandoned irrigation works of the Southwest, attested to Native Americans' land uses (4). Indians repeatedly used fire to renew grasses for grazing animals and left open woodland, where little underbrush hindered travel of the arriving Europeans (5).

MAKING LAND MORE PRODUCTIVE

Since there is a strong relationship between natural chemical elements in the soil and the ability to feed people (6, 7), early economists such as Malthus saw population as being limited by the amount of arable land. Through the centuries and especially in the 20th century, humans succeeded in producing more per acre, so that the correlation between amount of arable land and food production is no longer as strong as it once was. Various societies learned to rotate crops utilizing legumes. Irrigation and drainage, often on fertile soils, have increased production, as have plant selection and breeding and pest protection. As agricultural chemists began understanding more about plant growth in the 19th and 20th centuries, humans began mining guano and phosphorous for transportation around the world. The most far-reaching development affecting land use for agriculture has been the development of the Haber–Bosch process to synthesize ammonia. By one estimate these nitrogen compounds provide between 60 and 80% of the nutrients available to the most commonly produced grain crops, and are perhaps equal to half of the total nitrogen fixed by all bacteria in natural resource systems (8).

LEGACY OF COLONIALISM

In the last five decades, the creation of new nations from the former European colonies triggered vast socio-political changes that have altered land use. In places where traditional agricultural systems had been sufficient to feed a small population, western medical advances expanded the population (9, 10). Many new nations could not, or did not, build agricultural supply and production systems, including agricultural chemicals, to grow additional food. With more pressure on production, the result was too often indiscriminate exploitation of land in most developing countries. Considerations of ecosystem integrity and the need to maintain biodiversity are recent and are still not considered in many developing countries. Large-scale deforestation as in Southeast Asia and South America and expansion of cultivation into wildlife reserves became rampant for several reasons, including increasing population, greater number of people searching for land, and economics. In the richer western countries, application of soil and water conservation technology and financial support from governments initiated a land use system more dependent on science (7). In the centrally controlled economies of the Former Soviet Union and other socialist

(a)

(b)

Fig. 2 (a) Soil Conservation Service demonstration project at Coon Creek, La Crosse County, Wesconsin, U.S.A., 1934. Source of 1934 photograph is Records of the Natural Resources Conservation Service, negative 114-2501-167, National Archives. (b) Soil Conservation Service demonstration project at Coon Creek, La Crosse County, Wesconsin, U.S.A., 1967 (of approximately the same area of the 1934 photograph in a). Source of 1967 photographs is negative BII-2HH-291, Farm Services Agency, Salt Lake City, Utah., U.S.A.

countries, land use was strictly determined by the state. Although land use was designed according to landscape conditions, pressures to enhance productivity resulted in less care for the land, leading to severe land degradation.

GLOBAL LAND USE PATTERNS

In a recent modeling study, Ramankutty and Foley demonstrate the extent of natural ecosystems that were converted for agricultural use during the last three centuries (11). Fig. 1 is based on the empirical data they generated and shows the changing areas for forest, grassland, and cultivated land in the world and for several regions. The pattern of land use change in Africa (Fig. 1b) mimics the global pattern (Fig. 1a) due to the availability of land for agricultural expansion. Fig. 1c illustrates the situation of South Asia, a region where limits of available land for agriculture have been met. The remaining forest and savanna land in this region have many constraints for agricultural use. Fig. 1d illustrates the case of the former Soviet Union countries where major change in land use has been the conversion of savanna to agricultural use. The forested land of this region has low temperature as a major constraint for grain production and so less-forested land is converted for agricultural purposes.

Land use is not only the growing of crops or grazing of animals but also the architecture of fields. The layout and distribution of fields came about initially for efficiency of farm practices (access to the fields by the farmers) and later by concepts of soil and water conservation. Field layout had its best design in modern irrigated systems and these are evident even in developing countries. However, field size, shape, and patterns show a lack of organization in traditional systems. Many western countries went to great extent to reallocate land to enhance efficiency and economies of farming. In Belgium, for example, this process took place in the 1960s and was called "ruilverkaveling" or land reallocation. In other countries, economic pressures induced such changes. Figs. 2a and 2b show an example of change in field architecture over a 50-year period in the U.S.

The beginning of the new millennium is characterized by a number of conditions that affect or even control agricultural sustainability. The current world population of six billion people is projected to increase to nine billion in the next few decades and much of the increase is to take place in the countries that can least afford to enhance the quality of life for current population (7). Agricultural expansion has taken place at the expense of other kinds of land use (Fig. 1). To obtain an idea of the kinds of soils where conversion of land to agriculture was taking place, the authors combined a global land use/land cover map with the soil map of the world. Table 1 shows the distribution of land use as a function of soils.

About 21% of the global land is cultivated and the dominant soils exploited are the Entisols and Inceptisols (mostly for rice cultivation in Asia) Alfisols, Mollisols and Ultisols (for upland grain crops) and Oxisols (for plantation crops in the tropics). Some of the semiarid grasslands and shrublands in the developing countries are used for grazing, usually nomadic. About 658,000 km^2 of land (0.5%) is used

Table 1 Land use and land cover as a function of soils (data in 1000 km^2)

Soil orders	Urban	Cultivated land	Grass land	Shrub land	Shrub land–Grass land	Savanna	Forest	Wetland	Barren	Tundra
Alfisols	142	5708	652	454	24	2669	3296	26	97	11
Andisols	15	207	93	80	18	78	437	3	17	53
Aridisols	49	1286	1984	5851	216	1013	582	2	4551	66
Entisols	94	3365	1695	4125	133	2516	2156	62	7108	65
Gelisols	2	104	271	448	176	76	4204	787	246	4208
Histosols	3	240	48	24	4	66	1068	45	2	20
Inceptisols	89	3748	504	509	97	1865	2217	20	99	46
Mollisols	102	4811	2195	611	40	323	1003	3	19	47
Oxisols	13	2007	88	48	5	1923	5905	6	12	4
Spodosols	33	450	40	102	5	32	2487	19	3	77
Ultisols	86	3505	246	265	4	1834	5585	17	12	13
Vertisols	21	1112	365	493	2	949	189	14	15	0
Rocky land	7	736	2006	1019	539	579	3448	275	952	2901
Shifting sands	3	157	622	1535		427	13	0	2557	0
Total	658	27,436	10,809	15,564	1263	14,350	32,592	1279	15,690	7514

for urban purposes. The actual amounts vary between countries, and, as shown in the study of Nizeymana et al. the better quality lands succumb to urbanization faster. This study has shown that the highly productive soils of the U.S. have a higher level of urbanization (12).

CONCLUSION

There are a number of factors that determine land use patterns in a country. In a global perspective, historical land use reflects socio-cultural evolution, the impact of pests, diseases, conflicts between nations, developments in technologies, and the pressures introduced by population increases. Conversion of land for agricultural uses is currently at an increased pace particularly in developing countries although the reverse trend of restoring the original habitat is taking place in some countries such as the U.S. Recent developments in geographic information technology with appropriate models are now enabling the study of evolution of land use. These developments will provide a better analysis in the future.

REFERENCES

1. Diamond, J. *Guns, Germs, and Steel: The Fates of Human Societies*; W.W. Norton: New York, 1997.
2. Dyer, C. Rural Europe. In *The New Cambridge Medieval History*; Allmand, C., Ed.; Cambridge University Press: Cambridge, UK, 1998.
3. Crosby, A.W. *The Columbian Exchange: Biological and Cultural Consequences of 1492*; Greenwood Press: Westport, CT, 1973.
4. Hurt, R.D. *Indian Agriculture in America: Prehistory to the Present*; University Press of Kansas: Lawrence, 1987.
5. Helms, D. Soil and Southern History. Agric. History **2000**, *74* (Fall), 723–758.
6. Buol, S.W.; Eswaran, H. Assessment and Conquest of Poor Soils. *Proc. of Workshop on "Adaptation of Plants to Soil Stresses*"; INTSORMIL Pub. 94-2, University of Nebraska: Lincoln, NE, 1993; 17–27.
7. Eswaran, H.; Reich, P. Anthropic Landsystems of the World. 2001; In Press.
8. Smil, V. *Enriching the Earth: Fritz Haber, Carl Bosch, and the Transformation of World Food Production*; The MIT Press: Cambridge, MA, 2001.
9. Nye, P.H.; Greenland, D.J. Greenland. *The Soil Under Shifting Cultivation*; Technical Communication 51, Commonwealth Bureau of Soils: Bucks, UK, 1960.
10. Sanchez, P. *Properties and Management of Soils in the Tropics*; Wiley: New York, 1976.
11. Ramankutty, N.; Foley, J.A. Estimating Historical Changes in Global Land Cover: Croplands from 1700 to 1992. Global Biogeochem. Cycles **1999**, *13*, 997–1027.
12. Nizeyimana, E.L.; Peterson, G.W.; Imhoff, M.L.; Sinclair, H.R.; Waltman, S.W.; Reed-Margetan, D.S.; Levine, E.R.; Russo, J.M. Assessing the Impact of Land Conversion to Urban Use on Soils with Difference Productivity Levels in the USA. Soil Sci. Soc. Am. J. **2001**, *65*, 391–402.

LAND USE PLANNING, GIS IN

Egide Nizeyimana
Environmental Resources Research Institute, The Pennsylvania State University, University Park, Pennsylvania, U.S.A.

Jacob Opadeyi
The University of the West Indies, St. Augustine, Trinidad and Tobago

INTRODUCTION

Agricultural scientists are required to provide information needed to address land degradation and land use conflicts confronting the world today. As the population increases and land becomes a commodity in many parts of the world, careful planning of the use of land must be undertaken to accommodate conflicting people's needs and preserve and/or protect the environment. The decision about the use of land must be made based on analyses of each potential use in terms of its economic and biophysical suitability to the specific tract of land and possible impact to environmental degradation.

Geographic information system (GIS) and related technologies (e.g., remote sensing, global position system) have proven to be a valuable tool in land use planning activities. The GIS approach is important in this area because it provides functions to capture, store, organize, and analyze spatially referenced data. Moreover, GIS has been coupled to a variety of models (e.g., crop productivity, hydrology, and water quality simulations) and may be an important component of spatial decision support systems (SDSS) and land resource information systems (LRIS). GIS enhances model flexibility and efficiency in these systems where it is often regarded as a centralized data analysis, management, and planning system. As a result, decision makers and land use planners in private companies, universities, and at various levels of the government are using GIS to develop spatial environmental databases, perform land evaluations, and analyze and manage resources. There is no doubt that the use of GIS and GIS-based systems in land use planning activities will continue to increase in the future, as more detailed digital environmental datasets become available and the capability of computers to handle large volumes of data increases.

GIS USE IN LAND USE PLANNING ACTIVITIES

The goal of land use planning is to make decisions about the use of land and resources (1). Its implementation is often driven by future people's needs in terms of productivity and environmental sustainability. Land use planning is important in highly populated communities primarily due to conflicts between competing uses and interests of users (2). In this case, planning activities are tailored to make the optimal use of the limited land resources. In general, the land use planning involves sequentially an organization of thoughts and establishment of long-term goals, a land evaluation that includes appraising alternatives, and finally designing and implementing the plan. Land evaluation is the most important aspect of this process.

The information within a GIS consists of a spatial component represented by points (e.g., well locations), lines (e.g., streams, road networks), or polygons (e.g., soil delineations) and attribute data or information that describes characteristics of the spatial features. The spatial entity is referenced to a geographic coordinate system and is stored either in a vector or raster format (3). GIS is primarily used in the development of spatial databases and land evaluation, a procedure in the land use planning process that is aimed at determining the suitability of land units to current and alternative uses and the potential impact of each on the environment.

Development of Spatial Databases

GIS stores, retrieves, and allows efficient manipulation of database information. It provides powerful analysis and relational database facilities to modify and/or integrate spatial data from different sources and resolutions as well as advanced visualization functions to display output data in the form of interpretative maps. In this case, land attributes and qualities are derived from geographic databases and used to determine land suitability, limitations, or ratings for various land use types. The analysis results may be presented in tabular or graphical form and are intended to provide key information necessary for land users or planners to make meaningful decisions about land management and conservation and/or land use planning.

Site Suitability Analyses

A site suitability analysis typically involves the assessment of the level of affinity a specific land has for a particular use. Soil information available in soil databases is rarely enough for site evaluations. In addition to soils, the analysis often integrates local information on landforms, current land uses, the relative location of the land, and associated social and political restrictions. The proposed use may have additional limitations that should be taken into account. For example, an analysis for suitable sites for land application of sewage sludge should consider the physical, chemical, and biological properties of the waste in soil and water (4, 5).

GIS is efficient in identifying and ranking sites for various land use planning activities. First, site-specific analyses often require many and detailed data sources. Second, GIS overlay features, logical operations, and display functions are tailored to speed up data processing and therefore allow efficiently suitability class assignment and graphical display of results. A good example is the use of GIS for locating appropriate sites for forestland application of sewage waste (6). The authors derived physical site suitability ratings for an area in Vermont based on EPA guidelines (7) and merged them with social and political restrictions of the state and counties to derive a land applicability classification. Similar GIS-based approaches have been used to locate sites for solid waste disposals (8).

Linking GIS and Models

The spatial databases and associated attribute data described above may be part of a GIS/model graphical user interface (GUI). These models may be those that determine land suitabilities or those that simulate water quality for environmental impact assessments. As an entity of GIS/model interface, GIS allows easy access to database attributes by various algorithms, statistical software, and environmental models for a variety of land use analyses. In addition to model parameterization, a full integration of GIS and models also allows the user to interact with various modules, select data input and module utilities, and display graphical and/or tabular representations of modeling results (9). The use of this type of interface reduces significantly the processing time and resources required to develop input data and run the model. GIS/model linkage has been accomplished for field- and watershed-scale hydrology/water quality models such as the leaching and chemistry estimation (LEACHM) (10) and agricultural nonpoint source pollution (AGNPS) (11) models. Results of these analyses are used to support management and land use planning activities in the farm or watershed.

The GIS/model interface may also be part of an SDSS. In addition to modeling parameterization, SDSS offer the users with functions to evaluate different land use scenarios necessary for making management recommendations and/or land use planning decisions. Results may be used by farmers to adjust management practices of distinct fields in a given farm or may be used by field officers to help farmers set priorities while providing technical assistance for nutrient management planning. Fig. 1 shows a flow diagram indicating data integration in a GIS with models for water quality assessment in watersheds.

GIS has been incorporated in many land evaluation systems so that planners and public officials can take advantage of its spatial modeling and visualization capabilities. An example of such a system is the USDA–NRCS land evaluation and site assessment (LESA) (12). The land evaluation portion of LESA determines soil productivity levels, farm size and agricultural sales volume; the site assessment portion deals with factors such as location, amount of nonagricultural land, zoning restrictions, etc. (13). GIS, in GIS-based LESA systems, is used in both modules and at all levels of the analysis.

Development of Land Resource Information Systems

A number of land information delivery systems have been developed in recent years to make spatial data available to users for application in various aspects of land use planning. These systems, also called LRIS, are multipurpose systems that integrate geographic databases and GIS tools to analyze, record, report, and display spatial data relationships. Some of these systems have been embedded into the World Wide Web (WWW) for quick and easy access and analysis via Internet browsers.

RELIABILITY OF GIS-BASED ANALYSIS RESULTS

GIS and GIS-based systems used in spatial data analyses for land use planning have grown in recent years. GIS is

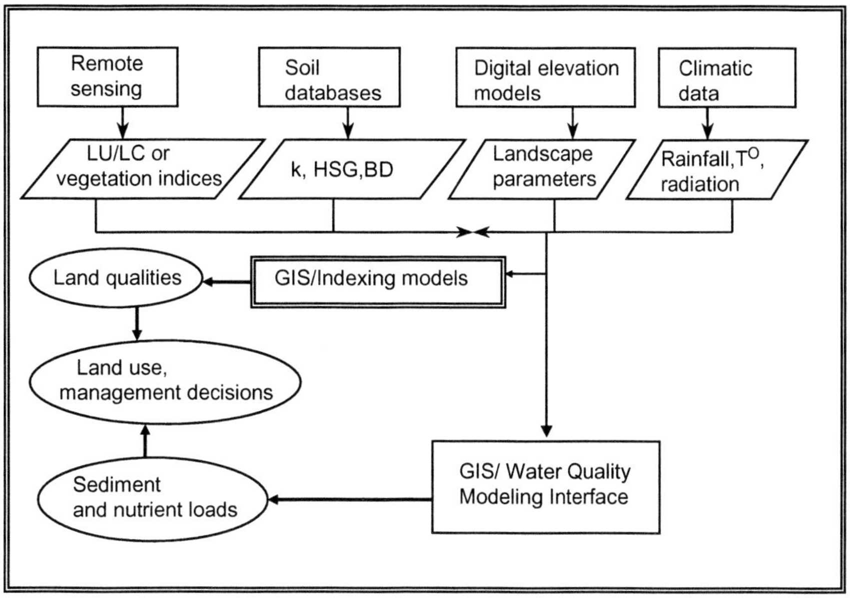

Fig. 1 Integration of GIS, spatial database, and models.

particularly attractive in these areas because it allows overlay of spatial data sets and the merging and analysis of attribute data from different sources. The resulting data are obtained using data from digital and paper maps of different scales or acquired at different resolutions such as in the case of digital elevation models (DEM) and remote sensing imagery. The combination of such data layers may produce unrealistic data and consequently lead to erroneous predictions. The question is how reliable are these results when used for developing land use and management plans. A discussion on the propagation of error and uncertainty in GIS-based systems is provided elsewhere (14).

The accuracy of GIS-based land evaluations is a function of the quality of attribute data and mapping delineations of databases, and the type of model or assessment scheme used in the analyses. Various algorithms for assessing the quality of GIS analysis results as affected by error and uncertainty in GIS layers, and their propagation are part of popular GIS and image analysis software (e.g., Arc/Info, ERDAS Imagine). The results, which are often in the form of reliability diagrams are not used by average GIS practitioners. This is probably due to the fact that they are not easy to understand or interpret. Similarly, models vary depending on how each represents various processes of the system. Lumped-parameter models and indexing/ranking schemes are mostly used in land evaluations because they are easy to parameterize. However, these models ignore spatial variations of parameters throughout the field, watershed, or region of study. Furthermore, models originally designed for fields and watersheds are often applied to regional analyses, thus adding some level of uncertainty in modeling predictions. For example, most land use planning programs use conventional methods of land evaluation. Each land parameter is given a range of values with corresponding ratings showing its suitability to crop production. These indexes are added or multiplied to create a single index that is to rank land units. The method is simple, but carries a high uncertainty because breaks between two ranges or ranks are subjective.

The effect of map scale and resolution on environmental assessment and modeling output data has been subject to many studies. Raster-based GIS systems require that a grid cell size be defined prior to the analysis. However, as pixel size increases above the resolution of the original data, the spatial variability decreases. This causes a decrease of the predictive power of generated input parameters particularly for small land areas.

SUMMARY AND CONCLUSION

GIS has been used primarily in land evaluation, a procedure in the land use planning process that deals with determining land suitability to current and alternative uses and the potential impact of each use on the environment. GIS and GIS-based systems are often considered to as integrated spatial information systems (ISIS) for data analysis needed in land use management and planning activities. Despite the advantages of GIS outlined above, the land use planner should be aware of the error and uncertainty in GIS and GIS-based analyses resulting from digitizing and scaling inaccuracies, data conversion between vector and raster formats, and others. Finally, GIS implementation in land use planning activities can be expensive. In addition to costs associated with hardware and software purchase and maintenance, a high level of technical expertise is required to perform complex modeling tasks in land evaluations for alternative uses and to sustain databases. Nonetheless, the demand for GIS and GIS-based analysis systems in land use planning is to increase in the future as more detailed spatial datasets become available.

REFERENCES

1. FAO, *Guidelines for Land Use Planning*; FAO Development Series 1, Soil Resources, Management and Conservation Service, FAO: Rome, 1993; 96.
2. Brinkman, R. Recent Developments in Land Use Planning, with Special Reference to FAO. In *The Future of the Land: Mobilizing and Integrating Knowledge for Land Use Options*; Fresco, L.O., Stroosnijder, L., Bouma, J., van Keulen, H., Eds.; Wiley: Chichester, 1994; 13–21.
3. Burrough, P.A. Matching Databases and Quantitative Models in Land Resource Assessment. Soil Use Mgmt **1989**, *5*, 3–8.
4. Nizeyimana, E.; Petersen, G.W.; Looijen, J.C. Land Use Planning and Environmental Impact Assessment Using GIS. In *Environmental Modeling with GIS and Remote Sensing*, Skidmore, A., Prins, H., Eds.; Taylor and Francis Ltd.: London, 2002; in press.
5. Petersen, G.W.; Nizeyimana, E.; Miller, D.A.; Evans, B.M. The Use of Soil Databases in Resource Assessments and Land Use Planning. In *Handbook of Soil Science*, Summes, M.E., Ed.; CRC Press: Boca Raton, 2000.
6. Hendrix, W.G.; Buckley, J.A. Use of a Geographic Information System for Selection of a Sites for Land Application of Sewage Waste. J. Soil Water Conserv. **1992**, *92*, 271–275.
7. U.S. *Environmental Protection Agency, Process Design Manual for Land Treatment of Municipal Waste*; PB 299655/1 Environmental Protection Agency: Cincinnati, OH, 1981.
8. Weber, R.S.; Jenkins, J.; Leszkiewicz, J.J. Application of Geographic Information System Technology to Landfill Site Selection. *Proceedings of the Application of Geographic Information Systems, and Knowledge-Based Systems for Land Use Management*; Nov. 12–14 VPI and State University: Blacksburg, VA, 1990.
9. Petersen, G.W.; Nizeyimana, E.; Evans, B.M. Application of GIS to Land Degradation Assessments. In *Methods for Assessment of Soil Degradation*; Lal, R., Blum, W.H., Valentine, C., Stewart, B.A., Eds.; CRC Press: Boca Raton, FL, 1997; 377–391.
10. Inskeep, P.P.; Wraith, J.M.; Wilson, J.P.; Snyder, R.D.; Macur, R.E.; Gaber, H.M. Input Parameter and Model Resolution Effects on Predictions of Solute Transport. J. Environ. Qual. **1996**, *25*, 453–462.
11. Tim, U.S.; Jolly, R. Evaluating Agricultural Nonpoint Source Pollution Using Integrated Geographic Information Systems and Hydrology/Water Quality Models. J. Environ. Qual. **1994**, *23*, 25–35.
12. Wright, L.E.; Zitzmann, W.; Young, K.; Googins, R. LESA—Agricultural Land Evaluation and Site Assessment. J. Soil Water Conserv. **1983**, *38*, 82–86.
13. Daniels, T. Using LESA in a Purchase of Development Rights Program. J. Soil Water Conserv. **1990**, *45*, 617–621.
14. Heuvelink, G.B.M. *Error Propagation in Environmental Modeling with GIS*; Taylor & Francis: London, 1998; 127.

LANDSCAPE

George Hall
The Ohio State University, Columbus, Ohio, U.S.A.

INTRODUCTION

The landscape is defined in several different ways. It has been defined as the portion of the land surface that the eye can comprehend in a single view (1). The glossary of soil science terms (2) states that the landscape is a collection of related landforms; usually the land surface which the eye can comprehend in a single view. Landforms are any physical, recognizable form or feature on the earth's surface (2). Thus, the scope of the landscape is variable depending on the position of the observer, whether in a deep valley or on an airplane flying over the valley. The landforms are the features that make up that landscape.

LANDSCAPE TERMINOLOGY

The present terminology used by soil scientists is based on the historic concepts of William Morris Davis and Walther Penck. Wood (3) and King (4) formulated models for the fully developed hillslope which included the elements of waxing slope, free face, debris slope, and pediment (Fig. 1). Ruhe (5, 6) developed a soil–geomorphic model in which the soil properties were integrated with the hillslope models. Ruhe (6, 7) modified the Wood and King models by proposing the hillslope elements of summit, shoulder, backslope, footslope, and toeslope (Fig. 1). Following the introduction of the hillslope positions, Ruhe and co-workers (7, 8) formalized terms for geomorphic slope components: headslope for the concave position, sideslope for the linear position, and noseslope for the convex position (Fig. 2).

A different approach was taken by Conacher and Dalrymple (9). They used the catena concept and defined the slopes as a three-dimensional unit extending from summit to the valley floor. The model consisted of nine units, all of which need not occur on a given landscape. Of particular importance in the model were the interactions among soil materials and mobilization, translocation, and redeposition of the material by water and gravity.

LANDSCAPE UNITS

The specific landscape position elements have differing processes and as a result have differing properties. Although these five positions are separately defined, as is the case with many natural phenomenon, the positions and thus the processes grade from one to the other. The following discussion is most applicable for humid climates.

Summit

Where the summit is greater than approximately 30 m in width, much of the water is retained on the surface. This position is considered the most stable, and the water movement is predominantly vertical except near the transition to the shoulder position or on small undulations on the summit.

Shoulder

Convexity is the rule for the shoulder positions. Surface runoff of water is maximized in this element resulting in a highly erosional and relatively unstable land surface. Lateral movement of surface material (soil creep) may become an important process on this part of the landscape. Lateral subsurface water movement or throughflow is often an important process in this landscape position.

Backslope

General linearity is characteristic of the backslope position. The dominant process on this position is the transport of material as well as water. Surface transport of material may be in the form of surface wash, creep, flow, or slump. Lateral subsurface water movement (throughflow) is important in this element. The position is considered to be relatively unstable depending on the degree of slope.

Footslope

Concavity is characteristic of this landscape position. This concavity results in deposition from upslope of

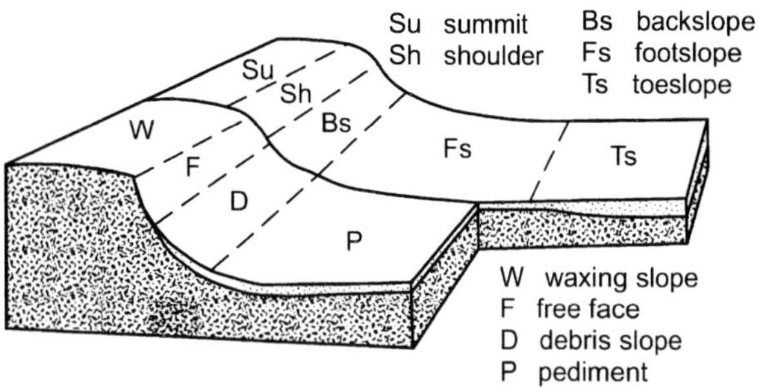

Fig. 1 Elements of a hillslope model. (From Ref. 6.)

particulate material as well as material carried in solution. Seepage zones may be present and water retention is high. Soils in footslope positions may be heterogeneous due to mass movement, irregular seepage, and nonuniform deposition. Paleosols often can be identified in these positions.

Toeslope

The toeslope element is unstable as a result of its dominantly constructional nature. Materials in this position are derived from up valley and to some extent from superjacent footslope and backslope positions.

CATENA

The catena concept was proposed by Milne (10, 11) as an interlocking of elements of the landscape. The term is derived from the latin word for chain. In U.S., catena is now used almost interchangeably with toposequence. Toposequence or local relief carries with it a morphological connotation that is related to a relative elevation and thus to changes in hydrology. Catena on the other hand carries with it a process–response connotation. The soils of a catena differ not only in morphology but also are considered to differ as a result of erosion, transport and deposition of surficial material as well as leaching, translocation and deposition of chemical and particulate constituents in the soil. The concept of catena has been used extensively to study the soil genesis (12–14).

The morphology and processes of each member of the catena is related to every other member of the catena. It is a dynamic process where in the members of the catena are continuously adjusting to the environmental changes of a given landscape (15).

LANDSCAPE MORPHOLOGY

Landforms are individually transformed by erosion and deposition during the process of landscape evolution.

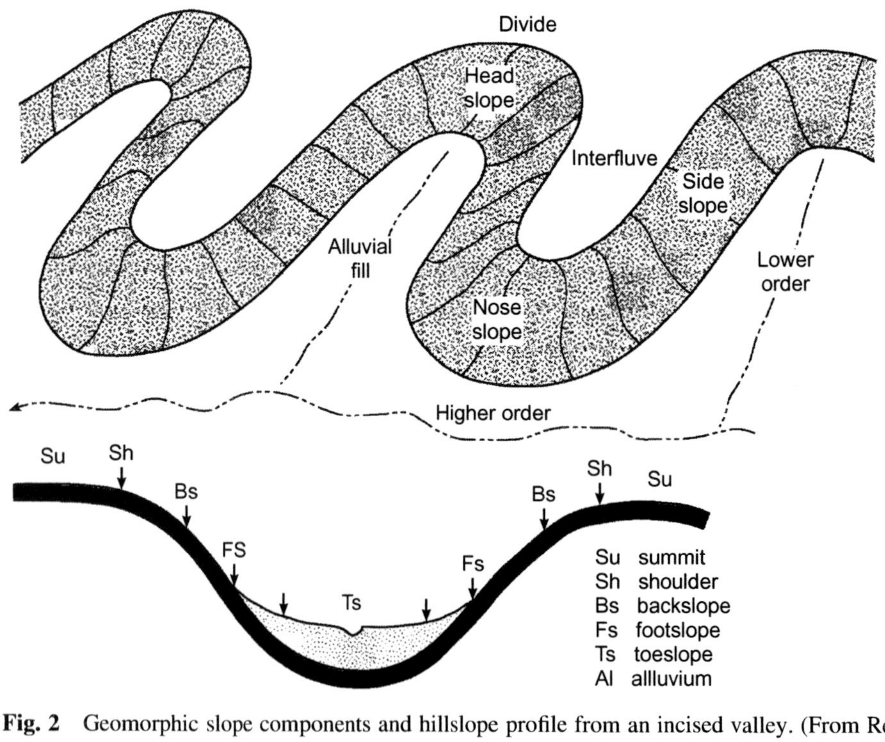

Fig. 2 Geomorphic slope components and hillslope profile from an incised valley. (From Ref. 8.)

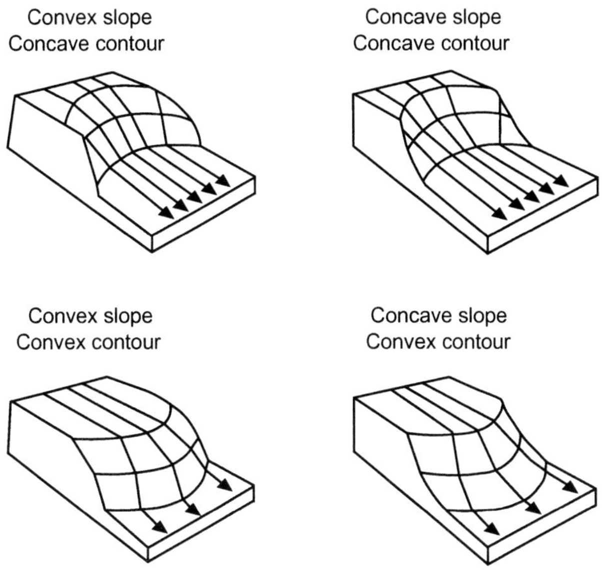

Fig. 3 Four basic slope shapes with flow lines. (From Ref. 13.)

Water movement is a major driving force that shapes and modifies the morphology of the landscape. To properly evaluate the processes occurring on the landscape, the morphology of the landscape must be determined. This evaluation should include at a minimum: gradient, aspect, and vertical and horizontal curvature.

Water movement on the slopes is largely determined by the planar curvature of the slopes; both the longitudinal and latitudinal profile. Huggett (13) added surface flow lines to the basic slope shapes (Fig. 3).

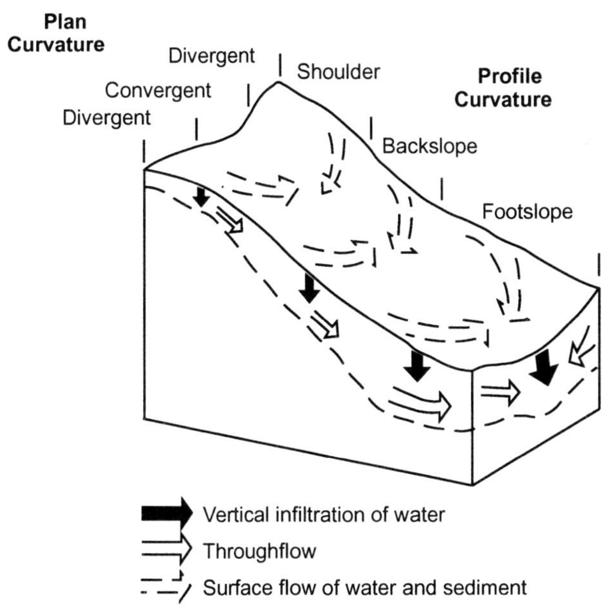

Fig. 4 Water movement in a hillslope system. (From Ref. 16.)

Pennock et al. (16) studied the complexity of water movement on various hillslope elements and curvatures. They used the terms convergent and divergent to identify the slope positions in which water accumulates or disperses. One combination of convergent and divergent slope positions is shown in Fig. 4. In this example the landscape element with the least moisture would be the divergent shoulder and the element with the most moisture would be the convergent footslope. Other elements would be expected to have intermediate moisture contents.

Landscape models have shown that landscapes are predictable; they have a large nonrandom variability component. The landscape must be defined in three dimensions, considering the surface configuration as well as the vertical changes in stratigraphic materials.

SLOPE ORIENTATION/ASPECT

Landscape modifying processes are influenced by the slope orientation or aspect. The orientation affects the amount of radiation reaching the soil surface which in turn influences the microclimate, and thus, the moisture and vegetation on slopes. In the areas of low relief, aspect has little influence while in the areas of high relief the influence becomes more important.

Global latitude is also a determinant because the angle of solar radiation increases with increased latitude. Thus, shading increases with increased latitude. Aspect becomes most important in the high relief areas between 40 and 60° latitude (17). In the northern hemisphere, south-facing slopes receive more radiation and are thus warmer and dryer than north-facing slopes (14). The differences in moisture and temperature are reflected in the vegetative and soil process differences for different slope orientations.

IMPLICATIONS FOR LAND USE AND MANAGEMENT

Landscape and Land Use

Most land use decisions are influenced by the landscape characteristics. In agriculture, cropping patterns and cultivation practices are conditioned by the gradient and configuration of the landscape. Conservation practices must be invoked as slope gradients increase. These practices include contour strip cropping and conservation tillage to control erosion. Increasing slope gradients also limit the use of agricultural equipment and increases the cost of production. In steeper portions of the landscape, the land must be utilized for grazing or forestry. Footslope and toeslope positions are more subject to seepage and flooding, and thus, may restrict the cropping options on these portions of the landscape.

In urban land-use planning, consideration must also be given to landscape. As in the case of agriculture, footslope and toeslope positions are subject to water seepage and flooding, and urban development of these areas must take these limitations into consideration when land-use decisions are made. Although housing developments on steep gradients are aesthetically pleasing, they are more expensive to build and maintain, and may be subject to slippage and slope failure.

Risks of Soil Erosion

Water, wind, and gravity are each considered agents of soil erosion. Erosion resulting from water and gravity are most commonly related to landscape positions. Wind erosion is more closely related to vegetative soil cover but may be modified by landscape conditions. Soil loss increases as the length of the slope increases up to some maximum. Soil loss increases exponentially as the slope gradient increases. Erosion as the result of gravity (mass wasting) is most common on very steep gradient slopes, but may also be found on lesser slopes where the surface and subsurface conditions such as a seep may cause slumps and slips to occur.

On-farm and In-situ Water Management

Water management is important in the control of erosion and to provide optimum moisture for plant growth. Diversions and waterways are often used to control water movement on the landscape. On steeper gradient slopes, contour and strip cropping may be used to slow water movement over the soil and thus encourage greater moisture infiltration and retention. Tile drainage may be used to intercept drainage from seeps in the footslope positions. In level or depressional areas where soil water accumulates at or near the surface, tile drainage must be installed to remove the excess water.

REFERENCES

1. Ruhe, R.V. *Quaternary Landscapes in Iowa*; Iowa State Univ. Press: Ames, IA, 1969; 255.
2. Roth, C.B. *Glossary of Soil Science Terms*; Soil Science Society of America: Madison, WI, 1997; 134.
3. Wood, A. The Development of Hillside Slopes. Geol. Assoc. Proc. **1942**, *53*, 128–138.
4. King, L.C. Canons of Landscape Evolution. Geol. Soc. Am. Bull. **1953**, *64*, 721–752.
5. Ruhe, R.V. Geomorphic Surfaces and the Nature of Soils. Soil Sci. **1956**, *82*, 441–455.
6. Ruhe, R.V. Elements of the Soil Landscape. Trans. Int. Congr. Soil Sci. 7th. **1960**, *4*, 165–170.
7. Ruhe, R.V. *Geomorphology*; Houghton Mifflin: Boston, MA, 1975; 246.
8. Ruhe, R.V.; Walker, P.H. Hillslope Model and Soil Formation: I. Open Systems. Trans. Int. Cong. Soil Sci. 9th. **1968**, *4*, 551–560.
9. Conacher, A.J.; Dalrymple, J.B. The Nine Unit Landscape Model: An Approach to Pedogeomorphic Research. Geoderma **1977**, *18*, 1–154.
10. Milne, G. A Provisional Soil Map of East Africa. Amani Memoirs No. 28, East African Agric. Res. Stn., Tanganyika Territory; 1936; 34.
11. Milne, G. Normal Erosion as a Factor in Soil Profile Development. Nature **1936**, *138*, 548.
12. Yaalon, D.H. Soil-Forming Processes in Time and Space. In *Paleopedology: Origin, Nature and Dating of Paleosols*; Yaalon, D.H., Ed.; Israel University Press: Jerusalem, Israel, 1971; 29–39.
13. Huggett, R.J. Soil Landscape Systems: A Model of Soil Genesis. Geoderma **1975**, *13*, 1–22.
14. Birkeland, P.W. *Soils and Geomorphology*; Oxford University Press: Oxford, 1999; 430.
15. Dan, J.; Yaalon, D.H. The Application of the Catena Concept in Studies of Pedogenesis in Mediterranean and Desert Fringe Regions. Trans. 8th Int. Cong. Soil Sci. **1964**, *5*, 751–758.
16. Pennock, D.J.; Zebarth, F.J.; deJong, E. Landform Classification and Soil Distribution in Hummocky Terrain, Saskatchewan, Canada. Geoderma **1987**, *40*, 297–315.
17. Hunckler, R.V.; Schaetzl, R.J. Spodosol Development as Affected by Geomorphic Aspect, Baraga County, Michigan. Soil Sci. Soc. Am. J. **1997**, *61*, 1105–1115.

LANDSCAPE RELATIONSHIPS

Timothy A. Quine
University of Exeter, Exeter, Devon, United Kingdom

L

INTRODUCTION

Tillage erosion is inextricably linked to landscape characteristics. The magnitude of tillage erosion rates is determined by the characteristics of the landscape and the spatial variation in soil redistribution by tillage contributes to the evolution of spatial variation in the landscape. The dynamic relationships between tillage erosion and landscape are, therefore, addressed under the headings *landscape sensitivity* and *landscape heterogeneity*. Sensitive landscapes are topographically complex, or have a high density of field boundaries, or display both characteristics. Tillage erosion contributes to the evolution of landscape heterogeneity through creation of distinctive landforms, such as lynchets, terraces and field boundary steps, and through progressive but relatively rapid redistribution of soil from spurs to hollows. The resultant variability in soil properties is an important influence on crop production and water erosion.

LANDSCAPE SENSITIVITY

Tillage erosion potential, described by the tillage transport coefficient is controlled by and may be predicted from cultivation practice; however, in order to determine rates of tillage erosion this erosion potential must be combined with information concerning landscape sensitivity to tillage erosion.

Tillage erosion occurs where the outflux of soil (usually downslope) due to tillage translocation exceeds the influx of soil (usually from upslope) due to tillage translocation—accumulation occurs under the opposite conditions. Net soil loss and gain due to tillage erosion are focussed on areas in the landscape with discontinuous or rapidly changing flux. Therefore, high landscape sensitivity to tillage erosion is found in landscapes with complex within-field topography or where there are numerous field boundaries located across steep slopes. In general terms, for landscapes of the same average slope angle, complexity and sensitivity to tillage erosion are inversely related to the distance between slope crest and valley; and for landscapes with similar crest to valley distance, relative sensitivity to tillage erosion will be proportional to average slope angle (Table 1).

The location of such sensitive landscapes is determined by lithology and long-term erosion history and, consequently, strong spatial variation in landscape sensitivity to tillage erosion exists even where there is relatively little variation in tillage erosion potential. In a study of topographic controls on medium-term soil redistribution on 5 agricultural fields in the UK, slope curvature (and, by implication, tillage erosion) dominated soil redistribution on the 4 fields underlain by sands, siltstone and chalk whereas on the field underlain by clay with flints no such relationship was identified (1). All the fields were subject to similar cultivation practice and the differences observed represent variation in landscape sensitivity due to lower topographic complexity. This is most clearly reflected in the range of slope profile curvatures: On the clay soils, the range was only 25% of the smallest range observed on the other fields.

High sensitivity landscapes include strongly rolling glacial till-lands, incised loess plains, and dry valley systems (Fig. 1).

LANDSCAPE HETEROGENEITY

The discussion of landscape sensitivity has addressed the control that large scale landscape heterogeneity exercises on spatial variation in tillage erosion rates. Changing the scale of examination to the field/sub-field scale sees a reversal in causation and the control of tillage erosion rates on the existence and degree of heterogeneity may be examined. Two aspects are considered: landform creation and landform modification.

Landform Creation

The creation of distinctive landforms provided some of the first evidence for the importance of tillage erosion

Table 1 Indicative mean soil redistribution rates due to tillage erosion

	Soil redistribution rates ($t\,ha^{-1}\,year^{-1}$)				
Distance (m)	**Maximum (slope 10°)**	**Maximum (slope 5°)**	**Maximum (slope 3°)**	**Slope change from 3° to 10° or from 10° to 3°**	**Slope change from 3° to 5° or from 5° to 3°**
50	18	9	5	12	4
40	22	11	7	15	4
30	29	15	9	21	6
20	44	22	13	31	9
10	88	44	26	62	18
5	176	87	52	124	35

Note: (1) An annual tillage transport coefficient of $500\,kg\,m^{-1}$ has been used — this is indicative of one typical mouldboard plough operation and two supplementary operations: e.g., discing and harrowing. (2) These are mean rates and mask significant smaller scale variability.
Where the distance refers to the land surface either from the upslope boundary to the location of maximum slope or over which slope angle increases, the rates are mean rates of erosion over the specified distances.
Where the distance refers to the land surface either from the location of maximum slope to the downslope boundary or over which slope angle decreases, the rates are mean rates of accumulation over the specified distance.

Fig. 1 A landscape of high sensitivity to tillage erosion—seasonally dry valleys and rolling hills near Cordoba in southern Spain. Note: This landscape is also highly sensitive to water erosion.

as a pedo-geomorphic process on agricultural land (2) and "monuments" of tillage erosion (Fig. 2) associated with boundaries of cultivation may be seen in a wide range of agricultural systems.

Field boundaries form lines of zero tillage flux. In landscapes where relatively steep slopes are divided by cross-slope boundaries these lines of zero flux represent *significant* flux discontinuities with high influx and zero outflux on the upslope side, and high outflux and zero influx on the downslope side. This flux discontinuity results in soil accumulation adjacent to the boundary in the upslope field and soil loss adjacent to the boundary in the downslope field with the consequent formation of a linear step along the boundary line (Fig. 3). Such steps are near universal features of agricultural landscapes:

Fig. 2 Dr. Dino Torri (shown) and Dr. Paolo Bazzoffi coined the phrase "monuments of tillage erosion" to describe features such as this relict landsurface surrounding a solitary pine on a strongly convex hill in Tuscany, Italy.

Fig. 3 Evidence of tillage erosion is widely seen in large banks at contour field boundaries on sloping agricultural land subject to mechanized agriculture. This example is from Aarslev in Denmark. (See also Ref. 2.)

1. Where fields are or have been narrow contour strips, bench terraces (3) or strip lynchets are created—the latter have provided landscape archaeologists with evidence of past land use.
2. When larger fields are located on steep slopes subject to mechanized agriculture then the steps may develop into large changes in slope—for example, steps 3–4 m high have been identified in the Palouse region of the Pacific Northwest of the USA (2).
3. Even in landscapes where water erosion processes dominate geomorphological evolution and cultivation is undertaken by hand, discontinuity in cultivation leads to step formation (4) and such steps can become unstable and act as foci for small-scale mass failure and pipe or gully initiation (Fig. 4).

The formation of these distinctive landforms is associated with local change in soil properties associated with accumulation and loss of plow soil. In landscapes in which rolling topography is divided into relatively large fields this may have relatively little impact on field-scale soil variation and productivity although boundary zones of depleted crop production are often visible. However, where moderately steep slopes are divided into narrow strips and rapid lynchet or terrace development takes place (Table 1), the landscape will become characterized by large variation in soil properties over small distances (5–7). The significance of this variation will be dependent on the nature of the sub-plow horizons. If the sub-plow material is nutrient rich and easily weathered the impact on soil productivity may be small (7) but if the sub-plow soil is significantly less productive than the plow soil, significant variation in crop production over distances of a few meters can be expected (6, 8) and a large proportion of the available agricultural land may become unproductive.

Landform Modification

The effect of differential flux on complex landforms is in marked contrast to the effect of zero flux on relatively simple landforms. While the latter leads to landform creation, the former leads to gradual landform obliteration although the rate is such that landform modification will usually be a more appropriate description. In essence, tillage erosion leads to a gradual smoothing of the topography and reduction in relief intensity. Convex slope elements are subject to greatest soil loss (Fig. 2) and concavities such as seasonal or relict channel-lines are subject to soil accumulation. Tillage acts like a conveyor belt transporting soil from crest to hollow leading to gradual infilling of the hollow and reduction in crest altitude. This creates important synergy with water erosion processes.

This homogenization of landform is, nevertheless, associated with an increase in landscape heterogeneity because of the associated increase in soil variability. Differences in soil properties between crest and hollow become more marked as (topsoil-rich) plowsoil accumulates in the hollows leading to the formation of overdeepened Ap horizons, while the plow soil on convexities becomes diluted by subsoil inclusion (Fig. 5) and depleted in surface-derived material (e.g., organic matter and surface applied nutrients) (9). Even where no net soil loss occurs on linear slopes below convexities, downslope tillage translocation of soil from the crest leads to soil depauperation (due to subsoil incorporation) (10, 11).

Systematic spatial variation in water movement and availability driven by topographic controls on hillslope hydrological processes will combine with this erosion-produced systematic soil variability to create a landscape of strong within-field spatial variation in plant growing conditions (Fig. 6) (12, 13) that may be economically very significant where water supply is a limiting factor for crop growth.

While landform modification is usually the most appropriate term, landform obliteration may take place where slope curvature is very high, slopes are short and implements are large. Under these conditions planing of landforms (14) can result in much more rapid landform evolution and strong variation in soil properties over very short distances.

Fig. 4 Water erosion is the dominant erosion process on agricultural fields in the rolling hills region of the Loess Plateau (4) but, even here, tillage steps develop at the lower boundaries of fields and these can be foci for mass failure and pipe or gully initiation.

Fig. 5 Tillage erosion of soil from the ridges and transport into the hollows in this field in Minnesota, USA, has led to the development of visible strong within-field spatial variation in soil properties. In this case the land-user is attempting to remediate the problem by manuring the convexities.

Fig. 6 Spatial variation in soil properties combines with hillslope hydrology to create significant spatial variation in conditions for crop growth seen in variation in time of flowering in this oilseed rape crop in Devon, UK (13).

REFERENCES

1. Quine, T.A.; Walling, D.E. Use of Caesium-137 Measurements to Investigate Relationships Between Erosion Rates and Topography. In *Landscape Sensitivity*; Thomas, D.S.G., Allison, R.J., Eds.; John Wiley: Chichester, 1993; 31–48.
2. Papendick, R.I.; Miller, D.E. Conservation Tillage in the Pacific Northwest. Journal of Soil and Water Conservation **1977**, *32*, 49–56.
3. Dabney, S.M.; Liu, Z.; Lane, M.; Douglas, J.; Zhu, J.; Flanagan, D.C. Landscape Benching from Tillage Erosion Between Grass Hedges. Soil and Tillage Research **1999**, *51*, 219–231.
4. Quine, T.A.; Govers, G.; Walling, D.E.; Zhang, X.; Desmet, P.J.J.; Zhang, Y. Erosion Processes and Landform Evolution on Agricultural Land—New Perspectives from Caesium-137 Data and Topographic-Based Erosion Modelling. Earth Surface Processes and Landforms **1997**, *22*, 799–816.
5. Agus, F.; Cassel, D.K.; Garrity, D.P. Soil-Water and Soil Physical Properties Under Contour Hedgerow Systems on Sloping Oxisols. Soil and Tillage Research **1997**, *40*, 185–199.
6. Lewis, L.A. Terracing and Accelerated Soil Loss on Rwandan Steeplands: A Preliminary Investigation on the Implications of Human Activities Affecting Soil Loss. Land Degradation and Rehabilitation **1992**, *3*, 241–246.
7. Quine, T.A.; Walling, D.E.; Zhang, X. Tillage Erosion, Water Erosion and Soil Quality on Cultivated Terraces at Xifeng in the Loess Plateau, China. Land Degradation and Development **1999**, *10*, 251–274.
8. Turkelboom, F.; Poesen, J.; Ohler, I.; Ongpraset, S. Reassessment of Tillage Erosion Rates by Manual Tillage on Steep Slopes in Northern Thailand. Soil and Tillage Research **1999**, *51*, 245–259.
9. Papendick, R.I.; Young, D.K.; McCool, D.K.; Krauss, H.A. Regional Effects of Soil Erosion on Crop Productivity—Pacific Northwest. In *Soil Erosion and Crop Productivity*; Follet, R.F., Stewart, B.A., Eds.; Amer. Soc. Agron. WI: USA, 1985; 305–320.
10. Van Oost, K.; Govers, G.; Van Muysen, W.; Quine, T.A. Modeling Translocation and Dispersion of Soil Constituents by Tillage on Sloping Land. Soil Science Society of America Journal **2000**, *64*, 1733–1739.
11. Quine, T.A.; Walling, D.E.; Chakela, Q.K.; Mandiringana, O.T.; Zhang, X. Rates and Patterns of Tillage and Water Erosion on Terraces and Contour Strips: Evidence from Caesium-137. Catena **1999**, *36*, 115–142.
12. Schumacher, T.E.; Lindstrom, M.J.; Schumacher, J.A.; Lemme, G.D. Modeling Spatial Variation in Productivity Due to Tillage and Water Erosion. Soil and Tillage Research **1999**, *51*, 331–339.
13. Quine, T.A.; Zhang, Y. An Investigation of Spatial Variation in Soil Erosion, Soil Properties and Crop Production within an Agricultural Field in Devon, UK. Journal of Soil and Water Conservation in press.
14. Lobb, D.A.; Kachanoski, R.G.; Miller, M.M. Tillage Translocation and Tillage Erosion in the Complex Upland Landscapes of Southwestern Ontario, Canada. Soil and Tillage Research **1999**, *51*, 189–209.

LANDSLIDE AND LANDCREEP

Hau-Jung Shieh
Jau-Yau Lu
National Chung-Hsing University, Taichung, Taiwan

INTRODUCTION

Landslide and landcreep are slope failures by mass movement. These types of slope failure cause costly damage to property, roads, utilities, structures, and stream channels, and many times, human life.

In geomorphology, there are many different classifications for slope failure. In general, "landslide" is used when the interface between the moving soil mass and the stationary underlayer is clear and the moving speed is fast. The term "creep" is used when the interface is unclear or gradually varied, and the moving speed is relatively slow. Slide can be further classified into fall (such as rock fall, debris fall, earth fall), topple, slump, and glide. Also, creep becomes flow, such as rock flow, debris flow, and earth flow, if the soil mass is deformed during the movement.

In nature, the distinction between landslide and landcreep sometimes is not very clear. The scale of the failure and movement speed can be used to distinguish them. Landslide is for rapid, small scale surface failure and landcreep for slow, large scale movement of soil mass. Table 1 shows the major differences between landcreep and landslide.

LANDSLIDE

Landslide refers to the abrupt failure of steep hill slope due to effects of seepage water and the weight of soil mass. The displacement of the soil mass varies from more than ten to several hundred meters. Landslide occurs when a rainstorm saturates the soil layer and one or more of the following mechanisms usually trigger it.

Surface Saturation

When there is a dense and impermeable subsurface layer, a saturated surface may form after a prolonged rainstorm. Water saturation increases the weight of the surface layer and decreases the soil strength. On a steep slope, the weakened surface layer can easily initiate a landslide. This is the most common type of landslide and it usually occurs near the peak of the storm. Although this mechanism is simple, the variability of the soil and geologic properties on the landscape, as well as the variability of the rainfall pattern makes it very difficult to predict when and where the landslide may occur.

Piping

This type of landslide occurs in sandy soils where interflows or piping are easily formed. First, piping causes the seepage flow to increase suddenly and a small landslide occurs at a certain point on the hill slope. Gradually, larger slope failure occurs above the initial point of piping-induced slope failure. This process may occur at many parts of a hill slope. This type of landslide starts during the early stage of a rainstorm.

Pressurized Groundwater

This type of landslide occurs in the places where surface soil layer is thick and the groundwater is easy to concentrate. The mass movement occurs in the valley-shaped soil layer with little disturbance. Due to the topographical and geological conditions, landslide occurs near bedrock due to the decrease of effective stress from the pressurized groundwater.

Surface Erosion

Initially a small landslide occurs at a certain point near the upper portion of the slope surface. When the soil mass of the small landslide flows down the slope, it may trigger additional down slope failure through surface erosion or failure of saturated surface soil. This type of landslide usually occurs on the shallow surface layer for a steep, valley-shaped slope.

Flow

When the coarse and loose surface soil layer, containing fine grains and organic material, is saturated, it loses its strength and starts to flow. Some of the material may move

Table 1 Differences between landcreep and landslide

	Landcreep	Landslide
Geology	Occurs in places with particular geology or geological formation	Seldom related to geology
Soils	Moves with cohesive soil as slip surface	Frequently occurs even in sandy soils
Topography	Occurs on gentle slopes of 5–20°, especially if the upper portion is plateau topography	Frequently occurs on the slopes steeper than 20°
Occurrence	Occurs gradually and continuously	Occurs suddenly
Moving speed	Usually low at 0.01–10 mm/day	High speed, greater than 10 mm/day
Surface deformation	Soil moves with original shape and little disturbance	Soil mass being disturbed
Provoking causes	Greatly affected by groundwater	Affected by rainfall, especially rainfall intensity
Scale	Large scale, between 1 and 100 ha	Small scale, less than 1 ha
Warning signs	Cracks, depressions, upheavals, groundwater fluctuations occur before the landcreep	Seldom has symptoms and slips down suddenly

(From Ref. 1.)

as clods, but most of the material tends to move as an earth flow. This type of landslide usually occurs on bare lands or grass slopes with gentle gradient.

Landslides may also be induced by an earthquake. According to the analysis of aerial photographs by the Soil and Water Conservation Bureau in Taiwan, landslides occurred at 2365 locations covering a total area of 14,347 ha in central Taiwan on September 21, 1999, due to a devastating earthquake (100-year return period), 7.3 on the Richter scale.

Several landslides occurred along the Chin-Shui River as a result of the September 21 earthquake, resulting in an upstream impoundment of water, below the village of Tsao-Ling, as shown in Fig. 1. The minimum crest of the landslide at the impoundment was estimated to be approximately 540 m, with an estimated total storage volume of about 46 million cubic meters. This location has a long history of landslides caused by either earthquakes or heavy rainfall, resulting in the impoundment of water and subsequent breach outflow causing downstream damage and loss of life. The massive slope failure covered an area of 620 ha (about 5 km along the river channel) and with an amount of 126 million cubic meters of deposited material.

LANDCREEP

Landcreep usually occurs continuously or intermittently on gentle slopes at a large scale (usually greater than 1 ha). Sometimes landcreep may shift to an abrupt slope failure, or even becomes a debris flow.

The characteristics of landcreep include the following:

1. It usually occurs in places with particular geology.
2. The slip surface usually occurs at layers with clay or clayey soils. The movement is mainly induced by the groundwater effect.
3. It usually occurs on gentle slopes. More than 50% of the landcreeps occur on slopes with slope angle less than 20°.
4. The large-scale soil mass (1–100 ha) usually moves with its original shape with little surface disturbance.

Fig. 1 Aerial photographs of landslides near Tsao-Ling, central Taiwan due to a devastating earthquake, resulting in a lake upstream of a natural dam. (Courtesy of Fourth River Basin Management Bureau, Water Conservancy Agency in Taiwan.)

Cracks, depressions, upheavals, and groundwater fluctuations have been observed before the landcreep, but in general landcreep is a long-term, continuous, and slow soil mass movement.
5. The moving speed of landcreep is usually low at 0.01–10 mm/day. The movement can be either continuous or intermittent. However, the final slope failure may occur rapidly.

The distribution of landcreep is highly influenced by the geological structure. In Japan, it usually occurs in (a) Neogene (new Tertiary); (b) tectonics; (c) areas affected by the volcanic activity induced upheavals, depressions, or areas influenced by strong geothermal gradients from volcanic gas or hot springs; (d) areas with the Eogene (old Tertiary) shale and tuff belt containing cap rock, or areas along the fault or with coal pit. But the areas with a low degree of consolidation for the Quaternary deposits seldom have this type of landcreep. In England and North European countries such as Sweden and Norway, landcreeps occurring in areas with thick clay deposits are of major concern.

Classifications of Landcreep

Landcreep can be classified according to the characteristics of geologic structure, formation features and history of landcreep development, type of movement, and shape of longitudinal profile for slip surface (2). Table 2 shows Varnes' classification of landcreep (3).

Causes of Landcreep

Besides geological factors, landcreep may be triggered by many causes but can be generally classified into two categories, i.e. from natural phenomena and human activities. The natural triggers of landcreep include rainfall, snowmelt, groundwater, earthquake, channel incision, and hill slope failure. The human-induced factors includes cut and fill of roadways, mining, excavation of tunnels, and construction of reservoirs.

Investigation of Landcreep

Investigations of landcreep include the collection of information on topography, soil, and geology; monitoring the weather, groundwater, and movement of the soil mass; and performance of slope stability assessment for slope failure hazard potential. Fig. 2 gives more detailed organization of landcreep investigations.

Landcreep Mitigation Measures

Landcreep mitigation measures are selected based on the provoking causes. In general, they can be classified into two categories, i.e. the precaution measures and the prevention measures. The precaution measures rely on a detailed geologic and hydrologic database to predict the potential for landcreep occurrence and a real-time monitoring system to provide information such that warning can be issued when a slope failure event becomes imminent. The prevention measures include the control works and the restraint works. The control works are used to eliminate the provoking causes of landcreeps. The restraint works are used to increase the shear strength of the soil layers, and to slow down the movement of the soil mass. Fig. 3 shows the detailed landcreep mitigation measures.

Table 2 Varnes' classification of landcreep

	Type of material		
		Engineering soils	
Type of movement	**Bedrock**	**Mainly coarse grains**	**Mainly fine grains**
Falls	Rock fall	Debris fall	Earth fall
Topples	Rock topple	Debris topple	Earth topple
Slides			
Rotational	Rock slump	Debris slump	Earth slump
Translational	Rock block slide	Debris block slide	Earth block slide
	Rock slide	Debris slide	Earth slide
Lateral spreads	Rock spread	Debris spread	Earth spread
Flows	Rock flow	Debris flow	Earth flow
	(deep creep)	(earth creep)	
Complex movement		Combination of two or more types of movement	

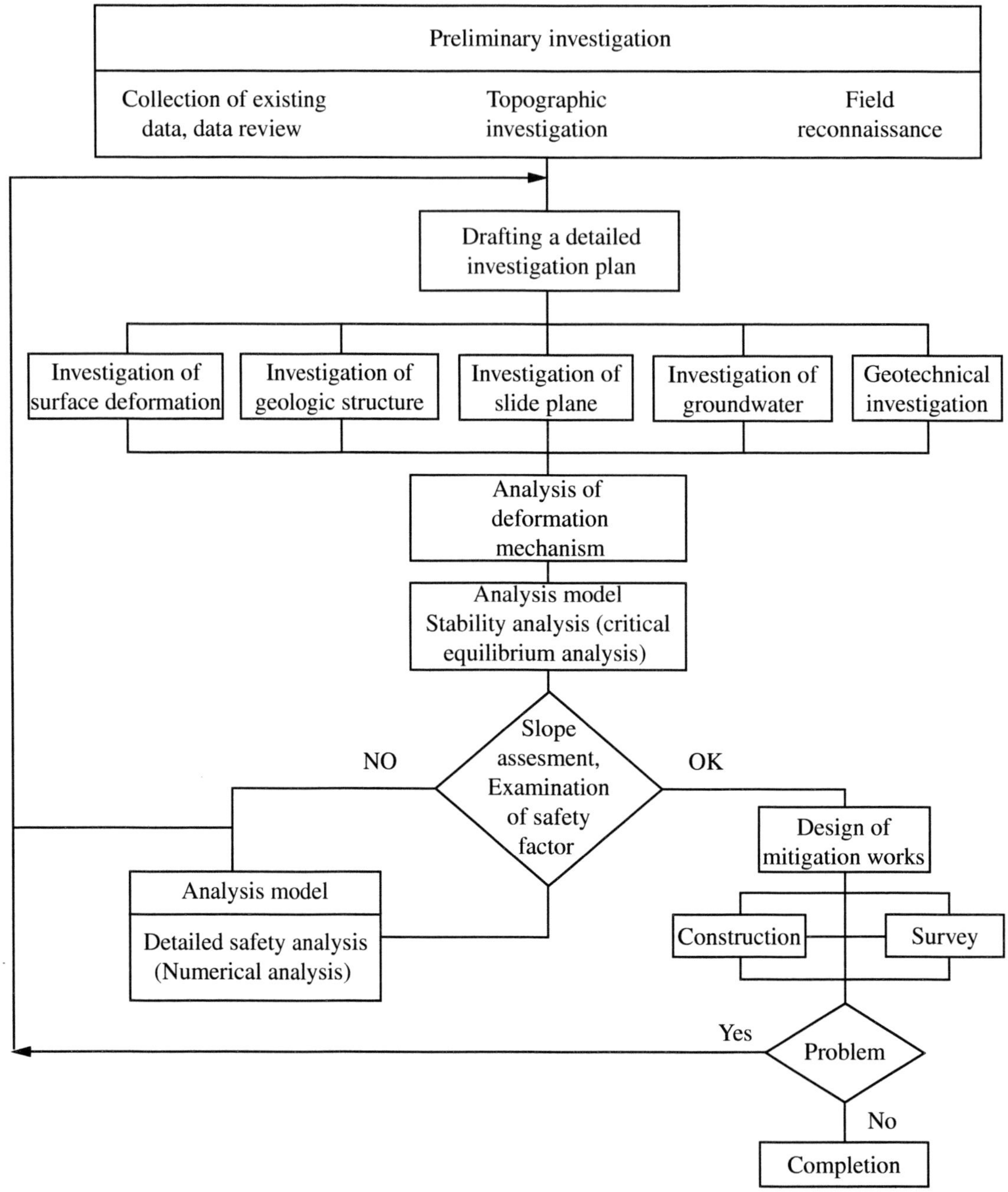

Fig. 2 Organization of landcreep investigation and analysis. (From Ref. 4.)

DEBRIS FLOW AND MUDFLOW

Debris flow is a mixture of water, sand, and large solid particles. It has strong impacting and destructive forces as it moves downstream a channel. Fig. 4 shows a severe debris flow disaster that occurred on October 9, 1973, near the Zhiban bridge, Zhiban Creek in eastern Taiwan. The debris flow was caused by the heavy rainfall from Typhoon Nora. The amount of deposited sediment was so large that the surface of the deposited material almost reached the lower surface of the bridge deck. As shown in the figure, the temple was almost completely buried.

Mudflow is an extreme case of the debris flow. It mainly carries sediment finer than coarse sand (2 mm), but

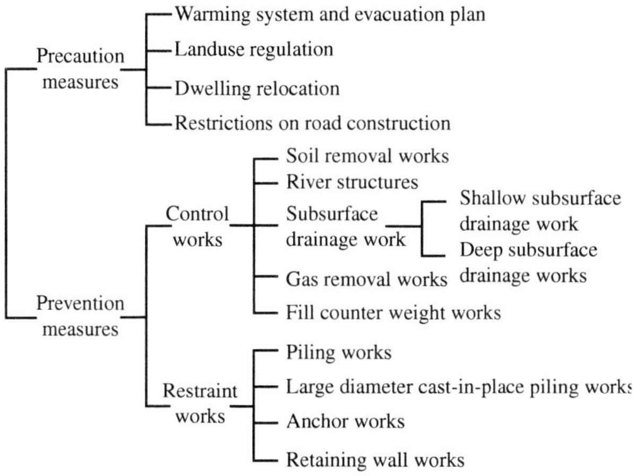

Fig. 3 Landcreep mitigation measures.

sometimes also a small amount of rock fragment. There are two types of mudflow (5). The first type occurs in the active volcano areas and it contains the volcanic ash. The second type occurs in the tertiary or dormant volcano areas and it contains the old volcanic ash or weathered tephra. The moving speed of the mudflow is usually at least two times faster than the ordinary debris flow, which contains more coarse particles.

Fig. 4 A severe debris flow disaster near the Zhiban bridge in Taiwan due to the heavy rainfall from Typhoon Nora (October 9, 1973).

Fig. 5 Panoramic view of the 1990 volcano eruption and mudflow in Wunsen, Nagasaki, Japan (aerial photograph taken on September 7, 1998).

Fig. 5 is an aerial photograph (6) showing the disaster due to the volcano eruption that occurred in Wunsen, Nagasaki, Japan in 1990. Enormous amount of ash ejected from the crater and the mudflow covered the mountain slope. The eruption lasted for five years and ceased in 1995.

In general, the initiation of debris flow is related to the following three conditions: (a) available loose solid material; (b) steep slope gradient; and (c) heavy rainfall. The loose solid material may be due to the prior landslides, landcreeps, typhoons, earthquakes, and volcanic eruptions. The gravitational force associated with steep slopes is an important driving force for the initiation of debris flow. An intensive rainstorm is usually the triggering factor for the formation of a debris flow.

REFERENCES

1. Watari, M.; Kobashi, S. *Prediction and Mitigation Measures of Landslides and Slope Failures*; Sankaido Co., Tokyo, Japan 1987.
2. Takai, A. *Soil Conservation*; Sankaido Co., Tokyo, Japan 1990.
3. Schuster, R.L.; Krizek, R.J., Eds. *Landslides, Analysis and Control*; National Academy of Sciences: Washington, DC, USA, 1978.
4. Japan Landslide Society. In: *Landslides in Japan*, National Conference of Landslide Control, 1996; 57pp.
5. Ikeya, H. *Debris Flow Disaster Investigation*; Sankaido Co., Tokyo, Japan 1980.
6. *Investigation of Watershed Management for Wunsen Mayo Mountain*; Kyushyu Forest Management Bureau, KFMB, 1999; 258pp.

LEACHING AND ILLUVIATION

L.E. Moody
California Polytechnic State University, San Luis Obispo, California, U.S.A.

INTRODUCTION

The dynamic system of soils includes various types of translocations, involving movements of materials in aqueous solution or suspension, from one part of the solum (leaching or eluviation) into another part of the solum (illuviation). During translocation, organic and inorganic materials may be deposited into the upper part of the soil profile, or produced or liberated in the upper part of the profile by physical, chemical, and biological weathering. The dominant translocation process in a soil profile depends partially on the form of the materials (inorganic or organic, solutes or particles). At a given stage of soil development, one process may be dominant or several may operate simultaneously or sequentially. In all cases, percolating water is the mechanism of translocation, and rates and intensities of translocation processes are controlled by the factors that regulate water percolation: climate, topography, and tortuosity of soil pores. The last is usually quantified and expressed as hydraulic conductivity.

LEACHING

Leaching is the translocation of solutes. Some authorities specify that leaching is the removal of solutes entirely out of the solum, representing a loss of materials from the soil profile (1), but according to many experts, leaching includes the translocation of solutes within the solum (2). Solutes arise mainly by chemical weathering, and include ions, complex ions, and ion pairs.

Mineral weatherability depends on several factors of mineral and soil chemistry. It also depends on particle size (smaller→more soluble). If chemical bonds between ions in a mineral are ionic in character, then the ions readily detach from the mineral surface and enter solution (3), as in the congruent dissolution of halite into $Na^+(aq)$ and $Cl^-(aq)$. Chemical bonds with some covalent character, as well as ionic character, are less easily disrupted. Because they generally require a strong polarizer such as H^+, dissolution proceeds less readily (3), as in the incongruent weathering of feldspar: $2KAlSi_3O_8 + 11H_2O + 2CO_2 \rightarrow 2K^+(aq) + Al_2Si_2O_5(OH)_4 + 2HCO_3^-(aq) + 4H_4SiO_4(aq)$. In this reaction, particulate kaolinite, as well as solutes, are produced.

The relative weatherability of the common silicates was tabulated by Goldich (4). Olivine, pyroxenes, and Ca-plagioclases are most weatherable, and quartz is most resistant (Fig. 1). Of the common nonsilicates, sodium salts (including sodium carbonates) and chlorides of Ca, Mg, and K are soluble, sulfates less so, and carbonates are slowly soluble. The solubility of Fe and Mn oxides depends on the oxidation state of the metals; reduction of Fe and Mn in saturated soils causes their oxides to dissolve. As minerals weather and elements are leached away, the soil unit decreases in mass and volume (5).

When chemical conditions in the soil are suitable, solutes precipitate, usually deeper than the origin of the solutes. One or more of several conditions, including the following, may initiate precipitation of solutes. Desiccation, the removal of water by evaporation or sorption by roots, may cause precipitation of soluble salts, gypsum, carbonates, or silica. Reduction in CO_2 partial pressure below the zone of maximum biological activity, is an additional cause of pedogenic carbonate precipitation. Solubility of many minerals is partially dependent on pH; therefore, a change in soil pH may also cause precipitation. Oxidation of Mn and Fe causes the oxides and oxyhydroxides of these metals to precipitate. As materials precipitate in the soil, mass is added and volume increases (5).

PODZOLIZATION

Podzolization is a specific term for such leaching and precipitation of Fe and Al chelates within the solum. Such leaching may occur in humid climates, acidic soil conditions, and characteristically under coniferous or mixed coniferous-deciduous forests. The types of soils formed by podzolization are podzols, generally equivalent to Spodosols in the U.S. taxonomy. The morphological expression of podzolization is a Bs, Bh, or Bhs (spodic)

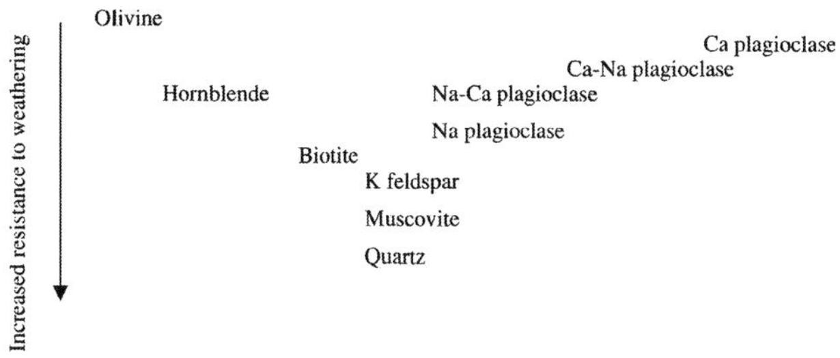

Fig. 1 Goldich's mineral stability series in weathering of the common rock-forming silicate minerals is displayed. Minerals least resistant to chemical weathering are at the top; resistance to weathering increases downward. Quartz is most resistant. (From Ref. 4.)

horizon, where the h symbolizes humus and s denotes Al and Fe sesquioxides. Often, the spodic horizon is overlain by a highly leached E (eluvial) horizon. A spodic horizon may be cemented by enrichment in humus, Al, Fe, and Si, in which case, it is known as *ortstein*. A placic horizon is a thin, wavy, cemented layer enriched in humus and Fe or Fe and Mn that is often associated with podzols.

Two models have been developed to describe the podzolization process. The first, favored by most authorities, involves the translocation of Al and Fe as organometallic complexes (6). Fulvic acids, components of humus, are formed during the decomposition of acidic leaf litter, especially litter from conifers and ericaceous (heath) plants. Humification is complex, and probably a combination of processes involving the breakdown of lignin and the synthesis of phenols and aromatics from carbohydrates and amino acids (7). Fulvic acids are water-soluble anions under acidic soil conditions, and, thus, mobile with percolating water. They are rich in phenols and carboxyls; the juxtaposition of these functional groups enables the organic matter to chelate trivalent cations Al and Fe. The organometallic complexes are carried downward in the percolating water. In an aerated, acidic environment, the metal cation:organic anion ratio mainly controls the solubility of the organometallic complex. If the cation:anion ratio is low, the complexes are soluble and, thus, mobile; they become less soluble by increases in the cation:anion ratio.

Precipitation of the organometallic complexes may be initiated at depth in the solum because of an increased ratio of the chelated trivalent cation to the organic anion. This shift in the cation:anion ratio may be brought about by microbial decomposition of the fulvic acids, or by saturation of the complexes by greater concentrations of Al and Fe released by weathering. As all available chelation sites are filled, the molecules lose polarity, become hydrophobic, and precipitate. Precipitation also may be initiated by desiccation, sorption of the complexes onto mineral surfaces, or flocculation in the presence of divalent cations. As the organic matter eventually decomposes, Al and Fe are released from complexation and are free to form oxyhydroxides or sesquioxides. With the addition of silica, Al can form imogolite and allophane.

A second, contrasting model of podzol formation holds that the dominant translocation process is inorganic, and that organic acids are involved mainly in weathering rather than in transport (8, 9). Weathering of primary minerals by carbonic and organic acids at the soil surface releases Al, Fe, and Si. At $pH < 5$, hydroxyaluminum cations react with silica to form stable and mobile, inorganic "proto-imogolite" complexes. The complexes precipitate in deeper horizons as imogolite and allophane, by desiccation or reaction to higher pH. Evidence supporting this model consists of podzols containing Al predominantly as imogolite, allophane, and proto-imogolite, and Fe as oxides, with Al-fulvic acid chelates sorbed and precipitated on allophane surfaces (10).

The two models may not be exclusive. Early podzolization in some Greenland soils involved translocation of Al and Fe by inorganic processes, whereas later stages involved their transport in complexes with organic acids (11). However, most studies of podzols favor the organic model (12, 13), even in early stages of podzolization (14).

ILLUVIATION OF CLAYS

As the finest particles in soils, clays (<2 mm diameter) are most susceptible to eluviation (removal) and illuviation (deposition). Fine clays (<0.2 mm), in particular, are small enough to be mobile in soil profiles. Larger particles, even silts and very fine sands, may be mobile in some soils, depending on size and geometry of pores and velocity of percolating water (15). Clays originate in soils by physical reduction of particles to clay size, by chemical transformation or precipitation, or by introduction as wind-transported dust. Once clays are present, their translocation in a soil profile involves mobilization, removal and transport, and deposition in a lower part of the profile.

Clay particles must be detached from the greater soil mass to enter into aqueous suspension. *Slaking* is physical detachment, by swelling during wetting, dislodgment by the shear of water flow, raindrop impact, or wind shear. *Dispersion* is chemical detachment that can result from the introduction of waters with low electrolyte concentration, dominance of Na compared to divalent cations on the zexchange complex, or high pH. Clay minerals differ in their mobility: of the phyllosilicates, smectites are most easily dispersed, and kaolinite and illite, less so (16). Once detached, particles are then free to move in percolating water.

Pore space in most soils consists of micropores and macropores. Macropores may be animal burrows, channels left by decayed roots, large interstitial pores between coarse fragments, or spaces between peds. Under saturated conditions or during infiltration of free water from the surface, percolating water carrying detached clays flows preferentially, and with greater velocity, through macropores than through micropores of the soil matrix. Water in the large pores is drawn into drier and microporous fabric by matric suction. The micropores serve a filtering function, and clay particles coat the walls of the large pores and the faces of peds as clay coatings (known as argillans, clay films, clay skins, or clay linings) (Figs. 2 and 3). Phyllosilicates assume a face-to-face configuration, with the c crystallographic axis approximately perpendicular to the ped face or pore wall. The orientation of phyllosilicate particles gives the argillan a strong optical orientation. To the unaided eye, and at low magnification, the argillan appears shiny or waxy. At high magnification of petrographic microscopy, the argillan displays undulatory extinction as the microscope stage is rotated under crossed polarizers.

Pore walls that do not allow appreciable water absorption, such as between rock fragments and sand grains, also develop clay coatings by illuviation. Drying of clay suspensions following downward flow of suspensions, draws in the water interface progressively closer to the grain surfaces and to contact points between grains. Clay

Fig. 2 Argillans (*arrows*) filling cracks between coarse prisms. This view is looking down on exposed tops of the prisms.

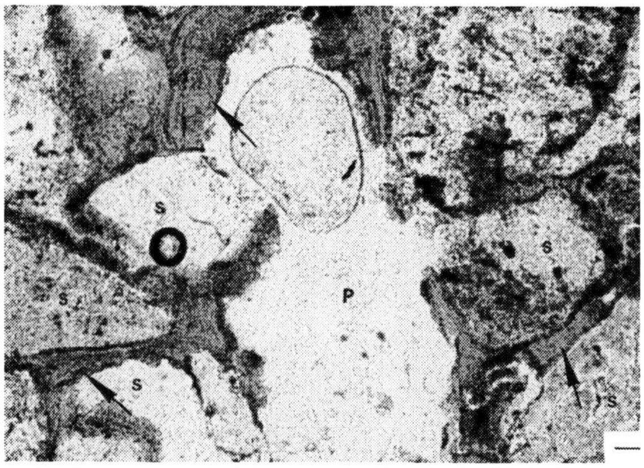

Fig. 3 Microscopic view under plane polarized light of loamy fine sand, with argillans (*arrows*) filling interstitial pores between sand grains (s) and lining a large pore (P).

particle deposition follows the pattern of drying, argillans are thicker at the contact points than away from them, and phyllosilicate particles orient with faces parallel and crystallographic axis perpendicular to the surface (17). Thus, clay coatings on grains, and as bridges between grains, have a strong optical orientation.

The depth to which the suspension percolates determines the depth at which deposition of clay particles occurs. The wetting front, driven by a potential gradient, stops where and when gravitational, pressure, and matric forces acting on the water reach equilibrium. The retarded water evaporates or is sorbed by roots, and suspended particles are deposited at the depth of water percolation (18, 19).

Soil scientists accept the strongly oriented argillan as evidence of clay illuviation. However, phyllosilicate clays, which are deposited by flocculation (upon encountering high electrolyte concentrations or in the presence of di- and trivalent cations), tend to assume edge-to-edge or edge-to-face configuration and may not develop strong optical orientation. Thus, their illuvial origin may be difficult to identify. Finally, after deposition, argillans may be modified or destroyed by shrink-swell or mixing of soil material by animal burrowing.

CONCLUSIONS

Because the processes of leaching and precipitation, podzolization, and illuviation are highly dependent on water movement into and through the soil, any activity or land use that interrupts water infiltration and percolation will disturb those processes. Physical disturbances of the soil, such as tilling, may mix soil materials in such a way that soil horizons will be disturbed, mixed, or effaced.

REFERENCES

1. Buol, S.W.; Hole, F.D.; McCracken, R.J.; Southard, R.J. *Soil Genesis and Classification*, 4th Ed.; Iowa State University Press: Ames, 1997.
2. Tan, Kim H. *Environmental Soil Science*; Marcel Dekker: New York, 1994.
3. Sposito, Garrison. *The Chemistry of Soils*; Oxford University Press: New York, 1989.
4. Goldich, Samuel S. A Study in Rock Weathering. Jour. Geology **1938**, *46*, 17–58.
5. Chadwick, O.A.; Nettleton, W.D. Quantitative Relationships Between Net Volume Change and Fabric Properties During Soil Evolution. In *Soil Micromorphology: Studies in Management and Genesis*; Ringrose-Voase, A.J., Humphreys, G.S. Eds.; Developments in Soil Science; Elsevier: Amsterdam, 1994; *22*, 353–359.
6. DeConinck, F. Major Mechanisms in Formation of Spodic Horizons. Geoderma **1980**, *24*, 101–123.
7. Oades, J.M. An Introduction to Organic Matter in Mineral Soils. In *Minerals in Soil Environments*; Dixon, J.B., Weed, S.B., Eds.; Soil Sci. Soc. Amer.: Madison, Wisconsin, 1989; 89–159.
8. Farmer, V.C.; Russell, J.D.; Berrow, M.L. Imogolite and Proto-Imogolite in Spodic Horizons: Evidence for a Mobile Aluminum Silicate Complex in Podzol Formation. Jour. Soil Sci. **1980**, *31*, 673–684.
9. Farmer, V.C. Significance of the Presence of Allophane and Imogolite in Podzol Bs Horizons for Podzolization Mechanisms: A Review. Soil Sci. Plant Nutrition **1982**, *28*, 571–578.
10. Anderson, H.A.; Berrow, M.L.; Farmer, V.C.; Hepburn, A.; Russell, J.D.; Walker, A.D. A Reassessment of Podzol Formation Processes. Jour. Soil Sci. **1982**, *33*, 125–136.
11. Jakobsen, B.H. Multiple Processes in the Formation of Subarctic Podzols in Greenland. Soil Sci. **1991**, *152*, 414–426.
12. Barrett, L.R.; Schaetzel, R.J. An Examination of Podzolization Near Lake Michigan Using Chronofunctions. Canadian Jour. Soil Sci. **1992**, *72*, 526–541.
13. Gustafsson, J.P.; Bhattacharya, P.; Bain, D.C.; Fraser, A.R.; McHardy, W.J. Podzolisation Mechanisms and the Synthesis of Imogolite in Northern Scandinavia. Geoderma **1995**, *66*, 167–194.
14. Certini, Giacomo; Ugolini, Fiorenzo C.; Corti, Guiseppe; Agnelli, Alberto. Early Stages of Podzolization Under Corsican Pine (*Pinus nigra* Arn. *ssp laricio*). Geoderma **1998**, *83*, 103.
15. Nettleton, W.D.; Brasher, B.R.; Baumer, O.W.; Darmody, R.G. Silt Flow in Soils. In *Soil Micromorphology: Studies in Management and Genesis*; Ringrose-Voase, A.J., Humphreys, G.S., Eds.; Developments in Soil Science; Elsevier: Amsterdam, 1994; *22*, 361–371.
16. Stern, R.; Ben-Hur, M.; Shainberg, I. Clay Mineralogy Effect on Rain Infiltration, Seal Formation, and Soil Losses. Soil Sci. **1991**, *152*, 455–462.
17. Sullivan, L.A. Clay Coating Formation on Impermeable Materials: Deposition by Suspension Retention. In *Soil Micromorphology: Studies in Management and Genesis*; Ringrose-Voase, A.J., Humphreys, G.S., Eds.; Developments in Soil Science; Elsevier: Amsterdam, 1994; *22*, 373–380.
18. Dijkerman, J.M.; Cline, M.G.; Olson, G.W. Properties and Genesis of Textural Subsoil Lamellae. Soil Sci. **1967**, *104*, 7–16.
19. Moody, L.E.; Graham, R.C. Pedogenic Processes in Thick Sand Deposits on a Marine Terrace, Central California. In *Whole Regolith Pedology*; Cremeens, David L., Brown, Randall B., Huddleston, J. Herbert, Eds.; Special Pub.; Soil Sci. Soc. Amer.: Madison, Wisconsin, 1994; *34*, 41–55.

LIMING AND LIME MATERIALS

M.K. Conyers

New South Wales Agriculture, Wagga Wagga, New South Wales, Australia

INTRODUCTION

"Liming" is the application of an alkaline compound of calcium (Ca) to soil. The main aim can be either or both to neutralize excessive acidity in the soil, or supply Ca. Liming is a major tool, along with tolerant plants, in the management of acidic soils.

ACIDITY

The quantity of acidity which is regarded as "excessive" varies with the soil and with the species and cultivar of plant. Direct hydrogen ion (H^+) toxicity is uncommon in mineral soils but can occur in peaty soils of low pH. A high concentration of H^+ also has an indirect effect on legume growth by inhibiting the symbiotic relationship between the host plant and rhizobia. In most mineral soils, low pH is associated with aluminum (Al) toxicity to plants. Plants vary in their tolerance to Al (e.g., medics are generally sensitive; tropical grasses are generally tolerant). In some soils, manganese (Mn) toxicity is also a limitation to plant growth. Plant tolerance to Al and to excessive Mn is independently inherited. Plant breeding for tolerance to Al and Mn toxicity is used either as an alternative to using limestone, where the cost of liming is uneconomic, or in conjunction with liming, where subsurface acidity or seasonal conditions limit the effectiveness of liming (1). The application of liming materials does not always overcome Mn toxicity, as it can occur at higher pH than Al toxicity and may be induced by hot, dry soil conditions or by waterlogging. On the other hand, too much alkali added to a soil can induce deficiencies of Mn and also of other trace elements such as iron (Fe), zinc (Zn), and boron (B). In view of the potential for imbalanced plant nutrition, what constitutes "excessive" soil acidity is best determined by trials on local soils with local species and cultivars.

Deficiencies of Ca and magnesium (Mg) also occur in some acidic soils. Calcium deficiency is more commonly found in tropical species such as the peanut. Calcium is also of benefit to soil structure; it is applied mainly to soils with large proportions of sodium (Na) on the cation exchange capacity, and also to noncracking clay soils with a large proportion of Mg on exchange sites. Where soil structure is a concern, gypsum is normally the preferred source of Ca, sometimes in combination with fine limestone. Gypsum is not an alkaline salt and hence does not add alkalinity to a soil.

While most liming materials can overcome Al toxicity and Ca deficiency, alleviation of Mg deficiency requires application of a liming material with a significant component of Mg. This may be either dolomite or a Mg rich limestone, commonly known as dolomitic limestone. There is in fact a continuum between pure limestone and pure dolomite. Magnesite may be used to provide Mg, but is rarely used as a liming material per se.

LIMING MATERIALS

Application rates of liming materials are best expressed relative to a standard such as calcium carbonate ($CaCO_3$). To overcome an existing problem of excessive acidity as opposed to a prophylactic dressing, application rates range from as little as 1 tonne $CaCO_3$ per hectare on weakly buffered soils, growing Al or Mn tolerant plants, to 10 tonne $CaCO_3$ per hectare on strongly buffered soils supporting sensitive plants. Weakly buffered soils include sandy textured soils and tropical oxisols, particularly with low concentrations of organic carbon (OC). Strongly buffered soils include acidified mollisols and soils with high contents of clay or OC. In high lime requirement situations, the $CaCO_3$, or its chemical equivalent, need not be applied in a single dose. Still higher application rates than outlined here may be required for the amelioration of acid sulphate soils; however, this requires a separate consideration from the present focus.

Some common lime materials are listed in Table 1. Most agricultural lime is some form of $CaCO_3$. This is generally in the mineral form of calcite, which occurs in marble, marl, and aged coralline deposits. Crushed shells and recently formed coralline material may be in the aragonite form of $CaCO_3$. The aragonite form is more soluble than the calcite form. Dolomite has a harder crystalline form than either of the $CaCO_3$ forms and is generally less reactive in soils (2). It has to be crushed finer than limestone to react in a similar time frame, but if Mg is required the only alternative to dolomite is to spread both crushed limestone and a separate source of Mg such as magnesite.

Industrial waste materials are also commonly used for liming. Precipitator dusts from clinker and cement manufacturing contain varying proportions of $CaCO_3$ and calcium oxide (CaO). The latter has an exothermic and caustic reaction with water and so should be handled with care (3). The risks increase as the proportion of oxide in the waste increases. Calcium silicates from steel manufacturing can also be good liming materials.

Table 1 Neutralizing value and calcium content of pure liming materials

Substance	Formula	%Neutralizing value[a]	% Calcium[a]
Limestone	$CaCO_3$	100	40
Dolomite	$(Ca_{0.5}\ Mg_{0.5})CO_3$	109	22
Burnt lime	CaO	178	71
Hydrated or slaked lime	$Ca(OH)_2$	135	54
Calcium silicate	CaH_2SiO_4	75	30
Magnesite	$MgCO_3$	119	0

[a] Rounded to nearest whole percent.

Liming materials can be broadly grouped as follows:

Natural deposits: Dolomite; calcite–limestone, coralline and earthy limestones, marble; aragonite–fresh shell and coralline materials.

Manufactured forms: Burnt lime; slaked or hydrated lime.

By-products: Precipitator dusts–mixtures of carbonates, oxides, etc.; slags–usually calcium silicate wastes from steel production.

Incidental forms, where liming is not the principal function of the material: reactive rock phosphate; dicalcic phosphate; sewage ash; manures; composts.

The reactions which neutralize acidity, other than for incidental liming materials, can be exemplified as follows:

$$CaO + 2H^+ = Ca^{2+} + H_2O$$
$$CaCO_3 + 2H^+ = Ca^{2+} + H_2O + CO_2$$
$$CaH_2SiO_4 + 2H^+ = Ca^{2+} + H_4SiO_4$$

In each case, 2 mole of acid are consumed per mole of liming compound. While the second reaction liberates the greenhouse gas CO_2, as does the increase in organic carbon mineralization which follows liming, in the longer term the removal of the constraint of soil acidity increases plant root and top growth and so net C fixation can be enhanced.

Purity and Fineness

The effectiveness of traditional liming materials such as limestone and dolomite depends on their purity and fineness. Clearly, the more pure the limestone or dolomite, the greater the quantity of acid which can be neutralized by each tonne of product. In more general terms, the ability of any type of liming material to neutralize acidity is termed Neutralizing Value (NV). It is a measure of the ability of a given mass of liming material to react with acidity, expressed relative to a standard such as CaO (4) or $CaCO_3$. The latter is adopted here. For pure minerals and compounds, NV is easily calculated by:

$$NV = \text{mole wt of } CaCO_3/\text{mole wt of compound}$$

For example, for CaO

$$NV = 100/56 = 178$$

Results for pure compounds are given in Table 1. For compounds of mixed or unknown composition, the NV is obtained by titration after dissolution of the liming material in excess acid. Details of the method can be obtained by reference to the latest edition of the Official Methods of Analysis of the Association of Official Analytical Chemists.

The fineness of a liming material is a more controversial issue, not because of differences in experimental results, but due to differences in opinion about the cost of production of fine lime and the ease of spreading. Fig. 1 shows the results of 3 studies on the ability of calcitic limestone to raise soil pH as a function of

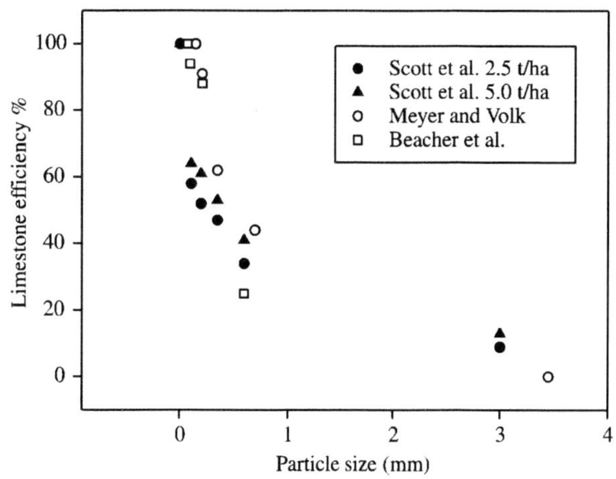

Fig. 1 The increase in limestone efficiency in raising soil pH with decrease in particle diameter. (From Refs. 5–7.)

particle size (5–7). The degree of agreement between these pot and field studies, conducted by different workers on different continents 50 years apart, is remarkable. In none of these studies was there an upper limit where fineness was of no further benefit in lifting pH. The concept of a "plateau" in particle size effectiveness, referred to in some texts, is a subjective assessment of firstly, the practical limitations of spreading fine limestone, and secondly, the cost of the liming material relative to its agricultural value.

Despite the clear demonstration of the value of fineness, there are two reasons why it is not necessarily helpful to define an "ideal" particle size distribution for all calcitic liming materials. First, the agricultural context has to be considered. The need for finer limestone is most apparent for crops grown in regions of relatively low rainfall. On one hand, for sandy soils receiving 2000 mm rainfall annually it may be unnecessary or wasteful to have a high proportion of fines as there may be losses of the limestone from surface soil through mass flow. Where cartage and spreading costs are a large proportion of the total cost of liming, it is more cost effective to cart and spread a product that is nearer to 100% effective. On the other hand, if the farm is near to a limestone quarry and if the farmer has his own lime spreader, the farmer may be able to regularly spread higher tonnages of relatively coarse limestone quite cheaply. A high ratio of lime cost to agricultural product value also means that limestone has to be used as cost effectively as possible. On the other hand in intensive horticulture, the cost of lime may be a minor input and a higher application rate of coarser limestone may be equally cost effective. Therefore, the agricultural context of the liming has to be considered.

Second, the lime crusher's industrial perspective needs to be considered. If the limestone crushing plant services industries such as glass, plastics, collieries, etc., there is more chance that the plant will have a ball mill, enabling the manufacture of fine limestone. If the main focus of the plant is to produce agricultural limestone, it may not be reasonable to expect the milling of very fine products. The cost of producing a fine limestone simply for soil amelioration may diminish the cost competitiveness of the crushing plant. On the other hand, the milling of fine limestone may give a lime producer a strong advantage over the competition. Local circumstances need to be considered. The same principle holds for by-product liming materials; the introduction of a processing step such as milling may in some instances make the by-product uneconomic. However, without milling, a coarse slag may be near useless. The way in which local by-products are best used should be determined by field trials and economic analysis.

CONCLUSION

The purpose of quantitative models of efficiency for liming materials should not be to dictate terms to crushers and other producers, but to enable consumers to assess the relative cost effectiveness of alternative products available to them. The concept of an "ideal" liming material is likely to vary with local circumstances.

REFERENCES

1. Cregan, P.D.; Hirth, J.R.; Conyers, M.K. Amelioration of Soil Acidity by Liming and Other Amendments. In *Soil Acidity and Plant Growth*; Robson, A.D., Ed.; Academic Press: Marrickville, Australia, 1989; 205–264.
2. Barber, S.A. Liming Materials and Practices. In *Soil Acidity and Liming*, 2nd Ed.; Adams, Fred., Ed.; Monograph 12, Agronomy; ASA, CSSA, SSSA: Madison, 1984; 171–209.
3. Boynton, R.S. *Chemistry and Technology of Lime and Limestone*, 2nd Ed.; John Wiley and Sons: New York, 1980.
4. Cooke, G.W. Calcium Fertilizers and Liming Materials. In *Fertilizing for Maximum Yield*, 3rd Ed.; Granada: London, 1982; 161–173.
5. Meyer, T.A.; Volk, G.W. Effect of Particle Size of Limestone on Soil Reaction, Exchangeable Cations and Plant Growth. Soil Science **1952**, *73*, 37–52.
6. Beacher, R.L.; Longnecker, D.; Merkle, F.G. Influence of Form, Fineness and Amount of Limestone on Plant Development and Certain Soil Characteristics. Soil Science **1952**, *73*, 75–82.
7. Scott, B.J.; Conyers, M.K.; Fisher, R.; Lill, W. Particle Size Determines the Efficiency of Calcitic Limestone in Amending Acidic Soil. Aust. J. Agric. Res. **1992**, *43*, 1175–1185.

MACROPOROSITY

M

Daniel Giménez
Rutgers, The State University of New Jersey, New Brunswick, New Jersey, U.S.A.

INTRODUCTION

Macroporosity refers to the areal or volumetric proportion of macropores in soils. The term *macropore* conveys connotations of rapid pathways for the movement of water and chemicals in soils. The glossary of soil science terms (1) defines macropores as pores with an equivalent cylindrical diameter larger than 75 μm, but in the literature the size range for defining macropores varies between 30 and 3000 μm (2). The hydrological properties of macropores are not only dependent on size but also on their continuity, tortuosity, and shape (3–5).

ORIGIN OF MACROPORES

Macropores can originate from soil fauna activity, root decay, shrinking of a soil matrix upon drying, or dissolution of chemical substances. Tillage results in the formation of structurally unstable macropores that subside during a growing season. Among soil fauna, earthworms have received much attention because of their abundance in soil, and because their burrows may have a significant impact in creating preferential flow pathways in soil (6, 7). Plant roots can also form an intricate network of macropores influencing water movement and chemical transport (8, 9). Taproot systems are particularly effective in providing stable and continuous pores that increase infiltration rates (10). Shrinking of a soil matrix upon drying forms planar macropores that are important for transport processes in initially dry soils (11). Surface cracks of cultivated soils and muds have similar geometry (12), but their development differs from that of a pore system caused by aggregation (13).

QUANTIFICATION OF MACROPORE PROPERTIES

Direct quantification of macropore size, number, and position is done using image analysis techniques. Images are obtained either in situ by tracing or photographing macropores on a plane parallel to a soil surface (7, 14), or from resin-impregnated soil blocks prepared in the laboratory (15). Dye stained patterns of macropores are used to quantify number, size, and shape of continuous macropores (16). Several studies were made using X- and gamma-ray computed tomography (17, 18). The number of these studies is expected to increase as the resolution of the tomographs is improved and they become more accessible to soil scientists.

The total number of macropores reported in the literature varies from 100 to 10,000 pores/m^2 depending on soil type, soil management and the methodology applied (7, 14). Distributions of macropore sizes are highly skewed, with number of pores decreasing exponentially as their size increases (7). For pores with a fractal surface, Pachepsky et al. (19) predicted that as pore sizes increase their shape becomes elongated, which would imply that a distribution of macropore sizes will be composed of a relatively small number of elongated or planar voids and a larger number of rounded pores. Tippkotter (20), working with a loess soil in West Germany, determined that macropores with sizes <1 mm were formed by roots. On the other hand, Edwards et al. (7) associated the presence of macropores >1 mm with earthworm activity. Macropore density and continuity decreases with depth and is usually interrupted by tillage (14).

Indirect quantification of macropores has been made through measurements of infiltration at various suctions and of air permeability (21). These methods require defining a lower limit of macropore radius, which is typically done based on the methodology used for the purpose of the study.

IMPACT OF MACROPORES ON WATER MOVEMENT AND CHEMICAL TRANSPORT

The effect of macropores on water movement through soil was recognized in the late 19th century, but received increased attention during the 1980s because of the risk macropores pose to contamination of subsurface water

(2, 22, 23). In the presence of macropores, matter can bypass a soil matrix reaching depths below a root zone where chances to be degraded and/or adsorbed are minimized. Flow regimes influenced by macropores have been studied in relation to the transport of chemicals, microorganisms, and colloidal particles through soils (24–26), using both undisturbed soil (26, 27), and packed soil containing artificial macropores (24, 28). Maximum flow is obtained under ponding conditions and with macropores connected to the surface, but macropore flow can also occur under a nonponding regime once the macropore walls are wet (28).

Typically, more than 70% of the total water moving through a soil is transmitted through pores that represent less than 1% of a soil volume (27). Heavy rainfalls occurring shortly after a chemical application can leach below the root zone up to 5% of the applied mass in sandy and loamy soils (29), but deep transport of chemicals has also been observed in heavy clay soils (30). This is consistent with the observation that in the U.S., groundwater contamination by pesticides is usually detected in areas with shallow groundwater tables regardless of soil type or land use, i.e., agricultural or urban (31).

Modeling Transport Processes in Macroporous Soils

Prediction of transport processes through macroporous soils is complicated by the intricate geometry of macropores. Laminar flow of water through pores with cylindrical and planar geometries can be theoretically estimated using Poiseuille's equation modified to account for the finite nature of planar voids (32), and the turbulent or transitional nature of flow through macropores (33, 34). Prediction of flow through natural macropores is, however, highly complicated by their irregular shape. The use in flow equations of measured diameter, length, and volume of macropores does not satisfactorily predict macropore hydraulic properties (33). Dunn and Phillips (5) showed that the smallest section in irregular macropores has a disproportionate effect on water flow through macropores, and, therefore, a potential to influence geometrical estimations from flow measurements.

A more pragmatic approach to modeling water movement in macroporous soils is to work with the concept of an *effective porosity* (i.e., the proportion of a soil volume occupied by macropores) coupled with relative simple relationships to estimate transport coefficients. Effective porosities are estimated from measurements of ponded and tension infiltration (27) or from information on an arbitrary lowest or highest limit of macropore radius, and on a fractal dimension defining the scaling of macropore size (35, 36). Data on areal macroporosity (14) were used by Rawls et al. (37) and Giménez et al. (38) to successfully predict macropore saturated hydraulic conductivity. This approach is promising and requires minimal data but it does not provide information on relationships between macropore morphology and water flow.

REFERENCES

1. *The Glossary of Soil Science Terms*; Soil Science Society of America: Madison, WI, 1997.
2. White, R.E. The Influence of Macropores on the Transport of Dissolved and Suspended Matter Through Soil. In *Advances in Soil Science*; Stewart, B.A., Ed.; Springer: New York, 1985; 95–120.
3. Skopp, J. Comment on "Micro-, Meso-, and Macroporosity of Soil". Soil Sci. Soc. Am. J. **1981**, *45*, 1246.
4. Groenevelt, P.H.; Kay, B.D.; Grant, C.D. Physical Assessment of a Soil with Respect to Rooting Potential. Geoderma **1984**, *34*, 101–114.
5. Dunn, G.H.; Phillips, R.E. Equivalent Diameter of Simulated Macropore Systems During Saturated Flow. Soil Sci. Soc. Am. J. **1991**, *55*, 1244–1248.
6. Ehlers, W. Observations on Earthworm Channels and Infiltration on Tilled and Untilled Loess Soil. Soil Sci. **1975**, *119*, 242–249.
7. Edwards, W.M.; Norton, L.D.; Redmond, C.E. Characterizing Macropores That Affect Infiltration into Nontilled Soils. Soil Sci. Soc. Am. J. **1988**, *52*, 483–487.
8. Gish, T.J.; Jury, W.A. Effect of Plant Roots and Root Channels on Solute Transport. Trans. ASAE **1983**, *26*, 440–444.
9. Li, Y.; Ghodrati, M. Preferential Transport of Nitrate Through Soil Columns Containing Root Channels. Soil Sci. Soc. Am. J. **1994**, *58*, 653–659.
10. Mitchell, A.R.; Ellsworth, T.R.; Meek, B.D. Effect of Root Systems on Preferential Flow in Swelling Soil. Commun. Soil Sci. Plant Anal. **1995**, *26*, 2655–2666.
11. Hoogmoed, W.B.; Bouma, J. A Simulation Model for Predicting Infiltration into Cracked Clay Soil. Soil Sci. Soc. Am. J. **1980**, *44*, 458–461.
12. Velde, B. Structure of Surface Cracks in Soil and Muds. Geoderma **1999**, *93*, 101–124.
13. Velde, B. Surface Cracking and Aggregate Formation Observed in a Rendzina Soil, La Touche (Vienne) France. Geoderma **2001**, *99*, 261–276.
14. Logsdon, S.D.; Allmaras, R.R.; Wu, L.; Swan, J.B.; Randall, G.W. Macroporosity and Its Relation to Saturated Hydraulic Conductivity Under Different Tillage Practices. Soil Sci. Soc. Am. J. **1990**, *54*, 1096–1101.
15. Giménez, D.; Allmaras, R.R.; Nater, E.A.; Huggins, D.R. Fractal Dimensions for Volume and Surface of Interaggregate Pores-Scale Effects. Geoderma **1997**, *77*, 19–38.
16. Droogers, P.; Stein, A.; Bouma, J.; de Boer, G. Parameters for Describing Soil Macroporosity Derived from Staining Patterns. Geoderma **1998**, *83*, 293–308.

M

17. Warner, G.S.; Nieber, J.L.; Moore, I.D.; Geise, R.A. Characterizing Macropores in Soil by Computer Tomography. Soil Sci. Soc. Am. J. **1989**, *53*, 653–660.
18. Perret, J.; Prasher, S.O.; Kantzas, A.; Langford, C. Three-Dimensional Quantification of Macropore Networks in Undisturbed Soil Cores. Soil Sci. Soc. Am. J. **1999**, *63*, 1530–1543.
19. Pachepsky, Ya.; Yakovchenko, V.; Rabenhorst, M.C.; Pooley, C.; Sikora, L.J. Fractal Parameters of Pore Surfaces Derived from Micromorphological Data: Effect of Long-Term Management Practices. Geoderma **1996**, *74*, 305–319.
20. Tippkotter, R. Morhology, Spatial Arrangement and Origin of Macropores in Some Hapludalfs, West Germany. Geoderma **1983**, *29*, 355–371.
21. Edwards, W.M.; Shipitalo, M.J.; Owens, L.B. Gas, Water, and Solute Transport in Soils Containing Macropores: A Review of Methodology. Geoderma **1993**, *57*, 31–49.
22. Beven, K.; Germann, P. Macropores and Water Flow in Soils. Water Resour. Res. **1982**, *18*, 1311–1325.
23. Bouma, J. Influence of Soil Macroporosity on Environmental Quality. Adv. Agron. **1991**, *46*, 1–37.
24. Czapar, G.F.; Horton, R.; Fawcett, R.S. Herbicide and Tracer Movement in Soil Columns Containing An Artificial Macropore. J. Environ. Qual. **1992**, *21*, 110–115.
25. Murphy, S.L.; Tate, R.L. III Bacterial Movement Through Soil. In *Soil Biochemistry*; Stotzky, G., Bollag, J.-M., Eds.; Marcel Dekker: New York, 1996; 253–286.
26. Jacobsen, O.H.; Moldrup, P.; Larsen, C.; Konnerup, L.; Petersen, L.W. Particle Transport in Macropores of Undisturbed Soil Columns. J. Hydrol. **1997**, *196*, 185–203.
27. Watson, K.W.; Luxmoore, R.J. Estimating Macroporosity in a Forest Watershed by Use of a Tension Infiltrometer. Soil Sci. Am. J. **1986**, *50*, 578–582.
28. Phillips, R.E.; Quisenberry, V.L.; Zeleznik, J.M.; Dunn, G.H. Mechanism of Water Entry into Simulated Macropores. Soil Sci. Soc. Am. J. **1989**, *53*, 1629–1635.
29. Flury, M. Experimental Evidence of Transport of Pesticides Through Field Soils—a Review. J. Environ. Qual. **1996**, *25*, 25.
30. Kelly, B.P.; Pomes, M.L. Preferential Flow and Transport of Nitrate and Bromide in Claypan Soil. Ground Water **1998**, *36*, 484–494.
31. Kolpin, D.W.; Barbash, J.E.; Gilliom, R.J. Pesticides in Ground Water of the United States, 1992–1996. Ground Water **2000**, *38*, 858–863.
32. Towner, G.D. Formulae for Calculating Water Flow in Macro-pores in Soil. Int. Agrophys. **1987**, *3*, 1–5.
33. Wang, D.; Norman, J.M.; Lowery, B.; McSweeney, K. Nondestructive Determination of Hydrogeometrical Characteristics of Soil Macropores. Soil Sci. Soc. Am. J. **1994**, *58*, 294–303.
34. Logsdon, S.D. Flow Mechanisms Through Continuous and Buried Macropores. Soil Sci. **1995**, *160*, 237–242.
35. Brakensiek, D.L.; Rawls, W.J.; Logsdon, S.D.; Edwards, W.M. Fractal Description of Macroporosity. Soil Sci. Soc. Am. J. **1992**, *56*, 1721–1723.
36. Rawls, W.J.; Brakensiek, D.L.; Logsdon, S.D. Estimation of Macropore Properties for No-till Soils. Trans. ASAE **1996**, *39*, 91–95.
37. Rawls, W.J.; Brakensiek, D.L.; Logsdon, S.D. Predicting Saturated Hydraulic Conductivity Utilizing Fractal Principles. Soil Sci. Soc. Am. J. **1993**, *57*, 1193–1197.
38. Giménez, D.; Perfect, E.; Rawls, W.J.; Pachepsky, Y. Fractal Models for Predicting Soil Hydraulic Properties: A Review. Eng. Geol. **1997**, *48*, 161–183.

MAGNESIUM

N.S. Bolan
Massey University, Palmerston North, New Zealand

K. Arulmozhiselvan
P. Paramasivam
Tamil Nadu Agricultural University, India

INTRODUCTION

Among the elements that constitute the solid surface of the Earth, magnesium (Mg) is the eighth most abundant and is an essential plant nutrient. Magnesium, the central component of chlorophyll, plays a major role in plant nutrition. Although Mg is regarded as a macronutrient, it is not always considered in fertilizer management, possibly because Mg is added to soil as an accessory element in many fertilizers. Increasing incidences of Mg deficiency are being observed mainly as a result of the use of Mg-free fertilizers. Furthermore, modern agricultural practices are inducing accelerated soil acidification. If uncorrected, this acidification will lead to a reduction in exchangeable basic nutrient cations, leading to the increasing incidence of Mg deficiency in soils. The role of Mg in plants and animals is discussed in relation to the supply of Mg in soils.

MAGNESIUM IN SOILS

Magnesium is a normal component of both igneous and sedimentary rocks and of the soils developed from such rocks (1). Soils developed from basic rocks (basalts and limestone) generally contain higher levels of Mg (0.27–2.86% Mg) than those developed from granite and sandstones (0.01–0.34% Mg) (2). The bulk of Mg in soil is usually in forms that are not readily available to the plant and exists in primary minerals and secondary silicate clay minerals (3). Some of the important silicate mineral species carrying Mg are shown in Table 1. The formulas presented represent the "ideal" composition of the minerals. In nature, however, the composition varies, depending on the extent to which elements substitute for each other in the crystalline structure of these mineral species.

Soil Mg is often subdivided into soluble, rapidly exchangeable, slowly exchangeable, and structural (mineral) forms (4). These arbitrary subdivisions account for differences in bioavailability (5). The availability of Mg in soils for plant uptake follows: solution>exchangeable> mineral. There is an equilibrium between the forms of Mg that allows release of Mg from less-available forms to the more-available forms. Only a small fraction of the total Mg is present in soil solution, and plants absorb Mg from soil solution, which is buffered by the readily exchangeable forms.

Most soils hold important reserves of Mg in primary and secondary minerals (5, 6). These reserves contribute a large proportion of the Mg needs of annual and perennial crops (7). The rate of release of Mg for plant uptake depends on the rate of weathering of these minerals. Studies on the kinetics of Mg release from soils and soil fractions using cation exchange resins, dilute salts, and organic acids showed that as weathering proceeds, the slowly exchangeable Mg originates from progressively coarser particle-size fractions (8). The rate of release of slowly exchangeable Mg increases with a decrease in particle-size. Further, the presence of surface coatings on a mineral surface acts as a semipermeable barrier, reducing the rate of mineral weathering and, consequently, the rate of Mg release to soil solution (9).

MAGNESIUM IN PLANTS

Magnesium forms the central unit of chlorophyll in plant leaves, and cannot be substituted by other metals. Magnesium is involved in the production of starch during photosynthesis. Because Mg is a component of chlorophyll, an insufficient supply reduces chlorophyll formation, which is likely to affect the photosynthetic ability of the plants. Magnesium has a controlling action on the swelling of plasma and plays a major role in many enzyme functions in plants (10).

Magnesium deficiency symptoms generally show up rather clearly in plant leaves, and Mg is readily mobile in plants. It moves from the older leaves to the younger leaves when Mg is deficient, and the deficiency symptoms

Encyclopedia of Soil Science

Table 1 Magnesium minerals in soils

Magnesium minerals	Chemical formula	Magnesium content (g Mg/kg)
Fosterite	Mg_2SiO_4	320–350
Pyrope	$3MgO{\cdot}Al_2O_3{\cdot}3SiO_2$	60–130
Iolite	$H_2(Mg,Fe)_4Al_8Si_{10}O_{37}$	50–80
Diopside	$CaMg(SiO_3)_2$	20–140
Augite	$CaMg(SiO_3)_2$	45–100
Enstatite	$MgSiO_3$	180–220
Actinolite	$Ca(Mg,Fe)_3Si_4O_{12}$	100–160
Hornblende	CaMg metasilicate	10–90
Serpentine	$H_4Mg_3Si_2O_9$	19–26
Talc	$H_2Mg_3Si_4O_{12}$	160–200
Phlogopite	$H_3Mg_3Al(SiO_4)_3$	130–180
Biotite	$(H,K)_2(Mg,Fe)_2Al_2Si_3O_{12}$	10–160
Clinochlore	$H_8(Mg,Fe)_5Al_2Si_3O_{18}$	100–120

(Adapted from Ref. 2.)

therefore show first on the older leaves as Mg is withdrawn from them. The deficiency is initially characterized by an interveinal chlorosis, although in acute stages, the leaf may be generally deficient in both green and yellow pigments (variegated coloration), and necrosis may occur in the areas of the leaf first affected by Mg deficiency. A deficiency of Mg also induces the formation of anthocyanins in some plant species, such as cotton (*Gossypium hirsutum*). Economically, one of the most important Mg deficiency diseases is identified in tobacco (*Solanum tobaccum*), commonly referred to as "sand drown." This occurs when tobacco leaves contain less than 0.25% Mg on a dry-weight basis. Similarly, Mg deficiency in pasture leads to "grass tetany" in grazing animals. The deficiency of Mg in pasture occurs when the concentration is less than approximately 0.2% (11).

Magnesium deficiency in plants generally occurs in soils that are low in exchangeable Mg, light textured, highly leached, acidic, and low in cation exchange capacity (12). In addition, induced deficiency may occur on some soils as a result of nutrient imbalances. Soils that are heavily fertilized, particularly with materials lacking in Mg or high in Ca, K, and NH_4, can also induce Mg deficiency (13). By growing crops that require high levels of Mg throughout the growing season (e.g., tobacco), Mg deficiency in soils may be exacerbated (12, 14).

MAGNESIUM IN ANIMALS

Magnesium plays an important role in the enzymatic metabolism of carbohydrates, lipids, proteins, and nucleic acids. Generally Mg is an activator for numerous enzymes, such as phosphatases. It is also involved in the nerve conduction and muscular contraction.

Deficiency of Mg in blood plasma causes a disorder known as hypomagnesemia (grass tetany or staggers), which usually occurs in dairy cows in the early part of lactation (15). Magnesium deficiency in animals can be overcome by regular use of Mg salts as drench or water trough treatment, as a lick, or in pasture after foliar application.

Application of excessive amounts of K fertilizer and K-rich farm effluents has been shown to increase the incidence of Mg deficiency leading to grass stagger. In pasture and fodder crops, grass stagger index (GSI) (Eq. 1) is used to predict the occurrence of Mg deficiency.

$$GSI = [K^+]/([Ca^{2+}] + [Mg^{2+}]) \quad (1)$$

where $[K^+, Ca^{2+}, Mg^{2+}]$ = milliequivalents/kg-DM.

It has been suggested that GSI values in the pasture in excess of 2.2 enhance the risk of grass stagger development. This condition is generally linked with animal serum Mg levels less than 1.0–1.5 mg/100 ml, compared with normal levels in cattle of 1.7–3.0 mg/100 ml. This condition arises in response to diets low in Mg or K-induced inhibition of Mg absorption in the rumen (15).

MAGNESIUM FERTILIZERS

Magnesium deficiency in soils can be overcome by adding Mg-containing compounds (Table 2) (16). Epsom salt ($MgSO_4$) is soluble in water and used as a fast-release Mg source. Other fertilizers are insoluble in water and are used as a slow-release source. The manufacture and use of serpentine superphosphate as a fertilizer material are declining. Dolomite, which contains both Ca and Mg, is more effective in acidic soils because the Mg is brought into solution by the acidic soil. Dolomite is the most widely used source of Mg, both as an ingredient of mixed fertilizers and as a separate amendment for liming.

All the slightly soluble Mg fertilizers listed in Table 2 are acid-neutralizing materials and would thus reduce the acid-forming potential of the fertilizer to which they might be added. Magnesium silicates, including magnesites, are ineffective sources of Mg for plants. Selectively calcined dolomite, in which the Mg component is oxidized to MgO, is more reactive than dolomite. Calcined magnesite can be added to soils or dusted onto pasture as a source of Mg.

Table 2 Magnesium fertilizers

Magnesium minerals	Chemical formula	Mg solubility (mmol/liter)	Solubility product (pK_{SP})	Magnesium content (g Mg/kg)
Dolomite	$MgCO_3{\cdot}CaCO_3$	0.038	17.09	100
Calcined dolomite	$MgO{\cdot}CaCO_3$			160
Hydrated dolomite	$MgO{\cdot}Ca(OH)_2$			170
Magnesite	$MgCO_3$	0.076	8.24	260
Brucite	$Mg(OH)_2$	0.091	11.41	360
Magnesia	MgO	0.150		560
Kieserite	$MgSO_4{\cdot}H_2O$	4943		160
Epsom salt	$MgSO_4{\cdot}7H_2O$	127.3		90
Kainite	$MgSO_4{\cdot}KCl{\cdot}3H_2O$			70
Langbeinite	$2MgSO_4{\cdot}K_2SO_4$			110
Fosterite	Mg_2SiO_4	0.067×10^{-3}	28.11	320–350

(Adapted from Ref. 16.)

FACTORS AFFECTING MAGNESIUM AVAILABILITY TO PLANTS

Plants differ markedly in their response to Mg deficiency in soil. The input and losses of Mg, forms of Mg in soils, and the interaction of Mg with other cations are some of the important properties affecting the plant availability of Mg (17). Magnesium concentration in plants is depressed by high concentrations of other cations, such as Ca, K, NH_4, and Al (18). Field trials have indicated that continuous use of ammonium-based fertilizers such as diammonium phosphate (DAP) along with muriate of potash (KCl) results in Mg deficiency in pasture (11). The addition of KCl decreases the concentration of exchangeable Mg in soils, which is attributed mainly to the displacement of Mg by K ions added through the fertilizer.

Lime-induced reductions in tissue Mg level and Mg uptake by plants have been observed by a number of researchers (19). This effect occurs following an increase in Mg adsorption resulting from an increase in pH-dependent adsorption sites in soils containing variable charge components. Liming has also been shown to increase the leaching potential of Mg, mainly because the Ca added through lime exchanges with Mg on the soil surfaces leading to the leaching of Mg in soil solution (20).

A number of soil test methods are used to predict Mg availability to plants, including 1N NH_4OAc-exchangeable Mg, percentage CEC saturated with Mg, and indices of relative Mg concentration in soil solution (pMg in soil solution, pMg–pCa in solution and half pMg–pK in solution) (21). The ability of these indices to predict the availability of Mg to plants varies depending on the relative concentration of Mg, Ca, and K in soil solution (22). Field calibrations for the Mg soil test are available that enable the soil testing service to predict the Mg status of the soils and to make necessary Mg fertilizer recommendations.

CONCLUSION

As the central unit of chlorophyll, magnesium is one of the most important plant nutrients. Magnesium is usually added as an accessory element along with other fertilizer materials. Plants take up Mg from soil solution, which is replenished through the continuous release from the exchange sites. Epsom salt is one of the most important soluble Mg sources added as a fertilizer material. Dolomite is a sparingly soluble material and is used extensively both as a source of Mg and as a liming material.

Excessive amounts of K in fertilizers and farm effluents generally decrease the uptake of basic cations, especially Ca and Mg by pasture. This is mainly the result of depletion of these cations in soil solution that results from leaching because of the competition from K ions for the cation exchange sites of the soil. To reduce the deficiency of Mg in grazing animals, it is important that this nutrient is added to soils, which receive regular effluent irrigation or K fertilizer application.

The timing of the application of K-based fertilizers to pasture soils is important in regulating the supply of Mg for dairy cattle. Increased Mg requirements at the end of pregnancy and in early lactation periods need to be supplemented by drenching with Mg salts. The magnesium requirement of dairy cattle can be regulated by adjusting the dietary cation–anion balance (DCAB). An acid DCAB in the animal ration is likely to overcome grass tetany disorders in dairy cattle caused by Mg deficiency.

REFERENCES

1. Aitken, R.L.; Scott, B.J. Magnesium. *Soil Analysis—An Interpretation Manual*; CSIRO Publishing: Collingwood, Australia, 1999; 255–262.
2. Beeson, K.C. Magnesium in Soils-Sources, Availability and Zonal Distribution. In *Magnesium and Agriculture, Proceedings of the Symposium*; Anderson, G.C., Jencks, E.M., Horvath, D.J., Eds. West Virginia University: Morgantown, 1959; 1–11.
3. Uzo Mokwunye, A.; Melsted, S.W. Magnesium Forms in Selected Temperate and Tropical Soils. Soil Sci. Soc. Am. Proc. **1972**, *36*, 762–764.
4. Metson, A.J.; Brookes, J.M. Magnesium in New Zealand Soils. II. Distribution of Exchangeable "Reserve" Magnesium in the Main Soil Groups. NZ. J. Expt. Agric. **1975**, *18*, 317–335.
5. Hailes, K.J.; Aitken, R.L.; Menzies, N.W. Magnesium in Tropical and Subtropical Soils from North-Eastern Australia. I. Magnesium Fractions and Interrelationships with Soil Properties. Aust. J. Soil Res. **1997**, *35*, 615–627.
6. Mayland, H.F.; Wilkinson, S.R. Soil Factors Affecting Magnesium Availability in Plant-Animal Systems: A Review. J. Ani. Sci. **1989**, *67*, 3437–3444.
7. Christenson, D.R.; Doll, E.C. Magnesium Uptake from Exchangeable and Non Exchangeable Sources in Soils as Measured by Intensive Cropping. Soil Sci. **1978**, *126*, 166–168.
8. Lombin, G.; Fayemi, A. Release of Exchangeable and Non-Exchangeable Magnesium from Nigerian Soils on Cropping with Maize or Chemical Extraction. J. Sci. Food Agric. **1976**, *27*, 101–108.
9. Courchesne, F.; Turmel, M.; Beauchemin, P. Magnesium and Potassium Release by Weathering in Spodosols: Grain Surface Coating Effects. Soil Sci. Soc. Am. J. **1996**, *60*, 1188–1196.
10. Krauss, R.W.; Gauch, H.G. Roles of Magnesium in Plants. In *Magnesium and Agriculture*; Proceedings of the Symposium, Anderson, G.C., Jencks, E.M., Horvath, D.J., Eds. West Virginia University: Morgantown, 1959; 39–61.
11. Bolan, N.S.; Horne, D.J.; Wilson, G.F.; Dawood, D.; Selvarajah, N. Composition and Nutritive Value of Pasture Irrigated with Farm Effluents. In *Soil Research: A Knowledge Industry for Land-Based Exporters*; Proceedings of Workshop, Currie, L.D., Loganathan, P., Eds. Massey University: Palmerston North, New Zealand, 2000; 157–164.
12. Pinkerton, A. The Fate of Magnesium Applied to Flue-cured Tobacco, and Its Effect on Leaf Quality and Magnesium Content. Australian J. Expt. Agric. Ani. Husbandry **1971**, *11*, 99–104.
13. Embleton, T.W. Magnesium. In *Diagnostic Criteria for Plants and Soils*; Chapman, H.D., Ed. University of California Press: California, 1966; 225–263.
14. Sinclair, A.H. Availability of Magnesium to Rye Grass from Soils During Intensive Cropping in the Glass House. J. Agric. Sci. **1980**, *96*, 635–642.
15. Grunes, D.L.; Rending, V.V. *Grass Tetany*; ASA Special Publication No 25, Am. Soc. Agron., Crop Sci. Soc. Am., Soil Science Society of America: Madison, Wisconsin, 1979; 1–23.
16. Augustin, S.; Mindrup, M.; Meiwes, K.J. Soil Chemistry. In *Magnesium Deficiency in Forest Ecosystem*; Huttl, R.F., Schaaf, W., Eds. Kluwer Academic Publishers: London, 1997; 1, 255–273.
17. Metson, A.J. Magnesium in New Zealand Soils. 1. Some Factors Governing the Availability of Soil Magnesium: A Review. NZ. J. Expt. Agric. **1974**, *2*, 277–319.
18. Ellis, R., Jr. *Influence of Soil, Liming, Magnesium, Potassium and Nitrogen on Magnesium Composition of Plants*; Am. Soc. Agron. Special publication, 1979; 35, 79–92.
19. Grove, J.H.; Sumner, M.E. Lime Induced Magnesium Stress in Corn: Impact of Magnesium and Phosphorus Availability. Soil Sci. Soc. Am. J. **1985**, *49*, 1192–1196.
20. Edmeades, D.C.; Judd, M.J. The Effects of Lime on the Magnesium Status and Equilibria in Some New Zealand Top Soils. Soil Sci. **1980**, *129*, 156–161.
21. Doll, E.C.; Lucas, R.E. Testing Soil for Potassium, Calcium and Magnesium. In *Soil Testing and Plant Analysis*; Walsh, L.M., Beaton, J.D., Eds. Soil Science Society of America: Madison, Wisconsin, 1973; 133–151.
22. Rahmatullah; Baker, D.E. Magnesium Accumulation by Corn (*Zea mays* L.) as a Function of Potassium-Magnesium Exchange in Soils. Soil Sci. Soc. Am. J. **1981**, *45* (5), 899–903.

MANURE, COMPOST, AND BIOSOLIDS

Bahman Eghball
United States Department of Agriculture-Agricultural Research Service (USDA-ARS), University of Nebraska–Lincoln, Lincoln, Nebraska, U.S.A.

Kenneth A. Barbarick
Colorado State University, Fort Collins, Colorado, U.S.A.

INTRODUCTION

Manure, compost, and biosolids (municipal sludge) are organic residuals that contain nutrients and organic matter. They are excellent substitutes for chemical fertilizers. The organic matter in these renewable organic residuals can significantly improve the chemical and physical properties of soil and enhance biological activities. Because of nutrients and organic matter in manure, compost, and biosolids, they can be used to improve degraded, eroded, or less productive soils as soil amendments. If not used properly, manure, compost, and biosolids can be sources of environmental pollution.

MANURE

Manure (animal waste) is generated in beef cattle feedlots, swine operations, dairy barns, poultry houses, and other livestock operations. The number of animals and the number of large production facilities in the U.S. have significantly increased in the past 10 years (1). Manure, as well as composts and biosolids, are renewable resources and are excellent sources of macro- (N, P, K, Ca, Mg, S) and micronutrients (Zn, Cu, Fe, Mn, etc.) that are essential for growing plants. For centuries manure was used throughout the world for improving soil fertility and enhancing crop productivity. However, with the advent of synthetic fertilizers after World War II, manure was considered more a liability than a nutrient resource for crop production.

When animals are grazing on pastures and rangelands, manure is dispersed across a large area and little management is needed because the material is not concentrated and decomposes rapidly. However, when animals are concentrated in small feeding areas, the quantity of manure requiring proper management increases greatly. Significant amounts of manure are generated each year in the U.S. from the confined feedlots of major livestock species (Table 1). The amount of N, P, and K in manure from these species would replace 25, 25, and 45% of the purchased N, P, and K fertilizers in the U.S. if utilized at agronomic application rates (Table 1). However, because of the high hauling cost, replacement of fertilizer is presently limited to specific areas in the country where the animal feeding operations are located. Crop producers are also reluctant to use manure because of factors such as hauling and spreading costs, potential introduction of weed seeds, availability of manure where needed, uncertainty about availability of manure nutrients to plants, and problems of odor and application uniformity. The global numbers of major livestock species are given in Table 2.

Even though manure is an excellent source of multiple nutrients and organic matter, it can also contribute to water, air, and land pollution because of the potential for environmental loading with excess phosphorus, nitrate, salts, undesirable microorganisms, pathogens (disease-causing organisms), and greenhouse gases. Manure application in excess of crop needs can cause a significant build-up of P, N, trace elements (As, Cd, Pb, Hg, Mo, Ni, Cu, Fe, Mn, Se, and Zn), and salts in soils. Trace-element limits in soil are given in Table 3. The elevated P and N levels in soil are of environmental concern when these nutrients are carried by runoff to streams and lakes and cause "eutrophication," which is nutrient enrichment of water that can promote algal growth and depletion of dissolved oxygen in water. This oxygen is essential for aquatic animals. Pathogens (such as bacteria, viruses, parasites) in runoff from fields receiving manure can be another source of water pollution. Pathogens and odorous materials can also be carried by wind from the feeding operations to neighboring areas. Excess manure application can contaminate the groundwater with nitrate–N. Nitrate is a water-soluble ion that moves with water into the soil and can reach the groundwater within a few days after application. The U.S. Environmental Protection Agency (USEPA) has set a 10 mg NO_3–N L^{-1} standard for drinking water.

Encyclopedia of Soil Science

Table 1 Annual manure, N, P, and K generated by animals confined in beef cattle feedlots, dairy barns, and poultry and swine operations, and fertilizer use in the U.S.

Animal species	Animals on feed[a] (million)	Manure (dry weight) (million Mg)	N[b] (Mg × 1000)	P[b] (Mg × 1000)	K[b] (Mg × 1000)
Beef cattle	13.22	31.67	602	206	633
Dairy cows	13.14	22.01	782	140	522
Chickens (broilers and layers)[c]	8263.00	13.26	544	186	278
Turkeys[c]	284.00	3.09	142	65	65
Swine	59.41	15.15	709	451	709
Total		85.18	2779	1048	2207
1999 Fertilizer use in U.S.[d]			12,436	4345	5016
1996 Global fertilizer use[d]			78,353	13,543	17,516
(Manure nutrient/U.S. fertilizer use) × 100 (%)			25	25	45

[a] From Ref. 1.
[b] Manure weight and N, P, and K contents taken from Ref. 2.
[c] Yearly production numbers.
[d] From Ref. 3.

COMPOST

Composting is the aerobic decomposition of organic materials in the thermophilic temperature range of 40–65°C. The composted material should be an odorless, fine textured, low moisture content material that can be bagged and sold for use in gardens, potting, and nurseries or used as a source of nutrients and organic matter on cropland with little fly breeding potential. Other advantages of composting include improving the handling characteristics of any organic residue by reducing its volume and weight. Composting also has the potential to kill pathogens and weed seeds. Disadvantages of composting organic residues include loss of N and other nutrients during composting, time for processing, cost for handling equipment, need for available land for composting, odors during composting, marketing, diversion of manure or residue from cropland, and slow release of available nutrients. Similar to manure or biosolids, composts can cause water, air, and land pollution if not used properly.

Table 2 Global numbers of major livestock species in 1997

Animal species	Animals[a] (million)
Cattle	1333
Chickens	14,156
Sheep and goats	1754
Swine	837

[a] From Ref. 4.

Temperature, water content, C:N ratio, pH level, aeration rate and the physical structure of organic materials are important factors influencing the rate and efficiency of the composting process. Ideal values for these factors include a temperature of 54–60°C, C:N ratio of 25:1–30:1, 50–60% moisture content, oxygen concentration >5%, pH of 6.5–8.0, and particle size of 3–13 mm. U.S. Environmental Protection Agency regulations for composting municipal waste require that the temperature be maintained at 55°C or above for at least 3 days in order to destroy pathogens. A temperature of 63°C within the compost pile is needed to destroy the weed seeds.

Homogeneous manure solids can be composted alone without mixing with bulk materials. Bulking agents are required to provide structural support when manure solids, or other organic residues, are too wet to maintain air space within the composting pile, and to reduce water content and/or to change the C:N ratio. Dry and fibrous materials, such as saw dust, leaves and finely chopped straw or peat moss are good bulking agents for composting wet manure or organic residues. Depending on the ambient temperature, a complete composting process may take 2–6 months.

There are a number of methods for composting organic materials. These include active windrow (with turning), passive composting piles, passively aerated windrow (supplying air through perforated pipes embedded in the windrow), active aerated windrow (forced air), bins, rectangular agitated beds, silos, rotating drums, containers, anaerobic digestion, and vermicompost (using

Table 3 United States Environmental Protection Agency (40CFR503.13, revised 1 July 1999) trace element limits, and concentrations in Littleton/Englewood, CO biosolids, 27 July 1999 (Concentration and quantities are on dry weight basis.)

Trace element	Agronomic rate concentration limit ($mg\,kg^{-1}$)	Ceiling concentration limit ($mg\,kg^{-1}$)	Annual soil loading limit ($kg\,ha^{-1}$)	Cumulative soil loading limit ($kg\,ha^{-1}$)	Littleton/Englewood biosolids ($mg\,kg^{-1}$)
Arsenic (As)	41	75	2.0	41	2.7
Cadmium (Cd)	39	85	1.9	39	5.6
Copper (Cu)	1500	4300	75	1500	256
Lead (Pb)	300	840	15	300	46
Mercury (Hg)	17	57	0.85	17	1.2
Molybdenum (Mo)	–	75	–	–	8.0
Nickel (Ni)	420	420	21	420	15
Selenium (Se)	100	100	5.0	100	4.6
Zinc (Zn)	2800	7500	140	2800	198

earth worms). Carcass composting can be used for all types of animals. Mortality composting can be accomplished in backyard-type bins, indicator composter bins, and in temporary open bins using layers of saw dust or chopped straw and dead animals. Water content is an important factor to be considered when composting dead animals, and should be maintained at about 40–50%.

BIOSOLIDS

Treatment of municipal wastewater results in a mostly organic by-product known as "biosolids." Land application of biosolids for beneficial use has been practiced since the early 20th century in the U.S. The USEPA announced requirements regarding beneficial use of biosolids with promulgation of the 40 CFR503 regulations in February 1993. The USEPA and the state agencies that control land application of biosolids encourage the judicious recycling of biosolids on crop- or rangeland, since they contain essential plant nutrients and organic matter.

A key aspect of USEPA and Colorado Department of Health (CDH) regulations requires application of biosolids at an agronomic rate. The CDH (5) defines agronomic rate as: "The rate at which biosolids are applied to land such that the amount of nitrogen required by the food crop, feed crop, fiber crop, cover crop or vegetation grown on the land is supplied over a defined growth period, and such that the amount of nitrogen in the biosolids which passes below the root zone of the crop or vegetation grown to groundwater is minimized." The USEPA trace-element limits for land application of biosolids are shown in Table 3. State agencies that control biosolids recycling on land are required to adopt these limits as minimum requirements to protect the environment and public health. Risk assessment of different biological pathways served as the foundation for establishing the trace-element restrictions. For example, the concentrations for Littleton/Englewood biosolids shown in Table 3 indicate that it meets the agronomic-rate limits and can, therefore, be applied at an agronomic rate with minimal restriction. New, aggressive pretreatment programs have significantly reduced trace-element concentration in biosolids since about 1970; therefore, environmental and public health risks are even more minimal.

The USEPA requires municipal wastewater treatment facilities to reduce pathogens and to reduce the attraction of insects and animals before applying biosolids to land. Most municipal wastewater treatment plants use heat and attack by beneficent microorganisms through anaerobic (without air) or aerobic (with air) digestion to kill potential pathogens and reduce odors that may reside in wastewater. Municipalities accomplish further reduction of pathogens and stabilization by composting, drying, or other techniques.

The major reason that the USEPA promotes land application of biosolids is that the plant nutrients and organic matter can benefit the soil–plant agroecosystem. For example, Littleton/Englewood biosolids used at two research locations contained up to 5.0% organic–N, 1.3% ammonium–N, 140 $mg\,kg^{-1}$ nitrate–N, 3.7% P, and 0.30% K. Biosolids can also provide plant micronutrients such as Fe and Zn. The organic carbon in biosolids can help to develop and stabilize soil structure with a concomitant increase in precipitation capture and decrease in soil erosion. Efficacious land application of biosolids changes the perspective from disposal of a waste (i.e., a nuisance) to recycling a valuable resource (i.e., a beneficial process).

REFERENCES

1. United States Department of Agriculture, *Agricultural Statistics*; Government Printing Office: Washington, DC, 2000.
2. United States Department of Agriculture, *Agricultural Uses of Municipal, Animal, and Industrial Byproducts*; Conserv. Research Report No. 44, Government Printing Office: Washington, DC, 1998.
3. Fertilizer Statistics, 2001, The Fertilizer Institute, Washington, DC (http://www.tfi.org).
4. Food and Agricultural Organization, 2001, Statistical database. United Nations, New York (http://apps.fao.org).
5. Colorado Department of Health, 1993, Biosolids Regulation 4.9.0.

METAL–CLAY INTERACTIONS

Dean Hesterberg
North Carolina State University, Raleigh, North Carolina, U.S.A.

INTRODUCTION

Soils are vital for regulating the biological effects and mobility of metals in nature. Some metals such as potassium and calcium are essential nutrients for plants and animals, while others such as lead and mercury are potentially toxic. Therefore, a detailed understanding of the chemistry of metals in soils is essential for managing their agricultural and ecological impacts. Two soil components that are particularly important for regulating metal impacts are clays and organic matter. Clays consist of inorganic minerals in the fine particle-size fraction of soils (<0.002 mm particles), and may include associated organic matter. Modern concepts of metal interactions within soils evolved from the discovery in the mid-1800s that soils retain (bind) cations. Since then, researchers have applied a wide variety of analytical techniques and chemical principles to understand metal interactions within soils and with soil components.

Most chemical elements are metals, and many of these occur naturally in soils. Metals usually occur as cations in soils. However, metal cations of higher charge such as chromium (Cr^{6+}) and molybdenum (Mo^{6+}) combine with oxygen to form stable molecules called oxyanions (CrO_4^{2-} and MoO_4^{2-}). The most abundant metals in mineral soils, aluminum (Al^{3+}) and iron (Fe^{3+} or Fe^{2+}), are mostly combined with oxygen anions (O^{2-}) into soil minerals such as aluminosilicates, aluminum hydroxides, and iron oxides. Potassium (K^+), sodium (Na^+), calcium (Ca^{2+}), and magnesium (Mg^{2+}) are abundant in all but highly acidic soils. Metals having a density greater 5 g cm^{-3} are called heavy metals. Except for manganese (Mn) and iron, concentrations of heavy metals in noncontaminated soils are typically <100 mg kg^{-1}. In soils that are impacted by nuclear wastes, metals such as uranium, plutonium, cesium, and strontium are of concern.

Clays are especially important for binding metal ions because they have a high amount of surface area and the surfaces can be charged. Because oxygen constitutes on an average 60% of the atoms in soils, the surfaces of clay particles are typically dominated by oxygen atoms (1). Phyllosilicate (layer silicate) clay surfaces have a permanent negative charge due to minor substitutions of equivalently sized, but low-charged metal cations for more abundant cations within the crystal structure. The surfaces on the edges of aluminosilicate clay particles, and surfaces of iron- and aluminum- and manganese-oxide minerals have a variable charge. This so-called pH-dependent surface charge is due to the loss or gain of hydrogen ions (H^+) on the surface oxygens with changes in the concentration of H^+ (i.e., pH) in the surrounding water.

MECHANISMS OF METAL RETENTION BY CLAYS

Metal ions are retained by clays through 1) adsorption, 2) surface cluster formation, 3) surface precipitation, 4) diffusion into an existing mineral structure, or 5) coprecipitation. The first three mechanisms occur at clay surfaces, the interface between clay particles and the surrounding water (aqueous solution). Diffusion results from a metal concentration gradient from the clay surface into the particle and occurs very slowly (2). Coprecipitation involves incorporation of metal ions into a clay structure as it forms.

Adsorption—Macroscopic Aspects

Adsorption is the net accumulation of a substance into a two-dimensional structure at the clay surface (3). Adsorption processes occur over time periods of seconds to days (4). The amount of a given type of metal ion adsorbed depends mainly on the total concentration of metal ions present, their affinity for the clay surfaces relative to the aqueous (water) solution, and factors such as pH and the presence of other (competing) ions. The concentration and affinity effects at constant pH and temperature are indicated by an adsorption isotherm, a plot of adsorbed vs. dissolved metal. Isotherms are classified according to their shape (Fig. 1). An H-curve reflects a high affinity of metal ions for the soil or clay, as is common for heavy metals at low concentrations. The H-curve can be fit mathematically with a Freundlich adsorption isotherm model (3). An L-curve suggests that the mineral

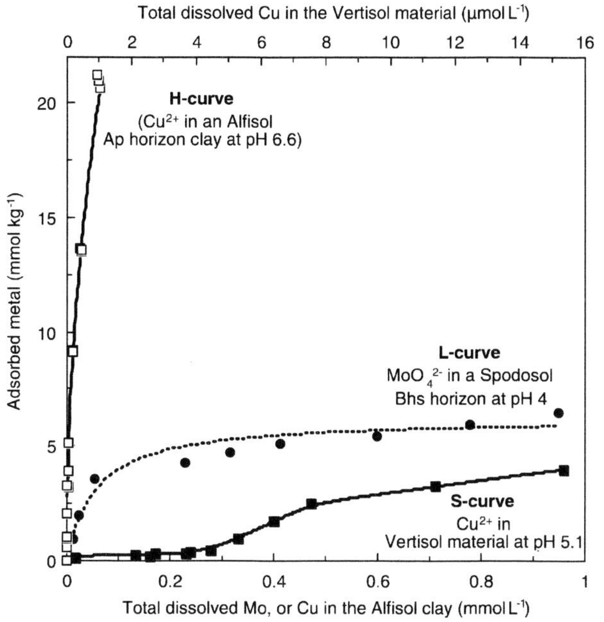

Fig. 1 Classes of adsorption isotherms for metal ions in soils or clays. H-curve data are for <0.001 mm soil clay (5); adsorbed Cu values are scaled by a factor of 0.1 and data are fit with a Freundlich isotherm model. L-curve data (6) are fit with a Langmuir isotherm model. S-curve data (3); adsorbed Cu values are scaled by a factor of 0.1 and fit with a smooth curve.

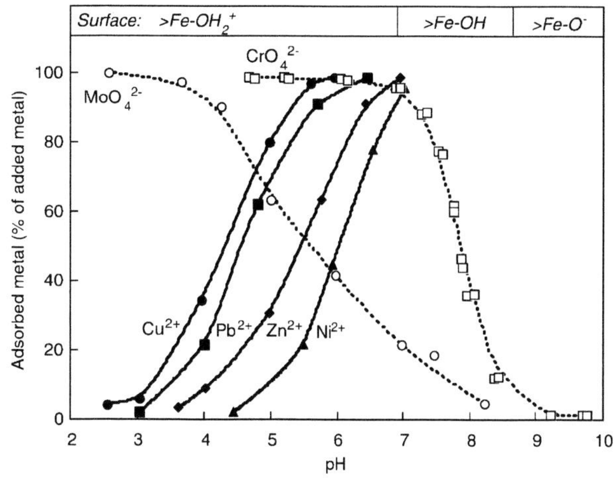

Fig. 2 Effects of pH on the adsorption of metal cations (Cu^{2+}, Pb^{2+}, Zn^{2+}, and Ni^{2+}) and oxyanions (MoO_4^{2-} and CrO_4^{2-}) on the iron oxide mineral goethite (α-FeOOH). The generalized number of protons and charge on reactive surface oxygens (e.g., $>Fe-OH_2^+$) at different pH ranges assumes that the goethite particles have no charge at pH 7.8 (7). Cation data (9) (added metal = 20 mmol kg^{-1} goethite); molybdate data (10) (added metal = 300 mmol kg^{-1} goethite); and chromate data (11) (added metal = 0.6 mmol kg^{-1} goethite).

particles have a high affinity for the metal ions at low concentration, and a decreasing affinity as the adsorbed metal concentration approaches the maximum adsorption capacity of the soil or clays (7). However, an isotherm does not reveal the molecular aspects of metal–clay associations, and metal retention mechanisms other than adsorption can also produce such a curve (8). The L-curve is fit mathematically with a Langmuir or Freundlich isotherm model (3). One explanation of an S-curve is that the presence of a metal-binding agent in the aqueous solution such as dissolved organic matter competes with the surfaces for the metal ions. As the total metal concentration increases, the capacity of the dissolved agent to bind metals is exceeded, and adsorption increases (3).

Typical pH dependence of metal-ion adsorption on oxide minerals is illustrated in Fig. 2. With increasing pH, metal cation adsorption increases and metal oxyanion adsorption decreases. These trends are related to the pH-dependent surface charge (Fig. 2). However, shifts in the adsorption curves along the *x*-axis for different metal cations added at the same level suggest that metal–clay interactions other than electrostatic attractions between the surface and oppositely charged ions are important. Heavy metal binding on such pH-dependent surfaces is strong under favorable pH conditions. Therefore, adsorption is not affected much by the presence of weakly bound cations such as Na^+ and K^+. Adsorption of heavy-metal cations on permanent charge sites of phyllosilicate minerals is usually weaker than on pH-dependent surfaces (2), so other cations more effectively compete for adsorption sites through a process called cation exchange.

Adsorption—Molecular Mechanisms

On a molecular scale, three mechanisms of metal ion adsorption are distinguished: inner-sphere surface complexes, outer-sphere surface complexes, and diffuse-layer ions (Fig. 3). Inner-sphere complexation, also called chemisorption, involves metal bonding directly to atoms at the clay surface as one or more water molecules that surround the dissolved ion are lost. A metal cation would bond directly to a deprotonated surface oxygen, and an oxyanion would bond through its own oxygen to the metal ion in the mineral that is one atomic layer behind the surface oxygens (2, 3, 7) (Fig. 3). As such, metal cations and oxyanions may compete for the same surface sites. Chemisorption involves a combination of covalent (electron sharing) and ionic (electrostatic) bonding. Outer-sphere surface complexes

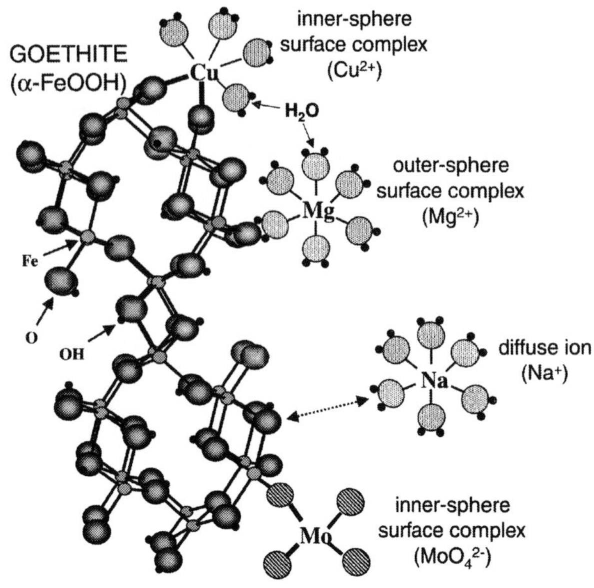

Fig. 3 Schematic illustration of molecular-scale mechanisms of metal cation or oxyanion adsorption on a goethite surface. (Adapted from Ref. 12.)

and diffuse ions are primarily bound to the surface through electrostatic attractions between the ions and oppositely charged surface sites. Outer-sphere surface complexes involve weaker binding of the metal ion directly on the clay surface, but with water molecules positioned between the ions and the surface. Diffuse-layer ions are not located directly on the surface, but are moving about within a region called the diffuse double layer that extends nanometer-scale distances from the surface. Conceptually, the diffuse double layer exists because of the opposing forces of electrostatic attraction of ions to the charged surface and diffusion of the accumulated ions away from the surface.

Metal cation adsorption on clays is modified by adsorbed organic matter. Adsorption involving interactions between clay surfaces, metal ions, and organic matter may consist of either the metal ion bridging between the clay and the organic matter (Type A ternary complex), or the metal ion adsorbing to the organic matter that is adsorbed on the clay surface (Type B ternary complex) (12).

Surface Clusters and Surface Precipitates

With increasing concentration of adsorbed metal and depending on pH, metal ions may cluster together on the clay surface through a three-dimensional structure consisting of metal cations and hydroxyl anions (OH^-) (8, 13). As metal concentration on the surface increases, these surface clusters may grow into a surface precipitate that entirely covers the surface.

Coprecipitation

Metal ions can be incorporated into the bulk clay structure by a process called coprecipitation, which is the simultaneous precipitation (formation) of one compound in conjunction with another (14). As a mineral forms, coprecipitation may involve the replacement of ions that are common to the mineral by metal ions of similar size and charge as the mineral forms. The substituted ions are dispersed as a minor component throughout the mineral. Another type of coprecipitation involves a metal accumulating as its own pure mineral phase encapsulated within a different pure mineral phase. A third type of coprecipitation involves an adsorption mechanism, whereby metal ions bind at surfaces of newly formed mineral particles, then become incorporated within the particles as they continue to grow.

CONCLUSION

Interactions between metals and clays in soils help regulate the plant availability of metals that are essential for agronomic crop growth, while affecting the mobility and environmental impacts of potentially toxic heavy metals. Because soils comprise a complex mixture of clays, organic matter, coarse-grained particles, water, air, and biological entities with properties that change both in space and time, accurately predicting metal reactions in a given soil remains a challenge.

REFERENCES

1. Schulze, D.G. An Introduction to Soil Mineralogy. In *Minerals in Soil Environments*, 2nd Ed.; SSSA Book Series 1; Dixon, J.B., Weed, S.B., Eds.; Soil Sci. Soc. Am.; Madison, WI, 1989; 1–34.
2. McBride, M.B. Chemisorption and Precipitation of Inorganic Ions. *Environmental Chemistry of Soils*; Oxford University Press: New York, 1994; 121–168.
3. Sposito, G. Inorganic and Organic Solute Adsorption in Soils. *The Surface Chemistry of Soils*; Oxford University Press: New York, 1984; 113–153.
4. Amacher, M.C. Methods of Obtaining and Analyzing Kinetic Data. In *Rates of Soil Chemical Processes*; SSSA Special Publication Number 27; Sparks, D.L., Suarez, D.L., Eds.; Soil Sci. Soc. Am.: Madison, WI, 1991; 19–59.

5. Weitzel, S.C. Flocculation of Soil Clay As Affected by Adsorbed Metals, Organic Matter, and PH. Master of Science Thesis, North Carolina State University, 1999.
6. Bibak, A.; Borggaard, O.K. Molybdenum Adsorption by Aluminum and Iron Oxides and Humic Acid. Soil Sci. **1994**, *158* (5), 323–328.
7. Sparks, D.L. Sorption Phenomena on Soils. *Environmental Soil Chemistry*; Academic Press: San Diego, CA, 1995; 99–139.
8. Manceau, A.; Charlet, L.; Boisset, M.C.; Didier, B.; Spadini, L. Sorption and Speciation of Heavy Metals on Hydrous Fe and Mn Oxides. From Microscopic to Macroscopic. Appl. Clay Sci. **1992**, *7*, 201–223.
9. McKenzie, R.M. The Adsorption of Lead and Other Heavy Metals on Oxides of Manganese and Iron. Aust. J. Soil Res. **1980**, *18*, 61–73.
10. Zhang, P. Chu; Sparks, D.L. Kinetics and Mechanisms of Molybdate Adsorption/Desorption At the Goethite/Water Interface Using Pressure-Jump Relaxation. Soil Sci. Soc. Am. J. **1989**, *53* (4), 1028–1034.
11. Ainsworth, C.C.; Girvin, D.C.; Zachara, J.M.; Smith, S.C. Chromate Adsorption on Goethite: Effects of Aluminum Substitution. Soil Sci. Soc. Am. J. **1989**, *53* (2), 411–418.
12. Alcacio, T.E.; Hesterberg, D.; Chou, J.W.; Martin, J.D.; Beauchemin, S.; Sayers, D.E. Molecular Scale Characteristics of Cu(II) Bonding in Goethite–Humate Complexes. Geochim. Cosmochim. Acta **2001**, *65*, 1355–1366.
13. McBride, M.B. Reactions Controlling Heavy Metal Solubility in Soils. Adv. Soil Sci. **1989**, *10*, 1–56.
14. Sposito, G. Mineral Solubility. *Chemical Equilibria and Kinetics in Soils*; Oxford University Press: Oxford, UK, 1994; 93–137.

METHANE EMISSIONS IN RICE, MITIGATION OPTIONS FOR

Kazuyuki Yagi
National Institute for Agro-Environmental Sciences, Tsukuba, Ibaraki, Japan

INTRODUCTION

The atmospheric concentration of methane (CH_4) has increased rapidly in recent years. Because it is a radiative trace gas and takes part in atmospheric chemistry, the rapid increase could be of significant environmental consequence. Of the wide variety of sources, rice fields are considered an important source of atmospheric CH_4, because the harvest area of rice has increased by about 70% during last 50 years and it is likely that CH_4 emission has increased proportionally. Recent estimates suggest that global emission rates of CH_4 from rice fields account for about 4–19% of the emission from all sources (1). Due to the large amount of the global emission from rice cultivation, reduction of CH_4 emission from this source is very important in order to stabilize atmospheric concentration. In addition, because of the possibility of controlling the emission by agronomic practices, rice cultivation must be one of the most hopeful sources for mitigating CH_4 emission.

PROCESSES CONTROLLING CH_4 EMISSIONS FROM RICE FIELDS

Table 1 provides a summary of measured methane emissions at a number of specific research sites around the world (2). It should be noted that methane fluxes from rice fields show pronounced diel and seasonal variations and vary substantially with different climate, soil properties, agronomic practices, and rice cultivars.

Processes involved in CH_4 emission from rice fields are illustrated in Fig. 1. Like other biogenic sources, CH_4 is produced by the activity of CH_4 producing bacteria, or methanogens, as one of the terminal products in the anaerobic food web in paddy soils. Methanogens are known as strict anaerobes that require highly reducing conditions. After soil is flooded, the redox potential of soil decreases rapidly by sequential biochemical reactions.

Flooded paddy soils have a high potential to produce CH_4, but part of CH_4 produced is consumed by CH_4 oxidizing bacteria, or methanotorophs. In rice fields, it is possible that a proportion of CH_4 produced in the anaerobic soil layer is oxidized in the aerobic layers, such as the surface soil–water interface and the rhizosphere of rice plants.

The emission pathways of CH_4 that is accumulated in flooded paddy soils is: diffusion into the flood water, loss through ebullition, and transport through the aerenchyma system of rice plants. In the temperate rice fields, more than 90% of CH_4 is emitted through plants (5), while significant amounts of CH_4 may evolve by ebullition, in particular during the early part of the season in the tropical rice fields (6). Therefore, it is concluded that possible strategies for mitigating CH_4 emission from rice cultivation can be made by controlling either production, oxidation, or transpor processes.

OPTIONS FOR MITIGATING CH_4 EMISSION

Water Management

Mid-season drainage (aeration) in flooded rice fields supplies oxygen into soil, resulting in a reduction of CH_4 production and a possible enhancement of CH_4 oxidation in soil (7, 8). A study using an automated sampling and analyzing system clearly showed that short-term drainage had a strong effect on CH_4 emission, as shown in Fig. 2. Total emission rates of CH_4 during the cultivation period were reduced by 42–45% by short-term drainage practices compared with continuously flooded treatment (9). These results indicate that improvement in water management can be one of the most promising mitigation strategies for CH_4 emission from rice fields. Increasing the rate of water percolation in rice fields by installing underground pipe drainage may also have an influence on CH_4 production and emission.

Soil Amendments and Mineral Fertilizers

The progress of soil reduction can be retarded by adding one of several electron acceptors in the sequential soil

Table 1 Methane emission from rice fields in various world locations[a]

Country	Daily average (g/m^2 day)	Flooding period (days)	Season total: Average (g/m^2)	Season total: Range (g/m^2)
China	0.19–1.39	75–150	13	10–22
India	0.04–0.46	60	10	5–15
Italy	0.10–0.68	130	36	17–54
Japan	0.01–0.39	110–130	11	3–19
Spain	0.10	120	12	
Thailand	0.04–0.77	80–110	16	4–40
U.S.A.	0.05–0.48	80–100	25	15–35

[a]The data are for the fields without organic fertilizer.

redox reactions. Sulfate is one of the most promising candidates for this strategy because it is commonly used as a component of mineral fertilizer and soil amendment. Field measurements have shown that CH_4 emission rate decreased by at most 55–70% by application of ammonium sulfate or gypsum (10, 11).

Additions of other oxidants, such as nitrate and iron-containing materials, may influence CH_4 emission from rice fields. As well as adding oxidants, dressing paddy fields with other soils that contain a large amount of free iron and manganese may decrease CH_4 emission. Other chemical candidates are nitrification inhibitors and acetylene releasing materials.

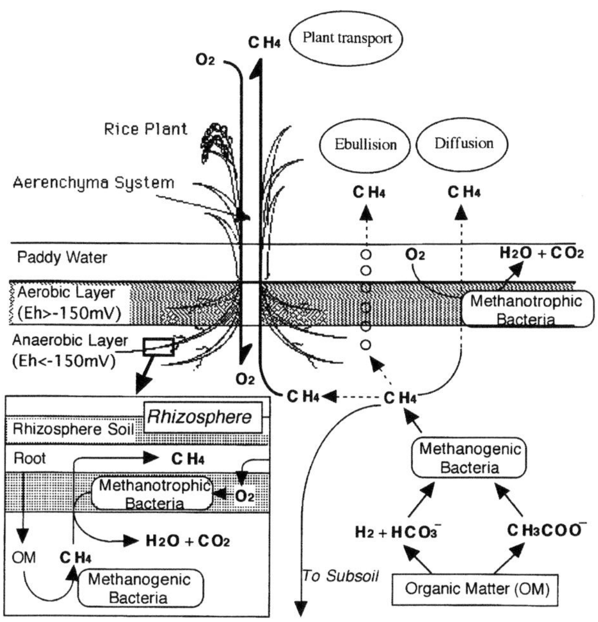

Fig. 1 Production, oxidation and emission of CH_4 in rice paddy fields. (Adapted from Refs. 3, 4.)

Organic Matter Management

In rice cultivation, fresh organic matter and animal wastes are often applied as fertilizers. In the fields, a proportion of the biomass of previous crops and weeds remains in soils at the start of rice cultivation. Such organic matter is decomposed in soils and acts as a substrate for fermentation reactions. Many researchers have demonstrated that incorporation of rice straw and green manure into rice paddy soils dramatically increases CH_4 emission

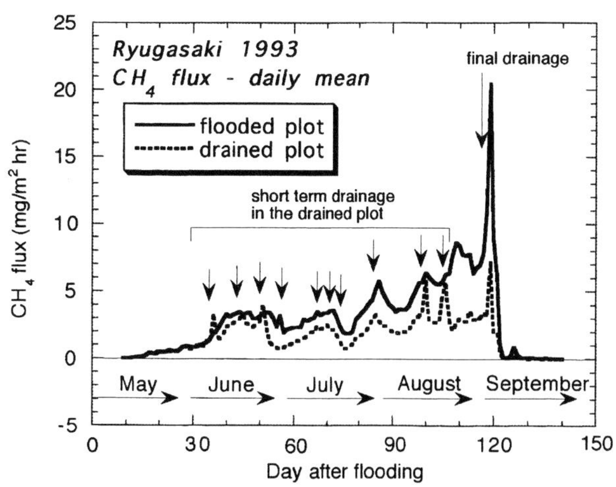

Fig. 2 Effect of water management on CH_4 emission from a rice paddy field. The arrows indicate period of midseason drainage in the intermittent irrigation plot and the timing of final drainage in both of the plots.

Table 2 Evaluation of the mitigation options for methane emission from fields

		Problem for application							
		Applicability		Economy		Effects on			
	CH_4 mitigation efficiency	Irrigated	Rainfed	Cost	Labor	Yield	Fertility	Time span	Other trade-off effects
Water management									
Midseason drainage	□	○	●	~	↑	+	~	○	May promote N_2O emission
Short flooding	□	○	●	~	~	−	−	○	May promote N_2O emission
High percolation	□	○	●	↑	↑	+	~	○	May promote nitrate leaching
Soil amendments									
Sulfate fertilizer	□	○	○	↑	~	Δ	−	○	May cause H_2S injury
Oxidants	□	○	○	↑	↑	Δ	−	○	
Soil dressing	○	○	○	↑	↑	−	−	○	
Organic matter									
Composting	□	○	○	↑	↑	+	+	○	
Aerobic decomposition	□	○	○	~	↑	~	~	○	
Burning	○	○	○	~	↑	~	~	○	Causing atmospheric pollution
Others									
Deep tillage	○	○	○	↑	↑	−	−	○	
No tillage	?	○	○	~	↓	−	~	○	
Rotation	○	○	Δ	~	↑	−	−	○	
Cultivar	○	○	○	~	~	~	~	●	

Key:
□ Very effective
○ Effective/applicable
Δ Case by case
● Not applicable/require long time
? No information
↑ Increase
↓ Decrease
~ About equal to previous situation
+ Positive
− Negative
(For further information see Refs. 13–15.)

(7, 10, 12). The impact of organic amendments on CH_4 emissions can be described by a dose–response curve which adopts correction factors for composted and fermented organic matter (6). Mitigation of CH_4 emission requires that the quantities of organic amendments be minimized.

Field experiments also indicated that composted or fermented organic matter increased CH_4 emission much less than fresh organic matter, due to a lower content of easily decomposable carbon (6, 7). Therefore, stimulation of composting organic amendments appears to be a promising mitigation option. Plowing the fields during the fallow period and promoting aerobic degradation of organic matter is also likely to reduce CH_4 emission.

Others

Different tillage and cropping practices change the physical, chemical, and microbiological properties of the plow layer soil and may reduce CH_4 emission. These include deep tillage, no tillage, and flooded rice-upland crop rotation.

Selecting and breeding rice cultivars that emit lower CH_4 is a desirable approach because it is easy to adopt. There are four points to consider for selecting cultivars: 1) they should exude low levels carbon from their roots; 2) they should have a low level of CH_4 transport and a high level of CH_4 oxidation in the rhizosphere; 3) they should have a higher harvest index, in order to reduce organic matter input into soil after harvest; and 4) they should be suitable and have a high productivity when other mitigation options are performed.

PROBLEMS AND FEASIBILITY OF THE OPTIONS

If the above mitigation options could be applied to world's rice cultivation, global CH_4 emission from rice fields could decrease significantly. However, there are several formidable obstacles to adopting the mitigation options into local rice farming. Table 2 summarizes the problems and feasibility of the individual mitigation options along with the efficiency of the options.

Application of some options is limited to specific types of rice fields. In particular, altering water management practices may be limited to rice paddy fields where the irrigation system is well equipped. Long midseason drainage and short flooding may cause possible negative effects on grain yield and soil fertility. Improving percolation by underground pipe drainage requires laborious engineering work. The increased water requirement is another problem in the water management options because water is a scarce commodity in many regions.

Cost and labor are serious obstacles for applying each option to local farmers. Most of the mitigation options will decrease profitability and the farmer net returns in the short run. To overcome these obstacles, an effort to maximize net returns by joining CH_4 mitigation and increased rice production will be needed, as well as political support.

It is recognized that the mitigation options should not have any significant trade-off effects, such as decreased rice yield, a decline in soil fertility, or increased environmental impact by nitrogen compounds. The development of anaerobic conditions in soil by flooding decreases decomposition rates of soil organic matter compared with aerobic soils, resulting in soil fertility being sustained for a long time. Flooded rice cultivation shows very little growth retardation by continuous cropping. Some mitigation options may reduce these advantages of rice fields. Application of sulfate-containing fertilizer may cause a reduction in rice yield due to the toxicity of hydrogen sulfide. Mid-season aeration and soil amendments may induce nitrogen transformation resulting in enhanced N_2O emissions (16, 17).

REFERENCES

1. Prather, M.; Derwent, R.; Ehhalt, D.; Fraser, P.; Sanhueza, E.; Zhou, X. Other Trace Gases and Atmospheric Chemistry. In *Climate Change 1994, Radiative Forcing of Climate Change and an Evaluation of the IPCC IS92 Emission Scenarios*; Houghton, J.T., Meira Filho, L.G., Bruce, J., Lee, H., Callander, B.A., Haites, E., Harris, N., Maskell, K., Eds.; Cambridge University Press: Cambridge, England, 1995; 73–126.
2. Intergovernmental Panel on Climate Change, *Greenhouse Gas Inventory Reference Manual*; IPCC Guidelines for National Greenhouse Gas Inventories, OECD: Paris, France, 1997; Vol. 3, 46–60.
3. Conrad, R. Control of Methane Production in Terrestrial Ecosystems. *Exchange of Trace Gases Between Terrestrial Ecosystems and the Atmosphere*; Andreae, M.O., Schimel, D.S., Eds.; John Wiley & Sons Ltd.: New York, 1989; 39–58.
4. Knowles, R. Processes of Production and Consumption. *Agricultural Ecosystem Effects on Trace Gases and Global Climate Change*; Harper, L.A., Mosier, A.R., Duxbury, J.M., Rolston, D.E., Eds.; American Society of Agronomy: Madison, WI, 1993; 145–156.
5. Cicerone, R.J.; Shetter, J.D. Sources of Atmospheric Methane: Measurements in Rice Paddies and a Discussion. J. Geophys. Res. **1981**, *86*, 7203–7209.
6. Denier van der Gon, H.A.C.; Neue, H.-U. Influence of Organic Matter Incorporation on the Methane Emission from a Wetland Rice Field. Global Biogeochem. Cycles **1995**, *9*, 11–22.
7. Yagi, K.; Minami, K. Effect of Organic Matter Application on Methane Emission from Some Japanese Paddy Fields. Soil Sci. Plant Nutr. **1990**, *36*, 599–610.
8. Sass, R.L.; Fisher, F.M.; Wang, Y.B.; Turner, F.T.; Jund, M.F. Methane Emission from Rice Fields: The Effect of Flood-Water Management. Global Biogeochem. Cycles **1992**, *6*, 249–262.
9. Yagi, K.; Tsuruta, H.; Kanda, K.; Minami, K. Effect of Water Management on Methane Emission from a Japanese Rice Paddy Field: Automated Methane Monitoring. Global Biogeochem. Cycles **1996**, *10*, 255–267.
10. Schütz, H.; Holzapfel-Pschorn, A.; Conrad, R.; Rennenberg, H.; Seiler, W. A 3-Year Continuous Record on the Influence of Daytime, Season, and Fertilizer Treatment on Methane Emission Rates from an Italian Rice Paddy. J. Geophys. Res. **1989**, *94*, 16405–16416.
11. Denier van der Gon, H.A.C.; Neue, H.-U. Impact of Gypsum Application on Methane Emission from a Wetland Rice Field. Global Biogeochem. Cycles **1994**, *8*, 127–134.
12. Sass, R.L.; Fisher, F.M.; Harcombe, P.A.; Turner, F.T. Mitigation of Methane Emission from Rice Fields: Possible

M

Adverse Effects of Incorporated Rice Straw. Global Biogeochem. Cycles **1991**, *5*, 275–287.
13. Ranganathan, R.; Neue, H.-U.; Pingali, P.L. Global Climate Change: Role of Rice in Methane Emission and Prospects for Mitigation. *Climate Change and Rice*; Peng, S., Ingram, K.T., Neue, H.-U., Ziska, L.H., Eds.; Springer-Verlag: Berlin, Germany, 1995; 122–135.
14. Neue, H.-U.; Wassmann, R.; Lantin, R.S. Mitigation Options for Methane Emissions from Rice Fields. In *Climate Change and Rice*; Peng, S., Ingram, K.T., Neue, H.-U., Ziska, L.H., Eds.; Springer-Verlag: Berlin, Germany, 1995; 137–144.
15. Yagi, K.; Tsuruta, H.; Minami, K. Possible Options for Mitigating Methane Emission from Rice Cultivation. Nutr. Cycling Agro-Ecosys. **1997**, *49*, 213–220.
16. Cai, Z.; Xing, G.; Yan, X.; Xu, H.; Tsuruta, H.; Yagi, K.; Minami, K. Methane and Nitrous Oxide Emissions from Rice Paddy Fields as Affected by Nitrogen Fertilisers and Water Management. Plant Soil **1997**, *196*, 7–14.
17. Bronson, K.F.; Neue, H.-U.; Singh, U.; Abao, E.B., Jr. Automated Chamber Measurements of Methane and Nitrous Oxide Flux in a Flooded Rice Soil: I. Residue, Nitrogen, and Water Management. Soil Sci. Soc. Am. J. **1997**, *61*, 981–987.

METHANE EMISSIONS FROM RICE PADDIES, FACTORS AFFECTING

M

Ronald L. Sass
Rice University, Houston, Texas, U.S.A.

INTRODUCTION

Methane (CH_4) is an important greenhouse gas with a global warming potential approximately 21 times that of carbon dioxide (CO_2). Of the agricultural sources that may account for as much as 40% of the total annual emission of atmospheric methane, a significant portion is contributed by rice cultivation. As one of the major crops of the world, rice accounts for approximately 23% of its per capita calorie intake. Recent estimates suggest that 50 ± 20 teragrams (Tg)/year (1 Tg = 1 million metric tons) of methane is emitted from global rice paddies. Methane is produced under anaerobic conditions generated in flooded rice fields. Irrigated rice fields account for 51%, rainfed fields for 27%, and deep-water fields for 8% of the world's total harvested ricelands. Dry upland ricelands (14%) are not flooded for any significant duration of time and do not produce methane (1).

Methane emission from rice fields is the result of soil bacterial processes; i.e., production in flooded anaerobic microsites and consumption (oxidation) in aerobic microsites. The flooding of rice fields promotes anaerobic fermentation of carbon (C) sources supplied by the rice plants and other incorporated organics, resulting in the formation of methane. Methane reaches the atmosphere primarily through a gas-conducting system in the rice plants and by ebullition of gas bubbles and diffusion. The process is governed by a complex set of parameters linking the physical and biological characteristics of flooded soil environments with specific agricultural management practices (1).

FACTORS AFFECTING METHANE EMISSIONS

Diel and Seasonal Patterns of Methane Emission

A positive dependence on temperature is observed in the diel pattern of methane emission in rice fields and in incubation studies (1). Seasonal variations in methane emission from rice paddies are more complex and differ among studies. Growth-seasonal methane fluxes observed in temperate rice fields show a general correlation with temperature, but respond primarily to seasonal trends in plant development and C inputs. From negligible values at the beginning of the season, methane emission gradually increases during the vegetative phase correlating with increasing plant biomass, and peaks near panicle differentiation, a period of rapid root development. Emission remains relatively constant during the reproductive stage, but may decrease during late grain filling because of root degradation. Prior to the end of the season, a second emission peak may be observed. This late-season increase in emission can be attributed to an increase in soil C substrate due to accelerating leaf and root senescence (1). The addition of readily degradable C, such as rice straw, before planting results in an increase in the early-season methane emission. The decomposition of the straw may also result in an enhanced or additional early-season emission maximum (1).

Photosynthetic Activity

In irrigated tropical rice paddies with double cropping, both methane emission and rice grain yields are consistently higher in the dry season than in the wet season (2). One interpretation of this finding is that higher photosynthetic rates during the sunnier days of the dry season lead to larger amounts of C available to methanogenic bacteria and, consequently, greater production and emission rates of methane. Seasonal rates of methane emission and rice grain yield have been positively correlated with accumulated solar radiation (3). A 1% increase in accumulative solar radiation is accompanied by a 1.1% increase in methane emission and a 1% increase in rice grain yield. These findings are consistent with the hypothesis that solar radiation, and hence photosynthetic activity of the rice plant, correlates with methane production and grain yield through the partitioning of nonstructural carbohydrates to the root system and grain panicle.

Organic Amendments

Organic amendments of flooded rice paddies increase both methane production and emission to the atmosphere. The

Encyclopedia of Soil Science

amount of methane emitted as a result of organic soil amendments depends greatly on the amount and condition of readily available decomposable C contained in the treatment. Methane emission with incorporated rice straw has been found to be higher than that from either compost or mineral-fertilized plots (4). The effect of incorporating green manure is even more dramatic (5). Schütz et al. (6) observed a 2.4-fold increase over control values with added rice straw. Cicerone et al. (7) observed increases from a control value of 1.4 g $CH_4\,m^{-2}\,yr^{-1}$ to up to 58.18 g $CH_4\,m^{-2}\,yr^{-1}$ with added straw, a factor of over 40 times. Wassmann et al. (8), investigating the effect of fertilizers on methane emission rates in Chinese rice fields (Hunan Province), found that the rate of increase in methane emission depended on the total amount of organic manure applied. A single dose of organic manure increased the emission rates by factors of 2.7 to 4.1, as compared with fields without organic manure. In field studies in the Philippines, Denier van der Gon and Neue (9) found that fields treated with green manure applied at a rate of 22 t ha^{-1} emitted over twice as much methane as fields in which the application rate was 11 t ha^{-1}. Studies indicate that methane emission enhancement due to added straw depends on soil type. In a study conducted in Bali (10), total amounts of methane emitted during the rice growth period were 2.6–3.3 g m^{-2} with applied chemical fertilizers compared with 3.9–6.8 g m^{-2} with rice straw application in Alfisol plots, and 4.2–5.8 g m^{-2} with applied chemical fertilizers compared with 6.9–10.7 g m^{-2} with rice straw application in Inceptisol plots. These findings suggest that although methane emissions from soils of volcanic origin are low, they show proportionate increases with added straw (C) comparable to those from other higher-emitting soils.

Inorganic Fertilization

Reports on the effects of mineral nitrogen (N) fertilizer application (i.e., source, method of application, and amount) on methane production and emission are difficult to interpret and inconsistent. Some form of fertilizer is necessary to ensure adequate plant development, which subsequently increases methane emission through increased plant biomass and root activity. On the other hand, the lack of adequate fertilization could result in poor plant growth and yield as well as premature death, which would increase methane emission through C made available by plant decay. The use of ammonium sulfate fertilizer may reduce methane emission through the competition by sulfate-reducing bacteria for hydrogen or the presence of hydrogen sulfide toxicity (1).

Water Management

Methane emission rates are highly sensitive to soil aeration controlled through water management. Periodic drainage of irrigated rice paddies, which significantly decreases methane emissions, may be the most effective method of mitigating them. In the Philippines, midseason drainage of 2 weeks' duration at either midtillering or panicle initiation was very successful in suppressing methane flux rates by up to 60%. However, the flux of nitrous oxide, a potent greenhouse gas, increased sharply during the drainage period (12). A single midseason drain reduced seasonal methane emission rates by 50% in fields in the United States. In addition, multiple short periods of drainage (2–3 days) approximately every 3 weeks during the growing season reduced methane emissions to an insignificant amount (11).

Soils

Rice soils prone to methane production belong mainly to the orders of Entisols, Inceptisols, Alfisols, Vertisols, and Mollisols. Oxisols, most of the Ultisols, and some of the Aridisols, Entisols, and Inceptisols are less favorable to methane production when flooded. After the flooding of calcareous or alkaline soils (high pH), methane production may occur almost immediately, whereas emission may take weeks to develop with acid soils and may not occur at all in highly acid soils. In general sandy soils high in organic C produce more methane than clay soils with similar C content (1). A strong positive correlation between seasonal methane emission and % sand has been observed in Vertisols in Texas (13). However, methane production may be limited in all soils if water percolation and the resultant oxygen levels are high.

Rice Cultivars

The effect of cultivar differences (14) contributes to variation in methane emissions from rice fields. A study of five rice cultivars in irrigated fields near Beijing, China indicated that methane emission during the tilling–flowering stage varied by a factor of two. Methane emission from eight different cultivars grown under comparable conditions near New Delhi, India differed by as much as one order of magnitude. Methane emissions from six different cultivars grown in Louisiana fields showed that semidwarf varieties evolved significantly less methane than tall varieties. Similarly, in an Indonesian paddy field, methane emissions from eight popular modern varieties of rice varied, depending on the cultivar, from 32.6–41.7 and from 51.3–64.6 g m^{-2} (15). Fields planted with the Italian rice cultivar Lido showed methane emissions 24–31% lower than fields planted with the cultivar Roma (16). The difference in methane emissions was attributed to a significantly lower gas transport capacity in the Lido cultivar. A study in Texas of ten rice cultivars (14) showed a cultivar-dependent variation in seasonal methane emissions, ranging from 17.95–41.0 g m^{-2}. Cultivars exhibiting higher seasonal methane emissions also exhibited higher soil acetate concentrations (17), suggesting that intervarietal differences in methane emissions may be the result of different soil organic substrate levels and, hence, different rates of methane production.

CONCLUSION

Significant progress has been made over the past quarter century in understanding processes of methane flux, their temporal and spatial variations, and their relationship to country level and global atmospheric trace-gas inventories. This progress has come through international and interdisciplinary research collaborations and cooperation. The known factors governing the levels of methane emissions from flooded rice fields, including climate and photosynthetic activity, added organic fertilizers, type of inorganic fertilizers, water management, soil type, and choice of rice cultivar, are largely understood. These factors have been incorporated into effective process-level models capable of calculating site-specific emission strengths throughout the world. Further integration of mechanistic models with geographic information systems of factors controlling emissions can provide improved estimates of the methane source strength of various rice ecologies and farming practices and lead to environmentally and socioeconomically sound management and mitigation policies.

REFERENCES

1. Neue, H.U.; Sass, R.L. Rice Cultivation and Trace Gas Exchange. In *Global Atmospheric-Biospheric Chemistry*; Prinn, R.G., Ed.; Plenum Press: New York, 1994; 119–147.
2. Neue, H.U.; Lantin, R.S.; Wassmann, R.; Aduna, J.B.; Alberto, M.C.R.; Andales, M.J.F. Methane Emission from Rice Soils of the Philippines. In *CH_4 and N_2O Global Emissions and Controls from Rice Fields and Other Agricultural and Industrial Sources*; Minami, K., Mosier, A., Sass, R.L., Eds.; NIAES Series 2; Yokendo Publishers: Tokyo, 1994; 55–63.
3. Sass, R.L.; Fisher, F.M.; Turner, F.T.; Jund, M.F. Methane Emission from Rice Fields as Influenced by Solar Radiation, Temperature, and Straw Incorporation. Global Biogeochem. Cy. **1991**, *5*, 335–350.
4. Yagi, K.; Minami, K. Effects of Organic Matter Applications on Methane Emission from Japanese Paddy Fields. Soil Sci. Plant Nutr. **1990**, *36*, 599–610.
5. Denier van der Gon, H.A.C.; Neue, H.U.; Lantin, R.S.; Wassmann, R.; Alberto, M.C.R.; Aduna, J.B.; Tan, M.J.P. Controlling Factors of Methane Emission from Rice Fields. In *World Inventory of Soil Emission Potentials*; Batjes, N.H., Bridges, E.M., Eds.; WISE Report 2; ISRIC: Wageningen, Netherlands, 1993; 81–92.
6. Schütz, H.; Holzapfel-Pschorn, A.; Conrad, R.; Rennenberg, H.; Seiler, W. A 3-year Continuous Record on the Influence of Daytime, Season and Fertilizer Treatment on Methane Emission Rates from an Italian Rice Paddy. J. Geophys. Res. **1989**, *94*, 16,405–16,416.
7. Cicerone, R.J.; Delwiche, C.C.; Tyler, S.C.; Zimmermann, P.R. Methane Emissions from California Rice Paddies with Varied Treatments. Global Biogeochem. Cy. **1992**, *6*, 233–248.
8. Wassmann, R.; Shangguan, X.J.; Cheng, D.X.; Wang, M.X.; Papen, H.; Rennenberg, H.; Seiler, W. Spatial and Seasonal Distribution of Organic Amendments Affecting Methane Emission from Chinese Rice Fields. Biol. Fert. Soils **1996**, *22*, 191–195.
9. Denier van der Gon, H.A.C.; Neue, H.U. Influence of Organic Matter Incorporation on the Methane Emission from a Wetland Rice Field. Global Biogeochem. Cy. **1995**, *9*, 11–23.
10. Subadiyasa, N.; Arya, N.; Kimura, M. Methane Emissions from Paddy Fields in Bali Island, Indonesia. Soil Sci. Plant Nutr. **1997**, *43*, 387–394.
11. Sass, R.L.; Fisher, F.M.; Wang, Y.B.; Turner, F.T.; Jund, M.F. Methane Emission from Rice Fields: The Effect of Flood Water Management. Global Biogeochem. Cy. **1992**, *6*, 249–262.
12. Bronson, K.F.; Neue, H.U.; Singh, U.; Abao, E.B. Automated Chamber Measurements of Methane and Nitrous Oxide Flux in a Flooded Rice Soil: I. Residue, Nitrogen, and Water Management. Soil Sci. Soc. Am. J. **1997**, *61*, 981–987.
13. Sass, R.L.; Fisher, F.M.; Lewis, S.T.; Turner, F.T.; Jund, M.F. Methane Emission from Rice Fields: Effect of Soil Properties. Global Biogeochem. Cy. **1994**, *8*, 135–140.
14. Sass, R.L.; Fisher, F.M., Jr. Methane from Irrigated Rice Cultivation. Curr. Top. Wetland Biogeochem. **1996**, *2*, 24–39.
15. Nugroho, S.G.; Sunyoto; Jumbanraja, J.; Suprapto, H.; Ardjasa, W.S.; Kimura, M. Effect of Rice Variety on Methane Emission from An Indonesian Paddy Field. Soil Sci. Plant Nutr. **1997**, *43*, 799–809.
16. Butterbachbahl, K.; Papen, H.; Rennenberg, H. Impact of Gas Transport Through Rice Cultivars on Methane Emission from Rice Paddy Fields. Plant Cell Environ. **1997**, *20*, 1175–1183.
17. Sigren, L.K.; Byrd, G.T.; Fisher, F.M.; Sass, R.L. Comparison of Soil Acetate Concentrations and Methane Production, Transport, and Emission in Two Rice Cultivars. Global Biogeochem. Cy. **1997**, *11*, 1–14.

METHANE IN FLOODED RICE

Ronald L. Sass
Rice University, Houston, Texas, U.S.A.

INTRODUCTION

Methane is a radiatively active trace gas with approximately 21 times more infrared sorbing capability per molecule than carbon dioxide as a source of potential global warming (1). Worldwide, agricultural sources of methane may account for as much as 40% of the total annual emission of atmospheric methane with a significant portion contributed by rice cultivation. To meet the rice supply of growing populations, rice cultivation will continue to increase at or beyond its current rate. Based on projected population growth rates in countries where rice is the main food crop, it is estimated that the world's annual rough rice production must increase from a 1990 value of 473 million tons to 564 million tons by 2000 and 781 million tons by 2020—a 65% increase (1.7% per year). In south Asia, for example, rice production is projected to double by the year 2020 (2). Because arable land is limited in major rice growing areas, increased production has to be achieved mainly by intensifying cropping (i.e., two or three crops per year) and developing new technologies rather than expanding the area of rice cultivation. Irrigated rice will continue to dominate production. Irrigated rice land now comprises about half the total harvested area but contributes more than two-thirds the total grain production. With present agronomic practices, this trend will lead to significantly increased methane emissions.

METHANE PRODUCTION AND EMISSION FROM RICE FIELDS

The processes involved in methane emission from flooded rice to the atmosphere include methane production in the soil by methanogens, methane oxidation within oxic zones of floodwater and the soil by methanotrophs, and vertical transport from soil to the atmosphere. Methane is produced in the terminal step of several anaerobic degradation chains. The amount of methane produced in flooded rice soils is primarily determined by the availability of methanogenic substrates and the influence of environ-mental factors. The sources of organic carbon for methanogenic substrates are derived from rice plants via root processes and biomass litter or added organic matter. The environmental factors affecting methane production include soil texture, climate, and agricultural practices, such as water regime and management. Plant-mediated transport is the primary mechanism for the emission of methane from rice paddies, with approximately 90% of CH_4 transported to the atmosphere through the aerechymal system of the rice plants. The rice aerenchymal system not only transports methane from the rhizosphere to the atmosphere but also promotes the movement of atmospheric oxygen into the rhizosphere supporting methane oxidation. More than 50% of the generated methane is oxidized during the rice growing season (3).

DISTRIBUTION OF METHANE EMISSIONS FROM RICE AGRICULTURE

Rice is grown under a variety of climatic, soil, and hydrological conditions in over 90 countries of the world and on all continents except Antarctica. The ecosystems within which rice is grown vary widely with respect to elevation, rainfall pattern, depth of flooding and drainage, soil type, and by the variety of rice planted. Fields of rice can be found from 50° North in regions of China and Japan to 40° South in southern regions of Australia. Rice is grown from sea level to altitudes of more than 2,500 meters. It is a unique crop in that it adapts to a broad range of soil water content. It grows well in flood prone areas of South and Southeast Asia (as much as 5 meters of floodwater) and in drought-prone upland areas of Asia, South America, and Africa (4).

A rich variation in local cultural patterns as well as in natural conditions can be found in the various regions of the world where rice agriculture is practiced. These variations are reflected in different regional and country values of methane emission rates (5). Rice agriculture can be conveniently divided into four categories based on water availability: upland or dry rice, rainfed lowland rice, floodprone rice and irrigated rice. The relative methane flux strength for different rice ecosystems follows the order: irrigated rice ≥ favorable rainfed rice

> flood prone rainfed rice ≥ deepwater rice > drought prone rainfed rice > tidal wetland rice. Upland rice is not a source of CH_4, because it is grown in well-aerated soils. Differences in crop residue treatment, types and amounts of organic amendments, scheduled or opportunistic aeration periods, soil classification and texture, fertilization practices, and choice of rice cultivars are major causes for variations in methane fluxes in irrigated rice systems. Table 1 gives scaling factors for methane emissions with different ecosystem water regimes and organic amendments relative to continuously flooded (irrigated) fields without organic amendments. An estimate for each of these different systems is obtained by multiplying the expected methane emission from a continuously flooded field without organic amendments by the two relevant scaling factors. Highest methane fluxes are observed in continuously flooded fields receiving organic amendments such as green or animal manure. Lowest methane fluxes are recorded in fields with low residual carbon content, multiple aeration periods, poor soils, low fertilization, and poor rice growth and low grain yields.

The methane source strength of rainfed rice is highly uncertain because of the high variability in all factors controlling methane emission. Methane fluxes are generally lower from rainfed rice because of drought periods and low rice yields. The availability and use of recycled soil carbon residues and available organic amendments are also generally less in areas where rainfed rice is grown compared to areas of irrigated rice. Methane fluxes may approach those from irrigated rice in highly flood-prone rain-fed areas. Some areas of deepwater rice may exhibit an even higher methane source strength than irrigated rice because of long flooded growing periods and high rice biomass production, even though grain yields are low. The source strength of tidal wetland rice is small, generally because it is affected by salt water high in sulphate (5, 6).

Table 1 Methane emission scaling factors for rice ecosystems relative to continuously flooded fields without organic amendments

	Scaling factor	Range
1. Rice ecosystem[a]		
Upland	0	
Irrigated		
Continuously flooded	1.0	
Intermittently flooded, single aeration	0.5	0.2–0.7
Intermittently flooded, multiple aeration	0.2	0.1–0.3
Rainfed		
Flood prone	0.8	0.5–1.0
Drought prone	0.4	0–0.5
Deep water		
Water depth 50–100 cm	0.8	0.6–1.0
Water depth >100 cm	0.6	0.5–0.8
2. Organic amendments[b]		
Straw, animal manure, green manure, agricutlural wastes, etc.:		
0 t ha^{-1} (inorganic fertilizer only)	1.0	
1–2 t ha^{-1}	1.5	1.0–2.0
2–4 t ha^{-1}	1.8	1.5–2.5
4–8 t ha^{-1}	2.5	11.5–3.5
8–15 t ha^{-1}	3.5	2.0–4.5
15+ t ha^{-1}	4.0	3.0–5.0
Fermented amendments, compost, biogas pit residue, etc.:		
1–6 t ha^{-1}	1.0	1.0–1.5
6–12 t ha^{-1}	1.5	1.0–2.0

[a] (From Ref. 17.)
[b] (From Ref. 18.)

RANGE OF GLOBALLY OBSERVED SEASONAL EMISSIONS

Observed methane emissions for a single rice cropping season reported from several countries around the world show large ranges, reflecting local as well as regional differences in agricultural, biological, and climatic factors. An average of all such reported median country methane emission values from irrigated rice is 27.23 g CH_4 m^{-2}, with a range from a minimum value of less than 1 g CH_4 m^{-2} to a maximum value of 155 g CH_4 m^{-2} (5). A data set from five countries (China, India, Indonesia, Thailand, and Philippines) obtained under a network coordinated by the International Rice Research Institute showed averages of less than 30 g CH_4 m^{-2} under mineral fertilisation (7). Emissions under organic amendments were generally higher than 20 g CH_4 m^{-2} except for a site in Northern India where average emissions remained as low as 2 g CH_4 m^{-2} (7). In a series of studies in Texas, conducted between 1991 and 1995, 32 seasonal methane emission rates were measured under conditions of continuous flooding, no organic amendments, in three different soil types with 11 different rice cultivars. These studies show a measured mean methane emission of 21.99 ± 8.53 g CH_4 m^{-2} and a median value of 19.84 g CH_4 m^{-2} (8). The situation is similar for rainfed and deepwater rice. For rainfed rice, the median of average observed value is 13.33 g CH_4 m^{-2} and reported ranges are from 1 to 68 g CH_4 m^{-2} (5). Emission values (9) for rainfed rice ranges from 8 g CH_4 m^{-2} (drought prone rice) to 16 g CH_4 m^{-2} (flood prone rice). The average methane emission value for deepwater rice ranges from 12 g CH_4 m^{-2} (50–100 cm water depth) to 16 g CH_4 m^{-2} (>100 cm water depth) (5).

Estimates of the global release of methane from rice paddies made over the past three decades range from 12 to 200 Tg yr^{-1} (8). Many early estimates are based on a limited number of field measurements and range widely. The IPCC suggested a most probable value of methane emitted from global rice paddies to be 60 ± 40 Tg yr^{-1} (10). Three recent estimates based on extensive field data and model calculations report ranges of 25–54 Tg yr^{-1} (11), 21–41 Tg yr^{-1} (12), and 33–49 Tg yr^{-1} (5). Considering these latest results, a current best estimate ranging between 26 and 50 Tg for the annual world methane emission from rice fields is posited (8). Asia accounts for approximately 90% of the world rice production (13) and over 90% of the annual methane emissions from rice fields. Seven Asian countries lead the world in rice production and methane emission: China, India, Indonesia, Thailand, Philippines, Japan and The Republic of Korea (8).

MITIGATION OF METHANE EMISSIONS FROM RICE FIELDS

The Intergovernmental Panel on Climate Change (14) calculated that to stabilize the concentration of atmospheric methane at the 1990 level would require a 15–20% reduction in total methane emission. Irrigated rice cultivation is currently among the larger global sources of atmospheric methane and will become an even larger source due to increasing food demands. Because it is one of the few sources where the management of methane emission is possible, it is also likely to be a critical focus of mitigation efforts. Several mitigation options to reduce atmospheric methane concentrations have been developed through recent research efforts (2, 6, 15, 16). These include water management, soil amendments, organic matter management, cultivar selection, and field preparation, with water management being the most promising. However because rice is also the world's most important wetland crop and the primary calorie source of a large fraction of the world's population, mitigation efforts must be based on sound agricultural practices as well as good scientific judgment.

REFERENCES

1. Rodhe, H. A Comparison of the Contribution of Various Gases to the Greenhouse Effect. Science **1990**, *248*, 1217–1219.
2. Sass, R.L.; Fisher, F.M.; Wang, Y.B.; Turner, F.T.; Jund, M.F. Methane Emission from Rice Paddies: The Effect of Floodwater Management. Global Biogeochem. Cycles **1992**, *6*, 249–262.
3. Huang, Y.; Sass, R.L.; Fisher, F.M. A Semi-Empirical Model of Methane Emission from Flooded Rice Paddy Soils. Global Change Biology **1998**, *4*, 247–268.
4. Neue, H.U.; Sass, R.L. Trace Gas Emissions from Rice Fields. In *Global Atmospheric-Biospheric Chemistry*; Environmental Science Res. 48, Prinn, R.G., Ed.; Plenum Press: New York, 1994; 119–148.
5. Neue, H-U.; Sass, R.L. The Budget of Methane from Rice Fields. IGACtivities NewsLetter **1998**, *12*, 3–11.
6. Neue, H.U. Fluxes of Methane from Rice Fields and Potential for Mitigation. Soil Use and Management **1997**, *13*, 258–267.
7. Wassmann, R.; Neue, H.U.; Lantin, R.S.; Buiendia, L.V.; Rennenberg, H. Characterization of Methane Emissions from Rice Fields in Asia. 1. Comparison Among Field Sites in Five Countries. Nutrient Cycling in Agroecosystems **2000**, *58*, 1–12.
8. Sass, R.L.; Fisher, F.M.; Ding, A.; Huang, Y. Exchange of Methane from Rice Fields: A Model System for Wetland Emission Modeling. Journal of Geophysical Research Atmospheres **1999**, *104*, 26943–26952.

9. Wassmann, R.; Neue, H.U.; Lantin, R.S. Characterization of Methane Emissions from Rice Fields in Asia. 2. Differences Among Irrigated, Rainfed and Deepwater Rice. Nutrient Cycling in Agroecosystems **2000**, *58*, 13–22.
10. *Climate Change 1992*; The IPCC Supplementary Report to the IPCC Scientific Assessment. Supporting Material, Working Group III, Response Strategies, Subgroup AFOS; WMO/UNEP, Geneva, 1992; 14–22.
11. Sass, R.L.; Fisher, F.M.; Lewis, S.T.; Jund, M.F.; Turner, F.T. Methane Emission from Rice Fields: Effect of Soil Properties. Global Biogeochem. Cycles **1994**, *8*, 135–140.
12. Yagi, K.; Minami, K. Methane Emission from Paddy Fields as Affected by Soil Properties and Its Implications to the Global Estimation. In *IRRI-UNDP Final Workshop*; Beijing, China, 10–13 August, 1998; 34–38.
13. *World Rice Statistics 1993–94*. International Rice Research Institute, P. O. Box 933, Manila 1099, Philippines, 1995; 2–5.
14. Houghton, J.T.; Callender, B.A.; Varney, S.K., Eds.; *Climate Change*; The Supplementary Report to the IPCC Scientific Assessment; IPCC-Intergovernmental Panel on Climate Change, Cambridge University Press: UK, 1992.
15. Sass, R.L.; Fishe, F.M. Methane Emissions from Texas Rice Fields: A Five-Year Study. In *Climate Change and Rice*; Peng, S., Ingram, K.T., Neue, H.U., Ziska, L.H., Eds.; Springer: New York, 1995; 46–59.
16. Yagi, K.; Tsuruta, H.; Minami, K. Possible Options for Mitigating Methane Emission from Rice Cultivation. Nutrient Cycling in Agroecosystems **1997**, *49*, 213–220.
17. Intergovernmental Panel on Climate Change Revised 1996 IPCC Guidelines for National Greenhouse Gas Inventories, J. T. Houghton, IPCC/OECD/IEA, Paris, France. **1997**, *3*, 4–12.
18. Denier van der Gon, H.A.C.; Neue, H.U. Influence of Organic Matter Incorporation on the Methane Emission from a Wetland Rice Field. Global Biogeochem. Cycles **1995**, *9*, 11–22.

MINELAND RECLAMATION AND SOIL CARBON SEQUESTRATION

Vasant A. Akala
Rattan Lal
School of Natural Resources, The Ohio State University, Columbus, Ohio, U.S.A.

INTRODUCTION

Revegetation of mined lands is one of the management options for mitigation of the negative impacts of mining. The area of minelands that have been reclaimed since 1970 is 1 Mha in the United States alone (1, 2). Land restoration measures can reverse the degradation trends, leading to positive changes in the ecosystem. These include improved water, air, and soil qualities and the attendant socio-economic benefits (3). An important ancillary benefit of mineland reclamation is its potential to sequester soil organic carbon (SOC) (4). The potential for sequestration of carbon (C) in the aboveground biomass and minesoils of reclaimed minelands may be high and merit serious consideration as a significant benefit of restoration of highly disturbed land.

MINESOIL DEVELOPMENT AND SOIL ORGANIC CARBON SEQUESTRATION

The rate of formation of minesoil or minesoil development over time may be the single most important edaphic factor determining the success of the reclamation plan. The rate of minesoil development is a function of reclamation plan management. Freshly exposed mine spoil may be thought of as soil at time zero, in respect to formation, because it consists only of rock and pulverized rock material (5). Mine spoils reflect the properties of the parent material more closely than natural soils because of the initial stage of pedogenic development (6).

Although there are different theories regarding the rate of minesoil development, there is a general consensus that soil weathering is rapid in the early stages of reclamation, and the rate of change decreases over time (7–9). Very rapid chemical reactions occur in spoil as the exposed material begins to come into equilibrium with its new environment (7). Also, minesoils are exposed to processes of physical weathering such as cycles of wetting and drying, freezing and thawing, and mechanical disruption by roots (9). Rapid changes in particle size distribution take place as spoil material weathers (10).

The potential of SOC enhancement in minelands depends on biomass productivity, root development in sub-soil, and changes in minesoil properties resulting from overburden weathering (11). Improvement in soil aggregation is an important factor influencing minesoil development (12, 13). The changes in minesoil properties impact on biomass productivity and pedogenic processes that lead to SOC sequestration in minesoils. Soil processes that are important in the C dynamics of minesoils and those that lead to SOC sequestration are weathering, stable structure formation, erosion, compaction, aeration, aggregation, nutrient recycling, humification, and mineralization. Table 1 shows how some of these processes relate to SOC sequestration in minesoils.

CASE STUDY

Akala and Lal (14) conducted a study on a chronosequence of reclaimed minelands, where the effects of reclamation duration (time), land use (pasture and forest, with and without topsoil application), and soil processes (aggregation and humification) on SOC sequestration were observed. The study showed that soil disturbance by mining resulted in high loss of the antecedent SOC pools and reclamation of mined lands regained SOC. Temporal changes in the SOC pool of reclaimed minesoils showed that the potential to sequester SOC was high. The SOC pool of the soil surface (0–15 cm depth) during the initial phases of reclamation was 10–15 $Mg\,ha^{-1}$, which increased to 45–55 $Mg\,ha^{-1}$ in 25–30 years (Fig. 1). The low SOC pool during the initial phases of reclamation was attributed to the drastic disturbance caused by mining. Soil that was removed and stored, and applied during reclamation lost significant amounts of SOC by decomposition and erosion during mining. Reclamation, minimum perturbation, and management of restored ecosystems,

Table 1 Minesoil processes and soil organic carbon (SOC) sequestration

Processes	Relation to SOC sequestration
Weathering	Formation of soils from the geologic material in the soil reduces the particle size of rock and coarse fragments, leading to the formation of stable peds and aggregates
Erosion	Probably the most important process to be addressed in the initial stages of reclamation because the topsoil is susceptible to erosion. Control of erosion in the initial phases assures establishment of vegetation and favorable impact on environmental quality.
Aggregation	Increasing clay and organic matter content favor the eventual formation of minesoil peds and minesoil structure. The presence and development of minesoil structure by aggregation leads to SOC sequestration. Indirectly, aggregation reduces erodibility and increases pore space, thereby increasing infiltration and permeability, and thus improving minesoil quality.
Compaction	Vehicular traffic involved in implementing the reclamation plan causes compaction. This result in increased density and soil strength and decreased aeration of the re-graded spoil material. As a result of these changes, permeability is reduced and root proliferation and rooting depth are reduced. Water erosion may also be accelerated.

over a period of 20–25 years, had an ameliorative effect on minesoils because of decreases in soil compaction (reduction in soil bulk density), improvement in soil structure (high aggregation), and increases in SOC accumulation (enhanced biomass productivity). The SOC pool for 15–30 cm depth also increased over the period of reclamation. Possible mechanisms of C sequestration in reclaimed mineland were development of an A horizon, increased aggregation through formation of organomineral complexes, and humification of soil organic matter (SOM).

Minesoil aggregation and C sequestration are closely related. Akala and Lal (14) observed a high correlation between water stable aggregates and SOC content. Land disturbance caused by mining not only decreased SOC content of the soil aggregates (secondary organomineral complexes) but also caused a drastic decrease in SOC content associated with the primary soil particles (primary organomineral complexes). The loss in SOC content (in comparison to the antecedent levels in control sites) was 65% in the clay fraction, 75% in the silt fraction, and 40% in the sand fraction. The temporal increase in SOC content in all particle size fractions over the reclamation duration reflected the increase in total SOC pool. The increase (difference in SOC contents between the end and beginning of reclamation period) was 3–6 times the antecedent level for the clay and silt fractions and 1–2 times for the sand fraction. A large increase in the clay fraction showed the reactive nature of minesoils and its ability to respond to changes in SOC pool. The reactive nature of the primary particles coupled with the increase in SOM input was the most important factors for SOC sequestration in the initial stages of mineland reclamation.

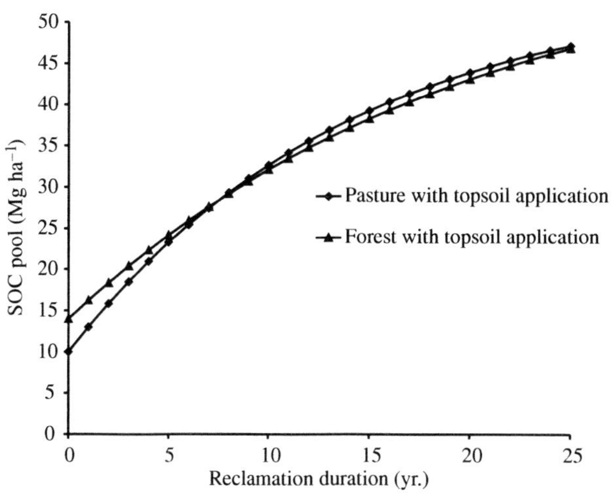

Fig. 1 Change in soil organic carbon (SOC) during reclamation over time.

DISCUSSION

The rates of SOC sequestration in reclaimed minelands can be 2–3 Mg ha^{-1} yr^{-1} for 0–15 cm depth and 1–2 Mg ha^{-1} yr^{-1} for 15–30 cm depth, and are higher than those observed in other land uses. For example, SOC sequestration rates by adoption of conservation tillage are in the range of 0.1–0.5 Mg ha^{-1} yr^{-1} in temperate regions and 0.05–0.2 Mg ha^{-1} yr^{-1} in tropical regions (15). Similarly, SOC sequestration rates are in the range of 0.3–0.5 Mg ha^{-1} yr^{-1} by restoration of wetlands (16). Rates of SOC sequestration reported above can be sustained for a period of 10–15 years. The rate of change of SOC pools or SOC sequestration in reclaimed minelands depends on the magnitude of loss by perturbation (mining) and how far the reclaimed sites are from attaining the equilibrium SOC level. Soils that initially have low SOC content and those that are converted to restorative land management have a potential SOC sink capacity. Reclaimed minelands are an example of such aggrading ecosystems.

The SOC pool that can be potentially attained in reclaimed minelands in the United States is 25 million metric tonnes carbon (15). The net emission of carbon dioxide (CO_2) increased from 1037 Tg (1 Tg = 1 Teragram = 1×10^{12} g = 1 million metric ton) C equivalent in 1990, to 1263 Tg C equivalent in 1996 in the United States (17). Assuming that the reclaimed minelands were at least 25 years old, they had a potential to offset 4% of CO_2 emissions for the United States. The importance of this potential may be realized when considered in combination with the restoration of all degraded soils and ecosystems and in the context of other soil management strategies (15).

CONCLUSION

1. The potential to sequester SOC in reclaimed minelands is high at 2–3 Mg ha^{-1} yr^{-1}.
2. SOC sequestration can be an environmentally friendly use of reclaimed minelands, and long-term C storage in such systems can offset part of the CO_2 emissions (2–4%).
3. The formation of primary organomineral complexes is the first step in the process of SOC sequestration.
4. The total SOC content of the 25- to 30-year-old reclaimed sites can be 30–35% higher than that of the undisturbed sites, but not all of the SOC is sequestered in the secondary organomineral complexes, thereby making it vulnerable to future land disturbance.
5. Soil aggregation and humification in reclaimed minelands can be higher than the undisturbed sites, resulting in high SOC sequestration.
6. Pasture with topsoil application can achieve the highest SOC sequestration in a short period of time.

REFERENCES

1. Office of Surface Mining (OSM). *1998 Annual Report*; Office of Surface Mining, Department of Interior: Washington D.C., 1998.
2. National Mining Association (NMA). *Facts About Coal*; National Mining Association: Washington D.C., 1998.
3. Hossner, L.R., Ed.; *Reclamation of Surface Mined Lands*; CRC Press: Boca Raton, FL, 1991; Vol. I and II.
4. United States Department of Energy (USDOE). *Energy Department Launches Thirteen New Research Projects to Capture and Store Greenhouse Gases*; DOE Fossil Energy Techline: Washinton D.C., 2000.
5. Kohnke, H. The Reclamation of Coal Minesoils. Adv. Agron. **1950**, *2*, 317–349.
6. Sobek, A.A.; Smith, R.M.; Schuller, W.A.; Freeman, J.R.; *Overburden Properties That Influence Minesoils*; NCA/BCR: 1976; 159.
7. Struthers, P.H. Chemical Weathering of Strip Minesoils. The Ohio J. of Sci. **1964**, *64*, 125–131.
8. Caspall, F.C. *Soil Development on Surface Mine Spoils in Western Illinois*; NCA/BCR, 1975; 228.
9. Schafer, W.M.; Nielsen, G.A.; Dollhopf, D.J.; Temple, K. *Soil Genesis, Hydrological Properties, Root Characteristics and Microbial Activity of 1 to 50 Year Old Strip Mine Spoils*, EPA-600/7-79-100; USEPA: Cincinnati, OH, 1979.
10. Van Lear, D.H. *Effects of Spoil Texture on Growth of K-31 Tall Fescue*, Research Note NE-141; USDA Forest Service, 1971.
11. Haering, K.C.; Daniels, W.L.; Roberts, J.A.; Changes in Minesoil Properties Resulting from Overburden Weathering. J. Environ. Qual. **1993**, *22* (1), 194–200.
12. Boerner, R.E.J.; Scherzer, A.J.; Brinkman, J.A.; Spatial Patterns of Inorganic N, P Availability, and Organic C in Relation to Soil Disturbances: A Chronosequence Analysis. Appld. Soil Ecology **1998**, *7* (2), 59–177.
13. Malik, A.; Scullion, J. Soil Development on Restored Opencast Coal Sites with Particular Reference to Organic Matter and Aggregate Stability. Soil Use Manage. **1998**, *14* (4), 234–239.
14. Akala, V.A.; Lal, R. Potential of Mineland Reclamation for Soil Organic Carbon Sequestration in Ohio. Land Degrad. Develop. **2000**, *11*, 289–297.
15. Lal, R.; Kimble, J.M.; Follett, R.F.; Cole, C.V.; *The Potential of U.S. Cropland to Sequester Carbon and Mitigate the Greenhouse Effect*; Ann Arbor Press: Chelsea, MI, 1998; 128.
16. Mitsch, W.J.; Wu, X. Wetlands and Global Change. In *Soil Management and Greenhouse Effect*; Lal, R., Kimble, J., Levine, E., Stewart, B.A., Eds.; CRC Press: Boca Raton, FL, 1995; 205–230.
17. Environmental Protection Agency (EPA). *Inventory of US Greenhouse Gas Emissions and Sinks: 1990–1996*; Environmental Protection Agency: Washington D.C., 1998.

MINERALS, PRIMARY

M

Nikolaos I. Barbayiannis
Vissarion Z. Keramidas
Aristotle University of Thessaloniki, Thessaloniki, Greece

INTRODUCTION

According to the Glossary of Soil Science Terms (1), primary minerals are those that have not been altered chemically since their deposition or crystallization from molten lava. Primary minerals identified in soils belong mainly to the classes of silicates, oxides of Fe, Zr and Ti and phosphates (apatite). They are inherited in soils from the parent material. Their presence is predicated by the type of the parent material and the intensity of weathering, which results in the formation of new phases in soils and the release of elements in the soil solution. Some of these elements are essential for plant life, e.g., K, Ca, Mg, Fe. Therefore, primary minerals play a significant role in soil formation and in the long run they serve as sources of plant nutrients. Primary minerals in soils are usually found in the sand and silt fraction, but certain minerals like quartz, feldspar, and Fe oxides can also be found in the clay fraction.

SILICATES

Of the primary minerals found in soils, silicates are the most common, comprising nearly 40% of the common minerals. The building unit of the silicates is the silicon tetrahedron. Silicate structures may consist of single tetrahedra (nesosilicates), double tetrahedra (sorosilicates), rings (cyclosilicates), single or double chains (inosilicates), sheets (phyllosilicates), or framework patterns (tectosilicates). Typical silicate minerals most likely to be found in soils are presented in Table 1.

Olivines

Olivines are olive-green nesosilicates in which divalent cations mainly Mg^{2+} and Fe^{2+} join the silicon tetrahedra by forming octahedra of Mg^{2+} or Fe^{2+}. The forsterite–fayalite series include the most abundant naturally occurring olivines, with Mg^{2+} and Fe^{2+} being the respective cation of the series end members.

Olivine is the first mineral to crystallize in the initial stages of magma solidification and it is among the most unstable mineral because the high ratio of the divalent cations to silicon in its structure renders it vulnerable to chemical attack. Therefore, olivines are very easily weathered in soils and for this reason are considered relatively rare soil constituents and can be found only in very young soils and in the coarser sand fractions.

Zircon

Zircon is another nesosilicate mineral. Its structure consists of alternating edge-sharing silicon tetrahedra that are held together with Zr^{4+} ions that are located in the center of triangular dodecahedra (3).

Zircon is found as residual grains in the sand and silt fractions of soils and because of its stability in pedogenic environments it is frequently related to the degree of soil development and soil age.

Pyroxenes and Amphiboles

Pyroxenes and amphiboles are "ferro-magnesian" minerals with single and double chain structures, respectively (Table 1). Pyroxene chains have silicon tetrahedra sharing two O atoms. In amphiboles the chains are formed by silicon tetrahedra sharing alternatively two and three O atoms. In both classes of minerals, the chains are held together mainly by divalent cations (Mg^{2+}, Fe^{2+} or Ca^{2+}) located at the center of an octahedron (4).

Augite and hornblende are the most important minerals of the pyroxenes and amphiboles, respectively. Pyroxenes and amphiboles are formed at lower temperatures and pressures during magma solidification compared to olivines and are more stable in the weathering environment. In soils, however, they weather relatively faster and are largely confined in the sand and silt fractions. Some of them may occur in the clay fraction of soils that are young and have not been subjected to intensive weathering (5).

Feldspars

Feldspars are tectosilicates, with three-dimensional framework of corner linked silicon and aluminum tetrahedra. The

Table 1 Common primary minerals of the silicate class found in soils

Silicon tetrahedra arrangement	Name	Ideal formula
Nesosilicates $(SiO_4)^{4-}$ (single)	Olivine	$(Mg,Fe)_2SiO_4$
	Zircon	$ZrSiO_4$
Sorosilicates $(Si_2O_7)^{6-}$ (double)	Epidote	$Ca_2(Al,Fe)Al_2O(SiO_4)_4(O,OH)_2$
Cyclosilicates $(Si_6O_{18})^{12-}$ (rings)	Tourmaline	$(Na,Ca)(Li,Mg,Al)(Al,Fe,Mn)_6(BO_3)_3(Si_6O_{18})(OH)_4$
Inosilicates		
(Single chains) $(SiO_3)^{2-}$	Pyroxenes	
	Augite	$(Ca,Na)(Mg,Fe,Al)(Si,Al)_2O_6(Mg,Fe)SiO_6$
	Hypersthene	$(Mg,Fe)SiO_3$
(Double chains) $(Si_4O_{11})^{6-}$	Amphiboles	
	Hornblende	$(Ca,Na)_{2-3}(Mg,Fe,Al)_5Si_6(Si,Al)_2O_{22}(OH)_2$
Phyllosilicates $(Si_4O_{10})^{2-}$ (sheets)	Micas	
	Muscovite	$KAl_2(AlSi_3O_{10})(OH)_2$
Tectosilicates $(SiO_2)^0$ (framework)	Feldspars	
	Orthoclase	$KAlSi_3O_8$
	Albite	$NaAlSi_3O_8$
	Anorthite	$CaAl_2Si_2O_8$
	Quartz	SiO_2

(From Ref. 2.)

charge deficit created from Al substitution for Si is compensated by Na, K and Ca. Feldspars are virtually present in all sediments and soils in quantities that vary with the nature of the parent material and the stage of weathering. The majority of feldspars belong to the ternary system $NaAlSi_3O_8$ (albite) – $KAlSi_3O_8$ (K-feldspar) – $CaAl_2Si_2O_8$ (anorthite). In K-feldspars one out of every four Si is replaced by Al, with K balancing the charge. Most common K-feldspars are sanidine, orhoclase, microcline and adularia. Sodium and calcium feldspars form a series called "plagioclase" with compositions ranging from pure albite (Na-feldspar) to pure anorthite (Ca-feldspar).

The presence of feldspars in soils is related to the overall mineralogical composition and the prevailing environmental conditions, including climate, topography, degree of leaching, presence of chelating agents, redox status and soil solution composition (6).

Feldspars usually follow the stability sequence: anorthite < albite < K-feldspars (3).

Weathering of plagioclase minerals increases the supply of Ca in soils in a manner similar to that of weathering of K-feldspars in the clay, and silt fractions provide an important source of K in soils.

Micas

Micas are 2:1 phyllosilicate minerals formed by two silicon tetrahedral sheets and an aluminum octahedral sheet in between. One out of four Si in the tetrahedral sheet is replaced by Al and the high negative layer charge developed is balanced by nonhydrated interlayer cations that hold the layers tightly together (7). The most abundant and important micas in soils (muscovite and biotite) have K as the interlayer cation which upon weathering is released in the soil solution and serves as a plant nutrient.

Micas are inherited in soils from the parent material and since they tend to weather to other minerals they are expected to prevail in younger, less weathered soils (8). In young soils they tend to occur as discrete sand and silt particles, while in more weathered soils they are usually found in the clay fraction. Biotite is uncommon in most soils since it is transformed very easily to other secondary minerals, even under mild weathering conditions. Muscovite is commonly found in soils of advanced weathering.

Like the other primary minerals in soils mica becomes unstable in the weathering conditions at the soil surface. This is demonstrated by the fact that biotite is almost absent from the surface horizons of most, except very young, soils. Micas are considered as the precursors of several secondary 2:1 minerals and clay minerals in soil like illites, vermiculites and smectites (8, 9).

Quartz

Quartz is one of the seven polymorphs of SiO_2, found in soils and sediments: quartz, cristobalite, tridymite, coesite, stishovite, lechatelerite (silica glass) and opal (10). Quartz and disordered cristobalite are the typical forms found in soils while tridymite is present mostly in soils developed from siliceous volcanic rocks.

Quartz belongs to the tectosilicate class of the silicate minerals (Table 1), which have each O of silicon tetrahedra linked to Si atoms of adjacent tetrahedra, forming a three-dimensional framework structure. Due to its resistance to weathering, quartz is generally concentrated in the sand and silt fractions, and generally elluvial (surface) horizons are enriched due to weathering and removal of less resistant minerals.

Quartz is considered as inert sceletic material and exhibits very small to nil anion and cation exchange properties. In certain horizons, however, poorly crystalline quartz and other silica minerals may serve as the cementing agent. In normal soils quartz particles are also considered important components of the soil structure because they are linked together or to clay domains through organic matter bridges (11). Quartz is also a suitable mineral to assess parent material uniformity and degree of weathering because of its abundance, resistance to weathering, and immobility.

OXIDES

Iron

Of the several iron oxides and hydroxides found in soils, magnetite (Fe_3O_4), is the only one of primary origin. The crystal structure of magnetite, like all iron oxides, consists of Fe ions surrounded by six O, in a six-fold coordination.

Magnetite occurs as black grains in the heavy fraction of the sand and silt fraction of soils and is of lithogenic origin (12). Since magnetite is found in the coarser soil fractions and is easily transformed to maghemite in the finer fractions, its influence on soil chemistry is minor.

Titanium

Rutile (TiO_2) and ilmenite ($FeTiO_3$) are the main primary titanium minerals found in soils. Rutile is tetragonal and consists of TiO_6 polyhedra that share their opposite edges and form chains along the *c* axis.

Titanium oxides are common accessory minerals of igneous and metamorphic rocks and have been identified in the sand and silt fractions of soils originating from such rocks.

Rutile develops a negative charge through adsorption of OH^- on surface Ti ions. This charge is pH dependent. Charge development plays an important role in sandy acid soils and influences physical and chemical properties such as formation of stable aggregates and retention of nutrients such as phosphate.

Titanium oxides, like zircon, are most frequently related to soil genesis studies. Their occurrence in the nonclay fraction in the different horizons of soils and the parent material provides information about the profile uniformity and soil development.

PHOSPHATES

Phosphate minerals comprise only a small part of the inorganic fraction of soils and their basic structural unit is the orthophosphate ion (PO_4^{3-}), which forms a tetrahedron where P is surrounded by four O^{2-}. The high affinity of PO_4^{3-} for cations, forms structures that are classified as framework, chain and layer phosphates, similar to silicates (13). Apatite [$Ca_{10}(PO_4)_6(Cl,F,OH)_2$] is the most frequently reported phosphate of igneous origin and has been identified with photographic microscopy as discrete grains in the sand and silt fraction of a number of soils. Along with their weathering products, apatites are considered a natural source of P for plants.

REFERENCES

1. Soil Science Society of America, *Glossary of Soil Science Terms*; SSSA: Madison, WI, 1997; 134.
2. Shultze, D.E. An Introduction to Soil Mineralogy. In *Minerals in Soil Environments*, 2nd Ed.; Series 1; Dixon, J.B., Weed, S.B., Eds.; SSSA: Madison, WI, 1989; 1–34.
3. Deer, W.A.; Howie, R.A.; Zussman, J. *An Introduction to the Rock Forming Minerals*; Longman Ltd. Essex, England, 1992; 696.
4. Huang, P.M. Feldspars, Olivines and Amphiboles. In *Minerals in Soil Environments*, 2nd Ed.; Series 1; Dixon, J.B., Weed, S.B., Eds.; SSSA: Madison, WI, 1989; 975–1050.
5. Churchman, G.J. The Alteration and Formation of Soil Minerals by Weathering. In *Handbook of Soil Science*; Sumner, M.E., Ed.; CRC Press: Boca Raton, FL, 2000; F3–F76.
6. Allen, B.L.; Hajek, B.F. Mineral Occurrence in Soil Environments. In *Minerals in Soil Environments*, 2nd Ed.; Series 1; Dixon, J.B., Weed, S.B., Eds.; SSSA: Madison, WI, 1989; 199–278.
7. Fanning, D.S.; Keramidas, V.Z.; El-Desoky, M.A. Micas. In *Minerals in Soil Environments*, 2nd Ed.; Series 1; Dixon, J.B., Weed, S.B., Eds.; SSSA: Madison, WI, 1989; 551–634.
8. von Reichenbach, H.G.; Rich, C.I. Fine Grained Micas in Soils. In *Soil Components. Part 2. Inorganic Com-*

ponents; Gieseking, J.E., Ed.; Springer: Berlin, 1975; 59–95.
9. Borchardt, G. Smectites. In *Minerals in Soil Environments*, 2nd Ed.; Series 1; Dixon, J.B., Weed, S.B., Eds.; SSSA: Madison, WI, 1989; 675–727.
10. Drees, L.R.; Wilding, L.P.; Smeck, N.E.; Senkeyi, Abu L. Silica in Soils: Quartz and Disordered Silica Polymorphs. In *Minerals in Soil Environments*, 2nd Ed.; Series 1; Dixon, J.B., Weed, S.B., Eds.; SSSA: Madison, WI, 1989; 913–974.
11. Emerson, W.W. The Structure of Soil Crumbs. J. Soil Sci. **1959**, *10*, 235–244.
12. Schwertman, U.; Taylor, R.M. Iron Oxides. In *Minerals in Soil Environments*, 2nd Ed.; Series 1; Dixon, J.B., Weed, S.B., Eds.; SSSA: Madison, WI, 1989; 379–438.
13. Lindsay, W.L.; Vlek, L.G.; Chien, S.H. Phosphate Minerals. In *Minerals in Soil Environments*, 2nd Ed.; Series 1; Dixon, J.B., Weed, S.B., Eds.; SSSA: Madison, WI, 1989; 1089–1130.

MINERALS, SECONDARY

M

Deborah A. Soukup
Komex H2O Science, Inc., Bakersfield, California, U.S.A.

INTRODUCTION

There are three possible origins for minerals in soil environments including inheritance from parent materials, weathering transformations of existing minerals, and neoformation or crystallization from solution (1). Secondary minerals are defined as *minerals formed later than the rock enclosing them, usually at the expense of an earlier-formed primary mineral, as a result of weathering, metamorphism, or solution* (2). Alternatively, secondary minerals may be defined as *recrystallized products of the chemical breakdown and/or alteration of primary minerals* (3). Secondary minerals are generally characterized by smaller particle size, because the particle size of primary minerals is decreased during weathering and the release of soluble materials. Therefore, secondary minerals are typically principal components of the silt and clay fraction of soils (4).

TYPES OF SECONDARY MINERALS IN SOIL

The major secondary minerals and the soil orders that they commonly occur in are listed in Table 1. Primary minerals including feldspars, pyroxenes, amphiboles, micas, and primary chlorite may be altered to secondary minerals such as illite, vermiculite, clay chlorite, smectites, kaolinite, halloysite, and oxides of Fe and Al through the removal of silica and bases, and the addition of water. Additional details and the significance of these secondary minerals in soil environments are discussed in the following sections. Selected physical and chemical characteristics of the secondary minerals are summarized in Table 2.

Iron and Aluminum Oxides

The most common oxides of Fe and Al typically reported in soils are goethite, hematite and gibbsite (4–7). These minerals, sometimes referred to as *sesquioxides*, generally occur in soils subject to intense weathering, e.g., Oxisols and Ultisols. The sesquioxide minerals are amphoteric in character and exhibit variable charges. These minerals are also characterized by high surface charge density, high specific surface area, and high cation and anion adsorption capacity (4–7). Additionally, sesquioxide minerals are reported to enhance soil aggregation. Finely divided iron oxides are believed to act as binding agents among other soil particles and may result in cementation of these particles into large units (7).

Even at low concentrations, the iron oxides goethite and hematite play an important role in influencing soil color because of their pigmenting power (7). Goethite minerals are naturally yellow, whereas hematite is bright red. In fact, the greater pigmenting power of hematite can mask the yellow color of higher concentrations of goethite. Gibbsite, in contrast, is colorless; however, its content is used in the oxidic ratio to indicate the relative degree of weathering:

$$\text{Oxidic ratio} = \frac{(\%\ \text{extractable}\ Fe_2O_3 + \%\ \text{gibbsite})}{\%\ \text{clay}}$$

If the oxidic ratio is ≥ 0.2, the soil is considered to be highly weathered.

Kaolinite and Halloysite

Kaolinite is one of the most common clay minerals in soils, particularly those of warm, moist climates. Kaolinite is a 1:1 aluminosilicate mineral composed of one octahedral sheet stacked above one tetrahedral sheet. The two crystal units comprising one kaolinite particle are held together by hydrogen bonds, and the space between the structural layers, therefore, has a fixed dimension (4, 8). Halloysite is also a 1:1 aluminosilicate mineral with the same composition as kaolinite, except that halloysite may be hydrated and contain water between the structural layers. Both kaolinite and halloysite are products of acid weathering, but halloysite is formed more rapidly in soils of volcanic origin.

Kaolinite and halloysite typically have low surface areas and low cation- and anion-exchange capacities. Isomorphous substitution within the crystal is limited, contributing to the low permanent charge. However, kaolinite and halloysite may develop variable or pH dependent negative charge because of the dissociation of protons for exposed OH groups (4, 8). The typical range in

Table 1 Ideal chemical formula of secondary minerals commonly occurring in soils

Mineral name	Chemical formula	Major occurrences
Oxides of Fe and Al		
Goethite	$FeOOH$	Oxisols, Ultisols
Hematite	Fe_2O_3	Oxisols, Ultisols
Gibbsite	$Al(OH)_3$	Oxisols, Ultisols
Aluminosilicate clays		
Kaolinite	$Al_2Si_2O_5(OH)_4$	Ultisols, Oxisols, Alfisols
Halloysite	$Al_2Si_2O_5(OH)_4.\ 2H_2O$	Andisols
Illite	$K(Si_3Al)(AlMg)_2O_{10}(OH)_2$	Alfisols, Mollisols
Chlorite[a]	$[(M^{2+},M^{3+})_3(SiAl)_4(Al_2)O_{10}(OH)_2]^{x-}[(M^{2+},M^{3+})_3(OH)_6]^{x+}$	Common accessory mineral in many soils
Vermiculite	$Mg(SiAl)_4(Al_2)O_{10}(OH)_2$	Common in many soils
Smectite	$Na(Si_4)(AlMg)_3O_{10}(OH)_2$	Vertisols, Mollisols, Alfisols
Hydroxy-interlayer Vermiculite and Smectite	Chemical composition highly variable	Wide geographic distribution but most abundant in Ultisols and Alfisols
Carbonate and sulfate minerals		
Dolomite	$CaMg(CO_3)_2$	Aridisols, Mollisols, Alfisols, arid Entisols
Calcite	$CaCO_3$	Aridisols, Mollisols, Alfisols, arid Entisols
Gypsum	$CaSO_4{\cdot}2H_2O$	Aridisols

[a] M^{2+} and M^{3+} represent cations with charges of +2 and +3, respectively.

Table 2 Selected physical and chemical characteristics of secondary minerals commonly occurring in soils

Mineral name	Surface area (m^2/g^{-1})	Cation exchange capacity ($cmol_c/kg^{-1}$)
Oxides of Fe and Al		
Goethite[a]		3
Hematite[a]		3
Gibbsite[b]		3
Aluminosilicate clays		
Kaolinite	7–30	2–15
Halloysite	10–45	10–40
Illite	80–150	20–40
Chlorite	25–150	10–40
Vermiculite	600–800	100–200
Smectite	600–800	80–150
Hydroxy-interlayer		25–150
Vermiculite and Smectite		CEC varies depending on degree of filling of interlayer space
Carbonate and Sulfate minerals		
Dolomite[c]		
Calcite[c]		
Gypsum[c]		

[a] See Refs. 5 and 7 for additional information.
[b] See Refs. 5 and 6 for additional information.
[c] See Ref. 14 for additional information.

cation exchange capacity (CEC) for kaolinite and halloysite is from 1 to 10 cmol(+)/kg.

Kaolinite-containing clays are used extensively in the production of brick, sewer pipes, and drain tiles. Kaolinite is also used in the ceramic industry, because of its low expansion and contraction capacity. Additionally, kaolinite is used for the production of false teeth, paper making, and in the pharmaceutical industry (4, 8).

Illites

Illites are micaceous clays and have been referred to as *hydrous micas* and *hydrobiotite* (4). However, illite differs from micas, such as muscovite, in that it contains more SiO_2 and less K. Illites are 2:1 aluminosilicate minerals consisting of one octahedral sheet sandwiched between two tetrahedral sheets. The 2:1 layers are bound together strongly by K ions, and thus, illites are non-expansive and do not exhibit shrink–swell behavior upon wetting and drying (4, 9). Illite typically weathers to smectite under conditions of high precipitation, because of the loss of interlayer K.

Vermiculites

Vermiculite is also a 2:1 aluminosilicate mineral with Mg occupying the octahedral positions between the two tetrahedral sheets (4, 10). Vermiculites are common weathering products of micas in well-drained soils and are characterized by high CECs ranging from 144 to 182 cmol(+)/kg (10). These minerals exhibit *wedge zones*, attributed to marginal curling of layers on the mineral surface (4). These wedge zones provide partially enlarged interlayer spacings in which organic matter may be entrapped, or for fixation of K, NH4+, and other cations. The high fixation capacity of many soils for K and NH4+ is primarily attributable to the presence of vermiculite. Vermiculite is also widely used as a potting medium in nurseries and in cat litter.

Smectites

Smectites are a group of expansive, 2:1 aluminosilicate minerals, commonly occurring in soils with impeded drainage and/or in soil environments characterized by high Si and basic cation activities (11). Minerals within the smectite group are classified based upon the location of the layer charge, either within the tetrahedral or octahedral sheets. In montmorillonite, substitution of Mg for Al in the octahedral sheet produces the layer charge. The layer charges of beidellite and nontronite, two other important smectite minerals, are concentrated in the tetrahedral sheets. The predominant octahedral cations in beidellite and nontronite are Al and Fe, respectively (11).

Smectites are generally responsible for the high shrink–swell conditions in soil and result in more damage annually than any other natural disaster including earthquakes and floods. Annual estimates of damage to residential structures in the U.S. were approximately 798.1 million in 1970 and projected to increase to 997 million by the year 2000 (12). These figures increase by a factor of 2–3 if damage to industrial and commercial buildings and transportation infrastructure are included.

Hydroxy-Interlayered Vermiculite and Smectite

Hydroxy-interlayered vermiculite (HIV) and smectite (HIS) are believed to occur in soils as weathering products derived from chlorite or micas or from the deposition of hydroxy-Al or -Fe polymeric components within the interlayer spaces of vermiculite and smectite (13). HIV and HIS are widely distributed geographically and are found in several soil orders, although they are most commonly reported in Ultisols and Alfisols (13).

There is significant variability in the composition of HIV and HIS, because the composition is dependent on the basic 2:1 mineral structure and the type and amount of hydroxy-interlayer material within the interlayer spaces. The presence of hydroxy-Al and -Fe interlayer components may have a profound effect on some of the physical and chemical properties of HIV and HIS, including swelling, CEC, and adsorption of metals and anions. Many investigators have reported that formation of hydroxy-Al and -Fe interlayers resulted in reduced swelling and dispersion of smectites. Hydroxy-Al polymers were more effective than hydroxy-Fe polymers in reducing swelling, presumably because the Al interlayer components are more uniformly distributed within the interlayer space, and their greater relative stability (13).

The CEC of HIV and HIS may also be reduced as the interlayer space is filled. The primary mechanisms proposed for the observed charge reduction in HIV and HIS compared to the noninterlayered end-members are precipitation of Al on surfaces and/or into interlayer spaces, sterically blocking exchange sites, and adsorption of positively charged hydroxy-Al polymers that are non-exchangeable (13). Hydroxy-Al interlayers can also prevent fixation of K^+, NH4+, Cs^+ and Rb^+ because the interlayer material inhibits collapse of the layers about these ions. Finally, the presence of hydroxy interlayers in 2:1 clay minerals significantly enhances the adsorption of anions, such as P (13). The increase in anion adsorption in hydroxy-interlayered clays occurs because anions are only adsorbed at edge sites in minerals not containing hydroxy-interlayers.

Chlorites

Chlorites are hydrated Mg and Al silicates that are commonly green in color and resemble micas in appearance (4, 13). Structurally, chlorite is a 2:2 aluminosilicate mineral consisting of one octahedral sheet and two tetrahedral sheets with the interlayer space occupied by $Mg(OH)_2$ or *brucite layers*. The replacement of Mg by Al in the brucite layers creates enough positive charge to nearly neutralize the negative charges. Thus, chlorite has little or no charge and a small CEC. Chlorites are less stable than most of the other clays in acidic environments and are subject to rapid weathering (4, 13).

Carbonate and Sulfate Minerals

Calcium and magnesium carbonate minerals, including calcite and dolomite, may originate from several sources or combinations of sources, either directly in the form of carbonates, or by a solution-precipitation mechanism (14). Dissolution of more soluble Ca-bearing minerals, such as gypsum, may result in the subsequent precipitation of calcite. Carbonates may also be formed by the reaction of Ca ions in rainwater and surface water with CO_2-charged H_2O in the soil. Coatings of calcium carbonate may eventually accumulate and result in cementation in some soils. Movement of clay in calcareous soils has also been reported to be restricted because of flocculation of the silicate clays by Ca^{2+} (14).

Gypsum may be precipitated in soil from surface or subsurface waters rich in Ca and SO_4. It may also be formed through the oxidation of pyrite and subsequent reaction of acid sulfate with $CaCO_3$ (14). Gypsum has frequently been used to reclaim sodic soils because of its high solubility. However, numerous cases of soil subsidence and deterioration of concrete due to the presence of gypsum have been reported.

SUMMARY AND CONCLUSIONS

In summary, it should be noted that the information presented in this chapter on the formation and characteristics of common secondary minerals in soils is a broad overview that simplifies many relevant details. The actual alteration pathways occurring in the soil environment are very complex and are influenced by a variety of abiotic and biotic factors. The secondary minerals discussed in this chapter may also occur in soils as a result of inheritance from parent materials. Additionally, it may be extremely difficult in some soil environments to determine the precise origin of a given mineral.

REFERENCES

1. Allen, B.L.; Hajek, B.F. Mineral Occurrence in Soil Environments. In *Minerals in Soil Environments*, 2nd Ed.; Dixon, J.B., Weed, S.B., Eds.; Soil Science Society of America: Madison, WI, 1989; 199–278.
2. Bates, R.L.; Jackson, J.A. *Dictionary of Geological Terms*, 3rd Ed.; Doubleday: New York, NY, 1984.
3. Brady, N.C.; Weil, R.R. *The Nature and Properties of Soils*, 12th Ed.; Prentice-Hall: Upper Saddle River, NJ, 1999.
4. Tan, K.H. *Environmental Soil Science*, 2nd Ed.; Marcel Dekker: New York, 2000; 28–79.
5. McBride, M.B. *Environmental Chemistry of Soils*; Oxford University Press: New York, 1994.
6. Hsu, P.H. Aluminum Hydroxides and Oxyhydroxides. In *Minerals in Soil Environments*; 2nd Ed.; Dixon, J.B., Weed, S.B., Eds.; Soil Science Society of America: Madison, WI, 1989; 331–378.
7. Schwertmann, U.; Taylor, R.M. Iron Oxides. In *Minerals in Soil Environments*, 2nd Ed.; Dixon, J.B., Weed, S.B., Eds.; Soil Science Society of America: Madison, WI, 1989; 379–438.
8. Dixon, J.B. Kaolin and Serpentine Group Minerals. In *Minerals in Soil Environments*, 2nd Ed.; Dixon, J.B., Weed, S.B., Eds.; Soil Science Society of America: Madison, WI, 1989; 467–525.
9. Fanning, D.S.; Keramidas, V.Z.; El-Desoky, M.A. Micas. In *Minerals in Soil Environments*, 2nd Ed.; Dixon, J.B., Weed, S.B., Eds.; Soil Science Society of America: Madison, WI, 1989; 551–634.
10. Douglas, L.A. Vermiculites. In *Minerals in Soil Environments*, 2nd Ed.; Dixon, J.B., Weed, S.B., Eds.; Soil Science Society of America: Madison, WI, 1989; 635–674.
11. Borchardt, G. Smectites. In *Minerals in Soil Environments*, 2nd Ed.; Dixon, J.B., Weed, S.B., Eds.; Soil Science Society of America: Madison, WI, 1989; 675–727.
12. Reid, D.A.; Ulery, A.L. Environmental Applications of Smectites. In *Environmental Soil Mineralogy*; Dixon, J.B., Schulze, D., Eds.; in press. Soil Science Society of America: Madison, WI.
13. Barnhisel, R.I.; Bertsch, P.M. Chlorites and Hydroxy-Interlayered Vermiculite and Smectite. In *Minerals in Soil Environments*, 2nd Ed.; Dixon, J.B., Weed, S.B., Eds.; Soil Science Society of America: Madison, WI, 1989; 729–788.
14. Doner, H.E.; Lynn, W.C. Carbonate, Halide, Sulfate, and Sulfide Minerals. In *Minerals in Soil Environments*, 2nd Ed.; Dixon, J.B., Weed, S.B., Eds.; Soil Science Society of America: Madison, WI, 1989; 279–330.

MITES

Hans Klompen
The Ohio State University, Columbus, Ohio, U.S.A.

M

INTRODUCTION

Mites (Acari) form the most diverse lineage of Chelicerata. They are generally small (0.25–1 mm for most taxa) and are characterized morphologically by the division of their bodies. Functionally there are two parts: the gnathosoma, containing the mouth, chelicera, and pedipalps; and the idiosoma, the body proper, containing the legs, eyes (if present), brain, and all remaining body systems. The nymphal and adult instars generally carry the 4 pairs of walking legs characteristic for Chelicerata, but the larval (and prelarval) instars carry only 3 pairs. An excellent review of mite morphology can be found in Evans (1).

The basal division within Acari is between the orders Parasitiformes (= Anactinotrichida) and Acariformes (= Actinotrichida). The former includes the Ixodida (ticks), and the Mesostigmata, a highly diverse lineage which is well-represented in soils. These mites tend to be relatively large, often with very discrete shields. Some Mesostigmata are fungal or pollen feeders, but most retain the ancestral, predaceous life-style. In terms of their impact on soil ecosystems, a basic distinction between Uropodina (slow and often fungivorous) and Gamasina (active and mostly predaceous) is important.

The bulk of mite diversity is in the Acariformes, with at least four rather distinct taxa in soil environments: Astigmata, oribatid mites, "Endeostigmata," and Prostigmata. Astigmata are often associated with ephemeral habitats. They have three nymphal instars, the second of which, the deutonymph, is highly modified for dispersal. It has no mouth, is well sclerotized, and generally carries an attachment organ (sucker plate) on the posterior venter. The other instars are often poorly sclerotized and can feed. The main group of interest for soil biologists are the oribatid mites. Like Astigmata, they are solid feeders, with generally well developed chelate (biting) chelicera. However, oribatid mites are usually well sclerotized in the adults, can live for a long time (1–3 year life cycles are not unusual in temperate regions), have generally low reproductive rates, and disperse poorly (2). In contrast, Astigmata disperse very well, tend to have very high reproductive rates, and live for much shorter periods of time (3).

The "Endeostigmata" form a paraphyletic assemblage of early derivative Acariformes. Poorly sclerotized and with diverse feeding behaviors, from fungivory to predation, these mites are particularly abundant in relatively poor quality habitats, including the deep soil. The final lineage of Acariformes is the Prostigmata sp. This group is enormously diverse in both morphology and life-history. In studies of soil ecosystems it tends to be ignored due to the difficulties in identifying these often very small and poorly sclerotized mites. Yet identification to at least major groups is desirable, as the main lineages, Anystina (including Parasitengona), Eupodina, Raphignathae, and Heterostigmae, are quite different in morphology, life history, and (most probably) ecosystem function. All of these mites are fluid feeders, with chelicerae modified as stylets. Prostigmata includes many fungivorous and predaceous taxa, and a few true plant feeders.

Comprehensive keys to the species of soil inhabiting mites have been published for the former Soviet Union (4). Keys to the families ("Endeostigmata," Prostigmata, oribatid mites) and genera (Mesostigmata, Astigmata) for North America can be found in Dindal (5). Comprehensive keys for soil mites in tropical areas are not yet available.

DISTRIBUTION AND ABUNDANCE

Mites are the most abundant group of arthropods within most soil and litter habitats, including deserts and arctic areas. Numbers of 10,000–500,000/m^2 are normal in temperate and tropical sites (6). Mesostigmata usually form about 20% of the total mite fauna (3,000–4,000/m^2) (6), with abundance and diversity generally being higher in forests than in more open habitats. In temperate regions, the number of Uropodina is generally lower than that of Gamasina, but in the tropics Uropodina abundance can rival that of oribatid mites. Overall, Uropodina are strongly associated with soils of high organic matter content.

Astigmata have often been ignored in studies on soil fauna, but can be common in disturbed sites, such as soil around isolated city streets, and cultivated fields (3, 5). A similar association with disturbance is found for most Heterostigmae (Prostigmata). The contrast with oribatid mites is once again strong. Oribatid mites are generally very common in relatively stable habitats ($10{,}000-300{,}000/m^2$) (6), and usually form the bulk of mite biomass in forest ecosystems. However, many oribatid mites are quite sensitive to soil disturbance, and the fauna of conventionally tilled agricultural fields is generally depauperate. Unlike most other groups of mites, they cannot take advantage of the resource flushes that are characteristic of many agricultural practices (e.g., manure applications). Overall, the group is highly diverse, and most of the 150+ families described do occur in soil and litter habitats.

The "Endeostigmata" are not very diverse, although many species have a broad geographic range. Among the 126+ described families of Prostigmata, about 50 occur mostly in soil environments, but representatives of many other families may be encountered during sampling (7). The number of Prostigmatid and Endeostigmatid mites (these groups are often combined) varies dramatically by general habitat. Numbers vary from $7{,}500/m^2$ in moss turf banks in Antarctica to over $100{,}000/m^2$ in pine forests in Northern Europe (7). Prostigmata plus Endeostigmata may form 40–50% of the total number of mites in deserts and grassland ecosystems, but this proportion drops sharply in temperate forests (less than 10%). The relative proportion of Prostigmata (+Endeostigmata) to oribatid mites is occasionally reported, but this number may not be very informative. The different lineages in Prostigmata are sufficiently diverse; to lump them together into a single value may not be useful.

FUNCTION

The role of mites in soil ecosystems is only partially understood. It is clear that their role in overall energy flow is limited (6), and most research is concentrating more on indirect effects. These include fragmentation of litter (comminution) and soil formation, nutrient cycling, dispersal of microbial spores, and stimulation of the microflora (bacteria and fungi) by grazing. A second level of indirect regulation results from predation on both other microarthropods and nematodes. It is useful to distinguish the two: the first group of processes mostly impact the fungal decomposition pathway (few mites feed directly on bacteria), while the second influences both the fungal and bacterial pathways (8). Most importantly from the standpoint of soil ecosystems, different groups of mites have distinctly different functions, and broad generalizations are nearly always obscuring, rather than illuminating (9).

Comminution, the breaking up of larger organic units such as dead leaves and wood into smaller pieces, is a process that is largely limited to oribatid mites and Astigmata. The other lineages are either predaceous, or feed by puncturing, not breaking up, fungal hyphae etc. Comminution greatly increases the surface areas on which bacteria can complete the actual process of decomposition. Through their feeding and production of faecal pellets, oribatid mites can alter the structure of the soil. As such, they are considered the most important group of arachnids from the standpoint of direct or indirect effects on the formation and maintenance of soil structure (2, 8). Many oribatid mites also sequester calcium and other minerals in their cuticle and thus their bodies may form important "sinks" for nutrients in some nutrient limited environments.

Oribatid soil mites disperse bacteria and fungi, both externally, on their body surface, and internally, by excreting undigested spores. Through this process they can enhance endomycorrhizal colonization (10). Such dispersal may be especially important in the early stages of saprophytic succession (11). Mites grazing on fungi influences both the composition and productivity of the microflora (11). What is still poorly understood are the effects of these actions on decomposition rates (8, 11). Compensatory growth of fungi in response to periodic grazing by mites and Collembola appears to be very important in the field and may significantly increase decomposition rates (11). In addition, comminution of litter and the disturbance of fungi by mite feeding may favor bacterial over fungal growth.

Mites are among the most important predators in the soil ecosystem, attacking most other components of the mesofauna. The two main taxa involved in this process are gamasine Mesostigmata and Prostigmata, but nematode feeding can be found in all taxa discussed (3). Strict specificity for microarthropods (most common among Prostigmata) or nematodes (some "Endeostigmata" and Mesostigmata) occurs, but the majority of mites are generalist predators, feeding on a variety of prey. So far, relatively little is known about the effect of arthropod predation in soil communities. In laboratory experiments (microcosms) predatory mites can suppress nematode numbers (12), but these results have not been translated to more realistic field conditions. In a broader sense, predation may alleviate grazing pressure by nematodes and oribatid mites on the primary decomposers, fungi and bacteria. Excessive grazing decreases, rather than increases, soil respiration and decomposition rates.

M

REFERENCES

1. Evans, G.O. *Principles of Acarology*; University Press: Cambridge, 1992; 563.
2. Norton, R.A. Aspects of the Biology and Systematics of Soil Arachnids, Particularly Saprophagous and Mycophagous Mites. Quaestiones Entomologicae **1985**, *21*, 523–541.
3. Walter, D.E.; Proctor, H.C. *Mites: Ecology, Evolution and Behaviour*; CABI Publishing: New York, 1999; 322.
4. Ghilyarov, M.S., Bregatova, N.G., Eds. *A Key to the Soil-Inhabiting Mites. Mesostigmata*; Zoologicheskogo Instituta Akademii Nauk SSSR: Leningrad, 1977; 1–718.
5. Dindal, D.L., Ed. *Soil Biology Guide*; John Wiley & Sons: New York, 1990; 1349.
6. Petersen, H.; Luxton, M. A Comparative Analysis of Soil Faunal Populations and Their Role in Decomposition Processes. Oikos **1982**, *39*, 288–388.
7. Kethley, J. Acarina: Prostigmata (Actinedida). In *Soil Biology Guide*; Dindal, D.L., Ed.; John Wiley & Sons: New York, 1990; 667–756.
8. Moore, J.C.; Walter, D.E.; Hunt, H.W. Arthropod Regulation of Micro- and Mesobiota in Below-Ground Detrital Foodwebs. Annual Review of Entomology **1988**, *33*, 419–439.
9. Walter, D.E.; Ikonen, E.K. Species, Guilds, and Functional Groups: Taxonomy and Behavior in Nematophagous Arthropods. Journal of Nematology **1989**, *21*, 315–327.
10. Klironomos, J.N.; Kendrick, W.B. Stimulative Effects of Arthropods on Endomycorrhizas of Sugar Maple in the Presence of Decaying Litter. Functional Ecology **1995**, *9*, 528–536.
11. Lussenhop, J. Mechanisms of Microarthropod-Microbial Interactions in Soil. Advances in Ecological Research **1992**, *23*, 1–33.
12. Laakso, J.; Setälä, H. Population- and Ecosystem-Level Effects of Predation on Microbial-Feeding Nematodes. Oecologia **1999**, *120*, 279–286.

MODERN CIVILIZATION, SOILS AND

K.R. Olson
University of Illinois, Urbana, Illinois, U.S.A.

INTRODUCTION

In many parts of the world, contemporary use of soils is not greatly different from ancient use of soils (1). The archeological record shows that some ancient people abused their soils, and that their civilizations were disrupted by the ecological and environmental consequences. Some ancient populations, such as the Middle Mississippian culture near Cahokia, Illinois, were larger in A.D. 1200 than their modern descendent populations (2). The archeological record gives our generation reasons to ponder whether contemporary societies are using their resources wisely (1). People should seek explanations for ancient population declines and shifts so they might avoid similar fates, since an important factor in the continuing prosperity of a nation is the nature of the husbandry of the soil and land resources through centuries of use and occupation.

The Middle Mississippian Indians settled in an area near Cahokia, in Southwestern Illinois (3–5) (adjacent to the current city of St. Louis, Missouri) in approximately A.D. 500. By A.D. 1350 the site was abandoned (2) after peaking in the Lohmann phase (A.D. 1050–1100) and Stirling phase (A.D. 1100–1200) (5–7). During the later years of settlement the population grew to 20,000 people, making it the largest city in the United States until the European settlement of Philadelphia, Pennsylvania exceeded 20,000 people in 1830. The Middle Mississippian people settled on stream and lacustrine deposits in the American Bottoms of the Mississippi River valley.

Previously it would not have been possible to feed such a large population by hunting and fishing alone. The Middle Mississippi Indians supplemented their diet by cultivating maize (corn) and garden crops on silty soils (Fluvaquents) (8) on floodplains. They could then store sufficient food for the winter season to feed the entire population. The people also cut down the adjacent forest for firewood and for making structures and fences. This deforestation caused accelerated erosion of steep, silty soils (Hapludalfs) on nearby hillsides. As a result, wood had to be transported greater and greater distances over time. These Middle Mississippian Indians were also the earthen mound builders. They created thousands of earthen burial and ceremonial (platform) mounds. The earthen mounds were built on clayey soils (Fluvaquents). They dug out the silty, clayey, and sandy stream deposits from the Mississippi River valley and carried the soil materials on their backs in sacks to the top of the mounds (2). Later the original borrow areas were filled to build a large plaza by digging and transporting more soil materials in sacks from borrow areas farther away.

Some of these mounds (almost 800 years old) still exist. The largest mound is called Monks Mound and is still 30 meters high with a rectangular base approximately 320 m by 294 m.

War, fire, food shortages, and disease were the most likely causes of the collapse of the Middle Mississippian civilization, but the final causes are still the subject of considerable debate. It is difficult to understand how the wastes from 20,000 people could be disposed of without polluting the adjacent water and land. Waste disposal problems could have contributed to the contamination of the water supplies, especially the water in the borrow areas. Over time, the demand for maize without crop rotation and nutrient additions could have resulted in soil depletion and reduced crop yields, which could have contributed to food shortages.

Overcutting of forests, overgrazing of pastures and ranges, and the overplowing of fields may have accelerated erosion in many ancient civilizations. Accelerated soil erosion many have diminished the ability of many farmers worldwide to produce satisfactory yields and maintain a reasonable standard of living (9). Tillage utilized in agriculture often accelerates soil loss from water and wind erosion (10). For survival and well-being, governments throughout the world need to respond to the challenge of stabilizing erodible soils.

MODERN CIVILIZATIONS

World demand for food climbs higher each year as a result of both population growth and rising expectations. As a nation's standard of living increases, diets often contain more meat. That change may entail growing feed crops, such as corn and soybeans, which lead to more erosion than small grains and forage (11). In the Third World, farmers are being pushed onto steeply sloping, erodible lands that are rapidly losing their topsoil. Even in the American Midwest, many farmers have abandoned more biologically diverse and soil-conserving rotations that include forage and small grains (oats and wheat) in favor of continuous corn, or a corn and soybean rotation. In other parts of the world, farming has extended into semiarid regions where land is vulnerable to soil loss from wind erosion when plowed (12).

The loss of topsoil affects the ability to grow food by decreasing the availability of nutrients. This results in a degrading of the soil's physical condition, increasing the need for fertilizers and irrigation, which, in turn, increase the cost of production. The loss of topsoil is a quiet crisis in the world economy, because unlike natural disasters such as earthquakes or volcanic eruptions, it is largely a human-made crisis that unfolds gradually (11).

Soil degradation is more severe in sub-Saharan Africa than anywhere else in the tropics (12). This is where soil degradation has severely reduced the soil's potential capacity to produce food, feed, and fiber. This is a result of physical, biological, and chemical deterioration due to erosion, desertification, laterization (formation of a stone-like layer high in iron), salt accumulation, excessive leaching and acidification, and nutrient imbalance.

Prior to European settlement, many of the northcentral U.S. soils were developed under prairie grass on nearly level plains and under forest on rolling hillslopes. European settlement and agricultural technology had significant impacts on northcentral U.S. soils (Table 1).

In the Dust Bowl of the 1930s, severe droughts in the Great Plains of the United States resulted in crop failures and major dust storms. In 1935, with dust settling in the congressional offices in Washington, D.C., Congress passed a law, creating the Soil Conservation Service to address the erosion problem. At that time, soil erosion of row cropland became so severe, and farm debt became so high, that lands were abandoned in many states such as Illinois, Indiana, Kentucky, Tennessee, Oklahoma, and Missouri. Numerous research stations and demonstration farms were established in the northcentral United States to test and disseminate information on how to restore soils and improve crop yields and farm income on hilly, highly eroded soils. Management techniques that helped achieve these goals included the removal, crushing, and application of limestone from nearby hills to highly eroded soils with acid subsoils. Rock phosphate was added to the phosphorus-poor soils and forage crops were seeded as pasture for beef and sheep, reducing the need for yearly tillage. Gullies were filled on the steep hillslopes as well.

In the 1950s, nitrogen fertilizer became available at low cost and the increasing acreage planted with soybeans and corn increased erosion. In the 1960s, the moldboard plow was partially replaced with a chisel plow, which reduced soil loss by leaving plant residue on the surface (conservation or mulch tillage). In the 1970s, no-till planting of limited acreages of corn and soybean further reduced tillage and soil erosion, but often with increased use of herbicides. In the 1980s, ridge tillage, which left residue between the rows at planting to reduce erosion, became more popular. Row cultivation was used to control weeds between the rows and to maintain ridges.

The U.S. Food Security Act of 1985 resulted in millions of hectares of highly erodible lands being put into forages

Table 1 Agricultural technology impacts of European settlement on northcentral U.S. soils

Timeline	Agricultural technology events related to European settlement
1814	The introduction of the cast-iron plow made it possible to plow the forest land but not the wet, sticky prairie soils. Consequently, settlement in the early 1800s along the Ohio, Mississippi, and Missouri Rivers resulted in the clearing of forest and cultivation of sloping land which was easily eroded.
1837	The John Deere Company started mass producing a steel plow which could be used to plow the prairie lands. Shortly thereafter, some of the agriculture from the highly eroded forest soils shifted or moved to the prairie soils.
1850s	The steam engine was introduced which helped with the wheat harvest and increased wheat production. In 1879 the U.S. Drainage Act and Levee Act allowed the drainage (initially surface ditches and later tile systems) and farming of swampy prairie areas in states like Illinois.
1890–1950	Gasoline tractors replaced horses resulting in a decline in the acreages of oats and hay.
1920s	Agronomic research resulted in better oil seed varieties of soybean. Crop rotations which had previously included corn planted in rows approximately 1 m wide, drilled wheat and oats in very narrow rows approximately 7.5 cm wide, and perennial hay was replaced by an intensive rotation of corn and soybeans. Over time the soybeans were grown in wide rows for grain instead of being closely grown for forage; consequently, accelerated soil erosion occurred.

or trees for 10 years or more, which further helped reduce soil movement from row cropland. Many states developed conservation programs in the 1980s that required soil loss to be reduced, over a specific time, to the tolerable soil loss level, which ranges from 2 to 11 metric tons per hectare per year.

EFFECTS OF AGRICULTURE ON SOILS

Sheet and rill erosion rates on cropland have been declining in the United States (13). Under current production agricultural systems, however, 23% of U.S. cropland is still eroding at rates higher than the tolerable soil loss (14, 15). The effects of erosion on cropland soils depend, in part, on the characteristics of the soils and the landscapes in which they occur. Even under favorable climatic conditions, topography, and parent materials, soil formation is a slow process, requiring hundreds of years to develop topsoils and subsoils. Recently adopted conservation management improvements, advances in conservation tillage technology, and land-use changes have contributed to the sheet and rill erosion-rate reductions; however, there still can be a net soil loss and reduced long-term soil productivity even under no-till systems. When considered in the context of long-term productivity, erosion rates alone are not the only indicator of soil degradation. Degradation of soil structure and aggregation or loss of irreplaceable soil attributes, is much more serious in certain soils than others when compared at the same erosion rates. Documented on-farm and off-site damages include loss of soil; sedimentation in streams, reservoirs, and fields; loss of nutrients; soil property changes; soil productivity losses; and air, soil, and water pollution.

Current government programs either provide incentives for or require soil loss to be held within tolerable levels. On average, total soil loss from wind and water activity has been declining, in part, as a consequence of government programs. However, erosion and sediment deposition are locally and regionally significant and, thus, still perceived by the scientific and nonscientific communities as a serious threat to our economy and environment. One of history's greatest problems has been the cultivating and erosion of sloping lands. Continued productivity losses on these soils would result in long-term costs to society since this natural resource (capital) would eventually be depleted. Sustaining the long-term productivity and quality of all soil resources needs to be a national priority and is essential to our nation's prosperity. It is in the national interest that management strategies and techniques continue to be developed to monitor and sustain soil productivity and quality.

EFFECTS OF MODERN URBANIZATION AND INDUSTRIALIZATION ON SOILS

For the past 200 years, as European settlement increased in the United States, the disturbance and alteration of the soils has accelerated. Many millions of hectares in the urban landscape have been altered or destroyed for buildings, transportation, or mineral extraction. Vast areas of soils near urban centers have been excavated or disturbed for the construction of buildings, roads, railroads, cemeteries, canals, utility lines, and landfills. In addition, there has been surface and subsurface mining for various minerals, including coal, limestone, clay, sand, and gravel; and waste disposal. Topsoil and even subsoils are often removed and sold from constructions sites. Topsoil has become a commodity that is of value and often transported and sold to the highest bidder. In the urban landscape, soils (often both topsoil and subsoil) have been removed, exposing the dense parent materials, which are often built on and then landscaped using a thin layer of sod. This sod has difficulty growing roots into the dense parent material. Often, soils are no longer present and cannot accept or hold the rainwater on the landscape. Runoff is much greater than it was prior to the alteration. The lack of soil water storage and the covering of other soils in the landscape by roofs, sidewalks, driveways, and roads has resulted in more runoff and increased flooding. As humankind has tried to clean up the air and water, the industrial and urban pollutants, urban and industrial waste, and sewage sludge have either been consolidated or buried under the soil materials, or transported and spread on the soils as amendments. Fortunately, environmental regulations no longer allow urban and industrial waste and sewage sludge to be transferred to the air, rivers, lakes, or oceans.

REFERENCES

1. Olson, G.W. *Soils and the Environment—A Guide to Soil Surveys and Their Applications*; Chapman and Hall: New York, 1981.
2. Fowler, M.L. *The Cahokia Atlas: A Historical Atlas of Cahokia Archaeology*, Revised Edition; University of Illinois at Urbana: Champaign, 1997.
3. Fowler, M.L. *The Cahokia Atlas: A Historical Atlas of Cahokia Archaeology*, Studies in Illinois Archaeology No. 6.; Illinois Historical Preservation Agency: Springfield, 1989.

4. Milner, G.R. The Later Prehistoric Cahokia Culture System of the Mississippi River Valley: Foundations, Florescence, and Fragmentation. Journal World Prehistory **1990**, *4*, 1–43.
5. Pauketat, T.R. *The Ascent of Chiefs: Cahokia and Mississippian Politics in Native North America*; University of Alabama Press: Tuscaloosa, 1994.
6. Bareis, C.J.; Porter, J.W. *American Bottom Archaeology: A Summary of the FAI-270 Project Contributions to the Culture History of the Mississippi River Valley*; University of Illinois Press: Urbana, 1984.
7. Hall, R.L. Cahokia Identity and Interaction Models of Cahokia Mississippian. In *Cahokia and the Hinterlands: Middle Mississippian Cultures of the Midwest*; Emerson, T.E., Lewis, R.B., Eds.; University of Illinois Press: Urbana, 1991.
8. Soil Survey Staff, *Soil Taxonomy, a Basic System of Soil Classification for Making and Interpreting Soil Surveys*, AH-436; U.S. Gov. Print. Office: Washington, DC, 1975.
9. Lowdermilk, W.C. *Conquest of the Land Through Seven Thousand Years*, Agricultural Information Bulletin No. 99, 2nd edition; 1975.
10. Troeh, F.R.; Hobbs, J.A.; Donahue, R.L. *Soil and Water Conservation for Productivity and Environmental Protection*; Prentice-Hall: Englewood Cliffs, New Jersey, 1980.
11. Brown, L.R.; Wolf, E.C. *Soil Erosion: Quiet Crisis in the World Economy*; Worldwatch Institute: Washington, DC, 1984.
12. Lal, R. Soil Degradation and the Future of Agriculture in Sub-Saharan Africa. J. Soil Water Conserv. **1988**, *43*, 444–451.
13. Lee, L.K. Land Use and Soil Loss: 1982 Update. J. Soil Water Conserv. **1984**, *39*, 226–228.
14. USDA Staff. *Soil and Water Resources Conservation Act Program Report and Environmental Impact Statement*; U.S. Govt. Print. Office: Washington, DC, 1981.
15. USDA Staff. *Preliminary Data, 1982*; National Resources Inventory, Executive Summary: Washington, DC, 1983.

MULCH FARMING

Rattan Lal
The Ohio State University, Columbus, Ohio, U.S.A.

INTRODUCTION

Mulching, the practice of leaving crop residues and other biomass on the soil surface to protect it from climatic elements, is an ancient practice. Its application, however, is even more relevant now than ever before because of the serious global issues relevant to human population growth, scarcity of prime agricultural land, ecological benefits of nutrient recycling and optimizing rates of fertilizer input, severe problems of soil degradation, adverse water quality, and the accelerated greenhouse effect. It is an ecological approach to addressing the problem of sustainability, soil degradation, environment quality, and nutrient cycling. Conventional agricultural practices, based on clean cultivation and intensive use of inorganic fertilizers and chemical amendments, make soils susceptible to degradative processes, e.g., erosion, acidification, and emission of greenhouse gases to the atmosphere. Stabilizing the atmospheric concentration of greenhouse gases is a major global challenge (1). The global release of soil organic carbon (SOC) from agriculture is estimated at 800 Tg C/yr ($T = tera = 10^{12} = 1$ million metric ton) (2). Soilrestoration (for enhancement of SOC content) is an important and feasible option for mitigating the greenhouse effect (3–5), and enhancing soil quality for increasing production and improving the environment (6, 7). These desirable properties and processes are achievable through frequent applications of crop residue mulch and other mulch farming techniques.

MULCH TYPES

Mulch is any material, other than soil, specifically established at the soil–air interface to manage soil and water and create favorable environments for plant growth. Most mulches are organic materials of plant, origin comprising crop residues, weed biomass, leaf litter, and by-products of agroindustries, e.g., saw dust, rice husk, corn cobs, etc. However, there is a wide variety of mulch materials ranging from gravels and plastic to crop residues and planted fallows. Mulches can be broadly classified into two categories: organic and inorganic mulches (Fig. 1).

1. *Organic mulches.* Most mulch materials used in agriculture are of organic origin, and comprise crop residue used both ex situ and in situ.
2. *Inorganic mulches.* Three most common inorganic mulches are plastic, gravels, and water.

MULCH PROCUREMENT FOR ARABLE LAND

Crop Residues

The most common practice of mulch procurement for use on agricultural land is the use of crop residues from the previous season's crops. Crop residue is a renewable resource, and annually a large quantity of residues are produced in the world (8). Residues produced in the world are estimated at 3460 million Mg, of which 1600 million Mg (46%) are produced in developed countries and 1860 million Mg (54%) in developing countries (Table 1). Properly used as mulch, these residues can cover the entire arable land of 1500 million ha at an average rate of 2.3 Mg/ha. Therefore, crop residues are an important resource that need to be carefully and judiciously used for sustainable use of soil, water and other natural resources.

Crop Residues Management

Management of crop residues is an important strategy of mulch procurement. Crop/plant residues, produced by annuals or perennials, in situ or ex situ, used as mulch can improve soil quality, increase productivity and sustainability, and enhance environmental quality. Principal management alternatives outlined in Fig. 2 indicate four options: 1) mulching, 2) using as animal feed, 3) composting, and 4) burning. Using residues as mulch is the best option for controlling soil erosion, conserving soil water, and at the same time replenishing plant nutrient reserves.

Planted Fallows and Cover Crops

Cover crops, legumes or grasses generally grown to produce biomass and provide ground cover, are a good source of mulch. Suitable cover crops are those that: 1) establish rapid

Encyclopedia of Soil Science

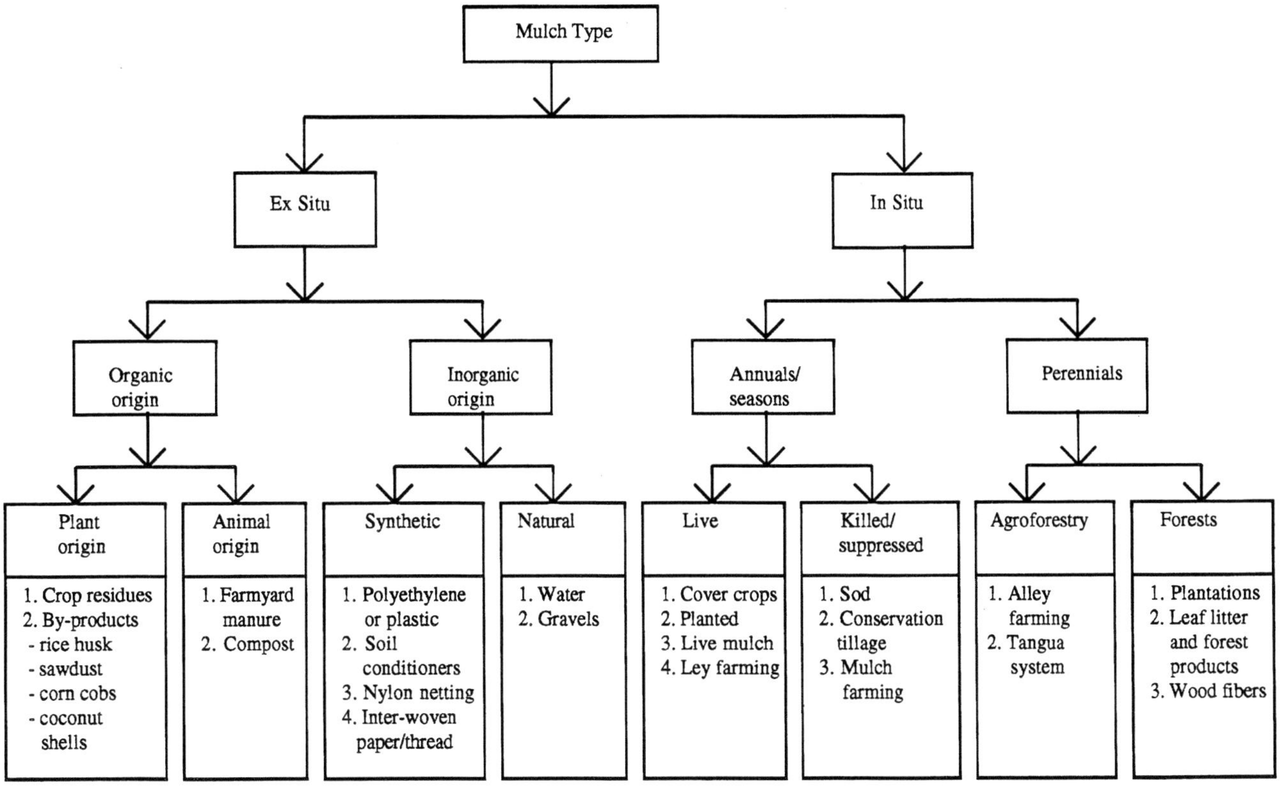

Fig. 1 Different types of mulch materials. (From Ref. 16.)

ground cover, 2) improve soil fertility through biological nitrogen fixation and recycling nutrients, 3) suppress weeds and other pests, and 4) produce a large quantity of biomass that can be used as mulch. Selected legumes and grasses, appropriate for the soil and environment, can ameliorate soil properties and produce the mulch material in a short time interval (9). Deep rooted cover crops are also effective in improving subsoil properties.

BENEFITS OF MULCH FARMING

Mulch farming is an ecological approach to agriculture. Properly used, it can improve and sustain agricultural production, enhance soil quality, and have minimal adverse effects on the environment (Fig. 3). Principal benefits of mulch farming are briefly described as follows.

Soil-Surface Management and Erosion Control

Soil erosion by water and wind is the principal degradative process (10), and it adversely affects soil quality (11, 12). Mulching, at the rate of 4–6 Mg/ha, is an effective erosion-control measure (13). Larson (14) recommended use of crop residue mulch for erosion control in the mid-western U.S.A. Moldenhauer and Wischmeier (15) observed significant reduction in soil erosion by mulch farming practices. Both runoff and soil erosion decrease exponentially with increase in surface area covered by mulch (Fig. 4). Structural improvement by mulching are due to increase in SOC or humus content that facilitates formation of organomineral complexes, and increase in soil biodiversity, notably the activity of earthworms, termites, and other beneficial soil fauna (16).

Water Management

There are four types of water: 1) surface water, 2) groundwater, 3) soil water, and 4) rainwater. Mulching affects all four types of water, but has a major impact on soil water and surface water. Mulching decreases runoff by improving infiltration rate, replenishes soil water by improving retention pores, and recharges groundwater by increasing percolation. Partitioning of rainwater into runoff, soil water and groundwater components is influenced by

Table 1 Estimates of crop residues produced by cereals, legumes, and other crops

Crop	Residues produced (10^6 ton/yr) U.S.A.	World
A. Cereals		
Maize	195.0	479.0
Wheat	81.0	826.5
Sorghum	23.0	87.0
Barley	15.0	253.5
Rice	11.0	780.0
Oats	4.0	34.0
Millet	0.3	43.5
Rye	0.3	40.5
Others	0.6	18.0
Subtotal	325.2	2562.0
B. Legumes		
Beans	1.5	23.0
Groundnut shells	2.2	23.0
Lentils	0.1	2.0
Peas	0.2	16.0
Soybeans	54.0	103.0
Pulses	–	71.0
Subtotal	58.0	238.0
C. Oil crops		
Linseed	0.2	3.0
Sesame	–	4.0
Rapeseed	0.2	40.5
Safflower	0.2	1.0
Seed cotton	15.0	90.0
Sunflower	23.0	23.0
Subtotal	17.1	161.5
GRAND TOTAL	400.3	2961.5

(From Ref. 8.)

mulching through its effect on the water balance. Mulching increases soil water and groundwater components by decreasing runoff and soil evaporation. Mulching improves soil water storage, and minimizes risks of drought. When the soil is wet at about field capacity level, as is the case soon after rainfall or irrigation, mulching decreases the evaporation rate and prolongs the duration during which soil remains moist (Fig. 5).

Soil Temperature Management

Crop residue mulch has a buffering or moderating effect on soil temperature regime. It influences the extremes by decreasing the maximum and increasing the minimum soil temperature. Therefore, soil temperature range is narrower in mulched than unmulched soil. Mulching increases soil temperature in the morning and decreases it in the afternoon (Fig. 6). Plastic mulches are often used to increase soil temperature in horticultural and other high-value plants. Soil temperature effects of mulching are both due to direct and indirect effects. Direct effects are due to reduction in insolation or energy load on the soil surface. Mulching shades the soil and prevents radiation reaching the soil surface. Indirectly, mulching influences soil temperature through its effect on soil's heat capacity and thermal conductivity.

Nutrient Cycling and Soil Fertility Enhancement

Severe nutrient depletion in sub-Saharan Africa (17, 18) and elsewhere can be controlled through nutrient cycling. Depending on plant species and management systems, crop residues contain a considerable amount of plant nutrients. Total amount of plant nutrients in crop residues range from 40 to 100 kg/Mg of residues (8). On a global scale, crop residues contain 22.6 million Mg N, 3.6 million Mg P, 47.4 million Mg K (Table 2). Estimates of global fertilizer consumption show annual use of 113 million tons of NPK vs. 74 million Mg contained in crop residues. In comparison, annual fertilizer consumption is 16 million Mg of NPK in U.S. vis-à-vis 9 million Mg contained in crop residues. It is feasible, therefore, that a considerable quantity of fertilizer can be saved by returning crop residues to the soil. Indeed, mulching and residues management are effective nutrient recycling mechanisms.

Soil Biodiversity

Applications of crop residues mulch improve activity of soil micro- and macrofauna and biomass carbon. The relative magnitude of biomass carbon is a good indicator of soil quality. Mulching improves population and activity of soil fauna especially those of earthworms and termites. These organisms improve soil structure by burrowing, mixing and soil turnover activities. They decompose and mineralize crop residues and make nutrients available to plants.

The Greenhouse Effect

The atmospheric CO_2 concentration has increased by about 30% from 280 ppm in 1850 to about 370 ppm in 2000 (19, 20). This increase is attributed to two principal

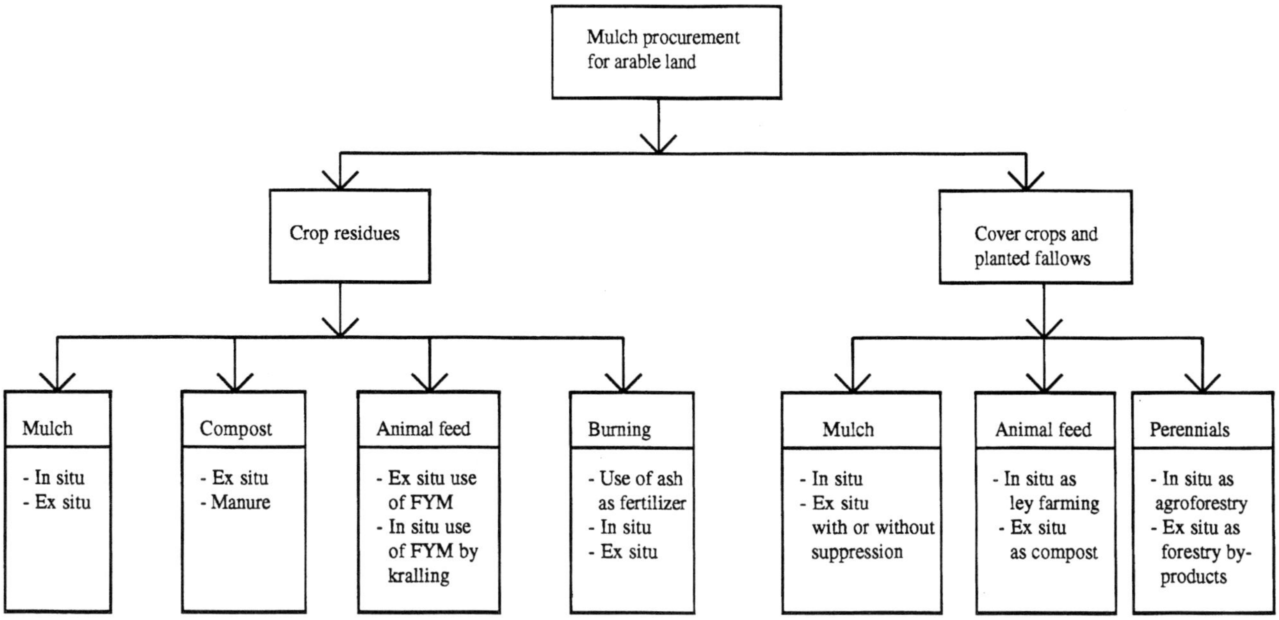

Fig. 2 Methods of procurement of mulch materials.

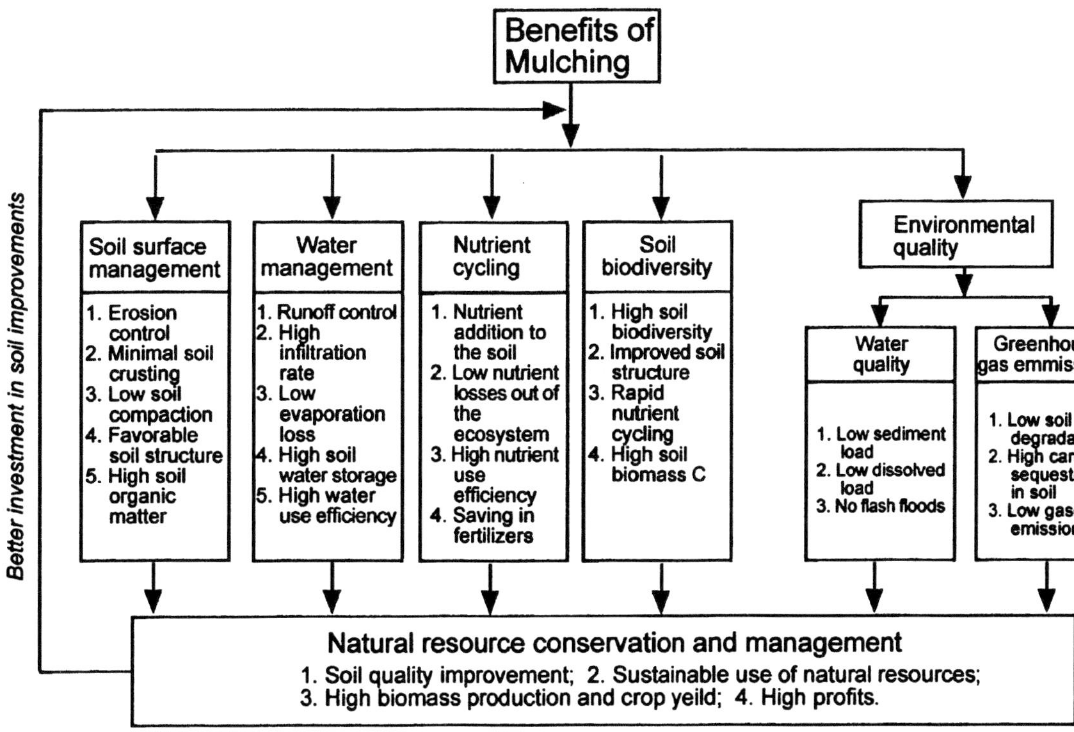

Fig. 3 Economic, ecologic, and natural resource improvements by mulch farming.

human activities: 1) land use change and soil management (1.6 Pg C/yr; 1 Pg = peta gram = 10^{15} g), and 2) fossil fuel consumption (6.8 Pg C/yr). The rate of increase of atmospheric CO_2 concentration is about 1.5 ppm or 3.5 Pg C/yr. Total crop residues production in the world estimated at 3460 million tons per annum (8) contains

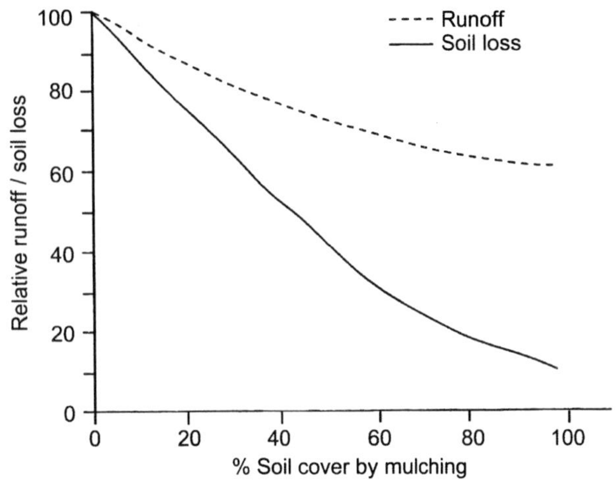

Fig. 4 A schematic depicting an exponential decline erosion with increasing soil cover by mulching.

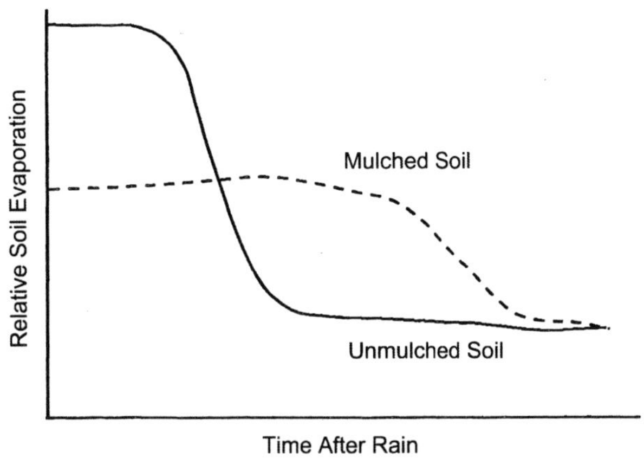

Fig. 5 A schematic depicting mulch effects on soil.

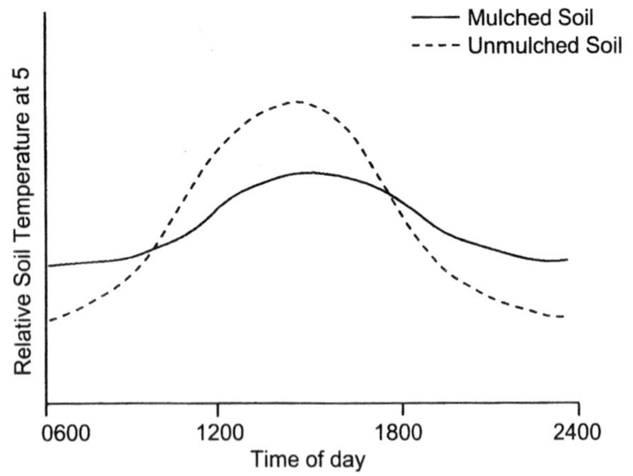

Fig. 6 A schematic showing diurnal fluctuations in mulched and unmulched soil during summer.

Table 2 Nutrients contained in crop residues of cereals and legumes produced in the world

	Plant nutrients in residues (million tons/yr)		Fertilizer use (million tons/yr)	
Nutrient	U.S.A.	World	U.S.A.	World
N	2.98	22.62	10	77
P	0.47	3.58	2	16
K	5.70	47.39	4	20
Ca	1.87	12.11	–	–
Mg	0.85	6.16	–	–
N+P+K	9.15	73.59	16	113

(From Ref. 8.)

1.5 Pg of C. This is about 43% of the annual atmospheric increase of 3.5 Pg/C, and cannot be ignored in the context of global C cycle. This C can have a major effect even if only 10% of the 1.5 Pg can be sequestered in soils as humus. In addition to the direct effect, mulching also sequesters C through erosion control, and increase in biomass production (21).

MULCHING AND AGRICULTURAL SUSTAINABILITY

Mulch farming affects agricultural sustainability through its impact on soil quality and productivity (Fig. 7). Mulch-induced changes in soil quality set-in-motion restorative processes leading to soil and water conservation, nutrient cycling and soil fertility improvement, improvement in soil structure and C sequestration, increase in soil biodiversity and soil structure enhancement. Mulch farming improves energy use efficiency by 1) reducing fertilizer use through decreasing nutrient losses in runoff and erosion, and recycling plant nutrients, 2) decreasing irrigation needs through increasing infiltration, and reducing losses due to runoff and evaporation, 3) decreasing herbicide use through suppressing weed growth, and 4) decreasing fuel consumption through reduction of tillage operations and vehicular traffic.

Crop residues are also a renewable source of energy that can be used in the farm household. Energy content of straw, depending on crop species and other factors, ranges from 3000 kcal/kg to 4000 kcal/kg (22). Energy value of 1 Mg of residues may be as much as 9 million BTU,

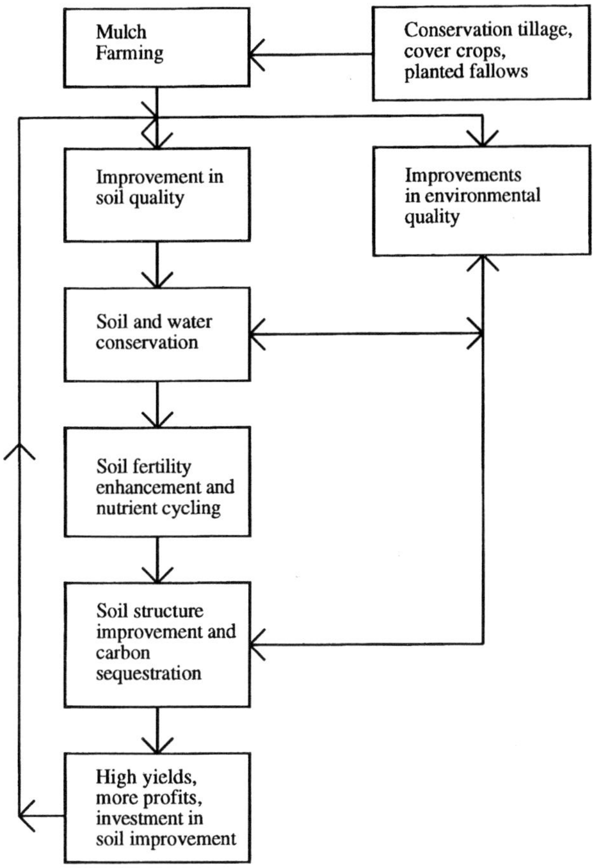

Fig. 7 Mulch farming and agricultural sustainability.

18.6×10^9 J, or equivalent to two barrels of oil (23). Estimates of fuel energy in crop residues show that cereal straw produced annually in the world is equivalent to 47×10^{18} J of energy (Table 3) (8).

Table 3 Fuel energy of cereal straw produced in the world

Crop	Energy equivalent (10^{18} J)
Barley	4.7
Maize	8.9
Millet	0.8
Oats	0.6
Rice	14.5
Rye	0.8
Sorghum	0.8
Wheat	15.4
Total	47.3

(From Ref. 8.)

LIMITATION OF MULCH FARMING

Biomass, the principal mulch material, has numerous alternative uses especially for resource-poor farmers of the tropics. Crop residues are used for feeding livestock, for building fences and homesteads, and as fuel for cooking and household energy use. It is a precious commodity, and little if any is left behind on the field as mulch on small farms in developing countries. Under some circumstances, mulches may also have negative effects on crop growth due to cool soil temperature during spring in temperate regions, high incidence of pests and pathogens, and N immobilization.

CONCLUSIONS

Mulch farming maintains or enhances soil quality by conserving soil and water resources, improving SOC content, increasing soil biodiversity, strengthening nutrient cycling mechanisms, and regulating soil temperature regime. Mulch farming increases probability of achieving agricultural sustainability through: 1) increasing production, 2) decreasing costs of fertilizers and water inputs, and 3) preserving the resource base and its productive potential.

Mulch farming has a potential to improve environmental quality by: 1) decreasing soil erosion and transport of chemicals in runoff, mulch farming decreases risks of eutrophication of surface waters and contamination of groundwater, and 2) reducing emission of radioactively active gases from soil to the atmosphere through increasing SOC content, stabilizing C within aggregates as organomineral complexes, decreasing losses due to soil erosion, and improving biomass production.

REFERENCES

1. Engelman, R. Stabilizing the Atmosphere: Population, Consumption, and Greenhouse Gases. *Population and Environment Program*; Population Action International: Washington, DC, 1994; 48.
2. Schlesinger, W.H. Evidence from Chronosequence Studies for a Low Carbon-Storage Potential of Soils. Nature **1990**, *348*, 232–234.
3. Lal, R., Kimble, J.M., Levine, E., Stewart, B.A., Eds. *Soils and Global Change*; CRC/Lewis Publishers: Boca Raton, FL, 1995; 440.
4. Lal, R., Kimble, J.M., Levine, E., Stewart, B.A., Eds. *Soil Management for Mitigating the Greenhouse Effect*; CRC/Lewis Publishers: Boca Raton, FL, 1995; 385.

5. Lal, R. Soil Management and Restoration for C Sequestration to Mitigate the Accelerated Greenhouse Effect. Prog. Environ. Sci. **1999**, *1*, 307–326.
6. Doran, J.W.; Parkin, T.B. Defining and Assessing Soil Quality. *Defining Soil Quality for a Sustainable Environment*; SSSA Special Publication No. 35, SSSA: Madison, WI, 1994; 3–21.
7. Lal, R. *Soil Quality and Agricultural Sustainability*; Lal, R., Ed.; Ann Arbor Press: Chelsea, 1998; 3–11.
8. Lal, R. The Role of Residues Management in Sustainable Agricultural Systems. J. Sustain. Agric. **1995**, *5*, 51–78.
9. Smith, M.S.; Frye, W.B.; Varco, J.J. Legume Winter Cover Crops. Adv. Soil Sci. **1987**, *7*, 95–140.
10. Oldeman, L.R. The Global Extent of Soil Degradation. In *Soil Resilience and Sustainable Land Use*; Greenland, D.J., Szabolcs, I., Eds.; CAB International: Wallingford, UK, 1994; 99–118.
11. Lal, R. Applying Soil Quality Concepts for Combating Soil Erosion. In *Soil Quality and Soil Erosion*; Lal, R., Ed.; CRC Press: Boca Raton, FL, 1999; 309–318.
12. Lal, R.; Mokma, D.; Lowery, B. Relation Between Soil Quality and Erosion. In *Soil Quality and Soil Erosion*; Lal, R., Ed.; CRC Press: Boca Raton, FL, 1999; 237–257.
13. Lal, R. *Soil Erosion in the Topics: Principles and Management*; McGraw-Hill: New York, 1990; 580.
14. Larson, W.E. Protecting the Soil Resource Base. J. Soil Water Conserv. **1981**, *36*, 13–16.
15. Moldenhauer, W.C.; Wischmeier, W.H. Soil and Water Losses and Infiltration Rates on Ida Silt Loam as Influenced by Cropping Systems, Tillage Practices, and Rainfall Characteristics. Soil Sci. Soc. Am. Proc. **1960**, *24*, 409–413.
16. Lal, R. Soil Structure and Sustainability. J. Sustain. Agric. **1991**, *1*, 67–92.
17. Stoorvogel, J.J.; Scaling, E.M. *Assessment of Soil Nutrient Depletion in Sub-Saharan Africa:* 1983–2000; Report No. 28, The World Staring Center: Wageningen, The Netherlands, 1990.
18. Stoorvogel, J.J.; Smaling, E.M.A.; Janssen, B.H. Calculating Soil Nutrient Balances in Africa at Different Scales. I. Supranational Scale. Fertil. Res. **1993**, *35*, 227–235.
19. IPCC, *Technical Summary*; Inter-Governmental Panel on Climate Change, WMO: Geneva, Switzerland, 1995; 44.
20. Kennel, C. *UC Revelle Program on Climate, Science and Policy*; Scripps Institute of Oceanography: La Jolla, CA, 2000.
21. Lal, R. Residue Management, Conservation Tillage and Soil Restoration for Mitigating Greenhouse Effect by CO_2^- Enrichment. Soil Tillage Res. **1997**, *43*, 81–107.
22. Stout, B.A. *Handbook of Energy for World Agriculture*; Elsevier: New York, 1990.
23. Anderson, A.W.; Anderson, J.F. On Finding a Use of Straw. In *Utilization and Recycles of Agricultural Wastes and Residues*; Shuler, M.L., Ed.; CRC Press: Boca Raton, FL, 1980; 237–278.

NEMATODES

Christien H. Ettema
Wageningen University, Wageningen, The Netherlands

INTRODUCTION

Nematodes are generally microscopic animals, belonging to the phylum Nematoda. They are very numerous, representing 80% of all multicellular animals. Nematodes live in soil, freshwater and marine environments. In addition, they occur as parasites of vertebrate (including humans) and invertebrate animals (such as insects).

Soil nematodes live in water-filled pore spaces or water-films surrounding soil particles. A handful of soil contains thousands of individuals, dozens of species, and several feeding groups. Plant disease caused by nematodes results in billions of dollars lost worldwide. However, most nematode species are beneficial: They enhance plant growth by promoting nutrient cycling and contribute to biological control of plant enemies, including insects, fungi and plant-parasitic nematodes.

HISTORY

Free-living (nonparasitic) nematodes were first discovered in 1656 A.D., when Borellus observed *Turbatrix aceti* in natural vinegar, using one of the first microscopes (1). In 1743, Needham gave the first account of a plant-parasitic nematode, *Anguina tritici*, from wheat galls. In the mid-nineteenth century, nematology developed as an independent branch of zoology. Since the study of soil nematodes has been driven by the economic impact of plant-parasites, free-living nematodes have received comparatively little attention. However, the most intensively studied nematode today is the free-living *Caenorhabditis elegans*, which is considered a model for genomic, developmental and neurobiological systems with relevance to all of biology (2).

CLASSIFICATION

Taxonomic Classification

As nematology is a relatively young science, nematode taxonomy is still in great flux, with no single classification agreed upon by all taxonomists. Taxonomic classification is generally based on morphological characteristics, but recent molecular phylogenetic research challenges the resulting higher-level classification (3). Usually, Nematoda are divided in two classes, the Adenophorea (Aphasmidia) and the Secernentea (Phasmidia), based primarily on the absence or presence of posterior sense organs, the phasmids. Examples of soil-dwelling orders are given in Fig. 1. A comprehensive overview of taxonomic literature is provided by (4).

Trophic Classification

The soil nematode fauna typically includes several trophic (feeding) groups. Nematodes may be bacterial feeders (predominantly Rhabditida, Araeolaimida, Monhysterida), fungal feeders (Aphelenchida, Dorylaimida, Tylenchida), plant parasites and plant associates (Tylenchida, Aphelenchida, Dorylaimida), carnivores (Mononchida, Enoplida), omnivores (Dorylaimida), algal feeders (Chromadorida, Desmodorida, Enoplida), and insect parasites (Rhabditida). Trophic groups are generally distinguished by the mouth structure and esophageal shape. However, as some taxa lack the typical anterior morphology, it is recommended that nematodes be identified to genus and classified trophically according to Ref. 5. A second note of caution is that unequivocal classification into trophic groups is not always possible, due to opportunistic feeding habits, life-stage dependent diets, or lack of dependable information (5).

Life History Classification

For environmental monitoring, soil nematodes are classified into five life history groups, ranging from *r*-selected "colonizers" (commonly members of Rhabditida) to K-selected "persisters" (mostly Dorylaimida). Their relative abundances are weighted using the Maturity Index, indicating the level of soil disturbance (6). However, the challenge remains to avoid ad hoc descriptions and establish clear cause-and-effect relationships between nematode community changes and soil disturbance (7). A bibliography on environmental monitoring with nematodes is provided by Ref. 8.

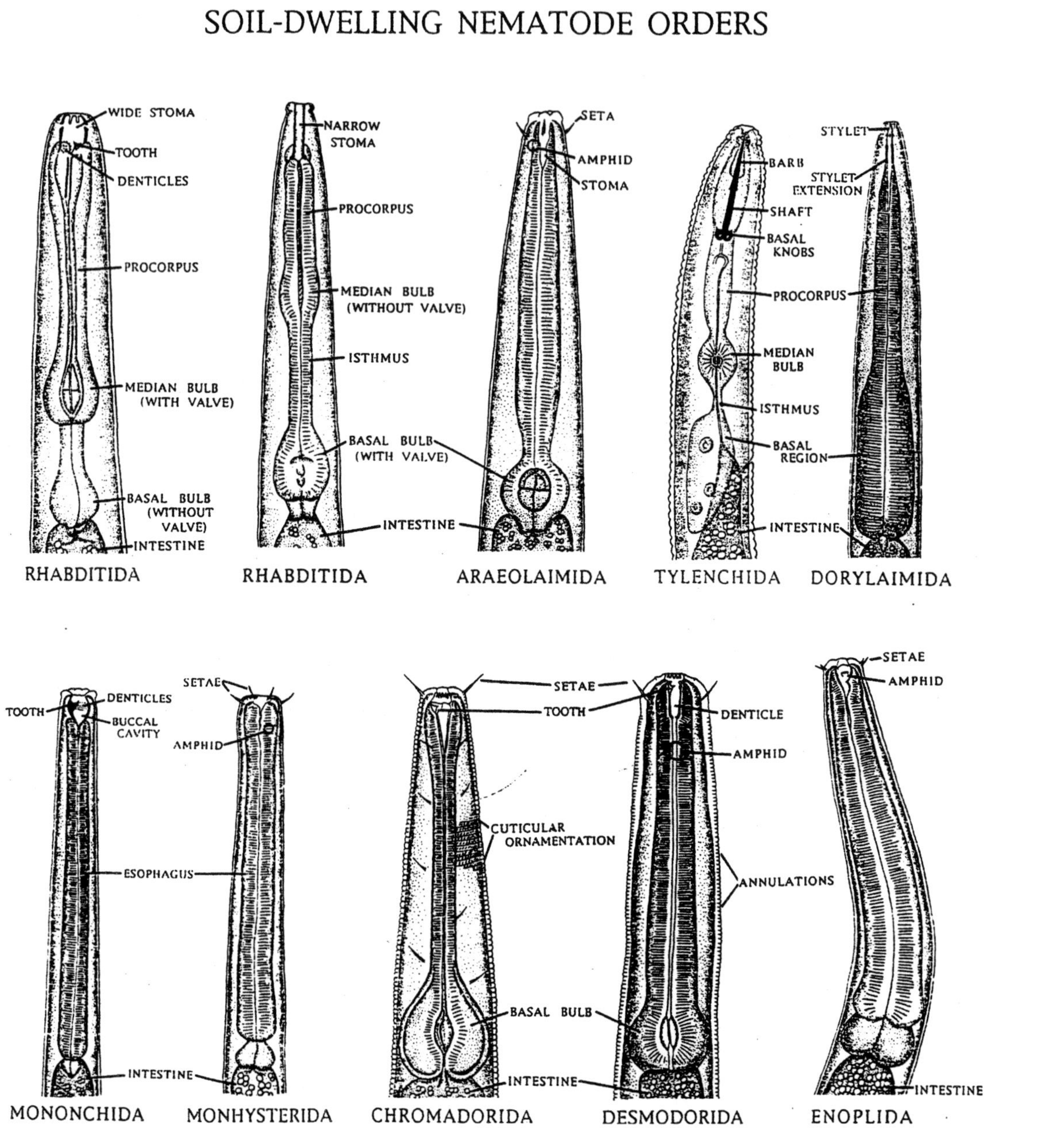

Fig. 1 The soil dwelling nematode orders (from Ref. 30). Note that the order Aphelenchida is missing, as the authors of Ref. 30 considered it a suborder (Aphelenchina) of the Tylenchida, following the classification used in Ref. 34.

MORPHOLOGY

Nematodes are bilaterally symmetrical, non-segmented pseudocoelomates. Their body plan is simply a "tube within a tube." The outer tube is the body wall, consisting of a flexible cuticle and longitudinal muscles, with which sinusoidal locomotion is achieved. The inner tube is the alimentary tract, which, starting at the anterior end (Fig. 1), consists of a stoma or stylet, esophagus, intestine, rectum and anus. Nematodes are normally bisexual, but for some species reproduction is parthenogenetic or hermaphroditic (1). Nematodes have excretory and nervous systems, but lack discrete circulatory or respiratory systems. Internal structures are easily observed with sub-stage lighting under a microscope. In the nematode life cycle, which depending on species and environment varies from 1 week to >1 year, the egg stage is followed by four juvenile stages before adulthood. In response to adverse environmental conditions (e.g., drought), nematodes may enter anabiosis, a state of metabolic dormancy which may last for years. Extensive information on nematode structure can be found in Ref. 9.

BIOGEOGRAPHY

Distribution and Abundance

Soil nematodes are found in all biomes, with topsoil abundances ranging from 50,000 m^{-2} in Antarctic deserts (10) up to tens of millions m^{-2} in temperate grasslands and deciduous forests (11). Generally, nematodes occur wherever there is moisture and food available. Since these resources have heterogeneous distributions, nematode populations are highly aggregated, which is a concern for quantitative soil sampling (12). Active migration is limited in range, but nematodes may be dispersed over long distances (notably in anabiotic form) by wind, water, insects, birds and mammals (1).

Species Diversity

Less than 6,000 terrestrial nematode species have been described, but it is estimated that there are least 100,000 species (13). Species richness is generally greatest in temperate broadleaf forest (62 species per soil sample) followed by cultivated soil, grassland, tropical rainforest, temperate coniferous forest and polar vegetation (14). The structural and functional implications of nematode biodiversity are still being debated (15).

HARMFUL NEMATODES

Plant-parasitic nematodes can cause great damage to agricultural crops. Ectoparasitic nematodes stay outside the root and use their stylet to puncture root cells (e.g., the sting nematode *Belonolaimus*; the stunt nematode *Tylenchorynchus*). Endoparasitic nematodes enter the root and move around (the lesion nematode *Pratylenchus*) or stay in one feeding site (the root-knot nematode *Meloidogyne*; the cyst nematodes *Globodera* and *Heterodera*). The resulting root damage can cause wilting, stunting,

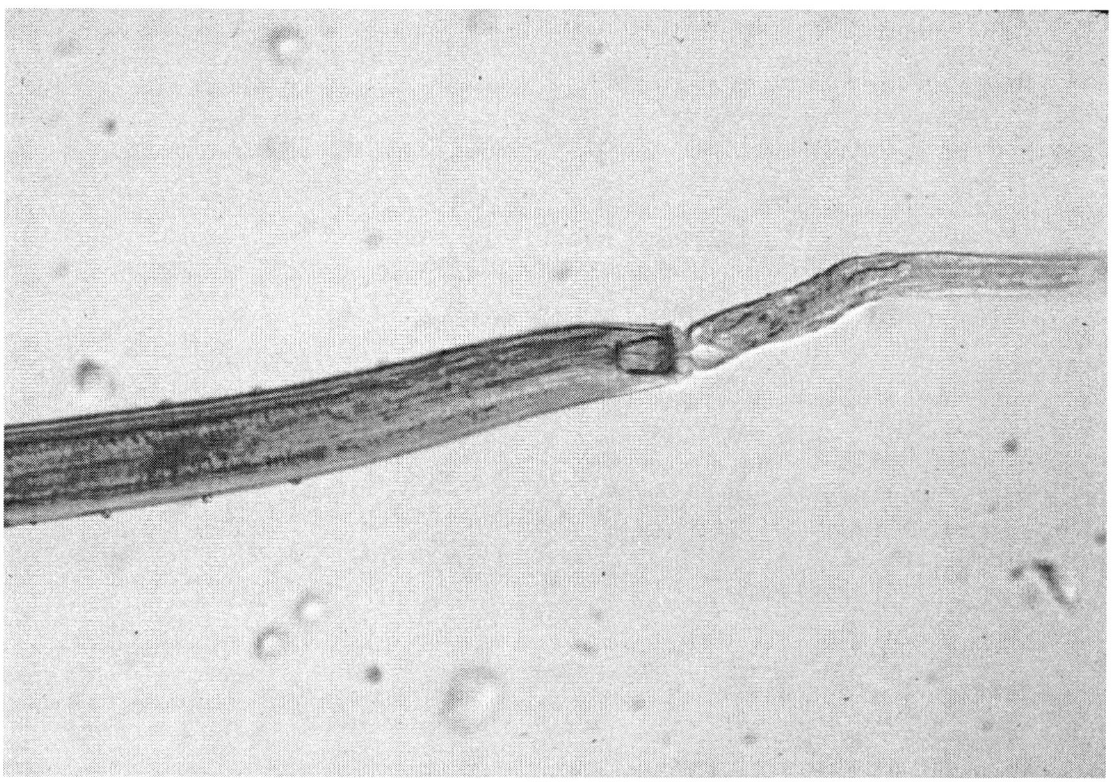

Fig. 2 A predacious nematode feeding on a gravid bacterivorous nematode. The predator itself is infected with *Pasteuria penetrans* (the "warts" attached to the predator's cuticle are bacterial endospores). (Photograph: C. Ettema.)

nutrient deficiencies, and yield losses. Some species transmit viruses (*Longidorus*, *Trichodorus*, and *Xiphinema*). In other cases, nematodes may cause damage to above ground plant parts (the stem nematode *Ditylenchus dipsaci*, the wheat gall nematode *Anguina tritici*). Worldwide, nematode damage may amount to $78 billion annually (16). Nematode problems of various crops and climates are reviewed by Refs. 17 and 18. An important observation is that low levels of root herbivory could *promote* plant growth by stimulating root branching and exudation, which in turn enhances rhizosphere microbial activity and nutrient availability (19).

Nematode control may include sanitary measures, crop rotation, use of resistant cultivars, application of nematicides and organic amendments, inter-cropping with nematode-antagonistic crops, and physical methods such as tillage, steaming and solarization (20). Nematodes have numerous natural enemies including soil bacteria, fungi, predatory nematodes, and insects (Fig. 2) (21, 22), but soil communities are not easily manipulated for biological control. Promising results have been obtained with *Verticillium chlamydosporium*, an egg-parasite of cyst and root knot nematodes (23). The nematode-trapping fungus *Arthrobotrys* is commercially available, and controls nematodes in cultivated mushroom and tomatoes (21).

BENEFICIAL NEMATODES

Microbial-Feeding Nematodes

Microbial-feeding nematodes (bacterivores and fungivores) may promote plant growth by enhancing nutrient mineralization, notably of nitrogen (24). As microbial-feeding nematodes have a higher carbon-to-nitrogen ratio than microbes, and low production efficiency, they mineralize the nitrogen immobilized in microbial biomass. In addition, they can stimulate microbial activity (and thus nitrogen mineralization) by reducing microbial competition, enhancing oxygen diffusion, and transporting microbial propagules to new substrates. Using knowledge of microbivorous nematode population dynamics, the application of organic amendments can be timed to optimize nitrogen availability to crops (25).

Predacious and Omnivorous Nematodes

By controlling microbivorous populations, predacious (Fig. 2) and omnivorous nematodes prevent "overgrazing" of microbes and thereby indirectly control nutrient cycling processes (26). In terms of biological control, it is unlikely that predacious nematodes could contribute significantly to plant-parasitic nematode control (27). Their generation times are too long for a timely response to plant-parasite population peaks, and there is evidence that predacious nematodes prefer microbial-feeding, not plant-feeding nematodes as diet. Predacious and omnivorous nematodes may be useful indicators of soil ecosystem health, because of their long life cycles and sensitivity to disturbance (6).

Entomopathogenic Nematodes

Entomopathogenic (insect-parasitic) nematodes are successfully used in biological control of pest insects, including white grubs, leaf miners and cockroaches (28). Entomopathogenic nematodes (order Rhabditida) carry a symbiotic bacterium. After penetrating the host insect, the nematode releases the bacterium, which kills the host within 72 hours, providing food to the nematode, which then multiplies inside the cadaver. Infective juveniles disperse to new hosts. (For detailed information on entomopathogenic nematodes see Ref. 29.)

METHODS

Methods for collection and extraction are reviewed by Refs. 17, 30, and 31. For ecological work with nematodes, useful methods are compiled by Ref. 32. A bibliography of nematode identification keys is provided by Ref. 4, and an interactive diagnostic key to plant-parasitic, free-living and predacious nematodes is found in Ref. 33.

REFERENCES

1. Poinar, George O., Jr. *The Natural History of Nematodes*; Prentice Hall: Englewood Cliffs, NJ, 1983.
2. Caenorhabditis Elegans http://helios.bto.ed.ac.uk/mbx/fgn/wow/celegans.html (accessed Aug 2000).
3. Blaxter, Mark L.; De Ley, Paul.; Garey, James R.; Liu, Leo X.; Scheldeman, Patsy; Vierstraete, Andy; Vanfleteren, Jaques R.; Mackey., Laura Y.; Dorris, Mark; Frisse, Linda M.; Vida, J.T.; Thomas, W. Kelley. A Molecular Evolutionary Framework for the Phylum Nematoda. Nature **1998**, *392*, 71–75.
4. Bongers, Tom. Nematode Identification Literature. http://www.spg.wau.nl/nema/ident_lit.html (accessed Aug 2000).
5. Yeates, Gregor W.; Bongers, Tom; De Goede, Ron G.M.; Freckman, Diana W.; Georgieva, S.S. Feeding Habits in Soil Nematode Families and Genera—An Outline for Soil Ecologists. J. Nematol. **1993**, *25*, 315–331.
6. De Goede, Ron G.M.; Bongers, Tom. The Nematode Maturity Index. In *Techniques in Nematode Ecology*; Wheeler, T., Forge, T., Caswell-Chen, E., LaMondia, J.A.,

N

Eds.; http://ianrwww.unl.edu/son/deGoede.htm (accessed Aug 2000). Society of Nematologists, 2000.

7. Niles, Robert K.; Freckman, Diana W. From the Ground Up: Nematode Ecology in Bioassessment and Ecosystem Health. In *Plant and Nematode Interactions*; Barker, K.R., Pederson, G.A., Windham, G.L., Eds.; Agronomy Monograph 36 American Society of Agronomy: Madison WI, 1998; 65–85.
8. Bongers, Tom. Maturity Index Literature. http://www.spg.wau.nl/nema/MI_lit.htm (accessed Aug 2000).
9. Bird, A.F.; Bird, J. *The Structure of Nematodes*; Academic Press: London, 1991.
10. Freckman, Diana W.; Virginia, Ross A. Low-Diversity Antarctic Soil Nematode Communities: Distribution and Response to Disturbance. Ecology **1997**, *78*, 363–369.
11. Sohlenius, Björn. Abundance, Biomass and Contribution to Energy Flow by Soil Nematodes in Terrestrial Ecosystems. Oikos **1980**, *34*, 186–194.
12. Webster, R.; Boag, Brian. Geostatistical Analysis of Cyst Nematodes in Soil. J. Soil Science **1992**, *43*, 583–595.
13. Andrássy, István. A Short Census of Free-Living Nematodes. Fund. Appl. Nematol. **1992**, *15*, 187–188.
14. Boag, Brian; Yeates, Gregor W. Soil Nematode Biodiversity in Terrestrial Ecosystems. Biodiv. Cons. **1998**, *7*, 617–630.
15. Ettema, Christien H. Soil Nematode Diversity: Species Coexistence and Ecosystem Function. J. Nematol. **1998**, *30*, 159–169.
16. Wheeler, T., Forge, T., Caswell-Chen, E., LaMondia, J.A., Eds. *Plant and Soil Nematodes: Societal Impact and Focus for the Future*, A Report by the Committee on National Needs and Priorities in Nematology. Society of Nematologists: Lawrence, KS, 1994.
17. Nickle, William R. *Manual of Agricultural Nematology*; Marcel Dekker: New York, 1991.
18. Luc, M.; Sikora, R.A.; Bridge, J. *Plant-Parasitic Nematodes in Subtropical and Tropical Agriculture*; CAB International: Wallingford, UK, 1990.
19. Denton, Crystal S.; Bardgett, Richard D.; Cook, Roger.; Hobbs, Philip J. Low Amounts of Root Herbivory Positively Influence the Rhizosphere Microbial Community in a Temperate Grassland Soil. Soil Biol. Biochem. **1999**, *31*, 155–165.
20. NEMABASE — A Database of the Host Status of Plants to Plant-Parasitic Nematodes. http://www.ipm.ucdavis.edu/NEMABASE/index.html (accessed Aug 2000).
21. Poinar, George O., Jr.; Jansson, Hans-Börje. *Diseases of Nematodes*; CRC Press: Boca Raton, FL, 1988; Vol. I and II.
22. Natural Enemies of Nematodes: Biological Control Images. http://sacs.cpes.peachnet.edu/nemabc/ (accessed Aug 2000).
23. Verticillium Chlamydosporium and Biological Control of Phytoparasitic Nematodes. http://www.area.ba.cnr.it/~e085ac01/bkfair3444.html (accessed Aug 2000).
24. Freckman, Diana W. Bacterivorous Nematodes and Organic-Matter Decomposition. Agric. Ecosys. Envir. **1988**, *24*, 195–217.
25. Ferris, Howard.; Venette, Robert C.; Lau, S.S. Dynamics of Nematode Communities in Tomatoes Grown in Conventional and Organic Farming Systems, and Their Impact on Soil Fertility. Appl. Soil. Ecol. **1996**, *3*, 161–175.
26. Hunt, H.W.; Coleman, David C.; Ingham, Elaine R.; Ingham, Russ E.; Elliott, Edward T.; Moore, John C.; Rose, S.L.; Reid, C.P.P.; Morley, C.R. The Detrital Food Web in a Short Grass Prairie. Biol. Fert. Soils **1987**, *3*, 57–68.
27. Yeates, Gregor W.; Wardle, David A. Nematodes as Predators and Prey: Relationships to Biological Control and Soil Processes. Pedobiologia **1996**, *40*, 43–50.
28. Smart, Grover C., Jr. Entomopathogenic Nematodes for the Biological Control of Insects. J. Nematol. **1995**, *4-S*, 529–534.
29. Gaugler, R.; Kaya, H. *Entomopathogenic Nematodes in Biological Control*; CRC Press: Boca Raton, 1990.
30. Ingham, Russ E. Nematodes *Methods of Soil Analysis. Part 2—Microbiological and Biochemical Properties*; Bigham, J.M., Ed.; Soil Science Society of America: Madison, WI, 1994; 459–490.
31. McSorley, Robert. Sampling and Extraction Techniques for Nematodes. In *Techniques in Nematode Ecology*; Wheeler, T., Forge, T., Caswell-Chen, E., LaMondia, J.A., Eds.; http://ianrwww.unl.edu/son/McSorley.htm (accessed Aug 2000). Society of Nematologists, 2000.
32. Wheeler, T., Forge, T., Caswell-Chen, E., LaMondia, J.A., Eds. *Techniques in Nematode Ecology*; http://ianrwww.unl.edu/son/Ecology_Manual_TOC.htm (accessed Aug 2000)., Society of Nematologists, 2000.
33. Interactive Diagnostic Key to Plant Parasitic, Freeliving and Predacious Nematodes. http://ianrwww.unl.edu/ianr/plntpath/nematode/key/nemakey.htm.
34. Golden, A.M. Classification of the Genera and Higher Categories of the Order Tylenchida (Nematoda). In *Plant Parasitic Nematodes*; Zuckerman, B.M., Mai, W.F., Rohde, R.A., Eds.; 1971; Vol. 1, 191–232.

NITROGEN AND ITS TRANSFORMATIONS

Oswald Van Cleemput
Pascal Boeckx
Ghent University, Ghent, Belgium

INTRODUCTION

Nitrogen (N) is essential to all life. It is the nutrient that most often limits biological activity. In agricultural and natural ecosystems, N occurs in many forms covering a range of valence states from -3 to +5. The change from one valence state to another depends primarily on environmental conditions. The transformations and flow from one form to another constitute the basics of the soil N cycle (Fig. 1). The use of N fertilizers has become essential to increase the productivity of agriculture, and has resulted in an almost doubling of the global food production in the past 50 years. However, this also implies that the natural N cycle has substantially been disturbed. In the following paragraphs an overview of the different N transformation processes in the soil is given.

THE NITROGEN CYCLE: GENERAL

Atmospheric N_2 gas (valence 0) can be converted by lightening to various oxides and finally to nitrate (NO_3^-) (valence +5), which can be deposited and taken up by growing plants. Also N_2 gas can be converted to ammonia (NH_3, valence -3) by biological N_2 fixation, with the NH_3 participating in a number of biochemical reactions in the plant. When plant residues decompose the N-compounds undergo a series of microbial conversions (mineralisation) leading first to the formation of ammonium (NH_4^+) (valence -3) and possibly ending up in NO_3^- (nitrification). Under anaerobic conditions NO_3^- can be converted to various N-oxides and finally to N_2 gas (denitrification). When mineral or organic N fertilizers are used they also undergo the same transformation processes and influence the rate of other N-transformations. In considering the soil compartment, there can be N gains (such as biological N_2 fixation) as well as N losses (such as leaching and denitrification). Furthermore N can be exported from the soil via harvest products, or immobilized in soil organic matter.

NITROGEN TRANSFORMATIONS IN THE SOIL

The principal forms of N in the soil are NH_4^+, NO_3^- or organic N-substances. At any moment, inorganic N in the soil is only a small fraction of the total soil N. Most of the N in a surface soil is present as organic N. It consists of proteins (20–40%), amino sugars, such as the hexosamines (5–10%), purine and pyrimidime derivates (1% or less), and complex unidentified compounds formed by reaction of NH_4^+ with lignin, polymerization of quinones with N compounds and condensation of sugars and amines. In the subsoil, an important fraction of the present N can be trapped in clay lattices (especially *illitic* clays) as nonexchangeable NH_4^+ and is consequently largely unavailable. Organic substances slowly mineralize by microorganisms to NH_4^+, which could be converted by other microorganisms to NO_3^- (see further).

The NH_4^+ can be adsorbed to negatively charged sites of clay minerals and organic compounds. This reduces its mobility in the soil compared to the more mobile NO_3^- ion.

Microorganisms can use both NH_4^+ and NO_3^- to satisfy their need for N. This type of N transformation is called microbial immobilization.

The ratio between carbon (C) and N (C:N ratio) in organic matter determines whether immobilization or mineralization is likely to occur. When utilizing organic matter with a low N content, the microorganisms need additional N, decreasing the mineral N pool of the soil. Thus, incorporation of organic matter with a high C:N ratio (e.g., cereal straw) results in immobilization. Incorporation of organic matter with a low C:N ratio (e.g., vegetable or legume residues) results in N-mineralization. A value of the C:N ratio of 25 to 30 is often taken as the critical point toward either immobilization or mineralization.

Nitrification is a two-step process. In the first step NH_4^+ is converted to nitrite (NO_2^-) (valence +3) by a group of obligate autotrophic bacteria known as *Nitrosomonas* species. The second step is carried out by another group of obligate autotrophic bacteria known as *Nitrobacter* species. Also a few heterotrophs can carry out nitrification, usually at much lower rates.

N

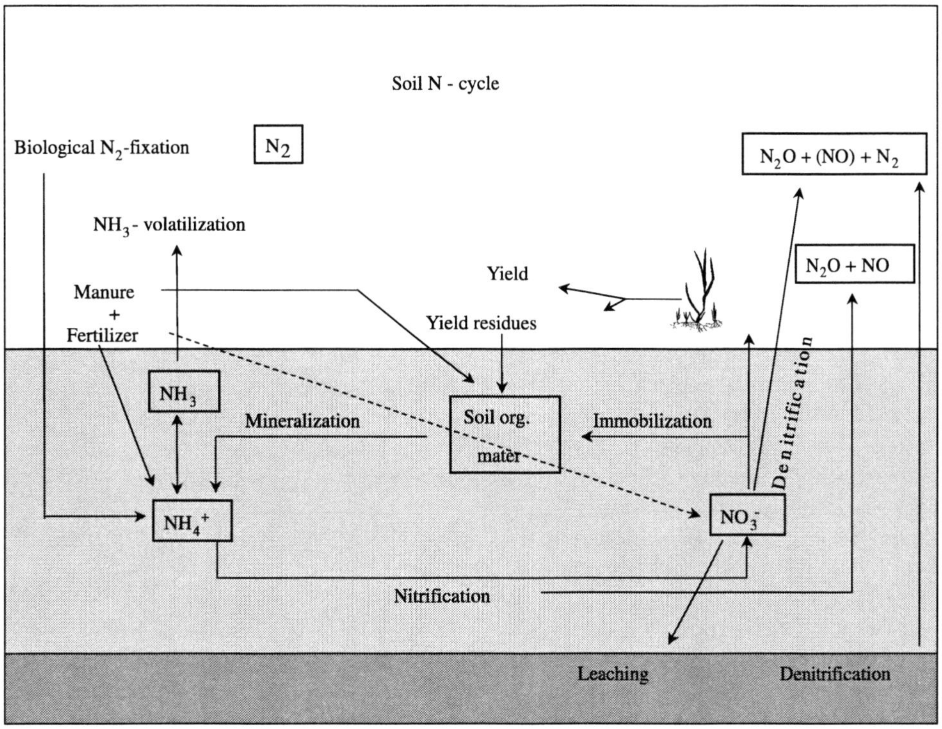

Fig. 1 The soil N cycle. The white compartment represents the atmosphere; the light gray compartment represents the biosphere and the dark gray compartment the subsoil.

Soil water and aeration are crucial factors for nitrification. At a water potential of 0 kPa (saturation), there is little air in the soil and nitrification stops, due to oxygen limitation; nitrification is greatest near field capacity (-33 kPa in medium- to heavy-textured soils, to 0 to -10 kPa in light sandy soils). Also in dry soils NH_4^+ and sometimes NO_2^- accumulate presumably because *Nitrobacter* species are more sensitive to water stress than the other microorganisms.

Nitrification is slow in acid conditions with an increasing rate at increasing pH. Mainly under alkaline conditions, nitrite is also accumulating, because *Nitrobacter* is known to be inhibited by ammonia, which is formed under alkaline conditions. Nitrification is a process that acidifies the soil as protons (H^+) are liberated:

$$NH_4^+ + 2O_2 \rightarrow NO_3^- + 2H^+ + H_2O$$

During nitrification minor amounts of nitrous oxide (N_2O) (valence +1) and nitric oxide (NO) (valence +2) are formed. Both compounds have environmental consequences, discussed below.

The effect of temperature on nitrification is climate dependent. There is a climatic selection of species of nitrifiers, with those from cooler regions having lower temperature optima and less heat tolerance than species from warmer regions. All above-mentioned factors influencing nitrification also influence the nitrifying population. The population and activity of nitrifiers can be reduced by the use of nitrification inhibitors, such as dicyanodiamide, nitrapirin and neem (*Azadirachta indica*) seed cake. They are used mostly to retard the nitrification of manure; otherwise their practicality is controversial and they are not extensively used. More details about nitrification and nitrification inhibitors can be found in McCarty (1) and Prosser (2).

NITROGEN INPUT PROCESSES

Atmospheric Nitrogen Deposition

The total atmospheric N (NH_4^+ and NO_3^-) deposition is in the order of 10–40 kg N ha^{-1} yr^{-1} in much of north-western and central Europe and some regions in North America. It ranges from 3–5 kg N ha^{-1} yr^{-1} in pristine areas (3). It is originating from previously emitted NH_3 and NO_x from agricultural and industrial activities or traffic.

Biological Nitrogen Fixation

Rhizobium species living in symbiotic relationship in root nodules of legumes, e.g. clover (*Trifolium*), lucerne (*Medicago*), peas (*Pisum*) and beans (*Faba*)—can convert atmospheric N_2 gas to NH_3, which is further converted to amino acids and proteins. Parallel to this process, the *rhizobium* species receive from the legume the energy they need to grow and to fix N_2. Photosynthetic cyanobacteria are also N-fixing organisms and are especially important in paddy rice (*Oryza*). The amount of N fixed varies greatly from crop to crop, ranging from a few kg to a few hundred $kg\,N\,ha^{-1}\,yr^{-1}$. The process is depressed by ample N supply from other sources, and it is sensitive to lack of phosphorus. The amount of globally fixed N is almost the double the amount of applied fertilizer N. Next to symbiotic N fixing bacteria also non-symbiotic species (e.g. *Azotobacter*) occur in soils. In general, free-living diazotrophs make a small but significant contribution to the soil N status. Some non-leguminous trees and plants (e.g. alder (*Alnus*), sugarcane (*Saccharum*) host N-fixing bacteria as well. Much uncertainty exists about the association of N fixing bacteria with non-legumes (so called associative N fixing bacteria).

Mineral and Organic Nitrogen Fertilization

Theoretically plants should prefer NH_4^+ above NO_3^-, because NH_4^+ does not need to be reduced before incorporation into the plant. In most well-drained soils oxidation of NH_4^+ is fairly rapid and therefore most plants have developed to grow better with NO_3^-. However, a number of studies have shown that plants better develop when both sources are available. Rice, growing under submerged conditions must grow in the presence of NH_4^+ as NO_3^- is not stable under flooded conditions. When urea is applied it rapidly hydrolyzes under well-drained conditions, unless a urease inhibitor is being added; under submerged conditions rice plants may also absorb N directly as molecular urea. Organic manure can be of plant or animal origin or a mixture of both. However, most comes from dung and urine from farm animals. It exists as farmyard or stable manure, urine, slurry or as compost. Because its composition is not constant and because plant material (catch or cover crops, legumes) is often added freshly (green manure) to the soil, less than 30% of its nutrients becomes available for the next crop.

NITROGEN UPTAKE BY PLANTS

Growing plants get their N from fertilizer N as well as from organic soil N upon mineralization. Plants take up N compounds both as NO_3^- and as NH_4^+. In general, NO_3^- is the major source of plant N. There is some evidence that small amounts of organic N (urea or amino acids) can be taken up by plants from the soils solution. Plant uptake of N can be studied through the use of mineral fertilizers or organic matter labeled with the stable N isotope ^{15}N. The proportion of applied N taken up by the crop is affected by many factors, including crop species, climate and soil conditions. Above ground parts of the crop can recover 40–60% of the fertilizer N applied.

NITROGEN LOSS PROCESSES

Ammonia Volatilization

Losses of N from the soil by NH_3 volatilization amount globally to 54 Mt (or 10^{12} g) NH_3-N yr^{-1} and 75% is of anthropogenic origin (4). According to the ECETOC (5), the dominant source is animal manure and about 30% of N in urine and dung is lost as NH_3. The other major source is surface application of urea or ammonium bicarbonate and to a lesser degree other ammonium-containing fertilizers. As urea is the most important N fertilizer in the world, it may lead to important NH_3 loss upon hydrolysis and subsequent pH rise in the vicinity of the urea till. The transformation of NH_4^+ to the volatile form NH_3 increases with increasing pH, temperature, soil porosity, and wind speed at the soil surface. It decreases with increasing water content and rainfall events following application. Ammonia losses from soils can be effectively reduced by fertilizer incorporation or injection instead of surface application.

Emission of Nitrogen Oxides (N_2O, NO) and Molecular Nitrogen (Nitrification and Denitrification)

Microbial nitrification and denitrification are responsible for the emission of NO and N_2O (6). They are by-products in nitrification and intermediates during denitrification. Probably about 0.5% of fertilizer N applied is emitted as NO (7) and 1.25% as N_2O (8). However, wide ranges have been reported. Intensification of arable agriculture and of animal husbandry has made more N available in the soil N cycle increasing the emission of N oxides. The relative percentage of NO and N_2O formation very much depends on the moisture content of the soil. At a water-filled pore space (WFPS, or the fraction of total soil pore space filled with water) below 40% NO is produced mainly from nitrification. Between a WFPS of 40% and 60% formation of NO and N_2O from nitrification occurs. Between a WFPS of 60% and 80% N_2O is predominantly produced from denitrification and the formation of NO is decreasing sharply. At a WFPS above 80% the formation of N_2 by denitrification is dominant. In practice these WFPS ranges will overlap and depend on the soil type (9). Next to water content, also temperature, land use and availability of N and decomposable organic matter are important determining factors for N_2O formation. Nitrous oxide is a greenhouse gas contributing 5–6% to the enhanced greenhouse effect. Increased concentrations are also detrimental for the stratospheric ozone layer (10). In the presence of sunlight, NO_x (NO and NO_2) react with volatile organic compounds from evaporated petrol and solvents and from vegetation and forms tropospheric ozone which is, even at low concentration, harmful to plants and human beings. The major gaseous end-product of denitrification is N_2. The ratio of N_2O to N_2 produced by denitrification depends on many environmental conditions. Generally the more anaerobic the environment the greater the N_2 production. Denitrification is controlled by three primary factors (oxygen, nitrate and carbon), which in turn are controlled by several physical and biological factors. Denitrification N loss can reach 10% of the fertilizer N input—more on grassland and when manure is also applied (11). Chemical denitrification is normally insignificant and is mainly related to the stability of NO_2^- and acid conditions (12). It is more difficult to reduce N_2O and NO from soils then NH_3 losses. A general principle is to minimize N surpluses in the soil profile via careful fertilizer adjustment, corresponding to the actual crop demands.

Leaching

Applied NO_3^- or NO_3^-, formed through nitrification from mineralized NH_4^+ or from NH_4^+ from animal manure, can leach out of the rooting zone. It is well possible that this leached NO_3^- can be denitrified at other places and returned into the atmosphere. The amount and intensity of rainfall, quantity and frequency of irrigation, evaporation rate, temperature, soil texture and structure, type of land use, cropping and tillage practices and the amount and form of fertilizer N are all parameters influencing the amount of NO_3^- leaching to the underground water. Nitrate leaching should be kept under control as it may influence the nitrate content in drinking water influencing human health and in surface water, causing eutrophication. Nitrate losses can be minimized by reducing the mineral N content in the soil profile during the winter period by careful fertilizer adjustment, growing of cover crops or riparian buffer areas.

REFERENCES

1. McCarty, G.W. Modes of Action of Nitrification Inhibitors. Biol. Fert. Soils **1999**, *29*, 1–9.
2. Prosser, J.I., Ed. *Nitrification, Special Publications of the Society of General Microbiology*; IRL Press: Oxford, 1986; 20.
3. Lagreid, M.; Bockman, O.C.; Kaarstad, O. *Agriculture, Fertilizers and the Environment*; CIBA Publishing: Oxon, UK, 1999; 294.
4. Sutton, M.A.; Lee, D.S.; Dollard, G.J.; Fowler, D. International Conference on Atmospheric Ammonia: Emission, Deposition and Environmental Impacts. Atmospheric Environment **1998**, *32*, 1–593.
5. ECETOC *Ammonia Emissions to Air in Western Europe*, (No. 62); European Centre for Ecotoxicology and Toxicology of Chemicals: Brussels, 1994; 196 p.
6. Bremner, J.M. Sources of Nitrous Oxide in Soils. Nutr. Cycl. Agroecosys. **1997**, *49*, 7–16.
7. Veldkamp, E.; Keller, M. Fertilizer-Induced Nitric Oxide Emissions from Agricultural Soils. Nutr. Cycl. Agroecosys. **1997**, *48*, 69–77.
8. Mosier, A.; Kroeze, C.; Nevison, C.; Oenema, O.; Seitsinger, S.; Van Cleemput, O. Closing the Global N_2O Budget: Nitrous Oxide Emissions Through the Agricultural N Cycle. Nutr. Cycl. Agroecosys. **1998**, *52*, 225–248.
9. Davidson, E.A. Fluxes of Nitrous Oxide and Nitric Oxide from Terrestrial Ecosystems. In *Microbial Production and Consumption of Greenhouse Gases: Methane, Nitrogen Oxides, and Halomethanes*; Rogers, J.E., Whitman, W.B., Eds.; American Society for Microbiology: Washington, DC, 1991; 219–235.
10. Crutzen, P.J. The Influence of Nitrogen Oxides on the Atmospheric Ozone Content. Quat. J. Royal Meteor. Soc. **1976**, *96*, 320–325.
11. von Rheinbaben, W. Nitrogen Losses from Agricultural Soils Through Denitrification—A Critical Evaluation. Z. Pflanzenern. Bodenk. **1990**, *153*, 157–166.
12. Van Cleemput, O. Subsoils: Chemo- and Biological Denitrification, N_2O and N_2 Emissions. Nutr. Cycl. Agroecosys. **1998**, *52*, 187–194.

NITROUS OXIDE FROM AGRICULTURAL SOILS, EMISSION OF

J.R. Freney
Commonwealth Scientific and Industrial Research Organisation (CSIRO) Plant Industry, Canberra, Australian Capital Territory, Australia

INTRODUCTION

Nitrous oxide is a gas that is produced naturally by many different micro-organisms in soils and waters, and as a result of human activity associated with agriculture, biomass burning, stationary combustion, automobiles, and the production of nitric and adipic acids for industrial purposes. According to the Intergovernmental Panel on Climate Change (IPCC) (1), ~23.1 million metric tons (Mt) of nitrous oxide is emitted each year, 14.1 Mt as a result of natural processes (~4.7 Mt from the oceans, ~6.3 Mt from tropical soils, and ~3.1 Mt from temperate soils); and ~9 Mt as a result of human activities (5.5 Mt from agricultural soils, 0.6 Mt from cattle and feedlots, 0.8 Mt from biomass burning, and 2.1 Mt from mobile sources and industry). While there is considerable uncertainty associated with each of these estimates, it is apparent that most nitrous oxide is derived from soils.

Because of the intimate connection between the Earth and the atmosphere, much of the nitrous oxide produced enters the atmosphere and affects its chemical and physical properties. Nitrous oxide contributes to the destruction of the stratospheric ozone layer that protects the Earth from harmful ultraviolet radiation, and is one of the more potent greenhouse gases that trap part of the thermal radiation from the Earth's surface. The atmospheric concentration of nitrous oxide is ~313 parts per billion. It is increasing at the rate of ~0.7 parts per billion each year, and its lifetime is ~166 years (2). It seems that the increased atmospheric concentration results from the increased use of synthetic fertilizer nitrogen, biologically fixed nitrogen, animal manure, crop residues, and human sewage sludge in agriculture to produce food and fiber for the rapidly increasing world population (3).

NITROUS OXIDE EMISSION FROM AGRICULTURE

All soils are deficient in nitrogen for the growth of plants, but the deficiency can be overcome by adding fertilizer nitrogen. When the fertilizer (e.g., urea or ammonia-based compounds) is applied to soil, it is transformed by micro-organisms as follows:

$$\begin{aligned}\text{Fertilizer nitrogen} &\xrightarrow{1} \text{Ammonium} \xrightarrow{2} \text{Nitrite} \\ &\xrightarrow{3} \text{Nitrate} \xrightarrow{4} \text{Nitrite} \\ &\xrightarrow{5} \text{Nitric oxide} \xrightarrow{6} \text{Nitrous oxide} \\ &\xrightarrow{7} \text{Dinitrogen}\end{aligned} \quad (1)$$

When the soil is aerobic (i.e., when oxygen is present) ammonium is oxidized to nitrite and nitrate (Steps 2 & 3). This process is called nitrification. After addition of irrigation water or rain, the soil may become anaerobic (devoid of oxygen). The nitrate is then reduced by soil organisms to nitrite and the gases nitric oxide, nitrous oxide, and dinitrogen (Steps 4–7) in a process termed denitrification (4).

When atmospheric scientists first expressed concern that nitrous oxide emission into the atmosphere, as a result of fertilizer use, would lead to destruction of the ozone layer, it was thought that nitrous oxide was produced mainly from the microbiological reduction of nitrate in poorly aerated soils. However, research in the latter part of the 1970s showed that significant nitrous oxide was emitted from aerobic soils during nitrification of ammonium, and subsequent work has shown that nitrification is a major source of nitrous oxide (4).

Nitrous Oxide from Denitrification

Certain micro-organisms in the absence of oxygen have the capacity to reduce nitrate (or other nitrogen oxides). Most denitrifying bacteria are heterotrophs—that is, they require a source of organic matter for energy—but denitrifying organisms that obtain their energy from light or inorganic compounds also occur in soils. The capacity to denitrify has been reported in more than 20 genera of bacteria, and almost all are aerobic organisms that can only grow anaerobically in the presence of nitrogen oxides. The dominant denitrifying organisms in soil are *Pseudomonas* and *Alcaligenes*. In addition to the free-living denitrifiers,

Rhizobia living symbiotically in root nodules of legumes are able to denitrify nitrate and produce nitrous oxide (4).

The general requirements for biological denitrification include the presence of micro-organisms with denitrifying capacity, nitrate (or other nitrogen oxides) and available organic matter, the absence of oxygen, and a suitable pH and temperature environment. In aerobic soils, denitrification can occur in anaerobic microsites in soil aggregates or in areas of high carbon content, where active microbial activity rapidly consumes all of the available oxygen (4).

Nitrous Oxide from Nitrification

The process of nitrification is normally defined as the biological oxidation of ammonium to nitrate with nitrite as an intermediate (4). The first step in the reaction, the oxidation of ammonium to nitrite, is carried out mainly by the micro-organism *Nitrosomonas*. The second step, oxidation of nitrite to nitrate, is carried out by *Nitrobacter*. It has been shown in a number of publications that *Nitrosomonas europaea* produces nitrous oxide during the oxidation of ammonium (4).

The possibility that significant nitrous oxide can be produced in soils by nitrifying organisms was indicated by studies that showed that soils incubated under aerobic conditions with ammonium produced more nitrous oxide than soils amended with nitrate (4). In addition, treatment of aerobic soils with nitrapyrin, which inhibits nitrification of ammonium but has little effect on denitrification, markedly reduced the emission of nitrous oxide (4). Production of nitrous oxide by nitrification in soils is increased by increasing temperature, pH, water content, available carbon, and the addition of ammonium-based fertilizers, plant residues, and animal manure.

Flooded Soils

In the past few years, increased attention has been given to nitrous oxide emission from paddy soils. The concern is that the introduction of management practices to reduce methane emissions from flooded soils may result in increased emissions of nitrous oxide.

Flooded soils are characterized by an oxygenated water column overlying an oxidized layer at the soil–water interface, an aerobic zone around each root, and anaerobic conditions in the remainder of the soil. This differentiation of the flooded soil into oxidized and reduced zones has a marked effect on the transformation of nitrogen (5). The resulting reactions are as follows:

1. Ammonium in the reduced zone diffuses to the oxidized zone;
2. Ammonium is oxidized to nitrate by nitrifying organisms;
3. The nitrate formed diffuses to the anaerobic zone;
4. Denitrification occurs with the production of nitrous oxide and dinitrogen;
5. The gaseous products diffuse through the soil and water layers to the atmosphere (5).

It is apparent that the rate of diffusion of ammonium to an oxidized layer and the rate of nitrification in the oxidized layer are factors controlling the production of nitrous oxide in flooded soils. The rate of diffusion of nitrous oxide through the soil and water layers will control its rate of emission to the atmosphere, or its further reduction to dinitrogen (5).

A number of mechanisms have been identified for the transfer of nitrous oxide from the soil to the atmosphere (3). Nitrous oxide may diffuse from the zone of production through the saturated soil and water layer to the atmosphere. It may also enter the roots of the rice plant and move by diffusion through the plant to the atmosphere in the same way as methane. Bhadrachalam et al. (6) studied the importance of the two pathways in intermittently flooded rice in the field in India and found that nitrogen gas fluxes were ~30% greater when transfer through the plants was included.

In the tropics, rice is usually transplanted and fertilized some time after flooding. Because of the anaerobic conditions that develop before fertilization and the slow rate of diffusion of nitrous oxide in flooded soils, most of the nitrous oxide is reduced to dinitrogen and very little escapes to the atmosphere. Nitrous oxide emission from intermittently flooded rice was relatively large compared with that from permanently flooded rice, reflecting the different oxidation states of intermittently and continuously flooded soils (6). Studies of nitrous oxide emission from rice fields from the time the soils were drained for harvest, through to flooding the soil in preparation for planting the next crop, showed that nitrous oxide was emitted continuously while the soil was not flooded. Overall, the rate of emission of nitrous oxide from flooded soils was less than that from upland soils after application of nitrogen fertilizer (3).

Biomass Burning

During combustion the nitrogen in the fuel can be converted into gaseous forms such as ammonia, nitric oxide, nitrous oxide, dinitrogen, and hydrogen cyanide. It is estimated that biomass burning contributes between

0.3 and 1.6 Mt nitrous oxide per year globally to the atmosphere (3). Most of the biomass burning (~90%) takes place in the tropics as a result of forest clearing, savanna and sugar cane fires, and burning of agricultural wastes and firewood (7).

Biomass burning is not only an instantaneous source of nitrous oxide, but it results in a longer-term enhancement of the production of this gas. Measurements of nitrous oxide emissions from soils, before and after burning showed that significantly more nitrous oxide was exhaled after the burn through alteration of the chemical, biological, and physical processes in soil (7).

Fertilizer Consumption and Nitrous Oxide Production

Nitrous oxide emissions from agricultural soils are generally greater and more variable than those from uncultivated land. Application of fertilizer nitrogen, animal manure, and sewage sludge usually results in enhanced emissions of nitrous oxide (7). Generally, there is a large emission of nitrous oxide immediately after the application of fertilizer. After about 6 weeks, the emission rate falls and fluctuates around a low value. Mosier (8) concluded that interactions between the physical, chemical, and biological variables are complex, that nitrous oxide fluxes are variable in time and space, and that soil management, cropping systems, and variable rainfall appear to have a greater effect on nitrous oxide emission than the type of nitrogen fertilizer. Consequently, Mosier et al. (9) recommend the use of one factor only for calculating the emission of nitrous oxide from different fertilizer types:

$$N_2O \text{ emitted} = 1.25\% \text{ of N applied (kg N/ha)} \quad (2)$$

This equation is based on data from long-term experiments with a variety of mineral and organic fertilizers, and encompasses 90% of the direct contributions of nitrogen fertilizers to nitrous oxide emissions.

Mosier et al. (3) developed a methodology to estimate agricultural emissions of nitrous oxide, taking into account all of the nitrogen inputs into crop production. They included direct emissions from agricultural soils as a result of synthetic fertilizer addition, animal wastes, increased biological nitrogen fixation, cultivation of mineral and organic soils through enhanced organic matter mineralization, and mineralization of crop residues returned to the field. Indirect nitrous oxide emissions resulting from deposition of ammonia and oxides of nitrogen, leaching of nitrate, and introduction of nitrogen into sewage systems were also included. They concluded that in 1989, 9.9 Mt of nitrous oxide was emitted into the atmosphere directly or indirectly, as a result of agriculture (Table 1).

Table 1 Calculated emission of nitrous oxide from agricultural activities

	Mt nitrous oxide per year
Direct soil emissions	
Synthetic fertilizer	1.4 (0.28–2.5)
Animal waste	0.9 (0.19–1.7)
Biological nitrogen fixation	0.16 (0.03–0.3)
Crop residue	0.6 (0.11–1.1)
Cultivation of Histosols	0.16 (0.03–0.3)
Total	3.3 (0.6–5.9)
Animal production	
Waste management systems	3.3 (0.9–4.9)
Indirect emissions	
Atmospheric deposition	0.47 (0.09–0.9)
Nitrogen leaching and runoff	2.5 (0.2–12.1)
Human sewage	0.3 (0.06–4.1)
Total	3.3 (0.35–17.1)
Total	9.9 (1.9–27.9)

(Modified from Ref. 3.)

MANAGEMENT PRACTICES TO DECREASE NITROUS OXIDE EMISSION

The low efficiency of fertilizer nitrogen in agricultural systems is primarily caused by the large losses of mineral nitrogen from those systems by gaseous loss: nitrous oxide emission is directly linked to the loss processes. It is axiomatic that any strategy that increases the efficiency of nitrogen fertilizer use will reduce emissions of nitrous oxide, and this has been directly demonstrated for a number of strategies (3).

The IPCC (1) reported that some combination of the following management practices, if adopted worldwide, would improve the efficiency of the use of synthetic fertilizer and manure nitrogen, and significantly reduce nitrous oxide emission into the atmosphere:

1. Match nitrogen supply with crop demand.
2. Tighten nitrogen flow cycles by returning animal wastes to the field and conserving residues instead of burning them.
3. Use controlled-release fertilizers, incorporate fertilizer to reduce volatilization, use urease and nitrification inhibitors, and match fertilizer type to precipitation.
4. Optimize tillage, irrigation, and drainage.

The potential decrease in nitrous oxide emissions from synthetic fertilizer, as a result of the mitigation techniques, could amount to 20% (1).

REFERENCES

1. IPCC (Intergovernmental Panel on Climate Change), *Climate Change 1995. Impacts, Adaptations and Mitigation of Climate Change: Scientific–Technical Analyses*; Watson, R.T., Zinyowera, M.C., Moss, R.H., Eds.; Cambridge University Press: Cambridge, England, 1996; 1–878.
2. Hengeveld, H.; Edwards, P. 1998. An Assessment of New Research Developments Relevant to the Science of Climate Change. Climate Change Newsletter **2000**, *12*, 1–52.
3. Mosier, A.; Kroeze, C.; Nevison, C.; Oenema, O.; Seitzinger, S.; van Cleemput, O. Closing the Global N_2O Budget: Nitrous Oxide Emissions Through the Agricultural Nitrogen Cycle. Nutri. Cycling Agroecosys. **1998**, *52*, 225–248.
4. Bremner, J.M. Sources of Nitrous Oxide in Soils. Nutri. Cycling Agroecosys. **1997**, *49*, 7–16.
5. Patrick, W.H., Jr. Nitrogen Transformation in Submerged Soils. *Nitrogen in Agricultural Soils*; Stevenson, F.J., Ed.; American Society of Agronomy: Madison, WI, 1982; 449–465.
6. Bhadrachalam, A.; Chakravorti, S.P.; Banerjee, N.K.; Mohanty, S.K.; Mosier, A.R. Denitrification in Intermittently Flooded Rice Fields and N-Gas Transport Through Rice Plants. Ecol. Bull. **1992**, *42*, 183–187.
7. Granli, T.; Bøckman, O.C. Nitrous Oxide from Agriculture. Norwegian J. Agric. Sci. **1994**, Supplement No. 12, 1–128.
8. Mosier, A.R. Chamber and Isotope Techniques. *Exchange of Trace Gas between Terrestrial Ecosystems and the Atmosphere*; Andreae, M.O., Schimel, D.S., Eds.; John Wiley & Sons: Chichester, England, 1989; 175–187.
9. Mosier, A.R.; Duxbury, J.M.; Freney, J.R.; Heinemeyer, O.; Minami, K. Nitrous Oxide Emissions from Agricultural Fields: Assessment, Measurement and Mitigation. Plant Soil **1995**, *181*, 95–108.

NITROUS OXIDE EMISSIONS: SOURCES, SINKS, AND STRATEGIES

Katsu Minami
National Institute for Agro-Environmental Sciences, Tsukuba, Japan

INTRODUCTION

Although it has been known for more than 50 years that nitrous oxide (N_2O) is a regular constituent of the atmosphere, it was not considered to be of any importance as an air constituent until the early 1970s. Atmospheric scientists hypothesized that N_2O released to the atmosphere through denitrification of nitrates in soils and natural waters may trigger reactions in the stratosphere leading to partial destruction of ozone layer protecting the earth from biologically harmful ultraviolet radiation from the sun (1). Nitrous oxide is also one of the natural components of Earth's atmosphere and contributes to the natural greenhouse effect therefore the increasing of N_2O in the atmosphere may be contributing to global warming (2).

The Fatmospheric concentration of N_2O has been increasing at an accelerated rate for several decades at a rate of 0.8 ppbv per year, and the lifetime of N_2O is 120 years. It has been estimated that doubling the concentration of N_2O in the atmosphere would result in a 10% decrease in the ozone layer which would increase the ultraviolet radiation reaching the earth by 20% (3), eventually leading to an increase in the occurrence of skin cancer and other health problems. The global warming potential (GWP) of each molecule of N_2O is about 210 times (20-year horizon) greater than each molecule of CO_2. Nitrous oxide currently accounts for 6% of total GWP (4).

THE BUDGET OF ATMOSPHERIC NITROUS OXIDE

Atmospheric Distribution of Nitrous Oxide

As a result of biotic and anthropogenic activities, the concentration of N_2O in the atmosphere is increasing at the rate of about 0.25% per year. The concentration of N_2O is about 0.75 ppbv higher in the Northern Hemisphere than in the Southern Hemisphere, suggesting the presence of greater source of N_2O in the Northern Hemisphere than in the Southern Hemisphere. Ice core measurements show that the pre-industrial value of N_2O was relatively stable at about 285 ppbv for most of the past 2000 years, and started to increase around the year 1700 associated with anthropogenic activity (4, 5).

Sinks for Nitrous Oxide

The major atmospheric loss process for N_2O is photochemical decomposition by sunlight (wavelengths 180–230 nm) in the stratosphere and is calculated to be 12.3 (range 9–16) Tg N per year. Tropospheric sinks such as surface loss in aquatic and soil systems are considered to be small as compared to atmospheric sink (5). However, the paucity of data does not enable us to estimate the importance of this sink on a global basis.

Sources of Nitrous Oxide

There are many sources of nitrous oxide, both natural and anthropogenic, which cannot be easily quantified. Undoubtedly, the Earth and oceans are significant N_2O sources. Nitrous oxide fluxes from upwelling regions of the Indian and Pacific Oceans clearly suggest that oceans may be a larger source. The ocean flux estimate is 3 (range 1–5) Tg N per year. Tropical forest soils are probably the single most important source of N_2O emission to the atmosphere. The total N_2O emission from tropical soils (forest, savannah) is estimated at 4 (range 2.7–5.7) Tg N per year. The magnitude of N_2O emissions from intensively fertilized tropical agricultural soils has not been quantified. Also, no attempt has been made to separate the tropical soil sources into natural and anthropogenic components (5).

The main anthropogenic sources are derived from agriculture and a number of industrial processes, such as adipic acid and nitric acid production. New research suggests that N_2O emissions from cropped, nitrogen-fertilized agricultural systems are significant on a global scale as shown in Table 1 (5).

Agricultural Fields

About 40% of N_2O sources are anthropogenic and, among them, fertilized soils account for about 60% (5). This figure could be an underestimate because tropical agricultural soil sources resulting from human activities have not been

N

Table 1 The amount of N_2O emission from agriculture (Mt/yr)

Mineral fertilizer	1.5 (0.5–2.5)
N-fixation	0.5 (0.25–0.75)
Soils after burning	0.1 (0.05–0.2)
Animal wastes	1.5 (0.5–2.5)
Biomass burning	0.2 (0.1–0.3)
Forest conversion	0.4 (0.1–1)
Total	4.2 (1.5–7.25)

separated from natural tropical soil sources. Agricultural N_2O emissions are considered to arise from fertilization of soils with mineral N and animal manures, from N derived from biological N fixation, and from enhanced soil N mineralization. Nitrous oxide is directly evolved during biomass burning, and produced in soil after burning, and enhanced emissions arise during the clearing of tropical forests for agricultural activities (4).

Bouwman (6) estimated the total emissions of N_2O from a regression equation: total annual direct field N_2O–N Loss = 1+0.0125 N-application (Kg N/ha). The value of 1 Kg N_2O-N/ha represents the background N_2O–N evolved and the 0.0125 factor expresses for the contribution from fertilization. This estimate includes N sources from a variety of mineral and organic N fertilizers and was based on long-term data sets. About 40% of the estimated N_2O production is derived from North and Central America, Europe and the former Soviet Union where about 20% of the world human population resides. Asian countries that hold about 55% of the global human population contribute about 40% of the estimated annual N_2O production (4).

MECHANISM OF NITROUS OXIDE PRODUCTION

Nitrification

Nitrification is the reaction whereby ammonium is oxidized to nitrate. In soils, autotrophic and heterotrophic bacteria mediate this process. The most common ammonium oxidizers are *Nitrosomonas* spp., which are involved in the formation of nitrite, while nitrite oxidation to nitrate is usually achieved by *Nitrobacter* spp. The overall nitrification sequence is as follows:

$$\begin{array}{c} \qquad\qquad N_2O \\ \qquad\qquad \uparrow \\ NH_4^+ \rightarrow NH_2OH \rightarrow (NOH) \rightarrow NO \rightarrow NO_2^- \rightarrow NO_3^- \end{array}$$

Two nitrogenous gases may evolve through nitrification, NO and N_2O. The nitrifiers are active over a wide range of temperatures (2–40°C). The overall nitrification process is controlled primarily by ammonium and oxygen concentrations. Because oxygen supply is moderated by soil moisture, the effect of soil water content on N transformation probably reflects its impact on oxygen diffusion in the normal soil moisture range. Nitrous oxide emissions from N-fertilizer applied agricultural fields have been detected by Bremner and Blackmer (7).

Denitrification

Biological denitrification refers to the dissimilatory reduction of nitrate and nitrite to produce NO, N_2O, and N_2 by a taxonomically diverse group of bacteria which synthesize a series of reductases that enable them to utilize successively more reduced N oxides as electron acceptors in the absence of oxygen. The general reductive sequence is as follows:

$$NO_3^- \rightarrow NO_2^- \rightarrow NO \rightarrow N_2O \rightarrow N_2$$

The most abundant denitrifiers are heterotrophs that require sources of electron-reducing equivalents contained in available organic matter. Soil factors that most strongly influence denitrification are oxygen, nitrate concentration, pH, temperature, and organic carbon. Nitrous oxide reductase appears to be more sensitive to oxygen than either nitrate or nitrite reductase. Therefore, N_2 production predominates in more anoxic sites and N_2O production may be higher under more aerobic conditions.

Chemical Decomposition of Nitrite (Chemodenitrification)

There is evidence that nitrite produced by nitrifying or denitrifying microorganisms can also react chemically to form N_2O via "chemodenitrification" (8). High nitrite concentrations have been attributed to the inhibition of nitrite oxidation, which is presumed to result from ammonia toxicity to *Nitrobacter*. Several investigators have noted that gaseous loss of nitrogen (via NO, N_2O or N_2) may accompany temporary nitrite accumulation. High concentration of nitrite is sometimes found in anaerobic soils where ammonium and ammonium type fertilizers are applied at high doses. Nitrite ions react chemically with organic molecules forming nitroso-groups (–N=O) that are unstable.

Other Mechanisms

Some of the N_2O evolved from soils may be formed by chemical decomposition of hydroxylamine (NH_2OH)

produced by nitrifying or nitrate-reducing microorganisms because NH_2OH has been identified as an intermediate in oxidation of ammonium to nitrate by *Nitrosmonas europeae* and has been postulated as an intermediate in microbial reduction of nitrate to ammonium, and several investigations have shown that NH_2OH is decomposed rapidly in soils with formation of N_2O and N_2 by processes that appear to be largely chemical. Other investigations indicated that Mn compounds are involved in the reactions leading to formation of N_2O and N_2 by chemical decomposition of NH_2OH in soils, and that $CaCO_3$ and Fe compounds are involved in the reactions leading to formation of N_2 in calcareous soils. The reaction is as follows:

$$2MnO_2 + 2NH_2OH \rightarrow 2MnO + N_2O + 3H_2O$$

Several workers (1) have postulated that N_2O is produced in soils through interaction of hydroxylamine and nitrite produced by soil microorganisms as follows:

$$NH_2OH + HNO_2 \rightarrow N_2O + 2H_2O$$

MEASUREMENT OF NITROUS OXIDE EMISSION

N_2O emissions from N-fertilized agricultural fields have been found to vary between 0.001 and 6.8% of the N applied to the field (9–11). A proportion of this variability in N_2O estimates, relative to the amount and type of fertilizer applied and to the type of cropping system such as grassland, upland crop, rice paddy and others, has been attributed to spatial and temporal change in the processes, which produce N_2O in soil.

MITIGATION STRATEGY

Many of the strategies were proposed by Mosier et al. (12): 1) match N supply with crop demand; 2) close N flow cycles; 3) use advanced fertilizer technologies; and 4) optimize tillage, irrigation and drainage, in which gaseous emissions deal primarily with cropping systems could be minimized. Although most of the practices listed are assumed to decrease N_2O emissions, there have been relatively few systematic studies in which various farming practices were compared as to their ability to conserve N and limit N_2O emissions. A number of field studies have been conducted with nitrification inhibitors that could decrease N_2O emissions when used. There are a few studies in which the potential of using controlled release fertilizer for decreasing N_2O emission was evaluated as follows.

Nitrification Inhibitors

Because ammonia and ammonium producing compounds are the main sources of fertilizer nitrogen, maintenance of the applied nitrogen in the ammonium form should result in a lower emission of N_2O from cultivated soils. One mechanism of maintaining added ammonium N is to use nitrification inhibitors with the fertilizer. Mosier et al. (12) summarized the effects of nitrification inhibitors on N_2O emission from fertilized soils both in laboratory and field studies. A number of field studies indicate that nitrification inhibitors do limit N_2O emissions from ammonium-based fertilizers. Several recent field tests also show that the utilization of a variety of nitrification inhibitors does significantly limit N_2O emissions from the application of ammonium-based fertilizers. To illustrates this point, a study to quantify the effect of nitrification inhibitors DCS (N-2, 5-dichlorophenyl succinamic acid) and the application of ammonium sulfate on N_2O emissions was conducted in field lysimeters using carrot (*Daucus carota* L.) as a test crop (11). The addition of DCS reduced about 30% of N_2O emission and leaching of nitrate.

Controlled Release Fertilizer

The use of controlled release fertilizer has the potential to improve N-use efficiency by matching nutrient release with crop demand, reducing NO_3^- leaching and denitrification losses. Polyolefin-coated fertilizers are a type of controlled release fertilizer where fertilizer granules are covered with a thermoplastic resin. The release of the N fertilizer is temperature dependent and is not controlled by hydraulic reactions or microbial attack of the coating. Greenhouse studies have revealed that controlled release fertilizer can increase yields with more N-fertilizer use efficiency and reduce NO_3^- leaching.

For example, Minami (11) observed that a controlled release fertilizer reduced N_2O emissions in lysimeter studies of carrot at Tsukuba, Japan, in which the fertilizer-induced emissions of N_2O-N during an 83 day-period of cultivation were 0.14 and 0.02% of the 250 Kg N applied following ammonium sulfate and a controlled release fertilizer, respectively.

REFERENCES

1. Bremner, J.M. Sources of Nitrous Oxide in Soils. Nutr. Cycling in Agroecosys **1997**, *49*, 7–16.
2. IPCC, In *Climate Change: The IPCC Scientific Assessment*; Houghton, J.T., Jenkins, G.J., Ephraume, J.J., Eds.; Cambridge University Press: Cambridge, 1990.

3. Crutzen, P.J.; Ehhalt, D.H. Effect of Nitrogen Fertilizers and Combustion on the Stratospheric Ozone Layer. Ambio **1977**, *6*, 112–117.
4. IPCC, In *Climate Change 1995, Impacts, Adaptations and Mitigation of Climate Change: Scientific-Technical Analyses*; Watson, R.T., Zinyowera, Mos R.H., Eds.; Cambridge University Press: Cambridge, 1996.
5. IPCC, In *Climate Change 1994: Radiative Forcing of Climate Change and an Evaluation of the IPCC IS92 Emission Scenarios*; Houghton, J.T., Meira Filho, L.G., Bruce, J., Lee, H., Callander, B.A., Haites, E., Harris, N., Maskell, K., Eds.; Cambridge University Press: Cambridge, 1995.
6. Bouwman, A.F. *Direct Emission of Nitrous Oxide from Agricultural Soils*, RIVM report No. 773004004; RIVM: Bilthoven, 1994; 1–28.
7. Bremner, J.M.; Blackmer, A.M. Effects of Acetylene and Soil Water Content on Emissions of Nitrous Oxide from Soils. Nature **1979**, *280*, 380–381.
8. Chalk, P.K.; Smith, C.J.; Chemodenitrification. In *Gaseous Loss of Nitrogen from Plant-Soil Systems*; Freney, J.R., Simpson, J.R., Eds.; Martinus Nijhof/Dr W. Junk Publishers: The Hague, 1983; 65–90.
9. Eichner, M.J. Nitrous Oxide Emissions from Fertilized Soils: Summary of Available Data. J. Environ. Qual. **1990**, *19*, 272–280.
10. Bouwman, A.F.; Exchange of Greenhouse Gases between Terrestrial Ecosystems and the Atmosphere. In *Soils and the Greenhouse Effect*; Bouwman, A.F., Ed.; John Wiley & Sons: New York, 1990; 61–127.
11. Minami, K. Nitrous Oxide Emissions from Agricultural Fields. In *Trace Gas Emissions and Plants*; Singh, S.N., Ed.; Kluwer Academic Publishers: Dordrecht, 2000; 215–230.
12. Mosier, A.R.; Duxbury, J.M.; Freney, J.R.; Heinemeyer, O.; Minami, K. Nitrous Oxide Emissions from Agricultural Fields: Assessment, Measurement and Mitigation. Plant and Soil **1996**, *131*, 95–108.

NO TILL

Paul W. Unger
United States Department of Agriculture (USDA), Bushland, Texas, U.S.A.

INTRODUCTION

Conservation tillage is any tillage or tillage and planting system that results in at least a 30% cover of crop residues on the soil surface after planting the next crop (1). It is used mainly to control soil erosion, but it also helps conserve water. In comparison, conventional tillage refers to tillage operations normally used for crop production that bury most residues and result in <30% cover after planting. Tillage that incorporates all residues into soil is clean tillage.

Tillage methods such as sweep, chisel, paraplow, subsoiling, slit, and strip rotary can usually qualify as conservation tillage. Even disk tillage may qualify, provided adequate residues are retained on the surface. The ultimate conservation tillage method is no tillage (or zero tillage) for which the next crop is planted without any soil disturbance since harvesting the previous crop. A special planter is usually needed to prepare a narrow, shallow seedbed for the seed being planted (1). Sometimes, no tillage is used in combination with a subsoiling operation that facilitates crop seeding and early plant root growth, but that leaves the surface residues virtually undisturbed, except for the slot caused by the subsoiling implement (1).

Adequate residues are not always produced to provide 30% cover (e.g., dryland (nonirrigated) crops). Also, a crop such as cotton (*Gossypium hirsutum* L.) may not produce enough residue under some conditions to satisfy the required ground cover for conservation tillage. Under such conditions, some conventional or even clean tillage methods can provide for soil and water conservation. Any tillage method that results in a rough or ridged surface helps reduce soil erosion by wind. Listing (ridge-forming tillage) is commonly used to help control wind erosion in the cotton-producing area of west Texas, U.S.A. where residue amounts are usually low (personal observation). Even plowing that brings erosion-resistant clods to the surface helps control wind erosion on some sandy soils (2). Any tillage method that impedes or prevents water flow across the surface helps reduce soil erosion by water and usually helps conserve water. Listing on the contour retains water on the surface, thus reducing erosion and conserving water. Furrow diking in conjunction with listing improves water retention where contour tillage is not used (3). Graded-furrow tillage allows excess water to flow slowly from land, thus reducing the potential for erosion; it also provides water conservation benefits (4).

ADVANTAGES AND DISADVANTAGES OF USING CONSERVATION TILLAGE AND NO TILLAGE

Advantages

Compared with clean tillage, advantages of different conservation tillage types, including no tillage, include improved erosion control, a cleaner environment, greater water conservation, equal or greater crop yields, less equipment and lower maintenance costs, lower energy and labor requirements, and greater net returns. Erosion control benefits from conservation tillage are due to more residues being retained on the soil surface. For controlling erosion by wind, residues shield the surface and reduce wind speed at the surface to below the threshold required for erosion to occur. Erosion by water is reduced because residues reduce the rate and amount of water flow across the surface. Residues also result in less soil particle detachment and transport due to raindrop splash and flowing water. The value of surface cover provided by crop residues for controlling erosion by wind and water is illustrated in Fig. 1 (5).

Greater water conservation with conservation tillage results from residues retarding the rate of water flow across the surface, thus providing more time for infiltration. Residues also shield the surface against raindrop impact, thus dissipating the energy of raindrops, reducing surface sealing, and maintaining favorable infiltration rates. Residues reduce soil water evaporation by shading the soil and slowing the wind at the soil surface. Of course, the soil must have adequate storage capacity for the water to be retained for later use by crops.

Use of conservation tillage reduces erosion, thus resulting in a cleaner environment. Erosion by wind damages crops, causes health and visibility problems, clogs

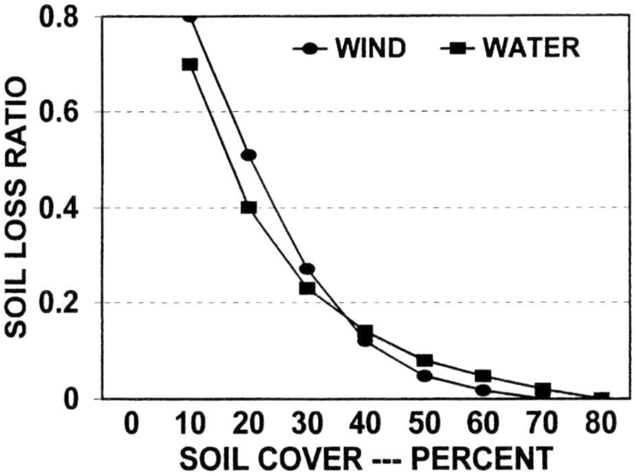

Fig. 1 Relationship of the soil loss ratio (soil loss with cover divided by soil loss from bare soil. (From Ref. 5.)

roads and waterways, damages machinery and homes, and pollutes the air. Erosion by water damages crops, roads, machinery, and homes. It also pollutes water with soil particles, chemicals adhering to the particles, and chemicals dissolved in water.

Crop yields are affected by numerous factors. Yields with conservation tillage systems are often greater than with clean tillage, provided no major problems are encountered. Yield increases, especially with no tillage, are usually attributable to greater soil water conservation, especially in subhumid and semiarid regions without irrigation. More favorable soil temperatures may also be involved. In warm or hot regions, high soil temperatures may injure plants, and surface residues with no tillage result in temperature decreases of up to 10°C (6, 7), which result in better crop performance. In cool regions, low temperatures with no tillage are usually detrimental to crop yields because planting is delayed beyond the optimal date (8).

Advantages of lower equipment inventories, equipment maintenance, and energy and labor requirements with conservation tillage are interrelated. With most conservation tillage methods, and especially no tillage, tillage frequency and intensity are lower than with clean tillage. As a result, less equipment may be needed, smaller tractors may be satisfactory (for no tillage), and the tractors and equipment are used less frequently. This results in less equipment maintenance and in lower fuel and labor requirements. Some fuel energy savings, however, may be partially offset by the energy required to produce herbicides and fertilizer, especially where no tillage is used. The no tillage system is based on using herbicides for weed control, and more nitrogen fertilizer is used under some conditions, especially when first converting to the system.

As for yields, many factors affect net returns for a crop production system. However, if production costs are not greater and yields are equal to or exceed those with clean tillage, then net returns should be equal or greater with conservation tillage, especially with no tillage, because equipment inventories and maintenance and labor and energy requirements are lower (9, 10).

Disadvantages

Problems with conservation tillage, especially no tillage, occur under some conditions (8, 11–13). A greater use of herbicides results in concern regarding the potential for polluting soil and water resources. Lower soil temperatures in cool regions delay crop planting, thereby potentially reducing crop yields. On poorly drained soils, additional water retained by using no tillage aggravates the excess soil water problem, thus generally reducing crop yields. Some weeds are difficult to control with herbicides, which, along with the high cost of some herbicides, may increase production costs. The possible need for new equipment may also increase production costs, especially when first converting to a no tillage system. Because crop residues are retained on the surface when a no-tillage system is used, there is the potential for increased pest problems (insects, diseases, rodents). Problems are greater with some insects and less with others, indicating that insect populations must be closely evaluated regardless of tillage system used. Organisms of some plant diseases are carried over to the next crop when residues are retained. Surface residues also provide shelter for rodents, which may be detrimental for the production of some crops. Other possible disadvantages include limited residue availability, greater soil compaction, and a need for greater managerial ability. Certainly, conservation tillage and no tillage are not suitable for all conditions. However, with good management, most problems (real or potential) can be minimized or avoided.

RESULTS ACHIEVED BY USING CONSERVATION OR NO TILLAGE

The value of conservation and no tillage farming methods for controlling erosion, conserving water, and increasing crop yields has been shown in numerous studies, but only a few examples will be given here. Probably the most dramatic example regarding the value of no tillage for controlling erosion occurred during a rainstorm on watersheds planted with corn (*Zea mays* L.) in Ohio (U.S.A.) (14). Treatments were clean tillage with sloping rows

N

(land slope 6.6%), clean tillage with contour rows (land slope 5.8%), and no tillage with contour rows (land slope 20.7%). On the respective treatment areas, rainfall was 140, 140, and 129 mm; runoff was 112, 58, and 64 mm; and sediment loss was 50.7, 7.2, and 0.07 Mg ha^{-1}. Even though the slope was much greater, soil loss was negligible from the no tillage area. Runoff was also low, which provided an opportunity to store more soil water, but soil water information was not given.

Soil losses determined by using a rainfall simulator averaged 2.1 and 5.7 Mg ha^{-1} where no tillage and conventional tillage, respectively, were used for wheat (*Triticum aestivum* L.) and soybean (*Glycine max* L.) production in Paraná, Brazil. Losses from bare soil averaged 100.2 Mg ha^{-1} during the same period (15). No tillage is widely used to control soil losses in Brazil (16) and Paraguay (personal observations). Crop residues retained on the soil surface by using no tillage was highly effective for reducing runoff and soil losses on Oxisols in Brazil (17).

Rainfall totaled 346 mm in July and August 1992 in Australia. Runoff from an Alfisol averaged 136 mm from bare soil where no, shallow, and deep tillage treatments were imposed, but only 4 mm with the same treatments when the soil was covered by straw (18). As a result, water storage was much greater in the straw-covered soil. Infiltration of water (simulated rainfall at 108 mm h^{-1} for 20 minutes) was greater (less runoff) in Queensland, Australia, when a clay soil was treated with gypsum or when a surface cover of wheat straw was present. With gypsum applied at 0 (check), 2, and 5 Mg ha^{-1}, cumulative infiltration was 12%, 37%, and 50%, respectively, with no surface cover and 45%, 68%, and 76%, respectively, with an 80% surface cover (19). The residue cover was more effective than gypsum for enhancing infiltration, thereby improving soil water conservation for crop production. Retaining crop residues on the soil surface by using no tillage was found to be effective for increasing water storage in soil in other studies in Australia (20, 21).

In India, soil water conservation and crop yield increases occurred when plant residues were placed on the soil surface as a mulch (22, 23). Applying the mulch soon after plant emergence was the most effective treatment (23). Similar results should occur by using no tillage, provided adequate residues are retained on the soil and weeds are effectively controlled. In some studies in India, however, weed control with no tillage was not as good as with plowing (24) or hand weeding (25), which resulted in lower crop yields with no tillage. Effective weed control is a prerequisite for successful crop production with no tillage.

After harvesting irrigated winter wheat, moldboard, rotary, disk, sweep, and no tillage treatments were imposed to manage the residues during the fallow period until planting dryland grain sorghum [*Sorghum bicolor* L. (Moench)] 10–11 months later at Bushland, Texas. U.S. Weed control was similar with all treatments. Plant available soil water contents averaged 149, 143, 158, 179, and 207 mm at sorghum planting and sorghum grain yields averaged 2.56, 2.19, 2.37, 2.77, and 3.34 Mg ha^{-1} with the respective treatments. Greater water contents and yields with conservation tillage (sweep and especially no tillage) resulted from the retention of more residues on the surface than with other treatments. The residues resulted in greater infiltration and lower evaporation, but the effect of the different processes could not be determined (26).

A field study at Akron, Colorado, U.S. clearly showed the value of surface residues with conservation tillage (minimum and no tillage) for reducing evaporation. Soil water content 1 day after 13.5 mm rain was similar to that at a depth of 15 cm where conventional, minimum, and no tillage treatments were imposed after harvesting winter wheat. The treatments resulted in 1.2, 2.2, and 2.7 Mg ha^{-1} of surface residues, respectively. After 34 rainless days, the soil had dried to a <0.1 m^3 m^{-3} water content at depths of 12, 9, and 5 cm, respectively (27). The value of surface residues for reducing evaporation was also shown under laboratory conditions (28, 29).

CONCLUSIONS

Conservation tillage and no tillage farming systems are based on retaining sufficient crop residues on the soil surface, mainly to control erosion. Other benefits include greater water conservation; improved environmental protection; equipment, energy, and labor savings; and often greater net returns to the producer. Some disadvantages occur under some conditions and the systems, especially no tillage, may not be suitable for all conditions. Most disadvantages, however, can be overcome or minimized by careful management.

REFERENCES

1. SSSA (Soil Science Society of America), *Glossary of Soil Science Terms, 1996*; Soil Science Society of America: Madison, WI, 1997.
2. Fryrear, D.W. Wind Erosion: Mechanics, Prediction, and Control. In *Dryland Agriculture, Strategies for Sustainability*; Advances in Soil Science, Singh, R.P., Parr, J.F., Stewart, B.A., Eds.; Springer-Verlag: New York, 1990; Vol. 13, 187–199.
3. Clark, R.N.; Jones, O.R. Furrow Dams for Conserving Rainwater in a Semiarid Climate. *Proceeding of Conference*

on Crop Production with Conservation in the 80's, Chicago, IL, December 1980; Am. Soc. Agric. Eng.: St. Joseph, MI, 1981; 198–206.
4. Richardson, C.W. Runoff, Erosion, and Tillage Efficiency on Graded-Furrow and Terraced Watersheds. J. Soil Water Conserv. **1973**, *28* (4), 162–164.
5. Papendick, R.I.; Parr, J.F.; Meyer, R.E. Managing Crop Residues to Optimize Crop/Livestock Production Systems for Dryland Agriculture. In *Dryland Agriculture, Strategies for Sustainability*; Advances in Soil Science, Singh, R.P., Parr, J.F., Stewart, B.A., Eds.; Springer-Verlag: New York, 1990; Vol. 13, 253–272.
6. Allen, R.R.; Musick, J.T.; Wiese, A.F. *No-Till Management of Furrow Irrigated Continuous Grain Sorghum*; Prog. Rpt. PR-3332 C, Texas Agric. Exp. Stn.: College Station, TX, 1975.
7. Rockwood, W.G.; Lal, R. Mulch Tillage: A Technique for Soil and Water Conservation in the Tropics. Span **1974**, *17*, 77–79.
8. Radke, J.K. Managing Early Season Soil Temperatures in The Northern Corn Belt Using Configured Soil Surfaces and Mulches. Soil Sci. Soc. Am. J. **1982**, *46* (5), 1067–1071.
9. Crosson, P.; Hanthorn, M.; Duffy, M. The Economics of Conservation Tillage. In *No-Tillage and Surface Tillage Agriculture*; Sprague, M.A., Triplett, G.B., Eds.; John Wiley and Sons: New York, 1986; 409–436.
10. Harman, W.L.; Martin, J.R. Economics of Conservation Tillage in Texas. In *Conservation Tillage: Today and Tomorrow*; Misc. Publ. MP-1634, Gerik, T.J., Harris, B.L., Eds.; Texas Agric. Exp. Stn.: College Station, TX, 1987; 24–37.
11. Phillips, R.E.; Phillips, S.H. *No-Tillage Agriculture, Principles and Practices*; Van Norstrand-Reinhold: New York, 1984.
12. Sprague, M.A.; Triplett, G.B. *No-Tillage and Surface-Tillage Agriculture*; John Wiley and Sons: New York, 1986.
13. Unger, P.W. Reduced Tillage. In *Semiarid Lands and Deserts, Soil Resource and Reclamation*; Skujiņš, J., Ed.; Marcel Dekker: New York, 1991; 387–422.
14. Harrold, L.L.; Edwards, W.M. A Severe Rainstorm Test of No-Till Corn. J. Soil Water Conserv. **1972**, *27* (1), 30.
15. Sidiras, N.; Derpsch, R.; Mondardo, A. Effect of Tillage Systems on Water Capacity, Available Moisture, Erosion, and Soybean Yield in Parana, Brazil. In *No-Tillage Crop Production in the Tropics*; Proceedings of a Symposium, Monrovia, Liberia, Akobunda, I.O., Deutsch, A.E., Eds.; International Plant Protection Center: Oregon State University: Corvallis, OR, 1983.
16. Muzilli, O. Soil Conservation Policies in the State of Parana, Brazil: The Role of Agricultural Research to Attain Sustainability. In *Abstracts: 10th International Soil Conservation Organization Conference*, West Lafayette, IN, May 1999; 96–97.
17. Derpsch, R.; Sidiras, N.; Roth, C.H. Results of Studies Made from 1977 to 1984 to Control Erosion by Cover Crops and No-Tillage Techniques in Paraná, Brazil. Soil Tillage Res. **1986**, *8*, 253–263.
18. Cogle, A.L.; Rao, K.P.C.; Reddy, M.V.; Srinivasan, S.T.; Megarry, D.; Smith, G.D.; Yule, D.F. The Impact of the Soil Biota and Cover on Runoff and Infiltration in a Hard Setting Alfisol in the SAT. In *Soil and Water Conservation, Challenges and Opportunities*; Proceedings 8th International Soil Conservation Conference, New Delhi, December 1994; Bhushan, L.S., Abrol, I.P., Rama Mohan Rao, M.S., Eds.; A.A. Balkema: Rotterdam, Brookfield, 1998; Vol. 2, 1546–1553.
19. Freebairn, D.M.; Woodruff, D.R.; Rowland, P.; Wockner, G.; Hamilton, A.; Radford, B.J. Options for Improving Water Conservation and Utilization in Semiarid Southern Queensland, Australia. In *Challenges in Dryland Agriculture—A Global Perspective*; Proceedings of the International Conference on Dryland Farming, Amarillo/Bushland, TX, August 1988; Unger, P.W., Sneed, T.V., Jordan, W.R., Jensen, R., Eds.; Texas Agric. Exp. Stn.: College Station, TX, 1988; 523–526.
20. Carter, M.R.; Mele, P.M.; Steed, G.R. The Effects of Direct Drilling and Stubble Retention on Water and Bromide Movement and Earthworm Species in a Duplex Soil. Soil Sci. **1994**, *157*, 224–231.
21. Carter, M.R.; Steed, G.R. The Effects of Direct-Drilling and Stubble Retention on Hydraulic Properties at the Surface of Duplex Soils in North-Eastern Victoria. Australian J. Soil Res. **1992**, *30*, 505–516.
22. Friesen, G.H.; Korwar, G.R. Conservation Tillage Systems for Sorghum Production Under Semi-Arid Conditions in India. Tropical Pest Manage. **1987**, *33* (4), 364–366.
23. Mahto, Deo Nandan; Sinha, Madan Prasad. Effect of Tillage on Weed Control and Yield of Maize. Indian J. Agron. **1980**, *25* (1), 146–148.
24. Agarwal, S.K.; De, Rajat. Effect of Application of Nitrogen, Mulching and Antitranspirants on the Growth and Yield of Barley Under Dryland Conditions. Indian J. Agric. Sci. **1977**, *47* (4), 191–194.
25. Mane, V.S.; Shingte, A.K. Use of Mulch for Conserving Moisture and Increasing the Yield of Sorghum in Dryland. Indian J. Agric. Sci. **1982**, *52* (7), 458–462.
26. Unger, P.W. Tillage and Residue Effects on Wheat, Sorghum, and Sunflower Grown in Rotation. Soil Sci. Soc. Am. J. **1984**, *48* (4), 885–891.
27. Smika, D.E. Seed Zone Soil Water Conditions with Reduced Tillage in the Semi-Arid Central Great Plains. Proceedings 7th Conference of the International Soil Tillage Research Organization, Sweden, 1976.
28. Unger, P.W.; Parker, J.J. Evaporation Reduction from Soil with Wheat, Sorghum, and Cotton Residues. Soil Sci. Soc. Am. J. **1976**, *40* (6), 938–942.
29. Ji, S.; Unger, P.W. Soil Water Accumulation Under Different Precipitation, Potential Evaporation, and Straw Mulch conditions. Soil Sci. Soc. Am. J. **2001**, *65* (2), 442–448.

NONPOINT SOURCE POLLUTION

Ravendra Naidu
Mallavarapu Megharaj
Peter Dillon
Rai Kookana
Commonwealth Scientific and Industrial Research Organisation (CSIRO) Land and Water, Glen Osmond, Adelaide, South Australia, Australia

Ray Correll
Commonwealth Scientific and Industrial Research Organisation (CSIRO) Mathematical and Information Sciences, Glen Osmond, Adelaide, South Australia, Australia

Walter Wenzel
University of Agricultural Sciences Vienna, Vienna, Austria

INTRODUCTION

Nonpoint source pollution (NPSP) has no obvious single point source discharge and is of diffuse nature (Table 1). An example of NPSP include aerial transport and deposition of contaminants such as SO_2 from industrial emissions leading to acidification of soil and water bodies. Rain water in urban areas could also be a source of NPSP as it may concentrate organic and inorganic contaminants. Examples of such contaminants include polycyclic aromatic hydrocarbons, pesticides, polychlorinated biphenyls that could be present in urban air due to road traffic, domestic heating, industrial emissions, agricultural treatments, etc. (1–3). Other examples of NPSP include fertilizer (especially Cd, N and P) and pesticide applications to improve crop yield. Use of industrial waste materials as soil amendments have been estimated to contaminate thousands of hectares of productive agricultural land in countries throughout the world.

CONTAMINANT INTERACTIONS

Nonpoint pollution is generally associated with low-level contamination spread at broad acre level. Under these circumstances, the major reaction controlling contaminant interactions are sorption–desorption processes, plant uptake, surface runoff, and leaching. However, certain contaminants, in particular organic compounds are also subjected to volatilization, chemical, and biological degradation. Sorption–desorption and degradation (both biotic and abiotic) are the two most important processes controlling organic contaminant behavior in soils. These processes are influenced by both soil and solution properties of the environment. Such interactions also determine the bioavailability and/or transport of contaminants in soils. Where the contaminants are bioavailable, risk to surface- and ground-water and soil, crop, and human health are enhanced.

IMPLICATIONS TO SOIL AND ENVIRONMENTAL QUALITY

Environmental contaminants can have a deleterious effect on nontarget organisms and their beneficial activities. These effects could include a decline in primary production, decreased rate of organic matter break-down, and nutrient cycling as well as mineralization of harmful substances that in turn cause a loss of productivity of the ecosystems. Certain pollutants, even though present in very small concentrations in the soil and surrounding water, have potential to be taken up by various microorganisms, plants, animals, and ultimately human beings. These pollutants may accumulate and concentrate in the food chain by several thousand times through a process referred to as biomagnification.

Urban sewage, because of its nutrient values and source of organic carbon in soils, is now increasingly being disposed to

N

Table 1 Industries, land uses, and associated chemicals contributing to nonpoint source pollution

Industry	Type of chemical	Associated chemicals
Agricultural activities	Metals/metalloid	Cadmium, mercury, arsenic, selenium
	Nonmetals	Nitrate, phosphate, borate
	Salinity/sodicity	Sodium, chloride, sulfate, magnesium, alkalinity
	Pesticides	Range of organic and inorganic pesticides including arsenic, copper, zinc, lead, sulfonylureas, organochlorine, organophosphates, etc., salt, geogenic contaminants (e.g., arsenic, selenium etc.)
	Irrigation	Sodium, chloride, arsenic, selenium
Automobile and industrial emissions	Dust	Lead, arsenic, copper, cadmium, zinc, etc.
	Gas	Sulfur oxides, carbon oxides
	Metals	Lead and lead organic compounds
Rain water	Organics	Polyaromatic hydrocarbons, polychlorbiphenyls, etc.
	Inorganic	Sulfur oxides, carbon oxides acidity, metals and metalloids

(From Barzi, F.; Naidu, R.; McLaughlin, M.J. Contaminants and the Australian Soil Environment. In *Contaminants and the Soil Environment in the Australasia-Pacific Region*; Naidu, R., Kookana, R.S., Oliver, D., Rogers, S., McLaughlin, M.J., Eds.; Kluwer Academic Publishers: Dordrecht, The Netherlands, 1996; 451–484.)

land. The contaminants present in sewage sludge (nutrients, heavy metals, organic compounds, and pathogens), if not managed properly, could potentially affect the environment adversely. Dumping of radioactive waste (e.g., radium, uranium, plutonium) onto soil is more complicated because these materials remain active for thousands of years in the soil and thus pose a continued threat to the future health of the ecosystem.

Industrial wastes, improper agricultural techniques, municipal wastes, and use of saline water for irrigation under high evaporative conditions result in the presence of excess soluble salts (predominantly Na and Cl ions) and metalloids such as Se and As in soils. Salinity and sodicity affect the vegetation by inhibiting seed germination, decreasing permeability of roots to water, and disrupting their functions such as photosynthesis, respiration, and synthesis of proteins and enzymes.

Some of the impacts of soil pollution migrate a long way from the source and can persist for some time. For example, suspended solids can increase water turbidity in streams, affecting benthic and pelagic aquatic ecosystems, filling reservoirs with unwanted silt, and requiring water treatment systems for potable water supplies. Phosphorus attached to soil particles, which are washed from a paddock into a stream, can dominate nutrient loads in streams and downstream water bodies. Consequences include increases in algal biomass, reduced oxygen concentrations, impaired habitat for aquatic species, and even possible production of cyanobacterial toxins, with serious impacts for humans and livestock consuming the water. Where waters discharge into estuaries, N can be the limiting factor for eutrophication; estuaries of some catchments where fertilizer use is extensive have suffered from excessive sea grass and algal growth.

More insidious is the leaching of nutrients, agricultural chemicals, and hydrocarbons to ground-water. Incremental increases in concentrations in ground-water may be observed over long periods of time resulting in initially potable water becoming undrinkable and then some of the highest valued uses of the resource may be lost for decades. This problem is most severe on tropical islands with shallow relief and some deltaic arsenopyrite deposits, where wells cannot be deepened to avoid polluted ground-water because underlying ground-water is either saline or contains too much As.

SAMPLING FOR NONPOINT SOURCE POLLUTION

The sampling requirements of NPSP are quite different from those of the point source contamination. Typically, the sampling is required to give a good estimate of the mean level of pollution rather than to delineate areas of pollution. In such a situation, sampling is typically carried out on a regular square or a triangular grid. Furthermore, gains may be possible by using composite sampling (4). However, if the pollution is patchy, other strategies may be used. One such strategy is to divide the area into remediation units, and to sample each of these. The possibility of movement of the pollutant from the soil to some receptor (or asset) is assessed, and the potential harm is quantified. This process requires an analysis of the bioavailability of the pollutant, pathway analysis, and the toxicological risk. The risk analysis is then assessed and decisions are then made as to how the risk should be managed.

MANAGEMENT AND OR REMEDIATION OF NONPOINT SOURCE POLLUTION

The treatment strategies used for managing NPSP are generally those that modify the soil properties to decrease the bioavailable contaminant fraction. This is particularly so in the rural agricultural environment where soil–plant transfer of contaminants is of greatest concern. Soil amendments commonly used include those that change the ion-exchange characteristics of the colloid particles and those that enhance the ability of soils to sorb contaminants. An example of NPSP management includes the application of lime to immobilize metals because the solubility of most heavy metals decreases with increasing soil pH. However, this approach is not applicable to all metals, especially those that form oxyanions—the bioavailability of such species increases with increasing pH. Therefore, one of the prerequisites for remediating contaminated sites is a detailed assessment of the nature of contaminants present in the soil. The application of a modified aluminosilicate to a highly contaminated soil around a zinc smelter in Belgium was shown to reduce the bioavailability of metals thereby reducing the Zn phytotoxicity (5). The simple addition of rock phosphates to form Pb phosphate has also been demonstrated to reduce the bioavailability of Pb in aqueous solutions and contaminated soils due to immobilization in the metal (6). Nevertheless, there is concern over the long-term stability of the processes. The immobilization process appears attractive currently given that there are very few cheap and effective in situ remediation techniques for metal-contaminated soils. A novel, innovative approach is using higher plants to stabilize, extract, degrade, or volatilize inorganic and organic contaminants for in situ treatment (cleanup or containment) of polluted topsoils (7).

PREVENTING WATER POLLUTION

The key to preventing water pollution from the soil zone is to manage the source of pollution. For example, nitrate pollution of ground-water will always occur if there is excess nitrate in the soil at a time when there is excess water leaching through the soil. This suggests that we should aim to reduce the nitrogen in the soil during wet seasons and the drainage through the soil. Local research may be needed to demonstrate the success of best management techniques in reducing nutrient, sediment, metal, and chemical exports via surface runoff and infiltration to ground-water. Production figures from the same experiments may also convince local farmers of the benefits of maintaining nutrients and chemicals where needed by a crop rather than losing them off site, and facilitate uptake of best management practices.

GLOBAL CHALLENGES AND RESPONSIBILITY

The biosphere is a life supporting system to the living organisms. Each species in this system has a role to play and thus every species is important and biological diversity is vital for ecosystem health and functioning. The detection of hazardous compounds in Antarctica, where these compounds were never used or no man has ever lived before, indicates how serious is the problem of long-range atmospheric transport and deposition of these pollutants. Clearly, pollution knows no boundaries. This ubiquitous pollution has had a global effect on our soils, which in turn has been affecting their biological health and productivity. Coupled with this, over 100,000 chemicals are being used in countries through out the world. Recent focus has been on the endocrine disruptor chemicals that mimic natural hormones and do great harm to animal and human reproductive cycles.

These pollutants are only a few examples of contaminants that are found in the terrestrial environment.

REFERENCES

1. Chan, C.H.; Bruce, G.; Harrison, B. Wet Deposition of Organochlorine Pesticides and Polychlorinated Biphenyls to the Great Lakes. J. Great Lakes Res. **1994**, *20*, 546–560.
2. Lodovici, M.; Dolara, P.; Taiti, S.; Del Carmine, P.; Bernardi, L.; Agati, L.; Ciappellano, S. Polycyclic Aromatic Hydrocarbons in the Leaves of the Evergreen Tree*Laurus Nobilis*. Sci. Total Environ. **1994**, *153*, 61–68.
3. Sweet, C.W.; Murphy, T.J.; Bannasch, J.H.; Kelsey, C.A.; Hong, J. Atmospheric Deposition of PCBs into Green Bay. J. Great Lakes Res. **1993**, *19*, 109–128.
4. Patil, G.P.; Gore, S.D.; Johnson, G.D. *Manual on Statistical Design and Analysis with Composite Samples, Technical Report No. 96-0501*; EPA Observational Economy Series, Center for Statistical Ecology and Environmental Statistics: Pennsylvania State University, 1996; Vol. 3.
5. Vangronsveld, J.; Van Assche, F.; Clijsters, H. Reclamation of a Bare Industrial Area, Contaminated by Non-Ferrous Metals:*In Situ* Metal Immobilisation and Revegetation. Environ. Pollut. **1995**, *87*, 51–59.
6. Ma, Q.Y.; Logan, T.J.; Traina, S.J. Lead Immobilisation from Aqueous Solutions and Contaminated Soils Using Phosphate Rocks. Environ. Sci. Technol. **1995**, *29*, 1118–1126.
7. Wenzel, W.W.; Adriano, D.C.; Salt, D.; Smith, R. Phytoremediation: A Plant–Microbe Based Remediation System. In *Bioremediation of Contaminated Soils*; Soil Science Society of America Special Monograph No. 37, Adriano, D.C., Bollag, J.M., Frankenberger, W.T., Jr, Sims, W.R. Eds.; Soil Science Society of America: Madison, USA, 1999; 772.

NUTRIENT DEFICIENCY AND TOXICITY SYMPTOMS

N

Noel J. Grundon
Atherton, Queensland, Australia

INTRODUCTION

When plants suffer from too much (a toxicity) or too little (a deficiency) of the mineral elements that are their food, they develop visual symptoms on their leaves, stems, fruit, and roots that are characteristic for the deficiency or toxicity. If we can understand these visual symptoms, we are better able to diagnose the causes of the disorders and to commence correcting the problem. In order to understand this visual language of sick plants, it is advantageous to understand the processes that underlie symptom development.

SYMPTOM DEVELOPMENT

Because mineral nutrients are associated with specific metabolic process, the development of symptoms and their location on the plant can be considered in relation to the function and mobility of the mineral elements within the plant.

In Relation to the Functions of Elements

The functions of mineral elements have been covered widely in the literature (see 1, 2) and include:

1. Being an integral constituent of an essential metabolite, complex or macro-molecular assembly, for example, amino acids; proteins (nitrogen and sulfur); enzymes and coenzymes (nitrogen, sulfur, iron, copper, zinc, manganese, nickel, molybdenum); nucleic acids (DNA, RNA), ADP, and ATP (nitrogen, phosphorus); cell walls (calcium); and chlorophyll (nitrogen, magnesium).
2. Being an activator of an enzyme or regulator of an enzyme-mediated process, for example, potassium, calcium, magnesium, zinc, manganese, and nickel.
3. Playing a nonstructural role in a physiological process, for example protein metabolism; pollen tube growth (boron); auxin metabolism (boron, zinc); membrane integrity and function (phosphorus, calcium, boron, zinc); photosynthesis (potassium, magnesium, chlorine, manganese); carbohydrate metabolism and partitioning (potassium, sulfur, magnesium, copper, zinc); stomatal movement (potassium, chlorine); and messenger in environmental signals (calcium).

Mineral elements may have nonspecific roles in addition to specific roles. For example, the ions of many elements play a role in cation–anion and pH balance, osmoregulation, and maintaining electrical neutrality within cells. Another example is the quality of harvested products: Calcium deficiency can lead to decreased quality of fruit causing bitter-pit in apple, blossom-end rot in tomato, and jelly-fruit in mango.

If an element is involved in the uptake or metabolism of another element, the symptoms may not be clearly defined and it can be difficult to identify the causal element. For example, nitrogen and sulfur are biochemically related in the formation of proteins. In forest trees there is no inorganic nitrogen in the foliage and the close biochemical linkage between organic nitrogen and organic (and total) sulfur causes the rate of nitrogen uptake to be limited by the rate of sulfur accumulation (3). Therefore, in many forest and horticultural tree species, the symptoms of mild nitrogen and sulfur deficiency are very similar (4).

In Relation to the Mobility of the Elements

Mineral elements differ markedly in their mobility and extent, and of retranslocation. All mineral elements are highly mobile within the xylem, but their mobility within the phloem can be divided into three classes: 1) readily mobile, 2) variably mobile, and 3) poorly mobile. Mineral elements that are readily mobile in the phloem (e.g., nitrogen, potassium, phosphorus, magnesium) are easily retranslocated from older to younger tissues when nutrient supply becomes limiting. Deficiency symptoms of these elements develop firstly in the older tissues and advance toward the younger tissues as the deficiency becomes more severe. Nutrients that have low phloem mobility (e.g., calcium, boron, iron) develop symptoms firstly in immature tissues. However, elements that are readily mobile under one set of conditions may appear to be only partly mobile, or even immobile under different conditions. For example, magnesium is normally classed as a mobile element and deficiency symptoms appear firstly on older leaves when a constant but inadequate supply occurs. However, when magnesium supply to the roots of subterranean clover was interrupted suddenly, symptoms developed firstly on the younger leaves (5).

Some elements show variable mobility depending on the degree of deficiency, and either the nitrogen or phosphorus status of the plant. For example, copper, zinc, and molybdenum are relatively immobile when nitrogen supply is adequate, and there is little or no movement of these elements from old leaves of plants deficient in them. Here, symptoms occur mainly in young tissues. However, should nitrogen become limited, copper, zinc, and molybdenum can be remobilized from older to younger leaves along with the retranslocation of nitrogen (6), and the symptoms of nitrogen deficiency are likely to dominate.

If a toxic nutrient element interacts with the metabolism of another element, symptoms of the toxicity occur on older tissues while symptoms of deficiency of the other element occur on those tissues characteristic for that nutrient. For example, on acidic soils, excessive levels of soluble manganese can induce iron deficiency in some plants, thereby causing the development of manganese toxicity symptoms on older leaves and iron deficiency symptoms on younger leaves (7).

DIAGNOSTIC SYMPTOMS

Descriptions and color photographs of mineral nutrient deficiencies and toxicities in a range of species have been published (see 8 for a comprehensive list). Symptom expression can vary greatly from one species to another, and also amongst cultivars within species. It is most important to check the specific symptoms described for a particular species. However, the first requirement in the use of visual symptoms is knowledge of the appearance of a healthy plant of the species concerned. Knowledge of the symptoms caused by insect pests and diseases also aids in the final diagnosis.

Deficiency

Because one element may have many functions, the characteristic symptoms of a deficiency are those associated with the function that is most sensitive or dominates. For example, zinc is associated with auxin metabolism, nucleotide synthesis, and membrane integrity, but it is impairment of auxin metabolism in zinc-deficient plants that leads to the characteristic symptoms of leaf distortion and failure of the internodes to elongate. General symptoms of deficiency of some nutrient elements are listed in Table 1.

Toxicity

Symptoms of nutrient toxicity occur more commonly in those tissues that have had a longer period of time to accumulate ions, e.g., the margins of older leaves in dicotyledons or the leaf-tips in monocotyledons and gymnosperms. Thus, marginal chlorosis and necrosis in older leaves are common symptoms of many nutrient toxicities (7, 9, 10). General symptoms of toxicity of some mineral elements are listed in Table 2.

Table 1 General symptoms of mineral element deficiency (Note that symptoms for specific crops and cultivars may differ)

Element	General symptoms
Nitrogen	Pale yellow chlorosis of the older leaves spreading to younger leaves, often with reddening in cold weather
Phosphorus	Dark green foliage; reddening or purpling of older leaves spreading to younger leaves
Potassium	Near-marginal to marginal chlorosis and necrosis of older leaves
Calcium	Death of the terminal meristems; distorted growth or death of leaf tips; disorders of fruit (e.g., blossom-end rot of tomato; bitter pit in apple; jelly-fruit in mango)
Magnesium	Marginal or interveinal chlorosis and necrosis of older leaves
Sulfur	Pale yellow chlorosis of immature leaves
Iron	Interveinal chlorosis or bleaching of immature leaves
Copper	Distorted growth of immature leaves or death of leaf tips; failure of fertilization and fruit set
Zinc	Little leaf, rosetting, and shortening of the internodes
Manganese	Interveinal chlorosis, necrotic spots, or streaks on older or middle leaves
Boron	Death of growing points; failure of fertilization and fruit set
Nickel	Interveinal chlorosis (mono-cotyledons) or leaf-tip necrosis (di-cotyledons)
Molybdenum	General chlorosis (legumes); mottled pale appearance (non-legumes)

(From Refs. 4, 7, 9, and 10.)

N

Table 2 General symptoms of mineral element toxicity (Note that symptoms for specific crops and cultivars may differ)

Nutrient	Visual symptoms
Nitrate-nitrogen	Marginal scorch of leaves followed by interveinal collapse
Ammonium-nitrogen	Necrosis and blackening of margins of leaves or leaf tips
Phosphorus	Marginal scorch and interveinal necrosis and death of older leaves
Sodium	Marginal scorch and necrosis of older leaves
Sulfate	Marginal scorch and necrosis of older leaves
Chloride	Bronzed chlorosis and marginal scorch of older leaves
Manganese	Marginal chlorosis and brown necrotic peppering on older leaves
Aluminium	Roots stunted, brown and multi-branched; symptoms on shoots may resemble phosphorus deficiency
Zinc	Interveinal necrosis on older leaves
Boron	Interveinal chlorosis becoming marginal necrosis on older leaves
Nickel	Transverse light and dark bands on immature leaves

(From Refs. 4, 7, 9, and 10.)

Advantages, Disadvantages, and Limitations

The use of visual symptoms to diagnose nutrient disorders has distinct advantages, including: 1) the technique can be applied in the field, and 2) it is not dependent on costly laboratory support services. However, even in the hands of experienced practitioners, there are a number of major disadvantages in relying on visual symptoms as the sole diagnostic tool, including:

1. When the disorder is mild or transient in nature, no foliar symptoms may be produced and clearly defined symptoms may occur only when the disorder is severe enough to depress yield.
2. For some micronutrient disorders, environmental factors (e.g., light and zinc deficiency symptoms) have a profound effect on the appearance or severity of the symptoms.
3. Under certain conditions, different nutrient disorders can produce similar symptoms.
4. By the time definitive symptoms become evident, it is often too late to correct the problem in that growing season.

The Final Diagnosis

Despite the limitations to its application, the use of visual symptoms remains a valuable diagnostic tool. It is important to recognize that the use of visual symptoms provides a preliminary diagnosis only. Confirmation by other methods such as plant and soil analyses, pot culture assays, field experiments, or test strips is an essential second step.

REFERENCES

1. Marschner, Horst. Functions of Mineral Nutrients: Macronutrients. *Mineral Nutrition of Higher Plants*, 2nd Ed.; Academic Press: London, 1995; 229–312.
2. Marschner, Horst. Functions of Mineral Nutrients: Micronutrients. *Mineral Nutrition of Higher Plants*, 2nd Ed.; Academic Press: London, 1995; 313–404.
3. Turner, J.; Lambert, M.J. Nutrition and Nutritional Relationships in *Pinus radiata*. Annual Review of Ecological Systems **1986**, *17*, 325–350.
4. Dell, B.; Malajczuk, N.; Grove, T.S. Atlas of Deficiencies. *Nutrient Disorders in Plantation Eucalypts*, ACIAR Monograph 31; Australian Centre for International Agricultural Research: Canberra, Australia, 1995; 9–92.
5. Scott, B.J.; Robson, A.D. Distribution of Magnesium in Subterranean Clover (*Trifolium subterraneum* L.) in Relation to Supply. Australian Journal of Agricultural Research **1990**, *41*, 499–510.
6. Hill, J.; Robson, A.D.; Loneragan, J.F. The Effects of Copper and Nitrogen Supply on the Retranslocation of Copper in Four Cultivars of Wheat. Australian Journal of Agricultural Research **1978**, *29*, 925–939.
7. Grundon, N.J.; Edwards, D.G.; Takkar, P.N.; Asher, C.J.; Clark, R.B. *Nutritional Disorders of Grain Sorghum*, ACIAR Monograph 2; Australian Centre for International Agricultural Research: Canberra, Australia, 1987; 99.
8. Grundon, N.J.; Robson, A.D.; Lambert, M.J.; Snowball, K. Nutrient Deficiency and Toxicity Symptoms. In *Plant Analysis: An Interpretation Manual*, 2nd Ed.; Reuter, D.J., Robinson, J.B., Eds.; CSIRO Publishing: Collingswood, Australia, 1997; 37–50.
9. Blamey, F.P.C.; Edwards, D.G.; Asher, C.J. *Nutritional Disorders of Sunflower*; Department of Agriculture, The University of Queensland: St. Lucia, Australia, 1987; 72.
10. Grundon, N.J. *Hungry Crops: A Guide to Nutrient Deficiencies in Field Crops*; Queensland Department of Primary Industries: Brisbane, Australia, 1987; 246.

NUTRIENT DIFFUSION, BIOAVAILABILITY AND PLANT UPTAKE

Niels Erik Nielsen
The Royal Veterinary and Agricultural University, Copenhagen, Denmark

INTRODUCTION

Entry of nutrient elements into plants and, therefore, into the food web of human beings, depends on the capability of the soil to release and maintain a concentration bigger than the minimum concentration (c_{min}) of elements in the soil solution at the root surface, and on the uptake capacity of the plant roots. Root-induced rhizosphere processes influence nutrient availability, maintenance of the nutrient concentration in the soil solution and plant uptake. The entry into plants and the associated flux of a nutrient element toward root surface may be controlled by a number of rate-limiting and/or rate-determining processes. One of the rate-limiting processes may be the flux by diffusion from the solid constituents in the soil via the soil solution to the nutrient-absorbing cell membrane in root tissue near the root surface. The goal of good agronomy is always to obtain a useful and sustainable modification of rate-limiting and/or rate-determining processes in the soil plant atmosphere system, aiming at high yields and qualities of the crops. This goal requires identification of the rate-limiting and/or rate-determining steps, an understanding of their dynamics, and knowledge of how to obtain appropriate modifications of these steps.

SOIL PLANT SYSTEM

The movement of any nutrient element, M, from solid soil constituents to the root surface, and its entry into plants, can be divided into a sequence of processes (steps), as illustrated in Fig. 1, which also indicates major agronomical actions used to improve nutrition and growth of crop plants. The ↔ denotes solid-phase processes slowly approaching equilibrium, or microbial-mediated net mineralization of N, S and P, for example. Also denoted are the source/sink processes by which diffusible nutrients are being produced or removed by chemical, physical, and biological transformation processes. In Fig. 1, L denotes ligands that are any dissolved solute reacting with M to form ML, which are organic complexes and ion pairs dissolved in the soil solution. The occurrence of L and, therefore, of ML, increases the total concentration (M + ML) and mobility of the nutrient element. The ⇆ denotes reversible processes which are spontaneously approaching equilibrium. Depending on ion species, ion concentration at the root surface, and plant age, the symbol ⇄ denotes processes that may be irreversible. The irreversible processes are always rate limiting and/or rate determining, whereas reversible processes may be rate limiting, only. Processes 2 and 3 are in the vicinity of the soil particle, whereas process 7 is in the vicinity of the root. Processes 4, 5, and 6 are transport processes by mass-flow and diffusion due to water uptake and nutrient uptake (process 8) by cell membranes of root cells near the root surface or root hairs. Process 9 is the nutrient translocation in the plant. Process 10 is the plant growth that also integrates the absorbed nutrient into the plant tissue. Processes 8, 9, and/or 10 create the concentration (electro-chemical) gradients for irreversible net flux of nutrients from the soil-soil solution system into the plants. Hence, at any time, the rate-determining processes (Fig. 1) are then either the release of the nutrient into the pool of plant available nutrient (source/sink processes in the soil), or the nutrient uptake into the roots, its translocation/circulation in the plant, and/or the rate by which the nutrient is built into new tissue. Mass-flow and diffusion may be rate limiting only, and not rate determining. Usually only a small fraction of the plant-available nutrients is dissolved in the soil solution. This implies that the bioavailability of nutrients to plant roots is governed by several soil properties including, for example, the characteristics of process 2 in Fig. 1 and the possibilities for movement via soil solution to the root surface by mass-flow and diffusion. The concept of a *bioavailable nutrient* can then be defined as a nutrient element that is present in a pool of diffusible (available) nutrients which are close enough to arrive at water- and/or nutrient-absorbing root surfaces during a period of 10 days, for example. This seems to fit with the observation that most of the depletion zones of slowly moving nutrients, such as phosphorus, are created during the first 10 days after root growth into a new soil volume unit (1). The bioavailable quantity of a nutrient in the soil is affected by at least five different groups of processes as indicated in Table 1.

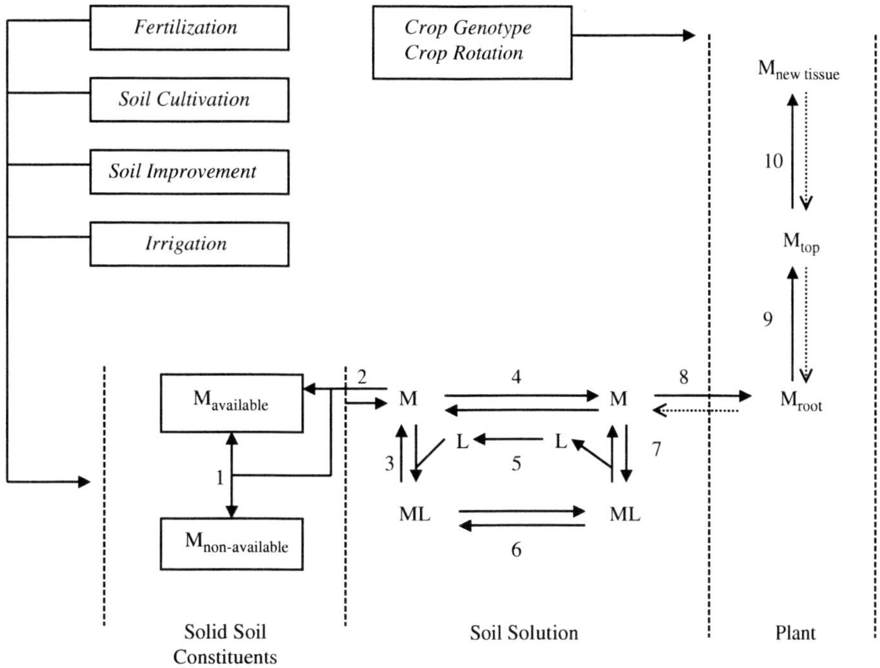

Fig. 1 Flux of a nutrient element in the soil plant atmosphere system and agronomy actions; nutrient element (M), ligands (L), reversible $\leftrightarrows$ and irreversible processes. (From Ref. 22.)

Table 1 Processes and factors involved in nutrient transfer from soil to plant roots

Process	Factors
Release, mineralization, and dissolvement of the nutrient in the soil solution	Chemical and physical properties of the solid phases, temperature, soil water content, and activity of the microbial biomass
Root development	Root length, distribution of roots in the root zone, root morphology, root hairs, rate of root growth, and root surface contact area with soil solution
Solute movement by mass flow and diffusion to roots	Transpiration rate (w_o), concentration of the nutrient in the soil solution (c_b), effective diffusion coefficient (D_e), nutrient buffer power of the soil (b)
Rhizosphere processes, increasing the rate of nutrient release in the soil	Depletion of the soil solution for nutrients by the roots; root exudates as protons, reducing agents, chelates, organic anions, enzymes; modification of microbial activity; mycorrhizal and Rhizobium symbioses.
Nutrient uptake	Concentration of the nutrient at the root surface (c_o); Transport kinetic parameters of nutrient uptake by the roots ($\bar{I}_{max}$, Km and c_{min})

(From Ref. 22.)

DIFFUSION

Diffusion is the net movement of a solute or a gas from a region that has a higher concentration, to an adjacent region that has a lower concentration. Diffusion is a result of the random thermal motions of molecules in the considered solids, solution, or air. The net movement caused by diffusion is a statistical phenomenon because the probability of the molecules' movement from the concentrated to the diluted region is greater than vice versa. Fick (1855) was one of the first to examine diffusion on a quantitative basis. The basic equation (Eq. 1) to express diffusion e.g., to a root is today known as Fick's first law of diffusion:

$$F = -D\frac{\partial C}{\partial r} \tag{1}$$

in which F is the diffusive flux (mol cm^{-2} s^{-1}) of a nutrient in the r-direction normal to the root cylinder. The driving force (the gradient in the electro-chemical potential) is, in most cases, approximated by the concentration gradient

$$\frac{\partial C}{\partial r}(\text{mol cm}^{-4})$$

and D is the diffusion coefficient (cm^2 s^{-1}). They way to describe diffusion processes mathematically under various conditions has been presented by Jost (2) and Crank (3). Great contributions to our understanding of solute movement in the soil root system by mass-flow and diffusion have been given or reviewed by Nye and Tinker (4, 6) and Barber (5). Recently, Willigen and colleagues (7) reviewed some aspects of the modeling of nutrient and water uptake by plant roots.

Nutrient Movement by Mass Flow and Diffusion from Soil to Plant Roots

Nutrients bound to the solid-soil phase are virtually immobile in the sense of its movements to roots. The nutrient has to be released into the soil solution as indicated in Fig. 1. Furthermore, contact between the root and nutrient-absorbing membranes in the root tissue near the root surface and the soil solution is a prerequisite for nutrient uptake. Contact to nutrient pools can be brought about by two means: 1) by growth of roots to the sites where nutrient pool are located (root interception), and 2) by movement of the nutrient from the bulk (the pool) of the soil to the root surface. Even so, nutrients may at any time move over a certain distance in the soil solution and cell wall before they reach the outside cell membrane of a root hair or a root cortical cell for uptake. The mechanisms for these transports are mass flow and diffusion (4, 8). The driving force for the net movement of nutrients (Fig. 1) is the water and the selective nutrient uptake by the plant root, creating a concentration gradient (dc/dr). The general equation of continuity (mass-balance) used to describe movements in a direction normal to a root cylinder at radial distance r and time t, may partly be developed from Eq. 1, extended and expressed as

$$\left[b\frac{\partial c}{\partial t}\right]_r = -\left[\frac{1}{r}\frac{\partial rF_T}{\partial r}\right]_t + U_{r,t} \tag{2}$$

in which

- U is the production/consumption term (mole cm^{-3} s^{-1}) at r (radial distance from the center of the root) and t (time).
- b is the buffer power (dC/dc) in which C is the total concentration of diffusible solute in the soil and c is the concentration of solute in the soil solution.
- F_T is total net flux of solute by mass flow and diffusion (mole cm^{-2} s^{-1}).

$$F_T = F_m + F_d$$

where $F_m = wc$ in which w is the flux of soil solution in the direction of the root (cm^3 cm^{-2} s^{-1}) and c is the nutrient concentration of soil solution (mole cm^{-3}). The expected rates of water flux at the root surface are 0.2–1 10^{-6} cm s^{-1} (5). The flux, F_d, by diffusion can be expressed by Flick's first law

$$F_d = -D_c b\frac{dc}{dr}$$

The $b = dC/dc$ is the soil buffer power defined previously. The C is the sum of the amount of nutrient in the soil solution and the amount of adsorbed nutrient that is able to replenish the nutrient in the soil solution spontaneously. Hence, b is the parameter mediating the effects of the soil chemical conditions on nutrient uptake by plants. D_e denotes the effective diffusion coefficient in the soil. D_e differs between media, but it can be related to the diffusion coefficient D_o for the nutrient in free soil solution. The influences of soil on diffusion, and thereby the relation between D_e and D_o, can be expressed by Eq. 3 (9).

$$D_e = D_0\theta f/b \tag{3}$$

where θ is the volumetric water content expressed as a fraction, and f is the impedance factor that essentially allows for the increase in the actual diffusion distance because of the tortuous pathway of water filled soil pores and water films. The volumetric water content that allows a reasonable root activity is between 0.1 and 0.4. The value of f increases with increase in water content (10), whereas the buffer power remains constant with changes in soil moisture at the same bulk density (11). It has been observed (12) that the relation between f and θ can be expressed empirically by $f = 1.58\theta - 0.17$ for θ>about 0.11. From this it may be estimated that D_e decreases about 18 times if θ decreases from 0.40 to 0.15. Hence D_e is the parameter mediating the effects of soil moisture, soil chemical, and soil physical conditions on diffusion in soil.

Almost all studies on solute movement in the soil plant system neglect U in Eq. 2 because our understanding of the biology caused by root-induced processes and its effects on production or consumption of available nutrients is incomplete as yet.

To solve Eq. 2 for a given soil plant system is a complicated process, and in most cases, difficult or even impossible because of the lack of information on the soil root interactions and root behavior. The method for obtaining analytical and numerical solutions of Eq. 2 under a number of often simplified soil plant conditions has been summarized (5–7). However, illustration of the importance of diffusion for the bioavailability of nutrients in soils may be based on Eq. 4

$$\Delta r = \sqrt{2D_e t} \tag{4}$$

in which Δr is the average distance of diffusion; e.g., in a direction normal to a root. The mathematics behind Eq. 4 has been presented by Jost (2). Based on Eq. 4, the equivalent soil volume (V in cm^3) of soil depleted for diffusible (available) nutrients—*the quantity of bioavailable nutrients*—can then be estimated as follows (Eq. 5) for roots without root hairs:

$$V = \pi(\Delta r + r_o)^2 L_v \tag{5}$$

in which Δr is estimated from Eq. 4, r_o is the root radius and L_v is the root density in cm cm^{-3} of soil. The data in Table 2 show the expected effective diffusion coefficient of a number of plant nutrients in soil and corresponding influences on the nutrient bioavailability; in addition, the data show how a decrease of the soil moisture from $\theta =$ 0.40 to $\theta = 0.15$ affects the bioavailability at a root density of 5 cm^{-2} of roots without root hairs. It may be calculated from $f = 1.58\,\theta - 0.17$ and Eq. 3 that the diffusive flux decreases by a factor of $18 = D_e^{\theta=0.40}/\ D_e^{\theta=0.15}$. At field capacity of water content, the expected, effective diffusion coefficient of nitrate in soil is $10^{-6}\,cm^2\,s^{-1}$. This is almost 10 times slower than in pure water. Hence, a pored media, such as soil, physically decreases the possibility for solute movement with a factor of nearly 10. As the soil dries out, this factor increases as illustrated in Fig. 2. Apart from nitrate and chlorine, nutrient elements are adsorbed more or less to the solid soil constituents. This is the main cause of the decrease of the diffusion coefficients below $10^{-6}\,cm^2\,s^{-1}$. The diffusion coefficient of phosphorus is as low as $10^{-9}\,cm^2\,s^{-1}$ mainly because approximately 0.1% of the diffusible (available) phosphorus is dissolved in the soil solution only. This has a large effect on the bioavailability of the plant-available quantities of the various nutrient elements in soil as illustrated in columns 3 and 4 of Table 2. The V-values ≥ 1 indicate that the root at

Table 2 Expected effective diffusion coefficients of some nutrients in soil at field capacity of water content (e.g, $\theta = 0.40$), and estimated bioavailability as a fraction of diffusible (available) nutrient (Eq. 4 and 5) at a root density ($L_v = 5\,cm^{-2}$) of roots without root hairs (Mean root radius, r = 0.01 cm; Time, t = 10 days)

Element	**D_e[a] ($cm^2\,s^{-1}$)**	**Bioavailable nutrient as a fraction of available nutrient (V, cm^3)**	
		$\theta = 0.40$	$\theta = 0.15$
Nitrate	$1*10^{-6}$	27.558	1.574
Potassium	$1*10^{-7}$	2.847	0.180
Boron	$1*10^{-7}$	2.847	0.180
Magnesium	$1*10^{-8}$	0.314	0.026
Calcium	$1*10^{-8}$	0.314	0.026
Phosphorus	$1*10^{-9}$	0.042	0.006
Manganese	$1*10^{-9}$	0.042	0.006
Molybdenum	$1*10^{-9}$	0.042	0.006
Zinc	$1*10^{-9}$	0.042	0.006
Iron	$1*10^{-10}$	0.008	0.003

[a] Values obtained from Refs. 5 and 22.

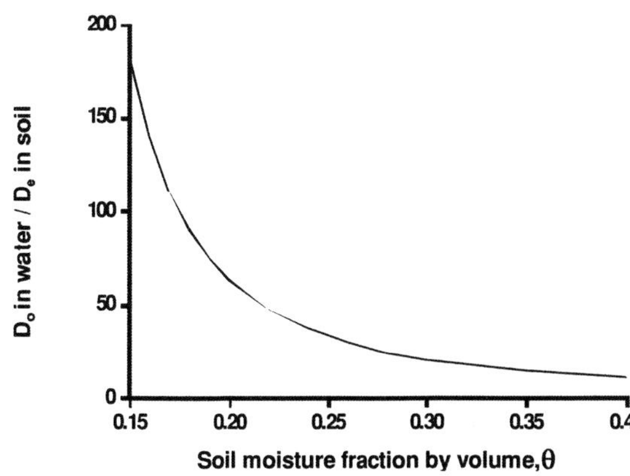

Fig. 2 Effects of soil and soil moisture on nutrient element mobility.

a density of $5\,cm^{-2}$ is able to deplete all the available nutrient as seen for nitrate, even under dry conditions, whereas only 4% of the available phosphorus is bioavailable inside a period of 10 days. If the soil dries out to $\theta = 0.15$, the bioavailability decreases to only 0.6%. This illustrates that the decrease of soil moisture may create nutrient deficiency even in soil with high phosphorus fertility. However, phosphorus uptake is increased by the activity of root hairs (discussed in the following).

IMPORTANCE OF ROOT HAIRS

Root hairs are outgrowths from specialized root epidermal cells (trichoblasts). Root hair length, diameter, and number per unit length of root, vary among plant species and among genotypes within the same species (5, 6, 13). Frequency and size of root hairs are affected by many environmental factors, as well. In nature, the length of root hairs vary from 0.01–0.15 cm, the radius varies from 0.0005–0.002 cm, and the number per unit of length varies from 100–1000 per cm root. The importance of root hairs for phosphorus uptake has been demonstrated directly in the laboratory (14, 15) and under field conditions (13). It is reasonable to assume that the clusters of root hairs' outer tips form a fairly well-defined cylinder to which phosphorus diffuses, on average, a distance Δr in 10 days, and that root hair density and its period of function are long enough to withdraw the entire available nutrient in the soil penetrated by root hairs. The bioavailability of phosphorus, for example, as affected by root hair length, can then be estimated from the following extension of Eq. 10:

$$V = \pi(\Delta r + \sigma + r_o)^2 L_v \tag{6}$$

in which σ is the root hair length in cm.

Fig. 3 illustrates that the bioavailability of phosphorus increases exponentially with the root hair length. Hence, root hairs play a very important role for the bioavailability of nutrients having a low effective diffusion coefficient (D_e) in soil.

BOUNDARY CONDITIONS AND NUTRIENT ENTRY

If depletion zones around the roots do not overlap, the solute concentration converges to the solute concentration c_b in the bulk solution, at which $F_T = F_m + F_d = 0$. The boundary conditions at the surface of the root are

$$F_T = F_m + F_d = \alpha c_o \tag{7}$$

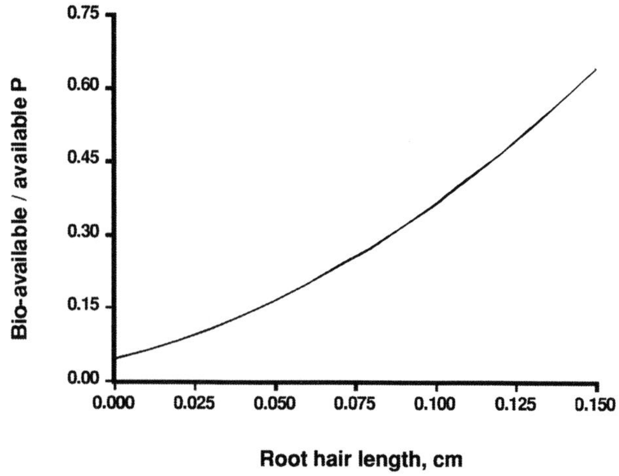

Fig. 3 The effect of root hair length on phosphorus bioavailability. Soil moisture ($\theta = 0.4$); root density $5\,cm^{-2}$; root radius 0.01 cm.

in which α ($cm\,s^{-1}$) is the root-absorbing power defined by Tinker and Nye (6) and c_o is the concentration of solute at the root surface. It can be learned from Eq. 7 that the actual concentration (c_o) at the root surface and, therefore, the rate of nutrient flow per unit length of root, is determined by the ratio F_T/α. The kinetics of net uptake of nutrients (5, 16–18) may be expressed by:

$$\overline{\text{In}}L^* = \frac{L^*\bar{I}_{max}(c_0 - c_{min})}{Km + c_0 - c_{min}} \tag{8}$$

in which L^* is the root length per unit of plant biomass, $\bar{I}_{max}$ ($mole\,cm^{-1}\,s^{-1}$) is the mean maximal net influx, Km ($mole\,cm^{-3}$) is the Michaelis–Menten factor, c_o is the concentration of the nutrient at the root surface, and c_{min} is the nutrient concentration at which $\overline{\text{In}} = 0$. The values of the parameters $\bar{I}_{max}$, Km and c_{min} vary according to the plant nutrient, temperature, and plant species/genotype and plant age. Furthermore, kinetics of nutrient uptake by roots may be influenced by ion interactions. Determined values of L^*, $\bar{I}_{max}$, Km and c_{min} for uptake of several nutrients by several plant species or genotypes, obtained under conditions in which the rate-determining step of nutrient uptake was located in the roots, has been noted (19). The data show that the values of L^*, $\bar{I}_{max}$, Km and c_{min} vary considerably among nutrients and among plant species and genotypes. This illustrates the efficiency by which these plants utilize soil as a source of nutrients. It is possible from $F_T = F_m + F_d = \alpha c_o$ and Eq. 8 to develop how α varies at varying solute concentration at the root surface by:

$$\alpha = \frac{\bar{I}_{max}(c_o - c_{min})}{2\pi r_0 c_o(Km + c_o - c_{min})} \tag{9}$$

Figs. 4A and 4B illustrate the variation of α for phosphorus uptake at low concentration (c_o) at the root surface of some plant species and barley genotypes. Hence, the root-absorbing power (α) varies also at low concentration (c_o) of solute at the root surface. This implies that phosphorus uptake at low P concentration is more under the control of the plant parameters determining the size of α than under the control of P diffusion in the soil, whereas at the range of c_o at which α has achieved its maximum, uptake is controlled by diffusion.

CONCLUSION

Even though the movement—and the main factors affecting the movement—of nutrient elements to root by mass-flow and diffusion is well known, the effect of soil conditions on crop growth is still not properly understood. It is obvious that the big variation (Table 2) of the effective diffusion coefficient (D_e), caused mainly by the variation of the soil chemistry of the various nutrient elements, has a large impact on the bioavailability of nutrient elements. The mobility of phosphorus and micronutrients is so low in most soils that the soil exploited by root hairs is the main source of these elements. The root-induced modifications to the soil in the rhizosphere would then have a considerable impact on the efficiency by which plants use the rhizosphere soil as a source of nutrients. The understanding of how root-induced processes accelerate solute movement and the transformation of non-available nutrients to bioavailable nutrients is increasing (6, 20, 21). Root hair length and root-induced processes appear to vary between genotypes of our crop plants (13). Hence, improvement of the efficiency by which plants use soil as a source of nutrients seems to be a possibility by targeted plant breeding.

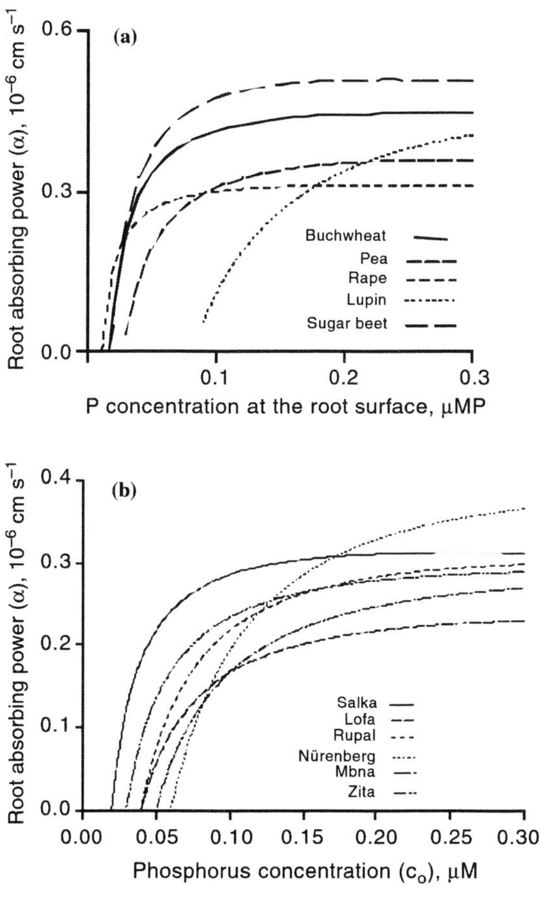

Fig. 4 Absorption power (α) at varying phosphorus concentration at the root surface of some plant species (A) and barley genotypes (B). (From Refs. 19 and 22.)

REFERENCES

1. Gahoonia, T.S.; Raza, S.; Nielsen, N.E. Phosphorus Depletion in the Rhizosphere as Influenced by Soil Moisture. Plant and Soil **1994**, *159*, 213–218.
2. Jost, W. *Diffusion in Solids, Liquids and Gasses*; Academic Press: New York, 1960.
3. Crank, J. *The Mathematics of Diffusion*, 2nd Ed.; Clarendon Press: Oxford, 1975.
4. Nye, P.H.; Tinker, P.B. *Solute Movement in the Soil-Root System*; Blackwell: Oxford, 1977.
5. Barber, S.A. *Soil Nutrient Bioavailability*, 2nd Ed.; Wiley: New York, 1995.
6. Tinker, P.B.; Nye, P.H. *Solute Movement in the Rhizosphere*; Oxford University Press: Oxford, 2000.
7. Willigen, P.; Nielsen, N.E.; Claassen, N.; Castrignanò, A.M. Modelling Water and Nutrient Uptake. In *Root Methods a Handbook*; Smit, A.L., Bengough, A.G., Engels, C., Noordwijk, M., Pellerin, S., Geijn, S.C., Eds.; Springer: Berlin, 2000.
8. Barber, S.A. A Diffusion and Mass-Flow Concept of Soil Nutrient Availability. Soil Science **1962**, *93*, 39–49.
9. Nye, P.H. The Measurement and Mechanism of Ion Diffusion in Soil. I. The Relation Between Self-Diffusion and Bulk Diffusion. J. Soil Sci. **1966**, *17*, 16–23.
10. Rowell, D.L.; Martin, M.W.; Nye, P.H. The Measurement and Mechanisms of Ion Diffusion in Soils. III. The Effect of Moisture Content and Soil Solution Concentration on the Self-Diffusion of Ions in Soils. J. Soil Sci. **1967**, *18*, 204–222.
11. Bhadoria, P.B.S.; Classen, J.; Jungk, A. Phosphate Diffusion Coefficient in Soil as Affected by Bulk Density and Water Content. Z. Pflanzenernaehr. Bodenk. **1991**, *154*, 53–57.

12. Barraclough, P.B.; Tinker, P.B. The Determination of Ionic Diffusion Coefficients in Field Soils. 1. Diffusion Coefficient in Sieved Soils in Relation to Water Content and Bulk Density. J. Soil Sci. **1981**, *32*, 225–236.
13. Gahoonia, T.S.; Nielsen, N.E.; Lyshede, O.B. Phosphorus Acquisition of Cereal Cultivars in the Field at Three Levels of P Fertilization. Plant and Soil **1999**, *211*, 269–281.
14. Barley, K.P.; Rovira, A.D. The Influence of Root Hairs on the Uptake of Phosphorus. Commun. Soil Sci. Plant Anal. **1970**, *1*, 287–292.
15. Gahoonia, T.S.; Nielsen, N.E. Direct Evidence on the Participation of Phosphorus (P^{32}) Uptake from Soil. Plant and Soil **1998**, *198*, 147–152.
16. Classen, N.; Barber, S.A. A Method for Characterizing the Relation Between Nutrient Concentration and Flux into Roots of Intact Plants. Plant Physiology **1974**, *54*, 564–568.
17. Nielsen, N.E. A Transport Kinetic Concept for Ion Uptake by Plants. III. Test of the Concept by Results from Water Culture and Pot Experiments. Plant and Soil **1976**, *45*, 659–677.
18. Nielsen, N.E.; Barber, S.A. Differences Among Genotypes of Corn in the Kinetics of P Uptake. Agronomy Journal **1978**, *70* (5), 695–698.
19. Nielsen, N.E. Bioavailability, Cycling and Balances of Nutrient in the Soil Plant System. In *Integrated Plant Nutrition Systems*; Dudal, R., Roy, R.N., Eds.; Fertilizer and Plant Nutrition Bulletin 12; FAO: Rome 1995; 333–348.
20. Jungk, A.; Classen, N. Ion Diffusion in the Soil-Root System. Adv. Agron. **1997**, *61*, 53–110.
21. Hensinger, P. How Do Plant Roots Acquire Mineral Nutrients? Chemical Processes Involved in the Rhizosphere. Adv. Agron. **1998**, *64*, 225–265.
22. Nielsen, N.E. Bioavailability of Nutrients in Soil. In *Roots and Nitrogen in Cropping Systems of the Semi-Arid Tropics*; Ito, O., Johansen, C., Adu-Gyamfi, J.J., Katayama, K., Kumar Rao, J.V.D.K., Rego, T.J., Eds.; International Crop Research Institute for the Semi-arid Tropics: India, 1996; 411–427.

NUTRIENT INTERACTIONS IN SOIL–PLANT SYSTEMS

N

F.S. Zhang
J. Shen
China Agricultural University, Beijing, People's Republic of China

Y-G. Zhu
Research Centre for Eco-environmental Sciences,
Chinese Academy of Sciences, Beijing, People's Republic of China

INTRODUCTION

The soil–plant system is one of the most important components in agricultural and natural ecosystems. Nutrient dynamics in soil–plant systems not only reflect the pattern of nutrient flow but also influence food production and quality, and the contaminant pathways in agricultural and natural ecosystems. In soil–plant systems, interactions between nutrients occur when the supply of one nutrient affects the movement, absorption, or utilization of another nutrient within the soil, soil–root interface (rhizosphere), or plant. There are several mechanisms of nutrient interaction in soil–plant systems. In this review, some specific examples are chosen to illustrate the mechanisms and effects of nutrient interactions, which include 1) nutrient interactions in soil, 2) nutrient interactions in the rhizosphere, and 3) nutrient interactions in plants.

NUTRIENT INTERACTIONS IN SOIL

Interactions between soil nutrients can occur through several soil physico-chemical processes, particularly adsorption/desorption, dissolution/precipitation, and mineralization/immobilization. In these processes, nutrient interactions affect plant growth by affecting nutrient bioavailability. Adsorption and precipitation are the two key processes affecting nutrient interactions in soil.

Soil pH can affect the adsorption of nutrient ions and then affect nutrient uptake. The application of $CaCO_3$ can markedly increase the uptake of Mo by plants because the adsorption of MoO_4^{2-} by soils and by Al and Fe oxides decreases as pH increases due to $CaCO_3$ application. Additionally interactions can also occur through the effects of one ion on the adsorption of another. For example, $CaSiO_3$ may displace $H_2PO_4^-$ adsorbed onto Al and Fe oxides in soil. The application of $H_2PO_4^-$ can increase Mo concentration in soil solutions by displacing MoO_4^{2-} from the surface of soil colloids. For cations, the Mn oxides can readily adsorb Co from soil solution. The phosphates adsorbed by hydrous oxides of Fe and Al can further adsorb Zn. Interaction of $H_2PO_4^-$ and Zn could occur through the formation of a complex between $H_2PO_4^-$ and Zn on the surface of the oxides (1).

For precipitation effects, the application of $CaCO_3$ can decrease the toxicity of Al to plants by increasing soil pH and thus forming precipitation of $Al(OH)_3$. Factors that decrease soil pH can result in Al toxicity. Aluminum toxicity can also be modified by P supply through forming a precipitate of $AlPO_4$. Similarly the concentration of Mn^{2+} in soil solution can be affected by the soil pH and oxidation–reduction potential. The availability of P for plants may be controlled by the dissolution and precipitation of Ca phosphates on calcareous soils and sparingly soluble Fe or Al phosphates on acid soils, which can also be modified by soil pH (2). Chloride or SO_4^{2-} can form stable complexes with Cd^{2+}, thus increasing the concentration of Cl^- or SO_4^{2-} in the growth medium can significantly increase plant uptake of Cd^{2+} (3).

NUTRIENT INTERACTIONS AND DYNAMICS IN THE RHIZOSPHERE

The rhizosphere is the narrow zone of soil adjacent to roots where root, soil, and microorganisms play joint roles. Therefore, the rhizosphere is regarded as a special habitat with intense nutrient interactions and biological activities. All mechanisms of nutrient interactions in bulk soil are also applicable to the rhizosphere, but some aspects of nutrient interactions in the rhizosphere are distinct from bulk soil due to specific rhizosphere processes. Only the effects of major rhizosphere processes on nutrient interactions are considered here (2).

Rhizosphere pH can affect adsorption–desorption and dissolution–precipitation of nutrient ions, thus affecting nutrient bioavailability to plants. Rhizosphere pH depends primarily on relative differences in root uptake of cations and anions. The supply of NO_3^--N normally increases soil pH in the rhizosphere, whereas the supply of NH_4^+-N decreases rhizosphere pH. Symbiotic nitrogen fixation by legume plants can decrease rhizosphere pH. Rhizosphere pH can also be decreased due to P deficiency for nongraminaceous species. This acidification can enhance the dissolution of sparingly soluble minerals containing P, K, Fe, Zn and Mn (2).

The oxidation–reduction status in the rhizosphere plays an important role in nutrient interactions. The heavy precipitation of Fe^{3+} oxide-hydrates can form an Fe film at the root surface due to the release of O_2 into the rhizosphere of rice, which severely hampers the uptake of other nutrients. An important factor for the decreased Mn^{2+} concentration in rhizosphere soil is adsorption of Mn^{2+} to the freshly precipitated Fe^{3+} oxide-hydrates. In calcareous soils with low availability of both Fe and Mn, the response to Fe deficiency may thus overcome Mn deficiency to some extent, whereas in calcareous soils with low Fe and high Mn availability, this effect may cause excessive mobilization of Mn in the rhizosphere and Mn toxicity in plants (2).

Roots can release considerable amounts of organic substances into the rhizosphere. Phosphorus deficiency can increase the exudation of citrate in white lupin growing on P-deficient calcareous soils. Citrate has considerable effects on the mobilization of micronutrients and P in the rhizosphere. Citrate may contribute to both desorption and chelation of Al and Fe from Fe and/or Al phosphates, and thus enhance P mobilization. Under Fe deficiency, phytosiderophores are released from roots in grasses, which can form stable chelates not only with Fe^{3+} but also with Cu^{2+}, Zn^{2+}, and less distinctly with Mn^{2+} (2, 4).

NUTRIENT INTERACTIONS IN PLANTS

Movement of nutrient ions from soil solution into the cell walls of individual cells or roots is the first step for nutrient uptake. The interaction between Si and Al may reduce the toxicity of Al by producing low-solubility aluminosilicate on the root cell wall (5, 6). Higher concentrations of Ca^{2+} in solution may depress the absorption of Zn^{2+} and Mn^{2+}, and enhance absorption of anions such as $H_2PO_4^-$ and SO_4^{2-}, possibly through the modification of the surface electrical potential of cell walls. Increased concentrations of Mg and Ca in solution may protect plants from Al toxicity by improving Mg or Ca nutrition and alleviating the toxic effect of Al on root development. Competition at absorption sites of root cell plasmamembranes is another mechanism for nutrient interactions. The absorption of MoO_4^{2-} by plant roots is strongly inhibited by SO_4^{2-}, but is associated with an increase in $H_2PO_4^-$ concentrations in solution. Nutrient interactions can also be operated through the regulation of efflux. For example, plants growing with higher external K^+ concentrations may have a higher efflux of Cs^+ than those growing with lower external K^+ concentrations, which contributes to the overall effect of K^+–Cs^+ interactions (7). Altered expression of gene(s) related to the ion transport system of root cell plasma membranes may also be involved in nutrient interactions, e.g., P/Zn.

Another source of interaction among ions occurs due to the regulation of the cation/anion balance in cells. Nitrate (NO_3-N) and ammonium (NH_4-N) account for more than 70% of the total cations and anions taken up by plants. Iron nutrition of plants can be significantly improved by the supply of NH_4^+-N (8). Moreover, NH_4-N supply could contribute to Fe translocation from primary leaves and roots to young leaves (2).

In nutrient utilization, the transport of one nutrient in xylem or phloem may be impaired by another nutrient. Furthermore, the function of the nutrient may be affected at this site (1). For example, Na^+ loading was larger than Mg^{2+} loading and was competed by K^+. Chloride loading competed with sulfate and phosphate. Additionally high concentration of Ca or Mg may form precipitates in phloem, particularly in the presence of phosphates. The supply of Si can alleviate Mn toxicity in cereals by altering the distribution of Mn within plant leaves or regulating the uptake rate of Mn. However, the mechanisms for interactions between Si and Mn are not yet understood.

Nutrient interactions may occur during functioning of nutrients. For example, Mo is a constituent of nitrate reductase. When MoO_4^{2-} is replaced by WO_4^{2-} the activity of NO_3-reductase in higher plants can be inhibited. Some species can use Na in place of K for some, but not all, physiological or metabolic functions. Additionally Mo, Ca, and Cu may be the components of the system of symbiotic N fixation.

Nutrient interactions can involve not only one mechanism but several processes from soil to plant. For example, the mechanisms of interactions between Zn and P may include: P depresses Zn bioavailability in soils and Zn absorption by roots and translocation of Zn from roots to shoots; P-induced growth response causes Zn dilution in plant tops; P–Zn imbalance leads to metabolic disorder; and P-activated interference affects Zn functioning. Additionally, P can also decrease the development of arbuscular mycorrhizas (AM), resulting in a decrease in plant Zn uptake mediated by AM (9). Overall, there is

N

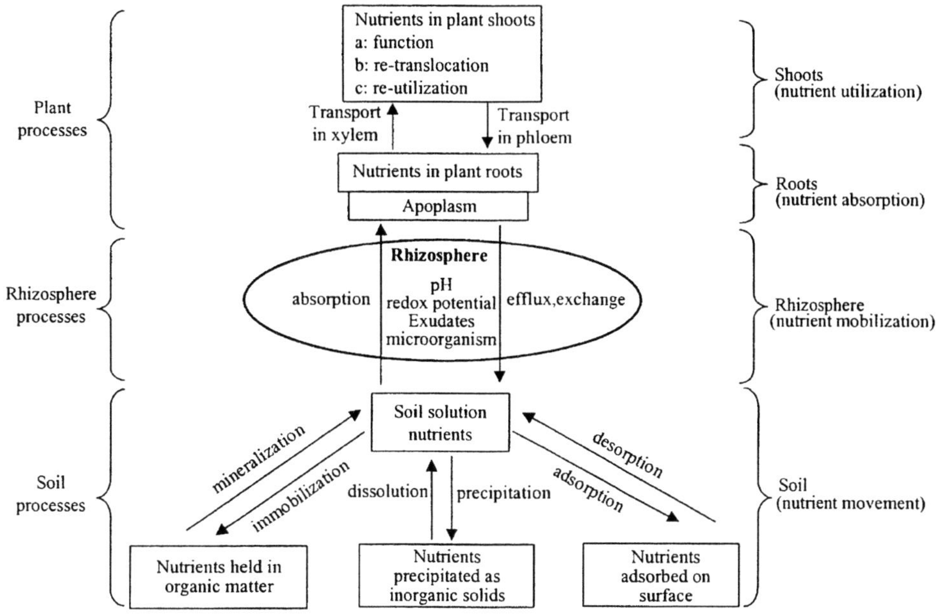

Fig. 1 Mechanisms and processes of nutrient interactions from soils to plants. (Adapted from Ref. 1.)

possibly a triangle-interaction between P, Zn, and AM; however, the mechanisms involved are not clear.

CONCLUSIONS

Interactions between nutrients can occur through several different mechanisms involving nutrient movement from soil to root surface; nutrient dynamics in the rhizosphere; nutrient absorption and utilization by plants; nutrient mobilization and retranslocation within plants; and the expression of genes involved in nutrient uptake, translocation, and function (Fig. 1). It is important to identify the limiting step(s) of the overall effects of the interactions, and what effects on plant growth are produced through nutrient interactions. Although some independent research on nutrient interactions in soils or plants has been done, systematic studies on nutrient interactions in the soil–plant system from molecular to ecosystem level should be emphasized in future research. Another dimension of nutrient interactions that needs attention is the control of plant uptake of trace elements (nutrients or contaminants) for food quality and environmental remediation.

ACKNOWLEDGMENTS

This project was supported by MSBRDPC (No. G1999011709), NSFC (No. 39790100 and 30000102) and the "Hundred-Talent Program" of the Chinese Academy of Sciences.

REFERENCES

1. Robson, A.D.; Pitman, M.G. Interactions Between Nutrients in Higher Plants. In *Inorganic Plant Nutrition*; Läuchli, A., Bieleski, R.L., Eds. Springer-Verlag: New York, 1983; 147–180.
2. Marschner, H. *Mineral Nutrition of Higher Plants*, 2nd Ed. Academic Press: London, England, 1995.
3. McLaughlin, M.J.; Lambrechts, R.M.; Smolders, E.; Smart, M.K. Effects of Sulfate on Cadmium Uptake by Swiss Chard: II. Effects Due to Sulfate Addition to Soil. Plant Soil **1998**, *202*, 217–222.
4. Zhang, F.; Römheld, V.; Marschner, H. Role of the Root Apoplasm for Iron Acquisition by Wheat Plants. Plant Physiol. **1991**, *97*, 1302–1305.
5. Hodson, M.J.; Evans, D.E. Aluminum/Silicon Interaction in Higher Plants. J. Exp. Bot. **1995**, *46*, 161–171.
6. Cocker, K.M.; Evans, D.E.; Hodson, M.J. The Amelioration of Aluminum Toxicity by Silicon in Higher Plants: Solution Chemistry or an in Plant Mechanism. Physiol. Plant. **1998**, *104*, 608–614.
7. Zhu, Y-G.; Shaw, G.; Nisbet, A.F.; Wilkins, B.T. Effects of External Potassium on Compartmentation and Flux Characteristics of Radiocaesium in Intact Spring Wheat Roots. Ann. Bot. **1999**, *84*, 639–644.
8. Mengel, K.; Planker, R.; Hoffman, B. Relationship Between Leaf Apoplast pH and Iron Chlorosis of Sunflower (*Helianthus annuus* L.). J. Plant Nutr. **1994**, *17*, 1053–1065.
9. Smith, S.E.; Read, D.J. *Mycorrhizal Symbioses*, 2nd Ed. Academic Press: London, England, 1997.

NUTRIENT–WATER INTERACTIONS

Ardell D. Halvorson
United States Department of Agriculture-Agricultural Research Service (USDA-ARS), Fort Collins, Colorado, U.S.A.

INTRODUCTION

Water is a major factor in nutrient availability to plants (1–4). It is the vehicle through which nutrients move through soil to access plant roots for uptake. Nutrients move via mass flow and diffusion in soil water to the root surface. Root interception is a third way in which plants obtain soil nutrients as root hairs develop and contact the soil particles and/or solution.

WATER AND NUTRIENT AVAILABILITY

In mass flow, nutrient ions are transported with water flow to the root as the plant absorbs water for transpiration. Many mobile nutrients, such as calcium (Ca), magnesium (Mg), nitrate-N (NO_3-N), and sulfate (SO_4), are transported to the root by mass flow. Diffusion of nutrients to the plant root occurs as ions move from high-concentration areas to low-concentration areas in the soil solution. Phosphorus (P) and potassium (K) are two nutrients that move by diffusion.

If soil water becomes limiting, as it frequently does under dryland or rainfed conditions, nutrient availability to plants can be affected (5). Water is held as a film around soil particles. As the water content of the soil decreases, the thickness of the film decreases. Most plant nutrients are most readily available when the soil is near field capacity, which is about the water content of the wet soil at 2 days after a rain has saturated the soil and free drainage has ceased. Nutrient availability is at a minimum as the soil water content approaches the permanent wilting point, which is the water content at which plant roots cannot extract water from the soil. As soil water content diminishes, some less-soluble nutrients may precipitate out of the soil solution and become unavailable to plants. However, these minerals will dissolve and become available once again as the soil is rewetted. Thus, soil water content influences nutrient availability and plant growth.

Micronutrients are generally supplied to plant roots by diffusion in soil. Therefore, low soil moisture conditions will reduce micronutrient uptake. Plants require smaller quantities of micronutrients to optimize productivity than macronutrients such as P; thus, drought stress effects on micronutrient deficiency are not as serious as for P. However, iron (Fe) and zinc (Zn) deficiencies are frequently associated with high soil moisture conditions (2).

Soil water content is an important factor in microbial activity in soils. Soil microbial activity is important in the breakdown of organic plant and animal residues, which release nutrients such as N and P for plant uptake. Microbial activity tends to be greatest when soil water is near field capacity with soil temperatures ranging from 25° to 35°C. As soils dry, microbial activity decreases and lowers the rate of nutrient release from soil organic matter (6, 7).

NUTRIENT AND WATER USE EFFICIENCY

Adequate levels of plant nutrients are needed to optimize rooting depth and water extraction from the soil (2, 3, 5). Healthy plants tend to root deeper into the soil profile, using more of the soil water in the root zone. Thus, plants not only need adequate water to optimize yield potential but they also require an adequate level of nutrients to allow the crop to take advantage of the available water supplies. Under dryland conditions, the crop will often use all of the available water (precipitation plus soil water in the root zone) during the growing season. Application of N and P fertilizers will frequently increase crop yields, thus increasing crop water use efficiency (WUE). WUE is the amount of crop produced per unit of available water from precipitation, soil, and irrigation. The influence of N fertilization on WUE of winter wheat, corn, and sorghum in a dryland wheat-corn or sorghum-fallow rotation is shown in Fig. 1.

When plant-available water is limited, overapplication of N can also result in reduced grain yields due to increased vegetative growth and water use in the early growth stage, with insufficient water remaining to maximize grain development and yield. Application of N will not increase yields without adequate plant-available water, and increasing plant-available water will not increase crop yield without adequate N supply. The percentage increase in response of crops such as wheat to P

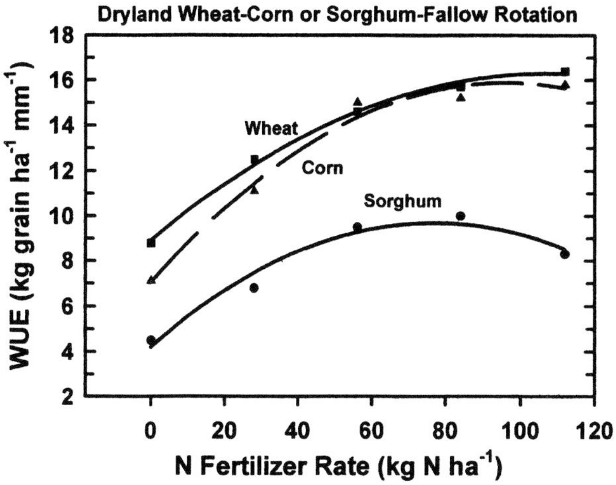

Fig. 1 Water use efficiency (WUE) of wheat, corn, and sorghum as a function of N fertilizer rate in a dryland wheat-corn or sorghum-fallow rotation near Akron, Colorado, U.S.A.

fertilization tends to be greater in dry years than in wet years on P-deficient soils, while both N and P are needed to optimize yields in wetter years.

Water is important for activation and movement of fertilizer nutrients applied to soils (1–3, 7, 8). Dry fertilizer granules must dissolve in the soil water before they become available to plants. When applied to dry soil, liquid fertilizers may become unavailable to plants until precipitation or irrigation water rewets the soil and they become part of the soil solution again. Rainfall affects the volatilization loss of N from ammonia-based fertilizers such as urea and urea ammonium nitrate (UAN). Rainfall received within 36 hours after surface applications of urea or UAN fertilizers will greatly reduce N volatilization losses and improve the N fertilizer use efficiency by crops. Rainfall moves the surface-applied N fertilizer into the soil where it can react and reduce NH_3 losses to the atmosphere. Excessive soil water, however, can result in anaerobic conditions and the loss of NO_3-N by denitrification. Nitrate-N is converted to various N gases, which are lost to the atmosphere under anaerobic conditions.

Water is essential for optimizing crop yields. Under irrigation, water is generally not a yield-limiting factor. Under dryland or rainfed conditions, crop yields are dependent on available soil water supplies and growing season precipitation. Adequate levels of essential plant nutrients are needed to optimize crop yields and WUE (i.e., kg grain produced per mm crop water use). Under rainfed conditions, crop water supplies during the growing season can vary weekly and annually. During periods of drought (i.e., low supply of plant-available water), less plant nutrients are needed to optimize crop yields than during years of average or above-average precipitation. In wetter years, both crop yield potential and the nutrients needed to optimize crop yield increase.

Soil management practices, such as reduced- and no-till systems, that increase soil organic matter and improve soil physical quality also improve soil aggregation and porosity. This, in turn, improves water infiltration into the soil and water availability for increased crop productivity and improved nutrient use efficiency.

IRRIGATION WATER QUALITY AND FERTILIZER APPLICATION

Irrigation water quality can affect the application of fertilizer nutrients through irrigation systems (3, 8). For example, the addition of anhydrous NH_3 or liquid ammonium polyphosphate fertilizers to irrigation waters high in Ca can result in the formation of lime and calcium phosphate precipitates. The precipitates can plug sprinkler and drip irrigation systems. In some instances, precipitation of the Ca can result in a higher sodium (Na) hazard of the irrigation water, which may subsequently reduce the water intake capacity of the soil.

Applying fertilizers with both flood and furrow irrigation systems requires that a uniform distribution of water be achieved throughout the field to obtain a uniform distribution of fertilizer nutrients to the crop. With flood and furrow irrigation systems, fertilizer should not be applied with the initial flush of irrigation water because of the generally non-uniform distribution of water during the initial wetting of the soil surface by the irrigation water. The reactions of fertilizers with the irrigation water and the fertilizer distribution to the crop are affected by (irrigation) water quality. If fertigation (i.e., application of fertilizer nutrients through an irrigation system) is to be used, the compatibility of fertilizers to be applied with the quality of irrigation water available must be examined to avoid poor distribution of fertilizer nutrients.

ENVIRONMENTAL QUALITY

Nitrogen is generally transported from soils into surface and groundwater by runoff, erosion, and leaching (7, 9). Runoff water from watersheds with high levels of soluble N and P sources on the soil surface can contribute to eutrophication of streams, lakes, ponds, bays, and estuaries. Placing or positioning applied N and P sources below the soil surface and using soil management practices to minimize runoff will help reduce agriculture's impact on eutrophication of water bodies. Water erosion of soil not only carries soluble plant nutrients from a watershed, but also carries soil particles with sorbed nutrients, such as P, into water bodies that can then contribute to degradation of water quality.

Soil management practices such as no-till and other conservation tillage practices can reduce soil erosion by water. Water moving through soil in excess of field capacity water content can move soluble nutrients, such as NO_3-N, below the root zone of crops and into groundwater. In summary, using cropping systems and an adequate fertility program to optimize crop WUE will help reduce loss of plant-available water and nutrients below the crop root zone.

REFERENCES

1. Engelstad, O.P., Ed. *Fertilizer Technology and Use*, 3rd Ed.; Soil Sci. Soc. Am.: Madison, Wisconsin, 1985.
2. Havlin, J.L.; Beaton, J.D.; Tisdale, S.L.; Nelson, W.L. *Soil Fertility and Fertilizers: An Introduction to Nutrient Management*, 6th Ed.; Prentice Hall: Upper Saddle River, New Jersey, 1999.
3. Mortvedt, J.J.; Murphy, L.S.; Follett, R.H. *Fertilizer Technology and Application*; Meister Publishing Co.: Willoughby, Ohio, 1999.
4. Troeh, F.R.; Thompson, L.M. *Soils and Soil Fertility*, 5th Ed.; Oxford University Press: New York, 1993.
5. Taylor, H.M.; Jordan, W.R.; Sinclair, T.R., Eds. *Limitations to Efficient Water Use in Crop Production*; Am. Soc. Agron., Crop Sci. Soc. Am., Soil Sci. Soc. Am.: Madison, Wisconsin, 1983.
6. Follett, R.F.; Stewart, J.W.B.; Cole, C.V., Eds. *Soil Fertility and Organic Matter as Critical Components of Production Systems*; Soil Sci. Soc. Am., Inc.: Madison, Wisconsin, 1987.
7. Pierzynski, G.M.; Sims, J.T.; Vance, G.F. *Soils and Environmental Quality*, 2nd Ed.; CRC Press: Boca Raton, Florida, 2000.
8. California Fertilizer Association. *Western Fertilizer Handbook*, 8th Ed.; Interstate Publishers: Danville, Illinois, 1995.
9. Follett, R.F., Ed. *Nitrogen Management and Ground Water Protection*; Elsevier: New York, 1989.

ORGANIC MATTER ACCUMULATION

Sylvie A. Quideau
University of Alberta, Edmonton, Alberta, Canada

O

INTRODUCTION

Chemical, biological, and physical processes continually transform organic matter added to soils in the form of plant or animal detritus. Soil organic matter includes residual components of original organic tissues, their degradation products, and products synthesized by soil fauna. Two major types of compounds are distinguished: 1) nonhumic substances, belonging to identifiable chemical classes such as carbohydrates, and 2) humic substances, a series of brown to dark-brown, high-molecular weight biopolymers distinctive to soil environments. The rate of organic matter (OM) accumulation varies greatly among soils, reflecting the influence of environmental factors on pedogenic processes (1):

$$OM = f\text{ (climate, organisms, parent material, topography, time)} \quad (1)$$

where f stands for "is a function of." Organic matter accumulation is used as a criterion for classifying soils into organic and mineral soil orders, and for identifying epipedons and diagnostic subsurface horizons.

PEDOGENIC PROCESSES

Additions to Soils

Littering refers to the deposition of dead plant and animal detritus on the soil surface (Table 1). As litter decays, soluble organic products leach into the mineral soil. Alternatively, particulate litter can be incorporated into the soil by faunal *pedoturbation*, which is soil mixing by animals, including termites, earthworms, ants, rodents, and even humans. Root turnover also may constitute an important addition of carbon into the mineral soil. In forests most carbon is added as surficial litter; in grassland ecosystems, up to two-thirds of carbon is added through the decay of roots.

Organic Matter Transformation

Decomposition refers to the chemical and biochemical reactions occurring during the decay of plant and animal remains as soil microorganisms colonize them (Fig. 1). Decomposition involves the fragmentation of biopolymers into smaller molecules, with carbon dioxide (CO_2) production as the ultimate step. Organic litter is composed of distinct chemical components that decompose at different rates in the following order, from easily decomposed to more resistant: simple sugars and amino acids>proteins>cellulose>hemicelluloses>fats, starches, and waxes>lignins and tannins. *Mineralization* releases soluble or gaseous inorganic constituents during decomposition processes.

Humification is a multistage process (3). Source materials for humus synthesis include residual components from incomplete decomposition of organic litter and the products of microbial anabolic activities. According to present concepts, polyphenols derived from lignin degradation, together with those synthesized by microorganisms, are oxidized to quinones, which undergo self-polymerization or combine with amino compounds to form nitrogen (N)-containing polymers. Sugar-amine condensation reactions may also participate in the formation of humic substances.

Accumulation in Organic Soils

Paludization can be considered a geogenic rather than pedogenic process because it involves deposition of initial parent material. Paludization occurs when conditions impede decomposition, enabling the buildup of a thick mass of organic deposits. Decomposition is hampered by poorly drained conditions, as in Histosols, and by extreme cold, as in Gelisols. Under anaerobic conditions, humic substances exhibit an accumulation of aromatic carbon (C) compounds arising from the absence of lignin-degrading fungi (4). Aromaticity can also develop in organic horizons from sources without lignin, such as detritus from algae and mosses. In contrast to paludization, *ripening* refers to the decomposition processes occurring in the organic horizon under oxidizing conditions after exposure to the air.

Surface Accumulation in Mineral Soils

Melanization produces thick, dark-colored surface horizons characteristic of Mollisols. The formation of mollic epipedons is promoted by the proliferation of grass roots

Table 1 Fundamental pedogenic processes associated with organic matter accumulation

Term	Definition	Representative horizon (soil order)
Littering	Accumulation of fresh organic detritus on the mineral soil surface to a depth of <30 cm	Oi horizon (all soil orders)
Decomposition	Break-down of organic molecules into simpler compounds	Oe, Oa, and A horizons (all soil orders)
Mineralization	Release of inorganic constituents of organic matter	
Humification	Formation of humic substances from raw organic materials	
Synthesis	Formation of new organic molecules by combination of elements or constituents	
Paludization	Accumulation of thick (>30 cm deep) organic materials on the mineral soil surface	Histic epipedon (Histosols and Gelisols)
Ripening	Changes in organic soil promoted by entry of air into previously waterlogged material	Histic epipedon (Histosols)
Melanization	Darkening of light-colored initial mineral soil by addition of organic matter	Mollic epipedon (Mollisols) Melanic epipedon (Andisols)
Podzolization	Translocation of organic matter in the soil profile associated with Al and Fe migration	Spodic horizon (Spodosols)

(From Ref. 2).

that constitute a considerable input of plant residues (5). Another key factor in melanization is the active faunal community, which contributes to the rapid incorporation of the residues into the mineral soil and favors high initial mineralization rates. Subsequent decomposition and humification processes result in the formation of chemically stable, dark-colored humic substances, characterized by a high proportion of high-molecular weight, highly aromatic, acid-insoluble humic acids. Multivalent cations, such as calcium, act as bridges between organic colloids and clay particles, and stabilize organic substances within the soil matrix. In allophanic soils derived from volcanic parent material (Andisols), organic matter accumulation is favored by the formation of resistant organic-alumina complexes.

Subsurface Accumulation

Podzolization results in the formation of subsurface horizons of organic matter accumulation characteristic of Spodosols. Subsurface organic matter accumulation occurs at the top of the spodic horizon due to the migration of water-soluble organic compounds from the mineral surface (6). Soluble organics are composed mainly of polyphenolics and lower molecular weight polymers (fulvic acids), originating from the decomposition processes of nutrient-poor, acidic organic residues, such as heath and coniferous litter. According to the traditional metal–fulvate theory, organometallic complexes form in the decomposing litter under conditions of low metal saturation, complex more iron (Fe) and aluminum (Al) as they flush down the soil profile with percolating water, and precipitate when the ratio of metal to C reaches a critical level. Other proposed mechanisms for immobilization include flocculation, polymerization, and sorption on mineral surfaces.

ENVIRONMENTAL CONTROLS

Climate

Climate influences organic matter accumulation by controlling the balance between litter production and decomposition rates. On a global basis, soil carbon content increases with increasing precipitation but decreases with increasing temperature (7). In situations in which moisture is not a controlling factor, decomposition increases with increasing temperature. Litter production follows the same trend. Therefore, the worldwide accumulation of soil organic matter is related more to factors controlling decomposition than to the productivity of ecosystems. On a local and regional scale, OM content decreases exponentially with rising annual temperature at any given level of precipitation (1):

$$\mathrm{OM} = C\mathrm{e}^{-kT} \tag{2}$$

where T is the mean annual air temperature (°C), and C and k are constants. For instance, in North American prairie soils, carbon content decreases two to three times for each 10°C increase in mean annual temperature when other

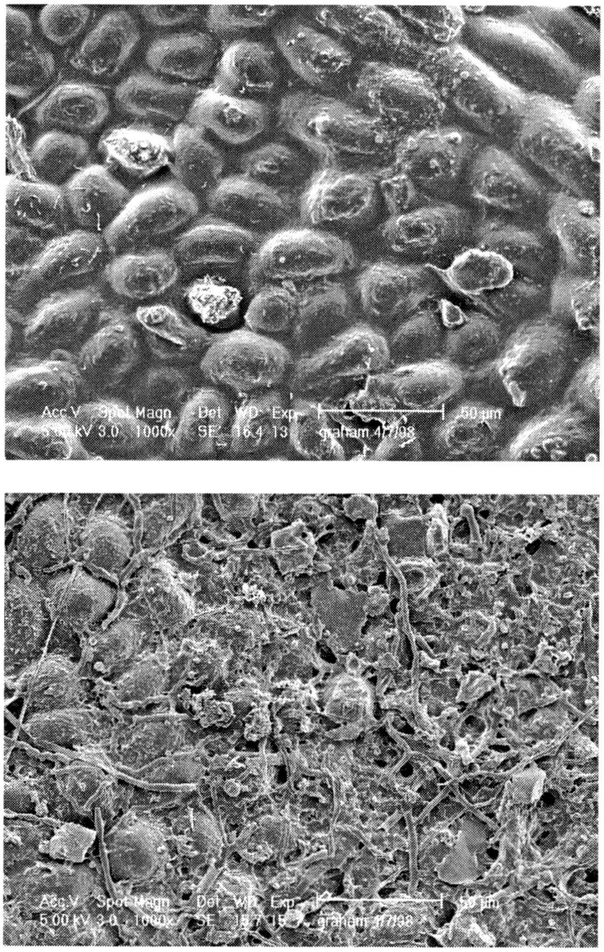

Fig. 1 Scanning electron micrographs of a fresh (*top*) and decomposing (*bottom*) manzanita leaf (magnification: 1000×).

factors are kept constant. On the other hand, organic matter accumulation does not follow a climatic pattern in poorly drained soils where anaerobic conditions impede decomposition processes.

In addition to its influence on total C content, climate may also affect the chemical composition of soil organic matter. High rainfall favors high leaching regimes and reduces the development of aromaticity in soil organic matter (4). Aromaticity also has been reported to be negatively correlated to the precipitation/temperature ratio. For prairie soils of the Great Plains, polysaccharides decrease with increasing temperature but increase with increasing precipitation (8).

Organisms

The amount, placement, and chemical composition of the organic residues of vegetation also affects organic matter accumulation (9). Litter production worldwide declines with increasing latitude from tropical to arctic forests, following the same distribution patterns as net primary productivity. The placement of organic residues affects the distribution of organic matter accumulation within the soil profile. In grassland soils, where belowground production is abundant, organic matter is more evenly distributed than it is in forest soils, where most accumulation occurs in the uppermost horizon. Coniferous litter, such as pine, tends to be low in nutrients but high in recalcitrant constituents, such as lignin and waxes, and typically decomposes at slower rates than deciduous litter. The chemical composition of litter also exerts a significant influence on the accumulation and turnover of soil organic matter by determining the palatability of the plant material, which in turn can alter the distribution and activity of soil fauna. Soil animals, such as earthworms, may accelerate decomposition rates by contributing to the rapid mixing of fresh plant residues into the mineral soil.

Parent Material

Parent material may influence organic matter accumulation through its effect on soil fertility. Soils formed from base cation-rich volcanic rocks (e.g., basalt) are typically more fertile, and thus experience more organic matter accumulation than soils with lower inherent mineral-derived nutrients, such as those formed from granitic materials. Parent material is also effective through its determination of soil texture. Soil clay content and organic matter accumulation are positively correlated. Clay content affects soil moisture and water availability, thereby modifying plant productivity and litter production. Additionally, a high clay content may induce organic matter accumulation by stabilizing humic substances formed during decomposition. Clay and organic matter form organomineral complexes that are resistant to further biodegradation.

Topography

Topography interacts with microclimate to influence organic matter distribution in soils. Organic matter accumulation is often favored at the bottom of hills where conditions are wetter than at mid- or upper-slope positions. In a similar fashion, organic matter accumulation is usually greater on north-facing slopes compared with south-facing slopes (in the Northern Hemisphere) because temperature is lower.

Time

Long-term rates of organic matter accumulation in Holocene-aged soils vary from about 1 to 12 g C m^{-2} yr^{-1} (10). Organic matter, however, does not accumulate indefinitely in soils. Depending on other soil forming factors, an equilibrium level is reached over time. Organic matter encompasses a series of pools with varying turnover rates. Amounts of the young, labile organic matter may level off in decades because plant biomass stabilizes, while amounts of the more recalcitrant fractions, composed of humic substances often complexed with clay minerals, may continue to increase for tens of thousands of years.

REFERENCES

1. Jenny, H. *Factors of Soil Formation*; McGraw-Hill: New York, 1941.
2. Buol, S.W.; Hole, F.D.; McCracken, R.J. Pedogenic Processes: Internal, Soil-Building Processes. *Soil Genesis and Classification*, 3rd Ed.; Iowa State University Press: Ames, 1989; 114–125.
3. Stevenson, F.J. Biochemistry of the Formation of Humic Substances. *Humus Chemistry: Genesis, Composition, Reactions*, 2nd Ed.; John Wiley and Sons: New York, 1994; 188–211.
4. Preston, C.M. Applications of NMR to Soil Organic Matter Analysis: History and Prospects. Soil Science **1996**, *161* (3), 145–166.
5. Oades, J.M. The Retention of Organic Matter in Soils. Biogeochemistry **1988**, *5*, 35–70.
6. Browne, B.A. Towards a New Theory of Podzolization. In *Carbon Forms and Functions in Forest Soils*; McFee, W.W., Kelly, J.M., Eds.; Soil Science Society of America: Madison, Wisconsin, 1995; 253–273.
7. Post, W.A.; Emanuel, W.R.; Zinke, P.J.; Stangenberger, A.G. Soil Carbon Pools and World Life Zones. Nature **1982**, *298* (8), 156–159.
8. Amelung, W.; Flach, K.; Zech, W. Climatic Effects on Soil Organic Matter Composition in the Great Plains. Soil Sci. Soc. Am. J. **1997**, *61*, 115–123.
9. Quideau, S.A.; Anderson, M.A.; Graham, R.C.; Chadwick, O.A.; Trumbore, S.E. Soil Organic Matter Processes: Characterization by ^{13}C NMR and ^{14}C Measurements. Forest Ecology and Management **2000**, *138*, 19–27.
10. Chadwick, O.A.; Kelly, E.F.; Merritts, D.M.; Amundson, R.G. Carbon Dioxide Consumption During Soil Development. Biogeochemistry **1994**, *24*, 115–127.

ORGANIC MATTER AND GLOBAL C CYCLE

O

Keith Paustian

Colarado State University, Fort Collins, Colorado, U.S.A.

INTRODUCTION

Soil organic matter (SOM) comprises an integral component of the global carbon (C) cycle, both as the largest overall repository of C within the terrestrial system and as a major source and sink for C exchanges between the atmosphere, terrestrial vegetation, and aquatic environments. Consequently, SOM plays a significant role in regulating the composition of the atmosphere, particularly with respect to carbon dioxide (CO_2). Thus, there is both concern that the effects of climate and land use change on soils will exacerbate the problem of increasing CO_2 in the atmosphere and hope that through better management, soils can play a part in mitigating increasing CO_2 levels.

CARBON COMPONENTS OF SOIL

Carbon is present in both inorganic and organic forms in soil. With the exception of some soils formed on carbonate-rich parent material or arid soils containing high levels of primary or secondary carbonates, organic forms of C predominate in soil. These organic compounds range from fresh plant residues—which are the primary source of SOM—to highly recalcitrant, amorphous humic substances. Plant residues are broken down by the soil biota, chiefly microorganisms, to derive energy (i.e., respiration) and C and nutrient elements needed to grow and reproduce. The soil biota comprise a complex food web in which microbial-, plant-, and animal-derived materials are continually consumed, decomposed, and reformed as soil biomass and other secondary compounds. Plant compounds that are not fully metabolized by the biota, e.g., lignin derivatives, microbial metabolites and other organic substances, can also recombine through chemical and physical reactions. Thus, the decomposition process can be viewed as a "cascade" of biophysiochemical reactions and products (1), resulting in the loss of C via respiration, together with the formation of a wide array of secondary SOM compounds including complex recalcitrant substances which can persist in soil for several hundreds to thousands of years.

CARBON STOCKS IN SOIL

Carbon comprises a relatively minor component of most soils, in terms of mass. Most soils contain from 1 to 10% C by mass in surface horizons, with the majority having C contents in the range of 1–3%. The most notable exceptions are soils formed under waterlogged conditions, which restricts the flow of oxygen to soil organisms, greatly reducing the rates of SOM decay. Such organic soils or "histosols" may contain 10–30% or more of their total dry mass as C. Despite this generally low concentration, on an area basis, the C contained in soils usually exceeds that contained in the living and dead vegetation (Table 1). Estimates of the global C stock in soils vary, although most recent estimates are on the order of 1400–1600 Pg (Petagram = 10^{15} g = billion metric tonnes) organic C (2, 3) and 700–900 Pg inorganic C (4). Concentrations of organic C are highest in wetlands and in cold, wet environments (e.g., boreal forest) where decomposition rates are suppressed, and are lowest in desert and tundra soils where plant productivity, and hence C additions to soil, is low. Stocks of inorganic C are greatest in desert and semiarid to subhumid grasslands and savannas, where carbonate leaching is restricted and the formation of secondary carbonate minerals is favored.

CARBON FLUXES

The CO_2 exchanges between the atmosphere, vegetation, and soils are among the largest annual fluxes in the global C cycle (Fig. 1), which are clearly detectable in the regular, seasonal variations in atmospheric CO_2 concentrations. Global estimates of net primary production, expressed as C assimilated by live plants (minus respiration) are on the order of 60 Pg C/yr (5). Much of the annually produced biomass is added to the soil and soil-surface each year as senesced leaves, roots, and woody debris. Emissions of CO_2 from soils through decomposition processes (via heterotrophic respiration) are thought to be on the order of 50 Pg/yr with additional losses of about 10 Pg/yr from disturbances such as fire (5). If terrestrial biome C stocks were at equilibrium, then

Table 1 Global carbon stocks in soil and vegetation by major biome types

Biome	Area (10^6 km^2)	Average C density (Mg ha^{-1})	Soil C stock (Pg)	Vegetation C stock (Pg)
Tropical forests	17.6	123	216	212
Temperate forests	10.4	96	100	59
Boreal forests	13.7	344	471	88
Tropical savannas	22.5	117	264	66
Temperate grasslands	12.5	236	295	9
Deserts and semi-deserts	45.5	42	191	8
Tundra	9.5	13	121	6
Wetlands	3.5	642	225	15
Croplands	16.0	80	128	3
Total	151.2	–	2011	466

(Adapted from Ref. 5.)

additions of C through net primary production would be fully balanced by losses through decomposition and other emission sources such as fire. However, overall estimates of the global C cycle suggest that a net accumulation of C presently occurs in terrestrial vegetation and soils, at a rate of about 2 Pg C/yr (Table 2). During the 1980s, tropical regions are believed to have been a net source of C emissions, largely through deforestation. Similar global estimates for the 1990s have not yet been reported, but if the tropics remain a net source of C from deforestation then the size of the present-day terrestrial sink outside the tropics would be correspondingly greater. In the global budget, the terrestrial sink term is calculated as a difference between the estimates of fossil C emissions, ocean uptake, and observed increases in atmospheric CO_2. The existence and relative size of the inferred terrestrial sink are broadly consistent with estimates of a northern hemisphere terrestrial sink, based on atmospheric transport models (6)—although considerable uncertainty as to the magnitude and geographic distribution of the sink remains (7).

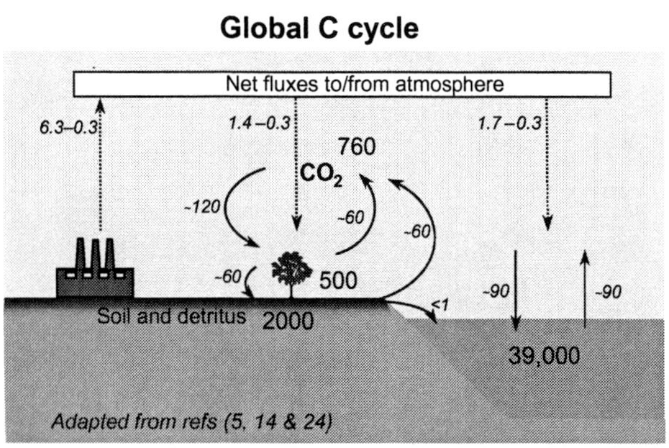

Fig. 1 Simplified depiction of the global carbon cycle. Total stocks of carbon in the atmosphere (as CO_2), oceans, vegetation and soils + detritus are shown in bold and approximate annual gross fluxes between the major biosphere pools are shown in italics. Dashed arrows denote direction and magnitude of the net flux of CO_2 from industrial sources (mainly fossil fuel combustion) and net sinks to terrestrial and marine environments, estimated as averages for 1990–1999. All stock values are in units of Pg (i.e., billion metric tonnes) and fluxes are Pg/yr.

SOIL CARBON FLUXES AND GLOBAL CHANGE

While the net terrestrial accumulation of C is small relative to the annual fluxes of C between the atmosphere and land, it is extremely significant in relation to the net change in CO_2 in the atmosphere. In other words, if the terrestrial C sink was to disappear, and everything else remained equal, the rate of growth of CO_2 in the atmosphere would increase by more than 50%, from about 3.2 to over 5 Pg/yr (Table 2).

The origin of and controls on this terrestrial C sink are not well understood. Model estimates (8) attribute much of the sink to increased plant growth rates due to higher CO_2 concentrations, although other factors such as N deposition and changing land use contribute as well. The contribution of soils to the overall C sink is still uncertain although it is hypothesized that most of the C accumulates as biomass and surface litter pools. In part, this is to be expected due to the lag time between C accumulation in biomass and fresh residue pools, and its subsequent appearance in more stabilized SOM pools, as well as

Table 2 Major net sources and sinks of C in the global budget for the 1980s and 1990s.

	1980–1989	1990–1999
Atmospheric increase in CO_2-C	3.3±0.1	3.2±0.1
Emission of C from fossil fuel and cement	5.4±0.3	6.3±0.3
Ocean–atmosphere flux	−1.9±0.6	−1.7±0.3
Land–atmosphere flux	−0.2±0.7	−1.4±0.3
(Net from land use change)	(1.7 [0.6–2.5])	NA[a]
(Net from other terrestrial sinks)	(−1.9 [−3.8–0.3])	NA

[a]Not available.
(Adapted from Ref. 24.)

the fact that much of the C added to soils is relatively rapidly decomposed and evolved as CO_2. Global accumulation of refractory humic compounds in mature native ecosystems has been estimated at 0.4 Pg/yr and it has been suggested that these increases roughly balance leaching losses and long-term average organic C transfers to oceans (9). Soil C increases attributed to global afforestation and forest regrowth since the 1950s are estimated to be on the order of 0.1 Pg/yr (10). Global estimates are still lacking for other ecosystems, but recent inventories for the U.S. (11) suggest that soils in cropland and grazing land currently represent a small sink, on the order of 10–30 Tg/yr (Teragram = 10^{12} g = million metric tonnes), which can be compared to estimates for C increases in U.S. forest biomass of about 210 Tg/yr (12).

The potential feedbacks of climate change on soil C stocks are currently being debated. Since both plant production (hence, C inputs) and decomposition (hence, C losses) are affected by changes in temperature, precipitation, and CO_2 concentrations, the interactions and feedbacks controlling the terrestrial C balance are complex and difficult to predict. Earlier estimates, assuming a stimulation of decomposition due to projected increases in temperature, suggested that soils could become a significant net source of CO_2, on the order of 40–60 Pg over a 50–100 yr period, under a climate regime predicted for double present-day CO_2 concentrations (13, 14). Other studies suggest that including positive effects of CO_2 on plant productivity, management, together with adaptations in land might largely offset climate-driven increases in decomposition potential (15, 16). Recent debates (17–19) have highlighted the complexity and continuing uncertainty of how temperature increase, and other changes in climate and CO_2, may impact soils and the global C cycle.

The other significant driver of soil C changes, both past and future, is land use and management. It is well established that the conversion of native ecosystems (e.g., forests, grasslands, wetlands), primarily to agricultural uses, has led to significant losses of C from terrestrial ecosystems, on the order of 120–170 Pg C or more over the past 150–300 yrs from vegetation and soils combined (20). From soils alone, estimates of historical losses over the same period are 50–100 Pg C (21, 22). However, this legacy of past land use offers an opportunity to reverse the historical trend and through better management of soils, exploit their potential to become a significant sink for CO_2. A variety of management options are available to increase soil C storage in agricultural (23) and other managed ecosystems (5). Recent estimates suggest that potential C sequestration through improvements in land use and management, globally, is on the order of 1 Pg/yr (5). Thus the role of SOM in the global C cycle, and how it will respond to future changes in climate and land use, will be determined by both the natural forces regulating the Earth's biosphere as well as the social, economic, and political actions of humankind.

REFERENCES

1. Swift, M.J.; Heal, O.W.; Anderson, J.M. *Decomposition in Terrestrial Ecosystems*; Blackwell Press: Oxford, UK, 1979; 372.
2. Post, W.M.; Emanuel, W.R.; Zinke, P.J.; Stangenberger, A.G. Soil Carbon Pools and World Life Zones. Nature **1982**, *298*, 156–159.
3. Eswaran, H.; Van Den Berg, E.; Reich, P. Organic Carbon in Soils of the World. Soil Sci. Soc. Am. J. **1993**, *57*, 192–194.
4. Eswaran, H.; Reich, P.F.; Kimble, J.M.; Beinroth, F.H.; Padmanabhan, E.; Moncharoen, P. Global Carbon Stocks. In *Global Climate Change and Pedogenic Carbonates*; Lal, R., Kimble, J.M., Eswaran, H., Stewart, B.A., Eds.; CRC Press: Boca Raton, FL, 2000; 15–25.
5. IPCC, *Land Use, Land Use Change, and Forestry*; Intergovernmental Panel on Climate Change Special Report, Cambridge University Press: Cambridge, UK, 2000; 377.
6. Tans, P.P.; Fung, I.Y.; Takahashi, T. Observational Constraints on the Global Atmospheric Carbon Dioxide Budget. Science **1990**, *247*, 1431–1438.
7. Field, C.B.; Fung, I.Y. The Not-So-Big US Carbon Sink. Science **1999**, *285*, 544–545.
8. McGuire, A.D.; Sitch, S.; Clein, J.S.; Dargaville, R.; Esser, G.; Foley, J.; Heimann, M.; Joos, F.; Kaplan, J.; Kicklighter, D.W.; Meier, R.A.; Melillo, J.M.; Moore, B. III; Prentice, I.C.; Ramankutty, N.; Reichenau, T.; Schloss, A.; Tian, H.; Williams, L.J.; Wittenberg, U. Carbon Balance of the Terrestrial Biosphere in the Twentieth Century: Analysis of CO_2, Climate and Land Use Effects with Four Process-Based Ecosystem Models. Glob. Biogeochem. Cycles **2001**, *15*, 183–206.

9. Schlesinger, W.H. Evidence from Chronosequence Studies for a Low Carbon-Storage Potential of Soils. Nature **1990**, *348*, 232–234.
10. Post, W.M.; Kwon, K.C. Soil Carbon Sequestration and Land-Use Change: Processes and Potential. Global Change Biol. **2000**, *6*, 317–327.
11. Eve, M.D.; Paustian, K.; Follett, R.; Elliott, E.T. A National Inventory of Changes in Soil Carbon from National Resources Inventory Data. In *Methods of Assessment of Soil Carbon*; Lal, R., Kimble, J.M., Follett, R.F., Stewart, B.A., Eds.; CRC Press: Boca Raton, FL, 2001; 593–610.
12. US Environmental Protection Agency, *Inventory of US Greenhouse Gas Emissions and Sinks: 1990–1998*; US EPA 236-R-00-001, US EPA: Washington, DC, 2000.
13. Jenkinson, D.S.; Adams, D.E.; Wild, A. Model Estimates of CO_2 Emissions from Soil in Response to Global Warming. Nature **1991**, *351*, 304–306.
14. Schlesinger, W.H. An Overview of the Carbon Cycle. In *Advances in Soil Science: Soils and Global Change*; Lal, R., Kimble, J., Levine, E., Stewart, B.A., Eds.; CRC Press: Boca Raton, FL, 1995; 9–25.
15. Prentice, K.C.; Fung, I.Y. The Sensitivity of Terrestrial Carbon Storage to Climate Change. Nature **1990**, *346*, 48–51.
16. Paustian, K.; Elliott, E.T.; Peterson, G.A.; Killian, K. Modelling Climate, CO_2 and Management Impacts on Soil Carbon in Semi-Arid Agroecosystems. Plant Soil **1996**, *187*, 351–365.
17. Giardina, C.P.; Ryan, M.G. Evidence That Decomposition Rates of Organic Carbon in Forest Mineral Soil do not Vary with Temperature. Nature **2000**, *404*, 858–861.
18. Davidson, E.A.; Trumbore, S.E.; Amundson, R. Soil Warming and Organic Carbon Content. Nature **2000**, *408*, 789–790.
19. Thornley, J.H.M.; Cannell, M.G.R. Soil Carbon Storage Response to Temperature: A Hypothesis. Ann. Bot. **2001**, *87*, 591–598.
20. Houghton, R.A. The Annual Net Flux of Carbon to the Atmosphere from Changes in Land Use 1850–1990. Tellus **1999**, *51B*, 298–313.
21. IPCC, Climate Change 1995: Chapter 23 – Agricultural Options for Mitigation of Greenhouse Gas Emission Impacts. *Adaptations and Mitigation of Climate Change: Scientific–Technical Analyses; Intergovernmental Panel on Climate Change (IPCC) Working Group II*; Cambridge University Press: Cambridge, UK, 1996, 745–771.
22. Lal, R. Soil Management and Restoration for Carbon Sequestration to Mitigate the Accelerated Greenhouse Effect. Prog. Environ. Sci. **1999**, *1*, 307–326.
23. Paustian, K.; Collins, H.P.; Paul, E.A. Management Controls on Soil Carbon. In *Soil Organic Matter in Temperate Agroecosystems; Long-Term Experiments of North America*; Paul, E.A., Paustian, K., Elliott, E.T., Cole, C.V., Eds.; CRC/Lewis Publishers: Boca Raton, FL, 1997; 15–49.
24. IPCC, Climate Change 2001: The Scientific Basis. In *Contribution of Working Group I to the Third Assessment Report of the Intergovernmental Panel on Climate Change (IPCC)*; Houghton, J.T., Yihui, D., Eds.; Cambridge University Press: Cambridge, UK, in press.

ORGANIC MATTER, GLOBAL DISTRIBUTION IN WORLD ECOSYSTEMS

O

W.M. Post
Oak Ridge National Laboratory, Oak Ridge, Tennessee, U.S.A.

INTRODUCTION

Globally, the amount of organic matter in soils, commonly represented by the mass of carbon, is estimated to be 1200–1500 Pg C (1 Pg C = 10^{15} g carbon) in the top 1 m of soil (1, 2). This is 2–3 times larger than the amount of organic matter in living organisms in all terrestrial ecosystems (1). The exact ratio between living and dead organic matter in terrestrial ecosystems varies, depending on the ecosystem. The amount of carbon stored in soil is determined by the balance of two biotic processes—the productivity of terrestrial vegetation and the decomposition of organic matter. Each of these processes has strong physical and biological controlling factors. These include climate; soil chemical, physical, and biological properties; and vegetation composition. Interactions among these controlling factors are of particular importance. These biological and physical factors are the same as the ones that influence the aboveground structure and composition of terrestrial ecosystems, so there are strong correspondences between soil organic matter content and ecosystem type.

ORGANIC MATTER INPUTS

Quantity

The amount of carbon stored in soils is to a great extent determined by the rate of organic matter input through litterfall, root exudates, and root turnover. The main factors that influence vegetation production are suitable temperatures for photosynthesis, available soil moisture for evapotranspiration, and rates of CO_2 and H_2O exchange. Dry and/or cold climates support low vegetation production rates and soils under such climates have low organic matter contents. Where climates are warm and moist, vegetation production is high and soil organic matter contents are correspondingly high. Fig. 1 shows the striking correspondence between soil organic matter content and general climate measurements that results from the relationship between vegetation production and suitable moisture and temperature conditions.

Vegetation production depends not only on climate but also on nutrient supply from decomposition and geochemical weathering. Walker and Adams (6) hypothesized that the level of available phosphorus during the course of soil development is the primary determinant of terrestrial net primary production. Numerous workers have examined this hypothesis. Tiessen et al. (7) and Roberts et al. (8) found that available phosphorus explained about one-fourth of the variance in soil organic matter in many different soil orders. The relationship between phosphorus and carbon is strongest during the aggrading stage of vegetation–soil system development (9). Initially, the production of acidic products by pioneer vegetation promotes the release of phosphorus by weathering of parent material. Organic matter builds up in the soil, increasing the storage of phosphorus in decomposing organic compounds. Nitrogen fixing bacteria populations, which depend on a supply of organic carbon and available phosphorus, can grow to meet ecosystem demands for nitrogen. Plant growth is enhanced by this increasing nitrogen and phosphorus cycling, resulting in increased rates of weathering. This process continues until the vegetation is constrained by other factors affecting phosphorus availability: Leaching losses become larger than the weathering inputs (10); or an increasing fraction of the phosphorus becomes unavailable by adsorption or precipitation with secondary minerals (11); or nitrogen availability (denitrification or leaching is affected) reaching or exceeding nitrogen inputs and fixation (12). In mature soils, net primary production is more likely to be limited by nitrogen. Availability of other nutrients that are largely derived from parent materials, such as most base cations, may also influence soil organic matter accumulation during early soil development (13). Soils derived from base cation rich volcanic parent materials (Andisols) have much higher carbon contents on average than soils from other parent materials (4).

Species Composition

Biotic factors, in particular plant species composition, also affect soil organic matter dynamics. Production and decomposition rates are to some degree controlled by

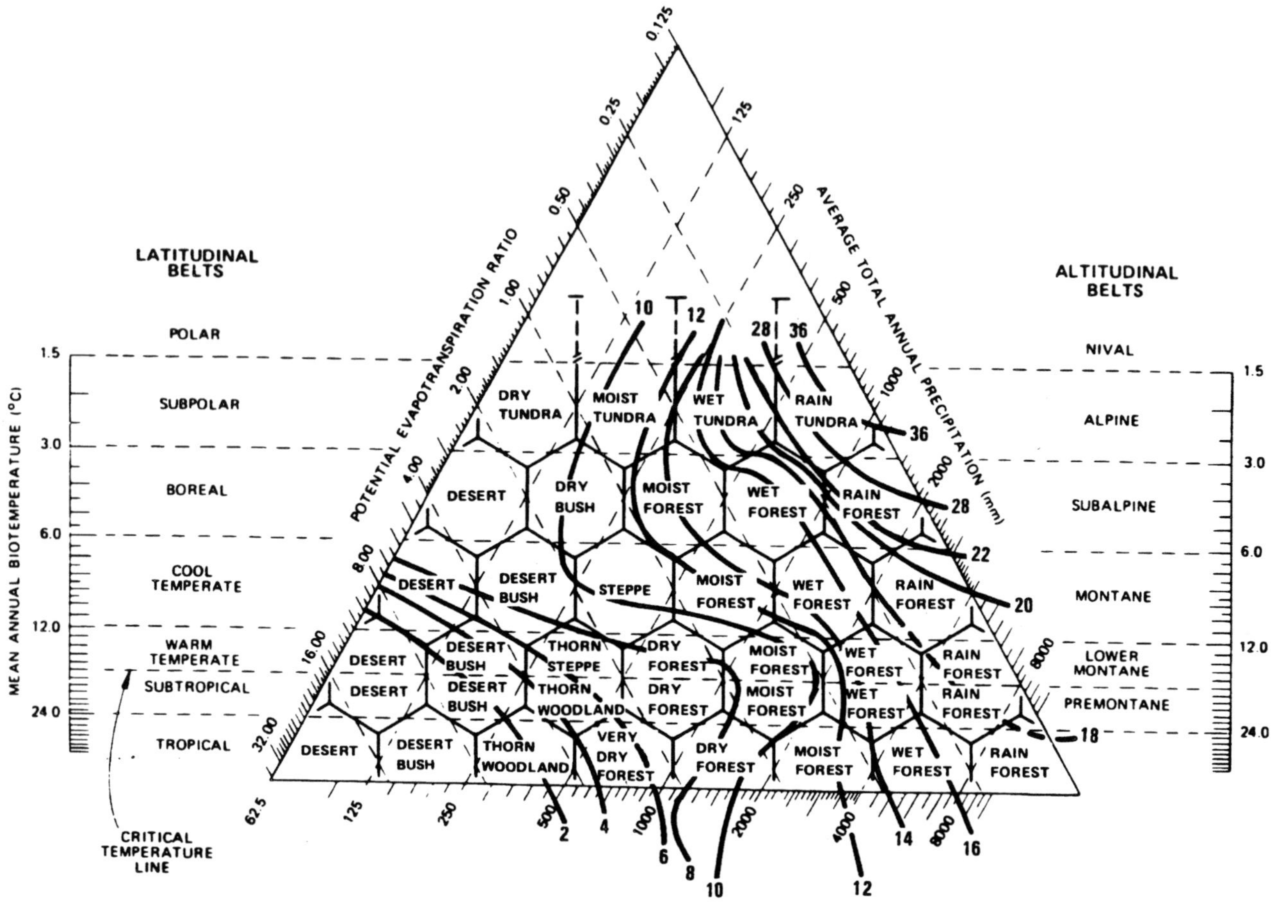

Fig. 1 Contours of soil carbon density ($kg\,m^{-2}$) plotted on Holdridge diagram (3) for world life-zone classification. Values of biotemperature and precipitation uniquely determine a life zone and associated vegetation. Contour lines for mean soil carbon content in the surface meter of soil are determined from data derived from over 3000 soil profiles (4, 5).

species composition. Each terrestrial plant species produces different amounts and chemical compositions of leaves, roots, branches, and wood of varying decomposability. This range of decomposability may be summarized by the lignin and nitrogen content of the organic material (14, 15). Litter decay rate is inversely related to C:N and lignin:N ratios and positively related to N content. Species with tissues that have low nutrient or high lignin content produce litter that is slow to decay. Nitrogen is made available to plants during the decomposition process. Nitrogen is a limiting element for productivity in most terrestrial ecosystems so the rate at which it is released during decomposition is an important factor in ecosystem production. Thus, the interactions between processes regulating plant populations and their productivity and microbial processes regulating nitrogen availability result in some of the observed variation in soil carbon and nitrogen storage (16–19).

Placement

The deeper that fresh detritus is placed in the soil, the slower it decomposes. This is a result of declining decomposer activity and increased protection from oxidation with depth in the soil. Prairies have a somewhat lower productivity than forests and produce no slowly decomposing woody material. Nevertheless, prairies have a very high soil organic matter content because prairie grasses allocate twice as much production to belowground roots and tillers than to aboveground leaves (20). The result is high soil organic matter contents with a uniform distribution in the upper 1 m of soil (Fig. 2). In contrast, a spruce–fir forest contains 50 percent of its soil organic matter in the top 10 cm. There are interesting exceptions to the rule that above-/belowground plant allocation determines soil organic matter distribution patterns in soil. Tropical moist forest soils show a uniform depth distribution similar to the depth distribution of temperate

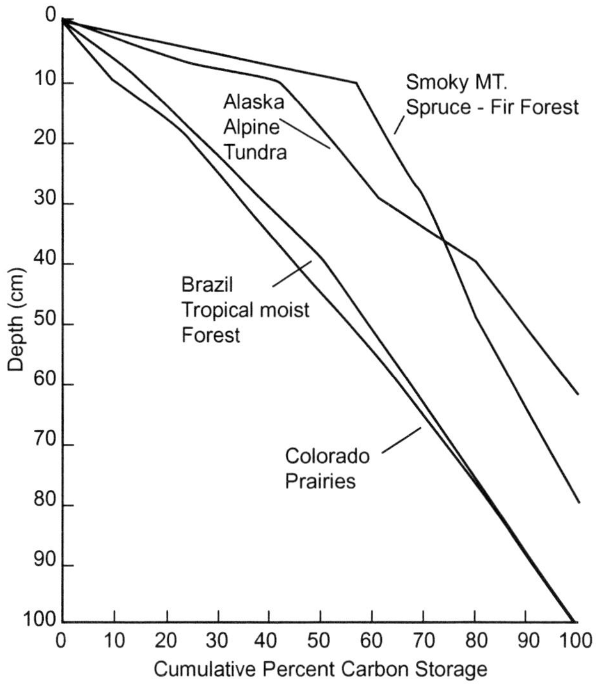

Fig. 2 Cumulative carbon storage as a function of depth for four ecosystems. Refer text for explanation of these patterns. (From Ref. 4.)

grasslands, however, in tropical forests this is largely due to a long-term accumulation of recalcitrant organic materials at lower depths in the soil rather than increased allocation to roots. Alpine tundra soils support a largely herbaceous flora but show a similar depth distribution as forest soils because of inhibition of surface litter decomposition by low temperatures and high water saturation.

DECOMPOSITION

Climate

Organic matter decay rates can be related to environmental parameters such as temperature and soil moisture. Climatic indices that correlate well with decay rates include plant moisture and temperature indices (21, 22), linear combinations of temperature and rainfall (23), and actual evapotranspiration (15). Warm temperatures and available soil moisture enhance microbial, and micro- and macro-invertebrate activity. These environmental conditions are also correlated with plant production. As a result, the amount of organic matter present in soil is highest in vegetation types with the highest rates of organic matter production. These are ones found in the warm, moist climate regions. The contours of soil carbon density displayed in Fig. 1 reflect the balance of input by vegetation production and loss from decomposition imposed by climate. Soil carbon content increases from lower left to the upper right in Fig. 1 as the temperature decreases in the cool temperate, boreal, and sub-polar life zones and as precipitation increases in the warm temperate, subtropical, and tropical life zones.

The combined influence of temperature and precipitation is presented by the third axis of the Holdridge diagram (Fig. 1) as the ratio of potential evapotranspiration (PET) to annual precipitation. When this ratio is less than 1.0, rainfall exceeds PET and vice versa. Life zones bordering the line with the PET is equal to precipitation (PET ratio = 1.0) have soil carbon contents around 10 kg m^{-2} except in warm temperate and subtropical zones where strong seasonality limits production, but decomposition conditions are favorable for most of the year. Soil carbon content increases as the PET ratio decreases indicating that productivity increases faster than the rate of decomposition with increasing moisture availability.

Organic Matter Quality

On global scale, climate may be the most important factor controlling decay rates, but within a given region, substrate chemistry is the more important factor (15, 24, 25). Decay rate is often negatively related to substrate C:N ratio. Litter C:N is initially much greater than microbial C:N but approaches microbial C:N as the microbes release the carbon as CO_2 while taking up nitrogen (nitrogen immobilization). The further the initial litter C:N is from microbial C:N, the slower the decay rate. Lignin content or lignin: N ratios may be better predictors of decay rates because lignin itself is difficult to decompose, and it shields nitrogen and other more easily degraded chemical fractions from microbes. Concise and simple models of decay rate are based on a combination of chemical and climatic indices.

The effect of litter quality on soil organic matter content is most dramatically expressed in Podzols (Spodosol in the United States Department of Agriculture classification). These occur over large areas in boreal zones dominated by evergreen conifers, but often occur in other regions on shallow or sandy soils. Low nitrogen content of organic matter inputs and cool temperatures reduce decomposition and soil animal activity. As a result, large surface organic matter accumulations occur over a thin A horizon. Low temperatures combined with

Table 1 Mean organic carbon contents ($kg\,m^{-2}$) by FAO–UNESCO soil units to 1 m depth

Soil unit	Mean C ($kg\,m^{-2}$)
Acrisols	9.4
Cambisols	9.6
Chernozems	12.5
Podzoluvisols	7.3
Ferrasols	10.7
Gleysols	13.1
Phaeozems	14.6
Fluvisols	9.3
Kastanozems	9.6
Luvisols	6.5
Greyzems	19.7
Nitosols	8.4
Histosols	77.6
Podzols	24.2
Arenosols	3.1
Regosols	5.0
Solonetz	6.2
Andisols	25.4
Vertisols	11.1
Planosols	7.7
Xerosols	4.8
Yermosols	3.0
Solochaks	4.2

These soil units generally span a wide range of climate conditions and therefore present a different view of soil organic matter content based on additional soil factors. In particular, the high C content of Podzols, Histosols, and Andisols is apparent. Refer text for additional explanation of biological, chemical and physical factors responsible.
(From Ref. 2.)

leaching of organic acids result in podsolization as the predominant soil-forming process. Leaching of iron, aluminum oxides, and organic matter result in a distinct E horizon near the surface where these materials are removed and deposited in the B horizon. If the surface organic layers are included, these soils can have substantial organic matter contents, exceeding the expected amount for the climate conditions. Batjes (2) gives an average value for Podzols of 24.2 $kg\,m^{-2}$ for the surface meter which is considerably above the mean for most other soil types (see Table 1).

SIGNIFICANT PHYSICAL AND CHEMICAL INFLUENCES

There are several notable exceptions to the climate-based explanation of variation in soil carbon content. There are two in particular that have lower rates of decomposition and therefore higher accumulations of organic matter than expected (Table 1). These include Histosols due to hydrological conditions and Andisols due to parent material chemical effects.

Histosols

In landscape positions where water accumulates at or above the surface of the soil for an appreciable part of the growing season, decomposition can be reduced to such an extent that large amounts of undecomposed organic matter can accumulate. This soil type is called a Histosol and can be found in any region in wetlands where decomposition is restricted. The soil-surface of mature or old-growth boreal forests over shallow water tables are often covered with *Sphagnum* moss which may also lead to development of Histosols. Histosols with the largest areas and thickest accumulations occur in lowland tundra where a mixture of sedges, lichens, and mosses grow at the northern limit of vegetation in the northern hemisphere. Production, decomposition, and evaporation are limited by low temperatures and water-saturated soils. In these cold regions, deeper layers may freeze and and not become thawed during the short growing season (permafrost). As a result, Histosols have carbon contents over 70 $kg\,m^{-2}$ in the surface meter (Table 1). Some regions have been accumulating organic matter since the last glacial period without any substantial decomposition. Histosols in such regions may be several meters thick and contain over 250 $kg\,C\,m^{-2}$ (2). Globally it is estimated that boreal and sub-arctic Histosols contain 455 Pg C that has accumulated during the postglacial period (26).

Andisols

Andisols form on young volcanic stone (basalt lava) rich in nutrients and alkaline. Andisols are weakly weathered soils associated with pyroclastic parent materials that are rich in allophane, ferrihydrite, and other minerals that readily form complexes with humus molecules. These chemical constituents provide conditions promoting high vegetation production and also the retention of organic matter in soil. As a result, Andisols typically have higher soil carbon contents (25.4 $kg\,m^{-2}$, Table 1) than soils with the same environmental conditions but different parent materials.

CONCLUSIONS

Over long periods of time, organic matter in soils is the result of climatic, biological, and geological factors. These factors are not independent. In particular there exists a strong relationship between climate and vegetation type. In Fig. 1, the Holdridge climate based life zones have names that depict the dominant vegetation of climates. Jobbágy and Jackson (27) provide a summary of soil data based on biomes that demonstrates similar soil carbon distribution as that based on climate (Table 2).

Over shorter periods of time soil carbon varies with vegetation disturbances and changes in land use patterns that affect rates of organic matter input and its decomposition. Various land uses result in very rapid declines in soil organic matter from the native condition (29–32). Losses of 50% in the top 20 cm and 30% for the surface 100 cm are average. Much of this loss in soil organic carbon can be attributed to erosion, reduced inputs of organic matter, increased decomposability of crop residues, and tillage effects that decrease the amount of physical protection to decomposition. Evidence from long-term experiments suggest that C losses due to oxidation and erosion can be reversed with soil management practices that minimize soil disturbance and optimize plant yield through fertilization. These experimental results are believed to apply to large regions and that organic matter is being restored as a result of establishment of perennial vegetation, increased adoption of conservation tillage methods, efficient use of fertilizers, and increased use of high yielding crop varieties (33, 34). Additionally, when agricultural land is no longer used for cultivation and allowed to revert to natural vegetation or replanted to perennial vegetation, soil organic carbon can accumulate by processes that essentially reversing some of the effects responsible for soil organic carbon losses initially—from when the land was converted from perennial vegetation—and return them to typical amounts for the climate, vegetation, landscape position, and parent material conditions (35, 36).

Table 2 Mean organic carbon content ($kg\,m^{-2}$) by biome to 1 m depth

Biome	Mean C ($kg\,m^{-2}$)
Boreal forest	9.3
Crops	11.2
Deserts	6.2
Sclerophyllous shrubs	8.9
Temperate deciduous forest	17.4
Temperate evergreen forest	14.5
Temperate grassland	11.7
Tropical deciduous forest	15.8
Tropical evergreen forest	18.6
Tropical grassland/savanna	13.2
Tundra	14.2

Biome classification is based on Whittaker (28).
(From Ref. 27.)

ACKNOWLEDGMENTS

Work sponsored by U.S. Department of Energy, Carbon Dioxide Research Program, Environmental Sciences Division, Office of Biological and Environmental Research and performed at Oak Ridge National Laboratory (ORNL). ORNL is managed by UT-Battelle, LLC, for the U.S. Department of Energy under contract DE-AC05-00OR22725.

REFERENCES

1. Post, W.M.; Peng, T.-H.; Emanuel, W.R.; King, A.W.; Dale, V.H.; DeAngelis, D.L. The Global Carbon Cycle. American Scientist **1990**, *78*, 310–326.
2. Batjes, N.H. Total Carbon and Nitrogen in the Soils of the World. European Journal of Soil Science **1996**, *47*, 151–163.
3. Holdridge, L.R. Determination of World Plant Formations from Simple Climatic Data. Science **1947**, *105*, 367–368.
4. Zinke, P.J.; Stangenberger, A.G.; Post, W.M.; Emanuel, W.R.; Olson, J.S. *Worldwide Organic Soil Carbon and Nitrogen Data*; ORNL/TM-8857, Oak Ridge National Laboratory: Oak Ridge, TN, 1984.
5. Post, W.M.; Pastor, J.; Zinke, P.J.; Stangenberger, A.G. Global Patterns of Soil Nitrogen Storage. Nature **1985**, *317*, 613–616.
6. Walker, T.W.; Adams, A.F.R. Studies on Soil Organic Matter: I. Influence of Phosphorus Content of Parent Materials on Accumulations of Carbon, Nitrogen, Sulfur, and Organic Phosphorus in Grassland Soils. Soil Science **1958**, *85*, 307–318.
7. Tiessen, H.J.; Stewart, W.B.; Cole, C.V. Pathways of Phosphorus Transformations in Soils of Differing Pedogenesis. Soil Science Society of America Journal **1984**, *48*, 853–858.
8. Roberts, T.L.; Stewart, J.W.B.; Bettany, J.R. The Influence of Topography on the Distribution of Organic and Inorganic Soil Phosphorus Across a Narrow Environmental Gradient. Canadian Journal of Soil Science **1985**, *65*, 651–665.
9. Anderson, D.W. The Effect of Parent Material and Soil Development on Nutrient Cycling in Temperate Ecosystems. Biogeochemistry **1988**, *5*, 71–97.
10. Jenny, H. *The Soil Resource*; Springer: Berlin, 1980.
11. Walker, T.W.; Syers, J.K. The Fate of Phosphorus During Pedogenesis. Geoderma **1976**, *15*, 1–19.
12. Schlesinger, W.H. *Biogeochemistry: An Analysis of Global Change*; Academic: New York, 1991.
13. Torn, M.S.; Trumbore, S.E.; Chadwick, O.A.; Vitousek, P.M.; Hendricks, D.M. Mineral Control of Soil Organic Carbon Storage and Turnover. Nature **1997**, *389*, 170–173.

14. Aber, J.D.; Melillo, J.M. Nitrogen Immobilization in Decaying Hardwood Leaf Litter as a Function of Initial Nitrogen and Lignin Content. Canadian Journal of Botany **1982**, *58*, 416–421.
15. Meentemeyer, V. Macroclimate and Lignin Control Of Litter Decomposition Rates. Ecology **1978**, *59*, 465–472.
16. Zinke, P.J. The Pattern of Influence of Individual Trees on Soil Properties. Ecology **1962**, *42*, 130–133.
17. Wedin, D.A.; Tilman, D. Species Effects on Nitrogen Cycling: A Test with Perennial Grasses. Oecologia **1990**, *84*, 433–441.
18. Hobbie, S.E. Effects of Plant Species on Nutrient Cycling. Trends in Ecology and Evolution **1992**, *7*, 336–339.
19. Hobbie, S.E. Temperature and Plant Species Control Over Litter Decomposition in Alaskan Tundra. Ecological Monographs **1996**, *66*, 503–522.
20. Sims, P.L.; Coupland, R.T. *Grassland Ecosystems of the World: Analysis of Grasslands and Their Uses*; Coupland, R.T., Ed.; Cambridge University Press: Cambridge, 1979.
21. Olson, J.S. Energy Storage and the Balance of Producers and Decomposers in Ecological Systems. Ecology **1963**, *44*, 322–331.
22. Fogel, R.; Cromack, K. Effect of Habitat and Substrate Quality on Douglas Fir Litter Decomposition in Western Oregon. Canadian Journal of Botany **1977**, *55*, 1632–1640.
23. Pandey, V.; Singh, J.S. Leaf-Litter Decomposition in an Oak–Conifer Forest in Himalaya: The Effects of Climate and Chemical Composition. Forestry **1982**, *55*, 47–59.
24. Flanagan, P.W.; VanCleve, K. Nutrient Cycling in Relation to Decomposition and Organic Matter Quality in Tiaga Ecosystems. Canadian Journal of Forest Research **1983**, *13*, 795–817.
25. McClaugherty, C.A.; Pastor, J.; Aber, J.D.; Melillo, J.M. Forest Litter Decomposition in Relation to Soil Nitrogen Dynamics and Litter Quality. Ecology **1984**, *66*, 266–275.
26. Gorham, E. Northern Peatlands: Role in the Carbon Cycle and Probable Responses to Climatic Warming. Ecological Applications **1991**, *1*, 182–195.
27. Jobbágy, E.G.; Jackson, R.B. The Vertical Distribution of Organic Carbon and Its Relation to Climate and Vegetation. Ecological Applications **2000**, *10* (2), 423–436.
28. Whittaker, R.H. *Communities and Ecosystems*; MacMillan: London, 1975.
29. Jenny, H. *Factors of Soil Formation*; McGraw-Hill: New York, 1941.
30. Davidson, E.A.; Ackerman, I.L. Changes in Soil Carbon Inventories Following Cultivation of Previously Untilled Soils. Biogeochemistry **1993**, *20*, 161–193.
31. Mann, L.K. Changes in Soil Carbon After Cultivation. Soil Science **1986**, *142*, 279–288.
32. Schlesinger, W.H. Changes in Soil Carbon Storage and Associated Properties with Disturbance and Recovery. In *The Changing Carbon Cycle: A Global Analysis*; Trabalka, J.R., Reichle, D.E., Eds.; Springer: New York, 1985.
33. Buyanovsky, G.A.; Wagner, G.H. Carbon Cycling in Cultivated Land and Its Global Significance. Global Change Biology **1998**, *4*, 131–142.
34. Lal, R.; Kimble, J.M.; Follett, R.F.; Cole, C.V. *The Potential of U.S. Cropland to Sequester Carbon and Mitigate the Greenhouse Effect*; Ann Arbor Press: Ann Arbor, 1998.
35. Post, W.M.; Kwon, K.C. Soil Carbon Sequestration and Land-Use Change: Processes and Potential. Global Change Biology **2000**, *6*, 317–328.
36. Silver, W.L.; Ostertag, R.; Lugo, A.E. The Potential for Carbon Sequestration Through Reforestation of Abandoned Tropical Agricultural and Pasture Lands. Restoration Ecology **2000**, *8* (4), 394–407.

ORGANIC MATTER IN THE LANDSCAPE

H.H. Janzen
B.H. Ellert
Agriculture and Agri-Food Canada, Lethbridge, Alberta, Canada

D.W. Anderson
Department of Soil Science, University of Saskatchewan, Saskatoon, Saskatchewan, Canada

O

INTRODUCTION

Soil organic matter is composed largely of plant tissues decomposing back to the simple molecules from which they were first formed: carbon dioxide, ammonium, water, and various salts. Although often present in large amounts, organic matter is transient—new litter is continually being added, and organic matter already present is always decomposing. So at any time, the amount present reflects the net balance between additions and losses.

Both inputs and losses at a site depend on conditions there. Litter inputs from plant growth and decomposition are both influenced by many factors: temperature, moisture, aeration, nutrients, plant community, and more. And wind and water move organic matter-laden soil about the landscape. So the amount and composition of organic matter differs from place to place in the landscape.

Organic matter varies in several dimensions: It varies vertically within the profile, and horizontally across the landscape. It changes across time, often by human influence. How organic matter is distributed over the landscape may be as important as its amount in affecting the way an ecosystem behaves (1). In this review, we describe how organic matter is distributed in "natural" landscapes, show how human activities can alter that pattern, and illustrate with a few examples how organic matter distribution (and redistribution) can affect ecosystem function.

ORGANIC MATTER IN LANDSCAPES UNAFFECTED BY HUMAN ACTIVITY

Vertical Distribution

In most soil profiles, organic matter (or organic carbon) concentration is highest near the surface, where most plant litter is added, and then declines with increasing depth (2, 3). In grasslands, globally, the surface 0.2 m of soil accounts for 42% of the organic carbon in the first 1 m; in shrublands, the proportion is 33% and in forests, it is 50% (4).

The rate of organic matter turnover also changes with depth. Compared to that deeper in the profile, surface soil usually has higher proportions of "young" organic matter from recent inputs of plant litter. Radiocarbon studies show that the mean turnover time or radiocarbon "age" of organic matter usually increases with depth (5–7).

Lateral Distribution

The amount of organic matter in the soil at a given spot is the result of a complex interaction, over time, of parent material, climate, vegetation, and topography (8). Among landscapes, at regional scales, organic matter content is controlled largely by precipitation, temperature, soil texture, and vegetation type (9, 10). But within a landscape, organic matter is also influenced by other factors. Topography, through effects on microclimate and water movement, can produce soils of widely different organic matter within meters (Fig. 1) (9, 11). Usually, the amount of organic matter is lowest near summits, and highest in toeslope positions (12–14). This pattern occurs for various reasons (13, 15, 16): higher moisture and nutrient status downslope increase litter production; organic matter in lower slope positions may decompose more slowly because of soil conditions (e.g., reduced aeration or accumulated clay); variations in microclimate produce different plant communities across topographical gradients; and erosion may move organic matter downhill.

Apart from topography, localized variations in texture, soil chemistry, vegetation (1, 17), and other properties may also cause organic matter to vary within landscapes. Even in landscapes that appear uniform, therefore, soil organic matter content varies significantly over scales of several meters. Tiessen and Santos (18) observed ccoefficients of

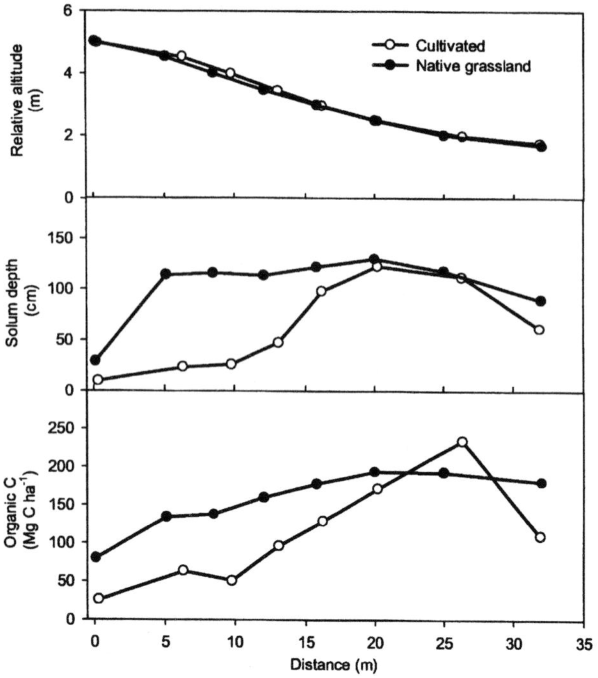

Fig. 1 An illustration of variability in soil organic carbon in transects across two toposequences, a native grassland and an agricultural field cultivated for about two decades. Organic matter is about 58% carbon, by weight, so it is often measured as organic carbon. (Adapted from Ref. 28.)

variation greater than 50% in organic carbon and total nitrogen concentrations in the surface soil of a tropical semiarid field ($65 \times 40\,m^2$) immediately after clearing.

Organic matter varies across landscapes not only in amount but also in form. For example, Paul et al. (5) observed that radiocarbon "age" of surface soil increased from toeslope to summit positions at two uncultivated grassland sites. And Schimel et al. (15) found that the relative mineralization rate of organic matter (N mineralized per unit of total nitrogen) decreased downslope, even though total mineralization increased, pointing to differences in organic matter quality.

Temporal Distribution

Organic matter in the landscape also changes with time. Soils in early stages of development accumulate organic matter, but the rate of build-up slows as soils approach a steady state, where decomposition roughly balances litter inputs (19). Even then, however, organic matter fluctuates during a year because litter additions follow a different pattern from decomposition. It also fluctuates from year to year with changes in weather: in some periods, when litter inputs exceed losses, organic matter accumulates; in others, it is depleted (20).

HUMAN INFLUENCE ON DISTRIBUTION

Human activities can profoundly alter soil organic matter content. Historically, disturbance of "natural" ecosystems has almost always resulted in losses (21, 22). Converting grass- and forestlands to arable agriculture, for example, typically results in the loss of about 30% of the organic carbon originally present in the solum (23). But with better land management, at least part of the organic matter lost can be restored (24).

Human influence on soil organic matter—whether it leads to losses or gains—is rarely uniform over the landscape. Thus, human intervention often alters not only the amount of organic matter, but also its distribution over the landscape. The rearrangement can occur in several ways.

Patchwork Application of Practices

Land practices are often not applied uniformly over the landscape. Forests may be cut in patches; diverse cropping practices may be used in scattered patterns in agricultural landscapes; organic materials such as manures, derived from large, far-flung areas, may be funneled into small areas (25); grazing animals may congregate near water or shade, depositing organic matter there (26). These, and other practices applied nonuniformly, can redistribute organic matter on the landscape.

Nonuniform Effects on C Balance

Even where the same management practice or land use change is applied uniformly over a large area, its effect on organic matter may vary from place to place because the landscape is not uniform to begin with. For example, Schimel et al. (27) observed that proportional losses of organic matter after cultivating grassland were higher in upper- than in lower-slope positions. And the mechanism of organic matter loss, whether by erosion or biological mineralization, may also vary among slope positions (28).

Tillage

Farmers cultivate soils to control pests, prepare land for seeding, and bury residues. This tillage dilutes organic matter-rich soil near the surface by mixing it with soil from deeper layers (14). It may also affect organic matter deeper in the profile by altering rooting patterns, leaching, faunal activity, and soil temperature (29, 30). Consequently, tillage changes organic matter distribution in the profile, even though total amount may not always be affected (31).

Tillage also alters lateral distribution of organic matter by physically "dragging" topsoil. Each tillage pass moves soil, but the extent of movement is greater downslope than upslope, so that over the years, tillage moves soil and organic matter to lower-slope positions (32).

Erosion

Globally, the predominant human influence on distribution of organic matter in the landscape is via erosion, especially in deforested and agricultural lands. Although erosion is a natural process and occurs also in undisturbed ecosystems, human disturbance may increase erosion by orders of magnitude (33, 34).

Erosion affects organic matter at a site within the landscape in three ways (11); it may: remove organic matter; result in mixing of subsoil into the surface layer by stripping surface soil away; or result in deposition of soil eroded from elsewhere in the landscape.

The effects of erosion are not uniform across the landscape. Highest losses usually occur from convex uplands or "shoulder elements" (11, 13). Much of the eroded soil removed from one location in the landscape may be deposited nearby, especially, in closed watersheds (35). But areas of net removal are usually larger than areas of deposition so that erosion often increases the variability of organic matter on the landscape. This is compounded by selective translocation of soil fractions rich in organic matter (14, 36). The net effect, therefore, is often a removal of organic matter from widespread areas in the landscape and its deposition in low-lying areas.

Erosion not only affects directly the distribution of organic matter on the landscape, but may also have long-lasting secondary consequences through effects on plant growth and litter input (14). While erosion can result in extensive translocation of organic matter, its net effect on total amount stored in the landscape is not always clear. If erosion suppresses productivity, thereby limiting replenishment of organic matter, the organic matter may spiral downwards over the long term. But when productivity of eroded areas can be restored, the eroded landscape might eventually contain more organic matter than before, because of the higher storage potential of eroded areas (37) and the accumulated and buried organic matter in depositional areas (14).

Disruption and management of ecosystems by humans often alter irreversibly the amount and distribution of organic matter in landscapes, often increasing variability on the landscape (e.g., Fig. 1) (38). Our understanding of this redistribution is still incomplete (39, 40).

IMPLICATIONS

The heterogeneity of organic matter in the landscape, and its further rearrangement by human activity, determine how ecosystems function. To illustrate, we present three examples.

First, the spatial variability of organic matter influences productivity on the landscape. Plant productivity is closely linked to organic matter (41); consequently, landscapes with variable organic matter usually show corresponding heterogeneity in productivity (whether high organic matter increases productivity or high productivity increases organic matter is not always clear). In farmlands, therefore, increasing attention has been devoted recently to "site-specific farming"—adjusting practices spatially to compensate for variability in organic matter and related soil properties (42).

Secondly, the way organic matter is distributed across the landscape influences the unintended release of nutrients into the broader environment. For example, organic matter affects transformations of nitrogen, both as a source of mineralized nitrogen and by effects on microbial activity. Consequently, nitrate leaching or nitrous oxide emissions may be linked to organic matter distribution, especially since sites where organic matter accumulates may also have high moisture (43, 44).

Thirdly, the heterogeneity of organic matter within the landscape makes it harder to measure changes in carbon storage. Soil organic matter has been proposed as a potential "sink" for carbon; widespread adoption of practices that build organic matter could increase carbon storage in soils, mitigating the increases in atmospheric CO_2 linked to global warming (45). But to quantify that "sink" precisely, organic matter distribution across the landscape would have to be taken into account; differences in organic matter among points on the landscape are usually much greater than the expected response to new management (46). At one location, Garten and Wullschleger (47) estimated that more than 100 samples would need to be taken to detect a change in soil organic carbon of about 1 Mg C ha^{-1}. The redistribution of organic matter by erosion makes measuring of the carbon sink even more complicated (48). When erosion occurs during the measurement interval, it may be hard to distinguish carbon exchanged with the atmosphere from that merely redistributed on the landscape.

The topsoil, enriched in organic matter, forms a veneer over the landscape, one that varies from place to place and year to year. The performance and persistence of ecosystems depend on this thin layer. And how that layer varies over the landscape, especially in response to management, therefore has long-lasting effects on how productive and resilient the ecosystem will be. In the past, human activity has often accentuated native variability, removing organic matter from sites where it was already in short supply and depositing it in areas of excess. New approaches may now favor the preservation of organic matter both in amount and distribution across the landscape. We know something about how past management has affected organic matter distribution; now, we need to learn how restorative practices will shape its future patterns across the landscape.

REFERENCES

1. Herrick, J.E.; Wander, M.M. Relationships Between Soil Organic Carbon and Soil Quality in Cropped and Rangeland Soils: The Importance of Distribution, Composition, and Soil Biological Activity. In *Soil Processes and the Carbon Cycle*; Lal, R., Kimble, J.M., Follett, R.F., Stewart, B.A., Eds.; CRC Press: Boca Raton, FL, 1997; 405–425.
2. Ajwa, H.A.; Rice, C.W.; Sotomayor, D. Carbon and Nitrogen Mineralization in Tallgrass Prairie and Agricultural Soil Profiles. Soil Sci. Soc. Am. J. **1998**, *62*, 942–951.
3. Zhao, Q.; Zhong, L.; Yingfei, X. Organic Carbon Storage in Soils of Southeast China. Nutr. Cycling Agroecosyst. **1997**, *49*, 229–234.
4. Jobbágy, E.; Jackson, R.B. The Vertical Distribution of Soil Organic Carbon and Its Relation to Climate and Vegetation. Ecol. Appl. **2000**, *10* (2), 423–436.
5. Paul, E.A.; Follett, R.F.; Leavitt, S.W.; Halvorson, A.; Peterson, G.A.; Lyon, D.J. Radiocarbon Dating for Determination of Soil Organic Matter Pool Sizes and Dynamics. Soil Sci. Soc. Am. J. **1997**, *61*, 1058–1067.
6. Trumbore, S. Age of Soil Organic Matter and Soil Respiration: Radiocarbon Constraints on Belowground C Dynamics. Ecol. Appl. **2000**, *10* (2), 399–411.
7. Scharpenseel, H.W.; Pfeiffer, E.M.; Becker-Heidmann, P. Ecozone and Soil Profile Screening for C-Residence Time, Rejuvenation, Bomb ^{14}C Photosynthetic $\delta^{13}C$ Changes. In *Assessment Methods for Soil Carbon*; Lal, R., Kimble, J.M., Follett, R.F., Stewart, B.A., Eds.; Lewis Publishers/CRC Press: Boca Raton, FL, 2001; 207–220.
8. Jenny, H. *The Soil Resource: Origin and Behavior*; Springer: New York, 1980; 377.
9. Arrouays, D.; Daroussin, J.; Kicin, J.L.; Hassika, P. Improving Topsoil Carbon Storage Prediction Using a Digital Elevation Model in Temperate Forest Soils of France. Soil Sci. **1998**, *163* (2), 103–108.
10. Burke, I.C.; Yonker, C.M.; Parton, W.J.; Cole, C.V.; Flach, K.; Schimel, D.S. Texture, Climate, and Cultivation Effects on Soil Organic Matter Content in U.S. Grassland Soils. Soil Sci. Soc. Am. J. **1989**, *53*, 800–805.
11. Pennock, D.J. Effects of Soil Redistribution on Soil Quality: Pedon, Landscape, and Regional Scales. In *Soil Quality for Crop Production and Ecosystem Health*; Gregorich, E.G., Carter, M.R. Eds.; Elsevier: Amsterdam, 1997; 167–185.
12. Burke, I.C.; Elliott, E.T.; Cole, C.V. Influence of Macroclimate, Landscape Position, and Management on Soil Organic Matter in Agroecosystems. Ecol. Appl. **1995**, *5* (1), 124–131.
13. Schimel, D.S.; Kelly, E.F.; Yonker, C.; Aguilar, R.; Heil, R.D. Effects of Erosional Processes on Nutrient Cycling in Semiarid Landscapes. In *Planetary Ecology*; Caldwell, D.E., Brierley, J.A., Brierley, C.L., Eds.; Van Nostrand Reinhold: New York, 1985; 571–580.
14. Gregorich, E.G.; Greer, K.J.; Anderson, D.W.; Liang, B.C. Carbon Distribution and Losses: Erosion and Deposition Effects. Soil Till. Res. **1998**, *47*, 291–302.
15. Schimel, D.; Stillwell, M.A.; Woodmansee, R.G. Biogeochemistry of C, N, and P in a Soil Catena of the Shortgrass Steppe. Ecology **1985**, *66* (1), 276–282.
16. Cheng, W.; Virginia, R.A.; Oberbauer, S.F.; Gillespie, C.T.; Reynolds, J.F.; Tenhunen, J.D. Soil Nitrogen, Microbial Biomass, and Respiration Along An Arctic Toposequence. Soil Sci. Soc. Am. J. **1998**, *62*, 654–662.
17. Mueller-Harvey, I.; Juo, A.S.R.; Wild, A. Soil Organic C, N, S and P After Forest Clearance in Nigeria: Mineralization Rates and Spatial Variability. J. Soil Sci. **1985**, *36*, 585–591.
18. Tiessen, H.; Santos, M.C.D. Variability of C, N, and P Content of a Tropical Semiarid Soil As Affected by Soil Genesis, Erosion and Land Clearing. Plant Soil **1989**, *119*, 337–341.
19. Chadwick, O.A.; Kelly, E.F.; Merritts, D.M.; Amundson, R.G. Carbon Dioxide Consumption During Soil Development. Biogeochemistry **1994**, *24*, 115–127.
20. Campbell, C.A.; Zentner, R.P.; Selles, F.; Biederbeck, V.O.; McConkey, B.G.; Blomert, B.; Jefferson, P.G. Quantifying Short-Term Effects of Crop Rotations on Soil Organic Carbon in Southwestern Saskatchewan. Can. J. Soil Sci. **2000**, *80*, 193–202.
21. Solomon, D.; Lehmann, J.; Zech, W. Land Use Effects on Soil Organic Matter Properties of Chromic Luvisols in Semiarid Northern Tanzania: Carbon, Nitrogen, Lignin and Carbohydrates. Agric. Ecosyst. Environ. **2000**, *78*, 203–213.
22. Tiessen, H.; Cuevas, E.; Chacon, P. The Role of Soil Organic Matter in Sustaining Soil Fertility. Nature **1994**, *371*, 783–785.
23. Davidson, E.A.; Ackerman, I.L. Changes in Soil Carbon Inventories Following Cultivation of Previously Untilled Soil. Biogeochemistry **1993**, *20*, 161–193.
24. Sampson, R.N.; Scholes, R.J. Additional Human-Induced Activities—Article 3.4. In *Land Use, Land Use Change, and Forestry. A Special Report of the IPCC*; Watson, R.T., Noble, I.R., Bolin, B., Ravindranath, N.H., Verardo, D.J., Dokken, D.J., Eds.; Cambridge University Press: Cambridge, 2000; 180–281.
25. Fernandes, E.C.M.; Motavalli, P.P.; Castilla, C.; Mukurumbira, L. Management Control of Soil Organic Matter

Dynamics in Tropical Land Use Systems. Geoderma **1997**, *79*, 49–67.
26. Franzluebbers, A.J.; Stuedemann, J.A.; Schomberg, H.H. Spatial Distribution of Soil Carbon and Nitrogen Pools Under Grazed Tall Fescue. Soil Sci. Soc. Am. J. **2000**, *64*, 635–639.
27. Schimel, D.S.; Coleman, D.C.; Horton, K.A. Soil Organic Matter Dynamics in Paired Rangeland and Cropland Toposequences in North Dakota. Geoderma **1985**, *36*, 201–214.
28. Gregorich, E.G.; Anderson, D.W. Effects of Cultivation and Erosion on Soils of Four Toposequences in the Canadian Prairies. Geoderma **1985**, *36*, 343–354.
29. Cihacek, L.J.; Ulmer, M.G. Effects of Tillage on Profile Soil Carbon Distribution in the Northern Great Plains of the U.S. In *Management of Carbon Sequestration in Soil*; Lal, R., Kimble, J.M., Follett, R.F., Stewart, B.A., Eds.; CRC Press: Boca Raton, FL, 1997; 83–91.
30. Mikhailova, E.A.; Bryant, R.B.; Vassenev, I.I.; Schwager, S.J.; Post, C.J. Cultivation Effects on Soil Carbon and Nitrogen Contents At Depth in the Russian Chernozem. Soil Sci. Soc. Am. J. **2000**, *64*, 738–745.
31. Yang, X.-M.; Wander, M.M. Tillage Effects on Soil Organic Carbon Distribution and Storage in a Silt Loam Soil in Illinois. Soil Till. Res. **1999**, *52*, 1–9.
32. Govers, G.; Lobb, D.A.; Quine, T.A. Tillage Erosion and Translocation: Emergence of a New Paradigm in Soil Erosion Research. Soil Till. Res. **1999**, *51*, 167–174.
33. Meade, R.H.; Yuzyk, T.R.; Day, T.J. Movement and Storage of Sediment in Rivers of the United States and Canada. In *The Geology of North America*; Wolman, M.G., Riggs, H.C., Eds.; Geological Society of America: Boulder, CO, 1990; Vol. O-1, 255–280.
34. Oldeman, L.R. The Global Extent of Soil Degradation. In *Soil Resilience and Sustainable Land Use*; Greenland, D.J., Szabolcs, I., Eds.; CAB International: Oxfordshire, 1994; 99–118.
35. Pennock, D.J.; De Jong, E. Rates of Soil Redistribution Associated with Soil Zones and Slope Classes in Southern Saskatchewan. Can. J. Soil Sci. **1990**, *70*, 325–334.
36. van Noordwijk, M.; Cerri, C.; Woomer, P.L.; Nugroho, K.; Bernoux, M. Soil Carbon Dynamics in the Humid Tropical Forest Zone. Geoderma **1997**, *79*, 187–225.
37. Izaurralde, R.C.; Nyborg, M.; Solberg, E.D.; Janzen, H.H.; Arshad, M.A.; Malhi, S.S.; Molina-Ayala, M. Carbon Storage in Eroded Soils After Five Years of Reclamation Techniques. In *Soil Processes and the Carbon Cycle*; Lal, R., Kimble, J.M., Follett, R.F., Stewart, B.A. Eds.; CRC Press: Boca Raton, FL, 1997; 369–385.
38. Beckett, P.H.T.; Webster, R. Soil Variability: A Review. Soils Fertil. **1971**, *34* (1), 1–15.
39. Starr, G.C.; Lal, R.; Kimble, J.M.; Owens, L. Assessing the Impact of Erosion on Soil Organic Carbon Pools and Fluxes. In *Assessment Methods for Soil Carbon*; Lal, R., Kimble, J.M., Follett, R.F., Stewart, B.A. Eds.; Lewis Publishers, CRC Press: Boca Raton, FL, 2001; 417–426.
40. Jacinthe, P.A.; Lal, R.; Kimble, J.M. Assessing Water Erosion Impacts on Soil Carbon Pools and Fluxes. In *Assessment Methods for Soil Carbon*; Lal, R., Kimble, J.M., Follett, R.F., Stewart, B.A., Eds.; Lewis Publishers, CRC Press: Boca Raton, FL, 2001; 427–449.
41. Bauer, A.; Black, A.L. Quantification of the Effect of Soil Organic Matter Content on Soil Productivity. Soil Sci. Soc. Am. J. **1994**, *58*, 185–193.
42. Beckie, H.J.; Moulin, A.P.; Pennock, D.J. Strategies for Variable Rate Nitrogen Fertilization in Hummocky Terrain. Can. J. Soil Sci. **1997**, *77*, 589–595.
43. van Kessel, C.; Pennock, D.J.; Farrell, R.E. Seasonal Variations in Denitrification and Nitrous Oxide Evolution At the Landscape Scale. Soil Sci. Soc. Am. J. **1993**, *57*, 988–995.
44. Pennock, D.J.; Corre, M.D. Development and Application of Landform Segmentation Procedures. Soil Till. Res. **2001**, *58*, 151–162.
45. Lal, R.; Bruce, J.P. The Potential of World Cropland Soils to Sequester C and Mitigate the Greenhouse Effect. Environ. Sci. Pol. **1999**, *2*, 177–185.
46. Ellert, B.H.; Janzen, H.H.; McConkey, B.G. Measuring and Comparing Soil Carbon Storage. In *Assessment Methods for Soil Carbon*; Lal, R., Kimble, J.M., Follett, R.F., Stewart, B.A., Eds.; Lewis Publishers/CRC Press: Boca Raton, FL, 2001; 131–146.
47. Garten, C.T., Jr.; Wullschleger, S.D., Jr. Soil Carbon Inventories Under a Bioenergy Crop (Switchgrass): Measurement Limitations. J. Environ. Qual. **1999**, *2*, 1359–1365.
48. Pennock, D.J.; Frick, A.H. The Role of Field Studies in Landscape-scale Applications of Process Models: An Example of Soil Redistribution and Soil Organic Carbon Modeling Using CENTURY. Soil Till. Res. **2001**, *58*, 183–191.

ORGANIC MATTER MANAGEMENT

R. César Izaurralde
Joint Global Charge Research Institute, College Park, Maryland, U.S.A.

Carlos C. Cerri
Universidade de São Paulo, CENA, Piracicaba, São Pãulo, Brazil

INTRODUCTION

Soil organic matter (SOM) consists of a complex array of living organisms such as bacteria and fungi, plant and animal debris in different stages of decomposition, and *humus*—a rather stable brown to black material showing no resemblance to the organisms from which it originates. Because SOM is or has been part of living tissues, its composition is dominated by carbon (C), hydrogen, oxygen and—in lesser abundance—by nitrogen, phosphorus, sulfur among other elements. Levels of SOM are expressed in terms of soil organic carbon (SOC) concentration (g kg^{-1}) or mass per unit area (g m^{-2}) to a given depth. The level of SOC in virgin soils reflects the action and interaction of the major factors of soil formation: climate, vegetation, topography, parent material, and age. These factors control SOC content by regulating the balance between C gains via photosynthesis and losses via autotrophic and heterotrophic respiration, as well as C losses in soluble and solid form. The SOC content usually ranges between 5 and 100 g kg^{-1} in mineral soils. These concentrations appear modest but at 1500 Pg, the amount of organic C stored globally in soils is second only to that contained in oceans and at least twice that found in either terrestrial vegetation or the atmosphere.

Cultivated soils usually contain less SOC than virgin soils (1) due to the magnification of two biophysical processes: 1) net nutrient mineralization accompanied by release of CO_2 due to microbial respiration, and 2) soil erosion. SOC losses of up to 50% have been reported within 30–70 years of land use conversions under temperate conditions (2–5). SOC losses reported in subtropical and tropical environments often match or even surpass those observed under temperate conditions (6–8). In subtropical and tropical environments, shifting cultivation systems appear to conserve more SOC than forestlands permanently cleared for cultivation (9).

This manuscript has been created by the Battelle Memorial Institute as operator of the Pacific Northwest National Laboratory under Contract No. DE-AC06-76RLO 1830 with the U.S.

MAJOR PROCESSES LEADING TO CARBON LOSSES FROM SOIL

Mineralization Processes

Depending on its frequency and kind, tillage changes the soil biophysical environment in ways that affect the net mineralization of nutrients and the release of carbon. These changes can be described in terms of increases or decreases in soil porosity, disruption of soil aggregates, and redistribution in the proportion of soil aggregate size, as well as alteration of energy and water fluxes. All these changes enhance, at least temporarily, the conversion of organic C into CO_2 (10) and the net release of nutrients from SOM. Much of the success of past agricultural practices relied heavily on the control of decomposition processes through tillage operations to satisfy plant nutrient demands. All this came at a price, however, for a heavy reliance on soil nutrients to feed crops without proper replenishment led to the worldwide declines of SOM (11).

Soil Erosion Processes

Agricultural ecosystems normally experience soil losses at rates considerably greater than natural ecosystems because of an incomplete plant or residue cover of the soil during rainy or windy conditions. When surface and environmental conditions are right (i.e., bare soil, sloping land, intense rain, windy weather), the kinetic energy embedded in wind and water is transferred to soil aggregates causing them to be detached and transported away from their original position across fields or downhill. Besides the physical loss of soil particles and on-site impact on soil productivity, the detachment and transport processes also cause aggregate breakdown, thereby exposing labile C to microbial activity. This aggregate breakdown also facilitates the preferential removal of soil materials comprised mainly of humus and clay or silt fractions. Consequently, water- and windborne sediments become enriched in C with respect to the contributing soil. Carbon enrichment ratios ranging from 3 to 360 have been reported (12, 13). The fate of these C-enriched sediments is not well known, for while transport and burial of C in eroded sediments may lead to "sequestration" (14), it may also result in part of it being emitted back to the atmosphere as CO_2 (15).

RESTORING SOIL ORGANIC MATTER: THE EMERGING SCIENCE OF SOIL CARBON SEQUESTRATION

The Role of Long-Term Field Experiments

SOM is an essential attribute of soil quality (16) and has an essential role in soil conservation and sustainable agriculture. Many practices—some involving land use changes—have been shown to increase SOM and thus received considerable attention for their possible role in climate change mitigation (17–19). Carbon sequestration in managed soils occurs when there is a net removal of atmospheric CO_2 because C inputs (nonharvestable net primary productivity) are greater than C outputs (soil respiration, C costs related to fossil fuels and fertilizers). Soil C sequestration has the additional appeal that all its practices conform to principles of sustainable agriculture (e.g., reduced tillage, erosion control, diversified cropping systems, improved soil fertility). Long-term field experiments have been instrumental to increase our understanding of SOM dynamics (20, 21). The first and longest standing experiment was started at Rothamsted, England, by J.B. Lawes and J.H. Gilbert who in 1843 began documenting the impact of nutrient manipulation on crop yields and soil properties (22). Other experiments were initiated thereafter in America, Europe, and Oceania with the goal of discovering interactions among climate, soil, and management practices. The knowledge that emerged from these experiments has been instrumental for the development and testing of agroecosystem and SOM models (23).

The Global Importance of Soil C Sequestration

There appears to be a significant opportunity for managed ecosystems to act as C sinks. For example, results from inverse modeling experiments suggest that during 1988–1992, terrestrial ecosystems may have been sequestering atmospheric C at rates of 1–2.2 $Pg\,y^{-1}$ (24). Some of the likely causes include the growth of new forest in previously cultivated land (25) and the "CO_2-fertilization effect" (26). Globally, agricultural soils have been estimated to have the capacity to sequester C at rates of 0.6 $Pg\,y^{-1}$ (11) during several decades. The realization of this potential C sequestration would not be trivial since it would offset roughly about one-tenth of the current emissions from fossil fuels. In the U.S., annual gains in soil C from improved agricultural practices have been estimated at 0.14 $Pg\,yr^{-1}$ (25). Whether or not soil C sequestration practices are widely adopted will depend on their value relative to other C capture and sequestration technologies.

Mechanisms of Soil C Sequestration

Recent reviews of experimental results have contributed to organize our understanding of the environmental and management controls of soil C sequestration in grassland (27, 28) and agricultural (29, 30) ecosystems. The use of C balance, soil fractionation, and isotope techniques have been instrumental to reveal how new C (from crop

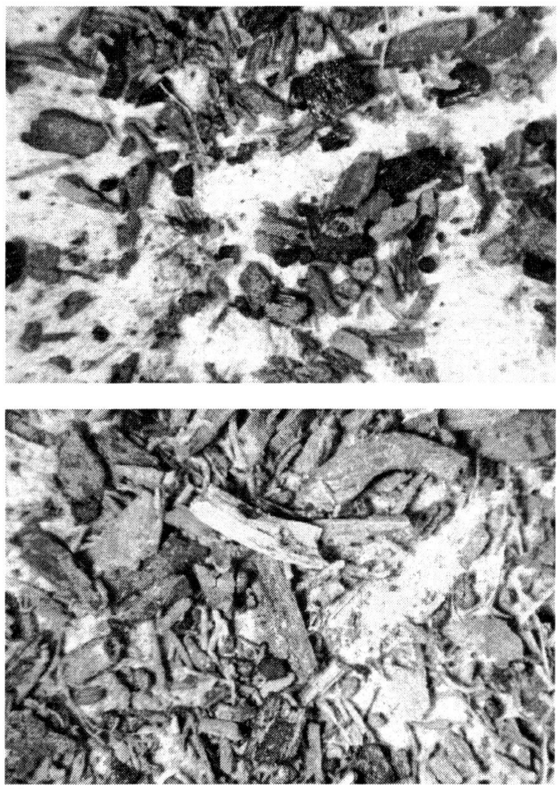

Fig. 1 Young organic matter (fine roots and other organic debris) extracted from two Cryoboralfs under cereal cropping for 13 years receiving N at annual rates of 0 (top) and 50 $kg\,ha^{-1}$ (bottom) (32). The black material is charcoal.

Table 1 Examples of worldwide land use and management impacts on SOC

Region/Country	Climate	Soil	Duration	Crop/land use	Treatment	Soil organic carbon ($kg\ m^{-2}$) Initial	Final	Depth (cm)	Reference
Argentina	Temperate humid	Argiudoll	17	Corn–wheat–soybean	Moldboard plow		4.95	20	(37)
					No tillage		5.46		
Chaco, Argentina	Subtropical semiarid	Alfisol	20	Highly restored			7.05	20	(6)
			10	Moderately restored			3.10		
			60	Highly degraded			1.50		
Rondonia, Brazil	Tropical humid			Forest			4.33	50	(38)
			5	Pasture			5.85		
			9	Pasture			5.26		
			20	Pasture			5.28		
			41	Pasture			6.56		
			81	Pasture			6.12		
Georgia, U.S.	Temperate humid	Hapludult	5	Bermudagrass	Unharvested	1.39	1.74	6	(39)
					Lightly grazed		2.01		
					Heavily grazed		2.00		
					Hayed		1.59		
Kentucky	Temperate humid	Paleudalf	20	Conventional tillage corn	0 kg N/ha^{-1}		4.89	30	(40)
					84 kg N/ha^{-1}		5.63		
					168 kg N/ha^{-1}		5.64		
					336 kg N/ha^{-1}		6.14		
				No till corn	0 kg N/ha^{-1}		5.54		
					84 kg N/ha^{-1}		5.84		
					168 kg N/ha^{-1}		5.89		
					336 kg N/ha^{-1}		6.63		
Kuztown, Pennsylvania	Temperate humid	Fragiudalf	15	Corn	Conventional	4.20	4.30	15	(35)
					Organic	4.40	5.00		
					Manure	4.10	5.30		
Michigan	Cool temperate humid	Hapludalf	7	No crops—natural succession	Tillage		2.06	15	(41)
					Control		2.22		
Swift Current, Canada	Cold semiarid	Haploboroll	10	Cont. wheat	Minimum tillage	3.05	3.52	15	(42)
				Fallow–wheat–wheat		2.99	3.34		
				Green manure–wheat–wheat		2.89	3.22		

Continued

Table 1 Examples of worldwide land use and management impacts on SOC *(Continued)*

Region/Country	Climate	Soil	Duration	Crop/land use	Treatment	Soil organic carbon ($kg\,m^{-2}$) Initial	Final	Depth (cm)	Reference
Breton, Canada	Cold subhumid	Cryoboralf	51	Wheat–fallow	Nil	2.64	1.81	15	(34)
					Fertilizer		2.13		
					Manure		3.11		
				Wheat–oat–barley–hay–hay	Nil		2.91		
					Fertilizer		3.37		
					Manure		4.32		
Russia	Cool temperate humid	Mollisol	300	Native grassland			2.07	50	(5)
			50	Hay			2.13		
			100	Continuous cropping			1.59		
			50	Continuous fallow			1.51		
Punjab, India	Subtropical subhumid	Alluvial	6	Corn–wheat	Minimum tillage, residue retained		0.48	15	(43)
					Minimum tillage, residue removed		0.48		
					Conventional tillage		0.50		
Morocco	Warm temperate semiarid	Calcixeroll	11	Continuous wheat and other rotations	Conventional tillage	3.20	3.73	20	(44)
					No tillage		3.39		
Western Nigeria			20	Bush fallow			2.77	15	(45)
			25	Bush fallow			2.96		
			10	Bermudagrass			3.90		
			10	Cultivation			1.29		

O

residues, roots, and organic amendments) enters soil, resides shortly (for a few years) in labile soil fractions, and finally becomes a long-time constituent (for hundreds of years) of recalcitrant organo-mineral complexes (31). Figure 1 contrasts young (labile) organic matter fractions extracted from two cultivated soils with and without N fertilizer (32). The amounts of labile organic matter—fine roots and other organic debris—present in each soil reflect differences in crop productivity induced by addition of N at annual rates of $50\,kg\,ha^{-1}$ for 13 years. 'Terra Preta' soil—in tropical regions of South America and West Africa—represents a prime example of ancient wisdom applied to develop sustainable agriculture through the improvement of soil fertility and SOM (33).

The quantity and quality of C entering soil as well as the interaction of this C with the soil biophysical environment are major factors determining the rate and duration of soil C sequestration. The quantity of C added to soil in the form of roots, crop residues, and organic amendments has been shown to play a dominant role in defining the trajectory of SOC over time (34). Management practices geared toward optimizing nutrient supply and building nutrient reserves (e.g., fertilization, use of legumes in crop rotations) are almost guaranteed to increase soil C stocks. The quality of crop residues and the timing of their incorporation to soil also have an influential role on C decomposition and, thus, on soil C storage (35). The degree of soil disturbance—through its impact on soil aggregation—constitutes another major factor regulating C decomposition and retention in soil (36). In this context, no tillage agriculture has come to represent one of the most significant technological innovations of the last 30 years because it allows farmers the possibility of growing crops economically while reducing erosion and improving both quantity and quality of SOM. A few examples of the management impacts on soil C sequestration from around the world are presented in Table 1.

Soil Organic Matter, Energy, and Full C Accounting

Land is the natural habitat of humans. Humans dwell on it and use it as a resource for the production of food, fiber, and other goods. Simply put, land is managed when there is a manipulation of energy and matter flows in order to meet certain economic and social objectives. Farm mechanization and fertilizers are two of the many technical innovations that—though they rely on the utilization of fossil energy—have brought dramatic increases in food production during the last century. Changes in management practices that include soil C sequestration as an objective require careful evaluation of their impact not only on soil C gains but also on C costs from the use of fossil energy (e.g., manufacture of fertilizers) (46, 47) and on the net greenhouse gas emissions (48).

THE ROLE OF SOIL ORGANIC MATTER IN THE 21ST CENTURY

SOM has played and will continue to play a central role in sustainable land management. The restoration of SOM at global scales offers a unique opportunity to mitigate global warming. As population levels and affluence increase, demands on land to produce food, fiber, biomass, and other products will remain high. Because land is finite, important decisions will have to be made in order to balance such demands with functional objectives such as the preservation of natural ecosystems. As part of any climate policy, the impact of land use changes and management on SOM storage should be included as a criterion for making these decisions. Depending on their degree of expansion, several evolving agricultural technologies—such as genetically modified crops, conservation tillage, organic farming, and precision farming—may have important implications for soil C sequestration (19). Their ultimate impact on C sequestration will depend not only on the economic benefit realized by individual producers but also on whether society recognizes the value of soil C storage to mitigate global warming.

REFERENCES

1. Davidson, E.A.; Ackerman, I.L. Changes in Soil Carbon Following Cultivation of Previously Untilled Soils. Biogeochemistry **1993**, *20*, 161–164.
2. Dalal, R.C.; Mayer, R.J. Long-Term Trends in Fertility of Soils Under Continuous Cultivation and Cereal Cropping in Southern Queensland. II. Total Organic Carbon and Its Rate of Loss from the Soil Profile. Aust. J. Soil Res. **1986**, *24*, 281–292.
3. Mann, L.K. Changes in Soil Carbon After Cultivation. Soil Sci. **1986**, *142*, 279–288.
4. Ellert, B.H.; Gregorich, E.G. Storage of Carbon, Nitrogen and Phosphorus in Cultivated and Adjacent Forested Soils of Ontario. Soil Sci. **1996**, *161*, 587–603.
5. Mikhailova, E.A.; Bryant, R.B.; Vassenev, I.I.; Schwager, S.J.; Post, C.J. Cultivation Effects on Soil Carbon and Nitrogen Contents at Depth in the Russian Chernozem. Soil Sci. Soc. Am. J. **2000**, *64*, 738–745.
6. Abril, A.; Bucher, E.H. Overgrazing and Soil Carbon Dynamics in the Western Chaco of Argentina. Appl. Soil Ecol. **2001**, *16*, 243–249.

O

7. Lal, R. Deforestation and Land Use Effects on Soil Degradation and Rehabilitation in Western Nigeria, II. Soil Chemical Properties. Land Degrad. Dev. **1996**, *7*, 87–98.
8. Lobe, I.; Amelung, W.; Du Preez, C.C. Losses of Carbon and Nitrogen with Prolonged Arable Cropping from Sandy Soils of the South African Highveld. Eur. J. Soil Sci. **2001**, *52*, 93–101.
9. Houghton, R.A. Changes in the Storage of Terrestrial Carbon Since 1850. In *Soils and Global Change*; Lal, R., Kimble, J., Levine, E., Stewart, B.A., Eds.; CRC/Lewis Publishers: Boca Raton, FL, 1995; 45–65.
10. Reicoski, D.C.; Lindstrom, M.J. Fall Tillage Method: Effect from Short-Term Carbon Dioxide Flux from Soil. Agron. J. **1993**, *85*, 1237–1243.
11. Cole, V.; Cerri, C.; Minami, K.; Mosier, A.; Rosenberg, N.J.; Sauerbeck, D. Agricultural Options for Mitigation of Greenhouse Gas Emissions. In *Climate Change 1995: Impacts, Adaptations and Mitigation of Climate Change*; Report of IPCC Working Group II, Watson, R.T., Zinowera, M.C., Moss, R.H., Eds.; Cambridge University Press: London, UK, 1996; 745–771.
12. Sterk, G.; Herrmann, L.; Bationo, A. Wind-Blown Nutrient Transport and Soil Productivity Changes in Southwest Niger. Land Degrad. Dev. **1996**, *7*, 325–336.
13. Zobeck, T.M.; Fryrear, D.W. Chemical and Physical Characteristics of Wind-Blown Sediment. Trans. Am. Soc. Agric. Eng. **1986**, *29*, 1037–1041.
14. Stallard, R.F. Terrestrial Sedimentation and the Carbon Cycle: Coupling Weathering and Erosion to Carbon Burial. Global Biogeochem. Cycles **1998**, *12*, 231–257.
15. Lal, R. Global Soil Erosion by Water and C Dynamics. In *Soils and Global Change*; Lal, R., Kimble, J., Levine, E., Stewart, B.A., Eds.; CRC/Lewis Publishers: Boca Raton, FL, 1995; 131–141.
16. Doran, J.W.; Coleman, D.C.; Bezdicek, D.F.; Stewart, B.A., Eds. *Defining Soil Quality for a Sustainable Environment*; Soil Science Society America Special Publication No. 35, SSSA: Madison, WI, 1994; 244.
17. Batjes, N.H. Mitigation of Atmospheric CO_2 Concentrations by Increased Carbon Sequestration in the Soil. Biol. Fert. Soils **1998**, *27*, 230–235.
18. Post, W.M.; Kwon, K.C. Soil Carbon Sequestration and Land-Use Change: Processes and Potential. Global Change Biol. **2000**, *6*, 317–327.
19. Izaurralde, R.C.; Rosenberg, N.J.; Lal, R. Mitigation of Climatic Change by Soil Carbon Sequestration: Issues of Science, Monitoring and Degraded Lands. Adv. Agron. **2001**, *70*, 1–75.
20. Powlson, D.S.; Smith, P., Smith, J.U., Eds. *Evaluation of Soil Organic Matter Models Using Existing Long-Term Datasets*; NATO ASI Series I, Springer: Heidelberg, 1996; Vol. 38, 429.
21. Paul, E.A., Paustian, K., Elliott, E.T., Cole, C.V., Eds. *Soil Organic Matter in Temperate Agroecosystems: Long-Term Experiments in North America*; CRC/Lewis Publishers: Boca Raton, FL, 1997; 414.
22. Jenkinson, D.S. The Rothamsted Long-Term Experiments: Are They Still of Use? Agron. J. **1991**, *83*, 2–10.
23. Smith, P.; Smith, J.U.; Powlson, D.S.; McGill, W.B.; Arah, J.R.M.; Chertov, O.; Coleman, K.W.; Franko, U.; Frolking, S.; Jenkinson, D.S.; Jensen, L.S.; Kelly, R.H.; Klein-Gunnewiek, H.; Komarov, A.S.; Li, C.; Molina, J.A.E.; Mueller, T.; Parton, W.J.; Thornley, J.H.M.; Whitmore, A.P. A Comparison of the Performance of Nine Soil Organic Matter Models Using Datasets from Seven Long-Term Experiments. Geoderma **1997**, *81*, 153–225.
24. Fan, S.; Gloor, M.; Mahlman, J.; Pacala, S.; Sarmiento, J.; Takahashi, T.; Tans, P. A Large Terrestrial Carbon Sink in North America Implied by Atmospheric and Oceanic Carbon Dioxide Data and Models. Science **1998**, *282*, 442–446.
25. Houghton, R.A.; Hackler, J.L.; Lawrence, K.T. The U.S. Carbon Budget: Contributions from Land Use Change. Science **1999**, *285*, 574–578.
26. Kimball, B.A. Carbon Dioxide and Agricultural Yield: An Assemblage and Analysis of 430 Prior Observations. Agron. J. **1983**, *75*, 779–782.
27. Conant, R.T.; Paustian, K.; Elliott, E.T. Grassland Management and Conversion into Grassland: Effects on Soil Carbon. Ecol. Appl. **2001**, *11*, 343–355.
28. Scott, N.A.; Tate, K.R.; Ford-Robertson, J.; Giltrap, D.J.; Smith, C.T. Soil Carbon Storage in Plantations and Pastures: Land-Use Implications. Tellus **1999**, *51B*, 326–335.
29. Janzen, H.H.; Campbell, C.A.; Izaurralde, R.C.; Ellert, B.H.; Juma, N.; McGill, W.B.; Zentner, R.P. Management Effects on Soil C Storage on the Canadian Prairies. Soil Till. Res. **1998**, *47*, 181–195.
30. Paustian, K.; Collins, H.P.; Paul, E.A. Management Controls on Soil Carbon. In *Soil Organic Matter in Temperate Ecosystems: Long-Term Experiments in North America*; Paul, E.A., Paustian, K., Elliott, E.T., Cole, C.V., Eds.; CRC/Lewis Publishers: Boca Raton, FL, 1997; 15–49.
31. Jastrow, J.D. Soil Aggregate Formation and the Accrual of Particulate and Mineral-Associated Organic Matter. Soil Biol. Biochem. **1996**, *28*, 665–676.
32. Solberg, E.D.; Nyborg, M.; Izaurralde, R.C.; Malhi, S.S.; Janzen, H.H.; Molina-Ayala, M. Carbon Storage in Soils Under Continuous Cereal Grain Cropping: N Fertilizer and Straw. In *Management of Carbon Sequestration in Soil*; Lal, R., Kimble, J., Follett, R., Stewart, B.A., Eds.; CRC/Lewis Publishers: Boca Raton, FL, 1998; 235–254.
33. Glaser, B.; Haumaier, L.; Guggenberger, G.; Zech, W. The 'Terra Preta' Phenomenon: A Model for Sustainable Agriculture in the Humid Tropics. Naturwissenschaften **2001**, *88*, 37–41.
34. Izaurralde, R.C.; McGill, W.B.; Robertson, J.A.; Juma, N.G.; Thurston, J.T. Carbon Balance of the Breton Classical Plots After Half a century. Soil Sci. Soc. Am. J. **2001**, *65*, 431–441.
35. Drinkwater, L.E.; Wagoner, P.; Sarrantonio, M. Legume-Based Cropping Systems Have Reduced Carbon and Nitrogen Losses. Nature **1998**, *396*, 262–265.
36. Six, J.; Elliott, E.T.; Paustian, K. Soil Macroaggregate Turnover and Microaggregate Formation: A Mechanism for C Sequestration Under No-Tillage Agriculture. Soil Biol. Biochem. **2000**, *32*, 2099–2103.
37. Alvarez, R.; Russo, M.E.; Prystupa, P.; Scheiner, J.D.; Blotta, L. Soil Carbon Pools Under Conventional and No-Tillage Systems in the Argentine Rolling Pampa. Agron. J. **1998**, *90*, 138–143.
38. Neill, C.; Cerri, C.C.; Melillo, J.M.; Feigl, B.J.; Steudler, P.A.; Moraes, J.F.L.; Piccolo, M.C. Stocks and

Dynamics of Soil Carbon Following Deforestation for Pasture in Rondônia. In *Soil Processes and the Carbon Cycle*; Lal, R., Kimble, J., Follett, R., Stewart, B.A., Eds.; CRC/Lewis Publishers: Boca Raton, FL, 1998; 9–28.
39. Franzluebbers, A.J.; Stuedemann, J.A.; Wilkinson, S.R. Bermudagrass Management in the Southern Piedmont USA: I. Soil and Surface Residue Carbon and Sulfur. Soil Sci. Soc. Am. J. **2001**, *65*, 834–841.
40. Ismail, I.; Blevins, R.L.; Frye, W.W. Long-Term No-Tillage Effects on Soil Properties and Continuous Corn Yields. Soil Sci. Soc. Am. J. **1994**, *58*, 193–198.
41. Richter, D.D.; Babbar, L.I.; Huston, M.A.; Jaeger, M. Effects of Annual Tillage on Organic Carbon in a Fine-Textured Udalf: The Importance of Root Dynamics to Soil Carbon Storage. Soil Sci. **1999**, *149*, 78–83.
42. Curtin, D.; Wang, H.; Selles, F.; McConkey, B.G.; Campbell, C.A. Tillage Effects on Carbon Fluxes in Continuous Wheat and Fallow–Wheat Rotations. Soil Sci. Soc. Am. J. **2000**, *64*, 2080–2086.
43. Ghuman, B.S.; Sur, H.S. Tillage and Residue Management Effects on Soil Properties and Yields of Rainfed Maize and Wheat in a Subhumid Subtropical Climate. Soil Till. Res. **2001**, *58*, 1–10.
44. Mrabet, R.; Saber, N.; El-Brahli, A.; Lahlou, S.; Bessam, F. Total, Particular Organic Matter and Structural Stability of a Calcixeroll Soil Under Different Wheat Rotations and Tillage Systems in a Semiarid Area of Morocco. Soil Till. Res. **2001**, *57*, 225–235.
45. Lal, R. Land Use and Soil Management Effects on Soil Organic Matter Dynamics on Alfisols in Western Nigeria. In *Soil Processes and the Carbon Cycle*; Lal, R., Kimble, J., Follett, R., Stewart, B.A., Eds.; CRC/Lewis Publishers: Boca Raton, FL, 1998; 109–126.
46. Schlesinger, W.H. Carbon and Agriculture: Carbon Sequestration in Soils. Science **1999**, *284*, 2095.
47. Izaurralde, R.C.; McGill, W.B.; Rosenberg, N.J. Carbon Cost of Applying Nitrogen Fertilizer. Science **2000**, *288*, 811–812.
48. Robertson, G.P.; Paul, E.A.; Harwood, R.R. Greenhouse Gases in Intensive Agriculture: Contributions of Individual Gases to the Radiative Forcing of the Atmosphere. Science **2000**, *289*, 1922–1925.

ORGANIC MATTER MODELING

Pete Smith
University of Aberdeen, Aberdeen, United Kingdom

O

INTRODUCTION

There are a number of approaches to modeling soil organic matter (SOM) turnover including: 1) process-based multi-compartment models; 2) models that consider each fresh addition of plant debris as a separate cohort which decays in a continuous way; and 3) models that account for C and N transfers through various trophic levels in a soil food web. These approaches are described in more detail below.

PROCESS-BASED, MULTI-COMPARTMENT SOM MODELS

Most models are process-based, i.e. they focus on the processes mediating the movement and transformations of matter or energy and usually assume first order rate kinetics (1). Early models simulated the SOM as one homogeneous compartment (2). Some years later two-compartment models were proposed (3, 4) and, as computers became more accessible, multi-compartment models were developed (5, 6). Of the 33 SOM models currently represented within the Global Change and Terrestrial Ecosystems (GCTE) Soil Organic Matter Network (SOMNET) database (7–9), 30 are multi-compartment, process-based models. Each compartment or SOM pool within a model is characterized by its position in the model's structure and its decay rate. Decay rates are usually expressed by first-order kinetics with respect to the concentration (C) of the pool

$$dC/dt = -kC$$

where t is the time. The rate constant k of first-order kinetics is related to the time required to reduce by half the concentration of the pool *when there is no input.* The pool's half life [$h = (\ln 2)/k$], or its turnover time ($\tau = 1/k$) are sometimes used instead of k to characterize a pool's dynamics: the lower the decay rate constant, the higher the half-life, the turnover time, and the stability of the organic pool.

The flows of C within most models represent a sequence of carbon going from plant and animal debris to the microbial biomass, then to soil organic pools of increasing stability. Some models also use feed-back loops to account for catabolic and anabolic processes and microbial successions. The output flow from an organic pool is usually split. It is directed to a microbial biomass pool, another organic pool, and, under aerobic conditions, to CO_2. This split simulates the simultaneous anabolic and catabolic activities and growth of a microbial population feeding on one substrate. Two parameters are required to quantify the split flow. They are often defined by a microbial (utilization) efficiency and stabilization (humification) factor which control the flow of decayed C to the biomass and humus pools, respectively. The sum of the efficiency and humification factors must be inferior to one to account for the release of CO_2. A thorough review of the structure and underlying assumptions of different process-based SOM models is available (6).

COHORT MODELS DESCRIBING DECOMPOSITION AS A CONTINUUM

Another approach to modeling SOM turnover is to treat each fresh addition of plant debris into the soil as a cohort (5). Such models consider one SOM pool that decays with a feed-back loop into itself. Q-SOIL (10), for example, is represented by a single rate equation. The SOM pool is divided into an infinite number of components, each characterized by its "quality" with respect to degradability as well as impact on the physiology of the decomposers. The rate equation for the model Q-SOIL represents the dynamics of each SOM component of quality q and is quality dependent. Exact solutions to the rate equations are obtained analytically (11).

FOOD-WEB MODELS

Another type of model simulates C and N transfers through a food web of soil organisms (1, 12); such models explicitly account for different trophic levels or functional groups of biota in the soil (13–18). Some models have been developed which combine an explicit description of the soil biota with a process-based approach (19). Food-web models require a detailed knowledge of the biology of the system to be simulated and are usually parameterized for application at specific sites.

Table 1 Overview of SOM models represented within GCTE-SOMNET in January 2001

Model	Inputs				Factors affecting decay rate constants	Soil outputs	Reference
	Timestep	Meteorology	Soil and plant	Management			
ANIMO	Day, week, month	P, AT, Ir, EvW	Des, Lay, Imp, Cl, OM, N, pH	Rot, Ti, Fert, Man, Res, Irr, AtN	T, W, pH, N, O	C, N, W, ST, gas	(26)
APSIM	Day	P, AT, Ir	Lay, W, C, N, BD, Wi, PG, PS	Rot, Ti, Fert, Irr	T, W, pH, N	C, N, W, ST, gas	(27)
Candy	Day	P, AT, Ir	D, Imp, W, N, C, Wi, PD, Nup	Rot, Ti, Fert, Man, Res, Irr, AtN	T, W, N, Cl	C, N, W, ST, gas	(28)
CENTURY	Month	P, AT	W, Cl, OM, pH, C, N	Rot, Ti, Fert, Man, Res, Irr, AtN	T, W, N, Cl, pH, Ti	C, BioC, 13C, 14C, N, W, ST, gas	(29)
Chenfang Lin Model	Day	ST	OM, BD, W	Man, Res	T, W, F	C, BioC, gas	(30)
DAISY	Hour, day	P, AT, Ir, EvG	Lay, Cl, C, N, PG, PS	Rot, Ti, Fert, Man, Res, Irr, AtN	T, W, N, Cl	C, BioC, N, W, ST, gas	(31)
DNDC	Hour, day, month	P, AT	Lay, Cl, OM, pH, BD	Rot, Ti, Fert, Man, Res, Irr, AtN	T, W, N, Cl, Ti	C, BioC, N, W, ST, gas	(32)
DSSAT	Hour, day, month, year	P, AT, Ir	Des, Lay, Imp, W, Cl, PS, OM, pH, C, N	Rot, Ti, Fert, Man, Res, Irr	T, W, N, Cl, Ti	C, BioC, N, W, ST	(33)
D3R	Day	P, AT	Y, PS	Rot, Ti, Res	T, W, N, Cv, Ti	Decomp. of surface and buried residue	(34)
Ecosys	Minute, hour	P, AT, Ir, WS, RH	Lay, W, Cl, CEC, PS, OM, pH, N, BD, PG, PS	Rot, Ti, Fert, Man, Res, Irr, AtN	T, W, N, O, Cl, Cv	C, BioC, N, W, ST, pH, Ph, EC, gas, ExCat	(35)
EPIC	Day	P, AT	Lay, Imp, W, Cl, OM, pH, C, BD, Wi	Rot, Ti, Fert, Man, Res, Irr,AtN	T, W, N, pH, Cl, Ce, Cv	C, BioC, N, W, ST	(36)

(*Continued*)

Table 1 Overview of SOM models represented within GCTE-SOMNET in January 2001 (*Continued*)

Model	Inputs: Timestep	Inputs: Meteorology	Inputs: Soil and plant	Inputs: Management	Factors affecting decay rate constants	Soil outputs	Reference
FERT	Day	P, AT, WS	Des, Lay, W, Cl, OM, pH, C, N, BD, W, Ph, K, Nup, Y, PS	Rot, Ti, Fert, Man, Res, Irr	T, W, N, pH, Cv	C, N, Ph, K	(37)
ForClim-D	Year	P, AT	W, AG	None	T, W	C	(38)
GENDEC	Day, month	ST, W	W, InertC, LQ	Can be used - not essential	T, W, N	C, BioC, N, gas, LQ	(39)
HPM/EFM	Day	P, AT, Ir, WS	W, Cl, PS	Rot, Fert, Irr, AtN	T, W, N	C, BioC, N, W, gas	(40)
ICBM	Day, year	Combination of weather & climate	Many desirable: none essential	C inputs to soil	T, W, Cl	C	(41)
KLIMAT-SOIL-YIELD	Day, year	P, AT, ST, Ir, EvG, EvS, VPD, SH	Des, Lay, Imp, W, Cl, PS, OM, pH, C, N	Fert, Man, Res, Irr	T, W, N, Cl	C, BioC, N, W, ST	(42)
CNSP Pasture Model	Day	P, AT, Ir	Lay, Inp, W, Cl, CEC, OM, pH, C, N, PS, AS	Fert	T, W, N, pH	C, N, W, ST	(43)
Humus Balance	Year	Climate based on P and AT	Des, Lay, PS, OM, pH, C, N	Rot, Fert, Man	N, H, Cl, Cv	C, N	(44)
MOTOR	User specified	P, AT, EvG	Des, OM	Rot, Ti, Fert, Man	T, W, N, Cl, Ti	C, BioC, 13C, 14C, gas	(45)
NAM SOM	Year	P, AT	Des, PS, OM, Ero	Man, Res	T, W, Cl, Cv	C, BioC	(46)
NCSOIL	Day	ST, (P, AT)	W, OM, C, N	Fert, Man, Res	T, W, N, pH, Cl, Ti	C, BioC, 14C, N, 15N, gas	(47)
NICCE	Hour, day	P, AT, Ir, WS	Imp, OM, C, N, W, TC, PG	Fert, Man, Res, Irr, AtN	T, W, Cl, N	C, BioC, 13C, 14C, N, 15N, W, ST, gas	(48)
O'Brien Model	Year	None	Lay, C, 14C	None	None	C, 14C	(49)
O'Leary Model	Day	P, AT	Lay, W, Cl, pH, N	Ti, Fert, Res	T, W, N, Cl, Ti	C, BioC, N, W, ST, gas, ResC, ResN	(50)
Q-Soil	Year	Optional	C, N	Rot, Fert, Man, Res, AtN	T, W, N	C, BioC, 13C, N	(10)
RothC	Month	P, AT, EvW	Cl, C, InertC (can be estimated)	Man, Res, Irr	T, W, Cl, Cv	C, BioC, gas, 14C	(51)
SOCRATES	Week	P, AT	CEC, Y	Rot, Fert, Res	T, W, N, Cv, Ce	C, BioC, gas	(52)

(*Continued*)

O

Table 1 Overview of SOM models represented within GCTE-SOMNET in January 2001 (*Continued*)

Model	Inputs				Factors affecting decay rate constants	Soil outputs	Reference
	Timestep	Meteorology	Soil and plant	Management			
SOMM	Day	P, ST	OM, N, AshL, NL	Man	T, W, N	C, N, gas	(53)
Sundial	Week	P, AT, EvG	Imp, Cl, W, Y	Rot, Fert, Man, Res, Irr, AtN	T, W, N, Cl	C, BioC, N, 15N, W, gas	(54)
Verberne	Day	P, AT, Ir, WS, EvS	Des, W, Cl, PS, OM, C, N	Man, AtN	T, W, N, Cl	C, BioC, N, W	(55)
VOYONS	Day, week, month	P, ST	Cl, OM, C, N	Fert, Man, Res, Irr, AtN	T, W, Cl	C, BioC, 13C, 14C, N, gas	(56)
Wave	Day	P, AT, Ir, EvG	Lay, OM, C, N, W, PG	Rot, Ti, Fert, Man, Res, Irr, AtN	T, W, N	C, N, W, ST, gas	(57)

Key:
Meteorology: P = Precipitation, AT = Air Temperature, ST = Soil Temperature, Ir = Irradiation, EvW = Evaporation over water, EvG = Evaporation over grass, EvS = Evaporation over bare soil, WS = Wind speed, RH = Relative humidity, VPD = Vapor Pressure deficit, SH = Sun hours.
Soil & plant inputs: Des = Soil description, Lay = Soil layers, Imp = Depth of impermeable layer, Cl = Clay content, OM = Organic matter content, N = Soil nitrogen content/dynamics, C = Soil Carbon content/dymanics, InertC = Soil inert carbon content, pH = pH, W = Soil water characteristics, Wi = Wilting point, PD = Soil particle size distribution, CEC = Cation exchenge capacity, Ero = Annual erosion losses, BD = Soil bulk density, TC = Thermal conductivity, PG = Plant growth characteristics, PS = Plant species composition, AS = Animal species present, AG = Animal growth characteristics, Y = Yield, Nup = Plant nitrogen uptake, LQ = Litter quality, AshL = Ash content of litter, NL = N content of litter.
Management input details: Rot = rotation, Ti = tillage practice, Fert = Inorganic fertilizer applications, Man = Organic manure applications, Res = Residue management, Irr = Irrigation, AtN = Atmospheric nitrogen inputs.
Factors affecting decay rate constants: T = temperature, W = water, pH = pH, N = nitrogen, O = oxygen, Cl = clay, Ce = cation exchange capacity, Cv = cover crop, Ti = tillage, F = Fauna.
Soil outputs: C, N, W, LQ and ST as above. BioC = Biomass carrbon, 13C = ^{13}C dynamics, 14C = ^{14}C dynamics, 15N = ^{15}N dynamics, gas = gaseous losses (e.g. CO_2, N_2O, N_2), ResC = Surface residue carbon, ResN = Surface residue nitrogen, Ph = Phosphorus dynamics, K = Potassium dynamics. EC = Electrical conductivity, ExCat = Exchangeable cations. NB: N in the soil inputs and outputs section is used to denote all aspects of the N cycle.
Further details regarding optimum decay conditions, SOM components, rate constants, methods of pool fitting and refractory SOM are given elsewhere (Refs. 6, 58). A metadatabase of all models is available (Ref. 7).

FACTORS AFFECTING SOM TURNOVER IN MODELS

Rate "constants" (k) are constant for a given set of biotic and abiotic conditions. For non-optimum environmental circumstances, the simplest way to modify the maximum value of k is by multiplication by a reduction factor μ—ranging from 0 to 1. Environmental factors considered by SOM models include temperature, water, pH, nitrogen, oxygen, clay content, cation exchange capacity, type of crop/plant cover and tillage.

Many studies show the effect of temperature on microbially mediated transformations in soil, either expressed as a reduction factor or the Arrhenius equation, but recently the assumption that SOM decomposition is temperature dependent has been challenged by showing that old SOM in forest soils does not decompose more rapidly in soils from warmer climates than in soils from colder regions (20). Water and oxygen have a major impact on the microbial physiology. Whilst some models simulate O_2 concentrations in soil explicitly (21, 22), many define the extent of anaerobiosis based on soil pore space filled with water (WFPS) (23, 24). Soil clay content and total SOM are correlated. Various schemes simulate the effect of clay on rate equations to obtain SOM accumulation. Nitrogen is an essential element for microbial growth which will be maximal when enough N is assimilated to maintain the microbial C:N ratio (25). Table 1 presents an overview of the 33 models represented in the GCTE-SOMNET (7, 8) including the factors affecting SOM turnover.

SOM MODEL EVALUATION

There are many reasons for evaluating the performance of a SOM model. Model evaluation shows how well a model can be expected to perform in a given situation, it can help to improve the understanding of the system (especially where the model fails), it can provide confidence in the models ability to predict changes in SOM in the future or where there are no data, and it can be used to assess the uncertainties associated with the model's predictions. Models can be evaluated at a number of different levels. They can be evaluated at the individual process level, at the level of a sub-set of processes (e.g., net mineralization), or the models' overall outputs (e.g., changes in total SOM over time) can be tested against measured laboratory and field data. Models can also be evaluated for their applicability in different situations, e.g. for scaling-up simulated net C storage from a site specific to a regional level (59). Many examples of different forms of SOM model evaluation are presented elsewhere (6).

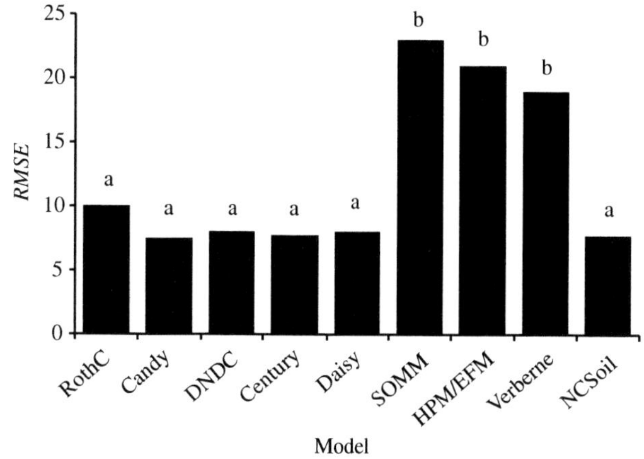

Fig. 1 Overall *RMSE* (Root Mean Square Error) value for nine SOM models when simulating changes in total soil organic carbon in up to 12 datasets from seven long term experiments. The *RMSE* values of the models with the same letter (a or b) do not differ significantly (two sample, two tailed t-test; $p > 0.05$), but the *RMSE* values of the two groups (a and b) do differ significantly (two sample, two tailed t-test; $p < 0.05$) (60).

In the most comprehensive evaluation of SOM models to date (60), nine models were tested against 12 data sets from seven long-term experiments representing arable rotations, managed and unmanaged grassland, forest plantations and natural woodland regeneration. The results showed that six models had significantly lower overall errors (RMSE) than another group of three models (Fig. 1).

The poorer performance of three of the models was related to failures in other parts of the ecosystem models, thus providing erroneous inputs into the SOM module (60).

SOM MODEL APPLICATION

SOM models are often used as research tools in that they are hypotheses of the dynamics of C and N in soil and can be used to distinguish between competing hypotheses (47). Another increasing application of SOM models is in agronomy; many SOM models are now being used to improve agronomic efficiency and environmental quality through incorporation into decision support systems; e.g., SUNDIAL-FRS (54), DSSAT (33), and APSIM (27).

SOM models are now used, more than ever, to extrapolate our understanding of SOM dynamics both temporally (in to the future) and spatially (to assess C fluxes from whole regions or continents). An early example of a regional scale application was the use of the CENTURY model to predict the effects of alternative management practices and policies in agroecosystems of the central United States (61). Since then, many studies have adopted similar methodologies to assess SOM dynamics at the regional (62), national (63, 64) and global scales (65–73). SOM models are increasingly being used by policy makers at the national, regional or global scales, for example in the post-Kyoto debate on the ability of the terrestrial biosphere to store carbon (74). With such an important role in society, it is important that SOM models are transparent, well-evaluated and well documented. There is still a variety of understanding and different hypotheses incorporated in our current SOM models. Future developments in SOM models will further improve our understanding and allow models to be used truly predictively, without the need for site-specific calibration. These developments will improve estimates of, and reduce, the uncertainty associated with SOM model predictions.

REFERENCES

1. Paustian, K. Modelling Soil Biology and Biochemical Processes for Sustainable Agricultural Research. In *Soil Biota. Management in Sustainable Farming Systems*; Pankhurst, C.E., Doube, B.M., Gupta, V.V.S.R., Grace, P.R., Eds.; CSIRO Information Services: Melbourne, 1994; 182–193.
2. Jenny, H. *Factors of Soil Formation. A System of Quantitative Pedology*; McGraw-Hill: New York, N.Y., 1941.
3. Beek, J.; Frissel, M.J. *Simulation of Nitrogen Behaviour in Soils*; Pudoc: Wageningen, The Netherlands, 1973.
4. Jenkinson, D.S. Studies on the Decomposition of Plant Material in Soil. V. Journal of Soil Science **1977**, *28*, 424–434.
5. McGill, W.B. Review and Classification of Ten Soil Organic Matter (SOM) Models. In *Evaluation of Soil Organic Matter Models Using Existing, Long-Term Datasets*; NATO ASI I38; Smith, J.U., Eds.; Powlson, D.S., Smith, P., Springer-Verlag: Berlin, 1996; 111–133.
6. Molina, J.A.E.; Smith, P. Modeling Carbon and Nitrogen Processes in Soils. Advances in Agronomy **1998**, *62*, 253–298.
7. Global Change and Terrestrial Ecosystems (GCTE) Soil Organic Matter Network (SOMNET) database Accessible at http://www.iacr.bbsrc.ac.uk/aen/somnet/index.htm
8. Smith, P.; Smith, J.U.; Powlson, D.S., Eds. *Soil Organic Matter Network (SOMNET): 1996 Model and Experimental Metadata*, GCTE Report 7; GCTE Focus 3 Office: Wallingford, U.K., 1996; 255.
9. Smith, P.; Powlson, D.S.; Smith, J.U.; Glendining, M.J. The GCTE SOMNET: A Global Network and Database of Soil Organic Matter Models and Long-Term Datasets. Soil Use and Management **1996**, *108*, 57.
10. Bosatta, E.; Ågren, G.I. Theoretical Analyses of the Interactions Between Inorganic Nitrogen and Soil Organic Matter. European J. Soil Science **1995**, *76*, 109–114.
11. Bosatta, E.; Ågren, G.I. Theoretical Analysis of Microbial Biomass Dynamics in Soils. Soil Biol. Biochem. **1994**, *26*, 143–148.
12. Smith, P.; Andrén, O.; Brussaard, L.; Dangerfield, M.; Ekschmitt, K.; Lavelle, P.; Tate, K. Soil Biota and Global Change at the Ecosystem Level: Describing Soil Biota in Mathematical Models. Global Change Biology **1998**, *4*, 773–784.
13. Hunt, H.W.; Coleman, D.C.; Cole, C.V.; Ingham, R.E.; Elliott, E.T.; Woods, L.E. Simulation Model of a Food Web with Bacteria, Amoebae, and Nematodes in Soil. In *Current Perspectives in Microbial Ecology*; Klug, M.J., Reddy, C.A., Eds.; American Society for Microbiology: Washington DC, 1984; 346–352.
14. Hunt, H.W.; Coleman, D.C.; Ingham, E.R.; et al. The Detrital Food Web in a Shortgrass Prairie. Biology and Fertility of Soils **1987**, *3*, 57–68.
15. Hunt, H.W.; Trlica, M.J.; Redente, E.F.; et al. Simulation Model for the Effects of Climate Change on Temperate Grassland Ecosystems. Ecological Modelling **1991**, *53*, 205–246.
16. de Ruiter, P.C.; Van Faassen, H.G. A Comparison Between an Organic Matter Dynamics Model and a Food Web Model Simulating Nitrogen Mineralization in Agroecosystems. European Journal of A gronomy **1994**, *3*, 347–354.
17. de Ruiter, P.C.; Van Veen, J.A.; Moore, J.C.; Brussaard, L.; Hunt, H.W. Calculation of Nitrogen Mineralization in Soil Food Webs. Plant and Soil **1993**, *157*, 263–273.
18. de Ruiter, P.C.; Neutel, A.-M.; Moore, J.C. Energetics and Stability in Belowground Food Webs. In *Food Webs, Integration of Patterns and Dynamics*; Polis, G.A., Winemiller, K.O., Eds.; Chapman & Hall: New York, 1995; 201–210.
19. McGill, W.B.; Hunt, H.W.; Woodmansee, R.G.; Reuss, J.O. PHOENIX, a Model of the Dynamics of Carbon and Nitrogen in Grassland Soil. In *Terrestrial Nitrogen Cycles. Processes, Ecosystem Strategies and Management Impacts*; Clark, F.E., Roswall, T.; Ecological Bulletins **1981**, *33*, 49–115.
20. Giardina, C.P.; Ryan, M.G. Evidence that Decomposition Rates of Organic Carbon in Mineral Soil Do Not Vary with Temperature. Nature **2000**, *393*, 249–252.
21. Grant, R.F. A Technique for Estimating Denitrification Rates at Different Soil Temperatures, Water Contents and Nitrate Concentrations. Soil Sci. **1991**, *152*, 41–52.
22. Sierra, J.; Renault, P. Respiratory Activity and Oxygen Distribution in Natural Aggregates in Relation to Anaerobiossis. Soil Sci. Soc. Am. J. **1996**, *60*, 1428–1438.
23. Skopp, J.; Jawson, M.D.; Doran, J.W. Steady-State Aerobic Microbial Activity as a Function of Soil Water Content. Soil Sci. Soc. Am. J. **1990**, *54*, 1619–1625.
24. Doran, J.W.; Mielke, L.N.; Stamatiadis, S. *Microbial Activity and N Cycling as Regulated by Soil Water-Filled Pore Space*. Proc. 11th Conf. Inter. Soil Tillage Res. Org. Edinburg, Scotland **1988**, *1*, 49–54.

O

25. Molina, J.A.E.; Clapp, C.E.; Shaffer, M.J.; Chichester, F.W.; Larson, W.E. NCSOIL, a Model of Nitrogen and Carbon Transformations in Soil: Description, Calibration, and Behavior. Soil Sci. Soc. Am. J. **1983**, *47*, 85–91.
26. Rijtema, P.E.; Kroes, J.G. Some Results of Nitrogen Simulations with the Model ANIMO. Fertilizer Research **1991**, *27*, 189–198.
27. McCown, R.L.; Hammer, G.L.; Hargreaves, J.N.G.; Holzworth, D.P.; Freebairn, D.M. APSIM: A Novel Software System for Model Development, Model Testing and Simulation in Agricultural Systems Research. Agricultural Systems **1996**, *50*, 255–271.
28. Franko, U. Modelling Approaches of Soil Organic Matter Turnover Within the CANDY System. In *Evaluation of Soil Organic Matter Models Using Existing, Long-Term Datasets*; Powlson, D.S., Smith, P., Smith, J.U., Eds.; NATO ASI I38; Springer-Verlag: Berlin, 1996; 247–254.
29. Parton, W.J.; Stewart, J.W.B.; Cole, C.V. Dynamics of C, N, P, and S in Grassland Soils: A Model. Biogeochemistry **1987**, *5*, 109–131.
30. Lin, C.; Liu, T.S.; Hu, T.L. Assembling a Model for Organic Residue Transformation in Soils. Proc. Natl. Council (Taiwan) Part B **1987**, *11*, 175–186.
31. Mueller, T.; Jensen, L.S.; Hansen, S.; Nielsen, N.E. Simulating Soil Carbon and Nitrogen Dynamics with the Soil-Plant-Atmosphere System Model DAISY. In *Evaluation of Soil Organic Matter Models Using Existing, Long-Term Datasets*; Powlson, D.S., Smith, P., Smith, J.U.; Eds.; NATO ASI I38; Springer-Verlag: Berlin, 1996; 275–281.
32. Li, C.; Frolking, S.; Harriss, R. Modelling Carbon Biogeochemistry in Agricultural Soils. Global Biogeochemical Cycles **1994**, *8*, 237–254.
33. Hoogenboom, G., Jones, J.W., Hunt L.A.; Thornton P.K.; Tsuji, G.Y. An Integrated Decision Support System for Crop Model Applications. Paper 94-3025 Presented at ASAE Meeting, Missouri, June, 1994; 23.
34. Douglas, C.L., Jr.; Rickman, R.W. Estimating Crop Residue Decomposition from Air Temperature, Initial Nitrogen Content, and Residue Placement. Soil Science Society of America Journal **1992**, *56*, 272–278.
35. Grant, R.F. Dynamics of Energy, Water, Carbon and Nitrogen in Agricultural Ecosystems: Simulation and Experimental Validation. Ecological Modelling **1995**, *81*, 169–181.
36. Williams, J.R. The Erosion-Productivity Impact Calculator (EPIC) Model: A Case History. Philosophical Transactions of the Royal Society, London B **1990**, *329*, 421–428.
37. Kan, N.A.; Kan, E.E. Simulation Model of Soil Fertility. Physiology and Biochemistry of Cultivated Plants **1991**, *23*, 3–16, (in Russian).
38. Perruchoud, D.O. *Modeling the Dynamic of Non-living Organic Carbon in a Changing Climate: a Case Study for Temperate Forests*. PhD Thesis, ETH Diss. No. 11900, 1996; 196.
39. Moorhead, D.L.; Reynolds, J.F. A General Model of Litter Decomposition in the Northern Chihuahuan Desert. Ecological Modelling **1991**, *56*, 197–219.
40. Thornley, J.H.M.; Verberne, E.L.J. A Model of Nitrogen Flows in Grassland. Plant, Cell and Environment **1989**, *12*, 863–886.
41. Andrén, O.; Kätterer, T. ICBM—The Introductory Carbon Balance Model for Exploration of Soil Carbon Balances. Ecological Applications **1997**, *7*, 1226–1236.
42. Sirotenko, O.D. The USSR Climate-Soil-Yield Simulation System. Meteorologia i Gidrologia **1991**, *4*, 67–73, (in Russian).
43. McCaskill, M.; Blair, G.J. A Model of S, P and N Uptake by a Perennial Pasture. I. Model Construction. Fertilizer Research **1990**, *22*, 161–172.
44. Schevtsova, L.K.; Mikhailov, B.G. *Control of Soil Humus Balance Based on Statistical Analysis of Long-Term Field Experiments Database*; (in Russian) VIUA: Moscow, 1992.
45. Whitmore, A.P.; Klein-Gunnewiek, H.; Crocker, G.J.; Klír, J.; Körschens, M.; Poulton, P.R. Simulating Trends in Soil Organic Carbon in Long-term Experiments Using the Verberne / MOTOR Model. Geoderma **1997**, *81*, 137–151.
46. Ryzhova, I.M. Analysis of Sensitivity of Soil-Vegetation Systems to Variations in Carbon Turnover Parameters Based on a Mathematical Model. Eurasian Soil Science **1993**, *25*, 43–50.
47. Molina, J.A.E.; Hadas, A.; Clapp, C.E. Computer Simulation of Nitrogen Turnover in Soil and Priming Effect. Soil Biol. Biochem. **1990**, *22*, 349–353.
48. Van Dam, D.; Van Breemen, N. NICCE—A Model for Cycling of Nitrogen and Carbon Isotopes in Coniferous Forest Ecosystems. Ecological Modelling **1995**, *79*, 255–275.
49. O'Brien, B.J. Soil Organic Carbon Fluxes and Turnover Rates Estimated from Radiocarbon Measurements. Soil Biology and Biochemistry **1984**, *16*, 115–120.
50. O'Leary, G.J. *Soil Water and Nitrogen Dynamics of Dryland Wheat in the Victorian Wimmera and Mallee*. PhD. Thesis, University of Melbourne, 1994, 332.
51. Coleman, K.; Jenkinson, D.S.; Crocker, G.J.; Grace, P.R.; Klír, J.; Körschens, M.; Poulton, P.R.; Richter, D.D. Simulating Trends in Soil Organic Carbon in Long-Term Experiments Using RothC-23.6. Geoderma **1997**, *81*, 29–44.
52. Grace, P.R., Ladd, J.N. *SOCRATES v2.00 User Manual.* Co-operative Research Centre for Soil and Land Management. PMB 2, Glen Osmond 5064, South Australia, 1995.
53. Chertov, O.G.; Komarov, A.S. SOMM—A Model of Soil Organic Matter and Nitrogen Dynamics in Terrestrial Ecosystems. In *Evaluation of Soil Organic Matter Models Using Existing, Long-Term Datasets*; Powlson, D.S., Smith, P., Smith, J.U., Eds.; NATO ASI I38; Springer-Verlag: Berlin, 1996; 231–236.
54. Smith, J.U.; Bradbury, M.J.; Addiscott, T.M. SUNDIAL: Simulation of Nitrogen Dynamics in Arable Land. A User-Friendly, PC-based Version of the Rothamsted Nitrogen Turnover Model. Agronomy J. **1996**, *88*, 38–43.
55. Verberne, E.L.J.; Hassink, J.; de Willigen, P.; Groot, J.R.R.; van Veen, J.A. Modelling Soil Organic Matter Dynamics in Different Soils. Netherlands Journal of Agricultural Science **1990**, *38*, 221–238.

56. André, M.; Thiery, J.M.; Courmac, L. ECOSIMP Model: Prediction of CO_2 Concentration Changes and Carbon Status in Closed Ecosystems. Adv. Space Research **1992**, *14*, 323–326.
57. Vanclooster, M.; Viaene, P.; Diels, J.; Feyen, J. A Deterministic Evaluation Analysis Applied to an Integrated Soil-Crop Model. Ecological Modelling **1995**, *81*, 183–195.
58. Falloon, P.; Smith, P. Modelling Refractory Organic Matter—A Review. Biology and Fertility of Soils **2000**, *30*, 388–398.
59. Izaurralde, R.C.; Haugen-Kozyra, K.H.; Jans, D.C.; McGill, W.B.; Grant, R.F.; Hiley, J.C. Soil Organic Carbon Dynamics: Measurement, Simulation and Site to Region Scale-up. In *Assessment Methods for Soil Carbon. Advances in Soil Science*; Lal, R., Kimble, J.M., Follett, R.F., Stewart, B.A., Eds.; Lewis Publishers: Boca Raton, FL, 2001; 553–575.
60. Smith, P.; Smith, J.U.; Powlson, D.S.; McGill, W.B.; Arah, J.R.M.; Chertov, O.G.; Coleman, K.; Franko, U.; Frolking, S.; Jenkinson, D.S.; Jensen, L.S.; Kelly, R.H.; Klein-Gunnewiek, H.; Komarov, A.; Li, C.; Molina, J.A.E.; Mueller, T.; Parton, W.J.; Thornley, J.H.M.; Whitmore, A.P. A Comparison of the Performance of Nine Soil Organic Matter Models Using Datasets from Seven Long-term Experiments. Geoderma **1997**, *81*, 153–225.
61. Donigian, A.S. Jr., Barnwell, T.O. Jr., Jackson, R.B. IV; Patwardhan, A.S., Weinrich, K.B., Rowell, A.L., Chinnaswamy, R.V., Cole, C.V. Assessment of Alternative Management Practices and Policies Affecting Soil Carbon in Agroecosystems of the Central United States. US EPA Report EPA/600/R-94/067, Athens, 1994, 194.
62. Falloon, P.; Smith, P.; Smith, J.U.; Szabó, J.; Coleman, K.; Marshall, S. Regional Estimates of Carbon Sequestration Potential: Linking the Rothamsted Carbon Model to GIS Databases. Biol. Fert. Soil **1998**, *27*, 236–241.
63. Lee, J.J.; Phillips, D.L.; Liu, R. The Effect of Trends in Tillage Practices on Erosion and Carbon Content of Soils in the US Corn Belt. Water, Air, and Soil Pollution **1993**, *70*, 389–401.
64. Parshotam, A.; Tate, K.R.; Giltrap, D.J. Potential Effects of Climate and Land-use Change on Soil Carbon and CO_2 Emissions from New Zealand's Indigenous Forests and Unimproved Grasslands. Weather and Climate **1996**, *15*, 3–12.
65. Post, W.M.; Emanuel, W.R.; Zinke, P.J.; Stangenberger, A.G. Soil Carbon Pools and World Life Zones. Nature **1982**, *298*, 156–159.
66. Post, W.M.; Pastor, J.; Zinke, P.J.; Staggenberger, A.G. Global Patterns of Soil Nitrogen Storage. Nature **1985**, *317*, 613–616.
67. Post, W.M.; King, A.W.; Wullschleger, S.D. Soil Organic Matter Models and Global Estimates of Soil Organic Carbon. In *Evaluation of Soil Organic Matter Models Using Existing, Long-Term Datasets*; Powlson, D.S., Smith, P., Smith, J.U., Eds.; NATO ASI I38; Springer-Verlag: Berlin, 1996; 201–222.
68. Potter, C.S.; Randerson, J.T.; Field, C.B.; Matson, P.A.; Vitousek, P.M.; Mooney, H.A.; Klooster, S.A. Terrestrial Ecosystem Production: A Process Model Based on Satellite and Surface Data. Global Biogeochemical Cycles **1993**, *7*, 811–841.
69. Schimel, D.S.; Braswell, B.H., Jr.; Holland, E.A.; McKeown, R.; Ojima, D.S.; Painter, T.H.; Parton, W.J.; Townsend, J.R. Climatic, Edaphic, and Biotic Controls Over Storage and Turnover of Carbon in Soils. Global Biogeochemical Cycles **1994**, *8*, 279–293.
70. Goto, N.; Sakoda, A.; Suzuki, M. Modelling Soil Carbon Dynamics as a Part of the Carbon Cycle in Terrestrial Ecosystems. Ecological Modelling **1993**, *74*, 183–204.
71. Esser, G. Modelling Global Terrestrial Sources and Sinks of CO_2 with Special Reference to Soil Organic Matter. In *Soils and the Greenhouse Effect*; Bouwman, A.F., Ed.; John Wiley & Sons: NewYork, 1990; 247–261.
72. Goldewijk, K.K.; van Minnen, J.G.; Kreileman, G.J.J.; Vloedbeld, M.; Leemans, R. Simulating the Carbon Flux Between the Terrestrial Environment and the Atmosphere. Water, Air and Soil Pollution **1994**, *76*, 199–230.
73. Melillo, J.M.; Kicklighter, D.W.; McGuire, A.D.; Peterjon, W.T.; Newkirk, K.M. Global Change and Its Effect on Soil Organic Carbon Stocks. In *Role of Nonliving Organic Matter in the Earth's Carbon Cycle*; Zepp, R.G., Sonntag, C.H., Eds.; John Wiley and Sons: New York, 1995; 175–189.
74. IPCC. *Land Use, Land-Use Change, and Forestry. A Special Report of the IPCC*; Cambridge University Press: Cambridge, UK, 2000; 377.

ORGANIC MATTER AND NUTRIENT DYNAMICS

O

Charles W. Rice
Kansas State University, Manhattan, Kansas, U.S.A.

INTRODUCTION

Soil organic matter (SOM) is a fundamental component of soil and the global carbon (C) cycle. Soil organic matter controls many of the chemical, physical, and biological properties of the soil (1). The estimated amount of organic C stored in world soils is about 1100–1600 Pg, more than twice the C in living vegetation (560 Pg) or in the atmosphere (750 Pg) (2). Hence, even relatively small changes in soil C storage per unit area could have a significant impact on the global C balance. Soil organic matter is derived mostly from plant residues. Plants convert CO_2 into tissue through photosynthesis. Upon their death, plant tissues decompose, primarily by soil microorganisms, and most of the C in the plant material is eventually released back into the atmosphere as CO_2. Between 10 and 20% of the C in plant residue forms SOM, sometimes referred to as "humus." Some of this C can persist in soils for hundreds and even thousands of years. Associated with the C in soil organic matter are many essential plant nutrients—primarily N, P, and S. Concentrations of soil organic matter range from 0.2 to over 80% in peat soils although the typical range for temperate soils is 0.4–10% (3). While it is a minor component of most soils, SOM is essential to the living component of soil. Soil organic matter provides the energy source for most soil microorganisms, and provides the nutrient for plants and the soil biological community. Thus, knowledge of SOM dynamics is crucial for the understanding of global C cycling and plant production.

The accumulation of SOM is dependent on the quantity and quality of organic residue inputs, largely as plant material, the rates of microbial decomposition, and the capacity of the soil to store organic matter. The quality of the plant residue affects both the extent and rate of decomposition. Labile C compounds such as simple sugars degrade relatively rapidly and more completely to CO_2. On the other end of the spectrum, lignin is more difficult to degrade. Most microorganisms do not have the capacity to completely degrade lignin to CO_2. Thus, many of the partial degradation products form the precursors to soil organic matter. Generally, the C:N ratio is a guide of decomposability. A ratio >30 slows decomposition and immobilizes N; a ratio <20 releases N and allows microbial decomposition to proceed. Lignin content or lignin:N ratios are also used to assess organic matter degradability.

The rate of microbial decomposition of plant material and soil organic matter also is a function of climate. As temperature increases, microbial activity increases; generally for every 10°C increase, microbial activity doubles ($Q_{10} = 2$). Soil water content also is important, optimum microbial activity occurs at near "field capacity"which is equivalent to 60% water-filled spore space (4). As soil becomes waterlogged, decomposition slows and becomes less complete. Peat soils often are a result of these waterlogged conditions. As soils dry, decomposition is also slowed. The third factor affecting the amount of soil organic matter is soil texture—primarily clay content and mineralogy. Clay content stabilizes soil organic matter by two mechanisms. First, organic molecules are chemically bound to clay surfaces, which retards degradation. Clays with high adsorption capacities, such as montmorillinitic clays, are effective in retention of organic molecules. Secondly, soils with greater clay content increase the potential for aggregate formation. Macro-aggregates physically protect organic matter trapped inside the aggregates from microbial degradation. Generally, aggregates >250 μm provide the greatest protection (5, 6).

COMPOSITION

Soil organic matter is not one definable entity. Since SOM is formed from plant material and microbial decomposition products, it is a myriad of organic compounds. There have been several theories on the formation of soil humus but the most widely accepted theory is that organic residues undergo decomposition by microorganisms (7). The altered compounds and new compounds synthesized by soil microbes polymerize through chemical or enzymatic reactions. Thus, SOM is undergoing constant transformation. Typically, most soil organic models define three pools of SOM (Fig. 1). The active pool, which is comprised of microbial biomass and labile organic compounds makes up less than 5% of the soil organic C. The slow pool usually makes up 20–40% of the total organic C and the recalcitrant pool makes up 60–70% of the soil C. These

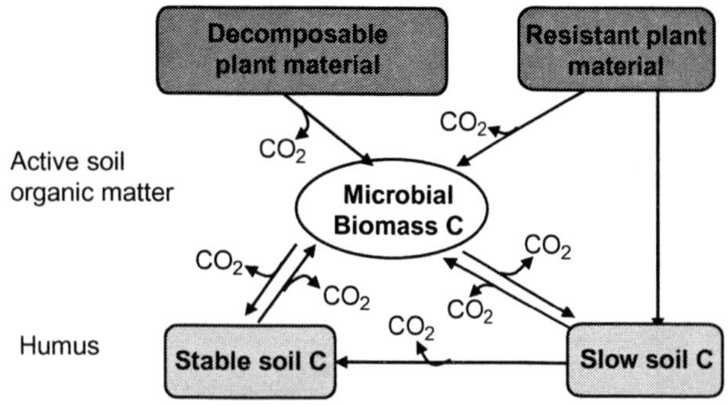

Fig. 1 Schematic of plant decomposition through microbial biomass in the formation of soil organic matter. (Adapted from Ref. 7.)

fractions are often defined kinetically based on laboratory mineralization (5, 8). Microbial biomass is the processor and the slow pool is the one in which much of the plant-associated nutrients reside for mineralization. The recalcitrant pool is material that is difficult to degrade and contains what in the older literature was known as humic and fulvic acids—fractions obtained by chemical-fractionation procedures. The active pool has turnover times on the order of months to years, the slow pool takes decades to turn over, while C in the recalcitrant pool takes from hundred to thousands of years to turn over completely. However, 2–5% of the recalcitrant pool is degraded annually. Since the recalcitrant pool is generally in equilibrium in natural systems, then the rate for formation equals the rate of degradation.

NUTRIENTS

Along with carbon, SOM contains important plant nutrients. Soil organic matter can be a source or sink of plant nutrients. Plant productivity is directly associated with SOM content and turnover. Approximately 90–95% of the soil N, 40% of the soil P, and 90% of the soil S is associated with SOM (3). Generally, the C:N:P:S ratio is 100:10:1:1. In agricultural soils, approximately 2–4% of the organic matter is rendered available for plant uptake on an annual basis. As discussed earlier, the more active pools of SOM are likely to be the major source of plant nutrients.

Nitrogen

The plant nutrient that is needed from soil in the greatest quantities is N. Approximately 90–95% of the soil N is in the organic form. The net release of organic N as inorganic N from the soil can be a significant source of N to satisfy plant needs. This process is called net N mineralization, which is the sum of two simultaneous processes, N mineralization and N immobilization. Nitrogen mineralization, or more correctly called "ammonification," is the conversion of organic N to ammonium. Nitrogen immobilization is the conversion of ammonium to organic N. Microorganisms control both these processes, thus factors that regulate microbial activity, as described earlier, will impact N availability (9). In addition to environmental factors, the quality of the organic substrate for microorganisms is important. Low-quality substrate, i.e., high C:N ratio, microorganisms degrading the residue require additional nutrients, primarily N. As a result, soil microorganisms assimilate or immobilize inorganic N. Plants may become deficient in N as the microbes fulfill their N need during decomposition of high C:N residue. Later as the organic material is processed, the N previously assimilated by microorganisms is re-released as inorganic N and can become available to plants. Often the release of organic N to the inorganic forms is in synchrony with plant uptake since favorable temperatures and water availability that promote microbial activity also promote plant growth (9, 10). In most native ecosystems and organic agriculture, N mineralization is the major source of plant N needs. In cropland as much as 11–300 kg N ha^{-1} can be supplied from organic matter (11). Further biological transformations of N occur in the soil, including nitrification (oxidation of NH_4^+ to NO_3^-) and denitrification (conversion of NO_3^- to N_2); however, these are not directly linked to SOM and will not be discussed in this section. Please refer to Sylvia et al. (12) for further discussion.

Phosphorus and Sulfur

Sulfur follows similar transformations as N. P is more controlled by soil chemistry and chemical transformations. Thus, SOM is not usually a major source of P by plants, except in very high organic soils. However, organic P can represent 80% of the total P in some soils. Significant amounts of P and S are contained within the microbial biomass. As a rule, a C:S or C:P ratio greater than 60 promotes immobilization of S and P into the microorganisms.

Because of the importance of SOM to the quality of the soil and plant productivity, an understanding of SOM dynamics is critical to preserve natural ecosystems and ensure the long-term productivity of managed ecosystems. Gains and losses of SOM have added significance because it is a reservoir for global C and the associated interaction with climate change.

Table 1 Land use for C sequestration

	Management strategies	
Land use	**Soil management**	**Crop management**
Cultivation	Tillage	Varieties
Rangeland	Residue management	Crop rotations
Forestry	Fertility	Cover crops
	Water management	
	Erosion control	

(Adapted from Ref. 13.)

MANAGEMENT OF SOM TO ENHANCE SOIL QUALITY

Agriculture in the 1800s and early 1900s relied upon the plowing the soil with low crop yields and crop residues were often removed. This combination of agricultural practices resulted in reducing the replenishment of organic C to the soil. Approximately, 50% of the SOM has been lost from soil over a period of 50–100 years of cultivation. In recent decades, higher yields, return of crop residues, and development of conservation tillage practices have increase SOM. Table 1 lists several practices affecting the soil's ability to sequester C (13). Examples of rates of soil C increases are summarized in Table 2 (14). Nitrogen management that increases crop productivity results in an increase in SOM (15). Nitrogen fertilizer applied at recommended rates for 10 years increase soil C approximately 2 MT C ha^{-1}. Grassland systems also can contribute to C sequestration when properly managed. Under elevated atmospheric CO_2, the soil contained 6% more C to a depth of 15 cm compared with ambient conditions (16). The increase in soil C was due to increased plant production followed by incorporation into the soil. The amount of C sequestered over the 8-year experimental period was equivalent to 4 Mg ha^{-1}. Proper fire management may also increase soil C (17).

Table 2 Estimates of C sequestration potential of agricultural practices of U.S. cropland

Agricultural practice	(MTC/ha/yr)
Conservation Reserve Program	0.3–0.7
Conservation tillage	0.24–0.40
Fertilizer management	0.05–0.15
Rotation with winter cover crops	0.1–0.3
Summer fallow elimination	0.1–0.3

(Adapted from Ref. 14.)

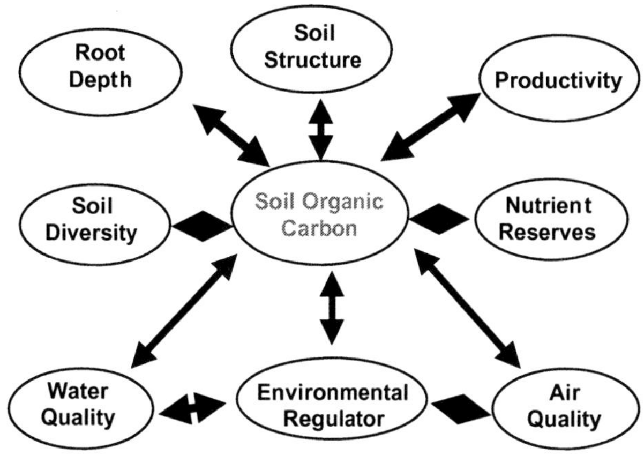

Fig. 2 Relationship between SOM and soil, water, and air quality. (Adapted from Ref. 14.)

Managing agricultural soils for sequestering C will result in additional benefits. Increasing soil organic C include increased crop productivity and enhanced soil, water, and air quality (Fig. 2). In addition, management practices that increase soil C also tend to reduce soil erosion, reduce energy inputs, and improve soil resources.

REFERENCES

1. Doran, J.W.; Parkin, T.B. Defining and Assessing Soil Quality. In *Defining Soil Quality for a Sustainable Environment*; Spec. Publ 35, Doran, J.W., Coleman, D.C., Bezdicek, D.F., Stewart, B.A., Eds.; Soil Sci. Soc. Am. Madison, WI, 1994; 3–24.
2. Sundquist, E.T. The Global Carbon Dioxide Budget. Science **1993**, *259*, 934–941.
3. Smith, J.L.; Lynch, J.M.; Bezdicel, D.F.; Papendick, R.I. Soil Organic Matter Dynamics and Crop Residue Management. In *Soil Microbial Ecology*; Metting, B., Ed.; Marcel Dekker: New York, NY, 1992; 65–94.
4. Linn, D.M.; Doran, J.W. Effect of Water-Filled Pore Space Carbon Dioxide and Nitrous Oxide Production in Tilled and Nontilled Soils. Soil Sci. Soc. Am. J. **1984**, *48*, 1267–1272.
5. van Veen, J.A.; Paul, E.A. Organic C Dynamics in Grassland Soils. 1. Background Information and Computer Simulations. Can. J. Soil Sci. **1981**, *61*, 185–201.
6. Jastrow, J.D.; Miller, R.M. Soil Aggregate Stabilization and Carbon Sequestration: Feed Backs Through Organo-Mineral Associations. In *Soil Process and the Carbon Cycle*; Lal, R., Kimble, J.M., Follett, R.F., Stewart, B.A., Eds.; CRC Press: New York, 1996; 207–223.
7. Paul, E.A.; Clark, F.E. *Soil Biology and Biochemistry*; Academic Press: San Diego, CA, 1996.
8. Rice, C.W.; Garcia, F.O. Biologically Active Pools of Soil C and N in Tallgrass Prairie. In *Defining Soil Quality for*

a Sustainable Environment; Spec. Publ 35, Doran, J.W., Coleman, D.C., Bezdicek, D.F., Stewart, B.A., Eds.; Soil Sci. Soc. Am. Madison, WI, 1994; 201–208.
9. Rice, C.W.; Havlin, J.L. Integrating Mineralizable N Indices into Fertilizer N Recommendations. In *Soil Testing: Prospects for Improving Nutrient Recommendations*; Spec. Pub. No. 40, Havlin, J.L., Jacobsen, J.S., Eds.; Soil Sci. Soc. Am. Madison, WI, 1994; 1–13.
10. McGill, W.B.; Meyers, R.J.K. Controls on Dynamics of Soil and Fertilizer Nitrogen. In *Soil Fertility and Organic Matter as Critical Components of Production Systems*; Spec. Pub. No. 19; Follett, R.F., Stewart, J.W.B., Cole, C.V., Eds.; Soil Sci. Soc. Am. Madison, WI, 1987; 73–99.
11. Smith, J.L.; Paul, E.A. The Significance of Soil Microbial Biomass Estimations. Soil Biochem. **1990**, *6*, 357–396.
12. Sylvia, D.M.; Fuhrman, J.F.; Hartel, P.G.; Zuberer, D.A. *Principles and Application of Soil Microbiology*; Prentice Hall: Upper Saddle River, NJ, 1998.
13. Lal, R.; Kimble, J.R.; Follett, R.F.; Cole, C.V. *The Potential of U.S. Cropland to Sequester Carbon and Mitigate the Greenhouse Effect*; Ann Arbor Press: Chelsa, MI, 1998.
14. Lal, R.; Follett, R.F.; Kimble, J.; Cole, C.V. Managing U.S. Cropland to Sequester Carbon in Soil. J. Soil Water Conserv. **1999**, *54*, 374–381.
15. Espinoza, Y. Dynamics and Mechanisms of Stabilization of C and N in Soil Ph.D. Dissertation Kansas State Univ.
16. Williams, M.A.; Rice, C.W.; Owensby, C.E. Carbon and Nitrogen Dynamics and Microbial Activity in Tallgrass Prairie Exposed to Elevated CO_2 For 8 Years. Plant and Soil **2000**, *227*, 127–137.
17. Rice, C.W.; Owensby, C.E. Effects of Fire and Grazing on Soil Carbon in Rangelands. In *The Potential of U.S. Grazing Lands to Sequester Carbon and Mitigate the Greenhouse Effect*; Follet, R., Ed.; Lewis Publishers: Boca Raton, 2000; 323–342.

ORGANIC MATTER STRUCTURE AND CHARACTERIZATION

O

Georg Guggenberger
Institute of Soil Science and Soil Geography, University of Bayreuth, Bayreuth, Germany

INTRODUCTION AND CONCEPTS

Soil organic matter (SOM) encompasses all biologically-derived organic material found in the soil or on its surface irrespective of 1) source, 2) whether it is living or dead, or 3) stage of decomposition, but excluding the aboveground portion of living plants (1). This implies large structural heterogeneity and close linkage to SOM functions. Litter composition and its changes during biotic and abiotic transformation are key variables in the processes and the size of carbon sequestration in soil. In the context of the Kyoto Protocol, analysis of SOM structure helps us to understand these processes and to predict changes in the sign and magnitude of terrestrial carbon fluxes in a changing environment.

To reduce SOM heterogeneity, different components of SOM need to be separated into entities that differ in terms of source, composition, and turnover. Recent evidence suggests that the classical chemical fractionation into fulvic acids, humic acids, and humin according to solubility characteristics in dilute acid and base (for standard procedure see Ref. 2) is not useful in this respect (3). Fractionations that are more promising rely mostly on physical fractionations according to particle size or density (4, 5), and the analysis of dissolved organic matter in the soil solution (6).

ANALYTICAL METHODS

The analytical approaches can be divided into degradative methods involving chemolysis, thermolysis, or thermochemolysis, and noninvasive spectroscopic techniques such as nuclear magnetic resonance (NMR) spectroscopy (Table 1). Degradative methods provide molecular-level information on specific organic compounds accessible to the degradative step whereas spectroscopic techniques inform on the bulk composition of SOM. Since all analytical approaches have their drawbacks, a comprehensive picture of the SOM structure can only be obtained using a combination of various methods (7).

Degradative Methods

Chemolysis of SOM involves various degradation procedures (hydrolysis, oxidation, extraction) that are selective in their attack on specific molecular structures (Table 1). The techniques are well suited to follow the often subtle changes in the composition of plant- and microbial-derived biopolymers during microbial transformation processes, in particular when different soil fractions are comparatively investigated (8, 9). But only a part of SOM present in specific structures—as determined by NMR spectroscopy—can be identified with these techniques, and information on the original macromolecular structure of SOM is limited (10).

Analytical pyrolysis uses thermal degradation to cleave bonds in the organic macromolecules and enables a sensitive and rapid characterization of organic constituents (11). Problems may arise due to the complicated pyrolysis behavior of many organic compounds, in particular in the presence of catalytic minerals. Thermal secondary reaction causes considerable modification of the organic compound, which may bias the interpretation of the pyrolysis products with respect to their mother compounds (12, 13). Some of the problems can be solved by thermochemolysis utilizing tetramethylammonium hydroxide (TMAH) (13, 14). Hydroxyl and carboxyl groups are converted into their respective methyl ethers and methyl esters to avoid fragmentation of aliphatic and benzenecarboxylic acids.

Spectroscopic Techniques

Nuclear magnetic resonance spectroscopy has a large potential to analyze the different chemical species of ^{1}H, ^{13}C, ^{15}N, and ^{31}P nuclei in solution and solid (not ^{1}H) state. With the cross-polarization magic angle spinning (CP MAS) pulse sequence, the chemical structure of C and N can be characterized *in situ*, nondestructively, and within a feasible time of analysis (15, 16). However, even with completion of detailed experiments on rates of signal generation and relaxation for each type of carbon (17), quantitative analysis of NMR data is not possible in the presence of paramagnetic species. Samples containing charred materials cannot be quantified by the CP technique; instead, the C nuclei must be directly excited using a Bloch decay sequence (18).

Table 1 Important degradative and nondegradative techniques to study SOM structure

Technique	Components addressed	Applications	Drawbacks	References
Chemolytic techniques	Address defined biomolecules released mostly as their monomer units from the organic matrix by an appropriate chemical treatment	Analysis of defined organic components of plant, microbial and pyrogenic origin Often well-suited as biomarkers	Only about 60% (fresh organic residues) to about 20% (transformed, mineral-associated compounds) of SOM is characterized	(7, 34)
Acid hydrolysis	Polysaccharides	Distinguishes different types of polysaccharides by sequential extraction Pattern of monosaccharides released gives information on the source (plant vs. microorganisms)	Not all polysaccharides are hydrolyzable Yield differs for different types of with organic matter	(8, 35)
	Proteinaceous compounds	Comprises the dominant form of organic nitrogen in soil Assessment of D/L ratios	Only a part of the proteinaceous compounds is hydrolyzable	(36, 37)
	Amino sugars	Traces microbial input to SOM Distinguishes between fungal and bacterial residues	Source of some amino sugars is not well defined	(22, 38)
Solvent extraction	Extractable lipids	Comprises hydrophobic aliphatic compounds of various origins	Complex and polymerized lipids are only scarcely accessible	(39, 40)
Saponification	Cutin and suberin components	Detects aliphatic biopolymers derived from vascular plants	Does not comprise nonester plant biomacromolecules	(25)
CuO oxidation	Lignin	Traces plant input to SOM Informs about type of plant Informs about lignin decomposition in soil	Only cleaves arylether bondings Yield is not defined and decreases with increasing lignin decomposition	(41, 42)
	Cutin and suberin components	Detects aliphatic biopolymers derived from vascular plants	Does not comprise nonester plant biomacromolecules	(43)
HNO_3 oxidation	Pyrogenic carbon	Traces the highly aromatic core of charred organic components	Yield varies with the type of the charred organic components	(44, 45)
Analytical pyrolysis	Products of SOM released by thermal degradation	Pattern of pyrolysis products allows detailed information on the chemical composition of the parent molecules Well suited for biomarker analysis	Volatilization of different types of pyrolysis fragments varies Pyrolysis fragments may have different sources Secondary pyrolysis reactions may occur	(11, 12)
Py-GC MS	Pyrolysis products that are volatile to be separated by gas chromatography	Separates pyrolysis products that have the same mass/charge (m/z) ratio	Nonvolatile pyrolysis products cannot be detected	(13)
Py-FIMS	Pyrolysis products are treated by soft ionization technique	Soft ionization produces predominantly molecular ions of the pyrolysis products	Volatilization decreases with increasing polarity of the fragments	(11)

(Continued)

Thermochemolysis with TMAH	In particular aliphatic and benzenecarboxylic acids as their methyl esters	Pyrolysis/methylation renders polar products volatile for gas chromatographic separation	Reproducibility needs to be improved	(46, 47)
NMR spectroscopy	Analysis of all C, H, N, or P species within a solid or soluble sample	Signal intensity relates to the the concentration of nuclei creating the signal	Most applications are not quantitative	(15, 16, 17, 48)
Solution ^{13}C, ^{1}H, ^{15}N, and ^{31}P NMR spectroscopy	Analysis of C, H, N, and P species in aqueous samples	Composition of C, H, and N species in the soil solution and in alkaline extracts (humic and fulvic acids)	Insoluble compounds are not detected	(49)
CP MAS ^{13}C, and ^{15}N spectroscopy	Analysis of C and N species in solid samples	Characterization of the chemical structure of organic C and N *in situ* and nondestructively	Limited quantitative interpretation caused by paramagnetic species, by different C and N relaxation, by spinning side bands, and by the cross-polarization technique	(15, 16)
BD ^{13}C NMR spectroscopy	Analysis of C species in solid samples	Direct excitation of ^{13}C spins enables analysis of charred materials	Very time consuming	(18)
IR spectroscopy (FTIR and DRIFT)	Analysis of C species in solid samples	Informs on the type of atoms to which C is bonded and on the nature of the bond	Reveals only a few well-resolved peaks Signals from minerals often dominate	(50) (19, 50)

Key: SOM = soil organic matter, Py-GC MS = pyrolysis-gas chromatography mass spectrometry, Py-FIMS = pyrolysis-field ionization mass spectrometry, TMAH = tetramethylammonium hydroxide, NMR spectroscopy = nuclear magnetic resonance spectroscopy, CP MAS NMR spectroscopy = cross-polarization magic angle spinning nuclear magnetic resonance spectroscopy, BD NMR spectroscopy = Bloch decay nuclear magnetic resonance spectroscopy, IR spectroscopy = infrared spectroscopy, FTIR = Fourier transform infrared spectroscopy, DRIFT = diffuse reflectance infrared Fourier transform spectroscopy.

Additional information on the structure of soil organic carbon (SOC) can be obtained by specific pulse sequences such as interrupted decoupling (ID), proton spin relaxation editing (PSRE), and mixing of proton spins (MOPS) (for literature see Ref. 1). Of the other spectroscopic techniques, infrared spectroscopy provides information on the type of atoms to which C is bonded and on the nature of the bond (19), whereas electron spin resonance spectroscopy gives information about free radicals in SOM (20).

CHEMICAL STRUCTURE OF SOIL ORGANIC MATTER

Individual Plant and Microbial Components

Individual components derived from plants (primary resources) and microorganisms (secondary resources) can be considered as the parent material of SOM. Within plant and microbial tissues, polysaccharides are generally the most important organic components. Analysis of sugar monomers released by hydrolysis or pyrolysis revealed that crystalline and noncrystalline plant polysaccharides are rapidly decomposed and microbial polysaccharides accumulate in soil (8, 21). Most amino sugars in soil are also of microbial origin, predominantly from fungal chitin and bacterial peptidoglycan (22). An important biopolymer in soil is vascular plant-derived lignin with its *p*-hydroxyphenyl, guaiacyl, and syringyl monomeric units. The pattern of lignin monomers can be used to identify the primary resource of SOM; in contrast to angiosperm lignin, gymnosperm lignin does not contain syringyl units. In soil, lignin is decomposed in aerobic environments via side-chain oxidation and ring opening (23, 24). This results in a shortening of alkyl side-chains and an increase in carbonyl and carboxyl groups. Thus, parts of the lignin

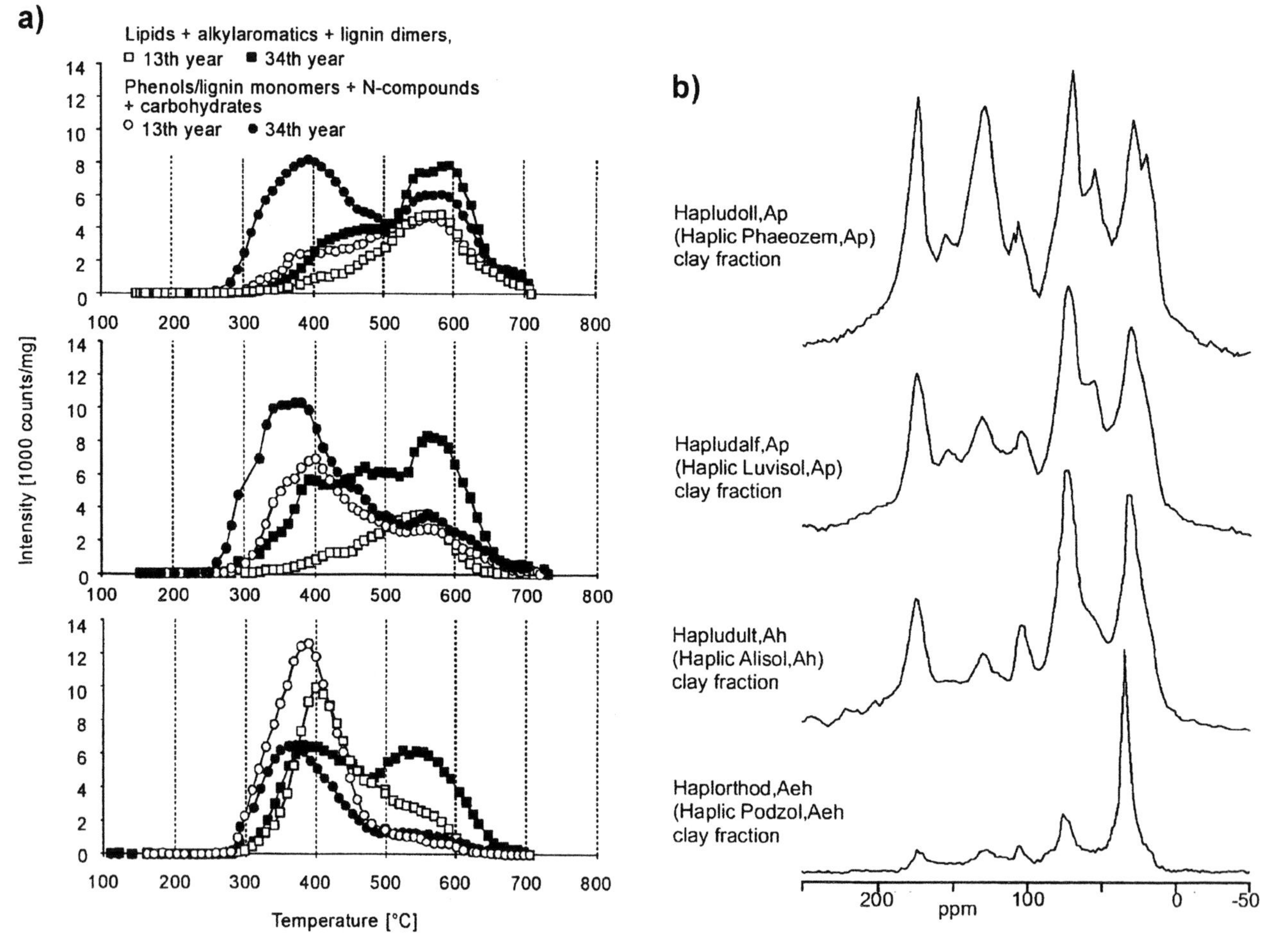

Fig. 1 (a) Thermograms for compound classes evolved by pyrolysis-field ionization from clay (above), fine silt (center), and medium silt (below) from a humus formation experiment after 13 and 34 years of soil development (From Ref. 11) and (b) solid-state ^{13}C NMR spectra of the clay fraction isolated from soils of different pedogenesis. (From Ref. 10.)

degradation products are water soluble (23) and represent, together with the polysaccharides, the majority of the dissolved organic matter in soil (6). The third major component is free and bound lipids of plant and microbial origin, including waxes, organic acids, steroids, glycerides, and phospholipids (3, 25). Plant cutin and suberin (insoluble polyesters) can be easily decomposed (26), while nonsaponifiable cutan and suberan (insoluble nonpolyesters) appear to be resistant to degradation (27). Some aliphatic compounds of microbial origin are also selectively preserved (28).

The Soil Organic Matter Continuum

Soil organic matter is generally composed of the above-mentioned plant and microbial residues, and their transformation products (7, 29). The latter refers largely to the term "humic substances" as frequently used in older references e.g., (30, 31). Recently, there has been a shift in paradigm away from humic substances. Organic matter degradation is considered to be predominantly a process of attrition, during which relatively resistant biomolecules are selectively concentrated (32). In surface soil horizons with large inputs of plant residues, the SOM continuum varying from fresh residues to highly degraded components is dominated by the former material (7). In contrast, SOM in subsoil horizons is enriched by highly-altered, recalcitrant organic materials that can only be characterized to a minor extent by chemolytic treatments (32). Accessibility of these compounds is limited to (chemo)thermolytic and spectroscopic methods.

In aging SOM, Schulten *et al.* (33) and Schulten & Leinweber (11) observed that increasing amounts of thermal energy applied in pyrolysis-field ionization mass spectrometry were required for volatilizing lignin dimers, alkylaromatics, and lipids (Fig. 1a). The authors suggested that the development of stronger chemical bonds was due to either formation of three-dimensional crosslinking by aryl–alkyl combinations or by formation of strong organic–mineral complexes. Kögel-Knabner (7) concluded from ID pulse sequences at ^{13}C NMR spectroscopy that cross-linked aliphatic compounds accumulate during SOM decomposition.

The pedogenic environment has a large impact on the structure of SOM. This can be particularly confined by solid-state ^{13}C NMR spectroscopy of mineral-associated organic matter, while spectra obtained on bulk samples are influenced by plant residues that are rather similar in composition (10). The Mollisol in Fig. 1b is characterized by about similar contributions of alkyl (0–50 ppm), O-alkyl (50–110 ppm), aromatic (110–160 ppm) and carboxyl/amide (160–200 ppm) carbon. In contrast, Alfisols and Ultisols show a high proportion of alkyl and O-alkyl C, and Spodosols are dominated by alkyl C.

Future work may benefit from the application of microscopic and nondestructive microspectroscopic techniques to investigate SOM within its mineral and microbiological soil environment. For in-depth information on modern methodological approaches and concepts on SOM structure, the excellent reviews of Kögel-Knabner (10), Baldock and Nelson (1), and Hedges *et al.* (32) are recommended.

REFERENCES

1. Baldock, J.A.; Nelson, P.N. Soil Organic Matter. In *Handbook of Soil Science*; Sumner, M.E., Ed.; CRC Press: Boca Raton, FL, 2000; B-25–B-84.
2. Swift, R.S. Organic Matter Characterisation. In *Methods of Soil Analysis. Part 3. Chemical Methods*; Sparks, D.L., Ed.; Soil Science Society of America: Madison, WI, 1996; 1011–1069.
3. Stevenson, F.J.; Elliott, E.T. Methodologies for Assessing the Quantity and Quality of Soil Organic Matter. In *Dynamics of Soil Organic Matter in Tropical Ecosystems*; Coleman, D.C., Oades, J.M., Uehara, G., Eds.; University of Hawaii Press: Honolulu, HI, 1989; 173–199.
4. Christensen, B.T. Carbon in Primary and Secondary Organomineral Complexes. In *Structure and Organic Matter Storage in Agricultural Soils*; Carter, M.R., Stewart, B.A., Eds.; CRC Press: Boca Raton, FL, 1996; 97–165.
5. Golchin, A.; Oades, J.M.; Skjemstad, J.O.; Clarke, P. 1994. Soil Structure and Carbon Cycling. Aus. J. Soil Res. **1994**, *32*, 1043–1068.
6. Guggenberger, G.; Zech, W.; Schulten, H.-R. Formation and Mobilization Pathways of Dissolved Organic Carbon: Evidence from Chemical Structural Studies of Organic Carbon Fractions in Acid Forest Floor Solutions. Org. Geochem. **1994**, *21*, 51–66.
7. Kögel-Knabner, I. Biodegradation and Humification Processes in Forest Soils. In *Soil Biochemistry*; Bollag, J.-M., Stotzky, G., Eds.; Marcel Dekker: New York, NY, 1993; Vol. 8, 101–137.
8. Guggenberger, G.; Christensen, B.T.; Zech, W. Land-Use Effects on the Composition of Organic Matter in Particle-Size Separates of Soils: I. Lignin and Carbohydrate Signature. Eur. J. Soil Sci. **1994**, *45*, 449–458.
9. Hedges, J.I.; Oades, J.M. Comparative Organic Geochemistries of Soils and Sediments. Org. Geochem. **1997**, *27*, 319–361.
10. Kögel-Knabner, I. Analytical Approaches for Characterizing Soil Organic Matter. Org. Geochem. **2000**, *31*, 609–625.
11. Schulten, H.-R.; Leinweber, P. Characterization of Humic and Soil Particles by Analytical Pyrolysis and Computer Modeling. J. Anal. Appl. Pyrol. **1996**, *38*, 1–53.

12. Saiz-Jimenez, C. Analytical Pyrolysis of Humic Substances: Pitfalls, Limitations and Possible Solutions. Environ. Sci. Technol. **1994**, *28*, 1773–1780.
13. van Bergen, P.; Flannery, M.B.; Poulton, P.R.; Evershed, R.P. Organic Geochemical Studies of Soils from Rothamsted Experimental Station: III. Nitrogen-Containing Organic Matter in Soil from Geescroft Wilderness. In *Fate of N-Containing Macromolecules in the Biosphere and Geosphere*; Symposium Series 707, Stankiewicz, B.A., van Bergen, P.F., Eds. American Chemical Society, Oxford University Press: England, 1998; 321–338.
14. Saiz-Jimenez, C. The Chemical Structure of Humic Substances: Recent Advances. In *Humic Substances in Terrestrial Ecosystems*; Piccolo, A., Ed. Elsevier: Amsterdam, The Netherlands, 1996; 1–44.
15. Knicker, H.; Nanny, M.A. Nuclear Magnetic Resonance Spectroscopy. Basic Theory and Background. In *NMR Spectroscopy in Environmental Science and Technology*; Nanny, M.A., Minear, R.A., Leenheer, J.A., Eds. Oxford University Press: London, England, 1997; 3–15.
16. Skjemstad, J.O.; Clarke, P.; Golchin, A.; Oades, J.M. Characterisation of Soil Organic Matter by Solid-State ^{13}C NMR Spectroscopy. In *Driven by Nature: Plant Litter Quality and Decomposition*; Giller, K.E., Ed. CAB International: Wallingford, England, 1997; 253–271.
17. Pfeffer, P.E.; Gerasimowicz, W.V. *Nuclear Magnetic Resonance in Agriculture*; CRC Press: Boca Raton, FL, 1989.
18. Skjemstad, J.O.; Clarke, P.; Taylor, J.A.; Oades, J.M.; McClure, S.G. The Chemistry and Nature of Protected Carbon in Soil. Aust. J. Soil Res. **1996**, *34*, 251–271.
19. Piccolo, A.; Conte, P. Advances in Nuclear Magnetic Resonance and Infrared Spectroscopies of Soil Organic Particles. In *Structure and Surface Reactions of Soil Particles*; Huang, P.M., Senesi, N., Buffle, J., Eds. John Wiley & Sons: Chichester, England, 1998; 183–250.
20. Cheshire, M.V.; Senesi, N. Electron Spin Resonance Spectroscopy of Organic and Mineral Soil Particles. In *Structure and Surface Reactions of Soil Particles*; Huang, P.M., Senesi, N., Buffle, J., Eds. Wiley: Chichester, England, 1998; 325–376.
21. Huang, Y.; Eglinton, G.; VanderHage, E.R.E.; Boon, J.J.; Bol, R.; Ineson, P. Dissolved Organic Matter in Grass Upland Soil Horizons Studied by Analytical Pyrolysis Techniques. Eur. J. Soil Sci. **1998**, *49*, 1–15.
22. Parsons, J.W. Chemistry and Distribution of Amino Sugars in Soils and Soil Organisms. In *Soil Biochemistry*; Paul, E.A., Ladd, J.N., Eds. Marcel Dekker: New York, NY, 1981; Vol. 5, 197–227.
23. Haider, K. Problems Related to the Humification Processes in Soils of the Temperate Climate. In *Soil Biochemistry*; Bollag, J.-M., Stotzky, G., Eds. Marcel Dekker: New York, NY, 1992; Vol. 7, 55–94.
24. Shevchenko, S.M.; Bailey, G.W. Life After Death: Lignin–Humic Relationships Reexamined. Critical Rev. Environ. Sci. Technol. **1996**, *26*, 95–153.
25. Kögel-Knabner, I.; Ziegler, F.; Riederer, M.; Zech, W. Distribution and Decomposition Pattern of Cutin and Suberin in Forest Soils. Z. Pflanzenernähr. Bodenk. **1989**, *152*, 409–413.
26. Riederer, M.; Matzke, K.; Ziegler, F.; Kögel-Knanber, I. Inventories and Decomposition of the Lipid Plant Biopolymers Cutin and Suberin in Temperate Forest Soils. Org. Geochem. **1993**, *20*, 1063–1076.
27. Tegelaar, E.W.; de Leeuw, J.W.; Saiz-Jimenez, C. Possible Origin of Aliphatic Moieties in Humic Substances. Sci. Total Environ. **1989**, *81/82*, 1–17.
28. Lichfouse, E.; Chenu, C.; Baudin, F.; Leblond, C.; da Silva, M.; Behar, F.; Derenne, S.; Largeau, C.; Wehrung, P.; Albrecht, P. A Novel Pathway of Soil Organic Matter Formation by Selective Preservation of Resistant Straight-Chain Biopolymers: Chemical and Isotopic Evidence. Org. Geochem. **1998**, *28*, 411–415.
29. Waksman, S.A. *Humus, Origin, Chemical Composition and Importance in Nature*; Balliére, Tindall & Cox: London, England, 1938.
30. Aiken, G.R.; McKnight, D.M.; Wershaw, R.L.; MacCarthy, P. *Humic Substances in Soil, Sediment and Water*; John Wiley & Sons: New York, NY, 1985.
31. Hayes, M.H.B.; MacCarthy, P.; Malcolm, R.; Swift, R.S. *Humic Substances II. In Search of Structure*; Wiley Interscience: Chichester, England, 1989.
32. Hedges, J.I.; Eglinton, G.; Hatcher, P.G.; Kirchman, D.L.; Arnosti, C.; Derenne, S.; Evershed, R.P.; Kögel-Knabner, I.; de Leeuw, J.W.; Littke, R.; Michaelis, W.; Rullkötter, J. The Molecularly-Uncharacterized Component of Nonliving Organic Matter in Natural Environments. Org. Geochem. **2000**, *31*, 945–958.
33. Schulten, H.-R.; Leinweber, P.; Reuter, G. Initial Formation of Soil Organic Matter from Grass Residues in a Long-Term Experiment. Biol. Fertil. Soils **1992**, *14*, 237–245.
34. Stevenson, F.J. *Humus Chemistry, Genesis, Composition, Reactions*, 2nd Ed. John Wiley & Sons: New York, NY, 1994.
35. Cheshire, M.V. *Nature and Origin of Carbohydrates*; Academic Press: London, England, 1979.
36. Chen, C.-N.; Shufeldt, R.C.; Stevenson, F.J. Amino Acid Analysis of Soils and Sediments: Extraction and Desalting. Soil Biol. Biochem. **1975**, *7*, 143–151.
37. Amelung, W.; Zhang, X. Determination of Amino Acid Enantiomers in Soils. Soil Biol. Biochem. **2001**, *33*, 553–562.
38. Amelung, W. Methods Using Amino Sugars As Markers for Microbial Residues in Soil. In *Assessment Methods for Soil Carbon*; Lal, R., Kimble, J.M., Follett, R.F., Stewart, B.A., Eds. Lewis Publishers: Boca Raton, FL, 2001; 233–272.
39. Dinel, H.; Schnitzer, M.; Mehuys, G.R. Soil Lipids Origin, Nature, Content, Decomposition and Effect on Soil Physical Properties. In *Soil Biochemistry*; Bollag, J.-M., Stotzky, G., Eds. Marcel Dekker: New York, NY, 1990; Vol. 5, 397–429.
40. Capriel, P.; Beck, T.; Borchert, H.; Härter, P. Relationship Between Soil Aliphatic Fraction Extracted With Supercritical Hexane, Soil Microbial Biomass, and Soil Aggregate Stability. Soil Sci. Soc. Am. J. **1990**, *54*, 415–420.
41. Ertel, J.R.; Hedges, J.I. The Lignin Component of Humic Substances: Distribution Among Soil and Sedimentary

Humic, Fulvic, and Base-Insoluble Fractions. Geochim. Cosmochim. Acta **1984**, *48*, 2065–2074.

42. Kögel-Knabner, I.; Zech, W.; Hatcher, P.G. Chemical Structural Studies of Forest Soil Humic Acids: Aromatic Carbon Fraction. Soil Sci. Soc. Am. J. **1991**, *55*, 241–247.
43. Goñi, M.A.; Hedges, J.I. Potential Applications of Cutin-derived CuO Reaction Products for Discriminating Vascular Plant Sources in Natural Environments. Geochim. Cosmochim. Acta **1990**, *54*, 23,073–23,081.
44. Glaser, B.; Haumaier, L.; Guggenberger, G.; Zech, W. Black Carbon in Soils: The Use of Benzencarboxylic Acids As Specific Markers. Org. Geochem. **1998**, *29*, 811–819.
45. Glaser, B.; Haumaier, L.; Guggenberger, G.; Zech, W. The Terra Preta Phenomenon: A Model for Sustainable Agriculture in the Humid Tropics. Naturwissenschaften **2001**, *88*, 37–41.
46. del Rio, J.C.; McKinney, D.E.; Knicker, H.; Nanny, M.A.; Minard, R.D.; Hatcher, P.G. Structural Characterization of Bio- and Geo-Macromolecules by Off-Line Thermochemolysis With Tetramethylammonium Hydroxide. J. Chromatogr. **1998**, *823*, 433–448.
47. Filley, T.R.; Minard, R.D.; Hatcher, P.G. Tetramethylammonium Hydroxide (TMAH) Thermochemolysis: Proposed Mechanisms Based Upon the Application of ^{13}C-Labeled TMAH to a Synthetic Model Lignin Dimer. Org. Geochem. **1999**, *30*, 607–621.
48. Preston, C.M. Applications of NMR to Soil Organic Matter Analysis: History and Prospects. Soil Sci. **1996**, *161*, 144–166.
49. Preston, C.M. Review of Solution NMR of Humic Substances. In *NMR of Humic Substances and Coal*; Wershaw, R.L., Mikita, M.A., Eds. Lewis Publishers: Chelsea, MI, 1987; 3–32.
50. Parfitt, R.L.; Fraser, A.R.; Farmer, V.C. Adsorption on Hydrous Oxides III. Fulvic Acid and Humic Acid on Goethite, Gibbsite and Imogolite. J. Soil Sci. **1977**, *28*, 289–296.

O

ORGANIC MATTER TURNOVER

J. Six
Colorado State University, Fort Collins, Colorado, U.S.A

J.D. Jastrow
Argonne National Laboratory, Argonne, Illinois, U.S.A

INTRODUCTION

Soil organic matter (SOM) is a dynamic entity. The amount (stock) of organic matter in a given soil can increase or decrease depending on numerous factors including climate, vegetation type, nutrient availability, disturbance, land use, and management practices. But even when stocks are at equilibrium, SOM is in a continual state of flux; new inputs cycle—via the process of decomposition—into and through organic matter pools of various qualities and replace materials that are either transferred to other pools or mineralized. For the functioning of a soil ecosystem, this "turnover" of SOM is probably more significant than the sizes of SOM stocks (1). An understanding of SOM turnover is crucial for quantifying C and nutrient cycles and for determining the quantitative and temporal responses of local, regional, or global C and nutrient budgets to perturbations caused by human activities or climate change (2).

DEFINITION OF SOIL ORGANIC MATTER TURNOVER

The turnover of an element (e.g., C, N, P) in a pool is generally determined by the balance between inputs (I) and outputs (O) of the element to and from the pool (Fig. 1). Turnover is most often quantified as the element's mean residence time (MRT) or its half-life ($T_{1/2}$). The MRT of an element in a pool is defined as 1) the average time the element resides in the pool at steady state or 2) the average time required to completely renew the content of the pool at steady state. The term half-life is adopted from radioisotope work, where it is defined as the time required for half of a population of elements to disintegrate. Thus, the half-life of SOM is the time required for half of the currently existing stock to decompose.

The most common model used to describe the dynamic behavior or turnover of SOM is the first-order model, which assumes constant zero-order input with constant proportional mass loss per unit time (3, 4)

$$\frac{\partial S}{\partial t} = I - kS, \tag{1}$$

where S is the SOM stock, t is the time, k is the decomposition rate, and kS is equivalent to output O. Assuming equilibrium ($I = O$), the MRT can then be calculated as

$$\mathrm{MRT} = \frac{1}{k} \tag{2}$$

and MRT and $T_{1/2}$ can be calculated interchangeably with the formula

$$\mathrm{MRT} = T_{1/2}/\ln 2 \tag{3}$$

MEASURING SOIL ORGANIC MATTER TURNOVER

Most often the turnover of SOM, more specifically the turnover of SOM-C, is estimated by one of four techniques:

1. Simple first-order modeling
2. ^{13}C natural abundance technique
3. ^{14}C dating technique
4. "bomb" ^{14}C technique.

This list does not include tracer studies where a substrate (e.g., plant material) enriched in ^{13}C, ^{14}C, and/or ^{15}N is added to soil, and its fate is followed over time. Most studies of this type (see Ref. 5 for a review) use the tracers to quantify the short-term (1–5 yr) decomposition rate of freshly added material rather than the long-term turnover of whole-soil C.

Eqs. 1 and 2 form the basis for estimates of SOM turnover derived from first-order modeling; the unknown k is calculated as

Encyclopedia of Soil Science

O

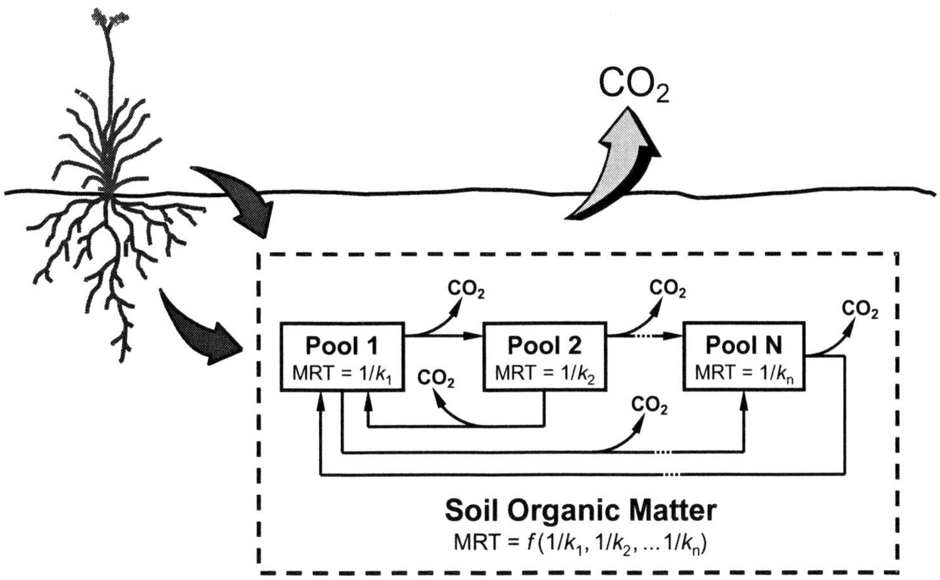

Fig. 1 The turnover of soil organic matter (SOM) is determined by the balance of inputs and outputs. Total SOM consists of many different pools that are turning over at different rates. The mean residence time (MRT) of total SOM is a function of the turnover rates of its constituent pools.

$$k = \frac{I}{S}$$

by assuming a steady state

$$\frac{\partial S}{\partial t} = 0.$$

This approach requires estimates of annual C input rates, which can be assumed to be continuous or discrete (3). The input can also be written as

$$I = hA$$

where A is the annual addition of C as fresh residue and h (the isohumification coefficient) represents the fraction that, after a rapid initial decomposition of A, remains as the actual annual input to S. An estimate of h is then necessary. A value of 0.3 is commonly used for agricultural crops, but the value can be higher for other materials such as grasses or peat (6, 7).

Another approach to estimate k by first-order modeling is "chronosequence modeling" (8). An increase (or decrease) in C across a chronosequence of change in vegetation, land use, or management practice can be fitted to a first-order model

$$S = S_e\left[1 - \left(\frac{S_e - S_0}{S_e}\right)e^{-kt}\right]$$

which is equivalent to

$$S = S_0 + (S_e - S_0)(1 - e^{-kt}) \quad (4)$$

where t is the time since the change, S_e is the C content at equilibrium, and S_0 is the initial C content before the change ($t = 0$). An average value of I can then be calculated

$$I = kS_e,$$

but in this case I represents annual inputs of new SOM (hA) rather than inputs of fresh litter or detritus. This approach is also used for chronosequences of primary succession (e.g., on glacial moraines, volcanic deposits, river terraces, dune systems), in this case $S_0 = 0$ (4).

The ^{13}C natural abundance technique relies on 1) the difference in ^{13}C natural abundance between plants with different photosynthetic pathways [Calvin cycle (C_3 plants) vs. Hatch–Slack cycle (C_4 plants)], and 2) the assumption that the ^{13}C natural abundance signature of SOM is identical to the ^{13}C natural abundance signature of the plants from which it is derived (9). Thus, where a change in vegetation type has occurred at some known point in time, the rate of loss of the C derived from the original vegetation and the incorporation of C derived from the new vegetation can be inferred from the resulting change in the ^{13}C natural abundance signature of the soil. The turnover of C derived from the original vegetation is then calculated by using the first-order decay model

$$\text{MRT} = \frac{1}{k} = \frac{t}{\ln(S_t/S_0)} \quad (5)$$

where t is the time since conversion, S_t is the C content derived from original vegetation at time t, and S_0 is the C content at $t = 0$. For further details on the technique see Refs. 9, 10.

The presence of ^{14}C with a half-life of 5570 yr in plants and the transformation of this ^{14}C into SOM with little isotopic discrimination allows the SOM to be dated, providing an estimate of the age of the SOM. The ^{14}C datingtech nique is applicable within a time frame of 200–40,000 yr; samples with an age less than 200 yr are designated as modern (See Ref. 11 for further details of the methodology.)

Thermonuclear bomb tests in the 1950s and 1960s caused the atmospheric ^{14}C content to increase sharply and then to fall drastically after the tests were halted. This sequence of events created an in situ tracer experiment; the incorporation of bomb-produced radiocarbon into SOM after the tests stopped allows estimates of the turnover of SOM. Further details of the technique are described in Refs. 2, 12, 13.

RANGE AND VARIATION IN ESTIMATES OF TOTAL SOIL ORGANIC MATTER TURNOVER

Comparisons of MRT values estimated by the four methods previously described (see, also, Table 1) reveal a wide range of MRTs. Although variations within each method are attributable to differences in vegetation, climate, soil type, and other factors, the largest variations in observed MRTs are method dependent. For example, MRTs estimated by simple first-order modeling and ^{13}C natural abundance are generally smaller by an order of magnitude than MRTs estimated by radiocarbon dating, because of the different time scales that the two methods measure. The ^{13}C method is generally used in medium-term observations or experiments (5–50 yr); hence, this method gives an estimate of turnover dominated by relatively recent inputs and C pools that cycle within the time frame of the experiment. In contrast, the oldest and most recalcitrant C pools dominate estimates by radiocarbon dating because of the long-term time frame (200–40,000 yr) that this method measures (11).

Table 1 Range and average mean residence times (MRTs) of total soil organic C in various ecosystem types as estimated by four different methods

Method and ecosystem	Sites and sources[a]	MRT (yr)		
		Low[b]	High[b]	Average ± SE[c]
First-order modeling				
Cultivated systems and recovering grassland or woodland systems	7/7	15 (14)	102 (15)	67 ± 12
^{13}C natural abundance				
Cultivated systems	20/10	18 (16)	165 (17)	61 ± 9
Pasture systems	12/10	17 (18)	102 (19)	38 ± 7
Forest systems	2/2	18 (20)	25 (21)	22 ± 4
Radiocarbon aging[d]				
Cultivated systems	21/8[e]	327 (22)	1770 (23)	880 ± 105
Grassland systems	4/3[f]	Modern (23)	1040 (24)	–[g]
Forest systems	4/3	422 (22)	1550 (25)	1005 ± 184
"Bomb" ^{14}C analysis				
Cultivated systems	1/1	1863 (13)	1863 (13)	1863[g]
Forest and grassland systems	14/12	36 (26)	1542 (27)	535 ± 134

[a] First value indicates the number of sites used to calculate average MRT values; second value indicates the number of literature sources surveyed (i.e., some sources provided data for multiple sites).
[b] Number in parentheses indicates reference to literature.
[c] SE, standard error.
[d] Values presented in MRT columns for this technique are radiocarbon ages in years B.P.
[e] Includes two sites dating as "modern."
[f] Includes three sites dating as "modern."
[g] Only one value available.

FACTORS CONTROLLING SOIL ORGANIC MATTER TURNOVER

Primary production (specifically, the rate of organic matter transfer below-ground) and soil microbial activity (specifically, the rates of SOM transformation and decay) are recognized as the overall biological processes governing inputs and outputs and, hence, SOM turnover. These two processes (and the balance between them) are controlled by complex underlying biotic and abiotic interactions and feedbacks, most of which can be tied in some way to the state factor model of soil formation (4). Climate (especially temperature and precipitation) constrains both production and decomposition of SOM. Vegetation type affects production rates and the types and quality of organic inputs (e.g., below- vs. above-ground, amounts of structural tissue, C/N and lignin/N ratios), as well as the rates of water and nutrient uptake—all of which, in turn, influence decomposition rates. The types, populations, and activities of soil biota control decomposition and nutrient cycling/availability and hence influence vegetative productivity. Parent material affects SOM turnover as soil type, mineralogy, texture, and structure influence pH, water and nutrient supply, aeration, and the habitat for soil biota, among other factors. Topography modifies climate, vegetation type, and soil type on the landscape scale and exerts finer-scale effects on temperature, soil moisture, and texture. Lastly, time affects whether inputs and outputs are at equilibrium, and temporal scale influences the relative importance of various state factor effects on production and decomposition.

Disturbance or management practices also exert considerable influence on SOM turnover via direct effects on inputs and outputs and through indirect effects on the factors controlling these fluxes. An example of management effects on MRT is illustrated in Table 2; in most cases, the MRT of whole-soil C is significantly longer under no tillage agriculture than under conventional tillage practices.

TURNOVER OF DIFFERENT SOIL ORGANIC MATTER POOLS

The previous discussion is focused on the turnover and MRT of whole-soil C; hence, it treats SOM as a single, homogeneous reservoir. But, in fact, SOM is a heterogeneous mixture consisting of plant, animal, and microbial materials in all stages of decay combined with a variety of decomposition products of different ages and levels of complexity. Thus, the turnover of these components varies continuously, and any estimate of MRT for SOM as a whole merely represents an overall average value (Fig. 1).

Although average MRTs are useful for general comparisons of sites or the effects of different management practices, they can be misleading because soils with similar average MRTs can have very different distri-

Table 2 Effect of tillage practices on mean residence time (MRT) of total soil organic C estimated by the ^{13}C natural abundance technique

Site (Ref.)	Cropping system[a]	Depth (cm)	t^b (yr)	MRT (yr)
Sidney, NE (28)	Wheat–fallow (NT)	0–20	26	73
	Wheat–fallow (CT)			44
Delhi, Ont. (29)	Corn (NT)	0–20	5	26
	Corn (CT)			14
Boigneville, France (16)	Corn (NT)	0–30	17	127
	Corn (CT)			55
Rosemount, MN (30)	Corn (NT, 200 kg N ha^{-1} yr^{-1})	0–30	11	118
	Corn (CT, 200 kg N ha^{-1} yr^{-1})			73
	Corn (NT, 0 kg N ha^{-1} yr^{-1})			54
	Corn (CT, 0 kg N ha^{-1} yr^{-1})			72
Average ± SE[c]		NT		80 ± 19
		CT		52 ± 11

[a] NT, no tillage; CT, conventional (moldboard plow) tillage.
[b] Time period of experiment.
[c] SE, standard error.

Table 3 Mean residence time (MRT) of macro- and microaggregate-associated C estimated by the ^{13}C natural abundance technique

Ecosystem (Ref.)	Aggregate size class[a]	μm	MRT (yr)
Tropical pasture (44)	M	>200	60
	m	<200	75
Temperate pasture grasses (19)	M	212–9500	140
	m	53–212	412
Soybean (45)	M	250–2000	1.3
	m	100–250	7
Corn (46)	M	>250	14
	m	50–250	61
Corn (47)	M	>250	42
	m	50–250	691
Wheat–fallow, no tillage (48)	M	250–2000	27
	m	53–250	137
Wheat–fallow, conventional tillage (48)	M	250–2000	8
	m	53–250	79
Average±SE[b]	M		42±18
	m		209±95

[a] M, macroaggregate; m, microaggregate.
[b] SE, standard error.

butions of organic matter among pools with fast, slow, and intermediate turnover rates (2, 31). Simulation models that account for variations in turnover rates for different SOM pools are now used to generate more realistic descriptions of SOM dynamics. A few models represent decomposition as a continuum, with each input cohort following a pattern of increasing resistance to decay (32), but most models are multicompartmental, with several organic matter pools (often 3–5) that are kinetically defined with differing turnover rates. For example, the CENTURY SOM model (33) divides soil C into active, slow, and passive pools, with MRTs of 1.5, 25, and about 1000 yr, respectively, and separates plant inputs into metabolic (readily decomposable; MRT of 0.1–1 yr) and structural (difficult to decompose; MRT of 1–5 yr) pools as a function of lignin:N ratio. Even though compartmental models are reasonably good at simulating changes in SOM, the compartments are conceptual in nature, and thus it has been difficult to relate them to functionally meaningful pools or experimentally verifiable fractions (34, 35).

The use of isotopic techniques to analytically determine the MRTs of physically and chemically separated SOM fractions has demonstrated the existence of various turnover rates for different pools. For example, low-density SOM (except for charcoal) invariably turns over faster than high-density, mineral-associated SOM, and hydrolyzable SOM turns over faster than nonhydrolyzable residues (36, 37). The MRTs of primary organomineral associations generally increase with decreasing particle size, although there are exceptions (particularly among fine gradations of silt- and clay-sized particles) that have been variously related to climate, clay mineralogy, and fractionation methodology (34, 38, 39).

For a given set of biotic and abiotic conditions, the turnover of different SOM pools depends mechanistically on the quality and biochemical recalcitrance of the organic matter and its accessibility to decomposers. With other factors equal, clay soils retain more SOM with longer MRTs than do sandy soils (40). Readily decomposable materials can become chemically protected from decomposition by association with clay minerals and by sorption to humic colloids (38, 41). Clay mineralogy also plays an important role. For example, montmorillonitic clays and allophanes generally afford more protection than illites and kaolinites (42). In addition, the spatial location of SOM within the soil matrix determines its physical accessibility to decomposers. Relatively labile material may become physically protected by incorporation into soil aggregates (43) or by deposition in micropores inaccessible even to bacteria. Studies of the average MRTs of organic matter in macroaggregates vs. microaggregates show consistently slower turnovers in microaggregates (Table 3). Thus, a much higher proportion of the SOM occluded in microaggregates consists of stabilized materials with relatively long MRTs.

REFERENCES

1. Paul, E.A. Dynamics of Organic Matter in Soils. Plant Soil **1984**, *76*, 275–285.
2. Trumbore, S.E. Comparison of Carbon Dynamics in Tropical and Temperate Soils Using Radiocarbon Measurements. Global Biogeochem. Cycles **1993**, *7*, 275–290.
3. Olson, J.S. Energy Storage and the Balance of Producers and Decomposers in Ecological Systems. Ecology **1963**, *44*, 322–331.
4. Jenny, H. *The Soil Resource—Origin and Behavior*; Springer: New York, 1980; 377.
5. Schimel, D.S. *Theory and Application of Tracers*; Academic Press: San Diego, CA, 1993; 119.
6. Buyanovsky, G.A.; Kucera, C.L.; Wagner, G.H. Comparative Analyses of Carbon Dynamics in Native and Cultivated Ecosystems. Ecology **1987**, *68*, 2023–2031.
7. Jenkinson, D.S. The Turnover of Organic Carbon and Nitrogen in Soil. Phil. Trans. R. Soc. Lond. Ser. B **1990**, *329*, 361–368.
8. Jastrow, J.D. Soil Aggregate Formation and the Accrual of Particulate and Mineral-Associated Organic Matter. Soil Biol. Biochem. **1996**, *28*, 665–676.
9. Cerri, C.; Feller, C.; Balesdent, J.; Victoria, R.; Plenecassagne, A. Application Du Tracage Isotopique Natural En ^{13}C a L'etude De La Dynamique De La Matiere Oganique Dans Les Sols. C.R. Acad. Sci. Paris Ser. II **1985**, *300*, 423–428.
10. Balesdent, J.; Mariotti, A. Measurement of Soil Organic Matter Turnover Using ^{13}C Natural Abundance. In *Mass Spectrometry of Soils*; Boutton, T.W., Yamasaki, S., Eds.; Marcel Dekker: New York, 1996; 83–111.
11. Goh, K.M. Carbon Dating. In *Carbon Isotope Techniques*; Coleman, D.C., Fry, B., Eds.; Academic Press: San Diego, CA, 1991; 125–145.
12. Goh, K.M. Bomb Carbon. In *Carbon Isotope Techniques*; Coleman, D.C., Fry, B., Eds.; Academic Press: San Diego, CA, 1991; 147–151.
13. Harrison, K.G.; Broecker, W.S.; Bonani, G. The Effect of Changing Land Use on Soil Radiocarbon. Science **1993**, *262*, 725–726.
14. Hendrix, P.F. Long-Term Patterns of Plant Production and Soil Carbon Dynamics in a Georgia Piedmont Agroecosystem. In *Soil Organic Matter in Temperate Agroecosystems: Long-Term Experiments in North America*; Paul, E.A., Paustian, K., Elliott, E.T., Cole, C.V., Eds.; CRC Press: Boca Raton, FL, 1997; 235–245.
15. Buyanovsky, G.A.; Brown, J.R.; Wagner, G.H. Sanborn Field: Effect of 100 Years of Cropping on Soil Parameters Influencing Productivity. In *Soil Organic Matter in Temperate Agroecosystems: Long-Term Experiments in North America*; Paul, E.A., Paustian, K., Elliott, E.T., Cole, C.V., Eds.; CRC Press: Boca Raton, FL, 1997; 205–225.
16. Balesdent, J.; Mariotti, A.; Boisgontier, D. Effect of Tillage on Soil Organic Carbon Mineralization Estimated from ^{13}C Abundance in Maize Fields. J. Soil Sci. **1990**, *41*, 587–596.
17. Vitorello, V.A.; Cerri, C.C.; Andreux, F.; Feller, C.; Victoria, R.L. Organic Matter and Natural Carbon-13 Distributions in Forested and Cultivated Oxisols. Soil Sci. Soc. Am. J. **1989**, *53*, 773–778.
18. Desjardins, T.; Andreux, F.; Volkoff, B.; Cerri, C.C. Organic Carbon and ^{13}C Contents in Soils and Soil Size-Fractions, and Their Changes Due to Deforestation and Pasture Installation in Eastern Amazonia. Geoderma **1994**, *61*, 103–118.
19. Jastrow, J.D.; Boutton, T.W.; Miller, R.M. Carbon Dynamics of Aggregate-Associated Organic Matter Estimated by Carbon-13 Natural Abundance. Soil Sci. Soc. Am. J. **1996**, *60*, 801–807.
20. Martin, A.; Mariotti, A.; Balesdent, J.; Lavelle, P.; Vuattoux, R. Estimate of Organic Matter Turnover Rate in a Savanna Soil by ^{13}C Natural Abundance Measurements. Soil Biol. Biochem. **1990**, *22*, 517–523.
21. Trouve, C.; Mariotti, A.; Schwartz, D.; Guillet, B. Soil Organic Carbon Dynamics Under *Eucalyptus* and *Pinus* Planted on Savannas in the Congo. Soil Biol. Biochem. **1994**, *26*, 287–295.
22. Paul, E.A.; Collins, H.P.; Leavitt, S.W. Dynamics of Resistant Soil Carbon of MidWestern Agricultural Soils Measured by Naturally-Occurring ^{14}C Abundance. Geoderma **2001**, *104*, 239–256.
23. Paul, E.A.; Follett, R.F.; Leavitt, S.W.; Halvorson, A.; Peterson, G.A.; Lyon, D.J. Radiocarbon Dating for Determination of Soil Organic Matter Pool Sizes and Dynamics. Soil Sci. Soc. Am. J. **1997**, *61*, 1058–1067.
24. Jenkinson, D.S.; Harkness, D.D.; Vance, E.D.; Adams, D.E.; Harrison, A.F. Calculating Net Primary Production and Annual Input of Organic Matter to Soil from the Amount and Radiocarbon Content of Soil Organic Matter. Soil Biol. Biochem. **1992**, *24*, 295–308.
25. Trumbore, S.E.; Bonani, G.; Wolfli, W. The Rates of Carbon Cycling in Several Soils from AMS ^{14}C Measurement of Fractionated Soil Organic Matter. In *Soils and the Greenhouse Effect*; Bouwman, A.F., Ed.; Wiley: London, 1990; 407–414.
26. O'Brien, B.J. Soil Organic Carbon Fluxes and Turnover Rates Estimated from Radiocarbon Enrichments. Soil Biol. Biochem. **1984**, *16*, 115–120.
27. Bol, R.A.; Harkness, D.D.; Huang, Y.; Howard, D.M. The Influence of Soil Processes on Carbon Isotope Distribution and Turnover in the British Uplands. Eur. J. Soil Sci. **1999**, *50*, 41–51.
28. Six, J.; Elliott, E.T.; Paustian, K.; Doran, J.W. Aggregation and Soil Organic Matter Accumulation in Cultivated and Native Grassland Soils. Soil Sci. Soc. Am. J. **1998**, *62*, 1367–1377.
29. Ryan, M.C.; Aravena, R.; Gillham, R.W. The Use of ^{13}C Natural Abundance to Investigate the Turnover of the Microbial Biomass and Active Fractions of Soil Organic Matter Under Two Tillage Treatments. In *Soils and Global Change*; Lal, R., Kimble, J., Levine, E., Stewart, B.A., Eds.; CRC Press: Boca Raton, FL, 1995; 351–360.
30. Clapp, C.E.; Allmaras, R.R.; Layese, M.F.; Linden, D.R.; Dowdy, R.H. Soil Organic Carbon and ^{13}C Abundance as Related to Tillage, Crop Residue, and Nitrogen Fertiliza-

tion Under Continuous Corn Management in Minnesota. Soil Till. Res. **2000**, *55*, 127–142.
31. Davidson, E.A.; Trumbore, S.E.; Amundson, R. Soil Warming and Organic Carbon Content. Nature **2000**, *408*, 789–790.
32. Ågren, G.I.; Bosatta, E. Theoretical Analysis of the Long-Term Dynamics of Carbon and Nitrogen in Soils. Ecology **1987**, *68*, 1181–1189.
33. Parton, W.J.; Schimel, D.S.; Cole, C.V.; Ojima, D.S. Analysis of Factors Controlling Soil Organic Matter Levels in Great Plains Grasslands. Soil Sci. Soc. Am. J. **1987**, *51*, 1173–1179.
34. Balesdent, J. The Significance of Organic Separates to Carbon Dynamics and Its Modeling in Some Cultivated Soils. Eur. J. Soil Sci. **1996**, *47*, 485–493.
35. Christensen, B.T. Matching Measurable Soil Organic Matter Fractions with Conceptual Pools in Simulation Models of Carbon Turnover: Revision of Model Structure. In *Evaluation of Soil Organic Matter Models*; Powlson, D.S., Smith, P., Smith, J.U., Eds.; Springer: Berlin, 1996; 143–159.
36. Martel, Y.A.; Paul, E.A. The Use of Radiocarbon Dating of Organic Matter in the Study of Soil Genesis. Soil Sci. Soc. Am. Proc. **1974**, *38*, 501–506.
37. Trumbore, S.E.; Chadwick, O.A.; Amundson, R. Rapid Exchange Between Soil Carbon and Atmospheric Carbon Dioxide Driven by Temperature Change. Science **1996**, *272*, 393–396.
38. Christensen, B.T. Physical Fractionation of Soil and Organic Matter in Primary Particle Size and Density Separates. Adv. Soil Sci. **1992**, *20*, 1–90.
39. Feller, C.; Beare, M.H. Physical Control of Soil Organic Matter Dynamics in the Tropics. Geoderma **1997**, *79*, 69–116.
40. Sorensen, L.H. The Influence of Clay on the Rate of Decay of Amino Acid Metabolites Synthesized in Soils During Decomposition of Cellulose. Soil. Biol. Biochem. **1974**, *7*, 171–177.
41. Jenkinson, D.S. Soil Organic Matter and Its Dynamics. In *Russell's Soil Conditions and Plant Growth*; Wild, A., Ed.; Wiley: New York, 1988; 564–607.
42. Dalal, R.C.; Bridge, B.J. Aggregation and Organic Matter Storage in Sub-Humid and Semi-Arid Soils. In *Structure and Organic Matter Storage in Agricultural Soils*; Carter, M.R., Stewart, B.A., Eds.; CRC Press: Boca Raton, FL, 1996; 263–307.
43. Tisdall, J.M.; Oades, J.M. Organic Matter and Water-Stable Aggregates in Soils. J. Soil Sci. **1982**, *33*, 141–163.
44. Skjemstad, J.O.; Le Feuvre, R.P.; Prebble, R.E. Turnover of Soil Organic Matter Under Pasture as Determined by ^{13}C Natural Abundance. Aust. J. Soil Res. **1990**, *28*, 267–276.
45. Buyanovsky, G.A.; Aslam, M.; Wagner, G.H. Carbon Turnover in Soil Physical Fractions. Soil Sci. Soc. Am. J. **1994**, *58*, 1167–1173.
46. Monreal, C.M.; Schulten, H.R.; Kodama, H. Age, Turnover and Molecular Diversity of Soil Organic Matter in Aggregates of a Gleysol. Can. J. Soil Sci. **1997**, *77*, 379–388.
47. Angers, D.A.; Giroux, M. Recently Deposited Organic Matter in Soil Water-Stable Aggregates. Soil Sci. Soc. Am. J. **1996**, *60*, 1547–1551.
48. Six, J.; Elliott, E.T.; Paustian, K. Aggregate and Soil Organic Matter Dynamics Under Conventional and No-Tillage Systems. Soil Sci. Soc. Am. J. **1999**, *63*, 1350–1358.

ORGANISMS AND SOIL FOOD WEBS

David C. Coleman
University of Georgia, Athens, Georgia, U.S.A.

INTRODUCTION

Soils may be viewed as the organizing centers for terrestrial ecosystems. This is largely the result of organismal activities in the soil. Major functions such as ecosystem production, respiration and nutrient recycling are controlled by the rates at which nutrients are released by decomposition in the soil and litter horizons. The array of biota, including microbes, microbe-feeding fauna, vegetation, and consumers are all influenced by soil processes, and the organisms in turn have an impact on the soil system.

Soils provide a wide range and variety of microhabitats, thus accommodating a very diverse biota. The enormous surface area (hundreds of m^2/g of soil) of soil particles, ranges in size classes from clays (0.1–2 μm in diameter), to silts (2–50 μm in diameter), and sands (0.05–2 mm in diameter). Numerous microbes and micro- and mesofauna (protozoa and nematodes) exist in water films on these particles, and in or on the surfaces of microaggregates formed from the primary particles (1). In turn, the more mobile fauna, from collembola and mites (larger mesofauna) to the macrofauna (earthworms, millipedes, ants, termites, and fossorial or earth-dwelling vertebrates) move through macro- and micropores in the soil. The macrofauna plays a role in moving parts of the soil profile around, and form many sorts of burrows and pores; they are often termed "ecological engineers."

SOIL FOOD WEBS

The initial breaking up or "comminution," of plant litter (above- and below- ground) results from the chewing and macerating action of both large and small animals. This comminution benefits the fauna, which derive nutritional benefit from the litter and/or microbes initially colonizing the plant material. The increased surface area and further inoculation of the smaller pieces enhances the microbial access to, and breakdown of, these tissues. The decomposition process drives complex food webs (2, 3) in the soil, with numerous interactions between the initial agents of decomposition, the bacteria and fungi, and the fauna that in turn feed upon them, which facilitates nutrient return in the soil matrix (Fig. 1). These feeding relationships have several trophic levels, with bacteria and fungi being fed upon by microbe-feeders, such as protozoa, nematodes and microarthropods, which are in turn preyed upon by predatory nematodes or mites, and these in turn fed upon by higher predators (Fig. 1). In forests or no tillage agroecosystems, where fungi dominate in the surface litter, the dominant flows of energy and nutrients will go via fungal pathways. In contrast, in conventional tilled fields where the organic matter is incorporated in the plow layer (usually 6–8 in. = 15–20 cm), the dominant flows of energy and nutrients may be more bacterially dominated, usually decomposing faster than in the no tillage system (1).

ZONES OF INFLUENCE

The heterogeneous distribution of food resources in the soil matrix makes it difficult to sample adequately for abundances and activities of the biota in a repeatable fashion. A useful approach is to consider soils as being comprised of zones of influence (ZOI), that can be targeted for further study. These ZOI, also termed "hot spots," are located in the root-rhizosphere, in regions of organic detritus accumulation, or detritusphere, and also in earthworm-influenced regions, such as burrows, which are termed a drilosphere (4) (Fig. 2). These ZOI may represent less than 10% of the volume of the surface A horizon, but account for up to 90% of the total biological activity in soils worldwide.

ROLES OF BIOTA IN SOIL FOOD WEBS

The functional roles of soil organisms can be compared most usefully in terms of body width. The microbes and microfauna inhabit soil water films, and are restricted to this aquatic milieu. In contrast, the meso- and macrofauna, from acari (mites) to earthworms, inhabit gas-filled pores, and move around in the soil matrix for considerable distances (Fig. 3) (5).

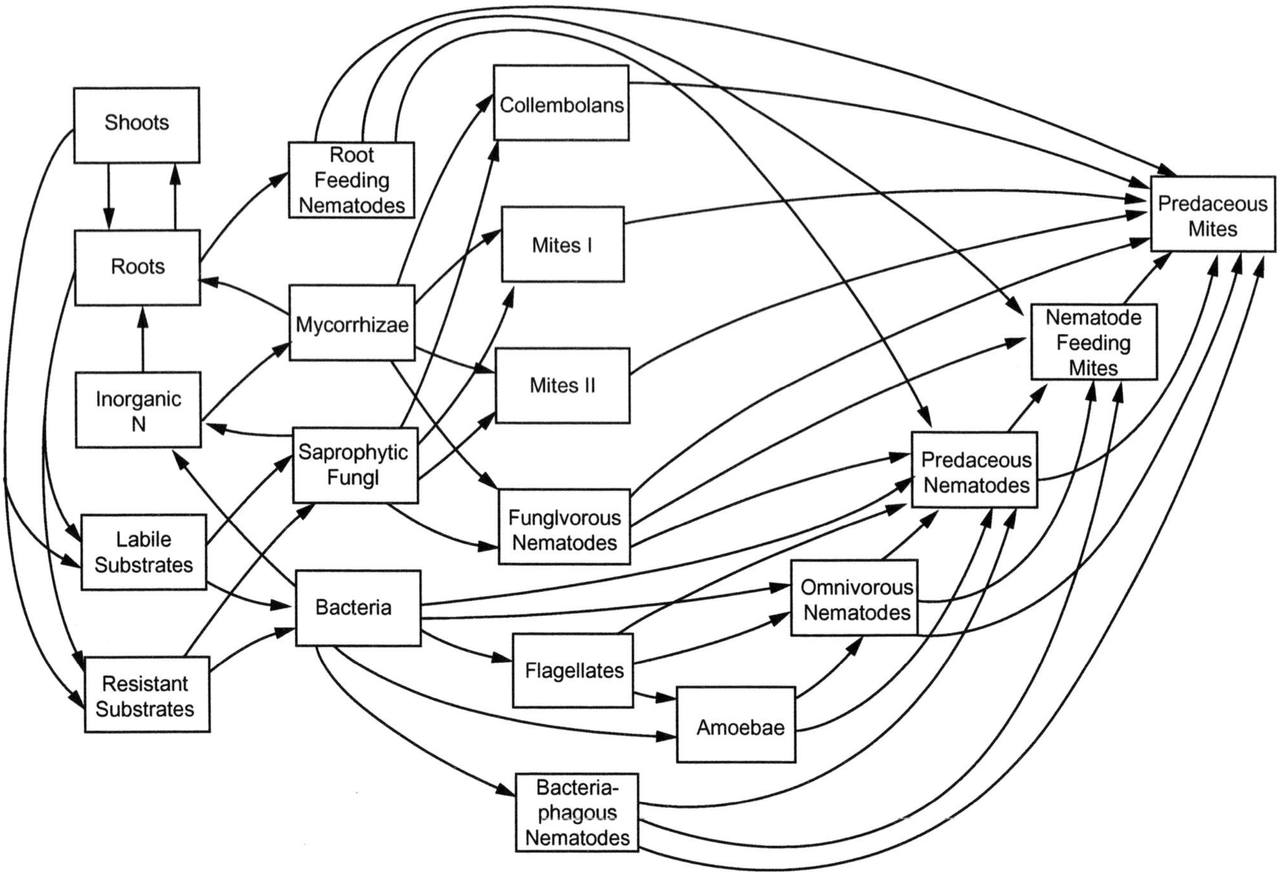

Fig. 1 Representation of detrital food web in shortgrass prairie. Fungal-feeding mites are separated into two groups (I and II) to distinguish the slow-growing oribatids from faster-growing taxa. Flows omitted from the figure for the sake of clarity include transfers from every organism to the substrate pools (death) and transfers from every animal to the substrate pools (defecation) and to inorganic N (ammonification). (From Ref. 2.)

Bacteria

Bacteria are unicellular prokaryotes (organisms lacking a unit membrane-bounded nucleus and other organelles), that are found in all habitats on earth. They are exceedingly numerous (more than 10^{30}, or one million trillion trillion (6)) and diverse, currently comprising over 35 kingdoms in two domains, the Archaea and Eubacteria. They are active in all aspects of elemental cycling, and needed for nitrogen cycling, both in nitrogen fixation (splitting N_2 and incorporating N into organic compounds), and subsequent transformational pathways as well. They are also primary agents of decomposition in many habitats, and are particularly active in rhizospheres (4).

Fungi

Fungi are multicellular eukaryotes that are found in many habitats worldwide. They have long, ramifying strands (hyphae) which can grow into and explore many microhabitats, and are used for obtaining water and nutrients. The hyphae secrete a considerable array of enzymes, such as cellulases, and even lignases in some specialized forms (useful in breaking down wood), decomposing substrates in situ, taking up the decomposed subunits, and translocating them back through the hyphal network. Fungi are very abundant, particularly in undisturbed forest floors, in which literally thousands of kilometers of hyphal filaments will occur per gram of leaf litter.

The roles of mycorrhizas (literally "fungus–root," or symbiotic fungi associated with many plants) in soil systems are being increasingly viewed as central to much of terrestrial ecosystem function. Mycorrhizas are essential to the growth and reproduction of numerous families of plants (1).

Microfauna

The unicellular eukaryotes, or Protoctista, are more often called protozoans. They include the flagellates, naked

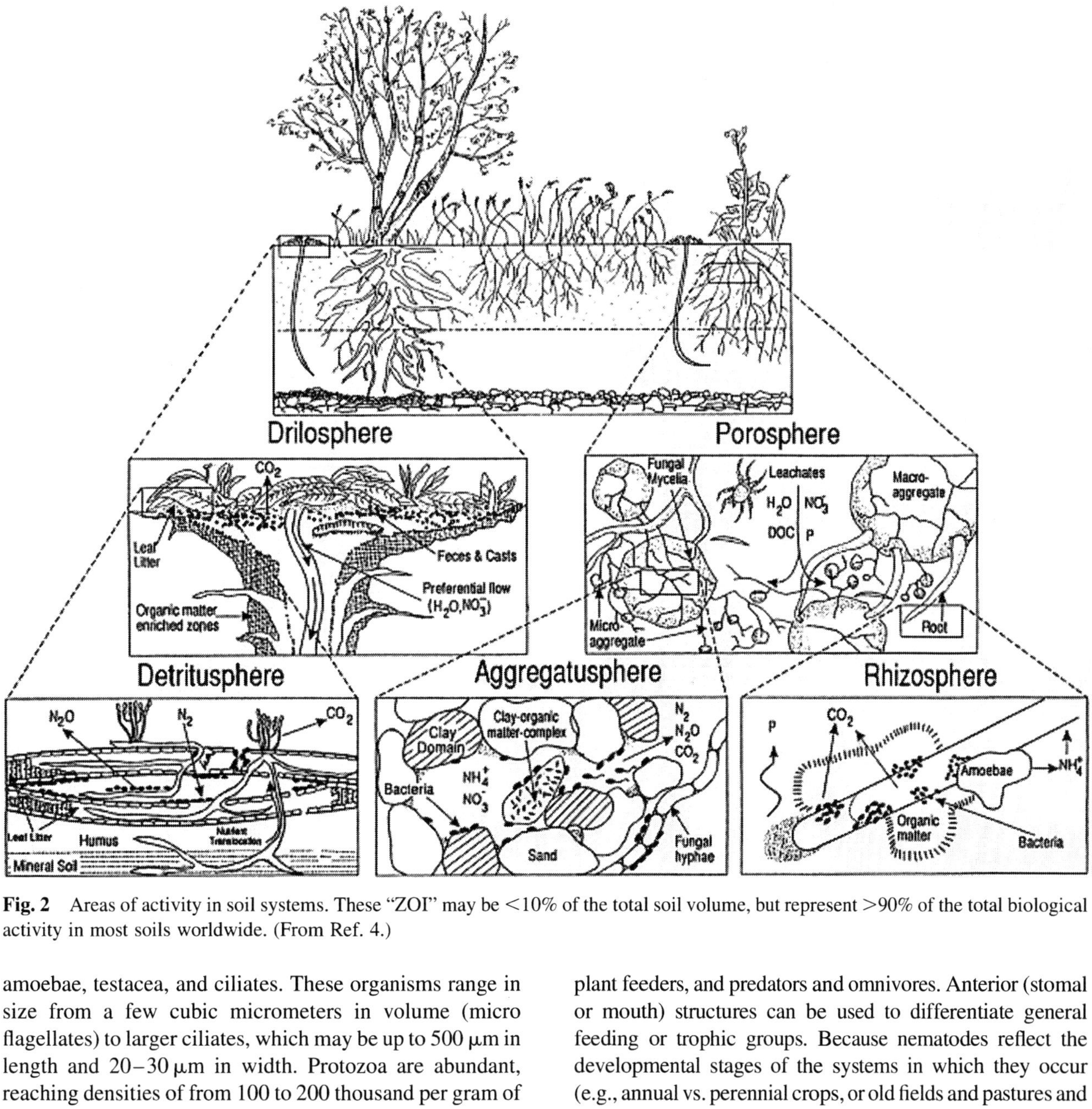

Fig. 2 Areas of activity in soil systems. These "ZOI" may be <10% of the total soil volume, but represent >90% of the total biological activity in most soils worldwide. (From Ref. 4.)

amoebae, testacea, and ciliates. These organisms range in size from a few cubic micrometers in volume (micro flagellates) to larger ciliates, which may be up to 500 μm in length and 20–30 μm in width. Protozoa are abundant, reaching densities of from 100 to 200 thousand per gram of soil. Bacteria, their principal prey, often exist in numbers up to one billion per gram of soil. All of these organisms are true water-film dwellers, and become dormant or inactive during episodes of drying in the soil. They can exist in inactive or resting stages literally for decades at a time in very xeric environments.

Mesofauna

Nematodes

Nematodes have a wide range of feeding preferences. A general trophic grouping is bacterial feeders, fungal feeders, plant feeders, and predators and omnivores. Anterior (stomal or mouth) structures can be used to differentiate general feeding or trophic groups. Because nematodes reflect the developmental stages of the systems in which they occur (e.g., annual vs. perennial crops, or old fields and pastures and more mature forests), they have been used as indicators of overall ecosystem condition (7, 8).

Collembola

Collembolans, or "springtails" are primitive apterygote (wingless) insects. They are called "springtails" because many of them have a spring-like lever, or furcula, which enables them to move many body-lengths away from predators in a springing fashion. Collembolans are ubiquitous members of the soil fauna, often reaching abundances on 100,000 or more per m^2. They occur

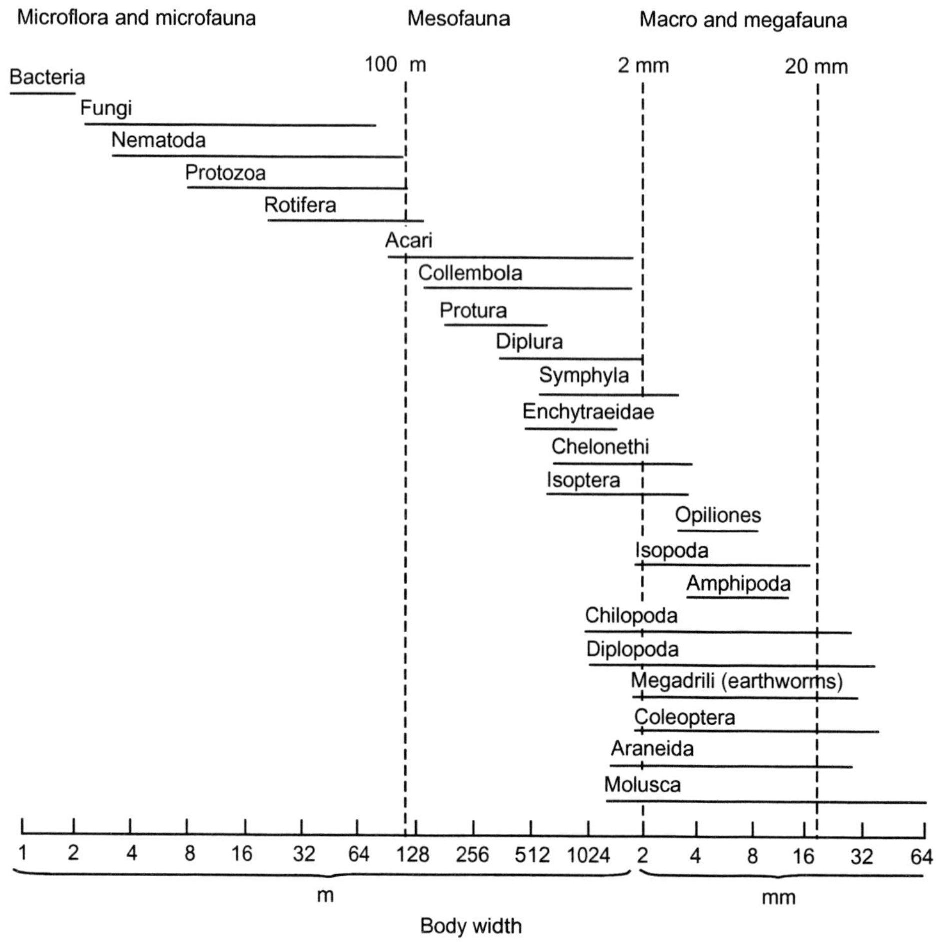

Fig. 3 Size classification of organisms in decomposer food webs by body width. (From Ref. 5.)

throughout the soil profile, where their major diet is decaying vegetation and associated microbes (usually fungi). However, like many members of the soil fauna, collembolans defy placement in exact trophic groups. Many collembolan species will eat nematodes when those are abundant. Some feed on live plants or their roots. One family (Onychiuridae) may feed in the rhizosphere and consume mycorrhizas or even plant pathogenic fungi (9).

Mites (Acari)

The soil mites, Acari, are chelicerate arthropods related to the spiders. They are often the most abundant microarthropods in soils. A 100-g sample may contain as many as 500 mites representing nearly 100 genera. This diverse array includes participants in three or more trophic levels, with varied strategies for feeding, reproduction, and dispersal.

The oribatid mites (Oribatei) are the characteristic mites of the soil and are usually fungivorous or detritivorous. Mesostigmatid mites are nearly all predators on other small fauna, although a few species are fungivores and may become numerous at times. Astigmatid mites are associated with rich, decomposing nitrogen sources and are rare except in agricultural soils. The Prostigmata contains a broad diversity of mites with several feeding habits (1).

Macrofauna

Termites (Isoptera) are one of the major ecosystem "engineers," particularly in tropical regions. Termites are social insects with a well-developed caste system. By their ability to digest wood, they have become economic pests of major importance in some regions of the world. The termites in a primitive family, the Kalotermitidae, possess a gut flora of protozoans, which enables them to digest cellulose. Their normal food is wood that has come into contact with soil. Many species of termites construct runways of soil, or along root channels, and some are

O

builders of large, spectacular mounds. Members of the phylogenetically advanced family Termitidae possess a formidable array of microbial symbionts (bacteria and fungi, but not protozoa), and elevated pH in their hindguts (10), which enable them to process and digest the humified organic matter in tropical soils and to thrive on it.

Termites parallel earthworms in ingestive and soil turnover functions. The principal difference is that earthworms egest much of what they ingest in altered form (that enriches microbial action), whereas termites transfer large amounts of soil/organic material into building nests and mounds (carbon sinks).

Earthworms

Much of the evidence for earthworm effects on soil processes comes from agroecosystems and involves a small group of European lumbricids (Lumbricidae family in the Oligochaeta order). Impacts of exotic earthworms on native species are not well understood, although there is evidence that when native habitat is destroyed and native earthworm species extirpated, exotic earthworms colonize the newly empty habitat. As more extensive studies are carried out, it is becoming clear that earthworms are present in a wide variety of tropical as well as temperate ecosystems.

Earthworms have important roles in the fragmentation, breakdown and incorporation of soil organic matter (SOM). This affects the distribution of SOM, and also its chemical and physical characteristics. Changes in any of these soil parameters may have significant effects on other soil biota, by changing their resource base (e.g., distribution and quality of SOM, microbes or microarthropods) or by changing the physical structure of the soil (11). Earthworm activities impact the communities of other soil biota through their effects on the chemical and physical characteristics of SOM, causing changes in the microbial and microarthropod communities, and also having impacts elsewhere in the soil food web (12).

SUMMARY

Soil biota are very interconnected with each other by a variety of trophic and nutrient-flow pathways. The possibility of enhancing biotic activity in agricultural systems via conservation or no tillage regimes is very great, and much research is now focusing on this area of interest. This approach permits a melding of interests between those who want to reduce fossil fuel inputs to agroecosystems and those who are concerned with enhancing carbon sequestration in soils as well.

REFERENCES

1. Coleman, D.C.; Crossley, D.A., Jr. *Fundamentals of Soil Ecology*; Academic Press: San Diego, CA, 1996; 205.
2. Hunt, H.W.; Coleman, D.C.; Ingham, E.R.; Ingham, R.E.; Elliott, E.T.; Moore, J.C.; Rose, S.L.; Reid, C.P.P.; Morley,C.R. The Detrital Food Web in a Shortgrass Prairie. Biol. Fert. Soils **1987**, *3*, 57–68.
3. Moore, J.C.; de Ruiter, P.C. Invertebrates in Detrital Food Webs Along Gradients of Productivity. In *Invertebrates as Webmasters in Ecosystems*; Coleman, D.C., Hendrix, P.F., Eds.; CABI: Wallingford, UK, 2000; 161–184.
4. Beare, M.H.; Coleman, D.C.; Crossley, D.A., Jr.; Hendrix, P.F.; Odum, E.P. A Hierarchical Approach to Evaluating the Significance of Soil Biodiversity to Biogeochemical Cycling. Plant Soil **1995**, *170*, 5–22.
5. Swift, M.J.; Heal, O.W.; Anderson, J.M. *Decomposition in Terrestrial Ecosystems*; Univ. of California Press: Berkeley, CA, 1979; 379.
6. Whitman, W.B.; Coleman, D.C.; Wiebe, W.J. Prokaryotes: The Unseen Majority. Proc. Natl. Acad. Sci. **1998**, *95*, 6578–6583.
7. Bongers, T. The Maturity Index: An Ecological Measure of Environmental Disturbance Based on Nematode Species Composition. Oecologia **1990**, *83*, 14–19.
8. Yeates, G.W.; Bongers, T.; de Goede, R.G.M.; Freckman, D.W.; Georgieva, S.S. Feeding Habits in Soil Nematode Families and Genera—An Outline for Soil Ecologists. J. Nematol. **1993**, *25*, 315–331.
9. Lartey, R.T.; Curl, E.A.; Peterson, C.M. Interactions of Mycophagous Collembola and Biological Control Fungi in the Suppression of *Rhizoctonia Solani*. Soil Biol. Biochem. **1994**, *26*, 81–88.
10. Bignell, D.E.; Eggleton, P. On the Elevated Intestinal pH of Higher Termites (Isoptera Termitidae). Insectes Sociaux **1995**, *42*, 57–69.
11. Hendrix, P.F., Ed. *Earthworm Ecology and Biogeography in North America*; Lewis Publishers: Boca Raton, FL, 1995; 335.
12. Fu, S.; Cabrera, M.L.; Coleman, D.C.; Kisselle, K.W.; Garrett, C.J.; Hendrix, P.F.; Crossley, D.A., Jr. Soil Carbon Dynamics of Conventional Tillage and No-Till Agroecosystems at Georgia Piedmont—HSB-C Models. Ecol. Model. **2000**, *131*, 229–248.

OXISOLS

S.W. Buol
North Carolina State University, Raleigh, North Carolina, U.S.A.

INTRODUCTION

The name Oxisol was introduced in 1960 to avoid connotations associated with laterite and associated terms (1). Oxisols are sandy loam or finer textured soils. Their subsoils have an apparent cation exchange capacity at pH 7 (CEC_7) of 16 cmol kg^{-1} clay or less, an apparent effective cation exchange capacity (ECEC) of 12 cmol kg^{-1} clay or less and contain less than 10 percent Ca, Mg or K bearing weatherable minerals in the 50- to 200-micron sand fraction.

If the surface 18 cm contains less than 40 percent clay, the amount of clay in the subsoil must be less than 20 percent greater than the clay content of the surface 18 cm. If the surface 18 cm contains 40 percent or more clay increased clay content with depth (i.e., a kandic horizon) is permitted providing it contains less than 10 percent weatherable minerals (2).

Reaction and base saturation are not criteria for defining Oxisols. Most Oxisols are acidic in reaction and have low base saturation percentages but some have high base saturation and neutral pH values. Most Oxisols have low silt content, low available water holding capacity, low bulk density, strong fine and very fine granular structure and a more rapid hydraulic conductivity than would be predicted by particle size class.

Almost all Oxisols are located within tropical latitudes and have average annual soil temperatures of 15°C or warmer. The mean soil temperature of June, July and August differs less than 6°C from that of December, January and February.

Oxisols are divided into five suborders according to their soil moisture regime (SMR):

1. Aquox (3 percent of all Oxisols) are poorly drained with aquic conditions and most often located in depressions and seepage areas in association with other Oxisols.
2. Torrox (0.3 percent of all Oxisols) have an aridic SMR with less than 90 consecutive days of plant available moisture in normal years.
3. Ustox (53 percent of all Oxisols) have an ustic SMR with from 90 to 270 consecutive days of plant available moisture in normal years. One or two crop-growing seasons are possible each year.
4. Udox (32 percent of all Oxisols) have less than 90 consecutive days that are too dry for reliable crop growth in normal years. Two and sometimes three consecutive crops can be grown each year.
5. Perox (12 percent of all Oxisols) have rainfall that exceeds potential evapotranspiration every month of normal years. With no dry season it is difficult to harvest grain crops and slash and burn management is difficult because of no reliable time to burn.

GEOLOGIC SETTINGS

The most extensive contiguous areas of Oxisols are related to mid- to late-Tertiary geomorphic surfaces composed of material that has been weathered, eroded, transported and re-deposited many times (3, 4). The materials within which Oxisols form on such surfaces are an accumulation of resistant minerals that have the properties of Oxisol subsoil at the time they are deposited. Such deposits, known as oxic soil material sometimes have stone-lines of quartz and/or oxide-cemented gravel. The inert nature of oxic soil material precludes many pedogenic processes and Oxisols are present on recent alluvial deposits of such material (5).

Oxisols also form on stable surfaces where easily weatherable materials, such as basic and ultra basic rocks and geologically old volcanic deposits are exposed to warm humid conditions (6, 7). These occurrences are generally of limited spatial extent.

PROCESSES OF FORMATION

Processes that concentrate aluminum and deplete silicon favor the formation of oxic soil material. Low Si/Al ratios may have their origin in the magma of the Earth's crust and persist when exposed to the soil environment. Silicon loss is to be expected in the surface of almost all soils as rainwater infiltrates and moves through the surface layer. The amount of silicon removed depends upon the type of silicate mineral and residence time of water around the silicate mineral. Net silicon loss results in destruction of 2:1 lattice clays and favors kaolinite, halloysite and gibbsite enrichment.

Iron released by the dissolution of iron bearing silicates is concentrated as iron oxides unless reducing conditions are present. Almost all of the iron in Oxisols is present as iron oxides. Iron oxides impart many red, red-yellow and yellow hues to soil. Red hues are indicative of hematite while the yellow colors indicate goethite. Mixtures of these minerals have intermediate colors (8). Hematite is more easily reduced, and thus more readily dissolved, than goethite (9, 10). Even short periods of reduction in micro sites around decaying roots will, over time, cause the removal of hematite leaving goethite as the dominate iron oxide in the more yellow colored Oxisols. If soil material is subjected to reducing conditions all iron oxides are chemically reduced to soluble ferrous ions and removed leaving only the gray color of the clays and sand (11).

Oxisols contain more organic carbon than most other mineral soils (12). The organic carbon in the subsoil is relatively unavailable for microbial oxidation and masked by red and yellow colored iron oxides (13).

LOCATIONS

It is estimated that about 981 million hectares of the Earth's surface is dominated by Oxisols. Of this 76 percent is in South America, 22 percent in Africa and 2 percent in Asia. In South America the central plateau between the Amazon and Parana rivers and lower Amazon basin is perhaps the most extensive area of Oxisols. Multiple erosion-deposition cycles have created deep sedimentary deposits of material rich in quartz sand, iron and aluminum oxides and 1:1 layer clays. Quartzipsamments are present in deposits too sandy to classify as Oxisols. The eastern part of the Amazon basin is filled with oxic materials eroded from adjacent plateaus and Oxisols predominate. In the western part of the Amazon basin materials eroded from the Andes enrich the parent materials with weatherable minerals and almost no Oxisols are present (14). In central Africa, most Oxisols are in the central Congo River basin but seldom present where relatively recent volcanic materials contribute material to the present soil.

The presence of Oxisols within all soil moisture regimes and on recently formed geomorphic surfaces argues against generalizations about climatic and time factors necessary for Oxisol formation. The close association of Oxisols with parent materials almost devoid of weatherable silicate minerals and the near absence of Oxisols in adjacent geologic materials clearly identify parent material composition as the controlling factor in Oxisol formation.

HUMAN USE

Relative to most other soils Oxisols are chemically infertile. Except where formed in basic and ultra basic material Oxisols are extremely poor in basic cations and exchangeable aluminum dominates the cation exchange sites. Size and density of trees in undisturbed "cerrado" vegetation growing on Oxisols in central Brazil is closely related to chemical fertility (15, 16). Even the more highly base saturated "Eutro" great groups of Oxisols contain small total quantities of exchangeable calcium, magnesium and potassium (1). Phosphorus contents are relatively low and phosphorus reacts with aluminum and iron oxides to such a degree that it is very slowly available to plant growth. Fast growing food-crop plants seldom sustain satisfactory growth on unfertilized Oxisols.

Many of the relatively base-rich Oxisols support natural forest growth. Forest stands slowly accumulate plant-essential nutrients in their biomass. When a sufficient quantity of forest vegetation is cut, allowed to dry and then burned, enough nutrients are rapidly released from the ashes to allow subsistence farmers to harvest one or two food crops. The site is then abandoned and a natural succession of woody vegetation that is able to grow with a slower rate of nutrient uptake than crop plants reestablishes. After several years the biomass of the natural vegetation again acquires enough essential nutrients that subsistence farmers can repeat the slash and burn process.

On the least fertile Oxisols only poor quality grass and shrub-type trees grow. An extensive area of this condition is the "cerrado" savanna vegetation in central Brazil. The "cerrado" vegetation is so nutrient poor that the bones of grazing cattle deteriorate from calcium and phosphorus deficiency unless supplemental mineral concentrates are supplied. Farmers attempting slash and burn techniques are discouraged by negligible yields. Inability to sustain indigenous human populations is a characteristic of the least fertile Oxisols.

With the advent of a clear understanding of chemical limitations and in those locations where economic stability enabled infrastructure necessary for commercial agriculture an entirely different relationship of human interaction with Oxisols is possible. In 1992, with only 10 million hectares of the estimated 204 million hectares of Oxisol dominate "cerrado" cultivated, 28 percent of the grain production in Brazil was from the "cerrado" area (17). Beef production has rapidly increased on fertilized pastures. With market access it is economically feasible to make initial applications of lime and phosphate fertilizer to overcome the natural acidity and phosphate fixation capacity of the most infertile Oxisols. Small amounts of zinc, boron, molybdenum and copper are needed on some sites. The initial application of lime and phosphate is mixed as deeply as possible to maximize a favorable rooting depth. Facilitated by the low cation capacity of the Oxisol subsoil calcium moves downward, replaces exchangeable aluminum, increases the exchangeable Ca/Al ratio and facilitates deeper root growth (18). The deeper rooting zone enhances available moisture supply during rainless periods during the growing season. Total amounts of exchangeable aluminum in the subsoil are low and relatively small amounts of calcium are required. Gypsum is especially favorable for calcium movement and supplies sulfur that is often needed in Oxisols. The initial investment in lime and phosphate fertilizer is large but not prohibitive. In subsequent years nitrogen, phosphorus and potassium fertilizers and lime are needed to replace nutrients exported in harvested crops and maintain desirable pH values but annual amounts are no greater than on other soils growing similar crops.

Many Oxisol dominated landscapes are nearly level and trafficability for large equipment and road construction are facilitated by physically stable low activity silicate clay and iron oxide. Low indigenous populations facilitate acquisition of large management units for efficient mechanized agriculture. Oxisols with ustic and udic soil moisture regimes and isothermic and isohyperthermic soil temperature regimes have reliable rains during at least one growing season each year followed by a period of dryness as grain crops mature. The reliable onset of a dry season decreases the risk of grain spoilage, decreases cost of drying harvested grain and permits maximum efficiency of harvest equipment and marketing infrastructure.

CONCLUSION

Oxisols are physically deep but nutrient poor and sparsely inhabited soils in the world. Most Oxisols are located in warm tropical areas with ustic or udic soil moisture regimes that permit the growth of one or two grain crops each year without irrigation. Without lime and fertilizer even subsistence slash and burn agriculture is not possible on the poorest Oxisols. With stable economic infrastructure, markets, chemical fertilizer and lime needed to initially correct natural chemical infertility and replace nutrients harvested intensive mechanized agriculture and grazing of improved pastures are now realities.

REFERENCES

1. Buol, S.W.; Eswaran, H. Oxisols. Advances in Agronomy **2000**, *68*, 151–195.
2. Soil Survey Staff. *Soil Taxonomy*, 2nd Ed.; Agriculture Handbook No. 436; US Department of Agriculture, Natural Resources Conservation Service: Washington, DC, 1999; 869 p.
3. Lepsch, I.F.; Buol, S.W.; Daniels, R.B. Soils-Landscape Relationships in the Occidental Plateau of Sao Paulo State, Brazil: I. Geomorphic Surfaces and Soil Mapping Units. Soil Sci. Soc. Am. J. **1977**, *41*, 104–109.
4. Lepsch, I.F.; Buol, S.W.; Daniels, R.B. Soils-Landscape Relationships in the Occidental Plateau of Sao Paulo State, Brazil: II. Soil Morphology, Genesis and Classification. Soil Sci. Soc. Am. J. **1977**, *41*, 109–115.
5. Odell, R.T.; Dijkerman, J.C.; Van Vuure, W.; Melsted, S.W.; Beavers, A.H.; Sutton, P.M.; Kurtz, L.T. *Characteristics, Classification and Adoption of Soils in Selected Areas of Sierra Leone, West Africa.* Bull. 748; Agric. Exp. Stn.: Univ. Ill., Urbana-Champaign, 1974; 194 p.
6. Buurman, P.; Soepraptohardjo, M. Oxisols and Associated Soils on Ultramafic and Felsic Volcanic Rocks in Indonesia. In *Red Soils in Indonesia*; Buurman, P., Ed.; Bull. No. 4, Soil Research Inst.: Bogor. Pub. by Centre for Agric. Pub. and Documentation, Wageningen, The Netherlands, 1980; 71–87.
7. Beinroth, F.H. Some Highly Weathered Soils of Puerto Rico; 1. Morphology, Formation and Classification. Geoderma **1982**, *27*, 1–73.
8. Bigham, J.M.; Golden, D.C.; Buol, S.W.; Weed, S.B.; Bowen, L.H. Iron Oxide Mineralogy of Well-drained Ultisols and Oxisols: II. Influence on Color, Surface Area and Phosphate Retention. Soil Sci. Soc. Am. J. **1978**, *42*, 825–830.
9. Macedo, J.; Bryant, R.B. Morphology, Mineralogy, and Genesis of a Hydrosequence of Oxisols in Brazil. Soil Sci. Soc. Am. J. **1987**, *51*, 690–698.
10. Macedo, J.; Bryant, R.B. Preferential Microbial Reduction of Hematite Over Goethite in Brazilian Oxisol. Soil Sci. Soc. Am. J. **1989**, *53*, 1114–1118.
11. Buol, S.W.; Camargo, M.N. Wet Oxisols. In *Proc. Eigth International Soil Correlation Meeting (VIII ISCOM): Characterization, Classification, and Utilization of Wet*

Soils; Kimble, J.M., Ed.; USDA, Soil Conservation Service, Nat. Soil Survey Center: Lincoln, NE, 1992; 41–49.

12. Eswaran, H.; Van Den Berg, E.; Reich, P. Organic Carbon in Soils of the World. Soil Sci. Soc. Am. J. **1993**, *57*, 194.
13. Couto, W.; Sanzonowicz, C.; De, O.; Barcellos, A. Factors Affecting Oxidation-Reduction Processes in an Oxisol with a Seasonal Water Table. Soil Sci. Soc. Am. J. **1985**, *49*, 1245–1248.
14. EMBRAPA. Mapa de Solos do Brasil. (Escala 1:5,000,000); Serviço Nacional de Levantamento e Conservação de Solos: EMBRAPA: Geocarta, SA, 1981.
15. Lopes, A.S.; Cox, F.R. A Survey of the Fertility Status of Surface Soils Under Cerrado Vegetation in Brazil. Soil Sci. Soc. Am. J. **1977**, *41*, 742–747.
16. Lopes, A.S.; Cox, F.R. Cerrado Vegetation in Brazil: An Edaphic Gradient. Agron. J. **1977**, *69*, 828–831.
17. Lopes, A.S. Soils Under Cerrado: A Success Story in Soil Management. Better Crops International **1996**, *10* (2), 9–15.
18. Ritchey, K.D.; Sousa, D.M.G.; Lobato, E.; Sousa, O.C. Calcium Leaching to Increase Rooting Depth in a Brazilian Savannah Oxisol. Agron. J. **1980**, *32*, 40–44.

PARTICLE DENSITY

Keith R.J. Smettem
The University of Western Australia, Nedlands, Western Australia, Australia

INTRODUCTION

Density is a standard physical term defined as weight (mass) per unit volume. In soils, the density of the solid particles in a particular sample is referred to as the particle density. SI units are $kg\,m^{-3}$, but the fractional units of $g\,m^{-3}$ are also used ($1\,g\,cm^{-3} = 1\,Mg\,m^{-3}$). In most soils the average density of particles, ρ_s, is in the range $2.6–2.7\,Mg\,m^{-3}$. This narrow range reflects the predominance of quartz and clay minerals in the soil matrix. An average particle density of $2.65\,Mg\,m^{-3}$ (the density of quartz) is often applied to soils comprised principally of silicate materials. Exceptions can occur if the soil is high in organic matter, which reduces the soil particle density. Humus, for example, has a density that is usually less than $1.5\,Mg\,m^{-3}$. Conversely, soils that are rich in iron can have high particle densities. Ferromagnesian minerals, for example, have densities ranging from 2.9 to $3.5\,Mg\,m^{-3}$. Density of iron oxides and other so-called heavy minerals can exceed $4\,Mg\,m^{-3}$.

Oven dry soil is essentially a two-phase porous medium comprising air and solids. The porosity can be determined if the bulk density, ρ_b, (the ratio of the mass of dry solids to the bulk volume of the soil) and the particle density are known. The porosity, ε, is then given by

$$\varepsilon = 1 - \rho_b/\rho_s$$

The particle density is also necessary for calculation of the sedimentation parameter required for estimation of particle size by sedimentation methods (5, 6). It is also used to determine the particle settling velocity in studies of sediment deposition from flowing water (7).

MEASUREMENT OF PARTICLE DENSITY

Widely accepted standard methods are available for measuring the weighted mean particle density of a sample (2) and are based on measuring the mass and volume. The mass is determined by weighing and the volume is then calculated from the mass and density of a fluid displaced by the sample. The two most common methods of measuring particle density are the pycnometer method (1) and the submersion method (3).

A pycnometer is a specific gravity flask that is sometimes fitted with a thermometer. For large samples, a volumetric flask may also be used with sufficient measurement accuracy for most applications. In brief, the method involves careful weighing to obtain the particle density from

$$\rho_s = \rho_w(W_s - W_a)/[W_{sw} - (W_s - W_a)]$$

where ρ_w is the density of water at the observed temperature, W_s is the pycnometer weight containing a soil sample at oven dry water content, W_a is the empty (air filled) pycnometer weight, W_{sw} is the pycnometer weight when filled with soil and water, and W_w is the pycnometer weight when filled with water at the observed temperature. Note that the denominator represents the volume of water displaced by the soil particles.

The soil in the pycnometer is covered with a layer of water and evacuated to ensure complete saturation of the soil before topping up the pycnometer. Note that de-aired distilled water is used throughout the procedure and organic matter is generally removed prior to use.

An air pycnometer can be used as an alternative method, particularly if it is important to retain organic matter in the sample. The method uses gas rather than water as the displacing fluid and the ideal gas law to calculate the volume of solids. Care must be taken to ensure that the procedure is performed at a constant temperature.

The submersion method involves measuring the volumetric displacement of water or a nonpolar liquid by a soil sample of known air-dry weight. The samples of air dried soil, usually about 25 g in weight, are formed into spaghetti-like threads by moistening to a plastic consistency and forcing through a 2-mm sieve. After oven drying at 105°C and cooling in a dessicator, the prepared sample is placed in a weighing dish attached to a weighing beam by a thin wire and both the dish and the sample are immersed in a container of liquid. The particle density then follows from the equation:

$$\rho_s = \rho_l(W_{sd} - W_d)/[(W_{sd} - W_d) - (W_{sdl} - W_{dl})]$$

where ρ_l is the density of water or organic liquid used, W_{sd} is the oven dried weight of both the soil and the weighing dish, W_d is the weight of the weighing dish, W_{sdl} is the weight of sample and weighing dish submerged in liquid and W_{dl} is the weight of the submerged dish without the soil sample.

The submersion method sacrifices some precision compared to the pycnometer method, but is easy to perform and has the advantage of being less laborious. The method cannot, however, be used with sandy soils because coherence is usually too low to allow the spaghetti-like sample threads to be prepared.

The particle density of finely divided active soil measured in water is usually greater than that measured in a nonpolar liquid (2). As the proportion of minerals with expanding lattices increases in a given sample, so too does the specific gravity (8). Water density is also known to be affected by the surfaces of finely divided particles, so a more accurate measure of particle density would be obtained using a nonpolar inorganic liquid in a pycnometer (2). However, if the particle density information is required in order to determine the volume of solids in a soil in contact with water, then it is logical to use water as the measurement liquid.

If the soil is high in organic matter then it is easier to achieve wetting using a nonpolar organic liquid. Organic particles also have a tendency to be buoyant in water whereas they sediment more readily in nonpolar liquids. This is obviously an advantage when attempting to measure the weight of a submerged sample.

The density of a suspension can also be measured very precisely using a vibrating tube (4) which is filled with a solution or suspension. The resonant frequency of the vibrating tube is directly related to the density of the suspension.

REFERENCES

1. American Society for Testing and Materials, *Am. Soc. Test. Mater. Procedures Testing Soils*; American Society for Testing and Materials: Philadelphia, 1958.
2. Blake, G.R.; Hartge, K.H. Particle Density. In *Methods of Soil Analysis 1: Physical and Mineralogical Methods*, 2nd Ed.; Klute, A., Ed.; American Society of Agronomy: Madison, WI, 1986; 377–382.
3. Capek, M.;1933; Cited by DiGleria, J.A.; Klimes-Szmik, A.; Dvoracsek, M.; Bodenphysik und Bodenkolloidik, 1962; German Edition Jointly by Akademiai Kiado, Budapest, and VEB Gustav Fischer Verlag, Jena.
4. Elder, J.G. Density Measurement by the Mechanical Oscillator. *Methods in Enzymology*; Academic Press: NY, 1979; 12–25.
5. Gee, G.W.; Bauder, J.W. Particle Size Analysis by Hydrometer: A Simplified Method for Routine Textural Analysis and a Sensitivity Test of Measurement Parameters. Soil Sci. Soc. Am. J. **1979**, *43*, 1004–1007.
6. Gee, G.W.; Bauder, J.W. Particle-Size Analysis. In *Methods of Soil Analysis 1: Physical and Mineralogical Methods*, 2nd Ed.; Klute, A., Ed.; American Society of Agronomy: Madison, WI, 1986; 383–411.
7. Julien, P.Y.; Simons, D.B. Sediment Transport Capacity of Overland Flow. Trans. Am. Soc. Agric. Eng. **1985**, *28*, 755–761.
8. Gradwell, M.W. The Determination of Specific Gravities of Soils as Influenced by Clay-Mineral Composition. N. Z. J. Sci. Technol. **1955**, *37B*, 283–289.

P

PARTICLE PACKING

S.C. Gupta
D.P. Thoma
University of Minnesota, St. Paul, Minnesota, U.S.A.

INTRODUCTION

Packing refers to the arrangement of solids in a given volume. Packing is important in many different fields, including aeronautics, agriculture, ceramics, chemical engineering, chemistry, foundation engineering, electrical engineering, foods, geology, mechanical engineering, medicine, pharmaceuticals, and polymer science (1). A few specific applications include packing of solid fuel for rockets and missiles, packing and storage of seeds, strength of powder compacts, packing of fluidized beds, packing of soil and roadway foundations, packing of pharmaceutical pills, and strengths of polymer structure. The underlying premise of packing is to identify the most efficient way of packing solids (whether it be particles, aggregates, powder, solid fuel, fruits, etc.) such that it provides maximum strength, uses minimum space, maximizes the efficiency of chemical reaction, and/or minimizes the shipping costs. In soils, geology, and hydrology, packing is important in understanding and quantifying the pathways of water and chemical transport and the sites of their retention.

INDICES OF PACKING

There are several indices that can be used to characterize particle packing. These include bulk density (ρ_b), void ratio (e), and porosity (f):

$$\rho_b = \frac{\text{mass of particles}}{\text{bulk volume}} \tag{1}$$

$$e = \frac{\text{volume of voids}}{\text{volume of particles}} \tag{2}$$

$$f = \frac{\text{volume of voids}}{\text{bulk volume}} \tag{3}$$

The above three indices are related to one another through their relationship with particle density (ρ_p).

$$f = \left(1 - \frac{\rho_b}{\rho_p}\right) = \frac{e}{1+e} \tag{4}$$

$$\rho_p = \frac{\text{mass of particles}}{\text{volume of particles}} \tag{5}$$

In some fields, packing density has also been expressed as fractional density—defined as a fraction of the theoretical density of the material. Another index that characterizes a packing arrangement is the coordination number, the number of neighboring particles touching a given particle.

TYPES OF PACKING

Particle packing can be geometrically systematic (regular) or random (disordered). There are five different ways of systematically packing spherical particles of one size (Fig. 1). These are cubic, cubical-tetrahedral, tetragonal-sphenoidal, pyramidal, and tetrahedral (2). Over the years, different terminology has been used to describe these packing arrangements. Graton and Fraser (3) defined these packing arrangements as cubical, orthorhombic, rhombohedral, tetragonal-sphenoidal, and rhombohedral, respectively. White and Walton (4) defined them as cubical, single stagger, double stagger, pyramidal, and tetrahedral, respectively.

In cubical packing, spheres are arranged at 90° angles (Fig. 2). Each sphere touches four spheres in the same layer and one sphere in the layer above and one sphere in the layer below. Therefore, the coordination number for the cubical packing is 6 (Table 1). For other packing arrangements, the coordination number varies from 8 to 12 (Table 1).

PACKING DENSITY

The packing density or porosity of a given packing arrangement can be calculated by knowing the geometry of packing. For example, in cubic packing, each sphere

Encyclopedia of Soil Science

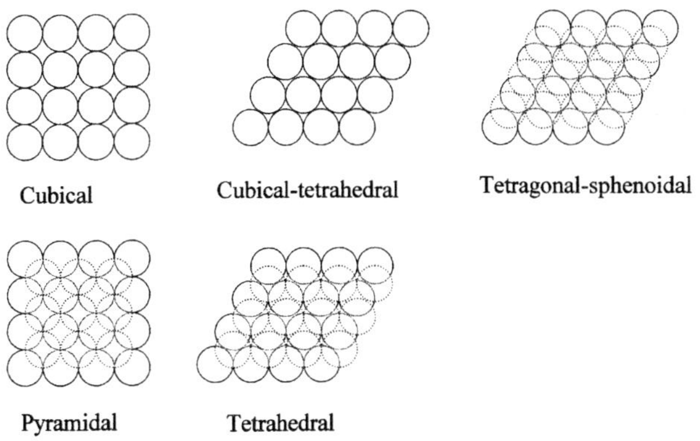

Fig. 1 The full circles represent spheres in one plane while the dotted circles represent the spheres in the layer above.

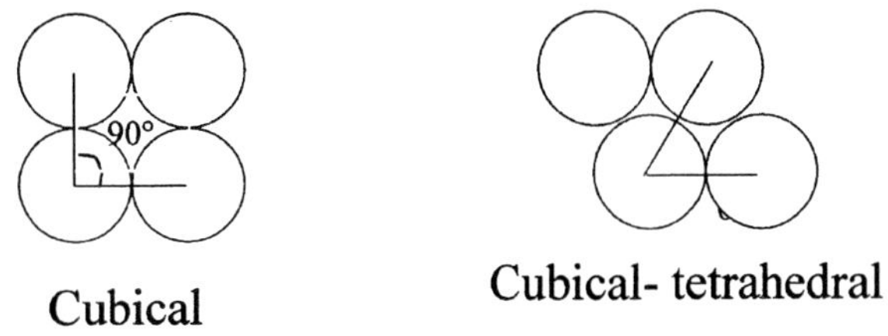

Fig. 2 Angle at which particles are aligned in cubical and cubical–tetrahedral packing.

(with diameter of $2r$) is surrounded by an imaginary cube (unit prism) with a unit volume equal to $8r^3$. Since the volume of the sphere is $(4/3)\pi r^3$, the porosity and void ratio of the cubical packing are:

$$f = \frac{8r^3 - (4/3)\pi r^3}{8r^3} = \left(1 - \frac{\pi}{6}\right) = 0.4764 \qquad (6)$$

$$e = \frac{8r^3 - (4/3)\pi r^3}{4/3(\pi)r^3} = \left(\frac{\pi}{6} - 1\right) = 0.9099 \qquad (7)$$

Similar calculations of porosity for other types of packing arrangements have been made by White and Walton (4) and are listed in Table 1.

Porosities in Table 1 are independent of sphere radius. Cubical is the loosest and pyramidal and tetrahedral and the densest packing of any spherical particles of a given size. Additional reduction in the above porosities can occur by filling the irregular void spaces between the primary particles with smaller secondary, tertiary, quaternary, and quinary spheres. White and Walton (4) showed that the porosity of tetrahedral packing can be reduced to as low as 3.9% (Table 2) through the introduction of smaller particles. Similar reductions in porosity are also possible for other packing arrangements.

MODELS OF PARTICLE PACKING

Over the years, attempts have been made to predict the packing density of a mixture of particles varying in diameter. Westman and Hugill (5) and later Bodman and Constantin (6) and Staples (7) showed that one could approximate the packing density of a mixture of two or more particle fractions using an analytical expression based on the packing density of individual fractions. The assumption underlying this method is that if larger particles are dominant, then the contribution of the finer particles to bulk volume is minimal. However, if the finer particles are dominant then the bulk volume of the mixture is equal to the bulk volume of finer particles plus the particle volume of larger particles. Extrapolating these two assumptions to various proportions of coarse and finer particles then identifies an optimum mixture of coarse and fine particles that can result in maximum density.

Fig. 3 illustrates the basis of the analytical model of Ref. 5. The abscissa refers to the volume fraction of coarse (X_1) and fine (X_2) particles whereas the ordinate refers to particle volume, void volume, and bulk volume of coarse and fine fractions. Points C and F refer to the bulk volume

Table 1 Various types of systematic packing and the corresponding coordination number, radius of the largest cavity, volume of unit prism, porosity, and void ratio

Type of packing	Coordination number	Radius of largest cavity	Volume of unit prism	Porosity	Void ratio
Cubical	6	$0.732r$[a]	$8r^3$	0.4764	0.9099
Cubical-tetrahedral	8	$0.528r$	$4r^3 \times 3$	0.3955	0.6540
Tetragonal-sphenoidal	10	$0.285r$	$6r^3$	0.3020	0.4324
Pyramidal	12	$0.414r$	$4r^3 \times 2$	0.2595	0.3504
Tetrahedral	12	$0.414r$	$4r^3 \times 2$	0.2595	0.3504

[a] r is the radius of the packing particles.

Table 2 Radius of secondary (B), tertiary (C), quaternary (D), and quinary (E) particles that can fit in the irregular spaces between the primary (A) particles in tetrahedral packing and the corresponding porosities

	A	B	C	D	E	Filler
Radius of sphere	r	$0.414r$	$0.225r$	$0.177r$	$0.116r$	Very small
Relative number of spheres	1	1	2	8	8	
Volume of a sphere	$4.189r^3$	$0.298r^3$	$0.0476r^3$	$0.0225r^3$	$0.0066r^3$	
Total solid volume of spheres	$4.189r^3$	$4.487r^3$	$4.582r^3$	$4.762r^3$	$4.815r^3$	$5.437r^3$
Porosity, %	25.95	20.7	19.0	15.8	14.9	3.9

(From Ref. 4.)

of coarse and fine fractions when packed individually. If these fractions are stacked one above the other without mixing then the bulk volume of the mixture is given by the line CF. Lines CE and HE refer to the particle and bulk volumes of the coarse fraction, respectively. Lines FD and GD refer to particle and bulk volumes of the fine fraction, respectively. If the coarse fraction dominates the mixture then the fine particles will be enclosed in the voids of the coarse fraction. The bulk volume describing this relationship is line CE. If on the other hand, the fine fraction is dominant then the addition of any coarse particles adds to the bulk volume the equivalent of the particle volume of the coarse fraction. The line FH then describes the bulk volume of the mixture. The intersection of lines CE and FH defines the volume fraction of coarse and fine particles with the minimum bulk volume or the maximum packing density.

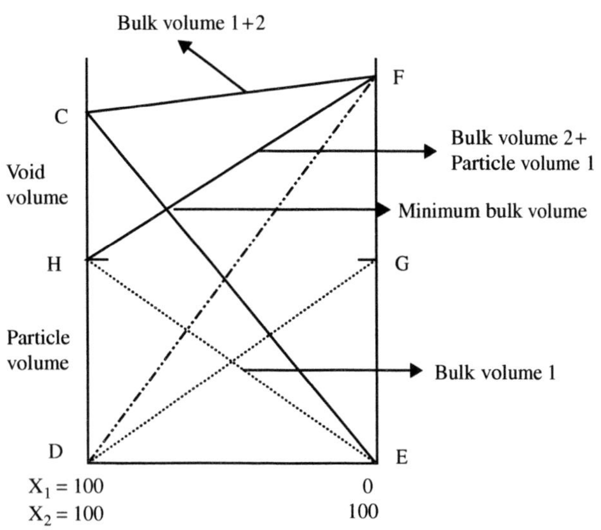

Fig. 3 Relationship between particle volume, void volume, and bulk volume of coarse and fine fractions in the packing model of Westman and Hugill. (Adapted from Ref. 5.) The intersection of lines CE and HF represents the minimal volume or maximum bulk density of a mixture.

In Fig. 3 we assumed that the particle density of both fractions was the same. If there is a difference in particle density of various fractions then the ordinate can be expressed in normalized volume by dividing the bulk volume by the particle volume. Equations of the two lines (CE and FH), along with the assumption that the sum of the volume fraction of the two components is 1.0, help define the minimum volume or the maximum density (5).

Gupta and Larson (8) presented a computer model that simulated maximum, random, and minimum bulk densities of a soil based on the particle size distribution and the packing density of individual fractions. The packing model is based on the concept that certain particles (donor) will be enclosed in the voids of the packing assemblages of other particles (acceptor) of larger radii. The decision to designate which particle fraction acts as the acceptor and which acts as the donor depends upon whether the maximum density or random density are to be predicted. The model uses the cavity radius of stable packing arrangements (tetragonal or pyramidal-tetrahedral) as a guide in deciding when donor fractions can be fitted into the voids of the acceptor fractions.

Maximum density refers to the conditions when all the voids of the coarse fractions are systematically filled with finer fractions. Random density refers to conditions when particle fractions are randomly selected as acceptor or donor particles. Minimum density refers to conditions when all the fractions are acceptor particles and there is no mixing between the fractions.

Gupta and Larson (8) tested the predictions of random packing from the packing model against the measured values for 22 mixtures of various size glass beads (7) and 43 soils and dredged sediment mixtures. In many cases, the predicted bulk densities were close to the measured values. Spivey et al. (9) showed that artificially compacted (at 2 MPa) bulk densities of coastal plain soils in the southeastern United States compared well with random bulk densities predicted from the packing model. Reinsch and Grossman (10) showed that mean bulk densities estimated from the packing model were an acceptable approximation of oven-dry bulk densities after inundation.

Gupta and Larson (11) also extended the packing model to predict the bulk density of artificially prepared aggregated soils consisting of aggregates that varied in diameter from 0.053 to 50.8 mm. For aggregate size distributions varying in geometric mean diameter from 0.52 to 9.06 mm, the random packing density of aggregated soils was within $0.17\,Mg\,m^{-3}$ of the measured values.

With the availability of high-speed computing, there have been several advances in modeling particle packing. These advances have been summarized in a review article by Davis (12).

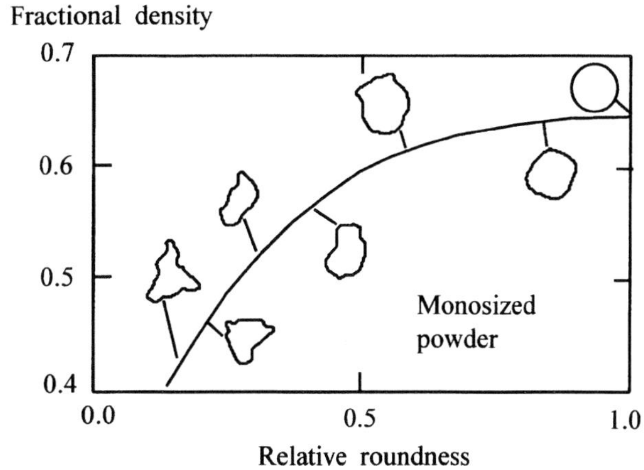

Fig. 5 Variation in fractional density versus roundness of randomly packed mono-sized particles. (Adapted from Ref. 1.)

PARTICLE SHAPE EFFECTS

Concepts of particle packing have generally been developed assuming the particles to be spherical. Next to the particle size distribution, the shape of the particles has a major influence on the packing density of granular materials. Particle shape can be described by parameters such as sphericity and roundness (13, 14). Sphericity refers to the degree to which particles approach spherical shape, whereas roundness describes the angularity of particle corners. Wadell (15) defined the sphericity index as the ratio of the surface area of a sphere (with the same volume as the test particle) to the surface area of the test particle. Riley (16) defined the sphericity as the ratio of the diameter of the smallest circumscribing circle to the largest inscribed circle. The closer the particles are to a perfect sphere, the larger the sphericity index. German (1) calculated the sphericity index of various shapes using Wadell's procedure (15, 17). The sphericity index varied from 0.87 for a cylinder of 1:1 diameter to length, 0.83 for a cylinder of 1:2 diameter to length, 0.47 for a disk of 5:1 diameter to thickness, 0.81 for a cube, and 0.76 for a rectangle of 2:2:1 length, width, and height ratio. German (1) showed that for monosized particles, the particle packing increases with an increase in the sphericity of the particles (Fig. 4).

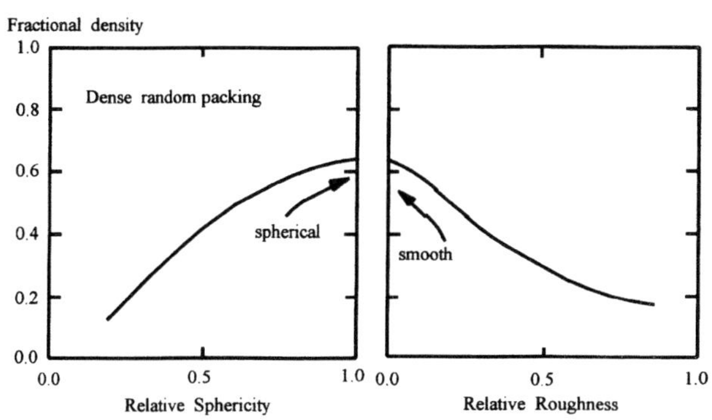

Fig. 4 Effect of particle shape and smoothness on random fractional packing density. (Adapted from Ref. 1.)

A common method of assessing particle roundness is to compare images of individual particles to standard charts showing particles with varying roundness (13, 14). German (1) showed that packing density increases with an increase in roundness of randomly packed monosized particles (Fig. 5).

Another particle shape parameter affecting packing density is the surface roughness. This parameter refers to interparticle friction due to irregularities on the particle surface (1). German (1) showed that the greater the surface roughness the lower the particle density of monosized particles (Fig. 4).

REFERENCES

1. German, R.M. *Particle Packing Characteristics*; Metal Powder Industries Federation: Princeton, NJ, 1989; 443.
2. Gray, W.A. *The Packing of Solid Particles*; Chapman and Hall: London, 1968.
3. Graton, L.C.; Fraser, H.J. Systematic Packing of Spheres with Particular Relation to Porosity and Permeability. J. Geol. **1935**, *43*, 785–909.
4. White, H.E.; Walton, S.F. Particle Packing and Particle Shape. J. Am. Ceram. Soc. **1937**, *20*, 155–166.
5. Westman, A.E.R.; Hugill, H.R. The Packing of Particles. J. Am. Ceram. Soc. **1930**, *13*, 767–779.

P

6. Bodman, G.B.; Constantin, G.K. Influence of Particle Size Distribution on Soil Compaction. Hilgardia **1965**, *36*, 567–591.
7. Staples, W.J. The Influence of Size Distribution on the Bulk Density of Uniformly Packed Glass Particles. Soil Sci. Soc. Am. Proc. **1975**, *39*, 404–408.
8. Gupta, S.C.; Larson, W.E. A Model for Predicting Packing Density of Soils Using Particle Size Distribution. Soil Sci. Soc. Am. J. **1979**, *43*, 758–764.
9. Spivey, L.D.; Busscher, W.J.; Campbell, R.B. The Effect of Texture on Strength of Southeastern Coastal Plain Soils. Soil Till. Res. **1986**, *6*, 351–363.
10. Reinsh, T.G.; Grossman, R.B. A Method to Predict Bulk Density of Tilled Ap Horizons. Soil Till. Res. **1995**, *34*, 95–104.
11. Gupta, S.C.; Larson, W.E. Modeling Soil Mechanical Behavior During Tillage. In *Predicting Tillage Effects on Soil Physical Properties and Processes*; Unger, P.W., Van Doren, D.M., Eds.; Special Pub. 44 ASA, SSSA: Madison, WI, 1982.
12. Lee, Davis I. Particle Pack Influence on Highly Filled Material Properties. Curr. Opin. Solid Stat. Mater. Sci. **1999**, *4*, 505–513.
13. Krumbein, W.C. Measurement and Geological Significance of Shape and Roundness of Sedimentary Particles. J. Sediment. Petrol. **1941**, *11*, 64–72.
14. Griffiths, J.C. *Scientific Methods in Analysis of Sediments*; McGraw-Hill: New York, 1967.
15. Wadell, H. Volume, Shape and Roundness of Rock Particles. J. Geol. **1932**, *40*, 443–451.
16. Riley, N.A. Projection Sphericity. J. Sediment. Petrol. **1941**, *11*, 94–97.
17. Wadell, H. Sphericity and Roundness of Rock Particles. J. Geol. **1933**, *41*, 310–311.

PARTICLE SHAPE

P

Brian M. Schafer
The University of Queensland, Gatton, Queensland, Australia

INTRODUCTION

Soil particle shape refers to the morphology of grains of essentially individual or composite mineral grains comprising the fine earth fraction of soil. The term grain has been used in geological literature to describe minerals ranging in size from 2 mm to submicroscopic dimensions whilst mineral grains larger than 2 mm are classed as phenocrysts (1). In soils, particulate material is generally described as particles smaller than 2 mm average diameter being measured as the various ratios of a particle's long, intermediate, and short axes (2). They are categorized into size classes that in turn are used to determine texture grades in soils. Generally the particles are in tangential contact with each other rather than in continuous contact with their neighbors as found in igneous rocks and collectively determine soil fabric classes by virtue of their packing arrangement.

The use of size and shape analysis in soil science has been extended from sedimentary petrology origins to express the frequency distribution of size and shape ranges of particulate matter (3). They form an integral part of both structure and mineral analysis used to describe quantitatively and interpret the origin and processes of formation of soil material. Cementing materials such as carbonates and iron oxides are excluded except where they are formed separately from the soil matrix as secondary formations. Thus most work in soil particle shape has been confined to primary minerals that are residual or transported from other sources.

Secondary clay minerals also play a significant role in structure and mineral analysis. However clay particle shape has not been subjected to the same analytical treatment as the coarser fractions.

Particle shape is related to size in that the size is reflected as volume expressed as true nominal diameter whereby the particles form an open three-dimensional network. This implies that the particle volume has a porosity component determined by a packing arrangement of different diameter particles but invariably leaving a residual pore space which can be filled with cementing agents, fluids, or air.

METHODS OF MEASUREMENT

Shape is based on the concept of sphericity and roundness (4–6). Sphericity is a description of the overall form of the particle irrespective of angularity of edges and corners whereas roundness refers only to sharpness (3). Both of these properties have been defined mathematically (7, 8).

Sphericity and roundness assessments are commonly made visually on individual grains and are tedious and imprecise due to the partly altered nature of minerals that makes it difficult to positively identify their origin in source material. They require the skills of optical mineralogy techniques and an appreciation of sedimentary petrology.

Particle shape analysis is usually performed on samples obtained from particle size analysis. This latter method results in separation of particles into grade scale classes based on a geometric scale. The broadness of the individual classes in the International grade scale tends to hide the continuum of sizes meaning that the nature of the size distribution cannot be determined from the grade scale classes. Consequently shape analysis conducted on each size grade scale class also cannot be expressed as a function of the whole and therefore measurement is semi-quantitative.

A number of grade scales have been proposed for shape analysis. Krumbein (8) used a sphericity classification proposed by Zingg (7) (Table 1) and Brewer (3) suggested the addition of planar, acicular, and acicular–planar shape classes to this classification. Roundness grade scales have been proposed (9–11) and are listed in Table 2. The scales are arbitrary divisions of a geometric scale of sizes or shapes (sphericity) so that each unit or grade serves as a class interval for analytical purposes. The units are mathematical derivations of particle diameter (millimeters) and phi scale (3) and alternatively a rho scale (11). Combining sphericity and roundness scales have not been attempted but estimation of the two parameters is used in defining particle shape.

PEDOGENESIS

Sphericity and roundness are properties of particulate matter that have been used to interpret the abrasion effect

Table 1 Shape classes. *a*, *b*, *c* = crystallographic axes

Class I	Disc	$b/a > 2/3$	$c/b < 2/3$
Class II	Spheroid	$b/a > 2/3$	$c/b > 2/3$
Class III	Blade	$b/a < 2/3$	$c/b < 2/3$
Class IV	Rod	$b/a < 2/3$	$c/b > 2/3$

(Adapted from Refs. 3, 7.)

of transport processes on specific detritus resulting from predominant physical comminution of source rocks and sedimentary bodies. The more spherical or rounder the particles, the greater was the effect of abrasion and hence weathering and transport.

These processes are common in diagenesis and weathering processes associated with sedimentary rocks and sedimentary bodies (alluvium, colluvium) and are integral in the rock cycle of igneous, sedimentary, and metamorphic rocks.

Table 2 Roundness grades

	Class limits		
Grade term	**Ref. (9)**	**Ref. (10)**	**Ref. (11) Rho Scale**[a]
Very angular		0.12–0.17	0.00–1.00
Angular	0–0.15	0.17–0.25	1.00–2.00
Subangular	0.15–0.25	0.25–0.35	2.00–3.00
Subrounded	0.25–0.40	0.35–0.49	3.00–4.00
Rounded	0.40–0.60	0.49–0.70	4.00–5.00
Well-rounded	0.60–1.00	0.70–1.00	5.00–6.00

[a] Based on Ref. 10 class limits.
(Adapted from Ref. 3.)

In soils, shape analysis is used to determine uniformity of parent material, origin and mode of soil formation and in mineral and structure analysis.

Soil forming processes at or near the Earth's surface can then be invoked to link parent material origin and genesis of soil materials and soil profiles.

REFERENCES

1. American Geological Institute, *Dictionary of Geological Terms*, 2nd Ed.; Dolphin Reference Books: USA, 1962.
2. Pettijohn, F.J.; Potter, P.E.; Siever, R. *Sand and Sandstone*; Springer: New York, 1987.
3. Brewer, R. *Fabric and Mineral Analysis of Soils*; Wiley: USA, 1964.
4. Wadell, H. Volume, Shape, and Roundness of Rock Particles. J. Geol. **1932**, *40*, 443–451.
5. Wadell, H. Sphericity and Roundness of Rock Particles. J. Geol. **1933**, *41*, 310–331.
6. Wadell, H. Volume, Shape and Roundness of Quartz Particles. J. Geol. **1935**, *43*, 250–279.
7. Zingg, Th. Beitragzur Schotteranalyse. Schweiz. Mineral. Petrog. Mitt. **1935**, *15*, 39–140.
8. Krumbein, W.C. Measurement and Geological Significance of Shape and Roundness of Sedimentary Particles. J. Sediment. Petrol. **1941**, *11*, 64–72.
9. Pettijohn, F.H. *Sedimentary Rocks*, 2nd Ed.; Harper Brothers: New York, 1957.
10. Powers, M. A New Roundness Scale for Sedimentary Particles. J. Sediment. Petrol. **1953**, *23*, 117–119.
11. Folk, R.L. Student Operated Error in Determination of Roundness, Sphericity and Grain Size. J. Sed. Petrol. **1955**, *25*, 297–301.

PEDOGENIC SILICA ACCUMULATION

Katherine J. Kendrick
U.S. Geological Survey, Pasadena, California, U.S.A.

P

INTRODUCTION

Pedogenic silica is defined as silica that has precipitated in the soil environment, taking the form of opal-A, opal-CT, or microcrystalline quartz. Opaline silica, which is amorphous to poorly crystalline, is the most common form of pedogenic silica (1, 2). Pedogenic silica is formed in arid, semiarid, and Mediterranean climates where there is adequate precipitation to mobilize the silica, but not so much as to leach the silica out of the profile. Pedogenic silica accumulation has been reported throughout the western U.S., Australia, Italy, South Africa, and New Zealand (3). In its most advanced form, it cements the soil fabric to form extremely hard horizons known as duripans or silcrete.

SOURCES OF PEDOGENIC SILICA

Pedogenic silica is derived from two main sources. The weathering and hydrolysis of primary silicate minerals provide for the slow release of silica into solution as silicic acid. Easily weathered silicates, such as olivine and Ca-plagioclase, are particularly likely to contribute silica into solution. The second source is from the weathering of primary amorphous silica, including volcanic glass and biogenic silica (e.g., phytoliths, diatoms). Such amorphous silica is the most soluble form of silica found in the soil environment. Weathering of glassy volcanic products, including ash and tephra, to release silicic acid into solution occurs readily in soils.

DISSOLUTION OF SILICA

Factors controlling the dissolution of silica include soil solution pH, the presence of organic matter, the particle size of the material, and the presence of coatings on grains. The solubility of silica is moderate throughout the pH range of most soils, but is particularly high at pH values ≥ 9, as found in sodium carbonate soil systems. Organic matter enhances the dissolution of silica, including quartz, and impedes precipitation (3). The dissolution of quartz has been shown to be highest in the root zone due to organic complexation of monosilicic acid (4). On the other hand, organic matter may actually form a coating on opal, impeding its dissolution (5). Because of surface area effects, smaller particles are more susceptible to dissolution. This phenomenon accounts for the fact that quartz does not persist in the clay fraction of most soils (6). Aluminum- and Fe-oxyhydroxides can form insoluble coatings on grains, thereby decreasing rates of dissolution from those grains (3).

MOVEMENT AND PRECIPITATION OF SILICA

Silica is moved in solution through the soil as silicic acid. Incomplete leaching allows for the precipitation of this silicic acid in the pedogenic zone. During drying induced by evapotranspiration, SiO_4 is adsorbed onto surfaces, forming amorphous opaline silica. This process led to the silica cementation of south-facing terrace edges on the central California coast (7). Alternatively, in situ alteration of eolian dust to opal-A, rather than precipitation from silica in soil solution, has been proposed for duripans in Idaho (8). The eolian dust comprised volcanic glass and feldspars.

Conditions favoring the precipitation of silica from solution include pH values less than 7, a high available surface area within the soil fabric, and high ionic strength of the soil solution. Although a high available surface area favors precipitation, silica cementation is most common in medium-textured parent materials with abundant skeletal grains. The grain-to-grain cementation is an important component in the formation of a duripan (6).

Aluminum- and Fe-oxyhydroxides specifically adsorb soluble silica. Since these compounds are nearly ubiquitous in soils, often as coatings on other grains, they form a significant sink for the precipitation of silica on surfaces (3).

SILICA MORPHOLOGY

Stages of silica cementation have been defined for coarse grained deposits in arid climatic regimes (9). The first stage is precipitation of silica on the undersides of clasts. These are termed *pendants*, and are equivalent to *opal beards* in Australian soil descriptions (10). Dramatic examples of opal pendants have been described in some central Californian soils (11). The second stage is precipitation within the matrix, where silica forms bridging contacts between grains. The third stage is silica precipitation on all sides of the clasts, and the final stage is a laminar cap above a plugged matrix. Apart from silica precipitation on gravels, silica can cement the finer matrix materials into nodules termed *durinodes*.

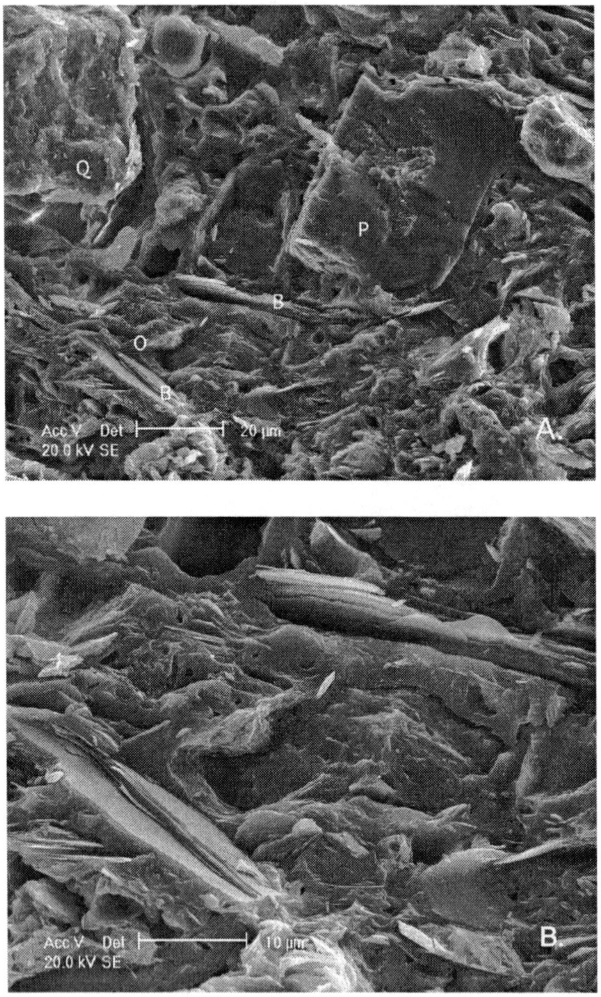

Fig. 1 Scanning electron micrograph of soil fabric in a duripan from southern California. (A) Primary mineral grains (P: plagioclase; B: biotite; Q:, quartz) cemented by opaline silica (O: opaline silica). (B) Close-up view of the lower left portion of Fig. 1a, showing biotite grains embedded in opaline silica. (Photo by Dr Krassimir Bozhilov, Central Facility for Advanced Microscopy and Microanalysis, UC Riverside.)

Duripans are defined as soil horizons that are irreversibly cemented by various forms of silica such that they do not slake after soaking in water or hydrochloric acid (12). Duripans take two general forms. In Mediterranean climates, duripans often have prismatic structure and pedogenic calcite is minimal or absent (1). Small opal flocs within the matrix of the prisms are the precursors to durinodes. Further development of the duripan yields extensive grain-to-grain cementation in the soil matrix (Fig. 1) and silica coatings on the tops and upper sides of the prisms. Ultimately, silica laminae with platy structure cap the prismatic horizon (1). This is in contrast to duripans that form in arid environments, which typically have a large component of pedogenic calcite and are characterized by a platy structure throughout, with plates 1–15 cm thick (13). As little as 10% Si as opaline silica is adequate to cement horizons effectively, and form a fully developed duripan (1).

Thorough cementation of a soil horizon to form a duripan requires geomorphic stability of long duration, particularly in the absence of readily soluble volcanic ash. The ancient landscapes of Australia fit this criterion. Silica cemented horizons in Australia are common and are referred to as silcrete, duricrust, and grey billy (10).

Silica cementation has also been reported in *hardsetting* soils. Hardsetting soils have one or more horizons with hard to very hard consistence when dry. This phenomenon is recognized in soils with alternating seasons of wetting and drying, primarily those in Australia. Hardsetting soils eventually slake on wetting, and are thus not irreversibly cemented (14). Amorphous particles and coatings of silica have been documented in these soils (14), suggesting that silica is an important constituent in the process of seasonal cementation (15). The easily mobilized silica might be ephemeral, and may or may not be related to incipient formation of a duripan.

REFERENCES

1. Flach, K.W.; Nettleton, W.D.; Gile, L.H.; Cady, J.G. Pedocementation: Induration by Silica, Carbonates, and Sesquioxides in the Quaternary. Soil Sci. **1969**, *107*, 442–453.
2. Chadwick, O.A.; Hendricks, D.M.; Nettleton, W.D. Silica in Duric Soils: 1. A Depositional Model. Soil Sci. Soc. Am. J. **1987**, *51*, 975–982.
3. Drees, L.R.; Wilding, L.P.; Smeck, N.E.; Senkayi, A.L. Silica to Soils: Quartz and Disordered Silica Polymorphs. In *Minerals in Soil Environments*, 2nd Ed.; Dixon, J.B., Weed, S.B., Eds. Soil Sci. Soc. Am. Madison, WI, 1989; 913–974.
4. Cleary, W.J.; Conolly, J.R. Embayed Quartz Grains in Soils and Their Significance. J. Sediment. Petrol. **1972**, *42*, 899–904.

5. Wilding, L.P.; Drees, L.R. Contributions of Forest Opal and Associated Crystalline Phases to Fine Silt and Clay Fractions of Soils. Clays Clay Min. **1974**, *22*, 295–306.
6. Moody, L.E.; Graham, R.C. Silica-Cemented Terrace Edges, Central California Coast. Soil Sci. Soc. Am. J. **1997**, *61*, 1723–1729.
7. Blank, R.R.; Fosberg, M.A. Duripans in Idaho, U.S.A. In Situ Alteration of Eolian Dust (loess) to an Opal-A/ X-Ray Amorphous Phase. Geoderma **1991**, *48*, 131–149.
8. Norton, L.D. Micromorphology of Silica Cementation in Soils. In *Soil Micromorphology: Studies in Management and Genesis*; A.J. Ringrose-Voase and G.S. Humphreys, Eds. Proc. IX Int. Working Meeting on Soil Micromorphology, Townsville, Australia, July 1992, Dev. Soil Sci. **1994**, *22*, 811–824.
9. Harden, J.W., Taylor, E.M., Reheis, M.C., McFadden, L.D. Calcic, Gypsic and Siliceous Soil Chronosequences in Arid and Semiarid Environments. In *Occurrence, Characteristics, and Genesis of Carbonate, Gypsum, and Silica Accumulations in Soils*; Nettleton, W.D., Ed. Soil Sci. Soc. of Am. Spec. Publ. Madison, WI, **1991**, *26*, 1–16.
10. Milnes, A.R., Wright, M.J., Thiry, M. Silica Accumulations in Saprolites and Soils in South Australia. In *Occurrence, Characteristics, and Genesis of Carbonate, Gypsum, and Silica Accumulations in Soils*; Nettleton, W.D., Ed. Soil Sci. Soc. of Am. Spec. Publ. Madison, WI **1991**, *26*, 121–149.
11. Munk, L.P.; Southard, R.J. Pedogenic Implications of Opaline Pendants in Some California Late-Pleistocene Palexeralfs. Soil Sci. Soc. Am. J. **1993**, *57*, 149–154.
12. Soil Survey Staff, *Soil Taxonomy*; U.S. Gov. Printing Office: Washington, DC, 1975.
13. Chadwick, O.A.; Graham, R.C. Pedogenic Processes. In *Handbook of Soil Science*; Sumner, M.E., Ed. CRC Press: Boca Raton, FL, 2000; E-41–E-75
14. Chartres, C.J.; Norton, L.D. Micromorphological and Chemical Properties of Australian Soils with Hardsetting and Duric Horizons. Dev. Soil Sci. **1994**, *22*, 825–834.
15. Chartres, C.J.; Kirby, J.M.; Raupach, M. Poorly Ordered Silica and Aluminasilicates as Temporary Cementing Agents in Hard-Setting Soils. Soil Sci. Soc. Am. J. **1990**, *54*, 1060–1067.

PEDOLOGICAL MODELING

Ronald Amundson
University of California, Berkeley, California, U.S.A.

INTRODUCTION

The fascinating impressiveness of rigorous mathematical analysis, with its atmosphere of precision and elegance, should not blind us to the defects of the premise that condition the whole process—T.C. Chamberlin, commenting on Lord Kelvin's (ultimately incorrect) calculation of the age of the Earth. (1)

In pedology, and in other sciences, *models* are increasingly used as tools for understanding natural phenomena. But what is a model, how are models used in pedology, and how are models developed and modified?

OVERVIEW

To begin, the term *model* has been used interchangeably in pedology with other concepts, sometimes leading to confusion or miscommunication. Very simply, a *model* has been described by some as "a form of highly complex scientific hypothesis" (2), that is "a simplified and idealized description or conception of a particular system, situation, or process (often in mathematical terms) that is put forward as a basis for calculations, predictions, or further investigation" (3). As briefly outlined below, the mathematical approaches to describing a phenomenon can be varied, but all must rest on a solid understanding of the soil, and the factors and processes which affect it. This empirical knowledge in turn constrains the *assumptions* which underlie any mathematical model development. A model based on an incorrect, or poorly developed, understanding of a soil and the processes that affect it will likely be unable to describe the processes of interest, but that inability may in turn inspire the modeler to better understand the soil. Therefore, modeling can help refocus attention to field work and to the type of data to be collected. A specific example of how assumptions affect the development of a model is given later in this entry.

The first step in modeling soils—or anything—is to define the object of interest. In applying models to pedology, it should be recognized that soil is, in reality, a continuum of objects distributed across the earth's surface—both in space and time. The exact lateral boundary between one "soil" and another, or the vertical boundary between soil and non-soil, is arguably impossible to determine. Jenny (4) first applied principles derived from the physical sciences to the conceptualization and modeling of soils. Jenny's approach was to divide the continuum of soils on the earth's surface into *systems*, which are arbitrarily defined, discrete, three-dimensional segments of the landscape that are amenable to mass or energy budgeting. The volume of these systems (both the chosen area and depth) is arbitrary, but it sets the stage for the mathematical formulations which are chosen to represent or describe it. A second important aspect of soils is the vast amount of time, and the array of unknown processes, that may have affected any soil system. This complex history in turn forces pedologists to develop tools, concepts, and modes of enquiry not always confronted by their experimental colleagues. Finally, soil formation is the result of an incompletely understood array of processes, and pedological models invariably are attempts to mathematically capture one, or at the most, a very restricted subset of the full suite of biogeochemical processes that affect a soil system.

In the pedological literature, more attention has possibly been devoted to classifying and discussing models (5) than actually developing them. Unfortunately, much discussion has focused on the "pros and cons" of the *Factors of Soil Formation* (4) in the realm of pedological models. Briefly, the factorial "model" discussed in a book-length treatise by Jenny (4) can be symbolically represented as:

$$s = f(cl, o, r, p, t, \ldots)$$

where s = soil properties, cl = climate, o = biota, r = topography, p = parent material, and t = time (4, 6). The general truthfulness of this statement is almost beyond dispute (virtually every pedologist would ultimately have to agree that soil forms in response to variations in these factors, and that soil properties can also be numerically correlated with variations in these factors). Indeed, soil is *defined* in terms of this statement:

Soil is the "collection of natural bodies occupying portions of the earth's surface that support plants and that have properties due to the integrated effect of climate and living matter, acting upon parent material, as conditioned by relief, over periods of time" (7).

Based on this definition, it would be more correct to define the state factor "model or theory," as it has been alternately called, as a pedological *law* given its universal truthfulness and common definitions of scientific laws (8). At the very least, it is a fundamental underlying theory of pedology in the sense of Kuhn (9). This definition would also allow the discussion of models in pedology to move beyond this fundamental truism (and a useful quantitative and mathematical tool in its own right) to the development of mathematical models as they are commonly considered and used in geochemistry, geophysics, and related fields. The remainder of this article deals with these more practical modeling issues.

MODELING PROCESSES

Typical approaches to mathematically modeling soil processes involve the development of a mass or energy balance model. The mathematics used will ultimately hinge upon one's understanding of the soil properties, the processes that presumably control them, and how these processes may vary over time. Some of the simplest models may be analytical models with time independent variables. Alternative modeling approaches may involve the abandonment of time invariant parameters and ultimately, the incorporation of relatively random changes in the rate of the process and factors that affect it.

As an example, I begin with possibly the first true mathematical model of a pedogenic process—a time dependent, analytical, mass-balance model of O horizon formation in forest soils developed by Jenny et al. (10). Jenny et al (10) defined the system of study, discussed the processes that affect it, and for the simplest cases they considered (tropical forests with nearly constant litter inputs with time), described the change in O horizon mass (F, in mass area^{-1}) with time:

$$dF = Adt - k(F + A)dt \tag{1}$$

where A = litter inputs (mass area^{-1} time^{-1}), and k = decomposition constant (time^{-1}). (See Ref. 10 for a fuller discussion of calculation and definition of k.) Upon integration, the solution to Eq. 1 provided by Jenny was:

$$F = \frac{A(1-k)}{k}(1 - e^{-kt}) \tag{2}$$

This model, and permutations of it, has served as the foundation for decades of research and modeling of the soil organic C budget. Today the model has been extended for use in stable isotopic studies of soil organic matter (11) and in modeling turnover times of soil organic carbon (12). However, the analytical model as developed by Jenny assumes constant inputs with time (a restriction that he noted does not occur in all situations) and constant decomposition rates with time. It also assumes that all organic matter is homogeneous (and by implication, has the same decomposition rates).

Work over the past 15 years in particular, has revealed that soil organic matter (and even litter layers), can best be viewed as multiple pools of soil organic matter, each with their own characteristic input and decomposition rates. In simplest terms, multiple pool models of soil organic matter can be expressed as:

$$dC/dt = \sum_{i=1}^{n}(I_i - k_iC_i) \tag{3}$$

where C = total mass of soil C (mass volume^{-1}), I = inputs of pool i (mass volume^{-1} time^{-1}), k_i = decomposition constant of pool i (time^{-1}) and C_i = mass of soil C in pool i (mass volume^{-1}) (11). Even these multiple pool models do not capture other aspects of soil formation: 1) the variation in soil C with depth, and 2) the likelihood that the process may have varied unpredictably over time. To address the first issue, models may include a downward transport term and depth dependent inputs. For a single pool of organic matter, a basic depth dependent model is:

$$dC/dt = I - \upsilon\frac{\partial C}{\partial Z} - kC \tag{4}$$

where υ = advection coefficient (distance time^{-1}) (11). Even these models can be made more complex, to include the process of diffusion for example (13).

Analytical solutions to all the aforementioned models require time invariant parameters. The inclusion of time-dependent parameters may be accomplished through numerical means. Yet, soils form over thousands to millions of years, with numerous unknown perturbations to the system that elude the fundamental framework of these simple models (14). It is these real world complications imposed by nature that weaken the utility of relatively simple, time invariant, deterministic models. Recently, Phillips (14) began the discussion of applying nonlinear dynamical systems theory to soil modeling, with the goal of explicitly incorporating the role that random differences in initial conditions and historical contingencies have on the processes of soil formation. Undoubtedly, work of this nature is one of the future challenges in the field of pedological modeling.

CONCLUSION

While models are powerful means of testing hypotheses and synthesizing contemporary knowledge in a concise way, the previous discussion serves to illustrate that models of all types are simplified, incomplete, mathematical descriptions of the "real" world. Increasingly, it is emphasized that models can never be fully verified (confirmed as the establishment of truth) (2). Experience shows that multiple models (or versions of the same model) may faithfully mimic empirical observations of interest, a dilemma illustrating a practical verification problem. In addition, mathematical models may be internally correct but they may poorly represent the phenomena they intend to describe because of incomplete knowledge of the system and, as a result, incorrect assumptions. Knowledge of the soil is essential in modeling, for as Baker has recently noted, "mathematics (is) the science that draws necessary conclusions without regard to facts" (15).

Given the ultimate simplicity and abstractness of models, and the ultimate complexity of nature in general, and soils in particular, the question may be asked "why model in the first place" (2). Pedological processes are first order controls on global atmospheric (16) and aquatic chemistry, and even relatively simple analytical models of soil processes have thus far proven useful to link these global reservoirs. The truly unique role for pedologists, in addition to applying mathematics on their own, is to provide the unique conceptual foundation peculiar to soils —one informed by extensive field observations guided by ideas generated during previous modeling attempts—that will make pedological models relevant to scientists and society.

REFERENCES

1. Chamberlin, T.C. On Lord Kelvin's Address on the Age of the Earth as an Abode Fitted for Life. Smithsonian Institution Annual Report. Washington, DC, 1899; 223–246.
2. Oreskes, N.; Shrader-Frechette, K.; Belitz, K. Verification, Validation, and Confirmation of Numerical Models in the Earth Sciences. Sci. **1994**, *263*, 641–646.
3. The Oxford English Dictionary. 2nd Ed.; Clarendon Press: Oxford, 1989.
4. Jenny, H. *Factors of Soil Formation. A System of Quantitative Pedology*; McGraw Hill Book Co.: New York, 1941; 281.
5. Hoosbeek, M.R.; Amundson, R.; Bryant, R.B. Pedological Modeling. In *Handbook of Soil Science*; Sumner, M.E., Ed.; CRC Press: Boca Raton, FL, 1999; E77–E116.
6. Amundson, R.; Jenny, H. On a State Factor Model of Ecosystems. Bioscience **1997**, *47*, 536–543.
7. Soil Survey Staff. *Soil Survey Manual*, U.S. Dep. Agric. Handbook No. 18; U.S. Govt. Printing Office: Washington, DC, 1951; 503.
8. Morris, C., Ed. *Academic Press Dictionary of Science and Technology*; Academic Press: New York, 1992.
9. Kuhn, T.S. *The Structure of Scientific Revolutions*, 2nd Ed.; University of Chicago Press: Chicago, 1970; 210.
10. Jenny, H.; Gessel, S.P.; Bingham, F.T. Comparative Decompostion Rates of Organic Matter in Temperate and Tropical Regions. Soil Sci. **1949**, *68*, 419–432.
11. Amundson, R.; Baisden, W.T. Stable Isotope Tracers and Models in Soil Organic Matter Studies. In *Methods in Ecosystem Science*; Sala, O., Mooney, H., Howarth, B., Jackson, R.B., Eds.; Springer Verlag: New York, 2000; 117–137.
12. Trumbore, S.E. Comparison of Soil Carbon Dynamics in Tropical and Temperate Soil Using Radiocarbon Measurements. Global Biogeochem. Cycles **1993**, *9*, 515–528.
13. Elzein, A.; Balesdent, J. Mechanistic Simulation of Vertical Distribution of Carbon Concentrations and Residence Times in Soils. Soil Sci. Soc. Am. J. **1995**, *59*, 1328–1335.
14. Phillips, J.D. On the Relations Between Complex Systems and the Factorial Model of Soil Formation (with Discussion). Geoderma **1998**, *86*, 1–42.
15. Baker, V.R. Geosemiosis. Geological Soc. Am. Bul. **1999**, *111*, 633–645.
16. Amundson, R.; Stern, L.; Baisden, T.; Wang, Y. The Isotopic Composition of Soil and Soil-Respired CO_2. Geoderma **1998**, *82*, 83–114.

PEDOTRANSFER FUNCTIONS

J.H.M. Wösten
Alterra Green World Research, Wageningen, The Netherlands

P

INTRODUCTION

Simulation models, which are indispensable tools in modeling water and solute movement into and through soil, require as key input parameters easily accessible and representative hydraulic characteristics. Techniques to measure these characteristics are relatively time-consuming and therefore costly. At the same time, good predictions of the characteristics instead of direct measurements may be accurate enough for many applications. Considering the desired accuracy and the available financial resources, it is rewarding to analyze existing databases containing measured hydraulic characteristics and to establish relationships that predict the characteristics from measured basic soil data. These predictive relationships are called "pedotransfer functions" (PTFs) (1) and they essentially translate data *we have* into data *we need.* Basically, PTFs relate soil characteristics being assembled during soil survey to more complex characteristics needed for simulation. Predicting soil hydraulic characteristics dominates the research field, although soil chemical and soil biological characteristics are also being predicted. Several reviews on PTF development and use have been published (2,3).

Large databases on measured hydraulic characteristics, such as UNSODA (4), HYPRES (5), WISE (6), and United States Department of Agriculture Natural Resource Conservation Service pedon database (7), form the essential, basic sources of information for the derivation of PTFs. In using PTFs, insight is needed to determine which input variables are to be included in a PTF, what technique is to be used to establish a PTF, and how accuracy and reliability of PTFs are to be quantified.

FUNCTIONS USED TO DESCRIBE THE WATER RETENTION AND HYDRAULIC CONDUCTIVITY CHARACTERISTICS

Describing hydraulic characteristics as functions rather than as tables has the clear advantage that they can be easily incorporated in simulation models. There exists a wide range of different equations for the description of the characteristics. The following equations to describe volumetric soil water content, θ, and hydraulic conductivity, K, as functions of pressure head, h, are widely used (8):

$$\theta(h) = \theta_r + \frac{\theta_s - \theta_r}{(l + |\alpha h|^n)^{1-1/n}} \tag{1}$$

$$K(h) = K_s \frac{((l + |\alpha h|^n)^{1-1/n} - |\alpha h|^{n-1})^2}{(l + |\alpha h|^n)^{(1-1/n)(l+2)}} \tag{2}$$

In these equations the subscripts r and s refer to residual and saturated values and θ, and n and l are parameters that determine the shape of the curve. The residual water content θ_r refers to the water content where the gradient $d\theta/dh$ becomes zero ($h \rightarrow -\infty$). The parameter α (cm^{-1}) approximately equals the inverse of the pressure head at the inflection point. The dimensionless parameter n reflects the steepness of the curve. The dimensionless parameter l determines the slope of the hydraulic conductivity curve in the range of more negative values of h.

PTFs to predict the model parameters θ_r, θ_s, K_s, α, l and n from basic soil data were built by many authors (9). Figure 1 shows the mean water retention and hydraulic conductivity characteristics, also called class PTFs, for the texture class "medium fine topsoil" (5).

SELECTION OF PEDOTRANSFER FUNCTION PREDICTOR VARIABLES

Soil properties affecting water retention and hydraulic conductivity are manifold (10). Table 1 lists the properties used most often as predictors because of their availability and because they proved to be the most promising ones.

Particle size distribution is used in almost any PTF. Particle size classes differ in different national and international classification systems, and so differ the number and the size of classes used in PTFs may also differ. Using sand, silt, and clay contents is a common approach.

Limited, measured water retention data at, for instance, two pressure heads may dramatically improve predictions of the complete water retention characteristic.

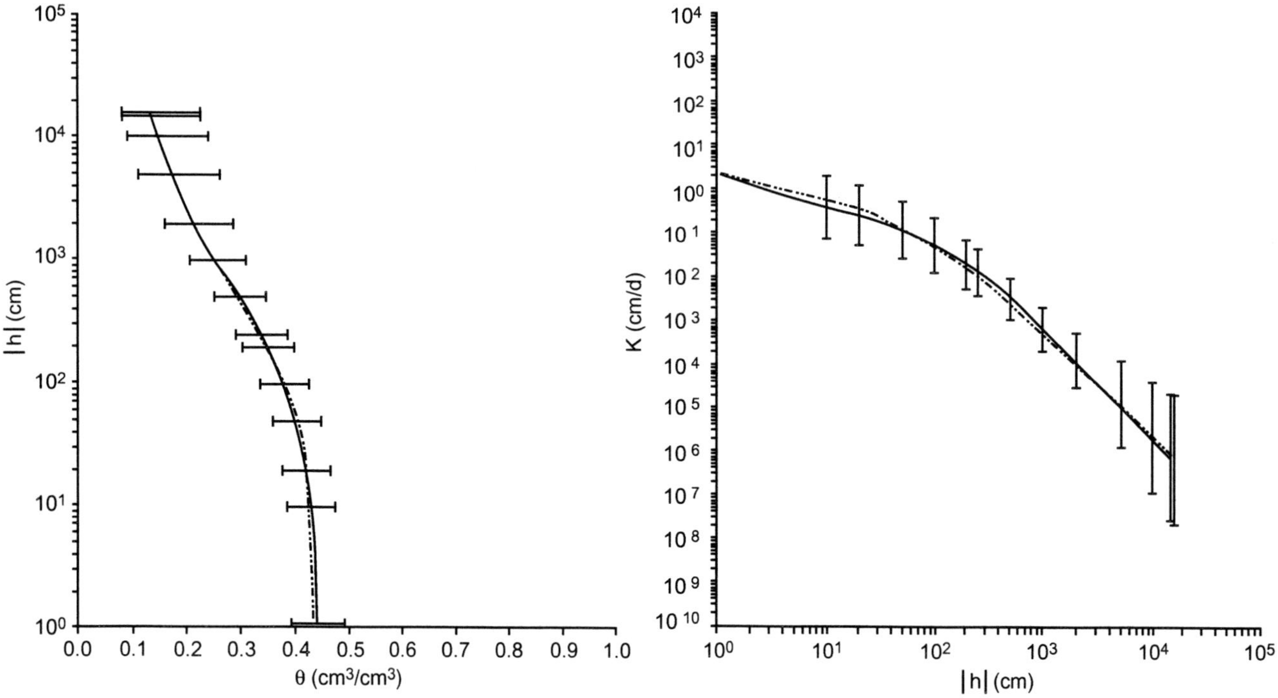

Fig. 1 Geometric mean water retention (left graph) and hydraulic conductivity (right graph) characteristic (solid lines), standard deviations (bars) and van Genuchten fits (dotted lines) for the texture class "medium fine topsoil."

Porosity or *bulk density* is an important variable in many PTFs.

Soil structure and *morphology descriptors* such as the parameter topsoil and subsoil are successfully included in PTFs.

Landscape position is used as a topographic variable in PTFs.

Organic matter content is often used as predictor, but because bulk density and organic matter content are correlated, bulk density may effectively substitute organic matter content.

Mechanical properties and shrink–swell parameters as characterized by the coefficient of linear extensibility (COLE) are used to estimate both water retention and K_s.

Table 1 Soil properties often used in PTFs

Particle size properties	Hydraulic characteristics	Morphological properties	Chemical/mineralogical properties	Mechanical properties
Sand, silt, clay	Water content at	Bulk density	Organic carbon	Penetration resistance
Fine sand	−33 kPa	Porosity	Organic matter	
	−1500 kPa	Horizon	CEC	
Very coarse sand, coarse fragments	Reference moisture retention curve	Structure	Clay type	
Median or geometric mean particle size		Grade	$CaCO_3$	
		Size	Iron	
		Shape		
Water-stable aggregates		Color		
		Consistence		
		Pedality		
		Landscape position		

Table 2 Continuous PTFs developed from the HYPRES database

$$\theta_s = 0.7919 + 0.001691C - 0.29619D - 0.000001491S^2 + 0.0000821OM^2 + 0.02427C^{-1} + 0.01113S^{-1} + 0.01472\ln(S) - 0.0000733OMC - 0.000619DC - 0.001183DOM - 0.0001664 \times \text{topsoil} \times S$$

$(R^2 = 76\%)$

θ_s is the saturated water content in the van Genuchten equations; C, percentage clay (i.e., percentage <2 μm); S, percentage silt (i.e., percentage between 2 and 50 μm); OM, percentage organic matter; D, bulk density; topsoil and subsoil are qualitative variables having the value of 1 or 0 and ln, natural logarithm.

METHODS TO DEVELOP PEDOTRANSFER FUNCTIONS

When the set of PTF input parameters is defined and the PTF output is decided upon, a method is selected to relate input and output. The most prominent methods to create this relationship are discussed below.

Regression Analysis

The oldest and most widely used method to build PTFs was exclusively linear regressions, but this method is now gradually being replaced by nonlinear regression. An advantage of regression techniques is that the essential predictor variables can be found automatically using stepwise regression. A drawback of regression is that it requires a model of the dependency of soil properties which is almost impossible now that large digitized databases containing numerous soil properties are widely available. One remedial approach is to use principal component analysis. It allows identification of a small number of new variables that are linear combinations of the original predictors and that can explain a large percentage of variability within samples. An example of this type of PTF predicting the saturated water content θ_s is presented in Table 2.

Artificial Neural Networks

Artificial neural networks (ANNs) are becoming a common tool for modeling complex "input–output" dependencies. The advantage of ANNs is that no *a priori* model is required that relates input to output. After establishing the network structure and finding coefficients to express the degree of influence of the network components on each other, an ANN generates a complex formula relating input with output values. This formula can be used like a regression formula. ANNs consisting of many interconnected simple computational elements, called nodes or neurons, are comparable to the best nonlinear statistical regression approaches.

Group Method of Data Handling

The group method of data handling (GMDH) has, unlike ANNs, a built-in algorithm to retain only essential input variables in a flexible net of regression equations, which relates inputs to outputs. GMDH is a technique of finding an approximate relationship between a set of input variables x_1, x_2,H, x_N and an output variable y. When the number of input variables is very large, or the relationship between inputs and output is very complex, GMDH successfully competes with statistical regression.

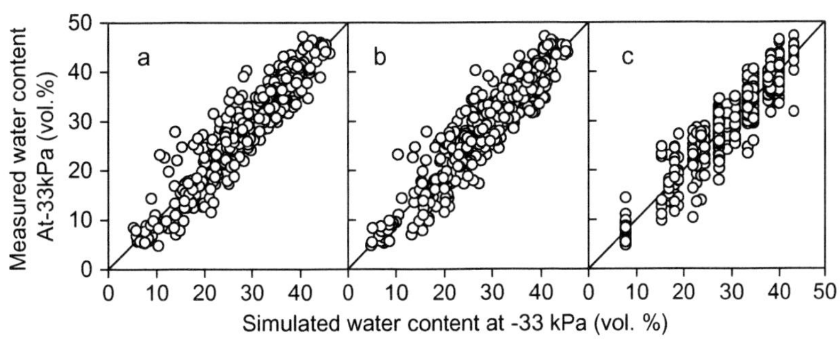

Fig. 2 One-to-one diagrams showing accuracy of predicting water content at −33 kPa suction with three methods: (a) backpropagation ANN, (b) group method of data handling, and (c) regression tree algorithm. As can be seen, the accuracy of the three methods is quite comparable. In all three cases the RMSE is around 3.4 vol%, the R^2 value of the regression is around 0.9, the slope of the regression is close to 1 and the intercept is about 0.001 vol%.

Regression Tree Algorithm

Regression tree algorithm is a recursive data partitioning algorithm that initially splits the dataset into two subsets based on a single best predictor variable (the variable that minimizes the variance in the response). It then does the same on each of the subsets and so on recursively. The output is a tree with branches and terminal nodes. The predicted value at each terminal node is the average at that node.

Fig. 2 shows the accuracy of predicting water content at −33 kPa pressure head with the last three methods.

ACCURACY AND RELIABILITY OF PEDOTRANSFER FUNCTIONS

As empirical equations, PTFs are routinely evaluated in terms of the correspondence between measured and estimated values. When measured values are those used to develop the equation, the accuracy of the equation is evaluated. When the measured values are different from the ones used to develop the equation, the reliability is evaluated. A multitude of statistics is used in PTF development including: multiple determination coefficient (R^2), root mean square error (RMSE), mean error (ME), and *t*-statistics to test the null hypothesis.

Accuracy of Pedotransfer Functions

Using the same statistics, the accuracy of existing PTFs varies appreciably. Accuracy estimates themselves may serve as benchmarks. However, they should be compared to variability of other measured input data. The model should not be more accurate than data used in the model development. Therefore, the PTF can be thought to be accurate if the variability of the PTF errors does not differ significantly from the variability in data, and if the average error does not significantly differ from zero.

Reliability of Pedotransfer Functions

The reliability of PTFs can be evaluated by cross-validation, i.e., splitting the available dataset in a development and validation subset, or by using an independent dataset.

Many studies assessed the reliability of PTFs by applying them to independent, regional datasets. From this it appears that PTFs developed from regional databases give good results in regions with a similar soil and landscape history.

CONCLUSIONS

A number of conclusions can be formulated.

1. PTFs are a powerful tool in estimating physical, chemical, and fertility properties of soils. Because PTFs predict properties that are difficult to obtain from already available basic soil properties, they have the clear advantage that they are relatively inexpensive and easy to derive and to use.
2. For application on a specific location, use of PTFs might not be appropriate. In this case direct measurement is the only option. PTFs should not be used to make predictions for soils that are outside the range of soils used to derive the PTFs. In other words, use of PTFs for interpolation purposes is safe but dangerous to use for extrapolation.
3. Most successful PTFs are developed from large and reliable databases. This implies that well structured and easily accessible national and international databases of measured soil hydraulic characteristics need to be created. PTFs should be periodically updated as more measured data become available from these databases.
4. Accuracy and reliability of PTFs will be appropriate for many applications on a regional and national scale. On these scales temporal and spatial variability of other than hydraulic characteristics most likely will also have an important impact on the modeling results.
5. The search for data mining tools to develop better, more flexible PTFs and the search for additional soil properties as inputs in PTFs are important directions for improving PTF accuracy and reliability.
6. Uncertainty in PTFs can be quantified. Its effects on calculated functional aspects of soil behavior will help to assess, if an input parameter needs further detailing, and which one, in order to arrive at more accurate results.

REFERENCES

1. Bouma, J.; van Lanen, J.A.J. Transfer Functions and Threshold Values: From Soil Characteristics to Land Qualities. In *Quantified Land Evaluation*; Proc. Workshop ISSS and SSSA, Washington, DC, 27 Apr.–2 May 1986; Beek, K.J., Ed.; Int. Inst. Acrospace Surv. Earth Sci. Publ. No. 6 ITC Publ.: Enschede, The Netherlands, 1987; 106–110.
2. Van Genuchten, M.Th.; Leij, F.J. On Estimating the Hydraulic Properties of Unsaturated Soils. In *Indirect Methods for Estimating the Hydraulic Properties of Unsaturated Soils*; van Genuchten, M.Th., Leij, F.J., Lund, L.J., Eds.; University of California: Riverside, CA, 1992; 1–14.

3. Pachepsky, Ya.A.; Rawls, W.J.; Timlin, D.J. The Current Status of Pedotransfer Functions: Their Accuracy, Reliability, and Utility in Field- and Regional-Scale Modelling. In *Assessment of Non-Point Source Pollution in the Vadose Zone*; Geophysical Monograph 108, Corwin, D.L., Loague, K., Ellsworth, T.R., Eds.; American Geophysical Union: Washington, DC, 1999; 223–234.
4. Leij, F.; Alves, W.J.; van Genuchten, M.Th.; Williams, J.R. *The UNSODA Unsaturated Soil Hydraulic Database*; User's Manual Version 1.0. EPA/600/R96/095, National Risk Management Laboratory, Office of Research and Development: Cincinnati, OH, 1996.
5. Wösten, J.H.M.; Lilly, A.; Nemes, A.; Le Bas, C. Development and Use of a Database of Hydraulic Properties of European Soils. Geoderma **1999**, *90*, 169–185.
6. Batjes, N.H. Development of a World Data Set of Soil Water Retention Properties Using Pedotransfer Rules. Geoderma **1996**, *71*, 31–52.
7. *Soil Survey Manual*; U.S. Dept. Agriculture Handbook No. 18, United States Department of Agriculture: Washington, DC, 1951.
8. Van Genuchten, M.Th. A Closed-Form Equation for Predicting the Hydraulic Conductivity of Unsaturated Soils. Soil Sci. Soc. Am. J. **1980**, *44*, 892–898.
9. Wösten, J.H.M.; van Genuchten, M.Th. Using Texture and Other Soil Properties to Predict the Unsaturated Soil Hydraulic Functions. Soil Sci. Soc. Am. J. **1988**, *52*, 1762–1770.
10. Rawls, W.J.; Gish, T.J.; Brakensiek, D.L. Estimating Soil Water Retention from Soil Physical Properties and Characteristics. Adv. in Soil Sci. **1991**, *16*, 213–234.

PERMAFROST: SOIL TEMPERATURE/SPECIAL PROBLEMS

Douglas L. Kane
Julia Boike
University of Alaska Fairbanks, Fairbanks, Alaska, U.S.A.

INTRODUCTION

On an extended south-to-north transect at northern latitudes, a transition from ground that never experiences seasonal freezing, to those that occasionally freeze during the winter, to those that freeze every year, to those that may remain frozen for an extended time could be encountered. The topic of discussion here is those surface soils and deeper geologic layers that remain at or below freezing for a duration of two years or more and how they impact people living in this environment. Such frozen ground, both unconsolidated and bedrock, are commonly referred to as permafrost (1). Although at or below the freezing point of bulk water (0°C), the term "permafrost" neither implies that water is present or that water, if present, is frozen. In fact, it is possible for significant amounts of water to remain unfrozen in permafrost; this is also true for water in seasonal frost.

PERMAFROST CHARACTERISTICS

Spatially extensive permafrost can be found in Russia and Canada as far south as 45°N, and even farther south on the elevated Tibetan Plateau and Himalayan Mountains (2). Approximately 25 million km^2 of permafrost exist in the northern hemisphere. In the higher latitudes, permafrost is continuous under the land surfaces. At intermediate latitudes permafrost is discontinuous or sporadic. Legget (3) reported that 20% of the land surface of the world is underlain by permafrost. More than 50% of Russia and Canada are underlain by permafrost. Alaska has continuous permafrost in the northern 1/3 of the state and discontinuous permafrost in the rest of the state, excluding the coastal areas from the Aleutian Islands to southeastern Alaska (Fig. 1). In the southern hemisphere, permafrost distribution is confined to Antarctica and high alpine or mountainous regions. Isolated permafrost is common at higher elevations, and evidence of past permafrost is common in areas that no longer have permafrost. The thickness of permafrost can vary from a thin lens of less than 1 m to greater than 1000 m (Fig. 1). Permafrost can also be found in coastal areas at the bottom of seas.

Permafrost ground is interesting to the engineer and scientist because the medium is usually composed of two solids (porous medium and ice) and two fluids (air and liquid water). The more components that are present in a mixture, the more difficult it is to predict the medium's response to an input of energy or mass. The amount of unfrozen water in saturated ground is strictly a function of the grain size (more specifically surface area) and the freezing temperature. A fine-grained soil such as clay has a very high surface area relative to coarse-grained soils such as sand; in frozen ground, this translates into much higher unfrozen water contents at the same temperature. The unfrozen water found in permafrost exists as a film of water around each soil particle. It is via these unfrozen films that water moves in permafrost. Frozen clay can have as much as 5–7% unfrozen water by volume at −15°C. As the temperature decreases the amount of unfrozen water also decreases. Thermal and hydraulic properties of permafrost are quite variable and depend upon the percentages of the various ground components. Most heat transfer in permafrost is by conduction and can be modeled by Fourier's law, although simpler methods have been developed.

A typical temperature profile of permafrost appears in Fig. 2. Since there is a geothermal flux outward from the center of the earth, this heat has to be successfully transferred to the ground surface or the permafrost will warm and melt. In order to maintain the thermal integrity of the permafrost, the soils above the top of the permafrost table must completely freeze during the winter so there is a continuously decreasing thermal gradient along which the geothermal heat can be transferred to the surface by conduction (Fig. 2).

The seasonally thawed soil layer at the ground surface that goes through freezing/thawing annually and mantles the permafrost is called the active layer (Fig. 2). This layer acts as a buffer to heat and mass transfer to the permafrost. A typical active layer in the continuous permafrost zone would typically thaw to a maximum depth of 60 cm. The top 15–25 cm of these soils are generally composed of organic material, with the deeper soils being mineral.

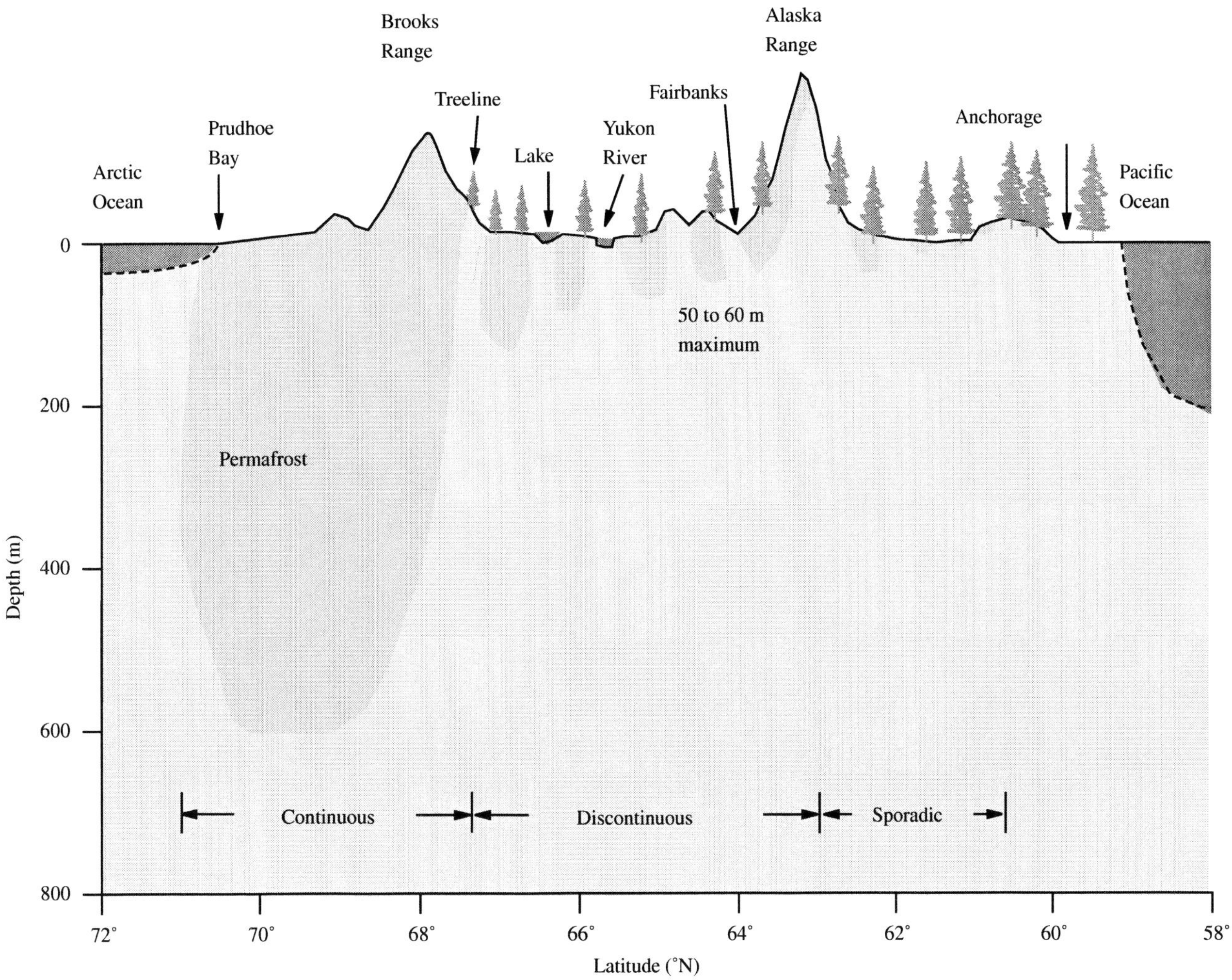

Fig. 1 Schematic of permafrost distribution in Alaska on a south-to-north transect.

Organic soils are good thermal insulators and, when coupled with snow cover during the winter months, they minimize heat loss from the ground. The thermal balance of the permanently frozen ground is maintained by heat loss to the atmosphere that occurs at high latitudes over the extended winter months.

Permafrost is usually considered to be impermeable to water movement. This is usually a good assumption for short durations of days or even a few months. For longer periods of time, the redistribution of water within permafrost can be appreciable. During snowmelt and major rainfall events, considerable water enters the active layer. Most of this water resides in the organic soils as the mineral soils are usually already near saturation. Since permafrost has relatively low hydraulic conductivity, there is no hydraulic connection between the perched water above the permafrost (suprapermafrost groundwater) and the subpermafrost groundwater below the permafrost (Fig. 2). In continuous permafrost, the subsurface hydrology is confined to the active layer. For areas of discontinuous permafrost, the subsurface hydrology is a combination of shallow flow over the frozen ground and deeper flow around and under it.

SURFACE ENERGY BALANCE

Any time the ground surface is disturbed, the surface energy balance that sustains permafrost is upset. This generally results in warming of the permafrost and thickening of the active layer. Much sporadic and discontinuous permafrost

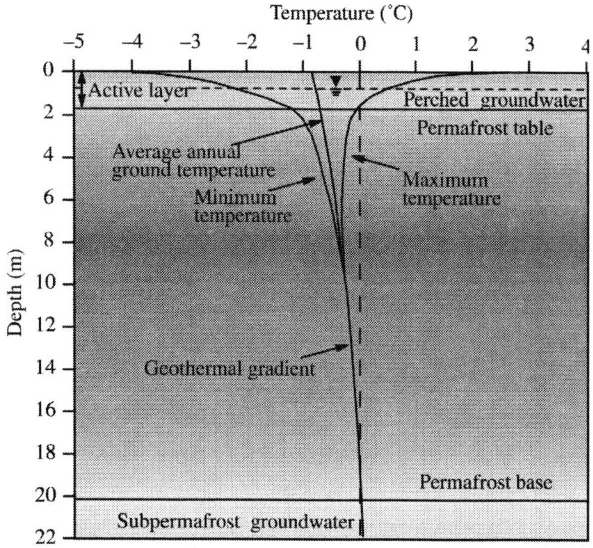

Fig. 2 Typical temperature profile for permafrost, with annual variation indicated near the ground surface.

is maintained at temperatures just below freezing; climatic warming of just a few degrees would result in the melting of this frozen ground and warming of colder permafrost. Also where permafrost exists, surface disturbances such as removing vegetation or surface soils and ponding of water are sufficient to alter the surface energy balance and cause permafrost degradation. Permafrost generally appears where the mean annual surface temperature is at least a few degrees C below freezing; for instance, Hay River, NWT, Canada has sporadic, shallow permafrost with a mean annual temperature of −3.4°C.

In areas of discontinuous permafrost, the south-facing slopes are generally permafrost-free (Fig. 1) while north-facing slopes have permafrost; the east- and west-facing slopes and valley bottoms may or may not have permafrost, depending upon subtle site-specific conditions that impact the energy balance (vegetation, slope, moisture content, etc.). Because it took thousands of years for permafrost to develop, deep permafrost is a particularly good recorder of past climates, and changes in the temperature profile will reflect these impacts. Many general atmospheric circulation models (GCM) predict variable global warming in the high latitudes, and in some areas there is already field evidence of atmospheric warming (4) and permafrost warming (5). To sustain permafrost, the geothermal heat radiating out from the core of the earth must be removed; this only occurs if the active layer is completely frozen during winter.

LIVING WITH PERMAFROST

Permafrost influences many facets of everyday living that are often taken for granted in warmer climates. Permafrost negatively impacts both the quality and quantity of the available groundwater and limits the potential for infiltration of wastewater. As a result, housing units with individual water and wastewater systems must be designed with innovative options to replace the traditional groundwater well and septic system for wastewater treatment. Utilities that are typically buried are often placed in aboveground utilidors. Also as roadways age, they become very bumpy as differential movement of the road surface occurs.

The significant property of permafrost from an engineering viewpoint is the ice content. The amount of ice can range from essentially none in well-drained ground to values that exceed the porosity of poorly drained ground. Within the permafrost, segregated ice with minimal entrained soil exists as ice lenses and wedges. As long as the permafrost stays frozen, the ground is relatively stable. However, thawing of ice-rich permafrost causes ground to settle or subside. This thaw settlement compromises the structural integrity of buildings, roadways, pipelines, and other structures built on it. For areas where construction must take place on frozen ground, special efforts—for example the use of insulation and thermosyphons—are necessary to maintain the structural integrity of the permafrost. Thermosyphons are vertical devices for removing heat from the ground, and they are used extensively under buildings and pipelines to maintain below freezing conditions in the ground.

Soil freezing causes additional engineering problems near the ground surface when strong thermal gradients induce water movement from warm ground to cold ground through unfrozen films of water on the surfaces of soil particles, regardless of whether it is frozen or not. This process causes "frost heave," as it results in an accumulation of ice in the form of lenses that can cause the ground to expand and deform. Surface heaving becomes significant over time periods of several years, and it is important for those engineering structures with design lives of tens of years to compensate for this additional stress.

CONCLUSIONS

Permafrost and seasonally frozen ground are very extensive in the high latitudes of Earth. Also, resource development and population in these areas is increasing. Initial efforts directed at building an infrastructure (roads, airports, water and sewer distribution, etc.) on permafrost assumed that the frozen ground was in thermal equilibrium. It is now obvious that because of climatic change that the permafrost is not in thermal equilibrium. Instead, it is warming in many areas and this needs to be considered in the present design criteria for all engineering structures built on permafrost. Warming, with subsequent thawing of permafrost, will affect the surface energy budget, water and gas fluxes (potential for release of greenhouse gases), and vegetation; hence, there is a direct feedback with the climate system. Just as it took a long time for permafrost to develop, it will take a long time for it to completely melt; but the consequences of thawing permafrost can impact our daily lives in the first or second season of melting.

REFERENCES

1. Muller, S.W. 1943. *Permafrost or Permanently Frozen Ground and Related Engineering Problems*. U.S. Engineers Office, Strategic Engineering Study, Special Report No. 62, 136.
2. Brown, J.; Ferrians, O.J.; Jr., Heginbottom, J.A.; Melnikov, E.S. 1997. *Circum-Arctic Map of Permafrost and Ground-Ice Conditions*. U.S. Dept. of Interior, Geological Survey, Map CP-45.
3. Legget, R.F. Permafrost Research. Arctic **1954**, *7* (3/4), 153–158.
4. Serreze, M.C.; Walsh, J.E.; Chapin, F.S.; Osterkamp, T.; Dyurgerov, M.; Romanovsky, V.; Oechel, W.C.; Morison, J.; Zhang, T.; Berry, R.G. Observational Evidence of Recent Change in the Northern High-Latitude Environment. Climate Change **2000**, *46*, 159–207.
5. Lachenbruch, A.H.; Marshall, B.V. Changing Climate: Geothermal Evidence from Permafrost in the Alaskan Arctic. Science **1986**, *234*, 689–696.

PETROCALCIC HORIZONS, SOILS WITH

Alain Ruellan
Montpellier, France

INTRODUCTION

In Mediterranean and tropical arid regions, B or C horizons of soils frequently contain, at shallow depths (some decimeters), more than 50% calcium carbonate. In general these horizons are layered (platy), hard, and 10–100 cm thick. These Bca and Cca horizons are called "lime crusts," "calcretes," or "petrocalcic horizons." Soils with petrocalcic horizons can be extensive on plateaus, slopes, and river terraces.

Petrocalcic horizons are strongly developed horizons that have accumulated calcium or other carbonates (1). They are indurated or cemented calcic horizons. Vertically, they are related to the A and B horizons above and to C horizons below that may or not be calcareous (2).

THE CALCIC HORIZONS

Calcic horizons (3, 4) are horizons of calcium carbonate accumulation. In the calcic horizons, the calcium carbonate:

- arrives from the top of the soil profile, or from the upper part of the pedological toposequence, or from the air (wind), or from the underground water;
- precipitates giving rise to very rich calcium carbonate forms which are either discontinuous (pseudomycelia, cutans, nodules, or veins), or continuous (petrocalcic horizons).

As development proceeds, the carbonate accumulates, first filling the pores and then replacing preexisting minerals (epigeny) (5–8).

THE PETROCALCIC HORIZONS (2–4)

When the calcium carbonate concentration in B or C horizons increases sufficiently to mask all or most of the pedological or lithological structure within which it develops, it becomes a petrocalcic horizon (or calcrete, or lime crust). When consolidated, the carbonate content of the horizon is 50–90%.

There are several types of petrocalcic horizons:

- Unlayered (nonplaty) petrocalcic horizons; the structure can be massive, angular, finely lamellar, or nodular. The colour is nearly white.
- Layered (platy) petrocalcic horizons:
  - Layered, noncompact, petrocalcic horizons are made up of superposed, clearly individualized layers of crusty material, hard, but not petrified (Fig. 1). The thickness of these layers varies from a few mm to several cm; they are not continuous vertically but are separated from each other by more or less horizontal fissures arranged in a braided manner. The internal basic structure can be massive, nodular, or finely lamellar; the color, generally white to creamy white, with frequent black spots. As the crusty material becomes harder, it tends to be pinker in color.
  - Layered compact petrocalcic horizons (slabs) (Fig. 2) are made up of one or several calcium carbonate layers, extremely hard, greyish- or, more often, pinkish-colored. Each layer can reach 10–20 cm in thickness. These layers are petrified, usually very continuous. The internal basic organization is very massive without fine lamellae.
  - Finely layered compact petrocalcic horizons are very hard formations with thicknesses of a few mm to a few cm. The horizons are clearly stratified, consisting of one or more series of very fine superposed lamellae. The general color is white or pink.

In addition to clalcium carbonate accumulation structures and hardness, other important data useful for characterizing and distinguishing different types of petrocalcic horizons are water pH and natural drainage:

- Basic types have a water pH between 8.0 and 8.7; ultrabasic types have a water pH >8.7. The very high pH indicates the presence of magnesium carbonate (9).
- Soils with petrocalcic horizons in environments with restricted drainage tend to form montmorillonite, attapulgite, and sepiolite clay minerals (6–8, 10).

Encyclopedia of Soil Science

Fig. 1 Below an A horizon (1): a layered noncompact petrocalcic horizon (crust) (2), on an unlayered (nonplaty) petrocalcic horizon (3), on a calcic horizon (4).

Fig. 2 A layered compact petrocalcic horizon (slab) (2) — below an A horizon (containing pebbles of slab) (1) — on a layered noncompact petrocalcic horizon (crust) (3), on a calcic horizon (4).

VERTICAL SUCCESSION OF THE CALCIC AND PETROCALCIC HORIZONS

Below A or B horizons, the calcium carbonate accumulation can be of three types:

- Slightly differentiated: the distribution of carbonates is diffuse, sometimes with pseudomycelia with very diffuse upper and lower limits.
- Moderately differentiated: the carbonates in the calcic horizon occur partly as diffuse distributions in the soil matrix and partly as concentrations, forming cutans, soft and hard nodules, or veins, with diffuse upper and lower limits.
- Highly differentiated (petrocalcic): the carbonates form laterally continuous concentrations resulting in one or more superposed petrocalcic horizons. The vertical complete succession from the top is: finely layered, over thicker compact layers (slabs), over thick noncompact layers, over a massive horizon. There are four main types of vertical successions of petrocalcic horizons:
  - single massive horizons;
  - layered noncompact horizons, over unlayered horizons;
  - finely layered horizons, over layered noncompact horizons, over unlayered horizons;
  - finely layered horizons, over layered compact horizons, over layered noncompact horizons, over unlayered horizons.

Petrocalcic horizons always have sharp upper boundaries and carbonate contents that decrease with depth. A gradual transition and discontinuous concentrations of carbonate mark the lower boundary of the petrocalcic horizon. The solum above the petrocalcic horizon is generally between 10 and 50 cm thick.

LATERAL DISTRIBUTION OF THE CALCIC AND PETROCALCIC HORIZONS

As with vertical transitions, progressive lateral transitions frequently occur between different forms of calcic and petrocalcic horizons. From this it can be concluded that the vertical and horizontal structures of calcium carbonate accumulation result from the same mechanisms.

In a landscape, the lateral distribution of the calcic and petrocalcic horizons is mainly a function of topography and age.

Along a slope or pediment, calcic horizon development increases downslope. In a complete toposequence (catena), differentiation progresses from a soil with

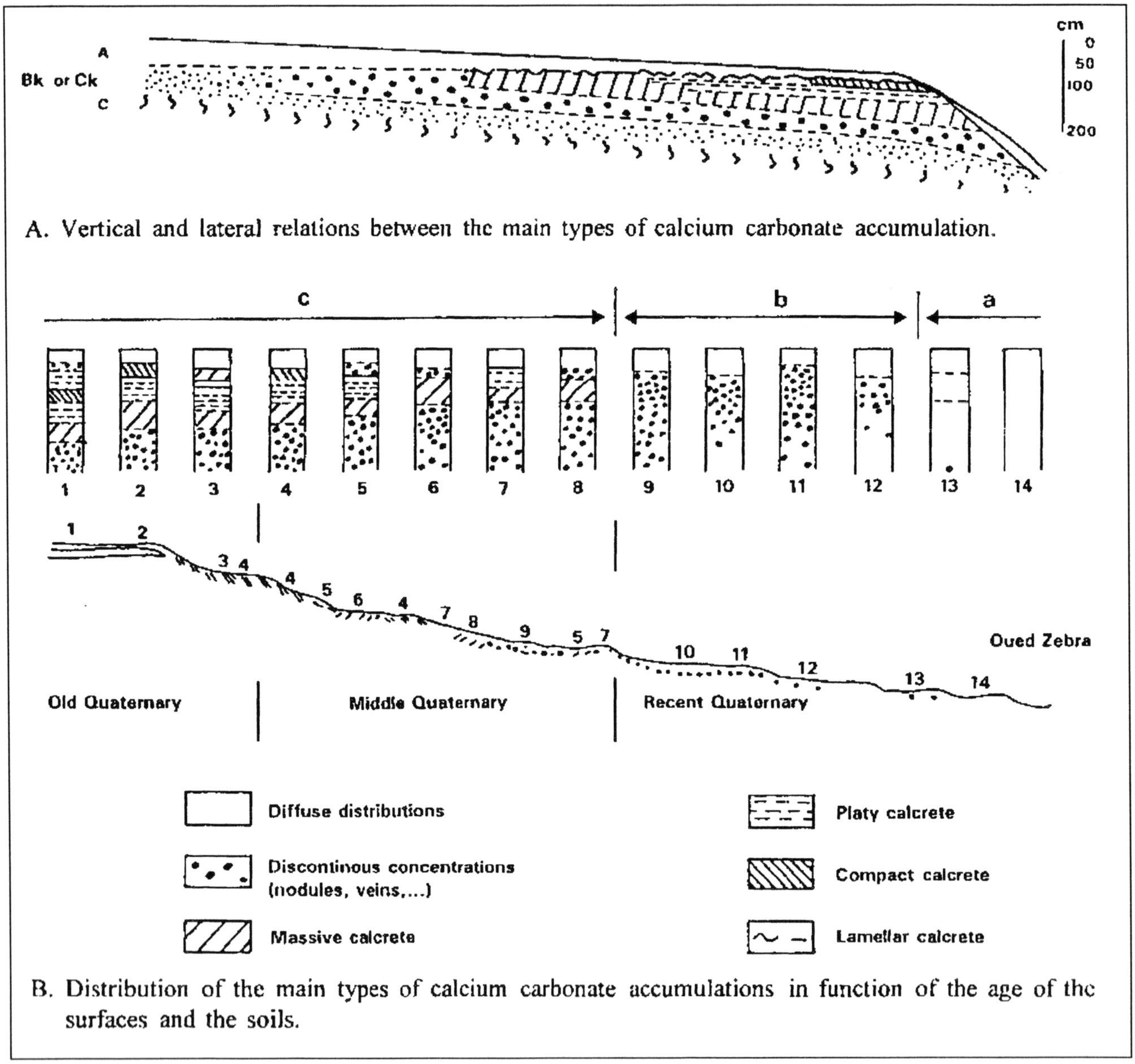

Fig. 3 Relationships between calcic and petrocalcic horizons in space and time (North Morocco). The length of the sequences may vary between some tens and several hundreds of meters; the difference in altitude, between the old and recent Quaternary surfaces, is some tens of meters. (From Ref. 1.)

minimal calcic horizon development upslope, to a soil with moderate calcic horizon development immediately downslope, and finally to soils with more and more strongly developed petrocalcic horizons, the different types and superposition of petrocalcic horizons appearing successively (Fig. 3A).

In time, calcic horizons evolve (Fig. 3B) from diffuse carbonate distributions to discontinuous accumulations (pseudomycelia, cutans, nodules, or veins) to unlayered petrocalcic horizons to layered noncompact petrocalcic horizons to slab. The finely layered (ribboned) horizons can exist as soon as noncompact layered horizons appear. On the other hand, decarbonation of the A and B horizons above the calcic and petrocalcic horizons, does not increase with age. It is only on the younger recent Quaternary surfaces that a small decarbonation can be observed as these surfaces age. However, this decarbonation does not increase on the older surfaces and, when petrocalcic horizons appear, the calcium carbonate content of the upper horizons may increase, due to erosion and subsequent formation of the upper horizons from the calcrete.

These facts confirm the following interpretations:

- There is a logical order of appearance of the calcic and petrocalcic accumulations and horizons; this logical order is the same in space, vertically and laterally, and in time; these accumulations and horizons are thus genetically linked, by toposequences and by chronosequences.
- The vertical leaching of the calcium carbonate, that impoverishes the upper horizons in favor of the calcic and petrocalcic horizons, is a limited phenomenon; the major part of the calcium carbonate that accumulates in the soils comes from lateral redistributions.

In arid and semiarid regions, which are the privileged domains of the petrocalcic horizons, calcic and petrocalcic horizons can occur in soils formed from noncalcareous or noncalcic rocks: this happens when landscapes upstream furnish calcium by lateral lixiviation. However, in very arid regions near the sea, very strong petrocalcic horizons occur in soils formed from noncalcareous, noncalcic rocks without possible upstream sources of calcium. So calcium carbonate can also arrive by air as calcareous dust and calcium from sea spray.

SOIL USE

Petrocalcic horizons in soils at depths of less than 50 cm limit the soil's agriculture uses, because of the physical (hardness and low porosity) and chemical (calcium carbonate, high pH) obstacles the horizon constitutes for the roots. It is possible to fracture the nonlayered and the layered noncompact petrocalcic horizons to facilitate penetration of water and roots for cotton, alfalfa, or fruit trees, but it is impossible to do that well when slabs are present. It costs too much money for an insufficient result. Chemically, the nonlayered, hard but noncompact petrocalcic horizons are more crop limiting because they contain too much active calcium and magnesium carbonate.

Layered petrocalcic horizons in some areas are used for construction of houses and building of roads.

REFERENCES

1. Soil Survey Staff. *Keys to Soil Taxonomy*. U.S. Department of Agriculture, Natural Resources Conservation Service, Second Edition, USDA-NRCS: Washington, 1999.
2. Ruellan, A. *Contribution à la Connaissance des Sols des Régions Méditerranéennes: les Sols à Profil Calcaire Différencié de la Basse Moulouya (Maroc Oriental)*; Mémoire ORSTOM n° 54, ORSTOM: Paris, France, 1971.
3. Ruellan, A. Calcisols. In *World Reference Base for Soil Resources, Draft*; Spargaren, O.C., Ed.; ISSS, ISRIC, FAO: Wageningen, Rome, Italy, 1994; 106–111.
4. Ruellan, A. Calcisols. In *World Reference Base for Soil Resources, Introduction*; Deckers, J.A., Nachtergaele, F.O., Spaargaren, O.C., Eds.; Acco: Leuven, Germany, 1998; 53–56.
5. Boulet, R. *Toposéquences de Sols Tropicaux en Haute-Volta: Equilibre et Déséquilibre Pédo-Bioclimatique*; Mémoire ORSTOM n° 62, ORSTOM: Paris, France, 1974.
6. Nahon, D.; Paquet, H.; Ruellan, A.; Millot, G. Encroûtements Calcaires Dans les Altérations des Marnes Eocènes de la Falaise de Thiés (Sénégal). Organisation Morphologique et Minéralogie. Sci. Géol. Bull. (Strasbourg) **1975**, *28*, 29–45.
7. Millot, G.; Nahon, D.; Paquet, H.; Ruellan, A.; Tardy, Y. L'Epigénie Calcaire des roches Silicatées Dans les Encroûtements Carbonatés en Pays Subaride, Anti-Atlas, Maroc. Sci. Géol. Bull. (Strasbourg) **1977**, *30* (3), 129–158.
8. Paquet, H.; Ruellan, A. Calcareous Epigenetic Replacement (Epigénie) in Soils and Calcrete Formation. In *Soils and Sediments, Mineralogy and Geochemistry*; Paquet, H., Clauer, N., Eds.; Springer: Berlin, Germany, 1997; 21–48.
9. Ruellan, A. Les Sols Salés et Alcalisés en Profondeur de la Plaine du Zebra (Basse Moulouya, Maroc); Premiers Résultats D'une Expérimentation Destinée à Etudier Leur Amélioration et Leur Evolution Sous Irrigation. *Proceedings of the 8th International Congress of Soil Science*, Bucharest, Romania, 31 August – 9 September, 1964; II, 937–948.
10. Millot, G.; Paquet, H.; Ruellan, A. Néoformation De l'Attapulgite Dans les Sols à Carapaces Calcaires de la Basse-Moulouya (Maroc Oriental). C.R. Ac. Sci. Paris **1969**, *268*, 2771–2774.

pH

Grant W. Thomas
Professor Emeritus, University of Kentucky, Lexington, Kentucky, U.S.A.

INTRODUCTION

pH and buffering (reserves) of soils are controlled by a number of soil components such as clay minerals, organic matter, oxides of aluminum and iron and compounds of calcium and sodium. In a few cases, the oxidation of sulfide overrides these, more common controllers of pH, causing exceedingly low pH values. Including the latter, a range of pH values from 2 to 10 can be encountered. In most soils, however, a pH range of 4 to 9 is more common. The chemical reactions that control the soil pH are usually what interest us. In other words, pH reveals the reactions which dominate the soils. As such, pH is very useful as a clue about what must be done to the soil to make it ideal for the purpose we desire. No other single soil measurement gives us so much information so easily.

WHAT SOIL pH VALUES MEAN

Although pH values are not adequate to determine lime requirement, they do indicate rather precisely what is going on in the soil in terms of the major chemical reactions. In addition, pH infers secondary reactions such as the availability of both primary and micronutrients. The major pH categories are described below, together with their significance.

PRESENCE OF FREE ACIDS

Soil pH values around 2 to 3 indicate the presence of a free mineral acid. In almost all cases, the acid is H_2SO_4 which generally arises from the oxidation of iron sulfides. The usual reason for their oxidation is exposure to the atmosphere during surface mining, excavation during construction, or drainage of submerged soils. Obviously, not all materials contain sulfides, but if they are present, exposure begins their oxidation, and the oxidation results in free sulfuric acid. This acid then begins to dissolve soil components, including the clay minerals. If pH in the range of 2 to 3 persists with time, this indicates that there is sufficient free acid present to dominate all soil reactions. The long-term effect is the virtual dissolution of clay minerals.

In addition to the long-term destruction of the soil, plant growth is almost totally inhibited and the cost of neutralizing the acidity will be very high. On a practical level, neutralization of free acid is too expensive to attain by depending of the income received from growing crops. Therefore, it usually is achieved only by high expenditures for agricultural limestone, commonly mandated by government regulation. For example, the use of 100 Mg of $CaCO_3$ per hectare is commonly required to neutralize the free acid. That amount of limestone commonly costs more than the value of the land treated (1).

PRESENCE OF Al^{3+} IONS

At pH values of 4 to 5, trivalent exchangeable aluminum is always present in soils unless they are quite high in organic matter. The pH of the soil is dominated by the hydrolysis reaction:

$$Al(H_2O)_6^{3+} \rightarrow Al(H_2O)_5OH + H^+$$

which has a minimum pH of 3.84 in montmorillonite and a maximum pH of 4.89 in hectorite (2).

This pH range indicates problems with growth of most crop plants because root growth is strongly inhibited by the presence of Al^{3+}. Nevertheless, the case is not nearly as serious as with free mineral acid for two reasons: First, plants are not killed outright and second, correction of the problem usually is not an economic impossibility. Instead of requiring 100 Mg of $CaCO_3$ per hectare, a range of 5 to 10 Mg per hectare is generally sufficient to correct the problem. And, this treatment will last for several years and will produce crop yield responses that will pay for the treatment.

P

PRESENCE OF HYDROXY-AI

At pH 5.5 and above, there is no longer any significant trival-ent, exchangeable Al (3). The chemistry is now dominated by complex, polymerized hydroxy-Al of the general composition:

$$[Al(OH)_{\sim 2.5}]_x^{\sim(0.5+)x}$$

These polymers are not exchangeable from either clay mine-rals or organic matter (4) so that they are reactive only with the H^+ ions formed by nitrogen fertilizer (or acid rain) or OH^- ions that come from application of lime. The adsorption of H^+ tends to reduce the OH^- ratio and the OH^- tends to raise it—to a maximum of three, where the charge becomes zero.

Because of these characteristics, hydroxy-Al, and in some soils, hydroxy-Fe, form the so-called pH-dependent charge in soils. In the case of clay minerals, the hydroxy-Al polymers lose positive charge by gaining hydroxyls. In the case of organic matter, the hydroxy-Al polymers block the strongest carboxylic groups effectively reducing the ionization of the remaining carboxylic groups at a given pH (5, 6). Thus, in both mineral and organic fractions, the effect of the hydroxy-Al polymers is to make the soil an apparently weaker acid that greatly increases the buffer capacity.

Although hydroxy-Al is not directly toxic to plants as trivalent-Al is, it does have a negative effect of short-term phosphorus availability because it represents a sink for soluble phosphorus. Whether this has a practical effect on long-term phosphorus availability is not at all clear because Al-phosphate is surely one of the dependable sources of phosphorus in acid soils (7).

NEUTRAL SOIL: pH 6 TO 7

A soil with pH 6 to 7 is free of trivalent exchangeable aluminum, has variable levels of non-exchangeable hydroxy-Al, and has Ca as the dominant exchangeable cation. In addition, there are no elements such as manganese present in toxic quantities. On the other hand, manganese, zinc, boron, and iron are usually available in amounts adequate for good crop growth. The 6 to 7 pH range increases molybdenum availability, which, in turn, favors crops that depend on nitrogen fixation, such as soybeans and alfalfa. This pH also favors mineralization of organic nitrogen with the eventual formation of nitrate-nitrogen.

Hence, the pH range of 6 to 7 generally indicates optimum soil conditions for crop growth, with the fewest negative consequences.

PRESENCE OF CALCIUM CARBONATES

Passing to alkaline soils, the presence of excess $CaCO_3$ in the soil is diagnostic in the pH range of 7.6 to 8.3, depending on the partial pressure of CO_2. In this pH range, a fizz test with HCl will reveal the presence of $CaCO_3$, which shows there is no need to worry about the use of lime because soil acidity will not develop. In addition, soil structure usually is superior because Ca so dominates the system. On the other hand, availability of certain elements such as zinc, manganese, and iron can be reduced enough to cause deficiency problems. On the whole, a calcareous soil is favorable but some problems can arise. Knowing whether a soil is calcareous or not can be very important in a landscape that has both calcareous and noncalcareous soils, a common occurrence. Treatments for soils and crop variety response will vary with the presence or absence of $CaCO_3$ in the soil.

PRESENCE OF SODIUM CARBONATE

As pH values rise above pH 8.3,the cause is invariably the presence of sodium carbonate. As the problem worsens, the pH can rise to 9 and even 10. In this pH range, most exchangeable calcium is precipitated as $CaCO_3$, leaving the soils dominated by exchangeable sodium. Soil structure gradually deteriorates because both clay minerals and organic matter are dispersed. As a rule, water infiltration is greatly reduced as is internal drainage in the soil. The result is that dry soils stay dry and wet soils stay wet so that crop growth is strongly inhibited and soil productivity declines radically. Rooting is poor and as an additional restrictive factor, iron deficiency is common.

Because this transformation to sodium-dominated soils does not occur uniformly, the usual situation is a normal landscape interspersed with sodium-affected circles. Crop growth is so much inferior within these circular zones as compared to growth in the normal soil areas that it is quite easy to determine the proportion of the land that is strongly affected by sodium. It is very difficult to reverse the soil degradation and make these circular areas productive again. This often results in abandonment of the fields because of poor crop production. Probably the exception to this is when there are very high value crops such as citrus where treatment with sulfuric acid and/or gypsum can recover production with time and expense.

Table 1 Diagnostic pH values and the dominant reaction at each one

pH	Dominant reaction
2–3	Free mineral acid, usually H_2SO_4
4–5	Presence of trivalent, exchangeable Al
≥5.5	Presence of hydroxy-Al
≥6.0–6.7	An ideal pH for many crops
7.6–8.3	Presence of $CaCO_3$
9–10	Presence of Na_2CO_3

A summary of the foregoing discussion on the significance of soil pH is given in Table 1. These values are extremely important in field diagnosis; a table such as this, carried in the clipboard or in the head is of constant value in making recommendations.

STRENGTH OF SOIL ACIDITY

As was shown clearly in 1931 (8), there is a wide difference in the strength of soil acidity from one soil to the other. Mehlich (9) showed similar results and attempted to relate the strength of soil acidity to the clay mineralogy present. Briefly, he showed that montmorillonite was strongly acid whereas kaolinite was much weaker.

Rich (10) showed that interlayers of hydroxy-Al greatly reduced apparent acid strength when added to clay minerals. Coleman & Thomas (11) showed the effect of adding both hydroxy-iron and aluminum to montmorillonite. Whereas pure montmorillonite had strongly acid titration curves, the addition of hydroxy iron or aluminum compounds to the clay gave titration curves of very weak acids (Fig. 1).

Both hydroxy Fe and Al compounds act as hydrogen ion sinks, which give exactly the same effect as if the acid were very slightly disassociated. They also act as OH^- sinks which makes the pH very difficult to raise. The positive side of this high buffering capacity is that pH remains quite stable even with the use of high nitrogen fertilizer rates or with a degree of overliming.

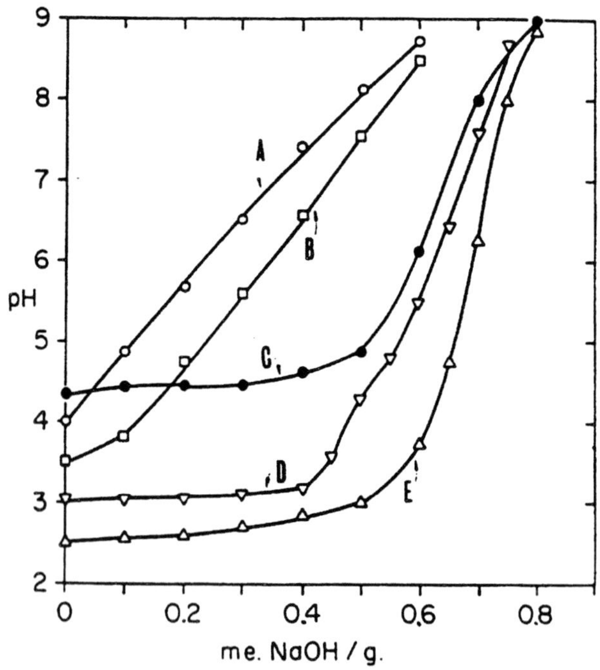

Fig. 1 Buffer curves (in 1*N* KCl) for hydroxy-Al and -Fe complexes with montmorillonite and for Al-, Fe-, and H-montmorillonites: A. hydroxy-Al complex; B. hydroxy-Fe^{3+} complex; C. Al-clays; D. Fe^{3+}-clay; E. H-clay. (From Ref. 11.)

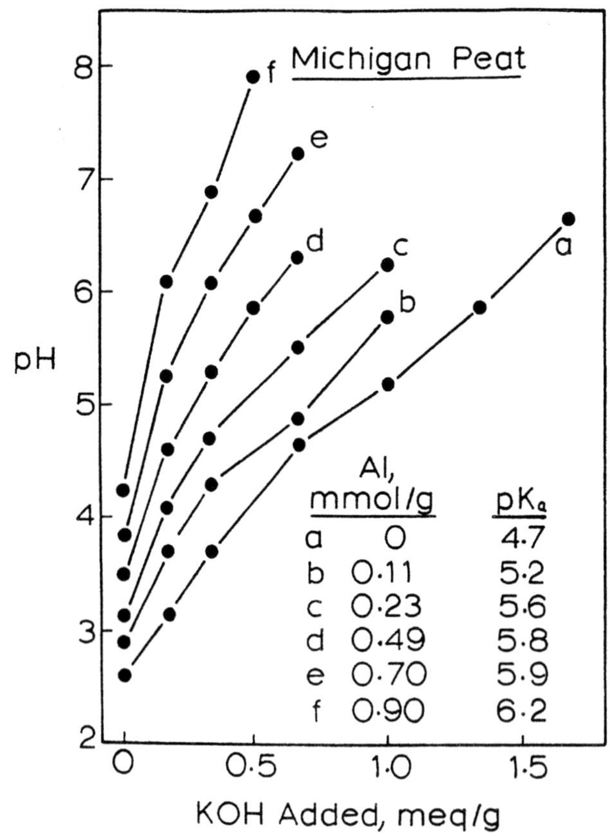

Fig. 2 The effect of adsorbed Al on Michigan peat titration curves and on pKa values. (From Ref. 6.)

ORGANIC MATTER

In the case of organic matter, the differences already mentioned in clay minerals—hydroxy Fe or Al are also observed. The strength of carboxylic acids in organic matter should be in the order of pKa~4, where, in fact, most soil organic matter has an apparent pKa of around 6. Martin and Reeve (5) showed why this was so. Organic matter contained Fe and Al compounds and as they were removed, the pKa of organic matter approached the expected pKa of 4. Schnitzer and Skinner (12) demonstrated how common the interaction between hydroxy Fe and Al and organic matter really was (6). Fig. 2 shows how strongly the adsorption of Al by organic matter affects apparent acid strength. The exchangeable Al in organic matter is almost insignificant. Instead, Al becomes hydroxylated and virtually non exchangeable. At pH values of ≥4, the average composition of Al adsorbed is $Al(OH)_2^+$.

The fact that organic matter essentially immobilizes exchangeable Al helps explain why it can be so beneficial in acid soils. In two artificial soils made of sand and montmorillonite and with 10% organic matter in one, the amount of exchangeable Al was greatly affected by organic matter, as shown in Table 2 (6).

Table 2 The effect of organic matter on exchangeable Al in montmorillonite + sand at given pH values

		Exchangeable Al cmol/kg
pH	**No organic matter**	**(+) 10% organic matter**
4.0	1.70	0.25
4.5	1.20	0.05
5.0	0.80	0.02

OTHER FACTORS THAT AFFECT SOIL pH

Salt Content

The salt content of the soil-water slurry greatly affects soil pH. Salt content tends to be higher in soils that have been fertilized heavily, soils that have been irrigated with fairly high salt water, and samples that have been taken in the fall vs those taken in the spring (13). Increased salt content tends to give lower pH values in temperate zone soils. This pH variation can be from one or two tenth's, up to perhaps a half unit. In tropical soils, this same salt effect can be exactly the opposite, causing a rise in pH of about the same amount (14).

This salt problem has led many researchers to add salts ranging from 0.001 M to 1 M to soil-water slurries in an attempt to "standardize" the pH values. The real effect is that the pH values all change, at the highest salt concentration up to 2 units (15) with the result that there is another set of relationships to record mentally for use in interpretations. In practical terms, adding salt is not generally worth the trouble. It certainly is true that some extreme conditions will give pH values that cause some misinterpretations. But, the vast majority are good enough without adding salt to make reasonable deductions and recommendations.

Dilution

Another factor that can affect pH is the soil:water ratio in the slurry. Diluting soils by 10 times (16) will raise pH by about 0.4 unit. Since most slurries range from 1:1 to 1:2.5, the change in pH is only about 0.1 unit, which is not a big problem.

Carbon Dioxide

Carbon dioxide content of calcareous soils can affect the pH greatly, pH 8.3 at the atmospheric partial pressure (0.0003 atm) to 6.2 at 1.0 atm. Because soil levels of CO_2 are almost always higher than those in the above ground atmosphere, there is a tendency for the reaction below to move toward the right.

$$CaCO_3 + H_2CO_3 \leftrightarrow Ca(HCO_3)_2$$

On dried and ground soil, the effect should be small but on pH measurements done in the field, this effect can be quite important.

Suspension Effect

In most soils, there will be a reduction in pH when the electrodes are placed in the suspension as compared to the supernatant solution. (In tropical soils, the effect may be exactly the reverse.) This suspension effect has been interpreted in two ways. First, according to Marshall (17), the higher activity of H^+ ions near the clay surface causes the lowest pH. Second, according to Coleman et al. (18) the suspension effect is mostly from the electrical charge on the soil which differentially affects the mobilities of K^+ and Cl^- in the calomel electrode, not the glass electrode. In a strongly negatively-charged soil, the mobility of K^+ would be high and that of Cl^- low while in a positively-charged tropical soil the opposite would occur.

It is generally accepted that the suspension effect is largely spurious (19). Nevertheless, there is evidence that H^+ activity is somewhat higher at the clay surface (20).

CONCLUSIONS

Soil pH is one of the most useful, economic and rapid measurements that can be made on soils. Remembering the reasons for unusually low and high pH values, it also interprets well the intermediate values rather precisely. Because of this, a rapid soil pH value can be one of the most useful measurements taken on the soil, whether the news is favorable or bad.

REFERENCES

1. Miller, W.L.; Godfrey, C.L.; McCully, W.G.; Thomas, G.W. Formation of Soil Acidity in Carbonaceous Soil Materials Exposed by Highway Excavations in East Texas. Soil Sci. **1976**, *121*, 162–169.
2. Frink, C.R.; Peech, M. Hydrolysis and Exchange Reactions of the Aluminum Ion in Hectorite and Montmorillonite Suspensions. Soil Sci. Soc. Am. Proc. **1963**, *27*, 527–530.
3. Thomas, G.W.; Hargrove, W.L. The Chemistry of Soil Acidity. In *Soil Acidity and Liming*, 2nd Ed.; Adams, F., Ed.; Agron Monogr. 12; ASA and SSSA: Madison, WI, 1984; 3–56.
4. Thomas, G.W. Beyond Exchangeable Aluminum: Another Ride on the Merry-Go-Round. Commun. in Soil Sci. Plant Anal. **1988**, *19* (7–12), 833–856.
5. Martin, A.E.; Reeve, R. Chemical Studies of Podzolic Illuvial Horizons. III. Titration Curves of Organic Matter Suspensions. J. Soil Sci. **1958**, *9*, 89–100.
6. Hargrove, W.L.; Thomas, G.W. Effect of Organic Matter on Exchangeable Aluminum and Plant Growth in Acid Soils. In *Chemistry in the Soil Environment*; Am. Soc. Agron., Soil Sci. Soc. Am.: Madison, WI, 1981.
7. Chu, C.R.; Moschler, W.W.; Thomas, G.W. Rock Phosphate Transformation in Acid Soils. Soil Sci. Soc. Am. Proc. **1962**, *26*, 476–478.
8. Pierre, W.H.; Scarseth, G.D. Determination of the Percentage Base Saturation of Soils and Its Value in Different Soils at Definite pH Values. Soil Sci. **1931**, *31*, 99–114.
9. Mehlich, A. The Significance of Percentage Base Saturation and pH in Relation to Soil Colloids. Soil Sci. Soc. Am. Proc. **1943**, *7*, 167–174.
10. Rich, C.I. Aluminum in Interlayers of Vermiculite. Soil Sci. Soc. Am. Proc. **1960**, *24*, 26–32.
11. Coleman, N.T.; Thomas, G.W. Buffer Curves of Acid Clays as Affected by the Presence of Ferric Iron and Aluminum. Soil Sci. Soc. Am. Proc. **1964**, *28*, 187–190.
12. Schnitzer, M.; Skinner, S.I.M. Organo-Metallic Interactions in Soils. 3. Properties of Iron- and Aluminum-Organic Matter Complexes, Prepared in the Laboratory and Extracted from a Soil. Soil Sci. **1964**, *98*, 197–203.
13. Baver, L.D. Factors Affecting the H^+-ion Concentration of Soils. Soil Sci. **1927**, *23*, 399–414.
14. Van Raij, B.; Peech, M. Electrochemical Properties of Some Oxisols and Alfisols of the Tropics. Soil Sci. Soc. Am. Proc. **1972**, *36*, 587–593.
15. Okwsami, T.A.; Rust, R.H.; Juo, A.S.R. Reactive Characteristics of Certain Soils from South Nigeria. Soil Sci. Soc. Am. J. **1987**, *51*, 1256–1262.
16. Davis, L.E. Measurements of pH with the Glass Electrode as Affected by Soil Moisture. Soil Sci. **1943**, *56*, 405–422.
17. Marshall, C.E. *The Physical Chemistry and Mineralogy of Soils*; John Wiley & Sons, Inc. New York, 1964; Vol. I.
18. Coleman, N.T.; Williams, D.E.; Nielsen, T.R.; Jenny, H. On the Validity of Interpretations of Potentiometrically Measured Soil pH. Soil Sci. Soc. Am. Proc. **1951**, *15*, 106–110.
19. Olsen, R.A.; Robbins, J.E. The Cause of the Suspension Effect in Resin-Water Systems. Soil Sci. Soc. Am. Proc. **1971**, *35*, 260–265.
20. Swobada, A.R.; Kunze, G.W. Reactivity of Montmorillonite Surface with Weak Organic Bases. Soil Sci. Soc. Am. Proc. **1968**, *32*, 806–811.

PLANET AND HUMAN SOCIETY, SOIL AS A HERITAGE OF

Fred P. Miller
The Ohio State University, Columbus, Ohio, U.S.A.

INTRODUCTION

The early Latin derivation of *Homo* (Latin for man) was *hemo*, meaning the earthly one; akin to the Latin word for earth or soil, namely, *humas*, and the Latin word for humans, namely, *humanus* (1). The history of humankind, and our ancestral Homo species, is the history of our relationship to the Earth and environment (2). Archeological evidence in many parts of the world reveals our penchant to alter the environment (3–5), be it to our boon or bane (6). Soil is a major component of the natural resource trinity (soil–sun–climate) from which humankind's sustenance is rooted. And soil is the common earthen parchment upon which its cultural signature is imprinted. Speech, tools, and fire (allowing for clustered habitation in inclement climates) formed the tripod of culture for Paleolithic man and his later counterparts (6). Today, the sophistication and capability of modern tools, the development of multiple energy sources, and the increase in population have resulted in the capacity for modern civilization to alter the environment in ways, and at scales, unprecedented in human history. As early as the mid-nineteenth century, human activity was recognized as a significant force in altering the environment (6–10). Even at the turn of the 20th century, geologists were calling humans the dominant geological force of the planet (9). Still, progress was measured in terms of our increasing control over nature, marching through the stages of our cultural evolution as if the Earth was simply the stage upon which this human drama was acted out (9).

The impact of progressive stages of civilization on the environment, and the tapping of the Earth's stocks of natural resources to sustain these civilizations, were largely ignored (9). Yet, there were those thinkers and writers who recognized that civilization was dependent upon soil and other natural resources for its preservation, and that humankind had failed in its appointed role as steward of the soil and natural resources heritage that sustains it and civilization itself (7, 8, 11).

Both nature and humans have left (and are still leaving) imprints on the Earth's soil resources. Soil is the Earth's equivalent to anatomical derma—a planetary vellum with the "signature" of pedogenic processes, such as climate, acting upon a variety of soil parent materials (e.g., sedimentary, igneous, or metamorphic bedrock; alluvium; glacial deposits; wind-blown silts or loess; and volcanic ash and lava). The morphology and character of the soils originating from these interactions are further modified by the topographic configuration of landscapes, the flora and fauna adapted to the resulting ecological niches, and the duration of these active processes. The spectrum of soils and ecosystems born of these pedogenic and ecological processes constitutes the natural resource heritage from which humankind, and civilization itself, are sustained. Agriculturally productive soils, occupying only a limited portion of the planetary soil resource pool, were precursors of the genesis of civilization. They are primary requisites for generating the necessary biomass to sustain the more than 6 billion humans now occupying this planet, with another 2 to 3+ billion projected to be here within the next four to five decades.

HISTORICAL PERSPECTIVE

The dawn of civilization was rooted in the soil. As Bradley (12) noted, "the fabric of human life has been woven on earthen looms. It everywhere smells of the clay." And an old saying, attributed to the Chinese, states that, "Man—despite his artistic pretensions, his sophistication, and his many accomplishments—owes his existence to a six-inch layer of topsoil and the fact that it rains."

The domestication of plants and animals, i.e., the beginning of agriculture, occurred nearly 10,000 years ago (2, 5, 13–17), and released the human species from its bondage to a hunting and gathering existence since its origin. Evidence of plant and animal domestication in the early Holocene has been documented on several continents (5, 15–17). The effect of this profound and transforming revolution in human history was to increase the carrying capacity (18) of a region and its ecosystem. Archeological evidence within the Fertile Crescent and areas around the Mediterranean suggests that their inhabitants were domesticators of animals and cultivators of the earliest founder cereals, such as einkorn wheat, emmer wheat, and barley, and founder legumes such as lentil, chick pea, pea, and bitter vetch (14). This allowed the accumulation of surplus food supplies that underwrote major population expansion, the division of labor, which released many from the task of gathering–producing food, and the rise of cities and urban states (13, 19).

Diamond (20) posits that the development of agriculture was "the worst mistake in the history of the human race." Most historians, anthropologists, and others see this agricultural transformation of 10,000 years ago as "the greatest single step forward in the history of mankind"—the most momentous turn in the progress of humankind (13, 21). In the words of Thomas Hobbes, life before agriculture was "nasty, brutish, and short" (21).

An axiom of agricultural geography, still valid today, holds that cultivation-based agriculture is predominantly located on soils derived from relatively young geologic parent materials such as alluvium, glacial deposits, loess, and volcanic ash. Soil provides the medium through which nutrient flows and energy conversions take place. Thus, it is not coincidental that early agriculture, and the civilizations that sprang from it, originated in broad alluvial flood plains and adjacent foothills. They were the Tigris and Euphrates rivers (the Fertile Crescent), the Jordan Valley, the Nile River, the Indus River, the major rivers of Asia, and intermontane (alluvial) valleys of the Middle East. Similarly, other early cultures were nurtured by agriculture rooted in soils derived from loess and other relatively young soil parent materials (5, 9, 16, 22). Contemporary agricultural equivalents include the North American Corn Belt (glacial deposits and loess), the San Joquin Valley of California (alluvium), the wheat region of the northwestern United States (volcanic deposits), the rice cultures of Asia (alluvium), and the Chernozem-Black Soils of the Russia grain belt (glacial materials).

The advent of the agricultural revolution was accompanied by increasing ecological manipulation. The development of the ox-drawn hoe, followed by the plow, occurred approximately 5,000 to 6,000 years ago throughout Mesopotamia, Egypt, and China (13, 23, 24). Then came plows of increasing sophistication and improved design, including seeder plows that simultaneously allowed the opened furrow to be planted, the design of which is still used today in parts of the Middle East (13). These developments ratcheted up the capacity to manipulate the environment and commonly resulted in deleterious environmental impacts, with soil erosion being the Achilles heel of cultivation-based agriculture. Hillel (13) has opined that, contrary to the prophet Isaiah, the plowshare became more destructive than the sword. Similarly, animal herding and overgrazing of hillsides exacerbated the environmental impacts of agriculture.

Another early technological innovation accompanied the advance of cultivation-based agriculture: irrigation. The early societies, cultures, and civilizations that developed in the arid and semiarid regions of the Middle East needed to manipulate the hydrologic cycle for crop production. The earliest evidence of irrigated farming was found in the Jordan River Valley, within which lie the ruins of the ancient city Jericho, dating back perhaps eight millennia (23). It is not surprising that irrigation was an early achievement, given the juxtaposition of the alluvial soils and the watercourses from which the alluvium was derived.

Canal systems were built throughout much of the Middle East to intercept portions of the adjacent river flows and distribute the water to crop fields. These elaborate irrigation networks relied on gravity flow, although there were gates and other features to accommodate the rise and fall of the rivers' seasonal discharges. Water-lifting devices, such as Archimedes' screw (tambour) and the animal-powered water wheel (sagia) were designed to lift water from irrigation canals and rivers to crop fields (25).

Irrigation came to be relied on to support civilizations throughout the history of Mesopotamia and much of the semiarid Middle East. Today these cultures are characterized as hydraulic civilizations (26), riverine (27, 28), or irrigation-based civilizations (25). But civilizations that try to sustain themselves in these rainfall-deficient environments are vulnerable to two soil-related problems: silt and salt. As these civilizations' populations expanded, and agriculture and grazing moved further up-slope into the watersheds supplying the rivers and irrigation systems, accelerated soil erosion began its insidious gnawing away at the soil resources of the uplands. The silts and sediments carried downstream eventually flowed into the irrigation canals and networks, clogging them and reducing their effectiveness. Likewise, the river channels themselves became silt-laden, raising the riverbed, rendering the river unstable and prone to flooding the adjacent fields. This situation also raised the water table, bringing with it salt. Thus, in addition to tending to the agriculture and irrigation systems, intensive diking and levee building were required—the classic example being China' s Yellow River (24). Removing this silt and diking river courses became labor-intensive. Some cultures resorted to enslaving others for labor, capturing them as the spoils of conflict between nation-states.

Without drainage systems and intensive water management in these water-deficient environments, irrigation caused water tables to slowly rise, exacerbated by the rising riverbed. These rising water tables brought with them salts that eventually wicked to the soil surface, rendering the soils sterile. Coupled with invasions from other peoples, competition for water, and internal conflicts and weaknesses, the collapse of civilizations throughout Mesopotamia and the Middle East reads like an historical casualty list in the "Graveyard of Empires," including once-mighty Babylon itself (24).

The scarred and soil-denuded landscapes of this region and many areas around the Mediterranean, attest to the ravages of soil erosion caused by cropping or overgrazing the hillsides. Many structures and entire cities of these early civilizations are entombed in the sediments and salts unleashed by their mismanagement. It is ironic that the soil resources that were the heritage of these civilizations eventually became the materials contributing largely to the collapse and burial of these empires. Indeed, humankind had deeply etched its early history into the soil that became the heritage of subsequent civilizations.

The rocky and barren skeletal remnants of the upland landscapes in these regions stand in stark contrast to the original soil resources and vegetative cover. The lush cedars of Lebanon, once covering more than one half million hectares, were clear-cut for the ships of Phoenicia, for King Solomon's temple, to make way for agriculture, and other uses. Today, only four small (few hectares) remnant groves remain. Standing on Mount Nebo overlooking the Jordan Valley, the site where Moses once saw the lush land of "milk and honey," one sees today a denuded and decimated land, capable of producing only a small fraction of its original potential (24, 29). But within these degraded ecosystems is a lesson that is still valid today. Those soils that were protected from erosion by terraces and other measures are still being cultivated as they had been for two or more millennia.

The Nile Valley has nurtured and sustained more than five uninterrupted millennia of civilizations (13, 30, 31). The duration of the Nile River civilizations and multiple collapses of Mesopotamian civilizations, are due to the different soil and water regimes of the two riverine ecosystems. The scourges of siltation and salinization were not as severe along the Nile during its annual pulses as they were in the Tigris–Euphrates plain. Thus, the land of Egypt could remain perennially cultivated and productive while the land of Mesopotamia suffered degradation (13). The building of the Aswan High Dam in the 1960s may yet be problematic for Egypt's agriculture as water tables are rising, requiring the country to invest in artificial drainage systems (13) to stem the scourge of salinity to which it was so long immune.

The influence of soil on civilizations was not lost on the Greek or Roman empires. Plato offered the reason that Attica, a region in southeast Greece, in former times could support a soldiery exempt from the toil of farming: Its soils, as is proved by the remnants now left, surpassed all others in fertility (9). The Greek poet Hesiod and Roman writers such as Virgil, Pliny, Varro, and Columella recognized the human impacts on nature and the importance of soil quality, particularly soil fertility and its conservation, to sustain civilization and its cultures (9).

The Book of Genesis in the Old Testament had a profound impact on the formation of conceptions about the relationship of humans to the Earth. The commandment was to be fruitful and multiply, and take possession of the Earth. In much of western culture, humankind, by divine authority, assumed a powerful control over nature that led to seeing its place in manipulating nature, which had been so wonderfully designed by the Creator (9). By the seventeenth century, the classical notion of a design in nature, the Old Testament ideas, and the new passion for science coalesced and provided the stimulus for the study of nature. The concept of civilization cooperating with nature began to unfold. By the nineteenth century, especially in the works of Marsh (7), it is recognized that humankind's unsteward-like manipulations of nature were upsetting its balance and harmony (9).

Population growth, the progressing of civilization, and increased pressure on the environment do not necessarily have to lead to accelerated environmental damage, although most of human history has proven otherwise. As Butzer (32) noted, we can learn much about the environmental successes and failures through the study of human history and the lessons of our human heritage encoded in the settlement and land-use histories.

CONTEMPORARY PERSPECTIVE

More than two centuries ago, Malthus (33) recognized the decreased "power of the land" resulting from human-induced land degradation. During the eighteenth century, European powers were still vying for dominance, control, and settlement of the North American continent. Settlement of the continent was accompanied by the same exploitive behavior that occurred throughout history. It was during this time that Marsh (7) and others (6, 8–10) recognized the significance of human impacts on the environment and the historical arrogance toward nature (34–37). White (34) argued that this attitude is dominant in western traditions stemming from westerners' religious beliefs. But Tuan (37) argued that the tendency toward environmental degradation, and the desire to maximize one' s well-being characterize all human existence. White (34) stated that "the emergence in widespread practice of the Baconian creed that scientific knowledge means technological power over nature can scarcely be dated before about 1850. Its acceptance as a normal pattern of action may mark the greatest event in human history since the invention of agriculture, and perhaps in nonhuman terrestrial history as well."

The early American settlements became the catalyst for mass migrations of Europeans to this new land, migrations that lasted for three centuries. The vast and resource-rich continent that lay before these settlers, and the speed with which it was populated, are unprecedented in history (29). As a testimonial to the resources that these settlers inherited, the dense forests were considered an obstacle to agriculture even though they were utilized for fuel and timber. The volume of this timber resource was so large that the center of the commercial lumbering industry did not move beyond western New York until after 1850 (36). Between 1850 and 1910, American farmers cleared more forest than in the previous 250 years—about 77 million hectares (190 million acres), equivalent to clearing 35 square kilometers (13.6 square miles) every day for 60 years (38).

During these westward migrations, the vastness of the resource base, and the open and cheap land areas still available to the West, became the settler's talisman, and worked against a psychology of permanence (36). The knowledge that these western lands were still available tended to salve the anxiety of failure, and was not conducive to fostering a conservation ethic or to promoting a sense of stability. So plentiful was the North American natural resource heritage that its vastness betrayed its vulnerability.

To illustrate this impermanence syndrome and the environmental impact of its perpetrators, Trimble (39) cited one wit of the late 1830s, who summarized the situation in the southern Piedmont by noting that "the scratching farmer's cares and anxieties are only relieved by his land soon washing away. As that goes down the rivers he goes over the mountains." Gray (40) also pointed out the tendency to deplete land and then migrate west by stating that: "Over the upland soils from Virginia to Texas the wave of migration passed like a devastating scourge. Especially in the rolling piedmont lands the planting of corn and cotton in hill and drill hastened erosion, leaving the hillsides gullied and bare." An 1853 appraisal of Laurens County, South Carolina (39) was written in apocalyptic prose: "The destroying angel has visited these once fair forests and limpid streams. The farms, the fields are washed and worn into unsightly gullies and barren slopes—everything everywhere betrays improvident and reckless management."

As northern and southern migrations into the semiarid portion of the new nation continued, the exploitation of the soil resources under the onslaught of these migrations was exacerbated by yet another human imprint on the soil—the great Dust Bowl era. The vast expanse of prairie encountered by the western migration was unparalleled in the history of human settlements. The European background of the encroaching settlers, and any experience that may have been acquired in the humid East, left unprepared those who first entered the tallgrass prairie and then the semiarid shortgrass region. The latter ecosystem proved the undoing of those who broke the sod. Tempted by their tradition, lured by the bait of immediate profits, and encouraged by financial and industrial interests (36), the settlers' breaking of the sod in this region, coupled with the vagaries of its climate, triggered one of the great ecological disasters of human history. On May 12, 1934, the droughts common to this marginal-cropping region resulted in the first of many dust storms. It was a tragic event in American history that scarred both the land and its people. The impacts of this Dust Bowl were felt from Texas to the Dakotas, damaging some 60 to 80 million ha of land (41).

The commentary of various witnesses to the erosive demise of many landscapes provides a unique description of an ecological problem as seen through the eyes of those who experienced it personally. Such accounts echo a common concern, if not resignation, among those who wrote about it. Recorded in these accounts is an intensity of feeling and despair reserved only to those who are stunned by the immediacy and personal touch of a tragedy. Its shock and impact have not been tempered by either time, healing the landscape, or by the muffled accounts of historians.

Sears (36) quotes the owner of a large tract in the shortgrass country that had been plowed for wheat in the 1930s: "We're through. It's worse than the papers say. Our fences are buried, the house is hidden to the eaves, and our pasture which was kept from blowing by the grass, has been buried and is worthless now. We see what a mistake it was to plow up all that land, but it's too late to do anything about it." Others were not so articulate about the problem. They simply packed their few belongings and headed West, much as they and their predecessors had done when the water-induced erosion to the East had rendered the land scarred and unyielding. These hardy but tragic people became the human pulp for Steinbeck's *The Grapes of Wrath*. They left behind the scars of a squandered heritage: a damaged ecosystem totaling many tens of millions of hectares (42).

In response to this ecological disaster, a new federal agency was formed in 1933, the Soil Erosion Service, which became the Soil Conservation Service in 1935. Soil conservation became a national policy, although its implementation was based on voluntary participation of land owner-operators baited with the incentives of technical and financial assistance. As the necessity to manage ecosystems more holistically became understood, and the concept of ecological sustainability entered the global lexicon, agencies and institutions took a more comprehensive view of ecosystem management. During the 1990s, the Soil Conservation Service was renamed the Natural Resources Conservation Service and many federal and state agencies adopted sustainable ecosystem management as their mantra.

CONCLUSION

Is modern society, with all of its technological sophistication, still bound by the limits of its environmental heritage as were previous societies and cultures? Or do science and technology allow humankind to pursue its prerogatives, unhindered from any obligation for stewardship and sustainable management of its natural resources and soil heritage? The answer to these questions is a resounding "no." The laws of nature and thermodynamics preclude any shortcuts to sustaining civilizations without the sustainable management of the ecosystems and ecological processes that undergird our sustenance and well-being. Even though humankind has opted out of the tyranny of natural selection (43), the inextricable interconnectedness of the environment and human well-being requires our understanding of the global environment. The physical sustainability of the biosphere is now seen to be inseparably bound with issues of economic development, social equity, and international peace and security (44).

One of the major challenges for the future is to continue to ratchet up the carrying capacity of our soil heritage while sustaining its inherent productive qualities and minimizing leakages of production inputs. Humans have already expanded the carrying capacity of their agricultural ecosystems a thousandfold (18) through such innovations as genetic manipulation of plants and animals and the development of synthetic nitrogen. There are only certain areas of our soil heritage that can accommodate the intensity of these ecological manipulations productively and sustainably. Our soil heritage is as vital to modern civilization as it was to our ancestral hunter–gatherer kin. As Smil (18) has noted, "our 'postmodern' civilization would do quite well without computers and software, without ATMs and the WWW but it would disintegrate in a matter of years without synthetic nitrogen fertilizers, and it would collapse in a matter of months without thriving (soil) bacteria. Our first duty is to take care of these true essentials."

Modern civilization is no less dependent upon the soil and ecological heritage than its foraging and pastoral forebearers. The causes of the Earth's major environmental problems are rooted in human behavior. Thus, our greatest challenge is to understand the limits and vulnerability of our soil and natural resources heritage, and behave accordingly.

REFERENCES

1. Thomas, William L., Jr., Ed.; *The Random House College Dictionary*; Random House: New York, 1988.
2. Hymas, Edward. *Soil and Civilization*; Harper and Row: New York, 1976.
3. Thomas, William L., Jr., Ed.; *Man's Role in Changing the Face of the Earth*; The University of Chicago Press: Chicago, 1956; 1–1152.
4. Redman, Charles L. *Human Impact on Ancient Environments*; Univ. of Arizona Press: Tucson, 1999; 1–255.
5. Roberts, N. *The Holocene: An Environmental History*; Second Edition; Blackwell: Oxford, 1998.
6. Sauer, Carl O. The Agency of Man on the Earth. In *Man's Role in Changing the Face of the Earth*; Thomas, William L., Jr., Ed.; The University of Chicago Press: Chicago, 1956; 49–69.
7. Marsh, George Perkins. *Man and Nature*; Charles Scribner and Co. New York, 1864; 1–577.
8. Buffon, George Louis Le Clerc. *A Natural History, General and Particular: Containing the History and Theory of the Earth, General History of Man, the Brute Creation, Vegetables, Minerals, etc.*, Translated by William Smellie Thomas Kelly and Co. London, 1866; Volume II, 151–186.
9. Glacken, Clarence J. Changing Ideas of the Habitable World. In *Man's Role in Changing the Face of the Earth*; Thomas, William L., Jr., Ed. The University of Chicago Press: Chicago, 1956; 70–92.
10. Humboldt, Alexander von. *Aspects of Nature in Different Lands and Different Climates; with Scientific Elucidations*; Translated from the French by Mrs. Sabine Lea and Blanchard: Philadelphia, 1849; 475.
11. Shaler, Nathaniel S. The Origin and Nature of Soils. *Twelfth Annual Report of the United States Geological Survey, 1890–1891*; Part I: Geology U. S. Government Printing Office: Washington, D.C., 1891; 213–345.
12. Bradley, John Hodgdon. *Autobiography of Earth*; Coward-McCann, Inc. New York, 1935.
13. Hillel, Daniel J. *Out of the Earth*; The Free Press, A Division of MacMillan: New York, 1991, 3–174.
14. Lev-Yadun, Simcha.; Gopher, Avi.; Abbo, Shahal. The Cradle of Agriculture. Science **2000**, *288*, 1602–1603.
15. Vasey, Daniel E. *An Ecological History of Agriculture, 10,000 B.C. – A.D. 10,000*; Iowa State University Press: Ames, Iowa, 1992; 23–43.
16. Harlan, J.R. Agricultural Origins: Centers and Non-Centers. Science **1971**, *174*, 468–474.

17. Brown, Kathryn. New Trips Through the Back Alleys of Agriculture. Science **2001**, *292*, 631–633.
18. Smil, Vaclav. *Feeding the World: A Challenge for the Twenty-First Century*; The MIT Press: Cambridge, Mass, 2000; ix-52.
19. Higham, C.F.W. Prehistoric Rice Cultivation in Southeast Asia. Scientific American **1984**, *250*, 138–146.
20. Diamond, J. The Worst Mistake in the History of the Human Race. Discovery **1987**, 64–66.
21. Russell, Emily W.B. *People and the Land Through Time; Linking Ecology and History*; Yale Univ. Press: New Haven, Conn., 1997; 111–130.
22. Narr, Karl J. Early Food-Producing Populations. In *Man's Role in Changing the Face of the Earth*; Thomas, William L., Jr., Ed.; The University of Chicago Press: Chicago, 1956; 134–151.
23. Ehrlich, Paul R.; Ehrlich, Ann H.; Holdren, J.P. *Ecoscience: Population, Resources, Environment*; Freeman: San Francisco, 1977.
24. Lowdermilk, W.C. *Conquest of the Land Through Seven Thousand Years*; Agri. Information Bull. No. 99, Department of Agriculture, Soil Conservation Service: Washington D.C., 1953; 1–30.
25. Hillel, Daniel J. *Rivers of Eden; The Struggle for Water and the Quest for Peace in the Middle East*; Oxford Univ. Press: New York, 1994; 1–231.
26. Wittfogel, K.A. The Hydraulic Civilizations. In *Man's Role in Changing the Face of the Earth*; Thomas, W. R., Ed. Univ. of Chicago Press: Chicago, 1956, 152–164.
27. Simmons, I. *Environmental History: A Concise Introduction*; Blackwell: Oxford, 1993; 2–3.
28. Goudie, Andrew. *The Human Impact on the Natural Environment*, 5th Ed.; The MIT Press: Cambridge, Mass, 2000, 21.
29. Carter, V.G.; Dale, T. *Topsoil and Civilization*; Revised Ed. Univ. of Oklahoma Press: Norman, OK, 1974; 3–62.
30. Butzer, Karl W. The Human Role in Environmental History. Science **2000**, *287*, 2427.
31. James, T.G.H. *An Introduction to Ancient Egypt*; Farrar Straus Giroux: New York, 1979; 17–96.
32. Butzer, Karl W. *Early Hydraulic Civilization in Egypt: A Study in Cultural Ecology*; Univ. of Chicago Press: Chicago, 1976; 1–134.
33. Malthus, Thomas R. *An Essay on the Principle of Population, as it Affects the Future Improvement of Society*; J. Johnson: London, 1798; 1.
34. White, L., Jr The Historical Roots of Our Ecological Crisis. Science **1967**, *155*, 1203–1207.
35. Bouillenne, R. Man the Destroying Biotype. Science **1962**, *135*, 706–712.
36. Sears, P.B. *Deserts on the March*; 4th; Oklahoma University Press: Norman, Oklahoma, 1980; 1.
37. Tuan, Y. Man and Nature. Landscape **1966**, *15* (3), 30–36.
38. Powell, Douglas S.; Joanne, L. Faulkner; David, R. Darr; Zhiliang, Zhu.; Douglas, W. MacCleery. *Forest Resources of the United States, 1992*; General Technical Report Rm 234, U.S. Department of Agriculture, Forest Service: Washington, DC, 1993; 1–20.
39. Trimble, Stanley W. *Man-Induced Soil Erosion on the Southern Piedmont, 1700–1970*; Soil Conservation Society of America: Ankeny, IA, 1974, 1.
40. Gray, L.C. *History of Agriculture in the Southern United States to 1860*; Publication 430, Carnegie Institution of Washington D.C. Waverly Press, Inc. Baltimore, MD, 1933; 1.
41. Bennett, H.H. *Soil Conservation*; McGraw-Hill Book Co: New York, 1939; 1.
42. Miller, Fred P.; Wayne, D. Rasmussen; L. Donald Meyer, Historical Perspective of Soil Erosion in the United States. In *Soil Erosion and Crop Productivity*; Follett, Ronald F., Stewart, Bobby A., Eds. American Society of Agronomy, Crop Science Society of America, and Soil Science Society of America: Madison, Wisc, 1985; 23–48.
43. Maddox, John. Positioning the Goalposts. Nature **2000**, *403*, 139.
44. Jasanoff, S.; Colwell, R.; Dressolhaus, M.S.; Goldman, R.D.; Greenwood, M.R.C.; Huange, A.S.; Lester, W.; Levin, S.A.; Linn, M.C.; Lubchenco, J.; Novacek, M.J.; Roosevelt, A.C.; Taylor, J.E.; Wexler, N. Conversations with the Community. Science **1997**, *228*, 2066–2067.

PLANT AVAILABLE WATER

Argyrios Gerakis
Arta, Greece

Joe T. Ritchie
Belton, Texas, U.S.A.

P

INTRODUCTION

Accurately evaluating the available soil water reservoir is vital to developing optimum management for rain-fed crop production in marginally dry regions. Often there is a linear relationship between plant available water and yield (1) and between plant available water and leaf growth, within limits (2). Phenology can be delayed due to pre-anthesis drought stress (2).

Two forces retain water in the soil: adhesion and cohesion. Adhesion is the attraction of water molecules to soil particles. Cohesion is the attraction of water molecules to each other. These two forces oppose the effect of gravity that removes soil water by drainage. The maximum storage of useful water to the plants depends on the equilibrium between soil water retention forces and gravity. The minimum storage depends on the retention forces, mostly. There is some water that is held so tightly in the soil that the plants cannot extract (Fig. 1).

DEFINING THE LIMITS OF PLANT AVAILABLE WATER

We define potentially plant available water or potentially plant extractable water (θ_p) as the difference between the drained upper limit (θ_d) and the lower limit (θ_l). The drained upper limit is the highest, field-measured water content of a soil after thorough wetting and draining until drainage becomes practically negligible. The θ_d corresponds closely to water content at "field capacity" and to matric suction in the range of 10 to 33 kPa. The lower limit, θ_l, is the lowest, field-measured, volumetric water content of a soil after plants stop extracting water due to premature death or dormancy as a result of water deficit. The lower limit corresponds closely to the water content at the "permanent wilting point" and to matric suction of 1.5 MPa. Potentially plant available water is similar across a wide range of soil textures (Fig. 2). The mean and standard deviation of field measured θ_p for a wide range of soils is in Table 1.

MEASURING THE LIMITS OF PLANT AVAILABLE WATER

The traditional approach to evaluating the limits of available water has been to measure the "permanent wilting point" and "field capacity" of soil samples removed from the field through use of pressure chambers. Usually, soil water contents are measured at suctions of 1.5 MPa for the wilting point and 10 to 33 kPa for field capacity. Samples are taken from the soil at various depths to the measured or estimated rooting depth. The difference between the upper and lower limits of availability are summed over the rooting depth to determine the total potentially plant available water.

Several criticisms have been made of the above rather static definitions of available soil water limits. Water above the upper limit can be taken up while drainage is occurring. Plant growth can be retarded before the lower limit is reached. Water extraction by roots may continue beyond the 1.5 MPa range in some instances. Where root density is low, especially in deep soils, it is difficult to separate the effects of root distribution from the effect of water flow through the soil on available water.

FIELD MEASUREMENT OF THE UPPER LIMIT

The drained upper limit is a necessary input in water balance models. The "tipping bucket" approach in water balance models such as in DSSAT (4) can be sensitive to the value chosen for θ_d (5). Because water can be taken up by plants while drainage is occurring, the drained upper limit is not always the appropriate upper limit of soil water availability. Therefore, it is necessary to measure or reasonably estimate the drainage rates while they are significant. There are functions to estimate drainage flux that are relatively simple to evaluate from field data (6).

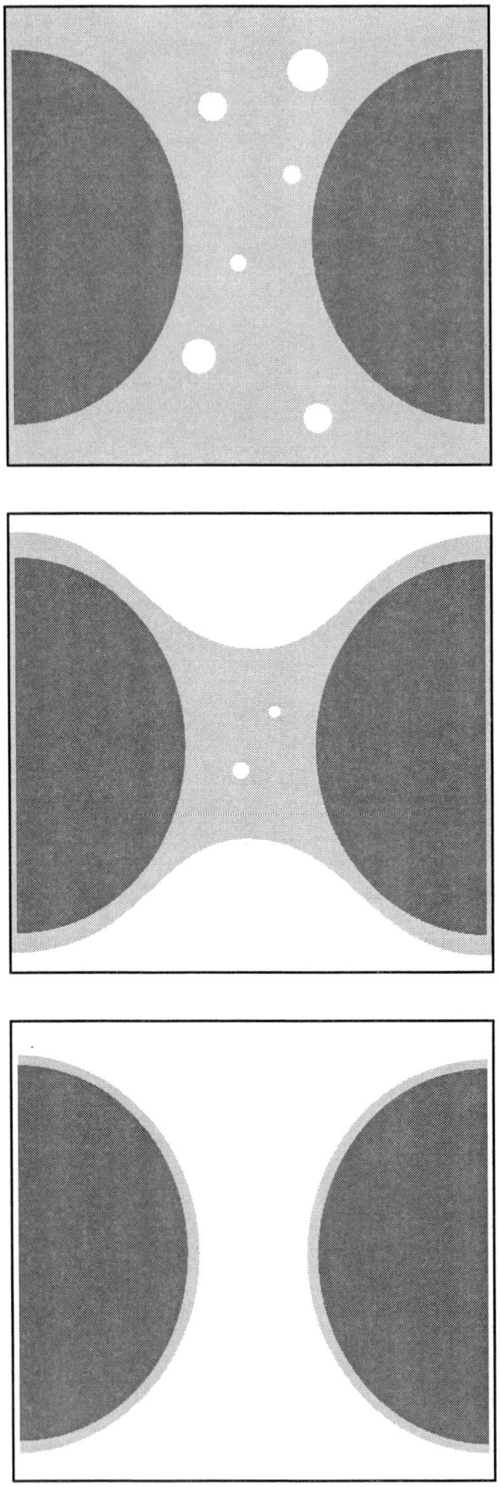

Fig. 1 Schematic of water in soil pores at three water contents: saturation (entrapped air bubbles persist), drained upper limit (larger pores empty, menisci form between soil particles), and lower limit (thin films around soil particles).

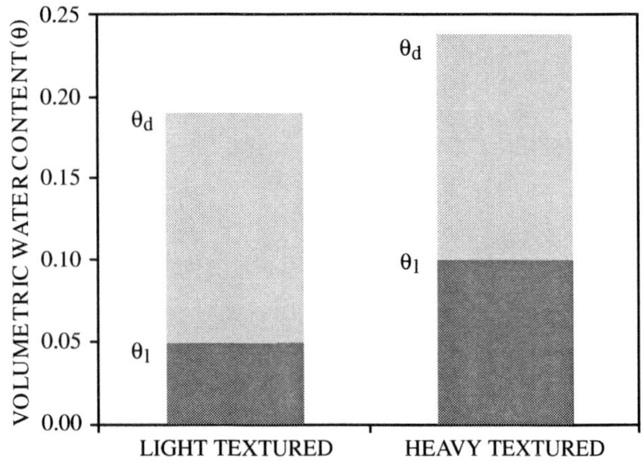

Fig. 2 Potentially extractable soil water ($\theta_d - \theta_l$) is similar among soils with a wide variety in texture. Heavy textured soils have higher proportion of smaller pores, therefore both the upper (θ_d) and the lower limit (θ_l) are higher than in light textured soils.

FIELD MEASUREMENT OF THE LOWER LIMIT

Field measurements of the lower limit of water availability are more difficult than measuring the upper limit because root growth dynamics are different between and within crops and also because conditions for the lower limit may not often naturally occur.

The best value for the lower limit is obtained when crops reach their maximum vegetative size without stress and then grow on stored soil water until plants are visibly severely distressed. Because leaf senescence and natural death of many crops occur at maturity, it is important that plants have undergone premature severe stress. On the other hand, any stress that causes a reduction in root growth before establishment of a "normal" rooting profile can cause incomplete root water extraction. Nutrient, drought, and aeration stresses, as well as chemical boundaries in soil, may restrict root growth and rooting depth.

Crop species and cultivars within a species may exhibit differing final water content values at crop maturity. Information on the lowest measured water content for several crops experiencing terminal drought is in Fig. 3. Results indicate that this soil is uniform with depth and the 1.5 MPa suction was quite close to the lowest measured water content when the crop had sufficient time and rooting density to remove the water down to the θ_l. The true θ_l was reached in the 20 to 80 cm depth for all crop situations but was not obtained below that depth for a short maturity soybean. In contrast, the long season sunflower Contiflor 3 extracted water to the θ_l to a depth of about 260 cm.

Table 1 Potentially plant available water for soils of different texture as measured in the field. The values range from a minimum of 8.0 ± 3.1% for sands to 14.8% for just one observation for the silt. The mean plant available water values for the remaining textural classes are within a narrow range of only 11.0 to 14.3%

Texture	No. samples	Mean and standard deviation for θ_p (volume percent)
Sand	76	8.0 ± 3.1
Loamy sand	7	12.9 ± 3.6
Sandy loam	31	13.2 ± 2.2
Loam	51	13.6 ± 3.0
Silt loam	83	14.3 ± 3.3
Silt	1	14.8
Silty clay loam	53	13.0 ± 2.1
Clay loam	41	12.5 ± 3.2
Sandy clay loam	24	11.0 ± 3.5
Sandy clay	0	
Silty clay	31	13.4 ± 3.0
Clay	3	12.9 ± 3.6

(From Ref. 3.)

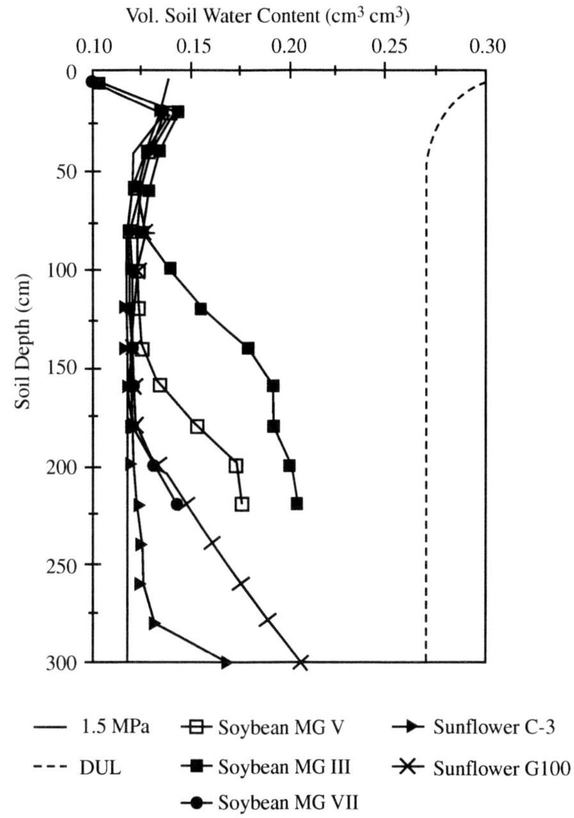

Fig. 3 Minimum soil water content reached by sunflower cultivar G 100, sunflower cultivar Contiflor 3, soybean cultivar Asgrow 3127, soybean cultivar Asgrow 5308, and soybean cultivar R.A. 702 planted on a silt loam soil (Entic Haplustoll) in Cordoba, Argentina, during the 1992–1993 growing season (7). The dashed line indicates the field measured drained upper limit, and the solid heavy line indicates water content measured in the laboratory at 1.5 MPa suction.

ESTIMATING THE LIMITS OF PLANT AVAILABLE WATER

When the cost and time of primary data collection is prohibitive, it is necessary to estimate reasonable values for the limits of the soil water reservoir. Many studies have found statistical relationships between plant available water and one or more soil properties such as texture, organic matter content, coarse fragment content, and bulk density (8–11). To achieve a good fit, statistical models can be very complex and may behave erratically. There are many soil properties that are related to the water limits. Yet the more predictors there are in a model, the more specific the model becomes to a particular data set and its generality is being lost. Also, the additional data required to use a complex model are not always available.

To derive a simple model, we used a database of field measured soil water limits (3). The database contained 401 soil samples from 15 states of the United States. Seven soil orders were represented, but over 60% of the soils were Mollisols or Alfisols. Histosols, Oxisols, and Spodosols were not represented. There were few data from the Midwest and southeast United States. The fact that these regions experience frequent precipitation during the crop growing season precludes the casual collection of field-measured lower limit.

We selected the gravimetric drained upper limit (w_d) as the first dependent variable to estimate, to make the model sensitive to bulk density, and to avoid the error associated with the measurement of bulk density. The model of w_d was

$$w_d = 0.186\ (\text{sand}/\text{clay})^{-0.141} \quad (1)$$

where w_d is the gravimetric water content and "sand" and "clay" are percent sand and clay. Then θ_d can be calculated from

$$\theta_d = w_d\ \rho_b/\rho_w \quad (2)$$

where θ_d is the volumetric water content, ρ_b is the soil bulk density, and ρ_w is the density of water.

The second dependent variable that we estimated was θ_p, the volumetric, potentially plant extractable water content. We picked sand content as the independent variable once we realized that θ_p is approximately the same across soils, except for soils with high sand content (3). The model was

$$\theta_p = 0.132 - 2.5 \times 10^{-6} e^{0.105\ \text{sand}} \quad (3)$$

where 0.132 is the mean θ_p for soils with less than 65% sand.

Then, we modified the estimates of θ_d and θ_l for coarse fragment content (particles larger than 2 mm in diameter) and for unusually high organic C content. The equations were tested with a limited independent data set with good results (12).

The θ_l can be calculated as the difference between θ_d and θ_p. The plant extractable water for the soil profile can be calculated as the product of θ_p times soil depth, in units of soil depth. If a profile consists of layers of different composition, then the plant extractable water of the profile is calculated as the sum of the extractable water of each layer. It is important to know the rooting depth and the root density of the crop. The concept of plant extractable water is meaningful only in that depth where roots are dense enough to effectively extract water.

PLANT AVAILABLE WATER AND SOIL–WATER MANAGEMENT

Irrigation scheduling is an obvious application of plant available water information. Through proper irrigation scheduling, it should be possible to apply only the water that matches the evapotranspiration or to meet some other desired criteria (13, 14). A rule of thumb is that irrigation is needed when the soil water content drops to 50% of potentially extractable soil water for such crops as alfalfa, corn, and spring grains (13). However, reduction in growth has been observed at a water content as low as 85% of potentially extractable soil water, in sandy soils (2).

Precision agriculture is the management of spatial and temporal variability in all aspects of agricultural production to improve crop performance and environmental quality (15). Ideal management variables are those with high spatial dependence and low temporal variance. Plant available water probably has high spatial dependence and high temporal variance, so it is not an ideal candidate for precision management, yet there are two possibilities for precision water management that are related to plant available water: variable rate irrigation and matching agronomic inputs to plant available water defined by soil and/or landscape properties.

Variable sprinkler irrigation systems exist, usually coupled with chemigation. Limited experience so far indicates that the benefits of such systems have not been fully realized. The potential usefulness of matching agronomic inputs to plant available water increases with spatial variability in plant available water. Mapping small-scale variability is expensive, so perhaps a more viable approach is to map relief or some landscape attribute that is indirectly related to plant available water (15).

Tillage seems an important factor in the conservation of plant available water in dryland production. In one case, no-tillage resulted in improved soil water conservation during fallow compared to stubble mulch tillage (16). No-tillage afforded substantially higher soil water contents during planting of the crop after fallow. This could be because no-tillage results in less evaporation.

SUMMARY AND CONCLUSIONS

Definitions of the upper and lower limits of plant available water are difficult to establish because of water flow into and out of the root zone and because of incomplete extraction by sparse roots at the lower boundaries of the root zone. Pressure extraction equipment used on soil samples removed from a field often does not provide reliable estimates of the limits of water availability when comparing it to observations in the field. Therefore, for accurate soil water balance, it is important to measure the upper and lower limits in the field.

If the limits of plant available water cannot be measured, they can be estimated based on routinely measured soil properties. We present a simple model that uses few parameters. The model can make it easier to approximate soil water limits in water balance simulation, irrigation scheduling, and watershed management. A challenge for the future will be to manage the spatial and temporal variability in plant available water in precision agriculture systems to improve crop performance and environmental quality.

REFERENCES

1. Jones, O.R.; Johnson, W.C. Cropping Practices: Southern Great Plains. In *Dryland Agriculture. Agron. Monograph 23*; Dregne, H.E., Willis, W.O., Eds.; ASA and SSSA: Madison, WI, 1983; 365–385.
2. NeSmith, D.S.; Ritchie, J.T. Short- and Long-Term Responses of Corn to a Pre-Anthesis Soil Water Deficit. Agron. J. **1992**, *84*, 107–113.

P

3. Ratliff, L.F.; Ritchie, J.T.; Cassel, D.K. Field-Measured Limits of Soil Water Availability as Related to Laboratory-Measured Properties. Soil Sci. Soc. Am. J. **1983**, *47*, 770–775.
4. Jones, C.A.; Ritchie, J.T.; Kiniry, J.R.; Godwin, D.C. Subroutine Structure. In *CERES-Maize: A Simulation Model of Maize Growth and Development*; Jones, C.A., Kiniry, J.R., Eds.; Texas A&M Press: College Station, TX, 1986, 49–105.
5. Buttler, I.W.; Riha, S.J. Water Fluxes in Oxisols: A Comparison of Approaches. Water Resour. Res. **1992**, *28* (1), 221–229.
6. Black, T.A.; Gardner, W.R.; Thurtell, G.W. The Prediction of Evaporation, Drainage, and Soil Water Storage for a Bare Soil. Soil Sci. Soc. of Am. Proc. **1969**, *33*, 655–660.
7. Dardanelli, J.L.; Bachmeier, O.A.; Sereno, R.; Gil, R. Rooting Depth and Soil Water Extraction Patterns of Different Crops in a Silty Loam Haplustoll. Field Crops Res. **1997**, *54*, 29–38.
8. Cassel, D.K.; Ratliff, L.F.; Ritchie, J.T. Models for Estimating In-Situ Potential Extractable Water Using Soil Physical and Chemical Properties. Soil Sci. Soc. Am. J. **1983**, *47*, 764–769.
9. Cosby, B.J.; Hornberger, G.M.; Clapp, R.B.; Ginn, T.R. A Statistical Exploration of the Relationships of Soil Moisture Characteristics to the Physical Properties of Soils. Water Resour. Res. **1984**, *20*, 682–690.
10. Ghosh, R.K. Estimation of Soil–Moisture Characteristics from Mechanical Properties of Soils. Soil Sci. **1980**, *130*, 60–83.
11. Gupta, S.C.; Larson, W.E. Estimating Soil Water Retention Characteristics from Particle Size Distribution, Organic Matter Percent, and Bulk Density. Water Resour. Res. **1979**, *15*, 1633–1635.
12. Ritchie, J.T.; Gerakis, A.; Suleiman, A. Simple Model to Estimate Field-Measured Soil Water Limits. Trans. ASAE **1999**, *42* (6), 1609–1614.
13. Hill, R.W. Irrigation Scheduling. In *Modeling Plant and Soil Systems. Agron. Monograph 31*; Hanks, J., Ritchie, J.T., Eds.; ASA and SSSA: Madison, WI, 1991, 491–509.
14. Ritchie, J.T.; Amato, M. Field Evaluation of Plant Extractable Soil Water for Irrigation Scheduling. Acta Horticulturae **1990**, *278*, 595–615.
15. Pierce, F.J.; Nowak, P. Aspects of Precision Agriculture. Advances in Agronomy **1999**, *67*, 1–85.
16. Jones, O.R.; Hauser, V.L.; Popham, T.W. No-Tillage Effects on Inflitration, Runoff, and Water Conservation on Dryland. Trans. ASAE **1994**, *37* (2), 473–479.

PLANT NUTRIENT AVAILABILITY—INTENSITY, QUANTITY AND BUFFER POWER

Ram C. Dalal
Department of Natural Resources, Indooroopilly, Queensland, Australia

INTRODUCTION

Plant nutrient availability in soil is dominated by soil solution phase processes such as hydrolysis, hydration, carboxylation, and oxidation–reduction reactions. In addition, microbes and plant roots, through their synthetic and metabolic activities, mediated by both intracellular and extracellular enzymes, affect these solution phase processes. These processes determine the activity of plant nutrient ions in the soil solution, and, in dilute soil solution, nutrient ion activity may be taken as concentration of nutrient or nutrient intensity factor (C_l). Sorption, desorption and precipitation and dissolution, and complexation and ion exchange with the nutrient on the solid phase provide quasi equilibrium with nutrient concentration in the soil solution. The concentration of nutrient in the entire soil mass is the quantity factor (C_s). The change in intensity factor with a change in quantity factor—that is, dC_s/dC_l—is the nutrient buffer power. Because growing plants continually remove nutrient from the soil solution (intensity), the measure of capacity of the solid (and solution) phase nutrient (quantity) to maintain nutrient concentration during its withdrawal by plants (buffer power) is important for continual nutrient supply for plant growth. This is presented in Fig. 1.

This article is concerned with the relative contribution of the intensity, quantity, and buffer power and their interactions in nutrient availability to plants.

INTENSITY FACTOR

The initial concentration of a nutrient in soil solution may be obtained by soil solution displacement (1), and by equilibrating in dilute salt solutions such as 0.01 M $CaCl_2$ or 0.02 M KCl.

The relationship between low concentration range of a nutrient and plant uptake is described by the following equation (2):

$$U = 2\pi r \ \alpha C_r \quad (1)$$

where U = nutrient uptake by a unit m root segment, r = root radius, C_r = nutrient concentration at the root surface, and α = root absorbing power. Diffusion and/or mass flow and root interception controls the nutrient concentration at the root surface.

The nutrient uptake by mass flow is closely related to plant water uptake. In most soils, plant uptake of NO_3^-, Ca^{2+}, Mg^{2+}, and Cl^- is primarily governed by mass flow (1). Thus, increasing water influx increases proportionately nutrient supply by mass flow. For example, by increasing plant water transpiration from 106 L/kg of plant shoot to 444 L/kg of shoot increased Ca^{2+} uptake from 72.8 mg to 177.6 mg/pot (1). However, nutrient ions such as $H_2PO_4^-$, K^+, and Zn^{2+}, which are strongly sorbed on soil surface, and usually low in soil solution, are primarily supplied to root surface by diffusion flow.

Nutrient concentration at the root surface cannot be measured directly. However, it can be estimated by using Fick's first law:

$$F = -DA \ (dC/dx) \quad (2)$$

where F = the nutrient flux, dC/dx = concentration gradient, A = area for diffusion, and D = the diffusion coefficient. The negative D means net nutrient movement from high concentration to low concentration. However, in dilute solutions D may be taken as constant over a range of concentrations in the soil solution (3).

For strongly sorbed ions such as P and Zn^{2+}, initial nutrient concentration in the soil solution rarely meets plant nutrient demand. Thus, intensity factor alone does not sufficiently explain the relationship between nutrient concentration and plant nutrient uptake. For example, $H_2PO_4^-$ activity or concentration in the soil solution is usually not related to grain yield or nutrient uptake (4) except in the early stages of plant growth. Quantity factor becomes an important consideration for plant availability of these nutrients.

QUANTITY FACTOR

The plant available quantity of most cations is that measured by displacing cations from the exchange

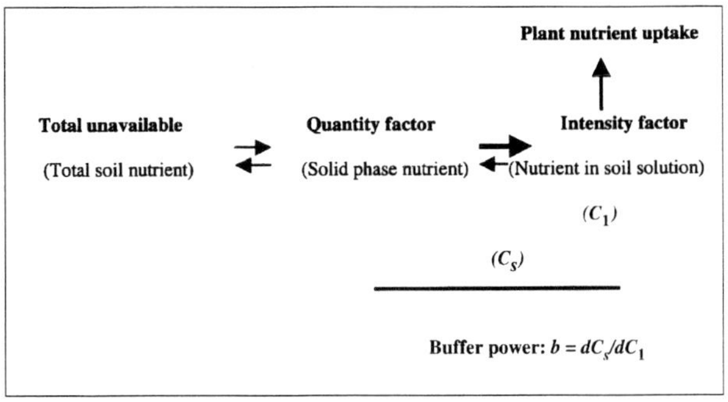

For highly mobile nutrients such as NO_3 and Cl, $b \leq 1$; for these ions, b is insignificant in plant nutrient availability. For relatively immobile nutrients such as P and Zn, $b \geq 1$; significance of b in plant nutrient availability increases as its value increases.

Fig. 1 A diagrammatic representation of intensity factor, quantity factor, and buffer power. (From R.C. Dalal, unpublished.)

complex with a strong concentration of salt solution such as ammonium acetate, sodium acetate, or silver thiourea (5). For micronutrient cations (Fe, Mn, Zn, Cu), complexation agents such as EDTA or DTPA have been used to reduce the precipitation and resorption of extracted cations from the soil solution (5).

Numerous methods are available for measuring the plant available quantity of phosphorus. These methods include extraction of P from soil by dilute and strong acids, dilute and strong alkali solutions, neutral salt solutions, and their combinations and iron oxide impregnated paper (6), isotopic exchangeable phosphate and ion-exchange resin (6), and electroultrafiltration (2).

The nutrient concentration of the quantity factor is 1–4 orders of magnitude greater than that of the intensity factor. For example, in 20 Australian soils, mean intensity of P was 0.03 mg/L (0.3 mg/kg), whereas mean quantity was 15 mg/kg (L value, Olsen P) or 30 mg/kg soil (Colwell P) (6). Also, quantity of nutrients such as P (7), K^+, and Zn^{2+} in soil often exceeds that taken up by plants in field conditions, although intensive cropping could remove as much quantities of P, K^+, and Zn^{2+} as measured by the quantity factor.

The quantity factor for P, K^+, and Zn^{2+} correlates better than the intensity factor with plant nutrient uptake. In 20 Australian soils, the intensity factor (0.01 M $CaCl_2$ P) accounted for only 19% each for the variation in P uptake at 35 days and 150 days of wheat growth while the quantity factor (Colwell P) accounted for 2–3 times more, 38% and 56% for the variation in P uptake at 35 days and 150 days of wheat growth, respectively (4). Similarly, in eight Australian soils, the intensity and quantity factors accounted for 29% and 81%, respectively, for wheat grain yield in the field (4).

This also applies to Zn^{2+} availability. For example, in 13 Vertisols, relative yield response to Zn application was better correlated with DTPA- extractable Zn (C_s), $r^2 = 0.69$, than soil solution Zn (C_l), $r^2 = 0.56$ (8).

BUFFER POWER

The availability of nutrients to growing plants is not only influenced by nutrient intensity and the quantity but also the ability of soil to replace the nutrient as plants take it up from soil solution. The effective diffusion of the nutrient to the root surface depends on the soil's buffer power for that nutrient, volumetric soil moisture, and the impedance factor. This is described by an adaptation of Eq. 2 to the soil system by the following equation (9):

$$D_e = D_l \ \Theta f_l \ (dC_l/dC_s) + R \tag{3}$$

where D_e = effective diffusion coefficient of nutrient ion in soil, D_l = diffusion coefficient of ion in water, Θ = volumetric soil water content, f_l = an impedance factor or tortuosity factor that is measured using nonadsorbed ions such as Cl^- ions, C_l = concentration of nutrient ion in the soil solution (intensity factor), C_s = quantity of nutrient ion in the entire mass of soil (quantity factor), and R = the residual term accounting for diffusion through the adsorbed phase, which is probably negligible in neutral soils for most ions.

The nutrient buffer power, b, is the reciprocal of dC_l/dC_s, that is,

$$b = dC_s/dC_l \tag{4}$$

The nutrient buffer power, b, is measured from either an adsorption isotherm or desorption isotherm. In adsorption isotherm, simulating fertilizer application, known amount of nutrient is added to soil and the soil incubated for periods from minutes (exchangeable cations) to several months to attain quasi equilibrium between the sorbed or exchangeable ions and those in the soil solution. At the end of the incubation period, the concen-tration of nutrient in the soil solution is measured, and C_l and C_s calculated. Desorption isotherms are constructed from desorption of nutrient by an electrolyte, electroultrafiltration, or ion exchange resin, simulating the nutrient uptake by roots.

For a narrow range in nutrient concentration, there is a linear relationship between C_s and C_l, that is,

$$C_s = b_l \ C_l + a_l \tag{5}$$

where b_l is the distribution coefficient and a_l is a constant. The buffer power, b, is calculated from the following equation (10):

Table 1 Percentage of variation in nutrient uptake accounted for by intensity factor, quantity factor, and a combination of intensity or quantity and buffer power

Nutrient	Intensity (C_l)	Quantity (C_s)	Buffer power (b)	C_l and b	C_s and b	Reference
P	17	75	27	49	86	(4)
P	15	51	–	79	89	(4)
K	–	6	–	–	19	(3)
Zn	13	–	–	38		(3)
Zn	27	27	0.08	27	29	(8)

$$b = \rho b_l + \theta \quad (6)$$

where ρ = bulk density, and θ = volumetric moisture content.

However, for a wide range in quantity and intensity, the relationship between C_s and C_l is frequently described by the Freundlich isotherm:

$$C_s = kC_l^n \quad (7)$$

where k and n are constants, k is related to nutrient sorption capacity of the soil, and n is related to nutrient bonding energy to soil or curvilinearity between the corresponding C_s and C_l values in a soil.

Langmuir isotherm has also been used to describe the relationship between C_s and C_l values in a soil (4, 11):

$$C_s = MBC_l/(1 + bC_l) \quad (8)$$

On differentiating with respect to C_l, Eq. 8 becomes

$$dC_s/dC_l = MB/(1 + bC_l)^2 \quad (9)$$

As $C_l \rightarrow 0$, Eq. 9 becomes

$$dC_s/dC_l = MB \quad (10)$$

where M is Langmuir adsorption maximum, B is bonding energy term, and the product, MB, is the maximum buffering capacity.

For 20 Australian soils, the P buffer power varied from 32 to 364, except in one soil, an Oxisol, which had the value of 3476 (6). For 33 U.S. and Canadian soils, the values varied from 33 to 299 (12).

The P buffer values are much lower, from 19 to 75, for central European soils (3). For 14 Australian Vertisols, Zn buffer power varied from 217 to 790 (8). The K buffer power varied from 142 to 592 for southern Indian soils (3). The relative importance of nutrient buffer power, intensity, and quantity factors is illustrated for a number of experiments in Table 1.

PERSPECTIVES

Other factors that affect nutrient availability from soil such as climate (temperature, moisture), pH, nutrient placement, soil biota, and plant type should also be considered along with the parameters of plant nutrient availability. For strongly sorbed ions, nutrient concentration in the soil solution is usually low. It is not determined routinely in soil testing laboratories. While quantity of the nutrient is routinely measured in soil testing services, accounting for the variation in plant nutrient uptake almost always improves by considering the soil's nutrient buffer power. Nutrient uptake models also include buffer power, so it should be widely measured. Optimal assessment of plant available nutrients from soil as well as fertilizer applications is essential to ensure the efficient use of nutrients for economic food and fiber production while minimizing nutrient loss/pollution to the environment. While techniques for measuring buffer power, quantity, and intensity factors should be further refined, consideration of buffer power along with quantity/intensity factors (13) will improve estimates and hence the efficient use of plant available nutrients from soil.

REFERENCES

1. Barber, S.A. *Soil Nutrient Bioavailability: A Mechanistic Approach*; 2nd Ed; John Wiley: New York, 1996; 1–414.
2. Mengel, K. Dynamics and Availability of Major Nutrients in Soils. In *Advances in Soil Science*; Stewart, B.A., Ed.; Springer-Verlag: New York, 1988; Vol. 2, 65–131.
3. Nair, K.P.P. The Buffering Power of Plant Nutrients and Effects on Availability. In *Advances in Agronomy*; Sparks, D.L., Ed.; Academic Press: Toronto, 1996; Vol. 57, 237–287.
4. Dalal, R.C.; Hallsworth, E.G. Evaluation of the Parameters of Soil Phosphorus Svailability Factors in Predicting Yield Response and Phosphorus Uptake. Soil Sci. Soc. Am. J. **1976**, *40*, 541–546.

5. Rayment, G.E.; Higginson, F.R. *Australian Laboratory Handbook of Soil and Water Chemical Methods*; Inkata Press: Melbourne, 1992; 1–330.
6. Dalal, R.C.; Hallsworth, E.G. Measurement of Isotopic Exchangeable Soil Phosphorus and Interrelationship Among Parameters of Quantity, Intensity, and Capacity Factors. Soil Sci. Soc. Am. J. **1977**, *41*, 81–86.
7. Moody, P.W.; Bolland, M.D.A. Phosphorus. In *Soil Analysis: An Interpretation Manual*; Peverill, K.I., Sparrow, L.A., Reuter, D.J., Eds.; CSIRO Publishing: Melbourne, 1999; 187–220.
8. Dang, Y.P.; Dalal, R.C.; Edwards, D.G.; Tiller, K.G. Understanding Zinc Reactions in Vertisols to Predict Zinc Responses by Wheat. Plant and Soil **1993**, *155/156*, 247–250.
9. Nye, P.H.; Tinker, P.B. *Solute Movement in the Soil-Root System*; Blackwell Scientific Publications: Oxford, 1977; 1–342.
10. van Rees, K.C.J.; Comerford, N.B.; Rao, P.S.C. Defining Soil Buffer Power: Implications for Ion Diffusion and Nutrient Uptake Modelling. Soil Sci. Soc. Am. J. **1990**, *54*, 1505–1507.
11. Holford, I.C.R. Soil Phosphorus: Its Measurement, and Its Uptake by Plants. Aust. J. Soil Res. **1997**, *35*, 227–239.
12. Kovar, J.L.; Barber, S.A. Phosphorus Supply Characteristics of 33 Soils as Influenced by Seven Rates of P Addition. Soil Sci. Soc. Am. J. **1988**, *52*, 160–165.
13. Probert, M.E.; Moody, P.W. Relating Phosphorus Quantity, Intensity, and Buffer Capacity to Phosphorus Uptake. Aust. J. Soil Res. **1998**, *36*, 389–393.

PLANT NUTRIENT SUFFICIENCY AND REQUIREMENTS

K.L. Sahrawat
West Africa Rice Development Association, Côte d'Ivoire (Ivory Coast), West Africa

INTRODUCTION

The need for food, feed, and fiber production is ever increasing because of continued rapid population growth. The need is most pressing in the developing world (1). Apart from water shortages, soil infertility is the major constraint to plant growth and yield throughout much of tropical regions. The increased crop production can be achieved through enhanced soil fertility (2). Soil fertility can only be sustained if the nutrients removed from the soil are replenished through addition. Also, crop production needs to be made efficient through optimal use of plant nutrients. Assessing the nutrient sufficiency and requirements are important components of research for efficient and rational use of nutrients from external sources. The objective of this communication is to provide a concise summary of the major developments in the research on plant nutrient sufficiency and requirements. The emphasis is placed on the salient research results from recent field studies. For detailed discussion and summary of the past research, the readers are referred to an excellent text by Black (2).

PLANT NUTRIENT SUFFICIENCY AND REQUIREMENT DEFINED

Nutrient Sufficiency

Nutrient sufficiency or critical nutrient concentration as more commonly termed, is a relative term as an absolute sufficiency cannot be determined. Nutrient sufficiency is a measure of nutrient concentration in the plant determined by plant analysis and is, preferably expressed as a range of concentrations rather than a single concentration. Sufficiency of a given nutrient lies between critical deficiency value and an excess or toxic concentration. Alternatively, nutrient sufficiency is the range of concentrations in which no yield reduction occurs and no nutrient stress symptoms appear (3). A critical deficiency concentration of a nutrient is the one that corresponds to 10% below the maximum yield on the low side and a critical toxicity concentration of a nutrient corresponds to 10% below the maximum yield on the high side (4). It is perhaps more useful to consider critical concentration as the concentration that separates plants that would respond to addition of more of a nutrient from those that would not respond (5). For practical purposes, economic criterion would appear more appropriate and the critical nutrient concentration may be termed as critical economic concentrations (6).

Nutrient Requirement

Nutrient requirement is the amount of nutrient required to achieve a yield target which is normally 90% of the maximum yield obtained under optimal nutrient regime. As for nutrient sufficiency, economic considerations are equally relevant for determining the nutrient requirements of various crop plants (2). Nutrient requirement comprises of external and internal nutrient requirements. External nutrient requirement is the concentration in the growing medium, usually soil while the internal requirement is the concentration in the plant tissue at a yield target, normally 90% of the maximum yield.

DETERMINING NUTRIENT SUFFICIENCY AND REQUIREMENTS

Soil and plant analysis are used for determining the nutrient requirements of crops while plant analysis is used to establishing nutrient sufficiency under well-defined conditions of experimentation and collection of soil and plant samples. Data on nutrient sufficiency range in plant tissue of selected grain crops (7) for macronutrients are summarized in Table 1.

Since early finding that the chemical composition of plants changes with the nutrient supplies, attempts have been made to use the elemental concentrations for evaluating nutrient requirements of crops. For historical developments covering the period up to 1990, the readers are referred to the excellent text entitled *Soil Fertility Evaluation and Control* by C.A. Black (2). The more recent literature is briefly covered in this paper. Evidently, much

Table 1 The range of sufficiency levels in plant tissue of N, P, and K for selected grain crops

		Sufficiency range concentration (%)		
Crop growth stage	**Plant part**	**N**	**P**	**K**
Corn or maize (*Zea mays*)				
Prior to tasseling	Leaf below the whorl	3.00–3.50	0.25–0.45	2.00–2.50
Rice (*Oryza sativa* L.)				
Maximum tillering	Newest, fully developed leaf	2.80–3.60	0.10–0.18	1.20–2.40
Barley (*Hordeum vulgare* and *Hordeum distichon*)				
Emergence of head from boot	Whole tops	1.75–3.00	0.20–0.50	1.50–3.00
Sorghum (*Sorghum bicolor*)				
37–56 days after planting	Newest, fully developed leaf	3.20–4.20	0.13–0.25	2.00–3.00
Winter wheat (*Triticum aestivum*)				
Just before heading	Top two leaves	1.75–3.00	0.20–0.50	1.5–3.00
Spring wheat (*Triticum aestivum*)				
Emergence of head from boot	Whole tops	2.00–3.00	0.20–0.50	1.50–3.00

research has been done in determining the concentrations of nutrient elements as indexes of the sufficiency of nutrient supplies for crop growth and for determining nutrient requirements (2, 7, 8).

A recent book entitled *The Plant Analysis Handbook* by H.A. Mills and J.B. Jones, Jr. (7) provides tables of interpretive plant analysis data for over 1000 individual horticultural, agronomic, and plantation crops. The publication provides data on sufficiency or critical concentration range for macro- and micronutrient elements for various crop plants where the deficiency, sufficiency, and toxicity concentrations have been established over a wide range of conditions in the solution culture to field. In addition, survey range and survey average values for various nutrient elements are also presented for growing conditions under which the upper and lower limits of nutrient concentrations have not been as clearly defined as in the case of sufficiency range. Survey average value refers to the nutrient level found in healthy, normal plants without the presence of visual symptoms on the foliage showing deficiency or toxicity. Another comprehensive treatise on *Plant Analysis: An Interpretation Manual* edited by Reuter and Robinson (8) is also an excellent source of information on nutrient levels for deficiency, toxicity, and sufficiency for temperate and tropical crops, pasture species, fruits, vines and nuts, vegetable crops, ornamentals, and forest plantations. The treatise discusses in detail the concepts and principles for interpreting plant analysis, describes nutrient deficiency and toxicity symptoms in the plant, apart from providing concentration ranges of nutrients covering the range from deficiency to toxicity.

The determination of nutrient sufficiency and requirements are based on the relationships with plant growth and yield. The nutrient elemental composition of the plant at optimal yield should approximate the nutrient sufficiency levels, expressed either as individual nutrient concentrations or ratios of the various nutrient elements (2, 8). Before soil and plant tests can be used for determining nutrient sufficiency and requirements, appropriate standards need to be developed through field experimentation (9). Field experiments are considered essential for calibration because standards developed under controlled conditions, e.g., greenhouse pots may not be representative of the real world situation and may not hold under field conditions. Using the primary standards developed under field conditions, the relative sufficiency and requirements of nutrients can be determined under nonexperimental conditions. Although it is generally accepted that the determination of sufficiency requirements for nutrients leads to improved prediction of yields, it is emphasized that these values depend on the growing conditions (2).

Several equations based on the response of crop yield or relative yield or growth to nutrient concentrations in soil or plant have been proposed for determining the nutrient requirements of crops. Black (2) provides an excellent historical perspective to the developments in research on the use of soil and plant tests for determining the nutrient requirements of crops. It is evident from the discussion that while qualitative aspects of nutrient status are useful and at times, especially in the past, it was considered acceptable, no generally satisfactory system of testing plants for nutrient deficiencies has emerged. This remains a major obstruction in the use of nutrient concentrations for determining requirements in a quantitative and consistent manner. The lack of consistency, however, is not entirely unexpected in view of the fact that the concentrations of various nutrient elements are determined by the total mass of the plant, which in turn is influenced by the growing conditions relative to environment, nutrient supply and interactions among various nutrient elements, and the genetics of the crop.

Soil characteristics greatly influence the use of soil tests for determining the nutrient sufficiency or critical concentration levels for various nutrient elements. For example, a recent study with grain sorghum under rainfed cropping on nearby Alfisol and Vertisol sites at the ICRISAT Center, Patancheru (India) showed that on the Vertisol, 90% relative grain yield was obtained at 2.8 $mg\,kg^{-1}$ Olsen extractable P while on the Alfisol, 90% relative grain yield was achieved at 5.0 $mg\,kg^{-1}$ Olsen P (10). These results demonstrate that a single critical limit of available P does not hold true for sorghum on the two soil types under similar agroclimatic conditions and that the critical limit is lower for the clayey Vertisol with higher phosphate buffering capacity than the sandy Alfisol with lower buffering capacity.

Field study with upland rice on an Ultisol in the humid forest zone of Ivory Coast showed that the critical limit of Bray 1 P in the soil at 90% relative rice grain yield varied from 12.5 and 15.0 $mg\,P\,kg^{-1}$ soil for the four cultivars tested. The critical or sufficiency value tended to be lower for the P-efficient rice cultivars (11). Using plant tissue P concentration as the criterion, the critical concentration of P in the whole rice plant tops at the tillering stage, at 90% relative grain yield, was found to be 2 $g\,P\,kg^{-1}$ for the four rice cultivars. It was indicated that while the rice cultivars differed in the external P requirement (concentration in the growing medium), they did not differ in their internal P requirement (concentration in the plant tissue) (11, 12). In these field experiments, a significant linear relationship was observed between total P uptake in the biomass and rice grain or rice straw yield (12).

Sahrawat et al. (13) provided results on the calibration of plant P test for predicting sorghum grain yield under rainfed cropping of a Vertisol (Typic Haplustert) in the semi-arid zone of India. Despite variability in the rainfall received during the three years of field study, the leaf (newest, fully developed leaf at 50 % flowering stage of the crop) P concentration was found to be linearly related to grain yield (r^2 varied from 0.724 to 0.993). The relationships between leaf P and sorghum yield during the three years of study were described by the following equations:

$$1987:\ \text{Grain yield } (t\,ha^{-1}) = 4.057 + 28.300 \text{ Leaf P } (\%) \quad r^2 = 0.993 \tag{1}$$

$$1988:\ \text{Grain yield } (t\,ha^{-1}) = -7.693 + 50.0085 \text{ Leaf P } (\%) \quad r^2 = 0.724 \tag{2}$$

$$1989:\ \text{Grain yield } (t\,ha^{-1}) = -3.414 + 21.787 \text{ Leaf P } (\%) \quad r^2 = 0.979 \tag{3}$$

The critical or the sufficiency leaf P concentration at 90% of the maximum yield was found to be about 2.5 $g\,P\,kg^{-1}$ (Fig. 1). Phosphorus content in the grain was not significantly correlated to yield. Two important points emerged from this study: 1) linear relationship between

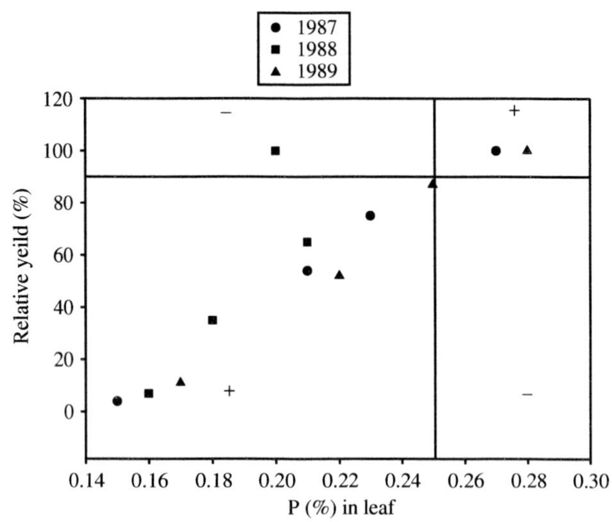

Fig. 1. Relationship between relative grain yield and P Content in the index leaf (new, fully developed leaf) at flowering in sorghum grown on a vertisol for three years (1987–1989). Each point represent an average value of four replications. (From Ref. 13.)

leaf P and yield over a range of applied P; and 2) the Cate and Nelson split method (14) of graphic presentation of relationship between leaf P and relative sorghum grain yield was found to be useful and had the practical advantage in dividing the data into responsive and non-responsive populations (Fig. 1) (13).

SIMPLE MODELS FOR DETERMINING NUTRIENT REQUIREMENTS

Sahrawat (15) evaluated two methods for determining the fertilizer phosphorus requirement (FPR) of sorghum. The first method used a simple method based on P uptake (16, 17) using the equation: $\mathrm{FPR} = (Up - U0)/\mathrm{PRF}$, where Up is P uptake at a given yield, $U0$ is P uptake from unfertilized soil, and PRF is the P recovery fraction of applied P. The parameters $U0$, Up, and PRF were determined in a three-year field experiment with sorghum grown on a calcareous Vertisol under rainfed conditions. The second method, proposed by the author, was based on the P applied, P uptake, and grain yield relationships. First, P uptake at a given sorghum yield was determined from the relationship between total P uptake and grain yield. The amount of fertilizer P applied for the given P uptake and grain yield was then, determined from the relationship between P applied and P uptake. Based on three years results, it was found that the fertilizer P applied and total P uptake by sorghum were significantly and linearly related, and described by regression Eqs. 4 and 5:

$$\mathrm{P\,uptake\,(kg\,ha^{-1})} = 1.67 + 0.200\,\mathrm{P\,applied\,(kg\,ha^{-1})} \quad (r^2 = 0.97;\ n = 12) \tag{4}$$

Total P uptake by the crop in turn, was closely linearly related to sorghum grain yield, and described by the following regression equation:

$$\mathrm{P\,uptake\,(kg\,ha^{-1})} = 0.51 + 2.62\ \mathrm{grain\,yield\ (t\,ha^{-1})} \quad (r^2 = 0.95) \tag{5}$$

There was a close agreement between the observed values of FPR and the predicted values determined by the two methods. These results suggested that the simple models based on P uptake can be utilized for determining the nutrient requirements of crops.

Evaluation of the two methods described, for determining the fertilizer P requirements of upland rice cultivars grown for two seasons (1992–1993) under a range of applied P on an Ultisol in the humid forest zone of Ivory Coast, showed that there was a good agreement between the observed value of FPR and the predicted values of FPR determined by the two methods, indicating the usefulness of these simple methods based on P uptake for determining the P requirements of crops (18). As observed for sorghum, fertilizer P applied and P uptake and grain yield and P uptake by four upland rice cultivars were significantly and linearly related. Overall, the results obtained with upland rice on an Ultisol in the humid forest zone and sorghum on a Vertisol in the semi-arid tropical zone indicate a wide applicability of these simple models for determining P requirements of crops and merit further evaluation.

It was further indicated that using the Cate and Nelson split method (14, 19) of graphic presentation of relationships between P applied and relative rice grain yield or between P uptake and relative yield provided a simple procedure for determining the P requirement of rice (18).

The author's unpublished results showed that a significant and linear relationship exists between fertilizer N (as urea) and N uptake, and between grain yield and N uptake for lowland rice cultivars grown under irrigated conditions on Alfisols in central Ivory Coast. It was further indicated that these relationships can be used for determining the N requirements of lowland rice cultivars.

PERSPECTIVE

The research on determining the nutrient sufficiency and nutrient requirements of plants is a prerequisite for efficient use of purchased inputs of nutrients and developing effective fertility management practices and thus forms an essential component of the strategy for making the overall agricultural production more efficient. However, there have been problems in consistency of the results which are greatly influenced by growing conditions and complex interactions among environmental, cultural, and genetic conditions (2). Despite these limitations, the knowledge of nutrient sufficiency and requirement is helpful in improving the prediction of plant growth and yield. A combination of soil and plant tests may be more effective in determining the nutrient sufficiency and requirements of crops. While soil tests are effective in determining the long-term nutrient requirement of crops, plant tests are necessary to diagnose and correct the nutrient deficiency of the current crops (13). The use of simple models based on nutrient uptake appear useful, as indicated by results with sorghum and rice, in determining N and P requirements and thus, merit further evaluation.

REFERENCES

1. IFPRI, *Feeding the World, Preventing Poverty and Protecting the Earth: A 2020 Vision*; International Food Policy Research Institute (IFPRI): Washington, DC, 1996.
2. Black, C.A. *Soil Fertility Evaluation and Control*; Lewis Publishers: Boca Raton, Florida, 1993; 155–269, 271–452.
3. Jones, J.B., Jr.; Eck, H.V.; Voss, R. Plant Analysis as an Aid in Fertilizing Corn and Grain Sorghum. In *Soil Testing and Plant Analysis*, 3rd Ed.; Westerman, R.L., Ed.; Soil Science Society of America Series 3, Soil Science Society of America: Madison, Wisconsin, 1990; 521–547.
4. Ohki, K. Zinc Nutrition Related to Critical Deficiency and Toxicity Levels for Sorghum. Agron. J. **1984**, *76*, 253–256.
5. Escano, C.R.; Jones, C.A.; Uehara, G. Nutrient Diagnosis in Corn Grown on Hydric Dystrandepts: I. Optimum Nutrient Concentration. Soil Sci. Soc. Am. J. **1981**, *45*, 1135–1139.
6. Malavolta, E.; Da Cruz, V.F. A Meaning for Foliar Diagnosis. In *Recent Advances in Plant Nutrition*; Samish, R.M., Ed.; Lavoisier Publishing: New York, 1971; 1, 1–3.
7. Mills, H.A.; Jones, J.B., Jr. *Plant Analysis Handbook: A Practical Sampling, Preparation, and Interpretation Guide*; MicroMacro Publishing, Inc.: Athens, Georgia, 1996.
8. Reuter, D.J., Robinson, J.B., Eds.; *Plant Analysis: An Interpretation Manual*, 2nd Ed.; CSIRO: Australia, 1997.
9. Westerman, R.L., Ed.; *Soil Testing and Plant Analysis*, 3rd Ed.; Soil Science Society of America Series 3, Soil Science Society of America: Madison, Wisconsin, 1990.
10. Sahrawat, K.L.; Pardhasaradhi, G.; Rego, T.J.; Rahman, M.H. Relationship Between Extracted Phosphorus and Sorghum Yield in a Vertisol and an Alfisol Under Rainfed Cropping. Fert. Res. **1996**, *44*, 23–26.
11. Sahrawat, K.L.; Jones, M.P.; Diatta, S. Extractable Phosphorus and Rice Yield in an Ultisol of the Humid Forest Zone in West Africa. Commun. Soil Sci. Plant Anal. **1997**, *28* (9&10), 711–716.
12. Sahrawat, K.L.; Jones, M.P.; Diatta, S. Plant Phosphorus and Rice Yield in an Ultisol of the Humid Forest Zone in West Africa. Commun. Soil Sci. Plant Anal. **1998**, *29* (7&8), 997–1005.
13. Sahrawat, K.L.; Rahman, M.H.; Rao, J.K. Leaf Phosphorus and Sorghum Yield Under Rainfed Cropping of a Vertisol. Nutrient Cycl. Agroecosyst. **1999**, *54*, 93–97.
14. Cate, R.B., Jr.; Nelson, L.A. *A Rapid Method for Correlation of Soil Test Analysis with Plant Response Data*; International Soil Testing Series Technical Bulletin 1, North Carolina Agricultural Experiment Station: Raleigh, North Carolina, 1965.
15. Sahrawat, K.L. Assessing the Fertilizer Phosphorus Requirement of Grain Sorghum. Commun. Soil Sci. Plant Anal. **1999**, *30* (11&12), 1593–1601.
16. Driessen, P.M. Nutrient Demand and Fertilizer Requirements. In *Modelling of Agricultural Production: Weather, Soils and Crops*; van Keulen, H., Wolf, J., Eds.; PUDOC, Wageningen: The Netherlands, 1986; 182–202.
17. Cornforth, I.S.; Metherell, A.K.; Sorn-Srivichai, P. Assessing Fertilizer Requirements. *Phosphorus Requirements for Sustainable Agriculture in Asia and Oceania*; International Rice Research Institute: Manila, Philippines, 1990; 157–166.
18. Sahrawat, K.L. Determining Fertilizer Phosphorus Requirement of Upland Rice. Commun. Soil Sci. Plant Anal. **2000**, *31* (9&10), 1195–1208.
19. Dahnke, W.C.; Olson, R.A. Soil Test Calibration, and the Recommendation. In *Soil Testing and Plant Analysis*, 3rd Ed.; Westerman, R.L. Ed.; Soil Science Society of America Series 3; Soil Science Society of America: Madison, Wisconsin, 1990; 45–71.

PLASTIC PROPERTIES

Gunnar Kirchhof
New South Wales Agriculture, Tamworth, New South Wales, Australia

P

INTRODUCTION

Application of a force to soil can result in an elastic, non-permanent, deformation that only lasts while the force is applied or a plastic, permanent deformation that persists after the load is removed. The latter may, or may not result in a decrease in the soil volume. How easily and how much the volume changes is determined by the soil's plasticity. It is an important factor influencing ease and outcome of soil tillage operations. Soil plasticity is caused by the lubricating film of water surrounding soil particles that allow the soil to change shape without rupturing upon application of forces. As an inherent soil property it is governed primarily by the surface area of the soil particles. Plasticity increases with increasing clay content, activity of clay minerals, position of the adsorbed cations in the lyotropic series and organic matter content.

DEFINITION

The adverb plastic is derived from the Latin noun *plasticus* [from Greek: plastikós (plássein)] meaning form or mould. Plasticity is the property of a solid body being moulded, receiving form or being brought into form. If the force leading to plastic deformation exceeds a yield value, permanent change in shape or size occurs. The definition of plasticity for soils includes the conditions that the deformation must not result in rupture and that the shape remains unchanged after water is removed during drying (1).

Although soil plasticity is closely related to soil consistence or consistency, there are two main differences: 1) Consistence and consistency usually refer to the resistance to deformation at a moisture content of the soil in the field (2) and are a manifestation of the forces of cohesion and adhesion within the soil at various water contents (3). Plasticity is usually described with the soil nearly saturated with water (4) or at water contents where the soil can easily be deformed. 2) Soil consistency or consistence is usually evaluated on undisturbed soil aggregates, fragments or peds. Plasticity is assessed on disturbed, often wetted and moulded soil material. The latter can therefore be regarded as an inherent soil property which does not change with the soil's macro structure.

The Atterberg limits (5) are most commonly used as parameters to describe soil plasticity, which is measured by the plasticity index (PI). PI is equal to the water content difference at the liquid limit (LL) and the plastic limit (PL). Although rarely used by the agriculturally oriented discipline of soil science, soil engineers frequently use the Casagrande Plasticity Chart (6) to describe soil plasticity. It is a plot of the plasticity index (PI) against the liquid limit (LL). Many fine-grained soils lie scattered along a line on the Casagrande Plasticity Chart:

$$PI = 0.73(LL - 20)$$

the units for the Plasticity Index (PI) and Liquid Limit (LL) are:

$$\text{g water}/100\,\text{g dry soil}$$

The line defined by this equation is called the Casagrande A-line. Engineering classification schemes use the position of soils on the chart in terms of their relation to the A-line for classification purposes. Soils with similar liquid limits tend to have greater soil toughness and dry strength when the plasticity index increases (7).

The plastic limit is generally regarded as the most suitable soil water content for seedbed preparation. Compared with wetter or drier soil water contents it tends to give the best tilth and least cloddiness. However, the soil's compactibilty tends to be greatest at soil water contents slightly wetter than the plastic limit.

THEORY OF SOIL PLASTICITY

Water Films

Water in soil is held inside pore spaces and is adsorbed along the surfaces of the soil particles as a thin water film. The latter determines the soil's plastic behavior because the water film acts as a lubricant between particles. The lubricating effect is most pronounced if particles are small, platy and are in face-to-face contact.

Cohesion

The same forces that determine the soil's behavior during application of stress govern the dynamics of the soil's plastic behavior: cohesion, adhesion, and angle of internal friction. Soil properties that determine the response to these forces, and therefore degree of plasticity, are texture, presence of cementation agents, cation exchange capacity and type of cations, salt concentration, soil water content, pH and organic matter content (8–10).

Cohesion can be separated into true and apparent cohesion. Apparent cohesion holds particles together through capillary forces and interlocking of rough particle surfaces (7, 11). It allows sands to be deformed without rupture, but this deformation is not plastic deformation since moulded sands fall apart when dry. True cohesion holds particles together through electrostatic and electromagnetic forces of attraction and cementing agents. True cohesion is the force field for plasticity.

FACTORS AFFECTING PLASTICITY

Clay Content and Clay Type

The specific surface area of soils, and thus their plasticity is largely determined by clay content. However, despite a highly significant empirical relationship between the Atterberg limits and clay content, reliable prediction of one from the other is often not possible, unless similar soil types are considered (12–16).

The plasticity of different types of clay minerals is related to the size of their surface area, shape of the clay lattice and type of exchangeable cations. Platy or sheet like structures are more plastic than tetrahedra type structures. An assessment of plasticity, or any other soil property, in relation to clay mineralogy must bear in mind that natural soils usually do not contain appreciable amounts of pure primary clay minerals. Therefore only a broad relationship can be indicated when relating clay mineralogy to plasticity.

Soils can be differentiated into three groups by the way in which their physical properties relate to plasticity (17):

1. Soils where the void volume changes little with changes in soil water content. These soils are generally governed by low activity or 1:1 clay minerals such as Kaolinite, Halloysite, Serpentine and Berthierine. Owing to their relatively low specific surface area ($10–150\,m^2\,g^{-1}$) and low cation exchange capacity ($< 15\,cmol + kg^{-1}$) they have low values of plasticity.
2. Soils where the void space changes reversibly with changes in soil water content. These soils are typically governed by high activity or 2:1 clay minerals. The most common types of clay minerals in soils are Illites, Vermiculites, Smectites and Chlorites. Their specific surface area is typically $50–800\,m^2\,g^{-1}$ and cation exchange capacity $20–200\,cmol{+}kg^{-1}$ thus having a much greater plasticity than the low activity clay minerals.
3. The last group are those soils where the void volume changes with soil water content, but is largely irreversible. These soils contain appreciable amounts of allophane. Upon drying the plasticity of these soils decreases irreversibly possibly due to the permanent collapse of a large number of dehydrated micro-pores (18). The degree with which allophane soils change their plastic behavior upon drying can be used for classification purposes.

The impact of other clay size particles in soils (sesquioxides, manganese and silicium oxides, calcium and magnesium carbonates) on soil plasticity is largely associated through their ability to cement particles together and thus tends to decrease plasticity or causes sub-plastic behavior.

Cations

The impact of different types of cations on the plasticity of soils is related to the size of the hydrated cation as well as to its charge, i.e., their position in the lyotropic series. Divalent cations (Ca and Mg) tend to increase both Atterberg limits and the plasticity of soils. Due to its larger effective ionic radius, Mg has a stronger effect on the Atterberg limits than Ca (19). Monovalent cations (K and Na) lower both Atterberg limits but the K-cation tends to decrease and the Na-cation increase plasticity (20).

pH

The effect of pH on plasticity is indirect and associated with the influence of pH on cation exchange capacity, amount of exchangeable bases and soil organic matter.

Organic Carbon

Soil organic matter has a high absorptive capacity for water. Lubricating water films tend not to form unless hydration of organic matter is fairly complete. Therefore organic matter increases the Atterberg limits but tends to have little effect on the plasticity index (1, 13–16). Besides total amount of organic matter, type or quality of organic matter influences plasticity. On soils with similar texture and organic matter content the presence of long-chained lipids tends to shift the Atterberg limit to higher soil water contents (21).

TYPES OF PLASTICITY

Three different types of plasticity, based on the apparent change in clayeyness when a standard field texture test is carried out, can be distinguished (2):

1. Soil with normal plasticity only shows minimal apparent change in texture during the field texture test.
2. Sub-plastic soils are those where the texture appears to become finer or heavier as the soil bolus is worked for the field texture test.
3. Super-plastic soils are those where clayeyness diminishes as the soil is worked.

Super plasticity may be of little practical importance. Soils that could fall into this category are very sticky clays found in buried soils formed through weathering of glacial drift. Most soils have normal plasticity. Although sub-plastic clay soils are mainly reported in Australia, they probably occur worldwide. Soils with sub-plastic properties appear to be associated with very strongly bonded silt size aggregates consisting largely of fine clay (22). Sub-plasticity of Oxisols was attributed to cementation by iron oxides while on Alfisols it was attributed to the presence of packets or domains of strongly oriented clay but not Ca-carbonates (23). Removal of the cementation agents in sub-plastic clays revealed the presence of highly active (2:1) clay minerals (24). Degree of dispersion of sub-plastic clays has a strong impact on the outcome of texture analysis but also on the outcome of many soil chemical analyses.

REFERENCES

1. Baver, L.D. *Soil Physics*, Third Edition. John Wiley & Sons, Inc.: New York, Chapman & Hall Limited: London, 1956.
2. Butler, B.E. A System for the Description of Soil Structure and Consistence in the Field. The Journal of the Australian Institute of Agricultural Science, **1955**, *18*, 239–249.
3. SSSA. Internet Glossary of Soil Science Terms. Soil Science Society of America, 677 South Segoe Road, Madison, WI 53711, 1997. http://www.soils.org/sssagloss/ (accessed Sept 2000).
4. Troeh, F.R.; Thompson, L.M. *Soils and Soil Fertility*; Oxford University Press: New York, Oxford, 1993.
5. Atterberg, A. Die Plastizität der Tone. Bodenkundliche Mitteilungen **1911**, *1*, 10–43.
6. Casagrande, A. Classification and Identification of Soils. Proceedings of the American Society of Civil Engineers **1947**, *73*, 783–810.
7. Lambe, T.W.; Whitman, R.V. *Soil Mechanics, SI Version*; John Wiley & Sons: New York, Chichester, Brisbane, Toronto, Singapore, 1976.
8. Means, R.E.; Parcher, J.V. *Physical Properties of Soils*; Constable and Company Ltd.: London, 1964.
9. Hartge, K.H. *Einführung in die Bodenphysik*; Ferdinand Enke Verlag: Stuttgart, 1978.
10. Zhang, H.Q.; Hargte, K.H. Die Wechselwirkung Zwischen Winkel der Inneren Reibung, Organischer Substanzen und Wasserspannung. Mitteilungen der Deutschen Bodenkundlichen Gesellschaft **1989**, *59*, 279–282.
11. Mitchell, J.K. *Fundamentals of Soil Behaviour*; John Wiley & Sons Inc.: New York, London, Sydney, Toronto, 1976.
12. Warkentin, B.P. Physical Properties Related to Clay Minerals in Soils of the Caribbean. Tropical Agriculture (Trinidad) **1974**, *51*, 279–287.
13. Lal, R. Physical Properties and Moisture Retention Characteristics of Some Nigerian Soils. Geoderma **1979**, *21*, 209–223.
14. Rosa de la, D. Relation of Seven Pedological Characteristics to Engineering Qualities of Soil. Journal of Soil Science **1979**, *30*, 793–799.
15. Larney, F.J.; Fortune, R.A.; Collins, J.F. Intrinsic Soil Physical Parameters Influencing Intensity of Cultivation Procedures for Sugar Beet Seedbed Preparation. Soil and Tillage Research **1988**, *12*, 253–267.
16. Jong de, E.; Acton, D.F.; Stonehouse, H.B. Estimating the Atterberg Limits of Southern Saskatchewan Soils from Texture and Carbon Contents. Canadian Journal of Soil Science **1990**, *70*, 543–554.
17. Maeda, T.; Takenaka, H.; Warkentin, B.P. Physical Properties of Allophane Soils. Advances in Agronomy **1977**, *29*, 229–264.
18. Ross, G.J. Mineralogical, Physical and Chemical Characteristics of Amorphous Constituents in Some Pozsolic Soils from British Columbia. Canadian Journal of Soil Science **1980**, *60*, 31–43.
19. Misopolinos, N.D.; Silleos, N.G.; Prodromou, K.P. The Influence of Exchangeable Mg on Certain Physical Soil Properties in a Number of Mg-Affected Soils. Catena **1988**, *15*, 127–136.
20. Baver, L.D. The Atterberg Consistency Constants: Factors Affecting Their Values and a New Concept of Their Significance. Journal of the American Society of Agronomy **1930**, *22*, 935–948.
21. Leinweber, P.; Kahle, P.; Schulten, H.R. Einfluss der Qualität der Organischen Substanz auf die Konsistenz Landwirtschaftlicher Böden. Zeitschrift für Pflanzenernährung und Bodenkunde **1991**, *154*, 169–170.
22. Walker, P.H.; Hutka, J. Sub-Plasticity in Australian Soils. III. Particle-Size Properties of Soil Materials of Varying Plasticity. Australian Journal of Soil Research **1976**, *14*, 249–260.
23. Brewer, R.; Blackmore, A.V. Sub-Plasticity in Australian Soils. II Relationship Between Sub-Plasticity Rating, Optically Oriented Clay, Cementation and Aggregate Stability. Australian Journal of Soil Research **1976**, *14*, 237–248.
24. Blackmore, A.V. Sub-Plasticity in Australian Soils. IV Plasticity and Structure Related to Cementation. Australian Journal of Soil Research **1976**, *14*, 261–272.

PLINTHITE AND PETROPLINTHITE

Eswaran Padmanabhan
University of Malaysia Sarawak, Sarawak, Malaysia

Hari Eswaran
United States Department of Agriculture-Natural Resources Conservation Service (USDA-NRCS), Washington, D.C., U.S.A.

INTRODUCTION

Laterite is a highly weathered material rich in secondary oxides of iron, aluminum, or both (1). The material has very low amounts of basic cations and primary silicates, but conversely, may be enriched in quartz and kaolinite. Laterites are either hard or capable of hardening on exposure to wetting and drying cycles. The term laterite originated in 1807 (2), but the first mention of it in formal scientific literature was in 1821 (3). Geologists, engineers, and pedologists consider laterite as some kind of sesquioxide accumulation that ranges from weathered rock to hard and cemented ironstone. The term Latosol (4) was introduced to distinguish it from laterite but was later suggested to be abandoned (5, 6).

These older terms were replaced with modern terms such as plinthite (7), petroplinthite (8), and petroferric contact (7). This paper describes the genesis and morphology of these features, as well as impact of their presence on use and management of the affected soils. Geologic deposits of iron stones and bauxite which have other modes of formation are not considered in this paper.

DEFINITIONS

The term plinthite (Fig. 1) is derived from the Greek word "plinthos" meaning brick. More recent definitions for this term have been introduced (6, 7, 9). Absolute amount of iron in plinthite is site dependent. Initially, diffuse mottles separated by bleached interconnected zones form in the soil. A continuous phase of plinthite is formed eventually, through the accretion of iron with time. Plinthite usually forms deep in the soil, and therefore, has low amounts of organic matter. Erosion and truncation of the soil could expose the plinthite. Subsequent to these processes, secondary enrichment with organic matter could occur.

Plinthite normally has a sharp upper boundary to surficial layers of the soil, but a diffuse lower boundary to reduced subsurface layers. The basic form of plinthite consists of quartz (silt and sand size) in a matrix of kaolinitic clay (with possibly some mica). In extreme cases, gibbsite, goethite and hematite can also occur in addition to quartz and kaolinitic clay (10). Manganese minerals such as birnessite and lithiophorite can also be present and consequently, color the system black.

Petroplinthite (formerly known as iron stone or lateritic gravel) either forms directly in a soil or from the hardening of plinthite. Petroplinthite usually occurs as loose or slightly cemented gravel (10, 11). Recementation of plinthites or petroplinthites by rapid and cyclic influx of iron forms the petroferric contact. Alternatively, the contact can also develop on exposed ironstone gravel brought to the surface through uplift and erosion.

MODE OF FORMATION AND MORPHOLOGY

The pedological environment in the tropics is conducive for the formation of plinthite and related sesquioxide accumulations (6). The chemical processes resulting in the formation of these sesquioxide forms are quite complicated (6). A large supply of iron is needed to form plinthite. The origin of iron is normally difficult to ascertain (12). Downslope migration and subsequent accumulation of iron at lower levels in the landscape constitutes one mode of formation (5). It has been suggested that iron can diffuse upwards (with fluctuating groundwater) towards oxidizing zones (13).

Cementation in soils may be due to two different chemical processes: 1) ionic chemical bonding within the compounds (in petrogypsic and petrocalcic horizons) 2) elements with intermediate electronegativity, such as Si, Al, Fe, Mn (14), that form covalent bonds within

Encyclopedia of Soil Science

Fig. 1 Plinthudult from Peninsular Malaysia (Plinthite is the combination of red and white mottles. The red blotches harden into hard petroplinthite when exposed and dried).

the cementing agents and with other soil constituents (15). Even in crystalline forms, these compounds have outer surfaces composed of OH_2–OH groups that can be used to form covalent bonds with the same elements (6). This results in the high density of the cement.

Cementation and hardening of petroplinthite is attributed to the close packing of crystals (16). Submicroscopy studies indicate that the crust of Fe-nodules are made up of densely packed Fe-oxides (10, 11, 17). Iron accumulates on this rigid surface, subsequently increasing the thickness of the crust. Interior regions of the nodules usually have higher porosities as a consequence of the random orientation of minerals (10, 11, 17). It has been suggested that formation of kaolinite is possible with Fe accretionary formations (18–20). Variations in ultramicroporosity and redox potentials appear to determine the type and geometry of minerals (17).

Alternating wet and dry conditions also favor hardening (6). Other factors that favor the hardening process include an ustic soil moisture regime (7) or geomorphic changes in the landscape such as uplift or lowering of the ground water table. Post-uplift erosion eventually exposes the hardened zone (most evident at the edges of uplifted zones). The resistance to further erosion causes the formation of escarpments and inversion of the landscape (21). Consequently, the altered local hydrology usually favors the development of petroferric contacts in the soils.

OCCURRENCE AND DISTRIBUTION

The various forms of sesquioxide accumulations discussed earlier can be found in several soil types. The sharp upper and lower boundaries to the petroplinthite represent discontinuities in the soil. Active plinthite formation is usually found in relatively young landscapes that have fluctuating water tables in the soil. Plinthite is distributed over large areas in India and petroplinthite gravel is present in the peneplains. West Africa has the largest contiguous extent of soils with plinthite. Such soils are present as a band from the West Coast of Gambia in the north to Cameroon in South and extend to southern Sudan and eastern Zaire (6). South of this band, soils with petroplinthite occur. The largest extent of soils with petroferric contacts also occurs in West Africa (Ivory Coast, eastern Ghana, Burkina Faso and Central African Republic) and sporadically in the eastern part of southern Africa.

MANAGEMENT IMPLICATIONS

Pure laterites and latosols cover the greater part of the humid tropics and are either agriculturally poor or virtually useless (22). Essentially, the six major agro-management-related problems for such soils are irreversible hardening upon exposure, restriction to water flow, reduction in soil volume for root development, soil degradation, high remediation costs and irreversible phosphorous fixation. The cause–effect relationships involving all these stresses and the subsequent impact on land productivity remains unclear. Corrective measures to overcome these problems are usually not undertaken since such projects are considered to be economically not viable. Economic and social factors have an impact on the use and management of such soils. However, these factors will not be discussed in this paper. Another effect of the presence of such soils would be the gradual reduction in C-sequestration capacity.

Irreversible hardening of plinthite occurs when the vegetative cover is removed (23). Presence of plinthite, petroplinthite or petroferric contacts indicates that some restriction to water (24, 25) and root penetration as well as natural soil degradation processes exists. Water flows faster above the plinthite but movement within this material is slow and restricted to the bleached zones (24, 25). Perched water above the plinthite layer restricts the potential use of such soils to crops that require saturated conditions, such as rice. Landscape evolution and climatic change can modify the hydrology (26).

Quality of such soils declines (degradation) for agriculture and engineering purposes. Storage capacities for moisture and nutrients reduces with decrease in effective soil volume. Plinthite by itself poses few constraints to agricultural use of the soil (6). Global climatic changes (increased aridity) enhance the development of petroplinthite or petroferric contacts. The Fe-nodules tend to have elevated P contents and higher P sorption capacities compared to the soil fines (27).

In the humid tropics (e.g., Malaysia), such soils are being used for rubber and oil palm cultivation. Apart from lowering the average yield, presence of such soils complicates site yield predictions, fertilizer recommendations and also leaf and soil sampling for nutrient analyses. Increasing demands for food has forced these soils to be cultivated. Mismanagement of such soils has resulted in erosion rates of up to 100 ton $ha^{-1} yr^{-1}$. Eventual exposure of the petroferric material causes such soils to be abandoned. Soil remediation is very expensive. In addition to this, revegetation techniques are extremely time-consuming. Reforestation may be the only viable option for such soils (6).

Laterite has always been a cheap raw material for the construction of roads and walls in the tropics. However, the raw material is not considered to be very durable. Methods to enhance the performance of the material include treatment with lime, bitumen or even cement (28). Stabilized soils have been used in road construction since the Roman times (29). Firing improves the durability of lateritic soil blocks (30). However, many of these methods are expensive and may not be viable options in developing nations.

CONCLUSIONS

The problem soils discussed in this paper are present in many developing countries in the tropics. Such soils can be easily mismanaged. Several agro-management and engineering related problems can be identified. The average far mer cannot manage these soils at high input levels. Zero- and low-input agricultural systems on such soils enhance soil degradation. The exact areal extent of such soils in many tropical countries is still unknown. There is also a lack of knowledge of the resilience of these soils to development. These issues merit rigorous and immediate attention.

REFERENCES

1. Alexander, L.T.; Cady, J.G. Genesis and Hardening of Laterite in Soils. *Tech. Bull. 1282*; Soil Conservation Service, U.S. Dept. Agric., 1962; 90 pp.
2. Buchanan, F. *A Journey from Madras Through the Countries of Mysore, Canara and Malabar*; East India Co: London, 1807; Vol. 2, 440–441.
3. Babington, B. Remarks on the Geology of the Country Between Tellicherry and Madras. Trans. Geol. Soc., London **1821**, *5*, 328–329.
4. Kellogg, C.E. Preliminary Suggestions for the Classification and Nomenclature of Great Soil Groups in Tropical and Equatorial Regions. Commonwealth Bur. Soil Sci., Tech Commun. **1949**, *46*, 76–85.
5. Maignien, R. *Review of Research on Laterites. Natural Resources Research IV*; UNESCO: Paris, 1966; 148.
6. Eswaran, H.; De Coninck, F.; Varghese, T. Role of Plinthite and Related Forms in Soil Degradation. In *Advances in Soil Science*; Lal, R., Stewart, B.A., Eds.; CRC Press: Boca Raton, USA, 1990; Vol. 11, 109–127.
7. Soil Survey Staff, *Soil Taxonomy: A Basic System of Soil Classification for Making and Interpreting Soil Surveys*, 2nd Ed.; Natural Resources Conservation Service. US Department of Agriculture, Handbook 436; US Govt. Print. Office: Washington, DC; 1999, 869.
8. Sys, C. Suggestions for the Classification of Tropical Soils with Lateritic Materials in the American Classification. Pedologie **1968**, *18*, 189–198.
9. Daniels, R.B.; Perkins, H.F.; Hajek, B.F.; Gamble, E.E. Morphology of Discontinuous Phase Plinthite and Criteria for Its Field Identification in the Southeastern United States. Soil Sci. Soc. Am. J. **1978**, *42*, 944–949.
10. Eswaran, H.; Comerma, J.; Sooryanarayanan, V. Scanning Electron Microscopic Observations on the Nature and Distribution of Iron Minerals in Plinthite and Petroplinthite. *Lateritisation Processes, Proceedings of the International Seminar on Laterisation Processes*; Oxford and IBH Publishing Co: New Delhi, 1980; 335–341.
11. Eswaran, H.; Raghumohan, N.G. The Microfabric of Petroplinthite. Soil Sci. Soc. Am. Proc. **1973**, *37*, 79–81.
12. Van Wambeke, A.; Eswaran, H.; Herbillon, A.J.; Comerma, J. Oxisols. In *Pedogenesis and Soil Taxonomy. II. The Soil Orders. Developments in Soil Science 11B*; Wilding, L.P., Smeck, N.E., Hall, G.F., Eds.; Elsevier: Amsterdam, The Netherlands, 1983; 325–354.
13. Lelong, F. Régime Des Nappes Phréatiques Continues Dans Les Formations D'altération Tropicale. Conséquence Pour La Pédogénèse. Sci. Terre. **1966**, *11*, 203–244.
14. Sanderson, R.T. *Chemical Periodicity*; Reinhold: New York, 1964.
15. Chadwick, F.; Nettleton, D. Micromorphologic Evidence of Adhesive and Cohesive Forces in Soil Cementation. In *Soil Micromorphology: A Basic and Applied Science. Developments in Soil Science* 19; Donglas, L. A., Ed.; Elsevier: Amsterdam, The Netherlands, 1990; 207–212.
16. Schwertmann, U. Some Properties of Soil and Synthetic Iron Oxides. In *Iron in Soils and Clay Minerals*; NATO ASI Series, Stucki, J.W., Goodman, B.A., Schwertmann, U., Eds.; D. Reidel Publishing Co: Dordrecht, 1985; 203–204.
17. Padmanabhan, E.; Mermut, A.R. Submicroscopic Structure of Fe-Coatings on Quartz Grains in Tropical Environments. Clays and Clay Minerals **1996**, Vol. *44* (6), 801–810.
18. Tardy, Y.; Nahon, D. Geochemistry of Laterites, Stability of Al-Goethite, Al-Hematite and Fe^{3+}-kaolinite in Bauxites and Ferricretes: An Approach to the Mechanism of Concretion Formation. Am. J. Sci. **1985**, *285*, 865–903.

19. Trolard, F.; Tardy, Y. A Model of Fe^{3+}-Kaolinite, Al^{3+}-goethite, Al^{3+}-Hematite Equilibria in Laterites. Clay Miner. **1989**, *24*, 1–21.
20. Nahon, D.B. Self-Organization in Chemical Lateritic Weathering. Geoderma **1991**, *51*, 5–13.
21. Sivarajasingham, S.; Alexander, L.T.; Cady, J.G.; Cline, M.G. Laterite. In *Advances in Agronomy*; Norman, A.G., Ed.; Academic Press: New York, 1962; 14, 1–60.
22. Kamarck, A.M. *Climate and Economic Development*; EDI Seminar Paper No. 2, Economic Development Institute, World Bank: Washington, DC, 1972.
23. Sombroek, W.G. Soils of the Amazon Region. In *The Amazon: Limnology and Landscape Ecology of a Mighty Tropical River and Its Basin*; Sioli, H., Ed.; Kluwer Press: Dordrecht, Netherlands, 1984; 521–535.
24. Carlan, W.L.; Perkins, H.F.; Leonard, R.A. Movement of Water in a Plinthic Paleudult Using a Bromide Tracer. Soil Sci. **1985**, *139*, 62–66.
25. Blume, L.J.; Perkins, H.F.; Hubbard, R.K. Subsurface Water Movement in an Upland Coastal Plain Soil as Influenced by Plinthite. Soil Sci. Soc. Am. J. **1987**, *51*, 774–779.
26. Macedo, J.; Bryant, R.B. Morphology, Mineralogy and Genesis of a Hydrosequence of Oxisols in Brazil. Soil Sci. Soc. Am. J. **1987**, *51*, 690–698.
27. Tiessen, H.; Frossard, E.; Mermut, A.R.; Nyamekye, A.L. Phosphorus Sorption and Properties of Ferruginous Nodules from Semiarid Soils from Ghana and Brazil. Geoderma **1991**, *48*, 373–389.
28. Okagbue, C.O.; Yakubu, J.A. Limestone Ash Waste as a substitute for Lime in Soil Improvement for Engineering Construction. Bull. Eng. Geol. Env. **2000**, *58*, 107–113.
29. Krebs, R.D.; Walker, R.B. *Highway Materials*; McGraw Hills: New York, 1971; 428.
30. Anifowose, A.Y.B. Stabilization of Lateritic Soils as a raw Material for Building Blocks. Bull. Eng. Geol. Env. **2000**, *58*, 151–157.

POINT SOURCE POLLUTION

Ravendra Naidu
Mallavarapu Megharaj
Peter Dillon
Rai Kookana
Commonwealth Scientific and Industrial Research Organisation (CSIRO) Land and Water, Adelaide, South Australia, Australia

Ray Correll
Commonwealth Scientific and Industrial Research Organisation (CSIRO) Mathematical and Information Sciences, Adelaide, South Australia, Australia

Walter Wenzel
University of Agricultural Sciences Vienna—BOKU, Vienna, Austria

INTRODUCTION

Environmental pollution is one of the foremost ecological challenges. Pollution is an offshoot of technological advancement and over-exploitation of natural resources. From the standpoint of pollution, the term environment primarily includes air, land, and water components including landscapes, rivers, parks, and oceans. Pollution can be generally defined as an undesirable change in the natural quality of the environment that may adversely affect the well-being of humans, other living organisms, or entire ecosystems either directly or indirectly. Although pollution is often the result of human activities (anthropogenic), it could also be due to natural sources such as volcanic eruptions emitting noxious gases, pedogenic processes or natural change in the climate. Where pollution is localized it is described as point source (PS). Thus, PS pollution is a source of pollution with a clearly identifiable point of discharge that can be traced back to the specific source such as leakage of underground petroleum storage tanks or an industrial site.

Some naturally occurring pollutants are termed geogenic contaminants and these include fluorine, selenium, arsenic, lead, chromium, fluoride and radionuclides in the soil and water environment. Significant adverse impacts of geogenic contaminants (e.g., As) on environmental and human health have been recorded in Bangladesh, West Bengal, India, Vietnam, and China. More recently reported is the presence of geogenic Cd and the implications to crop quality in Norwegian soils (1).

The terms contamination and pollution are often used interchangeably but erroneously. Contamination denotes the presence of a particular substance at a higher concentration than would occur naturally and this may or may not have harmful effects on human or the environment. Pollution refers not only to the presence of a substance at higher level than would normally occur but also is associated with some kind of adverse effect.

NATURE AND SOURCES OF CONTAMINANTS

The main activities contributing to PS pollution include industrial, mining, agricultural, and commercial activities as well as transport and services (Table 1). Uncontrolled mining, manufacturing, and disposal of wastes inevitably cause environmental pollution. Military land and land used for recreational shooting are also important sites of PS contamination. The contaminants associated with such activities are listed in Table 1. Contamination at many of these sites appears to have resulted because of lax regulatory measures prior to the establishment of legislation protecting the environment.

CONTAMINANT INTERACTIONS IN SOIL AND WATER

Inorganic Chemicals

Inorganic contaminant interactions with colloid particulates include: adsorption–desorption at surface sites, precipitation,

Encyclopedia of Soil Science

P

Table 1 Industries, land uses and associated chemicals contributing to points, nonpoint source pollution

Industry	Type of chemical	Associated chemicals
Airports	Hydrocarbons	Aviation fuels
	Metals	Particularly aluminum, magnesium, chromium
Asbestos production and disposal	Asbestos	
Battery manufacture and recycling	Metals	Lead, manganese, zinc, cadmium, nickel, cobalt, mercury, silver, antimony
	Acids	Sulfuric acid
Breweries/distilleries	Alcohol	Ethanol, methanol, esters
Chemicals manufacture and use	Acid/alkali	Mercury (chlor/alkali), sulfuric, hydrochloric and nitric acids, sodium and calcium hydroxides
	Adhesives/resins	Polyvinyl acetate, phenols, formaldehyde, acrylates, phthalates
	Dyes	Chromium, titanium, cobalt, sulfur and nitrogen organic compounds, sulfates, solvents
	Explosives	Acetone, nitric acid, ammonium nitrate, pentachlorophenol, ammonia, sulfuric acid, nitroglycerine, calcium cyanamide, lead, ethylene glycol, methanol, copper, aluminum, bis(2-ethylhexyl) adipate, dibutyl phthalate, sodium hydroxide, mercury, silver
	Fertilizer	Calcium phosphate, calcium sulfate, nitrates, ammonium sulfate, carbonates, potassium, copper, magnesium, molybdenum, boron, cadmium
	Flocculants	Aluminum
	Foam production	Urethane, formaldehyde, styrene
	Fungicides	Carbamates, copper sulfate, copper chloride, sulfur, chromium
	Herbicides	Ammonium thiocyanate, carbanates, organochlorines, organophosphates, arsenic, mercury
	Paints	
	Heavy metals	Arsenic, barium, cadmium, chromium, cobalt, lead, manganese, mercury, selenium, zinc
	General	Titanium dioxide
	Solvent	Toluene, oils natural (e.g., pine oil) or synthetic
	Pesticides	Arsenic, lead, organochlorines, organophosphates
	Active ingredients	Sodium, tetraborate, carbamates, sulfur, synthetic pyrethroids
	Solvents	Xylene, kerosene, methyl isobutyl ketone, amyl acetate, chlorinated solvents
	Pharmacy	Dextrose, starch
	General/solvents	Acetone, cyclohexane, methylene chloride, ethyl acetate, butyl acetate, methanol, ethanol, isopropanol, butanol, pyridine methyl ethyl ketone, methyl isobutyl ketone, tetrahydrofuran
	Photography	Hydroquinone, pheidom, sodium carbonate, sodium sulfite, potassium bromide, monomethyl paraaminophenol sulfates, ferricyanide, chromium, silver, thiocyanate, ammonium compounds, sulfur compounds, phosphate, phenylene diamine, ethyl alcohol, thiosulfates, formaldehyde
	Plastics	Sulfates, carbonates, cadmium, solvents, acrylates, phthalates, styrene
	Rubber	Carbon black
	Soap/detergent	
	General	Potassium compounds, phosphates, ammonia, alcohols, esters, sodium hydroxide, surfactants (sodium lauryl sulfate), silicate compounds
	Acids	Sulfuric acid and stearic acid
	Oils	Palm, coconut, pine, tea tree

(*Continued*)

Table 1 Industries, land uses and associated chemicals contributing to points, nonpoint source pollution *(Continued)*

Industry	Type of chemical	Associated chemicals
	Solvents	
	General	Ammonia
	Hydrocarbons	e.g., BTEX (benzene, toluene, ethylbenzene, xylene)
	Chlorinated organics	e.g., trichloroethane, carbon tetrachloride, methylene chloride
Defense works		See explosives under chemicals manufacture and use, foundries, engine works, service stations
Drum reconditioning		See chemicals manufacture and use
Dry cleaning		Trichlorethylene and ethane
		Carbon tetrachloride
		Perchlorethylene
Electrical		PCBs (transformers and capacitors), solvents, tin, lead, copper
Engine works	Hydrocarbons	
	Metals	
	Solvents	
	Acids/alkalis	
	Refrigerants	
	Antifreeze	Ethylene glycol, nitrates, phosphates, silicates
Foundries	Metals	Particularly aluminum, manganese, iron, copper, nickel, chromium, zinc, cadmium and lead and oxides, chlorides, fluorides and sulfates of these metals
	Acids	Phenolics and amines
		Coke/graphite dust
Gas works	Inorganics	Ammonia, cyanide, nitrate, sulfide, thiocyanate
	Metals	Aluminum, antimony, arsenic, barium, cadmium, chromium, copper, iron, lead, manganese, mercury, nickel, selenium, silver, vanadium, zinc
	Semivolatiles	Benzene, ethylbenzene, toluene, total xylenes, coal tar, phenolics and PAHs
Iron and steel works		Metals and oxides of iron, nickel, copper, chromium, magnesium and manganese, and graphite
Landfill sites		Methane, hydrogen sulfides, heavy metals, complex acides
Marinas		Engine works, electroplating under metal treatment
	Antifouling paints	Copper, tributyltin (TBT)
Metal treatments	Electroplating metals	Nickel, chromium, zinc, aluminum, copper, lead, cadmium, tin
	Acids	Sulfuric, hydrochloric, nitric, phosphoric
	General	Sodium hydroxide, 1,1,1-trichloroethane, tetrachloroethylene, toluene, ethylene glycol, cyanide compounds
	Liquid carburizing baths	Sodium, cyanide, barium, chloride, potassium chloride, sodium chloride, sodium carbonate, sodium cyanate
Mining and extractive industries		Arsenic, mercury and cyanides and also refer to Explosives under chemicals, manufacture and use
Power stations		Asbestos, PCBs, fly ash, metals
Printing shops		Acids, alkalis, solvents, chromium (see Photography under chemicals, manufacture and use)
Scrap yards		Hydrocarbons, metals, solvents
Service stations and fuel storage facilities		Aliphatic hydrocarbons
		BTEX (i.e., benzene, toluene, ethylbenzene, xylene)
		PAHs (e.g., benzo(a) pyrene)
		Phenols
		Lead

(Continued)

Table 1 Industries, land uses and associated chemicals contributing to points, nonpoint source pollution *(Continued)*

Industry	Type of chemical	Associated chemicals
Sheep and cattle dips		Arsenic, organochlorines and organophosphates, carbamates, and synthetic pyrethroids
Smelting and refining		Metals and the fluorides, chlorides and oxides of copper, tin, solver, gold, selenium, lead, aluminum
Tanning and associated trades	Metals	Chromium, manganese, aluminum
	General	Ammonium sulfate, ammonia, ammonium nitrate, phenolics (creosote), formaldehyde, tannic acid
Wood preservation	Metals	Chromium, copper, arsenic
	General	Naphthalene, ammonia, pentachlorophenol, dibenzofuran, anthracene, biphenyl, ammonium sulfate, quinoline, boron, creosote, organochlorine pesticides

(Adapted from Ref. 11.)

ion exchange with clay minerals, binding by organically coated particulate matter or organic colloidal material, or adsorption of contaminant ligand complexes. Depending on the nature of contaminants, these interactions are controlled by solution pH and ionic strength of soil solution, nature of the species, dominant cation, and inorganic and organic ligands present in the soil solution (2).

Organic Chemicals

The fate and behavior of organic compounds depend on a variety of processes including sorption–desorption, volatilization, chemical and biological degradation, plant uptake, surface runoff and leaching. Sorption–desorption and degradation (both biotic and abiotic) are perhaps the two most important processes as the bulk of the chemicals is either sorbed by organic and inorganic soil constituents, and chemically or microbially transformed/degraded. The degradation is not always a detoxification process. This is because in some cases the transformation or degradation process leads to intermediate products that are more mobile, more persistent or more toxic to nontarget organisms. The relative importance of these processes is determined by the chemical nature of the compound.

IMPLICATIONS TO SOIL AND ENVIRONMENTAL QUALITY

Considerable amount of literature is available on the effects of contaminants on soil microorganisms and their functions in soil. The negative impacts of contaminants on microbial processes are important from the ecosystem point of view and any such effects could potentially result in a major ecological perturbance. Hence it is most relevant to examine the effects of contaminants on microbial processes in combination with communities. The most commonly used indicators of metal effects on microflora in soil are 1) soil respiration, 2) soil nitrification, 3) soil microbial biomass, and 4) soil enzymes.

Contaminants can reach the food chain by way of water, soil, plants, animals. In addition to the food chain transfer, pollutants may also enter via direct consumption or dust inhalation of soil by children or animals. Accumulation of these pollutants can take place in certain target tissues of the organism depending on the solubility and nature of the compound. For example, DDT and PCBs accumulate in human adipose tissue. Consequently, several of these pollutants have the potential to cause serious abnormalities including cancer and reproductive impairments in animal and human systems.

SAMPLING FOR POINT SOURCE POLLUTION

The aims of the sampling system must be clearly defined before it can be optimized (3). The type of decision may be to determine land use, how much of an area is to be remediated, or what type of remediation process is required. Because sampling and the associated chemical and statistical analyses are expensive, careful planning of the sampling scheme is therefore a good investment. One of the best ways to achieve this is to use an ancillary data that is available. This data could be in the form of emission history from a stack, old photographs that give details of previous land uses, or agricultural records. Such data can at least give qualitative information.

As discussed before, PS pollution will typically be airborne from a stack, or waterborne from some effluent such as tannery waste, cattle dips, or mine waste. In many cases, the industry will have modified its emissions (e.g., cleaner production) or point of release (increased stack height), hence the current pattern of emission may not be closely related to the historic pattern of pollution. For example, liquid effluent may previously have been discharged into a bay, but that effluent may now be treated and perhaps discharged at some other point. Typically, the aim of a sampling scheme in these situations is to assess the maximum concentrations, the extent of the pollution, and the rate of decline in concentration from the PS. Often the sampling scheme will be used to produce maps of concentration isopleths of the pollutant.

The location of the sampling points would normally be concentrated towards the source of the pollution. A good scheme is to have sufficient samples to accurately assess the maximum pollution, and then space additional samples at increasing intervals. In most cases, the distribution of the pollutant will be asymmetric, with the maximum spread down the slope or down the prevailing wind. In such cases more samples should be placed in the direction of the expected gradient. This is a clear case of when ancillary data can be used effectively. A graph of concentration of the pollutant against the reciprocal of distance from the source is often informative (4). Sampling depths will depend on both the nature of the pollution and the reason for the investigation. If the pollution is from dust and it is unlikely to be leached, only surface sampling will be required. An example of this is pollution from silver smelting in Wales (5). In contrast, contamination from organic or mobile inorganic pollutants such as F compounds may migrate well down to the profile and deep sampling may be required (6, 7).

ASSESSMENT

In order to assess the impacts of pollution, reliable and effective monitoring techniques are important. Pollution can be assessed and monitored by chemical analyses, toxicity tests, and field surveys. Comparison of contaminant data with an uncontaminated reference site and available databases for baseline concentrations can be useful in establishing the extent of contamination. However, this may not be always possible in the field. Chemical analyses must be used in conjunction with biological assays to reveal site contamination and associated adverse effects. Toxicological assays can also reveal information about synergistic interactions of two or more contaminants present as mixtures in soil, which cannot be measured by chemical assays alone.

Microorganisms serve as rapid detectors of environmental pollution and are thus of importance as pollution indicators. The presence of pollutants can induce alteration of microbial communities and reduction of species diversity, inhibition of certain microbial processes (organic matter breakdown, mineralization of carbon and nitrogen, enzymatic activities, etc.). A measure of the functional diversity of the bacterial flora can be assessed using ecoplates (see http://www.biolog.com/section_4.html). It has been shown that algae are especially sensitive to various organic and inorganic pollutants and thus may serve as a good indicator of pollution (8). A variety of toxicity tests involving microorganisms, invertebrates, vertebrates, and plants may be used with soil or water samples (9).

MANAGEMENT AND/OR REMEDIATION OF POINT SOURCE POLLUTION

The major objective of any remediation process is to 1) reduce the actual or potential environmental threat and 2) reduce unacceptable risks to man, animals and the environment to acceptable levels (10). Therefore strategies to either manage and/or remediate contaminated sites have developed largely from application of stringent regulatory measures set up to safeguard ecosystem function as well as to minimize the potential adverse effects of toxic substances on animal and human health.

The available remediation technologies may be grouped into two categories: 1) ex situ techniques that require removal of the contaminated soil or groundwater for treatment either on-site or off-site, and 2) in situ techniques that attempt to remediate without excavation of contaminated soils. Generally, in situ techniques are favored over ex situ techniques because of 1) reduced costs due to elimination or minimization of excavation, transportation to disposal sites, and sometimes treatment itself; 2) reduced health impacts on the public or the workers; and, 3) the potential for remediation of inaccessible sites, e.g., those located at greater depths or under buildings. Although in situ techniques have been successful with organic contaminated sites, the success of in situ strategies with metal contaminants has been limited. Given that organic and inorganic contaminants often occur as a mixture, a combination of more than one strategy is often required to either successfully remediate or manage metal contaminated soils.

GLOBAL CHALLENGES AND RESPONSIBILITY

The last 100 years has seen massive industrialization. Indeed such developments were coupled with the rapid increase in world population and the desire to enhance economy and food productivity. While industrialization has led to increased economc activity and much benefit to human race, the lack of regulatory measures and appropriate waste management strategies until early 1980s (including the use of agrochemicals) has resulted in contamination of our biosphere. Continued pollution of the environment through industrial emissions is of global concern. There is, therefore, a need for politicians, regulatory organizations, and scientists to work together to minimize environmental contamination and to remediate contaminated sites. The responsibility to check this pollution lies with every individual and country although the majority of this pollution is due to the industrialized nations. There is a clear need of better coordination of efforts in dealing with numerous forms of PS pollution problems that are being faced globally.

REFERENCES

1. Mehlum, H.K.; Arnesen, A.K.M.; Singh, B.R. Extractability and Plant Uptake of Heavy Metals in Alum Shale Soils. Commun. Soil Sci. Plant Anal. **1998**, *29*, 183–198.
2. McBride, M.B. Reactions Controlling Heavy Metal Solubility in Soils. Adv. Soil Sci. **1989**, *10*, 1–56.
3. Patil, G.P.; Gore, S.D.; Johnson, G.D. *EPA Observational Economy Series Volume 3: Manual on Statistical Design and Analysis with Composite Samples, Technical Report No. 96-0501*. Center for Statistical Ecology and Environmental Statistics: Pennsylvania State University, 1996.
4. Ward, T.J.; Correll, R.L. Estimating Background Concentrations of Heavy Metals in the Marine Environment. In *Proceedings of a Bioaccumulation Workshop: Assessment of the Distribution, Impacts and Bioaccumulation of Contaminants in Aquatic Environments, Sydney, 1990*; Miskiewicz, A.G., Ed.; Water Board and Australian Marine Science Association: Sydney, 1992; 133–139.
5. Jones, K.C.; Davies, B.E.; Peterson, P.J. Silver in Welsh Soils: Physical and Chemical Distribution Studies. Geoderma **1986**, *37*, 157–174.
6. Barber, C.; Bates, L.; Barron, R.; Allison, H. Assessment of the Relative Vulnerability of Groundwater to Pollution: A Review and Background Paper for the Conference Workshop on Vulnerability Assessment. J. Aust. Geol. Geophys. **1993**, *14* (2–3), 147–154.
7. Wenzel, W.W.; Blum, W.E.H. Effects of Fluorine Deposition on the Chemistry of Acid Luvisols. Int. J. Environ. Anal. Chem. **1992**, *46*, 223–231.
8. Megharaj, M.; Singleton, I.; McClure, N.C. Effect of Pentachlorophenol Pollution Towards Microalgae and Microbial Activities in Soil from a Former Timber Processing Facility. Bull. Environ. Contam. Toxicol. **1998**, *61*, 108–115.
9. Juhasz, A.L.; Megharaj, M.; Naidu, R. Bioavailability: The Major Challenge (Constraint) to Bioremediation of Organically Contaminated Soils. In *Remediation Engineering of Contaminated Soils*; Wise, D., Trantolo, D.J., Cichon, E.J., Inyang, H.I., Stottmeister, U., Eds.; Marcel Dekker: New York, 2000; 217–241.
10. Wood, P.A. Remediation Methods for Contaminated Sites. In *Contaminated Land and Its Reclamation*; Hester, R.E., Harrison, R.M., Eds.; Royal Society of Chemistry, Thomas Graham House: Cambridge, UK, 1997; 47–73.
11. Barzi, F.; Naidu, R.; McLaughlin, M.J. Contaminants and the Australian Soil Environment. In *Contaminants and the Soil Environment in the Australasia–Pacific Region*; Naidu, R., Kookana, R.S., Oliver, D., Rogers, S., McLaughlin, M.J., Eds.; Kluwer Academic Publishers: Dordrecht, The Netherlands, 1996; 451–484.

POLAR REGIONS, SOILS OF THE

Charles Tarnocai
Agriculture and Agri-Food Canada, Ottawa, Canada

Iain Campbell
Land & Soil Consultancy Services, Nelson, New Zealand

INTRODUCTION

The Arctic polar region is the area of the Northern Hemisphere lying north of the tree line, while the Antarctic polar region in the Southern Hemisphere consists of the continent of Antarctica. Because these regions have the Earth's coldest climates, their soil materials are perennially frozen and their development is dependent on the amount of moisture in the soil and the stability of the landscape.

ENVIRONMENTAL SETTING

The Arctic has a land area of approximately 7.2×10^6 km^2, encompassing portions of Canada (33% of the area of the Arctic), Russia (32%), Greenland (31%), the United States/Alaska (3%) and Scandinavia (1%). Glaciers cover approximately 1.9×10^6 km^2 (26%) of this area, with most glaciers (92% or 1.7×10^6 km^2) occurring in Greenland.

Although Antarctica is large (approximately 14×10^6 km^2), 99.7% of it is covered by ice caps. The ice-free portion (0.05×10^6 km^2) (Table 1), which includes soils, occurs in many widely scattered small areas, the largest being the McMurdo Dry Valley region.

Short, cold summers and long, extremely cold winters characterize the climates of both polar regions. Mean daily temperatures above 0°C occur only during the warmest part of the summer. Total annual precipitation is generally low and occurs mostly as snow. In Antarctica the annual precipitation averages 50 mm, ranging from less than 10 mm in some parts of the Transantarctic Mountains to greater than 100 mm in northern coastal regions. Effective precipitation is low, however, since most snowfall either blows away or sublimes. Soil moistening from snow thaw persists for very few days, and vapor exchange is important in soil moisture transfers.

The Arctic vegetation is a nearly continuous cover of shrub-tundra in the south, grading to a sparse cover of dwarf shrubs, herbs, mosses and lichens in the north. Most of the Antarctic soils are unvegetated. Permafrost is continuous in both the Arctic and Antarctic and the terrain is associated with patterned ground. The dominant soils are Cryosols (1), Cryosols and Cryic Histosols (2), or Gelisols (3) (Table 1).

SOIL-FORMING PROCESSES

Soil formation in the Arctic is dominated by cryogenic processes, which are driven by the presence and mobility of unfrozen soil water as it migrates towards the frozen front along the thermal gradient in the frozen system. These processes, which produce a very dynamic landscape, include freeze–thaw, cryoturbation (frost churning) (Fig. 1), frost heave, cryogenic sorting, thermal cracking and ice segregation. Other soil-forming processes that can leave an imprint on Arctic soils include the gleyic process, brunification, eluviation and salinization.

In the Antarctic, however, soil formation is dominated by weathering processes because of the extraordinary landscape stability resulting from the exceedingly cold and arid climate. Patterned ground occurs widely in these dry Antarctic soils, especially on younger land surfaces, but the presence of volcanic ash in the cracks indicates that cryoturbic mixing is limited. Because almost all precipitation occurs as snow, which blows away or sublimes, little moisture is transferred to the soil and cryoturbation and ice build-up are minimal or nonexistent. As a result, cryogenic processes have little or no effect, and weathering processes, which operate over very long periods on this stable landscape, dominate even though they act extremely slowly. The soils thus produced are brunified (Fig. 2) and often contain salt accumulations.

Weathering processes operate extremely slowly in the Arctic as well as in the Antarctic. Physical weathering takes place mostly at the surface. In the Antarctic clast reduction occurs through abrasion, exfoliation, granular disintegration (salt weathering) and thermal fracture. Frost weathering (frost shattering) is minimal in the Antarctic

P

Table 1 Estimated areal distribution of soil groups in the polar regions

Soil classification		Area
US[a]	Canada[b]	($\times 10^3$ km^2)
Arctic:		
Turbels	Turbic cryosols	3351
Orthels	Static cryosols	456
Histels	Organic cryosols	264
Inceptisols	Brunisols	172
Entisols	Regosols	236
Spodosols	Podzols	50
Total Arctic soil area		4529
Antarctic:		
Orthels	Static cryosols	46
Total polar soil area		4575

Note: Using local terminology, the Antarctic Orthels can be separated into Arid Cryosols (30×10^3 km^2) and Humid Cryosols (16×10^3 km^2).
[a](From Ref. 3.)
[b](From Ref. 1.)

because of the absence of water, but it is more active in the Arctic, where the soil moisture content is higher. The dominant chemical weathering processes in Antarctic soils are oxidation and salinization (4, 5). Because evaporation exceeds precipitation, salts accumulate in the Antarctic soils in a variety of forms, commonly in distinct horizons. Although salt composition of the soil partially reflects that of the soil parent material, it typically mirrors atmospheric and climatic gradients, indicating that aerosolic transport is the main source. Clay mineral formation is limited in both Arctic and Antarctic soils.

SOIL CHARACTERISTICS

Arctic Soils

In Arctic soils cryoturbation results in irregular or broken soil horizons, involutions, organic intrusions (Fig. 1), organic matter build-up in the subsoil (often with concentration of this material along the top of the permafrost table), oriented rock fragments, silt-enriched layers and silt caps (6). Arctic soils contain various amounts of ice in the form of ice crystals, vein ice, ice wedges and massive ground ice several meters thick.

Arctic soils generally have high moisture content, especially near the permafrost table, so gleying, associated grayish colours and redoximorphic features occur on most landforms, including those on better-drained positions. These soils are very often saturated because the active layer (the upper layer of the soil that freezes and thaws annually) is shallow. Thin eluvial or leached horizons are common in Arctic soils, especially in the south. The brunification associated with coarse-textured soils is usually associated with this eluvial horizon.

The active layer not only supports biological activity, but also protects the underlying permafrost. The thickness

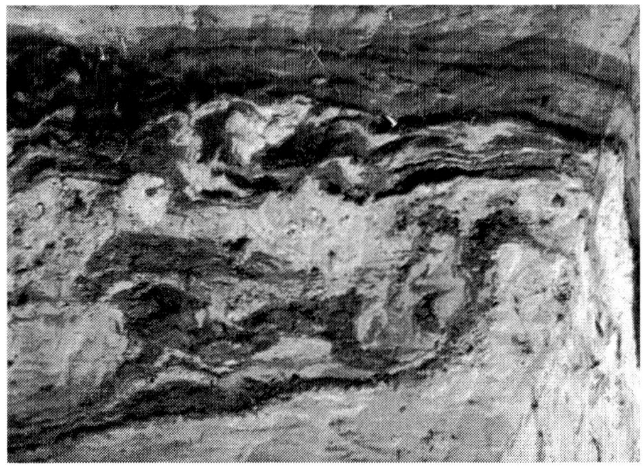

Fig. 1 Strongly cryoturbated permafrost-affected soil from Arctic Canada. Cryoturbation is displayed by the strongly contorted mineral and organic horizons.

Fig. 2 A Miocene-age soil weathering profile on till from near the head of Hatherton Glacier, Antarctica. There is a well-developed surface pavement and a salt horizon at 15 cm. The soil is dry frozen with the active layer depth at approximately 10 cm. Soil colors are reddish brown (5YR 4/4) at the surface passing to light yellowish brown (10YR 6/4) at depth.

of this layer is controlled by soil texture and moisture, thickness of the surface organic layer, vegetation cover, aspect and latitude. In Arctic soils the active layer normally extends 25 to 120 cm below the soil surface (7, 8).

Particle size distribution in Arctic Cryosols ranges from silty clay to coarse gravelly sand. The composition of the fine-earth fraction is commonly dependent on the mode of deposition of the parent materials. The pH of Arctic soils varies greatly and also depends on the chemistry of the parent materials. The similarity of pH to that of the parent material is caused, in part, by cryoturbation, which not only mixes parent material with soil materials, but also mixes soil material among the horizons.

The nitrogen, potassium and phosphorus contents of Arctic soils are generally low since most of these nutrients are locked into compounds contained in the surface organic layer (9). Salt crusts, which form by translocation of salts from the subsurface horizons, develop during dry periods in the summer because of increased evaporation from the soil surface. Such salt crusts are common on the surfaces of High Arctic soils. Calcareous crusts on the underside of rocks are also a common phenomenon, especially in coarse-textured soils. Soils developed on fine-textured marine sediments usually have both high conductivity and high salt content.

The total soil organic carbon concentration in mineral soils in Canada ranges from 3.0 to 100.9 kg/m^3 (10). Mineral soils that are cryoturbated contain the highest amount of carbon. Although the surface organic horizons become thinner and very patchy in the High Arctic, a large amount of carbon is still stored in both the surface and subsurface horizons. Perennially frozen organic soils (2–2.5 m thick) cover large areas in both Canada and Russia (Figs. 3 and 4).

Fig. 3 Lowland ice-wedge polygon landscape, Mackenzie River Delta area, Canada. The ice wedges are located in the polygonal trenches.

Fig. 4 Cross section of an ice-wedge polygon area associated with Cryosols, Mackenzie River Delta area, Canada. The ice wedge is approximately 3 m wide.

The diversity of microorganisms in Arctic soils is much greater than that in Antarctic soils because of the higher concentrations of available water and organic matter in Arctic soils. These microorganisms play an important role in the decomposition of the organic matter and in nutrient cycling (11).

Antarctic Soils

Although conditions for soil formation in the Antarctic are minimal, considerable spatial soil variation results from differences in soil-forming factors. The main physical features of Antarctic soils are:

1. a distinctive surface pebble or boulder pavement;
2. differing characteristics associated with differing rock types;
3. a soil form dominated by coarse, but highly variable, sandy to bouldery textures;
4. only a very small increase in fine particle size, associated with increasing age;
5. almost complete lack of cohesion or structural development;
6. increased reddening and increased oxidation depth with increasing soil age;
7. increased accumulation of soluble salts with increasing soil age;
8. an active layer from <5 cm to >70 cm, depending on climate; and
9. permafrost that is commonly hard and ice-cemented in coastal situations and on younger land surfaces, but dry frozen in areas of high aridity.

In the Transantarctic Mountains, the soils may thaw for only two months while moisture contents of <1–5% are common (12). Distinctions based on soil temperature and moisture differences are evident. In the Antarctic Peninsula and some northern coastal regions, the soils are more moist and may have organic accumulations associated with sporadic moss or grass occurrences in damp sites.

Biologically, Antarctic soils typically lack organic accumulations. Life forms are restricted to more favorable moist environments where small patches of mosses, lichens and algae may be present. Bacteria and algae occur within the soils, supporting grazers such as nematodes, tardigrades and rotifers (13).

CONCLUSIONS

Arctic and Antarctic soils are the coldest soils on earth. These soils have been studied intensively since the mid-1900s and there is now considerable knowledge of the processes that lead to the formation of these soils and of their role in relation to the unique ecosystems of these Polar Regions. Because of the extreme conditions and the high ice content of Arctic soils or the slow rate of soil processes in Antarctic soils, there is a very large potential for human impact in this environment. Although recovery rates from such impact vary according to the severity of the initial activity, recovery is extremely slow or even irreversible.

REFERENCES

1. Soil Classification Working Group. *The Canadian System of Soil Classification*; 3rd Ed.; Publication 1646; Research Branch, Agriculture and Agri-Food Canada: Ottawa, 1998; 187.
2. ISSS Working Group RB. *World Reference Base for Soil Resources*, World Soil Resources Report 84; International Society of Soil Science (ISSS), International Soil Reference and Information Centre (ISRIC) and Food and Agriculture Organization of the United Nations (FAO); FAO: Rome, 1998; 91.
3. Soil Survey Staff. *Soil Taxonomy: A Basic System of Soil Classification for Making and Interpreting Soil Surveys*, 2nd Ed.; Agriculture Handbook Number 436; United States Department of Agriculture, Natural Resources Conservation Service: Washington, DC, 1999; 869.
4. Campbell, I.B.; Claridge, G.G.C. *Antarctica: Soils, Weathering Processes and Environment*; Elsevier Science Publishers: Amsterdam, 1987; 368.
5. Bockheim, J.G. Relative Age and Origin of Soils in Eastern Wright Valley, Antarctica. Soil Science **1979**, *128*, 142–152.
6. Bockheim, J.G.; Tarnocai, C. Recognition of Cryoturbation for Classifying Permafrost-Affected Soils. Geoderma **1998**, *81*, 281–293.
7. Smith, C.A.S.; Swanson, D.K.; Moore, J.P.; Ahrens, R.J.; Bockheim, J.G.; Kimble, J.M.; Mazhitova, G.G.; Ping, C.L.; Tarnocai, C. A Description and Classification of Soils and Landscapes of the Lower Kolyma River, Northeastern Russia. Polar Geography and Geology **1996**, *19* (2), 107–126.
8. Tarnocai, C. Distribution of Soils in Northern Canada and Parameters Affecting Their Utilization. In *11th International Congress of Soil Science Transactions*; Symposia Papers: Edmonton, Canada, 1978; Vol. 3, 332–347.
9. Broll, G.; Tarnocai, C.; Müller, G. Interactions Between Vegetation, Nutrients and Moisture in Soils in the Pangnirtung Pass Area, Baffin Island, Canada. In *Cryosols and Cryogenic Environments*; Special Issue of Permafrost and Periglacial Processes; Tarnocai, C., King, R., Smith, S., Eds.; 1999; Vol. 10(3), 265–277.
10. Kimble, J.M.; Tarnocai, C.; Ping, C.L.; Ahrens, R.; Smith, C.A.S.; Moore, J.; Lynn, W. Determination of the Amount of Carbon in Highly Cryoturbated Soils. In *Post-Seminar Proceedings of the Joint Russian–American Seminar on Cryopedology and Global Change*, Pushchino, Russia, Nov 15–16, 1992; Gilichinsky, D.A., Chief Ed.; Russian Academy of Sciences: Pushchino, Russia, 1993; 277–291.
11. Robinson, C.H.; Wookey, P.A. Microbial Ecology, Decomposition and Nutrient Cycling. In *Ecology and Arctic Environment*; Woodin, S.J., Marquiss, M., Eds.; Special Publication 13, British Ecological Society; Blackwell Science: Oxford, 1997; 41–68.
12. Campbell, I.B.; Claridge, G.G.C.; Campbell, D.I.; Balks, M.R. The Soil Environment. In *Ecosystem Dynamics in a Polar Desert: The McMurdo Dry Valleys, Antarctica*; Priscu, J.C., Ed.; Antarctic Research Series; American Geophysical Union: Washington, DC, 1998; Vol. 72, 297–322.
13. Freckmann, D.W.; Virginia, R.A. Soil Biodiversity and Community Structure in the McMurdo Dry Valleys, Antarctica. In *Ecosystem Dynamics in a Polar Desert: The McMurdo Dry Valleys, Antarctica*; Priscu, J.C., Ed.; Antarctic Research Series; American Geophysical Union: Washington, DC, 1998; Vol. 72, 323–335.

POLLUTANTS, PERSISTENT ORGANIC

Mark Radosevich
E. Danielle Rhine
University of Delaware, Newark, Delaware, U.S.A.

INTRODUCTION

Biodegradation is the primary dissipation pathway for most organic pollutants in soils. Three prevalent factors related to chemical structure that commonly inhibit the biodegradation of organic pollutants are: 1) inherent susceptibility to enzymatic attack, 2) bioavailability, and 3) nutrient availability. Inherently recalcitrant xenobiotic chemicals (synthetic chemicals with no natural analogs) can persist in environments that lack sufficient numbers of microorganisms with the necessary catabolic pathways. However, the capacity of soil microbial communities to degrade toxic chemicals can increase with repeated chemical exposure. This suggests that the genetic potential to metabolize the chemical is initially present and the capacity to degrade the pollutant could be realized through community adaptation. Understanding the biological fate of these hazardous substances is of critical importance for the protection of human and ecosystem health.

MOLECULAR RECALCITRANCE

Many toxic organic chemicals are intrinsically stable (i.e., recalcitrant) in natural environments due to their fundamental structural properties. The structure of a compound can influence its stability by: 1) inhibiting its susceptibility to enzymatic attack or 2) causing favorable interactions (i.e., sorption) with the soil matrix that result in rate-limited mass transfer between the soil and aqueous phases thus reducing the amount of contaminant that is available for microbial metabolism.

Electrophillic Attack by Oxygenase Enzymes

Several generalizations can be made regarding the chemical structure and susceptibility to enzymatic attack involving two key mechanisms: steric hindrance and electron distribution effects. Under aerobic conditions (oxygen present), many xenobiotic compounds, especially aromatic and aliphatic hydrocarbons can be transformed by the electrophilic addition of oxygen by a class of enzymes known as oxygenases. These enzymes produce an electrophilic form of oxygen at the enzyme active site that subsequently attacks electron-dense sites on the substrate molecule (1). In the case of aromatic ring structures, electron-withdrawing substituents, such as halogens, lower the electron density in the ring, rendering it less susceptible to attack by oxygenase enzymes. Conversely, electron-donating substituents such as alkyl groups increase both the electron density in the ring and the oxygenase-mediated reactions. A classic example of the electronic effects on enzymatic reaction rates is the relative rate of transformation of the herbicide 2,4-D compared to its structural analog 2,4,5-T (Fig. 1). These two molecules are identical with the exception of an additional chlorine atom in position five on the aromatic ring of 2,4,5-T. Addition of this chlorine atom inhibits the electrophilic attack of the acetic acid side chain and dramatically increases the stability of this compound relative to 2,4-D, in natural environments. Halogen substitution of aromatic and aliphatic compounds in key positions can not only block metabolism, producing dead-end metabolites, but also can yield products (e.g., acyl halides) that are lethal to the microorganism (1). Another classic example is that of polychlorinated biphenyls (PCBs) and pentachlorophenol (PCP). Biological degradation of these compounds decreases dramatically as the degree of halogenation increases (Fig. 1).

Increasing the degree of halogenation generally stabilizes the organic compounds in aerobic environments. Under anaerobic conditions (oxygen absent), however, these chemicals can be transformed via reductive processes. Reductive reactions usually result in detoxification of the compound and in some cases are involved in the energy-yielding metabolism of some organisms. In such cases, the organism directly transfers electrons (reducing equivalents) from an electron donor such as lactate, acetate, or $H_2(g)$ to the halogenated compound. This terminal electron accepting process is a form of respiration and is analogous to oxidation of carbon substrates using oxygen as an electron acceptor.

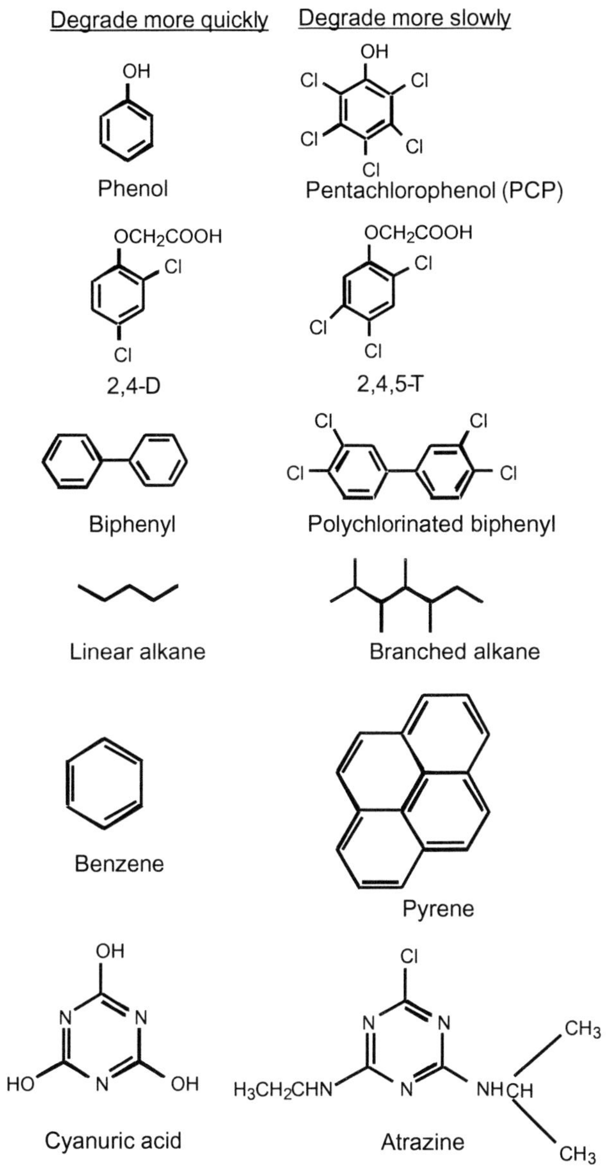

Fig. 1 Examples of persistent organic chemicals and their more degradable structural analogs illustrating the effect of various functional groups on biodegradability.

Hydrolytic Reactions

Microorganisms can degrade organic pollutants using hydrolytic reactions (1). Enzymes mediating these reactions are generally referred to as hydrolases and act by replacing substituents on the substrate with water. In these reactions, water acts as a nucleophile attacking centers of positive charge on the substrate molecule. Because oxygen does not participate in the reaction, hydrolytic transformations can occur in aerobic or anaerobic environments. Functional groups commonly occurring in many pesticides that are susceptible to hydrolytic attack include amides, esters, carbamates, C-halogen bonds, and sulfonyl ureas to name a few.

Steric Hindrance

Steric hindrance is another mechanism related to chemical structure that can influence enzymatic reactions. The addition of functional groups alters the structure in such a way that it is no longer recognized by the enzyme. Common examples of steric hindrance include highly branched aliphatic compounds or substituents, polycyclic aromatic hydrocarbons, and various polymers. As the molecular weight of the substrate increases by branching, polymerization, or addition of fused aromatic rings, susceptibility to enzymatic attack decreases (Fig. 1).

BIOAVAILABILITY

Sorption and Sequestration

Removal of contaminants from the bulk soil aqueous phase can limit their bioavailability to microorganisms (2), thus reducing the degradation rate. Reduction of aqueous phase contaminant concentrations can result from a variety of processes that are primarily a function of a chemical's properties and soil characteristics such as clay and organic matter content. These processes include slow dissolution of the contaminant solid or liquid phase (3), slow diffusion and sequestration in soil nanopores (4), soil sorption (either adsorption or partitioning) (5), and partitioning of the contaminant in a nonaqueous phase liquid (6) (Fig. 2). Ionizable functional groups on various organic acids and bases (e.g., 2,4-D, atrazine, and quinoline) result in pH-dependent charge of these molecules that can enhance or reduce interaction with sorption sites on soil-surfaces. Hydrophobic interactions of nonpolar organic compounds, primarily with soil organic matter, are greater for molecules with low water solubility (aromatic compounds with alkyl or halide substituents).

Data regarding sorption-limited biodegradation in soils and sediments suggest the existence of available and unavailable compartments. However, the concentration of contaminant in these two pools does not strictly correspond to the soil aqueous and solid phases, respectively. Substrate localized in the "unavailable" compartment(s) may be an exploitable resource for appropriately adapted microorganisms. For example, differences have been observed in the bioavailability of sorbed PAHs naphthalene- and

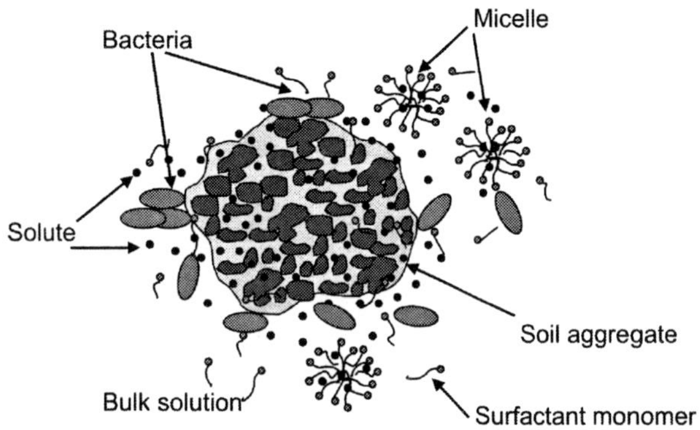

Fig. 2 Diagram illustrating interactions of solute molecules with the soil matrix resulting in a reduction in bioavailability to microorganisms. Rate limited processes depicted in the figure include sorption–desorption via electrostatic or hydrophobic interactions, entrapment and slow diffusion through nano/micropores, and release of solute from the soil matrix by interactions with surfactant micelles.

phenanthrene-degrading bacteria (7). Bacterial attachment at the interface between nonaqueous phase liquids (NAPL; a class of liquid organic contaminants such as solvents and fuels) and water is known to influence the uptake of NAPL and NAPL-partitioned organic compounds (8). Little is known about the role of attachment in the biodegradation of soil-sorbed pollutants. However, the ability of microorganisms to utilize surface accumulated substrates in dilute media has been recognized since the early 1940s (9). Biodegradation of soil sorbed phenanthrene varied substantially between two PAH-degrading bacteria with differing cell surface properties (10). These studies indicate that bioavailability may be species-dependent, and suggest that cellular attachment and as yet unidentified adaptations may allow some species greater access to sorbed pollutants.

Enhancing Bioavailability

Increasing the degradation of sorbed, trapped, or insoluble pollutants focused on enhancing the contaminant release from the soil matrix through the addition of cosolvents and surfactants. Enhancement and inhibition of biodegradation in the presence of surfactants have been reported. In general, enhanced degradation is attributed to enhanced solubility (3), an increase in the interfacial surface area between immiscible fluids (11), or enhanced desorption (12) (Fig. 2). Inhibition of biodegradation can result from surfactant toxicity, sequestration of the contaminant within surfactant micelles (13), or preferential utilization of the surfactant as a substrate (11). Understanding how microbes enhance desorption or gain access to sorbed substrates is paramount to reducing the persistence of sorbed organic pollutants.

NUTRIENT AVAILABILITY

Effect of Added Nutrients on Hydrocarbon Degradation

Most soil and subsurface environments are limited in available carbon for microbial growth. However, in hydrocarbon contaminated soils, the ratio of reduced C to other nutrients is very large and as a result, one or more nutrients other than C may become rate-limiting factors for contaminant biodegradation. Addition of inorganic nutrients usually enhances hydrocarbon biodegradation. This approach is frequently considered to stimulate bioremediation of petroleum contaminated environments. Optimal ranges of C:N:P:S ratios vary enormously but a C:N:P ratio of 100:10:1 is generally accepted as optimal for hydrocarbon biodegradation in most environments. Inorganic salts such as ammonium phosphate are commonly used as nutrient fertilizers in hydrocarbon contaminated environments. Inorganic salts are inexpensive and provide nutrients in bioavailable forms, but in some situations these fertilizers can become diluted or washed out of the contaminated zone and therefore do not provide a long term supply of the limiting nutrient(s). This is particularly a concern in near shore and saturated environments. In other cases, use of these fertilizers can lead to rather dramatic changes in pH or result in osmotic stress that can adversely affect biodegradation rates. These problems can sometimes be overcome by using organic nutrient sources with low C:N and C:P ratios such as urea, oleophillic forms of N and P, or a variety of other slow-release fertilizers (14).

Accelerated biodegradation in response to fertilization is frequently reported in hydrocarbon-contaminated soils and aquifer sediments (15). Some studies, however, have observed no increase or an inhibition in degradation rates (16). These contradictory results suggest that the response of the microbial community to manipulation of various environmental factors to optimize biodegradation is site specific and reflects the complex nature of heterogeneous systems such as soils (17).

Organic Chemicals that Are Macronutrient Sources

Ammonia is almost invariably the preferred N-source for most bacteria. However, in natural environments ammonia is not always available, and bacteria have evolved the ability to use a wide array of organic and inorganic compounds as N-sources. From a pragmatic viewpoint, coordinated regulation of degradative pathways and N-assimilatory pathways makes sense if the organic chemical in question is a poor carbon and energy source but potentially a good N-source.

Biodegradation of a N-Containing Herbicide

Unlike homocyclic aromatic compounds such as 2,4-D, atrazine (Fig. 1) can serve as a C and N source for some microorganisms. Hence the level of induction and expression of the atrazine-degrading genotype in soil microorganisms is likely to be affected by the C and N status in soils. Atrazine has a low C/N ratio (8:5) and its complete mineralization results in release of excess N. Therefore, atrazine is potentially a good N-source and its degradation is likely to be favored in N-limited soils. Addition of inorganic fertilizers to soils has been shown to inhibit atrazine mineralization (18, 19). The general suppression of atrazine mineralization in the presence of exogenous nitrogen suggests that regulation and expression of atrazine degradation genes in soil microbial communities is linked to the N status of soil. Atrazine and its N-heterocyclic derivatives are apparently substrates for a complex pathway that may be catabolic as well as linked to assimilatory pathways of C and N. A recent study with ^{15}N-ring-atrazine has definitively shown that atrazine-N is incorporated into the biomass of several bacterial cultures known to mineralize atrazine (19). Evidence for a regulatory linkage between atrazine catabolism and N assimilation has also been observed in an atrazine-mineralizing soil bacteria (19, 20). Given the apparent negative correlation between atrazine degradation rates and soil N-levels, perhaps the majority of soil bacteria capable of utilizing triazine-N can repress expression of the degradative pathway enzymes in favor of more readily available sources of N.

MICROBIAL ECOLOGY AND BIODEGRADATION

Soil is perhaps the most biologically diverse and abundant ecosystem in the biosphere. A few grams of topsoil may contain 5–15 thousand different microbial species. Only 0.1–1% of this biodiversity has been cultured and studied in the laboratory. Understanding of pollutant biodegradation is largely based on the study of microorganisms obtained via classical enrichment techniques. However, since these organisms were isolated under very selective conditions and represent less than 0.1% of the biodiversity of soil, they may not be representative of the full genetic biodegradation potential of soil communities and may play only a minor role in the detoxification of contaminated soil. To more fully understand environmental processes mediated by microorganisms, various culture-independent methodologies, based on the analysis of soil genomic DNA, are becoming widely used. Many of these approaches are based on the detection of 16S rRNA genes (16S rDNA) from community members that do not grow well in the laboratory.

One such technique amenable to ecological studies is denaturing gradient gel electrophoresis (DGGE). This DNA fingerprinting technique allows separation of DNA fragments of similar size but different sequence composition (21). When applied to microbial community analysis, conserved 16S rDNA sequences are used as primers to amplify community DNA. The resulting mixture of PCR products is then separated by DGGE. The result is a pattern of bands that roughly corresponds to the number of predominant members in a microbial community. Statistical analysis of banding patterns can then be used to monitor

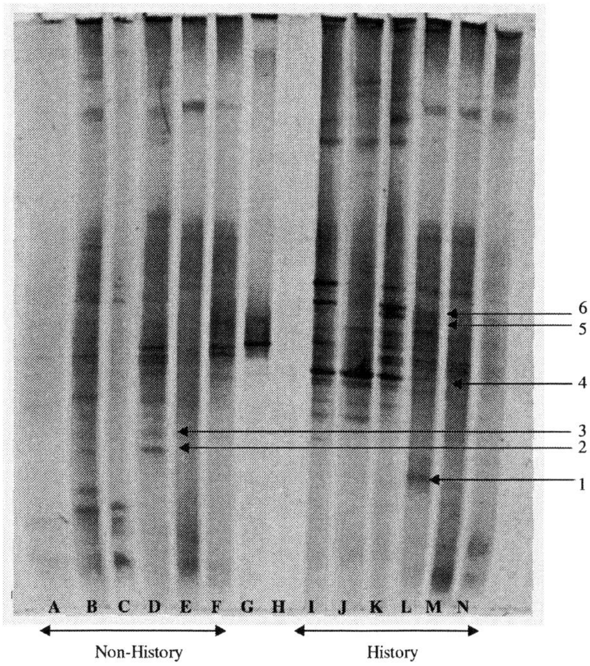

Fig. 3 PCR amplified 16S rDNA fingerprint of microbial communities obtained using DGGE analysis of DNA extracted from amended NH- (A–F) and H-soils (I–N). Soil amendments include: nonamended field soil (F and N), water control (E and M), atrazine (D and L), atrazine+NH_4NO_3 (C and K), atrazine+acetate (B and J), and atrazine+acetate+NH_4NO_3 (A and I). A pure bacterial strain, isolate P5-2 was used as a positive control (G) and a water control (H) was also included.

changes in the structure of microbial communities or to compare communities in different environments. The selection of the PCR primers used to amplify community DNA determines the relative breadth or narrowness of community members targeted for analysis. In some instances, DNA fragments isolated from DGGE bands can be sequenced to identify phylogenetically the organisms represented by that particular band.

This approach has been used to characterize microbial communities in soils with and without prior atrazine exposure histories (H- and NH-soil, respectively) and different C/N ratios. The DNA fingerprints were then related to atrazine mineralization capacity in the acclimated soils (soils exposed to three atrazine amendments in a lab acclimation study) (Fig. 3). The banding patterns revealed highly complex microbial communities. Although several shared bands were observed, the DNA fingerprints of the H- and NH-soils showed two distinctly different communities (lanes E and M). The bands (representing distinct phylotypes) that intensified in response to atrazine amendment in the H-soil were different from those that intensified in the NH-soil. For example, band 1 in the H-soil was not present in the NH-soil and bands 2 and 3 were present in the NH-soil amended with atrazine but absent in the H-soil. In soils amended with atrazine+NH_4NO_3 (low C/N), several bands in the H-soil became more intense (bands 4–6, lane K). These bands were absent in the corresponding NH-soil (lane C). Interestingly, band 4 was even more intense in the atrazine + acetate (high C/N) treated soil while bands 5 and 6 were absent. In general, the communities in the H- and NH-soils responded quite differently to the various amendments (i.e., variations in C/N ratio). Presumably the organisms represented by these bands in the DGGE gel play a very significant role in atrazine degradation in this soil.

REFERENCES

1. Hickey, W.J. Biochemistry and Metabolism of Xenobiotic Chemicals. In *Principles and Applications of Soil Microbiology*; Sylvia, D.M., Fuhrmann, J.J., Hartel, P.G., Zuberer, D.A., Eds.; Prentice Hall: Upper Saddle River, NJ, 1998; 447–468.
2. Ogram, A.V.; Jessup, R.E.; Ou, L.T.; Rao, P.S.C. Effects of Sorption on Biological Degradation Rates of (2,4-Dichlorophenoxy) Acetic Acid in Soils. Appl. Environ. Microbiol. **1985**, *49*, 582–587.
3. Tiehm, A. Degradation of Polycyclic Aromatic Hydrocarbons in the Presence of Synthetic Surfactants. Appl. Environ. Microbiol. **1994**, *60*, 258–263.
4. Nam, K.; Alexander, M. Role of Nanoporosity and Hydrophobicity in Sequestration and Bioavailability: Tests with Model Solids. Environ. Sci. Technol. **1997**, *32*, 71–74.
5. Smith, S.C.; Ainsworth, C.C.; Traina, S.J.; Hicks, R.J. The Effects of Sorption on the Transformation and Degradation of Quinoline. Soil Sci. Soc. Am. J. **1992**, *56*, 737–746.
6. Efroymson, R.A.; Alexander, M. Reduced Mineralization of Low Concentrations of Phenanthrene Because of Sequestering in Nonaqueous-Phase Liquids. Environ. Sci. Technol. **1995**, *29*, 515–521.
7. Guerin, W.F.; Boyd, S.A. Differential Bioavailability of Soil-Sorbed Naphthalene to Two Bacterial Species. Appl. Environ. Microbiol. **1992**, *58*, 1142–1152.
8. Stelmack, P.L.; Gray, M.R.; Pickard, M.A. Bacterial Adhesion to Soil Contaminants in the Presence of Surfactants. Appl. Environ. Microbiol. **1999**, *65*, 163–168.
9. Zobell, C.E. The Effect of Solid Surfaces Upon Bacterial Activity. J. Bact. **1943**, *46*, 39–56.
10. Dean, S.M.; Jin, Y.; Cha, D.K.; Radosevich, M. Phenanthrene Degradation in Soils Co-Inoculated with Phenanthrene-Degrading and Biosurfactant-Producing Bacteria. J. Environ. Qual. **2001**, *30* (4), 1126–1133.
11. Rouse, J.D.; Sabatini, D.A.; Suflita, J.M.; Harwell, J.H. Influence of Surfactants on Microbial Degradation of Organic Compounds. Crit. Rev. Environ. Sci. Technol. **1994**, *24*, 325–370.
12. Aronstein, B.N.; Alexander, M. Surfactants at Low Concentrations Stimulate Biodegradation of Sorbed Hydrocarbons in Samples of Aquifer Sands and Soil Slurries. Environ. Toxicol. Chem. **1992**, *11*, 1227–1233.
13. Guha, S.; Jaffe, P.R. Biodegradation Kinetics of Phenanthrene Partitioned into the Micellar Phase of Nonionic Surfactants. Environ. Sci. Technol. **1996**, *30*, 605–611.
14. Churchill, S.A.; Griffin, R.A.; Jones, L.P.; Churchill, P.F. Biodegradation Rate Enhancement of Hydrocarbons by an Oleophillic Fertilizer and a Rhamnolipid Biosurfactant. J. Environ. Qual. **1995**, *24*, 19–28.
15. Jackson, A.; Pardue, J.H. Seasonal Variability of Crude Oil Respiration Potential in Salt and Fresh Marshes. J. Eviron. Qual. **1997**, *26*, 1140–1146.
16. Swindoll, C.M.; Aelion, C.M.; Pfaender, F.K. Influence on Inorganinc and Organic Nutrients on Aerobic Biodegradation and on Adaptation Response of Subsurface Microbial Communities. Appl. Environ. Microbiol. **1988**, *54*, 212–217.
17. Leahy, J.G.; Colwell, R.R. Microbial Degradation of Hydrocarbons in the Environment. Microbiol. Rev. **1990**, *54*, 305–315.
18. Alvey, S.; Crowley, D.E. Influence of Organic Amendments on Biodegradation of Atrazine as a Nitrogen Source. J Enviro. Qual. **1995**, *24*, 1156–1162.
19. Bichat, F.; Sims, G.K.; Mulvaney, R.L. Microbial Utilization of Heterocyclic Nitrogen from Atrazine. Soil Sci. Soc. Am. J. **1999**, *63* (1), 100–110.
20. Gebendinger, N.; Radosevich, M. Inhibition of Atrazine Degradation by Cyanazine and Exogenous Nitrogen in Bacterial Isolate M91-3. Appl. Microbiol. Biotechnol. **1999**, *51*, 375–381.
21. Muyzer, G.; Smalla, K. Application of Denaturing Gradient Gel Electrophoresis (DGGE) and Temperature Gradient Gel Electrophoresis (TGGE) in Microbial Ecology. Antonie van Leewenhoek **1998**, *73*, 127–141.

POLLUTION

Zulkuf Kaya
University of Çukurova, Balcali, Adana, Turkey

P

INTRODUCTION

Soil—the medium of life and foundation of human existence, the contact of lithosphere, atmosphere, and biosphere—has been dominated by numerous physical (temperature and pressure changes, wind and water movements, etc.), chemical (oxidation, plant root, microorganism, excretions, solution, neosynthesis), and biological (humus formation) processes in the past and the present, ranging from hundreds to millions of years. Ensurance of sustainable life, i.e., bound to the preservation of the natural courses of the processes mentioned above, is ultimately secured by the "quality of soil" in physical, chemical, and biological terms.

SOIL POLLUTION: THE HUMAN IMPACT

The impact of urbanization and industrialization has been a major factor against the need of preserving the quality of soil, by reducing it via chemical contaminants, use of polluted waters for irrigation and deposition of harmful particulates to the atmosphere. Ecosystems are threatened by such contaminants and their interactions in the environment. Harmful chemicals can be studied via recently developed concepts of ecosystem/agroecosystem planning, keeping in mind the sources of chemical contamination, which are responsible for the degradation of the environment. These sources are: 1) the wastewaters, 2) the agricultural wastes, 3) the airborne pollutants, 4) the pesticides, 5) the urban wastes: a) sewage sludge, b) composts, c) fly ash, and 6) the industrial wastes: a) pesticides and fungicides, b) fertilizers, c) detergents, d) chlorinated solvents (1).

These sources may cause fatal effects and/or irreversible destructions for human health and the environment. The levels of contamination together with their, ability and mobility in decomposition and accumulation of the chemicals in the soil have been scientifically proven to be harmful to animals, plants, microorganisms via destroying the natural structure of water, soil and air, which is balanced by nature (2).

Inactivated enzymes are the measure of heavy metal (divalent) toxicity readily reacting with proteins, amines, and sulfohydryl (–SH) groups as well as Hg and Cd replacing Zn in metallic enzymes. Metal toxicity is known to decrease cell membrane permeabilities and change genetic characteristics of cells increasing risks of cancer. Mercury accumulates in fatty tissues as methyl-Hg, whereas cadmium replaces calcium in bones and kidneys destroying their excretory functions.

According to Forstner (2) the behavior of synthetic organic compounds in soils are not "sorptive" processes, but are related to dissolution mechanisms on lipid-like particle surfaces ousted from soil solutions. Functional groups varying in molecular size, form, and valency such as the cationic, basic, and acidic solutions together with nonpolar compounds are also responsible for the "sorption" of organic chemicals by solids.

Ecological studies have been carried out on the decomposition or hydrolysis of chemical bonds from water-soluble pesticides such as methylcarbamates or organophosphoric insecticides revealing shorter duration periods in soils than organochlorine compounds. However, weak bonding compounds of the chlorophenol group are easily mobilized by pH changes and other unknown environmental impacts, and expose higher risks for human and environmental health.

The highly persistent and biologically unsynthesizable "xenobiotica," i.e., the organic compounds with the highest poisoning risk such as 2,3,7,8-tetrachlorodibenz-dioxine with a lethal dose of $1 \mu g\ kg^{-1}$ body weight are 1000-fold more poisonous than sodium cyanide (2).

Metals such as lead, arsenic, mercury, chromium, cadmium, and copper are found at acceptable limits in nature. However, these limits might increase with contamination from agricultural, industrial, and infrastructural wastes (Table 1). For instance, concentrations of lead might increase by the gaseous emissions of vehicles and use of garbage compost as a fertilizer together with pesticides. Arsenic is naturally found at a level of 10 ppm in soils and might also increase to 500 ppm with industrial wastes and agricultural applications. Lead arsenate applications in orchards are determined to increase lead to 121 ppm (1). Mercury, which is determined to be

Table 1 Contents of source materials causing pollution in cultivated soils (mg kg^{-1}). Critical levels are in italics

Elements	Fertilizers[a]	Sewage sludge	Domestic compost	Fly ash	Precipitation[b]
Cd	P(*50*)	12	10	10	0.25
Cr	P(200)	250	120	280	1.4
Hg	Trace	*4.4*	–	–	*0.05*
Ni	Trace	80	120	270	7.3
Pb	P(100)	*700*	*1200*	330	*11.0*
Zn	P(150)	*3000*	*2000*	360	*29*

[a] Fertilizer containing highest amounts of heavy metals.
[b] The estimated accumulation in surface soils (top 20 cm) in 100 years by atmospheric precipitation.
P: Phosphate fertilizers, the figures in parentheses are normal levels.
(Adapted from Ref. 3.)

between 90 and 250 g ha^{-1} in normal soils, might increase at a rate of 5 g ha^{-1} yr^{-1} with the use of chemicals for cereal seeds. Additionally, about the same amount of mercury is added to the soil by rain and irrigation waters as well as garbage compost (Table 1) (1). The increase of Cd contents in the soils of Japan has been the main reason in establishing a law for the control of soil contaminants. The behavior as well as the qualitative and quantitative distributions and conversion mechanisms of the foreign organic materials in soils have always been obscure (Fig. 1). Figure 1 illustrates that the independent enzymes are responsible for the biotic conversions of organic chemicals occurring together with the nonbiotic. Soil quality on the long run is extensively affected by such conversions, forming nonextractable bonded residues (3), even by suitable agricultural practices (1).

A specific extraction method for each nonextractable residue should be determined for residues absorbed within clay mineral stacks, and bonded by noncovalent bonding in macro molecular spaces of humic substances, hydrogen bridges, van der Waals forces and attractions occurring via charge transfers as well as covalent bonding to monomeric humic substances and macro molecular structure of phenols and aromatic amines.

Chemical residues bonded by humic substances in soils should be carefully controlled due to their limited

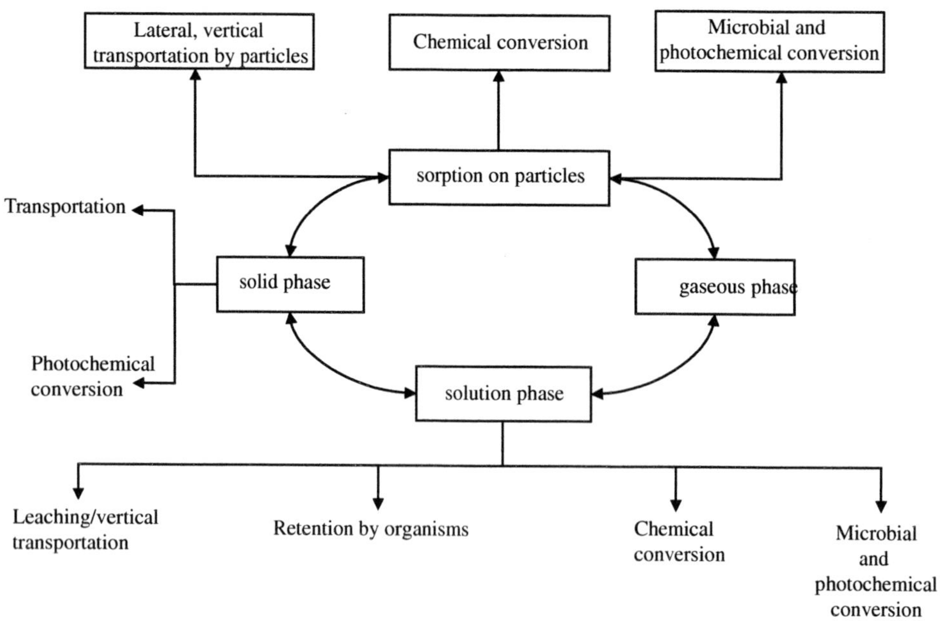

Fig. 1 Behavior of pesticides in soils. (Adapted from Ref. 1.)

availability to plants, and lack of mineralization preventing their inclusion to the C-cycle (1), thus algae, fungi and lichens, protozoa, and simple metazoa are used to isolate the anthropogenic chemicals from soils and soil solutions. However, information obtained on the conversion of environmental chemicals in laboratory conditions are not usually valid for field conditions, which are highly complex with different densities of populations, nutrient statuses, and conversion activities (4).

Microorganisms, insects, and larvae have yielded estimable results (5) at numerous experiments conducted in field or laboratory conditions on the conversion and transportation of chemicals in soils. In spite of being tedious to handle, lysimeters with soil columns, have also been documented to yield realistic results.

Cd, Pb, AND Cr CONTAMINATION IN SOILS

The natural Cd content of soils at the upper 20 cm depth is <0.5 mg kg^{-1}. Cd contents of soils in the European Union have significantly increased to 0.16–0.32 mg kg^{-1} Cd in the last decades by excess input (1). The sources of increasing Cd are the polluted (~1.5–35 g ha^{-1} yr^{-1}) sewage sludges (~1 g ha^{-1} yr^{-1}), phosphate fertilizers (~5–6 g ha^{-1} yr^{-1}), and others (~1 g ha^{-1} yr^{-1}) (1).

The major source of Cd contamination in agricultural soils is the excess use of fertilizers with varying amounts of Cd contents (0.1–90 mg kg^{-1}) depending on the phosphate rock materials utilized in production. Such contaminations in soils will cause a 0.1 mg kg^{-1} increase of Cd in 20–30 years. Other spots of high Cd contamination are adjacent to Zn mines. Another significant Cd contamination source is the sewage sludge used in cultivated soils. The Cd content in the dry matter of the sludge is 7–17 mg kg^{-1} (1). Cd adsorption by organic or inorganic soil compounds (minerals) depends on the soil texture, pH, and Ca together with other elements. The retention of low amounts of Cd in soil solutions in surface soils might take tens or hundreds of years depending on texture (6). The carryover of Cd to plants does not only depend on the type and growth period, but also on the availability of Cd in soils. Exchange, adsorption, fixation, and diffusion are the dominant mechanisms in soils controlling Cd uptake/carryover by plants.

Pb is less toxic than Cd in soils with contents of about 15 mg kg^{-1} in the crust of the Earth. This amount increases to 700 mg kg^{-1} in soils close to highway traffic. Pb is very immobile in soils compared to Cd and Zn, being slightly soluble in pH < 5, whereas it is more soluble and available to plants at pH = 4–4.5 (7).

Soil Pb contents generally vary from 1 to 20 mg kg^{-1}, (2, 6) with Pb solubility decreasing at pH > 6 and increasing in pH < 6 with increasing availability to plants. The decrease of solubility/availability has been shown in many cases and especially in Pb-free grains of corn grown in soils with high Pb (320 kg day^{-1}) (8). Tolerable limits of Pb in soils are given by Hock and Elstner (8) as 100 mg kg^{-1} and 50–300 mg kg^{-1} for EU standards by Forstner (2).

Chromium, the essential element for human and animal diets is not considered essential for plant nutrition with a tolerable level of 100 mg kg^{-1} in soils (1, 8). Cr^{3+} is irrevocable in its role in the induction of the effect of insulin and proteins as well as stabilizing nucleic acid structures and activating some enzymes in glucose metabolism (9).

Varying parent materials seem to be responsible for the wide range chromium contents of soils (5–3000 mg kg^{-1}). High pH conditions favor the formation of new insoluble compounds—$Cr(OH)_3$ and $Cr_2O_3 x H_2O$—with iron oxides. In strong acid conditions Cr is readily soluble, with a low uptake (carryover) by plants, whereas it is readily available even at very low amounts if retained by roots (7).

NICKEL CONTAMINATIONS IN SOILS

Soils contain as low as 5–50 mg kg^{-1} Ni (7), which may increase to 5000 mg kg^{-1} in soils developed from olivine and serpentine rocks. Ni reaches the soil from air, via burning coal (10–50 mg kg^{-1} Ni) and derivatives of petroleum (49–145 mg kg^{-1} Ni) transformed to Nickel-carbonyl, which is cancerogenous (9). Sewage sludge containing high amounts of Ni is also responsible for the pollution of the soil. Clay minerals, Mn, Fe, and Al oxides adsorb and/or bind-immobilize-soil Ni depending on pH (10). At pH < 5.5 the solubility and plant availability of Ni increases. In soils with pH >6—the threshold reaction for Ni—the extractable Ni content by (ethylene diamine tetra acetate) EDTA is 25 mg kg^{-1}. At pH < 6, lower values of Ni concentrations are needed to avoid toxicity.

MERCURY CONTAMINATION

Mercury compounds (esp. $CH_3{-}Hg^+$) are highly poisonous for humans and animals. Excess Hg affects the central nervous system, develops blindness, and causes mutagenous and teratogenous effects via $CH_3{\cdot}HgCl$. The Hg content of the Earth's crust is approximately 0.02 mg kg^{-1}, whereas slightly contaminated soils and sediments together with soils developed from Hg-rich parent materials (located near volcanic areas) may contain 0.5 mg kg^{-1} Hg (dominantly <0.1 mg kg^{-1}) and 40 mg kg^{-1} Hg, respectively, (8).

Soils of highly populated areas and ore production sites contain 0.1–0.4 and 1.8 mg kg^{-1} Hg, respectively. Similarly soils, irrigated by wastewaters, contain upto 1.6 mg kg^{-1} of Hg (7).

The following reaction, together with microorganisms taking part in the process, shows the conversion of Hg in soils as:

$$Hg_2^{2+} \rightarrow Hg^{2+} + Hg^0$$

The low solubility of Hg in soils and its immobilization caused by the effective groups (–S–H and –S–S) in soil organic matter, by inducing accumulation in the root zone, consequently reduce the carry over to the plants (11). Vegetables like lettuce and beans also seem to be suitable for growing in Hg-contaminated soils (7 mg kg^{-1} Hg) due to their low uptake rates (0.08–0.12 mg kg^{-1} Hg), in contrast to grass containing surprisingly high amounts of Hg at golf courses with 130 mg kg^{-1} Hg.

REFERENCES

1. Korte, F. Lehrbuch der Ökologischen Chemie. 3. Auflage; Georg Thieme: Stutgart, New York, 1992; Vol. 3, 373.
2. Förstner, U. *Umweltschutztechnik. 2. Auflage*; Georg Thieme: New York, 1992; 507.
3. Berrow, M.L. An Overview of Soil Contamination Problems. In *Proceedings of International Conference on Chemicals in the Environment*; Lester, J.N., Perry, R., Sterritt, R.M., Eds.; Jelper Ltd: London, 1986; 543–552.
4. Klein, W.; Scheunert, I. *Bound Pesticide Residues in Soil, Plants and Food with Particular Emphasis on the Application of Nuclear Techniques in Agrochemicals*; Fate in Food and the Environment, International Atomic Energy Agency: Wien, 1982.
5. Führ, F.; Steffens, W. Transport und Bilanzierung von Umweltchemikalien in Agroökosystemausschnitten. AGE Bericht Wege und Wirkungen von Umweltchemikalien Bonn, AGE: Bad Godesberg, 1985; 36–38.
6. Poelstra, P.; Friesel, M.J.; Bassam, N.E.L. Transport and Accumulation of Cd Ions in Soils and Plants. Z. Pflanzenern Bodenk **1979**, *142*, 848–864.
7. Schachtschabel, P.; Blume, P.; Brummer, G.; Hartge, K.H.; Schwertmann, U. *Lehrbuch der Bodenkunde 14. Auflage*; Ferdinand Enke: Mannheim, 1998; 494.
8. Hock, B.; Elstner, E.F. *Schadwirkungen auf Pflanzen. 2. Auflage*; Wissenschafts: Mannheim, 1988; 348.
9. Merian, E. *Metalle in der Umwelt*; Verlag Chemie: Weinheim, 1984; 722.
10. Tiller, K.G.; Geith, J.; Brummner, G. The Relative Affinities of Cd, Ni, and Zn for Different Soil Clay Fractions and Goethite. Geoderma **1984**, *34*, 17–35.
11. Fushtey, S.G.; Frank, R. Distribution of Mercury Residues from the Use of Mercurial Fungicides on Golf Course Greens. Can. J. Soil Sci. **1981**, *61*, 525–527.

POROSITY AND PORE SIZE DISTRIBUTION

P

Keith C. Cameron
Graeme D. Buchan
Lincoln University, Canterbury, New Zealand

INTRODUCTION

The relative arrangement of the soil units (primary particles and aggregates) defines the pore space of the soil. Pore space is essential as it provides the capacity to store water and air, as well as enabling drainage and root growth. For some soil processes, the total porosity is less important than the pore size distribution. For example, a sandy soil has a lower total porosity than a clayey soil, but usually drains faster. The ability of a pore to transmit water decreases dramatically with its size. The sandy soil has a higher proportion of large soil pores than the clay, and thus drains much faster. Small pores are however essential as they provide the ability to store water.

THE SOIL PORE SYSTEM

Structured or aggregated soils are effectively "dual porosity" systems. Their porosity can be separated approximately into two components.

1. "Intra-aggregate porosity" is the microscopic pore space created by the geometrical packing of individual soil particles.
2. "Inter-aggregate porosity" represents the pore space due to the arrangement of soil aggregates. It is created by biological activity, such as earthworm burrowing and root growth; shrinkage during soil drying; and by cultivation.

Definitions

Pore size

Soil particles are normally specified by a single size, the diameter of an equivalent idealized sphere. By contrast, we assume that the soil pore space can be partitioned into a system of idealized cylindrical tubes, each with an equivalent cylinder radius, *r*. This enables us to link the size of a pore with its ability to attract water by capillary action, as measured by the suction head *h* (cm) required to empty that pore:

$$h = (2\gamma/r)\cos\alpha \tag{1}$$

where γ is the surface tension, *r* the pore radius (cm), and α the contact angle.

Figure 1 shows how soil pore sizes can be classified using either soil micromorphology, or the hydraulic properties of soil.

Total porosity, ε

This is the ratio of the volume of pores to the total volume of soil:

$$\varepsilon = \text{volume of pores/total volume of soil} = V_p/V_t \tag{2}$$

Most soils have a porosity between 0.3 and 0.6, usually expressed as a percentage (i.e., 30 and 60%). Organic soils (e.g., peats) may have porosity up to 90%.

Air-filled porosity, ε_a

This is the ratio of the volume of air to the total volume of soil:

$$\varepsilon_a = \text{volume of air/total volume of soil} = V_a/V_t \tag{3}$$

For field soils, an approximate minimum value of ε_a required to maintain adequate aeration (especially oxygen supply to respiring roots and microbes) is about 10%, although a range of values from 10 to 25% may occur (1). If ε_a is less than 10% at "field capacity" then soil loosening or drainage is recommended. A desirable ε_a for horticultural potting mixes is in the range of 20–25%, which provides rapid drainage and low mechanical impedance for root growth, while the organic matter provides adequate water holding capacity (2).

Macroporosity

Unfortunately there is no single clear or agreed criterion for the definition of "macropores" in soil. The minimum equivalent diameter, *d*, for macropores reported in the literature ranges between 30 and 3000 μm. Here we adopt

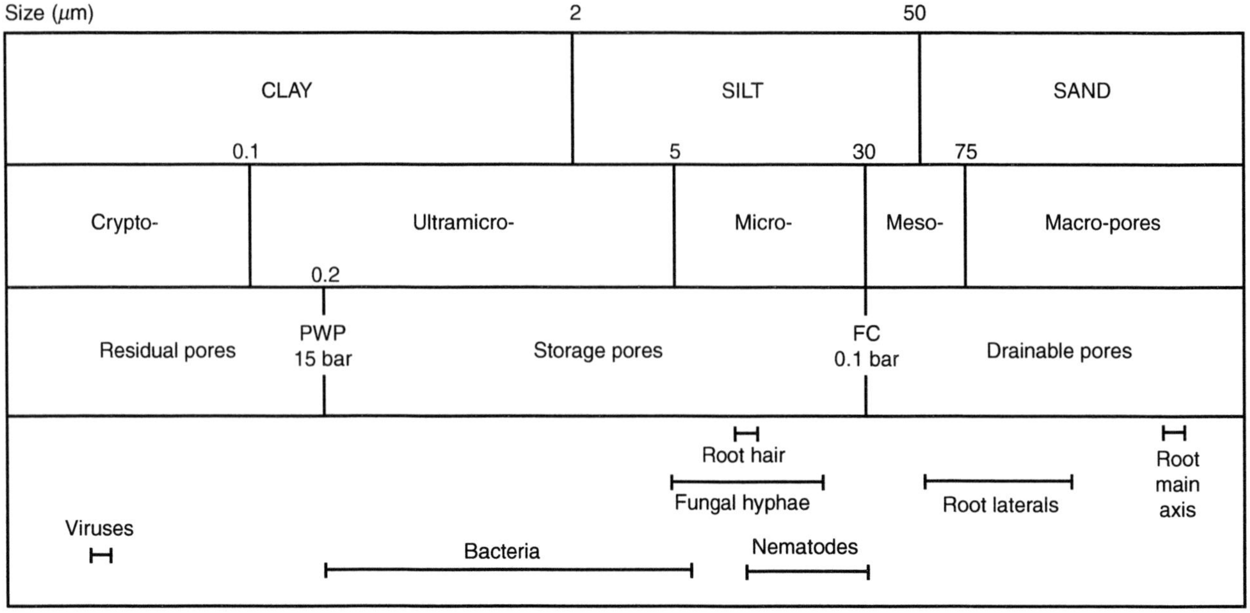

Fig. 1 Sizes of particles, pores and biota in soil. Top to bottom: Particle sizes (μm) in the USDA classification; pore sizes defined morphologically by Brewer (From Ref. 3); pore sizes defined hydraulically, with field capacity (FC, 0.1 bar) defining the lower limit of "drainable porosity", and permanent wilting point (PWP, 15 bar) defining the limit of "storage pores"; typical sizes (diameters) of roots and microbiota in soil. Root sizes are for cereal crops (From Ref. 4).

the lower limit, and define macroporosity as the fraction of soil volume occupied by pores with $d > 30\,\mu m$. This definition of macroporosity represents the air-filled porosity at field capacity, so that macropores are the "drainable pores" (Fig. 1).

Methods of Measurement

Total porosity can be measured by various techniques. The simplest method involves the collection and weighing of undisturbed soil cores of known volume (V_t). The ratio of the mass of dry soil to V_t gives the soil "bulk density", ρ_b. ε is then calculated as:

$$\varepsilon = 1 - \rho_b/\rho_p \tag{4}$$

where ρ_p is the particle density. For most common soil minerals, $\rho_p = 2.65\,g\,cm^{-3}$, but ρ_p decreases as organic matter increases. For example, a soil with $\rho_b = 1.3\,g\,cm^{-3}$ and $\rho_p = 2.65\,g\,cm^{-3}$ would have a porosity of 51%.

Porosity can also be measured using a gas pycnometer (5). The original method, developed by Torstensson and Eriksson (6), is based on Boyle's law of volume–pressure relationships. An oven dried soil core is sealed into a gas tight chamber of the pycnometer, and the air in the chamber is compressed by a known amount. The porosity is calculated from the difference between the volume of compressed air in the chamber when it contains a soil sample and when it is empty.

Macroporosity depends on the definition chosen for the lower limit of macropore diameter. It can be measured from the soil moisture characteristic specify where. Macroporosity may also be measured optically, using image analysis of soil thin sections. Bullock and Thomasson in their pioneering work (7) developed techniques for scanning soil thin sections with a TV camera and setting an image analyzer to detect differences in gray-scale indicating solids or pores. The volume and geometrical arrangement of the soil pores can be determined.

PORE SIZE DISTRIBUTION

Pores can be classified according to size or function, as shown in Fig. 1.

The *macropores* ($d > 30\,\mu m$) are usually air-filled and contain water only when the soil is saturated or partially drained. The smaller pores (*micropores*: $d < 30\,\mu m$) include the storage pores retaining plant-available water, and are generally found within, rather than between the soil aggregates.

Methods of Measurement

Lawrence (8) reviews techniques for measuring pore sizes in soil. The commonest method is to determine the "soil moisture characteristic" (SMC). The SMC is the relationship between soil water content θ, and the soil matric potential ψ_m (Fig. 2). Since ψ_m is negative in unsaturated soil, we drop the negative sign and refer instead to the *suction* h $(= -\psi_m)$.

The amount of water retained by the soil over a range of matric potentials can be measured by applying a suction to the soil to overcome the forces of capillarity and adhesion holding the water, using the apparatus shown in Fig. 3. An undisturbed core taken from the field is saturated with water and placed on the porous plate of the *tension table*. When the column of water is level with the soil, all the pores are water-filled. As the column is lowered, increasing suction removes water, first from the larger soil pores, followed by progressively smaller pores. The loss of water at each suction is measured by re-weighing the soil core once it has ceased to drain at each particular suction.

The suction (or tension table) technique is limited to suctions less than $\approx$10 kPa. Water held at lower matric potentials ($\psi_m < -10$ kPa) must be measured using a *pressure-plate* technique (5).

The clay soil in Fig. 2 has a higher water content at each matric potential than the sandy soil, because the clay has a finer texture with smaller pores. The sandy soil pores are larger and can be emptied at relatively low suctions. The shape of the curve for the sandy soil reflects the restricted size range of pores, most of which empty around the same suction (e.g., 10 kPa in Fig. 2). The clay soil has a more gradual curve reflecting the more uniform distribution of pore sizes.

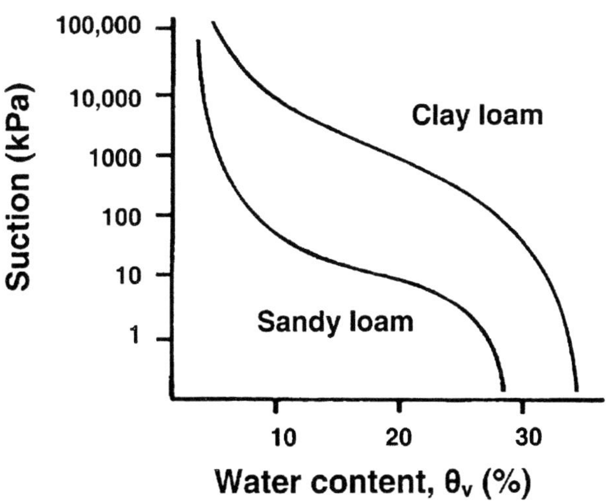

Fig. 2 Comparison of a soil moisture characteristic of a sandy vs. a clay soil. (Adapted from Ref. 9.)

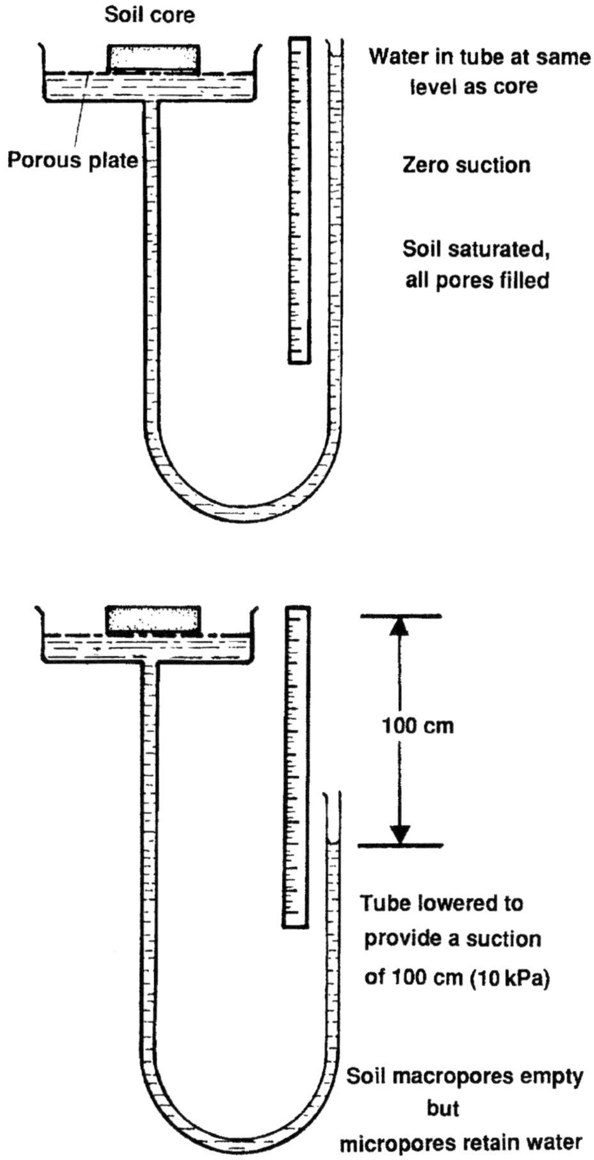

Fig. 3 Measurement of soil moisture release characteristic using tension table equipment. (Adapted from Ref. 9.)

Soil structure affects the shape of the moisture release curve at low suctions (e.g., 0 to about 10 kPa), because it is the shape and packing of the structural units which mainly define the volume of macropores. Differences between soils at low suctions provide information about the relative drainage and aeration capacities of the soils.

Pore size distribution can also be measured using mercury intrusion porosimetry (8). Pressure is used to force mercury *into* the soil pores. The relationship between applied pressure and the volume of intruded mercury provides the measure of pore-size distribution. Although the method is fast, errors may occur due to differences in the contact angle of the mercury on different mineral surfaces (5).

KEY FUNCTIONS OF THE PORE SYSTEM

Porosity and pore-size distribution control a wide range of soil phenomena, including:

(a) *Water storage in soil.* This is one of the critical soil properties that determine the ability of a soil to sustain plant growth. The amount of water that is stored in a soil depends not just on the total porosity but also on the pore-size distribution.

(b) *Transmission characteristics of the soil.* The hydraulic conductivity of the soil controls its drainage rate and its ability to support plant growth. Different size classes of pores perform different functions. Macropore flow controls drainage at or near saturation. Micropore flow regulates the rate of water supply to roots, and hence determines the onset of plant water stress. The soil pore system also affects the transport of solutes and particles with the soil solution. For saturated or near-saturated flow, macropores can provide a "preferential" or a "bypass" pathway for water, solutes or particles. Soil micropores provide a "storage volume" which can protect solutes against leaching. Diffusion of solutes out of soil micropores helps to supply nutrients to plant roots. Soil porosity and pore-size distribution control the rate of mass flow and diffusion of soil gases. Plant root respiration requires a constant supply of oxygen and produces carbon dioxide. The soil pore system influences the rate of supply of oxygen at the root surface of the plant and rate of diffusion of carbon dioxide away from the root.

(c) *Soil mechanical properties, including penetration resistance, soil consistency, compactibility and "bearing capacity."* Plant roots grow through the soil pore system in search of water and nutrients. Although roots are able to expand the diameter of a soil pore, the rate of root growth can be severely limited in a compact soil with low porosity. The greater the porosity of a soil, the more easily it can be compacted, and the more likely it is to yield or deform under stress or load.

(d) *Location of biological activity, including migration of worms, nematodes or pathogens.* Although soil organisms can create their own pores by burrowing through soil, they also utilise the existing pore system.

REFERENCES

1. da Silva, A.P.; Kay, B.D.; Perfect, E. Characterization of the Least Limiting Water Range of Soils. Soil Sci. Soc. Am. J. **1994**, *58*, 1775–1781.
2. Handreck, K. Gardening Down-Under. *Better Soils and Potting Mixes for Better Gardens*; CSIRO: Melbourne, Vic., 1993.
3. Brewer, R. *Fabric and Mineral Analysis of Soils*; Wiley: New York, 1964; 470.
4. Gregory, P.J. Growth and Functioning of Plant Roots. In *Soil Conditions and Plant Growth*; Chapter 4. Wild, A., Ed.; Longman: New York, 1988; 113–167.
5. Danielson, R.F.; Sutherland, P.L. Porosity. In *Methods of Soil Analysis. Part 1. Physical and Mineralogical Methods*, 2nd Ed.; Agronomy Monograph No. 9. Klute, A., Ed.; American Society of Agronomy: Madison, WI. 1986; 443–461.
6. Torstensson, G.; Eriksson, S. A New Method for Determining the Porosity of the Soil. Soil Sci. **1936**, *42*, 405–417.
7. Bullock, P.; Thomasson, A.J. Rothamsted Studies of Soil Structure II. Measurement and Characterisation of Macroporosity by Image Analysis and Comparison with Data from Water Retention Measurements. J. Soil Sci. **1979**, *30*, 391–413.
8. Lawrence, G.P. Measurement of Pore Sizes in Soils: A Review of Existing Techniques. J. Soil Sci. **1977**, *28*, 527–540.
9. McLaren, R.G.; Cameron, K.C. *Soil Science: Sustainable Production and Environmental Protection*; Oxford University Press: Auckland, 1996; 304.

POTASSIUM

Philippe Hinsinger
INRA UMR Sol et Environnement, Montpellier, France

P

INTRODUCTION

Potassium (atomic weight = 39.098) is an alkaline metal, as revealed by the etymology of its symbol (K) which derives from the Latin word "kalium" and from the Arabic word "qali" (alkali). K is thus strongly electropositive and always occurs as a monovalent cation. Consequently, its physico-chemistry and speciation are rather simple. K is an abundant alkaline metal cation, reaching a concentration of $26\,g\,kg^{-1}$ in the Earth's crust. It is a major nutrient for all living organisms.

POTASSIUM IN PLANTS

K is the major cation in most plants, occurring at concentrations ranging from 5 to $50\,g\,kg^{-1}$, twice as much as Ca and slightly less than N (1–3). The etymology of its name accounts for the abundance of K in plant-derived ash material (potash). K is involved in a large number of physiological processes: osmoregulation and cation–anion balance, protein synthesis and activation of enzymes (2, 3). K is often referred to as "a cation for anions" as it balances the abundant negative charges of inorganic (nitrate) and organic anions (carboxylates) in plant cells. It therefore occurs at large concentrations, 100–200 mM in the cytosol, 5–10 times less in the vacuole. Being a major inorganic solute, it plays a key role in the water balance of plants: maintenance of the osmotic potential and turgor pressure involved in cell extension. K-controlled changes in turgor pressure in guard cells is a key process of stomatal opening and closure and hence, of the regulation of plant transpiration. Many of these physiological roles are related to the high mobility of K at all levels in the plant. This unique, considerable mobility of K in the plant is essentially due to the large permeability of cell membranes to K-ions, which arises from the occurrence of a range of highly K selective, low and high affinity ion channels and transporters. These are now being increasingly characterized at a molecular level (4). Large rates of K uptake can thereby be achieved in plant roots. In addition, K-ions can easily be leached out of living plant tissues, as documented for tree foliage which contributes a large flux of K back to the soil via throughfall (3). K also rapidly leaves dead roots and other plant debris compared with N and P, which require hydrolysis of organic molecules. At an agronomic level, the demand for K largely varies with plant species and productivity. The uptake of K essentially occurs during the vegetative stage and can reach values of $10\,kg\,ha^{-1}\,day^{-1}$ and above. Depending also on the agricultural practices (removal of straw, for instance), the amount of K removed with the harvested material will range from $5–50\,kg\,ha^{-1}$ for cereal grains to $50–500\,kg\,ha^{-1}$ for forage, root and tuber or plantation crops.

POTASSIUM IN SOILS

Among major nutrients, K is usually the most abundant in soils as total K content ranges from 0.1 to 40, with an average of $14\,g\,kg^{-1}$ (1, 5). A major proportion of soil K occurs as structural K in feldspars and interlayer K in micaceous minerals (Fig. 1) (1, 6, 7). Some minor proportion of soil K (usually much less than 1%) is adsorbed on negatively charged soil constituents, namely clay minerals and organic matter. A marginal part is present as free K-ions in the soil solution. Bulk soil solution concentrations usually amount to 100–1000 μM (less than 0.01–0.1% of total K). The reason for this rather low concentration of K in the soil solution and hence restricted mobility of K in soils, compared to other metal cations such as Na or Ca is related to its selective adsorption onto some clay minerals. Because of its ionic radius and small hydration energy, K-ions indeed perfectly fit into the interlayer sites of micaceous minerals (micas, illites and mica-derived clays) (8). These sites and, to a lesser degree, the sites on the frayed edges of these minerals have thus a considerably larger affinity for K than for other cations, including divalent cations such as Ca or Mg (Fig. 2). Clay minerals also bear sites with larger affinity for divalent cations than for monovalent cations such as K. These sites are located on the planar faces of clay minerals and are thus dominant in clays such as kaolins and smectites. They also occur in organic compounds. K is

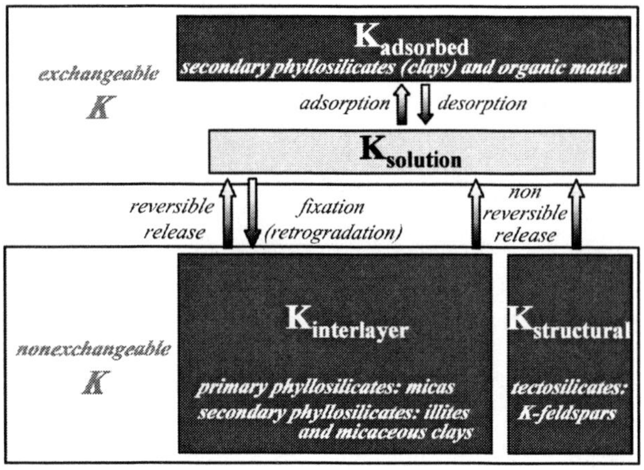

Fig. 1 The various forms of soil K and the chemical processes involved in soil K dynamics.

thus much less strongly held in soils dominated by kaolins (tropical soils), sand or organic matter than in soils dominated by illite–vermiculite clay minerals. Traces of mica-derived clay minerals can dramatically influence the dynamics of soil K as evidenced in tropical soils that are apparently dominated by kaolins (9). More generally, K dynamic is largely dependent on soil mineralogy which determines both ion exchange and release-fixation processes (6, 7), i.e., the dynamics of "nonexchangeable K." The latter is defined as that (major) portion of soil K which cannot be exchanged by NH_4-ions. NH_4-ions have the same charge and radius than K-ions and can successfully desorb K only from low K-affinity sites.

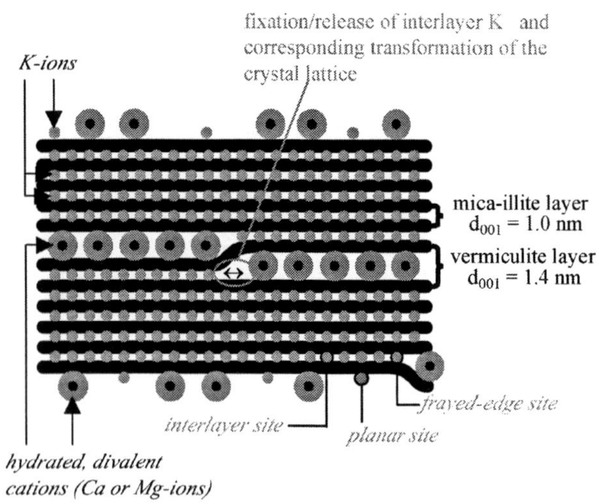

Fig. 2 The various sites of exchange of K-ions in micaceous clay minerals and the transition between mica and vermiculite layer that occurs when interlayer K is exchanged by hydrated, divalent cations.

Exchangeable K thus comprise soil solution and easily desorbable K-ions. Nonexchangeable K mostly comprise interlayer K (high K-affinity sites) of micaceous minerals and structural K of feldspars (Fig. 1), i.e., 90–99% of total K in many soils. The release of K from feldspars requires a complete and irreversible dissolution of the mineral and is enhanced under acidic conditions (6, 7). The release of interlayer K from micaceous minerals can proceed similarly or involve an ion exchange process (Fig. 1) leading to an expansion of the phyllosilicate (Fig. 2). This reversible release is essentially governed by the concentrations of K and competing cations in the outer solution (8, 10, 11). Cations which can be responsible for this release, such as Ca- and Mg-ions, have a large hydration energy, contrary to K- or NH_4-ions. Therefore, they remain hydrated when exchanging interlayer K-ions and expand the interlayer space (Fig. 2), making it possible for the release to proceed further, whereas NH_4-ions would block the reaction (12). However, because of the considerable affinity of these interlayer sites for K relative to Ca or Mg, the release can occur only for extremely low concentrations of K in the soil solution (Fig. 3), in the micromolar range (11, 12). Conversely, elevated concentrations of K are prone to the reverse reaction of fixation, i.e., the collapse of expanded layers and concomitant increase in nonexchangeable K at the expense of readily available K. As long as exchangeable K was assumed to be the only plant-available K fraction, K fixation, i.e., the poor recovery of applied K in the exchangeable fraction was seen as a poor efficiency of K fertilization. This apparent loss of applied K has therefore received considerably more interest than the reverse release process

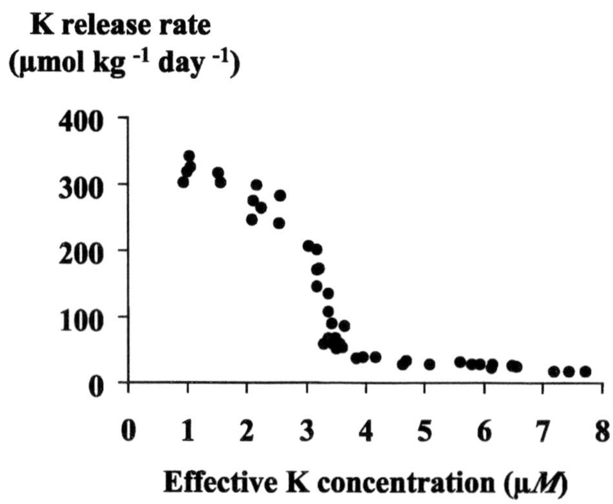

Fig. 3 Effect of solution K concentration on the rate of release of nonexchangeable K from a soil. (Adapted from Ref. 11.)

(7). In addition, the release of nonexchangeable K was considered unlikely to occur to any great extent, especially because K concentration in the bulk soil solution of fertilized soils is usually far above the critical concentrations that are prone to K release. However, numerous long-term fertilizer trials have shown that the release of nonexchangeable K contributes a major proportion of soil K supply in unfertilized and possibly fertilized plots too, although an overall net fixation is often found in the latter (Fig. 4) (5, 13). Nonexchangeable K can contribute up to $100\,kg\,ha^{-1}\,yr^{-1}$, 80–100% of soil K supply. Such annual fluxes of release are fairly large compared with K dissolution rates commonly estimated by geochemists (in the order of $5–15\,kg\,ha^{-1}\,yr^{-1}$). However, many geochemical models do not take into account the amount of K taken off in the vegetation, leading to large underestimates of the actual dissolution rates (14). Conversely, many K budgets provided by agronomists do not take into account atmospheric inputs and leaching of soil K, assuming that these terms are fairly negligible in most cases. This certainly holds true for atmospheric inputs which rarely exceed values in the order of a few $kg\,ha^{-1}\,yr^{-1}$. Leaching can, however, vary over a much wider range of values, from several $kg\,ha^{-1}\,yr^{-1}$ in most cases up to several tens (and up to 1–3 hundreds) of $kg\,ha^{-1}\,yr^{-1}$ in those situations that are the most prone to leaching: bare soil or poor soil coverage by the vegetation, excessive fertilizer rates, coarse-textured soils. Omitting the leaching term would, however, have led to underestimating the actual release rate (5). Hence, considerable amounts of nonexchangeable K can be released in agricultural soils and contribute a significant proportion of plant uptake, in contradiction to the widespread viewpoint shared by numerous agronomists and soil scientists. Reasons for this can be found when considering root–soil interactions occurring in the rhizosphere.

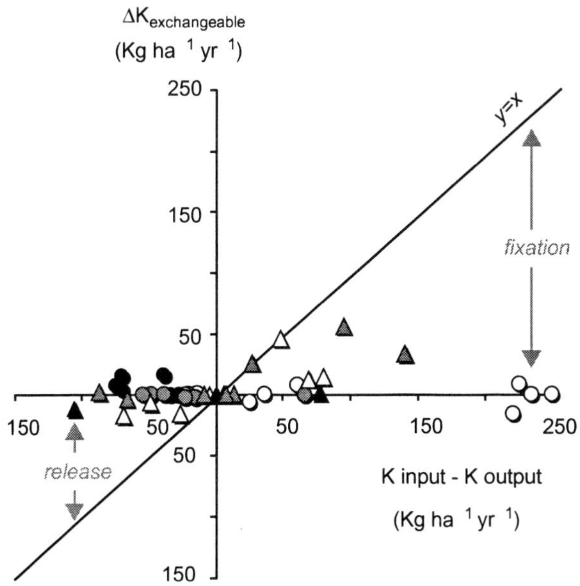

Fig. 4 Annual change in exchangeable K ($\Delta K_{exchangeable}$) as a function of the annual K budget in various K treatment plots of long-term fertilizer trials. The K input comprehends organic and inorganic, applied K fertilizers. The K output corresponds to the offtake of K in the harvested product. (Adapted from Refs. 5, 13.)

POTASSIUM IN PLANT–SOIL INTERACTIONS

K occurs at rather low concentrations in the soil solution, compared to other nutrient cations and to the large requirements of plants for K. The transfer of soil K via mass-flow toward plant roots (i.e., the convective flow of solute accompanying transpiration-driven water flow) contributes about 1–20% of plant demand (3, 15–17). A direct consequence is the rapid depletion of K-ions from the soil solution in the vicinity of plant roots, i.e., the rhizosphere. The resulting concentration gradient generates a diffusion of K-ions in the rhizosphere which plays a key role in the transport of K toward plant root (i.e., 80–99% of plant demand). Such depletion results in a shift of the cation exchange equilibria which rule the dynamics of both exchangeable and interlayer K. This ultimately results in a desorption of exchangeable K and eventually of interlayer K (7, 16, 17), as shown by their depletion in the rhizosphere (Fig. 5). The extent of the depletion of exchangeable K will depend on chemical parameters such as the initial level of exchangeable K and the K buffering capacity of the soil and on physical parameters that directly determine the diffusive transport of K-ions: soil texture and structure, soil water content (17). The K depletion zone will extend over several millimeters in clayey, dry soils up to several centimeters in wet, sandy soils (7, 16, 17). The intensity of the depletion will also depend on how far the K concentration of solution is decreased, which may vary among plant species according to the K uptake ability of their root. Plants with a lower external K efficiency, i.e., with a higher affinity transport system, will have the capability to take up K at lower K concentrations and may thus deplete soil K further (17). In the vicinity of roots, solution K concentration can indeed decrease by 2–3 orders of magnitude, down to as little as 2–3 μM (16). At such low K concentrations, the release of nonexchangeable K can occur at large rates, whereas it would be dramatically restricted at bulk soil K

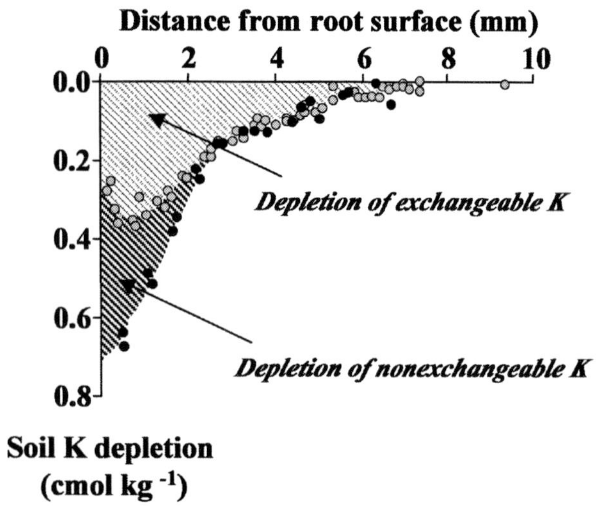

Fig. 5 Depletion of both exchangeable K (gray dots) and HCl-extractable K (black dots) as a function of the distance from rape roots. (Adapted from Ref. 17.)

concentrations of several hundreds of μM (Fig. 3) (11). Plants thus play a major role in the dynamic of interlayer K via the root-induced depletion of solution K (16). Measurements in pot experiments have indeed revealed that within several days of growth, the release of nonexchangeable K can amount up to 90% of K supplied to the plant (7, 16). Soil–root chemical interactions in the rhizosphere thus largely explain the unexpectedly large contribution of the release of nonexchangeable K to plant uptake that is found in many agricultural soils, including fertilized soils.

ASSESSING AND MANAGING POTASSIUM FERTILITY

Soil K fertility is most often evaluated by measuring exchangeable K (1, 7) most frequently with molar NH_4 acetate in batch conditions. However, the adequacy of exchangeable K to predict plant response, i.e., the actual bioavailability of soil K, is rather poor in many soils. This arises from the major contribution of the release of nonexchangeable K in some soils, especially when exchangeable K is low and/or when large reserves of nonexchangeable K are readily available as a consequence of: 1) soil mineralogical composition or 2) fertilization history (build-up of fixed K due to excessive K-fertilizer rates). In these situations, quantitative evaluation of the potential release of nonexchangeable K would be highly recommended for a better prediction of plant response and fertilizer needs (7). There are several methods for assessing nonexchangeable K but none of them is routinely used on a broad scale, because of their cost. These are either based on the use of 1) concentrated, strong acids that dissolve K-bearing minerals or 2) cationic resins or chemicals such as Na tetraphenylboron that can promote the release of interlayer K by removing K-ions from soil solution and by shifting the exchange equilibria (6, 7). Alternatively, correction factors can be used when interpreting exchangeable K values, which account either for the cationic exchange capacity (or clay content) or for the soil type and K release potential (7). Exchangeable K is nonetheless often used alone for fertilizer recommendations, resulting in frequently overestimated fertilizer needs to compensate for the expected large fixation and negligible release. Many long-term fertilizer trials have shown that adequate yields of crops can be obtained at fairly low rates of K fertilizer application, or even, for the least demanding crops such as cereals, without any K fertilizer for several years or decades (7, 13). Other more demanding crops, however, require the application of K fertilizer to achieve high yield and quality in the harvested products (18). The need for K - fertilizers will thus depend on the release potential of the soil and on the demand of the plant, the latter being now increasingly accounted for in fertilizer recommendations. Fertilizer trials have also shown that commonly used soluble K fertilizers and organic sources such as manure or crop residues have fairly comparable efficiencies. This is not surprising as K is highly mobile in organic compounds where it occurs as soluble or exchangeable K-ions. These sources are thus equally important as K-fertilizers and absolutely need to be accounted for in K budgets.

CONCLUSIONS

K is the major nutrient cation for plants and thus taken up at large rates by plant roots. These are achieved by both high and low affinity transport systems which explain the considerable mobility of K within the plant. In comparison, K is much less mobile in soils because of the strong affinity of some exchange sites of clays. The large K uptake rates achieved by roots result in a steep depletion of solution K in the rhizosphere, and hence in a shift of the equilibria of cation exchange. Exchangeable K and even nonexchangeable K can thereby be significantly depleted and contribute a substantial proportion of plant uptake. This is confirmed by K balance both in short-term pot experiments and long-term field trials. In addition to the desorption–adsorption of exchangeable K, release and fixation processes thus need to be accounted for when evaluating soil K fertility.

REFERENCES

1. Munson, R.D. *Potassium in Agriculture*; American Society of Agronomy, Crop Science Society of America, Soil Science Society of America: Madison, WI, 1985; 1223.
2. Mengel, K.; Kirkby, E.A.; Kosegarten, H.; Appel, T. *Principles of Plant Nutrition*, 5th Ed.; Kluwer Academic Publishers: Dordrecht, Netherlands, 2001; 673.
3. Marschner, H. *Mineral Nutrition of Higher Plants*, 2nd Ed.; Academic Press: London, 1995; 889.
4. Schachtman, D.P. Molecular Insights into the Structure and Function of Plant K^+ Transport Mechanisms. Biochim. Biophys. Acta **2000**, *1465*, 127–139.
5. Blake, L.; Mercik, S.; Koerschens, M.; Goulding, K.W.T.; Stempen, S.; Weigel, A.; Poulton, P.R.; Powlson, D.S. Potassium Content in Soil, Uptake in Plants and the Potassium Balance in Three European Long-Term Field Experiments. Plant Soil **1999**, *216*, 1–14.
6. Sparks, D.L. Potassium Dynamics in Soils. Adv. Soil Sci. **1987**, *6*, 1–63.
7. International Potash Institute, *Methodology in Soil-K Research*; Proceedings of the 20th Colloquium of the International Potash Institute, Baden bei Wien, Austria, International Potash Institute: Bern, Switzerland, 1987; 428.
8. Dixon, J.D.; Weed, S.B. *Minerals in Soil Environment*; Soil Science Society of America: Madison, WI, 1989; 1244.
9. Fontaine, S.; Delvaux, B.; Dufey, J.E.; Herbillon, A.J. Potassium Exchange Behaviour in Carribean Volcanic Ash Soils Under Banana Cultivation. Plant Soil **1989**, *120*, 283–290.
10. Schneider, A. Influence of Soil Solution Ca Concentration on Short-Term Release and Fixation of a Loamy Soil. Eur. J. of Soil Science **1997**, *48*, 513–522.
11. Springob, G.; Richter, J. Measuring Interlayer Potassium Release Rates from Soil Materials. II. A Percolation Procedure to Study the Influence of the Variable Solute K in the $<1 \ldots 10\,\mu M$ Range. Z. Pflanzen. Bodenk. **1998**, *161*, 323–329.
12. Springob, G. Blocking the Release of Potassium from Clay Interlayers by Small Concentrations of NH_4^+ And Cs^+. Eur. J. of Soil Sci. **1999**, *50*, 665–674.
13. Gachon, L. *Phosphore et Potassium dans les Relations Sol–Plante: Conséquences sur la Fertilisation*; Institut National de la Recherche Agronomique: Paris, France, 1988; 566.
14. Taylor, A.B.; Velbel, M.A. Geochemical Mass Balances and Weathering Rates in Forested Watersheds of the Southern Blue Ridge II. Effects of Botanical Uptake Terms. Geoderma **1991**, *51*, 29–50.
15. Barber, S.A. *Soil Nutrient Bioavailability. A Mechanistic Approach*, 2nd Ed.; Wiley: New York, 1995; 414.
16. Hinsinger, P. How Do Plant Roots Acquire Mineral Nutrients? Chemical Processes Involved in the Rhizosphere. Adv. Agron. **1998**, *64*, 225–265.
17. Jungk, A.; Claassen, N. Ion Diffusion in the Soil–Root System. Adv. Agron. **1997**, *61*, 53–110.
18. http://www.ipipotash.org/publications/publications.html (accessed October 2000).

PRECISION AGRICULTURE, ENGINEERING ASPECTS

Joel T. Walker
Reza Ehsani
Matthew O. Sullivan
The Ohio State University, Columbus, Ohio, U.S.A.

INTRODUCTION

Information technology is playing an increasingly important role in today's agricultural production systems of all sizes, commodities, and management philosophies. Precision agriculture (12, 14, 17) or site-specific management is an information-based management technique that has the potential to improve profitability (7) and reduce the environmental impact (19) of crop production. It also has the potential to improve the quality and nutrient content of the product. Precision agriculture, rather than the "one-size-fits-all" management strategy, provides for differential treatment of selected areas of a production field, called management zones, based upon expectation of increased yield, profit, or some other agronomic goal (2–4, 18). Management zones may be selected for differential treatment based upon various documented differences such as soil type, soil fertility or pH, yield history, presence of weeds, insects, or diseases, or other measures for which a differential treatment helps the producer achieve a selected goal. The ability to provide differential treatment to management zones, also called site-specific management, depends upon availability of both proper equipment and effective treatment algorithms.

WHAT MAKES IT POSSIBLE?

Precision agriculture techniques have been made possible by the advent of global positioning systems (GPS) and high speed computer processing. GPS provides real-time location information to a computer that, from stored information, determines the current management zone, selects appropriate treatment for that management zone, and controls mechanisms to provide the treatment. Figure 1 is a graphic representation of the precision agriculture paradigm. GPS provides position information for a variety of data gathering processes or for control of site-specific treatments. Information of various types (shown as layers) may be used in analysis of yield results or to develop an application map to control site-specific treatments. The whole system taken together is often called precision agriculture or site-specific agriculture. Note that a feedback loop is implied where results of the previous growing season (yield) become part of the information that influences current treatment practices. Given the proper treatment algorithms, current treatment practices may optimize the goal parameter. Maximum yield is not necessarily the best goal because the cost of treatments required to achieve that yield may be greater than the increased crop value (4, 18).

GPS consists of a minimum of 24 satellites circulating around the Earth sending signals to a local receiver for which location is desired (9). Each satellite broadcasts encoded information with particular timing. By measuring the time a signal travels (at the speed of light) to reach the receiver, the distance from a satellite to the receiver may be calculated. Determining distances from four or more satellites of known location may establish the receiver's location (latitude, longitude, and elevation). Even simple, inexpensive receivers are capable of accuracy better than 15 m, close enough to return to a favorite fishing spot. With specialized transmissions containing information to correct for known errors (called differential signals), accuracy better than 1 m may be achieved. Specialized local transmitters make possible real-time kinematic (RTK-GPS) systems with accuracy of better than 1 cm. RTK systems are used in surveying, guidance, and where the fine precision may justify a relatively high equipment cost.

High-speed computer processing systems have also played an important role in the advent of precision agriculture. Precision agriculture requires collection and storage of data, decision-making computation, and controlling of equipment by computers operating at billions of operations per second. GPS locations are recorded in real-time by the computer. Digital maps of field conditions and parameters are carried in memory or storage media. Digital data such as digital still or video images of weeds, insects, or disease damage, soil properties, crop spectral reflectance, or climactic conditions may also be collected in real-time by the computer. Using such information,

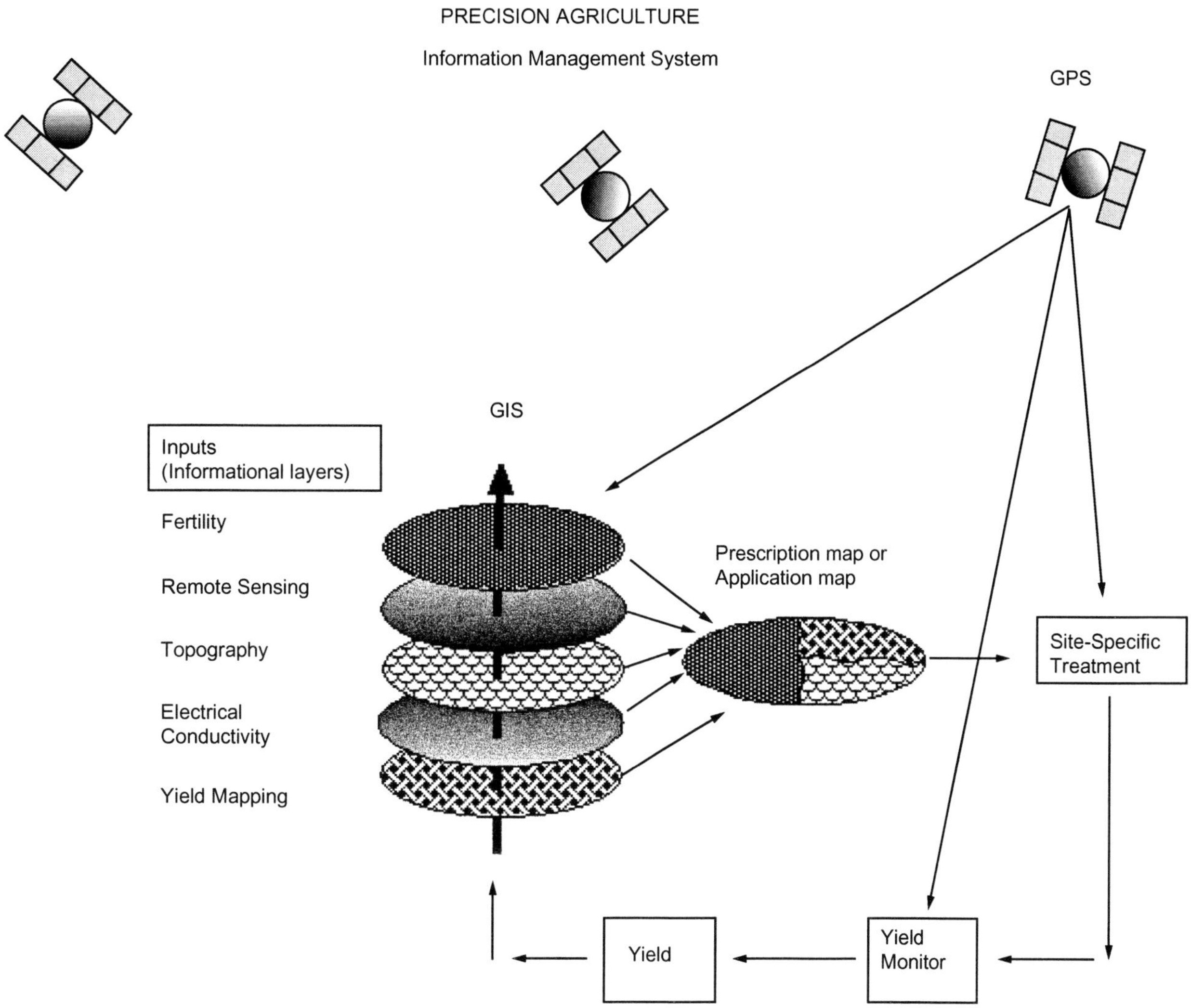

Fig. 1 Paradigm for precision agriculture.

the computer may work through a response model to decide what actions should be taken with current equipment. For example, a precision agriculture capable planter may be able to adjust planting rate and depth or change the seed variety on-the-go. Thus, the computer may decide that areas with a selected soil type and relative elevation will get a reduced population (investment of seeds) because less yield is expected. The computer may dictate that another soil type with a high yield potential and high existing moisture content will get a different variety planted to a shallower depth. Across an entire field, many combinations of the controlled variables will be chosen to optimize the desired result (yield, profit, or other goal). A large field would require constant decisions and adjustment of the equipment that would not be possible without computerized systems. Finally, volumes of data may be produced by precision agriculture techniques and computer methods are being developed to extract useful information; and models form this data (5, 6, 10, 11, 15, 16) and present it to the public in readily accessible form (13).

SENSING FOR PRECISION AGRICULTURE

Real-time sensors will be important to many precision agriculture systems (8). A real-time sensor provides data in a nearly constant stream as the machine traverses the field. For example, a camera may provide digital images from which the presence of weeds may be determined.

A sprayer may then be directed to spray only where weeds are present. Organic matter and moisture levels in soil affect the performance of some herbicides. Sensors that measure organic matter and/or moisture on-the-go allow for optimum rates of chemical application—adequate to control the weeds, but no more than necessary to preserve environmental quality.

One of the most popular real-time sensors measures grain yield on combines (12). Yield sensors are also available or are being developed for a variety of crops such as cotton, potatoes, tree fruits, and strawberries. Yield sensors provide a measure of yield over a whole field. Areas of the field with unusually high or low yield may be identified and corrective action (subject to some predictive model created by the producer or by computer) can be taken for the following year. This one aspect of precision agriculture has created a lot of activity in adjusting soil drainage and fertility and crop management decisions such as the relative value of fertilizer or chemical inputs.

For many agronomic parameters, site-specific soil sampling is more practical, either because a real-time sensor is not yet available, or because the spatial variation of the parameter is more gradual and can be estimated with a few site samples. Soil samples to determine fertility are common. The values of pH (acidity/alkalinity), nitrogen, potassium, and phosphorus are particularly important in prediction of yield levels. Soil type is another parameter that is often determined by a site visit.

Remote sensing is becoming increasingly important to precision agriculture. Early data that came from random satellite observations had low resolution and limited value. Now specially equipped satellites and aircraft may be hired to collect specific crop or soil information. Crop growth or health may be deduced from this data. Certain wavelengths of the electromagnetic spectrum are particularly helpful. The visible light frequencies provide some information. Unhealthy crops tend to reflect more yellow and red light. Various frequencies in the infrared range also have been correlated with plant health and soil moisture conditions. Multi-spectral systems provide data on the visible spectrum (often three primary colors) and a limited range of the infrared. More sophisticated equipment, called hyper-spectral, can provide data from a much broader range of the spectrum and from narrower sample bands. This type of data offers greater opportunity to correlate specific crop or soil conditions to measured spectral data.

Other types of remote sensing are becoming available. Light detection and ranging (LIDAR), for example, is a laser ranging technique that may be used to measure topography and plant height simultaneously (1). Currently applied to the forest industry and relatively expensive, this method holds promise for rapid feedback of information on crop growth problems to the producer, perhaps allowing solution of the problem before yield is permanently affected. For example, nematodes in soybeans are a common cause of reduced vigor and yield if left untreated. Conventional crop scouting or remote sensing may not detect small areas of infestation or provide feedback in time to take corrective action. With regular LIDAR imaging, a producer could detect and treat such trouble spots in a timely and efficient manner.

GEOGRAPHIC INFORMATION SYSTEMS

Geographic information systems (GIS) techniques are an integral part of precision agriculture. Basically, GIS is a storage system for geographically referenced digital data. Many of the data types discussed above can be digitized (if not already in digital form) and referenced to specific locations in a field. Each parameter or variable then may be represented as a layer of information. Geographic position of the information matches the position on a map of the field. The value of a parameter may be represented on a map as a shade of gray or a color. So a GIS information layer representing soil type can be drawn in the physical shape of the field with patches of color—each color representing a different soil type. Many such layers may exist in a GIS dataset for a particular field. Depending upon the need, these layers may be viewed as overlays—simultaneous presentation of several variables in one field. As many layers of information are added to the GIS system, it becomes apparent that human vision is inadequate to detect the important patterns. Computers, however, are infinitely more able to "see" data patterns with a potential to produce an economic advantage.

FUTURE

The future of precision agriculture depends on economic results. Can the cost of additional equipment and operations be more than offset by increased economic return and value of environmental protection? It seems possible that all agricultural operations could be monitored and recorded, linked by digital transmissions to databases containing weather, remote sensing, and historical data, and controlled through a general model of the crop's predicted response to specific inputs. Such a system may not only control specific agricultural operations, but also be involved in scheduling, ordering seed, fertilizer, and supplies, providing records to regulating agencies, and feeding valuable information back into the research system for further refinement of the control model. Already, manufacturers are producing prototype agricultural machines that perform without an operator. These machines are guided by GPS, are controlled by computer, and have onboard sensors to detect obstructions or people in the way. The advancement of agriculture will come with technology—technology to feed the world.

REFERENCES

1. Ackermann, F. Airborne Laser Scanning—Present Status and Future Expectations. ISPRS J. Photogram. Remote Sens. **1999**, *54* (2–3), 64–67.
2. Adams, M.L.; Cook, S.; Corner, R. Managing Uncertainty in Site-Specific Management: What is the Best Model? Precis. Agric. **2000**, *2*, 39–54.
3. Boote, K.J.; Jones, J.W.; Pickering, N.B. Potential Uses and Limitations of Crop Models. Agron. J. **1996**, *88*, 704–716.
4. Bullock, D.S.; Bullock, D.G. From Agronomic Research to Farm Management Guidelines: A Primer on the Economics of Information and Precision Technology. Precis. Agric. **2000**, *2*, 71–101.
5. Data Mining for Site-Specific Agriculture. http://www.gis.uiuc.edu/cfardatamining/ (accessed 4/16/2001).
6. Data Mining Puts Geographic Information to Work for Public, Private Sectors. http://csd.unl.edu/csd/resource/vol-13/datamine.htm (accessed 4/16/2001).
7. Erickson, K., Ed. *Precision Farming Profitability*; Agricultural Research Programs, Purdue University: West Lafayette, IN, 2000.
8. Gaultney, L.D. Prescription Farming Based on Soil Property Sensors. *ASAE Annual International Meeting*; ASAE Paper No. 89-1036, ASAE: St. Joseph, MI, 1989.
9. Kennedy, M. *The Global Positioning System and GIS: An Introduction*; Ann Arbor Press: Chelsea, MI, 1996.
10. Kerschberg, L.; Lee, S.W.; Tischer, L. *A Methodology and Life Cycle Model for Data Mining and Knowledge Discovery for Precision Agriculture*; Final Report, Center for Information Systems Integration and Evolution, George Mason University: Fairfax, VA, 1997.
11. Lazarevic, A.; Fiez, T.; Obradovic, Z.A. *Software System for Spatial Data Analysis and Modeling*, Report to INEEL University Research Consortium Project No. C94-175936.
12. Morgan, M.; Ess, D. *The Precision-Farming Guide for Agriculturalists*; John Deere Publishing: Moline, IL, 1997.
13. Ohio State Precision Agriculture. http://www.ag.ohio-state.edu/~precisfm/ (accessed 4/16/2001).
14. Pierce, F.J.; Sadler, E.J. (co-editors). *The State of Site-Specific Management for Agriculture*; ASA/CSSA/SSSA: Madison, WI, 1997.
15. Precision Farming Support. http://www.umac.org/farming/precision.html (accessed 4/16/2001).
16. Regression and Emerging Data Mining Tools Modeling. http://www/gis.uiuc.edu/cfardatamining/APRReport.htm (accessed 4/16/2001).
17. Sonka, S.T., Ed. *Precision Agriculture for the 21st Century: Geospatial and Information Technologies in Crop Management*; National Academy Press: Washington, DC, 1998.
18. Swinton, S.M.; Lowenberg-Deboer, J. *Site-Specific Management Guidelines—Profitability of Site-Specific Farming*; Publication SSMG-3, Phosphorus and Potash Institute, 1998.
19. Wang, X.; Tim, U.S. Problem-Solving Environment for Evaluating Environmental and Agronomic Implications of Precision Agriculture. http://www.esri.com/library/userconf/proc00/professional/papers/PAP102/p102.htm (accessed 4/16/2001).

PRECISION AGRICULTURE, NUTRIENT CYCLING AND

Robert J. Lascano
Texas A&M University, Lubbock, Texas, U.S.A.

INTRODUCTION

Precision farming, also known as site-specific management, refers to the practice of applying agronomic inputs, mainly fertilizers and other chemicals, across a farm at variable rates based on soil nutrients or chemical tests, soil textural changes, weed pressures, and/or yield maps for each field in the farm. In most large fields (e.g., >40 ha) crop yield is notoriously variable. The sources of this variation are related to the physical and chemical properties of the soil, pests, microclimate, genetic, and phenological responses of the crop, and their interactions. Currently, the technology for crop yield mapping is more advanced than current methodologies for determining and understanding causes of yield variability. Prevailing and traditional management practices treat fields uniformly as one unit. However, recent reports (1–3) show that to understand underlying soil processes that explain crop yield variability, research must be done at the landscape level and using appropriate statistical tools for large-scale studies (1, 3, 4).

There is a linear relation between crop yield and water use when the only limiting factor is water (5); however, root uptake of water is synergistically related to nutrient uptake and the two processes cannot be separated. Precision farming has the potential to improve water and nutrient use efficiency on large fields provided there is a quantitative understanding of what factors affect crop water and nutrient use and where in the field. It is known that crop water and nutrient use are a function of many biotic and abiotic variables—including managed inputs—and that harvestable yield is a manifestation of how these variables and inputs interact and are integrated during the growing season. However, it is difficult to determine a hierarchy of the contribution of each input and variable to the measured yield using classical statistics (2, 3). Often, variables that affect water and nutrient supply to the plant contribute to yield at a high level assuming an adequate plant stand and weed control. The cause and effect relation between a single state variable and crop yield is site specific and is difficult to establish without considerable sampling of the soil and/or crop. The establishment of response functions, i.e., crop water and nutrient use as a function of variable x_i, only gives a partial answer to explain crop water and nutrient use and yield based on inputs. The general idea of precision farming is to optimize input application to the measured crop yield at each sampling location. This is a simple premise; however, the decisions for variable rate application of any agronomic input must consider temporal and spatial variability of the soil's properties affecting crop growth, water and nutrient use, and yield. Soil factors that affect stored water, such as depth to root restricting layer and soil textural differences, must be considered in any precision farming operation that attempts to improve crop water use and yield related to agronomic inputs. Similarly, to improve the use of any soil macro- and micro-nutrient by the crop, the overall cycle of the nutrient must be considered, including its availability in the soil and demand by the crop.

Precision farming must incorporate the inherent spatial and temporal variability of soil physical, chemical, and biological factors within a field for input management. Accurate representation of spatial and temporal variability in a field requires taking and analyzing many samples. Sampling is normally done on a grid with a scale that can vary from one to several hundred meters (6). Once properties are measured, geostatisitical tools (e.g., semivariogram, kriging, cokriging, etc.) and other spatial statistical tools (e.g., autocorrelation, crosscorrelation, state-space analysis, etc.) can be used to establish statistical relations in space and to minimize the number of soil samples to characterize and map fields (2, 3, 7). The number of samples required *a priori* to determine spatial and temporal variability is perhaps the single largest deterrent in the application of precision farming practices to manage and improve crop water and nutrient use.

There is very little information published on crop water and nutrient use across large fields at the landscape level and in the context of precision farming (1, 8, 9). An exception is a study (1) where cotton water and total nitrogen use were measured along a 700 m transect with the objective to 1) illustrate the landscape pattern of cotton water and total nitrogen use and 2) determine the underlying soil processes governing cotton lint yield variability. In this study, state-space analysis (1, 3) is used to formulate management decisions that may improve crop water and nitrogen use and, thus, yield, using precision farming practices.

LANDSCAPE CROP WATER AND NITROGEN USE

The concept of crop water and nitrogen use in a large field is illustrated by the study of Li et al. (1). In 1999, a field experiment was conducted near Lamesa, Texas, on a research farm of Texas A&M University on the southern edge of the High Plains of Texas. The soil was classified as an Amarillo sandy loam. The field was 60 ha with slopes ranging between 0.3 and 6.3% (1). To assess the effect of soil water, nitrate-nitrogen (NO_3-N), and topography on cotton lint yield across the landscape, two irrigation levels were used. The irrigation treatments consisted of water applications at the 50 and 75% potential evapotranspiration (ET) with a center pivot LEPA irrigation system (10). At each irrigation level one transect was established following the circular pattern of the center pivot. The two transects were instrumented with 50 neutron access tubes each 15 m apart and volumetric water content (θ_v) was measured periodically throughout the growing season. At each point θ_v was measured in 0.3 m depth increments to a depth of 2.0 m using a neutron probe calibrated for this soil. In addition, at each transect point soil texture, soil and plant N-NO_3, leaf area index, lint yield, slope, plant density, and other parameters were measured (1).

It has been shown that the use of classical statistics, such as regression analysis and analysis of variance, fails to completely explain the cause and effect between, for example, crop yield and measured soil variables in precision farming experiments (1–4, 11). Instead, there are other more appropriate statistical tools for relating the variability of soil and plant parameters measured in space and time. For example, the structure of the spatial variance between measurements may be derived from the sample *semivariogram*, which is the average variance between neighboring measurements spatially separated by the same distance. Spatial structure between variables is often determined using *autocorrelation* and *crosscorrelation* functions. Autocorrelation measures the linear correlation of a variable in space along a transect. Crosscorrelation is the comparison of two variables measured along a transect and is used to describe the spatial correlation between two landscape variables, i.e., where one variable, the tail variable, lags behind the head variable by some distance. The spatial association between several variables can be described using *state-space* analysis, which is a multivariate autoregressive technique (1–4, 7, 11).

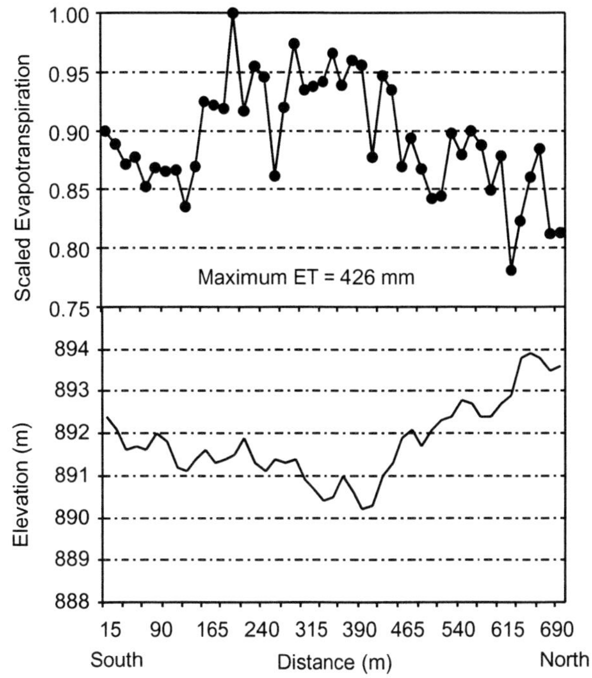

Fig. 1 Scaled evapotranspiration and elevation as a function of distance along a 700 m transect.

To illustrate the variability of crop water use or ET, values measured along the 50% irrigation transect were selected (1). In Fig. 1, the relation between the scaled ET and elevation, both as a function of distance along the transect, is shown. The ET data are scaled to the maximum of 426 mm of water measured 210 m from the south end of the transect. These results show that higher ET was measured at lower elevations and that ET decreased at higher elevations. Spatial crosscorrelation between lint yield and soil water, lint yield and site elevation, and soil water and site elevation are shown in Fig. 2. For a 95% confidence interval the cotton lint yield was positively crosscorrelated with soil θ_v across a lag distance of ± 30 m. Lint yield and θ_v were negatively crosscorrelated with elevation at a lag distance of ± 30 m. These results show the effect of topography on the θ_v and crop water use measured along the transect. Similar results are given in other reports (1, 8, 9, 11). In this example the crosscorrelation between θ_v and elevation shows the spatial structure of measured variables and further that more water was stored in lower elevations resulting in higher ET.

Linear regression analysis between θ_v and lint yield and relative site elevation is shown in Fig. 3, and the state-space analysis for the relation between lint yield and three measured parameters is shown in Fig. 4. Results in Fig. 3 show the shortcomings of using an inappropriate statistical

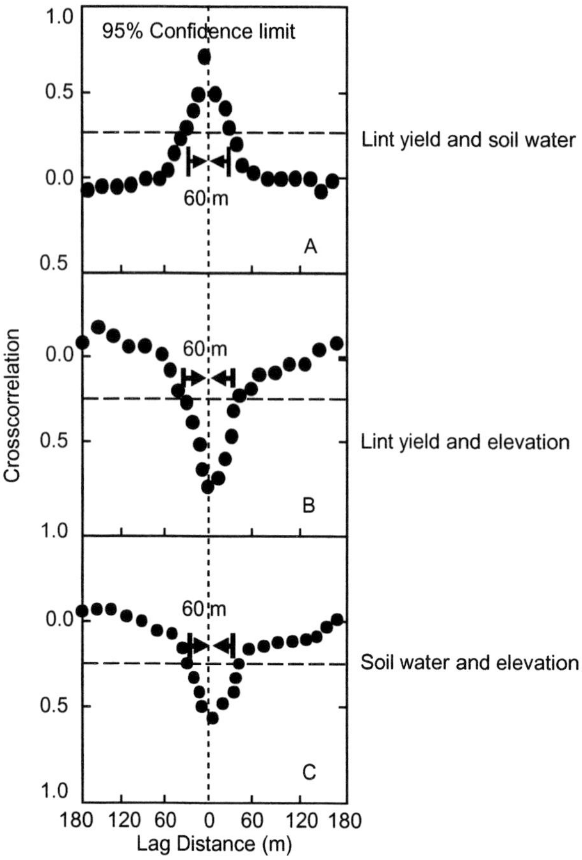

Fig. 2 Crosscorrelation as a function of lag distance. (A) Lint yield and soil water, (B) lint yield and elevation, and (C) soil water and elevation. Shown is the 95% confidence for the crosscorrelation distance. (From Ref. 1.)

tool to understand underlying processes explained with the state-space analysis. This analysis (Fig. 4) quantified how cotton lint yields varied as a function of distance and showed that by using θ_v, soil NO_3-N and elevation the variation in lint yield can be explained with a high level of confidence.

Benefits of precision farming to improve crop water and nutrient use may be obtained by an economic analysis of maximizing crop yield as a function of application of N fertilizer and irrigation water as given by the state-space equation. In the example given, the decision can be made to apply more N fertilizer to the lower areas of the field that also hold more water and increase crop water use and yield. With the introduction of variable rate planters it will be possible in the near future to discriminate site locations and plant more "drought" tolerant varieties or change the seeding rate in areas that are prone to have less soil water. This implies the delineation of management zones within a field that are defined based on potential crop water and

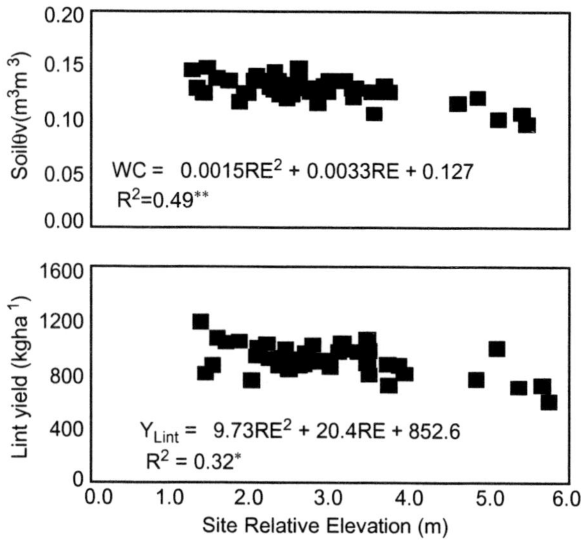

Fig. 3 Soil water content (θ_v) and cotton lint yield as a function of site relative elevation.

nutrient use and their interaction with other input variables to maximize economic yield across the field. This type of precision farming is not currently practiced but remains within the realm of possibilities that this type of farming has to offer.

A final consideration is the cost/benefit of precision agriculture practices and its impact on agriculture. Currently, hardware for variable rate application of agronomic inputs is relatively expensive and in many cases unavailable; however, with increased adaptation and use of these practices the cost will be reduced. Further, environmental concerns for a given area will probably place limits

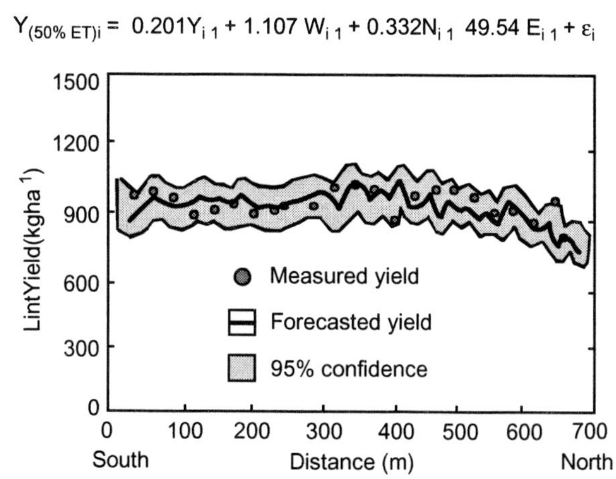

Fig. 4 State-space equation relating cotton lint yield (Y) to water content (W), nitrogen (N), and elevation (E) as a function of distance and location (i) along a 700 m transect. (From Ref. 1.)

on the amount of certain nutrients, e.g., N fertilizer, used for crop production. This will force producers to apply N and other nutrients across the field according to crop needs and position along the landscape. These practices will be beneficial from both an environmental and an economical point of view.

REFERENCES

1. Li, H.; Lascano, R.J.; Booker, Jill; Wilson, L.T.; Bronson, K.F. Cotton Lint Yield Variability in a Heterogeneous Soil at a Landscape Scale. Soil Tillage Res. **2001**, *58*, 245–258.
2. Nielsen, D.R.; Wendroth, O.; Pierce, F.J. Emerging Concepts for Solving the Enigma of Precision Farming Research. In *Precision Agriculture, Proceedings of the Fourth International Conference*, Minneapolis, MN; July 19–22, 1998; Robert, P.C., Rust, R.H., Larson, W. E., Eds.; 1999; pp. 303–318.
3. Wendroth, O.; Al-Oman, A.M.; Kirda, C.; Reichardt, K.; Nielsen, D.R. State-Space Approach to Spatial Variability of Crop Yield. Soil Sci. Soc. Am. J. **1992**, *56*, 801–807.
4. Cassel, D.K.; Wendroth, O.; Nielsen, D.R. Assessing Spatial Variability in an Agricultural Experiment Station Field: Opportunities Arising from Spatial Dependence. Agron. J. **2000**, *92* (4), 706–714.
5. Kramer, P.J.; Boyer, J.S. *Water Relations of Plants and Soils*; Academic Press: San Diego, 1995.
6. Sadler, E.J.; Busscher, W.J.; Baver, P.J.; Karlen, D.L. Spatial Requirements for Precision Farming: A Case Study in the Southern USA. Agron. J. **1998**, *90*, 191–197.
7. Shumway, R.H.; Stoffer, D.S. *Time Series Analysis and Its Application*; Springer Verlag: New York, 2000.
8. Halvorson, G.A.; Doll, E.C. Topographic Effects on Spring Wheat Yield and Water Use. Soil Sci. Soc. Am. J. **1991**, *55*, 1680–1685.
9. Hanna, A.Y.; Harlan, P.W.; Lewis, D.T. Soil Available Water as Influenced by Landscape Position and Aspect. Agron. J. **1982**, *74*, 999–1004.
10. Lyle, W.M.; Bordovsky, J.P. Low Energy Precision Application (LEPA) Irrigation System. Trans. ASAE. **1981**, *24*, 1241–1245.
11. Timlin, D.J.; Pachepsky, Ya.; Snyder, V.A.; Bryant, R.B. Spatial and Temporal Variability of Corn Grain Yield on a Hillslope. Soil Sci. Soc. Am. J. **1998**, *62*, 764–773.

PRODUCTIVITY, INFLUENCE OF NATURAL AND ANTHROPOGENIC DISTURBANCES ON

Douglas G. Maynard
Natural Resources Canada, Victoria, British Columbia, Canada

INTRODUCTION

Disturbances are relatively discrete events, either natural or anthropogenic (i.e., human-induced), that change the existing condition of an ecological system (1). Soil productivity refers to the capacity of a soil to support plant growth. Disturbances are an important process of virtually all forest ecosystems. It is not if a disturbance will happen, but when, where, and how (2). Natural disturbances such as fire, windthrow, insect and disease outbreaks, mass wasting (e.g., landslides), surface erosion, and catastrophic events (e.g., hurricanes, earthquakes, floods), control forest succession and affect soil productivity. The influence of human activity has increased anthropogenic disturbances such as harvesting, human caused fires, air pollution (e.g., acid deposition), and climate change which may affect forest succession and soil productivity. Human intervention (e.g., fire suppression) subsequently also has altered the frequency and degree of natural disturbances affecting soil productivity. How these disturbances alter soil processes and ultimately affect soil productivity depends on numerous factors, including the frequency, type, and degree of disturbances, soil properties, and forest ecosystem.

The most obvious effect associated with disturbance natural or anthropogenic, is the removal or destruction of above-ground biomass. However, changes as a result of disturbance may affect soil processes that result in either positive or negative changes to soil productivity. In northern temperate and boreal forests the effect of disturbances on soil processes may be the most important (3). The changes in soil productivity from disturbance may be direct or indirect. They include changes to the chemical, physical, and biological properties of the soil, such as loss of organic matter, soil compaction, soil displacement or mixing, exposure of subsurface material and bedrock, changes in nutrient supply, and soil organisms.

NATURAL DISTURBANCES

Fire

The most widely occurring and extensively studied natural disturbance is fire (4). Forest fires are a major large-scale disturbance in most forests of the world. These include temperate and boreal forests, most Australian ecosystems, Mediterranean-type ecosystems, and all but the wettest of tropical rainforests (5), including lowland rain forests of northwestern Amazonia (6). The intensity, type, and frequency of fire are important factors that affect soil productivity. High-intensity burns with high temperatures will result in volatilization of various nutrients. Carbon (C), nitrogen (N), and sulfur (S) will volatilize at relatively low temperatures (as low as 302 to <932°F). Phosphorus (P) and potassium (K) are less volatile but will volatilize at temperatures of more than 1112°F. Losses of other nonvolatile nutrients (>1832°F) such as calcium (Ca), magnesium (Mg), and manganese (Mn) may occur through blowing of fine ash. Similarly, in low temperature burns, P also can be lost through removal of fine ash. The accumulation of cations (e.g., Ca and Mg) has been observed in surface soils following high-intensity fires and results in increased soil pH.

Soil biological processes also may be influenced by fire, either directly by affecting soil organisms or indirectly as a result of chemical changes. It is difficult, however, to assess the relative influence of fire on soil biological processes. Some soil organisms are killed in ground fires, thus the species composition of the soil fauna and microbial population may change. This could indirectly influence soil productivity by affecting the microbial population of the soil (i.e., mycorrhizal fungi), thus altering the rate of nutrient release from the soil.

In the short-term, fires can increase nutrient availability by releasing those tied up in soil organic matter (humus). This is probably beneficial; however, under certain conditions (e.g., relatively infertile soils), organic matter and nutrients may be lost that could decrease the long-term productivity of the soil (7). Fire is rarely a uniform event, and generally, large burns have a variety of intensities across the landscape. Thus, generalizations are not entirely appropriate although fire disturbance is considered essential to the rejuvenation of most forest ecosystems through the release of nutrients found in humus and renewal of forest stands.

Encyclopedia of Soil Science

Mass Wasting

Mass wasting (e.g., landslides. snow avalanches) is a natural process in mountainous areas and steep terrain. Generally, mass wasting events tend to have longer-lasting effects on soil productivity, because surface soil is often removed or severely disturbed (4). The soils on upper portions of slides generally are shallower and coarser with more bedrock exposure than the original soils. The lower portion of slides consists mainly of materials deposited or mixed with original soil. Additionally, for several years after failure, slides are susceptible to surface erosion. Revegetation of slide areas generally occurs on modified rather than primary soil materials. Productivity of these areas is reduced although in many cases, colonizing plant species are similar to adjacent areas.

Surface Erosion

Surface erosion would be expected to be low in natural forests. Significant surface erosion may result from natural disturbances such as landslides, intense fires, and anthropogenic disturbances which remove vegetative cover and expose mineral soil. For example, many Amazonian soils are susceptible to erosion following deforestation, because rainfall is often intense and the terrain is steep (8).

Windthrow

Gaps are the most obvious result of windthrow, but soils can also be affected (9) by altering structure and mixing subsurface soil with surface material. In some tropical forests (e.g., seasonally dry, semi-deciduous), small-scale gaps are the dominant canopy disturbances between large-scale events such as hurricanes and fire (10).

Mixing of soil may affect nutrient availability, and soil-forming processes, as well as expose a variety of substrates that may negatively affect surface soil chemical and physical properties. In boreal forests long-term effects of windthrow may persist for 100 to 200 years.

Insects and Disease

Insects and disease indirectly accelerate or change soil nutrient dynamics. These processes include 1) altering the rate and amount of nutrients leached or deposited as litter; 2) changing light intensities; 3) reducing competition among plants; 4) altering plant species composition; and 5) stimulating translocation of nutrients from boles and branches to high turnover components such as leaves, buds, and flowers (11). Insect grazing occurs more slowly than obvious disturbances such as fire or forest harvesting. Similarly, pathogens (i.e., disease organisms) also differ from large-scale disturbances by selectively eliminating less vigorous individuals (12). The influence of insects and diseases, however, is similar to large-scale disturbances in that they recycle nutrients and renew forest stands within the landscape. Pathogens, and to some extent insects also can increase tree susceptibility to blow down which will affect soil productivity as outlined previously.

ANTHROPOGENIC (HUMAN-INDUCED) DISTURBANCES

Harvesting

The most common human-related activity is harvesting and its associated practices. The influence on soil productivity will vary greatly among species (forest type), time of harvest, and inherent soil properties.

Biomass removal is the main pathway for nutrient loss from harvesting. Losses will depend on the type of harvest (e.g., whole tree versus boles only), harvest system used (e.g., clear cut versus selective logging), forest type, and time of year. Generally, the influence of forestry practices on nutrient loss is long-term, and harvesting rotation times that are less than the time needed to replace lost nutrients will eventually reduce soil productivity. The primary N sources in forest soils are biological fixation or deposition from the atmosphere. Continuing availability of most nutrients other than N, however, is largely dependent on weathering of parent material. In most temperate and boreal forests, loss of N and soil processes related to N availability is the most critical. In some temperate and boreal forests, particularly those affected by acid deposition, the loss of cations, particularly calcium (Ca), also may reduce soil productivity (13).

Some tropical forests, such as tropical montane forests, appear to function similar to temperate and boreal forests with respect to N (14). However, in highly weathered clay soils of the tropics (e.g., oxisols and ultisols), there is little replacement of nutrients lost (15). In these forests (i.e., lowland rain forests), changes in the phosphorus P cycle (and possibly Ca) following harvesting may be the most critical process affecting soil productivity.

Increased leaching (less nutrient uptake and lower evapotranspiration) is another pathway for nutrient loss following harvest. In most studies of temperate and boreal forests, nutrient loss from increased leaching in harvesting does not appear to have a major influence on soil productivity in the short term (16).

Deforestation through anthropogenic fire (i.e., slash-and-burn agriculture) and land clearing is a special type of harvesting disturbance of particular concern in tropical forests. This phenomenon has contributed to reduced productivity across a number of tropical soil types and forest ecosystems including the Amazon (17). Following land clearing involving fire, loss of organic matter and nutrients is generally more rapid and complete in tropical forests than in temperate or boreal forests (18). This is believed to be caused by a combination of the climate, easily decomposable organic matter, and reduced organic inputs. In nutrient-poor ecosystems, disruption of the nutrient conservation processes will quickly result in reduced productivity following land clearing. Nutrient-rich sites do not lose their potential productivity in the short-term, but the long-term implications are still a concern (19).

Other important factors affecting soil productivity in harvesting are soil compaction and, displacement, and organic matter loss. Altered soil physical properties from forestry practices represent deviations from the natural range of soil conditions, usually with negative effects on soil productivity (16). Physical disturbances associated with harvesting, including roads, skid trails, and landings, may result in soil compaction, soil displacement and mixing, exposure of unfavorable subsurface material, and organic matter removal.

Soil compaction is a concern in virtually all forest ecosystems (20). Tracked areas generally have lower soil productivity because of soil compaction and possibly soil displacement (including removal of organic matter). Generally, the loss of productivity is related to the amount of compaction. Compaction affects soil structure, aeration, water infiltration, runoff, and surface erosion that will reduce soil productivity. Recovery of compacted soils is dependent on several factors and has been found to vary from several years to several decades in boreal, temperate, and tropical forests (16, 20). In contrast, tree growth has been found to be higher on displaced material such as berms and sidecasts compared with the harvested undisturbed soils. This has been related to improved edaphic conditions (e.g., aeration, porosity), increased soil organic matter, and lack of vegetative competition.

Site Preparation

Site preparation is used to produce more appropriate seedbeds for seedlings. In the short term, soil productivity may be improved. However, potential detrimental factors as a result of organic matter and nutrient loss could have long-term effects. The severity of disturbance is related to the degree of organic matter and nutrient losses as a result of mechanical site preparation (7).

Acid Deposition

Acid deposition is a large-scale (i.e., regional or greater) disturbance that may influence soil productivity. Acid deposition is decreasing in many parts of the world; however, it remains a concern in most of Europe and eastern North America, and is of increasing concern in many countries such as China. In high-deposition areas, soil acidification, reduced base cations in surface soils, increased Al, and nutrient imbalances, (particularly N saturation) have been observed. The loss in soil productivity will depend on the amount of soil acidification and the inherent physical and chemical properties. For example, the influence of acid deposition on the soil productivity of a sandy soil with low cation exchange capacity will probably be greater than on a fine-textured soil with high cation exchange capacity.

Climate Change

The spatial scale of climate change is similar to acid deposition; however, the potential influence of climate change on soil productivity is largely unknown. Changing climatic conditions may directly influence biological and chemical processes, and thus influence soil productivity. Indirect effects include changing the frequency and degree of natural disturbances. For example, if drought conditions increase, this may indirectly influence soil productivity by altering the frequency and severity of other disturbance agents such as fire and insect outbreaks (4).

SUMMARY

Disturbances are important processes that affect forest ecosystems. Increasingly, human activities have resulted in changes to the frequency, type, and degree of disturbance on forest ecosystems. Disturbances influence the physical, chemical, and biological properties of soil, and changes to these properties subsequently affect soil productivity (21). The difficulties in assessing changes associated with disturbances is that forest ecosystems are evolving and not static; available information is usually based on short-term results that may not be indicative of long-term effects (16, 22). Thus, a fundamental understanding of the soil physical, chemical, and biological processes is needed to assess the consequences of natural or anthropogenic disturbances on soil productivity.

P

REFERENCES

1. Pickett, S.T.A., White, P.S., Eds. *The Ecology of Natural Disturbances and Patch Dynamics*; Academic Press: Orlando, Florida, 1985.
2. Averill, Robert. D.; Larson, Louise; Saveland, Jim; Wargo, Philip; Williams, Jerry. Disturbance Processes and Ecosystem Management. USDA Forest Service. [Online] Available: http://www.fs.fed.us/research/disturb.html. [Accessed 07 April 1996].
3. Kimmins, J.P. Importance of Soil and Role of Ecosystem Disturbance for Sustained Productivity of Cool Temperate and Boreal Forests. Soil Sci. Soc. Am. J. **1996**, *60* (6), 1643–1654.
4. Rogers, Paul. *Disturbance Ecology and Forest Management: A Review of the Literature*, General Technical Report INT-GTR-336, USDA Forest Service Intermountain Research Station: Ogden, Utah, 1996.
5. Attiwill, Peter M. The Disturbance of Forest Ecosystems: The Ecological Basis for Conservation Management. For. Ecol. Manage. **1994**, *63*, 247–300.
6. Sanford, Robert L., Jr.; Saldarriaga, Juan.; Clark, Kathleen E.; Uhl, Christopher.; Herrera, Rafael. Amazon Rain-Forest Fires. Science **1985**, *227* (4682), 53–55.
7. Prescott, Cindy E.; Maynard, Doug G.; Laiho, Raija. Humus in Northern Forests: Friend or Foe? For. Ecol. Manage. **2000**, *133* (1-2), 23–36.
8. Richter, D.D.; Babbar, L.I. Soil Diversity in the Tropics. Adv. Ecol. Res. **1991**, *21*, 315–389.
9. Ulanova, Nina G. The Effects of Windthrow on Forests at Different Spatial Scales: A Review. For. Ecol. Manage. **2000**, *135* (1–3), 155–167.
10. Dickinson, M.B.; Whigham, D.F.; Hermann, S.M. Tree Regeneration in Felling and Natural Treefall Disturbances in a Semideciduous Tropical Forest in Mexico. For. Ecol. Manage. **2000**, *134* (1–3), 137–151.
11. Mattson, William J.; Addy, Norton D. Phytophagous Insects as Regulators of Forest Primary Production. Science **1975**, *190* (4214), 515–522.
12. Castello, John D.; Leopold, Donald J.; Smallidge, Peter J. Pathogens, Patterns and Processes in Forest Ecosystems. BioScience **1995**, *45* (1), 16–24.
13. Adams, M.B.; Burger, J.A.; Jenkins, A.B.; Zelazny, L. Impact of Harvesting and Atmospheric Pollution on Nutrient Depletion of Eastern US Hardwood Forests. For. Ecol. Manage. **2000**, *138* (1–3), 301–319.
14. Tanner, E.V.J.; Vitousek, P.M.; Cuevas, E. Experimental Investigation of Nutrient Limitations of Forest Growth on Wet Tropical Mountains. Ecology **1998**, *79* (1), 10–22.
15. Vitousek, P.M.; Sanford, R.L., Jr. Nutrient Cycling in Moist Tropical Forest. Ann. Rev. Ecol. Syst. **1986**, *17*, 137–167.
16. Grigal, David F. Effects of Extensive Forest Management on Soil Productivity. For. Ecol. Manage. **2000**, *138* (1–3), 167–185.
17. Laurance, William F.; Cochrane, Mark A.; Bergen, Scott.; Fearnside, Phillip M.; Delamônica, Patricia.; Barber, Christopher.; D'Angelo, Sammya.; Fernandes, Tito. The Future of the Brazilian Amazon. Science **2001**, *291* (5503), 438–439.
18. Scholes, R.J.; van Breemen, N. The Effects of Global Change on Tropical Ecosystems. Geoderma **1997**, *79* (1–4), 9–24.
19. Jordan, Carl F.; Herrera, Rafael. Tropical Rain Forests: Are Nutrients Really Critical? Am. Nat. **1981**, *117* (2), 167–180.
20. Kozlowski, T.T. Soil Compaction and Growth of Woody Plants. Scand. J. For. Res. **1999**, *14* (6), 596–619.
21. Grigal, David F.; Vance, Eric D. Influence of Soil Organic Matter on Forest Productivity. New Zealand J. For. Sci. **2000**, *30* (1/2), 169–205.
22. Worrell, R.; Hampson, A. The Influence of Some Forest Operations on the Sustainable Management of Forest Soils—A Review. Forestry **1997**, *70* (1), 61–85.

PROPERTIES AND PRODUCTIVITY OF SOIL: TILLAGE EROSION

Tom E. Schumacher
South Dakota State University, Brookings, South Dakota, U.S.A.

INTRODUCTION

The relation between soil productivity and soil erosion is complex. Soil erosion is defined as the wearing away of the land surface by natural or anthropogenic agents that abrade, detach, and remove soil from one point on the earth's surface and deposit it elsewhere (1). The capacity of a soil to produce a certain yield of crops or other plants within a specified system of management is defined as soil productivity (1). Soil productivity is dependent on the ability of the soil to provide water, nutrients, and oxygen to the plant root system as well as limiting exposure of the root system to toxic gases or solutions. The processes involved in the supply of beneficial and toxic compounds are dependent on storage capacities and transport rates that are determined by an array of soil properties. Soil and topographic properties that control the inherent productivity of a specific field location include slope, effective root zone depth, bulk density, available water holding capacity, plant available nutrient content, soil aeration, water infiltration and percolation, soil pH, aluminum and manganese concentration, root zone cation exchange capacity, aggregate stability, salinity, and nutrient cycling by soil biological organisms (2–4). As soil erodes, soil properties of the root zone often are altered. Consequently storage capacity and transport rates of compounds critical for plant growth also change. Several reviews have been written on various aspects of the subject of soil erosion, soil properties, and productivity (5–7).

Tillage erosion, the downslope displacement of soil through the action of tillage operations (1), is a specific form of soil erosion. The impact of tillage erosion on soil productivity is primarily related to soil removal from a specific landscape position and deposition in another part of the landscape. As a result many of the causes of change in soil productivity attributed to other forms of erosion also apply to tillage erosion. Lal (8) lists several direct effects of soil erosion on crop yield. These include a reduction in effective rooting depth, loss of plant nutrients, loss of available plant water, loss of land area (due to gully formation and unfavorable conditions in depositional areas), and damage to seedlings from the action of the erosion agent. Of these, tillage erosion acts on soil productivity through the first three: loss of effective rooting depth, loss of plant nutrients, and loss of plant available water.

TILLAGE EROSION AND SOIL PROPERTIES

Soil productivity changes due to tillage erosion at eroding landscape positions are dependent on properties of the underlying soil horizons and the severity of soil loss. Lal (8) describes three scenarios of erosion-induced change in soil productivity that depend on subsoil properties. Some soils may have no change in soil properties affecting crop growth. These soils usually are deep soils with relatively uniform soil properties in and below the root zone. A second scenario involves rare cases where soil properties in subsurface soil horizons are more favorable for plant growth than in the A horizon. In these soils, soil productivity actually increases with erosion. A third scenario, which is by far the most common in agricultural fields, are soils that drop in productivity as soil is removed from the surface. These soils have subsurface horizons with soil properties that are less beneficial to plant growth than soil in the A horizon.

Tillage erosion changes soil properties and thus soil productivity by at least two mechanisms. One mechanism involves the net loss of topsoil from field locations where the slope forms a convex shape such as the shoulder landscape position. A second mechanism involves the mixing of topsoil with subsoil exposed by tillage erosion as soil is translocated by the tillage operation beyond the eroding landscape position.

Over a period of time, tillage erosion can result in exposure of subsoil. Tillage operations spread and mix subsoil with topsoil as soil is translocated downslope even though there may be no net loss of soil in the translocation zone (9, 10).

This mixing of topsoil with subsoil spread by tillage equipment can result in a change in soil properties that reduces soil productivity in adjacent noneroding landscape positions. The subsoil materials over time will become mixed from the zone of soil loss (convex slopes) to the zone of deposition (concave slopes).

LANDSCAPE POSITION AND SOIL PRODUCTIVITY

The effects of different erosion agents on soil productivity can often be better differentiated by their location of action within the landscape than by unique changes in soil properties. Tillage erosion occurs in topographic positions that are convex such as shoulder positions (11). Water erosion primarily occurs in mid to lower portions of a backslope (12). Wind erosion occurs primarily on coarse textured soils and is most severe on shoulder positions facing into the prevailing wind (13). Tillage erosion appears to be responsible for most of the soil erosion occurring in shoulder positions where wind erosion is not a predominate agent of erosion (14). Therefore soil productivity losses observed in shoulder positions and on knolls in humid and semi-humid regions are likely the result of tillage erosion.

The shoulder position prior to agriculture often has lower solum thickness (15). As a result changes in soil productivity from tillage erosion may occur more rapidly and with greater intensity per unit of soil erosion because the soil tends to be naturally shallower in the shoulder position. Figure 1 illustrates a 32 ha field from central South Dakota showing variation in winter wheat (*Triticum aestivum* L.) yield, predicted soil gain/loss from moldboard plow tillage, and available phosphorous based on the Olsen P test (16). Lower yields in this field tended to be associated with field areas that had high rates of tillage erosion, lower plant available phosphorous, and exposed subsoil material.

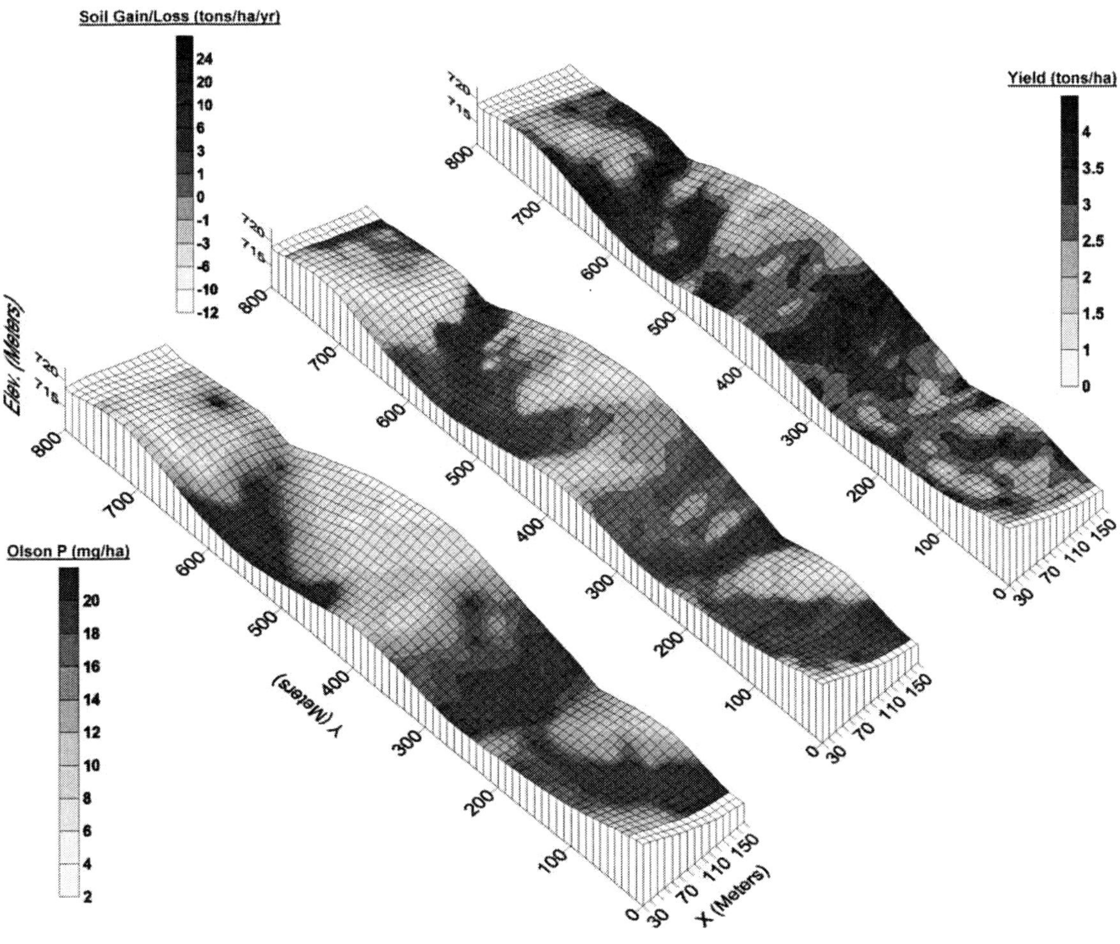

Fig. 1 Predicted soil loss from tillage erosion, winter wheat yield, and Olsen P test from a 32 ha field in central South Dakota. Soil gain (deposition) is represented by positive values in the middle graph, while soil loss (erosion) is represented by negative values.

MODELING TILLAGE EROSION AND SOIL PRODUCTIVITY

The complexity of the soil erosion-productivity relationship is such that few generalizations can be made. The mechanisms by which tillage erosion impacts soil productivity will change depending on soil and climate. Site-specific soil property measurements can be very expensive and time consuming if applied to large scale fields. Modeling erosion events, erosion-induced changes in soil properties within the root zone, and soil property-productivity relationships is one approach that may be used to evaluate fields. In one study, a tillage erosion model (17) and water erosion model (18) were used to evaluate the effects of erosion on soil property distribution in a closed depressional landscape (19). The erosion-induced change in soil properties was then coupled with a productivity index (20), a model that relates soil properties to soil productivity. Model results showed that considerable variability in soil productivity was introduced into the landscape by the combined action of tillage and water erosion.

In most cases soil erosion is due to an interaction of various erosive agents such as tillage and water, or tillage and wind, or a combination of all three agents. An evaluation of soil productivity within landscapes must take into account the interaction of all erosive agents on soil properties, and the specific effect of soil properties as related to supply of water, oxygen, and nutrients to the root systems of growing plants.

REFERENCES

1. Soil Science Society of America. *Glossary of Soil Science Terms*; SSSA: Madison, WI, 1997; 134.
2. Letey, J. Relationship Between Soil Physical Properties and Crop Production. Advances in Soil Science **1985**, *1*, 277–294.
3. Pierce, F.J.; Lal, R. Monitoring Soil Erosion's Impact on Crop Productivity. In *Soil Erosion Research Methods*; Lal, R., Ed.; Soil Water Conservation Society: Ankeny, IA, 1994; 235–264.
4. Shaffer, M.J.; Schumacher, T.E.; Ego, C.L. Simulating the Effects of Erosion on Corn Productivity. Soil Sci. Soc. Am. J. **1995**, *59*, 672–676.
5. Follet, R.F., Stewart, B.A., Eds. *Soil Erosion and Crop Productivity*; Am. Soc. Agron.: Madison, WI, 1985; 533.
6. Lal, R. Soil Erosion Impact on Agronomic Productivity and Environment Quality. Critical Reviews in Plant Sciences **1998**, *17*, 319–464.
7. Lal, R. Effects of Soil Erosion on Crop Productivity. CRC Critical Reviews in Plant Science **1987**, *5*, 303–367.
8. Lal, R. Monitoring Soil Erosion's Impact on Crop Productivity. In *Soil Erosion Research Methods*; Lal, R., Ed.; Soil Conservation Society: Ankeny, IA, 1988; 187–200.
9. Lobb, D.A.; Kachanoski, R.G. Modeling Tillage Translocation Using Step, Linear-Plateau, and Exponential Functions. Soil Tillage Research **1999**, *51*, 317–330.
10. Sibbesen, E.; Skjoth, F.; Rubaek, G.H. Tillage Caused Dispersion of Phosphorous and Soil in Four 16-year Old Field Experiments. Soil Tillage Research **2000**, *54*, 91–100.
11. Govers, G.; Vandaele, K.; Desmet, P.; Poesen, J.; Bunte, K. The Role of Tillage in Soil Redistribution on Hillslopes. European J. Soil Science **1994**, *45*, 469–478.
12. Young, R.A.; Mutchler, C.K. Effect of Slope Shape on Erosion and Runoff. Trans. ASAE **1969**, *12*, 231–233, 239.
13. Chepil, W.S.; Siddoway, F.H.; Armburst, D.V. Wind Erodibility of Knolly Terrain. J. Soil Water Conservation **1964**, *19*, 179–181.
14. Govers, G.; Lobb, D.A.; Quine, T.A. Tillage Erosion and Translocation: Emergence of a New Paradigm in Soil Erosion Research. Soil Tillage Research **1999**, *51*, 167–174.
15. Hall, G.F. Pedology and Geomorphology. Pedogenesis and Soil Taxonomy: Concepts and Interactions. In *Developments in Soil Science*; Wilding, L.P., Smeck, N.E., Hall, G.F., Eds.; 11A; Elsevier: Amsterdam, 1983; 117–138.
16. Bly, A.; Gerwing, J.; Gelderman, R.; Mitchel, V. Influence of Starter P and Zn Fertilizer on Yield of No-till Corn Across Strips with Residual P Rates. Soil Progress Report 97–16, Soil Water Science Research 1997 Annual Report. South Dakota State University: Brookings, SD, 1997; 1–3.
17. Lindstrom, M.J.; Schumacher, J.A.; Schumacher, T.E. TEP: A Tillage Erosion Prediction Model to Calculate Soil Translocation Rates from Tillage. J. Soil and Water Conservation **2000**, *55*, 105–108.
18. Flanagan, D.C.; Nearing, M.A. *USDA—Water Erosion Prediction Project Hillslope Profile and Watershed Model Documentation*, NSERL Report No. 10, USDA-ARS National Soil Erosion Research Laboratory: West Lafayette, Indiana, 1995.
19. Schumacher, T.E.; Lindstrom, M.J.; Schumacher, J.A.; Lemme, G.D. Modeling Spatial Variation in Productivity Due to Tillage and Water Erosion. Soil Tillage Research **1999**, *51*, 331–339.
20. Pierce, F.J.; Larson, W.E.; Dowdy, R.H.; Graham, W.A.P. Productivity of Soils: Assessing Long-Term Changes Due to Erosion. J. Soil Water Conservation **1983**, *38*, 39–44.

PROTOZOA

B.S. Griffiths
Scottish Crop Research Institute, Invergowrie, Dundee, United Kingdom

P

INTRODUCTION

Protozoa are single-celled eukaryotic organisms (1) commonly subdivided into four groups: flagellates, naked amoebae, shelled or testate amoebae (Testacea), and ciliates (Fig. 1). Naked amoebae are generally the most important group of protozoa in soil (2). Protozoa require a water film for locomotion and feeding, so their activity is limited to the water-filled pore space in soil. Most protozoa have adapted to fluctuating moisture by the ability to form resistant cysts, but ciliates require much higher soil moisture than flagellates or amoebae to be active (3, 4). Reproduction is usually asexual by means of binary fission, but sexual reproduction also occurs.

IMPORTANCE IN SOIL, PLANT, AND ENVIRONMENTAL PROCESSES

Protozoa play a substantial part in energy and nutrient flows of terrestrial ecosystems, consuming 22% of the C input to a beech-forest soil and 63% of the microbial biomass (5). Annual production is 39–98% of that of the total soil fauna (6), and their contribution to total net nitrogen mineralization in soil is 12–30% (7). Protozoa in soil systems regulate and modify microbial community activity and composition, and inoculate new substrates with micro-organisms (8, 9). Their effects result mainly from their feeding activities because they are too small to directly influence soil structure. Soil protozoa are heterotrophic, feeding on bacteria, although mycophagous (10), predatory, and saprophytic protozoa occur (1). Approximately 40% of ingested nutrients are used for the production of protozoan biomass, while 60% are excreted (11). The effects of this nutrient turnover are such that plants are significantly smaller and contain less nitrogen if protozoa are not present (2, 8, 9, 11). Protozoa are a key link in the terrestrial food chain as a major predator of bacteria and prey for larger organisms.

FACTORS AFFECTING PROTOZOAN ACTIVITY AND COMPOSITION

The impact of environmental and land management factors on protozoa results primarily from their effects on substrate supply, and secondarily on soil structure, moisture, and climatic factors. Protozoa decrease with soil depth, from approximately $10^6\,g^{-1}$ at 10 cm to $10^1\,g^{-1}$ below 1 m, resulting from decreasing levels of organic matter down the soil profile (12). The application of organic manure to soil is an obvious source of added substrate, and increases protozoan populations (13). Inorganic fertilizers also increase protozoan populations, because of the increased plant growth, leading to a greater substrate input to the soil (13). The application of elemental sulphur, however, increased acidification of a Canadian pasture and reduced numbers of soil micro-organisms and protozoa (14). Protozoa in soil are characterized by very rapid changes in density—amoebal numbers increase within two days of a rain due to a stimulation of bacterial prey (2). Seasonal maxima can be related to food supply, such as the spring and autumn peaks of *Dictyostelium mucoroides* (15) and the peak of Testacea following autumn leaf fall (16). Population minima are typically related to unfavorable environmental conditions, such as dry summers and low temperature (12). Protozoa can rapidly encyst/excyst (17), such that 100% of the protozoa were active six days after watering a desert soil, compared to 0% before (18). Protozoa have a clumped horizontal distribution, again linked to substrate availability in the rhizosphere and in decomposing organic matter (19). Rhizosphere populations are, on average, six-fold higher than in the surrounding soil (12). A 90-fold increase on decomposing grass residues in soil has been demonstrated (20). Protozoan species can be used as biological indicators of the environment (1). There were, for example, marked differences in the species of testate amoebae among mor, acid-mull, and calcareous-mull litter types (21), and the ratio of *Colpodea* to *Polyhymenophora* is an indicator of soil disturbance (22). Soil protozoa are also used in laboratory bioassays to monitor pesticides and heavy metals (23).

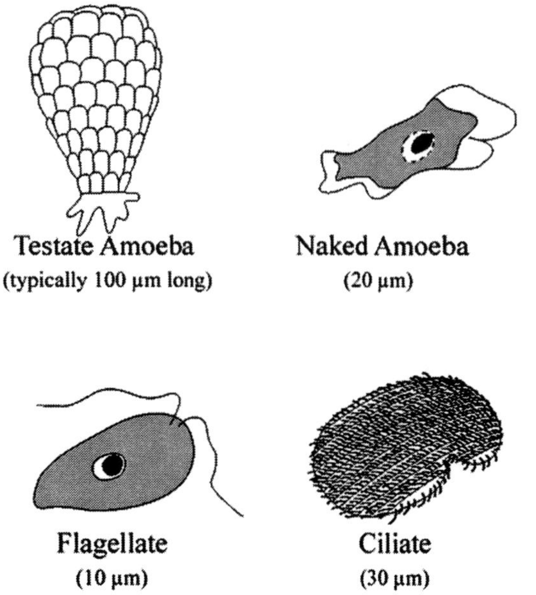

Fig. 1 Diagram of the four main groups of soil protozoa.

METHODS OF STUDYING PROTOZOA

The study of protozoa in soil is hampered by their relatively small size, low numbers, and variable morphology. Direct enumeration has been most successfully used for testate amoebae (24) and ciliates (25). Density-gradient centrifugation has enabled the direct observation of metabolically active protozoa (26). Culturing techniques rely on protozoan growth on laboratory media. A soil suspension is generally diluted using the most probable number (MPN) technique (27), with modifications to allow the enumeration of mycophagous and active protozoa (10, 28). Technological advances are being adapted to the study of protozoan populations in soil. Thus, a phospholipid fatty acid marker for protozoa has been identified (29), a β-N-acetyl-glucosaminiadase-like enzyme has been identified as a putative marker for protozoan bacterivory (30), while molecular probes (31) and analysis of 18S rRNA genes (32) allow examination of protozoan communities without the need for culturing.

REFERENCES

1. Foissner, W. Soil Protozoa: Fundamental Problems, Ecological Significance, Adaptations in Ciliates and Testaceans, Bioindicators, and Guide to the Literature. Prog. Protistol. **1987**, *2*, 69–212.
2. Clarholm, M. Effects of Plant–Bacterial–Amoebal Interactions on Plant Uptake of Nitrogen Under Field Conditions. Biol. Fertil. Soils **1989**, *8*, 373–378.
3. Zwart, K.B.; Brussaard, L. Soil Fauna and Cereal Crops. In *The Ecology of Temperate Cereal Fields*; Firbank, L.G., Carter, N., Darbyshire, J.F., Potts, J.R., Eds.; Blackwell Scientific Publications: Oxford, 1991; 139–168.
4. Darbyshire, J.F. Effect of Water Suctions on the Growth In soil of the Ciliate *Colpoda Steinii* and the Bacterium *Azotobacter Chroococcum*. J. Soil Sci. **1976**, *27*, 369–376.
5. Meisterfeld, R. The Importance of Protozoa in a Beech Forest Ecosystem. Symp. Biol. Hung. **1986**, *33*, 291–299.
6. Schönborn, W. The Role of Protozoan Communities in Freshwater and Soil Ecosystems. Acta Protozool. **1992**, *31*, 11–18.
7. Elliott, E.T.; Hunt, H.W.; Walter, D.E. Detrital Food–Web Interactions in North American Grassland Ecosystems. Agric. Ecosyst. Environ. **1988**, *24*, 41–56.
8. Darbyshire, J.F., Ed. *Soil Protozoa*; CAB International: Wallingford, England, 1994; 1–209.
9. Bardgett, R.D.; Griffiths, B.S. Ecology and Biology of Soil Protozoa, Nematodes, and Microarthropods. In *Modern Soil Microbiology*; Van Elsas, J.D., Trevors, J.T., Wellington, E.M.H., Eds.; Marcel Dekker: New York, 1997; 129–163.
10. Ekelund, F. Enumeration and Abundance of Mycophagous Protozoa in Soil, with Special Emphasis on Heterotrophic Flagellates. Soil Biol. Biochem. **1998**, *30*, 1343–1348.
11. Griffiths, B.S. Microbial-Feeding Nematodes and Protozoa in Soil: Their Effects on Microbial Activity and Nitrogen Mineralization in Decomposition Hotspots and the Rhizosphere. Pl. Soil **1994**, *164*, 25–33.
12. Schnürer, J.; Clarholm, M.; Rosswall, T. Fungi, Bacteria and Protozoa in Soil from Four Arable Cropping Systems. Biol. Ferti. Soils **1986**, *2*, 119–126.
13. Singh, B.N. The Effect of Artificial Fertilizers and Dung on the Numbers of Amoebae in Rothamsted Soils. J. Gen. Microbiol. **1949**, *3*, 204–210.
14. Gupta, V.V.S.R.; Germida, J.J. Populations of Predatory Protozoa in Field Soils After 5 Years of Elemental S Fertilizer Application. Soil Biol. Biochem. **1988**, *20*, 787–792.
15. Kuserk, F.T. The Relationship Between Cellular Slime Moulds and Bacteria in Forest Soils. Ecology **1980**, *61*, 1474–1485.
16. Lousier, J.D.; Parkinson, D. Annual Population Dynamics and Production Ecology of Testacea (Protozoa, Rhizopoda) in an Aspen Woodland Soil. Soil Biol. Biochem. **1984**, *16*, 103–114.
17. Bryant, R.J.; Woods, L.E.; Coleman, D.C.; Fairbanks, B.C.; Mclellan, J.F.; Cole, C.V. Interactions of Bacterial and Amoebal Populations in Soil Microcosms with Fluctuating Moisture Content. Appl. Environ. Microbiol. **1982**, *43*, 747–752.
18. Parker, L.W.; Freckman, D.W.; Steinberger, V.; Driggers, L.; Whitford, W.G. Effects of Simulated Rainfall and Litter Quantities on Desert Soil Biota: Soil Respiration, Microflora, and Protozoa. Pedobiologia **1984**, *27*, 185–195.
19. Feest, A. The Quantitative Ecology of Soil Mycetozoa. Prog. Protistol. **1987**, *2*, 331–361.
20. Griffiths, B.S.; Caul, S. Migration of Bacterial-Feeding Nematodes, But not Protozoa, to Decomposing Grass Residues. Biol. Fertil. Soils **1993**, *15*, 201–207.

21. Stout, J.D. Some Observations on the Protozoa of Some Beechwood Soils on the Chiltern Hills. J. Anim. Ecol. **1963**, *32*, 281–287.
22. Lüftenegger, G.; Foissner, W.; Adam, H. r-and K-Selection in Soil Ciliates: A Field and Experimental Approach. Oecologia **1985**, *66*, 574–579.
23. Forge, T.A.; Berrow, M.L.; Darbyshire, J.F.; Warren, A. Protozoan Bioassays of Soil Amended with Sewage Sludge and Heavy Metals, Using the Common Soil Ciliate *Colpodas Steinii*. Biol. Fertil. Soils **1993**, *16*, 282–286.
24. Lousier, J.D.; Parkinson, D. Evaluation of a Membrane-Filter Technique to Count Soil and Litter Testacea. Soil Biol. Biochem. **1981**, *13*, 209–213.
25. Lüftenegger, G.; Petz, W.; Foissner, W.; Adams, H. The Efficiency of Direct Counting the Numbers of Microscopic Soil Animals. Pedobiologia **1988**, *31*, 95–102.
26. Griffiths, B.S.; Ritz, K. A. Technique to Extract, Enumerate and Measure Protozoa from Mineral Soils. Soil Biol. Biochem. **1988**, *20*, 163–173.
27. Rønn, R.; Ekelund, F.; Christensen, S. Optimizing Soil Extract and Broth Media for MPN-Enumeration of Naked Amoebae and Heterotrophic Flagellates in Soil. Pedobiologia **1994**, *39*, 10–19.
28. Rutherford, P.M.; Juma, N.G. Influence of Soil Texture on Protozoa-Induced Mineralization of Bacterial Carbon and Nitrogen. Can. J. Soil Sci. **1992**, *72*, 183–200.
29. Frostegård, Å.; Petersen, O.; Bååth, E.; Nielsen, T. Dynamics of a Microbial Community Associated with Manure Hotspots as Revealed by Phospholipid Fatty Acid Analysis. Appl. Environ. Microbiol. **1997**, *63*, 2224–2231.
30. Vrba, J.; Ŝimek, K.; Nedoma, J.; Hartman, P. 4-Methylumbelliferyl-β-N-Acetylglucosaminide Hydrolysis by a High-Affinity Enzyme, a Putative Marker of Protozoan Bacterivory. Appl. Environ. Microbiol. **1993**, *59*, 3091–3101.
31. Zarda, B.; Mattison, G.; Hess, A.; Hahn, D.; Höhener, P.; Zeyer, J. Analysis of Bacterial and Protozoan Communities in an Aquifer Contaminated with Monoaromatic Hydrocarbons. FEMS Microbiol. Ecol. **1998**, *27*, 141–152.
32. Van Hannen, E.J.; Mooij, W.; Van Agterveld, M.P.; Gons, H.J.; Laanbroek, H.J. Detritus-Dependant Development of the Microbial Community in an Experimental System: Qualitative Analysis by Denaturing Gradient Gel Electrophoresis. Appl. Environ. Microbiol. **1999**, *65*, 2478–2484.

QUALITY

E.G. Gregorich
Agriculture and Agri-Food Canada, Central Experimental Farm, Ottawa, Ontario, Canada

INTRODUCTION

Concisely defined, soil quality is the degree of fitness of a soil for a specific use—its ability or capacity to function for a specific purpose (1, 2). Depending on this use, soil quality may mean different things to different people. For example, in the agricultural context, within which much of the discussion of soil quality has taken place, the farmer may think of soil quality in terms of productive land that maintains or increases farm profitability, as well as the conservation of farm resources so that future farming generations can make a living. To the consumer, agricultural soil quality may mean a dependable supply of wholesome, affordable food and the protection of drinking water quality. To the environmentalist, it may be defined in terms of global nutrient cycling, water regulation, climate change, and biodiversity. And for high-level policy- and decision-makers, soil quality may be seen in terms of a nation's ability to stay competitive in the international marketplace and to honor its international environmental commitments.

BASIC SOIL FUNCTIONS

Soil performs six main functions: sustaining plant and animal life, regulating water, regulating gases, regulating energy, buffering against or filtering out potential environmental contaminants, and providing physical support for human structures such as buildings and roads. How well a soil carries out these functions determines its quality.

INHERENT AND DYNAMIC SOIL QUALITY

Soil quality has both an inherent or natural element determined by geological materials and soil formation processes (such as chemical and physical weathering and a dynamic element determined by management practices. Natural processes, such as erosion and the subsequent loss of organic matter, compaction, and salinization can degrade the inherent quality of soil. Human activity, such as agriculture, can accelerate these processes through various land use and management practices, hastening the symptoms and effects of soil degradation. On the other hand, some agricultural land uses and practices (such as various tillage methods, cropping systems, and nutrient management plans) help to stabilize or improve soil quality.

SUSTAINABLE DEVELOPMENT

Current interest in soil quality is linked to sustainable development. A soil management system is considered sustainable only if it maintains or improves soil quality and does not compromise environmental quality beyond a degree acceptable to society. Only by understanding soil quality and promoting it now can we achieve environmental protection and preserve soils in a usable state for future generations. Regular monitoring of soil quality allows land managers to assess the sustainability of various land uses and management practices.

DEVELOPING A FRAMEWORK TO ASSESS SOIL QUALITY

Soils are the interface between terrestrial ecosystems and both aquatic ecosystems and the atmosphere. Based on the linkages between ecosystems, soil quality is normally defined using environmental and economic performance criteria, including those of plant and animal health and productivity. Soil quality and soil health are often considered synonymous, but some researchers make a distinction between these terms by considering soil health as a subset of soil quality and by stressing that the focus of soil health is on the organic and more dynamic soil properties. This latter concept expands to a larger view of soils as a key component for sustainably supporting all terrestrial ecosystems.

A useful framework for assessing soil quality begins with a description of the functions on which such quality is

Encyclopedia of Soil Science

to be based (i.e., those most likely to influence the performance criteria). For example, at the ecosystem or global scale, soil functions include regulating both biotic processes (e.g., supplying plants with mineral nutrients and water) and the flux of elements (e.g., the turnover and storage of C, N, P, and S); mitigating the accumulation of greenhouse gases (e.g., CO_2, NO_x, and CH_4); altering the chemical composition of precipitation and distributing water throughout the environment; contributing to the gas, water, and heat balance of the atmosphere; and serving as a biological habitat and genetic reserve (3, 4).

Once soil functions are known, soil characteristics or properties that support these functions, along with important interactions of these characteristics, must be identified (5). Much of the work to characterize soil quality has been carried out for agricultural soils. Table 1 lists some of the physical, chemical, and biological characteristics and processes of agricultural soils that contribute to the soil quality functions that support agricultural productivity and sustainability.

EVALUATING SOIL QUALITY

It is not feasible or even practical to measure all soil properties in order to assess the soil quality. Soil quality indicators are needed that encompass the whole range of physical, chemical, and physical properties of the soil; reflect soil function; are easy to measure for a variety of users and under various field conditions; and respond to changes in climate and management. The concepts underlying the criteria to select key indicators of soil quality, along with the interpretative framework needed to link the response of the measured indicator to management action and the issue of concern are described in another section of the encyclopedia. Assessment of the indicators is critical in determining whether a soil process is unrestricted or a soil is fit for use. The approaches used to set quantitative critical limits or ranges of soil indicators (which help determine whether the value of the indicator adequately describes soil performance) is discussed in another section of the encyclopedia.

Land management practices used in agricultural production usually result in a degradation of the structures, or a perturbation of the processes, that help to maintain natural ecosystems in equilibrium. The inherent and dynamic aspects of soil quality that help accurately assess sustainable crop and animal production, and the link between soil quality and development of sustainable management systems (i.e., those that foster reduction in the inputs of nonrenewable resources, maintain acceptable levels of productivity, and minimize impact on the environment) are described in other sections of the encyclopedia.

The framework used to interpret and assess soil quality must also be useful in evaluating effects in adjacent areas or systems. A description of soil properties important to the process of erosion and the consequences of erosion on soil functions, the relationship between soil quality and the quality of receiving waters, and the effects of both soil management and soil quality on the soil processes that produce and consume carbon and nitrogen gases are discussed other sections of the encyclopedia.

Table 1 Some soil quality functions that are important for crop production

Soil quality functions	Function characteristics and processes involved
Support plant growth	Provide suitable medium for seed germination and root growth
	Show absence of adverse chemical conditions (e.g., acidity, salinity, sodicity)
	Supply balance of nutrients
	Provide suitable medium for microorganisms (for decomposition, nutrient cycling)
	Promote root growth and development
Regulate water	Receive, store, and release water for plant use
	Retain water adequately to buffer and reduce effects of drought
	Provide adequate infiltration and storage capacity to reduce runoff
Regulate gases	Accept, hold, and release gases
	Have conditions for adequate air exchange with atmosphere
Regulate energy	Store energy-rich organic matter
	Decompose, mineralize, and recycle nutrients and energy
Buffer or filter	Accept, hold, and release nutrients
	Sequester biotoxic elements

(Adapted from Ref. 5.)

MONITORING SOIL QUALITY

In natural systems, soil quality is observed as a baseline value or set of values against which future changes in the system can be analyzed and compared. In agricultural systems soil quality is monitored with the view toward managing the system to enhance production while not degrading soils and the environment.

The selection and quantification of properties used to assess soil quality are often based on their spatial and temporal variability, on whether they are desirable for crop production, and on whether they can be controlled or managed in a time frame relevant for annual crop production (i.e., relatively short-term) or for long-term sustainability (6). A set of basic soil properties and characteristics has been proposed as indicators of soil quality (7). Table 2 lists a set of indicators that forms the core of most datasets used to characterize soil quality. From this dataset it is possible to derive other relevant indicators such as available water, microbial quotient, and respiratory quotient. It has been proposed that a suitable level of precision for soil quality indicators is the ability to detect a 10% change at the 90% confidence level (8).

Some indicators are less useful than others because of high temporal and/or spatial variability, because of strong correlation to other indicators, or because the response of the indicators to change is not readily interpreted (8). Selecting an appropriate suite of indicators or adopting surrogate indicators for a particular soil quality assessment is an important first step in monitoring soil quality, and there is a need to balance the cost of measuring particular indicators against the value of the information obtained.

Statistical quality control methods have been proposed to determine the range of variation and to set control limits for

Table 2 Some soil physical, chemical, and biological properties used as indicators of soil quality

Soil property	Method	Information
Physical properties		
Soil texture	Hydrometer method	Retention and transport of water, nutrients, and chemicals; susceptibility to erosion; stabilization of organic matter and soil structure
Topsoil and rooting depth	Soil coring, excavation	Water and nutrient availability, rooting volume for crop production
Soil bulk density	Soil core, neutron probe	Volume of pore space, compaction
Infiltration, hydraulic conductivity	Pressure, tension infiltrometer; permeameter; laboratory measurement on cores	Runoff and leaching potential, drainage
Water content	Gravimetric measurement by oven drying; volumetric measurement by TDR[a] or core	Available water
Water release curve, water-holding capacity	Water content at 33 and 1500 kPa tension	Availability of air and water, retention and transport of water and chemicals, drainage
Chemical properties		
pH	pH meter	Acidity or alkalinity of soil, nutrient availability
Electrical conductivity	Conductivity meter	Presence and quantity of soluble salts
Available N, P, and K	Laboratory extraction and analysis	Plant available nutrients
Organic C and N	Wet or dry combustion	Organic matter reserves, nutrient cycling, soil structure
Biological properties		
Potentially mineralizable N	Aerobic or anaerobic incubation	Potential to supply plant-available N
Microbial biomass	Chloroform fumigation followed by incubation or extraction	Size of microbiological population, pool of rapidly cycling organic matter and nutrients
Soil respiration	In situ and laboratory measurements by CO_2 gas analyzer or alkali traps	Availability of soil organic matter reserves, microbiological activity

[a] Time domain reflectometry.
(Adapted from Ref. 7.)

Table 3 Soil properties that can be estimated from input variables using pedotransfer functions

Soil property	Basic input variables[a]
Cation exchange capacity	Clay type and quantity + organic carbon
Water retention characteristic; hydraulic conductivity; volumetric water content	% Sand, silt, clay + organic carbon + bulk density
Aerobic and anaerobic microbial activity	Water-filled pore space (calculated from bulk density, particle density, and water content)
Soil strength	Bulk density and water content
Rooting depth	Bulk density, water-holding capacity, pH, electrical conductivity
Leaching potential	% Sand, silt, and clay; pH; organic carbon; hydraulic conductivity, cation exchange capacity; depth

[a] Variable used in pedotransfer function.
(Adapted from Ref. 11.)

properties (9) in order to provide the farmer with information to manage the soil in a sustainable manner. The theory behind these methods is that if upper and lower control limits, based on known or desired tolerances, can be established to represent minimum levels for sustainable production, then changes in critical parameters can be identified and corrective action taken to adjust the management system.

When data are unavailable, pedotransfer functions can be used to estimate values of soil characteristics (10, 11; Table 3). Pedotransfer functions are functional relationships (i.e., regression models) that transfer known soil properties (e.g., texture, bulk density, organic C content) into unknown soil properties (e.g., soil hydraulic conductivity, volumetric water content). The main advantage of pedotransfer functions is that they are relatively inexpensive and easy to derive and use. The disadvantage of these functions is that some may not be appropriate for application at a specific site and should not be used to make predictions for soils that are outside the range of soils that were used to derive the pedotransfer functions originally.

REFERENCES

1. Larson, W.E.; Pierce, F.J. The Dynamics of Soil Quality as a Measure of Sustainable Management. In *Defining Soil Quality for a Sustainable Environment*; Special Pub. No. 35, Doran, J.W., Coleman, D.C., Bezedick, D.F., Stewart, B.A. Eds.; American Society Agronomy: Madison, WI, 1994; 37B51.
2. Gregorich, E.G.; Carter, M.R.; Angers, D.A.; Monreal, C.M.; Ellert, B.H. Towards a Minimum Data Set to Assess Soil Organic Matter Quality in Agricultural Soils. Can. J. Soil Sci. **1994**, *74*, 367–385.
3. Ellert, B.H.; Clapperton, M.J.; Anderson, D.W. An Ecosystem Perspective of Soil Quality. In *Soil Quality for Crop Production and Ecosystem Health*; Gregorich, E.G., Carter, M.R., Eds.; Elsevier: Amsterdam, 1997; 115–141.
4. Blum, W.E.H.; Santelises, A.A. A Concept of Sustainability and Resilience Based on Soil Functions. In *Soil Resilience and Sustainable Land Use*; Greenland, D.J., Szabolcs, I., Eds.; CAB International: Wallingford, UK, 1994; 535–542.
5. Carter, M.R.; Gregorich, E.G.; Anderson, D.W.; Doran, J.W.; Janzen, H.H.; Pierce, F.J. Concepts of Soil Quality and Their Significance. In *Soil Quality for Crop Production and Ecosystem Health*; Gregorich, E.G., Carter, M.R., Eds.; Elsevier: Amsterdam, 1997; 1–9.
6. Singer, M.J.; Ewing, S. Soil Quality. In *Handbook of Soil Science*; Sumner, M.E., Ed.; CRC Press: Boca Raton, FL, 2000; G271–G298.
7. Doran, J.W.; Parkin, T.B. Defining and Assessing Soil Quality. In *Defining Soil Quality for a Sustainable Environment*; SSSA Special Publication No 35, 3-21, Doran, J.W., Coleman, D.C., Bezdicek, D.F., Stewart, B.A., Eds.; Soil Science Society of America: Madison, WI, 1994.
8. Schipper, L.A.; Sparling, G.P. Performance of Soil Condition Indicators Across Taxonomic Groups and Land Uses. Soil Sci. Soc. Am. J. **2000**, *64*, 300–311.
9. Pierce, F.J.; Gilliland, D.C. Soil Quality Control. In *Soil Quality for Crop Production and Ecosystem Health*; Gregorich, E.G., Carter, M.R., Eds.; Elsevier: Amsterdam, 1997; 203–219.
10. Wösten, J.H.M. Pedotransfer Functions to Evaluate Soil Quality. In *Soil Quality for Crop Production and Ecosystem Health*; Gregorich, E.G., Carter, M.R., Eds.; Elsevier: Amsterdam, 1997; 221–245.
11. Doran, J.W.; Parkin, T.B. Quantitative Indicators of Soil Quality: A Minimum Data Set. In *Methods for Assessing Soil Quality*; SSSA Special Publication No 49, 25-37, Doran, J.W., Jones, A.J., Eds.; Soil Science Society of America: Madison, WI, 1996.

QUALITY, CRITICAL LIMITS AND STANDARDIZATION

M.R. Carter

Agriculture and Agri-Food Canada, Charlottetown, Prince Edward Island, Canada

INTRODUCTION

Critical ("threshold," "trigger," "baseline," "reference") limits in soil quality assessment refer to the specific value or range of a soil property or indicator that is required to ensure that a soil process or function is not restricted or adversely influenced. Based on the primary concept of "fitness for use" for soil quality, critical limits denote the boundary values or margins of tolerance required for soil indicators that are associated with optimum soil functioning, and indicate if a soil is 'fit' for a specific use. Standardization refers to the technical protocol for the sampling, storage, analysis, and interpretation of soil properties, attributes, and indicators as a means to provide reliable and comparable methodology for soil quality assessment.

Concerns with soil degradation and sustainable soil management in agroecosystems have emphasized the need to define and evaluate soil and land quality (1–3). The basic idea of "fitness for use" in regard to agricultural and/or industrial use of soil, reflected in early attempts to classify "soil suitability" or "land capability," (4, 5) is the basic premise of soil quality. Ecological concepts such as function, processes, attributes, and indicators, provide a useful and logical framework to describe and evaluate soil quality (4). The framework is based on the following sequence: purpose, function, processes, properties/attributes (including critical values), indicators, and methods (including standardization) (Table 1). Selection of suitable attributes and indicators, the establishment of critical limits for these attributes; and standardization related to the protocol and methodology for attribute characterization are key aspects of soil quality evaluation.

CRITICAL LIMITS FOR SOIL QUALITY

A large range of attributes, such as chemical, physical, and biological properties, can be used to describe soil quality. A group of key soil properties/attributes or indicators is called a minimum data set (MDS) (1, 4, 5).

Concept of Critical Limits

The concept of "critical values" (also called "threshold," "trigger," "baseline" or "reference" values) is commonly used in soil pollution studies (6), and is analogous to the "degrees of limitation" for rating land qualities in land use requirements (3). Critical limits are needed for each soil quality property/attribute or indicator, within a MDS, so that the measured value of the property can be classed as "optimum" or "limited" for the specific land use, and to ensure that dependent soil processes and function are not restricted or adversely affected (1). If the value obtained for the selected indicator falls outside the desired range this would raise questions about the fitness of the soil for the specific use, and/or indicate the need for improved management (1).

Selecting Critical Soil Properties and Parameters

Past studies have noted the general relation between soil/land use and soil properties resulting in the development of land capability classes and soil productivity ratings (1, 4, 5). For example, the Canadian Land Suitability Rating System categorizes land into classes based on the limitations (e.g., climate, soil, and landscape restrictions) to productivity of a specific crop. The USDA Land Capability Classes based on the rating (eight classes) of six limitations/hazards (e.g., soil erosion, wetness, depth, salinity, sodicity, and climate) provides a meaningful assessment of soil productivity applicable to a wide range of soil types and agricultural crops (5). In other examples, a MDS of selected soil attributes (e.g., soil nutrient availability, organic matter, particle size, plant available water, structure, strength, rooting depth, pH, and salinity) is considered adequate to quantify and monitor soil quality (1).

Setting Critical Limits

Selection of critical soil properties is followed by the setting of critical limits, or ranges, that help characterize the value of the property as "optimum" or "limited." If

Q

Table 1 Sequential framework to evaluate soil quality for specific purpose or fitness of use

Sequence steps	Sequential framework	Questions implied by the framework
1	Purpose	What will the soil be used for?
2	Functions	What specific role is being asked of the soil?
3	Processes	What key soil processes support each function?
4	Properties/attributes	What are the critical soil properties for each process? What are their critical values or range?
5	Indicators/surrogates/pedotransfer function[a]	When the soil property/attribute is difficult to measure or not available, what indirect or related property or properties can be used in its place?
6	Methodology/standardization	What methods are available to measure the attribute? Technical rules and protocols for soil sampling, handling, storage, analysis, and interpretation of data.

[a] A pedotransfer function is a mathematical function that relates a soil property to another soil property (1, 4).
(Adapted from Ref. 4.)

information on critical values is limited, qualitative reference values can be selected based on soil resource inventories or general consensus (1, 5). An index can then be derived using scores (e.g., 0 to 1) for the value of each property, along with weighting values for the considered importance or priority of each property for soil quality (1). Combination of the scores for a group of selected properties provides a quantitative index as an overall measure of soil quality (5).

General Classes of Critical Limits

For many soil properties, a quantitative range of values (e.g., low, optimum, high) can be derived based on empirically derived relations between the property and soil processes or some aspect of soil function (e.g., plant growth). These values have general application to a wide range of situations. For example, soil hydraulic conductivity (7), macroporosity and aeration (8), bulk density (9), strength or penetration resistance (10), and the least limiting water range (11) have been characterized on the basis of their critical range for plant root development and growth (Table 2). Critical levels and ranges are also available for soil oxygen content (10), salinity and sodicity (7). Examples of desirable values are available for plant nutrients, soil pH, and soil rooting depth (1).

For soil properties that are multi-functional in nature (e.g., organic matter), the setting of critical limits can

Table 2 Critical ranges for specific soil properties in relation to root and plant growth applicable over a range of soil types and conditions

Range	Hydraulic conductivity ($m\,s^{-1}$)	Macropores (> 60 μm diameter) (% soil volume)	Relative bulk density (%)	Penetration resistance (MPa)	Least limiting water range ($m^3\,m^{-3}$)
Limiting (low)	$<10^{-7}$	<5	<80	<0.5	<0.10
Optimum	10^{-7}–10^{-4}	10–20	82–87	0.5–2.0	0.10–0.20
Limiting (high)	$>10^{-4}$	>25	>90	>2.5	—

The low range for hydraulic conductivity and macropore volume, and high range for relative bulk density and penetration resistance, are generally associated with compact soil conditions that inhibit root growth, while the contrasting limiting range is related to poor root-to-soil contact and poor retention of water. A limiting high range does not apply for least limiting water range.
(Adapted from Refs. 7–11.)

Table 3 Critical level or range of soil physical attributes or indicators for optimum root growth of irrigated fruit trees in duplex soils in southeastern Australia

Soil quality attribute	Indicator	Critical level
Soil volume	Soil depth	500 mm
Soil strength	Penetrometer resistance	<0.5 MPa
Soil water parameters	Macropores (>30 μm diam.)	>15%
	Storage pores (30–0.2 μm diam.)	>20%
	Unsaturated hydraulic conductivity (−100 kPa)	$>10^{-4}$ m day^{-1}
Soil aeration	Air-filled porosity (after 24 h drainage)	>15%
Soil temperature		18–25°C

(Adapted from Ref. 14.)

be problematic. Some studies have linked critical values of total soil organic matter for specific processes involved with soil fertility and erodibility (12), while others have attempted to link or elucidate cause–effect mechanisms between soil functions and soil organic matter, and soil microbiological indicators and macrofauna (13).

Site Specific Critical Limits

Although general classes or optimum ranges can be established for many soil attributes in soil quality assessment, it is recognized that the critical ranges must be determined and/or adapted and related to specific land use situations. Generally, critical values for soil attributes and indicators would be site or soil specific. For example, a high clay content may be favoured in a semi-arid region, where soil moisture retention is an advantage, but may be undesirable in humid conditions in which poor internal drainage may limit yields. In similar fashion, a certain soil bulk density can be optimal under a semi-arid moisture regime, but deleterious under a humid moisture regime due to changes in relative saturation and subsequent poor soil aeration (4). Table 3 provides an example of critical limits for some soil physical attributes selected for irrigated tree crops in Alfisols with a texture contrast in the A and B horizon in southeastern Australia (14). Other farm-based studies have developed critical limits for indicators to evaluate farmer satisfaction and resource conservation. Table 4 illustrates critical limits set for rice (*Oryza sativa* L.) culture in the Philippines based on consensus values at the farm level (15).

STANDARDIZATION

Once an attribute has been identified for a specific soil type or situation, information is needed in regard to soil quality standards for a given set of conditions. Soil quality standards are required to ensure that soil sampling, description, and analysis can set the limits for a quality soil and detect adverse changes in soil quality. Usually the critical

Table 4 Critical level for soil and associated non-soil indicators, based on improvements relative to community average, used to evaluate resource conservation and sustainability for rice culture on sloping land at Gubu, near Cebu City, Philippines

Indicator	Critical level (relative to community average)
Crop yield	More than 20% of average yield
Profit	Better than 20% of community average
Frequency of crop failure	Average frequency or 20%, whichever is lower
Soil depth	Average of similar soil types or 50 cm
Soil organic carbon	Average of community or 10 g kg^{-1}, whichever is higher
Permanent ground cover	Average of community or 15%, whichever is higher

(Adapted from Ref. 15.)

soil attribute is related to a specific methodology (1, 4). Standardization deals with the development and applications of technical rules, specification and protocols in regards to a measuring method (6). The International Organization for Standardization (ISO) has developed various standards for soil quality measurements that address the different phases (e.g., soil sampling, handling, storage, analysis) involved in soil characterization (6, 16).

REFERENCES

1. Larson, W.E.; Pierce, F.J. Conservation and Enhancement of Soil Quality. In *Evaluation for Sustainable Land Management in the Developing World*, IBSRAM Proceedings (12)2; International Board Soil Research Management: Bangkok, Thailand, 1991; Vol. 2, 175–203.
2. Doran, J.W.; Parkin, T.B. Defining and Assessing Soil Quality. In *Defining Soil Quality for a Sustainable Environment*; Doran, J.W., Coleman, D.C., Bezedick, D.F., Stewart, B.A., Eds.; Special Pub. No. 35; American Society Agronomy: Madison, Wisconsin, 1994; 3–21.
3. FAO. *Guidelines: Land Evaluation for Rainfed Agriculture*, Soils Bull. 52; Food and Agriculture Organization, United Nations: Rome, 1983; 237.
4. Carter, M.R.; Gregorich, E.G.; Anderson, D.W.; Doran, J.W.; Janzen, H.H.; Pierce, F.J. Concepts of Soil Quality and Their Significance. In *Soil Quality for Crop Production and Ecosystem Health*; Gregorich, E.G., Carter, M.R., Eds.; Elsevier: Amsterdam, 1997; 1–9.
5. Singer, M.J.; Ewing, S. *Handbook of Soil Science*; Sumner, M.E., Ed.; CRC Press: Boca Raton, FL, 1999; G271–298.
6. Nortcliff, S. Standardisation for Soil Quality Attributes. In *Soil Quality for Crop Production and Ecosystem Health*; Gregorich, E.G., Carter, M.R., Eds.; Elsevier: Amsterdam, 1997; 187–201.
7. Marshall, T.J.; Holmes, J.W. *Soil Physics*; Cambridge University Press: Cambridge, 1978; 102, 249–250, 257–258.
8. Thomasson, A.J. Towards an Objective Classification of Soil Structure. Journal Soil Science **1978**, *29*, 38–46.
9. Håkansson, I.; Lipiec, J. A Review of the Usefulness of Relative Bulk Density Values in Studies of Soil Structure and Compaction. Soil Tillage Res. **2000**, *53*, 71–85.
10. Gliński, J.; Lipiec, J. *Soil Physical Conditions and Plant Roots*; CRC Press: Boca Raton, FL, 1990; 250.
11. Kay, B.D.; Angers, D.A. Soil Structure. In *Handbook of Soil Science*; Sumner, M.E., Ed.; CRC Press: Boca Raton, FL, 1999; A229–276.
12. Feller, C.; Beare, M.H. Physical Control of Soil Organic Matter Dynamics in the Tropics. Geoderma **1997**, *79*, 69–116.
13. Carter, M.R. Organic Matter and Sustainability. *Sustainable Management of Soil Organic Matter*; CABI: Oxford, 2001; 9–22.
14. Cockroft, B.; Olsson, K.A. Case Study of Soil Quality in South-Eastern Australia: Management of Structure for Roots in Duplex Soils. In *Soil Quality for Crop Production and Ecosystem Health*; Gregorich, E.G., Carter, M.R., Eds.; Elsevier: Amsterdam, 1997; 339–350.
15. Gomez, A.A.; Swete Kelly, D.E.; Syers, J.K.; Coughlan, K.J. Measuring Sustainability of Agricultural Systems at the Farm Level. In *Methods for Assessing Soil Quality*; Doran, J.W., Jones, A.J., Eds.; Soil Science Society America Special Publication, Pub. No. 49; SSSA: Madison, WI, 1996; 401–410.
16. Hortensius, D.; Welling, R. International Standardization of Soil Quality Measurements. Commun. Soil Sci. Plant Anal. **1996**, *27*, 387–402.

QUALITY AND EROSION

Craig Ditzler
United States Department of Agriculture-Natural Resources Conservation Service (USDA-NRCS), Soil Quality Institute, Ames, Iowa, U.S.A.

INTRODUCTION

Soil erosion involves the detachment and removal of soil material from one site and its transport to another location. Soil erosion usually degrades soil quality and a soil of poorer quality is less able to withstand further erosion, thus creating a downward spiral of soil degradation. When the surface soil is removed through erosion, organic matter and clay particles may be lost, with consequent reductions in fertility, biological activity, aggregation and rooting depth. Other potential effects of erosion on soil quality include reduced porosity and infiltration, formation of crusts on the soil surface, changes in soil texture, and compaction. These changes in turn reduce the capacity of the soil to supply and cycle nutrients, filter and degrade toxic materials, store and supply moisture and sustain plant and biological productivity. They may also result in increased runoff, less biomass production and plant cover, and greater susceptibility to further erosion. Erosion increases the variability in soil quality across a field and, on a broad scale, is associated with widespread loss of agricultural productivity and declining water quality.

Soil erosion is a natural process by which soil particles are detached and moved by water, wind, gravity, or ice. All soils have an inherent erodibility, or natural susceptibility to erosion, based on soil features, topography, and climate. However, human activities, such as logging, livestock grazing, tillage, removal of vegetation, and urban development, can greatly accelerate natural rates of erosion. Cleared and managed as they are for crop and livestock production, agricultural soils are particularly susceptible to wind and water erosion (Figs. 1 and 2), and also to a recently recognized process known as tillage erosion—the loosening of soil by tillage equipment and its downslope movement under gravity (1).

Eroded soil may move only a few meters in a field and come to rest in lower positions, resulting in increased variability of surface soil properties across the field as subsoil becomes exposed in some places and surface layers are buried and over-thickened in others. It may also move great distances, being deposited in neighboring fields, roadside ditches, and water bodies. In some cases of wind erosion, fine soil particles may travel many kilometers before being deposited.

EFFECTS OF SOIL EROSION ON SOIL QUALITY

The loss of surface soil in a landscape may impair soil function, and thus soil quality, through adverse effects on many physical, chemical, and biological properties of the soil. Table 1 summarizes some of the effects of erosion on the soil quality functions listed by Gregorich.

Physical Effects

Erosion selectively removes the finer, lighter particles from the soil surface, leaving coarser particles behind. Depending on the severity of erosion, an eroded soil may become very coarse in texture, sometimes with a gravelly surface. The deposition of the eroded material in lower topographic areas may result in a thickening of the topsoil and an increase in rooting volume.

Tillage of eroded soils may result in a mixing of subsoil with the surface soil, altering its composition. For example, cultivation of eroded soils having clay-enriched subsoils may increase the clay content of the surface soil. As erosion progresses, plant roots must enter progressively deeper into the subsoil layer to obtain nutrients and water. Where subsoil layers restrict root growth because of their physical and/or chemical properties, the depth of rooting is reduced, along with the capacity of the soil to supply water and nutrients to plants.

Individual soil particles are held together in aggregates, which form the structural fabric of the soil. Soils with good structural arrangement of aggregates and the pore spaces between them provide good aeration for soil roots and microbes; allow ready movement and storage of water and plant nutrients in the pore spaces; and retain their structure when exposed to stresses such as cultivation and the impact of raindrops (2, 3). Abrasion by wind, rain, and tillage can disintegrate aggregates at the soil surface, and the

Fig. 1 Deposition of wind-blown soil material. (From USDA, Natural Resources Conservation Service—Soil Quality Institute.)

Fig. 2 Severe erosion by water has removed all of the surface layer and much of the subsoil. (From USDA, Natural Resources Conservation Service—Soil Quality Institute.)

resulting fine particles can plug larger soil pores and form a hard, thin crust on the surface. This crust seals the surface and limits the infiltration of air and water, as well as impedes the emergence of seedlings. Finer soil particles are also more easily compacted as the pore spaces between them are reduced under the pressure of farm machinery. Deteriorating soil structure further increases the risk of soil erosion—fine soil particles created by the breakdown of aggregates at the soil surface are especially vulnerable to wind and water erosion. Furthermore, compacted, crusted soils resist the infiltration of water, increasing the volume of surface runoff and compounding the effects of erosion. Thus, erosion reduces soil quality, making the soil prone to further erosion and further degradation

Chemical Effects

Organic matter and clay are important sites of cation exchange in the soil. Cation-exchange capacity is a measure of negatively charged sites on the soil particles that are capable of holding positively charged ions, including many plant nutrients. As organic matter and clay particles are lost from the soil surface through erosion, attached nutrients are relocated in the landscape, often to adjacent water bodies where they contribute to declining water quality. Loss of these fine soil particles also impairs the ability to store nutrients, reducing soil fertility. This effect is more pronounced in sandy soils containing small amounts of

Table 1 Some general effects of soil erosion on soil-quality functions

Erosion effects	Soil-quality functions affected
Reduced cation-exchange capacity	Sustaining plant growth and animal life Buffering or filtering Regulating energy
Formation of surface crusts	Regulating water, gases, and energy Sustaining plant growth and animal life
Changes in rooting volume	Sustaining plant growth and animal life
Changes in surface layer texture	Sustaining plant growth and animal life Regulating water and gases
Loss of organic matter	Sustaining plant growth and animal life Regulating energy
Compaction	Regulating water and gases Sustaining plant growth and animal life
Deterioration of soil structure	Regulating water and gases Sustaining plant growth and animal life
Exposure of subsoil material	Sustaining plant growth and animal life Regulating water, gases, and energy

clay, though it may be offset to some degree if erosion causes the surface layer to become more clayey. Where subsoils are more enriched in clay than the surface soil, the clay particles may form chemical bonds with phosphorus, fixing it into forms not easily available to plants and thus reducing fertility.

For some acid soils with subsoil pH of less than 5.0, concentrations of available aluminum may be at levels toxic to plant roots and removal of surface soil by erosion can effectively reduce the rooting depth of the soil.

Subsoil layers having pH of more than 8.5 often contain high levels of sodium. Surface soil removal and subsequent exposure of these subsoils to rainwater or irrigation waters with low ionic concentrations can lead to dispersion of clay particles, loss of soil structure, surface sealing, and greatly reduced water infiltration.

Biological Effects

Removal of organic matter and nutrients from the soil surface by erosion reduces the food and energy supply needed to sustain healthy populations of soil organisms and support plant growth. Soil organisms are the agents of organic matter decomposition and nutrient cycling in soil. In addition, they play an important role in the stabilization of soil aggregates through the production of binding agents such as roots, fungal hyphae, polysaccharides, gums, and complex molecules consisting of humic substances combined with iron, aluminum, or aluminosilicates (4, 5). So, as erosion proceeds, the food source for organisms is reduced, leading to declines in populations. Soil structure and stability then deteriorate, the soil becomes more susceptible to further erosion, and the cycle continues.

CONCLUSION

The interaction of accelerated soil erosion and soil quality is complex. Soil erosion usually reduces soil quality, and a soil of poorer quality is less able to withstand erosion, thus creating a downward spiral of soil degradation. However, many soil-conservation practices have been useful in mitigating the effects of erosion. Reduced tillage systems limit soil disturbance and build soil structure. Crop residue management, underseeding, cover cropping, and permanent cover (e.g., pasture) protect the soil from the action of wind and water. Contour cultivation, grassed waterways, and terracing alter the flow of surface water, curbing water erosion. Herbaceous wind barriers and woody windbreaks and shelterbelts help to control wind erosion. Nevertheless, large tracts of agricultural land throughout the world are still subject to the unsustainable loss of soil as a result of erosion, and continued adoption of preventive methods are needed to protect and restore soil quality.

REFERENCES

1. Govers, G.; Lobb, D.A.; Quine, T.A. Tillage Erosion and Translocation: Emergence of a New Paradigm in Soil Erosion Research. Soil Tillage Res. **1999**, *51*, 167–174.
2. Karlen, Douglas L.; Stott, Diane E. A Framework for Evaluating Physical and Chemical Indicators of Soil Quality. In *Defining Soil Quality for a Sustainable Environment*; Doran, J.W., Coleman, D.C., Bezdicek, D.F., Stewart, B.A., Eds. SSSA Special Publication Number 35; Soil Science Society of America: Madison, WI, 1994; 53–72.
3. Topp, G.C.; Wires, K.C.; Angers, D.A.; Carter, M.R.; Culley, J.L.B.; Holmstrom, D.A.; Kay, B.D.; Langille, D.R.; McBride, R.A.; Patterson, G.T.; Perfect, E.; Rasiah, V.; Rodd, A.V.; Webb, K.T. Changes in Soil Structure. In *The Health of Our Soil: Toward Sustainable Agriculture in Canada*; Acton, D.F., Gregorich, L.F., Eds. Agriculture Canada: Ottawa, Ont., 1995; 51–60.
4. Tisdall, J.M.; Oades, J.M. Organic Matter and Water-Stable Aggregation. J. Soil Sci. **1982**, *33* (2), 141–163.
5. Wright, S.F.; Starr, J.L.; Paltineanu, I.C. Changes in Aggregate Stability and Concentration of Glomalin During Tillage Management Transition. Soil Sci. Soc. Am. J. **1999**, *63* (6), 1825–1829.

QUALITY INDICATORS

Q

Graham Sparling
Landcare Research, Hamilton, New Zealand

INTRODUCTION

This chapter presents criteria to select soil properties for soil quality indicators. A minimum data set to measure soil quality will include key dynamic properties of the soil quality aspect being monitored, with regard for the spatial scale, and the time span. An interpretative framework is needed to link indicator response to management action. Interpretation will differ according to the soil and land use, and monitoring trends through time is useful where critical limits have not been defined. Multiple indicators can be combined to form a simple index, but in so doing, some soil quality information may be masked.

CRITERIA FOR SELECTING SOIL QUALITY INDICATORS

Many soil properties and characteristics have been proposed as indicators of soil quality (1, 2). In order to be considered a valuable indicator, a soil property must meet certain criteria: 1) it needs to get our attention, and 2) it needs to tell us something useful about the condition of the soil. Desirable attributes include being quantitative and measurable; responsive within the specified time scale; interpretable; cost effective; scientifically justifiable; socially acceptable; internationally recognized; and preferably a part of historical monitoring procedures (1–4).

There is some overlap between the categories of soil quality and land quality, of which soil quality is a subset (2, 5). Indicators for *land quality* generally describe the *inherent* and *intrinsic* characteristics of the soil, usually on a landscape scale. Inherent characteristics, such as soil mineralogy and texture, soil depth, and stoniness, are used to assess land capability or suitability for use, and generally are not greatly affected by land management. Other inherent characteristics, such as slope and aspect, overlap with soil quality indicators because they influence the more rapidly changing and dynamic soil properties that are responsive to land management. The variety of proposed indicators reflects the many ways in which soil quality has been defined (5) and the many components of a soil's chemical, physical, and biological attributes that contribute to its overall character. Of its many definitions, quality can be most briefly described as "fitness for use" (1, 5). Within a soil context, a quality rating depends on how well the soil characteristics match the suitability for a particular use. (M.R. Carter in "Soil Quality: Critical limits and standardization" (this volume) this definition.) Soil characteristics are not fixed—what is good soil quality for one land use may be deemed poor soil quality for another use. Thus, quality ratings are not absolute, but are instead values based on interpretation depending on the intended use of the soil (6). The soil itself does not change, only our perception of it in relation to human needs.

A Minimum Data Set

Because it is not feasible to measure all soil quality characteristics, indicators need to be selected to address the issues of greatest concern. There are three major categories of soil quality concerns: soil erosion and redistribution; chemical and biological contamination; and soil degradation and depletion. The examples of these categories given in Table 1 are of concern because they adversely influence the role of the soil in ecosystem functioning and services (5). Their degree of importance varies in different parts of the world.

Time and Spatial Aspects

Selected indicators must be appropriate to the scale being considered, within both spatial and time dimensions. We tend to classify these items based on human perceptions, and soil processes that geologists consider rapid may be considered slow or relatively static to soil biologists. Examples of time and spatial scales and some relevant soil quality characteristics are shown in Table 2. Human-induced changes to soil, including soil redistribution through erosion and deposition, organic matter depletion, and chemical and radioactive contamination, can take hundreds of years to remediate.

Table 1 Major aspects of soil quality requiring indicators

Soil quality aspect	Examples
Erosion and deposition	Soil redistribution by wind and water, mass movement, slips and slumps, and tillage displacement
Contamination/pollution	Presence of potentially toxic chemicals
	Pathogenic organisms
Degradation/depletion	Depletion of soil organic matter
	Loss of fertility
	Soil acidification
	Salinization
	Structural degradation

Interpretation

An interpretative framework is needed to make sense of the numeric values obtained for the various soil characteristics. We need to know what constitutes a high or low value, and what target value is desirable for each particular soil and land use. An interpretive framework should also indicate whether there might be off-site consequences for the wider ecosystem. For example, soluble fertilizer contents in soil may benefit crop production (good quality), but not matching the fertilizer to plant needs may result in an excess, increasing the risk of contaminating receiving waters (poor quality). In general, interpretative frameworks are better defined for contaminants than for other soil quality measures, because the presence of a contaminant can be related to an increased risk to plant and animal health. A critical value can be defined based on toxicity. Such dose–response curves are not available for most indicators used for soil loss, deposition, or soil depletion, and in many cases the response curve is an ill-defined continuum rather than a critical point (2). Much work still needs to be done to define justifiable targets for soil properties, and even the broad categories of "more is better," "less is better," or "optimum range" have been the subject of ongoing discussion (7). The units of expression can also influence interpretation—expression on a volume or area basis is generally preferred, particularly for comparisons between soils and land uses with differing bulk densities (1, 8).

Trends

In the absence of defined critical values for soil quality indicators, the importance of the trend over time has been emphasized (9). Trends away from a target range are interpreted as a decline in soil quality; trends toward the target are interpreted as improved soil quality. However, it can also be argued that short-term trends away from a target value could still be regarded as sustainable, provided the trend can be reversed within an acceptable time frame. Within the present context, recovery or restoration within a human generation of 20–25 years could be regarded as preserving intergenerational equity (10). Such exploitation–recovery cycles were much used during the shifting agriculture phase of human development. Soils differ in their resistance to change, and the rates at which they recover (11, 12) affects their soil quality rating. Generally, soils that are resistant to change, or do change but show a rapid rate of recovery, are regarded as having better quality (11).

Combining Indicators

Because multiple data sets are difficult to comprehend, it has been proposed that indicator sets be combined into a single index (1, 2). These single indices are usually additive or multiplicative combinations of the individual indicators, and a weighting factor is used to adjust for the relative importance of each factor. While a single number is easier to remember and communicate, its use has the disadvantage of possibly masking information. When too many indicators are combined, poor-quality soil aspects may be masked by those of modest or good quality. Reporting on any soil environmental quality should reflect the limiting quality rather than the average value. For example, excellent chemical fertility is of little use if the physical structure of the soil is badly degraded. This poor physical structure would therefore dominate the soil characteristics.

Table 2 Spatial and time scales of soil properties important for soil quality

Spatial scale	Time scale	Soil properties showing change	Soil quality aspects
Rapidly changing, highly dynamic characteristics:			
1–100 mm	Minutes and hours	Moisture content and temperature	Moisture and warmth for plant growth and biological activity
100 mm–1 m	Hours and days	Infiltration and drainage	Compaction due to treading or wheeled traffic, risk of run-off during rainfall, waterlogging in poorly draining soils, poor root environment
		Biological activity	Decomposition of fresh organic matter
Intermediate characteristics:			
1 m–100 m	Days and months	Available nutrients and mineralization	Nutrient depletion and supply, nutrient imbalances for plant growth
		Soil structure	Soil damage from compaction; influence on root environment, aeration, moisture retention, and drainage
100 m–1 km	1–5 years	Organic matter content	Turnover of moderately decomposable organic matter, nutrient supply
		Salinity	Saline water in root zone, toxic stress
		Acidification	Soil chemical process leading to acidification in root zone, nutrient imbalances, toxicity
		Erosion and redistribution	Loss of topsoil, soil deposition
Slowly changing characteristics:			
1 km–10 km	5–25 years	Soil loss and formation	Loss of topsoil and decrease in soil depth, soil deposition
		Humus formation or loss	Changes in soil structure and nutrient storage due to loss of organic matter
		Salinity	Salinity in the root zone, toxic stress, rising water tables
		Acidification	Increasing acidity due to leaching and cation/anion imbalance, toxic stress
		Weathering	Nutrient release and soil formation from rapidly weathering minerals
10 km–100 km	25–100 years	Stable organic matter	Humus provides exchange sites to retain ions in soil and contributes to soil aggregate formation
>100 km	>100 years	Soil formation	Nutrient release and soil formation from slowly weathering minerals
		Mineralogy	Proportions of sand, silt, and clay, exchange characteristics of the clays

REFERENCES

1. Doran, J.W.; Parkin, T.B. Defining and Assessing Soil Quality. In *Defining Soil Quality for a Sustainable Environment*; Doran, J.W., Coleman, D.C., Bezdicek, D.F., Stewart, B.A., Eds.; SSSA Special Publication No. 35, Soil Science Society of America: Madison, Wisconsin, 1994; 3–21.
2. Karlen, D.L.; Mausbach, M.J.; Doran, J.W.; Cline, R.G.; Harris, R.F.; Schuman, G.E. Soil Quality: A Concept, Definition, and Framework for Evaluation (Guest Editorial). Soil Sci. Soc. Am. J. **1997**, *61*, 4–10.
3. Doran, J.W.; Coleman, D.C.; Bezdicek, D.F.; Stewart, B.A., Eds. *Defining Soil Quality for a Sustainable Environment*, SSSA Special Publication Vol. 35; Soil Science Society of America: Madison, Wisconsin, 1994; 244.
4. Cornforth, I.S. Selecting Indicators for Assessing Sustainable Land Management. J. Environ. Manage. **1999**, *56*, 173–179.
5. Carter, M.R.; Gregorich, E.G.; Anderson, D.W.; Doran, J.W.; Janzen, H.H.; Pierce, F.J. Concepts of Soil Quality

and Their Significance. In *Soil Quality for Crop Production and Ecosystem Health*; Gregorich, E.G., Carter, M.R., Eds.; Elsevier: Amsterdam, 1997; 1–19.
6. Singer, M.J.; Ewing, S. Soil Quality. Interdisciplinary Aspects of Soil Science. In *Handbook of Soil Science*; Sumner, M.E., Ed.; CRC Press: New York, 1999; 272–298.
7. Sojka, R.E.; Upchurch, D.R. Reservations Regarding the Soil Quality Concept. Soil Sci. Soc. Am. J. **1999**, *63* (5), 1039–1054.
8. Reganold, J.P.; Palmer, A.S. Significance of Gravimetric Versus Volumetric Measurements of Soil Quality Under Biodynamic, Conventional, and Continuous Grass Management. J. Soil Water Conserv. **1995**, *50* (3), 298–305.
9. Larson, W.E.; Pierce, F.J. The Dynamics of Soil Quality as a Measure of Sustainable Management. In *Defining Soil Quality for a Sustainable Environment*; Doran, J.W., Coleman, D.C., Bezdicek, D.F., Stewart, B.A., Eds.; SSSA Special Publication No. 35; Soil Science Society of America: Madison, Wisconsin, 1994; 37–51.
10. Lunney, D.; Pressey, B.; Archer, M.; Hand, S.; Godhelp, H.; Curtin, A. Integrating Ecology and Economics: Illustrating the Need to Resolve the Conflicts of Space and Time. Ecol. Econ. **1997**, *23*, 135–143.
11. Greenland, D.J.; Szabolcs, I. *Soil Resilience and Sustainable Land Use*; CAB International: Wallingford, England, 1994.
12. Seybold, C.A.; Herrick, J.E.; Brejda, J.J. Soil Resilience: A Fundamental Concept of Soil Quality. Soil Sci. **1999**, *164*, 224–234.

QUALITY AND PRODUCTION OF CARBON AND NITROGEN GASES

Q

Philippe Rochette
Agriculture and Agri-Food Canada, Sainte-Foy, Québec, Canada

Reynald Lemke
Agriculture and Agri-Food Canada, Swift Current, Saskatchewan, Canada

Sean McGinn
Agriculture and Agri-Food Canada, Lethbridge, Alberta, Canada

INTRODUCTION

Several gases are either produced or consumed by biological and chemical transformations of carbon and nitrogen in soils. Some of these gases, including carbon dioxide (CO_2), methane (CH_4), and nitrous oxide (N_2O), play a role in the Earth's radiation balance (greenhouse effect) while others, like carbon monoxide (CO), ammonia (NH_3), and nitrogen oxides (NO_x, i.e., NO and NO_2) have negative impacts on atmospheric chemistry. In this article, we briefly describe how carbon and nitrogen gases are produced or consumed in soils and how soil management and soil quality can affect these processes.

CARBON GASES

Carbon is a major constituent of biomass and soil organic matter. However, more than 99% of global carbon is locked into sediments and fossil forms and is not available for biological processes. The small remaining active fraction of global carbon transits between atmospheric CO_2, biomass and soil organic matter, and detritus in the so-called carbon cycle (Fig. 1). The carbon cycle is driven by photosynthetic fixation of atmospheric CO_2 by plants. In global terrestrial ecosystems, it is estimated that plant photosynthesis fixes more than 200 Gt of CO_2 every year (1). Eventually, similar amounts are returned to the atmosphere by the respiration of animals and by the aerobic heterotrophic decomposition of soil organic matter and plant litter. In the absence of oxygen and other electron acceptors, CH_4 is the final product of soil organic matter decomposition. Human activities, including changes in land use and soil management, are contributing to unprecedented rapid increases in atmospheric CO_2 and CH_4 concentrations, which may result in important modifications of the Earth's climate. In dry soils, auto-oxidation of organic compounds can produce carbon monoxide (CO). Carbon monoxide can also be biologically oxidized to CO_2 in moist but well-aerated soils.

Carbon Dioxide

Soil-surface emitted CO_2, or soil respiration, is the sum of the CO_2 produced by root respiration and heterotrophic decomposition of root exudates, soil organic matter and plant litter. Decomposition processes, while dominated by soil microbes, are the result of complex interactions between soil fauna, fungi, actinomycetes, and bacteria. During decomposition, complex molecules like cellulose, hemicellulose, proteins, and lignin are broken down into low molecular weight substances and oxidized to CO_2 to produce energy and C for the growth of organisms (2, 3). The rate of decomposition is regulated by the quality and quantity of organic substrates and by physical/environmental properties of soil, such as temperature, moisture, and aeration (4).

When ecosystems are in equilibrium, soil CO_2 emissions are the result of the natural recycling of nutrients and equal the amount of atmospheric CO_2 fixed by plant photosynthesis. However, in the past 150 years, human activity has broken this equilibrium by burning fossil carbon reserves and by decreasing soil organic matter through land-use changes. During this period, the additional CO_2 that entered the carbon cycle could not be completely absorbed by increases in autotrophic activity and atmospheric CO_2 concentrations increased from 280 to 365 ppm.

Several measures have been identified to reduce net CO_2 emissions from soils. Among them, it is believed that modifications in agricultural and forest management

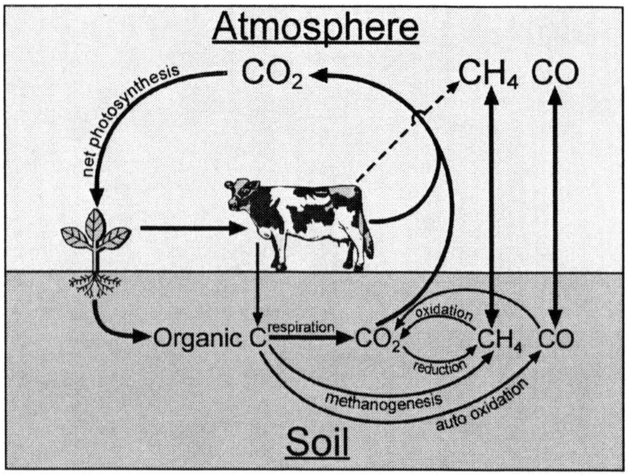

Fig. 1 The soil carbon cycle. In well-aerated soils, decomposition of soil organic matter produces carbon dioxide. Significant amounts of methane are produced only under highly anaerobic (waterlogged) soils.

practices could result in increased C storage in soils as organic matter. Converting cultivated land into natural ecosystems usually increases soil carbon content because of increased return of plant litter and decreased decomposition. Also, tillage breaks down aggregates and increases soil carbon losses by exposing organic matter to microbes that decompose it. Therefore, reducing tillage intensity, fallow frequency, and the amount of cultivated land (by increasing productivity), and increasing perennial crops in rotation and permanent grasslands are seen as potential management practices for mitigating global warming (5).

Methane

Methane is produced in soils by the decomposition of organic matter and by the reduction of CO_2 under highly anaerobic environments. Such conditions are found in wetlands and in rice paddies that, together with landfills, contribute about half of the total emissions of anthropogenic CH_4. Methane is stable in waterlogged soils and can be emitted to the atmosphere via diffusion, ebullition, and transport through plants. In the presence of oxygen, certain bacteria can oxidize CH_4 to CO_2 (6) (Fig. 1).

Human intervention can greatly influence CH_4 production or consumption in soils. Flooding of soils in natural or agricultural ecosystems usually results in increased emissions (7) while drainage of wetlands can turn a source of methane into a sink (8). Flooded rice fields are a major source of anthropogenic CH_4. Management practices have been proposed to reduce CH_4 emissions from rice paddies, including draining the fields during the growing season, replacing urea by other types of nitrogen fertilizers, and reducing the input of crop residues by using new cultivars and alternative cultural practices (5). Reduced rates of CH_4 oxidation in well-aerated soils have been observed following cultivation and addition of nitrogen fertilizers (9).

NITROGEN GASES

The passage of nitrogen (N) through ecosystems can be represented as a "loop-within-a-loop" (Fig. 2). Nitrogen enters ecosystems primarily through biotic or abiotic processes that "fix" [convert molecular nitrogen (N_2) to biologically available forms] N_2 from the atmosphere. Within the soil–plant system, N undergoes a complex series of transformations, resulting in a continual transfer of N between inorganic and organic forms—the inner loop (Fig. 2). Nitrous oxide and NO_x are both produced and consumed during these transformations. With time, most of the N entering ecosystems returns to the atmosphere as N_2, but an important fraction is emitted as gaseous NH_3, NO_x, and N_2O. More than 70% of the estimated 18 Tg N_2O–N entering the atmosphere each year is emitted by soils (1, 10). Above plant-canopy emissions of NO_x from soils are probably in the range of 3.3–21 Tg N yr^{-1} (11, 12). This exchange of N between the atmosphere and the soil–plant system—the outer

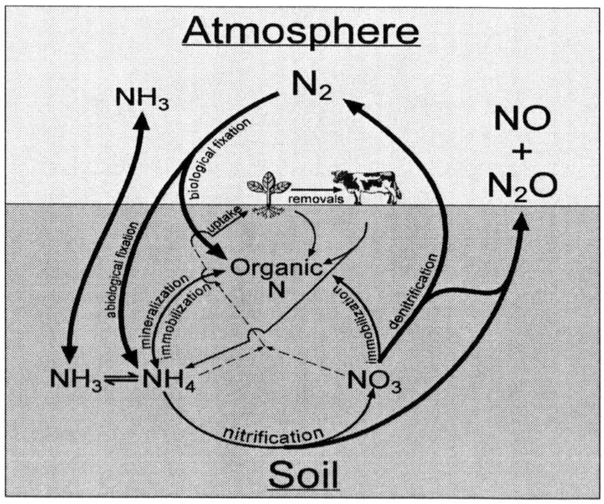

Fig. 2 Schematic representation of the nitrogen cycle depicted as a "loop-within-a-loop". The outer loop traces the passage of N from the atmosphere through ecosystems and back to the atmosphere, while the inner loop traces the interconversion of N between organic and inorganic forms within the soil–plant system.

loop—strongly mediates atmospheric concentrations of NH_3, NO_x, and N_2O.

Nitric and Nitrous Oxides

The majority of soil-emitted NO_x is nitric oxide (NO) (13); therefore, the rest of this discussion will focus on nitrogen gases other than nitrogen dioxide (NO_2). Nitric oxide is derived primarily via autotrophic nitrification (14, 15), whereas N_2O emissions can be a product of autotrophic nitrification (15), denitrification (16), or a combination of both (14). Nitrous oxide, and particularly NO, may also be emitted from various chemical reactions collectively known as chemodenitrification (17).

The magnitude of N_2O and NO emissions is strongly influenced by site-specific variables—most particularly soil attributes—and human intervention. Given that both gases are produced primarily during microbial transformations of inorganic N, the potential of a soil to produce and emit N_2O or NO increases with increasing N availability (18). Anthropogenic activities that increase the flow of N into the system also increase emissions of N_2O and NO. Such activities include, for example, the cultivation of legume grain or forage crops (19), and intensive use of nitrogen fertilizers (20).

Total N_2O emissions tend to increase as soil organic carbon (SOC) content increases. This is understandable, since N turnover is closely tied to SOC turnover, and the amount of carbon (C) available to drive microbial processes is directly related to the quantity and quality of SOC. The relative availability of C and N is of particular importance when livestock manure or sewage sludge is added to soils. The addition of both C and N tend to increase N_2O and NO emissions more than N alone, particularly in C limited systems (21, 22).

Agricultural management practices that minimize soil disturbance tend to increase soil water status and soil bulk density, favoring the development of anaerobic conditions and higher N_2O (4) but possibly lower NO emissions (23). Conversely, equal or lower emissions of N_2O have also been reported in no-tillage systems compared to conventional tillage systems (24). In this instance, reduced soil disturbance appeared to alter N cycling, resulting in lower NO_3^- availability—hence lower N_2O emissions—during the spring thaw period.

In general, soil conditions or management that lead to accumulations of NH_4^+, NO_3^-, and particularly NO_2^-, favor gaseous N losses. This may happen if N is released during soil C mineralization at a time when plant uptake is absent or minimal, an example being bare soil fallow periods commonly employed in the semiarid regions of North America, or if fertilizer or organic N is added at a time when crop growth is not vigorous. Low soil pH levels (<5) inhibit NO_2^- oxidizers more than NH_4^+ oxidizers, resulting in accumulations of NO_2^- and abiotic production of NO (25). The latter may occur even in near-neutral soils as a result of very high concentrations of NH_4^+ localized in urine patches, or concentrated in bands or nests of ammonium based fertilizers (26).

Ammonia

The production of ammonia (NH_3) in soils is closely linked to the accumulation of ammonium (NH_4^+) resulting from the mineralization of organic matter (ammonification) by soil microbes. In agricultural soils, the NH_4^+ content is enhanced where inorganic nitrogen and livestock manure fertilizers are used. In soil solution, NH_4^+ dissociates and forms an equilibrium with NH_3, which can volatilize to the soil air and eventually reach the atmosphere (Eq. 1). Ammonium is also bound to negative exchange sites on soil organic matter and clay minerals (exchangeable NH_4^+) and can be "fixed" within clays. The latter is released slower than exchangeable NH_4^+ (27). Ammonium is converted through biological oxidation (nitrification) by soil bacteria (nitrifiers) to nitrate (NO_3^-) which is taken up by plants. Atmospheric NH_3 and NH_4^+ can be recycled back to the soil through dry and wet deposition (Fig. 2).

$$\begin{array}{l} NH^+_{4(\text{solution})} \Leftrightarrow NH^+_{4(\text{exchangeable})} \Leftrightarrow NH^+_{4(\text{non-exchangeable})} \\ \Updownarrow \\ NH_{3(\text{solution})} \Leftrightarrow NH_{3(\text{soil air})} \Leftrightarrow NH_{3(\text{atmosphere})} \end{array} \quad (1)$$

The volatilized NH_3 from soil is not only a loss of valuable nitrogen for crop growth (28) but is a significant source of atmospheric NH_3 (29, 30) which can damage plants (31, 32) and contribute to odor nuisances that are associated with the application of manure to soils. In the atmosphere, NH_3 neutralizes atmospheric acids and the resulting aerosols contribute to atmospheric haze. Atmospheric NH_3 can also cause acidification of ecosystems by augmenting the capture of sulphur dioxide in clouds (33) and through nitrification of deposited NH_4^+ and NH_3 (34).

Most management techniques that relate to minimizing NH_3 emissions pertain to managing inorganic and livestock manure fertilizers, which are the most significant sources of this gas (30, 35). The key to these techniques is to reduce exposure of the amendment to the atmosphere either by limiting the amount applied or by injection/incorporation of the amendment. Manipulating the chemistry of the soil can also be used to control NH_3 emissions especially in amended soils (36, 37). For example, the addition of acids to fertilized soils can potentially reduce NH_3 emissions. Adding heavy metals or organic compounds with urea fertilizer can inhibit the conversion of urea to NH_4^+ by retarding urease enzyme activity. There is also the potential of adding Ca or Mg salts with urea fertilizers to reduce NH_3 emissions.

Meteorological and soil conditions also modify the emission of NH_3. Application of manure when soil moisture is initially high but subject to rapid drying can result in higher than normal emissions of NH_3 (36, 38). High soil moisture content would put more NH_4^+ into solution near the soil surface and increase volatilization. However, irrigation or rainfall following application can leach near-surface NH_4^+ and decrease volatilization.

CONCLUSION

Soil quality and carbon and nitrogen gas dynamics are closely linked. In general, high-quality soils have high nutrient availability and aeration, good infiltration and retention of water, and a stable structure, all of which are factors that control the production and consumption of gases in soils. Improving soil quality through higher soil organic C content or better soil structure often results in a reduction of net emissions of greenhouse gases into the atmosphere. Also, soil degradation increases emissions of greenhouse gases as a result of greater use of nitrogen fertilizers, erosion, denitrification, deforestation, and grassland cultivation (39). Therefore, adoption of most management practices aimed at improving soil quality also help to improve atmospheric quality.

REFERENCES

1. Duxbury, J.M.; Harper, L.A.; Mosier, A.R. Contributions of Agroecosystems to Global Climate Change. In *Agricultural Ecosystem Effects on Trace Gases and Global Climate Change*; ASA Special Publication No. 55, Harper, L.A., Mosier, A.R., Duxbury, J.M., Eds.; American Society of Agronomy: Madison, WI, 1993; 1–18.
2. Kilham, K. The Ecology of Soil Nutrient Cycling. *Soil Ecology*; Cambridge University Press: New York, 1994; 89–108.
3. Coleman, D.C.; Crossley, D.A. Decomposition and Nutrient Cycling. *Fundamentals of Soil Ecology*; Associated Press: New York, 1996; 109–140.
4. Linn, D.M.; Doran, J.W. Effect of Water-Filled Pore Space on Carbon Dioxide and Nitrous Oxide Production in Tilled and Nontilled Soils. Soil Sci. Soc. Am. J. **1984**, *48*, 1267–1272.
5. Intergovernmental Panel on Climate Change, In *Impacts, Adaptations and Mitigation of Climate Change: Scientific–Technical Analyses*; Watson, R.T., Zinyowera, M.C., Moss, R.H., Eds.; Cambridge University Press: New York, 1995.
6. Topp, E.; Pattey, E. Soils as Sources and Sinks for Atmospheric Methane. Can. J. Soil Sci. **1997**, *77*, 167–178.
7. Duchemin, E.; Lucotte, M.; Canuel, R.; Chamberland, A. Production of Greenhouse Gases CH_4 and CO_2 by Hydroelectric Reservoirs of the Boreal Region. Global Biogeochem. Cycles **1995**, *9*, 529–540.
8. Roulet, N.T.; Ash, R.; Quinton, W.; Moore, T.R. Methane Flux from Drained Northern Peatlands: Effect of a Persistent Water Table Lowering on Flux. Global Biogeochem. Cycles **1993**, *7*, 749–769.
9. Mosier, A.R.; Scimel, D.; Valentine, D.; Bronson, K.; Parton, W. Methane and Nitrous Oxide Fluxes in Native Fertilized and Cultivated Grasslands. Nature **1991**, *350*, 330–332.
10. Kroeze, C.; Mosier, A.; Bouwman, L. Closing the Global N_2O Budget: A Retrospective Analysis 1500–1994. Global Biogeochem. Cycles **1999**, *13* (1), 1–8.
11. Delmas, R.; Serca, D.; Jambert, C. Global Inventory of NO_x Sources. Nutr. Cycl. Agroecosyst. **1997**, *48*, 51–60.
12. Davidson, E.A.; Kingerlee, W. A Global Inventory of Nitric Oxide Emissions from Soils. Nutr. Cycl. Agroecosyst. **1997**, *48*, 37–50.
13. Hutchinson, G.L.; Davidson, E.A. Processes for Production and Consumption of Gaseous Nitrogen Oxides in Soil. *Agricultural Ecosystem Effects on Trace Gases and Global Climate Change*; ASA Spec. Public. No. 55, American Society of Agronomy: Madison, WI, 1993, 79–93.
14. Skiba, U.; Smith, K.A.; Fowler, D. Nitrification and Denitrification as Sources of Nitric Oxide and Nitrous Oxide in a Sandy Loam Soil. Soil Biol. Biochem. **1993**, *25* (11), 1527–1536.
15. Hutchinson, G.L.; Guenzi, W.D.; Livingston, G.P. Soil Water Controls on Aerobic Soil Emissions of Gaseous Nitrogen Oxides. Soil Biol. Biochem. **1993**, *25* (1), 1–9.
16. Williams, P.H.; Jarvis, S.C.; Dixon, E. Emission of Nitric Oxide and Nitrous Oxide from Soil Under Field and Laboratory Conditions. Soil Biol. Biochem. **1998**, *30* (14), 1885–1893.
17. Haynes, R.J.; Sherlock, R.R. Gaseous Losses of Nitrogen. In *Mineral Nitrogen in the Plant–Soil System*; Haynes, R.J., Ed.; Academic Press: New York, 1986; 242–286.
18. Matson, P.A.; Vitousek, P.M. Cross-System Comparisons of Soil Nitrogen Transformations and Nitrous Oxide Flux in Tropical Forest Ecosystems. Global Biogeochem. Cycles **1987**, *1*, 163–170.
19. Galbally, I.E. Biosphere–Atmosphere Exchange of Trace Gases Over Australia. In *Australia's Renewable Resources: Sustainability and Global Change*; Gifford, R.M., Barson, M.M., Eds.; Bureau of Rural Resources: Canberra, Australia, 1992; 117–149.
20. Smith, K.A.; McTaggart, I.P.; Tsuruta, H. Emissions of N_2O and NO Associated with Nitrogen Fertilization in Intensive Agriculture, and the Potential for Mitigation. Soil Use Management **1997**, *13*, 296–304.
21. Rochette, P.; Angers, D.A.; van Bochove, E.; Prevost, D.; Cote, D.; Bertrand, N. Soil C and N Dynamics Following Application of Pig Slurry for the 19th Consecutive Year: II—Mineral N and N_2O Fluxes. Soil Sci. Soc. Am. J. **2000**, *64*, 1396–1403.

22. Thornton, F.C.; Shurpali, H.J.; Bock, B.R.; Reddy, K.C. N_2O and NO Emissions from Poultry Litter and Urea Applications to Bermuda Grass. Atmos. Environ. **1998**, *32* (9), 1623–1630.
23. Civerolo, K.L.; Dickerson, R.R. Nitric Oxide Soil Emissions from Tilled and Untilled Cornfields. Agri. Forest Meteorol. **1998**, *90* (4), 307–311.
24. Lemke, R.L.; Izaurralde, R.C.; Nyborg, M.; Solberg, E.D. Tillage and N-Source Influence Soil-Emitted Nitrous Oxide in the Alberta Parkland Region. Can. J. Soil Sci. **1999**, *79*, 15–24.
25. Yamulki, S.; Harrison, R.M.; Goulding, K.W.T.; Webster, C.P. N_2O, NO and NO_2 Fluxes from a Grassland: Effect of Soil PH. Soil Biol. Biochem. **1997**, *29* (8), 1199–1208.
26. Burns, L.C.; Stevens, R.J.; Smith, R.V.; Cooper, J.E. The Occurrence and Possible Sources of Nitrite in a Grazed, Fertilized, Grassland Soil. Soil Biol. Biochem. **1995**, *27* (1), 47–59.
27. Mengel, K. Dynamics and Availability of Major Nutrients in Soils. Adv. Soil Sci. **1985**, *2*, 65–131.
28. McGinn, S.M.; Janzen, H.H. Ammonia Sources in Agriculture and Their Measurement. Can. J. Soil Sci. **1998**, *78*, 139–148.
29. Bouwman, A.F.; Lee, D.S.; Asman, W.A.H.; Dentener, F.J.; Van Der Hoek, K.W.; Olivier, J.G.J. A Global High-Resolution Emission Inventory for Ammonia. Global Biogeochem. Cycles **1997**, *11* (4), 561–587.
30. Duxbury, J. The Significance of Agricultural Sources of Greenhouse Gases. Fertilizer Res. **1994**, *38*, 151–163.
31. Van der Eerden, L.J.M. Toxicity of Ammonia to Plants. Agric. Environ. **1982**, *7*, 223–235.
32. Van der Eerden, L.J.M.; de Visser, P.H.B.; van Dijk, C.J. Risk of Damage to Crops in the Direct Neighbourhood of Ammonia Sources. Environ. Pollut. **1998**, *102*, 49–53.
33. ApSimon, H.M.; Kruse, M.; Bell, J.N.B. Ammonia Emission and Their Role in Acid Deposition. Atmos. Environ. **1987**, *21* (9), 1939–1946.
34. Fahey, T.J.; Wiliams, C.J.; Rooney-Varga, J.N.; Cleveland, C.C.; Postek, K.M.; Smith, S.D.; Bouldin, D.R. Nitrogen Deposition in and Around an Intensive Agricultural District in Central New York. J. Environ. Qual. **1999**, *28*, 1585–1600.
35. Buijsman, E.; Maas, H.F.M.; Asman, W.A.H. Anthropogenic NH_3 Emissions in Europe. Atmos. Environ. **1987**, *21* (5), 1009–1022.
36. Hargrove, W.L. Soil, Environmental, and Management Factors Influencing Ammonia Volatilization Under Field Conditions. In *Ammonia Volatilization from Urea Fertilizers*; Brock, B.R., Kissel, D.E., Eds.; National Fertilizer Development Center, Tennessee Valley Authority: Muscle Shoals, AL, 1988; 17–36.
37. Fenn, L.B.; Hossner, L.R. Ammonia Volatilization from Ammonium or Ammonium-Forming Nitrogen Fertilizers. Adv. Soil Sci. **1985**, *1*, 124–169.
38. Brunke, R.; Alvo, P.; Schuepp, P.; Gordon, R. Effect of Meteorological Parameters on Ammonia Loss from Manure in the Field. J. Environ. Qual. **1988**, *17* (3), 431–436.
39. Lal, R.; Kimble, J.; Stewart, B.A. World Soils as a Source or Sink of Radiatively-Active Gases. In *Soil Management and Greenhouse Effect*; Advances in Soil Science, Lal, R., Kimble, J., Stewart, B.A., Eds.; CRC Press: Boca Raton, 1995; 1–7.

Q

QUALITY AND PRODUCTIVITY

M.H. Beare
New Zealand Institute for Crop & Food Research Limited, Christchurch, New Zealand

INTRODUCTION

The capacity to nurture and sustain plant and animal productivity is a key function of high quality soils (1, 2). Indicators of soil quality reflect the key properties and processes that support this function and can be used to assess the fitness of soils for production. In addition to extrinsic factors (e.g., climate), productivity is influenced both by the intrinsic characteristics of a soil (i.e., inherent soil quality) and those processes or properties that are affected by its use and management (i.e., dynamic soil quality). Sustainable production depends on selecting land uses that are well-suited to the capability of the soil (and wider environment) and maintaining soil conditions that minimize the risk of productivity declines (3). The soil quality conditions (e.g., indicator optimum ranges) required to sustain agricultural productivity may be different from, and perhaps less stringent than, those needed to ensure environmental, economic, or social sustainability.

PRODUCTIVITY INDICATORS

Plant productivity depends directly or indirectly on a soil's ability to carry out a variety of functions. The fitness of soil to carry out these functions within acceptable boundaries is the basis of soil quality assessment. Soil quality indicators that can characterize this fitness are useful in assessing plant productivity. For example, a soil's ability to accommodate unrestricted plant emergence and root growth, and to permit effective infiltration and supply of air, water and nutrients may be described by physical indicators of soil quality (4). Chemical indicators of soil quality for plant productivity include properties associated with plant nutrition (e.g., nutrient availability, soil acidity) and properties that reflect an inhibitory or toxic effect on plant growth (e.g., presence of pesticide residues). Soil also supports a diverse community of organisms that both affect and are affected by its chemical and physical properties. Some biological indicators of soil quality reflect a risk to plant productivity (e.g., fungal pathogens) whereas others indicate a potential to enhance plant growth (e.g., plant growth-promoting bacteria). Recently there has been much interest in using biological indicators to describe a soil's capacity to regulate or maintain important soil physical or chemical properties (5). For example, the soil microbial biomass may be used as a biological indicator because it reflects a pool of potentially available nutrients (i.e., chemical function) or the ability of soil to maintain soil organic matter levels (i.e., chemical function) or soil structural stability (i.e., physical function).

Animal productivity depends to a significant extent on plant productivity. However, many other factors influence animal productivity, some of which are determined directly or indirectly by soil quality. These factors include forage nutrient content (e.g., selenium) and quality (e.g., protein content), chemical contaminants (e.g., DDT), and soil-borne diseases (e.g., parasites). This discussion focuses on soil quality for plant productivity, though many of the concepts also apply to animal productivity.

INHERENT SOIL QUALITY

Characteristics that define a soil's inherent capacity for plant production are usually static, changing little over relatively short time frames (years to decades) (1). They are the product of soil formation driven by climate, topography, parent material, vegetation and time. Soil mineralogy and particle size distribution are commonly included as properties of inherent soil quality for productivity (1). Other attributes, such as total soil C, cation exchange capacity, and sodicity may be defined as inherent properties where broad soil type or regional comparisons are of interest at one point in time, even though they may be altered by management over longer time frames (e.g., decades to centuries).

Inherent soil properties are the basis of many land use capability or suitability assessments (6, 7) that are key components of land use planning and policy development in many regions of the world. Most were undertaken with the primary aim of evaluating potential soil productivity (8, 9). These assessments are usually developed around a broad set of "land quality" criteria that integrate extrinsic factors (e.g., climate, topography, hydrology) affecting the composition and productivity of plant communities in an area with the inherent attributes of the soil. The interplay between these extrinsic factors and inherent soil quality determines the suitability of a land area for a particular agricultural use (e.g., arable cropping, extensive pastoral farming, forestry).

Thus, the inherent quality of a soil should be viewed in light of its intended agricultural use. For example, in warm humid climatic zones, free-draining, stony soils may be judged as low quality for arable cropping, but of high quality for viticultural production. Similar, but subtler differences in inherent soil quality have been applied to the selection of specific crops, cultivars, or rotations. The Canadian Land Suitability Rating System (10), for example, categorizes land into suitability classes based on the degree to which its environment limits the productivity of specific crops. Suitability classes are indexed from ratings that reflect limitations to productivity posed by climate (e.g., moisture, temperature), soil (e.g., topsoil depth, drainage) and landscape (e.g., slope, stoniness) factors.

DYNAMIC SOIL QUALITY

Soils of high inherent quality for a specific use may not necessarily function within their optimum range (i.e., exhibit maximum potential plant productivity) because of past use and management. This aspect of soil quality is determined by the dynamic properties of a soil.

Properties of dynamic soil quality are those that change in response to human use and management (1), normally over relatively short time frames (years to decades). Agricultural soils of high dynamic quality maintain high nutrient availability, permit adequate infiltration of air and water, have a relatively stable structure, and maintain a functionally diverse community of soil organisms that support relatively high levels of plant productivity. These processes are reflected in the specific physical, chemical and biological properties of soils. The soil properties of greatest importance to productivity in a particular agricultural system are often grouped into a minimum dataset (3, 5) defined by identifying the primary issues of soil management (e.g., nutrient deficiencies, risk to compaction, water storage) that influence crop production on the soils and in the production systems in which they are to be used (11).

The terms "dynamic soil quality" and "soil health" are often used interchangeably (1). Two soils may be equally "healthy" but achieve different levels of plant productivity

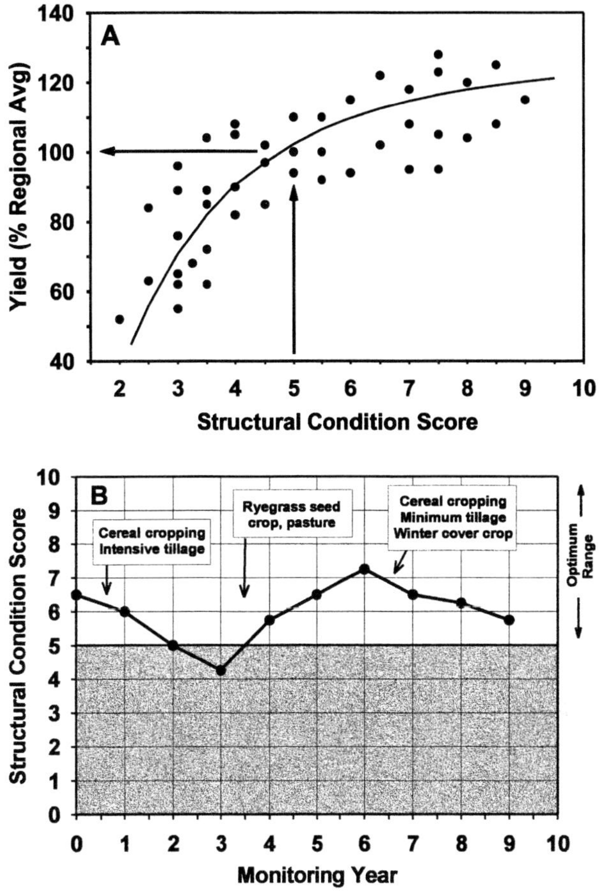

Fig. 1 (A) A relationship between soil structural condition and relative crop yield used to define the indicator's critical limit. (B) An example of how the critical limit for soil structural condition has been used to improve soil management decisions on the Canterbury Plains, New Zealand. (From Ref. 11.)

because of differences in their inherent quality. In terms of productivity, the optimum range or critical limit of a soil quality indicator is usually defined by applying established criteria to empirically derived relationships between the soil property and some measure of productivity (e.g., annual crop yield, root biomass at crop maturity) (Fig. 1) (11, 12). Critical limits for productivity have been established for a number of soil physical and chemical properties. In many cases, as in the example given (Fig. 1), the criteria used in defining critical limits are based on the objectives of the user (e.g., yield targets, profit margins). Where relationships between individual properties and productivity are less clear, some studies have succeeded in indexing data from several different soil properties (minimum dataset) into a single value that explains much more of the variation in productivity (13). While this approach has some advantages where complex interactions between factors may affect productivity, a single value can also mask information that is important to identify and manage the limitations to productivity.

Many factors influence the level of crop production in any one year (e.g., climatic conditions, cultivars, disease incidence, fertilizer rates and timing). For this reason the relationship between dynamic properties (or indicators) of soil quality and crop productivity is often difficult to define at a field scale. Where the relationship is known, loss of dynamic soil quality is probably best described as a risk to crop productivity. In this respect, the management of dynamic soil quality is often aimed at lowering the risk of productivity loss.

SOIL QUALITY AND SUSTAINABLE PRODUCTION

The use of soil quality information to evaluate the sustainability of soil management systems has involved two common approaches (3). With the *comparative assessment* approach, the performance of a system is determined in relation to alternative systems using one-off measures of soil properties. This approach is often used to evaluate potential impacts of land use change. With the *dynamic assessment* approach, the performance of a management system is assessed by measuring the changes in soil properties over time. For example, Fig. 2 shows the nutritive (N and P) benefits of different organic amendments (e.g., cattle manure, composted manure, barley straw, pea hay) for restoring the productivity (wheat yield) of a desurfaced (artificially eroded) cropping soil in southern Alberta, Canada (14). In this example, 87% of the restored yield could be attributed to N and P in the organic amendments while 13% was attributed to soil structural improvements.

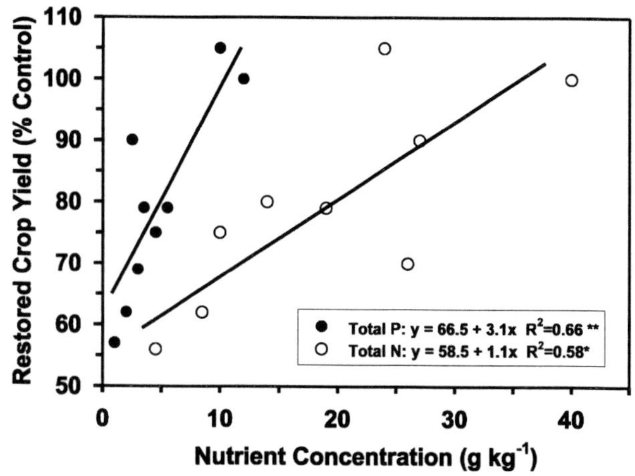

Fig. 2 Effects of total N and P concentrations in organic amendments on the restoration of wheat yields of a desurfaced (artificially eroded) cropping soil in southern Alberta, Canada. Restoration of crop yields is expressed as a percentage of the non-desurfaced (uneroded topsoil) control. (Adapted from Ref. 16.)

In agricultural systems, applications of the dynamic assessment approach usually involve both monitoring (i.e., repeated measurement) and control (i.e., regulation or management), and the focus shifts from describing soil conditions to maintaining soil conditions that sustain high productivity through improved management (Fig. 1). To this end, soil quality monitoring may be coupled with best management practice recommendations to develop soil management decision support systems that improve the sustainability of production systems.

Using the dynamic assessment approach, sustainable soil management practices are often defined as those that maintain or improve dynamic soil quality (3). However, from the standpoint of productivity alone, this represents a somewhat narrow definition of sustainable management. In the short term, agricultural production often results in a degradation of the ordered complex structures that characterize natural ecosystems in equilibrium (15). Sustainable management of soil for crop production may involve practices that approach a new steady state or balance degradative processes by restorative processes over relatively short time frames (Fig. 3). For example, in New Zealand and other regions of the world, farmers practice mixed cropping rotations in which the degradative effects of short-term arable cropping are balanced by the restorative effects of pastoral management under grazing (16). Where best management practices are employed, the degradation of soil properties is not usually sufficient to significantly impair crop production (Fig. 3). However, the critical limit (or optimum range) for soil properties (or indicators) that sustain high levels of crop production may be less restrictive than those needed to achieve environmental, economic, or social sustainability. Furthermore, a decline in soil quality can be compensated for by an increase in production inputs (e.g., fertilizer, tillage, pesticides) that may ultimately lower profitability or result in adverse environmental impacts (e.g., nitrate leaching).

The adoption of specific soil management practices is often driven by practical or economic considerations, rather than the potential benefits of improved soil quality for crop production and the environment. Agricultural profitability is difficult to achieve and sustain without a relatively high productivity, but high productivity does not necessarily ensure environmental, economic, or social sustainability. Nevertheless, there are many documented cases in which soil management practices designed to

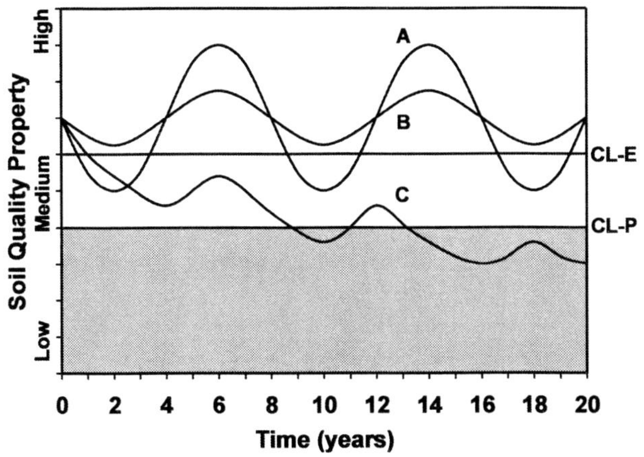

Fig. 3 A conceptual illustration of changes in soil quality under different types of management in relation to critical limits for productivity (CL-P) and environmental impacts (CL-E). Three types of mixed-cropping (degradative–restorative) management are shown: (A) intensive (e.g., cereal cropping with conventional tillage followed by grazed pasture), (B) extensive (e.g., cereal cropping with minimum tillage followed by grass seed crops and pasture), and (C) unbalanced (i.e., restorative practices insufficient to offset degradative management).

lower production costs and reduce environmental impacts have also resulted in improved crop performance and sustained economic performance.

REFERENCES

1. Carter, M.R.; Gregorich, E.G.; Anderson, D.W.; Doran, J.W.; Janzen, H.H.; Pierce, F.J. Concepts of Soil Quality and Their Significance. In *Soil Quality for Crop Production and Ecosystem Health*; Gregorich, E.G., Carter, M.R., Eds.; Elsevier: Amsterdam, 1997; 1–9.
2. Soil Science Society of America; Statement on Soil Quality. Agronomy News June, 1995; 7.
3. Larson, W.E.; Pierce, F.J. The Dynamics of Soil Quality as a Measure of Sustainable Management. In *Defining Soil Quality for a Sustainable Environment*; Special Pub. No. 35, Doran, J.W., Coleman, D.C., Bezedick, D.F., Stewart, B.A., Eds.; American Society Agronomy: Madison, WI, 1994; 37–51.
4. Doran, J.W.; Parkin, T.B. Defining and Assessing Soil Quality. In *Defining Soil Quality for a Sustainable Environment*; Special Pub. No. 35, Doran, J.W., Coleman, D.C., Bezedick, D.F., Stewart, B.A., Eds.; American Society Agronomy: Madison, WI, 1994; 3–21.
5. Gregorich, E.G.; Carter, M.R.; Angers, D.A.; Monreal, C.M.; Ellert, B.H. Towards a Minimum Data Set to Assess Soil Organic Matter Quality in Agricultural Soils. Can. J. Soil Sci. **1994**, *74*, 367–385.
6. Wösten, J.H.M.; Bouma, J.; Stoffelsen, G.H. Use of Soil Survey Data for Regional Soil Water Simulation Models. Soil Sci. Soci. Am. J. **1985**, *49*, 1238–1244.
7. Webb, T.H.; Claydon, J.J.; Harris, S.R. Quantifying Variability of Soil Physical Properties within Soil Series to Address Modern Land-Use Issues on the Canterbury Plains New Zealand. Austr. J. Soil Res. **2000**, *38*, 1115–1129.
8. Huddleston, J.H. Development and Use of Soil Productivity Ratings in the United States. Geoderma **1984**, *32*, 297–317.
9. Petersen, G.W.; Bell, J.C.; McSweeney, K.; Nielsen, G.A.; Robert, A.C. Geographic Information Systems in Agronomy. Adv. in Agron. **1995**, *55*, 67–111.
10. Pettapiece, W.W. Land Suitability Rating System for Agricultural Crops. 1. Spring-Seeded Small Grains. Technical Bulletin 1995-6E. Centre for Land and Biological Resources Research, Agric. Agri-food Canada. 1995.
11. Beare, M.H.; Williams, P.H.; Cameron, K.C. On-Farm Monitoring of Soil Quality of Sustainable Crop Production. *Best Soil Management Practices for Production*, Proceedings of the Fertiliser and Lime Research Centre Conference, Palmerston North, New Zealand. 1999; 81–90.
12. Pierce, F.J.; Gilliland, D.C. Soil Quality Control. In *Soil Quality for Crop Production and Ecosystem Health*; Gregorich, E.G., Carter, M.R., Eds.; Elsevier: Amsterdam, 1997; 203–219.
13. Hussain, I.; Olson, K.R.; Wander, M.M.; Karlen, D.L. Adaptation of Soil Quality Indices and Applications to Three Tillage Systems in Southern Illinois. Soil Tillage Res. **1999**, *50*, 237–249.
14. Larney, F.J.; Janzen, H.H. Restoration of Productivity to a Desurfaced Soil with Livestock Manure, Crop Residue, and Fertilizer Amendments. Agron. J. **1996**, *88*, 921–927.
15. Addiscott, T.M. Entropy and Sustainability. Eur. J. Soil Sci. **1995**, *46*, 161–168.
16. Haynes, R.J.; Swift, R.S.; Stephen, R.C. Influence of Mixed Cropping Rotations (Pasture-Arable) on Organic Matter Content, Water Stable Aggregation and Clod Porosity in a Group of Soils. Soil Tillage Res. **1991**, *19*, 77–87.

QUALITY, SOIL AND WATER

T.M. Nissen
M.M. Wander
University of Illinois, Urbana, Illinois, U.S.A.

INTRODUCTION

Sediments and chemicals originating on farms now impair more U.S. waters than any other source. Managing soil quality is an essential part of a strategy to reduce these forms of pollution and the costs they incur. Water quality concerns focus attention on soil functions related to regulating water flow and absorbing, buffering, and transforming chemical flows. Organic matter plays a significant role in a soil's ability to perform these functions and serves as one of several key indicators for measuring soil quality. Indicators, however, are scale dependent, and care must be taken in extending soil quality data—typically collected at the field level—to assess water quality at later temporal and larger physical scales. The use of interdisciplinary approaches to implement soil quality mirrors a trend already seen in the development of the water quality concept: after the technical details of indicator selection, sampling, and interpretation are addressed, disciplinary information is integrated into problem-solving frameworks for use by individuals, communities, municipalities, and nations.

SOILS AND WATER QUALITY

According to a U.S. EPA report of 1995, the leading source of water pollution in the United States is agriculture (Table 1). Declines in water quality worldwide have also been blamed chiefly on the increases in agricultural inputs and crop productivity that have occurred since the 1950's (1, 2). As efforts in industrialized countries have succeeded in mitigating municipal/industrial point-source pollution, the chemicals and sediments that originate on farms now impair more aquatic habitats and drinking water supplies than any other source.

In response, many land managers are becoming increasingly judicious in their use of chemical inputs. In the U.S. Corn Belt between 1982 and 1992, for example, commercial nitrogen and phosphorus consumption dropped by 3.5% and 21%, respectively (3, cited in 4). But crop productivity is not simply a function of the amount of chemicals applied—soils differ in their ability to release nutrients, maintain adequate moisture and oxygen levels, and allow deep rooting and other functions related to productivity that determine the need for, and efficiency of, chemical inputs. The same is true of water quality—many soil factors affect water quality in addition to, and in feedback with, chemical applications (Fig. 1). For example, increased soil compaction can result from excessive or poorly timed tillage operations, lowering the rate of water infiltration into the soil and leading to increased overland flow, soil erosion, and stream-channel erosion. These processes carry adsorbed nutrients, pesticides, and salts into surface waters; reduce clarity and aquatic-plant photosynthesis; and fill channels and reservoirs, limiting boat passage, water-treatment capacity, and hydroelectric capacity. In the United States, off-farm erosion costs were estimated to be $17 billion annually (5). That management practices cause differences in the ability of a soil to maintain or enhance water quality is one of the bases for evaluating soils in terms of their quality. Although soil quality is not limited to agricultural applications, it is most often considered in this context because of agriculture's large impact on water quality.

SOIL FUNCTIONS RELATED TO WATER QUALITY

One approach to soil quality evaluates a soil's ability to perform its essential *functions*. Soil functions specifically related to water quality include the ability to partition and regulate water flow and absorb, buffer, and transform chemical flows.

Partition and Regulate Water Flow

Soils are an important switching station in the global water cycle, partitioning precipitation into overland or infiltrated flow. A soil's water-holding capacity and the rate at which water infiltrates the soil help determine the proportion of rainfall that flows through the soil matrix, which in turn controls the volume of stream flow and the rate of groundwater recharge.

Encyclopedia of Soil Science

Table 1 Leading sources of pollution of assessed waters in the USA: 1992 and 1993

Sources	Rivers & streams (10^3 miles)[a]	Rank	Lakes & reservoirs (10^3 acres)[b]	Rank	Estuaries (sq mi)[c]	Rank
Agriculture	135	(1)	3350	(1)	3321	(3)
Hydro/habitat change	37	(4)	[d]		[d]	
Municipal/industrial point sources	53	(2)	2025	(2)	6436	(1)
Natural	42	(3)	965	(5)	2949	(4)
Unspecified nonpoint sources	[d]		989	(4)	991	(5)
Urban runoff/storm sewers	27	(5)	1200	(3)	4508	(2)

[a] Total river and stream mileage is 3.5 million miles.
[b] Total lake, reservoir, and pond area is 40.8 million acres.
[c] Total estuarine water area is 34.4 thousand square miles.
[d] Not among the top 5 sources.
(From Ref. 4.)

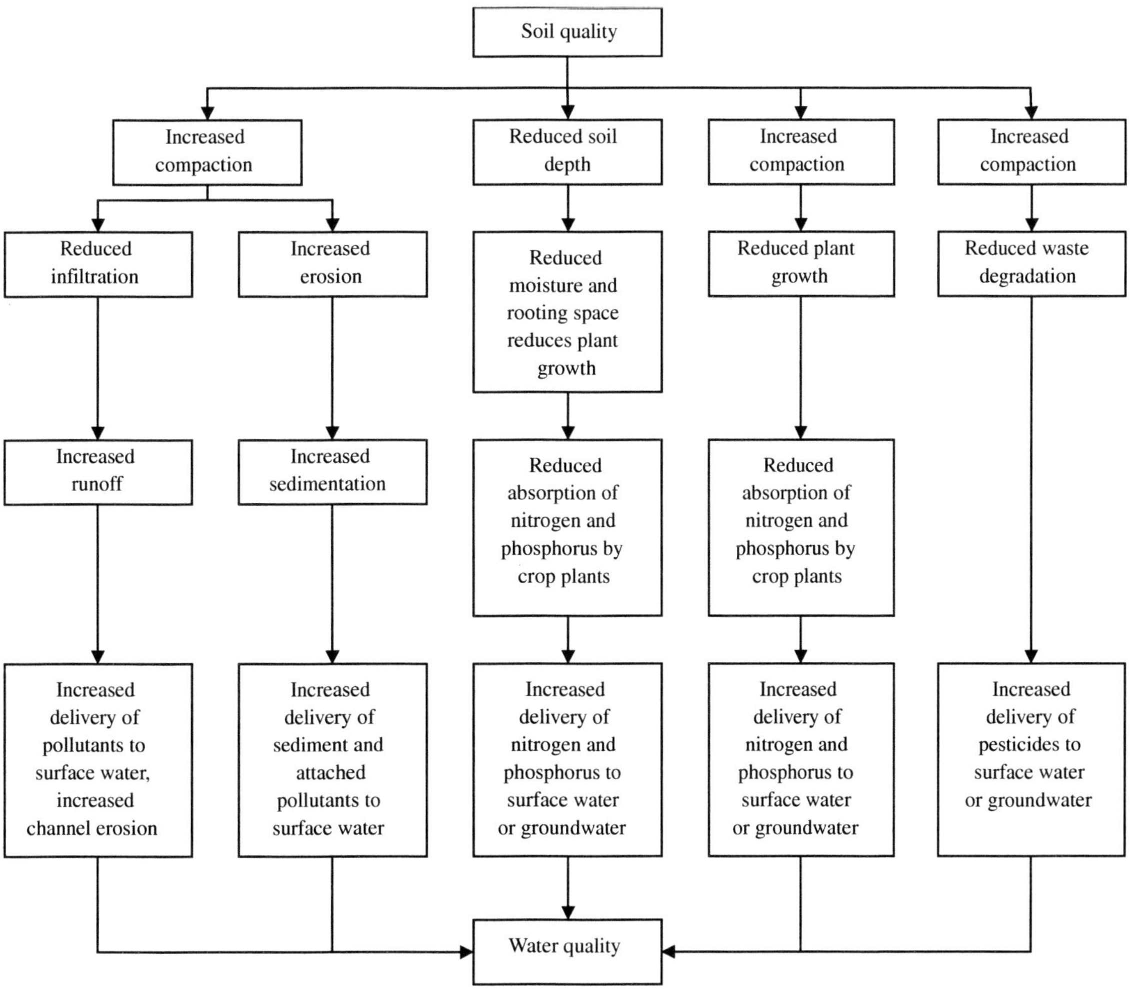

Fig. 1 Changes in soil quality affect water quality. (From National Academy of Sciences. *Soil and Water Quality: An Agenda for Agriculture*. National Academy Press: Washington, D.C., 1993.)

Absorb, Buffer, and Transform Chemical Flows

In natural settings, soils receive and store inputs of nitrogen, phosphorus, and organic wastes, transforming them into forms available for plant uptake at rates that rarely exceed plant demand. Losses to surface and groundwaters are minimal. In agriculture, water pollution occurs when these inputs exceed the capacity of a soil/plant system to process them or when the soil itself is transported through erosion. Excess nitrogen and phosphorus levels in lakes and estuaries are the main causes of eutrophication, depleting oxygen levels for aquatic life (1, 2). Elevated concentrations of nitrates, pesticides, and other organic compounds in waters pose direct risks to human health, although at current levels, these risks are less than to aquatic ecosystems (4).

Fig. 1 highlights the important feedback among productivity, soil quality, and water quality. As soil quality declines, so does productivity, reducing the uptake capacity of the plants and leaving more chemicals exposed to leaching and runoff. Profitability likewise declines as input-use efficiency is diminished. In general, farming profitability and water quality are complementary goals, as both are enhanced by improved input-use efficiency resulting from improved soil quality (6, 7).

INDICATORS OF SOIL QUALITY RELATED TO WATER QUALITY

So that these functions may be maintained or enhanced, research in soil quality has sought to identify the relative importance of management-affected soil indicators on multiple outcomes, the optimal levels of these indicators, and the extent to which these indicators are changed by management.

Selecting Indicators

The relative importance of an indicator depends on the outcome(s) of interest. As an example, Karlen and Stott (8) selected indicators for erosion based on a sensitivity analysis of a simulation model (Table 2). Weightings were assigned to indicators based on their relative influence on sediment loss, with infiltration rate and aggregate stability emerging as the leading indicators. If runoff and leaching concentrations of N and P had been the outcomes of interest, macronutrient concentrations would have been weighted more heavily than they were for sediment loss.

For soil quality monitoring, selected indicators are evaluated either against themselves in relation to a temporal baseline or against "benchmark" soils. Because water quality is generally high in natural systems such as forests and prairies, undisturbed soils are often used as benchmarks for managed soils with the same inherent properties (9–11).

Influence of Management on Indicators

As Table 2 indicates, maintaining soil cover and good soil physical conditions positively affects water quality. Both are related to organic matter management. On the surface, residues and cover crops protect against erosion by absorbing raindrop and flow energy. Together with roots, these additions become the young soil organic matter (SOM) that plays an important role in stable aggregation (12–14). Tillage and cropping regimes that favor organic matter accumulations generally maintain or improve water quality. For example, in data from Wisconsin where surface cover was not significantly different between conventional tillage and no-till, lower erosion in the no-till was associated with the enhancement of SOM-influenced characteristics (Table 3). Because of their contributions to SOM status, animal manures, municipal sludges, and some industrial by-products, if applied properly, can also have a beneficial impact on water quality, especially in the context of alternative disposal methods such as lagoons and landfills (15, 16). Management strategies to improve SOM status and reduce erosion may nevertheless have other water-quality tradeoffs. For example, a complete evaluation of no-till would include its reliance on pesticides compared to alternative weed control strategies such as cover crops (16).

SCALES OF SOIL AND WATER QUALITY

The influence of management on soil quality is typically measured at the field scale, yet its influence on water quality is expressed at the landscape or watershed level. The time scales over which practices affect soil and water properties also differ. Efforts to extrapolate soil quality information to larger scales are plagued by the fact that processes are not scale independent (17).

At larger physical and temporal scales, the connection between soil and water quality has been made indirectly by relating management practices to outcomes (18). The Universal Soil Loss Equation (USLE), which was developed and implemented before the soil quality concept was popularized, is the best example of this kind of soil and water management strategy (19). The USLE

Table 2 Soil quality functions and indicators related to water erosion, based on a sensitivity analysis of an erosion simulation model. Weightings reflect relative influence of function or indicator on the next highest level

		Indicator			
Function	**Weight**	**Level I**	**Weight I**	**Level II**	**Weight II**
Accommodate water entry	0.50	Infiltration rate	1.0	Surface crust	0.20
				Surface roughness	0.20
				Crop residue cover	0.50
				Macropores	0.10
Facilitate water transfer and absorption	0.10	Hydraulic conductivity	0.60	Soil texture	0.50
				Capillary water content	0.30
				Bulk density	0.20
		Porosity	0.15		
		Macropores	0.25	Plant roots	0.40
				Earthworms	0.60
Resist degradation	0.35	Aggregate stability	0.80	Mineralogy	0.20
				Soil carbohydrates	0.35
				Microbial biomass	0.30
				Available cations	0.15
		Shear strength	0.10		
		Soil texture	0.05		
		Heat transfer capacity	0.05		
Sustain plant growth	0.05	Rooting depth	0.25	Restrictive layer depth	0.70
				Texture	0.30
		Water relations	0.35	Available water capacity	0.70
				Drainage	0.20
				Organic C	0.10
		Nutrient relations	0.30	pH	0.15
				Organic C	0.25
				Macronutrients	0.40
				Micronutrients	0.20
		Chemical relations	0.10	Salinity	0.50
				Heavy metals	0.15
				Organics	0.25
				Radioactivity	0.10

(From Ref. 8.)

and related models have been used to estimate payments to farmers by U.S. government conservation programs, based in part on the amount of crop residue measured on individual farms. Those models have also provided the basis for ongoing efforts to scale up field-based information for use at the watershed level (20).

The successful experience with the USLE points to future opportunities for using spatially explicit models to extrapolate field-scale indicators to larger scales. Model sensitivity analysis (8), metamodeling techniques (21, 22), and spatial analyses of watershed land-use patterns (23) will be useful in relating soil quality indicators to desirable outcomes for water quality. Spatial analyses are also likely to increase the number of options for managing water quality. For example, watersheds may contain sinks for excess nutrients such as riparian zones to help mitigate plot-level problems (24). Some producers will find the construction of a wetland or riparian buffer-strips a more economical way to reduce nutrient loadings in streams than reducing application rates of fertilizers.

DECISION MAKING AND POLICY

Establishing the scientific links between soil quality and water quality will not be enough to affect soil management decisions. In order for soil quality to become a valued decision-making tool, methodologies need to be developed that can estimate the economic or environmental value of

Table 3 Soil-quality data from surface soils (0–7.5 cm) in Wisconsin after 12 years of tillage treatments. Measurements of runoff and sediment loss obtained from sprinkler-infiltration study

Parameter	No-till	Chisel	Moldboard	LSD
Bulk density (g cm^{-3})[a]	1.48	1.4	1.42	0.06
Turbidity (log% transmittance)[b]	1.36	1.04	0.97	0.21
Total C in aggregates (g kg^{-1})	24	16	11	4
Biomass C (Mg C kg^{-1} soil)	696	394	260	276
Respiration (Mg C kg^{-1} soil)	352	139	74	114
Earthworms (# 0.25 m^{-2})	78	52	53	18
Average surface cover (%)	68		58	NS
Soil quality index[c]	0.68	0.49	0.48	
Corn yield (t ha^{-1})	8.6	8.7	8.8	NS
Runoff amount (mm)	35 a[d]		42 a	
Sediment concentration (g L^{-1})	1.4 b		5.0 a	
Estimated soil loss (Mg ha^{-1})	0.5 b		2.1 a	

[a] Bulk density calculated from porosity where BD = (1-proportion porosity) × particle density.
[b] A measure of aggregate stability.
[c] Index value calculated by aggregating scores of individual parameters; scores assigned based on estimated optimal values.
[d] Values followed by the same letter within the same row are not significantly different at the $p < 0.05$ level.
NS = Not Significant.
(Adapted from Karlen, D.L.; et al. Long-term tillage effects on soil quality. Soil Till Res. 1994, 32(4), 313–327.)

marginal improvements in soil quality (25). Simulation models will be useful in assessing the dynamic biophysical tradeoffs for a wide range of what-if scenarios and management options relating soil and water quality (26). The tradeoffs that result from emphasizing one outcome over another—for example, productivity over nitrate concentration in an adjacent water body—are difficult to evaluate using traditional cost–benefit analyses. Interdisciplinary efforts to identify human preferences and assign economic value to nature, health, and other elusive human-welfare outcomes are being used to bridge this gap (27–30).

CONCLUSIONS

Several decades ago, the concept of water quality itself was unfamiliar and controversial (31). Evolving through this same interdisciplinary process, water quality has become the centerpiece, along with air quality, of resource-management strategy and policy. Now, the very dependence of water quality on soils ensures that farmers and developers, individuals and governments, will also need to consider soil quality in making resource-use decisions.

REFERENCES

1. Mannon, A.M. The Environmental Impact of Agriculture in Middle and High Latitudes. In *Agriculture and Environmental Change: Temporal and Spatial Dimensions*; John Wiley and Sons: Chichester, England, 1995; 227–263.
2. Tilman, D. Global Environmental Impacts of Agricultural Expansion: The Need for Sustainable and Efficient Practices. Proc. Natl. Acad. Sci. U.S.A. **1999**, *96* (11), 5995–6000.
3. Lander, C. *Minimizing Agricultural Nonpoint Source Impacts with Management of Nutrients for Crop and Forage Production*; NRCS: Washington, D.C., 1994.
4. National Resources Conservation Service, *Water Quality and Agriculture: Status, Conditions, and Trends. Working Paper #16*, 1997; 1–125.http://www.nhq.nrcs.usda.gov/land/pubs/wq.html (accessed March 2001).
5. Pimentel, D.; Harvey, C.; Resosudarmo, P.; Sinclair, K.; Kurz, D.; McNair, M.; Crist, S.; Shpritz, L.; Fitton, L.; Saffouri, R.; Blair, R. Environmental and Economic Costs of Soil Erosion and Conservation Benefits. Science **1995**, *267* (5201), 1117–1123.
6. Wander, M.M.; Walter, G.; Nissen, T.M.; Bollero, G.A.; Andrews, S.S.; Cavanaugh-Grant, D. Soil Quality: Science and Process. Agron. J. **(In press).**
7. Cassman, K.G. Ecological Intensification of Cereal Production Systems: Yield Potential, Soil Quality, and Precision Agriculture. Proc. Natl. Acad. Sci. U.S.A. **1999**, *96* (11), 5952–5959.
8. Karlen, D.L.; Stott, D.E. A Framework for Evaluating Physical and Chemical Indicators of Soil Quality. In

Q

Defining Soil Quality for a Sustainable Environment. SSSA Special Publication no. 35; Soil Science Society of America: Madison WI, 1994; 53–72.
9. Seybold, C.A.; Mausback, M.J.; Karlen, D.J.; Rogers, H.H. Quantification of Soil Quality. In *Carbon Sequestration in Soil. Advances in Agronomy*; Stewart, B.A., Lal, R., Eds.; Academic Press: San Diego, 1997; 387–404.
10. Brye, K.R.; Norman, J.M.; Bundy, L.G.; Gower, S.T. Water-Budget Evaluation of Prairie and Maize Ecosystems. Soil Sci. Soc. Am. J. **2000**, *64* (2), 715–724.
11. Wander, M.M.; Bollero, G.A. Soil Quality Assessment of Tillage Impacts in Illinois. Soil Sci. Soc. Am. J. **1999**, *63* (4), 961–971.
12. Chenu, C.; Le Bissonnais, Y.; Arrouays, D. Organic Matter Influence on Clay Wettability and Soil Aggregate Stability. Soil Sci. Soc. Am. J. **2000**, *64* (4), 1479–1486.
13. Six, J.; Elliott, E.T.; Paustian, K.; Doran, J.W. Aggregation and Soil Organic Matter Accumulation in Cultivated and Native Grassland Soils. Soil Sci. Soc. Am. J. **1998**, *62* (5), 1367–1377.
14. Beare, M.H.; Hendrix, P.F.; Coleman, D.C. Water-Stable Aggregates and Organic Matter Fractions in Conventional-Tillage and No-Tillage Soils. Soil Sci. Soc. Am. J. **1994**, *58* (3), 777–786.
15. Karlen, D.L., Wright, R.J., Kemper, W.D., Eds. *Agricultural Utilization of Urban and Industrial By-Products. ASA Special Publication 58*; ASA, CSSA, and SSSA: Madison, WI, 1995.
16. Kelly, T.C.; Lu, Y.C.; Teasdale, J. Economic-Environmental Tradeoffs Among Alternative Crop Rotations. Agric. Ecosyst. Environ. **1996**, *60* (1), 17–28.
17. Halvorson, J.J.; Smith, J.L.; Papendick, R.I. Issues of Scale for Evaluating Soil Quality. J. Soil Water Conserv. **1997**, *52* (1), 26–30.
18. Forster, D.L.; Richards, R.P.; Baker, D.B.; Blue, E.N. EPIC Modeling of the Effects of Farming Practice Changes on Water Quality in Two Lake Erie Watersheds. J. Soil Water Conserv. **2000**, *55* (1), 85–90.
19. Meyer, L.D. Evolution of the Universal Soil Loss Equation. J. Soil Water Conserv. **1984**, *March-April* 99–104.
20. Karlen, D.L. Opportunities and Challenges Associated with Watershed-Scale Assessments of Soil and Water Quality. J. Soil Water Conserv. **1999**, *54* (4), 626–627.
21. Bouzaher, A.; Lakshminarayan, P.G.; Cabe, R.; Carriquiry, A.; Gassman, P.W.; Shogren, J.F. Metamodels and Nonpoint Pollution Policy in Agriculture. Water Resour. Res. **1993**, *29* (6), 1579–1587.
22. Wu, J.J.; Babcock, B.A. Metamodeling Potential Nitrate Water Pollution in the Central United States. J. Environ. Qual. **1999**, *28* (6), 1916–1928.
23. Lovejoy, S.B.; Lee, J.G.; Randhir, T.O.; Engel, B.A. Research Needs for Water Quality Management in the 21st Century: A Spatial Decision Support System. J. Soil Water Conserv. **1997**, *52* (1), 18–22.
24. Groffman, P.M. Nitrogen in the Environment. In *Handbook of Soil Science*; Sumner, M.E., Ed.; CRC Press: Boca Raton, FL, 2000; C-190–C-200.
25. Jaenicke, E.C.; Lengnick, L.L. A Soil-Quality Index and Its Relationship to Efficiency and Productivity Growth Measures: Two Decompositions. Am. J. Agr. Econ. **1999**, *81* (4), 881–893.
26. Wagenet, R.J.; Hutson, J.L. Soil Quality and Its Dependence on Dynamic Physical Processes. J. Environ. Qual. **1997**, *26* (1), 41–48.
27. Smith, V.; Desvousges, W. *Measuring Water Quality Benefits*; Kluwer-Nijhoff: Boston, 1986.
28. Fox, G.; Umali, G.; Dickinson, T. An Economic Analysis of Targeting Soil Conservation Measures with Respect to Off-Site Water Quality. Can. J. Agric. Econ.-Rev. Can. Econ. Rural. **1995**, *43* (1), 105–108.
29. Bockstael, N.E.; Freeman, A.M.; Kopp, R.J.; Portney, P.R.; Smith, V.K. On Measuring Economic Values for Nature. Environ. Sci. Technol. **2000**, *34* (8), 1384–1389.
30. Andrews, S.S.; Lohr, L.; Cabrera, M.L. A Bioeconomic Decision Model Comparing Composted and Fresh Litter for Winter Squash. Agric. Syst. **1999**, *61* (3), 165–178.
31. Somlyody, L. Water-quality Management—Can We Improve Integration to Face Future Problems? Water Sci. Technol. **1995**, *31* (8), 249–259.

QUALITY AND SUSTAINABLE AGRICULTURE

John W. Doran
United States Department of Agriculture-Agricultural Research Service (USDA-ARS), University of Nebraska, Lincoln, Nebraska, U.S.A.

E.G. Gregorich
Agriculture and Agri-Food Canda Central Experimental Farm, Ottawa, Ontario, Canada

INDRODUCTION

Sustainable agriculture is a way of farming that can be carried out for generations to come. This long-term approach to agriculture combines efficient production with the wise stewardship of the Earth's resources, including land management to maintain or enhance the quality of agricultural soils. Over time, sustainable agriculture is expected to meet human needs for food and fiber; protect the natural resource base, preventing the degradation of soil, water, and air quality and conserving biodiversity; use nonrenewable resources efficiently; work with natural biological cycles and controls; and assure the economic survival of farming and the well-being of farmers and their families. Thus, the concept of sustainability embraces three distinct but overlapping areas of concern: social, economic, and environmental factors. Taking a holistic ecosystem approach to agriculture is the key to integrating these components of sustainability.

Growing human populations, diminishing nonrenewable resources, social and political instability, and environmental degradation threaten the natural processes that sustain the global ecosphere and life on Earth (1, 2). With losses of farmland to degradation and urbanization and little new agricultural land to develop, meeting the needs of future populations for food, fiber, and other agricultural products depends on a significant increase in crop yields. Under current food production practices, this yield increase spells greater use of inputs (e.g., energy, fertilizer, pesticides), and raises the question of the environmental costs of increased production.

AGRICULTURE AND ENVIRONMENTAL SUSTAINABILITY

To a large extent, the economic and social viability of agriculture depends on the sustainability of the resources that agriculture both depends on and affects. Farmers have always intuitively known that good soil quality underpins the success of their operations, and soil conservation programs in many countries throughout the last century supported this recognition. Since the late 1980s, interest has grown in a fuller accounting of the environmental costs of agricultural production. These costs include declining water quality and competition for finite water resources, declining air quality, emission of greenhouse gases, loss of wildlife habitat, and decline in species and genetic diversity (3). To a large extent, conserving and enhancing soil quality can mitigate these costs.

Soil

Agricultural land management in the past has taxed the soil system, often withdrawing more than it has returned. In particular, mechanical cultivation and the continuous production of row crops has resulted in the displacement and loss of soil by erosion, large decreases in soil organic matter, and a concomitant release of carbon as carbon dioxide into the atmosphere (4). Within the last decade, inventories of soil productive capacity indicate human-induced degradation on nearly 40% of the Earth's arable land as a result of land clearing, extensive soil cultivation, soil erosion, atmospheric deposition of pollutants, over-grazing, salinization, and desertification (5). The projected doubling of the human population in the next century threatens even greater degradation of soils and other natural resources (6).

Water

Intensive farming practices have, in many parts of the world, jeopardized the quality of surface water and groundwater through the addition of agriculturally derived nitrate nitrogen, phosphorus, pesticides, sediment, and pathogens (e.g., bacteria from manure). Agriculture is considered the most widespread contributor to nonpoint-source water pollution in the U.S. (7). The major agricultural water contaminant in North America and Europe is nitrate nitrogen derived from atmospheric deposition, livestock manures, and commercial fertilizers. Human alterations of the nitrogen cycle have almost doubled the rate of nitrogen input to terrestrial ecosystems over the past 30 years, resulting in large increases in the transfer of nitrogen from land to the atmosphere and to rivers, estuaries, and coastal waters (8–10).

Agriculture, competing as it does with many other uses of fresh water, is also at the heart of current discussions of water quantity and is a key consideration in the trend toward demand management, rather than supply management, of water resources. Global warming scenarios predict water shortages in some existing agricultural areas in the world, and water surpluses in others. Changing conditions will call for new ways of farming and of resolving conflict among water users.

Controlling soil erosion reduces the amount of sediments entering waterways, along with the chemicals (e.g., pesticides) and pathogens they carry. Nutrient management plans that account for all sources of nitrogen and phosphorus and match nutrient application more closely to plant needs reduce the risk of nitrogen and phosphorus entering groundwater and surface water. Maintaining soil quality improves the ability of soil to receive and partition water, making the most efficient use of this resource.

Air and Atmosphere

Atmospheric concentrations of greenhouse gases—particularly nitrous oxide, methane, and carbon dioxide—have been increasing dramatically over the past 20 years, enhancing the greenhouse effect by which the Earth's atmosphere is warmed. Agriculture is a significant contributor of these gases to the atmosphere; in Canada, for example, agriculture accounted for about 10% of total 1996 emissions of these gases, an increase of about 4% since 1981 (11). Soil management practices such as tillage, cropping patterns, and the use of manure and commercial fertilizers influence atmospheric quality through changes in the soil's capacity to produce or consume greenhouse gases (12, 13). Under the United Nations Framework Convention on Climate Change and its Protocols, signatory countries are considering ways to reduce emissions, including those from agriculture, and also to capitalize on agricultural sinks of greenhouse gases, such as by carbon sequestration in agricultural soils. The present threat of ozone depletion and global climate change necessitates a better understanding of the influence of land management on soil processes.

Agricultural emissions of ammonia (from fertilizer and manure) and particulate matter (dust released during soil tillage and subsequent erosion) have been linked to various environmental effects, such as acidification, eutrophication, and smog. Reducing the emission of these substances involves improving the quality of soils so that they are more resistant to erosion, reducing tillage and increasing the amount of soil cover, and practicing better nitrogen management (both manure and commercial fertilizer).

Agroecosystem Biodiversity

Agriculture benefits from biodiversity in many ways, but it has also reduced biodiversity over the years, mainly through the conversion of natural habitats, but also through effects on soil and water quality. Improving soil quality not only maintains a healthy biological community in the soil itself, along with the functions these organisms carry out, but also supports the other flora and fauna dependent on this community. Soil quality improvements that benefit water quality can also improve the viability of aquatic ecosystems.

ASSESSMENT OF SOIL QUALITY

Assessment of soil quality or health is invaluable in determining the sustainability of land management systems (14). Soil quality is the major link between the strategies of conservation management practices and the achievement of the major goals of agriculture (15, 16).

Assessment of soil quality can be used to identify problem production areas, make realistic estimates of food production, monitor changes in sustainability and environmental quality as related to agricultural management, and assist government agencies in formulating and evaluating sustainable agricultural and land use policies (Fig. 1). Using simple indicators of soil quality and soil health that have meaning to farmers and other land managers is likely the most fruitful means of linking science with practice in assessing the sustainability of land management practices (17).

Doran et al. (18) stressed the importance of holistic management approaches that optimize the multiple functions of soil, conserve soil resources, and support strategies for promoting soil quality. They proposed a basic set of indicators to assess soil quality and health in various agricultural management systems. Many of these key indicators are very useful to specialists (e.g., researchers, consultants, extension staff, and conservationists) but are beyond the expertise or time constraints of the producer to measure. Success in developing and implementing standards for assessing soil quality and sustainability hinges on partnership with agricultural producers, who are the primary

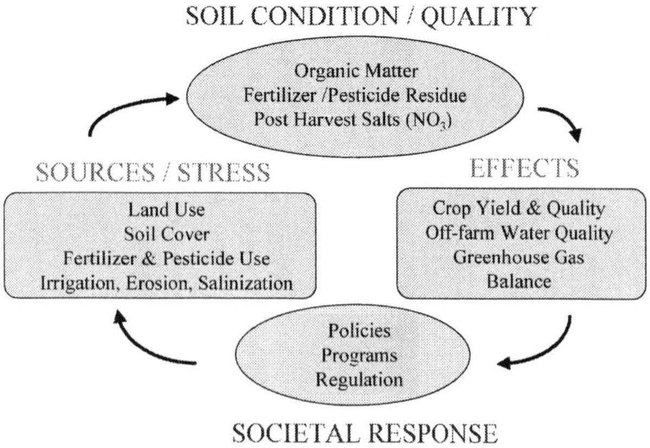

Fig. 1 The politics of soil health. The assessment of soil condition/quality is needed to monitor changes in sustainability and environmental quality as related to agriculture management and to assist governmental agencies in formulating realistic agricultural and land use policies.

stewards of the land and the chief decision makers regarding land use and management. In light of this consideration, Doran et al. (18) also presented strategies for sustainable management that include a more generic set of indicators to assess soil quality and health that are practical for producers (Table 1). A study conducted in the northern U.S. corn and dairy belt used a similar approach to determine how farmers assess soil quality and health and found that they ranked soil organic matter, crop appearance, and risk to erosion as the three most important properties for describing soil health and sustainable management (17). Such strategies maximize the benefits of natural cycles, reduce dependence on nonrenewable resources, and help producers identify long-term goals for sustainability that also meet short-term needs for production. Soil quality assessment must be directed at the financial survival of the farm as well as environmental preservation.

Although much remains to be done, useful models exist for translating soil science into practice. For example, Gomez et al. (19) provide a practical framework for determining the sustainability of hill country agriculture in the Philippines. It uses indicators that satisfy both the needs of the farmer (e.g., productivity, profitability, stability, and viability) and the conservation of soil and water resources. Threshold values for sustainability are identified relative to the average local conditions for crop yield, profit, risk of crop failure, soil depth, percent soil cover, and soil organic matter content. This conceptual framework for assessing sustainability could be expanded to include other needs of society and environmental conservation. In particular, adding a category for balancing energy input and output, as well as monetary costs, would better assess the short- and long-term sustainability of management and the value of greater reliance on renewable resources and less dependence on fossil fuels and petrochemicals in enhancing economic, ecological, and environmental resources.

Table 1 Strategies for sustainable agricultural management and proposed indicators of crop performance and soil and environmental health

Sustainability strategy	Indicators for producers
Conserve soil organic matter through maintaining soil C and N levels by reducing tillage, recycling plant and animal manures, and/or increasing plant diversity where C inputs C outputs	Direction/change in organic matter levels with time (visual or remote sensing by color or chemical analysis) Specific organic matter potential for climate, soil, and vegetation Soil water storage
Minimize soil erosion through conservation tillage and increased protective cover (residue, stable aggregates, cover crops, green fallow)	Visual (gullies, rills, dust, etc.) Surface soil properties (topsoil depth, organic matter content/texture, water infiltration, runoff, ponding, % cover)
Balance production and environment through conservation and integrated management systems (optimizing tillage, residue, water, and chemical use) and by synchronizing available N and P levels with crop needs during the year	Crop characteristics (visual or remote sensing of yield, color, nutrient status, plant vigor, and rooting characteristics) Soil physical condition/compaction Soil and water nitrate levels Amount and toxicity of pesticides used
Better use of renewable resources through relying less on fossil fuels and petrochemicals and more on renewable resources and biodiversity (e.g., crop rotations, legumes, manures, integrated pest management)	Input and output ratios of costs and energy Leaching losses/soil acidification Crop characteristics (as listed above) Soil and water nitrate levels

(From Ref. 2.)

Expanding the list of resource conservation variables to include leachable salts (especially nitrate), measured as soil electrical conductivity at the time of fertilization and after harvest, would permit land managers to better quantify the impact of agricultural practices on air and water quality.

Confirmation of the effectiveness of systems for residue management, organic matter formation, nitrogen and carbon cycling, soil structure maintenance, and biological control of pests and diseases will assist in discovering approaches that are both profitable and environmentally sound. The challenge in the future will be to better use the diversity and resiliency of the soil biological community to maintain a quality ecosystem, thus fostering sustainability. Sustainability strategies must be fine-tuned using such practices as crop rotation for greater crop diversity and tighter cycling of nutrients, reduction of soil disturbance to maintain soil organic matter and reduce erosion, and development of systems that make greater use of renewable biological resources. Ultimately, the indicators of soil quality and strategies for sustainable management must be linked to the development of management systems that foster reduction in the inputs of nonrenewable resources, maintain acceptable levels of productivity, and minimize impact on the environment.

REFERENCES

1. Bouma, J. Soil Environmental Quality: A European Perspective. J. Environ. Qual. **1997**, *26*, 26–31.
2. Postel, S.; et al. Carrying Capacity: Earth's Bottom Line. In *State of the World*; Brown, L.R., Ed.; W.W. Norton & Co.: New York, US, 1994; 3–21.
3. Lal, R. Basic Concepts and Global Issues: Soil Quality and Agricultural Sustainability. In *Soil Quality and Agricultural Sustainability*; Lal, R., Ed.; Ann Arbor Press: Chelsea, MI, 1998; 3–12.
4. Houghton, R.A.; Hobbie, J.E.; Melillo, J.M.; Moore, B.; Peterson, B.J.; Shaver, G.R.; Woodwell, G.M. Changes in the Carbon Content of Terrestrial Biota and Soils Between 1860 and 1980: A Net Release of CO_2 to the Atmosphere. Ecol. Monogr. **1983**, *53*, 235–262.
5. Doran, J.W.; Sarrantonio, M.; Liebig, M.A. Soil Health and Sustainability. In *Advances in Agronomy*; Vol. 56, Sparks, D.L., Ed.; Academic Press: San Diego, CA, 1996; 1–54.
6. Ruttan, V.W. The Transition to Agricultural Sustainability. Proc. Natl. Acad. Sci. **1999**, *96*, 5960–5967.
7. National Research Council, *Soil and Water Quality: An Agenda for Agriculture, Committee on Long-Range Soil and Water Conservation, Board on Agriculture, National Research Council*; National Academy Press: Washington, DC, 1993; 516.
8. Matson, P.A.; Parton, W.J.; Power, A.G.; Swift, M.J. Agricultural Intensification and Ecosystem Processes. Science **1997**, *277*, 504–509.
9. Socolow, R.H. Nitrogen Management and the Future of Food: Lessons from the Management of Energy and Carbon. Proc. Natl. Acad. Sci. **1999**, *96*, 6001–6008.
10. Vitousek, P.M.; Aber, J.D.; Howarth, R.W.; Likens, G.E.; Matson, P.A.; Schindler, D.W.; Schlesinger, W.H.; Tilman, D.G. Human Alteration of the Global Nitrogen Cycle: Sources and Consequences. Ecol. Appl. **1997**, *7*, 737–750.
11. McRae, T., Smith, C.A.S., Gregorich, L.J., Eds. *Environmental Sustainability of Canadian Agriculture*; Report of the Agri-Environmental Indicator Project, Agriculture and Agri-Food Canada: Ottawa, Ont., 2000; 224.
12. Mosier, A.R. Soil Processes and Global Change. Biol. Fert. Soils **1998**, *27*, 221–229.
13. Rolston, D.E.; Harper, L.A.; Mosier, A.R.; Duxbury, J.M. *Agricultural Ecosystem Effects on Trace Gases and Global Climate Change*; American Society of Agronomy Spec. Publ. 55, American Society of Agronomy: Madison, WI, 1993.
14. Karlen, D.L.; Mausbach, M.J.; Doran, J.W.; Cline, R.G.; Harris, R.F.; Schuman, G.E. Soil Quality: A Concept, Definition, and Framework for Evaluation. Soil Sci. Soc. Am. J. **1997**, *61*, 4–10.
15. Acton, D.F.; Gregorich, L.J. *The Health of Our Soils: Toward Sustainable Agriculture in Canada*; Agriculture and Agri-Food Canada: Ottawa, Ont, 1995.
16. Parr, J.F.; Papendick, R.I.; Hornick, S.B.; Meyer, R.E. Soil Quality: Attributes and Relationship to Alternative and Sustainable Agriculture. Am. J. Altern. Agric. **1992**, *7*, 5–11.
17. Romig, D.E.; Garlynd, M.J.; Harris, R.F.; McSweeney, K. How Farmers Assess Soil Health and Quality. J. Soil Water Conserv. **1995**, *50*, 229–236.
18. Doran, J.W.; Jones, A.J.; Arshad, M.A.; Gilley, J.E. Determinants of Soil Quality and Health. In *Soil Quality and Soil Erosion*; Lal, R., Ed.; CRC Press: Boca Raton, FL, 1999; 17–36.
19. Gomez, A.A.; Swete Kelly, D.E.; Seyers, J.K.; Coughlan, K.J. Measuring Sustainability of Agricultural Systems at the Farm Level. In *Methods for Assessing Soil Quality*; Soil Sci. Soc. Am. Spec. Publ. 49, Doran, J.W., Jones, A.J., Eds.; SSSA: Madison, WI, 1996; 401–410.

RADIONUCLIDES

Philip M. Jardine
Oak Ridge National Laboratory, Oak Ridge, Tennessee, U.S.A.

INTRODUCTION

Soil, the thin veneer of matter covering the Earth's surface and supporting a web of living diversity, is often abused through anthropogenic inputs of toxic waste. The disposal of radioactive waste generated at U.S. Department of Energy (DOE) facilities within the Weapons Complex has historically involved shallow land burial in unsaturated soils and sediments. Disposal methods from the 1940s to the 1980s ranged from unconfined pits and trenches to single- and double-shell buried steel tanks. Most of the below-ground burial strategies were deemed to be temporary (i.e., an average life span of several decades) until suitable technologies were developed to deal with the legacy waste issues. Technologies for retrieving and treating the below-ground radionuclide waste inventories have been slow to evolve and are often cost prohibitive or marginally effective. The scope of DOE's disposal problem is massive, with landfills estimated to contain more than 3 million cubic meters of radioactive and hazardous buried waste; a significant proportion of which migrated into surrounding soils and groundwater. It is estimated that the migration of these waste plumes contaminated over 600 billion gallons of water and 50 million cubic meters of soil.

The fate and transport of radionuclides through soil is controlled by coupled hydrologic, geochemical, and microbial processes. A multitude of complex soil processes are tightly linked that can both accelerate and impede the subsurface mobility of radioactive contaminants. Often the extent and magnitude of subsurface biogeochemical reactions is controlled by the spatial and temporal variability in soil hydrologic processes.

FATE AND TRANSPORT PROCESSES

Hydrologic Processes

Soil is a complex continuum of pore regions ranging from large macropores at the mm scale to small micropores at the sub-μm scale. It is the physical properties of the media (e.g., structured or layered), coupled with the duration and intensity of precipitation events that dictates the avenues of water and radionuclide movement through the subsurface. In humid environments where structured media is commonplace, transient storm events invariably result in the preferential migration of water (1–8). Highly conductive voids within the media (e.g., fractures, macropores) carry water around low-permeability, high-porosity matrix blocks or aggregates resulting in water bypass of the latter. In these humid regimes, recharge rates are very high with more than 50% of the infiltrating precipitation resulting in groundwater and surface water recharge. This condition promotes the formation of massive contaminant plumes in the soil since storm flow and groundwater interception with waste trenches is frequent and long-lasting. Even in semiarid environments, where recharge is typically small, subsurface preferential flow is a key mechanism controlling water and solute mobility (9, 10). Lithologic discontinuities and sediment layering promote perched water tables and unstable wetting fronts that drive both lateral and vertical subsurface preferential flow.

In both humid and semiarid regimes, water that is preferentially flowing through the soil media often remains in intimate contact with the porous matrix, and physical and hydrologic gradients drive the exchange of mass from one pore regime to another. Mass exchange is time-dependent and is often controlled by diffusion to and from the matrix. Thus, a significant inventory of radionuclide waste can reside within the soil matrix. This waste source is hydrologically linked to preferred flow paths which significantly enhances the extent and longevity of subsurface contaminant plumes. This scenario is commonplace at the Oak Ridge National Laboratory, located in eastern Tennessee, U.S., where thousands of underground disposal trenches and ponds have contributed to the spread of radionuclides such as ^{137}Cs, ^{60}Co, ^{90}Sr, and $^{235/238}U$ across tens of kilometers of landscape. Highly concentrated contaminant plumes move through soil and groundwater at time scales of meters per day (see Fig. 1) since the soils are highly structured and conducive to rapid preferential flow. However, the soil matrix, which has a high porosity and low permeability, serves as a source/sink for contaminants (5, 6, 11). The preferential movement of water and radionuclides through the subsurface also significantly impacts geochemical and microbial processes by controlling the extent and rate of various reactions with the solid phase. It imposes

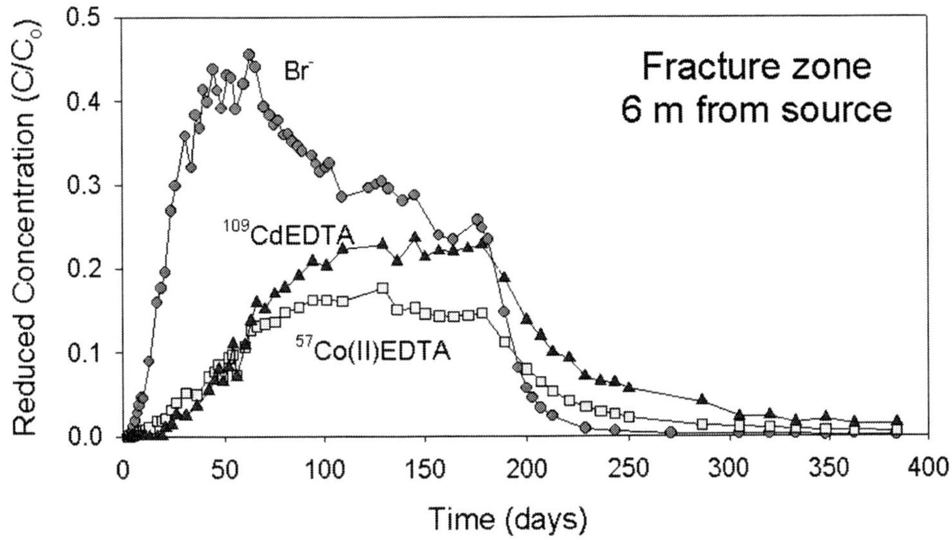

Fig. 1 Field-scale fate and transport of nonreactive Br^- and reactive $^{57}Co(II)EDTA^{2-}$ and $^{109}CdEDTA^{2-}$ in fractured subsurface media at the Oak Ridge National Laboratory. Although transport rates are rapid, geochemical reactions significantly impede the mobility of the chelated radionuclides as is indicated by their delayed breakthrough. (From Ref. 40.)

kinetic constraints on biogeochemical reactions and limits the surface area of interaction by partially excluding water and mass from the matrix porosity.

Geochemical Processes

Radionuclide fate and transport in soil and sediments is also controlled by interfacial reactions with the soil solid phase. Most soils are a complex mixture of variably charged phyllosilicates, redox reactive Fe- and Mn-oxides, organic matter, and mineral carbonates. Radionuclides interact with these solid phases through coulombic exchange, chemisorption, redox alterations, transformation processes such as polymerization, precipitation/dissolution, and complexation reactions. Both the extent and rate of these processes can be significantly influenced by variations in water content and the degree of pore regime connectivity. To make matters worse, radionuclide waste generated at the U.S. DOE facilities was often co-disposed with various chelating agents and organic acids. These synthetic organic constituents form highly stable, water-soluble complexes with a wide variety of radionuclides (12, 13). The presence of the complexing agent significantly alters the geochemical behavior of the disposed contaminants in soils and sediments through increased solubility, accelerated redox reactions, and ionic charge reversal.

The geochemical mechanism controlling the fate and transport of chelated radionuclides has been well characterized in numerous soils and subsurface materials (14–22). Typically, Fe(III) and Mn(IV) oxyhydroxides are the dominant subsurface mineral assemblages that catalyze co-contaminant oxidation/reduction and dissociation reactions (see Fig. 2). The mineral oxides have repeatedly been shown to catalyze the oxidation of $^{60}Co(II)EDTA^{2-}$ to $^{60}Co(III)EDTA^{-}$, thereby adversely enhancing the transport and persistence of ^{60}Co in a variety of subsurface environments ranging from aquifer sands to fractured weathered shale saprolites (15, 17, 18, 20, 22). Further, Fe(III)-oxides have also been shown to effectively dissociate a large number of chelated metal and radionuclide complexes (e.g., ^{60}Co–, ^{90}Sr–EDTA) through ligand competition (15, 20–22).

Certain radionuclides such as ^{137}Cs do not form strong bonds with many of the chelating agents and organic acids that were used during decontamination. Nevertheless, these radionuclides still interact aggressively with the soil solid phase. In the case of ^{137}Cs, 2:1 phyllosilicates and micas serve as excellent sorbents since the interlayer spaces of these mineral assemblages strongly attenuate the radionuclide. The migration tendency of ^{137}Cs in soils is often related to colloid mobility of contaminated sediments (23) or cation competition for surface sites in harsh environments such as those found beneath the Hanford tank farms in western Washington State, U.S. (24).

Microbial Processes

Radionuclides such as ^{60}Co and $^{235/238}U$ can exist in more than one oxidation state, and their behavior in

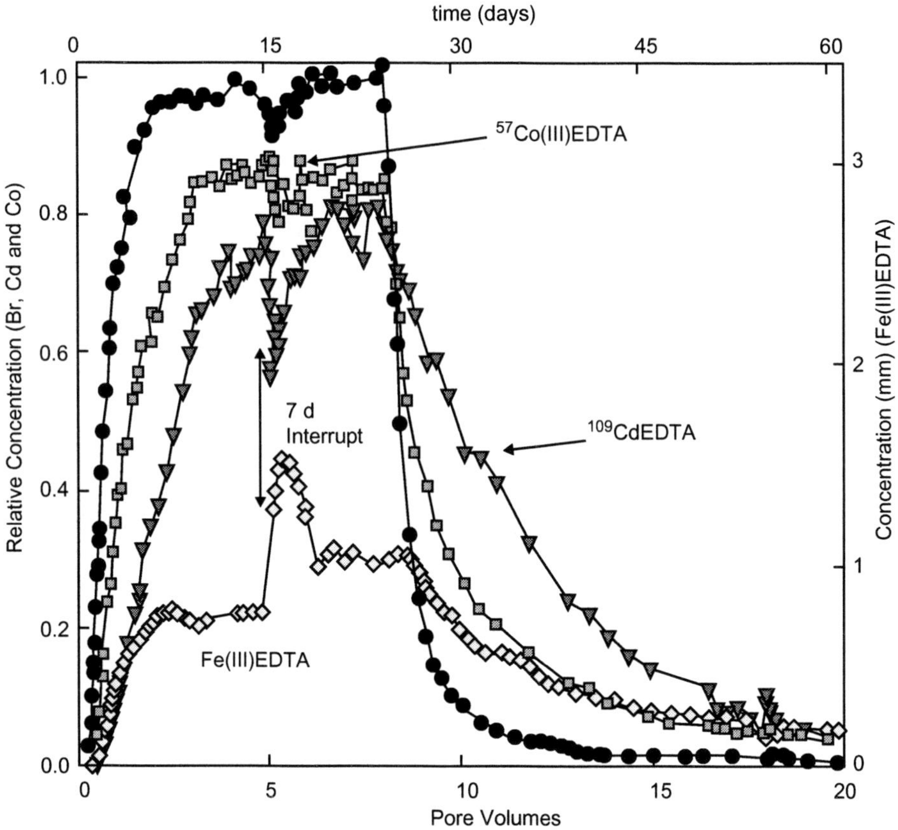

Fig. 2 Fate and transport of nonreactive Br^- and reactive $^{57}Co(II)EDTA^{2-}$ and $^{109}CdEDTA^{2-}$ in undisturbed soil columns of fractured weathered shale. Geochemical reactions impede the mobility of the chelated radionuclides. $Co(II)EDTA^{2-}$ is oxidized to $Co(III)EDTA^-$ where Mn-oxides serve as the oxidant. Fe-oxides effectively dissociate $CdEDTA^{2-}$ complexes resulting in the formation of free Cd and $Fe(III)EDTA^-$. A flow interruption technique was employed to quantify the presence of physical and geochemical nonequilibrium processes. (From Ref. 22.)

the environment depends on their oxidation state. For example, U(VI) is soluble and mobile in the environment whereas U(IV) is much less soluble and relatively immobile. Likewise, the oxidized ^{60}Co(III)EDTA complexes are much more stable and exhibit greater mobility in subsurface environments than the reduced ^{60}Co(II)EDTA (15, 17, 18). Subsurface Fe- and Al-oxides can effectively dissociate the Co(II)EDTA complex to Fe(III)EDTA (20) and Al(III)EDTA (16), respectively, and aqueous Co^{2+} is free to participate in sorption or precipitation reactions. Co(III) EDTA, on the other hand, is unaffected by Fe(III)- and Al-oxides. Therefore, the oxidized forms of these radionuclides and metals promote their undesirable enhanced migration through subsurface environments.

Numerous metal-reducing bacteria have been isolated that enzymatically reduce toxic metals and radionuclides to stable end-products. Microbial reduction of U(VI) to form the sparingly soluble U(IV) has been shown using chemostat experiments for a number of metal-reducing bacteria (25, 26). Gorby et al. (27) have also shown that certain metal-reducing bacteria can link the enzymatic reduction of $^{60}Co(III)EDTA^-$ to support cell growth. Recently, important advances have been made towards implementing field-scale microbially mediated metal reduction strategies in oxygen-deficient environments. Several studies have investigated contaminant reduction in the presence of solid phase material (27–29). Gorby et al. (27) have shown that the metal-reducing bacterium *Shewanella alga* preferentially reduced $Co(III)EDTA^-$ to $Co(II)EDTA^{2-}$ in the presence of Mn-oxides. Likewise, Wielinga et al. (29) documented the bioreduction of U(VI) by *Shewanella alga* in the presence of various Fe-oxide mineral phases. These authors noted that the rate of U(VI) bioreduction was unaffected in the presence of goethite and only slightly diminished in the presence of poorly crystalline Fe(III)-oxides, where the latter Fe solid phase effectively competed as a terminal electron acceptor. Recent studies by Brooks et al. (28) showed the sustained microbial reduction of Co(III)EDTA$^-$ under dynamic flow conditions. The net reduction of the Co(III)EDTA$^-$ dominated the fate and transport of the contaminant even in the presence of strong mineral oxidants such as Mn- and Fe-oxides that are known to effectively reoxidize Co(II)EDTA^{2-} back to Co(III)EDTA$^-$ (15, 18, 20). The research findings of Brooks et al. (28) provide new and important information on how to successfully implement a bioreduction strategy at the field scale. Their use of a dynamic flow system with sustained bacterial growth conditions in geochemically reactive media is consistent with contaminant migration scenarios in situ.

The studies of Brooks et al. (28), however, used uniformly packed media that contained little structure. Undisturbed subsurface soils and geologic material consist of a complex continuum of pore regions ranging from large macropores and fractures at the mm scale to small micropores at the sub-μm scale. Structured media, common to most subsurface environments throughout the world, accentuates this physical condition that often controls the geochemical and microbial processes affecting solute transport. Redox sensitive radionuclides such as U(VI), Co(III)EDTA, and Tc(VI) reside within nearly all of the pore structure of the subsurface media, with the greatest concentration of contaminants held within micropores (2, 3). Bacteria that are capable of reducing these contaminants are too big to reach a large fraction of the micropore regime and are largely restricted to macro- and mesopore domains (30, 31). Fortunately, the pore structure of the media is hydrologically interconnected, and contaminants move from one pore class to another via hydraulic and concentration gradients (4, 6, 7). This process is slow, however, and is often the rate-limiting factor governing the success of contaminant bioremediation. Thus, faster-flowing fracture-dominated regimes will most likely be physically more appealing for sustained bioreduction as long as a suitable electron donor can be supplied. In contrast, bioreduction processes in slower-flowing matrix regimes will most likely be limited by rate-dependent mass transfer of contaminants from smaller pores into larger pores.

Certain bacteria are also capable of degrading chelates and thus potentially immobilizing radionuclides in situ. The biodegradation of the commonly used aminopolycarboxylate chelates NTA, EDTA, and DTPA have been studied in soil and sediment systems for many years (32, 33, 35). Research has shown that NTA has the greatest potential for biodegradation in subsurface systems compared with the other aminopolycarboxylates (34, 35).

Bolton et al. (36) and Bolton and Girvin (37) have shown that the bacterial strain *Chelatobacter heintzii* (ATCC 29600) is capable of degrading NTA in the presence of many different toxic metals and radionuclides. Likewise, Payne et al. (38) and Liu et al. (39) have deciphered the mechanisms by which certain bacteria degrade radionuclide–EDTA complexes. These studies lend promise to the potential for using bacteria to biodegrade chelates and enhance the geochemical immobilization of radionuclides in situ.

SUMMARY

Radionuclide fate and transport in soils is controlled by coupled time-dependent hydrologic, geochemical, and microbial processes. Hydrologic processes such as preferential flow and matrix diffusion can serve to both accelerate and impede radionuclide migration, respectively. Preferential flow results in hydraulic, physical, and geochemical nonequilibrium conditions since differences in fluid velocities and solute concentrations in different-sized pores create hydraulic and concentration gradients that drive time-dependent inter-region advective and diffusive mass transfer. Thus, in soil systems with a large matrix porosity or a significant quantity of disconnected immobile water, radionuclide migration rates can be greatly retarded due to the slow transfer of mass to actively flowing preferential flow paths. Nevertheless, the prevalence of preferential flow can greatly accelerate the transport of mass in soil systems. Geochemical processes such as sorption, redox alterations, and dissociation reactions can also serve to both accelerate and impede radionuclide migration. Sorption and radionuclide-chelate dissociation reactions almost always result in retarded radionuclide migration rates, whereas oxidation reactions often result in more soluble, and thus more mobile, radionuclide species. Microbial processes can also potentially influence the fate and transport of radionuclides in soil. Metal-reducing bacteria and chelate degraders can alter the geochemical behavior of redox sensitive radionuclides which facilitates their immobilization via solid phase sorption and precipitation reactions.

Enhanced knowledge of the coupled hydrologic, geochemical, and microbial processes controlling radionuclide migration in soils will improve our conceptual understanding and predictive capability of the risks associated with spread of radioactive material in the subsurface environment. Too often risk assessment models treat soil and bedrock as inert media or assume that the media is in equilibrium with migrating contaminants.

Failure to consider the time-dependent coupled processes that control radionuclide migration will greatly over-predict the off-site contribution of contaminants from the primary waste source and thus provide an inaccurate assessment of pending risk. By recognizing the importance of soil processes on radionuclide migration, we can improve our decision-making strategies regarding the selection of effective remedial actions and improve our interpretation of monitoring results after remediation is complete.

REFERENCES

1. Shuford, J.W.; Fitton, D.D.; Baker, D.E. Nitrate-Nitrogen and Chloride Movement Through Undisturbed Field Soil. J. Environ. Qual. **1977**, *6*, 255–259.
2. Jardine, P.M.; Wilson, G.V.; Luxmoore, R.J. Unsaturated Solute Transport Through a Forest Soil During Rain Storm Events. Geoderma **1990**, *46*, 103–118.
3. Jardine, P.M.; Wilson, G.V.; McCarthy, J.F.; Luxmoore, R.J.; Taylor, D.L. Hydrogeochemical Processes Controlling the Transport of Dissolved Organic Carbon Through a Forested Hillslope. J. Contam. Hydrol. **1990**, *6*, 3–19.
4. Jardine, P.M.; O'Brien, R.; Wilson, G.V.; Gwo, J.P. Experimental Techniques for Confirming and Quantifying Physical Nonequilibrium Processes in Soils. In *Physical Nonequilibrium in Soils: Modeling and Application*; Selim, H.M., Ma, L., Eds.; Ann Arbor Press: Chelsea, MI, 1998; 243–271.
5. Jardine, P.M.; Wilson, G.V.; Luxmoore, R.J.; Gwo, J.P. Conceptual Model of Vadose-Zone Transport in Fractured Weathered Shales. *Conceptual Models of Flow and Transport in the Fractured Vadose Zone*; U.S. National Committee for Rock Mechanics. National Research Council, National Academy Press: Washington, DC, 2001; 87–114.
6. Wilson, G.V.; Jardine, P.M.; O'Dell, J.D.; Collineau, M. Field-Scale Transport from a Buried Line Source in Variable Saturated Soil. J. Hydrol. **1993**, *145*, 83–109.
7. Wilson, G.V.; Gwo, J.P.; Jardine, P.M.; Luxmoore, R.J. Hydraulic and Physical Nonequilibrium Effects on Multi-Region Flow and Transport. In *Physical Nonequilibrium in Soils: Modeling and Application*; Selim, H.M., Ma, L., Eds.; Ann Arbor Press: Chelsea, MI, 1998; 37–61.
8. Hornberger, G.M.; Germann, P.F.; Beven, K.J. Through-Flow and Solute Transport in an Isolated Sloping Soil Block in a Forested Catchment. J. Hydrol. **1991**, *124*, 81–97.
9. Porro, I.; Wierenga, P.J.; Hills, R.G. Solute Transport Through Large Uniform and Layered Soil Columns. Water Resour. Res. **1993**, *29*, 1321–1330.
10. Ritsema, C.J.; Dekker, L.W.; Nieber, J.L.; Steenhuis, T.S. Modeling and Field Evidence of Finger Formation and Finger Recurrence in a Water Repellent Sandy Soil. Water Resour. Res. **1998**, *34*, 555–567.
11. Jardine, P.M.; Sanford, W.E.; Gwo, J.P.; Reedy, O.C.; Hicks, D.S.; Riggs, R.J.; Bailey, W.B. Quantifying Diffusive Mass Transfer in Fractured Shale Bedrock. Water Resour. Res **1999**, *35*, 2015–2030.
12. Riley, R.G.; Zachara, J.M. *Chemical Contaminants on DOE Lands and Selection of Contaminant Mixtures for Subsurface Science Research*; DOE/ER-0547T, U.S. Govt. Print. Office: Washington, DC, 1992.
13. Toste, A.P.; Osborn, B.C.; Polach, K.J.; Lechner-Fish, T.J. Organic Analyses of an Actual and Simulated Mixed Waste: Hanford's Organic Complexant Site Revisited. J. Radioanal. Nucl. Chem. **1995**, *194*, 25–34.
14. Swanson, J.L. Effect of Organic Complexants on the Mobility of Low-Level Waste Radionuclides in Soils: Status Report PNL-3927, UC-70. 1981.
15. Jardine, P.M.; Jacobs, G.K.; O'Dell, J.D. Unsaturated Transport Processes in Undisturbed Heterogeneous Porous Media: II. Co-Contaminants. Soil Sci. Soc. Am. J. **1993**, *57*, 954–962.
16. Girvin, D.C.; Gassman, P.L.; Bolton, H. Adsorption of Aqueous Cobalt Ethylenediaminetetraacetate by δ-Al_2O_3. Soil Sci. Soc. Am. J. **1993**, *57*, 47–57.
17. Zachara, J.M.; Gassman, P.L.; Smith, S.C.; Taylor, D. Oxidation and Adsorption of $Co(II)EDTA^{2-}$ Complexes in Subsurface Materials with Iron and Manganese Oxide Grain Coatings. Geochim. Cosmochim. Acta. **1995**, *59*, 4449–4463.
18. Brooks, S.C.; Taylor, D.L.; Jardine, P.M. Reactive Transport of EDTA-Complexed Cobalt in the Presence of Ferrihydrite. Geochim. Cosmochim. Acta. **1996**, *60*, 1899–1908.
19. Read, D.; Ross, D.; Sims, R.J. The Migration of Uranium Through Clashach Sandstone: The Role of Low Molecular Weight Organics in Enhancing Radionuclide Transport. J. Contam. Hydrol. **1998**, *35*, 235–248.
20. Szecsody, J.E.; Zachara, J.M.; Chilakapati, A.; Jardine, P.M.; Ferrency, A.S. Importance of Flow and Particle-Scale Heterogeneity on Co(II/III)EDTA Reactive Transport. J. Hydrol. **1998**, *209*, 112–136.
21. Davis, J.A.; Kent, D.B.; Coston, J.A.; Hess, K.M.; Joye, J.L. Multispecies Reactive Tracer Test in an Aquifer with Spatially Variable Chemical Conditions. Water Resour. Res. **2000**, *36* (1), 119–134.
22. Mayes, M.A.; Jardine, P.M.; Larsen, I.L.; Brooks, S.C.; Fendorf, S.E. Multispecies Transport of Metal-EDTA Complexes and Chromate Through Undisturbed Columns of Weathered, Fractured Saprolite. J. Contam. Hydrol. **2000**, *45*, 243–265.
23. Solomon, D.K.; Marsh, J.D.; Larsen, I.L.; Wickliff, D.S.; Clapp, R.B. *Transport of Contaminants During Storms in the White Oak Creek and Melton Branch Watersheds*; ORNL/TM-11360, Oak Ridge National Laboratory: Oak Ridge, TN, 1991.
24. Serne, R.J.; Burke, D.S. *Chemical Information on Tank Supernatants, Cs Adsorption from Tank Liquids onto Hanford sediments, and Field Observations of Cs Migration from Past Tank Leaks*; Report No. PNNL-11495, Pacific Northwest National Laboratory: Richland, WA, 1997.
25. Gorby, Y.A.; Lovley, D.R. Enzymatic Uranium Precipitation. Environ. Sci. Technol. **1992**, *26*, 205–207.
26. Francis, A.J.; Dodge, C.J.; Lu, F.; Halada, G.P.; Clayton, C.R. XPS and XANES Studies of Uranium Reduction by *Clostridium* Sp. Environ. Sci. Technol. **1994**, *28*, 636–639.
27. Gorby, Y.A.; Caccavo, F.; Drektrah, D.B.; Bolton, H. Microbial Reduction of $Co(III)EDTA^{-}$ In the Presence and

Absence of Manganese(IV) Dioxide. Environ. Sci. Technol. **1998**, *32*, 244–250.

28. Brooks, S.C.; Carroll, S.L.; Jardine, P.M. Sustained Bacterial Reduction of Co(III)EDTA$^-$ In the Presence of Competing Geochemical Oxidation During Dynamic Flow. Environ. Sci. Technol. **1999**, *33*, 3002–3011.
29. Wielinga, B.; Bostick, B.; Rosenzweig, R.F.; Fendorf, S. Inhibition of Bacterially Promoted Uranium Reduction: Ferric (Hydr)Oxides as Competitive Electron Acceptors. Environ. Sci. Technol. **2000**, *34*, 2190–2195.
30. Smith, M.S.; Thomas, G.W.; White, R.E.; Ritonga, D. Transport of *Escherichia Coli* Through Intact and Disturbed Soil Columns. J. Environ. Qual. **1985**, *14*, 87–91.
31. McKay, L.D.; Cherry, J.A.; Bales, R.C.; Yahya, M.T.; Gerba, C.P. A Field Example of Bacteriophage as Tracers of Fracture Flow. Environ. Sci. Technol. **1993**, *27*, 1075–1079.
32. Tiedje, J.M. Microbial Degradation of Ethylenediaminetetraacetic Acid in Soils and Sediments. Appl. Environ. Microbiol. **1975**, *30*, 327–329.
33. Tiedje, J.M.; Mason, B.B. Biodegradation of Nitrilotriacetic Acid (NTA) in Soils. Soil Sci. Soc. Am. Proc. **1974**, *38*, 278–283.
34. Means, J.L.; Kucak, T.; Crerar, D.A. Relative Degradation Rates of NTA, EDTA, and DTPA and Environmental Implications. Environ. Pollut. Ser. B. **1980**, *1*, 45–60.
35. Bolton, H., Jr.; Li, S.E., Jr.; Workman, D.J.; Girvin, D.C. Biodegradation of Synthetic Chelates in Subsurface Sediments from the Southeast Coastal Plain. J. Environ. Qual. **1993**, *22*, 125–132.
36. Bolton, H.; Girvin, D.C.; Plymale, A.E.; Harvey, S.D.; Workman, D.J. Degradation of Metal-Nitrilotriacetate Complexes By *Chelatobacter Heintzii*. Environ. Sci. Technol. **1996**, *30*, 931–938.
37. Bolton, H., Jr.; Girvin, D.C., Jr. Effect of Adsorption on the Biodegradation of Nitrilotriacetate By *Chelatobacter Heintzii*. Environ. Sci. Technol. **1996**, *30*, 2057–2065.
38. Payne, J.W.; Bolton, H.; Campbell, J.A.; Xun, Y.L. Purification and Characterization of EDTA Monooxygenase from the EDTA-Degrading Bacterium BNC1. J. Bacteriology **1998**, *180*, 3823–3827.
39. Liu, Y.; Louie, T.M.; Payne, J.; Bohuslavek, J.; Bolton, H.; Xun, L.Y. Identification, Purification, and Characterization of Iminodiacetate Oxidase from the EDTA-Degrading Bacterium BNC1. Applied Environ. Microbiol. **2001**, *67*, 696–701.
40. Jardine, P.M.; Mehlhorn, T.L.; Larsen, I.L.; Bailey, W.B.; Brooks, S.C.; Roh, Y.; Gwo, J.P. Influence of Hydrological and Geochemical Processes on the Transport of Chelated metals and Chromate in Fractured Shale Bedrock. J. Contamin. Hydrol. **2001**, (in press).

REDOX PHENOMENA

Bruce R. James
University of Maryland, College Park, Maryland, U.S.A.

INTRODUCTION

Electron transfers from electron-rich reductants to electron-poor oxidants in soils change the valence state of the electron donor and acceptor, and thereby commonly affect their form, solubility, and mobility. These oxidation and reduction processes involving losses and gains of electrons, respectively, are known as redox reactions, and studying them provides a means to observe and quantify key properties governing the speciation of many elements important in plant nutrition, soil contamination, and soil genesis (1–4). Studying how redox conditions change in soils in response to natural and anthropogenic perturbations also provides a way to observe the sensitivity and resilience of soil chemical properties related to electron transfers. Soils are metastable, open systems in a remarkably stable state of nonequilibrium in which electron transfer reactions are controlled by abiotic and biotic processes involving gaseous, aqueous, and colloidal phases (5). Interfaces between these phases are centrally important to the energetics and rates of reactions involving redox of soils and natural waters (6–8). Many redox reactions in soils involve a phase change for the element transformed, thereby making quantification of the redox status of soils difficult and subject to innovation and new interpretation (5). Relatively new understandings of the kinetics of redox reactions in natural waters, particularly as influenced by light, will provide guidance for innovation in the field of redox theory and measurement in soils (7, 8).

THE NATURE OF THE ELECTRON AND THE PROTON AND THEIR ACTIVITIES IN SOILS

To understand and characterize "electron activity" and oxidation–reduction reactions in heterogeneous, multiphase soils and in soil solutions, an appreciation of the characteristics of electrons and closely allied protons is needed (4). An examination of the complementary nature of electrons and protons affirms the importance of hydrogen ion and electron activities as master variables in soils. The H atom, composed of one proton and one electron may be visualized and modeled as a spherical puff of cotton candy with a radius of approximately 10 cm and a proton nucleus with a radius of 5 μm—essentially an invisible fleck of unspun sugar in the center! The remaining volume of the atom is occupied by the electron: the density of the spun sugar represents the probability of finding the electron in any one location, and it becomes increasingly thinner (less likely) with distance away from the positively charged proton (9). The radius of the H atom (0.3 Å), therefore, is approximately 20,000 times that of the proton ($\sim 1.5 \times 10^{-5}$ Å). The proton also may be visualized as the size of a 0.1 μm colloidal clay particle, compared to a 2000 μm sand grain in a soil. In contrast to the large proportion of the volume of the H atom occupied by the negatively charged, wave-like electron, the electron has only neglible mass equal to approximately 550 $\mu g\ mol^{-1}$, 1/1836 of the mass of the H atom ($10^6\ \mu g\ mol^{-1}$).

In soil chemical calculations and theory, one considers the electron as a "species," designated "e^-" with neglible mass and thermodynamically as a ligand, reactant, and product. The electron is not ionic, but it is "negatively charged" as the carrier of negative electricity, as described by J.J. Thompson, who discovered the electron in 1897 (10). Its "activity" is conceptually analogous to that of H^+, but its concentration in "$mol\ L^{-1}$" is undefined. All these caveats about the electron require that one understands that electron activity in soils and natural waters should be regarded as related strictly to energy functions. Such functions can be described simply as "the ability to do work," "electrochemical potential," and more colloquially, "electron pressure," or quantified as a voltage (4).

Viewing the sibling concepts of "proton activity" (pH) and "electron activity" (pe) in soils, they must not be considered twins. Recognition must be given to similarities and differences in the formulation of conceptual and operational definitions for these key variables, and such comparisons are based on the differences in the nature of the proton and electron, as described above. In both cases, however, thermodynamic activity is defined as the ratio of e^- or H^+ in a given system relative to its activity under standard state conditions. In the case of the e^-, the standard state is at $(H^+) = 1$ M, a partial pressure of $H_2 = 1$ atm, and a temperature of 298 K, characteristics of the standard hydrogen electrode (SHE) with 0.00 V under these conditions (6, 11). For H^+ activity, the reference state is 1 M or pH 0. The measurement of pH and pe as analogous master variables in soils has been classically done with the glass and platinum electrodes, respectively. The potential difference between the sensing electrode and a calomel or silver–silver chloride reference electrode is corrected for the potential of the reference electrode relative to the SHE. In this way, an oxidation-reduction potential (ORP) is converted into a value known as Eh, or the platinum electrode potential relative to the SHE.

The familiar concept of buffer capacity for pH is defined as the change in acid or base added to effect a one unit change in pH (6). Similarly, the capacity factor in redox is referred to as poise and is defined as the change in added equivalents of reductant or oxidant to bring about a one unit change in pe (or Eh change of 59.2 mV) (1). Poise in soils and natural waters has been less intensively studied than pH buffering, but it is a central concept governed by reductant and oxidant activities and microbial processes. The above discussion of electron and proton activities in soils may be reviewed in more detail in recent publications (4, 6).

The U.S. Environmental Protection Agency has developed MINTEQ (Mineral Thermal Equilibria), a DOS-based computer program that calculates the distribution of myriad species based on minimizing the Gibbs free energy of a system of multiple chemical equilibria. It incorporates algorithms for redox reactions and allows the estimation of pe, pH, and reductant and oxidant activities in conjunction with dissolution and exchange equilibria. The use of MINTEQ also permits an easy way to conduct a sensitivity analysis for estimated pe, pH, or ion activities when just one value for a given parameter is varied (12). Other models have been developed and verified for redox-controlled, multi-species systems to predict transport of particular ions (13, 14).

APPLICATIONS OF REDOX PRINCIPLES IN SOIL ENVIRONMENTS

The thermodynamic and theoretical principles surrounding redox reactions have been used to predict and explain soil chemical phenomena in myriad environments and fields, including paddy rice production, wetland delineation, tidal marsh soil development, rhizosphere function, nutrient cycling, heavy metal remediation, synthetic organic compound clean-up, and other areas pertinent to environmental protection, agricultural production, and ecosystem function. The coupling of Eh-pH diagrams with empirical measurements of pH, Eh, and activities of oxidants and reductants has proved to be a powerful tool for assessing redox processes in heterogeneous, colloidal environments.

Given the importance of paddy rice (*Oryza sativa* L.) in worldwide production of food, early redox work in soil chemistry focused on the reduction reactions coupled with organic matter oxidation in flooded soils of Asia (15). This provided the foundation for more recent practical applications of microbial and chemical redox reactions to understand wetland function and delineation and to study tidal marsh soil development. These two areas have become worldwide concerns associated with human encroachment into these environments and with controversies over allowable uses of such "wet soils."

A current environmental concern linking soil redox processes and atmospheric chemistry is that emissions of nitrous oxide (N_2O) from waterlogged soils may contribute to global warming since N_2O absorbs infrared radiation, as do CO_2, CH_4, and H_2O in the troposphere. Nitrous oxide is an intermediate product in denitrification, the anaerobic microbial respiration process that converts nitrate (NO_3^-) to dinitrogen gas (N_2) as microbes oxidize reduced carbon compounds to organic acids or CO_2. The energetics of denitrification and its role in releasing N_2O are summarized in the following equations (4):

Reduction of nitrate to nitrous oxide by oxidation of carbohydrate to an organic acid:

$$1/2\ CH_2O + 1/4\ NO_3^- + 1/4\ H^+ \rightarrow 1/2\ HCOOH + 1/8\ N_2O + 1/8\ H_2O \quad (1)$$

Gibbs free energy (kcal per mole of electrons transferred) equals −21 at pH 7.

Reduction of nitrous oxide to dinitrogen by oxidation of carbohydrate to an organic acid:

$$1/2\ CH_2O + 1/2\ N_2O \Rightarrow 1/2\ HCOOH + 1/2\ N_2 \quad (2)$$

Gibbs free energy (kcal per mole of electrons transferred) equals −36 at pH 7.

Reduction of nitrate to dinitrogen by oxidation of carbohydrate to an organic acid:

$$1/2\ CH_2O + 1/5\ NO_3^- + 1/5\ H^+ \Rightarrow 1/2\ HCOOH + 1/10\ N_2 + 1/10\ H_2O \quad (3)$$

Gibbs free energy (kcal per mole of electrons transferred) equals −24 at pH 7.

The smaller Gibbs free energy of the reaction that produces N_2O from NO_3^- than for the conversion of N_2O to N_2 indicates that the first step in denitrification is less energetically favorable than is the second. This first step requires higher electron pressures (more reduced soil conditions or lower Eh values) to effect the reduction, and it may be less kinetically favorable also, even though both reactions are enzymatically catalyzed. Soil conditions, such as pH, Eh, temperature, and carbon availability will affect how much N_2O is emitted from soils to the atmosphere rather than being reduced further to harmless N_2. The overall reduction of NO_3^- to N_2 is energetically favorable, as indicated in Eq. 3, and is intermediate between the first and second steps in terms of energy released. The combination of thermodynamic calculations and kinetic experimentation can provide essential information to predict the nature of the linkage between soil redox processes and global warming.

Redox reactions in the rhizosphere also are important with respect to nutrient availability in this soil environment immediately adjacent to plant root surfaces (within one mm or so). On a broader scale, societal concerns about surface and groundwater chemical contamination and eutrophication of surface waters have focused attention on redox reactions with soil profiles as environments where nutrients, especially nitrogen, may undergo oxidation and reduction reactions leading to release of environmentally sensitive forms of nitrogen. For example, nitrification involves the oxidation of ammonium to nitrate, a form of nitrogen that is mobile in most soils and and is a common pollutant of ground and surface waters.

Soils contaminated with chromium(III,VI) or arsenic(III,V) may undergo transformations between the oxidation states of these elements and thereby change the mobility, toxicity, and potential bioavailability of each element. Hexavalent Cr is a soluble and toxic anion, in contrast to Cr(III), a sparingly soluble, nontoxic cation in most soils. Because of this difference in solubility and toxicity of the common valence states of Cr in soil environments, reduction of Cr(VI) is being used to remediate soils without changing the total concentration of Cr (16). Arsenic(III) oxidation to As (V) results in a decrease in the solubility of As, and thereby may prevent contamination of groundwater, as has occurred in recent years in Bangladesh (17).

Synthetic organic chemicals added to soils by humans in wastes or as pesticides may undergo oxidation reactions that usually result in a degradation and a decrease in toxicity. Examples are aromatic compounds used as organic solvents in industrial applications; and pesticides, such as herbicides and insecticides, that may be retained in soils or be leached to groundwater, depending on redox reactions. (18).

REFERENCES

1. Russell, E.W. The Chemistry of Waterlogged Soils. In *Soil Conditions and Plant Growth*, 10th Ed.; Longman: London, 1973; 670–695.
2. Rowell, D.L. Oxidation and Reduction. In *The Chemistry of Soil Processes*; Greenland, Dennis J., Hayes, Michael H.B., Eds.; John Wiley & Sons: London, 1981; 401–461.
3. Bartlett, Richmond J.; James, Bruce R. Redox Chemistry of Soils. In *Advances in Agronomy*; Sparks, Donald L., Ed.; Academic Press: San Diego, 1993; *50*, 151–208.
4. James, Bruce R.; Bartlett, Richmond J. Redox Phenomena. In *Handbook of Soil Science*; Sumner, Malcolm E., Ed.; CRC Press: Boca Raton, 2000; B169–B194.
5. Bartlett, Richmond J. Characterizing Soil Redox Behavior. In *Soil Physical Chemistry*, 2nd Ed.; Sparks, Donald L., Ed.; CRC Press: Boca Raton, 1999; 371–397.
6. Stumm, Werner; Morgan, James J. Oxidation and Reduction; Equilibria and Microbial Mediation. In *Aquatic Chemistry: Chemical Equilibria and Rates in Natural Waters*, 3rd Ed.; Wiley-Interface: New York, 1996; 425–515.
7. Stumm, Werner; Morgan, James J. Kinetics of Redox Processes. In *Aquatic Chemistry: Chemical Equilibria and Rates in Natural Waters*, 3rd Ed.; Wiley-Interface: New York, 1996; 672–725.
8. Stumm, Werner; Morgan, James J. Photochemical Processes. In *Aquatic Chemistry: Chemical Equilibria and Rates in Natural Waters*, 3rd Ed.; Wiley-Interface: New York, 1996; 726–759.
9. Castellan, G.W. *Physical Chemistry*, 3rd Ed.; Addison-Wesley Publishing Co: Reading, MA, 1983.
10. Thompson, J.J. *The Electron in Chemistry*; Franklin Institute Press: Philadelphia, 1923.
11. Lindsay, Willard L. *Chemical Equilibria in Soils*; Wiley-Interscience: New York, 1979.
12. Allison, J.D.; Brown, D.S. MINTEQA2/PRODEFA2–A Geochemical Speciation Model and Interactive Processor. In *Chemical Equilibria and Reaction Models*; Loeppart, R.H., et al., Ed.; Soil Sci. Soc. Am. Spec. Publ: Madison, WI, 1995; 241–252.
13. Liu, Chen W.; Narasimhan, T.N. Redox-Controlled Multiple-Species Reactive Chemical Transport 1. Model Development. Water Resources Research **1989**, *25* (5), 869–882.
14. Liu, Chen W.; Narasimhan, T.N. Redox-Controlled Multiple-Species Reactive Chemical Transport 2. Verification and Application. Water Resources Research **1989**, *25* (5), 883–910.
15. Ponnamperuma, F.N. The Chemistry of Submerged Soils. In *Advances in Agronomy*; Brady, Nyle C., Ed.; Academic Press: New York, 1972; 29.
16. James, B.R. The Challenge of Remediating Chromium-Contaminated Soils. Environmental Science and Technology **1996**, *30*, 248A–251A.
17. Bagla, P. India's Spreading Health Crisis Draws Global Arsenic Experts. Science **1996**, *274*, 174–175.
18. Schwarzenbach, R.P.; Gschwend, P.M.; Imboden, D.M. *Environmental Organic Chemistry*; John Wiley and Sons: New York, 1993.

REHABILITATION, INDICATORS AND MONITORING OF

R

David Jasper
The University of Western Australia, Crawley, Western Australia

INTRODUCTION

Measuring soil parameters to use as indicators of ecosystem sustainability or to define soil quality is an increasingly common theme in land management. While substantially more research has occurred in relation to agricultural soils (1), the principles developed are equally applicable to soils that are being restored after mining. In the agricultural context, Karlen et al. (2) listed four functions of soil which were integral to soil quality: 1) accommodating water entry; 2) retaining and supplying water to plants; 3) resisting degradation; and 4) supporting plant growth.

These four functions are equally applicable for rehabilitation of soils disturbed during mining, where re-establishment of a sustainable ecosystem requires vegetative diversity and productivity that are appropriate for the desired end-use. Plant productivity, which is a key driver of many processes in a sustainable ecosystem, directly reflects the soil conditions that have been created. Therefore constructing an appropriate soil environment is a crucial first step in rehabilitation.

The information that needs to be acquired in monitoring rehabilitated soils is a combination of measures which define the starting condition of the soil, together with those that may be more sensitive and can be used to demonstrate change over time (Table 1). For example, total soil carbon can be considered one of the most important indicators of soil quality and productivity (3), yet it is slow to respond to changes in management. As a result, microbial biomass and respiration have been used as a more sensitive indicator of long-term trends in organic matter dynamics in soils (4).

In general, indicators of successful ecosystem rehabilitation should have some key characteristics as follows:

1. Accommodate all land uses;
2. Describe generic factors and those that add resilience to the ecosystem;
3. Able to be repeatedly measured to indicate direction of change;
4. Absolute value needs to be comparable with:
   a. Premining environment (without specific expectations of replicating it), and
   b. Postmining/prerehabilitation conditions (with the on-going need to demonstrate appropriate development).

DEVELOPING SOIL BIOLOGICAL INDICATORS FOR ASSESSING REHABILITATION AFTER MINING

Key soil biological indicators may include soil microbial biomass and respiration, incidence of symbiotic microorganisms such as mycorrhizal fungi or rhizobia, and soil invertebrates. Soil microbial biomass is the living microbial fraction (bacteria, fungi, protozoa, microfauna, etc.) comprising 1–4% of soil organic matter (4). It is responsible for decomposition and mineralization of organic matter, contributes to soil structure development and is a source and sink of plant nutrients. The soil microbial biomass has potential as an indicator because it is always present at some level and is very responsive to management changes. The relative proportion of soil microbial C to total soil C has been shown to be a sensitive index of ecosystem recovery in land that was disturbed by coal mining and subsequently reforested (7). In general, low or decreasing soil microbial biomass implies low or declining plant productivity (4).

Mycorrhizal fungi assist in nutrient uptake for most plants, and therefore will contribute to the sustainability of revegetation. These fungi can be severely reduced by soil disturbance and stockpiling during mining (8). The occurrence of mycorrhizal fungi depends on which plant species are present, and this restricts their potential as a generic indicator of rehabilitation success.

Soil invertebrates are responsible for the breakdown of plant residues, which enables subsequent mineralisation, and through their activities enhance soil structure. Given their role and their widespread occurrence, soil animals do have potential as generic indicators (2, 9). In contrast, N_2-fixing microsymbionts have limited potential as indicators because they may have a specific host range

Table 1 Selected soil properties that are useful indicators of the capacity of a soil to sustain a restored ecosystem

Physical fertility	Chemical fertility	Biological fertility
Bulk density	pH	Organic carbon
Soil strength	Electrical conductivity	Microbial biomass & respiration
Aggregate stability	Plant-available & potentially-available nutrients	Soil invertebrates
Water infiltration	Heavy metal availability	
Plant-available water		

(Adapted from Refs. 1, 2, 4–6.)

and survive well without host plants in topsoils disturbed during mining.

All soil biological components and processes are driven ultimately by carbon inputs derived from plant production. Zak et al. (10) demonstrated, on a continental scale, that microbial biomass and organic matter pools were positively related to the annual net productivity of vegetation. This relationship has also been demonstrated at the local scale in bauxite rehabilitation (11). In the following case study, the author has explored the concept that the rate of recovery of biological processes in rehabilitated mine soils is likely to reflect the productivity of the vegetation that has been established.

CASE STUDY OF THE DEVELOPMENT AND APPLICATION OF INDICATORS

In the bauxite mining industry in south-western Australia, more than 700 ha of native eucalyptus forests are removed and a similar area replanted in previously mined areas every year. It is well recognized that the forest topsoil contains important reserves of viable seed, nutrients and microorganisms, therefore this soil is always salvaged prior to mining (12). Where possible the salvaged topsoil is directly returned to a previously mined area or it is stored in stockpiles until a suitable area becomes available. The relatively uniform nature of the soils and landscapes in the bauxite mining area, and the consistent rehabilitation strategies that are applied, make it feasible to use chronosequences of rehabilitation to estimate changes in the various soil biological components over time.

In addition to recent research in nutrient cycling (13) and soil development (12), there has been considerable research on soil organisms in the rehabilitated bauxite mines of south-western Australia. These organisms are involved in processes which are essential in nutrient supply and uptake including:

- mineralization of organic matter (soil invertebrates, soil microbial biomass);
- enhanced uptake of immobile nutrients (mycorrhizal fungi);
- supply of nitrogen through N_2-fixation (rhizobia, *Frankia* spp.).

Total C content in surface soils is substantially lower in topsoils spread again after bauxite mining and increases very slowly, even under highly productive revegetation (11) (Fig. 1). In contrast, soil microbial biomass increases rapidly and substantially from around 2 to 8 years (14) (Fig. 1). Its rate and magnitude of response parallels that of

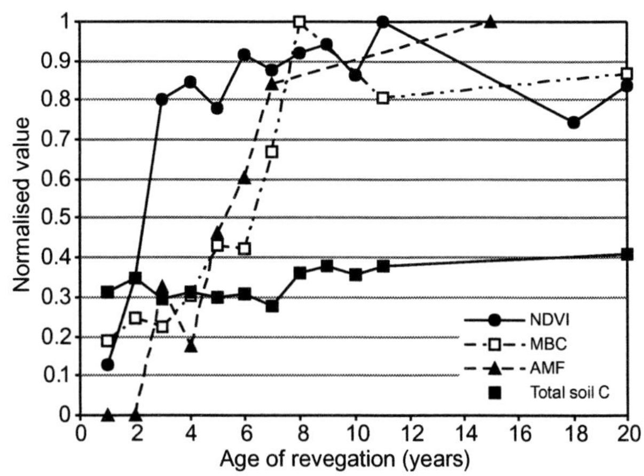

Fig. 1 The pattern of recovery of total soil carbon (From Ref. 11), infectivity of arbuscular mycorrhizal fungi (AMF) (Jasper, unpublished data), soil microbial carbon (MBC) (From Ref. 11) and the normalized difference vegetation index (NDVI) (From Ref. 14) with increasing age of revegetation following bauxite mining. (Total soil C is expressed as a proportion of the total C in premining forest soil (5.6%), while other parameters are expressed as a proportion of the maximum value reached. Values for total soil C and soil microbial C are derived from the same plots, but the other parameters are each from separate studies using different plots, within the same mining operation.)

the vegetation, and these attributes enhance its utility as an indicator.

VA mycorrhizal fungi follow a similar recovery pattern to that of microbial biomass (Fig. 1). However they are more difficult to measure and given their constraints described before, are less suitable as an indicator. Although equivalent time sequences are not available for total biomass of soil animals, Greenslade and Majer (9) found the total number of "decomposer" species of collembola, and the number of individuals, related very strongly to increasing plant and litter cover. Majer and Nichols (15) reported similar relationships of ant species richness with plant species diversity and vegetative cover, over an equivalent time-course to that in Fig. 1, although ant biomass was not measured.

Plant productivity is not easily measured on a large scale in bauxite rehabilitation. Relevant available data is restricted largely to estimates of cover (15). As an alternative, Jasper et al. (14) used remote sensing technology to estimate photosynthetic leaf area in bauxite rehabilitation, using the normalized difference vegetation index (NDVI) (Fig. 1). The pattern of recovery of NDVI was closely related to that derived from direct measures of plant cover (14, 15). NDVI values increased rapidly with increasing age of vegetation in rehabilitated bauxite mines, reaching maximum values after 4–6 years. This rate of increase was more rapid than that of soil microbial biomass or mycorrhizal fungi, which by comparison was delayed by 3–4 years.

SUMMARY AND CONCLUSIONS

The key role of soil microorganisms in processes of nutrient mineralization and acquisition by plants in sustainable ecosystems means that they are a relevant parameter to measure as an indicator of successful rehabilitation. However, it may be concluded from this integration of measurements from bauxite rehabilitation that plant productivity should be included as perhaps the most important indicator of revegetation success. This is not only because of its primary importance to recovery of soil processes, but also because it integrates all aspects of site fertility. Under the current protocols for bauxite mine rehabilitation, it can be argued that if diverse and productive vegetation is successfully established, then appropriate populations of soil microorganisms, and the processes to which they contribute, will subsequently develop.

The concept of using plant productivity as an integrated measure of rehabilitation performance is mirrored in current research in the Australian mining industry, which is investigating the application of "landscape function analysis" (LFA) (16) as a tool for assessing mine rehabilitation. In the LFA approach, the capacity of the landscape to retain essential resources of water and nutrients is assessed. This retention capacity reflects both the underlying characteristics of the constructed landform and the development of plant biomass and soil cover. Thus, it is an integrative measure of plant performance and the factors that affect the performance.

The importance of plant productivity, and its dependence on site fertility, reaffirms the importance of constructing a soil profile with physical and chemical characteristics that are appropriate for the particular vegetation to be established. If the right components of physical and chemical fertility are put in place, together with microbial and plant propagules, the key biological components, then the likelihood of successful rehabilitation is enhanced. In large-scale reestablishment of native ecosystems, such as in rehabilitation after bauxite mining, direct-returned fresh topsoil is the best source of those propagules.

ACKNOWLEDGMENTS

The data presented in this paper were derived from studies that were generously supported by Alcoa World Alumina—Australia.

REFERENCES

1. Karlen, D.L.; Mausbach, M.J.; Doran, J.W.; Cline, R.G.; Harris, R.F.; Schuman, G.E. Soil Quality: A Concept, Definition, and Framework for Evaluation. Soil Sci. Soc. Am. J. **1997**, *61*, 4–10.
2. Karlen, D.L.; Wollenhaupt, N.C.; Erbach, D.C.; Berry, E.C.; Swan, J.B.; Eash, N.S.; Jordahl, J.L. Crop Residue Effects On Soil Quality Following 10-Years of No-Till Corn. Soil Till. Res. **1994**, *31*, 149–167.
3. Reeves, D.W. The Role of Soil Organic Matter in Maintaining Soil Quality in Continuous Cropping Systems. Soil Till. Res. **1997**, *43*, 131–167.
4. Sparling, G.P. Soil Microbial Biomass, Activity and Nutrient Cycling as Indicators of Soil Health. In *Biological Indicators of Soil Health*; Pankhurst, C.E., Doube, B.M., Gupta, V.V.S.R., Eds.; CAB International: Wallingford, UK, 1997; 97–119.
5. Haigh, M.J. Soil Quality Standards for Reclaimed Coal-mine Disturbed Lands: A Discussion Paper. Int. J. Surf. Min., Reclam. Environ. **1995**, *9*, 187–202.
6. Doran, J.W.; Zeiss, M.R. Soil Health and Sustainability: Managing the Biotic Component of Soil Quality. Appl. Soil Ecol. **2000**, *15*, 3–11.

R

7. Insam, H.; Domsch, K.H. Relationship Between Soil Organic Carbon and Microbial Biomass on Chronosequences of Reclamation Sites. Microb. Ecol. **1988**, *15*, 177–188.
8. Jasper, D.A.; Abbott, L.K.; Robson, A.D. The Loss of VA Mycorrhizal Infectivity During Bauxite Mining May Limit the Growth Of *Acacia Pulchella* R.Br. Aust. J. Bot. **1989**, *37*, 33–42.
9. Greenslade, P.; Majer, J.D. Recolonization by Collembola of Rehabilitated Bauxite Mines in Western Australia. Aus. J. Ecol. **1993**, *18*, 385–394.
10. Zak, D.R.; Tilman, D.; Parmenter, R.R.; Rice, C.W.; Fisher, F.M.; Vose, J.; Milchunas, D.; Martin, C.W. Plant Production and Soil Microorganisms in Late-Successional Ecosystems: A Continental-Scale Study. Ecology **1994**, *75* (8), 2333–2347.
11. Sawada, Y. *Indices of Microbial Biomass and Activity to Assess Minesite Rehabilitation*; Proceedings Minerals Council of Australia 21st Annual Environmental Workshop, Newcastle, Australia, October 14–18, 1996; Minerals Council of Australia: Canberra, 223–236.
12. Ward, S.C. Soil Development on Rehabilitated Bauxite Mines in South-West Australia. Aust. J. Soil Res. **2000**, *38*, 453–464.
13. Todd, M.C.L.; Adams, M.A.; Grierson, P.F. Mineralisation of Nitrogen in a Chronosequence of Rehabilitated Bauxite Mines. Aust. J. Soil Res. **2000**, *38*, 435–451.
14. Jasper, D.A.; Sawada, Y.; Gaunt, E.; Ward, S.C. Indicators of Reclamation Success—Recovery Patterns of Soil Biological Activity Compared to Remote Sensing of Vegetation. In *Land Reclamation: Achieving Sustainable Benefits*; Fox, H.R., Moore, H.M., McIntosh, A.D., Eds.; A.A. Balkema: Rotterdam, 1998; 21–24.
15. Majer, J.D.; Nichols, O.G. Long-Term Recolonization Patterns of Ants in Western Australian Rehabilitated Bauxite Mines with Reference to Their Use as Indicators of Restoration Success. J. Appl. Ecol. **1998**, *35*, 161–182.
16. Ludwig, J.; Tongway, D.; Freudenberger, D.; Noble, J.; Hodgkinson, K. *Landscape Ecology, Function and Management: Principles from Australia's Rangelands*; CSIRO: Melbourne, 1997.

REHABILITATION OF MINERALS PROCESSING RESIDUE (TAILINGS)

L.R. Hossner
H. Shahandeh
Texas A&M University, College Station, Texas, U.S.A.

INTRODUCTION

Metals have been mined and exploited with the growth of world industry. In particular, iron, lead, zinc and copper have been mined extensively in regions all over the world. In the process of metal benefication, massive heaps of spoil (tailings and waste rock) have been created at mine sites or areas distant from mines (Fig. 1). Tailings and waste rock create esthetic problems in the landscape and affect water, soil, plant, and public health. Tailings is defined as the solid waste product of the milling and mineral concentration process (1). Mill tailings are the finely ground host rock materials from which the desired mineral values have been extracted during the concentration process. Generally, tailings are transported from the mill to their place of disposal as a water slurry containing 15–50% solids by weight and discharged by impoundment on land in settling ponds adjacent to the mills, used as backfill in the open pit or underground mine, disposed in deep lakes or offshore, or processed for secondary metal recovery followed by disposal. Tailings impoundments range in size from <10 to >2000 ha, and may be stacked as high as 50 m (2).

There are a large number of abandoned mine waste and tailings deposits from a wide variety of industries around the world. The total area of land disturbed by mining in China is estimated to be about 2 million ha (3). In the U.S. between 1930 and 1980 more than 2 million ha of land were affected by mining operations (4). The mine tailings produced in Malaysia, England, Thailand, and Canada are estimated in billions of tons occupying hundreds of thousands of hectares (2).

POTENTIAL ENVIRONMENTAL PROBLEMS

Abandoned spoil and tailings contain the waste products of both mining and ore processing operations. Chemical extractants such as sulfuric acid and sodium bicarbonate are used for uranium ores, cyanide for gold ores, sodium hydroxide for aluminum ores, and sulfuric acid and hydrochloric acids for copper, nickel, and cobalt ores. In addition to these toxic solvents and the dissolved heavy metals, tailings from uranium and phosphate operations can contain radionuclides such as thorium and radium. These materials are often a major source of pollution in the local environment due to dust blow and the potential leaching of the products of mineral weathering into water sources.

Tailings are also subject to the process of weathering and, over time, changes may occur in their properties which could be hazardous to the environment. For example, sulfides are associated minerals that are readily oxidized in the tailings when exposed to air, water, and iron oxidizing bacteria. One of the oxidation products is sulfuric acid. Tailings containing sulfide minerals may eventually have a pH of 1.5–3.5 (2). Toxic ions may also contaminate soils and waters adjacent to smelters through seepage, runoff waters, and eroded sediments. Effluents arising from tailings seepage could be toxic in varying degrees to man, animals, and plant life. Also, there is a concern for the amount of heavy metal uptake by plants growing on tailings and its effect on the food chain (5).

MINE WASTE REHABILITATION PRACTICES

Engineering Approach

Mine waste residue and soils contaminated with toxic metals and radionuclides can be remediated and stabilized using engineering approaches including chemical, physical, and thermal techniques (e.g., in situ mobilization, immobilization, degradation, and burial; or removal and reburial, vitrification, vacuum extraction, steam flooding, pumping and leaching, electroosmosis, and electroacoustic extraction) (6). The use of chemical and physical techniques to stabilize mineral wastes against wind and water erosion are limited because of the cost and maintenance (7). Physical stabilization of materials such as waste rock from strip mining can be used to reduce wind and water erosion.

Chemical stabilization requires a chemical agent, such as lignin sulfate or resinous adhesive, to react with mine waste to provide a crust resistant to wind and water erosion.

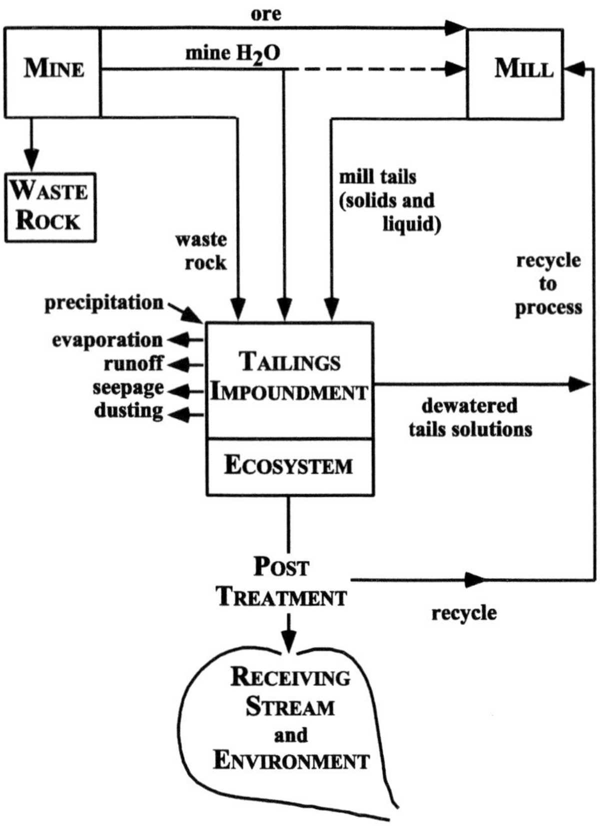

Fig. 1 Mine/mill environment. (From Ref. 2.)

Physical stabilization can be performed by protecting or isolating the waste from the environment with physical barriers such as liners and clay caps. Usually, the design and installation of a cover system consists of three zones: an erosion resistant layer, a moisture retention layer, and an underlying clay barrier to prevent entrance of water into the waste material. The design of the capping sequence and incorporation of a coarse layer to break capillarity are shown to be very important. When waste material poses a threat to groundwater quality or to plant growth, a layer of impermeable plastic or compacted clay may be placed between the waste and the soil cap to prevent movement of water into the toxic wastes. In other materials it may be necessary to place a diffusion barrier of very coarse material between the toxic waste and the cover soil to break a water column that might permit diffusion of the toxic materials from the waste into the cover soil. Soil caps placed over nontoxic materials may be thinner than soil caps placed over toxic materials (8).

Ecological Approach

According to the Center for Minesite Rehabilitation in Australia the strategies for rehabilitation of mine waste should be based on stabilization and sustainable revegetation of tailings and waste rock in a reconstructed ecosystem (9). Tailings must be stabilized before or concurrent with full-scale revegetation. The keys to successful revegetation of metalliferous mill tailings are: 1) obtain a complete understanding of the physical and chemical properties of the tailing material; 2) select an appropriate final land use of the area under consideration; 3) select the most cost effective combinations of amendments, nutrient supplements, cover materials, and plant species to be employed; and 4) monitor the performance of the vegetation and make adjustments as needed. It is now widely accepted that the establishment of vegetation is a desirable method for stabilization of mine wastes (10, 11). Vegetation reduces wind velocity at the surface, captures dust particles, reduces raindrop impact, reduces runoff by increasing infiltration, and reduces the overland flow of water and sediment. Vegetative stabilization also improves the chemical, biological, and physical properties of the mine wastes by increasing the organic matter content, nutrient level, cation exchange capacity, and biological activity.

Vegetation established on mine tailings has several advantages over physical and chemical methods of stabilization. Vegetation is esthetically pleasing and accepted by the public, relatively inexpensive, less disruptive of site remediation, creates a beneficial habitat for wildlife, plant roots and shoots can take up heavy metals, and plants may stimulate microbial immobilization of heavy metals in the rhizosphere. Thus, establishment of vegetation in mining areas has the potential of reducing contamination of adjacent soil, surface water, and ground water (8).

The main approaches to revegetation have been summarized in Table 1 (7). The waste characteristics of a site determine which approach is most suitable. Generally there are two approaches to revegetation which have been used in combination or separately. They are adaptive and ameliorative approaches. The adaptive approach to revegetation is to combat the toxicity of the waste by direct seeding with metal tolerant cultivars. The metal tolerant plants may be of great benefit to developing countries for low-cost revegetation. The other approach is ameliorative and the toxicity is avoided or diluted, rather than tolerated by using some form of covering system. This approach has been widely used and subjected to considerable research. The two types of covering material are ameliorants and inert amendments. Ameliorants are materials such as sewage sludge, compost, domestic refuse, peat, and topsoil. Inert ameliorants include materials such as clay, shale, or gypsum. Inert ameliorants are commonly wastes from other industry or mining activities. This approach has the

Table 1 Approaches to revegetation of minerals processing residue (tailings)

Waste characteristics	Reclamation technique	Problems encountered
Low toxicity: total metal content <0.1%. No major acidity or alkalinity problems	Amelioration and direct seeding with agricultural or amenity grasses and legumes. Using traditional or specialized techniques	Probable commitment to a medium/long-term maintenance program. Grazing management must be monitored
Low toxicity and climatic limitations: toxic metal content <0.1%. No major acidity or alkalinity problems, extreme of temperature, rainfall, etc.	Amelioration and direct seeding with native species. Seed or transplant ecologically adapted native species using amelioration treatments where appropriate	Irrigation often necessary at establishment. Expertise required on the characteristics of native flora
High toxicity: toxic metal content >0.1%. High salinity in some cases	1) Amelioration and direct seeding with tolerant ecotypes. Apply lime, fertilizer and organic matter, as necessary, before seeding	Regular fertilizer application. Few species have evolved tolerance and few are available commercially. Grazing management not possible
	2) Surface treatment and seeding with agricultural or amenity grasses and legumes. Amelioration with 10 50 cm of innocuous mineral waste and/or organic material. Apply lime and fertilizer as necessary	Regression will occur if depths of amendments are shallow or if upward movement of metals occurs. Availability and transport costs may be limiting
Extreme toxicity: very high toxic metal content. Intense salinity or acidity	Isolation: surface treatment with 30–100 cm of innocuous barrier and surface binding with 10–30 cm of a suitable rooting medium. Apply lime and fertilizer as necessary	Susceptibility to drought according to the nature and depth of amendments. High cost and potential limitations of materials availability

(From Ref. 7.)

added benefit of using one category of waste to overcome the problems of other forms of waste (12).

Although revegetation is desirable, metal wastes can present a very unfavorable environment for plants because of the presence of many growth limiting factors such as high salinity, metal toxicity, or nutrient deficiency in the mine tailings and soils (13).

CHARACTERISTICS OF MINE WASTE THAT LIMIT REHABILITATION

Characteristics of tailings from 43 selected mine waste sites are shown in Table 2 (14). Alleviation or modification of each chemical and physical limitation of the tailing is required prior to the establishment of vegetation.

Chemical Properties

The chemical composition of tailings depends on the original ore mineralogy, extraction techniques, and associated minerals. Among the minerals present in tailings, sulfides are often an important constituent that must be considered in tailings management.

Mining often exposes the sulfide-bearing minerals (pyrite, marcasite, pyrohotite, chalcopyrite, arsenopyrite, cobalite) to the atmosphere. In the presence of water, iron sulfide oxidation by weathering (Eq. 1) and by iron oxidizing bacteria catalysis (Eq. 2) will convert the sulfides to sulfuric acid (15):

$$2FeS_2 + 15/2O_2 + 7H_2O = 2Fe(OH)_3 + 4H_2SO_4 \quad (1)$$

$$FeS_2 + 14Fe^{3+} + 8H_2O = 15Fe^{2+} + 2SO_4^{2-} + 16H^+ \quad (2)$$

Large amounts of lime, between 10–150 t ha^{-1}, may be required to neutralize the acidity produced in sulfidic tailings (5).

Various salts may appear on the surface of tailings depending on the nature of the original mill process. The factors which are involved in the appearance of salts on the tailings surface include: excess concentrations of soluble salts in tailings materials, availability of shallow subsurface

Table 2 Mean and range of values for selected physical and chemical characteristics of tailings

Property	Unit	Mean	Range
Particle size distribution			
<2 mm	%	95	20–100
Sand	%	51	1–97
Silt	%	43	0–96
Clay	%	7	0–40
Moisture retention (bar)			
0.1	%	22	0–55
0.3	%	18	0–55
15	%	4	0–20
Available water holding capacity	%	16	0–35
Bulk density	$g\,cm^{-3}$	1.5	0.2–3.1
Particle density	$g\,cm^{-3}$	2.91	0.01–4.29
pH		6.2	1.8–9.4
Cation exchange capacity	cmolc/kg	2.63	0.19–46.5
Organic matter	%	2	0.02–25
Electrical conductivity	dS/m	2	0.1–22.4
Available nutrients			
Phosphorus (P)	mg/kg	10	1–400
Potassium (K)	mg/kg	63	1–564
Calcium (Ca)	mg/kg	11,930	40–52,480
Magnesium (Mg)	mg/kg	230	15–1328
Total analysis			
Nitrogen (N)	%	0.013	0.001–0.166
Sulfur (S)	%	4.02	0.001–38.87
Iron (Fe)	%	15.5	0.4–56.81
Aluminum (Al)	%	2.8	0.1–8
Calcium (Ca)	%	1.7	0.01–10.95
Magnesium (Mg)	%	1.2	0.04–5.0
Sodium (Na)	%	0.5	0.01–2.9
Potassium (K)	%	0.7	0.04–3.32
Manganese (Mn)	%	0.2	0.01–4.0
Silicon (Si)	%	22	4–37
Cadmium (Cd)	mg/kg	38	2–280
Chromium (Cr)	mg/kg	1000	70–7000
Cobalt (Co)	mg/kg	1140	100–9999
Molybdenum (Mo)	mg/kg	70	10–800
Nickel (Ni)	mg/kg	96	10–546
Lead (Pb)	mg/kg	340	0.3–2810
Titanium (Ti)	mg/kg	2500	200–10,000
Zinc (Zn)	mg/kg	510	1–5000
Copper (Cu)	mg/kg	130	1–750

water, salt concentrations within the tailing water, and temperature and chemical potential gradients between the surface and the interior of the tailing. The upward migration of salts and their inhibiting effect on root growth and revegetation have been major concerns in the reclamation of many mine tailings. For example, a common problem associated with the establishment of vegetation in asbestos tailings, bentonite tailings, and red mud from bauxite ore processing for alumina is salinity and/or sodicity. These wastes can be highly alkaline (pH > 10), saline (EC > 30 $dS\,m^{-1}$), and sodic. Vegetation establishment on these tailings is difficult because of salinity, alkalinity, clay dispersion and low hydraulic conductivity (5).

Mine wastes are usually deficient in major plant nutrients and almost universally deficient in nitrogen. Nutrient deficiencies, especially nitrogen, phosphorus and potassium, are often the principal limiting constraints to revegetation of kaolinitic china clay, iron, copper, gold, silver, and other heavy metal mine wastes (5). The success of establishing plants on tailings depends on providing an adequate supply of plant nutrients. High levels of the major essential nutrients may reduce the harmful effects of metal ions.

Toxic ions are frequently present in tailings of heavy metals in sufficient concentrations to prevent plant growth unless considerable amelioration is undertaken. Toxic ions decrease root respiration, limit water and nutrient uptake, reduce enzymatic activity and microbial populations, and inhibit cell mitosis in root meristematic regions (15). Toxic ions may also contaminate soils and waters adjacent to tailings through seepage, runoff waters, and eroded sediments. Another problem with heavy metal contaminated tailings is the possible uptake of metals by plants in quantities that could be toxic in the general food chain. Radioactivity associated with uranium and phosphate tailings and high concentrations of arsenic, mercury and cyanide in silver and gold mine tailings are other environmental concerns.

Physical Properties

The physical properties of tailings vary with the mineral being processed, the origin of the ore body, and the process used for mineral concentration. Physical properties are the most important in determining productivity of vegetated mine wastes because of the cost involved in trying to ameliorate particle size distribution and water-holding capacity. Nonuniform texture is the main physical problem of mine wastes which limits the availability of water to plants. Mining generates a wide range of particle size materials. This includes coarse mine wastes, fine clays, flotation tailings, chemical precipitates, and slimes. In the mineral industry the slime size fraction is $<5\ \mu m$.

Fine texture is a major problem in gold tailings. Gold tailing is finely crushed ($0.01-0.1\ \mu m$) because of the large surface area needed for fast reaction with processing chemicals. It is relatively easy to establish vegetation on these tailings but it is difficult to obtain full ground cover because young plants are sand blasted by finely crushed, wind blown materials. Laying large rock fragments on the slope of sand deposits and installing reed wind breaks on the entire sand deposit has helped the establishment of vegetation (2). A critical factor in reclaiming and vegetating a tailing is the availability of moisture for plant growth. Moisture curve characteristics of tailings derived from iron, nickel, copper, lead, zinc, and gold mines showed properties similar to a sandy loam soil, an indication of potential water deficiency for plant growth (2).

Vegetation establishment and maintenance on coarse tailings, where there will be little water retention and rapid drainage, is also related to water availability. Establishment of vegetation on very coarse tailings is difficult but it can be achieved by adding a soil cover followed by hydromulching and hydroseeding (15).

Crusting, cracking, and a general lack of structure are common characteristics of mine tailing brought about by differences in texture, lack of organic matter, and variable mineralogy. Structure determines the bulk density of tailings. Water infiltration is often limited in fine textured tailings due to poor structural characteristics. Root penetration and moisture stress of plants due to limited rooting generally becomes a problem with dry bulk density values above $1.5\ Mg\ m^{-3}$ in tailings (15).

Tailings have a high heat capacity. Tailings exposed to direct solar radiation can have temperatures of 55–65°C at 1–2 cm depth. Internal temperature in a silver mine in Canada reached 45°C during the oxidation process. Hay or straw mulch has been an effective insulator for stabilizing the tailings temperature (2).

SUMMARY

There are large areas of abandoned mine tailings from a variety of industrial operations around the world. Tailings contain a large proportion of the original ore that was mined. Tailings are deposited in impoundments with dimensions ranging from several square meters to some that are many square kilometers in area. In the past, these tailings have been largely abandoned and allowed to revegetate under natural conditions. In recent years there have been both public and scientific awareness of the possible consequences of long term disposal of tailings. Now many countries have enacted legislation to ensure reclamation of tailings of disposal sites once the mining and milling operations have ceased. A complete understanding of the physical and chemical properties of tailings is essential for planning successful rehabilitation programs.

REFERENCES

1. Richmond, T.C. The Revegetation of Metalliferous Tailings. In *Reclamation of Drastically Disturbed Lands*; Barnhisel, R.I., Darmody, R.G., Daniels, W.L., Eds.; American Society of Agronomy: Madison WI, 2000; 801–818.
2. Ritcey, G.M. *Tailings Management: Problems and Solutions in the Mining Industry*; Elsevier: New York, USA, 1989.

3. Ye, Z.H.; Wong, J.W.C.; Wong, M.H. Vegetation Response to Lime and Manure Compost Amendments on Acid Lead/Zinc Mine Tailings. Restor. Ecol. **2000**, *8*, 289–295.
4. Johnson, W.; Paone, J. *Land Utilization and Reclamation in the Mining Industry, 1930-1980*; U.S. Bureau of Mines Information Circular 8862, United States Printing Office: Washington, DC, 1982.
5. Hossner, L.R.; Shahandeh, H. Chemical and Physical Limitations to Mine Tailings Reclamation. Trends Soil Sci. **1991**, *1*, 291–305.
6. Francis, A.J.; Dodge, C.J. Remediation of Soils and Wastes Contaminated with Uranium and Toxic Metals. Environ. Sci. Technol. **1998**, *32*, 3993–3997.
7. Tordoff, G.M.; Baker, A.J.M.; Willis, A.J. Current Approaches to the Revegetation and Reclamation of Metalliferous Mine Wastes. Chemosphere **2000**, *41*, 219–228.
8. Menzies, N.W.; Mulligan, D.R. Vegetation Dieback on Clay-Capped Mine Waste. J. Environ. Qual. **2000**, *29*, 437–442.
9. Bell, L.C. The Australian Centre for Minesite Rehabilitation Research: An Initiative to Meet the Strategic Research Needs for Sustainable Mining Rehabilitations. Water, Air, Soil Pollut **1996**, *91*, 125–133.
10. Harris, J. A.; Birch, P.; Palmer, J. *Land Restoration and Reclamation: Principles and Practices*; Addison-Wesley/Longman: Harlow/Essex, England, 1996.
11. Munshower, F.F. *Practical Handbook of Disturbed Land Revegetation*; Lewis/CRC Press: London/Boca Raton, FL, 1994.
12. Bradshaw, A.D.; Johnson, M.S. Revegetation of Metalliferous Mine Waste: The Range of Practical Techniques Used in Western Europe. In *Minerals, Metals and the Environment*; Institute of Mining and Metallurgy: London, 1992.
13. Johnson, M.S.; Cooke, J.K.W.; Stevennson, J.K.W. Revegetation of Metalliferous Wastes and Land After Metal Mining. In *Mining and Environmental Impact*; Issues in Environmental Science and Technology, Hester, R.E., Harrison, R.M., Eds.; Royal Society of Chemistry: London, 1994; 31–48.
14. Murray, D.R. Pit Slope Manual. Supplement 10–1; Reclamation by Revegetation; Report No. 77–31, vol. 1. Mine Waste Description and Case Histories; CANMET (Can. Center Min. Energy Technol.); Toronto, Canada, 1977.
15. Hossner, L.R.; Hons, F.M. Reclamation of Mine Tailings. Adv. Soil Sci. **1992**, *17*, 311–349.

REHABILITATION AFTER OPEN CUT MINES

R

Douglas J. Dollhopf
Department of Land Resources and Environmental Science, Montana State University, Bozeman, Montana, U.S.A.

INTRODUCTION

Mining directly disturbs approximately 240,000 km^2 of the Earth's surface (1). Surface mining methods may be classified as: 1) open pit mining, 2) strip mining, 3) dredging, and 4) hydraulic mining (2). Open pit mining includes quarries used to produce limestone, sandstone, marble, and granite; pits used to produce sand, gravel, and bentonite; and large excavations used to produce talc, copper, gold, iron, silver, and other metals. This mining method is distinguished by having one large pit or numerous small pits across the landscape. State or federal governments establish regulations that require the operator to restore the land to an approved land use(s) that is equal to or better than the premine land use. The approved postmine land use(s) is instrumental in determining the final graded topography and vegetation community.

LANDSCAPE REGRADING

Reclamation of an open pit mine must address the open pit itself, waste rock removed from the pit to gain access to the ore, tailing impoundments, and access roads (Fig. 1). For open pit mines, complete backfilling of the pit is not usually economically feasible. Because the pit remaining after mineral extraction may be hundreds of feet deep, the cost of moving waste rock back into the depression is prohibitive. Surface and groundwater may flow into the depression after mining is terminated and an impoundment develops. Pit slopes should be reduced to the limits required for safe access for humans, livestock, and wildlife. Pit slopes should not be permitted at gradients that jeopardize the success of postmine reclamation.

COVERSOIL RESOURCES

Successful plant community establishment on an open pit mine site is a direct function of the coversoil quality applied. Coversoil suitability criteria vary in the US from state to state, but are similar to those presented in Table 1. The coversoil resource emanates from two procedures: 1) salvaging the natural soil resource from the area to be disturbed and 2) mining unconsolidated geologic stratum when the soil resource is absent. Under present US federal and state mine regulatory programs, all portions of the soil resource shall be salvaged in a project area if it meets physicochemical suitability criteria. Soil materials that do not meet these suitability criteria are generally not used since they may impair plant establishment and growth. Once the soil resource is salvaged, it should be directly hauled to an area that has been backfilled and is ready to receive coversoil. Direct hauled soil contains a viable seed bank, mycorrhizal associations, organic matter, and nutrients that aid in plant establishment. Conversely, if the soil resource is stockpiled, these resources will deteriorate with time.

Prior to implementation of mine land reclamation regulations during the 1970s and 1980s, open pit US mine operations often did not institute land reclamation. The soil resource was not salvaged, the disturbed landscape was not graded to the approximate original contour, and the site was not seeded. Consequently, the soil resource was lost. Today, these lands are being coversoiled using unconsolidated geologic stratum. Valley fill areas adjacent to the disturbed landscape often contain alluvial materials beneath the soil resource that meet all soil suitability criteria (Table 1), except organic matter is absent. This alluvial stratigraphy may be 10–30 m thick or more. Earth moving equipment is used to stockpile the soil resource, separating true topsoil (A-horizon) from subsoil (B- and C-horizons), to enable excavation of deeper alluvial material. Following placement of this alluvial coversoil on the disturbed landscape, the pit created is contoured and the stockpiled soil resource is replaced. Lands receiving coversoil emanating from a geologic stratum should have organic matter applied to expedite establishment of nutrient cycles in the plant root zone. Cattle manure and municipal compost are two sources of organic matter used in land reclamation projects.

Fig. 1 Large open pit mine in the U.S. with water impounded in bottom.

COVERSOIL THICKNESS REQUIREMENTS

The thickness of coversoil required for maximum vegetation performance is dependent on several factors including the 1) vegetation species present, 2) local climate, 3) coversoil quality, and 4) physicochemical quality of the substrate beneath the coversoil. The maximum rooting depth of many rangeland ecosystem plant species is less than 45 cm and generally does not exceed 100 cm (3). Investigators found that a minimum of 40 cm to a maximum of 150 cm of coversoil is needed for optimum plant growth (4–8). This wide variation of findings is a function of site specific conditions identified before. The Barth (4) investigation is representative of most research. This investigator found that generic spoil, defined as nonalkaline and nonsaline loams with sodium adsorption ratio (SAR) below seven, showed an increase in plant production up to a coversoil depth of approximately 50 cm. Sodic spoil, or material with SAR of 26 or higher, required a minimum of 70 cm of coversoil to reach maximum cool season grass production. Coversoiled acid spoils, defined as spoils exhibiting pH values between 3.6 and 4.3, revealed increased grass production up to the maximum coversoil depth used in this study (152 cm).

Table 1 Coversoil suitability criteria

Soil parameter	Suitability criteria
pH	>6.5 and <8.0 standard units
Electrical conductivity	<4.0 dS/cm
Exchangeable sodium percentage	<12%
USDA textural classes (12 types)	All suitable except clay, loamy sand, sand
Rock content (particles >2 mm diameter)	
Slope gradients <25%	<35%, weight basis
Slope gradients >25%	>35 and <60%, weight basis
As, Cd, Cu, Pb, Zn, other metals	Near back ground levels, i.e., no enrichment
Organic matter	>0.5%, weight basis

IN SITU SOIL RECLAMATION

In situ soil reclamation at open cut mine sites means no coversoil resource is used. Chemical and/or organic amendments are applied to the graded spoil, tailings, or waste rock landscape to enable plant establishment and growth.

In Situ Sodic Minesoil Remediation

Thousands of hectares of land in Wyoming and Montana were open pit mined for the mineral bentonite prior to passage of state regulations requiring land reclamation. The absence of a soil resource to reclaim these lands means the plant ecosystem must be established on graded spoils. Spoils are overburden materials cast aside to access the underlying mineral, bentonite. These overburden materials have a sodic condition (ESP 20–40), clay texture (60% clay) dominated by swelling clay minerals such as smectite, and preclude plant establishment and growth. In situ soil remediation requires a two-fold approach: 1) permanently reduce exchangeable sodium percentage (ESP) to less than 10 within the 0–30 cm soil profile and 2) provide an organic amendment to immediately prevent soil crust development and increase water infiltration. Crust development in these clayey-sodic soil systems is precluded with applications of wood chips from saw mill waste or manure (9). Applications of gypsum, calcium chloride, magnesium chloride, and sulfuric acid to these sodic soils have been shown to be effective at reducing the ESP to acceptable levels in minesoils and provide a root zone suitable for plant growth (10).

In Situ Acid Minesoil Remediation

Open pit mines associated with gold, silver, and metal extraction from sulfide ore bodies frequently produce tailings impoundments and waste rock areas that are acidic (pH 2.5–5.5) and fail to support plant growth. Soil acidity is formed when sulfide minerals, such as pyrite, are exposed to oxidizing conditions in the presence of water. In the absence of a coversoil resource, tailings and waste rock material are treated with alkaline amendments to permanently neutralize the soil acidity. Soil analysis of the acid base account and active acidity enable determination of the calcium carbonate requirement (11). Amendments used include calcium carbonate, calcium hydroxide, and calcium oxide. In situ treatment with amendments should be at least to the 45 cm soil depth and special engineered plowing equipment is required to attain this depth of incorporation (12).

ACID MINE DRAINAGE

Upon exposure of open pit mine wastes to water and oxygen, sulfide minerals oxidize to form acidic, sulfate rich drainage (13). Iron-oxidizing bacteria, e.g., *Thiobacillus ferrooxidans*, expedite the acid producing reaction rate up to a million times. Metal composition (Al, Cu, Fe, Mn, Pb, Zn, and others) and concentrations in acid mine drainage (AMD) depend on the type and quantity of sulfide minerals present. Approaches to prevent or treat AMD include: 1) removal of air and water from the sulfide body that is disturbed, 2) wetland construction to precipitate contaminants with oxidation ponds and anaerobic-microbial reactions in plant root zones, 3) underground anoxic- and above ground open-limestone drains to raise water pH, and 4) treatment of drainage with neutralizing chemicals such as hydrated lime, quicklime, soda ash, caustic soda, and ammonia (14).

STEEP SLOPE RECLAMATION

Open pit mines are frequently located in mountainous terrain and reconstructed slopes may have gradients as steep as 50%. Sediment yields tend to increase linearly with slope gradient. There is an inverse relationship between soil loss on slopes and rock cover (15). High rock content in soils increases infiltration rate and surface roughness decreasing runoff and soil loss. As erosion occurs on rocky soil, rock cover increases as the coarse fragments below the surface are exposed. This armoring of the soil surface can help reduce soil erosion. It has been reported that plant growth may be impaired when soil rock content exceeds 35% (16). High rock contents of 35–60% applied on steep slopes (Table 1) may impair plant growth, but are considered a best management practice to facilitate plant establishment and slope stability.

Construction of shallow pits across steep slopes is commonly practiced to increase water storage on the slope and minimize runoff. Pitting techniques using equipment referred to as Dammer–Diker, gouger, and dozer basin blade have been shown to be effective in controlling sediment loss (17).

CONCLUSION

During these past 30 years, land rehabilitation sciences at open pit mines developed at an exponential rate. Grading the mined landscape to the approximate original contour, use of adequate amounts of—and quality of—coversoil, and development of in situ soil treatments at sites where coversoil resources were absent have all provided means to establish a diverse plant community. Techniques have been developed to revegetate and stabilize slopes approaching a 50% gradient, which is an asset for mines located in mountainous terrain. Methods to control and treat AMD from open pit mines continue to improve, but this impact to surface water resources remains unresolved. Future research is required to develop better AMD treatment methods.

REFERENCES

1. Solomons, W. Mining Impacts Worldwide. *Proceedings of the International Hydrology Program*; UNESCO, Asian Institute of Technology: Bangkok, 1988; 7–9.
2. Paone, J.; Struthers, P.; Johnson, W. Extent of Disturbed Lands and Major Reclamation Problems in the United States. In *Reclamation of Drastically Disturbed Lands*; Schaller, F.W., Sutton, P., Eds.; American Society of Agronomy: Madison, WI, 1978; 11–22.
3. Wyatt, J.W.; Dollhopf, D.J.; Schafer, W.M. Root Distribution in 1- to 48-Year Old Strip-Mine Spoils in Southeastern Montana. J. Range Mgmt **1980**, *33*, 101–104.
4. Barth, R.C. *Soil-depth Requirements to Reestablish Perennial Grasses on Surface-Mined Areas in the Northern Great Plains*; No. 1, Publication 0192-6179/84/2701-0001 Mineral & Energy Resources Research Institute, Colorado School of Mines: Golden, 1984; Vol. 27.
5. Doll, E.C.; Merrill, S.D.; Halvorson, G.A. *Soil Replacement for Reclamation of Strip Mined Lands in North Dakota*; Agricultural Experiment Station Bulletin 514, North Dakota State University: Fargo, 1984; 23.
6. Halvorson, G.A.; Melsted, S.W.; Schroeder, S.A.; Smith, C.M.; Pole, M.W. Topsoil and Subsoil Thickness Requirements for Reclamation of Nonsodic Mined-land. Soil Sci. Soc. Am. J. **1986**, *50*, 419–422.
7. Power, J.F.; Sandoval, F.M.; Ries, R.E. Topsoil–Subsoil Requirements to Restore North Dakota Mined Lands to Original Productivity. Mining Engng **1979**, December, *23–27*.

8. Schuman, G.E.; Taylor, E.M.; Pinchak, B.A. Revegetation of Mined Land: Influence of Topsoil Depth and Mulching Method. J. Soil Water Conserv. **1985**, *40*, 249–252.
9. Schuman, G.E.; King, L.A.; Smith, J.A. Reclamation of Bentonite Mined Lands. In *Reclamation of Drastically Disturbed Lands*; Barnhisel, R.I., Darmody, R.G., Daniels, W.L., Eds.; American Society of Agronomy: Madison, WI, 2000; Vol. 41, 687–707.
10. Dollhopf, D.J.; Rennick, R.B.; Smith, S.C. *Long-Term Effects of Physicochemical Amendments on Plant Performance at a Bentonite Mine Site in the Northern Great Plains*; Reclamation Research Unit Publication 88-02, Montana State University: Bozeman, 1988; 126.
11. Sobek, A.A.; Schuller, J.R.; Freeman, J.R.; Smith, R.M. *Field and Laboratory Methods Applicable to Overburdens and Minesoils*; Publication 600/2-78-054, U.S. Environmental Protection Agency, Office of Research and Development: Cincinnati, Ohio, 1978; 47–67.
12. Dollhopf, D.J. *Deep Lime Incorporation Methods for Neutralization of Acidic Minesoils*; Reclamation Research Unit Publication 9201, Montana State University: Bozeman, 1992; 94.
13. Stumm, W.; Morgan, J.J. *Aquatic Chemistry*; Wiley: New York, 1970, 178.
14. Skousen, J.G.; Sexstone, A.; Ziemkiewicz, P.F. Acid Mine drainage Control and Treatment. In *Reclamation of Drastically Disturbed Lands*; Barnhisel, R.I., Darmody, R.G., Daniels, W.L., Eds.; American Society of Agronomy: Madison, WI, 2000; Vol. 41, 131–168.
15. Kapolka, N.M.; Dollhopf, D.J. Effect of Slope Gradient and Plant Growth on Soil Loss on Reconstructed Steep Slopes. Int. J. Surf. Mining, Reclam. Environ. **2001**, *15* (2), 86–89.
16. Munn, L.; Harrington, N.; McGirr, D.R. Rock Fragments. In *Reclaiming Mine Soils and Overburden in the Western United States; Analytical Parameters and Procedures*; Williams, R.D., Schuman, G.E., Eds.; Soil Conservation Society of America: Ankeny Iowa, 1987; 259–282.
17. Larson, J.E. *Revegetation Equipment Catalog*; Stock No. 001-00518-5 Forest Service, U.S. Department of Agriculture, U.S. Government Printing Office: Washington, DC, 1980.

REHABILITATION OF WASTE ROCK AND OVERBURDEN DUMPS

R

M.V. Fey
R.D. O'Brien
A.J. Mills
University of Stellenbosch, Stellenbosch, South Africa

INTRODUCTION

Improved mining technology over the last few decades has resulted in an increasing trend toward open-cast methods; causing large areas of land to be disturbed in order to access the natural resources beneath the surface. This article focuses on principles that would apply to the whole range of spoil materials—from those produced by conventional quarrying, to those from strip mining of vast ore deposits over hundreds of square kilometers. The nature of the spoil may vary from relatively hard, coarse, unweathered rock fragments to fine, soft, easily erodible earthy materials. In all cases, the common thread is the need first to remove whatever soil cover exists before mining can commence. This soil cover may either have been lost or, at least if stockpiled for later use in rehabilitation work, severely disturbed through mixing of topsoil with deeper horizons and sterilized to some degree as a result of stockpiling. A wide variety of practices have been applied to restore landscapes that have been affected by mining. The focus of this article will be on modern methods and recognized principles of rehabilitation.

DETERMINATION OF REHABILITATION GOALS BEFORE MINING COMMENCES

A prerequisite to mining should be the execution of a baseline survey that identifies and maps the existing soils and establishes the potential land use as a means of guiding the formulation of rehabilitation goals. While the planned rehabilitation program is unlikely to completely dictate mining methods, the greater the cooperation between the mine and rehabilitation planners, the better the chances of successful rehabilitation. Physical and chemical characterization of each overburden layer needs to be done prior to mine development. This will reveal the potential for compaction, erosion, and satisfaction of plant growth requirements in terms of water availability, nutrition, and toxicity problems. Soil scientists and mining engineers often disagree over what constitutes "topsoil"; the onus should be on the soil scientist to demonstrate the special character of true topsoil in terms of the contribution made by humus to soil structure, aeration, water holding capacity, plant nutrient supply, and toxicity buffering and also in terms of microbiological activity. Topsoil may also constitute a seed bank that may be of enormous value during revegetation.

The final land use must be decided upon before the development of the mine, and should consider factors such as stability and degree of permanence and not just the level of productivity to be achieved. The final land use will also influence the landscape design; fertilizer, liming, and maintenance requirements; and the cost of reclaiming the area. However, the final land use should be seen as the minimum requirement for rehabilitation and not inhibit research to develop improved methods that could lead to a more stable end land use. The aim of landscaping of waste dumps is to produce slopes that are not prone to erosion and are compatible with the proposed land use and the surrounding landscape. Erosion of waste dumps can have severe effects on the success of the reclamation, leading to exposure of encapsulated waste rock, high sediment loads, and lower water quality in receiving streams. Various computer models have been developed to aid the design of waste dumps (1) by predicting erosion and soil loss from waste rock dumps. These models range in complexity from short-term stability during the construction stage to topographic evolution models simulating erosion over centuries.

There is no typical design or shape of waste rock and overburden dumps. Economic circumstances often dictate, however, that the amount of landscaping be kept to a minimum, in which case there is still a considerable slope angle on the sides of the dump. This may have important implications for revegetation because of differential temperature and moisture regimes on north and south aspects, and the need for special measures such as terracing to trap water and reduce erosion. The latter can in some cases be counterproductive because of the tendency of

Fig. 1

contour terraces to concentrate the flow of water and to become breached if rainfall is too intense. The consequences can then be quite devastating in the form of deep gully erosion. Fig. 1 gives some indication of the difficulty of designing an acceptable shape for waste rock dumps.

RESTORING A GROWING MEDIUM FOR PLANTS

There are physical, chemical, and biological factors that need to be considered in restoring the capacity of overburden materials to sustain plant growth.

One of the biggest physical problems is compaction arising from the use of machinery during overburden placement after mining. Some degree of soil compaction is inevitable as soils are stripped and stored for later use. It is generally agreed that soil handling has a marked effect on the success of any reclamation scheme. In an attempt to remove the effects of compaction on the replaced soil material, various forms of subsoiling or ripping using tines drawn by crawler tractors are used. These methods are not always successful and alternative methods of soil placement may need to be developed. One such method, called loose tipping (2), involves the tipping of soil material from dump trucks, with the heaps being spread by an excavator operating from the overburden surface. With this method there is no need for vehicles to travel over newly restored soil. In some cases the heaps are left in place to serve as windbreaks, assist in recruitment of windblown seeds, and reduce both erosion and sand-blasting effects on new vegetation in areas exposed to strong winds. Eventually, natural forces reduce the relief of the mounds.

Another serious physical problem is the tendency of many spoils to disperse readily under raindrop impact, developing a surface crust that inhibits seed germination and seedling emergence and reducing the infiltration of water and diffusion of oxygen into underlying layers. Severe erosion can result from crust formation. Organic or gravel mulches, or aggregate-promoting chemicals applied on the surface, such as gypsum or polyacrylamide, are variously employed in such circumstances. Some spoils, even though finely divided and showing dispersive tendencies, may contain a certain proportion of coarse gravel and the value of this should be recognized: Initially severe erosion may rapidly abate as the lag concentrate of surface gravel develops into a stable mulch.

The placement of topsoil, although valuable, is not a requirement for reclamation success. The quality of topsoil that has been stockpiled may deteriorate rapidly (3) and the material often acidifies, requiring liming and fertilizer application to ensure optimal plant growth. The viability of seeds of many plants may decrease sharply after a few months of burial in a stockpile, and microbiological activity may also subside considerably. The importance of using freshly removed topsoil, therefore, needs to be emphasized by the rehabilitation specialist when dealing with mine planners.

In areas with insufficient topsoil, suitable waste rock or overburden is identified as a topsoil substitute. Such material will often be greatly improved by addition of organic matter and soil amendments such as lime (to neutralize acidity if the spoils are sulfidic) or gypsum (to improve permeability to water). Acidic rock drainage (ARD) can be particularly problematic in some mining wastes in view of the gradual nature of pyrite oxidation, leading to the slow release of acid, and the difficulty of accurately estimating the amount of lime needed for neutralization. Burial beneath 0.5 m or more of soil can be quite effective in minimizing ARD by cutting off a ready supply of oxygen from the pyritic soil.

The addition of fertilizers to topsoil substitutes may be needed for some years until the nutrient cycle between soil and plants has been established. Slow-release formulations of nitrogen fertilizers may be especially valuable, although in some cases cost may prohibit their use. A leguminous component in the vegetation cover can make a big difference to longer-term nitrogen supply by harnessing the nitrogen-fixing capacity of symbiotic bacteria. Stimulating the population of soil fauna (bacteria, fungi, earthworms, termites, and rodents, for example) is seldom pursued through direct inoculation or introduction of the desired species but can usefully be encouraged by understanding their special requirements (e.g., a suitable soil pH in the case of earthworms) and ensuring that these are met to some degree. The loosening and aerating effects of burrowing animal activity may well have been underestimated as prerequisites of successful rehabilitation. Even natural soils in their undisturbed state would tend, after long periods, to become increasingly compacted and structurally inferior were it not for the continual reworking of the soil by resident fauna.

REVEGETATION

The choice of vegetation cover depends on the end use, climatic conditions, and the potential productivity of the growing medium. Where cultivated crops, pastures, or forestry existed prior to mining, the goal is usually to restore productivity by reestablishing the same pattern of land use as far as possible. Often the goal is partly aesthetic in achieving harmony with the surrounding landscape.

In many rehabilitation projects, it is desirable to reinstate the natural plant and animal communities of the area. This may be because the land is required for conservation or wildlife habitat purposes. Alternatively, it may be decided that natural plant communities will constitute the most effective rehabilitation because of low maintenance requirements as well as resilience to drought and other extreme climatic events. Plant species in undisturbed ecosystems are an expression of the prevailing soil and climate. Consequently, if premining soil conditions can be approximated, plant species from surrounding natural areas are likely to be suitable for achieving revegetation of the disturbed area. Seeds should ideally be collected from as near to the rehabilitation site as possible in order to maintain the genetic integrity of surrounding ecosystems and to ensure that the reintroduced plants are adapted to the local conditions. The number of plant species and functional types on the rehabilitation site should preferably be maximized, given the ecological principle that ecosystem stability tends to increase with increasing biodiversity (4).

If topsoil is limited it may not be possible to reinstate the local flora in its entirety. The chemical status of rock wastes is often not conducive to plant growth and if roots escape the topsoil layer and come into contact with high metal concentrations, acidity, alkalinity, or salinity of the waste materials, revegetation efforts may be thwarted. In such a situation, plant species should be carefully selected for their tolerance to the particular chemical conditions of the dump. Plant species that have colonized old dumps in the area are often suitably adapted to the chemistry of the waste material and constitute a valuable seed source for rehabilitation.

The rehabilitation of a waste dump and the maintenance of a recreated ecosystem may not depend solely on the reintroduction of local plant species. Effective management of the vegetation may include prescribing suitable fire regimes or herbivore stocking densities. Besides their bioturbation role mentioned above, invertebrates may also play a role in key ecosystem processes such as pollination, seed dispersal, and nutrient cycling. Provision of organic mulch on the soil surface can assist colonization of the soil by these invertebrates. Monitoring populations of invertebrates such as ants, nematodes, and spiders is useful for comparing rehabilitated land with adjacent ecosystems. Such studies can be used for assessing the extent to which the rehabilitation process is moving toward a premining ecosystem state.

ENSURING SUSTAINABILITY THROUGH AN UNDERSTANDING OF SOIL-FORMING PROCESSES

A key factor, of which reclamation ecologists are often unaware, is the value of studying natural soil profiles in the undisturbed landscape for clues as to the intensity of soil-forming processes that have given rise to the development of soil horizons. Features such as cemented calcic or duripan horizons in arid landscapes, plinthite in the subhumid tropics, and gleying in hydromorphic soils often reveal subtleties of climatic and topographic effects that are key ingredients in the functioning of the ecosystem. Inability to recognize their significance may doom the rehabilitation effort to failure in the long run.

REFERENCES

1. Evans, K.G. Methods for Assessing Mine Site Rehabilitation Design for Erosion Impact. Aust. J. Soil Res. **2000**, *38*, 231–247.
2. Moffat, A.J.; Bending, N.A.D. Replacement of Soil and Soil-Forming Materials by Loose Tipping in Reclamation to Woodland. Soil Use and Management **2000**, *16*, 75–81.
3. Davies, R.; Hodgkinson, A.; Younger, A.; Chapman, R. Nitrogen Loss from a Soil Restored After Surface Mining. J. Environ. Qual. **1995**, *24*, 1215–1222.
4. Tilman, D.; Wedin, D.; Knops, J. Productivity and Sustainability Influenced by Biodiversity in Grassland Ecosystems. Nature **1996**, *379*, 718–720.

REMOTE SENSING AND GIS INTEGRATION

Egide Nizeyimana
The Pennsylvania State University, University Park, Pennsylvania, U.S.A.

INTRODUCTION

Remote sensing (RS) is often viewed as a technology for data acquisition on the Earth's surface. It was described in detail in one of the previous articles of this section. This article provides ways in which RS, including aerial photography, is integrated to a Geographic Information System (GIS) for environmental assessments needed in land use planning and management. RS is compatible with GIS because the data generated are geo-referenced to a known coordinate system and are spatial in nature. These data are generally stored and queried as grid, a data structure supported by the most popular commercial GIS software. Later, they are transferred to a GIS platform after processing for integration with other spatial land attributes from spatial data sources such as soil databases and digital elevation models. This RS/GIS integration and associated datasets can also be part of GIS-based systems such as GIS/model interfaces and spatial decision support systems (SDSS).

COMPATIBILITY ISSUES BETWEEN REMOTE SENSING AND GIS

A GIS can be defined as a set of computer tools for capturing, storing, analyzing, and displaying spatially referenced data. The data within a GIS consists of two elements: spatial entities represented by points (e.g., well locations), lines (e.g., streams, road networks), and polygons (e.g., soil delineations) and attribute data or information that describes characteristics of the spatial features. The spatial entity is referenced to a geographic coordinate system and is stored in either a vector or raster model. GIS is primarily a platform that integrates spatial information from variable data sources and provides tools to overlay and analyze it.

RS is compatible with GIS because the information acquired by imaging sensors carried aboard satellites and airplanes is geographic in nature, referenced to a known coordinate system, and in a grid layout, a raster data model commonly found in most popular GIS software packages. In a raster model environment, the analysis is performed pixel by pixel. Current commercial GIS-based software packages have vector and raster capability analyses and allow data conversion from one data model to the other. Nonimaging systems are also important to RS/GIS integration because they provide point data, which at geo-referenced locations, are often combined with other environmental data for site-specific assessments.

LINKING REMOTE SENSING AND GIS

Remote Sensing as a Source of Spatial Data

RS is one of the most important sources of land use/cover information used in GIS analyses. It provides information on the location and spatial and temporal distribution of land cover on the Earth's surface. Land use/cover distributions may be derived from aerial photographs after they are corrected for relief displacement and distortions caused by camera angle, and registered to a coordinate system. Delineations of different land use/cover types on photographs are made from visual interpretations of characteristics (e.g., tone, texture and color) aided by optical devices. Digital maps of these classes are created by digitizing and processing boundaries between land uses or by scanning photographs covering the area of interest and screen-digitizing their boundaries.

Land use distributions on the Earth's surface may also be obtained using RS imaging systems. In this case, RS provides opportunities in the area of land use planning which would not be otherwise available. While aerial photographs are effective and appropriate for analyses of small areas, RS offers tremendous advantages when planning for large areas such as river basins and regions. The fact that the land surface is observed from reflected/emitted energy for large areas and over a wide range of wavelengths allows for easy differentiation of existing land uses (e.g., wetlands, degraded landscapes) (1). Earth observation satellites also offer a repetitive coverage of the land, thus providing the possibility of monitoring land use pattern changes over time. Gathering information on land uses at several time intervals is particularly important in the monitoring and land evaluation stages of the land use planning process. Factors such as plant stress, crop growth and yields that serve as measures of agricultural productivity can be rapidly estimated following digital processing and analysis of RS imagery. RS has been used in many instances to monitor land surface conditions such as soil degradation and soil salinity (2). It is true that RS cannot replace field mapping of land use/land cover. However, RS supplements provide information that would not otherwise be available to land use planners and managers. There is no doubt that the development of sensors of high spatial resolution (1 m and higher) presently in orbits would increase the use of RS in land use planning and management.

As indicated before, RS and GIS technologies are highly compatible primarily because of the nature of RS as a source of spatial land use/land cover used in various environmental applications. Remote sensed imagery is a grid cell layer, a data format easily handled by GISs. In the RS/GIS integration, GIS appears as a platform that stores and integrates spatial data from different sources including remote sensing and output information needed for environmental analyses depending on the type of analysis sought. Land use /cover distributions derived from RS are input to GIS, GIS-based systems (SDSS), and models (e.g., hydrologic/water quality, crop yields, primary productivity) commonly used in land evaluation for land use planning and management. GIS has been proven to be a valuable tool in integrating RS-derived data with climate and land surface parameters (soils, terrain, etc.) to generate digital maps of ecosystem productivity or vulnerability to environmental factors (3). The parameterization of the Terrestrial Ecology Model (TEM) and a forest ecosystem model using GIS- and RS-derived data at regional scales has also been accomplished (4, 5). The RS/GIS approach has been adopted by many state and federal agencies involved in environmental assessments. For example, in the GAP Analysis Program, the U.S. Geological Survey (USGS) and collaborative institutions map and/or model potential natural habitats of native vertebrate species from remotely sensed data across the country, and use GIS map overlay tools to determine the degree of richness of habitats in these species (6).

Finally, GIS and RS applications have promoted the development of other high-resolution spatial technologies that enhance the land use assessment and planning. Some of these are the global positioning system (GPS) and digital orthophoto quadrangles (DOQ). GPS allows the user to record accurately and rapidly geographic coordinates of any location in the field with precisions ranging from several meters to a centimeter. Soil, terrain and land use attributes observed or measured can then be input to a GIS along with their precise locations and extent. GPS coupled with a GIS can improve the accuracy of land quality mapping by increasing the spatial variability of soil and landscape attributes. DOQs, on the other hand, are digital images of aerial photographs that were corrected to remove relief displacement and distortion caused by the camera angle. The USGS distributes single-band, 256-scale, gray-scaled DOQs at 1 m grid resolution. Although DOQs have not been used extensively in land use planning in the past, their high resolution and photograph-like characteristics make them a potential source of data in this area. Furthermore, repetitive DOQs acquired at different times can be used to monitor the magnitude of changes in land use over time. The relationship between GIS, RS, and other data sources is shown in Fig. 1.

Applications of Remote Sensing/GIS in Soil Science

The need for geospatial data for use in various spatial database development and analyses in industry, government, and universities have increased the demand for remotely sensed data. In the area of soil science, the remote sensing technology is used to acquire data, and digitally process and analyze it. Subsequent analyses may involve error assessment, and data conversion from raster to vector format and vice versa before the data is merged

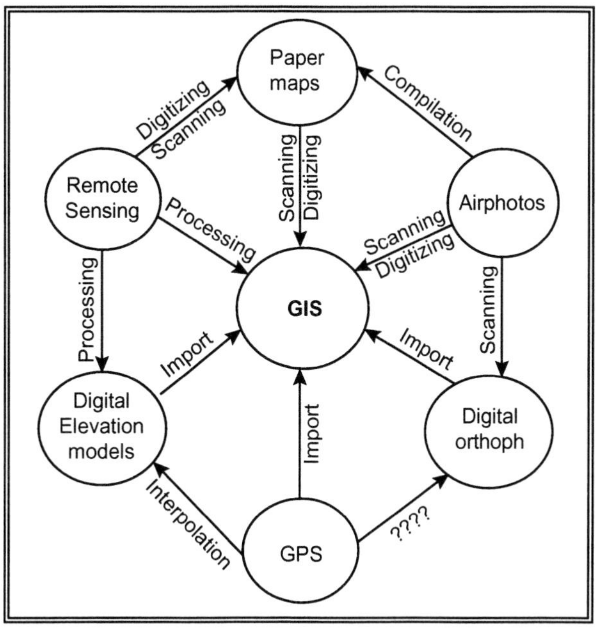

Fig. 1 Relationships between remote sensing and GIS. (Adapted from Ref. 7.)

Table 1 Documented uses of remote sensing in various areas of soil science

Application	Advantages	Potential limitations	References
Laboratory methods	Provides accurate measurements of reflectance values	Provide point data rather than areal extent of soil properties	(8)
Field methods	Easily related to on-site conditions	Provide limited data coverage; measurements are affected by soil conditions (soil roughness, moisture), sun angle, etc.	(9)
Airborn-/satellite-based methods	Provides areal extent and temporal coverage of soil properties	Radiance measurements affected by atmospheric and soil-surface conditions	(–)
a) Interpretation of digital images	Good results when data are well analyzed and interpreted	Image analysis can be costly; requires experienced technician for good results; field verification or prior knowledge of area required for best results	(10)
b) Interpretation of color composites	Easy to use and rapid interpretations	Intensive field verification is needed for good results due to the fact that interpretation is based on differences in tone and physical characteristics of objects; requires experienced interpreter for good results	(11)
c) Interpretation of radiances	Relatively easy, cheap, quantitative; data can be normalized to remove environmental effects	Results often unreliable	(12)

(Adapted from Ref. 13.)

with other datasets in GIS for specific analyses. Table 1 summarizes remote sensing techniques used in various aspects of soil science and potential limitations for use in GIS and advantages of each. These are laboratory approaches, field methods, and aircraft-/satellite-based methods. The latter involves interpretation and analyses of digital images, color composites, or radiances.

Remote Sensing/GIS Integration for Site-Specific Farming

The goal of site-specific farming (SSF) or precision farming is to optimize the profitability of a farm practicing variable management according to soil conditions found at each site. The concept is based primarily on the fact that soil chemical and physical properties that affect crop production (e.g., pH, nutrients, available water, impeding layers) vary spatially across agricultural landscapes. Management practices such as fertilizer and pesticide applications, irrigation water, and crop varieties should, therefore, be prescribed according to this soil variability. The profitability associated with variable rate applications of SSTs should avoid or reduce waste and the risk for environmental pollution because these agrochemicals are applied to the field only in amounts needed for optimal crop growth.

In an SST, the real-time detection of continuous soil variable is made possible by mobile devices including sensors and differential GPS units while agrochemicals are applied using variable rate field applicators mounted on farm equipments. These sensors are based on the same principles as RS and have been developed to detect, directly or indirectly, soil properties such as soil moisture, soil texture, nitrates, etc. The data acquired, along with their respective locations, are often integrated with spatial information from other sources such as soil databases, climate, landscape, and satellite/airborne remote sensing to delineate meaningful management zones within the farm.

REFERENCES

1. Petersen, G.W.; Nizeyimana, E.; Evans, B.M. Applications of Geographic Information Systems in Soil Degradation Assessments. In *Methods for Assessment of Soil Degradation*; Advances in Soil Science, Lal, R., Blum, W.H., Valentine, C., Rose, B.A., Eds.; CRC Press: Bacon Raton, FL, 1997; 377–391.
2. Raina, P.; Joshi, D.C.; Kolarkar, A.S. Mapping of Soil Degradation by Remote Sensing on Alluvial Plan, Rajasthan, India. Arid Soil Res. Rehab. **1993**, *7*, 145–161.
3. Parrish, D. A.; Townsend, L.; Saunders, J.; Carney, G.; Langston, C. USEPA *Region 6 Comparative Risk Project*: Evaluating Ecological Risk. EPA Unpublished Report, **1993**.
4. Pan, Y.; McGuire, A.D.; Kicklighter, D.W.; Melillo, J.M. The Importance of Climate and Soils for Estimates of Net Primary Production: A Sensitivity Analysis with the Terrestrial Ecosystem Model. Global Change Biol. **1996**, *2*, 5–23.
5. Lathrop, R.G., Jr.; Abler, J.D., Jr.; Bognar, J.A., Jr. Spatial Variability of Digital Soil Maps and Its Impact on Regional Ecosystem Modeling. Ecol. Model. **1995**, *82*, 1–10.
6. Scott, J.M.; Davis, F.; Csuti, B.; Noss, R.; Butterfield, B.; Groves, C.; Anderson, H.; Caicco, S.; D'Erchia, F.; Edwards, T.C., Jr.; Ulliman, J.; Wright, R.G. GAP Analysis: A Geographic Approach to Protection of Biological Diversity. Wildlife Monogr. **1993**, *123*, 1–41.
7. Nizeyimana, E.; Petersen, G.W. Land Use Planning and Environmental Impact Assessment Using GIS. Environmental Modeling Using GIS and Remote Sensing, 2001.
8. Stoner, E.R.; Baumgardner, M.F. Characteristic Variations in Reflectance Surface of Soils. Soil Sci. Soc. Am. J. **1981**, *45*, 1161–1165.
9. Gausman, H.W.; Leamer, R.W.; Noriega, J.R.; Rodriguez, R.R.; Wiegand, C.L. Field-Measured Spectrometric Reflectance of Disked and Non-Disked Soil with and without Wheat Straw. Soil Sci. Soc. Am. J. **1977**, *41*, 493–496.
10. Connors, K.F.; Gardner, T.W.; Petersen, G.W. Digital Analysis of the Hydrologic Components of Watersheds Using Simulated SPOT Imagery. *Proceedings of Workshop on Hydrologic Applications of Space Technology*; IAHS: Cocoa Beach, FL, 1985; Vol. 160, 355–365.
11. Bocco, G.; Palacio, J.; Valenzuela, C.R. Gully Erosion Modeling Using GIS and Geomorphologic Knowledge. ITC J. **1990**, *3*, 253–261.
12. Pickup, G.; Nelson, D.J. Use of Landsat Radiance Parameters to Distinguish Soil Erosion, Stability, and Deposition in Arid Central Australia. Remote Sens. Environ. **1984**, *16*, 195–209.
13. Nizeyimana, E.; Petersen, G.W. Remote Sensing Applications to Soil Degradation Assessments. In *Methods for Assessment of Soil Degradation*; Advances in Soil Science, Lal, R., Blum, W.H., Valentine, C., Rose, B.A., Eds.; CRC Press: Bacon Raton, FL, 1997; 393–405.

REMOTE SENSING OF SOIL MOISTURE, PASSIVE AND ACTIVE MICROWAVE

E.D. Warner
G.W. Petersen
The Pennsylvania State University, University Park, Pennsylvania, U.S.A.

INTRODUCTION

Soil moisture measurement with in situ methods over large areas and at frequent time intervals is not practical for most purposes. Remote sensing from air- and spaceborne platforms permits data collection from large areas and at frequent time intervals, matching the dynamic nature of soil moisture. Remote sensing with the microwave spectrum, wavelengths ranging from 1 mm to 1 m, is attractive as data can be collected day or night and through clouds. Terrestrial applications of microwave sensing typically use wavelengths from 3 to 75 cm, although the convention of microwave remote sensing is to refer to frequency rather than wavelength. The topic of microwave sensing is a broad one as it not only includes two general types of instruments, but also a variety of applications spanning fields such as agronomy, soil science, hydrology, meteorology, and astronomy.

This discussion will only include passive microwave sensing, synthetic aperture radar (SAR), and their use for detecting soil moisture. There are other types of radar instruments; however, only SAR systems have come to be widely accepted for terrestrial applications. Active and passive microwave systems are briefly described, excluding many critical details about instruments, data collection and processing, and applications. The reader should consult Refs. 1–3 for a comprehensive examination of many of these issues. More brief, but very informative, descriptions are available from Refs. 4–6.

THE PHYSICAL BASIS FOR MICROWAVE REMOTE SENSING OF SOIL MOISTURE

Microwave remote sensing is similar to sensing in other parts of the electromagnetic spectrum in that a sensor is used to detect the results of the interaction of energy from a part(s) of the electromagnetic spectrum with some type of media. Soil is a composite media of solid, water and air components. As the amount of water in the soil changes, the emission and reflection of energy from the soil also changes. The influence of soil moisture on emission and reflection is described numerically by the dielectric constant. Although the dielectric constant is referred to as a constant, it actually changes with the amount of water in the soil and is indicative of the differences in dielectric properties of dry soil and water. For example, the dielectric constant of water is around 80, while those of soils can vary from 3 (very dry soil) to around 30 (very wet soil).

Figure 1 displays the change in dielectric constant with volumetric soil moisture at different frequencies for one texture. Most applications of microwave sensing address soil moisture on a volumetric basis to minimize differences due to texture. As can be seen, the relationship between dielectric constant and volumetric moisture content is nonlinear. The rapid increase in dielectric constant with volumetric moisture content can be attributed to the amount of water in direct contact with the soil. At drier moisture contents the water is bound more tightly to the surface of the soil particles, which limits their ability to propagate energy. As the moisture content increases, the water is less bound and permits energy to propagate more freely resulting in an increase in dielectric constant. As can be seen, the dielectric constant is described by a real (e') and an imaginary (e'') component. The real dielectric constant is that which describes the propagation of the energy at a given wavelength or a range of wavelengths. The imaginary part of the dielectric constant describes losses, like friction in which energy is converted to heat, that occur during the interaction of a wave with a material.

While both passive and active microwave remote sensing of soil moisture are based on the dielectric properties of soil the physical mechanisms that are the basis for detection by these instruments are very different. A brief background to passive and active microwave sensing is provided as preparation for a description of the application of the instruments for soil moisture detection.

Encyclopedia of Soil Science

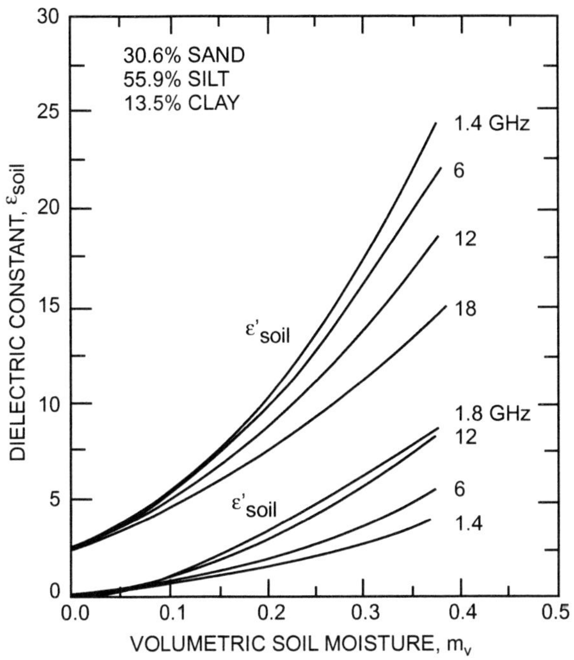

Fig. 1 Variation of dielectric constant with volumetric soil moisture. (Adapted from Ref. 3.)

SYNTHETIC APERTURE RADAR SYSTEMS AND DATA

Active microwave systems employ RADAR, an acronym for radio detection and ranging, technology for generating radio waves for illuminating areas or objects. The radio waves are directed away from a source antenna, which is also used to detect waves reflected from incident surfaces. Figure 2 depicts the general operation of a RADAR system, in this case a side looking airborne radar (SLAR). The detection of reflected waves does not have to be done by the same antenna that generated them, but nearly all commercial and government systems use this approach.

Pulses of known electromagnetic properties are generated by the RADAR hardware and propagated by the antenna. End-to-end knowledge of pulse electromagnetic properties are the foundation of SAR data processing, which can produce the fine azimuth resolution that differentiates SAR from real aperture RADAR systems. The pulse properties on reception provide information about an area from which they were reflected. An area is illuminated by many pulses as the antenna is moved along a trajectory. SAR processing takes the intensity of each received pulse and sums them. Pulses can contribute positively or negatively depending on the wave properties at detection. Ultimately an image of values is assembled, indicative of the intensity of the energy received at the antenna and the output of the summation operation executed during processing. Graphically an image is composed of pixels, giving SAR data an appearance similar to that acquired by the SPOT and Thematic Mapper optical and near-infrared instruments.

The magnitude of the intensity values, also called the backscatter cross-section, in an image is the result of the interaction of the transmitted pulses with the scattering and dielectric properties of the object(s) encompassed in the area defined by a pixel. Fig. 2 (4) depicts the difference between the intensity of pulses received from different features present in one image line. For example, the hill has the largest cross-section of all features imaged. The cross-section does not necessarily have to be related to a feature's size. There are conditions in which relatively few small scattering surfaces can generate a large cross-section. Factors influencing backscatter cross-section include surface conditions (soil roughness, soil moisture, and vegetation density and canopy properties) and illumination characteristics (incidence angle, azimuth angle, wavelength, and polarization).

Some additional processing is done to yield an image of values known as the normalized backscatter cross-section. Backscatter cross-section is a ratio of power received over that incident on a surface, and also includes antenna and illumination characteristics. For terrestrial applications backscatter cross-section is usually normalized by the area spanned by the pulses of power summed for each pixel in an image. Targets with area are referred to as distributed, due to the spread of transmitted power across the Earth's surface. The spread of power over an area matches the extensive arrangement of most terrestrial targets, such as crop and tree stands.

The normalized backscatter cross-section is a dimensionless quantity equal to the average cross-section per unit area. Normalized backscatter cross-section can vary over orders of magnitude, hence is usually calculated in decibels (10 times the log backscatter cross-section). Note that the normalized backscatter cross-section is the ratio of average scattered power density over the average incident power density over the scattering surface. It is not the total scattered power divided by the total incident power, which means that the normalized backscatter will vary with different incidence and azimuth angles, and can be greater than 1 for certain orientations.

PASSIVE MICROWAVE SYSTEMS AND DATA

Passive microwave sensing is based on the detection of radiation by an antenna that is constructed only to receive energy. The passive instrument then detects the intensity of microwave energy that originates from a source that is

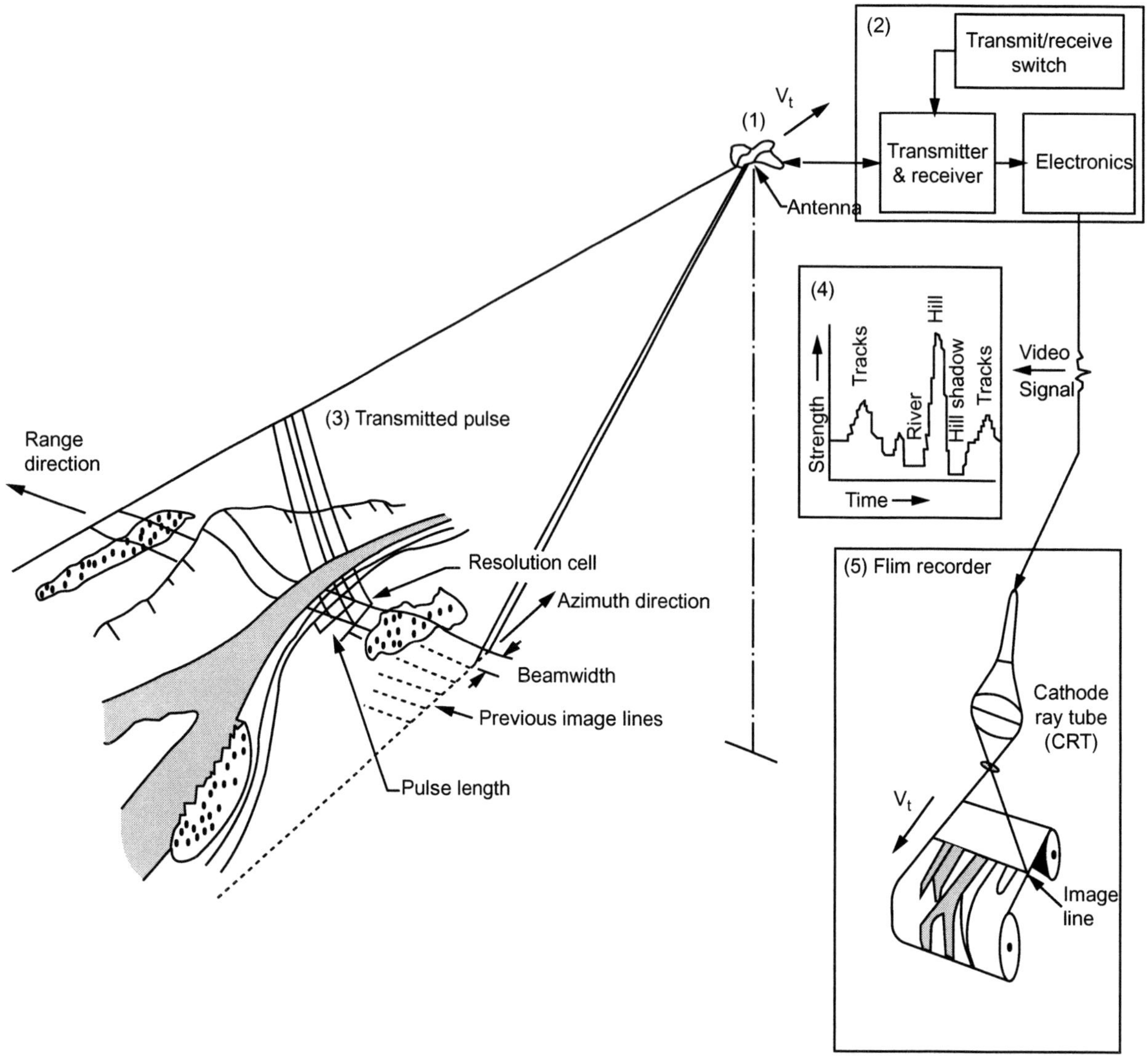

Fig. 2 SLAR instrument operation and data processing. (Adapted from Ref. 6.)

not the antenna. Passive instruments are referred to as radiometers, the same terminology used for thermal instruments sensitive to shorter wavelengths. Passive microwave and shorter wave thermal radiometry share many of the same physical and instrument concepts.

The basic structure of a microwave radiometer is shown in Figure 3. The passive microwave radiometer has three major components. The first component is an antenna that receives the incoming energy for a given instrument orientation. The second component is a receiver that detects and amplifies the signal from the antenna. The receiver is designed to be sensitive to energy of a selected frequency range and polarization. The third component of a microwave radiometer is a data handling system that has two major parts, a switch and a data recorder. The switch alternately passes data to the data recorder from the receiver and a thermal reference. Data from the receiver is in units of V. The thermal reference attains the temperature that occurs at the surface in order to achieve the detected voltage at the antenna. The nearly simultaneously acquired signals allow for the determination of the apparent antenna temperature, which is a conversion of the antenna acquired voltage to temperature, usually expressed in K. Other recorded data assist post-processing, including antenna pattern and atmospheric corrections. After post-processing, the apparent temperature of the antenna is converted to a quantity referred to as brightness temperature.

If a scanning radiometer is used to acquire data, then an image can be constructed from the brightness temperature

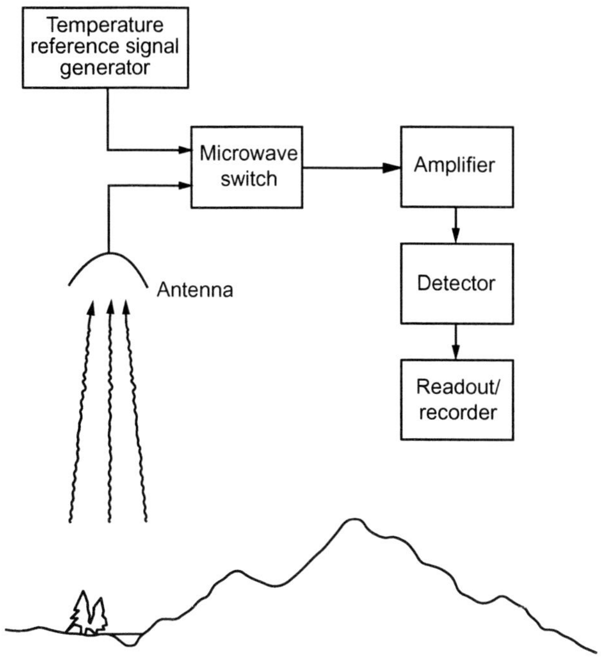

Fig. 3 Passive microwave radiometer sensing system. (Adapted from Ref. 6.) (© 1979, 1987, 1984 John Wiley and Sons, Inc.)

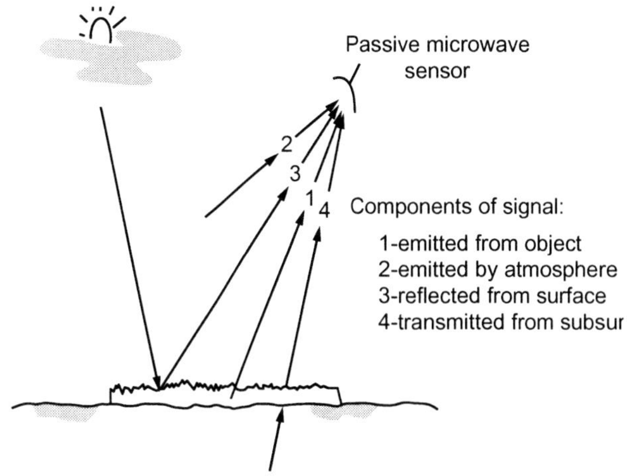

Fig. 4 Signal components for passive microwave sensing. (Adapted from Ref. 6.) (© 1979, 1987, 1994, by John Wiley and Sons, Inc.)

values. The passive microwave image is composed of pixels, the same structure as described for SAR images. Energy detected with passive microwave instruments is low in power and, hence, the spatial resolution of the acquired images is very coarse when compared to SAR systems, usually in the range of kilometers for orbiting instruments.

In the absence of instrument effects, brightness temperature is the product of radiation from sunlight reflected by the Earth and atmosphere, and energy emitted from the Earth's surface (Fig. 4). For frequencies used for terrestrial remote sensing, the contribution from the atmosphere is negligible. Reflection and emission from the Earth's surface are then the dominant sources of radiation detected by the antenna. Given the variety of sources of energy detected at the antenna, the application of brightness temperature for soil moisture sensing can be complicated. One similarity between brightness temperature and backscatter cross-section is that they are influenced by the same scene and illumination factors.

SOIL MOISTURE SENSING WITH SYNTHETIC APERTURE RADAR AND PASSIVE MICROWAVE INSTRUMENTS

Many investigations have established the sensitivity of normalized backscatter cross-section and brightness temperature to soil moisture. However, passive microwave brightness temperature and SAR normalized backscatter cross-section are also influenced by soil roughness and vegetation, which complicates the soil moisture detection. It is not possible to describe passive and active sensing of soil and vegetation thoroughly here, but more complete reviews of these sensitivities can be found in Refs. 7–10. A quick summary of these would imply that backscatter and brightness temperature increase with soil moisture and roughness, and vegetation canopy density.

However, the exact nature of the relationship between normalized backscatter cross-section and brightness temperature, and soil moisture and roughness, and vegetation canopy is also highly dependent on illumination factors (incidence angle, instrument frequency, and polarization, etc.). Also note that soil and vegetation vary greatly over short distances and are independent of each other, further complicating the interpretation of active and passive microwave data. The emphasis of the current research is to quantitatively estimate soil moisture from normalized backscatter cross-section and brightness temperature even given the difficulties presented by the soil moisture sensing problem. It is hoped that remotely sensed estimates of soil moisture can be used for applications such as crop management, flood prediction, and regional and global hydrologic research.

Quantitative Soil Moisture Estimation

A variety of modeling approaches have been used to quantitatively estimate soil moisture from SAR normalized backscatter and passive microwave brightness temperature. To date, no approach has been found to be universally applicable. The development and evaluation of quantitative soil moisture algorithms is hampered by the need for extensive supporting ground and sensor data.

The most rigorous investigations have been undertaken through the coordinated effort of a few organizations using airborne multi-instrument sensing of an area, which is simultaneously sampled for a number of soil and plant properties. The Wahita '92 and '94 programs, using a study area in south-central Oklahoma, are excellent examples of such efforts. Summary of the programs, references to published articles and access to generated data are available through the World Wide Web at http://hydrolab.arsusda.gov. More recently, Jackson et al. (11) realized promising results from an airborne passive microwave investigation to estimate soil moisture over a large area of the same section of Oklahoma. See Ref. 12 for a description of the use of airborne and orbiting SAR instruments to predict soil moisture of sites in Spain and the Sahel.

Progress in the quantitative estimation of soil moisture has also been hindered by the instruments used to collect data. For example, the only orbiting SAR instruments are single frequency/polarization, which restricts the information that can be extracted from the collected data. The lack of information can be compared to attempting to conduct land cover investigations from a single band instrument that only collects data in the green portion of the spectrum. Orbiting passive microwave instruments acquire multifrequency data however the spatial resolution of the data is coarse and the frequencies are not ideal for soil moisture detection. The estimation of soil moisture from spaceborne instruments will then rely on having the right instruments as much as having an appropriate method.

REFERENCES

1. Ulaby, F.T.; Moore, R.K.; Fung, A.K. *Microwave Remote Sensing: Active and Passive, Vol. I: Microwave Remote Sensing Fundamentals and Radiometry*; Artech House: Dedham, MA, 1981.
2. Ulaby, F.T.; Moore, R.K.; Fung, A.K. *Microwave Remote Sensing: Active and Passive, Vol. II: Radar Remote Sensing and Surface Scattering and Emission Theory*; Artech House: Dedham, MA, 1984.
3. Ulaby, F.T.; Moore, R.K.; Fung, A.K. *Microwave Remote Sensing: Active and Passive, Vol: III From Theory to Application*; Artech House: Dedham, MA, 1986.
4. Elachi, C. *Introduction to the Physics and Techniques of Remote Sensing*; Wiley: New York, 1987.
5. Rees, W.G. *Physical Principles of Remote Sensing*; Cambridge University Press: Cambridge, U.K., 1990.
6. Lillesand, T.M.; Kiefer, R.W. *Remote Sensing and Image Interpretation*; Wiley: New York, 1998.
7. Engman, E.T. Applications of Microwave Remote Sensing of Soil Moisture for Water Resources and Agriculture. Remote Sens. Environ. **1990**, *35*, 213–226.
8. Engman, E.T.; Chauhan, N. Status of Microwave Soil Moisture Measurements with Remote Sensing. Remote Sens. Environ. **1995**, *51*, 189–198.
9. Schmugge, T.J.; Jackson, T.J.; McKim, H.L. Survey of Methods for Soil Moisture Determination. Water Resour. Res. **1980**, *16* (6), 961–979.
10. Schmugge, T.J. Remote Sensing of Soil Moisture: Recent Advances. IEEE Trans. Geosci. Remote Sens. **1983**, GE-*21* (3), 336–344.
11. Jackson, T.J.; Le Vine, D.M.; Hsu, A.Y.; Oldak, A.; Starks, P.J.; Swift, C.T.; Isham, J.D.; Haken, M. Soil Moisture Mapping at Regional Scales Using Microwave Radiometry: The Southern Great Plains Hydrology Experiment. IEEE Trans. Geosci. Remote Sens. **1999**, *37* (5), 2136–2150.
12. van Oevelen, P.J.; Hoekman, D.H. Radar Backscatter Inversion Techniques for Estimation of Surface Soil Moisture: EFEDA-Spain and HAPEX-Sahel Case Studies. IEEE Trans. Geosci. Remote Sens. **1999**, *37* (1), 113–123.

RENDZINA

Thomas E. Fenton
Iowa State University, Ames, Iowa, U.S.A.

R

INTRODUCTION

Rendzina is a term that has a long history of use on a worldwide basis. In soil classification, it has always included soils high in carbonates. Most countries, except the U.S., have used the name to describe shallow to very shallow soils, gray to black in color that developed in limestone or chalk. Early usage in the U.S. differed in that. Rendzinas included deeper, dark soils with profiles ranging from 38 to 152 cm in thickness, with many, but not all containing carbonates (1).

The origin of the word "Rendzina" has a complex history. Simonson (2) described the original use of Rendzina as a folk term used to convey the clinking sound of a plow being drawn through certain stony soils. In the late 19th century, Polish scientists began using the name for soils formed in highly calcareous materials. Part of Poland was then in the Russian Empire and the name was used in some of the early Russian soil classifications.

HISTORY OF USE

De'Sigmond (3) described the conditions needed for the development of Rendzinas as parent rock rich in calcium carbonate or gypsum, that has the fineness of structure and solubility necessary to neutralize the podzolizing effect of the humid climate under which they developed the forest vegetation. He considered these soils to be azonal, that is, they did not have completely developed profiles.

Robinson (4) described the Rendzinas of Central Europe as humus-carbonate soils. They were typically dark colored soils with 3–12% organic matter and varying amounts of free calcium carbonate. In the Russia classification system proposed by Glinka (5), Rendzinas were considered to be typical endodynamomorphic soils, which had not yet reached mature development and in which the profile character was mainly determined by the nature of the parent material. Glinka (5) observed that these humus-carbonate soils usually contained considerable undecomposed rock, the amount of which decreased with soil development. All the Rendzina soils described by Glinka (5) are shallow soils similar to those described by De'Sigmond (3). As described by Glinka (5), the A horizon was 15–30 cm in thickness with varying amounts of limestone fragments and graded through a thin transition of lighter-colored material into fragmental limestone at about 46 cm or less. Spanish red soils regarded as terra rossa were thought to be formed by degradation of Rendzinas, consequent on deforestation under semiarid conditions. The placement of Rendzinas as endodynamomorphic soils was later criticized by Marbut (6) because he thought Glinka's (5) groupings were poorly balanced, in that they represented such a small part of the total soils of the world as compared to the more fully developed soils. Rode (7) used the term "turfy-carbonate" soils as a synonym for Rendzina for Russian soils within Podzol zones. The presence of the high carbonate content in the parent rock resulted in a high content of calcium-carbonate humus and was thought to have prevented these soils from becoming podozolized.

Marbut (8) in his final classification system included Rendzinas in Category III which he defined as those soils in which extreme youth, excessive erosion, wetness and excessive contents of carbonates, salt, or alkali prevented development of the normal soil. Rendzinas were the soils with excessive contents of carbonates.

Baldwin et al. retained the Rendzinas as intrazonal soils in their highest category order, calomorphic in the suborder level, and named Rendzina as a great soil group. A general description of the soils included as Rendzinas in this classification system was: an intrazonal group of soils, usually with brown or black friable surface horizons, underlain by light-gray or pale-yellow soft calcareous material, developed under grass vegetation or mixed grass and forest vegetation, in humid and semiarid regions from relatively soft, highly calcareous parent material (9).

Twelve years after the publication of the above soil classification system Oakes and Thorp (1) suggested that some of the soils included as Rendzinas such as Houston Black, were markedly different from the Rendzinas of Europe. The included soils they objected to, were high in clay, generally dark in color, free of stones, and were deep soils. Many, but not all were calcareous. The common property was shrink–swell that resulted in churning of the soil. They proposed that these soils be reclassified within a great soil group named Grumusols and developed a tentative definition. This proposal gradually gained acceptance both in the U.S. and other countries. The acceptance of the Grumusol concept substantially reduced Rendzina soils to a minor group in the U.S. Oakes and Thorp (1) observed that the Lithosols of the prairies of the Southern U.S. conformed more closely to the Rendzina of other countries rather than the medium depth and deep soils that had been included with Rendzinas in the U.S.

Papadakis (10) characterized the Rendzina group as those soils that contained active carbonates and gave effervescence with HCL in all their horizons. He attributed the deep, dark accumulation of humus to the carbonate environment. As climate became drier, the accumulations of humus were less and desert Rendzinas (serozems) lack humus.

Bridges (11) discussed Rendzinas in the category of intrazonal, calcimorphic soils. He described Rendzinas in the following way. A shallow soil rich in organic matter and biological activity. It has a stable crumb structure and is dark in color. It is a relatively simple soil with an A horizon directly overlying a C horizon, which is the limestone parent material. The humus is well-incorporated in the mull form and micromorphological evidence showed that these soils are largely composed of the fecal pellets of soil arthropods and the casts of earthworms.

PRESENT USE

Presently a part of the word Rendzina is used in the U.S. soil classification system—Soil Taxonomy (12). The "rend" part of the word was used to coin a new term for a suborder in the Mollisol order—Rendolls. The term was used to convey the concept of a high level of carbonates in these soils. Many of the soils previously called Rendzinas are included within the Mollisol Order and the Rendoll Suborder in Soil Taxonomy. Rendolls have mollic epipedons less than 50 cm in thickness that are underlain by calcareous parent material or a cambic horizon with a high carbonate content. The parent material is limestone, chalk, drift composed mainly of limestone, or shell bars. They occur in humid climates (udic moisture regimes) or cold climates with temperatures lower than 8°C (cryic soil temperature regime) or both and have a high calcium carbonate content (calcium equivalent of 40% or more) in the mollic epipedon or in the underlying mineral material. The native vegetation is thought to be forest or grass and shrubs. They are not extensive in the U.S. but on a worldwide basis occupy approximately 0.2% of the land area.

Rendzinas were recognized in the FAO–UNESCO System for soil shallow to limestone prior to 1988 (13). However, this term was dropped in the 1988 revision (14). These soils were then included with the leptosols or thin shallow soils. Presently the term Rendzic is an adjective in the FAO–UNESCO legend having been added in 1988 and is used for naming map units at the second level.

SUMMARY

The term Rendzina has a long history of use to describe shallow soils that are calcareous and are underlain by parent materials high in carbonates. Based on present classification systems, it appears that the term Rendzina will be used less frequently in the future. However, the use of the "Rend" root of the word will continue at least in Soil Taxonomy and the FAO–UNESCO soil classification systems.

REFERENCES

1. Oakes, H.; Thorp, J. Dark-Clay Soils of Warm Regions Variously Called Rendzina, Black Cotton Soils, Regur and Tirs. Soil Sci. Soc. Am. Proc. **1950**, *15*, 347–354.
2. Simonson, R.W. Rendzina—Origin, History, and Descendants. Soil Surv. Horizons **1997**, *38* (3), 71–92.
3. De'Sigmond, A.A.J. *The Principles of Soil Science*; Translation by Yolland, A.B. Thomas Murby: London, 1936; 236–244.
4. Robinson, G.W. *Soils—Their Origin, Constitution, and Classification*; Wiley: New York, 1949; 373.
5. Glinka, K.D. The Great Soil Groups of the World and Their Development, Translated from the German by C.F. Marbut, USDA, 1927; 146–149.
6. Marbut, C.F. Soils: Their Genesis and Classification. Soil Sci. Soc. Am. **1951**, 134, Lectures Given in the Graduate School of USDA, 1928.
7. Rode, A.A. *Soil Science*; Translated from Russian by Israel Program for Scientific Translations, Jerusalem National Science Foundation: Washington, DC, 1962; 517.
8. Marbut, C.F. Soils of the United States. *Atlas of American Agriculture*; USDA: USA, 1935.
9. Baldwin, M.; Kellogg, C.E.; Thorp, J. Soils and Men. *USDA Yearbook of Agriculture*; U.S. Govt. Printing Office: Washington, DC, 1938; 1232.
10. Papadakis, J. *Soils of the World*; 1964; 141.
11. Bridges, E.M. *World Soils*; Cambridge University Press: Cambridge, 1970; 89.
12. Soil Survey Staff, *Soil Taxonomy*, 2nd Ed. NRCS USDA U.S. Govt. Printing Office: Washington, DC, 1999; 869.
13. Food and Agriculture Organization, *FAO–UNESCO Soil Map of the World*; UNESCO: Paris, 1971–1981; Vol. I–X.
14. Food and Agriculture Organization, *FAO–UNESCO Soil Map of the World: Revised Legend*; FAO: Rome, 1988.

RESILIENCE

M.M. Tenywa
J.G.M. Majaliwa
A. Lufafa
Makerere University, Kampala, Uganda

INTRODUCTION

The soil resilience concept is relatively new (1, 2) and has emerged alongside the efforts to define and measure sustainability of agricultural systems (3). It describes the level of resistance to forces driving soil degradation, and provides insight into mechanisms that endow soils with favorable properties and beneficial processes from physical, chemical, and biological perspectives. Like sustainability, which has remained a difficult concept to elucidate, the soil resilience concept remains inexactly expressed (4, 5) and quantified yet it is recognized as the most relevant parameter for understanding and management of soil ecosystems (6). Various efforts to define soil resilience can broadly be grouped into three categories; as a process, nature, and soil state.

A Process

Soil resilience as a process focuses on the mechanical role of soil in countering degradative processes (e.g., rainsplash, detachment). It refers to the ability of the soil to resist or recover when subjected to degradative forces. This definition is primarily inspired by the Hooke's spring elasticity model. Using this model, a soil resilience coefficient was proposed as follows (3):

$$S_r = -2A/x^2 \tag{1}$$

where

S_r = resilience coefficient whose inverse indicates vulnerability to degradation;
A = the amount of work required to move soil from one state to another (e.g., 1 to 2); and
x = variable reflecting changes in soil state.

A resilient soil, therefore, exhibits a higher S_r and minimal change in soil state (x) as stress accumulates. Although the analogy is satisfactory when used to elaborate resilience of the physical soil system, its utility is limited on five major premises:

1. Lack of means to translate the chemical, biological, and other forms of work into "*unified mechanical form.*"
2. The absence of a lumped parameter adequate in capturing "wholly" the changes in soil state.
3. Assumes classical *resilience* to be characteristically independent of *stability*. Yet for soil systems, the two may not necessarily be mutually exclusive, in that alterations in resilience from equilibrium may result in deterioration of soil structural attributes.
4. Eq. 1 holds if the alterations are slight and temporary. Otherwise, if the soil system undergoes irreversible changes and the degree or duration of perturbations is drastic and intense, the relationship ceases to be valid (7).
5. The model describes the macroscopic behavior of the soil matrix and therefore does not give a complete picture of what is happening at finer levels of soil organization.

A Nature

This school of thought focuses on the functional integrity of the soil judged, based on its properties (e.g., soil formation stage, degree, and level of use/management), and is the most common approach found in literature (7). Soil resilience from this perspective is defined as the ability of the soil system to remain productive through efficient processing over an extended period of time (8, 4). Accordingly, it refers to the longevity of the soil resource and its health measured in terms of overall performance and quality (9). The mass flux balance approach (4) is typically used to express the functional integrity of the soil in form of a soil resilience factor:

$$S_r = S_a + \int (S_n - S_d + I_m)\, dt \tag{2}$$

where

S_r = soil resilience;
S_a = the antecedent soil condition;
S_n = the rate of new soil formation;

S_d = the rate of soil degradation/depletion; and
I_m = connotes the management inputs from outside the ecosystem.

The major limitation of this approach lies in its failure to account for the impact of restorative processes (buffering capacities of soil) on the prevailing soil properties; i.e., focuses on the "what" of the antecedent state has changed as a result of the interaction of soil formation, degradation, and management but overlooks the mechanisms of change—the "how."

A State

The state concept views soil resilience as a lumped, dynamic macroscopic parameter that is a consequence of fine-scale interactions between properties, processes, micro-climate, and management impacts within the soil body (10). Difficulties exist in the development of a single indicator of soil resilience because the soil consists of at least three "states," its physical, chemical, and biological properties (each of which consists of numerous sub-properties), which interact in a complex manner but are quantified through extremely different procedures. Seen as a combined effect of both soil-buffering capacities (physical, chemical, and biological) and soil mass flux balance due to pedological and anthropological actions, a resilience factor as described by the following equation was suggested (3):

$$S_r = BC_{ph} + BC_{ch} + BC_b + \int (r_p + r_a)\,dt \qquad (3)$$

where

S_r = soil resilience;
BC_{ph} = physical buffering;
BC_{ch} = chemical buffering;
BC_b = biological buffering;
r_p = rate of pedological soil fluxes;
r_a = rate of anthropological fluxes; and
dt = change over time.

Although this approach is process based, its major limitations emanate from the assumption that the named different soil "states" are independent, ignoring their known interactions (8). Nonetheless, this "state" approach to resilience combines the three fundamentally different components of the soil, and as such is worthy of attention by those who seek to quantify soil impacts in a holistic manner.

With respect to agronomy, soil productivity is a crucial indicator of soil resilience. Although soil productivity also remains difficult to quantify (11, 4), crop yield over time is a practical measure of productivity (1, 12, 13). Long-term crop performance under specific environmental and management conditions is mainly dependent upon Relevant Soil Depth (RSD) characteristics (14, 15). Assuming an infinitesimal pedon, a productivity index D_p of specific productivity level P ranging from $P_{\min}$ to $P_{\max}$ yield can be defined a manner similar to the degree of polarization (16a) by use of statistical mechanics on an RSD matrix, as:

$$D_p = 2(P - P_c)/g \qquad (4)$$

where: $g = (P_{\max} - P_{\min})$ is the maximum agro-ecological productivity gap; and $(P - P_c)$ is the gap between specific productivity level P and the critical productivity level $P_c = (P_{\max} + P_{\min})/2$ of the agro-ecological zone. D_p ranges from -1 for highly degraded (HD) land to +1 for highly productive (HP) land, insinuating the existence of three resilience phases: namely, high resilience (associated with HP), marginal resilience (associated with HD), and diminishing resilience (transitional phase) (5).

The major advantage of using D_p as an index of soil resilience is that it is a macroscopic state parameter derived from microscopic properties. Considering that the soil can exhibit more than one accessible state (configuration of pedon of specific parameters of state; e.g., temperature, energy, chemical potential, volume, and pressure) at different equilibria, D_p provides an excellent opportunity for capturing all of them (5).

Finally it is noteworthy that a similar index ($D_p = P/P_{\max}$) for crop productivity is used in land evaluation and suitability assessment (17), and Eq. 1 can be applied near the limit of the highly productive macrostate. However, a great challenge remains to establish a state parameter for nonagricultural soils.

FACTORS AFFECTING SOIL RESILIENCE

The factors that affect soil resilience fall in two broad groups, namely; endogenous and exogenous (4). Endogenous factors are related to the inherent soil properties (e.g., soil depth, relief, drainage status, structure, texture, SOM, humus quality, faunal and floral activities, nutrient level) and climate especially precipitation (18). Exogenous factors are largely a result of land use and management. Good management aims at providing a buffer against sudden fluctuations in the system, minimizing degradative effects of atmospheric agents (e.g., heat, wind, rain, runoff), enhancing restorative mechanisms, and providing adequate protection to soil against degradation. Several methods of management exist that impact on resilience factors: Agricultural

R

operations (e.g., irrigation, tillage, drainage, terracing, ridging, mulching, crop and water harvesting) influence weathering, soil formation, and synthesis and hence the direction of the degradative process.

The net balance of interaction between the two categories of factors (exogenous and endogenous) determines the predominant mechanism, rate, duration, and direction of beneficial processes in soil (19). Under unsuitable land use and/or poor management, the restoration capacity of the beneficial processes becomes undermined and the net force balance is tilted negatively toward soil resilience by endogenous factors, hence prevalence of the degradation mechanisms that include:

- Progressive removal of the top soil layer by soil erosion, tillage erosion, soil mining (e.g., brick making, sand quarrying)
- Destruction of soil structure/matrix by bad tillage
- Nutrient depletion through crop uptake, harvest, and burning
- Degradation of subsoil by leaching of Al and Si and formation of allophane and imogolite from residual silica and alumina (20).

On the other hand, when land is well managed, the positive effects of exogenous factors on soil resilience accrue through the following restorative mechanisms:

- Invasion of subsoil by micro-organisms
- Subsequent differentiation of sub-soil and plant succession (decomposition and humification)
- Weathering enhanced by soil and water conservation measures of stabilization (20).
- Silica absorbed by plants to form plant and Al-humus complex which makes up the thick A horizon peculiar to the black soil (21).

Notwithstanding the great progress of recent years in the development of soil resilience theory, there still exists a pressing need for development of a more adequate, single measure of soil resilience that can be used by soil conservation/land husbandry engineers who are commonly confronted with various combinations of land degradation forms.

REFERENCES

1. Beinroth, F.H.; Eswaran, H.; Reich, P.F.; Van der Berg, P.F. Land Related Stresses in Agroecosystems. In *Stressed Ecosystems and Sustainable Agriculture*; Virmani, S.M., Katyal, J.C., Abrol, I.P., Eds.; Oxford & IBH Publishing Co. PVT. Ltd.: New Delhi, Bombay, Calcutta, 1994; 131–148.
2. Eswaran, H. Role of Soil Information in Meeting the Challenges of Sustainable Land Management (18th Dr. R.V. Tamhane Memorial Lecture). Journal of the Indian Society for Science **1992**, *40*, 6–24.
3. Szabolcs, I. The Concept of Soil Resilience. In *Soil Resilience and Sustainable Land Use*; Greenland, D.J., Szabolcs, I., Eds.; CAB International Publishers, UK, 1994; 33–39.
4. Lal, R. Sustainable Lands Use Systems and Soil Resilience. In *Soil Resilience and Sustainable Land Use*; Greenland, D.J., Szabolcs, I., Eds.; CAB International Publishers, UK, 1994a; 41–67.
5. Tenywa, M.M.; Lal, R.; Majaliwa, M.J.G. Characterization of the Stages of Soil Resilience to Degradative Stresses: Soil Erosion. Stott, D. E.; Mohtar, R.H.; Steinhard, G.C. eds. Sustaining the Global Farm. Selected papers from the 10th International Soil Conservation Organization Meeting, May 24–29, 1999, West Lafayette, Purdue University and the USDA-ARS National Soil Erosion Research Laboratory, 1990; 606–610
6. Conway, G.R. Sustainability in Agricultural Development: Tradeoff with Productivity, Stability and Equitability. In *Proceedings of the 11th Annual AFSRE/E Symposium*, held 5–10 October in Michigan, USA, 1991; 1–22.
7. Lal, R. Soil Degradation and Agricultural Sustainability. In *Soil and Water Conservation Challenges and Opportunities, V1*; Bhushan, L.S., Abrol, I.P., Rama Mohan Rao, M.S., Eds.; Proc. 8th Int. Soil Cons. Conference, Dehra dun, India, 1994b; 191–207.
8. Lal, R. Effects of Macrofauna on Soil Properties in Tropical Ecosystems. Agric. Ecosystem Environ. **1988**, *24*, 101–115.
9. Hannam, I. Ecologically Sustainable Soil: The Role of Environmental Policy and Legislation. Paper Presented at the 10th International Soil Conservation Organization Conference, May 23–27, 1999; West Lafayette, Indiana-USA, 1999.
10. Kay, B.D.; Rasiah, V.; Perfect, E. The Structural Aspect of Soil Resilience. In *Soil Resilience and Sustainable Land-use*; Greenland, D.J., Szabolcs, I., Eds.; CAB International: Wallingford, UK, 1993; 449–468.
11. Douglas, M. *Sustainable Use of Agricultural Soils. A Review of the Prerequisites for Success or Failure*; Development and Environment reports No. 11, University of Berne: Switzerland, 1994.
12. Arnold, R.W., Szabolcs, I., Targulian, V.O., Eds. *Global Soil Change*; International Institute for Applied Systems Analysis: Laxenburg, Austria, 1990.
13. Uehara, G.; Tsuji, G.Y.; Beinrouth, F.H. Extrapolating Results of Long-Term Experiments. In *Soil Management Experimental Basis for Sustainability and Environmental Quality*; Lal, R., Stewart, B.A. Eds.; Lewis Publishers: 1995; 105–109.
14. Papendick, R.I.; Parr, J.F.; Van Schilgaarde, J. Soil Quality: New Perspective for a Sustainable Agriculture. In *Soil and Water Conservation Challenges and Opportunities, V1*; Bhushan, L.S., Abrol, I.P., Rama Mohan Rao, M.S., Eds.; Proc. 8th Int. Soil Cons. Conference, Dehra dun, India, 1994; 227–237.
15. Passioura, J.B. Soil Structure and Plant Growth. Aust. J. Soil. Res. **1991**, *29*, 717–728.
16. Reif, F. Ed. Fundamentals of Statistical and Thermal Physics. McGraw-Hill International Editions, Singapore, 1987; 651.

16a. Rozanov, B.G. Human Impacts on Evolution of Soils Under Various Ecological Conditions of the World,

14th International Congress of Soil Science, 12–18 August, 1990, Kyoto, Japan, Plenary papers, 1990; 53–62.

17. Sys, C.; Van Ranst, E.; Debaneye, J. *Land Evaluation. Part I. Methods in Land Evaluation*, Agricultural Publications No. 7.
18. Vlek, P.L.G.; Vielhauer, K. Nutrient Management Strategies in Stressed Environments. In *Stressed Ecosystems and Sustainable Agriculture*; Virmani, S.M., Katyal, J.C., Eswaran, H., Abrol, I.P., Eds.; Oxford & IBH Publishing Co. PVT. Ltd.: New Delhi, Bombay, Calcutta, 1994; 203–229.
19. Lal, R. Sustainable Land Use System and Soil Resilience, Soil Quality and Sustainability. Soil Tillage Res. **1993**, *27*, 1–8.
20. Zolcinski, J. A New Genetic Physio-Chemical Theory of the Formation of Humus, Peat and Coal. The Role and Significance of Biological Factors in These Processes. *Proc., First Intern. Congr. Soil Sci.*, Commission III and IV, 1930; 335–338.
21. Wada, K. Active Aluminium in Koroboko Soils and Non- and Para-Crystalline Clay Minerals. Clay Sci. **1977**, *17*, 143–151.

RESILIENCE, LAND USE, SOIL MANAGEMENT, AND

R

S.J. Park
University of Bonn, Bonn, Germany

INTRODUCTION

Most human activities associated with land use cause direct and indirect changes in soil-forming processes and soil properties. Urban and industrial development might be the most destructive form of soil disturbance, causing either total removal of existing soil layers, introduction of hazardous chemicals into soils, or drastic changes in soil-forming processes within a short time period. Even though the rate of change is slow and adverse effects are not clear in the short run, agricultural activities are also greatly contributing to soil disturbance and consequent loss of soil resilience. Land use practices and their influence on soil resilience are so diverse that it is virtually impossible to describe them in detail. Discussions here are mostly limited to agricultural land use in which soil-management practices are essential human interventions to maintain agricultural production. A more theoretical discussion on the influence of other land use systems on soils may be found elsewhere (1, 2). For agricultural land use, one should also note that the influence of certain soil-management practices on soil resilience varies widely depending on soil types formed as a result of highly interacting environmental factors. The ignorance of the intrinsic and spatial variability of soils and environmental conditions may result in significant reduction in crop production and accelerated degradation of soil resources.

BASIC CONCEPT

Soils are the product of the highly complex interactions of many interdependent variables that have been categorized as the five soil-forming factors: climate, parent material, organisms, topography and time. Under natural conditions, soil evolves along two developmental pathways: progressive and regressive (3). The progressive pathway of soil evolution includes conditions, processes and factors that promote horizonation, developmental upbuilding, and/or soil deepening. The regressive pathway promotes haploidization (destruction of soil horizons), retardant upbuilding, and/or subsurface removal. Either of these two pathways may be dominant for a soil in a given location, resulting in a steady state or equilibrium of a soil's physical, chemical and biological properties in a given environment. Any kind of soil management practices leads to changes of such a steady state and accelerates either progressive or regressive pathways.

It does not necessarily mean that soils in progressive pathways are more beneficial for agricultural and engineering use than soils in regressive pathways, but soils in progressive pathways are more likely restored after temporal disturbances induced by soil management by favorable soil-forming conditions. Continuous or sometimes catastrophic dominance of regressive pathways caused by both natural processes and poor soil management may lead to the loss of resilience and even irreversible degradation of soils. However, it might also be possible to reverse the regressive pathway of soils and to restore highly degraded soils with good management and appropriate technological input. The main question in maintaining soil resilience for sustainable agriculture or other human activities is how to minimize the regressive influence of human-induced soil disturbance within the range of progressive pathways of soil formation, which can be modeled:

$$S_r = S_a + \int (S_n - S_d \pm I_m)dt$$

where S_r is the soil's resilience, S_a is the antecedent soil condition, S_n is the rate of new soil formation (progressive condition), S_d is the rate of soil degradation/depletion (regressive condition), and I_m is the management inputs from outside the ecosystem (4).

AGRICULTURAL SOIL MANAGEMENT EFFECTS ON SOIL RESILIENCE

Most land use practices have largely been accelerating regressive pathways of soil formation, which is evidenced by rapid soil degradation worldwide (5). Agricultural land use is currently one of the main land uses contributing to such soil degradation processes along with deforestation and overgrazing. Table 1 illustrates progressive and regressive changes induced by soil cultivation. When the original vegetation is removed for cultivation, the soil

Table 1 Possible progressive and regressive alteration of soil forming factors by agricultural land use and soil management

Soil forming factor	Progressive influence	Regressive influence
Parent material	Adding mineral fertilizers	Removing through harvest more plants and animal nutrients than are replaced
	Accumulating shells and bones	Adding material in amounts toxic to plants or animals
	Accumulating ash locally	Altering soil constituents in a way to depress plant growth
	Removing excess amounts of substances such as salts	
Topography	Checking erosion through surface roughening	Causing subsidence by drainage of wetlands
	Land forming and structure building	Accelerating erosion
	Land leveling	Excavating
Climate	Adding water by irrigation	Subjecting soil to excessive insolation, to extended frost action, to expose to wind, to compaction
	Release of CO_2 to atmosphere with possible warming trend in climate	Altering aspect by land forming
	Changing color of soil surface to change albedo	Clearing and burning off organic cover
	Removing water by drainage	
	Diverting wind	
Organisms	Introducing and controlling populations of plants and animals	Removing plants and animals
	Adding organic matter to soil directly or indirectly through organisms	Reducing organic content of soil through burning; plowing, overgrazing, harvesting, accelerating oxidation, leaching
	Loosening soil by ploughing to admit more oxygen	Adding or fostering pathogenic organisms
	Removing pathogenic organisms, e.g, by controlled burning	
Time	Rejuvenating the soil by adding fresh parent material or through exposure of local parent material by soil erosion	Degrading the soil by accelerated removal of nutrients from soil and vegetation cover
	Reclaiming land from under water	Burying soil under solid fill or water

(Adapted from Ref. 12 in which progressive and regressive influences were presented as beneficial and detrimental effects, respectively.)

surface exposed to air becomes vulnerable to air and wind erosion. It also modifies the microclimate of the near-ground atmosphere. With continuous cultivation, organic matter levels in soil gradually decrease. The decrease of organic matter causes the deterioration of soil structure, reduction of microbial activities and cation exchange capacity, increase of nutrient leaching, and sensitivity to erosion. The use of heavy machinery for tillage and farm traffics destroys soil structures and increases soil compaction, which greatly accelerates erosion processes.

Agricultural soil management can be grouped into five major activities: 1) tillage, 2) maintaining organic material content and nutrients, 3) application of agricultural chemicals, 4) drainage and irrigation, and 5) mechanical control of erosion. In addition to the mechanical control of erosion, soil-erosion control is the most important consideration in selecting the different tillage types and maintaining the organic matter content. Tillage is frequently divided into two groups: conventional and conservation tillage. Conventional tillage turns over surface soils using a plow in order to remove weeds and to provide sufficient moisture and nutrient for planted seeds. Conservation tillage includes various tillage systems, such as no tillage; strip tillage; ridge-tillage; and

R

Table 2 Major soil-management practices and their possible influence on soil forming processes

Soil management	Management practice	Progressive influence on soil formation	Regressive influence on soil formation
Tillage	Conventional tillage	Increase aeration within plowed layers	Oxidative loss of organic matter
		Enhancing circulation of oxygen and water	Deterioration of soil structure
		Increase biological activity	Formation of subsurface plow pans
		Breaking surface and impermeable subsurface layers	Exposure of soil surface to raindrop impact, and water and wind erosion
		Increasing weathering rates	Reduction of plant diversity
	Conservation tillage (combined with residue management)	Increase of organic matter content in soil	Soil compaction (possibly in short term)
		Reducing erosion by increasing surface residue	Possible herbicide use
		Increase of biological activities	
		Reduced soil structure disruption (than conventional tillage)	
Maintaining organic matter and nutrients	Residue management (Crop residue mulching)	Protecting soil surface from raindrop impact	Reduction of soil temperature
		Reducing surface runoff	
		Increase organic matter content by the decomposition of residue	
		Increase biological activity	
		Less variation of soil temperature	
	Crop rotation (including cover and green manure crop)	Protecting soil surface by plant cover	Structural damage during harvesting of certain crops (e.g., potatoes, sugar beet)
		Increase organic matter input	
		Increase biodiversity in soils	
		Enhancing soil structure and stable aggregates	
		Enhancing nutrient balance	
	Organic fertilization (manure and compost)	Increase oganic matter content in a short time period	Introduction of toxic chemicals and heavy metals
		Increase water retention capacity	Introduction of animal pathogens
		Promoting stable soil aggregates	Gaseous damage (e.g., ammonia) for certain plant species and microorganisms

Table 2 Major soil-management practices and their possible influence on soil forming processes *(Continued)*

Soil management	Management practice	Progressive influence on soil formation	Regressive influence on soil formation
	Shifting cultivation	Addition of nutrients into soils Initial improvement of soil structure in heavy soils Initial soil pH increases	Exposing soil surface to erosion processes Decrease of soil fauna population by heat Deterioration of soil structure through cropping Nutrient loss and soil acidification with cropping Decrease of infiltration rate with cropping
	Soil acidity adjustment (lime and gypsum application)	Reducing soil acidity	Ionic imbalance of soil solution
		Enhance stable soil aggregate formation	Selective leaching for certain ions (e.g., K^+)
Application of agricultural chemicals	Inorganic fertilizer	Increase of plant growth and crop yield	Reduction of soil organic matter replacing traditional soil organic management
		Enhance balance of solutes in soil	Increase of soil acidity by certain fertilizers (e.g., superphosphate) Nutrient imbalance and toxicity Possible loss of vital soil organisms
	Pesticide application		Possible loss of vital soil organisms (e.g., earthworms, termites)
Irrigation and drainage	Irrigation	Supplying water in soils	Accumulation of salts or alkali compounds in soil (in arid and semi-arid climate)
		Removing excessive salts	Deterioration of soil structure and consequent reduction of infiltration rate, when sodium is present
		Encouraging microbial activities	Increase of toxic elements (e.g., boron, selenium) Poor aeration caused by water logging
	Drainage	Enhancing soil aeration Lowering the regional water table	Rapid oxidation of organic matter Soil shrinkage and consequent surface lowering
Mechanical erosion control	Contouring, terraces and waterways	Reducing erosion potential	Possible slope failure
			Removal of topsoil and expose of subsurface soil during construction

This table is not intended to be exhaustive, it is also highly controversial for any generalization for all soil types.

mulch-tillage, which aims to reduce soil and water losses (6). Crop residue management designed to retain all or a portion of the previous crop's stalls on the soil surface is an important component of most conservation tillage. Maintaining an adequate level of organic materials by means of residue management, crop rotation, and manure, is essential in soil-management practices not only for providing major nutrients for crop growth but also for improving soil structure and reducing soil erosion (7). Without regular addition of adequate amounts of organic materials, the potential of erosion and leaching, and gradual deterioration of soil physical properties increases. The strong positive correlation of organic matter content in soils with soil biodiversity has been well documented. Fertilizers and pesticides have led to a rapid increase of crop production, but wide use of agricultural chemicals also causes water pollution and soil degradation. Reduction of soil organic matter caused by replacing traditional organic fertilization through inorganic fertilizer is another negative influence on soil resilience. Table 2 presents a more detailed description on soil management practices and their possible influence on soil's progressive and regressive pathways.

PRINCIPLES AND PRACTICES OF IMPROVING SOIL RESILIENCE

The basic principles of soil management to improve soil resilience are already well known. For agricultural soil management, they can be categorized as 1) increase in soil organic matter content, 2) improvement in soil structure, 3) increase in soil biodiversity, 4) reduction of erosion rates, and 5) increase in nutrient capital and recycling mechanisms (8). Given the complexities of both soil-forming processes and socio-economic causes of soil degradation, however, there is no single strategy or management policy for maintaining soil resilience. The rapid increase of agricultural production in recent years has been achieved by the intensification of land use, the use of motorized farm equipment, and the application of agrochemicals. These technological inputs frequently induce the loss of soil resilience and environmental pollution in many agricultural areas (Table 2).

In recent years, new ideas have emerged for more sustainable soil management in both developed and developing countries, replacing high input agriculture (9). First, our improved understanding of the processes in the soil, water and plant ecosystem should be used for adapting management methods to the infinite variety of local conditions. Secondly, greater emphasis should be placed on efficiency of using soils, water, fertilizers and plant resources instead of opening new fragile lands or relying on higher inputs. Finally, the joint efforts of resource scientists and the knowledge and skills of the local people should be used to address issues of soil management.

The first and second principles require more scientific and quantitative understanding of relationships between different soil management systems and long-term soil resilience. Soils are extremely complex media with many intrinsic feedback mechanisms, and the magnitude and degree of a given human intervention on soil resilience vary widely. It is also frequently recognized that a progressive influence of a human activity in one soil might be a regressive influence in other soils or the same type of soil in a different environment. Much information is available on how different soil-management practices directly and indirectly regulate crop yields, but less is known about quantitative relationships between soil management and long-term soil processes that influence soil resilience characteristics. One important aspect of soil resilience studies is to develop appropriate restorative measures to improve degrading or already degraded soils. This is especially important for many developing countries where cultivation of new land and advanced technological inputs are limited due to high population pressure and the lack of socio-economic incentives.

Basic principles of enhancing soil resilience are known from many regions, but often those technical interventions are not available to farmers because of population pressures, limited natural resources, lack of secure land tenure systems, unfavorable macro-economic conditions, and poor national and international policies, etc. (10). The processes of soil degradation are too locally specific to be effectively combated with the best land management developed elsewhere. The introduction of western farming techniques in many developing countries has been proven largely inadequate and sometimes disastrous for soil quality (9). Preserving and improving traditional land-management techniques is increasingly considered the more sustainable land management system (11). Efforts to develop optimum sustainable soil management must include land users in both planning and implementation.

CONCLUSION

The rate of soil formation is slow, ranging from 0.001 to 0.4 cm a year (4), and the quality of such slowly forming soils is rapidly degrading due to inappropriate soil management in many parts of the world. Current soil-degradation trends give an alarming signal to the security of future food production. Historically, the rapid increase of agricultural production has been achieved by the intensification of land use and the use of agrochemicals, but more recently conservative soil management practices attract more attention than high input agriculture in order to reduce soil degradation and agriculture related pollution. The basic principles for conservative soil management are how to reduce the destructive influence of soil management practices within the resilience potentials of soils, and what is the most appropriate soil management to maintain or even improve soil resilience characteristics in given natural and socio-economic conditions. Our improved understanding of the processes in soil, water and plant ecosystems in relation to highly variable local environmental conditions is essential for these tasks. In addition, the joint efforts of resource scientists and the knowledge of the local people should be used to address issues of optimal local soil management. Soil resilience is a relatively new concept in soil science and land management studies, and current understanding of soil resilience and interaction with soil and land management remains rather qualitative and descriptive. A high research priority should be given to develop objective and quantitative criteria of soil resilience, especially in relation to different soil forming processes, and also to assess the cause–effect relationship between different soil management practices and resilience.

REFERENCES

1. Amundson, R.; Jenny, H. The Place of Humans in the State Factor Theory of Ecosystems and Their Soils. Soil Sci. **1991**, *151*, 99–109.
2. Seybold, C.A.; Herrick, J.E.; Brejda, J.J. Soil Resilience: A Fundamental Component of Soil Quality. Soil Sci. **1999**, *164*, 224–234.
3. Johnson, D.L.; Watson-Stegner, D. Evolution Model of Pedogenesis. Soil Sci. **1987**, *143*, 349–366.
4. Lal, R. Sustainable Land Use Systems and Soil Resilience. In *Soil Resilience and Sustainable Land Use*; Greenland, D.J., Szabolcs, I., Eds.; CAB International: Wallingford, 1994; 41–67.
5. Oldeman, L.R. Global Extend of Soil Degradation. In *Soil Resilience and Sustainable Land Use*; Greenland, D.J., Szabolcs, I., Eds.; CAB International: Wallingford, 1994; 99–118.
6. Braford, J.M.; Peterson, G.A. Conservation Tillage. In *Handbook of Soil Science*; Sumner, M.E., Ed.; CRC Press: Boca Raton, FL, 1999; G247–270.
7. Swift, M.J.; Sanchez, P.A. Biological Management of Tropical Soil Fertility for Sustained Productivity. Nat. Resour. **1984**, *20*, 2–10.
8. Lal, R. Soil Quality and Sustainability. In *Methods for Assessment of Soil Degradation*; Lal, R., Blum, W.H., Valentine, C., Stewart, B.A., Eds.; CRC Press: Boca Raton, FL, 1998; 17–30.
9. Young, A. *Land Resources, Now and for the Future*; Cambridge University Press: Cambridge, UK, 1998; 319.
10. Barrow, C.J. *Land Degradation*; Cambridge University Press: Cambridge, UK, 1994; 362.
11. Katyal, J.C.; Vlek, P.L.G. *Desertification—Concept, Causes and Amelioration*; ZEF Discussion Papers on Development Policy, No.33, ZEF: Bonn, 2000; 73.
12. Bidwell, O.W.; Hole, F.D. Man as a Factor of Soil Formation. Soil Sci. **1965**, *99*, 65–72.

RESILIENCE AND QUALITY

R

Rattan Lal
The Ohio State University, Columbus, Ohio, U.S.A.

INTRODUCTION

Soil resilience, soil's ability to restore its productivity and environmental moderation/buffering capacity (1), is important because of the ever increasing risks of soil and environmental degradation due to anthropogenic activities. It is important to understand the factors and processes governing soil resilience, and the techniques of measuring and modeling changes in it. The topic is more important and relevant now than ever before because of at least five major global issues of the 21st century. These are: 1) soil degradation, 2) greenhouse effect, 3) water scarcity, 4) land disposal of solid waste, and 5) agricultural sustainability. While all of these issues are influenced by soil resilience and quality, the global problem of soil degradation (2) is an important issue that needs to be addressed (3). Soil degradation is also closely linked with the emission of greenhouse gases (4–6), water disposal (6), and water quality. Thus, an in-depth understanding of the mechanisms involved in imparting resilience to soils is a necessary prerequisite to identifying appropriate techniques for averting degradation, enhancing or maintaining soil resilience.

Soil quality, productivity, and environmental moderation capacity (8) is the net balance between soil degradative processes and soil resilience (Fig. 1). In practice, soil quality is the capacity of the soil to perform a range of functions of value to humans. Important among these functions are: 1) producing biomass, 2) filtering pollutants out of water, 3) buffering emissions of greenhouse gases, 4) moderating cycles of C, N, P, S, H_2O and other elements, and 5) biodegrading contaminants, etc. Soil resilience maintains or enhances these functions while keeping degradative processes under check and restoring soil's life-supporting and environmental cleansing processes. Soil resilience can also decrease the risk of an accelerated greenhouse effect by C sequestration in stable microaggregates or deep incorporation in the soil profile; reduce eutrophication of surface water and contamination of groundwater by moderating its filtration and bioremediation processes; and facilitate achievement of agricultural sustainability by improving its life-support processes.

BASIC CONCEPTS AND DEFINITIONS

Applications of ecological concepts to soil science bring into focus several interrelated terms (9–13), some of these include the following:

- *Resilience*: the capacity to restore.
- *Soil stability*: the absence of change or constancy in soil quality.
- *Persistence*: the length of survival.
- *Elasticity*: the rate of restoration after perturbation
- *Amplitude*: the magnitude of change in soil quality from which restoration is possible.

These terms are subjective, and to be useful in soil management and restoration, it is important to operationalize them and develop methods to quantify their attributes.

There are two principal schools in relation to the concept of soil resilience (14). Per the first school, soil resilience refers to soil's capacity to restore itself (15, 16). If not excessively degraded beyond the point of no return, soil, like any natural ecosystem, has the capacity to self-regulation and self-maintenance (7). Soil resilience, as per this school, is related to its: 1) filtration, detoxification and transformation including bioremediation, abilities, and 2) its renewability, structural rejuvenation, and maintenance of a gene pool in soils (13, 17). Per the second school, soil resilience refers to its ability to resist change (2, 18), or to stability. According to this concept, therefore, soil resilience or stability is related to its buffering capacity. Although there may be some soils that can resist perturbation to a limit, most soils undergo change but may recover once the perturbation is removed.

Therefore, soil resilience may be defined as "soil's ability to resist change and restore its quality following stress alleviation and adoption of improved management systems." Land use and soil management are important to soil resilience (15). Similar to ecosystems, soils can also be grouped according to their resilience characteristics (7, 14) as follows:

High resilience: Soils which are not susceptible to degradation or are only slightly susceptible under extreme stresses. Some highly fertile and young soils come under

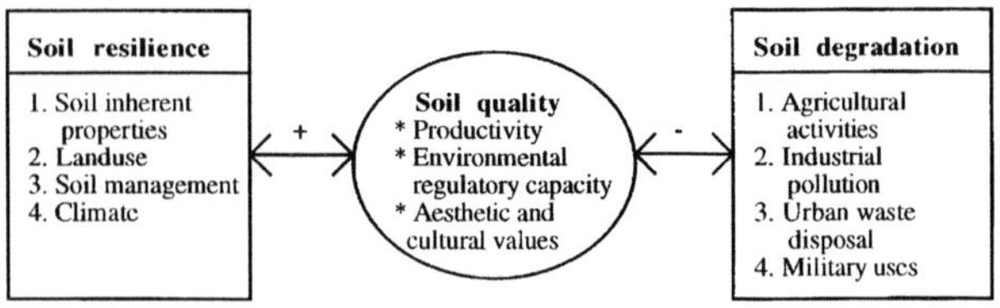

Fig. 1 Soil quality is the net effect of dynamic equilibrium between soil degradative and soil resilience processes.

this category, e.g., alluvial soil of the flood plains, loess soils and Andisols on flat terrain, Mollisols, etc. These are dynamically stable soils.

Moderate resilience: Soils are moderately susceptible to degradative processes with continuous and intensive cultivation.

Slight resilience: These soils are easily susceptible to degradation and have low capacity to recover and restore.

Nonresilient: Such soils are easily degraded beyond the point of no return and cannot restore themselves. Some shallow soils on steep slopes are nonresilient against accelerated erosion. Similarly, a plinthic soil with a hardened layer at a shallow depth can be nonresilient. Nonresilient soils may be also called fragile soils. These soils are easily destroyed in relation to

Table 1 Types of soil resilience in relation to agricultural land use

Class	Type	Description	Soil order
1	Resilient	Highly responsive to science-based inputs; prime agricultural land, with a wide range of soil characteristics. These soils have a high buffering capacity.	Mollisols, Histosols, Inceptisols, Andisols and other soils which are not easily degraded
2	Moderately resilient	These soils have fair to good agricultural potential and undergo changes in soil properties which can be restored with good soil management and science-based inputs. These soils respond to management. Changes in soil properties are modest, and soil properties can be restored by science-based management.	Oxisols, Ultisols, Alfisols, Entisol, Aridisols
3	Slightly resilient	These soils have marginal or low agricultural potential, and the range of soil properties for productive use is narrow. These soils have only modest response to input. Critical limits for irreversible degradation are readily achieved.	Vertisols, Spodosols, Psamments soils on steep terrain
4	Sensitive	These soils undergo drastic and adverse changes in soil properties, e.g., bulk density, infiltration, porosity, aggregation, nutrient depletion, and are not suitable for agriculture. Examples include acid sulfate soils, shallow soils, soils on steep terrain, and ecologically sensitive regions. Intensive cultivation of poorly structured soils, draining of acid sulfate soils, cultivation of marginal lands can lead to irreversible degradation by mismanagement or any intensive land use.	Sulfaquents, Psamments, Entisols

(Adapted from Ref. 19.)

their principal functions, e.g., productivity, environmental quality, social and aesthetic values, gene pool reserves, and support for engineering structures. Soil grouping according to this classification is shown in Table 1 (19).

Susceptibility to degradation is just the reverse of soil resilience. However, the concept of "stability" cannot be applied to soils for agricultural use. For example, plinthic (rocky) and gravelly soils may be highly stable but are not highly productive under intensive agricultural use. Such soils are more suited for engineering (construction) than agricultural use. Soil is in dynamic equilibrium with its environment and management. A productive soil of high quality is not stable; it responds to management intensity. The term "management" is a broad-based concept and "input" is a subset. The management may involve varying intensities of the same facets or different facets depending on the constraint to be alleviated. It is always undergoing a change, and attains equilibrium at a level determined by its inherent properties and management (Figs. 2 and 3). Another important aspect of resilience and management is the economics of production. If the management intensity or the input is not economic, the soil loses its resilience for the specific use.

FACTORS AFFECTING SOIL RESILIENCE

Soil resilience is affected by a range of intrinsic soil properties (or endogenous factors) and external conditions (or exogenous factors) (Fig. 4). Soil quality constitutes an important endogenous factor and includes physical, chemical, and biological processes and entities of soil quality. Soil resilience decreases with increasing soil

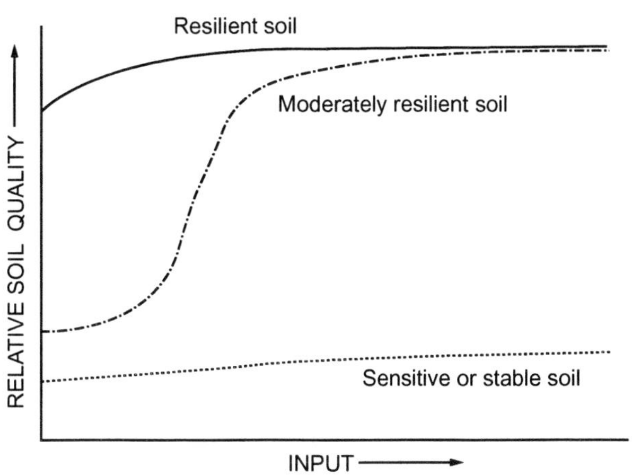

Fig. 2 Response to input of resilient and stable soils.

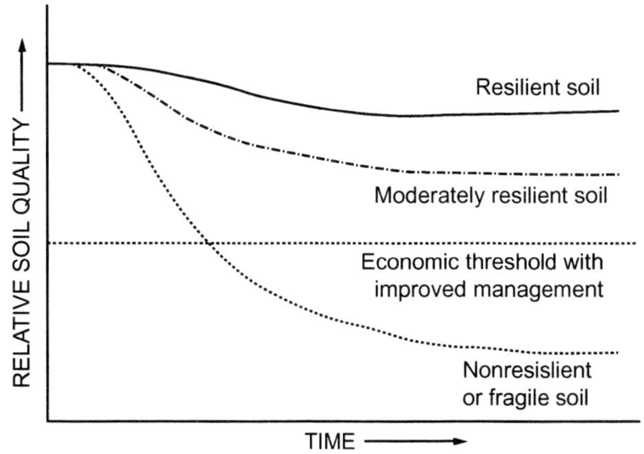

Fig. 3 Temporal changes in soil quality as affected by soil resilience and management.

capability class (19). Important among exogenous factors are land use and management, socioeconomic and institutional factors, and climate. Soils of arid regions are slightly to moderately resilient, all other factors remaining the same. Factors that affect productivity, profitability, and the farmer's ability to reinvest in soil improvement affect soil resilience. Productivity, sustainability and environments are linked through soil resilience (Eq. 1, Fig. 5).

$$A_s = f(S_r, P, E)_t \tag{1}$$

where A_s is the agricultural sustainability, S_r the soil resilience, P the productivity, E the environment, and t is the time. Resilient soils are productive, easy to manage, and have a wide range of management and farming systems options for sustainable use.

ASSESSMENT OF SOIL RESILIENCE

Soil resilience is an ecological concept applied to soil. Operationalization of the concept depends on the development of methods to assess and predict soil resilience. Lal (1, 15) proposed that soil resilience may be assessed by measuring components outlined in Eq. 2.

$$S_r = S_a + \int_0^t (S_n - S_d + I_m)\,dt \tag{2}$$

where S_a is the antecedent condition, S_n the renewability, S_d the degradation, and I_m is the input or management. S_a is in relative terms and refers to conditions prior to subjection to a stress. The magnitude and sign of the term

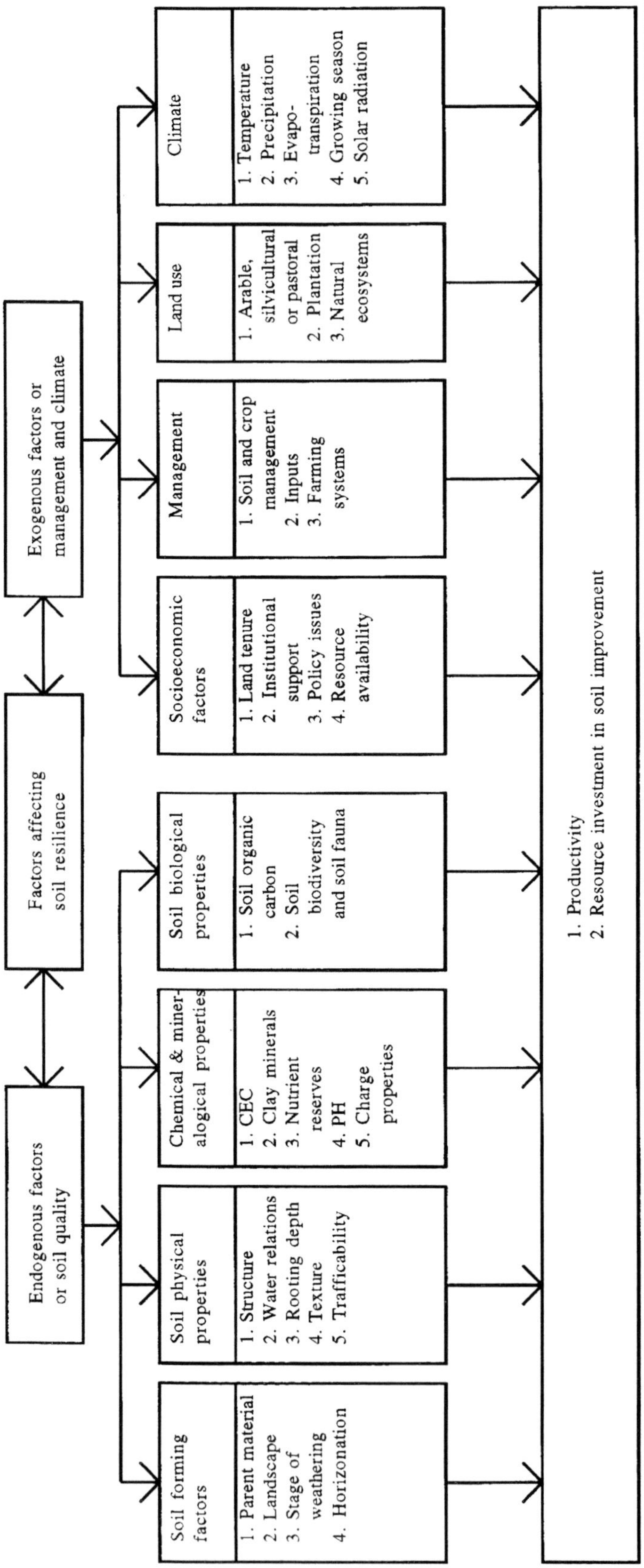

Fig. 4 Exogenous and endogenous factors and their interaction affecting soil resilience.

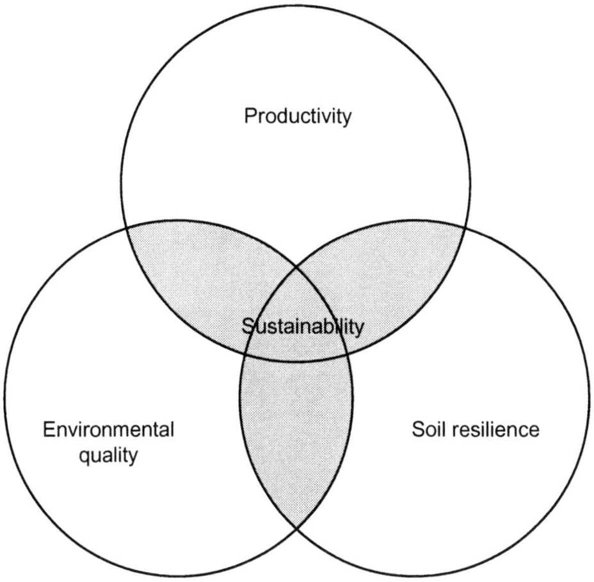

Fig. 5 Sustainability depends on productivity, environmental quality, and soil resilience.

$(S_n - S_d + I_m)$ is important in defining soil resilience class. Application of Eq. 2 involves a separate assessment of the individual soil properties (soil organic carbon or SOC content, top soil depth, available water capacity AWC, nutrient reserves etc.) and then combines these as per a rating system (20). Developing an aggregate or composite S_r index may also involve statistical techniques including principal component analysis and cluster analysis, etc. Soil resilience can also computed for individual soil properties, such as structural resilience in relation to erodibility and compactabilty (21). Quantitative assessment of structural resilience is important in developing and adopting soil and crop management systems that minimize the risks of physical degradation, e.g., erosion, compaction, and crusting. Soil resilience may also assessed for properties that affect other degradative processes, e.g., acidification and salinization.

Because it is a new concept in soil science, not much progress has been made in predicting soil resilience. Rozanov (14) proposed a physical model to predict soil resilience (Eqs. 3 and 4).

$$S_r = \frac{dI/dx}{x} \quad (3)$$

$$S_r = -2I/x^2 \quad (4)$$

where S_r is a coefficient of soil resilience, I is the work or input required to change the soil property, x. Soil resilience of two soils or two land use systems is equal if a similar level of input is required to bring about the desired change in soil property x.

SOIL RESILIENCE AND SOIL RESTORATION

Restoration of degraded soils and ecosystems is a high priority because global soil resources are finite and the world population is rapidly increasing at the rate of about 73 million per year. The soil quality depends on an optimal level of soil properties that a given soil needs to have in order to perform its functions satisfactorily. Therefore, identification of appropriate techniques of soil restoration requires knowledge of: 1) soil resilience, and 2) critical limits of key soil properties beyond which soil loses its resilience and becomes fragile and nonrestorable for a given function. Key soil properties and their limits vary among soils and land use functions. However, important properties are rooting depth, SOC content, AWC, nutrient reserves, and soil structure as measured by aggregation and stability.

Knowledge of critical limits of soil properties is important because it may enable the land user to take soil out of a given land use before it becomes nonresilient or fragile. Similarly, an appropriate land use and management system may be chosen that will maintain or enhance soil quality, soil resilience, and soil stability. There are numerous examples of soil quality and soil resilience enhancement by choice of an appropriate land use system. Soils of the Machaktos region of Kenya were severely degraded by early to mid-1930s, and yet these badlands have been converted to productive farm land (22). Lessons learned from this example of soil resilience and land restoration can be applied to restore degraded soils elsewhere, e.g., the Himalayan–Tibetan ecosystem, Andean region, Central America, and the Caribbeans. Some soils of volcanic origin are highly stable and resist degradative processes.

Methods and degree of soil restoration or reclamation also depend on the land use or functions. With a change in soil resilience class, land use may also change. For example, a soil capable of growing maize (*Zea mays*) can now grow only sorghum (*Sorghum bicolor*), or a soil capable of growing rice (*Oryza rativa*) can now only grow an upland crop. Therefore, the techniques and intensity of needed inputs depend on the land use or farming system intended.

There is a need to establish a link between the strategy of process of soil restoration and soil quality. This link is through key soil properties and the knowledge of their critical limits. Whether a degraded soil has been restored, through change in land use or management, can be determined through analyses of key soil properties and their dynamics.

CONCLUSIONS

Soil resilience is an ecological concept being used to understand soil processes and their dynamics under stress. It refers to a soil's capacity to resist change and restore itself when stress is removed and improved management systems are adopted. Soil resilience depends on inherent soil properties or soil quality, management and institutional factors, and climate. It can be measured for key soil properties and then aggregated to determine soil resilience class. Soil resilience has an important application to soil management for minimizing the risks of soil degradation, and for restoring degraded soils. In view of the severity of soil degradation worldwide, the economic importance of soil resilience in maintaining soil quality and the multifunctions of soil, cannot be overemphasized. However, additional research is needed to improve the concept and enhance its application to sustainable management of soil and water resources.

REFERENCES

1. Lal, R. Degradation and Resilience of Soils. Phil. Trans. R. Soc. Lond. B **1997**, *352*, 997–1010.
2. Oldeman, L.R. The Global Extent of Soil Degradation. In *Soil Resilience and Sustainable Land Use*; Greenland, D.J., Szabolcs, I., Eds.; CAB International: Wallingford, UK, 1994; 99–117.
3. Barrow, C.J. *Land Degradation: Development and Breakdown of Industrial Environments*; Cambridge University Press: Cambridge, UK, 1991; 295 pp.
4. Sombroek, W.G. *The Greenhouse Effect, Plant Growth and Soils*; Special Report on 25th Anniversary of ISRIC, ISRIC: Wageningen, The Netherlands, 1991.
5. Lal, R., Kimble, J.M., Levine, E., Stewart, B.A., Eds. *Soils and Global change*; CRC/Lewis Publishers: Boca Raton, FL, 1995; 544 pp.
6. Adger, W.N.; Brown, K. *Land Use and The Causes of Global Warming*; Wiley: Chichester, UK, 1994; 271 pp.
7. OECD, *The State of the Environment*; OECD: Paris, France, 1991; 295 pp.
8. Lal, R. Tillage Effects on Soil Degradation, Soil Resilience, Soil Quality, and Sustainability. Soil Till. Res. **1993**, *27*, 1–8.
9. Orians, G. Diversity, Stability and Maturity in Natural Ecosystems. In *Unifying Concepts in Ecology*; Van Dobben, W.H., Rowe-McConnell, R.H., Eds.; Junk Publishers: The Hague, 1975; 138–139.
10. Whitaker, R.H. *Communities and Ecosystems*, 2nd Ed.; MacMillan: New York, 1975.
11. Westman, W.E. Measuring the Inertia and Resilience of Ecosystems. Bioscience **1978**, *28*, 705–710.
12. Peters, R.H. *A Critique for Ecology*; Cambridge University Press: Cambridge, UK, 1991.
13. Blum, W.E.H. Soil Resilience: General Approches and Definitions. *Proceedings of the XVth International Congress Soil Science Acapulco, Mexico, 10–16 July*; 1994; Vol. 29, 233–237.
14. Rozanov, B. Stressed Soil Systems and Soil Resilience in Drylands. *Proceedings of the XVth International Congress Soil Science Acapulco, Mexico, 10–16 July*; 1994; Vol. 29, 238–245.
15. Lal, R. Sustainable Land Use Systems and Soil Resilience. In *Soil Resilience and Sustainable Land Use*; Greenland, D.J., Szabolcs, I., Eds.; CAB International: Wallingford, UK, 1994; 41–67.
16. Kay, B.D.; Rasiah, V.; Perfect, E. The Structural Aspects of Soil Resilience. In *Soil Resilience and Sustainable Land Use*; Greenland, D.J., Szabolcs, I., Eds.; CAB International: Wallingford, UK, 1994.
17. Szabolcs, I. Introduction to the Symposium on "Stressed Ecosystems and Soil Resilience". *Proceedings of the XVth International Congress Soil Science, Acapulco, Mexico, 10–16 July*; 1994; Vol. 29, 231–237.
18. Johnson, D.L.; Lewis, L.A. *Land Degradation: Creation and Destruction*; Blackwell: Cambridge, MA, 1995; 335.
19. Lal, R. Land Use and Soil Resilience. *Proceedings of the XVth International Congress Soil Science, Acapulco, Mexico, 10–16 July*; 1994; Vol. 29, 246–261.
20. Lal, R. *Methods and Guidelines for Assessing Sustainable Use of Soil and Water Resources in the Tropics*; SMSS Soil's Bulletin 21: Washington, DC, 1994; 78 pp.
21. Kay, B.D. Rates of Change of Soil Structure Under Different Cropping Systems. Adv. Soil Sci. **1990**, *12*, 1–30.
22. Tiffen, M.; Mortimore, M.; Gichuki, F. *More People, Less Erosion: Environmental Recovery in Kenya*; Wiley: Chichester, UK, 1994; 311 pp.

RESILIENCE AND RESTORATION

Frits W.T. Penning de Vries
Eric T. Craswell
*International Board for Soil Research and Management (IBSRAM),**
Jatujak, Bangkok, Thailand

R

INTRODUCTION

Soil is a substrate and a medium that serves many purposes both agricultural and nonagricultural. Degradation is the broad term that describes the loss of soil value for a particular function. Cultivating soil sometimes causes it to degrade or to improve, depending on the management, although improvements are relatively rare. Environmental conditions may also cause soils to change. Resilience, or resistance to change, varies greatly between soil types, climates, and management practices. The degree of resilience and hence the period over which degradation can occur is therefore also quite variable, and so is the time and effort required to restore degraded soils. Typically, virgin soils converted to annual cultivation can be used for 3–30 years before the first signs of reduced function become apparent, after which degradation can take place for another 5–50 years until the soil is no longer productive (1, 2, 17). While restoration is often technically feasible, the economic and sociological costs are often prohibitive (2, 19). Preferably, interventions should be made early to mitigate degradative forces, taking advantage of inherent soil resilience to offset the restoration costs.

DEFINITIONS AND EXAMPLES

Soils are dynamic because the amount of soil at a particular location can increase or decrease and also the absolute and relative amounts of their components can change nearly independent of each other. Some changes in soils are endogenous (e.g., weathering), and others are exogenous, caused by animals (e.g., termite hills) and humans (e.g., reinforced drainage, deposition).

Soils Are Used for Many Purposes

In many cases degradation implies the transfer of soil, organic matter, and nutrients to another location, where it may be beneficial (e.g., the Nile Valley clay) or detrimental (e.g., reservoir siltation).

*IBSRAM programs merged with those of the International Water Management Institute on 1 April, 2001.

Change and Degradation Are Not Identical

A soil may be changing considerably while its regular function is not clearly affected, particularly in economic terms, that is, erosion of a fertile, deep soil (21). In such cases, soil degradation may not appear to be a concern, and in the short term the need for soil and water conservation is not recognized. Yet if change continues, leading eventually to obvious degradation, it will be more difficult to retain the same function of the soil, the cost of restoration will escalate, and the loss of the natural resource base will reduce options for future populations. Even if the change is clear, but the benefits to accrue from conservation are not clear, the resources may be used elsewhere. In other cases, subtle changes in the soil may already have major consequences for its functions (for instance, the addition of *Rhizobium* inoculum and small amounts of micro-nutrients can increase crop productivity). Fig. 1 shows examples of the resilience of different soil types to erosion. Others are discussed elsewhere (5, 8, 21).

Like any object with mass, *soils are resilient to change*, meaning that their properties and components resist change in spite of external pressures that are constant or are sudden perturbations. An example is the alluvial soil of the Nile Valley, the agricultural function of which has remained unchanged for a thousand years or more. This example also shows that, in practice, resilience is a characteristic of the soil and of its environment combined. Another example of resilience is the China Loess Plateau, where soils are very deep, and erosion of top soil for hundreds of years has not diminished the productivity very much. Resilience is caused by 1) the natural generation of soil from deeper layers, which is generally quite slow ($<<$mm yr^{-1}) except for humid, tropical, volcanic soils and fluviatile sedimentation ($<$cm yr^{-1}); 2) by weathering of rock minerals that provide all the nutrients for crops (P, K, micronutrients; a few kg ha^{-1} yr^{-1}) and biological N-fixation (1–50 kg N ha^{-1} yr^{-1}); and 3) the buffering capacity of the soil mass.

Encyclopedia of Soil Science

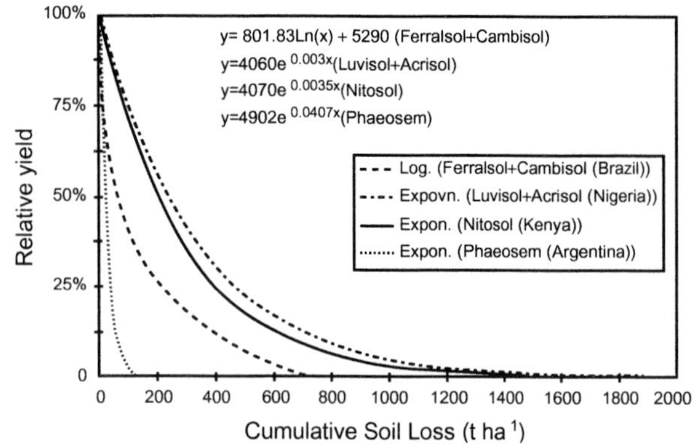

Fig. 1 Example of resilience of the function "relative crop yield" to erosion of different soil types. (From Ref. 21.)

When soils change too much and beyond their threshold of resilience, land users may be interested in *restoration of the prior status* of the soil. Restoration refers mainly to returning the capacity of the soil to support productive processes to the former level, but not necessarily to the same physical, chemical, and biological state. While restoration is often technically feasible, economic and sociological conditions may not allow the process, and the land becomes barren. This is particularly pertinent to areas with low crop cover, such as in semi-arid environments, where crops are grown for short periods, and on very infertile land where runoff relocates soil and nutrients.

Sustainable soil and land management refers to situations where all biophysical, agronomic, economic, and sociological conditions for long-term farming are met simultaneously at the farm level (6, 10). Soil resilience can help to overcome short periods of unsustainable land management, but it can also disguise an ongoing degradation process. Such short-term benefits of resilience offer the advantage that, if appropriate management actions are taken, soil changes can be limited and the threshold of soil resilience is not exceeded.

EXTENT AND RATE OF DEGRADATION

Soil degradation is widespread, to the extent that its productive uses have been diminished significantly. All continents suffer from loss in the capacity of soils to support agricultural crops, although some countries suffer more than others. An excellent description of the status of soil degradation was made by Van Lynden and Oldeman (12, 14). Yet it needs urgent updating, as the area is rapidly spreading where the resilience thresholds are surpassed ecologically and economically (2, 16, 17). For agricultural production, degradation implies that the yield ceiling is reduced, that efficiency of inputs may be less (and hence the cost of production is increased), and the risk of crop failure escalates (Fig. 2).

Some degradation and erosion processes that put soil resilience to the test are elaborated below. Many technologies for soil and water conservation are relatively well known, but their adoption and adaptation depends heavily on whether the farmers' socioeconomic circumstances allow them to adopt the technology concerned. For instance, while terracing and alley cropping are known to be very effective for controlling erosion, farmers often find such methods too expensive and too labor demanding to match their farm household resources. Willingness to follow soil conservation and restoration practices is particularly low when farmers do not own the land on which they farm and are not certain whether their investments will bring returns to their families (10). Therefore lack of knowledge and land tenure and distance from markets, compounded by opportunities for extensification and shifting cultivation, may lead small holders in the tropics to pursue exploitative and degrading land management practices that exceed the resilience capacity of their fragile environment.

Resilience, Degradation, and Restoration

The degree of resilience to the numerous processes of soil degradation (erosion, nutrient mining, salinization, compaction, surface sealing, etc.) varies widely, depending on

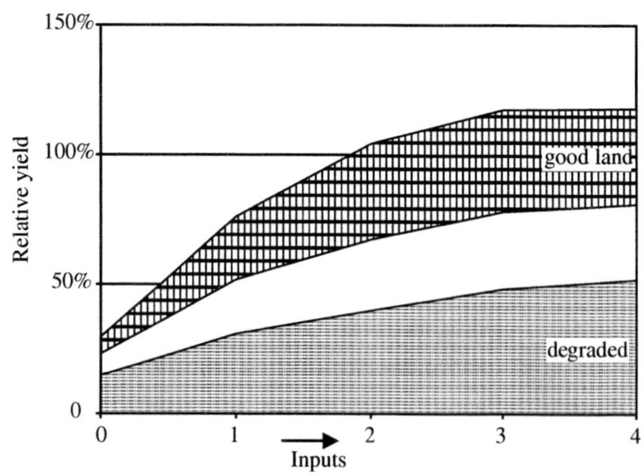

Fig. 2 Consequences of soil degradation for the response of the relative crop yield to management. (From Ref. 15.)

the nature of the process and the characteristics of the soil and environment. Some of the important dimensions are outlined below; the Reference section provides a comprehensive list of recommended literature on this subject.

Globally, *erosion* is one of the most important degradation processes (1, 2, 11). It involves transport of soil particles, organic matter, and soil nutrients, via flowing water or wind. Soil depth is a key factor determining resilience to these surface erosive processes. In the humid tropics, water erosion can cause all of the topsoil to be removed in one human generation. Wind erosion occurs in semi-arid environments and contributes to desertification. Eroded soil can be valuable downstream, but it can also silt up river bottoms and water reservoirs. The degree of resilience depends strongly on soil type and land management. Restoration is generally not feasible, but halting the process is possible in the right socioeconomic environment. WOCAT (21) provides good examples. The more the soil surface is covered by vegetation, particularly when early rainy season (erosive) rains arrive, the more erosion is prevented. This generally calls for the use of perennial crops and mulches.

Another key degradation process is *soil nutrient mining*. Mining typically occurs when cropping removes more nutrients than the soil generates (13). Resilience to this process depends on both capacity and intensity factors (4, 8). If only one crop is harvested in 5–20 years and the other years are fallow (the slash-and-burn management system), weathering may provide as many nutrients as removed by the single crop, and the system is sustainable. But when there is insufficient space or time for fallow, which is common these days, then mining reduces soil organic matter (SOM) and its nutrients. In many developing countries, SOM levels have declined by 50% over the past 50–100 years, with concomitant reduction in pH and increased Al toxicity. Conversely, nutrient levels have risen in some areas of developed countries among others because of soil fertility transfer. Even when SOM remains constant, potassium can be reduced (e.g., intensive rice cultivation) and productivity decreased. Resilience of some soils in terms of nutrient and organic reserves is generally significant. For example, Vertisols may be cultivated for 20 years with good yields but without fertilizer application (20). Soil mining is often reversible, provided that the process has not gone too far, by adding legumes to the crop cycle and greater use of organic and inorganic fertilizers (4, 18). Soil fertility can be viewed as a valuable natural resource that may be depleted to a certain degree and then replenished when conditions allow.

Erosion and loss of soil fertility *together* are processes that tend to reinforce each other: less soil nutrients > less crops and less SOM > less soil cover > more erosion, and so on.

Salinization is a common soil degradation process in irrigated areas (2, 17). Resilience to salinization is not a soil characteristic *per se*, but relates to the climate and the management of water, trees, and crops. In some areas, dryland salinity occurs when deep soil contains salts that rise to the surface due to insufficient rain infiltration. Large parts of Australia and areas such as Northeast Thailand currently suffer from this phenomenon. Restoration is possible with appropriate management of water and vegetation in the landscape and when the problem has not progressed too far.

Compaction is the increase in density of soils, particularly due to heavy mechanical traffic. When soils are too compact, roots cannot penetrate; the soils absorb less water and nutrients, and production is reduced. Resilience to compaction depends much on the particle size distribution and cropping pattern. Compaction generally increases runoff of rainwater. Some soils are much less resilient and more sensitive than others. Restoration under conventional tillage may require heavy equipment for deep plowing. *Sealing* of the surface has a similar effect; it may be caused by physical, chemical, and biological agents and can be eliminated with light equipment; however, plant cover is needed to prevent it from occurring again.

REFERENCES

1. Agassi, M., Ed. *Soil Erosion, Conservation and Rehabilitation*; Marcel Dekker Inc: New York, 1996; 402.
2. Bridges, M.; Oldeman, R. Food Production and Environmental Degradation. In *Response to Land Degradation*; Bridges, E.M., Ed.; Oxford: New Delhi, India, 2001; pp 36–40. Science Publishers: Plymouth, UK.
3. Bridges, M.; Hannam, I.; Oldeman, R.; Penning de Vries, F.W.T.; Scherr, S.; Sombatpanit, S., Eds. *Response to Land Degradation*; Oxford: New Delhi, India, 2001; Science Publishers: Plymouth, UK. 510 pp.
4. Buresh, R.J., Sanchez, P.A.; Calhoun, F. *Replenishing Soil Fertility in Africa*, Soil Science Society of America and ICRAF, Agronomy Society of America, 51; Madison, WI, 1997; 252.
5. Conway, G. *Agroecology and Farming Systems Research*; Remenyi, J.V., Ed.; ACIAR Proceedings 11; Canberra, 1987; 43–59.
6. Enters, T. *Methods for the Economic Assessment of the On- and Off-Site Impacts of Soil Erosion*, 2nd Ed; Issues in Sustainable Land Management No. 2; IBSRAM: Bangkok, 2000; 73.

7. Eswaran, H.; Lal, R.; Reich, P.; et al. Land Degradation: An Overview. In *Response to Land Degradation*; Bridges, E.M., Ed.; Oxford: New Delhi, India, 2001; pp 20–35. Science Publishers: Plymouth, UK.
8. Greenland, D.J.; Szabolcs, I. *Soil Resilience and Sustainable Land Use*; CABI: UK, 1994; 560.
9. Greenland, D.J.; Gregory, P.J.; Nye, P.H.; et al. Summary and Conclusions. In *Land Resources: On the Edge of the Malthusian Precipice?* Greenland, D.J., Ed.; CABI: UK, 1998; 1–7.
10. International Board for Soil Research and Management (IBSRAM). *Farmers' Adoption of Soil-Conservation Technologies*, IBSRAM Proceedings No. 17; IBSRAM: Bangkok, Thailand, 1997; 144.
11. Lal, R. Erosion-crop Productivity Relationships for the Soils of Africa. Soil Science Society of America Journal **1995**, *59*, 661–667.
12. Lynden, G.W.J. van.; Oldeman, L.R. *The Assessment of the Status of Human-Induced Soil Degradation in South and Southeast Asia*; UNEP/FAO/ISRIC: Nairobi/Rome/Wageningen, 1997.
13. Miwa, E. Global Nutrient Flow and Degradation of Soils and Environment. *Transactions 14th International Soils Science Society Congress*, Kyoto, Japan. 1990; Vol 5, 271–276.
14. Oldeman, L.R.; Hakkeling, R.T.A.; Sombroek, W.G. *World Map of the Status of Human-Induced Soil Degradation*, 2nd Ed.; ISRIC–UNEP: Wageningen, 1991.
15. Penning de Vries, F.W.T. Food Security? We are Losing Ground Fast; In *Crop Science: progress and prospects.* Noerberger, J., Geiger, H.H., Struik, P.C., Eds.; pp. 1–14.
16. Penning de Vries, F.W.T.; et al. Potential and Attainable Food Production and Food Security in Different Regions. Phil. Trans. R. Soc. London B. **1997**, *352*, 917–928.
17. Ponting, C. *Green History of the Earth*; Cambridge University Press: Cambridge, UK, 1991; 350.
18. Rey, Ch.; Scoones, I.; Toulmin, C. *Sustaining the Soil, Indigenous Soil and Water Conservation in Africa*; EARTHSCAN: London, UK, 1996; 260.
19. Scherr, S.J. *Soil Degradation. A Threat to Developing Country Food Security by 2020?* Food, Agriculture and Environment Discussion Paper 27; IFPRI: Washington, D.C., 1999; 60.
20. Syers, J.K.; Penning de Vries, F.W.T. *The Sustainable Management of Vertisols*; CABI and IBSRAM: Wallingford, UK, 2001; 400.
21. Tengberg, A.; Stocking, M. Erosion-Induced Loss in Soil Productivity and Its Impacts on Agricultural Production and Food Security. *FAO/Agritex Expert Consultation on Integrated Soil Management for Sustainable Agriculture and Food Security in Southern and Eastern Africa*, Harare. Land and Water Development Division, UN Food and Agriculture Organization, Rome. 1997; 42.
22. WOCAT 2000. FAO Land and Water Digital Media Series 9. circular dichroism–ROM. See also www.wocat.net.

RESOURCES, AGRICULTURAL PRODUCTIVITY, AND FOOD SECURITY

R

Keith Wiebe
United States Department of Agriculture (USDA), Washington, D.C., U.S.A.

INTRODUCTION

Soil resources and agricultural productivity affect food security both through their impact on food supply and through their impact on the incomes of that half of the world's people whose livelihoods depend directly on agriculture. Agricultural productivity depends in turn on a variety of factors, including the quality of soil and other natural resources. Resource quality has often received insufficient attention due to the scarcity of appropriate data, but recent improvements in data and methods allow better understanding of the impact of differences in the quality of land, labor, institutions, and infrastructure on agricultural productivity, and thus on food security.

WHAT IS FOOD SECURITY?

Food security has been defined and used in many ways, but the World Bank's 1986 definition, "access by all people at all times to sufficient food for an active, healthy life," is typical of most (1). Access to food derives from opportunities to produce food directly or to exchange other commodities or services for food (2). These opportunities are based in turn on resources, including soil, as well as production technologies, environmental and market conditions, and other factors (3).

In order to ensure food security, access to food must be sufficient under all possible circumstances in any particular period of time, because all sources of access are subject to variation. Food production varies with weather and other environmental factors, for example, while access to food via exchange depends on market factors such as wages and food prices. Access to sufficient food must also be sustainable over the long term. A household (or nation) can hardly be considered food secure in the long run if it is able to meet its current nutritional requirements only by depleting its endowment of resources. Appropriate use and conservation of soils are thus central to the concept of food security.

TRENDS IN GLOBAL FOOD SECURITY

The world's population recently passed the six-billion mark and continues to grow. Food production has grown even more rapidly in recent decades, increasing in per capita terms by 0.9 percent annually on a global scale, and even faster in China, India, and other populous developing countries (4). Production growth notwithstanding, over 800 million people, most of them in developing countries, suffer from chronic food insecurity because they cannot produce or purchase enough food (5). The Economic Research Service projects that the gap between what the poorest countries can produce or purchase and what they need to meet their nutritional requirements will grow by over 50 percent over the next decade, due in part to depletion and degradation of soil and water resources (6).

Population growth has slowed in recent years, but even so the world's population is projected to reach 7.5 billion by 2020. Almost all of this growth will occur in the developing world. Average incomes also continue to rise, and by 2005 over half of the world's population is expected to live in urban areas, with associated changes in dietary preferences (7). As a result of these factors, the International Food Policy Research Institute projects an increase of about 40 percent in world demand for grain over the period 1993–2020, primarily in the developing countries (8).

Such growth in food demand, representing an annual increase of 1.3 percent, is well within the range of growth in production over the past half century (9). Yet historic rates of growth in both cultivated area and yields have slowed in recent years. The Food and Agriculture Organization (FAO) reports that only a third of the estimated 2.8 billion hectares of land with potential for crop production in the developing countries is presently in cultivation, but the costs of expanded cultivation are often high in economic and environmental terms (10). As a result of these costs, the FAO projects that area expansion will contribute only about 20 percent of production growth over the next three decades. Increased multiple cropping and shorter fallow periods will contribute another 11 percent, while the majority of future production increases (69 percent) is projected to come from increased yields. Global average cereal yield growth slowed from 3.0 percent per year during the 1960s to 1.1 percent per year during the 1990s (11), due to reduced growth in input use, low cereal prices, and low levels of investment in agricultural research and technology. Poorly functioning markets and a lack of appropriate infrastructure and credit also contribute (8). Changes in soil quality may also play a role.

SOIL QUALITY AND AGRICULTURAL PRODUCTIVITY

Soils play a central role in food security, both in terms of food production and in terms of the livelihoods of farmers who produce crops other than food. Given limits on the expansion of agricultural land, increases in both food production and farmers' incomes depend critically on growth in agricultural output per unit of land, labor, and other inputs, i.e., agricultural productivity. While it has long been recognized that agricultural productivity depends directly on soil fertility, water-holding capacity, and other aspects of the quality of soil resources, these relationships have been difficult to quantify on global and regional scales due to limitations of data and spatial variations in climate, topography, management practices, and other factors (12). Recent improvements in spatially referenced data and analytical methods have allowed better understanding of the relationship between soil quality and agricultural productivity.

In order to isolate and control for the effects of differences between countries in land quality, Wiebe et al. (13) used spatially referenced soil and climate data in combination with new high-resolution land-cover data to measure the share of each country's cropland that is not limited by major soil or climate constraints to agricultural production. Econometric analysis of this measure, controlling for levels of labor, fertilizer, and other inputs to agricultural production, indicates that in most regions of the world over the period 1961–1997, agricultural productivity was 20–30 percent higher in countries with above-average soils and climates than it was in countries with below-average soils and climates. Soil quality also influences the impact of other inputs on agricultural productivity. For example, fertilizer response in Sub-Saharan Africa is significant and positive both in countries that have good soils and climate and in countries that do not, but the magnitude of the response is about twice as large in the latter countries (14). While these productivity impacts are estimated from differences in soil quality across countries, they suggest the importance of examining more closely the productivity impacts of changes in soil quality over time, i.e., soil degradation.

SOIL DEGRADATION AND FOOD SECURITY

Although data are limited, it is estimated that about 22 percent of the area under agriculture, permanent pasture, forest, or woodland worldwide has been degraded to some degree by physical, chemical, or biological processes over the past half century (15). Numerous studies have documented the impacts of such processes on crop yields in specific locations, but extrapolation to larger scales is difficult. Rough estimates suggest that soil degradation has reduced global average agricultural productivity by about 0.1 percent annually (16). Such losses have historically been masked by increased use of inputs and improvements in production technology, and as a result, it is generally agreed that they do not pose a threat to global food security. Degradation-induced losses may become more apparent in future, however, or more costly to mitigate, if yield growth continues to slow.

Yield effects may also be much more severe in particular areas. In a recent analysis of 90 original field studies in North America, den Biggelaar et al. (17) found that the impact of soil erosion on yields varied widely by crop, soil, and region. For example, wheat yields on Alfisols in the United States declined by an average of only 0.03 percent per year as a result of soil erosion, while soybean yields on Ultisols in the United States declined by an average of 0.27 percent per year. Lal (12) and Scherr (18) report similar variation in impacts across crops, soils, and regions elsewhere in the world, with corresponding variation in the potential impact of soil degradation on food security.

Actual impacts in any particular location depend not only on crop and soil but also on climate, topography, and a host of other factors, including management practices. Understanding the impact of soil degradation on agricultural productivity and food security thus requires an understanding of the incentives farmers face when making choices about management practices.

INCENTIVES FOR SOIL USE AND CONSERVATION

Some forms of soil degradation—such as nutrient depletion—can generally be feasibly reversed given current technologies and market conditions, but others—such as topsoil loss due to erosion—are effectively irreversible. How soils are actually managed to avoid, minimize, or reverse degradation depends on a variety of factors influencing farmer decisions, including the prices of agricultural inputs and outputs, farmer characteristics, and land tenure (19). Where markets function well and property rights are well-defined, farmers generally have an incentive to manage soils to protect their long-term productive potential (20). Where land tenure is uncertain or access to credit is limited, on the other hand, this incentive may be weak and farmers may not find it optimal to adopt costly soil conservation measures (21).

In addition to its effects on productivity, soil degradation may also affect water quality, siltation in reservoirs, flood frequency, and other outcomes (22). Since these impacts are felt off-site, farmers generally need additional incentives to address such concerns sufficiently. These concerns motivate policy measures to encourage increased adoption of conservation measures by farmers in many countries, including farmers in the United States.

CONCLUSION

Growth in global population, incomes, and food demand present continuing challenges for agriculture in the twenty-first century. New technologies offer considerable potential to maintain and improve soil quality, agricultural productivity, and food security, and evidence suggests that soil degradation has had a relatively small impact on global agricultural productivity to date. Actual resource management strategies and outcomes vary widely with circumstances, however, and depend critically on the incentives and institutions faced by farmers. The greater the spatial scale of potential costs and benefits, and the farther they are into the future, the greater is the challenge for policymakers to structure appropriate incentives and institutions, and the greater is the importance of research and improved understanding of the alternatives available.

REFERENCES

1. World Bank. *Poverty and Hunger: Issues and Options for Food Security in Developing Countries*; The World Bank: Washington, DC, 1986.
2. Sen, Amartya. *Poverty and Famines*; Clarendon Press: Oxford, 1981; 257.
3. Maxwell, Daniel; Wiebe, Keith. Land Tenure and Food Security: Exploring Dynamic Linkages. Development and Change **1999**, *30* (4), 825–849.
4. World Bank. *World Development Indicators*; The World Bank: Washington, DC, 1998.
5. Food and Agriculture Organization (FAO) of the United Nations. *The State of Food Insecurity in the World 1999*; FAO: Rome, 1999; 32.
6. Shapouri, Shahla; Rosen, Stacey. *Food Security Assessment: Why Countries are at Risk*, Agriculture Information Bulletin 754; U.S. Department of Agriculture Economic Research Service: Washington, DC, 1999; 21.
7. Food and Agriculture Organization (FAO) of the United Nations. *The State of Food and Agriculture 1998*; FAO: Rome, 1998; 371.
8. Pinstrup-Andersen, Per; Pandya-Lorch, Rajul; Rosegrant, Mark W. *World Food Prospects: Critical Issues for the Early Twenty-First Century*; International Food Policy Research Institute: Washington, DC, 1999; 32.
9. Byerlee, D.; Heisey, P.; Pingali, P. Realizing Yield Gains for Food Staples in Developing Countries in the Early Twenty-First Century: Prospects and Challenges; Presented at the Study Week on Food Needs of the Developing World in the Early Twenty-First Century, The Vatican, January 27–30, 1999.
10. Food and Agriculture Organization (FAO) of the United Nations. *Agriculture: Towards 2015/30: Technical Interim Report*; FAO: Rome, April 2000; 249.
11. Food and Agriculture Organization (FAO) of the United Nations. FAOSTAT Database http://www.faostat.fao.org, FAO: Rome, Accessed March 7, 2000.
12. Lal, R. Soil Erosion Impact on Agronomic Productivity and Environment Quality. Critical Reviews in Plant Sciences **1998**, *17* (4), 319–464.
13. Wiebe, Keith; Soule, Meredith; Narrod, Clare; Breneman, Vince. Resource Quality and Agricultural Productivity: A Multi-Country Comparison; Presented at the Annual Meeting of the American Agricultural Economics Association, Tampa, Florida, July 31, 2000.
14. Wiebe, Keith; Soule, Meredith; Narrod, Clare; Breneman, Vince. Resource Quality and Agricultural Productivity in Sub-Saharan Africa: A Multi-Country Comparison; Presented at the International Conference of Agricultural Economists, Berlin, August 14, 2000.
15. Oldeman, L.R. *An International Methodology for an Assessment of Soil Degradation and Georeferenced Soils and Terrain Database*, Working Paper and Preprint 93/06; International Soil Reference Information Centre: Wageningen, The Netherlands, 1993; 22.
16. Crosson, Pierre. *Soil Erosion and Its On-Farm Productivity Consequences: What do we Know?*, Discussion Paper 95–29; Resources for the Future: Washington, DC, 1995; 18.
17. Den Biggelaar, Christoffel; Lal, Rattan; Wiebe, Keith; Breneman, Vince. Impact of Soil Erosion on Crop Yields in North America. Advances in Agronomy, **2000**, *72*, 1–52.
18. Scherr, Sara. *Soil Degradation: A Threat to Developing-Country Food Security by 2020?* Food, Agriculture, and Environment Discussion Paper 27; International Food Policy Research Institute: Washington, DC, 1999; 63.

19. Soule, Meredith J; Tegene, Abebayehu; Wiebe, Keith D. Land Tenure and the Adoption of Conservation Practices. American Journal of Agricultural Economics, **2000**, *82* (4), 993–1005.
20. Hopkins, Jeffrey W.; Schnitkey, Gary D.; Sohngen, Brent L.; Miranda, Mario J.; Tweeten, Luther G. Optimal Cropland Degradation with Reversible and Irreversible Components; Presented at the Annual Meeting of the American Agricultural Economics Association, Salt Lake City, Utah, August 2–5, 1998.
21. Pagiola, Stefano. Economic Analysis of Incentives for Soil Conservation. In *Using Incentives for Soil Conservation: From Theory to Practice*; Sanders, D.W., Huszar, S., Sombatpanit, S., Enters, T., Eds.; Science Publishers, Inc: Enfield, New Hampshire, 1999, 41–56.
22. Crosson, Pierre. Future Supplies of Land and Water for World Agriculture. In *Population and Food in the Early Twenty-First Century*; Nurul, Islam, Ed.; International Food Policy Research Institute: Washington, DC, 1995; 143–159.

RESTORATION ECOLOGY

Richard Hobbs
Murdoch University, Murdoch, Western Australia, Australia

R

INTRODUCTION

Restoration ecology is the science behind ecological restoration, which covers a wide range of activities involved with the repair of damaged or degraded ecosystems, and is usually carried out for one of the following reasons:

1. To restore highly disturbed, but localized sites, such as mine sites
2. To improve productive capability in degraded production lands
3. To enhance nature conservation values in protected landscapes
4. To restore ecological processes over broad landscape-scale or regional areas.

Ecological restoration occurs along a continuum from the rebuilding of totally devastated sites to the limited management of relatively unmodified sites. Restoration aims to return the degraded system to some form of cover which is protective, productive, esthetically pleasing, or valuable in a conservation sense (1). A further tacit aim is to develop a system which is sustainable in the long term.

The conceptual basis of restoration ecology has developed rapidly in recent years (2–6). The term "ecological restoration" covers a wide range of activities involved with the repair of damaged or degraded ecosystems. An array of terms has been used to describe these activities including restoration, rehabilitation, reclamation, reconstruction, and reallocation. Generally, restoration is used to describe the complete reassembly of a degraded system to its undegraded state, while rehabilitation describes efforts to develop some sort of functional protective or productive system on a degraded site. In addition, some authors also use the term "reallocation" to describe the transfer of a site from one land use to a more productive or otherwise beneficial use. Here I will follow Ref. (1) and use the term restoration to refer broadly to activities which aim to repair damaged systems, although the other terms are used as above in particular examples. Restoration ecology involves a number of interconnected activities, as summarized in Fig. 1.

DYNAMIC SYSTEMS AND RESTORATION GOALS

Ecosystem characteristics which may be used when considering restoration goals include (1):

1. *Composition*: Species present and their relative abundances
2. *Structure*: Vertical arrangement of vegetation and soil components (living and dead)
3. *Pattern*: Horizontal arrangement of system components
4. *Heterogeneity*: A complex variable made up of components 1–3
5. *Function*: Performance of basic ecological processes (energy, water, nutrient transfers)
6. *Species interactions*: Includes pollination, seed dispersal etc.
7. *Dynamics and resilience*: Succession and state-transition processes, recovery from disturbance.

Ecosystems are naturally dynamic entities, and hence the setting of restoration goals in terms of static compositional or structural attributes is problematic. Often, past system composition or structure is unknown or partially known, and past data provide only static snapshots of system parameters. Current undegraded reference systems can therefore act as potential reference systems against which the success of restoration efforts in degraded systems can be measured. An alternative approach is to recognize explicitly the dynamic nature of ecosystems, and to accept that there are a range of potential short- and long-term outcomes of restoration projects. The aim should be to have a transparent and defensible method of setting goals for restoration which focuses on the desired characteristics for the system *in the future*, rather than in relation to what these were in the past. Where it is impossible or extremely expensive to restore composition and structure, alternative goals are appropriate. These may aim to repair damage to ecological function or ecosystem services, or to create a novel system using species not native to the region or those suited to changed environmental conditions.

Goal setting thus becomes an extremely important component of the restoration process. Goals for a particular site, or more broadly for a landscape, will need to be

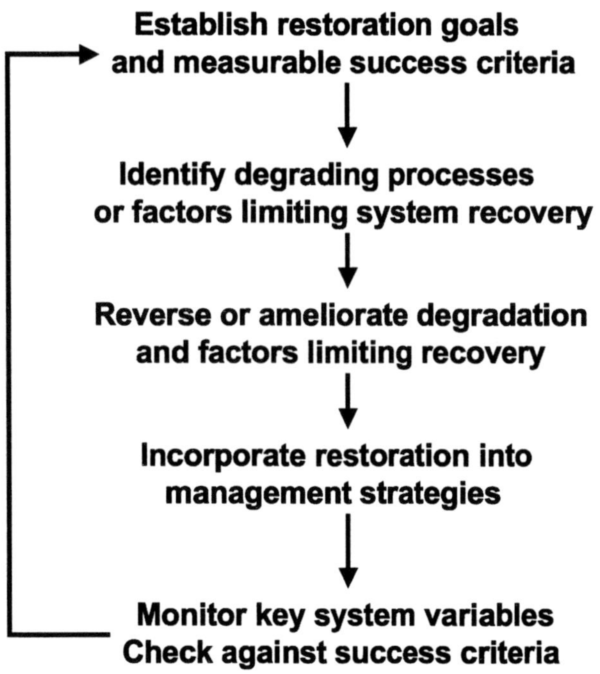

Fig. 1 Processes involved in ecological restoration. (From Ref. 9.)

determined iteratively by considering the ecological potential for restoration and matching this against societal desires. This argues for an adaptive approach to restoration, which garners ecological knowledge from as many sources as possible (including on-ground practitioners), and uses this knowledge to develop ecological response models which can indicate the likely outcomes of restoration activities. Which of the restoration options is taken up is decided on the basis of the stakeholder expectations and goals, and the extent to which it is implemented depends on the degree of financial and resource input from various sectors, including individual investment and public subsidy or incentives.

Fig. 2 Mine site in *Eucalyptus marginata* forest in Western Australia, following soil removal for extraction of alumina.

Fig. 3 Mine site in *Eucalyptus marginata* forest in Western Australia, following restoration involving soil ripping, topsoil replacement and seeding with native species.

RESTORATION OPTIONS

Arriving at clear restoration goals requires a clear picture of the restoration options available for a particular site, landscape or region. Restoration activities need to be prefaced by a rigorous assessment of the current state of the particular system or landscape, and the underlying factors leading to that state. Once this has been achieved, a clearer picture of the necessary restoration activities is possible, and a range of restoration options can be determined (Figs. 2 and 3).

Ecological restoration is required only where the system's resilience has been diminished in some way, or where the normal recovery processes are too slow to achieve management goals within a desirable time frame. Restoration requires that the stressors acting on the system are removed and may also involve replacing components that have been lost during the degradation of the system. Degrading processes can result in a variety of ecosystem responses, depending on the intensity, duration, and scale of the impact, and on which system components and processes are affected. Stressors which impair the processes of resource capture by plants (e.g., soil erosion, alteration of hydrology) are likely to have much greater impacts than stressors which remove or damage plants or consumers (e.g., pathogens or unsuitable fire regime).

A general feature of many systems seems to be the potential for the system to exist in a number of possible states, and the likelihood that restoration thresholds exist, which prevent the system from returning to a less-degraded state without the input of management effort (1). It has recently been suggested that two main types of such thresholds are likely, one which is caused by biotic interactions, and the other caused by abiotic limitations (6). If the system has degraded mainly due to biotic changes (such as grazing-induced changes in vegetation composition), restoration efforts need to focus on biotic manipulations which remove the degrading factor (e.g., the grazing animal) and adjust the biotic composition (e.g., replant desired species). If, on the other hand, the system has degraded due to changes in abiotic features (such as through soil erosion or contamination), restoration efforts need to focus first on removing the degrading factor and repairing the physical and/or chemical environment. In the latter case, there is little point in focusing on biotic manipulation without first tackling the abiotic problems.

The above argument is akin to ensuring that system functioning is corrected or maintained before questions of biotic composition and structure are considered. Considering system function provides a useful framework for the initial assessment of the state of the system and the subsequent selection of repair measures (7, 8). Where function is not impaired, restoration can legitimately focus on composition and structure as parameters to be considered when setting goals.

MEASUREMENTS OF SUCCESS

There have been numerous attempts to provide categories of assessment that will contribute to a picture of the "healthy ecosystem," which have varying degrees of ease of measurement. Biological potential inventory is probably the earliest form of ecosystem assessment, typified by the species list. Although this can be extremely useful for assessing conservation status, and is greatly improved by measurements over time, it often does not get to the basics of what is causing the degradation, rather simply reflecting the magnitude and direction of its effect.

More complex measurements of biological integrity can assess food-web complexity and the development of symbiotic relationships. However, difficulty of assessment increases greatly. Measurements relating to ecosystem function can include estimates of production, standing crop, mass balance and mineral cycling pools and rates. The problem with all these measures lies in determining what the target should be, in relation to the problems discussed above concerning reference systems. More generally, indicators of soil quality improvement with restoration can include factors such as pH, organic matter content, physical properties such as texture, erodibility, compaction, water holding capacity and drainage, and chemical properties such as toxicities associated with heavy metals and other contaminants, and the availability of nitrogen, potassium and phosphorus (10).

Measures of success have to be linked back to the clear definitions of goals for restoration. Assessment processes can be complicated and expensive, and if they are too complicated or expensive they will not be carried out. There is no point in assessing something unless it relates to specific goals. These goals can be set in relation to data on the pre-existing ecosystem, or in relation to adjacent systems, or can be settled on by discussion with stakeholders about what may be possible and desirable on the site.

REFERENCES

1. Hobbs, R.J.; Norton, D.A. Towards a Conceptual Frame-Work for Restoration Ecology. Restor. Ecol. **1996**, *4*, 93–110.
2. Allen, E.B.; Covington, W.W.; Falk, D.A. Developing the Conceptual Basis for Restoration Ecology. Restor. Ecol. **1997**, *5*, 275–276.
3. Aronson, J.; Floret, C.; Le Floc'h, E.; Ovalle, C.; Pontanier, R. Restoration and Rehabilitation of Degraded Ecosystems in Arid and Semiarid Regions. I. A View from the South. Restor. Ecol. **1993**, *1*, 8–17.
4. Aronson, J.; Le Floc'h, E.; Floret, C.; Ovalle, C.; Pontanier, R. Restoration and Rehabilitation of Degraded Ecosystems in Arid and Semiarid Regions. II. Case Studies in Chile, Tunisia and Cameroon. Restor. Ecol. **1993**, *1*, 168–187.
5. Urbanska, K.M., Webb, N.R., Edwards, P.J., Eds. *Restoration Ecology and Sustainable Development*; Cambridge University Press: Cambridge, 1997.
6. Whisenant, S.G. *Repairing Damaged Wildlands: A Process-Orientated, Landscape-Scale Approach*; Cambridge University Press: Cambridge, 1999.
7. Tongway, D.J.; Ludwig, J.A. Rehabilitation of Semi-Arid Landscapes in Australia. I. Restoring Productive Soil Patches. Restor. Ecol. **1996**, *4*, 388–397.
8. Ludwig, J., Tongway, D., Freudenberger, D., Noble, J., Hodgkinson, K., Eds. *Landscape Ecology, Function and Management: Principles from Australia's Rangelands*; CSIRO Publishing: Melbourne, 1997.
9. Hobbs, R.J. Restoration of Disturbed Ecosystems. In *Ecosystems of the World 16. Disturbed Ecosystems*; Walker, L., Ed.; Elsevier: Amsterdam, 1999; 673–687.
10. Harris, J.; Birch, P.; Palmer, J. *Land Restoration and Reclamation: Principles and Practice*; Longman: Harlow, 1996.

RESTORATION, SUCCESS AND COMPLETION CRITERIA FOR

S.C. Ward
Bicton, Australia

INTRODUCTION

Restoration of a disturbed area entails restoring the land so that the predisturbance conditions are replicated as closely as possible with all the area's environmental values intact (1). The term generally applies to the restoration of native ecosystems. Approaches to success and completion criteria vary in different countries. In the United States, the system relies on vigorously-enforced regulation, whereas in other countries, including Australia, a more flexible system of guidelines for rehabilitation or restoration is used.

Any project that aims to repair degraded land must have clear goals. Without goals, there is no way of determining whether a project is successful. This chapter refers specifically to success and completion criteria for restoration. It includes a case study of the restoration of a eucalyptus forest after mining to demonstrate the level of success and the completion criteria that have been used in a major project that is currently restoring around 500 hectares of forest each year.

Restoration infers that some aspects of the following characteristics of the area are restored (adapted from Ref. 2):

- composition: species presence and their relative abundances;
- structure: vertical arrangement of vegetation and soil components;
- pattern: horizontal arrangement of system components;
- heterogeneity: a complex variable made up from the three variables above plus heterogeneity of soil characteristics, litter distribution, etc.;
- function: performance of basic ecological processes (energy, water, and nutrient transfers);
- dynamics and resilience: successional processes, recovery from disturbance.

Debate surrounds the use of reference ecosystems against which restoration projects should be assessed. Pickett and Parker (3) consider that choosing a reference state or system is a "trap" and a "pitfall" to be avoided. They make the point that natural systems are dynamic, rather than static and predictable, and therefore contend that it is futile to try to recreate an ecosystem as it was at a particular time. However, Aronson et al. (4) argue that for any restoration project there must be a reference against which the results of the project can be assessed. Aronson et al. (5) define an "ecosystem of reference" as "some standard of comparison and evaluation, even if the choice is somewhat arbitrary." Restoration after coal mining in the United States uses land that is representative of the premining abiotic and biotic conditions and large enough to withstand natural and man-induced perturbations as reference areas (6). Whether or not an "ecosystem of reference" is adopted, some of the desirable characteristics of the target ecosystem must be defined and measured in the developing restored ecosystem before restoration can be considered successful.

ENVIRONMENTAL INDICATORS

It is obvious that not all of the characteristics of a restored ecosystem can be measured. What needs to be measured are a number of indicators that summarize the characteristics of systems and make it possible to gauge the general status of a system. Environmental indicators are physical, chemical, biological, or socio-economic measures that best represent the key elements of a complex ecosystem or environmental issues (7). The "Holy Grail" of restoration ecology is one measure that integrates all aspects of ecosystem function and structure into one or a small number of "one-off" measurements. Unfortunately, a single indicator or index (a mathematical combination of a number of indicators) is unlikely to be able to represent anything as complex as the successful restoration of an ecosystem.

For a general rehabilitation project, where the object is to return a stable, vegetated landform that doesn't replicate any existing predisturbance ecosystem, relatively simple methods to judge rehabilitation success may be appropriate. For example, a system called Ecosystem Function Analysis (EFA) has been developed in a project funded by the Australian mining industry (8). EFA consists of three modules: landscape function analysis, vegetation development, and habitat complexity. The system conceives

Encyclopedia of Soil Science

rehabilitation as being land degradation in reverse with successful rehabilitation becoming more able to regulate the movement, and minimize the losses, of vital ecosystem resources such as water, topsoil, nutrients, and organic matter. This method offers a simple methodology for assessing the sustainability of a rehabilitated ecosystem and is especially useful for small or "one-off" rehabilitation efforts where financial and other resources may be limited.

For large restoration projects, especially those where rehabilitation is progressive over a long period, a more intensive assessment effort is required that can assess progression toward a specific end. The assessment should also be able to give feedback so that restoration procedures on areas that are still undergoing rehabilitation can be continually improved.

General criteria for assessing all restoration projects have not been established. A large number of potential indicators of restoration success could be adopted. The challenge for the managers of any restoration project is to select the most relevant indicators for them to make decisions on the management of a restored area. The indicators ideally should be simple, repeatable, reliable, affordable, sensitive, and able to be aggregated with other measures. Indicators that can be measured repeatedly are valuable, as a time series of measurements allows the trajectory of the ecosystem to be visualized. "One-off" measurements often don't provide an accurate representation of a dynamic ecosystem.

Aronson et al. (5) proposed a number of "vital ecosystem attributes" that may serve as indicators of ecosystem structure and function. Examples of attributes of ecosystem structure are perennial species richness, annual species richness, total plant cover, above-ground phytomass, beta diversity, life form spectrum, keystone species, microbial biomass, and soil biota diversity. Examples of attributes of ecosystem function are biomass productivity, soil organic matter, maximum available soil water reserves, coefficient of rainfall efficiency, rain use efficiency, length of water availability period, nitrogen use efficiency, microsymbiont effectiveness, and cycling indices.

Hobbs and Norton (2) suggested these attributes were a good starting point for developing a list of parameters (indicators) to be measured to assess rehabilitation development and success. Smyth and Dearden (6) proposed a two-stage approach to measuring rehabilitation success. The first stage is successful landscaping and soil reconstruction. Soils hold some of the most important nonrenewable resources of an ecosystem (9) and the depth of soil available for root growth is a fundamental determinant of which plant species will establish and survive. The development of a stable soil with suitable physical and chemical characteristics is a basic requirement for successful restoration. The second stage is the evaluation of functional capacity and long-term successional trends. Ecosystems may not develop in an ordered and gradual manner but may undergo rapid transitions between different metastable states (10, 11). The dynamic nature and possible multiple metastable states of natural ecosystems means that the target values for "vital ecosystem attributes" will need to be defined with a mean and a variance. The values of the combined set of parameters in rehabilitated areas can then be compared against the target values or the natural variability found in "reference ecosystems" to provide a "scorecard" for the rehabilitation project (2).

COMPLETION CRITERIA

Completion criteria have been defined in a number of ways (12, 13) but can be considered as qualitative or quantitative indicators against which a restoration or rehabilitation project can be assessed so that responsibility for the area can be relinquished. They may include physical, biological, water quality, and safety indicators. It is generally accepted that completion criteria need to be developed on a site-specific basis (13, 14). This can be difficult and time-consuming and requires a detailed knowledge of the ecosystem being restored. For example, Smyth and Dearden (6) present a flow chart of reclamation planning, implementation, and monitoring for surface coal mine rehabilitation in the United States. Identifying the major impediments to restoration can help in selecting appropriate indicators.

Completion criteria for land disturbed by mining, construction, or other human-induced degradation serve two main purposes. First, they give both the landowner or land manager (e.g., private citizens or companies, or local, state or federal government departments) and the parties responsible for the disturbance (e.g., mining companies) a clear direction for restoration. Second, they describe the state of the restored area and surrounding lands that were influenced by the disturbance at which any financial obligation or legal responsibilities can be relinquished by those responsible for the disturbance.

CASE STUDY

An example of established completion criteria comes from the restoration of the jarrah (*Eucalyptus marginata* Donn ex Smith) after bauxite mining in southwest Australia,

Table 1 Criteria used by Alcoa World Alumina to assess the establishment, development, and growth of the flora on rehabilitated bauxite mines in southwest Australia

Criteria and intent	Timing	Guidelines for acceptance	Standard	Corrective action
Has the area been deep ripped?	During rehabilitation	All ripping must prevent water runoff and soil erosion Riplines must not discharge water into forest Ripping must be according to criteria established	No uncontrolled water runoff or soil erosion	Areas of unacceptable[a] erosion to be reworked and erosion control methods applied
Have areas of caprock been fractured by blasting or ripping?	During rehabilitation	All caprock should have been broken by ripping or blasting. Small areas less than 0.1 hectare are acceptable	No area greater than 0.1 hectare has unbroken caprock	Areas of caprock not broken over 0.1 of a hectare must be broken and revegetated
Is there an adequate cover of topsoil?	During rehabilitation	Topsoil should be spread across the whole rehabilitated area. Areas (less than 0.5 ha) not receiving topsoil are acceptable provided these areas do not exceed 10% of the rehabilitated area	Topsoil is spread over a minimum of 90% of the rehabilitated area	Topsoil or additional seed/fertilizer may be spread over the bare areas
Are there adequate numbers of both jarrah and marri?	9 months after rehabilitation	Rehabilitated areas must have a stocking rate which will meet proposed land use	An average of 1300 stems per hectare to be present at 9 months over 65% of the pit, of which at least 200 stems are marri to ensure that dieback does not devastate the site	Rehabilitated areas not meeting the standard will be re-planted as required
Is there an adequate legume content?	9 months after rehabilitation	Areas to have at least one legume/m^2. Areas up to 0.5 ha not meeting the standard are acceptable provided they are not greater than 10% of the pit	1 legume/m^2 based on 9 month establishment monitoring	Areas will be scarified and seeded with legumes
Are there any bare areas other than sumps greater than 0.5 ha?	9 months after rehabilitation	As above	There are no areas greater than 0.5 ha with less than 1 legume/m^2	Areas of 0.5 ha or greater not stocked at the rate of 1 legume/m^2 to be reseeded the following autumn
Is there an appropriate species richness?	15 months after rehabilitation	Areas to have a representative number of forest species present	Minimum of 50% of species in forest controls based on 15 month monitoring	Areas may need to receive additional seed. Sites to be scarified and seeded
Are there adequate stocking rates of eucalypts and under storey species capable of regenerating after a wildfire?	Approximately 10 years after rehabilitation	Guidelines for acceptance still being established based on fire recovery patterns	The site is capable of recovering from a wildfire	Treatments may be needed such as reseeding or thinning of areas
Are there adequate numbers of both jarrah and marri?	Approximately 15–20 years after rehabilitation	Numbers need to be adequate to meet the designated land use. Tree crowns need to be healthy, of suitable density and in proportion to their tree height	Minimum of 300 stems per ha with the potential to produce trees with sawn timber potential	A thinning may be required

[a]Unacceptable erosion is that which poses an accident hazard to a walking human or could lead to uncontrolled water discharge from a pit.
(From Ref. 17.)

where Alcoa World Alumina Australia mines and restores around 500 hectares of jarrah forest each year. Alcoa's rehabilitation objective is "to establish a self-sustaining jarrah forest ecosystem planned to enhance or maintain water, timber, recreation, conservation and/or other forest values." The restoration process involves landscaping, soil return, ripping, seeding, and fertilizing. Restoration is described more fully by Nichols et al. (15) and Ward et al. (16).

According to Elliott et al. (17) the restored areas must:

1. Meet land-use objectives.
2. Be integrated into the landscape.
3. Exhibit sustainable growth and development.
4. Be able to be integrated with forest management.
5. Vegetation must be resilient to disturbance.

Jarrah forests may take several centuries before attaining maturity, much longer than it is practical for Alcoa to maintain responsibility for the area. The company developed completion criteria, with input from all the stakeholders, that can be measured during restoration or relatively early in the development of the restored ecosystem. Alcoa's completion criteria consist of both assessments of the quality of the restoration procedures and indicators of restoration development and success, similar to the two-stage strategy proposed by Smyth and Dearden (6). Any impediments to successful rehabilitation need to be identified as early as possible. Early identification of problems may allow them to be remedied, or for rehabilitation procedures to be changed in subsequent years.

Restoration of the jarrah forest after mining demands that tree roots have access to water stored deep in the soil profile, propagules of a wide range of species are available, adequate nutrients are available, and soil conditions are suitable for the establishment and growth of the flora. It is impossible to measure all the indicators of ecosystem function or structure that may be of interest on every hectare of restored land. Different indicators are measured at different times after the restoration process. Indicators are measured at different intensities for a number of reasons, not the least of which is cost. Examples of the criteria used to assess the establishment, growth, and development of the flora are given in Table 1. The establishment of eucalypts and legumes is assessed each year on transects that cover all of that year's rehabilitation. Plant species richness is measured on randomly located plots within each 2–4 hectares of 15-month-old rehabilitation.

When all these criteria are met, monitoring and research that has been carried out in the past give a degree of confidence that the ecosystem is on a trajectory leading to a sustainable, diverse jarrah forest that does not require extra management costs, above those required for the native forest. Many examples of measures of ecosystem structure or function on Alcoa's bauxite mines of various ages, which are similar to, or trending toward, values found in the un-mined jarrah forest, have been published. Examples include soil nitrogen (18, 19), litter production (18, 19), litter nutrient content (18, 19), total ecosystem nitrogen (18–20), microbial biomass and microbial quotient (21, 22), botanical diversity (species richness) (23, 24), the abundance of VA mycorrhiza (25–27), depth to the water table (28), and the resilience to fire of the major plant species (29, 30).

REFERENCES

1. EPA, *Best Practice Environmental Management in Mining—Rehabilitation and Revegetation*; Commonwealth of Australia Environment Protection Agency: Canberra, 1995.
2. Hobbs, R.J.; Norton, D.A. Towards a Conceptual Framework for Restoration Ecology. Restoration Ecology **1996**, *4*, 93–110.
3. Pickett, S.T.A.; Parker, V.T. Avoiding the Old Pitfalls: Opportunities in a New Discipline. Restoration Ecology **1994**, *2*, 75–79.
4. Aronson, J.; Dhillion, S.; Le Floc'h, E. On the Need to Select An Ecosystem of Reference, However Imperfect: A Reply to Pickett and Parker. Restoration Ecol. **1995**, *3*, 1–3.
5. Aronson, J.; Floret, C.; Le Floc'h, E.; Ovalle, C.; Pontainier, R. Restoration and Rehabilitation of Degraded Ecosystems in Arid and Semi-Arid Lands. I. A View from the South. Restoration Ecol. **1993**, *1*, 8–17.
6. Smyth, C.R.; Dearden, P. Performance Standards and Monitoring Requirements of Surface Coal Mine Reclamation Success in Mountainous Jurisdictions of Western North America: A Review. Journal of Environmental Management **1998**, *53*, 209–215.
7. Saunders, D.; Margules, C.; Hill, B. Environmental Indicators for National State of the Environment Reporting—Biodiversity, Australia: State of the Environment (Environmental Indicator Reports). Department of the Environment, Canberra, 1998.
8. Tongway, D.; Hindley, N.; Ludwig, J.; Kearns, A.; Barnett, G. Early Indicators of Ecosystem Rehabilitation on Selected Minesites. In *Proceedings 22nd Annual Environmental Workshop*, Adelaide, South Australia, October, 1997; Minerals Council of Australia: Canberra, 1997.
9. Bradshaw, A.D. The Importance of Soil Ecology in Restoration Science. In *Restoration Ecology and Sustainable Development*; Urbanski, K.M., Webb, N.R., Edwards, P.J., Eds.; Cambridge University Press: Cambridge, UK, 1997; 33–65.
10. Westoby, M.; Walker, B.; Noy-Meir, I. Opportunistic Management for Rangelands Not At Equilibrium. Journal of Range Management **1989**, *42*, 266–274.

11. Hobbs, R.J. Dynamics of Vegetation Mosaics: Can We Predict Responses to Global Change. Ecoscience **1994**, *1*, 346–356.
12. Farrell, T.P. Some Considerations in Planning for Mine Decommissioning. In *Proceedings 18th Annual Environmental Workshop*, Burnie, Tasmania, October 24–29, 1993; Australian Mining Industry Council: Canberra, 1993, 235–247.
13. Tacey, W.; Treloar, J. What Do We Want Completion Criteria to Achieve? In *Proceedings 19th Annual Environmental Workshop*, Karratha, Western Australia, October 9–14, 1994; Australian Mining Industry Council: Canberra, 1994, 246–256.
14. Hollands, K. Lease Relinquishment in New South Wales—Completion Criteria. In *Proceedings 18th Annual Environmental Workshop*, Burnie, Tasmania, October 24–29, 1993; Australian Mining Industry Council: Canberra, 1993, 223–234.
15. Nichols, O.G.; Carbon, B.A.; Colquhoun, I.J.; Croton, J.T.; Murray, N.J. Rehabilitation After Bauxite Mining in South-western Australia. Landscape Planning **1985**, *12*, 75–92.
16. Ward, S.C.; Slessar, G.C.; Glenister, D.J. Environmental resource practices of alcoa of Australia Ltd. In *Australasian Mining and Metallurgy*; 2nd Ed.; Woodcock, J.T., Hamilton, J.K., Eds.; Australian Institute of Mining and Metallurgy: Melbourne, Victoria, 1993; Vol. 1, 104–108.
17. Elliott, P.; Gardner, J.H.; Allen, D.; Butcher, G. Completion Criteria for Alcoa of Australia Limited's Bauxite Mine Rehabilitation. In *Proceedings 21st Annual Environmental Workshop*, Newcastle, New South Wales, October 14–18, 1996; Minerals Council of Australia: Canberra, 1996, 79–89.
18. Ward, S.C.; Koch, J.M. Biomass and Nutrient Distribution In a 15-Year Old Forest Growing on a Rehabilitated Bauxite Mine. Australian Journal of Ecology **1996**, *21*, 309–315.
19. Ward, S.C.; Pickersgill, G.E. Nutrient Distribution in Two Eucalypt Plantations Growing on Rehabilitated Bauxite Mines. Australian Journal of Ecology **1985**, *10*, 111–124.
20. Koch, J.M. Nitrogen Accumulation in a Rehabilitated Bauxite Mined Area in the Darling Range, Western Australia. Australian Forestry Research **1987**, *17*, 59–72.
21. Sawada, Y.; Sparling, G.P.; Jasper, D.A. Use of Microbial Biomass and Activity Indices to Assess Mine-Site Rehabilitation. In *Proceedings, Soil 1994, Australian Society Soil Science Incorporated, Conference*, July 1994, Bunbury. Australian Society Soil Science: Perth.
22. Sawada, Y. Indices of Microbial Biomass and Activity to Assess Minesite Rehabilitation. In *Papers, Environmental Workshop 1996*, Newcastle, New South Wales, October 7–11 1996. Minerals Council of Australia: Canberra, 1996.
23. Koch, J.M.; Ward, S.C. Establishment of Understorey Vegetation for Rehabilitation of Bauxite-mined Areas in the Jarrah Forest of Western Australia. Journal of Environmental Management **1994**, *41*, 1–15.
24. Ward, S.C.; Koch, J.M.; Ainsworth, G.L. The Effect of Timing of Rehabilitation Procedures on the Establishment of a Jarrah Forest After Bauxite Mining. Restoration Ecology **1996**, *4*, 19–24.
25. Jasper, D.A.; Abbott, L.K.; Robson, A.D. The Loss of VA Mycorrhizal Infectivity During Bauxite Mining May Limit the Growth Of *Acacia Pulchella* R. Br. Australian Journal of Botany **1989**, *37*, 33–42.
26. Hutton, B.J.; Dixon, K.W.; Sivasithamparam, K.; Pate, J.S. Effect of Habitat Disturbance on Inoculum Potential of Ericoid Endophytes of Western Australian Heaths (Epacridaceae). New Phytologist **1997**, *135*, 739–744.
27. Gardner, J.H.; Malajczuk, N. Recolonisation of Rehabilitated Bauxite Mine Sites in Western Australia by Mycorrhizal Fungi. Forest Ecology and Management **1988**, *24*, 27–42.
28. Ruprecht, J.K.; Ainsworth, G.L.; Lareau, N.G.; Schofield, N.J. Groundwater and Vegetation Response to Mining and Subsequent Rehabilitation within Del Park Catchment, South-West Western Australia. Report No, WS 67; Water Authority of Western Australia: Perth, 1990.
29. Grant, C.D.; Koch, J.M.; Bell, D.T.; Loneragan, W.A. Tree Species Response to Prescribed Burns in Rehabilitated Bauxite Mines in Western Australia. Australian Forestry **1997**, *60*, 84–89.
30. Grant, C.D.; Koch, J.M.; Bell, D.T.; Loneragan, W.A. The Effect of Burning, Soil Scarification and Seeding on the Understorey Composition of 12 Year Old Rehabilitated Bauxite Mines in Western Australia. Australian Forestry **1997**, *60*, 16–23.

REVEGETATION OF LANDFILL SITES

R

G.Y.S. Chan
Department of Applied Biology and Chemical Technology, The Hong Kong Polytechnic University, Hung Hom, Hong Kong, China

M.H. Wong
Department of Biology and Institute of Natural Resources and Environmental Management, Hong Kong Baptist University, Kowloon Tong, Hong Kong, China

INTRODUCTION

The disposal of waste to land has been the prime means of waste disposal since the evolution of humans. In the past, people disposed of refuse at dumping points within a reasonable distance, for example, outside the city wall. As cities expanded, dumping points expanded and wastes were piled up until they reached an unacceptable level and were eventually covered by soil. These rudimentary open dumping methods are still common in small and remote villages. Landfill was developed from open dumping and is a method of codisposal of municipal solid waste and soil, on land with proper engineering consideration. Alternative names are sanitary landfill or controlled tipping; however, it has become more common simply to call it "landfill." It can be regarded as the only way for the ultimate disposal of municipal solid waste in a controlled manner that causes minimal nuisance to public health or safety. Hazardous or radioactive wastes are generally not accepted in ordinary landfill sites.

LANDFILL DESIGN AND OPERATION

The daily operation of a landfill site is to spread and compact the waste unloaded from vehicles by using waste-moving equipment. A better practice is to cover the waste with another layer of waste or with a temporary cover soil of about 0.2 m thick and compact with compactors to maximize the landfill capacity. The layers of waste and soil formed are called landfill cells. The final top cover is a layer of soil about 0.2 m to 1 m in depth. To prevent differential settlement, a high degree of compaction over the whole area of the cell is maintained. An *in situ* density of $1.2\,t\,m^{-3}$ can be achieved using a bulldozer (1). Where densities of only $0.6-0.7\,t\,m^{-3}$ are achieved, settlement of 20% or greater is expected; while on sites where a high density of $1.0-1.2\,t\,m^{-3}$ is achieved, the settlement may be 10% or less (2). A site is considered completed after its filling capacity is reached. Completed sites have to be redeveloped and special techniques have to be applied for vegetation establishment.

There are two contrasting approaches to landfill design: 1) to consider landfill as a bioreactor and try to maximize waste degradation; and 2) to attempt to isolate the embedded waste from the outside environment as far as technology can achieve and costs allow. The former principle relies upon attenuation of the leachate both within the waste and in the adjacent geology by biological and physicochemical processes. However, contamination of adjacent environments and groundwater has been experienced (3). The latter is also called containment landfill; where the rate of release of leachate into the environment is extremely low and landfill gas is either flared on-site or collected for energy recovery (4). In modern landfill design, the latter concept is commonly adopted to protect nearby environments and minimize pollutant discharge (5). To achieve this, an impermeable bottom liner is laid at the base of the site to avoid seepage of leachate into groundwater or adjacent water bodies. The very top of the landfill cells is covered by a thicker layer of inert soil, usually about 1–1.5 m thick, which is compacted by machinery (1). In order to further restrict gas migration, a costly synthetic impermeable layer may be laid together with the soil, but the efficiency of the composite liner (also called landfill cap) depends on the material used (6). Between the bottom liner and the landfill cap is a network of pipes to collect leachate and gas. The top liner not only prevents gas migration into the topsoil, it also prevents the infiltration of rainwater into the landfill.

Other criteria to be considered in the design, planning, and development of a landfill include engineering aspects, mode of landscaping, environmental impact, operational issues, and legal questions (1, 2, 7).

ANAEROBIC DEGRADATION OF MUNICIPAL SOLID WASTE

Anaerobic biodegradation of waste embedded in landfill cells generates landfill gas. The typical composition of pure landfill gas is about 60% CO_2 and 30% CH_4 ($v\ v^{-1}$); however, concentrations as high as 89.3% CO_2 and 77.1% CH_4 have been detected (1). In addition to the major gases present in landfill cover soil, other organics, such as alkanes, aromatic compounds, cycloalkanes, terpenes, alcohols, ketones, phosphine, and over 140 volatile organic compounds that are generally at levels of $<0.1\ mg\ m^{-3}$ have been identified in landfill gas (8, 9). The production of gases depends on climatic factors and the chemical nature of the waste.

The aqueous secondary waste percolating through embedded municipal solid waste and its degradation by-products is called landfill leachate. In most cases, rainwater contributes the greater portion of water that forms the leachate. Groundwater and other water sources from the adjacent environment may also supply moisture to the waste and produce leachate. Biodegradation of waste also produces water. The properties of leachate mainly depend on the nature of the embedded waste and the design of the site, which determines the infiltration rate of rainwater. High ammonium–nitrogen (NH_4^+-N) is a general characteristic of landfill leachate, as NH_4^+-N oxidation generally requires aerobic conditions. It is not uncommon for leachate to contain in excess of 1000 mg NH_4^+-N l^{-1} (10). The ratio of biochemical oxygen demand (BOD) to chemical oxygen demand (COD) of leachate collected from a new site is generally higher than that of leachate from an old site (11).

LANDFILL FACTORS AFFECTING PLANT GROWTH

Although the design and operation of landfills in different countries are not the same, the basic problems of landfill gas and leachate contamination are common to all (12).

Landfill Gas

Flower et al. (13) presented one of the early reports on how landfill gas affects tree growth and how its injurious effects may be prevented. Gas problems were experienced in sites where the clay layer was not formed properly or was cracked by uneven settlement of topsoil. Similarly, for landfill sites without an impermeable top liner, or for areas in larger sites where only temporary cover was laid, high methane (CH_4) and carbon dioxide (CO_2) and low oxygen (O_2) were detected in the cover soil and tree growth was affected (14). The tolerance of 10 subtropical woody plants to landfill gas was compared (15) and the results indicated that legumes were generally more resistant to landfill gas. For sensitive plants, such as *Liquidambar formosana* and *Castanopsis fissa* (both nonlegumes), chlorosis and stunted growth were observed after they had been fumigated with simulated landfill gas. Adverse effects of landfill gas on plants were caused by high levels of CO_2 (>20%) in the rhizosphere, rather than low partial pressure of O_2 (<10%). However, low levels of CO_2 (<10%) fumigation in the rhizosphere stimulated root growth, as low levels of gaseous CO_2 or bicarbonate stimulate cellular respiration in root cells. Therefore, stratification of CO_2 in landfill cover soil stimulates shallow root system development (Fig. 1). A landfill gas fumigation test indicated that differences in sensitivity of plants to gas level also caused differential stimulatory (*Lotus corniculatus* and *Trifolium pratense*) and inhibitory effects (*Vicia villosa* and *Trifolium repens*) on root growth subjected to landfill gas (CO_2) (16). However, the mechanisms by which CO_2, especially at high concentrations (e.g., >10%), differentially effect tissue respiration are still not fully understood.

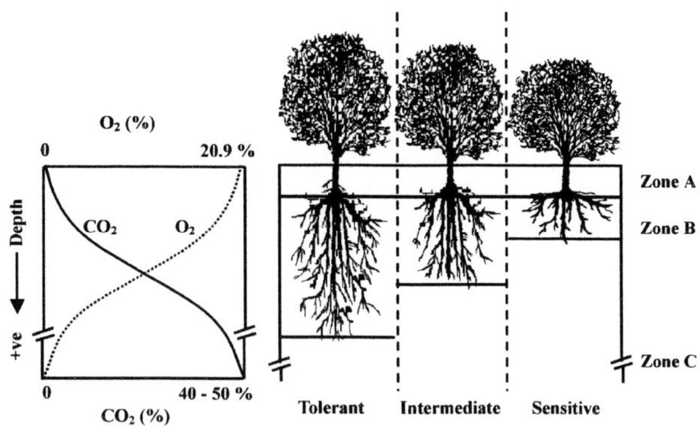

Fig. 1 Stratification of landfill gas in cover soil. Light penetration and occasional drying make Zone A unsuitable for root growth. In Zone B, root growth is stimulated by low levels of CO_2 (e.g., <10%). Zone C is restricted for root growth by high levels of CO_2 (e.g., >10%). Plants with roots sensitive to CO_2 have a shallow root system in Zone B. Plants with moderate sensitivity to CO_2 have an intermediate depth of Zone B for root growth. Plants tolerant to high CO_2 level have a deep root system.

When the O_2 partial pressure is low, leguminous N_2 fixation declines. Under a subambient partial pressure of O_2, N_2 fixation by a nodule is proportional to the O_2 concentration (17). The influence of CO_2 after landfill gas contamination on the growth and activity of nitrogen fixing nodules is complicated. Carbon dioxide at high levels as may exist in landfill cover soil was shown to be effective in inhibiting rhizobia-legume N_2 fixation and the effect was reversible within 1 h (18). However, it has been observed that rhizobia are effective in nodulating woody legumes (*Acacia confusa* and *Leucaena leucocephala*) under landfill gas fumigation conditions and are effective in nitrogen fixation (19).

Methane (CH_4), a major ingredient of landfill gas, is physiologically inert to plants. However, in the presence of CH_4, O_2 in the rhizosphere is depleted. Methane in cover soil can be converted to CO_2 by methanogenic bacteria (20). This process decreases the level of CH_4 and increases the level of CO_2 in cover soil.

Landfill Leachate

The problem of leachate seepage normally happens in confined areas and usually at the edge of a site. "Leachate breakout" is caused by high leachate levels within the waste and ingress into a site through weaknesses in the cap or on an uncapped site (21). The breakout can cause injury to trees due to leachate toxicity. In West Virginia, U.S.A., six tree species were irrigated with leachate for 4 years resulting in significant mortality (22). In Hong Kong, *Acacia confusa* irrigated with leachate exhibited growth depression to about 25% of the control after 50 days (23). The phytotoxicity of treated and untreated leachate samples collected from three landfills in Germany was evaluated using luminescent bactera and bioassays conducted using the aquatic plant *Lemna minor*, and terrestrial plants *Brassica rapa* and *Lepidium sativm* (24). In addition to high levels of NH_4^+-N, accumulation of chloride and increases in soil salinity are major causes of phytotoxicity of leachate (25). On the contrary, when leachate concentration and irrigation rate were properly controlled, leachate irrigation could benefit plant growth (26, 27).

Leachate recirculation on landfill sites can be regarded as an alternative approach to leachate treatment (28). It was found that leachate recycling increased the rate of biodegradation of organics and the landfill would become "stabilized" (29). When leachate recycling was practiced, the duration of CH_4 production could be shortened from 20–30 years to a few years (30). Application of leachate to the surface of a landfill can result in a significant decrease in leachate volume due to evapotranspiration. However, leachate recycling is not commonly adopted in cities demanding groundwater for their main water supply and for landfill without proper bottom liners. For example, leachate recycling has been banned in New Jersey (30) and is also not allowed by law in the U.K.

Other Landfill Factors

In addition to gas and leachate, a number of unique characteristics found in landfill final cover soil might affect plant growth. For landfill sites with a high-quality composite cap, leachate contamination of cover soil is unlikely. The physical and chemical properties of the cover soil being laid on top of the impermeable cap are therefore crucial for the performance of the vegetation established there. Nutrient deficiency, especially nitrogen, is a common characteristic of low-quality cover soil, such as demolition waste or unweathered soil (31). Application of fertilizer is therefore necessary.

Waterlogging is common in landfill sites where the evaporation rates are low or the rainfall is intense (21). On the contrary, the problem of drought was reported in subtropical landfill sites (32). Drought problems may be severe in landfills located in places with a high evaporation rate. The impermeable top liner imposes a constraint on plant growth because of the relatively shallow effective root zone for water storage.

Another common feature is the high bulk density in cover soil caused by compaction of soil and waste. Excessive compaction of soil might hinder the penetration of common plant roots. The restriction of root growth not only depends on bulk density but also on soil particle size. Plant roots will rarely penetrate into light textured soils with a bulk density of greater than $1.7-1.8\,g\,cm^{-3}$ or heavy textured soils with a bulk density greater than $1.5-1.6\,g\,cm^{-3}$. It was demonstrated that the growth of Austrian pine roots on a silt loam and sandy loam was restricted at bulk density of $1.4-1.8\,g\,cm^{-3}$ (33). A study conducted in Hamburg, Germany, indicated that the reduction of compaction during construction of landfill cover soil or the use of amelioration increased the water retention capacity in the topsoil layer (34).

Elevated temperature in cover soil, e.g., 45°C (15), caused by the heat-release process of the biodegradation of waste and solar radiation on bare lands, is another common feature of landfill sites. The adverse effects of elevated temperature in landfill cover soil on plant growth have been summarized (21).

Landfill cover soil may have elevated metal levels after contamination by leachate that has high metal levels. However, landfill cover soil does not normally accumulate metals to such high levels that it has harmful effects on vegetation. On old landfill sites, other landfill factors such as landfill gas are more significant than metals in their affect vegetation performance (35).

There is a lack of information related to the role of mycorrhizal fungi on tree growth on closed landfills. The need for mycorrhizal tree seedlings in site-restoration programs has been questioned (36, 37).

Suitable Tree Species for Landscaping Completed Landfill

There are two basic criteria that determine the suitability and success of a plant to be established on a landfill site: tolerance to landfill gas and/or leachate and drought tolerance. The superior performance of *Acacia confusa*, *A. magium*, and *A. auriculiformis* is mainly due to their high drought tolerance (38) and capablility to fix nitrogen when growing on nitrogen-deficient soil (15, 39). Due to their high drought tolerance, *Tristania conferta*, *Eucalyptus citriodora*, and *E. torelliana* are some of the nonlegumes that also show superior performance in landfill (31). Tree species suitable for woodland establishment in subtropical completed landfill are listed in (19, 31, 40), in which about one half of the trees are legumes. Suitable and unsuitable trees for planting in the U.K. are listed in (21); of which five of the 30 “likely to be tolerant” trees are legumes.

ECOLOGICAL RESTORATION AND END LAND USE

The number and size of landfill developments are increasing in most cities. As a direct consequence, more completed landfill will be sites awaiting redevelopment in the near future. Due to uneven settlement of the top profile and the potential accumulation of explosive landfill gas in infrastructure, building of estates or large civil constructions is not recommended on completed landfill sites. Therefore, closed landfill sites are commonly converted into parks, golf courses, or botanical gardens.

A sustainable landfill should be properly managed so that the environmental risk is acceptable (41). However, information about sustainable development of closed landfill is limited. To protect the environment, an assessment of the impact of landfill on ecosystems is required throughout the operation and postoperation stages of a site, both within the site boundaries and in the vicinity of the site (2).

The interaction between landfill factors and organisms inhabiting landfill sites is complex. In the rhizosphere, vegetated landfill caps may enhance methane oxidation, but the effects seem not to be significant or consistent (42). Introduction of earthworms to cover soil on landfills in the U.K. was successful but the physical soil conditions of the cover soil have not changed significantly after 3 years (43). *In situ* flaring of landfill gas may produce oxide deposits containing rather high concentrations of elements that form volatile species such as P, As, Sb, and Sn (44). However, adverse effects of such deposits on ecosystems established within landfill sites have not been reported.

On the contrary, all landfill sites are not necessarily unsuitable for ecosystem development. An ecological study was conducted on temporary cover soil in two ongoing landfill sites in Hong Kong, on which the cover soils were slightly contaminated by landfill gas and leachate. Higher plant coverage and diversity and higher densities of soil and litter animals were detected on these two sites, compared with normal sites. This was due to the seepage into landfill cover soil of leachate elevated in nutrient levels, especially nitrogen, which benefited plant growth. Indirectly, the presence of plant coverage and leaf litter benefited soil and litter fauna in the sites (45). The absence of plant cover at a completed landfill site limited the occurrence of some common soil animals such as Diplopoda, Hemiptera, Isopoda, and Ispotera (46).

A 4-year revegetation study was conducted on one of the world’s largest landfills in Hong Kong, constructed with composite impermeable top liners, and no landfill gas and leachate problems were detected. The performance of 24 tree species was tested and the results indicated that *Acacia confusa*, *A. mangium*, *A. auriculiformis* (legumes), *Eucalyptus citriodora*, and *E. torelliana* (nonlegumes) served as excellent pioneer species; while other species, such as *Castanopsis fissa*, *Cinnamomum camphora*, and *Machilus thunbergii* (native species to Hong Kong) had extremely high mortalities on the test site. After 3 years, the pioneer species provided shelter for native species. Animals, including insects and small vertebrates, started to inhabit the newly established woodland, with a species richness comparable to nearby natural landscapes (31). Tree planting has not been recommended in closed landfill sites in the U.K. as trees were suspected to damage the landfill top liner. However, evidence indicates that tree roots will not penetrate the top liner if it has a high bulk density of about 1.8 g g cm^{-3} (14, 47). The roles of trees in ecologically sound and sustainable reclamation of old landfill sites have been reviewed (35).

Wildlife colonization may also take place in woodlands established on landfill sites. The succession and maturation processes and the sustainability of the woodland rely on the physical and chemical properties of the man-made cover soil and also on climatic factors. Sustainable development of an ecosystem with vegetation comparable to ecosystems in the adjacent natural or less-disturbed lands has proved to be feasible on closed landfills. However, special technologies, management skill, and understanding of landfill design and operation are generally required.

R

REFERENCES

1. Department of the Environment (U.K.), *Landfilling Wastes*; Waste Management Paper No. 26, 6th Impression, HMSO: London, 1994.
2. Street, A.; Dumble, J.P. Above-Ground Landfill: The Engineered Approach. Journal of the Institution of Water and Environmental Management **1990**, *4*, 430–435.
3. Kuajara, O.; Sanchez, J.C.D.; Ballestrin, R.A.; Teixeira, E.C. Environmental Monitoring of the North Porto Alegre Landfill Brazil. Water Environment Research **1997**, *69*, 1170–1177.
4. Wuebben, P.; George, S.; Watkins, L.; Bonny, A. South Coast Air-Quality Management District (SCAQMD): The Permitting and Commercialization of Landfill Gas-Based Methanol Production Facilities. Applied Biochemistry and Biotechnology **1996**, 57/58, 729–740.
5. Gronow, J. The Role of Landfills in the United Kingdom. Waste Management and Research **1999**, *17*, 409–412.
6. Bachmann, J.; Von Felde, D. Comparison of Different Surface Sealing for Municipal Waste Landfills with Respect to Gas Emission and Recultivation. Zietschrift Fuer Kulturtechnik und Landentwicklung **1998**, *39*, 220–227.
7. Burton, S.A.Q.; Watson, C.I.A. Ammonia and Nitrogen Fluxes in Landfill Sites. Application to Sustainable Landfilling. Waste Management and Research **1998**, *16*, 41–53.
8. Allen, M.R.; Braithwaite, A.; Hills, C.C. Trace Organic Compounds in Landfill Gas at Seven U.K. Waste Disposal Sites. Environmental Science & Technology **1997**, *31*, 1054–1061.
9. Jenkins, R.O.; Morris, T.A.; Craig, P.J.; Ritchie, A.W.; Ostah, N. Phosphine Generation by Mixed- and Monoseptic-Culture of Anaerobic Bacteria. Science of the Total Environment **2000**, *250*, 73–81.
10. Robinson, H.; Gronow, J. Ground Water Protection in the U.K.: Assessment of Landfill Leachate Source-Term. Journal of the Institution of Water and Environmental Management **1992**, *6*, 229–236.
11. Obbard, J.P.; Barr, M.J.; Robinson-Howard, D.; Carville, M.S. Landfill Leachate: Characteristics and Biological Treatment in Hong Kong. Resource and Environmental Biotechnology **1999**, *2*, 235–248.
12. Lisk, D.J. Environmental Effects of Landfills. The Science of the Total Environment **1991**, *100*, 415–468.
13. Flower, F.B.; Gilman, E.F.; Leone, I.A. Landfill Gas, What it Does to Trees and How Its Injurious Effects May be Prevented? Journal of Arboriculture **1981**, *7*, 43–52.
14. Dobson, M.C.; Moffat, A.J. A Re-Evaluation of Objections to Tree Planting on Containment Landfills. Waste Management & Research **1995**, *13*, 579–600.
15. Chan, Y.S.G.; Wong, M.H.; Whitton, B.A. Effects of Landfill Gas on Subtropical Woody Plants. Environmental Management **1991**, *15*, 411–431.
16. Marchiol, L.; Cesco, S.; Pinton, R.; Zerbi, G. Germination and Initial Root Growth of Four Legumes as Affected by Landfill Biogas Atmosphere. Restoration Ecology **2000**, *8*, 93–98.
17. Heckmann, M.O.; Drevon, J.J. Inhibition of Symbiotic Nitrogen Fixation by Nitrate. In *Physiological Limitations and the Genetic Improvement of Symbiotic Nitrogen Fixation*; O'Gara, F., Manian, S., Drevon, J.J., Eds.; Kluwer Academic Publishers: The Netherlands, 1988; 97–106.
18. Zhang, J.; Liang, J.; Wong, M.H. The Effect of High CO_2 and Low O_2 Concentrations in Simulated Landfill Gas on the Growth and Nodule Activity of *Leucaena Leucocephala*. Plant and Cell Physiology **1995**, *36*, 1431–1438.
19. Chan, Y.S.G.; Wong, M.H.; Whitton, B.A. Effects of Landfill Gas on Growth and Nitrogen Fixation of Two Leguminous Trees (*Acacia Confusa*, *Leucaena Leucocephala*). Water, Air and Soil Pollution **1998**, *107*, 409–421.
20. Fielding, E.R.; Archer, D.B.; Macario, E.C.; MaCario, A.J.L. Isolation and Characteristization of Methanogenic Bacteria from Landfills. Applied and Environmental Microbiology **1988**, *54*, 835–836.
21. Dobson M.C. and Moffat A.J. (1993) The Potential for Woodland Establishment on Landfill Sites. Department of the Environment, HMSO, London.
22. Menser, H.A.; Winant, W.M.; Bennett, O.L. Spray Irrigation with Landfill Leachate. BioCycle **1983**, *24*, 22–25.
23. Wong, M.H.; Leung, C.K. Landfill Leachate as Irrigation Water for Tree and Vegetable Crops. Waste Management and Research **1989**, *7*, 311–324.
24. Devare, M.; Bahadir, M. Biological Monitoring of Landfill Leachate Using Plants and Luminescent Bacteria. Chemosphere **1994**, *28*, 261–271.
25. Stephens, W.; Tyrrel, S.F.; Tiberghien, J.E. Irrigation Short Rotation Coppice with Landfill Leachate: Constraints to Productivity Due to Chloride. Bioresource Technology **2000**, *75*, 227–229.
26. Ettala, M.O. Short Rotation Tree Plantations at Sanitary Landfills. Waste Management and Research **1988**, *6*, 291–302.
27. Cureton, P.M.; Groenvelt, P.H.; McBride, R.A. Landfill Leachate Recirculation: Effects on Vegetation Vigour and Clay Surface Cover Infiltration. Journal of Environmental Quality **1991**, *20*, 17–24.
28. Gordon, A.M.; McBride, R.A.; Fisken, A.J. Effect of Landfill Leachate Irrigation on Red Maple (*Acer Rubrum* L.) and Sugar Maple (*Acer Saccharum* Marsh.) Seedling Growth and on Foliar Nutrient Concentrations. Environmental Pollution **1989**, *56*, 327–336.
29. O'-Keefe, D.M.; Chynoweth, D.P. Influence of Phase Separation, Leachate Recycle and Aeration on Treatment of Municipal Solid Waste in Simulated Landfill Cells. Bioresource-Technology **2000**, *72*, 55–66.
30. Lee, G.F.; Jones, R.A.; Ray, C. Sanitary Landfill Leachate Recycle. Biocycle **1986**, *37*, 36–38.
31. Chan G.Y.S.; Wong M.H. Third Year Plant Performance Monitoring Report, South East New Territories (SENT) Landfill, Hong Kong, PRC, Submitted to Green Valley Landfill Ltd., 2001; 54 pp.

32. Wong, M.H.; Yu, C.T. Monitoring of Gin Drinkers' Bay Landfill, Hong Kong: II. Gas Contents, Soil Properties, and Vegetation Performance on the Side Slope. Environmental Management **1988**, *13*, 753–762.
33. Zisa R.P.; Halverson H.G.; Stout B.J. Stablishment and Conifers on Compact Soils in Urban Areas. USDA Forest Service Research Paper NE, **1980**, 451
34. Tresselt, K.; Groengroeft, A.; Leonhardt, T.; Miehlich, G. Soil Physical Properties of Topsoil Layers in Landfill Cover Systems and Their Influence on the Growth of Vegetation: Case Study Hamburg–Francop. Zeitschrift Fuer Kulturtechnik und Landentwicklung **1998**, *39*, 228–233.
35. Dickinson, N.M. Strategies for Sustainable Woodland on Contaminated Soils. Chemosphere **2000**, *41*, 259–263.
36. Tosh, J.E.; Senior, E.; Smith, J.E.; Watson-Craik, I.A. Ectomycorrhizal Seedling Response to Selected Components of Landfill Leachate. Mycological Research **1993**, *97*, 129–135.
37. Tosh, J.E.; Senior, E.; Smith, J.E.; Watson-Craik, I.A. The Role of Ectomycorrhizal Inoculations in Landfill Site Restoration Programmes. Letter in Applied Microbiology **1993**, *16*, 187–191.
38. Webb R. Tree Planting and Maintenance in Hong Kong Standing Interdepartmental Landscape Technicial Group, Hong Kong Government 1991.
39. Chan, Y.S.G.; Wong, M.H.; Whitton, B.A. Effects of Landfill Factors on Tree Cover—A Field Survey at 13 Landfill Sites in Hong Kong. Land Contamination & Reclamation **1996**, *4*, 115–128.
40. Chan, Y.S.G.; Wong, M.H.; Whitton, B.A. Effects of Landfill Leachate on Growth and Nitrogen Fixation of Two Leguminous Trees (*Acacia Confusa*, *Leucaena Leucocephala*). Water, Air and Soil Pollution **1999**, *111*, 29–40.
41. Westlake, K. Sustainable Landfill–Possibility of Pipe–Dream? Waste Management and Research **1997**, *15*, 453–461.
42. Hilger-Helen, A.; Wollum-Arthur, G.; Barlaz-Morton, A. Landfill Methane Oxidation Response to Vegetation, Fertilization, and Liming. Journal of Environmental Quality **2000**, *29*, 324–334.
43. Butt, K.R.; Frederickson, J.; Lowe, C.N. Colonisation, Survival and Spread of Earthworms on a Partially Restored Landfill Site. Pedobiologia **1999**, *43*, 684–690.
44. Glindemann, D.; Morgenstern, P.; Wennrich, R.; Stottmeister, U.; Bergmann, A. Toxic Oxide Deposits from the Combustion of Landfill Gas and Biogas. Environmental Science and Pollution Research International **1996**, *3*, 75–77.
45. Chan, Y.S.G.; Chu, L.M.; Wong, M.H. Influence of Landfill Factors on Plants and Soil Fauna—An Ecological Perspective. Environmental Pollution **1997**, *97*, 39–44.
46. Wong, M.H.; Cheung, K.C.; Lan, C.Y. Factors Related to the Diversity and Distribution of Soil Fauna on Gin Drinkers' Bay Landfill, Hong Kong. Waste Management & Research **1992**, *10*, 423–434.
47. Handel, S.N.; Robinson, G.R.; Parsons, W.F.J.; Mattei, J.H. Restoration of Woody Plants to Capped Landfills: Root Dynamics in an Engineered Soil. Restoration Ecology **1997**, *5*, 178–186.

ROCKS

Vsevolod V. Dobrovolsky
Moscow State Pedagogical University, Moscow, Russia

INTRODUCTION

Soil develops, as a rule, on a mineral substrate that plays the role of soil skeleton and accumulates plant residues and products of their biochemical transformation. The significance of soil-forming mineral substrate (parent rock) is very high, since it constitutes about 95% of the total mass of most soils. Soil inherits all peculiarities of mineral substrate connected with the particle size distribution, mineralogical and bulk chemical composition, and many physicochemical properties important for the agronomic technology.

Fully developed soil with the regular sequence of certain horizons determined by climatic factors can be formed only on those mineral substrates which are loose and porous enough to provide the free migration of water, air, mezofauna, and easy penetration by plant roots. These are the properties of late-tertiary and quaternary loose substrates covering nearly all the terrestrial surface till to date. The loose consistency is also characteristic of the volcanic eruptions, on which specific soils—Andosols—are formed.

Consolidated rocks play a less significant role in pedogenesis because their high density is unfavorable for an active interaction between biota, water, gases, and mineral matter. Soils formed on these rocks consist, as a rule, of a single horizon of plant residues decomposed into different degrees with an admixture of fine eolian material. These soils can also include a shallow transitional horizon consisting of rock debris, loose allochthonous material, and illuvial humus. The composition of hard rocks can strongly affect the possibility of full-profile soils' development and the trend of pedogenesis. On dense quartzites in case of moderate climate, the regolith cannot be formed even for $n = 10^4$ years. However, on clayey-calcareous rocks (marls) the soils with the complete sequence of soil horizons can be formed in a much shorter time. The high content of calcite in hard rocks and consequent neutralization of acid products of organic matter decomposition results in the occurrence of soils with a mollic horizon, even in the case of humid climate and forest vegetation. At the same time extremely high content of some chemical elements in parent material (e.g., Mg in serpentinites) may lead to low bioproductivity of ecosystems, and, consequently, to weak development of pedogenic features (1).

With respect to loose substrates, although mantle deposits differ by the mechanisms of transport and accumulation (glacial, fluvial, eolian, etc.), all of them consist of weakly consolidated products of rock weathering. The Earth's crust is composed mostly of compact crystalline rocks consisting of minerals of the silicate class. Being exposed on the surface, these rocks undergo mechanical disintegration, as the crystalline and chemical structures of minerals formed at a great depth become unstable. For this reason, a very slow but continuous processes of total decomposition (hydrolysis) and transformation into new mineral structures occur (2). Therein lies the essence of rock weathering (hypergenesis).

COARSE AND FINE MINERAL MATERIAL OF MANTLE DEPOSITS INHERITED BY SOILS

The mineral components of mantle deposits can be subdivided into three groups (3). The first one includes the coarse particles of prevailing bedrock minerals, i.e., quartz and silicates. The stability of these minerals is different and decreases along the following range: quartz (the most stable), feldspar, mica, amphibole, pyroxene, plagioclase, olivine (4). For this reason, redeposited products of weathering are characterized by the prevalence of quartz, lower contents of feldspars and micas, rare amphiboles, single plagioclases and pyroxenes, and very infrequent olivine.

Some of the minerals contained in widespread rocks in very small quantities are very stable. These are ilmenite, magnetite, garnet, epidote, zircon, rutile, disthene, staurolite, etc. They are insignificantly changed during the redeposition of weathering products and can be easily separated because of a considerable specific weight (2.9) as a single fraction of coarse particles (the so-called "heavy fraction".) Despite the mineralogical heterogeneity, this fraction cannot cause a significant variation in the bulk chemical composition of mantle deposits, since it usually constitutes less than 1% of their total mass. However, the composition of heavy fraction points to the original location of weathering products. Since microelements specific for certain territories are concentrated in this fraction, its study allows us to define the mineralogical–geochemical provinces of mantle deposits, the peculiarities of which are inherited by soils.

The second group of mineral components of mantle loams comprises minerals formed as a result of hypergene transformation of structures of primary silicates. Secondary (hypergene) silicates have a specific (layered) structure and are represented by very fine particles (<0.002 mm) that can be observed only with the use of electron microscope and with magnifications of several thousands (5).

The structure of hypergene silicates and their large specific surface account for the ability of these minerals to adsorb cations (which can be equivalently substituted by other cations) from solution. Hypergene silicates are represented by the group of clay minerals: kaolinite, halloysite, montmorillonite, hydrochlorides, hydromicas, etc. They differ by the cation exchange capacity, which is the lowest for kaolinite and the highest (about 10 times higher) for montmorillonite.

Soils formed on mantle deposits inherit the content and composition of clay minerals. The cation exchange capacity of many soils is related to the mineralogical composition of mantle deposits being prevalent parent rocks.

The composition of clay minerals may predetermine not only soil characteristics but also microrelief. For example, in the case of tropical and subtropical climates, with alternations of dry and humid seasons on heavy substrates with montmorillonite predominance Vertisols with well-pronounced microrelief gilgai are developed (6).

Another important property of mantle deposits is also connected with fine-earth minerals. This is the color of deposits inherited by soils. In loamy and clayey substrates clay particles serve as chromophores (coloring agents), because F(III) hydroxides are firmly adsorbed (nonexchangeable) on their surface. These iron hydroxides are neither destroyed during erosion and redeposition, nor are dissolved in acids and alkalis, and can be removed only in reduction conditions, as trivalent iron is reduced to an easily soluble bivalent iron. The latter is used for determining Fe(III) hydroxides in soils (7).

There are several modifications of iron oxides formed in different reduction–oxidation, acid–alkali, and temperature conditions. In mantle deposits beyond tropical areas, Fe(III) hydroxides (amorphous α-FeO*OH) on the

Fig. 1 Podzolized (bleached) soils on red-colored substrate.

surface of clay particles have yellow–brown colors: from 7.5 to 10YR according to the Munsell's Color Charts (8). In the tropics and partly in the subtropics, such iron substances are represented mainly by α-Fe_2O_3, and several modifications of FeO*OH and have bright red and orange colors (from 10R to 2.5YR), which could be changed only in upper horizons (Fig. 1).

The color of parent rock is well preserved in conditions of a free gas exchange between soil and atmosphere. Therefore, many tropical soils have red or orange colors, and soils of cold and temperate regions have yellow–brown colors. In conditions of hindered gas exchange (caused by overmoistening of loose deposits) and oxygen deficiency, parent rocks and soils acquire bleached bluish-gray color (from 5GY to 5BG) due to iron reduction and dissolution (Fig. 2).

In case of pedogenesis on light-colored limestones the color of deep soil horizons is inherited from the rock whereas the upper horizons turned black because of humus accumulation (Fig. 3).

Fig. 2 Gleyed soils with bluish-gray color due to iron reduction.

Fig. 3 Mollisol on limestone—inheritance of white color in deep soil horizons and black color of upper humus horizon.

CHEMICAL COMPOSITION AND PARTICLE SIZE DISTRIBUTION IN MANTLE DEPOSITS

The bulk chemical composition of mantle deposits and their soils depend on the proportion of fine and coarse mineral fractions. A high content of Si results from the predominance of quartz; the contents of Al, Fe, K, and Mg depend on the amount of clay minerals (fraction <0.002 mm). The average chemical composition of mantle deposits and soils differs from that of the Earth's crust by an increased share of Si, decreased shares of Al, Ca, Mg, K, and Na, and by the prevalence of Fe(III) over Fe(II).

The particle size distribution is the content of different particle size fractions in parent rocks and soils expressed in percent. This is a very important characteristic of parent rocks, because it results in mechanical and hydrophysical properties (stickness, swelling capacity, rupture resistance, porosity, permeability, water capacity, water availability for plants, etc.) of soils formed on these deposits.

The particle size distribution in mantle deposits varies considerably. It is connected with the chemical composition of mantle deposits and their soils: the higher the share of clay fraction, the lower is the Si content, and the higher are the contents of Al, Fe, Mg, and K.

THE TRANSFORMATION OF MANTLE DEPOSITS

The third group of mineral components of mantle loams includes epigenetic mineral neoformations appearing after the accumulation of loose deposits as a result of soil-hypergenic processes. These are concretions and cements of Fe(III) and Mn(IV) hydroxides; calcareous concretions, pore encrustation, and films on fissure walls; Ca, Mg, and Na sulfates and Na chlorides in the form of efflorescence, individual crystals, and druses.

Some epigenetic minerals are formed not by free crystallization in voids and fissures, but as a result of the selective metasomatic substitution of certain elements, which mainly affects very fine clay particles, while coarse mineral grains are only weakly corroded (9). The phenomenon of hypergenic metasomatosis is most typical for the oxides of Fe(III) and Mn(IV) and, to some extent, for calcium carbonate. Besides the neoformation of individual minerals, epigenetic processes play an important role in forming the micromorphological fabric, porosity, and bulk density of mantle deposits. Deposits with most common particle size distribution (silty loams) have a bulk density of 1.5 g/cm^3 and a porosity varying from 40 to 50% of the total volume.

From the aforesaid it follows that mantle deposits, the most widespread soil-forming substrates, are one of the most important factors of pedogenesis responsible for the particle size distribution, chemical composition, physico-mechanical, hydrophysical, and physicochemical properties of many soils.

REFERENCES

1. Graham, R.C. Soils and Mineral Weathering on Phyllite Colluvium and Serpentinite in Northwestern California. Soil Sci. Soc. Am. J. **1990**, *54*, 1682–1690.
2. Birkeland, P.W. *Pedology, Weathering, and Geomorphological Research*; Oxford University Press: New York, 1992; 286.
3. Dobrovolsky, V.V. *Biogeochemistry of the Worlds Land*; CRC Press: Boca Raton, FL, 1994; 362.
4. Goldish, S.S. A Study of Rock Weathering. J. Geol. **1938**, *46* (1), 17–58.
5. Grim, R.E. *Clay Mineralogy*; McGraw-Hill: New York, 1953.
6. Wilding, L., Puentes, R., Eds. *Vertisols: Their Distribution, Properties, Classification and Management*; Texas A & M University Printing Center: College Station, TX, 1988; 193.
7. Mehra, O.P.; Jackson, M.L. Iron Oxide Removal from Soils and Clays System Buffered with Sodium Bicarbonate by a Ditionite-Citrate. Clay Clay miner. **1960**, *5*, 317–327.
8. Pendleton, R.L.; Nickelson, D. Soil Colors and Special Munsell Soil Color Charts. Soil Sci. **1951**, *71*, 35.
9. Dobrovolsky, V.V. Micromorphological Effects of Metasomatic and Colloidal Phenomena During Hypergenesis. In *Soil Micromorphology*; Jongerius, A., Ed.; Elsevier: Amsterdam, 1964; 131–137.

SATELLITE MAPPING

Bruce E. Frazier
Washington State University, Pullman, Washington, U.S.A.

S

INTRODUCTION

Soil is that "ecstatic skin of the earth" (1), containing much life, formed in place over the face of the Earth as a three-dimensional skin. The whole soil is seen only by probing, digging, or otherwise cutting into it to expose its third dimension. For over a century we have organized the soil's properties according to classification schemes and mapped its geographic distribution. It has only been the last 30 years that we have had the opportunity to view the surface of the soil from a satellite. From such a perspective, we must infer what properties the soil has in three dimensions by convergence of several forms of evidence. Direct evidence of soil properties is communicated by their spectral nature and patterns of distribution. Indirect evidence derives from geomorphology, differences in types of vegetation growing on the soils, differences in vegetation caused by soil properties acting as a source of stress to the plants, and land cover created by man in accordance with soil properties. The indirect evidence is related to three of the state factors of soil formation (2): organisms, topography and parent material.

Collecting direct evidence of soil properties with an optical space-borne sensor usually requires that the soil be bare. The exception may be microwave sensing (non-optical) which has shown some success in mapping soil moisture. Low frequency microwaves partially penetrate vegetation and the soil surface, in addition to clouds, and can be used without light (3). Timing is critical for optical sensing except in those deserts where the soil is always bare. Some agricultural soils are exposed to a view from above at planting time, but with the adoption of cropping systems built around no till and minimum till practices, many agricultural soils are no longer bare at any time. As a result, indirect evidence of soil properties is crucial for mapping by satellite or any other optical airborne sensor. The ensuing discussion will first describe the available sensors, then research into direct evidence of soil properties, and finally, the indirect evidence and applications as described in the literature and the author's experience.

AVAILABLE SENSORS

The number of sensors, wavelengths, resolution, and satellites is wide as we enter the new millennium. Private satellites are available that can provide a range of resolutions in the optical wavelengths from high (1 m) to low (1 km) and a range of wavelengths from visible (400–700 nm), reflective infrared (700–2500 nm), to thermal infrared (8000–14,000 nm). Panchromatic (black and white) imagery that is similar to aerial photography is available from IKONOS (4), if necessary. Acquisition of images can be timed to capture the ground scene when soils are bare or when soil-response patterns are detectable in vegetation. Of course, no one has control over cloud cover; so for persistently cloudy areas, any clear image may dictate the choice of date, and all other factors become secondary.

A wealth of choices of satellite data inevitably leads to the realization that if one is planning to use satellite data to map soils, needs of the survey must be well-defined. Is high resolution data necessary? One meter panchromatic data may be useful where highly detailed surveys are conducted, but a synoptic view of a wide region will not be available. Would normal color or color infrared imagery show patterns of vegetation that relate to soil properties? These data are available in resolutions from 5 m to 1 km, providing some choices in detail and synoptic views. Is information about water absorption patterns of the landscape important? These data are available from the 30 m LANDSAT thematic mapper (TM) sensor which has two bands in the middle infrared region (1550–1750 and 2080–2350 nm). The future will witness more satellites launched by both government and private entities that will provide more choices of wavelength and resolution as dictated by economic feasibility.

DIRECT EVIDENCE

Direct evidence of soil properties in satellite data has been investigated and reported by several authors. Soil organic carbon (SOC), iron oxide, and water are constituents of soil that absorb energy. As long as the soil is bare and dry, information about constituents that appear at the surface can be collected by the satellite. This technique is especially adaptable to areas of the world that employ periods of bare fallow in multi-year cropping systems. Bowers and Hanks (5), Condit (6), Stoner et al. (7), and Baumgardner et al. (8) provide examples of basic research on energy and soil property interactions that were necessary to understand which wavelengths are important for satellite sensors and then to understand how to apply the satellite-acquired data. Taking into consideration these basic energy absorption features, Weismiller et al. (9, 10) were able to incorporate detailed soil patterns derived from Landsat multi-spectral scanner (MSS) data into the survey of Jasper County, IN. Spectral map overlays were used in the field with aerial photographs to help locate the hand-drawn boundaries. Spectral patterns were correlated with soil drainage classes. A few years later, Frazier and Cheng (11) investigated Landsat TM bands and found that ratios of several of the bands could be used to reliably map soil patterns in eastern Washington State. They used bands centered at 483 (TM1), 660 (TM3), 825 (TM4), and 1650 nm (TM5). The ratios TM 1:4, 3:4, 5:4, and 5:3 scaled to 8-bit range formed images that were related to SOC levels and to eroded soils. TM 5:4 pixel values were well correlated ($R^2 = 0.98$) to SOC (g/kg) in samples gathered from the test fields. Likewise TM 5:3 pixel values were well correlated ($R^2 = 0.96$) with the ratio of amorphous iron to SOC from the field samples. This meant that eroded soils could easily be mapped because elevated amorphous iron/SOC occurred where B horizons were exposed. Both of these sensors, MSS at 80 m resolution and TM at 30 m resolution, provided data of sufficient quality to benefit soil surveys published at 1:15,840 scale.

INDIRECT EVIDENCE

Indirect evidence of soil properties derived from processing satellite images has been added to actual soil surveys in numerous instances since the Landsat series of satellites was first launched in 1972. Usually, Landsat provides a background image to be used in a reconnaissance survey in an area where little ancillary data is available. Most examples described in the literature were spectral maps made from individual MSS bands or combinations of bands that covered the visible and near infrared spectra. Soil landscapes observed on images made from Landsat digital data were found to exhibit a characteristic surface geometry, type and density of vegetation, and hydrologic pattern (12). Frequent satellite overpasses were seen as an advantage over conventional aerial photography because comparison of multi-date images would show soil-specific vegetation changes. Multi-date capability also allowed surveyors to acquire low sun elevation images showing detailed hydrologic patterns (12).

Roudabush et al. (13), in surveys on Arizona rangeland, processed two dates of Landsat imagery. The resulting image consisted of clusters representing soil, landform, or geomorphic features. The clusters were related to soil map unit data within broad landform/parent material units. Upon completion of the survey it was concluded that Landsat spectral data added to the mapping of soils on about 35% of the mapping units.

In Washington State, a general soil map was produced with the intent of illustrating the wide diversity of soils in the state. Ten of the 12 soil orders that occur worldwide have been identified in Washington (14). Normally, a general soil map is made by condensing the detailed map units to achieve an appropriate level of abstraction. In some cases, as in the state of Washington, large blocks of land devoted to wilderness and parks had never been surveyed, and thus, had no prior map units to generalize from. For these areas, a schematic soil map was made using corroborating evidence from geology, climate, landforms, and vegetation. Patterns observed in Landsat MSS data served to define prominent geomorphic boundaries and vegetative indicators that could be incorporated into logical map units (15). The contribution of satellite data to the mapping process is shown in Fig. 1 of southeastern Washington, showing glacial outburst of flood channels cutting through the loess-covered Columbia Plateau (16). Soils within the channels are formed in less than a meter of recent loess over flood gravels and bedrock. Soils outside the channels are formed in deep loess. Flood channel boundaries are seen because of plant cover differences between the two landforms. Deep loess soils are cropped to small grains while shallow soils of the channels primarily support range grasses. The image underlying the general soil map polygons is a mid-February Landsat TM band 5 scene that captures the bright return of middle infrared energy from the dry range grasses in contrast to the dark return from green winter grain crops and bare, moist soils. A few apparent mismatches between the satellite image patterns and soil boundary line locations are evident upon close examination. These occur where irrigation has made cropping within the channels possible, where light toned residue from the fall grain crop is still on the field, and where mapping of channel width was exaggerated to preserve their continuity at a publication scale of 1/750,000.

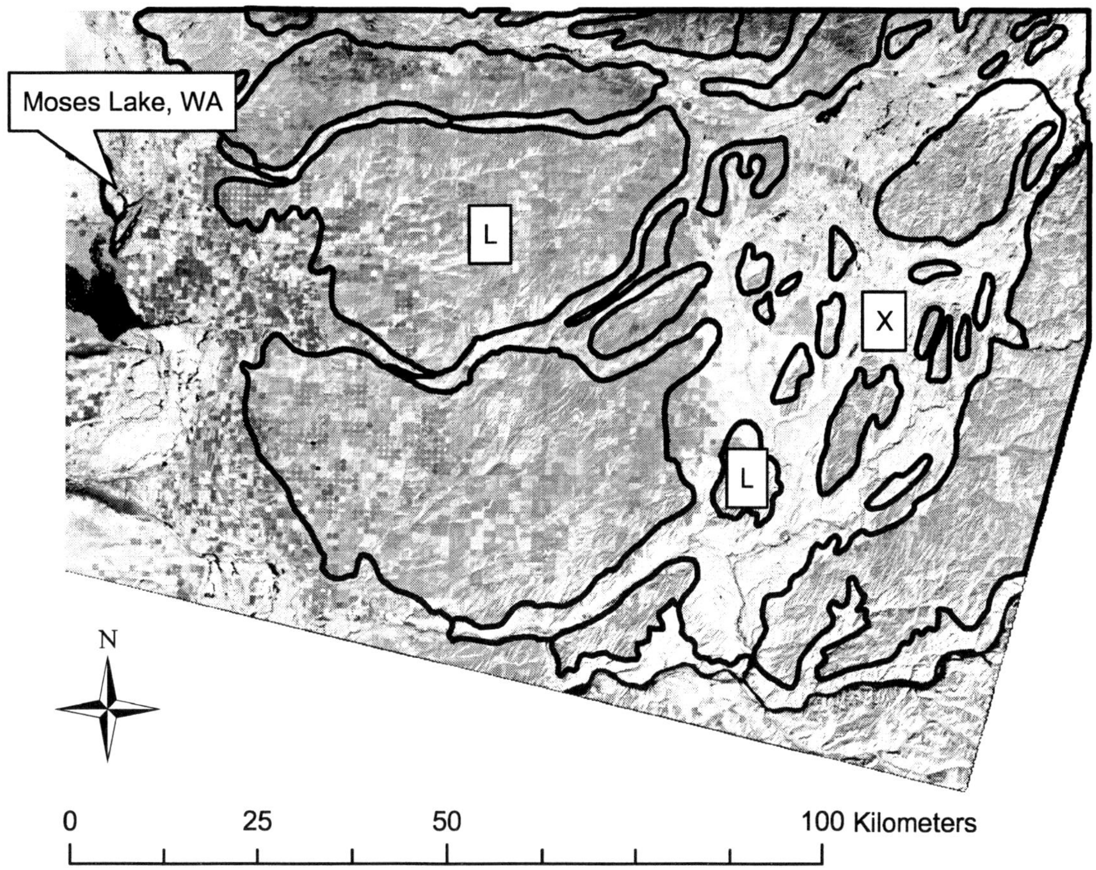

Fig. 1 Selected polygons from the Washington State general soil map superimposed on Landsat 7 TM band 5 (1550–1750 nm) scene, February 16, 2000. Light-toned X units are shallow soils of the channeled scablands. Dark-toned L units are deep soils of the loess uplands.

What is common to all these examples of satellite contributions to soil mapping is that the data provided information to delineate soil boundaries by traditional standard map production procedures, that is, lines drawn on photographic base, USGS quadrangle sheets, or mylar overlays. In the latest example (17), however, satellite data are one of the several mapped databases that are used by a model to predict the occurrence of certain soils in wilderness areas. The entire map, a result of a GIS modeling process, is a computer-drawn product based on elevation data, geology maps, current vegetation (from Landsat TM data), and climate data or models. Soil map units were classified at the subgroup level of soil taxonomy (18) and designed to be consistent with requirements of a fourth-order survey (19). Landsat TM data were used to map current vegetation classes, some of which include indicator species for certain soil properties, and to separate thin soils from bare rock and snow that occur in high elevation alpine areas. Verification was done by transecting the wilderness via existing trails and sampling soils within predicted soil groupings and landscape positions.

SUMMARY

During the past 30 years, land-observing satellites have provided data about soil that were beneficial to the making of soil maps. In some cases, these data gave information about soil properties, like color, that was used to infer erosion or drainage status. In other cases the evidence was about landform boundaries, vegetative indicators, or less well-understood spectral patterns expressed in terms of principal components that can be related to soil patterns. The launch of Landsat IV with the TM instrument brought the first wide-area access to the middle infrared bands. These data proved useful to map levels of SOC, iron oxide and moisture relationships in plants. Today's satellites now provide flexibility in dates of imagery, wavelength, resolution, and synoptic, allowing very focused views of land areas to be surveyed.

REFERENCES

1. Logan, W.B. Ecstatic Skin of the Earth. Conservation Voices **1999**, *2* (2), 14–17.
2. Jenny, H. *The Soil Resource: Origin and Behavior*; Ecological Studies 37, Springer: New York, 1980, 366–367.
3. Van de Griend, A.A.; Owe, M. The Influence of Polarization on Canopy Transmission Properties at 6.6 GHz and Implications for Large Scale Soil Moisture Monitoring in Semi-Arid Environments. IEEE Trans. Geosci. Remote Sens. **1994**, *32* (2), 409–415.
4. Space Imaging http://www.spaceimaging.com/level1/index23.htm.
5. Bowers, S.S.; Hanks, R.J. Reflection of Radiant Energy from Soils. Soil Sci. **1965**, *100*, 130–138.
6. Condit, H.R. The Spectral Reflectance of American Soils. Photogram. Eng. **1970**, *36*, 955–966.
7. Stoner, E.R.; Baumgardner, M.F.; Biehl, L.L.; Robinson, B.F. *Atlas of Soil Reflectance Properties*; Research Bull. 962, Agric. Exp. Sta.: Purdue Univ. West Lafayette, IN, 1980; 1–75.
8. Baumgardner, M.F.; Silva, L.F.; Biehl, L.L.; Stoner, E.R. Reflectance Properties of Soils. Adv. Agron. **1985**, *38*, 1–43.
9. Weismiller, R.A.; Van Scoyoc, G.E.; Pazar, S.E.; Latz, K.; Baumgardner, M.F. Use of Soil Spectral Properties for Monitoring Soil Erosion. In *Soil Erosion and Conservation*; El-Swaify, S.A., Moldenauer, W.C., Lo, A., Eds.; Soil Cons. Soc. Am. Ankeny, IA, 1985; 119–127.
10. Weismiller, R.A.; Kirschner, F.R.; Kaminsky, S.A.; Hinzel, E.J. *Spectral Classification of Soil Characteristics to Aid the Soil Survey of Jasper County, Indiana*; LARS Tech. Rpt. 040179, Lab. for Application of Remote Sensing, Purdue Univ. West Lafayette, IN, 1979.
11. Frazier, B.E.; Cheng, Y. Remote Sensing of Soils in the Eastern Palouse Region with Landsat Thematic Mapper. Remote Sens. Environ. **1989**, *28*, 317–325.
12. Westin, F.C.; Frazee, C.J. Landsat Data, Its Use in a Soil Survey Program. Soil Sci. Soc. Am. J. **1976**, *40*, 81–89.
13. Roudabush, R.D.; Herriman, R.C.; Barmore, R.L.; Schellentrager, G.W. Use of Landsat Multispectral Scanning Data for Soil Surveys on Arizona Rangeland. J. Soil Water Cons. **1985**, *40* (2), 242–245.
14. Boling, M.S.; Frazier, B.E.; Busacca, A.J. *General Soil Map of Washington*; Dept. Crop and Soil Sciences, Washington State Univ. Pullman, WA, 1998.
15. Boling, M.S. *General Soil Map of Washington State* Unpublished MS Thesis Dept. Agron. and Soils, Washington State Univ., Pullman, WA.
16. Baker, V.R.; Nummedal, D., Eds. *The Channeled Scabland: A Guide to the Geomorphology of the Columbia Basin, Washington*; NASA: Washington, DC, 1978; 1–186.
17. Rodgers, T.M. *Modeling Soils of the Sawtooth and Pasayten Wilderness Areas with a GIS* Unpublished MS Thesis Dept. Crop and Soil Sciences, Washington State Univ., Pullman, WA.
18. Soil Survey Staff, *Keys to Soil Taxonomy*, 8th Ed.; USDA, NRCS, USGPO: Washington, DC, 1998.
19. Soil Survey Staff, *Soil Survey Manual*; USDA Handbook No. 18, U.S. Government Printing Office: Washington, D.C., 1993.

SCALING EFFECTS

S

Yakov A. Pachepsky
Walter J. Rawls
United States Department of Agriculture-Agricultural Research Service (USDA-ARS), Hydrology and Remote Sensing Laboratory, Beltsville, Maryland, U.S.A.

John W. Crawford
University of Abertay, Dundee, Scotland, United Kingdom

INTRODUCTION

Soil moisture and soil hydraulic properties of soils are measured over temporal and spatial *measurement scales* that are different to the scale at which knowledge is required. For example, the measurement scale of the moisture release characteristic is equal to the volume of the core, whereas soil surface moisture measured from a satellite has the measurement scale equal to the pixel size of the imagery. Soil water balance calculations are based on subdividing soil into cells, the size of which defines the *model scale*. Both measurement and model scales define *supports*, i.e. minimum lengths, areas, volumes, or time intervals below which variation in properties is ignored.

Soil volume within a given measurement scale can be thought of as composed from soil volumes corresponding to smaller measurement scales. Averaging volumetric soil moisture content across the smaller volumes results in a correct value for volumetric moisture content for the larger scale. A similar averaging soil hydraulic properties, such as moisture retention and hydraulic conductivity, across small scales may not result in the correct values for the larger scale. The reason for that is that the larger scale represents levels of heterogeneity in soil structure that are absent in small-scale volumes. For the same reason, the spatial variability of soil moisture and soil hydraulic properties appears to be dependent on scale. Scaling as a noun means a relationship between soil moisture data at different scales, where as scaling as a gerund means relating data at different scales. Upscaling (downscaling) refers to estimating soil moisture or soil hydraulic properties at a scale that is coarser (finer) than the one at which data are available.

PHYSICS-BASED SCALING

Models of soil structure used for scaling assume preservation of some structure-related properties across a range of measurement scales or among samples of the same measurement scale.

Similitude-Based Scaling

The similitude model postulates that pore-size distributions in several soil samples of the same size can be converted into each other by simple multiplication of radii by a sample-specific constant or by adding a sample-specific constant to the logarithms of pore radii as shown in Fig. 1a. The Washburn equation

$$h = |\Psi| = \frac{a}{R} \tag{1}$$

is often used to replace values of pore radii R with values of the maximum matric potential Ψ at which a pore is filled during the drainage; h is the absolute value of matric potential Ψ, a is the capillary constant. Using (1), one can transform similarity in pore-size distributions into similarity in dependencies of h on saturation degree S as shown in Fig. 1b. Therefore, water retention curves of samples exhibiting the similitude should coincide in a 'lg$h$ vs. S' coordinate system if appropriately moved along the axis lg$h$. Let $h = f_i(S)$ are water retention curves in n samples, $i = 1, 2, 3, \ldots, n$. The coincidence of these curves following the shifts will be achieved if, for the same values of h,

$$\begin{aligned} \lg\lambda_1 + \lg f_1(S) &= \lg\lambda_2 + \lg f_2(S) = \cdots \\ &= \lg\lambda_n + \lg f_n(S) = \lg f_*(S) \end{aligned} \tag{2}$$

or

$$\lambda_1 f_1(S) = \lambda_2 f_2(S) = \cdots = \lambda_n f_n(S) = f_*(S) \tag{3}$$

where λ_i is the sample-specific scaling constant for the sample "i", $f_*(S)$ is the reference water retention curve, lg λ_i is the shift along the axis lg$h$ needed for the coalescence of the ith curve with the reference curve. The scaling allows the compression of variations in functions

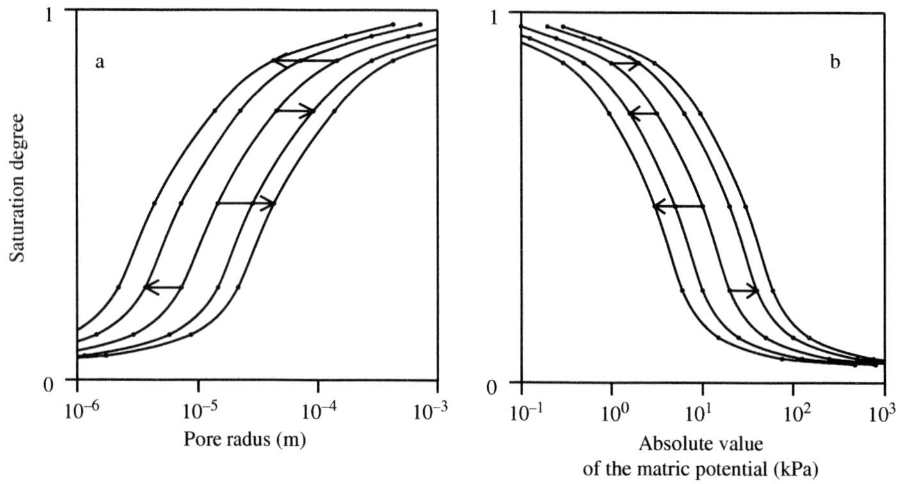

Fig. 1 Similarity of pore size distributions under the similitude hypothesis; a—distribution of pore volume expressed as saturated degree; b—pore radii transformed into absolute values of matric potentials. Arrows show shifts of the middle reference curve to transform it into other curves.

$f_i(S)$ into variations in single number λ_i (7). Figure 2 shows the application of this scaling. Although no ideal similitude is observed, a significant decrease in variability is achieved by applying scaling, i.e. shifting original water retention curves as shown in Fig. 2c. No large different groups of water retention curves are present, and the distribution of values of λ_i is close to the lognormal distribution (Fig. 2d). The similitude assumption can be combined with the hypothesis that the Kozeni-Karman law for the hydraulic conductivity K is applicable in soil. Then a coalescence of dependencies K(h) can be achieved.

Scale Invariance of Water Transport Equations

Soil water transport equations can be transformed into dimensionless forms that relate the dimensionless combination of the original dependent and independent variables. Dimensional or inspectional analysis is used for that purpose (1). Scaling invariance of the Richards' water transport equation requires, in general, water retention and hydraulic conductivity to be loglinear functions of volumetric water content:

$$\lg h = \lg h_0 - b_h \lg \theta \tag{4}$$

$$\lg K = \lg K_s + b_k \lg \theta \tag{5}$$

where the saturated hydraulic conductivity K_s and the air-entry pressure h_0 can be spatially variable whereas slopes b_k and b_h are the same across the spatial extent of the sampling. Where this scaling invariance applies, the volumetric water content at the depth, z at the time, t, can be obtained from the volumetric water content, θ', at another depth z' and at time t' as:

$$\frac{\theta}{\theta'} = \left(\frac{z}{z'}\right)^{-1/b_h}; \qquad \frac{t}{t'} = \left(\frac{z}{z'}\right)^{(b_k+b_h-1)/b_h} \tag{6}$$

Simplified models of water transport, such as the Green-Ampt infiltration model, the unit gradient drainage model, or the model with a negligible gravity also yield scaling relationships that compress spatio-temporal variations in water content profiles into spatial variability of one or two dimensionless parameters. To use the same hydraulic properties at different scales, it was suggested to incorporate the scaling directly into water transport equations by using fractional derivatives.

Fractal Scaling Using Self-Similarity, Self-Affinity, and Multiscaling Hypotheses

Self-similarity models are based on assumption of fractal geometry. Natural objects often have similar features when viewed at different scales. Measures of these features, e.g. the total number of features, the total lengths, the average roughness, the total surface area, etc., are dependent on the measurement scale. Fractal geometry assumes that this dependence is the same over a range of scales, i.e., scale-invariant within this range. Many soil properties were shown to obey fractal scaling (5).

To apply fractal geometry, one must have in mind a model of formation of the fractal features in soil. The Sierpinski

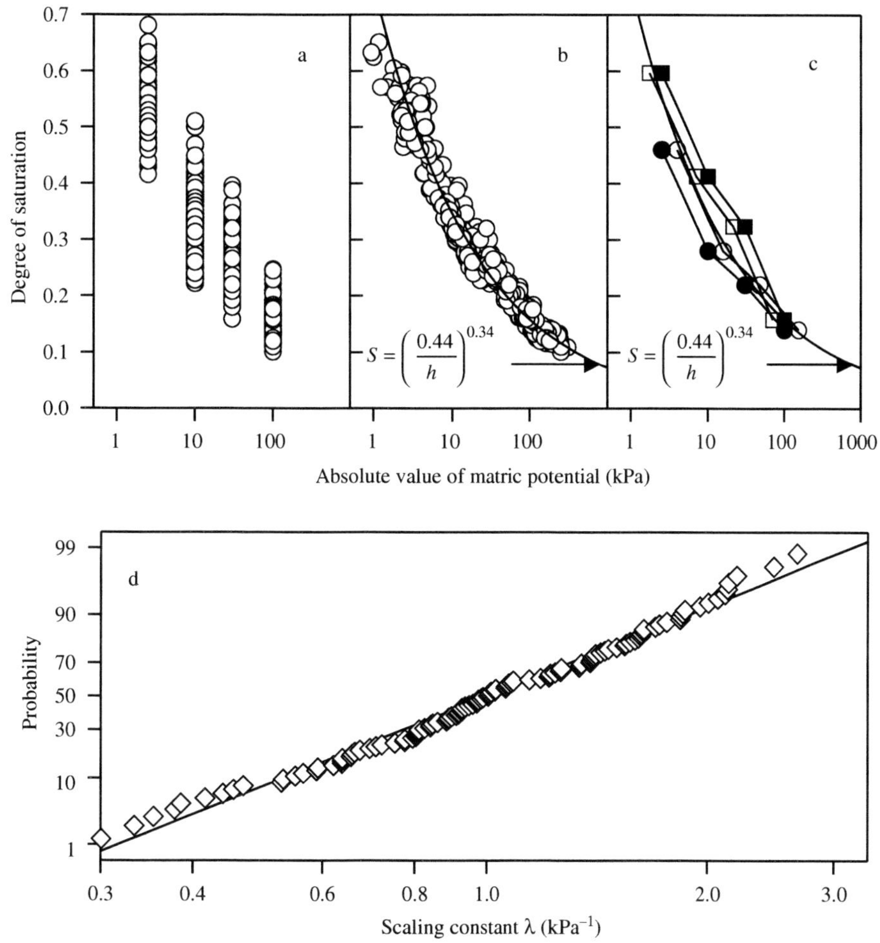

Fig. 2 Testing the similitude hypothesis with 120 of sandy siliceous, mesic Psammentic Hapludult across $170 \times 210\,m^2$ area; a—variability of the original water retention data; b—water retention curves shifted to be close to the reference water retention curves shown as a solid line with an equation; c—two examples of shifts of water retention curves; original and shifted curves are shown with filled and hollow symbols, respectively; d—probability distribution of the scaling constant; symbols-experimental data, line—the theoretical lognormal distribution.

carpet and the Menger sponge (Fig. 3) are examples of such models that are often used to simulate soil pore space structure. Those and other fractal models are obtained by an iterative extraction of similar pieces of progressively small sizes from an initially solid geometrical figure, square in case of the Sierpinski carpet and cube in case of the Menger sponge. Fractal objects can be also generated from aggregation processes.

Fractal modeling of pore volume based on the Sierpinski carpet and Menger sponge led to the widely used Brooks-Corey equation of soil water retention. The Sierpinski carpet model has been successfully used to develop equations for saturated hydraulic conductivity of macropores in soil (2). A fractal model of pore outlines has been useful in relating the unsaturated hydraulic conductivity k to soil water retention and the tortuosity caused by fractal surface of pores. Soils with water retention curves from fractal models satisfy the requirement (4) for scaling invariance of Richards equation, and also exhibit the similitude-based scaling (2) provided the spatial variability of the fractal dimension is not substantial (3).

The self-affinity model appears to be appropriate to simulate the spatial variability of soil moisture for large resolutions and extents (6). A self-affine surface remains statistically similar to itself when the resolution in horizontal plane increases μ times and the resolution in vertical direction increases μ^H times. The variances of self-affine elevations, z_1 and z_2, measured at two resolutions, d_1 and d_2, obey the scaling relationship $\text{Variance}(z_1) = (d_1/d_2)^{2H}\text{Variance}(z_2)$. Assuming spatial

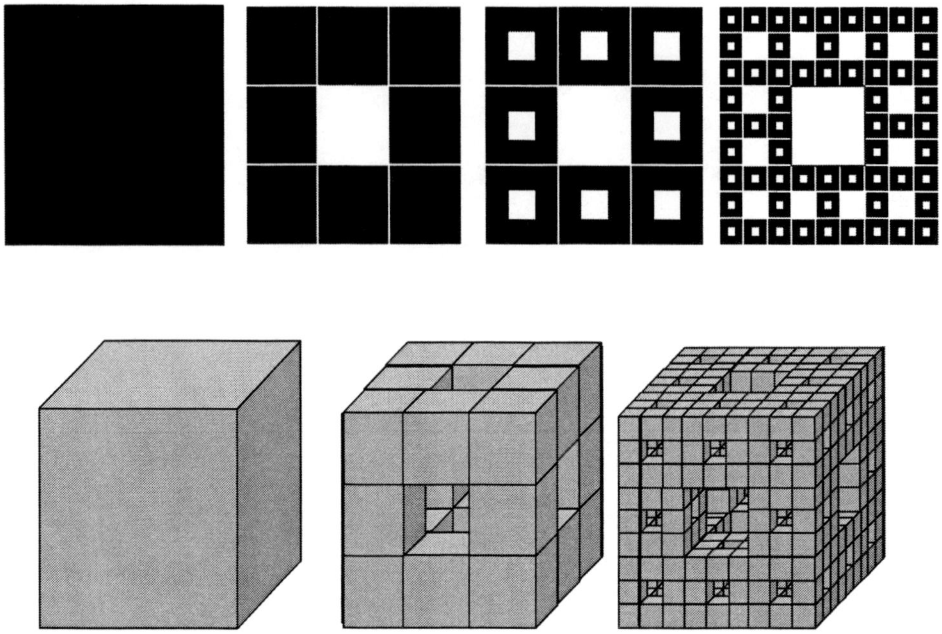

Fig. 3 Iterative construction of the fractal objects; top—first three stages of constructing fractal surface using the Sierpinski carpet model, bottom—first two stages of constructing fractal volume using the Menger sponge model.

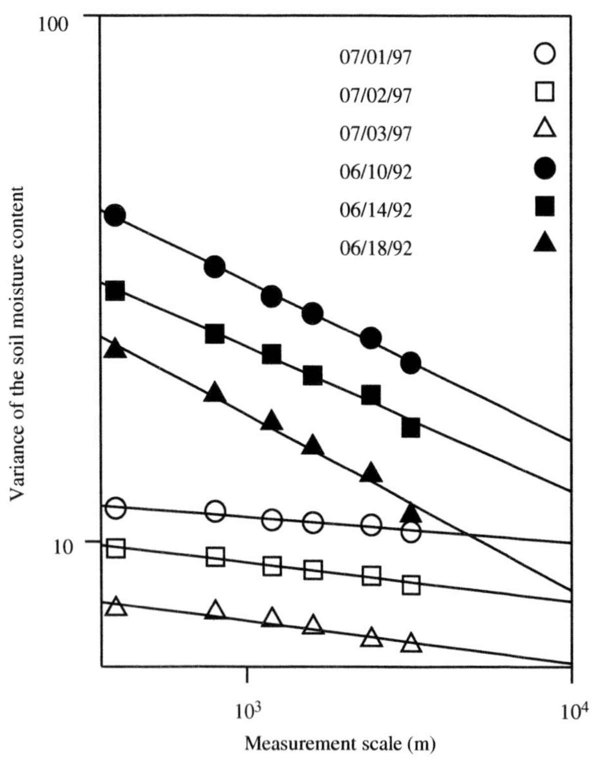

Fig. 4 Fractal scaling of the variance of soil surface moisture measured with passive microwave radiometry across Little Washita watershed in Oklahoma in 1992 and 1997 during drying periods. Observation dates are shown with different symbols, measurement scale is the pixel size.

dependencies of soil moisture contents to be self-affine, one can apply the self-affinity model as shown in Fig. 4.

It has been proposed that multiscaling, or multifractal, models expand the range of resolutions in which fractal scaling is applicable to soil moisture fields. Unlike the generation of Sierpinski carpet or Menger sponge fractals in which the similarity of structure persists across iterative steps, multifractals are generated by cascade processes that create more heterogeneity as the resolution decreases.

EMPIRICAL SCALING

Physics-based scaling is usually applicable within a range of scales covering one or two orders of magnitude. At larger scales, soil heterogeneity manifests itself by the appearance of new types of heterogeneity, and self-similar scaling ceases to be applicable. A further increase in the scale may reveal new scaling. Empirical observations are usually substituted to proceed from one level of physics-based scaling to another.

Transition from Laboratory to Plot Scale

Transition from the laboratory soil sample scale of 5–10 cm to the plot scale of 1–5 meters reveals a change in soil water retention. A regression equation has been developed to convert volumetric water content from the sample resolution to the plot resolution at the same soil matric potential $\theta_{\text{Plot}} = 1.287\theta_{\text{Lab}} - 0.977\theta_{\text{Lab}}^2$. Saturated hydraulic conductivity measured in soil cores increases according to the power law $K_{\text{sat}} \propto R^m$, where R is the core radius, $m = 0.3-0.4$ $(0.6-0.7)$ for coarse-textured (fine-textured) soils without macropores.

Using Pedotransfer Functions and Soil–Landscape Relationships

The use of pedotransfer functions, i.e., relationships between soil properties, is a common way to estimate effective soil hydraulic properties for large areas with data available from soil survey maps. Because the soil textural and other classes used in soil mapping are quite broad, nonlinear hydrologic models may give distorted answers if average class values are entered in pedotransfer functions. Soil–landscape relationships can be used to describe the variability within soil map units. In particular, spatial variations of soil hydraulic properties within extents of 10–1000 m correlate with terrain attributes (4).

Inverse modeling

Inverse modeling has been used to estimate effective soil hydraulic properties for large-area supports by matching measured and predicted soil surface moisture, runoff, or infiltration. Simulation studies show a good correspondence between weighted-average hydraulic parameters for individual soil units within the support and parameters found from the inverse modeling.

IMPORTANCE OF SCALING

Scaling relationships both enhance understanding of soil moisture dynamics and underlying processes and serve as a unique contrivance to overcome the mismatch of scales between measurements available in soil moisture studies, hydrological models, and applications of soil moisture parameters and values in environmental decision making. One should expect scaling of soil moisture to play a major role in the fast growing field of data assimilation, i.e. combining data obtained at different scales for environmental monitoring and prediction.

REFERENCES

1. Sposito, G., Eds. *Scale Dependence and Scale Invariance in Hydrology*; Cambridge University Press: Cambridge, MA, 1998; 423.
2. Giménez, D.; Perfect, E.; Rawls, W.J.; Pachepsky, Ya.A. Fractal Models for Predicting Soil Hydraulic Properties: A Review. Eng. Geol. **1997**, *48*, 161–183.
3. Hillel, D., Elrick, E., Eds. *Scaling in Soil Physics, Principles and Applications*; SSSA Spec. Publ. 25, Soil Science Society of America: Madison, WI, 1990; 122.
4. Kabat, P.; Hutjes, R.W.A.; Feddes, R.A. The Scaling Characteristics of Soil Parameters: From Plot Scale Heterogeneity to Subgrid Parameterization. J. Hydrol. **1997**, *190*, 363–396.
5. Pachepsky, Ya.A., Crawford, J.W., Rawls, W.J., Eds. *Fractals in Soil Science*; Elsevier: Amsterdam, 2000; 295.
6. Stewart, J.B., Engman, E.T., Feddes, R.A., Kerr, Y., Eds. *Scaling up in Hydrology Using Remote Sensing*; John Wiley & Sons: New York, 1996; 255.
7. Warrick, A.W.; Nielsen, D.R. Spatial Variability of Soil Physical Properties in the Field. In *Applications of Soil Physics*; Hillel, D., Ed.; Academic Press: New York, 1980; 319–344.

SEEPAGE

Patricia M. Gallagher
Drexel University, Philadelphia, Pennsylvania, U.S.A.

INTRODUCTION

The term *seepage* is used to describe the flow of water through the subsurface. Soil formations consist of soil particles with interconnected voids; seepage occurs through the voids. Seepage is important in engineering applications because it affects the stability of dams and levees, as well as dewatering of excavations during construction. Additionally, since groundwater is often used for water supply and irrigation, it has to be protected from overuse and contamination. This article provides a brief overview of seepage principles and applications. For more detailed information, see Refs. (1–8).

Water reaches the soil when precipitation percolates through the unsaturated zone under gravity flow until it reaches the *water table* or *phreatic surface*, which is the top of the zone of saturation and may vary seasonally. Once the water reaches the phreatic surface, seepage is typically horizontal through the soil. Eventually, the groundwater recharges streams and rivers or may be pumped from the ground as a water supply.

The quantity and speed of flow through a soil formation depend on the size of the voids through which the water flows. There are essentially two types of soil deposits: sand and clay. Sand deposits have bulky particles and larger voids than clay deposits so that more water tends to flow faster through sand than through clays. Deposits that readily transmit water are called aquifers, while aquicludes permit very little flow. Clays are often used as barriers to flow, e.g., in dams or as compacted liners while sand deposits often function as water-supply aquifers.

SEEPAGE PRINCIPLES

Water, both static and flowing, affects stresses and stability in soil. The static water pressure below the groundwater table increases linearly with depth according to the relationship

$$u = \gamma_w z$$

where u is the water pressure (lb/ft^2 or kN/m^2), γ_w is the unit weight of water (lb/ft^3 or kN/m^3), and z is the depth beneath the groundwater table (ft or m). When water pressure is described as $z = u/\gamma w$, it is called pressure head.

For water flowing through the subsurface the total energy potential is a combination of pressure head, elevation head and velocity head. Total head can be calculated using Bernoulli's equation

$$h = \frac{u}{\gamma_w} + z + \frac{v^2}{2g} = \text{constant}$$

where h is the total head, or the sum of the energy potentials, u is the water pressure (lb/ft^2 or kN/m^2), z is the elevation head (ft or m), v is the water velocity (ft/s or m/s), and g is the gravity (ft/s^2 or m/s^2). The velocity of groundwater is generally so small that velocity head can be considered to be zero for most flow problems.

While the total energy is constant at any point in the region of flow, the viscous resistance within the pores of the formation causes a loss of energy across the region of flow, described as *head loss*, Δh, as shown in Fig. 1. In Bernoulli's equation, the energy between points A and B can be written as

$$\frac{u_A}{\gamma_w} + z_A = \frac{u_B}{\gamma_w} + z_B + \Delta h$$

where Δh is the total head loss over the flow distance ΔL.

The ratio of head loss to the flow distance is called the *hydraulic gradient*, i:

$$i = \frac{\Delta h}{\Delta L}$$

While the velocity head can usually be considered zero in Bernoulli's equation, the flow velocity is very important for determining the quantity of flow moving through the soil formation. Darcy's law is used to relate the flow of water through soil to the hydraulic gradient.

$$v = ki$$

where v is the Darcy velocity or specific discharge (ft/s or m/s), and k is the hydraulic conductivity (ft/s or m/s).

The *Darcy velocity* is actually a quantity rather than a velocity, so it is more correctly called the *specific discharge* or *flux*. It accounts for the quantity of flow

Encyclopedia of Soil Science

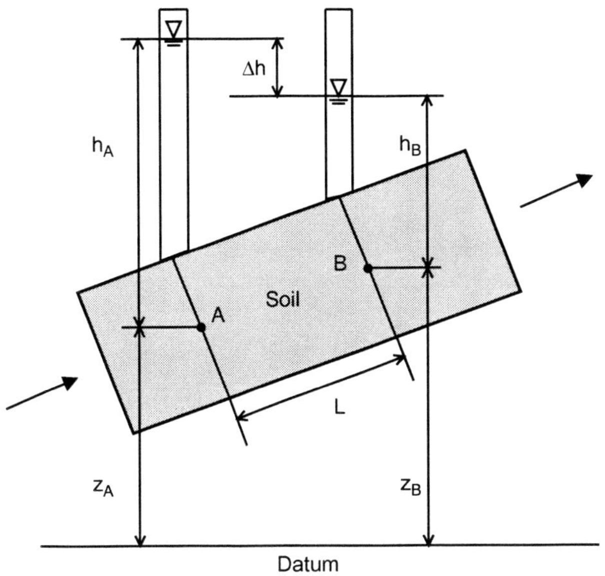

Fig. 1 Illustration of head loss between two points in a soil mass.

traveling through the entire cross-sectional area of flow, including both soil particles and voids. The *seepage velocity*, v_s, is the velocity at which water actually moves through the area occupied by the pore spaces. Therefore, it is always larger than the Darcy velocity. The area occupied by the pore spaces is typically quantified in terms of *porosity*, n, the ratio of the volume of voids to the total volume. Porosity ranges from about 25 to 50% for most soil deposits. The Darcy velocity and the seepage velocity are related by the porosity.

$$v = v_s n$$

Darcy's law, which is valid for most types of soil formations, is the basic equation used to describe groundwater flow.

HYDRAULIC CONDUCTIVITY

Hydraulic conductivity, a measure of water's ability to move through the soil, is the proportionality constant that relates the Darcy velocity to the hydraulic gradient. It describes the properties of both the formation and the fluid moving through it. A second parameter, *intrinsic permeability*, K, describes the properties of the media, such as the grain size and the tortuosity of the flow path. The hydraulic conductivity and the intrinsic permeability are related by the following equation

$$k = K\frac{\rho g}{\mu}$$

where ρ is the density of the permeant (kg/m^3) and μ is the viscosity of the permeant (Pa s or cP).

Hydraulic conductivity is highly variable spatially due to the heterogeneous nature of soil deposits and/or the presence of discontinuous layers or faults. Additionally, hydraulic conductivity varies with the direction of flow and can vary by orders of magnitude over the distance of a few inches or feet, even in soil deposits that appear to be fairly homogeneous. The hydraulic conductivity is a function of the void ratio and the grain size distribution of the formation. In general, sands have much higher hydraulic conductivity values than clays and can transmit large quantities of water. Typically, hydraulic conductivity values of coarse sands and gravels are in excess of $>10^{-3}$ m/s; fine sands range from 10^{-3} to 10^{-5} m/s; silty sands range from 10^{-5} to 10^{-7} m/s; silts range from 10^{-7} to 10^{-9} m/s; and clays tend to be $<10^{-9}$ m/s.

Hydraulic conductivity can be measured in either the laboratory or the field. Laboratory tests are done on small samples in a controlled environment, so while they may give accurate values, they cannot capture the variability of a deposit. Field methods may provide a better overall idea of the hydraulic conductivity of a formation, but they are expensive and can be difficult to interpret.

Two types of laboratory tests are commonly used to determine the hydraulic conductivity. The constant head test is used on highly permeable soils such as sand or gravel. The falling head test is more suitable for measuring the hydraulic conductivity of less permeable soils such as silt and clay. In both tests, the flow of water through a specimen of known area is measured and Darcy's law is used to calculate the hydraulic conductivity.

Field tests typically require several wells, including a central pumping well and surrounding observation wells. Water is pumped out of the pumping well and the water levels in the observation wells are measured when steady conditions are reached. The interpretation of pump tests is based on certain simplifying assumptions; if these assumptions are not valid for the formation in question, the results may be questionable. Single wells may also be used to measure the hydraulic conductivity. The tests are done by causing an instantaneous change in the water level, either through addition (a slug test) or removal (a bail test) of a known volume of water. The time taken by the water level to recover is measured and used to calculate the hydraulic conductivity. There are also indirect methods of assessing the hydraulic conductivity, including correlations based on the grain size distribution of the formation.

APPLICATIONS OF SEEPAGE THEORY

Dams and Levees

Dams are used for a variety of reasons, including flood control, water supply, and storage of liquid wastes. Levees are used primarily for flood control. Seepage through and beneath dams and levees is an important consideration in design. The primary concerns in seepage analysis are: 1) the amount of seepage that will go through or beneath the structure, 2) the potential for erosion and piping at the downstream end, and 3) the uplift pressure acting underneath the structure. If a dam is used for water supply, care must be taken to limit the amount of seepage through or beneath the dam, since water flowing beneath the dam cannot be used as water supply.

The simplest method for analyzing the flow through or beneath a dam is a *flow net.* An example of a flow net is shown in Fig. 2. A flow net is a graphical solution to the Laplace continuity equation, which describes steady flow in the x and y directions.

$$\frac{\partial^2 h}{\partial x^2} + \frac{\partial^2 h}{\partial y^2} = 0$$

As shown in Fig. 2, the solution consists of two families of mutually perpendicular lines: 1) *flow lines,* which show the path water travels along and 2) *equipotential* lines, which show the contours of total head. Rules for drawing flownets can be found in most textbooks on introductory soil mechanics. Computer programs may also be used to solve the Laplace equation, but the flow net is simple to use and, considering the uncertainty associated with the determination of the hydraulic conductivity, often accurate enough for most engineering analyses.

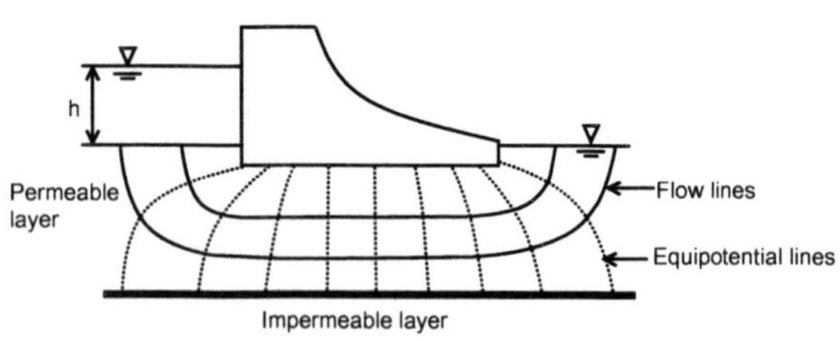

Fig. 2 Example flow net.

The flow net can be used to calculate the total head and the water pressure at any point beneath the structure, the seepage quantity and the risk of damage to the dam from erosion and piping. Piping is the loss of soil from beneath the dam. If enough soil is removed through erosion, channels can form in the foundation, allowing uncontrolled flow of water beneath the dam and often failure. Erosion is most likely to occur where the hydraulic gradient is highest, typically where the water exits from beneath the structure. If erosion is a problem, solutions include lowering the upstream water level, placing a filter at the downstream end of the dam, and increasing the length of the flow path by placing an impermeable layer at the upstream edge of the dam.

Water Supply and Groundwater Contamination

Groundwater must be properly managed as a resource. Aquifers are often the source of drinking and irrigation water. If over-pumped, resources could be depleted, causing loss of water supply, loss of base flow to streams or saltwater intrusion in areas near coastlines. Additionally, since groundwater contamination can affect water quality, it is necessary to protect against it.

Aquifers are either confined or unconfined. An unconfined aquifer is one in which the water level is below the top of the soil formation, so the water is under atmospheric pressure. If more water flows into the aquifer, the water level will rise. In contrast, a confined aquifer is one in which the top boundary is an aquiclude, confining the water in the aquifer and causing it to be under pressure. If a well is drilled into a confined aquifer, the water will rise above the top of the aquifer. If the water flows to the ground surface, the well is called artesian.

Although groundwater modeling is a complex three-dimensional problem, there are some simple analyses for modeling the flow of water to a single well. For a confined aquifer (Fig. 3), the flow rate (Q) from the well may be calculated using the following equation

$$Q = \frac{2\pi k H_a (h_0 - h_w)}{\ln\left(\frac{r_0}{r_w}\right)}$$

where H_a is the aquifer thickness, r_0 is the drawdown radius, h_0 is the water height at r_0, r_w is the well radius, and h_w is the water height at well.

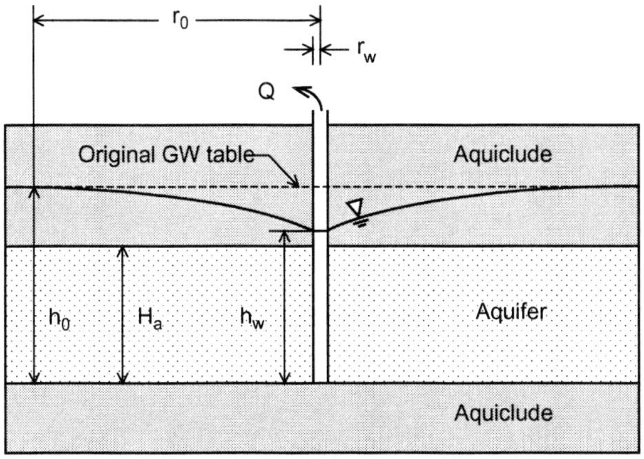

Fig. 3 Well in a confined aquifer.

For an unconfined aquifer (Fig. 4), the equation for water quantity is

$$Q = \frac{\pi k(h_0^2 - h_w^2)}{\ln\left(\frac{r_0}{r_w}\right)}$$

While these simple equations are useful to demonstrate the factors that influence flow to a well, most groundwater modeling requires the use of numerical codes. The complexity of subsurface conditions, the heterogeneity and anisotropy of the parameters involved in modeling and the uncertainty of the measurements make groundwater modeling extremely difficult.

As water resources dwindle and contamination affects the quality of drinking water supplies, proper management of groundwater resources becomes more important.

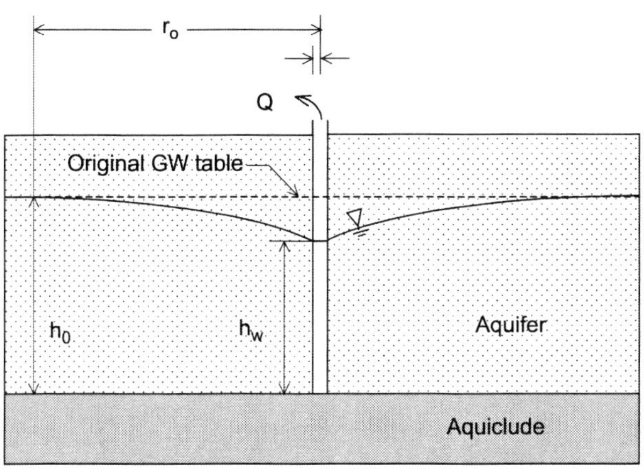

Fig. 4 Well in an unconfined aquifer.

Groundwater contamination can be caused by leakage from underground tanks, solid waste landfills or lagoons, and by deep-well injection of liquid wastes. In addition, animal wastes, fertilizers and pesticides may leach into the soil and contribute to groundwater contamination. The movement of contaminants through the subsurface depends on the type present.

Contaminants that dissolve in the groundwater are carried along with the flowing groundwater by the processes of advection, the bulk flow of contaminants along with the flowing groundwater, and dispersion, the spreading out of the contaminant in front of the advancing contaminant front. Dispersion is caused by the variation in pore size in the formation.

If contaminants are lighter than water and insoluble, they will float on top of the water table. These types of contaminants are called light nonaqueous phase liquids (LNAPLs). Contaminants that are heavier than water and insoluble (dense nonaqueous phase liquids, or DNAPLs) tend to sink to the bottom of the aquifer and get into larger pores and fractures. These types of contaminants can be very difficult to locate and treat.

Dewatering

Dewatering is another important application of seepage principles. When excavations are required below the groundwater table, dewatering is used to remove water from the excavation so that construction can occur in the dry. The basic types of construction dewatering methods include open pumping, predrainage, and cutoffs. Selecting an appropriate method of dewatering depends on numerous factors, including the hydraulic conductivity and soil conditions, the depth of the excavation below the groundwater table, and the size of the excavation. It can be a complex problem and is typically left to specialty dewatering contractors.

Open pumping consists of controlling the water from the excavation using trenches and sumps. The water is directed to the sumps through the trenches and removed from the excavation by pumping from the sumps. It is a simple method of dewatering and is most suited for use in excavations where the amount of flow is small. In cases where the hydraulic conductivity is high and large amounts of water could flow into an excavation, predrainage may be used. Predrainage is used to intercept the groundwater before it reaches the excavation by pumping from wells or wellpoints placed around the perimeter. It can involve pumping large quantities of water and has the potential to alter the regional groundwater level. In areas where the flow of water through the soil is significant, it can be difficult to keep an excavation dewatered. In cases like this, cutoffs may be used to prevent the water from entering the excavation. A cutoff, or seepage barrier, is designed to minimize the flow of water into an excavation. Typical examples include sheet pile walls and slurry walls. A slurry wall is a trench filled with a soil–bentonite backfill. It is excavated using a bentonite slurry to support the trench walls and then filled with a backfill that has a low hydraulic conductivity. Seepage barriers can be very effective in reducing the amount of water that can flow into the excavation.

CONCLUSION

Seepage is a complex process that can be difficult to quantify and analyze. In certain situations, such as flow through dams, simple analysis methods may be employed. More complex analysis is required for applications such as developing and managing groundwater resources. The difficulties associated with measuring the hydraulic conductivity accurately and accounting for the inherent variability of natural soil deposits remain the most challenging aspects of seepage analysis.

REFERENCES

1. Cedergren, H.R. *Seepage, Drainage, and Flow Nets*, 3rd Ed.; Wiley Classics in Ecology and Environmental Science; Wiley: New York, 1997; 496.
2. Coduto, D.P. *Geotechnical Engineering: Principles and Practices*; Prentice Hall: Upper Saddle River, NJ, 1999; 759.
3. Driscoll, F.G. *Groundwater and Wells*, 2nd Ed.; Johnson Division: St Paul, MN, 1986; 1089.
4. Duncan, J.M. *Seepage Course Notes*; Center for Geotechnical Practice and Research, The Charles, E. Via, Jr., Ed.; Department of Civil and Environmental Engineering, Virginia Polytechnic Institute and State University: Blacksburg, VA, 1997.
5. Fetter, C.W. *Contaminant Hydrogeology*; Prentice Hall: Upper Saddle River, NJ, 1993; 458.
6. Fetter, C.W. *Applied Hydrogeology*, 4th Ed.; Prentice Hall: Upper Saddle River, NJ, 2000; 691.
7. Freeze, R.A.; Cherry, J.A. *Groundwater*; Prentice Hall: Englewood Cliffs, NJ, 1979; 604.
8. Harr, M.E. *Groundwater and Seepage*; Dover Publications: New York, 1992; 315.

SELENIUM

S

Gunnar Gissel-Nielsen
Risoe National Laboratory, Roskilde, Denmark

INTRODUCTION

The traditional definition of a plant nutrient is an element essential for the growth and development of plants. However, with the increasing concern on the quality of agricultural crops as animal and human food, elements or compounds that improve the quality should be considered as plant nutrients, too. A clear example is selenium (Se).

Se has been recognized as an essential mineral in human and livestock nutrition for more than 30 years but never proved to be essential for plants. Lack of Se in food has often been viewed as an example of a mineral imbalance related to intensive crop production since the main source of Se for animal and humans is the soil–plant system.

This article explains why field treatment with Se is an inexpensive, safe, and easy way of ensuring a desirable Se intake in humans and animals.

GEOGRAPHICAL DISTRIBUTION OF Se

The lower and the upper limits for adequate Se supply are approximately 0.05 mg/kg dry matter and 2 mg/kg, respectively (1). Outside Europe, crops containing toxic Se concentrations are found in the Midwestern regions of the United States and Canadian prairies, in Venezuela, India, and China. Selenium deficiencies are more common and have been reported in both western and eastern coastal areas of North America, Venezuela, Australia, New Zealand, Japan, and China. However no information is available about the Se status of most countries of the world. In Europe, Scandinavia is a natural low-Se area, while central Europe ranges between deficiency and sufficiency. Selenium toxicity has only been observed in a few spots of Wales, Eire, and Russia (2).

Blood Se content is correlated with Se concentrations in the food for both humans and animals. Cereal and vegetables are dominant Se sources in the human diet and, along with pasture plants, in livestock fodder; the Se intake can be estimated to a great extent by evaluating the Se uptake by the fodder crops grown in the area in question. Consequently, an increase in the Se concentration of crops is an obvious way to remedy Se deficiency in humans or livestock (Fig. 1).

Se UPTAKE BY PLANTS

Low to moderate soil-Se is not correlated with plant-Se because of the large number of other factors influencing the availability of soil-Se to plants. In soils with high pH, inorganic Se occurs mainly as selenate, which is not fixed in the soil. Consequently, if precipitation and leaching are low, crops will be rich in Se. This is known in some places in India and in South Dakota, United States (3). On the other hand, a low pH favors the selenite form, which is fixed strongly to the soil clay particles and iron hydroxides. Consequently, soils with similar Se content will produce crops with much lower Se content in low pH soils than in high pH soils. Volatile Se is lost to the atmosphere through microbial activity, but Se also returns to the soil from the atmosphere through precipitation. This leaves only a minor part of Se in the cycle to pass through the plant–animal system.

The Se concentration in plants also depends on the time of sampling. The release of Se from clay and organic matter is a slow process. Selenium released during the winter is available for crops during spring when the yield of grass is low, giving a relatively high Se concentration. When the growth of grass increases during summer, less Se is available and it is distributed over a far greater yield. Therefore, the Se concentration drops as a result of dilution.

FIELD TREATMENT WITH Se

Different ways of raising Se concentrations of crops have been investigated over the years. Of all the different ways, application of selenate to the soil and foliar application of selenate or selenite are those of practical importance. Pre-sowing treatment of barley (*Hordeum vulgare* L.) and soybean (*Soya hispida* L.) seeds with selenite and selenate have also been tested.

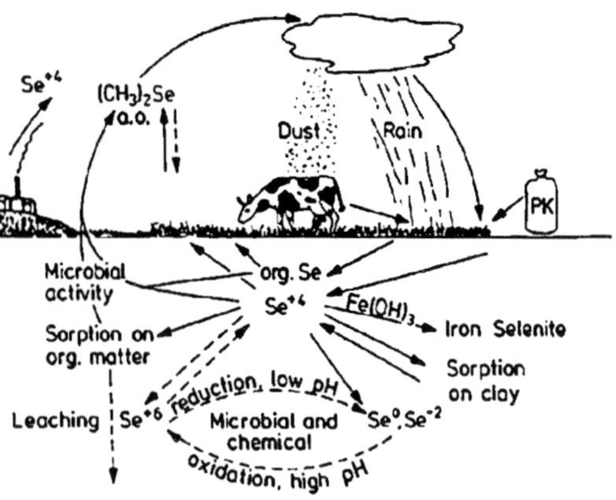

Fig. 1 Possible cycles of Se under field conditions with morainic soils. (From Ref. 1.)

Soil Application

The first studies on soil application of Se involved spraying of selenite or selenate solutions onto the soil surface in 1967 (4, 5). These relatively simple experiments showed the promising effects of Se addition, and they have been followed by comprehensive studies worldwide (2). Different Se salts have been tested in pot and field experiments. Selenates in general are 20 to 50 times more available than selenites, at least in the first year (1).

A large-scale field experiment involving an annual addition of 60 and 120 g Se/ha for 5 years as Na_2SeO_3 incorporated in a NPK compound fertilizer was carried out on 21 Danish farms in 1972–73 (3). The soils were all glacial deposit mineral soils with a pH of 5–7. The 120 g Se treatment raised the native Se concentration of 0.02–0.04 mg/kg of a number of fodder crops to 0.08–0.13 mg Se/kg, considered a sufficient and safe level for animal nutrition.

A strip of a paddock (5% of the area) was supplied with 340 g Se/ha, which is equivalent to 17 g Se/ha for the total area. The grazing sheep had the same blood Se content as sheep grazing a paddock with even distribution of the Se. The free movement of the grazing animals compensates for an uneven distribution of the Se (2). For perennial pastures, a single application of some hundred g Se per ha can give a desirable residual effect for 4–5 years.

Flue gas desulphurization waste, fly ash, and sewage sludge are examples of alternative sources of Se for field crops. They all resulted in varying increases in Se concentration of crops. In most cases, the amount of waste needed for remedying a Se deficiency situation was so high that other problems occurred, such as too high Cd concentrations or salinity problems. These waste products, therefore, could be acknowledged to have a positive effect on the Se concentration of crops when deposited in the field, but they should not be considered Se fertilizers as such (2).

Foliar Application

The complication of soil effect on the availability of added Se can be avoided by using foliar application. In a test of foliar applications of up to 50 g Se/ha as selenite in field experiments with spring barley, a linear correlation was found between added Se and Se concentration in the plants. Five to ten g Se/ha resulted in the desired minimum level of about 0.1 mg/kg Se in the grain. The increase in Se concentration was greater in the mature grain than in the straw, indicating an effective translocation of the absorbed Se to the edible part of the plant. Addition of a detergent to the Se solution improved the efficiency of foliar application, giving the desirable minimum concentration of about 0.1 mg/kg in the grain using 2–5 g Se per ha. The foliar application was performed at two growth stages (end of May and end of June). The late application also resulted in a two-fold Se increase. The likely explanation is a better plant cover of the soil and thereby a greater part of the solution wetting the leaves (2).

In another field experiment, foliar application in the spring of 10 g Se/ha as sodium selenite to an established pasture resulted in concentrations within the desirable range in all cuts during the growing season. The recovery of the added Se in crops was up to 35%.

Comprehensive Finnish field experiments included foliar applications of selenite and selenate. The results were in accordance with the earlier Danish studies: selenate is somewhat more effective than selenite, an amount of 5–10 g Se/ha is sufficient, and the effect depends on the stage of development of the crop at the time of spraying (6).

Seed Treatment with Selenium

Seed treatment with Se has been tested in barley and soybean (2, 7). The amount of selenite needed to obtain a desirable Se concentration in harvested barley grain was the same as when selenite was added through the fertilizer. The pre-sowing treatment of barley seed with selenite resulted in a concentration of several hundred mg Se/kg seed making it highly toxic to animals and humans.

Experience on a Larger Scale

In Finland, field treatment with Se is performed to an extent that enables study of overall long-term effects. Since 1985, all Finnish commercial fertilizers for food and feed crops are enriched with sodium selenate: 16 mg Se/kg fertilizer for cereals and horticulture crops; 6 mg Se/kg fertilizers for fodder beets and grass production. The Se concentrations in crops are monitored carefully and Se content in all crops are elevated, typically from about 0.01 to 0.1–0.3 mg/kg dry matter (8).

It was decided to remove fertilizers containing 16 mg Se/kg from the market, leaving only those having 6 mg/kg for all crops. This change stabilized the average Se concentrations of Finnish crops at a somewhat lower level: about 0.1–0.2 mg/kg in cereal grain, and about 0.13 and 0.17 in hay and grass silage, respectively. In 1998 the Finnish authorities decided to increase the Se content of fertilizers to 10 mg Se/kg. In any case, Finland has become a country producing crops with adequate Se levels for livestock and human nutrition instead of a Se-deficient area.

Main Guidelines

All these experiments leads to the following conclusion: Foliar application of about 5 g Se/ha as selenite or selenate, and soil fertilization using about 10 g Se/ha as selenate, or about 120 g Se/ha as selenite, are effective annual treatments for raising the Se content of annual crops to a desirable level for human and animal nutrition. The effect of Se is enhanced when it is used with a detergent for foliar application, and for all treatments the effect is greatest when carried out on a well-established crop. The residual effect of these treatments is very small. A somewhat higher amount is needed for pasture crops, but this gives a residual effect lasting 2–3 years.

BIOAVAILABILITY OF Se

Whether Se is offered in pre-mixed fodder, in concentrates, mineral supplements, water, lick-stones, or ruminant pellets, it occurs as inorganic selenite, whereas the Se in crops occurs predominantly as free seleno-amino acids or in proteins (2). Organic Se is often found to be more effective than inorganic Se to prevent Se deficiency in livestock. However, the significance of organic versus inorganic Se in animal nutrition is still subject to much discussion, so one cannot recommend a single Se compound as the best in all situations.

It is well established that the bioavailability of different Se compounds varies greatly and this has been shown in several experiments. Such variation stresses the importance of the form of Se in animal and human foodstuffs, and research has been carried out to evaluate the factors that influence the chemical form of Se in plants.

Earlier short-term experiments carried out with tomato roots (*Lycopérsicum esculéntum Mill.*) and other experiments with maize (*Zea mays* L.), showed that Se is translocated in the xylem sap as selenate when added to the nutrient solution as selenate, whereas selenite is metabolized immediately to Se-amino acids and translocated as such (2).

To evaluate the significance of these findings, rye grass (*Lolium perenne* L.) and barley were supplied Se as selenite or selenate through the roots or by foliar application in a pot experiment (2). A series of fractionations of Se compounds in grass and barley grain separated the compounds into total and water-soluble Se proteins, Se amino acids, selenite, and selenate. Neither the oxidation state of the added Se nor the method of application had any significant effect on the distribution of Se in the various compounds. In all four cases only about 10% of Se was present in an inorganic form. This implies that any method of supplementing crops with inorganic Se leads to the same Se compounds in the mature plants. Therefore, the effect of different methods can be evaluated purely on the basis of total uptake as a percentage of the total added.

CONCLUSION

Se toxicity in humans and livestock is rare, and only China has reported severe cases of deficiency leading to the death of many humans. Contrary to this, moderate Se deficiencies leading to problems in animals are seen in many countries, and these deficiencies might be more dangerous to human health than often considered. As described earlier, in view of the high rates of cardiovascular disease and certain forms of cancer in Finland at the time, the Finnish authorities legislated introduction of selenate in all compound fertilizers sold in Finland since 1985. In New Zealand selenized compound fertilizers are available, and in some countries (e.g., Sweden and the United Kingdom), microelement solutions including Se are on the market for foliar application. In many countries pre-mixed feed enriched with inorganic Se is available for livestock production. This reflects the growing concern for Se deficiency.

Selenium is available in tablet form for human consumption in most countries, and in general a much more varied diet including relatively Se-rich foodstuffs such as fish and some vegetables are consumed by humans, but far more attention is paid to the Se nutrition of animals than to the human population. In future it is likely that increased attention will be paid to the importance of sufficient Se in human diets, as well as in livestock feed.

This review demonstrates that a reasonable Se supplementation of crops is an inexpensive, safe, and easy way of ensuring a desirable Se intake in humans and animals.

REFERENCES

1. Gissel-Nielsen, G.; Gupta, U.C.; Lamand, M.; Westermarck, M. Selenium in Soils and Plants and Its Importance in Livestock and Human Nutrition. Advances in Agronomy **1984**, *37*, 397–460.
2. Gissel-Nielsen, G. Effects of Selenium Supplementation of Field Crops. In *Environmental Chemistry of Selenium*; Frankenberger, W.T., Jr.; Engberg, R.A., Eds.; Marcel Dekker, Inc.: New York, 1998; 99–112.
3. Gissel-Nielsen, G. *Control of Selenium in Plants*, Risø Report No. 370; Roskilde: Denmark, 1977; 1–42.
4. Cary, E.E.; Wiecorek, G.A.; Allaway, W.H. Reactions of Selenite-Selenium Added to Soils that Produce Low-Selenium Forage. Soil Sci. Soc. Am. Proc. **1967**, *31*, 21–26.
5. Watkinson, J.H.; Davies, E.B. Uptake of Native and Applied Selenium by Pasture Species. III. Uptake of Selenium from Various Carriers. N.Z.J. Agric. Res. **1967**, *10*, 116–121.
6. Yläranta, T. Effect of Selenite and Selenate Fertilization and Foliar Spraying on Selenium Content of Timothy Grass. Ann. Agric. Fenn. **1984**, *23*, 96–108.
7. Gupta, U.C.; MacLeod, J.A. Relationship Between Soybean Seed Selenium and Harvested Grain Selenium. Can. Soc. Soil Sci. **1999**, *79*, 221–223.
8. Jukola, E.; Hakkarainen, J.; Saloniemi, H.; Sankari, S. Effect of Selenium Fertilization on Selenium in Feedstuffs and Selenium, Vitamin E, and Beta-Carotene Concentrations in Blood of Cattle. J. Dairy Sci. **1996**, *79*, 831–837.

SERPENTINITIC SOILS

S

R. Burt
M.A. Wilson
United States Department of Agriculture (USDA), Lincoln, Nebraska, U.S.A.

INTRODUCTION

Serpentinite is a rock consisting primarily of serpentine minerals, e.g., chrysotile, lizardite, and antigorite, with the generalized chemical formula [$Mg_3Si_2O_5(OH)_4$], derived from the alteration (serpentinization) of such minerals as olivines, pyroxenes, and amphiboles in dunites, peridotites, pyroxenites, and other ultramafic rocks (1). Lizardite and antigorite have platy morphology, whereas chrysotile has tubular or fibrous morphology. Fibrous chrysotile is a mineral of the asbestos group (1) and is known to cause serious human-health problems (2, 3).

Currently, U.S. *Soil Taxonomy* (4) identifies serpentinitic soils at the family level as having magnesic mineralogy "any particle-size class, except for fragmental, with >40% by weight magnesium-silicate minerals, such as serpentine minerals (antigorite, chrysotile, lizardite) plus talc, olivines, Mg-rich pyroxenes, and Mg-rich amphiboles, in the fine-earth fraction (<2 mm)." With little substitution of Fe for Mg in serpentine, Fe in peridotite minerals is incorporated into common accessory minerals such as magnetite and hematite (5).

IDENTIFICATION OF SERPENTINE MINERALS

Serpentine minerals present quantification problems by X-ray diffraction (XRD) or optical analysis without labor-intensive pretreatments to remove Fe oxides from mineral surfaces. Oxides attenuate peaks during XRD analysis and mask mineral grains during optical analysis of sands or silts, preventing proper identification. Additionally, crystalline units of individual serpentine minerals are generally below the observational limit of the optical microscope, and two or more mineral species are often intermixed in sand-sized grains (6, 7). This characteristic thus requires the use of an electron microscope for exact mineral identification and limits the utility of the optical microscope in quantification of individual mineral species. Thermogravimetric analysis (dehydroxylation at 600 to 650°C) and total analysis are other methods used in the identification and quantification of serpentine minerals (8), typically requiring no pretreatments and allowing determination on the entire <2-mm fraction.

SOIL PROPERTIES

Ultramafic rocks occupy <1% of the land surface of the Earth but are found in many parts of the world, e.g., North America, Europe, continental Asia, Japan, Australia, New Zealand, Africa, Brazil, Puerto Rico, and Cuba (9). "Serpentine barrens" is a common term for soils developed from ultramafic (serpentinized) rock (10–12). Serpentinitic soils are recognized for their unusual, diverse, and sparse flora (9). Vegetation on these soils has been linked to nutrient deficiencies and imbalances (low N, P, K; adverse Ca/Mg ratios), non-anthropogenic metal enrichments (Ni, Cr, Co), low-water holding capacity, and susceptibility to erosion, with resulting slope instability and/or landsliding (13–18). Vegetation on serpentinitic soils in the Siskiyou-Klamath Mountains of southwest Oregon is reduced in kind and number, relative to surrounding landscapes, with broad-leaved trees typically absent and shrubs and pines of greater importance (12).

Fertility studies of serpentinitic soils documented Ca/Mg ratios <0.7 (9, 13, 19), considered unfavorable for the growth of most plants (20–22). Plant deficiencies of Ca can be either the result of low Ca and/or excess Mg, antagonistic to the uptake of Ca by plants (23). Low Ca^{2+} can also result in lowered ability to counteract the adverse effects of such ions as Na^+, Mg^{2+}, and H^+ (24, 25). In a study of 22 soils on serpentinized peridotite in California (14), productivity (timber-yield index) was related more to Ca/Mg ratio in surface horizons than to any combination of climatic and physical soil variables or Fe, Mn, Cr, Ni, or Co content. Research on associated mafic and serpentinitic soils in Maryland found extractable Ca/Mg ratios <0.10 have a 98% probability of having formed from serpentine rather than from nonserpentine mafic parent materials (26). A Ca/Mg ratio <0.1 is used in the Magnesic Great Group in the Australian Soil Classification System (27) but this ratio has yet to be incorporated into U.S. *Soil Taxonomy* (4, 12).

High levels of Ni, Cr, and Co in some serpentinitic soils have been linked to potential toxicity and/or antagonistic effects upon other ions (9, 22, 28, 29). Many studies have focused on Ni because of characteristic symptoms exhibited in plants growing on these soils (30). Ni substitutes for Mg in the serpentine crystal structure (5) and is generally considered a more likely candidate for chemical toxicity because of its greater availability to plants than Cr and because of its greater abundance than Cr and Co, respectively (9). A study of serpentinitic subalpine soils in Switzerland (31) showed exchangeable Ni $>0.1\ \mathrm{mmol\ kg^{-1}}$ and exchangeable Cr $<0.01\ \mathrm{mmol\ kg^{-1}}$. Most Cr was located in the structure of primary minerals (e.g., garnet, pyroxene, spinels) and secondary minerals (Fe oxides). However, research on serpentinitic soils in northwest Spain (32) showed the foliage of some agricultural crops (e.g., sugarbeet, cabbage, pasture) accumulates considerable Ni and Cr, despite low to moderate EDTA-extractable amounts of both elements in the soils.

Field observations of ultramafic rocks often indicate widely differing weathering rates within the same general area (19, 33). Elemental distribution (e.g., Mg, Ni, Cr) and mineral alteration have been used as indicators of pedogenesis of ultramafic soils. As Mg leaches from the soil profile, such elements as Fe, Al, and Si may concurrently be enriched (34). Ni, Co, and Mn generally increase with depth, except in the driest environments, whereas Cr is less mobile, tending to accumulate in upper parts of the soil profile, associated with high levels of Fe oxides (31, 35). Mineral transformations during pedogenesis of ultramafic soils are determined by such factors as topography and the climate. A study of serpentinitic soils in Oregon (36) shows smectite formation promoted on wetter landscape positions and concentrated in soils on footslopes as opposed to higher landscape positions. Similarly, Mg-rich smectite is found in poorly drained landscape positions of California serpentine terrain, and is attributed to neosynthesis in an environment with high pH and accumulations of Mg and Si by subsurface flow of soil water (37). A pedogenic study of ultramafic soils along a climatic gradient in southwestern British Columbia shows nearly equal proportions of hydroxyl-interlayered vermiculite/smectite, chlorite, serpentine, and talc in the wet-cool zones, whereas serpentine, talc, chlorite, and only small amounts of smectite are found in the dry-cool zones (38). In a soil catena on serpentinite in northwestern Italy, serpentine minerals appear to weather to low charge vermiculite in upper and drier horizons or to smectite in poorly drained conditions (39).

MANAGEMENT STRATEGIES

Timber on some serpentine soils (40, 41) is considered impractical to manage because of low productivity and sparse stands, due to inherent infertility and nutrient imbalance, complicated by susceptibility to landsliding and erosion. However, other studies (14) suggest that on some old peridotite terrains, which are at least partially serpentinized, the timber may be as productive as timber on some soils on more silicic (less mafic) rocks in the area. Little information exists on agricultural soils and the effect of their management on heavy metal availability (30, 32), due in part to the limited agronomic use of these soils (9, 32). Some studies (32) have shown however that despite moderate amounts of soil-extractable heavy metals, such crops as cabbage (*Brassica oleracea*), barley (*Secale cereale*), and pasture grown on managed serpentinite soils can accumulate significant amounts of Cr and Ni, with sugarbeet (*Beta vulgaris*) accumulating dangerous levels of Ni. Some management strategies that help reduce plant availability of Ni and Cr include liming and additions of organic matter, other plant nutrients (N, K, Ca), and phosphate fertilizers (42–45).

REFERENCES

1. Bates, R.L., Jackson, J.A., Eds. *Glossary of Geology*, 3rd Ed.; American Geological Institute: Alexandria, Virginia, 1987.
2. Skinner, H.C.W.; Ross, M.; Frondel, C. *Asbestos and Other Fibrous Minerals*; Oxoford Univ. Press: New York, 1988.
3. Schreier, H.E.; Omueti, J.A.; Lavkulich, L.M. Weathering Processes of Asbestos-Rich Serpentinitic Sediments. Soil Sci. Am. J. **1987**, *51*, 993–999.
4. Soil Survey Staff, *Soil Taxonomy. A Basic System of Soil Classification for Making and Interpreting Soil Surveys*; United States Department of Agriculture, Natural Resources Conservation Service Govt. Print. Office: Washington DC, 1999; 869.
5. Deer, W.A.; Howie, R.A.; Zussman, J. Serpentine. *An Introduction to the Rock-Forming Minerals*, 2nd Ed.; Longman Scientific and Technology: Essex, England, 1992; 344–352.
6. Kerr, P.F. *Optical Mineralogy*; 4th Ed.; McGraw-Hill Publ. Co.: New York, 1977.
7. Dixon, J.B. Kaolin and Serpentine Group Minerals. In *Minerals in Soil Environment*, 2nd Ed.; Dixon, J.B., Weed, S.B., Eds.; Soil Sci. Soc. Am. Madison, Wisconsin, 1989; 467–525.
8. Karathanasis, A.D.; Harris, W.G. Quantitative Thermal Analysis of Soil Materials. In *Quantitative Methods in Soil Mineralogy*; Amonette, J.E., Zelazny, L.W.; Eds.; Soil Sci. Soc. Am. Misc. Publ. Madison, Wisconsin, 1994; 360–411.
9. Brooks, R.R. *Serpentine and Its Vegetation. A Multidisciplinary Approach*; Ecology, Phytogeography, and

S

Physiology Series, Dioscorides Press: Portland, Oregon, 1987; Vol. 1, 454.
10. Kruckeberg, A.R. Plant Life on Serpentinite and Other Ferromagnesian Rocks in Northwestern North America. Syesis **1969**, *2*, 14–114.
11. Whittaker, R.H. The Ecology of Serpentine Soils: IV. The Vegetation Response to Serpentine Soils. Ecology **1954**, *35*, 275–288.
12. Whittaker, R.H. Vegetation of Siskiyou Mountains, Oregon and California. Ecology Monograph **1960**, *30*, 279–338.
13. Alexander, E.B.; Wildman, W.E.; Lynn, W.C. Ultramafic (Serpentinitic) Mineralogy Class. In *Mineral Classification of Soils*; Kittrick, J.A.; Ed.; Am. Soc. Agron. and Soil Sci. Soc.Am Spec. Publ. No. 16: Madison, Wisconsin, 1985; 135–146.
14. Alexander, E.B.; Adamson, C.; Graham, R.C.; Zinke, P.J. Soils and Conifer Forest Productivity on Serpentinized Peridotite of the Trinity Ophiolite, California. Soil Sci. **1989**, *148*, 412–423.
15. Jenny, H. *The Soil Resource: Origin and Behavior*; Ecol. Series. No. 37, Springer-Verlag: New York, 1980.
16. Kruckeberg, A.R. *California Serpentines: Flora, Vegetation, Geology, Soils, and Management Problems*; Univ. Calif. Publ. in Botany, Univ Calif. Press: Berkeley, 1984; Vol. 78.
17. Proctor, J. The Plant Ecology of Serpentine: II. Plant Response to Serpentine Soils. J. Ecol. **1971**, *59*, 397–410.
18. Alexander, E.B. Morphology, Fertility, and Classification of Productive Soils on Serpentinized Peridotite in California (U.S.A.). Geoderma **1988**, *41*, 337–351.
19. Johnston, W.R.; Proctor, J. Ecological Studies on the Lime Hill Serpentine, Scotland. Trans. Proc. Bot. Soc. Edinb **1979**, *43*, 145–150.
20. Proctor, J.; Woodell, S.R.J. The Ecology of Serpentine Soils. Adv. Ecol. Res. **1975**, *9*, 255–366.
21. Woodell, S.R.J.; Mooney, H.A.; Lewis, H. The Adaptation to Serpentine Soils in California of the Annual Species *Linanthus Androsaceus*(Polemoniaceae). Bull Torrey Bot. Club **1975**, *102*, 232–238.
22. Woolhouse, H.W. Toxicity and Tolerance in the Responses of Plant to Metals. In *Encyclopedia of Plant Physiology*; Lange, O.L. Ed.; Springer: New York, 1983; Vol. 12C.
23. Proctor, J. The Plant Ecology of Serpentine, III. The Influence of a high Calcium/Magnesium Ratio and High Nickel and Chromium Levels in Some British and Swedish Serpentine Soils. J. Ecol. **1971**, *59*, 827–842.
24. Wallace, A.; Frolich, E.; Lunt, O.R. Calcium Requirements of Higher Plants. Nature **1966**, *209*, 634.
25. Wyn Jones, R.G.; Lunt, O.R. The Function of Calcium in Plants. Bot. Rev. **1967**, *33*, 407–426.
26. Rabenhorst, M.C.; Foss, J.E. Soil and Geologic Mapping Over Mafic and Ultramafic Parent Materials in Maryland. Soil Sci. Soc. Am. **1981**, *45*, 1156–1160.
27. Isbell, R.F. *The Australian Soil Classification*; CSIRO Publ. Collingwood, Australia, 1996; 143.
28. Johnston, W.R.; Proctor, J. Growth of Serpentine and Non-Serpentine Races of *Festuca Rubra* in Solutions Simulating The Chemical Conditions in a Toxic Serpentine Soil. J. Ecol. **1981**, *69*, 855–869.
29. Wild, H. Indigenous Plants and Chromium in Rhodesia. Kirkia **1974**, *9*, 233–241.
30. Proctor, J.; Baker, A.J. The Importance of Nickel for Plant Growth in Ultramafic (Serpentine) Soils. In *Toxic Metals in Soil-Plant Systems*; Ross, S.M. Ed.; John Wiley & Sons: New York, 1994, 417–432.
31. Gasser, U.G.; Juchler, S.J.; Hobson, W.A.; Sticher, H. The Fate of Chromium and Nickel in Subalpine Soils Derived from Serpentinite. Can. J. Soil Sci. **1995**, *75*, 187–195.
32. Fernandez, S.; Seoane, S.; Merino, A. Plant Heavy Metal Concentrations and Soil Biological Properties Agricultural Serpentine Soils. Commun. Soil Sci. Plant Anal. **1999**, *30* (13&14), 1867–1884.
33. Proctor, J.; Woodell, S.R.J. The Plant Ecology of Serpentine Soil. Adv. Ecol. Res. **1971**, *9*, 255–366.
34. Wildman, W.E.; Jackson, M.L.; Whittig, L.D. Iron-Rich Montmorillonite Formation in Soils Derived from Serpentinite. Soil Soc. Am. J. **1968**, *32*, 787–794.
35. Bulmer, C.E.; Lavkulich, L.M. Pedogenic and Geochemical Processes of Ultramafic Soils Along a Climatic Gradient in Southwestern British Columbia. Can. J. Soil Sci. **1994**, *74*, 165–177.
36. Istok, J.D.; Edward, M.E. Influence of Soil Moisture on Smectite Formation in Soils Derived from Serpentinite. Soil Sci. Soc. Am. J. **1982**, *46*, 1106–1108.
37. Senkayi, A.L. Clay Mineralogy of Poorly Drained Soils Developing from Serpentinitic Rocks Ph.D Thesis Univ. Calif. Press: Davis. 1977.
38. Bulmer, C.E.; Lavkulich, L.M.; Shreier, H.E. Morphology, Chemistry, and Mineralogy of Soils Derived from Serpentinite and Tephra in Southwestern British Columbia. Soil Sci. **1992**, *154*, 72–82.
39. Bonifacio, E.; Zanini, E.; Boero, V.; Franchini-Angela, M. Pedogenesis in a Soil Catena on Serpentinite in North-Western Italy. Geoderma **1997**, *75*, 33–51.
40. Mason, R.S. *Geology, Mineral Resources and Rock Material of Curry County, Oregon*; Bulletin 93, Department of Geology and Mineral Industries: Oregon, 1977; 1–10.
41. Curry County Oregon Soil Survey; United States Department of Agriculture, Natural Resources Conservation Service, unpubl.
42. Halstead, R.L.; Finn, B.J.; MacLean, A.J. Extractability of Nickel Added to Soils and Its Concentration in Plants. Can. J. Soil Sci. **1969**, *49*, 335–342.
43. Hunter, G.R.; Vergnano, O. Nickel Toxicity in Plants. Ann. Appl. Biol. **1952**, 279–284.
44. Crooke, W.M. Effect of Soil Reaction on Uptake of Nickel from a Serpentine Soil. Soil Sci. **1956**, *81*, 269–276.
45. Pais, I.; Jones, J.B. *The Handbook of Trace Elements*; St. Lucie Press: Boca Raton, Florida, 1997; 223.

SESQUIOXIDES

Joey N. Shaw
Auburn University, Auburn, Alabama, U.S.A.

Larry T. West
University of Georgia, Athens, Georgia, U.S.A.

INTRODUCTION

Oxides, oxyhydroxides, or hydroxides (referred to collectively as oxides) of iron (Fe), aluminum (Al), and manganese (Mn), are termed sesquioxides. Oxide minerals largely affect soil chemical and physical properties, morphology, and classification. These minerals occur in soils as cations in four- (tetrahedral) and sixfold (octahedral) coordination with oxygen and hydroxyl groups. Substantial ionic substitution occurs in these minerals mostly due to the ability of many cations to reside in octahedral coordination. Most oxide minerals in soils form pedogenically after the release of these elements from primary mineral weathering, but a few Fe (e.g., magnetite) and Al (e.g., corundum) oxide minerals are predominantly of primary origin (inherited from the parent material). Some oxide minerals require an oxidizing (well-drained) soil environment for stability (e.g., most Fe and Mn oxides), while others do not (Al oxides). Virtually all soils possess sesquioxide minerals, with the possible exception of extremely young soils, extremely calcareous soils, or soils that possess seasonal high water tables for extended periods. In seasonally wet soils, transformation and translocation of the redox sensitive Fe and Mn results in soil wetness (redoximorphic) features that are used to evaluate depths to seasonally high water tables. Iron oxides in particular, and Mn oxides in some instances, impart much color to soils. Substantial coulombic and specific adsorption of organic molecules, cations, and anions occur on sesquioxides. Research suggests most sesquioxides possess no permanent charge, but develop surface charge dependent on the pH and ionic strength of the soil environment. Readers are encouraged to consult these following sources for additional information: Hsu (1), Schwertmann and Taylor (2), and McKenzie (3).

IRON OXIDES

Iron Oxide Minerals

Most Fe oxides in soils form by pedogenic processes. In an oxidizing environment, Fe released from the weathering of primary minerals such as hornblende and biotite coordinates octahedrally with oxygens and/or OHs to form Fe oxide minerals. Goethite (α-FeOOH) and hematite (α-Fe_2O_3) are common hexagonal close packed Fe oxides. Because Al^{3+} is similar in size to Fe^{3+}, substantial Al substitution occurs in both of these minerals (2). The cubic close packed lepidocrocite (γ-FeOOH), a polymorph of goethite found in periodically saturated soils, tolerates little Al substitution within its structure. Ferrihydrite, a para-crystalline Fe oxide, along with amorphous Fe forms, are generically called "soil Fe." No consensus on the formula and structure of ferrihydrite and other "soil Fe" phases exist, but it is believed these forms control Fe activities in soil solutions. Ferrihydrite is also considered a necessary precursor to hematite formation. Magnetite (Fe_3O_4) is a primary mineral, and most evidence suggests that it oxidizes to maghemite (Fe_2O_3) in soils. This oxidation may be facilitated in near-surface environments by fire (4). Although most Fe oxides possess some degree of magnetism, magnetite and maghemite exhibit the strongest magnetic properties.

Properties of Iron Oxides

Fe oxide amount and type dictates subsurface soil colors because their fine-grained nature allows them to coat other soil minerals. Red soil colors are due to the presence of hematite, yellow to brown colors indicate the presence of goethite, and orange colors arise from the presence of lepidocrocite (Table 1; Fig. 1a). Iron oxide color is also dependent on crystal size, degree of crystallinity, and the amount of ionic substitution (mostly Al) within the mineral

Encyclopedia of Soil Science

Table 1 General properties of oxide minerals commonly found in soils

Element	Common oxide form in soil	Color	Soil environments	Major effects on soil properties
Fe	Goethite, α-FeOOH	Yellow–brown	Oxidizing, many environments	Color, aggregation, sorption of cations, anions, and organics
Fe	Hematite, α-Fe_2O_3	Red to purple	Advanced weathering, oxidizing, warm and humid	Color, aggregation, sorption of cations, anions, and organics
Fe	Ferrihydrite, $Fe_5HO_8{\cdot}4H_2O$	Reddish-brown	Wet, rapid Fe release and oxidation, young soils	Color, aggregation, sorption of cations, anions, and organics
Al	Gibbsite, $Al(OH)_3$	Colorless	Advanced weathering or rapidly leached aluminous environments	Aggregation, sorption of anions and cations
Mn	Many minerals	Black to dark brown	Oxidizing	Color, sorption of anions and cations

(Adapted partly from Ref. 11.)

structure (2). Soil colors are often the result of mixtures of Fe oxide minerals, and certain Fe oxides (e.g., hematite) possess greater pigmenting ability than others (2). Many techniques have been developed that estimate soil Fe oxide content from either Munsell color notation or spectral reflectance values acquired via remote sensing. Soil color is also used as a differentiating characteristic at many levels of soil classification schemes, including soil taxonomy. For example, Rhodic subgroups of Ultisols and Alfisols possess dark red hematite enriched argillic horizons, while wet soil groups possess features related to Fe transformations in seasonally saturated soils.

Wet Soil Features

A discussion of Fe (and Mn) oxides is not complete without a reference to biochemical reduction affects on their solubility. Iron and Mn possess oxidation states that are dependent on the redox potential of the soil environment, which is largely controlled by soil saturation and microbial activity (5). When soils become saturated for significant time periods, aerobic microbes deplete the available O_2 and cause oxidized soil components such as Fe^{3+} and Mn^{4+} in oxide minerals to undergo chemical reduction induced by anaerobic microbial respiration and relatively high labile organic carbon content. As Fe and Mn are reduced, the oxide minerals bearing these elements become soluble. These elements can re-oxidize as water tables recede, precipitate as reduced forms in sulfide minerals, be adsorbed onto exchange sites, or move from one part of the soil to another by mass flow and diffusion. Reductive dissolution and loss of Fe and Mn oxide grain coatings exposes the silicate minerals that are commonly gray to white, resulting in the formation of redox depletions (see redox depletion in Fig. 1a). Oxidation of Fe and Mn in another part of the soil results in Fe and Mn oxide concentrations and a general reddening to blackening of the area. These depletions and concentrations, or redoximorphic features, are commonly used to interpret the depth to a seasonally high water table in soils (Fig. 1a and b).

Chemical and Physical Properties of Iron Oxides

Iron oxides greatly influence soil physical and chemical properties due to their charge characteristics, fine-particle size, and high surface area. Generally, Fe oxides possess surfaces that protonate (gain H^+), deprotonate (lose H^+), and develop charge based on soil solution pH and ionic strength. The pH at which the negative charge equals the positive charge has been termed the point of zero charge (PZC). Soil pH levels below this point result in (+) charged oxide surfaces, while the pH levels above the PZC result in (−) charged surfaces (Fig. 2a) (6). Fe and Al oxides tend to have higher PZCs than Mn oxides. Because the PZC of most soil Fe oxides occurs above pH 7, many soils with pH < 7 have Fe oxides with (+) charged surfaces. Therefore, many highly weathered soils with high Fe (and Al) oxide content have substantial ability to sorb anions on their surfaces, and possess appreciable anion exchange capacity (AEC). Phosphorus (P) and other anions are also sorbed through specific mechanisms involving coordination of the anion to the Fe within the oxide mineral (Fig. 2b). Because of this, P

(a) (b) (c)

Fig. 1 (a) Hematite (H), goethite (G), and redox depletions (D) in a Plinthic Paleudult; bar = 20 mm. (b) Photomicrograph (plane polarized light) of an Fe oxide concentration (C) on a pore wall of a Bw horizon in a Fluventic Dystrudept; bar = 0.2 mm. (c) Mn oxide (M) concentrations in the Bt horizon of an Oxyaquic Hapludalf; bar = 20 mm.

deficiency in crop and forest production is common on soils possessing high Fe (and Al) oxide quantities.

Iron oxides also play a large role in soil aggregation. Because many Fe oxide surfaces possess a (+) charge, they tend to facilitate aggregation by forming bridges between (−) charged phyllosilicate minerals. Poorly crystalline Fe oxides may promote increased aggregation compared to more crystalline forms (2).

ALUMINUM OXIDES

Aluminum Oxide Minerals

Gibbsite [$Al(OH)_3$] is the most common Al oxide found in soils (1) (Table 1). Corundum (Al_2O_3), boehmite (AlOOH), and diaspore (AlOOH) are primary minerals that are not typically found in soils, but are common in bauxite ore deposits (1). Unlike Fe oxides, poorly crystalline phases of Al oxide have not been readily identified in soils, but most likely exist. Aluminum oxides are essentially colorless, and their visual identification in soils is difficult.

Gibbsite Formation

Gibbsite is a common constituent of tropical and subtropical soils, and its presence generally suggests an advanced stage of soil development. Gibbsite formation is mostly due to the desilication (weathering and leaching of Si) of kaolinite or other aluminosilicate that occurs over long-term weathering under humid or extreme leaching environments. However, in highly leaching aluminous environments gibbsite can form rapidly (7). Because gibbsite can form rapidly as well as exist in transported

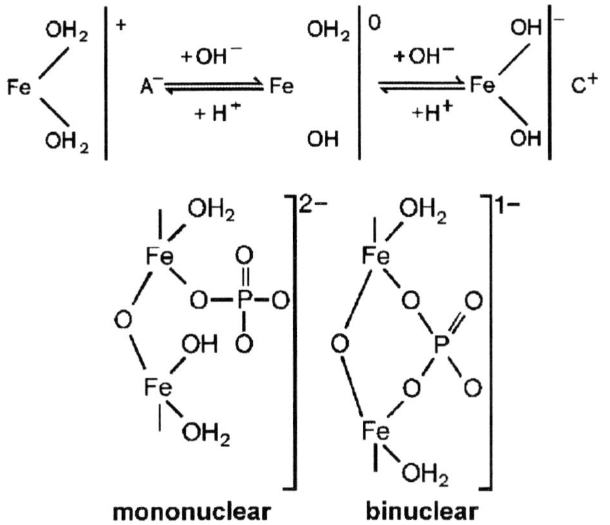

Fig. 2 Proposed mechanisms for development of pH dependent charge (a) and specific adsorption of oxyanions (PO_4) (b) by Fe oxides (Adapted from Ref. 2, chapter 8, p. 407 for (a) and p. 412 for (b), respectively.)

parent materials, its presence does not always indicate an advanced degree of soil development.

Aluminum Oxide Properties

Similar to Fe oxides, gibbsite develops surface charge based on soil solution pH and ionic strength. Most evidence suggests that Al oxides have a PZC ≈ pH 8 (1). Thus, gibbsite is (+) charged in most soils, and readily sorbs anions, cations, and organic molecules by both specific and coulombic attraction. Specific adsorption of ions (similar to Fe oxides) also occurs with gibbsite. Similar to Fe oxides, gibbsite contributes to soil aggregation by bridging (−) charged phyllosilicate surfaces.

MANGANESE OXIDES

Manganese Oxide Minerals

Manganese ions are similar in size to Fe, thus, Mn is found in several primary minerals. However, Mn oxides are not as abundant as Fe oxides because lower quantities of Mn typically exist in parent materials. Because Mn is essential to plant growth, but high concentrations sometimes found in wet or acid soils are toxic, an understanding of its mineral forms is critical for crop management (3). McDaniel and Buol (8) determined that approximately half of the Mn in soils developed from a gneiss parent material existed in oxide mineral form, while the other half was either organically bound or resided on exchange sites. Similar to Fe, most pedogenic Mn minerals exist as oxides with Mn in sixfold coordination with oxygens or OHs. Because Mn is redox sensitive, morphological features resulting from its transformations are also used as a soil wetness indicator. Manganese oxides mostly exist as black, hardened concentrations, called concretions, that violently effervesce upon treatment with H_2O_2. They also commonly exist as black coatings on ped faces (Table 1; Fig. 1c).

Manganese occurs in several oxidation states (Mn^{2+}, Mn^{3+}, and Mn^{4+}) within oxide minerals of varying crystallinity (8). Manganese oxides are classified as either layer or tunnel type, the difference depending on the presence or absence of tunnels within the mineral structure (3). Birnessite [$(Na_{0.7}Ca_{0.3})Mn_7O_{14}\cdot 16H_2O$], vernadite ($MnO_2$), and lithiophorite [$(Al,Li)MnO_2(OH)_2$] are layer-type Mn oxides that have been identified in soils, while hollandite ($Ba_2Mn_8O_{16}$) is a tunnel-type soil Mn oxide (3). Other primary Mn oxides sometimes found in soils include the tunnel-types pyrolusite (MnO_2) and todorokite [$(Na,Ca,K,Ba,Mn^{2+})_2Mn_4O_{12}\cdot 3H_2O$] (3).

Manganese Oxide Properties

Most evidence suggests that Mn oxides possess PZCs < pH 4.5 (3). This renders most Mn oxide surfaces (−) charged except under extremely acid conditions. Cations, particularly heavy metals, are adsorbed onto Mn oxide surfaces by both specific mechanisms and coulombic attractive forces (3). For example, in a study of metal-contaminated soils in Colorado, Zn, Cd, and Pb were preferentially adsorbed onto Mn and Fe oxide surfaces, while Cu was predominantly associated with the organic matter fraction (9). Manganese in oxide minerals is reduced to the soluble Mn^{2+} in saturated environments by similar mechanisms as described in the Fe oxide section earlier. Soluble Mn^{2+} is either leached, sorbed onto organic or clay surfaces, or reoxidizes to black concentrations.

GENERAL RELATIONSHIPS BETWEEN OXIDE MINERALS

The presence of high quantities of sesquioxides suggests an advanced degree of soil weathering. Although many soil orders possess these minerals, sesquioxides are concentrated in Oxisols and Ultisols. Because of the properties Fe and Al oxides impart to soils, these soils are often red, possess significant AEC, and are strongly aggregated. Phosphorus deficiency is also common.

Table 2 Oxide dominated soil mineralogical families in soil taxonomy and their occurrence

Mineralogical family	Criteria	Soil orders where commonly found
Ferritic	>40% Fe_2O_3 in the <2 mm fraction	Oxisols, Ultisols, Inceptisols
Gibbsitic	>40% gibbsite in the <2 mm fraction	Oxisols, Ultisols
Sesquic	18–40% Fe_2O_3 and 18–40% gibbsite in the <2 mm fraction	Oxisols
Ferruginous	18–40% Fe_2O_3 in the <2 mm fraction	Oxisols, Ultisols
Allitic	18–40% gibbsite in the <2 mm fraction	Oxisols, Ultisols (none currently correlated)
Ferrihydritic	(8× Oxalate extractable Si+2× oxalate extractable Fe)>5	Andisols, Spodosols
Parasesquic	>10% Fe_2O_3+gibbsite in the <2 mm fraction	Ultisols, Alfisols, Mollisols, Aridisols, Spodosols, Inceptisols

(From Ref. 10.)

Oxide minerals can also act as cementing agents in soils. Plinthite, a common constituent of southeastern U.S. coastal plain soils, is a hardened mass of mostly Fe oxides that also typically contains quartz, phyllosilicates, and gibbsite. In large quantities plinthite can have a major effect on soil properties and cause significant reductions in rooting depth and hydraulic conductivity.

Many soil mineralogy classes defined by oxide mineral quantities exist in soil taxonomy due to the influence these minerals have on soil properties (10). Most of the 12 soil orders have at least one soil series identified that has an oxide-dominated mineralogical class (Table 2).

REFERENCES

1. Hsu, P.H. Aluminum Oxides and Oxyhydroxides. In *Minerals in Soil Environments*; Soil Sci. Soc. Am. Book Series # 1, 2nd Ed.; Dixon, J.B., Weed, S.B., Eds.; Soil Sci. Soc. Am. Madison, WI, 1989; 1, 331–378.
2. Schwertmann, U.; Taylor, R.M. Iron Oxides. In *Minerals in Soil Environments*; Soil Sci. Soc. Am. Book Series #1, 2nd Ed.; Dixon, J.B., Weed, S.B., Eds.; Soil Sci. Soc. Am. Madison, WI, 1989; 379–438.
3. McKenzie, R.M. Manganese Oxides and Hydroxides. In *Minerals in Soil Environments*; Soil Sci. Soc. Am. Book Series #1, 2nd Ed.; Dixon, J.B., Weed, S.B., Eds.; Soil Sci. Soc. Am. Madison, WI, 1989; 439–466.
4. Schwertmann, U. Occurrence and Formation of Iron Oxides in Various Pedoenvironments. In *Iron in Soils and Clay Minerals*; Series C: Mathematical and Physical Sciences #217, Stucki, J.W., Goodman, B.A., Schwertmann, U., Eds.; D. Reidel Publ. Dordrecht, Holland, 1985.
5. Bohn, H.L.; McNeal, B.L.; O'Connor, G.A. *Soil Chemistry*; Wiley: New York, 1985.
6. McBride, M.B. Surface Chemistry of Soil Minerals. In *Minerals in Soil Environments*; Soil Sci. Soc. Am. Book Series #1, 2nd Ed.; Dixon, J.B., Weed, S.B., Eds.; Soil Sci. Soc. Am. Madison, WI, 1989; 35–88.
7. Norfleet, M.L.; Smith, B.R. Weathering and Mineralogical Classification of Selected Soils in the Blue Ridge Mountains of South Carolina. Soil Sci. Soc. Am. J. **1989**, *53*, 1771–1778.
8. McDaniel, P.A.; Buol, S.W. Manganese Distributions in Acid Soils of the North Carolina Piedmont. Soil Sci. Soc. Am. J. **1991**, *55*, 152–158.
9. Levy, D.B.; Barbarick, K.A.; Siemer, E.G.; Sommers, L.E. Distribution and Partitioning of Trace Metals in Contaminated Soils Near Leadville, Colorado. J. Environ. Qual. **1992**, *21*, 185–195.
10. Soil Survey Staff, *Soil Taxonomy. A Basic System of Soil Classification for Making and Interpreting Soil Surveys*; Handbook No. 436, USDA–NRCS: Washington, DC, 1999.
11. Schwertmann, U. Relations Between Iron Oxides, Soil Color, and Soil Formation. In *Soil Color*; Soil Sci. Soc. Am. Spec. Pub #31, Bigham, J.M., Ciolkosz, E.J., Eds.; Soil Sci. Soc. Am. Madison, WI, 1993; 51–70.

SHRINKAGE

D. McGarry
D.F. Yule
Natural Resource Sciences, Indooroopilly, Queensland, Australia

INTRODUCTION

Soil shrinkage is the process of soil material contracting to a lesser volume, with the loss of water on drying. In agriculture, soil shrinkage focused on two topics: the characterization and classification of soils containing clay, as the degree of shrinkage (and swelling) largely determines the potential for soil structure development (1), and the redevelopment of soil structure (soil resilience), compacted by wheels and implements (2). In engineering, soil shrinkage is a key determinant of material stability for both intrinsic and stress states associated with structures such as foundations, pipelines and roads, and soil conservation structures (3).

Negative aspects of soil shrinkage include overall volume change and uneven shrinkage of soil supporting loads in engineering, and in agriculture root breakage caused by soil cracking and the super-dehydration of soils from air thermals in cracks (4). Positive, agricultural aspects of soil shrinkage include soil cracking, providing aeration and facilitating water entry, and the formation of naturally, aggregated, fine seedbeds with wetting and drying cycles (5).

FACTORS AFFECTING SOIL SHRINKAGE

Many factors, singly and in combination, affect the degree of shrinkage in a soil, including soil fabric, mineralogy, saturating cation, electrolyte concentration and speciation, clay content, surface area, antecedent water content, frequency of wet–dry cycles, confining pressures, and soil thickness (6). Shrinkage is positively correlated with greater total and fine clay contents, and particularly greater proportions of clay composed of expanding 2:1 clay minerals, such as the smectites (6). Antecedent water content is positively correlated with clay percent, the proportion of expansible clay minerals and surface area, hence is also positively correlated with total shrinkage. Particle arrangement also affects shrinkage characteristics where random (edge to face) arrangements can trap larger volumes of water (or air) than stratified (face to face) arrangements; hence have greater capacity for shrinkage (4). Shrinkage is reduced by such factors as increased proportions of sand relative to clay as the interlocking of sand and silt grains increases the frictional forces (6); increased organic matter as the organic matter dilutes the shrinkage potential of the mineral components (7); and, cited by some studies, low cation exchange capacity and high exchangeable sodium levels.

MECHANISMS OF SOIL SHRINKAGE

Smectites, because of the charge deficiency, have strong ability to attract and adsorb water molecules onto clay domains (6). The mechanisms of soil shrinkage explain the loss of this water, there being two common representations (Figs. 1 and 2). The first involves the measurement of the volume of a soil sample as it slowly dries from a fully wet condition and provides a relationship between soil volume and water content (Fig. 1). This relationship is termed as a soil shrinkage characteristic curve (8) and consists of four zones, though all four zones are not always present: structural, basic or normal, residual, and zero shrinkage (9). In structural shrinkage, volume change is less than the volume of water lost, arising from the drainage of large pores that do not completely close with water loss. Basic shrinkage, traditionally termed normal shrinkage, is commonly depicted as equal volume and water loss. This zone accounts for most shrinkage in agricultural soils, generally between soil water potentials of -33 to <-1500 kPa (10). Water is lost from between clay domains (lumps of individual clay particles); the random orientation of these domains causing shrinkage to be equidimensional, resulting in crack formation. Residual shrinkage, like structural shrinkage, has volume change less than the volume of water lost. Water is lost from stable pores between clay particles, unable to completely close as the particles retain the clay matrix structure. Zero shrinkage, recognized by some authors, has no volume change with water loss (11).

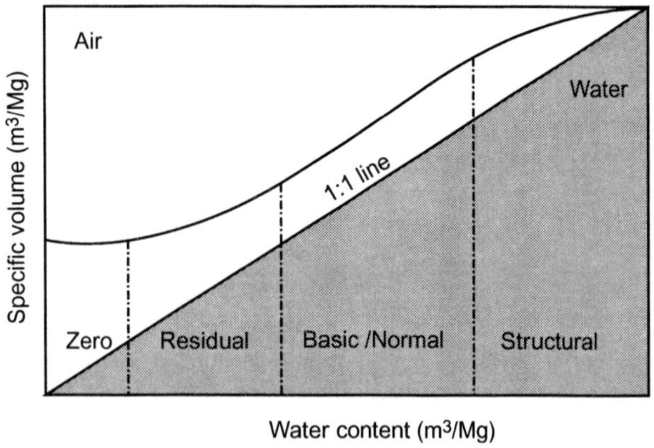

Fig. 1 A plot of a soil shrinkage characteristic curve showing the four shrinkage zones and the theoretical (1:1) shrinkage line.

On the soil shrinkage characteristic curve the location of experimental data is referred to a 1:1 line, termed the saturation line as it represents the shrinkage of a structureless (air-less) clay paste (Fig. 1). With reference to this theoretical line, the relative proportions of the specific volumes of the solid, liquid, and air for an experimental plotted value may be obtained.

The second representation of soil shrinkage aims to represent "whole soil" shrinkage behavior and is generally applicable to the shrinkage of large, intact soil units—either soil cores in the laboratory or field soil (Fig. 2). A measurable field result is a reduction in the vertical dimension, i.e., the soil surface moves down, and a visible field result is the formation of large cracks that form laterally at irregular intervals (Fig. 2). Importantly, the soil block between the cracks contains all the structural and residual porosity, the cracks (including the "crack" above the soil surface) contain only the basic (or normal) shrinkage porosity.

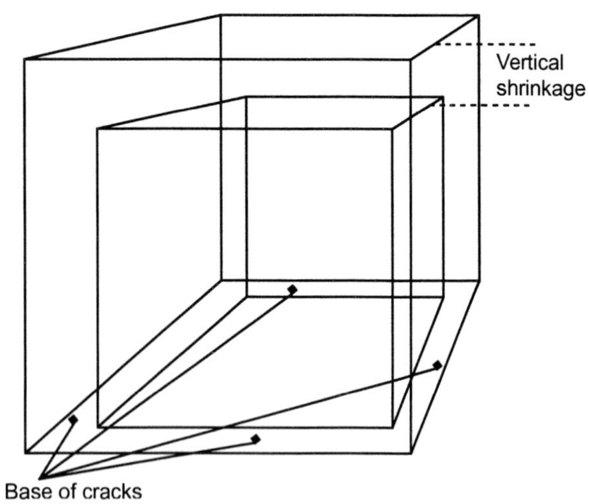

Fig. 2 Shrinkage of a soil cube showing equidimensional shrinkage, resulting in soil cracking.

THE MEASUREMENT OF SOIL SHRINKAGE

Soils may be characterized by measurement of total or partial shrinkage, of either linear dimensions or volume as an initially wet sample loses water. There is no unique measure of soil shrinkage; different techniques provide different aspects of a soil's shrinkage nature and potential.

Measurements may be made on soil in different states: 1) remolded samples of finely ground material, 2) natural soil aggregates, 3) intact blocks or cores, and 4) volume change in the field (12). A vital consideration is that for each of the tests, alteration of the start and end points in terms of water content will provide different results, hence negating the potential for soil comparisons.

Remolded Samples

The most common measure used in engineering applications, is linear shrinkage: LS_{std} (based on Ref. 13). Determination is made of the percent decrease in length of a subsample of ground soil (<425 μm) that has been wet to its liquid limit, mixed to a homogeneous paste, placed in a mold, air-dried (~20°C) for 24 hr, then oven-dried at 105°C. The LS_{std} of the specimen is:

$$LS_{std} = \frac{L_S}{L}$$

where, L is the length of the mold in mm and L_S is the longitudinal length of the oven-dry specimen in mm.

A modification to LS_{std}, towards providing a more agriculturally pertinent measure of soil shrinkage, is the modified linear shrinkage test: LS_{mod} (14), where linear shrinkage is determined on sieved (<2 mm), not remolded soil.

Natural Soil Aggregates

Several techniques are available, measuring soil volume by liquid displacement. Waterproofing coatings for the soil sample include paraffin wax, rubber balloons, and industrial resins. The most common test is the coefficient of linear extensibility: COLE (15), where determinations of volume change of intact soil clods are made between a moist and an oven-dry state. The test is now integral in defining the Vertisol soil group (16).

Intact soil clods of size ~0.05–0.08 m diameter are wetted on pressure plates to a specified matric potential (commonly, −33 kPa), before being coated in SARAN® resin. Two paired weighings are conducted—the first on the coated clod at the −33 kPa matric potential and then a second on the clod after oven-drying (105°C). The bulk density of the clod is calculated at each water content and COLE calculated as:

$$COLE_{std} = 100((BD_{OD}/BD_{33})^{1/3} - 1)$$

where, BD_{OD} is the bulk density at oven-dry and BD_{33} is the bulk density at −33 kPa.

Several indices of soil shrinkage can be derived from the shrinkage of coated soil clods by collecting data from the whole range of the soil shrinkage characteristic curve, not one part alone as with COLE. Ref. 17 specified 17 indices based on fitting straight lines to the shrinkage curve and Ref. 11 specified 13 parameters including consideration of the points of transition between the different shrinkage zones. Such indices have been used to quantify and rationalize differences in soils or treatments (e.g., Ref. 2).

Vertical Volume Change on Cores and in the Field

These measures of volume change are, most commonly, gained by measuring the physical dimensions of soil samples. Three types are recognized: volumetric and destructive (based on core bulk density measurements at various moisture states), nondestructive measures with dual gamma attenuation, and "one-dimensional" measures based on monitoring surface elevation (18). Disadvantages with each include: small cores may give nonrepresentative sampling in cracking soils, access tubes can initiate preferential cracking of soil around the tube giving biased measures, and there are problems with relating a local surface elevation measurement with an average value of water contents collected around the gauge. New techniques employ electronic displacement transducers with block kriging interpolation to improve the precision of the water content estimates (18).

THE DEFINITION OF SOIL CLASSES FROM CUT-OFF VALUES OF SHRINKAGE TEST DATA

An important aim of determining shrinkage values is to define soil classes in terms of volume change activity for either engineering or soil (pedological) classifications.

Table 1 Examples of classification schemes for the assessment of volume change potential based on (a) linear shrinkage LS_{std} (19) and (b) COLE (20)

		Volume change rating
(a) LS_{std}		
Arid to semi arid climate	Humid climate	
0–0.05	0–0.12	Low
0.05–0.12	0.12–0.18	Moderate
>0.12	>0.18	High
(b) COLE		
<0.03		Low
0.03–0.06		Medium
0.06–0.09		High
>0.09		Very high

Engineering classes define broad categories of soil in terms of their predicted soil behavior for conservation earthworks or building foundations. Examples of the use of LS_{std} and COLE classes are given (Table 1). Ref. 19 included climatic considerations in their classification, recognizing LS_{std} becomes limiting at lower values in drier climates, as the potential for desiccation (hence shrinkage) is greater. In pedological studies, (land survey and assessment) variable volume soils may be classed in terms of their potential for volume change. The aim is to provide a numeric, independent assessment of volume-change potential, to replace or supplement visual assessments of crack intensity toward classifying Vertisols. For example, in South Africa, $LS_{std} > 0.12$ is required for vertic horizons and $LS_{std} < 0.12$, but >0.08 for melanic horizons. There is also potential and a growing demand to use such classifications to express soil resiliency, i.e., a soil's potential to self-repair soil compaction through wet–dry cycles.

REFERENCES

1. Loveday, J. Field Aspects of Swelling and Shrinking. In *Physical Aspects of Swelling Clay Soils*; McGarity, J.W., Hoult, E.H., So, H.B., Eds.; University of New England: Armidale, NSW, Australia, 1972; 45–52.
2. Pillai, U.; McGarry, D. Structure Repair of a Compacted Vertisol with Wet/Dry Cycles and Rotation Crops. Soil Sci. Soc. Am. J. **1999**, *63*, 201–210.
3. Aitchison, G.D. The Quantitative Definitions of the Physical Behaviour of Expansive Soils—An Engineering Viewpoint. In *Physical Aspects of Swelling Clay Soils*;

McGarity, J.W., Hoult, E.H., So, H.B., Eds.; University of New England: Armidale, NSW, Australia, 1972; 63–74.
4. Yong, R.N.; Warkentin, B.P. *Soil Properties and Behaviour; Developments in Geotechnical Engineering 5*; Elsevier: Amsterdam, 1975.
5. Pillai-McGarry, U.P.P.; Collis-George, N. Laboratory Simulation of the Surface Morphology of Self-Mulching Vertisols. I—Materials, Methods and Preliminary Results. Austr. J. Soil Res. **1990**, *28*, 129–139.
6. Wilding, L.P.; Tessier, D. Genesis of Vertisols: Shrink–Swell Phenomena. In *Vertisols: Their Distribution, Properties, Classification and Management*; Technical Monograph No. 18, Wilding, L.P., Puentes, R., Eds.; Soil Management Support Services: Texas A & M, College Station, Texas, 1988; 55–81.
7. McGarry, D. The Structure and Grain Size Distribution of Vertisols. In *Vertisols and Technologies for Their Management*; Developments in Soil Science, Ahmad, N., Mermut, A., Eds.; Elsevier: Amsterdam, 1996; Vol. 24, 231–259.
8. Mitchell, A.R. Shrinkage Terminology: Escape from "Normalcy". Soil Sci. Soc. Am. Proc. **1992**, *56*, 993–994.
9. Tariq, A.U.R.; Durnford, D.S. Analytical Volume Change Model for Swelling Clay Soils. Soil Sci. Soc. Am. Proc. **1993**, *57*, 1183–1187.
10. Yule, D.F.; Ritchie, J.T. Soil Shrinkage Relationships of Texas Vertisols: I. Small Cores. Soil Sci. Soc. Am. Proc. **1980**, *44*, 1285–1291.
11. Braudeau, E.; Costantini, J.M.; Bellier, G.; Colleuille, H. New Device and Method for Soil Shrinkage Curve Measurement and Characterisation. Soil Sci. Soc. Am. Proc. **1999**, *63*, 525–535.
12. Warkentin, B.P. Soil Shrinkage. In *Soil Sampling and Methods of Analysis*; Carter, M.R., Ed.; Lewis Publishers: London, 1993; 513–518.
13. Standards Association of Australia. AS 1289. C4.1—1977. Determination of the Linear Shrinkage of a Soil (Standard Method), 1977.
14. McKenzie, N.J.; Jacquier, D.J.; Ringrose-Voase, A.J. A Rapid Method for Estimating Soil Shrinkage. Austr. J. Soil Res. **1994**, *32*, 931–938.
15. Grossman, R.B.; Brasher, B.R.; Franzmeier, D.P.; Walker, J.L. Linear Extensibility as Calculated from Natural-Clod Bulk Density Measurements. Soil Sci. Soc. Am. Proc. **1968**, *32*, 570–573.
16. Soil Survey Staff, *Soil Survey Laboratory Methods and Procedures for Collecting Soil Samples*; Soil Survey Investigations Report 1, Soil Conservation Service, USDA: Washington, DC, 1972.
17. McGarry, D. Quantification of the Effects of Zero and Mechanical Tillage on a Vertisol by Using Shrinkage Curve Indices. Austr. J. Soil Res. **1988**, *26*, 537–542.
18. Coquet, Y. In Situ Measurement of the Vertical Linear Shrinkage Curve of Soils. Soil Tillage Res. **1998**, *46*, 289–299.
19. Holland, J.E.; Richards, J. Road Pavements as Expansive Clays. Austr. Road Res. **1982**, *12*, 173–179.
20. Dasog, G.S.; Acton, D.F.; Mermut, A.R.; De Jong, E. Shrink–Swell Potential and Cracking in Clay Soils of Saskatchewan. Can. J. Soil Sci. **1988**, *68*, 251–260.

SILICON AND SODIUM

Jian Feng Ma
Kagawa University, Kagawa, Japan

S

INTRODUCTION

Silicon (Si) and sodium (Na) are both beneficial elements for higher plants. Beneficial elements are defined as those for which positive effects are expressed only in certain plant species or under specific growth conditions (1).

SILICON

Silicon is the second most abundant element in the Earth's crust. The concentration of Si in soil solution ranges between 3.5 and 40 mg l^{-1}. Silicon in soil solution is mainly present as the uncharged monomeric molecule, silicic acid [$Si(OH)_4$], when the solution pH is below 9.0. At a higher pH (>9.0), silicic acid dissociates into silicate ions [$(OH)_3SiO^{-1}$]. The solubility of silicic acid in water is 2.0 mM at 25°C, and polymerization of silicic acid into silica gel ($SiO_2 \cdot H_2O$) occurs when the concentration of silicic acid exceeds 2 mM.

Uptake, Translocation, and Accumulation of Si

All soil-grown plants contain Si. However, the Si concentration of plant shoots varies greatly between plant species, ranging from 0.1 to 10% Si on a dry weight basis. This variation is largely due to different capacities of Si uptake by plant roots. Three uptake modes have been suggested: active, passive, and rejective uptake (2).

Active uptake

Plant species which employ active uptake of Si, such as rice, take up Si faster than water (3). The Si concentration in rice leaf blades can reach more than 10% Si on a dry weight basis. The uptake of Si by rice roots is inhibited by metabolic inhibitors such as 2,4-DNP, iodo-acetate, and 2,4-D, suggesting that Si uptake is energy dependent (4). The mechanisms responsible for active uptake by rice are still unknown. A putative transporter for silicic acid uptake has been suggested, but neither the gene encoding it nor the transporter protein has been isolated.

Passive uptake

With the passive mode of uptake, the uptake rate of Si by plant roots is similar to that of water (3). The Si concentration in the shoot largely depends on Si concentration in the medium, but rarely exceeds 2.5% Si. Cucumber shows this type of uptake.

Rejective uptake

Some plants, such as tomato, take up Si more slowly than water (3). The Si concentration in the shoots is typically lower than 0.5% Si. A rejective mechanism seems to be involved in this uptake mode although it is not well understood.

Following uptake by the roots, silicic acid is translocated to the shoot where it is concentrated through loss of water and is polymerized. The process of Si polymerization converts silicic acid to colloidal silicic acid and finally to silica gel with increasing silicic acid concentration. In rice plants, more than 90% of total Si in the shoot is present in the form of silica gel, while the concentration of colloidal plus monomeric Si is kept below 140–230 mg Si l^{-1} (3). A similar pattern of accumulation is observed in cucumber leaves, although the total Si concentration of cucumber is much lower than that of rice.

Silicon is deposited as a 2.5 μm thick layer in the space immediately beneath the thin (0.1 μm) cuticle layer, forming a cuticle–Si double layer in leaf blades of rice (5). There are two types of silicified cells in rice leaf blades; silica cells, and silica bodies or silica motor cells (Fig. 1) (3). Silica cells are located on vascular bundles and are dumbell-like in shape, while silica bodies are in bulliform cells of rice leaves. The silicification of cells proceeds from silica cells to silica bodies. In addition to leaf blades, silicified cells are also observed in the epidermis and vascular tissues of the stem, leaf sheath, and hull of rice.

Currently, since there is no evidence that Si is involved in metabolic activities of plants, silica gel deposited on tissue surfaces is proposed to play an important role in alleviating biotic and abiotic stresses. Accumulation of Si in the shoots is a prerequisite for the beneficial functions of Si.

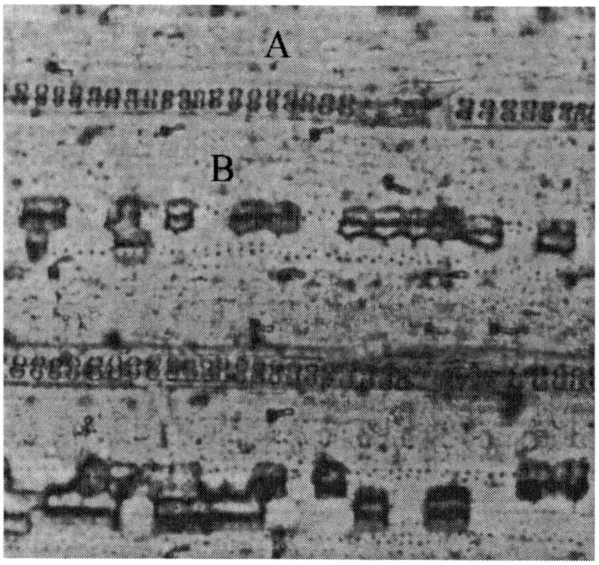

Fig. 1 Silica cell (A) and silica body (B) in rice leaf blades. (From Ref. 3.)

Beneficial Effects of Si

Positive effects of Si have been observed in a number of plant species, including rice, wheat, and barley. The beneficial effects of Si differ between plant species. Beneficial effects are usually obvious in plants which actively accumulate Si in their shoots (Fig. 2). The more the Si that accumulates in the shoots, the larger is the effect that is gained. This is because most effects of Si are expressed through the formation of silica gel, which is deposited on the surface of leaves, stems, and other organs of plants. On the other hand, the beneficial effects of Si vary with growth conditions. The effects are usually expressed more clearly when plants are under abiotic or biotic stresses. In addition, Si is the only element which does not damage plants when accumulated in excess. This is because silicic acid remains undissociated at physiological pH, and is also due to silica polymerization. Neither binding to cellular substances nor the creation of high osmotic pressure occurs in plants, even when accumulation is up to 10% Si on a dry weight basis. The beneficial effects of Si are outlined in Fig. 3.

Si-stimulated photosynthesis

Silicon deposited on the leaf blade keeps the leaf erect, thereby improving light interception by the leaf blade and indirectly stimulating photosynthesis. This is particularly important in dense plant stands and when nitrogen fertilizers are heavily applied, to minimize mutual shading.

Si-decreased susceptibility to disease and insect damage

It is well known that Si application reduces the severity of fungal diseases such as blast and sheath blight of rice, powdery mildew disease of barley and wheat, and vermin damage of rice by the plant hopper in the field. In addition to species with an active Si uptake mode, it was also reported that Si prevents powdery mildew disease of solution-cultured cucumber and musk melon (6).

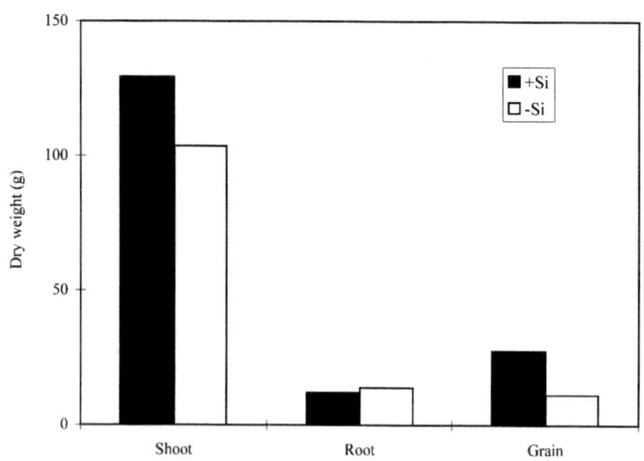

Fig. 2 Effect of Si application on the growth of rice. Rice was cultured in a nutrient solution with or without 1.67 mM Si as silicic acid.

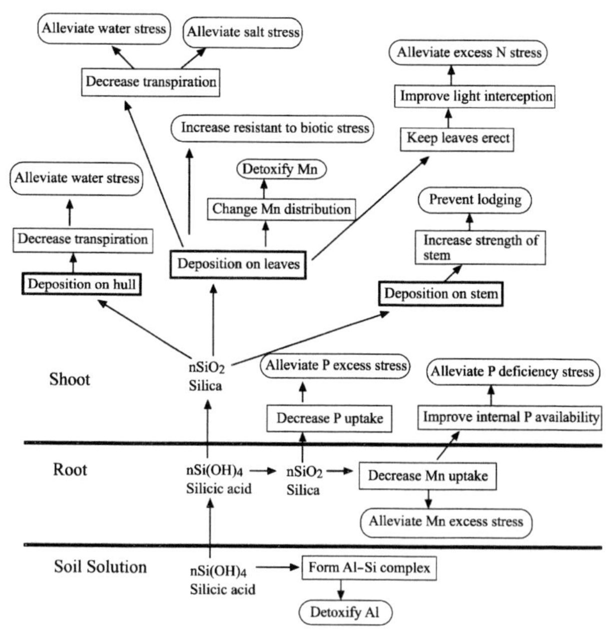

Fig. 3 Uptake, distribution, and accumulation of Si and its beneficial effects on crop growth in relation to biotic and abiotic stresses. (From Ref. 3.)

In recent years, it was reported that foliar application of Si is effective in inhibiting powdery mildew development on cucumber, musk melon, and grape leaves (7, 8). Foliar applied Si may deposit on the surface of leaves and play a role similar to Si transported from the roots. This approach may be particularly useful for crops with passive or rejective Si uptake modes.

Silicon deposited on the tissue surface is proposed to be responsible for the protective effects of Si against biotic stress. The deposited Si prevents physical penetration by insects and/or makes the plant cell less susceptible to enzymatic degradation by fungal pathogens. Recently, it was reported that Si application resulted in a stimulation of chitinase activity in cucumber and the rapid activation of peroxidases and polyphenoloxidases after infection with *Pythium* spp. (9). Glycosidically bound phenolics extracted from Si-treated plants when subjected to acid- or beta-glucosidase hydrolysis displayed a strong fungistatic activity.

Si-increased resistance to climatic stress

Si application reduces injury to rice caused by climatic stresses such as typhoons, low or high temperature, and low light. Typhoon damage usually results from lodging and sterility. Deposition of Si in rice increases the thickness of the culm wall and the size of the vascular bundle and can help to prevent lodging. Si deposited at the hull decreases the transpiration of panicles by about 30% at either milky or maturity stage (Table 1) (3) and prevents excess water loss. This is the means by which Si application can significantly increase the percentage of ripened grain (10).

Low temperature during summer can cause serious damage to rice crops. Low temperature decreases Si uptake by rice and insufficient sunshine lowers the Si:N ratio in rice, which can lead to an outbreak of blast disease.

Table 1 Effect of Si application on transpiration of the panicle at different stages. The excised panicles, with or without Si supply, were placed in an incubator at 30°C, relative humidity of 30% for 1 h

Treatment	Transpiration (mg H_2O/g fresh wt. lh)		Si conc. in hull (%)
	Milky stage	Maturity stage	
+Si	204	39.2	6.36
−Si	279	50.4	0.18
+Si/−Si	1.37	1.29	

(From Ref. 3.)

Application of Si under such conditions remarkably reduces the incidence of blast in rice.

The positive effects of Si under low-light conditions are particularly evident. Silicon effects on rice growth under shading treatment are greater than for a nonshaded treatment (3). However, the mechanism responsible for this phenomenon is unknown. It has been hypothesized that Si deposited on the leaf epidermal system might act as a "window" to facilitate the transmission of light through to photosynthetic mesophyll tissue. To date, however, evidence supporting this hypothesis is not available (11).

Si-alleviated water stress

Silicon can alleviate water stress by decreasing transpiration. Transpiration of the leaves is mediated through stomata and the cuticle. Rice plants have a thin cuticle and the formation of a cuticle–Si double layer significantly decreases cuticular transpiration. The transpiration of rice decreases with increasing Si concentration in the shoot (3). Water stress causes stomatal closure and consequently decreases photosynthetic rate. Therefore, Si stimulates the growth of rice more under water-stressed conditions (low humidity) compared to nonstressed conditions (high humidity).

Si-alleviated mineral stress

Mineral stress can be either a deficiency of an essential element or an excess of an essential or other element. Many reports have shown the beneficial effects of Si under various mineral stresses including P deficiency, excess of P, and toxicity of Al, Na, and Mn (3).

Beneficial effects of Si under P deficiency stress have been observed in many plants, including rice and barley. Silicon does not have any effect on P availability in soil (12). The beneficial effects of Si on growth under P deficiency stress appear to result from decreased Mn and Fe uptake, and thus increased P availability within P-deficient plants (13, 14).

Silicon decreases P uptake when the P supply is high (13). This phenomenon has been observed in rice and some non-Si-accumulators such as tomato, strawberry, cucumber, and soybean (3). In rice, the organic-P component is not affected by Si, but cellular inorganic-P is significantly decreased by Si when P is supplied at high concentrations (0.7 mM) (13). Therefore, stress caused by P excess can be alleviated by Si-induced decrease of P uptake.

Beneficial effects of Si under salt stress have also been observed in rice (15). The Na concentration in the shoot is decreased by about 50% with Si amendment. Translocation of Na to the shoot is partly regulated by transpiration. Since Si decreases transpiration, it is suggested that beneficial effects of Si during salt stress result from decreased transpiration.

Alleviative effects of Si on Mn toxicity have been reported in water-cultured rice, barley, bean, and pumpkin (3). Three different mechanisms seem to be involved, depending on the plant species. In rice (16), Si reduces Mn uptake by promoting the Mn-oxidizing power of the roots. In bean, it was found that Si does not reduce Mn uptake, but minimizes damage by effecting homogenous distribution of Mn in the leaf blade (17). In a cultivar of pumpkin, Si causes localized accumulation of Mn around the base of trichomes (18).

Alleviative effects of Si on Al-induced inhibition of root elongation have been reported in many crop species, including maize, cotton, rice, teosinte, sorghum, and wheat (19). Formation of nontoxic Al–Si complexes in the culture solution, and/or interaction between Al and Si within the plant have been suggested to be the mechanisms for this alleviative effect (20).

Silicate Fertilizers

In 1955, Si was first recognized as a fertilizer in Japan. Since then, 1.5–2.0 tons per ha of silicate fertilizers have been applied to paddy soils in Japan leading to a significant increase in rice production (2). Nowadays, Si fertilizers are also widely applied in Korea, China, and the U.S.A. Various Si fertilizers have been developed including slags, which are byproducts of the iron manufacturing industry (mainly containing calcium silicate), and sodium and potassium silicates. The solubility of these fertilizers, and their availability to plants vary greatly. They are used as a basal dressing or for topdressing. Potassium and sodium silicates are also used as foliar sprays for plants which are not able to take up Si via the roots.

SODIUM

Sodium (Na) is a common element in the environment. The content of this element in the Earth's crust is about 2.8% and Na concentrations in soil solutions range from 0.4 to 150 mM Na, depending on soil type. It is well known that excess Na inhibits plant growth, but small quantities of Na are essential and/or beneficial for some plant species.

Essentiality of Na

Certain plant species using the C_4 photosynthetic pathway (C_4 and CAM plants) require Na as an essential element. In these plants, Na deficiency causes chlorosis, necrosis, and decreased growth. Sodium is supposed to play a significant role in the conversion of pyruvate to phosphoenolpyruvate (PEP) in the mesophyll chloroplasts (21). Phosphoenolpyruvate is the first product of C fixation in C_4 plants. Failure to regenerate the PEP from pyruvate will result in decreased photosynthesis in C_4 plants. A co-transporter of pyruvate/Na was found in mesophyll chloroplasts of *Panicum miliaceum* (22), suggesting that Na is involved in the transport of pyruvate from bundle sheath cells to mesophyll chloroplasts. In some plant species such as *Amaranthus tricolor* L., Na enhances growth by stimulating nitrate uptake and reduction (23). Since only trace amounts of Na are required for its function in plants, Na deficiency rarely occurs in nature.

Beneficial Effects of Na

Beneficial effects of Na on plant growth are observed under both K deficiency and with optimal K supply. Under K deficiency, Na is able to act as a substitute for some functions of K such as charge compensation and osmoregulation. The K content required for optimal growth decreases from 4.3 to 1.0% in lettuce (24). However, the extent of beneficial effects of Na under low K stress differs between plant species. Sodium addition under low K stress is effective in rice, barley, and cotton, but not in corn, potato, or soybean (1). The substitution capacity of Na also differs between cultivars within a species. The high Na substitution capacity of some strains of tomato is associated with a high accumulation of Na in the top under low K stress (Table 2) (25). However, the beneficial effects of Na in these plant species are not apparent when K supply is sufficient.

On the other hand, in particular plant species, a beneficial effect of Na under optimal K supply is observed. This effect is not attributed to a partial substitution of Na for K, but to stimulation of cell elongation and

Table 2 Effect of Na addition on plant growth and Na accumulation in two strains of tomato cultivated under low K stress

	Plant dry weight (g)		
Strain	**−Na**	**+Na**	**Na accumulation (mg plant^{-1})**
576	1.06	1.67	9.9
546	0.80	0.85	2.9

(From Ref. 25.)

maintenance of water balance, although the exact mechanism is unknown. A typical example is sugar beet. Usually, Na concentration in plants is less than 10% of K levels, but in sugar beet, the Na concentration is similar to or greater than K and increased Na enhances sugar beet growth (1).

Sodium Fertilizers

Application of Na may be effective for some plant species (e.g., natrophilic species) under K deficiency conditions. Sodium chloride or sulfate has been applied to sugar beet stands (1). Sodium fertilizers are also used to increase the Na content of forage and pasture plants for animal nutrition.

REFERENCES

1. Marschner, H. Beneficial Mineral Elements. *Mineral Nutrition of Higher Plants*; Academic Press: San Diego, 1995; 405–426.
2. Takahashi, E.; Ma, J.F.; Miyake, Y. The Possibility of Silicon as an Essential Element for Higher Plants. Comments Agric. Food Chem. **1990**, *2*, 99–122.
3. Ma, J.F.; Miyake, Y.; Takahashi, E. Silicon as a Beneficial Element for Crop Plants. In *Silicon in Agriculture*; Datonoff, L., Ed.; Elsevier: Amsterdam, 2001, 17–39.
4. Okuda, A.; Takahashi, E. Effect of Metabolic Inhibitors on Si Uptake by Rice Roots. Jpn. J. Soil Sci. Plant Nutr. **1962**, *33*, 453–455.
5. Yoshida, S.; Ohnishi, Y.; Kitagishi, K. Histochemistry of Si in Rice Tissues. III. The Presence of Cuticle–Silica Double Layer in the Epidermal Tissue. Soil Sci. Plant Nutr. **1962**, *8*, 1–5.
6. Miyake, Y.; Takahashi, E. Effect of Silicon on the Resistance of Cucumber Plant to the Microbial Disease. Jpn. J. Soil Sci. Plant Nutr. **1982**, *53*, 106–110.
7. Menzies, J.; Bowen, P.; Ehret, D. Foliar Application of Potassium Silicate Reduces Severity of Powdery Mildew on Cucumber, Muskmelon, and Zucchini Squash. J. Am. Soc. Hort. Sci. **1992**, *117*, 902–905.
8. Bowen, P.; Menzies, J.; Ehret, D. Soluble Silicon Sprays Inhibit Powdery Mildew Development on Grape Leaves. J. Am. Soc. Hort. Sci. **1992**, *117*, 906–912.
9. Cherif, M.; Asselin, A.; Belanger, R.R. Defense Responses Induced by Soluble Silicon in Cucumber Roots Infected By *Pythium* Spp. Phytopathology **1994**, *84*, 236–242.
10. Ma, J.F.; Nishimura, K.; Takahashi, E. Effect of Silicon on the Growth of Rice Plant at Different Growth Stages. Soil Sci. Plant Nutr. **1989**, *35*, 347–356.
11. Agarie, S.; Agata, W.; Uchida, H.; Kubota, F.; Kaufman, P.B. Function of Silica Bodies in the Epidermal System of Rice (*Oryza Sativa* L.): Testing the Window Hypothesis. J. Exp. Bot. **1996**, *47*, 655–660.
12. Ma, J.F.; Takahashi, E. The Effect of Silicic Acid on Rice in a P-Deficient Soil. Plant Soil **1990**, *126*, 121–125.
13. Ma, J.F.; Takahashi, E. Effect of Silicate on Phosphate Availability of Rice in a P-Deficient Soil. Plant Soil **1991**, *133*, 151–155.
14. Ma, J.F.; Takahashi, E. Effect of Silicon on the Growth and Phosphorus Uptake of Rice. Plant Soil **1990**, *126*, 115–119.
15. Matoh, T.; Kairusmee, P.; Takahashi, E. Salt-Induced Damage to Rice Plants and Alleviation Effect of Silicate. Soil Sci. Plant Nutr. **1986**, *32*, 295–304.
16. Okuda, A.; Takahashi, E. Effect of Silicon Supply on the Injuries due to Excessive Amounts of Fe, Mn, Cu, As, Al, Co of Barley and Rice Plant. Jpn. J. Soil Sci. Plant Nutr. **1962**, *33*, 1–8.
17. Horst, W.J.; Marschner, H. Effect of Silicon on Manganese Tolerance of Bean Plants (*Phaseolus Vulgaris* L.). Plant Soil **1978**, *50*, 287–303.
18. Iwasaki, K.; Matsumura, A. Effect of Silicon on Alleviation of Manganese Toxicity in Pumpkin (*Cucurbita Moschata* Duch cv. Shintosa). Soil Sci. Plant Nutr. **1999**, *45*, 909–920.
19. Cocker, K.M.; Evans, D.E.; Hodson, M.J. The Amelioration of Aluminium Toxicity by Silicon in Higher Plants: Solution Chemistry or an in Planta Mechanism? Physiol. Plant **1998**, *104*, 608–614.
20. Ma, J.F.; Sasaki, M.; Matsumoto, H. Al-Induced Inhibition of Root Elongation in Corn, *Zea Mays* L. Is Overcome by Si Addition. Plant Soil **1997**, *188*, 171–176.
21. Brownell, P.E. Sodium in C_4 Photosynthesis. In *Plants in Action*; Atwell, B., Kriedemann, P., Turnbull, C., Eds.; Macmillan Education Australia: South Yarra, 1999, 518–520.
22. Ohnishi, J.; Kanai, M. Na^+-Induced Uptake of Pyruvate into Mesophyll Chloroplasts of a C_4 Plant, *Panicum Miliaceum* L. FEBS Letter **1987**, *219*, 347–350.
23. Ohta, D.; Yasuoka, S.; Matoh, T.; Takahashi, E. Sodium Stimulates Growth of *Amaranthus Tricolor* L. Plants Through Enhanced Nitrate Assimilation. Plant Physiol. **1989**, *89*, 1102–1105.
24. Costigan, P.A.; Mead, G.P. The Requirements of Cabbage and Lettuce Seedlings for Potassium in the Presence and Absence of Sodium. J. Plant Nutr. **1987**, *10*, 385–401.
25. Figdore, S.S.; Gerloff, G.C.; Gabelman, W.H. The Effect of Increasing NaCl Levels on the Potassium Utilization Efficiency of Tomatoes Grown Under Low-K Stress. Plant Soil **1989**, *119*, 295–303.

SLAKING, DISPERSION, AND CRUST FORMATION

Hwat Bing So
The University of Queensland, St. Lucia, Queensland, Australia

INTRODUCTION

Soil structure is generally defined as the arrangement, orientation, and organization of the primary particles of sand, silt, and clay into compound aggregates, which exhibit properties that are unequal to the properties of a mass of nonaggregated material with a similar texture (6). Therefore the nature of soil structure is that it conveys specific properties to the soil and any alteration, i.e., breakdown or structural development, to the soil structural units will affect the physical properties of the soil. The aggregation and organization of the soil particles tend to form a hierarchical order (4, 5) where the lower orders tend to have higher densities and greater internal strength than the higher orders. A schematic diagram of the hierarchical nature of soil structural elements in a clay soil is given in Fig. 1 (4). Clay particles tend to form domains (packets of parallel clay sheets, generally consisting of 5–7 sheets), in turn several domains form clusters, followed by several orders of clusters, micro- and macroaggregates. The hierarchical nature implies that the destruction of a lower order will result in the destruction of all higher hierarchical orders. An example is the dispersion of sodic clay domains which results in the destruction of all higher orders, resulting in a dense soil with low hydraulic conductivity. Hence the clay domains are the fundamental building blocks of the soil and its integrity may determine the soil's physical properties and behavior.

MECHANISMS OF SLAKING, DISPERSION, AND CRUST FORMATION

Soil aggregates breakdown in response to an external force or energy source such as tillage or raindrop energy, or internally generated forces such as when aggregates are subjected to rapid wetting, with the latter being the most common cause operating in the field. Upon rapid wetting, aggregates breakdown in two stages. The first is "slaking" where macroaggregates breakdown into microaggregates (20–250 μm in diameter). The second stage is "dispersion" where the microaggregates breakdown into individual particles of sand, silt, and clay. The result of these processes is the formation of a surface seal at the soil–air interface, associated with an increased packing density of the soil particles and a reduction in pore sizes. These, in turn, give rise to a reduced hydraulic conductivity or infiltration rates. Upon drying, the surface seal dries out into a surface crust. Surface sealing and crusting is a common phenomena with cultivated soils (Fig. 2). When slaking and dispersion occurs throughout the surface horizon following wetting, the whole surface horizon experienced an increased packing density and a reduction in pore sizes. In cases where the surface horizon hardens rapidly following drying, the soil is referred to as hardsetting (7, 8). These soils are associated with high contents of silt and fine sand which do not readily form aggregates.

Slaking of the surface soil is not necessarily a detrimental process. In its virgin state, the surface soil of self-mulching vertisols slakes upon wetting followed by surface sealing, yet these are highly productive. However, when slaking is followed by dispersion, the soil's productivity is generally reduced associated with the lower soil hydraulic conductivity, which in turn limits the supply of water to the root zone.

Mechanisms of Slaking

Slaking following rapid wetting is the result of three distinct processes:

1. softening of bonds;
2. differential swelling following wetting of the aggregates from the outside. The stresses formed between the swelling outer layers and the inner parts of the aggregates result in the successive disintegration of the outer layers of the aggregates, readily observed when an aggregate is dropped in water;
3. increased pressure within entrapped air pockets from the tensions exerted by the water meniscus. Combined with the softening of the bonds, this pressure results in miniexplosions of the soil aggregates when dropped in water.

Encyclopedia of Soil Science

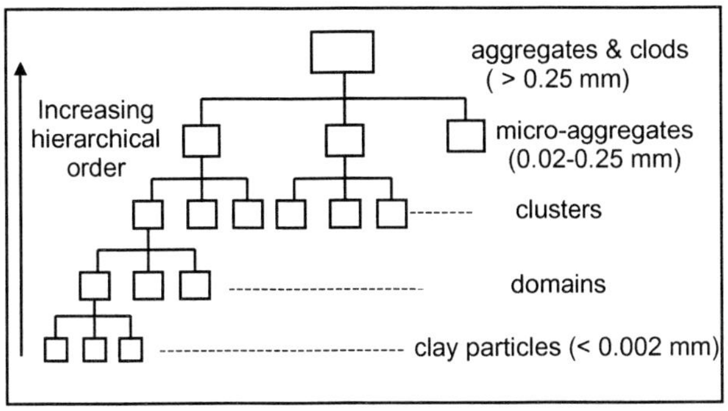

Fig. 1 Hierarchical organization of particles and structural elements. (Adapted from Ref. 4.)

Mechanisms of Dispersion

Dispersion is the result of swelling pressures derived from the osmotic pressure difference between the ion concentrations within the diffuse double layer between the clay particles and the external solution created by the addition of water. Dispersion occurs when the swelling pressure (repulsive force) is greater than the attractive forces

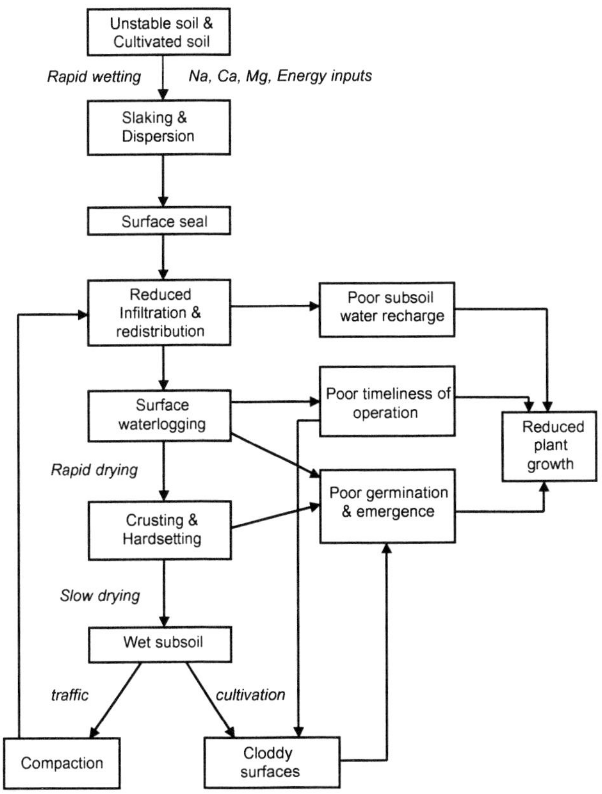

Fig. 2 A schematic diagram of how slaking and dispersion may affect the soil physical properties and plant growth. (Adapted from Ref. 13.)

between the clay particles. Dispersion is seen as the cloudy appearance around the aggregates when left in water. In the field it is seen as the cloudy and muddy appearance of water in puddles or in streams and rivers near construction sites or cultivated fields, after a rainstorm. Swelling of the clay domains is enhanced by the presence of monovalent cations such as sodium (Na) and the presence of low electrolyte concentrations in the water or soil solution. Swelling is suppressed by the presence of divalent and trivalent cations (1) and by the presence of high electrolyte concentrations.

CONSEQUENCES OF SLAKING AND DISPERSION

Slaking and dispersion are the fundamental processes of degradation of structurally unstable soils, and most cultivated soils exhibit some degree of instability. These degradation processes can be associated with many soil physical problems observed in the field that affect plant growth or soil behavior (12, 13). Waterlogging, reduced infiltration, crusting, hardsetting, poor timing of farming operations, and poor germination and establishment are generally perceived as distinctly different problems in the field, but as shown in Fig. 2, they have a common cause in slaking and dispersion and can therefore be corrected using the same treatment. Hence an ameliorative treatment applied to reduce or prevent dispersion and to correct one particular problem will also correct many of the other physical problems. The most important effect of slaking and dispersion is the reduction of the soil's hydraulic conductivity (K) associated with the blockage of pores between the larger particles or stable aggregates. A small increase in dispersion may result in large decreases in K (11, 14).

FACTORS AFFECTING SUSCEPTIBILITY TO SLAKING AND DISPERSION

Since slaking and dispersion are associated with swelling of the clay domains and the inability of the forces of attraction or bonding to hold the clay domains together, they are affected by the nature of the clay minerals, the adsorbed cations and the bonding mechanism.

Clay Type and Composition of Adsorbed Cations

Swelling clays (montmorillonites, vermiculites and illites) tend to confer a degree of instability on soil aggregates whereas nonswelling clays (kaolinites) are generally associated with high structural stability. Swelling is enhanced by the presence of sodium and low electrolyte concentrations. Sodium dominated clays continue to swell when the water content is increased and float apart in suspension (1), which is referred to as spontaneous dispersion. Swelling, slaking and dispersion will also increase in response to increasing proportion of Na or exchangeable sodium percentage (ESP). Although there is laboratory evidence for a threshold ESP (10), there is no clear evidence that a threshold ESP exists in field soils (13). However, for soil classification purposes, a soil is classified as sodic in Australia when the soil's ESP is equal to or greater than 6% (8), and in the U.S. when the ESP is equal to or greater than 15% (14), indicating the increased probability of physical problems occurring on these soils when used for cropping. On the other hand, dispersion is reduced when the electrical conductivity (EC) of the bulk solution increases. Therefore, a soil with a high ESP and high salinity may appear as well structured when dry, but will immediately disperse following rainfall or when irrigated with low EC water.

Calcium-dominated clays exhibit limited swelling. Ca-montmorillonite swells to a basal distance between two clays of 18 Å (9). At this distance, the electrostatic attractive forces between the clay plates are greater than the repulsive or swelling pressures between the plates. Calcium and magnesium dominated clays tend to flocculate whereas sodium dominated clays tend to disperse. However, Mg-clay does not have the same stability as Ca-clay and its tendency to disperse will depend on its combination with other factors, e.g., Mg-clay will disperse following cultivation at much lower water contents than Ca-clay (3). Soils with similar ESP tend to show greater dispersion when the other cation Mg is compared to Ca (2).

Trivalent cations such as Al^{3+} and Fe^{3+} confer the greatest stability onto clay domains. Hence the presence of Al and Fe oxides and hydroxides ensures a large proportion of these cations besides providing bonding materials to the aggregates. Highly acidic soils ($pH<5$) will have a high concentration of Al in solution and these will stabilize the clay domains. Hence oxisols are generally nondispersive soils and possess high structural stability.

Nature of Aggregate Bonding

Swelling and dispersion may be prevented if slaking of aggregates is prevented by strong bonds. The most common bonding materials for aggregates are organic matter, iron and aluminum oxides and hydroxides, Ca-carbonates and silica oxides, arranged in the order of abundance. The strength of bonding increases with the amount present in the soil. The material that is readily

Table 1 Strategies for managing slaking and dispersion

	Aggregate hierarchical level affected	Action required	Management options
Slaking	Macroaggregates	Prevent breakdown of bonds	1. Zero tillage or minimum tillage 2. Slow wetting during irrigation 3. Reduce droplet energy during spray irrigation
		Increased aggregate bonding and bonding strength	1. Incorporate organic matter 2. PVA 3. PAM 4. Lime or calcium carbonates 5. Introduce grasses in crop rotation
Dispersion	Microaggregates and clay domains	Increase flocculation: 1. Replace monovalent cations with di or polyvalent cations (permanent exchangeable cation effect)	1. Gypsum application 2. Lime application
		2. Increase electrolyte concentrations (temporary electrolyte effect)	3. Slaked lime ($Ca(OH)_2$) application

affected by human activity is soil organic matter. In the virgin and undisturbed state, soils with high ESP are generally nondispersive because of the accumulated soil organic matter in the surface soil. However, cultivation reduces the soil organic matter contents and when the bond strength from organic matter decreases below the swelling pressures from the sodic clay, dispersion occurs followed by the appearance of soil physical problems. Hence, the addition of adequate organic matter that will readily breakdown and produce suitable bonding material may prevent or reduce dispersion of the clay domains.

MANAGEMENT OF SLAKING AND DISPERSION

Managing slaking and dispersion of clay soils is the essence of managing soil structure. Table 1 lists a range of management options that have been used. Slaking can be managed at the macroaggregate level by avoiding the breakdown of existing bonds (organic matter) through minimizing soil disturbance (zero tillage) and/or through the addition of bonding material such as organic matter, polymers such as polyvinyl alcohol (PVA), polyacrylamide (PAM), and calcium carbonates (lime). It can also be reduced by incorporating grasses into the existing crop rotation as grass roots and associated fungal hyphae binds microaggregates into macroaggregates.

On the other hand, dispersion can be managed at the microaggregate level by using a combination of divalent cations and high EC in the soil solution, e.g., the use of Ca ameliorants such as gypsum ($CaSO_4 \cdot 2H_2O$) which provides Ca^{2+} and a continuous presence of adequate EC levels associated with its limited solubility (2.2 dS/m at 25°C). Gypsum is most useful on Na-dominated clays and other soils when dispersion occurs, e.g., following excessive cultivation. However, it may not be helpful on clays which are already strongly dominated by Ca ions. Ca^{2+} replaces Na^{+} from the clay exchange sites (exchangeable cation effect) which represent a permanent or residual effect of gypsum. The presence of a high electrolyte where free gypsum is present provides the dominant effect on sodic soils. It represents the temporary electrolyte effect of gypsum (11, 13). Lime will provide a similar effect but it is a slow-acting ameliorant.

REFERENCES

1. Aylmore, L.A.G.; Quirk, J.P. The Structural Status of Clay Systems. Clays Clay Miner **1962**, 9, 104–130.
2. Bakker, A.C.; Emmerson, W.W.; Oades, J.M. The Comparitive Effects of Exchangeable Calcium, Magnesium, and Sodium on Some Soil Physical Properties of Red–Brown Earth Subsoils 1. Exchange Reactions and Water Contents for Dispersion of Shepparton Soil. Aust. J. Soil Res. **1973**, *11*, 143–150.
3. Emmerson, W.W. Physical Properties and Structure. In *Soil Factors in Crop Production in a Semi-Arid Environment*; Russell, J.S., Greacen, E.L., Eds.; University of Queensland Press: St Lucia, Australia, 1977.
4. Dexter, A.R. Advances in Characterisation of Soil Structure. Soil Till. Res. **1988**, *11*, 199–238.
5. Hadas, A. Long-Term Tillage Practice Effects on Soil Aggregation Modes and Strength. Soil Sci. Soc. Am. J. **1987**, *51*, 191–197.
6. Hillel, D. *Introduction to Soil Physics*; Academic Press: Orlando, 1982.
7. Mullens, C.E.; McLeod, D.A.; Northcote, K.H.; Tisdall, J.M.; Young, I.M. Hardsetting Soils: Behaviour, Occurrence and Management. Adv. Soil Sci. **1990**, *11*, 37–108.
8. Northcote, K.H.; Skene, J.K.M. *Australian Soils with Saline and Sodic Properties*; Soil Publication No. 27, CSIRO: Australia, 1972.
9. Quirk, J.P. Interparticle Forces: A Basis for the Interpretation of Soil Physical Behaviour. Adv. Agron. **1994**, *53*, 121–183.
10. Shainberg, I.; Letey, J. Response of Soils to Sodic and Saline Conditions. Hilgardia **1984**, *52*, 1–57.
11. So, H.B.; Tayler, D.W.; Yates, W.J.; McGarity, J.W. Amelioration of Structurally Unstable Grey and Brown Clays. In *Modification of Soil Structure*; Emmerson, W.W., Bond, R.D., Dexter, A.R., Eds.; Wiley: Chichester, 1978.
12. So, H.B.; Smith, G.D.; Raine, S.R.; Schafer, B.M.; Loch, R.J. *Sealing, Crusting and Hardsetting Soils: Productivity and Conservation*; Australian Society of Soil Science Inc.: Queenland Branch, Brisbane, 1985.
13. So, H.B.; Aylmore, L.A.G. How do Sodic Soils Behave? the Effect of Sodicity on Soil Physical Behaviour. Aust. J. Soil Res. **1983**, 31.
14. So, H.B.; Cook, G.D. The Effect of Slaking and Dispersion on the Hydraulic Conductivity of Clay Soils. In *Soil Surface Sealing and Crusting*; Poesen, J.W.A., Nearing, M.A., Eds.; Catena Suppl., 1993; 24, 55–64.

SODIC SOILS

Pichu Rengasamy
Adelaide University, Glen Osmond, South Australia, Australia

INTRODUCTION

Farming land in regions where annual rainfall is below the level of evapotranspiration without the possibility of supplemental water input by irrigation is considered to be dryland. However, successful crop production is sustainable in regions with annual rainfall between 250 and 600 mm when soil management and agronomic practices are developed to suit rainfall pattern. Efficient capturing of rainfall and storing in soil layers, and subsequent water use efficiency by crops are critical for farming in dryland.

Sodic soils are widespread in arid and semiarid regions of the world extending up to 30% of the total land area (Table 1). Farming practices in these soils are mainly performed under dryland conditions. Sodicity is a latent problem in many salt-affected soils where deleterious effects on soil properties are evident only when salts are leached below a threshold level (1). While soil salinity reduces plant growth, directly affecting physiological functions through osmotic and toxicity effects on plants, sodicity causes deterioration of soil physical properties indirectly affecting plant growth and survival. Sodic soils are subjected to severe structural degradation and exhibit poor soil-water and soil-air relations; these properties (see Table 2) adversely affect root growth, restricting plant production and making the soils difficult to work when wet or dry.

YIELD DECLINE IN SODIC SOILS

Sodic soils make the paddocks prone to waterlogging, poor crop emergence and establishment, gully erosion, and in some instances tunnel erosion. Because of the heterogeneity in the accumulation of sodium by soil particles, these symptoms may only be observed in certain parts of the paddock. Generally, patchy growth and barren patches are visible in a number of spots in a paddock while the rest of the field may look normal. However, the effects of sodicity are fully realized in the harvested yield. The actual yield obtained in sodic soils is often less than half of the potential yield expected on the basis of climate, particularly rainfall and evapotranspiration (2, 3). Relative yield of cereals grown in dryland sodic soils in Australia in relation to average root zone exchangeable sodium percentage (ESP) is given in Fig. 1.

Swelling and dispersion of sodic aggregates destroy soil structure, reduce the porosity and permeability of soils, and increase the soil strength even at low suction (i.e., high water content). These adverse conditions restrict water storage and transport. Soils are, therefore, either too wet immediately after rain or too dry within a few days for optimal plant growth. Thus, the range of soil water content that does not limit plant growth and function (*nonlimiting water range*) is very small (4). Dense, slowly permeable sodic subsoils reduce the supplies of water, oxygen, and nutrients needed for obtaining maximum potential yield. During the rainy season, even with prolonged ponding of water on the surface, only a small increase in water content occurs in subsoil. The low porosity leads to slow internal drainage and water redistribution within the profile (5). This reduction in water storage causes crop water stress during prolonged dry periods. Subsoil as a source of water and nutrients becomes more important in dryland cropping regions than in irrigated soils.

SALT ACCUMULATION IN ROOT ZONES OF SODIC SOILS

Soils with sodic subsoils are characterized by moderate to high exchangeable sodium and, in many cases, with high pH (>8.5) where carbonate and bicarbonate minerals are present. Subsoil sodicity restricts drainage beyond the root zone and as a result salts accumulate in this zone. The concentration of accumulated salts fluctuates with rainfall pattern, input of salt from agronomic practices, and soil weathering, as schematically explained in Fig. 2. Dryland salinity or *seepage salinity* in many countries is associated with rising saline groundwater tables. However, the extent of subsoil salinity, also called *transient salinity*, not associated with saline groundwater is large in many landscapes dominated by subsoil sodicity. A relationship between rainfall, subsoil ESP, and EC_e for northeastern Australian soils has been reported (6). In dryland regions

Encyclopedia of Soil Science

Table 1 World distribution of sodic soils

Continent	Country	Area of sodic soils (000 ha)
North America	Canada	6,974
	U.S.	2,590
South America	Argentina	53,139
	Bolivia	716
	Brazil	362
	Chile	3,642
Africa	Algeria	129
	Angola	86
	Botswana	670
	Cameroon	671
	Chad	5,950
	Ethiopia	425
	Ghana	118
	Kenya	448
	Liberia	44
	Madagascar	1,287
	Namibia	1,751
	Niger	1,389
	Nigeria	5,837
	Somalia	4,033
	Sudan	2,736
	Tanzania	583
	Zimbabwe	26
South Asia	Bangladesh	538
	India	574
	Iran	686
North and Central Asia	China	437
	U.S.S.R.	119,628
Australasia	Australia	339,971

(From Ref. 7.)

with annual rainfall between 250 and 600 mm, sodic subsoils have an EC_e between 2 and 20 that can dramatically affect crop production through osmotic effects during dry periods. Laboratory measured EC_e will increase severalfold under field conditions as the soil layers dry in between rainy days. The combination of poor water storage and osmotic stress enhance water stress to crops under dryland cropping.

ROOT ZONE CONSTRAINTS IN DRYLAND SODIC SOILS

Multiple problems occur in soils with subsoil sodicity. Soil compaction, crusting, and induration of subsoil layers require "physical" reclamation. Sodicity, salt accumulation, and alkaline pH require "chemical' reclamation. All of these conditions cause, in addition to water stress, macro- and micronutrient deficiency and toxicity due to Na^+, Cl^-, HCO_3^-, CO_3^{2-}, B, $Al(OH)_4^-$, and others. Low organic matter and biological activity compound these problems encountered in sodic subsoils.

MANAGEMENT OF DRYLAND SODIC SOILS

Major criteria in increasing productivity in dryland sodic soil are improved water storage and transport in the root zone and crop water use efficiency. More information is

Table 2 Physical and chemical properties of a typical sodic soil profile in South Australia

Properties	0–20 cm	20–40 cm	40–100 cm
Chemical properties			
$pH_{1:5}$ (water)	7.9	8.9	9.2
EC_e (dS m^{-1})	0.4	3.8	4.9
Organic carbon (%)	1.2	0.6	0.3
Exchangeable sodium (%)	6.2	14.6	24.5
$CaCO_3$ (%)	0.1	2.8	4.5
Boron (mg kg^{-1})	1.2	22.0	38.5
Water soluble $Al(OH)_4^-$ (mg kg^{-1})	0.0	1.2	2.6
Physical properties			
Spontaneously dispersed clay (%)	1.2	8.6	9.4
Swelling (mm mm^{-1})	0.04	0.18	0.20
Hydraulic conductivity at saturation (mm day^{-1})	22.8	4.5	2.3
Penetrometer resistance at 100 kPa suction (MPa)	1.8	4.2	4.8
Aeration porosity (%)	9.7	4.8	3.9
Bulk density (Mg m^{-3})	2.0	2.2	2.3
Final infiltration rate in the field (mm h^{-1})	0.2		

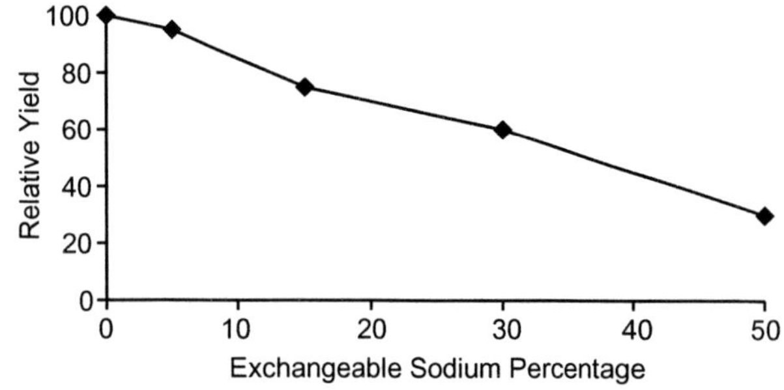

Fig. 1 Relative yield of cereals grown in Australian sodic soils in relation to average root zone ESP.

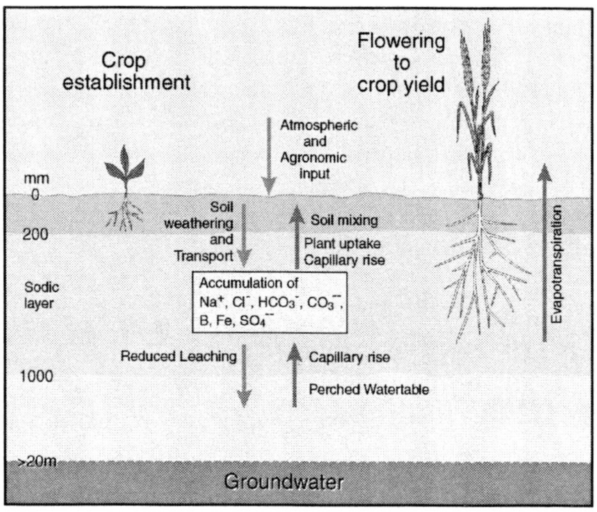

Fig. 2 Schematic diagram explaining salt accumulation in sodic subsoils.

available on agricultural management in sodic soils that is more relevant to irrigated lands (5). Reclamation procedures involving high costs are prohibitive in dryland regions because of low benefit/cost ratio.

Diagnosis of multiple problems with large variations, vertically and horizontally across the paddock, is primarily important. Gypsum is the most commonly used compound to reclaim sodic soils. Subsoil reclamation may involve higher rates of gypsum application or deep placement of gypsum by deep ripping or deep ploughing. Salt-tolerant plant species may alleviate subsoil salinity. Plants that can tolerate ion toxicity such as boron, carbonate, sodium, and chloride have also been identified. Strategies to improve subsoil fertility may include 1) mechanical means of placing nutrients deeper in the profile; 2) using nutrient sources of lower or higher mobility; 3) using deep-rooted legumes to fix nitrogen at depths; and 4) selection of plant species and genotypes better suited to acquiring nutrients from subsoils. Future research is needed on developing plants that modify the rhizosphere and adapt to edaphic conditions. Farming systems should be developed to prevent accumulation of salts and toxic elements in the root zone of sodic soils.

REFERENCES

1. Rengasamy, P.; Olsson, K.A. Sodicity and Soil Structure. Aust.J.Soil Res. **1991**, *31*, 821–837.
2. French, R.J.; Schultz, J.E. Water Use Efficiency of Wheat in a mediterranean-Type Environment. 1. The Relation Between Yield, Water Use and Climate. Aust. J. Agric. Res. **1984**, *35*, 765–775.
3. Rengasamy, P. Sodic Soils. In *Methods for Assessment of Soil Degradation*; Lal, R., Blum, W.H., Valentine, C., Stewart, B.A., Eds.; CRC Press: New York, 1997; 265–277.
4. Letey, J. The Study of Soil Structure: Science or Art. Aust.J.Soil Res. **1991**, *29*, 699–707.
5. Oster, J.D.; Jayawardane, N.S. Agricultural Management of Sodic Soils. In *Sodic Soils: Distribution, Properties, Management and Environmental Consequences*; Sumner,-M.E., Naidu, R., Eds.; Oxford University Press: New York, 1998; 125–147.
6. Shaw, R.J.; Coughlan, K.J.; Bell, L.C. Root Zone Sodicity. In *Sodic Soils: Distribution, Properties, Management and Environmental Consequences*; Sumner, M.E., Naidu, R., Eds.; Oxford University Press: New York, 1998; 95–106.
7. Bui, E.N.; Krogh, L.; Lavado, R.S.; Nachtergaele, F.O.; Toth, T.; Fitzpatrick, R.W. Distribution of Sodic Soils: The World Scene. In *Sodic Soils: Distribution, Properties, Management and Environmental Consequences*; Sumner, M.E., Naidu, R., Eds.; Oxford University Press: New York, 1998; 19–33.

SODIC SOILS, FORMATION, AND GLOBAL DISTRIBUTION OF

S

Brian Murphy
Centre for Natural Resources, Cowra, Australia

INTRODUCTION

Sodic soils, which are considered a subset of salt affected soils, have levels of exchangeable sodium that are sufficiently high to cause swelling and clay dispersion in water having low electrolyte concentrations, such as under rainfall and in many irrigation waters. These properties cause serious environmental and productivity problems.

A functional definition of a sodic soil is:

"Whenever exchangeable sodium is present in sufficient concentrations to influence soil behaviour, sodicity occurs, a soil is said to be sodic" (1).

Precise problems depend on where sodicity is in profile. Typical problems associated with sodic soils include (2, 3):

- dense impermeable subsoils causing waterlogging and restricting root growth;
- surface crusting and sealing restricting emergence and surface infiltration, hardsetting seedbeds;
- development of a toxic chemical environment for root growth;
- high susceptibility to erosion especially deep gully erosion (Fig. 1);
- contamination of surface waters by sediments and colloid assisted contaminants;
- instability in engineering structures piping or tunnelling can occur in structures associated with these soils (3).

ASSESSMENT AND RECOGNITION

The occurrence of any of the above problems is often sufficient to indicate at least the possibility that soils are sodic. From a classification viewpoint, sodic soils are traditionally considered to have dense impermeable subsoils with a coarse columnar to prismatic structure. Surface soils are often of a lighter texture than the subsoil, and may be bleached in the lower layer (A2 or E horizon) (Fig. 2). These soils are identified in the FAO/UNESCO, Russian, Canadian, and old Australian schemes by the terms solonetz or soloth (4, 5). In *Soil Taxonomy* this group of soils relates to the Natric Alfisols. More recently, in a new Australian soil classification, the order Sodosols is used to identify soils with sodic properties (6). However, soils with sodic layers are common in many other soil types, especially the Vertisols and Aridisols of *Soil Taxonomy* (7) and the Vertosols of the Australian Soil Classification System (6).

The level of exchangeable sodium percentage (ESP), which is the amount of exchangeable sodium as a percentage of the total cation exchange capacity of the soil, is used to confirm the identification of sodic soils. ESP is used as this gives a strong indication of the actual and potential dispersion behavior of the soil. Two of the most commonly used critical levels of ESP are 15% for U.S. soils (8, 9) and 6% for Australian soils (1, 6, 10). The level of ESP that begins to adversely affect soil behavior can vary with the level of soluble salts in the soil, the nature of soil clays, soluble minerals present in the soil, and other factors. Sumner (11) presents a more detailed history of the definition of sodic soils and the development of critical values of ESP.

As dispersion is the most critical property that causes the limitations of these soils, measures of dispersion have been used to identify sodic soils. Dispersion can be measured by aggregate tests or suspension tests (1). Spontaneous dispersion of soil aggregates in water is usually a strong indication of the presence of a sodic soil. However, the accuracy of dispersion tests to predict the presence of sodic soils is influenced by the level of electrolyte concentration as described in Fig. 3, with the critical level of ESP that causes dispersion being dependent on the electrical conductivity of the soil solution. As a general rule, classification systems have adopted the convention that the levels of ESP used to identify sodic soils are those that predict the dispersion behavior of the soil in water with a low electrolyte concentration (rainwater or fresh water) (11).

Sodic and Saline Soils

The close association between saline and sodic soils has led to some confusion when discussing these soils. Sodic soils are salt-affected soils but are not necessarily saline. Sodicity

Fig. 1 Severe gully erosion in a sodic subsoils formed on granite parent material.

Fig. 2 Example of a typical profile of a sodic soil profile—solonetzic soil or soloth–Sodosol–Natric Xeralf.

relates to the level of exchangeable sodium in the soil with the positive charge of the sodium ion being balanced by the negative charge on the clay minerals and organic matter. Soils may be sodic without being saline. Salinity relates to the level of free salt present in the soil solution. In saline soils the positive change of the soluble cations, such as sodium ions, is balanced by anions such as chloride, sulfate, or bicarbonate present in the soil solution.

In general usage now the following soils have come to be recognized (1, 11):

1. Sodic and saline soils with a high ESP (>15% in the U.S., >6% in Australia) and high soluble salts (EC sat > 4 dS/m), where EC sat is the electrical conductivity of the saturation extract.
2. Sodic and nonsaline soil with a high ESP (>15% in the U.S., >6% in Australia) and low soluble salts (EC sat < 4 dS/m).

ORIGIN AND FORMATION

Sodium, which on average is 2.64% of the atoms in the Earth's crust (13), is highly mobile on the landscape and tends to become the dominant cation as solutions move further from their original source. This has a significant effect on the origins and formation of sodic soils. Ultimately the only condition for sodic soils to form is that sodium becomes accumulated on the clay exchange complex. Therefore there are numerous mechanisms for sodic soils to form.

Traditionally sodic soils are thought to develop by the following pathway as described in (4):

1. Initial salinization of the soil from saline parent material or rising saline groundwater.
2. Leaching of salts, and dispersion and downward movement of clays (illuviation) to form an abrupt texture change at the top of the B horizon.

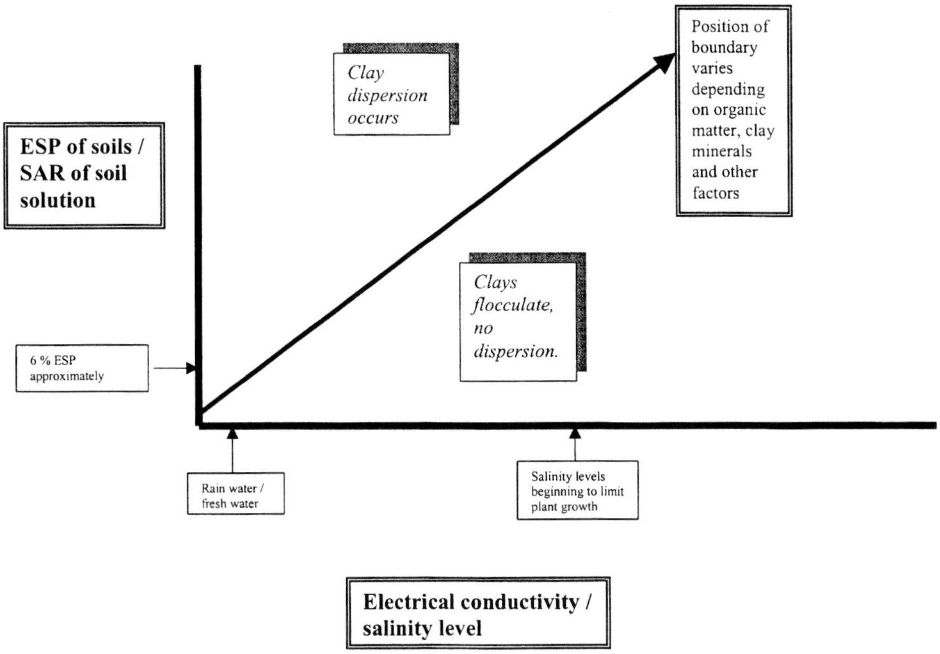

Fig. 3 Schematic diagram of effect of exchangeable sodium levels (ESP), Sodium Adsorption Ratio (SAR), and electrical conductivity levels on dispersion in sodic soils. (Adapted from Ref. 12.)

The link to saline and arid areas for sodic soils is evident.

Subsequent work has shown that this apparently simple process can, when combined with weathering and the pedological, geomorphological, and chemical processes occurring on the landscape, lead to a number of pathways for the development of sodic soils (4, 14).

The Formation of Sodic Soils

A brief summary of the possible pathways for formation of sodic soils is given below and is summarized in Fig. 4. At any location, more than one pathway for the formation of sodic soils can be operating.

Geological—Long-term development of sodic soils

1. Weathering of rocks with high levels of sodium relative to calcium and magnesium can release sodium, especially from sodic plagioclases and alkali feldspars, and lead to the development of sodic soils associated with these rocks (4).
2. Aeolian sources of salt and therefore sodium cations are a major source in many areas (14). The ocean is the main source of aeolian salt in some areas, but aeolian salt blown from arid terrestrial sources is highly significant for many areas of sodic soils.
3. The erosion and deposition of sodic clays is another possible pathway for sodic soils to develop on the landscape. Sodic clays are especially susceptible to water erosion and the erosion and deposition may result in the formation of sodic clays in areas of deposition. Saline sodic clays may be susceptible to wind erosion because of aggregation associated with their salinity.
4. The build up of sodium and salt in association with marine incursions of the sea in the Quaternary or Tertiary periods (5, 14).

Short-term development of sodic soils associated with human activities

1. Rising groundwater with high salinity can lead to salinization of the soil and the development of sodic soils.
2. Irrigation with water with high sodium levels relative to calcium and magnesium can lead to the development of sodic soils.

GLOBAL DISTRIBUTION OF SODIC SOILS

Sodic soils are widespread throughout the world, many of them occurring in important agricultural areas. Most of

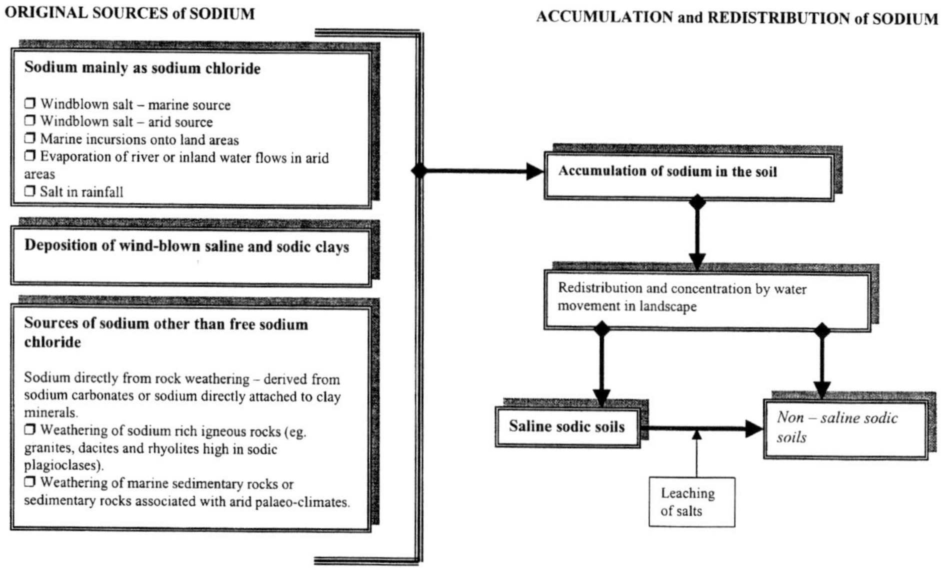

Fig. 4 Schematic diagram summarizing the origins and development of sodic soils.

them occur in drier areas but significant amounts also occur in more humid areas. The total area of sodic soils is estimated by Lubbock (15) to be more than 550 million hectares. Szabolcs (16) estimated that 932 million ha were affected by salinity/sodicity worldwide of which 581 million ha were sodic soils. Although there are difficulties in estimating such figures because of varying criteria for identifying sodic soils and the lack of specific on-the-ground data, this estimate does give an indication of the potential problem of sodic soils as a world environmental and productivity issue. Sodic soils occur on all the continents but the largest areas are in the driest continent, Australia.

The broad distribution of sodic soils on a global basis is given in Table 1 (4).

Table 1 Summary of the global distribution of sodic soils

Continent	Estimated area (million ha)
North America	9.6
South America	57.9
Africa	27.1
Europe	22.9
Northern and Central Asia	120.1
Southern Asia	1.8
Australasia	340

(From Refs. 4, 5, 10, 14, 16.)

CONCLUSION

Sodic soils are a major group of problem soils. They are potentially productive if managed properly, but can be severely degraded or cause severe environmental problems if managed inappropriately. It is likely that increasing salinization worldwide will increase the areas of these soils. Therefore, knowledge about the origins, formation, distribution, and management of these soils is seen as a valuable input to managing scarce and vulnerable natural resources.

REFERENCES

1. Rengasamy, P.; Churchman, G.J. Cation Exchange Capacity, Exchangeable Cations and Sodicity. In *Soil Analysis, an Interpretation Manual*; Peverill, K.I., Sparrow, L.A., Reuter, D.J. Eds.; CSIRO Publishing: Melbourne, Australia, 1999; 147–158.
2. So, H.B.; Aylmore, L.A.G. How Do Sodic Soils Behave? The Effects of Sodicity on Soil Behaviour. Australian Journal of Soil Research **1993**, *31*, 761–778.
3. Sumner, M.E.; Miller, W.P.; Kookana, R.S.; Hazelton, P. Sodicity, Dispersion and Environmental Quality. In *Sodic Soils, Distribution, Properties, Management and Environmental Consequences*; Sumner, M.E., Naidu, R., Eds.; Oxford University Press: New York, NY, 1998; 149–172.
4. Bui, E.N.; Krogh, L.; Lavado, R.S.; Nachtergaele, F.O.; Toth, T.; Fitzpatrick, R.W. Distribution of Sodic Soils: The World Scene. In *Sodic Soils, Distribution, Properties,*

Management and Environmental Consequences; Sumner, M.E., Naidu, R., Eds.; Oxford University Press: New York, NY, 1998; 19–33.

5. Gupta, R.K.; Abrol, I.P. Salt-Affected Soils: Their Reclamation and Management for Crop Production. Advances in Soil Science **1990**, *11*, 224–288.
6. Isbell, R.F. *The Australian Soil Classification*; CSIRO Publishing: Melbourne, Australia, 1996.
7. U.S. Soil Survey Staff, *Soil Taxonomy: A Basic System of Soil Classification for Making and Interpreting Soil Surveys*; USDA, Soil Conservation Service: Washington, DC, 1975.
8. USSL Staff, *Diagnosis and Improvement of Saline and Alkali Soils*; USDA, U.S. Government Printing Office: Washington, DC, 1954.
9. Sposito, G. *The Chemistry of Soils*; Oxford University Press: New York, NY, 1989.
10. Northcote, K.H.; Skene, J.K.M. *Australian Soils with Saline and Sodic Properties*. CSIRO Australia Soil Publication No. 27, 1972.
11. Sumner, M.E.; Rengasamy, P.; Naidu, R. Sodic Soils: A Reappraisal. In *Sodic Soils, Distribution, Properties, Management and Environmental Consequences*; Sumner, M.E., Naidu, R., Eds.; Oxford University Press: New York, NY, 1998; 3–17.
12. Rengasamy, P.; Greene, R.S.B.; Ford, G.W.; Mahanni, A.H. Identification of Dispersive Behaviour and the Management of Red-Brown Earths. Australian Journal of Soil Research **1984**, *22*, 413–431.
13. Paton, T.R. *The Formation of Soil Material*; George Allen and Unwin: London, UK, 1978.
14. Isbell, R.F.; Reeve, R.; Hutton, J.T. Salt and Sodicity. *Soils—An Australian Viewpoint*; Division of Soils, CSIRO, CSIRO Publishing: Melbourne, Australia, 1983; 107–117.
15. Lubbock in Bui, E.N.; Krogh, L.; Lavado, R.S.; Nachtergaele, F.O.; Toth, T.; Fitzpatrick, R.W. Distribution of Sodic Soils: The World Scene. In *Sodic Soils, Distribution, Properties, Management and Environmental Consequences*; Sumner, M.E., Naidu, R., Eds.; Oxford University Press: New York, NY, 1998; 19–33.
16. Szabolcs, I. *Salt-Affected Soils*; CRC Press: Boca Raton, FL, 1989.

SODIC SOILS, IRRIGATION FARMING AND

David Burrow
D.N.R.E. Agriculture Victoria, Tatura, Victoria, Australia

INTRODUCTION

Sodic soils, which occur naturally in many of the arid and semi-arid regions of the world, exhibit poor chemical and physical properties, particularly when surface waters of low total salt concentration are used for irrigation. The increased use of poorer quality saline–sodic waste-waters and groundwaters for irrigation, which are frequently blended with surface waters, has also led to increased land area affected by sodicity. With the increased use of saline–sodic irrigation water and its application to many soil types, there is an increasing need to better understand the processes underpinning management of sodic and saline–sodic soils for the maintenance or improvement of crop productivity.

PROPERTIES OF SODIC SOILS AFFECTING IRRIGATION FARMING

Hydraulic Properties

The movement of water into the soil following rainfall or surface irrigation, its redistribution within sub-surface layers, and drainage below the rootzone are important for productive irrigation farming. Water entry into the soil, measured by the infiltration rate (IR), and through soil layers, measured by the hydraulic conductivity (HC) both decrease with increasing exchangeable Na and decreasing soil salinity. For strongly sodic soils, heavy rainfall or low-salinity irrigation may lead to surface waterlogging on flat land or run-off on sloping ground (1). Similarly, freshwater irrigation of sodic soils with low HC may lead to shallow depths of wetting, uneven wetting, and low plant available water capacity (2). These hydraulic traits, together with lower air-filled porosity and excessive penetration resistance in dry soils, can restrict plant growth (3).

The low IR of a strongly sodic soil during irrigation is usually attributed to the formation of a surface seal or crust, which in turn derives from the physiochemical dispersion of clay-sized particles from unstable surface aggregates. These dispersed clay particles migrate with water flow and eventually block conducting pores.

Early studies showed that soil HC varies up to 2 orders of magnitude, depending on the salt concentration of a percolating solution and soil exchangeable sodium percentage (4). Reductions in HC were explained by swelling and dispersion of clay particles but these factors are modulated by clay mineralogy, sesquioxide and $CaCO_3$ content, soil texture, organic matter content, pH, soil weathering, and other exchangeable cations.

The effects of soil sodicity on IR and HC, resulting from saline–sodic irrigation are exacerbated in climatic zones with a marked rainfall season, due to the leaching of salts from surface soil horizons. Rainfall, with negligible salinity ($EC < 0.1\ dS\ m^{-1}$) infiltrates and reduces soil EC below a threshold electrolyte concentration (TEC) in these layers, resulting in clay swelling or dispersion. In the field, these processes are not necessarily detrimental to HC as they may be accommodated by pore geometry and slow wetting rates, or circumvented by preferential flow through large cracks. However, the maintenance of soil EC above TEC is an essential first step toward managing sodic soils under irrigation.

Nutrient Status

In sodic soils, high Na:Ca and Mg:Ca ratios reduce the amount of Ca available for plant uptake and can induce Ca deficiencies in crops. Data, obtained principally from greenhouse and laboratory studies, indicate that cereal species are sensitive to Ca deficiencies and particularly during seedling growth stages (5).

Poor productivity on sodic soils has also been attributed to lowered availability and plant uptake of nitrogen or micronutrients. Nitrogen availability in these soils is restricted under anaeorbic conditions (6), while micro-nutrient availability is controlled by adsorption and precipitation reactions (7).

The relative solubility of boron (B) and high Na/Ca ratios in some sodic soils also means that B and Na toxicity may contribute to crop production losses. However, the relative importance of nutrient availability and specific ion toxicity on crop production losses for sodic soils is difficult to judge, because of the compounding influence of poor soil physical properties on crop growth and yield.

Encyclopedia of Soil Science

S

SOIL MANAGEMENT

Specific tillage operations and cropping systems have been designed to overcome some of the poor physical properties of sodic soils (1, 8). Field grading and raised-bed cropping systems are used in landscapes of low gradient to encourage runoff and limit surface waterlogging. Deep ripping is used to break-up massive subsoil structures and for the incorporation of gypsum into subsurface horizons. Shallow tillage, at moisture contents close to the lower plastic limit, is used to generate seed-beds of fine tilth, rather than massive clods. Following crop establishment, zero tillage is typically advocated in order to minimize mechanical destruction of surface aggregates and subsequent dispersion of clays during irrigation.

Tillage practices, which aim to stabilize surface soil aggregates of an appropriate size-distribution, are aided by the use of cover crops and mulch applications (1). Soil cover crops, such as *Lolium rigidum*, tend to stabilize topsoil aggregates by means of entwining root systems and the incorporation of root mucilages within the rootzone. Mulches are used for the creation of a more favorable soil micro-climate, which encourages turnover in microbial populations, microbial detritus, and the generation of stabilizing, organic glues.

IRRIGATION MANAGEMENT

Water Quality

Crop production guidelines for the nonlimiting use of irrigation water, based on salinity (EC_{iw}) and sodicity (SAR_{iw}), have been developed for infiltration, crop salinity tolerance, and specific ion toxicity (8). In contrast, guidelines directed toward soil permeability (HC) and potential dispersion of clay tend to be based on soil chemical properties.

According to infiltration guidelines, calculated on paired EC_{iw} and SAR_{iw} ranges, the quality of irrigation waters used throughout the world (Table 1) should rarely limit their application to land (8). However, when the EC of infiltrating water falls below 0.2 dS m^{-1}, as occurs after rainfall, severe infiltration problems are predicted. Crop salinity tolerance guidelines predict severe Na toxicity for sensitive crops, based on $SAR_{iw} > 9$, and severe salinity effects on crop productivity, based on $EC_{iw} > 3$ dS m^{-1}. However, these predictions when applied to data in Table 1 only apply to irrigation of crops with drainage waters, due to their typically high SAR_{iw} and EC_{iw}.

The use of an adjusted SAR_{iw} (adj $R_{Na\,iw}$) in Table 1, accounts for the effects of high bicarbonate concentrations in some irrigation waters. Bicarbonate tends to precipitate Ca, and to a lesser extent Mg, as a carbonate crystal, with the net result of an increase in SAR_{iw}. Transformations of Ca (and Mg) between soluble, exchangeable, and crystal forms are modulated by pH and the partial pressure of CO_2, hence these two factors indirectly affect SAR_{iw}.

Irrigation Practices

Specific irrigation practices complement tillage and chemical additions to improve the production potential of sodic soils. These practices include increased frequency of irrigation, pre-planting irrigation, and adoption of alternative irrigation systems. Uniform applications of polyacrylamide (PAM) or finely powdered gypsum to topsoils, for the purpose of stabilizing fragile aggregates, is also best achieved through an initial irrigation.

Table 1 Mean and range of values for EC_{iw}, SAR_{iw}, and adj $R_{Na\,iw}$ of irrigation sources throughout the world. (Range values are shown in italics)

Measure	Canals	Drains	Ground waters	Rivers	Wadis	Wells	Waste waters	All sources
Samples	11	5	4	88	12	115	5	250
Countries	5	3	2	29	3	42	3	54
EC	1.37	4.38	1.45	0.80	2.74	1.78	0.84	1.47
	0.20–4.15	*1.06–6.20*	*0.46–3.80*	*0.03–7.42*	*0.30–8.01*	*0.14–7.93*	*0.69–1.11*	*0.03–8.01*
SAR	4.75	9.98	1.43	1.87	8.63	5.12	2.46	4.04
	0.60–14.0	*3.50–18.0*	*1.20–1.90*	*0.10–14.0*	*0.30–36.0*	*0.20–49.0*	*0.40–4.10*	*0.10–49.0*
adj R_{Na}	3.91	11.74	1.98	1.99	9.3	5.73	2.78	4.42
	0.60–11.2	*3.70–22.0*	*1.30–3.80*	*0.10–18.0*	*0.30–38.0*	*0.20–66.0*	*0.50–4.70*	*0.10–66.0*

Units EC_{iw} (dS m^{-1}), SAR_{iw}, and adj $R_{Na\,iw}$ [(mmol$_c$L^{-1})$^{0.5}$].
(Modified from Ref. 8.)

Increasing the frequency of short-duration irrigations on soils with high initial IR (e.g., cracking clays) and low final IR (due to sodicity), maximizes periods of high IR to wet the rootzone, and decreases the risk of waterlogging. Pre-planting irrigation wets the entire rooting depth and if timed correctly allows restoration of soil aeration before crop planting. This technique is especially effective on sodic soils of very low IR. A change from surface irrigation systems (e.g., flood or furrow) to sprinklers or drippers, can alleviate waterlogging on low IR soils by matching water delivery rates to the IR. The slow wetting of topsoils, under these alternative irrigation sytems, also decreases aggregate slaking and clay dispersion.

DRAINAGE OF SODIC SOILS

Drainage is essential for the continued practice of irrigation farming on sodic soils. Surface drainage prevents waterlogging and the development of toxic anaerobic conditions in the topsoil, while adequate subsurface drainage allows leaching of toxic or plant-dehydrating salts from the profile.

For some irrigated regions, the presence of a shallow watertable restricts drainage through a diminished hydraulic gradient. These watertables are frequently saline–sodic and contribute to sodification of soils via capillary transmission of Na-salts. Hence, the removal in these regions of subsurface water through drains (open or piped) and pumped bores is essential for sodium management. Since installation of open or piped drains is expensive, the use of cheaper mole drains within subsurface layers has been proposed. However, in unstable sodic soils, mole drains soon collapse and therefore are not generally used (9).

Periodic leaching of soil profiles is required for removal of salt or toxic ions. Analytical equations have been developed to predict the leaching requirement or soil SAR under saline–sodic irrigation (10). However, these equations do not fully account for Na-induced changes in IR or HC, which impinge on leaching capability.

Practically, intermittent irrigation is more efficient in salt removal than continuous ponding by promoting diffusion of salts into more favorable leaching pathways. Efficient leaching of sodic soils also requires irrigation with water of low SAR_{iw} and $EC_{iw} > TEC$, together with provision of sub-surface drainage for removal of salts.

REFERENCES

1. Oster, J.D.; Jayawardane, N.S. Agricultural Management of Sodic Soils. In *Sodic Soils: Distribution, Properties, Management, and Environmental Consequences*; Sumner, M.E., Naidu, R., Eds.; Oxford University Press: New York, 1998; 125–147.
2. Shaw, R.J.; Coughlan, K.J.; Bell, L.C. Root Zone Sodicity. In *Sodic Soils: Distribution, Properties, Management, and Environmental Consequences*; Sumner, M.E., Naidu, R., Eds.; Oxford University Press: New York, 1998; 95–106.
3. Jayawardane, N.S.; Blackwell, J. The Effects of Gypsum-Enriched Slots on Moisture Movement and Aeration in an Irrigated Swelling Clay. Aust. J. Soil Res. **1985**, *23*, 481–492.
4. Fireman, M. Permeability Measurements on Disturbed Soil Samples. Soil Sci. **1944**, *58*, 337–353.
5. Curtin, D.; Naidu, R. Fertility Constraints to Plant Production. In *Sodic Soils: Distribution, Properties, Management, and Environmental Consequences*; Sumner, M.E., Naidu, R., Eds.; Oxford University Press: New York, 1998; 107–123.
6. Cairns, R.R.; Bowser, W.E.; Milne, R.A.; Chang, P.C. Effect of Nitrogen Fertilization of Bromegrass on Solonetz Soil. Can. J. Soil Sci. **1967**, *47*, 1–6.
7. Naidu, R.; Rengasamy, P. Ion Interactions and Constraints to Plant Nutrition in Australian Sodic Soils. Aust. J. Soil. Res. **1993**, *31*, 801–819.
8. Ayers, R.S.; Westcot, D.W. Water Quality for Agriculture. *FAO Irrigation and Drainage Paper*; Food and Agricultural Organization of the United Nations: Rome, 174 pp, 1985; 29 Rev 1.
9. Muirhead, W.A.; Christen, E.; Moll, J. *Mole Drains Under Slots Reduce Waterlogging and Salinity in an Irrigated Clay Soil of Semi-Arid Australia*, XII World Congr. Agric. Eng. Rep. 94-A-018.
10. Jurinak, J.J.; Suarez, D.L. The Chemistry of Salt-Affected Soils and Waters. In *Agricultural Salinity Assessment and Management*; Tanji, K.K., Ed.; American Society of Civil Engineers: New York, 1990; 42–63.

SODIC SOILS, PROCESSES, CHARACTERISTICS AND CLASSIFICATION OF

Pichu Rengasamy
Department of Soil and Water, Adelaide University, Glen Osmond, S.A., Australia

INTRODUCTION

Land managers with paddocks that are prone to waterlogging, poor crop or pasture emergence, gully erosion, or tunnel erosion may be experiencing the effects of sodicity. Sodic soils are formed by the adsorption of Na^+ by the negatively charged sites on soil particles, particularly soil clays from soil solutions containing free sodium salts such as NaCl, Na_2CO_3, $NaHCO_3$, and Na_2SO_4. The soils are considered to be sodic when the free salts are leached from the soil layers and only exchangeable sodium remains adsorbed on soil particles. If free salts are also present, the soils become saline-sodic. Sodic soils are generally found in arid and semi-arid regions with high evapotranspiration and low rainfall associated with low leaching that are responsible for salt accumulation.

CHARACTERISTICS OF SODIC SOILS

The negative charge, which depends on the type and amount of clay materials, plays a vital role in the adsorption of sodium. Negative charge in soils increases with the content of clay minerals in the order smectites > illites > kaolinites. Soils dominant in positive charged sites such as Oxisols containing oxides of Al and Fe and low pH soils generally have negligible amounts of exchangeable sodium. Soils with pH-dependent charge minerals increase their negative charge with increasing pH. The more the negative charge, the more is the adsorption of sodium in soils.

Although soil sodicity is a result of chemical reaction (cation exchange) with the salts, it mainly affects soil physical properties and thus plant growth and productivity. It is possible that the release of high amounts of Na^+ in soil solutions can be toxic, which are encountered in saline-sodic soils. Sodic soils have an extremely poor physical condition (Table 1), leading to an inadequate balance between water and air regimes within the soil. The imbalance stems from restricted water acceptance and transmission properties, which result in the soil being too wet or dry for much of the time and leads to poor root development and crop growth. In addition, sodic soils are difficult to cultivate and have poor load-bearing characteristics. The lack of structural stability in these soils promotes soil hardening throughout soil profile and seal and crust formation at the soil surface, resulting in soil erosion and pollution of water bodies (more details are given in Refs. 1–3). Poor drainage in sodic soils also causes secondary salinity in subsoil layers (4).

SOIL PROCESSES

Swelling and dispersion of clay particles on wetting are the major processes responsible for the deterioration of physical behavior of sodic soils. In saline-sodic soils with high electrolyte concentration, swelling is minimal and clay dispersion is absent. Both swelling and dispersion behavior are governed by the balance between attractive and repulsive forces, arising from intermolecular and electrostatic interactions between solution and the solid phases in the soil. The distinction between saline and sodic soils arises because these forces vary, depending on whether the soil solution is concentrated (saline) or dilute with a high proportion of Na to divalent ions in the solid exchange phase to cause swelling and dispersion (sodic).

Soil scientists have used models based on pure clay minerals involving Lifshitz-van der Waals, ion correlation, hydration, and electrical diffuse double layer forces generated between colloidal particles suspended in water to explain sodic soil behavior (see Ref. 5). But soil clay systems that are complex, heterogenous intergrowths of different clay structures intimately associated with organic and biopolymers do not behave in the same way as their pure clay mineral counterparts. Thus, classical theories of colloidal behavior such as Derjaguin-Landau-Verwey-Overbeek (DLVO) theory (6) may not satisfactorily explain the behavior of sodic soils.

The processes that occur during the initial wetting of dry aggregates, resulting in swelling to the final stage of aggregate disintegration and leading to dispersion of soil clays when completely wet, are important in understanding sodic soil behavior. The polar nature of water molecules and solvation reactions with the solid phase are primary factors in causing swelling and dispersion. Clay particles in dry soil aggregates are bound together by inorganic and

Table 1 Example of properties of sodic soils compared to an ideal soil that is highly productive

Properties	Sodic soil	Ideal soil
$PH_{1:5}$ (water)	9.2	6.0–8.0
$EC_{1:5}$ ($dS\,m^{-1}$)	0.2	<0.4
Organic carbon (%)	0.3	>1.0
$SAR_{1:5}$	9.9	<3.0
Spontaneously dispersed clay (%)	8.7	0
Hydraulic conductivity at saturation ($mm\,day^{-1}$)	4.0	>80.0
Penetrometer resistance (MPa) at 100 kPa suction	3.8	<2.0
Aeration porosity (%)	5.6	>15
Bulk density ($Mg\,m^{-3}$)	2.2	<1.5
Nonlimiting water range ($mm^3\,mm^{-3}$)	0.38–0.42	0.1–0.5

(From Ref. 11.)

organic compounds involving several types of bonding, which produce strong, attractive pressures of the magnitude of megapascals (7). The water stability of an aggregate depends on the strength and persistence of these linkages in the presence of water molecules, which, in turn, are functions of the type of bonding. Bond strength in the presence of water generally decreases in the order: covalent, hydrophobic, Lifshitz-van der Waals, coordination complexing, hydrogen, and finally ionic bonds. In contrast to covalent bonds, ionic bonds are readily solvated by water molecules. Similar to pure sodium compounds, Na-clay linkage is easily solvated and ionic bond is broken, whereas calcium and magnesium ions are linked to clay particles by polar covalent (combination of covalent and ionic) bonds, with decreased ionicity compared to Na linkage. The degree of covalency in a bond involving metal cations depends on ionization and ionic potentials (3). The tendency to form covalent bonding increases in the order: Na^+, K^+, Mg^{2+}, Ca^{2+}, Fe^{3+}, and Al^{3+}, for example. Thus, both the type of cation and the nature of clay ligand determine the ionicity of the clay–cation–clay bonds.

Highly ionic bonding in sodic clays leads to extensive hydration and swelling with increasing water content, and finally, with separation of linkages on water saturation, clay particles disperse spontaneously in water, whereas in Ca, or Mg-saturated clays with polar covalent bonding, limited hydration leads to limited swelling without any separation (or dispersion) of clay particles. Only on mechanical agitation these clays can disperse. Flocculation of dispersed clay particles is brought about by the addition of electrolyte, which as result of its osmotic effect causes dehydration of clay water system, thereby bringing clay particles together. Therefore saline-sodic soils do not have severe deterioration of physical properties. The order of flocculating power is $Ca^{2+} > Mg^{2+} > K^+ > Na^+$. Thus, gypsum, a calcium compound, is very effective in reclaiming sodic soils. In addition to its flocculating effect, gypsum also promotes removal of sodium from clays by way of cation (calcium) exchange.

CLASSIFICATION

At present, there is a need for a uniform system of classification of salt-affected soils that distinguishes between saline and sodic conditions and is useful for soil management. The international nomenclature (8) includes the terms *sodic*, *alkali*, *solonchak*, *solonetz*, and *solodized solonetz*, and the complex interrelationship between them makes comparison between sodic soils difficult. Classification of saline and sodic soils devised by the U.S. Department of Agriculture, widely followed in many countries is as follows:

Saline, nonsodic soils – Exchangeable sodium percentage (ESP) < 15

Electrical conductivity in saturation paste (EC_e) $> 4\,dS/m$

Sodic, nonsaline – ESP > 15 and $EC_e < 4\,dS/m$

Saline, sodic – ESP > 15 and $EC_e > 4\,dS/m$

Nonsaline, nonsodic – ESP < 5 and $EC_e < 4\,dS/m$

The effects of exchangeable sodium on soil physical behavior varies from soil to soil and are influenced by several factors such as electrolyte concentration, pH, organic matter, biopolymers, and aggregate stability in water. Therefore, the definitions used on the basis of ESP vary according to practical experience. The following classes have been proposed (9) on the basis of physical behavior of sodic soils:

Class 1: *Dispersive Soils*. Soils that disperse spontaneously without shaking will have severe problems associated with crusting, reduced porosity, etc., even when subjected to minimum mechanical stress, e.g., under zero tillage.

Class 2: *Potentially Dispersive Soils*. Soils that require inputs of mechanical energy (raindrop impact, tillage) to bring about dispersion will experience soil physical problems when mechanically disturbed.

Class 3: *Flocculated Soils*. Soils that contain more than the minimum electrolyte concentration required for flocculation of clays (or prevention of dispersion of clays) will present few physical problems, but salts could be excessive and limit productivity.

Measurements of clay dispersion along with ESP or SAR (sodium adsorption ratio), EC, and pH will be necessary for managing saline and sodic soils. Based on these principles, a recent detailed proposal (10) on classification of salt-affected soils is available, but this has yet to be tested for practical applications.

REFERENCES

1. Quirk, J.P. Some Physico-Chemical Aspects of Soil Structural Stability: A Review. In *Modification of Soil Structure*; Emerson, W.W., Bond, R.D., Dexter, A.R., Eds.; John Wiley: New York, 1978; 3–16.
2. Shainberg, I.; Letey, J. Response of Soils to Sodic and Saline Conditions. Hilgardia **1984**, *52*, 1–57.
3. Rengasamy, P.; Sumner, M.E. Processes Involved in Sodic Behaviour. In *Sodic Soils: Distribution, Properties, Management, and Environmental Consequences*; Sumner, M.E., Naidu, R., Eds.; Oxford University Press: New York, 1998; 35–50.
4. Shaw, R.J.; Coughlan, K.J.; Bell, L.C. Root Zone Sodicity. In *Sodic Soils: Distribution, Properties, Management, and Environmental Consequences*; Sumner, M.E., Naidu, R., Eds.; Oxford University Press: New York, 1998; 95–106.
5. Quirk, J.P. Interparticle Forces: A Basis for the Interpretation of Soil Physical Behaviour. Adv. Agron. **1994**, *53*, 121–183.
6. van Olphen, H. *An Introduction to Clay Colloid Chemistry*; John Wiley: New York, 1977.
7. Rengasamy, P.; Olsson, K.A. Sodicity and Soil Structure. Aust. J. Soil. Res. **1991**, *31*, 821–837.
8. Szabolcs, I. *Salt-Affected Soils*; CRC Press: Boca Raton, Florida, 1989.
9. Rengasamy, P.; Greene, R.S.B.; Ford, G.W.; Mehanni, A.H. Identification of Dispersive Behaviour and the Management of Red-Brown Earths. Aust. J. Soil. Res. **1984**, *22*, 413–431.
10. Sumner, M.E.; Rengasamy, P.; Naidu, R. Sodic Soils: A Reappraisal. In *Sodic Soils: Distribution, Properties, Management, and Environmental Consequences*; Sumner, M.E., Naidu, R., Eds.; Oxford University Press: New York, 1998; 3–17.
11. Rengasamy, P. Sodic Soils. In *Methods for Assessment of Soil Degradation*; Lal, R., Blum, W.H., Valentine, C., Stewart, B.A., Eds.; CRC Press: Boca Raton, FL, 1998; 265–277.

S

SODIC SOILS, RECLAMATION OF

G.J. Churchman
Commonwealth Scientific and Industrial Research Organisation (CSIRO) Land and Water, Glen Osmond, South Australia, Australia

INTRODUCTION

Two main mechanisms are involved in the reclamation of sodic soils: 1) improvement in water flow, and 2) displacement of exchangeable Na ions.

Improvement in Water Flow

Sodic soils show a low permeability to water. Their permeability can be improved quite rapidly by increasing electrolyte concentrations (1, 2) (the "electrolyte effect"). Ions in solution tend to inhibit the swelling and dispersion of fine clay particles by compressing the electric double layers of adjacent particles (Fig. 1). The volume around the particle surfaces within which repulsion takes place is reduced, ultimately to zero.

It has been found (2) that there is no swelling and dispersion when the total electrolyte concentration exceeds a critical level, i.e., the "threshold electrolyte concentration" that depends on soil properties (1, 3).

Displacement of Exchangeable Na Ions

The deleterious effects of exchangeable sodium reflect the more extensive hydration of the sodium ion than that of other common exchangeable cations on soil particles (Fig. 2). Exchangeable sodium, which forms an ionic association with the clay surface, causes hydration, dissociation of particles, and hence swelling and dispersion (4). By contrast, exchangeable calcium forms polar covalent bonds with the clay surface (4). Calcium shows only limited hydration and dissociation of particles, and therefore causes only limited dispersion and swelling (4). Na^+ on exchange sites is replaced by Ca^{2+} for the longer term remediation of sodic soils.

CHEMICAL AMELIORANTS

Gypsum

Soluble sources of calcium ions are the most suitable sodic soil ameliorants. While gypsum is used most often, not all sources of gypsum are similarly effective or similarly suitable for each specific problem from sodicity. It was found (1, 5) that only "by-product" gypsum, as a surface application, could prevent crusting. Mined gypsum was too slowly soluble for the purpose (1). Gypsum is a by-product of many industrial processes (5). Furthermore, gypsums from different mines can differ quite markedly in purity (6). They can also differ in dissolution rate. A finer particle size generally provides a greater solubility (5). As well, sodium salts associated with gypsum (Table 1) (6) decrease its efficiency for amelioration (7).

Lime

Lime has also been used as an ameliorant for sodic soils (6, 8, 9). While generally less soluble than gypsum, lime may provide a useful incidental source of calcium ions when used to raise the pH of acid sodic soils. It may provide a cheaper alternative to gypsum or else be used in association with gypsum to extend the lifetime of the calcium reserves (6). Because of its increased solubility in the presence of sodium salts, it may be more effective than gypsum for the reclamation of saline sodic soils. Lime also contributes less to the salt load than gypsum (9).

Reclamation of Calcareous Sodic Soils

Calcareous sodic soils have an abundance of Ca, but it is present as $CaCO_3$, hence is insoluble at the high pH of these soils. Their sodicity is often overcome through acid additions (10, 11), either directly, or indirectly. Additions of elemental sulfur, pyrite (FeS_2), iron and aluminum sulfates (10, 12), and also organic matter (13) each provide acid through microbial activity for indirect acidification. Plant growth can also help dissolve native calcium carbonate (12).

Miscellaneous Chemical Ameliorants

Sodic soils may be reclaimed with soluble sources of Ca that include waste materials, e.g., acidic cottage cheese whey (1), acid resin by-products of the oil industry, and

Encyclopedia of Soil Science

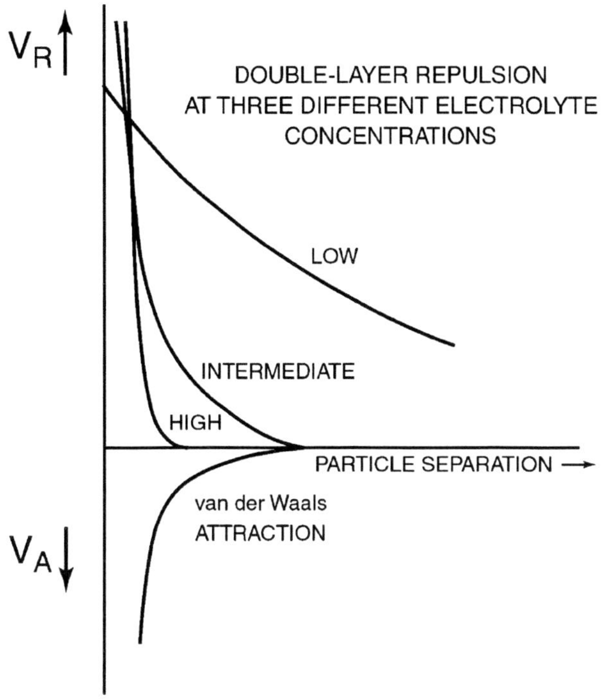

Fig. 1 Repulsive and attractive energy as a function of particle separation at three electrolyte concentrations. (From Ref. 19.)

also $CaCl_2$ as an industrial by-product (3). Addition of polymers can also increase the permeability of soils (1).

ROLE OF BIOLOGY AND ORGANIC MATTER

Effects of Plant Growth

Rice culture ameliorates sodic soils through the build-up of carbon dioxide and hence dissolution of calcium carbonate, among other effects (10). Among crops, rice is particularly tolerant to sodicity, and some grasses are also highly tolerant (10, 12) (Table 2).

Ameliorants are readily leached out of soils, so losing the benefit of the electrolyte effect. In addition, Na^+ from sodium salts often present in soils can replace exchangeable Ca^{2+}. Therefore long-term reclamation probably requires increased plant growth, hence enhanced biological activity (13) and stabilization of soil structure. This occurs partly through transient binding agents, e.g., roots, hyphae, polysaccharides, and hydrophobic binding (13) and partly through associations between soil organic matter and Ca^{2+} in the soil system. Organic matter has a stronger affinity for Ca than for Na (13), so is most effective when used with gypsum or lime to help restore a stable structure to soils (13). There is a resulting increase in stable pores for the transmission of water and nutrients, root growth, and soil stability.

Effects of Additions of Organic Matter

Due to its stabilizing effect on soil structure, organic matter can also help to reclaim sodic soils when added to soils, even though additions of organic matter can sometimes also enhance their dispersion, with resulting physical problems (13). The different effects vary with types of soils and also of organic matter (13). The addition of readily decomposable organic matter tends to reduce the pH of calcareous sodic soils, and hence aid their reclamation (13).

STRATEGIES FOR THE RECLAMATION OF SODIC SOILS

Reclamation Without the Addition of Ameliorants

Reclamation can be achieved slowly without ameliorants by many cycles of irrigation and cropping (10). As well, underlying calcareous or gypsiferous layers may be incorporated into upper parts of the soil profile through deep plowing (3).

Optimal Supply of Ameliorants

The effectiveness of gypsum can be increased by its placement in cracks for their stabilization so that they remain open to conduct water, air, and plant nutrients when soils become wet (1). Water flow to subsoil layers, and also aeration, hence higher crop yields, were achieved by incorporating gypsum in narrow tilled slots (12). More generally, gypsum is often incorporated into subsoil layers by tillage.

Ameliorants may also be used in association with other additions. Only a little gypsum applied with acidifying nitrogenous fertilizers led to effective reclamation of a saline-sodic soil (1). This combination confers immediate benefits through a higher electrolyte concentration and also medium-term benefits through an increased supply of exchangeable Ca ions. Enhanced plant growth from fertilizer use provides long-term benefits.

Supply of Water for Reclamation

Ameliorants are only effective when dissolved. A supply of water and its flow through the profile are required

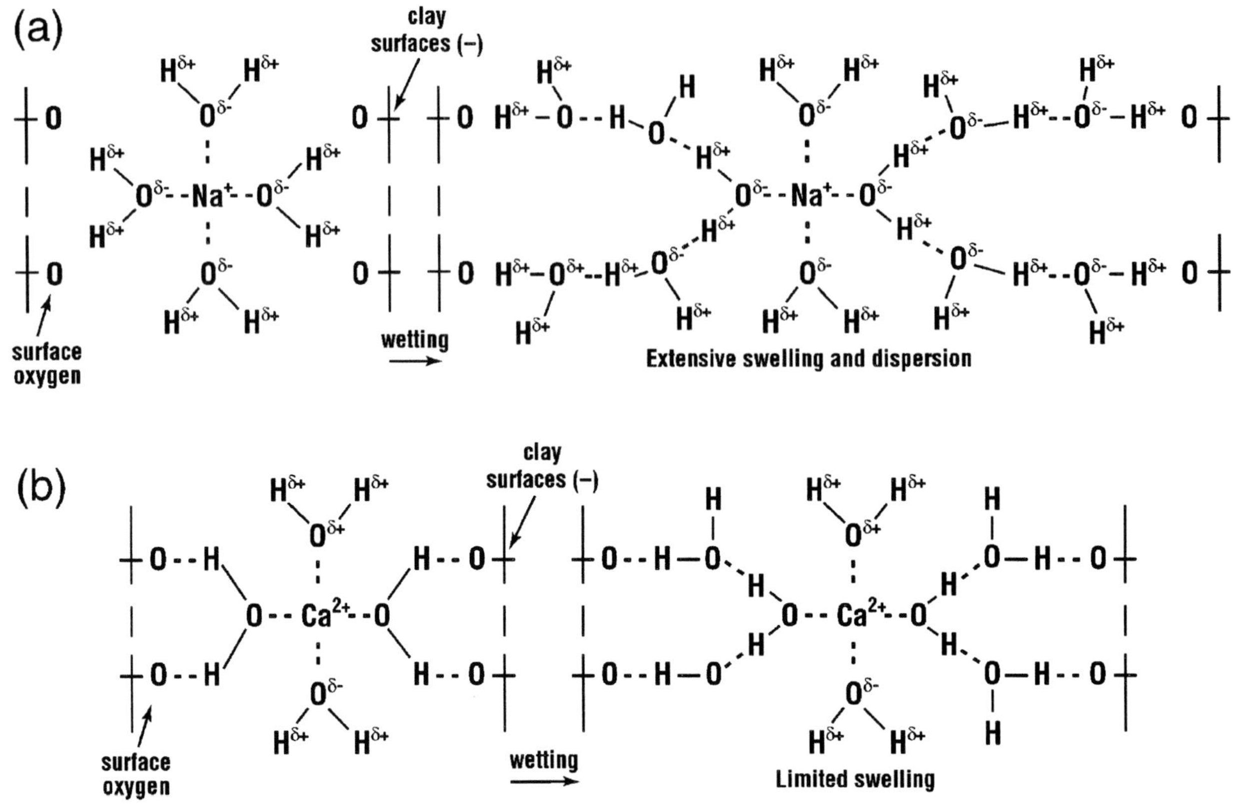

Fig. 2 Schematic representation of the effect of wetting and nature of bonding between cations and clay surfaces on swelling and dispersion: (a) Na^+ aquo ion linking clay particles by ionic bonding, (b) Ca^{2+} aquo ion linking clay particles by polar covalent bonding. Water molecules are linked to cations by hydrogen bonding. (From Ref. 4.)

for effective amelioration. Irrigation and/or rainfall may supply water but the quality of water available for irrigation can vary. Gypsum can be supplied in electrolyte-free irrigation water, as in parts of California (12). Saline-sodic irrigation water, including water commonly derived from groundwater sources, may exacerbate the deleterious

Table 1 Composition of some mined gypsum

Sample	Fe_2O_3 (%)	MnO (%)	TiO_2 (%)	CaO (%)	K_2O (%)	SO_3 (%)	P_2O_5 (%)	SiO_2 (%)	Al_2O_3 (%)	MgO (%)	Na_2O (%)	ZrO_2 (%)	Sr (%)	SUM (%)	Org C (%)	$CaCO_3$ (%)
Plant life	0.48	0.010	0.10	25.7	0.260	30.6	0.057	18.9	1.16	0.47	0.18	<0.001	0.271	78.1		
Neindorf	1.54	0.021	0.19	23.7	0.662	28.9	0.066	17.6	3.32	1.64	0.28	<0.001	0.300	78.3	0.41	1.8
Salt lake	0.03	<0.003	0.02	33.5	0.011	29.7	0.051	0.20	0.07	<0.009	0.05	<0.001	0.241	73.8	0.53	3.4
Austral Pacific	0.26	0.006	0.04	22.6	0.70	28.5	0.047	30.2	0.56	0.04	<0.03	<0.001	0.115	82.4	0.07	0.1
Agra unsieved	0.15	0.004	0.03	31.2	0.093	39.0	0.050	4.50	0.61	0.06	<0.03	<0.001	0.341	76.0		
Agra sieved (A)	0.12	0.010	0.03	31.0	0.067	39.5	0.049	3.90	0.40	0.04	<0.03	<0.01	0.845	75.9	0.02	0.1
Agra sieved (B)	0.11	0.012	0.03	31.0	0.067	39.4	0.053	3.80	0.28	0.06	<0.03	<0.001	0.847	75.6	0.1	
Cresco	0.15	0.004	0.02	31.0	0.089	39.2	0.071	4.20	0.37	0.07	0.04	<0.001	0.357	75.5	0.03	0.1
Top gypsum	0.32	0.006	0.10	23.4	0.340	28.5	0.057	26.3	1.12	0.35	0.19	0.001	0.194	80.9		
Jomoco (A)	0.46	0.010	0.08	28.0	0.237	33.9	0.286	11.8	1.02	0.68	0.17	<0.001	0.403	77.0	n.d.	n.d
Jomoco (B)	0.44	0.007	0.07	27.9	0.232	33.7	0.277	12.4	1.01	0.67	0.14	<0.001	0.402	77.2	n.d.	n.d.

(From Ref. 6.)

Table 2 Relative tolerance of crops to sodicity

ESP range	Crops[a]
10–15	Safflower, mash, pea, lentil, pigeon-pea, curd bean
16–20	Bengal gram, soybean
20–25	Groundnut, cowpea, onion, pearl millet
25–30	Linseed, garlic, guar
30–50	Indian mustard, wheat, sunflower, berseem, hybrid napier, guinea grass
50–60	Barley, sesbania, saftal, panicums
60–70	Rice, para grass
70+	Karnal, rhodes, and bermuda grasses

[a] Yields are about 50% of the potential yields in the respective sodicity ranges.
(From Ref. 12, following Ref. 10.)

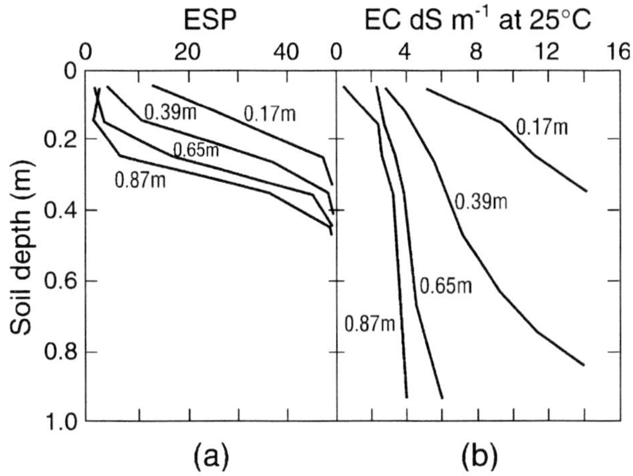

Fig. 3 Computer-model results (ESP and EC) for reclamation of a soil (initial ESP = 50; CEC = 200 $mmol_c\,kg^{-1}$ with gypsum and water (EC = 0). Numbers next to each line are depths of applied water. (From Ref. 9.)

effects of sodicity when its application during the dry season is followed by rains, as can occur in Israel (12). In this case, (by-product) gypsum is applied to the surface prior to the rainy season (12). Adequate drainage to beyond the root zone should accompany irrigation in order to leach displaced electrolyte but this should not cause increased salinity in underlying groundwater (11).

As long as there is provision for the disposal of the salt water, dilution with salt water (3, 11) is also useful in sodic soil reclamation. Its immediate effect is to increase the hydraulic conductivity of sodic soils by the electrolyte effect, but long-tern reclamation will occur provided it also contains sufficient Ca (Ca: total cations≥0.3). This can be added as gypsum.

Rate of Supply of Ameliorants

All additives increase costs so their optimal use is required. The cost of transporting gypsum often limits application rates. Recommendations from field trial results of the most effective application rates for ameliorants are necessarily specific to soil type and climate, among other factors. Nonetheless, both simulations (9, 14, 15) and laboratory studies (16) have shown that the solubility of gypsum increases as the proportion of exchangeable sodium on the soil increases. They have also shown that gypsum is more effective when mixed throughout the soil than when applied to the soil surface alone. Simulations (9) have shown that reclamation is primarily limited to the depth interval in which gypsum was applied (Fig. 3a). They have also shown (Fig. 3b) that the electrical conductivity EC of the solution within and below the gypsum-amended layer decreases as the exchangeable sodium percentage (ESP) within that layer decreases.

Nonetheless, calculations of gypsum requirements have been based on 1) exchange of Na by Ca, rather than the attainment of a sufficiently high electrolyte concentration EC to maintain a low permeability, and 2) assumptions of chemical equilibrium. The gypsum requirements for the achievement of a suitable permeability by the electrolyte effect also depend on the inherent EC of the soil solution (11). Because mineral dissolution can maintain a high EC throughout most of the soil profile, gypsum often needs to be added to the soil-surface only, mainly to prevent crusting. Assumptions of equilibrium, while strictly invalid, have probably been useful for the calculation of gypsum requirements because of low flow rates of percolating solutions and high surface areas of gypsum particles (10). Models of soil and water flow, e.g., UNSATCHEM (16) which consider equilibrium and kinetic expressions for ion exchange and the dissolution and precipitation of minerals and also ameliorants, may enable more robust calculations of requirements for ameliorants.

Ultimately, the main purpose of reclaiming sodic soils is to obtain the maximum possible improvement of yield of the particular crop in relation to the cost of ameliorant applied. Work in the Australian sugar industry (17) has established a quantitative relationship between percentage increase in ESP and consequent loss in sugarcane yield. Sugarcane yield decreased by 1.5–2.1 tonne ha^{-1} for every 1% increase in ESP (17). This relationship forms the basis of a cost–benefit analysis for calculating gypsum application rates from the ESP, the quality of the gypsum used, the quality of irrigation water and economic factors. These include the cost of gypsum and its application, the price for the product (sugarcane) and the discount rate (18). Clearly there are opportunities for more practical calculations of this kind to be carried out for other crops and in other soil and climatic environments.

REFERENCES

1. Sumner, M.E. Sodic Soils: New Perspectives. Austr. J. Soil Res. **1993**, *31* (6), 683–750.
2. Quirk, J.P.; Schofield, R.K. The Effect of Electrolyte Concentration on Soil Permeability. J. Soil Sci. **1955**, *6* (2), 163–178.
3. Loveday, J. Amendments for Reclaiming Sodic Soils. In *Soil Salinity Under Irrigation: Processes and Management*; Shainberg, I., Shalhevet, J., Eds.; Springer: Berlin, 1984; 353–374.
4. Rengasamy, P.; Sumner, M.E. Processes Involved in Sodic Behaviour. In *Sodic Soils*; Sumner, M.E., Naidu, R., Eds.; Oxford University Press: New York, 1998; 35–50.
5. Shainberg, I.; Sumner, M.E.; Miller, W.P.; Farina, M.P.W.; Pavan, M.A.; Fey, M.V. Use of Gypsum on Soils: A Review. Adv. Soil Sci. **1989**, *9*, 1–112.
6. Naidu, R.; Merry, R.H.; Churchman, G.J.; Wright, M.J.; Murray, R.S.; Fitzpatrick, R.W.; Zarcinas, B.A. Sodicity in South Australia. A Review. Austr. J. Soil Res. **1993**, *31* (6), 911–930.
7. Rengasamy, P.; Olsson, K.A. Irrigation and Sodicity. Austr. J. Soil Res. **1993**, *31* (6), 821–837.
8. Levy, G.J.; Shainberg, I.; Miller, W.P. Physical Properties of Sodic Soils. In *Sodic Soils*; Sumner, M.E., Naidu, R., Eds.; Oxford University Press: New York, 1998; 77–94.
9. Oster, J.D.; Frenkel, H. The Chemistry of the Reclamation of Sodic Soils with Gypsum and Lime. Soil Sci. Soc. Am. J. **1980**, *44* (1), 41–45.
10. Gupta, R.K.; Abrol, I.P. Salt-Affected Soils: Their Reclamation and Management for Crop Production. Adv. Soil Sci. **1990**, *11*, 223–288.
11. Oster, J.D.; Shainberg, I.; Abrol, I.P. Reclamation of Salt-Affected Soil. In *Soil Erosion, Conservation and Rehabilitation*; Agassi, M., Ed.; Marcel Dekker: New York, 1996; 315–352.
12. Oster, J.D.; Jayawardene, N.S. Agricultural Management of Sodic Soils. In *Sodic Soils*; Sumner, M.E., Naidu, R., Eds.; Oxford University Press: New York, 1998; 125–147.
13. Nelson, P.N.; Oades, J.M. Organic Matter, Sodicity and Soil Structure. In *Sodic Soils*; Sumner, M.E., Naidu, R., Eds.; Oxford University Press: New York, 1998; 51–75.
14. Tanji, K.K.; Deverel, S.J. Simulation Modeling for Rreclamation of Sodic Soils. In *Soil Salinity Under Irrigation: Processes and Management*; Shainberg, I., Shalhevet, J., Eds.; Springer: Berlin, 1984; 238–251.
15. Frenkel, H.; Gerstl, Z.; Alperovitch, N. Exchange Induced Dissolution of Gypsum and the Reclamation of Sodic Soils. J. Soil Sci. **1989**, *40* (3), 599–611.
16. Suarez, D.L.; Šimunek, J. UNSATCHEM: Unsaturated Water and Solute Transport Model with Equilibrium and Kinetic Chemistry. Soil Sci. Soc. Am. J. **1997**, *61*, 1633–1646.
17. Nelson, P.; Ham, G. Exploring the Response of Sugar Cane to Sodic and Saline Conditions Through Natural Variation in the Field. Field Crops Res. **2000**, *66* (3), 245–255.
18. Nelson, P.; Fitzgerald, T.; Swan, G.; Brennan, L. *Gypsy, Discounted Cash Flow Analysis for Application of Gypsum to Sodic Soils Under Sugarcane: Manual*; CRC for Sustainable Sugar Production: Townsville, Australia, 2000.
19. van Olphen, H. *An Introduction to Clay Colloid Chemistry*; Wiley: New York, 1963.

SODIUM-AFFECTED SOILS IN HUMID AREAS

Samuel J. Indorante
United States Department of Agriculture-Natural Resources Conservation Service (USDA-NRCS), Carbondale, Illinois, U.S.A.

S

INTRODUCTION

Sodium affected soils (SAS) or sodium soils are soils that have been adversely affected by sodium salts and/or exchangeable sodium. They usually occur in arid, semiarid, and subhumid climates where rainfall is insufficient to leach soluble salts from the soils, or where internal drainage is restricted. To a much lesser extent, SAS can occur in humid regions (Table 1) with a mean annual precipitation (MAP) >100 cm because of factors that restrict leaching of soluble salts from the soil. The main factors are a source of sodium, high water table, clayey dense subsoils, impermeable underlying geologic strata, and seasonal periods of high evapotranspiration.

SAS in humid areas tend to be leached of excess salts, but they contain enough exchangeable sodium to disperse the clay particles. The presence of exchangeable sodium, as measured by exchangeable sodium percentage (ESP), under low salt conditions, as measured by electrical conductivity of the saturation extract (ECe), causes the breakdown (peptization or dispersion) of soil aggregates (Fig. 1) (1). These high ESP, low ECe soils tend to be sticky when wet, highly erodible, and have poor tilth, low porosities, low infiltration rates, and low permeabilities.

Soil moisture recharge in these SAS profiles is slow and these soils also dry slowly. When these soils dry they become hard, cloddy, and crusty (Fig. 2); when wet, they are sticky and difficult to work. The adverse affect of exchangeable sodium on soil structure and associated soil–water and soil–plant relationships makes crop production difficult even in humid climates (MAP >100 cm).

GENESIS OF SAS IN HUMID AREAS

Sodium affected soils can occur over a wide range of climates, but generally under highly seasonally contrasted ones of arid to humid conditions (2, 3). The rate and degree of SAS genesis varies in each of the climates, and also varies with parent materials, topography, vegetation or land use, and the length of time sodium soil genesis has taken place. There are several main environmental factors that promote the genesis of SAS in all climates: the presence of shallow saline groundwater or a saline soil solution; textural discontinuities during the deposition of eolian, glacial, alluvial, or colluvial sediments; occurrence of perched water tables within 1 meter of the surface; low slope gradients; and impeded drainage (2). These environmental factors in a given climate control water movement over, into, and through the soil profile and the soil landscape. Water movement, in turn, controls the distribution of soluble sodium salts and exchangeable sodium in sodium soil profiles and across the soil landscape (4, 5).

Sodium affected soil genesis pathways have been described in arid climates (2), semiarid climates (3, 6), subhumid climates (3, 5), and humid climates (7–10). The sequence of pedogenic processes tends to follow, but is not restricted to the classic sodic soil genesis model as summarized by Miller and Pawluk (11):

Sodic Soil Genesis Model

Salinization—Salt accumulation due to sodium-rich parent material or capillary rise of shallow saline groundwater. Periodic alternation of dry and humid conditions is necessary, with high rates of evapotranspiration in the summer.

Desalinization—Leaching of salts by water moving through the soil profile.

Solonization—This step begins with the leaching of salt by rainwater, which leads to the dispersion of clay if ESP exceeds 10–15% and total soluble salt content is 0.1–1.5% or less, and the formation of a sodic soil. Translocation of dispersed clays will eventually lead to an abrupt textural change between A and B horizons.

Solodization—The final step, which is Na-induced hydrolysis and eluviation at the top of the slowly permeable B horizon.

The sources of sodium in humid climate SAS, include Na-bearing minerals in medium-textured loess (9); coarse-textured loess (10); alluvial deposits (12); and fluvial terraces (7). Humid climate SAS have formed under both forest (7), and mixed forest prairie and prairie vegetation

Table 1 Global distribution (km^2) of SAS with humid soil moisture regimes

Country or continent	Area (km^2)
Argentina	143,246
Australia	138,185
China	34,676
Asia, Africa, North and South America[a]	85,891
Global total	401,998

[a] The balance of humid region SAS are scattered in these continents.
(Data from United States Department of Agriculture-Natural Resources Conservation Service, Soil Survey Division, World Soil Resources, 2000.)

Fig. 2 A sodium affected soil spot intermingled with nonsodium affected soils in southern Illinois, USA. The climate is humid.

(9), and occur on level (9, 10, 12) and gently sloping (7, 9) landscapes.

NOMENCLATURE AND CLASSIFICATION OF SAS

SAS are primarily classified in terms of soil morphology, including soil texture, soil structure, and the presence of a diagnostic horizon. They are also classified in terms of chemical properties including pH, ECe, electrical conductivity of the 1:5 soil to solution ratio ($EC_{1:5}$), ESP and sodium adsorption ratio (SAR). The specific criterion used to define and classify sodic soils varies by geographic region and the goal of the classification system. Taxonomical classifications are designed to serve the multiple purposes of soil survey (13). Land management classifications group soils for SAS for the purpose of site-specific management of sodium soil areas for agronomic and related uses.

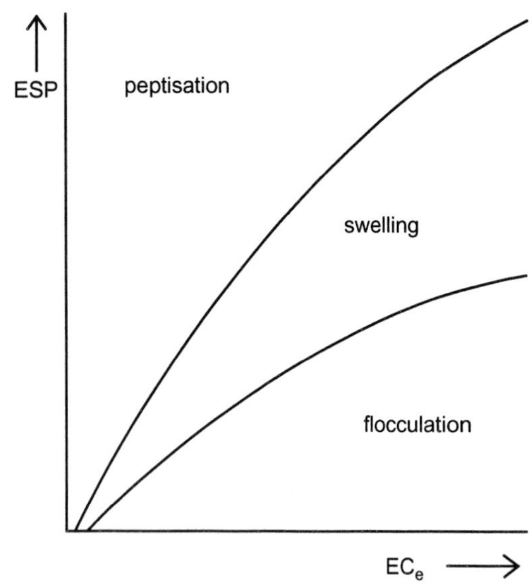

Fig. 1 Schematic diagram illustrating clay behavior as a function of exchangeable sodium percentage (ESP) and electrical conductivity of the saturation extract (ECe). (From Ref. 1.)

Taxonomical Classification

One of the most common taxonomical characteristics for classifying sodium soils in all climates is the presence of a natric horizon. That term is defined as a mineral soil horizon that satisfies the requirements of an argillic horizon, but also has prismatic, columnar, or blocky structure, and a subhorizon having an ESP >15% (13, 14). The Australian Soil Classification System (15) uses the presence of a sodic horizon to define its Sodosols (SAS). The Australian sodic horizon is a clear or abrupt textural B horizon which is sodic (ESP >6) in all or part of the B horizon, depending on the thickness of the horizon and a pH >5.5. Some taxonomical classifications of SAS soils in humid regions are Natraqualfs, Natrudalfs, Natralbolls, Natraquolls, Natraquerts, and Natrudolls (13); others are hydromorphical Solonetzes (16), Solodic Sodosol (15), and Gleyic and Mollic Solonetz (14).

Land-Management Classification of SAS

Common terms used to label, describe, and classify SAS in humid regions for management purposes are:

- Sodic—A nonsaline soil containing sufficient exchangeable sodium to adversely affect crop production and soil structure under most conditions of plant growth. The SAR of the saturation extract is >13 (17).

- Sodic Nonsaline—Soils that have an SAR of the 1:5 soil to soil solution ration ($SAR_{1:5}$) between about 3 and 10, but insufficient salt to maintain the clay in a flocculated condition (18).
- Very Sodic Nonsaline—Soils having an $SAR_{1:5}$ >10 and very low $EC_{1:5}$ values, which would be highly dispersive and have poor physical properties (18).
- Nonsaline-alkali—Soils for which the ESP is >15 and the ECe is <4 mmhos/cm. The pH usually ranges between 8.5 and 10 (19).
- Solonetz (in humid regions)—Soils that are leached of excess salts and have a well-developed structural B horizon with ESPs from 5 to 20% (3).

IDENTIFYING, MAPPING, AND MANAGING HUMID SAS

Sodium affected soils have primarily been identified and mapped using traditional soil survey methods, which include aerial photo interpretation, application of local soil-landscape models, soil sampling with the associated field tests, and laboratory analysis. Even when mapping at large scales (1:12,000 to 1:31,680), SAS are often mapped as complexes of SAS–nonSAS soils, because of the irregular shape, relatively small size, and the intricate geographic mix of SAS–nonSAS soils. Along with the traditional soil survey methods, geophysical tools, such as electromagnetic induction, have been applied successfully in the identification and detailed mapping of SAS in humid regions (20, 21) and subhumid regions (22) for both taxonomical classification and land management classification.

Once the SAS have been identified, mapped, and classified, the next step is the management of these areas for specific crops or land uses. An example of the importance of site-specific management and reclamation of SAS for crop production can be seen in a comparison of corn (*Zea mays* L.) and soybean (*Glycine max* [L.] Mer.) yields in SAS and corresponding nonSAS in the humid midwestern United States (23). The two soils are Cisne (fine, montmorillonitic, mesic Vertic Albaqualfs) and Huey (fine-silty, mixed, mesic Typic Natraqualfs) (13), and are typically mapped as a complex. Under the same climate and management factors, and under nonirrigated conditions, the Huey soils (with more than 15% ESP in the subsoil) had an average 18% yield reduction for soybean and an average 38% yield reduction for corn when compared to the Cisne soils.

Applications of gypsum ($CaSO_4.H_2O$), done in conjunction with water table management and tillage, are likely to produce the most success in reclamation of SAS in humid areas for crop production. These cultural practices should help prevent soil swelling and dispersion, and increase porosity, structural stability, hydraulic properties, soil tilth, drainage, and leaching, and reduce dry soil strength (24).

CONCLUSION

The adverse effect of exchangeable sodium on soil structure and associated soil–water and soil–plant relationships, make crop production difficult, even in humid climates. Crop production in these areas provides a unique challenge and opportunity for soil scientists and crop scientists. Through a better understanding of humid region SAS systems, scientists may find better ways to manage these areas to overcome the production problems caused by the high sodium concentration and poor physical makeup of SAS.

REFERENCES

1. Kamphorst, A.; Bolt, G.H. Saline and Sodic Soils. In *Soil Chemistry A. Basic Elements–Developments in Soil Science 5A*, 2nd Ed.; Bolt, G.H., Bruggenwert, M.G.M., Eds.; Elsevier: New York, 1978; 171–191.
2. Bui, E.N.; Krough, L.; Lavado, R.S.; Nachtergaele, F.O.; Toth, T.; Fitzpatrick, R.W.; Distribution of Sodic Soils: The World Scene. In *Sodic Soils—Distribution, Properties Management, and Environmental Consequences*; Sumner, M.E., Naidu, R., Eds.; Oxford University Press: New York, 1998; 19–33.
3. Szabolcs, I. Solonetz Soils in Europe. Their Formation and Properties with Particular Regard to Utilization. In *European Solonetz Soils and Their Reclamation*; Szabolcs, I., Ed.; Akademiai Kiado: Budapest, 1971; 9–34.
4. Richardson, J.L.; Wilding, L.P.; Daniels, R.B. Recharge and Discharge of Groundwater in Aquic Conditions Illustrated with Flownet Analysis. Geoderma **1992**, *53*, 65–78.
5. Seelig, B.D.; Richardson, J.L. Sodic Soil Toposequence Related to Focused Water Flow. Soil Sci. Soc. Am. J. **1994**, *58*, 156–163.
6. Heck, R.J.; Mermut, A.R. Genesis of Natriborolls (Solontzic) in a Closed Basin in Saskatchewan, Canada. Soil Sci. Soc. Am. J. **1992**, *56*, 842–848.
7. Pettry, D.E.; Brent, F.V.; Nash, V.E.; Koos, W.M. Properties of Natraqualfs in the Upper Coastal Plain of Mississippi. Soil Sci. Soc. Am. J. **1981**, *45*, 587–593.
8. Smith, G.D. Intrazonal Soils: A Study of Some Solonetz-like Soils Found Under Humid Conditions. Soil Sci. Soc. Am. Proc. **1937**, *2*, 461–469.
9. Wilding, L.P.; Odell, R.T.; Fehrenbacher, J.B.; Beavers, A.H. Source and Distribution of Sodium

Solonetzic Soils in Illinois. Soil Sci. Soc. Am. Proc. **1963**, *27*, 432–438.
10. Lavado, R.S.; Taboada, M.A. Water Salt, and Sodium Dynamics in a Natraquoll in Argentina. Catena **1988**, *15*, 577–594.
11. Miller, J.J.; Pawluk, S. Genesis of Solonetzic Soils as a Function of Topography and Seasonal Dynamics. Can. J. Soil Sci. **1994**, *74*, 207–217.
12. Horn, M.E.; Rutledge, E.M.; Dean, H.C.; Lawson, M. Classification and Genesis of Some Solonetz (Sodic) Soils in Eastern Arkansas. Soil Sci. Soc. Proc. **1964**, *28*, 688–692.
13. United States Soil Survey Staff, *Soil Taxonomy: A Basic System of Soil Classification for Making and Interpreting Soil Surveys*, Agriculture Handbook No. 436; U.S. Department of Agriculture–Natural Resources Conservation Service; U.S. Govt. Printing Office: Washington, DC, 1999.
14. FAO-UNESCO. *Soil Map of the World. Revised Legend*, World Soil Resources Report 60; Food and Agricultural Organization: Rome, Italy, 1988.
15. Isbell, R.F. *The Australian Soil Classification*, Australian Soil and Land Survey Handbook; CSIRO: Collingwood, Australia, 1996.
16. VASKhNIL (V.V. Dokuchaev Institute of Soil Science). *Classification and Diagnostics of Soils of the USSR*. Translated by S.Viswanathan; Amerind Publishing Co.: New Delhi, India, 1986.
17. *Glossary of Soil Science Terms*; Soil Science Society of America: Madison, Wisconsin, 1997.
18. Sumner, M.E.; Rengasamy, P.; Naidu, R. Sodic Soils: A Reappraisal. In *Sodic Soils–Distribution, Properties, Management, and Environmental Consequences*; Sumner, M.E., Naidu, R., Eds.; Oxford University Press: New York, 1998; 3–17.
19. United States Salinity Laboratory Staff. *Diagnosis and Improvement of Saline and Alkali Soils*, Agriculture Handbook No. 60; U.S. Department of Agriculture—Agriculture Research Service; U.S. Govt. Printing Office: Washington, DC, 1954.
20. Nettleton, W.D.; Bushue, L.J.; Doolittle, J.A.; Endres, T.J.; Indorante, S.J. Sodium-Affected Soil Identification in South-Central Illinois by Electromagnetic Induction. Soil Sci. Soc. Am. J. **1994**, *58*, 1190–1193.
21. Ammons, J.T.; Timpson, M.E.; Newton, D.L. Application of an Aboveground Electromagnetic Conductivity Meter to Separate Natraqualfs and Ochraqualfs in Gibson County, Tennessee. Soil Surv. Horiz. **1989**, *40*, 66–77.
22. Bennett, D.L.; George, R.J. Using the EM38 to Measure the Effect of Soil Salinity on Eucalyptus Globules in Southwestern Australia. Agric. Water Mgmt. **1995**, *27*, 69–86.
23. Olson, K.R.; Endres, T.J. Corn and Soybean Response to Exchangeable Sodium in South-Central Illinois. Soil Surv. Horiz. **1996**, *37*, 49–54.
24. Shainberg, I.; Sumner, M.E.; Miller, W.P.; Farina, M.P.W.; Pavan, M.A.; Fey, M.V. Use of Gypsum on Soils: A Review. In *Advances in Soil Science*; Stewart, B.A., Ed.; Springer–Verlag: New York, 1989; 1–112.

SOIL: THE ESSENCE OF LIFE AND ITS INTERACTION WITH OTHER NATURAL RESOURCES

S

D.J. Greenland
University of Reading, Reading, United Kingdom

INTRODUCTION

It is impossible to overestimate the importance of soils. All life on Earth, except for fish and other aquatic species, depends on the soil. The very first living things may have evolved in the oceans, but emergence onto the land followed and developed because of the unique properties of the soil. The relation of humans to the soil has long been recognized in religious beliefs. Hillel (1) traces these relationships in Jewish beliefs, in the sayings of Buddha, in the Koran, and in early Greek civilization.

The common theme is that Earth is the mother of life. In some places, the Earth had to be made fertile by water. In ancient Egypt, the waters of the Nile were worshipped, and in India the Ganges is known as the Mother of India. But in China, the Huanghe (formerly known as the Yellow River) is called the Sorrow of China, because of the huge quantities of silt eroded from the loessal soils in its catchment and the fearsome floods that have devastated the downstream areas for thousands of years.

It is not only the interaction of soil with water that creates its unique and essential properties but also its interaction with all natural resources. The soil has been rightly described as the attic of the earth sciences and the basement of the plant sciences. Soil is formed from rocks, and its interaction with air and water enables plants to grow. Forests and grasslands, as well as the crops on which we live, are all dependent on soil.

To grow, plants need air, water, and nutrients. Soil is the great storehouse of the nutrients essential for life. Not only does soil contain all essential nutrients, but it also exchanges them with soil solution in such a way that there is seldom an excess and normally a sufficient amount to allow development of a varied population of invertebrates and micro-organisms.

Nitrogen is an exception. It is seldom present in sufficient quantities within the rocks from which the soil was formed to support an active soil population. However, among the earliest organisms were those with the ability to process atmospheric nitrogen. The presence of air within the soil provides the nitrogen for these organisms; as these organisms die, their remains are attacked by other organisms and their nitrogen partially assimilated and partially added to the nitrogen in soil organic matter.

THE UNIQUE PROPERTIES OF SOILS

To support life, soil must enable water to be stored, so that it does not drain away under the pull of gravity but can be available to organisms in and on the soil and be retained for plant use. The soil must also allow air to move into it, so that oxygen is available and carbon dioxide may escape. This requires the soil to have a structure consisting of large cracks and pores; large pores will drain under gravity (transmission pores) and smaller pores will hold water against gravity but not hold it so strongly that plants are unable to use it (storage pores).

Such a structure is possible because of the wide range of soil particle sizes, from fine (<2 micron equivalent spherical diameter, esd) clay particles to silt particles (2 to 20 μ esd) and larger sand, gravel, and stones ($>20\ \mu$ esd). The physical forces associated with swelling and shrinkage—due to wetting and drying and freezing and thawing—create cracks and pores within the soil, between which are denser aggregates containing a wide range of smaller pores. Worms and other soil animals also create relatively large transmission pores, and fungi, bacteria, and plant roots release mucilaginous material to coat the sides of the pores and strengthen them against subsequent compression.

Thus, the special structure of soil essential to the development of the natural vegetation is built. This structure is able to provide water and nutrients, allow gases to move in and out, and provide channels through which plant roots can grow and seedlings can emerge. Once the vegetation is established, it contributes to the soil by adding organic matter as litter decays and nutrients are recycled. The organic matter feeds the soil population and contributes to the exchange of nutrients between soils and organisms. When the organic matter decomposes, it releases nitrogen, sulphur, and phosphorus contained within to feed other organisms.

Among the materials that make up soil organic matter are many acidic compounds (e.g., humic and fulvic acids), which adsorb important nutrient cations and exchange them with others in the soil solution.

When clays are formed by minerals weathering, they too retain nutrient cations at their surfaces. These cations are held either by negative charges—which develop from isomorphous replacement of higher-charged cations with lower-charged cations from within the clay lattice — or by dissociation from acidic groups on the surfaces of hydrous oxides (2).

THE FORMATION AND DEGRADATION OF SOILS

Soils tend to be formed slowly over thousands of years, as rocks weather, organic matter accumulates, and as eluviation, sedimentation, and erosion occur under the natural effects of climate and topography, modified by the influence of vegetation (3, 4). These processes which have given rise to the fertile soils of the world may be drastically modified by human intervention, which can both enhance and reduce the ability of soils to sustain life on Earth. Soils are used in many ways to improve human life, but their misuse or exploitation can also destroy the essential features that have made them the essence of life on Earth.

SOIL AND CIVILIZATION

Two hundred years ago, Malthus (5) warned the world of the dangers that arise as the demands of a growing population place increasing stress on world resources. More recently, Evans (6) has traced how the world population has grown and how it has produced the food to sustain it. Until the world population reached 3 billion in 1960, the main source of greater food production was cultivation of more land. Since 1960, the world population has doubled. The extra food needed to sustain the present population of 6 billion has come from increased crop yields, associated with higher-yielding plant varieties and the use of fertilizers and pesticides—techniques of the Green Revolution. There is scope for further production increases where good quality water is available, more land can be irrigated, and the number of crops grown per year increased. The Food and Agriculture Organization of the United Nations (FAO) believes that there is still scope for more land to be cultivated for arable production (7), but Young (8) and others believe that the FAO estimates are too high. Most good land was cultivated first; subsequently, agriculturists have been forced to develop poorer and more fragile land that is susceptible to degradation. Further, because the good land was close to cities and the cities have grown, alienation of land for buildings, roads, and factories has affected good land more than poor land.

Much of the essential resource for sustaining life has been lost by alienation of good land and degradation of poorer land. This is perhaps most vividly illustrated by erosion. The rate of natural erosion may be enormously accelerated by actions that expose the soil directly to rainfall, without the intervention of a forest or grass cover. Such actions mostly derive from deforestation to allow the land to be used for agriculture or animal production (8, 9). Although there are several sources of soil degradation, erosion is the most widespread and that which most commonly leads to irreversible losses of productivity (10).

The continued intensification of land use that has led to increasing degradation of land and water supplies has brought the world closer to the Malthusian Precipice (11), where the rate of population growth exceeds the rate of increase in food production.

To preserve life as we know it, it is essential that we conserve our soils and increase their productivity. In the past there have been several examples of the end of civilizations because of failure to control soil degradation (1, 12). Only by careful conservation of our soils, and technical innovations to increase their productivity while avoiding environmental damage, can a similar fate for our present civilization be prevented.

REFERENCES

1. Hillel, D. *Out of the Earth*; University of California Press: Berkeley, California, 1991.
2. Greenland, D.; Hayes, M. *Chemistry of Soil Constituents*; Wileys: Chichester, England: 1989.

3. Jenny, H. *Factors of Soil Formation*; McGraw Hill: New York, 1941.
4. Buol, S.W.; Hole, F.D.; McCracken, R.J. *Soil Genesis and Classification*; Iowa State University Press: Ames, Iowa, 1973.
5. Malthus, T.R. An Essay on the Principle of Population as It Affects the Future Improvement of Society. Johnson, St. Paul's Churchyard, London, England, 1798.
6. Evans, L.T. *Feeding the Ten Billion: Plants and Population Growth*; Cambridge University Press: Cambridge, England, 1998.
7. Alexandratos, N. *World Agriculture: Towards 2010. An FAO Study*; FAO, Rome, and John Wiley and Sons: Chichester, England, 1995.
8. Young, A. *Land Resources: Now and for the Future*; Cambridge University Press: Cambridge, England, 1998.
9. Lal, R. *Soil Erosion in the Tropics*; McGraw Hill: New York, 1994.
10. Oldeman, L.R. The Global Extent of Soil Degradation. In *Soil Resilience and Sustainable Land Use*; Greenland, D.J., Szabolcs, I., Eds.; CAB International: Wallingford, England, 1994.
11. Greenland, D.J.; Gregory, P.J.; Nye, P.H., Eds.; *Land Resources: On the Edge of the Malthusian Precipice?*; CAB International: Wallingford, England, 1998.
12. Hyams, E. *Soil and Civilisation*; Thames and Hudson: London, England, 1952.

SOIL HEAT AND WATER MOVEMENT

G.N. Flerchinger
United States Department of Agriculture-Northwest Watershed Research Center (USDA-NWRC), Boise, Idaho, U.S.A.

The interrelation between heat and water flow in soil is complex. Temperature gradients can induce vapor and liquid water transfer within the soil; in turn, water movement carries heat with it, thus altering the thermal regime of soil. These interactions are often neglected to simplify analysis of thermal and/or moisture regimes of soil. However, many situations and processes require that soil heat and water flow be analyzed simultaneously. Soil freezing during cold-season processes, and water vapor movement in conjunction with soil surface evaporation are two examples where thermal and moisture processes are tightly linked and require simultaneous evaluation. These and other less obvious interrelations between soil heat and water are discussed.

THEORY OF SOIL HEAT AND WATER MOVEMENT

Temperature distribution in soil with provisions for water movement on heat transfer is given by:

$$C_s \frac{\partial T}{\partial t} - \rho_i L_f \frac{\partial \theta_i}{\partial t} = \frac{\partial}{\partial z}\left[k_s \frac{\partial T}{\partial z}\right] - \rho_l c_l \frac{\partial q_l T}{\partial z} - L_v\left(\frac{\partial q_v}{\partial z} + \frac{\partial \rho_v}{\partial t}\right) \quad (1)$$

The terms (W m^{-3}) represent, respectively: specific heat for change in energy stored due to a temperature increase; latent heat required to freeze water; net thermal conduction into a layer; net thermal advection into layer due to water flow; and net latent heat of evaporation within the soil layer. In Eq. 1, C_s and T are volumetric heat capacity (J kg^{-1}C^{-1}) and temperature (C) of the soil, t is time (s), ρ_i is density of ice (kg m^{-3}), L_f is latent heat of fusion required to freeze water (J kg^{-1}), θ_i is volumetric ice content (m^3 m^{-3}), z is depth within the soil (m), k_s is soil thermal conductivity (W m^{-1} C^{-1}), ρ_l is density of water(kg m^{-3}), c_l is specific heat capacity of water (J kg^{-1} C^{-1}), q_l is liquid water flow (m s^{-1}), L_v is latent heat of vaporization required to evaporate water (J kg^{-1}), q_v is water vapor transfer (kg m^{-2} s^{-1}), and ρ_v is vapor density (kg m^{-3}) within the soil. Three specific processes addressed by Eq. 1 in which heat and water movement are most closely coupled are: soil freezing; water vapor movement, and thermal advection.

Soil Freezing

Due to negative water potentials, soil water exists in equilibrium with ice at temperatures below the normal freezing point of bulk water and over the entire range of soil freezing temperatures normally encountered. When ice is present, soil water potential is a function of temperature (1). This relation is expressed as:

$$\phi = \pi + \psi = \frac{L_f}{g}\left(\frac{T}{T_K}\right) \quad (2)$$

where φ is total water potential (m), π is soil water osmotic potential (m), ψ is soil matric potential, g is acceleration of gravity (m s^{-2}) and T_K is absolute temperature (K). As temperature at the freezing front decreases, more and more water freezes, water potential becomes more negative, and liquid water content continues to drop, creating a gradient in water potential and liquid water content. This drop in liquid water content at the freezing front has a similar effect as drying of the soil, and water will migrate from moist regions to the freezing front. This often results in elevated ice contents, ice lenses and frost heave.

Water Vapor Movement

Water vapor transfer can have a significant effect on the soil thermal environment due to the large latent heat of vaporization. Evaporation and subsequent vapor loss from the soil is very effective at cooling the soil. Vapor transfer in soil can be computed as the sum of flow due to a gradient in vapor density, q_{vp}, and that due to a temperature gradient, q_{vT} (2), where:

$$q_v = q_{vp} + q_{vT} = -D_v \rho_v \frac{dh_r}{dz} - \zeta D_v h_r s \frac{dT}{dz} \quad (3)$$

Here D_v is vapor diffusivity (m^2 s^{-1}) in soil, h_r is relative humidity within the soil based on water potential, s is the slope of the saturated vapor pressure curve (kg m^{-3}C^{-1}),

Encyclopedia of Soil Science

and ζ is an enhancement factor. Observed vapor transfer in response to a temperature gradient exceeds that predicted by Eq. 3; therefore, an enhancement factor is included, which can vary from around 10 at high water contents to unity under very dry conditions (3).

Thermal Advection

As water moves through soil, it carries thermal energy with it. Liquid water moving from warm soil to relatively cooler soil will tend to increase the temperature of the cooler soil. The heat transported is based on heat capacity of water and is proportional to the liquid water flow, q_l, and temperature gradient, as expressed in Eq. 1. Under most conditions, water movement through soil is sufficiently slow that this term can be ignored.

Thermally Induced Liquid Flow

Variation in water surface tension with temperature and the heat of wetting can cause liquid flow in response to thermal gradients. Dependence of surface tension (or water potential) on temperature, and the effect of thermal gradients on liquid water movement must be considered for nonisothermal conditions (4). As a result, the Richards equation for water flow can become complex when generalized for nonisothermal conditions. Studies comparing simulations for surface soil evaporation with and without thermal effects show that liquid flow in response to thermal gradients accounted for only 1% of the evaporation (5). Thus, thermally induced liquid flow can usually be ignored.

EXAMPLES OF COUPLED HEAT AND WATER FLOW

Soil heat transfer due to soil freezing, water vapor transfer, and thermal advection is described by Eq. 1. The effects of these processes are illustrated using data collected in southwestern Idaho, USA and simulated by the Simultaneous Heat and Water (SHAW) Model (6, 7). The site had two soil types: a loamy sand and a silt loam.

Soil Freezing

Migration of water due to soil freezing is controlled primarily by the rate of freezing front advance in relation to the unsaturated hydraulic conductivity. With rapid soil freezing, soil water is essentially frozen in place and there is little opportunity for water to migrate to the freezing front. Similarly, if the unsaturated conductivity is low, water migration to the freezing front will be slow. Thus, dry and/or coarse-textured soils exhibit much less frost-related water movement than moist, fine-textured soils.

Soil-water dynamics during soil freezing is illustrated for the silt loam soil in Fig. 1. For comparison, simulated water content without considering freeze/thaw processes is also plotted. Elevated total water content due to water migration to the freezing front can be observed for all three depths, while the agreement between simulated and

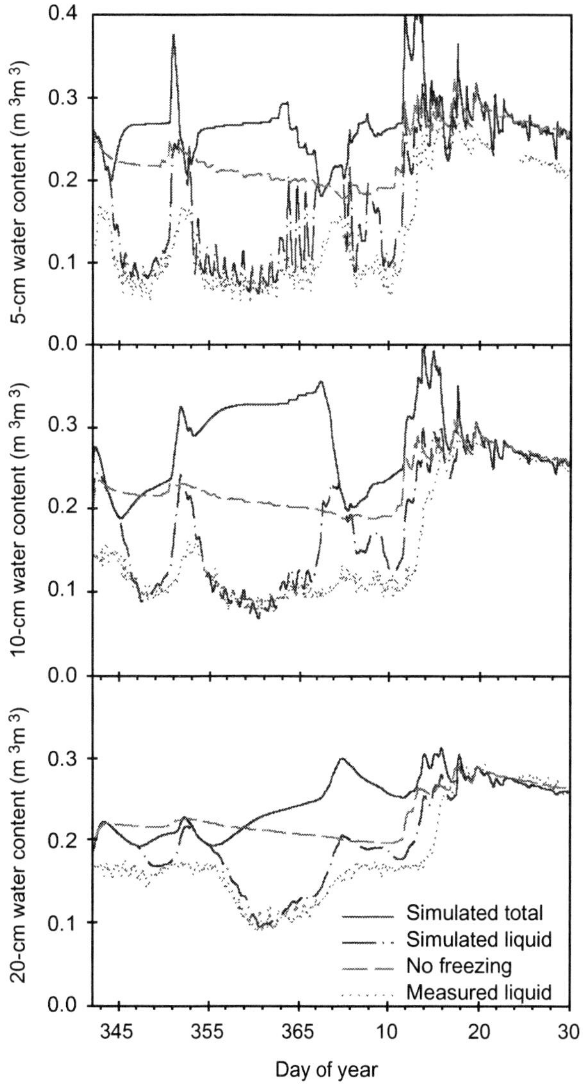

Fig. 1 Simulated total water content and simulated and measured liquid water content for a silt loam soil for the 5-, 10-, and 20-cm depths. Also plotted is simulated water content without considering freezing dynamics.

measured liquid water content was reasonable. Prior to day 344, simulated water content above 20 cm was decreasing due to drainage. After initiation of soil freezing on day 344, direction of flow reversed, and simulated water flow above 10 cm was upward toward the freezing front. As the 5-cm depth began to freeze on day 344, total water content began to increase, while simulated liquid water content continued to decrease. As the frost front advanced, the 5-cm total water content began to level off, and the 10-cm total water content increased upon freezing on day 345. Subsequently, the 20-cm depth began to freeze on day 347.

Soil-water dynamics for the loamy sand were considerably less responsive to freeze/thaw processes than the silt loam (Fig. 2). Due to the low unsaturated conductivity of the loamy sand, the simulated increase in total water content was much smaller compared to the silt loam. Even so, total water content in the frozen layer increased gradually, while water content of the simulation which ignored freezing processes continued to drain.

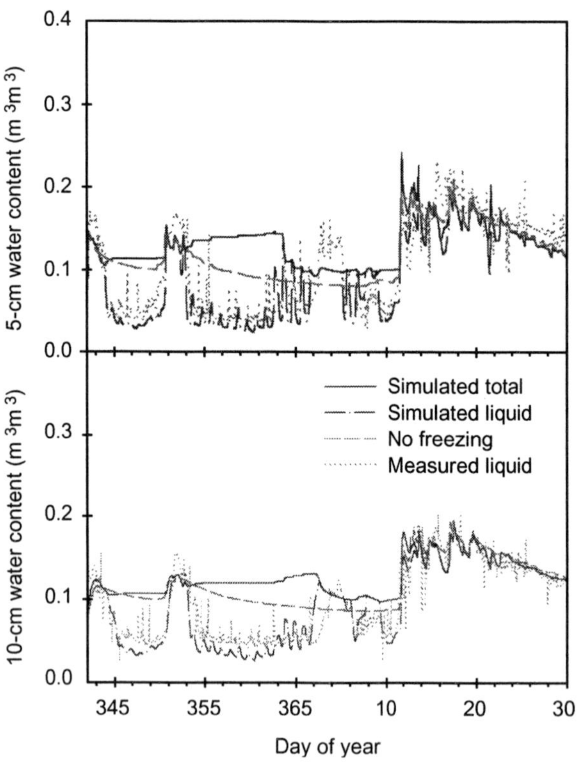

Fig. 2 Simulated total water content and simulated and measured liquid water content for a loamy sand soil for the 5- and 10-cm depths. Also plotted is simulated water content without considering freezing dynamics.

Soil-water dynamics during freezing and thawing are critical to predicting frozen-soil-related runoff. Although not measured, simulated runoff for the silt loam and loamy sand were 16 and 13 mm, respectively; no runoff was simulated when freezing was not considered.

Ignoring freeze/thaw processes had a minor, but not insignificant effect on simulated soil temperature. Compared to the full freeze/thaw simulation, ignoring effects of soil freezing resulted in simulated soil temperatures differing by typically 2.2°C for the 5- and 10-cm depths. Differences between the loamy sand simulations were somewhat less because there was less water to freeze.

Water Vapor Movement

Under moist conditions, vapor transfer within soil is usually negligible due to the lack of air-filled pore space. However, under drier conditions, vapor transfer can significantly contribute to soil water movement and has a large effect on soil thermal conditions. Water vapor movement is especially important for a situation where the surface soil is dry and underlain by moist soil.

Measured 2-cm soil water content is shown in Fig. 3 along with simulated water content for three simulations: vapor transfer included, vapor transfer ignored, and vapor transfer due to thermal gradients (q_{vT}) ignored. The simulation that ignored all water vapor transfer overestimated water content at the 2-cm depth because liquid flow through the dry ($<0.05\,m^3m^{-3}$) soil above 2 cm was insufficient to dry the soil beyond 1 cm.

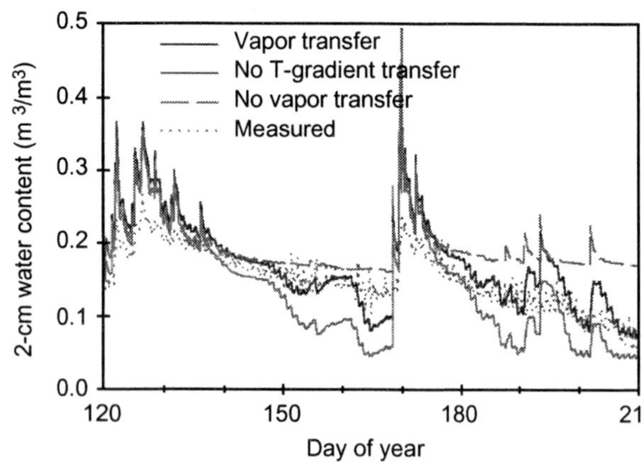

Fig. 3 Measured 2-cm water content for a silt loam soil and simulated 2-cm water contents 1) considering water vapor transfer, 2) assuming no water vapor transfer, and 3) neglecting the effect of thermal gradients on water vapor transfer.

Conversely, the simulation which ignored thermal gradient vapor transfer, q_{vT}, overestimated water vapor transfer and therefore resulted in excessive water loss. Typically daytime temperature gradients inhibit vapor movement toward the surface and reduce evaporation. Simulated evaporation for the 90-day period was 111 mm for the silt loam soil when vapor transfer was considered, 105 mm with no vapor transfer, and 117 mm when q_{vT} was ignored. Simulated evaporation for the loamy sand soil was 101, 96, and 105 mm, respectively.

Differences in evaporation between simulations had a moderate effect on simulated soil temperature. On average, the simulation without water vapor transfer had soil temperatures approximately 0.5°C warmer due to reduced evaporation. However, the absolute difference between simulations was as much as 1.2°C, because ignoring vapor transfer caused higher temperatures during the day and cooler temperatures at night.

Thermal Advection

Heat transfer by soil-liquid water flow is usually quite small compared to thermal conduction. Average absolute thermal gradient near the soil surface simulated for the bare soil plots was $1.6\,C\,cm^{-1}$. For typical values of thermal conductivity, it would require $1.3\,mm\,hr^{-1}$ of liquid water flow (q_l) to transfer an equivalent amount of heat within a 1-cm layer. Thus, the effect of thermal advection of heat becomes important only for rapid infiltration of water. Indeed, one-year simulations neglecting the advection term during infiltration resulted in near-surface soil temperatures differing by typically 0.3–0.4°C compared to simulations in which heat of advection during infiltration was considered. Maximum hourly differences of over 10°C occurred due to summer rainstorms on a hot soil surface (40°C). When advective heat transfer was considered during infiltration but ignored during redistribution, the temperature was typically within 0.03°C of the simulation considering advection. Thus, advective heat transfer can be significant during infiltration, but not generally important during redistribution of soil water.

SUMMARY AND PRACTICAL SIGNIFICANCE

Interrelation between water and heat flow within the soil can influence the soil temperature regime. This influence is most pronounced in the case of soil freezing and thawing, and water vapor movement during surface evaporation. With the exception of rapid infiltration of water, heat carried by liquid water flow can usually be ignored. Thermally induced liquid water flow is usually insignificant.

From a practical standpoint, although water vapor movement influences evaporation from the soil and in turn affects water availability for plant growth, ignoring this process usually provides evaporation estimates within 10% of actual values. Soil freezing, however, can significantly influence infiltration and runoff, causing severe flooding from relatively mild rainfall or snowmelt events.

REFERENCES

1. Fuchs, M.; Campbell, G.S.; Papendick, R.I. An Analysis of Sensible and Latent Heat Flow in a Partially Frozen Unsaturated Soil. Soil Sci. Soc. Amer. J. **1978**, *42* (3), 379–385.
2. Campbell, G.S. *Soil Physics with BASIC: Transport Models for Soilplant Systems*; Elsevier: Amsterdam, 1985; 150.
3. Cass, A.; Campbell, G.S.; Jones, T.L. Enhancement of Thermal Water Vapor Diffusion in Soil. Soil Sci. Soc. Amer. Jour. **1984**, *48*, 25–32.
4. Milly, P.C.D. Moisture and Heat Transport in Hysterectic, Inhomogeneous Porous Media: A Matric Head-Based Formulation and a Numerical Model. Water Resour. Res. **1982**, *18* (3), 489–498.
5. Milly, P.C.D. A Simulation Analysis of Thermal Effects on Evaporation from Soil. Water Resour. Res. **1984**, *20* (8), 1087–1098.
6. Flerchinger, G.N.; Saxton, K.E. Simultaneous Heat and Water Model of a Freezing Snow-Residue-Soil System I. Theory and Development. Trans. of Amer. Soc. Agric. Engr. **1989**, *32* (2), 565–571.
7. Flerchinger, G.N. *Simultaneous Heat and Water Model: Technical Documentation*, NWRC Technical Report 2000-09; http://www.nwrc.ars.usda.gov/models/shaw/index.html; USDA — Agricultural Research Service, Northwest Watershed Research Center: Boise, Idaho, 2000; 38.

SOIL WATER CONTENT

T.J. Jackson
United States Department of Agriculture-Agricultural Research Service (USDA-ARS), Beltsville, Maryland, U.S.A.

INTRODUCTION

Soil water content is a basic hydrologic state variable. It is used in applications that range in scale from irrigation management to flood forecasting. Soil water content has also increased in importance in weather forecasting and climate modelling as these disciplines have begun to recognize its importance in large-scale water cycles.

As important as soil water is in human activities and food supplies, it has not been as widely observed and monitored as other variables such as precipitation and temperature. This is caused by inadequate instrumentation as well as the numerous scales of variability that affect this state variable.

Soil water content generally refers to the volumetric soil water content. This is the percentage or fraction of a known volume occupied by water (θ_v). This is related to the gravimetric soil water (θ_g) and the bulk density of the soil (ρ_{soil}).

$$\theta_v = \frac{\theta_g \rho_{soil}}{\rho_{water}}$$

Soil bulk density is the mass of dry soil within a known volume. Typical values will range from 1 to 1.5 g cm^{-3}. The measurement and estimation of ρ_{soil} is described elsewhere in this volume as well as in Ref. (1). ρ_{water} is the density of water and is close to a value of 1 g cm^{-3} and is usually dropped from the equation.

VARIABILITY AND SAMPLING

A problem with both characterizing and sampling soil water content is that it exhibits variability that is related to physical characteristics and forcing variables that vary on scales of soil pores to continents. These sources and scales are discussed in further detail in Refs. (2–5). At the field scale important factors are the soil texture, vegetation, and topography. The next level is related primarily to precipitation and to a lesser degree evapotranspiration. Finally, larger scale climatology will influence soil water content over regions and continents. From these works it is evident that variations within a field can be as large as within a 100 km region.

METHODS FOR MEASURING SOIL WATER CONTENT

There is a wide range of techniques available for measuring soil water content. These are described in depth in Refs. (3, 6–9). Some of these are reviewed in the following section. It is important to note that there is no universal standard method (gravimetric with oven drying is the closest). This is because of the difficulties in capturing the variability, instrument design, and logistics. Table 1 summarizes advantages and disadvantages discussed in the following sections.

Gravimetric Sampling and Oven Drying

Gravimetric sampling combined with oven drying is generally accepted as the standard technique for characterizing soil water content. In this approach, a sample from a specific depth interval is removed in the field and placed in a container of known tare weight. The sealed sample is weighed, opened, and placed in an oven at 105°C for 24 hours. It is then weighed again. The gravimetric soil water content is computed as follows:

$$\theta_g = \frac{\text{Wet weight} - \text{Dry weight}}{\text{Dry weight} - \text{Tare}}$$

If the sample is collected using a tool of known volume, the volumetric soil moisture can be determined directly.

$$\theta_v = \frac{(\text{Wet weight} - \text{Dry weight})/\rho_{water}}{\text{Volume}}$$

Otherwise, an estimate of the bulk density must be used to convert θ_g to θ_v.

Coring tools can be used to extract a sample of known volume. This method works well for small-diameter samples from moderate depth intervals on compacted

Encyclopedia of Soil Science

Table 1 Comparison of some methods of measuring soil water content

Technique	Advantages	Disadvantages
Gravimetric	• Only method providing a direct measurement • Inexpensive equipment • Accurate	• Destructive • Labor required • Time delay for drying • Small sample size
TDR	• Most accurate indirect method • Rapid • Some spatial averaging	• Expensive • Salinity • Depth resolution for surface layer
Capacitance	• Low cost • Rapid • Probe design very flexible	• Sensitive to installation
Heat dissipation	• Low cost • Easy to install • Provides water potential	• Limited range of moisture • Calibration-hysteresis • Small sampling area • Slow response
Radioactive source	• Some spatial averaging	• Radiation hazard • Expensive • Calibration • Can't measure surface layer
Remote sensing	• Local, regional, and global mapping • Spatial averaging	• Poor spatial resolution • Only surface information

soils. For instance, it is difficult to sample shallow surface layers on tilled soils with this method. Scoops and augers can be used to collect gravimetric water content samples. Many of these devices can be seen at the following website http://www.soilsample.com/.

As noted in methods manuals, care must be exercised when using this method. Samples should be over 100 g, and consideration should be given to the general soil conditions and oven conditions when choosing a drying period. The major drawback to gravimetric sampling is that it is time consuming. In addition, it is destructive. Repetitive sampling at a location could modify conditions.

In Situ and Portable Measurement Techniques

A brief overview of these methods is presented here. An excellent source of additional information is the web site http://www.sowacs.com/, which includes links to manufacturers as well as comparisons.

Time domain reflectometry (TDR)

TDR instruments represent a fairly mature technology at this point in time. These devices are capable of measuring two properties of the soil, apparent dielectric constant and conductivity. The discussion here only considers the dielectric constant.

The soil dielectric constant is a function of the volume fractions and dielectric constants of its constituents: water, soil, and air. There is a very large difference between the dielectric constants of water and the other components, which makes these devices highly sensitive to soil water content.

A TDR, as typically used in soil water devices, measures the time it takes for an electrical wave to travel down parallel rods and reflect back. The rods are embedded in the soil and can be used in situ or as a portable sensor. The travel time will vary with the sensor configuration and the dielectric properties of the media. There are limitations on how short and how long the rods can be.

Topp et al. (10) and numerous other investigators have found that, for most field soils, a single calibration function can be used. When the soils have high clay contents (or organic matter), a separate calibration is required (11).

Therefore, TDR is a relatively robust method for measuring soil water content and can be incorporated as an in situ or as a portable sensor. Earlier system designs used low-cost sensors with a relatively expensive cable tester unit. Newer designs include all components in each individual sensor at a modest cost.

Capacitance devices

The dielectric constant of the soil media can also be measured using capacitance sensors. As described above, this dielectric constant is a function of the volumetric soil water content. A capacitance sensor consists of two or more electrodes with the soil media between them. An electronic oscillator is used to determine the oscillator frequency, which is a function of the dielectric constant (as well as the probe configuration).

These devices have become quite popular, and there are several commercially available systems. One difference between these systems is the capacitor design, which affects how the device can be used. Some devices utilize horizontal plates separated by a low dielectric material. This probe unit is placed in a low dielectric tube (PVC pipe) installed in the soil. Probes can be left in place for in situ measurement (12) or used in a fashion similar to neutron probes, carrying from site to site and logging the profile as the probe is lowered in the tube.

Another capacitance sensor design utilizes parallel rods (13). These can be used as in situ or portable sampling tools. One advantage of this device as opposed to the TDR is that shorter rods can be used (5–6 cm are typical). This means measurement closer to the soil surface is possible and that there are likely to be fewer problems when the device is used as a portable surface sampler. Fig. 1 shows a handheld portable version of the capacitance probes. Capacitance probes will be subject to the same calibration problems as TDR devices as related to soil texture and possible mineralogy. Most instruments are supplied with a general calibration.

Heat dissipation

The thermal conductivity of a soil is a function of its water content. Heat dissipation sensors are based on this relationship (14). A heater and a thermocouple are embedded in a ceramic probe and placed in the soil. The initial temperature is recorded, and then the heater is turned on for a short period of time and the final temperature is recorded. The change in temperature is related to the soil water content. The ceramic probe is assumed to be in matric potential equilibrium with the surrounding soil. This device is generally used to measure the matric potential and not the soil water content. Once calibrated for matric potential, the probe can be installed in any soil. However, in order to obtain soil water content, the probe must be either calibrated for the particular site or the matric potential must be related to soil water content using additional calibration or pedotransfer functions.

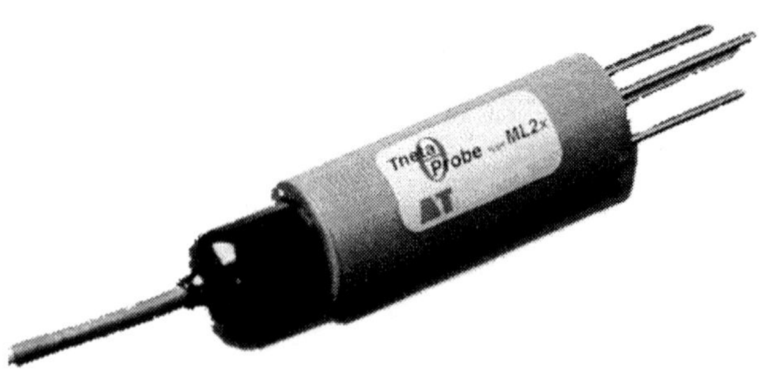

Fig. 1 A handheld capacitance type soil water content probe.

Radioactive source devices

These methods have been in use for a number of years. One of the most common is the neutron probe. Here a radioactive source and a detector are lowered down a tube in the soil. Water within the sphere of influence affects the response (neutron scattering). Another radioactive device is the gamma attenuation probe. A source is lowered down a tube while a detector is lowered down a parallel tube. The attenuation is related to the water content. Considering the range of low-cost and reliable alternatives available and the risks associated with radiation, radioactive source methods are not recommended.

Remote Sensing

Remote sensing refers to noncontact methods that generally utilize responses within portions of the electromagnetic spectrum. For soil water content measurements, these devices will always be viewing the soil from the surface. Sensors can be installed on towers, aircraft, and satellites. A key advantage of remote sensing is that systems can be designed to map large regions quickly on a frequent basis.

Although all regions of the electromagnetic spectrum have been explored for soil water measurement (15), only microwave techniques can provide a robust measurement. Microwave sensors appropriate for soil water content measurement can see through clouds and most vegetation. Observations can be made at any time of day since solar illumination isn't required. The microwave signal originates within a depth of the surface soil. The lower the frequency used, the deeper is the depth. However, there are restrictions on the frequencies that can be used, which limit this depth to approximately 5 cm.

Two methods can be used, active and passive. Active methods or radars send out a microwave signal, which interacts with the Earth's surface (soil and vegetation). A portion of this signal is reflected back to the sensor. The response depends upon the dielectric properties of the soil, among other factors. As described previously, the dielectric properties are functionally related to the soil water content.

Passive methods measure the Earth's natural emission at these frequencies. This emission is called the brightness temperature, and it depends upon the physical temperature, and the emissivity. Emissivity is a function of the dielectric constant and thus soil water content.

Fig. 2 shows examples of soil moisture maps derived over a portion of Oklahoma (16). These cover a watershed ($600\,km^2$) at a spatial resolution of 200 m. Here the patterns are related to soil type differences that affect the soil water content. The other image is of the same region ($10{,}000\,km^2$) at a spatial resolution of 800 m. Here the spatial patterns are dominated by rainfall features.

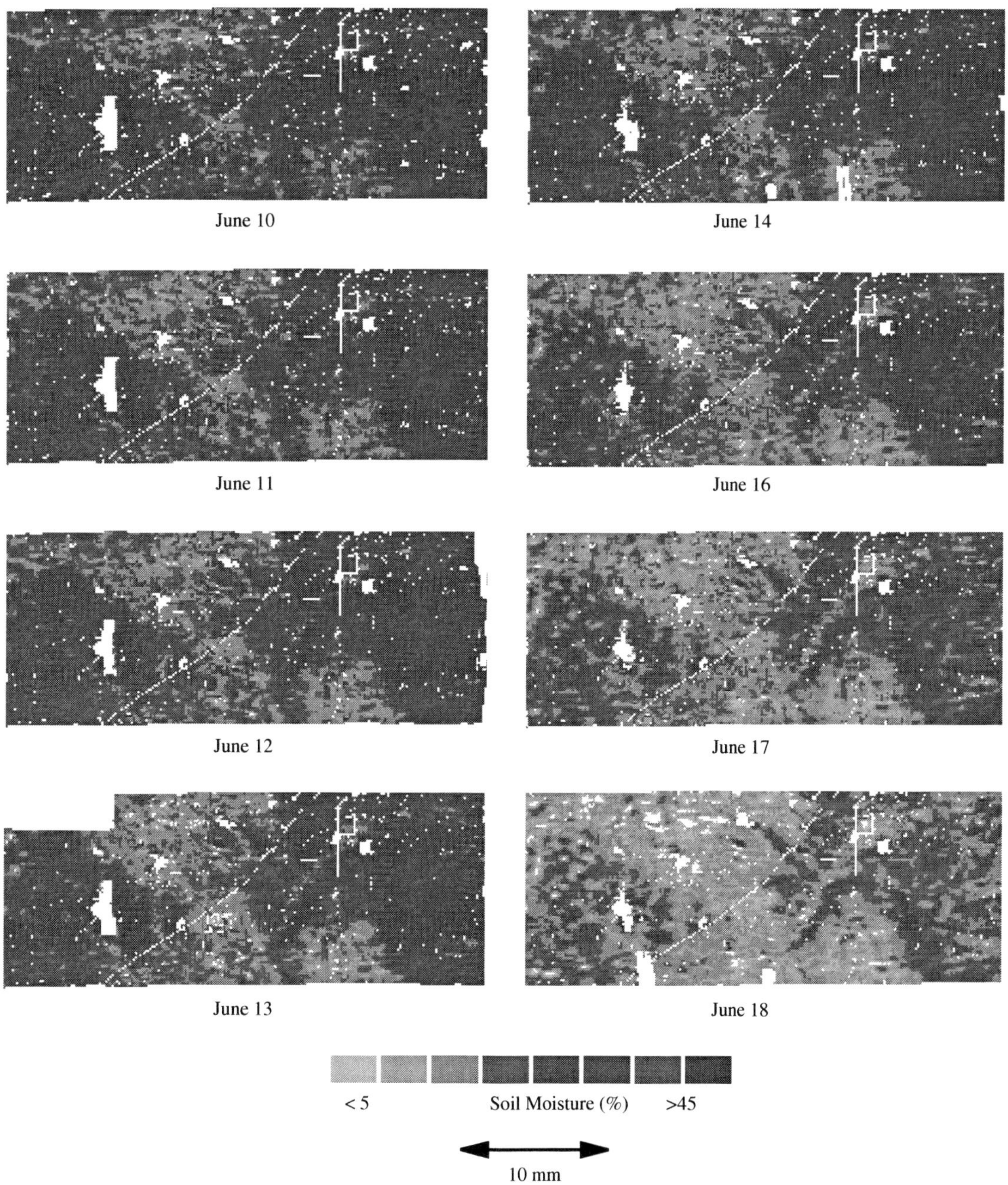

Fig. 2 Soil moisture maps for the Little Washita Watershed, Oklahoma, 1992.

DATA SOURCES

For the reasons discussed in the previous sections, there have been few attempts to measure soil water content on a routine basis such as is done with precipitation. Therefore, there are very few data resources and networks. Much of the historic records have been compiled in Ref. 17 as part of the Global Soil Moisture Data Bank. This information can be obtained through the following website http://climate.envsci.rutgers.edu/soil_moisture/.

The paper by Robock et al. (17) also describes most of the existing networks that collect soil moisture data. Of these, the Soil Climate Analysis Network (SCAN) operated by the U.S. Department of Agriculture is the most readily available. SCAN provides real time and historic (a few years at this point) hourly soil moisture and meteorological data at over thirty stations in the United States. These data can be obtained through the following website http://www.wcc.nrcs.usda.gov/scan/.

Remotely sensed soil moisture data sets from several previous experiments are available from the following website http://hydrolab.arsusda.gov/rsbarc/RSofSM.htm. In addition, there are new satellites due for launch in 2002 that will provide global surface soil moisture maps. More information on these is available at http://eos-pm.gsfc.nasa.gov/.

REFERENCES

1. Blake, G.R.; Hartge, K.H. Bulk Density. In *Methods of Soil Analysis. Part I*, 2nd Ed. Klute, A., Ed.; American Society of Agronomy: Madison, WI, 1986; 363–376.
2. Hawley, M.E.; McCuen, R.H.; Jackson, T.J. Soil Moisture Sampling Volume vs. Accuracy. Journal of the Irrigation and Drainage Division of the ASCE **1982**, *108*, 1–11.
3. Kutilek, M.; Nielsen, D.R. *Soil Hydrology*; Catena Verlag: Cremlingen-Destedt, Germany, 1994.
4. Petersen, R.G.; Calvin, L.D. Sampling. In *Methods of Soil Analysis. Part I*, 2nd Ed. Klute, A., Ed.; American Society of Agronomy: Madison, WI, 1986; 33–52.
5. Vinnikov, K.Y.; Robock, A.; Qiu, S.; Entin, J.K. Optimal Design of Surface Networks for Observation of Soil Moisture. J. Geophys. Res. **1999**, *104*, 19743–19749.
6. Campbell, G.S.; Mulla, D.J. Measurement of Soil Water Content and Potential. *Irrigation of Agricultural Crops*; Agronomy Monograph 30 American Society of Agronomy: Madison, WI, 1990; 127–142.
7. Gardner, W.H. Water Content. In *Methods of Soil Analysis. Part I*, 2nd Ed. Klute, A., Ed.; American Society of Agronomy: Madison, WI, 1986; 493–544.
8. Hillel, D. *Environmental Soil Physics*; Academic Press: San Diego, CA, 1998.
9. Or, D.; Wraith, J.M. Soil Water Content and Water Potential Relationships. In *Handbook of Soil Science*; Sumner, M.E., Ed.; CRC Press: Boca Raton, FL, 1999; A-53–A-85.
10. Topp, G.C.; Davis, J.L.; Annan, A.P. Electromagnetic Determination of Soil Water Content: Measurements in Coaxial Transmission Lines. Water Resources Research **1980**, *16*, 574–582.
11. Topp, G.C.; Reynolds, W.D. Time Domain Reflectometry: A Seminal Technique for Measuring Mass and Energy in Soil. Soil and Tillage Research **1998**, *47*, 125–131.
12. Paltineanu, I.C.; Starr, J.L. Real-time Soil Water Dynamics Using Multisensor Capaitance Probes: Laboratory Calibration. Soil Science Society of America **1997**, *61*, 1576–1585.
13. Robinson, M.; Dean, T.J. Measurement of Near Surface Soil Water Content Using a Capacitance Probe. Hydrological Processes **1993**, *77*, 77–86.
14. Reece, C.F. Evaluation of Line Heat Dissipation Sensor for Measuring Soil Matric Potential. Soil Science Society of America Journal **1996**, *60*, 1022–1028.
15. Schmugge, T.J.; Jackson, T.J.; McKim, H.L. Survey of Methods for Soil Moisture Determination. Water Resources Research **1980**, *16*, 961–979.
16. Jackson, T.J.; Le Vine, D.M.; Swift, C.T.; Schmugge, T.J. Large Scale Mapping of Soil Moisture Using the ESTAR Passive Microwave Radiometer. Remote Sensing of Environment **1995**, *53*, 27–37.
17. Robock, A.; Vinnikov, K.Y.; Srinivasan, G.; Entin, J.K.; Hollinger, S.E.; Speranskaya, N.A.; Liu, S.; Namkhai, A. The Global Soil Moisture Data Bank. Bull. Amer. Meteorol. Soc. **2000**, *81*, 1281–1299.

SOLUBLE SALTS, TRANSLOCATION AND ACCUMULATION OF

S

Jimmie L. Richardson
Department of Soil Science, North Dakota State University, Fargo, North Dakota, U.S.A.

INTRODUCTION

Translocation and accumulation of soluble salts in soils encompass both physical transport through a landscape to an accumulation area and soil profile processes of accumulation at that point. Soluble salts accumulate by sequences of mineral precipitation (evaporite sequences) that control the salt types of the resulting evaporite deposits either in or on the soil. These processes are driven by water that is invariably in the soil, not surface water. The water progressively becomes more concentrated in dissolved solids either by dissolving more salts as the water travels in the landscape or by losing water to the atmosphere via evapotranspiration. Either way the dissolved solids precipitate as salts in zones in discharge areas on the landscape.

TRANSLOCATION IN LANDSCAPES

In a landscape, step one of salt translocation results from removal of labile salts in the recharge and throughflow zones of the landscape and saturated, often-transient flow, to the discharge zones. Dissolution occurs in soils located in the recharge zone of landscapes when rainwater or snowmelt infiltrates and percolates through soil or geologic materials. The landscape sources of water can be divided into local, regional, and irrigation water. Local water moves as transient flow (inter-flow or throughflow) from nearby areas either in periods of wetness or from local water sources. The regional systems are larger scale created by fracture or aquifer flow to the accumulation areas, often via artesian aquifers that discharge under pressure. Irrigation water is gathered from distant streams, aquifers, or other source areas and transported in engineered channel or pipe systems to the accumulation sites. Landscape translocation removes salts from recharge areas and transfers the dissolved solids to discharge areas where accumulation can occur.

Accumulation, or step two of translocation within the landscape, occurs because inefficient water removal exists during discharge. In semiarid and arid climates, the lack of water and the high potential evaporation rates create conditions with poorly integrated intermittent streams. Thus, when water is present, it cannot be transported externally from the accumulation zones or saline soils. In young or constructional landscapes, such as a Pleistocene or Holocene lake or till plain, an integrated drainage network has usually not developed; without drainage channels water cannot be effectively removed. For example, in Grand Forks County, North Dakota, the subhumid climate with 0.5 m of annual precipitation at about 45 degrees north latitude, maintains relatively low evaporation potentials. Over 25% of the land area is variously impacted by saline or sodic soils. Saline soils have a saturated paste electrical conductivity that exceeds 4 dS/m and sodic soils have sodium absorption ratios exceeding 12. Some areas that are strong discharge areas or receive abundant water in deserts have very high evaporation potentials and ineffective drainage. These result in hypersalinity or salinity that exceeds an electrical conductivity of 20 dS/m. The Bonneville Salt Flats, which lie around the Great Salt Lake in Utah, are examples.

Accumulation sites often have a high water table either seasonally or periodically during pluvial times of drought-pluvial cycles. At these sites, increases in salinity correlate best to pluvial or wetter periods because this is when the water table is highest and evapotranspiration losses concentrate salts from the water table. Irrigation water creates a perpetually pluvial condition that often creates a high water table, resulting in persistent salinization. Salinization or sodification potentially occurs rapidly unless adequate drainage has been established before irrigation.

As a short synopsis of influence of the landscape, the translocation of dissolved solids relates closely to periods of saturated flow whereby salts increase as the path of transportation increases and water is lost to evapotranspiration.

TRANSLOCATION AND ACCUMULATION WITHIN SOILS

Translocation of salts within the soil involves a constant flux between upward water losses by matric and osmotic flow, and the downward infiltration and subsequent percolation of rain or irrigation water. The actual accumulation in the root zone of plants, which follows, relates to matric flow and to osmotic flow. In matric flow, which is often called capillarity, water moves slowly from wet soil to dry soil. The flow direction is independent of gravity, instead it is dictated by moisture content and the matric potential possessed by the soil. Soils with abundant surface area, such as those with clay textures, tend to have higher matric potentials at given water contents. In osmotic flow, water moves from low salt content to high salt content. Soil salinity increases hygroscopic water because of the increase in osmotic potential, often to the point where a soil looks moist but plants cannot obtain water.

Salt efflorescences often disappear after a rain event because the water that infiltrates carries the labile salts downward. Often, however, casts of less soluble minerals such as gypsum remain that are good indicators of saline condition. The salt efflorescence will return after a few rainless days with high evaporation stress. As evaporation stress occurs on the soil, the matric forces move water to the surface where the water vaporizes and the salts precipitate. Few really labile minerals form in the soil. Usually when somebody notes a "salt" in the soil it is gypsum. One exception is the winter precipitation of mirabilite. Mirabilite is a sodium sulfate that has 10 waters of hydration. With such a high amount of hydrogen bonding the ice-like crystals form easier at cool temperatures (1, 3). The usual events that tend to dominate saline soils are evaporative discharge driven by evaporation with flow reversals during rainfall, irrigation, or snowmelt events.

Saline seeps occur in western U.S. and Canada when an aquifer is activated during a pluvial time of the pluvial-drought cycle. The aquifer does not function unless enough water reaches it to create saturated flow. Once saturated flow occurs, the water moves quickly to points of discharge on the landscape and the seepage water dries out in the air or by plant usage, leaving the salts behind. In these cases the translocation to the soil is lateral from the landscape. A possible rule of thumb for the recharge area of most small saline seeps is about 100 m or less.

Precipitation of translocated salts occurs in sequence as evaporation or evapotranspiration concentrates water. First to form is calcite which removes 1 mole each of Ca and carbonate for every mole of calcite formed. Depending on which of these two ions is dominant, the next to form is either gypsum or Mg-calcite or dolomite with a rise in pH. In fact if the pH is over 8.5 one can almost be sure that the dominant carbonate is of sodium (2).

One mole of gypsum removes 1 mole of calcium and 1 mole of sulfate. If the sulfate is dominant over calcium as gypsum precipitates, the water and the salt efflorescence become high in Na and Mg with sulfate and chloride as the anions (2). Although the Eugster and Jones model was for a fully closed basin system, their sequence of evaporites is a good predictive tool in most saline situations because these situations represent groundwater discharge or ineffective drainage that act like a closed basin system or labile ion sink.

REFERENCES

1. Beke, G.J.; Palmer, C.J. Subsurface Occurrence of Mirabilite in a Mollisol of Southern Alberta, Canada: A Case Study. Soil Sci. Soc. Am. J. **1989**, *53*, 1611–1614.
2. Eugster, H.P.; Jones, B.F. Behavior of Major Solutes During Closed Basin Brine Evolution. Am. J. Sci. **1979**, *279*, 609–631.
3. Timpson, M.E.; Richardson, J.L.; Keller, L.P.; McCarthy, G.J. Evaporite Mineralogy Associated with Saline Seeps in Southwestern North Dakota. Soil Sci. Soc. Am. J. **1986**, *50*, 490–493.

SPECTRAL MIXTURES, SOIL–PLANT

S

A.R. Huete
Department of Soil, Water, and Environmental Science, University of Arizona, Tucson, Arizona, U.S.A.

INTRODUCTION

There exists considerable knowledge about soil optical properties based on extensive laboratory analyses studies (1–3). Information about soils are generally obtained by analyses of their spectral signatures, which are a function of the absorption properties of their biogeochemical constituents, moisture condition, and the size, shape, and geometry of the soil particles. At the landscape level, it is much more difficult to monitor and assess soil properties due to their extreme spatial variability and the masking of the soil surface by vegetation and litter. Airborne and satellite remote sensing are used to characterize and map soils in several ways: 1) direct remote sensing measurements of exposed soil surfaces in vegetated (e.g., canopy gaps) and nonvegetated areas (hyperarid zones); 2) employment of spectral mixture models to generate soil component images from remote sensing data over partially vegetated areas; 3) one can infer soil properties based on remote sensing measurements of the overlying vegetation.

SOIL–VEGETATION CANOPY MIXTURES

In arid, semiarid, and subhumid regions, soil surfaces are rarely spectrally pure and often contain significant quantities of litter and vegetation mixing with the soil signal (4). The remotely sensed, spectral responses from these surfaces are a function of the number and type of reflecting components, their optical properties, and their relative proportions. The optical properties and mixture proportions further vary seasonally and with land cover conversions (5). Resolving the extent of soil and vegetation cover is important to hydrologic and biogeochemical processes and crucial for functional analyses of ecosystems (6–8).

BASICS OF MIXTURE MODELING

Spectral mixture models are used in remote sensing to separate remote sensing measurements into their soil, nonphotosynthetically active vegetation (NPV), and vegetation components. Such models have utility in a variety of applications, including biogeochemical studies, leaf water content, land degradation, land cover conversions, fuelwood assessment, and soil and vegetation mapping (9–12). Soil–plant spectral mixtures may be modeled as linear or nonlinear mixtures (13, 14). In the linear case, also known as the "checkerboard" model, photons only interact with a single material and the fractions of each material are equivalent to their aerial proportions (15). Linear models are easy to use and work quite well over certain types of land cover conditions, such as desert shrub canopies (16). Mixture modeling generally involves three steps:

1. assess the dimensionality or number of spectral constituents in an image;
2. identify the physical nature of each of the constituents or "endmembers";
3. derive the amounts of each component for each pixel in the image.

The first step is generally accomplished with principal components analysis (PCA). The second step is achieved with an endmember analysis whereby various reference spectra are used to model the image. Several types of endmember signatures may be utilized in spectral mixture analyses, including image endmembers, bundled spectra endmembers, and laboratory reference endmembers, which are cataloged in "spectral libraries" (17) (Fig. 1). Typical endmembers used in remote sensing include green vegetation, soil, NPV, and "shade." Shade is used to model pixels not fully illuminated and vary with topography (18).

In matrix notation, a linear mixture model is expressed as

$$[\mathbf{D}] = [\mathbf{R}][\mathbf{C}] + [\varepsilon],$$

where $[\mathbf{D}]$ is the spectral data matrix, $[\mathbf{R}]$ is the reflectance matrix of the spectral constituents (endmembers), $[\mathbf{C}]$ is the component contributions or "loadings" matrix, and $[\varepsilon]$ is the residual errors. One can first model the spectral variability of a scene with several image endmembers followed by the use of reference endmembers to determine the composition of the image endmembers (19). Residual images are a quick way to find "outlying" pixels not adequately explained by the mixture model and may indicate the presence of unknown reflecting components. Residuals in the

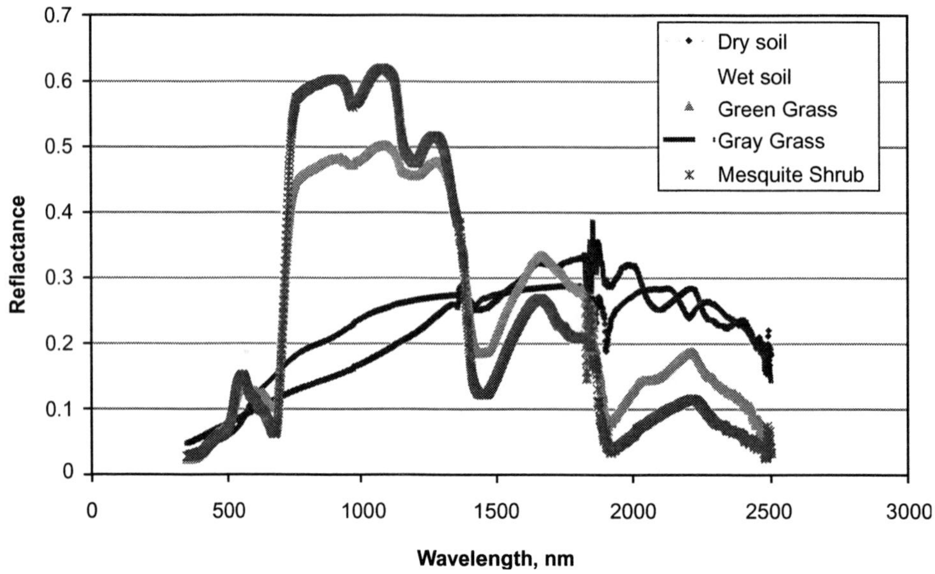

Fig. 1 Examples of reference spectral signatures of soil–plant components commonly found in semiarid canopies of the southwest U.S.

short-wave infrared (SWIR) region, attributable to cellulose and lignin in the vegetation, have been used to resolve and separate the NPV signal from soil (20). Residual images also contain the bulk of the noise and error in images. The products of mixture modeling include a set of component fraction images and a set of residual images containing the differences between modeled and measured data.

HYPERSPECTRAL SENSORS

Hyperspectral imaging sensors, with a large number of contiguous bands, carry valuable diagnostic information about soil and vegetation surfaces and have great potential for quantitative retrieval of their biophysical properties. The airborne visible–infrared imaging spectrometer (AVIRIS) sensor is of particular interest in soil–vegetation spectral mixture studies. AVIRIS operates in the 400–2450 nm region collecting 224 spectral bands with a nominal 10 nm spectral response function (21). This enables one to observe high-resolution spectra associated with unique absorption features, which are typically lost in coarser waveband data, such as from Landsat TM, SPOT, and AVHRR. The Hyperion Hyperspectral Imager onboard the Earth Observing-1 provides 220 contiguous bands covering the spectrum from 400 to 2500 nm (22). This sensor is a part of NASA's New Millennium Program, Earth Observing series designed to test new technologies for Earth System Science studies.

Okin et al. (23) used AVIRIS data in retrieving soil and vegetation information in arid and semiarid environments. They applied a multiple endmember spectral mixture analysis (MESMA) on AVIRIS imagery collected over the Mohave Desert in California and were able to reliably distinguish and map soil-surface types and vegetation cover. Palacios-Orueta et al. (24) used hyperspectral AVIRIS imagery to map the spatial variation of soil organic matter and soil iron contents over two watersheds in the Santa Monica Mountains in California. They used a sequential spectral mixture analysis technique known as hierarchical foreground/background analysis (HFBA) to limit the influence of the vegetation signal while extracting the variation in the soil signal with laboratory derived "training" vectors.

INFERRING SOIL PROPERTIES THROUGH VEGETATION

About 25–30% of the land area have significant quantities of forest and dense vegetation with leaf area index values exceeding 1, including tropical and temperate forests and cropland. Where the surface is strongly vegetated, soil properties must be inferred from measurements of the vegetated surface in conjunction with models and field measurements. Spectral vegetation indices (VI) as well as mixture models have been used to map vegetation characteristics to infer soil properties. Cwick et al. (25) mapped the forest soils in Manitoba through hyperspectral reflectance measurements of Black Spruce (*Picea mariana*) needles. They utilized the concentrations of potassium in the needles to infer the potassium distributions in the soil rooting zone with primary variations associated with either poorly or well-drained soils.

CONCLUSION

Approximately 70% of the Earth's terrestrial surface consists of open canopies (deserts, grasslands, savannas, and open forests) with vegetation overlying various proportions of exposed soil and litter background (4). The aim of remote sensing is to exploit and model these complex patterns of surface energy interactions for the purpose of mapping and extracting information about the biophysical and biochemical character of soils. Several approaches are used to analyze soil–plant spectral mixtures, including vegetation indices and mixture models. Vegetation indices are useful in assessing the spatial and temporal dynamics of the vegetation component, but yield little information on soil type and soil properties. Mixture models, on the other hand, have been successfully used to discriminate and monitor both soil and vegetation biophysical properties from broadband multispectral sensor systems (Landsat Thematic Mapper, SPOT, and AVHRR) to hyperspectral sensor imagery (AVIRIS, Hyperion). Hyperspectral sensors have greatly improved the identification of numerous soil and plant absorption features, from soil–plant mixtures, related to mineralogy, liquid water, chlorophyll, cellulose, and lignin contents (1, 7, 9).

REFERENCES

1. Ben-Dor, E.; Irons, J.R.; Epema, G. Soil Reflectance. *Remote Sensing for the Earth Sciences*; Rencz, A.N., Ed.; Wiley: New York, 1999; 111–188.
2. Irons, J.R.; Weismiller, R.A.; Petersen, G.W. Soil Reflectance. *Theory and Application of Optical Remote Sensing*; Asrar, G., Ed.; Wiley: New York, 1989; 66–106.
3. Baumgardner, M.F.; Silva, L.F.; Biehl, L.L.; Stoner, E.R. Reflectance Properties of Soils. Adv. Agron. **1985**, *38*, 1–44.
4. Graetz, R.D. Remote Sensing of Terrestrial Ecosystem Structure: An Ecologist's Pragmatic View. *Remote Sensing of Biosphere Functioning*; Hobbs, R.J., Mooney, H.A., Eds.; Springer: New York, 1990; 5–30.
5. Adams, J.B.; Sabol, D.E.; Kapos, V.; Almeida Filho, R.; Roberts, D.A.; Smith, M.O.; Gillespie, A.R. Classification of Multispectral Images Based on Fractions of Endmembers: Applications to Land-Cover Change in the Brazilian Amazon. Remote Sens. Environ. **1995**, *52*, 137–154.
6. Franklin, J.; Duncan, J.; Turner, D. Reflectance of Vegetation and Soil in Chihuahuan Desert Plant Communities from Ground Radiometry Using SPOT Wavebands. Remote Sens. Environ. **1993**, *46*, 291–304.
7. Asner, G.P.; Lobell, D.B. A Biogeophysical Approach for Automated SWIR Unmixing of Soils and Vegetation. Remote Sens. Environ. **2000**, *74*, 99–112.
8. Novo, E.M.; Shimabukuro, Y.E. Identification and Mapping of the Amazon Habitats Using a Mixing Model. Int. J. Remote Sens **1997**, *18*, 663–670.
9. Ustin, S.L.; Smith, M.O.; Adams, J.B. Remote Sensing of Ecological Processes: A Strategy for Developing and Testing Ecological Models Using Spectral Mixture Analysis. *Scaling Physiological Processes: Leaf to Globe*; Ehleringer, J.R., Field, C.B., Eds.; Academic Press: New York, 1993; 339–357.
10. Smith, M.O.; Ustin, S.L.; Adams, J.B.; Gillespie, A.R. Vegetation in Deserts. 1. A Regional Measure of Abundance from Multispectral Images. Remote Sens. Environ. **1990**, *31*, 1–26.
11. Gillespie, A.G. Spectral Mixture Analysis of Multispectral Thermal Infrared Images. Remote Sens. Environ. **1992**, *42*, 137–145.
12. Wessman, C.A.; Bateson, C.A.; Benning, T.L. Detecting Fire and Grazing Patterns in Tallgrass Prairies Using Spectral Mixture Analysis. Ecol. Appl. **1997**, *7*, 493–512.
13. Huete, A.R. Soil-Dependent Spectral Response in a Developing Plant Canopy. Agron. J. **1987**, *79*, 61–68.
14. Borel, C.C.; Gerstl, S.A.W. Nonlinear Spectral Mixing Models for Vegetative and Soil Surfaces. Remote Sens. Environ. **1994**, *47*, 403–416.
15. Huete, A.R. Separation of Soil–Plant Spectral Mixtures by Factor Analysis. Remote Sens. Environ. **1986**, *19*, 237–251.
16. McGwire, K.; Minor, T.; Fenstermaker, L. Hyperspectral Mixture Modeling for Quantifying Sparse Vegetation Cover in Arid Environments. Remote Sens. Environ. **2000**, *72*, 360–374.
17. Bateson, C.A.; Asner, G.P.; Wessman, C.A. Endmember Bundles: A New Approach to Incorporating Endmember Variability in Spectral Mixture Analysis. Trans. Geosci. Remote Sens. **2000**, *38*, 1083–1094.
18. Adams, J.B.; Smith, M.O.; Johnson, P. Spectral Mixture Modeling, A New Analysis of Rock and Soil Types at the Viking Lander 1 Site. J. Geophys. Res. **1986**, *91* (B8), B8098–B8112.
19. Farrand, W.H.; Singer, R.B.; Merenyi, E. Retrieval of Apparent Surface Reflectance from AVIRIS Data: A Comparison of Empirical Line, Radiative Transfer, and Spectral Mixture Methods. Remote Sens. Environ. **1994**, *47*, 311–321.
20. Roberts, D.A.; Smith, M.O.; Adams, J.B. Green Vegetation, Nonphotosynthetic Vegetation, and Soils in AVIRIS Data. Remote Sens. Environ. **1993**, *44*, 255–269.
21. Vane, G.; Green, R.O.; Vhrien, T.G.; Enmark, H.T.; Hansen, E.G.; Porter, W.M. The Airborne Visible/Infrared Imaging Spectrometer (AVIRIS). Remote Sens. Environ. **1993**, *44*, 127–143.
22. http://eo1.gsfc.nasa.gov/ (accessed October 2001).
23. Okin, G.S.; Roberts, D.; Murray, B.; Okin, W.J. Practical Limits on Hyperspectral Vegetation Discrimination in Arid and Semiarid Environments. Remote Sens. Environ. **2001**, *77*, 212–225.
24. Palacios-Orueta, A.; Pinzón, J.E.; Ustin, S.L.; Roberts, D. Remote Sensing of Soils in the Santa Monica Mountains: II. Hierarchical Foreground and Background Analysis. Remote Sens Environ. **1999**, *68*, 138–151.
25. Cwick, G.J.; Aide, M.T.; Bishop, M.P. Use of Hyperspectral and Biochemical Data from Black Spruce Needles to Map Soils at a Forest Site in Manitoba. Can. J. Remote Sens. **1998**, *24*, 187–193.

SPODOSOLS

Delbert Mokma
Michigan State University, East Lansing, Michigan, U.S.A.

INTRODUCTION

Spodosols are a group of soils that are characterized by the presence of a spodic horizon in which active amorphous organic matter and aluminum, with or without iron, have accumulated (1). The formative element of the word, "Spodosols," was derived from the Greek word, *spodos*, which means wood ashes. The name is connotative of bleached eluvial (E) horizons that overly distinctive reddish brown spodic (Bhs or Bs) horizons. Most Spodosols are sandy. These materials have many macropores and are rapidly permeable. Sandy properties give Spodosols low available water capacities. Available water capacity increases as the amount of humus and sesquioxides in the spodic horizons increases (2). Cation exchange capacity is related to amount of organic C rather than clay. Spodosols are acidic, thus they are naturally infertile.

A typical Spodosol has a distinctive profile with four master horizons (Table 1). O and/or A horizons have black or dark brown colors which result from accumulation of organic materials. E horizons have ashy gray colors, albic materials (1), as a result of the weathering of the non-resistant minerals and subsequent eluviation of amorphous materials. Bhs or Bs horizons have dark reddish brown colors resulting from illuvial accumulations of organo-metallic complexes. Reddest hues, darkest values, and lowest chromas usually occur in upper parts of spodic horizons. Dark reddish brown colors of spodic horizons are related to amounts of amorphous materials (3). Hues become more yellow and value and chroma increase with depth. Colors of C horizons tend to have yellower hues and higher values than spodic horizons.

Eluvial horizons have primarily uncoated sand grains. Immobilized, organo-metallic complexes coat most sand grains in B horizons. These coatings become thicker with time. As vegetation removes water from B horizons, desiccation causes cracks to form in the coatings. Thin aluminum and /or iron oxide coatings occur on sand grains in C horizons.

DISTRIBUTION

Spodosols are most extensive in areas of cool, humid, or perhumid climates (1). Some have formed in warm and hot, humid regions especially in areas of quartz-rich sand that have fluctuating water tables. Spodosols have not formed in arid areas.

FORMATION OF SPODOSOLS

Soil Forming Processes

Translocation of humus and sesquioxides

Formation of Spodosols involves the formation, translocation and immobilization of organo-metallic complexes. Organic acids form in forest canopies, canopy leachate and soil organic matter. These acids cause weathering of minerals and release of Al and Fe in upper mineral horizons. Soluble organo-metallic complexes form and move downward through O and A horizons. Al and Fe in several Swedish and Finnish Spodosols are translocated downward as organo-metallic complexes (4).

Immobilization of organo-metallic complexes occurs as a result of increases in pH, microbial decomposition of the organic ligand, and when the amount of Al and Fe taken up by organic molecules exceeds a threshold. Vanillic and *p*-hydroxybenzoic acids dominated surface horizons of Spodosols in Michigan, whereas, protocatechuic acid dominated spodic horizons (5). Distribution of protocatechuic acid paralleled those of Al and Fe suggesting it plays a role in the translocation process.

Ortstein is a spodic horizon in which sand grains are cemented together by organic matter and Al with minor amounts of Fe and little or no Si (6). Ortstein in somewhat poorly drained Spodosols usually has horizontal orientation, whereas that in well drained Spodosols usually has vertical orientation (7). Ortstein in the former soils is more continuous and more strongly cemented thereby more root restrictive than the later soils. Ortstein reduces forest and blueberry production.

Table 1 Selected properties of a typic haplorthod

Horizon	Depth (cm)	Texture	Color	Organic C (%)	ODOE[a]	Al[b] (%)	Fe[b] (%)
A	0–8	S	10YR 3/1	1.8	0.02	0.04	0.05
E	8–23	S	10YR 5/2	0.3	0.01	0.08	0.04
Bhs	23–28	S	5YR 3/2	1.8	0.24	0.40	0.45
Bs	28–41	S	7.5YR 4/4	0.7	0.07	0.43	0.15
BC	41–53	S	10YR 4/6	0.4	0.02	0.19	0.05
C	53–178	S	10YR 6/4	0.1	0.02	0.04	0.01

[a] Optical density of oxalate extract.
[b] Extracted with ammonium oxalate.

Silicon

Silicon may play a role in the translocation of Al through the formation of imogolite (8). Imogolite tends to be greater in lower Bs horizons than in upper Bhs or Bs horizons. It may form in spodic horizons after Al had been translocated as organo-metallic complexes (9) which may be broken down by soil microbes, thereby releasing ionic Al. The Al could then form imogolite in the B horizon.

Soil Forming Factors

Parent material

Spodosols developed in sandy to coarse loamy materials. They usually have little silicate clay. Quartz is the primary mineral in sand and silt fractions, with lesser amounts of orthoclase, plagioclase, mica, pyroxene, and amphibole (10). Accessory minerals are often highly weathered and not easily identifiable. Less resistant minerals tend to have weathered from E and B horizons.

As clay content of parent materials increases, the translocation of organic C, Al, and Fe decreases, thickness of E and B horizons decreases, and B horizon color is lighter and less red (11). Silicate clays inhibit translocation of organic C, Al, and Fe by adsorbing these materials.

Topography

Spodosols form on surfaces that have shallow to very deep water tables. Spodic horizon development is influenced by depth and duration of water saturation (12). Spodic horizons do not readily form in soils that are saturated with water near the soil surface for long periods of time. Water saturation inhibits translocation processes. Poorly drained soils in Spodosol landscapes usually have less organic C, Al and Fe in their B horizons than better drained Spodosols in the hydrosequence. Some poorly and somewhat poorly drained Spodosols, especially those in warmer climates, have little Fe because it is chemically reduced and leached from the soil. Somewhat poorly drained Spodosols usually have more humus and sesquioxides than other members of the hydrosequence. Saturated zones inhibit water movement, thus, organo-metallic complexes are carried from A and E horizons to B horizons but rarely through B horizons. In well drained Spodosols organo-metallic complexes can be carried by percolating waters to and through B horizons to lakes and streams.

Vegetation

Spodosols form under coniferous and deciduous forest but not under grasses. Spodic horizons formed under conifers have more Al and Fe than those under deciduous trees (13). Because organic compounds are involved with translocation of Al and Fe, more leaf or needle litter means more organic compounds available to form organo-metallic complexes.

Fire frequency influences spodic horizon development by affecting forest succession (14). When intervals between intense fires are 5 to 50 years, jack pine (*Pinus banksiana* L.) survives and regenerates. Much organic matter on the forest floor is destroyed by frequent fires, therefore, less organic compounds are available to form organo-metallic complexes with Al and Fe. Thus, E horizons are thin or non-existent and B horizons have small amounts of organic C, Al, and Fe. If fire intervals are 50 to 350 years, white (*Pinus strobus* L.) and/or red (*Pinus resinosa* Ait.) pine succeed jack pine. Less organic matter is destroyed by fewer fires and more organic compounds are available to form complexes with Al and Fe and translocate them from E horizons to B horizons. If fire intervals exceed 350 years, northern hardwoods (sugar maple, *Acer saccharum* Marsh; beech, *Fagus grandifolia*) succeed the pines. These trees are more shade tolerant and more fire resistant. Translocation of organic C, Al and Fe continues, but more slowly because less Al and Fe and more Ca are biocycled.

Climate

Spodosols usually form in areas with cool, humid climates. Spodosols of warmer climates have less Fe compared to those of cool climates. Spodosols dominate areas with cryic and frigid soil temperature regimes, but are less common in areas with a mesic soil temperature regime in New England (15). Strong spodic horizon development is associated with areas of deep snowpack before soils freeze in Michigan. Meltwater from snow can freely percolate through the soil in spring.

Time

Under optimum conditions Bs horizons may form in a few hundred years, but usually they form more slowly. Eluvial horizons form in 300 years or less (16). Weak Bs horizons form in about 2000 years (17). At least 8000 years are required for most soils on a surface to qualify as Spodosols.

REFERENCES

1. Soil Survey Staff. *Soil Taxonomy: A Basic System of Soil Classification for Making and Interpreting Soil Surveys*; USDA Agriculture Handbook 436, Second Edition; US Government Printing Office: Washington, DC, 1999.
2. Shetron, S.G. Distribution of Free Iron and Organic Carbon as Related to Available Water in Some Forested Sandy Soils. Soil Sci. Soc. Am. Proc. **1974**, *38*, 359–362.
3. Mokma, D.L. Color and Amorphous Materials in Spodosols from Michigan. Soil Sci. Soc. Am. J. **1993**, *57*, 125–128.
4. Lundstrom, U.S.; van Breeman, N.; Bain, D.C.; van Hees, P.A.W.; Giesler, R.; Gustafsson, J.P.; Ilvesniemi, H.; Karltun, E.; Melkerud, P.-A.; Olsson, M.; Riise, G.; Wahlberg, O.; Bergelin, A.; Bishop, K.; Finlay, R.; Jongmans, A.G.; Magnusson, T.; Mannerkoski, H.; Nordgren, A.; Nyberg, L.; Starr, M.; Tau Strand, L. Advances in Understanding the Podzolization Process Resulting from a Multidisciplinary Study of Three Coniferous Forest Soils in the Nordic Countries. Geoderma **2000**, *94*, 335–353.
5. Vance, G.F.; Mokma, D.L.; Boyd, S.A. Phenolic Compounds in Soils of Hydrosequences and Developmental Sequences of Spodosols. Soil Sci. Soc. Am. J. **1986**, *50*, 992–996.
6. Lee, F.Y.; Yuan, T.L.; Carlisle, V.W. Nature of Cementing Materials in Ortstein Horizons of Selected Florida Spodosols: I. Constituents of Cementing Materials. Soil Sci. Soc. Am. J. **1988**, *52*, 1411–1418.
7. Mokma, D.L.; Doolittle, J.A.; Tornes, L.A. Continuity of Ortstein in Sandy Spodic Horizons of Michigan. Soil Surv. Horiz. **1994**, *35*, 6–10.
8. Farmer, V.C.; Russell, J.D.; Berrow, M.L. Imogolite and Proto-Imogolite Allophane in Spodic Horizons: Evidence for a Mobile Aluminum Silicate Complex in Podzol Formation. J. Soil Sci. **1980**, *31*, 673–684.
9. Ugolini, F.C.; Dalgren, R. The Mechanism of Podzolization as Revealed by Soil Solution Studies. In *Podzols et Podzolisation*; Righi, D., Chauvel, A., Eds.; AFES et INRA: Paris, 1987; 195–203.
10. Haile-mariam, S.; Mokma, D.L. Mineralogy of Two Sandy Spodosol Hydrosequences in Michigan. Soil Surv. Horiz. **1995**, *36*, 121–132.
11. Gardner, D.R.; Whiteside, E.P. Zonal Soils in the Transitional Region Between the Podzol and Gray-Brown Podzolic Regions in Michigan. Soil Sci. Soc. Am. Proc. **1952**, *16*, 137–141.
12. Mokma, D.L.; Sprecher, S.W. Water Table Depths and Color Patterns in Spodosols of Two Hydrosequences in Northern Michigan, USA. Catena **1994**, *22*, 275–286.
13. Messenger, A.S.; Whiteside, E.P.; Wolcott, A.R. Climate, Time, and Organisms in Relation to Podzol Development in Michigan Sands: I. Site Descriptions and Microbiological Observations. Soil Sci. Soc. Am. Proc. **1972**, *36*, 633–638.
14. Mokma, D.L.; Vance, G.F. Forest Vegetation and Origin of Some Spodic Horizons, Michigan. Geoderma **1989**, *43*, 311–324.
15. Evans, C.V. Preliminary Investigations of Hydric Soil Hydrology and Morphology in New Hampshire. In *Preliminary Investigations of Hydric Soil Hydrology and Morphology in the United States*; Technical Report WRP-DE-13, Wakely, J.S., Sprecher, S.W., Lynn, W.C., Eds.; U.S. Army Engineer Waterways Experiment Station: Vicksburg, MS, 1996; 114–126.
16. Gjems, O. Some Notes on Clay Minerals in Podzol Profiles in Fennoscandia. Clay Miner. Bul. **1960**, *4*, 208–211.
17. Franzmeier, D.P.; Whiteside, E.P. A Chronosequence of Podzols in Northern Michigan. I. Ecology and Description of Pedons. II. Physical and Chemical Properties. Mich. State Univ. Agric. Exp. Sta. Quart. Bul. **1963**, *46*, 1–36.

STRUCTURE

Eileen J. Kladivko
Purdue University, West Lafayette, Indiana, U.S.A.

S

INTRODUCTION

Soil structure is an important characteristic of the soil. There is no one single definition of soil structure, but most descriptions of structure refer to the *arrangement* (size, shape, and orientation) of particles and the pores between them, or the *stability* of the current particle arrangement to some disruptive force (e.g. hand manipulation, water, wind, wheel traffic). Soil structure affects many properties and processes in the soil.

Soil aggregates are the backbone of soil structure. Aggregates are clusters of organic matter, sand, silt, and clay held together by a variety of forces. A hierarchical order of aggregates is sometimes described, with smaller groupings of particles comprising microaggregates and then groupings of microaggregates forming macroaggregates (1, 2). Pores within aggregates are smaller and are important for water retention, whereas pores between aggregates are larger and are important for water flow and aeration.

DESCRIPTION OF STRUCTURE

Descriptions of soil structure are often approached from the viewpoint of pedology, plant growth, or physics. The pedological view describes the shape, size, and relative strength of soil peds or aggregates. These descriptions reflect long-term soil forming processes as well as shorter-term human-induced changes due to soil management practices.

From an agronomic point of view, "good" soil structure promotes rapid water infiltration, adequate drainage of excess water, good aeration, water retention, resistance to erosion, and good germination, emergence, and rooting of plants (3). "Poor" soil structure may result in surface crusting, decreased infiltration and increased runoff and erosion, poor seedling emergence, poor rooting ability, or inadequate air or water for plants. Much of the soil management research conducted over the decades has been to devise ways to improve soil structure while enhancing cash crop yields.

The soil physics view of soil structure is often to assess the stability of aggregates to disruptive forces that simulate disruptive forces in the field (2). Size distribution and stability of aggregates after disruption are used to calculate quantitative indices of stability. Other ways to describe structure include pore size distribution and continuity. The particular method used to assess or describe soil structure should be matched with the intended use of the information.

AGGREGATE FORMATION

Soil aggregates are formed through biological, chemical, and physical processes. As soil microorganisms decompose plant residues and other organic materials, they exude polysaccharides which tend to glue particles together. Fungal hyphae also enmesh particles, thereby forming aggregates. Earthworms help aggregate soil through excretion of mucilages, which stick particles together, as well as bringing mineral and organic particles together in their castings. Plant roots also help to aggregate soil through several mechanisms, including root exudation and physical enmeshment by fibrous roots. Chemical conditions that impact aggregation include the concentrations and types of cations present in the soil solution, as they affect flocculation of the clay particles. Physical processes affecting aggregation include wetting/-drying cycles, freeze/thaw cycles, and mechanical tillage. These physical processes break down large clods and destroy aggregates, but they also bind particles together to form small aggregates. The net impact of these physical processes on formation and destruction of aggregation is highly dependent on initial soil conditions.

An important point to remember is that soil aggregation is dynamic (2, 3). Aggregates are not permanent, but are continuously being created and destroyed by natural means as well as by human management of the soil. Soil management practices must include regular "maintenance" of soil structure, to sustain productivity over the long-term. For example, soils maintained as native grasslands generally have higher organic matter contents and greater aggregate stability than similar soils used for row crop production (4). When soils are used for crop production, sustainable management practices often include periodic inputs of organic materials (animal manures, crop residues) to the soil to provide a foodsource for the soil organisms which in turn form aggregates. The relative stability of different organic binding agents in soil micro- and macroaggregates, is discussed in more detail in other sections of this encyclopedia Growing a fibrous-rooting plant as a cover crop after the main crop is harvested is a way to use the normally fallow time periods of the year to improve near-surface structure. Conservation tillage practices, especially no-tillage, are another way to improve soil aggregation and stability compared with conventional tillage practices (5). The increased aggregation results from both greater aggregate formation near the surface due to the presence of surface crop residues, and less physical disruption of aggregates by the tillage operation itself.

ENVIRONMENTAL SIGNIFICANCE OF SOIL STRUCTURE

Stable soil aggregates contain organic binding agents, as discussed earlier, and therefore are involved in carbon sequestration in soils. Within a given soil, increased soil aggregation is usually associated with increased soil carbon. Conversely, as aggregates are disrupted or destroyed, carbon mineralization and CO_2 evolution increase (6). Therefore increasing soil aggregation and aggregate stability can be an important method to increase carbon sequestration in soil. It must be remembered, however, that these higher aggregation levels must be continuously maintained, as previously discussed, or else the sequestered carbon will be released again. Soil organic matter associated with different aggregate size fractions appears to have different turnover times, and the interplay of organic matter chemistry, aggregate size, and physical location within or between aggregates is an active areas of research.

Although aggregates are naturally degraded with time as microorganisms feed on the organic binding materials, aggregates are more quickly and noticeably destroyed by tillage operations. Aggregates near the soil surface are also subjected to the disruptive impact of raindrop energy or rapid wetting, which can result in slaking, dispersion, and crust formation. Such surface structural breakdown can have major impacts on plant growth and overall soil functioning. Surface crusts can delay or inhibit seedling emergence, leading to low plant populations and yields. Surface seals can also reduce infiltration and increase runoff and erosion. The partitioning of less water into the soil profile may reduce water availability to the plant at a later date. The structure of soil below the surface may be degraded by tillage or wheel traffic, with potential consequences of poor rooting ability due to compaction, poor drainage and aeration of the rooting zone, and reduced water availability to the plant.

The structure of surface and subsurface horizons affects water flow into and through soil profiles and across landscapes. The partitioning of water between infiltration and surface runoff is controlled largely by near-surface conditions, although subsurface impeding layers may limit infiltration as well. Impeding layers may occur naturally or be formed as a result of management. In fields with a good subsurface drainage system, for example, water collects temporarily in depressions after a rain. If these depressions are tilled when the soil is too wet, near-surface soil horizons become compacted and impermeable. After the next rain, water stands in these depressions even longer, due to the compacted soil. Many soils have sufficient structural development below the surface to exhibit some degree of preferential flow of water, whereby water flows quite rapidly through large, inter-pedal voids while moving more slowly into and through smaller pores within aggregates.

CONCLUSION

In summary, it is well known that soil structure has a major influence on the ability of soil to perform different functions. Whether the soil is used to grow plants, accept wastes, partition rainfall into runoff, interflow, and groundwater recharge, or build houses and roads, the arrangement of particles and the stability of that arrangement are important factors to consider. Current research attempts to more quantitatively describe the in situ structure and the response of soil to various disruptive forces, as well as the relationships between soil structure, plant growth, and water flow.

REFERENCES

1. Tisdall, J.M.; Oades, J.M. Organic Matter and Water-Stable Aggregates in Soils. J. Soil Sci. **1982**, *33*, 141–163.
2. Hillel, D. *Fundamentals of Soil Physics*; Academic Press: San Diego, CA, 1980; 413.
3. Karlen, D.L.; Erbach, D.C.; Kaspar, T.C.; Colvin, T.S.; Berry, E.C.; Timmons, D.R. Soil Tilth: A Review of Past Perceptions and Future Needs. Soil Sci. Soc. Am. J. **1990**, *54*, 153–161.
4. Six, J.; Elliott, E.T.; Paustian, K.; Doran, J.W. Aggregation and Soil Organic Matter Accumulation in Cultivated and Native Grassland Soils. Soil Sci. Soc. Am. J. **1998**, *62*, 1367–1377.
5. Kladivko, E.J. Residue Effects on Soil Physical Properties. In *Managing Agricultural Residues*; Unger, P.W., Ed. Lewis Publishers: Boca Raton, FL, 1994; 123–141.
6. Elliott, E.T. Aggregate Structure and Carbon, Nitrogen and Phosphorus in Native and Cultivated Soils. Soil Sci. Soc. Am. J. **1986**, *50*, 627–633.

STRUCTURE, EARTHWORMS AND

M.J. Shipitalo
United States Department of Agriculture-Agricultural Research Service (USDA-ARS), Coshocton, Ohio, U.S.A.

INTRODUCTION

Earthworms are almost universally regarded by farmers and gardeners as a sign of healthy soil. Articles in farmer-oriented publications frequently include testimonials of how changes in crop and soil management result in the return of earthworms to fields where they were previously absent. The recent publication of the *Soil Biology Primer* and the proliferation of websites devoted to earthworms are further recognition of the perceived role of earthworms and other soil fauna in maintaining healthy soils (Table 1). From a scientific viewpoint, however, it is uncertain how much earthworms contribute to soil quality or if they are a consequence or cause of good soil health. Obviously, there are healthy soils that don't have earthworms.

Earthworms affect a number of soil processes including nutrient and carbon cycling, plant growth, and the activity and distribution of microorganisms. Perhaps the most noticeable impact of earthworms, however, is their effect on soil structure. Earthworms burrow into and ingest soil and in doing so modify soil porosity, aggregate size, and aggregate stability. The amount of soil ingested is highly dependent on the size, composition, and activity of the earthworm population and is hard to accurately measure because below-ground activity is difficult to monitor. Nevertheless, estimated ingestion rates for temperate-region soils are as high as 100 $Mg\,ha^{-1}\,yr^{-1}$. In tropical areas, where climatic conditions are less likely to inhibit activity, ingestion rates as high as 2600 $Mg\,ha^{-1}\,yr^{-1}$ have been reported (1). Similarly, the contributions of earthworms to soil porosity and aggregation and the benefits of earthworm-enhanced soil structure to plant growth (2), and effects on water quality are difficult to quantify (1, 3).

TYPES OF EARTHWORMS

Part of the problem in determining the effects of earthworms on soil structure comes from incomplete knowledge of their behavior. Worldwide there are about 3000 species of earthworms (4), few of which have been investigated in detail. A number of classification schemes have been proposed that group these species based on various aspects of their behavior. The most widely used system is that of Bouché in which earthworms are divided into three groups (4, 5). *Epigeic* earthworms are generally found beneath or within accumulations of organic matter and rarely burrow into or ingest much soil (Fig. 1). Typical habitats include forest litter or manure piles, thus they have little direct effect on the structure of mineral soils. *Endogeic* earthworms burrow extensively below ground and obtain their nutrition by ingesting a mixture of soil and organic matter (Fig. 1). They form extensively branched, sub-horizontal networks of burrows in search of food, but most of their activity is in the upper 10–15 cm where organic matter levels are generally highest. Portions of their burrows are often occluded with their excrement (casts) and they occasionally cast on the soil surface. *Anecic* earthworms normally live in permanent or semi-permanent burrows that can extend deep into the soil. They feed primarily on decaying surficial organic litter that they frequently pull into their burrows or mix with excrement to form a midden (Fig. 1). The midden blocks the burrow entrance and promotes further decay of the incorporated organic residues. These categories are not absolute, however, as the behavior of many species is intermediate to these groupings and can vary with environmental conditions (5).

EFFECTS ON SOIL POROSITY

Because they burrow extensively into mineral soil, endogeic and anecic earthworms can substantially alter soil porosity. Although earthworm burrows usually account for a small fraction of the soil volume, due to their continuity, stability, and relatively large size compared to pores formed by most other mechanisms, these macropores can greatly affect movement of air, water, and solutes. A number of investigators have demonstrated that burrows made by anecic and endogeic earthworms can effectively conduct water (1, 3, 6). Because most of their activity is

Table 1 Web-based resources on earthworms and soil structure

Source	Contents	Address
USDA-NRCS Soil Biology Primer	General information on soil fauna and their effects on soil	http://www.statlab.iastate.edu/survey/SQI/SoilBiologyPrimer/index.htm
Agriculture Canada	General information on earthworms includes FAQ and numerous links	http://res2.agr.ca/london/pmrc/faq/earthwor.html
Purdue University	Extension publication on earthworms and crop management	http://www.agcom.purdue.edu/Agcom/Pubs/AY/AY-279.html
University of California	Articles on earthworm biology and sustainable agriculture	http://www.sarep.ucdavis.edu/worms
Worm Digest	Commercial publication including general articles and numerous links	http://www.wormdigest.org/

confined to surficial soil horizons, however, endogeic earthworms probably do not directly influence movement deep into the profile (7). The fact that portions of their burrows are often occluded with casts, probably further limits their effectiveness.

On the other hand, anecic earthworms have the potential to influence gas, water, and solute movement throughout the profile. For example, burrows created by *Lumbricus terrestris* L. (a widespread anecic species) are normally single, nearly vertical channels up to 12 mm in diameter and 2.4 m deep (5). These burrows can have several entrances directly underneath the midden, but these usually coalesce into a single channel within the upper few cm of soil (Fig. 1). Although the midden would seem to inhibit entry of water, field studies conducted on burrows with

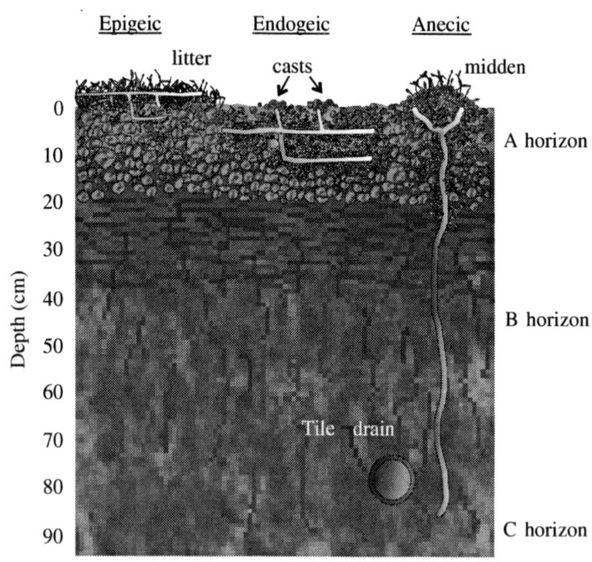

Fig. 1 Diagrammatic representation of the burrows made by the three ecological groups of earthworms as defined by Bouché.

undisturbed entrances indicate that they can transmit substantial amounts of water, up to 10% of rainfall (3, 8).

Both burrow types can increase infiltration thereby increasing plant available water and reducing surface runoff. For instance, when earthworms were eliminated from a pasture, a three-fold reduction in infiltration rate and a two-fold increase in runoff were noted (9). In cultivated soils earthworms can also reduce runoff by disrupting surface crusts that impede infiltration (10). The contribution of earthworm burrows to infiltration, however, is dependent on a number of factors. High intensity rainfall and dry soil can increase flow in earthworm burrows (3, 6, 8). Although it seems logical to assume that earthworms might block flow in their burrows, infiltration rates for burrows with the worm removed are similar to those for occupied burrows (11). In fact, occupied burrows are probably more effective in transmitting water than abandoned burrows because they are more likely to maintain near-surface continuity. Theoretically, it should be possible to model the contribution of earthworm burrows to infiltration based on their distribution and geometrical properties. This has proved difficult as not all burrows conduct water and their dimensions are not strongly correlated to their infiltration capacity (11, 12).

In rare circumstances, increased infiltration due to earthworm burrows can have negative consequences. Earthworm burrows can contribute to non-uniform distribution of water during furrow irrigation, loss of water through unlined irrigation ditches (7), and leakage of manure storage lagoons (13). Anecic earthworms can burrow close to tile lines (Figs. 1 and 2) and may increase losses of injected animal wastes in drainage waters (12). Earthworm burrows can also increase leaching of surface-applied agrochemicals, particularly when intense storms occur shortly after application on residue-covered no-till soils (3). The potential for this to occur is greatly reduced with time and low intensity, intervening, rainfalls.

Fig. 2 This *L. terrestris* burrow, impregnated with plastic and excavated in situ, passed within 2 cm of a buried tile and had an average infiltration rate of 353 mL min^{-1}.

Ingestion of herbicide-coated residues by earthworms can also reduce leaching losses (14). Once the water enters the burrows the organic matter-rich linings may further reduce herbicide transport by increasing sorption and degradation (1).

EFFECTS ON AGGREGATION

Although earthworms feed on decaying organic matter and the microorganisms that colonize it, the material ingested by endogeic and anecic species during feeding and burrowing is predominately mineral matter (1). This mixture is excreted as casts on the soil surface or below ground, depending of the species of earthworm, location of the food source, and soil bulk density (15). The casts usually contain more clay and less sand than the surrounding soil due to selective ingestion with the effect more prominent with endogeic species, which tend to be smaller than anecic earthworms (1). Moreover, earthworm casts are usually higher in pH, contain more available nutrients, and have higher levels of microbial activity than the uningested soil (4, 5).

Freshly excreted casts are initially less water stable than uningested soil because digestive secretions and the peristaltic action of the earthworm gut disrupt many of the existing interparticle bonds. During passage through the earthworm, however, the mineral matter is intimately mixed with ingested organic matter. If casts are allowed to age or dry before being subjected to disruption their stability can exceed that of the uningested soil, thereby enhancing formation of a desirable, water-stable, granular, soil structure (1).

A number of mechanisms can contribute to the increased stability of earthworm casts with aging or drying. These include chemical or mechanical stabilization by: 1) internal secretions of earthworms, 2) plant fibers incorporated into casts, 3) growth of fungal hyphae, 4) bacterially produced gums, 5) bonding by calcium humate or mucilage, 6) wetting and drying cycles, and 7) age-hardening/thixotropic effects combined with organic bonding (1). The mechanisms are not mutually exclusive and the relative contribution of a particular process is probably dependent on a number of factors. For the most part, however, incorporation of organic matter into casts is critical either as bonding agent or as promoter of microbial activity that leads to the production of bonding agents. For this reason a positive correlation between organic carbon content and cast stability is frequently noted. Thus, casts of earthworms that have higher organic matter ingestion rates are more stable than those that ingest more mineral-rich mixtures. Additionally, the distribution of the organic bonding agents is probably more important than the total amount of organic matter within the casts (1).

Because freshly deposited casts are initially of low stability, they are subject to dispersion and transport if not protected from raindrop impact or the action of flowing water. Thus, earthworm activity can increase infiltration and reduce runoff while increasing losses of soil and sediment associated nutrients from pastures (9) and cultivated fields (15). Furthermore, foraging and midden building by anecic earthworms can reduce surface residue cover thus exposing more soil and casts to raindrop impact with negative consequences for soil structure (16).

CONCLUSIONS

In general, earthworm activity improves soil structure by increasing macroporosity and enhancing aggregation, which in turn can reduce runoff and sediment loss and provide a better environment for plant growth. Under some circumstances, however, increased infiltration, deposition of casts on the soil surface, and excessive residue removal can have undesirable consequences. For the most part, these problems can be minimized by adopting modified management practices.

REFERENCES

1. Tomlin, A.D.; Shipitalo, M.J.; Edwards, W.M.; Protz, R. Earthworms and Their Influence on Soil Structure and Infiltration. In *Earthworm Ecology and Biogeography in North America*; Hendrix, P.F., Ed.; Lewis: Boca Raton, FL, 1995; 159–183.
2. Logsdon, S.D.; Linden, D.R. Interactions of Earthworms with Soil Physical Conditions Influencing Plant Growth. Soil Sci. **1992**, *154* (4), 330–337.
3. Shipitalo, M.J.; Dick, W.A.; Edwards, W.M. Conservation Tillage and Macropore Factors that Effect Water Movement and the Fate of Chemicals. Soil Tillage Res. **2000**, *53* (3-4), 167–183.
4. Lee, K.E. *Earthworms, Their Ecology and Relationships with Soils and Land Use*; Academic Press: New York, 1985; 411 pp.
5. Edwards, C.A.; Bohlen, P.J. *Biology and Ecology of Earthworms*; 3rd Ed.; Chapman & Hall: London, 1996; 426 pp.
6. Trojan, M.D.; Linden, D.R. Microrelief and Rainfall Effects on Water and Solute Movement in Earthworm Burrows. Soil Sci. Soc. Am. J. **1992**, *56* (3), 727–733.
7. Kemper, W.D.; Trout, T.J.; Segeren, A.; Bullock, M. Worms and Water. J. Soil Water Conserv. **1987**, *42* (6), 401–404.
8. Edwards, W.M.; Shipitalo, M.J.; Owens, L.B.; Norton, L.D. Water and Nitrate Movement in Earthworm Burrows Within Long-Term No-Till Cornfields. J. Soil Water Conserv. **1989**, *44* (3), 240–243.
9. Sharpley, A.N.; Syers, J.K.; Springett, J.A. Effect of Surface Casting Earthworms on the Transport of Phosphorous and Nitrogen in Surface Runoff from Pasture. Soil Biol. Biochem. **1979**, *11* (5), 459–462.
10. Kladivko, E.J.; Mackay, A.D.; Bradford, J.M. Earthworms As a Factor in the Reduction of Soil Crusting. Soil Sci. Soc. Am. J. **1986**, *50* (1), 191–196.
11. Shipitalo, M.J.; Butt, K.R. Occupancy and Geometrical Properties of *Lumbricus terrestris* L. Burrows Affecting Infiltration. Pedobiologia **1999**, *43* (6), 782–794.
12. Shipitalo, M.J.; Gibbs, F. Potential of Earthworm Burrows to Transmit Injected Animal Wastes to Tile Drains. Soil Sci. Soc. Am. J. **2000**, *64* (6), 2103–2109.
13. McCurdy, M.; McSweeney, K. The Origin and Identification of Macropores in an Earthen-Lined Dairy Manure Storage Basin. J. Environ. Qual. **1993**, *22* (1), 148–154.
14. Farenhorst, A.; Topp, E.; Bowman, B.T.; Tomlin, A.D. Earthworm Burrowing and Feeding Activity and the Potential for Atrazine Transport by Preferential Flow. Soil Biol. Biochem. **2000**, *32* (4), 479–488.
15. Binet, F.; Le Bayon, R.C. Space-Time Dynamics In Situ of Earthworm Casts Under Temperate Cultivated Soils. Soil Biol. Biochem. **1999**, *31* (1), 85–93.
16. Shuster, W.D.; Subler, S.; McCoy, E.L. Foraging by Deep-Burrowing Earthworms Degrades Surface Soil Structure of a Fluventic Hapludoll in Ohio. Soil Tillage Res. **2000**, *54* (3–4), 179–189.

STRUCTURE, MANAGING BELOWGROUND

S

K.A. Olsson
Tatura, Victoria, Australia

B. Cockroft
Kialla, Victoria, Australia

INTRODUCTION

Crop yield is strongly correlated with the production and functioning of roots. Soil management aims to produce and maintain conditions in the soil so that roots are provided with low mechanical resistance, non-limiting supplies of water, oxygen and nutrients and favorable soil temperature (1). Root growth and functioning are promoted by soil structure of low strength and of high porosity that facilitates the necessary transport of water, nutrients and gases through stable pores (2). While specifications for the necessary levels of soil strength and porosity are well documented (1, 3), they are not generally met in poorly structured subsoils underlying the surface (A) horizon(s). These include hardpans, claypans and compacted layers of high strength and low porosity. Such layers limit yield by physically restricting root proliferation, drainage and the storage and transport of water while making the management of otherwise well-structured surface horizons both difficult and inflexible (4, 5).

Researchers have long sought to improve the structure of difficult subsoils by deep tillage, chemical ameliorants and ameliorative crops, aiming for higher yields through faster infiltration of water, deeper wetting, increased water and nutrient availability to plants, increased drainage, leaching of salts and toxins, lower soil strength, and deeper rooting (6). Subsoil modification, however, has not consistently increased crop yields, while benefits are often transient and confounded (7–9). Many ameliorative measures do not modify the subsoil to a particular specification, while most do not adequately address the stabilization and subsequent management of the modified soil.

IMPROVING AND MANAGING SUBSOIL STRUCTURE

Principles guiding the improvement and management of subsoil structure for the growth and functioning of roots are documented in the literature (10). An initial survey of the soil profile provides information on soil physical and chemical properties including soil horizons, textures and clay mineralogy. Any chemical problems such as salinity, sodicity and acidity are corrected and nutrients are supplied at non-limiting rates. Where not limited by soil temperature, the main soil physical factors that directly influence root growth and functioning are soil strength, soil water supply and aeration. Specifications for each that are non-limiting for roots are given in the literature (1). Management to reduce soil strength and to increase porosity is then based on meeting each of these in a defined volume of soil within the root zone.

Soil Strength

For unrestricted elongation, roots require a soil strength, measured as penetrometer resistance, of less than 0.5 MPa (11). However, subsoils and compacted layers may have penetrometer resistances greater than 2 MPa, even when wet. In dense clay subsoils, strength increases rapidly on drying and roots may grow only during brief periods and over a small range of matric suction. This results in shallow, confined root systems and low root concentrations in the matrix (10). In expansive clays, roots tend to be restricted to cracks.

Soil strength is reduced by loosening the clay and fragmenting hardpans (4) while producing macropores (>30 μm diameter). The resulting fragments and aggregates assist root proliferation (10). Loosening the soil to about 0.5 m depth is considered a practical compromise in reducing the effects of dense soil (1, 3), but the depth of modification will depend on the position of restricting layers within the root zone. Ripping is effective in brittle soils, hardpans and compact sandy layers; it is less effective in loams and clays. Loosening of clay subsoils is best performed using rigid tines working in soil slightly drier than the Plastic Limit, aiming to achieve brittle failure (12). The tines fail the soil upwards by means of low angle tips operating at modest increments of depth and at low speed to avoid high confining stresses at the points. Wings behind each point shear and fracture many of the fragments under the confining pressure from the overburden. Deep tillage may be combined with additions of specified quantities of ameliorants, such as gypsum and lime to help stabilize the loosened soil (10).

An alternative technique mixes the soil layers and incorporates ameliorants, crop residues and fertilizers in parallel slots 100–150 mm wide at 1–2 m spacing tilled to the required depth (13).

The benefits to crops from changes in soil structure brought about by the activity of roots of previous crop and pasture species are limited in the present context. Crop responses are often confounded with improved nutrition and drainage of the surface soil and with reduced incidence of disease (14).

The loosened subsoil must be protected from compaction by traffic and from rapid wetting. An effective approach to the problem of soil compaction is to separate roots from traffic using beds (15). Management for stability in water and mechanical stability of the loosened subsoil also includes slow wetting and rapid drainage to avoid very low matric suctions to ensure an effective stress within aggregates and fragments.

Loosened subsoils under no traffic inevitably harden with time even though much of the improved macroporosity may be retained (16). While roots may grow in the macropores formed by tillage, or by the roots of earlier crops, they may find it difficult to penetrate the hard fragments >20 mm diameter where these retain the high strength of the original soil (17). A consequent slowing of root growth may be expressed physiologically as a slowing of the shoots, even where supplies of water, oxygen and nutrients are adequate (18). Research is needed to find ways to develop soft, porous aggregates that are stable both in water and to applied mechanical stresses. The process involves the penetration of unconfined fragments by roots and fungal hyphae and mellowing by successive cycles of wetting and drying (10). Stabilization requires additions of organic matter from roots, added calcium and electrolyte and a moist, well-drained environment.

Soil Water

In temperate crops, yield shows a close to linear increase with transpiration (19, 20). Poorly structured subsoils of low macroporosity slow the infiltration and penetration of water. Subsoils may show a predominance of fine pores (<30 μm diameter) which restrict both the storage and flow of water to crop roots. Storage pores (0.2 to 30 μm diameter) in subsoils commonly comprise <15% of the soil volume (21, 22). Water availability from the subsoil is further reduced where roots have difficulty penetrating the strong matrix. Where topsoils are shallow, potential transpiration rates may be sustained for only a few days after irrigation (23).

For rapid water entry, free drainage and gas exchange, the soil must be stable in water and maintain >15% macropores (15). Initially, the fragments produced by tillage may be little altered from the original and still limit water storage, gas exchange and water flow to roots, while mechanically restricting root entry. Should roots become largely confined to and clumped in macropores, they may be poorly functional in water uptake since theory predicts a major resistance to water flow in the soil (24, 25). Management should aim to develop soft, porous aggregates to enable improved hydraulic properties and a more even distribution of roots. Ideally, these aggregates should have >20% storage pores (26).

Soil Aeration

Diffusion of oxygen in air-filled pores continuous to the soil surface is necessary for aeration of the subsoil. Dense, slowly permeable subsoils, however, have few interconnected macropores over depth and may restrict oxygen supply to the root, particularly when the clay is swollen (27). The poor aeration of wet clays slows the growth and functioning of roots (28) and may also increase the incidence of soil-borne disease (29).

Management should aim for a stable air-filled porosity of >15% of the soil volume to be reached within 24 hours drainage throughout the modified solum (1). Roots require at least 10% oxygen for unrestricted growth (17). Surface drainage is essential. Deep tillage then produces the necessary macroporosity to depth. In some cases, the restructured subsoil in the root zone may connect hydraulically with a more permeable layer beneath (1). In other situations, artificial drainage using a pipe or mole drain is required (30). In irrigated soils, a slow application rate minimizes filling the macropores with water.

Where fragments and aggregates are not larger than about 10 mm or smaller than 1 mm (31), and where the oxygen diffusion rate is higher than about 35 μg/m^2/s (32), intra-aggregate aeration is unlikely to be limiting for root growth. An increase in root and microbial respiration in fragments undergoing structural change as described earlier should be met by the increased porosity within the fragment.

S

OUTLOOK

The management of difficult subsoils to achieve and sustain high productivity from crops must provide the necessary low soil strength together with complements of macropores and storage pores which will not limit the transfer of water and gases, the proliferation of roots or the supply of water to crops. Mechanical loosening of dense subsoil layers is an initial step in an ongoing process of development of the prescribed levels of soil strength and porosity. The process is facilitated by chemical ameliorants, the high initial porosity of the loosened soil, high levels of biological activity and closely controlled wetting under non-trafficked conditions. A quantitative understanding of root and top responses to profile modification in relation to plant production is an important area for research.

REFERENCES

1. Cockroft, B.; Olsson, K.A. Case Study of Soil Quality in South-Eastern Australia: Management of Structure for Roots in Duplex Soils. In *Soil Quality for Crop Production and Ecosystem Health*; Gregorich, E.G., Carter, M.R., Eds.; Elsevier: Amsterdam, 1997; 339–350.
2. Letey, J. The Study of Soil Structure: Science or Art. Aust. J. Soil Res. **1991**, *29*, 699–707.
3. Greenland, D.J. Soil Management and Soil Degradation. J. Soil Sci. **1981**, *32*, 301–322.
4. Wildman, W.E.; Meyer, J.L.; Neja, R.A. *Managing and Modifying Problem Soils*, Leaflet 2791; University of California, Division of Agricultural Sciences: California, 1979.
5. Chartres, C.J.; Cresswell, H.V.; Murphy, B.W.; Greeves, G.W.; Little, I.P.; Gessler, P.E. Distribution of Physical and Chemical Subsoil Constraints and Their Prediction in Landscapes. In *National Workshop on Subsoil Constraints to Root Growth and High Soil Water and Nutrient Use by Plants*, Proceedings of National Workshop, Tanunda, South Australia, Aug. 30–Sept. 2, 1992.
6. Unger, P.W. Effects of Deep Tillage and Profile Modification on Soil Properties, Root Growth and Crop Yields in the United States and Canada. Geoderma. **1979**, *22*, 275–295.
7. Eck, H.V.; Unger, P.W. Soil Profile Modification for Increasing Crop Production. Adv. Soil Sci. **1985**, *1*, 65–100.
8. Soane, G.C.; Godwin, R.J.; Marks, M.J.; Spoor, G. Crop and Soil Responses to Subsoil Loosening, Deep Incorporation of Phosphorus and Potassium Fertilizer and Subsequent Soil Management on a Range of Soil Types. Part 2: Soil Structural Conditions. Soil Use and Management. **1987**, *3*, 123–130.
9. Sojka, R.E.; Horne, D.J.; Ross, C.W.; Baker, C.J. Subsoiling and Surface Tillage Effects on Soil Physical Properties and Forage Oat Stand and Yield. Soil Tillage Res. **1997**, *40*, 125–144.
10. Olsson, K.A.; Cockroft, B.; Rengasamy, P. Improving and Managing Subsoil Structure for High Productivity from Temperate Crops on Beds. In *Subsoil Management Techniques*; Jayawardane, N.S., Stewart, B.A., Eds.; Lewis Publishers: Boca Raton, 1995; 35–65.
11. Cockroft, B.; Barley, K.P.; Greacen, E.L. The Penetration of Clays by Fine Probes and Root Tips. Aust. J. Soil Res. **1969**, *7*, 333–348.
12. Spoor, G.; Godwin, R.J. An Experimental Investigation into the Deep Loosening of Soil with Rigid Tines. J. Agr. Engng. Res. **1978**, *23*, 243–258.
13. Jawawardane, N.S.; Blackwell, J.; Kirchhof, G.; Muirhead, W.A. Slotting—A Deep Tillage Technique for Ameliorating Sodic, Acid and Other Degraded Subsoils and for Land Treatment of Waste. In *Subsoil Management Techniques*; Jayawardane, N.S., Stewart, B.A., Eds.; Lewis Publishers: Boca Raton, 1995; 109–146.
14. Cresswell, H.P.; Kirkegaard, J.A. Subsoil Amelioration by Plant Roots—the Process and the Evidence. Aust. J. Soil Res. **1995**, *33*, 221–239.
15. Cockroft, B.; Tisdall, J.M. Soil Management, Structure and Root Activity. In *Modification of Soil Structure*; Emerson, W.W., Bond, R.D., Dexter, A.R., Eds., Wiley: Chichester, 1978; 387–391.
16. Tisdall, J.M.; Olsson, K.A.; Willoughby, P. Soil Structural Management and Production in a Non-Cultivated Peach Orchard. Soil Tillage Res. **1984**, *4*, 165–174.
17. Dexter, A.R. Advances in Characterization of Soil Structure. Soil Tillage Res. **1988**, *11*, 199–238.
18. Masle, J.; Passioura, J.B. The Effect of Soil Strength on the Growth of Young Wheat Plants. Aust. J. Plant Physiol. **1987**, *14*, 643–656.
19. Penman, H.L. Water as a Factor in Productivity. In *Potential Crop Productivity*; Wareing, P.F., Cooper, J.P., Eds.; Heinemann: London, 1971; 89–99.
20. Taylor, H.M.; Jordon, W.R.; Sinclair, T.R., Eds. *Limitations to Efficient Water Use in Crop Production*; Amer. Soc. Agron., CSSA, Soil Sci. Soc. Amer: Madison, 1983.
21. Olsson, K.A.; Rose, C.W. Hydraulic Properties of a Red-Brown Earth Determined from In Situ Measurements. Aust. J. Soil. Res. **1978**, *16*, 169–180.
22. Williams, J. Soil Hydrology. *Soils: An Australian Viewpoint*; CSIRO: Melbourne and Academic Press: London, 1983; 507–530.
23. Olsson, K.A.; Rose, C.W. Patterns of Water Withdrawal Beneath an Irrigated Peach Orchard on a Red-Brown Earth. Irrig. Sci. **1988**, *9*, 89–104.
24. Passioura, J.B. Soil Structure and Plant Growth. Aust. J. Soil Res. **1991**, *29*, 717–728.
25. Tardieu, F.; Bruckler, L.; Lafolie, F. Root Clumping may Affect the Root Water Potential and the Resistance to Soil-root Water Transport. Plant and Soil. **1992**, *140*, 291–301.

26. Hall, D.G.M.; Reeve, M.J.; Thomasson, A.J.; Wright, V.F. *Water Retention, Porosity, and Density of Field Soils*, Soil Survey Technical Monograph, No. 9; Rothamsted Experiment Station: Harpenden, UK, 1977.
27. Greacen, E.L.; Gardner, E.A. Crop Behaviour on Clay Soils. Trop. Agric. (Trinidad) **1982**, *59* (2), 123–132.
28. Cannell, R.Q.; Jackson, M.B. Alleviating Aeration Stresses. In *Modifying the Root Environment to Reduce Crop Stress*; Arkin, G.F., Taylor, H.M., Eds.; ASAE: Michigan, 1981; 141–192.
29. Bergman, H.F. Oxygen Deficiency as a Cause of Disease in Plants. Bot. Rev. **1959**, *25*, 418–485.
30. MacEwan, R.J.; Gardner, W.K.; Ellington, A.; Hopkins, D.G.; Bakker, A.C. Tile and Mole Drainage for Control of Waterlogging in Duplex Soils of South-Eastern Australia. Aust. J. Exper. Agric. **1992**, *32*, 865–878.
31. Braunack, M.V.; Dexter, A.R. Compaction of Aggregate Beds. In *Modification of Soil Structure*; Emerson, W.W., Bond, R.D., Dexter, A.R., Eds.; John Wiley and Son: Chichester, 1978; 119–126.
32. Stolzy, L.H.; Letey, J.; Szuskiewicz, T.E.; Lunt, O.R. Root Growth and Diffusion Rates as Functions of Oxygen Concentrations. Soil Sci. Soc. Amer. Proc. **1961**, *25*, 463–467.

STRUCTURE, PEDOLOGICAL CONCEPTS AND WATER FLOW

Clinton C. Truman
United States Department of Agriculture (USDA), Tifton, Georgia, U.S.A.

Donald P. Franzmeier
Purdue University, West Lafayette, Indiana, U.S.A.

INTRODUCTION

Soil structure is the combination or arrangement of primary soil particles into secondary units or peds (1). Soil structure involves the geometric arrangement of particles and inter-particle forces that act between them. Soil structure influences how a soil behaves, including how water moves into (infiltration), across (runoff), and through (percolation or drainage) a given soil. This chapter discusses pedologic classification of soil structure, general mechanisms of ped formation, and the importance of soil structure for water flow and water quality.

CHARACTERIZATION OF STRUCTURE

Soil peds are characterized by their shape, size, and grade (2). The shape of peds is mainly spherical or rectilinear (Fig. 1). Spherical peds comprise *granular* structure. Rectilinear peds are *platy* if their vertical dimension is smaller than the horizontal dimension, *prismatic* if the vertical dimension is larger, or *blocky* if peds have about equal vertical and horizontal dimensions. *Columnar* is a special kind of prismatic structure where ped tops are rounded. Blocky structure can be subdivided into *angular blocky* if peds have sharp edges and corners or *subangular blocky* if edges and corners are rounded.

The size and structure of peds is described by the diameter of granules, the thickness of the plates, the width of prisms, or any dimension of the blocks. Size limits vary from <1 to >100 mm (Table 1).

Structure *grade* is described according to how evident peds appear in a soil exposure and how well they hold together when the soil is handled. Four classes of grade are *structureless*, *weak*, *medium*, and *strong* (Table 1). Structureless soils have no visible structure and are subdivided into two contrasting classes. *Single grain* soils are sandy and individual particles are not associated with other particles. In *massive* structure, particles are strongly bound to each other but in no recognizable pattern. Soil horizons that have undergone little or no soil formation have massive structure. Also, massive structure is often formed when a fine-textured soil is compacted while wet.

Some soils have compound structure in which large peds break up (part) into smaller peds of different shape. For example, a coarse prismatic structure may break into medium blocky structure (Fig. 2).

FORMATION OF SOIL STRUCTURE

Chemical, biological, and physical processes interact to form soil structure. Soil particles, such as clay, act separately in a *dispersed* system, or they group together to form domains in a *flocculated* system. Domains, in turn, group together to form aggregates, and aggregates to create peds. Flocculated systems form stronger structure than dispersed systems because of particle interaction. If the salt content of a soil (Na^+, Ca^{2+}, Mg^{2+}, Cl^-, SO_4^{2-}, etc., in soil solution) is high, as in the desert, the soil tends to be flocculated. In most soils, however, salt content is not high enough to cause flocculation. In these soils the *kind* of exchangeable cation affects flocculation. Sodium ions (Na^+) and to a lesser extent, Mg^{2+}, tend to disperse soils, whereas Ca^{2+} and Al^{3+} tend to flocculate soils. Also, Fe and Al oxides tend to flocculate soils.

Biological processes are mainly responsible for *granular* structure. Ants and termites secrete organic "glues" that hold particles together. Earthworms ingest soil, mix it with organic compounds and excrete aggregated soil material. Plants also secrete organic materials that bind soil particles together. These organisms mainly affect surface (A) horizons. Thus, these horizons generally have granular structure.

Blocky, platy, and prismatic structures are formed by intersecting horizontal and vertical fissures or planes

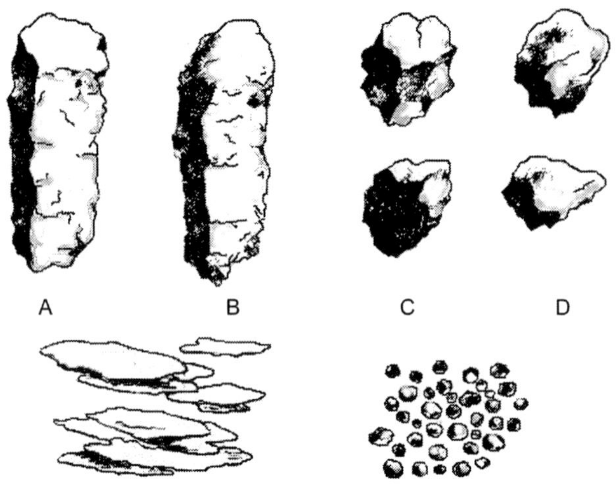

Fig. 1 Sketches of soil structure shapes: (A) prismatic, (B) columnar, (C) angular blocky, (D) subangular blocky, (E) platy, and (F) granular. (From Ref. 1.)

of weakness in soils, mainly subsoil horizons. Vertical fissures are created when a soil shrinks. For example, a water puddle often has a layer of fine soil particles at the bottom. When the water evaporates and the fine material dries, it breaks up into polygons separated from each other by cracks. The same process takes place in subsoils (3). When soil dries and shrinks, smaller sized peds in the east–west and north–south directions are balanced by creation of vertical fissures—the field does not get smaller in area. Soil shrinkage creates some, but fewer, horizontal fissures because soil elevation may actually decrease as soil shrinks. Horizontal fissures may also be caused by freezing and thawing. When soil freezes, liquid water moves to the freezing front where a layer of ice forms and displaces soil material. The freezing front tends to be at about the same horizontal level in the soil. When ice melts, a horizontal fissure is left.

Platy structure forms where horizontal fissures predominate, such as near-surface E horizons that undergo many freeze–thaw cycles. *Prismatic* structure forms where vertical fissures predominate, such as in clayey subsoil horizons that dry out due to evapotranspiration. *Blocky* structure forms when horizontal and vertical fissures intersect.

Soil structure is defined according to how individual particles (sand, silt, and clay) are joined together to form peds. Soil scientists, however, describe the end product—how the mass of soil in a profile breaks out into peds. Examples of soil structure descriptions from a published soil survey are "moderate medium granular structure" (A horizon) and "strong very coarse prismatic structure parting to moderate coarse angular blocky" (Bx horizon, fragipan).

Often ped surfaces are coated with clay, and the resulting clay films or skins make ped shape more apparent. Clay is mobilized in near-surface horizons, moves down the soil profile as a slurry, and is immobilized on a ped surface as water soaks into the ped and clay is filtered out on its surface. Chemical changes down the profile, such as a change in pH, may also help clay skin formation. Other mobile constituents (iron or manganese oxides) may also coat ped surfaces.

Table 1 Classes of grade and size of soil structure

Class	**Definition**			
Grade (strength of development):				
Structureless (0)	No observable structure (massive or single grain)			
Weak (1)	Structure barely observable in place; few peds survive digging			
Moderate (2)	Peds are well formed and evident in place; many peds survive digging			
Strong (3)	Structure distinct in place; most peds survive digging			
Class	Platy (mm)	Prismatic and columnar (mm)	Blocky (mm)	Granular (mm)
Size:				
Very fine (VF)	<1	<10	<5	<1
Fine (F)	1–2	10–20	5–10	1–2
Medium (M)	2–5	20–50	10–20	2–5
Coarse (C)	5–10	50–100	20–50	5–10
Very coarse (VC)	>10	>100	>50	>10

(From Ref. 2.)

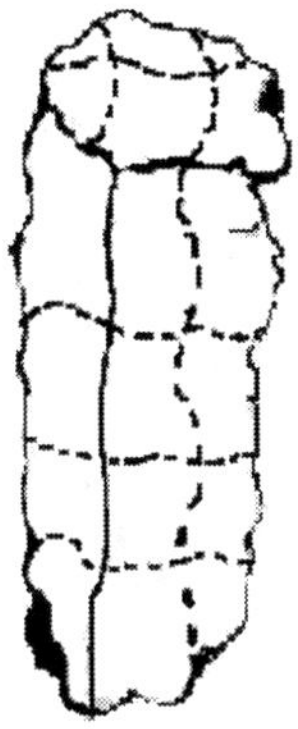

Fig. 2 Sketch illustrating compound structure, moderate coarse prismatic structure parting to weak medium angular blocky structure.

SIGNIFICANCE OF SOIL STRUCTURE

Characterizing soil structure is based on where soil particles are located, but the significance of soil structure relates to the space where there are no soil particles. Spaces between soil particles are called voids or pores. Pores are often divided into two types—macropores and micropores. Voids between peds are macropores and serve as pathways for roots and rapid water movement. Micropores are smaller pores between particles and aggregates. They hold water between rains for later use by plants.

Flat spaces between adjoining platy, blocky or prismatic peds are planar voids. The ease with which water moves and roots grow through a planar voids depends on ped shape (Fig. 3). The route through a platy structure is more tortuous than through a prismatic structure.

Tubular voids, such as root channels and earthworm burrows, are not usually included in the description of structure, but are important for water movement and root growth in soils. They act much like planar voids. Often, new roots follow old root channels in which old roots have decayed. Inactive worm channels are especially favorable environments for roots because they are often lined with highly-fertile worm castings.

Generally, permeability increases with stronger structural development and smaller peds. Soils with faster permeability are also conducive to better root growth and movement of soil air to roots.

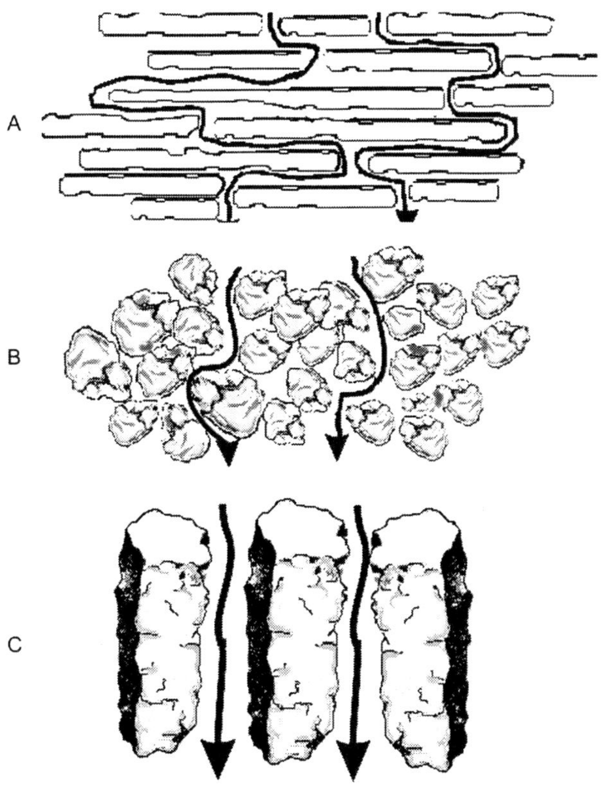

Fig. 3 Sketch illustrating paths of water flow through soils with different soil structure shapes: (A) platy, (B) blocky, and (C) prismatic.

WATER MOVEMENT AND SOIL STRUCTURE

Water moves slowly into and through ped interiors by *matrix flow*, and moves rapidly through larger voids as *preferential flow*. Interaction of these two kinds of flow is illustrated in Fig. 4. In the left part of the diagram, rectangles represent peds (prisms in this case), and the space between rectangles represents planar voids. The size of the arrow represents the magnitude of rainfall. Thin lines within a prism show matrix flow, and thick lines on the prism surface show preferential flow. Under a gentle rain, the water application, rate 1, all water soaks into (infiltrates) the prism as matrix flow, mainly from the top. With a moderate rain, rate 2, water starts to soak into the prism, but soon the intake rate of the prism is exceeded and preferential or nonmatrix flow begins. Then, water also soaks into the side of the prism. Under a hard rain, rate 3, water moves by matrix and preferential flow almost from the beginning of the storm. The same relations are shown in the graph, with the additional information that the infiltration rate into the prism decreases with time. Preferential flow begins when the horizontal line with the arrow crosses the infiltration curve.

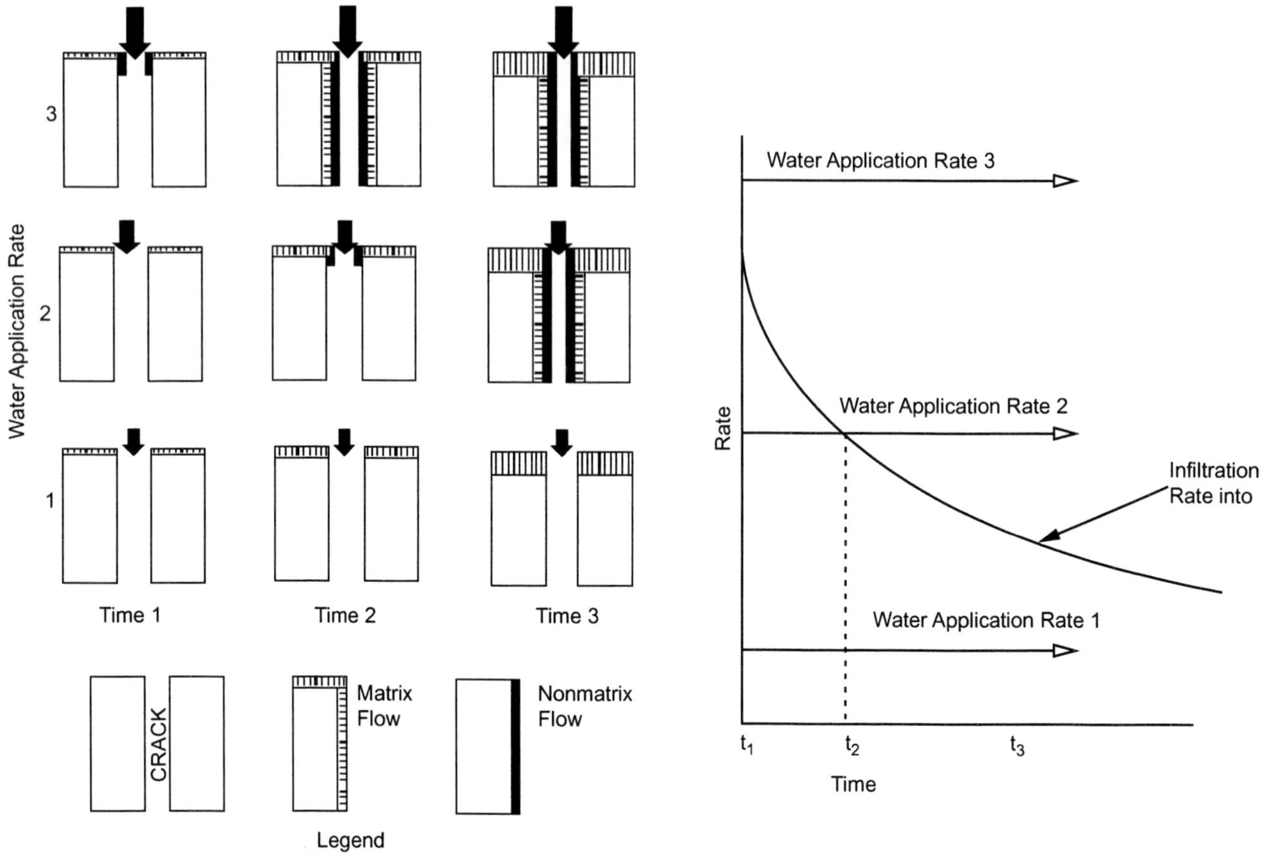

Fig. 4 Diagram illustrating water flow into and through a soil horizon with prismatic structure. Water application (rainfall) rate 1 is small enough that only matrix flow occurs. Application rate 2 initially produces matrix flow. As the soil wets, the infiltration rate into a prism decreases, application rate exceeds the infiltration rate, and nonmatrix flow begins. Application rate 3 is fast enough that nonmatrix (preferential) flow begins early and continues throughout the rain storm. (From Ref. 12.)

SOIL STRUCTURAL STABILITY

Soil structural stability describes the ability of a soil to retain its arrangement of soil and voids when exposed to different stresses and is one of the most important properties that affect a soil's ability to store and transmit water and agrichemicals (4). For example, compaction alters soil structure and hydrology by increasing bulk density, breaking down aggregates, decreasing porosity, aeration and infiltration, and by increasing soil strength and runoff.

SOIL MANAGEMENT FOR WATER FLOW

Runoff, infiltration, and subsequent drainage carry agrichemicals (nutrients and pesticides) to surface and ground waters. Therefore, soil structure affects rates, timing, and amounts of runoff, infiltration, and agrichemical fate and transport. In the subsurface, soil structure influences matrix or preferential flow (5).

Agrichemical transport by preferential flow means that water carrying the chemicals travels only through a small portion of the soil volume (not matrix flow) (6). When infiltration exceeds matrix flow, water and associated agrichemicals overflow into macrochannels causing preferential flow that transmits water and agrichemicals to deeper positions in relatively unsaturated soil materials (7).

Although coarse-textured sandy soils generally have little structure, they have good aeration and drainage, promote infiltration, limit runoff, yet lack the capacity to sorb and hold sufficient water and nutrients. They tend to be drought-prone and lack fertility. A management practice to improve the structural nature of these soils is to add organic matter, which binds soil particles and promotes water-holding capacity, and can be accomplished by managing residue or addition of amendments

Table 2 Selected topics and corresponding references associated with soil structure

Topic(s)	References
Methods of quantifying/studying soil structure	(13–15)
Modeling soil structure	(5, 9, 16, 17)
Improving soil structure	(18–20)
Biological effects on soil structure	(21, 22)
Tillage and cropping effects on soil structure	(4, 23–25)
Soil structure effects on chemical movement	(26–29)
Drainage and soil structure	(27, 30)
Soil structure, and Vertisols and cracking	(24, 31)
Soil classification, preferential flow, and soil structure	(5, 32)

such as manures or litter. Structural management of clayey soils is more difficult because potential plasticity and cohesion are high. Increasing aggregation in the surface of fine-textured soils will have a dramatic effect on rainfall partitioning into infiltration and runoff and sediment delivery, and subsequent agrichemical fate and transport.

Tillage has favorable and unfavorable effects on soil structure. Moldboard plowing destroys soil structure and aggregation in the Ap horizon, disrupts continuous macropores, increases compaction risk, and retards organic matter build up. One management practice that maintains or enhances soil structure and aggregate stability is reduced (or conservation) tillage. This type of tillage allows for residue management and subsequent organic matter build up. Organic matter is most important in modifying soil structural characteristics and can increase the resistance of the pore network to destructive forces. Because of enhanced soil structure, reduced tillage systems tend to decrease runoff and erosion; increase infiltration, soil water holding capacity, and hydraulic conductivity; and reduce agrichemical losses by runoff (8–10). Reduced tillage systems also conserve soil structure including preferential flow pathways (earthworm burrows, root channels, fissures, and/or cracks). Therefore, leaching by preferential flow can be enhanced by reduced tillage. Transport pathways are grouped into runoff and leaching, each a transporting agent for agrichemicals. Runoff and leaching potentials must be considered when evaluating management practices such as tillage on soil structure, water flow, and water quality. Additional literature on soil structure effects on water flow and water quality is given in Table 2.

REFERENCES

1. SSSA, *Glossary of Soil Science Terms*; Soil Sci. Soc. Am. Madison, WI 53711, 1997, 134.
2. Soil Survey Division Staff, *Soil Survey Manual*; USDA–Soil Conservation Service, Agric. Handbook 18, U.S. Govt Print. Office: Washington, DC, 1993.
3. Fanning, D.S.; Fanning, M.C.B. *Soil Morphology, Genesis, and Classification*; Wiley: New York, 1989; 395.
4. Gregorich, E.; Reynolds, W.; Culley, J.; McGovern, M.; Curnoe, W. Changes in Soil Physical Properties with Depth in a Conventionally Tilled Soil After No-Tillage. Soil Tillage Res. **1993**, *26*, 289–299.
5. Vervoort, R.; Radcliffe, D.; West, L. Soil Structure Development and Preferential Solute Flow. Water Resour. Res. **1999**, *35*, 913–928.
6. Flury, M. Experimental Evidence of Transport of Pesticides Through Field Soils—A Review. J. Environ. Qual. **1996**, *25*, 25–45.
7. McIntosh, J.; McDonnell, J.; Peters, N. Tracer and Hydrometric Study of Preferential Flow in Large Undisturbed Soil Cores from the Georgia Piedmont, USA. Hydrol. Process. **1999**, *13*, 139–155.
8. Leonard, R. Movement of Pesticides into Surface Waters. In *Pesticides in the Soil Environment*; Soil Science Society America Book Series 2, Cheng, H.H., Ed.; Soil Sci. Soc. Am. Madison, WI, 1990; 303–349.
9. Cresswell, H.; Smiles, D.; Williams, J. Soil Structure, Soil Hydraulic-Properties and the Soil-Water Balance. Austr. J. Soil Res. **1992**, *30*, 265–283.
10. Pagliai, M.; Raglione, M.; Panini, T.; Maletta, M.; La Marca, M. The Structure of Two Alluvial Soils in Italy After 10 Years of Conventional and Minimum Tillage. Soil Tillage Res. **1995**, *34*, 209–223.
11. Soil Survey Staff, *Soil Survey Manual*; USDA–Soil Conservation Service, Agric. Handbook 18, U.S. Govt Print. Office: Washington, DC, 1962.
12. Bouma, J.; Paetzold, R.F.; Grossman, R.B. *Measuring Hydraulic Conductivity for Use in Soil Survey*; Soil Survey Investigations Report No. 38, Soil Conservation Service, U.S. Dept. Agric., 1982.
13. Bartoli, F.; Bird, N.; Gomendy, V.; Vivier, H.; Niquet, S. The Relation Between Silty Soil Structures and Their Mercury Porosimetry Curve Counterparts: Fractals and Percolation. Eur. J. Soil Sci. **1999**, *50*, 9–22.
14. Gomendy, V.; Bartoli, F.; Burtin, G.; Doirisse, M.; Philippy, R.; Niquet, S.; Vivier, H. Silty Topsoil Structure and Its Dynamics: The Fractal Approach. Geoderma **1999**, *88*, 165–189.
15. Hakansson, I.; Lipiec, J. A Review of the Usefulness of Relative Bulk Density Values in Studies of Soil Structure and Compaction. Soil Tillage Res. **2000**, *53*, 71–85.
16. Gimenez, D.; Perfect, E.; Rawls, W.; Pachepsky, Y. Fractal Models for Predicting Soil Hydraulic Properties: A Review. Eng. Geol. **1997**, *48*, 161–183.
17. Connolly, R. Modelling Effects of Soil Structure on the Water Balance of Soil-Crop Systems: A Review. Soil Tillage Res. **1998**, *48*, 1–19.
18. Haynes, R.; Naidu, R. Influence of Lime, Fertilizer and Manure Applications on Soil Organic Matter Content and

Soil Physical Conditions: A Review. Nutr. Cycling Agroecosys. **1998**, *51*, 123–137.
19. Kort, J.; Collins, M.; Ditsch, D. A Review of Soil Erosion Potential Associated with Biomass Crops. Biomass Bioener. **1998**, *14*, 351–359.
20. Nemati, M.; Caron, J.; Gallichand, J. Using Paper De-inking Sludge to Maintain Soil Structural Form: Field Measurements. Soil Sci. Soc. Am. J. **2000**, *64*, 275–285.
21. Lee, K.; Foster, R. Soil Fauna and Soil Structure. Austr. J. Soil Res. **1991**, *29*, 745–775.
22. Shuster, W.; Subler, S.; McCoy, E. Foraging by Deep-burrowing Earthworms Degrades Surface Soil Structure of a Fluventic Hapludoll in Ohio. Soil Tillage Res. **2000**, *54*, 179–189.
23. Kay, B. Rates of Change of Soil Structure Under Different Cropping Systems. Adv. Soil Sci. **1990**, *12*, 1–51.
24. Oygarden, L.; Kvaerner, J.; Jenssen, P. Soil Erosion Via Preferential Flow to Drainage Systems in Clay Soils. Geoderma **1997**, *76*, 65–86.
25. Angers, D. Water-Stable Aggregation of Quebec Silty Clay Soils: Some Factors Controlling Its Dynamics. Soil Tillage Res. **1998**, *47*, 91–96.
26. Beck, A.; Johnston, A.; Jones, K. Movement of Nonionic Organic-Chemicals in Agricultural Soils. Critical Rev. Environ. Sci. Tech. **1993**, *23*, 219–248.
27. Ball, B.; Campbell, D.; Douglas, J.; Henshall, J.; O'Sullivan, M. Soil Structural Quality, Compaction and Land Management. Eur. J. Soil Sci. **1997**, *48*, 501–593.
28. Bergstrom, L.; Stenstrom, J. Environmental Fate of Chemicals in Soil. Ambio **1998**, *27*, 16–23.
29. Walton, R.; Volker, R.; Bristow, K.; Smettem, K. Experimental Examination of Solute Transport by Surface Runoff from Low-Angle Slopes. J. Hydro. **2000**, *233*, 19–36.
30. Collis-George, N. Drainage and Soil Structure—A Review. Austr. J. Soil Res. **1991**, *29* (6), 923–933.
31. Pillai, U.; McGarry, D. Structure Repair of a Compacted Vertisol with Wet–Dry Cycles and Crops. Soil Sci. Soc. Am. J. **1999**, *63*, 201–210.
32. Quisenberry, V.; Smith, B.; Phillips, R.; Scott, H.; Northcliff, S. A Soil Classification System for Describing Water and Chemical Transport. Soil Sci. **1993**, *156*, 306–315.

STRUCTURE, PLANT ESTABLISHMENT AND

D.L. Karlen
United States Department of Agriculture (USDA), Ames, Iowa, U.S.A.

INTRODUCTION

Soil structure influences plant germination, emergence, root development, growth, development, and yield. This section discusses how plants affect soil structure and how soil structure affects plants.

PLANT EFFECTS ON SOIL STRUCTURE

As a medium for plant growth and development, soil is a three-dimensional, dynamic, naturally occurring blanket on the surface of the earth (1). This medium has an inherent (naturally occurring) soil structure that reflects the forces of climate and living organisms acting on parent material, over time and within a specific topographic or landscape (slope or relief) position (2). The naturally occurring soil structure directly affects, and is affected by the type of plants (e.g., grass, forest, or agricultural crops) and how they are grown. Soils formed under grass (Fig. 1) generally have higher organic matter content and a structure that is conducive for rapid plant growth (3). This type of soil structure develops as perennial grasses replace most of their roots and top growth every year. As the litter layer dies, it falls on the soil-surface and may be decomposed or mixed into the upper part of the soil by earthworms and other soil organisms.

Grassland soils are generally less acid than soils that developed under forest. This occurs as grassland soils were generally formed with less annual precipitation than nearby forest soils and as grasses are more effective than trees in recycling nutrients that can be leached. Forest vegetation is more prevalent under a humid (higher precipitation) climate, which also contributes to greater leaching. Trees return organic matter to the soil-surface primarily in the form of fallen leaves, twigs, and eventually the entire trunk and its branches (3). As the litter layer decomposes, organic acids are formed, thus increasing the weathering of parent materials and subsequent leaching losses. Most organic matter under forest vegetation is concentrated in the litter layer and remains in the upper few inches of the surface or A horizon. Below that relatively shallow zone in forest-derived soils, organic matter is often more eluviated (leached) than under grassland. The net result of these natural processes is that A horizons in grassland soils (Mollisols) tend to have stronger, more granular structures and faster infiltration rates than in forest derived soils (Alfisols), but forest soils are more likely to have deeper profiles and more strongly developed B horizons than grassland soils.

HUMAN IMPACTS ON SOIL STRUCTURE

Human decisions regarding how to use the land for food, feed, and fiber production (i.e., agriculture) have a major impact on soil structure. The effect of choosing to cut native forests (especially by mechanical methods) or till [e.g., plowing, disking, rotary tillage, or even puddling (compaction) for rice] soils rather than to leave them in their natural condition can affect soil structure in both positive and negative ways depending upon the length of time for which the evaluation is being made. The short-term effects are generally positive with regard to crop production, especially on fine-textured soils. Tillage releases essential plant nutrients from plant residues or the soil organic matter and has traditionally provided for weed control. It can also improve the air and water relationships required for plant growth by breaking surface crusts and/or disrupting dense compacted layers that can restrict plant emergence and growth.

Negative effects of tillage or other aggressive physical disturbances of the soil (i.e., deforestation or over-grazing) become more evident when changes are observed over the long-term. These impacts include loss of soil organic matter, reductions in total aggregation, declines in the water-stability of the remaining aggregates, and decreases in the volume of macropores and biopores (Fig. 2). The long-term negative impacts actually begin immediately after cultivation as soil aggregates are broken apart and the existing structure is weakened or destroyed as the soil is loosened. However, the short-term benefits generally mask the long-term effects when observations are primarily focused on plant performance and above ground

Fig. 1 Well-aggregated soil showing plant root growth throughout the profile.

responses. The negative long-term effects begin to be evident after a short period of time as the soil particles gradually consolidate and create a denser soil than in nearby noncultivated sites. As soil aggregates are broken or dispersed, the fine particles are gradually filtered into the soil pores where they accumulate and block the narrow passageways, thus decreasing porosity and creating even stronger surface crusts. The soil degradation continues as the crusts restrict water infiltration, thus increasing runoff and subsequently soil erosion.

SOIL STRUCTURE EFFECTS ON PLANT GROWTH

Germination is affected by soil structure as the two primary factors affecting this plant process are soil temperature and water content. Soil temperature is affected by several factors including the angle of the sun's rays, soil color, surface residue or cover, and the depth and time of measurement. The most favorable temperature for seed germination depends on the plant species, but efforts to optimize those conditions, in addition to weed control prior to the development of herbicides, are among the primary reasons that agricultural soils have traditionally been tilled (4). The loose, friable soil created by cultivation at the expense of soil structure loses water through evaporation and warms more quickly than more-dense, nontilled soil. The effects of increased soil temperature and decreased water content are not always positive, especially in coarse-textured soils (5). If the seedbed is too dry, germination can be delayed by the lack of water. Under those conditions, it is possible to actually observe plant growth benefits from a slight

Fig. 2 Compacted soil with very little plant root growth throughout the soil profile.

amount of soil compaction. Also, if the effect of surface tillage on soil temperature below the seed zone is of primary concern, the increased surface porosity following tillage can actually insulate against temperature change deeper into the soil profile. This occurs as the air space created by the tillage operation conducts heat much less efficiently than either soil particles or water molecules.

Emergence of plant seedlings is affected by soil structure primarily when the surface soil consolidates and forms a crust. This occurs most frequently when soils have been tilled leaving very little surface residue prior to an intense rainfall or surface irrigation event. The thickness and strength of the crusts that can develop will be determined by several factors including the specific type and mineralogy of the soil, the amount of soil organic matter that is present, the stability of the soil aggregates, the intensity with which water strikes the soil-surface, and the air temperature or rate of drying after the wetting event (2).

Root development is another plant process that is strongly influenced by soil structure. Under normal conditions, plant roots during early stages of plant development will elongate and extend into the soil 5–10 times deeper than the above ground height. This occurs because after a seedling emerges, the demand for water increases tremendously and the root system must be sufficiently established to meet those demands. If the soil structure does not allow the roots to expand because of inherent fragipans or eluviated soil horizons (Fig. 3) or because of compaction caused by animal hooves or wheel-traffic (Fig. 4), plant growth, development, and ultimately yield or productivity will be reduced. The primary requirements for good root development are simply 1) adequate soil oxygen for physiological functioning of the root; 2) sufficient soil water for above and below ground plant needs; 3) soil temperature within the range suited to the plant species; 4) sufficient nutrients and sunlight for the plant to produce and transport carbohydrates (food and hormones) to the root system; and 5) relatively low or insignificant amounts of toxic chemicals, lethal gases, disease organisms, or insect damage (6). Keeping these factors in balance is a primary goal associated with agricultural soil management, when evaluated with regard to soil structure and plant emergence.

Fig. 4 Corn root growth in a U.S. Midwestern Mollisol confined to the nonwheel track area (right-hand side).

Fig. 3 Corn root growth in the surface horizons of a U.S. southeastern coastal plain soil confined to a mechanically disturbed area created by in-row subsoiling.

MANAGING SOIL STRUCTURE FOR PLANT ESTABLISHMENT

Soil management research around the world has shown that use of heavy machinery and frequent disk tillage can compact subsurface horizons in many irrigated soils. The result of such compaction in Morocco was a 12–23% decrease in wheat grain yield and a 9–20% decrease in straw yield (7). The decrease was accompanied by a consistent reduction in the number of shoots per unit area and significant changes in both root growth and distribution. In a U.S. study, improving aeration by subsoiling and avoiding wheel traffic significantly increased sugar beet yield (8). This study showed that regardless of the type of tillage used prior to planting, minimizing wheel traffic was one of the most important management decisions that could be made.

Finally, even with reduced-tillage or no-tillage practices, it is important to understand how subtle changes in soil structure can influence plant establishment. A specific example is known as seed furrow or sidewall smearing (9). This situation can occur if mechanical planters equipped with double-disk furrow openers are used when soils are too wet. When sidewall smearing occurs, penetration resistance adjacent to and below the seed is increased, pore size is reduced, and air permeability is decreased. One solution is to use different planter attachments, such as a triple-coulter, ahead of the double-disk openers. Another would be to delay planting until the soils were slightly drier, but that may also have negative consequences depending upon the crop, length of growing season, and other management factors.

REFERENCES

1. Doran, J.W.; Sarrantonio, M.; Liebig, M.A. Soil Health and Sustainability. In *Adv. Agron.* Sparks, D.L., Ed.; Academic Press: San Diego, CA, 1996; Vol. 56, 1–54.
2. Henry, D.; Fundamentals of Soil Science, 8th Ed.; Wiley: New York, 1990.
3. Troeh, F.R.; Thompson, L.M. *Soils and Soil Fertility*, 5th Ed.; Oxford University Press: New York, 1993.
4. Lal, R.; Stewart, B.A. *Soil Management: Experimental Basis for Sustainability and Environmental Quality*; Advances in Soil Science, CRC Press: Boca Raton, FL, 1995.
5. Karlen, D.L. Tillage and Planting System Effects on Corn Emergence from Norfolk Loamy Sand. Appl. Agric. Res. **1989**, *4*, 190–195.
6. Trouse, A.C., Jr. Root Tolerance to Soil Impediments. In *Crop Tolerance to Suboptimal Land Conditions*; ASA Spec. Publ. No. 32, Jung, G.A., Ed.; ASA/CSSA/SSSA: Madison, WI, 1978.
7. Oussible, M.; Crookston, R.K.; Larson, W.E. Subsurface Compaction Reduces the Root and Shoot Growth and Grain Yield of Wheat. Agron. J. **1992**, *84*, 34–38.
8. Johnson, B.S.; Erickson, A.E. Sugarbeet Response to Subsoiling and Wheel Traffic. Agron. J. **1991**, *83*, 386–390.
9. Iqbal, M.; Marley, S.J.; Erbach, D.C.; Kaspar, T.C. An Evaluation of Seed Furrow Smearing. Trans. ASAE **1998**, *41*, 1243–1248.

STRUCTURE, ROOTS AND

S

Alvin J.M. Smucker
Michigan State University, East Lansing, Michigan, U.S.A.

INTRODUCTION

Soil structure includes a vast array of heterogeneous and separable individual soil aggregates. Soil aggregates are the primary repositories of carbon, water, microbial communities, plant nutrients, and pollutants within the soil profile. These biophysical polymorph structures, ranging in size from submillimeter to many millimeters across, control the absorption, storage, and losses of most soil constituents. The dynamic properties of soil aggregates are controlled by soil type, anthropogenic inputs, daily changes in the weather, plant roots, and soil animal activities. Separable soil aggregates are formed during the combination of soil minerals, soil organic matter, plant residues, microbes and other soil biological constituents (9). Soil aggregates develop and function through complex biological, chemical and physical interactions with climate, water and ion activities occurring within and across interfaces of adjacent soil aggregates within the soil profile. Formation processes of soil aggregates include the cementation of adjacent smaller soil aggregates into larger aggregates (15) and/or accumulation and cementation of soil minerals, ions, and particulate organic matter onto surfaces of existing aggregates (3, 14). Repeated soil wetting and drying cycles develop stronger soil aggregates than the surrounding bulk soil, and create smaller internal pores than the surrounding macropores of the bulk soil.

PLANT ROOT CONTRIBUTIONS TO SOIL AGGREGATES

Plant species and root architectural contributions to the development of specific soil structural characteristics are well known. Root growth and exudation, fungal associations, and death contribute substantial quantities of carbon (C) into soil aggregates (6). Many of the mechanisms controlling soil aggregation processes and dynamics remain unknown. Current research suggests roots opportunistically invade pores between or within soil aggregates depositing soluble C substrates and residues that release C upon decomposition. These root products are energy-rich food sources that stimulate microbial growth and production of polysaccharides and other cementing agents that glue soil mineral particles into more stable soil aggregates (4, 12, 13). In relatively undisturbed soils, root derived C and subsequent decomposition products located within aggregates contribute more to the formation and stabilization of smaller soil aggregates than do plant residues located in the larger pores between aggregates (5). Greater numbers of aggregate wetting and drying cycles increase aggregate stability by promoting mineralization of root residues (16). Closer proximity of soil minerals, during drying, also increases the bridging strengths of adsorbed organic and inorganic compounds causing greater soil aggregate stabilities (10). These and other highly orchestrated plant genetic and environmental interactions govern the majority of these dynamic feed-forward and feed-back root and soil activities.

EVALUATING ROOT–SOIL AGGREGATION INTERACTIONS

Continuous rhizodeposition of C onto soil aggregates is one of the major contributions of crop management to soil structure formation and stability. There seems to be very little information on biogeochemical mechanisms associated with the improvement of C fixed by plants and retained for prolonged periods of time. Sissoko (13) demonstrated how plant root exudates increase the stability of soil aggregates. Using a seven-day root exudation, incubation, and soil wetting and drying protocol for nine cycles, he developed soil aggregates containing 70% more microbial biomass and produced 280% more stable aggregates than water-based controls.

Natural Isotopes of Carbon

Evaluating the complexities of C flow from plant roots to their various forms of cementation products within soil aggregates is complicated by the multiple biogeochemical processes of soil C mineralization, the spatial and temporal distributions of various C sources, and the heterogeneity of soil aggregate sizes and stabilities across a vast array of landscapes. Net accumulations of C, originating directly from plant residues or from soil organic matter (SOM), within soil aggregates can be estimated by mass spectrometric evaluations of the natural abundance of ^{13}C in bulk soil. Using soils that produced separate or combined populations of C3 or C4 plants whose metabolisms discriminate in their fixation and accumulation of $^{13}CO_2$ and $^{12}CO_2$ (6), identified SOM and plant root C contributions to the stabilities of multiple soil aggregate sizes. With natural carbon-13 isotopes it is possible to identify changes in bulk soil C sources after 6 years of continuous plant growth on the same soil (1).

Mechanical Removal of Concentric Layers

Plant-induced changes in soil C accumulations can be identified during shorter time periods by removing and analyzing the C isotope sources in 10–15 mg subsamples from concentric layers of soil aggregates. Santos (12) identified plant root C accumulations in peeled surface layers of soil aggregates as early as six weeks, for corn, using greenhouse containers, or 20 months, for alfalfa, in rotational field studies. Root depositions of ^{13}C and ^{15}N to specific concentric layers within soil aggregates are routinely being reported for aggregates sampled from the rhizosphere soils in surface horizons (7, 12). Smaller stainless steel soil aggregate erosion (SAE) peeling chambers (14) that more efficiently produce concentric layers of soil from individual and smaller aggregates have been developed for identifying C, N, P, pH (7, 8, 11), and microbial communities (2) at specific microsites within soil aggregates. These approaches provide numerous opportunities for identifying the rhizodeposition of additional plant compounds from specific C3 and C4 plant species.

Peeled soil layers from aggregates provide information on the deposition rates of both newly deposited labile C and older recalcitrant C within soil aggregates. Kinyangi (8) demonstrated interdependent and short-term relationships among soil C, N, and P. Using dry combustion analyses of small (10–20 mg) soil samples extracted from concentric layers of soil aggregates, he compared ratios of $^{13}C/^{12}C$ soil C, planted to C3 species of *Crotalaria grahamiana* plants whose roots produced a $\delta^{13}C$ signature of $-26.1‰$, and C4 cultivars of *Zea mays* plants whose roots produced a $\delta^{13}C$ signature of $-10.6‰$ with extractable P from the same peeled layers of aggregates from a Kenyan Oxisol. Using the equation below, he reported that roots of a contemporary maize crop sampled

$$\%\text{C from maize} = [(\delta^{13}\text{C layer}_{\text{final}} - \delta^{13}\text{C layer}_{\text{initial}}) / (\delta^{13}\text{C crop}_{\text{maize}} - \delta^{13}\text{C layer}_{\text{initial}})]$$

from both field (4 months) and container (5 weeks) studies contributed up to 50% of the measurable soil C in surface layers of Oxisol aggregates, 4–8 mm across. Carbon contents in surface layers of these same soil aggregates declined nearly 23% when fertilizer P was applied to the soil. This study demonstrated a high rate of C deposition by plant roots and transitory C mineralization that controls soil aggregate stability and P availability.

Observations that *Tithonia diversifolia* extracts P from P-deficient Oxisols when soil aggregates are penetrated by very fine roots were confirmed by identifying lower extractable P levels from the centers of peeled soil aggregates that had been penetrated by *Tithonia* roots (8). External soil layers of aggregates from P-deficient Oxisols were removed by the SAE method of peeling soil aggregates (14) and extractable P levels measured. Consequently, greater quantities of soil P can be gleaned from the centers of rigid soil aggregates of P-deficient Oxisols and transferred to food crops via decomposing *Tithonia* plant roots and aboveground residues (8).

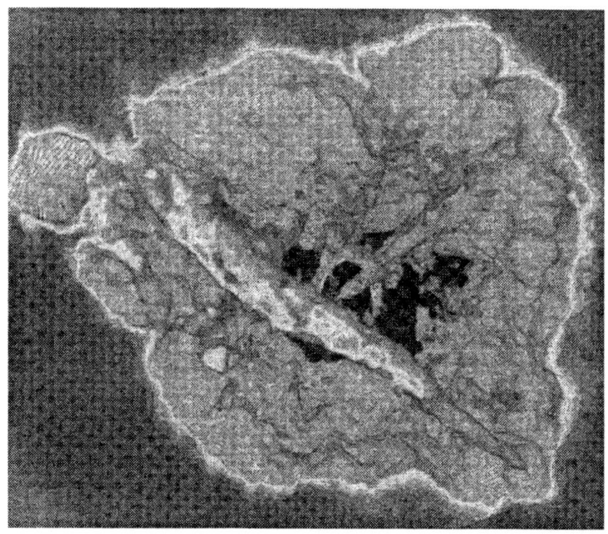

Fig. 1 Two-dimensional image of several electronic slices through microtomographic reconstructions of an undisturbed prairie grass soil aggregate, 1 mm across. Notice the diagonal root induced macropore (RIM) from top left to bottom right is the remnant of a dehydrated root that contributed root exudates and other plant carbon substrates to the soil aggregate. This soil aggregate image was produced by K. M. Kemner, B. Lai and H.-R. Lee at the x-ray beam in Sector 2 of the APS at ANL. Images were recorded by a CCD camera with a pixel unit cell size 7 x 6 microns.

S

MICRODENSITOMETRY OF SOIL AGGREGATES

Investigations of interior regions of soil aggregates could lead to a greater understanding of the complex porosities that sequester soil C, microbial communities, and ions. Greater knowledge of water and ion fluxes into and through soil aggregates will lead to knowledge-based opportunities for better managing these complex living centers of the soil. Microdensitometry evaluations of root and soil aggregate interactions are routine analytical options at the GeoSoilEnviro CARS (Consortium for Advanced Radiation Sources) beamline of Sector 13 at the Advanced Photon Source (APS), located within Argonne National Laboratory (ANL) near Chicago, Illinois. Computer algorithms rendered two- (Fig.1) and three-dimensional images for observations. This image and other synchrotron X-ray images of soil aggregates generate new spatial concepts that expand our knowledge of intra-aggregate and root interfaces. For example, root penetration of these very rigid soil aggregates, 1 mm across, of Kenyan Oxisols permitted root extraction of previously unavailable P from the internal regions of these aggregates (8). These types of root-induced macropores (RIMs) also increase macropore connectivities between adjacent soil aggregates, greatly increasing bypass flow through adjacent soil aggregates. Although the APS synchrotron used in these studies is quite large, immobile, and less expensive, new desktop models for routine microtomographic evaluations of porous materials are becoming commercially available. As microdensitometry evaluations of soil aggregates become less expensive, more specific internal porosity comparisons can be made across different soil types and management systems. These evaluations, although somewhat disruptive, when coupled with X-ray diffraction measurements should lead to greater understanding of the biogeochemical mechanisms controlling root uptake of ions as they are associated with the soil minerals at the root–soil interface.

REFERENCES

1. Balesdent, J.; Wagner, G.H.; Mariotti, A. Soil Organic Matter Turnover in Long-Term Field Experiments as Revealed by Carbon-13 Natural Abundance. Soil Sci. Soc. Am. J. **1988**, *52*, 118–124.
2. Blackwood, C.; Smucker, A.J.M.; Paul, E.A. *Microbial Distributions within Concentric Layers of Soil Aggregates*, Abstracts of Papers for the Ecol. Soc. Am., Snowbird, UT, July 2000.
3. Dexter, A.R.; Horn, R. Effects of Land Use and Clay Content on Soil Structure as Measured by Fracture Surface Analysis. Z. Pflanzenernahrung Bodenk. **1998**, *151*, 325–330.
4. Gale, W.J.; Cambardella, C.A.; Bailey, T.B. Surface Residue and Root Derived Carbon in Stable and Unstable Aggregates. Soil Sci. Soc. Am. J. **2000**, *64*, 196–201.
5. Gale, W.J.; Cambardella, C.A.; Bailey, T.B. Root Derived Carbon and the Formation and Stabilization of Aggregates. Soil Sci. Soc. Am. J. **2000**, *64*, 201–207.
6. Jastrow, J.D.; Boutton, W.; Miller, R.M. Carbon Dynamics of Aggregate-Associated Organic Matter Estimated by Carbon-13 Natural Abundance. Soil Sci. Soc. Am. J. **1996**, *60*, 801–807.
7. Kavdir, Y. Distribution of Cover Crop Nitrogen Retained by Soil Aggregates within a Rye-Corn Agroecosystem. Ph.D. Dissertation, Michigan State University, 1999; 157.
8. Kinyangi, J. Carbon, Nitrogen and Phosphorus Sequestration within Soil Aggregates of Maize and Tree-Based Agroforestry Systems. M.S. Thesis, Michigan State University, 2000; 101.
9. Oades, J.M. Soil Organic Matter and Structural Stability: Mechanisms and Implications for Management. Plant Soil **1984**, *76*, 319–337.
10. Reid, J.B.; Goss, M.J. Interactions Between Soil Drying due to Plant Water Use and Decreases in Aggregate Stability Caused by Maize Roots. J. Soil Sci. **1982**, *33*, 47–53.
11. Santos, D. Contributions of Roots and Organic Matter to Soil Aggregate Development and Stabilization. Ph.D. Dissertation, Michigan State University, 1998; 149.
12. Santos, D.; Murphy, S.L.S.; Taubner, H.; Smucker, A.J.M.; Horn, R. Uniform Separation of Concentric Surface Layers from Soil Aggregates. Soil Sci. Soc. Am. J. **1997**, *61*, 720–724.
13. Sissoko, F. Enhancement of Soil Aggregation by the Combined Influences of Soil Wetting and Drying and Root–Microbial Associations. M.S. Thesis, Michigan State University, East Lansing, 1997; 109.
14. Smucker, A.J.M.; Santos, D.; Kavdir, Y.; Paul, E.A.; Snider, R. *Concentric Gradients within Stable Soil Aggregates*, Proceedings of the 16th World Congress of Soil Science, France, August 1998.
15. Tisdall, J.M.; Oades, J.M. Organic Matter and Water-Stable Aggregates in Soils. J. Soil Sci. **1982**, *33*, 141–163.
16. Van Gestle, M.; Ladd, J.N.; Amato, M. Carbon and Nitrogen Mineralization from Two Soils of Contrasting Texture and Microaggregate Stability: Influence of Sequential Fumigation, Drying and Storage. Soil Biol. Biochem. **1991**, *23*, 313–322.

SUBIRRIGATION

Norman R. Fausey
United States Department of Agriculture (USDA), Columbus, Ohio, U.S.A.

INTRODUCTION

Water management has quite a broad scope, including all practices that influence any component of the hydrologic cycle. Within this broad scope are many practices, including things such as cloud seeding to increase precipitation, reservoir management to minimize flood events and to store water for municipal use, use of plastic mulches to reduce evaporation, and the use of infiltration basins to enhance recharge of groundwater aquifers. Municipal water supply; public safety issues related to flood forecasting, minimization, and urban storm water management; recreational needs; and agricultural production are examples of why water may need to be managed. Agricultural water management practices generally fall into one of five primary categories: irrigation, drainage, soil erosion control, water supply for animal needs, and waste water disposal. Irrigation involves adding water to assure an adequate supply for crop needs. Irrigation water may be applied on the soil surface or below the soil surface by various methods. One of the methods used to apply water below the soil surface is subirrigation.

DEFINITION AND DESCRIPTION OF SUBIRRIGATION

Subirrigation is the practice of adding water to the soil by means of subsurface drains that are also used to drain water from the soil during periods when the soil is too wet. The drains may be open drains (ditches) or closed drains (drainpipes). Water may be supplied from a surface or a subsurface source, and is delivered into the subsurface drains and allowed to redistribute within the soil from these subsurface drains. Control structures within the ditches or at the outlet of the closed drains are used to block the water from leaving through the outlet, and, thereby, to establish a pressure gradient to cause water to flow from the drains into the soil.

Subirrigation is only applicable to areas needing subsurface drainage and having an available adequate water supply. These areas typically have high water tables during some times of the year that can be lowered by subsurface drainage. Subirrigation depends upon being able to reestablish an elevated water table; this requires a substantial amount of water not only to raise the water table but also to meet the evapotranspiration demand of the crop in order to maintain the water table at a raised position within the soil.

THE PAST

Agricultural water management using subirrigation is not new. It seems reasonable to presume that once the idea to construct drains to remove excess water from the soil proved successful, the idea of putting water back into the soil through the drains could not have been too far behind. Providing an adequate source of water and a means to move the water against the gravitational gradient would certainly have limited the feasibility prior to the advent of efficient pumping systems. Although not well documented, certainly there were efforts and progress toward applying subirrigation as suggested by an anonymous (1) quote found in an early publication on drainage, stating: "I want the drains to irrigate with as much as to drain." An extension bulletin providing guidance for subirrigation in Florida was published in 1938 (2), indicating a recurring early demand for this information in this region. Renfro (3) summarized and discussed the use of subirrigation in the United States up to the mid-1950s. Most of the early applications were on very permeable organic or sandy soils, using open ditches, for high value crops (vegetables and citrus), and in areas with a readily available water supply including: the Sacramento–San Joaquin Delta in central California, the Everglades of southern Florida, the San Luis Valley in Colorado, the Flatwoods of the Florida coastal plain, the Cache Valley in northern Utah, the Egin Bench in southern Idaho, and the Great Lakes states.

THE PRESENT

During the past 30 to 40 years, agriculture has evolved rapidly. World population pressures, scientific advances,

Encyclopedia of Soil Science

and changing economic and social values have accelerated a shift in agriculture; diversity has given way to specialization. There are fewer farms and fewer farmers; production per unit of land is greater and is increasing. Risk reduction is a strong driving force in management decisions. Water management has become an important tool for reducing the risk of too much and too little water. This environment has led to a greater awareness of water management options and impacts, and, as a result, subirrigation has become a topic of increased interest for farmers and researchers.

In 1991, an international conference on subirrigation and controlled drainage water management was held in East Lansing, Michigan, U.S.A. The book (4) that resulted from this conference gives an excellent overview of the present status of subirrigation around the world. Reports of studies and experience from Canada, China, England, Finland, Italy, the Netherlands, and various locations within the United States are included and illustrate a high level of interest in subirrigation.

Subirrigation, when properly managed in concert with subsurface drainage, can produce consistently high yields every year regardless of the weather conditions during the growing season. The subsurface drainage function of the system is used to remove excess soil water to assure trafficability for early planting, lengthening the growing season, and to avoid flooding and lack of oxygen in the root zone, which causes root damage and stunting or death of the plants. The subirrigation function of the system is used to avoid deficit water conditions in the root zone, causing stunting and premature senescence of plants. An assured adequate supply of water also allows planning and management for high yields that would not occur when relying on natural rainfall. High yield management involves higher plant populations and more fertilizer application in order to take full advantage of the available water. With subirrigation, yield goals can be raised and still be reached consistently.

Subirrigation water management is also beneficial to the environment. An adequate supply of water to meet crop needs encourages maximum growth and, therefore, efficient use of applied nutrients. Nutrients are taken up rather than being left in the soil where they would be subject to being transported to surface water and ground water as non-point-source pollutants. Subirrigation promotes greater production and increases the amount of organic residue returned to the soil. Sustained soil quality and soil health depend upon maintaining or increasing the amount of organic matter in the soil. Present environmental goals also promote the capture and sequestration of carbon in the soil rather than the release of carbon dioxide into the atmosphere. Increasing the soil organic matter content stores more carbon in the soil.

THE FUTURE

Economic and social pressures will likely continue to have a significant impact on agriculture and, consequently, on agricultural water management practices. As society becomes more environmentally aware and sensitive overall, an ethic will emerge for the elimination, or certainly the reduction, of delivery of non-point-source contaminants from agriculture to surface and ground waters. Some of this has already begun and has resulted in the promotion of uncultivated vegetated corridors along streams designed to slow and filter runoff waters moving to streams. Cost share programs are available in some states to offset the annual loss of production for farmers willing to establish permanent vegetation in these corridors or to assist with installation of structures to control subsurface drainage discharge in order to reduce nitrate delivery to streams.

Realistically, it is difficult to control water quality during storm events. Subsurface drainage increases infiltration and decreases surface runoff. The use of subirrigation during the summer months would result in less available storage in the soil for rainwater and, therefore, increased runoff, unless the drains are opened quickly and the water table is allowed to fall rapidly ahead of the infiltrating water. Anything that increases runoff encourages the transport of sediments and other pollutants with the runoff water. Water that can be retained in the soil or on the land during storm events will not contribute to runoff and pollutant transport.

Historically the landscape included more wetland areas where runoff waters were retained or slowed and filtered before reaching streams. Agricultural and cultural development resulted in the loss of many of these wetlands that were also barriers to transportation or breeding grounds for diseases. Removal of the wetlands caused rapid delivery of runoff water to streams resulting in increased flooding and increased transport of sediments and other pollutants to the streams. One way to improve water quality would be to reestablish more wetland areas back into the landscape. Surface runoff and subsurface drainage waters could be directed to these wetlands for treatment and volume reduction before being discharged to streams. Some or all of the water leaving a wetland could be captured/harvested and stored on site in lieu of continuing offsite and downstream. The stored water could meet irrigation or other water supply needs.

Water supply is a critical need for subirrigation systems to be feasible and economical. The concept of capture of water during periods when excess water is available and its reuse to meet crop needs during deficit water periods offers an opportunity to make subirrigation affordable and practical in many areas. A holistic approach involving the integration of constructed wetlands, water storage facilities, and subirrigation of crops offers a unique opportunity to realize consistent and high crop yields. Such a system will generate more wetland habitat and protect water quality—two of society's current high priority goals. Therefore, the future for subirrigation is bright, considering the value to many segments of society to develop and use such integrated systems.

Subirrigation as part of an overall water management plan that protects water quality, increases wetland habitat, and stabilizes crop yields will become a water management best management practice.

REFERENCES

1. Anonymous; Tile to Irrigate. The Drainage Journal **1890**, *12* (10), 283.
2. Spencer, A.P. Subirrigation. Florida Agricultural Extension Bulletin 99, 1938.
3. Renfro, G., Jr. Applying Water Under the Surface of the Ground. *Water: The 1955 Yearbook of Agriculture*; U.S. Government Printing Office: Washington, DC, 1955; 273–278.
4. Belcher, H.W., D'Itri, F.M., Eds. Subirrigation and Controlled Drainage. Lewis Publishers: Boca Raton, FL, 1995.

SULFATE AND SULFIDE MINERALS

S

Delvin S. Fanning
Steven N. Burch
University of Maryland, College Park, Maryland, U.S.A.

INTRODUCTION

Sulfide and sulfate minerals (1) are rare to nonexistent in most soils of humid regions, but sulfate minerals commonly occur in some soils of arid regions. Where present, these minerals have strong effects upon soil chemical properties. It is important to know about the conditions under which they form in soils and effects they may have on soil properties. Sulfide minerals form in certain anaerobic soils, where there is a source of S as dissolved sulfate in the soil water, e.g., from sea water in tidal marsh soils classified as *Sulfaquents* and *Sulfihemists* by *Soil Taxonomy* (2–4).

Sulfate minerals, in contrast to sulfides, occur in soils that are more aerobic in nature. They may form from sulfides as part of the overall process of sulfuricization (1, 4). Generally, they are not present in soils of humid regions because most sulfate minerals readily dissolve and leach from such soils. However, they are present even in humid regions in most active acid sulfate soils, classified as *Sulfaquepts and Sulfudepts* by *Soil Taxonomy* (2, 3) and in some postactive acid sulfate soils. They are more common in soils of subhumid and arid regions, from which gypsum and other relatively soluble sulfates are not readily leached because of limited water supply.

Awareness of sulfur-bearing minerals in soils is important because the oxidation of sulfide minerals and the hydrolysis of certain sulfate minerals form sulfuric acid. When this acid is not sufficiently neutralized by carbonate or other minerals, the soils may become active acid sulfate soils with pH <3.5, which can be extremely detrimental to plant growth. The water that emanates from such soils has the properties of acid mine drainage (5) and can cause fish kills and other severe environmental problems. For more information about acid sulfate soils and associated problems, see other sections of this encyclopedia.

SULFIDES

The S in sulfide minerals occurs as the sulfide anion, S^{2-}, as the disulfide anion, S_2^{2-}, or less commonly as a polysulfide anion (1). The sulfide anion is large compared to the S^{6+} cation of the sulfate anion, to which sulfide can be converted by oxidation (Fig. 1). Iron sulfide minerals, highly insoluble under strong anaerobic conditions, are the only sulfides known to form and occur to any significant extent in soils. The sulfides of other heavy metals are also insoluble under strongly anaerobic conditions and may occur in very small quantities in some soils.

The iron disulfide pyrite, FeS_2, which crystallizes in the cubic crystal system, is the main sulfide mineral that forms and occurs in soils (1). Pyrite can occur in soils as a result of sulfidization in tidal environments (4), where it commonly occurs as framboids (see Fig. 7.2 of Ref. 1). Alternatively, it can occur in macroscopic crystals in soils in mine spoils from the mining of coal, lignite, and ore deposits. In such cases, it may also be present as marcasite, the less stable orthorhombic polymorph of pyrite. The Eh(pe)/pH conditions under which pyrite—the main iron sulfide mineral that occurs in soils, is stable relative to certain iron sulfates and geothite, the most stable FeOOH mineral in soils—is shown in Fig. 2.

Monosulfide minerals such as mackinawite, FeS, and polysulfide minerals such as griegite, Fe_3S_4, occur in small quantities in some soils undergoing sulfidization (1, 4), particularly where the levels of Fe in the soil materials occur in excess relative to the amount of S (6). In spite of their presence in low quantities, these minerals have a strong black pigmenting effect upon soil materials in which they occur, because of their occurrence in dispersed extremely fine particles. Their presence can be detected by the evolution of foul smelling hydrogen sulfide gas from the materials upon the addition of (dilute) hydrochloric acid, whereas no such evolution occurs with the disulfide minerals.

SULFATES

Many sulfate minerals are very soluble and behave as salts. However, others such as jarosite and barite are essentially insoluble and can remain in strongly leached soils for many millennia. Iron and aluminum sulfate salts form sulfuric acid upon hydrolysis to form minerals such as goethite, schwertmannite, and jarosite (1). For a current

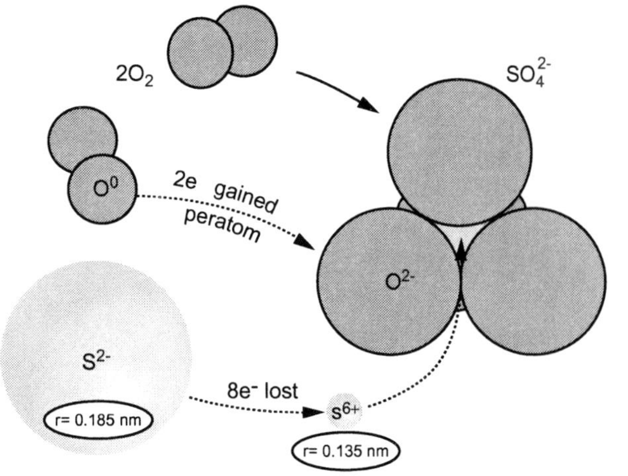

Fig. 1 Diagram illustrating idealized oxidation of sulfide anion to produce sulfate anion, where the sizes of the atoms and ions are shown in proportion to their expected sizes in nature. (Adapted from Ref. 1, Fig. 7.1, with the permission of the Soil Science Society of America.)

extensive review of sulfate minerals that occur in soils and related environments, see Ref. (7).

Gypsum

Gypsum, $CaSO_4{\cdot}4H_2O$, is essentially the only calcium sulfate mineral that occurs extensively in soils. In arid regions, the quantities in some soil horizons may be sufficient for the recognition of the *gypsic* or *petrogypsic* diagnostic horizons and of the *Gypsids* suborder of *Aridisols* of *Soil Taxonomy* (2, 3).

Gypsum forms in acid sulfate and other soils upon the reaction of sulfuric acid with $CaCO_3$ and is also a byproduct of industrial neutralization of waste sulfuric acid with $CaCO_3$ (1). It is also readily dissolved and reprecipitated, which enables it to move in response to wetting and drying of soils. Sometimes, gypsum is deliberately added to soils, as a calcium source, e.g., to enhance the growth of certain plants (peanuts) or for the reclamation of soils that contain high levels of exchangeable Na.

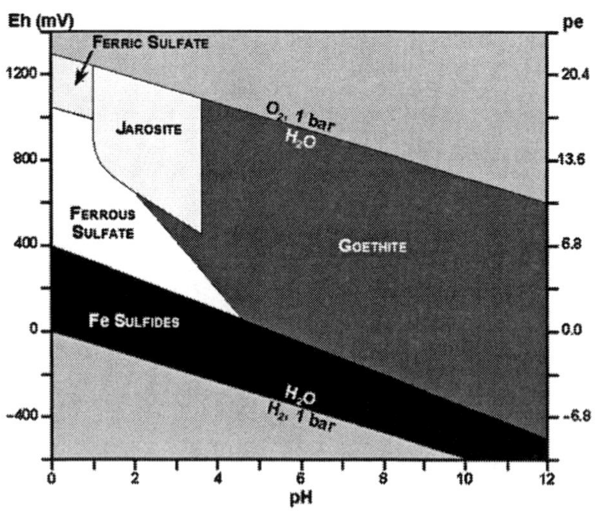

Fig. 2 Idealized Eh(pe)/pH diagram for the Fe–S–O system showing mineral phases that might be expected to be stable under various conditions and likely colors that would be expected to be associated with the minerals. (Adapted from Ref. 1, Fig. 7.3, with the permission of the Soil Science Society of America.)

Jarosite

Jarosite, $KFe_3(SO_4)_2(OH)_6$, and natrojarosite, $NaFe_3(SO_4)_2(OH)_6$, form in certain active acid sulfate soils. They are distinctive minerals in soils because of their pale yellow color. Jarosites show the present or past occurrence of acid sulfate soil conditions and, in combination with a soil pH measured in water of 3.5 or less, their presence can enable the identification of a *sulfuric horizon* as defined by *Soil Taxonomy* (2, 3). Jarosite forms in soils under acid sulfate conditions, when the soil pH is sufficiently low and the Eh/pe is sufficiently high (Fig. 2).

Jarosite in soils is highly insoluble as long as the Eh/pe remains sufficiently high for the ferric, Fe^{3+}, form of Fe to be stable. Because of its durability in soils, it may enable the dating of old postactive acid sulfate soils by K–Ar (potassium/argon) or $^{40}Ar/^{39}Ar$ (argon/argon) methods (1), although soil scientists have yet to employ this technique. Chemical reducing conditions can convert the ferric iron of jarosite to the ferrous form and make the mineral soluble, like with iron oxide and oxyhydroxide minerals.

Relatively Soluble Iron Sulfates

Ferrous sulfate minerals, e.g., rozenite, $FeSO_4{\cdot}4H_2O$ (usually white in color), or melanterite, $FeSO_4{\cdot}7H_2O$ (usually aqua blue in color), or other ferrous sulfate minerals depending on humidity conditions (1), can form in/on soils when ferrous sulfate-rich solutions (see ferrous sulfate stability field of Fig. 2) are desiccated. These minerals, or the dissolved ions of them, oxidize and hydrolyze readily to form sulfuric acid and iron (hydr)oxide minerals and are big contributors to the formation of acid drainage waters in some environments (1). Relatively soluble ferric sulfate minerals such as copiapite, $Fe_{14}O_3(SO_4)_{18}{\cdot}63H_2O$, and coquimbite, $Fe_2(SO_4)_3{\cdot}9H_2O$, form and behave in a similar way, although the Fe is already largely oxidized in these minerals (1). Halotrichite, a ferrous aluminum sulfate mineral, $FeAl_2(SO_4)_4 \cdot 22H_2O$, can also behave in a similar manner.

Sodium and Magnesium Sulfates

A number of sodium, e.g., mirabilite, $Na_2SO_4 \cdot 10H_2O$, and magnesium, e.g., hexahydrite, $MgSO_4 \cdot 6H_2O$, and mixed Na and Mg, e.g., bloedite, $Na_2Mg(SO_4)_2 \cdot 4H_2O$, minerals occur on certain salinized soils such as in North Dakota and Saskatchewan. It is suspected that the S for the formation of these minerals has been released from sulfide minerals deeper in the overall soil-geologic columns upon which these minerals occur (1).

Barite

Barite, $BaSO_4$, occurs in a few soils (1). The origin of this extremely white and very insoluble mineral in soils is poorly understood, although it occurs in some soils that have experienced sulfuricization (1). Color pictures and electron micrographs showing the appearance of this mineral in soils are given by Carson et al. (8).

REFERENCES

1. Fanning, D.S.; Rabenhorst, M.C.; Burch, S.N.; Islam, K.R.; Tangren, S.A. Sulfides and Sulfates. In *Soil Mineralogy with Environmental Implications*; Dixon, J.B., Schulze, D.G., Eds.; Soil Sci. Soc. Am. Madison, WI, 2002; 229–260.
2. Soil Survey Staff, *Soil Taxonomy*, 2nd Ed.; U.S. Dept. Agric. Handbook 436; U.S. Govt. Printing Office: Washington, DC, 1999.
3. Soil Survey Staff, *Keys to Soil Taxonomy*, 8th Ed.; U.S. Govt. Printing Office: Washington, DC, 1998.
4. Rabenhorst, M.C.; Fanning, D.S.; Burch, S.N. Acid Sulfate Soils, Formation. In *Encyclopedia of Soil Science*, Lal, R., Ed.; Marcel Dekker: New York, 2002; 1–4.
5. Gagliano, W.B.; Bigham, J.M. Acid Mine Drainage. In *Encyclopedia of Soil Science*, Lal, R., Ed.; Marcel Dekker: New York, 2001.
6. Rabenhorst, M.C.; James, B.R. Iron Sulfidization in Tidal Wetlands. In *Biomineralization Processes of Iron and Manganese*; Skinner, C.W., Fitzpatrick, R.W., Eds.; Catena Verlag: Cremlingen-Destedt, Germany, 1992; 203–217.
7. Alpers, C.N., Jambor, J.L., Nordstrom, D.K., Eds. *Sulfate Minerals: Crystallography, Geochemistry, and Environmental Significance*; Reviews in Mineralogy and Geochemistry, The Mineralogical Society of America: Washington, DC, 2000; Vol. 40.
8. Carson, C.D.; Fanning, D.S.; Dixon, J.B. Alfisols and Ultisols with Acid Sulfate Weathering Features in Texas. In *Acid Sulfate Weathering*; Soil Sci. Soc. Am. Spec. Publ. No. 10, Kittrick, J.A., Fanning, D.S., Hossner, L.R., Eds.; Soil Sci. Soc. Am. Madison, WI, 1982; 127–146.

SULFUR

Silvia Haneklaus
Elke Bloem
Ewald Schnug
Federal Agricultural Research Center, Braunschweig, Germany

INTRODUCTION

Soil fertility is generally defined as the ability of soils to yield crops and is therefore closely intertwined with plant nutrient cycles. The soil sulfur cycle is driven by biological and physico-chemical processes, which affect flora and fauna. The organic matter pool is an important source and sink for sulfur though its contribution to the mineral nutrition of high yielding crops has been overestimated in the past. Under temperate conditions it is rather the spatio-temporal variation of physico-chemical soil properties, which control the plant available sulfate-S content in the soil via the access of plant roots to sulfur rich groundwater or capillary ascending porous water. Therefore soil methods determining plant available sulfur status will always deliver only information for that instant ignoring relationships with the plant sulfur status or crop yield. A more promising way for the evaluation of the sulfur supply is a site-specific sulfur budget, which includes information about geomorphology, texture, climatic data and crop type and characteristics of the local soil water regime. The major soil properties and external sulfur sources affecting the amount of plant available sulfate in the soil are shown in Fig. 1.

For a comprehensive understanding of the relationship between sulfur and soil fertility it is necessary to identify the main factors and processes controlling the plant available sulfur pool and to appraise the ecotoxicological significance of sulfur within the content of a limiting factor rather than a pollutant. Detailed reviews about the role of sulfur in agro-ecosystems already exist (1–9) and will provide further information, not explicitly mentioned here.

SULFUR IN SOILS

The concentration of sulfur (S) in parent materials ranges from 0.026% to 1% S with igneous rocks $\leq$ metamorphic rocks $\leq$ magmatic rocks of upper continental crust $\ll$ limestones $<$ sedimantary rocks (sulphides) $<$ shales $<$ sedimentary rocks (sulfates) $\ll$ coal (10–12). The typical range of S in agricultural soils of humid and semi-humid regions is 100 to 500 $\mu g\,g^{-1}$, or 0.01 to 0.05 % S. This equals 224 to 1120 $kg\,ha^{-1}$ S in the A_h-horizon (8). The total S content of soils may be as low as 20 $\mu g\,g^{-1}$ (0.002%) in highly leached and weathered soils of humid regions or as high as 35,000 $\mu g\,g^{-1}$ (3.5%) in marine marsh soils and up to 50,000 $\mu g\,g^{-1}$ (5%) in calcareous and saline soils of arid and semiarid regions (8). Tropical soils generally contain low amounts of S due to their low organic matter content.

BIOLOGICAL ASPECTS OF SULFUR IN SOILS

Most of the S in terrestrial soils is bound in the organic fraction, which amounts normally to more than 95% of the total S content (2, 8). Organic S in soils is a heterogeneous mixture of soil organisms, partly decomposed plant material, animal and microbial residues. Little is, however, known about the composition of individual chemical compounds. Many different approaches have been developed to separate soil organic S into major fractions. The following approaches to identify distinct forms and properties of soil organic matter were made: chemical extraction followed by physical-chemical separation into humic acids, fulvic acids and humins reactivity with reducing agents in order to separate carbon-bonded S (C–S) and sulfate esters (C–O–S); physical separation into organo-mineral size fractions and molecular weight fractionation (13).

Sulfate-S is rapidly bound in the form of sulfate-esters, which is regarded as a short term source for S (14). The incorporation of S into high-molecular weight fractions such as humic acids prevents the rapid mineralization of S. In litter, S is predominately carbon-bonded S (C–S) to the humin fraction to a level of about 89%. The incorporation of organic S into complex organic substances followed by association with clay minerals can lead to the relative isolation of S from decayed soil micro-organisms. This process is considered as a physical protection of

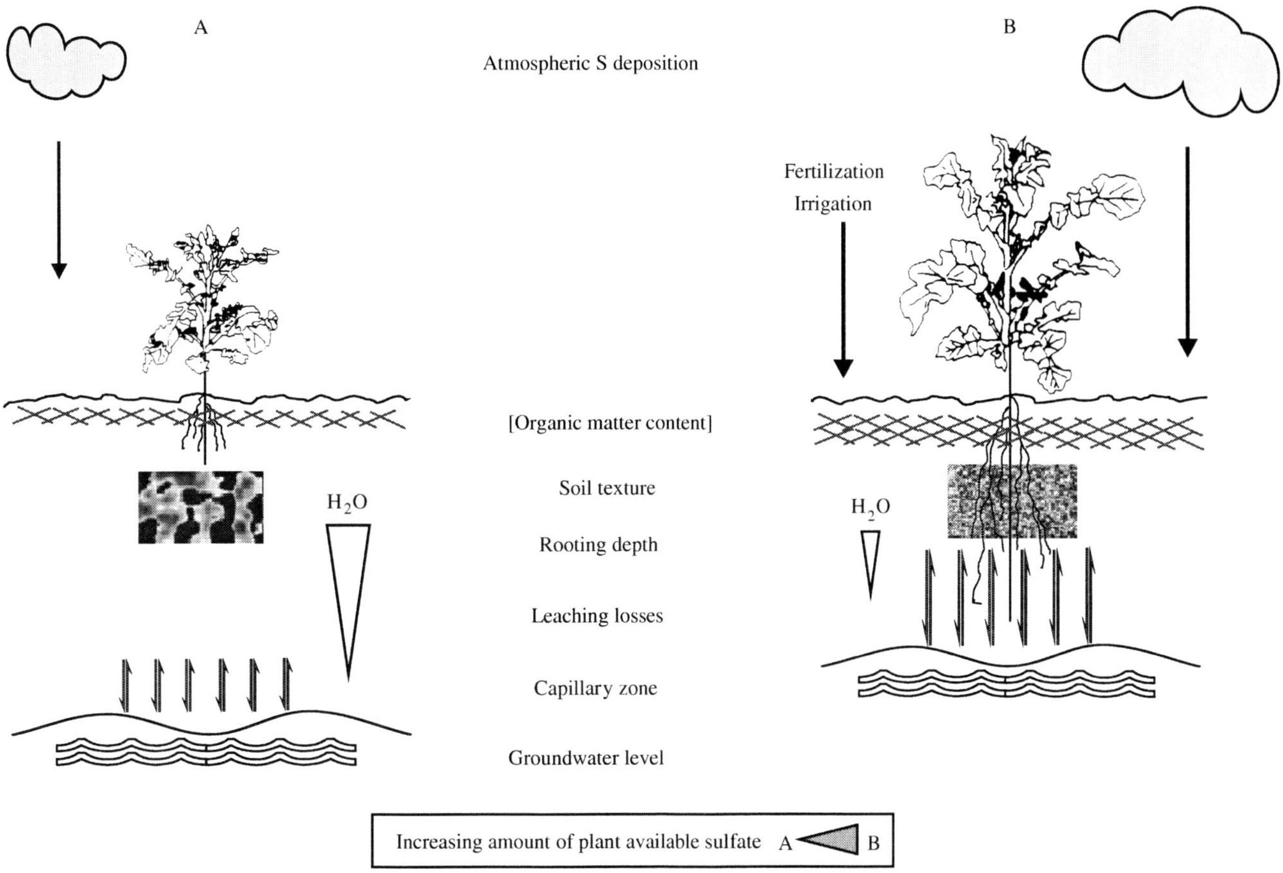

Fig. 1 Overview and efficacy of essential impact factors modifying plant available sulfate in soils.

organic S against degradation (15), and it results in a decreased availability of S to plants.

Sulfur Mineralization

In soils microbial mediated processes are mainly responsible for S transformations, so that the factors affecting the microbial activity, such as temperature, moisture, pH and substrate availability will also influence the process of mineralization, immobilization, oxidation and reduction. In aerobic agricultural soils the main factor in these processes is the release of inorganic, plant-available sulfate from organic matter.

Two types of processes are involved in the mineralization of S: biological and the biochemical mineralization (16). The biological mineralization is considered to be driven by the microbial need for organic C to provide energy, and S released as sulfate is a by-product of the oxidation of C to carbon-dioxide. This process is faster the more recently the organic matter was formed (17). Biochemical mineralization relies on the release of sulfate from the sulfate-ester pool through enzymatic hydrolysis.

Sulfur in the microbial biomass is actively turned over in soils whilst comprising only approximately 2–3% of the total soil S. The turnover of soil microbial bio-mass is fundamental to the incorporation of sulfate-S into soil organic matter but quantitative measures to assess this are unavailable.

The contribution of mineralization to the S supply of plants is only small (2), because mineralization, immobilization and possible leaching of S occur concurrently (14). The amount of S mineralized within the organic S pool in the soil ranges from 0.5 to 3.1% annually (2). The contribution of net mineralization accounts on average for 10–30 $kg\,ha^{-1}\,yr^{-1}$ S (18) in soils with carbon contents between 1 and 4% C. The studies of Eriksen et al. (2) and Bloem (18) reveal that mineralization is an important, however not cardinal S pool for plants. High yielding crops can not satisfy their S demand solely by mineralization and atmospheric S depositions (7).

Crops and Crop Rotation

Dead plant material contributes to the organic matter pool but the living plant organic matter which partly remains on the field may be regarded as a storage pool for S, too. Sulfur offtake and S demand differ depending on the plant species. Sugar cane has a very low S demand with only $0.3\,kg\,S\,Mg^{-1}$ dry matter yield (19), cereals about $1.5-2\,kg\,S\,Mg^{-1}$ of grain and $1-3\,kg\,S\,Mg^{-1}$ of straw yield (18), soybean between 4.3 and $8.8\,kg\,S\,Mg^{-1}$ (20, 21), mustard about $16\,kg\,S\,Mg^{-1}$ (22, 23) and oilseed rape up to $20\,kg\,S\,Mg^{-1}$ (24).

The S content of straw highly influences the mineralization of plant residues. Crops which leave residues with high S contents, may be considered as catch crops for S and are able to reduce S losses from soil by leaching. Cruciferous crops such as oilseed rape show a high S content of the residual straw (24, 25), which provide a large amount of rapidly decomposable organic material (26). In comparison, plants without a secondary metabolism such as cereals leave smaller amounts of S. Only if the S content of for example barley straw is above 0.13% S, can a release of S from the mineralization of straw be expected (27). If the value is below 0.13% S, a net immobilization of S takes place because the microbes need more S for the formation of their microbial biomass. This may exacerbate S deficiency of the following crop, for example oilseed rape, particularly on light soils, where macroscopic symptoms may become visible even before winter (25) because of the net immobilization of S. So far there is no evidence that winter crops benefit during their main growth from the previous crop's is plant organic matter residues with high S.

PHYSICO-CHEMICAL ASPECTS OF SULFUR IN SOILS

Sulfate has a high mobility in soils and can be delivered from subsoil or shallow groundwater. The water soluble fraction can be leached, adsorbed, immobilized or taken up by the plant. Availability of sulfate is therefore more a question of transfer between pools in terms of space and time rather than between biological or chemical systems. The insoluble, organic fraction is only slightly leached and is not directly available for plant uptake.

Soil Water Regime

Under temperate conditions groundwater is often an important S pool and there are three ways groundwater can contribute to the S nutrition of plants. Firstly, there is a direct S input if the groundwater level is only 1–2 m below the surface. This is sufficient to cover the S requirement of most crops, as plants can utilize the sulfate in the groundwater directly by their root system. An average yielding ($3\,Mg\,ha^{-1}$ seeds) oilseed rape crop covers more than 50% of its S demand by shallow groundwater or soil water (24). Secondly, groundwater contains between 5 and $100\,mg\,L^{-1}$ sulfate-S which, if used for irrigation can supply up to $100\,kg\,ha^{-1}$ S to the crop (18). Thirdly, the capillary rise of groundwater under conditions of a soil moisture deficit in the upper soil layers leads to a S input. The contribution of this process to the S supply of plants depends on the soil texture within the soil profile and climatic conditions. Generally heavier soils show less frequent S deficiency than lighter ones as they retain more S-rich pore water. This is proven by a significant relationship between groundwater level and clay content, and the sulfate-S content of soils (18).

Capillary Rise/Leaching

While rainfall water contains only about $2.5\,mg\,S\,L^{-1}$ (28), the S content of adsorbed soil water varies between 15 and $100\,mg\,S\,L^{-1}$ (18). Variation of the sulfate content of water in soil pores is caused by temporal changes of the groundwater level, soil depth and differences in soil texture (18). Sulfur-rich water in the soil pore ascends in the soil profile if the evapo-transpiration rate is higher than the amount of precipitation.

A higher groundwater level below the surface and a high soil moisture deficit at the lower periphery of the root zone will increase the extent of the capillary rise. These authors (29) calculated an average capillary rise of $0.3\,mm\,d^{-1}$ on a sandy soil, $3\,mm\,d^{-1}$ on a loamy sand and $5\,mm\,d^{-1}$ on a loamy silt with a groundwater level of 1.5 m below the surface. This means that under extreme conditions on sandy soils with a groundwater level >2 m below the surface virtually no sulfate will be available to the plant due to capillary rise, but on loamy and silty soils, the S supply of crops may be fully satisfied by capillary water, even if the sulfate concentration of the groundwater is low ($10\,mg\,L^{-1}$ S) (18). Spring crops particularly benefit from capillary ascending soil water as their main growth takes place when capillary rise dominates soil water movements.

The rate of leaching of S under temperate conditions depends on soil type, winter precipitation and the concentration of sulfate S in the soil water during the period of leaching (18). Sulfur losses through leaching are higher over winter due to higher precipitation rates and the lack of plant uptake. As a result, S stored in soils decreases with increasing precipitation and on sandy, free-draining soils all sulfate beneath the capillary zone and rooting depth of the crop may be leached before the beginning of the vegetative period (18) because of the rapid movement of the leachate through the soil and the low adsorption capacities of such soils. Average sulfate leaching depths are 15 cm with 50 mm of precipitation on a loamy soil and 25 cm on a sandy soil, respectively (30). Additionally leaching is higher on fallow than on cropped soils (31) and is enhanced by S fertilization. Average leaching losses under temperate conditions are in the range of 30–80 $kg\,ha^{-1}\,yr^{-1}$ S (32, 33). In addition to vertical water movements in the soil, lateral fluxes can occur particularly in the landscapes with pronounced differences in geomorphology. These fluxes are the major reason for the high spatio-temporal variability of sulfate in soils.

Soil Compaction

Soil compaction and tillage operations causing soil compaction will also decrease the amount of plant available S because of the reduced soil volume and consequent impact on soil pore space. Consequently rooting depth and density are decreased so that S rich capillary water can not be used (34, 35). This effect of soil compaction is supported by the fact that S deficiency regularly becomes visible first in the headland and along tramlines of fields (7) and because it is sulfate-S from the sub-soil that mainly contributes to the S nutrition of crops (18).

SULFUR EMISSIONS

Sulfur containing atmospheric constituents are principally SO_2, SO_4^{2-}, SO_3^{2-}, H_2S, COS, DMS, CS_2 and methylmercaptan. Natural S emissions from biogenic sources and from volcanoes add around 60 $Tg\,S\,yr^{-1}$ to the Earth's atmosphere with an upward tendency due to the increasing consumption of fossil fuels in South America, Africa and Asia (9).

Before the industrial revolution, atmospheric S depositions were, on an average, below 10 $kg\,S\,ha^{-1}\,yr^{-1}$. As a consequence of increased burning of fossil fuels with industrial development from 1890 to 1980 atmospheric S depositions, mainly as SO_2, increased steadily by 0.47 $kg\,S\,ha^{-1}\,yr^{-1}$ (1). In some rural areas of northern Scotland, however, SO_2 depositions were zero or below 2 $kg\,S\,ha^{-1}$ in the 1970s, while the S input in the midlands of England was about 160 $kg\,S\,ha^{-1}$ at the same time (36). At its peak the negative impacts of SO_2 emissions to humans, plants, soils and buildings were so serious that S became an unpopular nutrient and was called as the "yellow poison" (37). The political consequences resulted in clean air acts in European countries and North America and the desulfurization of emissions led to a drastic decrease of atmospheric S depositions.

Agriculture adapted and co-evolved to increasing S loads. Soil acidification caused by high S deposition required higher amounts of liming materials. Increasing S demands of agricultural crops due to higher yield potential as a result of plant breeding progress and production technology coincided with increased atmospheric S supply. Additionally the use of S-containing fertilizers went down nearly to zero (25). Consequently the reduction of atmospheric S deposition had a major effect on the productivity of crops. During the 1990s, macroscopic S deficiency became the most widespread nutrient disorder in northern Europe (7).

SULFUR FERTILIZERS

The worldwide demand of S fertilizers in 2010 is estimated to be about 11.3 million tons of S (38). Numerous S fertilizers and secondary raw material fertilizer products are available for either soil or foliar application in sulfate or elemental form (39). Arguments for and against the use of individual products depend on local farming conditions. Organic fertilizers such as manure and slurry contain about 1 $kg\,S\,Mg^{-1}$ and 0.5 $kg\,S\,Mg^{-1}$, respectively (40). This equals to 0.07 kg S per kg nitrogen.

AGRO-ECOLOGICAL ASPECTS OF SULFUR

Interaction of Agro-ecosystems with Other Ecosystems

The S input and off-take in agro-ecosystems may have a strong influence on neighboring eco-systems. Sulfur is often in excess in natural systems despite the low atmospheric S input because the turnover of S is much lower in these than in agricultural systems. Acid sub-soils, e.g., under forest vegetation, and peat soils show pH values <5, which are far below the acidity levels of fertile agricultural soils. Large storage capacities for adsorbed sulfate exist on such sites because the pH value of the soil is lower than the zero point of charge (ZPC) resulting in the hydration of metal oxides and thus a positive surface (41). Although adsorbed sulfate plays only a minor role in the direct S nutrition of agricultural crops it may contribute positively to the S balance of the whole surrounding landscape. Natural vegetation, land without plant production and forests show a positive S balance even when atmospheric inputs are low (2). Agricultural crops may benefit from this S pool, if groundwater reservoirs of both eco-systems are connected. This also means, however, that with an increasing share of agricultural farmland in landscapes, the risk of S deficiency overall also increases.

Sustainability

Sustainable agriculture should use soils in such a way that the present and future human needs for food or other agricultural goods are realized and the quality of the environment and the natural resources remain preserved (42). The contamination of groundwater with nitrates is a most serious problem. Nitrogen (N) and S are both involved in protein biosynthesis and a shortage in the S supply of crops also lowers the utilization of applied fertilizer N and thus deteriorates the crop quality. Non-protein N is accumulated in plant parts and besides poor efficiency for N fertilization, S deficiencies may increase the loss of N from agricultural soils through volatilization and leaching (43). On average each kg of S shortage to satisfy the S demand of the crop causes 15 kg of N to be lost to the environment. Such N inputs endanger strongly the stability of natural communities as for example the growth of algae in water bodies (44). Correcting S deficiency by fertilization is environmentally safe as sulfate is, in comparison with N, geogenously abundant.

Global Change

Climate is one of the major factors involved in pedogenesis. Soil formation is directly influenced mainly by temperature and water and indirectly via the climate-depending vegetation (45). Changes in climate may change soil types, increase erosion, affect element cycles and increase the release of greenhouse gases. Increased temperature and humidity accelerate pedogenesis. Global change thus would allow soils in the northern hemisphere to proceed faster through the individual stages of development and/or degradation associated with their individual soil series (46). Higher temperatures accelerate the decomposition of organic materials and thus decrease the organic matter content of soils. At the same time the net mobilization of S might increase while the organic S pool decreases. This effect is expected to be more pronounced in cultivated than in range soils (47). But at the same time under more humid climate conditions organic matter tends to accumulate in soils (47). Therefore an expected adverse effect of global warming on soil organic matter might be at least partially compensated in those areas where this coincides with increasing humidity (48). With increasing carbon accumulation, the other growth limiting elements, such as N, S, and P, may dilute relatively (49), which might result in decreased soil fertility.

Plant Health

Owing to the fact that higher anthropogenic S inputs in the past decades enabled plants to adapt to increasing environmental stress, the decline in the S supply within only one decade might have serious consequences for the stability of recent ecosystems (49). Sulfur metabolism provides several efficient mechanisms by which plants are able to tackle abiotic (e.g., xenobiotics and increasing surface ozone levels) and biotic (e.g., pests and diseases) stress, particularly via the glutathione metabolism which again is closely related to the S supply of the plants (3). Other mechanisms involved in response to plant pathogens include the production of S containing compounds in the secondary metabolism of the agriculturally important *Brassica* species, the release of volatile S compounds, the production of S rich proteins, localized deposition of elemental S and the production of phytochelatines, which detoxify heavy metals by forming complexes (50, 51).

Certain diseases (e.g., light leaf spot in oilseed rape) occur more frequently particularly in areas with low S input in Europe (52) and an improved knowledge of the significance of S metabolites in crop resistance to diseases will be beneficial for the improvement of S fertilizer strategies and could therefore minimize the input of pesticides.

ACKNOWLEDGMENTS

The authors cordially thank Dr. Kerr C. Walker (SAC, Aberdeen) for his linguistic efforts on our paper.

REFERENCES

1. Dämmgen, U.; Walker, K.C.; Grünhage, L.; Jäger, H.-J. The Atmospheric Sulphur Cycle. In *Sulphur in Agroecosystems*; Schnug, E., Ed.; Kluwer Academic Publ.: Dordrecht, 1998; 75–114.

S

2. Eriksen, J.; Murphy, M.D.; Schnug, E. The Soil Sulphur Cycle. In *Sulphur in Agroecosystems*; Schnug, E., Ed.; Kluwer Academic Publ.: Dordrecht, 1998; 39–74.
3. Hell, R.; Rennenberg, H. The Plant Sulphur Cycling. In *Sulphur in Agroecosystems*; Schnug, E., Ed.; Kluwer Academic Publ.: Dordrecht, 1998; 135–174.
4. Howarth, R.W.; Stewart, J.W.B. The Interactions of Sulphur with Other Element Cycles in Ecosystems. In *Sulphur Cycling on the Continents: Wetlands, Terrestrial Ecosystems and Associated Water Bodies, SCOPE 48*; Howarth, R.W., Stewart, J.W.B., Ivanov, M.V., Eds.; John Wiley & Sons: Chichester, 1992; 67–84.
5. Howarth, R.W.; Stewart, J.W.B.; Ivanov, M.V., Eds. *Sulphur Cycling on the Continents: Wetlands, Terrestrial Ecosystems and Associated Water Bodies, SCOPE 48*; John Wiley & Sons: Chichester, 1992; 345.
6. Janzen, H.H.; Ellert, B.H. Sulfur Dynamics in Cultivated, Temperate Agroecosystems. In *Sulfur in the Environment*; Maynard, D.G., Ed.; Marcel Dekker Inc.: New York, 1998; 11–44.
7. Schnug, E.; Haneklaus, S. Diagnosis of Sulphur Nutrition. In *Sulphur in Agroecosystems*; Schnug, E., Ed.; Kluwer Academic Publ.: Dordrecht, 1998; 1–38.
8. Stevenson, F.J. *Cycle of Soil. Carbon, Nitrogen, Phosphorus, Sulfur, Micronutrients*; John Wiley & Sons: New York, 1986; 285–320.
9. Whelpdale, D.M. An Overview of the Atmospheric Sulphur Cycle. In *Sulphur Cycling on the Continents: Wetlands, Terrestrial Ecosystems and Associated Water Bodies, SCOPE 48*; Howarth, R.W., Stewart, J.W.B., Ivanov, M.V., Eds.; John Wiley & Sons: Chichester, 1992; 5–26.
10. Bowen, H.J.M. *Trace Elements in Biochemistry*; Academic Press: London, 1966.
11. Friend, J.P. The Global Sulfur Cycle. In *Chemistry of the Lower Atmosphere*; Rasool, S.I., Ed.; Plenum Press: New York, 1973; 177–201.
12. Wedepohl, K.H. Chemical Fractionation in the Sedimentary Environment. In *Origin and Distribution of the Elements*; Ahrens, L.H., Ed.; Vol. 30 of Earth Sciences, Pergamon Press: London, 1968.
13. Anisimova, M.; Haneklaus, S.; Schnug, E. Significance of Sulfur for Soil Organic Matter. In *Sulfur Nutrition and Sulfur Assimilation in Higher Plants*; Brunold, C., Rennenberg, H., De Kok, L.J., Stulen, I., Davidian, J.-C., Eds.; Paul Haupt Publishers: Berne, 2000; 239–244.
14. Ghani, A.; McLaren, R.G.; Swift, R.S. The Incorporation and Transformations of 35S in Soil: Effect of Soil Conditioning and Glucose or Sulfate Addition. Soil Biol. Biochem. **1993**, *25*, 327–335.
15. Eriksen, J.; Lefroy, R.D.; Blair, G.J. Physical Protection of Soil Organic S Studied by Extraction and Fractionation of Soil Organic Matter. Soil Biol. Biochem. **1995**, *27*, 1011–1016.
16. McGill, W.B.; Cole, C.V. Comparative Aspects of Cycling of Organic C, N, S and P Through Soil Organic Matter. Geoderma **1981**, *26*, 267–286.
17. Ghani, A.; McLaren, R.G.; Swift, R.S. Mobilization of Recently-Formed Soil Organic Sulphur. Soil Biol. Biochem. **1993**, *25*, 1739–1744.
18. Bloem, E. Schwefel- Bilanz von Agrarökosystemen Unter Besonderer Berücksichtigung Hydrologischer und Bodenphysikalischer Standorteigenschaften. Landbauforschung Voelkenrode **1998**, *192*, 1–156.
19. Katyal, J.C.; Sharma, K.L.; Srinivas, K. In *Sulphur in Indian Agriculture*, Proc. TSI/FAI/IFA Symposium on Sulphur in Balanced Fertilisation, KS-2/1-KS-2/12, 1997.
20. Aulakh, M.S.; Pasricha, N.S.; Azad, A.S. Phosphorus-Sulphur Interrelationships for Soybeans on Phosphorus and Sulphur Deficient Soils. Soil Sci. **1990**, *150*, 705–709.
21. Nambiar, K.K.M.; Gosh, A.B. Highlights of Research of a Long-Term Fertilizer Experiment in India (1971–82), Techn. Bull. No.1, Longterm Fertilizer Experiment Project, IARI, 100, 1984.
22. Jain, G.L.; Sahu, M.P.; Somani, L.L. Balanced Fertilization Programme with Special Reference to Secondary and Micronutrients Nutrition of Crops Under Intensive Cropping, Proc. FAI/NR Seminar, Jaipur 147–174, 1984.
23. Aulakh, M.S.; Pastricha, N.S.; Sahota, N.S. Yield, Nutrient Concentration and Quality of Mustard Crops as Influenced by Nitrogen and Sulphur Fertilizers. J. Agric. Sci. **1980**, *94*, 545–549.
24. Schnug, E. *Quantitative und Qualitative Aspekte der Diagnose und Therapie der Schwefelversorgung von Raps (Brassica napus L.) Unter Besonderer Berücksichtigung Glucosinolatarmer Sorten*; DSC, Christian-Albrechts-University: Kiel, Germany, 1988.
25. Schnug, E.; Haneklaus, S. Sulphur Deficiency in Brassica Napus- Biochemistry- Symptomatology- Morphogenesis. Landbauforschung Völkenrode **1994**, *144*, 1–31.
26. Wu, J.A.G.; O'Donnell, Z.L.; Syers, J.K. Microbial Growth and Sulphur Immobilization Following the Incorporation of Plant Residues into Soil. Soil Biol. Biochem. **1993**, *25*, 1567–1573.
27. Chapman, S.J. Barley Straw Decomposition and S Immobilisation. Soil Biol. Biochem. **1997**, *29*, 109–114.
28. Dämmgen, U.; Grünhage, L.; Küsters, A.; Jäger, H.J. Konzentrationen von Luftinhaltsstoffen. I. Criteria Pollutants. Landbauforschung Voelkenrode **1996**, *170*, 196–221.
29. Giesel, W.; Renger, M.; Strebel, O. Berechnung des Kapillaren Aufstiegs aus dem Grundwasser in den Wurzelraum Unter Stationären Bedingungen. Z. Pflanzenern. und Bodenkde **1972**, *132*, 17–30.
30. Kumar, V.; Karwasra, S.P.S.; Singh, M.; Dhankar, J.S. An Evaluation of the Sulphur Status and Crop Responses in the Major Soils of Haryana, India. Sulphur in Agriculture **1994**, *18*, 23–26.
31. Kirchmann, H.; Pichlmayer, F.; Gerzabek, M.H. Sulfur Balance and Sulfur-34 Abundance in a Long-Term Fertilizer Experiment. Soil Sci. Amer. J. **1996**, *59*, 174–178.
32. Mansfeld, T. Schwefeldynamik von Böden des Dithmarscher Speicherkoogs und der Bornhoeveder Seenkette in Schleswig-Holstein. PhD, Kiel, Germany, 1994.
33. Preuschoff, M. Untersuchungen zur Schwefelversorgung von Weißkohl an Zwei Lößstandorten. PhD, University Hanover, Verlag Ulrich E. Grauer, Stuttgart, 1995.
34. Singh, B.R. Effect of Soil Compaction on S Availability to Crop Plants. Abstracts of the COST Action 829: Fundamental, Agronomical and Environmental Aspects of Sulfur Nutrition and Assimilation in Plants, Goslar, Germany, 1998.

35. Unger, P.W.; Kaspar, T.C. Soil Compaction and Root Growth: A Review. Agron. J. **1994**, *86*, 759–766.
36. Semb, A. Sulfur Emissions in Europe. Atmospheric Environm. **1978**, *12*, 455–460.
37. Boelsche, J. *Das gelbe Gift*; Rowohlt Verlag: Reinbek, 1994.
38. Cecotti, S.; Morris, R.J.; Messick, D. A Global Overview of the Sulphr Situation: Industry's Background, Market Trends, and Commercial Aspects of Sulphur Fertilisers. In *Sulphur in Agroecosystems*; Schnug, E., Ed.; Kluwer Academic Publ.: Dordrecht, 1998; 175–202.
39. Paulsen, H.M. Produktionstechnische und Ökologische Bewertung der Landwirtschaftlichen Verwertung von Schwefel aus Industriellen Prozessen. Landbauforschung Voelkenrode **1999**, *197*, 1–143.
40. Pedersen, C.A.; Knudsen, L.; Schnug, E. Sulphur Fertilisation. In *Sulphur in Agroecosystems*; Schnug, E., Ed.; Kluwer Academic Publ.: Dordrecht, 1998; 115–134.
41. Curtin, D.; Syers, J.K. Extractability and Adsorption of Sulfate in Soils. J. Soil Sci. **1990**, *41*, 305–312.
42. Anon. *Nachhaltiges Deutschland: Wege zu Einer Dauerhaft Umweltgerechten Entwicklung*; Umweltbundesamt Berlin, Erich Schmidt Verlag: Berlin, 1997.
43. Schnug, E. Sulphur Nutritional Status of European Crops and Consequences for Agriculture. Sulphur in Agriculture **1991**, *15*, 7–12.
44. Wild, A. *Umweltorientierte Bodenkunde*; Spektrum Akademischer Verlag: Heidelberg, 1993.
45. Hugget, R.J. *Geoecology*; Routledge: London, 1995.
46. Rogasik, J.; Daemmgen, U.; Luettich, M.; Obenauf, S. Wirkungen Physikalischer und Chemischer Klimaparameter auf Bodeneigenschaften und Bodenprozesse. Landbauforschung Voelkenrode **1994**, *148*, 107–139.
47. Burke, I.C.; Yonker, C.M.; Parton, W.J.; Cole, C.V.; Flach, K.; Schimel, D.S. Texture, Climate and Cultivation Effects on Soil Organic Matter Content in U.S. Grassland Soils. Soil Sci. Soc. Am. J. **1989**, *53*, 800–805.
48. Schnug, E. Response of Plant Metabolism to Air Pollution and Global Change Impact on Agriculture. In *Responses of Plant Metabolism to Air Pollution and Global Change*; DeKok, L.J., Stulen, I., Eds.; Backhys Publ.: Leiden, 1998; 15–22.
49. Schnug, E.; Haneklaus, S. Ecological Aspects of Plant Sulphur Supply. Proc. 15th Int. Congr. Soil Sci. Acapulco/Mexico, 1994; 5a: Comm.IV: Symposia, 364–371, 1994.
50. Resende, M.L.V.; Flood, J.; Ramsden, J.D.; Rowan, M.G.; Beale, M.H.; Cooper, R.M. Novel Phytoalexins Including Elemental Sulphur in the Resistance of Cocoa (*Theobroma cacao* L.) to Verticillium Wilt (*Verticillium dahliae Kleb.*). Physiol. Molec. Plant Pathol. **1996**, *48*, 347–359.
51. Schnug, E. Significance of Sulphur for the Nutritional and Technological Quality of Domesticated Plants. In *Sulfur Nutrition and Sulfur Assimilation in Higher Plants*; Cram, W.J., DeKok, L.J., Stulen, I., Brunold, C., Rennberg, H., Eds.; Backhys Publ.: Leiden, 1997; 109–130.
52. Thomas, J. Watch Out for Light Leaf Spot. NIAB Fellows Newsletter 1994.

SURFACE AREA, SPECIFIC

L.A. Graham Aylmore
The University of Western Australia, Western Australia, Australia

INTRODUCTION

Surface area is an extremely important soil property because a solid can establish contact with another solid or a liquid or a gas, only at its surface. The larger the specific surface area (SSA) (the surface area per unit mass of material), the larger is the surface energy per unit mass which is available for both physical and chemical interaction with the surroundings. From the agricultural or environmental point of view, most chemical reactions in soils take place at surfaces within the porous matrix. It is the interaction between the surfaces of the finely divided fraction and the soil water solution, which determines the mechanical or physical properties such as consistency, plasticity, swelling, and shrinking of the soil (i.e., the clay–water interaction). It is on the surfaces of the primary particles that the majority of plant nutrients, heavy metals, pesticides, etc. are held and hence any meaningful study of their retention must, of necessity, be referred back to a unit surface area basis for an understanding of the mechanisms involved.

RELATION WITH PARTICLE SIZE DISTRIBUTION

Surface area depends primarily on the state of subdivision of a particulate material. A system in which the particles are in a fine state of subdivision is termed a "disperse" system. Large surface per unit mass is a characteristic property of all disperse systems and the physical and chemical behavior of such systems depend greatly on the effects wich take place at the interface between two different phases of the system (e.g., between solid and liquid phases). Thus, the clay fraction containing finely divided crystalline and amorphous constituents plays a major role in this regard. For example, 1 g of <2 μm clay possesses approximately 50 times the SSA of the same quantity of very fine sand and 10 times that of the same quantity of silt; colloidal clay (100 mμ) possesses approximately 20 times the SSA of the same quantity of < 2 μm size clay and 1000 times that of the same quantity of very fine sand. It is for this reason that the particles of smallest size in the soil are frequently referred to as the "active fraction".

DETERMINATION OF SPECIFIC SURFACE AREA

The most universally accepted method for determining the specific surface area of finely divided porous materials is that of Brunauer et al. (1), the BET method. The theoretical description of the process of surface adsorption derived by Brunauer et al. can be written as

$$\frac{p}{v(p_0 - p)} = \frac{1}{v_m c} + \frac{c-1}{v_m c}\frac{p}{p_0} \qquad (1)$$

for adsorption on a free surface, where v is the quantity of gas adsorbed at pressure p; v_m is the amount required to form a monolayer and c is a constant related to the difference between the heat of adsorption of the first and subsequent layers. The main simplifying assumptions in the derivation of this equation is that the first layer of molecules is adsorbed at a particular energy and that all subsequent layers are adsorbed at an energy equal to that for the condensation of the pure liquid. This is a linear equation and a plot of $p/v(p_0 - p)$ against p/p_0 gives a straight line if the theory is obeyed. The intercept on the vertical axis is $1/v_m c$ and the slope is

$$(c-1)/v_m c$$

so that the two constants v_m and c can be obtained from experimental data. In practice this equation is found to be fairly closely obeyed by most adsorbents over the range

$$p/p_0 = 0.05\text{–}0.35.$$

By measuring v, the volume of gas adsorbed at several pressures within the range over which the equation is applicable, and knowing p_0, the saturation vapor pressure, we can obtain an estimate of the volume of gas required to form a monolayer over the surface of the soil from the appropriate plot. Knowing the area occupied by one molecule of gas (i.e., molecular area), the surface area of the sorbent material can be accurately estimated.

A number of gases have been utilized in the application of this method to the physical adsorption of gases at low temperature. When vapor adsorption methods were first developed, Emmett and Brunauer (2) advocated the use of nitrogen as adsorbate at 77 K (the boiling point of liquid nitrogen) and this gas has been generally accepted as the most satisfactory. The use of nitrogen as the primary standard has, however, been questioned (3, 4) there being suggestions that the quadrupole moment of nitrogen can interact with hydroxyl or other polar groups, to varying extents, leading to a change in the cross-sectional area of the adsorbed molecule.

Argon adsorption at 77 or 91 K (the boiling point of liquid oxygen) has been preferred by some workers, since this is a symmetrical nonpolar atom which should be less subject to specific interactions affected by changes in the chemical nature of surfaces. However, argon is less strongly adsorbed than nitrogen, raising some doubts as to whether the BET procedure provides a satisfactory estimate of the monolayer capacity ($c < 50$). Also there are strong suggestions that the effective saturation vapor pressure for argon at 77 K varies between that of solid argon above nonporous materials to that of the supercooled liquid within micro- (<20 Å) and transitional (20–200 Å) pores (5, 6).

Carbon dioxide adsorption at 196 K has been extensively used in studies on carbon blacks, charcoals and similar materials (7). The surface areas obtained using carbon dioxide on these materials frequently vastly exceed those obtained by nitrogen adsorption at 77 K, indicating the presence of large volumes of microporous regions inaccessible to the less energetic nitrogen adsorption. Similar effects have been noted with the expanding layer lattice aluminosilicates such as smectite and vermiculite (8).

For the measurement of very small surface areas ($<1\ m^2\ g^{-1}$), krypton sorption at 77 K has the advantage of providing a low saturation vapor pressure at the working temperature and hence a marked reduction in the volume of unadsorbed gas which has to be corrected for in the BET calculation (9). Krypton adsorption also offers the benefit that its inertness makes it unlikely that it will interact specifically with the solid surface. Because krypton and carbon dioxide sorption at 77 and 196 K respectively, like that of argon at 77 K, take place below the bulk triple point, the use of these two gases in SSA measurement is likewise subject to some uncertainty in the correct values for saturation vapor pressure to be used in the BET analysis.

The *t*-Method

Uncertainties in the validity of the BET approach when micro- and transitional pores are present, and in the absolute values of molecular areas and how these will vary with a substrate, led to the development of alternative methods for comparison of gas adsorption isotherms and surface area determination. Among these, the *t*-method of Lippens and De Boer (10) and similar variations (11, 12) have attracted attention as a means of interpreting vapor adsorption and characterizing the porosity of solid adsorbents. These methods involve the use of "reduced isotherms" in which the volume adsorbed V is plotted against the corresponding statistical thickness (t) of the adsorbed layer on a nonporous reference solid (10), or alternatively against some arbitrarily chosen parameter

$$V_s = V/V_x,$$

where V_x is the volume adsorbed at

$$p/p_0 = 0.4$$

(11). Where adsorption results entirely from monolayer–multilayer formation, straight line plots with slope proportional to the surface area of the sample will be obtained (13).

Microporosity and Accessibility of Surfaces

The nitrogen surface areas for the layer lattice aluminosilicate clay materials range from about $10–30\ m^2\ g^{-1}$ for the coarser kaolinite minerals through $50–180\ m^2\ g^{-1}$ for the illitic clays. If only the external surface area of the smectites is accessible to the measuring sorbate, as is generally the case for nonpolar gases, the surface area involved is frequently of the magnitude of $38\ m^2\ g^{-1}$ (e.g., Wyoming bentonite) to $99\ m^2\ g^{-1}$ (Redhill montmorillonite) (14). Even these values are, however, influenced by the nature of the dominant exchangeable cation on the clay surface suggesting differences in lamellar packing arrangement (8). The use of more strongly sorbed polar molecules such as water, ethylene glycol (15), ethylene glycol monoethyl ether (EGME) (16), carbon dioxide (8, 13), and adsorption from solution of cetyl pyridinium bromide (CPB) (17), methylene blue (MB) (18), or *para*-nitrophenol (*p*NP) (19, 20), provides access to the interlayer surfaces of the smectite clays to varying extents giving substantially higher values ($300–800\ m^2\ g^{-1}$). From chemical analysis and crystallographic parameters, the definitive total SSA can be calculated to be $750\ m^2\ g^{-1}$ (21).

The value of the constant c in the BET equation gives a measure of the energy with which the gas molecules are adsorbed by the porous material. For nonpolar gas (i.e., electrically neutral) molecules, the presence of the charged cations on the clay surface will have little, if any, effect on the sorption. On the other hand, conducting a similar experiment with polar molecules such as water, the differing effects of different cations upon the adsorption of water is illustrated by the quite significant differences in sorption energies calculated with different exchangeable cations in the soil. Using this method, Keenan et al. (22) demonstrated on kaolinitic clay the decreasing effect with increasing size of the monovalent alkali metal ions in the order Na>K>Rb>Cs, as would be expected, the smaller ions having a greater affinity for water of hydration. The obvious consequence of the presence of cations on the clay surfaces is dramatic variations in the apparent molecular area covered by one molecule as the sorbate clusters around the charged cations.

Similar difficulties are observed with other materials containing micropores, such as the tubular halloysites (23) and allophanes (SSA ranging from 400 to 900 $m^2 g^{-1}$) (24), where the possibility of enhanced sorption of gases through capillary condensation and the physical configuration and pore shape (25) make it difficult to interpret sorbed volumes in a meaningful way in terms of surface coverage. Measurement of the SSA of soils is frequently further complicated by the presence of organic matter, which can both significantly reduce the value measured by effectively blocking access of the sorbate to micropores (20, 26, 27) or, alternatively, increase it by virtue of its considerable capacity to absorb polar molecules into the structure (28, 29).

Molecular Areas

Values of molecular area have been determined with fair accuracy for a number of gases, e.g., 0.162 nm^2 for N_2, 0.138 nm^2 for Ar and 0.196 nm^2 for Kr at 77 K, 0.144 nm^2 for Ar at 90.1 K, and 0.221 nm^2 for CO_2 at 195 K. As could be expected, given the tendency of polar molecules to cluster around cation sites, the molecular areas estimated for most such sorbates vary substantially. Derived values include: 0.108 nm^2 for water (30); 0.33 nm^2 for ethylene glycol (15); 0.52 nm^2 for EGME (16); 0.54 nm^2 for CPB (using adsorption from solution) (17); 1.3 nm^2 for MB (18); and, 0.424 nm^2 for *p*NP (19, 20). These derived values remain the subject of some controversy (21, 31).

Negative Adsorption (Co-ion exclusion)

In 1947, Schofield (32) described a method for determining the SSA of clays based on the electrostatic repulsion of anions from the negatively charged clay surface. The volume from which anions (co-ions) are excluded from a negatively charged clay surface in a 1:1 electrolyte solution is given by (33)

$$V_{ex} = 2A(1 - e^{\Psi/2})/K \quad (2)$$

in which V_{ex} is the volume of exclusion, $2/K$ is the characteristic length in the Debye–Huckel theory of strong electrolytes and is given by $3.04/C^{1/2}$ expressed in Å, C is the concentration, A is the area of the interface and Ψ is the reduced electrical potential at the Gouy plane associated with the development of the diffuse layer of co-ions. For large electric potentials of the magnitude of −200 mV, the value of $(1 - e^{\Psi/2})$ is 0.98 so the slope of a plot of V_{ex} against $2/K$ would give a good estimate of the surface area. However, potentials of this magnitude are only attained in very dilute solutions. The layer lattice aluminosilicate clay surfaces are constant charge surfaces for which the Gouy plane potential increases with decrease in concentration. For Li-montmorillonite, Quirk and Marcelja (34) found that the surface potential varied from −36 to −225 mV in LiCl solutions ranging from 0.3 to 10^{-4} M. Edwards et al. (35, 36) measured the volume of exclusion for Wyoming bentonite and Fithian illite saturated with the alkali metal cations. The co-ion exclusion plot yielded an SSA of 560 $m^2 g^{-1}$ for Li-montmorillonite and 156 $m^2 g^{-1}$ for Cs-montmorillomite compared with the theoretical hydratable area of some 750 $m^2 g^{-1}$. For the illite saturated with these cations, the exclusion area varied from 80 to approximately 0 $m^2 g^{-1}$ compared with a nitrogen area of 110 $m^2 g^{-1}$. The decreasing slope of the co-ion exclusion plots with increasing radius of the ion is associated with the greater accommodation of ions in the Stern layer, thereby reduc-ing the Gouy plane potential and consequently co-ion exclusion. Since the basic assumption underlying Schofield's equation that the surface charge is large is not met, co-ion exclusion is clearly not appropriate for measuring the surface area of clays although it is still incorrectly cited in the literature.

SUMMARY

The surface area available to any given sorbate is determined essentially by surface reactivity and geometrical considerations. Hence their use should be governed by the intent and purpose of the surface area measurement. Despite the reservations with respect to saturation vapor pressures, application of the BET theory to the adsorption of N_2, Ar and Kr gases at low temperatures nowadays provides the standard procedures for the measurement of the external surface areas of finely divided materials using automated systems. More realistic measurement of the surface areas available in hydrated or solvated systems, where interlayer surfaces are accessible, can be obtained by the use of more strongly adsorbed polar molecules such as water, EG, EGME, or CPB. However, the affinity of charged molecules for cations and uncertainty in their surface orientation and extent of surface coverage severely limits confidence in the accuracy of these methods. For comparative purposes, the use of water vapor sorption at

$$p/p_0 = 0.19$$

on homoionic soils probably provides the most useful method of assessing the relative surface areas of soils.

REFERENCES

1. Brunauer, S.; Emmett, P.H.; Teller, E. Adsorption of Gases in Multimolecular Layers. J. Am. Chem. Soc. **1938**, *60*, 309–310.
2. Emmett, P.H.; Brunauer, S. The Use of Low Temperature van der Waals Adsorption Isotherms in Determining the Surface Area of Iron Synthetic Ammonia Catalysts. J. Am. Chem. Soc. **1937**, *59*, 1553–1564.
3. Aristov, B.C.; Kiselev, A.V. Effect of Dehydration of Silica Surface on the Adsorption Isotherms for Gaseous Nitrogen and Argon. Russian J. Phys. Chem. **1963**, *37*, 1359–1363.
4. Pierce, C.; Ewing, B. Area of Uniform Graphgite Surfaces. J. Phys. Chem. **1964**, *68*, 2562–2568.
5. Harris, M.R.; Sing, K.S.W. Use of Argon Adsorption for the Determination of Specific Surface Area. Chem. Ind. **1967**, 757–758.
6. Carruthers, J.D.; Payne, D.A.; Sing, K.S.W.; Stryker, L.J. Specific and Nonspecific Interactions in the Adsorption of Argon, Nitrogen and Water Vapour on Oxides. J. Coll. Interf. Sci. **1971**, *36*, 205–216.
7. Anderson, R.B.; Bayer, J.; Hofer, L.J.E. Determining Surface Areas from CO2 Isotherms. Fuel **1965**, *44*, 443–452.
8. Aylmore, L.A.G.; Sills, I.D.; Quirk, J.P. Surface Area of Homoionic Illite and Montmorillonite Clay Minerals by the Sorption of Nitrogen and Carbon Dioxide. Clays Clay Miner. **1970**, *18*, 91–96.
9. Beebe, R.A.; Beckwith, John B.; Honig, J.M. The Determination of Small Surface Areas by Krypton Adsorption at Low Temperatures. J. Am. Chem. Soc. **1945**, *67*, 1554–1558.
10. Lippens, B.C.; De Boer, J.H. Studies on Pore Systems in Catalysis.V. The *t* Method. J. Catal. **1965**, *4*, 319–323.
11. Sing, K.S.W. Empirical Method for Analysis of Adsorption Isotherms. Chem. Ind. **1968**, 1520–1521.
12. Pierce, C. The Universal Nitrogen Isotherm. J. Phys. Chem. **1968**, *72*, 3673–3676.
13. Aylmore, L.A.G. Gas Sorption in Clay Mineral Systems. Clays Clay Miner. **1974**, *22*, 175–183.
14. Aylmore, L.A.G.; Quirk, J.P. The Micropore Size Distributions of Clay Mineral Systems. J. Soil Sci. **1967**, *18*, 1–17.
15. Dyal, R.S.; Hendricks, S.B. Total Surface of Clays in Polar Liquids as a characteristic Index. Soil Sci. **1950**, *69*, 421–432.
16. Carter, D.L.; Mortland, M.M.; Kemper, W.D. Specific Surface. In *Methods of Soil Analysis. Part 1. Physical and Mineralogical Methods*, 2nd Ed.; Klute, A., Ed.; Soil Sci. Soc. Am. Madison, WI, 1986.
17. Greenland, D.J.; Quirk, J.P. Determination of Surface Area by Adsorption of Cetyl Pyridinium Bromide from Aqueous Solution. J. Phys. Chem. **1963**, *67*, 2886–2887.
18. Pham, T.H.; Brindley, G.W. Methylene Blue Adsorption By Clay Minerals. Determination of Surface Areas and Cation Exchange Capacities. Clays Clay Miner. **1970**, *18*, 203–212.
19. Ristori, G.G.; Sparvoli, E.; Quirk, J.P.; Martelloni, C. Measurement of Total and External Surface Area of Homoionic Smectites By *p*-Nitrophenol Adsorption. Mineral. Petrogr. Acta **1985**, *29A*, 137–143.
20. Theng, B.K.G.; Ristori, G.G.; Santi, C.A.; Percival, H.J. An Improved Method for Determining the Surface Areas of Topsoils with Varied Organic Matter Content, Texture and Clay Mineral Contents. Eur. J. Soil Sci. **1999**, *50*, 309–316.
21. Quirk, J.P.; Murray, R.S. Appraisal of the Ethylene Glycol Monoethyl Ether Method for Measuring Hydratable Surface Area of Clays and Soils. Soil Sci. Soc. Am. J. **1999**, *63*, 839–849.
22. Keenan, A.G.; Mooney, R.W.; Wood, L.A. The Relation Between Exchangeable Ions and Water Adsorption on Kaolinite. J. Phys. Coll. Chem. **1951**, *55*, 1462–1474.
23. Churchman, G.J.; Davy, T.J.; Aylmore, L.A.G.; Gilkes, R.G.; Self, P.G. Characteristics of Fine Pores in Some Halloysites. Clay Miner. **1995**, *30*, 89–98.
24. Hall, P.L.; Churchman, G.J.; Theng, B.K.G. Size Distribution of Allophane Unit Particles in Aqueous Suspensions. Clay Clay Miner. **1985**, *33*, 345–349.
25. Aylmore, L.A.G. Hysteresis in Gas Sorption Isotherms. J. Coll. Interf. Sci. **1974**, *46*, 410–416.
26. Burford, J.R.; Deshpande, T.L.; Greenland, D.J.; Quirk, J.P. Influence of Organic Materials on the Determination of the Specific Surface Areas of Soils. J. Soil Sci. **1964**, *15*, 192–201.
27. Sequi, P.; Aringhieri, R. Destruction of Organic Matter by Hydrogen Peroxide in the Presence of Pyrophosphate and Its Effect on Soil Specific Surface Area. Soil Sci. Soc. Am. **1977**, *41*, 340–342.
28. Chiou, C.T.; Lee, J.F.; Boyd, S.A. The Surface Area of Soil Organic Matter. Environ. Sci. Tech. **1990**, *24*, 1164–1166.
29. de Jonge, H.; de Jonge, L.W.; Mittelmeijer-Hazeleger, M.C. The Microporous Structure of Organic and Mineral Soil Materials. Soil Sci. **2000**, *165*, 99–108.
30. Quirk, J.P. Significance of Surface Areas Calculated from Water Vapour Isotherms by the Use of the B.E.T. Equation. Soil Sci. **1955**, *161*, 9–21.
31. Tiller, K.G.; Smith, L.H. Limitations of EGME Retention to Estimate the Surface Area of Soils. Aust. J. Soil Res. **1990**, *28*, 1–26.
32. Schofield, R. Kenworthy Calculation of Surface Areas from Measurements of Negative Adsorption. Nature **1947**, *160*, 408–410.
33. Chan, D.Y.C.; Pashley, R.M.; Quirk, J.P. Surface Potentials Derived from Co-Ion Exclusion Measurements on Homoionic Montmorillonite. Clays Clay Miner. **1984**, *32*, 131–138.
34. Quirk, J.P.; Marcelja, S. Application of Double-layer Theories to the Extensive Crystalline Swelling of Li-Montmolillonite. Langmuir **1997**, *13*, 6241–6248.
35. Edwards, D.G.; Posner, A.M.; Quirk, J.P. Repulsion of Chloride Ions by Negatively Charged Clay Surfaces. Part 1. Monovalent Cation Fithian Illites. Trans. Faraday Soc. **1965**, *61*, 2808–2815.
36. Edwards, D.G.; Posner, A.M.; Quirk, J.P. Repulsion of Chloride Ions by Negatively Charged Clay Surfaces. Part 2. Monovalent Montmorillonites. Trans. Faraday Soc. **1965**, *61*, 2816–2819.

SURFACE MANAGEMENT

Kwong Yin Chan
Wagga Wagga Agricultural Institute, Wagga Wagga, New South Wales, Australia

S

INTRODUCTION

Surface soil, the uppermost part of the soil profile ordinarily moved in tillage (i.e., the plow layer, Ap horizon), or its equivalent in uncultivated soils, ranges in depth from 7–25 cm (1). This is the part of the soil profile: 1) that is in intimate contact with the atmosphere, 2) that regulates and partitions water in the environment, 3) where most of plant roots and soil organic carbon are concentrated, and 4) where most of the biological and nutrient-cycling activities take place. All these processes, which are vital to the functioning of the soil, are directly controlled by the structure of the surface soil. At the same time, surface soil is also most affected by human activities associated with agriculture, mining, and urbanization. Clearing land of native vegetation for agricultural development and inappropriate management practices, both of which affect mostly the surface soil structure, are the causes of worldwide land degradation problems, namely, compaction, erosion, desertification, siltation, dust storms, eutrophication, etc. (2, 3).

IDEAL SOIL SURFACE STRUCTURAL CONDITIONS

Functionally, the surface soil should have a range of pore sizes suitable for transmission as well as storage, free drainage, and root proliferation (4). A system of classifying structural quality of surface soil based on these factors has been defined (Table 1) (5). Determined mainly by soil bulk density, this system is influenced by texture only for the extreme classes, namely, clays and loamy sands.

Besides having a suitable porosity (structural form), surface soil structure should also be assessed in terms of stability and resilience (6). Stability is a measure of the ability to resist structural breakdown when subjected to stresses, while resilience is a measure of the ability of the soil to recover its structural form after disturbance. Both stability and resilience are, to a varying extent, dependent on soil organic carbon levels (6).

DIFFERENT MANAGEMENT METHODS

Many management practices in agriculture and other land use affect soil structure, either directly or indirectly, because of their effects on the factors and processes involved in soil structure formation (3, 4, 6). While soil surface structural degradation problems (such as compaction, crusting, surface sealing, and hardsetting) are often caused by poor management practices such as excessive and inappropriate tillage, bare fallowing, and stubble burning/removal, management practices that are effective in improving/ameliorating soil surface structure also exist. These practices are described under the following categories:

Mechanical and Physical Methods

Tillage

Tillage is traditionally the most common practice used to modify soil surface structure. It can have large impacts, beneficial as well as adverse, on surface soil structural condition. Appropriately used, it is useful in providing suitable seedbeds, alleviating physical impedance (e.g., crusts), and managing crop residue.

On the other hand, inappropriate and excessive tillage has been the cause of widespread land degradation problems in many parts of the world (e.g., the dust bowls in the United States during the 1930s) (3, 7).

The worldwide trend is toward reducing the number of tillage operations (reduced tillage) and even eliminating them altogether in cropping (zero tillage), as advocated by the conservation tillage movement (8, 9). Significant reduction in soil erosion, and improvements in water infiltration and surface soil aggregation have been reported under reduced/zero tillage systems when compared with conventional tillage systems (3, 8, 9). There is a recognition of the need to rationalize tillage so as to maximize its benefits and minimize its adverse effects. Adopting controlled traffic systems of farming, in which the wheel tracks of all operations are confined to fixed paths, prevents recompaction of soil by traffic outside the selected paths. As the field is divided into cropping and

Table 1 A system for classification of surface soil structural quality

Class	Air-filled porosity at FC* % v/v	Water-holding capacity (mm/m)
Very good	>15%	>200
Good	10–15%	150–200
Moderate	5–10%	100–150
Poor	<5%	<100

* FC = soil water potential (−5 kPa).
(From Ref. 5.)

traffic zones, the undesirable effects of tillage on soil structure in the cropping zones can be avoided. Careful consideration of the type of tillage implement, as well as the frequency of tillage, is also important in crop-residue management (amount and orientation), which plays a vital role in wind- and water-erosion control (8).

Natural processes

These refer to the natural processes of wetting and drying, and freezing and thawing, which can influence soil surface structure (11). These processes can be influenced to improve surface soil structure. Many traditional farming practices use these natural processes to manage soil structure. For instance, one aim of autumn cultivation by European farmers is to put the soil into a condition where it will gain maximum benefits from the freezing and thawing of frost action by exposing the maximum clod surface at a suitable moisture content to produce frost tilth (10). Combined actions of wetting and drying are responsible for mellowing due to production of incipient cracks (11). Both processes of soil restructuring work better for clay soils than for sandy soils.

Chemical Methods

These refer to the application of natural or synthetic chemicals to improve soil structure. Generally, they act either by overcoming the chemical causes of instability (e.g., dispersion) or as stabilizing agents. For example, the application of gypsum to sodic soils can stop dispersion, improve physical condition of the surface soil, and increase crop yield (12).

Soil conditioners in forms of synthetic organic polymers [e.g., Polyvinyl alcohol (PVA) and Polyacrylamide (PAM)] can stabilize soil structure and improve soil physical conditions (4). Hitherto, their uses have been restricted to intensive horticulture because of the high costs involved. However, with the availability of new water-soluble polymers, the application rate, and therefore, the costs involved, have been greatly reduced (13).

Biological Methods

These are practices that involve using living organisms or their products for the improvement/maintenance of surface soil structure. The benefits of a pasture phase (ley) as part of the crop production system in maintaining favorable soil physical conditions have long been recognized (4, 14). The pasture phase improves soil structure because of the activities of the denser root network and higher organic carbon inputs when compared with the crops (3, 6, 15). The pasture roots create new pores as well as enlarge and stabilize existing ones. The increase in structural stability results from a combination of the direct effects of enmeshing and entanglement, as well as the indirect effects of associated microbial activities and the release of binding material by the roots. Improvement in soil structural stability due to increases in soil organic carbon under increasing duration of pasture has been documented (Fig. 1) (15). Retention of crop residue (stubble) provides surface protection against raindrop impact and wind in the

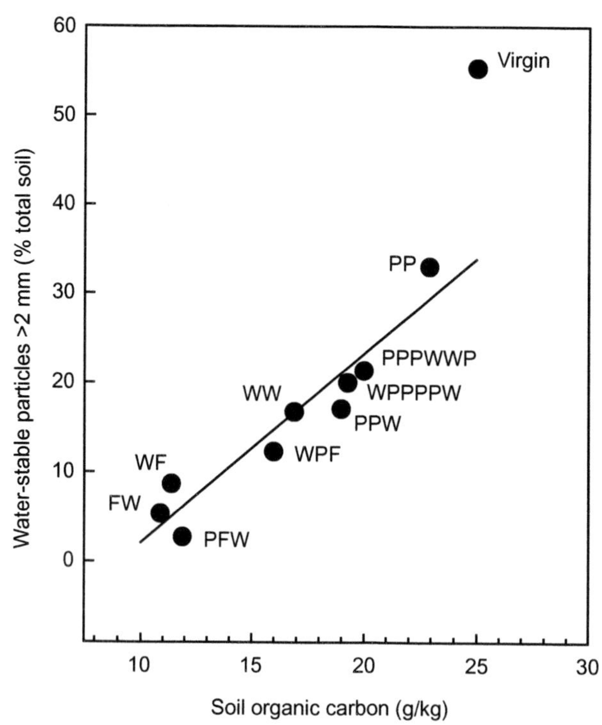

Fig. 1 The relationship between water-stable aggregation and organic carbon levels under different crop/pasture rotation (P = pasture; W = wheat; F = fallow). (Adapted from Ref. 15.)

S

Table 2 Beneficial effects on soil surface structure of different management practices

Management practices	Beneficial effects on soil structure	References
Mechanical and physical		
No tillage/reduced tillage	Increased water infiltration, soil water storage, aggregation; reduced soil erosion	3, 8, 9
Controlled traffic	Reduced soil compaction; increased biological activity of crop zone	18
Natural processes	Increased tilth, friability	10, 11
Chemical		
Use of ameliorants (e.g., gypsum, organic polymers)	Increased water infiltration, water-stable aggregation; reduced soil erosion, crusting	12, 13
Biological		
Pasture leys	Increased water-stable aggregation, water infiltration, soil organic matter	3, 4, 14, 15
Introduction of soil fauna	Increased macroporosity, water infiltration, water-holding capacity	17

short term and assists in maintaining/increasing the soil organic carbon levels, and hence the soil structural stability, in the long term. Application of organic matter as manure is a time-honored practice for maintaining/improving surface soil structure.

Introduction of soil fauna

Activities of many of the soil fauna, such as earthworms, ants, and termites, can modify soil structure. Earthworms can improve soil structure by creating burrows (macropores) and promoting surface aggregation because of their casting activities (16, 17). Significant improvements in surface soil structure and hence soil physical properties, such as increased water infiltration, porosity, and water-holding capacity, have been demonstrated as a consequence of introducing certain earthworm species to new locations (17).

IMPORTANCE OF AN INTEGRATED APPROACH

While many of the aforementioned methods, when implemented separately, have beneficial effects on soil structure and therefore on soil physical properties (Table 2), an integrated approach is required in practice. For effective improvement/maintenance of surface soil structure, these methods complement each other and usually more than one method is needed. For instance, in conservation tillage both reduction in tillage intensity and retention of crop residue are employed. Chemical/mechanical methods alone are merely short-term solutions that are ineffective, and may even result in adverse effects, in the longer term. Excessively tilling a soil that is already low in stability (because of low soil organic matter level) will only make the soil more tillage-dependent, vulnerable to complete structural collapse on wetting, and prone to crop failure (19). However, under certain circumstances they can serve as a starting point for overcoming the initial/inherent limiting soil structural conditions (e.g., the use of gypsum to overcome clay dispersion). Macropores, which are important components of the surface soil structure, can be effectively created/maintained only by biological agents (e.g., plant roots, earthworms, termites) as biopores (11). This highlights the importance of biological methods as part of the long-term soil surface management package.

REFERENCES

1. SSSA., *Glossary of Soil Science Terms*; Soil Science Society of America: Madison Wisconsin, 1997; 108.
2. Lal, R.; Stewart, B.A. Soil Degradation. In *Advances in Soil Science*; Springer-Verlag: New York, 1990; *11*, 8–12.
3. Chan, K.Y.; Pratley, J. Soil Structural Declines – How the Trend can be Reversed? In *Agriculture and the Environmental Imperative*; Pratley, J., Robinson, A., Eds.; CSIRO: Collingwood, Australia, 1998; 129–163.
4. Greenland, D.J. Soil Management and Soil Degradation. J. Soil Sci. **1981**, *32*, 301–322.
5. Hall, D.G.M.; Reeve, M.R.; Thomasson, A.J.; Wright, V.F. *Water Retention Porosity and Density of Field Soils*, Soil Survey Technical Monograph No. 9; Harpenden: United Kingdom, 1977, 55–60.
6. Kay, B.D. Rates of Change of Soil Structure Under Different Cropping Systems. In *Advances in Soil Science*; Stewart, B.A., Ed.; Springer-Verlag: Heidelberg, New York, 1990; *12*, 1–52.

7. Phillips, S.H.; Young, H.M. *No-Tillage Farming*; Reisman Associates: Milwaukee, Wisconsin, 1973; 11–16.
8. Unger, P.W.; McCalla, T.M. Conservation Tillage Systems. In *Advances in Agronomy*; Brady, N.C., Ed.; Academic Press: New York, 1980; *33*, 2–53.
9. Carter, M.R. *Conservation Tillage in Temperate Agroecosystem Development and Adaption to Soil Climate and Biological Constraints*; CRC: Boca Raton, Florida, 1994; 400.
10. Payne, D. Soil Structure, Tilth and Mechanical Behaviour. In *Russell's Soil Conditions & Plant Growth*; Wild, A., Ed.; Longman: Essex, England, 1988; 379–411.
11. Dexter, A.R. Amelioration of Soil by Natural Processes. Soil Tillage Res. **1991**, *20*, 87–100.
12. Shainberg, I.; Sumner, M.E.; Miller, W.P.; Farina, M.P.W.; Pavan, M.A.; Fey, M.V. Use of Gypsum on Soils: A Review. In *Advances in Soil Science*; Stewart, B.A., Ed.; Springer-Verlag: New York, 1989; *9*, 2–101.
13. Wallace, A.; Wallace, G.A. Soil and Crop Improvement with Water-Soluble Polymers. Soil Technol. **1990**, *3*, 1–8.
14. Low, A.J. The Effect of Cultivation on the Structure and Other Physical Characteristics of Grassland and Arable Soils (1945–1970). J. Soil Sci. **1972**, *23*, 363–380.
15. Tisdall, J.M.; Oades, J.M. The Effect of Crop Rotation on Aggregation in a Red-Brown Earth. Aust. J. Soil Res. **1980**, *18*, 423–433.
16. Lee, K.E. Physical Effects on Soils. In *Earthworms – Their Ecology and Relationships with Soils and Land Use*; Academic Press: Sydney, 1985; 33–64.
17. Edwards, C.A.; Bohlen, P.J. Role of Earthworms in Soil Structure, Fertility and Productivity. In *Biology and Ecology of Earthworms*; Chapman & Hall: London, 1996; 196–212.
18. Chamen, W.C.T.; Longstaff, D.J. Traffic and Tillage Effects on Soil Conditions and Crop Growth on a Swelling Clay Soil. Soil Use Manage. **1995**, *11*, 168–176.
19. Greacen, E.L. Tillage Research in a Semi-Arid Environment. Soil Tillage Res. **1983**, *3*, 107–109.

SUSTAINABILITY AND INTEGRATED NUTRIENT MANAGEMENT

S

G.S. Saroa
Rattan Lal
The Ohio State University, Columbus, Ohio, U.S.A.

INTRODUCTION

World population of 6 billion in 2000, increasing at the rate of 1.8 percent per year, is expected to reach 8 billion by the year 2025 and 9.4 billion by 2050 (1–3). Most of the population increase will occur in developing countries, which by 2050 will have increased by another 575 million in India, 300 million in China, 200 million in Nigeria, 200 million in Pakistan and 140 million in Ethiopia (4). About 780 million people are already food insecure in developing countries. The Asian farmers currently feed nearly 60 percent of the world population from only 25 percent of the world's land available for agriculture. The global demand for food may double over the period 1990-2030 with an increase of 2.5 to 3 times in developing countries.

The green revolution, which started in the 1970s, was perceived by some as a near miracle preventing starvation in many Asian countries. It was brought about by increased yields of irrigated rice (*Oryza sativa*) and wheat (*Triticum aestivum*) grown on fertile soils with intensive use of fertilizers. However, gains in agricultural production are not being sustained where inputs are insufficient or imbalanced, soil fertility is declining and other soil degradative processes are severe. Thus food insecurity in the Asian region is once again a matter of grave concern. The threat to food security in Asia is caused by the degradation of hitherto fertile soils, especially due to depletion of soil organic matter (SOM) content and associated plant nutrients. There is little scope for an increase in arable area, which is likely to decrease due to urbanization and soil degradation. Thus a pre-requisite to increasing food production is to develop, fine tune, and implement soil/crop management practices which enhance and maintain at appropriate levels the SOM content and plant nutrient reserves. Adopting a strategy of integrated nutrient management (INM) can increase nutrient use efficiencies, strengthen nutrient cycling, and increase and sustain food production.

The INM strategy involves meeting a crop's nutritional requirement to achieve the desired yield through judicious use of inorganic fertilizers, bio-fertilizers, bio-solids, biological nitrogen fixation, and other nutrient cycling mechanisms that enhance nutrient use efficiency, and optimize the use of most limited or non-renewable resources. This manuscript outlines the basic principle of INM in the context of resource-poor farmers of Asia.

SUSTAINABLE AGRICULTURE

Sustainable agriculture involves enhancing and sustaining agricultural production while preserving or improving soil fertility, maintaining a clean environment, and protecting or regenerating biological diversity and ecosystem health. The Technical Advisory Committee of the Consultative Group on International Agricultural Research defined sustainable agriculture as the "successful management of resources for agriculture to satisfy changing human needs while maintaining or enhancing the quality of the environment and conserving natural resources" (5). Traditional farming has been basically a sustainable agriculture, but at a low level of output supporting subsistence living. The transition from traditional to intensive agriculture is necessitated by the increase in global food demand and a wide spread problem of soil and environmental degradation.

FERTILIZER MANAGEMENT

The use of plant nutrients per unit of net-cropped area is more in Asia ($129\,kg\,ha^{-1}$) as compared to the world ($90\,kg\,ha^{-1}$). However the ratio of N: P: K use is quite balanced for the world (3.8:1.5:1) but not (7.9:2.7:1) for the Asian region (6). The wide ratio is not conducive to sustained agricultural production, and needs to be altered to achieve the desired ratio of about 4:2:1. The fertilizer consumption of China and India is about 70 percent of the total fertilizer consumption in Asia (6), and approximately 60 percent of the fertilizer consumed in Asia is used on rice (7). The increase in agricultural production in China during 1978 to 1995 was 32 percent by fertilizer use,

28 percent by irrigation, 17 percent by improved seed, 13 percent by machinery, and 6 percent by other factors (8). Furthermore, the use of chemical fertilizers will continue to play an important role in augmenting global food production. To enhance the use efficiency and maximize the benefits, fertilizers must be applied at the right amount, placed in the root zone at the right time, and applied in the right formulation and combination. High crop yields deplete the nutrient reserves (9) and, unless replenished judiciously and regularly, can lead to nutrient imbalance in soil. Indeed, crop yields in intensively cultivated irrigated areas of northwestern India are either stagnating or declining in the post green revolution period because of the degree of nutrient imbalance, severity of soil degradation, and intensity of cropping. Nutrient removal by different crops for a definite yield target is presented in Table 1 (9).

Nitrogen

Because of its large requirement for vigorous growth, crop response to N application is universal. Nitrogen represents approximately 70 percent of the total fertilizer nutrients (N+P+K) consumed in Asia (10). Nitrogen requirement for one Mg (metric ton) of cereal grains is 15–30 kg (11). Being a staple food group in Asia, an earnest effort has been made in the past for increasing cereal production leading to increased demand for fertilizer N. This trend is likely to continue in future, and fertilizer N will continue to be the most wanted nutrient.

Rice crop on an average needs about 20 kg N to produce one Mg of grain yield (12). Thus high yielding varieties producing 5 Mg of grain yield need 100 kg N both from soil and fertilizer sources. If grain yield without fertilizer (native soil N) is about 2.5 Mg ha^{-1}, it requires about 50 kg N ha^{-1} through fertilizer use. Assuming fertilizer N recovery of 40 percent, it involves applying 125 kg N ha^{-1} through fertilizer use for obtaining a target yield of 5 Mg ha^{-1}. The efficiency of applied N is usually low and ranges from 25 to 45 percent in rice and 50 to 70 percent in upland crops (13). In general, a 1 percent increase in recovery rate of N will normally result in an economy of 0.15 million Mg of urea/year in India, which in terms of additional crop production would be equivalent to one million Mg of food grains (14).

Efficiency of fertilizer N is also dependent on the form of N applied and the ecosystem in which they are used. Meelu and Morris (15) reported that amide (urea) and ammonical sources (ammonium sulphate) of N are superior to nitrate (calcium ammonium nitrate) for rice. But in case of hybrid rice there are contrary reports from China. The superiority of ammonical source, however, is due to its stability under flooded conditions compared to nitrate, which is highly unstable and lost through leaching and denitrification. The reported superiority of nitrate for hybrid rice is attributed to the physiological/genetic differences between the inbreds and hybrids (16).

Phosphorus

Asia consumes about 30 percent of the world's P fertilizer (10), and China and India are the main P users in Asia. In contrast to N, only 2–5 kg P is removed per Mg of cereal produced, and P use efficiency by single crop rarely exceeds 15–20 percent of added P. Because of its relative immobility, applied P is fixed in the soil and may be

Table 1 Nutrient requirement of principle crops

Crop	Scientific name	Yield (Mg ha^{-1})	Nutrients (kg Mg^{-1} of yield)		
			N	P_2O_5	K_2O
Wheat	*Triticum aestivum*	5.0	31	5	37
Rice	*Oryza sativa*	6.3	15	4	24
Corn	*Zea mays*	2.7	42	17	40
Sorghum	*Sorghum bicolor*	1.1	47	16	8
Millet	*Pennisetum typhoides*	0.6	50	18	170
Sugarcane	*Sachharum offinarum*	87.1	1.4	1	2.6
Potato (tubers)	*Solanum tuberosum*	17.6	5	2	8
Barley	*Hordeum vulgare*	2.5	22	12	60
Groundnut	*Erychus hypogea*	1.9	41	11	24
Mustard	*Brassica compestris*	0.7	31	16	40

(From Ref. 9.)

absorbed by subsequent crops. The magnitude of the residual effect depends upon the rate and kind of P fertilizer, cropping and management practices, and largely on soil type. Thus P fertilizers applied to a single crop often can meet the P requirement of subsequent crop/crops in a cropping system. Gill and Meelu (17) reported that when P is applied to wheat, at the recommended rate, the following rice crop can be grown on residual P. However, when P is applied only to rice, grain yield of wheat is considerably lowered. De Datta et al. (7) also reported that P use efficiency can be greatly improved by the choice of the crop to be fertilized in a sequence. The fertilization of a single crop in a sequence can lead to net negative P balance in the soil. Regular monitoring of nutrient balance, therefore, is important for greater precision in scheduling fertilizer use.

The use efficiency of P can vary widely depending on soil type, crop, and the source of fertilizer. For short duration and winter crops water-soluble phosphorus (WSP) sources and nitrophosphates with 70 percent WSP have higher efficiency than water insoluble or less water-soluble sources (18). For long duration and summer crops, the differences among sources with regard to P use efficiency are minimal. Method of P application plays an important role in P use efficiency, and about 50 percent of fertilizer can be saved by placement in row-crops as compared to the broadcast method (19), while the reverse may be true for rice.

The dynamics of P in upland soils is markedly different from that of lowland soils. Available P increases during the flooding phase and decreases during the drying phase. In general, wetland rice is less responsive than all dryland crops grown on the same soil. Soil phosphorus continues to be an *enigma* to soil scientists, and INM strategies can help in solving some of the problems of soil phosphorus management.

Potassium

Asia consumes only 15 percent of the world's K fertilizer (10), and China and India are the main K fertilizer consuming countries in Asia. The K use efficiency is about 80 percent. The neglect of K application is evident from the highly imbalanced fertilizer consumption ratio in most developing countries. Unlike N and P, response to K application is small on soils with illitic clay minerals. The response to K is usually observed in acidic and sandy soils during the summer season in the high rainfall areas where it is prone to leaching. The K response to corn (*Zea mays*) and rice is more than that to wheat (20). Production of one Mg of rice grains requires 17 kg K ha^{-1} (21) to 25 kg K ha^{-1} (22). Thus continuous appraisal of the crop response to K in relation to soil test is important.

Other Nutrients

Among the secondary nutrients, sulphur deficiency is wide spread with continuous use of high analysis S-free fertilizers. The S deficiency usually occurs in soils of humid regions, which are highly oxidized, sandy, and low in total S and SOM contents. Similarly, Zn deficiency is another disorder limiting the yield of crops, especially of rice. Other micronutrient deficiencies are also appearing because of low use of organic manures and increased use of high analysis fertilizers.

INTEGRATED NUTRIENT MANAGEMENT

The multilocational testing by All India Co-ordinated Rice Improvement Project has shown a 50 per cent decline in fertilizer response of rice from 15–16 kg grain kg^{-1} NPK nutrients applied in the1960s and 1970s, to 7–8 kg grain kg^{-1} NPK applied in the 1990s (23). Thus there is a strong need to assess both balanced fertilization and integrated plant nutrient management, which are not synonymous. The balanced fertilization implies balancing the ratio of applied nutrients to that of the nutrient requirements of the crop, and the INM implies combining organic and mineral methods of soil fertility management to reduce risks of nutrient imbalances and depletion of SOM reserves. The SOM is a biological buffer, which plays an important role in maintaining the balanced supply of available nutrients for plant growth. The higher SOM content reflects higher soil productivity (24). Therefore, application and management of biosolids as fertilizers are viable options to improving plant growth. Soils with low SOM content lose their buffering capacity and productivity, and have low fertilizer use efficiency. In areas of intensive cropping an increased level and better quality of SOM can contribute significantly to high productivity (25). Organic fertilizers vary widely in terms of their nutrient contents (Table 2), depending on the biosolids and their constituents (26).

The data from long term experiments under diversified agroclimatic conditions in India showed that neither organic manures nor mineral NPK fertilizers alone can produce high yields under intensive farming where nutrient turnover in soil-plant system is large (27). Only less than 40 percent of the nutrients present in bulky organic manures are available to the first crop following their application but may have notable residual effect. On the other hand, residues of high quality organic inputs such as green manures and legume tree pruning decompose quickly and may release 70–95 percent of their N within a season under tropical conditions (28), with a little impact

Table 2 Nutrient analysis of some common bio-fertilizers

Manures	Nutrient content (percent)		
	N	P_2O_5	K_2O
Fresh dung			
Cattle	0.3–0.4	0.1–0.2	0.1–0.3
Horse	0.3–0.4	0.3–0.4	0.3–0.4
Sheep	0.5–0.6	0.4–0.6	0.3–1.0
Poultry manure (fresh)	1.0–1.8	1.4–1.8	0.8–0.9
Farm yard manure (dry)	0.4–1.5	4.0–5.0	0.3–1.9
Crop residues			
Rice	0.36	0.08	0.71
Corn	0.42	1.57	1.65
Wheat	0.53	0.10	1.10
Green manure			
Sesbania aculeate	0.62	–	–
Cymposis tetragonoloba	0.34	–	–
Crotolaria juncea	0.75	0.12	0.61

(From Ref. 26.)

on SOM content. The recovery of this released N may be 10–30 percent by the first crop, and little if any for the second crop (29). The farm yard manure generally contains twice as much micronutrients as in *Sesbania* green manure on equal weight basis (30).

The crop residues contain appreciable amounts of plant nutrients, and can contribute to soil quality. Prolonged removal or burning of the crop residues usually decreases soil nutrient reserves and SOM content (31). Increasing SOM content under tropical conditions often requires frequent and high doses of biosolids. Approximately 7 Mg $ha^{-1} yr^{-1}$ drymatter of low quality residues (roots, stems), or 10 Mg $ha^{-1} yr^{-1}$ of high quality residues (green manure leaves) may be required to maintain a 1 percent organic carbon in soils of the sub-humid tropics (32).

CONCLUSIONS

There is a great scope for sustaining agricultural production through judicious combination of mineral, organic and microbial sources of plant nutrients. The dependence on chemical fertilizers can be reduced by supplementing the plant nutrients through the use of organic manures, green manures, crop residues, bio-fertilizers and inclusion of legume crops in cropping sequences. This will lead to maintenance of soil productivity and improvement of soil physical conditions. There is a need for development of simple, cost-effective, and easily adopted technologies for the production of organic manures. A judicious use of chemical fertilizers can reduce the risk of environment pollution because of reduction in use of fossil fuel for fertilizer production and low risks of contamination of natural waters. If the choice is to apply once within a rotation cycle, it may be applied to the most responsive crop in the sequence. Soil testing is an integral part of any INM program and regular monitoring of soil health and diagnosis of nutrient availability is essential for minimizing risks of soil and environmental degradation.

REFERENCES

1. Hulse, J.H. *Science Agriculture and Food Security*; NRC Research Press: Ottawa, Canada, 1995; 7–28.
2. Fischer, G.; Heling, G.K. Population Momentum and the Demand on Land and Water Resources. Phil. Trans. R. Soc. (London) **1997**, *352* (Ser. B), 869–889.
3. Litvin, D. Dirt poor. Economist, 21 March, 1998; 3–16.
4. Daily, C.; Dasgupta, P.; Bolin, B.; Crosson, P.; Guerry, J.D.; Ehrlich, C.; Folke, A.M.; Jansson, B.O.; Jansson, N.; Kautsky, A.; Kinzig, S.; Levin, K.G.; Maler, P.; Pinstrup-Andersen, D. Siniseal co.; Walker, B. Food Production, Population Growth, and the Environment. Science **1998**, *281*, 1291–1292.
5. TAC. *Sustainable Agricultural Production*; Implications for International Agricultural Research, CGIAR: Washington, DC, 1989; 1–15.
6. Kumar, V. Balanced Use of Plant Nutrients with Particular Reference to Integrated Plant Nutrient Systems in the Asian Region. In *Nutrient Management for Sustainable Crop Production in Asia*; Johnston, A.E., Syers, J.K., Eds.; Proceedings of an International Conference on Nutrient Management for Sustainable Crop Production, Bali, Indonesia, Dec 9–12, 1996; CAB International: New York, 1998; 85–93.
7. De Datta, S.K.; Buresh, R.J.; Mamaril, C.P. Increasing Nutrient Use Efficiency in Rice with Changing Needs. Fert. Res. **1990**, *26*, 157–167.
8. Xie, J.C.; Xing, W.Y.; Zhou, M.J. Current Use and Requirements for Nutrients for Sustainable Food Production in China. In *Nutrient Management for Sustainable Crop Production in Asia*; Johnston, A.E., Syers, J.K., Eds.; Proceedings of an International Conference on Nutrient Management for Sustainable Crop Production, Bali, Indonesia, Dec 9–12, 1996; CAB International: New York, 1998; 267–277.
9. *Fertilizers Statistics*. Fertilizer Association of India: 1974–75.
10. Food and Agriculture Organization. FAO Fertilizer Yearbook; Rome, Italy **1989**, *47*, 3–16.
11. Dobermann, A.; White, P.F. Strategies for Nutrient Management in Irrigated and Rainfed Lowland Rice System. Nutrient Cycling in Agroecosystems **1998**, *53*, 1–18.
12. Yoshida, S.; Oka, I.N. *Rice Research Strategies for the Future*; Int. Rice Research Inst. Los Banos, Phillipines, 1982; 51–78.
13. Hauck, R.D. *Nitrogen-15 in Soil Plant Studies*; Int. Atomic Energy Agency: Vienna, Austria, 1971; 65–80.

14. Abrol, I.P. Macronutrients in Soils and Crops. In *Proceedings National Symposium on Macronutrients in Soils and Crops*; Meelu, O.P., Bajwa, M.S., Singh, Ranjodh, Vig, A.C., Sidhu, P.S., Beri, V., Sadana, U.S., Eds.; PAU: Ludhiana, Punjab, India, 1991; 1–15.
15. Meelu, O.P.; Morris, R.A. Green Manure Management in Rice Based Cropping System. In *Green Manuring in Rice Farming*; Int. Rice Res. Inst. Manila, 1988; 209–222.
16. Reddy, M. Narayana; Krishnaiah, K. Current Status of Crop Response to Fertilizers in Different Agro-Climatic Regions. Fert. News **1999**, *44* (4), 113–126.
17. Gill, H.S.; Meelu, O.P. Studies on the Utilization of Phosphorus and Causes for Its Differential Response in Rice-Wheat Rotation. Plant Soil **1983**, *74*, 211–222.
18. Meelu, O.P.; Rana, D.S.; Sharma, K.N.; Singh, Raghbir Comparative Efficiency of Complex Phosphatic Fertilizers. J. Indian Soc. Soil Science **1977**, *25*, 374–378.
19. Saroa, G.S.; Vig, A.C. Response of Maize and Wheat to Nitrophosphates of Varying Water Solubility. Proceedings of the Third Agricultural Science Congress, NAAS, PAU, March 12–15 **1997**, *Vol. II*, 17–18.
20. Meelu, O.P.; Bishnoi, S.R.; Randhawa, N.S.; Sharma, K.N. Response of Wheat to Graded Doses of N and P on Different P and Organic Carbon Fertility Classes Under Rainfed Conditions. J. Agric. Science (Camb) **1976**, *86*, 425–426.
21. De Datta, S.K. *Principles and Practices of Rice Production*; John Wiley & Sons: New York, 1981; 348–419.
22. Velayutham, M.; Reddy, C.K.C. Balanced Fertilization for Increasing Fertilizer Use Efficiency. *Proceedings of FAI Regional Agriculture Committee Meeting*; 1987.
23. Anonymous. Quin-Quinneal Report of All India Co-ordinated Rice Improvement Project, Fertilizer Association of India, New Delhi, 1996.
24. Karama, A.S.; Marzuki, A.R.; Manwandan, I. Penggunaan Pupuk Organic Pada Tanaman Pangan. *Prosiding Seminar Nasional Penggunaan Pupuk V. Pusat Penelitian Tanah dan Agroklimat, Bogor. Hlm.* 1990; 395–425.
25. Finck, Arnold Integrated Nutrient Management: An Overview of Principles, Problems and Possibilities. Annals of Arid Zone **1998**, *37* (1), 1–24.
26. *Handbook of Agriculture*. Indian Council of Agricultural Research: New Delhi, 1961; 78–114.
27. Nambiar, K.K.M.; Abrol, I.P. Long Term Fertilizer Experiments in India—An Overview. Fert. News **1989**, *34* (4), 11–20.
28. Giller, K.E.; Cadisch, G. Future Benefits from Biological Nitrogen Fixation: An Ecological Approach to Agriculture. Plant Soil **1995**, *174*, 255–277.
29. Mafongoya, P.L.; Nair, P.K.P. Multipurpose Tree Pruning as a Source of Nitrogen to Maize Under Semi-Arid Conditions in Zimbabwe. Interaction of Pruning Quality and Time and Method of Application on Nitrogen Recovery by Maize in Two Soil Types. Agroforestry Systems **1997**, 1–14.
30. Katyal, J.C.; Sharma, B.D. Role of Micronutrients in Crop Production. Fert. News **1979**, *24* (7), 33–50.
31. Hooker, M.L.; Schepers, J.S. Effect of Residue Removal vs. Incorporation on N Uptake and Growth of Winter Wheat. *Agronomy Abstracts*; ASA: Madison, Wisconsin, 1984; 207.
32. Janssen, B.H. Integrated Nutrient Management: The Use of Organic and Mineral Fertilizers. In *The Role of Plant Nutrients for Sustainable Food Crop Production in Sub-Saharan Africa*; Van Reuler, H., Prins, W., Eds.; Leidschendam: VKP, 1993; 89–106.

TEMPERATURE MEASUREMENT

Kevin McInnes
Texas A&M University, College Station, Texas, U.S.A.

INTRODUCTION

There is no set design either for construction or procedure for installation of soil thermometers, other than the requirement that they operate in a continuously moist environment. Soil thermometers may be tailored for measurement of either surface or subsurface temperatures. Surface temperature measurements are desirable because they can be used to estimate subsurface temperatures. Measurement of subsoil temperature is difficult because installation disturbs the soil and alters the thermal regime.

VARIABILITY OF SOIL TEMPERATURE

Soil temperatures generally fall between −20 and 60°C, except in extreme environments. Mean diel soil surface temperature is usually the same as the mean diel air temperature. Similarly, the mean annual soil temperature is usually equal to the mean annual air temperature (1). Differences between extremes of the daily average surface temperatures vary from a few degrees near the equator to upward of 30°C or more in the mid-latitudes. For the diel cycle, differences between the periodic extreme surface temperatures depend on the nature of the soil cover. Bare soil produces the largest differences, with 15–30°C being common. Both diel and annual patterns in temperature at the soil surface travel down through the soil where each magnitude is increasingly attenuated and each phase increasingly retarded with increasing depth.

Water in soil helps buffer against extreme temperature fluctuation. In the case of soil freezing, declining soil temperatures are buffered near 0°C by the latent heat of fusion. Once water is frozen, soil temperatures may drop as low as −40°C in extremely cold environments (2). Likewise, as soil temperatures rise under extreme heating, they are buffered by the latent heat of vaporization at temperatures near 95°C (3). When all moisture is lost under prolonged heating, as might occur under a slow-moving fire, interfacial soil temperatures may rise hundreds of degrees (3).

MEASUREMENT OF SURFACE TEMPERATURE

Surface temperature may be used to calculate subsurface temperatures with a fair degree of accuracy (4). Surface temperatures can be measured with or without direct contact. Noncontact thermometers infer temperature from the amount of thermal radiation emitted from the soil. Infrared thermometers (IRTs) measure temperature by capturing and quantifying the amount of thermal radiation emitted in a narrow band of the infrared spectrum, typically 8–14 μm. Soil temperature is related to the quantity of radiation emitted through the Stefan–Boltzmann equation and knowledge of the soil's thermal emissivity (5). Since soil temperature is inferred from the amount of infrared radiation captured by the sensor, reflected background radiation introduces a source of uncertainty that must be considered. A soil's thermal emissivity changes with soil water content and surface roughness, whereas background radiation varies with air temperature, humidity, and cloud cover. Monitoring changes in emissivity and background radiation requires additional IRT measurements, adding to the complexity of the soil temperature measurements. A variation of the IRT used extensively in remote sensing is the infrared camera. An infrared camera provides detailed information on spatial variation in surface temperature that otherwise would be difficult to collect.

Contact thermometers used to measure surface temperature must have good thermal contact with the soil surface and minimally affect surface energy exchange. To satisfy both these requirements sensors may be coated with soil (6). The added soil provides weight that helps hold the thermometer in contact with the soil and gives the sensor similar emissivity and reflectivity as the soil.

TYPES OF SOIL THERMOMETERS

While non-electric contact sensors are still used to measure soil temperature, the advent of accurate and reliable electronics has removed nearly all advantages they held over electric thermometers. There are several types of

Encyclopedia of Soil Science

electric thermometers suited to measuring soil temperatures. Temperature probes made from thermocouples, resistance sensors, and integrated circuits are most common. All these sensors easily function over the normal range of soil temperature.

Thermocouples are popular soil temperature sensors because they are robust and easy to construct. Thermocouples consist of two dissimilar metal wires joined to each other at one end, the sensing junction, and to the terminals of a meter at the other end. A difference in temperature between the sensing junction and the meter terminals results in a voltage that can be measured and converted to temperature provided the temperature of the terminals is known. There are eight different combinations of metals and alloys, designated as letter types, for which there are internationally accepted values of the relationship between voltage and temperature difference (referenced to 0°C). Modern thermocouple meters have these relationships built into their software. Industry-accepted errors in wire composition allow a specified degree of uncertainty in voltage output at a given temperature (7). Type T wire consists of copper and constantan conductors and is the most popular type for constructing soil thermometers. Manufacture with standard type T wire allows an error ±0.8°C from the calibration equation used in modern thermocouple meter software. This allowable error for type T is better than that for other types of thermocouple wire by a factor of approximately 2. A limited selection of special wires with tolerances half those of the standard wires (e.g., ±0.4°C for type T) are available.

Thermocouples may be connected in a series configuration called a *thermopile* to measure minute temperature differences, such as those associated with gradients inducing heat fluxes in soil or they may be combined in parallel to measure a spatial average temperature (8). Caution must be exercised when using a parallel configuration of thermocouples because improper construction may cause serious errors (8–10).

Integrated circuit (IC) temperature transducers are a new class of thermometers. IC sensors are available with either a voltage or a current output. Both outputs are linearly related to temperature. Voltage output is typically $10\,mV\,K^{-1}$ (11) while current is usually regulated at $1\,\mu A\,K^{-1}$ (12). The IC sensors that regulate current proportional to temperature are particularly useful in measuring soil temperatures because they are insensitive to voltage drops over long lead wires. Any shielded and well-insulated twisted pair of conductors is sufficient for operation of the sensor hundreds of meters from the receiving meter. Current is calculated using Ohm's Law and the voltage drop across an inline resistor of known magnitude. Current-regulating sensors may be connected in parallel to measure a spatial average temperature.

Other soil temperature sensors take advantage of the sensitivity of electrical resistance of metals or semiconductors to temperature. Resistance temperature detectors (RTDs) are wire and metal film-type sensors that make accurate soil thermometers. Platinum RTDs are the most common. Resistances of platinum RTDs are usually $100\,\Omega$ at 0°C, although new thinfilm deposition technology has led to creation of RTDs with resistances $>1\,k\Omega$. The resistance of platinum increases almost linearly with temperature, approximately 0.4% per °C. Because of this linear relationship, RTDs may be connected in series and are used to measure an average soil temperature. The disadvantage to using RTDs as soil thermometers is that resistances of individual sensors are low enough that three- or four-conductor cables are required to account for the resistances of the lead wires (10).

Thermistors are also thermally sensitive resistors, but made from semiconductor materials. They have two distinct advantages over RTDs. First, their nominal resistance is greater so that only two conductors are usually required for lead wires. The second advantage is that their sensitivity to temperature is considerably greater. Meters are not required to measure with the precision necessary for RTDs. A disadvantage limiting the use of thermistors in averaging temperature is that their electrical resistance is usually a highly nonlinear function of temperature.

Nonelectric thermometers such as liquid-in-glass thermometers, dilatation thermometers, bimetallic thermometers, and manometric thermometers are sluggish and difficult to install in a manner that would minimize errors from heat conduction along the thermometer stem. In addition, thermometers that rely on thermal expansion as a measure of temperature are prone to serious errors in temperature measurement caused by external forces acting on the sensing portion of the device (13), such as forces present in soil prone to shrinking and swelling.

SOURCES OF ERROR

Spatial variability leads to uncertainty in measured soil temperature (8). The number of temperature sensors required to produce an average within a certain value of the mean must be determined from an in situ measure of variability (8). Fortunately, once the required number is known, then many thermocouples (8), IC temperature sensors (12), or RTDs sensors (8) are easily connected in a configuration that provides an average temperature. By their nature, IRTs measure an average surface temperature integrated over an area that depends on the distance of the thermometer from the surface and its field of view.

Heat conduction down the lead wires or stem of the thermometer to the sensing element may lead to error. Thermal conductivity of copper, for example, is two orders of magnitude greater than that of soil. To reduce conduction errors, a length of wire or portion of the thermometer stem should be buried along an isotherm of the temperature being measured. With electric thermometers, 0.2–0.5 m of wire buried along an isotherm is usually sufficient to reduce errors to within acceptable levels. Reducing the diameter of the lead wires near the sensor also will reduce heat conduction errors.

Lack of proper insulation and shielding of electric thermometers may also lead to serious errors. Insulation on sensors and cabling needs to be highly resistant to moisture. Adhesive-lined heat-shrinkable tubing or resin-filled stainless steel tubing is a commonly used insulator for a sensor. To reduce the potential from electromagnetically induced noise, sensor cabling should be shielded and properly grounded. Except for the self-powered thermocouple, most electric temperature sensors are also susceptible to self-heating errors. To prevent errors caused by Joule heating the current should be kept below the self-heating limit imposed by the heat dissipation coefficient (10), but large enough to give adequate resolution on the meter.

INSTALLATION OF SUBSURFACE THERMOMETERS

Soil thermometers may be buried in disturbed soil or they may be installed on supports driven into the soil. Burying is most efficient when the thermometers are to be placed within about 0.15 m of the surface. Small diameter pipe has been used to house soil thermometers installed at greater depth (13). Metal pipe may be driven to considerable depth in hard soil, but its considerably different thermal properties affect the local temperature regime. PVC pipe is a better support for soil thermometers because its ability to transfer heat is nearly equal to that of dry soil. Insertion is relatively easy, provided a pilot hole is created. Electric sensors should be embedded and sealed so they are flush with the outside of the pipe. The lead wires should be located inside the tube and, to reduce the potential for errors from heat conduction along the lead wires, a section of the wires should be bunched inside the pipe at the same depth as the sensor. Wooden stakes with embedded thermometers also have been used for support. Like PVC, wood transfers about the same amount of heat in response to a temperature gradient as does soil.

REFERENCES

1. Shul'gin, A.M. *The Temperature Regime of Soils.* Translated by A. Gourevitch, Israel Program for Scientific Translations. Sivan Press: Jerusalem, 1957.
2. Sharratt, B.S. Freeze-Thaw and Winter Temperature of Agricultural Soils in Interior Alaska. Cold Reg. Sci. Technol. **1993**, *22* (1), 105–111.
3. Campbell, G.S.; Jungbauer, J.D.; Bristow, K.L.; Hungerford, R.D. Soil Temperature and Water Content Beneath a Surface Fire. Soil Sci. **1995**, *159* (6), 363–374.
4. Campbell, G.S. *Soil Physics with BASIC: Transport Model for Soil-Plant Systems*; Elsevier: Amsterdam, 1985.
5. Fuchs, M.; Tanner, C.B. Surface Temperature Measurements of Bare Soils. J. Appl. Meteor. **1968**, *7* (2), 303–305.
6. Ham, J.M.; Senock, R.S. On the Measurement of Soil-Surface Temperature. Soil Sci. Soc. Am. J. **1992**, *56* (2), 370–377.
7. American Society for Testing and Materials, *Manual on the Use of Thermocouples in Temperature Measurement*, 4th Ed. ASTM Manual Series: MNL12 (PCN: 28-012093-40), ASTM: Philadelphia, 1993.
8. Tanner, C.B. *Basic Instrumentation and Measurements for Plant Environment and Micrometeorology*; Department of Soils, Bull. 6, University of Wisconsin: Madison, 1963.
9. Schooley, James F. *Thermometry*; CRC Press: Boca Raton, FL, 1986.
10. Michalski, L.; Eckersdorf, K.; McGhee, J. *Temperature Measurement*; John Wiley: New York, 1991.
11. National Semiconductor Corporation, *LM35/LM35A/LM35C/LM35CA/LM35D Precision Centigrade Temperature Sensors*; Datasheet 005516, National Semiconductor Corporation: Arlington, 1997.
12. Analog Devices, *Two-Terminal IC Temperature Transducer AD590*; Analog Devices: Norwood, 1997.
13. Garvitch, Z.S.; Probine, M.C. Soil Thermometers. Nature **1956**, *177* (4522), 1245–1246.

TERMITES

T

David E. Bignell
Queen Mary, University of London, London, United Kingdom

John A. Holt
James Cook University, Townsville, Australia

INTRODUCTION

Termites have a profound effect on soil properties, resulting both from their mound-building and gallery excavating activities, and from their modes of feeding. Although other soil animals can affect soil properties, the importance of termites lies in their huge abundance and biomass in the tropics and sub-tropics, and their unique associations with a diversity of microorganisms. Contrary to common assumptions, most termite species are beneficial to plant growth: they promote organic matter decomposition and nutrient cycling, and reduce erosion through a strong influence on soil physical, chemical, and biological processes. They also have long-term effects on soil development and the character of landscapes. For a modern synthesis of termite biology see Ref. 1, and for a review of economic impact and control methods, see Ref. 2.

TYPES

Termites are eusocial, polymorphic insects (order Isoptera) with generally cryptic behavior. They live in large family groups that are composed of reproductive forms (sometimes winged), together with numerous sterile soldiers and workers. Sometimes known inaccurately as "white ants," termites are derived forms of cockroach which depend on mutualistic intestinal bacteria and, in some higher forms, externally cultivated basidiomycete fungi, for assistance with energy metabolism and the provision of nitrogen. Some termites also contain populations of flagellate protozoans in the hindgut, a number of which have the ability to degrade cellulose and other plant polysaccharides. Termites are the only social insects with a true soldier caste having no role other than to defend the colony. More than 2300 species have been described in 270 genera and 6 families. Most of these are beneficial inhabitants of tropical and subtropical forests, savannas, and rangelands. A small number of relatively primitive species, loosely termed "drywood termites," "dampwood termites," and "subterranean termites" are able to colonize and/or consume timber used for buildings, furniture, transmission poles, and fencing; such pest species are inadvertently distributed by humans and become pantropical, with one or two species extending their range to include cities in temperate zones. Other species are agricultural or silvicultural pests, especially where non-indigenous crops are introduced or where land is cleared of natural plant detritus prior to planting, or not fertilised with mulch (3). The large majority of termite species are not pests and can be grouped heuristically into four broad functional categories based on their diets: grass-feeders, litter-feeders, wood-feeders, and soil-feeders. The last group dominate lowland tropical forests and the wetter savannas; grass-feeders and wood-feeders become relatively more common as aridity increases; almost all termite species are associated with the soil in some way, using it as either a habitat, a resource, or both. Under most global change scenarios, non soil-feeding species are likely to become more dominant, but overall termite biomass may decline (4, 5).

HABITAT

Termite diversity is strongly influenced by latitude (the numbers of naturally occurring taxa declining steadily north and south of the Equator), and by rainfall (species richness and abundance generally increasing with greater rainfall at the same latitude). Constructions made by termites include: numerous galleries tunnelled through wood and soil; underground chambers containing the queen, king, larval nurseries, and (in the sub-family *Macrotermitinae*) symbiotic fungus gardens; runways attached to the sides of trees, decaying wood and buildings; soil sheeting covering the surface of the ground; and arboreal or epigeal (i.e., emerging from the surface) mound-nests with a complex internal structure. Materials utilised, manufactured, or translocated by

Fig. 1 A termite-dominated landscape: mounds of *Amitermes vitiosus* in mulga country, West MacDonnell National Park, Northern Territory, Australia. (Photo, D.E. Bignell.)

termites include: surface soil; sub-soil; compacted feces; and carton (a lightweight organic-rich mixture of partially digested cellulose, saliva and soil). Epigeal mounds can attain a height of more than 7 m, but most are much smaller (Fig. 1). Many termites live underground with little indication of their presence at the soil surface. A large number of species are not primary mound builders, either living entirely in diffuse subterranean gallery networks or becoming established in pockets within the mounds of other species as secondary occupants (6, 7). Arboreal species are often conspicuous, but less important in biomass terms than subterranean forms, except where inundation is frequent. All termites are detritivores, consuming and digesting (in different species) a wide range of freshly dead or decaying plant material including dry grass, leaf litter, sound wood, decaying wood, dung, and humus-rich soil. Lichen and living roots may also be eaten, often without damage to plants. Termite abundance varies from less than 50 individuals m^{-2} in arid savannas to more than 7000 m^{-2} in some African forests. The corresponding biomass densities range from negligible to more than 100 g live weight m^{-2} in exceptional cases. In the tropics as a whole they are estimated to constitute 10% of all animal biomass (up to 95% of soil insect biomass), and impact C mineralization (decomposition) to roughly the same extent as all mammalian herbivores and natural fires. Overall, they contribute about 2% of the CO_2 flux to the atmosphere from all terrestrial sources. Earlier concerns about the possible role of CH_4 production by termites in global warming have now been shown to be unfounded (8).

EFFECTS ON SOILS

Through their activities as decomposers and constructors, termites have important roles in the genesis, morphology, and properties of soils throughout the tropics and sub-tropics (5). They also fix nitrogen on a significant scale, through mutualistic gut bacteria, and may be the primary providers of this nutrient in many soils and food chains, with termite mounds serving as foci for nutrient redistribution in some landscapes. It is highly probable that termites function as keystone species in ecosystems (creating niche opportunities for a large number of other organisms) and as ecosystem engineers (mediating biological function and maintaining soil structure, fertility and stability). Evidence is accumulating that termites (with earthworms) strongly affect C mineralization, the establishment of stable pools of complex organic material, nitrogen fixation, and denitrification. As a consequence, long-term exclusion or poisoning of termites changes the character of plant communities, as well as adversely affecting soil porosity, aeration, water-holding capacity, hydraulic conductivity, infiltration rates, and drainage (4, 7).

The physical effects of termites on soils range from micromorphological to soil profile evolution, and structure (7). Soil translocation is a major consequence of nest-building and the construction of foraging runways at the soil surface or above the ground. Estimates of the quantities moved vary, but may approach 5 tonnes ha^{-1} $year^{-1}$ in some systems; over thousands of years these activities strongly influence soil profile development, especially lateritic profiles and ferrisols (7, 9, 10). Micromorphological analysis of tropical soils frequently shows the presence of small (< 1 mm) pellets of biological origin, thought to be produced either by termite fecal deposits or the mixing of soil with termite saliva using the mandibles. The pellets, which may comprise up to 20% of the soil matrix, have a major influence on microporosity, moisture storage capacity, and infiltration rate. Subterranean chambers and galleries are also a conspicuous feature of soil profiles, almost universally detectable in soil cores and accounting for up to 2% of total soil volume, with effects on both infiltration rates and drainage. The behavior of termites in repacking soil particles, augmented with organic material derived from salivary and fecal products, while constructing mounds and galleries, is a key activity (7). It can be shown to affect bulk density (this is increased, restricting seed lodgement and growth), structural stability (this is increased when termites construct mounds that are highly organic, for example containing hemicellulose or glycoprotein) and hydraulic conductivity (both conductivity and flow rates are reduced if termite galleries are

Table 1 Soil properties significantly affected by termites

Soil property	Reference
Soil profile development	(7, 10–12)
Bulk density	(7)
Hydraulic conductivity/infiltration rates	(7, 13)
Soil chemical properties	(5)
Organic matter decomposition	(7)

reduced and, correspondingly, increased after mulching which attracts additional termite foraging).

The chemical properties of soils depend primarily on their parent materials, but with additional influence from climate, vegetation cover, and the activities of soil organisms. A notable contribution of termites is the incorporation of cation-rich clay sub-soils, together with organic-rich fecal and salivary materials, into constructions (5, 7, 9). In many cases, when these are subsequently eroded, the surface soil becomes enriched in Ca, Mg, Na, K, together with organic N, P and interstratified clay minerals. Several studies have shown that the inorganic nitrogen concentrations of mound soil or of soil reworked by termites are higher than adjacent soils that are unaffected by termites; this nitrogen apparently originates from symbiotic fixation within the termite gut. Leaching of such nitrogen represents a short-term redistribution and can be a significant input to the soil pool that is immediately available to plants (7). Non-mutualistic bacteria and fungi may also be present in larger numbers in mound soils; assessments of their activities suggest a vigorous metabolism which includes (in separate species) cellulose degradation, nitrification, and denitrification. Many evidences indicate that the presence of termite mounds promotes the accumulation of plant biomass on or around the structure (3, 7). In semi-arid systems, mulching of the soil to promote termite activity can lead to significant improvements in plant cover, plant species number, biomass production, and rainfall use efficiency. A summary of the effects of termites on soils is given in Table 1.

CONCLUSIONS

Termites are abundant soil invertebrates in much of the tropics and sub-tropics. The physical effects of termites on soils range from the micromorphological to soil profile evolution, and soil structure. There is evidence of enhancement of soil hydraulic conductivity and infiltration rates. Chemical effects include the promotion of litter decomposition, nutrient recycling, and the formation of stable pools of soil organic matter. In some landscapes termite mounds act as foci for nutrient redistribution and contribute inorganic nitrogen following symbiotic fixation in the insect's gut. Termite mounds are sites of activity for diverse bacteria and fungi.

REFERENCES

1. Abe, T.; Bignell, D.E.; Higashi, M., Eds. *Termites: Evolution, Sociality, Symbioses, Ecology*; Kluwer Academic Publishers: Dordrecht, Netherlands, 2000; 466.
2. Pearce, M.J. *Termites, Biology and Pest Management*; CAB International: Wallingford, UK, 1997; 172.
3. Wood, T.G. The Agricultural Importance of Termites in the Tropics. Agr. Zool. Rev. **1996**, *7*, 117–155.
4. Lavelle, P.; Bignell, D.E.; Lepage, M. Soil Function in a Changing World; the Role of Invertebrate Ecosystem Engineers. Eur. J. Soil Biol. **1997**, *33*, 159–193.
5. Black, H.I.J.; Okwakol, M.J.N. Agricultural Intensification, Soil Biodiversity and Agroecosystem Function in the Tropics: The Role of Termites. Appl. Soil Ecol. **1997**, *6*, 37–53.
6. Noirot, C.; Darlington, J.P.E.C. Termite Nests. Architecture, Regulation and Defence. In *Termites: Evolution, Sociality, Symbioses, Ecology*; Abe, T., Bignell, D.E., Higashi, M., Eds.; Kluwer Academic Publishers: Dordrecht, Netherlands, 2000; 121–139.
7. Holt, J.A.; Lepage, M. Termites and Soil Properties. In *Termites: Evolution, Sociality, Symbioses, Ecology*; Abe, T., Bignell, D.E., Higashi, M., Eds.; Kluwer Academic Publishers: Dordrecht, Netherlands, 2000; 389–407.
8. Sugimoto, A.; Bignell, D.E.; MacDonald, J.A. Global Impact of Termites on the Carbon Cycle and Atmospheric Trace Gases. In *Termites: Evolution, Sociality, Symbioses, Ecology*; Abe, T., Bignell, D.E., Higashi, M., Eds.; Kluwer Academic Publishers: Dordrecht, Netherlands, 2000; 409–435.
9. Lobry de Bruyn, L.; Conacher, A.J. The Role of Termites and Ants in Soil Modification: A Review. Aust. J. Soil Res. **1990**, *28*, 55–93.
10. Lal, R. *Tropical Ecology and Physical Edaphology*; John Wiley and Sons: Chichester, UK, 1987; 732.
11. Stoops, G. Relict Properties of Soils in Humid Tropical Regions with Special Reference to Central Africa. Catena Suppl. **1989**, *16*, 95–106.
12. Tardy, Y.; Roquin, C. Geochemistry and Evolution of Lateritic Landscapes. In *Weathering, Soils & Paleosols*; Martini, I.P., Chesworth, W., Eds.; Elsevier: London, 1992; 407–443.
13. Ouedraogo, P.; Lepage, M. Rôle des Termitières de *Macrotermes subhyalinus* dans une Brousse Tigrée (Yatenga, Burkina Faso). In *Fonctionnement et Gestion des Écosystèmes Forestiers Contractés Sahéliens*; Herbès, M., Ambouta, J.M.K., Peltier, R., Eds.; John Libby Eurotext: Paris, 1997; 91–94.

TERRACE RELATIONSHIPS: TILLAGE EROSION

Seth M. Dabney
United States Department of Agriculture, Oxford, Mississippi, U.S.A.

INTRODUCTION

All tillage moves soil. Implements drawn by animals or a tractor move soil in the direction of travel and, to a lesser extent, sideways. While a moldboard plow throws soil to only one side, most tillage implements—including tandem disks, chisel plows, harrows, and cultivators—throw soil to both sides. With such implements, a tillage operation along the contour moves some soil uphill, more soil downhill, and still more soil along the contour in the direction of travel. Net soil loss occurs where the tillage tool first engages the soil, as well as on slope convexities. Soil is deposited in depressions, along the upper and lower edges of the tilled area, and at points where the tillage tool is removed from the ground and turned around. With each tillage operation, the field becomes flatter and some soil leaves the tilled zone in the form of clods deposited along field borders. Over time, these clods along with stones removed from the tilled area and sediment deposited by runoff, coalesce to form terraces or *lynchets*. The formation of such terraces may or may not be desirable, but can hardly be avoided if field boundaries are fixed for long periods of time.

This section reviews research into terrace formation by tillage, identifies ways in which these terraces can be advantageous to soil and water conservation, and discusses problems that have been recognized with tillage terraces.

LYNCHETS

Archaeologists use the term *lynchet* to refer to soil banks that are considered the morphologic response on a hillslope to the presence of field boundaries in cultivated landscapes (1). Lynchets form at all field boundaries, whether bounded by untilled grass strips (balks), fences, or ditches. They frequently contain piles of stones at their core. The aggrading side of a field boundary is often termed a *positive lynchet* while the degrading side (the upslope edge of a tilled area) is termed a *negative lynchet* (Fig. 1). Positive lynchets have deeper, more fertile, and more productive soils than negative lynchets, where subsoil may be exposed (2–5).

Hand tillage with a hoe differs from that associated with draft animals or tractors because soil usually moves in a direction opposite of the workers' travel. Because hoeing is hard work, it is common for soil to move mainly downhill, with work starting at the bottom of a field and proceeding upslope (3). Again, the net result is gradual terrace formation.

Soil erosion by water generally increases with increasing slope length because of an accumulation of runoff. This should lead to the development of concave hillslope profiles with little change in elevation near hilltops (7). In contrast, the quantity of soil moved by tillage is independent of field size and leads to rapid soil loss at upslope field boundaries. Since erosion rates are generally considered in terms of mass per unit area, the amount of soil "eroded" by tillage is much greater on narrow fields than on long slopes (3, 8). Thus, the more strips into which a field is divided, the greater the contribution of tillage translocation to terrace formation.

ENGINEERED TERRACES

Engineered terraces are designed to manage runoff from a certain contributing area. Therefore, when a field is chosen to receive more than one terrace, it is common practice to start at the top of the field and work down so that the bottom terrace is not overloaded before upslope terraces are completed. While specialized equipment such as bulldozers and scrapers are generally used for terrace construction, ordinary farm machinery can also be used to create broad-base terraces (9). Proper tillage operations are needed as part of routine maintenance to preserve functionality of all terraces (10).

BENCH TERRACES

For deep soils, tillage translocation can be used to reduce the cost and increase the farmability of bench terraces. On irregular fields, starting terrace construction at the bottom of the slope may facilitate development of straight parallel

Encyclopedia of Soil Science

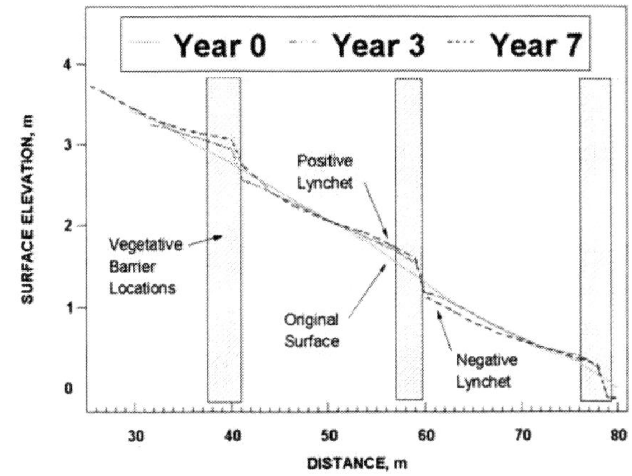

Fig. 1 Landscape change during seven years of tilled fallow between vegetated strips shows progression toward formation of bench terraces near Coffeeville, Mississippi.

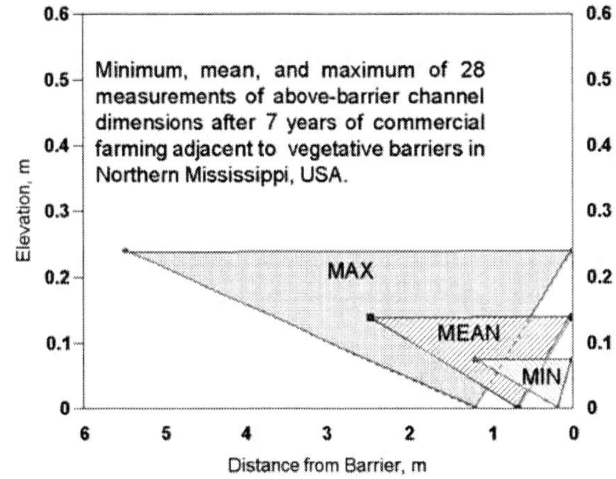

Fig. 2 When vegetated strips deviate slightly from the contour, small channels can act as small gradient terraces and divert an appreciable fraction of the runoff. (From Ref. 13.)

terrace systems. First, a bulldozer pushes up soil from below into a straight terrace that is extra high where it crosses existing gullies (11, 12). A tile outlet near the thalweg drains impounded runoff, leaving sediment trapped in the low area. When the thalweg area is sufficiently filled (up to 5 m of fill in 3 years), further sediment is cut off by construction of the next upslope terrace, parallel to the first. In one case, downhill moldboard plowing between terraces accounted for 50% of the soil moved in reducing a slope from 14% to 4% over a 15-year period.

A similar process occurs when fields are tilled between parallel vegetated strips. The vegetation conserves soil and water while benches gradually develop. The vertical interval between vegetated strips should not exceed 2 m and the planned vegetated backslope should not be steeper than 1 to 2, horizontal to vertical (10, 13).

GRADIENT TERRACES

Tillage adjacent to untilled, off-contour vegetated strips may create small berms and channels that can greatly alter runoff flow patterns. As an illustration, consider the asymmetrical triangular channels (Fig. 2) formed by farming with tandem disks and chisel plows between grass strips in a field with 6–9% slope steepness. After seven years of farming, microtopography was assessed at 28 locations along the upslope edge of several grass strips. In every case, a slope reversal that created a channel was observed. The hydraulic radius of the smallest channel was 0.04 m. At 0.2% grade and Manning's $n = 0.025$, this smallest channel could carry 25 mm/hr of runoff from a 15-m wide cropped strip more than 100 m long. Since the distance between local depressions in the field was less than 200 m, these tillage-induced channels could divert most of the field runoff to depressions where a stable outlet could be provided. However, note that the dimensions of such tillage-induced terraces are much smaller than those of standard terraces, and the channels could fill with sediment if upslope erosion rates were high and channel grades low.

The dimensions of the berm and channel combinations illustrated in Fig. 2 are probably a function of tillage implement type and speed, slope steepness, superposition of sequential tillage operations, and channel scour. A practical berm-height limit for modern disks is probably approximately 0.15 m. If berms get too high, clods cannot be thrown to their top, but will fall back into the channel. The same phenomenon limits upslope tillage translocation at the upper field border after a negative lynchet has formed.

TERRACE BENEFITS AND PROBLEMS

Bench terraces facilitate contour farming operations. On very steep slopes, mechanical cultivating is difficult and contour plowing can be dangerous because of the risk of overturning. The formation of bench terraces reduces the slope steepness and makes contour tillage easier and safer. When tractor or animal power is limited, it may be easier to plow downhill (8). However, when sufficient power is available, it is most efficient to plow on the contour because a uniform load can be matched to a tractor or a team of animals.

Contour tillage reduces runoff and soil erosion, thereby conserving plant nutrients and making more water available to grow crops (6). On the other hand, large fertility differences can develop between the negative and positive lynchets (2–4). This reduces uniformity of crop growth where fertilizer and liming amendments are not available. In these situations, farmers often cut soil from the terrace backslope and distribute it onto the degraded negative lynchet. If the upper field is held by a different landowner, an entire terrace may be undermined in this fashion (4). Manuring, fertilization with seaweed, and liming with marl were ancient remedies to similar problems (1).

REFERENCES

1. Bell, M. The Prehistory of Soil Erosion. In *Past and Present Soil Erosion: Archaeological and Geographical Perspectives*; Bell, M., Boardman, J., Eds. Oxbow Monogragh 22; Oxbow Books: Oxford, United Kingdom, 1992; 21–35.
2. Nyssen, J.; Haile, M.; Moeyersons, J.; Poesen, J.; Deckers, J. Soil and Water Conservation in Tigray (Northern Ethiopia): The Traditional Daget Technique and Its Integration with Introduced Techniques. Land Degrad. Develop. **2000**, *11*, 199–208.
3. Turkelboom, F.; Poesen, J.; Ohler, I.; Ongprasert, S. Reassessment of Tillage Erosion Rates by Manual Tillage on Steep Slopes in Northern Thailand. Soil Tillage Res. **1999**, *51*, 245–259.
4. Lewis, L.A. Terracing and Accelerated Soil Loss on Rwandian Steeplands: A Preliminary Investigation of the Implications of Human Activities Affecting Soil Movement. Land Degrad. Rehab. **1992**, *3*, 241–246.
5. Dabney, S.M.; Liu, Z.; Lane, M.; Douglas, J.; Flanagan, D.C. Landscape Benching from Tillage Erosion Between Grass Hedges. Soil Tillage Res. **1999**, *51*, 219–231.
6. Wadsworth, R.; Swetnam, R. Modeling the Impact of Climate Warming on the Landscape Scale: Will Bench Terraces Become Economically and Ecologically Viable Structures Under Changed Climate? Agr. Ecosys. Env. **1998**, *68*, 27–39.
7. Meyer, L.D.; Kramer, L.A. Erosion Equations Predict Land Slope Development. Ag. Eng. **1969**, *50*, 522–523.
8. Poesen, J.; van Wesemael, B.; Govers, G.; Martinez-Fernandez, J.; Desmet, P.; Vandaele, K.; Quine, T.; Degraer, G. Patterns of Rock Fragment Cover Generated by Tillage Erosion. Geomorphology **1997**, *18*, 183–197.
9. Samsel, L.G. Building and Maintaining Terraces with Ordinary Farm Machinery. Agr. Eng. **1943**, *24*, 337–338, 342.
10. USDA-Soil Conservation Service. *A Manual of Conservation of Soil and Water*, Agriculture Handbook No. 61; U.S. Dept. of Agriculture: Washington, DC, 1954; 79–107.
11. Jacobson, P. New Methods of Bench Terracing Steep Slopes. Transactions of the ASAE **1963**, *6*, 257–258, 261.
12. Jacobson, P. Improvements in Bench Terracing. Paper No. 67–233, 60th Annual Meeting of the ASAE, Saskatoon, Saskatchewan, Canada, June 27–30, 1967; American Society of Agricultural Engineers: St. Joseph, Michigan, 1967.
13. USDA-Natural Resources Conservation Service. Vegetative Barriers. *Conservation Buffers Workbook*, CORE4 Training Materials; USDA-NRCS: Lincoln, Nebraska, 1999; 117–137.

TESTING

Yash P. Kalra
J.S. Bhatti
Canadian Forest Service, Edmonton, Alberta, Canada

INTRODUCTION

Soil testing is generally defined as any chemical and physical measurement that is made on a soil. The term *soil testing* may be a rapid chemical analysis to assess the plant-available nutrient status, salinity, and elemental toxicity of a soil, or may represent a program that includes interpretations, evaluations, and fertilizer and amendment recommendations based on the results of chemical analysis (1, 2). As an agronomic tool, soil testing represents an index of measurement of nutrient availability. The objectives of soil testing are: a) to provide an index of nutrient supply in a soil; b) to predict the probability of obtaining a profitable response to fertilizers and/or lime; c) to provide a basis for recommendations on the amount of fertilizer and/or lime to apply; and d) to evaluate the fertility status of soils by the use of soil test summaries (3). Soil testing, combined with other agronomic information, helps determine the best way to manage nutrients. There is an interest currently to use soil testing to determine the pollution of surface and groundwater from fertilizer/waste material application. However, to identify the potential impact of pollutants, a comprehensive approach should be taken on the environment and soil testing. Different steps in soil testing follow.

SOIL SAMPLING

Soils are heterogeneous (both horizontally and vertically) and it is, therefore, important that the sample collected for analysis be representative of the field (1). Errors in soil sampling fall into three general categories: sampling error, selection error, and measurement error. Sampling error is caused by the inherent variation and can be avoided by including the total population in the sample. Selection error arises from any tendency to select some units of the population with greater or lesser probability than was intended. Measurement error is caused by the failure of the observed measurement to be true to value for the unit. It includes both the random errors of measurement and biases.

The most critical aspect of soil testing is obtaining a soil sample that is representative of the site (3). Statistical concepts determine which of the following three commonly used sampling strategies are used: a) simple random sampling; b) stratified random sampling; and c) systematic or grid sampling. Other important factors to consider are the depth of sampling, degree of field variability, number of cores needed to constitute a single composite sample, tools used to collect soil sample cores, type of containers, frequency of sampling, and the effect of season on sample collection (4).

PREPARATION OF SOIL SAMPLES FOR LABORATORY ANALYSIS

Sample preparation is critical in obtaining accurate results (4). The field-moist soil samples must be dried promptly to minimize microbial activity. The drying temperature can have a significant effect on the physio-chemical properties of soils. Generally the soils are air-dried. If an air-circulating oven is used to speed up the process, temperature should not exceed 38°C. The soils are ground to 2.00-mm size for routine analysis. For certain analyses, the samples need to be ground further to 1.00-mm size or finer.

APPARATUS FOR SOIL ANALYSIS

A soils laboratory would contain the following apparatus: soil grinder, sieves, auto-dispensers, funnel racks, titrator, time-controlled stirrer, time-controlled shaker, and balances. It would also have ovens, muffle furnace, pH meter, conductivity meter, sulfur analyzer, nitrogen analyzer, continuous flow analyzer (air-segmented and non-segmented stream instruments), UV-visible spectrophotometer, ion chromatograph (IC), atomic absorption spectrophotometer (AAS), flame emission spectrophotometer, and x-ray fluorescence spectrophotometer. Finally, the laboratory would have a potentiometric instrument, inductively coupled plasma-atomic emission spectrophotometer (ICP-AES) and glassware, plasticware, and other apparatus generally considered common to soil analysis laboratories. Computers, integral parts of modern laboratory instruments, are used for setting instrument parameters, controlling the operation of the instruments, and data capture and storage.

METHODS OF SOIL TESTING

Chemical soil tests are designed to estimate the amounts of plant nutrients available for plant growth. Different reagents are being used to extract different nutrients as listed in Table 1 and analyzed using the instruments given in the previous paragraph. Apart from nutrient contents, loss-on ignition or a potassium dichromate ($K_2Cr_2O_7$) wet oxidation procedure routinely determines organic matter. Soil texture is determined by sedimentation procedure in the laboratory and hand texturing method in the field (4).

Soil pH is determined in a soil-water slurry of 1:1 or 1:2 or in a soil slurry of 0.01 *M* $CaCl_2$ or 1 *M* KCl. Exchangeable acidity is determined by using the $BaCl_2$-TEA (triethanolamine) extractant buffered to pH 8.2. The Adams-Evans buffer (5) and the Shoemaker, Mclean, and Pratt (SMP) buffer (6) methods are used to determine the lime requirements of coarse-textured and fine-textured soils, respectively.

Fixed extraction ratios (e.g., 1:2, 1:5, etc.) and saturation extract methods are used for the measurement of electrical conductivity (EC). However, an estimation of the soluble salt content of soils is carried out using a 1:2 soil: water ratio. The Sodium Adsorption Ratio (SAR) is a useful index of soil sodicity based on the ratio of basic cations, which can be calculated from chemical analysis of the soil extract (4).

Soil structure, air porosity, water desorption characteristics, bulk density, saturated hydraulic conductivity, and infiltration rate are measured in situ and laboratory (4).

CORRELATION AND CALIBRATION

Correlation is the process of determining whether there is a relationship between the soil test results and plant uptake of a nutrient or yield. A soil test for a given nutrient has historically involved three steps: a) selecting an extractant; b) correlating the amount of nutrient

Table 1 Commonly used extractants for plant nutrients

Soil nutrient	Extractant
NO_3-N	2.0 *M* KCl
NH_4-N	2.0 *M* KCl
P	0.03 *M* NH_4F + 0.025 *M* HCl (Bray-P1)
	0.05 *M* HCl + 0.0125 *M* H_2SO_4 (Mehlich 1)
	0.5 *M* $NaHCO_3$, pH 8.5 (Olsen -P)
	0.2 *M* acetic acid (CH_3COOH) + 0.25 *M* NH_4NO_3 +0.015 *M* NH_4F + 0.013 *M* HNO_3 + 0.001 *M* EDTA (ethylene diamine tetraacetic acid) (Mehlich 3)
	0.25 *M* CH_3COOH + 0.15 *M* NH_4F (Kelowna)
Cation exchange capacity (CEC, Ca, Mg, K, Na)	1.0 *M* Ammonium acetate ($NH_4C_2H_3O_2$), pH 7.0
S	Ca $(H_2PO_4)_2$, containing 500 mg/L P
	0.01 *M* $CaCl_2$
	0.25 *M* CH_3COOH + 0.15 *M* NH_4F (Kelowna)
Fe, Mn, Cu, Zn	0.005 *M* DTPA (diethylene triamine pentaacetic acid) +0.01 *M* $CaCl_2$, and 0.1 *M* TEA (triethanolamine) $(HOCH_2CH_2)_3$ N adjusted to pH 7.30 ± 0.05 with 1:1 HCl
B	Hot water
Cl	Water

(From Ref. 4.)

extracted with the amount taken up by the plant; and c) calibrating the test values with crop yields. Nutrients extracted in soil testing procedures are frequently referred to as "available nutrients." The *Glossary of Soil Science Terms* (7) defines *available nutrients* as the amount of soil nutrient in chemical forms accessible to plant roots or compounds likely to be convertible to such forms during the growing season. For the purpose of soil testing correlation, the rate at which a nutrient is taken up by the plant becomes more important than the amount of the available nutrient. Thus, correlation of estimated nutrient uptake based on soil testing and other pertinent factors with total uptake by plant, seems to be a logical approach (8). This can be accomplished by greenhouse experiments and field trials, along with the use of empirical and mechanistic modeling approaches.

Calibration is the process of determining the probability of getting a growth response to applied nutrients. Calibration of a soil test is basic to a good soil-testing program. Soil test extractants were designed to assess rapidly and accurately the available nutrient status and/or elemental toxicity in soils. They were also designed to provide a quantitative basis for recommendations of the rates of plant nutrients that should be added as fertilizers, manures, or other materials to achieve economically optimum yield. A properly calibrated soil test should provide information on the degree of deficiency or sufficiency of an element, and identify how much of the element should be applied if it is deficient. There are additional criteria of a successful soil test. It should be easy to perform to ensure rapid turn-around time, cost effective to promote wide usage, and reproducible.

INTERPRETATION AND RECOMMENDATIONS

Soil testing interpretation is the process by which one assesses the fertility status of a field based upon a set of known factors. It involves an evaluation of chemical test results in terms of basic soil-plant relationship phenomena. More complex interpretation may include consideration of factors such as the crop to be grown, yield goal, individual soil characteristics, climate, tillage system, and environmental quality (9). At present, many fertilizer recommendations are made using interpretation of both the concepts and designs to fertilize the soil rather than the plants. Recommendations based on basic cation saturation ratio are designed to alter the existing ratio to achieve the ideal K:Ca:Mg ratio, regardless of existing and potential yield levels. The recommendations are based on soil test results and other agronomic conditions (9).

QUALITY CONTROL PROCEDURES FOR SOIL ANALYSIS

The quality of soil test results is evaluated on the basis of two components: bias and precision. Bias refers to the deviation of analytical result from the true value and measures accuracy of the result. Precision refers to the reproducibility of a given test value. A primary means to evaluating soil test quality procedures, as well as performance, is through a comparison of laboratories performing the same test procedure on the same soil. The use of appropriate reference soils to assess, monitor, maintain, and to ensure accuracy of analytical results, should be part of a soil testing procedure. Incorporation of reference materials is one cost-effective facet of a data quality program to ensure accuracy of analytical results (4).

It is vital that laboratories monitor and document the quality of their analytical results on a regular basis. For this purpose, many laboratories participate in proficiency testing programs. The Soil and Plant Analysis Council, Inc. (formerly the Council on Soil Testing and Plant Analysis) was founded in 1969 to address the issues of uniformity in the analytical procedures and fertilizer recommendations (10). The AOAC INTERNATIONAL is an association of scientists and organizations devoted to promoting methods validation and quality measurements in the analytical sciences. In 1990, the Soil Science Society of America and AOAC INTERNATIONAL established a joint committee to conduct validation studies for methods of soil analysis. The first validation was for soil pH methods (11).

FUTURE OF SOIL TESTING

The future of soil testing is bright. There will be greater automation in the laboratory and wider use of multi-element extractants. The search for better methods to evaluate bioavailability of elements in soils will continue to be an important concern, and will affect both the agronomic and analytical aspects of soil testing. No unusually drastic change is anticipated concerning laboratory procedures, but there is an expected change in the future use of soil tests, as less of a diagnostic procedure and more of a monitoring one.

ACKNOWLEDGMENTS

The authors are grateful to Drs. J.D. Beaton, J.B. Jones, Jr., T. D. Peck, and B. van Raij for their help in the preparation of the article.

REFERENCES

1. Westerman, R.L., Ed. *Soil Testing and Plant Analysis*, 3rd Ed.; SSSA Book Series Number 3, Soil Science Society of America: Madison, Wisconsin, 1990; 1–784.
2. Walsh, L.M.; Beaton, J.D. *Soil Testing and Plant Analysis*; Soil Science Society of America: Madison, Wisconsin, 1973; 1–491.
3. Havlin, J.L.; Beaton, J.D.; Tisdale, S.L.; Nelson, W.L. *Soil Fertility and Fertilizers: An Introduction to Nutrient Management*; Prentice Hall: Upper Saddle River, New Jersey, 1999; 1–499.
4. Carter, M.R., Ed. *Soil Sampling and Methods of Analysis*; CRC Press: Boca Raton, Florida, 1993; 1–823.
5. Adams, F.; Evans, C.E. A Rapid Method for Measuring the Lime Requirement of Red-Yellow Podzilic Soils. Soil Sci. Soc. Am. Proc. **1962**, *26*, 355–357.
6. Shoemaker, H.E.; Mclean, E.O.; Pratt, P.F. Buffer Methods for Determining the Lime Requirement of Soils with Appreciable Amounts of Extractable Aluminum. Soil Sci. Soc. Am. Proc. **1961**, *25*, 274–277.
7. *Glossary of Soil Science Society of America Terms*; Soil Science Society of America: Madison, Wisconsin, 1996; 1–134.
8. Jones, J.B., Jr. Soil Test Methods: Past, Present, and Future. Commun. Soil Sci. Plant Anal. **1998**, *29*, 1543–1552.
9. Brown, J.R., Ed. *Soil Testing: Sampling, Correlation, Calibration, and Interpretation*; SSSA Special Publication Number 21, Soil Science Society of America: Madison, Wisconsin, 1987; 1–244.
10. Jones, J.B., Jr.; Kalra, Y.P. Soil Testing and Plant Analysis Activities—The United States and Canada. Commun. Soil Sci. Plant Anal. **1992**, *23*, 2015–2027.
11. Kalra, Y.P. Determination of pH of Soils by Different Methods: Collaborative Study. JAOAC INTERNATIONAL **1995**, *78*, 310–324.

TEXTURE

Harold R. Geering
The University of Sydney, Sydney, New South Wales, Australia

Hwat Bing So
The University of Queensland, St. Lucia, Queensland, Australia

INTRODUCTION

Besides color, one of the principal descriptors of a mineral soil (i.e., less than 10% organic matter content) is its texture, which is defined as the particle size distribution of the fine earth fraction (<2 mm fraction). The distribution of particle size is conveniently partitioned by particle size analysis (PSA) into subfractions of sand, silt and clay, the proportions of which will have a dominant influence on many of the practical soil properties important to agriculture, the environment and engineering purposes, such as ease of cultivation, nutrient and water holding capacities and transmission characteristics, earth dam construction and susceptibility to erosion. However, with respect to particle size ranges the subdivision into sand, silt and clay fractions varies with country and professional institution (10). Examples of four of these particle size classifications are shown in Fig. 1 (2). In defining soil texture, the first three systems shown in Fig. 1 are amongst the most widely used in textural class designation.

For describing a soil and/or soil horizon portion of a soil profile, the soil is allocated to a textural class. The textural class descriptors provide an indication of which particle size fraction is dominant, with the exception of a loam where no single size range exerts a dominant influence. In the field the textural class is estimated by the hand-feel of a moist sample of fine earth fraction, when molded between the thumb and forefingers at just below the "sticky point," where the soil just begins to stick to the fingers (3). Initially, the soil is remolded several times to form a uniform moist bolus of about the size of a golf ball. Then, the bolus is deformed into threads (see Fig. 2) or ribbons approximately 3 mm thick (Fig. 3). The force necessary to deform and shear the bolus, and the length of these threads or ribbons tends to increase with clay content. At the same time, if the sample contains a particular size fraction that is readily identifiable by feel to the fingers, this is taken into account in deciding on the soil texture class, e.g., the grittiness of coarse sand or the slippery silky feel of silt. Alternatively, medium and coarse sand can also be seen with a hand lens, and fine sand can be heard by its crunchy grinding sound when remolding and shearing is done close to the ears. Finally, organic matter when present as humus at >10%, usually confers cohesion to sandy textures and greasiness to clayey textures (15), or as described by Clarke (4) tends to make sandy and clayey soils feel more loamy in texture.

Taking all of the above description and criteria into account, soil texturing by hand is essentially a consistency test (see entry on soil consistency) at the sticky point (5). It then follows that since the amount of work and cohesion increase in the order of the three principal texture class of sand, loam and clay, they are often well correlated with the texture classes derived from PSA (10).

PARTICLE SIZE ANALYSIS

When accurate data on the soils particle size distribution are desired, a PSA can be conducted in the laboratory on a sample (25–50 g of oven dry soil) of the fine earth fraction. The sample is treated to allow the individual primary soil particles to be freely suspended in water prior to the determination of the proportion of sand, silt and clay. As primary soil particles are generally found as conglomerates or bonded aggregates, they are treated with one or more of the following sequence of treatments: boiling to soften the bonds, 30% hydrogen peroxide to oxidize organic matter bonds and 1 M HCl acid to dissolve silica and iron-aluminum bonds. This is then followed by the addition of Na cations (as NaOH–Na hexa-*meta* phosphate mixture) and mechanical shaking using a milk shaker or an end over end shaker (for 16 h), to ensure that all clay particles are fully dispersed and remain dispersed. Alternatively, the ultrasonic probe is also routinely used to achieve effective separation of individual particles. Separation of the various size classes of particles is achieved using a combination of sieving, sedimentation and decantation. Each size fraction is expressed as an oven dry mass fraction or percentage of the original sample.

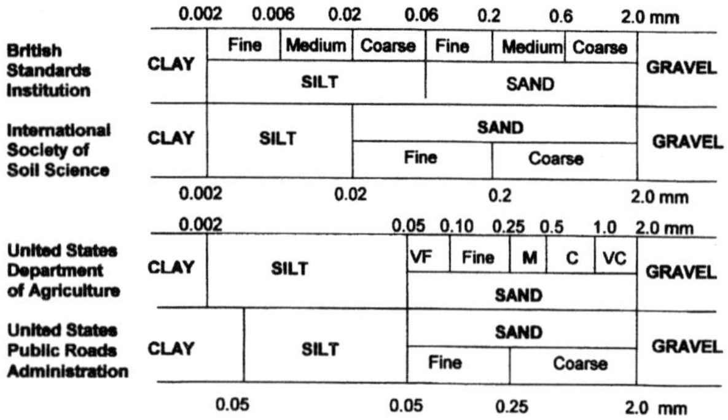

Fig. 1 Designation of particle size ranges to sand, silt and clay by four professional institutions. The British Standards Institution and Massachusetts Institute of Technology use the same classification. (From Ref. 2.)

Details can be found in a number of standard soils methodology texts (12–14). The percentages of sand, silt and clay fractions are then plotted on a texture triangle that is appropriate for the size fractions selected. Some of the most frequently used texture triangles are shown in Fig. 4. The selection of which texture diagram to use is dependent on the silt size fraction used.

RELEVANCE OF SOIL TEXTURE

The most fundamental property associated with particle size distribution is the specific surface area (SSA) of the soil expressed as surface area per unit mass or m^2g^{-1} (see entry on SSA). Smaller particles possess larger SSA, e.g., SSA for sand is approximately $0.01-0.1\,m^2g^{-1}$, for silt approximately $1\,m^2g^{-1}$ and for clay may range from 10 to $800\,m^2g^{-1}$. As most soils contain a considerable amount of clay, it is characterized by large SSA. All physical and chemical processes in the soil occur at the interface of the soil solids, hence the larger the SSA the greater will be the cumulative effect of these processes, and hence the clay fraction plays the major role in this regard.

Ease of Cultivation

The ease of cultivation for soils of different textures gives rise to the expressions of “light textured soils” and “heavy textured soils” being used for sandy and clayey soils, respectively. Coarse-grained, sandy soils tend to be loose, well aerated and easy or light to cultivate. Fine-textured soils tend to absorb much water and may become plastic and sticky when wet. Hence, they require much more energy and are heavy to cultivate. Upon drying, they tend to become dense and cohesive (9).

Nutrient- and Water-Holding Capacity, and Transmission Properties

The SSA and the magnitude of the electrical charges per unit area of clay and silt determine the cation-exchange capacity (CEC) of the soil, which is the amount of cations that a unit mass of soil can hold (see entry on CEC). The larger the SSA, the larger is the soil’s nutrient-holding capacity, e.g., CEC for kaolinite, illite and montmorillonite clays range 3–15, 10–40 and 60–120 mmole+/kg, respectively. Sandy soils tend to have higher infiltration rates and greater leaching, resulting in acid pH and low base saturation. On the other hand, clayey soils tend to have lower infiltration rates and less leaching, resulting in neutral to alkaline pH and high base saturation.

Particle size distribution and SSA are important determinant of the soil’s water-holding capacity. Water is adsorbed on soil-surfaces and water will accumulate in the pore space between the particles as capillary water. At field capacity and wilting point, sandy soils tends to have lower water contents than clayey soils. The plant available water capacity (PAWC) or the amount of water held between field capacity and wilting point is also lower in sandy soils. The approximate relationship between PAWC and texture is shown in Fig. 5 and typically the largest plant available water capacity tends to be highest in the intermediate texture range (6). Particle size distribution will affect the pore size distribution of a soil and consequently it will influence the soil’s transmission characteristic, which is termed the soils hydraulic conductivity (see entry on soil hydraulic conductivity).

Packing of Particles, Hardsetting and Engineering Properties of Soils

Particle size distribution also affects the potential packing density of the soil or soil aggregates (1, 8, 17) and the engineering properties of the soil, or its response to an applied force generally referred to as soil consistency. The presence of sufficient quantities of small particles that will fill the space between the larger particles will result in a massive soil. This generally occurs on soils with 30–35% sand content. One specific soil characteristic associated with high contents of silt and fine sand particles (silty or fine sandy textured soils) is hardsetting, where soil aggregates readily break down upon wetting and dry out rapidly into a hard consistency. It softens considerably upon rewetting (18). Hardsetting soils are characterized by low water-holding capacities and they alternate rapidly between soft and hard consistencies, often resulting in poor soil physical conditions for most plants.

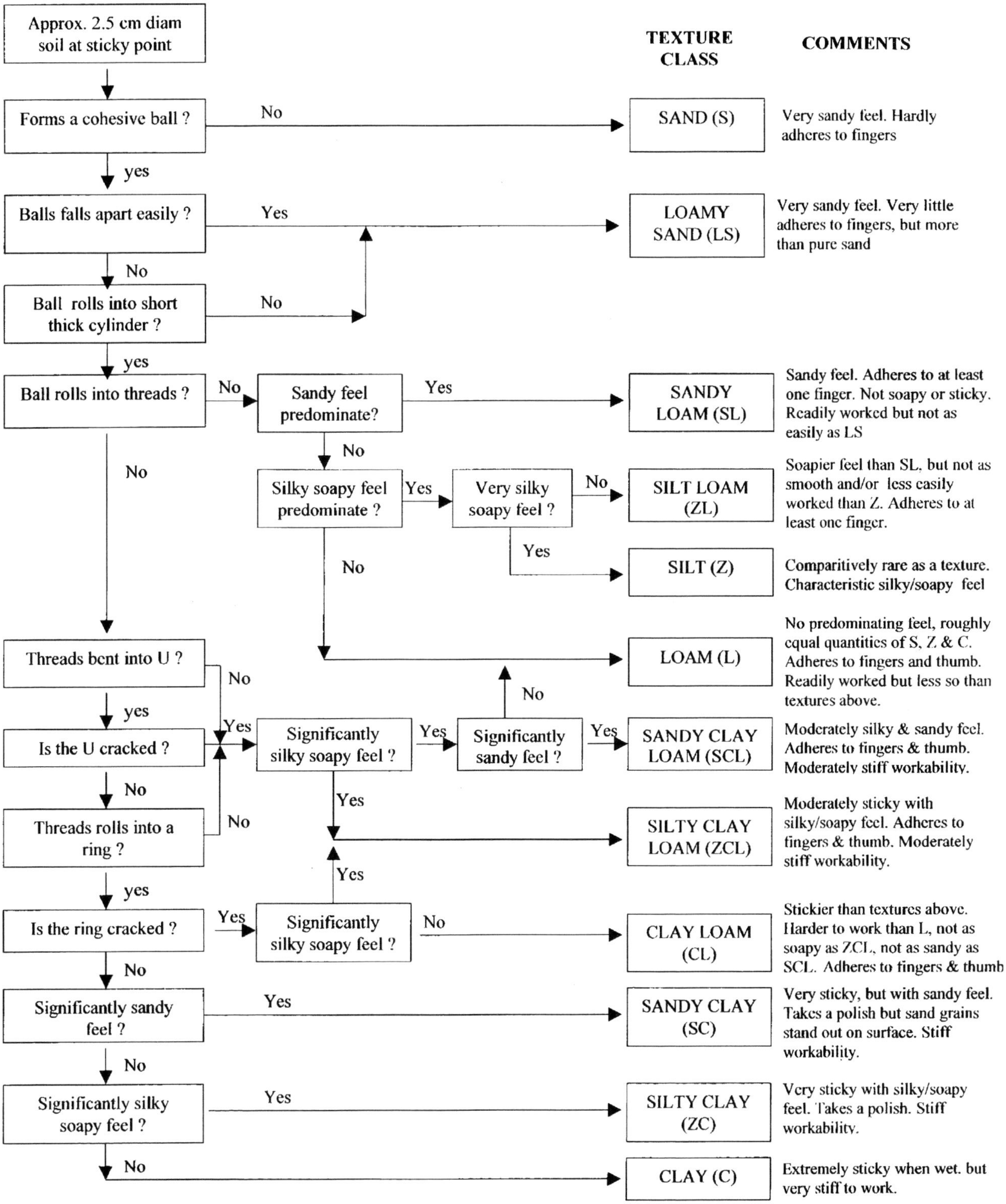

Fig. 2 A guide to field assessment of texture for mineral soils in the U.K. by S. Northcliff, Reading University and J.R. Landon, Booker Agricultural International. (From Ref. 16.)

Field Texture Groups	Ribbon length (mm)	Coherence of the bolus at sticky point	Feel	Other features	Texture Grade (Synonomous with Texture class)	Approx Clay %
1 ***The Sands***	Nil	Nil	Sandy	Single sand grains adhere to fingers	**1. Sand (S)**	Common-ly <5
	5	Slight	Sandy	Discolors fingers with an organic stain	**2. Loamy Sand (LS)**	5-10
	5 – 15	Slight	Sticky	Sand grains sticks to fingers and discolors with a clay stain	**3. Clayey Sand (CS)**	5-10
2 The Sandy Loams	15-25	Just Coherent	Very sandy	Medium Sand readily visible	**4. Sandy Loam (SL)**	10-20
	15-25	Just Coherent	Very sandy	Fine sand may be heard	**5. Fine Sandy Loam (FSL)**	10-20
	20-25	Strong	Sandy	Medium Sand easily visible	**6. Light Sandy Clay Loam (SCL-)**	15-20
3 The Loams	About 25	Coherent	Spongy & greasy	No obvious sandiness	**7. Loam (L)**	25
	About 25	Coherent	Slightly Spongy	Fine sand	**8. Loam Fine Sandy (Lfsy)**	25
	About 25	Coherent	Smooth	Silky; very smooth when manipulated	**9. Silt Loam (SiL)**	25 (>25%silt)
	25-40	Strong	sandy	Medium Sand in fine matrix	**10. Sandy Clay Loam (SCL)**	20-30
4 The Clay Loams	40-50	Strong	Smooth	No obvious sand grains	**11. Clay Loam (CL)**	30-35
	40-50	Coherent	Smooth	Silky feeling	**12. Silty Clay Loam (SiCL)**	30-35 (>25% silt)
	40-50	Coherent	Smooth & sandy	Fine sand can be felt and heard	**13. Fine Sandy Clay Loam (FSCL)**	30-35
5 The Light Clays	50-75	Coherent	Plastic	Fine to medium sand	**14. Sandy Clay (SC)**	35-40
	50-75	Coherent	Plastic	Smooth and silky	**15. Silty Clay (SiC)**	35-40 (>25% silt)
	50-75	Coherent	Plastic	Smooth with slight resistance to shearing	**16. Light Clay (LC)**	35-40
	> 75	Coherent	Plastic	Smooth with a little resistance to shearing	**17. Light Medium Clay (LMC)**	40-45
6 The Medium & Heavy Clays	> 75	Coherent	Plastic	Fair resistance to shearing	**18. Medium Clay (MC)**	45-55
	> 75	Coherent	Plastic	Firm resistance to shearing	**19. Heavy Clay (HC)**	>50

Fig. 3 A summary guide for assessment of soil texture for mineral soils in Australia, silt fraction is 0.02–0.002 mm. (From Ref. 14.)

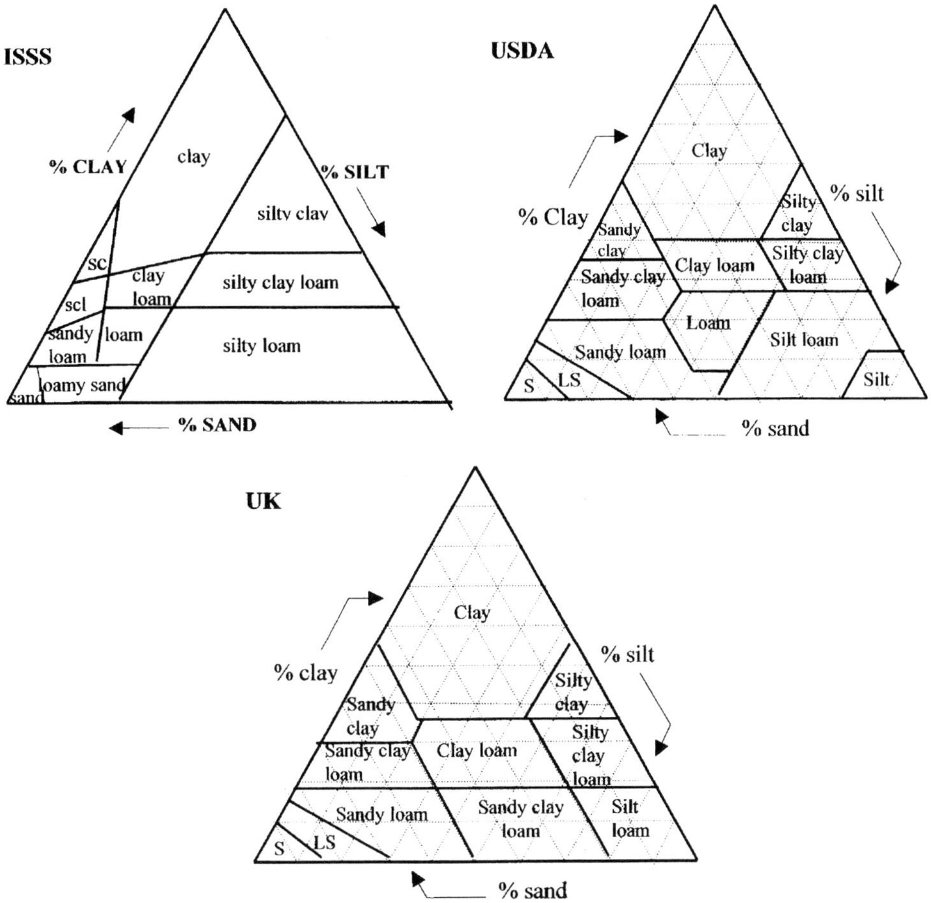

Fig. 4 Three of the most frequently used texture triangles. The ISSS is based on particle size of sand (2–0.02 mm), silt (0.02–0.0002 mm) and clay (<0.002 mm). The USDA triangle is based on silt fractions of 0.05–0.002 mm and the U.K. triangle on silt fractions of 0.06–0.002 mm. (From Ref. 7, 14, 16, 19.)

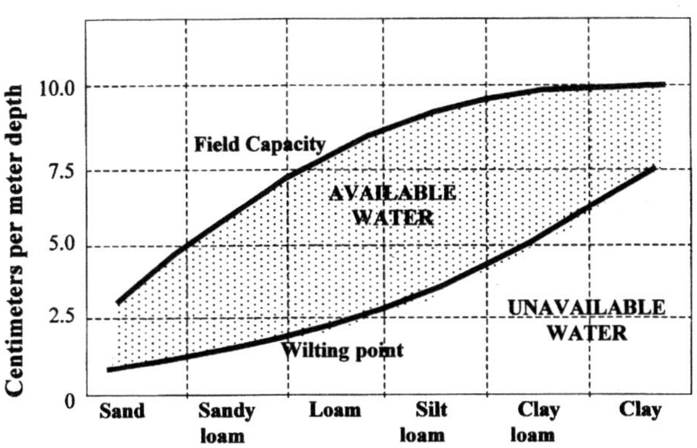

Fig. 5 Typical water-holding capacities of different textured soils. (From Ref. 6.)

REFERENCES

1. Bodman, G.B.; Constantin, G.K. Influence of Particle Size Distribution in Soil Compaction. Hilgardia **1965**, *36*, 567–591.
2. Brady, N.C. *The Nature and Properties of Soils*; 8th Ed. MacMillan: New York, 1974.
3. Cassel, D.K.; Nielson, D.R. Field Capacity and Available Water Capacity. In *Methods of Soil Analysis. Part 1. Physical and Mineralogical Methods*; Klute, A., Ed.; Chapter 36. ASA/SSSA: Madison, WI, 1986; 901–926.
4. Clark, G.R. *The Study of Soil in the Field*; Clarendon Press: Oxford, 1971; 46.
5. Collis-George, N. The Mechanical and Structural Properties of Soil. In *The Fundamentals of Modern Agriculture*; Blake, C.D., Ed.; Chapter 3. The University of Sydney Press: Sydney, Australia, 1967; 61–88.
6. Foth, H.D. *Fundamentals of Soil Science*, 6th Ed.; Wiley: New York, 1978.

7. Gee, G.W.; Bauder, J.W. Particle-Size Analysis. In *Methods of Soil Analysis. Part 1. Physical and Minerlogical Methods*; Klute, A., Ed.; Chapter 15. ASA/SSSA: Madison, WI, 1986; 383–412.
8. Gupta, S.C.; Larson, W.E. A Model for Predicting Packing Density of Soils Using Particle Size Distribution. Soil Sci. Soc. Am. Proc. **1979**, *19*, 758–764.
9. Hillel, D. *Introduction to Soil Physics*; Academic Press: New York, 1982.
10. Hodgson, J.M. *Soil Survey Field Handbook*; Technical Monograph No. 5, Soil Survey of England and Wales: Harpenden, 1974.
11. Hodgson, J.M. *Soil Sampling and Soil Description*; Oxford University Press: Oxford, 1978.
12. Klute, A., Ed.; *Methods of Soil Analysis. Part 1. Physical and Mineralogical Methods*; ASA/SSSA: Madison, WI, 1986.
13. Loveland, P.J.; Whalley, R.W. Particle Size Analysis. In *Soil Analysis, Physical Methods*; Smith, K.A., Mullins, C.E., Eds.; Chapter 6 Marcel Dekker: New York, 1991.
14. McDonald, R.C.; Isbell, R.F.; Speight, J.G.; Walker, P.J.; Hopkins, M.S. *Australian Soil and Field Handbook*, 2nd Ed.; Inkata Press: Melbourne, 1990; 115–123.
15. Northcote, K.H. *A Factual Key for the Recognition of Australian Soils*; CSIRO, Rellim Technical Publications: Adelaide, S. Australia, 1979.
16. zRowell, D.L. *Soil Science, Methods and Applications*; Chapters 1 and 2 Longman/Wiley: UK/New York, 1994.
17. Smith, G.D.; Coughlan, K.J.; Fox, W.E. The Role of Texture in Soil Structure. In *Modifications of Soil Structure*; Emmerson, W.W., Bond, R.D., Dexter, A.R., Eds.; Wiley: New York, 1978.
18. So, H.B.; Smith, G.D.; Raine, S.R.; Schafer, B.M.; Loch, R.J. *Sealing, Crusting and Hardsetting Soils: Productivity and Conservation*; Australian Society of Soil Science: Brisbane, 1995; 527.
19. Soil Survey Staff, *Soil Taxonomy. A Basic System of Soil Classification for Making and Interpreting Soil Surveys*; USDA Soil Conservation Service, R.E. Krieger: Malabar, Florida, 1988.

THERMAL ENVIRONMENT OF SEASONALLY FROZEN SOIL AFFECTED BY CROP AND SOIL MANAGEMENT

T

Brenton S. Sharratt
United States Department of Agriculture (USDA), Morris, Minnesota, U.S.A.

INTRODUCTION

The thermal regime of soils during winter and spring can have a profound influence on civil engineering designs, watershed hydrology, and agricultural productivity. Alfalfa (*Medicago sativa* L.) and winter wheat (*Triticum aestivum* L.), for example, can die as a consequence of heaving as soils freeze and thaw. Heaving may lift plants above the soil surface where there is a greater risk for exposure of crown and root tissues to lethal temperatures of about −20°C. In addition, the presence of frozen layers within the soil profile can impede water infiltration during snowmelt and rain events as well as retard drainage as the soil thaws in spring. Drainage, as well as soil temperatures, can also affect the timing of field operations such as planting. Agricultural productivity, therefore, depends on moderating soil temperatures during winter and rapidly thawing the soil profile in spring. Soil temperature and thaw can be regulated to some extent through crop residue and soil management.

The thermal regime of seasonally frozen soils is dependent on the interrelated processes of heat and water transfer at the soil surface and within the soil profile. The processes of heat and water transfer at the soil surface are governed by atmospheric conditions and physical properties of the soil at the soil–atmosphere interface. Atmospheric conditions such as air temperature, humidity, solar radiation, wind, and precipitation influence the energy available for processes such as soil heating and evaporation. In cold regions, snow cover will modify the atmospheric conditions at the soil surface. Texture, density, water content, and ice content are important physical properties of soils affecting the transmission of heat through a frozen soil. Ice formation typically occurs in soils at temperatures below 0°C due to capillary and adsorption forces that reduce the free energy status of water in soils. The process of freezing and thawing affects the quantity of heat transmitted through soils since heat ($334\,J\,g^{-1}$) is liberated as pore ice freezes and must be absorbed to melt pore ice. The process of ice formation and melting requires nearly 100 times the heat ($4.2\,J\,g^{-1}$) expended in warming or cooling water by 1°C. Thus, heat transfer processes in soils are influenced by phase transitions that occur as soils freeze and thaw. Atmospheric conditions and physical properties of the soil at the soil–atmosphere interface can also be modified by the quantity and orientation of crop residue on the soil surface as well as by the depth and type of tillage.

CROP RESIDUE

Crop residues can mitigate soil erosion, protect plants from winter temperature extremes, and provide a favorable environment for seed germination in the spring. Residues act as a barrier to heat and water transfer between the atmosphere and soil, thereby retarding heat loss from the soil during winter and hindering the warming of soil in spring. This thermal retardation is illustrated in Fig. 1 for a clear day in late autumn, winter, and spring. On warm autumn and spring days, soil without residue cover was warmer during the daytime and cooler at night than soil with residue cover. On cold winter days, however, soil with residue cover remained warmer throughout the day than soil without residue cover.

Residue management alters the amount and orientation of crop residue on the soil surface. Various techniques employed in managing crop residues include cutting stubble at various heights, burning residue, removing residue from the seed row, and altering the color of the residue. The winter thermal regime of soil can be dramatically affected by the height of stubble, especially in regions where strong winds redistribute snow. Taller stubble traps more snow and thereby provides additional insulation to the soil that can reduce frost penetration, hasten thawing of the soil profile, and elevate soil temperatures (Table 1). Burning crop residue in autumn or spring can hasten soil warming in the spring by removing residue from the soil surface and temporarily elevating soil temperatures by 100°C or more (1). These high temperatures, however, are only sustained for a few minutes during the burn and near the soil surface. In

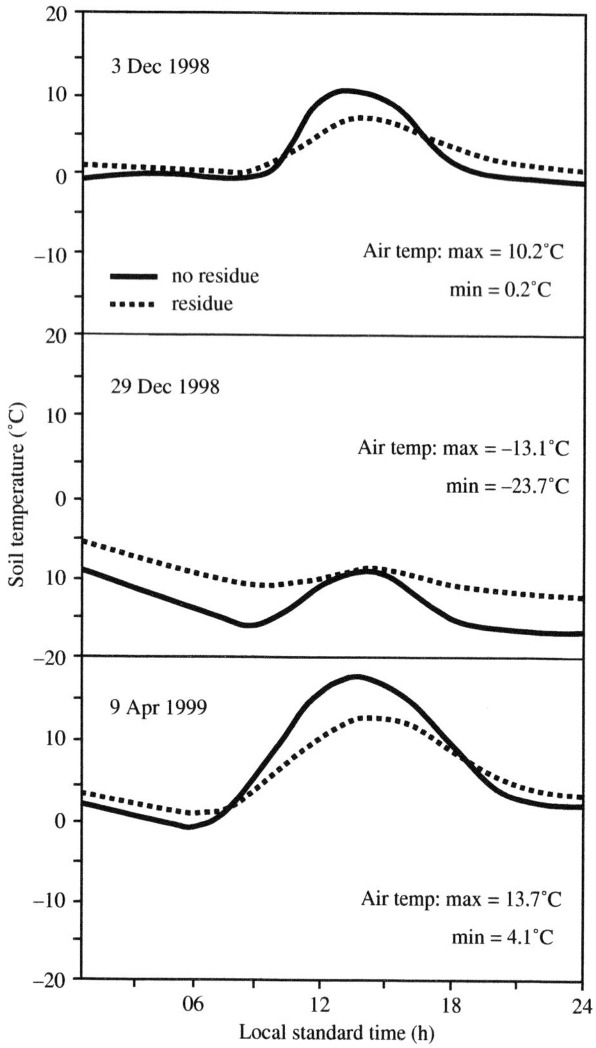

Fig. 1 Temperature at 1-cm depth of a soil with and without soybean (*Glycine max* L.) residue cover and without snow cover on a clear day in late autumn, winter, and spring near Morris, Minnesota.

addition, crop residue can be removed along the seed row to bolster soil temperatures during spring. Daily temperatures can rise by 2°C as the width of the band increases from 0 to 20 cm, but are unaffected by bands greater than 20 cm (2). Residue color can alter soil temperatures, but only on clear days without snow cover. Daily temperatures can be as much as 1–2°C higher for soils covered with black than with natural straw (3).

SOIL MANAGEMENT

Tillage is used to prepare a seedbed, alter the physical properties of soil to curtail erosion, and optimize the thermal regime of the seed zone. In cold regions, methods are sought that roughen, darken, and reduce the amount of residue on the soil surface to hasten soil warming in the spring. No tillage is advocated to conserve the soil resource, but this method often retards warming of the soil. Other methods such as strip tillage, ridge tillage, chisel plow, and moldboard plow are alternatives to managing the soil thermal regime. Strip tillage is accomplished in autumn or spring by cultivating in bands, thus resulting in a residue-free band. Little is known concerning spring temperatures achieved using strip tillage, but temperatures may be similar to those of residue-free bands. Ridge tillage can dramatically affect soil temperature in cold regions (4). Daily temperatures on a southerly slope can be elevated as much as 2°C over those on a level surface and 5°C over those on a northerly slope (Table 2). Spring soil temperatures are generally higher for moldboard and chisel plow owing to the rougher, darker, and smaller amount of residue on the soil surface compared with other tillage methods. Daytime temperatures can be 15°C higher, while nighttime temperatures can be 5°C lower, for moldboard plow than no tillage (5).

Crop and soil management practices that retard heat loss from soils in winter also slow the warming of soil in spring. Discovery of new residue management or tillage techniques that enhance soil warming during winter and spring is essential to the viability of agriculture in cold regions. These techniques will be identified only by

Table 1 Depth of frost penetration, day of year of complete soil thaw, and minimum soil (1-cm depth) temperature as influenced by corn (*Zea mays* L.) stubble height over three winters near Morris, Minnesota

	Stubble height			
	No stubble/residue	**15 cm**	**30 cm**	**60 cm**
Frost depth (cm)	91	78	46	22
Day of thaw	122	120	98	87
Soil temperature (°C)	−12.5	−10.0	−6.5	−6.0

Table 2 Average soil temperature at 1 cm depth on various aspects of a ridged soil surface for a clear, spring day at Fairbanks, Alaska[a] and Morris, Minnesota

		Aspect (°C)				
Location	**Date**	**North**	**South**	**West**	**East**	**Level**
Fairbanks	6 May 1990	4.9	9.0	7.5	7.3	8.0
Morris	22 March 1999	−2.0	2.5	0.6	0.4	0.7

[a] (Based on Ref. 4.)

improving our understanding of those physical properties of residue and soil that influence heat and water transfer between the soil and atmosphere.

REFERENCES

1. Rasmussen, P.E.; Rickman, R.W.; Douglas, Jr., C.L. Air and Soil Temperatures During Spring Burning of Standing Wheat Stubble. Agron. J. **1986**, *78*, 261–263.
2. Shinners, K.J.; Nelson, W.S.; Wang, R. Effects of Residue-Free Band Width on Soil Temperature and Water Content. Trans ASAE **1994**, *37*, 39–49.
3. Sharratt, B.S.; Flerchinger, G.N. Straw Color for Altering Soil Temperature and Heat Flux in the Subarctic. Agron. J. **1995**, *87*, 814–819.
4. Sharratt, B.S. Soil Temperature, Water Content, and Barley Development of Level vs. Ridged Subarctic Seedbeds. Soil Sci. Soc. Am. J. **1996**, *60*, 258–263.
5. Gupta, S.C.; Larson, W.E.; Linden, D.R. Tillage and Surface Residue Effects on Soil Upper Boundary Temperatures. Soil Sci. Soc. Am. J. **1983**, *47*, 1212–1218.

TILLAGE EROSION, DESCRIPTION AND PROCESS OF

Michael J. Lindstrom
United States Department of Agriculture-Agricultural Research Service (USDA-ARS), Morris, Minnesota, U.S.A.

INTRODUCTION

Tillage erosion is a problem that has been present since the dawn of cultivation. The problem has intensified with increased tillage speed, depth, and size of tillage tools, and with the tillage of steeper and more undulating lands. Evidence of tillage erosion is commonly observed as a difference in soil color between hilltops and adjacent lowerslope positions. Tillage erosion is defined by the Soil Science Society of America as the downslope displacement of soil through the action of tillage. It is easy to visualize that when tillage operations are conducted in the upslope direction, forward soil movement will be less than when conducted in the downslope direction (Fig. 1). This difference in soil translocation distance is a function of gravity. Assuming that tillage direction occurs equally often in the upslope and downslope directions, then a net downslope displacement of soil will take place. However, it is not an easy matter to move from this simple concept to one that suggests that soil loss from hilltops in undulating landscapes due to soil translocation by tillage can exceed levels that would be considered sustainable for crop production.

Tillage erosion has often been described in qualitative rather than quantitative terms. Evidence of the mass downslope movement of soil by tillage has been present for years. One example frequently cited comes from the Palouse region of the Pacific Northwest of the United States (1) where soil banks, 3 to 4 m high, have developed at fenceline locations on steep sideslope. These soil banks are the result of moldboard plowing, where the tillage above the fenceline turned the furrow slice toward the fenceline, and tillage below the fenceline turned the furrow slice away from the fenceline. In studies designed to measure the effect of soil variability across landscapes on crop production potentials, tillage has been implicated as the cause for observed downslope soil movement and an increase variability of soil properties (2, 3).

Examination of stereoscopic aerial photographs taken in 1947 and 1991 in the Loam Belt of Belgium showed a severe surface lowering on the top of the hillslopes and on hillslope convexities. Deposition occurred on the lowermost parts of the hillslope in hillslope concavities, and in topographic-defined convergence lines. The observed pattern differed markedly from that expected from water erosion processes, indicating that soil redistribution was dominated by tillage operations (4).

DETERMINATION OF TILLAGE EROSION

A simple linear regression of the form $Y = a + b(S)$ has been developed (5, 6), which describes the relationship between slope gradient (S) and mean soil translocation distance (Y) in the direction of tillage. Slope gradients were considered positive when tilling upslope, and negative when tilling downslope. Expanding on this relationship, it has been proposed (6) that tillage translocation could be considered a diffusion-type geomorphological process, similar to rainsplash and soil creep, and characterized by a single constant, the tillage transport coefficient (k).

$$\mathrm{k} = -D\rho_b B$$

where D is the depth of tillage (m), ρ_b is the soil bulk density (kg m^{-3}), and B is the slope of the linear regression equation of the relationship between soil displacement (m) and slope gradient (m m^{-1}). Using this relationship, the unit soil transport rate in the direction of tillage (Q_s) at any specific point in a field can be calculated as

$$Q_s = \mathrm{k}S$$

where S is the slope gradient (m m^{-1}). Representative tillage transport coefficients (k-value) for moldboard plow tillage have ranged between 230–330 kg m^{-1} (6). Commonly, agricultural fields undergo a series of tillage operation resulting in k-values of 400–600 kg m^{-1}.

It is not possible to directly calculate soil erosion using Q_s, since this value essentially represents a soil flux rate at a cross section for a specific tillage operation or series of operations. However, soil loss or gain will result when, for an elementary slope segment of unit width, the incoming flux is different than the outgoing flux:

$$E = (Q_s, \mathrm{in} - Q_s, \mathrm{out})/\mathrm{X}$$

Encyclopedia of Soil Science

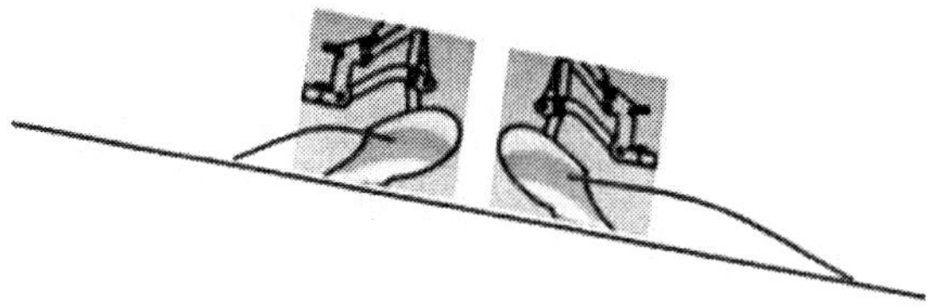

Fig. 1 Relative soil displacement distances when the thrust is upslope vs. downslope.

where E is the tillage erosion rate (kg m^{-2}) and X is the length (m) of the elementary slope segment under consideration. Since Q_s is directly proportional to the slope gradient, soil loss or gain will be proportional to the change in slope gradient. Soil translocation by tillage will result in soil loss on convex slope positions such as crests and shoulder slopes because there is an increase in slope gradient, thus an increase in soil transport rate. Conversely, soil deposition will take place in concave slope positions in the foot and toeslope positions. When slope gradients between adjacent elemental slope segments are equal, irrespective of their magnitude, no net soil loss or gain takes place because the Q_s in equals Q_s out. Thus, in backslope positions where slope gradients are commonly the greatest, exhibiting the greatest soil transport rate, net soil loss or gain will be minimal provided slope gradients remain constant. Therefore, the rate of soil gain or loss will depend on the unit transport rate and the degree of change in slope gradients:

$$E = \Delta Q_s / X$$

The magnitude of soil erosion rates by tillage vs. water is affected by many variables, i.e., topography, rainfall intensity, tillage intensity (depth and frequency), and land use. Examination of the relationship between a range of topographic parameters and ^{137}Cs-derived erosion rates, from fields in the United Kingdom (7), showed that the highest correlation was between erosion rate and landscape curvature at four of the five sites investigated. These results were not consistent with the dominance of water erosion, where slope angle and upslope lengths, or areas, are the primary influences. In a study comparing the roles of tillage and water erosion on landform development on agricultural land in Belgium (8), it was suggested that if water erosion were the dominant process, the landscape would be characterized by increased incision of the concavities and convergent waterways. A gradual increase in slope angles on upland convex slopes was also noted. In contrast, tillage produces maximum erosion on convex slopes, leading to reduced slope angles and infilling of concavities and hollows. The pattern of landform development observed was an infilling of the slope concavities and convergent waterways by sediment displaced through tillage that more then compensated for the lower-frequency, but more visible, rill and gully incision. The pattern indicates, that despite high susceptibility of the area to water erosion, landform development in this agricultural landscape is currently dominated by tillage erosion processes. These processes result in a reversal of the expected landscape evolution, with a gradual obliteration of topographic features. Other studies have indicated that tillage erosion rates are of the same order of magnitude as water erosion (9, 10).

EFFECTS OF TILLAGE EROSION

The tillage-transport coefficient (k-value) is a measure of the mean distance a mass of soil per unit width is moved by tillage, in a specified direction relative to the direction of tillage. The soil mass is translocated in the forward direction (parallel to the direction of tillage), but is also translocated in the lateral direction (perpendicular to the direction of tillage). Determination of k-values has mostly been in the forward direction. Using the mean displacement distances does not fully describe soil translocation, however. To illustrate, a single pass with a chisel plow may move 70 kg of soil forward per meter width of tillage. The mean forward displacement of this 70 kg of soil may be 40 cm, but significant quantities of soil may be moved as little as 5 cm or as much as 300 cm. Soil displacement will vary across the width of a tillage implement because of the spacing and arrangement of the individual tillage tools. This variation in distance over which soil is translocated is important since it affects the distance that soil constituents (amendments and contaminants) are dispersed or mixed by tillage.

The rate of soil loss by tillage erosion within topographically complex landscapes is several times more than is considered sustainable for crop production. Soil loss on a convex slope position in the Ontario Province of Canada was estimated to be 54 t ha^{-1} yr^{-1} (11). Estimates made using resident ^{137}Cs indicate that between 70% and 100% of soil lost on convex slope positions, is the direct result of tillage erosion (12). Crop yield reductions of 40% to 50% have been associated with these eroded landscape positions throughout southwestern Ontario (13).

Although tillage erosion can result in considerable soil loss and accumulation within fields, soil is not directly lost from fields by tillage erosion. However, tillage erosion exposes subsoil material on upperslope positions, which may become more susceptible to wind and water erosion. Furthermore, the redistribution of soil by tillage erosion delivers topsoil to areas of concentrated overland water flow on both the microtopographic scale (i.e., rills) and the macrotopographic scale (i.e., convergent landforms) (11, 14). As such, tillage erosion acts as a delivery mechanism of soil, which is then subject to water erosion.

Soil translocation by tillage produces maximum erosion at abrupt convex slope positions, causing a reduction in slope angles, and an infilling of hollows, resulting, over time, in a gradual obliteration of topographic features. As tillage erosion proceeds, the erosion process occurs over an increasingly larger area. In contrast, when water erosion is the dominant process, the landscape is characterized by increased incision of concavities and ephemeral gullies. A gradual increase in slope angle on convex slope positions also occurs.

CONCLUSIONS

Tillage erosion is directly proportional to the degree and scale of topographic complexity. Soil conservation measures that do not include a reduction in tillage erosion will not be effective in controlling soil loss on upperslope landscape positions of cultivated agricultural lands. To reduce soil loss caused by tillage erosion, frequency, tillage intensity (speed and depth), and the size of tillage implements must be reduced.

REFERENCES

1. Papandick, R.I.; Miller, D.E. Conservation Tillage in the Pacific Northwest. Journal Soil and Water Conservation **1977**, *32*, 40–56.
2. Kachanoski, R.G.; Rolston, D.E.; deJong, E. Spatial Variability of a Cultivated Soil as Affected by Past and Present Microtopography. Soil Science Society American Journal **1985**, *49*, 1082–1087.
3. Cao, Y.Z.; Coote, D.R.; Rees, H.W.; Wang, C.; Chow, T.L. Effects of Potato Production on Soil Quality and Yield at a Benchmark Site in New Brunswick. Soil Tillage Research **1994**, *29*, 23–34.
4. Vandaele, K.; Vanommeslaeghe, J.; Muylaert, R.; Govers, G. Monitoring Soil Redistribution Patterns Using Sequential Aerial Photographs. Earth Surface Processes and Landforms **1995**, *21*, 353–364.
5. Lindstrom, M.J.; Nelson, W.W.; Schumacher, T.E. Quantifying Tillage Erosion Rates due to Moldboard Plowing. Soil Tillage Research **1992**, *24*, 243–255.
6. Govers, G.; Vandaele, K.; Desmet, P.J.J.; Poesen, J.; Bunte, K. The Role of Tillage in Soil Redistribution on Hillslopes. European Journal of Soil Science **1994**, *45*, 469–478.
7. Quine, T.A.; Walling, D.E. Use of Caesium-137 Measurements to Investigate Relationships Between Erosion Rates and Topography. In *Landscape Sensitivity*; Thomas, D.S.G., Allison, R.J., Eds.; John Wiley: New York, 1993; 31–48.
8. Quine, T.A.; Desmet, P.J.J.; Govers, G.; Vandaele, K; Walling, D.E. A Comparison of the Role of Tillage and Water Erosion in Landform Development and Sediment Export on Agricultural Land Near Leuven, Belgium. In *Variability in Stream and Sediment Transport*, Proceedings of the Canberra Symposium, Canberra, Australia, December 1994; 77–86.
9. Quine, T.A.; Walling, D.E.; Chakela, G.K.; Mandiringana, O.T.; Zhang, X. Rates and Patterns of Tillage and Water Erosion on Terraces and Contour Strips: Evidence from Caesium-137. Catena **1999**, *36*, 115–142.
10. Govers, G.; Quine, T.A.; Walling, D.E. The Effect of Water Erosion and Tillage Movement on Hillslope Profile Development: A Comparison of Field Observations and Model Results. In *Farm Land Erosion: In Temperate Plains Environment and Hills*; Wicherek, S., Ed.; Elsevier: Amsterdam, 1993; 285–300.
11. Lobb, D.A.; Kachanoski, R.G.; Miller, M.H. Tillage Translocation and Tillage Erosion on Shoulder Slope Landscape Positions Measured Using ^{137}Cs as a Tracer. Canadian Journal of Soil Science **1995**, *75*, 211–218.
12. Lobb, D.A.; Kachanoski, R.G. Modeling Tillage Translocation Using Step, Linear-Plateau and Exponential Functions. Soil Tillage Research **1999**, *51*, 317–330.
13. Battison, L.A.; Miller, M.H.; Shelton, I.J. Soil Erosion and Corn Yield. I. Field Evaluation. Canadian Journal Soil Science **1987**, *67*, 731–745.
14. Govers, G.; Quine, T.A.; Desmet, P.J.J.; Walling, D.E. The Relative Contribution of Soil Tillage and Overland Flow Erosion to Soil Redistribution on Agricultural Land. Earth Surface Processes and Landforms **1996**, *21*, 929–946.

TILLAGE EROSION, MEASUREMENT TECHNIQUES OF

T

David A. Lobb
University of Manitoba, Winnipeg, Manitoba, Canada

INTRODUCTION

Tillage erosion is the redistribution of soil that occurs within a landscape as a direct result of tillage. Variations in the amount of soil moved by tillage cause loss and accumulation. Tillage translocation is the movement of soil by tillage; variability in translocation is affected by the design and operation of tillage implements, and by the topographic and soil properties of landscapes. Typically, tillage results in the progressive downslope movement of soil, causing severe soil loss on upperslope positions and accumulation in lowerslope positions (Fig. 1). A wide variety of materials and methods has been used to measure tillage erosion.

The earliest scientific study of tillage erosion took place in the United States in the 1930s (1). Since then, 30 to 40 studies have been conducted around the world, most undertaken in the last ten years (2). Very few of these studies used identical materials and methods, reflecting the evolution of experimental techniques and the diversity of experimental conditions. There are two general methods of measuring tillage erosion: measurement of soil loss and accumulation at points over the surface of a landscape; and measurement of tillage translocation at points over a landscape and of the difference between those points.

TRANSLOCATION MEASUREMENTS

Tillage translocation is normally measured with a tracer that is incorporated into the soil in plots. The distributions of the tracer before and after tillage are used to calculate translocation—forward translocation from the distribution parallel to the direction of tillage, and lateral translocation from the distribution perpendicular to the direction of tillage (Fig. 2).

There are two methods of calculating translocation from tracer distributions. For the more common method, the quantity of soil translocated per unit width of tillage is calculated directly from the tracer distributions (3). For the less common method, a summation curve is generated from a tracer distribution by employing convolution, and translocation per unit width of tillage is calculated from this curve (4, 5). Both methods provide accurate measures of translocation, but the latter provides additional information regarding the distance over which translocated soil is dispersed. Translocation can be expressed as a volume, a mass, or a depth-averaged length.

The calculation of tillage erosion from translocation measurements is simply the difference in translocation between two points, divided by the separating distance. Tillage erosion is expressed similarly to wind and water erosion; i.e., a change in soil mass per unit area or a change in elevation.

Within a complex soil landscape, numerous measurements of tillage translocation may be required. Plots should be placed so that the variability in translocation is characterized in detail. Where slope gradient has been assumed to be the dominant factor affecting soil movement, plots should be located over the full range of gradients. Greater attention should be paid to those areas where slope gradient is changing, (i.e., convex and concave areas), because changes in slope gradient result in soil loss and accumulation. A pair of plots positioned at the steepest point along a slope profile—one tilled upslope and the other tilled downslope—provides measures of the minimum and maximum translocation, net downslope translocation, and the total soil loss from the upper slope and the total soil accumulation on the lower slope. Although the minimum and maximum values of translocation occur at this point on the hillslope, if the gradient in this region of the slope is uniform, no soil loss or accumulation will occur at this point.

Tracers consist of plot tracers and point tracers. Point tracers are individually labeled tracers of various shapes and materials, such as steel nuts (6), or plastic spheres (7). Plot tracers are used to label a volume of soil; they can be physical, [e.g., gravel (8)], or chemical [e.g., ^{134}Cs (3), Cl (5)]. Point tracers also represent a volume of soil, but they have the distinct advantage that they can be used to characterize the complexity of soil movement in three dimensions. However, plot tracers provide a more accurate measure of bulk soil movement in two dimensions.

After tillage, soil is sampled and analyzed to determine the distribution of the tracer. Sampling usually involves

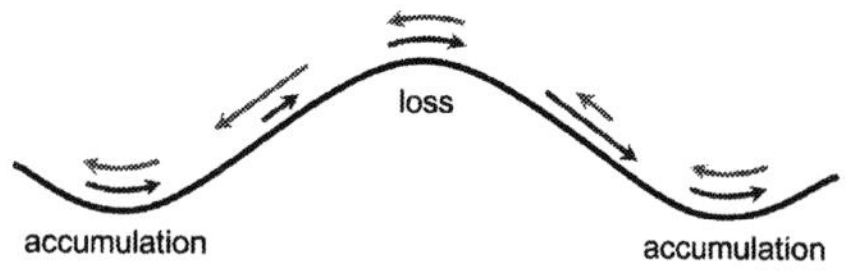

Fig. 1 Variability of translocation in hilly landscape (upslope–downslope tillage): Soil loss occurs on convex areas and accumulation occurs on concave areas. (Translocation is indicated by arrows.)

destruction of the site, eliminating the possibility of successive translocation measurements from recurrent tillage operations. Non-destructive sampling methods do exist (5). Sampling must be carried out as soon as possible after tillage to isolate the movement of soil by tillage from that by other processes. Other measures can also be taken to isolate soil movement by tillage (4). The precise relocation of plots and the recovery of the tracer are important considerations when sampling.

The design of plots is possibly the most important consideration when using tracers. Even under controlled operating conditions, the movement of soil during tillage is inherently variable due the nature of soil-tillage interactions. Soil strains and fails during tillage, and tools and implements flex and shift—the "flow" of soil during tillage is not steady. For accurate measurement of soil movement by tillage, plots must be sufficiently long and wide to filter out this "noise." The length of plots used in tillage erosion studies has ranged from a few centimeters to a few meters. As stated previously, a longer plot provides a better measure of the average movement of soil by tillage. The movement of soil varies across the width of tillage as a function of the tool type, arrangement, and spacing. For accurate measurement of soil movement by tillage, the width of the plot must be a multiple of the unit width of tillage. For a single pass of a simple tillage implement, the unit width is the distance between two adjacent tools. However, for an implement with multiple tool types or for a sequence of different implements, establishing the unit width of tillage may not be possible. To reduce measurement errors associated with plot width, the plot must be as wide as possible. The paths of a tillage implement overlap in the field (as much as 10%), with the exception of implements such as the moldboard plow, causing additional soil movement at the edges of each path. To account for this additional movement, plot widths must be equal to the spacing of implement paths; however, this is not practical in most cases. The depth of plots must exceed the maximum depth of the tillage operation(s) under examination. Tillage depth can vary greatly within a landscape, so caution must be used in estimating the maximum tillage depth—it is better to overestimate tillage depth than to underestimate.

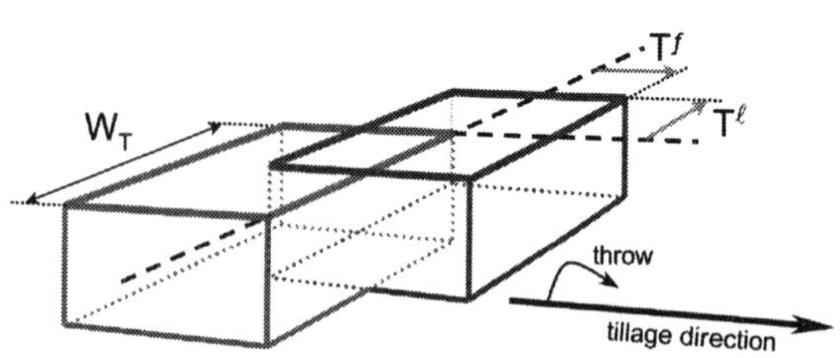

Fig. 2 Simplified illustration of soil translocation by tillage: T^f = forward translocation; T^l = lateral translocation; W_T = unit width of tillage.

Tillage translocation can also be measured using detailed elevation surveys of the soil surface. The redistribution of soil volume, along with soil bulk density, can be used to measure soil translocation (1, 8). There are two good examples of this technique. In the study of tillage erosion by contour moldboard plowing, the profiles of furrow slices turned upslope are compared to those of furrow slices turned downslope—the difference is a measure of net downslope translocation. In the study of tillage erosion by hoeing on steep slopes (soil is only hoed downslope), the profiles of the furrow created at the top of the slope and the ridge created at the bottom are measures of downslope translocation.

MEASUREMENTS OF LOSS AND ACCUMULATION

Soil loss and accumulation by tillage erosion can be estimated from changes in soil properties or surface elevation. In most cases, the loss or accumulation from a single tillage event or an annual sequence of events would be too small to measure accurately; consequently, measurements are usually made over a period of years. The distribution of these changes over the surface of the landscape allows the assessment of soil redistribution.

Soil surface elevation is commonly used as an indicator of recurrent tillage erosion. The elevation within a tilled portion of the landscape is compared to an adjacent non-eroding landscape feature, such as a fence-line or hedgerow. Decreases in surface elevation on upperslope landscape positions, and increases in elevation on lowerslope positions are classic evidence of tillage erosion. Using tillage history and elevation changes, it is possible to estimate the average annual rate of tillage erosion.

Soil properties, such as the amount of organic matter and carbonates, are commonly used as indicators of recurrent tillage erosion. Organic-poor, carbonate-rich subsoil exposed on the surface of upperslope landscape positions is classic evidence of soil loss by tillage erosion. Accumulation of organic rich soil in convergent slope positions is also classic evidence of tillage erosion. However, such familiar soil properties are not considered reliable estimators of soil erosion. Radioisotopes found in soil are used to accurately estimate soil erosion; the most frequently used are ^{137}Cs and ^{210}Pb (4, 9). Lead-210 naturally occurs; ^{137}Cs originates from human use of nuclear radiation. Both isotopes contact the soil through atmospheric deposition and remain concentrated in the surface soil. The spatial distribution of radioactivity is compared to a baseline distribution or to a single value based on a non-eroded reference site. This technique is suited to time-scales of 5–40 years for ^{137}Cs, and 50–150 years for ^{210}Pb. Soil magnetic susceptibility (10) and flyash concentrations have been used to estimate soil erosion.

The use of long-term soil loss and accumulation to estimate rates of tillage erosion is problematic in that it is not possible to isolate individual erosion processes. Furthermore, erosion processes interact. The contribution of tillage erosion to the total soil redistribution within a landscape must be determined based on the signature patterns of redistribution by individual erosion processes (11).

REFERENCES

1. Mech, S.J.; Free, G.R. Movement of Soil During Tillage Operations. Agriculture Engineering **1942**, *23*, 379–382.
2. Govers, Gerard; Lobb, David A.; Quine, Timothy A. Tillage Erosion and Translocation: Emergence of a New Paradigm in Soil Erosion Research. Soil Tillage Research **1999**, *51*, 167–174.
3. Quine, Timothy A.; Govers, Gerard; Poesen, Jean; Walling, Desmond E.; van Wesemael, Bas; Martinez-Fernandez, J. Fine Earth Translocation by Tillage in Stony Soils in the Guadalentin, Southeast Spain: An Investigation Using Caesium-134. Soil Till. Res. **1999**, *51*, 279–301.
4. Lobb, David A.; Kachanoski, R. Gary; Miller, Murray H. Tillage Translocation and Tillage Erosion in the Complex Upland Landscapes of Southwestern Ontario, Canada. Soil Till. Res. **1999**, *51*, 189–209.
5. Lobb, David A.; Kachanoski, R. Gary; Miller, Murray H. Tillage Translocation and Tillage Erosion on Shoulder Slope Landscape Positions Measured Using ^{137}Cs as a Tracer. Can. J. Soil Sci. **1995**, *75*, 211–218.
6. Lindstrom, Michael J.; Nelson, W.W.; Schumacher, Thomas. E. Quantifying Tillage Erosion Rates due to Moldboard Plowing. J. Soil Till. Res. **1992**, *24*, 243–255.
7. Govers, Gerard; Vandaele, K.; Desmet, Phillipe J.J.; Poesen, Jean; Bunte, K. The Role of Tillage in Soil Redistribution on Hillslopes. European J. Soil Sci. **1994**, *45*, 469–478.
8. Turkelboom, F.; Poesen, Jean; Ohler, I.; Ongprasert, S. Reassessment of Tillage Erosion Rates by Manual Tillage on Steep Slopes in Northern Thailand. Soil Till. Res. **1999**, *51*, 245–259.
9. Walling, Desmond E.; He, Q.; Quine, Timothy A. Use of Fallout Radionuclide Measurements in Sediment Budget Investigations. Géomorphologie **1996**, *3*, 17–28.
10. deJong, Eeltje; Nestor, P.A.; Pennock, Dan J. The Use of Magnetic Susceptibility to Measure Long-Term Soil Redistribution. Catena **1998**, *32*, 23–35.
11. Govers, Gerard; Quine, Timothy A.; Desmet, Phillipe J.J.; Walling, Desmond E. The Relative Contribution of Soil Tillage and Overland Flow Erosion to Soil Redistribution on Agricultural Land. Earth Surface Processes **1996**, *21*, 929–946.

TILLAGE EROSION, RELATIONSHIP TO WATER EROSION

Gerard Govers
Katholicke Universiteit Leuven, Leuven, Belgium

INTRODUCTION

Tillage and water erosion interact in various ways. It has long been known that tillage affects soil's resistance to detachment and transport by water. Soil tillage alters the soil structure, reduces soil's bulk density, breaks up the soil crust, and incorporates superficial residue in to the soil. However, these phenomena are not directly related to the fact that during tillage operations, soil displacement and tillage erosion occur. Tillage translocation and erosion have specific effects on the water erosion processes. Conversely, water erosion processes may affect tillage translocation and erosion. We discuss the interactions between tillage erosion and water erosion which occur at various time scales.

TILLAGE EROSION AND WATER EROSION

On a short timescale, tillage erosion acts as a delivery mechanism for water erosion. Water erosion is often strongest along the central axis of hill slope concavities (hollows); here, a large amount of surface runoff is concentrated, whereas slope gradients are often relatively high (1). This often leads to ephemeral gully erosion (2). At the same time, important amounts of soil are deposited here as a result of tillage. The delivery of sediment by tillage may exceed the average water erosion rate, in which case net deposition will occur.

Conversely, if soil delivery is lower than the average water erosion rate, the result will be net incision. The balance depends upon the relative intensity of both processes and the landscape morphology. Deposition by tillage will increase with increasing slope concavity. On the other hand, water erosion rates will increase with increasing slope gradient and upslope water contributing area. The interaction between tillage and water erosion is not one way; when ephemeral gullies form, farmers will often undertake special tillage operations to fill them in, which then leads to an increased soil displacement by tillage.

On a longer timescale, tillage translocation and erosion affects topography. Tillage translocation leads to the deposition of soil in concavities (hollows) as well as on the upslope side of field borders. It causes soil erosion on convexities (shoulders) as well as on the downslope side of field borders. Therefore, the process will lead to a diminution of the overall relief energy within a field and the water erosion risk, as the latter depends strongly on the slope gradient.

Accumulation and erosion of soil at field borders can be fairly rapid, leading to the development of soil banks of several meters high over a time period of a few decades if the soil is consistently thrown downslope during tillage. In intensive tillage systems in Western Europe, concavities are now filling at a rate of $5-20\,\mathrm{mm\,yr^{-1}}$ due to tillage deposition, while convexities are typically eroding at rates of $0.5-2\,\mathrm{mm\,yr^{-1}}$ (3). Considering a typical slope of 100 m having an average gradient of 10%, it will take 44 to 180 years to reduce average slope gradient by 1 percentage point, neglecting the effects of water erosion on topography. Changes in relief energy will be more rapid when convexities and concavities are more pronounced, as is the case on many moraine landscapes in northern Europe and America (4).

In many agricultural systems, slopes are interrupted by vegetative borders and/or water diversion structures to reduce the water erosion risk. In such systems, soil translocation by tillage will lead to much more rapid and important changes in topography. For example, tillage translocation will gradually decrease the slope of the field strips because of the deposition of soil at the downslope end of the field and erosion at the upslope end (Fig. 1) (5). Again, the speed at which these changes occur depends on the initial topography and the intensity of the tillage erosion process. Changes will be more rapid when initial slopes are high, tillage is intense, and field borders are closely spaced. Calculations for 6 m wide, slow-forming terraces in Ecuador show that tillage erosion will reduce the initial slope gradient of 30% to approximately 6% in 25 years using traditional tillage techniques. In some cases, this effect may be deliberately speeded up by consistently throwing the soil downslope when tilling the field strips.

Some positive effects of reduction of the within-field slope gradient may be counteracted by erosion processes specifically related to the soil banks. In some cases, gullies erosion may form on the terrace banks. Such bank gullies can be initiated by piping caused by animal activity,

Encyclopedia of Soil Science

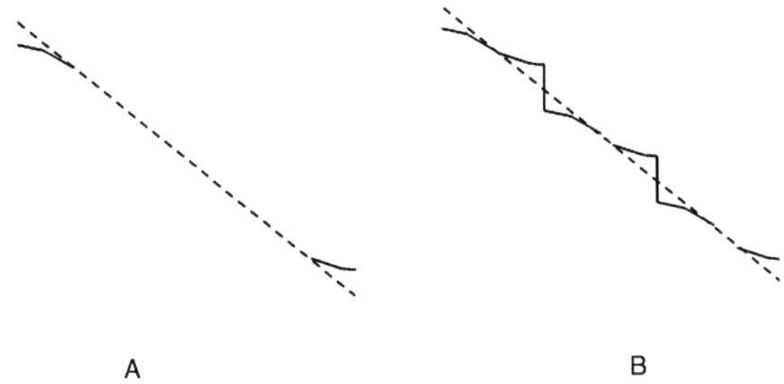

Fig. 1 Schematic representation of the evolution of the slope profile by tillage erosion with (B) and without (A) the presence of diversion structures or terraces.

cracking, and/or mass transport processes occurring on the banks. The risk of bank failure and the speed at which bank gullies will develop depends to a large extent upon the properties of the soil profile.

Evidently, water erosion may also lead to topographical changes affecting tillage erosion rates and patterns. Generally, water erosion will tend to incise the landscape, leading to steeper slope gradients and more pronounced convexities and concavities. This can be seen in its most extreme form in badlands. As a consequence, the potential for tillage erosion is increased. However, the reverse may also occur when within-field deposition of sediment by water is important. This deposition will lead to a reduction in relief energy and, eventually, a reduction in tillage erosion.

Tillage erosion also affects water erosion by changing soil properties. The extent and nature of such changes depend upon the initial properties of the soil. In general, such changes will be more important when the soil's A-horizon is shallow and when there is a large difference in properties between the A-horizon and lower soil horizons.

In a first phase, tillage erosion leads to the exposure of subsoil on convexities and at upslope field borders. If tillage erosion continues for a sufficiently long period, this subsoil material will be transported further downslope. During downslope transport, the subsoil material is mixed with the original topsoil. After a long time period, the properties of the plough layer will be determined predominantly by the properties of the original subsoil at convexities and upslope field borders. They will gradually change to those of the original topsoil in the downslope direction. The area where subsoil properties dominate will increase with time.

In a final stage, a soil profile inversion may occur over large parts of the landscape, whereby the original topsoil material is buried by a layer of subsoil material eroded further upslope. The effect of these changes on water erosion rates and patterns may vary. On stony soils, tillage may increase stone cover on convexities, thereby leading to a considerable decrease in water erosion risk from the protective effect of the stone (6). However, in most cases, the properties of the subsoil are such that tillage erosion will lead to an increase in the water erosion risk. Very often, the subsoil material has a lower structural stability than the original topsoil, which causes a higher sensitivity for crusting and more runoff generation in the areas of exposure.

Because the subsoil material is exposed in upslope landscape positions, this will increase the erosion risk over the whole landscape. The runoff generated in the upslope landscape positions increases soil erosion further downslope. This process may be further enhanced by reduced vegetative growth on the less-productive tillage eroded areas, and poorer soil cover and increased erosion rates. In concave landscape positions, the potential negative effect of water erosion on soil quality is to some extent compensated for by tillage deposition. Tillage deposition causes the depth of the A-horizon to either decrease more slowly or increase in these landscape positions, despite the removal of soil by water erosion. However, in the long term the original A-horizon may become buried by subsoil material.

Changes in soil properties like soil texture and organic matter content caused by water erosion are unlikely to affect tillage erosion to a great extent. Results of current tillage erosion experiments suggest that tillage speed and depth, tillage tool geometry, and dynamic soil properties, like bulk density and moisture content, have a far greater influence on tillage displacement and erosion (7).

Finally, different spatial signatures of tillage and water erosion have implications for the calculation of average soil erosion rates at the field scale and for the assessment of effects on soil quality. Tillage erosion causes only soil redistribution within a field so that the average soil erosion rate over the field is zero. The intensity of tillage erosion may be better assessed by calculating an average rate over the eroding part of the landscape, as is done for water erosion. To calculate an overall soil erosion rate, erosion and deposition rates for tillage and water erosion must be calculated using a spatially distributed model, so that soil losses and gains can be assessed at each landscape position. These may be combined to produce a map of the total soil erosion or deposition rate at each point (8). This rate can then be calculated by taking the average of the values for the eroding landscape positions.

Because of the compensating nature of water and tillage erosion, the total average soil erosion rate that is calculated using this procedure will generally be much lower but it will also more realistic than the sum of the separately calculated average water erosion rate and the average tillage erosion rates. Similarly, the assessment of the soil quality effects of erosion and deposition by tillage and water is only possible using a spatially distributed model that accounts for both processes.

REFERENCES

1. Govers, G.; Quine, T.A.; Desmet, P.J.J.; Walling, D.E. The Relative Contribution of Soil Tillage and Overland Flow Erosion to Soil Redistribution on Agricultural Land. Earth Surface Processes and Landforms **1996**, *2*, 929–946.
2. Vandaele, K.; Poesen, J.; Govers, G.; Van Wesemael, B. Geomorphic Threshold Conditions for Ephemeral Gully Erosion. Geomorphology **1996**, *16*, 161–173.
3. Govers, G.; Van Daele, K.; Desmet, P.J.J.; Poesen, J.; Bunte, K. The Role of Soil Tillage in Soil Redistribution on Hillslopes. European Journal of Soil Science **1994**, *45*, 469–478.
4. Lobb, D.A.; Kachanoski, R.G.; Miller, M.H. Tillage Translocation and Tillage Erosion on Shoulder Slope Landscape Positions Measured Using ^{137}Cs as a Tracer. Can. J. Soil Sci. **1995**, *75*, 211–218.
5. Dabney, S.M.; Liu, Z.; Lane, M.; Douglas, J.; Zhu, J.; Flanagan, D.C. Landscape Benching from Tillage Erosion Between Grass Hedges. Soil and Tillage Res. **1999**, *51*, 219–232.
6. Poesen, J.; Van Wesemael, B.; Govers, G.; Martinez-Fernandez, J.; Desmet, P.J.J.; Vandaele, K.; Quine, T.A.; Degraer, G. Patterns of Rock Fragment Cover Generated by Tillage Erosion. Geomorphology **1997**, *18*, 183–197.
7. Van Muysen, W.; Govers, G.; Bergkamp, G.; Roxo, M.; Poesen, J. Measurement and Modelling of the Effects of Initial Soil Conditions and Slope Gradient on Soil Translocation by Tillage. Soil and Tillage Res. **1999**, *51*, 303–316.
8. Van Oost, K.; Govers, G.; Desmet, P.J.J. Evaluating the Effects of Changes in Landscape Structure on Soil Erosion by Water and Tillage. Landscape Ecology **2000**, *15*, 577–589.

TILLAGE AND GAS EXCHANGE

T

D.C. Reicosky
United States Department of Agriculture-Agriculture Research Service (USDA-ARS), North Central Soil Conservation Research Laboratory, Morris, Minnesota, U.S.A.

INTRODUCTION

Agriculture is the economic foundation of rural America and has a major influence on components of industry, world trade, and global ecology. In traditional agricultural production, tillage of the soil has been an integral part of the production process. Tillage is the mechanical manipulation of soil and crop residue to prepare a seedbed where crop seeds are planted, sprout, take root, and grow into plants to produce grain. Intensive tillage loosens soil, buries crop residue, enables the soil to warm and dry, enhances release of soil nutrients for crop growth, kills the weeds that compete with crop plants for water and nutrients, facilitates root growth in compacted soil, and improves the flow of water and air within the soil. The enhanced gas exchange affects the processes that impact the accumulation and loss of soil carbon (C) in agricultural systems. Tillage increases water infiltration and increases the soil porosity, especially large pores, which allow greater movement of soil gases through the soil. Diffusion allows movement of gases into or out of the soil from higher to lower concentrations.

Concern for environmental quality and tillage-induced greenhouse gas emissions (carbon dioxide, methane, nitrous oxide) requires new knowledge to minimize agriculture's impact on the environment. In the United States, the moldboard plow has been a significant symbol of agriculture over the last 150 years and now is being reevaluated in many parts of the world as new conservation tillage techniques are developed and researched.

CROP PRODUCTION AND SOIL CARBON LOSS

The link between global warming and atmospheric carbon dioxide (CO_2), a greenhouse gas, has heightened interest in soil C storage, as soil organic matter, in agricultural production systems. Agricultural soils and agricultural production play an important role in C sequestration or storage and thus can help mitigate global warming (1). Intensive tillage has mineralized or oxidized 30% and 50% of the native soil C or soil organic matter since the pioneers brought the soils into cultivation. Tillage processes and mechanisms leading to C loss are directly linked to soil productivity, soil properties, and environmental issues (2). Soil C dynamics can have an indirect affect on climate change through net absorption or release of CO_2 from soil to the atmosphere in the natural C cycle. In agriculture, C comes into the system through photosynthesis and is returned to the atmosphere as CO_2 through human and microbial respiration. Good soil C management is vital because of its role in maintaining soil fertility, physical properties, and biological activity required for food production. Good soil C management is also needed to partially offset greenhouse gas emissions from manufacture and use of acid fertilizers, liming, fossil fuels, and the release of more potent nitrous oxide and methane from agricultural systems. Minimizing agriculture's impact on the global increase of CO_2 requires that we sequester and maintain high C levels in soil.

TILLAGE-INDUCED CO_2 LOSS

Tillage affects soil microbial activity, organic matter decomposition, and soil C loss in agricultural systems. Much of the C is lost as CO_2, which is the end product of microbial feeding on soil organic matter. Reicosky and Lindstrom (3) showed major short-term gaseous loss of C, immediately after tillage, which partially explains long-term C loss from tilled soils. Gas exchange was measured using a large, portable chamber to determine CO_2 loss from various types of tillage. Moldboard plow was the most intensive tillage and caused more CO_2 loss than less intensive tillage methods. No till or no soil disturbance lost the least amount of CO_2 suggesting minimal environmental impact. Moldboard plowing loosens and inverts soil and allows rapid CO_2 loss and oxygen entry. It also incorporates and mixes residues to enhance microbial decomposition and respiration (oxidation) (4). Stirring the soil in tillage is analogous to stirring the coals in a fire. Plowing accelerates microbial decomposition and soil aggregate breakdown to cause decreased soil C content in the surface layer. Ellert and Janzen (5) and Rochette and Angers (6) found similar results for different soils and less intensive tillage methods.

Reicosky (7) reported that average short-term CO_2 loss from four conservation tillage tools was 31% of the CO_2 from the moldboard plow. The moldboard plow lost 13.8 times more CO_2 as the soil not tilled while conservation tillage tools averaged about 4.3 times more CO_2 loss. The smaller CO_2 loss from conservation tillage tools was significant and suggests progress in equipment development for enhanced soil C management. Conservation tillage reduces the extent, frequency, and magnitude of mechanical disturbance caused by the moldboard plow and reduces the large air-filled pores or holes in the soil to slow the rate of gas exchange and C oxidation.

Strip tillage tools are designed to minimize soil disturbance. Different strip tillage tools and moldboard plow were compared to quantify short-term tillage-induced CO_2 loss relative to tillage intensity (8). Less intensive strip tillage reduced soil CO_2 losses. No till had the lowest CO_2 loss, and moldboard plow had the highest immediately after tillage. Forms of strip tillage had an initial soil CO_2 loss related to tillage intensity intermediate between the extremes of plowing and no till. The cumulative CO_2 losses for 24 hours were directly related to the soil volume disturbed by the tillage tool. Reducing the volume of soil disturbed by tillage should enhance soil and air quality by increasing the soil C content and suggests that soil and environmental benefits of strip tillage be considered in soil management decisions. The CO_2 released immediately after moldboard plowing suggests little C sequestration. Conservation tillage methods that leave most of the crop residue on the surface with limited soil contact yield better C sequestration to enhance environmental quality.

MECHANICS OF GAS EXCHANGE

Tillage affects all physical soil conditions, especially aeration (9). Tillage can drastically change the configuration, continuity, and size of soil pores. The moldboard plow is probably the most efficient implement to loosen a large volume of soil and to break up dense, massive soil clods into smaller units. After plowing, all plants and plant residues are buried and partially mixed into the soil, simplifying subsequent tillage and planting operations.

Several tillage-related factors affect soil gas exchange, especially the soil porosity and air permeability. The exchange of air between the soil and the atmosphere is bi-directional and can occur by two different mechanisms called diffusion and convection. In diffusion, the moving force is the gradient of partial pressure or concentration of the specific gas that causes the unevenly distributed molecules to randomly migrate from a zone of high concentration to low concentration. In convection, also called mass flow, the moving force consists of a gradient of total gas pressure and results in the entire mass of air streaming from a zone of high pressure to a zone of low pressure. Barometric pressure changes, soil temperature gradients, and wind gusts over the loosened soil surface can create pressure differences between soil air and the external atmosphere, thereby inducing convective flow into or out of the soil. Whether diffusion or mass flow is the dominant gas flow mechanism from soil depends on the total pressure gradient and pore size and pore continuity. When the soil is consolidated with only small pores, gas exchange is primarily by diffusion. When the soil has large pores, gas exchange also can occur by mass flow. Out in the field, both processes are occurring simultaneously.

The degree to which air pressure fluctuations and convective flow can exchange gas between soil and the atmosphere has long been debated among soil physicists. Most believe that diffusion, rather than convection, is the more important gas exchange mechanism. Recent evidence suggests that convection can, in certain circumstances following intensive tillage, contribute significantly to gas exchange and soil aeration, particularly at shallow depths and in soils with large pores (10). The tillage-induced CO_2 loss recently identified demonstrates the role of mass flow as a cause of C loss from tilled soils. The magnitudes of CO_2 fluxes were too large to be accounted for by simple diffusion from the soil (3). The tillage-induced change in soil–air permeability showed that convection contributes significantly to gas exchange (11).

While the effects of mass flow are intermittent and variable, they tend to be particularly significant immediately after a tillage event up to the time that the soil reconsolidates. External factors that cause soil reconsolidation may include secondary tillage, raindrop impact, or wheel track compaction. The real concern follows an intensive tillage operation, where the change in the soil physical properties increases soil–air permeability and changes the gaseous loss from a diffusion-controlled process to a convectively controlled process. Methods for measuring change in soil–air permeability on large scales have not been developed.

SOIL PRODUCTIVITY AND ENVIRONMENTAL BENEFITS

While moldboard plowing and other forms of intensive tillage have done much to increase U.S. crop production over the past 150 years, the increase in production has been accompanied by unseen costs of decreased soil quality from erosion and increased greenhouse gas emissions (1, 12). The organic matter of many of the prairie's soils has declined from that present under virgin conditions. The unseen, unmeasured costs that result from intensive tillage include loss in soil C due to enhanced oxidation and depletion of soil fertility reserves. The magnitude of these unseen costs depends primarily on the intensity of tillage, the quantity and quality of crop residue returned to the soil, and the crop rotation. Intensive tillage, primarily moldboard plowing, decreases soil C in virtually all crop production systems.

SUMMARY

Concern for soil productivity and greenhouse gas emissions requires new knowledge to minimize agriculture's impact on the environment. Soil C is the foundation of a healthy environment and sustainable agriculture. This is highly dependent on management decisions that influence intensity of tillage and the amount and placement of residues. Conservation tillage or no-till systems have shown increases in soil organic matter within 10 to 12 years of consistent use. The increase in soil organic matter depends on a delicate balance between the residue inputs of the previous crops and the tillage intensity associated with establishing the next crop. Farmers are faced with serious decisions with respect to environmental consequences of maintaining sustainable production and managing this delicate balance.

REFERENCES

1. Lal, R.; Kimball, J.; Follett, R.F.; Cole, C.V. *The Potential of U.S. Cropland to Sequester Carbon and Mitigate the Greenhouse Effect*; Sleeping Bear Press: Ann Arbor, MI, 1998; 128.
2. Paustian, K.; Collins, H.P.; Paul, E.A. Management Controls on Soil Carbon. In *Soil Organic Matter and Temperate Ecosystems: Long-Term Experiments in North America*; Paul, E.A., Paustian, K., Elliot, E.T., Cole, C.V., Eds.; CRC Press: Boca Raton, FL, 1997; 15–49.
3. Reicosky, D.C.; Lindstrom, M.J. Impact of Fall Tillage and Short-Term Carbon Dioxide Flux. In *Soil and Global Change*; Lal, R., Kimble, J., Levine, E., Stewart, B.A., Eds.; Lewis Publishers: Chelsea, MI, 1995; 177–187.
4. Reicosky, D.C.; Kemper, W.D.; Langdale, G.W.; Douglas, C.L., Jr.; Rasmussen, P.E. Soil Organic Matter Changes Resulting from Tillage and Biomass Production. J. Soil Water Conserv. **1995**, *50* (3), 253–261.
5. Ellert, B.H.; Janzen, H.H. Short-Term Influence of Tillage on CO_2 Fluxes from a Semi-Arid Soil on the Canadian Prairies. Soil Tillage Res. **1999**, *50*, 21–32.
6. Rochette, P.; Angers, D.A. Soil Surface Carbon Dioxide Fluxes Induced by Spring, Summer, and Fall Moldboard Plowing in a Sandy Loam. Soil Sci. Soc. Am. J. **1999**, *63*, 621–628.
7. Reicosky, D.C. Tillage-Induced CO_2 Emissions from Soil. Nutrient Cycling in Agroesystems **1997**, *49*, 273–285.
8. Reicosky, D.C. Strip Tillage Methods: Impact on Soil and Air Quality. *Proceedings of the Australian Society of Soil Science Incorporated National Soils Conference*; Brisbane, Australia, April 27–30, 1998; Australian Society of Soil Science Inc.: Brisbane, Australia, 1998; 56–60.
9. Erickson, A.E. Tillage Effects on Soil Aeration. In *Predicting Tillage Affects on Soil Physical Properties and Processes*; Unger, P.W., Van Doren, D.M., Eds.; ASA Special Publication No. 44; ASA: Madison, WI, 1982; 91–104.
10. Renault, P.; Mohrath, D.; Gaudu, J.C.; Fumanal, J.C. Air Pressure Fluctuations in a Prairie Soil. Soil Sci. Soc. Amer. J. **1998**, *62*, 553–563.
11. Reicosky, D.C.; Lindstrom, M.J. Fall Tillage Methods: Effect on Short-Term Carbon Dioxide Flux from Soil. Agron. J. **1993**, *85* (6), 1237–1243.
12. Schlesinger, W.H. Changes in Soil Carbon Storage and Associated Properties with Disturbance and Recovery. In *The Changing Carbon Cycle: A Global Analysis*; Trabalha, J.R., Reichle, D.E., Eds.; Springer-Verlag: New York, 1985; 194–220.

TIME AND SPACE, SOILS IN

Neil E. Smeck
The Ohio State University, Columbus, Ohio, U.S.A.

C. Lee Burras
Iowa State University, Ames, Iowa, U.S.A.

INTRODUCTION

Soil, the relatively thin organic-mineral entity at the Earth's surface, forms at the interface between the atmosphere and lithosphere by the assimilation of energy. The sun not only warms the Earth's surface which increases the rate of chemical reactions, but also drives the photosynthetic production of biomass. Death and decay of biomass results in the accumulation of humus in the upper horizon(s) of soil. Soil formation is also driven by the weathering of minerals in the lithosphere which releases energy utilized in soil development and nutrients that are essential for the growth of plants. Gravitational energy provides the driving force for the infiltration and percolation of water into and through the Earth's surface. The magnitude of humus accumulation, mineral weathering, and translocation or leaching of solutes and colloids by water moving through the lithosphere plays a major role in differentiating the great diversity of soils that cover the face of the Earth.

Soils vary greatly in their characteristics and potential to meet human needs. The most essential function of soil from a human perspective is that soil serves as the reservoir of water and nutrients essential for sustained plant growth. Plants serve as our food, either via direct production of fruit, vegetables, and grain, or indirectly as feed for livestock. Soils also serve as the foundation for societal infrastructure and as a medium for the disposal and amelioration of wastes. The regional and global differences in soil characteristics and potentials for use are due to differences in energy fluxes in the various ecosystems of the world. The objective of this chapter is to discuss the evolution of soils as a function of time and factors that result in the spatial distribution of the multitude of soils that mantle the Earth.

SOIL EVOLUTION AND DEGRADATION AS A FUNCTION OF TIME

Soil formation is initiated the moment a portion of the lithosphere is exposed to the atmosphere following major depositional or erosional geologic events. The geologic materials deposited by water, wind, ice, or gravity or exposed by erosion are referred to as "parent material" because they are the precursor of soil. Common parent materials are alluvium, glacial till, glacial outwash, lacustrine and marine sediments, loess (wind-blown silt), volcanic ash, pedisediment, colluvium, and residuum (weathered bedrock). Very youthful soils differ little from their parent material but as time progresses, soils with characteristics distinctive of the ecosystems in which they form will evolve (such soils are considered mature soils). The evolution of a youthful to a mature soil proceeds due to additions and losses of materials and translocation and transformation of components within the soil (1). Specific processes include the weathering of primary minerals to secondary minerals (clays, salts, and oxides) and soluble ions, leaching of soluble components from the soil, accumulation of organic matter in the soil, and the translocation and accumulation of secondary minerals within the soil. All of these processes contribute to the formation of unique sequences of soil horizons characteristic of the various mature soils.

Young soils exhibit weak expression of horizons, little accumulation of organic matter above that in the geologic parent material, and minimal mineral weathering. Such soils generally exhibit only A and C horizons (lack B horizons) and classify as Entisols in *Soil Taxonomy* (2). With additional time, such soils will develop weak B horizons, become more intensely weathered, and accumulate organic matter. These developments will qualify the soils as Inceptisols in *Soil Taxonomy* (2). Eventually as weathering and developmental processes continue, a mature soil will form. A mature soil can be defined as a soil that has attained a steady state with the energy fluxes in the ecosystem in which it occurs. This implies that there are sufficient energy influxes from the ecosystem to maintain the soil in its current state for an indefinite time, but insufficient energy for additional soil formation. Most of the soil orders in *Soil Taxonomy* represent such steady states. The distribution of these mature orders will be discussed in the second section of this chapter.

Encyclopedia of Soil Science

Rates of soil formation are a function of the ecosystem in which soil formation occurs and the current stage of soil development. Soil formation occurs more rapidly in high energy ecosystems (tropical) and in youthful, relatively shallow soils. As soil formation progresses toward maturity, increases in soil depth gradually slow and eventually cease because weathering, leaching, and energy inputs are reduced at depth. Buol et al. (3) list soil formation rates, obtained from the literature, ranging from 0.1 to 750 yr cm^{-1}; the former at the surface of a recent mudflow and the latter at a depth of 1 m in an Oxisol. Although rudimentary soils can form in <100 years, thousands of years are necessary for mature soils to form. Because more time is required to acquire the energy necessary for intensely weathered and/or highly developed soils to attain maturity, soil orders can be arrayed by the amount of time required to attain a steady state (Fig. 1). The steady state condition attained by mature soils necessitates that such soils are being regenerated at the same rate as they are being degraded by weathering or truncated by geologic erosion. Troeh et al. (5) suggest "a rate of 1 mt ac^{-1} yr^{-1} may be considered typical for geologic erosion from gently sloping soils." Using this geologic erosion rate, a steady-state condition could be maintained with a rate of soil formation of 100 yr cm^{-1} of soil.

Accelerated erosion caused by human activities on the landscape, however, can remove soil at rates that are many times that of the capacity for soil regeneration. Whereas maximum "tolerable" soil loss rates have been established at 11 mt ac^{-1} yr^{-1} by the USDA—Natural Resources Conservation Service (10 times the rate of geologic erosion), accelerated erosion rates exceeding 450 mt ac^{-1} yr^{-1} have been documented (5). Wischmeier and Smith (6) defined tolerable soil loss as "the maximum level of soil erosion that will permit a high level of crop productivity to be sustained economically and indefinitely." It is obvious that mismanagement of soils can result in the loss of soil at a rate greater than that of soil formation.

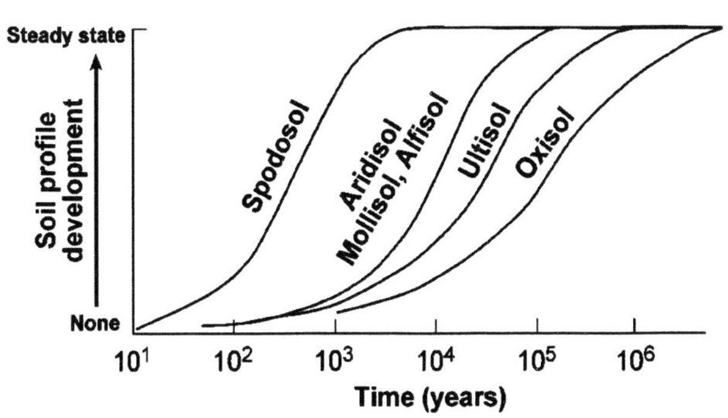

Fig. 1 Schematic diagram showing the variations in time to attain the steady state for various soil orders. (From *Soils and Geomorphology* by Peter W. Birkeland, copyright 1984 Oxford University Press, Inc. Used by permission of Oxford University Press, Inc.)

Reduction of the world's soil resources poses a serious threat for the longevity of humankind. Accelerated erosion is not the only process leading to soil degradation; compaction, oxidation of organic matter due to tillage and drainage, and salinity due to poorly managed irrigation systems all contribute to soil degradation. Based on the rate of soil formation relative to the length of a human lifetime, soil must be considered a nonrenewable resource. Soil loss due to accelerated erosion will not be regenerated in our lifetime. Degradation of the quality of soil due to inappropriate management practices, however, can be reversed by the use of best management practices; thus, soil can be considered a renewable resource with respect to soil quality. The most unique attribute of soil is that soil can be used indefinitely without deterioration if managed for sustainability.

SPATIAL DISTRIBUTION OF MATURE SOILS

Soil results from the interactions among climate, vegetation, geological parent material, topography, and time. These five "state factors," as formulated by Jenny (7), determine the nature and distribution of soils over the face of the Earth. The relative importance of a given state factor's influence on soil properties is a function of scale. Climate is generally the most important state factor at a global scale, whereas vegetation is the most important factor at a regional scale. At a local scale, topography and parent material separately or in combination often control the distribution of soils.

Precipitation and temperature are the most important climatic factors controlling the weathering and, ultimately, global distribution of mature soils (Fig. 2). Warm, wet climates generally result in highly weathered soils, whereas dry, cold climates result in minimally weathered soils. Climates between these extremes result in the formation of soils with intermediate degrees of weathering. This relationship can be illustrated by considering a few major climatic areas of the world. Wet tropical regions such as the rainforests of the Amazon Basin of South America and the Congo Basin of Africa are noted for deep, highly weathered, nutrient-poor soils classified as Oxisols in the U.S. soil classification system (2). Many Oxisols, however, occur in dry climates as relicts of paleo-tropical climates.

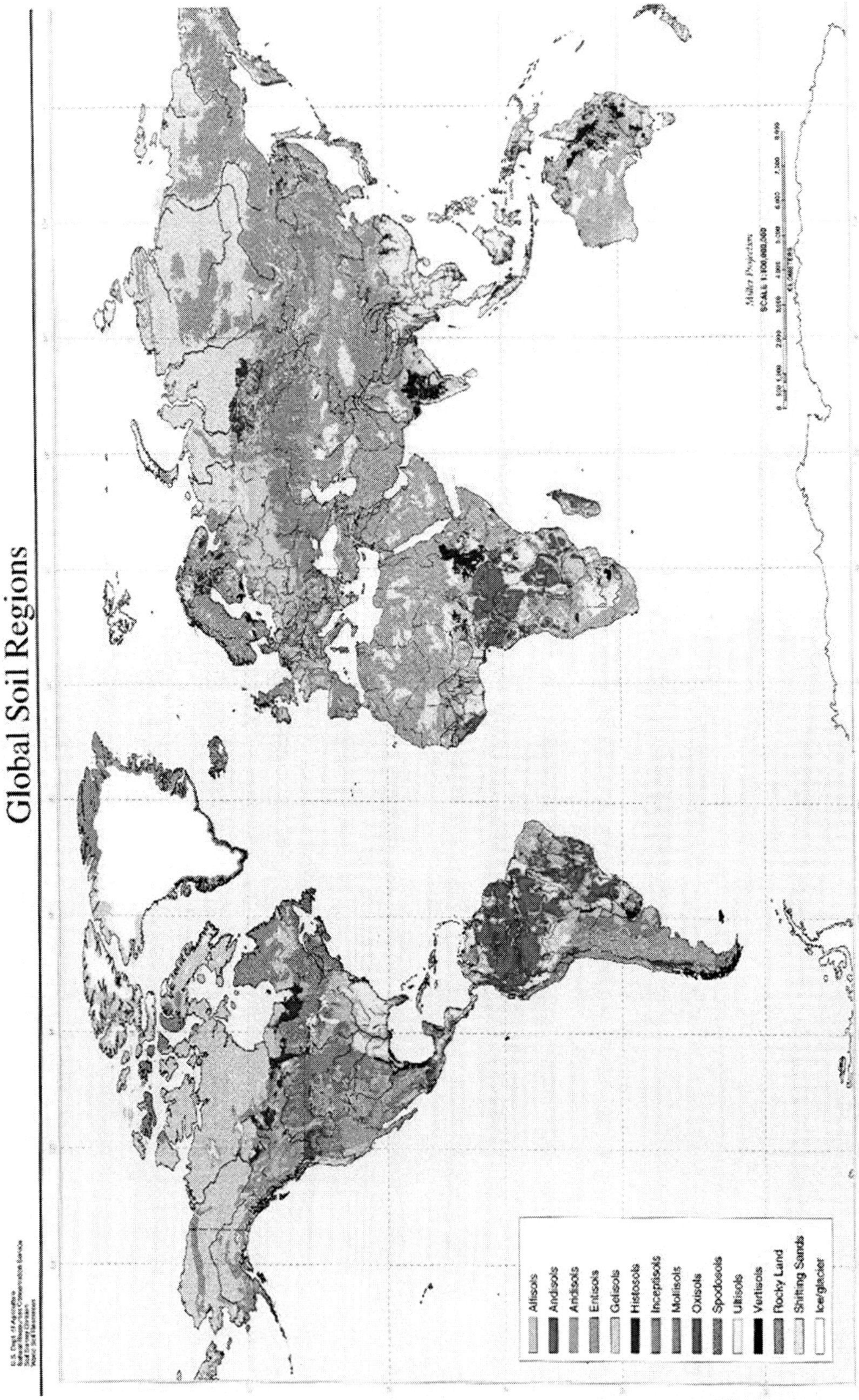

Fig. 2 World soil map (8).

Oxisols are often red in color due to the concentration of iron and aluminum oxides as other chemical components are released by weathering and leached from the soil. On the opposite end of this spectrum, the polar regions of North America, Europe, and Asia have large areas of Gelisols (2), which are permanently frozen soils. Because little weathering occurs when a soil is frozen, Gelisols are generally minimally weathered soils. Both Gelisols and Oxisols very readily degrade due to mismanagement but for contrasting reasons. Gelisols are fragile because they develop and regenerate at such an exceedingly slow rate. Oxisols are so highly weathered that they are depleted of primary minerals and nutrients; thus, there is little internal energy remaining for regeneration.

On progressing from humid-tropical to humid-temperate climates, soils generally become less weathered. The following soil orders can be considered a weathering continuum: Oxisols (tropical)–Ultisols–Alfisols (temperate). Ultisols generally occur in humid, temperate to semitropical climates, but may also occur in wet tropical locations where there has been insufficient time for Oxisols to form. Ultisols are concentrated in Central America, the southeastern quadrant of the U.S., Asia, and Africa. The key properties of Ultisols are low-nutrient and nutrient-bearing mineral contents, an A horizon with low organic matter content, and a thick, prominent B horizon. The B horizon is a zone of clay concentration due to both clay illuviation and formation. Alfisols are found in humid-temperate climates (with deciduous forests) that are too cool for Ultisols to form. These regions include much of eastern North America, central Europe and Great Britain, and China. Alfisols are more fertile and contain more nutrient-bearing minerals than Ultisols due to less intense weathering. Alfisols are also characterized by clay accumulation in B horizons, but the accumulation is less pronounced than in Ultisols. In addition, the clay accumulation in Alfisols is more dependent on clay illuviation and less dependent on clay formation than that of the B horizons in Ultisols.

Wet and/or cool climates such as in Scotland and Ireland result in the formation of Histosols (2). Histosols are organic soils developed in wet, anaerobic locations where plant production of organic matter far exceeds decomposition of the litter produced. Histosols can be tens of meters deep although they are generally 1–3 m deep. They are commonly referred to as peats and mucks. Historically, these soils have been mined for use as fuel. The arid and semiarid climates of the world result in the presence of either Aridisols or Entisols (2). Aridisols and Entisols in arid regions cover nearly 28% of the globe due to extensive dry areas in Africa, North America, South America, Asia, and Australia. Aridisols range from moderately developed soils that are relicts of past more humid climates to weakly developed soils commonly containing soluble salts due to minimal leaching. Such diverse soils are grouped together as Aridisols primarily due to sufficient moisture for only very limited vegetative growth without irrigation. Entisols, which commonly consist of shifting sands and consequently show very little soil development, dominate the driest portions of the Earth's surface. Entisols also occur in other climatic regions on very young or weathering resistant parent materials.

Vertisols (2) are very clayey soils associated with seasonally wet and dry climates. They are widespread in western India, eastern Australia, and along the northwestern coast of the Gulf of Mexico. During the dry season, Vertisols develop wide, deep cracks that swell shut during the wet season due to the expansive nature of the clays. The swelling pressures in the subsoil cause subsoil to be thrust toward the surface resulting in heaving of the soil-surface. Repetition of this cyclic process between seasons results in churning or inversion of the soil.

The distribution of vegetative communities is a function of climate, comparable to the impact of climate on the distribution of soils; however, vegetative communities superimpose unique soil forming processes and characteristics on those imposed by climate. This can be illustrated by considering the distribution of Spodosols, Alfisols, and Mollisols (2), which are strikingly different soils, Spodosols occur in ecosystems dominated by conifers (e.g., Maine and other northern New England states, Michigan, Wisconsin, Minnesota, and upper New York state), Alfisols occur in ecosystems dominated by deciduous trees (e.g., New York, Pennsylvania, Ohio) and Mollisols in prairie ecosystems common throughout the central U.S. (e.g., North Dakota to Texas, Illinois to Colorado). Spodosols are generally shallow, sandy, low fertility, acidic soils characterized by a prominent bleached E horizon over B horizons that have accumulated iron and aluminum oxides and/or humic materials due to the chelation and translocation of these components by compounds released by the coniferous vegetation or decomposition products of the litter. Alfisols are fertile soils characterized by an A horizon with low organic matter content and clayey B horizons. The translocation of clay from near surface horizons to the B horizon is attributed to the high degree of acidity produced by the decomposition of the leaves of deciduous trees. Mollisols are fertile soils characterized by thick, dark, humus-rich A horizons that form in prairie ecosystems. The annual production of a deep, fibrous root system by prairie grasses in conjunction with the Ca-rich chemistry of the soils results in the accumulation of stable humic compounds in the soil to considerable depth.

The distribution of soils on a landscape scale is generally related to the distribution of vegetation, topography, and

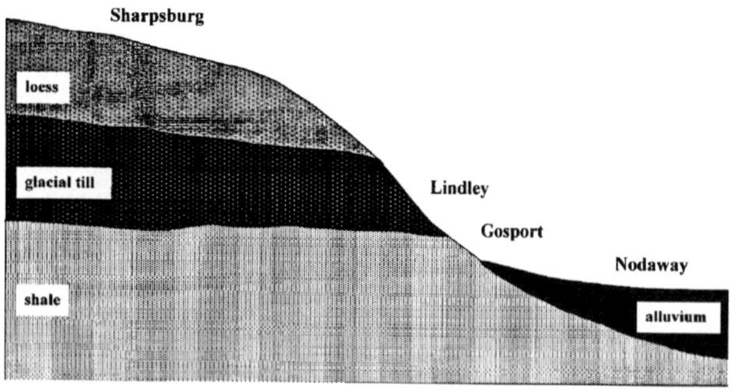

Fig. 3 Cross-section of a landscape showing the relationship between soils and geological parent materials from southern Iowa and northern Missouri.

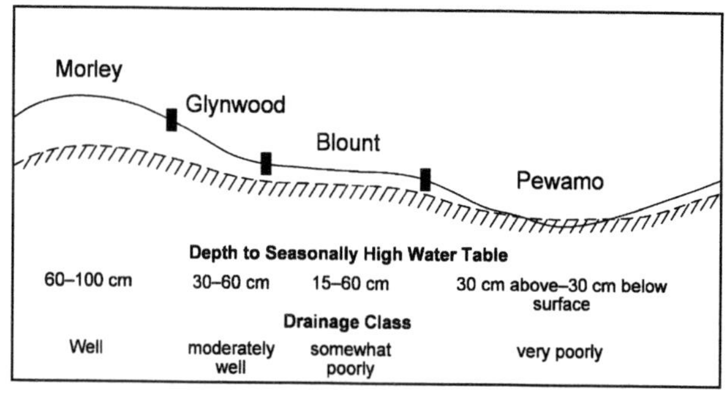

Fig. 4 Cross-section of landscape showing relationship between soil distribution and topography, drainage class, and depth to seasonally high water table (undrained condition) in Morley–Glynwood–Blount–Pewamo association of western Ohio.

parent materials or a combination thereof. An example of vegetation determining soil distribution occurs in some Iowa landscapes where soils formed in deciduous forest, savannah, and prairie ecosystems exist contiguously. Aifisols formed in the forested ecosystems, Mollisols in the prairie ecosystems, and soils with properties intermediate between Alfisols and Mollisols in the savannah ecosystem. Due to the causative relationship between ecosystem biology and soil pattern, such soil distributions are referred to as "biosequences."

The impact of parent materials on the distribution of soils occurs at any scale, but is most evident on large scale maps. Sharpsburg, Lindley, Gosport, and Nodaway soil series form an association in the various parent materials that occur in the hilly landscapes of southern Iowa and northern Missouri (Fig. 3). The soils comprising this association are quite different as indicated by their classification into different soil orders: Sharpsburg derived from loess is a Mollisol, Lindley derived from glacial till is an Alfisol, Gosport derived from shale is an Inceptisol, and Nodaway formed in alluvium is an Entisol. Although differences in age of the four parent materials influence classification of the soils at the order level, the soils derived from the different parent materials would classify as different soil series based solely on properties inherited from the parent materials.

Another parameter influencing soil distributions, especially at the field scale, is the impact of water flow, both surface and subsurface, in the landscape. Much of the precipitation that falls on sloping landscape components is removed by surface runoff, whereas depressional soils accumulate surface runon and retain all precipitation. Potential water infiltration and percolation through the soils is, therefore, least on sloping landscape components, intermediate on level areas, and greatest in the lowest portions of the landscape. The depth to a water table (zone of saturation in the lithosphere) also varies across landscapes. The shallowest depth to the water table generally occurs in the lowest landscape positions which serve as the primary point for aquifer recharge, whereas, the greatest depths to the water table occurs in the landscape components with the highest elevations (Fig. 4). Due to differential water distribution and depths to water tables within landscapes, soil wetness and drainage classes vary within landscapes. Wetness and drainage characteristics have a major impact on the suitability of soils for various uses, both agricultural and nonagricultural (construction, sewage deposal, home sites, recreational activities, etc.).

In summary, the world's soils are distributed in a highly systematic manner. Soil properties and distribution are predictable with knowledge of the five state factors and their impact on soil formation. Time is required for mature soils, those steady-state entities comprising the uppermost portion of the lithosphere, to evolve from geological parent material. The relative importance of the other four factors is a function of scale. Climate and vegetation are most important in explaining global soil patterns, whereas parent material, topography, and time are more important at landscape scales.

REFERENCES

1. Simonson, R.W. Outline of a Generalized Theory of Soil Genesis. Soil Sci. Soc. Am. Proc. **1959**, *23*, 152–156.
2. Soil Survey Staff. *Soil Taxonomy: A Basic System of Soil Classification for Making and Interpreting Soil Surveys*, 2nd Ed.; Agriculture Handbook Number 436, USDA/Natural Resources Conservation Service: Washington, DC, 1999; 869.

3. Buol, S.W.; Hole, F.D.; McCracken, R.J.; Southard, R.J. Time as a Factor of Soil Formation. *Soil Genesis and Classification*, 4th Ed.; Iowa State University Press: Ames, IA, 1997; 179–194.
4. Birkeland, P.W. *Soils and Geomorphology*; Oxford University Press: Oxford, 1984; 372.
5. Troeh, F.R.; Hobbs, J.A.; Donahue, R.L. *Soil and Water Conservation Productivity and Environmental Protection*, 3rd Ed.; Prentice Hall: Upper Saddle River, NJ, 1999; 610.
6. Wischmeier, W.H.; Smith, D.D. *Predicting Rainfall Erosion Losses: A Guide to Conservation Planning*; Agric. Handbook 537, USDA/Agricultural Research Service: Washington, DC, 1978.
7. Jenny, H. *Factors of Soil Formation*; McGraw-Hill: New York, 1941; 281.
8. USDA–NRCS. 2001. World Soil Resources. http://www.nhq.nrcs.usda.gov/WSR/mapindx/metadata/Maps/ORDERS.JPG (Reviewed August 22, 2001).

TROPICS AND SUB-TROPICS, SOILS OF THE

Hari Eswaran
United States Department of Agriculture-Natural Resources Conservation Service (USDA-NRCS), Washington, D.C., U.S.A.

Eswaran Padmanabhan
University Malaysia Sarawak, Kota Samarahan, Malaysia

INTRODUCTION

The tropics straddle the equator and by definition do not have large differences in temperature in the months corresponding to "winter or summer" of the higher latitudes. In this section, the tropics are defined as the area with a mean summer to winter soil temperature difference of less than 6°C as per the definition in *Soil Taxonomy* (1). The wide diversity of soils of the tropics is elaborated by Eswaran et al. (2). With increases in elevation, the mean annual temperature decreases resulting in temperature zones with corresponding differences in floristic composition. With a relatively uniform temperature, subdivisions of the biome may be made on the availability of moisture. The semiarid tropic borders the deserts and occupies about 15.5% of the area, while the humid tropics include the intertropical convergence zone and occupy 11.1% of the tropical biome zone.

Within the humid area, there is a subzone where the evapotranspiration does not exceed the precipitation in any month of the year. This is the extremely wet part of the tropics and typifies the "tropical rain-forest." In these perhumid areas, there are no dry months and annual precipitation frequently exceeds 4 m and may be as high as 12 m. Tree growth is luxuriant with several layers of vegetation. The canopy is closed and the under-storey layers are usually very high in humidity. High humidity promotes growth of epiphytes. Tropical rainforests provide a large amount of leaf litter but these are rapidly humified with insignificant accumulation of organic matter in the soil. Diversity of plants is large and most of the families occur on all the continents. The humid and perhumid areas support a wide variety of animal life that clusters around specific layers of the forest. As a rule, grass-eating animals are scarce but increase where the forest grades into the savanna.

A tropical dry forest occurs in the semiarid areas. Although this is the climax vegetation, it is quickly destroyed and does not have enough resilience to regenerate. Degeneration is evident in the semiarid parts of Africa where previous forest lands now lie barren after deforestation. Species diversity is also lower in the dry forests, which are generally more fragile. The dry forests give way to grasslands or savanna. In these landscapes, gallery forests, where the forests follow riparian courses are frequent. In the semiarid areas, there is also considerable plant adaptation to drought conditions. Long tap roots and thick barks characterize the shrubs. The branches of the shrubs are also generally twisted and gnawed due to frequent fires. Herbivores dominate but carnivores are also present to maintain a balance in the animal population.

Though climate is a major determinant of the biological character of the tropics, significant differences are also introduced by the soils that are present. A classical example is illustrated by areas with highly weathered Oxisols belonging to great groups such as the Acrudox and Acrustox. These soils have a very low nutrient and water holding capacity and consequently, even in an area with more than 5 m of rain and no dry season, the vegetation may comprise stunted trees or grass. In soils with excess water, as in some wetland areas, there is a completely different association of vegetation and animal life.

The dominant soils and the role they play in providing the character of the biomes is now illustrated with a few examples. Table 1 provides the area occupied by the suborders in tropical biomes.

OXISOLS

Oxisols (3) are present in both the wet and dry tropics. Large contiguous areas of Oxisols are on the Amazon Shield in South America and in the Congo Basin in Central Africa. Small areas occur on the volcanic islands of the Pacific and in South East Asia, where they form on weathering products of basic and ultrabasic rocks. On the

Encyclopedia of Soil Science

Table 1 Area ($\times$1,000 km^2) occupied by soil suborders in tropical biomes

		Tropical			
		Semiarid		Humid	
Soil-order	Sub-order	Area	%	Area	%
	Saprists	37	0.11	281	0.81
Histosols		37	0.11	281	0.81
	Aquods			13	0.04
	Cryods				
	Humods			29	0.08
	Orthods			18	0.05
Spodosols				61	0.17
	Vitrands	139	0.40	48	0.14
	Ustands	59	0.17		
	Udands			187	0.54
Andisols		198	0.57	235	0.68
	Aquox	300	0.86	20	0.06
	Ustox	3,087	8.87		
	Perox			1,010	2.90
	Udox			5,169	14.85
Oxisols		3,387	9.73	6,199	17.81
	Aquerts			1	0.00
	Usterts	1,170	3.36		
	Uderts			86	0.25
Vertisols		1,170	3.36	87	0.25
	Argids	44	0.13		
	Calcids			1	0.00
	Cambids				
Aridisols		44	0.13	1	0.00
	Aquults	714	2.05	331	0.95
	Humults	24	0.07	254	0.73
	Udults			2,656	7.63
	Ustults	3,633	10.44		
	Xerults				
Ultisols		4,371	12.56	3,240	9.31
	Aquolls			1	0.00
	Rendolls			121	0.35
	Ustolls	136	0.39		
	Udolls			54	0.16
Mollisols		136	0.39	176	0.51
	Aqualfs	387	1.11	20	0.06
	Ustalfs	3,777	10.85		
	Udalfs			618	1.77
Alfisols		4,165	11.96	638	1.83
	Aquepts	758	2.18	744	2.14
	Cryepts				
	Ustepts	2,807	8.06		
	Udepts			1,757	5.05
Inceptisols		3,566	10.24	2,502	7.19
	Aquents	74	0.21	29	0.08
	Psamments	1,947	5.59	558	1.60
	Fluvents	521	1.50	396	1.14
	Orthents	683	1.96	111	0.32
Entisols		3,225	9.26	1,093	3.14
Total		20,299	58.31	14,514	41.69

older surfaces of Africa and South America, peneplanation has flattened the land surface. The Oxisols are generally formed on preweathered and transported sediments. One feature they have is a stone-line, marking differential deposition.

The soils have low nutrient retention capacity and low water holding capacity, and in many countries, these features have earned them the name "droughty soils". The ability to retain anions such as phosphates, nitrates, and sulfates is high. Extremes in cation and anion retention are seen in Oxisols with a net positive charge. These properties are detrimental to plant growth and therefore vegetation production is very low and sometimes only adapted plants grow on these soils. The Oxisols of the high plateaus of Central Africa have a subsurface dark horizon called a sombric horizon. They are distinctive in these nondescript soils and have caught the attention of local farmers who prefer them due to their better fertility.

ULTISOLS

Occupying 7.6 million km^2 (21.87% of the tropics), the Ultisols are the second most extensive soils. Although there are a few members on recent erosion surfaces, most of the Ultisols are highly weathered with few weatherable minerals and occur on stable geomorphic surfaces (4). The landscape is generally undulating. Ultisols occur in association with Oxisols and also share many of the physico-chemical properties. Water and nutrient retention is also low and some, as the Oxisols on old landscapes, have stone-lines.

The Ultisols have an argillic or a kandic horizon and the latter soils are generally associated with older landscapes. On the old surfaces of Central Africa and the Amazon Shield, lithological discontinuities, sometimes with stone-lines, are common in the Kandiustults and Kandiudults. Clay content differentiation in the upper part of the soil, a defining characteristic for Ultisols, is generally due to geomorphic processes; clay translocation and accumulation to form an argillic horizon is a relict feature in many Ultisols or confined to the soils on younger surfaces.

Organic matter content is related to moisture saturation. Some Ultisols (Humults), which occur at high elevations in cooler environments, also have high organic matter. For low-input farming, these organic matter rich soils are frequently the first choice for food production. Ultisols are preferred to Oxisols for grain production, but the latter soils are generally highly rated for perennial crops such as rubber (*Hevea braziliensis*).

Many Ultisols (and some Oxisols and Alfisols) have plinthite (5). Plinthite is a sesquioxide rich, organic matter poor material, which has a reticulate pattern with reddish and white colors. It forms under aquic conditions and when the water table is lowered and the surface exposed, it can harden to iron stone. Soils with plinthite are considered as degraded because their productivity is low and they are also difficult to manage.

ALFISOLS

The Alfisols are the pedological equivalent of the Ultisols with a high pH or base saturation. They are preferred to Ultisols for their better nutrient status and easy workability. The Alfisols are more frequent in the semiarid tropics. In the humid tropics, they occur on specific rock types, such as on limestone and/or basic rocks, or at locations where there is an external source of bases. The dust from the Sahara (Harmattan winds) and Kalahari deserts add large quantities of carbonates to the soils of Central Africa and so Alfisols are extensive in Africa.

Alfisols, particularly the Ustalfs, which have a loner period of moisture deficit, are widely used for food crop production. In the wet season they grow maize, sorghum, and soybeans. As moisture stress increases, the variety of crops grown in the absence of irrigation includes chickpeas and millet. In the Ustalfs and in some of the Udalfs, sugarcane is an important crop. In the drier tropics, the Alfisols and other soils are used for grazing. Managed pastures are only recent and in many countries the carrying capacity for animals is frequently exceeded. With an adequate supply of water and nutrients, particularly nitrogen and phosphorous, these are one of the most productive soils of the tropics. The other very productive soil is the Mollisol but these are not very extensive.

INCEPTISOLS AND ENTISOLS

The alluvial soils of valleys and coastal plains have been the sites for agricultural development in Asia. These soils are naturally wet or can be puddled for rice cultivation. The Entisols occur on the recent flood plains while the better-developed Inceptisols are on older surfaces. In the semiarid tropics, rice is grown during the wet season while an upland crop is attempted, if adequate moisture is available, for much of the growing period. In Africa, wet Inceptisols and Entisols were traditionally avoided due to heavy infestation with pests and diseases. The African farmer prefers to cultivate the better-drained upland soils.

On the coastal plains of the tropics acid sulfate soils occur and they are more extensive in the tropics than in temperate areas. Formed under brackish water conditions they support unique vegetation and are also a unique ecosystem. These soils occur in association with the peats or Histosols. The characteristic feature of acid sulfate soils is the presence of a mineral called pyrite (FeS_2). In the Entisols (Sulfaquents), the process of pyrite formation is active. The Entisols commonly occur close to tidal flats and are generally continuously reduced. The bearing capacity of such soils is low and, due to the presence of pyrite, the soils are not cultivated. Drainage of such soils, or where the soils have been naturally drained by down-cutting of rivers, results in oxidation of the upper part of the soil with a concomitant ripening process. The pyrite oxidizes and alters to a yellow mineral called jarosite and in the process liberates sulfuric acid that lowers soil pH. This chemical transformation is considered to be the greatest acidity change in soils. The resultant soil is an Inceptisol (Sulfaquept) and has the dubious status of being the most acid soil. Many Sulfaquepts are used for rice production with low production and even crop failures.

Large areas of the tropics (Table 1) are covered with sandy soils (Psamments). The largest extent is in the desert margins of the Sahara and the Kalahari, in Africa. The Kalahari sands have been blown eastwards into parts of Botswana, Zimbabwe, and Zambia, where they bury former Ultisols and Oxisols. Towards the north, the Kalahari sands occur in Congo as large, broad tongues. Most of the soils on these broad deposits are Psamments but in the depressions, they are associated with Spodosols.

The upland Inceptisols and Entisols occur on steep sloping land. The soils are shallow due to continuous down wearing by water erosion. The tropical Udepts and Ustepts and the Orthents are commonly found under shrub or forests. Land conditions prevented their cultivation, although today, due to scarcity of land in many of the poorer countries, these soils are also being used for agriculture.

VERTISOLS

The Vertisols (6, 7) of the tropics have only recently been used for agriculture in low-input systems. These soils are sticky and plastic when wet and extremely hard when dry and so cannot be managed with the traditional bullock-drawn plows. In India, for example, until recently when mechanical power became available, the soils were only planted at the end of the rainy season.

Vertisols are very productive soils for agriculture (8). The pH is between 6 and 7.5 and the soil is well supplied with nutrients. Vertisols do not have adverse quantities of toxic elements and are not deficient in any nutrients, but they must be supplied with N and P for sustained productivity. They are derived from basic parent materials, and the largest extent is in India (on the Deccan Plateau) and bordering the Nile in Sudan.

OTHER SOILS

Mollisols, Aridisols, Histosols, Spodosols, and Andisols occur in small areas in the tropics. Mollisols are present in the cooler areas, with an isothermic soil temperature regime. Tropical Aridisols form a small band in Africa and in India. Their properties are similar to Aridisols of other biomes.

The Histosols (9) are very extensive in South East Asia and mainly in Sumatra and Kalimantan of Indonesia. They are found on the coastal plains of these islands. They also occur at similar physiographic settings in the other countries. Unlike the Histosols of northern latitudes, the tropical Histosols are woody and have buried timber. Some also have acid sulfate features. These soils are extremely difficult to manage due to their very acid nature and the many nutrient (specifically, trace element) deficiencies. The beach ridges on the coast create a basin inland where the Histosols form under aerobic conditions. Associated with the Histosols are the Spodosols of the coastal platforms in the tropics. Spodosols are formed on sands and have a bleached surface E horizon with an organic matter rich subsurface horizon. A lateral seepage of organic colloid rich water from the peat swamps into the sand result in the formation of a "spodic-like" B horizon.

The Andisols occur on the circum-Pacific belt of volcanoes. They are present in all countries of this orogenic zone. In general, they share the properties of the temperate Andisols but have a lower content of organic matter. Some of the Andisols with udic soil moisture regimes, particularly those occurring on the older surfaces, have a net positive charge, similar to the Acrudox. The upper part of the soil may have a net negative charge as long as there is organic matter, and so organic matter management is crucial to the productivity of these soils.

REFERENCES

1. Soil Survey Staff, *Soil Taxonomy: A Basic System of Soil Classification for Making and Interpreting Soil Surveys*, 2nd Ed.; U.S. Dept. Agric. Handbook 436, Government Printing Office: Washington, DC, 1999.
2. Eswaran, H.; Beinroth, F.H.; Kimble, J.; Cook, T. Soil Diversity in the Tropics: Implications for Agricultural Development. In *Myths and Science of Soils of the Tropics*; Publ. Soil Sci. Soc. AMer., Madison, Wisconsin, Lal, R., Sanchez, P.A., Eds.; Special Publ. No. 29, 1992; 1–16.
3. Buol, S.W.; Eswaran, H. Oxisols. Advances in Agronomy **2000**, *68*, 152–195.
4. Lim Jit Sai; Khalsi, M.S.; Eswaran, H. Alfisols and Ultisols with Low Activity Clays in Malaysia. In *Proceedings of Second International Soil Classification Workshop. Part II, Thailand*; Beinroth, F.H., Panichapong, S., Eds.; Publ. Dept. of Land Development: Bangkok, Thailand, 1979; 107–118.
5. Eswaran, H.; De Coninck, F.; Varghese, T. Role of Plinthite and Related Forms in Soil Degradation. Advances in Soil Science **1990**, *11*, 109–128.
6. Dudal, R.; Eswaran, H. Distribution, Properties and Classification of Vertisols. In *Vertisols: Their Distribution, Properties, Classification and Management*; Wilding, L.P., Puentes, R., Eds.; Publ. Soil Management Support Services US Department of Agriculture Natural Resources Conservation Service: Washington, DC, 1988; 1–22.
7. Mermut, A.R.; Padmanabhan, E.; Eswaran, H.; Dasog, G.S. Pedogenesis. In: *Vertisols and Technologies for Their Management. Developments in Soil Science*; Ahmad, N., Mermut, M., Eds.; 1997; 24, 43–61.
8. Eswaran, H.; Virmani, S.M.; Abrol, I.P. Issues and Challenges of Dryland Agriculture in Southern Asia. In *Agriculture and the Environment: Bridging Food Production and Environmental Protection in Developing Countries*; ASA Special Publication No. 60, Juo, A.S.R., Freed, R.D., Eds.; American Society of Agronomy: Madison, WI, 1995; 161–180.
9. Eswaran, H. Classification and Management of Histosols. In *Proc. of 2nd. Int. Symposium on Peat Land Soils., Thailand/Malaysia*; Eswaran, H., Panichapong, S.M., Eds.; Publ. Dept. of Land Development: Thailand, 1986; 15–26.

TURBATIONS

Robert C. Graham
Soil and Water Sciences Program, Department of Environmental Sciences, University of California, Riverside, California, U.S.A.

INTRODUCTION

Various processes act within soils to physically disturb and mix the soil material. These processes are in general referred to as *turbations*, and prefixes are added to more specifically categorize them as *bioturbations*, *cryoturbations*, and *pedoturbations*. Turbations of some kind operate in most soils. Often their effects are relatively minimal and the development of soil horizons is primarily under the influence of such processes as leaching and illuviation. In some soils, turbations are a dominant process that strongly impact the soil morphology and may even overwhelm the effects of other processes. The mechanisms responsible for the different kinds of turbations, and the resulting effects on the soils in which they operate, are explored below.

BIOTURBATION

Mixing of soil caused by organisms, either flora or fauna, is known as bioturbation. Soil is the basic rooting medium for plants and, while root growth itself disturbs soil fabric, the process is generally so slow and localized that a mixing effect is not obvious. A much greater impact results from tree throw, that is the uprooting of trees, usually by wind. When a tree falls, a root wad consisting of soil and rock enmeshed by coarse roots, is pulled out of the ground (Fig. 1). The soil material held in the root wad is eventually eroded and released in a mixed heap on the soil surface. Forest fires speed the release process by burning the fallen tree and its roots. Where tree throw is common, a hummocky landscape of relatively weakly developed soils results (1). In some cases, tree throw lifts soil horizons intact and lays them back as an inverted profile rather than fully mixing them (2).

Animals digging in or burrowing through soil can be very effective mixing agents. The mixing power of earthworms has long been recognized (3). Earthworms bring organic material from the surface down into the mineral soil. As they travel through the soil they ingest, and intimately mix, both organic and inorganic materials throughout the zone in which they are active. This generally produces a thick, dark, organic-rich A horizon (mollic or umbric epipedon), and can counteract the development of illuvial horizons (4). Bioturbation has also been attributed to other invertebrates, including ants, termites, and crayfish (5).

Burrowing mammals are powerful bioturbation agents. Pocket gophers are among the most active burrowers as they continually explore the upper several decimeters of soil for food. Other burrowers, such prairie dogs and ground squirrels, have more fixed tunnel systems but still effectively mix the soil, often digging into underlying soft bedrock, thereby deepening the soil–bedrock boundary (6). Some predators, such as badgers, coyotes, and bears, pursue these small burrowing animals by attempting to excavate them, and in the process they disrupt even larger volumes of soil. Bioturbation by mammals, as with earthworms, tends to promote the formation of thick, organic-rich A horizons and retard the strong differentiation of illuvial horizons. Bare soil thrown out of burrows is subject to raindrop impact and erosion, but thorough mixing of the surface soil, as accomplished by gophers, may improve aeration and water infiltration.

Bioturbation by mammals can have direct impacts on environmental quality and cultural resources. Burrowing can move environmental contaminants from the surface to deeper within the soil (7) and can break through subsurface barriers constructed to retain contaminants in landfills (8). Pocket gopher activity in archaeological sites can destroy the stratification of artifacts, confounding interpretations (9).

CRYOTURBATION

The churning and mixing of soils caused by freezing and thawing is known as cryoturbation. Freezing of water in the soil causes expansion, subsequent thawing causes collapse, and together the resulting forces mix the soil in complex patterns (Fig. 2). Cryoturbation dominates in many Gelisols, where an "active layer" overlies the

Encyclopedia of Soil Science

Fig. 1 An example of bioturbation: tree throw has lifted root wads of soil and rock that will eventually be deposited in a mixed mound on the soil surface.

permafrost. The active layer, usually < 2 m thick, is frozen during the cold season but thaws in the summer and is a zone of intense cryoturbation (10). During freezing, liquid water migrates to freezing fronts, both near the surface and at the underlying permafrost, and the soil mass expands at those locations when the water freezes. Cryoturbation orients rock fragments within the soil, transports organic-rich A horizon material into the subsoil, and disrupts the lateral continuity of soil horizons (11). Ice wedges in the permafrost also promote cryoturbation (12). When soil temperatures drop below −30°C in the winter, the already solid ice in the permafrost recrystallizes and shrinks by 3%, forming vertically oriented cracks. In the spring, water from the active layer migrates into these cracks and freezes, with a 9% increase in volume (at temperatures between 0 and −30°C). The resulting ice wedges slowly expand over the years and generate subsoil pressures that must be resolved in an upward direction, thereby contributing to cryoturbation. The effects of cryoturbation in permafrost areas can also be seen at the soil surface in the form of patterned ground, with polygonal nets of sorted stones, frost-heave ridges, and frost boils.

Although cryoturbation is most strongly expressed in permafrost regions, it also occurs in temperate zone soils where freezing is limited to the upper part of the soil. Frost heave churns the upper several centimeters of these soils, lifting rock fragments and uprooting seedlings. The effect is pronounced where the soil is bare, not insulated by snow cover or leaf litter.

PEDOTURBATION

Some soils undergo substantial swelling when they are wetted and shrink upon drying. The resulting disruption of the soil is called pedoturbation, and is generally associated with clayey soils that are rich in smectite, the most extreme expressions of which are Vertisols. When the soil swells, often with differential wetting patterns (13), its fabric is sheared upward along slickensides, bringing subsoil material toward the surface (Fig. 3) (14). When the soil dries and shrinks, cracks are formed that can be in the order of 10 cm wide at the soil surface and extend to over a meter in depth. These cracks promote mixing because surficial materials fall into them and are thus added directly to the subsoil (11). The morphology of Vertisols is dominated by the effects of pedoturbation, but the mixing is not

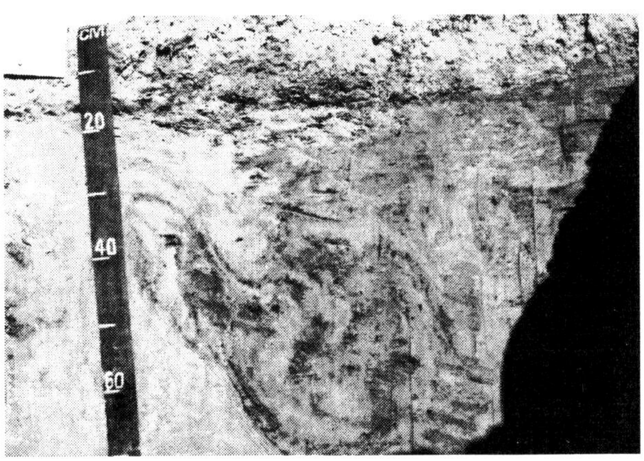

Fig. 2 An example of cryoturbation: freeze–thaw processes in a Gelisol have mixed organic matter into the subsoil in convoluted patterns. Scale is in centimeters. (Photograph provided by J.G. Bockheim.)

Fig. 3 An example of pedoturbation: shrink–swell processes in a Vertisol have forced lighter-colored subsoil material toward the surface. (Photograph provided by T.D. Cook.)

necessarily so rapid that it obliterates horizons formed by leaching and illuviation, such as calcic horizons (15). As with bioturbation, pedoturbation in Vertisols can rapidly mix anthropogenic contaminants to some depth (16).

Pedoturbation is not restricted to Vertisols and other clayey soils. Minor shrink–swell behavior is present in most soils, and to some degree it can alter the soil fabric. Narrow shrinkage cracks (<5 mm) under desert pavements allow eolian dust to infiltrate into shallow B horizons where it is incorporated into the soil fabric, slowly building the soil upward (17). In another example, the rock fabric of granodiorite was disrupted and converted to soil fabric by minor shrink–swell activity in the uppermost part of the weathered bedrock profile (18). Another type of pedoturbation happens in strongly developed petrocalcic horizons. Precipitation of calcium carbonate within voids in these cemented horizons causes expansion and internal shattering, whereupon the resulting cracks are sites for more calcite precipitation, and the process is continued. The result is a petrocalcic horizon composed of turbated and recemented angular fragments of the original petrocalcic horizon (19).

REFERENCES

1. Meyers, N.L.; McSweeney, K. Influence of Treethrow on Soil Properties in Northern Wisconsin. Soil Sci. Soc. Am. J. **1995**, *59*, 871–876.
2. Schaetzl, R.J. Complete Soil Profile Inversion by Tree Uprooting. Phys. Geogr. **1986**, *7*, 181–189.
3. Darwin, C. *The Formation of Vegetable Mould Through the Action of Worms, with Observations on Their Habits*; Appleton: New York, NY, 1881.
4. Graham, R.C.; Wood, H.B. Morphologic Development and Clay Redistribution in Lysimeter Soils Under Chaparral and Pine. Soil Sci. Soc. Am. J. **1991**, *55*, 1638–1646.
5. Hole, F.D. Effects of Animals on Soil. Geoderma **1981**, *25*, 75–112.
6. Munn, L.C. Effects of Prairie Dogs on Physical and Chemical Properties of Soils. In *Proceedings of the Symposium on the Management of Prairie Dog Complexes for the Reintroduction of the Black-Footed Ferret*; Oldemeyer, J.L., Biggins, D.E., Miller, B., Crete, R., Eds.; Biological Report 13 USDI-Fish and Wildlife Service: Washington, DC, 1993; 11–17.
7. Mace, J.E.; Graham, R.C.; Amrhein, C. Anthropogenic Lead Distribution in Rodent-Affected and Undisturbed Soils in Southern California. Soil Sci. **1997**, *162* (1), 46–50.
8. Suter, G.W.; Luxmoore, R.J.; Smith, E.D. Compacted Soil Barriers At Abandoned Landfill Sites are Likely to Fail in the Long Term. J. Environ. Qual. **1993**, *22*, 217–226.
9. Johnson, D.L. Biomantle Evolution and the Redistribution of Earth Materials and Artifacts. Soil Sci. **1990**, *149* (2), 84–102.
10. Bockheim, J.G.; Tarnocai, C. Gelisols. In *Handbook of Soil Science*; Sumner, M.E., Ed.; CRC Press: Boca Raton, FL, 2000; E-256–E-269.
11. Buol, S.W.; Hole, F.D.; McCracken, R.J.; Southard, R.J. *Soil Genesis and Classification*, 4th Ed.; Iowa State University Press: Ames, IA, 1997.
12. Rieger, S. *The Classification and Genesis of Cold Soils*; Academic Press: New York, NY, 1983.
13. Weitkamp, W.A.; Graham, R.C.; Anderson, M.A.; Amrhein, C. Pedogenesis of a Vernal Pool Entisol–Alfisol–Vertisol Catena in Southern California. Soil Sci. Soc. Am. **1996**, *60*, 316–323.
14. Wilding, L.P.; Tessier, D. Genesis of Vertisols: Shrink–Swell Phenomena. In *Vertisols: Their Distribution, Properties, Classification and Management*; Wilding, L.P., Puentes, R., Eds.; Texas A & M University Printing Center: College Station, TX, 1988; 55–81.
15. Southard, R.J.; Graham, R.C. Cesium-137 Distribution in a California Pelloxerert: Evidence of Pedoturbation. Soil Sci. Soc. Am. J. **1992**, *56*, 202–207.
16. Graham, R.C.; Ulery, A.L.; Neal, R.H.; Teso, R.R. Herbicide Residue Distributions in Relation to Soil Morphology in Two California Vertisols. Soil Sci. **1992**, *153* (2), 115–121.
17. Anderson, K.C., Wells, S.G., Graham, R.C., and McFadden, L.M. Processes of Vertical Accretion in the Stone-Free Zone Below Desert Pavement. In *Abstracts with Programs — Geological Society of America*, Boulder, CO, 1994, 26 (7), 87.
18. Frazier, C.S.; Graham, R.C. Pedogenic Transformation of Fractured Granitic Bedrock, Southern California. Soil Sci. Soc. Am. J. **2000**, *64*, 2057–2069.
19. Birkeland, P.W. *Soils and Geomorphology*, 3rd Ed.; Oxford University Press: Oxford, UK, 1999.

TURFGRASS, PROTECTION OF SOIL AND WATER WITH

T

James A. Murphy
Stephanie L. Murphy
Rutgers, The State University of New Jersey, New Brunswick, New Jersey, U.S.A.

INTRODUCTION

Turfgrasses, turfs, and lawns often bring to mind connotations of a society's affluence or playgrounds of the idle rich; however, the turfed areas of many landscapes have more humble beginnings and play an important role in the protection of the Earth's soil mantle. The origin of turfed areas in managed landscapes probably started along with man's effort to domesticate animals. Many centuries ago, human dwellings were close to the wilderness where predators roamed, and domesticated animals were herded and tethered close to dwellings to prevent escape and predation. Clearings eliminated cover for stalking predators, and grass provided incentive for the domesticated animals to stay put. Thus, a grazing system was developed that was necessary for survival of both man and his animals (1). Sheep and goats grazed the village green of ancient times, and the well-knit turf was esthetically inviting, encouraging residents to gather and socialize. These village greens became the squares or parks of many older towns and cities. Thus, this closely cropped system has become the basis for many landscapes around modern day homes, businesses, roadways, in parks and in other places of beauty. The total area of turfgrass in the U.S. has been estimated at 31 million acres, with more than 80% of the total in lawns (2).

FUNCTION OF TURF IN LANDSCAPES

The usefulness of closely cropped perennial grasses were probably recognized early in developing civilizations because of the important benefits in the stabilization and enhanced durability of soil under traffic. The ability to perform daily chores and recreational activities without interference from mud and dust would have been clearly recognized by early humans. The fibrous roots of perennial grasses are recognized today as the most effective means of stabilizing soil (3). This outstanding capacity of grasses to stabilize the soil mantle, particularly under repeated treading, enables humans to utilize the land in many ways, including parks, sports, playgrounds, and multiple other uses (4). The first airport runways were covered with turfgrasses until the aircraft became too large and heavy (5, 6). Turfed runways are still used for small aircraft and gliders in rural areas. The prehistoric Americans and homesteaders of the North American plains used sod to construct dwellings (7). Turf areas serve as recreational space for young and old. The grazed grounds around early human communities and homes probably provided for the first areas of play and sports. The game of lacrosse was modified from a sport played by Native Americans on grassed plains (7). Many sporting games including bowls, golf, and cricket have been played for centuries on turfgrass (8); football, soccer, and baseball are now the dominant sports that rely on the durability and safety of turfgrass fields. Many burial grounds are covered with turf, and many modern memorial gardens are designed into beautiful park-like settings with turf serving as the backdrop that ties together the landscape. Furthermore, a large proportion of the nearly 4 million miles of public roads in the U. S. (9) are bordered by low maintenance turf for reasons of safety as well as soil stabilization.

SOIL CONSERVATION

As the human population increases worldwide, the utilization of our land and soil resources continues to expand. Population growth tends to be greater in developing countries and puts intense pressure on these societies to intensively utilize much of the available soil resources in agricultural operations, housing, or infrastructure. Continued prosperity in developed countries fuels the expansion of suburban land development around redeveloping urban centers. These agricultural and construction activities of man result in extensive land disruption, which exposes soil to the erosive forces of wind and water. Soil erosion is a direct or contributing cause of declining soil productivity; destruction of land; sedimentation of streams, reservoirs and storm drains; damage to roadways; mounting floods; and degradation of air and water quality (10). Although many individuals were conscious of the problems of soil erosion earlier, it was not until the early 1900s that the agricultural community in U.S. began to regard soil conservation as fundamental to sound farming practice (11). Up to that time, vegetative methods of erosion control were given little attention and limited incidental use. In recent years, land development and use in urban centers has received greater attention for its impact on soil losses and the subsequent degradation of water quality, sedimentation of waterways, and severity of flooding (12, 13).

Turfgrasses are playing an increasingly important role in enhancing our physical environment, conserving our soil and water resources, and providing recreation (14, 15). Currently, there is considerable effort being expended to repair the landscape in China that was intentionally destroyed during the Cultural Revolution of the 1960s. Turf and trees are now being replanted in China to halt and eventually reverse the severe environmental degradation of smog and dust pollution, soil erosion, water pollution, and heat buildup in cities (16). Turf provides a relatively inexpensive durable ground cover that protects the valuable nonrenewable soil resource from water and wind erosion. Protection of soil by thick-growing vegetation takes place through several processes (17). One important way in which close-growing vegetation protects the land is through the interception of rainfall by the plant surfaces. This prevents the soil from the beating action of raindrops that dislodges soil particles and allows soil to move in water runoff. This positive effect can be frequently observed in runoff that is clear from turfed areas, whereas water runoff from land under cultivation or development is laden with silt, clay, organic matter, nutrients, and other materials potentially present in soil.

The maintenance of high humus or organic matter content in the soil is another benefit of keeping land in grass cover. Disruption of land is known to rapidly destroy soil organic matter. Organic matter is critical to creating or maintaining soil structure, the system of aggregates and pores in soil. Soils high in organic matter content typically have a rate of water infiltration that is considerably greater than soil low in organic matter (18). Greater infiltration of water into soil subsequently reduces the amount of runoff, further reducing the erosion hazard.

Another aspect of grass cover in erosion control is that it slows the movement of water over the land surface, thus reducing the erosive power of the water. Slower movement of water also provides greater time for water to infiltrate the soil, thus reducing runoff. The dense cover of vegetation provided by turfgrass also reduces the wind velocity at the soil surface compared to bare soil, and hence is effective in reducing erosion by wind; taller vegetation can reduce wind velocity to a greater degree, erosion control may depend on the extent of bare areas between plants. Grass roots are better than the roots of annual weeds and other plant species in binding soil particles together (3, 19). These principles of reducing soil erosion—including raindrop interception, improved infiltration, slowing runoff, and binding of soil particles—in grassland systems have proven to be just as effective in turfgrass settings (20–23).

WATER QUALITY

Nonpoint source pollution is now considered an important source of decreased water quality in U.S. (13). Soil erosion losses are a significant component of nonpoint source pollution. Erosion depletes land of important nutrients and organic matter content, and removes the most physically suitable soil for biologic sustenance. The eroded sediment that makes its way into streams, reservoirs, and other water bodies carries nutrients such as nitrogen and phosphorus that can significantly reduce the water quality. Close-growing vegetation, including turfgrasses, play an essential role in capturing water and filtering the sediment and nutrients in runoff (20, 24). Turf is recommended as the vegetation to enhance environmental quality in urbanized landscapes through the stabilization of exposed soil, slowing of runoff water, promotion of infiltration, and removal of pollutants such as sediment, organic matter, nutrients, and heavy metals from runoff water (13, 25).

SOIL ORGANIC CARBON

The loss of organic matter (carbon) from cultivated soils and other man-induced activities contribute to increased atmospheric CO_2, one of the so-called greenhouse gases (26). Perennial grasses are important components of ecosystems that conserve or enhance soil organic matter (27, 28), and it is thought that a grassland system can be managed to lessen the increases in atmospheric CO_2 by a process called carbon sequestration (29). Perennial grass cover on soil is more effective than cover by weedy species in maintaining high soil organic matter levels (27).

T

SUMMARY

When man alters a landscape, changes in the soil and vegetation take place. Perennial grasses often play an important role in the structural redevelopment of the altered ecosystems. Turf is established and maintained in landscapes to enhance the environment and quality of life through stabilization and accelerated restoration of soil. Turfgrasses are frequently a part of the secondary succession on abandoned farmland and other disturbed landscapes that may take hundreds or thousands of years to restore the soil, depending on the loss or degradation of soil. Other benefits of turfgrasses to humans and the environment include temperature moderation, reduction of noise and glare, reduction of pest populations, safety via firebreaks and increased visibility zones on roadsides, airfields and security-sensitive locations, and creation of surfaces for sport and leisure activities, as well as esthetic benefits (15).

REFERENCES

1. Hillel, D. *Out of the Earth: Civilization and the Life of the Soil*; Univ. of California Press: Berkeley, CA, 1991.
2. The Lawn Institute. *How the Environment Benefits from a Well-Maintained Lawn*; Online. http://www.turfgrasssod.org/lawninstitute/environmental_benefits.htm (Accessed September 2001).
3. Carter, M.R.; Angers, D.A.; Kunelius, H.T. Soil Structural Form and Stability and Organic Matter Under Cool-Season Perennial Grasses. Soil Sci. Soc. Am. J. **1994**, *58*, 1194–1199.
4. Asay, K.H.; Sleper, D.A. Contributions from Breeding Forage and Turf Grasses—An Overview. In *Contributions from Breeding Forage and Turf Grasses*; Sleper, D.A., Asay, K.H., Eds.; CSSA: Madison, WI, 1989; 15, 1–20.
5. Morrish, R.H.; Alton, A.E.; Cale, E.B. Airfields and Flight Strips. In *Grass: The Yearbook of Agriculture 1948*; USDA, Stefferud, A., Ed.; U.S. Gov. Print. Office: Washington, DC, 1948; 319–323.
6. Morrish, R.H.; Harrison, C.M. The Establishment and Comparative Wear Resistance of Various Grasses and Grass-Legumes Mixtures to Vehicular Traffic. Agron. J. **1948**, *40*, 168–179.
7. Roberts, E.C.; Huffine, W.W.; Grau, F.V.; Murray, J.J. Turfgrass Science—Historical Overview. In *Turfgrass*; Waddington, D.V., Carrow, R.N., Shearman, R.C., Eds.; Agron. Monogr ASA, CSSA and SSSA: Madison, WI, 1992; 32, 1–27.
8. Beard, J.B. *Turfgrass: Science and Culture*; Regents/Prentice Hall: Englewood Cliffs, NJ, 1973.
9. U.S. Department of Transportation/Federal Highway Administration. *Highway Statitistics 1999: Public Road Length – 1999*; Online. http://www.fhwa.dot.gov/ohim/hs99/tables/hm20.pdf (Accessed September 2001).
10. National Resource Conservation Service. *National Resource Inventory: Highlights*; 1997; Online. http://www.nhq.nrcs.usda.gov/land/pubs/97highlights.html (Accessed March 2001).
11. USDA, Bennett, H.H.; Lowdermilk, W.C. General Aspects of the Soil-Erosion Problem. *Soils and Men: Yearbook of Agriculture 1938*; USDA, U.S. Gov. Print. Office: Washington, DC, 1938.
12. U.S. Environmental Protection Agency. *Nonpoint Source Pointers (Factsheets)*, 1996; Online. http://www.epa.gov/OWOW/NPS/facts/ (Revised 21 January 1997) (Accessed March 2001).
13. U.S. Environmental Protection Agency. *National Water Quality Inventory: 1996 Report to Congress*, 1996; Online. http://www.epa.gov/305b/96index.html (Accessed March 2001).
14. Meyer, W.A.; Funk, C.R. Progress and Benefits to Humanity from Breeding Cool-Season Grasses for Turf. In *Contributions from Breeding Forage and Turf Grasses*; Spec. Publ. 15, Sleper, D.A., Asay, K.H., Eds.; CSSA: Madison, WI, 1989; 15, 31–48.
15. Beard, J.B.; Green, R.L. The Role of Turfgrasses in Environmental Protection and Their Benefits to Humans. J. Environ. Qual. **1994**, *23*, 452–460.
16. International Turf Producers Foundation, *Water Right — Conserving Our Water Preserving Our Environment*; http://www.turfgrasssod.org/Water.pdf (Accessed March 2001) International Turf Producers Foundation: Rolling Meadows, IL, 2001.
17. USDA, Enlow, C.R.; Musgrave, G.W. Grass and Other Thick-Growing Vegetation in Erosion Control. *Soils and Men: Yearbook of Agriculture 1938*; USDA, U.S. Gov. Print. Office: Washington, DC, 1938; 615–633.
18. Smith, F.B.; Brown, P.E.; Russell, J.A. The Effect of Organic Matter on the Infiltration Capacity of Clarion Loam. J. Am. Soc. Agron. **1937**, *29*, 521–525.
19. Weaver, J.E.; Harmon, G.W. *Quantity of Living Plant Materials in Prairie Soils in Relation to Runoff and Soil Erosion*; Bull. 8, Nebr. Univ. Conserv. Dept. Nebraska, 1935.
20. Gross, C.M.; Angle, J.S.; Welterlen, M.S. Runoff and Sediment Losses from Turfgrass. J. Environ. Qual. **1990**, *19*, 663–668.
21. Gross, C.M.; Angle, J.S.; Hill, R.L.; Welterlen, M.S. Runoff and Sediment Losses from Tall Fescue Under Simulated Rainfall. J. Environ. Qual. **1991**, *20*, 604–607.
22. Morton, T.G.; Gold, A.J.; Sullivan, W.M. Influence of Overwatering and Fertilization on Nitrogen Losses from Home Lawns. J. Environ. Qual. **1988**, *17*, 124–130.
23. Watschke, T.L.; Mumma, R.O. *The Effect of Nutrients and Pesticides Applied to Turf on the Quality of Runoff and Percolating Water*; ER 8904, Pennsylvania State University, Environmental Resources Res. Inst. University Park, PA, 1989.
24. Mendez, A. Sediment and Nitrogen Transport in Grass Filter Strips. J. Am. Water Resourc. Assoc. **1999**, *4*, 867–875.
25. U.S. Environmental Protection Agency, *Erosion, Sediment, and Runoff Control for Roads and Highways*; USEPA 841-

F-95-008d, http://www.epa.gov/OWOW/NPS/education/runoff.html (Accessed March 2001) U.S. Gov. Print. Office: Washington, DC, 1995.
26. Lal, R.; Kimble, J.M. Conservation Tillage for Carbon Sequestration. Nutr. Cycling Agro-Ecos. **1997**, *49*, 243–253.
27. McHenry, J.R. Influence of Some Perennial Grasses on the Organic Matter Content and Structure of An Eastern Nebraska Fine-Textured Soil. J. Am. Soc. Agron. **1947**, *39*, 981–994.
28. Paustian, K. Agricultural Soils As a Sink to Mitigate CO_2 Emissions. Soil Use Mgmt. **1997**, *13*, 230–244.
29. Malhi, S.S.; Nyborg, M.; Harapiak, J.T.; Heier, K.; Flore, N.A. Increasing Organic C and N in Soil Under Bromegrass with Long-Term N-Fertilization. Nutr. Cycling Agro-Ecosys. **1997**, *49*, 255–260.

21ST CENTURY, ROLE OF SOILS IN THE

T

Richard W. Arnold*
U.S. Department of Agriculture, Washington, D.C., U.S.A.

INTRODUCTION

Cultural attitudes determine the role of soils in today's world as our fragmented global community struggles to resolve the global issues of food security and environmental protection. A variety of world views influences the search for a sustainable, socially acceptable balance among soil functions that provides for viable economic growth and development, safe healthy environments, and intergenerational equity (1).

Linking entire social systems in a web of production, distribution, and consumption, agriculture often foreshadows the degree of economic well-being (2). Because agriculture operates simultaneously in the realms of ecology and economics, each of which marks time by different clocks, decisions affecting food security and environmental protection have become increasingly complex and variable over time and space.

THE PEDOSPHERE

Soils are a critical interface between society and natural resources. Thus, the basic principles of the organization and functioning of the Earth's soil cover, the *pedosphere*, can provide a scientific basis for global programs addressing food security and environmental protection throughout the 21st century (3).

Natural soils result from the interaction of processes taking place over time on the Earth's surface. Most involve gases and liquids that transform the solid phase of the surficial materials into features recognized as soils. These processes are influenced by soil-forming factors—namely, climate, biota, topography, parent material, and time—leading to great heterogeneity in the world's soil cover. Geomorphic processes that alter the landscape by erosion, transport, and deposition of rocks and sediments, and the interaction of biological systems with soils are all subject to major modifications associated with global and regional climate changes. The result is an intricate patchwork of soils, ranging in size from a few square meters to thousands of square kilometers.

For millennia, naturally evolving soils were dominant in the world, and determining the properties and distributions of major kinds of soils and their local geographic associates enabled societies to effectively tap these resources to satisfy their needs. Soils provided habitats for plants, animals, and microorganisms. Soils possessed the capacity for fertility and potential productivity due to water, physical support, and biological interactions that provided nutrients (4).

Mapping the pedosphere and deciphering the sequence of events and processes causing such complexity has revealed a multidimensional hierarchy that is meaningful for assessing many kinds of soils and predicting soil-related behavior.

ALTERED ECOSYSTEMS

For more than a century, scientific studies of soils as natural independent entities on the Earth's surface have contributed to a better understanding of the interconnectivity of the Earth's systems (5). As societies introduced more and more invasive procedures, they drastically altered ecosystem processes and biogeochemical cycles. Although some ecosystems have been enhanced, more have been degraded, and now it is relevant to predict and mediate the behavior and responses of both natural and artificial soil environments and landscapes.

With increases in the size of human population, the available natural resources have often been stressed, some beyond their limits of resilience. When soils are so stressed, they are unable to return to their former productive states without massive external inputs. Thus, sustainable integration of societal desires and natural resource capabilities is commonly jeopardized. The role of soils, consequently, can be viewed as the set of trade-offs among their various functions, as determined by current society.

FUNCTIONS OF SOILS

Functions of soils commonly include: 1) to promote biomass transformations; 2) to serve as the Earth's

*Retired

geomembrane to filter and buffer; 3) to provide biological habitats; 4) to provide usable materials; and 5) to serve as sources and repositories of our cultures (6).

Soils are used by most people primarily for the production of food, fuel, and fiber. The capacity to store and release water and the ability to renew, store, and release plant nutrients have dominated agronomic and forestry research and practical experimentation for many years. Biomass transformations are highly dependent on the microbiological populations that inhabit soils and facilitate the formation and use of beneficial compounds.

The pedosphere is a sensitive geomembrane, which mediates the transfer of air, water, and energy into, out of, and among the biosphere, atmosphere, hydrosphere, and the lithosphere (Fig. 1). Temperatures are moderated with depth, and moisture and associated compounds are filtered, retained, stored, and transferred in ways that contribute to clean, healthy environments.

Soils are the habitat for millions of organisms, ranging from cellular bacteria to burrowing animals. Communities of microorganisms decompose organic materials, facilitate the release of mineral elements, and produce the chemical and biological compounds essential for life on Earth (4).

Many soils are used directly as raw materials for constructing dams and foundations; others are source materials for landscaping industrial and urban sites; and some are ingredients for bricks and ceramic products.

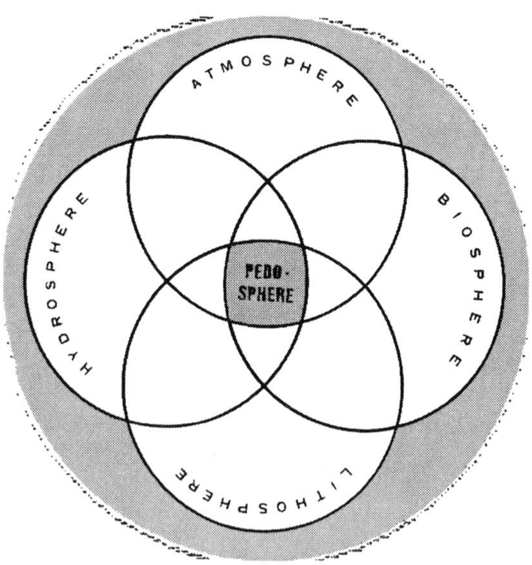

Fig. 1 The pedosphere results from the interaction throughout space and time of the atmosphere, biosphere, and hydrosphere with the lithosphere. Once developed, the pedosphere plays a significant role in supporting and maintaining life on the planet. (From Science Press, Beijing, China, with permission.)

Most transportation networks and urban communities rest on soils. In addition, soils are excavated to create disposal sites for society's numerous wastes. Soils that are disturbed or removed from their natural environments are commonly called "dirt." As such, displaced soil materials are generally considered a nuisance in our daily activities and need to be washed out, removed, and personal contact with them minimized. Historically, stigma was often attached to those associated with the "filth and dirt" of using and managing soils (7).

In the stories of most indigenous cultures, the Earth is sanctified and revered as a vital element of Nature, thereby reinforcing humankind's link with it. This sacredness remains in the use of soils as cemeteries, and as places where the spirits of ancestors reside. Archeological investigations generally involve deciphering the memories of times and events recorded in soil properties.

CONTINUING NEEDS

It is obvious that soils will continue to be used to maintain and improve bioproductivity for food, fuel, and fiber for a long time because the supply of elements and minerals necessary for life are derived, directly or indirectly, from soils. Healthy ecosystems and environments depend on soil-hosted microorganisms that facilitate filtering and purifying. There is still much to be learned about how soils behave in maintaining healthy anthropogenic landscapes. Unprecedented demands to safely handle wastes and provide major increases in food and feed are enormous challenges (8).

Perhaps less appreciated is the need of the human psyche for renewal through contact with nature. It appears that psychological well-being is, in part, related to our communion with the beauty, tranquility, and mysterious forces of nature. Although we are connected to the land and soils in ways that most of us do not readily comprehend, we do recognize the sense of belonging and the feeling of renewal associated with our contacts with the natural world. Is it a "dust thou art, and to dust thou shall return" syndrome in which our life cycle is subsumed in other natural life cycles? Whatever the explanation, the human-to-nature relationship is important, and soil is vital to our survival and growth.

CHANGING DEMOGRAPHICS

It has been suggested that sometime in this century more than our entire current population will be living in cities

(9). As cities evolve into megacities through the use of adjacent lands, the unique culture of urban dwellers, including their estrangement from rural landscapes and ecosystems, also develops.

Ecological functions of soils are common in extensive rural areas of the Earth, where the productive capacity of soils is easily recognized. In urban landscapes, which are very different in appearance, structure, and composition, the role of soils is commonly not observed, or even imagined, except in parks and residential lawns and gardens. Sewers and trash trucks remove numerous waste products, smoke stacks and vehicles exude particulates into the atmosphere, and streams and rivers wash away other debris. In rural landscapes, many of these functions are provided by the pedosphere—that covering of soils that seems to be ubiquitous and so common that it is taken-for-granted.

How much relatively undisturbed land would be required to provide the energy and products consumed in an urban environment, and to adequately handle its wastes? Such a measure could be thought of as a city's environmental footprint (8). At present we do not measure and monitor the enormous fluxes that occur in urban environments, yet they are becoming major stressors on our global habitat.

INFLUENCING THE ROLE

Is there a possibility of simply stretching current ecological theory to encompass urban ecosystems? It can be argued that our understanding of dynamics and processes of populations and soils can be extended to homeowners' associations and pavement (8). If people act as other organisms do, guided by individual self-interest, there is no basis for a moral or aesthetic call to environmental stewardship. Is there, perhaps, a spiritual or moral dimension that defies explanations offered by evolution or natural selection?

The challenge of understanding urban ecosystems requires specialists from many different disciplines, but it also requires that at least some individuals think in interdisciplinary and multidisciplinary ways—a task that may be difficult to accomplish.

The single most important force of landscape change in urban areas is land conversion driven by institutional decisions, population growth, and economic forces (8). A city's footprint, which indicates the dependence of an urban ecosystem on other ecosystems, may be tens or hundreds of times larger than the city itself.

Soil ecosystems are being altered, and even created, to meet the expectations of urban communities. Consequently, the kinds and patterns of adjacent soil landscapes are important to the monitoring and assessment of environmental health and sustainability.

THE FUTURE

Humans have made tremendous achievements in their mutual adjustment to accepted standards (10). For example, at a busy street corner clusters of pedestrians wait, voluntarily, for the light to change to green. When the light changes, surging individuals instinctively negotiate their way across the intersection, without bumping or colliding with those coming from the other direction. The voluntary acceptance of the established standard of traffic regulation by lights results in personal safety.

The adaptive flexibility observed in this "uncentralized" operation (10) suggests how the balance of the soil functions will be solved in the future. A global consensus must develop that will clearly define a minimum set of common norms and international standards for sustainable uses of various kinds of soils. Knowledge about the limitations of specific kinds of soils for particular uses is essential for informed decisions and agricultural policies (Table 1). For every conceivable use of soil there is a hypothetical ideal soil with the right set of properties and supporting processes to achieve a satisfactory level of success. Comparing local soils with a specified "ideal" soil can lead to recommended measures that minimize limitations and contribute toward attaining the expected

Table 1 Soil attributes that can become future constraints to achieving desired balances of soil functions if excessive demands are placed on soil resources

Soil attribute	Constraint
Resilience	Recovery from disturbance
Productivity	Capability for plant growth and yield
Responsiveness	Capacity for external enhancement
Sustainability	Dynamic equilibrium of interactions
Resistance	Stability to maintain current condition
Flexibility	Multiplicity of uses related to properties
Pedoclimate	Location and extent of suitable climate
Residence time	Capacity to store and release compounds
Geography	Availability due to location or intricacy of pattern

behavior of such an ideal soil. When local economics of managing individual soils are considered, it is possible to develop rankings of suitability and economic feasibility.

Understanding soil property-soil process relationships can provide the basis for creating and improving soil ecosystems needed to support the constantly changing global environment. International Organization for Standards (ISO) standards will serve as universal "rules of the road." Individuals and communities will adapt their actions and integration of needs with the available resources. Local flexibility will occur in the mutual adaptations applied to these global standards of ethical and environmental stewardship of the world's natural resources.

All of these conditions can be achieved in, and by, the world's diverse cultures through their own mutual adaptations. The future role of soils is truly in the hands and hearts of the people.

REFERENCES

1. Karlen, D.L.; Mausbach, M.J.; Doran, J.W.; Cline, R.G.; Harris, R.F.; Schuman, G.E. Soil Quality: A Concept, Definition, and Framework for Evaluation. Soil Sci. Soc. Amer. J. **1997**, *61*, 4–10.
2. De Soysa, I.; Gleditsch, N.P.; Gibson, M.; Sollenberg, M.; Westing, A.H. *To Cultivate Peace: Agriculture in a World of Conflict*, PRIO Report 1/99; International Peace Research Institute (PRIO): Oslo, 1999; 1–89.
3. Sokolov, I.A. Spatial-Temporal Organization of the Pedosphere and Its Evolutionary and Ecologic Causes. Eurasian Soil Sci. **1994**, *26* (4), 10–25.
4. Kovda, V.A. *The Role and Functions of the Soil Cover in the Earth's Biosphere*; Scientific Center of Biological Research, Academy of Science, USSR: Pushchino, 1985; 1–12.
5. Targulian, V.O.; Rode, A.A.; Dmitriev, N.A.; Armand, A.D. Soil as a Component of Natural Ecosystems and the Study of Its History, Modern Dynamics and Anthropogenic Changes. In *Selection, Management and Utilization of Biosphere Reserves*; Franklin, J.F., Krugman, S.L., Eds.; Proceedings of The United States-Union of Soviet Socialist Republics Symposium on Biosphere Reserves, Moscow, USSR, May 1976; USDA Pacific Northwest Forest and Range Experiment Station: Corvallis, Oregon, 1979; General Technical Report PNW 82, 186–197.
6. German Advisory Council on Global Change. *World in Transition: The Threat To Soils*, 1994, Annual Report; Economica Verlag: Bonn, 1995.
7. Yaalon, D.H.; Arnold, R.W. Attitudes Toward Soils and Their Societal Relevance; Then and Now. Soil Sci. **2000**, *165* (1), 5–12.
8. Collins, J.P.; Kinzig, A.; Grimm, N.B.; Fagan, W.F.; Hope, D.; Wu, J.; Borer, E.T. A New Urban Ecology. Am. Sci. **2000**, *88* (5), 416–425.
9. Brown, L.R.; Gardner, G.; Halweil, B. *Beyond Malthus: Sixteen Dimensions of the Population Problems*, Worldwatch Paper 143; Worldwatch Institute: Washington, DC, 1998.
10. Cleveland, H. Coming Soon: The Nobody-In-Charge Society. Futurist **2000**, *34* (5), 52–56.

ULTISOLS

U

Joey N. Shaw
Auburn University, Auburn, Alabama, U.S.A.

INTRODUCTION

Soils classified in the Ultisol order possess a subsurface horizon (argillic or kandic horizon) with more clay than overlying horizons, and low base saturation (percentage of exchangeable Ca, Mg, K, and Na) in the lower portion of the soil. In some regions, Ultisols commonly possess dense, root and water restrictive subsurface horizons (fragipans), and hardened accumulations of iron (Fe) oxides mixed with other minerals (plinthite and ironstone). Ultisols are found in all climates except deserts. Ultisols occupy approximately 9.2% of the Earth's land surface, are concentrated in tropical regions (80%), and are found in South America, Africa, China, Australia, the Pacific Islands, coastal India, and the southeastern U.S. (1, 2) (Fig. 1). Conceptually, the presence of Ultisols suggests: 1) in many cases, an advanced degree of soil weathering, 2) the land surface has been stable for a significant time, 3) long-term periodic wetting and drying of the soil, and 4) the soils are relatively infertile and have limited ability to replenish nutrients through inherent mechanisms. Although extensive cultivation of these soils occurs, Ultisols require that soil fertilizers and lime be added for sustainable crop production. Readers are encouraged to consult Refs. 1, 3–5 for detailed information.

MORPHOLOGY OF ULTISOLS

An idealized soil horizon sequence for a well-drained Ultisol consists of a darkened surface horizon where soil organic matter (SOM) has accumulated (A horizon), a lightened or bleached horizon (E) where materials have been removed, a subsurface horizon (B) where clay, iron (Fe), and aluminum (Al) oxides have concentrated, and an underlying horizon (C) that is relatively unweathered and comprised mostly of soil parent material. Transitional horizons exhibiting characteristics of both over- and underlying horizons are common.

Soil Organic Matter in Ultisols

Most Ultisols occur in temperate or tropical regions that possess a finite capacity to accumulate SOM due to aerobic microbial activity. Ultisols typically possess less SOM than Mollisols, Oxisols, or Vertisols, but often possess quantities comparable to Alfisols (1). Surface horizons (A) of poorly drained (wet) members (Aquults) as well as some Ultisols found either at high elevations or in portions of the tropics (Humults) contain relatively high SOM quantities.

Horizons of Ultisols

The A and E horizons comprise the eluvial portion of the soil where leaching and erosion have removed both soluble and colloidal soil constituents. The B and corresponding transitional horizons are considered the zone of illuviation where materials have accumulated. Besides possessing higher amounts of clay than overlying horizons, most B horizons of Ultisols possess translocated clay as evidenced by clay films, coatings on ped surfaces, or clay bridges between sand grains. These subsurface horizons qualify as argillic horizons, which in combination with a low base saturation, are diagnostic of Ultisols. Kandic horizons, the other subsurface diagnostic horizon commonly found in the southeastern U.S. Ultisols may not possess clay films.

Argillic and Kandic Horizons

Argillic and kandic horizons possess more clay than the overlying eluvial horizons. Argillic horizons are thought to form through soil wetting, colloid dispersion, and downward movement of clays from eluvial horizons, and truncation of movement in illuvial horizons induced by soil drying or other mechanism. Clay films are often better expressed in the lower portions of the argillic horizon due to either: 1) destruction of shallow films over long periods due to eluviation, or 2) mixing and destruction (bioturbation) of clay films by organisms (4). Because the destruction of clay films occurs in older soils, kandic horizons sometimes do not exhibit evidence of translocated clay. Although clay translocation is a formative process in Ultisol development, most soil clays likely move only short distances (few cm), and the majority of argillic horizon clay is derived in situ through mineral

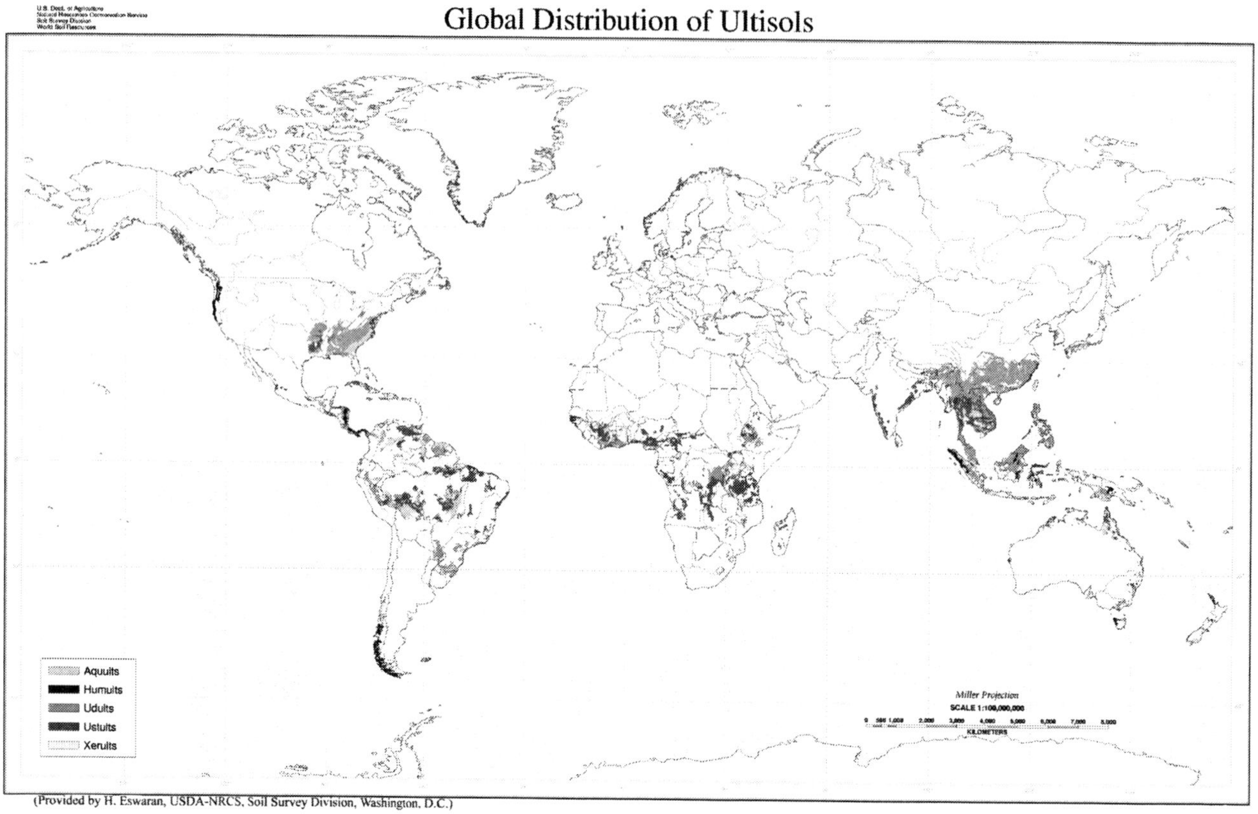

Fig. 1 Ultisol suborder distribution. (Provided by H. Eswaran, USDA–NRCS, Soil Survey Division, Washington, D.C.)

weathering (3). In addition, parent materials possessing high clay content (e.g., fine-textured fluvio-marine sediments, shales, argillaceous limestones) contribute significant subsurface clay. Argillic horizons of well-drained Ultisols also typically possess Fe and Al oxide minerals. Argillic horizons range in color from deep red (10R hues) to yellow (5Y hues) to gray depending mostly on the quantity, type, and degree of hydration of the Fe oxides present.

C horizons of Ultisols consist mostly of parent materials that include, but are not limited to, saprolite formed from igneous and metamorphic rocks, fluvio-marine sediments, wind-blown loess materials, local alluvium or colluvium, and thin, slightly weathered layers of sedimentary rocks (e.g., limestone, shale, or sandstone).

SOIL FORMING FACTORS

State Factor Approach

The state factor approach proposed by Jenny (6) surmises that soil formation is a function of climate, organisms, relief, and parent materials, acting over time. Parent materials on a local scale, climate on a regional scale, coupled with stable landscapes (function of relief), are considered the critical factors for Ultisol formation (3).

Parent material

Because Ultisols possess low amounts of exchangeable bases (primarily Ca, Mg, and K) by definition, parent materials both low in bases and weatherable minerals can induce their development (1). For example, Gamble and Daniels (7) suggested the weathered status of fluvio-marine parent materials of the North Carolina Coastal Plain mandates that Ultisols be the dominant order of that region. However, Ultisols also commonly develop on parent materials high in bases and weatherable minerals (4). In this situation, humid environments with seasonal surpluses of moisture enhance mineral weathering and base leaching. Therefore, seasonal drying and wetting of soils facilitate argillic horizon formation, base leaching, and subsequently, Ultisol development (3, 5).

SOIL CHEMICAL PROPERTIES

Base Saturation in Ultisols

The low-base criteria required for Ultisol placement [<35% base saturation in the lower portion of the solum by the pH 8.2 sum of cations method (2)] was originally designed to separate soils with argillic horizons that could sustain prolonged cultivation without amendment addition (Alfisols), and soils that require nutrient additions (Ultisols) (8). In reality, the base saturation criteria are largely for classification purposes, and the genetic and fertility significance of this value is suspect (4). Typically, base saturation decreases with depth in Ultisols, whereas it increases with depth in Alfisols (Fig. 2). Conceptually, base recycling by vegetation is thought to control base distribution within Ultisols, whereas base recycling coupled with mineral weathering is thought to sustain bases in Alfisols. The base saturation criteria have consistently separated soils formed in acid, felsic terrains (Ultisols), from soils derived from ultramafic parent materials (primarily Alfisols) in the Southeastern U.S. Piedmont (4).

Soil Acidity

Ultisols are typically acid (pH between 4.5 and 5.5) and possess relatively high exchangeable Al. In acid soils, the synergistic effects of Al, H, and Mn toxicity, coupled with Ca, Mg, and Mo deficiency, result in reduced root growth and a concomitant yield decrease (9). Therefore, increased yields upon lime application are common for Ultisols (Table 1). This is in part caused by the amelioration of Al toxicity due to the precipitation of Al forms at pH>5.5.

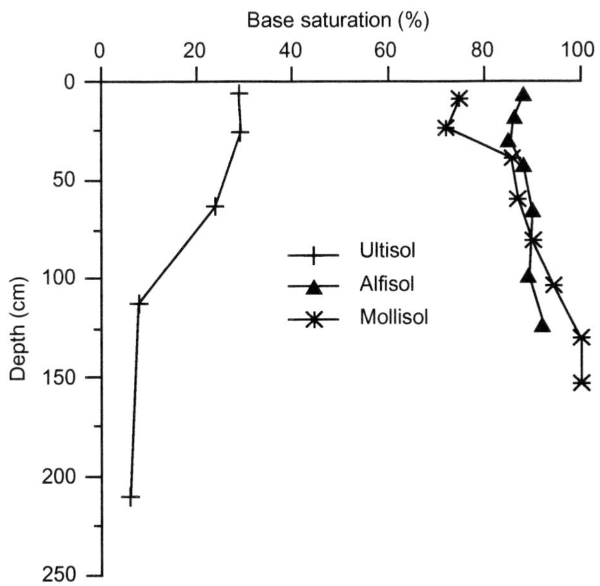

Fig. 2 Base saturation (%) for soils with argillic horizons representing the Ultisol (Typic Kandiudult), Alfisol (Typic Haplustalf), and Mollisol (Pachic Argiustoll) orders. (From Ref. 2.)

Table 1 Response of crop yields to lime application for select Ultisols. Typic Paleudult is from the Coastal Plain Province of Alabama; Typic Hapludult is from the Appalachian Plateau Province of Alabama

		Yield kg ha^{-1}	
Soil	**Crop**	**Unlimed**	**Limed**
Typic Paleudults	Seed-cotton[a]	990	1920
Typic Hapludults	Seed-cotton[a]	2430	2800
Typic Paleudults	Corn-grain[b]	2970	3920
Typic Hapludults	Corn-grain[b]	4700	5100
Typic Paleudults	Soybean-grain[c]	750	2030
Typic Hapludults	Soybean-grain[c]	1210	2550
Typic Paleudults	Wheat-grain[d]	1140	1610
Typic Hapludults	Wheat-grain[d]	1710	3020

(Adapted from Ref. 9.)
[a] (From Ref. 12.)
[b] (From Ref. 13.)
[c] (From Ref. 14.)
[d] (From Ref. 9.)

COMMON MINERALS

Low Activity Minerals

Kaolinite, hydroxy-interlayered minerals (HIM), and Fe and Al oxyhydroxides are commonly found in the clay fraction of Ultisols. Sand and silt fractions are often dominated by quartz, with lesser quantities of mica, feldspars, ferromagnesian, and other weatherable minerals. This results in soils with a relatively limited capacity (compared to most other orders) to: 1) sorb positively (+) charged nutrients [termed cation exchange capacity (CEC)], and 2) supply bases through mineral weathering. Similarly, kandic horizons are defined by having higher clay than eluvial horizons and an extremely low CEC. Conversely, Ultisols with high Fe and Al oxide content are capable of retaining negatively (−) charged species such as phosphate. The adsorption of phosphate on Fe oxides occurs both as a specific attraction and a ligand exchange governed mainly by pH.

Smectites in Ultisols

Although Ultisols with low activity clays are the most typical, areas of Ultisols containing appreciable smectite with relatively high CEC have been found in the southeastern U.S. Coastal Plain (10). In this situation, the smectite is inherited from the parent material.

SOIL MANAGEMENT AND USE

Cultivation of Ultisols

The southeastern U.S. Piedmont is an example of an anthropogenically degraded soil resource. The majority of this region was in intensive cotton cultivation subsequent to European settlement. Sandy loam textured surface horizons with appreciable SOM blanketed the region. However, monoculture cultivation and poor erosion control resulted in the destruction of these surface horizons and a degradation in this region's ability to produce row crops viably. Subsequently, this region is now predominantly in pulpwood production. Forest harvesting of Ultisols can also result in nutrient losses. Buol et al. (4) speculate that deforestation has accelerated the conversion of forest lands to savannas on Ultisols in the tropics due to the excessive removal of nutrients during timber harvesting.

Extensive acreage of cotton, corn, soybean, and peanuts in the Coastal Plain region of the U.S. illustrates the productivity potential of Ultisols. Cultivation of Ultisols is facilitated in this region by relatively long growing seasons, sufficient available moisture, good access to soil amendments, and favorable soil physical characteristics. Agronomic management systems that utilize reduced tillage, cover crops, and residue management that ultimately increase SOM can help sustain the productivity of cultivated Ultisols (5, 11).

Use of Ultisols

Well-drained Ultisols that are not on steep slopes or shallow to bedrock provide few limitations for many urban applications. However, Ultisols can range from excessively drained to very poorly drained, depending on landscape position, climate, and soil and parent material permeability. Excessively drained sandy Ultisols that allow rapid leaching of land-applied chemicals, soils on steep slopes, poorly drained soils that possess seasonal high water tables for significant portions of the year, and soils with high quantities of shrink–swell clay provide severe limitations for many urban and agronomic applications. Soil Survey reports produced by the Natural Resource Conservation Service (NRCS) should be consulted for evaluating soil for a particular use.

SUBORDERS

At the suborder level (the next level of soil classification below order), Ultisols are separated by soil climate. Aquults occupy 0.8% of the land surface area, can be found in most climates (except deserts), and possess seasonal high water tables for extended periods of the year that render suitability for most applications severe (2). These soils typically reside in depressions or other low lying positions. Their identification requires the interpretation of shallow-occurring soil wetness features (termed redoximorphic features). Humults possess high quantities of SOM in the upper portion of the soil, occupy 0.2% of the land surface area, and are mostly restricted to cooler, wetter environments (2). Humults occur in either mountainous terrains where aerobic decomposition of SOM is thought to be inhibited by temperature and moisture, or in the tropics, where extensive root growth may lead to excessive SOM accumulations (3). Ustults are found in semiarid regions that typically possess sufficient precipitation during the growing season for dryland cultivation. Xerults are found in Mediterranean climates characterized by cool, moist winters, and dry, warm summers. It is thought that the excessive winter-time precipitation during lower ET periods promotes leaching and subsequent Ultisol development in these climates. Udults are the most extensive Ultisol suborder (8.1%). These soils are found in humid environments, are fairly low in SOM, but possess sufficient subsoil moisture for dryland crop cultivation during most years. However, certain subsoil properties within Udults (e.g. plinthite, traffic or tillage pans, fragipans) may restrict root development as well as limit suitability for some urban uses.

CONCLUSIONS

The presence of Ultisols suggests appreciable pedogenic development, although pre-weathered parent materials often confound this relationship. Although not extensive, these soils can be found in many parts of the world. Unfortunately, some of these soils have been abused during the last two centuries, and have limited genetic ability to rebound. In some parts of the world, shifting agriculture is practiced on Ultisols because amendment application is not possible (2). This shifting cultivation is mostly caused by limitations in soil chemical properties, and not poor physical properties or climate. If managed properly, Ultisols are capable of providing and sustaining substantial food and other resources.

REFERENCES

1. West, L.T.; Beinroth, F.H. Ultisols. In *Handbook of Soil Science*; Sumner, M.E., Ed.; CRC Press: Boca Raton, FL, 1999,; E358–E372.
2. Soil Survey Staff, *Soil Taxonomy. A Basic System of Soil Classification for Making and Interpreting Soil Surveys*; Handbook No. 436, USDA–NRCS: Washington, DC, 1999.
3. Miller, B.J. Ultisols. In *Pedogenesis and Soil Taxonomy II. The Soil Orders*; Development in Soil Science 11B, Wilding, L.P., Smeck, N.E., Hall, G.F., Eds.; Elsevier: Amsterdam, Netherlands, 1983,; 283–323.
4. Buol, S.W.; Hole, F.D.; McCracken, R.J.; Southard, R.J. Ultisols: Low Base Status Soils. *Soil Genesis and Classification*, 4th Ed.; Iowa State University Press: Iowa, 1997,; 352–362.
5. West, L.T.; Beinroth, F.H.; Sumner, M.E.; Kang, B.T. Ultisols: Characteristics and Impacts on Society. Adv. Agron. **1997**, *63*, 179–236.
6. Jenny, H. *Factors of Soil Formation*; McGraw-Hill: New York, 1941.
7. Gamble, E.E.; Daniels, R.B. Parent Material of Upper and Middle Coastal Plain Soils in North Carolina. Soil Sci. Soc. Am. Proc. **1974**, *38*, 633–637.
8. Smith, G.D. Ultisols. In *The Guy Smith Interviews: Rationale for Concepts in Soil Taxonomy*; Tech. Monogr. No. 11, Forbes, T.R., Ed.; Soil Management Support Services, USDA: Washington, DC, 1986.
9. Adams, F. Crop Response to Lime in the Southern United States. In *Soil Acidity and Liming*, 2nd Ed.; Agronomy Number 12, Adams, F.; Ed.; American Society of Agronomy: Madison, WI, 1984, 211–266.
10. Karathanasis, A.D.; Hurt, G.W.; Hajek, B.F. Properties and Classification of Montmorillonite-Rich Hapludults in the Alabama Coastal Plains. Soil Sci. **1986**, *142*, 76–82.
11. Reeves, D.W. The Role of Organic Matter in Maintaining Soil Quality in Continuous Cropping Systems. Soil Till. Res. **1997**, *43*, 131–167.
12. Adams, F. *Response of Cotton to Lime in Field Experiments*; Bulletin #376, Alabama Agriculture Exp. Station: Auburn, AL, 1968.
13. Adams, F. *Response of Corn to Lime in Field Experiments*; Bulletin #391, Alabama Agriculture Exp. Station: Auburn, AL, 1969.
14. Rogers, H.T.; Adams, F.; Thurlow, D.L. *Lime Needs of Soybeans on Alabama Soils*; Bulletin #452, Alabama Agriculture Exp. Station: Auburn, AL, 1973.

URBAN SOILS AND CONSTRUCTION SITES

W. Burghardt
Soil Technology, University of Essen, Essen, Germany

INTRODUCTION

Land use will inevitably change soil properties. Disturbance of soils and the new substrates added to soils change the morphology of the soil profile and the processes in soil. Urban soils have, therefore, different quality characteristics than natural soils. In the urban ecosystem, a new soil landscape develops, but until recently, this was not the subject of serious study.

Cities in North and Latin America, Europe, and Asia are the home of more than 80% of the world's population. The demands placed on soils and therefore the targets for soil rehabilitation in urban areas must recognize the high degree of intervention by humans and the benefits of soils to the urban population and its economy.

To varying degrees (20–100%), the urban soils are covered by buildings, pavements and road surfaces, and communications and utility lines. Periodically new free spaces will be created and rehabilitated while others will disappear by conversion to another use. Thus, urban soils can change radically over time and acquire new properties.

URBAN SOILS

Types of Substrates

In the city, few soils still retain their natural state. Other soils are young and not yet differentiated into horizons by processes of soil development. Their properties are dominated by the characteristics of substrate of the soil. The majority of urban soils can be grouped in four ways (1) according to the anthropogenesis of the substrates involved:

Soils modified by treatment, or environmental exposure and use

The results will be compaction; stratification; mixing; reshaping of structure; enrichment by organic substances, sand, gasoline, oil, tar; contamination by toxic amounts of hazardous compounds, pathogenic microorganisms; acidification or alkalinisation; breaking of material; sorting and separating of substrate components.

Mode of deposition of soil substrates

Methods of deposition include: pouring of material (e.g., rail berth); methods of tipping materials (e.g., from shovel, wheel barrows, conveyor belt, or trucks); leveling by heavy machinery; lagoon deposits; spills (e.g., harbor sludge, fly ash); blast furnace sludge; casting waste (e.g., slag), dilapidation and demolition (e.g., of buildings); deposition of rubble from war raids; sealing (underground and surface cover); thorough mixing following repeated excavation and tipping.

Technogenic substrates (2, 3)

The technogenic substrates are from man-made material such as rubble, ashes and cinder, slag, foundry waste, industrial residues and waste, town refuse, sewage sludge, industrial sludge and dust, road paving materials, mine waste, material from soil remediation (e.g., biological, thermal, extractive treatment).

Reduction of fine Earth material by increase of skeletal soil fractions

Fine earth is often mixed with building material and other technogenic substrates, which are coarse sized. Hence, the mass of fine earth in the soil volume is reduced, as is its influence on the soil characteristics.

Degradation

Most of the changes in urban soils involve a decrease in the capability of soils to fulfill functions for the welfare of humans. The quality of urban soil can be defined by the degree to which it offers: protection against the occurrence and transfer to humans of hazardous compounds; improvement of the urban green space, climate, groundwater renewal, and concentration of chemicals; decay and safe storage of waste; and satisfactory features for economic and social purposes. The soil characteristics required to achieve these functions are very diverse and in part contradictory.

In the urban environment, soil degradation arises from a diverse array of processes. Soil degradation can occur by enrichment with compounds from emissions of households, industry, waste tips, traffic, and maintenance work.

Encyclopedia of Soil Science

Amendment of vegetable gardens with ash, sludge, and slag contribute to this too. Small service enterprises such as petrol stations, dry cleaners, car washes, garages, and diverse workshops are also sources of pollutants. Leakage and accidental spills from tanks, pipelines, gas pipes, sewage canals, and road tankers pollute the soils too. Fires can release harmful compounds to soils. Latrines not connected to the urban sewage canal system and the manure of dogs and other animals cause chemical soil degradation by eutrophication.

Soil degradation can be from the lack or imbalance of plant available nutrients of many urban substrates, which often results in deficits of phosphorus, nitrogen, and micronutrients. Atmospheric deposition of protons results in acidification. Similar effects arise from tipping of acid or acid producing substrates, e.g., of high sulfide content. Alkaline dust and substrates will raise the pH. Salinity can occur from uncontrolled irrigation in dry regions or use of de-icing salts in cold parts of the world.

Physical impacts are from increase of coarse particles (stone, gravel), which will reduce fine earth volume. Most of the man-made substrates are wastes or construction materials, which contain high amounts of coarse materials. Large areas of the city are degraded by sealing. In addition, unsealed ground is often strongly compacted. The compaction is not only confined to the topsoil as in rural sites. It also penetrates considerable depth. Soil structure can be destroyed by dispersing agents and may inhibit wetting. Decline of the soil structure can result in water and wind erosion and land sliding.

The loss of humus containing topsoil horizons and of subsoil by excavation in cities is one of the largest forms of soil loss. Excavation is the anthropogenic form of erosion.

Degradation can also mean a strong spatial heterogeneity of soil features, or the destruction and splitting of water catchment areas.

REHABILITATION OF URBAN SOILS

Hazardous Compounds

Rehabilitation of polluted urban soils has, besides others, the goal to avoid the transfer of hazardous compounds from soils to humans. The uptake of harmful compounds can result from direct consumption of polluted soil by dermal uptake through contact of the skin with soil; oral uptake by ingestion of dirty vegetables and fruit, drinking muddy water or direct ingestion of soil from dirty hands; pulmonary uptake by inhalation of fine dust from the fraction finer than 0.010 mm in diameter (PM10); uptake through open wounds to the skin.

The primary vehicle for the uptake of hazardous compounds will be food and drinking water and the release of volatile gases and vapor from polluted soils.

The most sensitive segment of the population is children. Direct soil uptake by children occurs in playgrounds, school yards, and sports grounds. All bare soil and sand pits should contain less than the limits for compounds dangerous to children. Therefore polluted soil and soil infected by pathogenic microorganisms (4) must be replaced. The depth of the new soil layer should be at least 35 cm, greater than the depth to which children will dig (5–7). Between the topsoil and the subsoil a geotextile could be placed to avoid deeper digging.

Earthen road materials are common in many suburban areas. They are sources of fine dust (PM10), which can be inhaled. This will be a large problem in arid and semiarid areas. Rehabilitation requires the stabilization of the road surface by binding agents or covering it by a fine gravel layer.

Polluted vegetable gardens are a source of contaminated food. There are several rehabilitation strategies practiced. The polluted soil can be replaced by clean soil. In case of heavy metal pollution, the mobility of metals can be reduced by liming. pH values above 5.0 for most metals and 6 for cadmium and nickel will minimize the amount of plant uptake. For vegetable gardens, the uptake of pollutants can be also diminished by reducing the cultivation area of vegetables in a garden or by the growth of vegetables, which accumulate low amounts of the hazardous compounds in the plant parts consumed by humans. A possible, but rarely practiced, strategy will be the addition of compounds to soils apart from lime, which decrease the plant availability of heavy metals. Iron minerals can have such an effect (8). Another novel, but still inadequately tested, strategy would be phytoremediation by the growth of crops that accumulate hazardous compounds and decrease the mobile fraction of the pollutant in the soil. Pollutant uptake by subsequent crops will be lowered (9).

Urban sites, which are degraded by the release of gases from aromatic polyhydrocarbons and mercury vapor, are generally rehabilitated by the removal of the polluted soil and its replacement by an unpolluted substrate.

From atmospheric deposition, soils of most areas of cities will be slightly polluted. Acute toxicity may not be the problem for the soils per se, but it might be of importance for long-term residents who gradually accumulate pollutants to harmful levels. Possibly the most effective rehabilitation measures for the whole city will be a dense vegetation cover that stabilizes the soil, prevents wind and water erosion of the soil and minimizes direct contact by humans with the pollutants.

Greenery and Street Trees

The quality of life in the cities will be enhanced by the greenery, which consists mainly of city forest, parks, lawns, and street trees. Rehabilitation of green areas can be achieved by optimizing the pH either by liming or adding sulfur. The nutrient content can be improved by fertilizing and manuring with compost. Restrictions to the root soil volume or to aeration and available water content can be corrected by soil amelioration before planting.

Street trees live under extreme conditions and the soil in which they grow will often be degraded (10). Street tree patches are often too small for aeration and water uptake. Sizes of at least $2 \times 2\,m^2$ are recommended for each street tree. When soils of street tree patches are stony or of man-made substrates of low soil quality, they can be replaced by suitable materials. For reducing compaction and improvement of aeration, the soil around the tree can be loosened by devices that release compressed air from lances pushed into the soil. Aeration will be also improved by inserting perforated drainage pipes near the tree into the rooting zone.

Salinity

Salinity problems occur in two major forms in cities. In cold climates, streets are kept free from ice and snow by spreading salts. De-icing salts often contaminate street tree patches. The increases in sodicity may damage the soil structure of loam, clay, and organic soils. Rehabilitation measures are liming and gypsum addition, loosening or replacing loam, clay and organic soil by sandy soil.

Soils of cities located in arid areas will often be irrigated. From groundwater and perched water near the soil surface evapotranspiration will be increased. In arid areas an accumulation of salts dissolved in the irrigation water will result. The secondary salinity by irrigation has to be reclaimed by loosening dense soil layers, destroying water percolation barriers caused by stone layers under fine textured horizons and controlling groundwater depth by drainage.

In arid and semiarid areas, gypsum- and carbonate-containing soils are common. Gypsum and carbonates can form compacted layers. Excess irrigation will dissolve gypsum and also carbonates creating voids. Voids are prone to collapse and may damage the stability of buildings and streets. When irrigation water is given in small doses, and only the topsoil is wetted, the consequences can be minimized or avoided.

Sealing

A serious problem of urbanization is the sealing of soils, which limits water infiltration and green area. To a degree, rehabilitation can be effected by changing the paved surface from impermeable to permeable materials, e.g., from tarmac to paving stone or lawn stone.

Much of the city landscape is sealed by buildings. Soil rehabilitation by unsealing is not possible. But above the sealed area a soil layer can be established. Examples are soil covers on roofs for rooftop plantings (11), or dense arrangements of potted plants on sealed surfaces.

Compaction

Compaction reduces water infiltration and water storage capacity of soils which are important land characteristics in cities for determining the volumes of drainage water as well as the ground water levels and the prevalence of floods in rivers flowing through cities. It determines also the storage capacity for storm water and influences the occurrence of surface water ponding. Topsoil compaction can be ameliorated by loosening the soil before establishment of the vegetation cover. On vegetated sites, deep rooting plants and liming to improve the soil structure and to favor the growth of earthworms are effective measures for improving soil porosity. However, deep compaction can only be rehabilitated by deep excavation and refilling the substrate again without compacting it.

Aeration

A significant function of soils in cities is as burial grounds. For the decay of the corpse, the soil must be well-aerated. Therefore, soils from silt and clay, of very high organic matter content ($>8\%$) and compacted soils have generally low suitability as burial grounds.

To achieve good aeration the soil must also be free of groundwater and stagnant water approximately to a depth of 2.5 m: alternatively burial grounds may need drainage or a raised surface by in-filling.

REHABILITATION OF CONSTRUCTION SITES

Rehabilitation of construction sites begins with demolishing buildings and ends with cleaning and improving the ground for a new use. Rehabilitation of these sites will reverse the sealing of soils.

One principle of rehabilitation of construction sites is to leave as much of the material as possible at the site.

This will minimize transport cost and will not damage other sites by transferring the problems of the rehabilitation site.

Construction materials will however need to be inspected to separate clean materials from those that are contaminated. The different kinds of the material will also be separated for the purpose of recycling. What is not sold can be crushed to sizes, that can be reused at the site either for construction, filling material or as a substrate for revegetation. The soil substrate will therefore often be stony, dry, calcareous, and poor in nutrients.

Management of contamination at construction sites requires documentation on individual layers of the profile and mapping of their spatial distribution. Contamination will occur in patches of variable size depending on the characteristics of the layers. The layers can be of materials that already have a certain content of pollutants, (e.g., ashes, cinder, rubble, dust, and slag from nonferrous metals) or they may have a high capacity for reducing solubility or for sorption, [e.g., well-buffered near-neutral pH, high carbonate, organic matter, and iron content (12)] or high pore volume (13) for the storage of compounds. Another significant property is the gravel content which reduces the amount of fine earth. Compounds moving through the reduced amount of fine earth will be concentrated and will increase their content in the fine earth mass (12).

The amount of contaminated material, which must be either cleaned or treated for immobilization of the pollutants (14) or removed to a waste tip, should be minimized. This will be achieved by a substrate survey (2) and excavating the individual substrate layers very carefully. For example at a textile plant at Nordhorn, Germany, 69% of the excavated soil was separated as unpolluted or only slightly polluted (15). For a 65 ha area of a marshalling yard near Ruhr, Germany, the amount of aromatic polyhydrocarbon contaminated substrate was determined to be less than 10% of the total amount of the excavated substrate of the site (16).

Mixing of the material of different layers raises the risk of uncontrolled distribution of hazardous compounds and the creation of future unforeseen hazards. Therefore, mixing contaminated materials must be strictly avoided.

Excess material, especially if treated for immobilization of pollutants, may be heaped to create a green area at the rehabilitated site. Necessary security constructions and control measures to prevent any release of residual pollutants can be established at the heap.

For recreation areas on only slightly polluted soil substrate or of substrates with low potential for a vigorous vegetation cover a 0.5–1 m covering of soil is usually sufficient. The soil cover should be applied without soil compaction. Therefore, use of heavy machinery should be avoided. Alternatively, green area planning can use the different kinds of substrates for the establishment of varied vegetation types on rocky regions, raw soils (soils at an early stage of pedogenesis), acid or carbonate containing soils, soils of low and high nutrient content, dry, wet, and peaty patches.

For substrates that are highly polluted, to minimize diffusion of hazardous compounds surface or subsoil sealing is required. The soils under the impermeable sealed layer will not be exposed to leaching by precipitation. This will only work when the groundwater is well below the sealed layer.

Rehabilitation of soils polluted by organic compounds may use the potential of soils for natural attenuation due to the in situ degradation by microorganisms. This will probably only work for soils freshly contaminated and with low to medium contamination levels. For soils contaminated many years earlier it can be assumed that the easily degraded compounds are already decomposed. An open question is whether the access of microorganisms to the residual polluted soil particles can be improved by in situ treatment. This would be an asset for the remediation of waste tips, slightly polluted soils under a soil cover and the more intensively polluted soil under sealed layers.

REFERENCES

1. Burghardt, W. The German Double Track Concept of Classifying Soils by Their Substrate and Their Anthroponatural Genesis: The Adoption to Urban Areas. In *Proceedings of the First International Conference on Soils of Urban, Industrial, Traffic and Mining Areas, University Essen, Germany,* July 12–18, 2000; Burghardt, W., Dornauf, Chr., Eds.; Working Group SUITMA/SU of IUSS: co. Fb9, Angewandte Bodenkunde/Soil Technology, University of Essen: Essen, Germany, 2000; Vol. 1, 217–222.
2. Burghardt, W. Substrate der Bodenbildung Urban, Gewerblich und Industriell Überformten Flächen (Substrates of the Soil Formation of the Urban, Commercial and Industrial Transformed Sites). In *Urbaner Bodenschutz (Urban Soil Protection)*; Arbeitskreis Stadtböden der Deutschen Bodenkundlichen Gesellschaft, Ed.; Springer: Berlin, 1996; 25–44.
3. Hiller, D.A.; Meuser, H. *Urbane Böden (Urban Soils)*; Springer: Berlin, 1998; 161.
4. Giammanco, G.; Marranzano, M.; Giannino, L.R. The Soil of Public Gardens as Salmonella Reservoir. Igiene Moderna **1984**, *82*, 762–765.
5. MAGS NRW, *Metalle auf Kinderspielplätzen (Metals on Playgrounds of Children)*; Erlaß VB4-0292.5.3; Ministerium

für Arbeit, Gesundheit und Soziales Nordrhein Westfalen: Düsseldorf, 1990.

6. Bachmann, G., Ed. *Fachliche Eckpunkte zur Ableitung von Bodenwerten im Rahmen des Bundesbodenschutzgesetzes (Professional Principles for the Deduction of Soil Characteristic Values in the Frame of Federal Soil Protection Law)*; Erich Schmidt: Berlin, 1988; 121.
7. FRG—Federal Republic of Germany; Bundes-Bodenschutz- und Altlastenverordnung (BbodschV) (Soil Protection and Contaminated Site Decree of the Federal Republic of Germany). Bundesgesetzblatt Jahrgang **1999**, I *(36)*.
8. Alloway, B.; Warren, G.; Lepp, N.; Singh, B.; Penny, C.; Bocherau, F.; Gregory, L.; Dourado, A. Minimising the Risk of Heavy Metal Exposure from Contaminated Soil by Reducing Bioavailability Using Adsorbtive Minerals and Selected Garden Vegetables. In *Proceedings of the First International Conference on Soils of Urban, Industrial, Traffic and Mining Areas, University Essen, Germany,* July 12–18, 2000; Burghardt, W., Dornauf, Chr., Eds.; Working Group SUITMA/SU of IUSS: co. Fb9, Angewandte Bodenkunde/Soil Technology, University of Essen: Essen, Germany, 2000; Vol. 3, 765–769.
9. Schwartz, Chr.; Perronnet, K.; Morel, J.L. Phytoremediation of Urban and Industrial Soils Contaminated by Heavy Metals. In *Proceedings of the First International Conference on Soils of Urban, Industrial, Traffic and Mining Areas, University Essen, Germany,* July 12–18, 2000; Burghardt, W., Dornauf, Chr., Eds.; Working Group SUITMA/SU of IUSS: co. Fb9, Angewandte Bodenkunde/Soil Technology, University of Essen: Essen, Germany, 2000; Vol. 3, 771–776.
10. Freie und Hansestadt Hamburg; Untersuchung Im Öffentlichen Grün (Investigation in the Public Green—Rehabilitation of Environmental Damaged Street Trees and Park Trees). Naturschutz und Landschaftspflege in Hamburg **1988**, *22*, 320.
11. Liesecke, H.-J., Ed. *Dachbegrünung—Beiträge zur Extensivbegrünung (Rooftop Planting—Contributions to Extensive Greening)*; Patzer: Berlin, Hannover, 1985; 145.
12. Burghardt, W.; Böhm, B.; Dornauf, Chr.; Rabearisoa, R. Verteilung von Stoffen aus Einträgen in Stadtböden (Distribution of Compounds from Deposits in Urban Soils). Bodenschutz **1998**, *3*, 92–97.
13. Baedjer, N.; Burghardt, W. The Influence of Man-Made Materials on the Solute Concentration of Percolating Water. In *Proceedings of the First International Conference on Soils of Urban, Industrial, Traffic and Mining Areas, University Essen, Germany,* July 12–18, 2000; Burghardt, W., Dornauf, Chr., Eds.; Working Group SUITMA/SU of IUSS: co. Fb9, Angewandte Bodenkunde/Soil Technology, University of Essen: Essen, Germany, 2000; Vol. 2, 471–476.
14. Testa, St. M. *The Reuse and Recycling of Contaminated Soil*; Lewis Publishers: Boca Raton, FL, 1997; 243.
15. Strasser, H.; Holland, K.; Schuller, D.; Rongen, P. *Bewertungskriterien für die Folgenutzung Eines Altstan-Dortes am Beispiel des Sanierungsfalles—Nordhorn Povel (Valuation Criteria for the Future Use of a Soil Contaminated Former Industrial Site Exemplified by a Case of Restoration—(Textile Factory) Nordhorn Povel). UBA Texte* 89-078; Umweltbundesamt: Berlin, 1989; 131.
16. Hiller, D.; Jessen, V.; Koeppner, Th.; Stenpass, R.; Burghardt, W. Characteristics of Soils of a Railway Marshalling Yard in the Heavy Industry Area of Duisburg (Ruhr Area, Germany). In *Proceedings of the First International Conference on Soils of Urban, Industrial, Traffic and Mining Areas, University Essen, Germany,* July 12–18, 2000; Burghardt, W., Dornauf, Chr., Eds.; Working Group SUITMA/SU of IUSS: co. Fb9, Angewandte Bodenkunde/Soil Technology, University of Essen: Essen, Germany, 2000; Vol. 3, 943–946.

VALUE TO HUMANS

V

W.E. Larson
University of Minnesota, St. Paul, Minnesota, U.S.A.

INTRODUCTION

The soil plays a major role in nearly all human activities. Plants growing in the soil provide most of our food, material for clothing and building, as well as medicinal and industrial chemicals. The soil is a storehouse for many chemicals. Carbon is stored in the soil from decaying plant residues, both cations and anions (organic and inorganic) are held on exchange sites within the soil and held from leaching or releasing to the atmosphere. The soil is also the home for millions of microorganisms, which degrade organic wastes, and chemicals such as pesticides.

The soil is the foundation upon which buildings, roadways, and other structures are formed. The soil provides recreational sites for playing fields, golf courses, outdoor camping and picnics, wildlife habitat, and biological diversity.

The soil is inextricably linked to water and air. Water running off the soil or percolating into aquifers is used by animals, and is cleansed and made fit for direct human consumption. It is also used for navigation and recreation. Water remaining in the soil can later provide water for plant use. The soil interacts with water to influence the atmosphere. Carbon from decaying plant residues is stored in the soil and thus does not contribute to atmospheric buildup of CO_2. The soil (and the water therein) interacts with atmospheric radiation and mitigates surface temperatures.

THE SOIL AS A PRODUCER OF FOOD AND FIBER

Approximately 11% of the land area, or 1461 million ha of the world, are presently cultivated . In the United States, nearly 186 million ha are cultivated (1). Crop and rangeland produce most of the food and fiber for the nation's human needs and for export. The United States' agricultural exports in recent years have exceeded $50 billion.

The quality of the soil and landscape varies greatly. Nearly 19.8 million ha of cropland are irrigated, mostly in the western states. Soil and landscape characteristics interact with weather to create a number of hazards for continued use of the land without serious degradation. Of all Nature's hazards, erosion from wind and water is probably the most serious, although salinity, compaction, acidity, and nutrient and organic matter depletion are also important.

Of the crops produced in the United States, those with the largest area include corn (29.6 million ha), wheat (25.4 million ha), soybeans (25.7 million ha), hay (24.7 million ha), sorghum (4.8 million ha), cotton (5.2 million ha), barley (2.7 million ha), and oats (1.1 million ha) (1). Grain is used for direct human consumption, as feed for animals, and for production of alcohol, sugar, and industrial chemicals. Cotton and other fiber crops are used for the making of cloth, and some seed crops are used for oil extraction and as feed for livestock. Many other crops are grown, including sugar beets, sugar cane, potatoes, canola, vegetables, fruits, nuts, and other products, for direct human consumption, animal feed, and specialty uses.

In the western United States about 239 million ha of land is used for rangeland and pastures (1), producing much of the nation's beef, mutton, and dairy products. As with cropland, rangeland can be damaged from overgrazing, erosion, and other causes. Some of the rangeland is interspersed with forestland and cropland.

THE SOIL AS A PRODUCER OF FORESTS

The United States has almost 262 million ha of forests (1), which are owned by the federal, state, and local governments, and by private individuals and institutions. These forests are a source of wood and chemicals, and are used for recreation and wildlife habitat.

Worldwide forests cover much of the landscape from the tropical rainforests to the boreal forests of the North and South hemispheres. Harvest and clearing procedures have been decreasing forestland in the tropics as well as in the United States.

Forests contribute a number of important functions for humans in addition to the production of wood and industrial products. Forests are a large sink for carbon, which is tied up in the vegetation and in the forest litter. Forests mitigate water balance, slowing runoff and releasing water slowly to surface and groundwaters. In mountainous areas, the forests store snow in the winter and the melt water is slowly released to streams and groundwater in the summer.

THE SOIL AND ITS INTERACTION WITH WATER

Nearly all atmospheric precipitation falls on the soil. As it strikes the surface, it either runs off the land and enters surface water bodies (lakes, streams, wetlands) or enters the soil. If it enters the soil, it may be stored for later use by plants, or may continue downward to enter groundwater. Thus, to a great extent, the condition of the soil (particularly at the surface) determines the ultimate disposition of water. Water moving across soil often carries with it eroded soil particles that may damage the soil for plant growth and may cause off-site damage leading to undesirable consequences. Sediment may be deposited in water supplies used for human consumption, or in road ditches, reservoirs, lakes, and other locations.

In 1992 the 48 contiguous states of the United States had approximately 50 million ha of wetlands, down from 89 million ha in 1780 (1).

The soil is also an excellent sorbant for chemicals. Exchange sites for both cations and anions can remove chemicals from water as it percolates downward. After moving through the mantle of soil and underlying Earth, water is usually purified, making it useable for human consumption. However, the capacity of soils to sorb chemicals varies greatly. Soils high in clay and/or organic matter usually have a greater sorbing capacity than sandy soils. Likewise, chemicals differ in their ability to attach to soil particles.

In the United States, as well as many other parts of the world, water supplies have sometimes been contaminated with excessive amounts of chemicals. Nitrates, phosphorus, and pesticides have been of most concern, although chemicals from industry and human wastes are also of concern.

THE SOIL AND ITS INTERACTION WITH THE ATMOSPHERE

The soil has an important bearing on atmospheric conditions, particularly the atmosphere close to the ground. The color of the surface soil affects the balance between radiation sorbed by the soil or reflected back into the atmosphere. The water content of the soil determines how much of the radiant energy is used for heating the soil and air, or evaporating water from the soil.

In addition, the soil is a sink for carbon, nitrogen, and other gaseous elements that may be converted to volatile forms and released into the atmosphere. Carbon in the form of plant residues is continually added to the soil. Whether it remains in the soil as organic matter or is decomposed to carbon dioxide and released depends on the characteristics of the soil and its management. When the native soil was first brought into cultivation, the organic matter within the soil was usually drastically reduced. Now with better management, the organic matter is in many cases being increased. This has important global implications on the reduction of atmospheric carbon dioxide and lessening possible climate change (2).

THE SOIL AS A BASE FOR BUILDINGS, LAWNS, AND GARDENS

Many different types of structures are built on soils and the underlying materials. Included are: 1) shallow excavations consisting of trenches and holes usually dug to a maximum of 1 to 2 m; 2) dwellings and small commercial buildings; 3) roads and streets; and 4) lawns and landscaping. In addition, several types of sanitary facilities are built in the soil including: 1) septic tank absorption fields; 2) sewage lagoons; 3) trench sanitary landfills; 4) area sanitary landfills; and 5) daily cover for landfills. In 1992 the United States had nearly 21 million ha of urban and built-up land.

Soil properties and site features determine, to a large extent, the performance of these structures. Type and content of clay, mineralogy of the silt and sand fraction, the kinds of adsorbed cations, erodibility, permeability, corrosivity, shrink–swell potential, available water capacity, and other behavioral characteristics affect engineering uses. Important landscape characteristics include depth to bedrock, soil wetness, depth to a seasonal high water table, slope, likelihood of flooding, natural soil structure, and soil density.

Of the 21 billion ha of urban and built-up land, gardens and lawns comprise a considerable share. This land is usually highly managed for the production of fruits, vegetables, lawns, and ornamentals.

THE SOIL FOR USE IN RECREATION AND SPORTS

The soil is the material upon which most outdoor sports and recreational facilities are built. Included are: 1) fields upon which a wide variety of sports are played (e.g., baseball, football, soccer); 2) picnic grounds and hiking trails; 3) camping sites; 4) golf courses; and 5) parks and open spaces. The United States has 16,743 golf courses occupying almost 1 million ha of land. Several million ha are probably devoted to picnic and local park areas.

Each of the recreation and sport uses requires either natural or engineered soil and/or landscape conditions. Sport fields usually require good drainage and the ability of the soil to support a grass cover. The soil is often graded to achieve the desired landscape features.

THE SOIL AS A MEDIUM FOR DIVERSITY

The soil is a medium in which a multitude of flora and fauna grow (3). Flora range from simple microorganisms to higher herbaceous plants, such as agricultural plants and trees. There are millions of microorganisms harbored in each gram of soil. Included are bacteria, actinomycetes, fungi, and algae. Many fauna (protozoa, worms, and arthropods) live in the soil, including higher animals such as mice, gophers, ground squirrels, and prairie dogs. Birds, such as the burrowing owl, access the soil through holes made by animals. Soil insects are an important source of food for many birds (e.g., worms for robins).

Because the soil is home to so many species, it is a repository for a great deal of genetic material. This material is a valuable resource that needs to be preserved. Under agricultural and highly managed forests the genetic diversity of species is often limited and changed.

THE SOIL AS AN ARCHIVE OF NATURAL AND HUMAN HISTORY

Soil science can contribute to the understanding of natural and human history, and thus provide valuable lessons for current management of natural resources. Detection of buried organic rich horizons (paleosols) and carbon dating of these horizons in relation to other horizons, as well as other chemical and physical properties, geomorphic, and landscape studies, are methods soil scientists use to reveal the past. Lowdermilk (4) documents deterioration of the land through erosion, siltation, salinity, and nutrient depletion in the Middle East region. Lowdermilk estimated that at least 11 empires have risen and fallen in the 7000-year history of the Tigris and Euphrates Valleys. While the area supported large populations in the past, today it depends upon oil exports to sustain its people. The decline of the Mayan Empire in Central America more than 1000 years ago has been related to soil deterioration from poor management (5). Other examples of the decline of civilization as a result of soil deterioration are in Turkey and the Negev Desert in Israel.

V

REFERENCES

1. *Agricultural Resources and Environmental Indicators, 1996–1997*; Agricultural Handbook Number 712, U.S. Department of Agriculture, Economic Research Service: Washington, DC, 1997.
2. Lal, R.; Kimble, J.M.; Follett, R.F.; Cole, C.V. *The Potential of U.S. Cropland to Sequester Carbon and Mitigate the Greenhouse Effect*; Sleeping Bear Press: Chelsea, Michigan, 1998.
3. Russell, E.W. *Soil Conditions and Plant Growth*, 10th Ed.; Longman Group Limited: London, 1973.
4. Lowdermilk, W.C. *Conquest of the Land Through Seven Thousand Years*; Agriculture Information Bulletin 99, U.S. Department of Agriculture – SCS: Washington, DC, 1953.
5. Olson, G.W. Archaeology: Lessons on Future Soil Use. Journal Soil and Water Conservation **1981**, *36*, 261–264.

VERTISOLS

Wouter A. Blokhuis
Agricultural University, Wageningen, The Netherlands

INTRODUCTION

Vertisols are dark-colored clay soils with a specific structure and with surface characteristics and deep cracks when dry. The soil swells upon wetting and shrinks when drying. Clay percentages, by definition 30 or higher (1), are often over 60. Smectite minerals (montmorillonite and others) dominate the clay fraction.

Soil parent materials are either silty/clayey deposits, rich in smectite minerals, or weathering material from mainly basic rock types (marl, limestone, basalt, basic igneous rock) that produce swelling clay minerals upon further weathering. Both the formation of the parent materials and the formation of the typical Vertisol profile are favored by a warm climate with alternating wet and dry seasons.

Preservation of basic ions during weathering and soil formation is essential for the formation of smectites, and this explains the occurrence of most Vertisols in poorly drained, often level terrain. Vertisols developed from basic rocks generally occupy sloping positions; external drainage is normal, but internal drainage sufficiently slow and leaching restricted, so that smectite clays can form.

Most Vertisols are found between 45° north and south latitudes in a semiarid to subhumid environment. The largest areas are in India (79 mha), Australia (70.5 mha) and Sudan (50 mha). Vertisols cover extensive areas in several other African countries, the Americas and China and smaller areas in Southern Asia and Southern Europe (Table 1). They are also described in arctic semiarid regions, e.g., in Canada and the U.S. (2). Vertisols and other soils with vertic properties may cover up to 320 mha worldwide (3). Because of their physical properties and the resulting problems with soil and water management, Vertisols are correctly labeled "problem soils." This should not, however, distract from the fact that with proper management Vertisols are amongst the most productive soils in semiarid agriculture.

Problems are also met in the transfer of research results and technologies from other soils to Vertisols. The applications of techniques that are effective in nonswelling soils have often produced misleading results. Concepts such as field capacity, permanent wilting point, available water capacity (AWC), and bulk density have a different or more restricted meaning and sometimes require specific laboratory or field investigations.

SOIL PHYSICAL AND CHEMICAL PROPERTIES RELEVANT TO AGRICULTURAL USE

Soil Structure

Most structural features of Vertisols have seasonal aspects. This implies nonpermanency of their porous system, a striking difference contrasted with rigid soils (4). When dry, most Vertisols have subangular and wedge-shaped angular peds, increasing in size with depth. Cracks, some centimeters wide at the soil-surface, extend to depths between 60 and 100 cm. Cracks are commonly obscured at the surface by fine crumb to granular soil aggregates (Fig. 1). This surface mulch develops due to repeated wetting and drying (5). Other Vertisols develop a hard surface crust, 2–3 cm thick.

Factors that Influence Soil Structure

Clay percentage and amount of swelling clay minerals

In general there is a positive correlation between clay percentage and amount of smectite clay minerals and degree of structure development (6). Vertisols with over 60% of mainly smectite clay are well structured: they have a surface mulch, a finely aggregated surface soil, a blocky structure with wedge-shaped peds, and abundant slickensides in the subsoil. Vertisols with relatively low clay and high sand contents are poorly structured. They have a thin surface crust and a coarse blocky structure with few large slickensides in the subsoil.

Organic matter

Organic matter in Vertisols ranges from 5 to 100 g/kg, and is commonly between 5 and 50 g/kg (7). Organic matter functions, such as binding nutrients and water, are, in Vertisols, provided by active clay surfaces. Still, organic matter usually increases the porosity of the soil (9); it has a positive effect on structure development and aggregate

Table 1 Distribution of dark clay soils

Countries and areas	**Estimated extent (mha)**
Angola: valleys of the Cunene and Cunbango, region of Catete and southwestern part of the country	0.5
Argentina: mainly in the northeastern part of the country in the province of Entre Rios, in the northeastern department of Buenos Aires, in south Corrientes, Santa Fe and the Eastern Chaco	6.0
Australia: mainly in Queensland (Darling Downs), northern plains, east–central part and coastal areas of the Northern Territories, patches in south Australia (near Adelaide), northwestern Australia and in Tasmania	70.5
Benin: Northern part of the country and in the region of Divo	0.1
Bolivia: mainly in the eastern part of the country, Livingstone and Tuli; probably large areas around the Okovambo and Makarikari swamps	2.0
Botswana: West of Livingstone and Tuli; probably large areas around Makarikari swamps	0.5
Brazil: Southwestern and western regions of Rio Grande do Sul (Bage, Uruguaiana, Alegrese and d. Pedrito districts); in the western parts of the country bordering Uruguay, Paraguay and Bolivia; in north-eastern Brazil	4.5
Burkina Faso: Souron Valley; poorly-drained basin deposits spread over the country	0.4
Cameroon: Logone Chari basin and part of the Chad Basin; peneplain of Kaele and Marona region	1.2
Chile: depressional areas associated with noncalcic Brown soils in Santiago and O'Higgins provices and also associated with Prairie and Chestnut soils in Magallanes Province	0.5
China: Central China	12.0
Ecuador: hilly lowland and valley bottoms in western part of the country (provinces of Guayas, Manabi, Esmeraldas)	1.0
Egypt: Nile delta	1.0
Ethiopia: rift valley and Ethiopia plateau	13.0
Ghana: mainly Accra, Ho-Keta and Wineba Plains; scattered patches near Kpandu, Kwamen and Kwesi	0.2
India: central and south-central Deccan platcau (parts of Bombay, Hyderabad and Madya Pradesh states)	79.0
Indonesia: mainly in central Java (Semarang, Demak, Bodjonegoro, Surabaya area), East Java and Lesser Sunda Islands (Lombok, Timor, Sumbawa and Flores)	1.8
Ivory Coast: in northern part of the country and in the region of Divo	2.8
Kenya: Athi Plains near Nairobi and other areas	?.?
Lesotho: Drakensberg	2.3
Madagascar: some valley depressions in the western part of the country and on the uplands in the western and northwestern part	0.8
Malawi: Chyre valley and areas around Nyasa and Chilwa lakes	1.6
Mali: Niger valley, and a large area along the borders with Mauritania	0.7
Marocco: mainly in the northwest (Gharb) and in the doukkalas (south of Casablanca); scattered spots	0.2
Mozambique: alluvial plains of the Limpopo and Inkomati rivers and surrounding uplands; Zambezi valley upstream of Tete	1.1
Namibia: probable occurrence in the Caprivi Strip along the Cubango river and around the Etosha Pan	0.7
Niger: Central Niger valley and several scattered patches	0.1
Nigeria: northeast Bornu and Benoue river basin	4.0
Paraguay: depressional areas in basaltic and limestone plateaus of the eastern region; in the Paraguay river basin and large areas in the Chaco region	1.5
Portugal: depressional area on diorite and limestone in Alentejo	0.1
Senegal: Senegal valley, lowland of the northwest	0.2
Somalia: in the plains extending between the Juba and Shebeli rivers	0.8
South Africa: Bush Veld and Springbok Flats (Transvaal)	2.1
Sudan: region between the White and Blue Nile extending east of the Blue Nile into Ethiopia and covering an area west of the White Nile: widespread in South Sudan, Bahr el Gasal, Upper Nile and Equatoria Province	50.0
Swaziland: mainly in the Middelveld, eastern Low-veld and Lebombo areas	0.2
Syria: Jezireh and basaltic plateaus south of Damascus	0.6
Tanzania: valley of the Mayowsi and Malagarasi, Great Ruaha valley areas near Tendigo swamps and Rukwa lake; scattered upland areas	7.0
Togo: in the northern part of the country and in the Mono valley	0.1
Uganda: valley of the Semliki, areas near Geroge, Albert and Edward Lakes, areas in the eastland northeast of the country	1.7

(Continued)

Table 1 Distribution of dark clay soils (*Continued*)

Countries and areas	Estimated extent (mha)
Union of Soviet Socialist Republics: compact chernozems in the Caucasus between Krasnodar and Groznyy	0.8
United States of America: "Blackland" in central Texas from the Red River bottomland on the north and northeast to the Rio Grande plain in the San Antonio area on the southwest; belt extending from Lowndes Country in Missori to Perry Country in Alabama; basaltic plateaus in Arizona; scattered spots in California, N. Dakota, and Montana; islands of Oahu, Kauai Molokai in Hawaii	18.0
Uruguay: in the southwest, northwest and south central part of the country and along the border with Brazil (Department of Cerro Largo)	1.0
Venezuela	1.5
Zaire: Ruizizi Plain, valley of the Semliki and Lufira valley	0.3
Zambia: valley of the Luangwa, Lukushi and Zambezi; Kafue flats	5.0
Zimbabwe: part of the valley of the Zambezi; Livingstone area, the southern part of the country	1.8

(Adapted from Refs. 3 and 56.)

stability (6, 8) and on soil–water relations. Coulombe et al. (7) consider clay-organic complexes to be fundamental to the formation and stability of soil aggregates. Research in Australia, summarized by McGarry (6), has shown that an improved structural stability was linked with larger amounts of the "active fraction", i.e. the more labile types of organic matter rather than with the total amount.

Cation exchange capacity, exchangeable, cations, and electrolyte concentration of the soil solution

Vertisols generally have high cation exchange capacities and high base saturations. Commonly, calcium or magnesium dominates, depending on the parent material (10). In Vertisols with a high proportion of divalent cations, the clays are flocculated and the soils well-structured. Vertisols with high exchangeable sodium (sodic soils) have surface crusts and widely spaced cracks enclosing compact large aggregates. They tend to disperse upon wetting (7).

Fig. 1 Cracks and surface mulch in a Vertisol.

Coupling electrical conductivity (EC) with ESP or sodium absorption ratio (SAR) is necessary to predict clay dispersion (11). Crescimanno et al. (12) found that at low cationic concentration physical soil degradation occurs in a 2–5 ESP range, and that at increasing ESP soil behavior appears as a continuum. Hebser et al. (13) observed that hydraulic conductivity decreased and dispersion increased with increasing ESP and decreasing electrolyte concentration. The presence of some minerals such as gypsum can overcome the dispersing action of sodium and flocculate soil particles (14).

Climate and landform

The length of the rainy season, the amount and intensity of annual rainfall, and the frequency of wet–dry cycles impact soil structure of Vertisols. In the Sudan, the strongly self-mulching soils with fine aggregation are in the northern arid part of the Vertisol belt (15–18). Generally, these soils have free carbonates and gypsum; calcium is the dominant cation. In higher-rainfall regions soil leaching lowered pH's, removed free carbonates and gypsum, and reduced mulch development by forming coarser aggregates. Crusty Vertisols are restricted to sodic spots in irrigated fields and topographic low situations with standing water after heavy rains (19).

SOIL CONSISTENCE

Soil consistence directly impacts tillage operations. Wet Vertisols are sticky and very plastic, dry Vertisols hard to very hard, and moist Vertisols friable. Moist consistence is the optimum for tillage operations, but unfortunately the time span that the soil is moist is usually short and its occurrence uncertain beacuse it depends on often erratic rainfall.

Sodic Vertisols are smearier when wet, more easily dispersed, and extremely hard when dry. Sandy Vertisols have lower plasticity and stickiness as compared to modal Vertisols, whereas the large aggregates—or large clods resulting from tillage—are extremely hard and do not crumble into a mulch upon wetting and drying.

SOIL–WATER RELATIONS

Infiltration and Hydraulic Conductivity

Infiltration into a dry surface mulch or prepared seedbed is rapid. Subsurface moistening is rapid as long as water can enter open cracks. When cracks close after wetting, infiltration is greatly reduced. Water in cracks passes through fissures between slickensides and ped surfaces and through biopores into the soil matrix (20). Infiltration is promoted by the retention of crop residue or fallow weeds, either as a cover or plowed into the soil.

It was observed in furrow irrigation (21) that water in cracks was 10 m ahead of water on the surface. Shrinkage fissures connect cracks, so that irrigation water passes through quickly. If a rainy season starts with heavy storms, wetting through cracks and connecting fissures is rapid. Lighter rains give a more gradual surface wetting whereby crack entrances are closed and cracks fail to function as preferential pathways (4).

The hydraulic conductivity in a moist Vertisol is slow; in a wet Vertisol, very slow. Measuring hydraulic conductivity in Vertisols remains problematic as there is no horizontal wetting front and pore size changes with the moisture content (22).

Moisture Retention

Vertisols have a relatively high moisture retention at all water potentials. The AWC—defined as the difference between the water contents at 0.033 MPa (0.33 bar) (field capacity) and 1.5 MPa (15 bar) (permanent wilting point)—has been reported as 110–250 mm for the surface 1 m depth (23). At both field capacity and wilting point a positive correlation was found between water content and both clay percentage and amount of smectite minerals in the clay fraction (24, 25). Table 2 gives chemical and physical characteristics, including water contents at field capacity and permanent wilting point and the AWC of a typical Vertisol in the Sudan.

Kutilek (4) found that Vertisols with ESP between 10 and 20 showed a drastic increase of the lower boundary of available water and suggested to abandon the concept of the 1.5 MPa wilting point in sodic Vertisols and use the direct experiment on plant wilting instead. He concluded that AWC in Vertisols cannot be measured by procedures empirically derived in nonvertic soils. Farbrother (26)

Table 2 Chemical and physical characteristics of a Typic Pellustert (S81FN 835–9) from near Damazin, Sudan (11°45'N and 32°47'E)

Depth (cm)	pH[a]	ECe (mmhos cm^{-1})	Colour moist	Clay (%)	Structure[b]	Water content FC[e] (%)	Water content PWP[f] (%)	Bilk[c] density (g cm^{-3})	Available[d] water (cm cm^{-1})
0–5	7.6	0.96	10YR 3/1	64.2	3cpl/3msbk	46.7	33.3	1.06	0.14
5–25	8.2	0.59	10YR 3/1	57.6	3cpl/2msbk	45.0	32.0	1.07	0.14
25–55	8.6	0.69	10YR 3/1	68.7	3cpl/2mabk		32.2		
55–90	8.8	0.94	10YR 3/1	71.8	3cpl/2mabk	47.6	33.3	1.09	0.16
90–125	8.5	1.45	10YR 3/1	74.7	massive		34.3		
125–150	8.3	1.52	10YR 3/1	77.4	massive	52.8	35.1	1.08	0.19

[a] – Measured in a 1:1 soil to water ratio.
[b] – Structure codes are: 3 = strong, 2 = moderate, c = coarse, m = medium, pl = platy, /= breaking to, sbk = subangular blocky, abk = angular blocky.
[c] – Bulk density at 0.3 bar tension.
[d] – Available water on a volume basis measured between FC and PWP.
[e] – FC = water content on a weight basis at 0.3 bar.
[f] – PWP = water content at 15 bar.
Source: Soil Management Support Services, USDA-SCS and Soil Survey Administration, Democratic Republic of the Sudan, 1982. Tour Guide, Fifth International Soil Classification Workshop. Soil Survey Administration. Khartoum, Sudan.

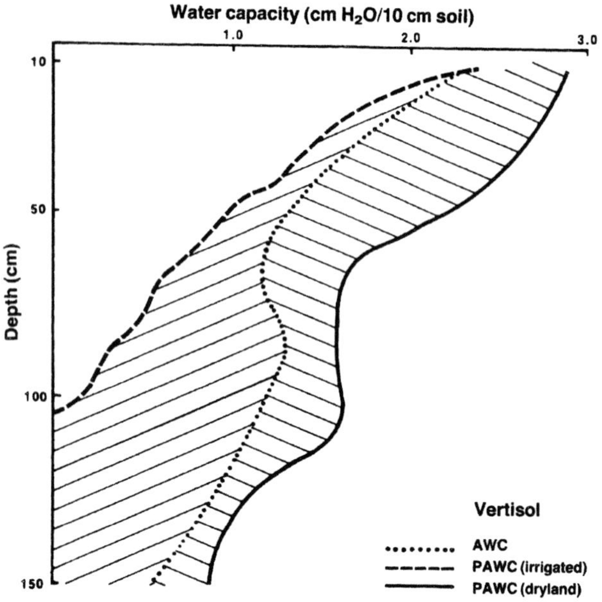

Fig. 2 Distribution with soil depths of the AWC and the PWAC at two levels of plant stress for a Typic Pellustert. (From Ref. 57.)

proposed to consider the saturated water content as the upper boundary of AWC in Vertisols of the semiarid zone.

The concept of plant available water capacity (PAWC) was introduced by Gardner et al. (27). PAWC is defined as the amount of water that can be extracted from an initially fully wet soil by a crop before a chosen stress symptom is observed (28). PAWC varies with crop, root development, depth of rooting, and other soil properties and gives a more realistic figure for the storage of soil water that is available to the plant roots in a given situation than does AWC (Fig. 2).

Groundwater in Vertisols?

At the bottom of cracks infiltration may stagnate and form transient perched water tables. Groundwater tables in Vertisols cannot be defined as lateral movement of water is negligible (29). However, a groundwater level could develop in a more permeable substratum underlying a Vertisol. Dowling et al. (30) report on groundwater tables at the interface of Vertisols and the underlying basalt.

SOIL PROPERTIES AND THE AGRICULTURAL USE OF VERTISOLS

Tillage Operations

Deep tillage of dry Vertisols requires high draft. Plowing or harrowing wet Vertisols is almost impossible. A moist soil can be tilled by hand or by animal-drawn and machine-drawn implements.

Plowing, especially deep plowing, has several disadvantages for soil quality: soil structure is destroyed (40), soil faunal activity (earthworms, termites) is reduced and the continuity of pores is reduced. This has a negative effect on water and air permeability. The use of heavy machinery compacts wet soils and plowing them forms shear planes or plow pans (5, 21, 32, 33).

The surface mulch that develops in many Vertisols after repeated wetting and drying is a great asset becasue it forms a natural seedbed and promotes infiltration of rain or irrigation water. Structure regenerates by the process of self-mulching. Pillai et al. (31) studied clod bulk density, images of soil structure, and evaporation in samples from compacted wheel furrows. A distinct repair of soil structure was shown. Rapidity of repair differed between crops and increased with the number of wet–dry cycles.

Soil tillage is technically difficult and costly in Vertisols, and several studies have been done of reduced tillage. Minimum, or zero, tillage and retention of stubble or fallow weeds show promise to improve soil structure, increase organic matter (34) and soil water storage (37), improve chemical fertility (35), optimize operation timeliness, and conserve water and soil in irrigated cotton (36).

Turpin et al. (39) found that the combination of zero tillage and stubble retention increased the stability of soil aggregates and so increased the storage of rainfall during the fallow period. Minimum tillage in irrigated cotton produced a less dispersive soil with lower ESP and greater organic matter content in comparison with traditional plowing (38). McGarry et al. (41) found that no tillage changes pore size distribution and specifically increased the amount of continuous soil pores and counts of earthworms and termites in the upper 24 cm of soil. They related the higher hydraulic conductivity and infiltration in the no tillage trials to this higher faunal activity.

Water Management

In arid to semiarid areas, rainfed cropping is strongly dependent on an efficient use of rain water. Water management in these regions must aim at maximizing transpiration and minimizing evaporation, runoff, and drainage (42).

In subhumid and semiarid regions after heavy rainfall, excess water on flat terrain may cause flooding, and prevent timeliness in seedbed preparations. On sloping terrain, runoff (often accompanied by soil erosion) is a danger.

Artificial internal drainage is generally not an option because of the low hydraulic activity and the internal soil movements (churning or pedoturbation) that will disrupt mole drains and break tile drains. However, successful mole and tile draining has been reported (33, 43, 44). Subsurface trenches were effective in yield increase (45).

Water use regulation, or budgeting, in irrigation schemes can only be approximate: the very low hydraulic conductivity and the absence of a horizontal wetting front make the calculation of the necessary volumes of water and timing of irrigations difficult (21). The final moisture addition at any specific depth approximates the original cross section of the cracks at that depth (26). This controls of the amount of water taken by the soil at each watering and has been referred to as the "self-regulating" characteristic of the clay.

Soil Erosion

Vertisol surface soils are usually well-aggregated, or there is a surface crust. Wind erosion of the dry surface soil is, therefore, unlikely. However, under extreme conditions wind erosion does occur. Harris (33) mentions wind erosion in southeastern Texas in Vertisols with calcareous surface soils tending to aggregate into sand-sized particles. Mermut et al. (2) report wind erosion from cold Vertisols in the U.S.A. and Canada. Surface granulation of the soil is probably encouraged by freeze/thaw cycles in addition to drying and wetting.

Yule and Willcocks (21) considered water erosion as the main limitation to sustain the production from Vertisols worldwide. Water erosion occurs on slopes as small as 1%. Some of the factors that make Vertisols highly susceptible to water erosion are:

- Much of the sediment erodes as aggregated material that has a lower bulk density than particles of a dispersed soil.
- Wet Vertisols have low infiltration rates causing high energy runoff and soil particle removal and transport (46).
- Crop or fallow weed cover, loose stubble, and roughing by harrowing help control soil erosion by water. Slopes may also be reduced in length and percentage by the layout of graded, free-draining furrows, thus disposing of runoff at noneroding velocities.

Gilgai

Gilgai is a repetitive pattern of small mounds and depressions found in many Vertisols. In India and Sudan it is restricted to the higher-rainfall regions (15, 47). There are various forms of gilgai microrelief (48, 49). Marked differences exist in morphological, chemical, and physical properties of soils associated with mounds and depressions, respectively, and with intervening soils between them (50). With repeated leveling, the microrelief will eventually disappear. However, the soil profile differences in a gilgai complex (the area with a specific type of gilgai) are permanent features that continue to affect the growth of plants (48). Uneven growth is likely to be a consequence.

LAND USE SYSTEMS

Raingrown Annual Crops

The specific properties of Vertisols have prompted cultivators to devise appropriate practices. Harris (33) considers timing of operations and water management the most important keys to successful production systems. In Australia, erosion control was the major innovation in the last several decades (21). Much of the cultivated land now is contoured. Residue retention and reduced tillage have become standard practices.

A commonly used land layout is a system of alternating small ridges and furrows, or cambered beds separated by larger furrows (Fig. 3). The furrows collect rainwater, but also serve as small irrigation ditches and surface drains. Examples are given by Ahmad (5).

In traditional cultivation, the deep Vertisols in India are fallowed during the rainy season and cropped on stored soil moisture in the post-rainy season. Weeds are controlled by tillage during the fallow period. Runoff and erosion may cause high water and soil losses. Towards the end of the rainy season the land is prepared for planting (5).

In the broadbed-and-furrow (BBF) system, developed at ICRISAT, Hyderabad, India, there is a rainy season crop and a post-rainy season crop (51, 52). A cropping calendar is developed from a detailed study of the rainfall data over the years. Commonly, an early start of the rainy-season crop allows a second crop, often by means of intercropping, using residual moisture. The land is laid out in semi-permanent graded BBFs on gradual slopes (usually 0.4–0.8%) (53). Beds are slightly raised, acting as "minibunds" for moisture conservation and erosion control. The furrow is shallow but provides good surface drainage to prevent waterlogging of the crops growing on the bed. Excess water is drained through a system of field drains and grassed waterways. The excess water collects in ponds and is used for complimentary irrigation of lower-lying fields.

Opportunity cropping refers to planting a crop whenever surface moisture conditions are satisfactory for seedling emergence, provided the soil moisture store equals or exceeds some specified fraction of PAWC. Carroll et al.

Fig. 3 Typical land layout for the cultivation of vegetable and small root crops. The beds are 3–4 m wide, and the drains serve for drainage or irrigation as needed. (From Ref. 5.)

(36) showed that opportunity cropping, in combination with zero or reduced tillage to control episodic soil erosion, was the most appropriate system to manage the variable rainfall in central Queensland, Australia.

Irrigated Crops

In Australia, Sudan, and other countries, cotton and several other annual crops are grown under irrigation. Fields are often laid out as cambered beds for crops alternating with furrows for flood irrigation. Irrigation gives full control of the cropping calendar and of water management in the dry season.

The quality of irrigation water is of particular importance in Vertisols. Soluble salts are not easily leached to deeper layers, and thus sodicity and salinity can develop. The Sudan Gezira irrigation scheme owes its success to high quality irrigation water from the Blue Nile.

The watercourses and embankments of irrigation schemes are vulnerable to slumping and cracking. However, unlined irrigation canals and channels function well in the impermeable Gezira clays, provided the embankments are kept moist most of the time.

Pasture

Worldwide, most Vertisols are used for extensive rearing of livestock. The full productive potential of the soils is not realized in this land use system (5). The unavailability of forage, the water shortage in the dry season, and the inaccessibility during the rainy season are serious constraints for exploitation of the land, even in this nonintensive way. Overgrazing, sometimes triggering soil erosion, can be a problem near settlements.

Forestry and Agroforestry

The shortage of timber and fuelwood in many traditional farming areas on Vertisols has received increasing attention in the last decades. Shrubs and trees, which provide a crop or forage as well as wood, have been introduced in agroforestry systems. Larger woodlands have been established with *Acacia*, *Eucalyptus*, *Casuarina* and other species in Africa and Australia. Teak forests (*Tectona grandis*) are successful in higher-rainfall areas in Trinidad (5) and Indonesia. Young (54) gives an outline of agroforestry practices that can be effective in the control of erosion and maintenance of soil fertility.

GEOTECHNICAL PROBLEMS

Buildings and roads built on Vertisols need special precautions against the effects of swell–shrink cycles and the associated movements of the soil. Walls of houses crack, and roads of any kind become uneven and intraffickable. Fredlund (55) discusses the geotechnical problems in urban land use on Vertisols.

REFERENCES

1. Soil Survey Staff, *Keys to Soil Taxonomy*, 8th Ed.; U.S. Department Agriculture, Natural Resources Conservation Service: Washington, DC, 1998; 325.
2. Mermut, A.R.; Patterson, D.D.; McDaniel, P.A. Cold Vertisols and Their Management. In *Vertisols and Technologies for Their Management*; Developments in Soil Science 24; Ahmad, N., Mermut, A., Eds.; Elsevier: Amsterdam, 1996; 479–497.
3. Dudal, R.; Eswaran, H. Distribution, Properties and Classification of Vertisols. In *Vertisols: Their Distribution, Properties, Classification and Management*; Wilding, L.P., Puentes, R., Eds.; Texas A & M University Printing Center, College Station: Texas, 1988; 1–22.
4. Kutilek, M. Water Relations and Water Management of Vertisols. In *Vertisols and Technologies for Their Management*; Developments in Soil Science 24; Ahmad, N., Mermut, A., Eds.; Elsevier: Amsterdam, 1996; 201–230.
5. Ahmad, N. Management of Vertisols in Rainfed Conditions. In *Vertisols and Technologies for Their Management*; Developments in Soil Science 24; Ahmad, N., Mermut, A., Eds.; Elsevier: Amsterdam, 1996; 363–428.
6. McGarry, D. The Structure and Grain Size Distribution of Vertisols. In *Vertisols and Technologies for Their Management*; Developments in Soil Science 24; Ahmad, N., Mermut, A., Eds.; Elsevier: Amsterdam, 1996; 231–259.
7. Coulombe, C.E.; Wilding, L.P.; Dixon, J.B. Overview of Vertisols: Characteristics and Impacts on Society. In *Advances in Agronomy*; Sparks, D.L., Ed.; Academic Press: San Diego, 1996; Vol. 75, 289–375.
8. Ohu, J.O.; Ekwue, E.J.; Folorunsa, O.A. The Effect of Addition of Organic Matter on the Compaction of a Vertisol from Northern Nigeria. Soil Technol. **1994**, *7* (2), 155–162.
9. Bellakki, M.A.; Badanus, V.P.; Setty, R.A. Effect of Long-term Integrated Nutrient Management on Some Important Properties of a Vertisol. J. Indian Soc. Soil Sci. **1998**, *46* (2), 176–180.
10. Ahmad, N. Vertisols. In *Pedogenesis and Soil Taxonomy II. The Soil Orders*; Developments in Soil Science 11B, Wilding, L.P., Smeck, N.E., Hall, G.F., Eds.; Elsevier: Amsterdam, 1983; 91–123.
11. Wilding, L.P.; Tessier, D. Genesis of Vertisols: Shrink–Swell Phenomena. In *Vertisols: Their Distribution, Properties, Classification and Management*; Wilding, L.P., Puentes, R., Eds.; Texas A & M University Printing Center, College Station: Texas, 1988; 55–81.
12. Crescimanno, G.; Iovino, M.; Provenzano, G. Influence of Salinity and Sodicity on Soil Structural and Hydraulic Characteristics. Soil Sci. Soc. Am. J. **1995**, *59* (6), 1701–1708.
13. Hebser, N.S.; Shedaksharappa, G.S.; Patil, C.V. Interacting Influence of SAR and Electrolyte Concentration on the Hydraulic Conductivity and Dispersion of Soils. Madras Agric. J. **1993**, *80* (4), 202–206.
14. Coulombe, C.E.; Wilding, L.E.; Dixon, J.B. Vertisols. In *Handbook of Soil Science*; Sumner, M.E., Ed.; CRC Press: Boca Raton, 2000; E 269–E 286.
15. De Vos, t.N.C., J.H.; Virgo, K.J. Soil Structure in Vertisols of the Blue Nile Clay Plains. Sudan J. Soil Sci. **1969**, *20*, 189–206.
16. Abdulla, H.H. Vertisols in the Semiarid Region of Sudan. In *Vertisol Management in Africa*, Proceedings Workshop in Harare, Zambia, 16–21 January; IBSRAM Proceedings no. 9, 1989; 283–296.
17. Jewitt, T.N.; Law, R.D.; Virgo, K.J. Vertisols of the Tropics and Subtropics: Their Management and Use. Outlook Agric. **1979**, *10*, 33–40.
18. Blokhuis, W.A. Morphology and Genesis of Vertisols. In *Vertisols and Rice Soils of the Tropics*; Symposium Papers II Trans. 12th Intern. Congr. Soil Sci. New Delhi, 1982; Vol. 3, 23–45.
19. Blokhuis, W.A. Vertisols in the Central Clay Plain of the Sudan. Agricultural University: Wageningen, 1993; XVI + 418 pp. Doctoral thesis.
20. Lin, H.S.; McInnes, K.J. Water Flow in Clay Soil Beneath a Tension Infiltrometer. Soil Sci. **1995**, *159* (6), 375–382.
21. Yule, D.F.; Willcocks, T.J. Tillage and Cultural Practices. In *Vertisols and Technologies for Their Management*; Developments in Soil Science 24; Ahmad, A., Mermut, A., Eds.; Elsevier: Amsterdam, 1996; 261–302.
22. Bouma, J.; Loveday, J. Characterizing Soil Water Regimes in Swelling Clay Soils. In *Vertisols: Their Distribution, Properties, Classification and Management*; Wilding, L.P., Puentes, R., Eds.; Texas A & M University Printing Center, College Station: Texas, 1988; 83–96.
23. Virmani, S.M.; Sahrawat, K.L.; Burford, J.R. Physical and Chemical Properties of Vertisols and Their Management. In *Vertisols and Rice Soils of the Tropics*; Symposium Papers II Trans. 12th Intern. Congr. Soil Sci. New Delhi, 1982; Vol. 3, 80–93.
24. Haider, G.; Hordofa, T.; Bekele, E. Irrigation Water Management for Cotton on Vertisols in the Middle Awash Region of Ethiopia. In *Management of Vertisols in Sub-Saharan Africa*; Proceedings of a Conference Held at ILCA, Addis Ababa, Ethiopia, 31 August–4 September 1987; Jutzi, S.C., Haque, I., McIntire, J., Stares, J.E.S., Eds.; ILCA: Addis Ababa, 1988; 166–182.
25. Padole, V.R.; Kalane, R.L.; Agarkar, S.C. Studies on Relationships of Moisture Retention with Soil Texture and Clay Mineralogy of Black Soils. Ann. Plant Physiol. **1996**, *10* (2), 212–214.
26. Farbrother, H.G. Supplementary Irrigation. In *Management of Vertisols Under Semiarid Conditions*, Proceedings of the First Regional Seminar, Kenya, 1–6 December 1986; IBSRAM; 1987; 267–282.
27. Gardner, E.A.; Shaw, R.J.; Smith, G.D.; Coughlan, K.J. Plant Available Water Capacity: Concept, Measurement

and Prediction. In *The Properties and Utilization of Cracking Clay Soils*; Proceedings of a Symposium Held at the University of New England, Armidale, New South Wales, Australia, 24–28 August 1981; McGarity, J.W., Hoult, E.H., So, H.B., Eds.; Reviews in Rural Science 5 University of New England, 1984; 164–175.

28. Coughlan, K.J.; Smith, G.D.; Yule, D.F. Soil Physical Research for Improved Dryland Crop Production on Vertisols in Queensland, Australia. In *Management of Vertisols for Improved Agricultural Production*; Proceedings of an IBSRAM Inaugural Workshop, 18–22 February 1985; ICRISAT Center, Patancheru, A.P. 502 324, India, ICRISAT: India, 1989; 87–99.
29. Blokhuis, W.A. Relationships Between Morphological Properties and Drainage in Vertisols. In *Taxonomy and Management of Vertisols and Aridisols*; Proc. 5th Intern. Soil Classification Workshop, Sudan, 2–11 November 1982; Part I: Papers, Soil Survey Administration: Khartoum, Sudan, 1982; 231–242.
30. Dowling, A.J.; Yule, D.F.; Lisle, A.T. The Salinization of Vertisols with Shallow Water Tables in the Emerald Irrigation Area. In *The Properties and Utilization of Cracking Clay Soils*; Proceedings of a Symposium Held at the University of New England, Armidale, New South Wales, Australia, 24–28 August 1981; McGarity, J.W., Hoult, E.H., So, H.B., Eds.; Reviews in Rural Science 5 University of New England, 1984; 285–291.
31. Pillai, U.P.; McGarry, D. Structure Repair of a Compacted Vertisol with Wet-Dry Cycles and Crops. Soil Sci. Soc. Am. Proc. **1999**, *63* (1), 201–210.
32. Mahmoud, M.A. Rainfed Agriculture and Cropping Systems on Vertisols in Sudan. In *Management of Vertisols in Sub-Saharan Africa*; Proceedings of a Conference Held at ILCA, Addis Ababa, Ethiopia, 31 August–4 September 1987; Jutzi, S.C., Haque, I., McIntire, J., Stares, J.E.S., Eds.; ILCA: Addis Ababa, 1988; 335–336.
33. Harris, B.L. Management of Cropland Vertisols in Texas. In *Vertisol Management in Africa*, Proceedings of a Workshop in Harare, Zambia, 16–21 January 1989; IBSRAM Proceedings, no. 9, 1989; 253–267.
34. Blair, N.; Crocker, G.J. Crop Rotation Effects on Soil Carbon and Physical Fertility of Two Australian Soils. Aust. J. Soil Res. **2000**, *38* (1), 71–84.
35. Chan, K.Y.; Hulugalle, N.R.; Arshad, M.A. Changes in Some Soil Properties Due to Tillage Practices in Rainfed Hardsetting Alfisols and Irrigated Vertisols of Eastern Australia. Soil Till. Res. **1999**, *53* (1), 49–57, Special Issue: Tillage and soil quality.
36. Carroll, C.; Halpin, M.; Burger, P.; Bell, K.; Sallaway, M.M.; Yule, D.F. The Effect of Crop Type, Crop Rotation and Tillage Practice on Runoff and Soil Loss on a Vertisol in Central Queensland. Aust. J. Soil Res. **1997**, *35* (4), 925–939.
37. O'Leary, G.J.; Connor, D.J. Stubble Retention and Tillage in a Semiarid Environment: 1. Soil Water Accumulation During Fallow. Field Crops Res. **1997**, *52* (3), 209–219.
38. Hulugalle, N.R.; Entwistle, P. Soil Properties, Nutrient Uptake and Crop Growth in an Irrigated Vertisol After Nine Years of Minimum Tillage. Soil Till. Res. **1997**, *42* (1–2), 15–32.
39. Turpin, J.E.; Bridge, B.J.; Orange, D.; Thompson, D.J. Water and Bromide Movement in a Vertisol Under Four Fallow Management Systems. Aust. J. Soil Res. **1999**, *37* (1), 75–89.
40. Daniells, I.; Larsen, D.L.; McKenzie, D.C.; Anthony, D.T.W. SOILpak: A Successful Decision Support System for Managing the Structure of Vertisols Under Irrigated Cotton. Aust. J. Soil Res. **1996**, *34* (6), 879–889.
41. McGarry, D.; Bridge, B.J.; Radford, B.J. Contrasting Soil Physical Properties After Zero and Traditional Tillage of an Alluvial Soil in the Semiarid Subtropics. Soil Till. Res. **1999**, *53* (2), 105–115.
42. Yule, D.F. Water Management of Vertisols in the Semiarid Tropics. In *Management of Vertisols Under Semi-Arid Conditions*, Proceedings of the First Regional Seminar, Kenya, 1–6 December 1986; IBSRAM 1987, 107–123.
43. Muirhead, W.A.; Humphreys, E.; Jaywardane, N.S.; Moll, J.L. Shallow Subsurface Drainage in an Irrigated Vertisol with a Perched Water Table. Agri. Water Mgmt **1996**, *30* (3), 261–282.
44. Holsambre, D.G.; Varade, S.B.; Acharya, H.S.; Rapte, S.L. Drainage Characteristics of Vertisols. J. Ind. Soc. Soil Sci. **1982**, *40*, 116–121.
45. Beyene, D., Regassa, H. Research Work on the Management of Vertisols in Ethiopia: Experience of the Institute of Agricultural Research. In *Vertisol Management in Africa*, Proc. Workshop in Harare, Zambia, 16–21 January 1989; IBSRAM Proceedings, no. 9; 1989; 165–171.
46. Freebairn, D.M.; Loch, R.J.; Silburn, D.M. Soil Erosion and Soil Conservation for Vertisols. In *Vertisols and Technologies for Their Management*; Developments in Soil Science 24; Ahmad, N., Mermut, A., Eds.; Elsevier: Amsterdam, 1996; 303–362.
47. Sehgal, J.L.; Bhattacharjee, J.C. Typic Vertisols of India and Iraq, Their Characterization and Classification. Pédologie **1988**, *38*, 67–95.
48. Probert, M.F.; Fergus, I.F.; Bridge, B.J.; McGarry, D.; Thompson, C.H.; Russell, J.S. The Properties and Management of Vertisols; CAB International (IBSRAM): Oron, UK, 2001; 36.
49. Hubble, G.D.; Isbell, R.F.; Northcote, K.H. Features of Australian Soils. In *Soils: An Australian Viewpoint, Division of Soils, CSIRO, Melbourne*; Academic Press: London, 2001; 17–47.
50. Thompson, C.H.; Beckmann, G.G. Gilgai in Australian Black Earth and Some of Its Effects on Plants. Trop. Agri. (Trinidad) **1982**, *59*, 149–156.
51. Virmani, S.M.; Rao, M.R.; Srivastava, K.L. Approaches to the Management of Vertisols in the SemiArid Tropics: The ICRISAT Experience. In *Management of Vertisols for Improved Agricultural Production*; Proceedings of an IBSRAM Inaugural Workshop, 18–22 February 1985; ICRISAT Center, Patancheru, A.P. 502 324, India, ICRISAT: India, 1989; 17–33.
52. Kampen, J. An Approach to Improved Productivity on Deep Vertisols. ICRISAT Information Bulletin No. 11, ICRISAT, Patancheru P.O., A.P. 502 324, India, July 1982; 14 pp.

53. Swindale, L.D. Developing, Testing and Transferring Improved Vertisol Technology: The Indian Experience. In *Management of Vertisols in Sub-Saharan Africa*; Proceedings of a conference held at ILCA, Addis Ababa, Ethiopia, 31 August–4 September 1987, Jutzi, S.C., Haque, I., McIntire, J., Stares, J.E.S., Eds.; ILCA: Addis Ababa, 1988; 13–43.
54. Young, A. The Potential of Agroforestry for Soil Conservation—with Special Reference to Vertisols. In *Management of Vertisols Under Semiarid Conditions*, Proceedings of the First Regional Seminar, Kenya, 1–6 December, 1986; IBSRAM; 1987; 187–199.
55. Fredlund, D.G. Geotechnical Problems Associated with Swelling Clays. In *Vertisols and Technologies for Their Management*; Developments in Soil Science 24, Ahmad, N., Mermut, A., Eds.; Elsevier: Amsterdam, 1996; 499–524.
56. Ahmad, N. Occurrence and Distribution of Vertisols. In *Vertisols and Technologies for Their Management*; Developments in Soil Science 24; Ahmed, N., Mermut, A., Eds.; Elsevier: Amsterdam, 1996; 1–42.
57. Gardner, E.A., Coughlan, K.J., Silburn, D.M. Soil Water Measurement and Management on Vertisols in Queensland, Australia. In *Management of Vertisols in Sub-Saharan Africa*; Proceedings of a Conference Held at ILCA, Addis Ababa, Ethiopia, 31 August–4 September 1987, Jutzi, S.C., Haque, I., McIntire, J., Stares, J.E.S., Eds.; ILCA: Addis Ababa, 1988; 131–165.

VOLATILIZATION OF AMMONIA FROM AGRICULTURAL SOILS

P.G. Saffigna
The University of Queensland-Gatton, Gatton, Queensland, and Saffcorp, Pty., Ltd., Tweed Heads, New South Wales, Australia

J.R. Freney
Commonwealth Scientific and Industrial Research Organisation (CSIRO) Plant Industry, Canberra, Australian Capital Territory, Australia

INTRODUCTION

The exchange of ammonia between soils, plants, waters, and the atmosphere is an important part of the terrestrial nitrogen cycle. The process by which ammonia is lost from the Earth's surface to the atmosphere is termed volatilization. The primary source of ammonia for loss is the natural microbial decomposition of amino acids and proteins in dead plants, animals and microorganisms in soils and waters, but substantial amounts come from the excreta of animals and the use of nitrogen fertilizers (1, 2). Ammonia (NH_3) has a strong affinity for water and it readily dissolves in it to form ammonium (NH_4^+) hydroxide, viz.

$$\begin{aligned} \text{ammonia(gas)} &\Leftrightarrow \text{ammonia (dissolved)} + \text{water} \\ &\Leftrightarrow \text{ammonium hydroxide} \\ &\Leftrightarrow \text{ammonium ions} + \text{hydroxyl ions} \end{aligned}$$

The ammonia and ammonium ions are in equilibrium and the reaction may be displaced to the left or right depending on the conditions. For example, if the pH of the system is increased by addition of alkali (hydroxyl ions), the reaction is displaced to the left and ammonia gas is formed and lost to the atmosphere. As ammonia is a gas at normal temperatures and pressures, and as the concentration in the atmosphere is usually low, it can be readily lost to the atmosphere. However, volatilization is a complex process affected by a combination of biological, chemical, and physical factors and the loss process may be hindered (1, 2).

THE MECHANISM OF AMMONIA VOLATILIZATION

Before volatilization can occur there must be a source of ammonia. This can be in the form of native organic matter, which decomposes to release ammonia, or fertilizer such as anhydrous ammonia, ammonium salts, or urea. Urea, either from animal urine or fertilizer, is rapidly hydrolyzed to ammonium carbonate in soil. This urease catalyzed reaction results in localized areas of high pH. Apart from the application of anhydrous ammonia, these sources tend to add ammonium ions rather than ammonia to the soil. Therefore, the conversion of ammonium ions to ammonia controls the loss of ammonia (1, 2).

The relative concentrations of ammonium and ammonia in solution are strongly affected by pH (acidity or alkalinity) and temperature. For example, the percentage of ammonia present at pH 6, 7, 8, and 9 is approximately 0.1, 1, 10, and 50 (7). Thus the higher the pH, the greater is the potential for ammonia loss from soil. The loss of ammonia from an application of ammonium sulfate increased from nil at pH 7, to 87% at pH 10.5.

The main driving force for ammonia volatilization is the difference in concentration between ammonia gas in the soil and ammonia in the atmosphere. Increasing windspeed increases the rate of volatilization by promoting more rapid transport of ammonia away from the soil surface. A four-fold increase in wind speed resulted in a ten-fold increase in ammonia loss. Ammonia volatilization and water loss from soils are directly related; no ammonia is emitted until evaporation commences.

Any factor which affects the ammonium ion concentration in soil will also affect the ammonia gas concentration and the loss process. Consequently, plant uptake, immobilization by microorganisms, nitrification, and leaching will reduce the amount of nitrogen available for volatilization, whereas increasing the rate of application of ammonium or ammonium producing fertilizers or organic residue will increase the potential for volatilization (1, 2).

Other factors which control ammonia volatilization are cation exchange capacity, buffer capacity, presence of calcium carbonate, water content of soil, soil texture, plant residues, fertilizer form, radiation, and atmospheric ammonia concentration (7). A number of models incorporating these factors have been developed to describe the volatilization of ammonia (3).

MEASUREMENT

A number of workers have used canopies over field crops and pastures, combined with acid traps, to measure ammonia loss, but the canopies affect the temperature, moisture and wind speed in the immediate environment of the plant, and thus the result obtained may not reflect ammonia loss from the natural environment. Simplified micrometeorological techniques have been developed to measure ammonia volatilization in the unconfined field situation, thus allowing the assessment of the importance of ammonia loss, and the factors controlling loss in different agricultural systems with minimum labor, equipment and skills (1, 3).

AMMONIA EMISSIONS

Animals and Their Wastes

Waste from farm animals is the principal source of atmospheric ammonia. Ammonia concentrations in the air range from as low as 1 $\mu g\,N\,m^{-3}$ over oceans to 5 in rural areas, 15 in urban areas, 50 in areas of intense animal husbandry, and 1000 over a field shortly after spreading animal waste as a slurry. The importance of animals as a source of ammonia is well illustrated by the situation in Europe, where nearly three quarters of the total emissions are from animals and their wastes (stables 34%, surface spreading 32%, grazing 8%), and only one quarter comes from a combination of fertilizer application (12%), industry (0.5%), crops (5%) and miscellaneous (e.g., treatment of waste water and sludge, pets, humans, refrigeration, 8%) (1).

The large loss of ammonia from livestock systems is due to the low conversion of dietary nitrogen into animal protein. More than 75% of the nitrogen intake is excreted in forms that give rise to ammonia emissions. Nitrogen is excreted mostly in urine with some present in droppings as microbial cell constituents and undigested food. Urine contains 70–90% urea which is rapidly hydrolyzed to ammonia by naturally occurring urease. The amount excreted depends on the feed composition—the better the quality of the food the less nitrogen excreted. Factors influencing ammonia emissions include manure properties, weather, soil attributes, and application measures (incorporation, dilution, soil preparation). Where animals are grazing pastures, about 10% of the excreted nitrogen is lost as ammonia—mostly from urine (1).

Cropping Systems

There has been a widespread move to urea as the major form of nitrogen fertilizer used in cropping systems because of its relatively low manufacturing cost and low transportation cost per unit of nitrogen. However, large losses of ammonia have been detected from rainfed and irrigated crops, and flooded rice in many countries following applications of urea (Table 1). Ammonia loss from cropping systems is affected by many of the factors discussed above, fertilizer composition, rate, time and method of application, and factors unique to the crop.

In flooded rice, up to 56% of the applied nitrogen is lost from the system by ammonia volatilization as a result of the growth of photosynthetic algae, which markedly increases the pH of the floodwater during daytime. The new practice of retaining tops and leaves of cut sugar cane plants on the soil surface following green cane harvesting has created problems for farmers when they apply urea. So as not to disturb the residue cover, which has many advantages, including weed control, many farmers apply urea by broadcasting onto the surface of the residue. The sugar cane residue has high urease activity and low ammonia retention capacity, so when the residue layer is moistened by dewfall, rainfall or condensation of evaporated soil water, some of the urea dissolves, is hydrolyzed, and when the water evaporates, between 30 and 40% of the applied nitrogen is lost as ammonia. Bananas have a

Table 1 Ammonia volatilized (% of N applied) from different cropping systems fertilized with urea

Plant	Treatment	Loss
Bananas	Surface applied	20
Corn	Surface applied	22
Rice	Broadcast into floodwater	10–56
	Broadcast into floodwater and incorporated	10–43
	Incorporated before flooding	5–16
	Broadcast 12 d after transplanting	21
	Broadcast at panicle initiation	3
Sugar Cane	Broadcast on trash	30–40
Wheat	Surface applied	36
	Buried	7

(Adapted from Ref. 7.)

very high requirement for nitrogen, frequent applications are made and more than 500 kg N $ha^{-1} yr^{-1}$ is applied. Direct measurements of ammonia volatilization from a banana crop in tropical Australia showed that, when urea was applied onto wet soil, 20% of the applied N was lost even though 90 mm of rain fell during the study. The extensive canopy of banana plants restricted rainwater from falling on the fertilized area and washing the urea into the soil.

Plants can absorb ammonia from the air or release it to the atmosphere. It has been established that plants have an ammonia compensation point, which is a finite ammonia concentration in the intercellular air spaces of plant leaves. Plants absorb or lose ammonia depending on whether the ambient ammonia concentration is above or below the compensation point. Ammonia losses of about 1 kg N $ha^{-1} yr^{-1}$ have been determined from fields of barley, maize, and wheat (1).

Biomass Burning

During combustion, considerable plant nitrogen is converted into gaseous forms including ammonia. Most biomass burning (~90%) occurs in the tropics as a result of forest clearing, savanna and sugar cane fires, and burning of agricultural wastes and firewood. According to IPCC (4), 8700 Mt of biomass is burned every year, and agriculture accounts for half of this. About 4% of the biomass nitrogen is released as ammonia during combustion, with the result that biomass burning contributes between 4.5 and 5.9 Mt N per yr^{-1} globally to the atmosphere (5, 6).

GLOBAL SIGNIFICANCE

The global emission of ammonia was estimated to be 54 Mt N yr^{-1} in 1990 (6). The contributions from the major sources were given as: 1) excreta from animals, 21.7 Mt, 2) synthetic fertilizers, 9.0 Mt, 3) oceans, 8.2 Mt, 4) biomass burning, 5.9 Mt, 5) crops, 3.6 Mt, 6) human population and pets, 2.6 Mt, 7) soils under natural vegetation, 2.4 Mt, 8) industrial processes, 0.2 Mt, and 9) fossil fuels, 0.1 Mt. About half of the global emission originates in Asia, and approximately 70% is associated with food production.

When ammonia is emitted into the atmosphere, some is absorbed by vegetation, some is dissolved in atmospheric water, converted to aerosols and transported long distances (>1000 km), and some is deposited nearby (3, 8). Model estimates indicate that about 50% of the emitted ammonia is deposited within 50 km of the source (8). High ammonia concentrations close to point sources such as cattle feedlots can damage vegetation, and deposition of ammonia and ammonium can result in acidification of soils and lakes, increased carbon storage in pristine areas, and increased emission of the greenhouse gas nitrous oxide (9); measures need to be instituted to reduce losses.

MITIGATION

Techniques proposed for reducing loss of ammonia from animal slurries applied to soils include incorporation or injection of the slurry into the soil, application with trail hoses, acidification before application, applying during rainfall, at night, or in winter, and matching nitrogen supply to the demand of the crop. Decreasing the water content of the slurry, and delaying application until a substantial canopy has developed (to reduce wind speeds) would also appear to have a large impact on ammonia loss (10). A logical option for limiting ammonia volatilization from cropping systems is to drill the fertilizer into the soil. Other recommendations include spreading the fertilizer just prior to rain, application in irrigation water, optimizing split application schemes, changing the fertilizer type to suit the conditions, and better matching of nitrogen supply with crop demand. Controlled release fertilizers provide an opportunity to match supply and demand while protecting the remainder of the fertilizer from release. Adoption of some or all of these practices will allow the farmer to reduce inputs and reduce the impact on the environment (4).

REFERENCES

1. ECETOC (European Centre for Ecotoxicology and Toxicology of Chemicals). Ammonia Emissions to Air in Western Europe. Technical Report No. 62. ECETOC, Brussels, Belgium. 1994.
2. Nelson, D.W. Gaseous Losses of Nitrogen Other than Through Denitrification. In *Nitrogen in Agricultural Soils*; Stevenson, F.J., Ed.; American Society of Agronomy: Madison, WI, 1982; 327–363.
3. Denmead, O.T. Progress and Challenges in Measuring and Modelling Gaseous Nitrogen Emissions from Grasslands: An Overview. In *Gaseous Nitrogen Emissions from Grasslands*; Jarvis, S.C., Pain, B.F., Eds.; CABI: Wallingford, UK, 1997; 423–438.
4. IPCC (Intergovernmental Panel on Climate Change), In *Climate Change 1995. Impacts, Adaptations and Mitigation of Climate Change: Scientific–Technical Analyses*;

Watson, R.T., Zinyowera, M.C., Moss, R.H., Eds.; Cambridge University Press: Cambridge, 1996; 1–878.
5. Schlesinger, W.H.; Hartley, A.E. A Global Budget For Atmospheric NH_3. Biogeochemistry 1992, *15*, 191–211.
6. Bouwman, A.F.; Lee, D.S.; Asman, W.A.H.; Dentener, F.J.; Van der Hoek, K.W.; Olivier, J.G.J. A Global High-Resolution Emission Inventory for Ammonia. Global Biogeochem. Cycles 1997, *11*, 561–587.
7. Peoples, M.B.; Freney, J.R.; Mosier, A.R. Minimizing Gaseous Losses of Nitrogen. In *Nitrogen Fertilization in the Environment*; Bacon, P.E., Ed.; Marcel Dekker: New York, 1995; 565–602.
8. Ferm, M. Atmospheric Ammonia and Ammonium Transport in Europe and Critical Loads—A Review. Nutr. Cycl. Agroecosyst. 1998, *51*, 5–17.
9. Smil, V. Nitrogen in Crop Production: An Account of Global Flows. Global Biogeochem. Cycles 1999, *13*, 647–662.
10. Jarvis, S.C.; Pain, B.F. *Gaseous Nitrogen Emissions from Grasslands*; CABI: Wallingford, UK, 1997.

V

WATER CONSERVATION IN SOIL

Paul W. Unger
United States Department of Agriculture-Agricultural Research Service (USDA-ARS), Bushland, Texas, U.S.A.

INTRODUCTION

Although this article emphasizes water conservation for agriculture, it is important also for residential, industrial, and recreational uses. Hence, all users should conserve water, especially where supplies are limited.

Adequate water is essential for optimum growth and development of all crops. For terrestrial crops, water is stored in soil between successive precipitation or irrigation events. Precipitation usually is adequate for crops in humid regions, but becomes increasingly limited and erratic when going through subhumid and semiarid regions into arid regions. Periods without effective precipitation (beneficial to crops) often occur during the growing season of most crops in the drier regions. Ratios of precipitation to potential evapotranspiration are >0.75, 0.50 to <0.75, 0.20 to <0.50, and <0.20 in humid, subhumid, semiarid, and arid regions, respectively (1).

In humid regions, plants seldom experience water deficiency because precipitation usually is frequent enough. However, even 7–10 days without rain may cause severe plant water deficiencies on soils having a low water holding capacity. Major droughts cause severe damage regardless of a soil's water holding capacity. Therefore, soil-water conservation is important in humid regions, especially where irrigation is not practiced.

Precipitation provides most water for crops in subhumid regions, with some water being available for irrigation. In semiarid regions where precipitation is more limited, crops often are irrigated where water is available. Precipitation usually is ineffective (of little or no benefit to crops) in arid regions and irrigation is the primary water source. Soil water storage is essential for successful crop production under all climatic conditions, whether the water is derived from precipitation or irrigation, except with drip or daily sprinkler irrigation for which small amounts of water are applied frequently. Water conservation is important under irrigated conditions because of increasing competition for water between agricultural, urban, industrial, and recreational users (2), especially where water supplies are limited or being exhausted.

SOIL–WATER CONSERVATION PRINCIPLES

Soil–water conservation is influenced by water infiltration into soil, evaporation from soil, use by non-crop plants (weeds), and deep drainage. To improve soil–water conservation, water infiltration must be increased and losses due to evaporation, use by weeds, and deep drainage must be reduced.

Increasing Infiltration

Effective infiltration depends on soil conditions being favorable for adequate water flow into soil and on sufficiently slow runoff to provide adequate time for infiltration. Runoff water does not benefit crops unless captured downstream and subsequently used for irrigation. Runoff is eliminated when the infiltration rate equals or exceeds the precipitation or irrigation rate. Conditions favoring infiltration include the presence of a surface cover to intercept and dissipate the energy of raindrops, thus reducing soil aggregate dispersion and surface seal development that promote runoff; a rough soil surface that temporarily detains water and retards runoff; a soil profile free of horizons that impede water flow; and a low antecedent soil water content. Once a soil is filled to its storage capacity, no additional storage can occur. A soil's water storage capacity depends on its texture (sand, silt, and clay content), organic matter content, profile depth, and horizon characteristics. Water infiltrating a soil in excess of its storage capacity is lost by deep percolation unless an impermeable layer is present. When a soil becomes saturated, runoff will be greater and some crops may be damaged.

Growing plants or crop residues (straw, stover, etc.), retained on soil through use of conservation tillage (reduced- or no-tillage), can provide a surface cover. Besides reducing the adverse effects of raindrops striking the surface, such cover also retards the rate of runoff across the surface, thus providing more time for infiltration. Growing plants also keep the soil drier, which increases infiltration. Other means to reduce runoff include tillage to disrupt surface seals caused by previous precipitation, create depressions for temporary water storage on the surface (surface roughness, contour tillage, furrow diking), reduce the rate of runoff from the surface (graded furrow), and disrupt restrictive layers in the soil profile (deep plowing or chiseling). Infiltration may be increased also by applying phosphogypsum to the soil surface (3) and injecting an anionic polymer into irrigation water (4), which stabilizes the soil surface.

Reducing Evaporation

Soil water losses by evaporation are greatest during first-stage evaporation, which depends on the net effect of water flow to the surface and environmental conditions (wind speed, temperature, relative humidity, and radiant energy). Losses decrease rapidly during second-stage evaporation when the soil water supply decreases and the rate depends mainly on soil conditions that control water flow to the surface. Third-stage evaporation is low and controlled mainly by adsorptive forces at the solid–liquid interface. Water moves to the surface as vapor in the third stage. The potential for decreasing evaporation is greatest during the first stage (5). Practices that decrease evaporation include decreasing turbulent water vapor transfer into the atmosphere, decreasing capillary continuity or water flow to the surface, and decreasing the water-holding capacity of the surface soil layer. A reduction of turbulent transfer of soil water to the atmosphere is most easily achieved by applying a surface mulch. Many materials can serve as a mulch (6), but crop residues and plastic films are used most commonly. Shallow tillage can reduce capillary water flow to the surface.

Reducing Use by Noncrop Plants (Weeds)

Soil–water transpired by weeds (including volunteer crop plants) hinders crop production under most conditions. Weeds present before planting decrease the soil water supply for later crop use while those present during the growing season compete directly with crops for soil water, space, light, and nutrients. Effective weed control is important especially for dryland (nonirrigated) crops (7) and efficient use of irrigation water (2). Weeds are controlled with tillage or applications of appropriate herbicides.

Reducing Deep Drainage

Deep drainage occurs when infiltrated precipitation or irrigation water exceeds a soil's water storage capacity. Such drainage reduces the amount that crops can use because it penetrates to depths beyond crop rooting depths. The water may also move some nutrients beyond the rooting depth, thus resulting in inefficient use of applied nutrients and possibly polluting underground water supplies.

To reduce drainage losses, crop growing seasons should closely match the season when the potential for excess water is greatest. Encouraging deeper plant rooting, as with deep tillage, and planting deep-rooting crop species also reduces deep drainage. Precipitation cannot be controlled, but irrigations can be applied in timely amounts to minimize losses. Other practices include deep plowing to bring materials to the surface that retain more water (8), installing subsurface barriers (9), and adding organic matter to increase the water holding capacity (10, 11).

SOIL–WATER CONSERVATION ACHIEVED UNDER DIFFERENT CONDITIONS

Extensive soil–water conservation research has been conducted; but only a few examples will be given because of space limitations. Under semiarid conditions at Bushland, Texas, moldboard-, rotary-, disk-, sweep-, and no-tillage treatments were imposed after harvesting irrigated winter wheat (*Triticum aestivum* L.) to manage the residues during the fallow period until planting dryland grain sorghum [*Sorghum bicolor* L. (Moench)] 10 to 11 months later (12). Plant available soil water contents averaged 149, 143, 158, 179, and 207 mm at sorghum planting and sorghum grain yields averaged 2.56, 2.19, 2.37, 2.77, and 3.34 Mg ha^{-1} with the respective treatments. Greater water contents and yields with sweep-tillage and especially no-tillage resulted from more residues retained on the surface than with other treatments during fallow and the sorghum growing season. The residues resulted in greater infiltration and lower evaporation, but the individual contribution of these processes could not be determined. Weed control was similar with all treatments and deep drainage was too small to affect the results.

Soil water contents were similar to the 15 cm depth one day after a 13.5 mm rain at Akron, Colorado, where conventional-, minimum-, and no-tillage treatments were imposed after harvesting winter wheat. Residue amounts were 1.2, 2.2, and 2.7 Mg ha^{-1} with the respective treatments. After 34 rainless days, water contents were $<0.1\ m^3\ m^{-3}$ to 12, 9, and 5 cm depths with conventional-, minimum-, and no-tillage, respectively (13). This field study clearly showed the value of surface residues for reducing evaporation, which also was shown under laboratory conditions (14, 15). Under no-tillage conditions in Colorado, increasing the croppingintensity from a wheat–fallow to a wheat–corn (*Zea mays* L.), or sorghum–fallow system resulted in a drier soil root zone. This, in turn, resulted in storing more water from precipitation in the soil because of greater infiltration and lower soil water evaporation and deep drainage (16).

When phosphogypsum (PG) was applied at $10\,\mathrm{Mg\,ha^{-1}}$ to a ridged sandy soil in Israel, runoff was reduced six-fold as compared with that from untreated soil (3). This showed the value of a soil stabilizing material for reducing runoff and increasing infiltration. When PG was applied at $3\,\mathrm{Mg\,ha^{-1}}$ to a clay loam under laboratory conditions, runoff was less than from bare soil, but greater than where wheat straw was applied at $2.2\,\mathrm{Mg\,ha^{-1}}$ (17). Injecting anionic polymers into water used for furrow irrigation on a silt loam in Idaho reduced soil loss 70% when applied at $0.7\,\mathrm{kg\,ha^{-1}}$ per irrigation and 97% when applied at $10\,\mathrm{g\,m^{-3}}$ of water. Infiltration was increased, probably because surface sealing and sediment transport were reduced (4, 18).

SUMMARY AND CONCLUSIONS

Soil–water conservation is important for crop production in all climatic regions, but is critical in subhumid, semiarid, and arid regions. Increasing demands for water for agricultural and nonagricultural purposes make water conservation increasingly important in some regions. Improved soil–water conservation depends on increasing infiltration and reducing evaporation, transpiration (by weeds), and deep drainage losses.

Practices for increasing infiltration include keeping the soil covered with plants or a mulch to reduce surface seal development and to impede runoff; applying materials to stabilize the soil surface; and using tillage to disrupt sealed soil surfaces, provide for temporary depressional water storage on the surface, and disrupt subsurface layers that impede water movement. Evaporation can be reduced by a surface mulch (crop residues or plastic films) or by shallow tillage. Weeds can be controlled by tillage or herbicides. Deep drainage can be minimized by matching crop growing seasons to seasons when water excesses are most likely, increasing soil water storage capacity, loosening the soil to allow deeper plant rooting, planting deep-rooting crops, and installing subsurface barriers that impede deep drainage.

REFERENCES

1. UNESCO (United Nations Educational, Scientific and Cultural Organization). *World Map of Desertification*, United Nations Conference on Desertification; FAO: Rome, Italy, 1977; A/Conf. 74/2.
2. Unger, P.W.; Howell, T.A. Agricultural Water Conservation—A Global Perspective. J. Crop Prod. **1999**, *2* (2), 1–36.
3. Agassi, M.; Shainberg, I.; Warrington, D.; Ben-Hur, M. Runoff and Erosion Control in Potato Fields. Soil Sci. **1989**, *148*, 149–154.
4. Lentz, R.D.; Shainberg, I.; Sojka, R.E.; Carter, D.L. Preventing Irrigation Furrow Erosion with Small Applications of Polymers. Soil Sci. Soc. Am. J. **1992**, *56* (6), 1926–1932.
5. Lemon, E.R. The Potentialities for Decreasing Soil Moisture Evaporation Loss. Soil Sci. Soc. Am. Proc. **1956**, *20* (1), 120–125.
6. Unger, P.W. Role of Mulches in Dryland Agriculture. In *Production and Improvement of Crops for Drylands*; Gupta, U.S., Ed.; Oxford & IBH Publishing Co. PVT. LTD.: New Delhi, Bombay, Calcutta, 1995; 241–270.
7. Lavake, D.E.; Wiese, A.F. Influence of Weed Growth and Tillage Interval During Fallow on Water Storage, Soil Nitrates, and Yield. Soil Sci. Soc. Am. J. **1979**, *43* (3), 565–569.
8. Miller, D.E.; Aarstad, J.S. *Effect of Deep Plowing on the Physical Characteristics of Hezel Soil,* Circ. 556; Washington Agric. Exp. Stn: Pullman, WA, 1972.
9. Saxena, G.K.; Hammond, L.C.; Robertson, W.K. Effects of Subsurface Asphalt Layers on Corn and Tomato Root Systems. Agron. J. **1973**, *65* (2), 191–194.
10. Jamison, V.C. Changes in Air-Water Relationships Due to Structural Improvement of Soils. Soil Sci. **1953**, *76* (2), 143–151.
11. Unger, P.W. *Relationships between Water Retention, Texture, Density and Organic Matter Content of West and South Central Texas Soils*, Misc. Publ. MP-1192C; Texas Agric. Exp. Stn.: College Station, 1975.
12. Unger, P.W. Tillage and Residue Effects on Wheat, Sorghum, and Sunflower Grown in Rotation. Soil Sci. Soc. Am. J. **1984**, *48* (4), 885–891.
13. Smika, D.E. *Seed Zone Soil Water Conditions with Reduced Tillage in the Semi-Arid Central Great Plains.* Proceedings 7th Conference of the International Soil Tillage Research Organization, Sweden, 1976.
14. Unger, P.W.; Parker, J.J.; Evaporation Reduction from Soil with Wheat, Sorghum, and Cotton Residues. Soil Sci. Soc. Am. J. **1976**, *40* (6), 938–942.
15. Ji, Shangning; Unger, P.W. Soil Water Accumulation Under Different Precipitation, Potential Evaporation, and Straw Mulch Conditions. Soil Sci. Soc. Am. J. **2001**, *40* (2), 442–448.
16. Farahani, H.J.; Peterson, G.A.; Westfall, D.G.; Sherrod, L.A.; Ahuja, L.R. Soil Water Storage in Dryland Cropping Systems: The Significance of Cropping Intensification. Soil Sci. Soc. Am. J. **1998**, *62* (4), 984–991.
17. Benyamini, Y.; Unger, P.W. Crust Development Under Simulated Rainfall on Four Soils. Agron **1984**, 243–244.
18. Trout, T.J.; Sojka, R.E.; Lentz, R.D. Polyacrilamide Effect on Furrow Erosion and Infiltration. Transactions ASAE **1995**, *38* (3), 761–765.

WATER HARVESTING

Gary Frasier
United States Department of Agriculture (USDA), Fort Collins, Colorado, U.S.A.

W

INTRODUCTION

Without water, land has no value and man cannot live. Yet with few exceptions, in most places on Earth, there is sufficient water available on an annual basis for man not only to survive but also to live. This water comes in the form of precipitation, and all that is required is a means of collecting and storing this precipitation from the sky. Accessing this type of water supply—collecting precipitation for reuse—is commonly called *water harvesting*. In more technical terms, water harvesting can be defined as "the process of collecting and storing precipitation (rain or snow) for beneficial use from an area that has been treated or modified to increase precipitation runoff" (1).

Water harvesting is an ancient practice dating back at least 9000 years when it was used in the Edom Mountains of Southern Jordan (2). It is probable that the first water harvesting system was nothing more than a simple depression that filled with water runoff from an uphill area. Evidence of extensive water harvesting systems used 3000–4000 years ago have been found in various areas of the Middle East. Many of these ancient systems were located in areas where the annual precipitation was less than 200 mm per year (3). Often in arid climates, small rain-filled depressions can be found at the bottom of rock outcroppings which provide drinking water for wildlife (4). Rain-filled depressions are being used for domestic drinking water supplies (Fig. 1). Currently water harvesting techniques are used in areas such as Israel, Egypt, Jordan, Mexico, Australia, and throughout parts of the United States.

GENERAL DESCRIPTION

Water harvesting systems are frequently grouped into two categories, depending upon the intended use of the collected water. Collected water can be used for domestic purposes (e.g., drinking by humans and animals) or crop growing. All water harvesting systems, whether used for drinking water supplies or crop growing, have two major components: a precipitation collection area (i.e., catchment or runoff area) and a water storage facility. Surfaces of the catchment areas range from rock outcroppings or smoothed–compacted soil, to chemically treated or membrane-covered soil, and building rooftops. The storage facility can be a pond, cistern, tank or, in the instance of crop growing, the soil profile (5).

RUNOFF FARMING

Runoff farming is a form of water harvesting used specifically for growing crops. Essentially, runoff farming maximizes the effect of limited precipitation over a large area by collecting the surface runoff and applying it to a smaller area. In its simplest form, a portion of land is dedicated to precipitation collection, and the collected water is diverted onto a crop growing area where it is allowed to infiltrate into the soil for use by plants. The crop area is usually smaller than the water collection area. The collected water can be applied directly to the crop area from the catchment area during the precipitation event or stored for later application using an irrigation system. Runoff farming is not necessarily a technique for increasing crop production, but rather a method for sustaining crop production or reducing the risk of crop failure caused by drought.

Types of Systems

Runoff farming systems can be grouped into three general types, depending upon the physical arrangement of the catchment area and the crop-growing area (Table 1). One method, called *floodwater farming*, diverts runoff flowing down a channel during a storm event directly onto a crop area (Fig. 2).

The second method is called *micro-catchment farming*. With this method, each plant or group of plants is assigned a small runoff area situated directly upslope of the growing area. The runoff water is held in the growing area by small earthen berms until it has infiltrated into the soil. This technique has been used extensively for tree crops such as pistachio, olive, and almond.

Fig. 1 A water harvesting system used for drinking water supplies on First Mesa of the Hopi Indian Reservation in Northern Arizona. The catchment surface is a natural rock outcropping and the collected water is stored in an excavated cistern at the lower end.

Fig. 2 Floodwater farming near Matrouh, Egypt.

A third method of runoff farming involves a combination of both onsite runoff water and irrigation with excess runoff water from a stored source. The land is formed into a series of large ridges and furrows, and crops, such as fruit trees or grapes, are planted in the furrow bottom. Water runs from the side slopes of the ridges into the crop area. Excess runoff water not directly infiltrated into the planted area continues down the center of the furrow into a storage facility. At a later date, the water is pumped back onto the crop area as needed, using an irrigation system (3).

Site Characteristics

The soil in the area of a runoff farming system must serve two diametrically opposed purposes. On the runoff collecting area, the soil must have a low rate of water infiltration and be resistant to soil erosion by water. The soil in the cropping area should have a high infiltration rate with a high water-holding capacity. Many loess soils are ideally suited for runoff farming applications because they have a high water-holding capacity yet form surface crusts that promote precipitation runoff (6).

The relative sizes of the catchment area to the cropping (runon) area depends upon the size and intensities of the precipitation events, the infiltration rate on the catchment area, and the quantity of water required. Large storms ($>25\,\text{mm}$) of moderate to high intensity can provide significant quantities of water from relatively small areas. Conversely, low-intensity storms on soils with high infiltration rates will require larger runoff areas to provide the same quantity of runoff. Typically, the runoff area is 5 to 20 times larger than the cropping area.

Catchment area slopes should be steep enough to maximize runoff and minimize surface storage in depressions, yet flat enough to prevent soil erosion (7). To minimize potential soil erosion on the runoff collecting area, overland flow distances should be minimized and the flow maintained as shallow sheet flow. Concentrated water flows should only be used in areas where channels are protected from erosion and designed to carry the flow (8). Landform areas with undulating, gently sloping land are best suited for runoff farming. The steeper portions of the land provide the water collection function, while the flatter areas are used for growing crops. In some instances, the crop area is shaped into a series of small terraces with essentially no slope in the planted area. This allows for a better distribution of the collected water in the soil profile.

Table 1 Types of runoff farming systems

Type	Technology level	Crops	Organization
Floodwater farming	Moderate to high	Annual grains and trees	Community
Micro-catchments	Low to moderate	Trees and shrubs	Family
Combination	High	Trees and vines	Community

Successful runoff farming systems include both adequate water management and cropping practices suited to the local area. Good water management includes reducing seepage and evaporative losses of the collected water. For maximum long-term effectiveness, water harvesting systems must have scheduled, timely maintenance and repair. Many systems have been adequately designed and constructed, yet have failed to supply the anticipated quantities of water within a relatively short time interval because of little or no maintenance. Usually, maintenance or repair can be accomplished in a short period of time without great expense.

CROPPING CONSIDERATIONS

Cropping practices must include plants that are capable of efficiently using available water yet withstanding prolonged time intervals of limited or non-existent water. Good cropping practices also must recognize that water requirements for plant establishment frequently differ from water requirements for mature plants. During the plant establishment phase, rooting systems are usually shallow, which necessitates that water be available in the upper layers of the soil profile. Under these conditions, there is also the potential for significant loss of soil water by evaporation from the unprotected (non-shaded) soil surface. Meeting these water requirements usually means applying water at a frequent interval during the plant establishment phase. Once the plant has achieved some above-ground biomass which shades the soil surface, the direct loss of water from the soil surface is lessened. Also, the deeper rooting system of mature plants allows a larger volume of soil to be used to retain the collected water until it is needed.

A feature that is not controlled but is important to the success of runoff farming systems is the precipitation characteristics of the area. The maximum net effectiveness of the water harvesting system is achieved if the precipitation occurs during cooler weather when evapotranspiration rates are lowest. Precipitation intensities must be greater than the infiltration rate of the catchment area. The timing of the precipitation compared with the water needs of the crops must be considered. Less extensive systems—especially water storage—are required if precipitation occurs during the cropping season. This reduces the length of time necessary to store the collected water. Selecting crops that are suited for or immediately following the normal rainy periods will improve the chances for a successful runoff farming system.

REFERENCES

1. Frasier, G.W. Water Harvesting for Rangeland Water Supplies: A Historic Perspective. In *Proc. A Symposium on Environmental, Economic, and Legal Issues Related to Rangeland Water Development*. Tempe, Arizona, November 13–15, 1997; The Center for the Study of Law, Science, and Technology, Arizona State University: Tempe, Arizona, 1998; 17–24.
2. Bruins, H.M.; Evenari, M.; Nessler, U. Rainwater-Harvesting Agriculture for Food Production in Arid Zones. The Challenges of the African Famine. Appl. Geogr. **1986**, *6*, 13–33.
3. Frasier, G.W. Water Harvesting for Crops. In *The Encyclopedia of Water Science*; Stewart, B.A., Howell, T., Eds.; Marcel Dekker: New York, **2002** *in press*.
4. Frasier, G.W. Water Supply for Arid and Semiarid Regions. J. Arid Land Studies, **2000**, Special Issue, Desert Technology V, *10* (S), 17–20.
5. Frasier, G.W. Water for Animals, Man and Agriculture by Water Harvesting. In *Rainfall Collection for Agriculture in Arid and Semiarid Regions*; Dutt, G.R., Hutchinson, C.F., Anaya Garduno, M., Eds.; Commonwealth Agricultural Bureaux, Farnham House: Farnham Royal, United Kingdom, 1981; 83–86.
6. Fink, Dwayne H.; Ehrler, William L. The Runoff Farming Agronomic System: Applications and Design Concepts. In *Proc. 1984 Arizona Section, American Water Resources Association and the Hydrology Section, Arizona-Nevada Academy of Science Tucson*, Arizona, April 7, 1984; 1984, 14, 33–40.
7. Flug, Marshall. Production of Annual Crops on Microcatchments. In *Rainfall Collection for Agriculutre in Arid and Semiarid Regions*; Dutt, G.R., Hutchinson, C.F., Anaya Garduno, M., Eds.; Commonwealth Agricultural Bureaux, Farnham House: Farnham Royal, United Kingdom, 1981; 39–42.
8. Reij, C.; Mulder, P.; Begemann, L. *Water Harvesting for Plant Production*, World Bank Technical Paper No. 91; The World Bank: Washington, DC, 1948; 1–120.

WATER RELATIONS IN FROZEN SOIL

John M. Baker
United States Department of Agriculture (USDA), University of Minnesota, St. Paul, Minnesota, U.S.A.

INTRODUCTION

In general, soils freeze from the surface downward, due to radiative and convective heat loss at the surface. Water contained in soil does not freeze at a single temperature; rather, it freezes gradually over a temperature range. Within this range, water and ice coexist in thermodynamic equilibrium, with the relative proportions of each dependent on temperature, solute content of the water, and retention properties of the medium. As the temperature decreases and more ice is formed, the water potential of the remaining liquid decreases as well. The decline in water potential accompanying freezing creates a gradient favoring water flow toward the freezing front. The extent of freezing-induced water redistribution depends on the balance between heat flow and water flow. If the delivery of latent heat (the product of the water flow rate and the latent heat of fusion) to the freezing front matches the (sensible) heat flow rate away from it, the downward movement of the freezing front will stall as ice accumulates, filling available pore space. Under the proper circumstances ice can continue to form even after all pore space is filled, resulting in the formation of lenses of pure ice and displacement of the soil above, a process known as frost heave (1).

RELATIONSHIPS

Once ice nucleation has occurred in a freezing soil, the pressure in the liquid phase is related to the temperature through the Clapeyron equation:

$$\frac{p_w - \pi}{\rho_w} - \frac{p_i}{\rho_i} = L_f \frac{T - T_0}{T_0}$$

where p_i and p_w are the gauge pressures within the ice and water phases, ρ_i and ρ_w are the densities of the respective phases, and π is the osmotic pressure of the soil solution. L_f is the latent heat of fusion ($334\,\text{kJ}\,\text{kg}^{-1}$), T is the temperature, and T_0 is the temperature at which bulk water freezes, both in K.

Because water also exists as a vapor, the relationship between phase pressures and temperature can also be expressed psychrometrically:

$$\frac{p_w - \pi}{\rho_w} - \frac{p_i}{\rho_i} = \frac{RT}{M} \ln \frac{e_i(T)}{e_w(T)}$$

The saturation vapor pressures over ice and water, e_i and e_w, are tabled functions of temperature, while R is the gas constant ($8.314\,\text{J}\,\text{mol}^{-1}\text{K}^{-1}$), and M is the partial molar mass of water ($0.018\,\text{kg}\,\text{mol}^{-1}$). Evaluation of either equation shows that the quantity on the left side has a temperature dependence of approximately $1.2\,\text{kJ}\,\text{kg}^{-1}\,\text{K}^{-1}$. The osmotic pressure depends on the solute concentration of the soil water, but in any event its temperature dependence is quite small, on the order of π/T. In many cases, and particularly in unsaturated soil, the gauge pressure within the ice phase, p_i, should be negligible. Hence the change in p_w (more commonly known as matric potential) with respect to T in a freezing soil will be about $1.2\,\text{MPa}\,\text{K}^{-1}$. The relationship between the temperature of a frozen soil and its liquid water content is graphically expressed in a freezing characteristic curve, analogous to the moisture characteristic curve that describes water retention in unfrozen soil (2, 3).

APPLICATIONS

Can this information be used to estimate the liquid water content of a frozen soil? Yes, but with some important restrictions and caveats (2). It has been shown that the moisture characteristic and the freezing characteristic are superimposable for porous media that are completely colloidal, i.e. clay suspensions, where surface tension effects are negligible. For such materials, the liquid water content corresponding to a specific pore water pressure (matric potential) should be the same whether the cause of the gauge pressure is drying or freezing (ignoring the issue of hysteresis). For media that are devoid of colloids, i.e. pure sands and silts, the rules for similarity are also clear, but different. Here the ratio of the surface tensions of an air–water interface (σ_{aw}) and an ice–water interface (σ_{iw})

Encyclopedia of Soil Science

must be taken into account. For such materials it has been shown experimentally for a given soil that for similar water contents during drying and freezing the pore water pressure will be more negative in the drying soil by a factor of 2.2, the ratio of σ_{aw} to σ_{iw}. In other words, at a specific pore water pressure there will be less liquid water in the frozen soil than in the drying soil. Unfortunately, most soils contain both colloidal and noncolloidal particles, so *a priori* scaling of a freezing characteristic curve from known moisture characteristic data is not possible. Recent developments in instrumentation and methodology offer means to directly measure soil freezing characteristic curves (3, 4).

Internal water movement within a frozen soil is a difficult process to study. Because thermodynamic similarity exists between freezing and drying, that is, both are functions of pore size, it is expected that for similar *liquid* water contents, the hydraulic conductivity of a frozen soil and an unfrozen soil will also be similar, but experimental verification data are scant. Coarse-textured, sandy soils, when unfrozen, generally have high saturated hydraulic conductivities, but since their pores drain at gauge pressures close to zero, their conductivities decrease dramatically with desaturation. For the reasons mentioned above, the decline in liquid water content (as a function of p_w) upon freezing is even steeper, so such soils rapidly lose their ability to conduct water as they freeze. Finer textured soils generally have lower saturated hydraulic conductivities, but since the decrease in water content upon freezing is more gradual, their conductivities decrease more slowly, so that they can often sustain more water movement in the frozen state than sandy soils. For this reason they are more prone to frost heave. Frost heave can cause tremendous structural damage to buildings and roadways and can also harm plants and trees.

Despite the substantial decrease in water potential accompanying freezing, there often is minimal movement of water as the freezing front penetrates. Unless there is a ready supply of water close to the plane of freezing, the soil beneath will soon become desiccated, causing a sharp decrease in hydraulic conductivity, to the point that the delivery of latent heat cannot match the rate of sensible heat loss, so the freezing front moves downward. Thus initially dry soils may freeze with little or no redistribution of moisture. Consistent with the Clapeyron equation, the largest water-filled pores freeze first, at temperatures closest to 0°C, and as the temperature decreases the water in progressively smaller pores freezes. Even in relatively moist soils, the hydraulic conductivity is often insufficient to support anything more than local redistribution of moisture. This is manifested in ice crystal formation in large pores and cracks, without significant change in profile water distribution at a scale detectable by traditional methods of soil moisture measurement.

The infiltration of water into frozen soil has received considerable attention, due to the sometimes catastrophic flooding that can occur following snowmelt or rainfall on frozen soil. It is commonly accepted that freezing dramatically lowers the infiltration capacity of a soil. This is generally true, for reasons alluded to earlier. In wet soils, particularly those with water tables near the surface, water movement during freezing fills large pores and in extreme cases creates lenses of pure ice. Also, just as the largest water-filled pores are the first to freeze, they are also the last to melt, at temperatures closest to 0°C. These are the pores that are the most important in infiltration; consequently, infiltration rates are much lower if they are ice-filled. Even in well-drained, unsaturated soils, local redistribution during freezing is often sufficient to fill large pores at the soil surface with ice, retarding subsequent infiltration. However, in drier soils, particularly if they freeze rapidly, the large pores can remain air-filled, and infiltration rates may approach those measured under unfrozen conditions (5). Though initial infiltration rates into frozen soil are usually quite low, often resulting in ponding, there have been observations of abrupt increases in infiltration prior to the complete disappearance of frozen soil below (6, 7). A number of possible explanations have been suggested, but none have been experimentally demonstrated.

CONCLUSION

Heat and water relations in frozen soil are difficult to study. Researchers are hampered by the inability to separately measure water and ice contents at similar spatial scales and by the complications arising from simultaneous, coupled transport of heat and water in a deformable medium. Formal theory describing phase equilibrium in ideal, uniform porous media is well-established, but application of that theory to transient phenomena in structured, yet plastic, soils with complex mineralogy remains elusive.

REFERENCES

1. Miller, R.D. Freezing Phenomena in Soils. In *Applications of Soil Physics*; Hillel, D., Ed.; Academic Press: New York, NY, 1980; 254–299.
2. Koopmans, R.W.R.; Miller, R.D. Soil Freezing and Soil Water Characteristic Curves. Soil Sci. Soc. Am. Proc. **1966**, *30*, 680–685.

3. Spaans, E.J.A.; Baker, J.M. The Soil Freezing Characteristic: Its Measurement and Similarity to the Soil Moisture Characteristic. Soil Sci. Soc. Am. J. **1996**, *60*, 13–19.
4. Spaans, E.J.A.; Baker, J.M. Examining the Use of Time Domain Reflectometry for Liquid Water Content Measurement in Frozen Soil. Water Resour. Res. **1995**, *31*, 2917–2925.
5. Granger, R.J.; Gray, D.M.; Dyck, G.E. Snowmelt Infiltration to Frozen Prairie Soils. Can. J. Earth Sci. **1984**, *21*, 669–677.
6. Baker, J.M.; Spaans, E.J.A. Mechanisms of Meltwater Movement Above and Within Frozen Soil. In *Proc. Intl. Symp. on Physics, Chemistry, and Ecology of Seasonally Frozen Soils*; Iskandar, I.K., Wright, E.A., Radke, J.K., Sharratt, B.S., Groenevelt, P.H., Hinzman, L.D., Eds.; Spec. Rept. 97-10, U.S. Army Cold Regions Research and Engineering Laboratory: Hanover, NH, 1997; 31–36.
7. Stahli, M.; Lundin, L.C. Soil Moisture Redistribution and Infiltration in Frozen Sandy Soils. Water Resour. Res. **1999**, *35*, 95–103.

WATER RETENTION IN SOIL

W

S.C. Gupta
D. Wang
University of Minnesota, St. Paul, Minnesota, U.S.A.

INTRODUCTION

Soil water retention refers to the mechanisms and processes related to changes in soil water content vs. its energy status. There are two components in the soil water retention: 1) the amount of water held in soil, and 2) the potential energy with which the water is held. The relationship between the amount of water and its potential energy is called the soil water retention characteristic curve (Fig. 1). The water retention characteristic curves are different for different soil types (Fig. 2). These differences are caused by the variation in soil particle-size distribution and their packing arrangement. Both these factors influence the soil water retention by affecting the pore-size distribution and the number of a given size pore in each size class.

BACKGROUND INFORMATION

The maximum amount of water held in a soil depends on the surface area of soil particles and the fraction of pores in a given volume of soil (porosity). The potential energy of the soil water, on the other hand, depends on the size distribution of the junction between pores. The junction between pores can be viewed as a small capillary tube, the radius of which determines the force needed to drain the pore (Fig. 3).

The energy with which the water is held in a capillary tube can be described by the capillary rise equation (Fig. 4):

$$h = \frac{2\sigma}{\rho g r} \cos \vartheta \qquad (1)$$

where h is the height of rise of water in a capillary tube, σ is the surface tension force of water, ρ is the density of water, g is the acceleration due to gravity, r is the radius of the capillary tube, and ϑ is the contact angle of water with the capillary tube. The smaller the radius of a capillary tube, the greater is the rise in water height. Applying this principle to soil water retention, the smaller the radius of the junction between two pore cavities, the greater is the energy needed to drain the pore restricted by the pore junction.

The energy with which water is held in soil is also called the soil matrix potential energy or simply the soil matrix potential. Because of the analogy between the capillary tube and the junction between soil pores, this energy is also called the capillary potential energy or simply the capillary potential. There are several ways to measure the potential energy of the soil. These include the tensiometer and the thermocouple psychrometer. The tensiometer consists of a small ceramic cup that is connected to a PVC tube on which a measuring device (vacuum gauge or mercury manometer) is mounted for recording the potential energy (Fig. 5).

In field applications of tensiometers, the ceramic cup and the PVC tube are first filled with de-aired water and sealed with a cap. The tensiometer is then installed in the soil and allowed to reach equilibrium with soil water. The negative pressure at equilibrium corresponds to the energy with which water is held in a soil.

The thermocouple psychrometer (Fig. 6) measures the relative humidity (RH) of the soil air that is in equilibrium with liquid soil water and its solutes. In the thermocouple psychrometer, water is first condensed at the thermocouple junction and then allowed to evaporate. The rate of evaporation is controlled by the relative humidity of the chamber that is in equilibrium with the potential energy of the soil water (11).

The water potential (ψ_w) is related to relative humidity by the relationship:

$$\psi_w = \frac{RT}{M} \ln(RH) \qquad (2)$$

where R is the ideal gas constant, T is absolute temperature, and M is the molecular weight of soil solution (water and solutes). Tensiometers are useful for measuring soil matric potential from 0 to -80 kPa, whereas thermocouple psychrometers are effective for measuring soil water potential from -1500 kPa to air-dry soil conditions.

The quantity of water held in the soil is expressed as the mass of water per unit mass of soil or the volume

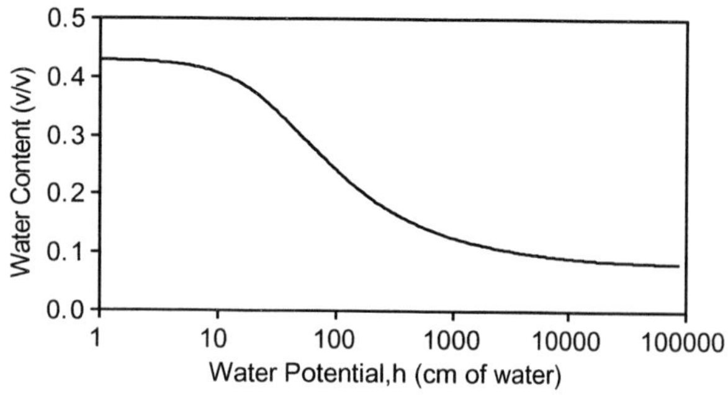

Fig. 1 Schematic of soil water retention characteristic curve.

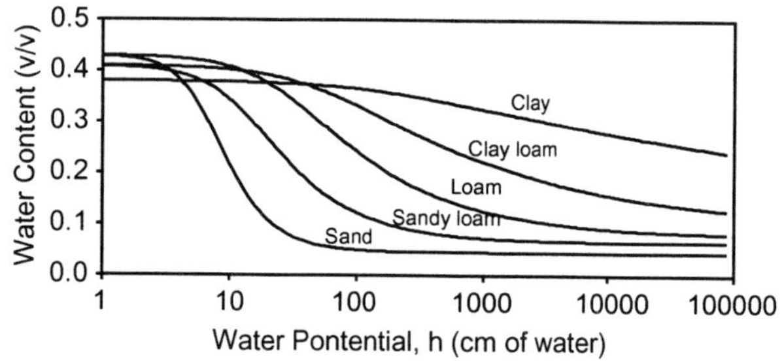

Fig. 2 Water retention of five different textured soils. (From Ref. 3.)

of water per unit volume of soil. The standard method of measuring soil water is by the gravimetric approach. In this case, the wet soil sample is weighed both before and after oven drying at 105°C for 24 hr or to a constant weight. The loss of water per unit weight of soil or per unit volume of soil thus quantifies the amount of water in the soil. The other methods include

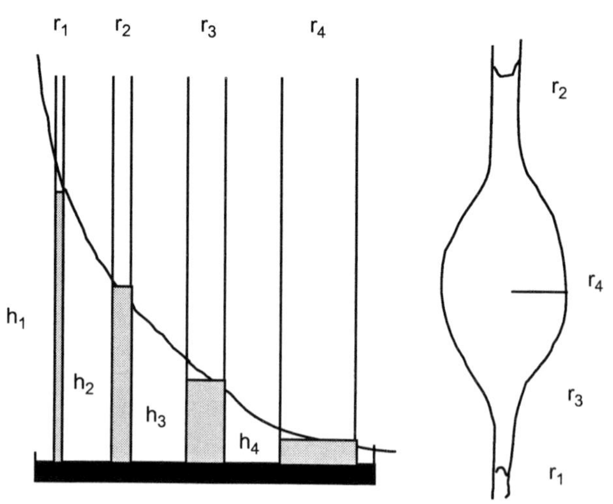

Fig. 3 Height in rise of water in capillary tubes corresponding to various neck radii in a pore.

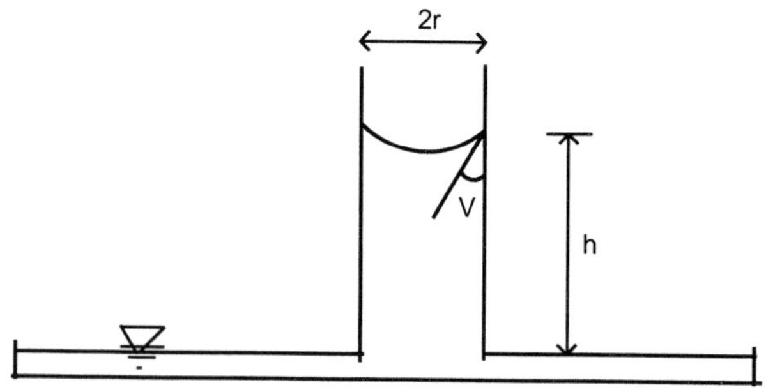

Fig. 4 Relationship between water rise, surface tension and capillary radius.

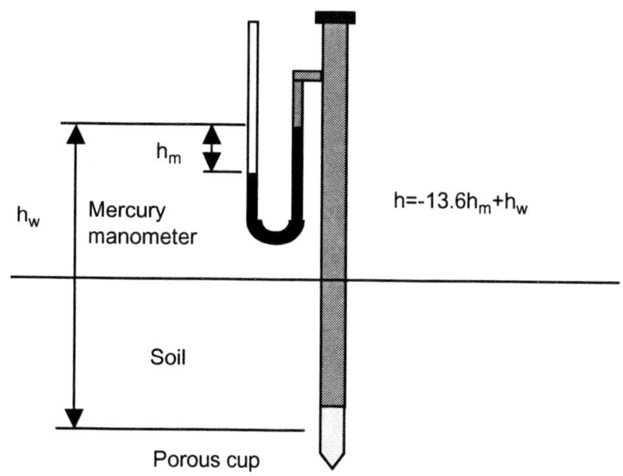

Fig. 5 Schematic of tensiometry for soil water potential measurement.

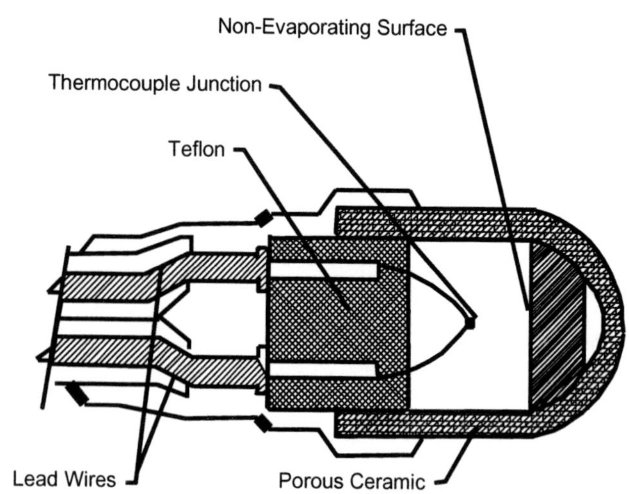

Fig. 6 Schematic of thermocouple psychrometer. (Adapted from Ref. 7.)

the neutron probe, the gamma ray attenuation, the time domain reflectometry (TDR), and the heat dissipation unit. The neutron probe is based on the principle that a fast neutron, when released in soil, will collide with hydrogen atoms thus resulting in slowed neutrons. Since most of the hydrogen atoms in soil are from water, the number of slowed neutron thus determines the amount of water in soil (4). Gamma ray attenuation is based on the principle that an increase in soil water will decrease the number of gamma particles passing through the soil and vice versa (6). TDR measures the dielectric constant of the moist soil. Since the dielectric constant of liquid water is 80 and that of the soil matrix is 4, changes in the soil dielectric constant reflect the changes in soil water content (16). The heat dissipation method works on the principle that at higher soil water content there is faster dissipation of heat and thus lower rise in soil temperature (10).

MODELS OF SOIL WATER RETENTION

In the literature, there are several methods available for estimating the water retention characteristics of soils. These methods are generally based on the premise that there exists some relationship between the soil water retention characteristics and the soil particle-size distribution, soil packing density, and soil organic matter content. These methods are either empirical or semiphysical. The empirical relationships are also called the "Pedotransfer" functions. Among the semiphysical methods, the most notable is the method of (1). The basis of this method is that the shape of the particle-size distribution curve is similar to the pore-size distribution curve. Then, it is a matter of translating the particles of a given size to the amount to water held by those particles, and the particle size to a corresponding pore size.

$$V_i = \left(\frac{w_i}{\rho_s}\right) e \tag{3}$$

$$r_i = -\frac{\sigma \cos \vartheta \sqrt{6}}{\rho g R_i \sqrt{e n_i^{(1-\alpha)}}} \tag{4}$$

where V_i and w_i are the pore volume and weight of ith fraction, respectively; ρ_s is the particle density; e is the void ratio; r_i and R_i are the mean pore and particle radii of the ith fraction, respectively; n_i is the number of particles in ith fraction, and α is an empirical constant ranging in value from 1.35 to 1.40.

Pedotransfer Functions

The pedotransfer functions, or the empirical methods of predicting soil water retention characteristics, use the easily measurable soil properties such as particle-size distribution (sand, silt, clay contents), bulk density, organic matter content, etc., as surrogates of water retention in soil (5). Using the measured database, an empirical equation that predicts the water retention at each soil matric potential is developed. Among the most widely used empirical method of predicting water retention characteristic is the method proposed by (12). This empirical method is based on water retention measurements of over 2000 soil horizons. In this method, the empirical relationships predict soil water content at 12 matric potentials (Table 1). The relationships are provided for three levels of input parameters. The first level assumes that the only inputs available are the percent sand, percent silt, percent clay, bulk density, and the organic matter content. The second level of relationships assumes that in addition to the input parameters for first level of regression equation, the water content of the soil at −1500 kPa matric potential is also known. The third level of regression further assumes that the soil water content at −33 kPa soil matrix potential is also known. Over the years, the method has been modified and improved by inclusions of additional data sets and/or other input variables (13–15).

Mathematical Description

Various mathematical functions have been used to describe the complete soil water retention characteristic curve. Notable among those are the functions suggested by Brooks and Corey (2) and by Van Genuchten (17). The Brooks and Corey function is:

$$\Theta = \frac{\theta - \theta_r}{\theta_s - \theta_r} = \left(\frac{h_e}{h}\right)^{-b} \tag{5}$$

whereas the van Genuchten function is:

$$\Theta = [1 + \alpha(-h)^n]^{-m} \tag{6}$$

$$m = (1 - n^{-1}) \tag{7}$$

where Θ is the normalized water content; θ_r is the residual water content; θ_s is the saturated water content; h is the apparent soil matric potential; h_e is the air entry matric potential; and α, b, n, and m are constants for a given soil. Because of their mathematical superiority, these two functions have been widely used in many soils and hydrologic modeling studies. Pedotransfer functions have

Table 1 Coefficient of linear regressions for predicting soil water contents at 12 matric potentials

Matric potential (kPa)	Regression coefficients *a*	*b*	*c*	*d*	*e*	*f*	*g*	*h*	Correlation coefficient (*R*)
	0.7899	−0.0037			0.0100	−0.1315			0.58
−4	0.6275	−0.0041			0.0239			−0.08	0.57
	0.1829				−0.0246	−0.0376	1.89	−1.38	0.77
	0.7135	−0.0030		0.0017		−0.1693			0.74
−7	0.4829	−0.0035			0.0263			0.25	0.74
	0.0888	−0.0003			−0.0107		1.53	−0.81	0.91
	0.4118	−0.0030		0.0023	0.0317				0.81
−10	0.4103	−0.0031			0.0260			0.41	0.81
	0.0619	−0.0002			−0.0067		1.34	−0.51	0.95
	0.3121	−0.0024		0.0032	0.0314				0.86
−20	0.3000	0.0024			0.0235			0.61	0.89
	0.0319	−0.0002					1.01	−0.06	0.99
	0.2576	−0.0020		0.0036	0.0299				0.87
−33	0.2391	−0.0019			0.0210			0.72	0.92
	0.2065	−0.0016		0.0040	0.0275				0.87
−60	0.1814	−0.0015			0.0178			0.80	0.94
	0.0136					−0.0091	0.66	0.39	0.99
	0.0349		0.0014	0.0055	0.0251				0.87
−100	0.1417	−0.0012			0.0151			0.85	0.96
	−0.0034				0.0022		0.52	0.54	0.99
	0.0281		0.0011	0.0054	0.0200				0.86
−200	0.0986	−0.0009			0.0116			0.9	0.97
	−0.0043				0.0026		0.36	0.69	0.99
	0.0238		0.0008	0.0052	0.0190				0.84
−400	0.0649	−0.0006			0.0085			0.93	0.98
	−0.0038				0.0026		0.24	0.79	0.99
	0.0216		0.0006	0.0050	0.0167				0.81
−700	0.0429	−0.0004			0.0062			0.94	0.98
	−0.0027				0.0024		0.16	0.86	0.99
	0.0205		0.0005	0.0049	0.0154				0.81
−1000	0.0309	−0.0003			0.0049			0.95	0.99
	−0.0019				0.0022		0.11	0.89	0.99
−1500	0.0260			0.0050	0.0158				0.80

$\theta_x = a + b$ sand(%) $+ c$ silt(%) $+ d$ clay (%) $+ e$ organic matter (%) $+ f$ bulk density (Mg m^{-3})$+g$ (water content at −33 kPa)$+h$ (water content at −1500 kPa) where θ_x is predicted water content (m^3 m^{-3}) at a given potential, sand = 0.5–2 mm, silt = 0.02–0.05, clay =< 0.002 mm. Sand + silt + clay = 100%.
(From Ref. 12.)

also been developed for predicting parameters of the Brooks and Corey and van Genuchten functions based on easily measurable soil properties such as percent sand, silt, clay, organic matter, and bulk density (13–15).

METHODS OF CHARACTERIZING SOIL WATER RETENTION

Measurement of soil water retention requires simultaneous determination of soil water content and potential. In the laboratory, the traditional methods of measuring soil water retention involve establishing a series of equilibrium between water in the soil sample and known potentials (9). Two of the most common methods that use this principle are the hanging water column and the pressure plate apparatus. In case of hanging water column, an initially saturated soil sample placed on a ceramic plate is allowed to drain by gravity until there is no more drainage. The soil water content at equilibrium thus corresponds to the gravitational forces equivalent to the distance between the soil sample and the water reservoir. The pressure plate apparatus consists of a ceramic plate and a pressure chamber. The soil sample is placed on a ceramic plate, the plate and the soil sample are allowed to saturate overnight, and then are desorbed by applying a known air pressure to the chamber. When no more water comes out of the soil sample, the sample is in equilibrium with the applied air pressure, and thus, the energy of the soil water is equivalent to the applied air pressure. The soil sample is taken out and its water content measured gravimetrically.

Soil water retention characteristics can also be characterized in situ using tensiometers and various water content measuring devices. A widely used approach is the instantaneous profile method where a flooded area is allowed to drain and then the soil water content and matric potential are measured simultaneously using a neutron probe and tensiometers (19). Wang et al. (18) used a combination of small tensiometers and TDR probes to measure in situ water retention. In this set-up, the authors used a tension infiltrometer as a source of water supply.

HYSTERESIS

For a given soil, the soil water retention characteristic curve varies depending upon whether the curve was obtained by desorbing a saturated soil or by wetting a dry soil (Fig. 7).

This behavior in water retention characteristics is called the "hysteresis". The main reason for the hysteresis is the so-called "ink-bottle" effect, which refers to narrow points

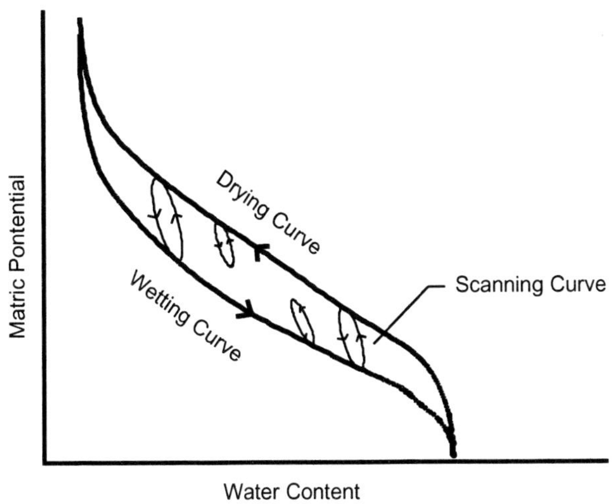

Fig. 7 Hysteresis in soil water retention. (Adapted from Ref. 8.)

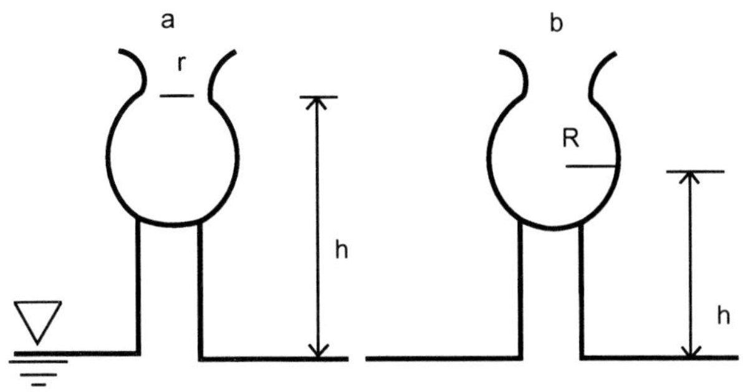

Fig. 8 "Ink-bottle" effect on water imbibition and drainage in large soil pores. (Adapted from Ref. 8.)

of connection between large cavities of the adjoining pores (Fig. 8).

During drying, the water content in the soil is controlled by the narrow points between pores (Fig. 8a), whereas during wetting water retention is controlled by the widest point in the pore cavity (Fig. 8b). As a result of this, the soil at any given potential retains more water while drying than wetting. The wetting and drying curves in Fig. 7 are called the primary curves whereas the curves connecting the primary curves are called the scanning curves. Scanning curves define the path that a soil takes when switching from wetting to drying, and vice versa.

PLANT AVAILABLE WATER

From the plant growth perspective, two points on the soil water retention curve are significantly important. These are the water retention at soil matric potential of -33 kPa

Table 2 Available water capacity of soils of various textural groups

Soil texture	Available water ($m^3 m^{-3}$)
Clay	0.076
Clay loam	0.121
Loam	0.077
Loamy sand	0.003
Silt	0.168
Sandy loam	0.136
Silty clay	0.071
Silty clay loam	0.142
Sand	0.001
Sandy clay	0.097
Sandy clay loam	0.058
Sandy loam	0.019

(Adapted from Ref. 3.)

(field capacity water content) and the water retention at soil matric potential of −1500 kPa (permanent wilting point, or PWP). The field capacity water content refers to soil water content after a saturated soil has been allowed to drain for 24–48 hr without any significant evaporation. Field measurements have shown that for most soils, the matric potential at field capacity water content corresponds to −33 kPa. If the water retention was measured on a disturbed soil then the field capacity water content corresponds to a soil matric potential of −10 kPa. PWP refers to soil water content when plants fail to recover its turgidity on watering. Field measurements have shown that this water content corresponds to about −1500 kPa for most plants. Water retention between soil matric potentials of −33 and −1500 kPa is referred to as the available water for plant growth.

As expected, available water capacity of soil varies with soil texture (Table 2). Generally, medium texture soils have higher available water capacity than either the fine or the coarse textured soils. This is due to both relatively higher field capacity and lower PWP water content of medium texture soils than the coarse or the fine textured soils. In coarse textured soil, water content at PWP is low but so is the water content at field capacity. In fine textured soils, water content at field capacity is higher but so is the water content at PWP.

REFERENCES

1 Arya, L.R.; Paris, J.F. A Physicoempirical Model to Predict the Soil Moisture Characteristics from Particle Size Distribution and Bulk Density Data. Soil Sci. Soc. Am. J. **1986**, *45*, 1023–1030.
2 Brooks, R.H.; Corey, A.T. Properties of Porous Media Affecting Fluid Flow. J. Irrig. Drain. ASCE **1966**, *92*, 61–88.
3 Carsel, R.F.; Parrish, R.S. Developing Joint Probability Distributions of Soil Water Retention Characteristics. Water Resour. Res. **1988**, *24*, 755–769.
4 Gardner, W.R.; Kirkham, D. Determination of Soil Moisture by Neutron Scattering. Soil Sci. **1952**, *73*, 392–401.
5 Gupta, S.C.; Larson, W.E. Estimating Soil Water Retention Haracteristics from Particle Size Distribution, Organic Matter, and Bulk Density. Water Resour. Res. **1979**, *15*, 1633–1635.
6 Gurr, C.G. Use of Gamma Rays in Measuring Water Content and Permeability in Unsaturated Columns of Soil. Soil Sci. **1962**, *94*, 224–229.
7 Hanks, R.J.; Ashcroft, G.L. *Applied Soil Physics*; Springer: Berlin, 1980.
8 Hillel, D. *Fundamental of Soil Physics*; Academic Press: New York, 1980.
9 Klute, A. Water Retention: Laboratory Methods. In *Methods of Soil Analysis, Part 1, Physical and Mineralogical Methods*; 2nd Ed. Klute, A. Ed.; ASA-SSSA: Madison, WI, 1986, 635–662.
10 Phene, C.J.; Hoffman, G.J.; Rawlins, S.L. Measuring Soil Matric Potential In Situ by Sensing Heat Dissipation within a Porous Body I: Theory and Sensor Construction. Soil Sci. Soc. Am. Proc. **1971**, *35*, 27–32.
11 Rawlins, S.L.; Campbell, G.S. Water Potential: Thermocouple Psychrometry. In *Methods of Soil Analysis, Part 1, Physical and Mineralogical Methods*; 2nd Ed.; Klute, A., Ed.; ASA-SSSA: Madison, WI, 1986; 597–618.
12 Rawls, W.J.; Brakensiek, D.L. Estimating Soil Water Retention from Soil Properties. J. Irrig. Drain. Div. Proc. ASCE **1982**, *108* (IR2), 167–171.
13 Rawls, W.J.; Brakensiek, D.L. 1985. Prediction of Soil Water Properties for Hydrologic Modeling. 293–299. *In* Jones, E.B., Ward, T.J., Ed.; Proc. Symp. Watershed Management in the Eighties. April 30–May 1, 1985, Denver, CO. Am. Soc. Civil Eng., New York, NY.
14 Rawls, W.J.; Ahuja, L.R.; Brakensiek, D.L. Estimating Soil Hydraulic Properties from Soil Data. In *Indirect Methods for Estimating the Hydraulic Properties of Unsaturated Soils*; van Genuchten, Leij & Lun, Eds.; Proceeding of the International Workshop, October 11–13, 1989 University of California: Riverside, CA, 1992.
15 Rawls, W.J.; Gish, T.J.; Brakensiek, D.L. Estimating Soil Water Retention from Soil Physical Properties and Characteristics. Adv. Soil Sci. **1991**, *16*, 213–234.
16 Topp, G.C.; Davis, J.L.; Annan, A.P. Electromagnetic Determination of Soil Water Content: Measurement in Coaxial Transmission Lines. Water Resour. Res. **1980**, *16*, 574–582.
17 van Genuchten, M.Th. A Closed Form Equation for Predicting the Hydraulic Conductivity of Unsaturated Soils. Soil Sci. Soc. Am. J. **1980**, *44*, 892–898.
18 Wang, D.; Yates, S.R.; Ernst, F.F. Determining Soil Hydraulic Properties Using Tension Infiltrometers, TDR, and Tensiometers. Soil Sci. Soc. Am. J. **1998**, *62*, 318–325.
19 Watson, K.K. An Instantaneous Profile Method for Determining the Hydraulic Conductivity of Unsaturated Porous Materials. Water Resour. Res. **1966**, *2*, 709–715.

WATER USE EFFICIENCY, ENHANCING

W

David C. Nielsen
United States Department of Agriculture-Agricultural Research Service (USDA-ARS), Akron, Colorado, U.S.A.

INTRODUCTION

In semiarid regions of the world, water is generally the most limiting factor to crop production. Therefore, a primary objective of cropping systems in these regions is to improve water use efficiency, i.e., get the most production of marketable yield for each unit of water used by the cropping system. There are many definitions of water use efficiency (1, 2), but this discussion uses the following: water use efficiency is the ratio of crop grain yield to the evapotranspiration used to produce that yield. The following discussion will deal mostly with enhancements to water use efficiency that are a result of altering and improving the cropping environment in semiarid regions.

An extremely important method for enhancing water use efficiency is to get more water to be used for plant transpiration and less for evaporation. This can take two forms in a dryland cropping system: increasing precipitation storage efficiency, and growing a crop in place of a fallow period.

REDUCING TILLAGE/MAINTAINING CROP RESIDUES

Increasing precipitation storage efficiency through soil-surface water evaporation reduction during the non-crop period of a crop rotation can be accomplished by: 1) reducing tillage to minimize the number of times moist soil is brought to the surface, and 2) maintaining residues to shade the soil surface, reduce convective exchange of water vapor at the soil–atmosphere interface, reduce runoff by preventing surface crusting, and catch snow.

Precipitation storage efficiency increases as tillage is reduced during the summer fallow period (3, 4) (Fig. 1a). The increased soil water storage is a result of both maintaining crop residues on the soil-surface and reducing the number of times that moist soil is brought to the surface. Data from western Kansas showed water use efficiency increasing by 28% for corn, 17% for sunflower, 10% for soybean, and 7% for grain sorghum. This efficiency was found when a no till production system was used in place of a conventional tillage system, (5) using three or four sweep plow tillage operations to control weeds in the fallow period prior to crop establishment.

The effect of residue amount on precipitation storage efficiency using data from three Great Plains sites is demonstrated in Fig. 1b (6). Increased precipitation storage efficiency during the 14-month fallow period of a winter wheat–fallow system resulted from increasing amount of surface wheat residue. Crop residues reduce soil water evaporation by shading the soil-surface and reducing convective exchange of water vapor at the soil–atmosphere interface (7, 8). Additionally, reducing tillage and maintaining surface residues reduce precipitation runoff and increase infiltration, thereby increasing precipitation storage efficiency (9).

The reduction in convective exchange of water vapor is more effective in standing crop residues (compared to flat residues) because wind speed is reduced near the soil-surface by the standing residue (10–13). Reductions of up to 50% in potential surface soil water evaporation have been demonstrated when wheat stubble height after harvest was increased from 0.1 to 0.5 m, or when stem populations of short (0.1 m) standing wheat stems were increased from 100 to 600 stems m^{-2} (14).

Standing crop residues also increase the soil water available for crop production in the central and northern Great Plains by increasing snow deposition during the overwinter period (15–19). Reduction in wind speed within the standing crop residue allows snow to drop out of the moving air stream. The greater the silhouette area index (SAI = height × diameter × number of stalks per unit ground area) through which the wind must pass, the greater the snow deposition. Typical populations and heights of standing sunflower stalks after harvest (SAI = 0.03–0.05 $m^2 m^{-2}$) can result in an overwinter soil water increase of about 10–12 cm when winter precipitation is near normal in amount and number of blizzards in northeast Colorado (20) (Fig. 2). Overwinter soil water storage in North Dakota increased 0.24 cm for each cm increase in wheat stubble height because of greater snow trapping (21). Because of the drifting and packing of snow that occurs in standing crop residues, precipitation storage efficiencies can be very high. Overwinter precipitation

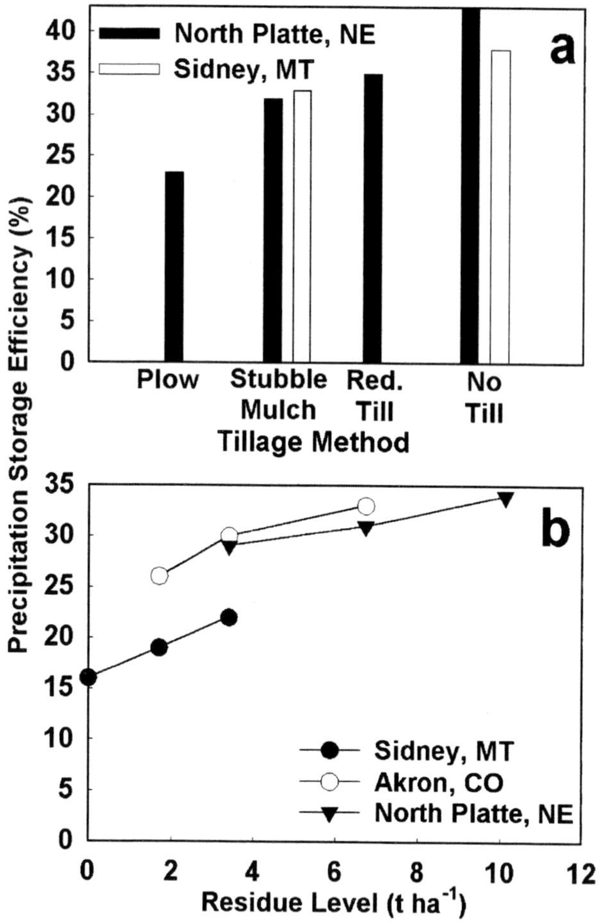

Fig. 1 Precipitation storage efficiency as influenced by a) tillage method during the 14-month fallow period in a winter wheat–fallow production system (data from North Platte, NE, from Ref. 3; data from Sidney, MT, from Ref. 4, and b) crop residue level on the soil surface. (From Ref. 6.)

storage efficiencies in excess of 100% in standing sunflower stalks have been reported in some years (20) as stalks collect snow blown in from adjacent areas without standing residues. Average overwinter precipitation storage efficiency of 55% has been reported for wheat stubble in northeast Colorado (22).

The increased precipitation storage efficiency from maintaining residues on the soil-surface will result in greater amounts of stored soil water. These higher soil water contents at planting have been shown to lead to higher crop yields for winter wheat and proso millet (23). The water use efficiency also increases with increased available soil water at planting, but only in years with normal or below normal growing season precipitation. In above-average precipitation years, when growing season precipitation makes up a higher percentage of total water

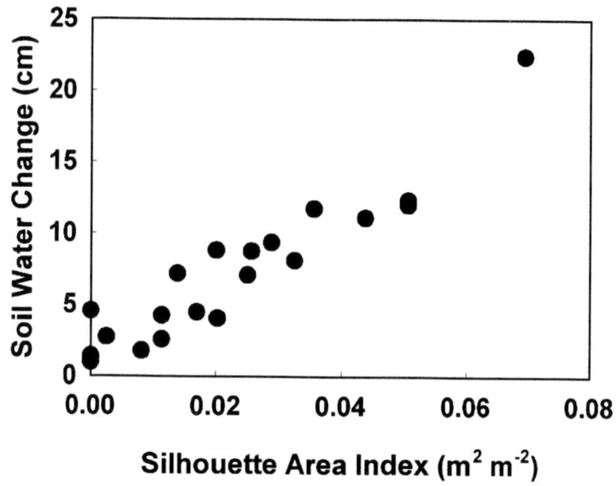

Fig. 2 Influence of sunflower silhouette area index on overwinter soil water change at Akron, CO. (From Ref. 20.)

use, there is no increase in water use efficiency with increasing water content at planting.

SHIFTING WATER USE TO CRITICAL GROWTH STAGES

Another method for enhancing water use efficiency is to shift growing season water use so that a greater proportion of the total is occurring during the more critical growth stages of flowering and grain filling, and less in the not-so-sensitive vegetative growth stage. When a growth retardant was applied to restrict vegetative development and leaf area in corn, water use during the vegetative stage was reduced, increasing the amount of water available for use during reproductive stages, resulting in a 16% increase in water use efficiency (24). Dryland winter wheat water use efficiency in the southern High Plains ranged widely from 0 to 36 kg ha^{-1} cm^{-1} due to the wide range in amount and seasonal distribution of precipitation (25). Winter wheat water use efficiency at Akron, CO increased linearly as the amount of precipitation falling between 15 May and 25 June (jointing through grain-filling) became a higher percentage of the total growing season precipitation.

Similar results are reported for experiments with limited irrigation in which water is withheld during vegetative growth and applied during flowering and grain-filling. When a set amount of limited irrigation was applied to corn, sorghum, and wheat during flowering and grain-filling, as opposed to applications throughout the entire growing season, water use efficiency increased by 19, 42, and 29%, respectively (26). Similarly, water applied to

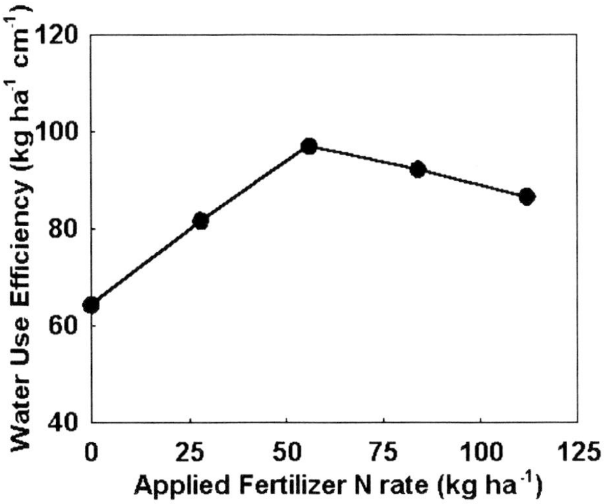

Fig. 3 Influence of nitrogen fertilizer application rate on winter wheat water use efficiency at Akron, Colorado. (From Ref. 28.)

winter wheat during the boot to anthesis stages averaged twice the yield response of water applied during tillering through jointing, without substantially increasing seasonal evapotranspiration (25).

APPROPRIATE FERTILIZER APPLICATION

In most cases when water supply is fixed, any management factor that increases yield will increase water use efficiency because evapotranspiration is little affected by the management (2, 27). For example, dryland winter wheat in northeast Colorado had increased water use efficiency with increased level of applied nitrogen fertilizer (Fig. 3) (28). However, water use efficiency can decline when high levels of nitrogen fertilization produce excessive vegetative growth which uses up the limited soil water supply early in the growing season (29). When water availability is low, plants grown under extremely high nitrogen fertility conditions may undergo more water stress during the more critical flowering and grain-filling stages than plants grown under lower fertility. Consequently, grain yield was reduced at the two highest fertilizer application rates while water use remained nearly the same for the highest N application levels.

REDUCING OR ELIMINATING FALLOW

One of the most productive ways of improving cropping system water use efficiency is to eliminate fallow periods. Cropping systems in the Great Plains have traditionally relied on the practice of summer fallowing in which one crop of winter wheat is grown every other year. The purpose of this was primarily to reduce the erratic yields (and sometimes total crop failure) associated with annual cropping. However, this practice is extremely inefficient in its storage and use of precipitation (30). Most of the precipitation received during the summer of fallow just preceding wheat planting in the wheat–fallow system is lost to evaporation. Intensifying cropping systems by reducing or eliminating fallow periods increases water use efficiency of the system.

Data averaged over 6 years at Akron, CO, show an 18% increase in water use efficiency through reduction of tillage in the wheat–fallow system, going from conventional tillage (35.5 kg ha^{-1} cm^{-1}) to no tillage (41.9 kg ha^{-1} cm^{-1}). When corn is added to the system such that two crops are produced in 3 years (wheat–corn–fallow), the water use efficiency increased by 55% (to 54.9 kg ha^{-1} cm^{-1}). With the continuous wheat–corn–millet system, the water use efficiency increased by 82% (to 64.4 kg ha^{-1} cm^{-1}). An even higher system water use efficiency (87.3 kg ha^{-1} cm^{-1}) was reported in a continuous dryland cropping experiment (8 years) with barley, corn, and winter wheat, also at Akron, CO (31).

CONCLUSION

In semiarid regions where lack of water is limiting to crop yield, water use efficiency needs to be maximized to get the most from this scarce resource. This can be accomplished through several management techniques to improve the cropping environment, including: reducing tillage to maintain surface crop residues, minimize evaporation, and trap snow; applying irrigations only during critical growth stages; matching fertilizer application rates to the expected available water condition; and employing crop rotations that reduce or eliminate fallow periods.

REFERENCES

1. Tanner, C.B.; Sinclair, T.R. Efficient Water Use in Crop Production: Research or Re-Search? In *Limitations to Efficient Water Use in Crop Production*; Taylor, H.M., Jordan, W.R., Sinclair, T.R., Eds.; ASA–CSSA–SSSA: Madison, WI, 1983; 1–27.
2. Stewart, B.A.; Steiner, J.L. Water Use Efficiency. In *Dryland Agriculture: Strategies for Sustainability*; Advances in Soil Science, Singh, R.P., Parr, J.F., Stewart, B.A., Eds.; Springer: New York, 1990; 13, 151–173.

3. Smika, D.E.; Wicks, G.A. Soil Water Storage During Fallow in the Central Great Plains as Influenced by Tillage and Herbicide Treatments. Soil Sci. Soc. Am. Proc. **1968**, *32*, 591–595.
4. Tanaka, D.L.; Aase, J.K. Fallow Method Influences on Soil Water and Precipitation Storage Efficiency. Soil Till. Res. **1987**, *9*, 307–316.
5. Norwood, C.A. Water Use and Yield of Dryland Row Crops as Affected by Tillage. Agron. J. **1999**, *91*, 108–115.
6. Greb, B.W.; Smika, D.E.; Black, A.L. Effect of Straw Mulch Rates on Soil Water Storage During Summer Fallow in the Great Plains. Soil Sci. Soc. Am. Proc. **1967**, *31*, 556–559.
7. Aiken, R.M.; Flerchinger, G.N.; Farahani, H.J.; Johnsen, K.E. Energy Balance Simulation for Surface Soil and Residue Temperatures with Incomplete Cover. Agron. J. **1997**, *89*, 405–416.
8. Van Doren, D.M., Jr.; Allmaras, R.R. Effect of Residue Management Practices on the Soil Physical Environment, Microclimate, and Plant Growth. In *Crop Residue Management Systems*; ASA Spec. Publ. 31, Oschwald, W.R., Ed.; ASA–CSSA– SSSA: Madison, WI, 1978; 49–83.
9. Unger, P.W.; Stewart, B.A. Soil Management for Efficient Water Use: An Overview. In *Limitations to Efficient Water Use in Crop Production*; Taylor, H.M., Jordan, W.R., Sinclair, T.R., Eds.; ASA–CSSA–SSSA: Madison, WI, 1983; 419–460.
10. Bilbro, J.D.; Fryrear, D.W. Wind Erosion Losses as Related to Plant Silhouette and Soil Cover. Agron. J. **1994**, *86*, 550–553.
11. Siddoway, F.H.; Chepil, W.S.; Armbrust, D.V. Effect of Kind, Amount, and Placement of Residue on Wind Erosion Control. Trans. ASAE **1965**, *8*, 327–331.
12. Nielsen, D.C.; Aiken, R.M. Wind Speed Above and within Sunflower Stalks Varying in Height and Population. J. Soil Water Conserv. **1998**, *53*, 347–352.
13. Smika, D.E. Soil Water Changes as Related to Position of Wheat Straw Mulch on the Soil Surface. Soil Sci. Soc. Am. J. **1983**, *47*, 988–991.
14. McMaster, G.S.; Aiken, R.M.; Nielsen, D.C. Optimizing Wheat Cutting Height for Harvest Efficiency and Soil and Water Conservation. Agron. J. **2000**, *92*, 1104–1108.
15. Steppuhn, H.; Erickson, D.; Zentner, R.P.; Nicholaichuk, W. Benefit/Cost Ratios for Snow Management Techniques in Semiarid Climates. In *Great Plains Agric. Council Publ. No. 120*; Proceedings of the Symp. on Snow Management for Agric., Swift Current, Saskatchewan, July, 1985, Steppuhn, H., Nicholaichuk, W., Eds.; Univ. of Nebraska: Lincoln, NE, 1986; 613–656.
16. Caprio, J.M. Potentials for Harvesting Water from Snow. In *Great Plains Agric. Council Publ. No. 120*; Proceedings of the Symp. on Snow Management for Agric., Swift Current, Saskatchewan, July, 1985; Steppuhn, H., Nicholaichuk, W., Eds.; Univ. of Nebraska: Lincoln, NE, 1986; 109–131.
17. Bauer, A.; Tanaka, D. Stubble Height Effects on Non-Growing Season Water Conservation. In *Great Plains Agric. Council Publ. No. 120*; Proceedings of the Symp. on Snow Management for Agric., Swift Current, Saskatchewan, July, 1985; Steppuhn, H., Nicholaichuk, W., Eds.; Univ. of Nebraska: Lincoln, NE, 1986; 255–272.
18. Smika, D.E.; Page, A.B.; Mickelson, R.H. Snow Water Management for Crop Production in the Central Great Plains. In *Great Plains Agric. Council Publ. No. 120*; Proceedings of the Symp. on Snow Management for Agric., Swift Current, Saskatchewan, July, 1985; Steppuhn, H., Nicholaichuk, W., Eds.; Univ. of Nebraska: Lincoln, NE, 1986; 335–344.
19. Campbell, C.A.; McConkey, B.G.; Zentner, R.P.; Selles, F.; Dyck, F.B. Benefits of Wheat Stubble Strips for Conserving Snow in Southwestern Saskatchewan. J. Soil Water Conserv. **1992**, *47*, 112–115.
20. Nielsen, D.C. Snow Catch and Soil Water Recharge in Standing Sunflower Residue. J. Prod. Agric. **1998**, *11*, 476–480.
21. Reis, R.F.; Power, J.F. Increased Soil Water Storage and Herbage Production from Snow Catch in North Dakota. J. Range Manag. **1981**, *34*, 359–362.
22. Greb, B.W. *Snowfall and Its Potential Management in the Semiarid Central Great Plains*; USDA-SEA Agric. Rev. Manage ARM-W-18, U.S. Govt. Print. Office: Washington, DC, 1980; 16–17.
23. Nielsen, D.C.; Anderson, R.L.; Bowman, R.A.; Aiken, R.M.; Vigil, M.F.; Benjamin, J.G. Winter Wheat and Proso Millet Yield Reduction Due to Sunflower in Rotation. J. Prod. Agric. **1999**, *12*, 193–197.
24. Shanahan, J.F.; Nielsen, D.C. Influence of Growth Retardants (Anti-Gibberellins) on Corn Vegetative Growth, Water Use, and Grain Yield Under Different Levels of Water Stress. Agron. J. **1987**, *79*, 103–109.
25. Musick, J.T.; Jones, O.R.; Stewart, B.A.; Dusek, D.A. Water-Yield Relationships for Irrigated and Dryland Wheat in the U.S. Southern Plains. Agron. J. **1994**, *86*, 980–986.
26. Shawcroft, R.W. Limited Irrigation May Drop Yield, Up Profit. Colorado Rancher Farmer **1983**, *37*, 35–38.
27. Viets, F.G., Jr. Fertilizers and Efficient Use of Water. Adv. Agron. **1962**, *14*, 223–264.
28. Nielsen, D.C.; Halvorson, A.D. Nitrogen Fertility Influence on Water Stress and Yield of Winter Wheat. Agron. J. **1991**, *83*, 1065–1070.
29. Power, J.F. Soil Management for Efficient Water Use: Soil Fertility. In *Limitations to Efficient Water Use in Crop Production*; Taylor, H.M., Jordan, W.R., Sinclair, T.R., Eds.; ASA–CSSA–SSSA: Madison, WI, 1983; 461–470.
30. Farahani, H.J.; Peterson, G.A.; Westfall, D.G. Dryland Cropping Intensification: A Fundamental Solution To Efficient Use of Precipitation. Adv. Agron. **1998**, *64*, 197–223.
31. Halvorson, A.D.; Reule, C.A. Nitrogen Fertilizer Requirements in an Annual Dryland Cropping System. Agron. J. **1994**, *86*, 315–318.

WATERSHED APPROACH

Timothy R. Green
Jan van Schilfgaarde
United States Department of Agriculture (USDA), Fort Collins, Colorado, U.S.A.

INTRODUCTION

Soils are open systems with natural and human-induced complexities. Water is an important change agent and state variable in soil systems that interacts over a broad range of space-time scales. Thus, soil and water management must be understood and practiced in the appropriate context of the landscape and hydrologic system.

THE HYDROLOGIC CYCLE

The hydrologic cycle is defined as the movement of water from the ocean and other sources into the atmosphere, then to the land surface, yielding flow and storage over and through soils and other geologic media to groundwater aquifers, streams, and rivers, and finally returning to the ocean. The water's physical state (liquid, solid, or vapor), energy state, and chemical composition (dissolved constituents) may be altered at various points along its journey through the cycle. Its particular state as it enters, transits, and exits the soil system plays an important role in the physical, chemical, and biological processes that occur in the soil, including their rates. The hydrologic processes of a particular system may vary significantly in space and time, and much of the local variability is linked to the global climate system. Various aspects of the global climate are not well understood (and poorly predicted), but multidisciplinary and international efforts are aimed at improving this aspect of the science and policy (1).

The hydrologic cycle may be considered a closed system (in terms of conservation of mass) that encompasses open systems of soils (Fig. 1). Water moves into and out of a soil system in multiple directions, even though one may tend to think of (vertical) surface infiltration and evapotranspiration dominating the water balance. Water is also stored in the soil–plant system for various periods of time. The residence time of this water and its constituents is important for biochemical transformations, and the rate of water flow affects the transport of key chemicals within and through soils.

A watershed, in turn, is defined as a region of land that contributes to a river system or reservoir. It is a compartment of the full hydrologic cycle that may cover an area of a few hectares to many thousands of square kilometers. Water flow and storage within a watershed can be related to its geology, landscape topography, and soil hydraulic properties. The surface topography generally determines the energy gradients for movement of incipient precipitation over and through the ground. The flow paths may be confounded by complex geology and development of the parent material, along with the partitioning of water between infiltration and runoff based on the soil hydraulic properties. The location of a soil system within a watershed will determine the soil physical properties and the potential for inflow and outflow of water (e.g., inflow of water from upland areas, exchange of soil water with shallow groundwater, and the degree of natural and human-induced drainage). The landscape position also affects the chemical status of the soil that controls processes such as plant growth and biological toxicity. Thus, the location of an open soil system within a watershed is paramount to how the soil system and processes of interest should be managed.

NATURAL VARIABILITY

Soil and water management on a watershed scale must consider the high degree of variability in soils and related factors within a watershed. Soil formation and development have been studied extensively, and the foundational work of Jenny (2) qualitatively identified five factors of soil formation: parent material, topography, climate, biota, and time. Additional factors include human activities (see below). While some soil attributes vary slowly compared with the time period of interest, the dynamics of other attributes cannot be ignored. From a practical point of view, some fraction of the natural variability cannot be explained; thus predictions must be assigned some level of uncertainty that the producers, land managers, or other stakeholders must accept.

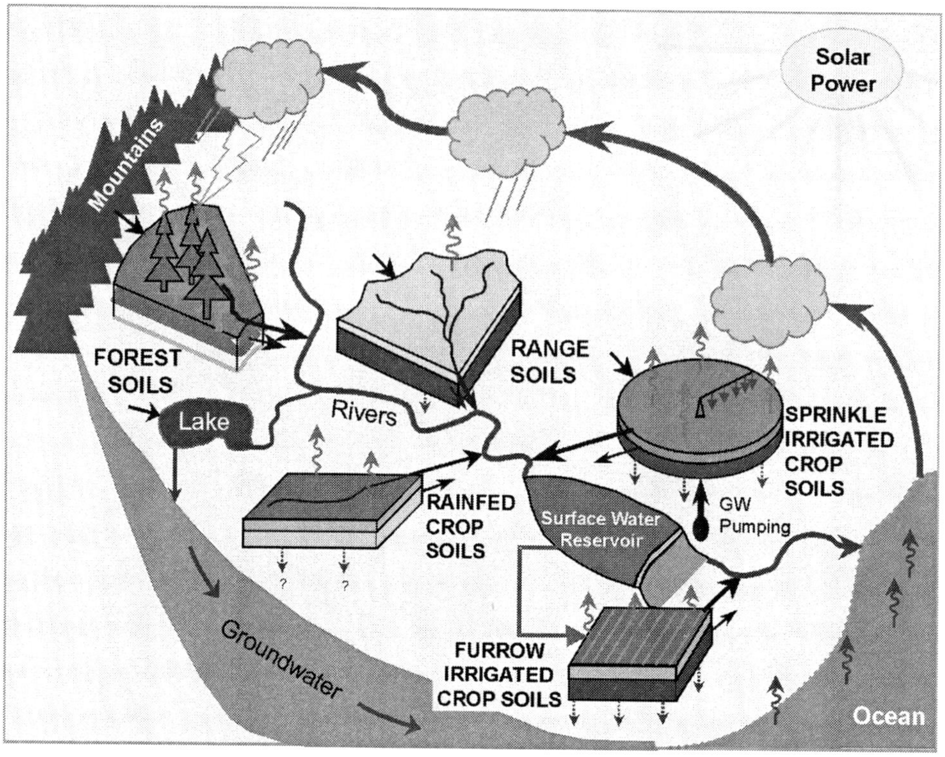

Fig. 1 A schematic view of open soil systems in the context of a watershed and hydrological cycle: upland forests, rangelands, dryland cropping systems, and pivot- and furrow-irrigated cropping systems.

SCALE: MOVING UP FROM POINT TO PLOT TO FIELD TO WATERSHED

Various properties of a soil may be measured or characterized on a small scale that is essentially a point in comparison with the variables or processes of interest. Many research studies in agronomy have been performed on experimental plots, typically on the order of a few tens of square meters. One of the most important challenges in the earth, soil, and agronomic sciences today is transferring information from one spatial scale to another (i.e., scaling) for management purposes. There are two problems with scaling within a watershed: aggregation of land units (upscaling) and disaggregation to individual land unit responses (downscaling). Upscaling provides important information about cumulative (off-site) effects on water quantity and quality over a watershed, while downscaling provides the necessary information to prescribe spatial management and predict on-site effects. The theoretical challenge involves both statistical scaling—based on measurements over a broad range of scales—and process-based, spatially distributed prediction. By addressing these issues and developing new computational methods to combine with measurements on real landscapes, practical prediction tools (with estimated uncertainties) for land management at appropriate space and time scales can be provided.

MANAGEMENT EFFECTS

The effects of historical land management must be borne in mind when devising management plans on a large scale. Most of the land masses of the Earth have been impacted to some degree by human land management practices. These include burning and clearing for agriculture, forestry, water diversion and storage, urbanization, transportation infrastructure, mining, and various agricultural and industrial land uses. In agriculture, widespread irrigation and drainage have converted deserts and wetlands, respectively, into croplands. Overirrigation and lack of adequate drainage have resulted in salinity problems and waterlogging. Less dramatic, yet significant, impacts on rainfed lands can be related to tillage practices, crop rotation, and grazing rates.

Recent advances in satellite-based remote sensing and geographical positioning have led to high-resolution mapping of landscape properties and state variables. This can be applied to spatial soil water management practices in the hope that average agricultural production will increase while adverse environmental effects decrease. For example, the practice of varying agronomic input rates within a farm field, known as "precision agriculture," touts such technology-dependent promises [(3); www.precision.agri.umn.edu]. Although the technology has improved dramatically in recent years, there is a need for improved scientific prediction to aid management decisions. Furthermore, the long-term feedback between landscape variability and spatial management will need to be addressed.

STRATEGY

A strategy for soil and water management should be based on a watershed approach to account for the various inputs that control the soil system, as well as to determine off-site impacts of the management practices used. The watershed approach recognizes multiple land uses within a given watershed such that management practices on one land area affect another. These interactions in space and time have been explored in some situations, but generally require further consideration.

Potential strategies are discussed below for croplands (irrigated and rainfed), rangelands, and forests. The soil scientist must recognize all important (or potentially important) uses and devise a means for accommodating all existing and planned uses. Diverse uses are likely to interact and often cause conflict. This requires developing technically sound alternative proposals, but even more importantly, engaging in conflict resolution by involving all interested parties in meaningful negotiation. Thus, community acceptance is added to the physical problem.

Irrigated Croplands

Some 37% of global crop production comes from the 12% of arable land that is irrigated, much of it in arid climates. Irrigation uses a large portion of available water resources and consequently may deprive cities or industries of water resources. Reduced water in streams may affect fish and wildlife adversely, and return flow from irrigation always degrades the quality of the remaining water resource. Thus, the strategy in managing irrigated agriculture must be to find a balance between crop production and other uses. One component of such a strategy may be to increase irrigation efficiency so as to make more water available for other purposes; another is to concentrate polluting salts in a smaller volume of drainage water or return flow to simplify separate disposal. In a report on agricultural salinity, van Schilfgaarde (4) addressed the question looming over irrigated agriculture: "Is it sustainable?" Due to the inevitable degradation of water quality in streams and terminal basins and the associated costs, the question of sustainability was restated as, "Are we prepared to maintain irrigated agriculture indefinitely?"

Rainfed Croplands

Many rainfed agricultural areas have been (or need to be) drained to enhance production. Reducing wetlands, however, deprives migrating birds of nesting and resting places, disturbing the natural order. Fish and wildlife specialists need to work with farm interests to find a balance. This may include the preservation of some wetlands, restoring fish habitat in streams and drainage channels, and using crop rotations that encourage diverse fauna. Income from hunting may offset possible losses from reduced crop production, and ecological criteria may supersede economic benefits.

Rangelands

Rangelands serve many functions: They store snow and modulate runoff to enhance stream flow, they support wildlife and diverse flora, and of course they provide grazing lands for cattle, among others. Overgrazing can affect the vegetation, hydrology, and soil erosion. Cattle on stream banks may also cause erosion and pollute the water. A strategy for sustainable management goes beyond providing the maximum number of grazing days for a herd of cattle.

Forests

Issues of forest management are not to be overshadowed by the above issues for agricultural lands. A disadvantage in terms of soil water management is that forest soils tend to be overshadowed by the trees—literally and figuratively. See www.soilslab.cfr.washington.edu/S-7/HistForSoilNA.html for a historical perspective. From the present perspective, however, forests have the advantage of being readily viewed in a watershed context. This is due in large part to the scale of operations and management of a forest system. The primary environmental indicators of success or failure are stream water quality and associated aquatic habitat. This has led to an emphasis on riparian zones and buffer strips to retain sediment and nutrients. Such an approach automatically integrates land area responses, but it is complicated by multiple users and stakeholders.

Multiple Use Conflicts

When water resources are managed to achieve a specific purpose, such management is likely to preclude, or at least interfere with, alternative uses. Irrigation, drainage, and erosion control are examples of water management practices that support sustainable agriculture. However, irrigation may exhaust water resources desired for urban or industrial uses, or reduce water quality and thus adversely affect fish and wildlife. Similarly, drainage may remove wetlands and eliminate nesting areas for migrating birds. Thus, major conflicts can arise. Whereas soil and hydrologic sciences can often explain or even predict the consequences of specific actions, resolving the conflicts generally requires input from the social sciences.

Many countries, states, and regions are adjusting their soil water practices to account for diversity in the opinions of stakeholders. Some of the policy direction is implicitly or explicitly directed by a watershed approach, simply because one cannot neglect interactions between land areas related to geomorphic and hydrologic factors. Development of appropriate strategies and specific practices is a site-specific endeavor that will lead to balanced resource use, not only to protect the environment but also to satisfy the people depending on the resource. Policy decisions will vary by region according to political circumstances, social desires, and economics, as well as the physical and biological situation. To be effective, nature demands that an integrated watershed approach be used.

EXAMPLES OF HISTORICAL PRACTICES

There are many examples of soil and water management practices from around the world that could illustrate the importance of a watershed approach. If space permitted, one could explore case studies on drainage of the Nile River Delta in Egypt, soil erosion of the Loess Plateau in China, historical land clearing and intensive agriculture throughout Europe, salinity and phosphorus issues in landscapes across Australia—including the Murray–Darling river basin and the wheat belt of Western Australia—and a variety of problems requiring (in hindsight, at least) an integrated watershed management approach. This article limited to three important concepts briefly illustrated by the following case studies.

Colorado River, U.S.A.

The Colorado River Basin includes parts of seven U.S. states, as well as a part of Mexico. The river rises as a pristine series of streams in the high Rocky Mountains and eventually drains into the Gulf of California. The average annual flow of the river is about $1.7 \times 10^{10}\,m^3$. Based on a relatively wet period of record, however, the courts have decreed that the seven U.S. states are entitled to $1.8 \times 10^{10}\,m^3$, and by treaty it has been agreed that Mexico is entitled to $1.8 \times 10^9\,m^3$. Furthermore, water quality standards have been imposed in terms of salinity (total dissolved solids). The origin of the salts is dissolution of minerals from the geologic formations and soils through which the water flows. The concentration of salts is increased when irrigated crops transpire pure water and thus pass on all the salts in a smaller volume of return flow.

Colorado River water is used for irrigation, as a municipal supply in cities like Los Angeles, and for many other uses. When the river finally reaches Mexico, the flow is minimal and the salt concentration tends to be high. Clearly, the legal entitlements cannot be met with the water nature provides. This water shortage and its distribution have been litigated for three-quarters of a century.

An example of an effort to reduce the problem somewhat is pertinent here. In the Grand Valley of Colorado, some 28 million ha were irrigated with Colorado River water when the Colorado River Salinity Control Program was started in the late 1970s. It was considered the largest single source of salt to the Colorado River. To maintain the agriculture and at the same time protect the river, a large project was implemented to line delivery canals, replace some of the ditches with pipe, and improve on-farm irrigation efficiency by the introduction of a number of up-to-date practices. These local practices integrated over the Grand Valley had a significant impact on the quantity and quality of water contributing to the Colorado River. The U.S. Bureau of Reclamation provides additional information at dataweb.usbr.gov/html/grandvalley2.html.

San Joaquin Valley, California, U.S.A.

In the 1920s, plans were developed to transport substantial amounts of water from northern California to the San Joaquin Valley, primarily for irrigation. Although engineers warned that drainage needed to be provided to make the irrigation enterprise viable, political and financial obstacles prevented the building of a drainage facility for decades. When parts of the nearly 20 million ha Westlands Irrigation District started to suffer substantially from waterlogging and salinity, one-half of a large drain (the upper half only!) was installed, with its water discharged "temporarily" into a low area euphemistically named the Kesterson Wildlife Refuge. Whereas the drain appeared to be successful as a salt-removal system, even if not completed, it soon became evident that selenium originating in the soils was transported into the Kesterson

Reservoir at concentrations sufficient to greatly damage, or kill, aquatic life and birds landing on the water. Finding a solution acceptable to all parties has been difficult. The interests involved include naturalists, sports fishermen and hunters, and farmers. Space does not permit further exploration other than to note that similar adverse effects from trace elements leached from open soil systems have been discovered in other watersheds.

Kissimee River, Florida, U.S.A.

Humid regions also present challenges for soil and water management. One such environment in south Florida is a system of streams and wetlands flowing into Lake Okeechobee, just to the north of the well-known Everglades National Park. Florida is very flat, and in its natural state, the shallow Kissimee River meandered from the south end of Lake Kissimee to Lake Okeechobee through wetlands that absorbed and cycled nutrients from agricultural runoff. Occasional heavy rains caused the river to overflow its natural banks and flood adjacent lands.

In the 1960s, the U.S. Army Corps of Engineers was commissioned for half a million 1962 U.S. dollars to dredge a straight canal (called C-38) with water-control gates and boat locks. The dimensions of the river were changed from approximately 166 km to 90 km long, from 30 m to 91 m wide, and from 3 m to 10 m deep. This provided for navigation and flood control, but dramatically changed the water quality and ecology. As a result, large-mouth bass and water fowl diminished in the Kissimee River ecosystem, and nutrients transported to Lake Okeechobee increased eutrophication problems.

In 1990, the State of Florida adopted the South Florida Water Management District's restoration plan (see www.sfwmd.gov for more details), and the U.S. Congress funded 50% of the estimated $372 million cost through the 1992 Water Resources Development Act. The plan includes acquisition of approximately 360 km^2 of land, backfilling large portions of the canal, and attempting to return the Kissimee River to its natural state. Further details of this project can be found at fcn.state.fl.us/eog/govdocs/opbenv/saveglades/everglades/html/kissimee.htm.

What Lessons Have We Learned?

The discussion here gives the reader an overview of soil and water management as an integrated process using the watershed approach. While it is true that the watershed approach integrates several processes and landscape factors, it also involves disaggregation of the watershed into its various geographical parts. Both issues remain challenging, primarily due to natural and human-induced variability and uncertainty in space and time. However, the advent of readily available spatial data is making research on these topics feasible, and in time sound management practices should follow.

The watershed approach to soil and water management also addresses political and economic issues of the stakeholders involved in regional, national, and international policy and decision making. Downstream water users have long recognized the importance of upstream land use practices, and it seems policy makers are increasingly facing the challenges of integrated watershed management for water quantity and quality. As history shows, practices of soil and water management are based primarily on social concerns and politics with varying degrees of import being placed on scientific understanding and engineering considerations. Regardless of the technical and political reasoning, some compromises or trade-offs will be required.

REFERENCES

1. IPCC- *Land Use, Land Use Change and Forestry. Special Report of the Intergovernmental Panel on Climate Change*; Watson, R.T., Ed.; Cambridge University Press: UK, 2000.
2. Jenny, H. *Factors of Soil Formation*; McGraw-Hill Book Company Inc.: New York, 1941.
3. van Schilfgaarde, J. Is Precision Agriculture Sustainable? Amer. J. of Alternative Agric. **1999**, *14* (1), 43–46.
4. van Schilfgaarde, J. Irrigated Agriculture: Is it Sustainable? In *Agricultural Salinity Assessment and Management*; ASCE Manual & Report on Engineering Practice, No. 71, Tanji, K.K., Ed.; American Society of Civil Engineers: New York, NY, 1990; 584–595.

WEATHERING

Dominique Righi
CNRS, Poitiers, France

INTRODUCTION

Weathering is the alteration of materials at, or near, the Earth's surface in response to the conditions that prevail there (1). It results from a complex set of interactions among the lithosphere, atmosphere, hydrosphere, and biosphere.

The minerals that constitute the lithosphere are conventionally divided into primary and secondary phases (Fig. 1). *Primary* phase minerals, produced at depth by high-temperature and/or high-pressure processes, are found in igneous and metamorphic rocks. *Secondary* phase minerals are transformations (secondary minerals that result from partial breakdown and reorganization of the primary phase minerals) or neoformations (secondary minerals that precipitate directly from solutions in contact with the rock) produced at the expense of primary phase minerals as a result of weathering. In most Earth-surface systems, secondary phase minerals include various clay minerals (e.g., aluminosilicates), oxides, and hydroxides.

The aqueous phase (hydrosphere) enters the weathering system as atmospheric precipitations, whose chemistry and abundance are determined by climatic and other atmospheric factors. Within the weathering system, the aqueous phase is rapidly modified by interaction with solids, including biomass. Dissolution of the most soluble minerals in the system (weathering phases), and precipitation of the least soluble minerals (new phases) tend to control the chemical composition of the aqueous phase. Residence time of the water in the weathering system is another important factor, so that a complex set of interactions connects the aqueous phase with other weathering factors.

Biomass is a major influence on the chemistry and behavior of water in weathering systems (2). The major effect of organic compounds is concentrated in the upper meter of the soil. Biomass yields important organic products containing chemical functional groups, such as carboxyl and phenol, that provide protons to the weathering system. Many of these products can attack a mineral surface directly to release ions into solution. Moreover, complex organic acids, such as oxalic and citric acids, are efficient in mineral breakdown because they hold cations (e.g., Al^{3+}, Fe^{3+}) in solution that would otherwise precipitate.

Weathering is a form of reaction that may be written schematically (3) as:

$$\text{Primary minerals} + \text{Reacting solution} = \text{Secondary minerals (clays)} + \text{Leachate} \quad (1)$$

This process is strongly influenced by the chemical composition of the rock (i.e., the nature of its primary minerals) and environmental factors (e.g., climate, living organisms, hydrodynamics). The environmental factors act through the aqueous phase, which is the first requirement of weathering.

MAIN WEATHERING MECHANISMS

Weathering mechanisms are determined by the highly variable physicochemical characteristics of the reacting solutions. Therefore, it is helpful to classify them as a function of the ionic concentration and the pH of these solutions (2, 3).

Acidolysis and Acidocomplexolysis

The low pH of <5 of this weathering alteration is obtained through the dissociation of organic acids. However, the effect on the mineral fraction is determined by the type of acid involved. Acidolysis, caused by noncomplexing organic acids, acts similarly to hydrolysis (see below), where-as acidocomplexolysis, which involves strongly complexing acids (e.g., oxalic and citric), leads to the destruction of all minerals except quartz and others that are highly resistant. Aluminum (Al) and iron (Fe) are solubilized as metalorganic complexes. Acidocomplexolysis is typical for the pedogenetic process called podzolization, in which organic acidity and complexing ability are the important factors.

Hydrolysis

In this alteration, weathering is by pure water or water with dissolved carbon dioxide (CO_2), which induces a pH of 4.5 to 6. Hydrolysis may be complete or only partial. Complete hydrolysis leads to the formation of Al

Encyclopedia of Soil Science

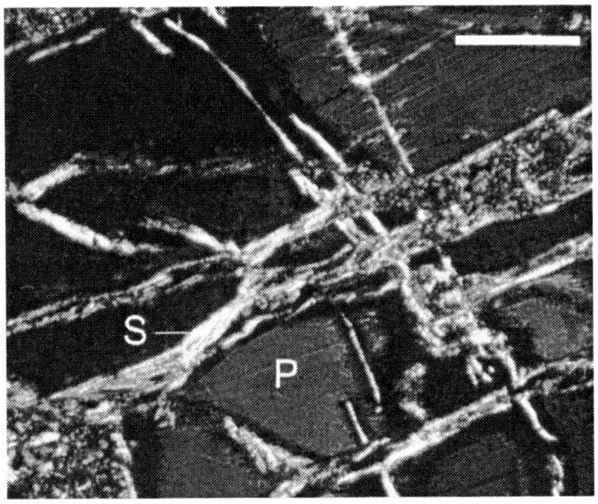

Fig. 1 Photograph from a thin section of a weathered rock (amphibolite) showing remnants of primary mineral (P) which is amphibole. Indicated by (S) are the secondary minerals (clay minerals) produced by reaction between the primary mineral and fluid percolating through the network of pores and fissures. Bar is 50 μm. (Photograph courtesy of Dr. Dominique Proust, Faculté des Sciences, Poitiers Cedex, France.)

hydroxide (gibbsite), and Fe oxides and oxyhydroxides (hematite, goethite), whereas partial hydrolysis leads to the formation of kaolinite or smectite clays as secondary minerals.

Salinolysis

Solutions in this weathering alteration are slightly alkaline and have a high ionic strength. This type of alteration is common in semiarid climatic zones, where there is a strong tendency for the aqueous phase to evaporate. The result is the precipitation of evaporite minerals (sulphates, chlorides) and the formation of various smectite clays.

Alkalinolysis

The aqueous solution in this alteration is salty and strongly alkaline, with a pH of 9 or more. As with salinolysis, evaporitic minerals form, but Al and silicon (Si) may reach relatively high concentrations in solution. This allows the formation of sodium (Na)-smectites and alkaline zeolites.

Secondary minerals produced through salinolysis and alkalinolysis mechanisms are essentially by precipitation from solution processes (neoformation). However, the required high Si and magnesium (Mg) concentrations in the soil solution are most often reached by alteration of preexisting silicate minerals. For example, dissolution of quartz at a high pH ensures that substantial soluble Si is available for neoformation of secondary phyllosilicate phases.

BASIC STRUCTURE OF A WEATHERING PROFILE

Over time, weathering processes lead to the development of a weathering profile (4). The basic structure of the weathering profile of a basement rock shows successive zones or horizons (Fig. 2). In the crystalline rock, which is considered the basic starting unit of this rock, all the minerals become unstable within the earth's surface conditions and are replaced by clays (secondary minerals), and ions are lost to the percolating aqueous solution.

Rock

The zone in which water–rock interactions begin is the initial unaltered rock. All rocks contain many cracks and fissures caused by the release of thermal and pressure constraints produced at depth. These cracks and fissures are the initial paths of penetration of the altering fluid. In the first stages of weathering, zones near cracks show intense alteration, whereas the zones furthest from the cracks remain unaltered. In the upper horizon of the bedrock, initial formation of clay minerals is localized around the initial cracks in the rock. Thus, in the initial stages of weathering, a rock sample is heterogeneously altered.

Saprock

The first zone of generalized alteration is called saprock. In this zone the minerals are partially or totally transformed into clays (secondary minerals), but the petrographic structure of the rock (i.e., the location and size of the initial mineral grains), is largely maintained. The overall quantity of clays is greatly increased over that of the basement rock through the alteration and transformation of the original minerals. The most important feature of this zone is the extreme heterogeneity of the mineralogy, with many types of clays and large portions of unaltered or little-altered rock minerals.

Saprolite

The next zone is that of saprolite, where there is little or no rock structure left and clays predominate. The effect of chemical transformation is greatest here; most of the original minerals are greatly altered. Due to a high degree of mineral transformation in the upper part of the saprolite zone, the rock structure often collapses and the initial fractures and fissures are lost.

Fig. 2 Schematic representation of a weathering profile showing the successives zones or horizons. (From Ref. 4.)

Organomineral Soil

The uppermost weathering zone commonly contains the soil. This is where, in a natural and undisturbed state, the biogeochemical processes become predominant. The chemical and physical properties of the system are dominated by pore structure. This is a secondary structure due to the weathering transformations which have totally effaced any rock structure. When porosity is sufficiently advanced to allow the establishment of organisms, their secretions and breakdown products induce a more rigorous weathering and the formation of soil types characteristic of the prevailing environmental conditions and vegetation.

BASIC FACTORS IN WEATHERING

Weathering is the result of a complex set of processes, which is controlled by a set of interacting factors. Soil-forming factors, including rock material, climate, vegetation, topography, and time, are relevant for weathering (5). In some cases, human activity disturbs natural soils so much that it could be considered an additional factor of soil formation.

Four variables govern weathering processes (6). The first is *rock type*, which is a chemical factor. The second, *climate*, is composed of rainfall (a chemical factor) and temperature (a physical factor). The third, *flow rate*, is a chemical factor because it determines, through drainage, the ratio of aqueous phase to rock and its contact time with rock. The fourth variable is the *age* of a weathering sequence, which is a physical parameter (time). The chemical factors of rock composition and flow rate are probably the most important in determining the type of secondary minerals (clay minerals) that will be produced.

On the *continental scale*, flow rate is determined by climate. In free-flowing water systems, drainage can be considered the driving force of weathering. Removing the leachate creates a mass action effect favoring the breakdown of the primary phase minerals. In humid zones, where drainage is free, such substractive weathering processes lead to the accumulation of Si, Al, and Fe, and the elimination of all other major cations. Kaolinite (an Al-rich clay mineral), gibbsite (Al hydroxide), and goethite and hematite (Fe oxyhydroxide and oxide) are the secondary minerals that form. Thick weathering profiles (20 m or more) characterized by this simple clay mineral assemblage, often called laterite (7), cover much of the intertropical zone. Under drier conditions the aqueous phase may develop high ionic concentrations, leading to precipitation of salts and/or formation of smectite or fibrous clays.

On the *watershed scale*, flow rate is determined by slope gradient. Similar to the effect of climate, the slope controls the water residence time and, hence, the chemistry of the reactions. The greater the slope, the more rapid the displacement of the aqueous solutions over the weathering zone. Moreover, material leached out of the upper part of the slope may deposit in the lower part. Therefore, the general type of secondary minerals changes with the slope. In the upper areas where water residence time is short, oxides of Fe and Al tend to form. As one proceeds downslope, kaolinite is more abundant and, in the lower regions, minerals such as smectites form.

On the *rock fragment* or *mineral scale*, porosity induces similar effects. In the larger pores of the system (1 mm or more in diameter), where time of contact between minerals and aqueous phase is short, kaolinite or Al and Fe oxides may form. In smaller pores ($<1\ \mu m$ diameter), longer residence time for the aqueous phase leads to higher concentrations of cations, which explains why smectites form at grain boundaries in solid rocks. In these systems of low water–rock ratio and long contact time, the chemical composition of the rock or mineral is very important in determining the secondary minerals (type of clay minerals) that will form. Acid rocks having a large proportion of K-feldspar and quartz will produce kaolinite, aluminous smectite, or illite (mica-like) clays. On the other hand, basic and ultrabasic rocks, which contain a large proportion of ferromagnesian minerals, will produce saponite, vermiculite, and chlorites (Mg- and/or Fe-rich clays) as weathering products. The parent rock effect is greater during the early stages of weathering or pedogenetic evolution. As weathering progresses (time increases), chemical leaching removes all but the most refractory components.

CONCLUSION

Clay minerals are the principal products of weathering and soil formation, which are essentially water–rock interaction processes. Clay formation occuring at the Earth's surface is a *slow reaction process*, meaning that some initial materials are only partially reacted and are not in equilibrium with the medium in which they are located. This creates the potential for further transformation of altered products. If the chemistry of the ambient soil or surface materials is changed, for example by pollution containing chemicals, there will be a high potential for change in the clay mineralogy of the surface material. Thus, the type of clay minerals produced by weathering reactions and their rate of formation are of the greatest importance in environmental problems.

REFERENCES

1. Chesworth, W. Weathering Systems. In *Weathering, Soils & Paleosols*; Martini, I.P., Chesworth, W., Eds.; Developments in Earth Surface Processes II; Elsevier: Amsterdam, 1992; 19–40.
2. Macias, F.; Chesworth, W. Weathering in Humid Regions, with Emphasis on Igneous Rocks and Their Metamorphic Equivalent. In *Weathering, Soils & Paleosols*; Martini, I.P., Chesworth, W., Eds.; Developments in Earth Surface Processes II; Elsevier: Amsterdam, 1992; 283–306.
3. Pédro, G. Caractérisation Générale des Processus de l'Altération Hydrolytique. Base des Méthodes Géochimique et Thermodynamique. Science Du Sol. **1979**, *2* (3), 93–105.
4. Righi, D.; Meunier, A. Origin of Clays by Rock Weathering and Soil Formation. In *Origin and Mineralogy of Clays, Clays and the Environment*; Velde, B., Ed.; Springer-Verlag: Berlin, 1995; 43–161.
5. Jenny, H. *Factors of Soil Formation*; McGraw-Hill: New York, 1941; 281.
6. Velde, B. Origin of Clays. In *Introduction to Clay Minerals*; Chapman & Hall: London, 1992; 101–154.
7. Tardy, Y. Diversity and Terminology of Lateritic Profiles. In *Weathering, Soils & Paleosols*; Martini, I.P., Chesworth, W., Eds.; Developments in Earth Surface Processes II; Elsevier: Amsterdam, 1992; 379–405.

WETLANDS AND BIODIVERSITY

Jean-Claude Lefeuvre
University of Rennes, Rennes, France

Virginie Bouchard
The Ohio State University, Columbus, Ohio, U.S.A.

INTRODUCTION

Wetlands are highly complex ecosystems subjected to a variety of hydrologic regimes, climatic conditions, soil formation processes, and geomorphologic settings. They include swamps, bogs, marshes, mire, fens, salt marshes, mangroves, and other types of ecosystems saturated by water during all or part of the growing season. They are found on every continent, except Antarctica, and in various climes, from the tropics to the tundra. The extent of the world's wetland is generally estimated to be from 7 to 9 million km^2 or about 4–6% of the earth. The variety of seasonal and perennial wetlands provides environmental conditions for highly distinctive fauna and flora.

ROLE OF HYDROLOGY

Wetlands differ significantly in their water source and seasonal hydrologic regime. Hydrological patterns (i.e., flooding frequency, duration and hydroperiod) influence physical and chemical characteristics (e.g., salinity, oxygen and other gas diffusion rates, reduction–oxidation potential, nutrient solubility) of a wetland. In return, these internal parameters and processes control flora and fauna distribution as well as ecosystem functions. Plants, animals and microbes are often oriented in predictable ways along the hydrological gradient (Fig. 1). Conversely, the biotic component affects the hydrology by eventually modifying flow or water level in a wetland (1, 2). Species also influence nutrient cycles and other ecosystem functions (3).

A LANDSCAPE PERSPECTIVE

Although the hydrology is part of the ecological signature of an individual wetland, wetlands are neither considered as aquatic nor terrestrial systems. They have characteristics from both systems and are defined as ecotones placed under this dual influence (4). Because wetlands are located at the interface of multiple systems, they assure vital functions (e.g., wildlife habitats) beneficial at the landscape level. Reduction of wetland area often reduces biodiversity in the landscape (2, 5). Increases in biodiversity occur when wetlands are created or restored in a disturbed landscape (6).

WETLAND FLORA

Wetland plants are adapted to a variety of stressful abiotic conditions (e.g., immersion, wave abrasion, water level fluctuation, low oxygen conditions). Identical adaptations to common environmental features have led taxonomically distinct species to sometimes look similar in terms of morphology, life cycle and life forms (7). Traditionally, wetland plants have been classified into groups of different life forms, primarily in relation with hydrological conditions. Helophytes are defined as plant species with over-wintering buds in water or in the submerged bottom (8). They are differentiated from hydrophytes in that their vegetative organs are partially raised above water level (8). Hejny's classification is based on relatively stable vegetative features that determine the ability of wetland plants to survive two unfavorable conditions, cold and drought (7). This classification uses the types of photosynthetic organs present in both the growth and flowering phases. Other classifications include both life form and growth form. When a species has a range of growth form, it is classified under the form showing the greatest achievement of its potential (7).

Plant communities in wetlands can be more or less homogeneous, mosaic-like, or distributed along a gradient resulting in a clear zonation of species. A gradient exists if one or several habitat parameters change gradually in space. This phenomenon is common in fresh water marshes that present a gradient of water depth and water saturated soils (Fig. 1). Such a gradient is often accompanied by differences in peat accumulation that is influenced by

Encyclopedia of Soil Science

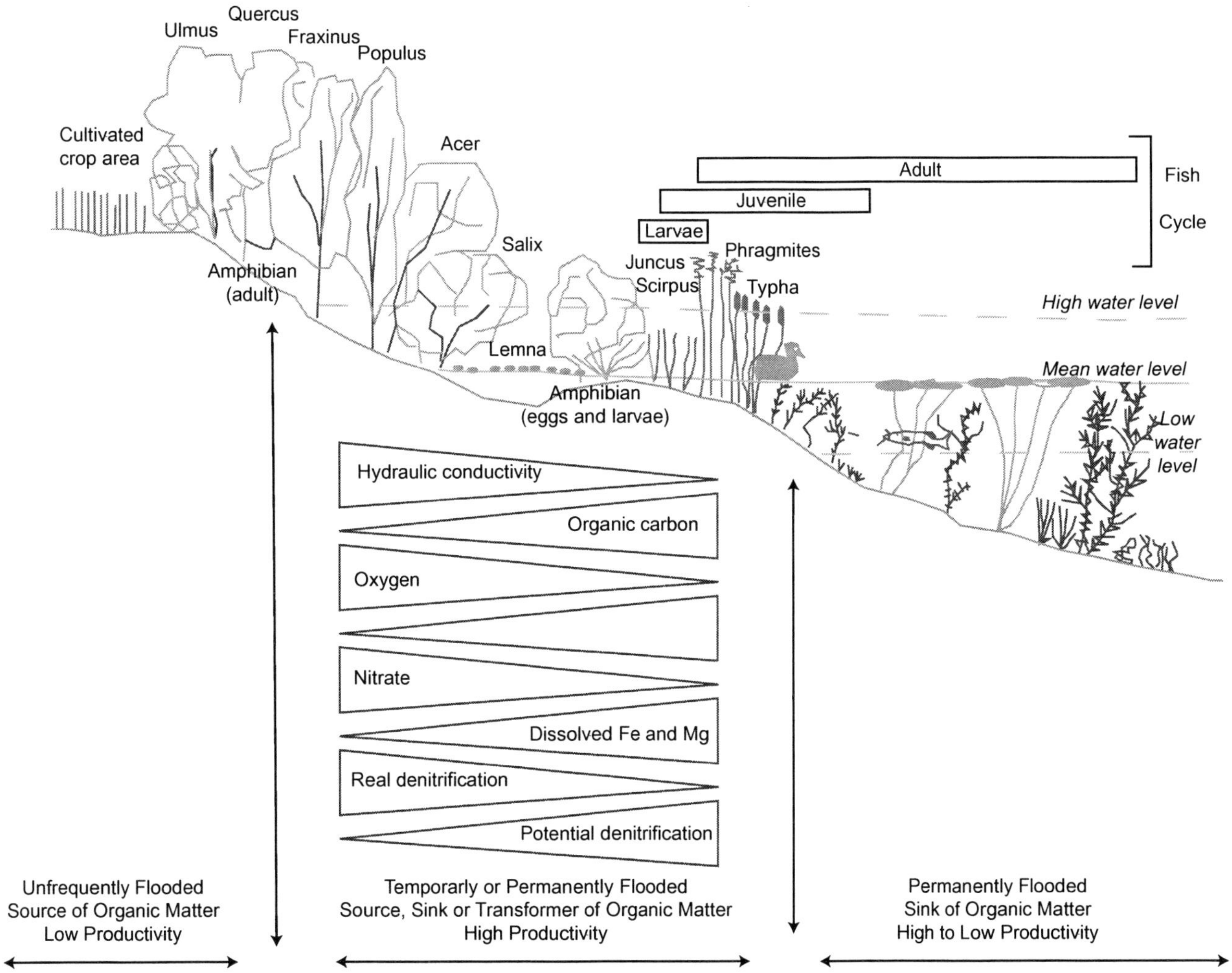

Fig. 1 Species distribution along the hydrological gradient in a freshwater marsh.

waves or currents. General principles of the zonation of aquatic plants have been largely described (7, 9, 10). Littoral vegetation can belong to several types of communities, which derived from the general principle that, from deeper water to the shore, we may expect successively submergent, floating, and emergent macrophytes. The most important habitat factor is water depth, depending on slope and peat accumulation (10). Other factors may be poor irradiance caused by high turbidity or exposure to waves or flow (7, 10).

Riparian ecosystems are found along streams and rivers that occasionally flood beyond confined channels or where riparian sites are created by channel meandering in the stream network. Riparian or bottomland hardwood forests contain unique tree species that are flood tolerant. Species distribution is associated with floodplain topography, flooding frequency, and flooding duration (11, 12). In southeastern U.S. bottomland, seasonally flooded forests are colonized by *Platanus occidentalis* (sycamore), *Ulmus americana* (American elm), *Populus deltoides* (cottonwood) and are flooded between 2 and 25% of the growing season (Fig. 1). Other species such as *Fraxinus pennsylvanica* (green ash), *Celtis laevigate* (sugarberry) and *Carya aquatica* (water hickory) colonized bottomlands that are flooded by less than 2% of the growing season (11). Freshwater marshes are dominated with emergent macrophytes rooted in the bottom with aerial leaves (i.e., helophytes). Species such as *Typha* (cattail), *Phragmites* (reed grass) and *Scirpus* (bulrush) are often clonal. A plant community is usually organized in sequence of patches that are dominated by one species. The second plant groups are the rooted plants with floating leaves (*Nymphaeid*). Lotus (*Nelumbo*) and water lilies (*Nymphea*) have very identical morphology (i.e., similar leaves and flowers) but a genetic analysis showed that lotus is more closely related with plane-tree than with water lilies (10). Submerged plants include elodeids (i.e., cauline species whose whole life cycle can be completed below the water surface or where only the flowers are emergent) and isoetids (i.e., species growing on the bottom whose whole life cycle can be completed without contact with the surface). Submerged species include species such as coontail (*Ceratophyllum demersum*) and water milfoil (*Myriophyllum* spp.). Plant species found in salt marshes are called halophytes (i.e., plants which complete their cycle in saline environments). A saltmarsh can be divided into low, middle, and upper marsh, according to flooding frequency and duration. Each zone is dominated by different plant species according to their tolerance to saline immersion (10).

A dominant competitive species—often a clonal species—can modify the theoretical zonation. Change in water chemistry (i.e., eutrophication) or hydrology may favor a particular species over the natural plant community. Highly competitive species are often invasive and aggressive in displacing native species. The expansion of *Phragmites australis* into tidal wetlands of North America causes a reduction in biodiversity as many native species of plants are replaced by a more cosmopolitan species (13). In riparian ecosystems, biodiversity is usually higher in the intermediate zones, whereas it is lower upstream and downstream (Fig. 2). The percentage of exotic species is low in upstream areas, but can represent up to 40 in downstream zones (Fig. 1) (14).

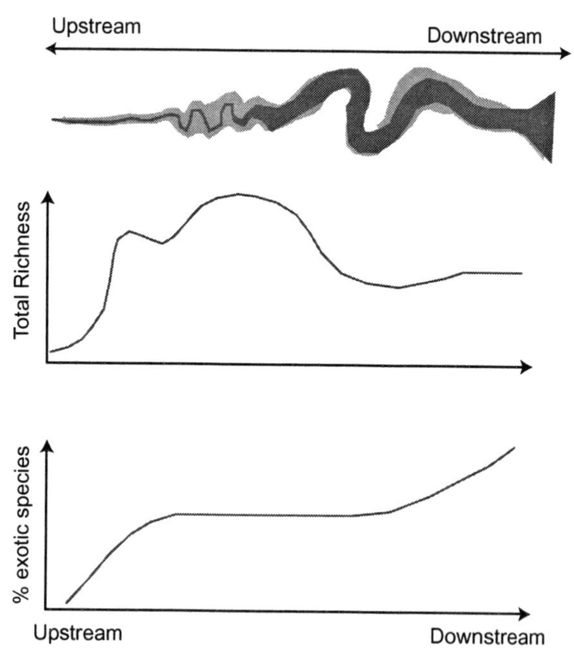

Fig. 2 Total richness and invasive species distribution along a stream longitudinal gradient. (From Ref. 14.)

WETLAND FAUNA

Diversity of vertebrate and invertebrate species is the result of a diverse community composed of resident and transient species, which use the space differently and at various times of day and year. The density and variety of animal populations at a particular wetland site is also explained by climatic events that affect geographic areas on a large scale. For example, the population of waterfowl during winter is largely dependent on climate variations in the northern part of the continents (15).

Resident species are often dependent on the type of vegetation. Animal communities are generally distributed along a zonation pattern, parallel to the plant communities, which is driven by the hydrological gradient. A few species depend entirely on a single plant species for their survival. One beetle (*Donacia* spp.) may depend on reeds, at least during its larval stage where another beetle (*Galerucella* spp.) uses only water lilies as their habitat and diet. For many other residents, their habitats extend to several plant communities during their life span. It is generally the case for many vertebrates such as amphibians, rodents, passerines, and waterfowl.

Many amphibian species depend on wetland or riparian zones for reproduction and larval stage. Likewise, wet meadows are necessary as a reproduction zone and nursery for a number of freshwater fishes. About 220 animals and 600 plant species are threatened by a serious reduction of wetlands in California, and the state's high rate of wetland loss (91% since the 1780s) is partly responsible (15). Many waterfowl species are sensitive to areas of reduction, patch size and distribution, wetland density, and proximity to other wetlands (16). When the Marais Poitevin (France), one of the principal wintering and passages sites for waterfowl in the Western Europe, underwent agricultural intensification in the 1980s, the population of ducks and waders declined tremendously. This decline was partly due to a 50% reduction of wet meadows between 1970 and 1995, primarily caused by the conversion to arable farmlands (17). Thus, maintenance of biodiversity depends on the existence of inter connections between wetlands, and between aquatic and terrestrial ecosystems. In fact, some authors have pointed out that increase in biodiversity occurs when wetlands are created or restored in a disturbed landscape (6, 18).

W

CONCLUSION

Despite the importance of wetlands, conservation efforts have ignored them for a long time. It is urgent to conserve the existing wetlands, and also to restore and create wetland ecosystems. A wide range of local, state, federal, and private programs are available to support the national policies of wetland "No Net Loss" in the U.S., and around the world. From a biodiversity perspective, on-going wetland protection policies may not be working because restored or created wetlands are often very different from natural wetlands (19). Created wetlands often result in the exchange of one type of wetland for another, and result in the loss of biodiversity and functions at the landscape level (19). We know now that it takes more than water to restore a wetland (20), even if an important place should be given to the ability of self-design of wetland ecosystems (21).

REFERENCES

1. Naiman, R.J.; Johnston, C.A.; Kelley, J.C. Alteration of North American Streams by Beaver. Bioscience **1988**, *38*, 753–762.
2. National Research Council, *Wetlands: Characteristics and Boundaries*; National Academy Press: Washington, DC, 1995; 306.
3. Naiman, R.J.; Pinay, G.; Johnston, C.A.; Pastor, J. Beaver Influences on the Long-Term Biogeochemical Characteristics of Boreal Forest Drainage Networks. Ecology **1994**, *75*, 905–921.
4. Hansen, A.J., Di Castri, F., Eds. Landscape Boundaries. *Consequences for Biotic Diversity and Ecological Flows*; Ecological Studies 92, Springer: Berlin, 1983; 452.
5. Gibbs, J.P. Wetland Loss and Biodiversity Conservation. Conser. Biol. **2000**, *14*, 314–317.
6. Weller, M.W. *Freshwater Marshes*, 3rd Ed.; University of Minnesota Press: Minneapolis, US, 1994; 192.
7. Westlake, D.F., Kvet, J., Szczepanski, A., Eds. *The Primary Ecology of Wetlands*; Cambridge University Press, 1998; 568.
8. Raunkiaer, C. *The Fife Forms of Plants and Statistical Plant Geography*; Oxford University Press: Oxford, England, 1934; 632.
9. Grevilliot, F.; Krebs, L.; Muller, S. Comparative Importance and Interference of Hydrological Conditions and Soil Nutrient Gradients in Floristic Biodiversity in Flood Meadows. Biodivers. Conserv. **1998**, *7*, 1495–1520.
10. Mitsch, W.J.; Gosselink, J.G. *Wetlands*, 3rd Ed; Wiley: New York, 2000; 920.
11. Clark, J.R., Benforado, J., Eds. *Wetlands of Bottomland Hardwood Forest*; Elsevier, Amsterdam: The Netherlands, 1981; 401.
12. Keogh, T.M.; Keddy, P.A.; Fraser, L.H. Patterns of Tree Species Richness in Forested Wetlands. Wetlands **1999**, *19*, 639–647.
13. Chambers, R.M.; Meyerson, L.A.; Saltonstall, K. Expansion of *Phragmites Australis* into Tidal Wetlands of North America. Aquat. Bot. **1999**, *64*, 261–273.
14. Planty-Tabacchi, A.M. Invasions Des Corridors Riverains Fluviaux Par Des Espèces Végétales D'origine Étrangère. These University Paul Sabatier Toulouse III, 1993; France, 177.
15. Hudson, W.E., Ed. *Landscape Linkages and Biodiversity*; Island Press: Washington, DC, 1991.
16. Leibowitz, S.C.; Abbruzzese, B.; Adams, P.R.; Hughes, L.E.; Frish, J. A Synoptic Approach to Cumulative Impact Assessment: A Proposed Methodology, 1992; EPA/600/R-92/167.
17. Duncan, P.; Hewison, A.J.M.; Houte, S.; Rosoux, R.; Tournebize, T.; Dubs, F.; Burel, F.; Bretagnolle, V. Long-Term Changes in Agricultural Practices and Wildfowling in an Internationally Important Wetland, and Their Effects on the Guild of Wintering Ducks. J. Appl. Ecol. **1999**, *36*, 11–23.
18. Hickman, S. Improvement of Habitat Quality for Resting and Migrating Birds at the Des Plaines River Wetlands Demonstration Project. Ecol. Engng. **1994**, *3*, 485–494.
19. Whigham, D.F. Ecological Issues Related to Wetland Preservation, Restoration, Creation and Assessment. Sci. Total Environ. **1999**, *240*, 31–40.
20. Zedler, J.B. Progress in Wetland Restoration Ecology. Trends Ecol. Evol. **2000**, *15* (10), 402–407.
21. Mitsch, W.J.; Wu, X.; Nairn, R.W.; Weihe, P.E.; Wang, N.; Deal, R.; Boucher, C.E. Creating and Restoring Wetlands: A Whole Ecosystem Experiment in Self-Design. BioScience **1998**, *48*, 1019–1030.

WETLAND AND CARBON SEQUESTRATION

Virginie Bouchard
Matthew Cochran
The Ohio State University, Columbus, Ohio, U.S.A.

INTRODUCTION

The increase in the concentration of "greenhouse gases" in the troposphere and its relation to human activities is now well-documented (1). Carbon fixation via photosynthesis and carbon release during decomposition have always been two important processes regulating the concentrations of CO_2 and CH_4 in the atmosphere, even in prehistoric and pre-industrial times. In addition to these internal processes, wetland ecosystems receive and release organic carbon through hydrologically-driven mass fluxes. Thus, the carbon pool within any wetland ecosystem is in balance between primary production, microbial decomposition, and carbon fluxes within interconnected ecosystems (Fig. 1). Wetlands play a particularly complex role in controlling greenhouses gases, as these ecosystems are intimately associated with all aspects of the production and consumption of both CO_2 and CH_4 (Fig. 2).

CARBON SEQUESTRATION WITH PRIMARY PRODUCTION

The concept of primary productivity is directly related to the ideas of energy flow in ecosystems. A portion of the photosynthetically active radiation (PAR), which is radiation in the 400–700 nm wave band, received by an ecosystem is absorbed by autotrophic organisms (photosynthetic plants and microorganisms). The absorbed energy is reradiated, lost as latent heat, or stored by the activity of photosynthesis in organic substances. This last flow of energy corresponds to net primary production (NPP). Fundamental ecological questions relating to the global carbon budget, the location of the missing carbon sink, and predictions of global climate change rely on obtaining good estimates of NPP.

Wetland ecosystems can have very high standing biomass values and correspondingly high NPP (2, 3). The annual aboveground NPP of macrophytes is reported to be up to 5 kg dry matter in the most productive sites (2). This production varies according to species, wetland type, and latitude (2, 3) and is often well correlated with the maximum aboveground standing crop. Belowground production is much more difficult to estimate because roots and rhizomes grow and die at different rates and times and because materials are translocated to and from shoots. The ratio of belowground to aboveground production can vary from 0.2 to 2.5 according to various studies (2). The connectivity of a wetland to hydrological fluxes (i.e., saltmarshes flooded by tides, coastal freshwater marshes flooded by seiche, riparian bottomland forests flooded by floods) is one of the most important factors enhancing primary productivity (4).

Net carbon capture from herbaceous vegetation is minimal compared to the long-term accumulation of carbon in bottomland hardwood forests. At the end of the growing season, most of the carbon trapped in the biomass of herbaceous macrophytes is found in dead plant litter and is released as CO_2 or CH_4 during decomposition or exported as dissolved or particulate organic matter to adjacent systems. Bottomland hardwood forests store carbon in tree biomass for a much longer period of time. The productivity of bottomland forest ranges between 200 and 2000 g dry organic matter per year (5).

CARBON SEQUESTRATION IN SOILS

Carbon fixed in photosynthesis either remains in the sediment as carbon accretion or is decomposed to CO_2 and CH_4 by a suite of fermentative microbes involved in soil organic matter decomposition. A diversity of biological, chemical, and physical mechanisms is also known to selectively "protect" different pools of soil organic matter from decomposition by soil microorganism. Most of the recalcitrant carbon destined for sediment accretion is derived from heavily lignified biomass. Organic soil is a result of the anaerobic conditions created by standing water or poorly drained conditions.

Carbon is even better protected in acidic environments, in marine sediments and under low temperatures. Peat accumulation is a result of reduced oxidation of the biomass produced in wetlands. The northern peatlands have accumulated 25–38 Pmol C since the last glaciation, equivalent to

Encyclopedia of Soil Science

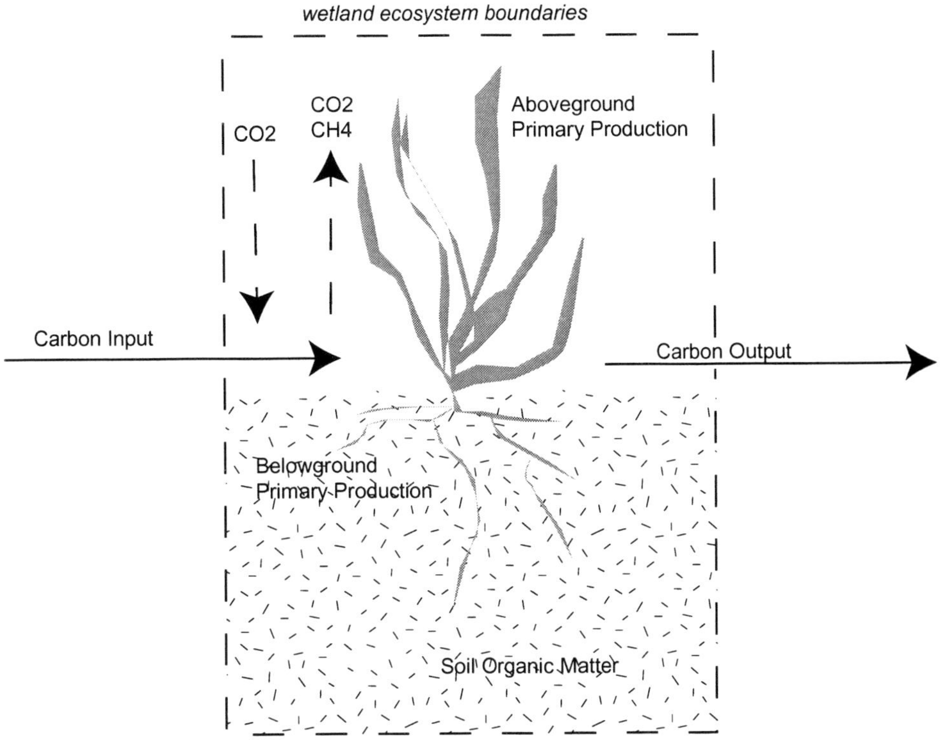

Fig. 1 Conceptual model showing the fluxes of organic carbon in wetlands.

50–70% of the total amount of carbon currently present in the atmosphere (6). Peat accumulation is greater in bogs than in fens due to differences in nutrient availability. Decomposition rates are enhanced in minerotrophic fens receiving nutrients from adjacent mineral soils relative to ombrotrophic bogs, which are fed only by rainwater.

Canadian wetland soils store about 154 Gt C, representing an average of 60% more carbon than what is stored in

Fig. 2 View of a freshwater wetland.

Canadian forests (95 Gt C in biomass and soils), which in turn is two orders of magnitude larger than the agricultural soil carbon pool (7) (Table 1). Peatlands are a sink for between 20 and 30 $g\,C\,m^{-2}\,yr^{-1}$, which means that Canadian peatlands sequester about 0.03 $Gt\,C\,yr^{-1}$ (Table 1). The sequestration rate of peatlands is currently smaller than the rate computed for boreal forests. However, the major difference between forests and peatlands is that a forest sink is transient and cannot be sustained continuously. If peatlands also have a theoretical limit to growth, it is reached only after many thousands of years (8). However, the fluxes of CO_2 and CH_4 from peatlands vary highly according to seasons, years, species and sites. Inter-annual measurement schemes—such as those that have been developed for forest ecosystems with the Ameriflux and Euroflux programs—are needed are needed to establish more precise fluxes of CO_2.

IMPACT OF HUMAN ACTIVITIES ON CARBON STORAGE

Rates of plant production are limited by phosphorus supply in freshwater systems and by nitrogen supply in saltwater and terrestrial systems (9, 10). An excess of

Table 1 Summary of carbon stocks and fluxes in Canadian wetlands/peatlands, forests and agricultural fields

Land use	Area (km^2)	C stock (Gt C)	Fluxes (Gt C yr^{-1})	Potential duration (years)
Peatlands/wetlands soils	1.24×10^6	154	−0.03	1000 years (?)
Forests	5.15×10^6	95	−0.2	50
Agricultural fields	0.45×10^6	6	0.0034	50

(Adapted from Ref. 7.)

nutrients leads to an increase of both primary production and decomposition in terrestrial systems (11). Decomposition rates increase, however, at a slower rate than primary production, leading to a potential increase of carbon storage in soil (11). In wetlands, little is known about the effect of excess nutrients on the denitrification process, which is directly linked to the amount of soil organic matter (12).

Wetlands have been drained on all continents due to human development. Rapid changes occur in organic soils after drainage as aerobic microbial decomposition is enhanced, releasing large amounts of CO_2 into the atmosphere. Carbon losses can lead to subsidence of the soil profile, changes in bulk density, and decreases in carbon content of the remaining soil.

CREATION OF WETLANDS TO SEQUESTER CARBON

Wetlands—particularly bottomland hardwood forests—are considered as potentially excellent "carbon sinks" because they take carbon dioxide out of the atmosphere and store it in living plant tissues and soil organic matter. Wetland creation might provide an opportunity to positively address wetland habitat losses while also addressing global warming (8, 13). However, a question, still remains: are we capable of constructing wetlands that function in similar ways as natural wetlands? How well these created and constructed wetlands mimic natural wetlands is still being debated (14, 15).

Various studies have indicated that created wetlands might not function in the same capacity as adjacent reference wetlands (16). In created *Spartina alterniflora* salt marshes, vegetation rapidly achieves 100% cover, although soil nitrogen and organic matter are slow to accumulate. Salt marshes constructed in North Carolina 25 years ago have lower soil organic carbon (SOC) and lower total N reservoirs than a 2000-year-old natural marsh (16). Their C accumulation rates are similar to those of reference sites, but N accumulation rates are higher, thus C/N ratios have declined over time. In Oregon, 95 restored freshwater marshes had lower soil organic matter than natural marshes and no evidence of accumulation (17).

Because bottomland hardwood forests combine both a long-term carbon storage in tree biomass and a slow release of soil organic carbon under flooded conditions, they are considered as excellent ecosystems that could potentially be used to sequester carbon. Restoration of bottomland hardwood forests on marginal agricultural lands in the Mississippi River Valley offers significant net carbon sequestration in the south-central region of the U.S. (13). However hardwood plantings have problems with slow initial growth and excessive early mortality rates. The use of herbaceous weed control and bedding may offer the potential to overcome these difficulties (13). Under the initiative of the U.S. Department of Energy, an 80 acre pilot study is currently underway in Louisiana. Abandoned marginally productive agricultural fields were planted in January 1997 with seven bottomland hardwood species. Following planting, the site will be monitored for planting success and survival. When the trees attain heights >4.5 ft, growth data and carbon sequestration per acre will be calculated (13).

CONCLUSION

The primary production in wetland ecosystems is greatly enhanced by hydrological fluxes, which leads to high biomass, primary production and accretion rates. The carbon that is either fixed stored in standing biomass, released by soil decomposing microorganisms, or stored in soil sediments. With the continual destruction of wetlands worldwide, more of this sequestered carbon is released to the atmosphere. The protection of wetlands will preserve the amount of carbon already stored in these ecosystems.

The creation and restoration of new wetlands will certainly contribute to sequestration of carbon and should be considered as one way to mitigate greenhouse emissions.

REFERENCES

1. IPCC (Intergovernemental Panel on Climate Change), Climate Change 1994. In *Radiative Forcing of Climate Change and an Evaluation of the IPCC IS92 Emission Scenarios*; Houghton, J.T., Meira Filho, L.G., Bruce, J., Hoesung, L., Callander, B.A., Haites, E., Harris, N., Maskell, K., Eds.; Cambridge University Press: Cambride, UK, 1995.
2. Westlake, D.F., Kvet, J., Szczepanski, A., Eds. *The Primary Ecology of Wetlands*; Cambridge University Press: Cambridge, 1998; 568.
3. Mitsch, W.J.; Gosselink, J.G. *Wetlands*, 3rd Ed.; Wiley: New York, 2000; 920.
4. Odum, W.E.; Odum, E.P.; Odum, H.T. Natures Pulsing Paradigm. Estuaries **1995**, *18* (4), 547–555.
5. Conner, J. Effects of Forest Management Practices on Southern Forested Wetland Productivity. Wetland **1994**, *14*, 27–40.
6. Gorham, E. Northern Peatlands: Role in the Carbon Cycle and Probable Responses to Climatic Warming. Ecol. Appl. **1991**, *1*, 182–195.
7. Roulet, N.T. Peatlands, Carbon Storage, Greenhouse Gases, and the Kyoto Protocol: Prospects and Significance for Canada. Wetlands **2000**, *20* (4), 605–615.
8. Clymo, R.S. Models of Peat Growth. SUO **1993**, *43*, 127–136.
9. Tamm, C.O. *Nitrogen in Terrestrial Ecosystems*; Springer: Berlin, 1991.
10. Vitousek, P.M.; Howarth, R.W. Nitrogen Limitation on Land and in the Sea: How Can it Occur? Biogeochemistry **1991**, *13*, 87–115.
11. Vitousek, P.M.; Aber, J.D.; Howarth, R.W.; Likens, G.E.; Matson, P.A.; Schindler, D.W.; Schlesinger, W.H.; Tilman, D.G. Human Alterations of the Global Nitrogen Cycle: Sources and Consequences. Ecol. Appl. **1997**, *7* (3), 737–750.
12. Davidsson, T.E.; Stahl, M. The Influence of Organic Carbon on Nitrogen Transformation in Five Wetland Soils. Soil Sci. Soc. Am. J. **2000**, *64*, 1129–1136.
13. Williams, J.R. Addressing Global Warming and Biodiversity Through Forest Restoration and Coastal Wetlands Creation. Sci. Total Environ. **1999**, *240*, 1–9.
14. Mitsch, W.J.; Wu, X.; Nairn, R.W.; Weihe, P.E.; Wang, N.; Deal, R.; Boucher, C.E. Creating and Restoring Wetlands: A Whole Ecosystem Experiment in Self-Design. BioScience **1998**, *48*, 1019–1030.
15. Zedler, J.B. Progress in Wetland Restoration Ecology. Trends Ecol. Evol. **2000**, *15* (10), 402–407.
16. Craft, C.; Reader, J.; Sacco, J.N.; Broome, S.W. Twenty-Five Years of Ecosystem Development of Constructed Spartina Alterniflora (Loisel) Marshes. Ecol. Appl. **1999**, *9* (4), 1405–1419.
17. Shaffer, P.; Ernst, T. Distribution of Soil Organic Matter in Freshwater Emergent/Open Water Wetlands in the Portland, Oregon Metropolitan Area. Wetlands **1999**, *19*, 505–516.

WETLAND CONSERVATION, POLICY FOR

Clayton Rubec
Habitat Conservation Division, Canadian Wildlife Service, Environment Canada, Ottawa, Ontario, Canada

INTRODUCTION

Meeting the challenge of conserving wetlands requires comprehensive national policies to provide a foundation for domestic action and a framework for international and national cooperation. Such policy is valuable as countries seek to address the management and habitat requirements for wildlife and other natural resources, such as soil and water, as well as human needs. Implementation of a national wetland policy is a key feature of the wise use principles of the Convention on Wetlands of International Importance (the "Ramsar Convention"). However, such policy remains an elusive goal for many of the 128 nations that, as of October 2001, are Contracting Parties to this global environmental treaty.

Responding to recommendations by the Convention, in 1998–1999 the author led a team of writers in preparing *Guidelines for Developing and Implementing National Wetland Policies* (1). These policy guidelines complement the Convention's guidance on wetland legislation (2). Highlights of the *Guidelines* and reports on the status of wetland policy development around the World, effective mid-1999, are presented here.

RELATIONSHIP BETWEEN POLICY AND WISE USE

The *Wise Use Principles* are a hallmark of the Ramsar Convention. "Wise use" applies not only to sites listed as Wetlands of International Importance (as of October 2001 covering over 87.2 million hectares at 1096 sites), but to all wetlands in the territory of Contracting Parties. These *Principles* help Contracting Parties improve institutional and organizational arrangements, address legislative and policy needs, increase knowledge and awareness of wetland values, inventory and monitor the status of wetlands, identify program priorities, and develop action plans for specific sites as components of a national wetland policy.

The formulation of national wetland policies sometimes involves a lengthy and complex process. Political, jurisdictional, institutional, legal, and financial constraints affect policy formulation, in addition to social and economic factors that continue to contribute to wetland loss while the policy process is underway.

WHAT ARE WETLAND POLICIES AND ARE THEY NEEDED?

A policy can simply be a document. However, policy making is also a process involving consensus-building, encapsulation of ideas and commitments, implementation, accountability, and review complemented by legislation, strategies, and operational programs. It is a mechanism for an administration to capture the public will or mandate on an issue and refine it with its own vision.

A national wetland policy is nationwide in scope, but it may be developed at several levels of government. In Australia and Canada, for example, both the federal government and state/provincial governments have developed wetland conservation policies. This reflects the federal nature of these two nations, wherein constitutional authority for natural resources management (including wetlands) is divided between the levels of government.

Wetlands are seldom explicitly covered in other natural resource management policies such as for water, soil, forest, land, or agriculture at a national level. Development of a "stand alone" wetland policy and/or strategy can be an important step in recognizing and solving wetland problems. A wetland policy recognizes wetlands as ecosystems requiring different management approaches not masked under other sectoral management objectives. Articulation of goals and objectives for these ecosystems identifies clear responsibilities of the Government and a public expectation that the Government will deliver these commitments.

Wetland policy objectives need to focus on a variety of themes as they become the image of the policy. However, practical implementation of the policy may result in only one or two of these objectives receiving the greatest public attention. For example, Canada's announcement of its

Encyclopedia of Soil Science

federal wetland policy in 1992 contained seven objectives, but "no net loss of wetland functions" has proven to be its catch phrase.

THE GUIDELINES

The *Guidelines* review the key steps and issues that may arise in both developing and implementing a National Wetland Policy. These include over 20 detailed sections, including defining the purpose of such an initiative, organizing a suitable process, deciding how to present the content of the policy document, and developing strategies for implementation and monitoring. The text is complemented by seven Wetland Policy Essays: 1) *Defining Stakeholders*, 2) *Consultations*, 3) *Wetland Policies Within a Federal State*, 4) *Sectoral Policies and Legislation*, 5) *Compliance Strategies*, 6) *Role of Non-Government Organizations*, and 7) *Development and Coordination Process*. The *Guidelines* provide a reference against which all nations can review their wetland action plans and strategies at the national level.

A team of contributors with governmental or non-governmental work experience and expertise in wetland policy development prepared the *Guidelines*. The team included writers from Ramsar national authorities in Australia, Canada, Trinidad and Tobago, Uganda, and the United States of America. Contributors from BirdLife International, University of Massachusetts, IUCN Environmental Law Centre, and Wetlands International were also involved.

Implementation Strategies

A national wetland policy includes specific implementation strategies that demonstrate the priorities of the Government but also foster the cooperation and involvement of other interests. Linkages between these strategies and national water, soil, biodiversity, and sustainable development policy initiatives are explored in the *Guidelines*.

An analysis of the strategies used in selected national wetland policies is summarized in Table 1. These include the Policies/Action Plans of Australia, Canada, Costa Rica, France, Jamaica, Malaysia, Peru, Trinidad and Tobago, and Uganda. These initiatives have many common strategic approaches including: a) ensuring public awareness and education, b) developing cooperation and partnerships between levels of government from national to local, c) developing supporting legislation and

Table 1 Implementation strategies in selected national wetland policies

Country	1	2	3	4	5	6	7	8	9	10	11	12
Australia	X	X	X	X	X	X						
Canada	X		X	X	X	X	X					
Costa Rica	X		X		X			X	X	X		X
France		X	X		X						X	X
Jamaica		X					X	X	X			
Malaysia		X	X		X	X	X	X	X	X		X
Peru			X		X	X		X	X			X
Trinidad and Tobago	X		X	X	X		X	X				
Uganda		X	X		X	X		X	X		X	X

Policy strategies:
1. Management of national wetland networks
2. Integration with other policies such as water, soil, forests
3. Public awareness and education
4. Partnerships
5. Science, monitoring, assessment, and research
6. International commitments
7. Managing special sites
8. Administration and institutions, capacity building
9. Enforcement, regulation, and legislation
10. Financial mechanisms
11. Restoration of degraded sites
12. Sustainable use and conservation

interrelated land and water use policies and programs, d) implementing wetland site management responsibilities, e) developing a sound basis for the policy through scientific research and expertise, f) developing institutional and financial capacity for policy implementation, and g) meeting international commitments. These strategies have been drafted to evoke a clear vision and acceptance across the nation.

Global Review of the Status of Wetland Policies and Strategies

Significant progress is evident globally in the development of national wetland policies since the Ramsar Convention focused attention on this issue in 1987 (1, 3). Meetings of the Contracting Parties every three years allow regular review of the status of wetland policies. The Contracting Parties last met in May 1999 when there were 114 Contracting Parties, while now (October 2001) there are 128 Contracting Parties.

Policies

As of May 1999, 42 of the 114 Ramsar Contracting Parties indicated they were engaged in development or implementation of a national wetland policy. Over the 1987 to 1999 period, the number of nations with an officially adopted national wetland policy grew from zero to 12. An additional 30 nations indicated that a National Wetland Policy was in draft or under consideration. However, 72 Ramsar Contracting Parties did not yet report any actions being taken in support of national wetland policy development. A number of nations, particularly those with a commonwealth or federal make-up, reported wetland policies and strategies at the subnational level also.

Strategies

National wetland strategies were considered separately from national wetland policies. As of May 1999, some 94 of 114 nations were engaged in national wetland strategy initiatives. Over the 1987 to 1999 period, the number of nations that indicated they have implemented a national wetland strategy (also called "action plans") had grown from four to 48. An additional 46 nations had strategies in draft or under consideration. Only 20 Contracting Parties did not report any steps toward developing such a strategy.

Table 2 summarizes the status of the development and adoption of national wetland policies or strategies from 1987 through May 1999. This table was developed by reviewing the reports and conference papers that summarize Convention activities every three years by each country. In 1987, only five nations indicated they were involved in any sort of national wetland policy or strategy; by 1999, this had grown to at least 94 nations. Table 3 summarizes the countries by Ramsar's regions (Africa, Asia, Europe, Neotropics, North America, and Oceania) that have either a national wetland policy or

Table 2 Evolution of Ramsar Convention national wetland policies and strategies[a]

Status of national wetland policies or strategies	1987 Regina COP3	1990 Montreux COP4	1993 Kushiro COP5	1996 Brisbane COP6	1999 San José COP7
National wetland policies:					
Policy adopted	0	0	3	6	12
Policy in draft or under consideration	0	2	12	21	30
No policy activities reported	17	43	36	65	72
National wetland strategies:					
Strategy adopted	4	9	9	35	48
Strategy in draft or under consideration	1	1	9	20	36
No strategy activities reported	12	35	33	37	20
Number of national reports tabled out of total contracting parties at this COP	17 of 35	45 of 60	51 of 76	92 of 92	98 of 114

[a] As of October 2001, the Ramsar Convention had 128 Contracting Parties and will next report on this issue in late 2002.

W

Table 3 Summary of national wetland policies and strategies by Ramsar Regions

Ramsar region	National wetland policy or strategy adopted by government	National wetland policy or strategy in preparation or under consideration
Africa	Congo Rep., Côte d'Ivoire, Senegal, Uganda	Algeria, Botswana, Burkina Faso, Chad, Comoros, Dem. Rep. Congo, Egypt, The Gambia, Ghana, Guinea, Kenya, Malawi, Mali, Morocco, Namibia, Niger, South Africa, Togo, Zambia
Asia	Indonesia, Japan, Thailand	Bangladesh, Cambodia, P.R. China, Georgia, India, Rep. Korea, Malaysia, Mongolia, Philippines, Russia,[a] Turkey, Vietnam
Europe	Belgium, Bulgaria, Denmark, Estonia, Finland, France, Greece, Iceland, Malta, Monaco, Netherlands, Norway, Romania, Sweden, Switzerland, United Kingdom	Austria, Belarus, Croatia, Czech Rep., Germany, Hungary, Ireland, Italy, Latvia, Lithuania, Poland, Portugal, Russia,[a] Slovak Rep., Slovenia, Ukraine, F.R. Yugoslavia
Neotropics (Central and South America and Caribbean)	Costa Rica, Peru	Argentina, Bahamas, Chile, Colombia, Ecuador, Guatemala, Honduras, Jamaica, Nicaragua, Panama, Paraguay, Trinidad and Tobago, Venezuela
North America	Canada, United States	Mexico
Oceania (Australia and Pacific)	Australia, New Zealand	–

[a] Russia straddles Asia and Europe.

national wetland strategy adopted, being drafted, or considered up to May 1999.

CONCLUSION

Development and implementation of national wetland policy in the 128 countries that have acceded to the Ramsar Convention is proceeding throughout the world. The Convention's Wise Use Principles and Guidelines on National Wetland Policy (1), complemented by guidelines for wetland legislation (2), are effective tools in fostering the completion and use of national wetland policies and strategies as important cornerstones of this Convention.

One of the most interesting aspects of the Ramsar Convention is its capacity to foster sharing of experience. Interchanges by wetland policy experts are now occurring internationally that involve short-term invited visits or sabbaticals, informal exchange of documents, confidential advice, and review of draft policy. Consultation workshops, working with nongovernment groups, meeting with senior government officials, exploring funding mechanisms, and assistance in drafting of text have also been involved. Experience gained in one nation's development of these policies is able to be shared and local expertise enhanced, filling a need among the Ramsar family.

REFERENCES

1. Rubec, C.D.A.; Pritchard, D.; Mafabi, P.; Nathai-Gyan, N.; Phillips, B.; Mahy, M.; Lynch-Stewart, P.; Chew, R.; Cintron, G.; Larson, J.; Ramakrishna, S. *Guidelines for Developing and Implementing National Wetland Policies*, Report to COP7 Ramsar Convention. Handbook No. 2; Ramsar Bureau; Gland: Switzerland, 2000.
2. Shine, C.; Glowka, L. *Reviewing Laws and Institutions to Promote the Conservation and Wise Use of Wetlands*, IUCN Environmental Law Centre Report to COP7 Ramsar Convention. Handbook No. 3; Ramsar Bureau; Gland: Switzerland, 2000.
3. Rubec, C.D.A. Status of National Wetland Policy Development in Ramsar Nations. In *Proceedings, Sixth Meeting of the Conference of the Contracting Parties*; Convention on Wetlands: Brisbane, Australia, 1996; Vol. 10/12A, 22–29.

WETLANDS, ECONOMIC VALUE OF

Gayatri Acharya
The World Bank, Washington, D.C., U.S.A.

INTRODUCTION

Wetlands have been centers of human civilization throughout history. For example, the Mekong and Red River deltas of Vietnam together comprise 16% of the country but support 42% of the population (1). Yet, wetlands have been historically transformed or converted and their capacity diminished in some manner by the impacts of land use changes. OECD (2) noted that: "The world may have lost 50% of the wetlands that existed since 1900; whilst much of this occurred in the northern countries during the first 50 years of the century, increasing pressure for conversion to alternative land use has been put on tropical and subtropical wetlands since the 1950s. By 1985, it was estimated that 56–65% of the available wetland had been drained for intensive agriculture in Europe and North America; the figures for tropical and subtropical regions were 27% for Asia, 6% for South America, and 2% for Africa, making a total of 26% worldwide."

The conversion of multiple use ecosystems to single land uses may signify a welfare loss through reduced access to, or availability of, valuable goods and services. Whether or not such conversions are merited can be assessed only with sufficient information on the relative efficiency of converting or conserving the wetland ecosystem vis-à-vis alternative land uses. Economic values are a measure of how much people are willing to pay for, and how much better or worse off they would consider themselves to be as a result of changes in the supply of different goods and services. Valuation provides a means of quantifying the benefits that people receive from wetlands. The net benefits of investing in land uses that are compatible with wetlands conservation, relative to those economic activities, which contribute to wetlands degradation, can then be assessed. Values generated through economic analysis are limited however by: a) their inability to capture intrinsic values associated with wetlands; and b) the level of understanding, both economists and natural scientists have with regard to the level of use of goods and services, and the variation in their availability with fluxes in the natural system. Furthermore, economic values measure preferences and welfare impacts associated with changes in the natural system or the availability of a service or product; they do not, in themselves, suggest anything about inter- or intra-generational distributional effects of conservation or development investments.

Wetland goods and services are also particularly difficult to value. This is because: a) many goods are not marketed but traded or consumed directly; b) wetland services, such as water quality or groundwater recharge often occur in areas away from the physical location of the wetland and may not be easily attributable to the wetland; c) many wetlands are transboundary resources and data on the use and consumption of goods and services are difficult to obtain; and d) many wetlands are public property. For these and other related reasons, the economic benefits generated by wetlands, and the economic costs associated with wetlands degradation or loss, are frequently unknown and omitted in project or policy analysis. As a result, the potential of wetlands to be used as contributors to economic growth, income generating activities, and as sources of goods and services, has been underestimated in many parts of the world, resulting in the loss of valuable species, services, and livelihoods.

ECONOMIC VALUES OF WETLANDS

Wetlands cover an estimated 6% of the world's land surface and may be of various types such as marshes, floodplains, freshwater ponds, lakes, swamps, etc. While these ecosystems are associated with a diverse and complex array of direct and indirect uses, the precise functions and benefits of a specific wetland can be determined only after careful study of the hydrological, physical, and biological characteristics of the site, over a period of time (3). Since the economic value of an ecosystem service (or good) is determined by the contribution it makes to maintaining the present level of human well being, it is calculated as a measure of a change in well being, or *social welfare*. Therefore, valuing a good or service requires us to study the change in a person's welfare due to a change in the availability of the resource. Economic value expressed in currency units is the monetary expression of the tradeoff between identified alternative land uses (see Ref. 3, among others, for more on valuation theory).

To fully assess these tradeoffs, the ultimate goal of valuation should be to assess the total economic value (TEV) of a system. Total economic value is defined as the sum of use value+existence value where use value is comprised of a) direct use value (consumptive or nonconsumptive); b) indirect use value; and c) option value. Box 1 defines each of these values while Fig. 1 shows how direct and indirect benefits can be identified in a wetland system. While valuation attempts to take account of all the components of the total economic benefit of wetlands, it is generally constrained by data availability and, therefore, partial valuation studies are more common (see Ref. 7 for more on the differences in total and partial valuation).

Direct uses include the use of the wetland for water supply and harvesting of wetland products such as fish and plant resources. Commonly recognized values of wetlands include those captured from the use of raw materials and physical products that are generally used in economic activities such as fisheries, water supply, and agriculture. Where wild foods, grasses, bird life, and other naturally occurring species are exploited for significant economic returns, such uses are also easily identifiable and may be valued in existing markets.

Direct uses however represent only a small proportion of the total value of wetlands, which generate economic benefits in excess of just physical products. *Indirect benefits* are derived from environmental functions that wetlands may provide, depending on the type of wetland, soil, and water characteristics, and associated biotic influences (5). Wetland functions are defined as a process or series of processes that take place within a wetland, including habitat and hydrological and water quality functions. Such functions have an economic value if they contribute to economic activity and enhance human welfare.

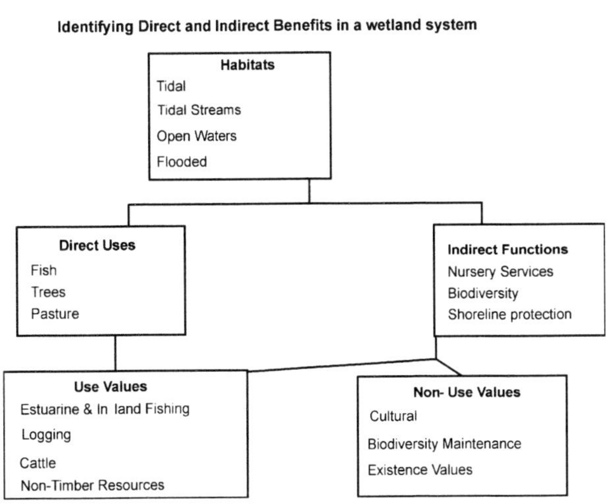

Fig. 1 Identifying direct and indirect benefits in a wetland system.

Box 1: Types of benefits valued by economic valuation techniques.

Direct benefits: These include the raw materials and physical products that are used directly for production, consumption, and sale including those providing energy, shelter, foods, water supply, transport, and recreation.

Indirect benefits: These include ecological functions which maintain, protect, and support natural and human systems through services such as maintenance of water quality, flow and storage, flood control and storm protection, nutrient retention, and micro-climate stabilization, and other productive and consumptive activities.

Option benefits: These refer to the premium placed on maintaining a pool of wetlands species and genetic resources for future uses such as for leisure or in commercial, industrial, agricultural and pharmaceutical applications and water-based developments, some of which may not be used or known in the present but may be used in the future.

Existence benefits: The intrinsic value of wetlands species and areas regardless of their current or future use possibilities, such as cultural, aesthetic, heritage and bequest significance.

Valuing Wetlands Goods—Techniques and Examples

A combination of the available techniques used to value environmental goods and services can be used to approximate a partial or total value of any one ecosystem, taking care that we can identify the various components of the system and have adequate economic and ecological data to do so. The following methods have been used in valuing wetland resources:

- Market prices: to estimate economic values for ecosystem products or services that are bought and sold in markets.
- Change in productivity: to estimate economic values for ecosystem products or services that contribute to the production of marketed goods or services.
- Travel cost: to estimate economic values associated with ecosystems/sites used for recreation based on the assumption that the value of a site is reflected in how much people are willing to pay to travel to visit the site.
- Hedonic pricing: to estimate economic values for ecosystem or environmental services that directly affect market prices of some other good, e.g., housing prices may reflect the value of local environmental attributes or amenities.

- Contingent valuation method: to estimate economic values for products and ecosystem services. This can be used to estimate nonuse values by asking people to directly state their willingness to pay for specific environmental services, based on hypothetical scenarios or contingent markets.
- Contingent choice, contingent ranking, contingent behavior: to estimate economic values for products and for ecosystem services by asking people to make tradeoffs between sets of ecosystem or environmental services or characteristics.
- Benefit transfer: to infer economic values by transferring benefit estimates from localized studies to similar areas under similar market conditions.
- Damage cost avoided, replacement cost, and substitute costs: to estimate economic values based on costs of avoided damages resulting from lost or diminished ecosystem services, replacing ecosystem services, or providing substitute services.

Examples of how these techniques have been applied to elicit values of various wetland benefits are given in Boxes 2 and 3 and in Table 1. These studies illustrate the range of values associated with different types of goods and services, wetlands provide.

In some parts of the U.S., the benefits of protecting natural watersheds to assure safe and plentiful drinking water supplies are being recognized. The city of New York recently found that it could avoid spending U.S. $6–8 billion in constructing new water treatment plants by protecting the upstate watershed that has traditionally

Box 2. Using contingent valuation and travel cost methods to assess the recreational value of Lake Nakuru, Kenya.

Lake Nakuru National Park is an important international tourist destination in Kenya. Fees currently charged to enter the park underestimate the total value that tourists place on the wetland and its component species, especially flamingos. A travel cost survey of visitors elicited information about length of stay, travel costs, place of origin, and visitation rates, distinguishing between resident and nonresident tourists. The contingent valuation survey used in this study asked visitors what their personal total costs of travel were; how much they would be willing to increase their expenditures to visit the park; how much they would contribute to a fund to clean-up and control urban pollution affecting the park; and how much they would contribute to a project to conserve flamingos. The survey also asked respondents for the minimum reduction in trip costs that they would be willing to accept should there be no flamingos. Results suggests that the annual recreational value of wildlife viewing in Lake Nakuru National Park was between U.S. $7.5 and $15 million and over a third of this value was accounted for by the willingness to pay for viewing flamingos.

(Adapted from Ref. 6.)

Table 1 Economic values of wetlands: some examples

Valuation approach	Good or service valued	Reference	Value	Location
Revealed willingness to pay (WTP)	Commercial fishing and trapping	(18)	$486.25 per acre	Louisiana, U.S.
Revealed WTP (mitigation costs)	Freshwater forested wetlands Freshwater emergent wetlands	(23)	$78,000 per acre $49,000 per acre	Maryland, U.S.
Contingent valuation, replacement cost	Nitrogen abatement	(21)	U.S. $59/kg N reduction capacity	Gotland, Sweden
Loss in productivity, market prices	Agriculture, forestry fishing from floodplain;	(7)	N109 (U.S. $15)/$10^3$ m^3 N381 (U.S. $51)/h	Hadejia–Nguru
Production function, contingent behavior	Groundwater recharge function (drinking water supply)	(9)	U.S. $13,029 per day for the wetlands	Wetlands, Nigeria
Production function	Groundwater recharge function (agriculture)	(8)	$62/h	
Market Prices	Artisanal fisheries, aquarium fish, sea cucumbers, shells, tourism	(24)	U.S. $2.14 million/yr	Kisite Marine National Park and Mpunguti Marine National Reserve, Kenya
Preventive expenditure	Coastal zone/shoreline protection	(25)	U.S. $603,248/yr	Mahe, Seychelles

W

Box 3. Valuing the Hadeija Nguru Wetlands, Nigeria.

In the Hadeija–Jama' are floodplain region in northern Nigeria, more than one half of the wetlands have already been lost to drought and upstream dams. Ecosystem valuation has been used in this area to weigh the costs and benefits of development projects that would divert some more water away from the floodplain for irrigated agriculture in upstream areas. The net benefits of such a diversion are estimated at U.S. $29 per hectare. In comparison, the floodplain, under the present flooding regime, provides U.S. $167 per hectare in benefit to a wider range of local people engaged in farming, fishing, grazing livestock, or gathering fuelwood and other wild products—benefits that would be greatly diminished by the project.

A study of the groundwater recharge function of the wetlands confirms furthermore that the wetlands play an important role by maintaining groundwater recharge in the floodplain. Groundwater recharge supports irrigated agricultural production in the floodplain. Irrigated agriculture using water from the shallow groundwater aquifer has a value of 36,308 Naira (U.S. $413) per hectare for the study area. A value of at least 2863 Naira or U.S. $32.5 per farmer per dry season or U.S. $62 per hectare is attributable to the present rate of groundwater recharge. In terms of maintaining water supply resources, the value of the recharge function is 1,146,588 Naira or U.S. $13,029 per day for the wetlands.

(Adapted from Refs. 7–9.)

provided these purification services. Based on this assessment, the city invested U.S. $1.5 billion in protecting its watershed by buying land around reservoirs and providing other incentives for land use (10). This decision ensures that in addition to getting a relatively cheap treatment system for its water supply, the city also maintains the watershed for recreation, wildlife habitat, and other ecological benefits derived from a healthy ecosystem.

CONCLUSIONS

Some caveats to valuation also apply. In a number of recent papers (11–13), researchers have attempted to value the world's ecosystems, some by extrapolating localized studies of specific resource losses to capture the impact of a global loss of such resources. Even if local values are calculated using the best possible data, extrapolating these values to all wetland resources on the planet creates problem not least because there is wide variation in the types of wetlands and their component products. Heimlich et al. (14) presents the results of over 30 studies carried out on wetlands. After accounting for geographical variation and other differences across the studies by calculating the present value over a 50-year period discounted at 6% for all the studies and then converting these values to an annual basis, they find significant variation among the different studies. For example, general recreational values range from $105 to $9859 per acre; waterfowl hunting from $108 to $3101 per acre; and recreational fishing from $95 to $28,845 per acre. Table 1 gives a few examples of various studies and types of goods or services valued. As the figures show, there is a wide variation amongst the types of wetlands valued and the values attributable to them. Great caution must therefore be exercised when extrapolating localized studies to other geographical areas. Resource values are influenced by local consumptive and productive use patterns, market conditions, and the institutional arrangements affecting the use of the resources. The good news is that there are now hundreds of studies of wetland values [See Ref. (15) for an extensive review] that we can refer to for some guidance on the range of values associated with wetlands (16–22).

Benefits of some wetlands will always be difficult to quantify and measure primarily because the required scientific, technical, or economic data is difficult to obtain and also that certain intrinsic values are not measurable by existing economic valuation methods. However, as the studies reported in this paper suggest, various wetland goods and services are extremely valuable and in many cases, a measurable economic value can be obtained. The economic value of ecological functions in sustaining the livelihood and cultures of human societies clearly cannot be disregarded in development and conservation policy since these values allow us to make informed decisions about tradeoffs and help in the conservation of natural resources to enhance human welfare. Hence, valuation studies need to be carried out wherever possible and with collaboration between economists and natural scientists to fully capture the economic and ecological linkages that make wetlands valuable.

REFERENCES

1. Dugan, P.J. *Wetland Conservation: A Review of Current Issues and Required Action*; IUCN: Gland, Switzerland, 1990.
2. OECD/IUCN, *Guidelines for Aid Agencies for Improved Conservation in Sustainable Use of Tropical and Subtropical Wetlands*; OECD: Paris, 1996.
3. Mitsch, W.J.; Gosselink, J.G. *Wetlands*, 2nd Ed. Van Nostrand Reinhold: New York, 1993.
4. Freeman, A.M. *The Measurement of Environmental and Resource Values: Theory and Methods*; Resources for the Future: Washington, DC, 1993.
5. Maltby, E. *Waterlogged Wealth*; Earthscan: London, 1986.

6. Navrud, S.; Mungatana, E. Environmental Valuation in Developing Countries: The Recreation Value of Wildlife Viewing. Ecological Economics **1994**, *11*, 135–151.
7. Barbier, E.B.; Adams, W.; Kimmage, K. Economic Valuation of Wetland Benefits. In *The Hadejia–Nguru Wetlands*; Hollis Ed.; IUCN: Gland, 1993.
8. Acharya, G.; Barbier, E.B. Valuing Groundwater Recharge Through Agricultural Production in the Hadejia–Nguru Wetlands in Northern Nigeria. Agricultural Economics **2000**, *22*, 247–259.
9. Acharya, G. Valuing the Hidden Hydrological Services of Wetland Ecosystems. Ecological Economics **2000**, *35*, 63–74.
10. National Research Council, (NRC), Committee to Review the New York City Watershed Management Strategy. *Watershed Management for Potable Water Supply: Assessing the New York City Strategy*; National Academy Press: Washington, DC, 2000.
11. Costanza, R.; d'Arge, R.; de Groot, R.; Farber, S.; Grasso, M.; Hannon, B.; Limburg, K.; Naeem, S.; O'Neill, R.V.; Paruelo, J.; Raskin, R.G.; Sutton, P.; van den Belt, M. The Value of the World's Ecosystem Services and Natural Capital. Nature **1997**, *387*, 253–260.
12. Pimentel, D.; Wilson, C.; McCullum, C.; Huang, R.; Dwen, P.; Flack, J.; Tran, Q.; Saltman, T.; Cliff, B. Economic and Environmental Benefits of Biodiversity. BioScience **1997**, *47* (11).
13. Ehrlich, P.R.; Ehrlich, A.H. *Betrayal of Science and Reason: How Anti-Environmental Rhetoric Threatens Our Future*; Putnam: New York, 1996.
14. Heimlich, R.E.; Wiebe, K.D.; Claassen, R.; Gadsby, D.; House, R.M. *Wetlands and Agriculture: Private Interests and Public Benefits*; Agricultural Economic Report No. 765, Resource Economics Division, Economic Research Service, USDA, 1998.
15. Bardecki, M. Wetlands and Economics: An Annotated Review of the Literature, 1988–1998 with Special Reference to the Wetlands of the Great Lakes. *Report Prepared for Environment Canada—Ontario Region*, 1998; http://www.on.ec.gc.ca/glimr/data/wetland-valuation/intro.html.
16. Hammack, J.; Brown, G.M. *Waterfowl and Wetlands: Toward Bioeconomic Analysis*; Resources for the Future: Washington, DC, 1974.
17. Lynne, G.D.; Conroy, P.; Prochaska, F.J. Economic Valuation of Marsh Areas for Marine Production Processes. Journal of Environmental Economics and Management **1981**, *8*, 175–186.
18. Farber; Costanza The Economic Value of Wetland Systems. Journal of Environmental Management **1987**, *24*, 41–51.
19. Bell, F. The Economic Valuation of Saltwater Marshes Supporting Marine Recreational Fishing in South-Eastern United States. Ecological Economics **1997**, *21*, 243–254.
20. Bergstrom, J.C.; Stoll, J.R.; Titre, J.P.; Wright, V.L. Economic Value of Wetlands Based Recreation. Ecological Economics **1990**, *2*, 129–147.
21. Gren, I.M.; Folke, C.; Turner, K.; Bateman, I. Primary and Secondary Values of Wetland Ecosystems. Environmental and Resource Economics **1994**, *4* (1), 55–74.
22. Morrison, M.D.; Bennett, J.W.; Blamey, R.K. *Valuing Improved Wetland Quality Using Choice Modelling*; Choice Modelling Research Report No. 6, University College, The University of New South Wales: Canberra, 1998.
23. Bohlen, C.; King, D. In *Towards a Sustainable Coastal Watershed: The Chesapeake Experiment*; Proceedings of a Conference, CRC Publication No. 149, Nelson, S., Hiss, P., Eds. Chesapeake Research Consortium: Edgewater, MD, 1995.
24. Emerton, L.; Tessema, Y. *Economic Constraints to the Management of Marine Protected Areas: The Case of Kisite Marine National Park and Mpunguti Marine National Reserve, Kenya*; IUCN—The World Conservation Union, 2001.
25. Emerton, L. *Economic Tools for Valuing Wetlands in Eastern Africa*; IUCN—The World Conservation Union, 1998.

WETLANDS AND METHANE EMISSION

W

Anna Joabsson
Torben Røjle Christensen
Climate Impacts Group, Department of Ecology, Lund University, Lund, Sweden

INTRODUCTION

Methane (CH_4) is a radiatively active trace gas that plays an important role for atmospheric chemistry and the energy balance of the Earth. At present, its contribution to the greenhouse effect is about 22%, which can be compared to carbon dioxide (CO_2) that contributes approximately 65% to the climate forcing of all long-lived greenhouse gases (excluding water vapour) (1). In addition to the direct warming effects, where infrared radiaton is absorbed and returned to the Earth's surface, chemical and photochemical reactions with CH_4 in the troposphere and stratosphere indirectly cause greenhouse warming. The preindustrial atmospheric concentration of CH_4 was about 0.75 ppmv (parts per million by volume) and since then it has more than doubled to approximately 1.73 ppmv (1). Anthropogenic sources of CH_4 include cattle (about 15% of the annual CH_4 release), rice paddies (20%), coal mining and oil production (14%), biomass burning (10%), natural gas leaks, and landfills and sewage disposal, while the major natural sources include wetlands (20–25%) and termites (5%) (2). Because of the substantial contribution from natural wetlands, knowledge about the dynamics leading to CH_4 formation and emission is of great importance to understand how these ecosystems interact with the climate and how they would respond to climate change.

METHANE PRODUCTION AND CONSUMPTION

Methane is produced by strictly anaerobic archaebacteria that are limited to the use of only a few simple substrates for biosynthesis and energy production (3). If available carbon compounds are not directly supplied, they depend on other groups of microorganisms for the initial breakdown of more complex organic structures into simpler molecules. The reduction of CO_2 with hydrogen is a common pathway to CH_4 formation, but acetate and formate are also important precursors of CH_4 in natural environments. Growth of methanogenic bacteria further requires a very low redox potential (E_h below -400 mV) and the ideal environmental conditions are often met in permanently waterlogged wetlands. However, CH_4 efflux from wetlands to the atmosphere depends not only on the rate of production (methanogenesis), but also on the extent of CH_4 consumption (methanotrophy) that may occur in oxic surface layers and in the close vicinity of plant roots (4–6). In this process, CH_4 is oxidized to CO_2, and the rate of CH_4 emission that can be measured at the surface is hence the net result of two counteracting processes—methanogenesis and methanotrophy.

ENVIRONMENTAL CONTROLS ON METHANE EMISSIONS

Water Table Position

Water table position in relation to the surface is considered to be the most important factor controlling CH_4 flux, because it indicates the boundary between anaerobic CH_4 production and aerobic CH_4 consumption (Fig. 1). Negative correlations between water table depth and rates of CH_4 emission have frequently been reported (7–9). It has also been found that the lowering of the water table may result in increased CH_4 emissions when episodic releases of CH_4 trapped in pore water occur and the diffusivity of gases in air-filled pore spaces increases (10, 11). Field experiments in peat-forming wetlands reveal a high spatial variability of CH_4 emissions caused by distinct microtopographical features on the wetland surface (12). Interactions between vascular plants and mosses create differences in elevation, and therefore distance to the water table within a radius of less than a metre (13). Scaling up of spot CH_4 emission measurements to larger areas without careful consideration of experimental plot location may therefore be misleading. Laboratory incubations have shown that there is a potential for CH_4 production both above and below the water table (14), but the actual production in the field peaks at approximately 5–15 cm below the water table (8). For methanotrophic CH_4 consumption to occur, a simultaneous supply of both CH_4 and oxygen is necessary. The abundance and activity of methanotrophic bacteria are therefore highest at the interface

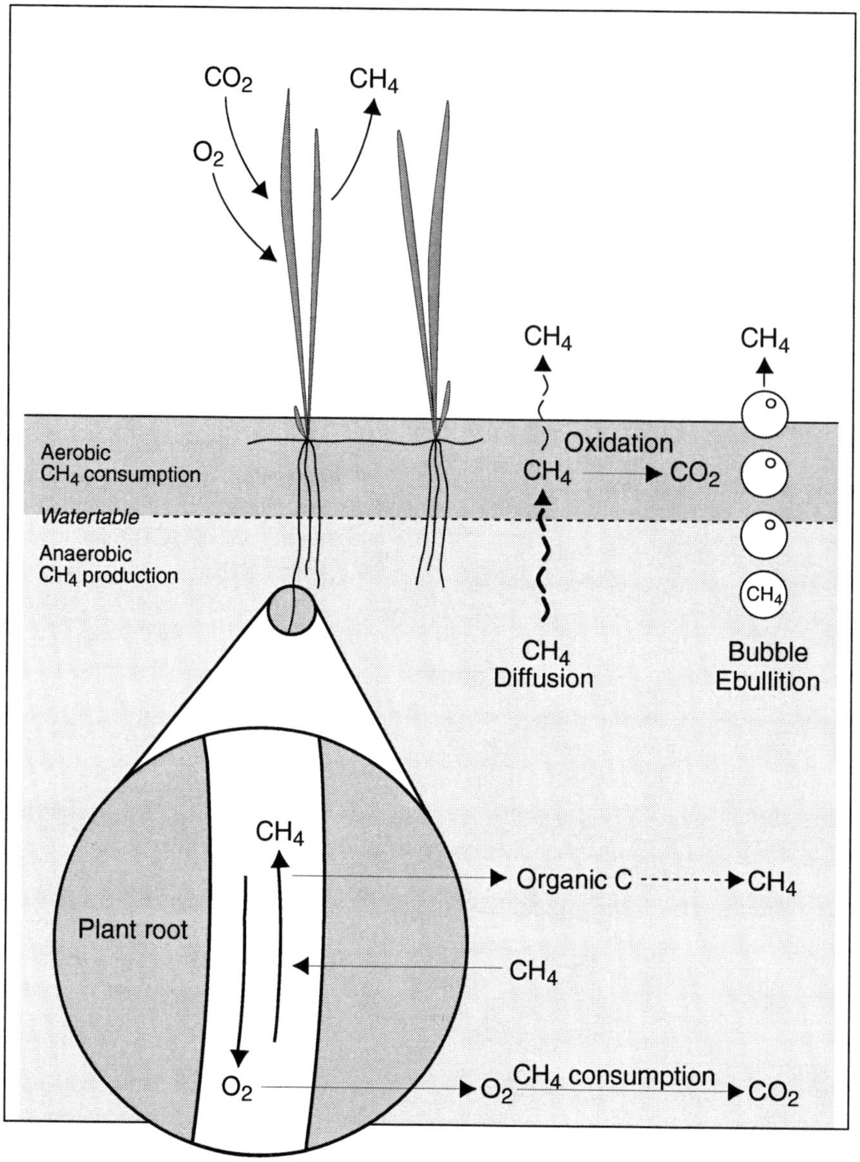

Fig. 1 The depth of the water table is important in determining net CH_4 emissions from natural wetlands because it indicates the boundary between anaerobic CH_4 production and aerobic CH_4 consumption. Methane may be transported from the soil to the atmosphere via diffusion through the soil profile, by bubble ebullition or by plant-mediated transport. In the first case, CH_4 is subjected to methanotrophic oxidation as it passes through the oxic surface layer and this acts to decrease net CH_4 emissions. Bubbles are quite stable, meaning that little oxidation takes place, and the same is also true when CH_4 is transported through vascular plants. Vascular plants are sources of methanogenic substrate since they release labile carbon compounds into the soil via root exudation and root turnover. Leakage of O_2 from roots may lead to inhibition of methanogenesis and methanotrophic CH_4 consumption in the otherwise anoxic soil.

between anoxic and oxic conditions, that is, at the depth of the water table (5, 14). Watson et al. (1997) (5) used a model to show a high capacity for CH_4 oxidation in peat, where methanotrophy accounted for 85% of the oxygen uptake potential when both oxygen and CH_4 were present in excess. In the absence of CH_4, maximum oxygen uptake potential was near the surface.

Soil Temperature

Correlations between soil temperature and CH_4 production and emission have been found at scales ranging from laboratory incubations and monolith experiments (8, 11, 15) to field investigations at the plot (16) and landscape scale (9). There is a direct effect of temperature on the metabolic rate of CH_4-producing microorganisms, but the sensitivity to temperature change is different for methanogens and methanotrophs (11, 17). In a study carried out with peat slurries under laboratory conditions, Dunfield et al. (1993) (17) found that CH_4 production in peat samples from temperate and subarctic areas was more sensitive to changes in temperature as compared to CH_4 consumption in the same samples. Increasing or decreasing temperatures also interact with other parameters with potential to control CH_4 production and emission to the atmosphere. One such interactive effect would be the relationship between temperature and water table position, where increasing temperatures lead to enhanced evaporation and plant transpiration, with lowering of the water table as a consequence (19). It has also been suggested that the temperature dependence of CH_4 production is constrained by substrate limitation and that temperature has an effect on methanogenesis mainly via its influence on substrate availability (18). In peat samples from a Swedish acid mire, CH_4 production was stimulated by increased temperatures only when substrate (glucose) concentrations were simultaneously increased (15). It has further been found that decomposition of labile material is more strongly controlled by temperature than decomposition of more recalcitrant components (19). From the same study, it was also concluded that substrate availability lagged behind rapid temperature changes and that the thermal/hydrological history of the soil was important to determine the rate of CH_4 production.

Methane Transport: Diffusion, Bubble Ebullition, and Transport Through Plants

The solubility of CH_4 in water is low, and it escapes through waterlogged soil to the atmosphere by diffusion, bubble ebullition, or by transport through vascular plants (Fig. 1). The rate of CH_4 emission is largely controlled by the mode of transport because the different transportation pathways are associated with more or less extensive CH_4 consumption in the soil profile. Methane diffusing through oxic environments is subjected to methanotrophic oxidation to CO_2, and a water table depth of only a few centimetres may cause considerable reductions of CH_4 emissions to the atmosphere. Bubbles form when the rate of CH_4 production causes the sum of the partial pressures of dissolved gases to exceed the value of the hydrostatic pressure in the soil. However, the concentration of CH_4 in bubbles will be at equlibrium with dissolved pore water CH_4. At low rates of methanogenesis and with no vascular plants present, the main release of CH_4 to the atmosphere is likely to be by diffusion, but bubbles begin to form when the production rate exceeds the capacity for diffusive loss (20). Once formed, bubbles are quite stable, meaning that little methanotrophic CH_4 oxidation occurs even as they pass through oxic environments. Studies conducted in rice paddies have shown that bubble formation tends to decrease as the plants mature because the progressively better developed rooting system acts as gas conduits and efficiently transports gases out of the soil (21). However, a close relationship between vascular plant production and the availability of methanogenic substrate (see below) may also speed up the rate of methanogenesis to such an extent that bubble formation is promoted. In order to supply oxygen for respiration to submerged structures, certain vascular plant species develop lacunae in stems, roots, and rhizomes (22). Depending on the species, the ventilation of growing roots is carried out either by pressurized bulk flow or by simple diffusion that follows concentration gradients between the atmosphere and the soil. Leakage of oxygen may give rise to CH_4 oxidation and inhibition of methanogenesis in the close vicinity of the roots (5, 6). However, net CH_4 emission is generally enhanced by the presence of vascular plants with a deep rooting system because of a "chimney effect," where CH_4 transported from the soil to the atmosphere within plant lacunae is withdrawn from consumption in oxic conditions (12, 23, 24).

Vascular Plants and Substrate for Methane Production

Vascular plants are sources of substrates for CH_4 production because they release easily degradable carbon compounds into the soil via root exudation and root turnover (Fig. 1) (20, 24, 25). In peat-forming wetlands, the organic matter becomes increasingly recalcitrant with depth (26) and several studies have pointed out that plant-derived inputs of labile carbon compounds could significantly contribute to increased CH_4 production (20, 23, 27). This "loading" of substrates into the soil may further be correlated to the photosynthetic rate of the vascular plants in the ecosystem (24, 28), since the amount of carbon allocated to belowground plant structures and eventually released to the soil likely is proportional to the CO_2 fixation rate (24). The overall primary productivity in the ecosystem may therefore be of great importance for the CH_4 emission rates from a wetland area.

REFERENCES

1. Lelieveld, J.; Crutzen, P.J.; Dentener, F.J. Changing Concentration, Lifetime and Climate Forcing of Atmospheric Methane. Tellus **1998**, *50B*, 128–150.

2. Chappelaz, J.A.; Fung, I.Y.; Thompson, A.M. The Atmospheric CH_4 Increase Since the Last Glacial Maximum (1). Source Estimates. Tellus **1993**, *45B*, 228–241.
3. Oremland, R.S. Biogeochemistry of Methanogenic Bacteria. In *Biology of Anaerobic Microorganisms*; Zehnder, A.J.B., Ed.; John Wiley: New York, 1988; 641–703.
4. King, G.M. Ecological Aspects of Methane Oxidation, a key Determinant of Global Methane Dynamics. Adv. Microb. Ecol. **1998**, *12*, 431–468.
5. Watson, A.; Stephen, K.D.; Nedwell, D.B.; Arah, J.R.M. Oxidation of Methane in Peat: Kinetics of CH_4 And O_2 Removal and the Role of Plant Roots. Soil. Biol. Biochem. **1997**, *29* (8), 1257–1267.
6. Popp, T.J.; Chanton, J.P.; Whiting, G.J.; Grant, N. Evaluation of Methane Oxidation in the Rhizosphere of a *Carex* Dominated Fen in North Central Alberta, Canada. Biogeochemistry **2000**, *51*, 259–281.
7. Christensen, T.R.; Friborg, T.; Sommerkorn, M.; Kaplan, J.; Illeris, L.; Soegaard, H.; Nordstroem, C.; Jonasson, S. Trace Gas Exchange in a High-Arctic Valley 1. Variations in CO_2 And CH_4 Flux Between Tundra Vegetation. Glob. Biogeochem. Cycles **2000**, *14* (3), 701–713.
8. Daulat, W.E.; Clymo, R.S. Effects of Temperature and Watertable on the Efflux of Methane from Peatland Surface Cores. Atmos. Environ. **1998**, *32* (19), 3207–3218.
9. Hargreaves, K.J.; Fowler, D. Quantifying the Effects of Water Table and Soil Temperature on the Emission of Methane from Peat Wetland at the Field Scale. Atmos. Environ. **1998**, *32* (19), 3275–3282.
10. Kettunen, A.; Kaitala, V.; Alm, J.; Silvola, J.; Nykänen, H.; Martikainen, P.J. Cross-Correlation Analysis of the Dynamics of Methane Emissions from a boreal Peatland. Glob. Biogeochem. Cycles **1996**, *10* (3), 457–471.
11. Moore, T.R.; Dalva, M. The Influence of Temperature and Water Table Position on Carbon Dioxide and Methane Emissions from Laboratory Columns of Peatland Soils. J. Soil Sci. **1993**, *44*, 651–664.
12. Frenzel, P.; Karofeld, E. CH_4 Emission from a Hollow-Ridge Complex in a raised Bog: The Role of CH_4 Production and Oxidation. Biogeochemistry **2000**, *51*, 91–112.
13. Waddington, J.M.; Roulet, N.T. Carbon Balance of a boreal Patterned Peatland. Glob. Change Biol. **2000**, *6*, 87–97.
14. Moore, T.R.; Dalva, M. Methane and Carbon Dioxide Exchange Potentials of Peat Soils in Aerobic and Anaerobic Laboratory Incubations. Soil. Biol. Biochem. **1997**, *29* (8), 1157–1164.
15. Bergman, I.; Svensson, B.H.; Nilsson, M. Regulation of Methane Production in a Swedish Acid Mire by PH, Temperature and Substrate. Soil Biol. Biochem. **1998**, *30* (6), 729–741.
16. Verville, J.H.; Hobbie, S.E.; Chapin, F.S., III; Hooper, D.U. Response of Tundra CH_4 And CO_2 Flux to Manipulation of Temperature and Vegetation. Biogeochem. **1998**, *41*, 215–235.
17. Dunfield, P.; Knowles, R.; Dumont, R.; Moore, T.R. Methane Production and Consumption in Temperate and Subarctic Peat Soils: Response to Temperature and pH. Soil Biol. Biochem. **1993**, *25* (3), 321–326.
18. Valentine, D.W.; Holland, E.A.; Schimel, D.S. Ecosystem and Physiological Controls Over Methane Production in Northern Wetlands. J. Geophys. Res. **1994**, *99* (D1), 1563–1571.
19. Updegraff, K.; Bridgham, S.D.; Pastor, J.; Weishampel, P. Hysteresis in the Temperature Response of Carbon Dioxide and Methane Production in Peat Soils. Biogeochemistry **1998**, *43*, 253–272.
20. Chanton, J.P.; Whiting, G.J. Trace Gas Exchange in Freshwater and Coastal Marine Environments: Ebullition and Transport by Plants. In *Biogenic Trace Gases: Measuring Emissions from Soil and Water*; Matson, P.A., Harriss, R.C., Eds.; Blackwell Science: Oxford, 1995; 98–125.
21. Holzapfel-Pschorn, A.; Conrad, R.; Seiler, W. Effects of Vegetation on the Emission of Methane from Submerged Paddy Soils. Plant Soil **1986**, *92*, 223–233.
22. Armstrong, W.; Justin, S.H.F.W.; Beckett, P.M.; Lythe, S. Root Adaptation to Soil Waterlogging. Aquat. Bot. **1991**, *39*, 57–73.
23. Greenup, A.L.; Bradford, M.A.; McNamara, N.P.; Ineson, P.; Lee, J.A. The Role of *Eriophorum Vaginatum* In CH_4 Flux from an Ombrotrophic Peatland. Plant Soil **2000**, *227*, 265–272.
24. Joabsson, A.; Christensen, T.R.; Wallén, B. Vascular Plant Controls on Methane Emissions from Northern Peatforming Wetlands. Trends Ecol. Evol. **1999**, *14*, 385–388.
25. Jones, D.L. Organic Acids in the Rhizosphere – A critical Review. Plant Soil **1998**, *205*, 25–44.
26. Christensen, T.R.; Jonasson, S.; Callaghan, T.V.; Havström, M. On the Potential CO_2 Releases from Tundra Soils in a Changing Climate. Appl. Soil Ecol. **1999**, *11*, 127–134.
27. Joabsson, A.; Christenesen, T.R. Methane Emissions from Wetlands and Their Relationship with Vascular Plants: an Arctic example. Glob. Change Biol., accepted.
28. Whiting, G.J.; Chanton, J.P. Plant-Dependent CH_4 Emission in a Subarctic Canadian Fen. Glob. Biogeochem. Cycles **1992**, *6* (3), 225–231.

WETLANDS SEDIMENTATION AND ECOLOGICAL ENGINEERING

W

Timothy C. Granata
Jay Martin
The Ohio State University, Columbus, Ohio, U.S.A.

INTRODUCTION

Wetlands are highly efficient at storing and transforming chemicals. The primary input of these chemicals are from overland flow. The characteristics of these chemicals are dependent on the sources in the watershed. The rate of input to riparian zones, however, is dependent on landscape features, such things as slope and surface resistance and soil properties, including grain size and cohesive strength. For large flows and loose soils, a significant portion of these chemical inputs can be a result of sediment transport.

Wetlands are intimately tied to soils through sedimentation processes. Accretion of sediments accelerates the aging of wetlands, reduces infiltration through bottom substrates increasing water heights, and removes phosphorous, a limiting nutrient in many freshwater ecosystems. In this paper, the mechanics of sedimentation are discussed in relation to natural and constructed wetlands, and ecological engineering principles are suggested as a way to mitigate the effects of excessive sedimentation.

PROBLEMS OF SEDIMENTATION IN WETLANDS

Sediment Accretion

The accretion of particulate material in wetlands limits the lifetime of these systems, advancing their succession to a terrestrial ecosystem as a function of sedimentation rates. Three sources of particulate material can be identified: suspended particles derived from the inflow, production of microbial particles in the wetland, and the production of detritus from macro-vegetation. The latter two increase with increased primary production in wetlands. The former is a function of hydraulic loading, defined as the flow, Q, per unit area of wetland, A, that is, QA^{-1}. Natural wetlands are prominent components of riparian ecosystems that exist either as floodplains, receiving periodic submersion, or as part of the riparian corridor, receiving continuous, though varying, inflow. In both cases, flow transports suspended sediments to wetlands. Suspended material is often deposited as current velocities decrease in broad, shallow expanses of wetlands. Sediment transport is a function of total flow (and thus rainfall intensity) in a watershed but also depends on the slope of the landscape and the susceptibility of soils to erosion. Steep slopes with highly unconsolidated soils will contribute to large suspended sediment loads during high run-off events. For low-flow events suspended sediment concentrations do not exceed $20\,\mathrm{mg\,L^{-1}}$ while for high flows they can exceed $80\,\mathrm{mg\,L^{-1}}$ (1). In extreme cases of accretion, sedimentation rates of $26.67\,\mathrm{cm\,yr^{-1}}$ can bury emergent grasses since sediment accumulation outpaces plant growth (2). In contrast, constructed wetlands are designed to treat wastewater having steady and high suspended solids concentrations, ranging from $20-75\,\mathrm{mg\,L^{-1}}$, and volumetric ratios of settable solids from $5-20\,\mathrm{ml\,L^{-1}}$ (3).

Reduced Conductivity of the Bed

Wetlands can experience reduced infiltration rates as particulate matter clogs bottom substrates. Generally, a reduction of hydraulic conductivity occurs with increasing hydraulic loading to the wetland. As a consequence of lower current velocities and higher sedimentation rates with increasing distance from the inlet, the hydraulic conductivity is lowest near the inlet and highest near the outlet. A general design criteria for wetlands is a hydraulic conductivity $1500\,\mathrm{m\,d^{-1}}$ based on bed of $1.25-2.5\,\mathrm{cm}$ diameter gravel (4).

Phosphorus Accumulation

The accretion of organic material on the bed results in wetlands with a high ion exchange capacity. The beds are partially responsible for removal of phosphates by sorption until they become saturated. The movement of phosphorus though the bed is slower than the hydraulic conductivity as a result of the storage (exchange) capacity of the bed (5). Plant uptake of phosphates is limited since a reduced hydraulic conductivity decreases infiltration and thus transport of phosphates to plant roots (6). A more consistent process for phosphate removal is by adsorption onto suspended sediments and their subsequent deposition (7). Phosphorus (P) loading is directly proportional to suspended sediment concentrations and hydraulic loading. It is estimated that 40 mg P accompanies each gram of suspended solids entering wetlands (1). Higher loading rates occur for flows over landscapes rich in P, such as agricultural lands. During periods of inundation, wetlands are characterized by aerobic and anaerobic soil zones. In anaerobic zones the decreased redox potential leads to an increase in the solubility of particulate phosphorus, which may be discharged from the wetlands. This dynamic feedback between removal of P by sedimentation and release of P after particulate accretion often leads to variable retention of P by wetlands. Despite mineralization of particulate P, wetlands are generally sinks for P, with removal as high as 97% (8). The retention of P, however, decreases as P loading increases (9) and often decreases through the life of the wetland (10).

PARTICLE TYPES AND SEDIMENTATION PROCESSES IN WETLANDS

The size distribution of particles in a wetland varies from nanometer-sized colloidal material to micrometer phytoplankton and heterotrophic organisms to millimeter, and centimeter-sized leaf detritus and sediments. For discrete particles, settling results from the force balance between particle weight (F_w) and drag on the particle such that

$$F_w = F_D = \rho C_D A_{pr} \frac{w^2}{2} = (\rho_p - \rho) V g \tag{1}$$

where C_D is the drag coefficient on a 2-D surface, A_{pr} is the projected area of the particle, w is the settling speed, ρ_p and ρ are the particle and water densities, respectively, V is the volume of the particle and g is gravitational acceleration ($9.8\,m\,s^{-2}$). Solving for settling rate, w, gives

$$w = \sqrt{2 \frac{(\rho_p - \rho)\, Vg}{\rho\, A_{pr}\, C_D}} \tag{2}$$

Thus, the settling rate of a particle is proportional to the excess density of the particles over that of water and the length scale of the particle defined by VA_{pr}^{-1}. Providing flow is not turbulent, specified as a Reynolds number $Re < 0.1$, and particle shape is unimportant to the drag, C_D can be approximated as

$$C_D = \frac{24}{Re} = \frac{24\nu}{2Rw} \tag{3}$$

where R is the radius of the particle. Assuming a spherical particle, the above equation reduces to

$$w = \frac{2\rho' g R^2}{9\nu} \tag{4}$$

the celebrated Stokes' Law for sinking particles. Here the excess density is written as

$$\rho' = \frac{\rho_p - \rho}{\rho} \tag{5}$$

For biological material, excess density is less than 0.1%, and size is the determining factor. For this reason small microbial organisms will sink slower than larger detrital material. The excess density of suspended sediments can reach 260%, indicating that density is responsible for the high settling rates of suspended sediments.

Sedimentation in a wetland will occur if the settling speed is greater than the surface loading rate per area of wetland, or $w > QA^{-1}$. Given the large spectrum of particle types in natural flows, w is usually calculated based on the size and density of the particle type to be removed. Dividing the depth by the sinking rate gives the retention time in the wetland. Particles that are vertically well mixed as they enter a wetland will have a uniform concentration distribution comprised of the spectrum of particle sizes. If the settling rate of the particle, w_p, is greater than w, the particle will be deposited in the wetland cell. For particles with $w_p < w$, the fraction of particles that will be removed is X_p. Thus, the total particle removal by a wetland can be calculated as

$$FR = (1 - X_c) + QA^{-1} \Sigma X_r \Delta x \tag{6}$$

where $1 - X_c$ is fraction of particles removed with sinking speeds $w_p > w$ and $QA^{-1}\Sigma X_r \Delta x$ is the fraction with a settling speed $w_p < w$ that are removed over a distance Δx, along the path of the flow (3). The crucial step in determining the sedimentation rate in wetlands is first determining the settling rate of each concentration fraction, either by settling columns or size-concentration measurements.

Often aggregations of particles form in wetlands (11), a process called flocculation. Sedimentation rates of aggregates are dominated by organic particles of fractal dimension (12). Aggregates tend to settle faster than their discrete component particles, and the only way to determine removal rate is by carefully transferring flocs to a settling column and determining the percent removal as a function of the height of the column and time (3).

In wetlands with a high concentration of suspended solids, settling is affected by the contact between particles. This would occur in the bottom sediments of most wetlands where two settling processes can be identified: hindered and compression settling (3). The hindered zone is marked by a large gradation in particle concentrations, which is less than the total concentrated in the compression zone. By plotting the concentration of particles over the height of a settling column as a function of time, a break-point can be found in the time from hindered settling to compression settling (3).

Because large, heavy particles settle out first, the highest sedimentation rates in wetlands are near the inlets. As mentioned above, the hydraulic conductivity through the bottom sediment decreases as clogging occurs. Since clogging is also a function of distance from the inlet, the most accurate estimate of the hydraulic conductivity is to measure the change in water surface elevation with increasing distance from the inlet. However, as Kadlec and Watson (11) show, this is not a simple function of Darcy's Law but must include evapotranspiration in the wetland. For newly constructed wetlands, the hydraulic conductivity of the bottom substrate (gravel, sand, mud) cannot be used since clogging will result from sediment accretion (4).

ECOLOGICAL ENGINEERING SOLUTIONS TO SEDIMENTATION

To keep suspended particles entrained by flows from accruing in wetlands and clogging bottom sediments, one or more settling basins can be included between the inlet channel and the wetland cell. This would have two effects: first, to collect all but the finest and least dense suspended particles and, second, to remove phosphorus from the inflow.

To overcome the problem of detrital accumulation in wetlands, woody plants could be substituted for grasses and periodically harvested. This would not only reduce the amount of biomass accrued in the wetland but would increase the efficiency of nutrient uptake in the unclogged root zone and provide a potentially marketable resource. Since phytoplankton are the most abundant nutrient filters in a wetland and have intrinsically low sinking rates, a wetland could be designed with low retention of suspended plankton to further reduce sedimentation while improving nutrient removal. A settling basin could then be sized to accumulate these nutrient-rich particles for harvesting before nutrients are remineralized. To keep flocs of particles from forming in wetlands, cells could be mixed with aerators or inlet pumps. This would break up flocs, which would be exported as discrete particles.

W

CONCLUSIONS AND FUTURE ISSUES

Since wetlands are always shallow to promote macrophtye growth, they essentially act as flat plate collectors of sinking particles. The dominant particles in wetlands are suspended particles in the inflow and biomass of vegetation resulting from growth in the wetlands. The processes of sedimentation can reduce a wetland's storage capacity, its efficiency to retain nutrients, and its lifetime. Future efforts should be aimed at enhancing nutrient removal while reducing sedimentation. This can be done with engineered structures, such as settling basins and grit chambers, or by less costly technologies such as stilling wells. More innovative Green solutions are on the horizon.

REFERENCES

1. Porter, K.S. Nitrogen and Phosphorous. *Food Production, Waste, and the Environment*; Ann Arbor Science Publishers Inc: Ann Arbor MI, 1975; 372.
2. Chung, C-H. Ecological Engineering of Coastlines with Salt-Marsh Plantations. In *Ecological Engineering: An Introduction to Ecotechnology*; Mitsch, W.J., Jørgenson, S.E., Eds.; John Wiley & Sons: New York, 1989; 472.
3. Metcalf and Eddy. *Wastewater Engineering: Treatment, Disposal, Reuse*; McGraw Hill: New York, 1991; 1334.
4. Watson, T.J.; Choate, K.D. Hydraulic Conductivity of Onsite Constructed Wetlands. In 9th National Symposium of Individual and Small Community Sewage Systems; Fort Worth, Texas, March 11–14, 2001.
5. Kadlec, R.H.; Knight, R.L. *Treatment Wetlands*; CRC Press/Lewis Publishers: Boca Raton, FL, 1996; 839.
6. Davies, T.H.; Cottingham, P.D. Phosphorus Removal from Wastewater in a Constructed Wetland. In *Constructed Wetlands for Water Quality Improvement*; Moshiri, G.A., Ed.; Lewis Publishers: Boca Raton, FL, 1993; 632.
7. Richardson, C.J.; Craft, C.B. Efficient Phosphorus Retention in Wetlands: Fact or Fiction? In *Constructed Wetlands for Water Quality Improvement*; Moshiri, G.A., Ed.; Lewis Publishers: Boca Raton, FL, 1993; 632.
8. Mitsch, W.J.; Gosselink, J.G. *Wetlands*; Van Nostrand Reinhold Co.: New York, 2000; 920.

9. Mitsch, W.J.; Reeder, B.C.; Klarer, D.M. Wetlands for Nutrient Control: Western Lake Erie. In *Ecological Engineering: An Introduction to Ecotechnology*; Mitsch, W.J., Jørgenson, S.E., Eds.; John & Wiley Sons: New York, 1989; 472.
10. White, J.S.; Bayley, S.E.; Curtis, P.J. Sediment Storage of Phosphorus in a Northern Prairie Wetland Receiving Municipal and Agro-Industrial Wastewater. Ecological Engineering **2000**, *14*, 127–138.
11. Kadlec, R.H.; Watson, J.T. Hydraulics and Solids Accumulation in a Gravel Bed Treatment Wetland. In *Constructed Wetlands for Water Quality Improvement*; Moshiri, G.A., Ed.; Lewis Publishers: Boca Raton, FL, 1993; 632.
12. Kilps, J.R.; Logan, B.E.; Alldregde, A.L. Fractal Dimensions of Marine Snow Determined from Image Analysis of *in situ* Photographs. Deep-Sea Research **1994**, *41* (8), 1159–1169.

WETLANDS

Ralph W. Tiner
U.S. Fish and Wildlife Service, National Wetlands Inventory Program, Hadley, Massachusetts, U.S.A.

INTRODUCTION

Wetland is a universal term used to describe the collection of flooded or saturated environments that have been referred to as marshes, swamps, bogs, fens, salinas, pocosins, mangroves, wet meadows, sumplands, salt flats, varzea forests, igapo forests, bottomlands, sedgelands, moors, mires, potholes, sloughs, mangals, palm oases, playas, muskegs, and other regional and local names. It has been defined as a basis for inventorying these natural resources, for conducting scientific studies, and, in some countries, for regulating uses of these areas. Given that wetlands include a diverse assemblage of ecosystems, classification schemes have been developed to separate and describe these different systems and to group similar habitats. Wetlands provide a number of functions that are considered valuable to society (e.g., surface water storage to minimize flood damages, sediment retention and nutrient transformation to improve water quality, shoreline stabilization, streamflow maintenance, and provision of vital habitat for fish, shellfish, wildlife, and plants that yield food and fiber for people). Because of these values and the widespread recognition of wetlands as important natural resources, numerous wetland definitions and classification systems have been developed to inventory these resources around the globe. The purpose of this article is to provide readers with an understanding of what wetlands are (wetland definition), how they vary globally (wetland types), and their extent as determined by various inventories. This article should serve as a starting point for learning about wetlands, with the listed references being sources of more detailed information.

WETLAND DEFINITIONS

Wetlands are aquatic to semiaquatic ecosystems where permanent or periodic inundation or prolonged waterlogging creates conditions favoring the establishment of aquatic life. Wetlands are often located between land and water and have, therefore, been referred to as ecotones (i.e., transitional communities). However, many wetlands are not ecotones between land and water, since they are not associated with a river, lake, estuary, or stream (1). Wetlands may derive water from many sources, including groundwater, river overflow, surface water runoff, precipitation, snowmelt, tides, melting permafrost, and seepage from impoundments or irrigation projects.

While the term "wetland" has many definitions, all definitions have common elements. (See Table 1; some definitions even include deepwater habitats.) The presence of water in wetlands may be permanent or temporary. Their water may be salty or fresh. Wetlands may be natural habitats or artificially created. They range from shallow water environments to temporarily wet (i.e., flooded or saturated) areas. All are wet long enough and often enough to, at least, periodically support hydrophytic vegetation and other aquatic life (including anaerobic microbes), to create hydric soils or substrates, and to activate biogeochemical processes associated with wet environments.

WETLAND TYPES

Differences in climate, soils, vegetation, hydrology, water chemistry, nutrient availability, and other factors have led to the formation of a multitude of wetland types around the globe. In general, wetlands are characterized by their hydrology (e.g., tidal vs. nontidal, inundation vs. soil saturation, frequency and duration of wetness), the presence or absence of vegetation (vegetated vs. nonvegetated), the type of vegetation (forested or treed, shrub, emergent, or aquatic bed), and soil type (e.g., organic vs. inorganic, peatland vs. nonpeatland). Table 2 presents brief descriptions of some North American types and Fig. 1 shows examples of vegetated wetlands.

Various countries have devised classification systems for describing differences among their wetlands and for categorizing wetlands for natural resources inventories. Scientists have created systems to organize certain wetlands into meaningful groups for analysis and management

Table 1 Examples of wetland definitions used for inventories

Country/organization	Wetland definition (source)
International/Ramsar	"areas of marsh, fen, peatland, or water, whether natural or artificial, permanent or temporary, with water that is static or flowing, fresh, brackish, or salt, including areas of marine water the depth of which at low tide does not exceed 6 m.... ...may incorporate riparian and coastal zone adjacent to wetlands, and islands or bodies of marine water deeper than 6 m at low tide lying within the wetlands" (2).
Australia	"areas of seasonally, intermittently, or permanently waterlogged soils or inundated land, whether natural or artificial, fresh or saline, e.g., waterlogged soils, ponds, billabongs, lakes, swamps, tidal flats, estuaries, rivers, and their tributaries " (12).
Canada	"land that is saturated with water long enough to promote wetland or aquatic processes as indicated by poorly drained soils, hydrophytic vegetation, and various kinds of biological activity which are adapted to a wet environment" (3).
U.S.	"lands transitional between terrestrial and aquatic systems where the water table is usually at or near the surface or the land is covered by shallow water." Wetland attributes include hydrophytic vegetation, undrained hydric soil, or saturated or flooded substrates (4).

Table 2 Brief nontechnical descriptions of some wetland types in North America

Wetland type	General description
Marsh	Herb-dominated wetland with standing water through all or most of the year, often with organic (muck) soils
Tidal marsh	Herb-dominated wetland subject to periodic tidal flooding
Salt marsh	Herb-dominated wetland occurring on saline soils, typically in estuaries and interior arid regions
Swamp	Wetland dominated by woody vegetation and usually wet for extended periods during the growing season
Mangrove swamp (Mangal)	Tidal swamp dominated by mangrove species
Peatland, mire, moor, or muskeg	Peat-dominated wetland
Bog	Nutrient-poor peatland, typically characterized by ericaceous shrubs, other woody species, and peat mosses
Fen	More or less nutrient-rich peatland, often represented by sedges and/or calciphilous herbs and woody species
Wet meadow	Herb-dominated wetland that may be seasonally flooded or saturated for extended periods, often with mineral hydric soils
Bottomland	Riverside or streamside wetland, usually on floodplain
Flatwood	Forested wetland with poorly drained mineral hydric soils located on broad flat terrain of interstream divides, common on coastal plains and glaciolacustrine plains
Farmed wetland	Wetland cultivated for rice, cranberries, sugar cane, mints, or other crops

These types may be defined differently in other regions.

Fig. 1 (*Facing page*) Some examples of North American wetlands (top to bottom, left to right): tidal salt marsh, inland marsh, pothole marsh, wet meadow/shrub swamp, northern peatland, bottomland swamp, hardwood swamp, and flatwood wetland. (Photos courtesy of U.S. Fish and Wildlife Service.)

(a)
(b)
(c)
(d)
(e)
(f)
(g)
(h)

(e.g., peatland classifications; see Ref. 1 for details). In 1998, the Ramsar Convention Bureau published a multinational classification system to provide consistency for inventorying wetlands and designating wetlands of international importance (2). This system includes 11 types of marine or coastal wetlands (i.e., shallow water and intertidal habitats: permanent shallow marine waters; marine subtidal aquatic beds; coral reefs; rocky marine shores; sand; shingle or pebble shores; estuarine waters; intertidal mud, sand, or salt flats; intertidal marshes; intertidal forests; coastal brackish/saline lagoons; coastal freshwater lagoons). This system also includes 19 inland wetland types (i.e., permanently flooded aquatic habitats to intermittently flooded sites are represented: permanent inland deltas; permanent rivers/streams/creeks; seasonal, intermittent, or irregular rivers/streams/creeks; permanent freshwater lakes; seasonal or intermittent freshwater lakes; seasonal or intermittent saline/brackish/alkaline lakes and flats; permanent saline/brackish/alkaline marshes and pools; seasonal or intermittent saline/brackish/alkaline marshes and pools; permanent freshwater marshes and pools; seasonal or intermittent freshwater marshes and pools; nonforested peatlands; alpine wetlands including meadows and temporary snowmelt waters; tundra wetlands; shrubby-dominated wetlands; freshwater tree-dominated wetlands on inorganic soils; forested peatlands; freshwater springs and oases; geothermal wetlands; and subterranean karst and cave hydrological systems). Lastly, nine man-made wetland types (aquaculture ponds; ponds; irrigated land including rice paddies; seasonally flooded agricultural land; salt exploitation sites; water storage areas including impoundments generally more than 8 ha; excavations; wastewater treatment areas; and canals and drainage channels) are also included in the system.

In North America, the Canadian and United States wetland classification systems were developed by government agencies interested in wetland conservation and management. The Canadian system emphasizes wetland origin (class), form, and vegetation in describing the wetland types (3). Five wetland classes are recognized: bog, fen, marsh, swamp, and shallow water. Within each class, different forms and types are characterized. Eight general vegetation types are defined by the presence or absence of vegetation (treed, shrub, forb, graminoid, moss, lichen, aquatic bed, and nonvegetated). These types may be subdivided into other types (e.g., treed into coniferous or deciduous, shrub into tall, low, and mixed, graminoids into grass, reed, tall rush, low rush, and sedge, aquatic bed into floating and submerged). The U.S. Fish and Wildlife Service's wetland and deepwater habitat classification (4) is the official federal system used for mapping wetlands and for reporting the status and trends of wetlands in the U.S. The features separating wetlands include general ecological and physical factors and specific features such as vegetation, soil/substrate composition, hydrology, water chemistry, and human alterations. Classification follows a hierarchical approach with five main levels designated: ecological system (marine, estuarine, lacustrine, riverine, and palustrine), subsystem, class (vegetated: forested, scrub-shrub, emergent, and aquatic bed; nonvegetated: unconsolidated shore, rocky shore, streambed, and reef), subclass, and modifiers. The modifiers are used to describe a wetland's hydrology (water regime), pH and salinity (water chemistry), soils, and the influence of humans and beaver (special modifiers). Common types include estuarine intertidal emergent wetlands (e.g., salt and brackish marshes), estuarine intertidal unconsolidated shore (e.g., tidal flats and beaches), palustrine emergent wetlands (e.g., marshes, fens, and wet meadows), palustrine forested wetlands, and palustrine scrub-shrub wetlands (e.g., shrub bogs and shrub swamps).

A hydrogeomorphic approach (HGM) to wetland classification has also been developing in the U.S. (5). The HGM system emphasizes abiotic features important for assessing wetland functions. Seven hydromorphic classes are identified: riverine, depressional, slope, mineral soil flats, organic soil flats, lacustrine fringe, and estuarine fringe. The U.S. Fish and Wildlife Service has adapted the HGM approach to provide additional modifiers to its classification system on a pilot basis. These HGM-type descriptors include landscape position (i.e., lotic, lentic, terrene, estuarine, and marine), landform (i.e., slope, basin, interfluve, floodplain, flat, island, and fringe), and water flow path (i.e., inflow, outflow, throughflow, bidirectional flow, isolated, and paludified) (6). These descriptors provide the required information to aid the evaluation of functions of wetlands across watersheds and large geographic areas.

EXTENT OF WETLANDS

Comprehensive wetland inventories do not exist in most countries. There are many inconsistencies among the inventories (e.g., different levels of effort, focus on particular types, and artificial wetlands such as rice paddies are often not included in wetland inventories) (7). Consequently, comparative analysis is fraught with problems. Nonetheless, Table 3 provides some perspective on the extent of wetlands in many regions. Most of the data came from a series of reports produced for the Bureau of the Ramsar Wetlands Convention (8). Globally, estimates

W

Table 3 Estimates of the current extent of wetlands in different regions of the world

Region/country	Wetland extent (ha)	Source
Africa	121,321,683–124,686,189	(13)
Asia	211,501,790–224,117,790	(14)
Central America		
Mexico	3,318,500 (very incomplete)	(15)
Europe		
Eastern	225,849,930	(16)
Western	28,821,979	(17)
Middle East	7,434,790	(18)
Neotropics	414,996,613	(19)
North America		
Canada	127,199,000–150,000,000	(20)
U.S.	114,544,800	(1)
Oceania	35,748,853	(21)
South America		
Tropical region	200,000,000	(22)

for wetlands range from about 750 million ha (7) to about 1.5 billion ha. Ten countries have over 2 million ha of peatlands alone, with Canada leading at nearly 130 million ha (represents about 18% of the country) followed by the former U.S.S.R. at 83 million ha (9, 10). About a third of Finland is covered by peatlands (10 million ha). The Pantanal of South America, perhaps the largest wetland in the world, reportedly covers about 200,000 km^2 (or 2 million ha) during the wet season (11).

REFERENCES

1. Tiner, R.W. *Wetland Indicators: A Guide to Wetland Identification, Delineation, Classification, and Mapping*; Lewis Publishers, CRC Press: Boca Raton, FL, 1999; 392.
2. Ramsar Convention Bureau. Information Sheet on Ramsar Wetlands: Gland, Switzerland, 1998.
3. National Wetlands Working Group, *Wetlands of Canada*; Ecological Land Classification Series No. 21, Land Conservation Branch, Canadian Wildlife Service, Environment Canada: Ottawa, Ont., 1988; 452.
4. Cowardin, L.M.; Carter, V.; Golet, F.C.; LaRoe, E.T. *Classification of Wetlands and Deepwater Habitats of the United States*; FWS/OBS-79/31; U.S. Department of the Interior, Fish and Wildlife Service: Washington, DC, 1979; 131.
5. Brinson, M.M. *A Hydrogeomorphic Classification for Wetlands*; Wetlands Research Program Tech. Rep. WRP-DE-4; U.S. Army Waterways Expt. Station: Vicksburg, MS, 1993; 103.
6. Tiner, R.W. *Keys to Waterbody Type and Hydrogeomorphic-Type Wetland Descriptors for U.S. Waters and Wetlands, Operational Draft*; U.S. Department of the Interior, Fish and Wildlife Service, Northeast Region: Hadley, MA, 2000; 20.
7. Finlayson, C.M.; Davidson, N.C. Global Review of Wetland Resources and Priorities for Wetland Inventory: Summary Report. In *Global Review of Wetland Resources and Priorities for Wetland Inventory*; Finlayson, C.M., Spiers, A.G., Eds.; Supervising Scientist Report 144: Canberra, Australia, 1999; 9.
8. Finlayson, C.M., Spiers, A.G., Eds. *Global Review of Wetland Resources and Priorities for Wetland Inventory*; Supervising Scientist Report 144: Canberra, Australia, 1999.
9. Taylor, J.A. Peatlands of the British Isles. In *Mires: Swamp, Bog, Fen, and Moor; Regional Studies*; Gore, A.J.P., Ed.; Elsevier: Amsterdam, The Netherlands; 1–46.
10. Botch, M.S.; Massing, V.V. Mire Ecosystems in the U.S.S.R. In *Mires: Swamp, Bog, Fen, and Moor; Regional Studies*; Gore, A.J.P., Ed.; Elsevier: Amsterdam, The Netherlands; 95–152.
11. Swarts, F.A., Ed. *The Pantanal of Brazil, Bolivia, and Paraguay*; Waterland Research Institute, Hudson MacArthur Publishers: Gouldsboro, PA, 2000; 287.
12. Wetland Advisory Committee. *The Status of Wetlands Reserves in System Six*; Report of the Wetland Advisory Committee to the Environmental Protection Authority: Australia, 1977.
13. Stevenson, N.; Frazier, S. Review of Wetland Inventory Information in Africa. In *Global Review of Wetland Resources and Priorities for Wetland Inventory*; Finlayson, C.M., Spiers, A.G., Eds.; Supervising Scientist Report 144: Canberra, Australia, 1999; 94.
14. Watkins, D.; Parish, F. Review of Wetland Inventory Information in Asia. In *Global Review of Wetland Resources and Priorities for Wetland Inventory*; Finlayson, C.M., Spiers, A.G., Eds.; Supervising Scientist Report 144: Canberra, Australia, 1999; 26.
15. Olmstead, I. Wetlands of Mexico. In *Wetlands of the World: Inventory, Ecology, and Management Volume I*; Whigham, D.F., Dykyjova, D., Heiny, S., Eds.; Kluwer Academic Publishers: Dordrecht, The Netherlands, 1993; 637–677.
16. Stevenson, N.; Frazier, S. Review of Wetland Inventory Information in Eastern Europe. In *Global Review of Wetland Resources and Priorities for Wetland Inventory*; Finlayson, C.M., Spiers, A.G., Eds.; Supervising Scientist Report 144: Canberra, Australia, 1999; 53.
17. Stevenson, N.; Frazier, S. Review of Wetland Inventory Information in Western Europe. In *Global Review of Wetland Resources and Priorities for Wetland Inventory*; Finlayson, C.M., Spiers, A.G., Eds.; Supervising Scientist Report 144: Canberra, Australia, 1999; 57.
18. Frazier, S.; Stevenson, N. Review of Wetland Inventory Information in the Middle East. In *Global Review of Wetland Resources and Priorities for Wetland Inventory*; Finlayson, C.M., Spiers, A.G., Eds.; Supervising Scientist Report 144: Canberra, Australia, 1999; 19.
19. Davidson, I.; Vanderkam, R.; Padilla, M. Review of Wetland Inventory Information in the Neotropics. In *Global Review of Wetland Resources and Priorities for Wetland Inventory*; Finlayson, C.M., Spiers, A.G., Eds.; Supervising Scientist Report 144: Canberra, Australia, 1999; 35.

20. Davidson, I.; Vanderkam, R.; Padilla, M. Review of Wetland Inventory Information in the North America. In *Global Review of Wetland Resources and Priorities for Wetland Inventory*; Finlayson, C.M., Spiers, A.G., Eds.; Supervising Scientist Report 144: Canberra, Australia, 1999; 35.
21. Watkins, D. Review of Wetland Inventory Information in Oceania. In *Global Review of Wetland Resources and Priorities for Wetland Inventory*; Finlayson, C.M., Spiers, A.G., Eds.; Supervising Scientist Report 144: Canberra, Australia, 1999; 26.
22. Junk, W.J. Wetlands of Tropical South America. In *Wetlands of the World: Inventory, Ecology, and Management Volume I*; Whigham, D.F., Dykyjova, D., Heiny, S., Eds.; Kluwer Academic Publishers: Dordrecht, The Netherlands, 1993; 679–739.

WINDBLOWN DUST

Dale Gillette
Air Resources Laboratory, National Oceanic and Atmospheric Administration, Research Triangle Park, North Carolina, U.S.A.

Windblown dust may be understood in terms of generation, transport, and deposition. By mass, windblown dust is probably the largest single contributor to particles in the Earth's atmosphere. The dominant mechanism for dust emission is "sandblasting," wherein sand-sized grains are initially made airborne; these sand grains either blast dust particles into the air from coatings of aggregated fine particles on larger particles, or splash fine particles from a reservoir of loose, fine particles. Size distributions of dust particles are broadly similar, despite differences in parent soil material, wind power, and other environmental factors.

The composition of dust is similar to the composition of its parent material. However, enrichments of certain elements or minerals are explained by the size distribution of parent soil minerals and the activity of soluble material. Dust residence time in the atmosphere is approximately less than 30 days, but windblown dust can be transported over phenomenal distances (e.g., across the Atlantic from the Sahara Desert to the Caribbean and North America).

Windblown dust rarely rises above the troposphere, and wet and dry deposition act to remove it from the atmosphere. Long-term mean deposition rates of windblown dust vary widely: approximately $10\,g\,m^{-2}\,yr^{-1}$ for undisturbed desert areas in the United States; 10 to $100\,g\,m^{-2}\,yr^{-1}$ for disturbed land areas and urban locations; several hundred $g\,m^{-2}\,yr^{-1}$ for locations within a few hundred kilometers of strong dust areas such as the Sahara.

EMISSION OF DUST

Although sufficiently strong winds can entrain dust directly, the predominant mechanism for emission of dust particles smaller than $10\,\mu m$ (PM_{10}) is abrasion of larger particles that previously became airborne. There is a minimum threshold friction velocity (u_{*t}) for individual particles to be entrained into the airstream from a smooth surface (1). This minimum u_{*t} occurs for a particle diameter of approximately $100\,\mu m$. Dust particles require much larger wind forces to remove them from the surface compared with sand-sized particles with diameters of approximately $100\,\mu m$. Because loose particles of $100\,\mu m$ are often found on soil surfaces, a "sandblasting" mechanism is almost always available for unprotected soils exposed to high winds. Physically, sand grains localize kinetic energy onto small target areas compared with that of the fluid energy transfer with the surface. Experiments (2, 3) have confirmed that particle-to-particle interaction (sandblasting) on an erodible surface is an important and probably dominating mechanism that produces suspension particle flux.

A model (2) for the particle flux of dust produced by the impacts of saltating particles showed that the vertical mass flux of dust particles (F_a) is proportional to the horizontal sand flux (q_{tot}) integrated over the total depth of the sand layer, multiplied by the mass per particle and divided by a binding energy Ω. The quantity Ω is characteristic of the strengths that hold dust particles in the soil. In other words, the model says that the ratio F_a/q_{tot} is related to the particle size of the saltating grains along with the binding energy of the dust particles.

Simultaneous data (4) on horizontal fluxes of sand (q_{tot}) and vertical fluxes of PM_{10} (F_a) on agricultural soils in Texas formed a body of experimental work with which to evaluate the above model for F_a/q_{tot}. These results (Fig. 1) show that the ratio F_a/q_{tot} for "sand" textures is highly variable but has a consistency regardless of friction velocity (which is proportional to wind speed). That is, for sandy-soil textures, the sandblasting model (2) for dust emission seems consistent with observations.

Sources of windblown dust are natural and anthropogenic. Natural sources include deserts, such as the Sahara, where there is lack of vegetation and (in some areas) soil crusting, coarse gravel and boulders, and presence of fine and sandy material in the soil. In the United States and other locations, deserts have more protection by vegetation, soil crusting, and coarse elements in the soil so that total dust emissions are smaller per unit area of desert. Highly active wind elements such as dust devils produce intense but highly local emissions when they encounter bare, loose soil.

Disturbances by man that leave large areas of loose soil exposed to high winds are some of the largest causes of wind blown dust emissions. Examples of these are agriculture, diversion of rivers, and vehicular travel in arid and semiarid lands.

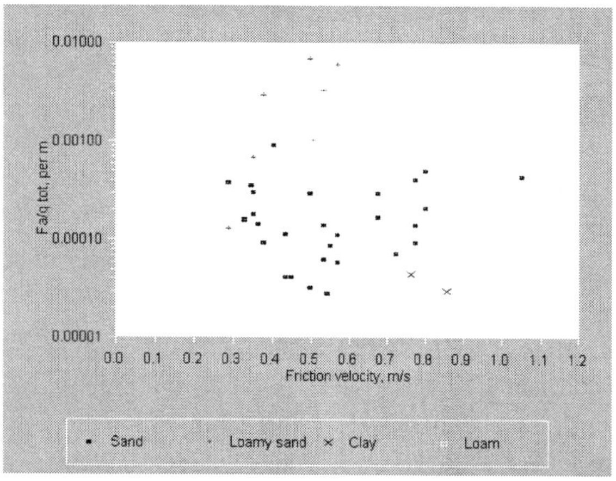

Fig. 1 Ratio of vertical flux of dust to horizontal flux of sand (per meter) versus friction velocity (meters per second). (From Ref. 4.)

SIZE DISTRIBUTIONS AND OPTICAL PROPERTIES OF DUST

Distributions of dust mass concentration have been observed to be normal in logarithm of particle size (i.e., lognormal) with two modes. For many different wind speeds above the threshold (lower limit), researchers (5) have found that there is a constant shape of dust mass size distribution for particles smaller than 10 μm. Scanning electron microscope (SEM) analysis of typical dust samples shows that many particles are aggregated and that individual particles have differing mineralogies.

An explanation (6) of the constant shape for dust size distributions is that the kinetic energy of individual saltating grains, along with the kinetic energy required to release suspendable particles, determines the quantities of different sizes of emitted dust. The huge variation in particle bond strengths and particle sizes explains the variability of F_a/q_{tot} as seen in Fig. 1, despite the overall similarity of dust size distributions produced over widely different soil locations and wind strengths.

Optical properties of the dust sampled in the United States may be summarized as a typical complex index of refraction for U.S. soil dust for visible light (488 nm); the real part of the index of refraction = 1.525, and the imaginary part of the index of refraction = 0.005i. For Saharan dust, the index of refraction was 1.55 + 0.008i. Visibility studies using the formula $M = C/V$, where M is the mass concentration of soil aerosols ($g\,m^{-3}$) and V is horizontal visibility in kilometers, measured the constant C as $0.011\,g\,m^{-3}\,km$.

COMPOSITION OF DUST

Mineralogically (chemically), a general model (7) of dust for particles 1 to 10 μm consists of (in order of the most abundant): quartz, mica, kaolinite, mixed-layer phyllosilicates, and feldspars. For particles smaller than 1 μm, the order is: mica, kaolinite, quartz, and mixed-layer phyllosilicates. Particles from 10 to 100 μm are usually almost exclusively quartz grains characteristically coated with clay. The finer mode particles (1–10 μm) are similar to clay platelets that stick to the surface of larger quartz particles. Collisions of sand-sized particles with other particles (sandblasting) act to release portions of aggregated particles and break up the crystalline structure of mineral particles. One soil containing more clay than another "sandier" soil had the proportions of the 1 to 10 μm mode increase with wind speed relative to the 10 to 100 μm mode (8). The sandier soil did not.

Usually, desert soils contain higher percentages of calcium carbonates than soils receiving greater rainfall. Because surface calcium carbonates are largely contained as fine particles, coatings, and interstitial material, the mechanisms of dust emission increase (enrich) the amount of calcium present in dust compared with that of the soil from which it was derived. Similar enrichments are found for other trace elements, such as the aluminum associated with clay composition. Soluble material, such as sodium chloride, is enriched in some barren desert soils. This concentration takes place through transport by water to low places, where the water is then lost largely through evaporation. Enrichment of salts in dust is also observed, especially at playas (drainage basins). Because windblown dust is rich in calcium carbonates, the percentage of carbonate may range from 5% to 25% by mass in deposited dust. Also, because soluble materials are concentrated at the surface in arid regions, salt content ranges from 0% to 20% by mass in deposited windblown dust.

TRANSPORT OF WINDBLOWN DUST

After being emitted from the surface to the air, windblown dust has the potential for transport. An approximate size limit for dust having the potential for long-range transport was defined as the size at which the fall (sedimentation) velocity is less than or equal to 0.1 times the friction velocity of the wind (9). Since the smallest friction velocity for the threshold of particle movement is approximately $20\,cm\,s^{-1}$, the size limit corresponds to a particle having a sedimentation velocity of $2\,cm\,s^{-1}$. For soil-derived particles with densities of approximately 2500 kg m^{-3}, this corresponds roughly to particle diameters between 10 and 20 μm. Particles of this size with potential for long-range transport could be depleted by deposition within a few minutes following their emission (10).

Although most dust is deposited near its point of origin, a small fraction of the dust emitted can be transported great distances. For example, dust has been shown to have transported from the Sahara Desert to Miami, Florida (11).

DEPOSITION OF WINDBLOWN DUST

Recent studies in desert regions that are not considered major global contributors of dust have sampled dust deposition, taking great care to exclude bird contributions and sand particles that are probably of local origin. The results showed that typical values of deposition for southern Nevada and southeastern California are 4.3 to 15.7 g m^{-2} yr^{-1} (12). Typical values for south-central New Mexico (northern Chihuahuan desert) are 7.6 to 28.1 g m^{-2} yr^{-1} (13). Depositions in urban desert areas of the United States are typically higher (e.g., 48 g m^{-2} yr^{-1}) (14); these values reflect dust contributions from anthropogenic disturbances of the soil followed by wind erosion on the disturbed soil (15)

Deposition rates downwind of agricultural areas, such as in western Texas, range from 10.8 to 13.7 g m^{-2} yr^{-1} (16), whereas in the Great Plains of the United States rates range from 20.2 to 90.8 g m^{-2} yr^{-1} (17). These higher deposition rates reflect sporadic large depositions from dust storms eroding bare farm fields. Deposition rates near major global atmospheric dust contributors, like the Sahara, are higher than other locations. Deposition in the Negev (Israel) ranges from 120 to 300 g m^{-2} yr^{-1} (18). Rates at Niger, West Africa range from 115 to 138 g m^{-2} yr^{-1} (19). However, far downwind from major dust sources, deposition rates of 1.3 g m^{-2} yr^{-1} were reported for dust from Africa reaching Miami, Florida (11).

REFERENCES

1. Bagnold, R. *The Physics of Blown Sand and Desert Dunes*; Methuen and Co. Ltd: London, England, 1941; 265.
2. Shao, Y.; Raupach, M.R.; Findlater, P.A. The Effect of Saltation Bombardment on the Entrainment of Dust by Wind. J. Geophys. Res. **1993**, *98* (D), 12719–12726.
3. Houser, C.A.; Nickling, W.G. The Emission and Vertical Flux of PM_{10} From a Disturbed Clay-Crusted Surface. Sedimentology **2001**, *48*, 255–268.
4. Gillette, D.A.; Fryrear, D.W.; Gill, T.E.; Ley, T.; Cahill, T.A.; Gearheart, E.A. Relation of Vertical Flux of Particles Smaller than 10 μm to Total Aeolian Horizontal Mass Flux At Owens Lake. J. Geophys. Res. **1997**, *102* (D), 26009–26015.
5. Gillette, D.A. Fine Particulate Emissions Due to Wind Erosion. Trans. Am. Soc. Agric. Eng. **1977**, *29*, 890–897.
6. Alfaro, S.C.; Gaudichet, A.; Gomes, L.; Maille, M. Modeling the Size Distribution of a Soil Aerosol Produced by Sandblasting. J. Geophys. Res. **1997**, *102* (D10), 11239–11249.
7. Gillette, D.A.; Clayton, R.N.; Mayeda, T.K.; Jackson, M.L.; Sridhar, K. Tropospheric Aerosols from Some Major Dust Storms of the Southwestern United States. J. Appl. Meteor. **1978**, *17*, 832–845.
8. Gillette, D.A.; Walker, T.R. Characteristics of Airborne Particles Produced by Wind Erosion of Sandy Soil, High Plains of West Texas. Soil Sci. **1977**, *123*, 97–110.
9. Gillette, D.A.; Blifford, I.H., Jr.; Fryrear, D.W. The Influence of Wind Velocity on Size Distributions of Soil Wind Erosion Aerosols. J. Geophys. Res. **1974**, *79*, 4068–4075.
10. Johnson, T.; Gillette, D.; Schwiesow, R. Fate Of Dust Particles from Unpaved Roads Under Various Atmospheric Conditions. In *Precipitation Scavenging and Atmosphere-Surface Exchange*; Schwartz, S., Slinn, W.G.N., Eds.; Hemisphere Publishing Co.: Washington, DC, 1992, 933–948.
11. Prospero, J.; Nees, R.; Uematsu, M. Deposition Rate of Particulate and Dissolved Aluminum Derived from Saharan Dust in Precipitation At Miami, Florida. J. Geophys. Res. **1987**, *92* (14), 723–14731.
12. Reheis, M.; Kihl, R. Dust Deposition in Southern Nevada and California, 1984–1989: Relation to Climate, Source Area, and Source Lithology. J. Geophys. Res. **1995**, *100*, 8893–8918.
13. Gile, L.; Hawley, J.; Grossman, R. Soils and Geomorphology on the Basin and Range Area of Southern New Mexico–Guidebook to the Desert Project, New Mexico Bureau of Mines Mineralogical. Research Memorandum **1981**, *39*, 222.
14. Pe'we', T.; Pe'we', E.; Pe'we', R.; Journaux, A.; Slatt, R. Desert Dust: Characteristics and Rates of Deposition in Central Arizona, U.S.A. In *Desert Dust: Origin, Characteristics, and Effect on Man*; Pe'we', T., Ed.; Geological Society of America: Boulder, Colorado, 1981, 169–190.
15. Holcombe, T.L.; Ley, T.; Gillette, D. Effects of Prior Precipitation and Source Area Characteristics on Threshold Wind Velocities for Blowing Dust Episodes, Sonoran Desert, 1948–1978. J. Appl. Meteor. **1997**, *36*, 1160–1175.
16. Rabenhorst, M.; Wilding, L.; Girdner, C. Airborne Dusts in the Edwards Plateau Region of Texas. Soil Sci. Soc. Am. **1984**, *48*, 621–627.
17. Smith, R.; Twiss, P.; Krauss, R.; Brown, M. Dust Deposition in Relation to Site, Season, and Climatic Variables. Soil Sci. Soc. Amer. Proc. **1970**, *34*, 112–117.
18. Offer, Z.; Goosens, D. Ten Years of Aeolian Dust Dynamics in a Desert Region (Negev Desert, Israel): Analysis of Airborne Dust Concentration, Dust Accumulation and the High Magnitude Dust Events. J. Arid Environ **2001**, *47*, 211–249.
19. Drees, L.; Manu, A.; Wilding, L. Characteristics of Aeolian Dusts in Niger, West Africa. Geoderma **1993**, *59*, 213–233.

WORLD REFERENCE BASE FOR SOIL RESOURCES

Jozef Deckers
Institute for Land and Water Management, Leuven, Belgium

Paul Driessen
Agricultural University, Wageningen, The Netherlands

Freddy Nachtergaele
FAO, Rome, Italy

Otto Spaargaren
International Soil Reference and Information Centre (ISRIC), Wageningen, The Netherlands

INTRODUCTION

In 1998, the International Union of Soil Sciences (IUSS) officially adopted the world reference base for soil resources (WRB) as the Union's system for soil correlation. The structure, concepts, and definitions of the WRB are strongly influenced by the FAO–UNESCO legend of the soil map of the world (1, 2). At the time of its inception, the WRB proposed 30 "Soil Reference Groups" accommodating more than 200 ("second level") soil units.

WRB (3–5) was endorsed by the IUSS in 1998 and provides an opportunity to create and refine a common and global language for soil classification. WRB aims to serve as a framework through which ongoing soil classification throughout the world can be harmonized. The ultimate objective is to reach international agreement on the major soil groups to be recognized at a global scale as well as on the criteria and methodology to be applied for defining and separating them. Such an agreement is needed to facilitate the exchange of information and experience, to provide a common scientific language, to strengthen the applications of soil science, and to enhance the communication with other disciplines and make the major soil names into household names (6).

ELEMENTS OF WRB: DIAGNOSTIC HORIZONS, PROPERTIES, AND MATERIALS

The taxonomic units of the WRB are defined in terms of measurable and observable "diagnostic horizons," the basic identifiers in soil classification. Diagnostic horizons are defined by combinations of characteristic "soil properties" and/or "soil materials." A sample of diagnostic horizons, properties, and materials used by the WRB to differentiate between reference soil groups are summarized in Tables 1–3.

Note that the generalized descriptions of diagnostic horizons, properties, and soil materials given in Tables 1–4 are solely meant as a first introduction to WRB terminology. For the exact concepts and full definitions reference is made to the FAO Soil Resources Reports No. 84 (5).

TWO-TIER APPROACH IN WRB

WRB comprises two tiers of categorical detail (7):

1. The "reference base" which is limited to the first level only, having 30 reference soil groups; and
2. The "WRB Classification System" consisting of combinations of a set of prefixes as unique qualifiers added to the reference soil groups, thus allowing very precise characterization and classification of individual soil profiles.

The Reference Base: An Overview of the World Soil Cover into 30 Reference Groups

In the present text, the 30 reference soil groups are aggregated into 10 "sets" of Major Soil Groups each allocated to one of the sets on the basis of "dominant identifiers," i.e., those soil forming factor(s) which most

Encyclopedia of Soil Science

Table 1 Sample description of diagnostic horizons in WRB

Surface horizons and subsurface horizons at shallow depth	
Anthropogenic horizons	Surface and subsurface horizons resulting from long-continued "anthropedogenic processes," notably deep working, intensive fertilization, addition of earthy materials, irrigation or wet cultivation
Histic horizon	(Peaty) Surface horizon, or subsurface horizon occurring at shallow depth, consisting of organic soil material
Umbric horizon	Well-structured, dark surface horizon with low base saturation and moderate to high organic matter content
Yermic horizon	Surface horizon of rock fragments ("desert pavement") usually, but not always, embedded in a vesicular crust and covered by a thin aeolian sand or loess layer
Subsurface horizons	
Cryic horizon	Perennially frozen horizon in mineral or organic soil materials
Duric horizon	Subsurface horizon with weakly cemented to indurated nodules cemented by silica (SiO_2) known as "durinodes"
Fragic horizon	Dense, noncemented subsurface horizon that can only be penetrated by roots and water along natural cracks and streaks
Vertic horizon	Subsurface horizon rich in expanding clays and having polished and grooved ped surfaces ("slickensides"), or wedge-shaped or parallelepiped structural aggregates formed upon repeated swelling and shrinking

Table 2 Descriptive summary of some diagnostic properties

Albeluvic tonguing	Iron-depleted material penetrating into an argic horizon along ped surfaces
Alic properties	Very acid soil material with a high level of exchangeable aluminum
Permafrost	Indicates that the soil temperature is perennially at or below $0°C$ for at least two consecutive years
Stagnic properties	Visible evidence of prolonged waterlogging by a perched water table

Table 3 Descriptive summary of some diagnostic materials

Anthropogenic soil material	Unconsolidated mineral or organic material produced largely by human activities and not significantly altered by pedogenetic processes
Organic soil material	Organic debris, which accumulates at the surface and in which the mineral component does not significantly influence soil properties
Sulfidic soil material	Waterlogged deposit containing sulfur, mostly sulfides, and not more than moderate amounts of calcium carbonate
Tephric soil material	Unconsolidated, non or only slightly weathered products of volcanic eruptions, with or without admixtures of material from other sources

clearly conditioned soil formation. These sets are discussed below.

Set #1 holds all soils with more than a defined quantity of *organic soil materials*. These organic soils are brought together in only one reference soil group: the Histosols.

Set #2 contains all *man-made soils*. These soils vary widely in properties and appearance and can occur in any environment but have in common that their properties are strongly affected by human intervention. They are aggregated to only one reference soil group: the Anthrosols.

Set #3 includes mineral soils whose formation is conditioned by the particular properties of their *parent material*. The set includes three reference soil groups:

Table 4 Sample of WRB qualifiers and their definitions

Examples of the unique qualifier definitions	
Carbi-	Having a cemented spodic horizon which does not contain enough amorphous iron to turn redder on ignition (in Podzols only)
Carbonati-	Having a soil solution with pH >8.5 (1:1 in water) and $HCO_3 > SO_4 \gg Cl$ (in Solonchaks only)
Chloridi-	Having a soil solution (1:1 in water) with $Cl \gg SO_4 > HCO_3$ (in Solonchaks only)
Cryi-	Having a cryic horizon within 100 cm of the soil surface

1. The Andosols of volcanic regions.
2. The sandy Arenosols of desert areas, beach ridges, inland dunes, areas with highly weathered sandstone, etc.
3. The swelling and shrinking heavy clayey Vertisols of backswamps, river basins, lake bottoms, and other areas with a high content of expanding 2:1 lattice clays.

Set #4 accommodates mineral soils whose formation was markedly influenced by their topographic/physiographic setting. This set holds soils in low terrain positions associated with recurrent floods and/or prolonged wetness, but also soils in elevated or steep terrain positions where soil formation is hindered by low temperatures or erosion.

The set holds four reference soil groups:

In low terrain positions:

1. Young alluvial Fluvisols, which show stratification or other evidence of recent sedimentation, and
2. Nonstratified Gleysols in waterlogged areas that do not receive regular additions of sediment.

In elevated and/or eroding areas:

1. Shallow Leptosols over hard rock or highly calcareous material, and
2. Deeper Regosols, which occur in unconsolidated materials and which have only surfacial profile development, e.g., because of low soil temperatures, prolonged dryness or erosion.

Set #5 holds soils that are only moderately developed on account of their limited pedogenetic age or because of rejuvenation of the soil material. Moderately developed soils occur in all environments, from sea level to the highlands, from the equator to the boreal regions, and under all kinds of vegetation. They have not more in common than "signs of beginning soil formation" so that there is considerable diversity among the soils in this set. Yet, they all belong to only one reference soil group: the Cambisols.

Set #6 accommodates the typical red and yellow soils of wet tropical and subtropical regions. High soil temperatures and ample moisture promote rock weathering and rapid decay of soil organic matter. The reference soil groups in this set have in common a long history of dissolution and transport of weathering products has produced deep and genetically mature soils:

1. Plinthosols on old weathering surfaces; these soils are marked by the presence of a mixture of clay and quartz ("plinthite") that hardens irreversibly upon exposure to the atmosphere.
2. Deeply weathered Ferralsols that have a very low cation exchange capacity and are virtually devoid of weatherable minerals.
3. Alisols with high cation exchange capacity and much exchangeable aluminum.
4. Deep Nitisols in relatively rich parent material and marked by shiny, nutty structural elements.
5. Strongly leached, red and yellow Acrisols on acid parent rock, with a clay accumulation horizon, low cation exchange capacity and low base saturation.
6. Lixisols with a low cation exchange capacity but high base saturation percentage.

Set #7 accommodates reference soil groups in arid and semiarid regions. Redistribution of calcium carbonate and gypsum is an important mechanism of horizon differentiation in soils in the dry zone. Soluble salts may accumulate at some depth or, in areas with shallow groundwater, near the soil-surface. The reference soil groups assembled in Set #7 are:

1. Solonchaks with a high content of soluble salts.
2. Solonetz with a high percentage of adsorbed sodium ions.
3. Gypsisols with a horizon of secondary gypsum enrichment.
4. Durisols with a layer or nodules of soil material that is cemented by silica.
5. Calcisols with secondary carbonate enrichment.

SET #8 holds soils that occur in the steppe zone between the dry climates and the humid temperate zone. This transition zone has a climax vegetation of ephemeral grasses and dry forest. Its location corresponds roughly with the transition from a dominance of accumulation processes in soil formation to a dominance of leaching processes. Set #8 includes three reference soil groups:

1. Chernozems with deep, very dark surface soils and carbonate enrichment in the subsoil.
2. Kastanozems with less deep, brownish surface soils and carbonate and/or gypsum accumulation at some depth (these soils occur in the driest parts of the steppe zone).
3. Phaeozems, the dusky red soils of prairie regions with high base saturation but no visible signs of secondary carbonate accumulation.

Set #9 holds the brownish and greyish soils of humid temperate regions. The soils in this set show evidence of redistribution of clay and/or organic matter. The cool climate and short genetic history of most soils in this zone explain why some soils are still relatively rich in bases despite a dominance of eluviation over enrichment processes. Eluviation and illuviation of metal-humus complexes produce the grayish (bleaching) and brown to black coating colors of soils of this set. Set #9 contains five reference soil groups:

1. Acid Podzols with a bleached eluviation horizon over an accumulation horizon of organic matter with aluminum and/or iron.
2. Planosols with a bleached topsoil over dense, slowly permeable subsoil.
3. Base-poor Albeluvisols with a bleached eluviation horizon tonguing into a clay-enriched subsurface horizon.
4. Base-rich Luvisols with a distinct clay accumulation horizon.
5. Umbrisols with a thick, dark, acid surface horizon that is rich in organic matter.

Set #10 holds the soils of permafrost regions. These soils show signs of "cryoturbation" (i.e., disturbance by freeze–thaw sequences and ice segregation) such as irregular or broken soil horizons and organic matter in the subsurface soil, often concentrated along the top of the permafrost table. Cryoturbation also results in oriented stones in the soil and sorted and nonsorted patterned ground features at the surface. All permafrost soils are assembled in one reference soil group: the Cryosols.

WRB Soil Correlation System

The most important innovation in the world reference base is the building-block approach. As described earlier, the building blocks are the uniquely defined qualifiers. There are 121 of these, which compares favorably with the 152 different soil units in the revised legend. A sample of modifiers and their definitions is given in Table 4.

Simplified Key to the WRB Reference Groups

The following successive steps have to be undertaken to classify a soil:

- Identify soil characteristics through observation in the field, supported by laboratory analyses;
- Determine the presence and kind of horizons;
- Identify specific vertical successions of horizons;
- Apply the key to the WRB Reference Groups to determine the soil group in terms of a specific combination of horizons;
- Consult the ranking list of the qualifiers to fully classify the Reference Group as described above for the Vertisols.

Once the Reference Group is identified, the building blocks assembled in Table 5 are used to define individual soil units as illustrated in the following example worked out to classify a Vertisol:

In Vertisols the following qualifiers have been recognized, in ranking order:

1. Thionic intergrade with acid sulphate Gleysols and Fluvisols.
2. Salic intergrade with the Solonchaks.
3. Sodic intergrade with the Solonetzes.
4. Gypsic intergrade with the Gypsisols.
5. Calcic intergrade with the Calcisols.
6. Alic intergrade with the Alisols.
7. Gypsiric containing gypsum.
8. Pellic dark colored, often poorly drained.
9. Grumic a mulched surface horizon.
10. Mazic very hard surface horizon; workability problems.
11. Chromic reddish colored.
12. Mesotrophic having less than 75% base saturation.
13. Hyposodic an ESP of 6–15.
14. Eutric base saturation over 50%.
15. Haplic the most common.

To classify a reddish colored Vertisol with a calcic horizon one would follow the priority list and note that qualifiers 5 and 11 apply. Therefore, the soil is classified as Chromi-Calcic Vertisol.

When more than two qualifiers can be used, they can be added within brackets after the standard name. If, for instance, the Vertisol discussed also has a very hard surface horizon (qualifier 10), the soil would be named Mazi-Calcic Vertisol (Chromic).

In addition to the unique qualifiers, an opportunity is created to indicate depth (from shallow to deep: Epi, Endo, Bathi) and intensity (from weak to strong: Proto, Para, Hypo, Ortho and Hyper) of features, important for management interpretations. In the example above, one may indicate the occurrence of the calcic horizon within 50 cm from the surface by classifying the soil as Chromi-Epicalcic Vertisol. In cases of polysequential soil profiles, the qualifiers Cumuli or Thapto can be used to indicate accumulation or burial.

For each reference soil group there is a defined list of which qualifier is available, which suggests qualifier rankings as in the example given above. For a comprehensive list of the qualifiers, reference is made to the FAO World Soil Resources Reports No. 84 (5).

Table 5 Simplified key to the WRB Reference Groups

No.	Criterion	If yes →	No.	Criterion	If yes →	No.	Criterion	If yes →
1	Organic matter > 40 cm deep	HISTOSOLS	11	Plinthite or petroplinthite within 50 cm	PLINTHOSOLS	21	Argic horizon and albeluvic tonguing	ALBELUVISOLS
2	Cyric horizon	CRYOSOLS	12	Ferralic horizon	FERRALSOLS	22	Argic horizon with $CEC_c > 24$, $Al_{sat} > 60\%$	ALISOLS
3	Human modifications	ANTHROSOLS	13	Natric horizon	SOLONETZ	23	Argic and nitic horizons	NITISOLS
4	Depth < 25 cm	LEPTOSOLS	14	Abrupt textural change	PLANOSOLS	24	Argic horizon with $CEC_c < 24$, BS < 50%	ACRISOLS
5	> 35% clay vertic horizon	VERTISOLS	15	Chernic or blackish mollic horizon	CHERNOZEMS	25	Argic horizon with $CEC_c > 24$, BS > 50%	LUVISOLS
6	Fluvic materials	FLUVISOLS	16	Brownish mollic horizon and secondary $CaCO_3$	KASTANOZEMS	26	Argic horizon with $CEC_c < 24$, BS > 50%	LIXISOLS
7	Salic horizon	SOLONCHAKS	17	Mollic horizon	PHAEOZEMS	27	Umbric horizon	UMBRISOLS
8	Gleyic properties	GLEYSOLS	18	Gypsic or petrogypsic horizon	GYPSISOLS	28	Cambic horizon	CAMBISOLS
9	Andic or vitric horizon	ANDOSOLS	19	Duric or petroduric horizon	DURISOLS	29	Coarse texture > 100 cm	ARENOSOLS
10	Spodic horizon	PODZOLS	20	Calcic or petrocalcic horizon	CALCISOLS	30	Other soils	→ REGOSOLS

(Each criterion: yes → reference group; ↓ no → next criterion. No at 10 → 11; no at 20 → 21.)

W

CORRELATION TABLE OF WRB (5) WITH USDA SOIL TAXONOMY (8)

WRB 1998 (5)	Soil taxonomy (8)	WRB 1998 (5)	Soil taxonomy (8)
Histosols	Histosols pp.	Phaeozems	Mollisols—Udolls pp.
Cryosols	Gelisols pp.	Gypsisols	Aridisols—Gypsids pp.
Anthrosols	Inceptisols pp., plaggepts	Durisols	Aridisols—Durids pp.
Leptosols	Entisols, lithic subgroups pp.	Calcisols	Aridisols—Calcids pp.
Vertisols	Vertisols	Albeluvisols	Alfisols—Fraglossudalfs
Fluvisols	Entisols—Fluvents	Alisols	Ultisols—Udults pp.
Solonchaks	Aridisols—Salorthids pp.	Nitisols	Oxisols—Kandiudox pp., Ultisols—Kandiudults pp.
Gleysols	Inceptisols—Aquepts pp., Entisols—Aquents pp.	Acrisols	Ultisols—Kandiudults, Kandiustults, Kanhaplustalfs pp.
Andosols	Andisols	Luvisols	Alfisols pp.
Podzols	Spodosols	Lixisols	Alfisols—Paleustalfs, Kandiustalfs, Kandiudalfs, Kanhaplustalfs pp.
Plinthosols	Oxisols—Plinthaquox pp.	Umbrisols	Inceptisols pp.
Ferralsols	Oxisols pp.	Cambisols	Inceptisols pp.
Solonetz	Aridisols—Natrargids pp.	Arenosols	Entisols—Psamments pp.
Planosols	Alfisols—Abruptic Albaqualf pp., Ultisols—Abruptic Albaquults pp.	Regosols	Entisols pp.
Chernozems	Mollisols—Ustolls pp.		
Kastanozems	Mollisols—Ustols, Xerolls pp.		

pp = pro paste = partial agreement only.

REFERENCES

1. FAO–UNESCO, *Soil Map of the World*; Legend FAO: Rome, Italy, 1974; Vol. 1, 59.
2. FAO–UNESCO–ISRIC, 1988, Revised Legend of the Soil Map of the World. *World Soil Resources Report No. 60*; FAO: Rome, 1990.
3. ISSS Working Group RB, In *World Reference Base for Soil Resources. Introduction*, 1st Ed.; Deckers, J.A., Nachtergaele, F.O., Spaargaren, O.C., Eds.; ISSS/ISRIC/FAO: Acco Leuven, 1998.
4. ISSS Working Group RB, In *World Reference Base for Soil Resources. Atlas*, 1st Ed.; Bridges, E.M., Batjes, N.H., Nachtergaele, F.O., Eds.; ISSS/ISRIC/FAO: Acco Leuven, 1998.
5. FAO/ISRIC/ISSS, World Reference Base for Soil Resources. *World Soil Resources Report No. 84*; FAO: Rome, 1998.
6. Dudal, R.A. World Reference Base for Soil Resources: Background, Principles and Challenges. *Proceedings International WRB Workshop*; Pretoria, 1996, 9.
7. Nachtergaele, F.O.; Spaargaren, O.; Deckers, J.A.; Ahrens, B. New Developments in Soil Classification World Reference Base for Soil Resources. Geoderma **2000**, *96*, 345–357.
8. Soil Survey Staff, *Keys to Soil Taxonomy*; USDA Natural Resources Conservation Service: Washington, DC, 1998; 326.

Index

Bold page numbers indicate article entry terms